FUNDAMENTAL AND PHYSICAL CONST

quantity	symbol	U.S.	
Charge			
electron	e		
proton	p		$+1.6022 \times 10^{-19}$ C
Density			
air [STP] [32°F (0°C)]		0.0805 lbm/ft^3	1.29 kg/m^3
air [70°F (20°C), 1 atm]		0.0749 lbm/ft^3	1.20 kg/m^3
earth [mean]		345 lbm/ft^3	5520 kg/m^3
mercury		849 lbm/ft^3	1.360×10^4 kg/m^3
seawater		64.0 lbm/ft^3	1025 kg/m^3
water [mean]		62.4 lbm/ft^3	1000 kg/m^3
Distance [mean]			
earth radius		2.09×10^7 ft	6.370×10^6 m
earth-moon separation		1.26×10^9 ft	3.84×10^8 m
earth-sun separation		4.89×10^{11} ft	1.49×10^{11} m
moon radius		5.71×10^6 ft	1.74×10^6 m
sun radius		2.28×10^9 ft	6.96×10^8 m
first Bohr radius	a_0	1.736×10^{-10} ft	5.292×10^{-11} m
Gravitational Acceleration			
earth [mean]	g	32.174 (32.2) ft/sec^2	9.8067 (9.81) m/s^2
moon [mean]		5.47 ft/sec^2	1.67 m/s^2
Mass			
atomic mass unit	u	3.66×10^{-27} lbm	1.6606×10^{-27} kg
earth		1.32×10^{25} lbm	6.00×10^{24} kg
electron [rest]	m_e	2.008×10^{-30} lbm	9.109×10^{-31} kg
moon		1.623×10^{23} lbm	7.36×10^{22} kg
neutron [rest]	m_n	3.693×10^{-27} lbm	1.675×10^{-27} kg
proton [rest]	m_p	3.688×10^{-27} lbm	1.673×10^{-27} kg
sun		4.387×10^{30} lbm	1.99×10^{30} kg
Pressure, atmospheric		14.696 (14.7) lbf/in^2	1.0133×10^5 Pa
Temperature, standard		32°F (492°R)	0°C (273K)
triple point, water		32.02°F, 0.0888 psia	0.01109°C, 0.6123 kPa
Velocity			
earth escape		3.67×10^4 ft/sec	1.12×10^4 m/s
light [vacuum]	c	9.84×10^8 ft/sec	2.9979 (3.00) $\times 10^8$ m/s
sound [air, STP]	a	1090 ft/sec	331 m/s
[air, 70°F (20°C)]		1130 ft/sec	344 m/s
Volume, molal ideal gas [STP]		359 ft^3/lbmol	22.41 m^3/kmol
Specific Heat			
air, constant pressure [100°F (38°C)]	c_p	0.240 Btu/lbm-F	1.005 kJ/kg·°C
air, constant volume [100°F (38°C)]	c_v	0.171 Btu/lbm-F	0.718 kJ/kg·°C
ice [32°F (0°C)]	$c(c_p)$	0.4896 Btu/lbm-F	2.050 kJ/kg·°C
water [60°F (18°C)]	$c(c_p)$	1.000 Btu/lbm-F	4.184 kJ/kg·K
Fundamental Constants			
Avogadro's number	N_A		6.022×10^{23} mol^{-1}
Bohr magneton	μ_B		9.2740×10^{-24} J/T
Boltzmann constant	k	5.65 × 10-24 ft-lbf/°R	1.38065×10^{-23} J/K
Faraday constant	F		96485 C/mol
gravitational constant	g_c	32.174 lbm-ft/lbf-sec^2	n.a.
gravitational constant	G	3.440 × 10–8 ft^4/lbf-sec^4	6.674×10^{-11} N·m^2/kg^2
nuclear magneton	μ_N		5.050×10^{-27} J/T
permeability of a vacuum	μ_0		1.2566×10^{-6} N/A^2 (H/m)
permittivity of a vacuum	ϵ_0		8.8542×10^{-12} C^2/N·m^2 (F/m)
Planck's constant	h		6.6260×10^{-34} J·s
Rydberg constant	R_∞		1.0974×10^7 m^{-1}
specific gas constant, air	R	53.35 ft-lbf/lbm-°R	287.03 J/kg·K
Stefan-Boltzmann constant		1.71×10^{-9} Btu/ft^2-hr-°R^4	5.670×10^{-8} W/m^2·K^4
universal gas constant	R^*	1545.35 ft-lbf/lbmol-°R	8314.47 J/kmol·K
	R^*	1.986 Btu/lbmol-°R	0.08206 atm·L/mol·K

PE
Mechanical Engineering Reference Manual

Fourteenth Edition

Michael R. Lindeburg, PE

PPI2PASS.COM
A KAPLAN COMPANY

Report Errors for This Book

PPI is grateful to every reader who notifies us of a possible error. Your feedback allows us to improve the quality and accuracy of our products. Report errata at **ppi2pass.com**.

MECHANICAL ENGINEERING REFERENCE MANUAL FOR THE PE EXAM
Fourteenth Edition

Current release of this edition: 2

Release History

date	edition number	revision number	update
Jan 2019	13	5	Minor corrections.
Jan 2020	14	1	New edition.
Mar 2021	14	2	Minor corrections.

Printed in the United States of America.

PPI
ppi2pass.com

ISBN: 978-1-59126-663-1

Topics

Where do I find practice problems to test what I've learned in this Reference Manual?

The *Mechanical Engineering Reference Manual* provides a knowledge base that will prepare you for the PE Mechanical exam. But there's no better way to exercise your skills than to practice solving problems. To simplify your preparation, please consider *Mechanical Engineering Practice Problems,* which provides you with more than 600 practice problems, each with a complete, step-by-step solution.

Mechanical Engineering Practice Problems may be purchased from PPI at **ppi2pass.com** or from your favorite retailer.

Table of Contents

Topic X: Dynamics and Vibrations

Topic XI: Control Systems

Topic XII: Plant Engineering

Topic XIII: Economics

Topic XIV: Law and Ethics

Topic XV: Support Material

Appendix Table of Contents

Preface

In 2020, the National Council of Examiners for Engineering and Surveying (NCEES) changed the Professional Engineering (PE) mechanical engineering licensing exams from a pencil-and-paper exam to a computer-based one. This in itself is not a big change. However, in the pencil-and-paper exam you were permitted to bring your own reference material, including *Mechanical Engineering Reference Manual* (*Reference Manual*), into the exam room. In the computer-based exam, you may not. Instead, you will have on-screen access to a searchable electronic copy of the *NCEES PE Mechanical Reference Handbook* (*NCEES Handbook*). This is the only reference you may consult during the exam.

This drastically changes how you must study for the exam. It is no longer enough to learn how to solve exam problems using the *Reference Manual* and other familiar reference books that you may annotate, highlight, and glue tabs to as you study. Now you must learn how to quickly find the equations and data you need in one specific source, an unmarked electronic copy of the *NCEES Handbook*.

That is the reason for this new edition. Most of the chapters in this *Reference Manual* have been revised to show you not just how to solve exam problems, but how to solve them using the particular equations and data that are found in the *NCEES Handbook*. Equations and data are labeled with blue pointers to their locations in the *NCEES Handbook*, so that as you review the material you need to know, you are also learning to use the *NCEES Handbook* efficiently.

This book is also a broad review of mechanical engineering principles, as well as a compendium of useful and typical data to support those principles. Because of its emphasis on undergraduate engineering subjects (the only prerequisite course of study required of most state licensing examinees), this book is an efficient method of preparing for the standardized PE exams in mechanical engineering. Tens of thousands of engineers have used the previous 13 editions for just that purpose.

This book presents each major topic as a standalone section subdivided into chapters. The choice of what to cover (i.e., topics) in sections and chapters has always been based on what I think a mechanical engineering graduate should know, only loosely influenced by the major subdivisions of the PE exams' specifications. Because of that, the scope of this book exceeds the scope of the PE exams. If you study everything in this book in preparation for your PE exam you will end up being overprepared.

We have used this new edition as an opportunity to make thousands of improvements, most too small to be noticed except through a side-by-side comparison. These improvements include updates to terminology, descriptions, and references; clarifications and rewording of explanations; additional typical data; updating or removing obsolete information; updated chapter nomenclature; improved consistency between chapters; and, the inevitable correction of author's errata.

Coincident with this new edition is a significant new edition of its companion book, *Mechanical Engineering Practice Problems*. Each chapter of *Practice Problems* is designed to provide practice for the fundamental concepts presented in the same chapter of the *Reference Manual*. More than 600 problems were reviewed and updated to demonstrate how to use *NCEES Handbook* equations to solve exam-like problems. The changes to both books go hand in hand.

The journey to "here" has been an adventure, and I'm looking forward to future editions.

Best career wishes.

Michael R. Lindeburg, PE

ABOUT THE AUTHOR

Michael R. Lindeburg, PE, is one of the best-known authors of engineering textbooks and references. His books and courses have influenced millions of engineers around the world. Since 1975, he has authored over 40 engineering reference and exam preparation books. He has spent thousands of hours teaching engineering to students and practicing engineers. He holds bachelor of science and master of science degrees in industrial engineering from Stanford University.

Acknowledgments

It's hard to believe that over 39 years ago, I wrote, typed, edited, and illustrated the first edition of the *Mechanical Engineering Reference Manual.* That process describes production of the first six editions of this book, which consisted of three-ring binders of loose pages printed from camera-ready copy prepared on my Smith-Corona typewriter.

This new edition was created using PPI's XML-based content management system and demanded a gigantic effort from dozens of people. I am grateful beyond words for their enormous contributions to this book.

New content lead contributor: David W. Burris, PE

New content contributors: Nathan R. Palmer, PE; Jared Anna, PE

Technical reviewers: Keith E. Elder, PE; N. S. Nandagopal, MS, PE; Nebojsa Sebastijanovic, PhD, PE

Calculation checker: Anil Acharya, PhD, PE

Content team: Bonnie Conner, Meghan Finley, Anna Howland

Editorial team: Martin Averill, Tyler Hayes, Scott Marley, Scott Rutherford

Editorial operations director: Grace Wong

Project manager: Beth Christmas

Product management team: Ellen Nordman, Megan Synnestvedt

Production team: Tom Bergstrom, Bradley Burch, Kim Burton-Weisman, Nikki Capra-McCaffrey, Robert Genevro, Richard, Iriye, Teresa Trego, Kim Wimpsett, and Stan Info Solutions

Publishing systems team: Jeri Jump, Sam Webster

Technical illustrations and cover design: Tom Bergstrom

Marketing team: John Golden, Jared Schulze

This edition incorporates the comments, suggestions, and errata submitted by many people who used the previous editions for their own preparations. As an author, I am humbled to know that these individuals read the previous edition in such detail as to notice typos, illogic, and other errata, and that they subsequently took the time to share their observations with me. Their suggestions have been incorporated into this edition, and their attention to detail will benefit you and all future readers.

There are hundreds and hundreds of additional contributors mentioned by name in the acknowledgments of those earlier editions. I haven't forgotten them, and their names will live on in the tens of thousands of old editions that remain in widespread circulation.

Near the end of the acknowledgments, after mentioning a lot of people who contributed to a book and, therefore, could be blamed for a variety of types of errors, it is common for an author to say something like, "I take responsibility for all of the errors you find in this book." Or, "All of the mistakes are mine." This is certainly true, given the process of publishing, since the author sees and approves the final version before his/her book goes to the printer. You would think that after 39 years of writing, I would have figured out (1) how to write without making mistakes, and (2) how to proofread without missing those mistakes that are so obvious to readers. However, such perfection continues to elude me. So, yes. The finger points straight at me.

In the absence of perfection, all I can say is that I'll do my best to incorporate to any suggestions and errata that you report through PPI's website, **ppi2pass.com**.

Thank you, everyone! This edition wouldn't exist without you.

Michael R. Lindeburg, PE

Codes Used to Prepare This Book

The documents, codes, and standards that I used to prepare this new edition were the most current available at the time. In the absence of any other specific need, that was the best strategy for this book.

Engineering practice is often constrained by law or contract to using codes and standards that have already been adopted or approved. However, newer codes and standards might be available. For example, the adoption of building codes by states and municipalities often lags publication of those codes by several years. By the time the 2018 codes are adopted, the 2020 codes have been released. Federal regulations are always published with future implementation dates. Contracts are signed with designs and specifications that were "best practice" at some time in the past. Nevertheless, the standards are referenced by edition, revision, or date. All of the work is governed by unambiguous standards.

All standards produced by ASME, ASHRAE, ANSI, ASTM, and similar organizations are identified by an edition, revision, or date. In mechanical engineering, NCEES does not specify "codes and standards" in its lists of PE exam topics. My conclusion is that the NCEES mechanical engineering PE exams are not sensitive to changes in codes, standards, regulations, or announcements in the Federal Register. That is the reason that I referred to the most current documents available as I prepared this new edition.

The relationship of the exams to specific codes is discussed in more detail in the Introduction.

Introduction

PART 1: HOW YOU CAN USE THIS BOOK

The main purpose of *Mechanical Engineering Reference Manual* (*Reference Manual*) is to get you ready for the NCEES PE mechanical exams. Use it along with the other PPI PE mechanical study tools to assess, review, and practice until you pass your exam.

Assess

To pinpoint the subject areas where you need more study, use the diagnostic exams on the PPI Learning Hub (**ppi2pass.com**). How you perform on these diagnostic exams will tell you which topics you need to spend more time on and which you can review more lightly.

Review

PPI offers a complete solution to help you prepare for exam day. Our mechanical engineering prep courses and *Reference Manual* offer a thorough review for the PE mechanical exams. *Mechanical Engineering Practice Problems* (*Practice Problems*), and the PPI Learning Hub quiz generator offer extensive practice in solving exam-like problems. *Mechanical Engineering HVAC and Refrigeration Practice Exam, Mechanical Engineering Machine Design and Materials Practice Exam*, and *Mechanical Engineering Thermal and Fluid Systems Practice Exam* provide practice exams that simulate the exam-day experience and let you hone your test-taking skills.

Practice

Learn to Use the *NCEES PE Mechanical Reference Handbook*

Download a PDF of the *NCEES PE Mechanical Reference Handbook* (*NCEES Handbook*) from the NCEES website. As you study, take the time to find out where important equations and tables are located in the *NCEES Handbook*. Although you could print out the *NCEES Handbook* and use it that way, it will be better for your preparations if you use it in PDF form on your computer. This is how you will be referring to it and searching in it during the actual exam.

A searchable electronic copy of the *NCEES Handbook* is the only reference you will be able to use during the exam, so it is critical that you get to know what it includes and how to find what you need efficiently. Even if you know how to find the equations and data you need more quickly in other references, take the time to search for them in *NCEES Handbook*. Get to know the terms and section titles used in the *NCEES Handbook* and use these as your search terms.

In this book, each equation from the *NCEES Handbook* is given in blue and annotated with the title of the section the equation is found in, also in blue. Whenever data are taken from a figure or table in the *NCEES Handbook*, the title of the figure or table is given in blue. Get to know these titles as you study; they will give you search terms you can use to quickly find the equations and data you need, saving valuable time during the exam.

Using steam tables, h_1 389.0 Btu/lbm, $s_1 = 1.567$ Btu/lbm-°R, and $p_2 = 4$ psia. h_2 represents the enthalpy for a turbine that is 100% efficient. Since the turbine is isentropic, $s_1 = s_2$. Using steam tables, find the appropriate enthalpy and entropy values at state 2′ where 2′ = 4 psia. **[Properties of Saturated Water and Steam (Temperature) - I-P Units]**

$$\begin{aligned} h_f &= 120.87 \text{ Btu/lbm} \\ s_f &= 0.2198 \text{ Btu/lbm-°R} \\ h_{fg} &= 1006.4 \text{ Btu/lbm} \\ s_{fg} &= 1.6424 \text{ Btu/lbm-°R} \end{aligned}$$

The steam quality at the turbine exhaust (state 2) for a 100% efficient turbine is found from the entropy relationship.

Properties for Two-Phase (Vapor-Liquid) Systems

$$\begin{aligned} s &= s_f + x s_{fg} \\ x &= \frac{s - s_f}{s_{fg}} \\ &= \frac{1.567\ \frac{\text{Btu}}{\text{lbm-°R}} - 0.2198\ \frac{\text{Btu}}{\text{lbm-°R}}}{1.6424\ \frac{\text{Btu}}{\text{lbm-°R}}} \\ &= 0.82 \end{aligned}$$

Some equations given in blue in this book may have a variable or two that is different from the equation as it appears in the *NCEES Handbook*. There are a small number of variables that are treated inconsistently in the *NCEES Handbook*; to minimize possible confusion while studying, in this book these variables have been made consistent.

For example, pressure is represented by both p and P in different sections of the *NCEES Handbook*; in this book pressure is always represented by p. Similarly, in this book heat is always represented by Q; heat rate is $\dot{Q}$ in reference to thermodynamic cycles and q otherwise; velocity is always v; and elevation is always z. All the variables and subscripts used in a chapter are defined in the nomenclature list at the end of each chapter.

Equations in blue may also differ from their presentation in the *NCEES Handbook* because of the presence of the gravitational constant, g_c. The *NCEES Handbook* generally does not indicate whether an equation requires g_c when used with U.S. customary units. On the PE exam, then, you will need to know when and how to include g_c in a calculation without any help from the *NCEES Handbook*.

To show the correct use of g_c, equations in this book are given in two versions where appropriate, one for use with SI units and one for use with U.S. customary units, with g_c correctly included in the U.S. version. When you solve practice problems, however, you should use the *NCEES Handbook* as your only reference, identifying when and how to use g_c on your own. This is more trouble than looking up the equations in this book, but it will better prepare you for the actual exam.

Access the PPI Learning Hub

Although the *Reference Manual*, *Practice Problems*, and the three mechanical engineering *Practice Exams* can be used on their own, they are designed to work with the PPI Learning Hub. At the PPI Learning Hub, you can access

- a personal study plan, keyed to your exam date, to help keep you on track
- diagnostic exams to help you identify the subject areas where you are strong and where you need more review
- a quiz generator containing hundreds of additional exam-like problems that cover all knowledge areas on the PE mechanical exams
- two full-length NCEES-like, computer-based practice exams for each of the PE mechanical engineering disciplines, to familiarize you with the exam day experience and let you hone your time management and test-taking skills
- electronic versions of *Mechanical Engineering HVAC and Refrigeration Practice Exam, Mechanical Engineering Machine Design and Materials Practice Exam, Mechanical Engineering Thermal and Fluid Systems Practice Exam, Mechanical Engineering Reference Manual*, and *Mechanical Engineering Practice Problems*

For more about the PPI Learning Hub, visit PPI's website at **ppi2pass.com**.

Be Thorough

Really do the work.

Time and again, customers ask us for the easiest way to pass the exam. The short answer is pass it the first time you take it. Put the time in. Take advantage of the problems provided and practice, practice, practice! Take the practice exams and time yourself so you will feel comfortable during the exam. When you are prepared you will know it. Yes, the reports in the PPI Learning Hub will agree with your conclusion but, most importantly, if you have followed the PPI study plan and done the work, it is more likely than not that you will pass the exam.

Some people think they can read a problem statement, think about it for 10 seconds, read the solution, and then say, "Yes, that's what I was thinking of, and that's what I would have done." Sadly, these people find out too late that the human brain makes many more mistakes under time pressure and that there are many ways to get messed up in solving a problem even if you understand the concepts. It may be in the use of your calculator, like using log instead of ln or forgetting to set the angle to radians instead of degrees. It may be rusty math, like forgetting exactly how to factor a polynomial. Maybe you can't find the conversion factor you need, or don't remember what joules per kilogram is in SI base units.

For real exam preparation, you'll have to spend some time with a stubby pencil. You have to make these mistakes during your exam prep so that you do not make them during the actual exam. So do the problems—all of them. Do not look at the solutions until you have sweated a little.

IF YOU ARE AN INSTRUCTOR

If you are teaching a prep course for the PE examination, you can use the material in this book as a guide to prepare your lectures. The first two editions of this book consisted of a series of handouts prepared for the benefit of my PE prep courses. These editions were intended to be compilations of all the long formulas, illustrations, and tables of data that I did not have time to put on the chalkboard. You can use this edition in the same way.

"Capacity assignment" is the goal in my prep courses. If you assign 20 hours of homework and a student is able to put in only 10 hours that week, that student will have worked to his or her capacity. After the PE examination, that student will honestly say that he or she could not have prepared any more than he or she did in your course. For that reason, you have to assign homework on the basis of what is required to become proficient in the subjects of your lecture. You must resist assigning only the homework that you think can be completed in an arbitrary number of hours.

Homework assignments in my prep courses are not individually graded. Instead, students are permitted to make use of existing solutions to learn procedures and techniques to the problems in their homework set, such as those in the companion *Practice Problems*, which contains solutions to all practice problems. However, each student must turn in a completed set of problems for credit each week. Though I don't correct the homework problems, I address comments or questions emailed to me, posted on the course forum, or noted on the assignments.

I believe that students should start preparing for the PE exam at least six months before the examination date. However, most wait until three or four months before getting serious. Because of that, I have found that a 13- or 14-week format works well for a live PE prep course. It's a little rushed, but the course is over before everyone gets bored with my jokes. Each week, there is a three-hour meeting, which includes lecture and a short break. If you can add more course time, your students will appreciate it. However, I don't think you can cover the full breadth of material in much less time or in many fewer weeks.

Lecture coverage of some examination subjects is necessarily brief; other subjects are not covered at all. These omissions are intentional; they are not the result of scheduling omissions. Why? First, time is not on our side in a prep course. Second, some subjects rarely contribute to the examination. For example, I have found that very few people study modeling and systems analysis, material handling, and manufacturing methods. Unless you have six months in which to teach your PE review, your students' time can be better spent covering other subjects.

All the skipped chapters and any related practice problems are presented as floating assignments to be made up in the students' "free time."

I strongly believe in exposing my students to a realistic sample examination, but I no longer administer an in-class mock exam. Since the prep course usually ends only a few days before the real PE examination, I hesitate to make students sit for several hours in the late evening to take a "final exam." Rather, I distribute and assign a sample exam at the first meeting of the prep course.

If the practice test is to be used as an indication of preparedness, caution your students not to even look at the sample exam prior to taking it. Looking at the sample examination, or otherwise using it to direct their review, will produce unwarranted specialization in subjects contained in the sample examination.

There are many ways to organize a PE prep course, depending on your available time, budget, intended audience, facilities, and enthusiasm. However, all good course formats have the same result: The students struggle with the workload during the course, and then they breeze through the examination after the course.

PART 2: EVERYTHING YOU EVER WANTED TO KNOW ABOUT THE PE EXAM

WHAT IS THE FORMAT OF THE PE EXAM?

The NCEES PE examination in mechanical engineering consists of two four-hour sessions separated by a one-hour lunch period. The examinee may choose one of three disciplines: HVAC and refrigeration, machine design and materials, or thermal and fluid systems. You must be approved by your state licensing board before you can register for the exam using the "My NCEES" system on the NCEES website. You select your discipline when you register for the exam. Switching disciplines is not possible at the exam appointment.

Both the morning and afternoon sessions contain 40 questions in multiple-choice (i.e., "objective") or alternative-item-type (AIT) format. As this is a "no-choice" exam, you must answer all questions in each session correctly to receive full credit. There are no optional questions.

WHAT SUBJECTS ARE ON THE PE EXAM?

NCEES has published a description of subjects on the exams. Irrespective of the published examination structure, the exact number of questions that will appear in each subject area cannot be predicted reliably.

There is no guarantee that any single subject will occur in any quantity. One of the reasons for this is that some of the questions span several disciplines. You might categorize and solve a steam flow question as a fluids problem, while NCEES might categorize it as a thermodynamics (ideal gas or compressible flow) problem.

Table 1 describes the subjects in detail and gives the approximate number of problems for each topic. Most examinees find the list to be formidable in appearance. NCEES adds,

> The examination is developed with questions that require a variety of approaches and methodologies including design, analysis, application, and operations. Some questions may require knowledge of engineering economics. These areas are examples of the kinds of knowledge that will be tested but are not exclusive or exhaustive categories.

WHAT IS THE TYPICAL PROBLEM FORMAT?

Almost all of the problems are stand-alone—that is, they are complete and independent. Problem types include traditional multiple-choice problems, as well as alternative item types (AITs). AITs include, but are not limited to

- *multiple correct*, which allows you to select multiple answers
- *point and click*, which requires you to click on a part of a graphic to answer
- *drag and drop*, which requires you to click on and drag items to match, sort, rank, or label
- *fill in the blank*, which provides a space for you to enter a response to the problem

Although AITs are a recent addition to the PE mechanical exams and may take some getting used to, they are not inherently difficult to master. For your reference, additional AIT resources are available on the PPI Learning Hub (**ppi2pass.com**).

Traditional multiple-choice problems will have four answer options, labeled A, B, C, and D. If the four answer options are numerical, they will be displayed in increasing value. One of the answer options is correct (or "most nearly correct"). The remaining answer options will consist of three "logical distractors," the term used by NCEES to designate options that are incorrect but look plausibly correct.

HOW MUCH "LOOK-UP" IS REQUIRED ON THE EXAM?

Since most of the questions are multiple choice in design, all required data will appear in the situation statement. Since the examination would be unfair if it was possible to arrive at an incorrect answer after making valid assumptions or using plausible data, you will not generally be required to come up with numerical data that might affect your success on the problem. Friction factors and pipe roughness, thermal conductivities, U- and R- heat transfer factors, most pipe sizes, material strengths and other properties, and relevant assumptions will be given in the question statement. There will also be superfluous information in the majority of questions.

WHAT DOES "MOST NEARLY" REALLY MEAN?

One of the more disquieting aspects of these questions is that the available answer choices are seldom exact. Answer choices generally have only two or three significant digits. Exam questions ask, "Which answer choice is most nearly the correct value?" or they instruct you to complete the sentence, "The value is approximately . . ."

A lot of self-confidence is required to move on to the next question when you don't find an exact match for the answer you calculated, or if you have had to split the difference because no available answer choice is close.

NCEES describes it like this:

> Many of the questions on NCEES exams require calculations to arrive at a numerical answer. Depending on the method of calculation used, it is very possible that examinees working correctly will arrive at a range of answers. The phrase "most nearly" is used to accommodate answers that have been derived correctly but that may be slightly different from the correct answer choice given on the exam. You should use good engineering judgment when selecting your choice of answer. For example, if the question asks you to calculate an electrical current or determine the load on a beam, you should literally select the answer option that is most nearly what you calculated, regardless of whether it is more or less than your calculated value. However, if the question asks you to select a fuse or circuit breaker to protect against a calculated current or to size a beam to carry a load, you should select an answer option that will safely carry the current or load. Typically, this requires selecting a value that is closest to but larger than the current or load.

The difference is significant. Suppose you were asked to calculate "most nearly" the volumetric pure airflow required to dilute a contaminated air stream to an acceptable concentration. Suppose, also, that you calculated the answer to be 823 cfm. If the answer choices were (A) 600 cfm, (B) 800 cfm, (C) 1000 cfm, and (D) 1200 cfm, you would go with answer choice (B), because it is most nearly what you calculated. If, however, you were asked to select a fan or duct with the same rated capacities, you would have to go with choice (C), because an 800 cfm fan wouldn't be sufficient. Got it?

HOW MUCH MATHEMATICS IS NEEDED FOR THE EXAM?

There are no pure mathematics questions (algebra, geometry, trigonometry, etc.) on the exam. However, you will need to apply your knowledge of these subjects to the exam questions.

Generally, only simple algebra, trigonometry, and geometry are needed on the PE exam. You will need to use the trigonometric, logarithm, square root, exponentiation, and similar buttons on your calculator. There is no need to use any other method for these functions.

Except for simple quadratic equations, you will probably not need to find the roots of polynomial equations. For second-order (quadratic) equations, the exam does not care if you find roots by factoring, completing the square, using the quadratic equation, graphing, or using

Table 1 Detailed Analysis of Tested Subjects[a,b]

HVAC and Refrigeration

I. Principles (28–43 problems)

A. Basic engineering practice (units and conversions, economic analysis, electrical concepts)

B. Thermodynamics (cycles, properties, compression processes)

C. Psychrometrics (heating/cooling processes, humidification/dehumidification processes)

D. Heat transfer

E. Fluid mechanics

F. Energy/mass balances

II. Applications (42–64 problems)

A. Heating/cooling loads

B. Equipment and components (cooling towers and fluid coolers, boilers and furnaces, heat exchangers, condensers/evaporators, pumps/compressors/fans, cooling/heating coils, control systems components, refrigerants, refrigeration components)

C. Systems and components (air distribution, fluid distribution/piping, refrigeration, energy recovery, basic control concepts)

D. Supportive knowledge (codes and standards, air quality and ventilation, vibration control, acoustics)

Machine Design and Materials Exam

I. Principles (35–55 problems)

A. Basic engineering practice (engineering terms and symbols, interpretation of technical drawings, quality assurance/quality control, project management and economic analysis, units and conversions, design methodology)

B. Engineering science and mechanics (statics, kinematics, dynamics)

C. Material properties (physical, chemical, mechanical)

D. Strength of materials (stress/strain, shear, bending, buckling, torsion, fatigue, failure theories)

E. Vibration (natural frequencies, damping, forced vibrations)

II. Applications (35–55 problems)

A. Mechanical components (pressure vessels and piping; bearings; gears; springs; dampers; belt, pulley, and chain drives; clutches and brakes; power screws; shafts and keys; mechanisms; basic mechatronics; hydraulic and pneumatic components; motors and engines)

B. Joints and fasteners (welding and brazing; bolts, screws, and rivets; adhesives)

C. Supportive knowledge (manufacturing processes, fits and tolerances, codes and standards, computational methods and their limitations, testing and instrumentation)

Thermal and Fluid Systems Exam

I. Principles (28–44 problems)

A. Basic engineering practice (engineering terms, symbols, and technical drawings; economic analysis; units and conversions)

B. Fluid mechanics (fluid properties, compressible flow, incompressible flow)

C. Heat transfer principles

D. Mass balance principles

E. Thermodynamics (thermodynamic properties, thermodynamic cycles, energy balances, combustion)

F. Supportive knowledge (pipe system analysis, joints, psychrometrics, codes and standards)

II. Hydraulic and Fluid Applications (21–33 problems)

A. Hydraulic and fluid equipment (pumps and fans, compressors, pressure vessels, control valves, actuators, connections)

B. Distribution systems

III. Energy/Power System Applications (21–33 problems)

A. Energy/power equipment (turbines, boilers and steam generators, internal combustion engines, heat exchangers, cooling towers, condensers)

B. Cooling/heating

C. Energy recovery

D. Combined cycles

[a]Considerable overlap, duplication, and flexibility exists in each topic.

[b]NCEES may occasionally revise exam subjects somewhat. For the most current information, visit **ppi2pass.com.**

your calculator's root finder. Occasionally, it will be convenient to use the equation-solving capability of your calculator. However, other solution methods will always exist.

There is essentially no use of calculus on the exam. Rarely, you may need to take a simple derivative to find a maximum or minimum of some simple algebraic function. Even rarer is the need to integrate to find an average, moment of inertia, statical moment, or shear flow.

There is essentially no need to solve differential equations. Questions involving radioactive decay, seismic vibrations, control systems, chemical reactions, and fluid mixing have appeared from time to time. However, these applications are extremely rare, have usually been first-order, and could usually be handled without having to solve differential equations.

Basic statistical analysis of observed data may be necessary. Statistical calculations are generally limited to finding means, medians, standard deviations, variances, percentiles, and confidence limits. Since the problems are multiple choice, you won't have to draw a histogram, although you might have to interpret one. Usually, the only population distribution you need to be familiar with is the normal curve. Probability, reliability, hypothesis testing, and statistical quality control are not explicit exam subjects, though their concepts may appear peripherally in some problems. You will not have to use linear or nonlinear regression and other curve fitting techniques to correlate data.

Quantitative optimization methods, such as linear, dynamic, and integer programming, generally associated with the field of operations research are not exam subjects.

The PE exam is concerned with numerical answers, not with proofs or derivations. You will not be asked to prove or derive formulas, use deductive reasoning, or validate theorems, corollaries, or lemmas.

Inasmuch as first assumptions can significantly affect the rate of convergence, problems requiring trial-and-error solutions are unlikely. Rarely, a calculation may require an iterative solution method. Generally, there is no need to complete more than two iterations. You will not need to program your calculator to obtain an "exact" answer. Nor will you generally need to use complex numerical methods.

HOW ABOUT ENGINEERING ECONOMICS?

For most of the early years of engineering licensing, questions on engineering economics appeared frequently on the examinations. This is no longer the case. What this means is that engineering economics concepts might appear in several questions on the exam, or the subject might be totally absent. While the degree of engineering economics knowledge has decreased somewhat, the basic economic concepts (e.g., time value of money, present worth, non-annual compounding, comparison of alternatives, etc.) are still valid test subjects.

If engineering economics is incorporated into other questions, its "disguise" may be totally transparent. For example, you might need to compare the economics of buying and operating two blowers for remediation of a hydrocarbon spill—blowers whose annual costs must be calculated from airflow rates and heads. Also, you may need to use engineering economics concepts and tables in problems that don't even mention "dollars" (e.g., when you need to predict future water demand, population, or traffic volume).

WHAT ABOUT FIRE PROTECTION ENGINEERING?

At one time, fire protection was a topic on the mechanical engineering PE exam. Numerical questions dealt with sprinkler capacity, sprinkler layout, fire pumps, hydrants, standpipes, hose and nozzle flow rate, and occupancy categories. This topic disappeared when the mechanical engineering PE exam adopted the breadth-and-depth format. However, piping, pumps, valve, and controls for fire protection are easily categorized into other exam topics. The fire protection chapter in this book covers basic material that might still be useful on the exam.

WHAT ABOUT NUCLEAR ENGINEERING?

At one time, nuclear engineering problems appeared regularly on the mechanical engineering PE exam. These problems dealt with shielding, health safety, core power development, decay, liquid metal flow and heat transfer, and core design. Such problems disappeared when the nuclear engineering PE exam became available. Problems involving nuclear reactor environments continue to appear, but these can always be solved with "traditional" heat transfer, thermodynamic, power cycle, and fluid machinery concepts.

WHAT ABOUT PROFESSIONALISM AND ETHICS?

For many decades, NCEES has considered adding professionalism and ethics questions to the PE exam. However, these subjects are not part of the test outline, and there has yet to be an ethics question in the exam. Professional practice questions dealing with obligations related to contracts, bidding, estimating, inspection, and regulations sometimes get pretty close. However, you won't encounter the phrase "ethical obligation" in the exam.

WHAT ABOUT CODES AND STANDARDS?

NCEES does not specify "codes and standards" in its lists of exam topics. For that reason, at least for the mechanical engineering PE exams, "codes and standards" seems to imply "knowledge about codes and standards," as opposed to "possession of and reference to the codes and standards" during the exam. The distinction is significant, because (without a specific list) it would be unreasonably expensive to purchase every code and standard affecting mechanical engineers. Among others, ASME, ASTM, ANSI, ASHRAE, SAE, NFPA, NEC, AGMA, EPA, OSHA, and other U.S. organizations publish numerous documents, as do Canada and the European Union (EU).

There are a few noteworthy exceptions: ASME Y14.5 (*Dimensioning and Tolerancing*); ASME *Boiler and Pressure Vessel Code* (BPVC) Sec. VIII, Div. 1; ASHRAE Standard 62.1 (*Ventilation for Acceptable Indoor Air Quality*); TEMA's *Standards of the Tubular Exchanger Manufacturers Association*; and, OSHA CFR 29. Depending on your discipline, one or more of these publications could be valuable.

Inasmuch as fire protection is no longer a specific topic on the mechanical PE exams, none of the NFPA publications should be needed. A useful standard for non-fatigue applications, ASME's *Code for Design of Transmission Shafting* (ASA-B17c), is obsolete and is unlikely to be needed during the exam.

IS THE EXAM TRICKY?

Other than providing superfluous data, the PE exam is not a "tricky exam." The exam does not overtly try to get you to fail. Examinees manage to fail on a regular basis with perfectly straightforward questions. The exam questions are difficult in their own right. NCEES does not need to provide misleading or conflicting statements. However, you will find that commonly made mistakes are represented in the available answer choices. Thus, the alternative answers (known as distractors) will be logical.

Questions are generally practical, dealing with common and plausible situations that you might experience in your job. You will not be asked to determine the radiated heat transfer from the side of a spacecraft at night after it has landed on a Jovian moon with a methane atmosphere.

WHAT MAKES THE QUESTIONS DIFFICULT?

Some questions are difficult because the pertinent theory is not obvious. There may be only one acceptable procedure, and it may be heuristic (or defined by a code) such that nothing else will be acceptable. For example, if you don't know the AGMA procedure for designing gears, no other knowledge of gear design is going to be helpful for an AGMA question.

Some questions are difficult because the data needed are hard to find. Solving some HVAC problems depends on having climatological data for a specific location and performance characteristics of specific construction types.

Some questions are difficult because they defy the imagination. Problems involving epicyclical gear trains can be like this. If you cannot visualize the operation of the mechanism, if you cannot get an intuitive feeling about what is going on, you probably cannot analyze it.

Some questions are difficult because the computational burden is high, and they just take a long time. Convective heat transfer, HVAC, and pipe networks analyzed with the Hardy-Cross method fall into this category.

Some questions are difficult because the terminology is obscure, and you just don't know what the terms mean. This can happen in almost any subject.

DOES THE PE EXAM USE SI UNITS?

The PE mechanical machine design and materials exam and the PE mechanical thermal and fluid systems exam primarily use customary U.S. units (also known as "English units," "inch-pound units," and "British units"), although SI and a variety of other metric systems are also used. The PE mechanical HVAC and refrigeration exam only uses customary U.S. units.

Questions use the units that correspond to commonly accepted industry standards. Metric units are used in chemical-related subjects, including electrical power (watts) and water concentration (mg/L) questions. Either system can be used for fluids, stress analysis, and thermodynamics.

Unlike this book, the exam does not differentiate between lbf and lbm (pounds-force and pounds-mass). Similarly, the exam does not follow this book's practice of meticulously separating the concepts of mass and weight, density and specific weight, and gravity, g, and the gravitational constant, g_c.

WHY DOES NCEES REUSE SOME QUESTIONS?

NCEES reuses some of the more reliable questions from each exam. The percentage of repeat questions isn't high—no more than 25% of the exam. NCEES repeats questions in order to equate the performance of one group of examinees with the performance of an earlier group. The repeated questions are known as *equaters*, and together, they are known as the *equating subtest*.

Occasionally, a new question appears on the exam that very few of the examinees do well on. Usually, the reason for this is that the subject is too obscure or the question is too difficult. Questions on control systems and some engineering management subjects (e.g., linear programming) fall into this category. Also, there have been cases where a low percentage of the examinees get the answer correct because the question was inadvertently stated in a poor or confusing manner. Questions that everyone gets correct are also considered defective.

NCEES tracks the usage and "success" of each of the exam questions. "Rogue" questions are not repeated without modification. This is one of the reasons historical analysis of question types shouldn't be used as the basis of your review.

DOES NCEES USE THE EXAM TO PRE-TEST FUTURE QUESTIONS?

NCEES does not use the PE exam to "pre-test" or qualify future questions. (It does use this procedure on the FE exam, however.) All of the questions you work will contribute toward your final score.

ARE THE EXAMPLE PROBLEMS IN THIS BOOK REPRESENTATIVE OF THE EXAM?

The example problems in this book are intended to be instructional and informative. They were written to illustrate how their respective concepts can be implemented. Example problems are not intended to represent exam problems or provide guidance on what you should study.

ARE THE PRACTICE PROBLEMS REPRESENTATIVE OF THE EXAM?

The practice problems in the companion *Practice Problems* book were chosen to cover the most likely exam subjects. Some of the practice problems are multiple choice, and some are alternative item types. Some may be more comprehensive and complex than actual exam problems.

Practice problems in the companion book were selected to complement subjects in the *Mechanical Engineering Reference Manual.* Over the many editions of both books, the practice problems have developed into a comprehensive review of the most important mechanical engineering subjects covered on the exams.

All of the practice problems are original. Since NCEES does not release old exams, and since examinees are sworn to secrecy before taking the exam, none of the practice problems are actual exam problems.

WHAT REFERENCE MATERIAL IS PERMITTED IN THE EXAM?

The PE Mechanical exam is a closed-book exam. You will be provided with a searchable electronic copy of the *NCEES Handbook.* This is the only reference material you can use during the exam. The PPI Learning Hub (**ppi2pass.com**) simulates exam day experience by allowing you to upload your *NCEES Handbook* PDF to use as you prepare with its Practice Exams.

WHAT ABOUT CALCULATORS?

The exam requires use of a scientific calculator. However, it may not be obvious that you should bring a spare calculator with you and leave it in your locker during the examination. It is always unfortunate when an examinee is not able to finish because his or her calculator stopped working for some unknown reason.

To protect the integrity of its exams, NCEES has banned communicating and text-editing calculators from the exam site. NCEES provides a list of calculator models acceptable for use during the exam. Calculators not included in the list are not permitted. Check the current list of permissible devices at the PPI website (**ppi2pass.com**).

The exam has not been optimized for any particular brand or type of calculator. In fact, for most calculations, a \$15 scientific calculator will produce results as satisfactory as those from a \$200 calculator. There are definite benefits to having built-in statistical functions, graphing, unit-conversion, and equation-solving capabilities. However, these benefits are not so great as to give anyone an unfair advantage.

It is essential that a calculator used for the mechanical PE examination have the following functions.

- trigonometric and inverse trigonometric functions
- hyperbolic and inverse hyperbolic functions
- π
- $\sqrt{x}$ and x^2
- both common and natural logarithms
- y^x and e^x

For maximum speed and utility, your calculator should also have or be programmed for the following functions.

- interpolation
- extracting roots of quadratic and higher-order equations
- calculating factors for economic analysis questions

You may not share calculators with other examinees. Be sure to take your calculator with you whenever you leave the examination room for any length of time.

Laptop and tablet computers (including the iPad®), and electronic readers (e.g., Nook® and Kindle™), are not permitted in the examination.

ARE CELL PHONES PERMITTED?

You may not possess or use a walkie-talkie, cell phone, or other communications or text-messaging device during the exam, regardless of whether it is on.

HOW YOU SHOULD GUESS

There is no deduction for incorrect answers, so guessing is encouraged. However, since NCEES produces defensible licensing exams, there is no pattern to the placement of correct responses. Since the quantitative responses are sequenced according to increasing values, the placement of a correct answer among other numerical distractors is a function of the distractors, not of some statistical normalizing routine. Therefore, it is irrelevant whether you choose all "A," "B," "C," or "D" when you get into guessing mode during the last minute or two of the exam period.

The proper way to guess is as an engineer. You should use your knowledge of the subject to eliminate illogical answer choices. Illogical answer choices are those that violate good engineering principles, that are outside normal operating ranges, or that require extraordinary assumptions. Of course, this requires you to have some basic understanding of the subject in the first place.

Otherwise, it's back to random guessing. That's the reason that the minimum passing score is higher than 25%.

You won't get any points using the "test-taking skills" that helped you in college—the skills that helped with tests prepared by amateurs. You won't be able to eliminate any [verb] answer choices from "Which [noun] . . ." questions. You won't find problems with options of the "more than 50" and "less than 50" variety. You won't find one answer choice among the four that has a different number of significant digits, or has a verb in a different tense, or has some singular/plural discrepancy with the stem. The distractors will always match the stem, and they will be logical.

HOW IS THE EXAM GRADED AND SCORED?

The maximum number of points you can earn on the mechanical engineering PE exam is 80. The minimum number of points for passing (referred to by NCEES as the *cut score*) varies from exam to exam. The cut score is determined through a rational procedure, without the benefit of knowing examinees' performance on the exam. That is, the exam is not graded on a curve. The cut score is selected based on what you are expected to know, not based on passing a certain percentage of engineers.

Each of the questions is worth one point. Grading is straightforward—either you get the question right or you don't. If you mark two or more answers for the same problem, no credit is given for the problem.

Your score is based on the number of correct answers you selected. It is converted to a scaled score which represents your ability compared to the minimum ability established for the exam.

Within 7 to 10 days of taking the exam, you will receive an email from NCEES notifying you to view your results in your MyNCEES account.

If you fail, you will also receive a diagnostic report showing your performance in each subject area.

HOW IS THE CUT SCORE ESTABLISHED?

The raw cut score may be established by NCEES before or after the exam is administered. Final adjustments may be made following the exam date.

NCEES uses a process known as the modified *Angoff procedure* to establish the cut score. This procedure starts with a small group (the cut score panel) of professional engineers and educators selected by NCEES. Each individual in the group reviews each problem and makes an estimate of its difficulty. Specifically, each individual estimates the number of minimally qualified engineers out of a hundred examinees who should know the correct answer to the problem. (This is equivalent to predicting the percentage of minimally qualified engineers who will answer correctly.)

Next, the panel assembles, and the estimates for each problem are openly compared and discussed. Eventually, a consensus value is obtained for each. When the panel has established a consensus value for every problem, the values are summed and divided by 100 to establish the cut score.

Various minor adjustments can be made to account for examinee population (as characterized by the average performance on any equater questions) and any flawed problems. Rarely, security breaches result in compromised problems or examinations. How equater questions, examination flaws, and security issues affect examinee performance is not released by NCEES to the public.

WHAT IS THE PASSING RATE?

Within a few percentage points, 69–77% of first-time takers pass the mechanical engineering PE exams. The passing rate for repeat exam takers is two-thirds of the first-time taker passing rate.

CHEATING AND EXAM SUBVERSION

There aren't very many ways to cheat on a computer-based test. You shouldn't try to smuggle your cell phone, camera, or notebook into the exam, or anything else that could be used to capture a record of the exam problems.

NCEES regularly reuses good problems that have appeared on previous exams. Therefore, examination integrity is a serious issue with NCEES, which goes to great lengths to make sure nobody copies the questions. You may not keep your scratch paper or enter text of questions into your calculator.

NCEES has become increasingly unforgiving about loss of its intellectual property. NCEES routinely prosecutes violators and seeks financial redress for loss of its examination problems, as well as invalidating any engineering license you may have earned by taking one of its examinations while engaging in prohibited activities. Your state board may impose additional restrictions on your right to retake any examination if you are convicted of such activities. In addition to tracking down the sources of any examination problem compilations that it becomes aware of, NCEES is also aggressive in pursuing and prosecuting examinees who disclose the contents of the exam in internet forum and "chat" environments. Your constitutional rights to free speech and expression will not protect you from civil prosecution for violating the nondisclosure agreement that NCEES requires you to sign before taking the examination. If you wish to participate in a dialog about a particular exam subject, you must do so in such a manner that does not violate the essence of your nondisclosure agreement. This requires decoupling your discussion from the examination and reframing the question to avoid any examination particulars.

PART 3: HOW TO PREPARE FOR AND PASS THE PE EXAM IN MECHANICAL ENGINEERING

WHAT SHOULD YOU STUDY?

The exam covers many diverse subjects. Strictly speaking, you don't have to study every subject on the exam in order to pass. However, the more subjects you study, the more you'll improve your chances of passing. You should decide early in the preparation process which subjects you are going to study. The strategy you select will depend on your background. Following are the four most common strategies.

A broad approach is the key to success for examinees who have recently completed their academic studies. This strategy is to review the fundamentals in a broad range of undergraduate subjects (which means studying all or most of the chapters in this book). The examination includes enough fundamentals problems to make this strategy worthwhile. Overall, it's the best approach.

Engineers who have little time for preparation tend to concentrate on the subject areas in which they hope to find the most problems. By studying the list of examination subjects, some have been able to focus on those subjects that will give them the highest probability of finding enough problems that they can answer. This strategy works as long as the examination cooperates and has enough of the types of questions they need. Too often, though, examinees who pick and choose subjects to review can't find enough problems to complete the exam. The PPI Mechanical Learning Hub offers diagnostic exams, hundreds of exam-like practice problems, and realistic practice exams to help you to assess, review, and practice.

Engineers who have been away from classroom work for a long time tend to concentrate on the subjects in which they have had extensive experience, in the hope that the exam will feature lots of problems in those subjects. This method is seldom successful.

Some engineers plan on modeling their solutions from similar problems they have found in textbooks, collections of solutions, and old exams. These engineers often spend a lot of time compiling and indexing the example and sample problem types in all of their books. This is not a legitimate preparation method, and it is almost never successful.

DO YOU NEED A CLASSROOM PREP COURSE?

Approximately 60% of first-time PE examinees take an instructor-led prep course of some form. Live online courses, as well as previously recorded lessons of various types, are available for some or all of the exam topics. Live courses and instructor-moderated courses provide several significant advantages over self-directed study, some of which may apply to you.

- A course structures and paces your review. It ensures that you keep going forward without getting bogged down in one subject.
- A course focuses you on a limited amount of material. Without a course, you might not know which subjects to study.
- A course provides you with the questions you need to solve. You won't have to spend time looking for them.
- A course spoon-feeds you the material. You may not need to read the book!
- The course instructor can answer your questions when you are stuck.

You probably already know if any of these advantages apply to you. A prep course will be less valuable if you are thorough, self-motivated, and highly disciplined.

HOW LONG SHOULD YOU STUDY?

We've all heard stories of the person who didn't crack a book until the week before the exam and still passed it with flying colors. Yes, these people really exist. However, I'm not one of them, and you probably aren't either. In fact, after having taught thousands of engineers in my own classes, I'm convinced that these people are as rare as the ones who have taken the exam five times and still can't pass it.

A thorough review takes approximately 300 hours. Most of this time is spent solving problems. Some of it may be spent in class; some is spent at home. Some examinees spread this time over a year. Others try to cram it all into two months. Most classroom prep courses last for three or four months. The best time to start studying will depend on how much time you can spend per week.

ADDITIONAL REVIEW MATERIAL

In addition to this book and its accompanying *Mechanical Engineering Practice Problems*, PPI can provide you with many targeted references and study aids, some of which are listed here. All of the books have stood the test of time, which means that examinees continually report their usefulness and that PPI keeps them up-to-date.

- *Mechanical Engineering HVAC and Refrigeration Practice Exam*
- *Mechanical Engineering Machine Design and Materials Practice Exam*
- *Mechanical Engineering Thermal and Fluid Systems Practice Exam*
- *PPI Learning Hub*, **ppi2pass.com**
- *Engineering Unit Conversions*
- *Thermal and Fluids Systems Reference Manual for the Mechanical PE Exam*

SHOULD YOU LOOK FOR OLD EXAMS?

The traditional approach to preparing for standardized tests includes working sample tests. However, NCEES does not release old tests or questions after they are used. Therefore, there are no official questions or tests available from legitimate sources. NCEES publishes booklets of sample questions and solutions to illustrate the format of the exam. However, these questions have been compiled from various previous exams, and the resulting publication is not a true "old exam." Furthermore, NCEES sometimes constructs its sample questions books from questions that have been pulled from active use for various reasons, including poor performance. Such marginal questions, while accurately reflecting the format of the examination, are not always representative of actual exam subjects.

WHAT SHOULD YOU MEMORIZE?

In theory, everything you need will be provided in the *NCEES Handbook*, so you can get by without memorizing anything.

In practice, you may find there are some very common, basic equations, such as the formula for the area of a circle, that are not given anywhere in the *NCEES Handbook*. You can recognize these equations in this book because they are not given in blue. You probably already have these memorized, but if you don't, make a note to know them by exam day.

You can speed up your problem solving significantly if you don't have to look up the conversion from ft-lbf/sec to horsepower, the definition of the sine of an angle, and the chemical formula for carbon dioxide. But you don't really have to memorize these simple things. As you work practice problems, you will automatically memorize the things that you come across more than a few times.

DO YOU NEED A STUDY PLAN?

The PPI Mechanical Learning Hub allows you to create a personalized study plan to plan your time and topics. It takes the guesswork out of what to study and for how long. PPI also offers a prep course for each of the three PE mechanical exams, which paces the review and practice of the important exam topics.

It is important that you develop and adhere to a study outline and schedule. Once you have decided which subjects you are going to study, you can allocate the available time to those subjects in a manner that makes sense to you. If you are not taking a classroom prep course (where the order of preparation is determined by the lectures), you should make an outline of subjects for self-study to use for scheduling your preparation.

HOW YOU CAN MAKE YOUR REVIEW REALISTIC

In the exam, you must be able to quickly recall solution procedures, formulas, and important data. You must remain sharp for eight hours or more. When you played a sport back in school, your coach tried to put you in game-related situations. Preparing for the PE exam isn't much different from preparing for a big game. Some part of your preparation should be realistic and representative of the examination environment.

There are several things you can do to make your review more representative. The most important is to refer to the *NCEES Handbook* frequently as you study and use it as your sole reference as you solve practice problems. Being able to find what you need quickly in the *NCEES Handbook* is crucial to performing well on the exam.

Learning to use your time wisely is one of the most important lessons you can learn during your review. You will undoubtedly encounter questions that end up taking much longer than you expected. In some instances, you will cause your own delays by spending too much time looking through books for things you need (or just by looking for the books themselves!). Other times, the questions will entail too much work. Learn to recognize these situations so that you can make an intelligent decision about skipping such questions in the exam.

WHAT TO DO A FEW DAYS BEFORE THE EXAM

There are a few things you should do a week or so before the examination. You should arrange for childcare and transportation. Since the examination does not always start or end at the designated time, make sure that your childcare and transportation arrangements are flexible.

If it's convenient, visit the exam location in order to find the building, parking areas, examination room, and restrooms. If it's not convenient, you may find driving directions and/or site maps on the web.

Take the battery cover off your calculator and check to make sure you are bringing the correct size replacement batteries. Some calculators require a different kind of battery for their "permanent" memories. Put the cover back on and secure it with a piece of masking tape. Write your name on the tape to identify your calculator.

If your spare calculator is not the same as your primary calculator, spend a few minutes familiarizing yourself with how it works. In particular, you should verify that your spare calculator is functional.

WHAT TO DO THE DAY BEFORE THE EXAM

Take the day before the examination off from work to relax. Do not cram the last night. A good night's sleep is the best way to start the examination. If you live a considerable distance from the examination site, consider getting a hotel room in which to spend the night.

Calculate your wake-up time and set the alarms on two bedroom clocks. Select and lay out your clothing items. (Dress in layers.) Select and lay out your breakfast items.

If it's going to be hot on exam day, put your (plastic) bottles of water in the freezer.

Make sure you have gas in your car and money in your wallet.

WHAT TO DO THE DAY OF THE EXAM

You should arrive at least 30 minutes before the examination starts. This will allow time for finding a convenient parking place, making room and seating changes, and calming down. Be prepared, though, to find that the examination room is not open or ready at the designated time.

You also will need to check in and verify your identity. Any items that you bring that are not approved for the testing room, including cell phone, watch, wallet, food, or drink, will need to be stored in a test center locker.

WHAT TO DO DURING THE EXAM

All of the procedures typically associated with timed, proctored, computer-graded assessment tests will be in effect when you take the PE examination.

Listen carefully to everything the proctors say. Do not ask your proctors any engineering questions. Even if they are knowledgeable in engineering, they will not be permitted to answer your questions.

If there are any questions that you think were flawed, in error, or unsolvable, ask a proctor for a "reporting form" on which you can submit your comments. Follow your proctor's advice in preparing this document.

WHAT ABOUT EATING AND DRINKING IN THE EXAM ROOM?

No food or beverages are allowed in the exam room. You may, however, leave them in your locker to have during the break in the middle of the exam.

You may also take an unscheduled break during the exam by raising your hand to notify a proctor. You'll be allowed to visit your locker and consume any food or beverage you stored there. However, any break time you take during the exam is lost; you won't get to stop the clock until you get back to the computer.

HOW TO SOLVE MULTIPLE-CHOICE QUESTIONS

When you begin each session of the exam, observe the following suggestions:

- Do not spend an inordinate amount of time on any single question. If you have not answered a question in a reasonable amount of time, make a note of it and move on.
- Five minutes before the end of each four-hour session, use the remaining time to guess at all of the remaining questions. Odds are that you will be successful with about 25% of your guesses, and these points will more than make up for the few points

that you might earn by working during the last five minutes.

- Make mental notes about any questions for which you cannot find a correct response, that appears to have two correct responses, or that you believe have some technical flaw. Errors in the exam are rare, but they do occur. Such errors are usually discovered during the scoring process and discounted from the examination, so it is not necessary to tell your proctor, but be sure to mark the one best answer before moving on.

SOLVE QUESTIONS CAREFULLY

Many points are lost to carelessness. Keep the following items in mind when you are solving the end-of-chapter questions. Hopefully, these suggestions will be automatic in the exam.

[ ] Did you recheck your mathematical equations?

[ ] Do the units cancel out in your calculations?

[ ] Did you convert between radius and diameter?

[ ] Did you convert between feet and inches?

[ ] Did you convert from gage to absolute pressures?

[ ] Did you convert between pounds and kips, or kPa and Pa?

[ ] Did you use the universal gas constant that corresponds to the set of units used in the calculation?

[ ] Did you recheck all data obtained from other sources, tables, and figures? (In finding the friction factor, did you enter the Moody diagram at the correct Reynolds number?)

SHOULD YOU TALK TO OTHER EXAMINEES AFTER THE EXAM?

The jury is out on this question. People react quite differently to the examination experience. Some people are energized. Most are exhausted. Some people need to unwind by talking with other examinees, describing every detail of their experience, and dissecting every examination question. Other people need lots of quiet space, and prefer to just get into a hot tub to soak and sulk. Most engineers, apparently, are in this latter category.

Since everyone who took the exam has seen it, you will not be violating your "oath of silence" if you talk about the details with other examinees immediately after the exam. It's difficult not to ask how someone else approached a question that had you completely stumped. However, keep in mind that it is very disquieting to think you answered a question correctly, only to have someone tell you where you went wrong.

To ensure you do not violate the nondisclosure agreement you signed before taking the exam, make sure you do not discuss any exam particulars with people who have not also taken the exam.

AFTER THE EXAM

Yes, there is something to do after the exam. Most people return home, throw their exam "kits" into the corner, and collapse. A week later, when they can bear to think about the experience again, they start integrating their exam kits back into their normal lives and all of the miscellaneous stuff you brought with you to the exam is put back wherever it came from.

Here's what I suggest you do as soon as you get home, before you collapse.

[ ] Thank your family for helping you during your preparation.

[ ] Take any paperwork you received on exam day out of your pocket, purse, or wallet. Put this inside your *Mechanical Engineering Reference Manual*.

[ ] Reflect on any statements regarding exam secrecy to which you signed your agreement in the exam.

[ ] Call your employer and tell him/her that you need to take a mental health day off on Monday.

A few days later, when you can face the world again, do the following.

[ ] Make notes about anything you would do differently if you had to take the exam over again.

[ ] Consolidate all of your application paperwork, correspondence to/from your state, and any paperwork that you received on exam day.

[ ] If you took a live prep course, call or email the instructor (or write a note) to say "Thanks."

[ ] Return any books you borrowed.

[ ] Write thank-you notes to all of the people who wrote letters of recommendation or reference for you.

[ ] Find and read the chapter in this book that covers ethics. There were no ethics questions on your PE exam, but it doesn't make any difference. Ethical behavior is expected of a PE in any case. Spend a few minutes reflecting on how your performance (obligations, attitude, presentation, behavior, appearance, etc.) might be about to change once you are licensed. Consider how you are going to be a role model for others around you.

[ ] Put all of your review books, binders, and notes someplace where they will be out of sight.

FINALLY

By the time you've "undone" all of your preparations, you might have thought of a few things that could help future examinees. If you have any sage comments about how to prepare, any suggestions about what to do in or bring to the exam, any comments on how to improve this book, or any funny anecdotes about your experience, I hope you will share these with me. By this time, you'll be the "expert," and I'll be your biggest fan.

AND THEN, THERE'S THE WAIT...

Waiting for the exam results is its own form of mental torture.

Yes, I know the exam grading should be almost instantaneous. But, you are going to wait, nevertheless. There are many reasons for the delay.

Although the actual machine grading "only takes seconds," consider the following facts: (a) NCEES prepares multiple exams for each administration, in case one becomes unusable (i.e., is inappropriately released) before the exam date. (b) Since the actual version of the exam used is not known until after it is finally given, the cut-score determination occurs after the exam date.

I wouldn't be surprised to hear that NCEES receives dozens, if not hundreds, of claims from well-meaning examinees who were 100% certain that the exams they took were seriously flawed to some degree—that there wasn't a correct answer for such-and-such question—that there were two answers for such-and-such question—or even, perhaps, that such-and-such question was missing from their exam booklet altogether. Each of these claims must be considered as a potential adjustment to the cut score.

After the individual exams are scored, the results are analyzed in a variety of ways. Some of the analysis looks at passing rates by such delineators as degree, major, university, site, and state. Part of the analysis looks for similarities between physically adjacent examinees (to look for cheating). Part of the analysis looks for exam sites that have statistically abnormal group performance. And, some of the analysis looks for exam questions that have a disproportionate fraction of successful or unsuccessful examinees. All of these steps have to be completed for 100% of the examinees before any results can go out.

NCEES releases the results electronically 7 to 10 days after the exam.

There is no pattern to the public release of results. None. The exam results are not released to all states simultaneously. (The states with the fewest examinees often receive their results soonest.) They are not released by discipline. They are not released alphabetically by state or examinee name. The people who failed are not notified first (or last). Your coworker might receive his or her notification today, and you might be waiting another three weeks for yours.

Some states post the names of the successful examinees, or unsuccessful examinees, or both on their official state websites before the results go out. Others update their websites after the results go out. Some states don't list much of anything on their websites.

AND WHEN YOU PASS...

- [ ] Celebrate.
- [ ] Notify the people who wrote letters of recommendation or reference for you.
- [ ] Ask your employer for a raise.
- [ ] Tell the folks at PPI (who have been rootin' for you all along) the good news.

Topic I: Background and Support

Chapter

1 Systems of Units

Content in blue refers to the *NCEES Handbook*.

NCEES EXAM SPECIFICATIONS AND RELATED CONTENT

HVAC AND REFRIGERATION EXAM

I.A.1. Basic Engineering Practice: Units and conversions

- 2. Common Units of Mass
- 3. Mass and Weight
- 6. The English Engineering System
- 8. Weight and Specific Weight
- 9. The English Gravitational System

MACHINE DESIGN AND MATERIALS EXAM

I.A.5. Basic Engineering Practice: Units and conversions

- 3. Mass and Weight
- 4. Acceleration of Gravity
- 5. Consistent Systems of Units
- 6. The English Engineering System
- 7. Other Formulas Affected by Inconsistency
- 8. Weight and Specific Weight
- 9. The English Gravitational System
- 10. The Absolute English System
- 11. Metric Systems of Units
- 12. SI Units (The mks System)
- 15. Dimensionless Groups

1. INTRODUCTION

The purpose of this chapter is to eliminate some of the confusion regarding the many units available for each engineering variable. In particular, an effort has been made to clarify the use of the so-called English systems, which for years have used the *pound* unit both for force and mass—a practice that has resulted in confusion even for those familiar with it.

2. COMMON UNITS OF MASS

The choice of a mass unit is the major factor in determining which system of units will be used in solving a problem. It is obvious that one will not easily end up with a force in pounds if the rest of the problem is stated in meters and kilograms. Actually, the choice of a mass unit determines more than whether a conversion factor will be necessary to convert from one system to another (e.g., between SI and English units). An inappropriate choice of a mass unit may actually require a conversion factor *within* the system of units.

The common units of mass are the gram, pound, kilogram, and slug.[1] There is nothing mysterious about these units. All represent different quantities of matter, as Fig. 1.1 illustrates. In particular, note that the pound and slug do not represent the same quantity of matter.[2] **[Measurement Relationships]**

Figure 1.1 *Common Units of Mass*

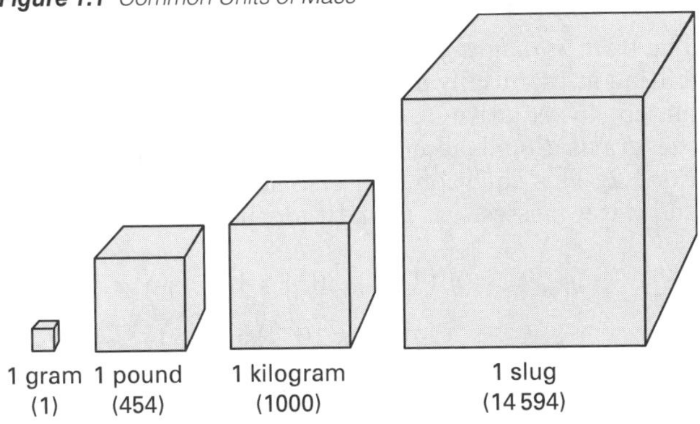

[1]Normally, one does not distinguish between a unit and a multiple of that unit, as is done here with the gram and the kilogram. However, these two units actually are bases for different consistent systems.
[2]A slug is approximately equal to 32.1740 pounds-mass.

3. MASS AND WEIGHT

In SI, *kilograms* are used for mass and *newtons* for weight (force). The units are different, and there is no confusion between the variables. However, for years the term *pound* has been used for both mass and weight. This usage has obscured the distinction between the two: mass is a constant property of an object; weight varies with the gravitational field. Even the conventional use of the abbreviations *lbm* and *lbf* (to distinguish between pounds-mass and pounds-force) has not helped eliminate the confusion.

It is true that an object with a mass of one pound will have an earthly weight of one pound, but this is true only on the earth. The weight of the same object will be much less on the moon. Therefore, care must be taken when working with mass and force in the same problem.

The relationship that converts mass to weight is familiar to every engineering student.

Concept of Weight

$$W = mg \qquad 1.1$$

Equation 1.1 illustrates that an object's weight will depend on the local acceleration of gravity as well as the object's mass. The mass will be constant, but gravity will depend on location. Mass and weight are not the same.

4. ACCELERATION OF GRAVITY

Gravitational acceleration on the earth's surface is usually taken as 32.2 ft/sec^2 or 9.81 m/s^2. These values are rounded from the more precise values of 32.1740 ft/sec^2 and 9.806 65 m/s^2. However, the need for greater accuracy must be evaluated on a problem-by-problem basis. Usually, three significant digits are adequate, since gravitational acceleration is not constant anyway but is affected by location (primarily latitude and altitude) and major geographical features.

The term *standard gravity*, g_0, is derived from the acceleration at essentially any point at sea level and approximately 45° N latitude. If additional accuracy is needed, the gravitational acceleration can be calculated from Eq. 1.2. This equation neglects the effects of large land and water masses. ϕ is the latitude in degrees.

$$\begin{aligned} g_{\text{surface}} &= g'\left(1 + (5.3024 \times 10^{-3})\sin^2\phi - (5.8 \times 10^{-6})\sin^2 2\phi\right) \\ g' &= 32.08769 \text{ ft/sec}^2 \\ &= 9.780327 \text{ m/s}^2 \end{aligned} \qquad 1.2$$

If the effects of the earth's rotation are neglected, the gravitational acceleration at an altitude h above the earth's surface is given by Eq. 1.3. r_e is the earth's radius.

$$\begin{aligned} g_h &= g_{\text{surface}}\left(\frac{r_e}{r_e + h}\right)^2 \\ r_e &= 3959 \text{ mi} \\ &= 6.3781 \times 10^6 \text{ m} \end{aligned} \qquad 1.3$$

5. CONSISTENT SYSTEMS OF UNITS

A set of units used in a calculation is said to be *consistent* if no conversion factors are needed.[3] For example, a moment is calculated as the product of a force and a lever arm length.

$$M = Fr \qquad 1.4$$

A calculation using Eq. 1.4 would be consistent if M was in newton-meters, F was in newtons, and r was in meters. The calculation would be inconsistent if M was in ft-kips, F was in kips, and r was in inches (because a conversion factor of 1/12 would be required).

The concept of a consistent calculation can be extended to a system of units. A *consistent system of units* is one in which no conversion factors are needed for any calculation. For example, Newton's second law of motion can be written without conversion factors. Newton's second law simply states that the force required to accelerate an object is proportional to the acceleration of the object. The constant of proportionality is the object's mass.

Newton's Second Law (Equations of Motion)

$$F = ma \qquad 1.5$$

Equation 1.5 is $F = ma$, not $F = Wa/g$ or $F = ma/g_c$. Equation 1.5 is consistent: it requires no conversion factors. This means that in a consistent system where conversion factors are not used, once the units of m and a have been selected, the units of F are fixed. This has the effect of establishing units of work and energy, power, fluid properties, and so on. The conversion between units of mass and units of weight is discussed in detail in Sec. 1.5.

The decision to work with a consistent set of units is desirable but not necessary. Problems in fluid flow and thermodynamics are routinely solved in the United States with inconsistent units. This causes no more of a problem than working with inches and feet when calculating a moment. It is necessary only to use the proper conversion factors.

[3]The terms *homogeneous* and *coherent* are also used to describe a consistent set of units.

6. THE ENGLISH ENGINEERING SYSTEM

Through common and widespread use, pounds-mass (lbm) and pounds-force (lbf) have become the standard units for mass and force in the *English Engineering System*. (The English Engineering System is used in this book along with the SI system.)

There are subjects in the United States in which the use of pounds for mass is firmly entrenched. For example, most thermodynamics, fluid flow, and heat transfer problems have traditionally been solved using the units of lbm/ft^3 for density, Btu/lbm for enthalpy, and Btu/lbm-°F for specific heat. Unfortunately, some equations contain both lbm-related and lbf-related variables, as does the steady flow conservation of energy equation, which combines enthalpy in Btu/lbm with pressure in lbf/ft^2.

The units of pounds-mass and pounds-force are as different as the units of gallons and feet, and they cannot be canceled. A mass conversion factor, g_c, is needed to make the equations containing lbf and lbm dimensionally consistent. This factor is known as the *gravitational constant* and has a value of 32.1740 lbm-ft/lbf-sec^2. The numerical value is the same as the standard acceleration of gravity, but g_c is not the local gravitational acceleration, g.[4] g_c is a conversion constant, just as 12.0 is the conversion factor between feet and inches.

The English Engineering System is an inconsistent system as defined according to Newton's second law. $F = ma$ cannot be written if lbf, lbm, and ft/sec^2 are the units used. The g_c term must be included.

Units

$$F_{\text{lbf}} = \frac{m_{\text{lbm}} a_{\frac{\text{ft}}{\text{sec}^2}}}{g_{c,\frac{\text{lbm-ft}}{\text{lbf-sec}^2}}} \qquad 1.6$$

In Eq. 1.6, g_c does more than "fix the units." Since g_c has a numerical value of 32.1740, it actually changes the calculation numerically. A force of 1.0 pound will not accelerate a 1.0-pound mass at the rate of 1.0 ft/sec^2.

In the English Engineering System, work and energy are typically measured in ft-lbf (mechanical systems) or in British thermal units (thermal and fluid systems). One Btu is equal to 778.17 ft-lbf.

Example 1.1

Calculate the weight in lbf of a 1.00 lbm object in a gravitational field of 27.5 ft/sec^2.

Solution

From Eq. 1.6,

Units

$$F = \frac{ma}{g_c} = \frac{(1.00\ \text{lbm})\left(27.5\ \frac{\text{ft}}{\text{sec}^2}\right)}{32.2\ \frac{\text{lbm-ft}}{\text{lbf-sec}^2}} = 0.854\ \text{lbf}$$

7. OTHER FORMULAS AFFECTED BY INCONSISTENCY

It is not a significant burden to include g_c in a calculation, but it may be difficult to remember when g_c should be used. Knowing when to include the gravitational constant can be learned through repeated exposure to the formulas in which it is needed, but it is safer to carry the units along in every calculation.

The following is a representative (but not exhaustive) listing of formulas that require the g_c term. In all cases, it is assumed that the standard English Engineering System units will be used.

Units

- kinetic energy

$$KE = \frac{mv^2}{2g_c} \quad \text{(in ft-lbf)} \qquad 1.7$$

- potential energy

$$PE = \frac{mgh}{g_c} \quad \text{(in ft-lbf)} \qquad 1.8$$

- shear stress

$$\tau = \left(\frac{\mu}{g_c}\right)\left(\frac{dv}{dy}\right) \quad \text{(in lbf/ft}^2\text{)} \qquad 1.9$$

- fluid pressure (pressure at a depth)

$$p = \frac{\rho g h}{g_c} \quad \text{(in lbf/ft}^2\text{)} \qquad 1.10$$

[4]It is acceptable (and recommended) that g_c be rounded to the same number of significant digits as g. Therefore, a value of 32.2 for g_c would typically be used.

Example 1.2

A rocket that has a mass of 4000 lbm travels at 27,000 ft/sec. What is its kinetic energy in ft-lbf?

Solution

From Eq. 1.7,

Units

$$KE = \frac{m\mathrm{v}^2}{2g_c} = (4000\ \text{lbm})\left(27{,}000\ \frac{\text{ft}}{\text{sec}}\right)^2 = 4.53 \times 10^{10}\ \text{ft-lbf}$$

8. WEIGHT AND SPECIFIC WEIGHT

Weight, W, is a force exerted on an object due to its placement in a gravitational field while specific weight is the force per unit volume. If a consistent set of units is used, Eq. 1.1 can be used to calculate the weight of a mass. In the English Engineering System, however, Eq. 1.11 must be used.

$$W = \frac{mg}{g_c} \quad 1.11$$

Both sides of Eq. 1.11 can be divided by the volume of an object to derive the specific weight, γ, of the object. Equation 1.12 illustrates that the specific weight (in units of lbf/ft^3) can also be calculated by multiplying the mass density (in units of lbm/ft^3) by g/g_c. Since g and g_c usually have the same numerical values, the only effect of Eq. 1.13 is to change the units of density.

$$\frac{W}{V} = \left(\frac{m}{V}\right)\left(\frac{g}{g_c}\right) \quad 1.12$$

Density, Specific Weight, and Specific Gravity

$$\gamma = \frac{W}{V} = \left(\frac{m}{V}\right)\left(\frac{g}{g_c}\right) = \frac{\rho g}{g_c} \quad 1.13$$

Weight does not occupy volume. Only mass has volume. The concept of specific weight has evolved to simplify certain calculations, particularly fluid calculations. For example, pressure at a depth is calculated from Eq. 1.14. (Compare this with Eq. 1.10.)

$$p = \gamma h \quad 1.14$$

9. THE ENGLISH GRAVITATIONAL SYSTEM

Not all English systems are inconsistent. Pounds can still be used as the unit of force as long as pounds are not used as the unit of mass. Such is the case with the consistent *English Gravitational System.*

If acceleration is given in ft/sec^2, the units of mass for a consistent system of units can be determined from Newton's second law. The combination of units in Eq. 1.15 is known as a *slug.* g_c is not needed at all since this system is consistent. It would be needed only to convert slugs to another mass unit.

$$\text{units of } m = \frac{\text{units of } F}{\text{units of } a} = \frac{\text{lbf}}{\frac{\text{ft}}{\text{sec}^2}} = \frac{\text{lbf-sec}^2}{\text{ft}} \quad 1.15$$

Slugs and pounds-mass are not the same, as Fig. 1.1 illustrates. However, both are units for the same quantity: mass. Equation 1.16 will convert between slugs and pounds-mass.

$$\text{no. of slugs} = \frac{\text{no. of lbm}}{g_c} \quad 1.16$$

The number of slugs is not derived by dividing the number of pounds-mass by the local gravity. g_c is used regardless of the local gravity. The conversion between feet and inches is not dependent on local gravity; neither is the conversion between slugs and pounds-mass.

Since the English Gravitational System is consistent, Eq. 1.17 can be used to calculate weight. The local gravitational acceleration is used.

$$W \text{ in lbf} = (m \text{ in slugs})\left(g \text{ in } \frac{\text{ft}}{\text{sec}^2}\right) \quad 1.17$$

10. THE ABSOLUTE ENGLISH SYSTEM

The obscure *Absolute English System* takes the approach that mass must have units of pounds-mass (lbm) and the units of force can be derived from Newton's second law. The units for F cannot be simplified any more than they are in Eq. 1.18. This particular combination of units is known as a *poundal.*[5] A poundal is not the same as a pound.

$$\begin{aligned} \text{units of } F &= (\text{units of } m)(\text{units of } a) \\ &= (\text{lbm})\left(\frac{\text{ft}}{\text{sec}^2}\right) \\ &= \frac{\text{lbm-ft}}{\text{sec}^2} \end{aligned} \quad 1.18$$

[5] A poundal is equal to 0.03108 pounds-force.

Poundals have not seen widespread use in the United States. The English Gravitational System (using slugs for mass) has greatly eclipsed the Absolute English System in popularity. Both are consistent systems, but there seems to be little need for poundals in modern engineering. Figure 1.2 shows the poundal in comparison to other common units of force.

Figure 1.2 *Common Force Units*

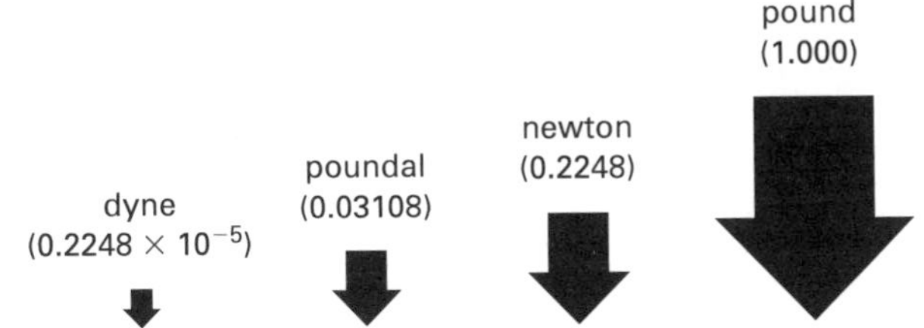

11. METRIC SYSTEMS OF UNITS

Strictly speaking, a *metric system* is any system of units that is based on meters or parts of meters. This broad definition includes *mks systems* (based on meters, kilograms, and seconds) as well as *cgs systems* (based on centimeters, grams, and seconds).

Metric systems avoid the pounds-mass versus pounds-force ambiguity in two ways. First, a unit of weight is not established at all. All quantities of matter are specified as mass. Second, force and mass units do not share a common name.

The term *metric system* is not explicit enough to define which units are to be used for any given variable. For example, within the cgs system there is variation in how certain electrical and magnetic quantities are represented (resulting in the ESU and EMU systems). Also, within the mks system, it is common practice in some industries to use kilocalories as the unit of thermal energy, while the SI unit for thermal energy is the joule. This shows a lack of uniformity even within the metricated engineering community.[6]

The "metric" parts of this book use SI, which is the most developed and codified of the so-called metric systems.[7] There will be occasional variances with local engineering custom, but it is difficult to anticipate such variances within a book that must itself be consistent.

12. SI UNITS (THE mks SYSTEM)

SI units comprise an *mks system* (so named because it uses the meter, kilogram, and second as base units). All other units are derived from the base units, which are completely listed in Table 1.1. This system is fully consistent, and there is only one recognized unit for each physical quantity (variable).

Table 1.1 *SI Base Units*

quantity	name	symbol
length	meter	m
mass	kilogram	kg
time	second	s
electric current	ampere	A
temperature	kelvin	K
amount of substance	mole	mol
luminous intensity	candela	cd

Two types of units are used: base units and derived units. The *base units* are dependent only on accepted standards or reproducible phenomena. The *derived units* (Table 1.2 and Table 1.3) are made up of combinations of base units. Prior to 1995, radians and steradians were classified as *supplementary units.*

Table 1.2 *Some SI Derived Units with Special Names*

quantity	name	symbol	expressed in terms of other units
frequency	hertz	Hz	1/s
force	newton	N	$kg·m/s^2$
pressure, stress	pascal	Pa	N/m^2
energy, work, quantity of heat	joule	J	N·m
power, radiant flux	watt	W	J/s
quantity of electricity, electric charge	coulomb	C	
electric potential, potential difference, electromotive force	volt	V	W/A
electric capacitance	farad	F	C/V
electric resistance	ohm	Ω	V/A
electric conductance	siemens	S	A/V
magnetic flux	weber	Wb	V·s
magnetic flux density	tesla	T	Wb/m^2
inductance	henry	H	Wb/A
luminous flux	lumen	lm	
illuminance	lux	lx	lm/m^2
plane angle	radian	rad	–
solid angle	steradian	sr	–

[6]In the "field test" of the metric system conducted over the past 200 years, other conventions are to use kilograms-force (kgf) instead of newtons and kgf/cm^2 for pressure (instead of pascals).
[7]SI units are an outgrowth of the *General Conference of Weights and Measures*, an international treaty organization that established the *Système International d'Unités* (*International System of Units*) in 1960. The United States subscribed to this treaty in 1975.

Table 1.3 *Some SI Derived Units*

quantity	description	symbol
area	square meter	m^2
volume	cubic meter	m^3
speed—linear	meter per second	m/s
speed—angular	radian per second	rad/s
acceleration—linear	meter per second squared	m/s^2
acceleration—angular	radian per second squared	rad/s^2
density, mass density	kilogram per cubic meter	kg/m^3
concentration (of amount of substance)	mole per cubic meter	mol/m^3
specific volume	cubic meter per kilogram	m^3/kg
luminance	candela per square meter	cd/m^2
absolute viscosity	pascal second	Pa·s
kinematic viscosity	square meters per second	m^2/s
moment of force	newton meter	N·m
surface tension	newton per meter	N/m
heat flux density, irradiance	watt per square meter	W/m^2
heat capacity, entropy	joule per kelvin	J/K
specific heat capacity, specific entropy	joule per kilogram kelvin	J/kg·K
specific energy	joule per kilogram	J/kg
thermal conductivity	watt per meter kelvin	W/m·K
energy density	joule per cubic meter	J/m^3
electric field strength	volt per meter	V/m
electric charge density	coulomb per cubic meter	C/m^3
surface density of charge, flux density	coulomb per square meter	C/m^2
permittivity	farad per meter	F/m
current density	ampere per square meter	A/m^2
magnetic field strength	ampere per meter	A/m
permeability	henry per meter	H/m
molar energy	joule per mole	J/mol
molar entropy, molar heat capacity	joule per mole kelvin	J/mol·K
radiant intensity	watt per steradian	W/sr

In addition, there is a set of non-SI units that may be used. This concession is primarily due to the significance and widespread acceptance of these units. Use of the non-SI units listed in Table 1.4 will usually create an inconsistent expression requiring conversion factors. [**Measurement Relationships**]

Table 1.4 *Acceptable Non-SI Units*

quantity	unit name	symbol or abbreviation	relationship to SI unit
area	hectare	ha	1 ha = 10 000 m^2
energy	kilowatt-hour	kW·h	1 kW·h = 3.6 MJ
mass	metric ton[a]	t	1 t = 1000 kg
plane angle	degree (of arc)	°	1° = 0.017 453 rad
speed of rotation	revolution per minute	r/min	1 r/min = $2\pi/60$ rad/s
temperature interval	degree Celsius	°C	1°C = 1K
time	minute	min	1 min = 60 s
	hour	h	1 h = 3600 s
	day (mean solar)	d	1 d = 86 400 s
	year (calendar)	a	1 a = 31 536 000 s
velocity	kilometer per hour	km/h	1 km/h = 0.278 m/s
volume	liter[b]	L	1 L = 0.001 m^3

[a]The international name for metric ton is *tonne*. The metric ton is equal to the *megagram* (Mg).
[b]The international symbol for liter is the lowercase l, which can be easily confused with the numeral 1. Several English-speaking countries have adopted the script ℓ or uppercase L (as does this book) as a symbol for liter in order to avoid any misinterpretation.

The SI unit of force can be derived from Newton's second law. This combination of units for force is known as a *newton*.

$$\text{units of force} = (m \text{ in kg})\left(a \text{ in } \frac{\text{m}}{\text{s}^2}\right) = \text{kg·m/s}^2 \qquad 1.19$$

Energy variables in SI units have units of N·m or, equivalently, $kg·m^2/s^2$. Both of these combinations are known as a *joule*. The units of power are joules per second, equivalent to a *watt*.

The various prefixes in SI units come with conversion factors in terms of powers of 10. [**Units**]

Example 1.3

A 10 kg block hangs from a cable. What is the tension in the cable? (Standard gravity equals 9.81 m/s^2.)

Solution

Newton's Second Law (Equations of Motion)

$$\begin{aligned}\sum F &= ma \\ &= mg \\ &= (10\,\text{kg})\left(9.81\ \frac{\text{m}}{\text{s}^2}\right) \\ &= 98.1\,\text{kg}\cdot\text{m/s}^2 \quad (98.1\,\text{N})\end{aligned}$$

Example 1.4

A 10 kg block is raised vertically 3 m. What is the change in potential energy?

Solution

Potential Energy

$$\begin{aligned}\text{PE} &= mgh \\ \Delta\text{PE} &= mg\Delta h \\ &= (10\ \text{kg})\left(9.81\ \frac{\text{m}}{\text{s}^2}\right)(3\ \text{m}) \\ &= 294\ \text{kg}\cdot\text{m}^2/\text{s}^2 \quad (294\ \text{J})\end{aligned}$$

13. RULES FOR USING SI UNITS

In addition to having standardized units, the set of SI units also has rigid syntax rules for writing the units and combinations of units. Each unit is abbreviated with a specific symbol. The following rules for writing and combining these symbols should be adhered to.

- The expressions for derived units in symbolic form are obtained by using the mathematical signs of multiplication and division. For example, units of velocity are m/s. Units of torque are N·m (not N-m or Nm).
- Scaling of most units is done in multiples of 1000.
- The symbols are always printed in roman type, regardless of the type used in the rest of the text. The only exception to this is in the use of the symbol for liter, where the use of the lower case "el" (l) may be confused with the numeral one (1). In this case, "liter" should be written out in full, or the script ℓ or L used. (L is used in this book.)
- Symbols are not pluralized: 45 kg (not 45 kgs).
- A period after a symbol is not used, except when the symbol occurs at the end of a sentence.
- When symbols consist of letters, there must always be a full space between the quantity and the symbols: 45 kg (not 45kg). However, for planar angle designations, no space is left: 42°12′45″ (not 42 ° 12 ′ 45 ″).
- All symbols are written in lowercase, except when the unit is derived from a proper name: m for meter, s for second, A for ampere, Wb for weber, N for newton, W for watt.
- Prefixes are printed without spacing between the prefix and the unit symbol (e.g., km is the symbol for kilometer).
- In text, when no number is involved, the unit should be spelled out. Example: Carpet is sold by the square meter, not by the m^2.
- Where a decimal fraction of a unit is used, a zero should always be placed before the decimal marker: 0.45 kg (not .45 kg). This practice draws attention to the decimal marker and helps avoid errors of scale.
- A practice in some countries is to use a comma as a decimal marker, while the practice in North America, the United Kingdom, and some other countries is to use a period as the decimal marker. Furthermore, in some countries that use the decimal comma, a period is frequently used to divide long numbers into groups of three. Because of these differing practices, spaces must be used instead of commas to separate long lines of digits into easily readable blocks of three digits with respect to the decimal marker: 32 453.246 072 5. A space (half-space preferred) is optional with a four-digit number: 1 234 or 1234.
- The word *ton* has multiple meanings. In the United States and Canada, the *short ton* of 2000 lbm (907.18 kg; 8896.44 N) is used. Previously, for commerce within the United Kingdom, a *long ton* of 2240 lbm (1016.05 kg) was used. A *metric ton* (or, *tonne*) is 1000 kg (10^6 Mg; 2205 lbm). In air conditioning industries, a *ton of refrigeration* is equivalent to a cooling rate of 200 Btu/min (12,000 Btu/hr). For explosives, a *ton of explosive power* is approximately the energy given off by 2000 lbm of TNT, standardized by international convention as 10^9 cal (1 Gcal; 4.184 GJ) for both "ton" and "tonne" designations. Various definitions are used in shipping, where *freight ton* (*measurement ton* or MTON) refers to volume, usually 40 ft^3. Many other specialty definitions are also in use.

14. PRIMARY DIMENSIONS

Regardless of the system of units chosen, each variable representing a physical quantity will have the same *primary dimensions.* For example, velocity may be expressed in miles per hour (mph) or meters per second (m/s), but both units have dimensions of length per unit time. Length and time are two of the primary dimensions, as neither can be broken down into more basic dimensions. The concept of primary dimensions is useful when converting little-used variables between different systems of units, as well as in correlating experimental results (i.e., dimensional analysis).

There are three different sets of primary dimensions in use.[8] In the $ML\theta T$ *system*, the primary dimensions are mass (M), length (L), time (θ), and temperature (T). All symbols are uppercase. In order to avoid confusion between time and temperature, the Greek letter theta is used for time.[9]

All other physical quantities can be derived from these primary dimensions.[10] For example, work in SI units has units of N·m. Since a newton is a kg·m/s^2, the primary dimensions of work are ML^2/θ^2. The primary dimensions for many important engineering variables are shown in Table 1.5. If it is more convenient to stay with traditional English units, it may be more desirable to work in the $FML\theta TQ$ system (sometimes called the *engineering dimensional system*). This system adds the primary dimensions of force (F) and heat (Q). Work (ft-lbf in the English system) has the primary dimensions of FL. (Compare this with the primary dimensions for work in the $ML\theta T$ system.) Thermodynamic variables are similarly simplified.

Dimensional analysis will be more conveniently carried out when one of the four-dimension systems ($ML\theta T$ or $FL\theta T$) is used. Whether the $ML\theta T$, $FL\theta T$, or $FML\theta TQ$ system is used depends on what is being derived and who will be using it, and whether or not a consistent set of variables is desired. Conversion constants such as g_c and J will almost certainly be required if the $ML\theta T$ system is used to generate variables for use in the English systems. It is also much more convenient to use the $FML\theta TQ$ system when working in the fields of thermodynamics, fluid flow, heat transfer, and so on.

15. DIMENSIONLESS GROUPS

A *dimensionless group* is derived as a ratio of two forces or other quantities. Considerable use of dimensionless groups is made in certain subjects, notably fluid mechanics and heat transfer. For example, the Reynolds number, Mach number, and Froude number are used to distinguish between distinctly different flow regimes in pipe flow, compressible flow, and open channel flow, respectively.

Table 1.6 contains information about the most common dimensionless groups used in fluid mechanics and heat transfer.

Table 1.5 *Dimensions of Common Variables*

variable	dimensional system		
(common symbol)	$ML\theta T$	$FL\theta T$	$FML T\theta Q$
mass (m)	M	$F\theta^2/L$	M
force (F)	ML/θ^2	F	F
length (L)	L	L	L
time (θ or t)	θ	θ	θ
temperature (T)	T	T	T
work (W)	ML^2/θ^2	FL	FL
heat (Q)	ML^2/θ^2	FL	Q
acceleration (a)	L/θ^2	L/θ^2	L/θ^2
frequency (n or f)	$1/\theta$	$1/\theta$	$1/\theta$
area (A)	L^2	L^2	L^2
coefficient of thermal expansion (β)	$1/T$	$1/T$	$1/T$
density (ρ)	M/L^3	$F\theta^2/L^4$	M/L^3
dimensional constant (g_c)	1.0	1.0	ML/θ^2F
specific heat at constant pressure (c_p); at constant volume (c_v)	L^2/θ^2T	L^2/θ^2T	Q/MT
heat transfer coefficient (h); overall (U)	M/θ^3T	$F/\theta LT$	$Q/\theta L^2T$
power (P)	ML^2/θ^3	FL/θ	FL/θ
heat flow rate ($\dot{Q}$)	ML^2/θ^3	FL/θ	Q/θ
kinematic viscosity (ν)	L^2/θ	L^2/θ	L^2/θ
mass flow rate ($\dot{m}$)	M/θ	$F\theta/L$	M/θ
mechanical equivalent of heat (J)	–	–	FL/Q
pressure (p)	$M/L\theta^2$	F/L^2	F/L^2
surface tension (σ)	M/θ^2	F/L	F/L
angular velocity (ω)	$1/\theta$	$1/\theta$	$1/\theta$
volumetric flow rate ($\dot{m}/\rho = \dot{V}$)	L^3/θ	L^3/θ	L^3/θ
conductivity (k)	ML/θ^3T	$F/\theta T$	$Q/L\theta T$
thermal diffusivity (α)	L^2/θ	L^2/θ	L^2/θ
velocity (v)	L/θ	L/θ	L/θ
viscosity, absolute (μ)	$M/L\theta$	$F\theta/L^2$	$F\theta/L^2$
volume (V)	L^3	L^3	L^3

[8] One of these, the $FL\theta T$ system, is not discussed here.

[9] This is the most common usage. There is a lack of consistency in the engineering world about the symbols for the primary dimensions in dimensional analysis. Some writers use t for time instead of θ. Some use H for heat instead of Q. And, in the worst mix-up of all, some have reversed the use of T and θ.

[10] A *primary dimension* is the same as a *base unit* in the SI set of units. The SI units add several other base units, as shown in Table 1.1, to deal with variables that are difficult to derive in terms of the four primary base units.

Table 1.6 Common Dimensionless Groups

name	symbol	formula	interpretation
Biot number	Bi	Transient Conduction Using the Lumped Capacitance Model $\frac{hV}{kA_s}$	$\frac{\text{surface conductance}}{\text{internal conduction of solid}}$
Euler number	Eu	$\frac{\Delta p}{\rho \nu^2}$	$\frac{\text{pressure force}}{\text{inertia force}}$
Fourier number	Fo	$\frac{kt}{\rho c_p L^2} = \frac{\alpha t}{L^2}$	$\frac{\text{rate of conduction of heat}}{\text{rate of storage of energy}}$
Grashof number*	Gr	$\frac{g\beta \Delta T L^3}{\nu^2}$	Grashof Number $\frac{\text{buoyancy force}}{\text{viscous force}}$
Mach number	Ma	Mach Number $\frac{V}{c}$	$\frac{\text{fluid velocity}}{\text{speed of sound}}$
Nusselt number	Nu	$\frac{hL}{k}$	$\frac{\text{temperature gradient at wall}}{\text{overall temperature difference}}$
Prandtl number	Pr	Convection: Terms $\frac{c_p \mu}{k} = \frac{\nu}{\alpha}$	$\frac{\text{diffusion of momentum}}{\text{diffusion of heat}}$
Reynolds number	Re	Reynolds Number $\frac{\mathrm{v} D \rho}{\mu} = \frac{\mathrm{v} D}{\nu}$	$\frac{\text{inertia force}}{\text{viscous force}}$

*Multiple definitions exist, most often the square or square root of the formula shown.

16. DIMENSIONAL ANALYSIS

Dimensional analysis is a means of obtaining an equation that describes some phenomenon without understanding the mechanism of the phenomenon. The most serious limitation is the need to know beforehand which variables influence the phenomenon. Once these are known or assumed, dimensional analysis can be applied by a routine procedure.

The first step is to select a system of primary dimensions. (See Sec. 1.14.) Usually the $ML\theta T$ system is used, although this choice may require the use of g_c and J in the final results.

The second step is to write a functional relationship between the dependent variable and the independent variables, x_i.

$$y = f(x_1, x_2, \ldots, x_m) \qquad 1.20$$

This function can be expressed as an exponentiated series. The C_1, a_i, b_i, ..., z_i in Eq. 1.21 are unknown constants.

$$y = C_1 x_1^{a_1} x_2^{b_1} x_3^{c_1} \cdots x_m^{z_1} + C_2 x_1^{a_2} x_2^{b_2} x_3^{c_2} \cdots x_m^{z_2} + \cdots \qquad 1.21$$

The key to solving Eq. 1.21 is that each term on the right-hand side must have the same dimensions as y. Simultaneous equations are used to determine some of the a_i, b_i, c_i, and z_i. Experimental data are required to determine the C_i and remaining exponents. In most analyses, it is assumed that the $C_i = 0$ for $i \geq 2$.

Since this method requires working with m different variables and n different independent dimensional quantities (such as M, L, θ, and T), an easier method is desirable. One simplification is to combine the m variables into dimensionless groups called *pi-groups*. (See Table 1.5.)

If these dimensionless groups are represented by π_1, π_2, π_3, ..., π_k, the equation expressing the relationship between the variables is given by the *Buckingham π-theorem*.

$$f(\pi_1, \pi_2, \pi_3, \ldots, \pi_k) = 0 \qquad 1.22$$

$$k = m - n \qquad 1.23$$

The dimensionless pi-groups are usually found from the m variables according to an intuitive process. Common dimensionless groups are given in Table 1.6.

Example 1.5

A solid sphere rolls down a submerged incline. Find an equation for the velocity, v.

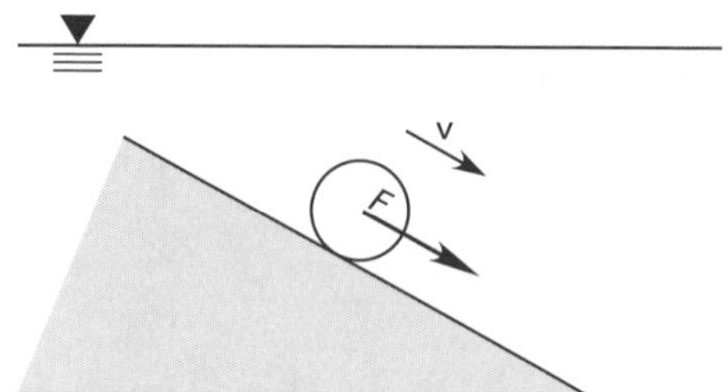

Solution

Assume that the velocity depends on the force, F, due to gravity, the diameter of the sphere, D, the density of the fluid, ρ, and the viscosity of the fluid, μ.

$$\mathrm{v} = f(F, D, \rho, \mu) = CF^aD^b\rho^c\mu^d$$

This equation can be written in terms of the primary dimensions of the variables.

$$\frac{L}{\theta} = C\left(\frac{ML}{\theta^2}\right)^a L^b \left(\frac{M}{L^3}\right)^c \left(\frac{M}{L\theta}\right)^d$$

Since L on the left-hand side has an implied exponent of one, a necessary condition is

$$1 = a + b - 3c - d \quad (L)$$

Similarly, the other necessary conditions are

$$-1 = -2a - d \quad (\theta)$$
$$0 = a + c + d \quad (M)$$

Solving simultaneously yields

$$b = -1$$
$$c = a - 1$$
$$d = 1 - 2a$$
$$\mathrm{v} = CF^aD^{-1}\rho^{a-1}\mu^{1-2a}$$
$$= C\left(\frac{\mu}{D\rho}\right)\left(\frac{F\rho}{\mu^2}\right)^a$$

C and a must be determined experimentally.

2 Engineering Drawing Practice

Content in blue refers to the *NCEES Handbook*.

NCEES EXAM SPECIFICATIONS AND RELATED CONTENT

MACHINE DESIGN AND MATERIALS EXAM

I.A.2. Basic Engineering Practice: Interpretation of technical drawings

4. Principal (Orthographic) Views
5. Auxiliary (Orthographic) Views
11. Surface Finish

II.C.2. Supportive Knowledge: Fits and tolerances

10. Tolerances

THERMAL AND FLUID SYSTEMS EXAM

I.A.1. Engineering terms, symbols, and technical drawings

1. Normal Views of Lines and Planes
3. Types of Views
4. Principal (Orthographic) Views
5. Auxiliary (Orthographic) Views
9. Sections

1. NORMAL VIEWS OF LINES AND PLANES

A *normal view* of a line is a perpendicular projection of the line onto a viewing plane parallel to the line. In the normal view, all points of the line are equidistant from the observer. Therefore, the true length of a line is viewed and can be measured.

Generally, however, a line will be viewed from an oblique position and will appear shorter than it actually is. The normal view can be constructed by drawing an auxiliary view from the orthographic view.[1] (See Sec. 2.5.)

Similarly, a normal view of a plane figure is a perpendicular projection of the figure onto a viewing plane parallel to the plane of the figure. All points of the plane are equidistant from the observer. Therefore, the true size and shape of any figure in the plane can be determined.

2. INTERSECTING AND PERPENDICULAR LINES

Figure 2.1 *Intersecting and Non-Intersecting Lines*

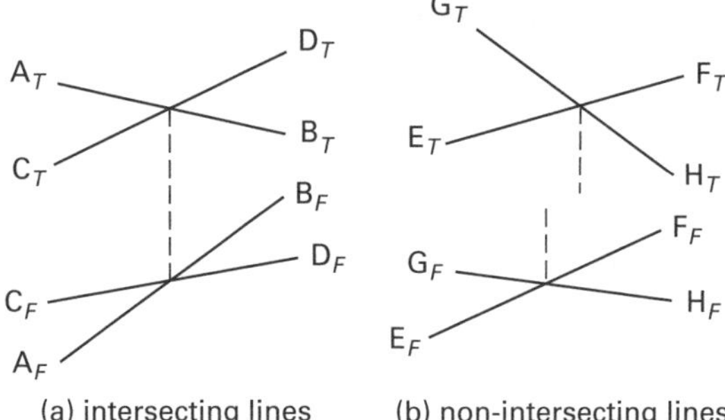

A single orthographic view is not sufficient to determine whether two lines intersect. However, if two or more views show the lines as having the same common point (i.e., crossing at the same position in space), then the lines intersect. In Fig. 2.1, the subscripts F and T refer to front and top views, respectively.

According to the *perpendicular line principle*, two perpendicular lines appear perpendicular only in a normal view of either one or both of the lines. Conversely, if two lines appear perpendicular in any view, the lines are perpendicular only if the view is a normal view of one or both of the lines.

[1]The technique for constructing a normal view is covered in engineering drafting texts.

3. TYPES OF VIEWS

Objects can be illustrated in several different ways depending on the number of views, the angle of observation, and the degree of artistic latitude taken for the purpose of simplifying the drawing process.[2] Table 2.1 categorizes the types of views.

***Table 2.1** Types of Views of Objects*

- orthographic views
  - principal views
  - auxiliary views
  - oblique views
    - cavalier projection
    - cabinet projection
    - clinographic projection
    - axonometric views
      - isometric
      - dimetric
      - trimetric
- perspective views
  - parallel perspective
  - angular perspective

The different types of views are easily distinguished by their *projectors* (i.e., projections of parallel lines on the object). For a cube, there are three sets of projectors corresponding to the three perpendicular axes. In an *orthographic* (*orthogonal*) *view*, the projectors are parallel. In a *perspective* (*central*) *view*, some or all of the projectors converge to a point. Figure 2.2 illustrates the orthographic and perspective views of a block.

***Figure 2.2** Orthographic and Perspective Views of a Block*

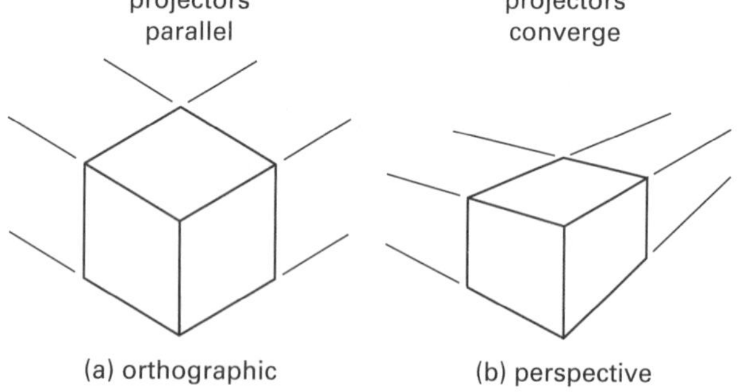

4. PRINCIPAL (ORTHOGRAPHIC) VIEWS

In a *principal view* (also known as a *planar view*), one of the sets of projectors is normal to the view. That is, one of the planes of the object is seen in a normal view. The other two sets of projectors are orthogonal and are usually oriented horizontally and vertically on the paper. Because background details of an object may not be visible in a principal view, it is necessary to have at least three principal views to completely illustrate a symmetrical object. At most, six principal views will be needed to illustrate complex objects. [**ANSI - Orthographic Projection Following Third-Angle Projection**][**ISO - Orthographic Projection Following First-Angle Projection**][**American National Standard for Engineering Drawings**]

The relative positions of the six views have been standardized and are shown in Fig. 2.3, which also defines the *width* (also known as *depth*), *height*, and *length* of the object. The views that are not needed to illustrate features or provide dimensions (i.e., *redundant views*) can be omitted. The usual combination selected consists of the top, front, and right side views.

***Figure 2.3** Positions of Standard Orthographic Views*

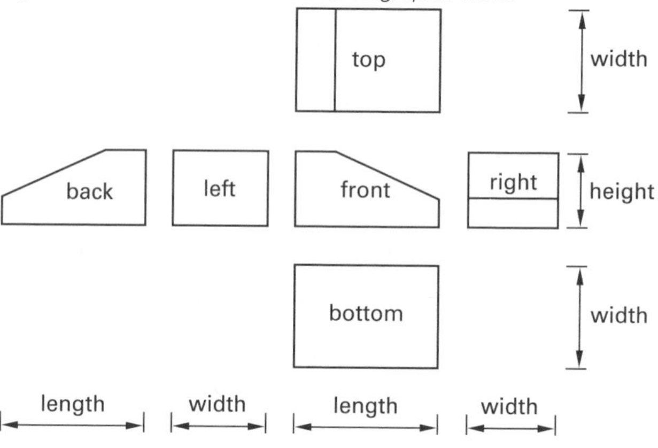

It is common to refer to the front, side, and back views as *elevations* and to the top and bottom views as *plan views*. These terms are not absolute since any plane can be selected as the front.

5. AUXILIARY (ORTHOGRAPHIC) VIEWS

An *auxiliary view* is needed when an object has an inclined plane or curved feature or when there are more details than can be shown in the six principal views. As with the other orthographic views, the auxiliary view is a normal (face-on) view of the inclined plane. Figure 2.4 illustrates an auxiliary view. [**ANSI - Orthographic Projection Following Third-Angle Projection**][**ISO - Orthographic Projection Following First-Angle Projection**][**American National Standard for Engineering Drawings**]

The projectors in an auxiliary view are perpendicular to only one of the directions in which a principal view is observed. Accordingly, only one of the three dimensions of width, height, and depth can be measured (scaled). In a *profile auxiliary view*, the object's width can be

[2]The omission of perspective from a drawing is an example of a step taken to simplify the drawing process.

measured. In a *horizontal auxiliary view* (*auxiliary elevation*), the object's height can be measured. In a *frontal auxiliary view*, the depth of the object can be measured.

Figure 2.4 *Auxiliary View*

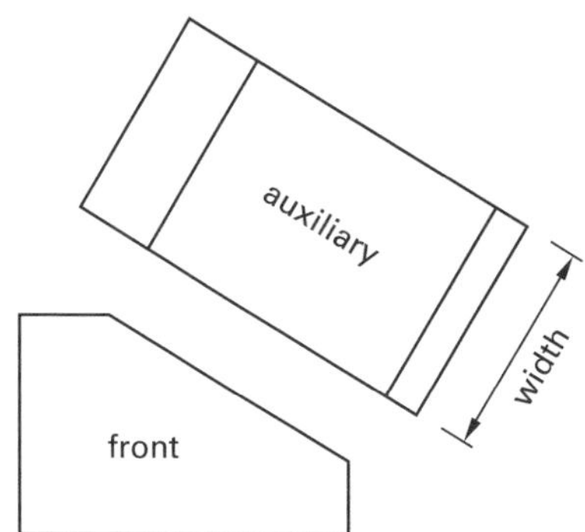

6. OBLIQUE (ORTHOGRAPHIC) VIEWS

If the object is turned so that three principal planes are visible, it can be completely illustrated by a single *oblique view*.[3] In an oblique view, the direction from which the object is observed is not (necessarily) parallel to any of the directions from which principal and auxiliary views are observed.

In two common methods of oblique illustration, one of the view planes coincides with an orthographic view plane. Two of the drawing axes are at right angles to each other; one of these is vertical, and the other (the *oblique axis*) is oriented at 30° or 45° (originally chosen to coincide with standard drawing triangles). The ratio of scales used for the horizontal, vertical, and oblique axes can be 1:1:1 or 1:1:½. The latter ratio helps to overcome the visual distortion due to the absence of perspective in the oblique direction.

Cavalier (45° oblique axis and 1:1:1 scale ratio) and *cabinet* (45° oblique axis and 1:1:½ scale ratio) *projections* are the two common types of oblique views that incorporate one of the orthographic views. If an angle of 9.5° is used (as in illustrating crystalline lattice structures), the technique is known as *clinographic projection*. Figure 2.5 illustrates cavalier and cabinet oblique drawings of a perfect cube.

Figure 2.5 *Cavalier and Cabinet Oblique Drawings*

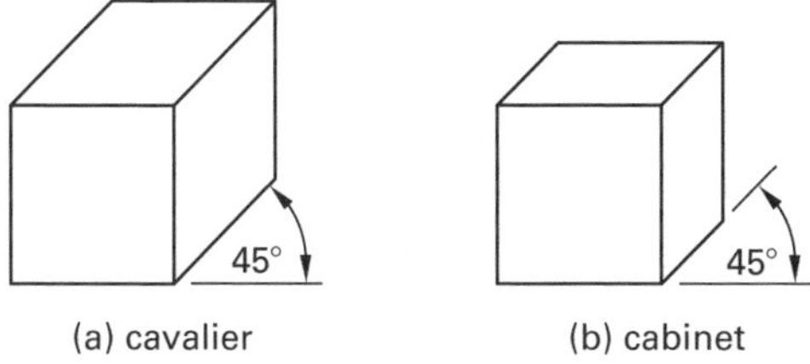

7. AXONOMETRIC (ORTHOGRAPHIC OBLIQUE) VIEWS

In axonometric views, the view plane is not parallel to any of the principal orthographic planes. Figure 2.6 illustrates types of axonometric views for a perfect cube. Axonometric views and axonometric drawings are not the same. In a *view* (*projection*), one or more of the face lengths is foreshortened. In a *drawing*, the lengths are drawn full length, resulting in a distorted illustration. Table 2.2 lists the proper ratios that should be observed.

In an *isometric view*, the three projectors intersect at equal angles (120°) with the plane. This simplifies construction with standard 30° drawing triangles. All of the faces are foreshortened by the same amount, to $\sqrt{2/3}$, or approximately 81.6%, of the true length. In a *dimetric view*, two of the projectors intersect at equal angles, and only two of the faces are equally reduced in length. In a *trimetric view*, all three intersection angles are different, and all three faces are reduced by different amounts.

Table 2.2 *Axonometric Foreshortening*

view	projector intersection angles	proper ratio of sides
isometric	120°, 120°, 120°	0.82:0.82:0.82
dimetric	131°25′, 131°25′, 97°10′	1:1:½
	103°38′, 103°38′, 152°44′	¾:¾:1
trimetric	102°28′, 144°16′, 113°16′	1:⅔:⅞
	138°14′, 114°46′, 107°	1:¾:⅞

8. PERSPECTIVE VIEWS

In a *perspective view*, one or more sets of projectors converge to a fixed point known as the *center of vision*. In the *parallel perspective*, all vertical lines remain vertical in the picture; all horizontal frontal lines remain horizontal. Therefore, one face is parallel to the observer and only one set of projectors converges. In the *angular perspective*, two sets of projectors converge. In the little-used *oblique perspective*, all three sets of projectors converge. Figure 2.7 illustrates types of perspective views.

[3]Oblique views are not unique in this capability—perspective drawings share it. Oblique and perspective drawings are known as *pictorial drawings* because they give depth to the object by illustrating it in three dimensions.

Figure 2.6 *Types of Axonometric Views*

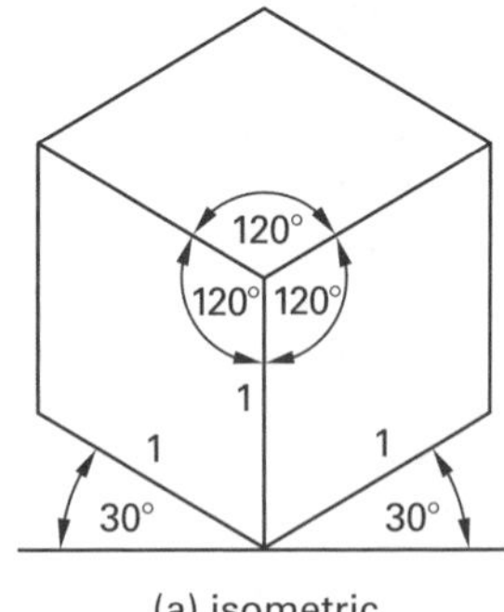

(a) isometric

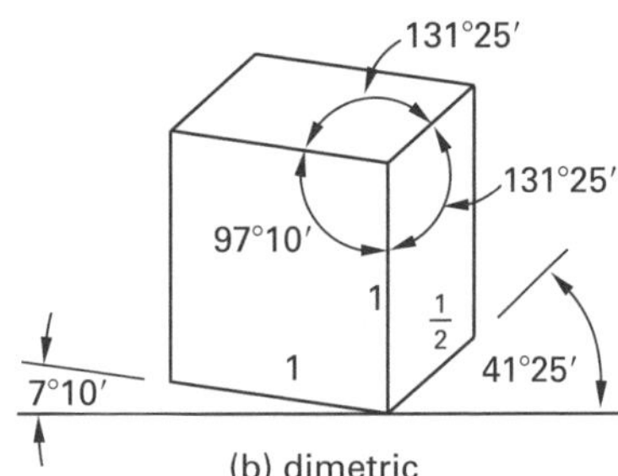

(b) dimetric

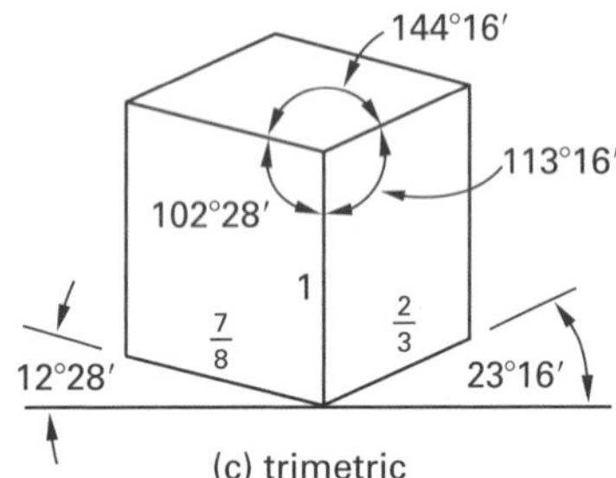

(c) trimetric

Figure 2.7 *Types of Perspective Views*

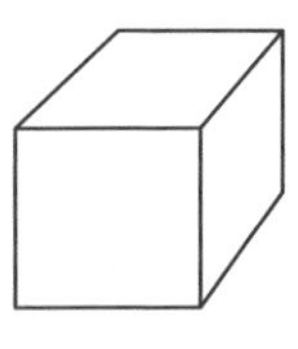

(a) parallel

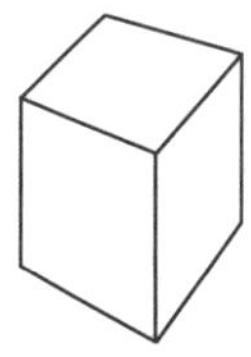

(b) angular

9. SECTIONS

A *section* is an imaginary cut taken through an object to reveal the shape or interior construction.[4] Figure 2.8 illustrates the standard symbol for a *sectioning cut* and the resulting sectional view. Section arrows are perpendicular to the cutting plane and indicate the viewing direction.

Figure 2.8 *Sectioning Cut Symbol and Sectional View*

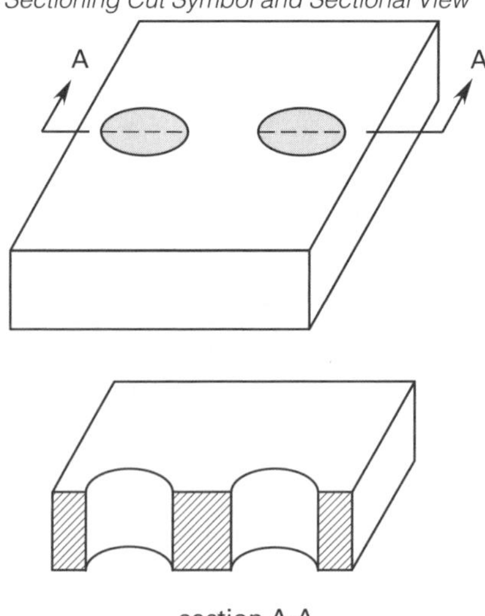

section A-A

10. TOLERANCES

The *tolerance* for a dimension is the total permissible variation or difference between the acceptable limits. The tolerance for a dimension can be specified in two ways: either as a general rule in the title block (e.g., ±0.001 in unless otherwise specified) or as specific limits that are given with each dimension (e.g., 2.575 in ± 0.005 in). [**ANSI B4.1 Fit Designations**][**ISO Symbols and Descriptions**]

11. SURFACE FINISH

ANSI B46.1 specifies surface finish by a combination of parameters.[5] The basic symbol for designating these factors is shown in Fig. 2.9. In the symbol, A is the maximum *roughness height index*, B is the optional minimum *roughness height*, C is the peak-to-valley *waviness height*, D is the optional peak-to-valley *waviness spacing* (*width*) rating, E is the optional *roughness width cutoff* (*roughness sampling length*), F is the *lay*, and G is the *roughness width*. Unless minimums are specified, all parameters are maximum allowable values, and all lesser values are permitted. [**Surface Texture Symbols and Construction**]

[4]The term *section* is also used to mean a *cross section*—a slice of finite but negligible thickness that is taken from an object to show the cross section or interior construction at the plane of the slice.

[5]Specification does not indicate appearance (i.e., color or luster) or performance (i.e., hardness, corrosion resistance, or microstructure).

Figure 2.9 *Surface Finish Designations*

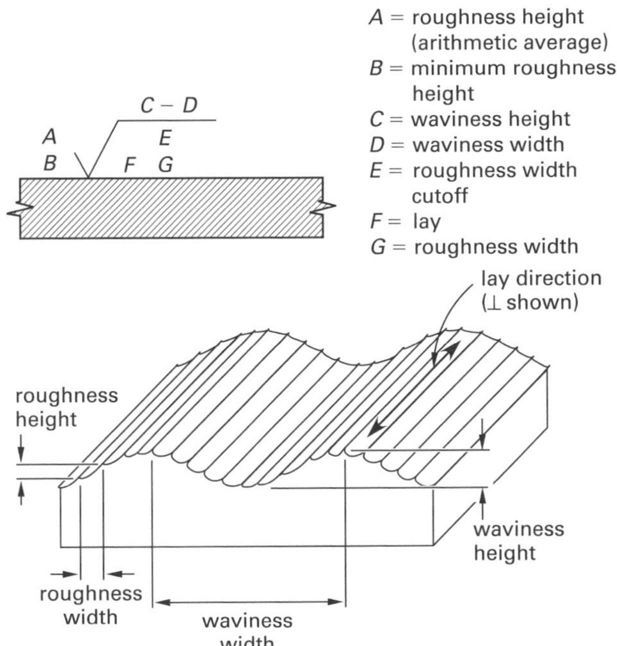

Since the roughness varies, the waviness height is an arithmetic average within a sampled square, and the designation A is known as the *roughness weight*, R_a.[6] Values are normally given in micrometers (μm) or microinches (μin) in SI or customary U.S. units, respectively. A value for the roughness width cutoff of 0.80 mm (0.03 in) is assumed when E is not specified. Other standard values in common use are 0.25 mm (0.010 in) and 0.08 mm (0.003 in). The lay symbol, F, can be = (parallel to indicated surface), ⊥ (perpendicular), C (circular), M (multidirectional), P (pitted), R (radial), or X (crosshatch).

If a small circle is placed at the A position, no machining is allowed and only cast, forged, die-cast, injection-molded, and other unfinished surfaces are acceptable.

[6]The symbol R_a is the same as the AA (arithmetic average) and CLA (centerline average) terms used in other (and earlier) standards.

3 Algebra

Content in blue refers to the *NCEES Handbook*.

1. INTRODUCTION

Engineers working in design and analysis encounter mathematical problems on a daily basis. Although algebra and simple trigonometry are often sufficient for routine calculations, there are many instances when certain advanced subjects are needed. This chapter and the following, in addition to supporting the calculations used in other chapters, consolidate the mathematical concepts most often needed by engineers.

2. SYMBOLS USED IN THIS BOOK

Many symbols, letters, and Greek characters are used to represent variables in the formulas used throughout this book. These symbols and characters are defined in the nomenclature section of each chapter. However, some of the other symbols in this book are listed in Table 3.2.

3. GREEK ALPHABET

Table 3.1 lists the Greek alphabet.

Table 3.1 *The Greek Alphabet*

uppercase	lowercase	Greek name	uppercase	lowercase	Greek name
A	α	alpha	N	ν	nu
B	β	beta	Ξ	ξ	xi
Γ	γ	gamma	O	o	omicron
Δ	δ	delta	Π	π	pi
E	ϵ	epsilon	P	ρ	rho
Z	ζ	zeta	Σ	σ	sigma
H	η	eta	T	τ	tau
Θ	θ	theta	Υ	υ	upsilon
I	ι	iota	Φ	ϕ	phi
K	κ	kappa	X	χ	chi
Λ	λ	lambda	Ψ	ψ	psi
M	μ	mu	Ω	ω	omega

4. TYPES OF NUMBERS

The *numbering system* consists of three types of numbers: real, imaginary, and complex. *Real numbers*, in turn, consist of rational numbers and irrational numbers. *Rational real numbers* are numbers that can be written as the ratio of two integers (e.g., 4, $^2/_5$, and $^1/_3$).[1] *Irrational real numbers* are nonterminating, nonrepeating numbers that cannot be expressed as the ratio of two integers (e.g., π and $\sqrt{2}$). Real numbers can be positive or negative.

Imaginary numbers are square roots of negative numbers. The symbols i and j are both used to represent the square root of -1.[2] For example, $\sqrt{-5} = \sqrt{5}\sqrt{-1} = \sqrt{5}i$. *Complex numbers* consist of combinations of real and imaginary numbers (e.g., $3 - 7i$).

[1] Notice that 0.3333333... is a nonterminating number, but as it can be expressed as a ratio of two integers (i.e., 1/3), it is a rational number.
[2] The symbol j is used to represent the square root of -1 in electrical calculations to avoid confusion with the current variable, i.

Table 3.2 *Symbols Used in This Book*

symbol	name	use	example
$\sum$	sigma	series summation	$\sum_{i=1}^{3} x_i = x_1 + x_2 + x_3$
π	pi	3.1415926...	$P = \pi D$
e	base of natural logs	2.71828...	
Π	pi	series multiplication	$\prod_{i=1}^{3} x_i = x_1 x_2 x_3$
Δ	delta	change in quantity	$\Delta h = h_2 - h_1$
$\bar{}$	over bar	average value	$\overline{x}$
$\dot{}$	over dot	per unit time	$\dot{m}$ = mass flowing per second
!	factorial[a]		$x! = x(x-1)(x-2)\cdots(2)(1)$
$\lvert\ \rvert$	absolute value[b]		$\lvert -3 \rvert = +3$
$\sim$	similarity		$\Delta ABC \sim \Delta DEF$
$\approx$	approximately equal to		$x \approx 1.5$
$\cong$	congruency		$ST \cong UV$
$\propto$	proportional to		$x \propto y$
$\equiv$	equivalent to		$a + bi \equiv re^{i\theta}$
∞	infinity		$x \to \infty$
log	base-10 logarithm		log 5.74
ln	natural logarithm		ln 5.74
exp	exponential power		$\exp(x) \equiv e^x$
rms	root-mean-square		$V_{\text{rms}} = \sqrt{\frac{1}{n}\sum_{i=1}^{n} V_i^2}$
$\angle$	phasor or angle		$\angle 53°$
$\in$	element of	set/membership	$Y = \sum_{j \in \{0,1,2,\ldots,n\}} y_j$

[a]*Zero factorial* (0!) is frequently encountered in the form of $(n-n)!$ when calculating permutations and combinations. Zero factorial is defined as 1.
[b]The notation abs(x) is also used to indicate the absolute value.

5. SIGNIFICANT DIGITS

The significant digits in a number include the leftmost, nonzero digits to the rightmost digit written. Final answers from computations should be rounded off to the number of decimal places justified by the data. The answer can be no more accurate than the least accurate number in the data. Of course, rounding should be done on final calculation results only. It should not be done on interim results.

There are two ways that significant digits can affect calculations. For the operations of multiplication and division, the final answer is rounded to the number of significant digits in the least significant multiplicand, divisor, or dividend. So, $2.0 \times 13.2 = 26$ since the first multiplicand (2.0) has two significant digits only.

For the operations of addition and subtraction, the final answer is rounded to the position of the least significant digit in the addenda, minuend, or subtrahend. So, $2.0 + 13.2 = 15.2$ because both addenda are significant to the tenth's position; however, $2 + 13.4 = 15$ since the 2 is significant only in the ones' position.

The multiplication rule should not be used for addition or subtraction, as this can result in strange answers. For example, it would be incorrect to round 1700 + 0.1 to 2000 simply because 0.1 has only one significant digit. Table 3.3 gives examples of significant digits.

Table 3.3 Examples of Significant Digits

number as written	number of significant digits	implied range
341	3	340.5–341.5
34.1	3	34.05–34.15
0.00341	3	0.003405–0.003415
341×10^7	3	340.5×10^7–341.5×10^7
3.41×10^{-2}	3	3.405×10^{-2}–3.415×10^{-2}
3410	3	3405–3415
3410*	4	3409.5–3410.5
341.0	4	340.95–341.05

*It is permitted to write "3410." to distinguish the number from its three-significant-digit form, although this is rarely done.

6. EQUATIONS

An *equation* is a mathematical statement of equality, such as 5 = 3 + 2. *Algebraic equations* are written in terms of *variables*. In the equation $y = x^2 + 3$, the value of variable y depends on the value of variable x. Therefore, y is the *dependent variable* and x is the *independent variable*. The dependency of y on x is clearer when the equation is written in *functional form*: $y = f(x)$.

A *parametric equation* uses one or more independent variables (*parameters*) to describe a function.[3] For example, the parameter θ can be used to write the parametric equations of a unit circle.

$$x = \cos\theta \tag{3.1}$$

$$y = \sin\theta \tag{3.2}$$

A unit circle can also be described by a *nonparametric equation*.[4]

$$x^2 + y^2 = 1 \tag{3.3}$$

7. FUNDAMENTAL ALGEBRAIC LAWS

Algebra provides the rules that allow complex mathematical relationships to be expanded or condensed. Algebraic laws may be applied to complex numbers, variables, and real numbers. The general rules for changing the form of a mathematical relationship are given as follows.

- *commutative law for addition*

$$A + B = B + A \tag{3.4}$$

- *commutative law for multiplication*

$$AB = BA \tag{3.5}$$

- *associative law for addition*

$$A + (B + C) = (A + B) + C \tag{3.6}$$

- *associative law for multiplication*

$$A(BC) = (AB)C \tag{3.7}$$

- *distributive law*

$$A(B + C) = AB + AC \tag{3.8}$$

8. POLYNOMIALS

A *polynomial* is a rational expression—usually the sum of several variable terms known as *monomials*—that does not involve division. The *degree of the polynomial* is the highest power to which a variable in the expression is raised. The following *standard polynomial forms* are useful when trying to find the roots of an equation.

$$(a + b)(a - b) = a^2 - b^2 \tag{3.9}$$

$$(a \pm b)^2 = a^2 \pm 2ab + b^2 \tag{3.10}$$

$$(a \pm b)^3 = a^3 \pm 3a^2b + 3ab^2 \pm b^3 \tag{3.11}$$

$$(a^3 \pm b^3) = (a \pm b)(a^2 \mp ab + b^2) \tag{3.12}$$

$$(a^n - b^n) = (a - b)\begin{pmatrix} a^{n-1} + a^{n-2}b + a^{n-3}b^2 \\ + \cdots + b^{n-1} \end{pmatrix} \tag{3.13}$$

[n is any positive integer]

$$(a^n + b^n) = (a + b)\begin{pmatrix} a^{n-1} - a^{n-2}b + a^{n-3}b^2 \\ - \cdots + b^{n-1} \end{pmatrix} \tag{3.14}$$

[n is any positive odd integer]

The *binomial theorem* defines a polynomial of the form $(a + b)^n$.

$$(a + b)^n = \underset{[i=0]}{a^n} + \underset{[i=1]}{na^{n-1}b} + \underset{[i=2]}{C_2a^{n-2}b^2} + \cdots + C_ia^{n-i}b^i + \cdots + nab^{n-1} + b^n \tag{3.15}$$

$$C_i = \frac{n!}{i!\,(n - i)!} \quad [i = 0, 1, 2, \ldots, n] \tag{3.16}$$

The coefficients of the expansion can be determined quickly from *Pascal's triangle*—each entry is the sum of the two entries directly above it. (See Fig. 3.1.)

[3]As used in this section, there is no difference between a parameter and an independent variable. However, the term *parameter* is also used as a descriptive measurement that determines or characterizes the form, size, or content of a function. For example, the radius is a parameter of a circle, and mean and variance are parameters of a probability distribution. Once these parameters are specified, the function is completely defined.
[4]Since only the coordinate variables are used, this equation is also said to be in *Cartesian equation form*.

The values r_1, r_2, ..., r_n of the independent variable x that satisfy a polynomial equation $f(x) = 0$ are known as *roots* or *zeros* of the polynomial. A polynomial of degree n with real coefficients will have at most n real roots, although they need not all be distinctly different.

Figure 3.1 *Pascal's Triangle*

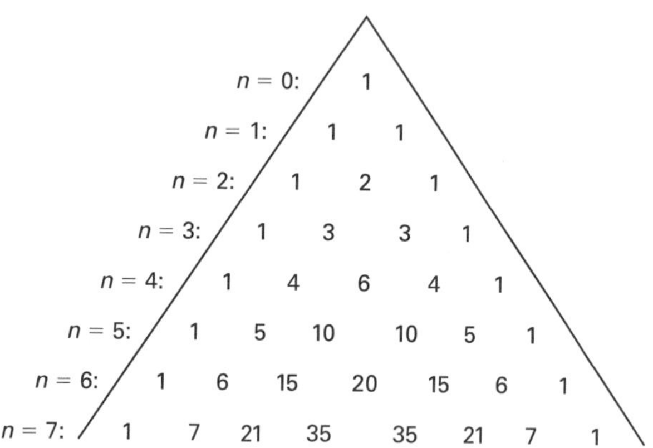

9. ROOTS OF QUADRATIC EQUATIONS

A *quadratic equation* is an equation of the general form $ax^2 + bx + c = 0$ $[a \neq 0]$. The *roots*, x_1 and x_2, of the equation are the two values of x that satisfy it.

$$x_1, x_2 = \frac{-b \pm \sqrt{b^2 - 4ac}}{2a} \qquad 3.17$$

$$x_1 + x_2 = -\frac{b}{a} \qquad 3.18$$

$$x_1 x_2 = \frac{c}{a} \qquad 3.19$$

The types of roots of the equation can be determined from the *discriminant* (i.e., the quantity under the radical in Eq. 3.17).

- If $b^2 - 4ac > 0$, the roots are real and unequal.
- If $b^2 - 4ac = 0$, the roots are real and equal. This is known as a *double root.*
- If $b^2 - 4ac < 0$, the roots are complex and unequal.

10. ROOTS OF GENERAL POLYNOMIALS

It is difficult to find roots of cubic and higher-degree polynomials because few general techniques exist. *Cardano's formula (method)* uses closed-form equations to laboriously calculate roots for general cubic (3rd degree) polynomials. Compromise methods are used when solutions are needed on the spot.

- *inspection and trial and error:* Finding roots by inspection is equivalent to making reasonable guesses about the roots and substituting into the polynomial.
- *graphing:* If the value of a polynomial $f(x)$ is calculated and plotted for different values of x, an approximate value of a root can be determined as the value of x at which the plot crosses the x-axis.
- *factoring:* If at least one root (say, $x = r$) of a polynomial $f(x)$ is known, the quantity $x - r$ can be factored out of $f(x)$ by long division. The resulting quotient will be lower by one degree, and the remaining roots may be easier to determine. This method is particularly applicable if the polynomial is in one of the standard forms presented in Sec. 3.8.
- *special cases:* Certain polynomial forms can be simplified by substitution or solved by standard formulas if they are recognized as being special cases. (The standard solution to the quadratic equation is such a special case.) For example, $ax^4 + bx^2 + c = 0$ can be reduced to a polynomial of degree 2 if the substitution $u = x^2$ is made.
- *numerical methods:* If an approximate value of a root is known, numerical methods (bisection method, Newton's method, etc.) can be used to refine the value. The more efficient techniques are too complex to be performed by hand.

11. EXTRANEOUS ROOTS

With simple equalities, it may appear possible to derive roots by basic algebraic manipulations.[5] However, multiplying each side of an equality by a power of a variable may introduce *extraneous roots.* Such roots do not satisfy the original equation even though they are derived according to the rules of algebra. Checking a calculated root is always a good idea, but is particularly necessary if the equation has been multiplied by one of its own variables.

Example 3.1

Use algebraic operations to determine a value that satisfies the following equation. Determine if the value is a valid or extraneous root.

$$\sqrt{x - 2} = \sqrt{x} + 2$$

[5]In this sentence, *equality* means a combination of two expressions containing an equal sign. Any two expressions can be linked in this manner, even those that are not actually equal. For example, the expressions for two nonintersecting ellipses can be equated even though there is no intersection point. Finding extraneous roots is more likely when the underlying equality is false to begin with.

Solution

Square both sides.

$$x - 2 = x + 4\sqrt{x} + 4$$

Subtract x from each side, and combine the constants.

$$4\sqrt{x} = -6$$

Solve for x.

$$x = \left(\frac{-6}{4}\right)^2 = \frac{9}{4}$$

Substitute $x = 9/4$ into the original equation.

$$\sqrt{\frac{9}{4} - 2} = \sqrt{\frac{9}{4}} + 2$$

$$\frac{1}{2} = \frac{7}{2}$$

Since the equality is not established, $x = 9/4$ is an extraneous root.

12. DESCARTES' RULE OF SIGNS

Descartes' rule of signs determines the maximum number of positive (and negative) real roots that a polynomial will have by counting the number of sign reversals (i.e., changes in sign from one term to the next) in the polynomial. The polynomial $f(x) = 0$ must have real coefficients and must be arranged in terms of descending powers of x.

- The number of positive roots of the polynomial equation $f(x) = 0$ will not exceed the number of sign reversals.
- The difference between the number of sign reversals and the number of positive roots is an even number.
- The number of negative roots of the polynomial equation $f(x) = 0$ will not exceed the number of sign reversals in the polynomial $f(-x)$.
- The difference between the number of sign reversals in $f(-x)$ and the number of negative roots is an even number.

Example 3.2

Determine the possible numbers of positive and negative roots that satisfy the following polynomial equation.

$$4x^5 - 5x^4 + 3x^3 - 8x^2 - 2x + 3 = 0$$

Solution

There are four sign reversals, so up to four positive roots exist. To keep the difference between the number of positive roots and the number of sign reversals an even number, the number of positive real roots is limited to zero, two, and four.

Substituting $-x$ for x in the polynomial results in

$$-4x^5 - 5x^4 - 3x^3 - 8x^2 + 2x + 3 = 0$$

There is only one sign reversal, so the number of negative roots cannot exceed one. There must be exactly one negative real root in order to keep the difference to an even number (zero in this case).

13. RULES FOR EXPONENTS AND RADICALS

In the expression $b^n = a$, b is known as the *base* and n is the *exponent* or *power*. In Eq. 3.20 through Eq. 3.33, a, b, m, and n are any real numbers with limitations listed.

$$b^0 = 1 \quad [b \neq 0] \tag{3.20}$$

$$b^1 = b \tag{3.21}$$

$$b^{-n} = \frac{1}{b^n} = \left(\frac{1}{b}\right)^n \quad [b \neq 0] \tag{3.22}$$

$$\left(\frac{a}{b}\right)^n = \frac{a^n}{b^n} \quad [b \neq 0] \tag{3.23}$$

$$(ab)^n = a^n b^n \tag{3.24}$$

$$b^{m/n} = \sqrt[n]{b^m} = \left(\sqrt[n]{b}\right)^m \tag{3.25}$$

$$(b^n)^m = b^{nm} \tag{3.26}$$

$$b^m b^n = b^{m+n} \tag{3.27}$$

$$\frac{b^m}{b^n} = b^{m-n} \quad [b \neq 0] \tag{3.28}$$

$$\sqrt[n]{b} = b^{1/n} \tag{3.29}$$

$$\left(\sqrt[n]{b}\right)^n = \left(b^{1/n}\right)^n = b \tag{3.30}$$

$$\sqrt[n]{ab} = \sqrt[n]{a}\sqrt[n]{b} = a^{1/n}b^{1/n} = (ab)^{1/n} \tag{3.31}$$

$$\sqrt[n]{\frac{a}{b}} = \frac{\sqrt[n]{a}}{\sqrt[n]{b}} = \left(\frac{a}{b}\right)^{1/n} \quad [b \neq 0] \tag{3.32}$$

$$\sqrt[m]{\sqrt[n]{b}} = \sqrt[mn]{b} = b^{1/mn} \tag{3.33}$$

14. LOGARITHMS

Logarithms can be considered to be exponents. For example, the exponent n in the expression $b^n = a$ is the logarithm of a to the base b. Therefore, the two expressions $\log_b a = n$ and $b^n = a$ are equivalent.

The base for *common logs* is 10. Usually, "log" will be written when common logs are desired, although "$\log_{10}$" appears occasionally. The base for *natural* (*Napierian*) *logs* is 2.71828..., a number that is given the symbol e. When natural logs are desired, usually "ln" is written, although "$\log_e$" is also used.

Most logarithms will contain an integer part (the *characteristic*) and a decimal part (the *mantissa*). The common and natural logarithms of any number less than one are negative. If the number is greater than one, its common and natural logarithms are positive. Although the logarithm may be negative, the mantissa is always positive. For negative logarithms, the characteristic is found by expressing the logarithm as the sum of a negative characteristic and a positive mantissa.

For common logarithms of numbers greater than one, the characteristics will be positive and equal to one less than the number of digits in front of the decimal. If the number is less than one, the characteristic will be negative and equal to one more than the number of zeros immediately following the decimal point.

If a negative logarithm is to be used in a calculation, it must first be converted to *operational form* by adding the characteristic and mantissa. The operational form should be used in all calculations and is the form displayed by scientific calculators.

The logarithm of a negative number is a complex number.

Example 3.3

Use logarithm tables to determine the operational form of $\log_{10} 0.05$.

Solution

Since the number is less than one and there is one leading zero, the characteristic is found by observation to be -2. From a book of logarithm tables, the mantissa of 5.0 is 0.699. Two ways of expressing the combination of mantissa and characteristic are used.

method 1: $\bar{2}.699$

method 2: $8.699 - 10$

The operational form of this logarithm is $-2 + 0.699 = -1.301$.

15. LOGARITHM IDENTITIES

Prior to the widespread availability of calculating devices, logarithm identities were used to solve complex calculations by reducing the solution method to table lookup, addition, and subtraction. Logarithm identities are still useful in simplifying expressions containing exponentials and other logarithms. In Eq. 3.34 through Eq. 3.45, $a \neq 1$, $b \neq 1$, $x > 0$, and $y > 0$.

$$\log_b b = 1 \tag{3.34}$$

$$\log_b 1 = 0 \tag{3.35}$$

$$\log_b b^n = n \tag{3.36}$$

$$\log x^a = a \log x \tag{3.37}$$

$$\log \sqrt[n]{x} = \log x^{1/n} = \frac{\log x}{n} \tag{3.38}$$

$$b^{n \log_b x} = x^n = \text{antilog}(n \log_b x) \tag{3.39}$$

$$b^{\log_b x/n} = x^{1/n} \tag{3.40}$$

$$\log xy = \log x + \log y \tag{3.41}$$

$$\log \frac{x}{y} = \log x - \log y \tag{3.42}$$

$$\log_a x = \log_b x \log_a b \tag{3.43}$$

$$\ln x = \ln 10 \log_{10} x \approx 2.3026 \log_{10} x \tag{3.44}$$

$$\log_{10} x = \log_{10} e \ln x \approx 0.4343 \ln x \tag{3.45}$$

Example 3.4

The surviving fraction, f, of a radioactive isotope is given by $f = e^{-0.005t}$. For what value of t will the surviving percentage be 7%?

Solution

$$f = 0.07 = e^{-0.005t}$$

Take the natural log of both sides.

$$\ln 0.07 = \ln e^{-0.005t}$$

From Eq. 3.36, $\ln e^x = x$. Therefore,

$$-2.66 = -0.005t$$
$$t = 532$$

16. PARTIAL FRACTIONS

The method of *partial fractions* is used to transform a proper polynomial fraction of two polynomials into a sum of simpler expressions, a procedure known as *resolution*.[6,7] The technique can be considered to be the act of "unadding" a sum to obtain all of the addends.

Suppose $H(x)$ is a proper polynomial fraction of the form $P(x)/Q(x)$. The object of the resolution is to determine the partial fractions u_1/v_1, u_2/v_2, and so on, such that

$$H(x) = \frac{P(x)}{Q(x)} = \frac{u_1}{v_1} + \frac{u_2}{v_2} + \frac{u_3}{v_3} + \cdots \quad 3.46$$

The form of the denominator polynomial $Q(x)$ will be the main factor in determining the form of the partial fractions. The task of finding the u_i and v_i is simplified by categorizing the possible forms of $Q(x)$.

case 1: $Q(x)$ factors into n different linear terms.

$$Q(x) = (x - a_1)(x - a_2)\cdots(x - a_n) \quad 3.47$$

Then,

$$H(x) = \sum_{i=1}^{n} \frac{A_i}{x - a_i} \quad 3.48$$

case 2: $Q(x)$ factors into n identical linear terms.

$$Q(x) = (x - a)(x - a)\cdots(x - a) \quad 3.49$$

Then,

$$H(x) = \sum_{i=1}^{n} \frac{A_i}{(x - a)^i} \quad 3.50$$

case 3: $Q(x)$ factors into n different quadratic terms, $x^2 + p_i x + q_i$.

Then,

$$H(x) = \sum_{i=1}^{n} \frac{A_i x + B_i}{x^2 + p_i x + q_i} \quad 3.51$$

case 4: $Q(x)$ factors into n identical quadratic terms, $x^2 + px + q$.

Then,

$$H(x) = \sum_{i=1}^{n} \frac{A_i x + B_i}{(x^2 + px + q)^i} \quad 3.52$$

Once the general forms of the partial fractions have been determined from inspection, the *method of undetermined coefficients* is used. The partial fractions are all cross multiplied to obtain $Q(x)$ as the denominator, and the coefficients are found by equating $P(x)$ and the cross-multiplied numerator.

Example 3.5

Resolve $H(x)$ into partial fractions.

$$H(x) = \frac{x^2 + 2x + 3}{x^4 + x^3 + 2x^2}$$

Solution

Here, $Q(x) = x^4 + x^3 + 2x^2$ factors into $x^2(x^2 + x + 2)$. This is a combination of cases 2 and 3.

$$H(x) = \frac{A_1}{x} + \frac{A_2}{x^2} + \frac{A_3 + A_4 x}{x^2 + x + 2}$$

Cross multiplying to obtain a common denominator yields

$$\frac{(A_1 + A_4)x^3 + (A_1 + A_2 + A_3)x^2 + (2A_1 + A_2)x + 2A_2}{x^4 + x^3 + 2x^2}$$

Since the original numerator is known, the following simultaneous equations result.

$$\begin{aligned} A_1 + A_4 &= 0 \\ A_1 + A_2 + A_3 &= 1 \\ 2A_1 + A_2 &= 2 \\ 2A_2 &= 3 \end{aligned}$$

The solutions are $A_1 = 0.25$; $A_2 = 1.50$; $A_3 = -0.75$; and $A_4 = -0.25$.

$$H(x) = \frac{1}{4x} + \frac{3}{2x^2} - \frac{x + 3}{4(x^2 + x + 2)}$$

17. SIMULTANEOUS LINEAR EQUATIONS

A *linear equation* with n variables is a polynomial of degree 1 describing a geometric shape in n-space. A *homogeneous linear equation* is one that has no constant term, and a *nonhomogeneous linear equation* has a constant term.

[6]To be a *proper polynomial fraction*, the degree of the numerator must be less than the degree of the denominator. If the polynomial fraction is improper, the denominator can be divided into the numerator to obtain whole and fractional polynomials. The method of partial fractions can then be used to reduce the fractional polynomial.

[7]This technique is particularly useful for calculating integrals and inverse Laplace transforms in subsequent chapters.

A solution to a set of simultaneous linear equations represents the intersection point of the geometric shapes in n-space. For example, if the equations are limited to two variables (e.g., $y = 4x - 5$), they describe straight lines. The solution to two simultaneous linear equations in 2-space is the point where the two lines intersect. The set of the two equations is said to be a *consistent system* when there is such an intersection.[8]

Simultaneous equations do not always have unique solutions, and some have none at all. In addition to crossing in 2-space, lines can be parallel or they can be the same line expressed in a different equation format (i.e., dependent equations). In some cases, parallelism and dependency can be determined by inspection. In most cases, however, matrix and other advanced methods must be used to determine whether a solution exists. A set of linear equations with no simultaneous solution is known as an *inconsistent system.*

Several methods exist for solving linear equations simultaneously by hand.[9]

- *graphing:* The equations are plotted and the intersection point is read from the graph. This method is possible only with two-dimensional problems.
- *substitution:* An equation is rearranged so that one variable is expressed as a combination of the other variables. The expression is then substituted into the remaining equations wherever the selected variable appears.
- *reduction:* All terms in the equations are multiplied by constants chosen to eliminate one or more variables when the equations are added or subtracted. The remaining sum can then be solved for the other variables. This method is also known as *eliminating the unknowns.*
- *Cramer's rule:* This is a procedure in linear algebra that calculates determinants of the original coefficient matrix **A** and of the n matrices resulting from the systematic replacement of column **A** by the constant matrix **B**.

Example 3.6

Solve the following set of linear equations by (a) substitution and (b) reduction.

$$2x + 3y = 12 \quad \text{[Eq. I]}$$

$$3x + 4y = 8 \quad \text{[Eq. II]}$$

Solution

(a) From Eq. I, solve for variable x.

$$x = 6 - 1.5y \quad \text{[Eq. III]}$$

Substitute $6 - 1.5y$ into Eq. II wherever x appears.

$$\begin{aligned}(3)(6 - 1.5y) + 4y &= 8\\ 18 - 4.5y + 4y &= 8\\ y &= 20\end{aligned}$$

Substitute 20 for y in Eq. III.

$$x = 6 - (1.5)(20) = -24$$

The solution $(-24, 20)$ should be checked to verify that it satisfies both original equations.

(b) Eliminate variable x by multiplying Eq. I by 3 and Eq. II by 2.

$$\begin{aligned}3 \times \text{Eq. I:}\ 6x + 9y &= 36 \quad \text{[Eq. I']}\\ 2 \times \text{Eq. II:}\ 6x + 8y &= 16 \quad \text{[Eq. II']}\end{aligned}$$

Subtract Eq. II′ from Eq. I′.

$$y = 20 \quad \text{[Eq. I' - Eq. II']}$$

Substitute $y = 20$ into Eq. I′.

$$\begin{aligned}6x + (9)(20) &= 36\\ x &= -24\end{aligned}$$

The solution $(-24, 20)$ should be checked to verify that it satisfies both original equations.

18. COMPLEX NUMBERS

A *complex number*, **Z**, is a combination of real and imaginary numbers. When expressed as a sum (e.g., $a + bi$), the complex number is said to be in *rectangular* or *trigonometric form.* The complex number can be plotted on the real-imaginary coordinate system known as the *complex plane*, as illustrated in Fig. 3.2.

[8] A homogeneous system always has at least one solution: the *trivial solution*, in which all variables have a value of zero.
[9] Other matrix and numerical methods exist, but they require a computer.

Figure 3.2 *A Complex Number in the Complex Plane*

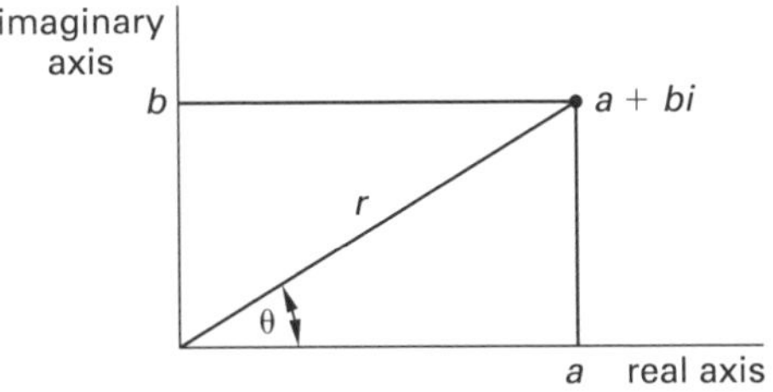

The complex number $\mathbf{Z} = a + bi$ can also be expressed in *exponential form*.[10] The quantity r is known as the *modulus* of $\mathbf{Z}$; θ is the *argument*.

$$a + bi \equiv re^{i\theta} \quad 3.53$$

$$r = \operatorname{mod} \mathbf{Z} = \sqrt{a^2 + b^2} \quad 3.54$$

$$\theta = \arg \mathbf{Z} = \arctan \frac{b}{a} \quad 3.55$$

Similarly, the *phasor form* (also known as the *polar form*) is

$$\mathbf{Z} = r\angle\theta \quad 3.56$$

The *rectangular form* can be determined from r and θ.

$$a = r\cos\theta \quad 3.57$$

$$b = r\sin\theta \quad 3.58$$

$$\begin{aligned}\mathbf{Z} &= a + bi = r\cos\theta + ir\sin\theta \\ &= r(\cos\theta + i\sin\theta)\end{aligned} \quad 3.59$$

The *cis form* is a shorthand method of writing a complex number in rectangular (trigonometric) form.

$$a + bi = r(\cos\theta + i\sin\theta) = r \operatorname{cis}\theta \quad 3.60$$

Euler's equation, as shown in Eq. 3.61, expresses the equality of complex numbers in exponential and trigonometric form.

$$e^{i\theta} = \cos\theta + i\sin\theta \quad 3.61$$

Related expressions are

$$e^{-i\theta} = \cos\theta - i\sin\theta \quad 3.62$$

$$\cos\theta = \frac{e^{i\theta} + e^{-i\theta}}{2} \quad 3.63$$

$$\sin\theta = \frac{e^{i\theta} - e^{-i\theta}}{2i} \quad 3.64$$

Example 3.7
What is the exponential form of the complex number $\mathbf{Z} = 3 + 4i$?

Solution

$$\begin{aligned} r &= \sqrt{a^2 + b^2} = \sqrt{3^2 + 4^2} = \sqrt{25} \\ &= 5 \end{aligned}$$

$$\theta = \arctan\frac{b}{a} = \arctan\frac{4}{3} = 0.927 \text{ rad}$$

$$\mathbf{Z} = re^{i\theta} = 5e^{i(0.927)}$$

19. OPERATIONS ON COMPLEX NUMBERS

Most algebraic operations (addition, multiplication, exponentiation, etc.) work with complex numbers, but notable exceptions are the inequality operators. The concept of one complex number being less than or greater than another complex number is meaningless.

When adding two complex numbers, real parts are added to real parts, and imaginary parts are added to imaginary parts.

$$(a_1 + ib_1) + (a_2 + ib_2) = (a_1 + a_2) + i(b_1 + b_2) \quad 3.65$$

$$(a_1 + ib_1) - (a_2 + ib_2) = (a_1 - a_2) + i(b_1 - b_2) \quad 3.66$$

Multiplication of two complex numbers in rectangular form is accomplished by the use of the algebraic distributive law and the definition $i^2 = -1$.

Division of complex numbers in rectangular form requires the use of the *complex conjugate*. The complex conjugate of the complex number $(a + bi)$ is $(a - bi)$. By multiplying the numerator and the denominator by the complex conjugate, the denominator will be converted to the real number $a^2 + b^2$. This technique is known as *rationalizing* the denominator and is illustrated in Ex. 3.8(c).

Multiplication and division are often more convenient when the complex numbers are in exponential or phasor forms, as Eq. 3.67 and Eq. 3.68 show.

$$(r_1 e^{i\theta_1})(r_2 e^{i\theta_2}) = r_1 r_2 e^{i(\theta_1 + \theta_2)} \quad 3.67$$

$$\frac{r_1 e^{i\theta_1}}{r_2 e^{i\theta_2}} = \left(\frac{r_1}{r_2}\right) e^{i(\theta_1 - \theta_2)} \quad 3.68$$

Taking powers and roots of complex numbers requires *de Moivre's theorem*, Eq. 3.69 and Eq. 3.70.

$$\mathbf{Z}^n = \left(re^{i\theta}\right)^n = r^n e^{in\theta} \quad 3.69$$

$$\sqrt[n]{\mathbf{Z}} = \left(re^{i\theta}\right)^{1/n} = \sqrt[n]{r}\, e^{i(\theta + k360^\circ/n)} \quad [k = 0, 1, 2, \ldots, n-1] \quad 3.70$$

[10]The terms *polar form*, *phasor form*, and *exponential form* are all used somewhat interchangeably.

Example 3.8

Perform the following complex arithmetic.

(a) $(3+4i)+(2+i)$

(b) $(7+2i)(5-3i)$

(c) $\dfrac{2+3i}{4-5i}$

Solution

(a)
$$\begin{aligned}(3+4i)+(2+i) &= (3+2)+(4+1)i\\ &= 5+5i\end{aligned}$$

(b)
$$\begin{aligned}(7+2i)(5-3i) &= (7)(5)-(7)(3i)+(2i)(5)\\ &\quad -(2i)(3i)\\ &= 35-21i+10i-6i^2\\ &= 35-21i+10i-(6)(-1)\\ &= 41-11i\end{aligned}$$

(c) Multiply the numerator and denominator by the complex conjugate of the denominator.

$$\begin{aligned}\frac{2+3i}{4-5i} &= \frac{(2+3i)(4+5i)}{(4-5i)(4+5i)} = \frac{-7+22i}{(4)^2+(5)^2}\\ &= \frac{-7}{41}+i\frac{22}{41}\end{aligned}$$

20. LIMITS

A *limit* (*limiting value*) is the value a function approaches when an independent variable approaches a target value. For example, suppose the value of $y=x^2$ is desired as x approaches 5. This could be written as

$$\lim_{x\to 5} x^2 \qquad 3.71$$

The power of limit theory is wasted on simple calculations such as this but is appreciated when the function is undefined at the target value. The object of limit theory is to determine the limit without having to evaluate the function at the target. The general case of a limit evaluated as x approaches the target value a is written as

$$\lim_{x\to a} f(x) \qquad 3.72$$

It is not necessary for the actual value $f(a)$ to exist for the limit to be calculated. The function $f(x)$ may be undefined at point a. However, it is necessary that $f(x)$ be defined on both sides of point a for the limit to exist. If $f(x)$ is undefined on one side, or if $f(x)$ is discontinuous at $x=a$ (as in Fig. 3.3(c) and Fig. 3.3 (d)), the limit does not exist.

Figure 3.3 *Existence of Limits*

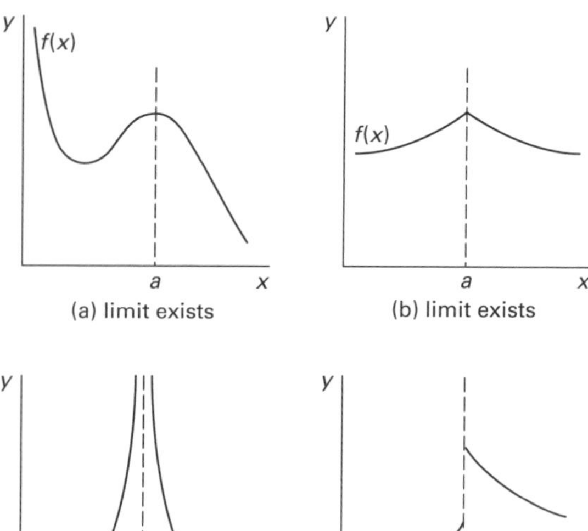

The following theorems can be used to simplify expressions when calculating limits.

$$\lim_{x\to a} x = a \qquad 3.73$$

$$\lim_{x\to a}(mx+b) = ma+b \qquad 3.74$$

$$\lim_{x\to a} b = b \qquad 3.75$$

$$\lim_{x\to a}\big(kF(x)\big) = k\lim_{x\to a}F(x) \qquad 3.76$$

$$\begin{aligned}&\lim_{x\to a}\left(F_1(x)\begin{Bmatrix}+\\-\\\times\\\div\end{Bmatrix}F_2(x)\right)\\ &= \lim_{x\to a}\big(F_1(x)\big)\begin{Bmatrix}+\\-\\\times\\\div\end{Bmatrix}\lim_{x\to a}\big(F_2(x)\big)\end{aligned} \qquad 3.77$$

The following identities can be used to simplify limits of trigonometric expressions.

$$\lim_{x\to 0}\sin x = 0 \qquad 3.78$$

$$\lim_{x\to 0}\left(\frac{\sin x}{x}\right) = 1 \qquad 3.79$$

$$\lim_{x\to 0}\cos x = 1 \qquad 3.80$$

The following standard methods (tricks) can be used to determine limits.

- If the limit is taken to infinity, all terms can be divided by the largest power of x in the expression. This will leave at least one constant. Any quantity divided by a power of x vanishes as x approaches infinity.
- If the expression is a quotient of two expressions, any common factors should be eliminated from the numerator and denominator.
- *L'Hôpital's rule*, Eq. 3.81, should be used when the numerator and denominator of the expression both approach zero or both approach infinity.[11] $P^k(x)$ and $Q^k(x)$ are the kth derivatives of the functions $P(x)$ and $Q(x)$, respectively. (L'Hôpital's rule can be applied repeatedly as required.)

$$\lim_{x\to a}\left(\frac{P(x)}{Q(x)}\right) = \lim_{x\to a}\left(\frac{P^k(x)}{Q^k(x)}\right) \qquad 3.81$$

Example 3.9

Evaluate the following limits.

(a) $\lim_{x\to 3}\left(\frac{x^3-27}{x^2-9}\right)$

(b) $\lim_{x\to\infty}\left(\frac{3x-2}{4x+3}\right)$

(c) $\lim_{x\to 2}\left(\frac{x^2+x-6}{x^2-3x+2}\right)$

Solution

(a) Factor the numerator and denominator. (L'Hôpital's rule can also be used.)

$$\begin{aligned}\lim_{x\to 3}\left(\frac{x^3-27}{x^2-9}\right) &= \lim_{x\to 3}\left(\frac{(x-3)(x^2+3x+9)}{(x-3)(x+3)}\right)\\ &= \lim_{x\to 3}\left(\frac{x^2+3x+9}{x+3}\right)\\ &= \frac{(3)^2+(3)(3)+9}{3+3}\\ &= 9/2\end{aligned}$$

(b) Divide through by the largest power of x. (L'Hôpital's rule can also be used.)

$$\begin{aligned}\lim_{x\to\infty}\left(\frac{3x-2}{4x+3}\right) &= \lim_{x\to\infty}\left(\frac{3-\frac{2}{x}}{4+\frac{3}{x}}\right)\\ &= \frac{3-\frac{2}{\infty}}{4+\frac{3}{\infty}} = \frac{3-0}{4+0}\\ &= 3/4\end{aligned}$$

(c) Use L'Hôpital's rule. (Factoring can also be used.) Take the first derivative of the numerator and denominator.

$$\begin{aligned}\lim_{x\to 2}\left(\frac{x^2+x-6}{x^2-3x+2}\right) &= \lim_{x\to 2}\left(\frac{2x+1}{2x-3}\right)\\ &= \frac{(2)(2)+1}{(2)(2)-3} = \frac{5}{1}\\ &= 5\end{aligned}$$

21. SEQUENCES AND PROGRESSIONS

A *sequence*, $\{A\}$, is an ordered *progression* of numbers, a_i, such as 1, 4, 9, 16, 25, ... The *terms* in a sequence can be all positive, all negative, or of alternating signs. a_n is known as the *general term* of the sequence.

$$\{A\} = \{a_1, a_2, a_3, \ldots, a_n\} \qquad 3.82$$

A sequence is said to *diverge* (i.e., be *divergent*) if the terms approach infinity or if the terms fail to approach any finite value, and it is said to *converge* (i.e., be *convergent*) if the terms approach any finite value (including zero). That is, the sequence converges if the limit defined by Eq. 3.83 exists.

$$\lim_{n\to\infty} a_n \begin{cases}\text{converges if the limit is finite}\\ \text{diverges if the limit is infinite}\\ \quad\text{or does not exist}\end{cases} \qquad 3.83$$

The main task associated with a sequence is determining the next (or the general) term. If several terms of a sequence are known, the next (unknown) term must usually be found by intuitively determining the pattern of the sequence. In some cases, though, the method of *Rth-order differences* can be used to determine the next term. This method consists of subtracting each term

[11] L'Hôpital's rule should not be used when only the denominator approaches zero. In that case, the limit approaches infinity regardless of the numerator.

from the following term to obtain a set of differences. If the differences are not all equal, the next order of differences can be calculated.

Example 3.10
What is the general term of the sequence $\{A\}$?

$$\{A\} = \left\{3, \frac{9}{2}, \frac{27}{6}, \frac{81}{24}, \ldots\right\}$$

Solution

The solution is purely intuitive. The numerator is recognized as a power series based on the number 3. The denominator is recognized as the factorial sequence. The general term is

$$a_n = \frac{3^n}{n!}$$

Example 3.11
Find the sixth term in the sequence $\{A\} = \{7, 16, 29, 46, 67, a_6\}$.

Solution

The sixth term is not intuitively obvious, so the method of Rth-order differences is tried. The pattern is not obvious from the first order differences, but the second order differences are all 4.

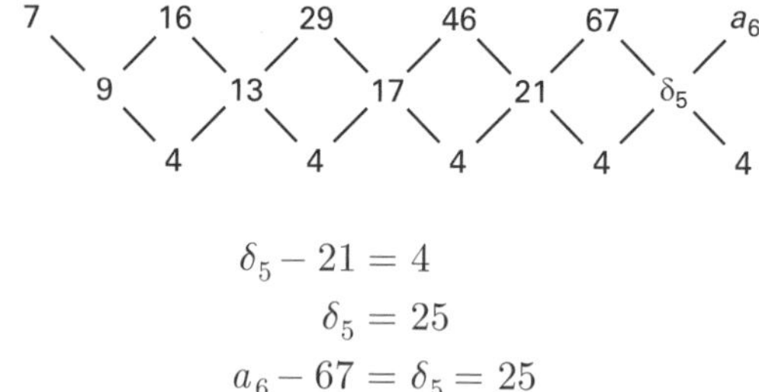

$$\delta_5 - 21 = 4$$
$$\delta_5 = 25$$
$$a_6 - 67 = \delta_5 = 25$$
$$a_6 = 92$$

Example 3.12
Does the sequence with general term e^n/n converge or diverge?

Solution

See if the limit exists.

$$\lim_{n\to\infty}\left(\frac{e^n}{n}\right) = \frac{\infty}{\infty}$$

Since ∞/∞ is inconclusive, apply L'Hôpital's rule. Take the derivative of both the numerator and the denominator with respect to n.

$$\lim_{n\to\infty}\left(\frac{e^n}{1}\right) = \frac{\infty}{1} = \infty$$

The sequence diverges.

22. STANDARD SEQUENCES

There are four standard sequences: geometric, arithmetic, harmonic, and p-sequence.

- The *geometric sequence* converges for $-1 < r \leq 1$ and diverges otherwise. a is known as the *first term*; r is known as the *common ratio*.

$$a_n = ar^{n-1} \quad \left[\begin{array}{l} a \text{ is a constant} \\ n = 1, 2, 3, \ldots, \infty \end{array}\right] \qquad 3.84$$

example: $\{1, 2, 4, 8, 16, 32\}$ $(a = 1,\ r = 2)$

- The *arithmetic sequence* always diverges.

$$a_n = a + (n-1)d \quad \left[\begin{array}{c} a \text{ and } d \text{ are constants} \\ n = 1, 2, 3, \ldots, \infty \end{array}\right] \qquad 3.85$$

example: $\{2, 7, 12, 17, 22, 27\}$ $(a = 2,\ d = 5)$

- The *harmonic sequence* always converges.

$$a_n = \frac{1}{n} \quad [n = 1, 2, 3, \ldots, \infty] \qquad 3.86$$

example: $\{1, 1/2, 1/3, 1/4, 1/5, 1/6\}$

- The *p-sequence* converges if $p \geq 0$ and diverges if $p < 0$. (This is different from the *p*-series, whose convergence depends on the sum of its terms.)

$$a_n = \frac{1}{n^p} \quad [n = 1, 2, 3, \ldots, \infty] \qquad 3.87$$

example: $\{1, 1/4, 1/9, 1/16, 1/25, 1/36\}$ $(p = 2)$

23. APPLICATION: GROWTH RATES

Models of *population growth* commonly assume arithmetic or geometric growth rates over limited periods of time.[12] *Arithmetic growth rate*, also called *constant growth rate* and *linear growth rate*, is appropriate when a population increase involves limited resources or occurs at specific intervals. Means of subsistence, such as areas of farmable land by generation, and budgets and enrollments by year, for example, are commonly assumed to increase arithmetically with time. Arithmetic growth is equivalent to simple interest compounding. Given a starting population, P_0, that increases every period by a constant *growth rate amount*, R, the population after t periods is

$$P_t = P_0 + tR \quad \text{[arithmetic]} \qquad 3.88$$

The *average annual growth rate* is conventionally defined as

$$r_{\text{ave},\%} = \frac{P_t - P_0}{tP_0} \times 100\% = \frac{R}{P_0} \times 100\% \qquad 3.89$$

Geometric growth (*Malthusian growth*) is appropriate when resources to support growth are infinite. The population changes by a fixed *growth rate fraction*, r, each period. Geometric growth is equivalent to discrete period interest compounding, and $(F/P, r\%, t)$ economic interest factors can be used. The population at time t (i.e., after t periods) is

$$P_t = P_0(1+r)^t \equiv P_0(P/F, r\%, t) \quad \text{[geometric]} \qquad 3.90$$

Geometric growth can be expressed in terms of a time constant, τ_b, associated with a specific base, b. Commonly, only three bases are used. For $b = 2$, τ is the *doubling time*, T. For $b = 1/2$, τ is the *half-life*, $t_{1/2}$. For $b = e$, τ is the *e-folding time*, or just *time constant*, τ. Base-e growth is known as *exponential growth* or *instantaneous growth*. It is appropriate for continuous growth (not discrete time periods) and is equivalent to continuous interest compounding.

$$\begin{aligned} P_t &= P_0(1+r)^t = P_0 e^{t/\tau} = P_0(2)^{t/T} \\ &= P_0\left(\tfrac{1}{2}\right)^{t/t_{1/2}} \quad \text{[geometric]} \end{aligned} \qquad 3.91$$

Taking the logarithm of Eq. 3.90 results in a *log-linear form*. Log-linear functions graph as straight lines.

$$\log P_t = \log P_0 + t\log(1+r) \qquad 3.92$$

24. SERIES

A *series* is the sum of terms in a sequence. There are two types of series. A *finite series* has a finite number of terms, and an *infinite series* has an infinite number of terms.[13] The main tasks associated with series are determining the sum of the terms and whether the series converges. A series is said to *converge* (be *convergent*) if the sum, S_n, of its term exists.[14] A finite series is always convergent.

The performance of a series based on standard sequences (defined in Sec. 3.22) is well known.

- *geometric series:*

$$S_n = \sum_{i=1}^{n} ar^{i-1} = \frac{a(1-r^n)}{1-r} \quad \text{[finite series]} \qquad 3.93$$

$$S_n = \sum_{i=1}^{\infty} ar^{i-1} = \frac{a}{1-r} \quad \left[\begin{array}{c}\text{infinite series} \\ -1 < r < 1\end{array}\right] \qquad 3.94$$

- *arithmetic series:* The infinite series diverges unless $a = d = 0$.

$$\begin{aligned} S_n &= \sum_{i=1}^{n}\big(a + (i-1)d\big) \\ &= \frac{n\big(2a + (n-1)d\big)}{2} \quad \text{[finite series]} \end{aligned} \qquad 3.95$$

- *harmonic series:* The infinite series diverges.
- *p-series:* The infinite series diverges if $p \leq 1$. The infinite series converges if $p > 1$. (This is different from the *p*-sequence whose convergence depends only on the last term.)

[12] Another population growth model is the logistic (S-shaped) curve.
[13] The term *infinite series* does not imply the sum is infinite.
[14] This is different from the definition of convergence for a sequence where only the last term was evaluated.

25. TESTS FOR SERIES CONVERGENCE

It is obvious that all finite series (i.e., series having a finite number of terms) converge. That is, the sum, S_n, defined by Eq. 3.96 exists.

$$S_n = \sum_{i=1}^{n} a_i \qquad 3.96$$

Convergence of an infinite series can be determined by taking the limit of the sum. If the limit exists, the series converges; otherwise, it diverges.

$$\lim_{n\to\infty} S_n = \lim_{n\to\infty} \sum_{i=1}^{n} a_i \qquad 3.97$$

In most cases, the expression for the general term a_n will be known, but there will be no simple expression for the sum S_n. Therefore, Eq. 3.97 cannot be used to determine convergence. It is helpful, but not conclusive, to look at the limit of the general term. If the limit, as defined in Eq. 3.98, is nonzero, the series diverges. If the limit equals zero, the series may either converge or diverge. Additional testing is needed in that case.

$$\lim_{n\to\infty} a_n \begin{cases} = 0 \text{ inconclusive} \\ \neq 0 \text{ diverges} \end{cases} \qquad 3.98$$

Two tests can be used independently or after Eq. 3.98 has proven inconclusive: the ratio and comparison tests. The *ratio test* calculates the limit of the ratio of two consecutive terms.

$$\lim_{n\to\infty} \frac{a_{n+1}}{a_n} \begin{cases} < 1 \text{ converges} \\ = 1 \text{ inconclusive} \\ > 1 \text{ diverges} \end{cases} \qquad 3.99$$

The *comparison test* is an indirect method of determining convergence of an unknown series. It compares a standard series (geometric and *p*-series are commonly used) against the unknown series. If all terms in a positive standard series are smaller than the terms in the unknown series and the standard series diverges, the unknown series must also diverge. Similarly, if all terms in the standard series are larger than the terms in the unknown series and the standard series converges, then the unknown series also converges.

In mathematical terms, if A and B are both series of positive terms such that $a_n < b_n$ for all values of n, then (a) B diverges if A diverges, and (b) A converges if B converges.

Example 3.13
Does the infinite series A converge or diverge?

$$A = 3 + \frac{9}{2} + \frac{27}{6} + \frac{81}{24} + \cdots$$

Solution

The general term was found in Ex. 3.10 to be

$$a_n = \frac{3^n}{n!}$$

Since limits of factorials are not easily determined, use the ratio test.

$$\lim_{n\to\infty}\left(\frac{a_{n+1}}{a_n}\right) = \lim_{n\to\infty}\left(\frac{\frac{3^{n+1}}{(n+1)!}}{\frac{3^n}{n!}}\right) = \lim_{n\to\infty}\left(\frac{3}{n+1}\right) = \frac{3}{\infty}$$

$$= 0$$

Since the limit is less than 1, the infinite series converges.

Example 3.14
Does the infinite series A converge or diverge?

$$A = 2 + \frac{3}{4} + \frac{4}{9} + \frac{5}{16} + \cdots$$

Solution

By observation, the general term is

$$a_n = \frac{1+n}{n^2}$$

The general term can be expanded by partial fractions to

$$a_n = \frac{1}{n} + \frac{1}{n^2}$$

However, $1/n$ is the harmonic series. Since the harmonic series is divergent and this series is larger than the harmonic series (by the term $1/n^2$), this series also diverges.

26. SERIES OF ALTERNATING SIGNS[15]

Some series contain both positive and negative terms. The ratio and comparison tests can both be used to determine if a series with alternating signs converges. If a series containing all positive terms converges, then the same series with some negative terms also converges. Therefore, the all-positive series should be tested for convergence. If the all-positive series converges, the original series is said to be *absolutely convergent.* (If the all-positive series diverges and the original series converges, the original series is said to be *conditionally convergent.*)

Alternatively, the ratio test can be used with the absolute value of the ratio. The same criteria apply.

$$\lim_{n\to\infty}\left|\frac{a_{n+1}}{a_n}\right| \begin{cases} <1 & \text{converges} \\ =1 & \text{inconclusive} \\ >1 & \text{diverges} \end{cases} \qquad 3.100$$

[15]This terminology is commonly used even though it is not necessary that the signs strictly alternate.

Linear Algebra

Content in blue refers to the *NCEES Handbook*.

1. MATRICES

A *matrix* is an ordered set of *entries* (*elements*) arranged rectangularly and set off by brackets.[1] The entries can be variables or numbers. A matrix by itself has no particular value—it is merely a convenient method of representing a set of numbers.

The size of a matrix is given by the number of rows and columns, and the nomenclature $m \times n$ is used for a matrix with m rows and n columns. For a *square matrix*, the number of rows and columns will be the same, a quantity known as the *order* of the matrix.

Bold uppercase letters are used to represent matrices, while lowercase letters represent the entries. For example, a_{23} would be the entry in the second row and third column of matrix **A**.

$$\mathbf{A} = \begin{bmatrix} a_{11} & a_{12} & a_{13} \\ a_{21} & a_{22} & a_{23} \\ a_{31} & a_{32} & a_{33} \end{bmatrix}$$

A *submatrix* is the matrix that remains when selected rows or columns are removed from the original matrix.[2] For example, for matrix **A**, the submatrix remaining after the second row and second column have been removed is

$$\begin{bmatrix} a_{11} & a_{13} \\ a_{31} & a_{33} \end{bmatrix}$$

An *augmented matrix* results when the original matrix is extended by repeating one or more of its rows or columns or by adding rows and columns from another matrix. For example, for the matrix **A**, the augmented matrix created by repeating the first and second columns is

$$\left[\begin{array}{ccc|cc} a_{11} & a_{12} & a_{13} & a_{11} & a_{12} \\ a_{21} & a_{22} & a_{23} & a_{21} & a_{22} \\ a_{31} & a_{32} & a_{33} & a_{31} & a_{32} \end{array}\right]$$

2. SPECIAL TYPES OF MATRICES

Certain types of matrices are given special designations.

- *cofactor matrix:* the matrix formed when every entry is replaced by the cofactor (see Sec. 4.4) of that entry
- *column matrix:* a matrix with only one column
- *complex matrix:* a matrix with complex number entries
- *diagonal matrix:* a square matrix with all zero entries except for the a_{ij} for which $i = j$
- *echelon matrix:* a matrix in which the number of zeros preceding the first nonzero entry of a row increases row by row until only zero rows remain. A *row-reduced echelon matrix* is an echelon matrix in which the first nonzero entry in each row is a 1 and all other entries in the columns are zero.
- *identity matrix:* a diagonal (square) matrix with all nonzero entries equal to 1, usually designated as **I**, having the property that $\mathbf{AI} = \mathbf{IA} = \mathbf{A}$
- *null matrix:* the same as a zero matrix
- *row matrix:* a matrix with only one row

[1] The term *array* is synonymous with *matrix*, although the former is more likely to be used in computer applications.
[2] By definition, a matrix is a submatrix of itself.

- *scalar matrix:*[3] a diagonal (square) matrix with all diagonal entries equal to some scalar k
- *singular matrix:* a matrix whose determinant is zero (see Sec. 4.10)
- *skew symmetric matrix:* a square matrix whose transpose (see Sec. 4.9) is equal to the negative of itself (i.e., $\mathbf{A} = -\mathbf{A}^t$)
- *square matrix:* a matrix with the same number of rows and columns (i.e., $m = n$)
- *symmetric(al) matrix:* a square matrix whose transpose is equal to itself (i.e., $\mathbf{A}^t = \mathbf{A}$), which occurs only when $a_{ij} = a_{ji}$
- *triangular matrix:* a square matrix with zeros in all positions above or below the diagonal
- *unit matrix:* the same as the identity matrix
- *zero matrix:* a matrix with all zero entries

Figure 4.1 shows examples of special matrices.

Figure 4.1 *Examples of Special Matrices*

$$\begin{bmatrix} 9 & 0 & 0 & 0 \\ 0 & -6 & 0 & 0 \\ 0 & 0 & 1 & 0 \\ 0 & 0 & 0 & 5 \end{bmatrix}$$

(a) diagonal

$$\begin{bmatrix} 2 & 18 & 2 & 18 \\ 0 & 0 & 1 & 9 \\ 0 & 0 & 0 & 9 \\ 0 & 0 & 0 & 0 \end{bmatrix}$$

(b) echelon

$$\begin{bmatrix} 1 & 9 & 0 & 0 \\ 0 & 0 & 1 & 0 \\ 0 & 0 & 0 & 1 \\ 0 & 0 & 0 & 0 \end{bmatrix}$$

(c) row-reduced echelon

$$\begin{bmatrix} 1 & 0 & 0 & 0 \\ 0 & 1 & 0 & 0 \\ 0 & 0 & 1 & 0 \\ 0 & 0 & 0 & 1 \end{bmatrix}$$

(d) identity

$$\begin{bmatrix} 3 & 0 & 0 & 0 \\ 0 & 3 & 0 & 0 \\ 0 & 0 & 3 & 0 \\ 0 & 0 & 0 & 3 \end{bmatrix}$$

(e) scalar

$$\begin{bmatrix} 2 & 0 & 0 & 0 \\ 7 & 6 & 0 & 0 \\ 9 & 1 & 1 & 0 \\ 8 & 0 & 4 & 5 \end{bmatrix}$$

(f) triangular

3. ROW EQUIVALENT MATRICES

A matrix $\mathbf{B}$ is said to be *row equivalent* to a matrix $\mathbf{A}$ if it is obtained by a finite sequence of *elementary row operations* on $\mathbf{A}$:

- interchanging the ith and jth rows
- multiplying the ith row by a nonzero scalar
- replacing the ith row by the sum of the original ith row and k times the jth row

However, two matrices that are row equivalent as defined do not necessarily have the same determinants. (See Sec. 4.5.)

Gauss-Jordan elimination is the process of using these elementary row operations to row-reduce a matrix to echelon or row-reduced echelon forms, as illustrated in Ex. 4.8. When a matrix has been converted to a row-reduced echelon matrix, it is said to be in *row canonical form.* The phrases *row-reduced echelon form* and *row canonical form* are synonymous.

4. MINORS AND COFACTORS

Minors and cofactors are determinants of submatrices associated with particular entries in the original square matrix. The *minor* of entry a_{ij} is the determinant of a submatrix resulting from the elimination of the single row i and the single column j. For example, the minor corresponding to entry a_{12} in a 3×3 matrix $\mathbf{A}$ is the determinant of the matrix created by eliminating row 1 and column 2.

$$\text{minor of } a_{12} = \begin{vmatrix} a_{21} & a_{23} \\ a_{31} & a_{33} \end{vmatrix} \quad 4.1$$

The *cofactor* of entry a_{ij} is the minor of a_{ij} multiplied by either $+1$ or -1, depending on the position of the entry. (That is, the cofactor either exactly equals the minor or it differs only in sign.) The sign is determined according to the following positional matrix.[4]

$$\begin{bmatrix} +1 & -1 & +1 & \cdots \\ -1 & +1 & -1 & \cdots \\ +1 & -1 & +1 & \cdots \\ \vdots & \vdots & \vdots & \end{bmatrix}$$

For example, the cofactor of entry a_{12} in matrix $\mathbf{A}$ (described in Sec. 4.4) is

$$\text{cofactor of } a_{12} = -\begin{vmatrix} a_{21} & a_{23} \\ a_{31} & a_{33} \end{vmatrix} \quad 4.2$$

Example 4.1

What is the cofactor corresponding to the -3 entry in the following matrix?

$$\mathbf{A} = \begin{bmatrix} 2 & 9 & 1 \\ -3 & 4 & 0 \\ 7 & 5 & 9 \end{bmatrix}$$

Solution

The minor's submatrix is created by eliminating the row and column of the -3 entry.

$$\mathbf{M} = \begin{bmatrix} 9 & 1 \\ 5 & 9 \end{bmatrix}$$

[3]Although the term *complex matrix* means a matrix with complex entries, the term *scalar matrix* means more than a matrix with scalar entries.
[4]The sign of the cofactor a_{ij} is positive if $(i + j)$ is even and is negative if $(i + j)$ is odd.

The minor is the determinant of **M**.

$$|\mathbf{M}| = (9)(9) - (5)(1) = 76$$

The sign corresponding to the -3 position is negative. Therefore, the cofactor is -76.

5. DETERMINANTS

A *determinant* is a scalar calculated from a square matrix. The determinant of matrix **A** can be represented as $D\{\mathbf{A}\}$, Det($\mathbf{A}$), $\Delta\mathbf{A}$, or $|\mathbf{A}|$.[5] The following rules can be used to simplify the calculation of determinants.

- If **A** has a row or column of zeros, the determinant is zero.
- If **A** has two identical rows or columns, the determinant is zero.
- If **B** is obtained from **A** by adding a multiple of a row (column) to another row (column) in **A**, then $|\mathbf{B}| = |\mathbf{A}|$.
- If **A** is triangular, the determinant is equal to the product of the diagonal entries.
- If **B** is obtained from **A** by multiplying one row or column in **A** by a scalar k, then $|\mathbf{B}| = k|\mathbf{A}|$.
- If **B** is obtained from the $n \times n$ matrix **A** by multiplying by the scalar matrix k, then $|\mathbf{kA}| = k^n|\mathbf{A}|$.
- If **B** is obtained from **A** by switching two rows or columns in **A**, then $|\mathbf{B}| = -|\mathbf{A}|$.

Calculation of determinants is laborious for all but the smallest or simplest of matrices. For a 2×2 matrix, the formula used to calculate the determinant is easy to remember.

$$\mathbf{A} = \begin{bmatrix} a & b \\ c & d \end{bmatrix}$$

$$|\mathbf{A}| = \begin{vmatrix} a & b \\ c & d \end{vmatrix} = ad - bc \qquad 4.3$$

Two methods are commonly used for calculating the determinant of 3×3 matrices by hand. The first uses an augmented matrix constructed from the original matrix and the first two columns (as shown in Sec. 4.1).[6] The determinant is calculated as the sum of the products in the left-to-right downward diagonals less the sum of the products in the left-to-right upward diagonals.

$$\mathbf{A} = \begin{bmatrix} a & b & c \\ d & e & f \\ g & h & i \end{bmatrix}$$

$$\text{augmented } \mathbf{A} = \left[\begin{array}{ccc|cc} a & b & c & a & b \\ d & e & f & d & e \\ g & h & i & g & h \end{array}\right] \qquad 4.4$$

$$|\mathbf{A}| = aei + bfg + cdh - gec - hfa - idb \qquad 4.5$$

The second method of calculating the determinant is somewhat slower than the first for a 3×3 matrix but illustrates the method that must be used to calculate determinants of 4×4 and larger matrices. This method is known as *expansion by cofactors*. One row (column) is selected as the base row (column). The selection is arbitrary, but the number of calculations required to obtain the determinant can be minimized by choosing the row (column) with the most zeros. The determinant is equal to the sum of the products of the entries in the base row (column) and their corresponding cofactors.

$$\mathbf{A} = \begin{bmatrix} a & b & c \\ d & e & f \\ g & h & i \end{bmatrix} \qquad 4.6$$

$$|\mathbf{A}| = a\begin{vmatrix} e & f \\ h & i \end{vmatrix} - d\begin{vmatrix} b & c \\ h & i \end{vmatrix} + g\begin{vmatrix} b & c \\ e & f \end{vmatrix}$$

Example 4.2

Calculate the determinant of matrix **A** (a) by cofactor expansion, and (b) by the augmented matrix method.

$$\mathbf{A} = \begin{bmatrix} 2 & 3 & -4 \\ 3 & -1 & -2 \\ 4 & -7 & -6 \end{bmatrix}$$

Solution

(a) Since there are no zero entries, it does not matter which row or column is chosen as the base. Choose the first column as the base.

$$\begin{aligned} |\mathbf{A}| &= 2\begin{vmatrix} -1 & -2 \\ -7 & -6 \end{vmatrix} - 3\begin{vmatrix} 3 & -4 \\ -7 & -6 \end{vmatrix} + 4\begin{vmatrix} 3 & -4 \\ -1 & -2 \end{vmatrix} \\ &= (2)(6 - 14) - (3)(-18 - 28) + (4)(-6 - 4) \\ &= 82 \end{aligned}$$

[5]The vertical bars should not be confused with the square brackets used to set off a matrix, nor with absolute value.
[6]It is not actually necessary to construct the augmented matrix, but doing so helps avoid errors.

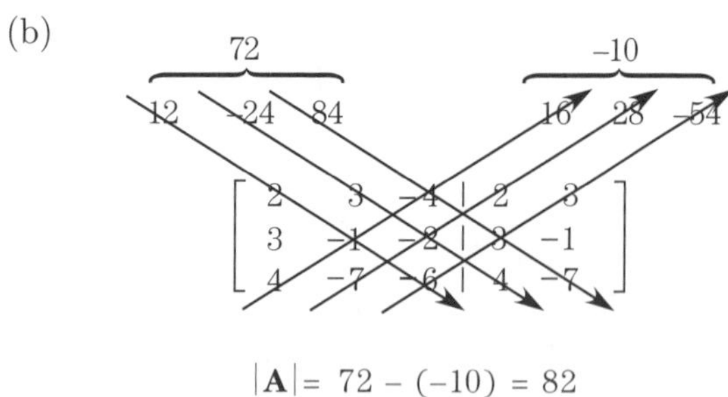

6. MATRIX ALGEBRA[7]

Matrix algebra differs somewhat from standard algebra.

- *equality:* Two matrices, **A** and **B**, are equal only if they have the same numbers of rows and columns *and* if all corresponding entries are equal.
- *inequality:* The $>$ and $<$ operators are not used in matrix algebra.
- *commutative law of addition:*

$$\mathbf{A}+\mathbf{B}=\mathbf{B}+\mathbf{A} \quad 4.7$$

- *associative law of addition:*

$$\mathbf{A}+(\mathbf{B}+\mathbf{C})=(\mathbf{A}+\mathbf{B})+\mathbf{C} \quad 4.8$$

- *associative law of multiplication:*

$$(\mathbf{AB})\mathbf{C}=\mathbf{A}(\mathbf{BC}) \quad 4.9$$

- *left distributive law:*

$$\mathbf{A}(\mathbf{B}+\mathbf{C})=\mathbf{AB}+\mathbf{AC} \quad 4.10$$

- *right distributive law:*

$$(\mathbf{B}+\mathbf{C})\mathbf{A}=\mathbf{BA}+\mathbf{CA} \quad 4.11$$

- *scalar multiplication:*

$$k(\mathbf{AB})=(k\mathbf{A})\mathbf{B}=\mathbf{A}(k\mathbf{B}) \quad 4.12$$

Except for trivial and special cases, matrix multiplication is not commutative. That is,

$$\mathbf{AB}\neq\mathbf{BA}$$

7. MATRIX ADDITION AND SUBTRACTION

Addition and subtraction of two matrices is possible only if both matrices have the same numbers of rows and columns (i.e., order). They are accomplished by adding or subtracting the corresponding entries of the two matrices.

8. MATRIX MULTIPLICATION AND DIVISION

A matrix can be multiplied by a scalar, an operation known as *scalar multiplication*, in which case all entries of the matrix are multiplied by that scalar. For example, for the 2×2 matrix **A**,

$$k\mathbf{A}=\begin{bmatrix} ka_{11} & ka_{12} \\ ka_{21} & ka_{22} \end{bmatrix}$$

A matrix can be multiplied by another matrix, but only if the left-hand matrix has the same number of columns as the right-hand matrix has rows. *Matrix multiplication* occurs by multiplying the elements in each left-hand matrix row by the entries in each right-hand matrix column, adding the products, and placing the sum at the intersection point of the participating row and column.

Matrix division can only be accomplished by multiplying by the inverse of the denominator matrix. There is no specific division operation in matrix algebra.

Example 4.3

Determine the product matrix **C**.

$$\mathbf{C}=\begin{bmatrix} 1 & 4 & 3 \\ 5 & 2 & 6 \end{bmatrix}\begin{bmatrix} 7 & 12 \\ 11 & 8 \\ 9 & 10 \end{bmatrix}$$

Solution

The left-hand matrix has three columns, and the right-hand matrix has three rows. Therefore, the two matrices can be multiplied.

The first row of the left-hand matrix and the first column of the right-hand matrix are worked with first. The corresponding entries are multiplied, and the products are summed.

$$c_{11}=(1)(7)+(4)(11)+(3)(9)=78$$

The intersection of the top row and left column is the entry in the upper left-hand corner of the matrix **C**.

The remaining entries are calculated similarly.

$$c_{12}=(1)(12)+(4)(8)+(3)(10)=74$$
$$c_{21}=(5)(7)+(2)(11)+(6)(9)=111$$
$$c_{22}=(5)(12)+(2)(8)+(6)(10)=136$$

The product matrix is

$$\mathbf{C}=\begin{bmatrix} 78 & 74 \\ 111 & 136 \end{bmatrix}$$

[7]Since matrices are used to simplify the presentation and solution of sets of linear equations, matrix algebra is also known as *linear algebra*.

9. TRANSPOSE

The *transpose*, $\mathbf{A}^t$, of an $m \times n$ matrix $\mathbf{A}$ is an $n \times m$ matrix constructed by taking the ith row and making it the ith column. The diagonal is unchanged. For example,

$$\mathbf{A} = \begin{bmatrix} 1 & 6 & 9 \\ 2 & 3 & 4 \\ 7 & 1 & 5 \end{bmatrix}$$

$$\mathbf{A}^t = \begin{bmatrix} 1 & 2 & 7 \\ 6 & 3 & 1 \\ 9 & 4 & 5 \end{bmatrix}$$

Transpose operations have the following characteristics.

$$(\mathbf{A}^t)^t = \mathbf{A} \qquad 4.13$$

$$(k\mathbf{A})^t = k(\mathbf{A}^t) \qquad 4.14$$

$$\mathbf{I}^t = \mathbf{I} \qquad 4.15$$

$$(\mathbf{AB})^t = \mathbf{B}^t\mathbf{A}^t \qquad 4.16$$

$$(\mathbf{A}+\mathbf{B})^t = \mathbf{A}^t + \mathbf{B}^t \qquad 4.17$$

$$|\mathbf{A}^t| = |\mathbf{A}| \qquad 4.18$$

10. SINGULARITY AND RANK

A *singular matrix* is one whose determinant is zero. Similarly, a *nonsingular* matrix is one whose determinant is nonzero.

The *rank* of a matrix is the maximum number of linearly independent row or column vectors.[8] A matrix has rank r if it has at least one nonsingular square submatrix of order r but has no nonsingular square submatrix of order more than r. While the submatrix must be square (in order to calculate the determinant), the original matrix need not be.

The rank of an $m \times n$ matrix will be, at most, the smaller of m and n. The rank of a null matrix is zero. The ranks of a matrix and its transpose are the same. If a matrix is in echelon form, the rank will be equal to the number of rows containing at least one nonzero entry. For a 3×3 matrix, the rank can either be 3 (if it is nonsingular), 2 (if any one of its 2×2 submatrices is nonsingular), 1 (if it and all 2×2 submatrices are singular), or 0 (if it is null).

The determination of rank is laborious if done by hand. Either the matrix is reduced to echelon form by using elementary row operations, or exhaustive enumeration is used to create the submatrices and many determinants are calculated. If a matrix has more rows than columns and row-reduction is used, the work required to put the matrix in echelon form can be reduced by working with the transpose of the original matrix.

Example 4.4
What is the rank of matrix $\mathbf{A}$?

$$\mathbf{A} = \begin{bmatrix} 1 & -2 & -1 \\ -3 & 3 & 0 \\ 2 & 2 & 4 \end{bmatrix}$$

Solution

Matrix $\mathbf{A}$ is singular because $|\mathbf{A}| = 0$. However, there is at least one 2×2 nonsingular submatrix.

$$\begin{vmatrix} 1 & -2 \\ -3 & 3 \end{vmatrix} = (1)(3) - (-3)(-2) = -3$$

Therefore, the rank is 2.

Example 4.5
Determine the rank of matrix $\mathbf{A}$ by reducing it to echelon form.

$$\mathbf{A} = \begin{bmatrix} 7 & 4 & 9 & 1 \\ 0 & 2 & -5 & 3 \\ 0 & 4 & -10 & 6 \end{bmatrix}$$

Solution

By inspection, the matrix can be row-reduced by subtracting two times the second row from the third row. The matrix cannot be further reduced. Since there are two nonzero rows, the rank is 2.

$$\begin{bmatrix} 7 & 4 & 9 & 1 \\ 0 & 2 & -5 & 3 \\ 0 & 0 & 0 & 0 \end{bmatrix}$$

11. CLASSICAL ADJOINT

The *classical adjoint* is the transpose of the cofactor matrix. (See Sec. 4.2.) The resulting matrix can be designated as $\mathbf{A}_{\text{adj}}$, $\text{adj}\{\mathbf{A}\}$, or $\mathbf{A}^{\text{adj}}$.

Example 4.6
What is the classical adjoint of matrix $\mathbf{A}$?

$$\mathbf{A} = \begin{bmatrix} 2 & 3 & -4 \\ 0 & -4 & 2 \\ 1 & -1 & 5 \end{bmatrix}$$

[8] The *row rank* and *column rank* are the same.

Solution

The matrix of cofactors is determined to be

$$\begin{bmatrix} -18 & 2 & 4 \\ -11 & 14 & 5 \\ -10 & -4 & -8 \end{bmatrix}$$

The transpose of the matrix of cofactors is

$$\mathbf{A}_{\text{adj}} = \begin{bmatrix} -18 & -11 & -10 \\ 2 & 14 & -4 \\ 4 & 5 & -8 \end{bmatrix}$$

12. INVERSE

The product of a matrix **A** and its inverse, $\mathbf{A}^{-1}$, is the identity matrix, **I**. Only square matrices have inverses, but not all square matrices are invertible. A matrix has an inverse if and only if it is nonsingular (i.e., its determinant is nonzero).

$$\mathbf{AA}^{-1} = \mathbf{A}^{-1}\mathbf{A} = \mathbf{I} \qquad 4.19$$

$$(\mathbf{AB})^{-1} = \mathbf{B}^{-1}\mathbf{A}^{-1} \qquad 4.20$$

The inverse of a 2×2 matrix is most easily determined by formula.

$$\mathbf{A} = \begin{bmatrix} a & b \\ c & d \end{bmatrix}$$

$$\mathbf{A}^{-1} = \frac{\begin{bmatrix} d & -b \\ -c & a \end{bmatrix}}{|\mathbf{A}|} \qquad 4.21$$

For any matrix, the inverse is determined by dividing every entry in the classical adjoint by the determinant of the original matrix.

$$\mathbf{A}^{-1} = \frac{\mathbf{A}_{\text{adj}}}{|\mathbf{A}|} \qquad 4.22$$

Example 4.7

What is the inverse of matrix **A**?

$$\mathbf{A} = \begin{bmatrix} 4 & 5 \\ 2 & 3 \end{bmatrix}$$

Solution

The determinant is calculated as

$$|\mathbf{A}| = (4)(3) - (2)(5) = 2$$

Using Eq. 4.22, the inverse is

$$\mathbf{A}^{-1} = \frac{\begin{bmatrix} 3 & -5 \\ -2 & 4 \end{bmatrix}}{2} = \begin{bmatrix} \frac{3}{2} & -\frac{5}{2} \\ -1 & 2 \end{bmatrix}$$

Check.

$$\mathbf{AA}^{-1} = \begin{bmatrix} 4 & 5 \\ 2 & 3 \end{bmatrix}\begin{bmatrix} \frac{3}{2} & -\frac{5}{2} \\ -1 & 2 \end{bmatrix} = \begin{bmatrix} 6-5 & -10+10 \\ 3-3 & -5+6 \end{bmatrix}$$

$$= \begin{bmatrix} 1 & 0 \\ 0 & 1 \end{bmatrix}$$

$$= \mathbf{I} \quad \text{[OK]}$$

13. WRITING SIMULTANEOUS LINEAR EQUATIONS IN MATRIX FORM

Matrices are used to simplify the presentation and solution of sets of simultaneous linear equations. For example, the following three methods of presenting simultaneous linear equations are equivalent.

$$a_{11}x_1 + a_{12}x_2 = b_1$$
$$a_{21}x_1 + a_{22}x_2 = b_2$$

$$\begin{bmatrix} a_{11} & a_{12} \\ a_{21} & a_{22} \end{bmatrix}\begin{bmatrix} x_1 \\ x_2 \end{bmatrix} = \begin{bmatrix} b_1 \\ b_2 \end{bmatrix}$$

$$\mathbf{AX} = \mathbf{B}$$

In the second and third representations, **A** is known as the *coefficient matrix*, **X** as the *variable matrix*, and **B** as the *constant matrix*.

Not all systems of simultaneous equations have solutions, and those that do may not have unique solutions. The existence of a solution can be determined by calculating the determinant of the coefficient matrix. Solution-existence rules are summarized in Table 4.1.

- If the system of linear equations is homogeneous (i.e., **B** is a zero matrix) and $|\mathbf{A}|$ is zero, there are an infinite number of solutions.
- If the system is homogeneous and $|\mathbf{A}|$ is nonzero, only the trivial solution exists.
- If the system of linear equations is nonhomogeneous (i.e., **B** is not a zero matrix) and $|\mathbf{A}|$ is nonzero, there is a unique solution to the set of simultaneous equations.

- If $|\mathbf{A}|$ is zero, a nonhomogeneous system of simultaneous equations may still have a solution. The requirement is that the determinants of all substitutional matrices (see Sec. 4.14) are zero, in which case there will be an infinite number of solutions. Otherwise, no solution exists.

Table 4.1 Solution-Existence Rules for Simultaneous Equations

	$\mathbf{B}=0$	$\mathbf{B}\neq 0$
$\lvert\mathbf{A}\rvert=0$	infinite number of solutions (linearly dependent equations)	either an infinite number of solutions or no solution at all
$\lvert\mathbf{A}\rvert\neq 0$	trivial solution only ($x_i=0$)	unique nonzero solution

14. SOLVING SIMULTANEOUS LINEAR EQUATIONS

Gauss-Jordan elimination can be used to obtain the solution to a set of simultaneous linear equations. The coefficient matrix is augmented by the constant matrix. Then, elementary row operations are used to reduce the coefficient matrix to canonical form. All of the operations performed on the coefficient matrix are performed on the constant matrix. The variable values that satisfy the simultaneous equations will be the entries in the constant matrix when the coefficient matrix is in canonical form.

Determinants are used to calculate the solution to linear simultaneous equations through a procedure known as *Cramer's rule.*

The procedure is to calculate determinants of the original coefficient matrix $\mathbf{A}$ and of the n matrices resulting from the systematic replacement of a column in $\mathbf{A}$ by the constant matrix $\mathbf{B}$. For a system of three equations in three unknowns, there are three substitutional matrices, $\mathbf{A}_1$, $\mathbf{A}_2$, and $\mathbf{A}_3$, as well as the original coefficient matrix, for a total of four matrices whose determinants must be calculated.

The values of the unknowns that simultaneously satisfy all of the linear equations are

$$x_1=\frac{|\mathbf{A}_1|}{|\mathbf{A}|} \qquad 4.23$$

$$x_2=\frac{|\mathbf{A}_2|}{|\mathbf{A}|} \qquad 4.24$$

$$x_3=\frac{|\mathbf{A}_3|}{|\mathbf{A}|} \qquad 4.25$$

Example 4.8

Use Gauss-Jordan elimination to solve the following system of simultaneous equations.

$$\begin{aligned}2x+3y-4z&=1\\3x-y-2z&=4\\4x-7y-6z&=-7\end{aligned}$$

Solution

The augmented matrix is created by appending the constant matrix to the coefficient matrix.

$$\left[\begin{array}{rrr|r}2&3&-4&1\\3&-1&-2&4\\4&-7&-6&-7\end{array}\right]$$

Elementary row operations are used to reduce the coefficient matrix to canonical form. For example, two times the first row is subtracted from the third row. This step obtains the 0 needed in the a_{31} position.

$$\left[\begin{array}{rrr|r}2&3&-4&1\\3&-1&-2&4\\0&-13&2&-9\end{array}\right]$$

This process continues until the following form is obtained.

$$\left[\begin{array}{rrr|r}1&0&0&3\\0&1&0&1\\0&0&1&2\end{array}\right]$$

$x=3$, $y=1$, and $z=2$ satisfy this system of equations.

Example 4.9

Use Cramer's rule to solve the following system of simultaneous equations.

$$\begin{aligned}2x+3y-4z&=1\\3x-y-2z&=4\\4x-7y-6z&=-7\end{aligned}$$

Solution

The determinant of the coefficient matrix is

$$|\mathbf{A}|=\begin{vmatrix}2&3&-4\\3&-1&-2\\4&-7&-6\end{vmatrix}=82$$

The determinants of the substitutional matrices are

$$|\mathbf{A}_1| = \begin{vmatrix} 1 & 3 & -4 \\ 4 & -1 & -2 \\ -7 & -7 & -6 \end{vmatrix} = 246$$

$$|\mathbf{A}_2| = \begin{vmatrix} 2 & 1 & -4 \\ 3 & 4 & -2 \\ 4 & -7 & -6 \end{vmatrix} = 82$$

$$|\mathbf{A}_3| = \begin{vmatrix} 2 & 3 & 1 \\ 3 & -1 & 4 \\ 4 & -7 & -7 \end{vmatrix} = 164$$

The values of x, y, and z that will satisfy the linear equations are

$$x = \frac{246}{82} = 3$$
$$y = \frac{82}{82} = 1$$
$$z = \frac{164}{82} = 2$$

15. EIGENVALUES AND EIGENVECTORS

Eigenvalues and eigenvectors (also known as *characteristic values* and *characteristic vectors*) of a square matrix **A** are the scalars k and matrices **X** such that

$$\mathbf{AX} = k\mathbf{X} \qquad 4.26$$

The scalar k is an eigenvalue of **A** if and only if the matrix $(k\mathbf{I} - \mathbf{A})$ is singular; that is, if $|k\mathbf{I} - \mathbf{I}| = 0$. This equation is called the *characteristic equation* of the matrix **A**. When expanded, the determinant is called the *characteristic polynomial*. The method of using the characteristic polynomial to find eigenvalues and eigenvectors is illustrated in Ex. 4.10.

If all of the eigenvalues are unique (i.e., nonrepeating), then Eq. 4.27 is valid.

$$[k\mathbf{I} - \mathbf{A}]\mathbf{X} = 0 \qquad 4.27$$

Example 4.10

Find the eigenvalues and nonzero eigenvectors of the matrix **A**.

$$\mathbf{A} = \begin{bmatrix} 2 & 4 \\ 6 & 4 \end{bmatrix}$$

Solution

$$k\mathbf{I} - \mathbf{A} = \begin{bmatrix} k & 0 \\ 0 & k \end{bmatrix} - \begin{bmatrix} 2 & 4 \\ 6 & 4 \end{bmatrix} = \begin{bmatrix} k-2 & -4 \\ -6 & k-4 \end{bmatrix}$$

The characteristic polynomial is found by setting the determinant $|k\mathbf{I} - \mathbf{A}|$ equal to zero.

$$(k-2)(k-4) - (-6)(-4) = 0$$
$$k^2 - 6k - 16 = (k-8)(k+2) = 0$$

The roots of the characteristic polynomial are $k = +8$ and $k = -2$. These are the eigenvalues of **A**.

Substituting $k = 8$,

$$k\mathbf{I} - \mathbf{A} = \begin{bmatrix} 8-2 & -4 \\ -6 & 8-4 \end{bmatrix} = \begin{bmatrix} 6 & -4 \\ -6 & 4 \end{bmatrix}$$

The resulting system can be interpreted as the linear equation $6x_1 - 4x_2 = 0$. The values of x that satisfy this equation define the eigenvector. An eigenvector **X** associated with the eigenvalue +8 is

$$\mathbf{X} = \begin{bmatrix} x_1 \\ x_2 \end{bmatrix} = \begin{bmatrix} 4 \\ 6 \end{bmatrix}$$

All other eigenvectors for this eigenvalue are multiples of **X**. Normally **X** is reduced to smallest integers.

$$\mathbf{X} = \begin{bmatrix} 2 \\ 3 \end{bmatrix}$$

Similarly, the eigenvector associated with the eigenvalue −2 is

$$\mathbf{X} = \begin{bmatrix} x_1 \\ x_2 \end{bmatrix} = \begin{bmatrix} +4 \\ -4 \end{bmatrix}$$

Reducing this to smallest integers gives

$$\mathbf{X} = \begin{bmatrix} +1 \\ -1 \end{bmatrix}$$

5 Vectors

Content in blue refers to the *NCEES Handbook*.

1. INTRODUCTION

Some characteristics can be described by scalars, vectors, or tensors. A *scalar* has only magnitude. Knowing its value is sufficient to define the characteristic. Mass, enthalpy, density, and speed are examples of scalar characteristics.

Force, momentum, displacement, and velocity are examples of vector characteristics. A *vector* is a directed straight line with a specific magnitude and that is specified completely by its direction (consisting of the vector's *angular orientation* and its *sense*) and magnitude. A vector's *point of application* (*terminal point*) is not needed to define the vector.[1] Two vectors with the same direction and magnitude are said to be *equal vectors* even though their *lines of action* may be different.[2]

A vector can be designated by a boldface variable (as in this book) or as a combination of the variable and some other symbol. For example, the notations $\mathbf{V}$, $\overline{V}$, $\widehat{V}$, $\vec{V}$, and $\underline{V}$ are used by different authorities to represent vectors. In this book, the magnitude of a vector can be designated by either $|\mathbf{V}|$ or V (italic but not bold).

Stress, dielectric constant, and magnetic susceptibility are examples of tensor characteristics. A *tensor* has magnitude in a specific direction, but the direction is not unique. Tensors are frequently associated with *anisotropic materials* that have different properties in different directions. A tensor in three-dimensional space is defined by nine components, compared with the three that are required to define vectors. These components are written in matrix form. Stress, σ, at a point, for example, would be defined by the following tensor matrix.

$$\sigma \equiv \begin{bmatrix} \sigma_{xx} & \sigma_{xy} & \sigma_{xz} \\ \sigma_{yx} & \sigma_{yy} & \sigma_{yz} \\ \sigma_{zx} & \sigma_{zy} & \sigma_{zz} \end{bmatrix}$$

2. VECTORS IN *n*-SPACE

In some cases, a vector, $\mathbf{V}$, will be designated by its two endpoints in n-dimensional vector space. The usual vector space is three-dimensional force-space. Usually, one of the points will be the origin, in which case the vector is said to be "based at the origin," "origin-based," or "zero-based."[3] If one of the endpoints is the origin, specifying a terminal point P would represent a force directed from the origin to point P.

If a coordinate system is superimposed on the vector space, a vector can be specified in terms of the n coordinates of its two endpoints. The magnitude of the vector $\mathbf{V}$ is the distance in vector space between the two points, as given by Eq. 5.1. Similarly, the direction is defined by the angle the vector makes with one of the axes. Figure 5.1 illustrates a vector in two dimensions.

$$|\mathbf{V}| = \sqrt{(x_2 - x_1)^2 + (y_2 - y_1)^2} \quad \text{5.1}$$

$$\phi = \arctan \frac{y_2 - y_1}{x_2 - x_1} \quad \text{5.2}$$

The *components* of a vector are the projections of the vector on the coordinate axes. (For a zero-based vector, the components and the coordinates of the endpoint are the same.) Simple trigonometric principles are used to resolve a vector into its components. A vector reconstructed from its components is known as a *resultant vector*.

[1] A vector that is constrained to act at or through a certain point is a *bound vector* (*fixed vector*). A *sliding vector* (*transmissible vector*) can be applied anywhere along its line of action. A *free vector* is not constrained and can be applied at any point in space.

[2] A distinction is sometimes made between equal vectors and equivalent vectors. *Equivalent vectors* produce the same effect but are not necessarily equal.

[3] Any vector directed from P_1 to P_2 can be transformed into a zero-based vector by subtracting the coordinates of point P_1 from the coordinates of terminal point P_2. The transformed vector will be equivalent to the original vector.

Figure 5.1 Vector in Two-Dimensional Space

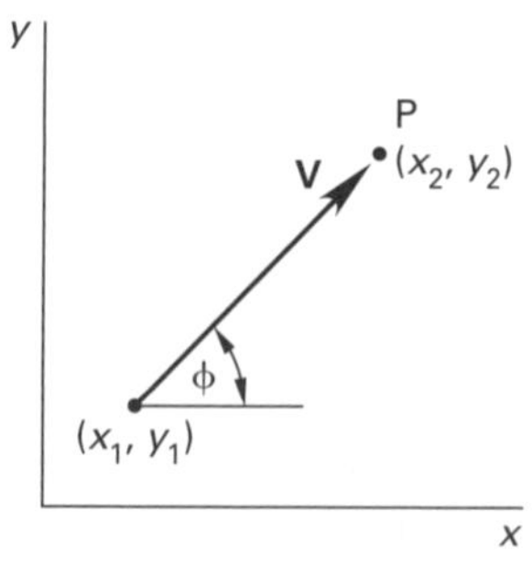

$$V_x = |\mathbf{V}|\cos\phi_x \qquad 5.3$$

$$V_y = |\mathbf{V}|\cos\phi_y \qquad 5.4$$

$$V_z = |\mathbf{V}|\cos\phi_z \qquad 5.5$$

$$|\mathbf{V}| = \sqrt{V_x^2 + V_y^2 + V_z^2} \qquad 5.6$$

In Eq. 5.3 through Eq. 5.5, ϕ_x, ϕ_y, and ϕ_z are the *direction angles*—the angles between the vector and the x-, y-, and z-axes, respectively. Figure 5.2 shows the location of direction angles. The cosines of these angles are known as *direction cosines*. The sum of the squares of the direction cosines is equal to 1.

$$\cos^2\phi_x + \cos^2\phi_y + \cos^2\phi_z = 1 \qquad 5.7$$

Figure 5.2 Direction Angles of a Vector

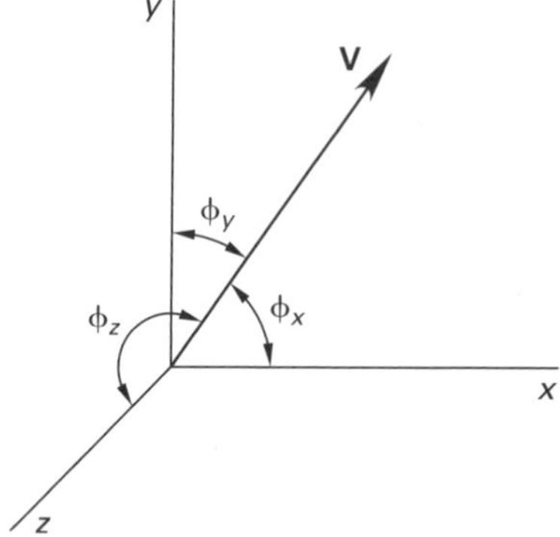

3. UNIT VECTORS

Unit vectors are vectors with unit magnitudes (i.e., magnitudes of 1). They are represented in the same notation as other vectors. (Unit vectors in this book are written in boldface type.) Although they can have any direction, the standard unit vectors (the *Cartesian unit vectors* **i**, **j**, and **k**) have the directions of the x-, y-, and z-coordinate axes and constitute the *Cartesian triad*, as illustrated in Fig. 5.3.

A vector **V** can be written in terms of unit vectors and its components.

$$\mathbf{V} = |\mathbf{V}|\mathbf{a} = V_x\mathbf{i} + V_y\mathbf{j} + V_z\mathbf{k} \qquad 5.8$$

Figure 5.3 Cartesian Unit Vectors

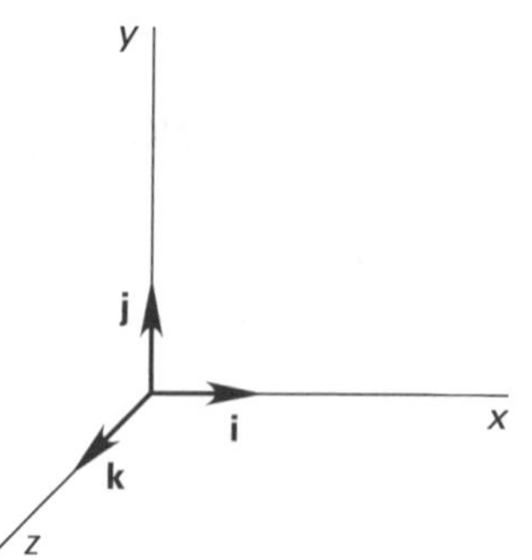

The unit vector, **a**, has the same direction as the vector **V** but has a length of 1. This unit vector is calculated by dividing the original vector, **V**, by its magnitude, |**V**|.

$$\mathbf{a} = \frac{\mathbf{V}}{|\mathbf{V}|} = \frac{V_x\mathbf{i} + V_y\mathbf{j} + V_z\mathbf{k}}{\sqrt{V_x^2 + V_y^2 + V_z^2}} \qquad 5.9$$

4. VECTOR REPRESENTATION

The most common method of representing a vector is by writing it in *rectangular form*—a vector sum of its orthogonal components. In rectangular form, each of the orthogonal components has the same units as the resultant vector.

$$\mathbf{A} \equiv A_x\mathbf{i} + A_y\mathbf{j} + A_z\mathbf{k} \quad \text{[three dimensions]}$$

However, the vector is also completely defined by its magnitude and associated angle. These two quantities can be written together in *phasor form*, sometimes referred to as *polar form*.

$$\mathbf{A} \equiv |\mathbf{A}|\angle\phi = A\angle\phi$$

5. CONVERSION BETWEEN SYSTEMS

The choice of the **ijk** triad may be convenient but is arbitrary. A vector can be expressed in terms of any other set of orthogonal unit vectors, **uvw**.

$$\mathbf{V} = V_x\mathbf{i} + V_y\mathbf{j} + V_z\mathbf{k} = V_x'\mathbf{u} + V_y'\mathbf{v} + V_z'\mathbf{w} \qquad 5.10$$

The two representations are related.

$$V_x' = \mathbf{V}\cdot\mathbf{u} = (\mathbf{i}\cdot\mathbf{u})\,V_x + (\mathbf{j}\cdot\mathbf{u})\,V_y + (\mathbf{k}\cdot\mathbf{u})\,V_z \qquad 5.11$$

$$V'_y = \mathbf{V}\cdot\mathbf{v} = (\mathbf{i}\cdot\mathbf{v})V_x + (\mathbf{j}\cdot\mathbf{v})V_y + (\mathbf{k}\cdot\mathbf{v})V_z \quad 5.12$$

$$V'_z = \mathbf{V}\cdot\mathbf{w} = (\mathbf{i}\cdot\mathbf{w})V_x + (\mathbf{j}\cdot\mathbf{w})V_y + (\mathbf{k}\cdot\mathbf{w})V_z \quad 5.13$$

Equation 5.11 through Eq. 5.13 can be expressed in matrix form. The dot products are known as the *coefficients of transformation*, and the matrix containing them is the *transformation matrix*.

$$\begin{pmatrix} V'_x \\ V'_y \\ V'_z \end{pmatrix} = \begin{pmatrix} \mathbf{i}\cdot\mathbf{u} & \mathbf{j}\cdot\mathbf{u} & \mathbf{k}\cdot\mathbf{u} \\ \mathbf{i}\cdot\mathbf{v} & \mathbf{j}\cdot\mathbf{v} & \mathbf{k}\cdot\mathbf{v} \\ \mathbf{i}\cdot\mathbf{w} & \mathbf{j}\cdot\mathbf{w} & \mathbf{k}\cdot\mathbf{w} \end{pmatrix} \begin{pmatrix} V_x \\ V_y \\ V_z \end{pmatrix} \quad 5.14$$

6. VECTOR ADDITION

Addition of two vectors by the *polygon method* is accomplished by placing the tail of the second vector at the head (tip) of the first. The sum (i.e., the *resultant vector*) is a vector extending from the tail of the first vector to the head of the second, as shown in Fig. 5.4. Alternatively, the two vectors can be considered as the two sides of a parallelogram, while the sum represents the diagonal. This is known as addition by the *parallelogram method*.

Figure 5.4 *Addition of Two Vectors*

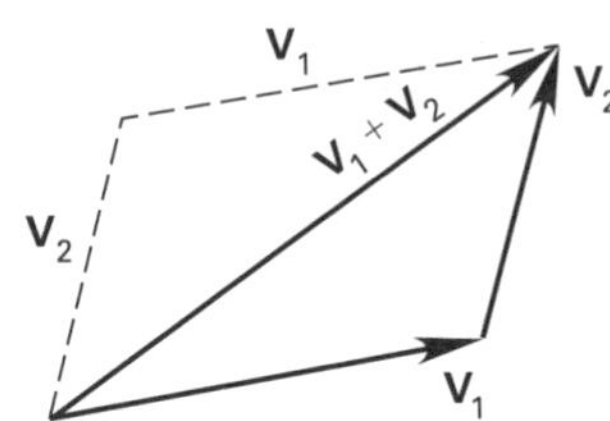

The components of the resultant vector are the sums of the components of the added vectors (that is, $V_{1x} + V_{2x}$, $V_{1y} + V_{2y}$, $V_{1z} + V_{2z}$).

Vector addition is both commutative and associative.

$$\mathbf{V}_1 + \mathbf{V}_2 = \mathbf{V}_2 + \mathbf{V}_1 \quad 5.15$$

$$\mathbf{V}_1 + (\mathbf{V}_2 + \mathbf{V}_3) = (\mathbf{V}_1 + \mathbf{V}_2) + \mathbf{V}_3 \quad 5.16$$

7. MULTIPLICATION BY A SCALAR

A vector, $\mathbf{V}$, can be multiplied by a scalar, c. If the original vector is represented by its components, each of the components is multiplied by c.

$$c\mathbf{V} = c|\mathbf{V}|\,\mathbf{a} = cV_x\mathbf{i} + cV_y\mathbf{j} + cV_z\mathbf{k} \quad 5.17$$

Scalar multiplication is distributive.

$$c(\mathbf{V}_1 + \mathbf{V}_2) = c\mathbf{V}_1 + c\mathbf{V}_2 \quad 5.18$$

8. VECTOR DOT PRODUCT

The *dot product* (*scalar product*), $\mathbf{V}_1\cdot\mathbf{V}_2$, of two vectors is a scalar that is proportional to the length of the projection of the first vector onto the second vector, as illustrated in Fig. 5.5.[4]

Figure 5.5 *Vector Dot Product*

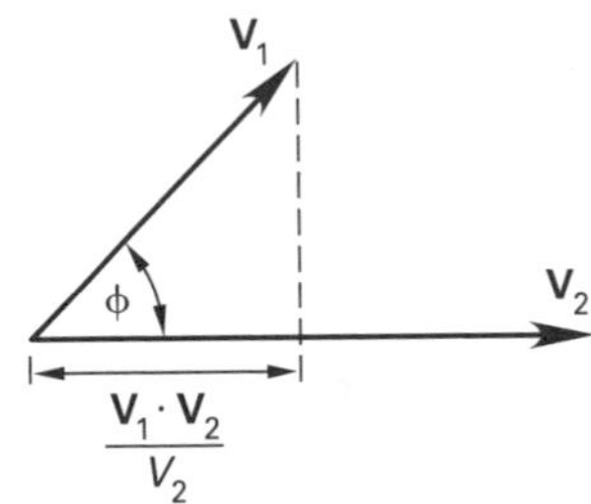

The dot product is commutative and distributive.

$$\mathbf{V}_1\cdot\mathbf{V}_2 = \mathbf{V}_2\cdot\mathbf{V}_1 \quad 5.19$$

$$\mathbf{V}_1\cdot(\mathbf{V}_2 + \mathbf{V}_3) = \mathbf{V}_1\cdot\mathbf{V}_2 + \mathbf{V}_1\cdot\mathbf{V}_3 \quad 5.20$$

The dot product can be calculated in two ways, as Eq. 5.21 indicates. ϕ is the acute angle between the two vectors and is less than or equal to 180°.

$$\begin{aligned}\mathbf{V}_1\cdot\mathbf{V}_2 &= |\mathbf{V}_1|\,|\mathbf{V}_2|\cos\phi \\ &= V_{1x}V_{2x} + V_{1y}V_{2y} + V_{1z}V_{2z}\end{aligned} \quad 5.21$$

When Eq. 5.21 is solved for the angle between the two vectors, ϕ, it is known as the *Cauchy-Schwartz theorem*.

$$\cos\phi = \frac{V_{1x}V_{2x} + V_{1y}V_{2y} + V_{1z}V_{2z}}{|\mathbf{V}_1|\,|\mathbf{V}_2|} \quad 5.22$$

The dot product can be used to determine whether a vector is a unit vector and to show that two vectors are orthogonal (perpendicular). For any unit vector, $\mathbf{u}$,

$$\mathbf{u}\cdot\mathbf{u} = 1 \quad 5.23$$

For two non-null orthogonal vectors,

$$\mathbf{V}_1\cdot\mathbf{V}_2 = 0 \quad 5.24$$

Equation 5.23 and Eq. 5.24 can be extended to the Cartesian unit vectors.

$$\mathbf{i}\cdot\mathbf{i} = 1 \quad 5.25$$

$$\mathbf{j}\cdot\mathbf{j} = 1 \quad 5.26$$

$$\mathbf{k}\cdot\mathbf{k} = 1 \quad 5.27$$

[4]The dot product is also written in parentheses without a dot; that is, $(\mathbf{V}_1\mathbf{V}_2)$.

$$\mathbf{i}\cdot\mathbf{j} = 0 \quad (5.28)$$

$$\mathbf{i}\cdot\mathbf{k} = 0 \quad (5.29)$$

$$\mathbf{j}\cdot\mathbf{k} = 0 \quad (5.30)$$

Example 5.1

What is the angle between the zero-based vectors $\mathbf{V}_1 = (-\sqrt{3}, 1)$ and $\mathbf{V}_2 = (2\sqrt{3}, 2)$ in an x-y coordinate system?

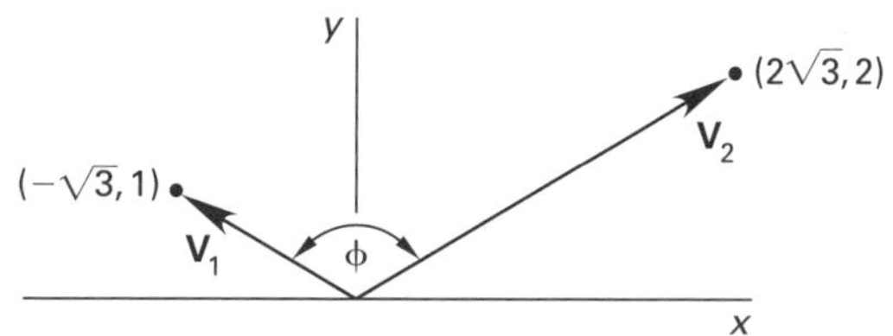

Solution

From Eq. 5.22,

$$\begin{aligned}\cos\phi &= \frac{V_{1x}V_{2x} + V_{1y}V_{2y}}{|\mathbf{V}_1|\,|\mathbf{V}_2|} = \frac{V_{1x}V_{2x} + V_{1y}V_{2y}}{\sqrt{V_{1x}^2 + V_{1y}^2}\sqrt{V_{2x}^2 + V_{2y}^2}}\\ &= \frac{(-\sqrt{3})(2\sqrt{3}) + (1)(2)}{\sqrt{(-\sqrt{3})^2 + (1)^2}\sqrt{(2\sqrt{3})^2 + (2)^2}}\\ &= \frac{-4}{8}\quad\left(-\frac{1}{2}\right)\\ \phi &= \arccos\left(-\frac{1}{2}\right) = 120^\circ\end{aligned}$$

9. VECTOR CROSS PRODUCT

The *cross product* (*vector product*), $\mathbf{V}_1 \times \mathbf{V}_2$, of two vectors is a vector that is orthogonal (perpendicular) to the plane of the two vectors.[5] The unit vector representation of the cross product can be calculated as a third-order determinant. Figure 5.6 illustrates the vector cross product.

$$\mathbf{V}_1 \times \mathbf{V}_2 = \begin{vmatrix} \mathbf{i} & V_{1x} & V_{2x} \\ \mathbf{j} & V_{1y} & V_{2y} \\ \mathbf{k} & V_{1z} & V_{2z} \end{vmatrix} \quad (5.31)$$

The direction of the cross-product vector corresponds to the direction a right-hand screw would progress if vectors $\mathbf{V}_1$ and $\mathbf{V}_2$ are placed tail-to-tail in the plane they define and $\mathbf{V}_1$ is rotated into $\mathbf{V}_2$. The direction can also be found from the *right-hand rule*.

***Figure 5.6** Vector Cross Product*

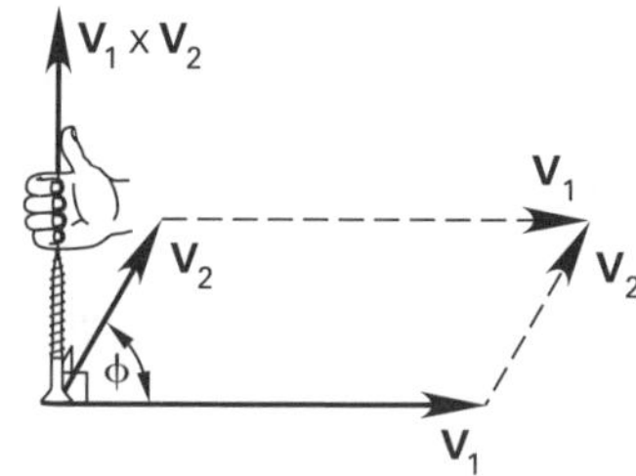

The magnitude of the cross product can be determined from Eq. 5.32, in which ϕ is the angle between the two vectors and is limited to 180°. The magnitude corresponds to the area of a parallelogram that has $\mathbf{V}_1$ and $\mathbf{V}_2$ as two of its sides.

$$|\mathbf{V}_1 \times \mathbf{V}_2| = |\mathbf{V}_1|\,|\mathbf{V}_2|\,\sin\phi \quad (5.32)$$

Vector cross multiplication is distributive but not commutative.

$$\mathbf{V}_1 \times \mathbf{V}_2 = -(\mathbf{V}_2 \times \mathbf{V}_1) \quad (5.33)$$

$$c(\mathbf{V}_1 \times \mathbf{V}_2) = (c\mathbf{V}_1) \times \mathbf{V}_2 = \mathbf{V}_1 \times (c\mathbf{V}_2) \quad (5.34)$$

$$\mathbf{V}_1 \times (\mathbf{V}_2 + \mathbf{V}_3) = \mathbf{V}_1 \times \mathbf{V}_2 + \mathbf{V}_1 \times \mathbf{V}_3 \quad (5.35)$$

If the two vectors are parallel, their cross product will be zero.

$$\mathbf{i}\times\mathbf{i} = \mathbf{j}\times\mathbf{j} = \mathbf{k}\times\mathbf{k} = 0 \quad (5.36)$$

Equation 5.31 and Eq. 5.33 can be extended to the unit vectors.

$$\mathbf{i}\times\mathbf{j} = -\mathbf{j}\times\mathbf{i} = \mathbf{k} \quad (5.37)$$

$$\mathbf{j}\times\mathbf{k} = -\mathbf{k}\times\mathbf{j} = \mathbf{i} \quad (5.38)$$

$$\mathbf{k}\times\mathbf{i} = -\mathbf{i}\times\mathbf{k} = \mathbf{j} \quad (5.39)$$

Example 5.2

Find a unit vector orthogonal to $\mathbf{V}_1 = \mathbf{i} - \mathbf{j} + 2\mathbf{k}$ and $\mathbf{V}_2 = 3\mathbf{j} - \mathbf{k}$.

Solution

The cross product is a vector orthogonal to $\mathbf{V}_1$ and $\mathbf{V}_2$.

$$\begin{aligned}\mathbf{V}_1 \times \mathbf{V}_2 &= \begin{vmatrix} \mathbf{i} & 1 & 0 \\ \mathbf{j} & -1 & 3 \\ \mathbf{k} & 2 & -1 \end{vmatrix}\\ &= -5\mathbf{i} + \mathbf{j} + 3\mathbf{k}\end{aligned}$$

Check to see whether this is a unit vector.

$$|\mathbf{V}_1 \times \mathbf{V}_2| = \sqrt{(-5)^2 + (1)^2 + (3)^2} = \sqrt{35}$$

[5]The cross product is also written in square brackets without a cross, that is, $[\mathbf{V}_1\mathbf{V}_2]$.

Since its length is $\sqrt{35}$, the vector must be divided by $\sqrt{35}$ to obtain a unit vector.

$$\mathbf{a} = \frac{-5\mathbf{i} + \mathbf{j} + 3\mathbf{k}}{\sqrt{35}}$$

10. MIXED TRIPLE PRODUCT

The *mixed triple product* (*triple scalar product* or just *triple product*) of three vectors is a scalar quantity representing the volume of a parallelepiped with the three vectors making up the sides. It is calculated as a determinant. Since Eq. 5.40 can be negative, the absolute value must be used to obtain the volume in that case. Figure 5.7 shows a mixed triple product.

Figure 5.7 *Vector Mixed Triple Product*

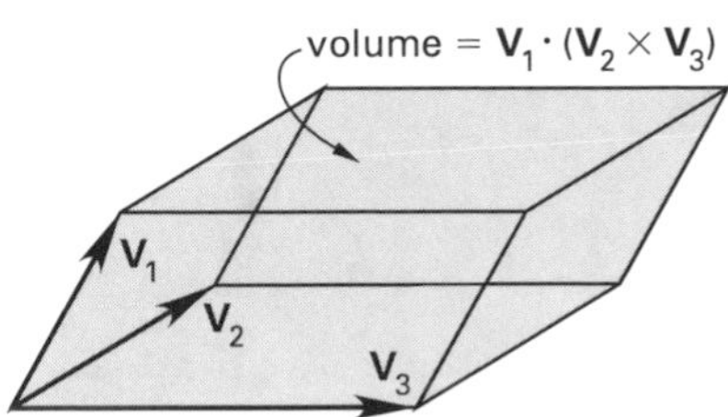

$$\mathbf{V}_1 \cdot (\mathbf{V}_2 \times \mathbf{V}_3) = \begin{vmatrix} V_{1x} & V_{1y} & V_{1z} \\ V_{2x} & V_{2y} & V_{2z} \\ V_{3x} & V_{3y} & V_{3z} \end{vmatrix} \qquad 5.40$$

The mixed triple product has the property of *circular permutation*, as defined by Eq. 5.41.

$$\mathbf{V}_1 \cdot (\mathbf{V}_2 \times \mathbf{V}_3) = (\mathbf{V}_1 \times \mathbf{V}_2) \cdot \mathbf{V}_3 \qquad 5.41$$

11. VECTOR TRIPLE PRODUCT

The *vector triple product* (also known as the *vector triple cross product*, *BAC-CAB identity* (spoken as "BAC minus CAB"), and *Lagrange's formula*) is a vector defined by Eq. 5.42.[6] The cross products in the parentheses on the right-hand side are scalars, so the vector triple product is the scaled difference between $\mathbf{V}_2$ and $\mathbf{V}_3$. Geometrically, the resultant vector is perpendicular to $\mathbf{V}_1$ and is in the plane defined by $\mathbf{V}_2$ and $\mathbf{V}_3$.

$$\mathbf{V}_1 \times (\mathbf{V}_2 \times \mathbf{V}_3) = (\mathbf{V}_1 \cdot \mathbf{V}_3)\mathbf{V}_2 - (\mathbf{V}_1 \cdot \mathbf{V}_2)\mathbf{V}_3 \qquad 5.42$$

12. VECTOR FUNCTIONS

A vector can be a function of another parameter. For example, a vector $\mathbf{V}$ is a function of variable t when its V_x, V_y, and V_z components are functions of t. For example,

$$\mathbf{V}(t) = (2t - 3)\mathbf{i} + (t^2 + 1)\mathbf{j} + (-7t + 5)\mathbf{k}$$

When the functions of t are differentiated (or integrated) with respect to t, the vector itself is differentiated (integrated).[7]

$$\frac{d\mathbf{V}(t)}{dt} = \frac{dV_x}{dt}\mathbf{i} + \frac{dV_y}{dt}\mathbf{j} + \frac{dV_z}{dt}\mathbf{k} \qquad 5.43$$

Similarly, the integral of the vector is

$$\int \mathbf{V}(t)\,dt = \mathbf{i}\int V_x\,dt + \mathbf{j}\int V_y\,dt + \mathbf{k}\int V_z\,dt \qquad 5.44$$

[6]This concept is usually encountered only in electrodynamic field theory.
[7]This is particularly valuable when converting among position, velocity, and acceleration vectors.

6 Trigonometry

Content in blue refers to the *NCEES Handbook*.

1. DEGREES AND RADIANS

Degrees and *radians* are two units for measuring angles. One complete circle is divided into 360 degrees (written 360°) or 2π radians (abbreviated *rad*).[1] The conversions between degrees and radians are [**Measurement Relationships**]

multiply	by	to obtain
radians	$\frac{180}{\pi}$	degrees
degrees	$\frac{\pi}{180}$	radians

The number of radians in an angle, θ, corresponds to twice the area within a circular sector with arc length θ and a radius of one, as shown in Fig. 6.1. Alternatively, the area of a sector with central angle θ radians is $\theta/2$ for a *unit circle* (i.e., a circle with a radius of one unit).

2. PLANE ANGLES

A *plane angle* (usually referred to as just an *angle*) consists of two intersecting lines and an intersection point known as the *vertex*. The angle can be referred to by a capital letter representing the vertex (e.g., B in Fig. 6.2), a letter representing the angular measure (e.g., B or β), or by three capital letters, where the middle letter is the vertex and the other two letters are two points on different lines, and either the symbol ∠ or ∢ (e.g., ∢ ABC).

Figure 6.1 *Radians and Area of Unit Circle*

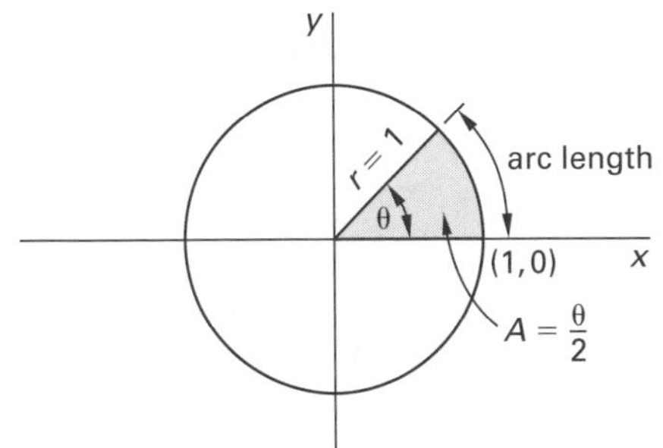

Figure 6.2 *Angle*

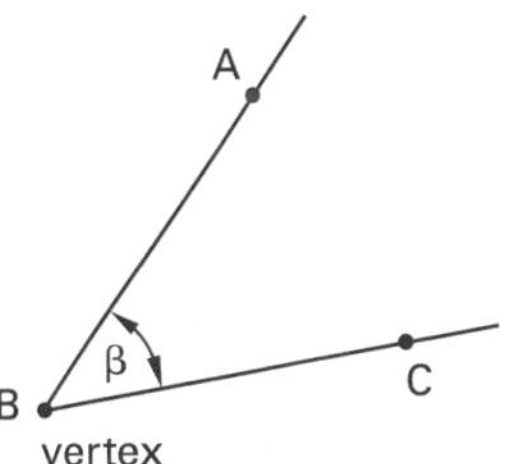

The angle between two intersecting lines generally is understood to be the smaller angle created.[2] Angles have been classified as follows.

- *acute angle:* an angle less than 90° ($\pi/2$ rad)
- *obtuse angle:* an angle more than 90° ($\pi/2$ rad) but less than 180° (π rad)
- *reflex angle:* an angle more than 180° (π rad) but less than 360° (2π rad)
- *related angle:* an angle that differs from another by some multiple of 90° ($\pi/2$ rad)
- *right angle:* an angle equal to 90° ($\pi/2$ rad)
- *straight angle:* an angle equal to 180° (π rad); that is, a straight line

Complementary angles are two angles whose sum is 90° ($\pi/2$ rad). *Supplementary angles* are two angles whose sum is 180° (π rad). *Adjacent angles* share a common

[1] The abbreviation *rad* is also used to represent *radiation absorbed dose*, a measure of radiation exposure.

[2] In books on geometry, the term *ray* is used instead of *line*.

vertex and one (the interior) side. Adjacent angles are supplementary if, and only if, their exterior sides form a straight line.

Vertical angles are the two angles with a common vertex and with sides made up by two intersecting straight lines, as shown in Fig. 6.3. Vertical angles are equal.

Figure 6.3 *Vertical Angles*

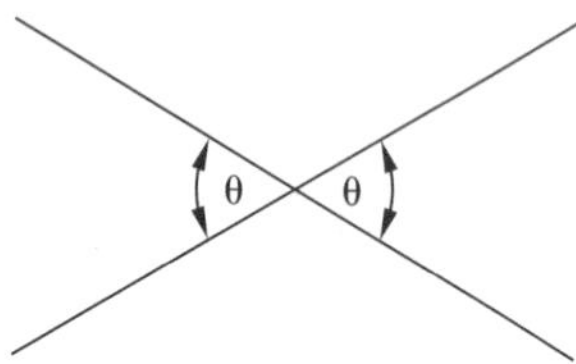

Angle of elevation and *angle of depression* are surveying terms referring to the angle above and below the horizontal plane of the observer, respectively.

3. TRIANGLES

A *triangle* is a three-sided closed polygon with three angles whose sum is 180° (π rad). Triangles are identified by their vertices and the symbol Δ (e.g., ΔABC in Fig. 6.4). A side is designated by its two endpoints (e.g., AB in Fig. 6.4) or by a lowercase letter corresponding to the capital letter of the opposite vertex (e.g., c).

In *similar triangles*, the corresponding angles are equal and the corresponding sides are in proportion. (Since there are only two independent angles in a triangle, showing that two angles of one triangle are equal to two angles of the other triangle is sufficient to show similarity.) The symbol for similarity is ~. In Fig. 6.4, ΔABC ~ ΔDEF (i.e., ΔABC is similar to ΔDEF).

Figure 6.4 *Similar Triangles*

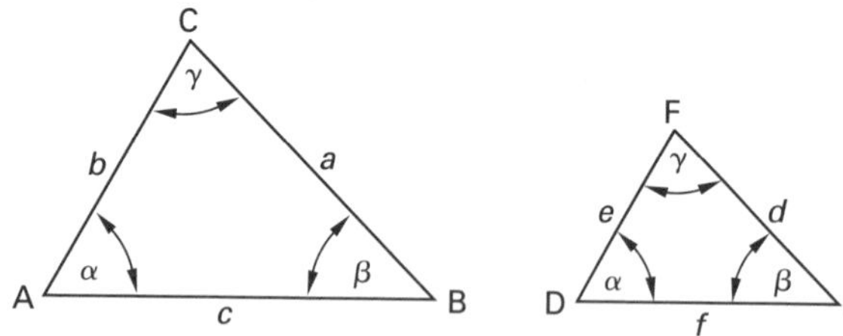

4. RIGHT TRIANGLES

A *right triangle* is a triangle in which one of the angles is 90° ($\pi/2$ rad). The remaining two angles are complementary. If one of the acute angles is chosen as the reference, the sides forming the right angle are known as the *adjacent side*, x, and the *opposite side*, y. The longest side is known as the *hypotenuse*, r. The *Pythagorean theorem* relates the lengths of these sides.

$$x^2 + y^2 = r^2 \qquad 6.1$$

In certain cases, the lengths of unknown sides of right triangles can be determined by inspection.[3] This occurs when the lengths of the sides are in the ratios of 3:4:5, 1:1:$\sqrt{2}$, 1:$\sqrt{3}$:2, or 5:12:13. Figure 6.5 illustrates a 3:4:5 triangle.

Figure 6.5 *3:4:5 Right Triangle*

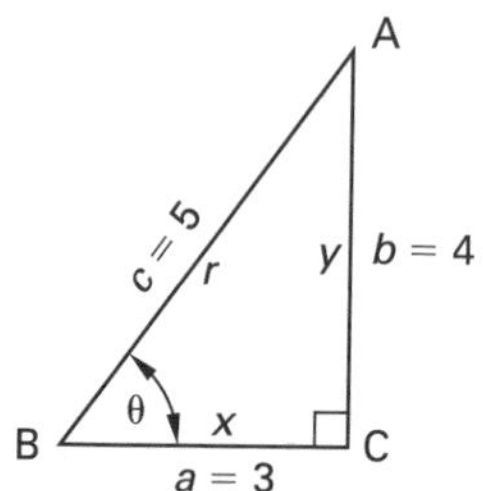

5. TRIGONOMETRIC FUNCTIONS

The *trigonometric functions* (also referred to as the *circular transcendental functions*, *transcendental functions*, or *functions of an angle*) are calculated from the sides of a right triangle. Equation 6.2 through Eq. 6.7 refers to Fig. 6.6.

Figure 6.6 *Right Triangle*

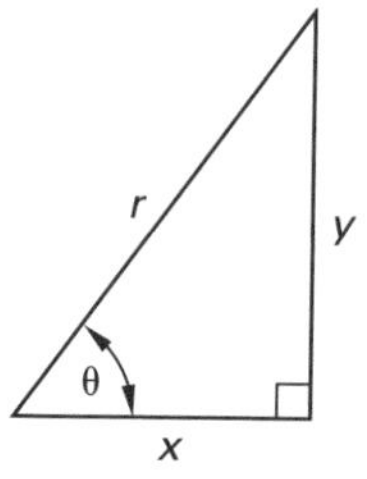

Trigonometry: Basics

$$\sin\theta = \frac{y}{r} = \frac{\text{opposite}}{\text{hypotenuse}} \qquad 6.2$$

$$\cos\theta = \frac{x}{r} = \frac{\text{adjacent}}{\text{hypotenuse}} \qquad 6.3$$

[3]These cases are almost always contrived examples. There is nothing intrinsic in nature to cause the formation of triangles with these proportions.

$$\tan\theta = \frac{y}{x} = \frac{\text{opposite}}{\text{adjacent}} \qquad 6.4$$

$$\cot\theta = \frac{x}{y} = \frac{\text{adjacent}}{\text{opposite}} \qquad 6.5$$

$$\sec\theta = \frac{r}{x} = \frac{\text{hypotenuse}}{\text{adjacent}} \qquad 6.6$$

$$\csc\theta = \frac{r}{y} = \frac{\text{hypotenuse}}{\text{opposite}} \qquad 6.7$$

Three of the trigonometric functions are reciprocals of the others. However, while the tangent and cotangent functions are reciprocals of each other, the sine and cosine functions are not.

Identities

$$\cot\theta = \frac{1}{\tan\theta} \qquad 6.8$$

$$\sec\theta = \frac{1}{\cos\theta} \qquad 6.9$$

$$\csc\theta = \frac{1}{\sin\theta} \qquad 6.10$$

The trigonometric functions correspond to the lengths of various line segments in a right triangle with a unit hypotenuse. Figure 6.7 shows such a triangle inscribed in a unit circle.

Figure 6.7 *Trigonometric Functions in a Unit Circle*

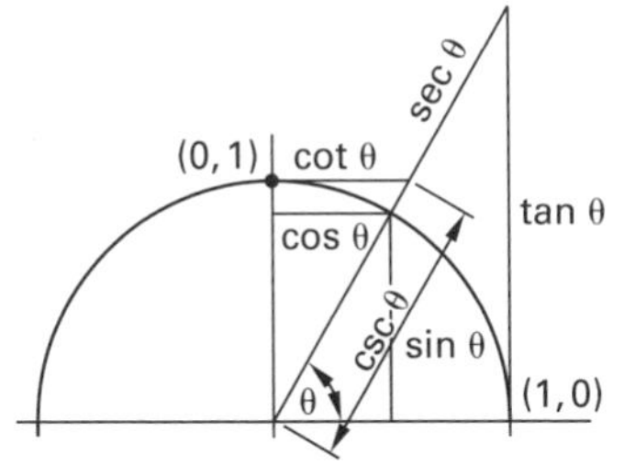

6. SMALL ANGLE APPROXIMATIONS

When an angle is very small, the hypotenuse and adjacent sides are essentially equal in length, and certain approximations can be made. (The angle θ must be expressed in radians in Eq. 6.11 and Eq. 6.12.)

$$\sin\theta \approx \tan\theta \approx \theta|_{\theta < 10^\circ\ (0.175\ \text{rad})} \qquad 6.11$$

$$\cos\theta \approx 1|_{\theta < 5^\circ\ (0.0873\ \text{rad})} \qquad 6.12$$

7. GRAPHS OF THE FUNCTIONS

Figure 6.8 illustrates the periodicity of the sine, cosine, and tangent functions.[4]

Figure 6.8 *Graphs of Sine, Cosine, and Tangent Functions*

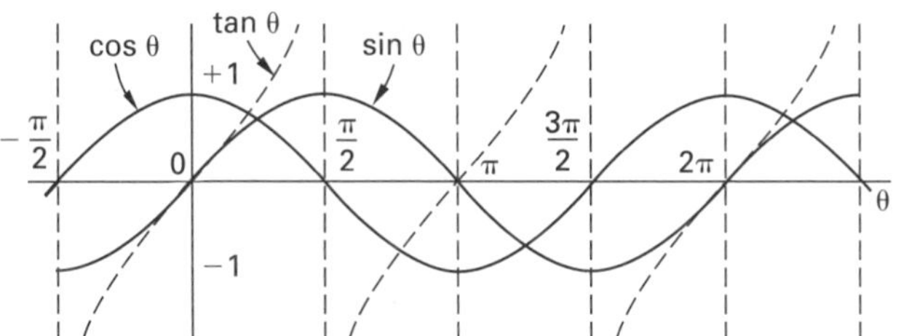

8. SIGNS OF THE FUNCTIONS

Table 6.1 shows how the sine, cosine, and tangent functions vary in sign with different values of θ. All three functions are positive for angles $0^\circ < \theta < 90^\circ$ ($0 < \theta < \pi/2$ rad), but only the sine is positive for angles $90^\circ < \theta < 180^\circ$ ($\pi/2$ rad $< \theta < \pi$ rad). The concept of quadrants is used to summarize the signs of the functions: angles up to 90° ($\pi/2$ rad) are in quadrant I, angles between 90° and 180° ($\pi/2$ and π rad) are in quadrant II, and so on.

Table 6.1 *Signs of the Functions by Quadrant*

	quadrant			
function	I	II	III	IV
sine	+	+	−	−
cosine	+	−	−	+
tangent	+	−	+	−

[4]The remaining functions, being reciprocals of these three functions, are also periodic.

Background and Support

9. TRIGONOMETRIC IDENTITIES

There are many relationships between trigonometric functions. For example, Eq. 6.13 through Eq. 6.15 are well known.

Identities

$$\sin^2\theta + \cos^2\theta = 1 \quad (6.13)$$

$$1 + \tan^2\theta = \sec^2\theta \quad (6.14)$$

$$1 + \cot^2\theta = \csc^2\theta \quad (6.15)$$

Other relatively common identities are listed as follows.[5]

- *double-angle formulas*

Identities

$$\sin 2\theta = 2\sin\theta\cos\theta = \frac{2\tan\theta}{1+\tan^2\theta} \quad (6.16)$$

$$\cos 2\theta = \cos^2\theta - \sin^2\theta = 1 - 2\sin^2\theta = 2\cos^2\theta - 1 = \frac{1-\tan^2\theta}{1+\tan^2\theta} \quad (6.17)$$

$$\tan 2\theta = \frac{2\tan\theta}{1-\tan^2\theta} \quad (6.18)$$

$$\cot 2\theta = \frac{\cot^2\theta - 1}{2\cot\theta} \quad (6.19)$$

- *two-angle formulas*

Identities

$$\sin(\theta \pm \phi) = \sin\theta\cos\phi \pm \cos\theta\sin\phi \quad (6.20)$$

$$\cos(\theta \pm \phi) = \cos\theta\cos\phi \mp \sin\theta\sin\phi \quad (6.21)$$

$$\tan(\theta \pm \phi) = \frac{\tan\theta \pm \tan\phi}{1 \mp \tan\theta\tan\phi} \quad (6.22)$$

$$\cot(\theta \pm \phi) = \frac{\cot\phi\cot\theta \mp 1}{\cot\phi \pm \cot\theta} \quad (6.23)$$

- *half-angle formulas* ($\theta < 180°$)

Identities

$$\sin\frac{\theta}{2} = \sqrt{\frac{1-\cos\theta}{2}} \quad (6.24)$$

$$\cos\frac{\theta}{2} = \sqrt{\frac{1+\cos\theta}{2}} \quad (6.25)$$

$$\tan\frac{\theta}{2} = \sqrt{\frac{1-\cos\theta}{1+\cos\theta}} = \frac{\sin\theta}{1+\cos\theta} = \frac{1-\cos\theta}{\sin\theta} \quad (6.26)$$

- *miscellaneous formulas* ($\theta < 90°$)

$$\cos\theta = \sin\left(\theta + \frac{\pi}{2}\right) = -\sin\left(\theta - \frac{\pi}{2}\right) \quad (6.27)$$

$$\sin\theta = \cos\left(\theta - \frac{\pi}{2}\right) = -\cos\left(\theta + \frac{\pi}{2}\right) \quad (6.28)$$

$$\sin\theta = 2\sin\frac{\theta}{2}\cos\frac{\theta}{2} \quad (6.29)$$

$$\sin\theta = \sqrt{\frac{1-\cos 2\theta}{2}} \quad (6.30)$$

$$\cos\theta = \cos^2\frac{\theta}{2} - \sin^2\frac{\theta}{2} \quad (6.31)$$

$$\cos\theta = \sqrt{\frac{1+\cos 2\theta}{2}} \quad (6.32)$$

$$\tan\theta = \frac{2\tan\frac{\theta}{2}}{1-\tan^2\frac{\theta}{2}} = \frac{2\sin\frac{\theta}{2}\cos\frac{\theta}{2}}{\cos^2\frac{\theta}{2} - \sin^2\frac{\theta}{2}} \quad (6.33)$$

$$\tan\theta = \sqrt{\frac{1-\cos 2\theta}{1+\cos 2\theta}} = \frac{\sin 2\theta}{1+\cos 2\theta} = \frac{1-\cos 2\theta}{\sin 2\theta} \quad (6.34)$$

$$\cot\theta = \frac{\cot^2\frac{\theta}{2} - 1}{2\cot\frac{\theta}{2}} = \frac{\cos^2\frac{\theta}{2} - \sin^2\frac{\theta}{2}}{2\sin\frac{\theta}{2}\cos\frac{\theta}{2}} \quad (6.35)$$

[5]It is an idiosyncrasy of the subject that these formulas are conventionally referred to as *formulas*, not *identities*.

$$\cot\theta = \sqrt{\frac{1+\cos 2\theta}{1-\cos 2\theta}} = \frac{1+\cos 2\theta}{\sin 2\theta} = \frac{\sin 2\theta}{1-\cos 2\theta} \quad 6.36$$

10. INVERSE TRIGONOMETRIC FUNCTIONS

Finding an angle from a known trigonometric function is a common operation known as an *inverse trigonometric operation.* The inverse function can be designated by adding "inverse," "arc," or the superscript −1 to the name of the function. For example,

$$\text{inverse}\sin 0.5 \equiv \arcsin 0.5 \equiv \sin^{-1} 0.5 = 30°$$

The range of each inverse trigonometric function is limited to only 180°. For instance, arctan will only return values between −90° and 90°. The result of a function must be checked against which quadrant the desired angle falls in to determine whether or not the output makes sense. If a correction is needed, arcsin and arccos will be off by a factor of 90°, and arctan will be off by a factor of 180°.

11. HYPERBOLIC TRANSCENDENTAL FUNCTIONS

Hyperbolic transcendental functions (normally referred to as *hyperbolic functions*) are specific equations containing combinations of the terms e^θ and $e^{-\theta}$. These combinations appear regularly in certain types of problems (e.g., analysis of cables and heat transfer through fins) and are given specific names and symbols to simplify presentation.[6]

- *hyperbolic sine*

$$\sinh\theta = \frac{e^\theta - e^{-\theta}}{2} \quad 6.37$$

- *hyperbolic cosine*

$$\cosh\theta = \frac{e^\theta + e^{-\theta}}{2} \quad 6.38$$

- *hyperbolic tangent*

$$\tanh\theta = \frac{e^\theta - e^{-\theta}}{e^\theta + e^{-\theta}} = \frac{\sinh\theta}{\cosh\theta} \quad 6.39$$

- *hyperbolic cotangent*

$$\coth\theta = \frac{e^\theta + e^{-\theta}}{e^\theta - e^{-\theta}} = \frac{\cosh\theta}{\sinh\theta} \quad 6.40$$

- *hyperbolic secant*

$$\operatorname{sech}\theta = \frac{2}{e^\theta + e^{-\theta}} = \frac{1}{\cosh\theta} \quad 6.41$$

- *hyperbolic cosecant*

$$\operatorname{csch}\theta = \frac{2}{e^\theta - e^{-\theta}} = \frac{1}{\sinh\theta} \quad 6.42$$

Hyperbolic functions cannot be related to a right triangle, but they are related to a rectangular (equilateral) hyperbola, as shown in Fig. 6.9. For a *unit hyperbola* ($a^2 = 1$), the shaded area has a value of $\theta/2$ and is sometimes given the units of *hyperbolic radians.*

Figure 6.9 *Equilateral Hyperbola and Hyperbolic Functions*

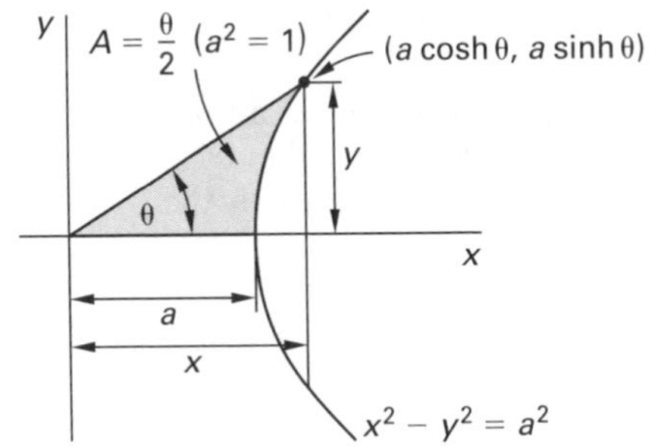

$$\sinh\theta = \frac{y}{a} \quad 6.43$$

$$\cosh\theta = \frac{x}{a} \quad 6.44$$

$$\tanh\theta = \frac{y}{x} \quad 6.45$$

$$\coth\theta = \frac{x}{y} \quad 6.46$$

$$\operatorname{sech}\theta = \frac{a}{x} \quad 6.47$$

$$\operatorname{csch}\theta = \frac{a}{y} \quad 6.48$$

[6]The hyperbolic sine and cosine functions are pronounced (by some) as "sinch" and "cosh," respectively.

12. HYPERBOLIC IDENTITIES

The hyperbolic identities are different from the standard trigonometric identities. Some of the most important identities are presented as follows.

$$\cosh^2\theta - \sinh^2\theta = 1 \qquad 6.49$$

$$1 - \tanh^2\theta = \operatorname{sech}^2\theta \qquad 6.50$$

$$1 - \coth^2\theta = -\operatorname{csch}^2\theta \qquad 6.51$$

$$\cosh\theta + \sinh\theta = e^{\theta} \qquad 6.52$$

$$\cosh\theta - \sinh\theta = e^{-\theta} \qquad 6.53$$

$$\sinh(\theta \pm \phi) = \sinh\theta\cosh\phi \pm \cosh\theta\sinh\phi \qquad 6.54$$

$$\cosh(\theta \pm \phi) = \cosh\theta\cosh\phi \pm \sinh\theta\sinh\phi \qquad 6.55$$

$$\tanh(\theta \pm \phi) = \frac{\tanh\theta \pm \tanh\phi}{1 \pm \tanh\theta\tanh\phi} \qquad 6.56$$

13. GENERAL TRIANGLES

A *general triangle* (also known as an *oblique triangle*) is one that is not specifically a right triangle, as shown in Fig. 6.10. Equation 6.57 calculates the area of a general triangle.

$$\text{area} = \tfrac{1}{2}ab\sin C = \tfrac{1}{2}bc\sin A = \tfrac{1}{2}ca\sin B \qquad 6.57$$

Figure 6.10 *General Triangle*

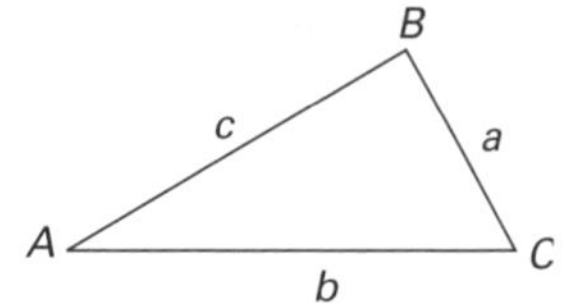

The *law of sines*, as shown in Eq. 6.58, relates the sides and the sines of the angles.

Trigonometry: Basics

$$\frac{\sin A}{a} = \frac{\sin B}{b} = \frac{\sin C}{c} \qquad 6.58$$

The *law of cosines* relates the cosine of an angle to an opposite side. (Equation 6.59 can be extended to the two remaining sides.)

Trigonometry: Basics

$$a^2 = b^2 + c^2 - 2bc\cos A \qquad 6.59$$

The *law of tangents* relates the sum and difference of two sides. (Equation 6.60 can be extended to the two remaining sides.)

$$\frac{a-b}{a+b} = \frac{\tan\left(\frac{A-B}{2}\right)}{\tan\left(\frac{A+B}{2}\right)} \qquad 6.60$$

Analytic Geometry

Content in blue refers to the *NCEES Handbook*.

1. MENSURATION OF REGULAR SHAPES

The dimensions, perimeter, area, and other geometric properties constitute the *mensuration* (i.e., the measurements) of a geometric shape. Appendixes 7.A and 7.B contain formulas and tables used to calculate these properties. [**Mensuration of Areas and Volumes: Nomenclature**]

Example 7.1

In the study of open channel fluid flow, the hydraulic radius is defined as the ratio of flow area to wetted perimeter. What is the hydraulic radius of a 6 in inside diameter pipe filled to a depth of 2 in?

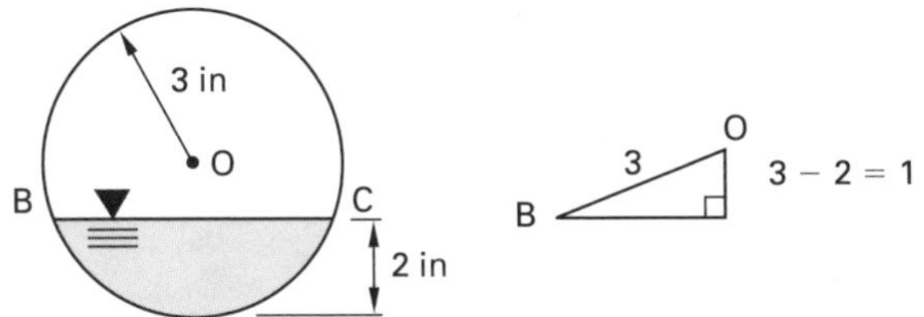

Solution

Points O, B, and C constitute a circular segment and are used to find the central angle of the circular segment.

$$\tfrac{1}{2}\angle\text{BOC} = \arccos\tfrac{1}{3} = 70.53°$$

$$\phi = \angle\text{BOC} = (2)(70.53°)\left(\frac{\pi}{180}\right) = 2.462\text{ rad}$$

From App. 7.A, the area in flow and arc length are

$$\begin{aligned} A &= \tfrac{1}{2}r^2(\phi - \sin\phi) \\ &= \left(\tfrac{1}{2}\right)(3\text{ in})^2\left(2.462\text{ rad} - \sin(2.462\text{ rad})\right) \\ &= 8.251\text{ in}^2 \\ s &= r\phi = (3\text{ in})(2.462\text{ rad}) \\ &= 7.386\text{ in} \end{aligned}$$

The hydraulic radius is

$$\begin{aligned} R_H &= \frac{A}{s} = \frac{8.251\text{ in}^2}{7.386\text{ in}} \\ &= 1.12\text{ in} \end{aligned}$$

2. AREAS WITH IRREGULAR BOUNDARIES

Areas of sections with irregular boundaries (such as creek banks) cannot be determined precisely, and approximation methods must be used. If the irregular side can be divided into a series of cells of equal width, either the trapezoidal rule or Simpson's rule can be used. Figure 7.1 shows an example of an irregular area.

Figure 7.1 *Irregular Areas*

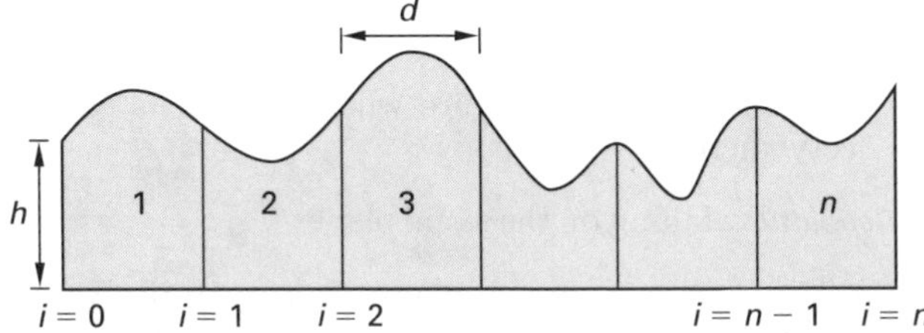

If the irregular side of each cell is fairly straight, the *trapezoidal rule* is appropriate.

$$A = \frac{d}{2}\left(h_0 + h_n + 2\sum_{i=1}^{n-1} h_i\right) \qquad 7.1$$

If the irregular side of each cell is curved (parabolic), *Simpson's rule* should be used. (n must be even to use Simpson's rule.)

$$A = \frac{d}{3}\left(h_0 + h_n + 4\sum_{\substack{i\,\text{odd}\\ i=1}}^{n-1} h_i + 2\sum_{\substack{i\,\text{even}\\ i=2}}^{n-2} h_i\right) \qquad 7.2$$

3. GEOMETRIC DEFINITIONS

The following terms are used in this book to describe the relationship or orientation of one geometric figure to another. Figure 7.2 illustrates some of the following geometric definitions.

Figure 7.2 Geometric Definitions

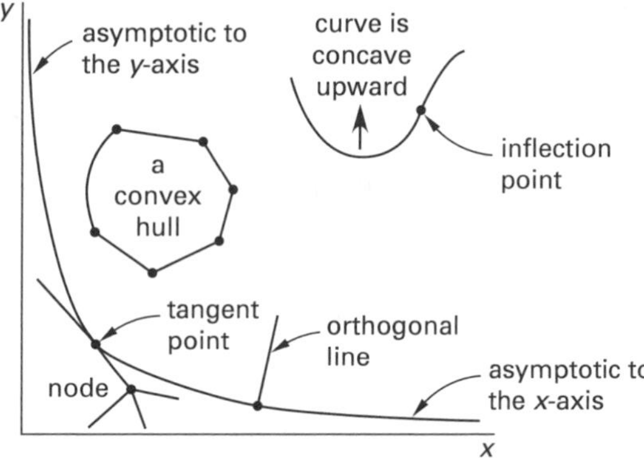

- *abscissa:* the horizontal coordinate, typically designated as x in a rectangular coordinate system
- *asymptote:* a straight line that is approached but not intersected by a curved line
- *asymptotic:* approaching the slope of a line; attaining the slope of a line in the limit
- *center:* a point equidistant from all other points
- *collinear:* falling on the same line
- *concave:* curved inward (in the direction indicated)
- *convex:* curved outward (in the direction indicated)
- *convex hull:* a closed figure whose surface is convex everywhere
- *coplanar:* falling on the same plane
- *inflection point:* a point where the second derivative changes sign or the curve changes from concave up to concave down or vice versa; also known as a *point of contraflexure*
- *locus of points:* a set or collection of points having some common property and being so infinitely close together as to be indistinguishable from a line
- *node:* a point on a line from which other lines enter or leave
- *normal:* rotated 90°; being at right angles
- *ordinate:* the vertical coordinate, typically designated as y in a rectangular coordinate system
- *orthogonal:* rotated 90°; being at right angles
- *saddle point:* a point on a three-dimensional surface where all adjacent points are higher than the saddle point in one direction (the direction of the saddle) and are lower than the saddle point in an orthogonal direction (the direction of the sides)
- *tangent point:* having equal slopes at a common point

4. CONCAVE CURVES

Concavity is a term that is applied to curved lines. A *concave up curve* is one whose function's first derivative increases continuously. Straight lines drawn tangent to concave up curves are all below the curve. The graph of such a function may be thought of as being able to "hold water."

The first derivative of a *concave down curve* decreases continuously. A graph of a concave down function may be thought of as "spilling water." (See Fig. 7.3.)

Figure 7.3 Concave Curves

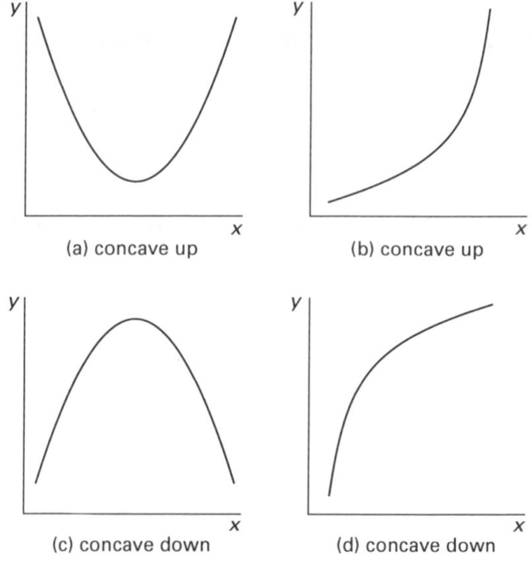

5. CONGRUENCY

Congruence in geometric figures is analogous to *equality* in algebraic expressions. Congruent line segments are segments that have the same length. Congruent angles have the same angular measure. Congruent triangles have the same vertex angles and side lengths.

In general, *congruency*, indicated by the symbol $\cong$, means that there is one-to-one correspondence between all points on two objects. This correspondence is defined by the *mapping function* or *isometry*, which can be a translation, rotation, or reflection. Since the identity function is a valid mapping function, every geometric shape is congruent to itself.

Two congruent objects can be in different spaces. For example, a triangular area in three-dimensional space can be mapped onto a triangle in two-dimensional space.

6. COORDINATE SYSTEMS

The manner in which a geometric figure is described depends on the coordinate system that is used. A three-dimensional system (also known as the *rectangular coordinate system* and *Cartesian coordinate system*) with its x-, y-, and z-coordinates is the most commonly used in engineering. Table 7.1 summarizes the components needed to specify a point in the various coordinate systems. Figure 7.4 illustrates the use of and conversion between the coordinate systems.

Figure 7.4 *Different Coordinate Systems*

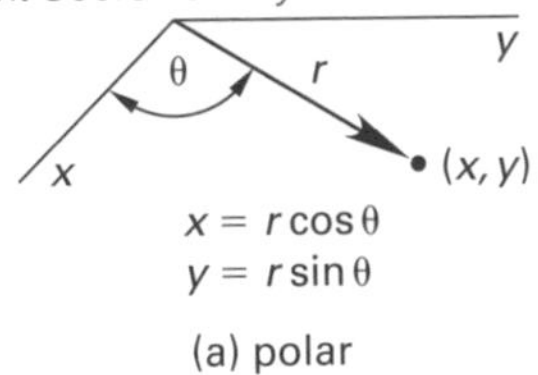

(a) polar

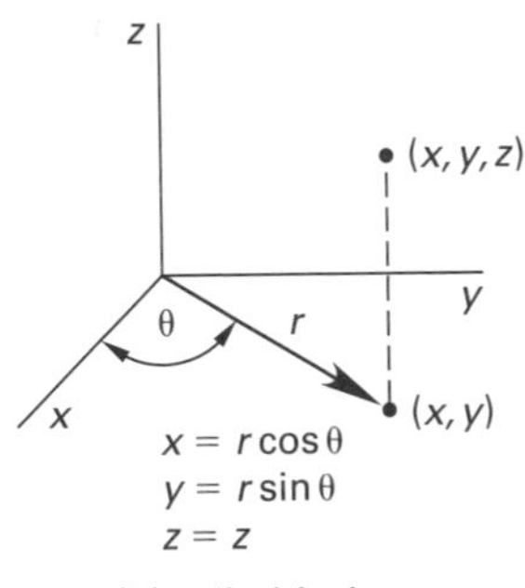

(b) cylindrical

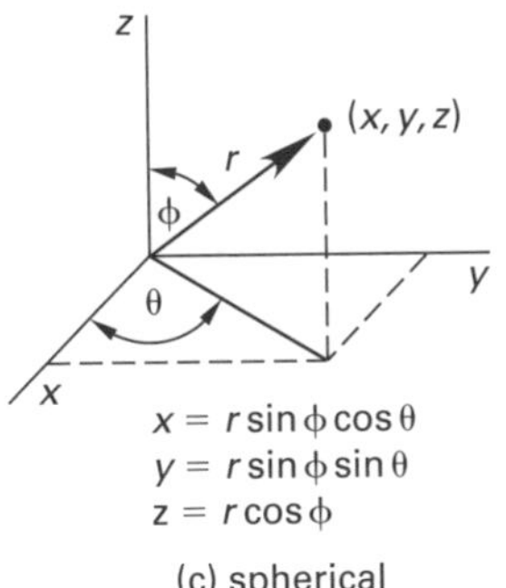

(c) spherical

Table 7.1 *Components of Coordinate Systems*

name	dimensions	components
rectangular	2	x, y
rectangular	3	x, y, z
polar	2	r, θ
cylindrical	3	r, θ, z
spherical	3	r, θ, ϕ

7. STRAIGHT LINES

Figure 7.5 illustrates a straight line in two-dimensional space. The *slope* of the line is m, the *y-intercept* is b, and the *x-intercept* is a. The equation of the line can be represented in several forms. The procedure for finding the equation depends on the form chosen to represent the line. In general, the procedure involves substituting one or more known points on the line into the equation in order to determine the coefficients.

Figure 7.5 *Straight Line*

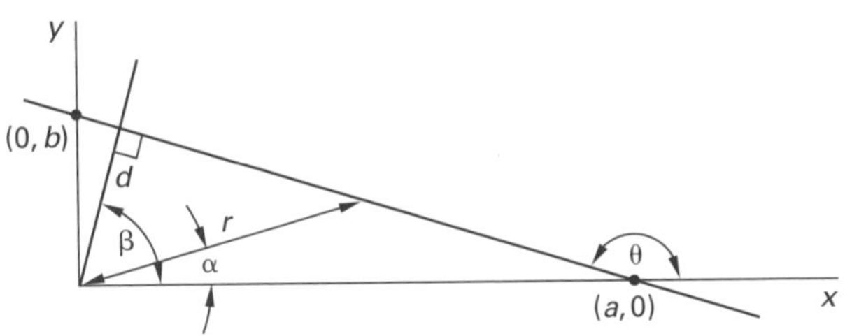

- *general form*

$$Ax + By + C = 0 \tag{7.3}$$

$$A = -mB \tag{7.4}$$

$$B = \frac{-C}{b} \tag{7.5}$$

$$\begin{aligned} C &= -aA \\ &= -bB \end{aligned} \tag{7.6}$$

- *slope-intercept form*

$$y = mx + b \tag{7.7}$$

$$\begin{aligned} m &= \frac{-A}{B} \\ &= \tan\theta \\ &= \frac{y_2 - y_1}{x_2 - x_1} \end{aligned} \tag{7.8}$$

$$b = \frac{-C}{B} \tag{7.9}$$

$$a = \frac{-C}{A} \quad 7.10$$

- *point-slope form*

$$y - y_1 = m(x - x_1) \quad 7.11$$

- *intercept form*

$$\frac{x}{a} + \frac{y}{b} = 1 \quad 7.12$$

- *two-point form*

$$\frac{y - y_1}{x - x_1} = \frac{y_2 - y_1}{x_2 - x_1} \quad 7.13$$

- *normal form*

$$x \cos\beta + y \sin\beta - d = 0 \quad 7.14$$

(d and β are constants; x and y are variables.)

- *polar form*

$$r = \frac{d}{\cos(\beta - \alpha)} \quad 7.15$$

(d and β are constants; r and α are variables.)

8. DIRECTION NUMBERS, ANGLES, AND COSINES

Given a directed line or vector, **R**, from (x_1, y_1, z_1) to (x_2, y_2, z_2), the *direction numbers* are

$$L = x_2 - x_1 \quad 7.16$$

$$M = y_2 - y_1 \quad 7.17$$

$$N = z_2 - z_1 \quad 7.18$$

The distance between two points is

$$d = \sqrt{L^2 + M^2 + N^2} \quad 7.19$$

The *direction cosines* are

$$\cos\alpha = \frac{L}{d} \quad 7.20$$

$$\cos\beta = \frac{M}{d} \quad 7.21$$

$$\cos\gamma = \frac{N}{d} \quad 7.22$$

The sum of the squares of direction cosines is equal to 1.

$$\cos^2\alpha + \cos^2\beta + \cos^2\gamma = 1 \quad 7.23$$

The *direction angles* are the angles between the axes and the lines. They are found from the inverse functions of the direction cosines.

$$\alpha = \arccos\frac{L}{d} \quad 7.24$$

$$\beta = \arccos\frac{M}{d} \quad 7.25$$

$$\gamma = \arccos\frac{N}{d} \quad 7.26$$

The direction cosines can be used to write the equation of the straight line in terms of the unit vectors. The line **R** would be defined as

$$\mathbf{R} = d(\mathbf{i}\cos\alpha + \mathbf{j}\cos\beta + \mathbf{k}\cos\gamma) \quad 7.27$$

Similarly, the line may be written in terms of its direction numbers.

$$\mathbf{R} = L\mathbf{i} + M\mathbf{j} + N\mathbf{k} \quad 7.28$$

Example 7.2

A directed line, **R**, passes through the points (4, 7, 9) and (0, 1, 6). Write the equation of the line in terms of its (a) direction numbers and (b) direction cosines.

Solution

(a) The direction numbers are

$$L = 4 - 0 = 4$$
$$M = 7 - 1 = 6$$
$$N = 9 - 6 = 3$$

Using Eq. 7.28,

$$\mathbf{R} = 4\mathbf{i} + 6\mathbf{j} + 3\mathbf{k}$$

(b) The distance between the two points is

$$d = \sqrt{(4)^2 + (6)^2 + (3)^2} = 7.81$$

From Eq. 7.27, the line in terms of its direction cosines is

$$\mathbf{R} = (7.81)\left(\frac{4\mathbf{i} + 6\mathbf{j} + 3\mathbf{k}}{7.81}\right)$$
$$= (7.81)(0.512\mathbf{i} + 0.768\mathbf{j} + 0.384\mathbf{k})$$

9. INTERSECTION OF TWO LINES

The intersection of two lines is a point. The location of the intersection point can be determined by setting the two equations equal and solving them in terms of a common variable. Alternatively, Eq. 7.29 and Eq. 7.30 can be used to calculate the coordinates of the intersection point.

$$x = \frac{B_2C_1 - B_1C_2}{A_2B_1 - A_1B_2} \qquad 7.29$$

$$y = \frac{A_1C_2 - A_2C_1}{A_2B_1 - A_1B_2} \qquad 7.30$$

10. PLANES

A *plane* in three-dimensional space, as shown in Fig. 7.6, is completely determined by one of the following.

- three noncollinear points
- two nonparallel vectors $\mathbf{V}_1$ and $\mathbf{V}_2$ and their intersection point P_0
- a point P_0 and a vector, $\mathbf{N}$, normal to the plane (i.e., the *normal vector*)

Figure 7.6 *Plane in Three-Dimensional Space*

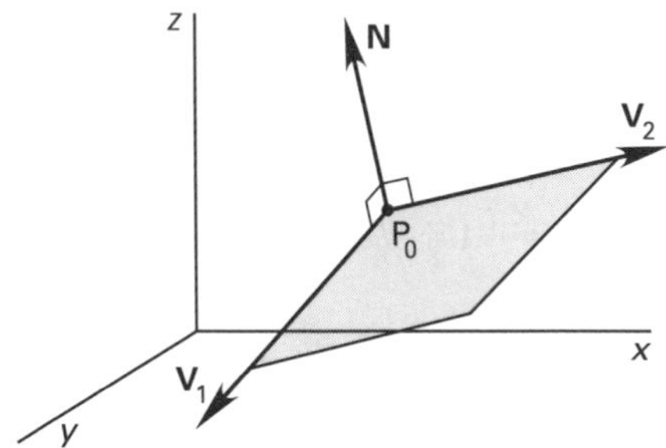

The plane can be specified mathematically in one of two ways: in rectangular form or as a parametric equation. The general form is

$$A(x - x_0) + B(y - y_0) + C(z - z_0) = 0 \qquad 7.31$$

x_0, y_0, and z_0 are the coordinates of the intersection point of any two vectors in the plane. The coefficients A, B, and C are the same as the coefficients of the normal vector, $\mathbf{N}$.

$$\mathbf{N} = \mathbf{V}_1 \times \mathbf{V}_2 = A\mathbf{i} + B\mathbf{j} + C\mathbf{k} \qquad 7.32$$

Equation 7.31 can be simplified as follows.

$$Ax + By + Cz + D = 0 \qquad 7.33$$

$$D = -(Ax_0 + By_0 + Cz_0) \qquad 7.34$$

The following procedure can be used to determine the equation of a plane from three noncollinear points, P_1, P_2, and P_3, or from a normal vector and a single point.

step 1: (If the normal vector is known, go to step 3.) Determine the equations of the vectors $\mathbf{V}_1$ and $\mathbf{V}_2$ from two pairs of the points. For example, determine $\mathbf{V}_1$ from points P_1 and P_2, and determine $\mathbf{V}_2$ from P_1 and P_3. Express the vectors in the form $A\mathbf{i} + B\mathbf{j} + C\mathbf{k}$.

$$\mathbf{V}_1 = (x_2 - x_1)\mathbf{i} + (y_2 - y_1)\mathbf{j} + (z_2 - z_1)\mathbf{k} \qquad 7.35$$

$$\mathbf{V}_2 = (x_3 - x_1)\mathbf{i} + (y_3 - y_1)\mathbf{j} + (z_3 - z_1)\mathbf{k} \qquad 7.36$$

step 2: Find the normal vector, $\mathbf{N}$, as the cross product of the two vectors.

$$\mathbf{N} = \mathbf{V}_1 \times \mathbf{V}_2 = \begin{vmatrix} \mathbf{i} & (x_2 - x_1) & (x_3 - x_1) \\ \mathbf{j} & (y_2 - y_1) & (y_3 - y_1) \\ \mathbf{k} & (z_2 - z_1) & (z_3 - z_1) \end{vmatrix} \qquad 7.37$$

step 3: Write the general equation of the plane in rectangular form using the coefficients A, B, and C from the normal vector and any one of the three points as P_0. (See Eq. 7.31.)

The parametric equations of a plane also can be written as a linear combination of the components of two vectors in the plane. Referring to Fig. 7.6, the two known vectors are

$$\mathbf{V}_1 = V_{1x}\mathbf{i} + V_{1y}\mathbf{j} + V_{1z}\mathbf{k} \qquad 7.38$$

$$\mathbf{V}_2 = V_{2x}\mathbf{i} + V_{2y}\mathbf{j} + V_{2z}\mathbf{k} \qquad 7.39$$

If s and t are scalars, the coordinates of each point in the plane can be written as Eq. 7.40 through Eq. 7.42. These are the parametric equations of the plane.

$$x = x_0 + sV_{1x} + tV_{2x} \qquad 7.40$$

$$y = y_0 + sV_{1y} + tV_{2y} \qquad 7.41$$

$$z = z_0 + sV_{1z} + tV_{2z} \qquad 7.42$$

Example 7.3

A plane is defined by the following points.

$$\begin{aligned} P_1 &= (2, 1, -4) \\ P_2 &= (4, -2, -3) \\ P_3 &= (2, 3, -8) \end{aligned}$$

Determine the equation of the plane in (a) general form and (b) parametric form.

Solution

(a) Use the first two points to find a vector, $\mathbf{V}_1$.

$$\begin{aligned} \mathbf{V}_1 &= (x_2 - x_1)\mathbf{i} + (y_2 - y_1)\mathbf{j} \\ &\quad + (z_2 - z_1)\mathbf{k} \\ &= (4-2)\mathbf{i} + (-2-1)\mathbf{j} \\ &\quad + \big(-3 - (-4)\big)\mathbf{k} \\ &= 2\mathbf{i} - 3\mathbf{j} + 1\mathbf{k} \end{aligned}$$

Similarly, use the first and third points to find $\mathbf{V}_2$.

$$\begin{aligned} \mathbf{V}_2 &= (x_3 - x_1)\mathbf{i} + (y_3 - y_1)\mathbf{j} \\ &\quad + (z_3 - z_1)\mathbf{k} \\ &= (2-2)\mathbf{i} + (3-1)\mathbf{j} \\ &\quad + \big(-8 - (-4)\big)\mathbf{k} \\ &= 0\mathbf{i} + 2\mathbf{j} - 4\mathbf{k} \end{aligned}$$

From Eq. 7.37, determine the normal vector as a determinant.

$$\mathbf{N} = \begin{vmatrix} \mathbf{i} & 2 & 0 \\ \mathbf{j} & -3 & 2 \\ \mathbf{k} & 1 & -4 \end{vmatrix}$$

Expand the determinant across the top row.

$$\begin{aligned} \mathbf{N} &= \mathbf{i}(12 - 2) - 2(-4\mathbf{j} - 2\mathbf{k}) \\ &= 10\mathbf{i} + 8\mathbf{j} + 4\mathbf{k} \end{aligned}$$

The rectangular form of the equation of the plane uses the same constants as in the normal vector. Use the first point and write the equation of the plane in the form of Eq. 7.31.

$$(10)(x-2) + (8)(y-1) + (4)(z+4) = 0$$

The three constant terms can be combined by using Eq. 7.34.

$$\begin{aligned} D &= -\big((10)(2) + (8)(1) + (4)(-4)\big) \\ &= -12 \end{aligned}$$

The equation of the plane is

$$10x + 8y + 4z - 12 = 0$$

(b) The parametric equations based on the first point and for any values of s and t are

$$\begin{aligned} x &= 2 + 2s + 0t \\ y &= 1 - 3s + 2t \\ z &= -4 + 1s - 4t \end{aligned}$$

The scalars s and t are not unique. Two of the three coordinates can also be chosen as the parameters. Dividing the rectangular form of the plane's equation by 4 to isolate z results in an alternate set of parametric equations.

$$\begin{aligned} x &= x \\ y &= y \\ z &= 3 - 2.5x - 2y \end{aligned}$$

11. DISTANCES BETWEEN GEOMETRIC FIGURES

The smallest distance, d, between various geometric figures is given by the following equations.

- Between two points in (x, y, z) format,

$$d = \sqrt{(x_2 - x_1)^2 + (y_2 - y_1)^2 + (z_2 - z_1)^2} \qquad 7.43$$

- Between a point (x_0, y_0) and a line $Ax + By + C = 0$,

$$d = \frac{|Ax_0 + By_0 + C|}{\sqrt{A^2 + B^2}} \qquad 7.44$$

- Between a point (x_0, y_0, z_0) and a plane $Ax + By + Cz + D = 0$,

$$d = \frac{|Ax_0 + By_0 + Cz_0 + D|}{\sqrt{A^2 + B^2 + C^2}} \qquad 7.45$$

- Between two parallel lines $Ax + By + C = 0$,

$$d = \left| \frac{|C_2|}{\sqrt{A_2^2 + B_2^2}} - \frac{|C_1|}{\sqrt{A_1^2 + B_1^2}} \right| \qquad 7.46$$

Example 7.4

What is the minimum distance between the line $y = 2x + 3$ and the origin $(0, 0)$?

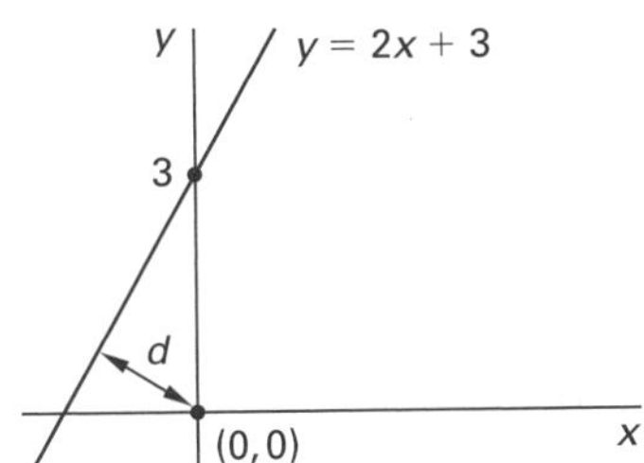

Solution

Put the equation in general form.

$$\begin{aligned} Ax + By + C &= 2x - y + 3 \\ &= 0 \end{aligned}$$

Use Eq. 7.44 with $(x, y) = (0, 0)$.

$$\begin{aligned} d &= \frac{|Ax + By + C|}{\sqrt{A^2 + B^2}} \\ &= \frac{(2)(0) + (-1)(0) + 3}{\sqrt{(2)^2 + (-1)^2}} \\ &= \frac{3}{\sqrt{5}} \end{aligned}$$

12. ANGLES BETWEEN GEOMETRIC FIGURES

The angle, ϕ, between various geometric figures is given by the following equations.

- Between two lines in $Ax + By + C = 0$, $y = mx + b$, or direction angle formats,

$$\phi = \arctan\left(\frac{A_1B_2 - A_2B_1}{A_1A_2 + B_1B_2}\right) \quad 7.47$$

$$\phi = \arctan\left(\frac{m_2 - m_1}{1 + m_1m_2}\right) \quad 7.48$$

$$\phi = |\arctan m_1 - \arctan m_2| \quad 7.49$$

$$\phi = \arccos\left(\frac{L_1L_2 + M_1M_2 + N_1N_2}{d_1d_2}\right) \quad 7.50$$

$$\phi = \arccos\left(\begin{array}{c}\cos\alpha_1\cos\alpha_2 + \cos\beta_1\cos\beta_2 \\ + \cos\gamma_1\cos\gamma_2\end{array}\right) \quad 7.51$$

If the lines are parallel, then $\phi = 0$.

$$\frac{A_1}{A_2} = \frac{B_1}{B_2} \quad 7.52$$

$$m_1 = m_2 \quad 7.53$$

$$\alpha_1 = \alpha_2;\ \beta_1 = \beta_2;\ \gamma_1 = \gamma_2 \quad 7.54$$

If the lines are perpendicular, then $\phi = 90°$.

$$A_1A_2 = -B_1B_2 \quad 7.55$$

$$m_1 = \frac{-1}{m_2} \quad 7.56$$

$$\begin{aligned} \alpha_1 + \alpha_2 &= \beta_1 + \beta_2 \\ &= \gamma_1 + \gamma_2 \\ &= 90° \end{aligned} \quad 7.57$$

- Between two planes in $A\mathbf{i} + B\mathbf{j} + C\mathbf{k} = 0$ format, the coefficients A, B, and C are the same as the coefficients for the normal vector. (See Eq. 7.32.) ϕ is equal to the angle between the two normal vectors.

$$\cos\phi = \frac{|A_1A_2 + B_1B_2 + C_1C_2|}{\sqrt{A_1^2 + B_1^2 + C_1^2}\sqrt{A_2^2 + B_2^2 + C_2^2}} \quad 7.58$$

Example 7.5

Use Eq. 7.47, Eq. 7.48, and Eq. 7.49 to find the angle between the lines.

$$\begin{aligned} y &= -0.577x + 2 \\ y &= +0.577x - 5 \end{aligned}$$

Solution

Write both equations in general form.

$$\begin{aligned} -0.577x - y + 2 &= 0 \\ 0.577x - y - 5 &= 0 \end{aligned}$$

(a) From Eq. 7.47,

$$\begin{aligned} \phi &= \arctan\left(\frac{A_1B_2 - A_2B_1}{A_1A_2 + B_1B_2}\right) \\ &= \arctan\left(\frac{(-0.577)(-1) - (0.577)(-1)}{(-0.577)(0.577) + (-1)(-1)}\right) \\ &= 60° \end{aligned}$$

(b) Use Eq. 7.48.

$$\begin{aligned}\phi &= \arctan\left(\frac{m_2 - m_1}{1 + m_1 m_2}\right) \\ &= \arctan\left(\frac{0.577 - (-0.577)}{1 + (0.577)(-0.577)}\right) \\ &= 60°\end{aligned}$$

(c) Use Eq. 7.49.

$$\begin{aligned}\phi &= |\arctan m_1 - \arctan m_2| \\ &= |\arctan(-0.577) - \arctan(0.577)| \\ &= |-30° - 30°| \\ &= 60°\end{aligned}$$

13. CONIC SECTIONS

A *conic section* is any one of several curves produced by passing a plane through a cone as shown in Fig. 7.7. If α is the angle between the vertical axis and the cutting plane and β is the cone generating angle, Eq. 7.59 gives the *eccentricity*, ϵ, of the conic section. Values of the eccentricity are given in

***Figure 7.7** Conic Sections Produced by Cutting Planes*

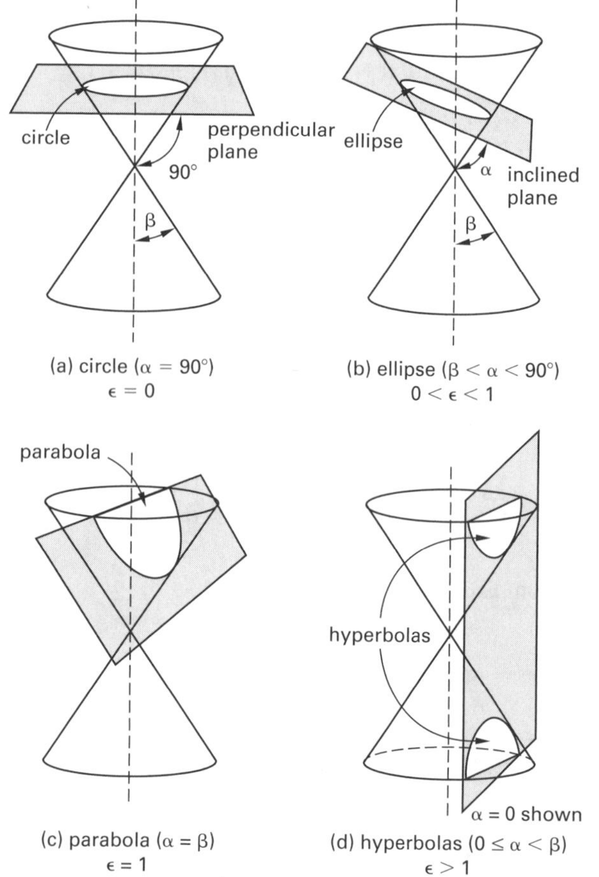

$$\epsilon = \frac{\cos\alpha}{\cos\beta} \tag{7.59}$$

All conic sections are described by second-degree polynomials (i.e., *quadratic equations*) of the following form.[1]

$$Ax^2 + Bxy + Cy^2 + Dx + Ey + F = 0 \tag{7.60}$$

This is the *general form*, which allows the figure axes to be at any angle relative to the coordinate axes. The *standard forms* presented in the following sections pertain to figures whose axes coincide with the coordinate axes, thereby eliminating certain terms of the general equation.

Figure 7.8 can be used to determine which conic section is described by the quadratic function. The quantity $B^2 - 4AC$ is known as the *discriminant*. Figure 7.8 determines only the type of conic section; it does not determine whether the conic section is degenerate (e.g., a circle with a negative radius).

***Figure 7.8** Determining Conic Sections from Quadratic Equations*

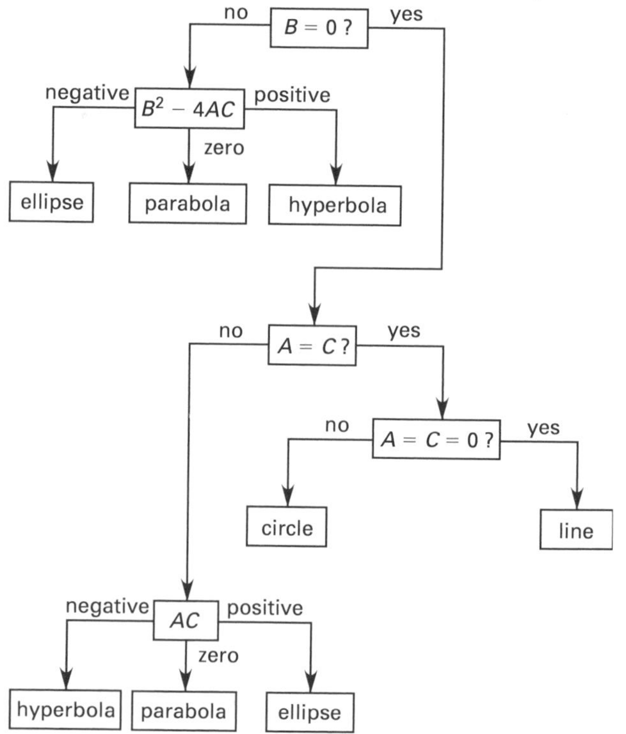

[1]One or more straight lines are produced when the cutting plane passes through the cone's vertex. Straight lines can be considered to be quadratic functions without second-degree terms.

Example 7.6

What geometric figures are described by the following equations?

(a) $4y^2 - 12y + 16x + 41 = 0$

(b) $x^2 - 10xy + y^2 + x + y + 1 = 0$

(c) $x^2 + 4y^2 + 2x - 8y + 1 = 0$

(d) $x^2 + y^2 - 6x + 8y + 20 = 0$

Solution

(a) Referring to Fig. 7.8, $B = 0$ since there is no xy term, $A = 0$ since there is no x^2 term, and $AC = (0)(4) = 0$. This is a parabola.

(b) $B \neq 0$; $B^2 - 4AC = (-10)^2 - (4)(1)(1) = +96$. This is a hyperbola.

(c) $B = 0$; $A \neq C$; $AC = (1)(4) = +4$. This is an ellipse.

(d) $B = 0$; $A = C$; $A = C = 1$ ($\neq 0$). This is a circle.

14. CIRCLE

The general form of the equation of a circle, as illustrated in Fig. 7.9, is

$$Ax^2 + Ay^2 + Dx + Ey + F = 0 \qquad 7.61$$

Figure 7.9 *Circle*

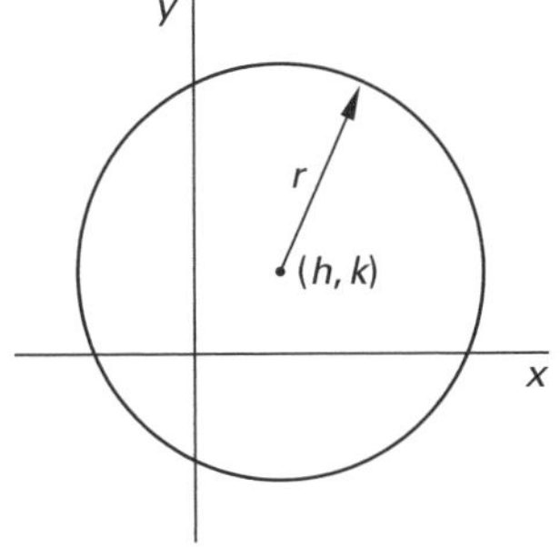

The *center-radius form* of the equation of a circle with radius r and center at (h, k) is

$$(x - h)^2 + (y - k)^2 = r^2 \qquad 7.62$$

The two forms can be converted by use of Eq. 7.63 through Eq. 7.65.

$$h = \frac{-D}{2A} \qquad 7.63$$

$$k = \frac{-E}{2A} \qquad 7.64$$

$$r^2 = \frac{D^2 + E^2 - 4AF}{4A^2} \qquad 7.65$$

If the right-hand side of Eq. 7.65 is positive, the figure is a circle. If it is zero, the circle shrinks to a point. If the right-hand side is negative, the figure is imaginary. A *degenerate circle* is one in which the right-hand side is less than or equal to zero.

15. PARABOLA

A *parabola* is the locus of points equidistant from the *focus* (point F in Fig. 7.10) and a line called the *directrix*. A parabola is symmetric with respect to its *parabolic axis*. The line normal to the parabolic axis and passing through the focus is known as the *latus rectum*. The eccentricity of a parabola is 1.

Figure 7.10 *Parabola*

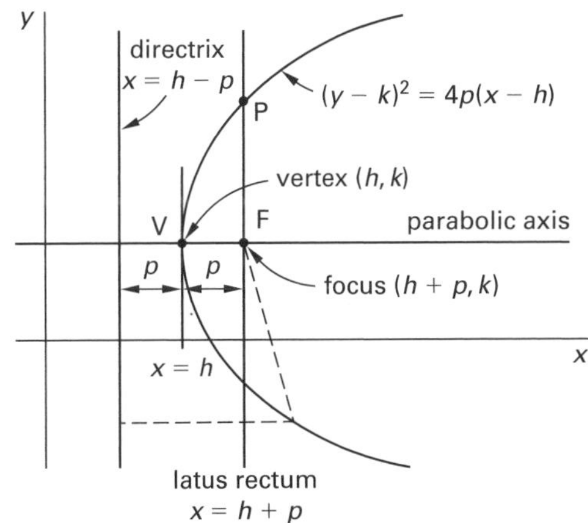

There are two common types of parabolas in the Cartesian plane—those that open right and left, and those that open up and down. Equation 7.60 is the general form of the equation of a parabola. With Eq. 7.66, the parabola points horizontally to the right if $CD > 0$ and to the left if $CD < 0$. With Eq. 7.67, the parabola points vertically up if $AE > 0$ and down if $AE < 0$.

$$Cy^2 + Dx + Ey + F = 0\Big|_{\substack{C,\,D \neq 0 \\ \text{opens horizontally}}} \qquad 7.66$$

$$Ax^2 + Dx + Ey + F = 0\Big|_{\substack{A,\,E \neq 0 \\ \text{opens vertically}}} \qquad 7.67$$

The *vertex form* of the equation of a parabola with vertex at (h, k), focus at $(h + p, k)$, and directrix at $x = h - p$, and that opens to the right or left is given by Eq. 7.68. The parabola opens to the right (points to the left) if $p > 0$ and opens to the left (points to the right) if $p < 0$.

$$(y - k)^2 = 4p(x - h)\Big|_{\text{opens horizontally}} \qquad 7.68$$

$$y^2 = 4px \Big|_{\substack{\text{vertex at origin} \\ h=k=0}} \qquad 7.69$$

The *vertex form* of the equation of a parabola with vertex at (h, k), focus at $(h, k+p)$, and directrix at $y = k - p$, and that opens up or down is given by Eq. 7.70. The parabola opens up (points down) if $p > 0$ and opens down (points up) if $p < 0$.

$$(x-h)^2 = 4p(y-k)\Big|_{\text{opens vertically}} \qquad 7.70$$

$$x^2 = 4py \Big|_{\text{vertex at origin}} \qquad 7.71$$

The general and vertex forms of the equations can be reconciled with Eq. 7.72 through Eq. 7.74. Whether the first or second forms of these equations are used depends on whether the parabola opens horizontally or vertically (i.e., whether $A = 0$ or $C = 0$), respectively.

$$h = \begin{cases} \dfrac{E^2 - 4CF}{4CD} & \text{[opens horizontally]} \\ \dfrac{-D}{2A} & \text{[opens vertically]} \end{cases} \qquad 7.72$$

$$k = \begin{cases} \dfrac{-E}{2C} & \text{[opens horizontally]} \\ \dfrac{D^2 - 4AF}{4AE} & \text{[opens vertically]} \end{cases} \qquad 7.73$$

$$p = \begin{cases} \dfrac{-D}{4C} & \text{[opens horizontally]} \\ \dfrac{-E}{4A} & \text{[opens vertically]} \end{cases} \qquad 7.74$$

16. ELLIPSE

An *ellipse* has two foci separated along its *major axis* by a distance $2c$, as shown in Fig. 7.11. The line perpendicular to the major axis passing through the center of the ellipse is the *minor axis.* The two lines passing through the foci perpendicular to the major axis are the *latera recta.* The distance between the two vertices is $2a$. The ellipse is the locus of those points whose distances from the two foci add up to $2a$. For each point P on the ellipse,

$$F_1P + PF_2 = 2a \qquad 7.75$$

Figure 7.11 *Ellipse*

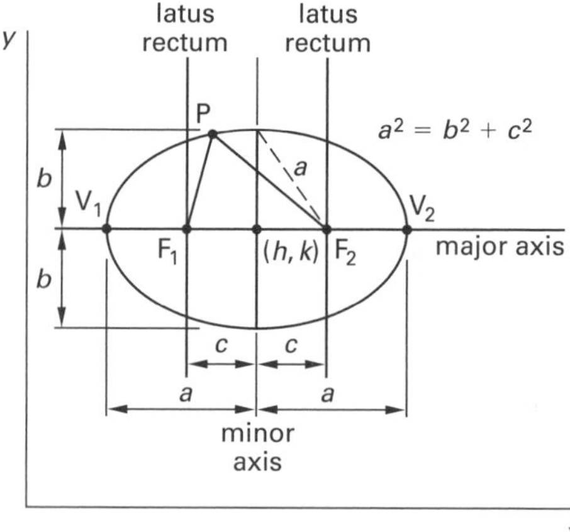

Equation 7.76 is the standard equation used for an ellipse with axes parallel to the coordinate axes, while Eq. 7.60 is the general form. F is not independent of A, C, D, and E for the ellipse.

$$Ax^2 + Cy^2 + Dx + Ey + F = 0\Big|_{\substack{AC>0 \\ A \neq C}} \qquad 7.76$$

Equation 7.77 gives the center form of the equation of an ellipse centered at (h, k). The larger of a or b is known as the *semi-major distance*, and the smaller is known as the *semi-minor distance.*

$$\frac{(x-h)^2}{a^2} + \frac{(y-k)^2}{b^2} = 1 \qquad 7.77$$

The distance between the two foci is $2c$.

$$2c = 2\sqrt{a^2 - b^2} \qquad 7.78$$

The *aspect ratio* of the ellipse is

$$\text{aspect ratio} = \frac{a}{b} \qquad 7.79$$

The *eccentricity*, ϵ, of the ellipse is always less than 1. If the eccentricity is zero, the figure is a circle (another form of a *degenerative ellipse*).

$$\epsilon = \frac{\sqrt{a^2 - b^2}}{a} < 1 \qquad 7.80$$

The standard and center forms of the equations of an ellipse can be reconciled by using Eq. 7.81 through Eq. 7.84.

$$h = \frac{-D}{2A} \qquad 7.81$$

$$k = \frac{-E}{2C} \qquad 7.82$$

$$a = \sqrt{C} \quad 7.83$$

$$b = \sqrt{A} \quad 7.84$$

17. HYPERBOLA

A *hyperbola* has two foci separated along its *transverse axis* (*major axis*) by a distance $2c$, as shown in Fig. 7.12. The line perpendicular to the transverse axis and midway between the foci is the *conjugate axis* (*minor axis*). The distance between the two vertices is $2a$. If a line is drawn parallel to the conjugate axis through either vertex, the distance between the points where it intersects the asymptotes is $2b$. The hyperbola is the locus of those points whose distances from the two foci differ by $2a$. For each point P on the hyperbola,

$$\text{F}_2\text{P} - \text{PF}_1 = 2a \quad 7.85$$

Figure 7.12 *Hyperbola*

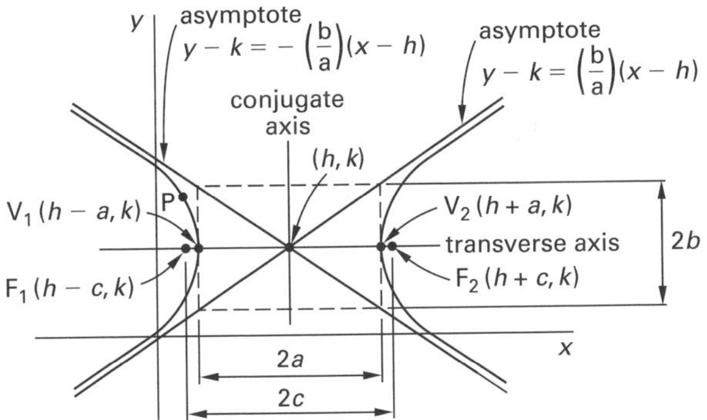

Equation 7.86 is the standard equation of a hyperbola. Coefficients A and C have opposite signs.

$$Ax^2 + Cy^2 + Dx + Ey + F = 0\Big|_{AC<0} \quad 7.86$$

Equation 7.87 gives the center form of the equation of a hyperbola centered at (h, k) and opening to the left and right.

$$\frac{(x-h)^2}{a^2} - \frac{(y-k)^2}{b^2} = 1\Big|_{\text{opens horizontally}} \quad 7.87$$

Equation 7.88 gives the center form of the equation of a hyperbola that is centered at (h, k) and is opening up and down.

$$\frac{(y-k)^2}{a^2} - \frac{(x-h)^2}{b^2} = 1\Big|_{\text{opens vertically}} \quad 7.88$$

The distance between the two foci is $2c$.

$$2c = 2\sqrt{a^2 + b^2} \quad 7.89$$

The *eccentricity*, ϵ, of the hyperbola is calculated from Eq. 7.90 and is always greater than 1.

$$\epsilon = \frac{c}{a} = \frac{\sqrt{a^2 + b^2}}{a} > 1 \quad 7.90$$

The hyperbola is asymptotic to the lines given by Eq. 7.91 and Eq. 7.92.

$$y = \pm\frac{b}{a}(x - h) + k\Big|_{\text{opens horizontally}} \quad 7.91$$

$$y = \pm\frac{a}{b}(x - h) + k\Big|_{\text{opens vertically}} \quad 7.92$$

For a *rectangular* (*equilateral*) *hyperbola*, the asymptotes are perpendicular, $a = b$, $c = \sqrt{2a}$, and the eccentricity is $\epsilon = \sqrt{2}$. If the hyperbola is centered at the origin (i.e., $h = k = 0$), then the equations are $x^2 - y^2 = a^2$ (opens horizontally) and $y^2 - x^2 = a^2$ (opens vertically).

If the asymptotes are the x- and y-axes, the equation of the hyperbola is simply

$$xy = \pm\frac{a^2}{2} \quad 7.93$$

The general and center forms of the equations of a hyperbola can be reconciled by using Eq. 7.94 through Eq. 7.98. Whether the hyperbola opens left and right or up and down depends on whether M/A or M/C is positive, respectively, where M is defined by Eq. 7.94.

$$M = \frac{D^2}{4A} + \frac{E^2}{4C} - F \quad 7.94$$

$$h = \frac{-D}{2A} \quad 7.95$$

$$k = \frac{-E}{2C} \quad 7.96$$

$$a = \begin{cases} \sqrt{-C} & \text{[opens horizontally]} \\ \sqrt{-A} & \text{[opens vertically]} \end{cases} \quad 7.97$$

$$b = \begin{cases} \sqrt{A} & \text{[opens horizontally]} \\ \sqrt{C} & \text{[opens vertically]} \end{cases} \quad 7.98$$

18. SPHERE

Equation 7.99 is the general equation of a sphere. The coefficient A cannot be zero.

$$Ax^2 + Ay^2 + Az^2 + Bx + Cy + Dz + E = 0 \qquad 7.99$$

Equation 7.100 gives the center form of the equation of a sphere centered at (h, k, l) with radius r.

$$(x-h)^2 + (y-k)^2 + (z-l)^2 = r^2 \qquad 7.100$$

The general and center forms of the equations of a sphere can be reconciled by using Eq. 7.101 through Eq. 7.104.

$$h = \frac{-B}{2A} \qquad 7.101$$

$$k = \frac{-C}{2A} \qquad 7.102$$

$$l = \frac{-D}{2A} \qquad 7.103$$

$$r = \sqrt{\frac{B^2 + C^2 + D^2}{4A^2} - \frac{E}{A}} \qquad 7.104$$

19. HELIX

A *helix* is a curve generated by a point moving on, around, and along a cylinder such that the distance the point moves parallel to the cylindrical axis is proportional to the angle of rotation about that axis. (See Fig. 7.13.) For a cylinder of radius r, Eq. 7.105 through Eq. 7.107 define the three-dimensional positions of points along the helix. The quantity $2\pi k$ is the *pitch* of the helix.

$$x = r\cos\theta \qquad 7.105$$

$$y = r\sin\theta \qquad 7.106$$

$$z = k\theta \qquad 7.107$$

Figure 7.13 Helix

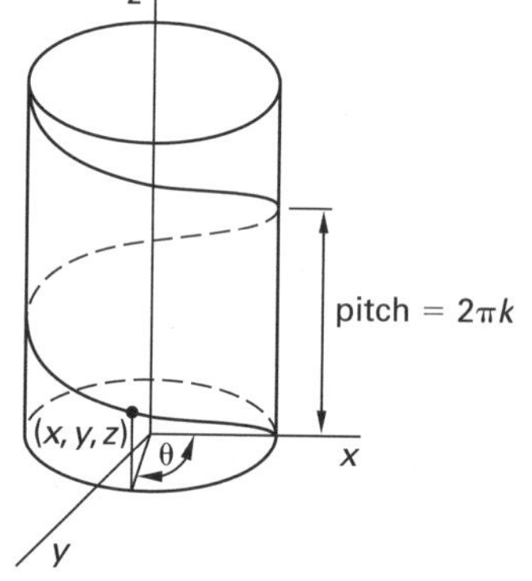

8 Differential Calculus

Content in blue refers to the *NCEES Handbook*.

1. DERIVATIVE OF A FUNCTION

In most cases, it is possible to transform a continuous function, $f(x_1, x_2, x_3, \ldots)$, of one or more independent variables into a derivative function.[1] In simple cases, the *derivative* can be interpreted as the slope (tangent or rate of change) of the curve described by the original function. Since the slope of the curve depends on x, the derivative function will also depend on x. The derivative, $f'(x)$, of a function $f(x)$ is defined mathematically by Eq. 8.1. However, limit theory is seldom needed to actually calculate derivatives.

$$f'(x) = \lim_{\Delta x \to 0} \frac{\Delta f(x)}{\Delta x} \qquad 8.1$$

The derivative of a function $f(x)$, also known as the *first derivative*, is written in various ways, including

$$f'(x), \frac{df(x)}{dx}, \frac{df}{dx}, \mathbf{D}f(x), \mathbf{D}_x f(x), \dot{f}(x), sf(s)$$

A *second derivative* may exist if the derivative operation is performed on the first derivative—that is, a derivative is taken of a derivative function. This is written as

$$f''(x), \frac{d^2f(x)}{dx^2}, \frac{d^2f}{dx^2}, \mathbf{D}^2f(x), \mathbf{D}_x^2f(x), \ddot{f}(x), s^2f(s)$$

Newton's notation (e.g., $\dot{m}$ and $\ddot{x}$) is generally only used with functions of time.

A *regular* (*analytic* or *holomorphic*) *function* possesses a derivative. A point at which a function's derivative is undefined is called a *singular point*, as Fig. 8.1 illustrates.

Figure 8.1 *Derivatives and Singular Points*

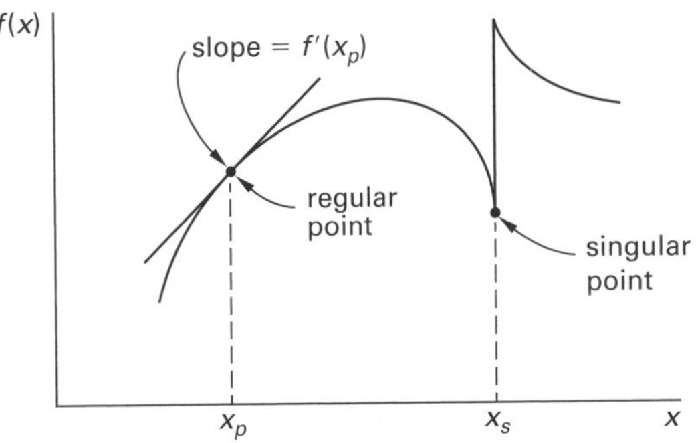

2. ELEMENTARY DERIVATIVE OPERATIONS

Equation 8.2 through Eq. 8.5 summarize the elementary derivative operations on polynomials and exponentials. Equation 8.2 and Eq. 8.3 are particularly useful. (a, n, and k represent constants. $f(x)$ and $g(x)$ are functions of x.)

$$\mathbf{D}\,k = 0 \qquad 8.2$$

$$\mathbf{D}\,x^n = nx^{n-1} \qquad 8.3$$

$$\mathbf{D}\,\ln x = \frac{1}{x} \qquad 8.4$$

$$\mathbf{D}\,e^{ax} = ae^{ax} \qquad 8.5$$

Equation 8.6 through Eq. 8.17 summarize the elementary derivative operations on transcendental (trigonometric) functions.

$$\mathbf{D}\,\sin x = \cos x \qquad 8.6$$

$$\mathbf{D}\,\cos x = -\sin x \qquad 8.7$$

$$\mathbf{D}\,\tan x = \sec^2 x \qquad 8.8$$

$$\mathbf{D}\,\cot x = -\csc^2 x \qquad 8.9$$

[1]A function, $f(x)$, of one independent variable, x, is used in this section to simplify the discussion. Although the derivative is taken with respect to x, the independent variable can be anything.

$$\mathbf{D}\sec x = \sec x \tan x \quad 8.10$$

$$\mathbf{D}\csc x = -\csc x \cot x \quad 8.11$$

$$\mathbf{D}\arcsin x = \frac{1}{\sqrt{1-x^2}} \quad 8.12$$

$$\mathbf{D}\arccos x = -\mathbf{D}\arcsin x \quad 8.13$$

$$\mathbf{D}\arctan x = \frac{1}{1+x^2} \quad 8.14$$

$$\mathbf{D}\operatorname{arccot} x = -\mathbf{D}\arctan x \quad 8.15$$

$$\mathbf{D}\operatorname{arcsec} x = \frac{1}{x\sqrt{x^2-1}} \quad 8.16$$

$$\mathbf{D}\operatorname{arccsc} x = -\mathbf{D}\operatorname{arcsec} x \quad 8.17$$

Equation 8.18 through Eq. 8.23 summarize the elementary derivative operations on hyperbolic transcendental functions. Derivatives of hyperbolic functions are not completely analogous to those of the regular transcendental functions.

$$\mathbf{D}\sinh x = \cosh x \quad 8.18$$

$$\mathbf{D}\cosh x = \sinh x \quad 8.19$$

$$\mathbf{D}\tanh x = \operatorname{sech}^2 x \quad 8.20$$

$$\mathbf{D}\coth x = -\operatorname{csch}^2 x \quad 8.21$$

$$\mathbf{D}\operatorname{sech} x = -\operatorname{sech} x \tanh x \quad 8.22$$

$$\mathbf{D}\operatorname{csch} x = -\operatorname{csch} x \coth x \quad 8.23$$

Equation 8.24 through Eq. 8.29 summarize the elementary derivative operations on functions and combinations of functions.

$$\mathbf{D}kf(x) = k\mathbf{D}f(x) \quad 8.24$$

$$\mathbf{D}\big(f(x) \pm g(x)\big) = \mathbf{D}f(x) \pm \mathbf{D}g(x) \quad 8.25$$

$$\mathbf{D}\big(f(x)\cdot g(x)\big) = f(x)\mathbf{D}g(x) + g(x)\mathbf{D}f(x) \quad 8.26$$

$$\mathbf{D}\left(\frac{f(x)}{g(x)}\right) = \frac{g(x)\mathbf{D}f(x) - f(x)\mathbf{D}g(x)}{\big(g(x)\big)^2} \quad 8.27$$

$$\mathbf{D}\big(f(x)\big)^n = n\big(f(x)\big)^{n-1}\mathbf{D}f(x) \quad 8.28$$

$$\mathbf{D}f\big(g(x)\big) = \mathbf{D}_g f(g)\mathbf{D}_x g(x) \quad 8.29$$

Example 8.1

What is the slope at $x = 3$ of the curve $f(x) = x^3 - 2x$?

Solution

The derivative function found from Eq. 8.3 determines the slope.

$$f'(x) = 3x^2 - 2$$

The slope at $x = 3$ is

$$f'(3) = (3)(3)^2 - 2 = 25$$

Example 8.2

What are the derivatives of the following functions?

(a) $f(x) = 5\sqrt[3]{x^5}$

(b) $f(x) = \sin x \cos^2 x$

(c) $f(x) = \ln(\cos e^x)$

Solution

(a) Using Eq. 8.3 and Eq. 8.24,

$$\begin{aligned} f'(x) &= 5\mathbf{D}\sqrt[3]{x^5} = 5\mathbf{D}(x^5)^{1/3} \\ &= (5)\left(\tfrac{1}{3}\right)(x^5)^{-2/3}\mathbf{D}x^5 \\ &= (5)\left(\tfrac{1}{3}\right)(x^5)^{-2/3}(5)(x^4) \\ &= 25x^{2/3}/3 \end{aligned}$$

(b) Using Eq. 8.26,

$$\begin{aligned} f'(x) &= \sin x\mathbf{D}\cos^2 x + \cos^2 x\mathbf{D}\sin x \\ &= (\sin x)(2\cos x)(\mathbf{D}\cos x) + \cos^2 x \cos x \\ &= (\sin x)(2\cos x)(-\sin x) + \cos^2 x \cos x \\ &= -2\sin^2 x \cos x + \cos^3 x \end{aligned}$$

(c) Using Eq. 8.29,

$$\begin{aligned} f'(x) &= \left(\frac{1}{\cos e^x}\right)\mathbf{D}\cos e^x \\ &= \left(\frac{1}{\cos e^x}\right)(-\sin e^x)\mathbf{D}e^x \\ &= \left(\frac{-\sin e^x}{\cos e^x}\right)e^x \\ &= -e^x \tan e^x \end{aligned}$$

3. CRITICAL POINTS

Derivatives are used to locate the local *critical points* of functions of one variable—that is, *extreme points* (also known as *maximum* and *minimum* points) as well as the *inflection points* (*points of contraflexure*). The plurals *extrema*, *maxima*, and *minima* are used without the word "points." These points are illustrated in Fig. 8.2. There is usually an inflection point between two adjacent local extrema.

Figure 8.2 Extreme and Inflection Points

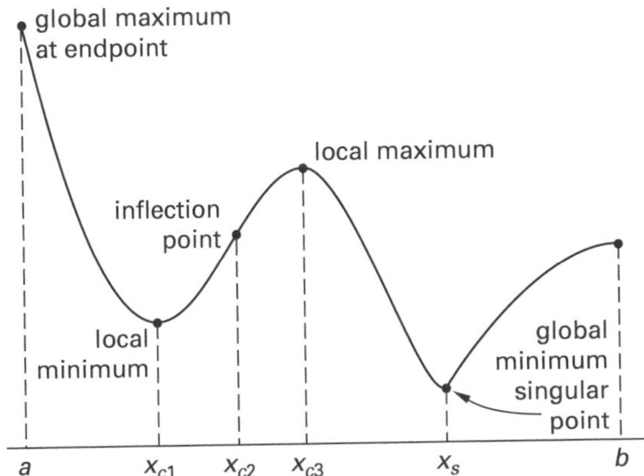

The first derivative is calculated to determine the locations of possible critical points. The second derivative is calculated to determine whether a particular point is a local maximum, minimum, or inflection point, according to the following conditions. With this method, no distinction is made between local and global extrema. Therefore, the extrema should be compared with the function values at the endpoints of the interval, as illustrated in Ex. 8.3.[2] Generally, $f'(x) \neq 0$ at an inflection point.

$$f'(x_c) = 0 \text{ at any extreme point, } x_c \qquad 8.30$$

$$f''(x_c) < 0 \text{ at a maximum point} \qquad 8.31$$

$$f''(x_c) > 0 \text{ at a minimum point} \qquad 8.32$$

$$f''(x_c) = 0 \text{ at an inflection point} \qquad 8.33$$

Example 8.3

Find the global extrema of the function $f(x)$ on the interval $[-2, +2]$.

$$f(x) = x^3 + x^2 - x + 1$$

Solution

The first derivative is

$$f'(x) = 3x^2 + 2x - 1$$

Since the first derivative is zero at extreme points, set $f'(x)$ equal to zero and solve for the roots of the quadratic equation.

$$3x^2 + 2x - 1 = (3x - 1)(x + 1) = 0$$

The roots are $x_1 = 1/3$, $x_2 = -1$. These are the locations of the two local extrema.

The second derivative is

$$f''(x) = 6x + 2$$

Substituting x_1 and x_2 into $f''(x)$,

$$f''(x_1) = (6)\left(\frac{1}{3}\right) + 2 = 4$$

$$f''(x_2) = (6)(-1) + 2 = -4$$

Therefore, x_1 is a local minimum point (because $f''(x_1)$ is positive), and x_2 is a local maximum point (because $f''(x_2)$ is negative). The inflection point between these two extrema is found by setting $f''(x)$ equal to zero.

$$f''(x) = 6x + 2 = 0 \text{ or } x = -\tfrac{1}{3}$$

Since the question asked for the global extreme points, it is necessary to compare the values of $f(x)$ at the local extrema with the values at the endpoints.

$$f(-2) = -1$$

$$f(-1) = 2$$

$$f\left(\tfrac{1}{3}\right) = 22/27$$

$$f(2) = 11$$

Therefore, the actual global extrema are at the endpoints.

4. DERIVATIVES OF PARAMETRIC EQUATIONS

The derivative of a function $f(x_1, x_2, \ldots, x_n)$ can be calculated from the derivatives of the parametric equations $f_1(s), f_2(s), \ldots, f_n(s)$. The derivative will be expressed in terms of the parameter, s, unless the derivatives of the parametric equations can be expressed explicitly in terms of the independent variables.

[2]It is also necessary to check the values of the function at singular points (i.e., points where the derivative does not exist).

Example 8.4

A circle is expressed parametrically by the equations

$$x = 5\cos\theta$$

$$y = 5\sin\theta$$

Express the derivative dy/dx (a) as a function of the parameter θ and (b) as a function of x and y.

Solution

(a) Taking the derivative of each parametric equation with respect to θ,

$$\frac{dx}{d\theta} = -5\sin\theta$$

$$\frac{dy}{d\theta} = 5\cos\theta$$

Then,

$$\frac{dy}{dx} = \frac{\frac{dy}{d\theta}}{\frac{dx}{d\theta}} = \frac{5\cos\theta}{-5\sin\theta} = -\cot\theta$$

(b) The derivatives of the parametric equations are closely related to the original parametric equations.

$$\frac{dx}{d\theta} = -5\sin\theta = -y$$

$$\frac{dy}{d\theta} = 5\cos\theta = x$$

$$\frac{dy}{dx} = \frac{\frac{dy}{d\theta}}{\frac{dx}{d\theta}} = \frac{-x}{y}$$

5. PARTIAL DIFFERENTIATION

Derivatives can be taken with respect to only one independent variable at a time. For example, $f'(x)$ is the derivative of $f(x)$ and is taken with respect to the independent variable x. If a function, $f(x_1, x_2, x_3, \ldots)$, has more than one independent variable, a *partial derivative* can be found, but only with respect to one of the independent variables. All other variables are treated as constants. Symbols for a partial derivative of f taken with respect to variable x are $\partial f/\partial x$ and $f_x(x, y)$.

The geometric interpretation of a partial derivative $\partial f/\partial x$ is the slope of a line tangent to the surface (a sphere, ellipsoid, etc.) described by the function when all variables except x are held constant. In three-dimensional space with a function described by $z = f(x, y)$, the partial derivative $\partial f/\partial x$ (equivalent to $\partial z/\partial x$) is the slope of the line tangent to the surface in a plane of constant y. Similarly, the partial derivative $\partial f/\partial y$ (equivalent to $\partial z/\partial y$) is the slope of the line tangent to the surface in a plane of constant x.

Example 8.5

What is the partial derivative $\partial z/\partial x$ of the following function?

$$z = 3x^2 - 6y^2 + xy + 5y - 9$$

Solution

The partial derivative with respect to x is found by considering all variables other than x to be constants.

$$\frac{\partial z}{\partial x} = 6x - 0 + y + 0 - 0 = 6x + y$$

Example 8.6

A surface has the equation $x^2 + y^2 + z^2 - 9 = 0$. What is the slope of a line that lies in a plane of constant y and is tangent to the surface at $(x, y, z) = (1, 2, 2)$?[3]

Solution

Solve for the dependent variable. Then, consider variable y to be a constant.

$$z = \sqrt{9 - x^2 - y^2}$$

$$\begin{aligned}\frac{\partial z}{\partial x} &= \frac{\partial(9 - x^2 - y^2)^{1/2}}{\partial x} \\ &= \left(\tfrac{1}{2}\right)(9 - x^2 - y^2)^{-1/2}\left(\frac{\partial(9 - x^2 - y^2)}{\partial x}\right) \\ &= \left(\tfrac{1}{2}\right)(9 - x^2 - y^2)^{-1/2}(-2x) \\ &= \frac{-x}{\sqrt{9 - x^2 - y^2}}\end{aligned}$$

[3] Although only implied, it is required that the point actually be on the surface (i.e., it must satisfy the equation $f(x, y, z) = 0$).

At the point (1, 2, 2), $x = 1$ and $y = 2$.

$$\left.\frac{\partial z}{\partial x}\right|_{(1,2,2)} = \frac{-1}{\sqrt{9-(1)^2-(2)^2}} = -\frac{1}{2}$$

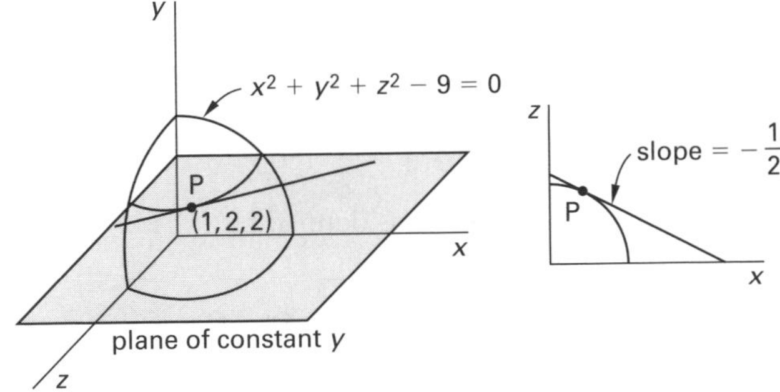

6. IMPLICIT DIFFERENTIATION

When a relationship between n variables cannot be manipulated to yield an explicit function of $n-1$ independent variables, that relationship implicitly defines the nth variable. Finding the derivative of the implicit variable with respect to any other independent variable is known as *implicit differentiation.*

An implicit derivative is the quotient of two partial derivatives. The two partial derivatives are chosen so that dividing one by the other eliminates a common differential. For example, if z cannot be explicitly extracted from $f(x, y, z) = 0$, the partial derivatives $\partial z/\partial x$ and $\partial z/\partial y$ can still be found as follows.

$$\frac{\partial z}{\partial x} = -\frac{\frac{\partial f}{\partial x}}{\frac{\partial f}{\partial z}} \qquad 8.34$$

$$\frac{\partial z}{\partial y} = -\frac{\frac{\partial f}{\partial y}}{\frac{\partial f}{\partial z}} \qquad 8.35$$

Example 8.7

Find the derivative dy/dx of

$$f(x, y) = x^2 + xy + y^3$$

Solution

Implicit differentiation is required because x cannot be extracted from $f(x, y)$.

$$\frac{\partial f}{\partial x} = 2x + y$$

$$\frac{\partial f}{\partial y} = x + 3y^2$$

$$\frac{dy}{dx} = \frac{-\frac{\partial f}{\partial x}}{\frac{\partial f}{\partial y}} = \frac{-(2x+y)}{x+3y^2}$$

Example 8.8

Solve Ex. 8.6 using implicit differentiation.

Solution

$$f(x, y, z) = x^2 + y^2 + z^2 - 9 = 0$$

$$\frac{\partial f}{\partial x} = 2x$$

$$\frac{\partial f}{\partial z} = 2z$$

$$\frac{\partial z}{\partial x} = \frac{-\frac{\partial f}{\partial x}}{\frac{\partial f}{\partial z}} = \frac{-2x}{2z} = -\frac{x}{z}$$

At the point (1, 2, 2),

$$\frac{\partial z}{\partial x} = -\frac{1}{2}$$

7. TANGENT PLANE FUNCTION

Partial derivatives can be used to find the equation of a plane tangent to a three-dimensional surface defined by $f(x, y, z) = 0$ at some point, P_0.

$$\begin{aligned} T(x_0, y_0, z_0) = (x - x_0)\left.\frac{\partial f(x,y,z)}{\partial x}\right|_{P_0} \\ + (y - y_0)\left.\frac{\partial f(x,y,z)}{\partial y}\right|_{P_0} \\ + (z - z_0)\left.\frac{\partial f(x,y,z)}{\partial z}\right|_{P_0} \\ = 0 \end{aligned} \qquad 8.36$$

The coefficients of x, y, and z are the same as the coefficients of **i**, **j**, and **k** of the normal vector at point P_0. (See Sec. 8.10.)

Example 8.9

What is the equation of the plane that is tangent to the surface defined by $f(x, y, z) = 4x^2 + y^2 - 16z = 0$ at the point (2, 4, 2)?

Solution

Calculate the partial derivatives and substitute the coordinates of the point.

$$\left.\frac{\partial f(x, y, z)}{\partial x}\right|_{P_0} = 8x\Big|_{(2,4,2)} = (8)(2) = 16$$

$$\left.\frac{\partial f(x, y, z)}{\partial y}\right|_{P_0} = 2y\Big|_{(2,4,2)} = (2)(4) = 8$$

$$\left.\frac{\partial f(x, y, z)}{\partial z}\right|_{P_0} = -16\Big|_{(2,4,2)} = -16$$

$$\begin{aligned} T(2,4,2) &= (16)(x-2) + (8)(y-4) - (16)(z-2) \\ &= 16x + 8y - 16z - 32 \end{aligned}$$

Substitute into Eq. 8.36, and divide both sides by 8.

$$2x + y - 2z - 4 = 0$$

8. GRADIENT VECTOR

The slope of a function is the change in one variable with respect to a distance in a chosen direction. Usually, the direction is parallel to a coordinate axis. However, the maximum slope at a point on a surface may not be in a direction parallel to one of the coordinate axes.

The *gradient vector function* $\nabla f(x, y, z)$ (pronounced "del f") gives the maximum rate of change of the function $f(x, y, z)$.

$$\nabla f(x, y, z) = \frac{\partial f(x, y, z)}{\partial x}\mathbf{i} + \frac{\partial f(x, y, z)}{\partial y}\mathbf{j} + \frac{\partial f(x, y, z)}{\partial z}\mathbf{k} \qquad 8.37$$

Example 8.10

A two-dimensional function is defined by

$$f(x, y) = 2x^2 - y^2 + 3x - y$$

(a) What is the gradient vector for this function? (b) What is the direction of the line passing through the point (1, −2) that has a maximum slope? (c) What is the maximum slope at the point (1, −2)?

Solution

(a) It is necessary to calculate two partial derivatives in order to use Eq. 8.37.

$$\frac{\partial f(x, y)}{\partial x} = 4x + 3$$

$$\frac{\partial f(x, y)}{\partial y} = -2y - 1$$

$$\nabla f(x, y) = (4x + 3)\mathbf{i} + (-2y - 1)\mathbf{j}$$

(b) Find the direction of the line passing through (1, −2) with maximum slope by inserting $x = 1$ and $y = -2$ into the gradient vector function.

$$\begin{aligned} \mathbf{V} &= \big((4)(1) + 3\big)\mathbf{i} + \big((-2)(-2) - 1\big)\mathbf{j} \\ &= 7\mathbf{i} + 3\mathbf{j} \end{aligned}$$

(c) The magnitude of the slope is

$$|\mathbf{V}| = \sqrt{(7)^2 + (3)^2} = 7.62$$

9. DIRECTIONAL DERIVATIVE

Unlike the gradient vector (covered in Sec. 8.8), which calculates the maximum rate of change of a function, the *directional derivative*, indicated by $\nabla_u f(x,y,z)$, $D_u f(x,y,z)$, or $f_u'(x,y,z)$, gives the rate of change in the direction of a given vector, **u** or **U**. The subscript u implies that the direction vector is a unit vector, but it does not need to be, as only the direction cosines are calculated from it.

$$\nabla_u f(x,y,z) = \left(\frac{\partial f(x,y,z)}{\partial x}\right)\cos\alpha + \left(\frac{\partial f(x,y,z)}{\partial y}\right)\cos\beta + \left(\frac{\partial f(x,y,z)}{\partial z}\right)\cos\gamma \qquad 8.38$$

$$\mathbf{U} = U_x\mathbf{i} + U_y\mathbf{j} + U_z\mathbf{k} \qquad 8.39$$

$$\cos\alpha = \frac{U_x}{|\mathbf{U}|} = \frac{U_x}{\sqrt{U_x^2 + U_y^2 + U_z^2}} \qquad 8.40$$

$$\cos\beta = \frac{U_y}{|\mathbf{U}|} \qquad 8.41$$

$$\cos\gamma = \frac{U_z}{|\mathbf{U}|} \qquad 8.42$$

Example 8.11

What is the rate of change of $f(x,y) = 3x^2 + xy - 2y^2$ at the point $(1,-2)$ in the direction $4\mathbf{i} + 3\mathbf{j}$?

Solution

The direction cosines are given by Eq. 8.40 and Eq. 8.41.

$$\cos\alpha = \frac{U_x}{|\mathbf{U}|} = \frac{4}{\sqrt{(4)^2 + (3)^2}} = \frac{4}{5}$$

$$\cos\beta = \frac{U_y}{|\mathbf{U}|} = \frac{3}{5}$$

The partial derivatives are

$$\frac{\partial f(x,y)}{\partial x} = 6x + y$$

$$\frac{\partial f(x,y)}{\partial y} = x - 4y$$

The directional derivative is given by Eq. 8.38.

$$\nabla_u f(x,y) = \left(\frac{4}{5}\right)(6x + y) + \left(\frac{3}{5}\right)(x - 4y)$$

Substituting the given values of $x = 1$ and $y = -2$,

$$\nabla_u f(1,-2) = \left(\frac{4}{5}\right)\big((6)(1) - 2\big) + \left(\frac{3}{5}\right)\big(1 - (4)(-2)\big) = \frac{43}{5} \quad (8.6)$$

10. NORMAL LINE VECTOR

Partial derivatives can be used to find the vector normal to a three-dimensional surface defined by $f(x,y,z) = 0$ at some point P_0. The coefficients of **i**, **j**, and **k** are the same as the coefficients of x, y, and z calculated for the equation of the tangent plane at point P_0. (See Sec. 8.7.)

$$\mathbf{N} = \left.\frac{\partial f(x,y,z)}{\partial x}\right|_{P_0}\mathbf{i} + \left.\frac{\partial f(x,y,z)}{\partial y}\right|_{P_0}\mathbf{j} + \left.\frac{\partial f(x,y,z)}{\partial z}\right|_{P_0}\mathbf{k} \qquad 8.43$$

Example 8.12

What is the vector normal to the surface of $f(x,y,z) = 4x^2 + y^2 - 16z = 0$ at the point $(2,4,2)$?

Solution

The equation of the tangent plane at this point was calculated in Ex. 8.9 to be

$$T(2,4,2) = 2x + y - 2z - 4 = 0$$

A vector that is normal to the tangent plane through this point is

$$\mathbf{N} = 2\mathbf{i} + \mathbf{j} - 2\mathbf{k}$$

11. DIVERGENCE OF A VECTOR FIELD

The *divergence*, div **F**, of a vector field $\mathbf{F}(x, y, z)$ is a scalar function defined by Eq. 8.45 and Eq. 8.46.[4] The divergence of **F** can be interpreted as the *accumulation* of flux (i.e., a flowing substance) in a small region (i.e., at a point). One of the uses of the divergence is to determine whether flow (represented in direction and magnitude by **F**) is compressible. Flow is incompressible if div $\mathbf{F} = 0$, since the substance is not accumulating.

$$\mathbf{F} = P(x, y, z)\mathbf{i} + Q(x, y, z)\mathbf{j} + R(x, y, z)\mathbf{k} \quad 8.44$$

$$\operatorname{div}\mathbf{F} = \frac{\partial P}{\partial x} + \frac{\partial Q}{\partial y} + \frac{\partial R}{\partial z} \quad 8.45$$

It may be easier to calculate divergence from Eq. 8.46.

$$\operatorname{div}\mathbf{F} = \nabla \cdot \mathbf{F} \quad 8.46$$

The vector del operator, ∇, is defined as

$$\nabla = \frac{\partial}{\partial x}\mathbf{i} + \frac{\partial}{\partial y}\mathbf{j} + \frac{\partial}{\partial z}\mathbf{k} \quad 8.47$$

If there is no divergence, then the dot product calculated in Eq. 8.46 is zero.

Example 8.13

Calculate the divergence of the following vector function.

$$\mathbf{F}(x, y, z) = xz\mathbf{i} + e^x y\mathbf{j} + 7x^3 y\mathbf{k}$$

Solution

From Eq. 8.45,

$$\begin{aligned}\operatorname{div}\mathbf{F} &= \frac{\partial}{\partial x}xz + \frac{\partial}{\partial y}e^x y + \frac{\partial}{\partial z}7x^3 y = z + e^x + 0\\ &= z + e^x\end{aligned}$$

12. CURL OF A VECTOR FIELD

The *curl*, curl **F**, of a vector field $\mathbf{F}(x, y, z)$ is a vector field defined by Eq. 8.49 and Eq. 8.50. The curl **F** can be interpreted as the *vorticity* per unit area of flux (i.e., a flowing substance) in a small region (i.e., at a point). One of the uses of the curl is to determine whether flow (represented in direction and magnitude by **F**) is rotational. Flow is irrotational if curl $\mathbf{F} = 0$.

$$\mathbf{F} = P(x, y, z)\mathbf{i} + Q(x, y, z)\mathbf{j} + R(x, y, z)\mathbf{k} \quad 8.48$$

$$\begin{aligned}\operatorname{curl}\mathbf{F} = &\left(\frac{\partial R}{\partial y} - \frac{\partial Q}{\partial z}\right)\mathbf{i} + \left(\frac{\partial P}{\partial z} - \frac{\partial R}{\partial x}\right)\mathbf{j}\\ &+ \left(\frac{\partial Q}{\partial x} - \frac{\partial P}{\partial y}\right)\mathbf{k}\end{aligned} \quad 8.49$$

It may be easier to calculate the curl from Eq. 8.50. (The vector del operator, ∇, was defined in Eq. 8.47.)

$$\begin{aligned}\operatorname{curl}\mathbf{F} &= \nabla \times \mathbf{F}\\ &= \begin{vmatrix}\mathbf{i} & \mathbf{j} & \mathbf{k}\\ \frac{\partial}{\partial x} & \frac{\partial}{\partial y} & \frac{\partial}{\partial z}\\ P(x, y, z) & Q(x, y, z) & R(x, y, z)\end{vmatrix}\end{aligned} \quad 8.50$$

If the velocity vector is **V**, then the vorticity is

$$\omega = \nabla \times \mathbf{V} = \omega_x\mathbf{i} + \omega_y\mathbf{j} + \omega_z\mathbf{k} \quad 8.51$$

The *circulation* is the line integral of the velocity, **V**, along a closed curve.

$$\Gamma = \oint V \cdot ds = \oint \omega \cdot dA \quad 8.52$$

Example 8.14

Calculate the curl of the following vector function.

$$\mathbf{F}(x, y, z) = 3x^2\mathbf{i} + 7e^x y\mathbf{j}$$

Solution

Using Eq. 8.50,

$$\operatorname{curl}\mathbf{F} = \begin{vmatrix}\mathbf{i} & \mathbf{j} & \mathbf{k}\\ \frac{\partial}{\partial x} & \frac{\partial}{\partial y} & \frac{\partial}{\partial z}\\ 3x^2 & 7e^x y & 0\end{vmatrix}$$

Expand the determinant across the top row.

$$\begin{aligned}&\mathbf{i}\left(\frac{\partial}{\partial y}(0) - \frac{\partial}{\partial z}(7e^x y)\right) - \mathbf{j}\left(\frac{\partial}{\partial x}(0) - \frac{\partial}{\partial z}(3x^2)\right)\\ &\quad + \mathbf{k}\left(\frac{\partial}{\partial x}(7e^x y) - \frac{\partial}{\partial y}(3x^2)\right)\\ &= \mathbf{i}(0 - 0) - \mathbf{j}(0 - 0) + \mathbf{k}(7e^x y - 0)\\ &= 7e^x y\mathbf{k}\end{aligned}$$

[4]A bold letter, **F**, is used to indicate that the vector is a function of x, y, and z.

13. TAYLOR'S FORMULA

Taylor's formula (series) can be used to expand a function around a point (i.e., approximate the function at one point based on the function's value at another point). The approximation consists of a series, each term composed of a derivative of the original function and a polynomial. Using Taylor's formula requires that the original function be continuous in the interval $[a, b]$ and have the required number of derivatives. To expand a function, $f(x)$, around a point, a, in order to obtain $f(b)$, Taylor's formula is

$$f(b) = f(a) + \frac{f'(a)}{1!}(b-a) + \frac{f''(a)}{2!}(b-a)^2 + \cdots + \frac{f^n(a)}{n!}(b-a)^n + R_n(b) \quad 8.53$$

In Eq. 8.53, the expression f^n designates the nth derivative of the function $f(x)$. To be a useful approximation, two requirements must be met: (1) point a must be relatively close to point b, and (2) the function and its derivatives must be known or easy to calculate. The last term, $R_n(b)$, is the uncalculated remainder after n derivatives. It is the difference between the exact and approximate values. By using enough terms, the remainder can be made arbitrarily small. That is, $R_n(b)$ approaches zero as n approaches infinity.

It can be shown that the remainder term can be calculated from Eq. 8.54, where c is some number in the interval $[a, b]$. With certain functions, the constant c can be completely determined. In most cases, however, it is possible only to calculate an upper bound on the remainder from Eq. 8.55. M_n is the maximum (positive) value of $f^{n+1}(x)$ on the interval $[a, b]$.

$$R_n(b) = \frac{f^{n+1}(c)}{(n+1)!}(b-a)^{n+1} \quad 8.54$$

$$|R_n(b)| \le M_n \frac{|(b-a)^{n+1}|}{(n+1)!} \quad 8.55$$

14. MACLAURIN POWER APPROXIMATIONS

If $a = 0$ in the Taylor series, Eq. 8.53 is known as the *Maclaurin series*. The Maclaurin series can be used to approximate functions at some value of x between 0 and 1. The following common approximations may be referred to as Maclaurin series, Taylor series, or power series approximations.

$$\sin x \approx x - \frac{x^3}{3!} + \frac{x^5}{5!} - \frac{x^7}{7!} + \cdots + (-1)^n \frac{x^{2n+1}}{(2n+1)!} \quad 8.56$$

$$\cos x \approx 1 - \frac{x^2}{2!} + \frac{x^4}{4!} - \frac{x^6}{6!} + \cdots + (-1)^n \frac{x^{2n}}{(2n)!} \quad 8.57$$

$$\sinh x \approx x + \frac{x^3}{3!} + \frac{x^5}{5!} + \frac{x^7}{7!} + \cdots + \frac{x^{2n+1}}{(2n+1)!} \quad 8.58$$

$$\cosh x \approx 1 + \frac{x^2}{2!} + \frac{x^4}{4!} + \frac{x^6}{6!} + \cdots + \frac{x^{2n}}{(2n)!} \quad 8.59$$

$$e^x \approx 1 + x + \frac{x^2}{2!} + \frac{x^3}{3!} + \cdots + \frac{x^n}{n!} \quad 8.60$$

$$\ln(1+x) \approx x - \frac{x^2}{2} + \frac{x^3}{3} - \frac{x^4}{4} + \cdots + (-1)^{n+1}\frac{x^n}{n} \quad 8.61$$

$$\frac{1}{1-x} \approx 1 + x + x^2 + x^3 + \cdots + x^n \quad 8.62$$

Integral Calculus

Content in blue refers to the *NCEES Handbook*.

1. INTEGRATION

Integration is the inverse operation of differentiation. For that reason, *indefinite integrals* are sometimes referred to as *antiderivatives*.[1] Although an expression can be a function of several variables, an integral can only be taken with respect to one variable at a time.[2] The *differential term* (dx in Eq. 9.1) indicates that variable. In Eq. 9.1, the function $f'(x)$ is the *integrand*, and x is the variable of integration.

$$\int f'(x)\,dx = f(x) + C \qquad 9.1$$

While most of a function, $f(x)$, can be "recovered" through integration of its derivative, $f'(x)$, a constant term will be lost, because the derivative of a constant term vanishes (i.e., is zero), leaving nothing from which to recover. A *constant of integration*, C, is added to the integral to recognize the possibility of such a constant term.

2. ELEMENTARY OPERATIONS

Equation 9.2 through Eq. 9.8 summarize the elementary integration operations on polynomials and exponentials.[3]

Equation 9.2 and Eq. 9.3 are particularly useful. (C and k represent constants. $f(x)$ and $g(x)$ are functions of x.)

$$\int k\,dx = kx + C \qquad 9.2$$

$$\int x^m\,dx = \frac{x^{m+1}}{m+1} + C \quad [m \neq -1] \qquad 9.3$$

$$\int \frac{1}{x}\,dx = \ln|x| + C \qquad 9.4$$

$$\int e^{kx}dx = \frac{e^{kx}}{k} + C \qquad 9.5$$

$$\int xe^{kx}dx = \frac{e^{kx}(kx-1)}{k^2} + C \qquad 9.6$$

$$\int k^{ax}dx = \frac{k^{ax}}{a\ln k} + C \qquad 9.7$$

$$\int \ln x\,dx = x\ln x - x + C \qquad 9.8$$

Equation 9.9 through Eq. 9.20 summarize the elementary integration operations on transcendental functions.

$$\int \sin x\,dx = -\cos x + C \qquad 9.9$$

$$\int \cos x\,dx = \sin x + C \qquad 9.10$$

$$\begin{aligned}\int \tan x\,dx &= \ln|\sec x| + C \\ &= -\ln|\cos x| + C\end{aligned} \qquad 9.11$$

$$\int \cot x\,dx = \ln|\sin x| + C \qquad 9.12$$

[1]The difference between an indefinite and definite integral (covered in Sec. 9.7) is simple: An *indefinite integral* is a function, while a *definite integral* is a number.
[2]This is not actually true for some higher mathematics, but it is true for engineering purposes.
[3]More extensive listings, known as *tables of integrals*, are widely available. (See App. 9.A.)

$$\int \sec x\, dx = \ln|\sec x + \tan x| + C = \ln\left|\tan\left(\frac{x}{2}+\frac{\pi}{4}\right)\right| + C \qquad 9.13$$

$$\int \csc x\, dx = \ln|\csc x - \cot x| + C = \ln\left|\tan\frac{x}{2}\right| + C \qquad 9.14$$

$$\int \frac{dx}{k^2+x^2} = \frac{1}{k}\arctan\frac{x}{k} + C \qquad 9.15$$

$$\int \frac{dx}{\sqrt{k^2-x^2}} = \arcsin\frac{x}{k} + C \quad [k^2 > x^2] \qquad 9.16$$

$$\int \frac{dx}{x\sqrt{x^2-k^2}} = \frac{1}{k}\operatorname{arcsec}\frac{x}{k} + C \quad [x^2 > k^2] \qquad 9.17$$

$$\int \sin^2 x\, dx = \tfrac{1}{2}x - \tfrac{1}{4}\sin 2x + C \qquad 9.18$$

$$\int \cos^2 x\, dx = \tfrac{1}{2}x + \tfrac{1}{4}\sin 2x + C \qquad 9.19$$

$$\int \tan^2 x\, dx = \tan x - x + C \qquad 9.20$$

Equation 9.21 through Eq. 9.26 summarize the elementary integration operations on hyperbolic transcendental functions. Integrals of hyperbolic functions are not completely analogous to those of the regular transcendental functions.

$$\int \sinh x\, dx = \cosh x + C \qquad 9.21$$

$$\int \cosh x\, dx = \sinh x + C \qquad 9.22$$

$$\int \tanh x\, dx = \ln|\cosh x| + C \qquad 9.23$$

$$\int \coth x\, dx = \ln|\sinh x| + C \qquad 9.24$$

$$\int \operatorname{sech} x\, dx = \arctan(\sinh x) + C \qquad 9.25$$

$$\int \operatorname{csch} x\, dx = \ln\left|\tanh\frac{x}{2}\right| + C \qquad 9.26$$

Equation 9.27 through Eq. 9.31 summarize the elementary integration operations on functions and combinations of functions.

$$\int kf(x)\,dx = k\int f(x)\,dx \qquad 9.27$$

$$\int \big(f(x)+g(x)\big)\,dx = \int f(x)\,dx + \int g(x)\,dx \qquad 9.28$$

$$\int \frac{f'(x)}{f(x)}\,dx = \ln|f(x)| + C \qquad 9.29$$

$$\int f(x)\,dg(x) = f(x)\int dg(x) - \int g(x)\,df(x) + C = f(x)g(x) - \int g(x)\,df(x) + C \qquad 9.30$$

Example 9.1

Find the integral with respect to x of

$$3x^2 + \tfrac{1}{3}x - 7 = 0$$

Solution

This is a polynomial function, and Eq. 9.3 can be applied to each of the three terms.

$$\int \left(3x^2 + \tfrac{1}{3}x - 7\right)dx = x^3 + \tfrac{1}{6}x^2 - 7x + C$$

3. INTEGRATION BY PARTS

Equation 9.30, repeated here as Eq. 9.31, is known as *integration by parts*. $f(x)$ and $g(x)$ are functions. The use of this method is demonstrated in Ex. 9.2.

$$\int f(x)\,dg(x) = f(x)g(x) - \int g(x)\,df(x) + C \qquad 9.31$$

Example 9.2

Find the following integral.

$$\int x^2 e^x dx$$

Solution

x^2e^x is factored into two parts so that integration by parts can be used.

$$f(x) = x^2$$
$$dg(x) = e^x dx$$
$$df(x) = 2x\,dx$$

$$g(x) = \int dg(x) = \int e^x dx = e^x$$

From Eq. 9.31, disregarding the constant of integration (which cannot be evaluated),

$$\int f(x)\,dg(x) = f(x)g(x) - \int g(x)\,df(x)$$

$$\int x^2e^x dx = x^2e^x - \int e^x(2x)\,dx$$

The second term is also factored into two parts, and integration by parts is used again. This time,

$$f(x) = x$$
$$dg(x) = e^x dx$$
$$df(x) = dx$$

$$g(x) = \int dg(x) = \int e^x dx = e^x$$

From Eq. 9.31,

$$\int 2xe^x dx = 2\int xe^x dx = 2\left(xe^x - \int e^x dx\right)$$
$$= 2(xe^x - e^x)$$

Then, the complete integral is

$$\int x^2 e^x dx = x^2 e^x - 2(xe^x - e^x) + C$$
$$= e^x(x^2 - 2x + 2) + C$$

4. SEPARATION OF TERMS

Equation 9.28 shows that the integral of a sum of terms is equal to a sum of integrals. This technique is known as *separation of terms*. In many cases, terms are easily separated. In other cases, the technique of *partial fractions* can be used to obtain individual terms. These techniques are used in Ex. 9.3 and Ex. 9.4.

Example 9.3
Find the following integral.

$$\int \frac{(2x^2+3)^2}{x} dx$$

Solution

Expand the numerator to gain access to all of its terms.

$$\int \frac{(2x^2+3)^2}{x} dx = \int \frac{4x^4 + 12x^2 + 9}{x} dx$$
$$= \int \left(4x^3 + 12x + \frac{9}{x}\right) dx$$
$$= x^4 + 6x^2 + 9\ln|x| + C$$

Example 9.4
Find the following integral.

$$\int \frac{3x+2}{3x-2} dx$$

Solution

The integrand is larger than 1, so use long division to simplify it.

$$\begin{array}{r} 1 \text{ rem } 4 \quad \left(1 + \dfrac{4}{3x-2}\right) \\ 3x-2 \,\big)\overline{\,3x+2\,} \\ \underline{3x-2} \\ 4 \text{ remainder} \end{array}$$

$$\int \frac{3x+2}{3x-2} dx = \int \left(1 + \frac{4}{3x-2}\right) dx$$
$$= \int dx + \int \frac{4}{3x-2} dx$$
$$= x + \tfrac{4}{3}\ln|(3x-2)| + C$$

5. DOUBLE AND HIGHER-ORDER INTEGRALS

A function can be successively integrated. (This is analogous to successive differentiation.) A function that is integrated twice is known as a *double integral*; if integrated three times, it is a *triple integral*; and so on. In engineering, double and triple integrals are generally used to calculate areas and volumes, respectively.

The successive integrations do not need to be with respect to the same variable. Variables not included in the integration are treated as constants.

There are several notations used for a multiple integral, particularly when the product of length differentials represents a differential area or volume. A double integral (i.e., two successive integrations) can be represented by one of the following notations.

$$\int\int f(x,y)\,dx\,dy, \int_{R^2} f(x,y)\,dx\,dy,$$
$$\text{or} \int\int_{R^2} f(x,y)\,dA$$

A triple integral can be represented by one of the following notations.

$$\int\int\int f(x,y,z)\,dx\,dy\,dz,$$
$$\int_{R^3} f(x,y,z)\,dx\,dy\,dz,$$
$$\text{or} \int\int\int_{R^3} f(x,y,z)\,dV$$

Example 9.5
Find the following double integral.

$$\int\int (x^2 + y^3 x)\,dx\,dy$$

Solution

Integrate the function twice, once with respect to x and once with respect to y.

$$\int (x^2 + y^3x)\,dx = \tfrac{1}{3}x^3 + \tfrac{1}{2}y^3x^2 + C_1$$

$$\int \left(\tfrac{1}{3}x^3 + \tfrac{1}{2}y^3x^2 + C_1\right) dy = \tfrac{1}{3}yx^3 + \tfrac{1}{8}y^4x^2 + C_1y + C_2$$

So,

$$\iint (x^2 + y^3x)\,dx\,dy = \tfrac{1}{3}yx^3 + \tfrac{1}{8}y^4x^2 + C_1y + C_2$$

6. INITIAL VALUES

The constant of integration, C, can be found only if the value of the function $f(x)$ is known for some value of x_0. The value $f(x_0)$ is known as an *initial value* or *initial condition*. To completely define a function, as many initial values—$f(x_0)$, $f'(x_0)$, $f''(x_0)$, and so on—as there are integrations are needed.

Example 9.6

It is known that $f(x) = 4$ when $x = 2$ (i.e., the initial value is $f(2) = 4$). Find the original function.

$$\int (3x^3 - 7x)\,dx$$

Solution

The function is

$$\begin{aligned} f(x) &= \int (3x^3 - 7x)\,dx \\ &= \tfrac{3}{4}x^4 - \tfrac{7}{2}x^2 + C \end{aligned}$$

Substituting the initial value determines C.

$$\begin{aligned} 4 &= \left(\frac{3}{4}\right)(2)^4 - \left(\frac{7}{2}\right)(2)^2 + C \\ 4 &= 12 - 14 + C \\ C &= 6 \end{aligned}$$

The function is

$$f(x) = \tfrac{3}{4}x^4 - \tfrac{7}{2}x^2 + 6$$

7. DEFINITE INTEGRALS

A *definite integral* is restricted to a specific range of the independent variable. (Unrestricted integrals of the types shown in all preceding examples are known as *indefinite integrals*.) A definite integral restricted to the region bounded by *lower* and *upper limits* (also known as *bounds*), x_1 and x_2, is written as

$$\int_{x_1}^{x_2} f(x)\,dx \qquad 9.32$$

Equation 9.33 indicates how definite integrals are evaluated. It is known as the *fundamental theorem of calculus*.

$$\begin{aligned} \int_{x_1}^{x_2} f'(x)\,dx &= f(x)\Big|_{x_1}^{x_2} \\ &= f(x_2) - f(x_1) \end{aligned} \qquad 9.33$$

A common use of a definite integral is the calculation of work performed by a force, F, that moves an object from position x_1 to x_2.

$$W = \int_{x_1}^{x_2} F\,dx \qquad 9.34$$

Example 9.7

Evaluate the following definite integral.

$$\int_{\pi/4}^{\pi/3} \sin x\,dx$$

Solution

From Eq. 9.33,

$$\begin{aligned} \int_{\pi/4}^{\pi/3} \sin x\,dx &= -\cos x\,\Big|_{\pi/4}^{\pi/3} \\ &= -\cos\tfrac{\pi}{3} - \left(-\cos\tfrac{\pi}{4}\right) \\ &= -0.5 - (-0.707) \\ &= 0.207 \end{aligned}$$

8. AVERAGE VALUE

The average value of a function $f(x)$ that is integrable over the interval $[a, b]$ is

$$\text{average value} = \frac{1}{b-a}\int_a^b f(x)\,dx \qquad 9.35$$

9. AREA

Equation 9.36 calculates the area, A, bounded by $x = a$, $x = b$, $f_1(x)$ above and $f_2(x)$ below. ($f_2(x) = 0$ if the area is bounded by the x-axis.) The area between two curves is illustrated in Fig. 9.1.

$$A = \int_a^b \big(f_1(x) - f_2(x)\big)\,dx \qquad 9.36$$

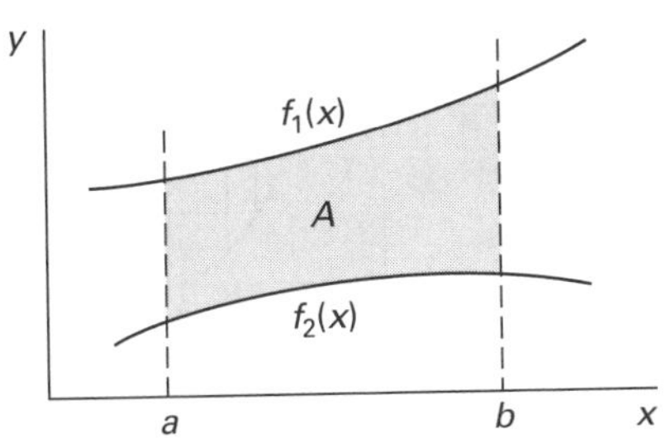

Figure 9.1 *Area Between Two Curves*

Example 9.8

Find the area between the x-axis and the parabola $y = x^2$ in the interval $[0, 4]$.

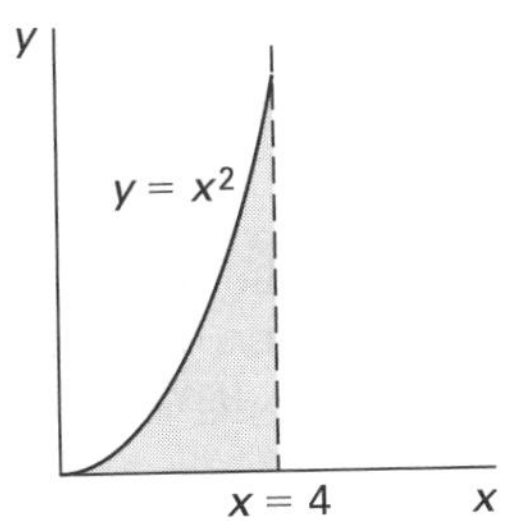

Solution

Referring to Eq. 9.36,

$$f_1(x) = x^2$$
$$f_2(x) = 0$$

$$\begin{aligned} A &= \int_a^b \big(f_1(x) - f_2(x)\big)\,dx \\ &= \int_0^4 x^2 dx \\ &= \left.\frac{x^3}{3}\right|_0^4 \\ &= 64/3 \end{aligned}$$

10. ARC LENGTH

Equation 9.37 gives the length of a curve defined by $f(x)$, whose derivative exists in the interval $[a, b]$.

$$\text{length} = \int_a^b \sqrt{1 + \big(f'(x)\big)^2}\;dx \qquad 9.37$$

11. PAPPUS' THEOREMS[4]

The first and second theorems of Pappus are as follows.[5]

- *First Theorem:* Given a curve, C, that does not intersect the y-axis, the area of the *surface of revolution* generated by revolving C around the y-axis is equal to the product of the length of the curve and the circumference of the circle traced by the centroid of curve C.

$$\begin{aligned} A &= \text{length} \times \text{circumference} \\ &= \text{length} \times 2\pi \times \text{radius} \end{aligned} \qquad 9.38$$

- *Second Theorem:* Given a plane region, R, that does not intersect the y-axis, the *volume of revolution* generated by revolving R around the y-axis is equal to the product of the area and the circumference of the circle traced by the centroid of area R.

$$\begin{aligned} V &= \text{area} \times \text{circumference} \\ &= \text{area} \times 2\pi \times \text{radius} \end{aligned} \qquad 9.39$$

12. SURFACE OF REVOLUTION

The surface area obtained by rotating $f(x)$ about the x-axis is

$$A = 2\pi \int_{x=a}^{x=b} f(x)\sqrt{1 + \big(f'(x)\big)^2}\;dx \qquad 9.40$$

The surface area obtained by rotating $f(y)$ about the y-axis is

$$A = 2\pi \int_{y=c}^{y=d} f(y)\sqrt{1 + \big(f'(y)\big)^2}\;dy \qquad 9.41$$

Example 9.9

The curve $f(x) = \frac{1}{2}x$ over the region $x = [0, 4]$ is rotated about the x-axis. What is the surface of revolution?

Solution

The surface of revolution is

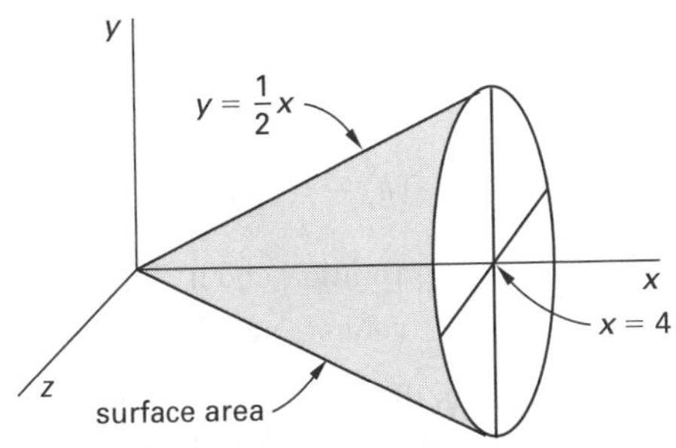

[4]This section is an introduction to surfaces and volumes of revolution. It does not involve integration.

[5]Some authorities call the first theorem the second and vice versa.

Since $f(x) = \frac{1}{2}x$, $f'(x) = \frac{1}{2}$. From Eq. 9.40, the area is

$$\begin{aligned} A &= 2\pi \int_{x=a}^{x=b} f(x)\sqrt{1+\left(f'(x)\right)^2}\,dx \\ &= 2\pi \int_0^4 \frac{1}{2}x\sqrt{1+\left(\frac{1}{2}\right)^2}\,dx \\ &= \frac{\sqrt{5}}{2}\pi \int_0^4 x\,dx \\ &= \frac{\sqrt{5}}{2}\pi \frac{x^2}{2}\Big|_0^4 \\ &= \frac{\sqrt{5}}{2}\pi\left(\frac{(4)^2-(0)^2}{2}\right) \\ &= 4\sqrt{5}\,\pi \end{aligned}$$

13. VOLUME OF REVOLUTION

The volume between $x = a$ and $x = b$ obtained by rotating $f(x)$ about the x-axis can be calculated from the *method of discs*, given by Eq. 9.42 and Eq. 9.44, or the *method of shells*, given by Eq. 9.43 and Eq. 9.45. In Eq. 9.42, $f^2(x)$ is the square of the function, not the second derivative. In Eq. 9.43, $h(y)$ is obtained by solving $f(x)$ for x (i.e., $x = h(y)$), $c = f(a)$, and $d = f(b)$.

$$V = \pi \int_{x=a}^{x=b} f^2(x)\,dx \quad \left[\begin{array}{l}\text{rotation about the } x\text{-axis;}\\ \text{volume includes the } x\text{-axis}\end{array}\right] \qquad 9.42$$

$$V = 2\pi \int_{y=c}^{y=d} y\left(h(b)-h(y)\right)dy \quad \left[\begin{array}{l}\text{rotation about the } x\text{-axis;}\\ \text{volume includes the } x\text{-axis}\end{array}\right] \qquad 9.43$$

The method of discs and the method of shells can also be used to determine the volume between $x = a$ and $x = b$, which is obtained by rotating $f(x)$ about the y-axis.

$$V = \pi \int_{y=c}^{y=d} h^2(y)\,dy \quad \left[\begin{array}{l}\text{rotation about the } y\text{-axis;}\\ \text{volume includes the } y\text{-axis}\end{array}\right] \qquad 9.44$$

$$V = 2\pi \int_{x=a}^{x=b} x\left(f(b)-f(x)\right)dx \quad \left[\begin{array}{l}\text{rotation about the } y\text{-axis;}\\ \text{volume includes the } y\text{-axis}\end{array}\right] \qquad 9.45$$

The method of discs is illustrated in Fig. 9.2(a), and the method of shells is illustrated in Fig. 9.2(b).

Whether or not a function, $f(x)$ (or $h(y)$), or its complement, $f(b) - f(x)$ (or $h(b) - h(y)$), is used for either method depends on whether or not the volume excludes (i.e., merely surrounds or encloses) or includes the axis of rotation. For example, for the volume below (i.e., bounded above by) a function, $f(x)$, rotated around the y-axis, the method of shells would integrate $xf(x)$ because the y-axis of rotation does not pass through the volume of rotation. (See Eq. 9.45.)

Figure 9.2 *Volume of a Parabolic Bowl*

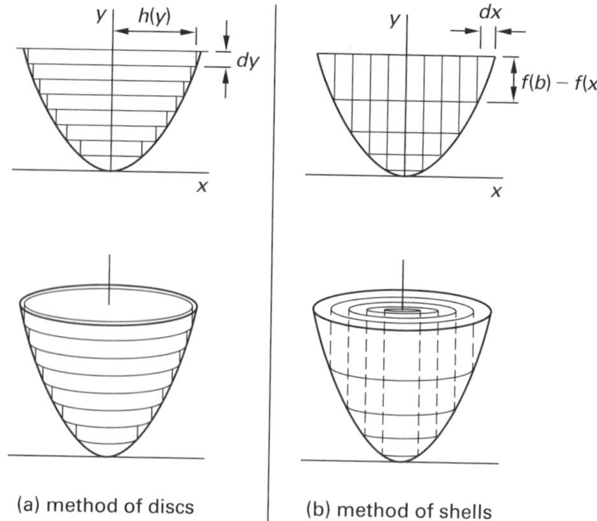

Example 9.10

The curve $f(x) = x^2$ over the region $x = [0, 4]$ is rotated about the x-axis. What is the volume of revolution that includes the x-axis?

Solution

Use the method of discs. The volume of revolution is

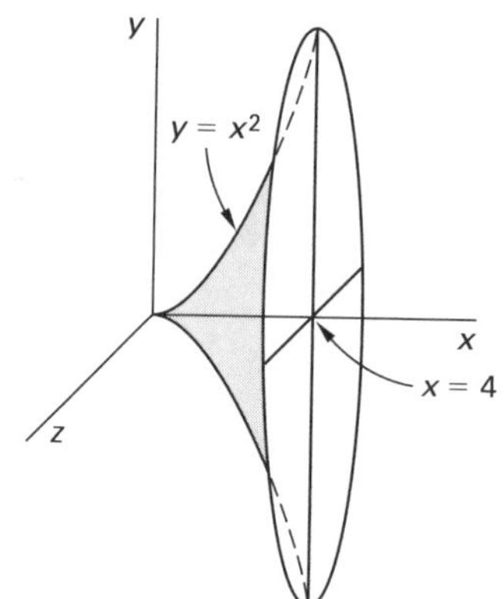

$$\begin{aligned} V &= \pi \int_a^b f^2(x)\,dx = \pi \int_0^4 \left(x^2\right)^2 dx \\ &= \pi \frac{x^5}{5}\Big|_0^4 \\ &= \pi\left(\frac{1024}{5} - 0\right) \\ &= 204.8\pi \end{aligned}$$

14. MOMENTS OF A FUNCTION

The *first moment of a function* is a concept used in finding centroids and centers of gravity. Equation 9.46 and Eq. 9.47 are for one- and two-dimensional problems, respectively. It is the exponent of x (1 in this case) that gives the moment its name.

$$\text{first moment} = \int x f(x)\,dx \qquad 9.46$$

$$\text{first moment} = \iint x f(x,y)\,dx\,dy \qquad 9.47$$

The *second moment of a function* is a concept used in finding moments of inertia with respect to an axis. Equation 9.48 and Eq. 9.49 are for two- and three-dimensional problems, respectively. Second moments with respect to other axes are analogous.

$$(\text{second moment})_x = \iint y^2 f(x,y)\,dy\,dx \qquad 9.48$$

$$(\text{second moment})_x = \iiint (y^2 + z^2) \times f(x,y,z)\,dy\,dz\,dx \qquad 9.49$$

15. FOURIER SERIES

Any periodic waveform can be written as the sum of an infinite number of sinusoidal terms, known as *harmonic terms* (i.e., an infinite series). Such a sum of terms is known as a *Fourier series*, and the process of finding the terms is *Fourier analysis*. (Extracting the original waveform from the series is known as *Fourier inversion*.) Since most series converge rapidly, it is possible to obtain a good approximation of the original waveform with a limited number of sinusoidal terms.

Fourier's theorem is Eq. 9.50.[6] The object of a Fourier analysis is to determine the coefficients a_n and b_n. The constant a_0 can often be determined by inspection since it is the average value of the waveform.[7]

$$f(t) = a_0 + a_1 \cos \omega t + a_2 \cos 2\omega t + \cdots + b_1 \sin \omega t + b_2 \sin 2\omega t + \cdots \qquad 9.50$$

ω is the *natural (fundamental) frequency* of the waveform. It depends on the actual waveform period, T.

$$\omega = \frac{2\pi}{T} \qquad 9.51$$

To simplify the analysis, the time domain can be normalized to the radian scale. The normalized scale is obtained by dividing all frequencies by ω. Then the Fourier series becomes

$$f(t) = a_0 + a_1 \cos t + a_2 \cos 2t + \cdots + b_1 \sin t + b_2 \sin 2t + \cdots \qquad 9.52$$

The coefficients a_n and b_n are found from the following relationships.

$$a_0 = \frac{1}{2\pi}\int_0^{2\pi} f(t)\,dt = \frac{1}{T}\int_0^T f(t)\,dt \qquad 9.53$$

$$a_n = \frac{1}{\pi}\int_0^{2\pi} f(t)\cos nt\,dt = \frac{2}{T}\int_0^T f(t)\cos nt\,dt \quad [n \geq 1] \qquad 9.54$$

$$b_n = \frac{1}{\pi}\int_0^{2\pi} f(t)\sin nt\,dt = \frac{2}{T}\int_0^T f(t)\sin nt\,dt \quad [n \geq 1] \qquad 9.55$$

While Eq. 9.54 and Eq. 9.55 are always valid, the work of integrating and finding a_n and b_n can be greatly simplified if the waveform is recognized as being symmetrical. Table 9.1 summarizes the simplifications.

Example 9.11

Find the first four terms of a Fourier series that approximates the repetitive step function illustrated.

$$f(t) = \begin{Bmatrix} 1 & 0 < t < \pi \\ 0 & \pi < t < 2\pi \end{Bmatrix}$$

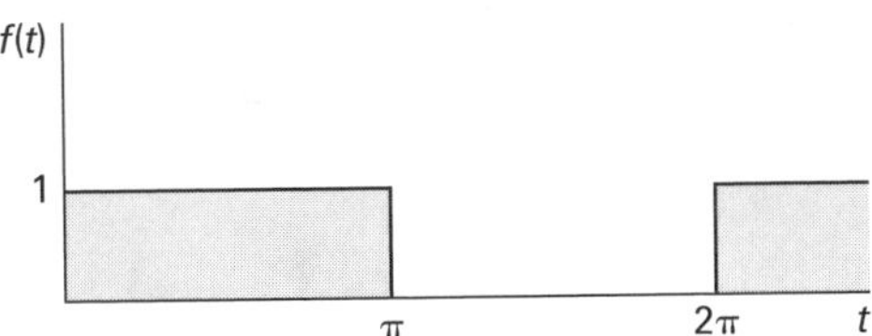

[6]The independent variable used in this section is t, since Fourier analysis is most frequently used in the time domain.

[7]There are different representations of the Fourier series, some with "$1/2a_0$" and some with "a_0," as shown in Eq. 9.50. The latter is used because it is consistent with the form used in NCEES publications. Both are correct, but using "$1/2a_0$" also changes Eq. 9.53. Regardless, the first term in the Fourier series represents the average value.

Table 9.1 Fourier Analysis Simplifications for Symmetrical Waveforms

	even symmetry $f(-t) = f(t)$	odd symmetry $f(-t) = -f(t)$
full-wave symmetry* $f(t + 2\pi) = f(t)$ $\lvert A_2\rvert = \lvert A_1\rvert$ $\lvert A_{\text{total}}\rvert = \lvert A_1\rvert$ *any repeating waveform	$b_n = 0$ [all n] $a_n = \frac{1}{\pi}\int_0^{2\pi} f(t)\cos nt\,dt$ [all n]	$a_0 = 0$ $a_n = 0$ [all n] $b_n = \frac{1}{\pi}\int_0^{2\pi} f(t)\sin nt\,dt$ [all n]
half-wave symmetry* $f(t + \pi) = -f(t)$ $\lvert A_2\rvert = \lvert A_1\rvert$ $\lvert A_{\text{total}}\rvert = 2\lvert A_1\rvert$ *same as rotational symmetry	$a_n = 0$ [even n] $b_n = 0$ [all n] $a_n = \frac{2}{\pi}\int_0^{\pi} f(t)\cos nt\,dt$ [odd n]	$a_0 = 0$ $a_n = 0$ [all n] $b_n = 0$ [even n] $b_n = \frac{2}{\pi}\int_0^{\pi} f(t)\sin nt\,dt$ [odd n]
quarter-wave symmetry $f(t + \pi) = -f(t)$ $\lvert A_2\rvert = \lvert A_1\rvert$ $\lvert A_{\text{total}}\rvert = 4\lvert A_1\rvert$	$a_0 = 0$ $a_n = 0$ [even n] $b_n = 0$ [all n] $a_n = \frac{4}{\pi}\int_0^{\frac{\pi}{2}} f(t)\cos nt\,dt$ [odd n]	$a_0 = 0$ $a_n = 0$ [all n] $b_n = 0$ [even n] $b_n = \frac{4}{\pi}\int_0^{\frac{\pi}{2}} f(t)\sin nt\,dt$ [odd n]

Solution

From Eq. 9.53,

$$a_0 = \frac{1}{2\pi}\int_0^{\pi}(1)\,dt + \frac{1}{2\pi}\int_{\pi}^{2\pi}(0)\,dt = \tfrac{1}{2}$$

This value of ½ corresponds to the average value of $f(t)$. It could have been found by observation.

$$\begin{aligned} a_1 &= \frac{1}{\pi}\int_0^{\pi}(1)\cos t\,dt + \frac{1}{\pi}\int_{\pi}^{2\pi}(0)\cos t\,dt \\ &= \frac{1}{\pi}\sin t\Big|_0^{\pi} + 0 \\ &= 0 \end{aligned}$$

In general,

$$a_n = \frac{1}{\pi}\,\frac{\sin nt}{n}\Bigg|_0^{\pi} = 0$$

$$\begin{aligned} b_1 &= \left(\frac{1}{\pi}\right)\int_0^{\pi}(1)\sin t\,dt + \frac{1}{\pi}\int_{\pi}^{2\pi}(0)\sin t\,dt \\ &= \frac{1}{\pi}(-\cos t)\Big|_0^{\pi} \\ &= \frac{2}{\pi} \end{aligned}$$

In general,

$$b_n = \left(\frac{1}{\pi}\right)\left(\frac{-\cos nt}{n}\right)\Bigg|_0^{\pi} = \begin{cases} 0 & \text{for } n \text{ even} \\ \dfrac{2}{\pi n} & \text{for } n \text{ odd} \end{cases}$$

The series is

$$f(t) = \frac{1}{2} + \tfrac{2}{\pi}\left(\sin t + \tfrac{1}{3}\sin 3t + \tfrac{1}{5}\sin 5t + \cdots\right)$$

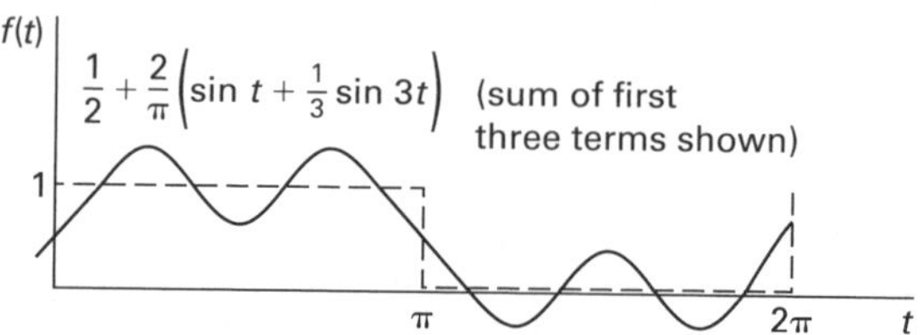

16. FAST FOURIER TRANSFORMS

Many mathematical operations are needed to implement a true Fourier transform. While the terms of a Fourier series might be slowly derived by integration, a faster method is needed to analyze real-time data. The *fast Fourier transform* (FFT) is a computer algorithm implemented in *spectrum analyzers* (*signal analyzers* or *FFT analyzers*) and replaces integration and multiplication operations with table lookups and additions.[8]

[8]*Spectrum analysis*, also known as *frequency analysis*, *signature analysis*, and *time-series analysis*, develops a relationship (usually graphical) between some property (e.g., amplitude or phase shift) versus frequency.

Since the complexity of the transform is reduced, the transformation occurs more quickly, enabling efficient analysis of waveforms with little or no periodicity.[9]

Using a spectrum analyzer requires choosing the frequency band (e.g., 0–20 kHz) to be monitored. (This step automatically selects the sampling period. The lower the frequencies sampled, the longer the sampling period.) If they are not fixed by the analyzer, the numbers of time-dependent input variable samples (e.g., 1024) and frequency-dependent output variable values (e.g., 400) are chosen.[10] There are half as many frequency lines as data points because each line contains two pieces of information—real (amplitude) and imaginary (phase). The *resolution* of the resulting frequency analysis is

$$\text{resolution} = \frac{\text{frequency bandwidth}}{\text{no. of output variable values}} \qquad 9.56$$

17. INTEGRAL FUNCTIONS

Integrals that cannot be evaluated as finite combinations of elementary functions are called *integral functions*. These functions are evaluated by series expansion. Some of the more common functions are listed as follows.[11]

- *integral sine function*

$$\begin{aligned}\text{Si}(x) &= \int_0^x \frac{\sin x}{x}\,dx \\ &= x - \frac{x^3}{3\cdot 3!} + \frac{x^5}{5\cdot 5!} - \frac{x^7}{7\cdot 7!} + \cdots\end{aligned} \qquad 9.57$$

- *integral cosine function*

$$\begin{aligned}\text{Ci}(x) &= \int_{-\infty}^x \frac{\cos x}{x}\,dx = -\int_x^\infty \frac{\cos x}{x}\,dx \\ &= C_E + \ln x - \frac{x^2}{2\cdot 2!} + \frac{x^4}{4\cdot 4!} - \cdots\end{aligned} \qquad 9.58$$

- *integral exponential function*

$$\begin{aligned}\text{Ei}(x) &= \int_{-\infty}^x \frac{e^x}{x}\,dx = -\int_{-x}^\infty \frac{e^{-x}}{x}\,dx \\ &= C_E + \ln x + x + \frac{x^2}{2\cdot 2!} + \frac{x^3}{3\cdot 3!} + \cdots\end{aligned} \qquad 9.59$$

C_E in Eq. 9.58 and Eq. 9.59 is *Euler's constant.*

$$\begin{aligned}C_E &= \int_{+\infty}^0 e^{-x}\ln x\,dx \\ &= \lim_{m\to\infty}\left(1 + \frac{1}{2} + \frac{1}{3} + \cdots + \frac{1}{m} - \ln m\right) \\ &= 0.577215665\end{aligned}$$

- *error function*

$$\begin{aligned}\text{erf}(x) &= \frac{2}{\sqrt{\pi}}\int_0^x e^{-x^2}\,dx \\ &= \left(\frac{2}{\sqrt{\pi}}\right)\left(\frac{x}{1\cdot 0!} - \frac{x^3}{3\cdot 1!} + \frac{x^5}{5\cdot 2!} - \frac{x^7}{7\cdot 3!} + \cdots\right)\end{aligned} \qquad 9.60$$

[9] Hours and days of manual computations are compressed into milliseconds.

[10] Two samples per time-dependent cycle (at the maximum frequency) is the lower theoretical limit for sampling, but the practical minimum rate is approximately 2.5 samples per cycle. This will ensure that *alias components* (i.e., low-level frequency signals) do not show up in the frequency band of interest.

[11] Other integral functions include the Fresnel integral, gamma function, and elliptic integral.

10 Differential Equations

Content in blue refers to the *NCEES Handbook*.

1. TYPES OF DIFFERENTIAL EQUATIONS

A *differential equation* is a mathematical expression combining a function (e.g., $y = f(x)$) and one or more of its derivatives. The *order* of a differential equation is the highest derivative in it. *First-order differential equations* contain only first derivatives of the function, *second-order differential equations* contain second derivatives (and may contain first derivatives as well), and so on.

A *linear differential equation* can be written as a sum of products of multipliers of the function and its derivatives. If the multipliers are scalars, the differential equation is said to have *constant coefficients.* If the function or one of its derivatives is raised to some power (other than one) or is embedded in another function (e.g., y embedded in $\sin y$ or e^y), the equation is said to be *nonlinear.*

Each term of a *homogeneous differential equation* contains either the function (y) or one of its derivatives; that is, the sum of derivative terms is equal to zero. In a *nonhomogeneous differential equation*, the sum of derivative terms is equal to a nonzero *forcing function* of the independent variable (e.g., $g(x)$). In order to solve a nonhomogeneous equation, it is often necessary to solve the homogeneous equation first. The homogeneous equation corresponding to a nonhomogeneous equation is known as a *reduced equation* or *complementary equation.*

The following examples illustrate the types of differential equations.

$y' - 7y = 0$	homogeneous, first-order linear, with constant coefficients
$y'' - 2y' + 8y = \sin 2x$	nonhomogeneous, second-order linear, with constant coefficients
$y'' - (x^2 - 1)y^2 = \sin 4x$	nonhomogeneous, second-order, nonlinear

An *auxiliary equation* (also called the *characteristic equation*) can be written for a homogeneous linear differential equation with constant coefficients, regardless of order. This auxiliary equation is simply the polynomial formed by replacing all derivatives with variables raised to the power of their respective derivatives.

The purpose of solving a differential equation is to derive an expression for the function in terms of the independent variable. The expression does not need to be explicit in the function, but there can be no derivatives in the expression. Since, in the simplest cases, solving a differential equation is equivalent to finding an indefinite integral, it is not surprising that *constants of integration* must be evaluated from knowledge of how the system behaves. Additional data are known as *initial values*, and any problem that includes them is known as an *initial value problem.*[1]

[1]The term *initial* implies that time is the independent variable. While this may explain the origin of the term, initial value problems are not limited to the time domain. A *boundary value problem* is similar, except that the data come from different points. For example, additional data in the form $y(x_0)$ and $y'(x_0)$ or $y(x_0)$ and $y'(x_1)$ that need to be simultaneously satisfied constitute an initial value problem. Data of the form $y(x_0)$ and $y(x_1)$ constitute a boundary value problem. Until solved, it is difficult to know whether a boundary value problem has zero, one, or more than one solution.

Most differential equations require lengthy solutions and are not efficiently solved by hand. However, several types are fairly simple and are presented in this chapter.

Example 10.1

Write the complementary differential equation for the following nonhomogeneous differential equation.

$$y'' + 6y' + 9y = e^{-14x}\sin 5x$$

Solution

The complementary equation is found by eliminating the forcing function, $e^{-14x}\sin 5x$.

$$y'' + 6y' + 9y = 0$$

Example 10.2

Write the auxiliary equation to the following differential equation.

$$y'' + 4y' + y = 0$$

Solution

Replacing each derivative with a polynomial term whose degree equals the original order, the auxiliary equation is

$$r^2 + 4r + 1 = 0$$

2. HOMOGENEOUS, FIRST-ORDER LINEAR DIFFERENTIAL EQUATIONS WITH CONSTANT COEFFICIENTS

A homogeneous, first-order linear differential equation with constant coefficients will have the general form of Eq. 10.1.

$$y' + ky = 0 \qquad 10.1$$

The auxiliary equation is $r + k = 0$ and it has a root of $r = -k$. Equation 10.2 is the solution.

$$y = Ae^{rx} = Ae^{-kx} \qquad 10.2$$

If the initial condition is known to be $y(0) = y_0$, the solution is

$$y = y_0 e^{-kx} \qquad 10.3$$

3. FIRST-ORDER LINEAR DIFFERENTIAL EQUATIONS

A first-order linear differential equation has the general form of Eq. 10.4. $p(x)$ and $g(x)$ can be constants or any function of x (but not of y). However, if $p(x)$ is a constant and $g(x)$ is zero, it is easier to solve the equation as shown in Sec. 10.2.

$$y' + p(x)y = g(x) \qquad 10.4$$

The *integrating factor* (which is usually a function) to this differential equation is

$$u(x) = \exp\left(\int p(x)\,dx\right) \qquad 10.5$$

The closed-form solution to Eq. 10.4 is

$$y = \frac{1}{u(x)}\left(\int u(x)g(x)\,dx + C\right) \qquad 10.6$$

For the special case where $p(x)$ and $g(x)$ are both constants, Eq. 10.4 becomes

$$y' + ay = b \qquad 10.7$$

If the initial condition is $y(0) = y_0$, then the solution to Eq. 10.7 is

$$y = \frac{b}{a}(1 - e^{-ax}) + y_0 e^{-ax} \qquad 10.8$$

Example 10.3

Find a solution to the following differential equation.

$$y' + -y = 2xe^{2x} \qquad y(0) = 1$$

Solution

This is a first-order linear equation with $p(x) = -1$ and $g(x) = 2xe^{2x}$. The integrating factor is

$$u(x) = \exp\left(\int p(x)\,dx\right) = \exp\left(\int -1\,dx\right) = e^{-x}$$

The solution is given by Eq. 10.6.

$$\begin{aligned} y &= \frac{1}{u(x)}\left(\int u(x)g(x)\,dx + C\right) \\ &= \frac{1}{e^{-x}}\left(\int e^{-x}2xe^{2x}\,dx + C\right) \\ &= e^{x}\left(2\int xe^{x}\,dx + C\right) \\ &= e^{x}(2xe^{x} - 2e^{x} + C) \\ &= e^{x}\left(2e^{x}(x-1) + C\right) \end{aligned}$$

From the initial condition,

$$\begin{aligned} y(0) &= 1 \\ e^0\big((2)(e^0)(0-1)+C\big) &= 1 \\ 1\big((2)(1)(-1)+C\big) &= 1 \end{aligned}$$

Therefore, $C = 3$. The complete solution is

$$y = e^x\big(2e^x(x-1)+3\big)$$

4. FIRST-ORDER SEPARABLE DIFFERENTIAL EQUATIONS

First-order separable differential equations can be placed in the form of Eq. 10.9. For clarity and convenience, y' is written as dy/dx.

$$m(x) + n(y)\frac{dy}{dx} = 0 \qquad \text{10.9}$$

Equation 10.9 can be placed in the form of Eq. 10.10, both sides of which are easily integrated. An initial value will establish the constant of integration.

$$m(x)\,dx = -n(y)\,dy \qquad \text{10.10}$$

5. FIRST-ORDER EXACT DIFFERENTIAL EQUATIONS

A *first-order exact differential equation* has the form

$$f_x(x,y) + f_y(x,y)y' = 0 \qquad \text{10.11}$$

$f_x(x,y)$ is the exact derivative of $f(x,y)$ with respect to x, and $f_y(x,y)$ is the exact derivative of $f(x,y)$ with respect to y. The solution is

$$f(x,y) - C = 0 \qquad \text{10.12}$$

6. HOMOGENEOUS, SECOND-ORDER LINEAR DIFFERENTIAL EQUATIONS WITH CONSTANT COEFFICIENTS

Homogeneous second-order linear differential equations with constant coefficients have the form of Eq. 10.13. They are most easily solved by finding the two roots of the auxiliary equation, Eq. 10.14.

$$y'' + k_1y' + k_2y = 0 \qquad \text{10.13}$$

$$r^2 + k_1r + k_2 = 0 \qquad \text{10.14}$$

There are three cases. If the two roots of Eq. 10.14 are real and different, the solution is

$$y = A_1e^{r_1x} + A_2e^{r_2x} \qquad \text{10.15}$$

If the two roots are real and the same, the solution is

$$y = A_1e^{rx} + A_2xe^{rx} \qquad \text{10.16}$$

$$r = \frac{-k_1}{2} \qquad \text{10.17}$$

If the two roots are imaginary, they will be of the form $(\alpha + i\omega)$ and $(\alpha - i\omega)$, and the solution is

$$y = A_1e^{\alpha x}\cos\omega x + A_2e^{\alpha x}\sin\omega x \qquad \text{10.18}$$

In all three cases, A_1 and A_2 must be found from the two initial conditions.

Example 10.4

Solve the following differential equation.

$$y'' + 6y' + 9y = 0$$
$$y(0) = 0 \qquad y'(0) = 1$$

Solution

The auxiliary equation is

$$\begin{aligned} r^2 + 6r + 9 &= 0 \\ (r+3)(r+3) &= 0 \end{aligned}$$

The roots to the auxiliary equation are $r_1 = r_2 = -3$. Therefore, the solution has the form of Eq. 10.16.

$$y = A_1e^{-3x} + A_2xe^{-3x}$$

The first initial condition is

$$\begin{aligned} y(0) &= 0 \\ A_1e^0 + A_2(0)e^0 &= 0 \\ A_1 + 0 &= 0 \\ A_1 &= 0 \end{aligned}$$

To use the second initial condition, the derivative of the equation is needed. Making use of the known fact that $A_1 = 0$,

$$y' = \frac{d}{dx}A_2xe^{-3x} = -3A_2xe^{-3x} + A_2e^{-3x}$$

Using the second initial condition,

$$\begin{aligned} y'(0) &= 1 \\ -3A_2(0)e^0 + A_2e^0 &= 1 \\ 0 + A_2 &= 1 \\ A_2 &= 1 \end{aligned}$$

The solution is

$$y = xe^{-3x}$$

7. NONHOMOGENEOUS DIFFERENTIAL EQUATIONS

A nonhomogeneous equation has the form of Eq. 10.19. $f(x)$ is known as the *forcing function.*

$$y'' + p(x)y' + q(x)y = f(x) \qquad \textit{10.19}$$

The solution to Eq. 10.19 is the sum of two equations. The *complementary solution*, y_c, solves the complementary (i.e., homogeneous) problem. The *particular solution*, y_p, is any specific solution to the nonhomogeneous Eq. 10.19 that is known or can be found. Initial values are used to evaluate any unknown coefficients in the complementary solution *after* y_c and y_p have been combined. (The particular solution will not have any unknown coefficients.)

$$y = y_c + y_p \qquad \textit{10.20}$$

Two methods are available for finding a particular solution. The *method of undetermined coefficients*, as presented here, can be used only when $p(x)$ and $q(x)$ are constant coefficients and $f(x)$ takes on one of the forms in Table 10.1.

The particular solution can be read from Table 10.1 if the forcing function is of one of the forms given. Of course, the coefficients A_i and B_i are not known—these are the *undetermined coefficients.* The exponent s is the smallest nonnegative number (and will be 0, 1, or 2), which ensures that no term in the particular solution, y_p, is also a solution to the complementary equation, y_c. s must be determined prior to proceeding with the solution procedure.

Once y_p (including s) is known, it is differentiated to obtain y_p' and y_p'', and all three functions are substituted into the original nonhomogeneous equation. The resulting equation is rearranged to match the forcing function, $f(x)$, and the unknown coefficients are determined, usually by solving simultaneous equations.

If the forcing function, $f(x)$, is more complex than the forms shown in Table 10.1, or if either $p(x)$ or $q(x)$ is a function of x, the method of *variation of parameters* should be used. This complex and time-consuming method is not covered in this book.

Table 10.1 *Particular Solutions**

form of $f(x)$	form of y_p
$P_n(x) = a_0x^n + a_1x^{n-1} + \cdots + a_n$	$x^s(A_0x^n + A_1x^{n-1} + \cdots + A_n)$
$P_n(x)e^{\alpha x}$	$x^s(A_0x^n + A_1x^{n-1} + \cdots + A_n)e^{\alpha x}$
$P_n(x)e^{\alpha x}\begin{Bmatrix}\sin\omega x\\ \cos\omega x\end{Bmatrix}$	$x^s\left(\begin{array}{l}(A_0x^n + A_1x^{n-1} + \cdots + A_n)\times(e^{\alpha x}\cos\omega x)\\ +(B_0x^n + B_1x^{n-1} + \cdots + B_n)\times(e^{\alpha x}\sin\omega x)\end{array}\right)$

*$P_n(x)$ is a polynomial of degree n.

Example 10.5

Solve the following nonhomogeneous differential equation.

$$y'' + 2y' + y = e^x\cos x$$

Solution

step 1: Find the solution to the complementary (homogeneous) differential equation.

$$y'' + 2y' + y = 0$$

Since this is a differential equation with constant coefficients, write the auxiliary equation.

$$r^2 + 2r + 1 = 0$$

The auxiliary equation factors in $(r+1)^2 = 0$ with two identical roots at $r = -1$. Therefore, the solution to the homogeneous differential equation is

$$y_c(x) = C_1e^{-x} + C_2xe^{-x}$$

step 2: Use Table 10.1 to determine the form of a particular solution. Since the forcing function has the form $P_n(x)e^{\alpha x}\cos\omega x$ with $P_n(x) = 1$ (equivalent to $n = 0$), $\alpha = 1$, and $\omega = 1$, the particular solution has the form

$$y_p(x) = x^s(Ae^x\cos x + Be^x\sin x)$$

step 3: Determine the value of s. Check to see if any of the terms in $y_p(x)$ will themselves solve the homogeneous equation. Try $Ae^x \cos x$ first.

$$\frac{d}{dx}(Ae^x \cos x) = Ae^x \cos x - Ae^x \sin x$$

$$\frac{d^2}{dx^2}(Ae^x \cos x) = -2Ae^x \sin x$$

Substitute these quantities into the homogeneous equation.

$$\begin{aligned} y'' + 2y' + y &= 0 \\ -2Ae^x \sin x + 2Ae^x \cos x & \\ -2Ae^x \sin x + Ae^x \cos x &= 0 \\ 3Ae^x \cos x - 4Ae^x \sin x &= 0 \end{aligned}$$

Disregarding the trivial (i.e., $A = 0$) solution, $Ae^x \cos x$ does not solve the homogeneous equation.

Next, try $Be^x \sin x$.

$$\frac{d}{dx}(Be^x \sin x) = Be^x \cos x + Be^x \sin x$$

$$\frac{d^2}{dx^2}(Be^x \sin x) = 2Be^x \cos x$$

Substitute these quantities into the homogeneous equation.

$$\begin{aligned} y'' + 2y' + y &= 0 \\ 2Be^x \cos x + 2Be^x \cos x & \\ +2Be^x \sin x + Be^x \sin x &= 0 \\ 3Be^x \sin x + 4Be^x \cos x &= 0 \end{aligned}$$

Disregarding the trivial ($B = 0$) case, $Be^x \sin x$ does not solve the homogeneous equation.

Since none of the terms in $y_p(x)$ solve the homogeneous equation, $s = 0$, and a particular solution has the form

$$y_p(x) = Ae^x \cos x + Be^x \sin x$$

step 4: Use the method of unknown coefficients to determine A and B in the particular solution. Drawing on the previous steps, substitute the quantities derived from the particular solution into the nonhomogeneous equation.

$$\begin{aligned} y'' + 2y' + y &= e^x \cos x \\ -2Ae^x \sin x + 2Be^x \cos x & \\ +2Ae^x \cos x - 2Ae^x \sin x & \\ +2Be^x \cos x + 2Be^x \sin x & \\ +Ae^x \cos x + Be^x \sin x &= e^x \cos x \end{aligned}$$

Combining terms,

$$\begin{aligned} (-4A + 3B)e^x \sin x + (3A + 4B)e^x \cos x & \\ = e^x \cos x & \end{aligned}$$

Equating the coefficients of like terms on either side of the equal sign results in the following simultaneous equations.

$$-4A + 3B = 0$$

$$3A + 4B = 1$$

The solution to these equations is

$$A = \frac{3}{25}$$

$$B = \frac{4}{25}$$

A particular solution is

$$y_p(x) = \tfrac{3}{25}e^x \cos x + \tfrac{4}{25}e^x \sin x$$

step 5: Write the general solution.

$$\begin{aligned} y(x) &= y_c(x) + y_p(x) \\ &= C_1 e^{-x} + C_2 x e^{-x} + \tfrac{3}{25}e^x \cos x \\ &\quad + \tfrac{4}{25}e^x \sin x \end{aligned}$$

The values of C_1 and C_2 would be determined at this time if initial conditions were known.

8. NAMED DIFFERENTIAL EQUATIONS

Some differential equations with specific forms are named after the individuals who developed solution techniques for them.

- *Bessel equation of order ν*

$$x^2 y'' + xy' + (x^2 - \nu^2)y = 0 \qquad 10.21$$

- *Cauchy equation*

$$\begin{aligned} a_0 x^n \frac{d^n y}{dx^n} + a_1 x^{n-1} \frac{d^{n-1} y}{dx^{n-1}} + \cdots & \\ + a_{n-1} x \frac{dy}{dx} + a_n y = f(x) & \end{aligned} \qquad 10.22$$

- *Euler's equation*

$$x^2 y'' + \alpha x y' + \beta y = 0 \qquad 10.23$$

- *Gauss' hypergeometric equation*

$$x(1-x)y'' + \big(c - (a+b+1)x\big)y' - aby = 0 \qquad 10.24$$

- *Legendre equation of order* λ

$$(1-x^2)y'' - 2xy' + \lambda(\lambda+1)y = 0 \quad [-1 < x < 1] \qquad 10.25$$

9. LAPLACE TRANSFORMS

Traditional methods of solving nonhomogeneous differential equations by hand are usually difficult and/or time consuming. *Laplace transforms* can be used to reduce many solution procedures to simple algebra.

Every mathematical function, $f(t)$, for which Eq. 10.26 exists has a Laplace transform, written as $\mathcal{L}(f)$ or $F(s)$. The transform is written in the s-domain, regardless of the independent variable in the original function.[2] (The variable s is equivalent to a derivative operator, although it may be handled in the equations as a simple variable.) Equation 10.26 converts a function into a Laplace transform.

$$\mathcal{L}\big(f(t)\big) = F(s) = \int_0^\infty e^{-st} f(t)\,dt \qquad 10.26$$

Equation 10.26 is not often needed because tables of transforms are readily available. (Appendix 10.A contains some of the most common transforms.)

Extracting a function from its transform is the *inverse Laplace transform* operation. Although other methods exist, this operation is almost always done by finding the transform in a set of tables.[3]

$$f(t) = \mathcal{L}^{-1}\big(F(s)\big) \qquad 10.27$$

Example 10.6

Find the Laplace transform of the following function.

$$f(t) = e^{at} \qquad [s > a]$$

Solution

Applying Eq. 10.26,

$$\mathcal{L}(e^{at}) = \int_0^\infty e^{-st} e^{at}\,dt = \int_0^\infty e^{-(s-a)t}\,dt$$

$$= -\frac{e^{-(s-a)t}}{s-a}\Bigg|_0^\infty$$

$$= \frac{1}{s-a} \qquad [s > a]$$

10. STEP AND IMPULSE FUNCTIONS

Many forcing functions are sinusoidal or exponential in nature; others, however, can only be represented by a step or impulse function. A *unit step function*, u_t, is a function describing the disturbance of magnitude 1 that is not present before time t but is suddenly there after time t. A step of magnitude 5 at time $t = 3$ would be represented as $5u_3$. (The notation $5u(t-3)$ is used in some books.)

The *unit impulse function*, δ_t, is a function describing a disturbance of magnitude 1 that is applied and removed so quickly as to be instantaneous. An impulse of magnitude 5 at time 3 would be represented by $5\delta_3$. (The notation $5\delta(t-3)$ is used in some books.)

Example 10.7

What is the notation for a forcing function of magnitude 6 that is applied at $t = 2$ and that is completely removed at $t = 7$?

Solution

The notation is $f(t) = 6(u_2 - u_7)$.

Example 10.8

Find the Laplace transform of u_0, a unit step at $t = 0$.

$$f(t) = 0 \text{ for } t < 0$$

$$f(t) = 1 \text{ for } t \geq 0$$

Solution

Since the Laplace transform is an integral that starts at $t = 0$, the value of $f(t)$ prior to $t = 0$ is irrelevant.

$$\mathcal{L}(u_0) = \int_0^\infty e^{-st}(1)\,dt = -\frac{e^{-st}}{s}\Bigg|_0^\infty$$

$$= 0 - \frac{-1}{s}$$

$$= \frac{1}{s}$$

11. ALGEBRA OF LAPLACE TRANSFORMS

Equations containing Laplace transforms can be simplified by applying the following principles.

- *linearity theorem* (c is a constant.)

$$\mathcal{L}\big(cf(t)\big) = c\mathcal{L}\big(f(t)\big) = cF(s) \qquad 10.28$$

[2] It is traditional to write the original function as a function of the independent variable t rather than x. However, Laplace transforms are not limited to functions of time.

[3] Other methods include integration in the complex plane, convolution, and simplification by partial fractions.

- *superposition theorem* ($f(t)$ and $g(t)$ are different functions.)

$$\mathcal{L}\big(f(t) \pm g(t)\big) = \mathcal{L}\big(f(t)\big) \pm \mathcal{L}\big(g(t)\big) = F(s) \pm G(s) \quad \text{10.29}$$

- *time-shifting theorem (delay theorem)*

$$\mathcal{L}\big(f(t-b)u_b\big) = e^{-bs}F(s) \quad \text{10.30}$$

- *Laplace transform of a derivative*

$$\mathcal{L}\big(f^n(t)\big) = -f^{n-1}(0) - sf^{n-2}(0) - \cdots - s^{n-1}f(0) + s^nF(s) \quad \text{10.31}$$

- *other properties*

$$\mathcal{L}\left(\int_0^t f(u)\,du\right) = \frac{1}{s}F(s) \quad \text{10.32}$$

$$\mathcal{L}\big(tf(t)\big) = -\frac{dF}{ds} \quad \text{10.33}$$

$$\mathcal{L}\left(\frac{1}{t}f(t)\right) = \int_0^\infty F(u)\,du \quad \text{10.34}$$

12. CONVOLUTION INTEGRAL

A complex Laplace transform, $F(s)$, will often be recognized as the product of two other transforms, $F_1(s)$ and $F_2(s)$, whose corresponding functions $f_1(t)$ and $f_2(t)$ are known. Unfortunately, Laplace transforms cannot be computed with ordinary multiplication. That is, $f(t) \neq f_1(t)f_2(t)$ even though $F(s) = F_1(s)F_2(s)$.

However, it is possible to extract $f(t)$ from its *convolution*, $h(t)$, as calculated from either of the *convolution integrals* in Eq. 10.35. This process is demonstrated in Ex. 10.9. χ is a dummy variable.

$$\begin{aligned} f(t) &= \mathcal{L}^{-1}\big(F_1(s)F_2(s)\big) \\ &= \int_0^t f_1(t-\chi)f_2(\chi)\,d\chi \\ &= \int_0^t f_1(\chi)f_2(t-\chi)\,d\chi \end{aligned} \quad \text{10.35}$$

Example 10.9

Use the convolution integral to find the inverse transform of

$$F(s) = \frac{3}{s^2(s^2+9)}$$

Solution

$F(s)$ can be factored as

$$F_1(s)F_2(s) = \left(\frac{1}{s^2}\right)\left(\frac{3}{s^2+9}\right)$$

As the inverse transforms of $F_1(s)$ and $F_2(s)$ are $f_1(t) = t$ and $f_2(t) = \sin 3t$, respectively, the convolution integral from Eq. 10.35 is

$$\begin{aligned} f(t) &= \int_0^t (t-\chi)\sin 3\chi\,d\chi \\ &= \int_0^t (t\sin 3\chi - \chi\sin 3\chi)\,d\chi \\ &= t\int_0^t \sin 3\chi\,d\chi - \int_0^t \chi\sin 3\chi\,d\chi \end{aligned}$$

Expand using integration by parts.

$$\begin{aligned} f(t) &= -\tfrac{1}{3}t\cos 3\chi + \tfrac{1}{3}\chi\cos 3\chi - \tfrac{1}{9}\sin 3\chi\Big|_0^t \\ &= \frac{3t - \sin 3t}{9} \end{aligned}$$

13. USING LAPLACE TRANSFORMS

Any nonhomogeneous linear differential equation with constant coefficients can be solved with the following procedure, which reduces the solution to simple algebra. A complete table of transforms simplifies or eliminates step 5.

step 1: Put the differential equation in standard form (i.e., isolate the y'' term).

$$y'' + k_1y' + k_2y = f(t) \quad \text{10.36}$$

step 2: Take the Laplace transform of both sides. Use the linearity and superposition theorems. (See Eq. 10.28 and Eq. 10.29.)

$$\mathcal{L}(y'') + k_1\mathcal{L}(y') + k_2\mathcal{L}(y) = \mathcal{L}\big(f(t)\big) \quad \text{10.37}$$

step 3: Use Eq. 10.38 and Eq. 10.39 to expand the equation. (These are specific forms of Eq. 10.31.) Use a table to evaluate the transform of the forcing function.

$$\mathcal{L}(y'') = s^2\mathcal{L}(y) - sy(0) - y'(0) \quad \text{10.38}$$

$$\mathcal{L}(y') = s\mathcal{L}(y) - y(0) \quad \text{10.39}$$

step 4: Use algebra to solve for $\mathcal{L}(y)$.

step 5: If needed, use partial fractions to simplify the expression for $\mathcal{L}(y)$.

step 6: Take the inverse transform to find $y(t)$.

$$y(t) = \mathcal{L}^{-1}\big(\mathcal{L}(y)\big) \qquad 10.40$$

Example 10.10

Find $y(t)$ for the following differential equation.

$$y'' + 2y' + 2y = \cos t$$
$$y(0) = 1 \qquad y'(0) = 0$$

Solution

step 1: The equation is already in standard form.

step 2: $\mathcal{L}(y'') + 2\mathcal{L}(y') + 2\mathcal{L}(y) = \mathcal{L}(\cos t)$

step 3: Use Eq. 10.38 and Eq. 10.39. Use App. 10.A to find the transform of cos t.

$$s^2\mathcal{L}(y) - sy(0) - y'(0) + 2s\mathcal{L}(y) - 2y(0) + 2\mathcal{L}(y) = \frac{s}{s^2+1}$$

But, $y(0) = 1$ and $y'(0) = 0$.

$$s^2\mathcal{L}(y) - s + 2s\mathcal{L}(y) - 2 + 2\mathcal{L}(y) = \frac{s}{s^2+1}$$

step 4: Combine terms and solve for $\mathcal{L}(y)$.

$$\mathcal{L}(y)(s^2 + 2s + 2) - s - 2 = \frac{s}{s^2+1}$$

$$\mathcal{L}(y) = \frac{\frac{s}{s^2+1} + s + 2}{s^2 + 2s + 2} = \frac{s^3 + 2s^2 + 2s + 2}{(s^2+1)(s^2+2s+2)}$$

step 5: Expand the expression for $\mathcal{L}(y)$ by partial fractions.

$$\mathcal{L}(y) = \frac{s^3 + 2s^2 + 2s + 2}{(s^2+1)(s^2+2s+2)} = \frac{A_1 s + B_1}{s^2+1} + \frac{A_2 s + B_2}{s^2+2s+2}$$
$$= \frac{s^3(A_1 + A_2) + s^2(2A_1 + B_1 + B_2) + s(2A_1 + 2B_1 + A_2) + (2B_1 + B_2)}{(s^2+1)(s^2+2s+2)}$$

The following simultaneous equations result.

$$\begin{aligned} A_1 + A_2 &= 1 \\ 2A_1 + B_1 + B_2 &= 2 \\ 2A_1 + A_2 + 2B_1 &= 2 \\ 2B_1 + B_2 &= 2 \end{aligned}$$

These equations have the solutions $A_1 = 1/5$, $A_2 = 4/5$, $B_1 = 2/5$, and $B_2 = 6/5$.

step 6: Refer to App. 10.A and take the inverse transforms. The numerator of the second term is rewritten from $(4s + 6)$ to $((4s + 4) + 2)$.

$$y = \mathcal{L}^{-1}\big(\mathcal{L}(y)\big)$$
$$= \mathcal{L}^{-1}\left(\frac{\left(\frac{1}{5}\right)(s+2)}{s^2+1} + \frac{\left(\frac{1}{5}\right)(4s+6)}{s^2+2s+2}\right)$$
$$= \left(\tfrac{1}{5}\right)\left(\begin{aligned} &\mathcal{L}^{-1}\left(\frac{s}{s^2+1}\right) + 2\mathcal{L}^{-1}\left(\frac{1}{s^2+1}\right) \\ &+ 4\mathcal{L}^{-1}\left(\frac{s-(-1)}{\big(s-(-1)\big)^2+1}\right) \\ &+ 6\mathcal{L}^{-1}\left(\frac{1}{\big(s-(-1)\big)^2+1}\right) \end{aligned}\right)$$
$$= \left(\tfrac{1}{5}\right)(\cos t + 2\sin t + 4e^{-t}\cos t + 6e^{-t}\sin t)$$

14. THIRD- AND HIGHER-ORDER LINEAR DIFFERENTIAL EQUATIONS WITH CONSTANT COEFFICIENTS

The solutions of third- and higher-order linear differential equations with constant coefficients are extensions of the solutions for second-order equations of this type. Specifically, if an equation is homogeneous, the auxiliary equation is written and its roots are found. If the equation is nonhomogeneous, Laplace transforms can be used to simplify the solution.

Consider the following homogeneous differential equation with constant coefficients.

$$y^n + k_1 y^{n-1} + \cdots + k_{n-1}y' + k_n y = 0 \qquad 10.41$$

The auxiliary equation to Eq. 10.41 is

$$r^n + k_1 r^{n-1} + \cdots + k_{n-1}r + k_n = 0 \qquad 10.42$$

For each real and distinct root r, the solution contains the term

$$y = Ae^{rx} \qquad 10.43$$

For each real root r that repeats m times, the solution contains the term

$$y = (A_1 + A_2x + A_3x^2 + \cdots + A_mx^{m-1})e^{rx} \qquad 10.44$$

For each pair of complex roots of the form $r = \alpha \pm i\omega$ the solution contains the terms

$$y = e^{\alpha x}(A_1 \sin \omega x + A_2 \cos \omega x) \qquad 10.45$$

15. APPLICATION: ENGINEERING SYSTEMS

There is a wide variety of engineering systems (mechanical, electrical, fluid flow, heat transfer, and so on) whose behavior is described by linear differential equations with constant coefficients.

16. APPLICATION: MIXING

A typical mixing problem involves a liquid-filled tank. The liquid may initially be pure or contain some solute. Liquid (either pure or as a solution) enters the tank at a known rate. A drain may be present to remove thoroughly mixed liquid. The concentration of the solution (or, equivalently, the amount of solute in the tank) at some given time is generally unknown. (See Fig. 10.1.)

Figure 10.1 *Fluid Mixture Problem*

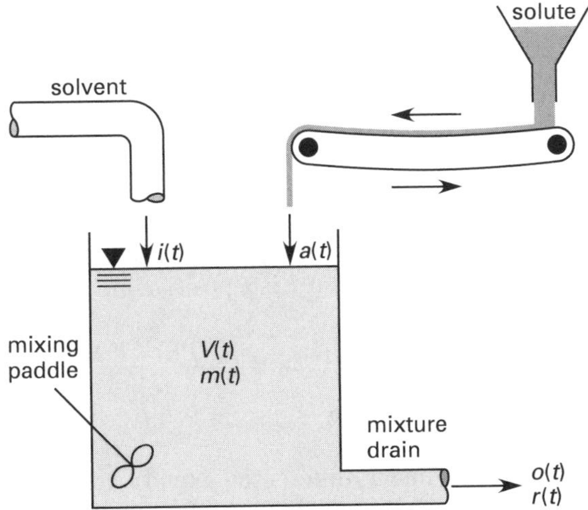

If $m(t)$ is the mass of solute in the tank at time t, the rate of solute change will be $m'(t)$. If the solute is being added at the rate of $a(t)$ and being removed at the rate of $r(t)$, the rate of change is

$$\begin{aligned} m'(t) &= \text{rate of addition} - \text{rate of removal} \\ &= a(t) - r(t) \end{aligned} \qquad 10.46$$

The rate of solute addition $a(t)$ must be known and, in fact, may be constant or zero. However, $r(t)$ depends on the concentration, $c(t)$, of the mixture and volumetric flow rates at time t. If $o(t)$ is the volumetric flow rate out of the tank, then

$$r(t) = c(t)o(t) \qquad 10.47$$

However, the concentration depends on the mass of solute in the tank at time t. Recognizing that the volume, $V(t)$, of the liquid in the tank may be changing with time,

$$c(t) = \frac{m(t)}{V(t)} \qquad 10.48$$

The differential equation describing this problem is

$$m'(t) = a(t) - \frac{m(t)o(t)}{V(t)} \qquad 10.49$$

Example 10.11

A tank contains 100 gal of pure water at the beginning of an experiment. Pure water flows into the tank at a rate of 1 gal/min. Brine containing ¼ lbm of salt per gallon enters the tank from a second source at a rate of 1 gal/min. A perfectly mixed solution drains from the tank at a rate of 2 gal/min. How much salt is in the tank 8 min after the experiment begins?

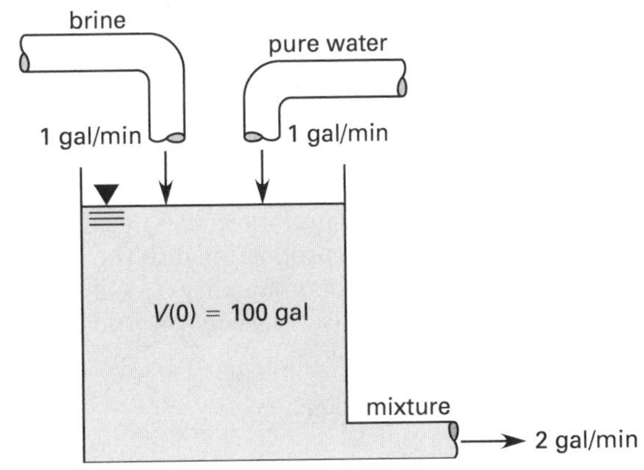

Solution

Let $m(t)$ represent the mass of salt in the tank at time t. 0.25 lbm of salt enters the tank per minute (that is, $a(t) = 0.25$ lbm/min). The salt removal rate depends on the concentration in the tank. That is,

$$\begin{aligned} r(t) = o(t)c(t) &= \left(2\ \frac{\text{gal}}{\text{min}}\right)\left(\frac{m(t)}{100\ \text{gal}}\right) \\ &= \left(0.02\ \frac{1}{\text{min}}\right)m(t) \end{aligned}$$

From Eq. 10.46, the rate of change of salt in the tank is

$$m'(t) = a(t) - r(t)$$
$$= 0.25\ \frac{\text{lbm}}{\text{min}} - \left(0.02\ \frac{1}{\text{min}}\right)m(t)$$

$$m'(t) + \left(0.02\ \frac{1}{\text{min}}\right)m(t) = 0.25\ \text{lbm/min}$$

This is a first-order linear differential equation of the form of Eq. 10.7. Since the initial condition is $m(0) = 0$, the solution is

$$m(t) = \left(\frac{0.25\ \frac{\text{lbm}}{\text{min}}}{0.02\ \frac{1}{\text{min}}}\right)\left(1 - e^{-\left(0.02\ \frac{1}{\text{min}}\right)t}\right)$$
$$= (12.5\ \text{lbm})\left(1 - e^{-\left(0.02\ \frac{1}{\text{min}}\right)t}\right)$$

At $t = 8$,

$$m(t) = (12.5\ \text{lbm})\left(1 - e^{-\left(0.02\ \frac{1}{\text{min}}\right)(8\ \text{min})}\right)$$
$$= (12.5\ \text{lbm})(1 - 0.852)$$
$$= 1.85\ \text{lbm}$$

17. APPLICATION: EXPONENTIAL GROWTH AND DECAY

Equation 10.50 describes the behavior of a substance (e.g., radioactive and irradiated molecules) whose quantity, $m(t)$, changes at a rate proportional to the quantity present. The constant of proportionality, k, will be negative for decay (e.g., *radioactive decay*) and positive for growth (e.g., compound interest).

$$m'(t) = km(t) \qquad 10.50$$

$$m'(t) - km(t) = 0 \qquad 10.51$$

If the initial quantity of substance is $m(0) = m_0$, then Eq. 10.51 has the solution

$$m(t) = m_0 e^{kt} \qquad 10.52$$

If $m(t)$ is known for some time t, the constant of proportionality is

$$k = \frac{1}{t}\ln\left(\frac{m(t)}{m_0}\right) \qquad 10.53$$

For the case of a decay, the *half-life*, $t_{1/2}$, is the time at which only half of the substance remains. The relationship between k and $t_{1/2}$ is

$$kt_{1/2} = \ln\frac{1}{2} = -0.693 \qquad 10.54$$

18. APPLICATION: EPIDEMICS

During an epidemic in a population of n people, the density of sick (contaminated, contagious, affected, etc.) individuals is $\rho_s(t) = s(t)/n$, where $s(t)$ is the number of sick individuals at a given time, t. Similarly, the density of well (uncontaminated, unaffected, susceptible, etc.) individuals is $\rho_w(t) = w(t)/n$, where $w(t)$ is the number of well individuals. Assuming there is no quarantine, the population size is constant, individuals move about freely, and sickness does not limit the activities of individuals, the rate of contagion, $\rho_s'(t)$, will be $k\rho_s(t)\rho_w(t)$, where k is a proportionality constant.

$$\rho_s'(t) = k\rho_s(t)\rho_w(t) = k\rho_s(t)\big(1 - \rho_s(t)\big) \qquad 10.55$$

This is a separable differential equation that has the solution

$$\rho_s(t) = \frac{\rho_s(0)}{\rho_s(0) + \big(1 - \rho_s(0)\big)e^{-kt}} \qquad 10.56$$

19. APPLICATION: SURFACE TEMPERATURE

Newton's law of cooling states that the surface temperature, T, of a cooling object changes at a rate proportional to the difference between the surface and ambient temperatures. The constant k is a positive number.

$$T'(t) = -k\big(T(t) - T_{\text{ambient}}\big) \quad [k > 0] \qquad 10.57$$

$$T'(t) + kT(t) - kT_{\text{ambient}} = 0 \quad [k > 0] \qquad 10.58$$

This first-order linear differential equation with constant coefficients has the following solution (from Eq. 10.8).

$$T(t) = T_{\text{ambient}} + \big(T(0) - T_{\text{ambient}}\big)e^{-kt} \qquad 10.59$$

If the temperature is known at some time t, the constant k can be found from Eq. 10.60.

$$k = \frac{-1}{t}\ln\left(\frac{T(t) - T_{\text{ambient}}}{T(0) - T_{\text{ambient}}}\right) \qquad 10.60$$

20. APPLICATION: EVAPORATION

The mass of liquid evaporated from a liquid surface is proportional to the exposed surface area. Since quantity, mass, and remaining volume are all proportional, the differential equation is

$$\frac{dV}{dt} = -kA \qquad 10.61$$

For a spherical drop of radius r, Eq. 10.61 reduces to

$$\frac{dr}{dt} = -k \qquad 10.62$$

$$r(t) = r(0) - kt \qquad 10.63$$

For a cube with sides of length s, Eq. 10.61 reduces to

$$\frac{ds}{dt} = -2k \qquad 10.64$$

$$s(t) = s(0) - 2kt \qquad 10.65$$

11 Probability and Statistical Analysis of Data

Content in blue refers to the *NCEES Handbook*.

NCEES EXAM SPECIFICATIONS AND RELATED CONTENT

MACHINE DESIGN AND MATERIALS EXAM

I.A.3. Basic Engineering Practice: Quality assurance/quality control
- 23. Measures of Central Tendency
- 24. Measures of Dispersion
- 32. Application: Statistical Process Control

1. SET THEORY

A *set* (usually designated by a capital letter) is a population or collection of individual items known as *elements* or *members*. The *null set*, $\varnothing$, is empty (i.e., contains no members). If A and B are two sets, A is a *subset* of B if every member in A is also in B. A is a *proper subset* of B if B consists of more than the elements in A. These relationships are denoted as follows.

$$A \subseteq B \quad \text{[subset]}$$

$$A \subset B \quad \text{[proper subset]}$$

The *universal set*, U, is one from which other sets draw their members. If A is a subset of U, then A' (also designated as A^{-1}, $\tilde{A}$, $-A$, and $\bar{A}$) is the *complement* of A and consists of all elements in U that are not in A. This is illustrated by the *Venn diagram* in Fig. 11.1(a).

The *union of two sets*, denoted by $A \cup B$ and shown in Fig. 11.1(b), is the set of all elements that are in either A or B or both. The *intersection of two sets*, denoted by $A \cap B$ and shown in Fig. 11.1(c), is the set of all elements that belong to both A and B. If $A \cap B = \varnothing$, A and B are said to be *disjoint sets*.

Figure 11.1 *Venn Diagrams*

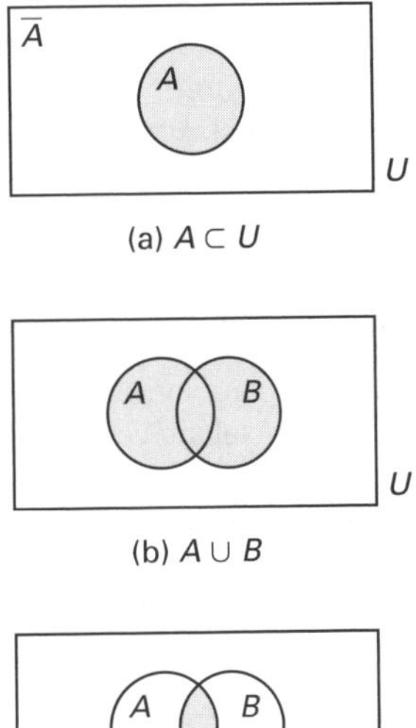

(a) $A \subset U$

(b) $A \cup B$

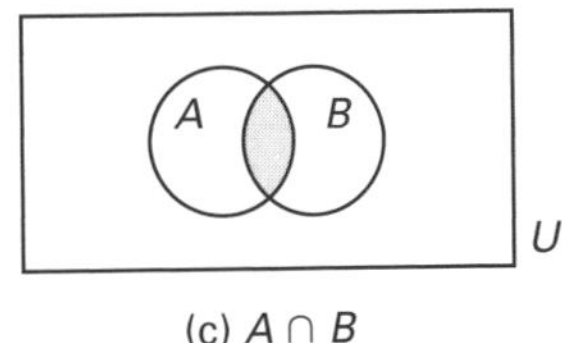

(c) $A \cap B$

(d) Graph the cumulative frequency distribution.

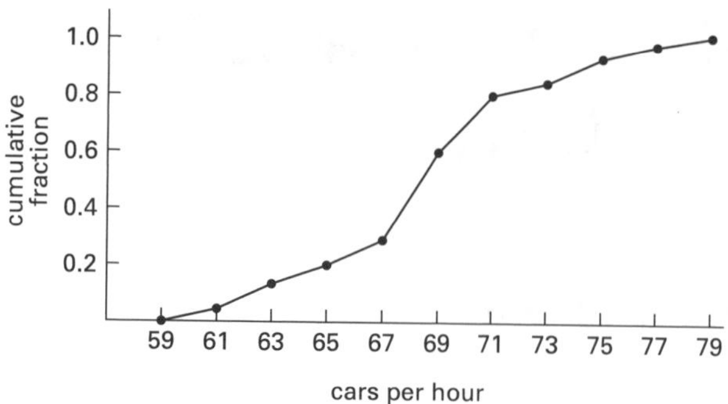

22. MEASURES OF EXPERIMENTAL ADEQUACY

An experiment is said to be *accurate* if it is unaffected by experimental error. In this case, *error* is not synonymous with *mistake*, but rather includes all variations not within the experimenter's control.

For example, suppose a gun is aimed at a point on a target and five shots are fired. The mean distance from the point of impact to the sight in point is a measure of the alignment accuracy between the barrel and sights. The difference between the actual value and the experimental value is known as *bias*.

Precision is not synonymous with accuracy. Precision is concerned with the repeatability of the experimental results. If an experiment is repeated with identical results, the experiment is said to be precise.

The average distance of each impact from the centroid of the impact group is a measure of the precision of the experiment. It is possible to have a highly precise experiment with a large bias.

Most of the techniques applied to experiments in order to improve the accuracy (i.e., reduce bias) of the experimental results (e.g., repeating the experiment, refining the experimental methods, or reducing variability) actually increase the precision.

Sometimes the word *reliability* is used with regard to the precision of an experiment. In this case, a "reliable estimate" is used in the same sense as a "precise estimate."

Stability and *insensitivity* are synonymous terms. A stable experiment will be insensitive to minor changes in the experimental parameters. For example, suppose the centroid of a bullet group is 2.1 in from the target point at 65°F and 2.3 in away at 80°F. The sensitivity of the experiment to temperature change would be

$$\text{sensitivity} = \frac{\Delta x}{\Delta T} = \frac{2.3\text{ in} - 2.1\text{ in}}{80°\text{F} - 65°\text{F}} = 0.0133\text{ in/°F}$$

23. MEASURES OF CENTRAL TENDENCY

It is often unnecessary to present the experimental data in their entirety, either in tabular or graphical form. In such cases, the data and distribution can be represented by various parameters. One type of parameter is a measure of *central tendency*. Mode, median, and mean are measures of central tendency.

The *mode* is the observed value that occurs most frequently. The mode may vary greatly between series of observations. Therefore, its main use is as a quick measure of the central value since little or no computation is required to find it. Beyond this, the usefulness of the mode is limited.

The *median* is the point in the distribution that partitions the total set of observations into two parts containing equal numbers of observations. It is not influenced by the extremity of scores on either side of the distribution. The median is found by counting up (from either end of the frequency distribution) until half of the observations have been accounted for.

When n is odd, the median is the $\left(\frac{n+1}{2}\right)^{\text{th}}$ item value.

For even numbers of observations, the median is estimated as some value (i.e., the average) between the two center observations—$\left(n/2\right)^{\text{th}}$ and $\left(n/2+1\right)^{\text{th}}$. **[Dispersion, Mean, Median, and Mode Values]**

Similar in concept to the median are *percentiles* (*percentile ranks*), *quartiles*, and *deciles*. The median could also have been called the *50th percentile* observation. Similarly, the 80th percentile would be the observed value (e.g., the number of cars per hour) for which the cumulative frequency was 80%. The quartile and decile points on the distribution divide the observations or distribution into segments of 25% and 10%, respectively.

The *arithmetic mean* is the arithmetic average of the observations. The sample mean, $\bar{x}$, can be used as an *unbiased estimator* of the population mean, μ. The *mean* may be found without ordering the data (as was necessary to find the mode and median). The mean can be found from Eq. 11.68. There are additional mean formulas for varying applications. The weighted arithmetic mean is used when certain data points are considered more influential to the overall dataset and can be determined through Eq. 11.69.

Dispersion, Mean, Median, and Mode Values

$$\bar{x} = \left(\frac{1}{n}\right)(x_1 + x_2 + \cdots + x_n) = \frac{\sum x_i}{n} \qquad 11.68$$

Dispersion, Mean, Median, and Mode Values

$$\bar{x}_W = \frac{\sum w_i x_i}{\sum w_i} \qquad 11.69$$

Solution

The groups are order conscious. From Eq. 11.20,

$$P(4,3) = \frac{n!}{(n-r)!} = \frac{4!}{(4-3)!} = \frac{4\cdot 3\cdot 2\cdot 1}{1} = 24$$

Example 11.3

Seven diplomats from different countries enter a circular room. The only furnishings are seven chairs arranged around a circular table. How many ways are there of arranging the diplomats?

Solution

All seven diplomats must be seated, so the groups are permutations of seven objects taken seven at a time. Since there is no head chair, the groups are ring permutations. From Eq. 11.22,

$$\begin{aligned} P_{\text{ring}}(7,7) &= (7-1)! = 6\cdot 5\cdot 4\cdot 3\cdot 2\cdot 1 \\ &= 720 \end{aligned}$$

4. PROBABILITY THEORY

The act of conducting an experiment (trial) or taking a measurement is known as *sampling*. *Probability theory* determines the relative likelihood that a particular event will occur. An *event*, e, is one of the possible outcomes of the *trial*. Taken together, all of the possible events constitute a finite *sample space*, $E = [e_1, e_2, \ldots, e_n]$. The trial is drawn from the *population* or *universe*. Populations can be finite or infinite in size.

Events can be numerical or nonnumerical, discrete or continuous, and dependent or independent. An example of a nonnumerical event is getting tails on a coin toss. The number from a roll of a die is a discrete numerical event. The measured diameter of a bolt produced from an automatic screw machine is a numerical event. Since the diameter can (within reasonable limits) take on any value, its measured value is a continuous numerical event.

An event is *independent* if its outcome is unaffected by previous outcomes (i.e., previous runs of the experiment) and *dependent* otherwise. Whether or not an event is independent depends on the population size and how the sampling is conducted. Sampling (a trial) from an infinite population is implicitly independent. When the population is finite, *sampling with replacement* produces independent events, while *sampling without replacement* changes the population and produces dependent events.

The terms *success* and *failure* are loosely used in probability theory to designate obtaining and not obtaining, respectively, the tested-for condition. "Failure" is not the same as a *null event* (i.e., one that has a zero probability of occurrence).

The *probability* of event e_1 occurring is designated as $p\{e_1\}$ and is calculated as the ratio of the total number of ways the event can occur to the total number of outcomes in the sample space.

Example 11.4

There are 380 students in a rural school—200 girls and 180 boys. One student is chosen at random and is checked for gender and height. (a) Define and categorize the population. (b) Define and categorize the sample space. (c) Define the trials. (d) Define and categorize the events. (e) In determining the probability that the student chosen is a boy, define success and failure. (f) What is the probability that the student is a boy?

Solution

(a) The population consists of 380 students and is finite.

(b) In determining the gender of the student, the sample space consists of the two outcomes $E = [\text{girl, boy}]$. This sample space is nonnumerical and discrete. In determining the height, the sample space consists of a range of values and is numerical and continuous.

(c) The trial is the actual sampling (i.e., the determination of gender and height).

(d) The events are the outcomes of the trials (i.e., the gender and height of the student). These events are independent if each student returns to the population prior to the random selection of the next student; otherwise, the events are dependent.

(e) The event is a success if the student is a boy and is a failure otherwise.

(f) From the definition of probability,

$$p\{\text{boy}\} = \frac{\text{no. of boys}}{\text{no. of students}} = \frac{180}{380} = 0.47$$

5. JOINT PROBABILITY

Joint probability rules specify the probability of a combination of events. If n mutually exclusive events from the set E have probabilities $p\{e_i\}$, the probability of any one of these events occurring in a given trial is the sum of the individual probabilities. The events in Eq. 11.23 come from a single sample space and are linked by the word *or*.

$$\begin{aligned} p\{e_1 \text{ or } e_2 \text{ or } \cdots \text{ or } e_k\} &= p\{e_1\} + p\{e_2\} \\ &\quad + \cdots + p\{e_k\} \end{aligned} \qquad 11.23$$

When given two independent sets of events, E and G, Eq. 11.24 will give the probability that events e_i and g_i will both occur. The events in Eq. 11.24 are independent and are linked by the word *and.*

$$p\{e_i \text{ and } g_i\} = p\{e_i\}p\{g_i\} \qquad 11.24$$

When given two independent sets of events, E and G, Eq. 11.25 will give the probability that either event e_i or g_i will occur. The events in Eq. 11.25 are mutually exclusive and are linked by the word *or.*

$$p\{e_i \text{ or } g_i\} = p\{e_i\} + p\{g_i\} - p\{e_i\}p\{g_i\} \qquad 11.25$$

Example 11.5

A bowl contains five white balls, two red balls, and three green balls. What is the probability of getting either a white ball or a red ball in one draw from the bowl?

Solution

Since the two possible events are mutually exclusive and come from the same sample space, Eq. 11.23 can be used.

$$\begin{aligned} p\{\text{white or red}\} &= p\{\text{white}\} + p\{\text{red}\} = \frac{5}{10} + \frac{2}{10} \\ &= 7/10 \end{aligned}$$

Example 11.6

One bowl contains five white balls, two red balls, and three green balls. Another bowl contains three yellow balls and seven black balls. What is the probability of getting a red ball from the first bowl and a yellow ball from the second bowl in one draw from each bowl?

Solution

Equation 11.24 can be used because the two events are independent.

$$\begin{aligned} p\{\text{red and yellow}\} &= p\{\text{red}\}p\{\text{yellow}\} \\ &= \left(\frac{2}{10}\right)\left(\frac{3}{10}\right) \\ &= 6/100 \end{aligned}$$

6. COMPLEMENTARY PROBABILITIES

The probability of an event occurring is equal to one minus the probability of the event not occurring. This is known as *complementary probability.*

$$p\{e_i\} = 1 - p\{\text{not } e_i\} \qquad 11.26$$

Equation 11.26 can be used to simplify some probability calculations. Specifically, calculation of the probability of numerical events being "greater than" or "less than" or quantities being "at least" a certain number can often be simplified by calculating the probability of the complementary event.

Example 11.7

A fair coin is tossed five times.[1] What is the probability of getting at least one tail?

Solution

The probability of getting at least one tail in five tosses could be calculated as

$$\begin{aligned} p\{\text{at least 1 tail}\} = p\{\text{1 tail}\} &+ p\{\text{2 tails}\} \\ &+ p\{\text{3 tails}\} + p\{\text{4 tails}\} \\ &+ p\{\text{5 tails}\} \end{aligned}$$

However, it is easier to calculate the complementary probability of getting no tails (i.e., getting all heads).

From Eq. 11.24 and Eq. 11.26 (for calculating the probability of getting no tails in five successive tosses),

$$\begin{aligned} p\{\text{at least 1 tail}\} &= 1 - p\{\text{0 tails}\} \\ &= 1 - (0.5)^5 \\ &= 0.96875 \end{aligned}$$

7. CONDITIONAL PROBABILITY

Given two dependent sets of events, E and G, the probability that event e_k will occur given the fact that the dependent event g has already occurred is written as $p\{e_k|\,g\}$ and given by *Bayes' theorem*, Eq. 11.27.

$$p\{e_k|g\} = \frac{p\{e_k \text{ and } g\}}{p\{g\}} = \frac{p\{g|e_k\}p\{e_k\}}{\sum_{i=1}^{n} p\{g|e_i\}p\{e_i\}} \qquad 11.27$$

8. PROBABILITY DENSITY FUNCTIONS

A *density function* is a nonnegative function whose integral taken over the entire range of the independent variable is unity. A *probability density function* is a mathematical formula that gives the probability of a discrete numerical event occurring. A *discrete numerical event* is an occurrence that can be described (usually) by an integer. For example, 27 cars passing through a bridge toll booth in an hour is a discrete numerical event. Fig. 11.2 shows a graph of a typical probability density function.

[1]It makes no difference whether one coin is tossed five times or five coins are each tossed once.

A probability density function, $f(x)$, gives the probability that discrete event x will occur. That is, $p\{x\} = f(x)$. Important discrete probability density functions are the binomial, hypergeometric, and Poisson distributions.

Figure 11.2 *Probability Density Function*

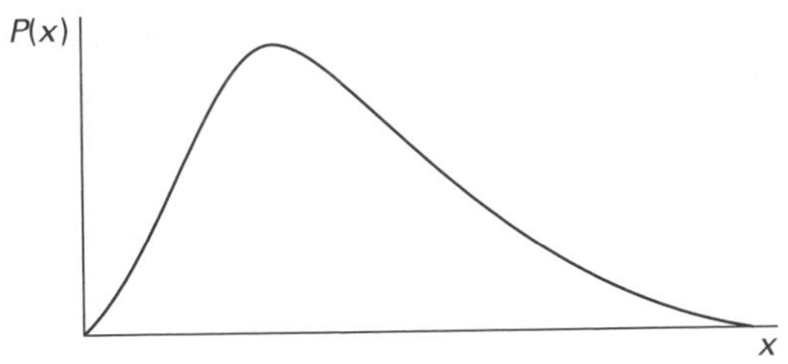

9. BINOMIAL DISTRIBUTION

The *binomial probability density function* (*binomial distribution*) is used when all outcomes can be categorized as either successes or failures. The probability of success in a single trial is $\widehat{p}$, and the probability of failure is the complement, $\widehat{q} = 1 - \widehat{p}$. The population is assumed to be infinite in size so that sampling does not change the values of $\widehat{p}$ and $\widehat{q}$. (The binomial distribution can also be used with finite populations when sampling with replacement.)

Equation 11.28 gives the probability of x successes in n independent *successive trials*. The quantity $\binom{n}{x}$ is the *binomial coefficient*, identical to the number of combinations of n items taken x at a time.

$$p\{x\} = f(x) = \binom{n}{x}\widehat{p}^{\,x}\widehat{q}^{\,n-x} \quad 11.28$$

$$\binom{n}{x} = \frac{n!}{(n-x)!\,x!} \quad 11.29$$

Equation 11.28 is a discrete distribution, taking on values only for discrete integer values up to n. The mean, μ, and variance, σ^2, of this distribution are

$$\mu = n\widehat{p} \quad 11.30$$

$$\sigma^2 = n\widehat{p}\,\widehat{q} \quad 11.31$$

Example 11.8

Five percent of a large batch of high-strength steel bolts purchased for bridge construction are defective. (a) If seven bolts are randomly sampled, what is the probability that exactly three will be defective? (b) What is the probability that two or more bolts will be defective?

Solution

(a) The bolts are either defective or not, so the binomial distribution can be used.

$$\widehat{p} = 0.05 \quad [\text{success} = \text{defective}]$$

$$\widehat{q} = 1 - 0.05 = 0.95 \quad [\text{failure} = \text{not defective}]$$

From Eq. 11.28,

$$\begin{aligned} p\{3\} = f(3) &= \binom{n}{x}\widehat{p}^{\,x}\widehat{q}^{\,n-x} = \binom{7}{3}(0.05)^3(0.95)^{7-3} \\ &= \left(\frac{7\cdot 6\cdot 5\cdot 4\cdot 3\cdot 2\cdot 1}{4\cdot 3\cdot 2\cdot 1\cdot 3\cdot 2\cdot 1}\right)(0.05)^3(0.95)^4 \\ &= 0.00356 \end{aligned}$$

(b) The probability that two or more bolts will be defective could be calculated as

$$p\{x \geq 2\} = p\{2\} + p\{3\} + p\{4\} + p\{5\} + p\{6\} + p\{7\}$$

This method would require six probability calculations. It is easier to use the complement of the desired probability.

$$p\{x \geq 2\} = 1 - p\{x \leq 1\} = 1 - \big(p\{0\} + p\{1\}\big)$$

$$p\{0\} = \binom{n}{x}\widehat{p}^{\,x}\widehat{q}^{\,n-x} = \binom{7}{0}(0.05)^0(0.95)^7 = (0.95)^7$$

$$\begin{aligned} p\{1\} &= \binom{n}{x}\widehat{p}^{\,x}\widehat{q}^{\,n-x} = \binom{7}{1}(0.05)^1(0.95)^6 \\ &= (7)(0.05)(0.95)^6 \end{aligned}$$

$$\begin{aligned} p\{x \geq 2\} &= 1 - \big((0.95)^7 + (7)(0.05)(0.95)^6\big) \\ &= 1 - (0.6983 + 0.2573) \\ &= 0.0444 \end{aligned}$$

10. HYPERGEOMETRIC DISTRIBUTION

Probabilities associated with sampling from a finite population without replacement are calculated from the *hypergeometric distribution*. If a population of finite size M contains K items with a given characteristic (e.g., red color, defective construction), then the probability of finding x items with that characteristic in a sample of n items is

$$p\{x\} = f(x) = \frac{\binom{K}{x}\binom{M-K}{n-x}}{\binom{M}{n}} \quad [\text{for } x \leq n] \quad 11.32$$

11. MULTIPLE HYPERGEOMETRIC DISTRIBUTION

Sampling without replacement from finite populations containing several different types of items is handled by the *multiple hypergeometric distribution.* If a population of finite size M contains K_i items of type i (such that $\Sigma K_i = M$), the probability of finding x_1 items of type 1, x_2 items of type 2, and so on, in a sample size of n (such that $\Sigma x_i = n$) is

$$p\{x_1, x_2, x_3, \ldots\} = \frac{\binom{K_1}{x_1}\binom{K_2}{x_2}\binom{K_3}{x_3}\cdots}{\binom{M}{n}} \quad \text{11.33}$$

12. POISSON DISTRIBUTION

Certain discrete events occur relatively infrequently but at a relatively regular rate. The probability of such an event occurring is given by the *Poisson distribution.* Suppose an event occurs, on the average, λ times per period. The probability that the event will occur x times per period is

$$p\{x\} = f(x) = \frac{e^{-\lambda}\lambda^x}{x!} \quad [\lambda > 0] \quad \text{11.34}$$

λ is both the mean and the variance of the Poisson distribution.

$$\mu = \lambda \quad \text{11.35}$$

$$\sigma^2 = \lambda \quad \text{11.36}$$

Example 11.9

The number of customers arriving at a hamburger stand in the next period is a Poisson distribution having a mean of eight. What is the probability that exactly six customers will arrive in the next period?

Solution

$\lambda = 8$ and $x = 6$. From Eq. 11.34,

$$p\{6\} = \frac{e^{-\lambda}\lambda^x}{x!} = \frac{e^{-8}(8)^6}{6!} = 0.122$$

13. CONTINUOUS DISTRIBUTION FUNCTIONS

Most numerical events are *continuously distributed* and are not constrained to discrete or integer values. For example, the resistance of a 10% 1 Ω resistor may be any value between 0.9 Ω and 1.1 Ω. The probability of an exact numerical event is zero for continuously distributed variables. That is, there is no chance that a numerical event will be *exactly* x. It is possible to determine only the probability that a numerical event will be less than x, greater than x, or between the values of x_1 and x_2, but not exactly equal to x.

Since an expression, $f(x)$, for a probability density function cannot always be written, it is more common to specify the *continuous distribution function,* $F(x_0)$, which gives the probability of numerical event x_0 or less occurring, as illustrated in Fig. 11.3.

$$p\{X < x_0\} = F(x_0) = \int_0^{x_0} f(x)\,dx \quad \text{11.37}$$

$$f(x) = \frac{dF(x)}{dx} \quad \text{11.38}$$

Figure 11.3 *Continuous Distribution Function*

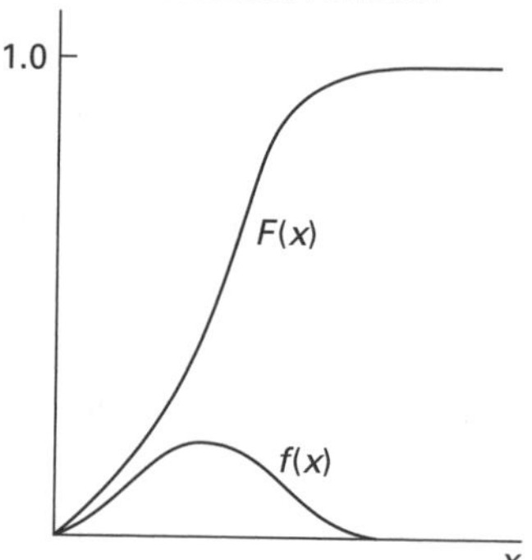

14. EXPONENTIAL DISTRIBUTION

The continuous *exponential distribution* is given by its probability density and continuous distribution functions.

$$f(x) = \lambda e^{-\lambda x} \quad \text{11.39}$$

$$p\{X < x\} = F(x) = 1 - e^{-\lambda x} \quad \text{11.40}$$

The mean and variance of the exponential distribution are

$$\mu = \frac{1}{\lambda} \quad \text{11.41}$$

$$\sigma^2 = \frac{1}{\lambda^2} \quad \text{11.42}$$

15. NORMAL DISTRIBUTION

The *normal distribution* (*Gaussian distribution*) is a symmetrical distribution commonly referred to as the *bell-shaped curve*, which represents the distribution of outcomes of many experiments, processes, and phenomena. (See Fig. 11.4.) The probability density and continuous distribution functions for the normal distribution with mean μ and variance σ^2 are

$$f(x) = \frac{e^{-\frac{1}{2}\left(\frac{x-\mu}{\sigma}\right)^2}}{\sigma\sqrt{2\pi}} \quad [-\infty < x < +\infty] \qquad 11.43$$

$$p\{\mu < X < x_0\} = F(x_0) = \frac{1}{\sigma\sqrt{2\pi}} \int_0^{x_0} e^{-\frac{1}{2}\left(\frac{x-\mu}{\sigma}\right)^2} dx \qquad 11.44$$

Figure 11.4 *Normal Distribution*

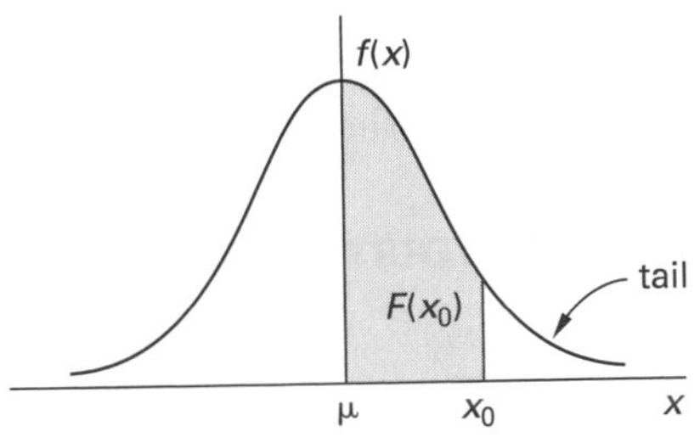

Since $f(x)$ is difficult to integrate, Eq. 11.43 is seldom used directly, and a *standard normal table* is used instead. (See App. 11.A.) The standard normal table is based on a normal distribution with a mean of zero and a standard deviation of 1. Since the range of values from an experiment or phenomenon will not generally correspond to the standard normal table, a value, x_0, must be converted to a *standard normal value*, z. In Eq. 11.45, μ and σ are the mean and standard deviation, respectively, of the distribution from which x_0 comes. For all practical purposes, all normal distributions are completely bounded by $\mu \pm 3\sigma$.

$$z = \frac{x_0 - \mu}{\sigma} \qquad 11.45$$

Numbers in the standard normal table, as given by App. 11.A, are the probabilities of the normalized x being between zero and z and represent the areas under the curve up to point z. When x is less than μ, z will be negative. However, the curve is symmetrical, so the table value corresponding to positive z can be used. The probability of x being greater than z is the complement of the table value. The curve area past point z is known as the *tail of the curve*.

Example 11.10

The mass, m, of a particular hand-laid fiberglass (Fiberglas™) part is normally distributed with a mean of 66 kg and a standard deviation of 5 kg. (a) What percent of the parts will have a mass less than 72 kg? (b) What percent of the parts will have a mass in excess of 72 kg? (c) What percent of the parts will have a mass between 61 kg and 72 kg?

Solution

(a) The 72 kg value must be normalized, so use Eq. 11.45. The standard normal variable is

$$z = \frac{x-\mu}{\sigma} = \frac{72\,\text{kg} - 66\,\text{kg}}{5\,\text{kg}} = 1.2$$

Reading from App. 11.A, the area under the normal curve is 0.3849. This represents the probability of the mass, m, being between 66 kg and 72 kg (i.e., z being between 0 and 1.2). However, the probability of the mass being less than 66 kg is also needed. Since the curve is symmetrical, this probability is 0.5. Therefore,

$$p\{m < 72\text{ kg}\} = p\{z < 1.2\} = 0.5 + 0.3849 = 0.8849$$

(b) The probability of the mass exceeding 72 kg is the area under the tail past point z.

$$p\{m > 72\text{ kg}\} = p\{z > 1.2\} = 0.5 - 0.3849 = 0.1151$$

(c) The standard normal variable corresponding to $m =$ 61 kg is

$$z = \frac{x-\mu}{\sigma} = \frac{61\,\text{kg} - 66\,\text{kg}}{5\,\text{kg}} = -1$$

Since the two masses are on opposite sides of the mean, the probability will have to be determined in two parts.

$$\begin{aligned} p\{61 < m < 72\} &= p\{61 < m < 66\} + p\{66 < m < 72\} \\ &= p\{-1 < z < 0\} + p\{0 < z < 1.2\} \\ &= 0.3413 + 0.3849 \\ &= 0.7262 \end{aligned}$$

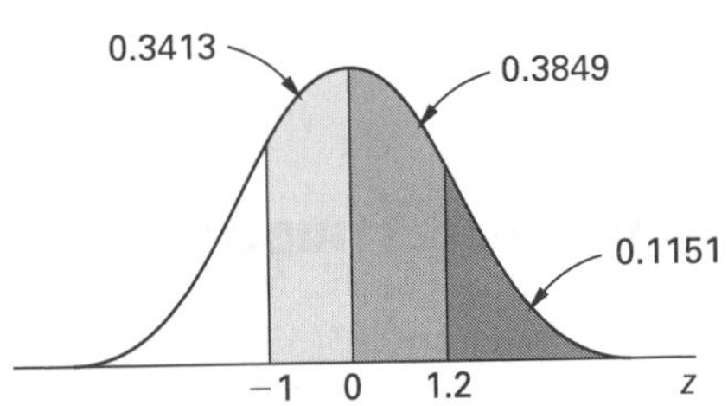

16. STUDENT'S *t*-DISTRIBUTION

In many cases requiring statistical analysis, including setting of confidence limits and hypothesis testing, the true population standard deviation, σ, and mean, μ, are not known. In such cases, the sample standard deviation, s, is used as an estimate for the population standard deviation, and the sample mean, $\overline{x}$, is used to estimate the population mean, μ. To account for the additional uncertainty of not knowing the population parameters exactly, a *t-distribution* is used rather than the normal distribution. The *t*-distribution essentially relaxes or expands the confidence intervals to account for these additional uncertainties.

The t-distribution is actually a family of distributions. They are similar in shape (symmetrical and bell-shaped) to the normal distribution, although wider, and flatter in the tails. t-distributions are more likely to result in (or, accept) values located farther from the mean than the normal distribution. The specific shape is dependent on the sample size, n. The smaller the sample size, the wider and flatter the distribution tails. The shape of the distribution approaches the standard normal curve as the sample size increases. Generally, the two distributions have the same shape for $n > 50$.

An important parameter needed to define a t-distribution is the *degrees of freedom*, df. (The symbol, ν, Greek nu, is also used.) The degrees of freedom is usually 1 less than the sample size.

$$\text{df} = n - 1 \quad \text{11.46}$$

Student's t-distribution is a standardized t-distribution, centered on zero, just like the standard normal variable, z. With a normal distribution, the population standard deviation would be multiplied by $z = 1.96$ to get a 95% confidence interval. When using the t-distribution, the sample standard deviation, s, is multiplied by a number, t, coming from the t-distribution.

$$t = \frac{\bar{x} - \mu}{\frac{s}{\sqrt{n}}} \quad \text{11.47}$$

That number is designated as $t_{\alpha,\text{df}}$ or $t_{\alpha,n-1}$ for a one-tail test/confidence interval, and would be designated as $t_{\alpha/2,\text{df}}$ or $t_{\alpha/2,n-1}$ for a two-tail test/confidence interval. For example, for a two-tail confidence interval with confidence level $C = 1 - \alpha$, the upper and lower confidence limits are

$$\text{UCL, LCL} = \bar{x} \pm t_{\alpha/2,n-1}\frac{s}{\sqrt{n}} \quad \text{11.48}$$

17. CHI-SQUARED DISTRIBUTION

The *chi-squared* (*chi square*, χ^2) *distribution* is a distribution of the sum of squared standard normal deviates, z_i. It has numerous useful applications. The distribution's only parameter, its *degrees of freedom*, ν or df, is the number of standard normal deviates being summed. Chi-squared distributions are positively skewed, but the skewness decreases and the distribution approaches a normal distribution as degrees of freedom increases. Appendix 11.B tabulates the distribution. The mean and variance of a chi-squared distribution are related to its degrees of freedom.

$$\mu = \nu \quad \text{11.49}$$

$$\sigma^2 = 2\nu \quad \text{11.50}$$

After taking n samples from a standard normal distribution, the *chi-squared statistic* is defined as

$$\chi^2 = \frac{(n-1)s^2}{\sigma^2} \quad \text{11.51}$$

The sum of chi-squared variables is itself a chi-squared variable. For example, after taking three measurements from a standard normal distribution, squaring each term, and adding all three squared terms, the sum is a chi-squared variable. A chi-squared distribution with three degrees of freedom can be used to determine the probability of that summation exceeding another number. This characteristic makes the chi-squared distribution useful in determining whether or not a population's variance has shifted, or whether two populations have the same variance. The distribution is also extremely useful in categorical hypothesis testing.

18. LOG-NORMAL DISTRIBUTION

With a *log-normal* (*lognormal* or *Galton's*) *distribution*, the logarithm of the independent variable, $\ln(x)$ or $\log(x)$, is normally distributed, not x. Log-normal distributions are rare in engineering; they are more common in social, political, financial, biological, and environmental applications.[2] The log-normal distribution may be applicable whenever a normal-appearing distribution is skewed to the left, or when the independent variable is a function (e.g., product) of multiple positive independent variables. Depending on its parameters (i.e., skewness), a log-normal distribution can take on different shapes, as shown in Fig. 11.5.

The symbols μ and σ are used to represent the *parameter values* of the transformed (logarithmitized) distribution; μ is the *location parameter* (*log mean*), the mean of the transformed values; and σ is the *scale parameter* (*log standard deviation*), whose unbiased estimator is the sample standard deviation, s, of the transformed values. These parameter values are used to calculate the standard normal variable, z, in order to determine event probabilities (i.e., areas under the normal curve). The *expected value*, $E\{x\}$, and standard deviation, σ_x, of the nontransformed distribution, x, are

$$E\{x\} = e^{\mu + \frac{1}{2}\sigma^2} \quad \text{11.52}$$

$$\sigma_x = \sqrt{(e^{\sigma^2} - 1)e^{2\mu + \sigma^2}} \quad \text{11.53}$$

$$z = \frac{x_0 - \mu}{\sigma} \quad \text{11.54}$$

[2]Log-normal is a popular reliability distribution. In electrical engineering, some transmission losses are modeled as log-normal. Some geologic mineral concentrations follow the log-normal pattern.

Figure 11.5 Log-Normal Probability Density Function

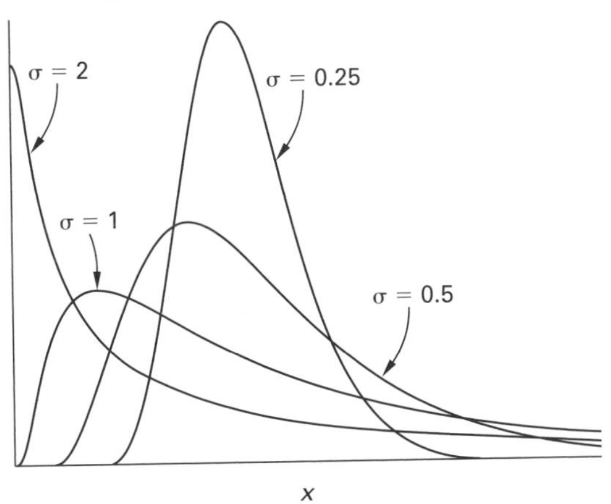

19. ERROR FUNCTION

The *error function*, erf(x), and its complement, the *complementary error function*, erfc(x), are defined by Eq. 11.55 and Eq. 11.56. The functions can be used to determine the probable error of a measurement, but they also appear in many engineering formulas. Values are seldom calculated from Eq. 11.55 or Eq. 11.56. Rather, approximation and tabulations (e.g., App. 11.D) are used.

$$\text{erf}(x_0) = \frac{2}{\sqrt{\pi}} \int_0^{x_0} e^{-x^2} dx \qquad 11.55$$

$$\text{erfc}(x_0) = 1 - \text{erf}(x_0) \qquad 11.56$$

The error function can be used to calculate areas under the normal curve. Combining Eq. 11.44, Eq. 11.45, and Eq. 11.55,

$$\frac{1}{\sqrt{2\pi}} \int_0^{z} e^{-u^2/2} du = \tfrac{1}{2}\text{erf}\left(\frac{z}{\sqrt{2}}\right) \qquad 11.57$$

The error function has the following properties.

$$\begin{aligned} \text{erf}(0) &= 0 \\ \text{erf}(+\infty) &= 1 \\ \text{erf}(-\infty) &= -1 \\ \text{erf}(-x_0) &= -\text{erf}(x_0) \end{aligned}$$

20. APPLICATION: RELIABILITY

Introduction

Reliability, $R\{t\}$, is the probability that an item will continue to operate satisfactorily up to time t. The *bathtub distribution*, Fig. 11.6, is often used to model the probability of failure of an item (or, the number of failures from a large population of items) as a function of time. Items initially fail at a high rate, a phenomenon known as *infant mortality*. For the majority of the operating time, known as the *steady-state operation*, the failure rate is constant (i.e., is due to random causes). After a long period of time, the items begin to deteriorate and the failure rate increases. (No mathematical distribution describes all three of these phases simultaneously.)

Figure 11.6 Bathtub Reliability Curve

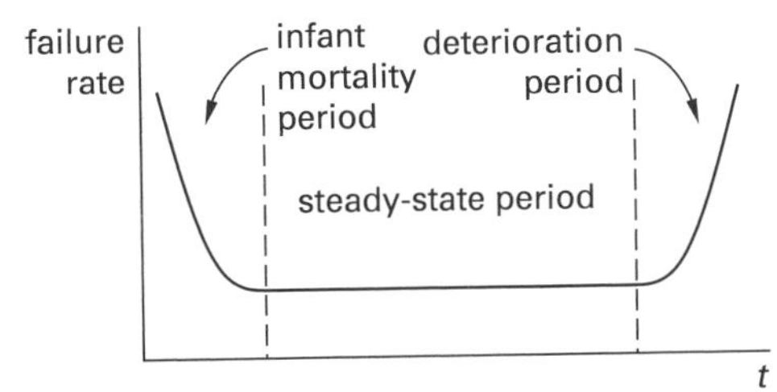

The *hazard function*, $z\{t\}$, represents the *conditional probability of failure*—the probability of failure in the next time interval, given that no failure has occurred thus far.[3]

$$z\{t\} = \frac{f(t)}{R(t)} = \frac{\frac{dF(t)}{dt}}{1 - F(t)} \qquad 11.58$$

A *proof test* is a comprehensive validation of an item that checks 100% of all failure mechanisms. It tests for faults and degradation so that the item can be certified as being in "as new" condition. The *proof test interval* is the time after initial installation at which an item must be either proof tested or replaced. If the policy is to replace rather than test an item, the item's lifetime is equal to the proof test interval and is known as *mission time*.

Exponential Reliability

Steady-state reliability is often described by the *negative exponential distribution*. This assumption is appropriate whenever an item fails only by random causes and does not experience deterioration during its life. The parameter λ is related to the *mean time to failure* (MTTF) of the item.[4]

$$R\{t\} = e^{-\lambda t} = e^{-t/\text{MTTF}} \qquad 11.59$$

$$\lambda = \frac{1}{\text{MTTF}} \qquad 11.60$$

Equation 11.59 and the exponential continuous distribution function, Eq. 11.40, are complementary.

$$R\{t\} = 1 - F(t) = 1 - (1 - e^{-\lambda t}) = e^{-\lambda t} \qquad 11.61$$

[3]The symbol $z\{t\}$ is traditionally used for the hazard function and is not related to the standard normal variable.
[4]The term "mean time *between* failures" is improper. However, the term *mean time before failure* (MTBF) is acceptable.

The hazard function for the negative exponential distribution is

$$z\{t\} = \lambda \quad \text{11.62}$$

Therefore, the hazard function for exponential reliability is constant and does not depend on t (i.e., on the age of the item). In other words, the expected future life of an item is independent of the previous history (length of operation). This lack of memory is consistent with the assumption that only random causes contribute to failure during steady-state operations. And since random causes are unlikely discrete events, their probability of occurrence can be represented by a Poisson distribution with mean λ. That is, the probability of having x failures in any given period is

$$p\{x\} = \frac{e^{-\lambda}\lambda^x}{x!} \quad \text{11.63}$$

Serial System Reliability

In the analysis of system reliability, the binary variable X_i is defined as 1 if item i operates satisfactorily and 0 if otherwise. Similarly, the binary variable Φ is 1 only if the entire system operates satisfactorily. Therefore, Φ will depend on a *performance function* containing the X_i.

A *serial system* is one for which all items must operate correctly for the system to operate. Each item has its own reliability, R_i. For a serial system of n items, the performance function is

$$\Phi = X_1X_2X_3 \cdots X_n = \min(X_i) \quad \text{11.64}$$

The probability of a serial system operating correctly is

$$p\{\Phi = 1\} = R_{\text{serial system}} = R_1R_2R_3 \cdots R_n \quad \text{11.65}$$

Parallel System Reliability

A *parallel system* with n items will fail only if all n items fail. Such a system is said to be *redundant* to the nth degree. Using redundancy, a highly reliable system can be produced from components with relatively low individual reliabilities.

The performance function of a redundant system is

$$\begin{aligned}\Phi &= 1 - (1 - X_1)(1 - X_2)(1 - X_3) \cdots (1 - X_n) \\ &= \max(X_i)\end{aligned} \quad \text{11.66}$$

The reliability of the parallel system is

$$\begin{aligned}R &= p\{\Phi = 1\} \\ &= 1 - (1 - R_1)(1 - R_2)(1 - R_3) \cdots (1 - R_n)\end{aligned} \quad \text{11.67}$$

With a fully redundant, parallel k-out-of-n system of n independent, identical items, any k of which maintain system functionality, the ratio of redundant MTTF to single-item MTTF $(1/\lambda)$ is given by Table 11.1.

Table 11.1 *MTTF Multipliers for k-out-of-n Systems*

	n				
k	1	2	3	4	5
1	1	3/2	11/6	25/12	137/60
2		1/2	5/6	13/12	77/60
3			1/3	7/12	47/60
4				1/4	9/20
5					1/5

Example 11.11

The reliability of an item is exponentially distributed with mean time to failure (MTTF) of 1000 hr. What is the probability that the item will not have failed before 1200 hr of operation?

Solution

The probability of not having failed before time t is the reliability. From Eq. 11.60 and Eq. 11.61,

$$\begin{aligned}\lambda &= \frac{1}{\text{MTTF}} = \frac{1}{1000 \text{ hr}} \\ &= 0.001 \text{ hr}^{-1}\end{aligned}$$

$$\begin{aligned}R\{1200\} &= e^{-\lambda t} = e^{(-0.001 \text{ hr}^{-1})(1200 \text{ hr})} \\ &= 0.3\end{aligned}$$

Example 11.12

What are the reliabilities of the following systems?

(a)

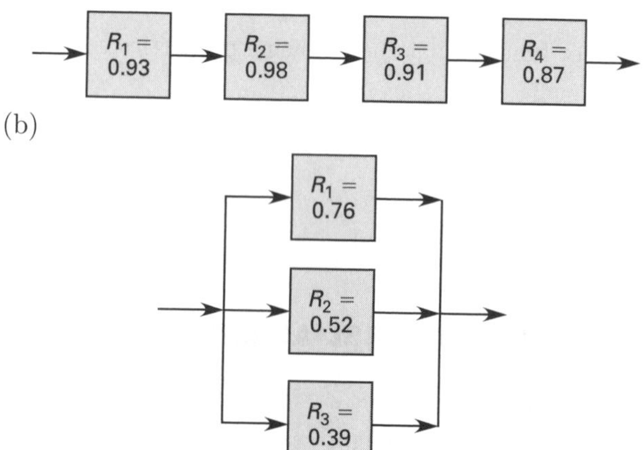

Solution

(a) This is a serial system. From Eq. 11.65,

$$R = R_1R_2R_3R_4 = (0.93)(0.98)(0.91)(0.87) = 0.72$$

(b) This is a parallel system. From Eq. 11.67,

$$\begin{aligned} R &= 1-(1-R_1)(1-R_2)(1-R_3) \\ &= 1-(1-0.76)(1-0.52)(1-0.39) \\ &= 0.93 \end{aligned}$$

21. ANALYSIS OF EXPERIMENTAL DATA

Experiments can take on many forms. An experiment might consist of measuring the mass of one cubic foot of concrete or measuring the speed of a car on a roadway. Generally, such experiments are performed more than once to increase the precision and accuracy of the results.

Both systematic and random variations in the process being measured will cause the observations to vary, and the experiment would not be expected to yield the same result each time it was performed. Eventually, a collection of experimental outcomes (observations) will be available for analysis.

The *frequency distribution* is a systematic method for ordering the observations from small to large, according to some convenient numerical characteristic. The *step interval* should be chosen so that the data are presented in a meaningful manner. If there are too many intervals, many of them will have zero frequencies; if there are too few intervals, the frequency distribution will have little value. Generally, 10 to 15 intervals are used.

Once the frequency distribution is complete, it can be represented graphically as a *histogram*. The procedure in drawing a histogram is to mark off the interval limits (also known as *class limits*) on a number line and then draw contiguous bars with lengths that are proportional to the frequencies in the intervals and that are centered on the midpoints of their respective intervals. The continuous nature of the data can be depicted by a *frequency polygon*. The number or percentage of observations that occur, up to and including some value, can be shown in a *cumulative frequency table*.

Example 11.13

The number of cars that travel through an intersection between 12 noon and 1 p.m. is measured for 30 consecutive working days. The results of the 30 observations are

79, 66, 72, 70, 68, 66, 68, 76, 73, 71, 74, 70, 71, 69, 67, 74, 70, 68, 69, 64, 75, 70, 68, 69, 64, 69, 62, 63, 63, 61

(a) What are the frequency and cumulative distributions? (Use a distribution interval of two cars per hour.) (b) Draw the histogram. (Use a cell size of two cars per hour.) (c) Draw the frequency polygon. (d) Graph the cumulative frequency distribution.

Solution

(a) Tabulate the frequency, cumulative frequency, and cumulative percent distributions.

cars per hour	frequency	cumulative frequency	cumulative percent
60–61	1	1	3
62–63	3	4	13
64–65	2	6	20
66–67	3	9	30
68–69	8	17	57
70–71	6	23	77
72–73	2	25	83
74–75	3	28	93
76–77	1	29	97
78–79	1	30	100

(b) Draw the histogram.

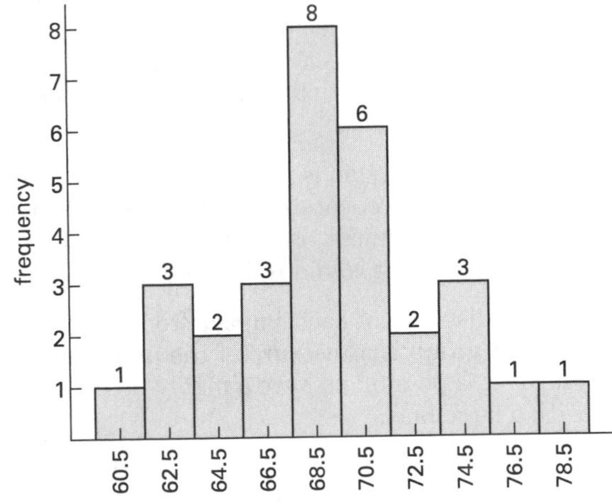

(c) Draw the frequency polygon.

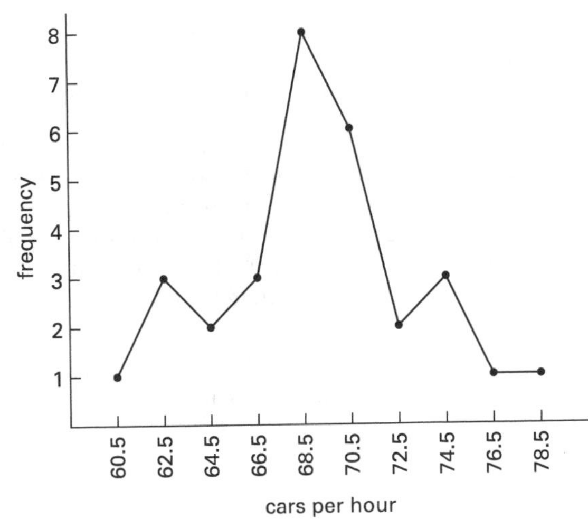

(d) Graph the cumulative frequency distribution.

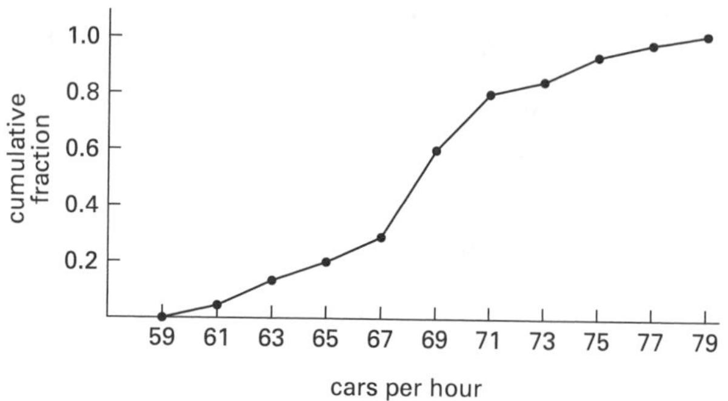

22. MEASURES OF EXPERIMENTAL ADEQUACY

An experiment is said to be *accurate* if it is unaffected by experimental error. In this case, *error* is not synonymous with *mistake*, but rather includes all variations not within the experimenter's control.

For example, suppose a gun is aimed at a point on a target and five shots are fired. The mean distance from the point of impact to the sight in point is a measure of the alignment accuracy between the barrel and sights. The difference between the actual value and the experimental value is known as *bias*.

Precision is not synonymous with accuracy. Precision is concerned with the repeatability of the experimental results. If an experiment is repeated with identical results, the experiment is said to be precise.

The average distance of each impact from the centroid of the impact group is a measure of the precision of the experiment. It is possible to have a highly precise experiment with a large bias.

Most of the techniques applied to experiments in order to improve the accuracy (i.e., reduce bias) of the experimental results (e.g., repeating the experiment, refining the experimental methods, or reducing variability) actually increase the precision.

Sometimes the word *reliability* is used with regard to the precision of an experiment. In this case, a "reliable estimate" is used in the same sense as a "precise estimate."

Stability and *insensitivity* are synonymous terms. A stable experiment will be insensitive to minor changes in the experimental parameters. For example, suppose the centroid of a bullet group is 2.1 in from the target point at 65°F and 2.3 in away at 80°F. The sensitivity of the experiment to temperature change would be

$$\text{sensitivity} = \frac{\Delta x}{\Delta T} = \frac{2.3\ \text{in} - 2.1\ \text{in}}{80°\text{F} - 65°\text{F}} = 0.0133\ \text{in}/°\text{F}$$

23. MEASURES OF CENTRAL TENDENCY

It is often unnecessary to present the experimental data in their entirety, either in tabular or graphical form. In such cases, the data and distribution can be represented by various parameters. One type of parameter is a measure of *central tendency*. Mode, median, and mean are measures of central tendency.

The *mode* is the observed value that occurs most frequently. The mode may vary greatly between series of observations. Therefore, its main use is as a quick measure of the central value since little or no computation is required to find it. Beyond this, the usefulness of the mode is limited.

The *median* is the point in the distribution that partitions the total set of observations into two parts containing equal numbers of observations. It is not influenced by the extremity of scores on either side of the distribution. The median is found by counting up (from either end of the frequency distribution) until half of the observations have been accounted for.

When n is odd, the median is the $\left({}^{n}\!/_{2}+1\right)^{\text{th}}$ item value. For even numbers of observations, the median is estimated as some value (i.e., the average) between the two center observations—$\left({}^{n}\!/_{2}\right)^{\text{th}}$ and $\left({}^{n}\!/_{2}+1\right)^{\text{th}}$. [Dispersion, Mean, Median, and Mode Values]

Similar in concept to the median are *percentiles* (*percentile ranks*), *quartiles*, and *deciles*. The median could also have been called the *50th percentile* observation. Similarly, the 80th percentile would be the observed value (e.g., the number of cars per hour) for which the cumulative frequency was 80%. The quartile and decile points on the distribution divide the observations or distribution into segments of 25% and 10%, respectively.

The *arithmetic mean* is the arithmetic average of the observations. The sample mean, $\bar{x}$, can be used as an *unbiased estimator* of the population mean, μ. The *mean* may be found without ordering the data (as was necessary to find the mode and median). The mean can be found from Eq. 11.68. There are additional mean formulas for varying applications. The weighted arithmetic mean is used when certain data points are considered more influential to the overall dataset and can be determined through Eq. 11.69.

Dispersion, Mean, Median, and Mode Values

$$\bar{x} = \left(\frac{1}{n}\right)(x_1 + x_2 + \cdots + x_n) = \frac{\sum x_i}{n} \qquad 11.68$$

Dispersion, Mean, Median, and Mode Values

$$\bar{x}_W = \frac{\sum w_i x_i}{\sum w_i} \qquad 11.69$$

The *geometric mean* is used occasionally when it is necessary to average ratios. The geometric mean is calculated as

Dispersion, Mean, Median, and Mode Values

$$\text{geometric mean} = \sqrt[n]{x_1 x_2 x_3 \cdots x_n} \quad [x_i > 0] \qquad 11.70$$

The *harmonic mean* is defined as

$$\text{harmonic mean} = \frac{n}{\frac{1}{x_1} + \frac{1}{x_2} + \cdots + \frac{1}{x_n}} \qquad 11.71$$

The *root-mean-squared (rms) value* of a series of observations is defined as

Dispersion, Mean, Median, and Mode Values

$$x_{\text{rms}} = \sqrt{\frac{\sum x_i^2}{n}} \qquad 11.72$$

The *ratio of exceedance* of a single measurement, x_i, from the mean (or, any other value), μ, is

$$\text{ER} = \left| \frac{x_i - \mu}{\mu} \right| \qquad 11.73$$

The exceedance of a single measurement from the mean can be expressed in standard deviations. For a normal distribution, this exceedance is usually given the variable z and is referred to as the *standard normal value, variable, variate,* or *deviate.*

$$z = \left| \frac{x_i - \mu}{\sigma} \right| \qquad 11.74$$

Example 11.14

Find the mode, median, and arithmetic mean of the distribution represented by the data given in Ex. 11.13.

Solution

First, resequence the observations in increasing order.

61, 62, 63, 63, 64, 64, 66, 66, 67, 68, 68, 68, 68, 69, 69, 69, 69, 70, 70, 70, 70, 71, 71, 72, 73, 74, 74, 75, 76, 79

The mode is the interval 68–69, since this interval has the highest frequency. If 68.5 is taken as the interval center, then 68.5 would be the mode.

The 15th and 16th observations are both 69, so the median is

$$\frac{69 + 69}{2} = 69$$

The mean can be found from the raw data or from the grouped data using the interval center as the assumed observation value. Using the raw data,

Dispersion, Mean, Median, and Mode Values

$$\begin{aligned} \overline{x} &= \frac{\sum x}{n} \\ &= \frac{2069}{30} \\ &= 68.97 \end{aligned}$$

24. MEASURES OF DISPERSION

The simplest statistical parameter that describes the variability in observed data is the *range.* The range is found by subtracting the smallest value from the largest. Since the range is influenced by extreme (low probability) observations, its use as a measure of variability is limited.

The *population standard deviation* is a better estimate of variability because it considers every observation. That is, in Eq. 11.75, N is the total population size, not the sample size, n.

Dispersion, Mean, Median, and Mode Values

$$\sigma = \sqrt{\frac{1}{N} \sum (x_i - \mu)^2} \qquad 11.75$$

Equation 11.76 can also be used to find the standard deviation.

$$\sigma = \sqrt{\frac{\sum x_i^2}{N} - \mu^2} \qquad 11.76$$

The standard deviation of a sample (particularly a small sample) is a biased (i.e., not a good) estimator of the population standard deviation. An *unbiased estimator* of the population standard deviation is the *sample standard deviation, s,* also known as the *standard error* of a sample.[5]

Dispersion, Mean, Median, and Mode Values

$$s = \sqrt{\left(\frac{1}{n-1}\right) \sum_{i=1}^{n} (x_i - \overline{x})^2} \qquad 11.77$$

If the sample standard deviation, s, is known, the standard deviation of the sample, σ_{sample}, can be calculated.

$$\sigma_{\text{sample}} = s\sqrt{\frac{n-1}{n}} \qquad 11.78$$

[5]There is a subtle yet significant difference between *standard deviation of the sample*, σ (obtained from Eq. 11.75 for a finite sample drawn from a larger population), and the *sample standard deviation*, s (obtained from Eq. 11.77). While σ can be calculated, it has no significance or use as an estimator. It is true that the difference between σ and s approaches zero when the sample size, n, is large, but this convergence does nothing to legitimize the use of σ as an estimator of the true standard deviation. (Some people say "large" is 30, others say 50 or 100.)

The *variance* is the square of the standard deviation. Since there are two standard deviations, there are two variances. The *variance of the sample* is σ^2, and the *sample variance* is s^2.

Dispersion, Mean, Median, and Mode Values

$$\sigma^2 = \left(\frac{1}{N}\right)\left((x_1-\mu)^2+(x_2-\mu)^2 + \cdots + (x_N-\mu)^2\right) = \frac{1}{N}\sum_{i=1}^{N}(x_i-\mu)^2 \quad 11.79$$

Dispersion, Mean, Median, and Mode Values

$$s^2 = \frac{1}{n-1}\sum_{i=1}^{n}(x_i-\overline{x})^2 \quad 11.80$$

The *relative dispersion* is defined as a measure of dispersion divided by a measure of central tendency. The *coefficient of variation* is a relative dispersion calculated from the sample standard deviation and the mean.

$$\text{coefficient of variation} = \frac{s}{\overline{x}} \quad 11.81$$

Example 11.15

For the data given in Ex. 11.13, calculate (a) the sample range, (b) the standard deviation of the sample, (c) an unbiased estimator of the population standard deviation, (d) the variance of the sample, and (e) the sample variance.

Solution

Calculate the quantities used in Eq. 11.75 and Eq. 11.77.

$$\sum x_i = 2069$$

$$\left(\sum x_i\right)^2 = (2069)^2 = 4{,}280{,}761$$

$$\sum x_i^2 = 143{,}225$$

$$n = 30$$

$$\overline{x} = \frac{2069}{30} = 68.967$$

(a) The sample range is

$$R = x_{\text{max}} - x_{\text{min}} = 79 - 61 = 18$$

(b) Normally, the standard deviation of the sample would not be calculated. In this case, it was specifically requested. From Eq. 11.75, using n for N and $\overline{x}$ for μ,

Dispersion, Mean, Median, and Mode Values

$$\sigma = \sqrt{\frac{\sum x_i^2}{n} - (\overline{x})^2} = \sqrt{\frac{143{,}225}{30} - \left(\frac{2069}{30}\right)^2} = 4.215$$

(c) From Eq. 11.77,

Dispersion, Mean, Median, and Mode Values

$$s = \sqrt{\frac{\sum x_i^2 - \frac{\left(\sum x_i\right)^2}{n}}{n-1}} = \sqrt{\frac{143{,}225 - \frac{4{,}280{,}761}{30}}{29}} = 4.287$$

(d) The variance of the sample is

$$\sigma^2 = 17.77$$

(e) The sample variance is

$$s^2 = 18.38$$

25. SKEWNESS

Skewness is a measure of a distribution's lack of symmetry. Distributions that are pushed to the left have negative skewness, while distributions pushed to the right have positive skewness. Various formulas are used to calculate skewness. *Pearson's skewness*, sk, is a simple normalized difference between the mean and mode. (See Eq. 11.82.) Since the mode is poorly represented when sampling from a distribution, the difference is estimated as three times the deviation from the mean. *Fisher's skewness*, γ_1, and more modern methods are based on the *third moment about the mean*. (See Eq. 11.83.) Normal distributions have zero skewness, although zero skewness is not a sufficient requirement for determining normally distributed data. Square root, log, and reciprocal transformations of a variable can reduce skewness.

$$\text{sk} = \frac{\mu - \text{mode}}{\sigma} \approx \frac{3(\overline{x} - \text{median})}{s} \quad 11.82$$

$$\gamma_1 = \frac{\sum_{i=1}^{N}(x_i-\mu)^3}{N\sigma^3} \approx \frac{n\sum_{i=1}^{n}(x_i-\overline{x})^3}{(n-1)(n-2)s^3} \quad 11.83$$

26. KURTOSIS

While skewness refers to the symmetry of the distribution about the mean, *kurtosis* refers to the contribution of the tails to the distribution, or alternatively, how flat the peak is. A *mesokurtic distribution* ($\beta_2 = 3$) is statistically normal. Compared to a normal curve, a *leptokurtic distribution* ($\beta_2 < 3$) is "fat in the tails," longer-tailed, and more sharp-peaked, while a *platykurtic distribution* ($\beta_2 > 3$) is "thin in the tails," shorter-tailed, and more flat-peaked. Different methods are used to calculate kurtosis. The most common, *Fisher's kurtosis*, β_2, is defined as the *fourth standardized moment* (*fourth moment of the mean*).

$$\beta_2 = \frac{\sum_{i=1}^{N}(x_i - \mu)^4}{N\sigma^4} \qquad 11.84$$

Some statistical analyses, including those in Microsoft Excel, calculate a related statistic, *excess kurtosis* (*kurtosis excess* or *Pearson's kurtosis*), by subtracting 3 from the kurtosis. A normal distribution has zero excess kurtosis; a peaked distribution has positive excess kurtosis; and a flat distribution has negative excess kurtosis.

$$\begin{aligned}\gamma_2 &= \beta_2 - 3 \\ &\approx \frac{n(n+1)\sum_{i=1}^{n}(x_i - \bar{x})^4}{(n-1)(n-2)(n-3)s^4} - \frac{3(n-1)^2}{(n-2)(n-3)}\end{aligned} \qquad 11.85$$

27. CENTRAL LIMIT THEOREM

Measuring a sample of n items from a population with mean μ and standard deviation σ is the general concept of an experiment. The sample mean, $\bar{x}$, is one of the parameters that can be derived from the experiment. This experiment can be repeated k times, yielding a set of averages ($\bar{x}_1, \bar{x}_2, \ldots, \bar{x}_k$). The k numbers in the set themselves represent samples from distributions of averages. The average of averages, $\bar{\bar{x}}$, and sample standard deviation of averages, $s_{\bar{x}}$ (known as the *standard error of the mean*), can be calculated.

The *central limit theorem* characterizes the distribution of the sample averages. The theorem can be stated in several ways, but the essential elements are the following points.

1. The averages, $\bar{x}_i$, are normally distributed variables, even if the original data from which they are calculated are not normally distributed.
2. The grand average, $\bar{\bar{x}}$ (i.e., the average of the averages), approaches and is an unbiased estimator of μ.

$$\mu \approx \bar{\bar{x}} \qquad 11.86$$

The standard deviation of the original distribution, σ, is much larger than the standard error of the mean.

$$\sigma \approx \sqrt{n}\, s_{\bar{x}} \qquad 11.87$$

28. CONFIDENCE LEVEL

The results of experiments are seldom correct 100% of the time. Recognizing this, researchers accept a certain probability of being wrong. In order to minimize this probability, experiments are repeated several times. The number of repetitions depends on the desired level of confidence in the results.

If the results have a 5% probability of being wrong, the *confidence level*, C, is 95% that the results are correct, in which case the results are said to be *significant*. If the results have only a 1% probability of being wrong, the confidence level is 99%, and the results are said to be *highly significant*. Other confidence levels (90%, 99.5%, etc.) are used as appropriate.

The complement of the confidence level is α, referred to as *alpha*. α is the *significance level* and may be given as a decimal value or percentage. Alpha is also known as *alpha risk* and *producer risk*, as well as the probability of a type I error. A *type I error*, also known as a *false positive error*, occurs when the null hypothesis is incorrectly rejected, and an action occurs that is not actually required. For a random sample of manufactured products, the null hypothesis would be that the distribution of sample measurements is not different from the historical distribution of those measurements. If the null hypothesis is rejected, all of the products (not just the sample) will be rejected, and the producer will have to absorb the expense. It is not uncommon to use 5% as the producer risk in noncritical business processes.

$$\alpha = 100\% - C \qquad 11.88$$

β (*beta* or *beta risk*) is the *consumer risk*, the probability of a type II error. A *type II error* occurs when the null hypothesis (e.g., "everything is fine") is incorrectly accepted, and no action occurs when action is actually required. If a batch of defective products is accepted, the products are distributed to consumers who then suffer the consequences. Generally, smaller values of α coincide with larger values of β because requiring overwhelming evidence to reject the null increases the chances of a type II error. β can be minimized while holding α constant by increasing sample sizes. The *power of the test* is the probability of rejecting the null hypothesis when it is false.

$$\text{power of the test} = 1 - \beta \qquad 11.89$$

29. NULL AND ALTERNATIVE HYPOTHESES

All statistical conclusions involve constructing two mutually exclusive hypotheses, termed the *null hypothesis* (written as H_0) and *alternative hypothesis* (written as H_1). Together, the hypotheses describe all possible outcomes of a statistical analysis. The purpose of the analysis is to determine which hypothesis to accept and which to reject.

Usually, when an improvement is made to a program, treatment, or process, the change is expected to make a difference. The null hypothesis is so named because it refers to a case of "no difference" or "no effect." Typical null hypotheses are:

H_0: There has been no change in the process.

H_0: The two distributions are the same.

H_0: The change has had no effect.

H_0: The process is in control and has not changed.

H_0: Everything is fine.

The alternative hypothesis is that there has been an effect, and there is a difference. The null and alternative hypotheses are mutually exclusive.

30. APPLICATION: CONFIDENCE LIMITS

As a consequence of the central limit theorem, sample means of n items taken from a normal distribution with mean μ and standard deviation σ will be normally distributed with mean μ and variance σ^2/n. The probability that any given average, $\bar{x}$, exceeds some value, L, is given by Eq. 11.90.

$$p\{\bar{x} > L\} = p\left\{z > \left|\frac{L-\mu}{\frac{\sigma}{\sqrt{n}}}\right|\right\} \qquad 11.90$$

L is the *confidence limit* for the confidence level $1 - p\{\bar{x} > L\}$ (normally expressed as a percent). Values of z are read directly from the standard normal table. As an example, $z = 1.645$ for a 95% confidence level since only 5% of the curve is above that value of z in the upper tail. This is known as a *one-tail confidence limit* because all of the probability is given to one side of the variation. Similar values are given in Table 11.2.

Table 11.2 *Values of z for Various Confidence Levels*

confidence level, C	one-tail limit, z	two-tail limit, z
90%	1.28	1.645
95%	1.645	1.96
97.5%	1.96	2.17
99%	2.33	2.575
99.5%	2.575	2.81
99.75%	2.81	3.00

With *two-tail confidence limits*, the probability is split between the two sides of variation. There will be upper and lower confidence limits, UCL and LCL, respectively.

$$p\{\text{LCL} < \bar{x} < \text{UCL}\} = p\left\{\left|\frac{\text{LCL}-\mu}{\frac{\sigma}{\sqrt{n}}}\right| < z < \left|\frac{\text{UCL}-\mu}{\frac{\sigma}{\sqrt{n}}}\right|\right\} \qquad 11.91$$

31. APPLICATION: BASIC HYPOTHESIS TESTING

A *hypothesis test* is a procedure that answers the question, "Did these data come from [a particular type of] distribution?" There are many types of tests, depending on the distribution and parameter being evaluated. The simplest hypothesis test determines whether an average value obtained from n repetitions of an experiment could have come from a population with known mean, μ, and standard deviation, σ. A practical application of this question is whether a manufacturing process has changed from what it used to be or should be. Of course, the answer (i.e., "yes" or "no") cannot be given with absolute certainty—there will be a confidence level associated with the answer.

The following procedure is used to determine whether the average of n measurements can be assumed (with a given confidence level) to have come from a known population.

step 1: Assume random sampling from a normal population.

step 2: Choose the desired confidence level, C.

step 3: Decide on a one-tail or two-tail test. If the hypothesis being tested is that the average has or has not *increased* or *decreased*, choose a one-tail test. If the hypothesis being tested is that the average has or has not *changed*, choose a two-tail test.

step 4: Use Table 11.2 or the standard normal table to determine the z-value corresponding to the confidence level and number of tails.

step 5: Calculate the actual standard normal variable, z'.

$$z' = \left|\frac{\bar{x}-\mu}{\frac{\sigma}{\sqrt{n}}}\right| \qquad 11.92$$

step 6: If $z' \geq z$, the average can be assumed (with confidence level C) to have come from a different distribution.

Example 11.16

When it is operating properly, a cement plant has a daily production rate that is normally distributed with a mean of 880 tons/day and a standard deviation of 21 tons/day. During an analysis period, the output is measured on 50 consecutive days, and the mean output is found to be 871 tons/day. With a 95% confidence level, determine whether the plant is operating properly.

Solution

step 1: The production rate samples are known to be normally distributed.

step 2: $C = 95\%$ is given.

step 3: Since a specific direction in the variation is not given (i.e., the example does not ask whether the average has decreased), use a two-tail hypothesis test.

step 4: The population mean and standard deviation are known. The standard normal distribution may be used. From Table 11.2, $z = 1.96$.

step 5: From Eq. 11.92,

$$z' = \left| \frac{\overline{x} - \mu}{\frac{\sigma}{\sqrt{n}}} \right| = \left| \frac{871 - 880}{\frac{21}{\sqrt{50}}} \right| = 3.03$$

Since $3.03 > 1.96$, the distributions are not the same. There is at least a 95% probability that the plant is not operating correctly.

32. APPLICATION: STATISTICAL PROCESS CONTROL

All manufacturing processes contain variation due to random and nonrandom causes. Random variation cannot be eliminated. *Statistical process control* (SPC) is the act of monitoring and adjusting the performance of a process to detect and eliminate nonrandom variation. [**Tests for Out of Control, for Three-Sigma Control Limits**]

Statistical process control is based on taking regular (hourly, daily, etc.) samples of n items and calculating the mean, $\overline{x}$, and range, R, of the sample. To simplify the calculations, the range is used as a measure of the dispersion. These two parameters are graphed on their respective x-bar and *R-control charts*, as shown in Fig. 11.7.[6] Confidence limits are drawn at $\pm 3\sigma/\sqrt{n}$. From a statistical standpoint, the control chart tests a hypothesis each time a point is plotted. When a point falls outside these limits, there is a 99.75% probability that the process is out of control. Until a point exceeds the control limits, no action is taken.[7]

Figure 11.7 Typical Statistical Process Control Charts

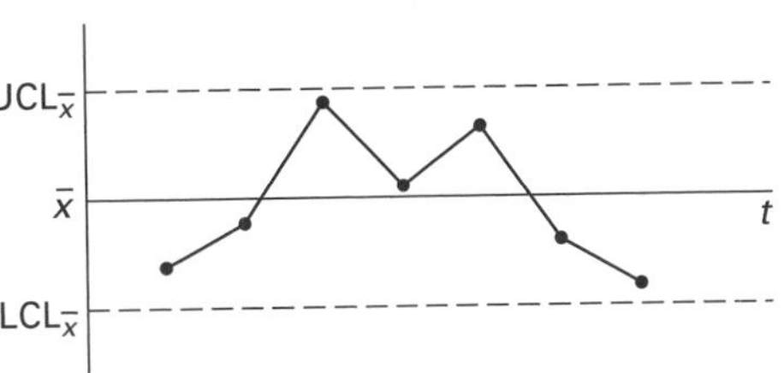

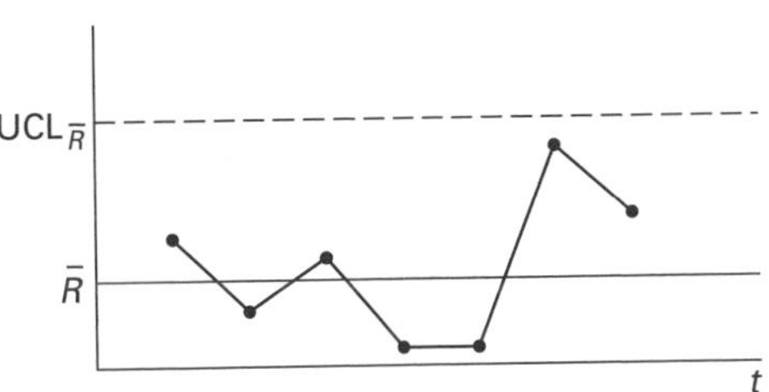

33. LINEAR REGRESSION

If it is necessary to draw a straight line ($y = mx + b$) through n data points $(x_1, y_1), (x_2, y_2), \ldots, (x_n, y_n)$, the following method based on the *method of least squares* can be used.

step 1: Calculate the following nine quantities.

$$\sum x_i \quad \sum x_i^2 \quad \left(\sum x_i\right)^2 \quad \overline{x} = \frac{\sum x_i}{n} \quad \sum x_i y_i$$

$$\sum y_i \quad \sum y_i^2 \quad \left(\sum y_i\right)^2 \quad \overline{y} = \frac{\sum y_i}{n}$$

step 2: Calculate the slope, m, of the line.

$$m = \frac{n\sum x_i y_i - \sum x_i \sum y_i}{n\sum x_i^2 - \left(\sum x_i\right)^2} \qquad 11.93$$

step 3: Calculate the y-intercept, b.

$$b = \overline{y} - m\overline{x} \qquad 11.94$$

[6]Other charts (e.g., the *sigma chart*, *p-chart*, and *c-chart*) are less common but are used as required.

[7]Other indications that a correction may be required are seven measurements on one side of the average and seven consecutively increasing measurements. Rules such as these detect shifts and trends.

step 4: To determine the goodness of fit, calculate the *correlation coefficient*, *r*.

$$r = \frac{n\sum x_i y_i - \sum x_i \sum y_i}{\sqrt{\left(n\sum x_i^2 - \left(\sum x_i\right)^2\right) \times \left(n\sum y_i^2 - \left(\sum y_i\right)^2\right)}} \qquad 11.95$$

If m is positive, r will be positive; if m is negative, r will be negative. As a general rule, if the absolute value of r exceeds 0.85, the fit is good; otherwise, the fit is poor. r equals 1.0 if the fit is a perfect straight line.

A low value of r does not eliminate the possibility of a nonlinear relationship existing between x and y. It is possible that the data describe a parabolic, logarithmic, or other nonlinear relationship. (Usually this will be apparent if the data are graphed.) It may be necessary to convert one or both variables to new variables by taking squares, square roots, cubes, or logarithms, to name a few of the possibilities, in order to obtain a linear relationship. The apparent shape of the line through the data will give a clue to the type of variable transformation that is required. The curves in Fig. 11.8 may be used as guides to some of the simpler variable transformations.

Figure 11.9 illustrates several common problems encountered in trying to fit and evaluate curves from experimental data. Fig. 11.9(a) shows a graph of clustered data with several extreme points. There will be moderate correlation due to the weighting of the extreme points, although there is little actual correlation at low values of the variables. The extreme data should be excluded, or the range should be extended by obtaining more data.

Figure 11.9(b) shows that good correlation exists in general, but extreme points are missed, and the overall correlation is moderate. If the results within the small linear range can be used, the extreme points should be excluded. Otherwise, additional data points are needed, and curvilinear relationships should be investigated.

Figure 11.9(c) illustrates the problem of drawing conclusions of cause and effect. There may be a predictable relationship between variables, but that does not imply a cause and effect relationship. In the case shown, both variables are functions of a third variable, the city population. But there is no direct relationship between the plotted variables.

Figure 11.8 *Nonlinear Data Curves*

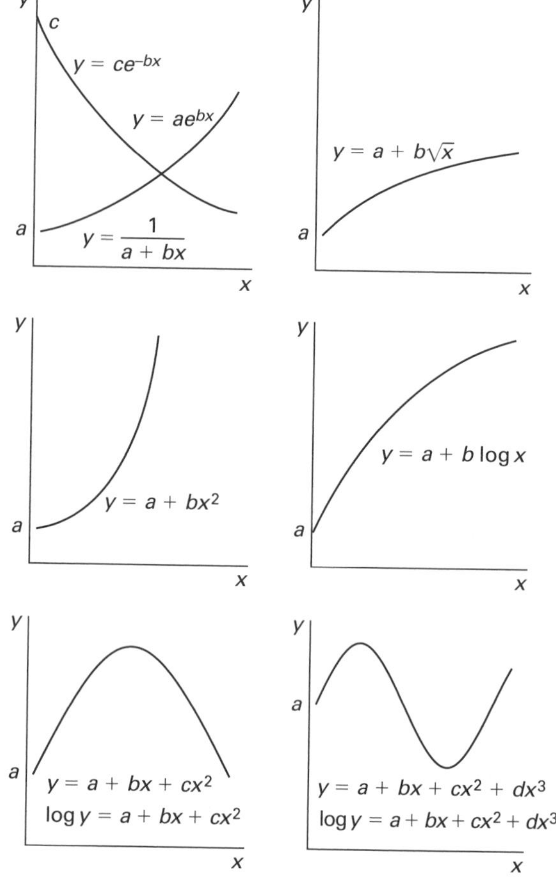

Figure 11.9 *Common Regression Difficulties*

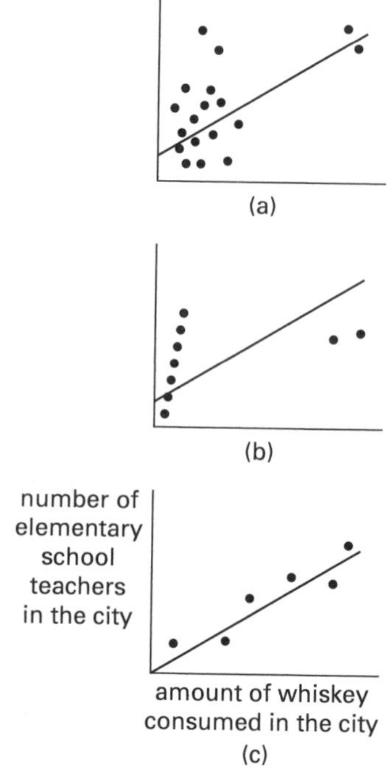

Example 11.17

An experiment is performed in which the dependent variable, y, is measured against the independent variable, x. The results are as follows.

x	y
1.2	0.602
4.7	5.107
8.3	6.984
20.9	10.031

(a) What is the least squares straight line equation that best represents this data? (b) What is the correlation coefficient?

Solution

(a) Calculate the following quantities.

$$\begin{aligned}\sum x_i &= 35.1\\ \sum y_i &= 22.72\\ \sum x_i^2 &= 529.23\\ \sum y_i^2 &= 175.84\\ \left(\sum x_i\right)^2 &= 1232.01\\ \left(\sum y_i\right)^2 &= 516.38\\ \overline{x} &= 8.775\\ \overline{y} &= 5.681\\ \sum x_i y_i &= 292.34\\ n &= 4\end{aligned}$$

From Eq. 11.93, the slope is

$$\begin{aligned}m &= \frac{n\sum x_i y_i - \sum x_i \sum y_i}{n\sum x_i^2 - \left(\sum x_i\right)^2} = \frac{(4)(292.34) - (35.1)(22.72)}{(4)(529.23) - (35.1)^2}\\ &= 0.42\end{aligned}$$

From Eq. 11.94, the y-intercept is

$$\begin{aligned}b &= \overline{y} - m\overline{x} = 5.681 - (0.42)(8.775)\\ &= 2.0\end{aligned}$$

The equation of the line is

$$y = 0.42x + 2.0$$

(b) From Eq. 11.95, the correlation coefficient is

$$\begin{aligned}r &= \frac{n\sum x_i y_i - \sum x_i \sum y_i}{\sqrt{\left(n\sum x_i^2 - \left(\sum x_i\right)^2\right)\left(n\sum y_i^2 - \left(\sum y_i\right)^2\right)}}\\ &= \frac{(4)(292.34) - (35.1)(22.72)}{\sqrt{\begin{array}{l}\left((4)(529.23) - 1232.01\right)\\ \quad\times\left((4)(175.84) - 516.38\right)\end{array}}}\\ &= 0.914\end{aligned}$$

Example 11.18

Repeat Ex. 11.17 assuming the relationship between the variables is nonlinear.

Solution

The first step is to graph the data. Since the graph has the appearance of the fourth case in Fig. 11.8, it can be assumed that the relationship between the variables has the form of $y = a + b\log x$. Therefore, the variable change $z = \log x$ is made, resulting in the following set of data.

z	y
0.0792	0.602
0.672	5.107
0.919	6.984
1.32	10.031

If the regression analysis is performed on this set of data, the resulting equation and correlation coefficient are

$$y = 7.599z + 0.000247$$

$$r = 0.999$$

This is a very good fit. The relationship between the variable x and y is approximately

$$y = 7.599\log x + 0.000247$$

12 Numbering Systems

Content in blue refers to the *NCEES Handbook*.

1. POSITIONAL NUMBERING SYSTEMS

A *base-b number*, N_b, is made up of individual *digits*. In a *positional numbering system*, the position of a digit in the number determines that digit's contribution to the total value of the number. Specifically, the position of the digit determines the power to which the *base* (also known as the *radix*), b, is raised. For *decimal numbers*, the radix is 10, hence the description *base-10 numbers*.

$$(a_n a_{n-1} \cdots a_2 a_1 a_0)_b = a_n b^n + a_{n-1} b^{n-1} + \cdots + a_2 b^2 + a_1 b + a_0 \qquad 12.1$$

The leftmost digit, a_n, contributes the greatest to the number's magnitude and is known as the *most significant digit* (MSD). The rightmost digit, a_0, contributes the least and is known as the *least significant digit* (LSD).

2. CONVERTING BASE-*b* NUMBERS TO BASE-10

Equation 12.1 converts base-b numbers to base-10 numbers. The calculation of the right-hand side of Eq. 12.1 is performed in the base-10 arithmetic and is known as the *expansion method*.[1]

Converting base-b numbers (i.e., decimals) to base-10 is similar to converting whole numbers and is accomplished by Eq. 12.2.

$$(0.a_1 a_2 \cdots a_m)_b = a_1 b^{-1} + a_2 b^{-2} + \cdots + a_m b^{-m} \qquad 12.2$$

3. CONVERTING BASE-10 NUMBERS TO BASE-*b*

The *remainder method* is used to convert base-10 numbers to base-b numbers. This method consists of successive divisions by the base, b, until the quotient is zero. The base-b number is found by taking the remainders in the reverse order from which they were found. This method is illustrated in Ex. 12.1 and Ex. 12.3.

Converting a base-10 fraction to base-b requires multiplication of the base-10 fraction and subsequent fractional parts by the base. The base-b fraction is formed from the integer parts of the products taken in the same order in which they were determined. This is illustrated in Ex. 12.2(d).

4. BINARY NUMBER SYSTEM

There are only two *binary digits* (*bits*) in the *binary number system*: zero and one.[2] Thus, all binary numbers consist of strings of bits (i.e., zeros and ones). The leftmost bit is known as the *most significant bit* (MSB), and the rightmost bit is the *least significant bit* (LSB).

As with digits from other numbering systems, bits can be added, subtracted, multiplied, and divided, although only digits 0 and 1 are allowed in the results. The rules of bit addition are

$$0 + 0 = 0$$
$$0 + 1 = 1$$
$$1 + 0 = 1$$
$$1 + 1 = 0 \text{ carry } 1$$

[1]Equation 12.1 works with any base number. The *double-dabble* (*double and add*) *method* is a specialized method of converting from base-2 to base-10 numbers.

[2]Alternatively, the binary states may be called *true* and *false*, *on* and *off*, *high* and *low*, or *positive* and *negative*.

Example 12.1

(a) Convert $(1011)_2$ to base-10. (b) Convert $(75)_{10}$ to base-2.

Solution

(a) Using Eq. 12.1 with $b = 2$,

$$(1)(2)^3 + (0)(2)^2 + (1)(2)^1 + 1 = 11$$

(b) Use the remainder method. (See Sec. 12.3.)

$$\begin{aligned} 75 \div 2 &= 37 \text{ remainder } 1 \\ 37 \div 2 &= 18 \text{ remainder } 1 \\ 18 \div 2 &= 9 \text{ remainder } 0 \\ 9 \div 2 &= 4 \text{ remainder } 1 \\ 4 \div 2 &= 2 \text{ remainder } 0 \\ 2 \div 2 &= 1 \text{ remainder } 0 \\ 1 \div 2 &= 0 \text{ remainder } 1 \end{aligned}$$

The binary representation of $(75)_{10}$ is $(1001011)_2$.

5. OCTAL NUMBER SYSTEM

The *octal* (*base-8*) *system* is one of the alternatives to working with long binary numbers. Only the digits 0 through 7 are used. The rules for addition in the octal system are the same as for the decimal system except that the digits 8 and 9 do not exist. For example,

$$\begin{aligned} 7+1 = 6+2 = 5+3 &= (10)_8 \\ 7+2 = 6+3 = 5+4 &= (11)_8 \\ 7+3 = 6+4 = 5+5 &= (12)_8 \end{aligned}$$

Example 12.2

Perform the following operations.

(a) $(2)_8 + (5)_8$

(b) $(7)_8 + (6)_8$

(c) Convert $(75)_{10}$ to base-8.

(d) Convert $(0.14)_{10}$ to base-8.

(e) Convert $(13)_8$ to base-10.

(f) Convert $(27.52)_8$ to base-10.

Solution

(a) The sum of 2 and 5 in base-10 is 7, which is less than 8 and, therefore, is a valid number in the octal system. The answer is $(7)_8$.

(b) The sum of 7 and 6 in base-10 is 13, which is greater than 8 (and, therefore, needs to be converted). Using the remainder method (see Sec. 12.3),

$$\begin{aligned} 13 \div 8 &= 1 \text{ remainder } 5 \\ 1 \div 8 &= 0 \text{ remainder } 1 \end{aligned}$$

The answer is $(15)_8$.

(c) Use the remainder method. (See Sec. 12.3.)

$$\begin{aligned} 75 \div 8 &= 9 \text{ remainder } 3 \\ 9 \div 8 &= 1 \text{ remainder } 1 \\ 1 \div 8 &= 0 \text{ remainder } 1 \end{aligned}$$

The answer is $(113)_8$.

(d) Refer to Sec. 12.3.

$$\begin{aligned} 0.14 \times 8 &= 1.12 \\ 0.12 \times 8 &= 0.96 \\ 0.96 \times 8 &= 7.68 \\ 0.68 \times 8 &= 5.44 \\ 0.44 \times 8 &= \text{etc.} \end{aligned}$$

The answer, $(0.1075...)_8$, is constructed from the integer parts of the products.

(e) Use Eq. 12.1.

$$(1)(8) + 3 = (11)_{10}$$

(f) Use Eq. 12.1 and Eq. 12.2.

$$\begin{aligned} &(2)(8)^1 + (7)(8)^0 \\ &\quad + (5)(8)^{-1} + (2)(8)^{-2} = 16 + 7 + \frac{5}{8} + \frac{2}{64} \\ &\quad = (23.656)_{10} \end{aligned}$$

6. HEXADECIMAL NUMBER SYSTEM

The *hexadecimal* (*base-16*) *system* is a shorthand method of representing the value of four binary digits at a time.[3] Since 16 distinctly different characters are needed, the capital letters A through F are used to represent the decimal numbers 10 through 15. The progression of hexadecimal numbers is illustrated in Table 12.1.

Example 12.3

(a) Convert $(4E3)_{16}$ to base-10. (b) Convert $(1475)_{10}$ to base-16. (c) Convert $(0.8)_{10}$ to base-16.

[3]The term *hex number* is often heard.

Table 12.1 Binary, Octal, Decimal, and Hexadecimal Equivalents

binary	octal	decimal	hexadecimal
0	0	0	0
1	1	1	1
10	2	2	2
11	3	3	3
100	4	4	4
101	5	5	5
110	6	6	6
111	7	7	7
1000	10	8	8
1001	11	9	9
1010	12	10	A
1011	13	11	B
1100	14	12	C
1101	15	13	D
1110	16	14	E
1111	17	15	F
10000	20	16	10

Solution

(a) The hexadecimal number D is 13 in base-10. Using Eq. 12.1,

$$(4)(16)^2 + (13)(16)^1 + 3 = (1235)_{10}$$

(b) Use the remainder method. (See Sec. 12.3.)

$$\begin{aligned} 1475 \div 16 &= 92 \text{ remainder } 3 \\ 92 \div 16 &= 5 \text{ remainder } 12 \\ 5 \div 16 &= 0 \text{ remainder } 5 \end{aligned}$$

Since $(12)_{10}$ is $(C)_{16}$, (or hex C), the answer is $(5C3)_{16}$.

(c) Refer to Sec. 12.3.

$$\begin{aligned} 0.8 \times 16 &= 12.8 \\ 0.8 \times 16 &= 12.8 \\ 0.8 \times 16 &= \text{etc.} \end{aligned}$$

Since $(12)_{10} = (C)_{16}$, the answer is $(0.CCCCC\ldots)_{16}$.

7. CONVERSIONS AMONG BINARY, OCTAL, AND HEXADECIMAL NUMBERS

The octal system is closely related to the binary system since $(2)^3 = 8$. Conversion from a binary to an octal number is accomplished directly by starting at the LSB (right-hand bit) and grouping the bits in threes. Each group of three bits corresponds to an octal digit. Similarly, each digit in an octal number generates three bits in the equivalent binary number.

Conversion from a binary to a hexadecimal number starts by grouping the bits (starting at the LSB) into fours. Each group of four bits corresponds to a hexadecimal digit. Similarly, each digit in a hexadecimal number generates four bits in the equivalent binary number.

Conversion between octal and hexadecimal numbers is easiest when the number is first converted to a binary number.

Example 12.4

(a) Convert $(5431)_8$ to base-2. (b) Convert $(1001011)_2$ to base-8. (c) Convert $(1011111101111001)_2$ to base-16.

Solution

(a) Convert each octal digit to binary digits.

$$\begin{aligned} (5)_8 &= (101)_2 \\ (4)_8 &= (100)_2 \\ (3)_8 &= (011)_2 \\ (1)_8 &= (001)_2 \end{aligned}$$

The answer is $(101100011001)_2$.

(b) Group the bits into threes starting at the LSB.

1 001 011

Convert these groups into their octal equivalents.

$$\begin{aligned} (1)_2 &= (1)_8 \\ (001)_2 &= (1)_8 \\ (011)_2 &= (3)_8 \end{aligned}$$

The answer is $(113)_8$.

(c) Group the bits into fours starting at the LSB.

1011 1111 0111 1001

Convert these groups into hexadecimal equivalents.

$$\begin{aligned} (1011)_2 &= (B)_{16} \\ (1111)_2 &= (F)_{16} \\ (0111)_2 &= (7)_{16} \\ (1001)_2 &= (9)_{16} \end{aligned}$$

The answer is $(BF79)_{16}$.

8. COMPLEMENT OF A NUMBER

The *complement*, N^*, of a number, N, depends on the machine (computer, calculator, etc.) being used. Assuming that the machine has a maximum number, n, of digits per integer number stored, the b's and $(b-1)$'s complements are

$$N_b^* = b^n - N \quad 12.3$$

$$N_{b-1}^* = N_b^* - 1 \quad 12.4$$

For a machine that works in base-10 arithmetic, the *tens* and *nines complements* are

$$N_{10}^* = 10^n - N \quad 12.5$$

$$N_9^* = N_{10}^* - 1 \quad 12.6$$

For a machine that works in base-2 arithmetic, the *twos* and *ones complements* are

$$N_2^* = 2^n - N \quad 12.7$$

$$N_1^* = N_2^* - 1 \quad 12.8$$

The binary ones complement is easily found by switching all of the ones and zeros to zeros and ones, respectively.

9. APPLICATION OF COMPLEMENTS TO COMPUTER ARITHMETIC

Equation 12.9 and Eq. 12.10 are the practical applications of complements to computer arithmetic.

$$(N^*)^* = N \quad 12.9$$

$$M - N = M + N^* \quad 12.10$$

The binary ones complement can be combined with a technique known as *end-around carry* to perform subtraction. End-around carry is the addition of the *overflow bit* to the sum of N and its ones complement.

Example 12.5

(a) Simulate the operation of a base-10 machine with a capacity of four digits per number and calculate the difference $(18)_{10} - (6)_{10}$ with tens complements.

(b) Simulate the operation of a base-2 machine with a capacity of five digits per number and calculate the difference $(01101)_2 - (01010)_2$ with twos complements.

(c) Solve part (b) with a ones complement and end-around carry.

Solution

(a) The tens complement of 6 is

$$(6)_{10}^* = (10)^4 - 6 = 10{,}000 - 6 = 9994$$

Using Eq. 12.10,

$$18 - 6 = 18 + (6)_{10}^* = 18 + 9994 = 10{,}012$$

However, the machine has a maximum capacity of four digits. Therefore, the leading 1 is dropped, leaving 0012 as the answer.

(b) The twos complement of $(01010)_2$ is

$$\begin{aligned} N_2^* &= (2)^5 - N = (32)_{10} - N \\ &= (100000)_2 - (01010)_2 \\ &= (10110)_2 \end{aligned}$$

From Eq. 12.10,

$$\begin{aligned} (01101)_2 - (01010)_2 &= (01101)_2 + (10110)_2 \\ &= (100011)_2 \end{aligned}$$

Since the machine has a capacity of only five bits, the leftmost bit is dropped, leaving $(00011)_2$ as the difference.

(c) The ones complement is found by reversing all the digits.

$$(01010)_1^* = (10101)_2$$

Adding the ones complement,

$$(01101)_2 + (10101)_2 = (100010)_2$$

The leading bit is the overflow bit, which is removed and added to give the difference.

$$(00010)_2 + (1)_2 = (00011)_2$$

10. COMPUTER REPRESENTATION OF NEGATIVE NUMBERS

On paper, a minus sign indicates a negative number. This representation is not possible in a machine. Hence, one of the n digits, usually the MSB, is reserved for sign representation. (This reduces the machine's capacity to represent numbers to $n - 1$ bits per number.) It is arbitrary whether the sign bit is 1 or 0 for negative numbers as long as the MSB is different for positive and negative numbers.

The ones complement is ideal for forming a negative number since it automatically reverses the MSB. For example, $(00011)_2$ is a five-bit representation of decimal 3. The ones complement is $(11100)_2$, which is recognized as a negative number because the MSB is 1.

Example 12.6

Simulate the operation of a six-digit binary machine that uses ones complements for negative numbers.

(a) What is the machine representation of $(-27)_{10}$? (b) What is the decimal equivalent of the twos complement of $(-27)_{10}$? (c) What is the decimal equivalent of the ones complement of $(-27)_{10}$?

Solution

(a) $(27)_{10} = (011011)_2$. The negative of this number is the same as the ones complement: $(100100)_2$.

(b) The twos complement is one more than the ones complement. (See Eq. 12.7 and Eq. 12.8.) Therefore, the twos complement is

$$(100100)_2 + 1 = (100101)_2$$

This represents $(-26)_{10}$.

(c) From Eq. 12.9, the complement of a complement of a number is the original number. Therefore, the decimal equivalent is -27.

13 Numerical Analysis

Content in blue refers to the *NCEES Handbook*.

1. NUMERICAL METHODS

Although the roots of second-degree polynomials are easily found by a variety of methods (by factoring, completing the square, or using the quadratic equation), easy methods of solving cubic and higher-order equations exist only for specialized cases. However, cubic and higher-order equations occur frequently in engineering, and they are difficult to factor. Trial and error solutions, including graphing, are usually satisfactory for finding only the general region in which the root occurs.

Numerical analysis is a general subject that covers, among other things, iterative methods for evaluating roots to equations. The most efficient numerical methods are too complex to present and, in any case, work by hand. However, some of the simpler methods are presented here. Except in critical problems that must be solved in real time, a few extra calculator or computer iterations will make no difference.[1]

2. FINDING ROOTS: BISECTION METHOD

The *bisection method* is an iterative method that "brackets" ("straddles") an interval containing the *root* or *zero* of a particular equation.[2] The size of the interval is halved after each iteration. As the method's name suggests, the best estimate of the root after any iteration is the midpoint of the interval. The maximum error is half the interval length. The procedure continues until the size of the maximum error is "acceptable."[3]

The disadvantages of the bisection method are (a) the slowness in converging to the root, (b) the need to know the interval containing the root before starting, and (c) the inability to determine the existence of or find other real roots in the starting interval.

The bisection method starts with two values of the independent variable, $x = L_0$ and $x = R_0$, which straddle a root. Since the function passes through zero at a root, $f(L_0)$ and $f(R_0)$ will almost always have opposite signs. The following algorithm describes the remainder of the bisection method.

Let n be the iteration number. Then, for $n = 0, 1, 2, \ldots$, perform the following steps until sufficient accuracy is attained.

step 1: Set $m = \frac{1}{2}(L_n + R_n)$.

step 2: Calculate $f(m)$.

step 3: If $f(L_n)f(m) \le 0$, set $L_{n+1} = L_n$ and $R_{n+1} = m$. Otherwise, set $L_{n+1} = m$ and $R_{n+1} = R_n$.

step 4: $f(x)$ has at least one root in the interval $[L_{n+1}, R_{n+1}]$. The estimated value of that root, x^*, is

$$x^* \approx \frac{1}{2}(L_{n+1} + R_{n+1})$$

The maximum error is $\frac{1}{2}(R_{n+1} - L_{n+1})$.

Example 13.1

Use two iterations of the bisection method to find a root of

$$f(x) = x^3 - 2x - 7$$

Solution

The first step is to find L_0 and R_0, which are the values of x that straddle a root and have opposite signs. A table can be made and values of $f(x)$ calculated for random values of x.

[1]Most advanced hand-held calculators have "root finder" functions that use numerical methods to iteratively solve equations.

[2]The equation does not have to be a pure polynomial. The bisection method requires only that the equation be defined and determinable at all points in the interval.

[3]The bisection method is not a closed method. Unless the root actually falls on the midpoint of one iteration's interval, the method continues indefinitely. Eventually, the magnitude of the maximum error is small enough not to matter.

x	−2	−1	0	+1	+2	+3
$f(x)$	−11	−6	−7	−8	−3	+14

Since $f(x)$ changes sign between $x=2$ and $x=3$, $L_0=2$ and $R_0=3$.

First iteration, $n=0$:

$$m=\tfrac{1}{2}(L_n+R_n)=\left(\frac{1}{2}\right)(2+3)=2.5$$

$$f(2.5)=x^3-2x-7=(2.5)^3-(2)(2.5)-7=3.625$$

Since $f(2.5)$ is positive, a root must exist in the interval [2, 2.5]. Therefore, $L_1=2$ and $R_1=2.5$. Or, using step 3, $f(2)f(2.5)=(-3)(3.625)=-10.875\leq 0$. So, $L_1=2$, and $R_1=2.5$. At this point, the best estimate of the root is

$$x^*\approx\left(\frac{1}{2}\right)(2+2.5)=2.25$$

The maximum error is $\left(\frac{1}{2}\right)(2.5-2)=0.25$.

Second iteration, $n=1$:

$$m=\left(\frac{1}{2}\right)(2+2.5)=2.25$$

$$f(2.25)=(2.25)^3-(2)(2.25)-7=-0.1094$$

Since $f(2.25)$ is negative, a root must exist in the interval [2.25, 2.5]. Or, using step 3, $f(2)f(2.25)=(-3)(-0.1096)=0.3288>0$. Therefore, $L_2=2.25$, and $R_2=2.5$. The best estimate of the root is

$$x^*\approx\tfrac{1}{2}(L_{n+1}+R_{n+1})\approx\left(\frac{1}{2}\right)(2.25+2.5)=2.375$$

The maximum error is $\left(\frac{1}{2}\right)(2.5-2.25)=0.125$.

3. FINDING ROOTS: NEWTON'S METHOD

Many other methods have been developed to overcome one or more of the disadvantages of the bisection method. These methods have their own disadvantages.[4]

Newton's method is a particular form of *fixed-point iteration*. In this sense, "fixed point" is often used as a synonym for "root" or "zero." Fixed-point iterations get their name from functions with the characteristic property $x=g(x)$ such that the limit of $g(x)$ is the fixed point (i.e., is the root).

All fixed-point techniques require a starting point. Preferably, the starting point will be close to the actual root.[5] And, while Newton's method converges quickly, it requires the function to be continuously differentiable.

Newton's method algorithm is simple. At each iteration ($n=0, 1, 2$, etc.), Eq. 13.1 estimates the root. The maximum error is determined by looking at how much the estimate changes after each iteration. If the change between the previous and current estimates (representing the magnitude of error in the estimate) is too large, the current estimate is used as the independent variable for the subsequent iteration.[6]

$$x_{n+1}=g(x_n)=x_n-\frac{f(x_n)}{f'(x_n)} \qquad 13.1$$

Example 13.2

Solve Ex. 13.1 using two iterations of Newton's method. Use $x_0=2$.

Solution

The function and its first derivative are

$$f(x)=x^3-2x-7$$

$$f'(x)=3x^2-2$$

First iteration, $n=0$:

$$x_0=2$$

$$f(x_0)=f(2)=(2)^3-(2)(2)-7=-3$$

$$f'(x_0)=f'(2)=(3)(2)^2-2=10$$

$$x_1=x_0-\frac{f(x_0)}{f'(x_0)}=2-\frac{-3}{10}=2.3$$

Second iteration, $n=1$:

$$x_1=2.3$$

$$f(x_1)=(2.3)^3-(2)(2.3)-7=0.567$$

[4]The *regula falsi (false position) method* converges faster than the bisection method but is unable to specify a small interval containing the root. The *secant method* is prone to round-off errors and gives no indication of the remaining distance to the root.

[5]Theoretically, the only penalty for choosing a starting point too far away from the root will be a slower convergence to the root.

[6]Actually, the theory defining the maximum error is more definite than this. For example, for a large enough value of n, the error decreases approximately linearly. Therefore, the consecutive values of x_n converge linearly to the root as well.

$$f'(x_1) = (3)(2.3)^2 - 2 = 13.87$$

$$x_2 = x_1 - \frac{f(x_1)}{f'(x_1)} = 2.3 - \frac{0.567}{13.87} = 2.259$$

4. NONLINEAR INTERPOLATION: LAGRANGIAN INTERPOLATING POLYNOMIAL

Interpolating between two points of known data is common in engineering. Primarily due to its simplicity and speed, straight-line interpolation is used most often. Even if more than two points on the curve are explicitly known, they are not used. Since straight-line interpolation ignores all but two of the points on the curve, it ignores the effects of curvature.

A more powerful technique that accounts for the curvature is the *Lagrangian interpolating polynomial.* This method uses an nth degree parabola (polynomial) as the interpolating curve.[7] This method requires that $f(x)$ be continuous and real-valued on the interval $[x_0, x_n]$ and that $n+1$ values of $f(x)$ are known corresponding to x_0, x_1, x_2, ..., x_n.

The procedure for calculating $f(x)$ at some intermediate point, x, starts by calculating the Lagrangian interpolating polynomial for each known point.

$$L_k(x^*) = \prod_{\substack{i=0\\ i\neq k}}^{n} \frac{x^* - x_i}{x_k - x_i} \qquad 13.2$$

The value of $f(x)$ at x^* is calculated from Eq. 13.3.

$$f(x^*) = \sum_{k=0}^{n} f(x_k) L_k(x^*) \qquad 13.3$$

The Lagrangian interpolating polynomial has two primary disadvantages. The first is that a large number of additions and multiplications are needed.[8] The second is that the method does not indicate how many interpolating points should be used. Other interpolating methods have been developed that overcome these disadvantages.[9]

Example 13.3

A real-valued function has the following values.

$$f(1) = 3.5709$$

$$f(4) = 3.5727$$

$$f(6) = 3.5751$$

Use the Lagrangian interpolating polynomial to estimate the value of the function at 3.5.

Solution

The procedure for applying Eq. 13.2, the Lagrangian interpolating polynomial, is illustrated in tabular form. Notice that the term corresponding to $i = k$ is omitted from the product.

$k = 0$: $\quad i = 0 \quad i = 1 \quad i = 2$

$$L_0(3.5) = \left(\frac{3.5-1}{1-1}\right)\left(\frac{3.5-4}{1-4}\right)\left(\frac{3.5-6}{1-6}\right) = 0.08333$$

(first factor crossed out)

$k = 1$:

$$L_1(3.5) = \left(\frac{3.5-1}{4-1}\right)\left(\frac{3.5-4}{4-4}\right)\left(\frac{3.5-6}{4-6}\right) = 1.04167$$

(second factor crossed out)

$k = 2$:

$$L_2(3.5) = \left(\frac{3.5-1}{6-1}\right)\left(\frac{3.5-4}{6-4}\right)\left(\frac{3.5-6}{6-6}\right) = -0.12500$$

(third factor crossed out)

Equation 13.3 is used to calculate the estimate.

$$\begin{aligned} f(3.5) &= (3.5709)(0.08333) + (3.5727)(1.04167) \\ &\quad + (3.5751)(-0.12500) \\ &= 3.57225 \end{aligned}$$

5. NONLINEAR INTERPOLATION: NEWTON'S INTERPOLATING POLYNOMIAL

Newton's form of the interpolating polynomial is more efficient than the Lagrangian method of interpolating between known points.[10] Given $n+1$ known points for $f(x)$, the *Newton form of the interpolating polynomial* is

$$f(x^*) = \sum_{i=0}^{n}\left(f[x_0, x_1, \ldots, x_i] \prod_{j=0}^{i-1}(x^* - x_j)\right) \qquad 13.4$$

$f[x_0, x_1, \ldots, x_i]$ is known as the ith *divided difference.*

$$f[x_0, x_1, \ldots, x_i] = \sum_{k=0}^{i}\left(\frac{f(x_k)}{\begin{array}{c}(x_k - x_0)\cdots \\ \times(x_k - x_{k-1}) \\ \times(x_k - x_{k+1}) \\ \times \cdots (x_k - x_i)\end{array}}\right) \qquad 13.5$$

[7] The Lagrangian interpolating polynomial reduces to straight-line interpolation if only two points are used.

[8] As with the numerical methods for finding roots previously discussed, the number of calculations probably will not be an issue if the work is performed by a calculator or computer.

[9] Other common methods for performing interpolation include the *Newton form* and *divided difference table.*

[10] In this case, "efficiency" relates to the ease in adding new known points without having to repeat all previous calculations.

It is necessary to define the following two terms.

$$f[x_0] = f(x_0) \qquad 13.6$$

$$\prod (x^* - x_j) = 1 \quad [i = 0] \qquad 13.7$$

Example 13.4

Repeat Ex. 13.3 using Newton's form of the interpolating polynomial.

Solution

Since there are $n + 1 = 3$ data points, $n = 2$. Evaluate the terms for $i = 0$, 1, and 2.

$i = 0$:

$$f[x_0] \prod_{j=0}^{-1} (x^* - x_j) = f[x_0](1) = f(x_0)$$

$i = 1$:

$$f[x_0, x_1] \prod_{j=0}^{0} (x^* - x_j) = f[x_0, x_1](x^* - x_0)$$

$$f[x_0, x_1] = \frac{f(x_0)}{x_0 - x_1} + \frac{f(x_1)}{x_1 - x_0}$$

$i = 2$:

$$f[x_0, x_1, x_2] \prod_{j=0}^{1} (x^* - x_j) = f[x_0, x_1, x_2](x^* - x_0)(x^* - x_1)$$

$$\begin{aligned} f[x_0, x_1, x_2] = &\frac{f(x_0)}{(x_0 - x_1)(x_0 - x_2)} \\ &+ \frac{f(x_1)}{(x_1 - x_0)(x_1 - x_2)} \\ &+ \frac{f(x_2)}{(x_2 - x_0)(x_2 - x_1)} \end{aligned}$$

Use Eq. 13.4. Substitute known values.

$$\begin{aligned} f(3.5) &= \sum_{i=0}^{n} \left(f[x_0, x_1, \ldots, x_i] \prod_{j=0}^{i-1} (x^* - x_j) \right) \\ &= 3.5709 + \left(\frac{3.5709}{1 - 4} + \frac{3.5727}{4 - 1} \right)(3.5 - 1) \\ &\quad + \left(\frac{3.5709}{(1 - 4)(1 - 6)} + \frac{3.5727}{(4 - 1)(4 - 6)} \right. \\ &\quad \left. + \frac{3.5751}{(6 - 1)(6 - 4)} \right) \\ &\quad \times (3.5 - 1)(3.5 - 4) \\ &= 3.57225 \end{aligned}$$

This answer is the same as that determined in Ex. 13.3.

Topic II: Fluids

Chapter

14 Fluid Properties

Content in blue refers to the *NCEES Handbook*.

NCEES EXAM SPECIFICATIONS AND RELATED CONTENT

THERMAL AND FLUID SYSTEMS EXAM

I.B.1 Fluid Mechanics: Fluid Properties
- 3. Fluid Pressure and Vacuum
- 4. Density
- 5. Specific Volume
- 6. Specific Gravity
- 7. Specific Weight
- 8. Mole Fraction
- 9. Viscosity
- 10. Kinematic Viscosity
- 18. Bulk Modulus
- 19. Speed of Sound

1. CHARACTERISTICS OF A FLUID

Liquids and gases can both be categorized as fluids, although this chapter is primarily concerned with incompressible liquids. There are certain characteristics shared by all fluids, and these characteristics can be used, if necessary, to distinguish between liquids and gases.[1]

- *compressibility:* Liquids are only slightly compressible and are assumed to be incompressible for most purposes. Gases are highly compressible.
- *shear resistance:* Liquids and gases cannot support shear, and they deform continuously to minimize applied shear forces.
- *shape and volume:* As a consequence of their inability to support shear forces, liquids and gases take on the shapes of their containers. Only liquids have free surfaces. Liquids have fixed volumes, regardless of their container volumes, and these volumes are not significantly affected by temperature and pressure. Unlike liquids, gases take on the volumes of their containers. If allowed to do so, gas densities will change as temperature and pressure are varied.
- *resistance to motion:* Due to viscosity, liquids resist instantaneous changes in velocity, but the resistance stops when liquid motion stops. Gases have very low viscosities.
- *molecular spacing:* Molecules in liquids are relatively close together and are held together with strong forces of attraction. Liquid molecules have low kinetic energy. The distance each liquid molecule travels between collisions is small. In gases, the molecules are relatively far apart and the attractive forces are weak. Kinetic energy of the molecules is high. Gas molecules travel larger distances between collisions.
- *pressure:* The pressure at a point in a fluid is the same in all directions. Pressure exerted by a fluid on a solid surface (e.g., container wall) is always normal to that surface.

2. TYPES OF FLUIDS

For computational convenience, fluids are generally divided into two categories: ideal fluids and real fluids. (See Fig. 14.1.) *Ideal fluids* are assumed to have no viscosity (and therefore, no resistance to shear), be incompressible, and have uniform velocity distributions when flowing. In an ideal fluid, there is no friction between moving layers of fluid, and there are no eddy currents or turbulence.

[1] The differences between liquids and gases become smaller as temperature and pressure are increased. Gas and liquid properties become the same at the critical temperature and pressure.

Figure 14.1 *Types of Fluids*

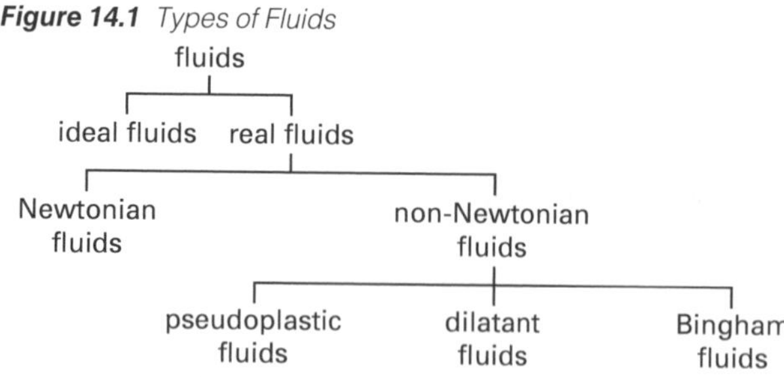

Real fluids exhibit finite viscosities and nonuniform velocity distributions, are compressible, and experience friction and turbulence in flow. Real fluids are further divided into *Newtonian fluids* and *non-Newtonian fluids*, depending on their viscous behavior. The differences between Newtonian and non-Newtonian fluids are described in Sec. 14.9.

For convenience, most fluid problems assume real fluids with Newtonian characteristics. This is an appropriate assumption for water, air, gases, steam, and other simple fluids (alcohol, gasoline, acid solutions, etc.). However, slurries, pastes, gels, suspensions, and polymer/electrolyte solutions may not behave according to simple fluid relationships.

3. FLUID PRESSURE AND VACUUM

In the English system, fluid pressure is measured in pounds per square inch (lbf/in^2 or psi) and pounds per square foot (lbf/ft^2 or psf), although tons (2000 pounds) per square foot (tsf) is occasionally used. In SI units, pressure is measured in pascals (Pa). Because a pascal is very small, kilopascals (kPa) and megapascals (MPa) are usually used. Other units of pressure include bars, millibars, atmospheres, inches and feet of water, torrs, and millimeters, centimeters, and inches of mercury. (See Fig. 14.2.)

Fluid pressures are measured with respect to two pressure references: zero pressure and atmospheric pressure. Pressures measured with respect to a true zero pressure reference are known as *absolute pressures.* Pressures measured with respect to atmospheric pressure are known as *gage pressures.*[2] Most pressure gauges read the excess of the test pressure over atmospheric pressure (i.e., the gage pressure). To distinguish between these two pressure measurements, the letters "a" and "g" are traditionally added to the unit symbols in the English unit system (e.g., 14.7 psia and 4015 psfg). For SI units, the actual words "gage" and "absolute" can be added to the measurement (e.g., 25.1 kPa absolute). Alternatively, the pressure is assumed to be absolute unless the "g" is used (e.g., 15 kPag or 98 barg).

Absolute and gage pressures are related by Eq. 14.1. It should be mentioned that $p_{\text{patmospheric}}$ in Eq. 14.1 is the actual atmospheric pressure existing when the gage measurement is taken. It is not standard atmospheric pressure, unless that pressure is implicitly or explicitly applicable. Also, since a barometer measures atmospheric pressure, *barometric pressure* is synonymous with atmospheric pressure. Table 14.1 lists standard atmospheric pressure in various units.

Figure 14.2 *Relative Sizes of Pressure Units*

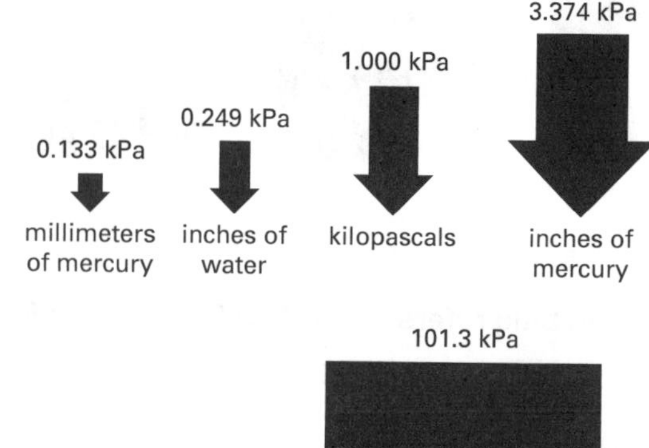

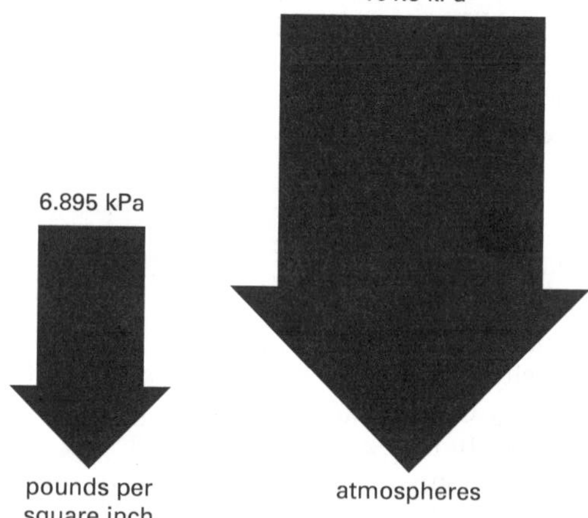

Pressure Field in a Static Liquid

$$p_{\text{absolute}} = p_{\text{atmospheric}} + p_{\text{gage}} \qquad 14.1$$

Table 14.1 *Standard Atmospheric Pressure*

1.000 atm	(atmosphere)
14.696 psia	(pounds per square inch absolute)
2116.2 psfa	(pounds per square foot absolute)
407.1 in wg	(inches of water, inches water gage)
33.93 ft wg	(feet of water, feet water gage)
29.921 in Hg	(inches of mercury)
760.0 mm Hg	(millimeters of mercury)
760.0 torr	
1.013 bars	
1013 millibars	
1.013×10^5 Pa	(pascals)
101.3 kPa	(kilopascals)

A *vacuum* measurement is implicitly a pressure below atmospheric (i.e., a negative gage pressure). It must be assumed that any measured quantity given as a vacuum is a quantity to be subtracted from the atmospheric pressure. For example, when a condenser is operating

[2]The spelling *gage* persists even though pressures are measured with *gauges.* In some countries, the term *meter pressure* is used instead of gage pressure.

with a vacuum of 4.0 in Hg (4 in of mercury), the absolute pressure is 29.92 in Hg − 4.0 in Hg = 25.92 in Hg. Vacuums are generally stated as positive numbers.

Pressure Field in a Static Liquid

$$p_{\text{absolute}} = p_{\text{atmospheric}} - p_{\text{vacuum}} \qquad 14.2$$

A difference in two pressures may be reported with units of *psid* (i.e., a *differential* in psi).

4. DENSITY

The *density*, ρ, of a fluid is its mass per unit volume.[3]

Density, Specific Weight, and Specific Gravity

$$\rho = \frac{m}{V} \qquad 14.3$$

In SI units, density is measured in kg/m^3. In a consistent English system, density would be measured in slugs/ft^3, even though fluid density is typically reported in lbm/ft^3.

The density of a fluid in a liquid form is usually given, known in advance, or easily obtained from tables in any one of a number of sources. (See Table 14.2.) Most English fluid data are reported on a per pound basis, and the data included in this book follow that tradition. To convert pounds to slugs, divide by g_c.

$$\rho_{\text{slugs}} = \frac{\rho_{\text{lbm}}}{g_c} \qquad 14.4$$

The density of an ideal gas can also be found from the specific gas constant and the ideal gas law.

$$\rho = \frac{p}{RT} \qquad 14.5$$

Table 14.2 *Approximate Densities of Common Fluids*

fluid	lbm/ft^3	kg/m^3
air (STP)	0.0807	1.29
air (70°F, 1 atm)	0.075	1.20
alcohol	49.3	790
ammonia	38	602
gasoline	44.9	720
glycerin	78.8	1260
mercury	848	13600
water	62.4	1000

(Multiply lbm/ft^3 by 16.01 to obtain kg/m^3.)

Example 14.1

The density of water is typically taken as $62.4\ \text{lbm/ft}^3$ for engineering problems where greater accuracy is not required. What is the value in (a) slugs/ft^3 and (b) kg/m^3?

Solution

(a) Equation 14.4 can be used to calculate the slug-density of water.

$$\rho = \frac{\rho_{\text{lbm}}}{g_c} = \frac{62.4\ \frac{\text{lbm}}{\text{ft}^3}}{32.2\ \frac{\text{lbm-ft}}{\text{lbf-sec}^2}} = 1.94\ \text{lbf-sec}^2/\text{ft-ft}^3$$
$$= 1.94\ \text{slugs/ft}^3$$

(b) The conversion between lbm/ft^3 and kg/m^3 is approximately 16.0, derived as follows.

$$\rho = \left(62.4\ \frac{\text{lbm}}{\text{ft}^3}\right)\left(\frac{35.31\ \frac{\text{ft}^3}{\text{m}^3}}{2.205\ \frac{\text{lbm}}{\text{kg}}}\right)$$
$$= \left(62.4\ \frac{\text{lbm}}{\text{ft}^3}\right)\left(16.01\ \frac{\text{kg-ft}^3}{\text{m}^3\text{-lbm}}\right)$$
$$= 999\ \text{kg/m}^3$$

In SI problems, it is common to take the density of water as $1000\ \text{kg/m}^3$.

5. SPECIFIC VOLUME

Specific volume, v, is the volume occupied by a unit mass of fluid.[4]

Properties of Single-Component Systems: Definitions

$$v = \frac{V}{m} \qquad 14.6$$

Specific volume is the reciprocal of density, so typical units will be ft^3/lbm, ft^3/lbmol, or m^3/kg.[5]

$$v = \frac{1}{\rho} \qquad 14.7$$

[3]Mass is an absolute property of a substance. Weight is not absolute, since it depends on the local gravity. The equations using γ that result (such as Bernoulli's equation) cannot be used with SI data, since the equations are not consistent. Thus, engineers end up with two different equations for the same thing.

[4]Care must be taken to distinguish between the italic v used for specific volume and the nonitalic v used for velocity.

[5]Units of ft^3/slug are also possible, but this combination of units is almost never encountered.

As with density, the specific volume of an ideal gas can be found from the specific gas constant and the ideal gas law.

$$v = \frac{RT}{p} \qquad 14.8$$

6. SPECIFIC GRAVITY

Specific gravity, SG, is a dimensionless ratio of a fluid's density to some standard reference density.[6] For liquids and solids, the reference is the density of pure water. There is some variation in this reference density, however, since the temperature at which the water density is evaluated is not standardized. Temperatures of 39.2°F (4°C), 60°F (16.5°C), and 70°F (21.1°C) have been used.[7]

Fortunately, the density of water is the same to three significant digits over the normal ambient temperature range: 62.4 lbm/ft^3 or 1000 kg/m^3. However, to be precise, the temperature of both the fluid and water should be specified (e.g., "... the specific gravity of the 20°C fluid is 1.05 referred to 4°C water ...").

Density, Specific Weight, and Specific Gravity

$$\text{SG}_{\text{liquid}} = \frac{\gamma_{\text{liquid}}}{\gamma_{\text{water}}} = \frac{\rho_{\text{liquid}}}{\rho_{\text{water}}} \qquad 14.9$$

Since the SI density of water is very nearly 1.000 g/cm^3 (1000 kg/m^3), the numerical values of density in g/cm^3 and specific gravity are the same. Such is not the case with English units.

The standard reference used to calculate the specific gravity of gases is the density of air. Since the density of a gas depends on temperature and pressure, both must be specified for the gas and air (i.e., two temperatures and two pressures must be specified). While STP (standard temperature and pressure) conditions are commonly specified, they are not universal.[8] Table 14.3 lists several common sets of standard conditions.

$$\text{SG}_{\text{gas}} = \frac{\gamma_{\text{gas}}}{\gamma_{\text{air}}} = \frac{\rho_{\text{gas}}}{\rho_{\text{air}}} \qquad 14.10$$

Table 14.3 *Commonly Quoted Values of Standard Temperature and Pressure*

system	temperature	pressure
SI	273.15K	101.325 kPa
scientific	0.0°C	760 mm Hg
U.S. engineering	32°F	14.696 psia
natural gas industry (U.S.)	60°F	14.65, 14.73, or 15.025 psia
natural gas industry (Canada)	60°F	14.696 psia

If it is known or implied that the temperature and pressure of the air and gas are the same, the specific gravity of the gas will be equal to the ratio of molecular weights and the inverse ratio of specific gas constants. The density of air evaluated at STP is listed in Table 14.2. At 70°F (21.1°C) and 1.0 atm, the density is approximately 0.075 lbm/ft^3 (1.20 kg/m^3).

$$\text{SG}_{\text{gas}} = \frac{M_{\text{gas}}}{M_{\text{air}}} = \frac{M_{\text{gas}}}{29.0} = \frac{R_{\text{air}}}{R_{\text{gas}}} = \frac{53.3\ \frac{\text{ft-lbf}}{\text{lbm-°R}}}{R_{\text{gas}}} \qquad 14.11$$

Specific gravities of petroleum liquids and aqueous solutions (of acid, antifreeze, salts, etc.) can be determined by use of a *hydrometer*. (See Fig. 14.3.) In its simplest form, a hydrometer is constructed as a graduated scale weighted at one end so it will float vertically. The height at which the hydrometer floats depends on the density of the fluid, and the graduated scale can be calibrated directly in specific gravity.[9]

There are two standardized hydrometer scales (i.e., methods for calibrating the hydrometer stem).[10] Both state specific gravity in degrees, although temperature is not being measured. The *American Petroleum Institute* (API) scale (°API) may be used with all liquids, not only with oils or other hydrocarbons. For the specific gravity value, a standard reference temperature of 60°F (15.6°C) is implied for both the liquid and the water.

$$\text{°API} = \frac{141.5}{\text{SG}} - 131.5 \qquad 14.12$$

$$\text{SG} = \frac{141.5}{\text{°API} + 131.5} \qquad 14.13$$

[6] The symbols S.G., sp.gr., S, and G are also used. In fact, petroleum engineers in the United States use γ, a symbol that civil engineers use for specific weight. There is no standard engineering symbol for specific gravity.

[7] Density of liquids is sufficiently independent of pressure to make consideration of pressure in specific gravity calculations unnecessary.

[8] The abbreviation "SC" (standard conditions) is interchangeable with "STP."

[9] This is a direct result of the buoyancy principle of Archimedes.

[10] In addition to °Be and °API mentioned in this chapter, the *Twaddell scale* (°Tw) is used in chemical processing, the *Brix* and *Balling scales* are used in the sugar industry, and the *Salometer scale* is used to measure salt (NaCl and $CaCl_2$) solutions.

Figure 14.3 Hydrometer

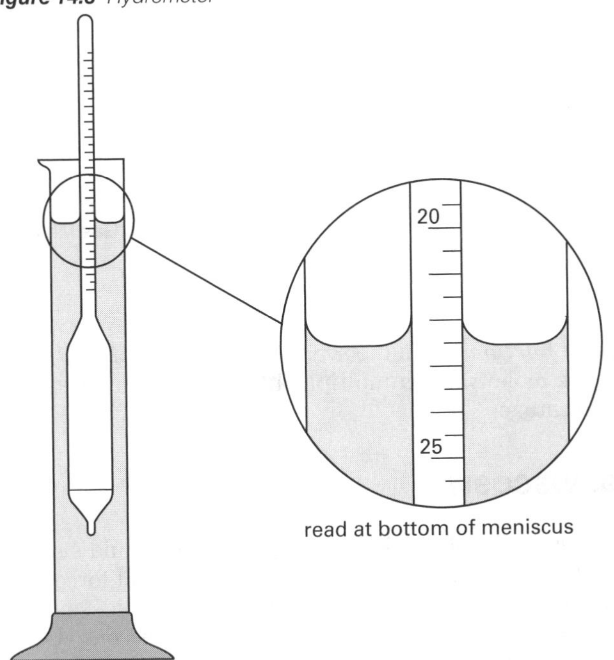

The *Baumé scale* (°Be) is used in the wine, honey, and acid industries. It is somewhat confusing because there are actually two Baumé scales—one for liquids heavier than water and another for liquids lighter than water. (There is also a discontinuity in the scales at SG = 1.00.) As with the API scale, the specific gravity value assumes 60°F (15.6°C) is the standard temperature for both scales.

$$\text{SG} = \frac{140.0}{130.0 + {}^\circ\text{Be}} \quad [\text{SG} < 1.00] \qquad 14.14$$

$$\text{SG} = \frac{145.0}{145.0 - {}^\circ\text{Be}} \quad [\text{SG} > 1.00] \qquad 14.15$$

Example 14.2

Determine the specific gravity of carbon dioxide gas (molecular weight = 44) at 66°C (150°F) and 138 kPa (20 psia) using STP air as a reference. The specific gas constant of carbon dioxide is 35.1 ft-lbf/lbm-°R.

SI Solution

Since the specific gas constant for carbon dioxide was not given in SI units, it must be calculated from the universal gas constant.

$$R = \frac{\overline{R}}{M} = \frac{8314.47\ \frac{\text{J}}{\text{kmol·K}}}{44\ \frac{\text{kg}}{\text{kmol}}} = 189\ \text{J/kg·K}$$

$$\rho = \frac{p}{RT} = \frac{1.38 \times 10^5\ \text{Pa}}{\left(189\ \frac{\text{J}}{\text{kg·K}}\right)(66^\circ\text{C} + 273^\circ)} = 2.15\ \text{kg/m}^3$$

From Table 14.2, the density of STP air is 1.29 kg/m^3. From Eq. 14.10, the specific gravity of carbon dioxide at the conditions given is

$$\text{SG} = \frac{\rho_{\text{gas}}}{\rho_{\text{air}}} = \frac{2.15\ \frac{\text{kg}}{\text{m}^3}}{1.29\ \frac{\text{kg}}{\text{m}^3}} = 1.67$$

Customary U.S. Solution

Since the conditions of the carbon dioxide and air are different, Eq. 14.11 cannot be used. Therefore, it is necessary to calculate the density of the carbon dioxide from Eq. 14.5. Absolute temperature (degrees Rankine) must be used. The density is

$$\rho = \frac{p}{RT} = \frac{\left(20\ \frac{\text{lbf}}{\text{in}^2}\right)\left(12\ \frac{\text{in}}{\text{ft}}\right)^2}{\left(35.1\ \frac{\text{ft-lbf}}{\text{lbm-}^\circ\text{R}}\right)(150^\circ\text{F} + 460^\circ)} = 0.135\ \text{lbm/ft}^3$$

From Table 14.2, the density of STP air is 0.0807 lbm/ft^3. From Eq. 14.10, the specific gravity of carbon dioxide at the conditions given is

$$\text{SG} = \frac{\rho_{\text{gas}}}{\rho_{\text{air}}} = \frac{0.135\ \frac{\text{lbm}}{\text{ft}^3}}{0.0807\ \frac{\text{lbm}}{\text{ft}^3}} = 1.67$$

7. SPECIFIC WEIGHT

Specific weight (*unit weight*), γ, is the weight of fluid per unit volume. The use of specific weight is most often encountered in civil engineering projects in the United States, where it is commonly called "density." The usual units of specific weight are lbf/ft^3.[11] Specific weight is not an absolute property of a fluid, since it depends not only on the fluid, but on the local gravitational field as well.

[11]Notice that the units are lbf/ft^3, not lbm/ft^3. Pound-mass (lbm) is a mass unit, not a weight (force) unit.

Density, Specific Weight, and Specific Gravity

$$\gamma = \frac{W}{V} = \rho g \quad \text{[SI]} \qquad 14.16(a)$$

$$\gamma = \frac{W}{V} = \frac{\rho g}{g_c} \quad \text{[U.S.]} \qquad 14.16(b)$$

If the gravitational acceleration is 32.2 ft/sec^2, as it is almost everywhere on earth, the specific weight in lbf/ft^3 will be numerically equal to the density in lbm/ft^3. This concept is demonstrated in Ex. 14.3.

Example 14.3

What is the sea level (g = 32.2 ft/sec^2) specific weight (in lbf/ft^3) of liquids with densities of (a) 1.95 slug/ft^3 and (b) 58.3 lbm/ft^3?

Solution

(a) Equation 14.16(a) can be used with any consistent set of units, including densities involving slugs.

Density, Specific Weight, and Specific Gravity

$$\begin{aligned} \gamma &= \rho g \\ &= \left(32.2\ \frac{\text{ft}}{\text{sec}^2}\right)\left(1.95\ \frac{\text{slug}}{\text{ft}^3}\right) \\ &= \left(32.2\ \frac{\text{ft}}{\text{sec}^2}\right)\left(1.95\ \frac{\text{lbf-sec}^2}{\text{ft-ft}^3}\right) \\ &= 62.8\ \text{lbf/ft}^3 \end{aligned}$$

(b) From Eq. 14.16(b),

$$\begin{aligned} \gamma &= \frac{\rho g}{g_c} \\ &= \left(58.3\ \frac{\text{lbm}}{\text{ft}^3}\right)\left(\frac{32.2\ \frac{\text{ft}}{\text{sec}^2}}{32.2\ \frac{\text{lbm-ft}}{\text{lbf-sec}^2}}\right) \\ &= 58.3\ \text{lbf/ft}^3 \end{aligned}$$

8. MOLE FRACTION

Mole fraction is an important parameter in many practical engineering problems, particularly in chemistry and chemical engineering. The composition of a fluid consisting of two or more distinctly different substances, A, B, C, and so on, can be described by the mole fractions, x_A, x_B, x_C, and so on, of each substance. (There are also other methods of specifying the composition.) The mole fraction of component A is the number of moles of that component, n_A, divided by the total number of moles in the combined fluid mixture, N.

$$x_A = \frac{N_A}{N_A + N_B + N_C + \cdots} \qquad 14.17$$

Ideal Gas Mixtures

$$x_i = \frac{N_i}{N} \qquad 14.18$$

Mole fraction is a number between 0 and 1. *Mole percent* is the mole fraction multiplied by 100%, expressed as a percentage.

9. VISCOSITY

The *viscosity* of a fluid is a measure of that fluid's resistance to flow when acted upon by an external force such as a pressure differential or gravity. Some fluids, such as heavy oils, jellies, and syrups, are very viscous. Other fluids, such as water, lighter hydrocarbons, and gases, are not as viscous. [**Absolute Viscosity (Left) and Kinematic Viscosity (Right) of Common Fluids at 1 atm**]

Most common liquids will flow more easily when their temperatures are raised. However, the behavior of a fluid when temperature, pressure, or stress is varied will depend on the type of fluid. The different types of fluids can be determined with a *sliding plate viscometer test*.[12]

Consider two plates of area A separated by a fluid with thickness y_0, as shown in Fig. 14.4. The bottom plate is fixed, and the top plate is kept in motion at a constant velocity, v_0, by a force, F.

Figure 14.4 *Sliding Plate Viscometer*

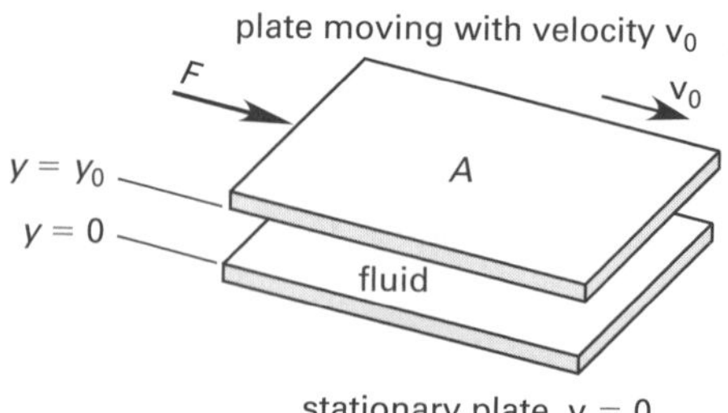

Experiments with water and most common fluids have shown that the force, F, required to maintain the

[12] This test is conceptually simple but is not always practical, since the liquid leaks out between the plates. In research work with liquids, it is common to determine viscosity with a *concentric cylinder viscometer*, also known as a *cup-and-bob viscometer*. Viscosities of perfect gases can be predicted by the kinetic theory of gases. Viscosity can also be measured by a *Saybolt viscometer*, which is essentially a container that allows a given quantity of fluid to leak out through one of two different-sized orifices.

velocity, v_0, is proportional to the velocity and the area and is inversely proportional to the separation of the plates. That is,

$$\frac{F}{A} \propto \frac{dv}{dy} \qquad 14.19$$

The constant of proportionality needed to make Eq. 14.19 an equality is the *absolute viscosity*, μ, also known as the *coefficient of viscosity*.[13] The reciprocal of absolute viscosity, $1/\mu$, is known as the *fluidity*.

$$\frac{F}{A} = \mu \frac{dv}{dy} \qquad 14.20$$

F/A is the *fluid shear stress*, τ. The quantity dv/dy (v_0/y_0) is known by various names, including *rate of strain*, *shear rate*, *velocity gradient*, and *rate of shear formation*. Equation 14.20 is known as *Newton's law of viscosity*, from which *Newtonian fluids* get their name. Sometimes Eq. 14.21 is written with a minus sign to compare viscous behavior with other behavior. However, the direction of positive shear stress is arbitrary. Equation 14.21 is simply the equation of a straight line.

Stress, Pressure, and Viscosity

$$\tau = \mu \frac{dv}{dy} \qquad 14.21$$

Not all fluids are Newtonian (although most common fluids are), and Eq. 14.21 is not universally applicable. Figure 14.5 (known as a *rheogram*) illustrates how differences in fluid shear stress behavior (at constant temperature and pressure) can be used to define Bingham, pseudoplastic, and dilatant fluids, as well as Newtonian fluids.

Gases, water, alcohol, and benzene are examples of Newtonian fluids. In fact, all liquids with a simple chemical formula are Newtonian. Also, most solutions of simple compounds, such as sugar and salt, are Newtonian. For a more viscous fluid, the straight line will be closer to the τ axis (i.e., the slope will be higher). (See Fig. 14.5.) For low-viscosity fluids, the straight line will be closer to the dv/dy axis (i.e., the slope will be lower).

Pseudoplastic fluids (muds, motor oils, polymer solutions, natural gums, and most slurries) exhibit viscosities that decrease with an increasing velocity gradient. Such fluids present no serious pumping problems.

Plastic materials, such as tomato catsup, behave similarly to pseudoplastic fluids once movement begins; that is, their viscosities decrease with agitation. However, a finite force must be applied before any fluid movement occurs.

Bingham fluids (Bingham plastics), typified by toothpaste, jellies, bread dough, and some slurries, are capable of indefinitely resisting a small shear stress but move easily when the stress becomes large—that is, Bingham fluids become pseudoplastic when the stress increases.

Figure 14.5 *Shear Stress Behavior for Different Types of Fluids*

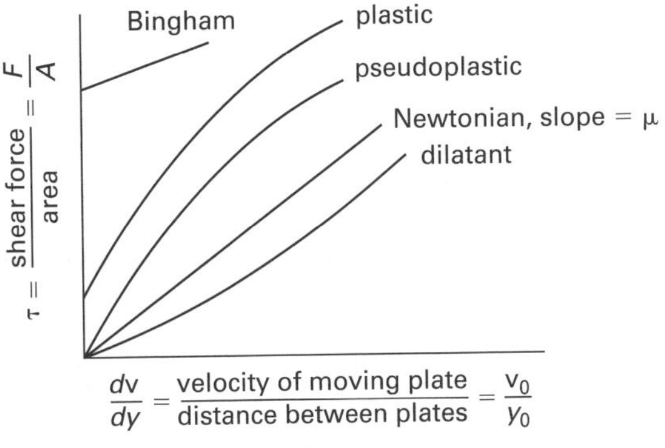

(a)

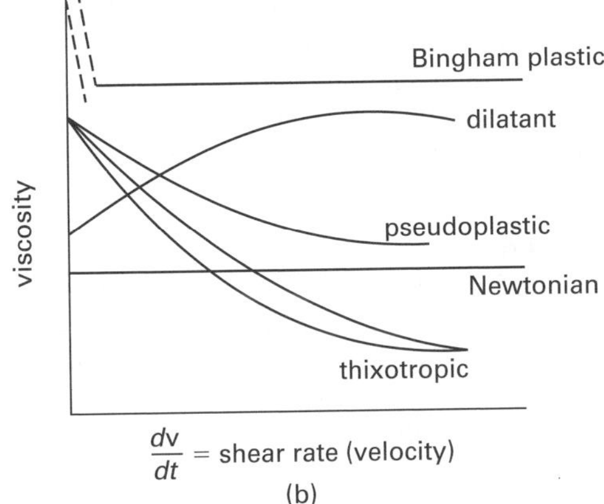

(b)

Dilatant fluids are rare but include clay slurries, various starches, some paints, milk chocolate with nuts, and other candy compounds. They exhibit viscosities that increase with increasing agitation (i.e., with increasing velocity gradients), but they return rapidly to their normal viscosity after the agitation ceases. Pump selection is critical for dilatant fluids because these fluids can become almost solid if the shear rate is high enough.

Viscosity can also change with time (all other conditions being constant). If viscosity decreases with time during agitation, the fluid is said to be a *thixotropic fluid*. If viscosity increases (usually up to a finite value) with time during agitation, the fluid is a *rheopectic fluid*. Viscosity does not change in time-independent fluids. *Colloidal materials*, such as gelatinous compounds, lotions, shampoos, and low-temperature solutions of soaps in water and oil, behave like *thixotropic liquids*—their viscosities decrease as the agitation continues. However, viscosity does not return to its original state after the agitation ceases.

Molecular cohesion is the dominating cause of viscosity in liquids. As the temperature of a liquid increases, these cohesive forces decrease, resulting in a decrease in viscosity.

[13]Another name for absolute viscosity is *dynamic viscosity*. The name *absolute viscosity* is preferred, if for no other reason than to avoid confusion with *kinematic viscosity*.

In gases, the dominant cause of viscosity is random collisions between gas molecules. This molecular agitation increases with increases in temperature. Therefore, viscosity in gases increases with temperature.

Although viscosity of liquids increases slightly with pressure, the increase is insignificant over moderate pressure ranges. Therefore, the absolute viscosity of both gases and liquids is usually considered to be essentially independent of pressure.[14]

The units of absolute viscosity, as derived from Eq. 14.21, are lbf-sec/ft². Such units are actually used in the English engineering system.[15] Absolute viscosity is measured in pascal-seconds (Pa·s) in SI units. Another common unit used throughout the world is the *poise* (abbreviated P), equal to a dyne·s/cm². These dimensions are the same primary dimensions as in the English system, $F\theta/L^2$ or $M/L\theta$, and are functionally the same as a g/cm·s. Since the poise is a large unit, the *centipoise* (abbreviated cP) scale is generally used. The viscosity of pure water at room temperature is approximately 1 cP.

Example 14.4

A liquid ($\mu = 5.2 \times 10^{-5}$ lbf-sec/ft²) is flowing in a rectangular duct. The equation of the symmetrical facial velocity distribution (in ft/sec) is approximately $\text{v} = 3y^{0.7}$ ft/sec, where y is measured in inches from the wall. (a) What is the velocity gradient at $y = 3.0$ in from the duct wall? (b) What is the shear stress in the fluid at that point?

Solution

(a) The velocity is not a linear function of y, so $d\text{v}/dy$ must be calculated as a derivative.

$$\frac{d\text{v}}{dy} = \frac{d}{dy}3y^{0.7} = (3)(0.7y^{-0.3}) = 2.1y^{-0.3}$$

At $y = 3$ in,

$$\frac{d\text{v}}{dy} = (2.1)(3)^{-0.3} = 1.51 \text{ ft/sec-in}$$

(b) From Eq. 14.21, the shear stress is

Stress, Pressure, and Viscosity

$$\tau = \mu\frac{d\text{v}}{dy} = \left(5.2 \times 10^{-5}\ \frac{\text{lbf-sec}}{\text{ft}^2}\right)\left(1.51\ \frac{\text{ft}}{\text{sec-in}}\right) \times \left(12\ \frac{\text{in}}{\text{ft}}\right) = 9.42 \times 10^{-4} \text{ lbf/ft}^2$$

10. KINEMATIC VISCOSITY

Another quantity with the name *viscosity* is the ratio of absolute viscosity to mass density. This combination of variables, known as *kinematic viscosity*, ν, appears sufficiently often in fluids and other problems as to warrant its own symbol and name. Typical units are ft²/sec and cm²/s (the *stoke*, St). It is also common to give kinematic viscosity in *centistokes*, cSt. The SI units of kinematic viscosity are m²/s.

Stress, Pressure, and Viscosity

$$\nu = \frac{\mu}{\rho} \quad \text{[SI]} \qquad 14.22(a)$$

$$\nu = \frac{\mu g_c}{\rho} = \frac{\mu g}{\gamma} \quad \text{[U.S.]} \qquad 14.22(b)$$

It is essential that consistent units be used with Eq. 14.22. The following sets of units are consistent.

$$\text{ft}^2/\text{sec} = \frac{\text{lbf-sec/ft}^2}{\text{slugs/ft}^3}$$

$$\text{m}^2/\text{s} = \frac{\text{Pa·s}}{\text{kg/m}^3}$$

$$\text{St (stoke)} = \frac{\text{P (poise)}}{\text{g/cm}^3}$$

$$\text{cSt (centistokes)} = \frac{\text{cP (centipoise)}}{\text{g/cm}^3}$$

Unlike absolute viscosity, kinematic viscosity is greatly dependent on both temperature and pressure, since these variables affect the density of the fluid. Referring to Eq. 14.22, even if absolute viscosity is independent of temperature or pressure, the change in density will change the kinematic viscosity.

The higher a fluid's kinematic viscosity, the more time will be required for the fluid to leak out of a container. *Saybolt Seconds Universal* (SSU) and *Saybolt Seconds Furol* (SSF) are scales of such viscosity measurement based on the smaller and larger orifices, respectively. Seconds can be converted (empirically) to viscosity in other units. The following relations are approximate conversions between SSU and stokes.

- For SSU < 100 sec,

$$\nu_{\text{stokes}} = 0.00226(\text{SSU}) - \frac{1.95}{\text{SSU}} \qquad 14.23$$

---

[14] This is not true for kinematic viscosity, however.

[15] Units of lbm/ft-sec are also used for absolute viscosity in the English system. These units are obtained by multiplying lbf-sec/ft² units by g_c.

- For SSU > 100 sec,

$$\nu_{\text{stokes}} = 0.00220(\text{SSU}) - \frac{1.35}{\text{SSU}} \qquad 14.24$$

11. VISCOSITY CONVERSIONS

The most common units of absolute and kinematic viscosity are listed in Table 14.4. Table 14.5 contains conversions between the various viscosity units.

Table 14.4 *Common Viscosity Units*

	absolute, μ	kinematic, ν
English	lbf-sec/ft² (slug/ft-sec)	ft²/sec
conventional metric	dyne·s/cm² (poise)	cm²/s (stoke)
SI	Pa·s (N·s/m²)	m²/s

Example 14.5

Water at 60°F has a specific gravity of 0.999 and a kinematic viscosity of 1.12 cSt. What is the absolute viscosity in lbf-sec/ft²?

Solution

The density of a liquid expressed in g/cm³ is numerically equal to its specific gravity.

$$\rho = 0.999 \text{ g/cm}^3$$

The centistoke (cSt) is a measure of kinematic viscosity. Kinematic viscosity is converted first to the absolute viscosity units of centipoise. From Table 14.5,

$$\begin{aligned}\mu_{\text{cP}} &= \nu_{\text{cSt}}\rho_{\text{g/cm}^3}\\ &= (1.12 \text{ cSt})\left(0.999 \frac{\text{g}}{\text{cm}^3}\right)\\ &= 1.119 \text{ cP}\end{aligned}$$

Next, centipoise is converted to lbf-sec/ft².

$$\begin{aligned}\mu_{\text{lbf-sec/ft}^2} &= \mu_{\text{cP}}(2.0885 \times 10^{-5})\\ &= (1.119 \text{ cP})(2.0885 \times 10^{-5})\\ &= 2.34 \times 10^{-5} \text{ lbf-sec/ft}^2\end{aligned}$$

Table 14.5 *Viscosity Conversions**

multiply	by	to obtain
absolute viscosity, μ		
dyne·s/cm²	0.10	Pa·s
lbf-sec/ft²	478.8	P
lbf-sec/ft²	47,880	cP
lbf-sec/ft²	47.88	Pa·s
slug/ft-sec	47.88	Pa·s
lbm/ft-sec	1.488	Pa·s
cP	1.0197×10^{-4}	kgf·s/m²
cP	2.0885×10^{-5}	lbf-sec/ft²
cP	0.001	Pa·s
Pa·s	0.020885	lbf-sec/ft²
Pa·s	1000	cP
reyn	144	lbf-sec/ft²
reyn	1.0	lbf-sec/in²
kinematic viscosity, ν		
ft²/sec	92,903	cSt
ft²/sec	0.092903	m²/s
m²/s	10.7639	ft²/sec
m²/s	1×10^6	cSt
cSt	1×10^{-6}	m²/s
cSt	1.0764×10^{-5}	ft²/sec
absolute viscosity to kinematic viscosity		
cP	$1/\rho$ in g/cm³	cSt
cP	$6.7195 \times 10^{-4}/\rho$ in lbm/ft³	ft²/sec
lbf-sec/ft²	$32.174/\rho$ in lbm/ft³	ft²/sec
kgf·s/m²	$9.807/\rho$ in kg/m³	m²/s
Pa·s	$1000/\rho$ in g/cm³	cSt
kinematic viscosity to absolute viscosity		
cSt	ρ in g/cm³	cP
cSt	$0.001 \times \rho$ in g/cm³	Pa·s
cSt	$1.6 \times 10^{-5} \times \rho$ in lbm/ft³	Pa·s
m²/s	$0.10197 \times \rho$ in kg/m³	kgf·s/m²
m²/s	$1000 \times \rho$ in g/cm³	Pa·s
m²/s	ρ in kg/m³	Pa·s
ft²/sec	$0.031081 \times \rho$ in lbm/ft³	lbf-sec/ft²
ft²/sec	$1488.2 \times \rho$ in lbm/ft³	cP

*cP: centipoise; cSt: centistoke; kgf: kilogram-force; P: poise

Fluids

12. VISCOSITY GRADE

The ISO *viscosity grade* (VG) as specified in ISO 3448, is commonly used to classify oils. (See Table 14.6.) Viscosity at 104°F (40°C), the approximate temperature of machinery, in centistokes (same as mm^2/s), is used as the index. Each subsequent viscosity grade within the classification has approximately a 50% higher viscosity, whereas the minimum and maximum values of each grade range ±10% from the midpoint.

Table 14.6 *ISO Viscosity Grade*

ISO 3448 viscosity grade	kinematic viscosity at 40°C (cSt)		
	minimum	midpoint	maximum
ISO VG 2	1.98	2.2	2.42
ISO VG 3	2.88	3.2	3.52
ISO VG 5	4.14	4.6	5.06
ISO VG 7	6.12	6.8	7.48
ISO VG 10	9.0	10	11.0
ISO VG 15	13.5	15	16.5
ISO VG 22	19.8	22	24.2
ISO VG 32	28.8	32	35.2
ISO VG 46	41.4	46	50.6
ISO VG 68	61.2	68	74.8
ISO VG 100	90	100	110
ISO VG 150	135	150	165
ISO VG 220	198	220	242
ISO VG 320	288	320	352
ISO VG 460	414	460	506
ISO VG 680	612	680	748
ISO VG 1000	900	1000	1100
ISO VG 1500	1350	1500	1650

Another method of classifying oil by viscosity is the Society of Automotive Engineers (SAE) J300 specification for engine oil. [**SAE J300 (1999) Motor Oil Grades—Low Temperature Specifications**]

13. VISCOSITY INDEX

Viscosity index (VI) is a measure of a fluid's viscosity sensitivity to changes in temperature. It has traditionally been applied to crude and refined oils through use of a 100-point scale.[16] The viscosity is measured at two temperatures: 100°F and 210°F (38°C and 99°C). These viscosities are converted into a viscosity index in accordance with standard ASTM D2270.

14. VAPOR PRESSURE

Molecular activity in a liquid will allow some of the molecules to escape the liquid surface. Strictly speaking, a small portion of the liquid vaporizes. Molecules of the vapor also condense back into the liquid. The vaporization and condensation at constant temperature are equilibrium processes. The equilibrium pressure exerted by these free molecules is known as the *vapor pressure* or *saturation pressure*. (Vapor pressure does not include the pressure of other substances in the mixture.) Typical values of vapor pressure are given in Table 14.7.

Table 14.7 *Typical Vapor Pressures*

fluid	lbf/ft^2, 68°F	kPa, 20°C
mercury	0.00362	0.000173
turpentine	1.115	0.0534
water	48.9	2.34
ethyl alcohol	122.4	5.86
ether	1231	58.9
butane	4550	218
Freon-12	12,200	584
propane	17,900	855
ammonia	18,550	888

(Multiply lbf/ft^2 by 0.04788 to obtain kPa.)

Some liquids, such as propane, butane, ammonia, and Freon, have significant vapor pressures at normal temperatures. Liquids near their boiling points or that vaporize easily are said to be *volatile liquids*.[17] Other liquids, such as mercury, have insignificant vapor pressures at normal temperatures. Liquids with low vapor pressures are used in accurate barometers.

The tendency toward vaporization is dependent on the temperature of the liquid. *Boiling* occurs when the liquid temperature is increased to the point that the vapor pressure is equal to the local ambient pressure. Therefore, a liquid's boiling temperature depends on the local ambient pressure as well as on the liquid's tendency to vaporize.

Vapor pressure is usually considered to be a nonlinear function of temperature only. It is possible to derive correlations between vapor pressure and temperature, and such correlations usually involve a logarithmic transformation of vapor pressure.[18] Vapor pressure can also be graphed against temperature in a (logarithmic) *Cox*

[16]Use of the *viscosity index* has been adopted by other parts of the chemical process industry (CPI), including in the manufacture of solvents, polymers, and other synthetics. The 100-point scale may be exceeded (on both ends) for these uses. Refer to standard ASTM D2270 for calculating extreme values of the viscosity index.

[17]Because a liquid that vaporizes easily has an aroma, the term *aromatic liquid* is also occasionally used.

[18]The *Clausius-Clapeyron equation* and *Antoine equation* are two such logarithmic correlations of vapor pressure with temperature.

chart when values are needed over larger temperature extremes. Although there is also some variation with external pressure, the external pressure effect is negligible under normal conditions.

15. SURFACE TENSION

The membrane or "skin" that seems to form on the free surface of a fluid is due to the intermolecular cohesive forces and is known as *surface tension*, σ. Surface tension is the reason that insects are able to walk and a needle is able to float on water. Surface tension also causes bubbles and droplets to take on a spherical shape, since any other shape would have more surface area per unit volume.

Data on the surface tension of liquids is important in determining the performance of heat-, mass-, and momentum-transfer equipment, including heat transfer devices.[19] Surface tension data is needed to calculate the nucleate boiling point (i.e., the initiation of boiling) of liquids in a pool (using the *Rohsenow equation*) and the maximum heat flux of boiling liquids in a pool (using the *Zuber equation*).

Surface tension can be interpreted as the tension between two points a unit distance apart on the surface or as the amount of work required to form a new unit of surface area in an apparatus similar to that shown in Fig. 14.6. Typical units of surface tension are lbf/ft (ft-lbf/ft^2), dyne/cm, and N/m.

Figure 14.6 *Wire Frame for Stretching a Film*

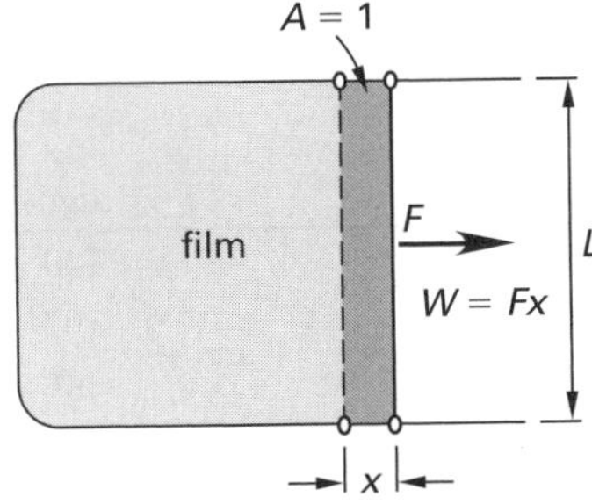

The apparatus shown in Fig. 14.6 consists of a wire frame with a sliding side that has been dipped in a liquid to form a film. Surface tension is determined by measuring the force necessary to keep the sliding side stationary against the surface tension pull of the film.[20] (The film does not act like a spring, since the force, F, does not increase as the film is stretched.) Since the film has two surfaces (i.e., two surface tensions), the surface tension is

$$\sigma = \frac{F}{2L} \tag{14.25}$$

Alternatively, surface tension can also be determined by measuring the force required to pull a wire ring out of the liquid, as shown in Fig. 14.7.[21] Since the ring's inner and outer sides are in contact with the liquid, the wetted perimeter is twice the circumference. The surface tension is

$$\sigma = \frac{F}{4\pi r} \tag{14.26}$$

Figure 14.7 *Du Nouy Ring Surface Tension Apparatus*

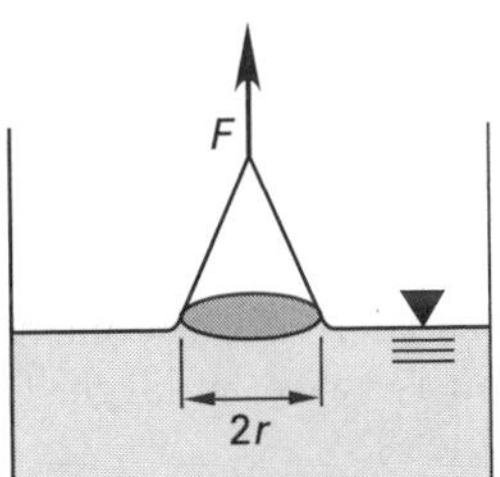

Surface tension depends slightly on the gas in contact with the free surface. Surface tension values are usually quoted for air contact. Typical values of surface tension are listed in Table 14.8.

At temperatures below freezing, the substance will be a solid, so surface tension is a moot point. As the temperature of a liquid is raised, the surface tension decreases because the cohesive forces decrease. Surface tension is zero at a substance's critical temperature. If a substance's critical temperature is known, the *Othmer correlation*, Eq. 14.27, can be used to determine the surface tension at one temperature from the surface tension at another temperature.

$$\sigma_2 = \sigma_1\left(\frac{T_c - T_2}{T_c - T_1}\right)^{11/9} \tag{14.27}$$

Surface tension is the reason that the pressure on the inside of bubbles and droplets is greater than on the outside. Equation 14.28 gives the relationship between the surface tension in a hollow bubble surrounded by a gas and the difference between the inside and outside pressures. For a spherical droplet or a bubble in a liquid,

[19]Surface tension plays a role in processes involving dispersion, emulsion, flocculation, foaming, and solubilization. It is not surprising that surface tension data are particularly important in determining the performance of equipment in the chemical process industry (CPI), such as distillation columns, packed towers, wetted-wall columns, strippers, and phase-separation equipment.

[20]The force includes the weight of the sliding side wire if the frame is oriented vertically, with gravity acting on the sliding side wire to stretch the film.

[21]This apparatus is known as a *Du Nouy torsion balance*. The ring is made of platinum with a diameter of 4.00 cm.

Table 14.8 *Approximate Values of Surface Tension (air contact)*

fluid	lbf/ft, 68°F	N/m, 20°C
n-octane	0.00149	0.0217
ethyl alcohol	0.00156	0.0227
acetone	0.00162	0.0236
kerosene	0.00178	0.0260
carbon tetrachloride	0.00185	0.0270
turpentine	0.00186	0.0271
toluene	0.00195	0.0285
benzene	0.00198	0.0289
olive oil	0.0023	0.034
glycerin	0.00432	0.0631
water	0.00499	0.0728
mercury	0.0356	0.519

(Multiply lbf/ft by 14.59 to obtain N/m.)

(Multiply dyne/cm by 0.001 to obtain N/m.)

where in both cases there is only one surface in tension, the surface tension is twice as large. (r is the radius of the bubble or droplet.)

$$\sigma_{\text{bubble}} = \frac{r(p_{\text{inside}} - p_{\text{outside}})}{4} \qquad 14.28$$

$$\sigma_{\text{droplet}} = \frac{r(p_{\text{inside}} - p_{\text{outside}})}{2} \qquad 14.29$$

16. CAPILLARY ACTION

Capillary action (*capillarity*) is the name given to the behavior of a liquid in a thin-bore tube. Capillary action is caused by surface tension between the liquid and a vertical solid surface.[22] In the case of liquid water in a glass tube, the adhesive forces between the liquid molecules and the surface are greater than (i.e., dominate) the cohesive forces between the water molecules themselves.[23] The adhesive forces cause the water to attach itself to and climb a solid vertical surface. It can be said that the water "reaches up and tries to wet as much of the interior surface as it can." In so doing, the water rises above the general water surface level. The surface is *hydrophilic* (*lyophilic*). This is illustrated in Fig. 14.8.

Figure 14.8 also illustrates that the same surface tension forces that keep a droplet spherical are at work on the surface of the liquid in the tube. The curved liquid surface, known as the *meniscus*, can be considered to be an incomplete droplet. If the inside diameter of the tube is less than approximately 0.1 in (2.5 mm), the meniscus is essentially hemispherical, and $r_{\text{meniscus}} = r_{\text{tube}}$.

Figure 14.8 *Capillarity of Liquids*

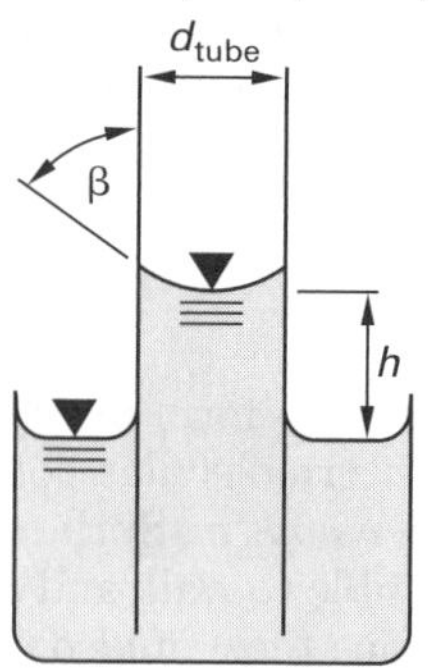

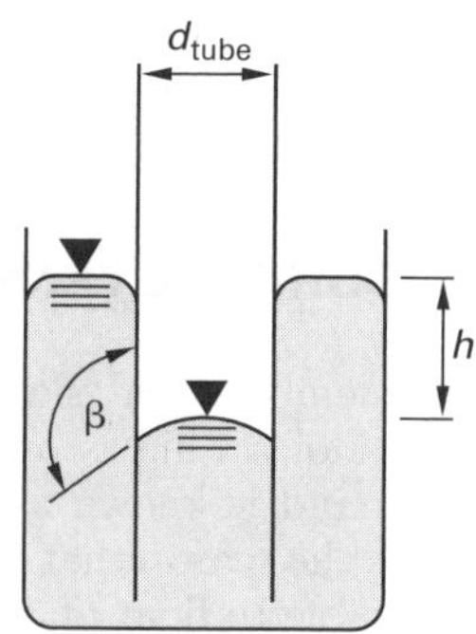

(a) adhesive force dominates (b) cohesive force dominates

For a few other liquids, such as mercury, the molecules have a strong affinity for each other (i.e., the cohesive forces dominate). The liquid avoids contact with the tube surface. The surface is *hydrophobic* (*lyophobic*). In such liquids, the meniscus in the tube will be below the general surface level.

The *angle of contact*, β, is an indication of whether adhesive or cohesive forces dominate. For contact angles less than 90°, adhesive forces dominate. For contact angles greater than 90°, cohesive forces dominate.

Equation 14.30 can be used to predict the capillary rise in a small-bore tube. Surface tension and contact angles can be obtained from Table 14.8 and Table 14.9, respectively.

Table 14.9 *Contact Angles, β*

materials	angle
mercury–glass	140°
water–paraffin	107°
water–silver	90°
silicone oil–glass	20°
kerosene–glass	26°
glycerin–glass	19°
water–glass	0°
ethyl alcohol–glass	0°

$$h = \frac{4\sigma\cos\beta}{\rho D_{\text{tube}} g} \quad \text{[SI]} \qquad 14.30(a)$$

$$h = \frac{4\sigma\cos\beta}{\rho D_{\text{tube}}} \times \frac{g_c}{g} \quad \text{[U.S.]} \qquad 14.30(b)$$

[22]In fact, observing the rise of liquid in a capillary tube is another method of determining the surface tension of a liquid.

[23]*Adhesion* is the attractive force between molecules of different substances. *Cohesion* is the attractive force between molecules of the same substance.

$$\sigma = \frac{h\rho D_{\text{tube}} g}{4\cos\beta} \quad \text{[SI]} \qquad 14.31(a)$$

$$\sigma = \frac{h\rho D_{\text{tube}}}{4\cos\beta} \times \frac{g}{g_c} \quad \text{[U.S.]} \qquad 14.31(b)$$

$$r_{\text{meniscus}} = \frac{D_{\text{tube}}}{2\cos\beta} \qquad 14.32$$

If it is assumed that the meniscus is hemispherical, then $r_{\text{meniscus}} = r_{\text{tube}}$, $\beta = 0°$, and $\cos\beta = 1.0$, and the above equations can be simplified. (Such an assumption can only be made when the diameter of the capillary tube is less than 0.1 in.)

Example 14.6

Ethyl alcohol's density is 49 lbm/ft^3 (790 kg/m^3). To what height will 68°F (20°C) ethyl alcohol rise in a 0.005 in (0.127 mm) internal diameter glass capillary tube?

SI Solution

From Table 14.8 and Table 14.9, respectively, the surface tension and contact angle are

$$\sigma = 0.0227 \text{ N/m}$$
$$\beta = 0°$$

From Eq. 14.30, the height is

$$h = \frac{4\sigma\cos\beta}{\rho D_{\text{tube}} g} = \frac{(4)\left(0.0227 \ \frac{\text{N}}{\text{m}}\right)(1.0)\left(1000 \ \frac{\text{mm}}{\text{m}}\right)}{\left(790 \ \frac{\text{kg}}{\text{m}^3}\right)(0.127 \text{ mm})\left(9.81 \ \frac{\text{m}}{\text{s}^2}\right)}$$
$$= 0.0923 \text{ m}$$

Customary U.S. Solution

From Table 14.8 and Table 14.9, respectively, the surface tension and contact angle are

$$\sigma = 0.00156 \text{ lbf/ft}$$
$$\beta = 0°$$

From Eq. 14.30, the height is

$$h = \frac{4\sigma\cos\beta g_c}{\rho D_{\text{tube}} g}$$
$$= \frac{(4)\left(0.00156 \ \frac{\text{lbf}}{\text{ft}}\right)(1.0)\left(32.2 \ \frac{\text{lbm-ft}}{\text{lbf-sec}^2}\right)\left(12 \ \frac{\text{in}}{\text{ft}}\right)}{\left(49 \ \frac{\text{lbm}}{\text{ft}^3}\right)(0.005 \text{ in})\left(32.2 \ \frac{\text{ft}}{\text{sec}^2}\right)}$$
$$= 0.306 \text{ ft}$$

17. COMPRESSIBILITY[24]

Compressibility (also known as the *coefficient of compressibility*), β, is the fractional change in the volume of a fluid per unit change in pressure in a constant-temperature process.[25] Typical units are in^2/lbf, ft^2/lbf, 1/atm, and 1/kPa. (See Table 14.10.) Equation 14.33 is written with a negative sign to show that volume decreases as pressure increases.

Stress, Pressure, and Viscosity

$$\beta = -\frac{1}{V}\frac{dV}{dp} \qquad 14.33$$

Compressibility can also be written in terms of partial derivatives.

$$\beta = \left(\frac{-1}{V_0}\right)\left(\frac{\partial V}{\partial p}\right)_T = \left(\frac{1}{\rho_0}\right)\left(\frac{\partial \rho}{\partial p}\right)_T \qquad 14.34$$

A fluid's compressibility is the reciprocal of its bulk modulus, a quantity that is more commonly tabulated than compressibility.

$$\beta = \frac{1}{B} \qquad 14.35$$

Compressibility changes only slightly with temperature. The small compressibility of liquids is typically considered to be insignificant, giving rise to the common understanding that liquids are incompressible.

The density of a compressible fluid depends on the fluid's pressure. For small changes in pressure, Eq. 14.36 can be used to calculate the density at one pressure from the density at another pressure.

$$\rho_2 \approx \rho_1\big(1 + \beta(p_2 - p_1)\big) \qquad 14.36$$

[24]Compressibility should not be confused with the *thermal coefficient of expansion*, $(1/V_0)(\partial V/\partial T)_p$, which is the fractional change in volume per unit temperature change in a constant-pressure process (with units of 1/°F or 1/°C), or the dimensionless *compressibility factor*, Z, which is used with the ideal gas law.

[25]Other symbols used for compressibility are c, C, and K.

Gases, of course, are easily compressed. The compressibility of an ideal gas depends on its pressure, p, its ratio of specific heats, k, and the nature of the process.[26] Depending on the process, the compressibility may be known as *isothermal compressibility* or (*adiabatic*) *isentropic compressibility*. Of course, compressibility is zero for constant-volume processes and is infinite (or undefined) for constant-pressure processes.

$$\beta_T = \frac{1}{p} \quad \text{[isothermal ideal gas processes]} \qquad 14.37$$

$$\beta_s = \frac{1}{kp} \quad \text{[adiabatic ideal gas processes]} \qquad 14.38$$

***Table 14.10** Approximate Compressibilities of Common Liquids at 1 atm*

liquid	temperature	β (in²/lbf)	β (1/atm)
mercury	32°F	0.027×10^{-5}	0.39×10^{-5}
glycerin	60°F	0.16×10^{-5}	2.4×10^{-5}
water	60°F	0.33×10^{-5}	4.9×10^{-5}
ethyl alcohol	32°F	0.68×10^{-5}	10×10^{-5}
chloroform	32°F	0.68×10^{-5}	10×10^{-5}
gasoline	60°F	1.0×10^{-5}	15×10^{-5}
hydrogen	20K	11×10^{-5}	160×10^{-5}
helium	2.1K	48×10^{-5}	700×10^{-5}

(Multiply 1/psi by 14.696 to obtain 1/atm.)

(Multiply in²/lbf by 0.145 to obtain 1/kPa.)

Example 14.7

Water at 68°F (20°C) and 1 atm has a density of 62.3 lbm/ft³ (997 kg/m³). The bulk modulus has a constant value of 320,000 lbf/in² (2.2×10^6 kPa). What is the new density if the pressure is isothermally increased from 14.7 lbf/in² to 400 lbf/in² (100 kPa to 2760 kPa)?

SI Solution

Compressibility is the reciprocal of the bulk modulus. From Eq. 14.35,

$$\beta = \frac{1}{B} = \frac{1}{2.2 \times 10^6 \text{ kPa}} = 4.55 \times 10^{-7} \text{ 1/kPa}$$

From Eq. 14.36,

$$\begin{aligned}\rho_2 &= \rho_1\big(1 + \beta(p_2 - p_1)\big)\\ &= \left(997 \ \frac{\text{kg}}{\text{m}^3}\right)\left(\begin{array}{l}1 + \left(4.55 \times 10^{-7} \ \dfrac{1}{\text{kPa}}\right)\\ \quad \times (2760 \text{ kPa} - 100 \text{ kPa})\end{array}\right)\\ &= 998.2 \text{ kg/m}^3\end{aligned}$$

Customary U.S. Solution

Compressibility is the reciprocal of the bulk modulus. From Eq. 14.35,

$$\beta = \frac{1}{B} = \frac{1}{320{,}000 \ \frac{\text{lbf}}{\text{in}^2}} = 0.3125 \times 10^{-5} \text{ in}^2\text{/lbf}$$

From Eq. 14.36,

$$\begin{aligned}\rho_2 &= \rho_1\big(1 + \beta(p_2 - p_1)\big)\\ &= \left(62.3 \ \frac{\text{lbm}}{\text{ft}^3}\right)\left(\begin{array}{l}1 + \left(0.3125 \times 10^{-5} \ \dfrac{\text{in}^2}{\text{lbf}}\right)\\ \quad \times \left(400 \ \dfrac{\text{lbf}}{\text{in}^2} - 14.7 \ \dfrac{\text{lbf}}{\text{in}^2}\right)\end{array}\right)\\ &= 62.38 \text{ lbm/ft}^3\end{aligned}$$

18. BULK MODULUS

The *bulk modulus*, B, of a fluid is analogous to the modulus of elasticity of a solid. Typical units are lbf/in², atm, and kPa. The term dp in Eq. 14.39 represents an increase in stress. The term dV/V_0 is a *volumetric strain*. Analogous to Hooke's law describing elastic formation, the *bulk modulus* of a fluid (liquid or gas) is given by Eq. 14.39.

Stress, Pressure, and Viscosity

$$B = \frac{\text{stress}}{\text{strain}} = -\frac{dp}{\frac{dV}{V}} \qquad 14.39$$

The bulk modulus can also be written in terms of partial derivatives.

$$B = -V\left(\frac{\partial p}{\partial V}\right)_T \qquad 14.40$$

The term *secant bulk modulus* is associated with Eq. 14.39 (the average slope), while the terms *tangent bulk modulus* and *point bulk modulus* are associated with Eq. 14.40 (the instantaneous slope).

[26]For air, $k = 1.4$.

The bulk modulus is the reciprocal of compressibility.

$$B = \frac{1}{\beta} \qquad 14.41$$

The bulk modulus changes only slightly with temperature. The bulk modulus of water is usually taken as 300,000 lbf/in^2 (2.1×10^6 kPa) unless greater accuracy is required, in which case Table 14.11 or App. 14.A can be used.

***Table 14.11** Approximate Bulk Modulus of Water*

pressure (lbf/in^2)	32°F	68°F	120°F	200°F	300°F
	(thousands of lbf/in^2)				
15	292	320	332	308	–
1500	300	330	340	319	218
4500	317	348	362	338	271
15,000	380	410	420	405	350

(Multiply lbf/in^2 by 6.8948 to obtain kPa.)

Reprinted with permission from Victor L. Streeter, *Handbook of Fluid Dynamics*, © 1961, by McGraw-Hill Book Company.

19. SPEED OF SOUND

The *speed of sound* (*acoustic velocity* or *sonic velocity*), c, in a fluid is a function of its bulk modulus (or, equivalently, of its compressibility).[27] Equation 14.42 gives the speed of sound through a liquid.

Mach Number

$$c = \sqrt{\frac{B}{\rho}} = \sqrt{\frac{1}{\beta\rho}} \qquad \text{[SI]} \qquad 14.42(a)$$

$$c = \sqrt{\frac{Bg_c}{\rho}} = \sqrt{\frac{g_c}{\beta\rho}} \qquad \text{[U.S.]} \qquad 14.42(b)$$

For an ideal gas, $B = kp$ (from (from Eq. 14.35 and Eq. 14.38)), and from the ideal gas law, $p/\rho = RT$; substituting these into Eq. 14.42 gives an expression for the speed of sound in an ideal gas.

Mach Number

$$c = \sqrt{kRT} = \sqrt{\frac{k\bar{R}T}{M}} \qquad \text{[SI]} \qquad 14.43(a)$$

$$c = \sqrt{kg_cRT} = \sqrt{\frac{kg_c\bar{R}T}{M}} \qquad \text{[U.S.]} \qquad 14.43(b)$$

In Eq. 14.43, T must be absolute temperature, in degrees Rankine or in kelvins. For air, the ratio of specific heats is $k = 1.4$, and the molecular weight is 28.967. The universal gas constant is $\bar{R} = 1545.35$ ft-lbf/lbmol-°R (8314.47 J/kmol·K).

Since k and R are constant for an ideal gas, the speed of sound is a function of temperature only. Equation 14.44 can be used to calculate the new speed of sound when temperature is varied.

$$\frac{c_1}{c_2} = \sqrt{\frac{T_1}{T_2}} \qquad 14.44$$

The *Mach number*, Ma, of an object is the ratio of the object's speed to the speed of sound in the medium through which it is traveling. (See Table 14.12.)

Mach Number

$$\text{Ma} = \frac{\text{v}}{c} \qquad 14.45$$

The term *subsonic travel* implies Ma < 1.[28] Similarly, *supersonic travel* implies Ma > 1, and usually Ma < 5. Travel above Ma = 5 is known as *hypersonic travel.* Travel in the transition region between subsonic and supersonic (i.e., 0.8 < Ma < 1.2) is known as *transonic travel.* A *sonic boom* (a shock wave phenomenon) occurs when an object travels at supersonic speed.

***Table 14.12** Approximate Speeds of Sound (at one atmospheric pressure)*

material	speed of sound (ft/sec)	(m/s)
air	1130 at 70°F	330 at 0°C
aluminum	16,400	4990
carbon dioxide	870 at 70°F	260 at 0°C
hydrogen	3310 at 70°F	1260 at 0°C
steel	16,900	5150
water	4880 at 70°F	1490 at 20°C

(Multiply ft/sec by 0.3048 to obtain m/s.)

[27] The symbol c is also used for the speed of sound.

[28] In the language of compressible fluid flow, this is known as the *subsonic flow regime.*

Example 14.8

What is the speed of sound in 150°F (66°C) water? The density is 61.2 lbm/ft^3 (980 kg/m^3), and the bulk modulus is 328,000 lbf/in^2 (2.26 × 10^6 kPa).

SI Solution

From Eq. 14.42(a),

Mach Number

$$c = \sqrt{\frac{B}{\rho}} = \sqrt{\frac{(2.26\times 10^6 \text{ kPa})\left(1000\ \frac{\text{Pa}}{\text{kPa}}\right)}{980\ \frac{\text{kg}}{\text{m}^3}}} = 1519 \text{ m/s}$$

Customary U.S. Solution

From Eq. 14.42(b),

Mach Number

$$c = \sqrt{\frac{Bg_c}{\rho}} = \sqrt{\frac{\left(328{,}000\ \frac{\text{lbf}}{\text{in}^2}\right)\left(12\ \frac{\text{in}}{\text{ft}}\right)^2\left(32.2\ \frac{\text{lbm-ft}}{\text{lbf-sec}^2}\right)}{61.2\ \frac{\text{lbm}}{\text{ft}^3}}}$$

$$= 4985 \text{ ft/sec}$$

Example 14.9

What is the speed of sound in 150°F (66°C) air at standard atmospheric pressure?

SI Solution

The specific gas constant, R, for air is

Mach Number

$$R = \frac{\overline{R}}{M} = \frac{8314.47\ \frac{\text{J}}{\text{kmol}\cdot\text{K}}}{28.967\ \frac{\text{kg}}{\text{kmol}}} = 287.03 \text{ J/kg}\cdot\text{K}$$

The absolute temperature is [**Temperature Conversions**]

$$T = 66°\text{C} + 273° = 339\text{K}$$

From Eq. 14.43(a),

Mach Number

$$c = \sqrt{kRT} = \sqrt{(1.4)\left(287.03\ \frac{\text{J}}{\text{kg}\cdot\text{K}}\right)(339\text{K})}$$

$$= 369 \text{ m/s}$$

Customary U.S. Solution

The specific gas constant, R, for air is

Mach Number

$$R = \frac{\overline{R}}{M} = \frac{1545.35\ \frac{\text{ft-lbf}}{\text{lbmol-°R}}}{28.967\ \frac{\text{lbm}}{\text{lbmol}}}$$

$$= 53.35 \text{ ft-lbf/lbm-°R}$$

The absolute temperature is [**Temperature Conversions**]

$$T = 150°\text{F} + 460° = 610°\text{R}$$

From Eq. 14.43(b),

Mach Number

$$c = \sqrt{kg_cRT}$$

$$= \sqrt{(1.4)\left(32.2\ \frac{\text{lbm-ft}}{\text{lbf-sec}^2}\right)\left(53.35\ \frac{\text{ft-lbf}}{\text{lbm-°R}}\right)(610°\text{R})}$$

$$= 1211 \text{ ft/sec}$$

20. NOMENCLATURE

A	area	ft^2	m^2
B	bulk modulus	lbf/ft^2	Pa
c	speed of sound	ft/sec	m/s
D	diameter	ft	m
F	force	lbf	N
g	gravitational acceleration, 32.2 (9.81)	ft/sec^2	m/s^2
g_c	gravitational constant, 32.2	lbm-ft/lbf-sec^2	n.a.
h	height	ft	m
k	ratio of specific heats	–	–
L	length	ft	m
m	mass	lbm	kg
M	molecular weight	lbm/lbmol	kg/kmol
Ma	Mach number	–	–
N	number of moles	–	–
p	pressure	lbf/ft^2	Pa
r	radius	ft	m
R	specific gas constant	ft-lbf/lbm-°R	J/kg·K
$\bar{R}$	universal gas constant, 1545.35 (8314.47)	ft-lbf/lbmol-°R	J/kmol·K
SG	specific gravity	–	–
T	absolute temperature	°R	K
v	specific volume	ft^3/lbm	m^3/kg
v	velocity	ft/sec	m/s
V	volume	ft^3	m^3

W	weight	lbf	N
x	mole fraction	–	–
y	distance	ft	m
Z	compressibility factor	–	–

Symbols

β	compressibility	ft^2/lbf	Pa^{-1}
β	contact angle	deg	deg
γ	specific weight	lbf/ft^3	N/m^3
μ	absolute viscosity	$lbf\text{-}sec/ft^2$	Pa·s
ν	kinematic viscosity	ft^2/sec	m^2/s
ρ	density	lbm/ft^3	kg/m^3
σ	surface tension	lbf/ft	N/m
τ	shear stress	lbf/ft^2	Pa

Subscripts

0	zero velocity (wall face)
c	critical
p	constant pressure
s	constant entropy
T	constant temperature

Fluids

15 Fluid Statics

Content in blue refers to the *NCEES Handbook*.

NCEES EXAM SPECIFICATIONS AND RELATED CONTENT

HVAC AND REFRIGERATION EXAM

I.E. Fluid Mechanics
- 2. Manometers
- 4. Fluid Height Equivalent to Pressure
- 5. Pressure on a Horizontal Plane Surface

MACHINE DESIGN AND MATERIALS EXAM

II.C.5. Supportive Knowledge: Testing and instrumentation
- 2. Manometers
- 4. Fluid Height Equivalent to Pressure

1. PRESSURE-MEASURING DEVICES

There are many devices for measuring and indicating fluid pressure. Some devices measure gage pressure; others measure absolute pressure. The effects of non-standard atmospheric pressure and nonstandard gravitational acceleration must be determined, particularly for devices relying on columns of liquid to indicate pressure. Table 15.1 lists the common types of devices and the ranges of pressure appropriate for each.

Table 15.1 *Common Pressure-Measuring Devices*

device	approximate range (in atm)
water manometer	0–0.1
mercury barometer	0–1
mercury manometer	0.001–1
metallic diaphragm	0.01–200
transducer	0.001–15,000
Bourdon pressure gauge	1–3000
Bourdon vacuum gauge	0.1–1

The *Bourdon pressure gauge* is the most common pressure-indicating device. (See Fig. 15.1.) This mechanical device consists of a C-shaped or helical hollow tube that tends to straighten out (i.e., unwind) when the tube is subjected to an internal pressure. The gauge is referred to as a *C-Bourdon gauge* because of the shape of the hollow tube. The degree to which the coiled tube unwinds depends on the difference between the internal and external pressures. A Bourdon gauge directly indicates *gage pressure*. Extreme accuracy is generally not a characteristic of Bourdon gauges.

Figure 15.1 *C-Bourdon Pressure Gauge*

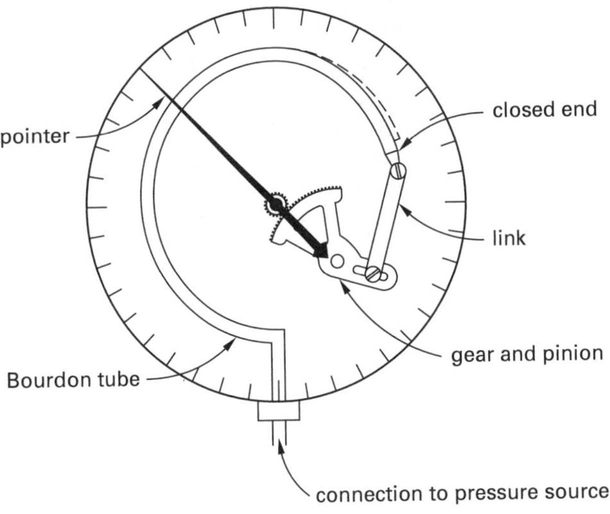

In non-SI installations, gauges are always calibrated in psi (or psid). Vacuum pressure is usually calibrated in inches of mercury. SI gauges are marked in kilopascals (kPa) or bars, units of kg/cm^2 are also found in some older gauges. Negative numbers are used to indicate vacuum. The gauge dial will be clearly marked if other units are indicated.

The *barometer* is a common device for measuring the absolute pressure of the atmosphere.[1] It is constructed by filling a long tube open at one end with mercury (or alcohol, or some other liquid) and inverting the tube so that the open end is below the level of a mercury-filled container. If the vapor pressure of the mercury in the tube is neglected, the fluid column will be supported only by the atmospheric pressure transmitted through the container fluid at the lower, open end.

Strain gauges, *diaphragm gauges*, *quartz-crystal transducers*, and other devices using the *piezoelectric effect* are also used to measure stress and pressure, particularly when pressure fluctuates quickly (e.g., as in a rocket combustion chamber). With these devices, calibration is required to interpret pressure from voltage generation or changes in resistance, capacitance, or inductance. These devices are generally unaffected by atmospheric pressure or gravitational acceleration.

Manometers (*U-tube manometers*) can also be used to indicate small pressure differences, and for this purpose they provide great accuracy. (Manometers are not suitable for measuring pressures much larger than 10 lbf/in^2 (70 kPa), however.) A difference in manometer fluid surface heights is converted into a pressure difference. If one end of a manometer is open to the atmosphere, the manometer indicates gage pressure. It is theoretically possible, but impractical, to have a manometer indicate absolute pressure, since one end of the manometer would have to be exposed to a perfect vacuum.

A *static pressure tube* (*piezometer tube*) is a variation of the manometer. (See Fig. 15.2.) It is a simple method of determining the static pressure in a pipe or other vessel, regardless of fluid motion in the pipe. A vertical transparent tube is connected to a hole in the pipe wall.[2] (No part of the tube projects into the pipe.) The static pressure will force the contents of the pipe up into the tube. The height of the contents will be an indication of gage pressure in the pipe.

The device used to measure the pressure should not be confused with the method used to obtain exposure to the pressure. For example, a static pressure *tap* in a pipe is merely a hole in the pipe wall. A Bourdon gauge, manometer, or transducer can then be used with the tap to indicate pressure.

Figure 15.2 *Static Pressure Tube*

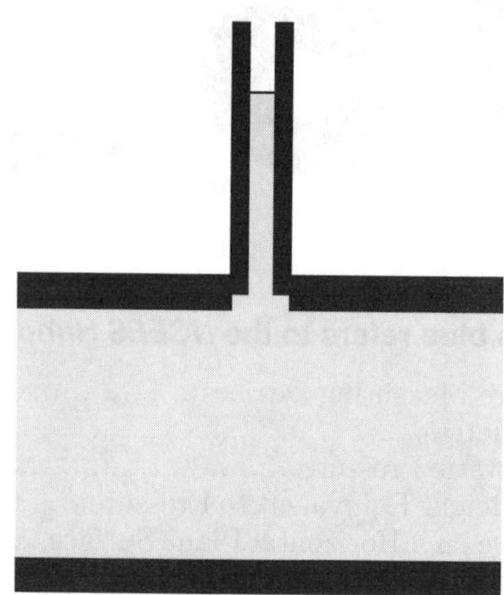

Tap holes are generally 1/8–1/4 in (3–6 mm) in diameter, drilled at right angles to the wall, and smooth and flush with the pipe wall. No part of the gauge or connection projects into the pipe. The tap holes should be at least 5 to 10 pipe diameters downstream from any source of turbulence (e.g., a bend, fitting, or valve).

2. MANOMETERS

Figure 15.3 illustrates a simple U-tube manometer used to measure the difference in pressure between two vessels. When both ends of the manometer are connected to pressure sources, the name *differential manometer* is used. If one end of the manometer is open to the atmosphere, the name *open manometer* is used.[3] The open manometer implicitly measures gage pressures.

Figure 15.3 *Simple U-Tube Manometer*

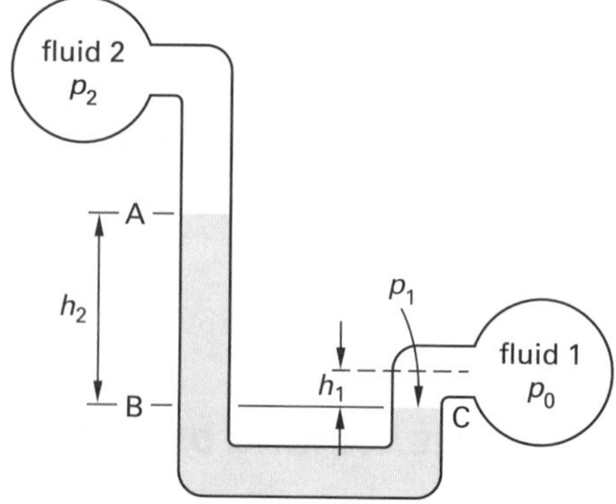

[1] A barometer can be used to measure the pressure inside any vessel. However, the barometer must be completely enclosed in the vessel, which may not be possible. Also, it is difficult to read a barometer enclosed within a tank.

[2] Where greater accuracy is required, multiple holes may be drilled around the circumference of the pipe and connected through a manifold (*piezometer ring*) to the pressure-measuring device.

[3] If one of the manometer legs is inclined, the term *inclined manometer* or *draft gauge* is used. Although only the vertical distance between the manometer fluid surfaces should be used to calculate the pressure difference, with small pressure differences it may be more accurate to read the inclined distance (which is larger than the vertical distance) and compute the vertical distance from the angle of inclination.

Since the pressure at point B in Fig. 15.3 is the same as at point C, the pressure differential produces the vertical fluid column between points A and B. The weight of this fluid column, in other words, balances the pressure differential (Eq. 15.1). In Eq. 15.2, A is the cross-sectional area of the tube. In the absence of any capillary action, the inside diameters of the manometer tubes are irrelevant.

$$F_{\text{net}} = F_{\text{C}} - F_{\text{A}} = \text{weight of fluid column AB} \quad 15.1$$

$$(p_0 - p_2)A = \rho_m g h_2 A \quad 15.2$$

$$p_0 - p_2 = \rho_m g h_2 \quad \text{[SI]} \quad 15.3(a)$$

$$p_0 - p_2 = \rho_m h_2 \times \frac{g}{g_c} = \gamma_m h_2 \quad \text{[U.S.]} \quad 15.3(b)$$

Equation 15.3(a) and Eq. 15.3(b) assume that the manometer fluid height is small, or that only only low-density gases fill the tubes above the manometer fluid.

The following equations hold true when the manometer (measuring) fluid is fluid 2 in an open manometer.

Manometers

$$\begin{aligned} p_0 &= p_2 + \gamma_2 h_2 - \gamma_1 h_1 \\ &= p_2 + g(\rho_2 h_2 - \rho_1 h_1) \end{aligned} \quad \text{[SI]} \quad 15.4(a)$$

$$\begin{aligned} p_0 &= p_2 + \gamma_2 h_2 - \gamma_1 h_1 \\ &= p_2 + (\rho_2 h_2 - \rho_1 h_1) \times \frac{g}{g_c} \end{aligned} \quad \text{[U.S.]} \quad 15.4(b)$$

When h_1 and h_2 are equal, letting $h = h_1 = h_2$,

Manometers

$$\begin{aligned} p_0 &= p_2 + (\gamma_2 - \gamma_1)h \\ &= p_2 + (\rho_2 - \rho_1)gh \end{aligned} \quad \text{[SI]} \quad 15.5(a)$$

$$\begin{aligned} p_0 &= p_2 + (\gamma_2 - \gamma_1)h \\ &= p_2 + (\rho_2 - \rho_1)h \times \frac{g}{g_c} \end{aligned} \quad \text{[U.S.]} \quad 15.5(b)$$

The quantity g/g_c has a value of 1.0 lbf/lbm in almost all cases, so γ is numerically equal to ρ, but with units of lbf/ft^3 instead of lbm/ft^3.

If a high-density fluid (such as water) is present above the measuring fluid, or if the columns h_1 or h_2 are very long, corrections will be necessary. (See Fig. 15.4.)

***Figure 15.4** Manometer Requiring Corrections*

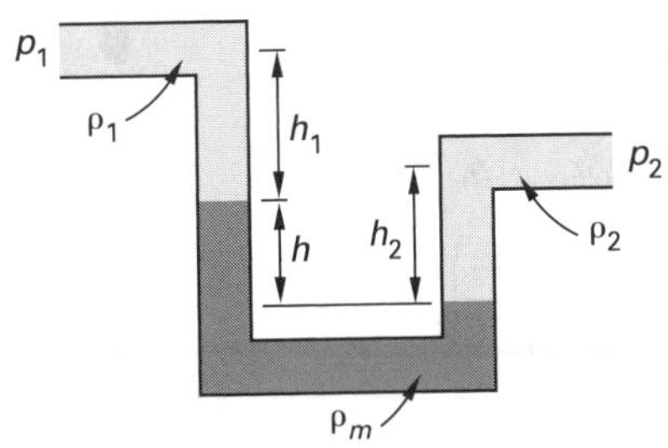

Fluid column h_2 "sits on top" of the manometer fluid, forcing the manometer fluid to the left. This increase must be subtracted out. Similarly, the column h_1 restricts the movement of the manometer fluid. The observed measurement must be increased to correct for this restriction.

$$p_2 - p_1 = g(\rho_m h + \rho_1 h_1 - \rho_2 h_2) \quad \text{[SI]} \quad 15.6(a)$$

$$\begin{aligned} p_2 - p_1 &= (\rho_m h + \rho_1 h_1 - \rho_2 h_2) \times \frac{g}{g_c} \\ &= \gamma_m h + \gamma_1 h_1 - \gamma_2 h_2 \end{aligned} \quad \text{[U.S.]} \quad 15.6(b)$$

Example 15.1

The pressure at the bottom of a water tank (ρ = 62.4 lbm/ft^3; ρ = 998 kg/m^3) is measured with a mercury manometer located below the tank bottom, as shown. (The density of mercury is 848 lbm/ft^3; 13 575 kg/m^3.) What is the gage pressure at the bottom of the water tank?

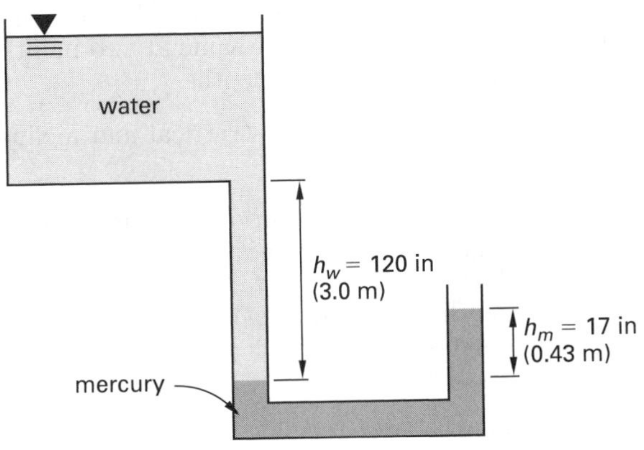

SI Solution

From Eq. 15.6(a),

$$\begin{aligned} \Delta p &= g(\rho_m h_m - \rho_w h_w) \\ &= \left(9.81\ \frac{\text{m}}{\text{s}^2}\right)\left(\begin{array}{c}\left(13\,575\ \frac{\text{kg}}{\text{m}^3}\right)(0.43\ \text{m}) \\ -\left(998\ \frac{\text{kg}}{\text{m}^3}\right)(3.0\ \text{m})\end{array}\right) \\ &= 27\,892\ \text{Pa} \quad (27.9\ \text{kPa gage}) \end{aligned}$$

Customary U.S. Solution

From Eq. 15.6(b),

$$\Delta p = (\rho_m h_m - \rho_w h_w) \times \frac{g}{g_c}$$

$$= \frac{\left(848 \ \frac{\text{lbm}}{\text{ft}^3}\right)(17 \text{ in}) - \left(62.4 \ \frac{\text{lbm}}{\text{ft}^3}\right)(120 \text{ in})}{\left(12 \ \frac{\text{in}}{\text{ft}}\right)^3} \times \left(\frac{32.2 \ \frac{\text{ft}}{\text{sec}^2}}{32.2 \ \frac{\text{lbm-ft}}{\text{lbf-sec}^2}}\right)$$

$$= 4.01 \text{ lbf/in}^2 \quad (4.01 \text{ psig})$$

3. HYDROSTATIC PRESSURE

Hydrostatic pressure is the pressure a fluid exerts on an immersed object or container walls.[4] Pressure is equal to the force per unit area of surface.

$$p = \frac{F}{A} \qquad 15.7$$

Hydrostatic pressure in a stationary, incompressible fluid behaves according to the following characteristics.

- Pressure is a function of vertical depth and density only. The pressure will be the same at two points in the same fluid with identical depths.
- Pressure varies linearly with (vertical and inclined) depth.
- Pressure is independent of an object's area and size and the weight (mass) of water above the object. Figure 15.5 illustrates the *hydrostatic paradox*. The pressures at depth h are the same in all four columns because pressure depends on depth, not volume.

Figure 15.5 Hydrostatic Paradox

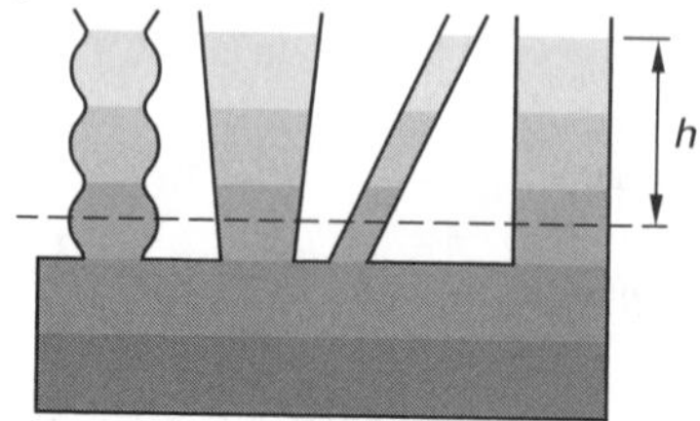

- Pressure at a point has the same magnitude in all directions (*Pascal's law*). Therefore, pressure is a scalar quantity.
- Pressure is always normal to a surface, regardless of the surface's shape or orientation. (This is a result of the fluid's inability to support shear stress.)
- The resultant of the pressure distribution acts through the *center of pressure.*
- The center of pressure rarely coincides with the average depth.

4. FLUID HEIGHT EQUIVALENT TO PRESSURE

Pressure varies linearly with depth. The relationship between pressure and depth (i.e., the *hydrostatic head*) for an incompressible fluid is given by Eq. 15.8.

$$p = \rho g h \qquad \text{[SI]} \qquad 15.8(a)$$

$$p = \frac{\rho g h}{g_c} = \gamma h \qquad \text{[U.S.]} \qquad 15.8(b)$$

Since ρ and g are constants, Eq. 15.8 shows that p and h are linearly related. Knowing one determines the other.[5] For example, the height of a fluid column needed to produce a pressure is

$$h = \frac{p}{\rho g} \qquad \text{[SI]} \qquad 15.9(a)$$

$$h = \frac{p g_c}{\rho g} = \frac{p}{\gamma} \qquad \text{[U.S.]} \qquad 15.9(b)$$

The difference in pressure between points 1 and 2, where z is the elevation of a point and h is the difference in elevation between the points, is

Pressure Field in a Static Liquid

$$p_2 - p_1 = -\gamma(z_2 - z_1) = -\gamma h = -\rho g h \qquad \text{[SI]} \qquad 15.10(a)$$

$$p_2 - p_1 = -\gamma(z_2 - z_1) = -\gamma h = -\frac{\rho g h}{g_c} \qquad \text{[U.S.]} \qquad 15.10(b)$$

Table 15.2 lists six common fluid height equivalents.[6]

[4]The term *hydrostatic* is used with all fluids, not only with water.

[5]In fact, pressure and height of a fluid column can be used interchangeably. The height of a fluid column is known as *head*. For example: "The fan developed a static head of 3 in of water," or "The pressure head at the base of the water tank was 8 m." When the term "head" is used, it is essential to specify the fluid.

[6]Of course, these values are recognized to be the approximate specific weights of the liquids.

Table 15.2 *Approximate Fluid Height Equivalents at 68°F (20°C)*

liquid	height equivalents	
water	0.0361 psi/in	27.70 in/psi
water	62.4 psf/ft	0.01603 ft/psf
water	9.81 kPa/m	0.1019 m/kPa
water	0.4329 psi/ft	2.31 ft/psi
mercury	0.491 psi/in	2.036 in/psi
mercury	133.3 kPa/m	0.00750 m/kPa

A barometer is a device that measures atmospheric pressure by the height of a fluid column. If the vapor pressure of the barometer liquid is neglected, the atmospheric pressure will be given by Eq. 15.11.

$$p_a = \rho g h \quad \text{[SI]} \qquad 15.11(a)$$

$$p_a = \frac{\rho g h}{g_c} = \gamma h \quad \text{[U.S.]} \qquad 15.11(b)$$

If the vapor pressure of the barometer liquid is significant (as it would be with alcohol or water), the vapor pressure effectively reduces the height of the fluid column, as Eq. 15.12 illustrates.

$$p_a - p_v = \rho g h \quad \text{[SI]} \qquad 15.12(a)$$

$$p_a - p_v = \frac{\rho g h}{g_c} = \gamma h \quad \text{[U.S.]} \qquad 15.12(b)$$

Example 15.2

A vacuum pump is used to drain a flooded mine shaft of 68°F (20°C) water.[7] The vapor pressure of water at this temperature is 0.34 lbf/in² (2.34 kPa). The pump is incapable of lifting the water higher than 400 in (10.16 m). What is the atmospheric pressure?

SI Solution

From Eq. 15.12,

$$\begin{aligned} p_a &= p_v + \rho g h \\ &= 2.34\ \text{kPa} + \frac{\left(998\ \frac{\text{kg}}{\text{m}^3}\right)\left(9.81\ \frac{\text{m}}{\text{s}^2}\right)(10.16\ \text{m})}{1000\ \frac{\text{Pa}}{\text{kPa}}} \\ &= 101.8\ \text{kPa} \end{aligned}$$

(Alternate SI solution, using Table 15.2)

$$\begin{aligned} p_a &= p_v + \rho g h = 2.34\ \text{kPa} + \left(9.81\ \frac{\text{kPa}}{\text{m}}\right)(10.16\ \text{m}) \\ &= 102\ \text{kPa} \end{aligned}$$

Customary U.S. Solution

From Table 15.2, the height equivalent of water is approximately 0.0361 psi/in. The unit psi/in is the same as lbf/in³, the units of γ. From Eq. 15.12, the atmospheric pressure is

$$\begin{aligned} p_a &= p_v + \rho g h = p_v + \gamma h \\ &= 0.34\ \frac{\text{lbf}}{\text{in}^2} + \left(0.0361\ \frac{\text{lbf}}{\text{in}^3}\right)(400\ \text{in}) \\ &= 14.78\ \text{lbf/in}^2 \quad (14.78\ \text{psia}) \end{aligned}$$

5. PRESSURE ON A HORIZONTAL PLANE SURFACE

The pressure on a horizontal plane surface is uniform over the surface because the depth of the fluid is uniform. (See Fig. 15.6.) The resultant of the pressure distribution acts through the center of pressure of the surface, which corresponds to the centroid of the surface.

The uniform pressure at depth h, where h is nonnegative and p_0 is atmospheric pressure, is given by Eq. 15.13.[8]

Figure 15.6 *Hydrostatic Pressure on a Horizontal Plane Surface*

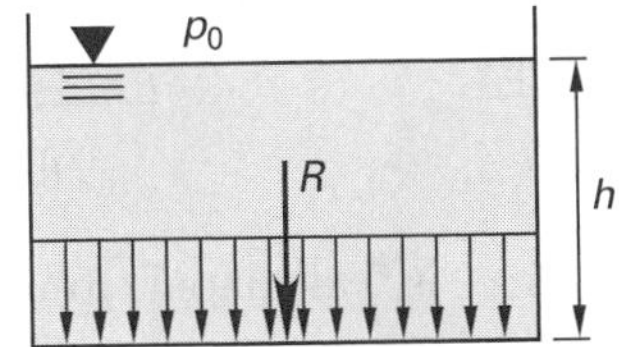

Forces on Submerged Surfaces and the Center of Pressure

$$p = p_0 + \rho g h \quad \text{[SI]} \qquad 15.13(a)$$

$$p = p_0 + \frac{\rho g h}{g_c} = p_0 + \gamma h \quad \text{[U.S.]} \qquad 15.13(b)$$

The total vertical force on the horizontal plane of area A is given by Eq. 15.14.

$$R = pA \qquad 15.14$$

[7]A reciprocating or other direct-displacement pump would be a better choice to drain a mine.

[8]The phrase *pressure at a depth* is universally understood to mean the *gage pressure*, as given by Eq. 15.13.

It is tempting, but not always correct, to calculate the vertical force on a submerged surface as the weight of the fluid above it. Such an approach works only when there is no change in the cross-sectional area of the fluid above the surface. This is a direct result of the *hydrostatic paradox*. (See Sec. 15.3.) Figure 15.7 illustrates two containers with the same pressure distribution (force) on their bottom surfaces.

Figure 15.7 *Two Containers with the Same Pressure Distribution*

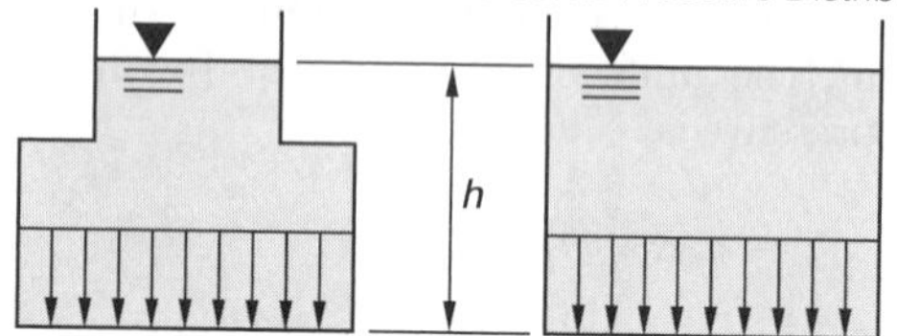

6. PRESSURE ON A RECTANGULAR VERTICAL PLANE SURFACE

The pressure on a vertical rectangular plane surface increases linearly with depth. If the plane surface extends to the surface, the pressure distribution will be triangular, as in Fig. 15.8(a); otherwise, the distribution will be trapezoidal, as in Fig. 15.8(b).

Figure 15.8 *Hydrostatic Pressure on a Vertical Plane Surface*

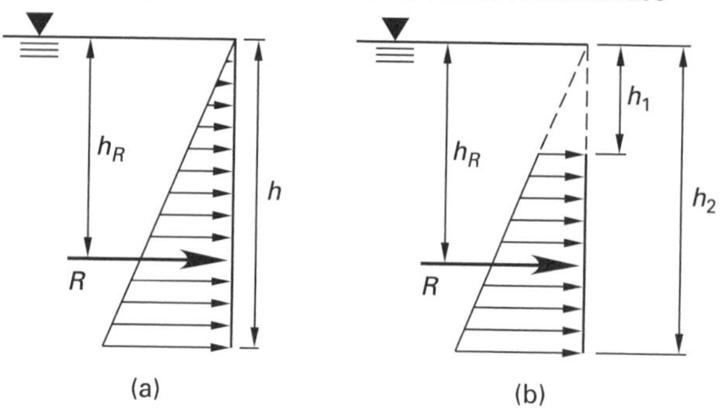

The resultant force is calculated from the *average pressure*.

$$\bar{p} = \tfrac{1}{2}(p_1 + p_2) \qquad 15.15$$

$$\bar{p} = \tfrac{1}{2}\rho g(h_1 + h_2) \qquad \text{[SI]} \qquad 15.16(a)$$

$$\bar{p} = \frac{\tfrac{1}{2}\rho g(h_1 + h_2)}{g_c} = \tfrac{1}{2}\gamma(h_1 + h_2) \qquad \text{[U.S.]} \qquad 15.16(b)$$

$$R = \bar{p}A \qquad 15.17$$

Although the resultant is calculated from the average depth, it does not act at the average depth. The resultant of the pressure distribution passes through the centroid of the pressure distribution. For the triangular distribution of Fig. 15.8(a), the resultant is located at a depth of $h_R = {}^2\!/_3 h$. For the more general case of Fig. 15.8(b), the resultant is located from Eq. 15.18.

$$h_R = \tfrac{2}{3}\left(h_1 + h_2 - \frac{h_1 h_2}{h_1 + h_2}\right) \qquad 15.18$$

7. PRESSURE ON A RECTANGULAR INCLINED PLANE SURFACE

The average pressure and resultant force on an inclined rectangular plane surface are calculated in a similar fashion as that for the vertical plane surface. (See Fig. 15.9.) The pressure varies linearly with depth. The resultant is calculated from the average pressure, which, in turn, depends on the average depth.

Figure 15.9 *Hydrostatic Pressure on an Inclined Rectangular Plane Surface*

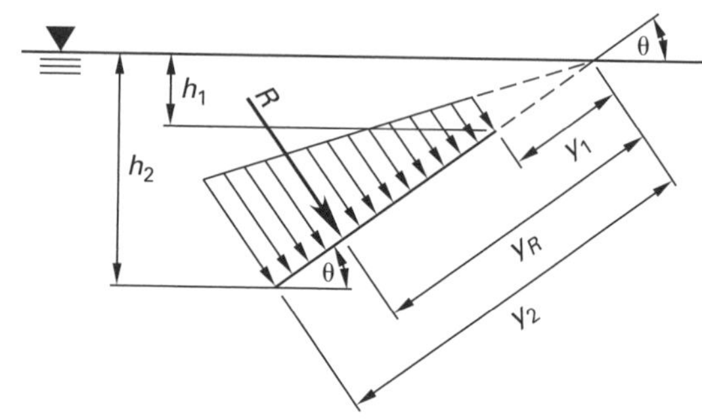

The average pressure and resultant are found using Eq. 15.19 through Eq. 15.21.

$$\bar{p} = \tfrac{1}{2}(p_1 + p_2) \qquad 15.19$$

$$\begin{aligned}\bar{p} &= \tfrac{1}{2}\rho g(h_1 + h_2) \\ &= \tfrac{1}{2}\rho g(y_1 + y_2)\sin\theta\end{aligned} \qquad \text{[SI]} \qquad 15.20(a)$$

$$\begin{aligned}\bar{p} &= \frac{\tfrac{1}{2}\rho g(h_1 + h_2)}{g_c} \\ &= \frac{\tfrac{1}{2}\rho g(y_1 + y_2)\sin\theta}{g_c} \\ &= \tfrac{1}{2}\gamma(h_1 + h_2) = \tfrac{1}{2}\gamma(y_1 + y_2)\sin\theta\end{aligned} \qquad \text{[U.S.]} \qquad 15.20(b)$$

$$R = \bar{p}A \qquad 15.21$$

As with the vertical plane surface, the resultant acts at the centroid of the pressure distribution, not at the average depth. Equation 15.18 is rewritten in terms of inclined depths.[9]

$$y_R = \left(\frac{\frac{2}{3}}{\sin\theta}\right)\left(h_1 + h_2 - \frac{h_1 h_2}{h_1 + h_2}\right) \qquad 15.22$$
$$= \frac{2}{3}\left(y_1 + y_2 - \frac{y_1 y_2}{y_1 + y_2}\right)$$

Example 15.3

The tank shown is filled with water ($\rho = 62.4$ lbm/ft^3; $\rho = 1000$ kg/m^3). (a) What is the total resultant force on a 1 ft (1 m) width of the inclined portion of the wall?[10] (b) At what depth (vertical distance) is the resultant force on the inclined portion of the wall located?

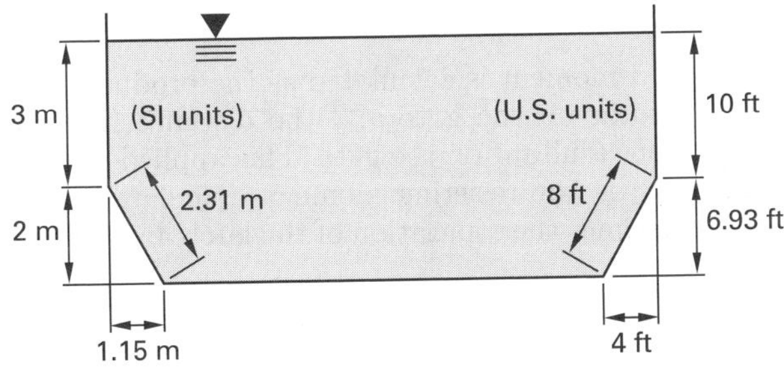

SI Solution

(a) The depth of the tank bottom is

$$h_2 = 3 \text{ m} + 2 \text{ m} = 5 \text{ m}$$

From Eq. 15.20, the average gage pressure on the inclined section is

$$\bar{p} = \left(\frac{1}{2}\right)\left(1000 \frac{\text{kg}}{\text{m}^3}\right)\left(9.81 \frac{\text{m}}{\text{s}^2}\right)(3 \text{ m} + 5 \text{ m})$$
$$= 39\,240 \text{ Pa} \quad \text{(gage)}$$

The total resultant force on a 1 m section of the inclined portion of the wall is

$$R = \bar{p}A = (39\,240 \text{ Pa})(2.31 \text{ m})(1 \text{ m})$$
$$= 90\,644 \text{ N} \quad (90.6 \text{ kN})$$

(b) θ must be known to determine y_R.

$$\theta = \arctan\frac{2 \text{ m}}{1.15 \text{ m}} = 60°$$

From Eq. 15.22, the location of the resultant can be calculated once y_1 and y_2 are known.

$$y_1 = \frac{3 \text{ m}}{\sin 60°} = 3.464 \text{ m}$$

$$y_2 = \frac{5 \text{ m}}{\sin 60°} = 5.774 \text{ m}$$

$$y_R = \left(\frac{2}{3}\right)\left(3.464 \text{ m} + 5.774 \text{ m} - \frac{(3.464 \text{ m})(5.774 \text{ m})}{3.464 \text{ m} + 5.774 \text{ m}}\right)$$
$$= 4.715 \text{ m} \quad \text{[inclined]}$$

The vertical depth at which the resultant force on the inclined portion of the wall acts is

$$h = y_R \sin\theta = (4.715 \text{ m})\sin 60°$$
$$= 4.08 \text{ m} \quad \text{[vertical]}$$

Customary U.S. Solution

(a) The water density is given in traditional U.S. mass units. The specific weight, γ, is

$$\gamma = \frac{\rho g}{g_c} = \frac{\left(62.4 \frac{\text{lbm}}{\text{ft}^3}\right)\left(32.2 \frac{\text{ft}}{\text{sec}^2}\right)}{32.2 \frac{\text{lbm-ft}}{\text{lbf-sec}^2}}$$
$$= 62.4 \text{ lbf/ft}^3$$

The depth of the tank bottom is

$$h_2 = 10 \text{ ft} + 6.93 \text{ ft} = 16.93 \text{ ft}$$

From Eq. 15.20, the average gage pressure on the inclined section is

$$\bar{p} = \left(\frac{1}{2}\right)\left(62.4 \frac{\text{lbf}}{\text{ft}^3}\right)(10 \text{ ft} + 16.93 \text{ ft})$$
$$= 840.2 \text{ lbf/ft}^2 \quad \text{(gage)}$$

The total resultant force on a 1 ft section of the inclined portion of the wall is

$$R = \bar{p}A = \left(840.2 \frac{\text{lbf}}{\text{ft}^2}\right)(8 \text{ ft})(1 \text{ ft}) = 6722 \text{ lbf}$$

---

[9] y_R is an inclined distance. If a vertical distance is wanted, it must usually be calculated from y_R and $\sin\theta$. Equation 15.22 can be derived simply by dividing Eq. 15.18 by $\sin\theta$.

[10] Since the width of the tank (the distance into and out of the illustration) is unknown, it is common to calculate the pressure or force per unit width of tank wall. This is the same as calculating the pressure on a 1 ft (1 m) wide section of wall.

(b) θ must be known to determine y_R.

$$\theta = \arctan\frac{6.93 \text{ ft}}{4 \text{ ft}} = 60^\circ$$

From Eq. 15.22, the location of the resultant can be calculated once y_1 and y_2 are known.

$$y_1 = \frac{10 \text{ ft}}{\sin 60^\circ} = 11.55 \text{ ft}$$

$$y_2 = \frac{16.93 \text{ ft}}{\sin 60^\circ} = 19.55 \text{ ft}$$

$$\begin{aligned} y_R &= \left(\frac{2}{3}\right)\left(11.55 \text{ ft} + 19.55 \text{ ft} - \frac{(11.55 \text{ ft})(19.55 \text{ ft})}{11.55 \text{ ft} + 19.55 \text{ ft}}\right) \\ &= 15.89 \text{ ft} \quad \text{[inclined]} \end{aligned}$$

The vertical depth at which the resultant force on the inclined portion of the wall acts is

$$\begin{aligned} h &= y_R \sin\theta = (15.89 \text{ ft})\sin 60^\circ \\ &= 13.76 \text{ ft} \quad \text{[vertical]} \end{aligned}$$

8. SPECIAL CASES: VERTICAL SURFACES

Several simple wall shapes and configurations recur frequently. Figure 15.10 indicates the depths, h_R, of their hydrostatic pressure resultants (*centers of pressure*). In all cases, the surfaces are vertical and extend to the liquid's surface.

Figure 15.10 *Centers of Pressure for Common Configurations*

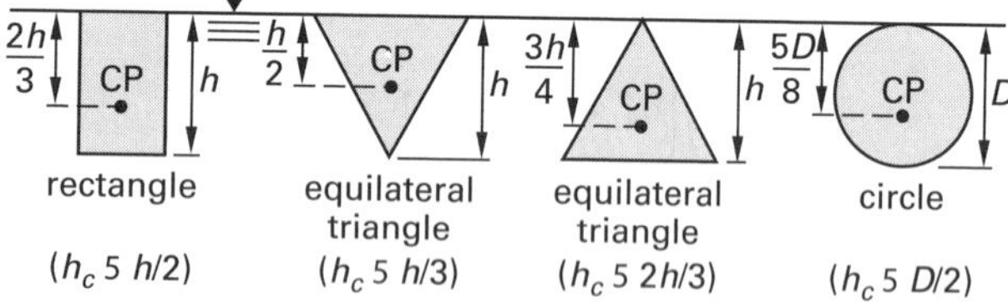

9. TORQUE ON A GATE

When an openable gate or door is submerged in such a manner as to have unequal depths of liquid on its two sides, or when there is no liquid present on one side, the hydrostatic pressure will act to either open or close the door. If the gate does not move, this pressure is resisted, usually by a latching mechanism on the gate itself.[11]

The magnitude of the resisting latch force can be determined from the *hydrostatic torque* (*hydrostatic moment*) acting on the gate. (See Fig. 15.11.) The moment is almost always taken with respect to the gate hinges.

Figure 15.11 *Torque on a Hinge (Gate)*

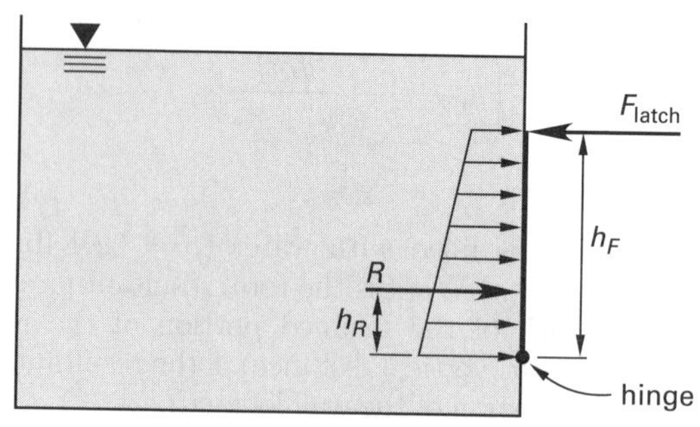

The applied moment is calculated as the product of the resultant force on the gate and the distance from the hinge to the resultant on the gate. This applied moment is balanced by the resisting moment, calculated as the latch force times the separation of the latch and hinge.

$$M_{\text{applied}} = M_{\text{resisting}} \qquad 15.23$$

$$Rh_R = F_{\text{latch}}h_F \qquad 15.24$$

10. PRESSURE DUE TO SEVERAL IMMISCIBLE LIQUIDS

Figure 15.12 illustrates the nonuniform pressure distribution due to two immiscible liquids (e.g., oil on top and water below).

Figure 15.12 *Pressure Distribution from Two Immiscible Liquids*

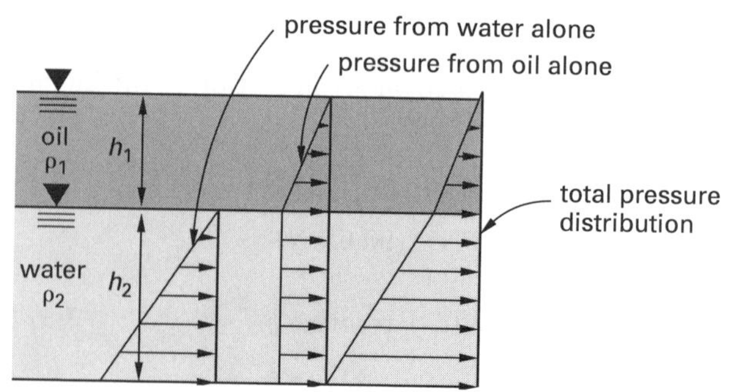

[11]Any contribution to resisting force from stiff hinges or other sources of friction is typically neglected.

The pressure due to the upper liquid (oil), once calculated, serves as a *surcharge* to the liquid below (water). The pressure at the tank bottom is given by Eq. 15.25. (The principle can be extended to three or more immiscible liquids as well.)

$$p_{\text{bottom}} = \rho_1 g h_1 + \rho_2 g h_2 \quad \text{[SI]} \qquad 15.25(a)$$

$$p_{\text{bottom}} = \frac{\rho_1 g h_1}{g_c} + \frac{\rho_2 g h_2}{g_c} = \gamma_1 h_1 + \gamma_2 h_2 \quad \text{[U.S.]} \qquad 15.25(b)$$

11. PRESSURE FROM COMPRESSIBLE FLUIDS

Fluid density, thus far, has been assumed to be independent of pressure. In reality, even "incompressible" liquids are slightly compressible. Sometimes, the effect of this compressibility cannot be neglected.

The familiar $p = \rho g h$ equation is a special case of Eq. 15.26. (It is assumed that $h_2 > h_1$. The minus sign in Eq. 15.26 indicates that pressure decreases as elevation (height) increases.)

$$\int_{p_1}^{p_2} \frac{dp}{\rho g} = -(h_2 - h_1) \quad \text{[SI]} \qquad 15.26(a)$$

$$\int_{p_1}^{p_2} \frac{g_c dp}{\rho g} = -(h_2 - h_1) \quad \text{[U.S.]} \qquad 15.26(b)$$

If the fluid is a perfect gas, and if compression is an isothermal (i.e., constant temperature) process, then the relationship between pressure and density is given by Eq. 15.27. The isothermal assumption is appropriate, for example, for the earth's *stratosphere* (i.e., above 35,000 ft (11 000 m)), where the temperature is assumed to be constant at approximately −67°F (−55°C).

$$pv = \frac{p}{\rho} = RT = \text{constant} \qquad 15.27$$

In the isothermal case, Eq. 15.27 can be rewritten as Eq. 15.28. (For air, R equals 53.35 ft-lbf/lbm-°R or 287.03 J/kg·K.) Of course, the temperature, T, must be in degrees absolute (i.e., in °R or K). Equation 15.28 is known as the *barometric height relationship*[12] because knowledge of atmospheric temperature and the pressures at two points is sufficient to determine the elevation difference between the two points.

$$h_2 - h_1 = \frac{RT}{g} \ln \frac{p_1}{p_2} \quad \text{[SI]} \qquad 15.28(a)$$

$$h_2 - h_1 = \frac{g_c RT}{g} \ln \frac{p_1}{p_2} \quad \text{[U.S.]} \qquad 15.28(b)$$

The pressure at an elevation (height) h_2 in a layer of perfect gas that has been isothermally compressed is given by Eq. 15.29.

$$p_2 = p_1 e^{g(h_1 - h_2)/RT} \quad \text{[SI]} \qquad 15.29(a)$$

$$p_2 = p_1 e^{g(h_1 - h_2)/g_c RT} \quad \text{[U.S.]} \qquad 15.29(b)$$

If the fluid is a perfect gas, and if compression is an *adiabatic process*, the relationship between pressure and density is given by Eq. 15.30,[13] where k is the *ratio of specific heats*, a property of the gas. ($k = 1.4$ for air, hydrogen, oxygen, and carbon monoxide, among others.)

$$pv^k = p\left(\frac{1}{\rho}\right)^k = \text{constant} \qquad 15.30$$

The following three equations apply to adiabatic compression of an ideal gas.

$$h_2 - h_1 = \left(\frac{k}{k-1}\right)\left(\frac{RT_1}{g}\right)\left(1 - \left(\frac{p_2}{p_1}\right)^{(k-1)/k}\right) \quad \text{[SI]} \qquad 15.31(a)$$

$$h_2 - h_1 = \left(\frac{k}{k-1}\right)\left(\frac{g_c}{g}\right)RT_1 \times \left(1 - \left(\frac{p_2}{p_1}\right)^{(k-1)/k}\right) \quad \text{[U.S.]} \qquad 15.31(b)$$

$$p_2 = p_1\left(1 - \left(\frac{k-1}{k}\right)\left(\frac{g}{RT_1}\right)(h_2 - h_1)\right)^{k/(k-1)} \quad \text{[SI]} \qquad 15.32(a)$$

$$p_2 = p_1\left(1 - \left(\frac{k-1}{k}\right)\left(\frac{g}{g_c}\right)\left(\frac{h_2 - h_1}{RT_1}\right)\right)^{k/(k-1)} \quad \text{[U.S.]} \qquad 15.32(b)$$

[12]This is equivalent to the work done in an isothermal compression process. The elevation (height) difference, $h_2 - h_1$ (with units of feet), can be interpreted as the work done per unit mass during compression (with units of ft-lbf/lbm).

[13]There is no heat or energy transfer to or from the ideal gas in an adiabatic process. However, this is not the same as an isothermal process.

$$T_2 = T_1\left(1 - \left(\frac{k-1}{k}\right)\left(\frac{g}{RT_1}\right)(h_2 - h_1)\right) \quad \text{[SI]} \qquad 15.33(a)$$

$$T_2 = T_1\left(1 - \left(\frac{k-1}{k}\right)\left(\frac{g}{g_c}\right)\left(\frac{h_2 - h_1}{RT_1}\right)\right) \quad \text{[U.S.]} \qquad 15.33(b)$$

The three adiabatic compression equations can be used for the more general *polytropic compression* case simply by substituting the *polytropic exponent*, n, for k.[14] Unlike the ratio of specific heats, the polytropic exponent is a function of the process, not of the gas. The polytropic compression assumption is appropriate for the earth's *troposphere*.[15] Assuming a linear decrease in temperature along with an altitude of −0.00356°F/ft (−0.00649°C/m), the polytropic exponent is $n = 1.235$.

Example 15.4

The air pressure and temperature at sea level are 1.0 standard atmosphere and 68°F (20°C), respectively. Assume polytropic compression with $n = 1.235$. What is the pressure at an altitude of 5000 ft (1525 m)?

SI Solution

The absolute temperature of the air is 20°C + 273° = 293K. From Eq. 15.32 (substituting $k = n = 1.235$ for polytropic compression), the pressure at 1525 m altitude is

$$p_2 = (1.0\ \text{atm}) \left(1 - \left(\frac{1.235 - 1}{1.235}\right)\left(9.81\ \frac{\text{m}}{\text{s}^2}\right) \times \left(\frac{1525\ \text{m}}{\left(287.03\ \frac{\text{J}}{\text{kg·K}}\right)(293\text{K})}\right)\right)^{1.235/(1.235-1)}$$
$$= 0.834\ \text{atm}$$

Customary U.S. Solution

The absolute temperature of the air is 68°F + 460° = 528°R. From Eq. 15.32 (substituting $k = n = 1.235$ for polytropic compression), the pressure at an altitude of 5000 ft is

$$p_2 = (1.0\ \text{atm}) \left(1 - \left(\frac{1.235 - 1}{1.235}\right)\left(\frac{32.2\ \frac{\text{ft}}{\text{sec}^2}}{32.2\ \frac{\text{lbm-ft}}{\text{lbf-sec}^2}}\right) \times \left(\frac{5000\ \text{ft}}{\left(53.35\ \frac{\text{ft-lbf}}{\text{lbm-°R}}\right)(528°\text{R})}\right)\right)^{1.235/(1.235-1)}$$
$$= 0.835\ \text{atm}$$

12. EXTERNALLY PRESSURIZED LIQUIDS

If the gas above a liquid in a closed tank is pressurized to a gage pressure of p_t, this pressure will add to the hydrostatic pressure anywhere in the fluid. The pressure at the tank bottom illustrated in Fig. 15.13 is given by Eq. 15.34.

$$p_{\text{bottom}} = p_t + \rho g h \quad \text{[SI]} \qquad 15.34(a)$$

$$p_{\text{bottom}} = p_t + \frac{\rho g h}{g_c} = p_t + \gamma h \quad \text{[U.S.]} \qquad 15.34(b)$$

Figure 15.13 *Externally Pressurized Liquid*

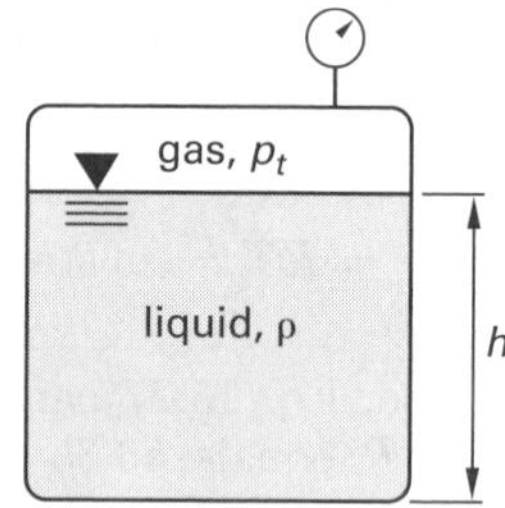

13. BUOYANCY

The *buoyant force* is an upward force that acts on all objects that are partially or completely submerged in a fluid. The fluid can be a liquid, as in the case of a ship floating at sea, or the fluid can be a gas, as in a balloon floating in the atmosphere.

[14] Actually, polytropic compression is the general process. Isothermal compression is a special case ($n = 1$) of the polytropic process, as is adiabatic compression ($n = k$).

[15] The *troposphere* is the part of the earth's atmosphere we live in and where most atmospheric disturbances occur. The *stratosphere*, starting at approximately 35,000 ft (11 000 m), is cold, clear, dry, and still. Between the troposphere and the stratosphere is the *tropopause*, a transition layer that contains most of the atmosphere's dust and moisture. Temperature actually increases with altitude in the stratosphere and decreases with altitude in the troposphere, but is constant in the tropopause.

There is a buoyant force on all submerged objects, not just those that are stationary or ascending. There will be, for example, a buoyant force on a rock sitting at the bottom of a pond. There will also be a buoyant force on a rock sitting exposed on the ground, since the rock is "submerged" in air. For partially submerged objects floating in liquids, such as icebergs, a buoyant force due to displaced air also exists, although it may be insignificant.

Buoyant force always acts to counteract an object's weight (i.e., buoyancy acts against gravity). The magnitude of the buoyant force is predicted from *Archimedes' principle* (*the buoyancy theorem*): The buoyant force on a submerged object is equal to the weight of the displaced fluid.[16] An equivalent statement of Archimedes' principle is: A floating object displaces liquid equal in weight to its own weight.

$$F_{\text{buoyant}} = \rho g V_{\text{displaced}} \quad \text{[SI]} \qquad 15.35(a)$$

$$F_{\text{buoyant}} = \frac{\rho g V_{\text{displaced}}}{g_c} = \gamma V_{\text{displaced}} \quad \text{[U.S.]} \qquad 15.35(b)$$

In the case of stationary (i.e., not moving vertically) objects, the buoyant force and object weight are in equilibrium. If the forces are not in equilibrium, the object will rise or fall until equilibrium is reached. That is, the object will sink until its remaining weight is supported by the bottom, or it will rise until the weight of displaced liquid is reduced by breaking the surface.[17]

14. BUOYANCY AND SPECIFIC GRAVITY

The specific gravity (SG) of an object submerged in water can be determined from its dry and submerged weights. Neglecting the buoyancy of any surrounding gases,

$$\text{SG} = \frac{W_{\text{dry}}}{W_{\text{dry}} - W_{\text{submerged}}} \qquad 15.36$$

Figure 15.14 illustrates an object floating partially exposed in a liquid. Neglecting the insignificant buoyant force from the displaced air (or other gas), the fractions, x, of volume exposed and submerged are easily determined.

$$x_{\text{submerged}} = \frac{\rho_{\text{object}}}{\rho_{\text{liquid}}} = \frac{(\text{SG})_{\text{object}}}{(\text{SG})_{\text{liquid}}} \qquad 15.37$$

$$x_{\text{exposed}} = 1 - x_{\text{submerged}} \qquad 15.38$$

Figure 15.14 *Partially Submerged Object*

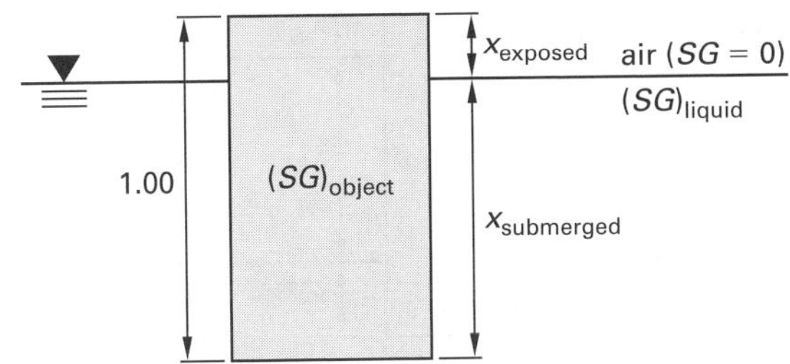

Figure 15.15 illustrates a somewhat more complicated situation—that of an object floating at the interface between two liquids of different densities. The fractions of immersion in each liquid are given by the following equations.

$$x_1 = \frac{(\text{SG})_2 - (\text{SG})_{\text{object}}}{(\text{SG})_2 - (\text{SG})_1} \qquad 15.39$$

$$x_2 = 1 - x_1 = \frac{(\text{SG})_{\text{object}} - (\text{SG})_1}{(\text{SG})_2 - (\text{SG})_1} \qquad 15.40$$

Figure 15.15 *Object Floating in Two Liquids*

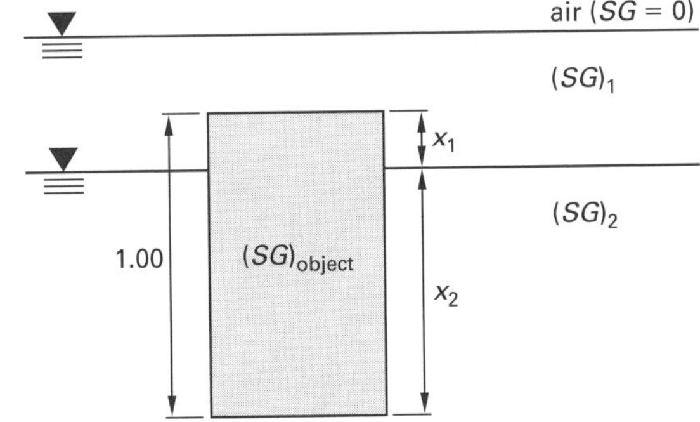

A more general case of a floating object is shown in Fig. 15.16. Situations of this type are easily evaluated by equating the object's weight with the sum of the buoyant forces.

In the case of Fig. 15.16 (with two liquids), the following relationships apply, where x_0 is the fraction, if any, extending into the air above.

$$(\text{SG})_{\text{object}} = x_1(\text{SG})_1 + x_2(\text{SG})_2 \qquad 15.41$$

$$(\text{SG})_{\text{object}} = (1 - x_0 - x_2)(\text{SG})_1 + x_2(\text{SG})_2 \qquad 15.42$$

$$(\text{SG})_{\text{object}} = x_1(\text{SG})_1 + (1 - x_0 - x_1)(\text{SG})_2 \qquad 15.43$$

[16]The volume term in Eq. 15.35 is the total volume of the object only in the case of complete submergence.

[17]An object can also stop rising or falling due to a change in the fluid's density. The buoyant force will increase with increasing depth in the ocean due to an increase in density at great depths. The buoyant force will decrease with increasing altitude in the atmosphere due to a decrease in density at great heights.

Figure 15.16 *General Two-Liquid Buoyancy Problem*

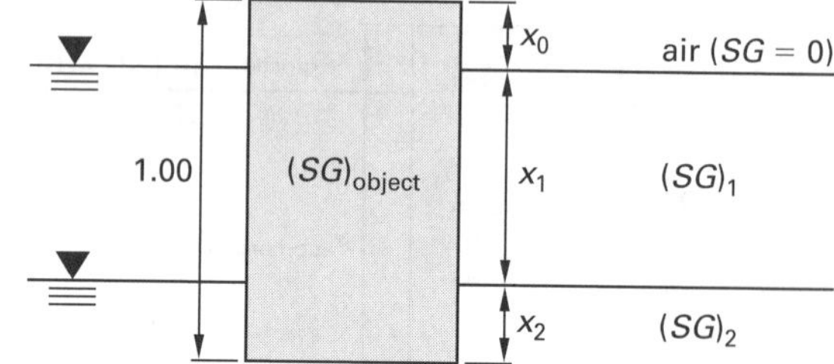

Example 15.5

An empty polyethylene telemetry balloon and payload have a mass of 500 lbm (225 kg). The balloon is filled with helium when the atmospheric conditions are 60°F (15.6°C) and 14.8 psia (102 kPa). The specific gas constant of helium is 2079 J/kg·K (386.3 ft-lbf/lbm-°R). What minimum volume of helium is required for lift-off from a sea-level platform?

SI Solution

$$\rho_{\text{air}} = \frac{p}{RT} = \frac{1.02 \times 10^5 \text{ Pa}}{\left(287.03 \ \frac{\text{J}}{\text{kg}\cdot\text{K}}\right)(15.6^\circ\text{C} + 273^\circ)}$$
$$= 1.231 \text{ kg/m}^3$$

$$\rho_{\text{helium}} = \frac{1.02 \times 10^5 \text{ Pa}}{\left(2079 \ \frac{\text{J}}{\text{kg}\cdot\text{K}}\right)(288.6\text{K})} = 0.17 \text{ kg/m}^3$$

$$m = 225 \text{ kg} + \left(0.17 \ \frac{\text{kg}}{\text{m}^3}\right) V_{\text{He}}$$

$$m_b = \left(1.231 \ \frac{\text{kg}}{\text{m}^3}\right) V_{\text{He}}$$

$$225 \text{ kg} + \left(0.17 \ \frac{\text{kg}}{\text{m}^3}\right) V_{\text{He}} = \left(1.231 \ \frac{\text{kg}}{\text{m}^3}\right) V_{\text{He}}$$

$$V_{\text{He}} = 212.1 \text{ m}^3$$

Customary U.S. Solution

The gas densities are

$$\rho_{\text{air}} = \frac{p}{RT} = \frac{\left(14.8 \ \frac{\text{lbf}}{\text{in}^2}\right)\left(12 \ \frac{\text{in}}{\text{ft}}\right)^2}{\left(53.35 \ \frac{\text{ft-lbf}}{\text{lbm-}^\circ\text{R}}\right)(60^\circ\text{F} + 460^\circ)}$$
$$= 0.07682 \text{ lbm/ft}^3$$

$$\gamma_{\text{air}} = \rho \times \frac{g}{g_c} = 0.07682 \text{ lbf/ft}^3$$

$$\rho_{\text{helium}} = \frac{\left(14.8 \ \frac{\text{lbf}}{\text{in}^2}\right)\left(12 \ \frac{\text{in}}{\text{ft}}\right)^2}{\left(386.3 \ \frac{\text{ft-lbf}}{\text{lbm-}^\circ\text{R}}\right)(520^\circ\text{R})}$$
$$= 0.01061 \text{ lbm/ft}^3$$

$$\gamma_{\text{helium}} = 0.01061 \text{ lbf/ft}^3$$

The total weight of the balloon, payload, and helium is

$$W = 500 \text{ lbf} + \left(0.01061 \ \frac{\text{lbf}}{\text{ft}^3}\right) V_{\text{He}}$$

The buoyant force is the weight of the displaced air. Neglecting the payload volume, the displaced air volume is the same as the helium volume.

$$F_b = \left(0.07682 \ \frac{\text{lbf}}{\text{ft}^3}\right) V_{\text{He}}$$

At lift-off, the weight of the balloon is just equal to the buoyant force.

$$W = F_b$$
$$500 \text{ lbf} + \left(0.01061 \ \frac{\text{lbf}}{\text{ft}^3}\right) V_{\text{He}} = \left(0.07682 \ \frac{\text{lbf}}{\text{ft}^3}\right) V_{\text{He}}$$
$$V_{\text{He}} = 7552 \text{ ft}^3$$

15. BUOYANCY OF SUBMERGED PIPELINES

Whenever possible, submerged pipelines for river crossings should be completely buried at a level below river scour. This will reduce or eliminate loads and movement due to flutter, scour and fill, drag, collisions, and buoyancy. Submerged pipelines should cross at right angles to the river. For maximum flexibility and ductility, pipelines should be made of thick-walled mild steel.

Submerged pipelines should be weighted to achieve a minimum of 20% negative buoyancy (i.e., an average density of 1.2 times the environment, approximately 72 lbm/ft^3 or 1200 kg/m^3). Metal or concrete clamps can be used for this purpose, as well as concrete coatings. Thick steel clamps have the advantage of a smaller lateral exposed area (resulting in less drag from river flow), while brittle concrete coatings are sensitive to pipeline flutter and temperature fluctuations.

Due to the critical nature of many pipelines and the difficulty in accessing submerged portions for repair, it is common to provide a parallel auxiliary line. The auxiliary and main lines are provided with crossover and mainline valves, respectively, on high ground at both sides of the river to permit either or both lines to be used.

16. INTACT STABILITY: STABILITY OF FLOATING OBJECTS

A stationary object is said to be in *static equilibrium*. However, an object in static equilibrium is not necessarily stable. For example, a coin balanced on edge is in static equilibrium, but it will not return to the balanced position if it is disturbed. An object is said to be *stable* (i.e., in *stable equilibrium*) if it tends to return to the equilibrium position when slightly displaced.

Stability of floating and submerged objects is known as *intact stability*.[18] There are two forces acting on a stationary floating object: the buoyant force and the object's weight. The buoyant force acts upward through the centroid of the displaced volume. This centroid is known as the *center of buoyancy*. The gravitational force on the object (i.e., the object's weight) acts downward through the object's center of gravity.

For a totally submerged object (as in the balloon and submarine shown in Fig. 15.17) to be stable, the center of buoyancy must be above the center of gravity. The object will be stable because a righting moment will be created if the object tips over, since the center of buoyancy will move outward from the center of gravity.

Figure 15.17 *Stability of a Submerged Object*

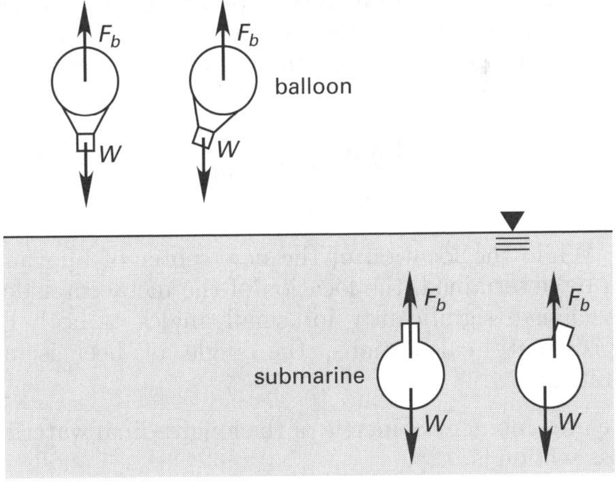

The stability criterion is different for partially submerged objects (e.g., surface ships). If the vessel shown in Fig. 15.18 heels (i.e., lists or rolls), the location of the center of gravity of the object does not change.[19] However, the center of buoyancy shifts to the centroid of the new submerged section 123. The centers of buoyancy and gravity are no longer in line.

This righting couple exists when the extension of the buoyant force, F_b, intersects line O–O above the center of gravity at M, the *metacenter*. For partially submerged objects to be stable, the metacenter must be above the center of gravity. If M lies below the center of gravity, an

Figure 15.18 *Stability of a Partially Submerged Floating Object*

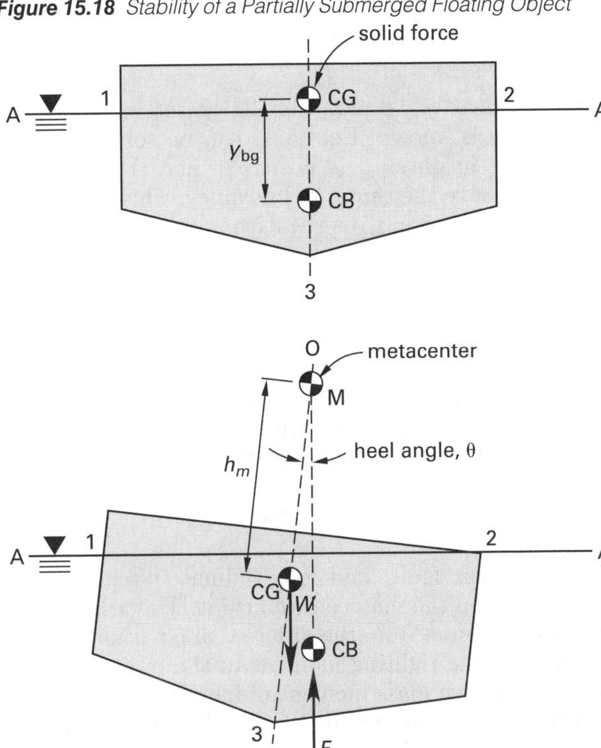

overturning couple will exist. The distance between the center of gravity and the metacenter is called the *metacentric height*, and it is reasonably constant for heel angles less than 10°. Also, for angles less than 10°, the center of buoyancy follows a locus for which the metacenter is the instantaneous center.

The metacentric height is one of the most important and basic parameters in ship design. It determines the ship's ability to remain upright as well as the ship's roll and pitch characteristics.

"Acceptable" minimum values of the metacentric height have been established from experience, and these depend on the ship type and class. For example, many submarines are required to have a metacentric height of 1 ft (0.3 m) when surfaced. This will increase to approximately 3.5 ft (1.2 m) for some of the largest surface ships. If an acceptable metacentric height is not achieved initially, the center of gravity must be lowered or the keel depth increased. The beam width can also be increased slightly to increase the waterplane moment of inertia.

For a surface vessel rolling through an angle less than approximately 10°, the distance between the vertical center of gravity and the metacenter can be found from Eq. 15.44. Variable I is the centroidal area moment of

[18]The subject of intact stability, being a part of naval architecture curriculum, is not covered extensively in most fluids books. However, it is covered extensively in basic ship design and naval architecture books.
[19]The verbs *roll*, *list*, and *heel* are synonymous.

inertia of the original waterline (free surface) cross section about a longitudinal (fore and aft) waterline axis; V is the displaced volume.

If the distance, y_{bg}, separating the centers of buoyancy and gravity is known, Eq. 15.44 can be solved for the metacentric height. y_{bg} is positive when the center of gravity is above the center of buoyancy. This is the normal case. Otherwise, y_{bg} is negative.

$$y_{bg} + h_m = \frac{I}{V} \qquad 15.44$$

The *righting moment* (also known as the *restoring moment*) is the stabilizing moment exerted when the ship rolls. Values of the righting moment are typically specified with units of foot-tons (MN·m).

$$M_{righting} = h_m \gamma_w V_{displaced} \sin\theta \qquad 15.45$$

The transverse (roll) and longitudinal (pitch) *periods* also depend on the metacentric height. The roll characteristics are found from the differential equation formed by equating the righting moment to the product of the ship's transverse mass moment of inertia and the angular acceleration. Larger metacentric heights result in lower roll periods. If r is the radius of gyration about the roll axis, the roll period is

$$T_{roll} = \frac{2\pi r}{\sqrt{g h_m}} \qquad 15.46$$

The roll and pitch periods must be adjusted for the appropriate level of crew and passenger comfort. A "beamy" ice-breaking ship will have a metacentric height much larger than normally required for intact stability, resulting in a short, nauseating roll period. The designer of a passenger ship, however, would have to decrease the intact stability (i.e., decrease the metacentric height) in order to achieve an acceptable ride characteristic. This requires a moderate metacentric height that is less than approximately 6% of the beam length.

Example 15.6

A 600,000 lbm (280 000 kg) rectangular barge has external dimensions of 24 ft width, 98 ft length, and 12 ft height (7 m × 30 m × 3.6 m). It floats in seawater ($\gamma_w = 64.0\ \text{lbf/ft}^3$; $\rho_w = 1024\ \text{kg/m}^3$). The center of gravity is 7.8 ft (2.4 m) from the top of the barge as loaded. Find (a) the location of the center of buoyancy when the barge is floating on an even keel, and (b) the approximate location of the metacenter when the barge experiences a 5° heel.

SI Solution

(a) Refer to the following diagram. Let dimension y represent the depth of the submerged barge.

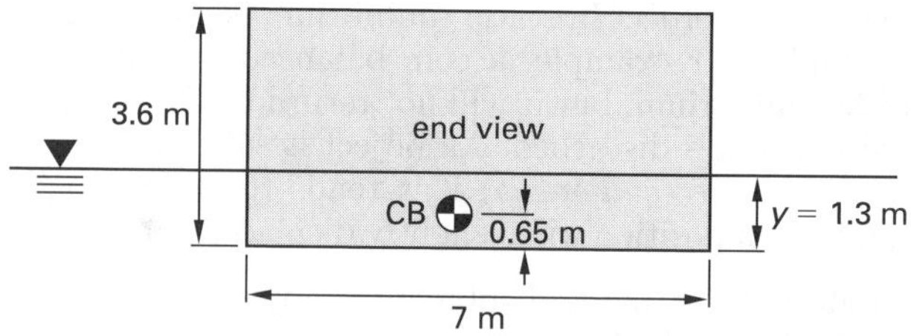

From Archimedes' principle, the buoyant force equals the weight of the barge. This, in turn, equals the weight of the displaced seawater.

$$F_b = W = V\rho_w g$$

$$(280\,000\ \text{kg})\left(9.81\ \frac{\text{m}}{\text{s}^2}\right) = y\big((7\ \text{m})(30\ \text{m})\big)\left(1024\ \frac{\text{kg}}{\text{m}^3}\right) \times \left(9.81\ \frac{\text{m}}{\text{s}^2}\right)$$

$$y = 1.3\ \text{m}$$

The center of buoyancy is located at the centroid of the submerged cross section. When floating on an even keel, the submerged cross section is rectangular with a height of 1.3 m. The height of the center of buoyancy above the keel is

$$\frac{1.3\ \text{m}}{2} = 0.65\ \text{m}$$

(b) While the location of the new center of buoyancy can be determined, the location of the metacenter does not change significantly for small angles of heel. For approximate calculations, the angle of heel is not significant.

The area moment of inertia of the longitudinal waterline cross section is

$$I = \frac{Lw^3}{12} = \frac{(30\ \text{m})(7\ \text{m})^3}{12} = 858\ \text{m}^4$$

The submerged volume is

$$V = (1.3\ \text{m})(7\ \text{m})(30\ \text{m}) = 273\ \text{m}^3$$

The distance between the center of gravity and the center of buoyancy is

$$y_{bg} = 3.6\ \text{m} - 2.4\ \text{m} - 0.65\ \text{m} = 0.55\ \text{m}$$

The metacentric height measured above the center of gravity is

$$h_m = \frac{I}{V} - y_{\text{bg}} = \frac{858 \text{ m}^4}{273 \text{ m}^3} - 0.55 \text{ m} = 2.6 \text{ m}$$

Customary U.S. Solution

(a) Refer to the following diagram. Let dimension y represent the depth of the submerged barge.

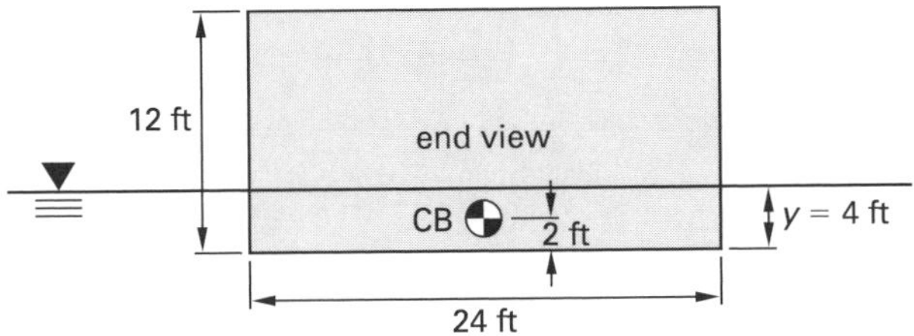

From Archimedes' principle, the buoyant force equals the weight of the barge. This, in turn, equals the weight of the displaced seawater.

$$F_b = W = V\gamma_w$$

$$600{,}000 \text{ lbf} = y\big((24 \text{ ft})(98 \text{ ft})\big)\left(64 \ \frac{\text{lbf}}{\text{ft}^3}\right)$$

$$y = 4 \text{ ft}$$

The center of buoyancy is located at the centroid of the submerged cross section. When floating on an even keel, the submerged cross section is rectangular with a height of 4 ft. The height of the center of buoyancy above the keel is

$$\frac{4 \text{ ft}}{2} = 2 \text{ ft}$$

(b) While the location of the new center of buoyancy can be determined, the location of the metacenter does not change significantly for small angles of heel. Therefore, for approximate calculations, the angle of heel is not significant.

The area moment of inertia of the longitudinal waterline cross section is

$$I = \frac{Lw^3}{12} = \frac{(98 \text{ ft})(24 \text{ ft})^3}{12} = 112{,}900 \text{ ft}^4$$

The submerged volume is

$$V = (4 \text{ ft})(24 \text{ ft})(98 \text{ ft}) = 9408 \text{ ft}^3$$

The distance between the center of gravity and the center of buoyancy is

$$y_{\text{bg}} = 12 \text{ ft} - 7.8 \text{ ft} - 2 \text{ ft} = 2.2 \text{ ft}$$

The metacentric height measured above the center of gravity is

$$h_m = \frac{I}{V} - y_{\text{bg}} = \frac{112{,}900 \text{ ft}^4}{9408 \text{ ft}^3} - 2.2 \text{ ft}$$
$$= 9.8 \text{ ft}$$

17. NOMENCLATURE

A	area	ft^2	m^2
D	diameter	ft	m
F	force	lbf	N
g	gravitational acceleration, 32.17 (9.807)	ft/sec^2	m/s^2
g_c	gravitational constant, 32.17	lbm-ft/lbf-sec^2	n.a.
h	vertical depth or height	ft	m
I	moment of inertia	ft^4	m^4
k	ratio of specific heats	–	–
L	length	ft	m
M	moment	ft-lbf	N·m
n	polytropic exponent	–	–
p	pressure	lbf/ft^2	Pa
r	radius of gyration	ft	m
R	resultant force	lbf	N
R	specific gas constant	ft-lbf/lbm-°R	J/kg·K
SG	specific gravity	–	–
T	absolute temperature	°R	K
v	specific volume	ft^3/lbm	m^3/kg
V	volume	ft^3	m^3
W	weight	lbf	n.a.
x	fraction	–	–
y	distance	ft	m
y	inclined distance	ft	m
z	elevation	ft	m

Symbols

γ	specific weight	lbf/ft^3	N/m^3
θ	angle	deg	deg
ρ	density	lbm/ft^3	kg/m^3
μ	coefficient of friction	–	–
ρ	density	lbf/ft^3	kg/m^3
ω	angular velocity	rad/sec	rad/s

Subscripts

a	atmospheric
b	buoyant
bg	between CB and CG
c	centroidal

Fluids

F	force
He	helium
m	manometer fluid or metacentric
R	resultant
t	tank
v	vapor
w	water

16 Fluid Flow Parameters

Content in blue refers to the *NCEES Handbook*.

NCEES EXAM SPECIFICATIONS AND RELATED CONTENT

MACHINE DESIGN AND MATERIALS EXAM

II.C.5. Supportive Knowledge: Testing and instrumentation
- 5. Bernoulli Equation
- 7. Pitot Tube

THERMAL AND FLUID SYSTEMS EXAM

I.B.3. Fluid Mechanics: Incompressible flow
- 9. Hydraulic Diameter
- 10. Reynolds Number

III.A.2. Energy/Power Equipment: Boilers and steam generators
- 23. Steam Traps

1. INTRODUCTION TO FLUID ENERGY UNITS

Several important fluids and thermodynamics equations, such as Bernoulli's equation and the steady-flow energy equation, are special applications of the *conservation of energy* concept. However, it is not always obvious how some formulations of these equations can be termed "energy." For example, elevation, z, with units of feet, is often called *gravitational energy*.

Fluid energy is expressed per unit mass, as indicated by the name, *specific energy*. Units of fluid specific energy are commonly ft-lbf/lbm in the traditional English system and J/kg (or m^2/s^2, which is equivalent) in the SI system.[1]

2. KINETIC ENERGY

Energy is needed to accelerate a stationary body. Therefore, a moving mass of fluid possesses more energy than an identical, stationary mass. The energy difference is the *kinetic energy* of the fluid.[2] If the kinetic energy is evaluated per unit mass, the term *specific kinetic energy* is used. Equation 16.1 gives the specific kinetic energy corresponding to a fluid flow with average velocity, v.

$$E_\mathrm{v} = \frac{\mathrm{v}^2}{2} \quad \text{[SI]} \qquad 16.1(a)$$

$$E_\mathrm{v} = \frac{\mathrm{v}^2}{2g_c} \quad \text{[U.S.]} \qquad 16.1(b)$$

3. POTENTIAL ENERGY

Work is performed in elevating a body. Therefore, a mass of fluid at a high elevation will have more energy than an identical mass of fluid at a lower elevation. The energy difference is the *potential energy* of the fluid.[3]

[1]The ft-lbf/lbm unit may be thought of as just feet, although lbf and lbm do not really cancel out. The combination of variables $E \times (g_c/g)$ is required in order for the units to resolve into "ft."

[2]The terms *velocity energy* and *dynamic energy* are used less often.

[3]The term *gravitational energy* is also used.

Like kinetic energy, potential energy is usually expressed per unit mass. Equation 16.2 gives the specific potential energy of fluid at an elevation z.[4]

$$E_z = zg \quad \text{[SI]} \quad 16.2(a)$$

$$E_z = \frac{zg}{g_c} \quad \text{[U.S.]} \quad 16.2(b)$$

z is the elevation of the fluid. The reference point (i.e., zero elevation point) is entirely arbitrary and can be chosen for convenience. This is because potential energy always appears in a difference equation (i.e., ΔE_z), and the reference point cancels out.

4. PRESSURE ENERGY

Work is performed and energy is added when a substance is compressed. Therefore, a mass of fluid at a high pressure will have more energy than an identical mass of fluid at a lower pressure. The energy difference is the *pressure energy* of the fluid.[5] Pressure energy is usually found in equations along with kinetic and potential energies and is expressed as energy per unit mass. Equation 16.3 gives the specific pressure energy of fluid at pressure p.

$$E_p = \frac{p}{\rho} \quad 16.3$$

5. BERNOULLI EQUATION

The *Bernoulli equation* is an ideal energy conservation equation based on several reasonable assumptions. The equation assumes the following.

- The fluid is incompressible.
- There is no fluid friction.
- Changes in thermal energy are negligible.[6]

The Bernoulli equation states that the *total specific energy* of a fluid flowing without friction losses in a pipe is constant.[7] The total specific energy possessed by the fluid is the sum of its pressure, kinetic, and potential specific energies. Drawing on Eq. 16.1, Eq. 16.2, and Eq. 16.3, the Bernoulli equation is written as

$$E_t = E_p + E_v + E_z = \text{constant} \quad 16.4$$

Duct Design: Bernoulli Equation

$$E_t = \frac{p}{\rho} + \frac{v^2}{2} + zg = \text{constant} \quad \text{[SI]} \quad 16.5(a)$$

$$E_t = \frac{p}{\rho} + \frac{v^2}{2g_c} + \frac{zg}{g_c} = \text{constant} \quad \text{[U.S.]} \quad 16.5(b)$$

The Bernoulli equation may be expressed in head instead of specific energy by dividing both sides by g/g_c (in U.S. units) or by g (in SI units).

$$h_t = \frac{p}{\rho g} + \frac{v^2}{2g} + z = \text{constant} \quad \text{[SI]} \quad 16.6(a)$$

$$h_t = \frac{p g_c}{\rho g} + \frac{v^2}{2g} + z = \text{constant} \quad \text{[U.S.]} \quad 16.6(b)$$

In Eq. 16.6(a) and Eq. 16.6(b), the three terms added together are *static head* (also called *pressure head*), *velocity head*, and *elevation head*, in that order.

The Bernoulli equation may be stated in terms of the properties at two points in the same flow, with point 1 upstream of point 2. An additional term, h_f, is often added to account for the *head loss* (also called *friction head*) due to frictional forces acting against the fluid's motion between points 1 and 2.

Bernoulli Equation

$$\frac{p_1}{\rho g} + \frac{v_1^2}{2g} + z_1 = \frac{p_2}{\rho g} + \frac{v_2^2}{2g} + z_2 + h_f$$

$$\frac{p_1}{\gamma} + \frac{v_1^2}{2g} + z_1 = \frac{p_2}{\gamma} + \frac{v_2^2}{2g} + z_2 + h_f \quad \text{[SI]} \quad 16.7(a)$$

$$\frac{p_1 g_c}{\rho g} + \frac{v_1^2}{2g} + z_1 = \frac{p_1 g_c}{\rho g} + \frac{v_2^2}{2g} + z_2 + h_f$$

$$\frac{p_1 g_c}{\gamma} + \frac{v_1^2}{2g} + z_1 = \frac{p_1 g_c}{\gamma} + \frac{v_2^2}{2g} + z_2 + h_f \quad \text{[U.S.]} \quad 16.7(b)$$

The Bernoulli equation is valid for both laminar and turbulent flows. Since the original research by Bernoulli assumed laminar flow, when the Bernoulli equation is used for turbulent flow, it is sometimes referred to as the "steady-flow energy equation." The Bernoulli equation can also be used for gases and vapors if the incompressibility assumption is valid.[8]

[4]Since $g = g_c$ (numerically), it is tempting to write $E_z = z$. In fact, many engineers in the United States do just that.

[5]The terms *static energy* and *flow energy* are also used. The name *flow energy* results from the need to push (pressurize) a fluid to get it to flow through a pipe. However, flow energy and kinetic energy are not the same.

[6]In thermodynamics, the fluid flow is said to be *adiabatic*.

[7]This is the total specific energy because the energy is per unit mass. However, the word "specific," being understood, is often omitted. Of course, "the total energy of the system" means something else and requires knowing the fluid mass in the system.

[8]A gas or vapor can be considered to be incompressible as long as its pressure does not change by more than 10% between the entrance and exit, and its velocity is less than Mach 0.3 everywhere.

The *total head*, h_t, and *total pressure*, p_t, can similarly calculated from the total specific energy.

$$h_t = \frac{E_t}{g} \quad \text{[SI]} \qquad 16.8(a)$$

$$h_t = E_t \times \frac{g_c}{g} \quad \text{[U.S.]} \qquad 16.8(b)$$

$$p_t = \rho g h_t \quad \text{[SI]} \qquad 16.9(a)$$

$$p_t = \rho h_t \times \frac{g}{g_c} \quad \text{[U.S.]} \qquad 16.9(b)$$

Example 16.1

A pipe draws water from the bottom of a reservoir and discharges it freely at point C, 100 ft (30 m) below the surface. The flow is frictionless. (a) What is the total specific energy at an elevation 50 ft (15 m) below the water surface (i.e., point B)? (b) What is the velocity at point C?

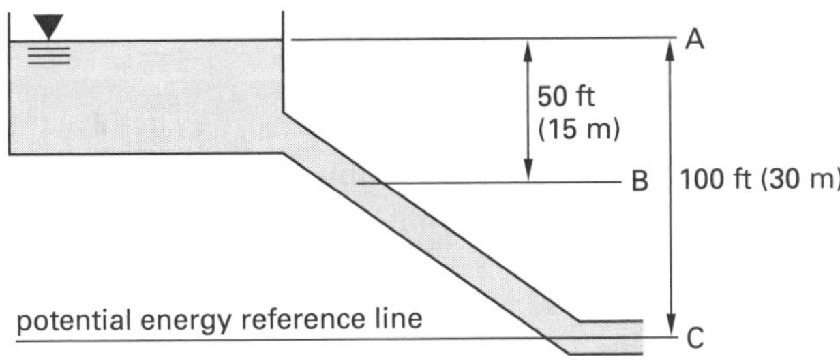

SI Solution

(a) At point A, the velocity and gage pressure are both zero. Therefore, the total energy consists only of potential energy. Choose point C as the reference ($z = 0$) elevation.

$$\begin{aligned} E_{\text{A}} &= z_{\text{A}} g \\ &= (30\ \text{m})\left(9.81\ \frac{\text{m}}{\text{s}^2}\right) \\ &= 294.3\ \text{m}^2/\text{s}^2 \quad (\text{J/kg}) \end{aligned}$$

At point B, the fluid is moving and possesses kinetic energy. The fluid is also under hydrostatic pressure and possesses pressure energy. These energy forms have come at the expense of potential energy. (This is a direct result of the Bernoulli equation.) Also, the flow is frictionless. Therefore, there is no net change in the total energy between points A and B.

$$\begin{aligned} E_{\text{B}} &= E_{\text{A}} \\ &= 294.3\ \text{m}^2/\text{s}^2 \quad (\text{J/kg}) \end{aligned}$$

(b) At point C, the gage pressure and pressure energy are again zero, since the discharge is at atmospheric pressure. The potential energy is zero, since $z = 0$. The total energy of the system has been converted to kinetic energy. From Eq. 16.5,

$$\begin{aligned} E_t &= 294.3\ \frac{\text{m}^2}{\text{s}^2} \\ &= 0 + \frac{\text{v}^2}{2} + 0 \\ \text{v} &= 24.3\ \text{m/s} \end{aligned}$$

Customary U.S. Solution

(a)

$$\begin{aligned} E_{\text{A}} &= \frac{z_{\text{A}} g}{g_c} \\ &= \frac{(100\ \text{ft})\left(32.2\ \frac{\text{ft}}{\text{sec}^2}\right)}{32.2\ \frac{\text{lbm-ft}}{\text{lbf-sec}^2}} \\ &= 100\ \text{ft-lbf/lbm} \\ E_t &= E_{\text{B}} \\ &= E_{\text{A}} \\ &= 100\ \text{ft-lbf/lbm} \end{aligned}$$

(b)

$$\begin{aligned} E_t &= E_{\text{C}} \\ &= 100\ \frac{\text{ft-lbf}}{\text{lbm}} \\ &= 0 + \frac{\text{v}^2}{2g_c} + 0 \\ \text{v}^2 &= 2g_c E_t \\ &= (2)\left(32.2\ \frac{\text{lbm-ft}}{\text{lbf-sec}^2}\right)\left(100\ \frac{\text{ft-lbf}}{\text{lbm}}\right) \\ &= 6440\ \text{ft}^2/\text{sec}^2 \\ \text{v} &= \sqrt{6440\ \frac{\text{ft}^2}{\text{sec}^2}} \\ &= 80.2\ \text{ft/sec} \end{aligned}$$

Example 16.2

Water (62.4 lbm/ft^3; 1000 kg/m^3) is pumped up a hillside into a reservoir. The pump discharges water with a velocity of 6 ft/sec (2 m/s) and a pressure of 150 psig (1000 kPa). Disregarding friction, what is the maximum elevation (above the centerline of the pump's discharge) of the reservoir's water surface?

SI Solution

At the centerline of the pump's discharge, the potential energy is zero. The atmospheric pressure at the pump inlet is the same as (and counteracts) the atmospheric pressure at the reservoir surface, so gage pressures may be used. The pressure and velocity energies are

$$E_p = \frac{p}{\rho} = \frac{(1000\ \text{kPa})\left(1000\ \frac{\text{Pa}}{\text{kPa}}\right)}{1000\ \frac{\text{kg}}{\text{m}^3}} = 1000\ \text{J/kg}$$

$$E_\text{v} = \frac{\text{v}^2}{2} = \frac{\left(2\ \frac{\text{m}}{\text{s}}\right)^2}{2} = 2\ \text{J/kg}$$

The total energy at the pump's discharge is

$$E_{t,1} = E_p + E_\text{v} = 1000\ \frac{\text{J}}{\text{kg}} + 2\ \frac{\text{J}}{\text{kg}} = 1002\ \text{J/kg}$$

Since the flow is frictionless, the same energy is possessed by the water at the reservoir's surface. Since the velocity and gage pressure at the surface are zero, all of the available energy has been converted to potential energy.

$$E_{t,2} = E_{t,1}$$

$$z_2 g = 1002\ \text{J/kg}$$

$$z_2 = \frac{E_{t,2}}{g} = \frac{1002\ \frac{\text{J}}{\text{kg}}}{9.81\ \frac{\text{m}}{\text{s}^2}} = 102.1\ \text{m}$$

The volumetric flow rate of the water is not relevant since the water velocity was known. Similarly, the pipe size is not needed.

Customary U.S. Solution

The atmospheric pressure at the pump inlet is the same as (and counteracts) the atmospheric pressure at the reservoir surface, so gage pressures may be used.

$$E_p = \frac{p}{\rho} = \frac{\left(150\ \frac{\text{lbf}}{\text{in}^2}\right)\left(12\ \frac{\text{in}}{\text{ft}}\right)^2}{62.4\ \frac{\text{lbm}}{\text{ft}^3}} = 346.15\ \text{ft-lbf/lbm}$$

$$E_\text{v} = \frac{\text{v}^2}{2g_c} = \frac{\left(6\ \frac{\text{ft}}{\text{sec}}\right)^2}{(2)\left(32.2\ \frac{\text{lbm-ft}}{\text{lbf-sec}^2}\right)} = 0.56\ \text{ft-lbf/lbm}$$

$$E_{t,1} = E_p + E_\text{v} = 346.15\ \frac{\text{ft-lbf}}{\text{lbm}} + 0.56\ \frac{\text{ft-lbf}}{\text{lbm}} = 346.71\ \text{ft-lbf/lbm}$$

$$E_{t,2} = E_{t,1}$$

$$\frac{z_2 g}{g_c} = 346.71\ \text{ft-lbf/lbm}$$

$$z_2 = \frac{E_{t,2} g_c}{g} = \frac{\left(346.71\ \frac{\text{ft-lbf}}{\text{lbm}}\right)\left(32.2\ \frac{\text{lbm-ft}}{\text{lbf-sec}^2}\right)}{32.2\ \frac{\text{ft}}{\text{sec}^2}} = 346.71\ \text{ft}$$

6. IMPACT ENERGY

Impact energy, E_0 (also known as *stagnation energy* and *total energy*), is the sum of the kinetic and pressure energy terms.[9] Equation 16.11 is applicable to liquids and gases flowing with velocities less than approximately Mach 0.3.

$$E_0 = E_p + E_\text{v} \qquad 16.10$$

[9]It is confusing to label Eq. 16.10 *total* when the gravitational energy term has been omitted. However, the reference point for gravitational energy is arbitrary, and in this application the reference coincides with the centerline of the fluid flow. In truth, the effective pressure developed in a fluid which has been brought to rest adiabatically does not depend on the elevation or altitude of the fluid. This situation is seldom ambiguous. The application will determine which definition of total head or total energy is intended.

$$E_0 = \frac{p}{\rho} + \frac{v^2}{2} \quad \text{[SI]} \quad 16.11(a)$$

$$E_0 = \frac{p}{\rho} + \frac{v^2}{2g_c} \quad \text{[U.S.]} \quad 16.11(b)$$

Impact head, h_0, is calculated from the impact energy in a manner analogous to Eq. 16.8. Impact head represents the height the liquid will rise in a piezometer-pitot tube when the liquid has been brought to rest (i.e., stagnated) in an adiabatic manner. Such a case is illustrated in Fig. 16.1. If a gas or high-velocity, high-pressure liquid is flowing, it will be necessary to use a mercury manometer or pressure gauge to measure stagnation head.

Figure 16.1 *Pitot Tube-Piezometer Apparatus*

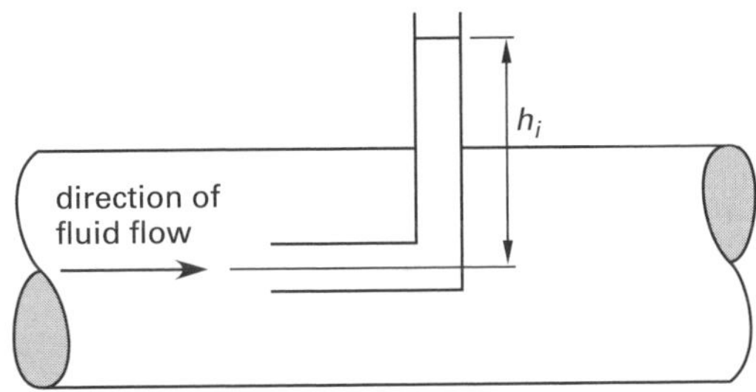

7. PITOT TUBE

A *pitot tube* (also known as an *impact tube* or *stagnation tube*) is simply a hollow tube that is placed longitudinally in the direction of fluid flow, allowing the flow to enter one end at the fluid's *velocity of approach*. (See Fig. 16.2.) It is used to measure velocity of flow and finds uses in both subsonic and supersonic applications.

Figure 16.2 *Pitot Tube*

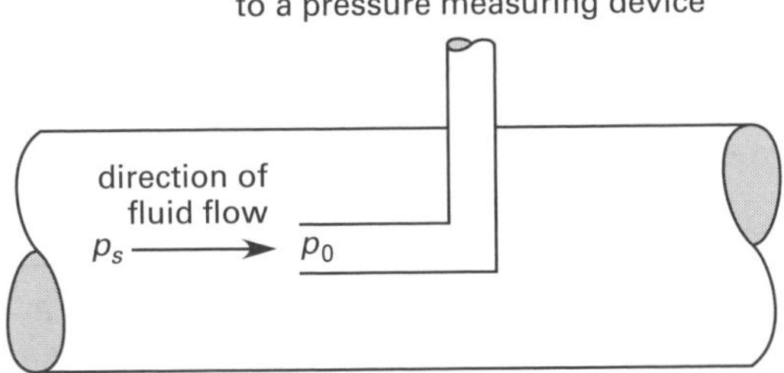

When the fluid enters the pitot tube, it is forced to come to a stop (at the *stagnation point*), and the velocity energy is transformed into pressure energy. If the fluid is a low-velocity gas, the stagnation is assumed to occur without compression heating of the gas. If there is no friction (the common assumption), the process is said to be adiabatic.

The Bernoulli equation can be used to predict the static pressure at the stagnation point. Since the velocity of the fluid within the pitot tube is zero, the upstream velocity, v, can be calculated if the static and stagnation pressures are known.

$$\frac{p_1}{\rho} + \frac{v_1^2}{2} = \frac{p_2}{\rho} \quad 16.12$$

Pitot Tube

$$v = \sqrt{\left(\frac{2}{\rho}\right)(p_0 - p_s)} \quad \text{[SI]} \quad 16.13(a)$$

$$v = \sqrt{\left(\frac{2g_c}{\rho}\right)(p_0 - p_s)} \quad \text{[U.S.]} \quad 16.13(b)$$

In reality, both friction and heating occur, and the fluid may be compressible. These errors are taken care of by a correction factor known as the *impact factor*, C_0, which is applied to the derived velocity. C_0 is usually very close to 1.00 (e.g., 0.99 or 0.995).

$$v_{\text{actual}} = C_0 v_{\text{indicated}}$$

Since accurate measurements of fluid velocity are dependent on one-dimensional fluid flow, it is essential that any obstructions or pipe bends be more than 10 pipe diameters upstream from the pitot tube.

Example 16.3

The static pressure of air (0.075 lbm/ft^3; 1.20 kg/m^3) flowing in a pipe is measured by a precision gauge to be 10.00 psig (68.95 kPa). A pitot tube-manometer indicates 20.6 in (0.523 m) of mercury. The density of mercury is 0.491 lbm/in^3 (13 600 kg/m^3). Losses are insignificant. What is the velocity of the air in the pipe?

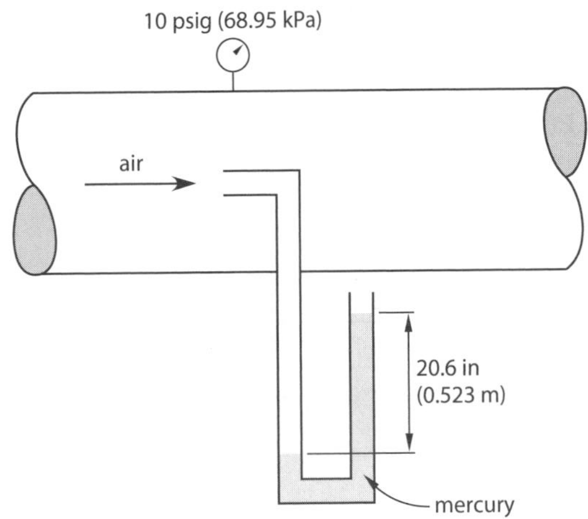

Fluids

Fluids

SI Solution

The impact pressure is

$$\begin{aligned} p_0 &= \rho g h \\ &= \frac{\left(13\,600\ \frac{\text{kg}}{\text{m}^3}\right)\left(9.81\ \frac{\text{m}}{\text{s}^2}\right)(0.523\ \text{m})}{1000\ \frac{\text{Pa}}{\text{kPa}}} \\ &= 69.78\ \text{kPa} \end{aligned}$$

From Eq. 16.13, the velocity is

$$\begin{aligned} \text{v} &= \sqrt{\frac{2(p_0 - p_s)}{\rho}} \\ &= \sqrt{\frac{(2)(69.78\ \text{kPa} - 68.95\ \text{kPa})\left(1000\ \frac{\text{Pa}}{\text{kPa}}\right)}{1.20\ \frac{\text{kg}}{\text{m}^3}}} \\ &= 37.2\ \text{m/s} \end{aligned}$$

Customary U.S. Solution

The impact pressure is

$$\begin{aligned} p_0 &= \frac{\rho g h}{g_c} = \frac{\left(0.491\ \frac{\text{lbm}}{\text{in}^3}\right)\left(32.2\ \frac{\text{ft}}{\text{sec}^2}\right)(20.6\ \text{in})}{32.2\ \frac{\text{lbm-ft}}{\text{lbf-sec}^2}} \\ &= 10.11\ \text{lbf/in}^2 \quad (10.11\ \text{psig}) \end{aligned}$$

Since impact pressure is the sum of the static and kinetic (velocity) pressures, the kinetic pressure is

$$\begin{aligned} p_\text{v} &= p_0 - p_s \\ &= 10.11\ \text{psig} - 10.00\ \text{psig} \\ &= 0.11\ \text{psi} \end{aligned}$$

From Eq. 16.13, the velocity is

$$\begin{aligned} \text{v} &= \sqrt{\frac{2g_c(p_0 - p_s)}{\rho}} \\ &= \sqrt{\frac{(2)\left(32.2\ \frac{\text{lbm-ft}}{\text{lbf-sec}^2}\right)\left(0.11\ \frac{\text{lbf}}{\text{in}^2}\right)\left(12\ \frac{\text{in}}{\text{ft}}\right)^2}{0.075\ \frac{\text{lbm}}{\text{ft}^3}}} \\ &= 116.6\ \text{ft/sec} \end{aligned}$$

8. HYDRAULIC RADIUS

The *hydraulic radius*, R_H, is defined as the area in flow divided by the *wetted perimeter*.[10] (The hydraulic radius is not the same as the radius of a pipe.) The area in flow is the cross-sectional area of the fluid flowing. When a fluid is flowing under pressure in a pipe (i.e., *pressure flow* in a *pressure conduit*), the area in flow will be the internal area of the pipe. However, the fluid may not completely fill the pipe and may flow simply because of a sloped surface (i.e., *gravity flow* or *open channel flow*).

The wetted perimeter, P, is the length of the line representing the interface between the fluid and the pipe or channel. It does not include the *free surface* length (i.e., the interface between fluid and atmosphere).

$$R_H = \frac{\text{area in flow}}{\text{wetted perimeter}} = \frac{A}{P} \qquad 16.14$$

Consider a circular pipe flowing completely full. The area in flow is πr^2. The wetted perimeter is the entire circumference, $2\pi r$. The hydraulic radius is therefore half the pipe's radius, or one-fourth its diameter.

$$R_H = \frac{\pi r^2}{2\pi r} = \frac{r}{2} = \frac{D}{4} \quad \left[\begin{array}{c}\text{circular pipe}\\ \text{flowing full}\end{array}\right] \qquad 16.15$$

The hydraulic radius of a pipe flowing half full is also $r/2$, because the flow area and wetted perimeter are both halved. However, it is time-consuming to calculate the hydraulic radius for pipe flow at any intermediate depth, due to the difficulty in evaluating the flow area and wetted perimeter. Appendix 16.A greatly simplifies such calculations.

Example 16.4

A pipe (internal diameter = 6) carries water with a depth of 2 flowing under the influence of gravity. (a) Calculate the hydraulic radius analytically. (b) Verify the result by using App. 16.A.

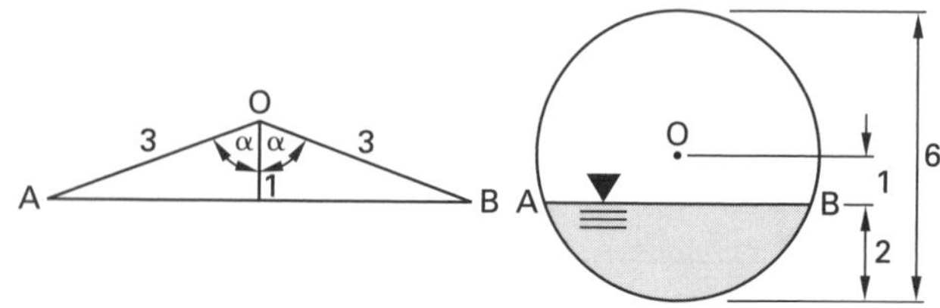

Solution

(a) Use App. 7.A. The equations for a circular segment must be used. The radius is $6/2 = 3$.

[10]The hydraulic radius can also be calculated as one-fourth of the hydraulic diameter of the pipe or channel, as will be subsequently shown. That is, $R_H = \frac{1}{4}D_H$.

Points A, O, and B are used to find the central angle of the circular segment.

$$\phi = 2\alpha = 2\arccos\frac{1}{3} = (2)(70.53^\circ)$$
$$= 141.06^\circ$$

ϕ must be expressed in radians.

$$\phi = 2\pi\left(\frac{141.06^\circ}{360^\circ}\right) = 2.46 \text{ rad}$$

The area of the circular segment (i.e., the area in flow) is

$$A = \tfrac{1}{2}r^2(\phi - \sin\phi) \quad [\phi \text{ in radians}]$$
$$= \left(\frac{1}{2}\right)(3)^2\big(2.46 \text{ rad} - \sin(2.46 \text{ rad})\big)$$
$$= 8.235$$

The arc length (i.e., the wetted perimeter) is

$$P = r\phi = (3)(2.46 \text{ rad}) = 7.38$$

The hydraulic radius is

$$R_H = \frac{A}{P} = \frac{8.235}{7.38} = 1.12$$

(b) The ratio of depth to diameter is needed to use App. 16.A.

$$\frac{d}{D} = \frac{2}{6} = 0.333$$

From App. 16.A,

$$\frac{R_H}{D} \approx 0.186$$

$$R_H = (0.186)(6)$$
$$= 1.12$$

9. HYDRAULIC DIAMETER

Many fluid, thermodynamic, and heat transfer processes are dependent on the physical length of an object. This controlling variable is generally known as the *characteristic dimension*. The characteristic dimension in evaluating fluid flow is the *hydraulic diameter* (also known as the *equivalent hydraulic diameter*).[11] The hydraulic diameter for a full-flowing circular pipe is simply its inside diameter. The hydraulic diameters of other cross sections in flow are given in Table 16.1. If the hydraulic radius is known, it can be used to calculate the hydraulic diameter.

Flow in Noncircular Conduits

$$D_H = \frac{4 \times \text{cross-sectional in flow}}{\text{wetted perimeter}} = \frac{4A}{P} = 4R_H \qquad 16.16$$

Table 16.1 *Hydraulic Diameters for Common Conduit Shapes*

conduit cross section	D_H
flowing full	
circle	D
annulus (outer diameter D_o, inner diameter D_i)	$D_o - D_i$
square (side L)	L
rectangle (sides L_1 and L_2)	$\frac{2L_1L_2}{L_1 + L_2}$
flowing partially full	
half-filled circle (diameter D)	D
rectangle (h deep, L wide)	$\frac{4hL}{L + 2h}$
wide, shallow stream (h deep)	$4h$
triangle, vertex down (h deep, L broad, s side)	$\frac{hL}{s}$
trapezoid (h deep, a wide at top, b wide at bottom, s side)	$\frac{2h(a + b)}{b + 2s}$

[11]The engineering community is very inconsistent, but the three terms *hydraulic depth*, *hydraulic diameter*, and *equivalent diameter* do not have the same meanings. Hydraulic depth (flow area divided by exposed surface width) is a characteristic length used in Froude number and other open channel flow calculations. Hydraulic diameter (four times the area in flow divided by the wetted surface) is a characteristic length used in Reynolds number and friction loss calculations. Equivalent diameter ($1.3(ab)^{0.625}/(a + b)^{0.25}$) is the diameter of a round duct or pipe that will have the same friction loss per unit length as a noncircular duct. Unfortunately, these terms are often used interchangeably.

Example 16.5

Determine the hydraulic diameter and hydraulic radius for the open trapezoidal channel shown.

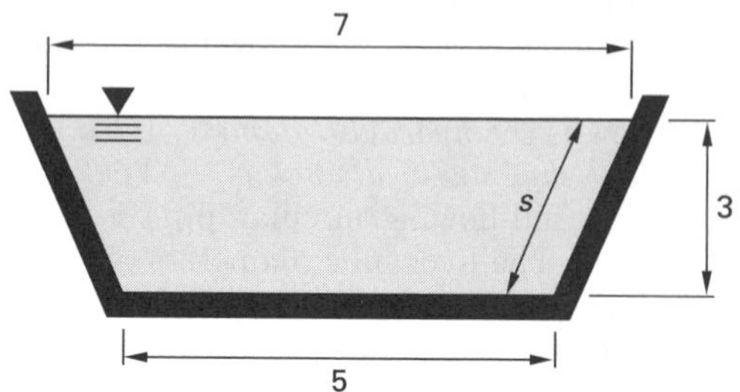

Solution

The batter of the inclined walls is $(7 - 5)/2$ walls = 1.

$$s = \sqrt{(3)^2 + (1)^2} = 3.16$$

Using Table 16.1,

$$D_H = \frac{2h(a+b)}{b+2s} = \frac{(2)(3)(7+5)}{5+(2)(3.16)} = 6.36$$

From Eq. 16.16,

Flow in Noncircular Conduits

$$R_H = \frac{D_H}{4} = \frac{6.36}{4} = 1.59$$

10. REYNOLDS NUMBER

The *Reynolds number*, Re, is a dimensionless number interpreted as the ratio of inertial forces to viscous forces in the fluid.[12]

$$\text{Re} = \frac{\text{inertial forces}}{\text{viscous forces}} \qquad 16.17$$

The inertial forces are proportional to the flow diameter, velocity, and fluid density. (Increasing these variables will increase the momentum of the fluid in flow.) The viscous force is represented by the fluid's absolute viscosity, μ. Thus, the Reynolds number is calculated as

$$\text{Re} = \frac{\text{v}D_H\rho}{\mu} \qquad \text{[SI]} \qquad 16.18(a)$$

$$\text{Re} = \frac{\text{v}D_H\rho}{g_c\mu} \qquad \text{[U.S.]} \qquad 16.18(b)$$

Since μ/ρ is defined as the *kinematic viscosity*, ν, Eq. 16.18 can be simplified.

$$\text{Re} = \frac{\text{v}D_H}{\nu} \qquad 16.19$$

For a circular pipe flowing full (or exactly half full), the hydraulic diameter is equal to the inside diameter of the pipe, so the Reynolds number is

Reynolds Number

$$\text{Re} = \frac{\text{v}D\rho}{\mu} = \frac{\text{v}D}{\nu} \qquad \text{[SI]} \qquad 16.20(a)$$

$$\text{Re} = \frac{\text{v}D\rho}{g_c\mu} = \frac{\text{v}D}{\nu} \qquad \text{[U.S.]} \qquad 16.20(b)$$

11. LAMINAR FLOW

Laminar flow gets its name from the word *laminae* (layers). If all of the fluid particles move in paths parallel to the overall flow direction (i.e., in layers), the flow is said to be *laminar*. (The terms *viscous flow* and *streamline flow* are also used.) This occurs in pipeline flow when the Reynolds number is less than (approximately) 2300. Laminar flow is typical when the flow channel is small, the velocity is low, and the fluid is viscous. Viscous forces are dominant in laminar flow.

In laminar flow, a stream of dye inserted in the flow will continue from the source in a continuous, unbroken line with very little mixing of the dye and surrounding liquid. The fluid particle paths coincide with imaginary *streamlines*. (Streamlines and velocity vectors are always tangent to each other.) A "bundle" of these streamlines (i.e., a *streamtube*) constitutes a complete fluid flow.

12. TURBULENT FLOW

A fluid is said to be in *turbulent flow* if the Reynolds number is greater than (approximately) 4000.[13] Turbulent flow is characterized by a three-dimensional movement of the fluid particles superimposed on the overall direction of motion. A stream of dye injected into a turbulent flow will quickly disperse and uniformly mix with the surrounding flow. Inertial forces dominate in turbulent flow. At very high Reynolds numbers, the flow is said to be *fully turbulent*.

[12] Engineering authors are not in agreement about the symbol for the Reynolds number. In addition to Re (used in this book), engineers commonly use **Re**, R, $\Re$, N_{Re}, and N_R.

[13] The *NCEES Handbook* characterizes flow as turbulent when the Reynolds number is greater than 10,000. This is not typical.

13. CRITICAL FLOW

The flow is said to be in a *critical zone* or *transition region* when the Reynolds number is between 2300 and 4000. These numbers are known as the lower and upper *critical Reynolds numbers* for fluid flow, respectively. (Critical Reynolds numbers for other processes are different.) It is difficult to design for the transition region because fluid behavior is not consistent, and few processes operate in the critical zone. In the event a critical zone design is required, the conservative assumption of turbulent flow will result in the greatest value of friction loss.

14. FLUID VELOCITY DISTRIBUTION IN PIPES

With laminar flow, the viscosity makes some fluid particles adhere to the pipe wall. The closer a particle is to the pipe wall, the greater the tendency will be for the fluid to adhere to the pipe wall. The following statements characterize laminar flow.

- The velocity distribution is parabolic.
- The velocity is zero at the pipe wall.
- The velocity is maximum at the center and equal to twice the average velocity.

$$v_{ave} = \frac{Q}{A} = \frac{v_{max}}{2} \quad \text{[laminar]} \qquad 16.21$$

With turbulent flow, there is generally no distinction made between the velocities of particles near the pipe wall and particles at the pipe centerline.[14] All of the fluid particles are assumed to have the same velocity. This velocity is known as the *average* or *bulk velocity*. It can be calculated from the volume flowing.

$$v_{ave} = \frac{Q}{A} \quad \text{[turbulent]} \qquad 16.22$$

Laminar and turbulent velocity distributions are shown in Fig. 16.3. In actuality, no flow is completely turbulent, and there is a difference between the *centerline velocity* and the average velocity. The error decreases as the Reynolds number increases. The ratio v_{ave}/v_{max} starts at approximately 0.75 for Re = 4000 and increases to approximately 0.86 at Re = 10^6. Most problems ignore the difference between v_{ave} and v_{max}, but care should be taken when a centerline measurement (as from a pitot tube) is used to evaluate the average velocity.

Figure 16.3 *Laminar and Turbulent Velocity Distributions*

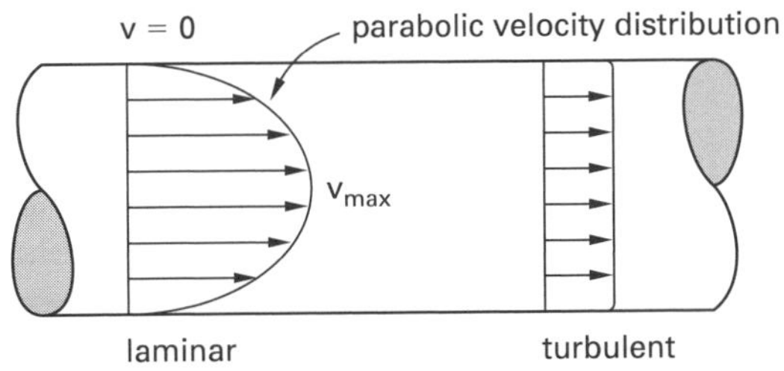

The ratio of the average velocity to maximum velocity is known as the *pipe coefficient* or *pipe factor*. Considering all the other coefficients used in pipe flow, these names are somewhat vague and ambiguous. Therefore, they are not in widespread use.

For turbulent flow (Re ≈ 10^5) in a smooth, circular pipe of radius r_o, the velocity at a radial distance r from the centerline is given by the $^1/_7$*-power law*.

$$v_r = v_{max}\left(\frac{r_o - r}{r_o}\right)^{1/7} \quad \text{[turbulent flow]} \qquad 16.23$$

The fluid's *velocity profile*, given by Eq. 16.23, is valid in smooth pipes up to a Reynolds number of approximately 100,000. Above that, up to a Reynolds number of approximately 400,000, an exponent of $^1/_8$ fits experimental data better. For rough pipes, the exponent is larger (e.g., $^1/_5$).

Equation 16.23 can be integrated to determine the average velocity.

$$\begin{aligned} v_{ave} &= \left(\frac{49}{60}\right)v_{max} \\ &= 0.817v_{max} \quad \text{[turbulent flow]} \end{aligned} \qquad 16.24$$

When the flow is laminar, the velocity profile within a pipe will be parabolic and of the form of Eq. 16.25. (Equation 16.23 is for turbulent flow and does not describe a parabolic velocity profile.) The velocity at a radial distance r from the centerline in a pipe of radius r_o is

$$v_r = v_{max}\left(\frac{r_o^2 - r^2}{r_o^2}\right) \quad \text{[laminar flow]} \qquad 16.25$$

When the velocity profile is parabolic, the flow rate and pressure drop can easily be determined. The average velocity is half of the maximum velocity given in the velocity profile equation.

$$v_{ave} = \tfrac{1}{2}v_{max} \quad \text{[laminar flow]} \qquad 16.26$$

[14]This disregards the *boundary layer*, a thin layer near the pipe wall, where the velocity goes from zero to v_{ave}.

The average velocity is used to determine the flow quantity and friction loss. The friction loss is determined by traditional means.

$$Q = A\mathrm{v}_{\text{ave}} \qquad 16.27$$

The kinetic energy of laminar flow can be found by integrating the velocity profile equation, resulting in Eq. 16.28.

$$E_{\mathrm{v}} = \mathrm{v}_{\text{ave}}^2 \quad \text{[laminar flow]} \quad \text{[SI]} \qquad 16.28(a)$$

$$E_{\mathrm{v}} = \frac{\mathrm{v}_{\text{ave}}^2}{g_c} \quad \text{[laminar flow]} \quad \text{[U.S.]} \qquad 16.28(b)$$

15. PIPE MATERIALS AND SIZES

Many materials are used for pipes. The material used depends on the application. Water supply distribution, wastewater collection, and air conditioning refrigerant lines place different demands on pipe material performance. Pipe materials are chosen on the basis of tensile strength to withstand internal pressures, compressive strength to withstand external loads from backfill and traffic, smoothness, corrosion resistance, chemical inertness, cost, and other factors.

The following are characteristics of the major types of legacy and new installation commercial pipe materials that are in use.

- *asbestos cement:* immune to electrolysis and corrosion, light in weight but weak structurally; environmentally limited
- *concrete:* durable, watertight, low maintenance, smooth interior
- *copper and brass:* used primarily for water, condensate, and refrigerant lines; in some cases, easily bent by hand, good thermal conductivity
- *ductile cast iron:* long lived, strong, impervious, heavy, scour resistant, but costly
- *plastic* (PVC, CPVC, HDPE, and ABS):[15] chemically inert, resistant to corrosion, very smooth, lightweight, low cost
- *steel*: high strength, ductile, resistant to shock, very smooth interior, but susceptible to corrosion
- *vitrified clay:* resistant to corrosion, acids (e.g., hydrogen sulfide from septic sewage), scour, and erosion

The required wall thickness of a pipe is proportional to the pressure the pipe must carry. However, not all pipes operate at high pressures. Therefore, pipes and tubing may be available in different wall thicknesses (*schedules*, *series*, or *types*). Steel pipe, for example, is available in schedules 40, 80, and others.[16]

For initial estimates, the approximate schedule of steel pipe can be calculated from Eq. 16.29. p is the operating pressure in psig; S is the allowable stress in the pipe material; and, e is the *weld-joint efficiency*, also known as the *joint quality factor* (typically 1.00 for seamless pipe, 0.85 for electric resistance-welded pipe, 0.80 for electric fusion-welded pipe, and 0.60 for furnace butt-welded pipe). For seamless carbon steel (A53) pipe used below 650°F (340°C), the allowable stress is approximately between 12,000 psi and 15,000 psi. So, with butt-welded joints, a value of 6500 psi is often used for the product Se.

$$\text{schedule} \approx \frac{1000p}{Se} \qquad 16.29$$

Steel pipe is available in black (i.e., plain *black pipe*) and galvanized (inside, outside, or both) varieties. Steel pipe is manufactured in plain-carbon and stainless varieties. AISI 316 stainless steel pipe is particularly corrosion resistant.

The actual dimensions of some pipes (concrete, clay, some cast iron, etc.) coincide with their *nominal dimensions*. For example, a 12 in concrete pipe has an inside diameter of 12 in, and no further refinement is needed. However, some pipes and tubing (e.g., steel pipe, copper and brass tubing, and some cast iron) are called out by a nominal diameter that has nothing to do with the internal diameter of the pipe. For example, a 16 in schedule-40 steel pipe has an actual inside diameter of 15 in. In some cases, the nominal size does not coincide with the external diameter, either.

PVC (polyvinyl chloride) pipe is used extensively as water and sewer pipe due to its combination of strength, ductility, and corrosion resistance. Manufactured lengths range approximately 10–13 ft (3–3.9 m) for sewer pipe and 20 ft (6 m) for water pipe, with integral gasketed joints or solvent-weld bells. Infiltration is very low (less than 50 gal/in-mile-day), even in the wettest environments. The low Manning's roughness constant (0.009 typical) allows PVC sewer pipe to be used with flatter grades or smaller diameters. PVC pipe is resistant to corrosive soils and sewerage gases and is generally resistant to abrasion from pipe-cleaning tools.

Truss pipe is a double-walled PVC or ABS pipe with radial or zigzag (diagonal) reinforcing ribs between the thin walls and with lightweight concrete filling all voids. It is used

[15]PVC: polyvinyl chloride; CPVC: chlorinated polyvinyl chloride; HDPE: high-density polyethylene; ABS: acrylonitrile-butadiene-styrene.
[16]Other schedules of steel pipe, such as 30, 60, 120, and so on, also exist, but in limited sizes. Schedule-40 pipe roughly corresponds to the standard weight (S) designation used in the past. Schedule-80 roughly corresponds to the extra-strong (X) designation. There is no uniform replacement designation for double-extra-strong (XX) pipe.

primarily for underground sewer service because of its smooth interior surface ($n = 0.009$), infiltration-resistant, impervious exterior, and resistance to bending.

Prestressed concrete pipe (PSC) or *reinforced concrete pipe* (RCP) consists of a concrete core that is compressed by a circumferential wrap of high-tensile strength wire and covered with an exterior mortar coating. *Prestressed concrete cylinder pipe* (PCCP), or *lined concrete pipe* (LCP), is constructed with a concrete core, a thin steel cylinder, prestressing wires, and an exterior mortar coating. The concrete core is the primary structural, load-bearing component. The prestressing wires induce a uniform compressive stress in the core that offsets tensile stresses in the pipe. If present, the steel cylinder acts as a water barrier between concrete layers. The mortar coating protects the prestressing wires from physical damage and external corrosion. PSC and PCCP are ideal for large-diameter pressurized service and are used primarily for water supply systems. PSC is commonly available in diameters from 16 in to 60 in (400 mm to 1500 mm), while PCCP diameters as large as 144 in (3600 mm) are available.

It is essential that tables of pipe sizes be used when working problems involving steel and copper pipes, since there is no other way to obtain the inside diameters of such pipes.[17] **[Pipe and Tube Data]**

16. MANUFACTURED PIPE STANDARDS

There are many different standards governing pipe diameters and wall thicknesses. A pipe's nominal outside diameter is rarely sufficient to determine the internal dimensions of the pipe. A manufacturing specification and class or category are usually needed to completely specify pipe dimensions.

Modern cast-iron soil (i.e., sanitary) pipe is produced according to ASTM A74. Ductile iron (CI/DI) pipe is produced to ANSI/AWWA C150/A21.50 and C151/A21.51 standards. Gasketed PVC sewer pipe up to 15 in inside diameter is produced to ASTM D3034 standards. Gasketed sewer PVC pipe from 18 in to 48 in is produced to ASTM F679 standards. PVC pressure pipe for water distribution is manufactured to ANSI/AWWA C900 standards. Reinforced concrete pipe (RCP) for culvert, storm drain, and sewer applications is manufactured to ASTM/AASHTO C76/M170 standards.

17. PIPE CLASS

The term *pipe class* has several valid meanings that can be distinguished by designation and context. For plastic and metallic (i.e., steel, cast and ductile iron, cast bronze, and wrought copper) pipe and fittings, pressure class designations such as 25, 150, and 300 refer to the pressure ratings and dimensional design systems as defined in the appropriate ASME, ANSI, AWWA, and other standards. Generally, the class corresponds roughly to a maximum operating pressure category in psig.

ASME pipe classes 1, 2, and 3 refer to the maximum stress categories allowed in pipes, piping systems, components, and supports, as defined in the ASME *Boiler and Pressure Vessel Code* (BPVC), Sec. III, Div. 1, "Rules for Construction of Nuclear Facility Components."

For precast concrete pipe manufactured according to ASTM C76, the pipe classes 1, 2, 3, 4, and 5 correspond to the minimum vertical loading (D-load) capacity as determined in a *three-edge bearing test* (ASTM C497). (See Table 16.2.[18]) Each pipe has two D-load ratings: the pressure that induces a crack 0.01 in (0.25 mm) wide at least 1 ft (100 mm) long ($D_{0.01}$), and the pressure that results in structural collapse (D_{ult}).[19]

$$\text{D-load} = \frac{F_{\text{lbf}}}{D_{\text{ft}}L_{\text{ft}}} \quad \text{[ASTM C76 pipe]} \qquad 16.30$$

Table 16.2 *ASTM C76 Concrete Pipe D-Load Equivalent Pipe Class**

ASTM C76 pipe class	$D_{0.01}$ load rating (lbf/ft^2)	D_{ult} load rating (lbf/ft^2)
class 1 (I)	800	1200
class 2 (II)	1000	1500
class 3 (III)	1350	2000
class 4 (IV)	2000	3000
class 5 (V)	3000	3750

(Multiply lbf/ft^2 by 0.04788 to obtain kPa.)
*As defined in ASTM C76 and AASHTO M170.

18. PAINTS, COATINGS, AND LININGS

Various materials have been used to protect steel and ductile iron pipes against rust and other forms of corrosion. *Red primer* is a shop-applied, rust-inhibiting primer applied to prevent short-term rust prior to shipment and the application of subsequent coatings. *Asphaltic coating* ("tar" coating) is applied to the exterior of underground pipes. *Bituminous coating* refers to a similar coating made from tar pitch. Both asphaltic and bituminous coatings should be completely removed or sealed with a synthetic resin prior to the pipe being finish-coated, since their oils may bleed through otherwise.

[17]It is a characteristic of standard steel pipes that the schedule number does not affect the outside diameter of the pipe. An 8 in schedule-40 pipe has the same exterior dimensions as an 8 in schedule-80 pipe. However, the interior flow area will be less for the schedule-80 pipe.

[18]The D-load rating for ASTM C14 concrete pipe is D-load $= F/L$ in lbf/ft.

[19]This D-load rating for concrete pipe is analogous to the *pipe stiffness* (PS) rating (also known as *ring stiffness*) for PVC and other plastic pipe, although the units are different. The PS rating is stated in lbf/in^2 (pounds per inch of pipe length per inch of pipe diameter).

Though bituminous materials (i.e., asphaltic materials) continue to be cost effective, epoxy-based products are now extensively used. Epoxy products are delivered as a two-part formulation (a polyamide resin and liquid chemical hardener) that is mixed together prior to application. *Coal tar epoxy*, also referred to as *epoxy coal tar*, a generic name, sees frequent use in pipes exposed to high humidity, seawater, other salt solutions, and crude oil. Though suitable for coating steel penstocks of hydroelectric installations, coal tar epoxy is generally not suitable for potable water delivery systems. Though it is self-priming, appropriate surface preparation is required for adequate adhesion. Coal tar epoxy has a density range of 1.9–2.3 lbm/gal (230–280 g/L).

19. CORRUGATED METAL PIPE

Corrugated metal pipe (CMP, also known as *corrugated steel pipe*) is frequently used for culverts. Pipe is made from corrugated sheets of galvanized steel that are rolled and riveted together along a longitudinal seam. Aluminized steel may also be used in certain ranges of soil pH. Standard round pipe diameters range from 8 in to 96 in (200 mm to 2450 mm). Metric dimensions of standard diameters are usually rounded to the nearest 25 mm or 50 mm (e.g., a 42 in culvert would be specified as a 1050 mm culvert, not 1067 mm).

Larger and noncircular culverts can be created out of curved steel plate. Standard section lengths are 10–20 ft (3–6 m). Though most corrugations are transverse (i.e., annular), helical corrugations are also used. Metal gauges of 8, 10, 12, 14, and 16 are commonly used, depending on the depth of burial.

The most common corrugated steel pipe has transverse corrugations that are ½ in (13 mm) deep and 2⅔ in (68 mm) from crest to crest. These are referred to as "2½ inch" or "68×13" corrugations. For larger culverts, corrugations with a 2 in (25 mm) depth and 3 in, 5 in, or 6 in (76 mm, 125 mm, or 152 mm) pitches are used. Plate-based products using 6 in × 2 in (152 mm × 51 mm) corrugations are known as *structural plate corrugated steel pipe* (SPCSP) and *multiplate* after the trade-named product "Multi-Plate™."

The flow area for circular culverts is based on the nominal culvert diameter, regardless of the gage of the plate metal used to construct the pipe. Flow area is calculated to (at most) three significant digits.

A Hazen-Williams coefficient, C, of 60 is typically used with all sizes of corrugated pipe. Values of C and Manning's coefficient, n, for corrugated pipe are generally not affected by age. *Design Charts for Open Channel Flow* (U.S. Department of Transportation) recommends a Manning constant of $n = 0.024$ for all cases. The U.S. Department of the Interior recommends the following values.

For standard (2⅔ in × ½ in or 68 mm × 13 mm) corrugated pipe with the diameter ranges given: 12 in (457 mm), 0.027; 24 in (610 mm), 0.025; 36–48 in (914–1219 mm), 0.024; 60–84 in (1524–2134 mm), 0.023; 96 in (2438 mm), 0.022.

For (6 in × 2 in or 152 mm × 51 mm) multiplate construction with the diameter ranges given: 5–6 ft (1.5–1.8 m), 0.034; 7–8 ft (2.1–2.4 m), 0.033; 9–11 ft (2.7–3.3 m), 0.032; 12–13 ft (3.6–3.9 m), 0.031; 14–15 ft (4.2–4.5 m), 0.030; 16–18 ft (4.8–5.4 m), 0.029; 19–20 ft (5.8–6.0 m), 0.028; 21–22 ft (6.3–6.6 m), 0.027.

If the inside of the corrugated pipe has been asphalted completely smooth 360° circumferentially, Manning's n ranges from 0.009 to 0.011. For culverts with 40% asphalted inverts, $n = 0.019$. For other percentages of paved invert, the resulting value is proportional to the percentage and the values normally corresponding to that diameter pipe. For field-bolted corrugated metal pipe arches, $n = 0.025$.

It is also possible to calculate the Darcy friction loss if the corrugation depth, 0.5 in (13 mm) for standard corrugations and 2.0 in (51 mm) for multiplate, is taken as the specific roughness.

20. TYPES OF VALVES

Valves used for *shutoff service* (e.g., gate, plug, ball, and some butterfly valves) are used fully open or fully closed. *Gate valves* offer minimum resistance to flow. They are used in clean fluid and slurry services when valve operation is infrequent. Many turns of the handwheels are required to raise or lower their gates. *Plug valves* provide for tight shutoff. A 90° turn of their handles is sufficient to rotate the plugs fully open or closed. *Eccentric plug valves*, in which the plug rotates out of the fluid path when open, are among the most common wastewater valves. *Plug cock valves* have a hollow passageway in their plugs through which fluid can flow. Both eccentric plug valves and plug cock valves are referred to as "plug valves." *Ball valves* offer an unobstructed flow path and tight shutoff. They are often used with slurries and viscous fluids, as well as with cryogenic fluids. A 90° turn of their handles rotates the balls fully open or closed. *Butterfly valves* (when specially designed with appropriate seats) can be used for shutoff operation. They are particularly applicable to large flows of low-pressure (vacuum up to 200 psig (1.4 MPa)) gases or liquids, although high-performance butterfly valves can operate as high as 600 psig (4.1 MPa). Their straight-through, open-disk design results in minimal solids build-up and low pressure drops.

Other valve types (e.g., globe, needle, Y-, angle, and some butterfly valves) are more suitable for *throttling service*. *Globe valves* provide positive shutoff and precise metering on clean fluids. However, since the seat is parallel to the direction of flow and the fluid makes two right-angle turns, there is substantial resistance and pressure drop through them, as well as relatively fast erosion of the seat. Globe valves are intended for frequent operation. *Needle valves* are similar to globe valves, except that the plug is a tapered, needlelike cone. Needle valves provide accurate metering of small flows of clean fluids. Needle valves are applicable to cryogenic fluids. *Y-valves* are similar to globe valves in operation, but their seats are inclined to the direction of flow, offering more of a straight-through passage and unobstructed flow than the globe valve. *Angle valves* are essentially globe valves where the fluid makes a 90° turn. They can be used for throttling and shutoff of clean or viscous fluids and slurries. *Butterfly valves* are often used for throttling services with the same limitations and benefits as those listed for shutoff use.

Other valves are of the *check* (*nonreverse-flow, antireversal*) variety. These react automatically to changes in pressure to prevent reversals of flow. Special check valves can also prevent excess flow. Figure 16.4 illustrates *swing*, *lift*, and *angle lift check valves*, and Table 16.3 gives typical characteristics of common valve types.

21. SAFETY AND PRESSURE RELIEF VALVES

A *pressure relief device* typically incorporates a disk or needle that is held in place by spring force, and that is lifted off its seat (i.e., opens) when the static pressure at the valve inlet exceeds the opening pressure. A *relief valve* is a pressure relief device used primarily for liquid service. It has a gradual lift that is approximately proportional (though not necessarily linear) to the increase in pressure over opening pressure. A *safety valve* is used for compressible steam, air, and gas services. Its performance is characterized by rapid opening ("pop action"). In a *low-lift safety valve*, the discharge area depends on the disc position and is limited by the lift amount. In a *full-lift* (*high-lift*) *safety valve*, the discharge area is not determined by the position of the disc. Since discs in low-lift valves have lift distances of only approximately 1/24th of the bore diameter, the discharge capacities tend to be much lower than those of high-lift valves.

A *safety relief valve* is a dual-function valve that can be used for either a liquid or a compressible fluid. Its operation is characterized by rapid opening (for gases) and by slower opening in proportion to the increase in inlet pressure over the opening pressure (for liquids), depending on the application.

22. AIR RELEASE AND VACUUM VALVES

Small amounts of air in fluid flows tend to accumulate at high points in the line. These air pockets reduce flow capacities, increase pumping power, and contribute to pressure surges and water hammer. *Air release valves* should be installed at high points in the line, particularly with line diameters greater than 12 in (3000 mm).[20] They are float operated and open against the line's internal pressure to release accumulated air and gases. Air release valves are essential when lines are being filled. Valves for sewage lines prevent buildups of sewage gases (e.g., hydrogen sulfide) and are similar in functionality, though they contain features to prevent clogging and fouling.

Vacuum valves are similar in design, except that they permit air to enter a line when the line is drained, preventing pipeline collapse. *Combination air valves* (CAV), also known as *double orifice air valves*, combine the functions of both air release valves and vacuum valves. Vacuum valves should not be used in drinking water supply distribution systems due to the potential for contamination.

If unrestricted, air moving through a valve orifice will reach a maximum velocity of approximately 300 ft/sec (91 m/s) at 6.7 lbf/in^2 (46 kPa) differential pressure. It is good practice to limit the flow to 10 ft/sec (3 m/s) (occurring at about 1 lbf/in^2 (7 kPa)), and this can be accomplished with slow-closing units.

Siphon air valves (*make-or-break valves*) are a type of air vacuum valve that includes a paddle which hangs down in the flow. As long as the flow exists, the valve remains closed. When the flow stops or reverses, the valve opens. In a typical application, a pump is used to initiate siphon flow, and once flow is started, the pump is removed.

23. STEAM TRAPS

Steam traps are installed in steam lines. Since the heat content of condensate is significantly less than that of the steam, condensate is not as useful in downstream applications. Traps collect and automatically release condensed water while permitting vapor to pass through. They may also release accumulated noncondensing gases (e.g., air). **[Steam Trap]**

Inverted bucket steam traps are simple, mechanically robust, and reliable. They are best for applications with water hammer. *Thermostatic steam traps* operate with either a bimetallic or a balanced pressure design to detect the difference in temperature between live steam and condensate or air. They are suitable for cold start-up conditions, but less desirable with variable loadings. The *float steam trap* is the best choice for variable loadings, but it is less robust due to its mechanical complexity and is less resistant to water hammer. A *thermodynamic disc steam trap* uses flash steam as the stream passes through the trap and is both simple and robust. It is suitable for intermittent, not continuous, operation only.

[20]Smaller diameter lines may develop flow velocities sufficiently high (e.g., greater than 5 ft/sec (1.5 m/s)) to flush gas accumulations.

Figure 16.4 *Types of Valves*

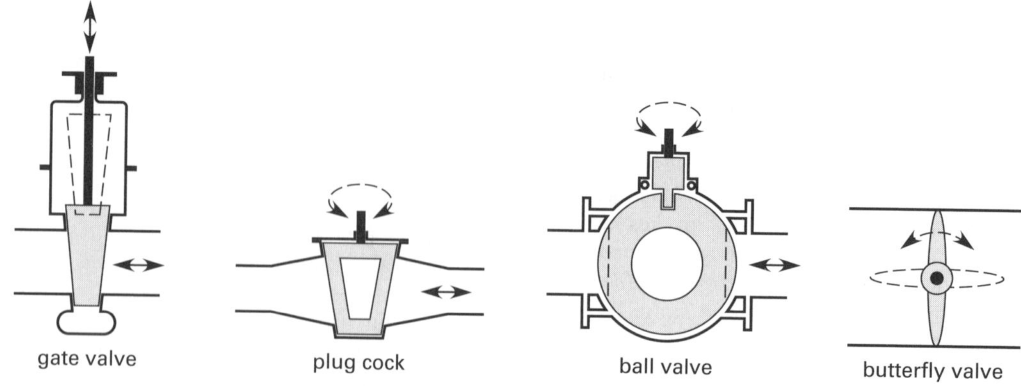

(a) valves for shutoff service

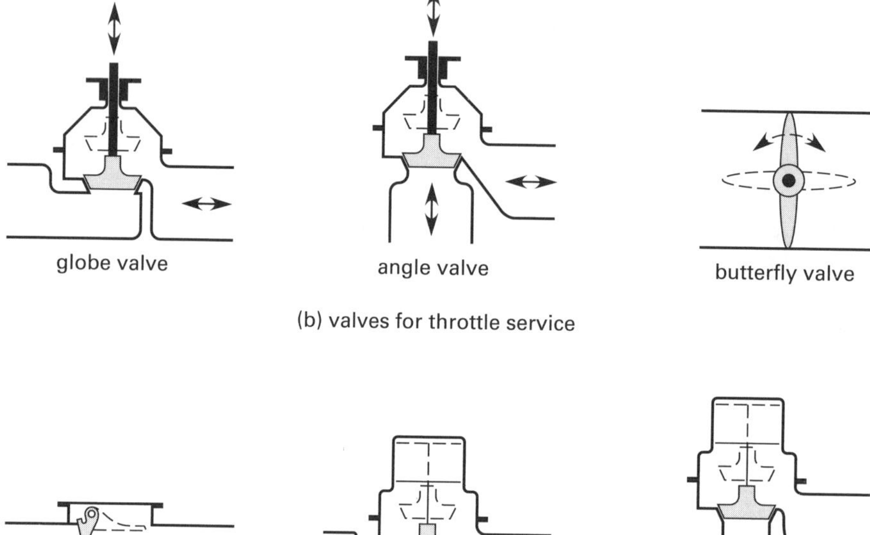

(b) valves for throttle service

swing check valve

lift check valve

angle lift check valve

(c) valves for antireversal service

Table 16.3 *Typical Characteristics of Common Valve Types*

valve type	fluid condition	switching frequency	pressure drop (fully open)	typical control response	typical maximum pressure (atm)	typical maximum temperature (°C)
ball	clean	low	low	very poor	160	300
butterfly	clean	low	low	poor	200	400
diaphragm* (not shown)	clean to slurried	very high	low to medium	very good	16	150
gate	clean	low	low	very poor	50	400
globe	clean	high	medium to high	very good	80	300
plug	clean	low	low	very poor	160	300

(Multiply atm by 101.33 to obtain kPa.)

*Diaphragm valves use a flexible diaphragm to block the flow path. The diaphragm may be manually or pneumatically actuated. Such valves are suitable for both throttling and shutoff service.

24. NOMENCLATURE

A	area	ft^2	m^2
C	correction factor	–	–
C	Hazen-Williams coefficient	–	–
d	depth	ft	m
D	diameter	ft	m
D-load	ASTM C497 load rating	lbf/ft^2	N/m^2
e	joint efficiency	–	–
E	specific energy	ft-lbf/lbm	J/kg
F	force	lbf	N
g	gravitational acceleration, 32.17 (9.807)	ft/sec^2	m/s^2
g_c	gravitational constant, 32.17	lbm-ft/lbf-sec^2	n.a.
h	head, head loss, or height	ft	m
L	length	ft	m
n	Manning's roughness coefficient	–	–
p	pressure	lbf/ft^2	Pa
P	wetted perimeter	ft	m
Q	volumetric flow rate	ft^3/sec	m^3/s
r	radius	ft	m
R_H	hydraulic radius	ft	m
Re	Reynolds number	–	–
s	batter (slope)	–	–
S	allowable stress	lbf/ft^2	Pa
v	velocity	ft/sec	m/s
z	elevation	ft	m

Symbols

α	angle	rad	rad
γ	specific weight	lbf/ft^3	N/m^3
μ	absolute viscosity	lbf-sec/ft^2	Pa·s
ν	kinematic viscosity	ft^2/sec	m^2/s
ρ	density	lbm/ft^3	kg/m^3
ϕ	angle	rad	rad

Subscripts

0	impact (stagnation)
ave	average
f	friction
i	inner
max	maximum
o	outer
p	pressure
r	radius
s	static
t	total
v	velocity
z	potential

17 Fluid Dynamics

Content in blue refers to the *NCEES Handbook*.

NCEES EXAM SPECIFICATION AND RELATED CONTENT

HVAC AND REFRIGERATION EXAM

- I.E. Fluid Mechanics
  - 2. Conservation of Mass
  - 6. Friction Factor
  - 7. Energy Loss Due to Friction: Laminar Flow
  - 8. Energy Loss Due to Friction: Turbulent Flow
  - 14. Valve Flow Coefficients
  - 25. Pitot-Static Gauge
- II.B.5. Equipment and Components: Pumps/compressors/fans
  - 13. Minor Losses

MACHINE DESIGN AND MATERIALS EXAM

- II.C.5. Supportive Knowledge: Testing and instrumentation
  - 25. Pitot-Static Gauge

THERMAL AND FLUID SYSTEMS EXAM

- I.B.3. Fluid Mechanics: Incompressible flow
  - 6. Friction Factor
  - 7. Energy Loss Due to Friction. Laminar Flow
  - 8. Energy Loss Due to Friction. Turbulent Flow
  - 13. Minor Losses
  - 14. Valve Flow Coefficients
  - 16. Discharge from Tanks
  - 29. Impulse-Momentum Principle
  - 30. Jet Propulsion
  - 31. Open Jet on a Single Stationary Blade
  - 32. Open Jet on a Single Moving Blade
  - 36. Water Hammer
  - 38. Drag

1. HYDRAULICS AND HYDRODYNAMICS

This chapter covers fluid moving through pipes, measurements with venturis and orifices, and other motion-related topics such as model theory, lift and drag, and pumps. In a strict interpretation, any fluid-related phenomenon that is not hydro*statics* should be hydro *dynamics*. However, tradition has separated the study of moving fluids into the fields of hydraulics and hydrodynamics.

In a general sense, *hydraulics* is the study of the practical laws of fluid flow and resistance in pipes and open channels. Hydraulic formulas are often developed from experimentation, empirical factors, and curve fitting, without an attempt to justify why the fluid behaves the way it does.

On the other hand, *hydrodynamics* is the study of fluid behavior based on theoretical considerations. Hydrodynamicists start with Newton's laws of motion and try to develop models of fluid behavior. Models developed in this manner are complicated greatly by the inclusion of viscous friction and compressibility. Therefore, hydrodynamic models assume a perfect fluid with constant density and zero viscosity. The conclusions reached by hydrodynamicists can differ greatly from those reached by hydraulicians.[1]

2. CONSERVATION OF MASS

Fluid mass is always conserved in fluid systems, regardless of the pipeline complexity, orientation of the flow, and fluid. This single concept is often sufficient to solve simple fluid problems.

$$\dot{m}_1 = \dot{m}_2 \qquad 17.1$$

When applied to fluid flow, the conservation of mass law is known as the *continuity equation.*

Continuity Equation

$$Q = A\text{v} \qquad 17.2$$

$$\dot{m} = \rho Q = \rho A\text{v} \qquad 17.3$$

Applying Eq. 17.3 to Eq. 17.1 gives

$$\rho_1 A_1 \text{v}_1 = \rho_2 A_2 \text{v}_2 \qquad 17.4$$

If the fluid is incompressible, then $\rho_1 = \rho_2$.

$$Q_1 = Q_2 \qquad 17.5$$

$$A_1 \text{v}_1 = A_2 \text{v}_2 \qquad 17.6$$

Various units and symbols are used for *volumetric flow rate*. This book uses Q, but the symbol $\dot{V}$ is often used as well. MGD (millions of gallons per day) and MGPCD (millions of gallons per capita day) are units commonly used in municipal waterworks problems. MMSCFD (millions of standard cubic feet per day) may be used to express gas flows.

Calculation of flow rates is often complicated by the interdependence between flow rate and friction loss. Each affects the other, so many pipe flow problems must be solved iteratively. Usually, a reasonable friction factor is assumed and used to calculate an initial flow rate. The flow rate establishes the flow velocity, from which a revised friction factor can be determined.

3. TYPICAL VELOCITIES IN PIPES

Fluid friction in pipes is kept at acceptable levels by maintaining reasonable fluid velocities. Table 17.1 lists typical maximum fluid velocities. Higher velocities may be observed in practice, but only with a corresponding increase in friction and pumping power.

4. HEAD LOSS DUE TO FRICTION

The original Bernoulli equation was based on an assumption of frictionless flow. In actual practice, friction occurs during fluid flow. This friction "robs" the fluid of energy, E, so that the fluid at the end of a pipe section has less energy than it does at the beginning.[2]

$$E_1 > E_2 \qquad 17.7$$

Most formulas for calculating friction loss use the symbol h_f to represent the *head loss due to friction.*[3] This loss is added into the original Bernoulli equation to restore the equality. Of course, the units of h_f must be the same as the units for the other terms in the Bernoulli equation. If the Bernoulli equation is written in terms of energy, the units will be ft-lbf/lbm or J/kg.

$$E_1 = E_2 + E_f \qquad 17.8$$

Consider the constant-diameter, horizontal pipe in Fig. 17.1. An incompressible fluid is flowing at a steady rate. Since the elevation of the pipe, z, does not change, the potential energy is constant. Since the pipe has a constant area, the kinetic energy (velocity) is constant. Therefore, the friction energy loss must show up as a decrease in pressure energy. Since the fluid is incompressible, this can only occur if the pressure, p, decreases in the direction of flow.

Figure 17.1 *Pressure Drop in a Pipe*

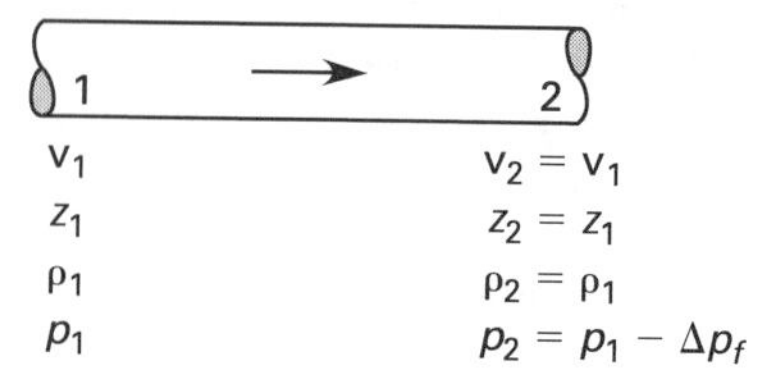

[1]Perhaps the most disparate conclusion is *D'Alembert's paradox*. In 1744, D'Alembert derived theoretical results "proving" that there is no resistance to bodies moving through an ideal (nonviscous) fluid.
[2]The friction generates minute amounts of heat. The heat is lost to the surroundings.
[3]Other names and symbols for this friction loss are *friction head loss* (h_L), *lost work* (LW), *friction heating* ($\mathcal{F}$), *skin friction loss* (F_f), and *pressure drop due to friction* (Δp_f). All terms and symbols essentially mean the same thing, although the units may be different.

Table 17.1 *Typical Full-Pipe Bulk Fluid Velocities*

fluid and application	velocity ft/sec	velocity m/s
water: city service	2–10	0.6–2.1
3 in diameter	4	1.2
6 in diameter	5	1.5
12 in diameter	9	2.7
water: boiler feed	8–15	2.4–4.5
water: pump suction	4	1.2
water: pump discharge	4–8.5	1.2–2.5
water, sewage: partially filled sewer	2.5 (min)	0.75 (min)
brine, water: chillers and coolers	6–8 typ (3–10)	1.8–2.4 typ (0.9–3)
air: compressor suction	75–200	23–60
air: compressor discharge	100–250	30–75
air: HVAC forced air	15–25	5–8
natural gas: overland pipeline	< 150 (60 typ)	< 45 (18 typ)
steam, saturated: heating	65–100	20–30
steam, saturated: miscellaneous	100–200	30–60
50–100 psia	< 150	< 45
150–400 psia	< 130	< 39
400–600 psia	< 100	< 30
steam, superheated: turbine feed	160–250	50–75
hydraulic fluid: fluid power	7–15	2.1–4.6
liquid sodium ($T > 525°C$): heat transfer	10 typ (0.3–40)	3 typ (0.1–12)
ammonia: compressor suction	85 (max)	25 (max)
ammonia: compressor discharge	100 (max)	30 (max)
oil, crude: overland pipeline	4–12	1.2–3.6
oil, lubrication: pump suction	< 2	< 0.6
oil, lubrication: pump discharge	3–7	0.9–2.1

(Multiply ft/sec by 0.3048 to obtain m/s.)

5. RELATIVE ROUGHNESS

It is intuitive that pipes with rough inside surfaces will experience greater friction losses than smooth pipes.[4] *Specific roughness*, ϵ, is a parameter that measures the average size of imperfections inside the pipe. Table 17.2 lists values of ϵ for common pipe materials. [**Moody Diagram (Stanton Diagram)**]

Table 17.2 *Values of Specific Roughness, ϵ, for Common Pipe Materials*

material	ϵ ft	ϵ m
plastic (PVC, ABS)	0.000005	1.5×10^{-6}
copper and brass	0.000005	1.5×10^{-6}
steel	0.0002	6.0×10^{-5}
plain cast iron	0.0008	2.4×10^{-4}
concrete	0.004	1.2×10^{-3}

(Multiply ft by 0.3048 to obtain m.)

However, an imperfection the size of a sand grain will have much more effect in a small-diameter hydraulic line than in a large-diameter sewer. Therefore, the *relative roughness*, ϵ/D, is a better indicator of pipe roughness. Both ϵ and D have units of length (e.g., feet or meters), and the relative roughness is dimensionless.

6. FRICTION FACTOR

The *Darcy friction factor*, f, is one of the parameters used to calculate friction loss.[5] The friction factor is not constant but decreases as the Reynolds number (fluid velocity) increases, up to a certain point known as *fully turbulent flow* (or *rough-pipe flow*). Once the flow is fully turbulent, the friction factor remains constant and depends only on the relative roughness and not on the Reynolds number. (See Fig. 17.2.) For very smooth pipes, fully turbulent flow is achieved only at very high Reynolds numbers.

Figure 17.2 *Friction Factor as a Function of Reynolds Number*

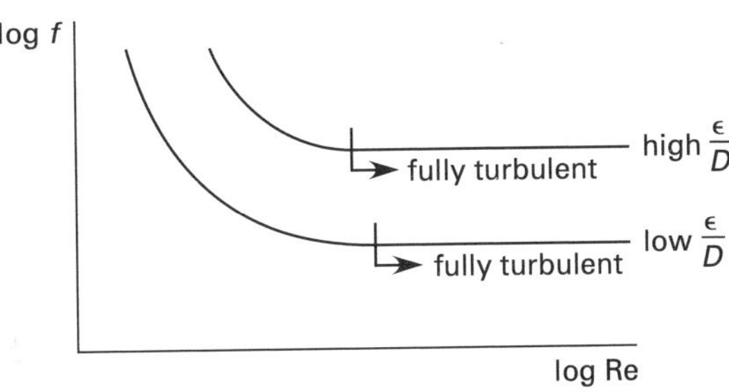

The friction factor is not dependent on the material of the pipe but is affected by the roughness. For example, for a given Reynolds number, the friction factor will be the same for any smooth pipe material (glass, plastic, smooth brass and copper, etc.).

The friction factor is determined from the relative roughness, ϵ/D, and the Reynolds number, Re, by various methods. These methods include explicit and

[4]Surprisingly, this intuitive statement is valid only for turbulent flow. The roughness does not (ideally) affect the friction loss for laminar flow.
[5]There are actually two friction factors: the Darcy friction factor and the *Fanning friction factor*, f_{Fanning}, also known as the *skin friction coefficient* and *wall shear stress factor*. Both factors are in widespread use, sharing the same symbol, f. Civil and (most) mechanical engineers use the Darcy friction factor. The Fanning friction factor is encountered more often by chemical engineers. One can be derived from the other: $f_{\text{Darcy}} = 4f_{\text{Fanning}}$.

implicit equations, the Moody diagram, and tables. The values obtained are based on experimentation, primarily the work of J. Nikuradse in the early 1930s.

When a moving fluid initially encounters a parallel surface (as when a moving gas encounters a flat plate or when a fluid first enters the mouth of a pipe), the flow will generally not be turbulent, even for very rough surfaces. The flow will be laminar for a certain *critical distance* before becoming turbulent.

Friction Factors for Laminar Flow

The easiest method of obtaining the friction factor for laminar flow (Re < 2300) is to calculate it. Equation 17.9 illustrates that roughness is not a factor in determining the frictional loss in ideal laminar flow.

Moody Diagram (Stanton Diagram)

$$f = \frac{64}{\text{Re}} \quad \text{[circular pipe]} \qquad 17.9$$

Table 17.3 gives friction factors for laminar flow in various cross sections.

Friction Factors for Turbulent Flow: by Moody Chart

The *Moody friction factor chart*, presents the friction factor graphically as a function of Reynolds number and relative roughness. There are different lines for selected discrete values of relative roughness. Due to the complexity of this graph, it is easy to mislocate the Reynolds number or use the wrong curve. Nevertheless, the Moody chart remains the most common method of obtaining the friction factor.

Example 17.1

Use the Moody diagram to find the friction factor for a Reynolds number of Re = 400,000 and a relative roughness of $\epsilon/D = 0.004$.

Solution

From the Moody diagram, the friction factor is approximately 0.028. [**Moody Diagram (Stanton Diagram)**]

Friction Factors for Turbulent Flow: by Formula

One of the earliest attempts to predict the friction factor for turbulent flow in smooth pipes resulted in the *Blasius equation* (claimed "valid" for 3000 < Re < 100,000).

$$f = \frac{0.316}{\text{Re}^{0.25}} \qquad 17.10$$

The *Nikuradse equation* can also be used to determine the friction factor for smooth pipes (i.e., when $\epsilon/D = 0$). Unfortunately, this equation is implicit in f and must be solved iteratively.

$$\frac{1}{\sqrt{f}} = 2.0\log_{10}(\text{Re}\sqrt{f}) - 0.80 \qquad 17.11$$

***Table 17.3** Friction Factors for Laminar Flow in Various Cross Sections**

tube geometry	D_H (full)	c/d or θ	friction factor, f
circle	D	–	64.00/Re
rectangle	$\frac{2cd}{c+d}$	1	56.92/Re
		2	62.20/Re
		3	68.36/Re
		4	72.92/Re
		6	78.80/Re
		8	82.32/Re
		∞	96.00/Re
ellipse	$\frac{cd}{\sqrt{\frac{1}{2}(c^2+d^2)}}$	1	64.00/Re
		2	67.28/Re
		4	72.96/Re
		8	76.60/Re
		16	78.16/Re
isosceles triangle	$\frac{d\sin\theta}{2\left(1+\sin\frac{\theta}{2}\right)}$	10°	50.80/Re
		30°	52.28/Re
		60°	53.32/Re
		90°	52.60/Re
		120°	50.96/Re

*$\text{Re} = v_{\text{bulk}}D_H/\nu$, and $D_H = 4A/P$.

The *Karman-Nikuradse equation* predicts the fully turbulent friction factor (i.e., when Re is very large).

$$\frac{1}{\sqrt{f}} = 1.74 - 2\log_{10}\frac{2\epsilon}{D} \qquad 17.12$$

The most widely known method of calculating the friction factor for any pipe roughness and Reynolds number is another implicit formula, the *Colebrook equation.* Most other equations are variations of this equation. (Notice that the relative roughness, ϵ/D, is used to calculate f.)

$$\frac{1}{\sqrt{f}} = -2\log_{10}\left(\frac{\frac{\epsilon}{D}}{3.7} + \frac{2.51}{\text{Re}\sqrt{f}}\right) \qquad 17.13$$

A suitable approximation would appear to be the *Swamee-Jain equation*, which claims to have less than 1% error (as measured against the Colebrook equation) for relative roughnesses between 0.000001 and 0.01, and for Reynolds numbers between 5000 and 100,000,000.[6] Even with a 1% error, this equation produces more accurate results than can be read from the Moody friction factor chart.

$$f = \frac{0.25}{\left(\log_{10}\left(\frac{\frac{\epsilon}{D}}{3.7} + \frac{5.74}{\mathrm{Re}^{0.9}}\right)\right)^2} \qquad 17.14$$

7. ENERGY LOSS DUE TO FRICTION: LAMINAR FLOW

The most common method of calculating frictional energy loss in a fluid experiencing laminar flow is the *Darcy-Weisbach equation* (also known as the *Darcy equation* or the *Weisbach equation*), which can be used for both laminar and turbulent flow.[7] One advantage of using the Darcy equation is that if f is known, the assumption of laminar flow does not need to be confirmed.

Darcy-Weisbach Equation

$$h_f = f\left(\frac{L}{D}\right)\left(\frac{\mathrm{v}^2}{2g}\right) \qquad 17.15$$

If necessary, h_f can be converted to an actual pressure drop in lbf/ft^2 or Pa by multiplying by the fluid density.

$$\Delta p = h_f \times \rho g \qquad \text{[SI]} \qquad 17.16(a)$$

$$\Delta p = h_f \times \rho\left(\frac{g}{g_c}\right) = h_f\gamma \qquad \text{[U.S.]} \qquad 17.16(b)$$

Values of the Darcy friction factor, f, are often quoted for new, clean pipe. The friction head losses and pumping power requirements calculated from these values are minimum values. Depending on the nature of the service, scale and impurity buildup within pipes may decrease the pipe diameters over time. Since the frictional loss is proportional to the fifth power of the diameter, such diameter decreases can produce dramatic increases in the friction loss.

$$\frac{h_{f,\mathrm{scaled}}}{h_{f,\mathrm{new}}} = \left(\frac{D_{\mathrm{new}}}{D_{\mathrm{scaled}}}\right)^5 \qquad 17.17$$

Equation 17.17 accounts for only the decrease in diameter. Any increase in roughness (i.e., friction factor) will produce a proportional increase in friction loss.

Because the "new, clean" condition is transitory in most applications, an uprating factor of 10–30% is often applied to either the friction factor, f, or the head loss, h_f. Of course, even larger increases should be considered when extreme fouling is expected.

Another approach eliminates the need to estimate the scaled pipe diameter. This simplistic approach multiplies the initial friction loss by a factor based on the age of the pipe. For example, for schedule-40 pipe between 4 in and 10 in (10 cm and 25 cm) in diameter, the multipliers of 1.4, 2.2, and 5.0 have been proposed for pipe ages of 5, 10, and 20 years, respectively. For larger pipes, the corresponding multipliers are 1.3, 1.6, and 2.0. Obviously, use of these values should be based on a clear understanding of the method's limitations.

8. ENERGY LOSS DUE TO FRICTION: TURBULENT FLOW

For turbulent flow, the *Darcy-Weisbach equation* is used almost exclusively to calculate the head loss due to friction.

Darcy-Weisbach Equation

$$h_f = f\left(\frac{L}{D}\right)\left(\frac{\mathrm{v}^2}{2g}\right) \qquad 17.18$$

The head loss can be converted to pressure drop.

Bernoulli Equation

$$p_1 - p_2 = \gamma h_f = \rho g h_f \qquad \text{[SI]} \qquad 17.19(a)$$

$$p_1 - p_2 = \gamma h_f = \frac{\rho g h_f}{g_c} \qquad \text{[U.S.]} \qquad 17.19(b)$$

In problems where the pipe size is unknown, it will be impossible to obtain an accurate initial value of the friction factor, f (since f depends on velocity). In such problems, an iterative solution will be necessary.

Civil engineers commonly use the *Hazen-Williams equation* to calculate head loss. This method requires knowing the Hazen-Williams *roughness coefficient*, C, values of which are widely tabulated.[8] **[Values of Hazen-Williams Coefficient C]**

The advantage of using this equation is that C does not depend on the Reynolds number. The Hazen-Williams equation is empirical and is not dimensionally

[6]American Society of Civil Engineers. *Journal of Hydraulic Engineering* 102 (May 1976): 657. This is not the only explicit approximation to the Colebrook equation in existence.
[7]The difference is that the friction factor can be derived by hydrodynamics: $f = 64/\mathrm{Re}$. For turbulent flow, f is empirical.
[8]An approximate value of $C = 140$ is often chosen for initial calculations for new water pipe. $C = 100$ is more appropriate for water pipe that has been in service for some time. For sludge, C values are 20–40% lower than the equivalent water pipe values.

consistent. The length, L, must be in feet, the flow rate, Q, in gallons per minute, and the inner diameter, D, in inches. It is then taken as a matter of faith that the units of h_f are feet of head.

Hazen-Williams Equation

$$h_{f,\text{ft}} = 0.002083 L_{\text{ft}} \left(\frac{100}{C}\right)^{1.85} \left(\frac{Q_{\text{gpm}}^{1.85}}{D_{\text{in}}^{4.8655}}\right) \qquad 17.20$$

When the Hazen-Williams equation is used to calculate head loss in a circular pipe expressed as a pressure drop per foot of pipe, the equation is

Circular Pipe Head Loss Equation (Head Loss Expressed as Pressure)

$$\Delta p_{\text{psi/ft}} = \frac{4.52 Q_{\text{gpm}}^{1.85}}{C^{1.85} D_{\text{in}}^{4.87}} \qquad 17.21$$

The Hazen-Williams equation should be used only for turbulent flow. It gives good results for liquids that have kinematic viscosities around 1.2×10^{-5} ft²/sec (1.1×10^{-6} m²/s), which corresponds to the viscosity of 60°F (16°C) water. At extremely high and low temperatures, the Hazen-Williams equation can be 20% or more in error for water.

Example 17.2

50°F water is pumped through 1000 ft of 4 in, schedule-40 welded steel pipe at the rate of 300 gpm. What friction loss (in feet of head) is predicted by the Darcy-Weisbach equation?

Solution

From a table of water properties, the kinematic viscosity of 50°F water is 1.410×10^{-5} ft²/sec. [**Properties of Water (I-P Units)**]

From a list of roughness coefficients, the range of coefficients for steel pipe is 0.0001 ft to 0.0003 ft; use an intermediate value of 0.0002 ft. [**Moody Diagram (Stanton Diagram)**]

From a table of pipe data, schedule-40 steel pipe with a nominal size of 4 in has an inside diameter of 4.026 in and an area of flow of 12.724 in². [**Schedule 40 Steel Pipe**]

$$D = \frac{4.026 \text{ in}}{12 \frac{\text{in}}{\text{ft}}} = 0.3355 \text{ ft}$$

$$A = \frac{12.724 \text{ in}^2}{\left(12 \frac{\text{in}}{\text{ft}}\right)^2} = 0.08835 \text{ ft}^2$$

Convert the flow to cubic feet per second.

$$Q = \frac{300 \frac{\text{gal}}{\text{min}}}{\left(7.481 \frac{\text{gal}}{\text{ft}^3}\right)\left(60 \frac{\text{sec}}{\text{min}}\right)} = 0.6684 \text{ ft}^3/\text{sec}$$

Use the continuity equation to find the velocity.

Continuity Equation

$$Q = A\text{v}$$

$$\text{v} = \frac{Q}{A} = \frac{0.6684 \frac{\text{ft}^3}{\text{sec}}}{0.08835 \text{ ft}^2} = 7.565 \text{ ft/sec}$$

Calculate the Reynolds number.

Reynolds Number

$$\text{Re} = \frac{\text{v}D}{\nu} = \frac{\left(7.565 \frac{\text{ft}}{\text{sec}}\right)(0.3355 \text{ ft})}{1.410 \times 10^{-5} \frac{\text{ft}^2}{\text{sec}}} = 180{,}000$$

The relative roughness is

$$\frac{\epsilon}{D} = \frac{0.0002 \text{ ft}}{0.3355 \text{ ft}} = 0.0006$$

From the Moody diagram, for a Reynolds number of 180,000 and a relative roughness of 0.0006, the friction factor is 0.0195.

Use the Darcy-Weisbach equation to calculate the friction loss.

Darcy-Weisbach Equation

$$h_f = f\left(\frac{L}{D}\right)\left(\frac{\text{v}^2}{2g}\right)$$

$$= (0.0195)\left(\frac{1000 \text{ ft}}{0.3355 \text{ ft}}\right) \times \left(\frac{\left(7.565 \frac{\text{ft}}{\text{sec}}\right)^2}{(2)\left(32.17 \frac{\text{ft}}{\text{sec}^2}\right)}\right)$$

$$= 51.70 \text{ ft}$$

Example 17.3

For the pipe in Ex. 17.2, the Hazen-Williams coefficient is 100. Use the Hazen-Williams equation to calculate the head loss due to friction (a) in feet of head, and (b) as a pressure drop in psi per foot.

Solution

(a) Substitute the values given in Ex. 17.2 into Eq. 17.20.

Hazen-Williams Equation

$$\begin{aligned} h_{f,\text{ft}} &= 0.002083 L_{\text{ft}} \left(\frac{100}{C}\right)^{1.85} \left(\frac{Q_{\text{gpm}}^{1.85}}{D_{\text{in}}^{4.8655}}\right) \\ &= (0.002083)(1000\ \text{ft}) \left(\frac{100}{100}\right)^{1.85} \left(\frac{\left(300\ \frac{\text{gal}}{\text{min}}\right)^{1.85}}{(4.026\ \text{in})^{4.8655}}\right) \\ &= 90.86\ \text{ft} \end{aligned}$$

(b) Substitute the values given in Ex. 17.2 into the equation for pressure drop in a circular pipe, Eq. 17.21.

Circular Pipe Head Loss Equation (Head Loss Expressed as Pressure)

$$\begin{aligned} \Delta p_{\text{psi/ft}} &= \frac{4.52 Q_{\text{gpm}}^{1.85}}{C^{1.85} d_{\text{in}}^{4.87}} \\ &= \frac{(4.52)\left(300\ \frac{\text{gal}}{\text{min}}\right)^{1.85}}{(100)^{1.85}(4.026\ \text{in})^{4.87}} \\ &= 0.03909\ \text{psi per foot} \end{aligned}$$

9. FRICTION LOSS FOR WATER FLOW IN STEEL PIPES

Since water's specific volume is essentially constant within the normal temperature range, tables and charts can be used to determine water velocity. Friction loss and velocity for water flowing through steel pipe (as well as for other liquids and other pipe materials) in table and chart form are widely available. (Appendix 17.B is an example of such a table.) Tables and charts almost always give the friction loss per 100 ft or 10 m of pipe. The pressure drop is proportional to the length, so the value read can be scaled for other pipe lengths. Flow velocity is independent of pipe length.

These tables and charts are unable to compensate for the effects of fluid temperature and different pipe roughness. Unfortunately, the assumptions made in developing the tables and charts are seldom listed. Another disadvantage is that the values can be read to only a few significant figures. Friction loss data should be considered accurate to only ±20%. Alternatively, a 20% safety margin should be established in choosing pumps and motors.

10. FRICTION LOSS IN NONCIRCULAR DUCTS

The frictional energy loss by a fluid flowing in a rectangular, annular, or other noncircular duct can be calculated from the Darcy-Weisbach equation by using the *hydraulic diameter*, D_H, in place of the diameter, D. The hydraulic diameter is defined as four times the cross-sectional area of flow divided by the wetted perimeter.[9]

The *hydraulic radius*, R_H, is defined as the cross-sectional area of flow divided by the wetted perimeter. The hydraulic diameter is therefore four times the hydraulic radius.

Flow in Noncircular Conduits

$$D_H = \frac{4 \times \text{cross-sectional area in flow}}{\text{wetted perimeter}} = \frac{4A}{P} = 4R_H$$

The friction factor, f, can be determined in any of the conventional manners.

11. FRICTION LOSS FOR OTHER LIQUIDS, STEAM, AND GASES

The Darcy equation can be used to calculate the frictional energy loss for all incompressible liquids, not just for water. Alcohol, gasoline, fuel oil, and refrigerants, for example, are all handled well, since the effect of viscosity is considered in determining the friction factor, f.[10]

In fact, the Darcy equation is commonly used with noncondensing vapors and compressed gases, such as air, nitrogen, and steam.[11] In such cases, reasonable accuracy will be achieved as long as the fluid is not moving too fast (i.e., less than Mach 0.3) and is incompressible. The fluid is assumed to be incompressible if the pressure (or density) change along the section of interest is less than 10% of the starting pressure.

[9]Although it is used for both, this approach is better suited for turbulent flow than for laminar flow. Also, the accuracy of this method decreases as the flow area becomes more noncircular. The friction drop in long, narrow slit passageways is poorly predicted, for example. However, there is no other convenient method of predicting friction drop. Experimentation should be used with a particular flow geometry if extreme accuracy is required.

[10]Since viscosity is not an explicit factor in the formula, it should be obvious that the Hazen-Williams equation is primarily used for water.

[11]Use of the Darcy equation is limited only by the availability of the viscosity data needed to calculate the Reynolds number.

If possible, it is preferred to base all calculations on the average properties of the fluid as determined at the midpoint of a pipe.[12] Specifically, the fluid velocity would normally be calculated as

$$\text{v} = \frac{\dot{m}}{\rho_{\text{ave}} A} \qquad 17.22$$

However, the average density of a gas depends on the average pressure, which is unknown at the start of a problem. The solution is to write the Reynolds number and Darcy equation in terms of the constant mass flow rate per unit area, G, instead of velocity, v, which varies.

$$G = \text{v}_{\text{ave}} \rho_{\text{ave}} \qquad 17.23$$

$$\text{Re} = \frac{DG}{\mu} \qquad \text{[SI]} \qquad 17.24(a)$$

$$\text{Re} = \frac{DG}{g_c \mu} \qquad \text{[U.S.]} \qquad 17.24(b)$$

$$\Delta p_f = p_1 - p_2 = \rho_{\text{ave}} h_f g = \frac{fLG^2}{2D\rho_{\text{ave}}} \qquad \text{[SI]} \qquad 17.25(a)$$

$$\begin{aligned} \Delta p_f = p_1 - p_2 = \gamma_{\text{ave}} h_f &= \rho_{\text{ave}} h_f \times \frac{g}{g_c} \\ &= \frac{fLG^2}{2D\rho_{\text{ave}} g_c} \end{aligned} \qquad \text{[U.S.]} \qquad 17.25(b)$$

Assuming a perfect gas with a molecular weight of M, the ideal gas law can be used to calculate ρ_{ave} from the absolute temperature, T, and $p_{\text{ave}} = (p_1 + p_2)/2$.

$$p_1^2 - p_2^2 = \frac{fLG^2 \overline{R} T}{DM} \qquad \text{[SI]} \qquad 17.26(a)$$

$$p_1^2 - p_2^2 = \frac{fLG^2 \overline{R} T}{Dg_c M} \qquad \text{[U.S.]} \qquad 17.26(b)$$

To summarize, use the following guidelines when working with compressible gases or vapors flowing in a pipe or duct. (a) If the pressure drop, based on the entrance pressure, is less than 10%, the fluid can be assumed to be incompressible, and the gas properties can be evaluated at any point known along the pipe. (b) If the pressure drop is between 10% and 40%, use of the midpoint properties will yield reasonably accurate friction losses. (c) If the pressure drop is greater than 40%, the pipe can be divided into shorter sections and the losses calculated for each section, or exact calculations based on compressible flow theory must be made.

Calculating a friction loss for steam flow using the Darcy equation can be frustrating if steam viscosity data are unavailable. Generally, the steam viscosities listed in compilations of heat transfer data are sufficiently accurate. Various empirical methods are also in use. For example, the *Babcock formula*, given by Eq. 17.27, for pressure drop when steam with a specific volume, v, flows in a pipe of diameter D is

$$\Delta p_{\text{psi}} = 0.470 \left(\frac{D_{\text{in}} + 3.6}{D_{\text{in}}^6} \right) \dot{m}_{\text{lbm/sec}}^2 L_{\text{ft}} v_{\text{ft}^3/\text{lbm}} \qquad 17.27$$

Use of empirical formulas is not limited to steam. Theoretical formulas (e.g., the *complete isothermal flow equation*) and specialized empirical formulas (e.g., the *Weymouth*, *Panhandle*, and *Spitzglass formulas*) have been developed, particularly by the gas pipeline industry. Each of these provides reasonable accuracy within their operating limits. However, none should be used without knowing the assumptions and operational limitations that were used in their derivations.

Example 17.4

0.0011 kg/s of 25°C nitrogen gas flows isothermally through a 175 m section of smooth tubing (inside diameter = 0.012 m). The viscosity of the nitrogen is 1.8 × 10^{-5} Pa·s. The pressure of the nitrogen is 200 kPa originally. At what pressure is the nitrogen delivered?

SI Solution

The flow area of the pipe is

$$A = \frac{\pi D^2}{4} = \frac{\pi (0.012 \text{ m})^2}{4} = 1.131 \times 10^{-4} \text{ m}^2$$

The mass flow rate per unit area is

$$G = \frac{\dot{m}}{A} = \frac{0.0011 \ \frac{\text{kg}}{\text{s}}}{1.131 \times 10^{-4} \text{ m}^2} = 9.73 \text{ kg/m}^2\text{·s}$$

The Reynolds number is

$$\text{Re} = \frac{DG}{\mu} = \frac{(0.012 \text{ m}) \left(9.73 \ \frac{\text{kg}}{\text{m}^2\text{·s}} \right)}{1.8 \times 10^{-5} \text{ Pa·s}} = 6487$$

The flow is turbulent, and the pipe is said to be smooth. From a Moody diagram, the friction factor is about 0.035.

[12]Of course, the entrance (or exit) conditions can be used if great accuracy is not needed.

Since two atoms of nitrogen form a molecule of nitrogen gas, the molecular weight of nitrogen is twice the atomic weight, or 28 kg/kmol. The temperature must be in degrees absolute: $T = 25°\text{C} + 273° = 298\text{K}$. The universal gas constant is 8314.47 J/kmol·K.

From Eq. 17.26, the final pressure is

$$p_2^2 = p_1^2 - \frac{fLG^2\overline{R}T}{DM}$$

$$= (200\,000\ \text{Pa})^2 - \frac{(0.035)(175\ \text{m})\left(9.73\ \frac{\text{kg}}{\text{m}^2\text{·s}}\right)^2 \times \left(8314.47\ \frac{\text{J}}{\text{kmol·K}}\right)(298\text{K})}{(0.012\ \text{m})\left(28\ \frac{\text{kg}}{\text{kmol}}\right)}$$

$$= 3.576 \times 10^{10}\ \text{Pa}^2$$

$$p_2 = \sqrt{3.576 \times 10^{10}\ \text{Pa}^2} = 1.89 \times 10^5\ \text{Pa} \quad (189\ \text{kPa})$$

The percentage drop in pressure should not be more than 10%.

$$\frac{200\ \text{kPa} - 189\ \text{kPa}}{200\ \text{kPa}} = 0.055 \quad (5.5\%) \quad [\text{OK}]$$

Example 17.5

Superheated steam at 140 psi and 500°F enters a 200 ft long steel pipe (Darcy friction factor of 0.02) with an internal diameter of 3.826 in. The pipe is insulated so that there is no heat loss. (a) Use the Babcock formula to determine the maximum velocity and mass flow rate such that the steam does not experience more than a 10% drop in pressure. (b) Verify the velocity by calculating the pressure drop with the Darcy-Weisbach equation.

Solution

(a) From a superheated steam table, interpolating as needed, the specific volume of the steam is 4.062 ft³/lbm. **[Properties of Superheated Steam - I-P Units]**

The maximum pressure drop is 10% of 140 psi or 14 psi.

From Eq. 17.27,

$$\Delta p_{\text{psi}} = 0.470\left(\frac{D_{\text{in}} + 3.6}{D_{\text{in}}^6}\right)\dot{m}_{\text{lbm/sec}}^2 L_{\text{ft}} v_{\text{ft}^3/\text{lbm}}$$

$$14\ \text{psi} = (0.470)\left(\frac{3.826\ \text{in} + 3.6}{(3.826\ \text{in})^6}\right)\dot{m}^2 \times (200\ \text{ft})\left(4.062\ \frac{\text{ft}^3}{\text{lbm}}\right)$$

$$\dot{m} = 3.935\ \text{lbm/sec}$$

The steam velocity is

Continuity Equation

$$\text{v} = \frac{Q}{A} = \frac{\dot{m}}{\rho A} = \frac{\dot{m}v}{\left(\frac{\pi}{4}\right)D^2}$$

$$= \frac{\left(3.935\ \frac{\text{lbm}}{\text{sec}}\right)\left(4.062\ \frac{\text{ft}^3}{\text{lbm}}\right)}{\left(\frac{\pi}{4}\right)\left(\frac{3.826\ \text{in}}{12\ \frac{\text{in}}{\text{ft}}}\right)^2}$$

$$= 200.2\ \text{ft/sec}$$

(b) Use the Darcy-Weisbach equation to find the steam friction head.

Darcy-Weisbach Equation

$$h_f = f\left(\frac{L}{D}\right)\left(\frac{\text{v}^2}{2g}\right)$$

$$= (0.02)\left(\frac{200\ \text{ft}}{\left(\frac{3.826\ \text{in}}{12\ \frac{\text{in}}{\text{ft}}}\right)}\right)\left(\frac{\left(200.2\ \frac{\text{ft}}{\text{sec}}\right)^2}{(2)\left(32.2\ \frac{\text{ft}}{\text{sec}^2}\right)}\right)$$

$$= 7808\ \text{ft of steam}$$

From Eq. 17.25,

$$\Delta p = \rho h_f \times \frac{g}{g_c} = \frac{h_f}{v} \times \frac{g}{g_c}$$

$$= \frac{7808\ \text{ft}}{\left(4.062\ \frac{\text{ft}^3}{\text{lbm}}\right)\left(12\ \frac{\text{in}}{\text{ft}}\right)^2} \times \left(\frac{32.2\ \frac{\text{ft}}{\text{sec}^2}}{32.2\ \frac{\text{lbm-ft}}{\text{lbf-sec}^2}}\right)$$

$$= 13.3\ \text{lbf/in}^2 \quad (13.3\ \text{psi})$$

12. EFFECT OF VISCOSITY ON HEAD LOSS

Friction loss in a pipe is affected by the fluid viscosity. For both laminar and turbulent flow, viscosity is considered when the Reynolds number is calculated. When viscosities substantially increase without a corresponding decrease in flow rate, two things usually happen: (a) the friction loss greatly increases, and (b) the flow becomes laminar.

It is sometimes necessary to estimate head loss for a new fluid viscosity based on head loss at an old fluid viscosity. The estimation procedure used depends on the flow regimes for the new and old fluids.

For laminar flow, the friction factor is directly proportional to the viscosity. If the flow is laminar for both fluids, the ratio of new-to-old head losses will be equal to the ratio of new-to-old viscosities. Therefore, if a flow is already known to be laminar at one viscosity and the fluid viscosity increases, a simple ratio will define the new friction loss.

If both flows are fully turbulent, the friction factor will not change. If flow is fully turbulent and the viscosity decreases, the Reynolds number will increase. Theoretically, this will have no effect on the friction loss.

There are no analytical ways of estimating the change in friction loss when the flow regime changes between laminar and turbulent or between semiturbulent and fully turbulent. Various graphical methods are used, particularly by the pump industry, for calculating power requirements.

13. MINOR LOSSES

In addition to the frictional energy lost due to viscous effects, friction losses also result from fittings in the line, changes in direction, and changes in flow area. These losses are known as *minor losses* or *local losses*, since they are usually much smaller in magnitude than the pipe wall frictional loss.[13] Two methods are used to calculate minor losses: equivalent lengths and loss coefficients.

With the *method of equivalent lengths*, each fitting or other flow variation is assumed to produce friction equal to the pipe wall friction from an *equivalent length* of pipe. For example, a 2 in globe valve may produce the same amount of friction as 54 ft (its equivalent length) of 2 in pipe. The equivalent lengths for all minor losses are added to the pipe length term, L, in the Darcy equation. The method of equivalent lengths can be used with all liquids, but it is usually limited to turbulent flow by the unavailability of laminar equivalent lengths, which are significantly larger than turbulent equivalent lengths.

$$L_t = L + \sum L_{\text{eq}} \qquad 17.28$$

Equivalent lengths are simple to use, but the method depends on having a table of equivalent length values. The actual value for a fitting will depend on the fitting manufacturer, as well as the fitting material (e.g., brass, cast iron, or steel) and the method of attachment (e.g., weld, thread, or flange).[14] Because of these many variations, it may be necessary to use a "generic table" of equivalent lengths during the initial design stages.

An alternative method of calculating the minor loss for a fitting is to use the *method of loss coefficients*. Each fitting has a *loss coefficient*, K, associated with it, which, when multiplied by the kinetic energy, gives the loss. (See Table 17.4.) Therefore, a loss coefficient is the minor loss, $h_{f,\text{fitting}}$, expressed in fractions (or multiples) of the velocity head.

Minor Losses in Pipe Fittings, Contractions, and Expansions

$$h_{f,\text{fitting}} = K\left(\frac{v^2}{2g}\right) \qquad 17.29$$

The minor loss can also be expressed as a drop in pressure.

Valve and Fittings Losses

$$\Delta p = K\left(\frac{\rho}{g}\right)\left(\frac{v^2}{2}\right) \qquad 17.30$$

The loss coefficient for any minor loss can be calculated if the equivalent length is known. However, there is no advantage to using one method over the other, other than convention and for consistency in calculations.

$$K = \frac{fL_{\text{eq}}}{D} \qquad 17.31$$

Exact friction loss coefficients for bends, fittings, and valves are unique to each manufacturer. Furthermore, except for contractions, enlargements, exits, and entrances, the coefficients decrease fairly significantly (according to the fourth power of the diameter ratio) with increases in valve size. Therefore, a single K value is seldom applicable to an entire family of valves. Nevertheless, generic tables and charts have been developed. These compilations can be used for initial estimates as long as the general nature of the data is recognized. [**K-Factors—Threaded Pipe Fittings**]

Loss coefficients for specific fittings and valves must be known in order to be used. They cannot be derived theoretically. However, the loss coefficients for certain changes in flow area can be calculated from the following equations.[15]

[13]Example and practice problems often include the instruction to "ignore minor losses." In some industries, valves are considered to be "components," not fittings. In such cases, instructions to "ignore minor losses in fittings" would be ambiguous, since minor losses in valves would be included in the calculations. However, this interpretation is rare in examples and practice problems.
[14]In the language of pipe fittings, a *threaded fitting* is known as a *screwed fitting*, even though no screws are used.
[15]No attempt is made to imply great accuracy with these equations. Correlation between actual and theoretical losses is fair.

- *sudden enlargements* (D_1 is the smaller of the two diameters)

$$K = \left(1 - \left(\frac{D_1}{D_2}\right)^2\right)^2 \qquad 17.32$$

- *sudden contractions* (D_1 is the smaller of the two diameters)

$$K = \tfrac{1}{2}\left(1 - \left(\frac{D_1}{D_2}\right)^2\right) \qquad 17.33$$

- *pipe exit* (projecting exit, sharp-edged, or rounded)

$$K = 1.0 \qquad 17.34$$

- *pipe entrance*

reentrant: $K = 0.78$
sharp-edged: $K = 0.50$
rounded:

bend radius / D	K
0.02	0.28
0.04	0.24
0.06	0.15
0.10	0.09
0.15	0.04

- *tapered diameter changes*

$$\beta = \frac{\text{small diameter}}{\text{large diameter}} = \frac{D_1}{D_2}$$

$$\phi = \text{wall-to-horizontal angle}$$

For enlargement, $\phi \leq 22°$:

$$K = 2.6 \sin\phi (1 - \beta^2)^2 \qquad 17.35$$

For enlargement, $\phi > 22°$:

$$K = (1 - \beta^2)^2 \qquad 17.36$$

For contraction, $\phi \leq 22°$:

$$K = 0.8 \sin\phi (1 - \beta^2) \qquad 17.37$$

For contraction, $\phi > 22°$:

$$K = 0.5\sqrt{\sin\phi}\,(1 - \beta^2) \qquad 17.38$$

Example 17.6

A pipeline contains five regular 90° elbows and 228 ft of straight pipe. The elbows are 1 in threaded steel pipe. Water travels through the pipeline at 6 ft/sec. Disregard entrance and exit losses. Determine the total equivalent length of the pipeline.

Solution

From a table of equivalent lengths, the equivalent length for a 90° elbow of 1 in threaded pipe carrying water at 6 ft/sec is 3.0 ft. The total equivalent length of the pipeline is

$$\begin{aligned} L_t &= L + \sum L_{\text{eq}} \\ &= 228 \text{ ft} + (5)(3.0 \text{ ft}) \\ &= 243 \text{ ft} \end{aligned}$$

14. VALVE FLOW COEFFICIENTS

Valve flow capacities depend on the geometry of the inside of the valve. The *flow coefficient*, C_v, for a valve (particularly a control valve) relates the flow quantity (in gallons per minute) of a fluid with specific gravity to the pressure drop (in pounds per square inch). (The flow coefficient for a valve is not the same as the coefficient of flow for an orifice or venturi meter.) As Eq. 17.39 shows, the flow coefficient is not dimensionally homogeneous. Calculations are simplified for water flow where SG = 1.0.

Metricated countries use a similar concept with a different symbol, K_v, (not the same as the loss coefficient, K) to distinguish the valve flow coefficient from customary U.S. units. K_v is defined[16] as the flow rate in cubic meters per hour of water at a temperature of 16°C with a pressure drop across the valve of 1 bar. To further distinguish it from its U.S. counterpart, K_v may also be referred to as a *flow factor*. C_v and K_v are linearly related by Eq. 17.40.

Valve Flow Coefficient

$$K_v = Q_{\text{m}^3/\text{h}}\sqrt{\frac{\text{SG}}{\Delta p_{\text{bars}}}} \quad \text{[SI]} \qquad 17.39(a)$$

$$C_v = Q_{\text{gpm}}\sqrt{\frac{\text{SG}}{\Delta p_{\text{psi}}}} \quad \text{[U.S.]} \qquad 17.39(b)$$

$$K_v = 0.86\,C_v \qquad 17.40$$

When selecting a control valve for a particular application, the value of C_v is first calculated. Depending on the application and installation, C_v may be further modified by dividing by *piping geometry* and *Reynolds*

[16]Several definitions of both C_v and K_v are in use. A definition of C_v based on Imperial gallons is used in Great Britain. Definitions of K_v based on pressure drops in kilograms-force and volumes in liters per minute are in use. Other differences in the definition include the applicable temperature, which may be given as 5–30°C or 5–40°C instead of 16°C.

number factors. (These additional procedures are often specified by the valve manufacturer.) Then, a valve with the required value of C_v is selected.

Although the flow coefficient concept is generally limited to control valves, its use can be extended to all fittings and valves. The relationship between C_v and the loss coefficient, K, is

$$C_v = \frac{29.9D_{\text{in}}^2}{\sqrt{K}} \quad \text{[U.S.]} \qquad 17.41$$

Additional adjustments are required when sizing steam control valves. Use Eq. 17.42 when steam pressure is less than or equal to 15 psig. Inlet and outlet pressures are absolute, given in psia.

Valve Flow Coefficient

$$C_v = \frac{Q_{\text{lbm/hr}}}{2.11\sqrt{p_i^2 - p_o^2}} \qquad 17.42$$

When the inlet pressure is greater than 15 psig, Eq. 17.42 can be simplified to Eq. 17.43.

Valve Flow Coefficient

$$C_v = \frac{Q_{\text{lbm/hr}}}{1.6p_i} \qquad 17.43$$

15. EXTENDED BERNOULLI EQUATION

The original Bernoulli equation assumes frictionless flow and does not consider the effects of pumps and turbines.

A pump adds energy to the fluid flowing through it. The amount of energy added by the pump, E_A, can be determined from the difference between the total energies on the two sides of the pump. In most situations, a pump will add primarily pressure energy.

A turbine extracts energy from the fluid flowing through it. The amount of energy extracted by the turbine, E_E, can be similarly determined from the difference between the total energies on the two sides of the turbine.

When friction is present and when there are minor losses such as fittings and other energy-related devices in a pipeline, the energy balance is affected. The *extended Bernoulli equation* takes these additional factors into account.

$$\begin{aligned} &(E_p + E_{\text{v}} + E_z)_1 + E_A \\ &= (E_p + E_{\text{v}} + E_z)_2 + E_E + E_f + E_m \end{aligned} \qquad 17.44$$

$$\begin{aligned} &\frac{p_1}{\rho} + \frac{\text{v}_1^2}{2} + z_1 g + E_A \\ &= \frac{p_2}{\rho} + \frac{\text{v}_2^2}{2} + z_2 g + E_E + E_f + E_m \end{aligned} \quad \text{[SI]} \qquad 17.45(a)$$

$$\begin{aligned} &\frac{p_1}{\rho} + \frac{\text{v}_1^2}{2g_c} + \frac{z_1 g}{g_c} + E_A \\ &= \frac{p_2}{\rho} + \frac{\text{v}_2^2}{2g_c} + \frac{z_2 g}{g_c} + E_E + E_f + E_m \end{aligned} \quad \text{[U.S.]} \qquad 17.45(b)$$

As defined, E_A, E_E, and E_f are all positive terms. None of the terms in Eq. 17.44 and Eq. 17.45 are negative.

The concepts of sources and sinks can be used to decide whether the friction, pump, and turbine terms appear on the left or right side of the Bernoulli equation. An *energy source* puts energy into the system. The incoming fluid and a pump contribute energy to the system. An *energy sink* removes energy from the system. The leaving fluid, friction, and a turbine remove energy from the system. In an energy balance, all energy must be accounted for, and the energy sources just equal the energy sinks.

$$\sum E_{\text{sources}} = \sum E_{\text{sinks}} \qquad 17.46$$

Therefore, the energy added by a pump always appears on the entrance side of the Bernoulli equation. Similarly, the frictional energy loss always appears on the discharge side.

Equation 17.47 shows the extended Bernoulli equation expressed in terms of head.

$$\begin{aligned} &\frac{p_1}{\rho} + \frac{\text{v}_1^2}{2} + z_1 g + h_{\text{pump}} \\ &= \frac{p_2}{\rho} + \frac{\text{v}_2^2}{2} + z_2 g + h_{\text{turbine}} \\ &\quad + h_f + h_m \end{aligned} \quad \text{[SI]} \qquad 17.47(a)$$

$$\begin{aligned} &\frac{p_1}{\rho} + \frac{\text{v}_1^2}{2g_c} + \frac{z_1 g}{g_c} + h_{\text{pump}} \\ &= \frac{p_2}{\rho} + \frac{\text{v}_2^2}{2g_c} + \frac{z_2 g}{g_c} + h_{\text{turbine}} \\ &\quad + h_f + h_m \end{aligned} \quad \text{[U.S.]} \qquad 17.47(b)$$

16. DISCHARGE FROM TANKS

The velocity of a jet issuing from an orifice in a tank can be determined by comparing the total energies at the free fluid surface and the jet itself. (See Fig. 17.4.) At the fluid surface, $p_1 = 0$ (atmospheric) and $v_1 = 0$. (v_1 is known as the *velocity of approach.*) The only energy the fluid has is potential energy. At the jet, $p_2 = 0$. All of the potential energy difference ($z_1 - z_2$) has been converted to kinetic energy. The theoretical velocity of the jet can be derived from the Bernoulli equation. Equation 17.48 is known as the equation for *Torricelli's speed of efflux.*

Jet Propulsion

$$v_2 = \sqrt{2gh} \qquad 17.48$$

$$h = z_1 - z_2 \qquad 17.49$$

Figure 17.3 *EGL and HGL for Minor Losses*

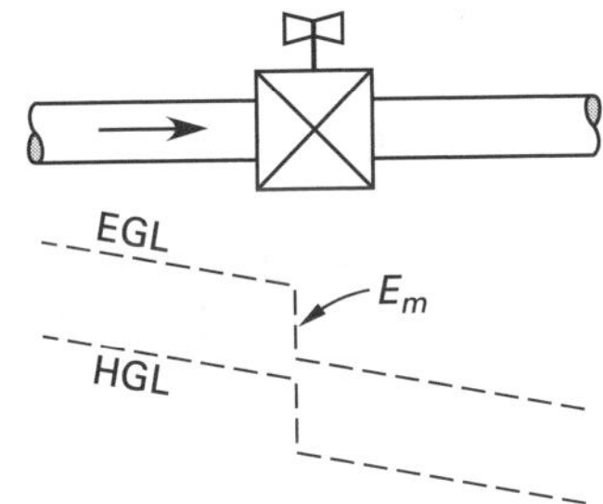

(a) valve, fitting, or obstruction

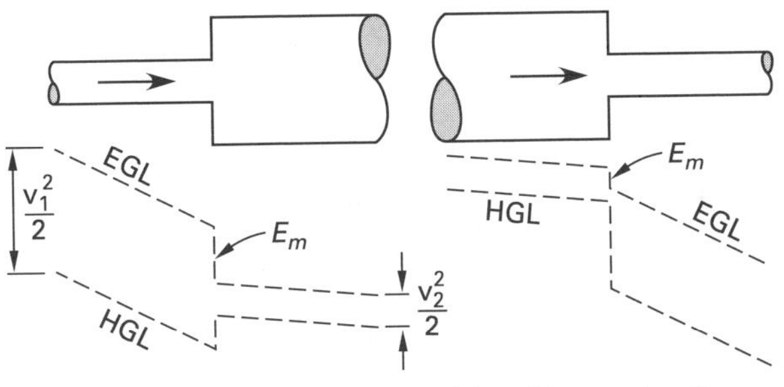

(b) sudden enlargement (c) sudden contraction

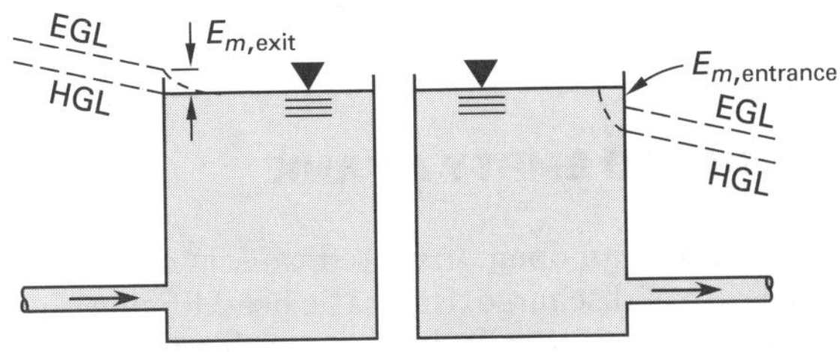

(d) transition to reservoir (e) transition to pipeline

The actual jet velocity is affected by the orifice geometry. The *coefficient of velocity*, C_v, is an empirical factor that accounts for the friction and turbulence at the orifice. [Orifices]

Figure 17.4 *Discharge from a Tank*

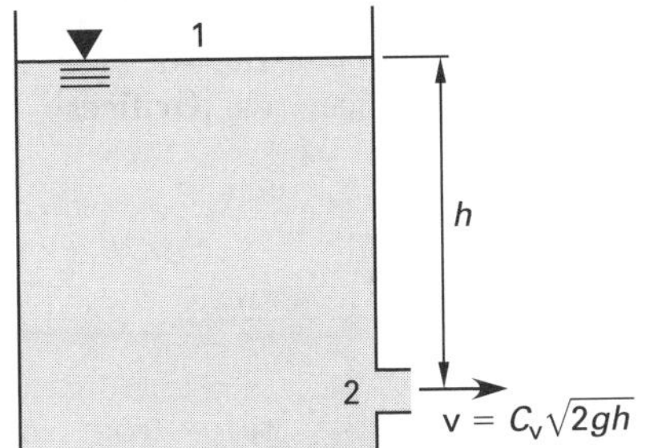

Orifice Discharging Freely into Atmosphere

$$v_o = C_v\sqrt{2gh} \qquad 17.50$$

$$C_v = \frac{\text{actual velocity}}{\text{theoretical velocity}} = \frac{v_o}{v_t} \qquad 17.51$$

The specific energy loss due to turbulence and friction at the orifice is calculated as a multiple of the jet's kinetic energy.

$$E_f = \left(\frac{1}{C_v^2} - 1\right)\frac{v_o^2}{2} = (1 - C_v^2)gh \quad \text{[SI]} \qquad 17.52(a)$$

$$E_f = \left(\frac{1}{C_v^2} - 1\right)\frac{v_o^2}{2g_c} = (1 - C_v^2)h \times \frac{g}{g_c} \quad \text{[U.S.]} \qquad 17.52(b)$$

The total head producing discharge (*effective head*) is the difference in elevations that would produce the same velocity from a frictionless orifice.

$$h_{\text{effective}} = C_v^2 h \qquad 17.53$$

The orifice guides quiescent water from the tank into the jet geometry. Unless the orifice is very smooth and the transition is gradual, momentum effects will continue to cause the jet to contract after it has passed through. The velocity calculated from Eq. 17.50 is usually assumed to be the velocity at the *vena contracta*, the section of smallest cross-sectional area. (See Fig. 17.5.)

Figure 17.5 *Vena Contracta of a Fluid Jet*

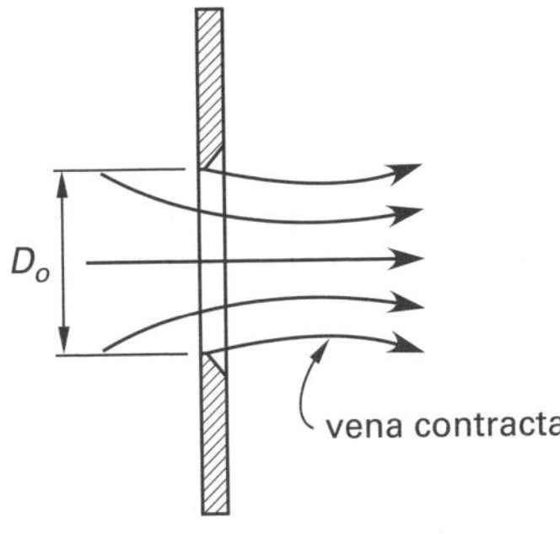

For a thin plate or sharp-edged orifice, the vena contracta is often assumed to be located approximately one half an orifice diameter past the orifice, although the actual distance can vary from $0.3D_o$ to $0.8D_o$. The area of the vena contracta can be calculated from the orifice area and the *coefficient of contraction*, C_c. For water flowing with a high Reynolds number through a small sharp-edged orifice, the contracted area is approximately 61–63% of the orifice area. **[Orifices]**

$$A_{\text{vena contracta}} = C_c A_o \qquad 17.54$$

$$C_c = \frac{A_{\text{vena contracta}}}{A_o} \qquad 17.55$$

The theoretical discharge rate from a tank is $Q_t = A_o\sqrt{2gh}$. However, this relationship needs to be corrected for friction and contraction by multiplying by C_v and C_c. The *coefficient of discharge*, C_d, is the product of the coefficients of velocity and contraction.

$$Q = C_c \text{v}_o A_o = C_d \text{v}_t A_o = C_d A_o \sqrt{2gh} \qquad 17.56$$

$$C_d = C_v C_c = \frac{Q}{Q_t} \qquad 17.57$$

For a submerged orifice, the volumetric flow rate is

Submerged Orifice Operating under Steady-flow Conditions

$$\begin{aligned} Q &= A_o \text{v}_o = C_c C_v A_o \sqrt{2g(h_1 - h_2)} \\ &= C_d A_o \sqrt{2g(h_1 - h_2)} \end{aligned}$$

17. DISCHARGE FROM PRESSURIZED TANKS

If the gas or vapor above the liquid in a tank is at gage pressure p, and the discharge is to atmospheric pressure, the head causing discharge will be

$$h = z_1 - z_2 + \frac{p}{\rho g} \qquad \text{[SI]} \qquad 17.58(a)$$

$$h = z_1 - z_2 + \frac{p}{\rho} \times \frac{g_c}{g} = z_1 - z_2 + \frac{p}{\gamma} \qquad \text{[U.S.]} \qquad 17.58(b)$$

The discharge velocity can be calculated from Eq. 17.50 using the increased discharge head. (See Fig. 17.6.)

Figure 17.6 *Discharge from a Pressurized Tank*

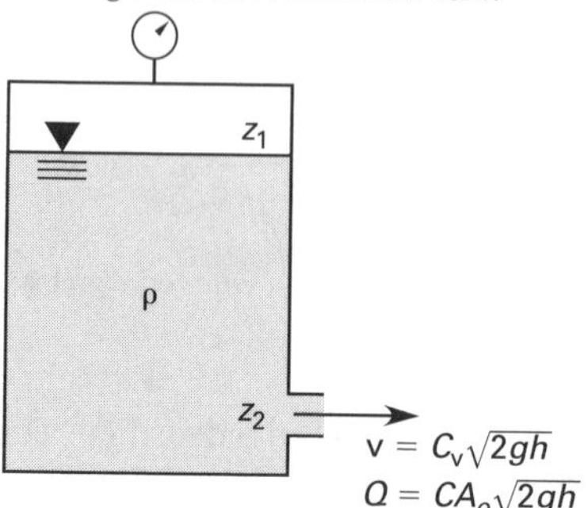

18. COORDINATES OF A FLUID STREAM

Fluid discharged from an orifice in a tank gets its initial velocity from the conversion of potential energy. After discharge, no additional energy conversion occurs, and all subsequent velocity changes are due to external forces. (See Fig. 17.7.)

Figure 17.7 *Coordinates of a Fluid Stream*

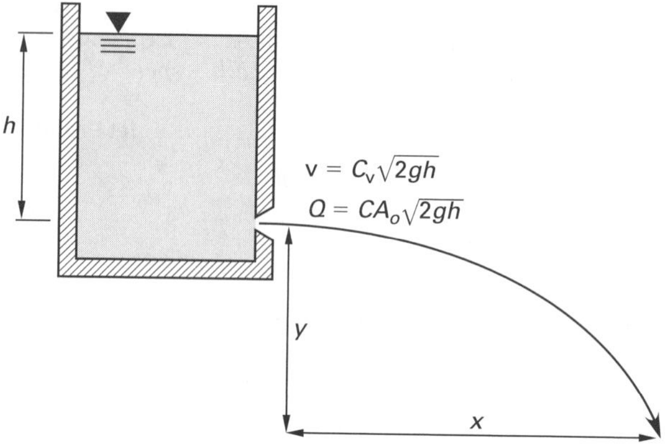

In the absence of air friction (drag), there are no decelerating or accelerating forces in the x-direction on the fluid stream. The x-component of velocity is constant. Projectile motion equations can be used to predict the path of the fluid stream.

$$\text{v}_x = \text{v}_o \quad \text{[horizontal discharge]} \qquad 17.59$$

$$x = \text{v}_o t = \text{v}_o \sqrt{\frac{2y}{g}} = 2C_v\sqrt{hy} \qquad 17.60$$

After discharge, the fluid stream is acted upon by a constant gravitational acceleration. The y-component of velocity is zero at discharge but increases linearly with time.

$$\text{v}_y = gt \qquad 17.61$$

$$y = \frac{gt^2}{2} = \frac{gx^2}{2\text{v}_o^2} = \frac{x^2}{4hC_v^2} \qquad 17.62$$

19. TIME TO EMPTY A TANK

If the fluid in an open or vented tank is not replenished at the rate of discharge, the static head forcing discharge through the orifice will decrease with time. If the tank has a varying cross section, A_t, Eq. 17.63 specifies the basic relationship between the change in elevation and elapsed time. (The negative sign indicates that z decreases as t increases.)

$$Q\,dt = -A_t\,dz \qquad 17.63$$

If A_t can be expressed as a function of h, Eq. 17.64 can be used to determine the time to lower the fluid elevation from z_1 to z_2.

$$t = \int_{z_1}^{z_2} \frac{-A_t\,dz}{C_d A_o \sqrt{2gz}} \qquad 17.64$$

For a tank with a constant cross-sectional area, A_t, the time required to lower the fluid elevation is

$$t = \frac{2A_t(\sqrt{z_1} - \sqrt{z_2})}{C_d A_o \sqrt{2g}} \qquad 17.65$$

If a tank is replenished at a rate of Q_{in}, Eq. 17.66 can be used to calculate the discharge time. If the tank is replenished at a rate greater than the discharge rate, t in Eq. 17.66 will represent the time to raise the fluid level from z_1 to z_2.

$$t = \int_{z_1}^{z_2} \frac{A_t\,dz}{C_d A_o \sqrt{2gz} - Q_{\text{in}}} \qquad 17.66$$

If the tank is not open or vented but is pressurized, the elevation terms, z_1 and z_2, in Eq. 17.65 must be replaced by the total head terms, h_1 and h_2, that include the effects of pressurization.

Example 17.7

A vertical, cylindrical tank 15 ft in diameter discharges 150°F water ($\rho = 61.20$ lbm/ft^3) through a sharp-edged, 1 in diameter orifice ($C_d = 0.62$) in the tank bottom. The original water depth is 12 ft. The tank is continually pressurized to 50 psig. How long does it take, in seconds, to empty the tank?

Solution

The area of the orifice is

$$A_o = \frac{\pi D^2}{4} = \frac{\pi (1 \text{ in})^2}{(4)\left(12 \ \frac{\text{in}}{\text{ft}}\right)^2} = 0.00545 \text{ ft}^2$$

The tank area is constant with respect to depth.

$$A_t = \frac{\pi D^2}{4} = \frac{\pi (15 \text{ ft})^2}{4} = 176.7 \text{ ft}^2$$

The total initial head includes the effect of the pressurization. Use Eq. 17.58.

$$\begin{aligned} h_1 &= z_1 - z_2 + \frac{p}{\rho} \times \frac{g_c}{g} \\ &= 12 \text{ ft} + \frac{\left(50 \ \frac{\text{lbf}}{\text{in}^2}\right)\left(12 \ \frac{\text{in}}{\text{ft}}\right)^2}{61.2 \ \frac{\text{lbm}}{\text{ft}^3}} \times \frac{32.2 \ \frac{\text{lbm-ft}}{\text{lbf-sec}^2}}{32.2 \ \frac{\text{ft}}{\text{sec}^2}} \\ &= 12 \text{ ft} + 117.6 \text{ ft} \\ &= 129.6 \text{ ft} \end{aligned}$$

When the fluid has reached the level of the orifice, the fluid potential head will be zero, but the pressurization will remain.

$$h_2 = 117.6 \text{ ft}$$

The time needed to empty the tank is given by Eq. 17.65.

$$\begin{aligned} t &= \frac{2A_t(\sqrt{z_1} - \sqrt{z_2})}{C_d A_o \sqrt{2g}} \\ &= \frac{(2)(176.7 \text{ ft}^2)(\sqrt{129.6 \text{ ft}} - \sqrt{117.6 \text{ ft}})}{(0.62)(0.00545 \text{ ft}^2)\sqrt{(2)\left(32.2 \ \frac{\text{ft}}{\text{sec}^2}\right)}} \\ &= 7036 \text{ sec} \end{aligned}$$

20. SIPHONS

A *siphon* is a bent or curved tube that carries fluid from a fluid surface at a high elevation to another fluid surface at a lower elevation. Normally, it would not seem difficult to have a fluid flow to a lower elevation. However, the fluid seems to flow "uphill" in a portion of a siphon. Figure 17.8 illustrates a siphon.

Figure 17.8 *Siphon*

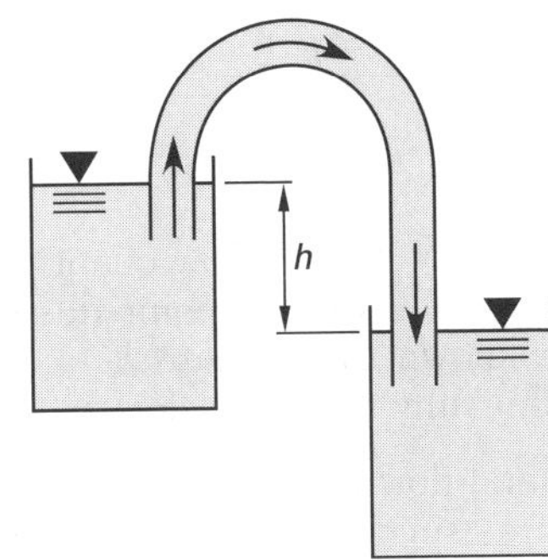

Starting a siphon requires the tube to be completely filled with liquid. Then, since the fluid weight is greater in the longer arm than in the shorter arm, the fluid in the longer arm "falls" out of the siphon, "pulling" more liquid into the shorter arm and over the bend.

Operation of a siphon is essentially independent of atmospheric pressure. The theoretical discharge is the same as predicted by the Torricelli equation. A correction for discharge is necessary, but little data is available on typical values of C_d. Therefore, siphons should be tested and calibrated in place.

$$Q = C_d A \mathrm{v} = C_d A\sqrt{2gh} \qquad 17.67$$

21. SERIES PIPE SYSTEMS

A system of pipes in series consists of two or more lengths of different-diameter pipes connected end-to-end. In the case of the series pipe from a reservoir discharging to the atmosphere shown in Fig. 17.9, the available head will be split between the velocity head and the friction loss.

$$h = h_{\mathrm{v}} + h_f \qquad 17.68$$

Figure 17.9 *Series Pipe System*

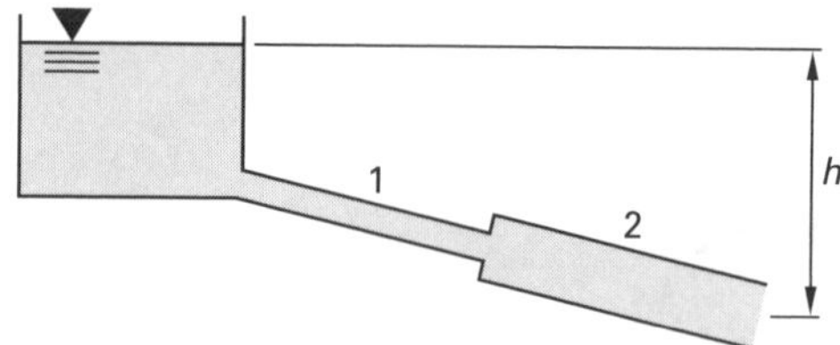

If the flow rate or velocity in any part of the system is known, the friction loss can easily be found as the sum of the friction losses in the individual sections. The solution is somewhat more simple than it first appears to be, since the velocity of all sections can be written in terms of only one velocity.

$$h_{f,t} = h_{f,a} + h_{f,b} \qquad 17.69$$

$$A_a \mathrm{v}_a = A_b \mathrm{v}_b \qquad 17.70$$

If neither the velocity nor the flow quantity is known, a trial-and-error solution will be required, since a friction factor must be known to calculate h_f. A good starting point is to assume fully turbulent flow.

When velocity and flow rate are both unknown, the following procedure using the Darcy friction factor can be used.[17]

step 1: Calculate the relative roughness, ϵ/D, for each section. Use the Moody diagram to determine f_a and f_b for fully turbulent flow (i.e., the horizontal portion of the curve).

step 2: Write all of the velocities in terms of one unknown velocity.

$$Q_a = Q_b \qquad 17.71$$

$$\mathrm{v}_b = \left(\frac{A_a}{A_b}\right)\mathrm{v}_a \qquad 17.72$$

step 3: Write the total friction loss in terms of the unknown velocity.

$$\begin{aligned} h_{f,t} &= \frac{f_a L_a \mathrm{v}_a^2}{2D_a g} + \left(\frac{f_b L_b}{2D_b g}\right)\left(\frac{A_a}{A_b}\right)^2 \mathrm{v}_a^2 \\ &= \left(\frac{\mathrm{v}_a^2}{2g}\right)\left(\frac{f_a L_a}{D_a} + \left(\frac{f_b L_b}{D_b}\right)\left(\frac{A_a}{A_b}\right)^2\right) \end{aligned} \qquad 17.73$$

step 4: Solve for the unknown velocity using the Bernoulli equation between the free reservoir surface ($p = 0$, $\mathrm{v} = 0$, $z = h$) and the discharge point ($p = 0$, if free discharge; $z = 0$). Include pipe friction, but disregard minor losses for convenience.

$$\begin{aligned} h &= \frac{\mathrm{v}_b^2}{2g} + h_{f,t} \\ &= \left(\frac{\mathrm{v}_a^2}{2g}\right)\left(\left(\frac{A_a}{A_b}\right)^2\left(1 + \frac{f_b L_b}{D_b}\right) + \frac{f_a L_a}{D_a}\right) \end{aligned} \qquad 17.74$$

step 5: Using the value of v_a, calculate v_b. Calculate the Reynolds number, and check the values of f_a and f_b from step 4. Repeat steps 3 and 4 if necessary.

22. PARALLEL PIPE SYSTEMS

Adding a second pipe in parallel with a first is a standard method of increasing the capacity of a line. A *pipe loop* is a set of two pipes placed in parallel, both originating and terminating at the same junction. The two pipes are referred to as *branches* or *legs*. (See Fig. 17.10.)

[17]If Hazen-Williams constants are given for the pipe sections, the procedure for finding the unknown velocities is similar, although considerably more difficult since v^2 and $\mathrm{v}^{1.85}$ cannot be combined. A first approximation, however, can be obtained by replacing $\mathrm{v}^{1.85}$ in the Hazen-Williams equation for friction loss. A trial-and-error method can then be used to find velocity.

Figure 17.10 Parallel Pipe System

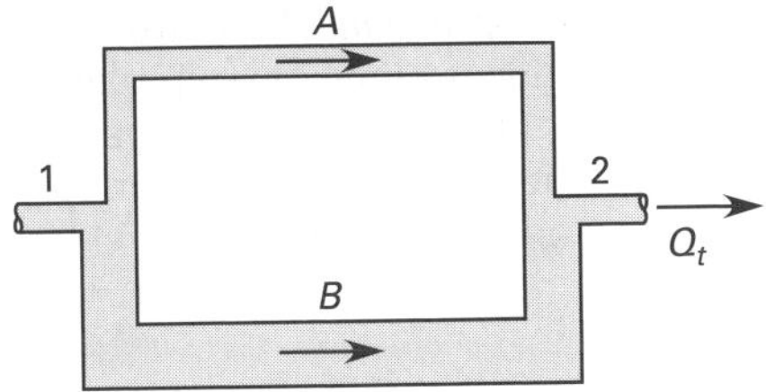

There are three principles that govern the distribution of flow between the two branches.

- The flow divides in such a manner as to make the head loss in each branch the same.

$$h_{f,1} = h_{f,2} \qquad 17.75$$

- The head loss between the junctions A and B is the same as the head loss in branches 1 and 2.

$$h_{f,\text{A-B}} = h_{f,1} = h_{f,2} \qquad 17.76$$

- The total flow rate is the sum of the flow rates in the two branches.

$$Q_t = Q_1 + Q_2 \qquad 17.77$$

If the pipe diameters are known, Eq. 17.75 and Eq. 17.77 can be solved simultaneously for the branch velocities. For first estimates, it is common to neglect minor losses, the velocity head, and the variation in the friction factor, *f*, with velocity.

If the system has only two parallel branches, the unknown branch flows can be determined by solving Eq. 17.78 and Eq. 17.80 simultaneously.

$$\frac{f_1 L_1 v_1^2}{2D_1 g} = \frac{f_2 L_2 v_2^2}{2D_2 g} \qquad 17.78$$

$$Q_1 + Q_2 = Q_t \qquad 17.79$$

$$\frac{\pi}{4}(D_1^2 v_1 + D_2^2 v_2) = Q_t \qquad 17.80$$

However, if the system has three or more parallel branches, it is easier to use the following iterative procedure. This procedure can be used for problems (a) where the flow rate is unknown but the pressure drop between the two junctions is known, or (b) where the total flow rate is known but the pressure drop and velocity are both unknown. In both cases, the solution iteratively determines the friction coefficients, *f*.

step 1: Solve the friction head loss, h_f, expression (either Darcy-Weisbach or Hazen-Williams) for velocity in each branch. If the pressure drop is known, first convert it to friction head loss.

$$v = \sqrt{\frac{2Dgh_f}{fL}} \quad \text{[Darcy-Weisbach]} \qquad 17.81$$

$$v = \frac{0.355\,CD^{0.63}h_f^{0.54}}{L^{0.54}} \quad \text{[Hazen-Williams] [SI]} \qquad 17.82(a)$$

$$v = \frac{0.550\,CD^{0.63}h_f^{0.54}}{L^{0.54}} \quad \text{[Hazen-Williams] [U.S.]} \qquad 17.82(b)$$

step 2: Solve for the flow rate in each branch. If they are unknown, friction factors, *f*, must be assumed for each branch. The fully turbulent assumption provides a good initial estimate. (The value of k' will be different for each branch.)

$$Q = Av = A\sqrt{\frac{2Dgh_f}{fL}} = k'\sqrt{h_f} \quad \text{[Darcy-Weisbach]} \qquad 17.83$$

step 3: Write the expression for the conservation of flow. Calculate the friction head loss from the total flow rate. For example, for a three-branch system,

$$Q_t = Q_1 + Q_2 + Q_3 = (k_1' + k_2' + k_3')\sqrt{h_f} \qquad 17.84$$

step 4: Check the assumed values of the friction factor. Repeat as necessary.

Example 17.8

3 ft^3/sec of water enter the parallel pipe network shown at junction A. All pipes are schedule-40 steel with the nominal sizes shown. Minor losses are insignificant. What is the total friction head loss between junctions A and B?

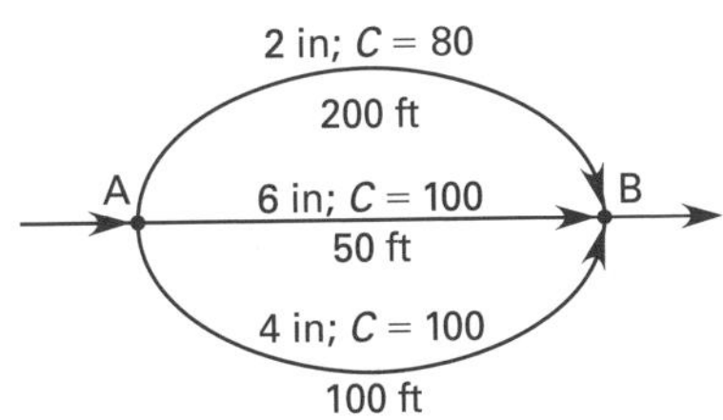

Fluids

Fluids

Solution

step 1: From tables of pipe data, find the flow areas and inner diameters of the pipes. If needed, convert to feet and square feet. [**Pipe and Tube Data**]

	2 in	4 in	6 in
flow area	0.0233 ft^2	0.0884 ft^2	0.2006 ft^2
diameter	0.1723 ft	0.3355 ft	0.5054 ft

Rearrange the Hazen-Williams equation to solve for the velocity (Eq. 17.82(b)).

$$\mathrm{v} = \frac{0.550\,CD^{0.63}h_f^{0.54}}{L^{0.54}}$$

The velocity (expressed in ft/sec) in the 2 in diameter pipe branch is

$$\begin{aligned}\mathrm{v}_{2\,\text{in}} &= \frac{(0.550)(80)(0.1723\ \text{ft})^{0.63}h_f^{0.54}}{(200\ \text{ft})^{0.54}} \\ &= 0.831h_f^{0.54}\end{aligned}$$

The friction head loss is the same in all parallel branches. The velocities in the other two branches are

$$\begin{aligned}\mathrm{v}_{6\,\text{in}} &= 4.327h_f^{0.54} \\ \mathrm{v}_{4\,\text{in}} &= 2.299h_f^{0.54}\end{aligned}$$

step 2: The flow rates are

$$\begin{aligned}Q &= A\mathrm{v} \\ Q_{2\,\text{in}} &= (0.0233\ \text{ft}^2)0.831h_f^{0.54} \\ &= 0.0194h_f^{0.54} \\ Q_{6\,\text{in}} &= (0.2006\ \text{ft}^2)4.327h_f^{0.54} \\ &= 0.8680h_f^{0.54} \\ Q_{4\,\text{in}} &= (0.0884\ \text{ft}^2)2.299h_f^{0.54} \\ &= 0.2032h_f^{0.54}\end{aligned}$$

step 3: The total flow rate is

$$\begin{aligned}Q_t &= Q_{2\,\text{in}} + Q_{6\,\text{in}} + Q_{4\,\text{in}} \\ 3\ \frac{\text{ft}^3}{\text{sec}} &= 0.0194h_f^{0.54} + 0.8680h_f^{0.54} + 0.2032h_f^{0.54} \\ &= (0.0194 + 0.8680 + 0.2032)h_f^{0.54} \\ h_f &= 6.5\ \text{ft}\end{aligned}$$

23. MULTIPLE RESERVOIR SYSTEMS

In the *three-reservoir problem*, there are many possible choices for the unknown quantity (pipe length, diameter, head, flow rate, etc.). In all but the simplest cases, the solution technique is by trial and error based on conservation of mass and energy. (See Fig. 17.11.)

Figure 17.11 *Three-Reservoir System*

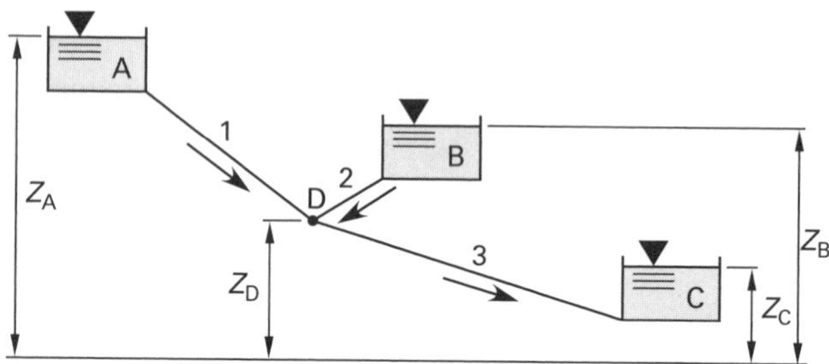

For simplification, velocity heads and minor losses are usually insignificant and can be neglected. However, the presence of a pump in any of the lines must be included in the solution procedure. This is most easily done by adding the pump head to the elevation of the reservoir feeding the pump. If the pump head is not known or depends on the flow rate, it must be determined iteratively as part of the solution procedure.

Case 1: Given all lengths, diameters, and elevations, find all flow rates.

Although an analytical solution method is possible, this type of problem is easily solved iteratively. The following procedure makes an initial estimate of a flow rate and uses it to calculate the pressure at the junction, p_D. Since this method may not converge if the initial estimate of Q_1 is significantly in error, it is helpful to use other information (e.g., normal pipe velocities) to obtain the initial estimate. (See Sec. 17.3.) An alternate procedure is simply to make several estimates of p_D and calculate the corresponding values of flow rate.

step 1: Assume a reasonable value for Q_1. Calculate the corresponding friction loss, $h_{f,1}$. Use the Bernoulli equation to find the corresponding value of p_D. Disregard minor losses and velocity head.

$$\mathrm{v}_1 = \frac{Q_1}{A_1} \qquad 17.85$$

$$z_A = z_D + \frac{p_D}{\gamma} + h_{f,1} \qquad 17.86$$

step 2: Use the value of p_D to calculate $h_{f,2}$. Use the friction loss to determine v_2. Use v_2 to determine Q_2. If flow is out of reservoir B, $h_{f,2}$ should be added. If $z_D + (p_D/\gamma) > z_B$, flow will be into reservoir B. In this case, $h_{f,2}$ should be subtracted.

$$z_B = z_D + \frac{p_D}{\gamma} \pm h_{f,2} \qquad 17.87$$

$$Q_2 = \text{v}_2 A_2 \qquad 17.88$$

step 3: Similarly, use the value of p_D to calculate $h_{f,3}$. Use the friction loss to determine v_3. Use v_3 to determine Q_3.

$$z_\text{C} = z_\text{D} + \frac{p_\text{D}}{\gamma} - h_{f,3} \qquad 17.89$$

$$Q_3 = \text{v}_3 A_3 \qquad 17.90$$

step 4: Check if $Q_1 \pm Q_2 = Q_3$. If it does not, repeat steps 1 through 4. After the second iteration, plot $Q_1 \pm Q_2 - Q_3$ versus Q_1. Interpolate or extrapolate the value of Q_1 that makes the difference zero.

Case 2: Given Q_1 and all lengths, diameters, and elevations except z_C, find z_C.

step 1: Calculate v_1.

$$\text{v}_1 = \frac{Q_1}{A_1} \qquad 17.91$$

step 2: Calculate the corresponding friction loss, $h_{f,1}$. Use the Bernoulli equation to find the corresponding value of p_D. Disregard minor losses and velocity head.

$$z_\text{A} = z_\text{D} + \frac{p_\text{D}}{\gamma} + h_{f,1} \qquad 17.92$$

step 3: Use the value p_D to calculate $h_{f,2}$. Use the friction loss to determine v_2. Use v_2 to determine $Q_2 A_2$. If flow is out of reservoir B, $h_{f,2}$ should be added. If $z_\text{D} + (p_\text{D}/\gamma) > z_\text{B}$, flow will be into reservoir B. In this case, $h_{f,2}$ should be subtracted.

$$z_\text{B} = z_\text{D} + \frac{p_\text{D}}{\gamma} \pm h_{f,2} \qquad 17.93$$

$$Q_2 = \text{v}_2 A_2 \qquad 17.94$$

step 4: From the conservation of mass, the flow rate into reservoir C is

$$Q_3 = Q_1 \pm Q_2 \qquad 17.95$$

step 5: The velocity in pipe 3 is

$$\text{v}_3 = \frac{Q_3}{A_3} \qquad 17.96$$

step 6: Calculate $h_{f,3}$.

step 7: Disregarding minor losses and velocity head, the elevation of the surface in reservoir C is

$$z_\text{C} = z_\text{D} + \frac{p_\text{D}}{\gamma} - h_{f,3} \qquad 17.97$$

Case 3: Given Q_1, all lengths, all elevations, and all diameters except D_3, find D_3.

step 1: Repeat step 1 from case 2.

step 2: Repeat step 2 from case 2.

step 3: Repeat step 3 from case 2.

step 4: Repeat step 4 from case 2.

step 5: Calculate $h_{f,3}$ from

$$z_\text{C} = z_\text{D} + \frac{p_\text{D}}{\gamma} - h_{f,3} \qquad 17.98$$

step 6: Calculate D_3 from $h_{f,3}$.

Case 4: Given all lengths, diameters, and elevations except z_D, find all flow rates.

step 1: Calculate the head loss between each reservoir and junction D. Combine as many terms as possible into constant k'.

$$\begin{aligned} Q = A\text{v} &= A\sqrt{\frac{2Dgh_f}{fL}} \\ &= k'\sqrt{h_f} \quad \text{[Darcy-Weisbach]} \end{aligned} \qquad 17.99$$

$$\begin{aligned} Q = A\text{v} &= \frac{A(0.550)\,CD^{0.63}h_f^{0.54}}{L^{0.54}} \\ &= k'h_f^{0.54} \quad \text{[Hazen-Williams; U.S.]} \end{aligned} \qquad 17.100$$

step 2: Assume that the flow direction in all three pipes is toward junction D. Write the conservation equation for junction D.

$$Q_{\text{D},t} = Q_1 + Q_2 + Q_3 = 0 \qquad 17.101$$

$$\begin{aligned} &k_1'\sqrt{h_{f,1}} + k_2'\sqrt{h_{f,2}} \\ &\quad + k_3'\sqrt{h_{f,3}} = 0 \\ &\quad \text{[Darcy-Weisbach]} \end{aligned} \qquad 17.102$$

$$\begin{aligned} &k_1'h_{f,1}^{0.54} + k_2'h_{f,2}^{0.54} \\ &\quad + k_3'h_{f,3}^{0.54} = 0 \\ &\quad \text{[Hazen-Williams]} \end{aligned} \qquad 17.103$$

Fluids

step 3: Write the Bernoulli equation between each reservoir and junction D. Since $p_A = p_B = p_C = 0$, and $v_A = v_B = v_C = 0$, the friction loss in branch 1 is

$$h_{f,1} = z_A - z_D - \frac{p_D}{\gamma} \qquad 17.104$$

However, z_D and p_D can be combined since they are related constants in any particular situation. Define the correction, δ_D, as

$$\delta_D = z_D + \frac{p_D}{\gamma} \qquad 17.105$$

Then, the friction head losses in the branches are

$$h_{f,1} = z_A - \delta_D \qquad 17.106$$

$$h_{f,2} = z_B - \delta_D \qquad 17.107$$

$$h_{f,3} = z_C - \delta_D \qquad 17.108$$

step 4: Assume a value for δ_D. Calculate the corresponding h_f values. Use Eq. 17.99 to find Q_1, Q_2, and Q_3. Calculate the corresponding Q_t value. Repeat until Q_t converges to zero. It is not necessary to calculate p_D or z_D once all of the flow rates are known.

24. FLOW MEASURING DEVICES

A device that measures flow can be calibrated to indicate either velocity or volumetric flow rate. There are many methods available to obtain the flow rate. Some are indirect, requiring the use of transducers and solid-state electronics, and others can be evaluated using the Bernoulli equation. Some are more appropriate for one variety of fluid than others, and some are limited to specific ranges of temperature and pressure.

Table 17.4 categorizes a few common flow measurement methods. Many other methods and variations thereof exist, particularly for specialized industries. Some of the methods listed are so basic that only a passing mention will be made of them. Others, particularly those that can be analyzed with energy and mass conservation laws, will be covered in greater detail in subsequent sections.

The utility meters used to measure gas and water usage are examples of *displacement meters*. Such devices are cyclical, fixed-volume devices with counters to record the numbers of cycles. Displacement devices are generally unpowered, drawing on only the pressure energy to overcome mechanical friction. Most configurations for positive-displacement pumps (e.g., reciprocating piston, helical screw, and nutating disk) have also been converted to measurement devices.

Table 17.4 *Flow Measuring Devices*

I direct (primary) measurements
- positive-displacement meters
- volume tanks
- weight and mass scales

II indirect (secondary) measurements
- obstruction meters
  - flow nozzles
  - orifice plate meters
  - variable-area meters
  - venturi meters
- velocity probes
  - direction sensing probes
  - pitot-static meters
  - pitot tubes
  - static pressure probes
- miscellaneous methods
  - hot-wire meters
  - magnetic flow meters
  - mass flow meters
  - sonic flow meters
  - turbine and propeller meters

The venturi nozzle, orifice plate, and flow nozzle are examples of *obstruction meters*. These devices rely on a decrease in static pressure to measure the flow velocity. One disadvantage of these devices is that the pressure drop is proportional to the square of the velocity, limiting the range over which any particular device can be used.

An obstruction meter that somewhat overcomes the velocity range limitation is the *variable-area meter*, also known as a *rotameter*, illustrated in Fig. 17.12.[18] This device consists of a float (which is actually more dense than the fluid) and a transparent sight tube. With proper design, the effects of fluid density and viscosity can be minimized. The sight glass can be directly calibrated in volumetric flow rate, or the height of the float above the zero position can be used in a volumetric calculation.

It is necessary to be able to measure static pressures in order to use obstruction meters and pitot-static tubes. In some cases, a *static pressure probe* is used. Figure 17.13 illustrates a simplified static pressure probe. In practice, such probes are sensitive to burrs and irregularities in the tap openings, orientation to the flow (i.e., *yaw*), and interaction with the pipe walls and other probes. A *direction-sensing probe* overcomes some of these problems.

[18] The rotameter has its own disadvantages, however. It must be installed vertically; the fluid cannot be opaque; and it is more difficult to manufacture for use with high-temperature, high-pressure fluids.

Figure 17.12 Variable-Area Rotameter

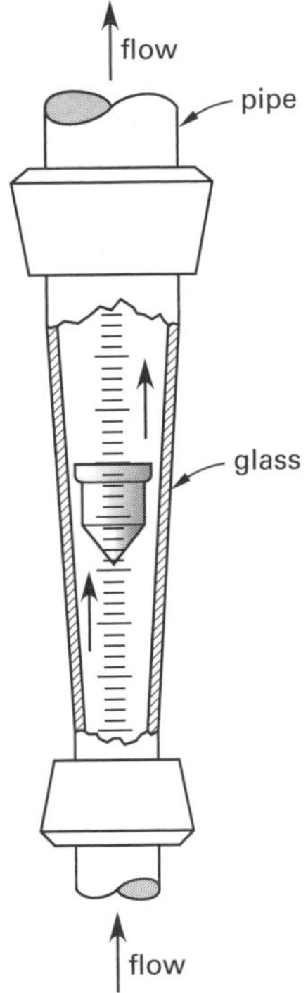

Figure 17.13 Simple Static Pressure Probe

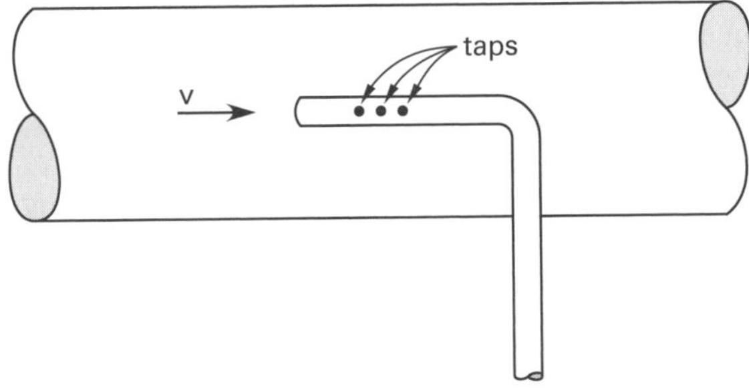

A weather station *anemometer* used to measure wind velocity is an example of a simple *turbine meter*. Similar devices are used to measure the speed of a stream or river, in which case the name *current meter* may be used. Turbine meters are further divided into cup-type meters and propeller-type meters, depending on the orientation of the turbine axis relative to the flow direction. (The turbine axis and flow direction are parallel for propeller-type meters; they are perpendicular for cup-type meters.) Since the wheel motion is proportional to the flow velocity, the velocity is determined by counting the number of revolutions made by the wheel per unit time.

More sophisticated turbine flowmeters use a reluctance-type pickup coil to detect wheel motion. The permeability of a magnetic circuit changes each time a wheel blade passes the pole of a permanent magnet in the meter body. This change is detected to indicate velocity or flow rate.

A *hot-wire anemometer* measures velocity by determining the cooling effect of fluid (usually a gas) flowing over an electrically heated tungsten, platinum, or nickel wire. Cooling is primarily by convection; radiation and conduction are neglected. Circuitry can be used either to keep the current constant (in which case, the changing resistance or voltage is measured) or to keep the temperature constant (in which case, the changing current is measured). Additional circuitry can be used to compensate for thermal lag if the velocity changes rapidly.

Modern *magnetic flowmeters* (*magmeter*, *electromagnetic flowmeter*) measure fluid velocity by detecting a voltage (potential difference, electromotive force) that is generated in response to the fluid passing through a magnetic field applied by the meter. The magnitude of the induced voltage is predicted by *Faraday's law of induction*, which states that the induced voltage is equal to the negative of the time rate of change of a magnetic field. The voltage is not affected by changes in fluid viscosity, temperature, or density. From Faraday's law, the induced voltage is proportional to the fluid velocity. For a magmeter with a dimensionless instrument constant, k, and a magnetic field strength, B, the induced voltage will be

$$V = kB\mathrm{v}D_{\mathrm{pipe}} \qquad 17.109$$

Since no parts of the meter extend into the flow, magmeters are ideal for corrosive fluids and slurries of large particles. Normally, the fluid has to be at least slightly electrically conductive, making magmeters ideal for measuring liquid metal flow. It may be necessary to dope electrically neutral fluids with precise quantities of conductive ions in order to obtain measurements by this method. Most magmeters have integral instrumentation and are direct-reading.

In an *ultrasonic flowmeter*, two electric or magnetic transducers are placed a short distance apart on the outside of the pipe. One transducer serves as a transmitter of ultrasonic waves; the other transducer is a receiver. As an ultrasonic wave travels from the transmitter to the receiver, its velocity will be increased (or decreased) by the relative motion of the fluid. The phase shift between the fluid-carried waves and the waves passing through a stationary medium can be measured and converted to fluid velocity.

25. PITOT-STATIC GAUGE

Measurements from pitot tubes are used to determine total (stagnation) energy. Piezometer tubes and wall taps are used to measure static pressure energy. The difference between the total and static energies is the kinetic energy of the flow. Figure 17.14 illustrates a comparative method of directly measuring the velocity head for an incompressible fluid.

$$\frac{v^2}{2} = \frac{p_t - p_s}{\rho} = hg \quad \text{[SI]} \qquad 17.110(a)$$

$$\frac{v^2}{2g_c} = \frac{p_t - p_s}{\rho} = h \times \frac{g}{g_c} \quad \text{[U.S.]} \qquad 17.110(b)$$

Jet Propulsion

$$v = \sqrt{2gh} \qquad 17.111$$

An alternative form of Eq. 17.111 is given by Eq. 17.112, where the velocity pressure p_w is equal to $p_t - p_s$. C is a unit conversion factor equal to 136.8. The calculated velocity, v, is in feet per minute.

Pitot-Static Tubes

$$v = C\sqrt{\frac{2p_w g_c}{\rho}} \qquad 17.112$$

Figure 17.14 *Comparative Velocity Head Measurement*

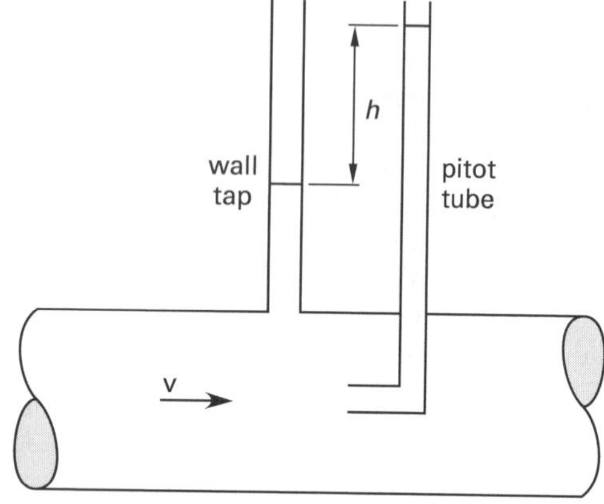

The pitot tube and static pressure tap shown in Fig. 17.14 can be combined into a *pitot-static gauge*. (See Fig. 17.15.) In a pitot-static gauge, one end of the manometer is acted upon by the static pressure (also referred to as the *transverse pressure*). The other end of the manometer experiences the total (stagnation or impact) pressure. The difference in elevations of the manometer fluid columns is the velocity head. This distance must be corrected if the density of the flowing fluid is significant.

$$\frac{v^2}{2} = \frac{p_t - p_s}{\rho} = \frac{h(\rho_m - \rho)g}{\rho} \quad \text{[SI]} \qquad 17.113(a)$$

$$\frac{v^2}{2g_c} = \frac{p_t - p_s}{\rho} = \frac{h(\rho_m - \rho)}{\rho} \times \frac{g}{g_c} \quad \text{[U.S.]} \qquad 17.113(b)$$

$$v = \sqrt{\frac{2gh(\rho_m - \rho)}{\rho}} \qquad 17.114$$

Another correction, which is seldom made, is to multiply the velocity calculated from Eq. 17.114 by C_I, the *coefficient of the instrument*. Since the flow past the pitot-static tube is slightly faster than the free-fluid velocity, the static pressure measured will be slightly lower than the true value. This makes the indicated velocity slightly higher than the true value. C_I, a number close to but less than 1.0, corrects for this.

Figure 17.15 *Pitot-Static Gauge*

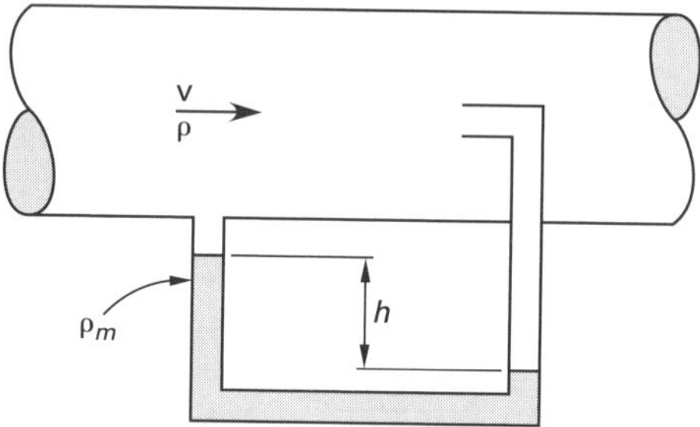

Pitot tube measurements are sensitive to the condition of the opening and errors in installation alignment. The *yaw angle* (i.e., the acute angle between the pitot tube axis and the flow streamline) should be zero.

A pitot-static tube indicates the velocity at only one point in a pipe. If the flow is laminar, and if the pitot-static tube is in the center of the pipe, v_{max} will be determined. The average velocity, however, will be only half the maximum value.

$$v = v_{max} = v_{ave} \quad \text{[turbulent]} \qquad 17.115$$

$$v = v_{max} = 2v_{ave} \quad \text{[laminar]} \qquad 17.116$$

Example 17.9

Water (62.4 lbm/ft^3; $\rho = 1000$ kg/m^3) is flowing through a pipe. A pitot-static gauge with a mercury manometer registers 3 in (0.076 m) of mercury. What is the velocity of the water in the pipe?

SI Solution

The density of mercury is $\rho = 13\,580$ kg/m^3. The velocity can be calculated directly from Eq. 17.114.

$$v = \sqrt{\frac{2gh(\rho_m - \rho)}{\rho}}$$

$$= \sqrt{\frac{(2)\left(9.81\ \frac{\text{m}}{\text{s}^2}\right)(0.076\ \text{m}) \times \left(13\,580\ \frac{\text{kg}}{\text{m}^3} - 1000\ \frac{\text{kg}}{\text{m}^3}\right)}{1000\ \frac{\text{kg}}{\text{m}^3}}}$$

$$= 4.33\ \text{m/s}$$

Customary U.S. Solution

The density of mercury is $\rho = 848.6$ lbm/ft^3.

From Eq. 17.114,

$$\mathrm{v} = \sqrt{\frac{2gh(\rho_m - \rho)}{\rho}}$$

$$= \sqrt{\frac{(2)\left(32.2\ \frac{\text{ft}}{\text{sec}^2}\right)(3\text{ in})\times\left(848.6\ \frac{\text{lbm}}{\text{ft}^3} - 62.4\ \frac{\text{lbm}}{\text{ft}^3}\right)}{\left(62.4\ \frac{\text{lbm}}{\text{ft}^3}\right)\left(12\ \frac{\text{in}}{\text{ft}}\right)}}$$

$$= 14.24 \text{ ft/sec}$$

26. VENTURI METER

Figure 17.16 illustrates a simple *venturi*. (Sometimes the venturi is called a *converging-diverging nozzle*.) This flow-measuring device can be inserted directly into a pipeline. Since the diameter changes are gradual, there is very little friction loss. Static pressure measurements are taken at the throat and upstream of the diameter change. These measurements are traditionally made by a manometer.

Figure 17.16 Venturi Meter

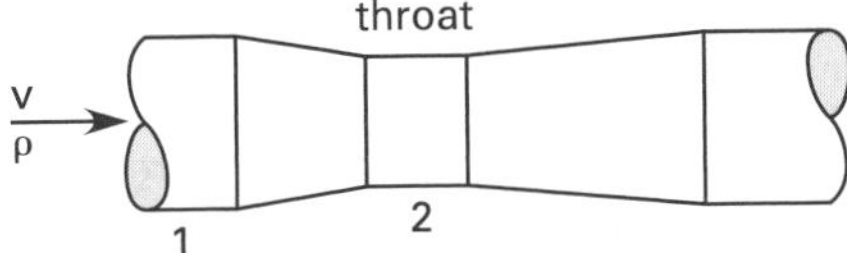

The analysis of *venturi meter performance* is relatively simple. The traditional derivation of upstream velocity starts by assuming a horizontal orientation (and therefore no change in elevation) and frictionless, incompressible, and turbulent flow. Then, the Bernoulli equation is written for points 1 and 2. Equation 17.117 shows that the static pressure decreases as the velocity increases. This is known as the *venturi effect*.

$$\frac{\mathrm{v}_1^2}{2} + \frac{p_1}{\rho} = \frac{\mathrm{v}_2^2}{2} + \frac{p_2}{\rho} \quad \text{[SI]} \quad 17.117(a)$$

$$\frac{\mathrm{v}_1^2}{2g_c} + \frac{p_1}{\rho} = \frac{\mathrm{v}_2^2}{2g_c} + \frac{p_2}{\rho} \quad \text{[U.S.]} \quad 17.117(b)$$

The two velocities are related by the continuity equation.

$$A_1\mathrm{v}_1 = A_2\mathrm{v}_2 \quad 17.118$$

Combining Eq. 17.117 and Eq. 17.118, assuming no elevation differences and eliminating the unknown v_1 produces an expression for the throat velocity. A *coefficient of velocity* is used to account for the small effect of friction. (C_v is very close to 1.0, usually 0.98 or 0.99.)

$$\mathrm{v}_2 = C_\mathrm{v}\mathrm{v}_{2,\text{ideal}} = \left(\frac{C_\mathrm{v}}{\sqrt{1-\left(\frac{A_2}{A_1}\right)^2}}\right)\sqrt{\frac{2(p_1-p_2)}{\rho}} \quad \text{[SI]} \quad 17.119(a)$$

$$\mathrm{v}_2 = C_\mathrm{v}\mathrm{v}_{2,\text{ideal}} = \left(\frac{C_\mathrm{v}}{\sqrt{1-\left(\frac{A_2}{A_1}\right)^2}}\right)\sqrt{\frac{2g_c(p_1-p_2)}{\rho}} \quad \text{[U.S.]} \quad 17.119(b)$$

The *velocity of approach factor*, F_va, also known as the *meter constant*, is the reciprocal of the first term of the denominator of the first term of Eq. 17.119. The *beta ratio* can be incorporated into the formula for F_va.

$$\beta = \frac{D_2}{D_1} \quad 17.120$$

$$F_\text{va} = \frac{1}{\sqrt{1-\left(\frac{A_2}{A_1}\right)^2}} = \frac{1}{\sqrt{1-\beta^4}} \quad 17.121$$

If a manometer is used to measure the pressure difference directly, Eq. 17.119 can be rewritten in terms of the manometer fluid reading. (See Fig. 17.17.)

$$\mathrm{v}_2 = C_\mathrm{v}\mathrm{v}_{2,\text{ideal}} = \left(\frac{C_\mathrm{v}}{\sqrt{1-\beta^4}}\right)\sqrt{\frac{2g(\rho_m-\rho)h}{\rho}} = C_\mathrm{v}F_\text{va}\sqrt{\frac{2g(\rho_m-\rho)h}{\rho}} \quad 17.122$$

Figure 17.17 Venturi Meter with Manometer

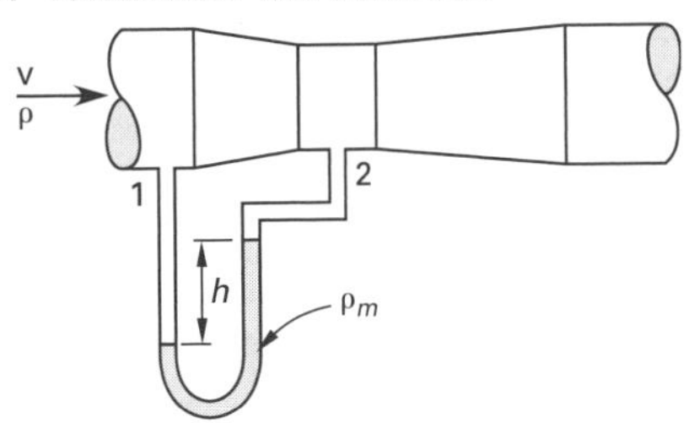

The flow rate through a venturi meter can be calculated from the throat area. There is an insignificant amount of contraction of the flow as it passes through the throat, and the *coefficient of contraction* is seldom encountered in venturi meter work. The *coefficient of discharge* ($C_d = C_c C_v$) is essentially the same as the coefficient of velocity. Values of C_d range from slightly less than 0.90 to over 0.99, depending on the Reynolds number. C_d is seldom less than 0.95 for turbulent flow. (See Fig. 17.18.)

$$Q = C_d A_2 \mathrm{v}_{2,\mathrm{ideal}} \qquad 17.123$$

$$Q = C_d A_2 \mathrm{v}_2 = \frac{C_v A_2}{\sqrt{1-\left(\frac{A_2}{A_1}\right)^2}} \sqrt{2g\left(\frac{p_1}{\gamma} + z_1 - \frac{p_2}{\gamma} - z_2\right)} \quad \text{[SI]} \qquad 17.124(a)$$

$$Q = C_d A_2 \mathrm{v}_2 = \frac{C_v A_2}{\sqrt{1-\left(\frac{A_2}{A_1}\right)^2}} \sqrt{2g\left(\frac{p_1}{\gamma}g_c + z_1 - \frac{p_2}{\gamma}g_c - z_2\right)} \quad \text{[U.S.]} \qquad 17.124(b)$$

Figure 17.18 *Typical Venturi Meter Discharge Coefficients (long radius venturi meter)*

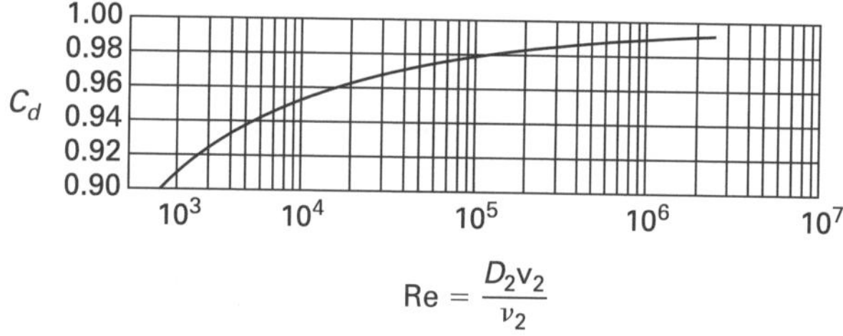

The product $C_d F_{va}$ is known as the *coefficient of flow* or *flow coefficient*, not to be confused with the coefficient of discharge.[19] This factor is used for convenience, since it combines the losses with the meter constant.

$$C_f = C_d F_{va} = \frac{C_d}{\sqrt{1-\beta^4}} \qquad 17.125$$

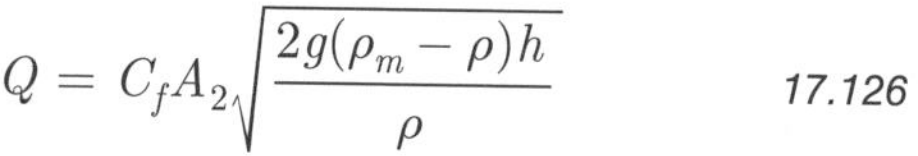

$$Q = C_f A_2 \sqrt{\frac{2g(\rho_m - \rho)h}{\rho}} \qquad 17.126$$

27. ORIFICE METER

The *orifice meter* (or *orifice plate*) is used more frequently than the venturi meter to measure flow rates in small pipes. It consists of a thin or sharp-edged plate with a central, round hole through which the fluid flows. Such a plate is easily clamped between two flanges in an existing pipeline. (See Fig. 17.19.)

Figure 17.19 *Orifice Meter with Differential Manometer*

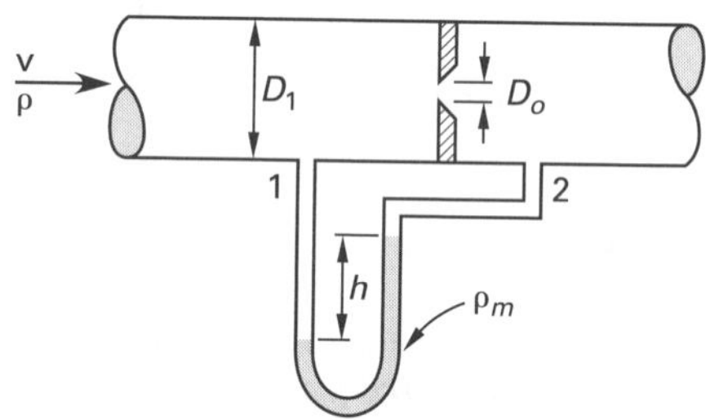

While (for small pipes) the orifice meter may consist of a thin plate without significant thickness, various types of bevels and rounded edges are also used with thicker plates. There is no significant difference in the analysis procedure between "flat plate," "sharp-edged," or "square-edged" orifice meters. Any effect that the orifice edges have is accounted for in the discharge and flow coefficient correlations. Similarly, the direction of the bevel will affect the coefficients but not the analysis method.

As with the venturi meter, pressure taps are used to obtain the static pressure upstream of the orifice plate and at the *vena contracta* (i.e., at the point of minimum pressure).[20] A differential manometer connected to the two taps conveniently indicates the difference in static pressures.

Although the orifice meter is simpler and less expensive than a venturi meter, its discharge coefficient is much less than that of a venturi meter. C_d usually ranges from 0.55 to 0.75, with values of 0.60 and 0.61 often being quoted. (The coefficient of contraction has a large effect, since $C_d = C_v C_c$.) Also, its pressure recovery is poor (i.e., there is a permanent pressure reduction), and it is susceptible to inaccuracies from wear and abrasion.[21]

[19] Some writers use the symbol K for the flow coefficient.

[20] Calibration of the orifice meter is sensitive to tap placement. Upstream taps are placed between one-half and two pipe diameters upstream from the orifice. (An upstream distance of one pipe diameter is often quoted and used.) There are three tap-placement options: flange, vena contracta, and standardized. Flange taps are used with prefabricated orifice meters that are inserted (by flange bolting) in pipes. If the location of the vena contracta is known, a tap can be placed there. However, the location of the vena contracta depends on the diameter ratio $\beta = D_o/D$ and varies from approximately 0.4 to 0.7 pipe diameters downstream. Due to the difficulty of locating the vena contracta, the standardized $1D$-$\frac{1}{2}D$ configuration is often used. The upstream tap is one diameter before the orifice; the downstream tap is one-half diameter after the orifice. Since approaching flow should be stable and uniform, care must be taken not to install the orifice meter less than approximately five diameters after a bend or elbow.

[21] The actual loss varies from 40% to 90% of the differential pressure. The loss depends on the diameter ratio $\beta = D_o/D_1$, and is not particularly sensitive to the Reynolds number for turbulent flow. For $\beta = 0.5$, the loss is 73% of the measured pressure difference, $p_1 - p_2$. This decreases to approximately 56% of the pressure difference when $\beta = 0.65$ and to 38%, when $\beta = 0.8$. For any diameter ratio, the pressure drop coefficient, K, in multiples of the orifice velocity head is $K = (1-\beta^2)/C_f^2$.

The derivation of the governing equations for an orifice meter is similar to that of the venturi meter. (The obvious falsity of assuming frictionless flow through the orifice is corrected by the coefficient of discharge.) The major difference is that the coefficient of contraction is taken into consideration in writing the mass continuity equation, since the pressure is measured at the vena contracta, not the orifice.

$$A_2 = C_c A_o \quad 17.127$$

$$\mathrm{v}_o = \left(\frac{C_d}{\sqrt{1 - \left(\frac{C_c A_o}{A_1} \right)^2}} \right) \sqrt{\frac{2(p_1 - p_2)}{\rho}} \quad \text{[SI]} \quad 17.128(a)$$

$$\mathrm{v}_o = \left(\frac{C_d}{\sqrt{1 - \left(\frac{C_c A_o}{A_1} \right)^2}} \right) \sqrt{\frac{2g_c(p_1 - p_2)}{\rho}} \quad \text{[U.S.]} \quad 17.128(b)$$

If a manometer is used to indicate the differential pressure $p_1 - p_2$, the velocity at the vena contracta can be calculated from Eq. 17.129.

$$\mathrm{v}_o = \left(\frac{C_d}{\sqrt{1 - \left(\frac{C_c A_o}{A_1} \right)^2}} \right) \sqrt{\frac{2g(\rho_m - \rho)h}{\rho}} \quad 17.129$$

The *velocity of approach factor*, F_{va}, for an orifice meter is defined differently than for a venturi meter, since it takes into consideration the contraction of the flow. However, the velocity of approach factor is still combined with the coefficient of discharge into the flow coefficient, C_f. Figure 17.20 illustrates how the flow coefficient varies with the area ratio and the Reynolds number.

$$F_{\text{va}} = \frac{1}{\sqrt{1 - \left(\frac{C_c A_o}{A_1} \right)^2}} \quad 17.130$$

$$C_f = C_d F_{\text{va}} \quad 17.131$$

The flow rate through an orifice meter is given by Eq. 17.132.

Orifices

$$Q = C A_o \sqrt{2g\left(\frac{p_1}{\gamma} + z_1 - \frac{p_2}{\gamma} - z_2 \right)} \quad \text{[SI]} \quad 17.132(a)$$

$$Q = C A_o \sqrt{2g\left(\frac{p_1 g_c}{\gamma} + z_1 - \frac{p_2 g_c}{\gamma} - z_2 \right)} \quad \text{[U.S.]} \quad 17.132(b)$$

In Eq. 17.132, the coefficient C is

Orifices

$$C = \frac{C_v C_c}{\sqrt{1 - C_c^2 \left(\frac{A_o}{A_1} \right)^2}} \quad 17.133$$

If a manometer is used to measure the pressure difference directly, Eq. 17.133 can be rewritten in terms of the manometer fluid reading.

$$Q = C_f A_o \sqrt{\frac{2g(\rho_m - \rho)h}{\rho}} = C_f A_o \sqrt{\frac{2(p_1 - p_2)}{\rho}} \quad \text{[SI]} \quad 17.134(a)$$

$$Q = C_f A_o \sqrt{\frac{2g(\rho_m - \rho)h}{\rho}} = C_f A_o \sqrt{\frac{2g_c(p_1 - p_2)}{\rho}} \quad \text{[U.S.]} \quad 17.134(b)$$

Figure 17.20 *Typical Flow Coefficients for Orifice Plates*

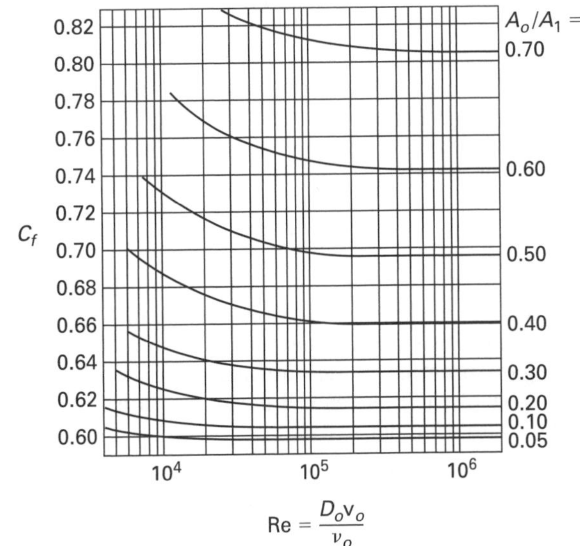

Example 17.10

150°F water ($\rho = 61.2$ lbm/ft³) flows in an 8 in schedule-40 steel pipe at the rate of 2.23 ft³/sec. A sharp-edged orifice with a 7 in diameter hole is placed in the line. A mercury differential manometer is used to record the pressure difference. (Mercury has a density of 848.6 lbm/ft³.) If the orifice has a flow coefficient, C_f, of 0.62, what deflection in inches of mercury is observed?

Solution

The orifice area is

$$A_o = \frac{\pi D_o^2}{4} = \frac{\pi\left(\frac{7\text{ in}}{12\ \frac{\text{in}}{\text{ft}}}\right)^2}{4} = 0.2673\text{ ft}^2$$

Equation 17.134 is solved for h.

$$\begin{aligned} h &= \frac{Q^2\rho}{2gC_f^2A_o^2(\rho_m-\rho)} \\ &= \frac{\left(2.23\ \frac{\text{ft}^3}{\text{sec}}\right)^2\left(61.2\ \frac{\text{lbm}}{\text{ft}^3}\right)\left(12\ \frac{\text{in}}{\text{ft}}\right)}{(2)\left(32.2\ \frac{\text{ft}}{\text{sec}^2}\right)(0.62)^2 \times(0.2673\text{ ft}^2)^2\left(848.6\ \frac{\text{lbm}}{\text{ft}^3}-61.2\ \frac{\text{lbm}}{\text{ft}^3}\right)} \\ &= 2.62\text{ in} \end{aligned}$$

28. FLOW MEASUREMENTS OF COMPRESSIBLE FLUIDS

Volume measurements of compressible fluids (i.e., gases) are not very meaningful. The volume of a gas will depend on its temperature and pressure. For that reason, flow quantities of gases discharged should be stated as mass flow rates.

$$\dot{m} = \rho_2 A_2 \text{v}_2 \qquad 17.135$$

Equation 17.135 requires that the velocity and area be measured at the same point. More importantly, the density must be measured at that point as well. However, it is common practice in flow measurement work to use the density of the upstream fluid at position 1. (This is not the stagnation density.)

The significant error introduced by this simplification is corrected by the use of an *expansion factor*, Y. For venturi meters and flow nozzles, values of the expansion factor are generally calculated theoretical values. Values of Y are determined experimentally for orifice plates. (See Fig. 17.21.)

$$\dot{m} = Y\rho_1 A_2 \text{v}_2 \qquad 17.136$$

Figure 17.21 Approximate Expansion Factors, k = 1.4 (air)

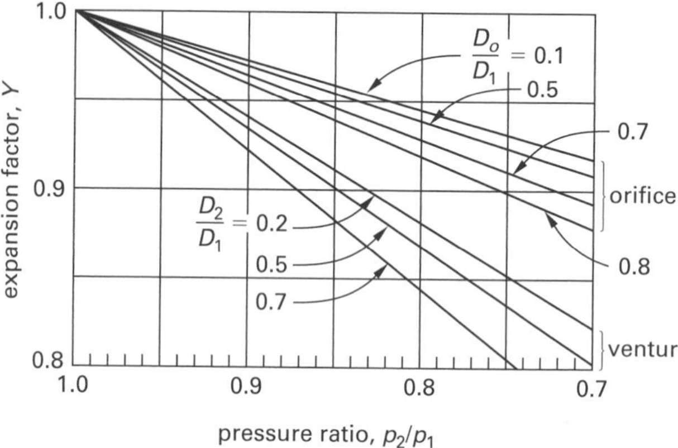

Geankoplis, Christie J., *Transport Processes and Unit Operations*, 3rd, © 1993. Printed and electronically reproduced by permission of Pearson Education, Inc., New York, New York.

Derivation of the theoretical formula for the expansion factor for venturi meters and flow nozzles is based on thermodynamic principles and an assumption of adiabatic flow.

$$Y = \sqrt{\frac{(1-\beta^4)\left(\left(\frac{p_2}{p_1}\right)^{2/k}-\left(\frac{p_2}{p_1}\right)^{(k+1)/k}\right)}{\left(\frac{k-1}{k}\right)\left(1-\frac{p_2}{p_1}\right)\left(1-\beta^4\left(\frac{p_2}{p_1}\right)^{2/k}\right)}} \qquad 17.137$$

Once the expansion factor is known, it can be used with the standard flow rate, Q, equations for venturi meters, flow nozzles, and orifice meters. For example, for a venturi meter, the mass flow rate would be calculated from Eq. 17.138.

$$\dot{m} = Y\dot{m}_{\text{ideal}} = \left(\frac{YC_dA_2}{\sqrt{1-\beta^4}}\right)\sqrt{2\rho_1(p_1-p_2)} \qquad \text{[SI]} \qquad 17.138(a)$$

$$\dot{m} = Y\dot{m}_{\text{ideal}} = \left(\frac{YC_dA_2}{\sqrt{1-\beta^4}}\right)\sqrt{2g_c\rho_1(p_1-p_2)} \qquad \text{[U.S.]} \qquad 17.138(b)$$

29. IMPULSE-MOMENTUM PRINCIPLE

(The convention of this section is to make F and x positive when they are directed toward the right. F and y are positive when directed upward. Also, the fluid is assumed to flow horizontally from left to right, and it has no initial y-component of velocity.)

The *momentum*, **P** (also known as *linear momentum* to distinguish it from *angular momentum*, which is not considered here), of a moving object is a vector quantity defined as the product of the object's mass and velocity.[22]

$$\mathbf{P} = m\mathbf{v} \quad \text{[SI]} \qquad 17.139(a)$$

$$\mathbf{P} = \frac{m\mathbf{v}}{g_c} \quad \text{[U.S.]} \qquad 17.139(b)$$

The *impulse*, **I**, of a constant force is calculated as the product of the force and the length of time the force is applied.

$$\mathbf{I} = \mathbf{F}\Delta t \qquad 17.140$$

The *impulse-momentum principle* (*law of conservation of momentum*) states that the impulse applied to a body is equal to the change in momentum. Equation 17.141 is one way of stating Newton's second law of motion.

$$\mathbf{I} = \Delta\mathbf{P} \qquad 17.141$$

$$F\Delta t = m\Delta \text{v} = m(\text{v}_2 - \text{v}_1) \quad \text{[SI]} \qquad 17.142(a)$$

$$F\Delta t = \frac{m\Delta \text{v}}{g_c} = \frac{m(\text{v}_2 - \text{v}_1)}{g_c} \quad \text{[U.S.]} \qquad 17.142(b)$$

For fluid flow, there is a mass flow rate, $\dot{m}$, but no mass per se. Since $\dot{m} = m/\Delta t$, the impulse-momentum equation can be rewritten as follows.

$$F = \dot{m}\Delta \text{v} \quad \text{[SI]} \qquad 17.143(a)$$

$$F = \frac{\dot{m}\Delta \text{v}}{g_c} \quad \text{[U.S.]} \qquad 17.143(b)$$

Equation 17.143 calculates the constant force required to accelerate or retard a fluid stream. This would occur when fluid enters a reduced or enlarged flow area. If the flow area decreases, for example, the fluid will be accelerated by a wall force up to the new velocity. Ultimately, this force must be resisted by the pipe supports.

As Eq. 17.143 illustrates, fluid momentum is not always conserved, since it is generated by the external force, F. Examples of external forces are gravity (considered zero for horizontal pipes), gage pressure, friction, and turning forces from walls and vanes. Only if these external forces are absent is fluid momentum conserved.

Since force is a vector, it can be resolved into its x- and y-components of force.

$$F_x = \dot{m}\Delta \text{v}_x \quad \text{[SI]} \qquad 17.144(a)$$

$$F_x = \frac{\dot{m}\Delta \text{v}_x}{g_c} \quad \text{[U.S.]} \qquad 17.144(b)$$

$$F_y = \dot{m}\Delta \text{v}_y \quad \text{[SI]} \qquad 17.145(a)$$

$$F_y = \frac{\dot{m}\Delta \text{v}_y}{g_c} \quad \text{[U.S.]} \qquad 17.145(b)$$

If the flow is initially at velocity v but is directed through an angle α with respect to the original direction, the x- and y-components of velocity can be calculated from Eq. 17.146 and Eq. 17.147.

$$\Delta \text{v}_x = \text{v}(\cos\alpha - 1) \qquad 17.146$$

$$\Delta \text{v}_y = \text{v}\sin\alpha \qquad 17.147$$

Since F and v are vector quantities and Δt and m are scalars, F must have the same direction as $\text{v}_2 - \text{v}_1$. (See Fig. 17.22.) This provides an intuitive method of determining the direction in which the force acts. Essentially, one needs to ask, "In which direction must the force act in order to push the fluid stream into its new direction?" (The force, F, is the force on the fluid. The force on the pipe walls or pipe supports has the same magnitude but is opposite in direction.)

Figure 17.22 *Force on a Confined Fluid Stream*

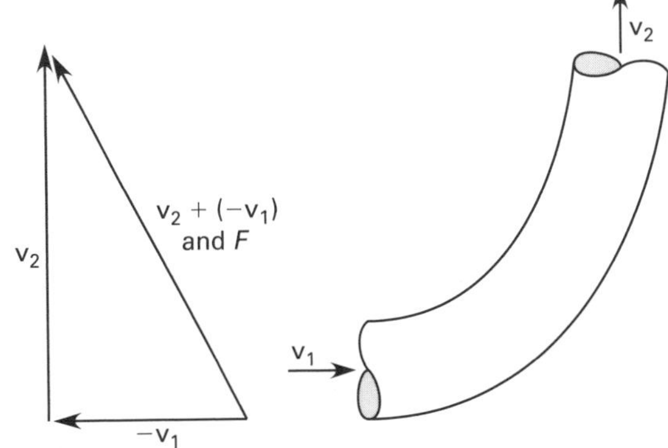

If a jet is completely stopped by a flat plate placed perpendicular to its flow, then $\alpha = 90°$ and $\Delta \text{v}_x = -\text{v}$. If a jet is turned around so that it ends up returning to where it originated, then $\alpha = 180°$ and $\Delta \text{v}_x = -2\text{v}$. A positive Δv indicates an increase in velocity. A negative Δv indicates a decrease in velocity.

Another approach is to use volumetric flow and pressure.

[22]The symbol B is also used for momentum. In many texts, however, momentum is given no symbol at all.

Impulse-Momentum Principle

$$\sum F = \sum Q_2\rho_2 \mathrm{v}_2 = \sum Q_1\rho_1 \mathrm{v}_1 \qquad 17.148$$

This can be broken down into the force components as

Impulse-Momentum Principle

$$p_1 A_1 - p_2 A_2 \cos\alpha - F_x = Q\rho(\mathrm{v}_2\cos\alpha - \mathrm{v}_1) \qquad 17.149$$

$$F_y - W - p_2 A_2 \sin\alpha = Q\rho(\mathrm{v}_2 \sin\alpha - 0) \qquad 17.150$$

30. JET PROPULSION

A basic application of the impulse-momentum principle is the analysis of jet propulsion. Air enters a jet engine and is mixed with fuel. The air and fuel mixture is compressed and ignited, and the exhaust products leave the engine at a greater velocity than was possessed by the original air. The change in momentum of the air produces a force on the engine. (See Fig. 17.23.)

Figure 17.23 *Jet Engine*

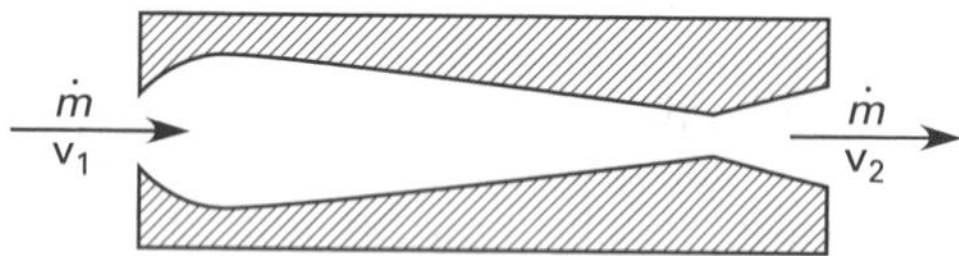

The governing equation for a jet engine is Eq. 17.151. The mass of the jet fuel is small compared with the air mass, and the fuel mass is commonly disregarded.

$$F_x = \dot{m}(\mathrm{v}_2 - \mathrm{v}_1) \qquad 17.151$$

$$F_x = Q_2\rho_2\mathrm{v}_2 - Q_1\rho_1\mathrm{v}_1 \quad \text{[SI]} \qquad 17.152(a)$$

$$F_x = \frac{Q_2\rho_2\mathrm{v}_2 - Q_1\rho_1\mathrm{v}_1}{g_c} \quad \text{[U.S.]} \qquad 17.152(b)$$

A slightly different approach is to take F as the force resulting from the flow through a nozzle opening in the side of a tank. This is very similar to an orifice discharging an incompressible fluid to atmosphere. With this approach, the force can be calculated as

Jet Propulsion

$$\begin{aligned} F &= Q\rho(\mathrm{v}_2 - 0) \\ &= 2\gamma h A_2 \end{aligned} \qquad 17.153$$

In Eq. 17.153, the volumetric flow rate and velocity are calculated as

Jet Propulsion

$$Q = A_2\sqrt{2gh} \qquad 17.154$$

$$\mathrm{v}_2 = \sqrt{2gh} \qquad 17.155$$

31. OPEN JET ON A SINGLE STATIONARY BLADE

Figure 17.24 illustrates a fluid jet being turned through an angle α by a stationary blade (also called a *vane*). It is common to assume that $|\mathrm{v}_2| = |\mathrm{v}_1|$, although this will not be strictly true if friction between the blade and fluid is considered. Since the fluid is both decelerated (in the x-direction) and accelerated (in the y-direction), there will be two components of force on the fluid.

$$\Delta\mathrm{v}_x = \mathrm{v}_2\cos\alpha - \mathrm{v}_1 \qquad 17.156$$

$$\Delta\mathrm{v}_y = \mathrm{v}_2 \sin\alpha \qquad 17.157$$

Fixed Blade

$$\begin{aligned} -F_x &= Q\rho(\mathrm{v}_2\cos\alpha - \mathrm{v}_1) \\ &= \dot{m}(\mathrm{v}_2\cos\alpha - \mathrm{v}_1) \end{aligned} \quad \text{[SI]} \qquad 17.158(a)$$

$$\begin{aligned} -F_x &= \frac{Q\rho(\mathrm{v}_2\cos\alpha - \mathrm{v}_1)}{g_c} \\ &= \frac{\dot{m}(\mathrm{v}_2\cos\alpha - \mathrm{v}_1)}{g_c} \end{aligned} \quad \text{[U.S.]} \qquad 17.158(b)$$

$$\begin{aligned} F_y &= Q\rho(\mathrm{v}_2\sin\alpha - 0) \\ &= \dot{m}(\mathrm{v}_2\sin\alpha - 0) \end{aligned} \quad \text{[SI]} \qquad 17.159(a)$$

$$\begin{aligned} F_y &= \frac{Q\rho(\mathrm{v}_2\sin\alpha - 0)}{g_c} \\ &= \frac{\dot{m}(\mathrm{v}_2\sin\alpha - 0)}{g_c} \end{aligned} \quad \text{[U.S.]} \qquad 17.159(b)$$

$$F = \sqrt{F_x^2 + F_y^2} \qquad 17.160$$

Figure 17.24 *Open Jet on a Stationary Blade*

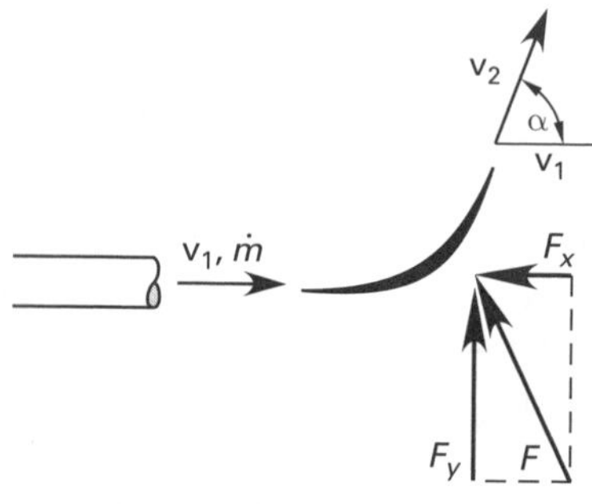

32. OPEN JET ON A SINGLE MOVING BLADE

If a blade is moving away at velocity v_b from the source of the fluid jet, only the *relative velocity difference* between the jet and blade produces a momentum change. Furthermore, not all of the fluid jet overtakes the moving blade. The equations used for the single stationary blade can be used by substituting $(\mathrm{v}-\mathrm{v}_b)$ for v and by using the effective mass flow rate, $\dot{m}_{\text{eff}}$. (See Fig. 17.25.)

$$\Delta \mathrm{v}_x = (\mathrm{v} - \mathrm{v}_b)(\cos\alpha - 1) \quad 17.161$$

$$\Delta \mathrm{v}_y = (\mathrm{v} - \mathrm{v}_b)\sin\alpha \quad 17.162$$

$$\dot{m}_{\text{eff}} = \left(\frac{\mathrm{v} - \mathrm{v}_b}{\mathrm{v}}\right)\dot{m} \quad 17.163$$

Moving Blade

$$\begin{aligned} -F_x &= Q\rho(\mathrm{v} - \mathrm{v}_b)(1 - \cos\alpha) \\ &= \dot{m}_{\text{eff}}(\mathrm{v} - \mathrm{v}_b)(1 - \cos\alpha) \end{aligned} \quad \text{[SI]} \quad 17.164(a)$$

$$\begin{aligned} -F_x &= \frac{Q\rho(\mathrm{v} - \mathrm{v}_b)(1 - \cos\alpha)}{g_c} \\ &= \frac{\dot{m}_{\text{eff}}(\mathrm{v} - \mathrm{v}_b)(1 - \cos\alpha)}{g_c} \end{aligned} \quad \text{[U.S.]} \quad 17.164(b)$$

$$\begin{aligned} F_y &= Q\rho(\mathrm{v}-\mathrm{v}_b)\sin\alpha \\ &= \dot{m}_{\text{eff}}(\mathrm{v}-\mathrm{v}_b)\sin\alpha \end{aligned} \quad \text{[SI]} \quad 17.165(a)$$

$$\begin{aligned} F_y &= \frac{Q\rho(\mathrm{v}-\mathrm{v}_b)\sin\alpha}{g_c} \\ &= \frac{\dot{m}_{\text{eff}}(\mathrm{v}-\mathrm{v}_b)\sin\alpha}{g_c} \end{aligned} \quad \text{[U.S.]} \quad 17.165(b)$$

Figure 17.25 *Open Jet on a Moving Blade*

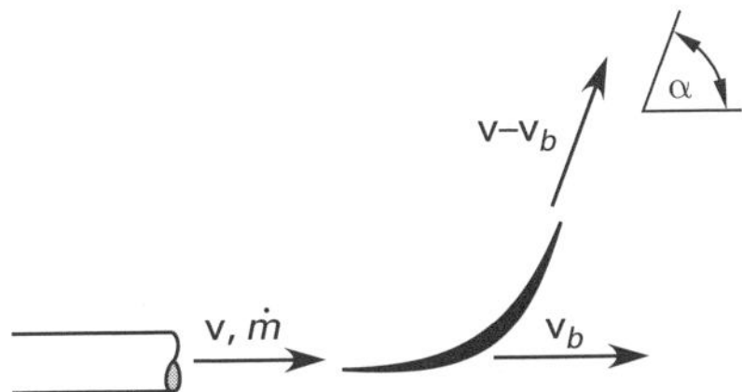

33. OPEN JET ON A MULTIPLE-BLADED WHEEL

An *impulse turbine* consists of a series of blades (buckets or vanes) mounted around a wheel. (See Fig. 17.26.) The tangential velocity of the blades is approximately parallel to the jet. The effective mass flow rate, $\dot{m}_{\text{eff}}$, used in calculating the reaction force is the full discharge rate, since when one blade moves away from the jet, other blades will have moved into position. All of the fluid discharged is captured by the blades. Equation 17.164 and Eq. 17.165 are applicable if the total flow rate is used. The tangential blade velocity is

$$\mathrm{v}_b = \frac{2\pi r n_{\text{rpm}}}{60} = \omega r \quad 17.166$$

Figure 17.26 *Impulse Turbine*

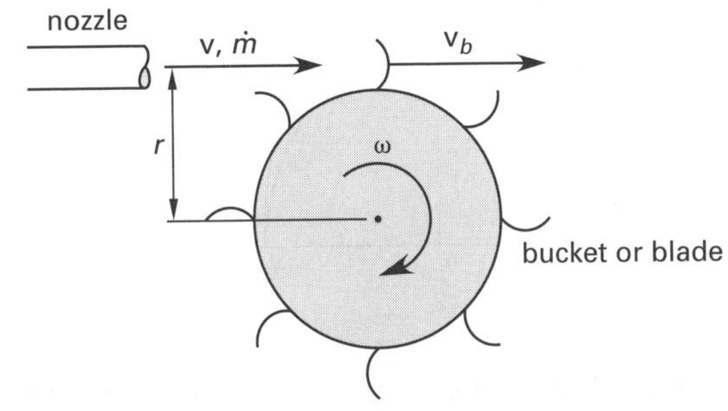

34. IMPULSE TURBINE POWER

The total power potential of a fluid jet can be calculated from the kinetic energy of the jet and the mass flow rate.[23] (This neglects the pressure energy, which is small by comparison.)

$$P_{\text{jet}} = \frac{\dot{m}\mathrm{v}^2}{2} \quad \text{[SI]} \quad 17.167(a)$$

$$P_{\text{jet}} = \frac{\dot{m}\mathrm{v}^2}{2g_c} \quad \text{[U.S.]} \quad 17.167(b)$$

The power transferred from a fluid jet to the blades of a turbine is calculated from the x-component of force on the blades. The y-component of force does no work.

$$P = F_x \mathrm{v}_b \quad 17.168$$

Impulse Turbine

$$\begin{aligned} P &= Q\rho(\mathrm{v} - \mathrm{v}_b)(1 - \cos\alpha)\mathrm{v}_b \\ &= \dot{m}(\mathrm{v} - \mathrm{v}_b)(1 - \cos\alpha)\mathrm{v}_b \end{aligned} \quad \text{[SI]} \quad 17.169(a)$$

$$\begin{aligned} P &= \frac{Q\rho(\mathrm{v} - \mathrm{v}_b)(1 - \cos\alpha)\mathrm{v}_b}{g_c} \\ &= \frac{\dot{m}(\mathrm{v} - \mathrm{v}_b)(1 - \cos\alpha)\mathrm{v}_b}{g_c} \end{aligned} \quad \text{[U.S.]} \quad 17.169(b)$$

The maximum theoretical blade velocity is the velocity of the jet: $\mathrm{v}_b = \mathrm{v}$. This is known as the *runaway speed* and can only be achieved when the turbine is unloaded. If Eq. 17.169 is maximized with respect to v_b, the maximum power will occur when the blade is traveling at

[23]The full jet discharge is used in this section. If only a single blade is involved, the effective mass flow rate, $\dot{m}_{\text{eff}}$, must be used.

half of the jet velocity: $v_b = v/2$. The power (force) is also affected by the deflection angle of the blade. Power is maximized when $\alpha = 180°$. Figure 17.27 illustrates the relationship between power and the variables α and v_b.

Figure 17.27 *Turbine Power*

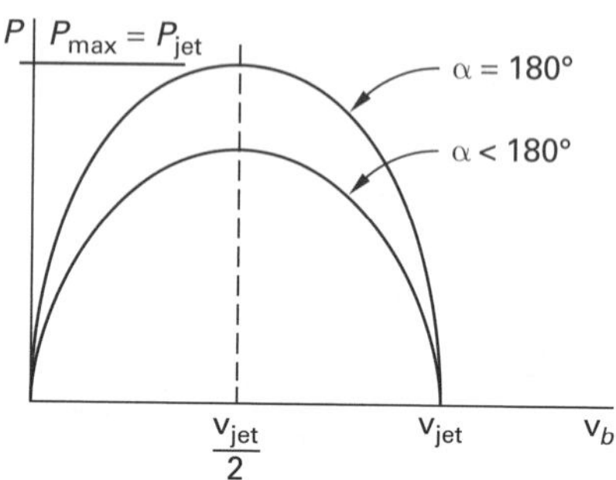

Putting $\alpha = 180°$ and $v_b = v/2$ into Eq. 17.169 results in $P_{max} = \dot{m}v^2/2$, which is the same as P_{jet} in Eq. 17.167. If the machine is 100% efficient, 100% of the jet power can be transferred to the machine.

35. CONFINED STREAMS IN PIPE BENDS

As presented in Sec. 17.29, momentum can also be changed by pressure forces. Such is the case when fluid enters a pipe fitting or bend. (See Fig. 17.28.) Since the fluid is confined, the forces due to static pressure must be included in the analysis. (The effects of gravity and friction are neglected.)

$$F_x = p_2A_2\cos\alpha - p_1A_1 + \dot{m}(v_2\cos\alpha - v_1) \quad \text{[SI]} \quad 17.170(a)$$

$$F_x = p_2A_2\cos\alpha - p_1A_1 + \frac{\dot{m}(v_2\cos\alpha - v_1)}{g_c} \quad \text{[U.S.]} \quad 17.170(b)$$

$$F_y = (p_2A_2 + \dot{m}v_2)\sin\alpha \quad \text{[SI]} \quad 17.171(a)$$

$$F_y = \left(p_2A_2 + \frac{\dot{m}v_2}{g_c}\right)\sin\alpha \quad \text{[U.S.]} \quad 17.171(b)$$

Figure 17.28 *Pipe Bend*

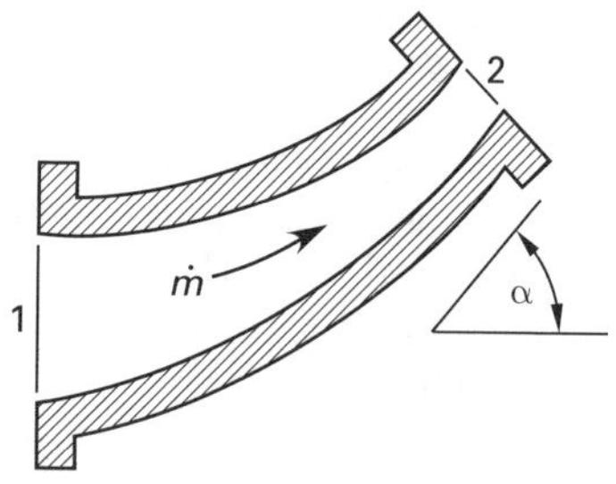

Example 17.11

60°F water (ρ = 62.4 lbm/ft³) at 40 psig enters a 12 in × 8 in reducing elbow at 8 ft/sec and is turned through an angle of 30°. Water leaves 26 in higher in elevation. (a) What is the resultant force exerted on the water by the elbow? (b) What other forces should be considered in the design of supports for the fitting?

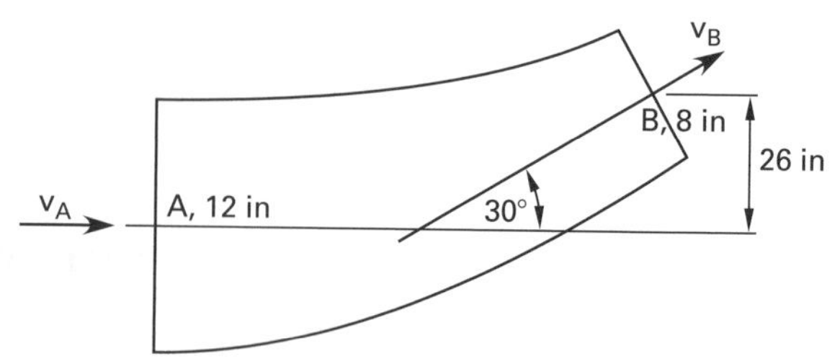

Solution

(a) The velocity and pressure at point B are both needed. The velocity is easily calculated from the continuity equation.

$$A_A = \frac{\pi D_A^2}{4} = \frac{\pi\left(\frac{12 \text{ in}}{12 \frac{\text{in}}{\text{ft}}}\right)^2}{4} = 0.7854 \text{ ft}^2$$

$$A_B = \frac{\pi D_B^2}{4} = \frac{\pi\left(\frac{8 \text{ in}}{12 \frac{\text{in}}{\text{ft}}}\right)^2}{4} = 0.3491 \text{ ft}^2$$

$$v_B = \frac{v_A A_A}{A_B} = \left(8 \frac{\text{ft}}{\text{sec}}\right)\left(\frac{0.7854 \text{ ft}^2}{0.3491 \text{ ft}^2}\right) = 18 \text{ ft/sec}$$

$$p_A = \left(40 \frac{\text{lbf}}{\text{in}^2}\right)\left(12 \frac{\text{in}}{\text{ft}}\right)^2 = 5760 \text{ lbf/ft}^2$$

The Bernoulli equation is used to calculate p_B. (Gage pressures are used for this calculation. Absolute pressures could also be used, but the addition of p_{atm}/ρ to both sides of the Bernoulli equation would not affect p_B.)

$$\frac{p_A}{\rho} + \frac{v_A^2}{2g_c} = \frac{p_B}{\rho} + \frac{v_B^2}{2g_c} + \frac{zg}{g_c}$$

$$\frac{5760 \ \frac{\text{lbf}}{\text{ft}^2}}{62.4 \ \frac{\text{lbm}}{\text{ft}^3}} + \frac{\left(8 \ \frac{\text{ft}}{\text{sec}}\right)^2}{(2)\left(32.2 \ \frac{\text{lbm-ft}}{\text{lbf-sec}^2}\right)}$$

$$= \frac{p_B}{62.4 \ \frac{\text{lbm}}{\text{ft}^3}} + \frac{\left(18 \ \frac{\text{ft}}{\text{sec}}\right)^2}{(2)\left(32.2 \ \frac{\text{lbm-ft}}{\text{lbf-sec}^2}\right)}$$

$$+ \frac{26 \text{ in}}{12 \ \frac{\text{in}}{\text{ft}}} \times \frac{32.2 \ \frac{\text{ft}}{\text{sec}^2}}{32.2 \ \frac{\text{lbm-ft}}{\text{lbf-sec}^2}}$$

$$p_B = 5373 \text{ lbf/ft}^2$$

The mass flow rate is

$$\dot{m} = Q\rho = vA\rho = \left(8 \ \frac{\text{ft}}{\text{sec}}\right)(0.7854 \text{ ft}^2)\left(62.4 \ \frac{\text{lbm}}{\text{ft}^3}\right)$$

$$= 392.1 \text{ lbm/sec}$$

From Eq. 17.170,

$$F_x = p_2A_2\cos\alpha - p_1A_1 + \frac{\dot{m}(v_2\cos\alpha - v_1)}{g_c}$$

$$= \left(5373 \ \frac{\text{lbf}}{\text{ft}^2}\right)(0.3491 \text{ ft}^2)(\cos 30°)$$

$$-\left(5760 \ \frac{\text{lbf}}{\text{ft}^2}\right)(0.7854 \text{ ft}^2)$$

$$+ \frac{\left(392.1 \ \frac{\text{lbm}}{\text{sec}}\right)\left(\left(18 \ \frac{\text{ft}}{\text{sec}}\right)(\cos 30°) - 8 \ \frac{\text{ft}}{\text{sec}}\right)}{32.2 \ \frac{\text{lbm-ft}}{\text{lbf-sec}^2}}$$

$$= -2807 \text{ lbf}$$

From Eq. 17.171,

$$F_y = \left(p_2A_2 + \frac{\dot{m}v_2}{g_c}\right)\sin\theta$$

$$= \left(\left(5373 \ \frac{\text{lbf}}{\text{ft}^2}\right)(0.3491 \text{ ft}^2) + \frac{\left(392.1 \ \frac{\text{lbm}}{\text{sec}}\right)\left(18 \ \frac{\text{ft}}{\text{sec}}\right)}{32.2 \ \frac{\text{lbm-ft}}{\text{lbf-sec}^2}}\right)\sin 30°$$

$$= 1047 \text{ lbf}$$

The resultant force on the water is

$$R = \sqrt{F_x^2 + F_y^2} = \sqrt{(-2807 \text{ lbf})^2 + (1047 \text{ lbf})^2}$$

$$= 2996 \text{ lbf}$$

(b) In addition to counteracting the resultant force, R, the support should be designed to carry the weight of the elbow and the water in it. Also, the support must carry a part of the pipe and water weight tributary to the elbow.

36. WATER HAMMER

Water hammer in a long pipe is an increase in fluid pressure caused by a sudden velocity decrease. (See Fig. 17.29.) The sudden velocity decrease will usually be caused by a valve closing. Analysis of the water hammer phenomenon can take two approaches, depending on whether or not the pipe material is assumed to be elastic.

Figure 17.29 *Water Hammer*

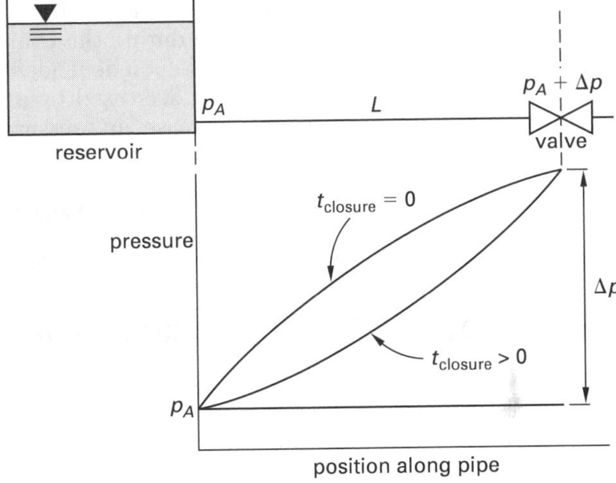

The *water hammer wave speed*, c, is given by Eq. 17.172. The first term on the right-hand side represents the effect of fluid compressibility, and the second term represents the effect of pipe elasticity.

$$\frac{1}{c^2} = \frac{d\rho}{dp} + \frac{\rho}{A}\frac{dA}{dp} \qquad 17.172$$

For a compressible fluid within an inelastic (rigid) pipe, $dA/dp = 0$, and the wave speed is simply the speed of sound in the fluid.

$$c = \sqrt{\frac{dp}{d\rho}} = \sqrt{\frac{E}{\rho}} \quad \left[\begin{array}{c}\text{inelastic pipe;}\\ \text{consistent units}\end{array}\right] \qquad 17.173$$

If the pipe material is assumed to be inelastic (such as cast iron, ductile iron, and concrete), the time required for the water hammer shock wave to travel from the suddenly closed valve to a point of interest depends only on the velocity of sound in the fluid, c, and the distance, L, between the two points. This is also the time required to bring all of the fluid in the pipe to rest.

$$t = \frac{L}{c} \qquad 17.174$$

When the water hammer shock wave reaches the original source of water, the pressure wave will dissipate. A rarefaction wave (at the pressure of the water source) will return at velocity c to the valve. The time for the compression shock wave to travel to the source and the rarefaction wave to return to the valve is given by Eq. 17.175. This is also the length of time that the pressure is constant at the valve.

$$t = \frac{2L}{c} \qquad 17.175$$

The fluid pressure increase resulting from the shock wave is calculated by equating the kinetic energy change of the fluid with the average pressure during the compression process. The pressure increase is independent of the length of pipe. If the velocity is decreased by an amount Δv instantaneously, the increase in pressure will be

Water Hammer

$$\Delta p_h = \rho c \Delta v \quad \text{[SI]} \qquad 17.176(a)$$

$$\Delta p_h = \frac{\rho c \Delta v}{g_c} \quad \text{[U.S.]} \qquad 17.176(b)$$

It is interesting that the pressure increase at the valve depends on Δv but not on the actual length of time it takes to close the valve, as long as the valve is closed when the wave returns to it. Therefore, there is no difference in pressure buildups at the valve for an "instantaneous closure," "rapid closure," or "sudden closure."[24] It is only necessary for the closure to occur rapidly.

Water hammer does not necessarily occur just because a return shockwave increases the pressure. When the velocity is low—less than 7 ft/sec (2.1 m/s), with 5 ft/sec (1.5 m/s) recommended—there is not enough force generated to create water hammer. The cost of using larger pipe, fittings, and valves is the disadvantage of keeping velocity low.

Having a very long pipe is equivalent to assuming an instantaneous closure. When the pipe is long, the time for the shock wave to travel round-trip is much longer than the time to close the valve. The valve will be closed when the rarefaction wave returns to the valve.

If the pipe is short, it will be difficult to close the valve before the rarefaction wave returns to the valve. With a short pipe, the pressure buildup will be less than is predicted by Eq. 17.176. (Having a short pipe is equivalent to the case of "slow closure.") The actual pressure history is complex, and no simple method exists for calculating the pressure buildup in short pipes.

Installing a *surge tank*, *accumulator*, *slow-closing valve* (e.g., a gate valve), or *pressure-relief valve* in the line will protect against water hammer damage. (See Fig. 17.30.) The surge tank (or *surge chamber*) is an open tank or reservoir. Since the water is unconfined, large pressure buildups do not occur. An accumulator is a closed tank that is partially filled with air. Since the air is much more compressible than the water, it will be compressed by the water hammer shock wave. The energy of the shock wave is dissipated when the air is compressed.

Figure 17.30 *Water Hammer Protective Devices*

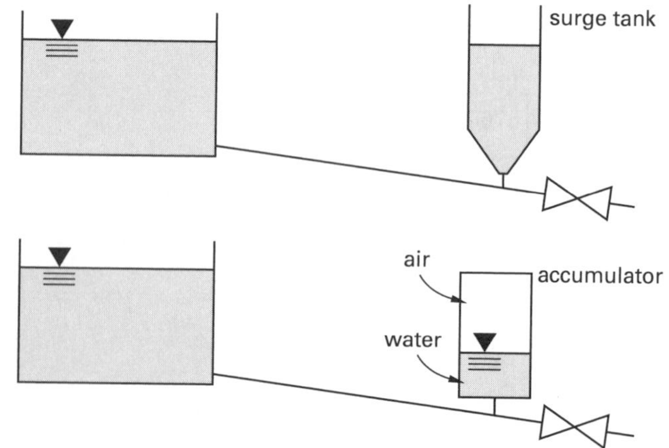

[24]The pressure elsewhere along the pipe, however, will be lower for slow closures than for instantaneous closures.

For an incompressible fluid within an elastic pipe, $d\rho/dp$ in Eq. 17.173 is zero. In that case, the wave speed is

$$c = \sqrt{\frac{A}{\rho}\frac{dp}{dA}} \quad \begin{bmatrix}\text{elastic pipe;}\\ \text{consistent units}\end{bmatrix} \qquad 17.177$$

For long elastic pipes (the common assumption for steel and plastic), the degree of pipe anchoring and longitudinal stress propagation affect the wave velocity. Longitudinal stresses resulting from circumferential stresses (i.e., Poisson's effect) must be considered by including a Poisson's effect factor, c_P. In Eq. 17.178, for pipes anchored at their upstream ends only, $c_P = 1 - 0.5\nu_{\text{pipe}}$; for pipes anchored throughout, $c_P = 1 - \nu^2_{\text{pipe}}$; for anchored pipes with expansion joints throughout, $c_P = 1$ (i.e., the commonly used *Korteweg formula*). Pipe sections joined with gasketed integral bell ends and gasketed ends satisfy the requirement for expansion joints. Equation 17.178 shows that the wave velocity can be reduced by increasing the pipe diameter.

$$c = \sqrt{\frac{\dfrac{E_{\text{fluid}}t_{\text{pipe}}E_{\text{pipe}}}{t_{\text{pipe}}E_{\text{pipe}} + c_P D_{\text{pipe}}E_{\text{fluid}}}}{\rho}} \qquad 17.178$$

At room temperature, the modulus of elasticity of ductile steel is approximately 2.9×10^7 lbf/in^2 (200 GPa); for ductile cast iron, 2.2–2.5 $\times$ 10^7 lbf/in^2 (150–170 GPa); for PVC, 3.5–4.1 $\times$ 10^5 lbf/in^2 (2.4–2.8 GPa); for ABS, 3.2–3.5 $\times$ 10^5 lbf/in^2 (2.2–2.4 GPa).

Example 17.12

Water ($\rho = 1000$ kg/m^3, $E = 2 \times 10^9$ Pa), is flowing at 4 m/s through a long length of 4 in schedule-40 steel pipe ($D_i = 0.102$ m, $t = 0.00602$ m, $E = 2 \times 10^{11}$ Pa) with expansion joints when a valve suddenly closes completely. What is the theoretical increase in pressure?

Solution

For pipes with expansion joints, $c_P = 1$. From Eq. 17.178, the modulus of elasticity to be used in calculating the speed of sound is

$$\begin{aligned} c &= \sqrt{\frac{\dfrac{E_{\text{fluid}}t_{\text{pipe}}E_{\text{pipe}}}{t_{\text{pipe}}E_{\text{pipe}} + c_P D_{\text{pipe}}E_{\text{fluid}}}}{\rho}} \\ &= \sqrt{\frac{\dfrac{(2\times10^9\ \text{Pa})(0.00602\ \text{m})(2\times10^{11}\ \text{Pa})}{(0.00602\ \text{m})(2\times10^{11}\ \text{Pa}) + (1)(0.102\ \text{m})(2\times10^9\ \text{Pa})}}{1000\ \frac{\text{kg}}{\text{m}^3}}} \\ &= 1307.75\ \text{m/s} \end{aligned}$$

From Eq. 17.176, the pressure increase is

Water Hammer

$$\begin{aligned} \Delta p_h &= \rho c \Delta \text{v} \\ &= \left(1000\ \frac{\text{kg}}{\text{m}^3}\right)\left(1307.75\ \frac{\text{m}}{\text{s}}\right)\left(4\ \frac{\text{m}}{\text{s}}\right) \\ &= 5.23 \times 10^6\ \text{Pa} \end{aligned}$$

37. LIFT

Lift is an upward force that is exerted on an object (flat plate, airfoil, rotating cylinder, etc.) as the object passes through a fluid. Lift combines with drag to form the resultant force on the object, as shown in Fig. 17.31.

Figure 17.31 *Lift and Drag on an Airfoil*

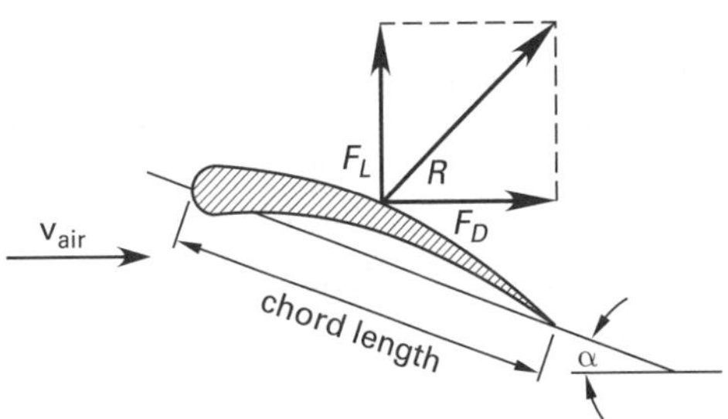

The generation of lift from air flowing over an airfoil is predicted by Bernoulli's equation. Air molecules must travel a longer distance over the top surface of the airfoil than over the lower surface, and, therefore, they travel faster over the top surface. Since the total energy of the air is constant, the increase in kinetic energy comes at the expense of pressure energy. The static pressure on the top of the airfoil is reduced, and a net upward force is produced.

Within practical limits, the lift produced can be increased at lower speeds by increasing the curvature of the wing. This increased curvature is achieved by the use of *flaps*. (See Fig. 17.32.) When a plane is traveling slowly (e.g., during takeoff or landing), its flaps are extended to create the lift needed.

Figure 17.32 *Use of Flaps in an Airfoil*

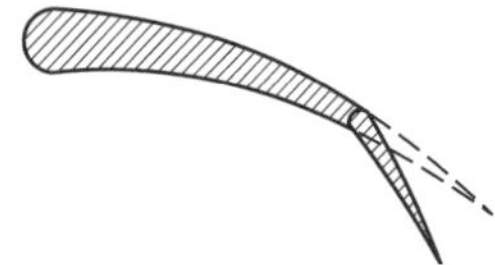

The lift produced can be calculated from Eq. 17.179, whose use is not limited to airfoils.

$$F_L = \frac{C_L A \rho \text{v}^2}{2} \qquad \text{[SI]} \qquad 17.179(a)$$

$$F_L = \frac{C_L A \rho \text{v}^2}{2g_c} \qquad \text{[U.S.]} \qquad 17.179(b)$$

The dimensions of an airfoil or wing are frequently given in terms of chord length and aspect ratio. The *chord length* (or just *chord*) is the front-to-back dimension of the airfoil. The *aspect ratio* is the ratio of the *span* (wing length) to chord length. The area, A, in Eq. 17.179 is the airfoil's area projected onto the plane of the chord.

$$A = \text{chord length} \times \text{span} \quad \begin{bmatrix}\text{rectangular} \\ \text{airfoil}\end{bmatrix} \qquad 17.180$$

The dimensionless *coefficient of lift*, C_L, modifies the effectiveness of the airfoil. The coefficient of lift depends on the shape of the airfoil, the Reynolds number, and the angle of attack. The theoretical coefficient of lift for a thin flat plate in two-dimensional flow at a low angle of attack, α, is given by Eq. 17.181. Actual airfoils are able to achieve only 80–90% of this theoretical value.

$$C_L = 2\pi \sin\alpha \quad [\text{flat plate}] \qquad 17.181$$

The coefficient of lift for an airfoil cannot be increased without limit merely by increasing α. Eventually, the *stall angle* is reached, at which point the coefficient of lift decreases dramatically. (See Fig. 17.33.)

Figure 17.33 *Typical Plot of Lift Coefficient*

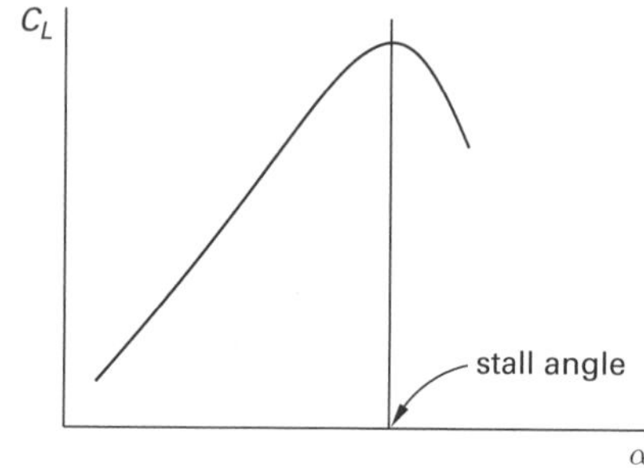

38. DRAG

Drag is a frictional force that acts parallel but opposite to the direction of motion. The total drag force is made up of *skin friction* and *pressure drag* (also known as *form drag*). These components, in turn, can be subdivided and categorized into *wake drag*, *induced drag*, and *profile drag*. (See Fig. 17.34.)

Figure 17.34 *Components of Total Drag*

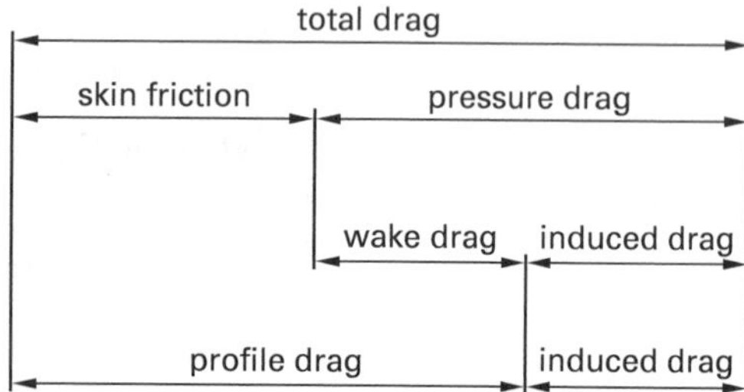

Most aeronautical engineering books contain descriptions of these drag terms. However, the difference between the situations where either skin friction drag or pressure drag predominates is illustrated in Fig. 17.35.

Figure 17.35 *Extreme Cases of Pressure Drag and Skin Friction*

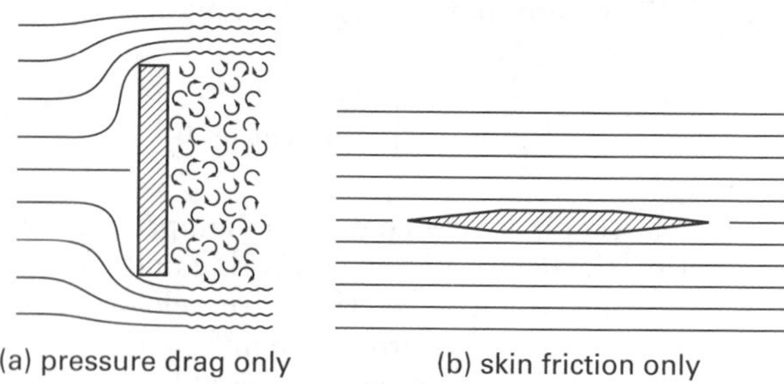

Total drag is most easily calculated from the dimensionless *drag coefficient*, C_D. The drag coefficient depends only on the Reynolds number.

Drag Force

$$F_D = \frac{C_D \rho \mathrm{v}^2 A}{2} \quad [\text{SI}] \qquad 17.182(a)$$

$$F_D = \frac{C_D \rho \mathrm{v}^2 A}{2 g_c} \quad [\text{U.S.}] \qquad 17.182(b)$$

In most cases, the area, A, in Eq. 17.182 is the projected area (i.e., the *frontal area*) normal to the stream. This is appropriate for spheres, cylinders, and automobiles. In a few cases (e.g., for airfoils and flat plates) as determined by how C_D was derived, the area is a projection of the object onto a plane parallel to the stream.

Typical drag coefficients for production cars vary from approximately 0.25 to approximately 0.70, with most modern cars being nearer the lower end. By comparison, other low-speed drag coefficients are approximately 0.05 (aircraft wing), 0.10 (sphere in turbulent flow), and 1.2 (flat plate). (See Table 17.5.)

Table 17.5 *Typical Ranges of Vehicle Drag Coefficients*

vehicle	low	medium	high
experimental race	0.17	0.21	0.23
sports	0.27	0.31	0.38
performance	0.32	0.34	0.38
’60s muscle car	0.38	0.44	0.50
sedan	0.34	0.39	0.50
motorcycle	0.50	0.90	1.00
truck	0.60	0.90	1.00
tractor-trailer	0.60	0.77	1.20

Used with permission from *Unit Operations*, by George Granger Brown, et al., John Wiley & Sons, Inc., © 1950.

In aerodynamic studies performed in *wind tunnels*, accuracy in lift and drag measurement is specified in *drag counts*. One drag count is equal to a C_D (or, C_L) of 0.0001. Drag counts are commonly used when describing the effect of some change to a fuselage or wing geometry. For example, an increase in C_D from 0.0450 to 0.0467 would be reported at an increase of 17 counts. Some of the best wind tunnels can measure C_D down to 1 count;

most can only get to about 5 counts. Outside of the wind tunnel, an accuracy of 1 count is the target in computational fluid dynamics.

39. DRAG ON SPHERES AND DISKS

The drag coefficient varies linearly with the Reynolds number for laminar flow around a sphere or disk. In laminar flow, the drag is almost entirely due to skin friction. For Reynolds numbers below approximately 0.4, experiments have shown that the drag coefficient can be calculated from Eq. 17.183.[25] In calculating the Reynolds number, the sphere or disk diameter should be used as the characteristic dimension.

$$C_D = \frac{24}{\text{Re}} \qquad \textit{17.183}$$

Substituting this value of C_D into Eq. 17.182 results in *Stokes' law* (see Eq. 17.184), which is applicable to laminar slow motion (ascent or descent) of spherical particles and bubbles through a fluid. Stokes' law is based on the assumptions that (a) flow is laminar, (b) Newton's law of viscosity is valid, and (c) all higher-order velocity terms (v^2, etc.) are negligible.

$$F_D = 6\pi\mu \text{v} R = 3\pi\mu \text{v} D \quad \text{[laminar]} \qquad \textit{17.184}$$

The drag coefficients for disks and spheres operating outside the region covered by Stokes' law have been determined experimentally. In the turbulent region, pressure drag is predominant. Figure 17.36 can be used to obtain approximate values for C_D for spheres and flat disks.

Figure 17.36 shows that there is a dramatic drop in the drag coefficient around $\text{Re} = 10^5$. The explanation for this is that the point of separation of the boundary layer shifts, decreasing the width of the wake. (See Fig. 17.37.) Since the drag force is primarily pressure drag at higher Reynolds numbers, a reduction in the wake reduces the pressure drag. Therefore, anything that can be done to a sphere (scuffing or wetting a baseball, dimpling a golf ball, etc.) to induce a smaller wake will reduce the drag. There can be no shift in the boundary layer separation point for a thin disk, since the disk has no depth in the direction of flow. Therefore, the drag coefficient for a thin disk remains the same at all turbulent Reynolds numbers.

Figure 17.36 *Drag Coefficients for Spheres and Circular Flat Disks*

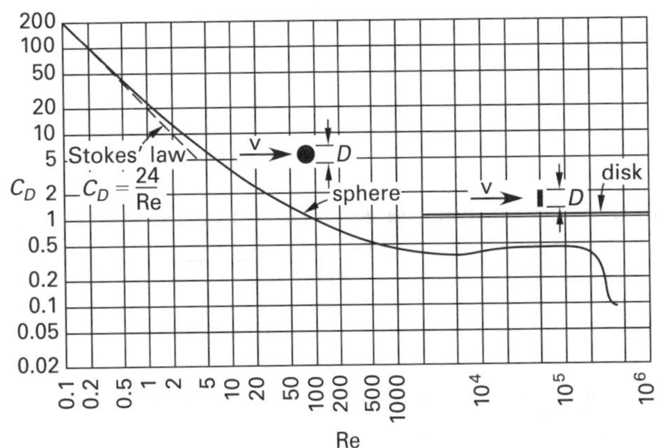

Binder, *Fluid Mechanics*, 5th, © 1973. Printed and electronically reproduced by permission of Pearson Education, Inc., New York, New York.

Figure 17.37 *Turbulent Flow Around a Sphere at Various Reynolds Numbers*

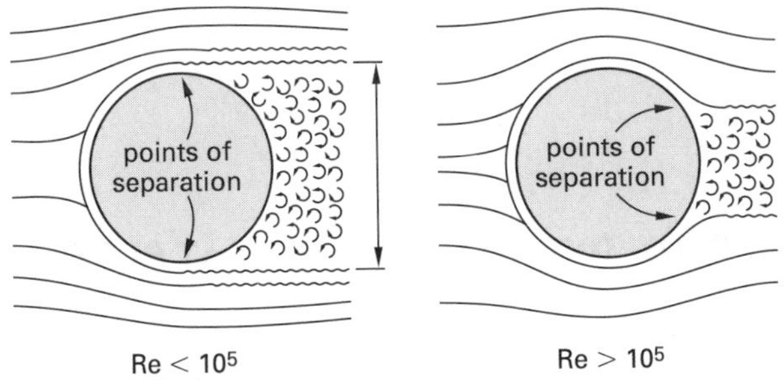

40. TERMINAL VELOCITY

The velocity of an object falling through a fluid will continue to increase until the drag force equals the net downward force (i.e., the weight less the buoyant force). The maximum velocity attained is known as the *terminal velocity* (*settling velocity*).

$$F_D = mg - F_b \quad \begin{bmatrix}\text{at terminal}\\ \text{velocity}\end{bmatrix} \quad \text{[SI]} \qquad \textit{17.185(a)}$$

$$F_D = \frac{mg}{g_c} - F_b \quad \begin{bmatrix}\text{at terminal}\\ \text{velocity}\end{bmatrix} \quad \text{[U.S.]} \qquad \textit{17.185(b)}$$

If the drag coefficient is known, the terminal velocity can be calculated from Eq. 17.186. For small, heavy objects falling in air, the buoyant force (represented by the ρ_{fluid} term) can be neglected.[26]

[25] Some sources report that the region in which Stokes' law applies extends to Re = 1.0.

[26] A skydiver's terminal velocity can vary from approximately 120 mi/hr (54 m/s) in horizontal configuration to 200 mi/hr (90 m/s) in vertical configuration.

Fluids

$$\text{v}_{\text{terminal}} = \sqrt{\frac{2(mg - F_b)}{C_D A \rho_{\text{fluid}}}} = \sqrt{\frac{2Vg(\rho_{\text{object}} - \rho_{\text{fluid}})}{C_D A \rho_{\text{fluid}}}} \qquad 17.186$$

For a sphere of diameter D, the terminal velocity is

$$\text{v}_{\text{terminal,sphere}} = \sqrt{\frac{4Dg(\rho_{\text{sphere}} - \rho_{\text{fluid}})}{3C_D \rho_{\text{fluid}}}} \qquad 17.187$$

If the spherical particle is very small, Stokes' law may apply. In that case, the terminal velocity can be calculated from Eq. 17.188.

$$\text{v}_{\text{terminal}} = \frac{D^2 g(\rho_{\text{sphere}} - \rho_{\text{fluid}})}{18\mu} \quad \text{[laminar]} \quad \text{[SI]} \qquad 17.188(a)$$

$$\text{v}_{\text{terminal}} = \frac{D^2(\rho_{\text{sphere}} - \rho_{\text{fluid}})}{18\mu} \times \frac{g}{g_c} \quad \text{[laminar]} \quad \text{[U.S.]} \qquad 17.188(b)$$

41. FLOW OVER A PARALLEL FLAT PLATE

The drag experienced by a flat plate oriented parallel to the direction of flow is almost totally skin friction drag. (See Fig. 17.38.) Prandtl's *boundary layer theory* can be used to evaluate the frictional effects. Such an analysis predicts that the shape of the boundary layer profile is a function of the Reynolds number. The characteristic dimension used in calculating the Reynolds number is the chord length, L (i.e., the dimension of the plate parallel to the flow).

Figure 17.38 *Flow Over a Parallel Flat Plate (one side)*

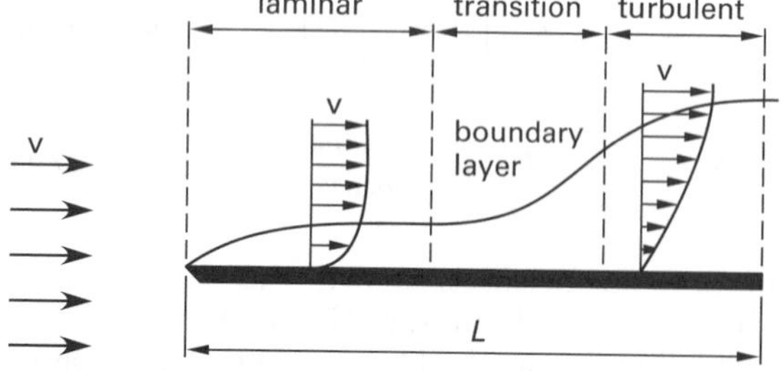

For laminar flow over a smooth, flat plate, the drag coefficient based on the boundary layer theory is given by Eq. 17.189, which is known as the *Blasius solution*.[27] The critical Reynolds number for laminar flow over a flat plate is often reported to be 530,000.

Drag Force

$$C_D = \frac{1.33}{\text{Re}^{0.5}} \quad [\text{laminar flow, } 10^4 < \text{Re} < 5 \times 10^5] \qquad 17.189$$

The drag coefficient for turbulent flow is

$$C_D = \frac{0.031}{\text{Re}^{1/7}} \quad [\text{laminar flow, } 10^6 < \text{Re} < 10^9] \qquad 17.190$$

The drag force is calculated from Eq. 17.191. The expression for drag force is multiplied by two because there is friction on both sides of the flat plate. The area, A, is for one side only.

Drag Force

$$F_D = 2\left(\frac{C_D \rho \text{v}^2 A}{2}\right) = C_D \rho \text{v}^2 A \quad \text{[SI]} \qquad 17.191(a)$$

$$F_D = 2\left(\frac{C_D \rho \text{v}^2 A}{2g_c}\right) = \frac{C_D \rho \text{v}^2 A}{g_c} \quad \text{[U.S.]} \qquad 17.191(b)$$

42. NOMENCLATURE

A	area	ft²	m²
B	magnetic field strength	T	T
c	speed of sound	ft/sec	m/s
c_P	Poisson's effect factor	–	–
C	coefficient or constant	–	–
C	Hazen-Williams coefficient	–	–
D	diameter	ft	m
E	modulus of elasticity	lbf/ft²	Pa
E	specific energy	ft-lbf/lbm	J/kg
f	Darcy friction factor	–	–
F	force	lbf	N
g	gravitational acceleration, 32.17 (9.807)	ft/sec²	m/s²
g_c	gravitational constant, 32.17	lbm-ft/lbf-sec²	n.a.
G	mass flow rate per unitarea	lbm/ft²-sec	kg/m²·s
h	height, head, or head loss	ft	m
I	impulse	lbf-sec	N·s
k	magmeter instrument constant	–	–
k	ratio of specific heats	–	–
K	coefficient	–	–
K	minor loss coefficient	–	–
L	length	ft	m
m	mass	lbm	kg

[27]Other correlations substitute the coefficient 1.44, the original value calculated by Prandtl, for 1.328 in Eq. 17.189. The Blasius solution is considered to be more accurate.

$\dot{m}$	mass flow rate	lbm/sec	kg/s
M	molecular weight	lbm/lbmol	kg/kmol
n	rotational speed	rev/min	rev/min
p	pressure	lbf/ft^2	Pa
p_v	velocity pressure	in wg	cm wg
P	momentum	lbm-ft/sec	kg·m/s
P	power	ft-lbf/sec	W
P	wetted perimeter	ft	m
Q	flow rate	gal/min	n.a.
r	radius	ft	m
R	resultant force	lbf	N
$\bar{R}$	universal gas constant, 1545.35 (8314.47)	ft-lbf/lbmol-°R	J/kmol·K
R_H	hydraulic radius	ft	m
Re	Reynolds number	–	–
SG	specific gravity	–	–
t	time	sec	s
T	absolute temperature	°R	K
v	specific volume	ft^3/lbm	m^3/kg
v	velocity	ft/sec	m/s
V	induced voltage	V	V
y	y-coordinate of position	ft	m
Y	expansion factor	–	–
z	elevation	ft	m

Symbols

α	angle	deg	deg
β	diameter ratio	–	–
γ	specific weight	lbf/ft^3	N/m^3
δ	flow rate correction	ft^3/sec	m^3/s
ϵ	specific roughness	ft	m
μ	absolute viscosity	lbf-sec/ft^2	Pa·s
ν	kinematic viscosity	ft^2/sec	m^2/s
ρ	density	lbm/ft^3	kg/m^3
ϕ	angle	deg	deg
ω	angular velocity	rad/sec	rad/s

Subscripts

ave	average
A	added (by pump)
b	blade or buoyant
c	contraction
d	discharge
D	drag
eff	effective
eq	equivalent
E	extracted (by turbine)
f	flow or friction
h	hammer
H	hydraulic
i	inside or inlet
I	instrument
L	lift
m	manometer fluid or minor
max	maximum
o	orifice, outlet, or outside
p	pressure
s	static
t	tank, theoretical, or total
v	valve flow
v	velocity
va	velocity of approach
x	x-component
z	potential

Fluids

18 Hydraulic Machines and Fluid Distribution

Content in blue refers to the *NCEES Handbook*.

NCEES EXAM SPECIFICATIONS AND RELATED CONTENT

HVAC AND REFRIGERATION EXAM

I.E. Fluid Mechanics

17. Net Positive Suction Head

MACHINE DESIGN AND MATERIALS EXAM

II.A.12. Mechanical Components: Hydraulic and pneumatic components

9. Terminology of Hydraulic Machines and Pipe Networks
10. Pumping Power
11. Pumping Efficiency
17. Net Positive Suction Head
26. Affinity Laws

THERMAL AND FLUID SYSTEMS EXAM

II.A.1 Hydraulic and Fluid Equipment: Pumps and fans

9. Terminology of Hydraulic Machines and Pipe Networks
17. Net Positive Suction Head

1. HYDRAULIC MACHINES

Pumps and turbines are the two basic types of hydraulic machines discussed in this chapter. Pumps convert mechanical energy into fluid energy, increasing the energy possessed by the fluid. Turbines convert fluid energy into mechanical energy, extracting energy from the fluid.

2. TYPES OF PUMPS

Pumps can be classified according to how energy is transferred to the fluid: intermittently or continuously.

Positive displacement pumps (*PD pumps*) transfer energy to the fluid continuously. The most common types are *reciprocating action pumps* (which use pistons, plungers, diaphragms, or bellows) and *rotary action pumps* (which use vanes, screws, lobes, or progressing cavities). Such pumps discharge a fixed volume for each stroke or revolution.

Kinetic pumps transfer energy to the fluid continuously by transforming fluid kinetic energy into fluid static pressure energy. The pump imparts the kinetic energy; the pump mechanism or housing is constructed in a manner that causes the transformation. *Jet pumps* and *ejector pumps* fall into the kinetic pump category, but centrifugal pumps are the primary examples.

In the operation of a *centrifugal pump*, liquid flowing into the *suction side* (the *inlet*) is captured by the *impeller* and thrown to the outside of the pump casing. Within the casing, the velocity imparted to the fluid by the impeller is converted into pressure energy. The fluid leaves the pump through the *discharge line* (the *exit*). It is a characteristic of most centrifugal pumps that the fluid is turned approximately 90° from the original flow direction. (See Table 18.1.)

***Table 18.1** Generalized Characteristics of Positive Displacement and Kinetic Pumps*

characteristic	positive displacement pumps	kinetic pumps
energy transfer	continuous	intermittent
flow rate	low	high
pressure rise per stage	high	low
constant quantity over operating range	flow rate	pressure rise
self-priming	yes	no
discharge stream	pulsing	steady
works with high viscosity fluids	yes	no

3. RECIPROCATING POSITIVE DISPLACEMENT PUMPS

Reciprocating positive displacement (PD) pumps can be used with all fluids, and are useful with viscous fluids and slurries (up to about 8000 Saybolt seconds universal (SSU), when the fluid is sensitive to shear, and when a high discharge pressure is required.[1] By entrapping a volume of fluid in the cylinder, reciprocating pumps provide a fixed-displacement volume per cycle. They are self-priming and inherently leak-free. Within the pressure limits of the line and pressure relief valve and the current capacity of the motor circuit, reciprocating pumps can provide an infinite discharge pressure.[2]

There are three main types of reciprocating pumps: power, direct-acting, and diaphragm. A *power pump* is a *cylinder-operated pump*. It can be single-acting or double-acting. A *single-acting pump* discharges liquid (or takes suction) only on one side of the piston, and there is only one transfer operation per crankshaft revolution. A *double-acting pump* discharges from both sides, and there are two transfers per revolution of the crank.

Traditional reciprocating pumps with pistons and rods can be either single-acting or double-acting and are suitable up to approximately 2000 psi (14 MPa). *Plunger pumps* are only single-acting and are suitable up to approximately 10,000 psi (70 MPa).

Simplex pumps have one cylinder, *duplex pumps* have two cylinders, *triplex pumps* have three cylinders, and so forth. *Direct-acting pumps* (sometimes referred to as *steam pumps*) are always double-acting. They use steam, unburned fuel gas, or compressed air as a motive fluid.

PD pumps are limited by both their NPSH_R characteristics, acceleration head, and (for rotary pumps) slip.[3] Because the flow is unsteady, a certain amount of energy, the *acceleration head*, h_{ac}, is needed to accelerate the fluid flow each stroke or cycle. If the acceleration head needed is too large, the NPSH_R requirements may not be attainable. Acceleration head can be reduced by increasing the pipe diameter, shortening the suction piping, decreasing the pump speed, or placing a *pulsation damper* (*stabilizer*) in the suction line.[4]

Generally, friction losses with pulsating flows are calculated based on the maximum velocity attained by the fluid. Since this is difficult to determine, the maximum velocity can be approximated by multiplying the average velocity (calculated from the rated capacity) by the factors in Table 18.2.

***Table 18.2** Typical v_{max}/v_{ave} Velocity Ratios[a,b]*

pump type	single-acting	double-acting
simplex	3.2	2.0
duplex	1.6	1.3
triplex	1.1	1.1
quadriplex	1.1	1.1
quintuplex and up	1.05	1.05

[a]Without stabilization. With properly sized stabilizers, use 1.05–1.1 for all cases.
[b]Multiply the values by 1.3 for metering pumps where lost fluid motion is relied on for capacity control.

When the suction line is "short," the acceleration head can be calculated from the length of the suction line, the average velocity in the line, and the rotational speed.[5]

[1]For viscosities of SSU greater than 240, multiply SSU viscosity by 0.216 to get viscosity in centistokes.
[2]For this reason, a relief valve should be included in every installation of positive displacement pumps. Rotary pumps typically have integral relief valves, but external relief valves are often installed to provide easier adjusting, cleaning, and inspection.
[3]Manufacturers of PD pumps prefer the term *net positive inlet pressure* (NPIP) to NPSH. NPIPA corresponds to NPSH_A; NPIPR corresponds to NPSH_R. Pressure and head are related by $p = \gamma h$.
[4]Pulsation dampers are not needed with rotary-action PD pumps, as the discharge is essentially constant.
[5]With a properly designed pulsation damper, the effective length of the suction line is reduced to approximately 10 pipe diameters.

In Eq. 18.1, C and K are dimensionless factors. K represents the relative compressibility of the liquid. (Typical values are 1.4 for hot water; 1.5 for amine, glycol, and cold water; and 2.5 for hot oil.) Values of C are given in Table 18.3.

$$h_{\text{ac}} = \frac{C}{K}\left(\frac{L_{\text{suction}}\text{v}_{\text{ave}}N}{g}\right) \quad 18.1$$

*Table 18.3 Typical Acceleration Head C-Values**

pump type	single-acting	double-acting
simplex	0.4	0.2
duplex	0.2	0.115
triplex	0.066	0.066
quadriplex	0.040	0.040
quintuplex and up	0.028	0.028

*Typical values for common connecting rod lengths and crank radii.

4. ROTARY PUMPS

Rotary pumps are *positive displacement* (PD) pumps that move fluid by means of screws, progressing cavities, gears, lobes, or vanes turning within a fixed casing (the *stator*). Rotary pumps are useful for high viscosities (up to 4×10^6 SSU for screw pumps). The rotation creates a cavity of fixed volume near the pump input; atmospheric or external pressure forces the fluid into that cavity. Near the outlet, the cavity is collapsed, forcing the fluid out. Figure 18.1 illustrates the external circumferential piston rotary pump.

Figure 18.1 External Circumferential Piston Rotary Pump

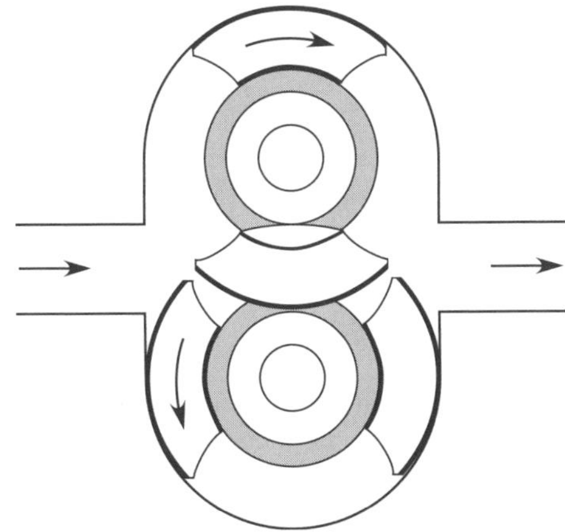

Discharge from rotary pumps is relatively smooth. Acceleration head is negligible. Pulsation dampers and suction stabilizers are not required.

Slip in rotary pumps is the amount (sometimes expressed as a percentage) of each rotational fluid volume that "leaks" back to the suction line on each revolution. Slip reduces pump capacity. It is a function of clearance, differential pressure, and viscosity. Slip is proportional to the third power of the clearance between the rotating element and the casing. Slip decreases with increases in viscosity; it increases linearly with increases in differential pressure. Slip is not affected by rotational speed. The *volumetric efficiency* is defined in terms of volumetric flow rate, Q, by Eq. 18.2. Figure 18.2 illustrates the relationship between flow rate, speed, slip, and differential pressure.

$$\eta_v = \frac{Q_{\text{actual}}}{Q_{\text{ideal}}} = \frac{Q_{\text{ideal}} - Q_{\text{slip}}}{Q_{\text{ideal}}} \quad 18.2$$

Figure 18.2 Slip in Rotary Pumps

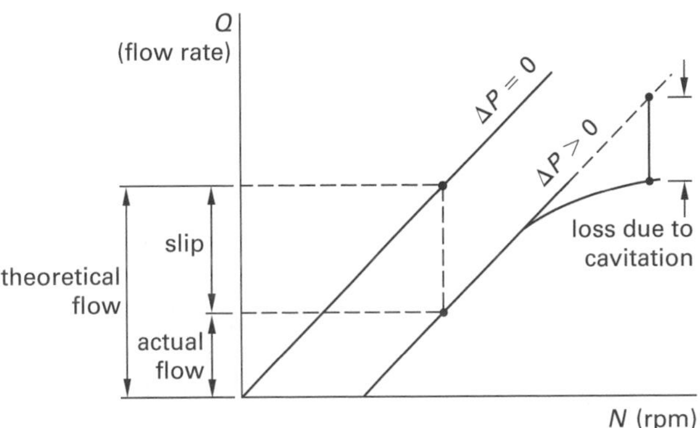

Except for screw pumps, rotary pumps are generally not used for handling abrasive fluids or materials with suspended solids.

5. DIAPHRAGM PUMPS

Hydraulically operated *diaphragm pumps* have a diaphragm that completely separates the pumped fluid from the rest of the pump. A reciprocating plunger pressurizes and moves a hydraulic fluid that, in turn, flexes the diaphragm. Single-ball check valves in the suction and discharge lines determine the direction of flow during both phases of the diaphragm action.

Metering is a common application of diaphragm pumps. Diaphragm metering pumps have no packing and are essentially leakproof. This makes them ideal when fugitive emissions are undesirable. Diaphragm pumps are suitable for pumping a wide range of materials, from liquefied gases to coal slurries, though the upper viscosity limit is approximately 3500 SSU. Within the limits of their reactivities, hazardous and reactive materials can also be handled.

Diaphragm pumps are limited by capacity, suction pressure, and discharge pressure and temperature. Because of their construction and size, most diaphragm pumps are limited to discharge pressures of 5000 psi (35 MPa) or less, and most high-capacity pumps are limited to 2000 psi (14 MPa). Suction pressures are similarly

Fluids

limited to 5000 psi (35 MPa). A pressure range of 3–9 psi (20–60 kPa) is often quoted as the minimum liquid-side pressure for metering applications.

The discharge is inherently pulsating, and dampers or stabilizers are often used. (The acceleration head term is required when calculating $NPSH_R$.) The discharge can be smoothed out somewhat by using two or three (i.e., duplex or triplex) plungers.

Diaphragms are commonly manufactured from stainless steel (type 316) and polytetrafluorethylene (PTFE) or other elastomers. PTFE diaphragms are suitable in the range of −50°F to 300°F (−45°C to 150°C) while metal diaphragms (and some ketone resin diaphragms) are used up to approximately 400°F (200°C) with life expectancy being reduced at higher temperatures. Although most diaphragm pumps usually operate below 200 spm (strokes per minute), diaphragm life will be improved by limiting the maximum speed to 100 spm.

6. CENTRIFUGAL PUMPS

Centrifugal pumps and their impellers can be classified according to the way energy is imparted to the fluid. Each category of pump is suitable for a different application and (specific) speed range. (See Table 18.9 later in this chapter.) Figure 18.3 illustrates a typical centrifugal pump and its schematic symbol.

Figure 18.3 *Centrifugal Pump and Symbol*

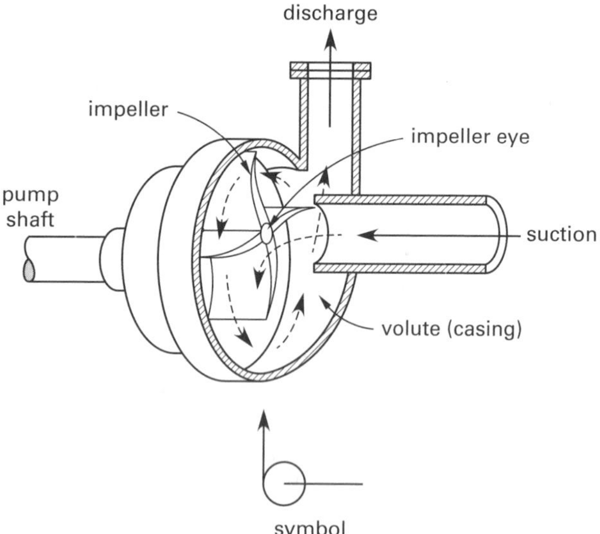

Radial-flow impellers impart energy primarily by centrifugal force. Fluid enters the impeller at the hub and flows radially to the outside of the casing. Radial-flow pumps are suitable for adding high pressure at low fluid flow rates. *Axial-flow impellers* impart energy to the fluid by acting as compressors. Fluid enters and exits along the axis of rotation. Axial-flow pumps are suitable for adding low pressures at high fluid flow rates.[6]

Radial-flow pumps can be designed for either single- or double-suction operation. In a *single-suction pump*, fluid enters from only one side of the impeller. In a *double-suction pump*, fluid enters from both sides of the impeller.[7] (That is, the impeller is two-sided.) Operation is similar to having two single-suction pumps in parallel. (See Fig. 18.4.)

Figure 18.4 *Radial- and Axial-Flow Impellers*

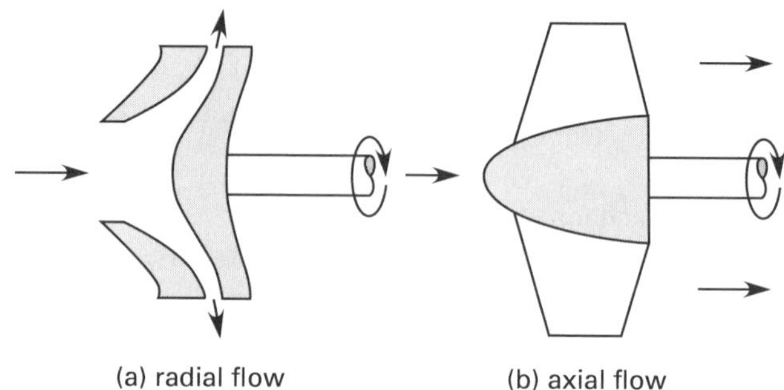

A *multiple-stage pump* consists of two or more impellers within a single casing. The discharge of one stage feeds the input of the next stage, and operation is similar to having several pumps in series. In this manner, higher heads are achieved than would be possible with a single impeller.

The *pitch* (*circular blade pitch*) is the impeller's circumference divided by the number of impeller vanes. The impeller *tip speed* is calculated from the impeller diameter and rotational speed. The impeller "tip speed" is actually the tangential velocity at the periphery. Tip speed is typically somewhat less than 1000 ft/sec (300 m/s).

$$\mathrm{v}_{\mathrm{tip}} = \frac{\pi D N}{60\,\frac{\mathrm{sec}}{\mathrm{min}}} = \frac{D\omega}{2} \quad 18.3$$

7. SEWAGE PUMPS

The primary consideration in choosing a pump for sewage and large solids is resistance to clogging. Centrifugal pumps should always be the single-suction type with nonclog, open impellers. (Double-suction pumps are prone to clogging because rags catch and wrap around the shaft extending through the impeller eye.) Clogging can be further minimized by limiting the number of impeller blades to two or three, providing for large passageways, and using a bar screen ahead of the pump.

[6]There is a third category of centrifugal pumps known as *mixed flow pumps*. Mixed flow pumps have operational characteristics between those of radial flow and axial flow pumps.

[7]The double-suction pump can handle a greater fluid flow rate than a single-suction pump with the same specific speed. Also, the double-suction pump will have a lower $NPSH_R$.

Though made of heavy construction, nonclog pumps are constructed for ease of cleaning and repair. Horizontal pumps usually have a split casing, half of which can be removed for maintenance. A hand-sized cleanout opening may also be built into the casing. A sewage pump should normally be used with a grit chamber for prolonged bearing life.

The solids-handling capacity of a pump may be specified in terms of the largest sphere that can pass through it without clogging, usually about 80% of the inlet diameter. For example, a wastewater pump with a 6 in (150 mm) inlet should be able to pass a 4 in (100 mm) sphere. The pump must be capable of handling spheres with diameters slightly larger than the bar screen spacing.

Figure 18.5 shows a simplified wastewater pump installation. Not shown are instrumentation and water level measurement devices, baffles, lighting, drains for the dry well, electrical power, pump lubrication equipment, and access ports. (Totally submerged pumps do not require dry wells. However, such pumps without dry wells are more difficult to access, service, and repair.)

Figure 18.5 *Typical Wastewater Pump Installation (greatly simplified)*

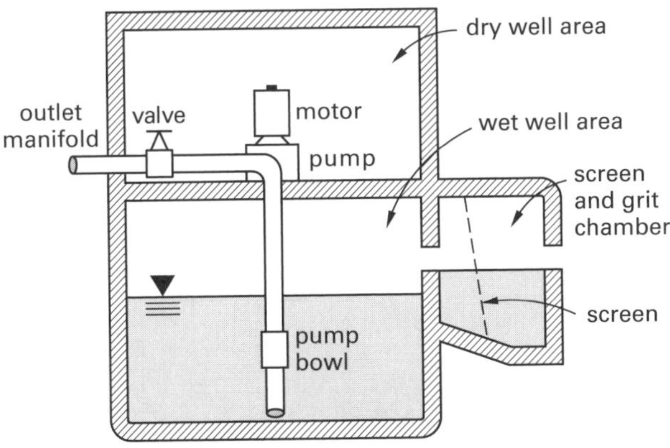

The multiplicity and redundancy of pumping equipment is not apparent from Fig. 18.5. The number of pumps used in a wastewater installation largely depends on the expected demand, pump capacity, and design criteria for backup operation. It is good practice to install pumps in sets of two, with a third backup pump being available for each set of pumps that performs the same function. The number of pumps and their capacities should be able to handle the peak flow when one pump in the set is out of service.

8. SLUDGE PUMPS AND GRAVITY FLOW

Centrifugal and reciprocating pumps are extensively used for pumping sludge. Progressive cavity screw impeller pumps are also used.

As further described in Sec. 18.28, the pumping power is proportional to the specific gravity. Accordingly, pumping power for dilute and well-digested sludges is typically only 10–25% higher than for water. However, most sludges are non-Newtonian fluids, often flow in a laminar mode, and have characteristics that may change with the season. Also, sludge characteristics change greatly during the pumping cycle; engineering judgment and rules of thumb are often important in choosing sludge pumps. For example, a general rule is to choose sludge pumps capable of developing at least 50–100% excess head.

One method of determining the required pumping power is to multiply the power required for pumping pure water by a service factor. Historical data is the best method of selecting this factor. Choice of initial values is a matter of judgment. Guidelines are listed in Table 18.4.

Table 18.4 *Typical Pumping Power Multiplicative Factors*

solids concentration	digested sludge	untreated, primary, and concentrated sludge
0%	1.0	1.0
2%	1.2	1.4
4%	1.3	2.5
6%	1.7	4.1
8%	2.2	7.0
10%	3.0	10.0

Derived from *Wastewater Engineering: Treatment, Disposal, Reuse*, 3rd ed., by Metcalf & Eddy, et al., © 1991, with permission from The McGraw-Hill Companies, Inc.

Generally, sludge will thin out during a pumping cycle. The most dense sludge components will be pumped first, with more watery sludge appearing at the end of the pumping cycle. With a constant power input, the reduction in load at the end of pumping cycles may cause centrifugal pumps to operate far from the desired operating point and to experience overload failures. The operating point should be evaluated with high-, medium-, and low-density sludges.

To avoid cavitation, sludge pumps should always be under a positive suction head of at least 4 ft (1.2 m), and suction lifts should be avoided. The minimum diameters of suction and discharge lines for pumped sludge are typically 6 in (150 mm) and 4 in (100 mm), respectively.

Not all sludge is moved by pump action. Some installations rely on gravity flow to move sludge. The minimum diameter of sludge gravity transfer lines is typically 8 in (200 mm), and the recommended minimum slope is 3%.

To avoid clogging due to settling, sludge velocity should be above the transition from laminar to turbulent flow, known as the *critical velocity*. The critical velocity for most sludges is approximately 3.5 ft/sec (1.1 m/s). Velocity ranges of 5–8 ft/sec (1.5–2.4 m/s) are commonly quoted and are adequate.

9. TERMINOLOGY OF HYDRAULIC MACHINES AND PIPE NETWORKS

A pump has an inlet (designated the *suction*) and an outlet (designated the *discharge*). The subscripts *s* and *d* refer to the inlet and outlet of the pump, not of the pipeline.

All of the terms that are discussed in this section are *head* terms and, as such, have units of length. When working with hydraulic machines, it is common to hear such phrases as "a pressure head of 50 feet" and "a static discharge head of 15 meters." The term *head* is often substituted for pressure or pressure drop. Any head term (*pressure head, atmospheric head, vapor pressure head,* etc.) can be calculated from pressure by using Eq. 18.4.[8]

$$h = \frac{p}{g\rho} \quad \text{[SI]} \qquad 18.4(a)$$

$$h = \frac{p}{\rho} \times \frac{g_c}{g} = \frac{p}{\gamma} \quad \text{[U.S.]} \qquad 18.4(b)$$

Some of the terms used in the description of pipe networks appear to be similar (e.g., suction head and total suction head). The following general rules will help to clarify the meanings.

Rule 1: The word *suction* or *discharge* limits the quantity to the suction line or discharge line, respectively. The absence of either word implies that both the suction and discharge lines are included. Example: discharge head.

Rule 2: The word *static* means that static head only is included (not velocity head, friction head, etc.). Example: static suction head.

Rule 3: The word *total* means that static head, velocity head, and friction head are all included. (Total does not mean the combination of suction and discharge.) Example: total suction head.

The following terms are commonly encountered.

- *friction head,* h_f: The head needed to overcome resistance to flow in the pipes, fittings, valves, entrances, and exits.

Head Loss Due to Flow: Darcy-Weisbach Equation

$$h_f = f\frac{L}{D}\frac{\mathrm{v}^2}{2g} \qquad 18.5$$

- *velocity head,* h_v: The specific kinetic energy of the fluid. Also known as *dynamic head.*[9]

Minor Losses in Pipe Fittings, Contractions, and Expansions

$$h_\mathrm{v} = \frac{\mathrm{v}^2}{2g} \qquad 18.6$$

- *static suction head,* h_s: The vertical distance above the centerline of the pump inlet to the free level of the fluid source. If the free level of the fluid is below the pump inlet, h_s will be negative and is known as *static suction lift.* (See Fig. 18.6.)

Figure 18.6 *Static Suction Lift*

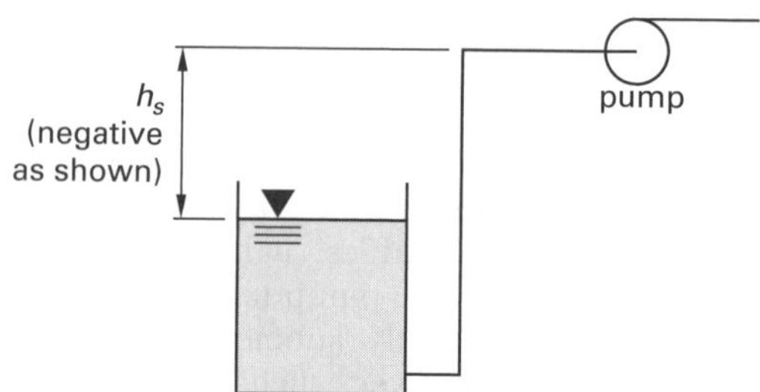

- *static discharge head,* h_d: The vertical distance above the centerline of the pump inlet to the point of free discharge or surface level of the discharge tank. (See Fig. 18.7.)

Figure 18.7 *Static Discharge Head*

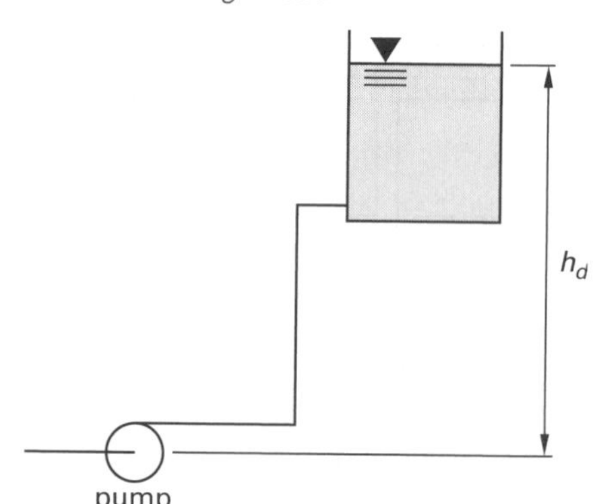

The ambiguous term *effective head* is not commonly used when discussing hydraulic machines, but when used, the term most closely means *net head* (i.e., starting head less losses). Consider a hydroelectric turbine that is fed by water with a static head of *H*. After frictional and other losses, the net head available to the turbine will be less than *H*. The turbine output will coincide with an ideal turbine being acted upon by the net or effective head. Similarly, the actual increase in pressure across a pump will be the effective head added (i.e., the head net of internal losses and geometric effects).

[8]Equation 18.4 can be used to define *pressure head, atmospheric head,* and *vapor pressure head,* whose meanings and derivations should be obvious.
[9]The term *dynamic* is not as consistently applied as are the terms described in the rules. In particular, it is not clear whether dynamic head includes friction head.

Example 18.1

Write the symbolic equations for the following terms: (a) the total suction head, (b) the total discharge head, and (c) the total head added.

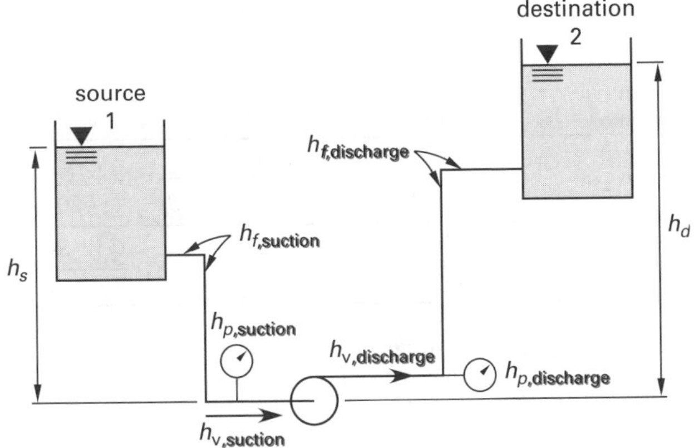

Solution

(a) The *total suction head* at the pump inlet is the sum of static (pressure) head and velocity head at the pump suction.

$$h_{t,\text{suction}} = h_{p,\text{suction}} + h_{v,\text{suction}}$$

Total suction head can also be calculated from the conditions existing at the source (1), in which case suction line friction would also be considered. (With an open reservoir, $h_{p(1)}$ will be zero if gage pressures are used, and $h_{v(1)}$ will be zero if the source is large.)

$$h_{t,\text{suction}} = h_{p(1)} + h_s + h_{v(1)} - h_{f,\text{suction}}$$

(b) The *total discharge head* at the pump outlet is the sum of the static (pressure) and velocity heads at the pump outlet. Friction head is zero since the fluid has not yet traveled through any length of pipe when it is discharged.

$$h_{t,\text{discharge}} = h_{p,\text{discharge}} + h_{v,\text{discharge}}$$

The total discharge head can also be evaluated at the destination (2) if the friction head, h_f, between the discharge and the destination is known. (With an open reservoir, $h_{p(2)}$ will be zero if gage pressures are used, and $h_{v(2)}$ will be zero if the destination is large.)

$$h_{t,\text{discharge}} = h_{p(2)} + h_d + h_{v(2)} + h_{f,\text{discharge}}$$

(c) The *total head added* by the pump is the total discharge head less the total suction head.

$$\Delta h = h_{t,\text{discharge}} - h_{t,\text{suction}}$$

Assuming suction from and discharge to reservoirs exposed to the atmosphere and assuming negligible reservoir velocities,

$$\Delta h \approx h_d - h_s + h_{f,\text{discharge}} + h_{f,\text{suction}}$$

10. PUMPING POWER

The energy (head) added by a pump can be determined from the difference in total energy on either side of the pump. Writing the Bernoulli equation for the discharge and suction conditions produces Eq. 18.7, an equation for the *total dynamic head*, often abbreviated TDH.

$$\Delta h = h_{t,\text{discharge}} - h_{t,\text{suction}} \qquad 18.7$$

$$\Delta h = \frac{p_d - p_s}{\rho g} + \frac{\mathrm{v}_d^2 - \mathrm{v}_s^2}{2g} + z_d - z_s \quad \text{[SI]} \qquad 18.8(a)$$

$$\Delta h = \frac{(p_d - p_s)g_c}{\rho g} + \frac{\mathrm{v}_d^2 - \mathrm{v}_s^2}{2g} + z_d - z_s \quad \text{[U.S.]} \qquad 18.8(b)$$

In most applications, the change in velocity and potential heads is either zero or small in comparison to the increase in pressure head. Equation 18.8 then reduces to Eq. 18.9.

$$\Delta h = \frac{p_d - p_s}{\rho g} \quad \text{[SI]} \qquad 18.9(a)$$

$$\Delta h = \frac{p_d - p_s}{\rho} \times \frac{g_c}{g} \quad \text{[U.S.]} \qquad 18.9(b)$$

It is important to recognize that the variables in Eq. 18.8 and Eq. 18.9 refer to the conditions at the pump's immediate inlet and discharge, not to the distant ends of the suction and discharge lines. However, the total dynamic head added by a pump can be calculated in another way. For example, for a pump raising water from one open reservoir to another, the total dynamic head would consider the total elevation rise, the velocity head (often negligible), and the friction losses in the suction and discharge lines.

The head added by the pump can also be calculated from the impeller and fluid speeds. Equation 18.10 is useful for radial- and mixed-flow pumps for which the incoming fluid has little or no rotational velocity component (i.e., up to a specific speed of approximately 2000 U.S. or 40 SI). In Eq. 18.10, $\mathrm{v}_{\text{impeller}}$ is the tangential impeller velocity at the radius being considered, and $\mathrm{v}_{\text{fluid}}$ is the average tangential velocity imparted to the fluid by the impeller. The impeller efficiency, η_{impeller}, is typically 0.85–0.95. This is much higher than the total pump efficiency because it does not include mechanical and fluid friction losses. (See Sec. 18.11.)

$$\Delta h = \frac{\eta_{\text{impeller}} v_{\text{impeller}} v_{\text{fluid}}}{g} \qquad 18.10$$

Head added can be thought of as the energy added per unit mass. The total pumping power depends on the head added, Δh, and the mass flow rate. For example, the product $\dot{m}\Delta h$ has the units of foot-pounds per second (in customary U.S. units), which can be easily converted to horsepower. Pump output power is known as *hydraulic power* or *water power*. Hydraulic power is the net power actually transferred to the fluid.

Horsepower is a common unit of power, which results in the terms *water horsepower* and *hydraulic horsepower*, whp, being used to designate the power that is transferred into the fluid. (This is also called *fluid power*.) Water horsepower can be calculated using the formula

Pump Power Equation

$$\text{whp} = \frac{\dot{m}\Delta h}{33{,}000} \qquad 18.11$$

In Eq. 18.11, $\dot{m}$ must be in lbm/min. 33,000 is the conversion factor changing ft-lbm/min to horsepower.

When flow is known in gallons per minute, water horsepower is

Pump Power Equation

$$\text{whp} = \frac{Q\Delta h}{3960} \qquad 18.12$$

In Eq. 18.12, 3960 converts ft-gal/min to horsepower.

When the added head is expressed as added pressure in psi, water horsepower is

Pump Power Equation

$$\text{whp} = \frac{Q\Delta p}{1714} \qquad 18.13$$

In Eq. 18.13, 1714 converts ft-lbf/in^2 to horsepower.

Various other relationships for finding the hydraulic horsepower are given in Table 18.5.

The unit of power in SI units is the watt (kilowatt). Table 18.6 can be used to determine *hydraulic kilowatts*, WkW.

***Table 18.5** Hydraulic Horsepower Equations*

	Q (gal/min)	$\dot{m}$ (lbm/min)	Q (ft^3/sec)
Δh in feet	$\frac{\Delta h Q(\text{SG})}{3956}$	$\frac{\dot{m}\Delta h}{550} \times \frac{g}{g_c}$	$\frac{Q\Delta h(\text{SG})}{8.814}$
Δp in psi	$\frac{Q\Delta p}{1714}$	$\frac{\dot{m}\Delta p}{238.3(\text{SG})} \times \frac{g}{g_c}$	$\frac{Q\Delta p}{3.819}$
Δp in psf[b]	$\frac{Q\Delta p}{2.468 \times 10^5}$	$\frac{\dot{m}\Delta p}{34{,}320(\text{SG})} \times \frac{g}{g_c}$	$\frac{Q\Delta p}{550}$
W in $\frac{\text{ft-lbf}}{\text{lbm}}$	$\frac{QW(\text{SG})}{3960}$	$\frac{\dot{m}W}{550}$	$\frac{QW(\text{SG})}{8.814}$

(Multiply horsepower by 0.7457 to obtain kilowatts.)
[a]based on $\rho_{\text{water}} = 62.4$ lbm/ft^3 and $g = 32.2$ ft/sec^2
[b]Velocity head changes must be included in Δp.

***Table 18.6** Hydraulic Kilowatt Equations[a]*

	Q L/s	$\dot{m}$ (kg/s)	Q (m^3/s)
Δh in meters	$\frac{9.81 Q\Delta h(\text{SG})}{1000}$	$\frac{9.81\dot{m}\Delta h}{1000}$	$9.81 Q\Delta h(\text{SG})$
Δp in kPa[b]	$\frac{Q\Delta p}{1000}$	$\frac{\dot{m}\Delta p}{1000(\text{SG})}$	$Q\Delta p$
W in $\frac{\text{J}}{\text{kg}}$[b]	$\frac{QW(\text{SG})}{1000}$	$\frac{\dot{m}W}{1000}$	$QW(\text{SG})$

(Multiply kilowatts by 1.341 to obtain horsepower.)
[a]based on $\rho_{\text{water}} = 1000$ kg/m^3 and $g = 9.81$ m/s^2
[b]Velocity head changes must be included in Δp.

Example 18.2

A pump adds 550 ft of pressure head to 100 lbm/sec of water ($\rho = 62.4$ lbm/ft^3 or 1000 kg/m^3). (a) Complete the following table of performance data. (b) What is the hydraulic power in horsepower and kilowatts?

item	customary U.S.	SI
$\dot{m}$	100 lbm/sec	_kg/s
h	550 ft	_m
Δp	_lbf/ft^2	_kPa
Q	_ft^3/sec	_m^3/s
W	_ft-lbf/lbm	_J/kg
P	_hp	_kW

Solution

(a) Work initially with the customary U.S. data.

$$\Delta p = \rho h \times \frac{g}{g_c} = \left(62.4\ \frac{\text{lbm}}{\text{ft}^3}\right)(550\ \text{ft}) \times \frac{g}{g_c}$$
$$= 34{,}320\ \text{lbf/ft}^2$$

$$Q = \frac{\dot{m}}{\rho} = \frac{100\ \frac{\text{lbm}}{\text{sec}}}{62.4\ \frac{\text{lbm}}{\text{ft}^3}} = 1.603\ \text{ft}^3/\text{sec}$$

$$W = h \times \frac{g}{g_c} = 550\ \text{ft} \times \frac{g}{g_c}$$
$$= 550\ \text{ft-lbf/lbm}$$

Convert to SI units.

$$\dot{m} = \frac{100\ \frac{\text{lbm}}{\text{sec}}}{2.201\ \frac{\text{lbm}}{\text{kg}}} = 45.43\ \text{kg/s}$$

$$h = \frac{550\ \text{ft}}{3.281\ \frac{\text{ft}}{\text{m}}} = 167.6\ \text{m}$$

$$\Delta p = \left(34{,}320\ \frac{\text{lbf}}{\text{ft}^2}\right)\left(\frac{1}{\left(12\ \frac{\text{in}}{\text{ft}}\right)^2}\right)\left(6.895\ \frac{\text{kPa}}{\frac{\text{lbf}}{\text{in}^2}}\right)$$
$$= 1643\ \text{kPa}$$

$$Q = \left(1.603\ \frac{\text{ft}^3}{\text{sec}}\right)\left(0.0283\ \frac{\text{m}^3}{\text{ft}^3}\right) = 0.0454\ \text{m}^3/\text{s}$$

$$W = \left(550\ \frac{\text{ft-lbf}}{\text{lbm}}\right)\left(1.356\ \frac{\text{J}}{\text{ft-lbf}}\right)\left(2.201\ \frac{\text{lbm}}{\text{kg}}\right)$$
$$= 1642\ \text{J/kg}$$

(b) From Table 18.5, the hydraulic horsepower is

$$\text{whp} = \frac{\dot{m}\Delta h}{550} \times \frac{g}{g_c} = \frac{\left(100\ \frac{\text{lbm}}{\text{sec}}\right)(550\ \text{ft})}{550\ \frac{\text{ft-lbf}}{\text{hp-sec}}} \times \frac{g}{g_c}$$
$$= 100\ \text{hp}$$

From Table 18.6, the power is

$$\text{WkW} = \frac{\dot{m}\Delta p}{1000(\text{SG})} = \frac{\left(45.43\ \frac{\text{kg}}{\text{s}}\right)(1643\ \text{kPa})}{\left(1000\ \frac{\text{W}}{\text{kW}}\right)(1.0)}$$
$$= 74.6\ \text{kW}$$

11. PUMPING EFFICIENCY

Fluid power (or water horsepower or hydraulic horsepower) is the net energy actually transferred to the fluid per unit time.

Centrifugal Pump Characteristics

$$P_{\text{fluid}} = \text{whp} = \rho g h Q \qquad 18.14$$

The input power delivered by the motor to the pump is known as the *brake pump power*. The term *brake horsepower*, bhp, is also often used to designate both the quantity and its units. Due to frictional losses between the fluid and the pump and mechanical losses in the pump itself, the brake pump power will be greater than the fluid power.

The ratio of fluid power to brake pump power is the pump efficiency, η_{pump}. Figure 18.8 gives typical pump efficiencies as a function of the pump's specific speed. The difference between the brake and hydraulic powers is known as the *friction power* (or *friction horsepower*, fhp).

Centrifugal Pump Characteristics

$$P_{\text{brake}} = \frac{P_{\text{fluid}}}{\eta_{\text{pump}}} = \frac{\rho g h Q}{\eta_{\text{pump}}} \qquad 18.15$$

$$\text{fhp} = P_{\text{brake}} - P_{\text{fluid}} \qquad 18.16$$

Figure 18.8 *Average Pump Efficiency Versus Specific Speed*

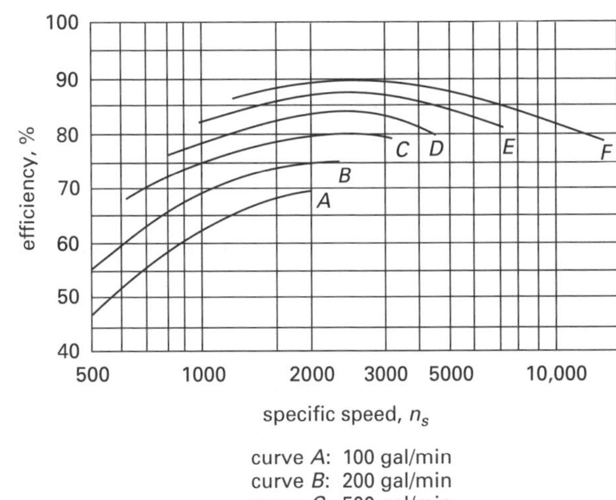

Pumping efficiency is not constant for any specific pump; rather, it depends on the operating point and the speed of the pump. (See Sec. 18.23.) A pump's characteristic efficiency curves will be published by its manufacturer.

With pump characteristic curves given by the manufacturer, the efficiency is not determined from the intersection of the system curve and the efficiency curve. (See Sec. 18.22.) Rather, the efficiency is a function of only the flow rate. Therefore, the operating efficiency is read from the efficiency curve directly above or below the operating point. (See Sec. 18.23 and Fig. 18.9.)

Figure 18.9 *Typical Centrifugal Pump Efficiency Curves*

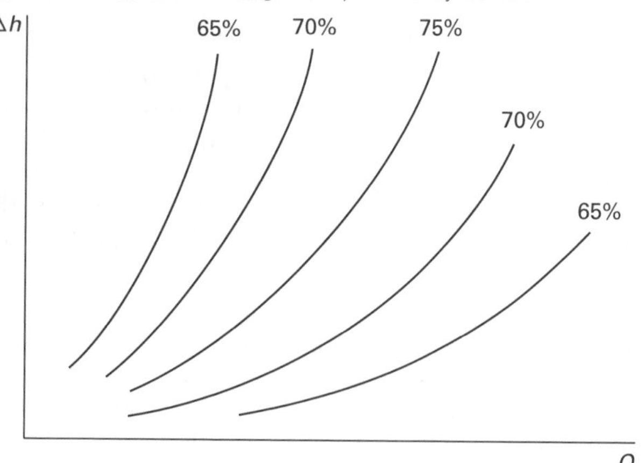

Efficiency curves published by a manufacturer will not include such losses in the suction elbow, discharge diffuser, couplings, bearing frame, seals, or pillow blocks. Up to 15% of the motor horsepower may be lost to these factors. Upon request, the manufacturer may provide the pump's installed *wire-to-water efficiency* (i.e., the fraction of the electrical power drawn that is converted to hydraulic power).

The pump must be driven by an engine or a motor.[10] The power delivered to the motor (the *purchased power*) is greater than the power delivered to the pump, as accounted for by the motor efficiency, η_{motor}. If the pump motor is electrical, its input power requirements will be stated in kilowatts.[11]

Centrifugal Pump Characteristics

$$P_{\text{purchased}} = \frac{P_{\text{brake}}}{\eta_{\text{motor}}} \qquad 18.17$$

$$P_{\text{purchased,kW}} = \frac{0.7457 P_{\text{brake}}}{\eta_{\text{motor}}} \qquad 18.18$$

The *total efficiency* of the pump installation is the product of the pump and motor efficiencies.

Pump Power Equation

$$\eta_t = \eta_{\text{pump}}\eta_{\text{motor}} = \frac{P_{\text{fluid}}}{P_{\text{purchased}}} \qquad 18.19$$

The pump power equation is

Pump Power Equation

$$P_{\text{purchased}} = \frac{Q\gamma h}{\eta_t} = \frac{Q\rho g h}{\eta_t} = \frac{P_{\text{fluid}}}{\eta_{\text{pump}}\eta_{\text{motor}}} \qquad 18.20$$

12. COST OF ELECTRICITY

The power utilization of pump motors is usually measured in kilowatts. The kilowatt usage represents the rate that energy is transferred by the pump motor. The total amount of work, W, done by the pump motor is found by multiplying the rate of energy usage (i.e., the delivered power, P), by the length of time the pump is in operation.

$$W = Pt \qquad 18.21$$

Although the units horsepower-hours are occasionally encountered, it is more common to measure electrical work in *kilowatt-hours* (kWh). Accordingly, the cost of electrical energy is stated per kWh (e.g., \$0.10 per kWh).

$$\text{cost} = \frac{W_{\text{kWh}} \times (\text{cost per kWh})}{\eta_{\text{motor}}} \qquad 18.22$$

13. STANDARD MOTOR SIZES AND SPEEDS

An effort should be made to specify standard motor sizes when selecting the source of pumping power. Table 18.7 lists NEMA (National Electrical Manufacturers Association) standard motor sizes by horsepower *nameplate rating*.[12] Other motor sizes may also be available by special order.

[10]The source of power is sometimes called the *prime mover*.
[11]A *watt* is a joule per second.
[12]The nameplate rating gets its name from the information stamped on the motor's identification plate. Besides the horsepower rating, other nameplate data used to classify the motor are the service class, voltage, full-load current, speed, number of phases, frequency of the current, and maximum ambient temperature (or, for older motors, the motor temperature rise).

Table 18.7 *NEMA Standard Motor Sizes (brake horsepower)*

1/8,*	1/6,*	1/4,*	1/3,*	1/2,*	3/4,*		
1,	1.5,	2,	3,	5,	7.5	10,	15
20,	25,	30,	40,	50,	60,	75,	100
125,	150,	200,	250,	300,	350,	400,	450
500,	600,	700,	800,	900,	1000,	1250,	1500
1750,	2000,	2250,	2500,	2750,	3000,	3500,	4000
4500,	5000,	6000,	7000,	8000			

*fractional horsepower series

For industrial-grade motors, the rated (nameplate) power is the power at the output shaft that the motor can produce continuously while operating at the nameplate ambient conditions and without exceeding a particular temperature rise dependent on its wiring type. However, if a particular motor is housed in a larger, open frame that has better cooling, it will be capable of producing more power than the nameplate rating. NEMA defines the *service factor*, SF, as the amount of continual overload capacity designed into a motor without reducing its useful life (provided voltage, frequency, and ambient temperature remain normal). Common motor service factors are 1.0, 1.15, and 1.25. When a motor develops more than the nameplate power, efficiency, power factor, and operating temperature will be affected adversely.

$$\text{SF} = \frac{P_{\text{max continuous}}}{P_{\text{rated}}} \qquad 18.23$$

Larger horsepower motors are usually three-phase induction motors. The *synchronous speed*, N in rpm, of such motors is the speed of the rotating field, which depends on the number of poles per stator phase and the frequency, f. The number of poles must be an even number. The frequency is typically 60 Hz, as in the United States, or 50 Hz, as in European countries. Table 18.8 lists common synchronous speeds.

$$N = \frac{120f}{\text{no. of poles}} \qquad 18.24$$

Table 18.8 *Common Synchronous Speeds*

number of poles	N (rpm) 60 Hz	50 Hz
2	3600	3000
4	1800	1500
6	1200	1000
8	900	750
10	720	600
12	600	500
14	514	428
18	400	333
24	300	250
48	150	125

Induction motors do not run at their synchronous speeds when loaded. Rather, they run at slightly less than synchronous speed. The deviation is known as the *slip*. Slip is typically around 4% and is seldom greater than 10% at full load for motors in the 1–75 hp range.

$$\text{slip (in rpm)} = \text{synchronous speed} - \text{actual speed} \qquad 18.25$$

$$\text{slip (in percent)} = \frac{\text{synchronous speed} - \text{actual speed}}{\text{synchronous speed}} \times 100\% \qquad 18.26$$

Induction motors may also be specified in terms of their kVA (kilovolt-amp) ratings. The kVA rating is not the same as the power in kilowatts, although one can be derived from the other if the motor's *power factor* is known. Such power factors typically range from 0.8 to 0.9, depending on the installation and motor size.

$$\text{kVA rating} = \frac{\text{motor power in kW}}{\text{power factor}} \qquad 18.27$$

Example 18.3

A pump driven by an electrical motor moves 25 gal/min of water from reservoir A to reservoir B, lifting the water a total of 245 ft. The efficiencies of the pump and motor are 64% and 84%, respectively. Electricity costs $0.08/kWh. Neglect velocity head, friction, and minor losses. (a) What minimum size motor is required? (b) How much does it cost to operate the pump for 6 hr?

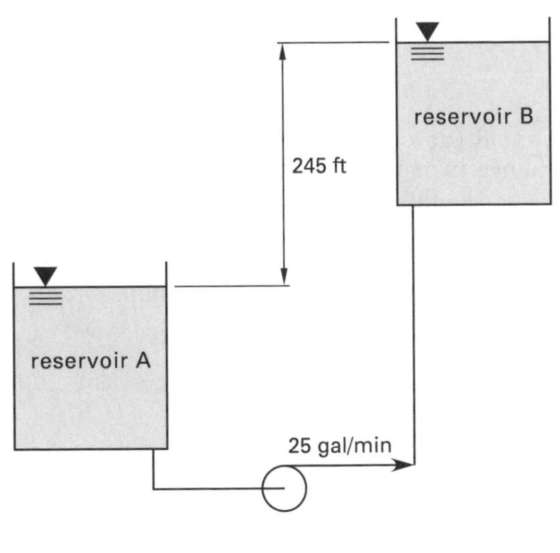

Fluids

Solution

(a) The head added is 245 ft. Using Table 18.5 and incorporating the pump efficiency, the motor power required is

$$P = \frac{hQ(\text{SG})}{3956\eta_p} = \frac{(245\ \text{ft})\left(25\ \frac{\text{gal}}{\text{min}}\right)(1.0)}{\left(3956\ \frac{\text{ft-gal}}{\text{hp-min}}\right)(0.64)}$$
$$= 2.42\ \text{hp}$$

From Table 18.7, select a 3 hp motor.

(b) The developed power, not the motor's rated power, is used. From Eq. 18.21 and Eq. 18.22,

$$\begin{aligned}\text{cost} &= (\text{cost per kWh})\left(\frac{Pt}{\eta_{\text{motor}}}\right)\\ &= \left(0.08\ \frac{\$}{\text{kWh}}\right)\left(\frac{(2.42\ \text{hp})(6\ \text{hr})}{0.84}\right)\\ &\quad\times\left(0.7457\ \frac{\text{kW}}{\text{hp}}\right)\\ &= \$1.03\end{aligned}$$

14. PUMP SHAFT LOADING

The torque on a pump or motor shaft can be calculated from the brake power and rotational speed.

$$T_{\text{in-lbf}} = \frac{63{,}025P_{\text{hp}}}{N} \quad \text{18.28}$$

$$T_{\text{ft-lbf}} = \frac{5252P_{\text{hp}}}{N} \quad \text{18.29}$$

$$T_{\text{N·m}} = \frac{9549P_{\text{kW}}}{N} \quad \text{18.30}$$

The actual (developed) torque can be calculated from the change in momentum of the fluid flow. For radial impellers, the fluid enters through the eye and is turned 90°. The direction change is related to the increase in momentum and the shaft torque. When fluid is introduced axially through the eye of the impeller, the tangential velocity at the inlet (eye), $\text{v}_{t,\text{suction}}$, is zero.[13]

$$T_{\text{actual}} = \dot{m}(\text{v}_{t,\text{discharge}}r_{\text{impeller}} - \text{v}_{t,\text{suction}}r_{\text{eye}}) \quad [\text{SI}] \quad \text{18.31(a)}$$

$$T_{\text{actual}} = \frac{\dot{m}}{g_c}(\text{v}_{t,\text{discharge}}r_{\text{impeller}} - \text{v}_{t,\text{suction}}r_{\text{eye}}) \quad [\text{U.S.}] \quad \text{18.31(b)}$$

Centrifugal pumps can be driven directly from a motor, or a speed changer can be used. Rotary pumps generally require a speed reduction. *Gear motors* have integral speed reducers. V-belt drives are widely used because of their initial low cost, although timing belts and chains can be used in some applications.

When a belt or chain is used, the pump's and motor's maximum overhung loads must be checked. This is particularly important for high-power, low-speed applications (such as rotary pumps). *Overhung load* is the side load (force) put on shafts and bearings. The overhung load is calculated from Eq. 18.32. The empirical factor K is 1.0 for chain drives, 1.25 for timing belts, and 1.5 for V-belts.

$$\text{overhung load} = \frac{2KT}{D_{\text{sheave}}} \quad \text{18.32}$$

If a direct drive cannot be used and the overhung load is excessive, the installation can be modified to incorporate a jack shaft or outboard bearing.

Example 18.4

A centrifugal pump delivers 275 lbm/sec (125 kg/s) of water while turning at 850 rpm. The impeller has straight radial vanes and an outside diameter of 10 in (25.4 cm). Water enters the impeller through the eye. The driving motor delivers 30 hp (22 kW). What are the (a) theoretical torque, (b) pump efficiency, and (c) total dynamic head?

SI Solution

(a) From Eq. 18.3, the impeller's tangential velocity is

$$\text{v}_t = \frac{\pi DN}{60\ \frac{\text{s}}{\text{min}}} = \frac{\pi(25.4\ \text{cm})\left(850\ \frac{\text{rev}}{\text{min}}\right)}{\left(60\ \frac{\text{s}}{\text{min}}\right)\left(100\ \frac{\text{cm}}{\text{m}}\right)}$$
$$= 11.3\ \text{m/s}$$

Since water enters axially, the incoming water has no tangential component. From Eq. 18.31, the developed torque is

$$\begin{aligned}T &= \dot{m}\text{v}_{t,\text{discharge}}r_{\text{impeller}}\\ &= \frac{\left(125\ \frac{\text{kg}}{\text{s}}\right)\left(11.3\ \frac{\text{m}}{\text{s}}\right)\left(\frac{25.4\ \text{cm}}{2}\right)}{100\ \frac{\text{cm}}{\text{m}}}\\ &= 179.4\ \text{N·m}\end{aligned}$$

[13]The tangential component of fluid velocity is sometimes referred to as the *velocity of whirl.*

(b) From Eq. 18.30, the developed power is

$$P_{\text{kW}} = \frac{NT_{\text{N·m}}}{9549} = \frac{\left(850\ \frac{\text{rev}}{\text{min}}\right)(179.4\ \text{N·m})}{9549\ \frac{\text{N·m}}{\text{kW·min}}}$$
$$= 15.97\ \text{kW}$$

The pump efficiency is

$$\eta_{\text{pump}} = \frac{P_{\text{developed}}}{P_{\text{input}}} = \frac{15.97\ \text{kW}}{22\ \text{kW}}$$
$$= 0.726 \quad (72.6\%)$$

(c) Use Table 18.6 to find the total dynamic head.

$$P_{\text{kW}} = \frac{9.81\dot{m}\Delta h}{1000}$$

$$\Delta h = \frac{P_{\text{kW}}(1000)}{\dot{m}(9.81)} = \frac{(15.97\ \text{kW})\left(1000\ \frac{\text{W}}{\text{kW}}\right)}{\left(125\ \frac{\text{kg}}{\text{s}}\right)\left(9.81\ \frac{\text{m}}{\text{s}^2}\right)}$$
$$= 13.0\ \text{m}$$

Customary U.S. Solution

(a) From Eq. 18.3, the impeller's tangential velocity is

$$\text{v}_t = \frac{\pi DN}{60\ \frac{\text{sec}}{\text{min}}} = \frac{\pi(10\ \text{in})\left(850\ \frac{\text{rev}}{\text{min}}\right)}{\left(60\ \frac{\text{sec}}{\text{min}}\right)\left(12\ \frac{\text{in}}{\text{ft}}\right)}$$
$$= 37.09\ \text{ft/sec}$$

Since water enters axially, the incoming water has no tangential component. From Eq. 18.31, the developed torque is

$$T = \frac{\dot{m}\text{v}_{t,\text{discharge}}r_{\text{impeller}}}{g_c}$$
$$= \frac{\left(275\ \frac{\text{lbm}}{\text{sec}}\right)\left(37.09\ \frac{\text{ft}}{\text{sec}}\right)\left(\frac{10\ \text{in}}{2}\right)}{\left(32.2\ \frac{\text{lbm-ft}}{\text{lbf-sec}^2}\right)\left(12\ \frac{\text{in}}{\text{ft}}\right)}$$
$$= 132.0\ \text{ft-lbf}$$

(b) From Eq. 18.29, the developed power is

$$P_{\text{hp}} = \frac{T_{\text{ft-lbf}}N}{5252} = \frac{(132.0\ \text{ft-lbf})\left(850\ \frac{\text{rev}}{\text{min}}\right)}{5252\ \frac{\text{ft-lbf}}{\text{hp-min}}}$$
$$= 21.36\ \text{hp}$$

The pump efficiency is

$$\eta_{\text{pump}} = \frac{P_{\text{developed}}}{P_{\text{input}}} = \frac{21.36\ \text{hp}}{30\ \text{hp}}$$
$$= 0.712 \quad (71.2\%)$$

(c) From Table 18.5, the total dynamic head is

$$\Delta h = \frac{\left(550\ \frac{\text{ft-lbf}}{\text{hp-sec}}\right)P_{\text{hp}}}{\dot{m}} \times \frac{g_c}{g}$$
$$= \frac{\left(550\ \frac{\text{ft-lbf}}{\text{hp-sec}}\right)(21.35\ \text{hp})}{275\ \frac{\text{lbm}}{\text{sec}}} \times \frac{32.2\ \frac{\text{lbm-ft}}{\text{lbf-sec}^2}}{32.2\ \frac{\text{ft}}{\text{sec}^2}}$$
$$= 42.7\ \text{ft}$$

15. SPECIFIC SPEED

The capacity and efficiency of a centrifugal pump are partially governed by the impeller design. For a desired flow rate and added head, there will be one optimum impeller design. The quantitative index used to optimize the impeller design is known as *specific speed*, N_s, also known as *impeller specific speed*. Table 18.9 lists the impeller designs that are appropriate for different specific speeds.[14]

Highest heads per stage are developed at low specific speeds. However, for best efficiency, specific speed should be greater than 650 (13 in SI units). If the specific speed for a given set of conditions drops below 650 (13), a multiple-stage pump should be selected.[15]

Specific speed is a function of a pump's capacity, head, and rotational speed at peak efficiency, as shown in Eq. 18.33. For a given pump and impeller configuration,

[14]Specific speed is useful for more than just selecting an impeller type. Maximum suction lift, pump efficiency, and net positive suction head required (NPSH_R) can be correlated with specific speed.

[15]*Partial emission, forced vortex centrifugal pumps* allow operation down to specific speeds of 150 (3 in SI). Such pumps have been used for low-flow, high-head applications, such as high-pressure petrochemical cracking processes.

***Table 18.9** Specific Speed Versus Impeller Design*

	approximate range of specific speed (rpm)	
impeller type	customary U.S. units	SI units
radial vane	500 to 1000	10 to 20
Francis (mixed) vane	2000 to 3000	40 to 60
mixed flow	4000 to 7000	80 to 140
axial flow	9000 and above	180 and above

(Divide customary U.S. specific speed by 51.64 to obtain SI specific speed.)

the specific speed remains essentially constant over a range of flow rates and heads. (Q in Eq. 18.33 is half of the full flow rate for double-suction pumps.)

$$N_s = \frac{N\sqrt{Q_{\text{m}^3/\text{s}}}}{(\Delta h_{\text{m}})^{0.75}} \quad \text{[SI]} \qquad 18.33(a)$$

$$N_s = \frac{N\sqrt{Q_{\text{gpm}}}}{(\Delta h_{\text{ft}})^{0.75}} \quad \text{[U.S.]} \qquad 18.33(b)$$

A common definition of specific speed is the speed (in rpm) at which a *homologous pump* would have to turn in order to deliver 1 gal/min at 1 ft total added head.[16] This definition is implicit to Eq. 18.33 but is not very useful otherwise. While specific speed is not dimensionless, the units are meaningless. Specific speed may be assigned units of rpm, but most often it is expressed simply as a pure number.

The numerical range of acceptable performance for each impeller type is redefined when SI units are used. The SI specific speed is obtained by dividing the customary U.S. specific speed by 51.64.

Specific speed can be used to determine the type of impeller needed. Once a pump is selected, its specific speed and Eq. 18.33 can be used to determine other operational parameters (e.g., maximum rotational speed). Specific speed can be used with Fig. 18.8 to obtain an approximate pump efficiency.

Example 18.5

A centrifugal pump powered by a direct-drive induction motor is needed to discharge 150 gal/min against a 300 ft total head when turning at the fully loaded speed of 3500 rpm. What type of pump should be selected?

Solution

From Eq. 18.33, the specific speed is

$$N_s = \frac{N\sqrt{Q_{\text{gpm}}}}{(\Delta h_{\text{ft}})^{0.75}} = \frac{\left(3500\ \frac{\text{rev}}{\text{min}}\right)\sqrt{150\ \frac{\text{gal}}{\text{min}}}}{(300\ \text{ft})^{0.75}} = 595\ \text{rpm}$$

From Table 18.9, the pump should be a radial vane type. However, pumps achieve their highest efficiencies when specific speed exceeds 650. (See Fig. 18.8.) To increase the specific speed, the rotational speed can be increased, or the total added head can be decreased. Since the pump is direct-driven and 3600 rpm is the maximum speed for induction motors, the total added head should be divided evenly between two stages, or two pumps should be used in series. (See Table 18.8.)

In a two-stage system, the specific speed would be

$$N_s = \frac{\left(3500\ \frac{\text{rev}}{\text{min}}\right)\sqrt{150\ \frac{\text{gal}}{\text{min}}}}{\left(\frac{300\ \text{ft}}{2}\right)^{0.75}} = 1000\ \text{rpm}$$

This is satisfactory for a radial vane pump.

Example 18.6

An induction motor turning at 1200 rpm is to be selected to drive a single-stage, single-suction centrifugal water pump through a direct drive. The total dynamic head added by the pump is 26 ft. The flow rate is 900 gal/min. What minimum size motor should be selected?

Solution

From Eq. 18.33, the specific speed is

$$N_s = \frac{N\sqrt{Q_{\text{gpm}}}}{(\Delta h_{\text{ft}})^{0.75}} = \frac{\left(1200\ \frac{\text{rev}}{\text{min}}\right)\sqrt{900\ \frac{\text{gal}}{\text{min}}}}{(26\ \text{ft})^{0.75}} = 3127\ \text{rpm}$$

From Fig. 18.8, the pump efficiency will be approximately 82%.

[16] *Homologous pumps* are geometrically similar. This means that each pump is a scaled up or down version of the others. Such pumps are said to belong to a *homologous family*.

From Table 18.5, the minimum motor horsepower is

$$P_{\text{hp}} = \frac{Q\Delta h(\text{SG})}{3956\eta_{\text{pump}}} = \frac{(26 \text{ ft})\left(900 \ \frac{\text{gal}}{\text{min}}\right)(1.0)}{\left(3956 \ \frac{\text{ft-gal}}{\text{hp-min}}\right)(0.82)}$$

$$= 7.2 \text{ hp}$$

From Table 18.7, select a 7.5 hp or larger motor.

Example 18.7

A single-stage pump driven by a 3600 rpm motor is currently delivering 150 gal/min. The total dynamic head is 430 ft. What would be the approximate increase in efficiency per stage if the single-stage pump is replaced by a double-stage pump?

Solution

From Eq. 18.33, the specific speed is

$$N_s = \frac{N\sqrt{Q_{\text{gpm}}}}{(\Delta h_{\text{ft}})^{0.75}} = \frac{\left(3600 \ \frac{\text{rev}}{\text{min}}\right)\sqrt{150 \ \frac{\text{gal}}{\text{min}}}}{(430 \text{ ft})^{0.75}}$$

$$= 467 \text{ rpm}$$

From Fig. 18.8, the approximate efficiency is 45%.

In a two-stage pump, each stage adds half of the head. The specific speed per stage would be

$$N_s = \frac{N\sqrt{Q_{\text{gpm}}}}{(\Delta h_{\text{ft}})^{0.75}} = \frac{\left(3600 \ \frac{\text{rev}}{\text{min}}\right)\sqrt{150 \ \frac{\text{gal}}{\text{min}}}}{\left(\frac{430 \text{ ft}}{2}\right)^{0.75}} = 785 \text{ rpm}$$

From Fig. 18.8, the efficiency for this configuration is approximately 60%.

The increase in stage efficiency is 60% − 45% = 15%. Whether or not the cost of multistaging is worthwhile in this low-volume application would have to be determined.

16. CAVITATION

Cavitation is a spontaneous vaporization of the fluid inside the pump, resulting in a degradation of pump performance. Wherever the fluid pressure is less than the vapor pressure, small pockets of vapor will form. These pockets usually form only within the pump itself, although cavitation slightly upstream within the suction line is also possible. As the vapor pockets reach the surface of the impeller, the local high fluid pressure collapses them. Noise, vibration, impeller pitting, and structural damage to the pump casing are manifestations of cavitation.

Cavitation can be caused by any of the following conditions.

- discharge head far below the pump head at peak efficiency
- high suction lift or low suction head
- excessive pump speed
- high liquid temperature (i.e., high vapor pressure)

17. NET POSITIVE SUCTION HEAD

The occurrence of cavitation is predictable. Cavitation will occur when the net pressure in the fluid drops below the vapor pressure. This criterion is commonly stated in terms of head: Cavitation occurs when the available head is less than the required head for satisfactory operation. (See Eq. 18.36.)

The minimum fluid energy required at the pump inlet for satisfactory operation (i.e., the required head) is known as the *net positive suction head required*, NPSH_R.[17] NPSH_R is a function of the pump and will be given by the pump manufacturer as part of the pump performance data.[18] NPSH_R is dependent on the flow rate. However, if NPSH_R is known for one flow rate, it can be determined for another flow rate from Eq. 18.36.

$$\frac{\text{NPSH}_{R,2}}{\text{NPSH}_{R,1}} = \left(\frac{Q_2}{Q_1}\right)^2 \qquad 18.34$$

Net positive suction head available, NPSH_A, is the actual total fluid energy at the pump inlet. Equation 18.35 is based on the conditions at the fluid surface at the top of an open fluid source (e.g., tank or reservoir). If the source is pressurized instead of being open to the atmosphere, the actual pressure can replace p_{atm} in Eq. 18.35.

Centrifugal Pump Characteristics

$$\text{NPSH}_A = \frac{p_{\text{atm}}}{\rho g} \pm h_s - h_f - \frac{\text{v}^2}{2g} - \frac{p_{\text{vapor}}}{\rho g} \qquad 18.35$$

[17]If NPSH_R (a head term) is multiplied by the fluid specific weight, it is known as the *net inlet pressure required*, NIPR. Similarly, NPSH_A can be converted to NIPA.

[18]It is also possible to calculate NPSH_R from other information, such as suction specific speed. However, this still depends on information provided by the manufacturer.

If $NPSH_A$ is less than $NPSH_R$, the fluid will cavitate. The criterion for cavitation is given by Eq. 18.36. (In practice, it is desirable to have a safety margin.)

$$NPSH_A < NPSH_R \quad \begin{bmatrix}\text{criterion for} \\ \text{cavitation}\end{bmatrix} \qquad 18.36$$

Example 18.8

2.0 ft^3/sec (50 L/s) of 60°F (15°C) water is pumped from an elevated feed tank to an open reservoir through 6 in (15.2 cm), schedule-40 steel pipe as shown. The friction loss for the piping and fittings in the suction line is 2.6 ft (0.9 m). The friction loss for the piping and fittings in the discharge line is 13 ft (4.3 m). The atmospheric pressure is 14.7 psia (101 kPa). What is the $NPSH_A$?

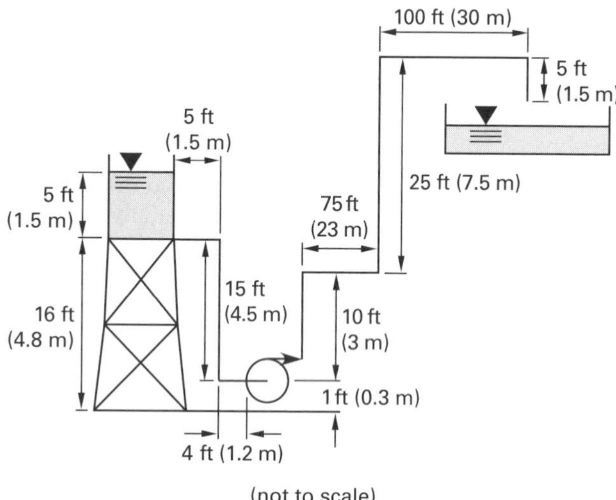

(not to scale)

SI Solution

From a table of water properties, at 15°C, the density of water is 999.1 kg/m^3, and the vapor pressure is 1.7 kPa. [**Properties of Water (SI Units)**]

The elevation difference, h_s, between the fluid reservoir surface and the centerline of the pump suction inlet is 1.5 m + 4.5 m = 6 m.

Use the continuity equation to find the fluid velocity. From a table of pipe data, the cross-sectional area of nominal 6 in schedule-40 pipe is 28.876 in^2. [**Schedule 40 Steel Pipe**]

$$A = \frac{(28.876\ \text{in}^2)\left(2.540\ \frac{\text{cm}}{\text{in}}\right)^2}{\left(100\ \frac{\text{cm}}{\text{m}}\right)^2} = 0.0186\ \text{m}^2$$

Find the fluid velocity.

Continuity Equation

$$Q = A\text{v}$$

$$\text{v} = \frac{Q}{A} = \frac{\left(\frac{50\ \frac{\text{L}}{\text{s}}}{1000\ \frac{\text{L}}{\text{m}^3}}\right)}{0.0186\ \text{m}^2} = 2.68\ \text{m/s}$$

Calculate the $NPSH_A$.

Centrifugal Pump Characteristics

$$\begin{aligned} NPSH_A &= \frac{p_{\text{atm}}}{\rho g} \pm h_s - h_f - \frac{\text{v}^2}{2g} - \frac{p_{\text{vapor}}}{\rho g} \\ &= \frac{(101\ \text{kPa})\left(1000\ \frac{\text{Pa}}{\text{kPa}}\right)}{\left(999.1\ \frac{\text{kg}}{\text{m}^3}\right)\left(9.807\ \frac{\text{m}}{\text{s}^2}\right)} + 6\ \text{m} \\ &\quad - 0.9\ \text{m} - \frac{2.68\ \frac{\text{m}^2}{\text{s}}}{(2)\left(9.807\ \frac{\text{m}}{\text{s}^2}\right)} \\ &\quad - \frac{(1.7\ \text{kPa})\left(1000\ \frac{\text{Pa}}{\text{kPa}}\right)}{\left(999.1\ \frac{\text{kg}}{\text{m}^3}\right)\left(9.807\ \frac{\text{m}}{\text{s}^2}\right)} \\ &= 14.86\ \text{m} \end{aligned}$$

Customary U.S. Solution

From a table of water properties, at 60°F, the density of water is 1.938 lbf-sec^2/ft^4, and the vapor pressure is 0.26 psi. [**Properties of Water (I-P Units)**]

The elevation difference, h_s, between the fluid reservoir surface and the centerline of the pump suction inlet is 5 ft + 15 ft = 20 ft.

Use the continuity equation to find the fluid velocity. From a table of pipe data, the cross-sectional area of nominal 6 in schedule-40 pipe is 28.876 in^2. [**Schedule 40 Steel Pipe**]

$$A = \frac{28.876\ \text{in}^2}{\left(12\ \frac{\text{in}}{\text{ft}}\right)^2} = 0.2005\ \text{ft}^2$$

Find the fluid velocity.

Continuity Equation

$$Q = A\text{v}$$

$$\text{v} = \frac{Q}{A} = \frac{2.0\ \frac{\text{ft}^3}{\text{sec}}}{0.2005\ \text{ft}^2} = 9.975\ \text{ft/sec}$$

Calculate the NPSH_A.

Centrifugal Pump Characteristics

$$\text{NPSH}_A = \frac{p_{\text{atm}}}{\rho g} \pm h_s - h_f - \frac{\text{v}^2}{2g} - \frac{p_{\text{vapor}}}{\rho g}$$

$$= \frac{\left(14.7\ \frac{\text{lbf}}{\text{in}^2}\right)\left(12\ \frac{\text{in}}{\text{ft}}\right)^2}{\left(1.938\ \frac{\text{lbf-sec}^2}{\text{ft}^4}\right)\left(32.17\ \frac{\text{ft}}{\text{sec}^2}\right)} + 20\ \text{ft}$$

$$-2.6\ \text{ft} - \frac{\left(9.975\ \frac{\text{ft}}{\text{sec}}\right)^2}{(2)\left(32.17\ \frac{\text{ft}}{\text{sec}^2}\right)}$$

$$- \frac{\left(0.26\ \frac{\text{lbf}}{\text{in}^2}\right)\left(12\ \frac{\text{in}}{\text{ft}}\right)^2}{\left(1.938\ \frac{\text{lbf-sec}^2}{\text{ft}^4}\right)\left(32.17\ \frac{\text{ft}}{\text{sec}^2}\right)}$$

$$= 49.21\ \text{ft}$$

18. PREVENTING CAVITATION

Cavitation is prevented by increasing NPSH_A or decreasing NPSH_R. NPSH_A can be increased by

- increasing the height of the fluid source
- lowering the pump
- reducing friction and minor losses by shortening the suction line or using a larger pipe size
- reducing the temperature of the fluid at the pump entrance
- pressurizing the fluid supply tank
- reducing the flow rate or velocity (i.e., reducing the pump speed)

NPSH_R can be reduced by

- placing a throttling valve or restriction in the discharge line[19]
- using an oversized pump
- using a double-suction pump
- using an impeller with a larger eye
- using an inducer

High NPSH_R applications, such as boiler feed pumps needing 150–250 ft (50–80 m), should use one or more booster pumps in front of each high-NPSH_R pump. Such booster pumps are typically single-stage, double-suction pumps running at low speed. Their NPSH_R can be 25 ft (8 m) or less.

Throttling the input line to a pump and venting or evacuating the receiving tank both increase cavitation. Throttling the input line increases the friction head and decreases NPSH_A. Evacuating the receiving tank increases the flow rate, increasing NPSH_R while simultaneously increasing the friction head and reducing NPSH_A.

19. CAVITATION COEFFICIENT

The *cavitation number* (or *cavitation coefficient*), σ, is a dimensionless number that can be used in modeling and extrapolating experimental results. The actual cavitation coefficient is compared with the *critical cavitation number* obtained experimentally. If the actual cavitation number is less than the critical cavitation number, cavitation will occur. Absolute pressure must be used for the fluid pressure term, p.

$$\sigma < \sigma_{\text{cr}} \quad \text{[criterion for cavitation]} \qquad 18.37$$

$$\sigma = \frac{2(p - p_{\text{vapor}})}{\rho \text{v}^2} = \frac{\text{NPSH}_A}{\Delta h} \qquad \text{[SI]} \qquad 18.38(a)$$

$$\sigma = \frac{2g_c(p - p_{\text{vapor}})}{\rho \text{v}^2} = \frac{\text{NPSH}_A}{\Delta h} \qquad \text{[U.S.]} \qquad 18.38(b)$$

The two forms of Eq. 18.38 yield slightly different results. The first form is essentially the ratio of the net pressure available for collapsing a vapor bubble to the velocity pressure creating the vapor. It is useful in model experiments. The second form is applicable to tests of production model pumps.

[19]This will increase the total head, Δh, added by the pump, thereby reducing the pump's output and driving the pump's operating point into a region of lower NPSH_R.

Fluids

20. SUCTION SPECIFIC SPEED

The formula for *suction specific speed*, N_{ss}, can be derived by substituting NPSH_R for added head in the expression for specific speed. Q is halved for double-suction pumps.

$$N_{ss} = \frac{N\sqrt{Q_{\text{m}^3/\text{s}}}}{(\text{NPSH}_{R,\text{m}})^{0.75}} \quad \text{[SI]} \qquad 18.39(a)$$

$$N_{ss} = \frac{N\sqrt{Q_{\text{gpm}}}}{(\text{NPSH}_{R,\text{ft}})^{0.75}} \quad \text{[U.S.]} \qquad 18.39(b)$$

Suction specific speed is an index of the suction characteristics of the impeller. Ideally, it should be approximately 8500 (165 in SI) for both single- and double-suction pumps. This assumes the pump is operating at or near its point of optimum efficiency.

Suction specific speed can be used to determine the maximum recommended operating speed by substituting 8500 (165 in SI) for N_{ss} in Eq. 18.39 and solving for N.

If the suction specific speed is known, it can be used to determine the NPSH_R. If the pump is known to be operating at or near its optimum efficiency, an approximate NPSH_R value can be found by substituting 8500 (165 in SI) for N_{ss} in Eq. 18.39 and solving for NPSH_R.

Suction specific speed available, SA, is obtained when NPSH_A is substituted for added head in the expression for specific speed. The suction specific speed available must be less than the suction specific speed required to prevent cavitation.[20]

21. PUMP PERFORMANCE CURVES

For a given impeller diameter and constant speed, the head added will decrease as the flow rate increases. This is shown graphically on the *pump performance curve* (*pump curve*) supplied by the pump manufacturer. Other operating characteristics (e.g., power requirement, NPSH_R, and efficiency) also vary with flow rate, and these are usually plotted on a common graph, as shown in Fig. 18.10.[21] Manufacturers' pump curves show performance over a limited number of calibration speeds. If an operating point is outside the range of published curves, the affinity laws can be used to estimate the speed at which the pump gives the required performance.

Figure 18.10 *Pump Performance Curves*

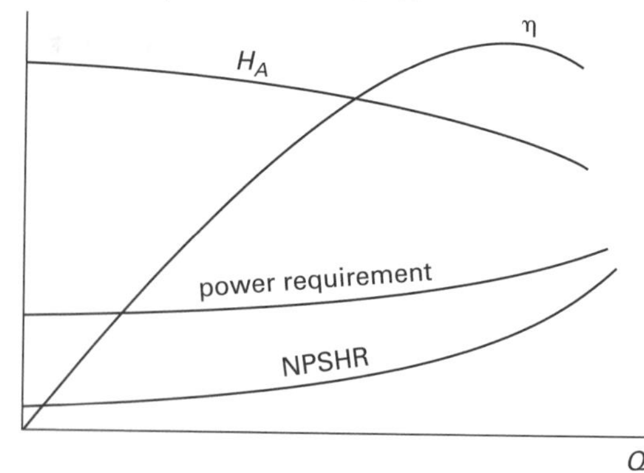

On the pump curve, the *shutoff point* (also known as *churn*) corresponds to a closed discharge valve (i.e., zero flow); the *rated point* is where the pump operates with rated 100% of capacity and head; the *overload point* corresponds to 65% of the rated head.

Figure 18.10 is for a pump with a fixed impeller diameter and rotational speed. The characteristics of a pump operated over a range of speeds or for different impeller diameters are illustrated in Fig. 18.11.

22. SYSTEM CURVES

A *system curve* (or *system performance curve*) is a plot of the static and friction energy losses experienced by the fluid for different flow rates. Unlike the pump curve, which depends only on the pump, the system curve depends only on the configuration of the suction and discharge lines. The following equations assume equal pressures at the fluid source and destination surfaces, which is the case for pumping from one atmospheric reservoir to another. The velocity head is insignificant and is disregarded.

$$\Delta h = h_z + h_f \qquad 18.40$$

$$h_z = h_d - h_s \qquad 18.41$$

$$h_f = h_{f,\text{suction}} + h_{f,\text{discharge}} \qquad 18.42$$

If the fluid reservoirs are large, or if the fluid reservoir levels are continually replenished, the net static suction head ($h_d - h_s$) will be constant for all flow rates. The friction loss, h_f, varies with v^2 (and, therefore, with Q^2)

[20] Since speed and flow rate are constants, this is another way of saying NPSH_A must equal or exceed NPSH_R.

[21] The term *pump curve* is commonly used to designate the h_A versus Q characteristics, whereas *pump characteristics curve* refers to all of the pump data.

Figure 18.11 Centrifugal Pump Characteristics Curves

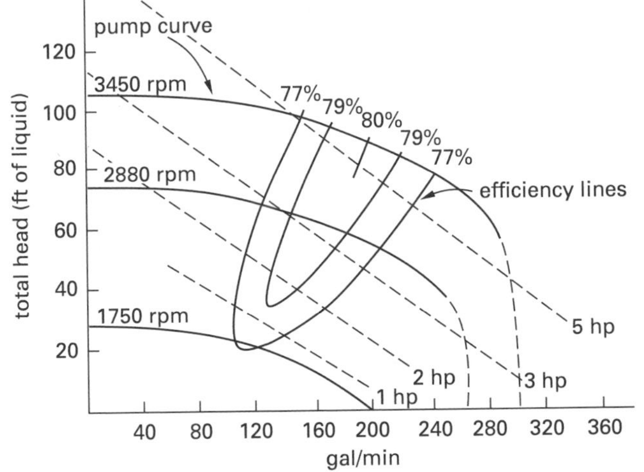

(a) variable speed

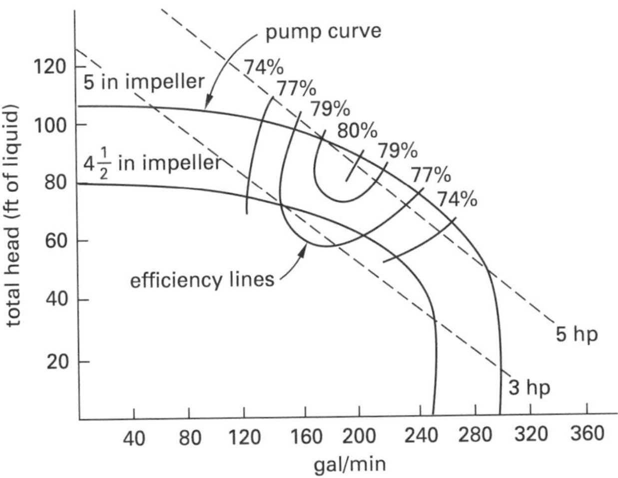

(b) variable impeller diameter

in the Darcy-Weisbach friction formula. This makes it easy to find friction losses for other flow rates (subscript 2) once one friction loss (subscript 1) is known.[22]

$$\frac{h_{f,1}}{h_{f,2}} = \left(\frac{Q_1}{Q_2}\right)^2 \qquad 18.43$$

Figure 18.12 illustrates a system curve following Eq. 18.40 with a positive added head (i.e., a fluid source below the fluid destination). The system curve is shifted upward, intercepting the vertical axis at some positive value of Δh.

Figure 18.12 System Curve

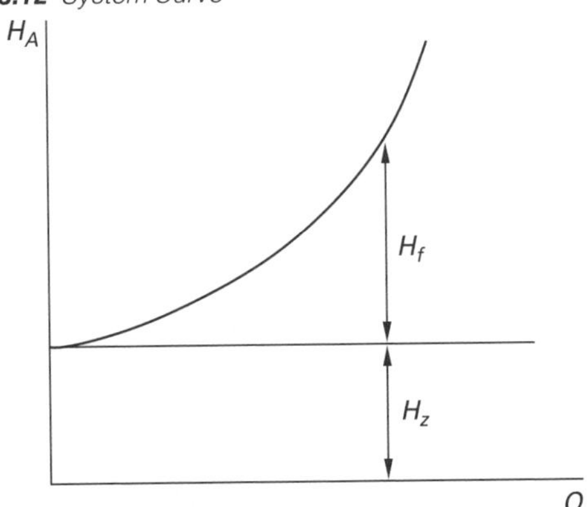

23. OPERATING POINT

The intersection of the pump curve and the system curve determines the *operating point*, as shown in Fig. 18.13. The operating point defines the system head and system flow rate.

Figure 18.13 Extreme Operating Points

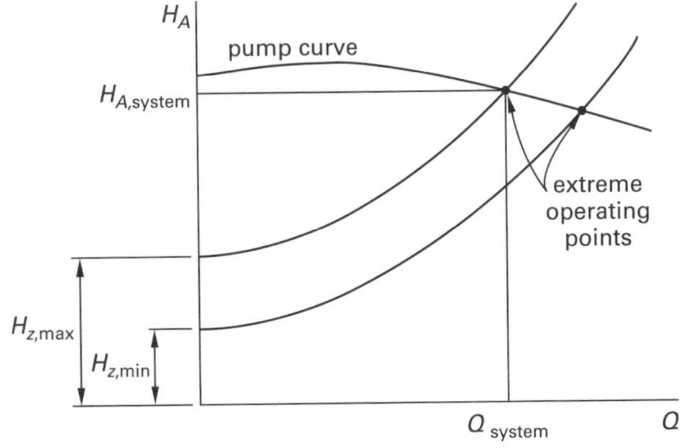

When selecting a pump, the system curve is plotted on manufacturers' pump curves for different speeds and/or impeller diameters (i.e., Fig. 18.11). There will be several possible operating points corresponding to the various pump curves shown. Generally, the design operating point should be close to the highest pump efficiency. This, in turn, will determine speed and impeller diameter.

In some systems, the static head varies as the source reservoir is drained or as the destination reservoir fills. The system head is then defined by a pair of matching system friction curves intersecting the pump curve. The two intersection points are called the *extreme operating points*—the maximum and minimum capacity requirements.

[22]Equation 18.43 implicitly assumes that the friction factor, f, is constant. This may be true over a limited range of flow rates, but it is not true over large ranges. Nevertheless, Eq. 18.43 is often used to quickly construct preliminary versions of the system curve.

After a pump is installed, it may be desired to change the operating point. This can be done without replacing the pump by placing a throttling valve in the discharge line. The operating point can then be moved along the pump curve by partially opening or closing the valve, as is illustrated in Fig. 18.14. (A throttling valve should never be placed in the suction line since that would reduce $NPSH_A$.)

Figure 18.14 *Throttling the Discharge*

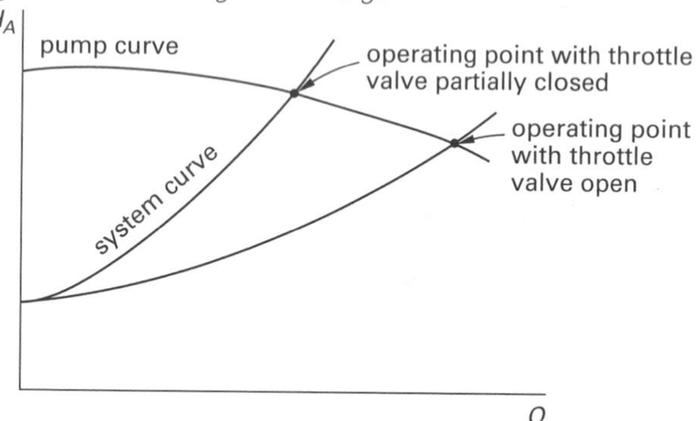

Figure 18.15 *Pumps Operating in Parallel*

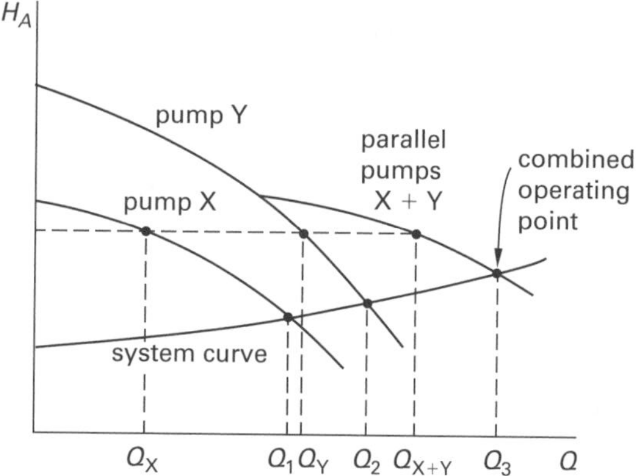

24. PUMPS IN PARALLEL

Parallel operation is obtained by having two pumps discharging into a common header. This type of connection is advantageous when the system demand varies greatly or when high reliability is required. A single pump providing total flow would have to operate far from its optimum efficiency at one point or another. With two pumps in parallel, one can be shut down during low demand. This allows the remaining pump to operate close to its optimum efficiency point.

Figure 18.15 illustrates that parallel operation increases the capacity of the system while maintaining the same total head.

The performance curve for a set of pumps in parallel can be plotted by adding the capacities of the two pumps at various heads. A second pump will operate only when its discharge head is greater than the discharge head of the pump already running. Capacity does not increase at heads above the maximum head of the smaller pump.

When the parallel performance curve is plotted with the system head curve, the operating point is the intersection of the system curve with the X + Y curve. With pump X operating alone, the capacity is given by Q_1. When pump Y is added, the capacity increases to Q_3 with a slight increase in total head.

25. PUMPS IN SERIES

Series operation is achieved by having one pump discharge into the suction of the next. This arrangement is used primarily to increase the discharge head, although a small increase in capacity also results. (See Fig. 18.16.)

The performance curve for a set of pumps in series can be plotted by adding the heads of the two pumps at various capacities.

Figure 18.16 *Pumps Operating in Series*

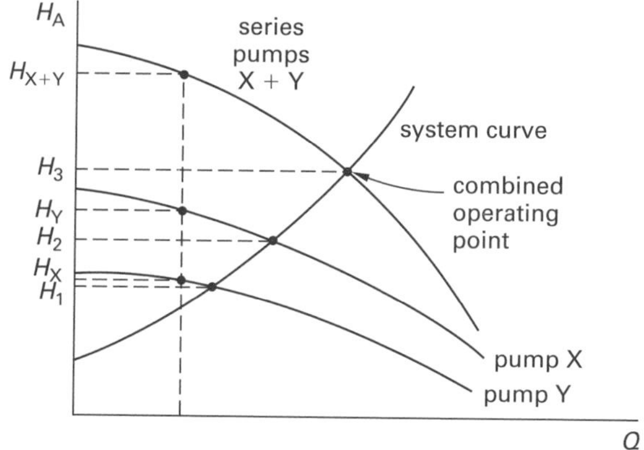

26. AFFINITY LAWS

Most parameters determining a specific pump's performance can be modified, including impeller diameter, speed, flow rate, specific gravity, head, and horsepower. If the impeller diameter is held constant and the speed is varied, and with an unchanged system curve, the following relationships are maintained with no change in efficiency.[23]

Pump Affinity Laws

$$Q_2 = Q_1\left(\frac{N_2}{N_1}\right) \qquad 18.44$$

[23]See Sec. 18.27 if the entire pump is scaled to a different size.

$$h_2 = h_1\left(\frac{N_2}{N_1}\right)^2 = h_1\left(\frac{Q_2}{Q_1}\right)^2 \quad 18.45$$

$$\text{bhp}_2 = \text{bhp}_1\left(\frac{N_2}{N_1}\right)^3 = \text{bhp}_1\left(\frac{Q_2}{Q_1}\right)^3 \quad 18.46$$

If the speed is held constant and the impeller size is reduced (i.e., the impeller is trimmed), while keeping the pump body, volute, shaft diameter, and suction and discharge openings the same, the following relationships exist.[24]

Pump Affinity Laws

$$Q_2 = Q_1\left(\frac{D_2}{D_1}\right) \quad 18.47$$

$$h_2 = h_1\left(\frac{D_2}{D_1}\right)^2 \quad 18.48$$

$$\text{bhp}_2 = \text{bhp}_1\left(\frac{D_2}{D_1}\right)^3 \quad 18.49$$

$$p_2 = p_1\left(\frac{D_2}{D_1}\right)^3 \quad 18.50$$

If speed and impeller diameter are kept constant and the specific gravity of the fluid is changed,

Pump Affinity Laws

$$\text{bhp}_2 = \text{bhp}_1\left(\frac{\text{SG}_2}{\text{SG}_1}\right) \quad 18.51$$

$$p_2 = p_1\left(\frac{\text{SG}_2}{\text{SG}_1}\right) \quad 18.52$$

The affinity laws are based on the assumption that the efficiency stays the same. In reality, larger pumps are somewhat more efficient than smaller pumps, and extrapolations to greatly different sizes should be avoided. Equation 18.53 can be used to estimate the efficiency of a differently sized pump. The dimensionless exponent, n, varies from 0 to approximately 0.26, with 0.2 being a typical value.

$$\frac{1-\eta_{\text{smaller}}}{1-\eta_{\text{larger}}} = \left(\frac{D_{\text{larger}}}{D_{\text{smaller}}}\right)^n \quad 18.53$$

Example 18.9

A pump operating at 1770 rpm delivers 500 gal/min against a total head of 200 ft. It is desired to have the pump deliver a total head of 375 ft. At what speed should this pump be operated to achieve this new head at the same efficiency?

Solution

Use the affinity law relating head and speed, and solve for the new speed.

Pump Affinity Laws

$$h_2 = h_1\left(\frac{N_2}{N_1}\right)^2$$

$$N_2 = N_1\sqrt{\frac{h_2}{h_1}} = \left(1770\ \frac{\text{rev}}{\text{min}}\right)\sqrt{\frac{375\ \text{ft}}{200\ \text{ft}}} = 2424\ \text{rpm}$$

Example 18.10

A pump is needed to pump 500 gal/min against a total dynamic head of 425 ft. The hydraulic system has no static head change. Only the 1750 rpm performance curve is known for the pump. At what speed must the pump be turned to achieve the desired performance with no change in efficiency or impeller size?

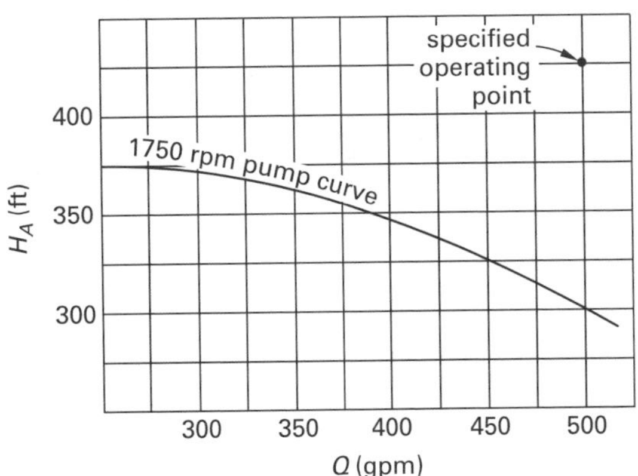

Solution

A flow of 500 gal/min with a head of 425 ft does not correspond to any point on the 1750 rpm curve.

[24]One might ask, "How is it possible to change a pump's impeller diameter?" In practice, a different impeller may be available from the manufacturer, but more often the impeller is taken out and shaved down on a lathe. Equation 18.47, Eq. 18.48, and Eq. 18.49 are limited in use to radial flow machines, and with reduced accuracy, to mixed-flow impellers. Changing the impeller diameter significantly impacts other design relationships, and the accuracy of performance prediction decreases if the diameter is changed much more than 20%.

From Eq. 18.45, the quantity h/Q^2 is constant.

$$\frac{h}{Q^2} = \frac{425 \text{ ft}}{\left(500 \ \frac{\text{gal}}{\text{min}}\right)^2} = 1.7 \times 10^{-3} \text{ ft-min}^2/\text{gal}^2$$

In order to use the affinity laws, the operating point on the 1750 rpm curve must be determined. Random values of Q are chosen and the corresponding values of h are determined such that the ratio h/Q^2 is unchanged.

Q	h
475	383
450	344
425	307
400	272

These points are plotted as the system curve. The intersection of the system and 1750 rpm pump curve at 440 gal/min defines the operating point at 1750 rpm. From Eq. 18.44, the needed pump speed is

$$N_2 = \frac{N_1 Q_2}{Q_1} = \frac{\left(1750 \ \frac{\text{rev}}{\text{min}}\right)\left(500 \ \frac{\text{gal}}{\text{min}}\right)}{440 \ \frac{\text{gal}}{\text{min}}}$$

$$= 1989 \text{ rpm}$$

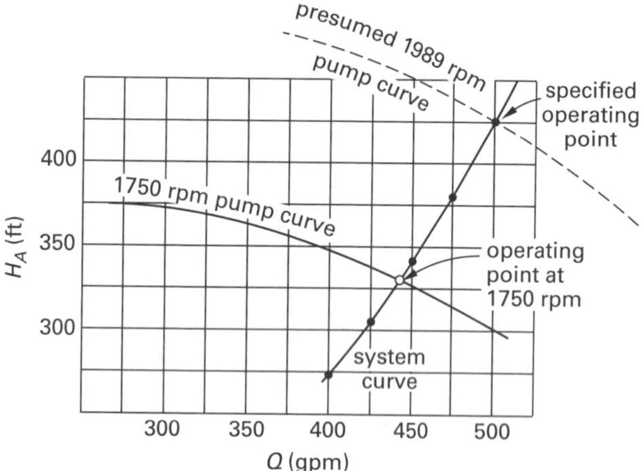

27. PUMP SIMILARITY

The performance of one pump can be used to predict the performance of a *dynamically similar* (*homologous*) pump. This can be done by using Eq. 18.54 through Eq. 18.59.

$$\frac{N_1 D_1}{\sqrt{h_1}} = \frac{N_2 D_2}{\sqrt{h_2}} \qquad 18.54$$

$$\frac{Q_1}{D_1^2 \sqrt{h_1}} = \frac{Q_2}{D_2^2 \sqrt{h_2}} \qquad 18.55$$

$$\frac{p_1}{\rho_1 D_1^2 h_1^{1.5}} = \frac{p_2}{\rho_2 D_2^2 h_2^{1.5}} \qquad 18.56$$

$$\frac{Q_1}{N_1 D_1^3} = \frac{Q_2}{N_2 D_2^3} \qquad 18.57$$

$$\frac{p_1}{\rho_1 N_1^3 D_1^5} = \frac{p_2}{\rho_2 N_2^3 D_2^5} \qquad 18.58$$

$$\frac{N_1 \sqrt{Q_1}}{h_1^{0.75}} = \frac{N_2 \sqrt{Q_2}}{h_2^{0.75}} \qquad 18.59$$

These *similarity laws* (also known as *scaling laws*) assume that both pumps

- operate in the turbulent region
- have the same pump efficiency
- operate at the same percentage of wide-open flow

Similar pumps also will have the same specific speed and cavitation number.

As with the affinity laws, these relationships assume that the efficiencies of the larger and smaller pumps are the same. In reality, larger pumps will be more efficient than smaller pumps. Therefore, extrapolations to much larger or much smaller sizes should be avoided.

Example 18.11

A 6 in pump operating at 1770 rpm discharges 1500 gal/min of cold water (SG = 1.0) against an 80 ft head at 85% efficiency. A homologous 8 in pump operating at 1170 rpm is being considered as a replacement. (a) What total head and capacity can be expected from the new pump? (b) What would be the new motor horsepower requirement?

Solution

(a) From Eq. 18.54,

$$h_2 = \left(\frac{D_2 N_2}{D_1 N_1}\right)^2 h_1$$

$$= \left(\frac{(8 \text{ in})\left(1170 \ \frac{\text{rev}}{\text{min}}\right)}{(6 \text{ in})\left(1770 \ \frac{\text{rev}}{\text{min}}\right)}\right)^2 (80 \text{ ft})$$

$$= 62.14 \text{ ft}$$

From Eq. 18.57,

$$\begin{aligned} Q_2 &= \left(\frac{N_2 D_2^3}{N_1 D_1^3}\right) Q_1 \\ &= \left(\frac{\left(1170\ \frac{\text{rev}}{\text{min}}\right)(8\ \text{in})^3}{\left(1770\ \frac{\text{rev}}{\text{min}}\right)(6\ \text{in})^3}\right)\left(1500\ \frac{\text{gal}}{\text{min}}\right) \\ &= 2350\ \text{gpm} \end{aligned}$$

(b) From Eq. 18.12, the water horsepower is

Pump Power Equation

$$\begin{aligned} \text{whp} &= \frac{Q\Delta h}{3960} \\ &= \frac{\left(2350\ \frac{\text{gal}}{\text{min}}\right)(62.14\ \text{ft})}{3960\ \frac{\text{ft-gal}}{\text{hp-min}}} \\ &= 36.88\ \text{hp} \end{aligned}$$

From Eq. 18.53,

$$\begin{aligned} \eta_{\text{larger}} &= 1 - \frac{1-\eta_{\text{smaller}}}{\left(\frac{D_{\text{larger}}}{D_{\text{smaller}}}\right)^n} = 1 - \frac{1-0.85}{\left(\frac{8\ \text{in}}{6\ \text{in}}\right)^{0.2}} \\ &= 0.858 \end{aligned}$$

$$\text{bhp}_2 = \frac{\text{whp}_2}{\eta_p} = \frac{36.92\ \text{hp}}{0.858} = 43.0\ \text{hp}$$

28. PUMPING LIQUIDS OTHER THAN COLD WATER

Many liquid pump parameters are determined from tests with cold, clear water at 85°F (29°C). The following guidelines can be used when pumping water at other temperatures or when pumping other liquids.

- Head developed is independent of the liquid's specific gravity. Pump performance curves from tests with water can be used with other Newtonian fluids (e.g., gasoline, alcohol, and aqueous solutions) having similar viscosities.
- The hydraulic horsepower depends on the specific gravity of the liquid. If the pump characteristic curve is used to find the operating point, multiply the horsepower reading by the specific gravity. Table 18.5 and Table 18.6 incorporate the specific gravity term in the calculation of hydraulic power where required.
- Efficiency is not affected by changes in temperature that cause only the specific gravity to change.
- Efficiency is nominally affected by changes in temperature that cause the viscosity to change. Equation 18.60 is an approximate relationship suggested by the Hydraulics Institute when extrapolating the efficiency (in decimal form) from cold water to hot water. n is an experimental exponent established by the pump manufacturer, generally in the range of 0.05–0.1.

$$\eta_{\text{hot}} = 1 - (1-\eta_{\text{cold}})\left(\frac{\nu_{\text{hot}}}{\nu_{\text{cold}}}\right)^n \qquad 18.60$$

- NPSH_A depends significantly on liquid temperature.
- NPSH_R is not significantly affected by variations in the liquid temperature.
- When hydrocarbons are pumped, the NPSH_R determined from cold water can usually be reduced. This reduction is apparently due to the slow vapor release of complex organic liquids. If the hydrocarbon's vapor pressure at the pumping temperature is known, Fig. 18.17 will give the percentage of the cold-water NPSH_R.

Figure 18.17 *Hydrocarbon NPSH_R Correction Factor*

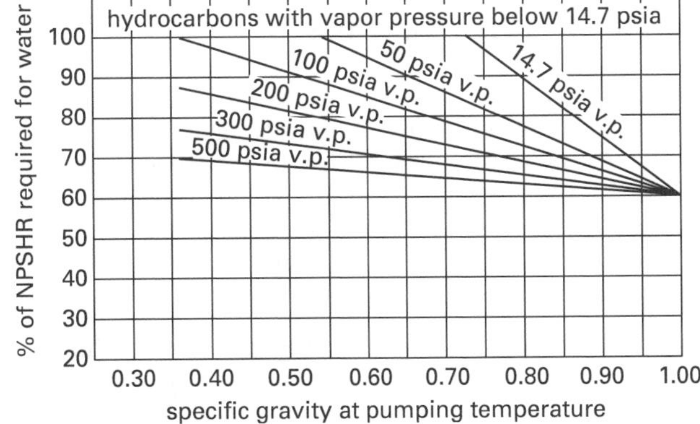

- Pumping many fluids requires expertise that goes far beyond simply extrapolating parameters in proportion to the fluid's specific gravity. Such special cases include pumping liquids containing abrasives, liquids that solidify, highly corrosive liquids, liquids with vapor or gas, highly viscous fluids, paper stock, and hazardous fluids.
- Head, flow rate, and efficiency are all reduced when pumping highly viscous non-Newtonian fluids. No exact method exists for determining the reduction factors, other than actual tests of an installation using both fluids. Some sources have published charts of correction factors based on tests over limited viscosity and size ranges.[25]

[25]A chart published by the Hydraulics Institute is widely distributed.

Example 18.12

A centrifugal pump has an $NPSH_R$ of 12 psi based on cold water. 10°F liquid isobutane has a specific gravity of 0.60 and a vapor pressure of 15 psia. What $NPSH_R$ should be used with 10°F liquid isobutane?

Solution

From Fig. 18.17, the intersection of a specific gravity of 0.60 and 15 psia is above the horizontal 100% line. The full $NPSH_R$ of 12 psi should be used.

29. TURBINE SPECIFIC SPEED

Like centrifugal pumps, turbines are classified according to the manner in which the impeller extracts energy from the fluid flow. This is measured by the turbine-specific speed equation, which is different from the equation used to calculate specific speed for pumps.

$$N_s = \frac{N\sqrt{P_{\text{kW}}}}{h_t^{1.25}} \quad \text{[SI]} \quad 18.61(a)$$

$$N_s = \frac{N\sqrt{P_{\text{hp}}}}{h_t^{1.25}} \quad \text{[U.S.]} \quad 18.61(b)$$

30. REACTION TURBINES

Reaction turbines (also known as *Francis turbines* or *radial-flow turbines*) are essentially centrifugal pumps operating in reverse. (See Fig. 18.18.) They are used when the total available head is small, typically below 600–800 ft (183–244 m). However, their energy conversion efficiency is higher than that of impulse turbines, typically in the 85–95% range.

In a reaction turbine, water enters the turbine housing with a pressure greater than atmospheric pressure. The water completely surrounds the turbine runner (impeller) and continues through the draft tube. There is no vacuum or air pocket between the turbine and the tailwater.

All of the power, affinity, and similarity relationships used with centrifugal pumps can be used with reaction turbines.

Example 18.13

A reaction turbine with a draft tube develops 500 hp (brake) when 50 ft³/sec water flow through it. Water enters the turbine at 20 ft/sec with a 100 ft pressure head. The elevation of the turbine above the tailwater level is 10 ft. Disregarding friction, what are the (a) total available head and (b) turbine efficiency?

Figure 18.18 Reaction Turbine

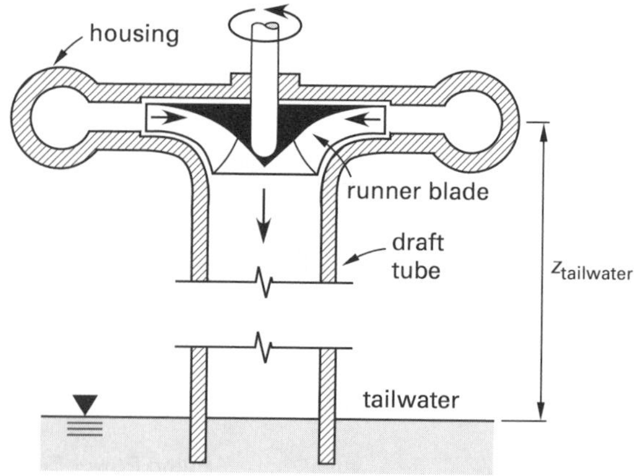

Solution

(a) The available head is the difference between the forebay and tailwater elevations. The tailwater depression is known, but the height of the forebay above the turbine is not known. However, at the turbine entrance, this unknown potential energy has been converted to pressure and velocity head. The total available head (exclusive of friction) is

$$\begin{aligned} h_t &= z_{\text{forebay}} - z_{\text{tailwater}} = h_p + h_v - z_{\text{tailwater}} \\ &= 100 \text{ ft} + \frac{\left(20 \ \frac{\text{ft}}{\text{sec}}\right)^2}{(2)\left(32.2 \ \frac{\text{ft}}{\text{sec}^2}\right)} - (-10 \text{ ft}) \\ &= 116.2 \text{ ft} \end{aligned}$$

(b) From Table 18.5, the theoretical hydraulic horsepower is

$$\begin{aligned} P_{\text{th}} &= \frac{Q\Delta h(\text{SG})}{8.814} \\ &= \frac{(116.2 \text{ ft})\left(50 \ \frac{\text{ft}^3}{\text{sec}}\right)(1.0)}{8.814 \ \frac{\text{ft}^4}{\text{hp-sec}}} \\ &= 659.2 \text{ hp} \end{aligned}$$

The efficiency of the turbine is

$$\eta = \frac{P_{\text{brake}}}{P_{\text{th}}} = \frac{500 \text{ hp}}{659.2 \text{ hp}} = 0.758 \quad (75.8\%)$$

31. TYPES OF REACTION TURBINES

Each of the three types of turbines is associated with a range of specific speeds.

- *Axial-flow reaction turbines* (also known as *propeller turbines*) are used for low heads, high rotational speeds, and large flow rates. (See Fig. 18.19.) These propeller turbines operate with specific speeds in the 70–260 range (266–988 in SI). Their best efficiencies, however, are produced with specific speeds between 120 and 160 (460 and 610 in SI).

Figure 18.19 *Axial-Flow Turbine*

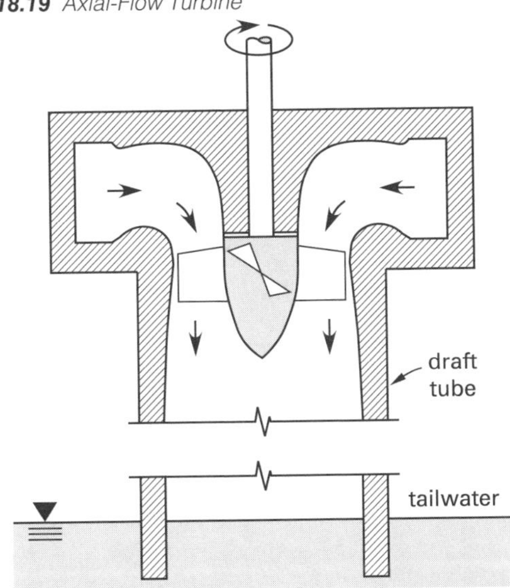

- For *mixed-flow reaction turbines*, the specific speed varies from 10 to 90 (38 to 342 in SI). Best efficiencies are found in the 40 to 60 (150 to 230 in SI) range with heads below 600 ft to 800 ft (180 m to 240 m).
- *Radial-flow reaction turbines* have the lowest flow rates and specific speeds but are used when heads are high. These turbines have specific speeds between 1 and 20 (3.8 and 76 in SI).

32. NOMENCLATURE

A	area	ft^2	m^2
bhp	brake horsepower	hp	n.a.
C	factor	–	–
D	diameter	ft	m
f	Darcy friction factor	–	–
f	frequency	Hz	Hz
fhp	friction horsepower	hp	n.a.
g	gravitational acceleration, 32.17 (9.807)	ft/sec^2	m/s^2
g_c	gravitational constant, 32.17	lbm-ft/lbf-sec^2	n.a.
h	height or head	ft	m
K	compressibility factor	–	–
L	length	ft	m
$\dot{m}$	mass flow rate	lbm/sec	kg/s
n	dimensionless exponent	–	–
N	rotational speed	rev/min	rev/min
$NPSH_A$	net positive suction head available	ft	m
$NPSH_R$	net positive suction head required	ft	m
p	pressure	lbf/ft^2	Pa
P	power	ft-lbf/sec	W
Q	volumetric flow rate	gal/min	L/s
r	radius	ft	m
SA	suction specific speed available	rev/min	rev/min
SF	service factor	–	–
SG	specific gravity	–	–
t	time	sec	s
T	torque	ft-lbf	N·m
v	velocity	ft/sec	m/s
W	work	ft-lbf	kW·h
whp	water horsepower	hp	n.a.
z	elevation	ft	m

Symbols

γ	specific weight	lbf/ft^3	N/m^3
η	efficiency	–	–
ν	kinematic viscosity	ft^2/sec	m^2/s
ρ	density	lbm/ft^3	kg/m^3
σ	cavitation number	–	–
ω	angular velocity	rad/sec	rad/s

Subscripts

ac	acceleration
atm	atmospheric
ave	average
cr	critical
d	discharge
f	friction
max	maximum
p	pressure or pump
s	specific or suction
ss	suction specific
t	tangential or total
th	theoretical
v	velocity
v	volumetric

Fluids

19 Hydraulic and Pneumatic Systems

Content in blue refers to the *NCEES Handbook*.

NCEES EXAM SPECIFICATIONS AND RELATED CONTENT

MACHINE DESIGN AND MATERIALS EXAM

II.A.12. Mechanical Components: Hydraulic and pneumatic components
5. Control Valves

THERMAL AND FLUID SYSTEMS EXAM

I.B.1 Fluid Mechanics: Fluid Properties
9. Pressure Drop

I.B.2. Fluid Mechanics: Compressible Flow
15. Pneumatic Systems

II.A.5. Hydraulic and Fluid Equipment: Actuators
14. Linear Actuators and Cylinders

II.A.6. Hydraulic and Fluid Equipment: Connections
7. Pressure Rating of Pipe and Tubing

1. INTRODUCTION TO FLUID POWER

Fluid power (*hydraulic power* or *power hydraulic*) *equipment* is hydraulically operating equipment that generates hydraulic pressure at one point in order to perform useful tasks at another. The equipment typically consists of a power source (i.e., an electric motor or internal combustion engine), pump, actuator cylinders or rotary fluid motors, control valves, high-pressure tubing or hose, fluid reservoir, and hydraulic fluid. (See Fig. 19.1.) During operation, the power source pressurizes the hydraulic fluid, and the control valves direct the fluid to the cylinders or hydraulic motors.

Figure 19.1 *Typical Fluid Power Circuit*

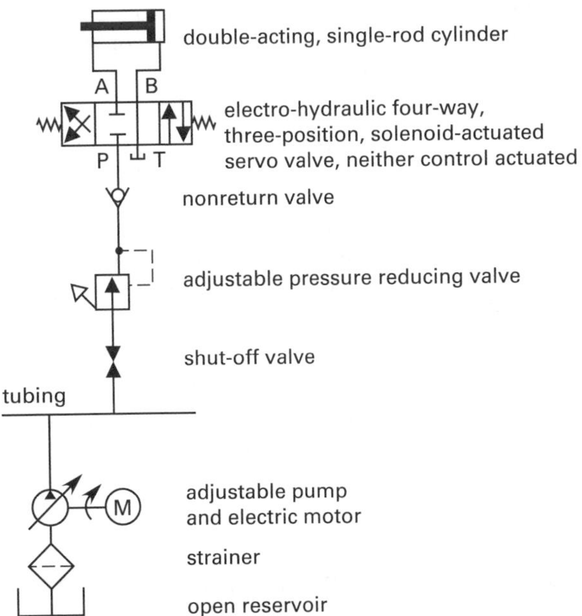

2. FLUID POWER SYMBOLS

Most symbols for fluid power equipment are simplified representations of their physical counterparts. Symbols have been standardized by the ANSI and ISO.[1] **[ANSI Symbols for Hydraulic Power]**

[1]Refer to ANSI Y32.10 and ISO 1219.

3. HYDRAULIC FLUIDS

Petroleum-based oils and fire-resistant fluids are the two general categories of *hydraulic fluids*. Petroleum oils are enhanced with additives intended to inhibit or prevent rust, foam, wear, and oxidation. The largest drawback to petroleum oils, however, is flammability. *Fire-resistant fluids* are categorized as water and oil emulsions,[2] water-glycol mixtures, or straight synthetic fluids (e.g., silicone or phosphate esters, ester blends, and chlorinated hydrocarbon-based fluids).

In addition to cost, the properties most relevant in selecting hydraulic fluids are lubricity (i.e., the ability to reduce friction and prevent wear), viscosity, viscosity index,[3] pour point, flash point, rust resistance, oxidation resistance, and foaming resistance. Relative to petroleum oils, fire resistance is usually achieved to the detriment of the other properties.

Water-oil emulsions are the lowest-cost fire-resistant fluids. They generally perform as well as, or better than, most petroleum fluids.

Due to their similarity to standard antifreeze solutions, water-glycol mixtures are a good choice for low-temperature use. Periodic checking is required to monitor alkalinity and water evaporation. The water content should not be allowed to drop below approximately 35% to 50%.

Synthetic hydraulic fluids are the most costly of the fire-resistant fluids. They have high lubricity. Special formulations may be needed for low-temperature use, and their viscosity indexes are generally lower than those of petroleum oils. A significant factor is that they are not chemically consistent with the seal materials in use for petroleum oils.[4] Therefore, synthetics cannot be used in all existing systems.

The temperature of the hydraulic fluid entering the pump is typically 100°F to 120°F (38°C to 49°C). The temperature is commonly limited by most specifications to 120°F (49°C) for water-based fluids and 130°F (54°C) for all other fluids.

4. DESIGNATIONS OF HYDRAULIC OILS

In hydraulic applications, the type of fluid required is based on many factors, such as type of pump, type of center system (e.g., open/closed), accumulator and cylinder requirements, and use of oil coolers. Hydraulic fluids are identified by both Society of Automotive Engineers (SAE) and/or International Organization for Standardization (ISO) designations.[5] Both designations indicate the oil's useful operating range and simplify the selection and identification of an oil with a given viscosity index. The higher the viscosity number, the more viscous is the fluid. The SAE designation is known as the grade or weight.[6] A recommendation of SAE 10W, 20, or 20W is typical in hydraulic applications. An SAE 20 oil (grade 20 or 20 weight) has a viscosity index of approximately 100. Higher viscosity indexes, such as 120 or 160, are used in applications such as manual transmissions, gear driven transfer cases, and front/rear drive axles, but not as hydraulic fluid.

The *ISO viscosity grade* (VG) designation is similar in purpose and is recognized internationally. Designations 32, 46, and 68 are common specifications for hydraulic applications with vane, piston, and gear-type pumps. As shown in Table 19.1, an ISO 32 oil is equivalent to an SAE 10W oil. An ISO 68 oil is equivalent to an SAE 20W oil.

Table 19.1 *Typical Properties of ISO Hydraulic Oils*

		absolute (dynamic) viscosity*					
		(cSt)		reyns × 10^6 (lbf-sec/in^2)		density	
ISO viscosity grade	equivalent SAE grade	40°C	100°C	104°F	212°F	(kg/m^3)	(lbm/ft^3)
32	10W	32	5.4	4	0.6	857	53.6
46	20	46	6.8	5.7	0.8	861	53.7
68	20W	68	8.7	8.5	1.1	865	54.1
100	30	100	11.4	12.6	1.4	869	54.3
150	40	150	15	19	1.8	872	54.4
220	50	220	19.4	27.7	2.4	875	54.6

(Multiply cSt by 1.0764×10^{-5} to obtain ft^2/sec.)
(Multiply lbm/ft^3 by 16.018 to obtain kg/m^3.)

*For convenience, viscosity in reyns has been increased by 10^6. Multiply table values by 10^{-6}.

5. CONTROL VALVES

Control valves are used to direct the flow of gases and fluids in a variety of HVAC and industrial applications. *Pressure reduction valves* are used to reduce the pressure to components that cannot tolerate or do not need the full system pressure. *Pressure relief valves* are used to protect the system from dangerously high pressures.

Although seated poppet (globe, gate, and plunger) valves can be used, the *spool-type valve* is most common for channel selection and directional control.

[2]An emulsion of water in oil (as opposed to oil in water) is known as an *invert emulsion*.
[3]*Viscosity index* is the relative rate of change in viscosity with temperature. The viscosities of fluids with high viscosity indexes change less with variations in temperature than those of fluids with low viscosity indexes.
[4]Some of the seal materials that are suitable for use with synthetics include butyl rubber, ethylene-propylene rubber, silicone, Teflon™, and nylon.
[5]In the United States, oils are also designated by their military performance numbers (e.g., MLL-PRF-87257).
[6]The designation "W" in SAE grades stands for "winter," not weight.

Control valves are described according to their number of ports, their normal configuration, and their number of positions. The "way" of a control valve is equal to the number of ports. Thus, a three-way diverting valve will have one port for pressurized fluid and two possible discharge ports. A four-way valve will have two ports for pressurized fluid and two discharge ports. If a valve prevents through-flow when de-energized (i.e., when off), it is designated as "normally closed" or "NC." If the valve permits through-flow when de-energized, it is designated as "normally open" or "NO."

In some valves, the fluid can be infinitely split between two discharge ports. In others, there are two distinct *positions*. Thus, a four-way, two-position valve (designated as a "4/2 valve") could switch the discharges completely. [**Two-Way Control Valves**]

A *three-way control valve* (*single-acting valve*) is commonly used to control a single-acting circuit. A *four-way valve* (*double-acting valve*) is typically used to control a double-acting circuit. There are several ways that a control valve in its neutral position can be designed to function. An *open-center valve* (*tandem-center valve*) is typically used with a fixed-displacement pump. In the neutral position, it allows hydraulic fluid to free-flow back to the tank. Shifting the spool position directs hydraulic fluid to the selected port. An *open-center-power-beyond valve* (*high-pressure carryover valve*) is similar to an open-center valve except that in the neutral position, hydraulic fluid flows to the downstream circuit instead of to the tank. A *closed-center valve* is typically used with a variable-displacement pump. Hydraulic fluid is blocked at the valve until the spool is moved out of the neutral position. [**Two-Way Control Valves**]

In the neutral position, a *motor-spool valve* (*free-flow valve*) allows hydraulic fluid to flow back to the tank. The operator is able to run a hydraulic motor under load and, when the valve is shifted back to neutral, the motor is allowed to coast to a stop. A *cylinder-spool valve* should be used in applications where a load is to be raised and held aloft with a hydraulic cylinder. With a cylinder-spool valve in its neutral position, fluid is blocked from flowing to the tank. This effectively locks the load in place. [**Valve Normal Position**]

6. TUBING, HOSE, AND PIPE FOR FLUID POWER

Carbon, alloy, and stainless steel fluid power tubings are available, particularly in the smaller sizes (e.g., up to approximately 2 in in diameter).[7] Fluid power tubing is available in two different pressure ratings: 0 psi to 1000 psi (0 MPa to 6.89 MPa) and 1000 psi to 2500 psi (6.89 MPa to 17.2 MPa).

Types S (seamless) and F (furnace butt-welded) varieties of A53 and A106 grade B black steel pipe can be used when larger diameter pipes are required.[8] The dimensions are the same as for normal steel pipe of the same schedule.

There are numerous types of flexible hose. Most are reinforced with steel wire, steel braid, or other high strength fiber. Fluid compatibility, bend radius, and operating pressure are selection criteria. The internal diameter of flexible hose is the hose's nominal size. When pressurized to more than 250 psi (1.7 MPa), the working pressure of hose is taken as 25% of the burst pressure.

7. PRESSURE RATING OF PIPE AND TUBING

Allowable working pressure (*pressure rating*) for pipe in fluid power systems is calculated in the same manner as that for pressure piping in other types of power piping systems.[9] The maximum working pressure is given by Eq. 19.1.

$$p_{\max} = \frac{2S(t'-C)}{D-2y(t'-C)} \quad \text{[U.S. only]} \qquad 19.1$$

In Eq. 19.1, S is the maximum allowable stress. The maximum allowable stress depends on the metal composition and temperature, but since most fluid power systems run at approximately 100°F (38°C), the maximum stress corresponding to that temperature should be used. By convention, the maximum allowable stress is calculated as the ultimate tensile strength divided by a factor of safety (referred to as a *design factor*) of 3 or, occasionally, 4.[10] Values of 12,500 psi (86.1 MPa) and 17,000 psi (117 MPa) are commonly used for initial studies for A285 carbon steel pipes and tubes, and these values correspond to approximate factors of safety over the ultimate strength of 4 and 3, respectively.

In Eq. 19.1, t' is the *design wall thickness* and is calculated as 87.5% of the nominal wall thickness listed in the pipe tables. C is an allowance for bending, production

[7]Copper pipe may react with some hydraulic fluids and is rarely used.
[8]The "grade" of a steel is usually its tensile strength in ksi. For example, the tensile strength of A516 grade 60 steel is 60 ksi. The materials, grades, specifications, classes, and types of steel pipes and tubes are easily confused. A285, A515, and A516 are common designations for the carbon steel used in pipes and tubes. The grade of the material may relate to its tensile strength, ductility, or other property. Additional specifications may apply to the manufacturer of the pipe. A285 pipe, for example, is often specified according to the additional specifications A53 and A106. The type (e.g., F or S) relates to how seams are formed.
[9]ASA B31.1.1
[10]The abbreviations SMYS and SMTS stand for "standard minimum yield strength" and "standard minimum tensile strength," respectively.

variations, corrosion, threading, and variations in mechanical strength. A value of $C = 0.05$ in is appropriate for threaded steel tubes up to 3/8 in diameter. For larger threaded steel tubes, C is equal to the depth of the thread in inches. $C = 0.05$ in for unthreaded steel tubes up to 0.5 in. $C = 0.065$ in for unthreaded steel tubes 1 1/4 in and larger. For both ferritic and austenitic steels, y is 0.4 for operation at temperatures common to fluid power systems, up to 900°F (480°C). For operation at 950°F (510°C), $y = 0.5$. For 1000°F (540°C and above), $y = 0.7$.[11]

Equation 19.1 is a code-based approach to calculating the working pressure. Three other theoretical methods are also used, primarily with steel hydraulic tubing connected with flared fittings, to calculate the working pressure.[12] Dimensions used in Eq. 19.2, Eq. 19.3, and Eq. 19.4 are nominal, tabulated values. The *Barlow formula* is the standard thin-wall cylinder formula.

Cylindrical Pressure Vessel

$$p = \frac{2St}{D} \tag{19.2}$$

The *Boardman formula* is

$$p = \frac{2St}{D - 0.8t} \tag{19.3}$$

The *Lamé formula* is

$$p = \frac{S(D^2 - d^2)}{D^2 + d^2} \tag{19.4}$$

Since power fluid systems are subject to rapid valve closures, the pressures calculated from Eq. 19.1 through Eq. 19.4 must be reduced for the effect of water hammer. The amount of reduction for water hammer is calculated from the *water hammer factor*, WHF, the ratio of pressure (in psi) to flow rate (in gpm).[13] The derating in working stress due to water hammer is

$$\Delta p_{\text{psi}} = (\text{WHF})Q_{\text{gpm}} \quad \text{[U.S. only]} \tag{19.5}$$

The working pressure after water hammer may be further reduced for connections and fittings. The amount of the reduction is approximately 25%.

8. BURST PRESSURE

The *burst pressure* is calculated from Eq. 19.6, which is derived from the thin-walled cylinder theory. S_{ut} is the ultimate tensile strength.

$$p_{\text{burst}} = \frac{2S_{\text{ut}}t}{D} \tag{19.6}$$

Example 19.1

A 2 in (nominal) schedule-80 A53 grade B (ultimate tensile strength of 60 ksi) unthreaded steel pipe is used in a fluid power system. The pipe carries 100 gpm of 100°F hydraulic fluid. The water hammer factor for this pipe is 6.52 psi/gpm. Considering a 25% reduction for connections and fittings, what is the working pressure?

Solution

From tables of pipe data, find the dimensions of 2 in nominal schedule-80 steel pipe. [**Schedule 80 Steel Pipe**]

$$\begin{aligned} d &= 1.939 \text{ in} \\ t_{\text{nominal}} &= 0.218 \text{ in} \\ D &= d + 2t \\ &= 1.939 \text{ in} + (2)(0.218 \text{ in}) \\ &= 2.375 \text{ in} \end{aligned}$$

The design wall thickness is

$$t_{\text{design}} = 0.875t_{\text{nominal}} = (0.875)(0.218 \text{ in}) = 0.191 \text{ in}$$

The allowable stress is

$$S = \frac{S_{\text{ut}}}{3} = \frac{60{,}000 \ \frac{\text{lbf}}{\text{in}^2}}{3} = 20{,}000 \text{ lbf/in}^2$$

From Eq. 19.1 using $S = 20{,}000$ lbf/in^2, $C = 0.065$ in, and $y = 0.4$,

$$\begin{aligned} p_{100^\circ\text{F}} &= \frac{2S(t' - C)}{D - 2y(t' - C)} \\ &= \frac{(2)\left(20{,}000 \ \frac{\text{lbf}}{\text{in}^2}\right)(0.191 \text{ in} - 0.065 \text{ in})}{2.375 \text{ in} - (2)(0.4)(0.191 \text{ in} - 0.065 \text{ in})} \\ &= 2216 \text{ lbf/in}^2 \end{aligned}$$

[11] Values of C and y are specified by Part 2 of ASME's *Code for Power Piping* (ASME B31.1).
[12] Tubing must conform to SAE J524, J525, and J356.
[13] Including an allowance for water hammer should be based on the type of fluid system, not just on the material used for the pipe or tube. Some sources suggest that the allowance for water hammer should be included only with cast-iron pipes. While it is true that most cast irons experience brittle (not ductile) failure, omitting the pressure increase with ductile pipes denies that water hammer actually occurs.

The derating due to water hammer is

$$\Delta p_{\text{psi}} = (\text{WHF})Q_{\text{gpm}} = \left(6.52\ \frac{\frac{\text{lbf}}{\text{in}^2}}{\frac{\text{gal}}{\text{min}}}\right)\left(100\ \frac{\text{gal}}{\text{min}}\right)$$

$$= 652\ \text{lbf/in}^2$$

The working pressure before fitting allowance is

$$2216\ \frac{\text{lbf}}{\text{in}^2} - 652\ \frac{\text{lbf}}{\text{in}^2} = 1564\ \text{lbf/in}^2$$

The fitting derating is 25%. The working pressure is

$$p = (1 - 0.25)\left(1564\ \frac{\text{lbf}}{\text{in}^2}\right) = 1173\ \text{lbf/in}^2$$

9. PRESSURE DROP

Flows below 15 ft/sec (4.5 m/s) in small-diameter (i.e., less than 1 in (2.5 cm) nominal) tubing are almost always laminar or transitional.[14] Using traditional units, the Reynolds number and friction factor are given by Eq. 19.7 and Eq. 19.8. Tabulations of friction losses for hydraulic oil often assume a viscosity of 33.1 cS (equivalent to 155 SSU) and a density of 62.4 lbm/ft³ (1000 kg/m³).

$$\text{Re} = \frac{3162 Q_{\text{gpm}}}{D_{\text{in}}\nu_{\text{cS}}} \qquad 19.7$$

Moody Diagram (Stanton Diagram)

$$f = \frac{64}{\text{Re}} \quad \text{[laminar flow]} \qquad 19.8$$

The friction loss is given by Eq. 19.9. For an adequate factor of safety, the value calculated should be multiplied by 2 or 3. This will allow for inaccuracies in density and viscosity.

$$\Delta p_{f,\text{psi}} = \frac{2.15 \times 10^{-4} f\rho_{\text{lbm/ft}^3} L_{\text{ft}} Q^2_{\text{gpm}}}{D^5_{\text{in}}} \qquad 19.9$$

The pressure drop through a valve or fitting can be determined in the conventional manner if the equivalent length of the valve or fitting is known. If the *valve flow factor*, C_v, is known, Eq. 19.10 can be used.

$$\Delta p_{v,\text{psi}} = \frac{Q^2_{\text{gpm}}(\text{SG})}{C^2_v} \qquad 19.10$$

10. FLUID POWER PUMPS

Most fluid power pumps are positive displacement pumps. Rotary and reciprocating designs are both used. All of the standard formulas for pump performance (horsepower, torque, etc.) from Chap. 18 apply. For example, the horsepower needed to drive the pump is given by Eq. 19.11. η_{pump} is commonly taken as 0.85.

$$P_{\text{hp}} = \frac{p_{\text{psi}} Q_{\text{gpm}}}{1714\eta_{\text{pump}}} \qquad 19.11$$

The torque on the pump shaft is

$$T_{\text{in-lbf}} = p_{\text{psi}}\left(\frac{\text{displacement in}\ \frac{\text{in}^3}{\text{rev}}}{2\pi}\right) = \frac{63{,}025 P_{\text{hp}}}{N_{\text{rpm}}} = \frac{36.77 Q_{\text{gpm}} p_{\text{psi}}}{N_{\text{rpm}}} \qquad 19.12$$

The flow rate in pumps is related to the displacement per revolution by Eq. 19.13.

$$Q_{\text{gpm}} = \frac{N_{\text{rpm}}\left(\text{displacement in}\ \frac{\text{in}^3}{\text{rev}}\right)}{231} \qquad 19.13$$

To prevent cavitation, the flow velocity in suction lines is generally limited by specification to 1 ft/sec to 5 ft/sec (0.3 m/s to 1.5 m/s). The flow velocity in discharge lines is limited to 10 ft/sec to 15 ft/sec (3 m/s to 4.5 m/s).[15] The actual fluid velocity can easily be calculated from Eq. 19.14.

$$\text{v}_{\text{ft/sec}} = \frac{0.3208 Q_{\text{gpm}}}{A_{\text{in}^2}} \qquad 19.14$$

When starting a hydraulic pump, the fluid viscosity should be 4000 SSU (870 cS) or less. During steady operation at higher temperaures, viscosity should be above 70 SSU (13 cS).

Pump life is a term that refers to bearing life in hours. Most bearings have a *rated life* at some specific speed and pressure. The rated (bearing) life can be modified for other speeds and pressures with Eq. 19.15. Bearing life predictions for pump applications are the same as for other applications.

$$\text{life} = (\text{rated life})\left(\frac{\text{rated speed}}{\text{actual speed}}\right) \times \left(\frac{\text{rated pressure}}{\text{actual pressure}}\right)^3 \qquad 19.15$$

[14]Since the friction factor is larger for laminar flow than for transitional and turbulent flows, laminar flow is the conservative assumption.
[15]Velocities could be faster—up to 25 ft/sec (7.5 m/s)—but are limited to the lower values in order to prevent excessive friction and noise.

11. ELECTRIC MOTORS

The power needed to drive a hydraulic pump is given by Eq. 19.11. Since motors are rated by their developed power, Eq. 19.11 also specifies the minimum motor size. Most fixed (i.e., not mobile) fluid power pumps are driven by single- or three-phase induction motors. Table 19.2 can be used to solve for typical electrical parameters describing the performance of an induction motor.[16]

Table 19.2 *Electric Motor Variables**

		formula	
find	given	single-phase	three-phase
I_{amps}	P_{hp}	$\frac{746P_{hp}}{V\eta(pf)}$	$\frac{746P_{hp}}{\sqrt{3}\,V\eta(pf)}$
I_{amps}	P_{kW}	$\frac{1000P_{kW}}{V(pf)}$	$\frac{1000P_{kW}}{\sqrt{3}\,V(pf)}$
I_{amps}	P_{kVA}	$\frac{1000P_{kVA}}{V}$	$\frac{1000P_{kVA}}{\sqrt{3}\,V}$
P_{kW}		$\frac{IV(pf)}{1000}$	$\frac{\sqrt{3}IV(pf)}{1000}$
P_{kVA}		$\frac{IV}{1000}$	$\frac{\sqrt{3}IV}{1000}$
P_{hp}		$\frac{IV\eta(pf)}{746}$	$\frac{\sqrt{3}IV\eta(pf)}{746}$

(Multiply hp by 0.7457 to obtain kW.)

*η is the motor efficiency.

12. STRAINERS AND FILTERS

Although fluid power systems are theoretically closed, they are never free from dirt, grit, and metal particles. Although cold-formed tubes are essentially free of internal scale, hot-rolled tubes will always have some inside and outside scale. Power systems in daily use should be protected by a 1 μm filter. Backup and occasional-use systems may be able to use a 25 μm filter.

A filter installed in the suction line will protect all components but will contribute to suction pressure loss. For that reason, it is common practice to install the filter after the pump, protecting all components except the pump. The pump is protected by a coarse screen in the suction line.

The maximum permissible pressure drop across a new and clean suction strainer or filter installed below the fluid level (i.e., submerged) is commonly limited by specification to 0.25 psi (1.7 kPa) for fire-resistant fluids and 0.50 psi (3.4 kPa) for all others.

13. ACCUMULATORS

Accumulators store potential energy in the form of pressurized hydraulic fluid. They are commonly used with intermittent duty cycles or to provide emergency power.[17] However, they can also be used to compensate for leakage, act as shock absorbers, and dampen pulsations.

The three basic types of accumulators are weight loaded, mechanical spring loaded, and gas loaded (i.e., hydropneumatic). *Hydropneumatic accumulators*, where pistons, diaphragms, or bladders separate the hydraulic fluid from the gas, are (by far) the most common.[18,19] Most accumulators are high-pressure tanks and, as such, should conform to ASME's *Code for Unfired Pressure Vessels.*

The volume of hydraulic fluid released by or captured in an accumulator is equal to the change in volume of the compressed gas. Compression and expansion of the gas in an accumulator is governed by standard thermodynamic principles. However, calculation of volumetric changes is complicated by the speed of the process, since the gas may heat or cool during the volume change.

The expression for polytropic processes, Eq. 19.16, is the most useful. (In Eq. 19.16, pressures should be strictly absolute pressures. However, the accumulator pressures are so high that use of gage pressures is a common practice.) $V_{charged}$ is the volume of the gas in the accumulator when it is charged with hydraulic fluid; $V_{discharged}$ is the gas volume after the fluid discharge.

$$p_{charged}V^n_{charged} = p_{discharged}V^n_{discharged} \quad \textit{19.16}$$

$$V_{discharged} = V_{charged} + \text{discharge} \quad \textit{19.17}$$

Table 19.3 illustrates the variation in the polytropic exponent, n, with the charge and discharge rates. If the discharge is rapid, the process will be adiabatic, and $n = k$ (the ratio of specific heats). If it is very slow, then the process will be isothermal, and $n = 1$.

Table 19.3 *Typical Polytropic Exponents for Accumulator Sizing*

time for discharge or charging (min)	diatomic gas[a]	monatomic gas[b]
< 1	1.4	1.7
2	1.3	1.5
3	1.15	1.25
3	1.0	1.0

[a]Nitrogen, air, etc.
[b]Helium, argon, etc.

[16]"Line voltage" and "line-to-line voltage" are synonymous terms. There is also a "phase voltage" in three-phase systems. Wye-connected sources are commonly in use, so phase and line voltage are related by $V_{line} = \sqrt{3}\,V_{phase}$.
[17]Use of accumulators is becoming less common.
[18]Reactive gases, such as hydrogen and oxygen, should never be used as accumulator gases.
[19]Spring and weight-loaded accumulators are much rarer.

The *precharge gas pressure* is the pressure in the accumulator when it is completely filled with gas and is empty of hydraulic fluid. Ideally, the precharge pressure would be the minimum system pressure. However, rapid discharge of the gas is an adiabatic (or semi-adiabatic) process, and the gas will cool to below its original system temperature, decreasing the pressure. Some of the accumulator fluid would not discharge. Therefore, the precharge pressure must be higher than the minimum system pressure.

After discharge, the accumulator will slowly warm back up to the system temperature in a constant-volume process.

$$\frac{p_{\text{discharged,cold}}}{T_{\text{discharged,cold}}} = \frac{p_{\text{discharged,warm}}}{T_{\text{discharged,warm}}} \qquad 19.18$$

The slow increase in temperature can also be considered to be an isothermal compression from the charged condition.

$$p_{\text{charged,warm}} V_{\text{charged,warm}} = p_{\text{discharged,warm}} V_{\text{discharged,warm}} \qquad 19.19$$

Since the accumulator has a rigid body, the discharged volume is equal to the total accumulator volume, regardless of the temperature or pressure in the discharged accumulator. Equation 19.17 and Eq. 19.19 can be combined to give a direct expression for the precharge pressure.

$$\begin{aligned} p_{\text{precharge}} &= p_{\text{precharged,warm}} \\ &= \frac{p_{\text{charged,warm}} V_{\text{charged,warm}}}{V_{\text{accumulator}}} \\ &= \frac{p_{\text{charged,warm}} \times (V_{\text{accumulator}} - \text{discharge})}{V_{\text{accumulator}}} \end{aligned} \qquad 19.20$$

Example 19.2

An accumulator must supply 250 in^3 of fluid at 2000 psi minimum pressure. When charged with air, the maximum accumulator pressure cannot exceed 3000 psi. Discharge is assumed to be adiabatic. What are (a) the accumulator size, and (b) the required precharge pressure?

Solution

(a) From Eq. 19.17, the charged accumulator gas volume is

$$V_{\text{charged}} = V_{\text{discharged}} - 250 \text{ in}^3$$

The gas starts out compressed at 3000 psi with volume $V_{\text{warm,charged}}$. It expands to volume $V_{\text{cold,discharged}}$ in an adiabatic process. Since the process is adiabatic, $n = k = 1.4$. The pressure after the discharge is 2000 psi. Use Eq. 19.16 and Eq. 19.17.

$$p_{\text{charged}} V^n_{\text{charged}} = p_{\text{discharged}} V^n_{\text{discharged}}$$

$$V_{\text{discharged}} = (V_{\text{discharged}} - 250 \text{ in}^3)\left(\frac{3000 \ \frac{\text{lbf}}{\text{in}^2}}{2000 \ \frac{\text{lbf}}{\text{in}^2}}\right)^{1/1.4}$$

Rearranging terms and solving, the total accumulator volume is

$$V_{\text{accumulator}} = V_{\text{discharged}} = 994 \text{ in}^3$$

(b) Use Eq. 19.20 to calculate the precharge pressure.

$$\begin{aligned} p_{\text{precharge}} &= \frac{p_{\text{charged,warm}}(V_{\text{accumulator}} - \text{discharge})}{V_{\text{accumulator}}} \\ &= \frac{\left(3000 \ \frac{\text{lbf}}{\text{in}^2}\right)(994 \text{ in}^3 - 250 \text{ in}^3)}{994 \text{ in}^3} \\ &= 2245 \text{ lbf/in}^2 \end{aligned}$$

14. LINEAR ACTUATORS AND CYLINDERS

Linear actuators (e.g., *hydraulic rams* and *hydraulic cylinders*) are the most common type of fluid power actuators.[20] The force exerted by a hydraulic cylinder is easily calculated.

Force and Pressure to Extend Cylinder

$$\begin{aligned} F_R &= \left(\frac{\pi D_1^2}{4}\right) p_R \\ &= A_1 p_R \end{aligned} \qquad 19.21$$

In Eq. 19.21, F_R is the force to extend the cylinder, and p_R is the applied pressure. D_1 is the piston diameter, and A_1 is the piston's cross-sectional area.

In Eq. 19.22, η_{cylinder} accounts for cylinder friction, and it is commonly taken as 0.85. Hydraulic rams are sized in pounds and tons (2000 lbf/ton) of force.

$$F_R = \left(\frac{\pi D_1^2}{4}\right) p_R \eta_{\text{cylinder}} = A_1 p_R \eta_{\text{cylinder}} \qquad 19.22$$

[20]Strictly speaking, a hydraulic motor is a rotary actuator or "linear motor."

Actuating speed of a cylinder rod depends on the fluid flow rate, Q, into the cylinder and the effective cylinder area, A. In most cases, retraction speed is less than the actuating speed, since the fluid acts on the piston area less the rod area.

$$\mathrm{v}_{\text{ft/sec}} = 0.3208\left(\frac{Q_{\text{gpm}}}{A_{\text{in}^2}}\right) \qquad 19.23$$

When fully extended, cylinder rods are like long columns and are subject to buckling failure. Therefore, piston rod sizing should include a check for column loading.

A *hydraulic ram* (*hydraulic jack*, *hydraulic press*, *fluid press*, etc.) is illustrated in Figure 19.2. This is a force-multiplying device. A force, F_p, is applied to the *plunger*, and a useful force, F_r, appears at the *ram*. Even though the pressure in the hydraulic fluid is the same on the ram and plunger, the forces on them will be proportional to their respective cross-sectional areas.

Figure 19.2 *Hydraulic Ram*

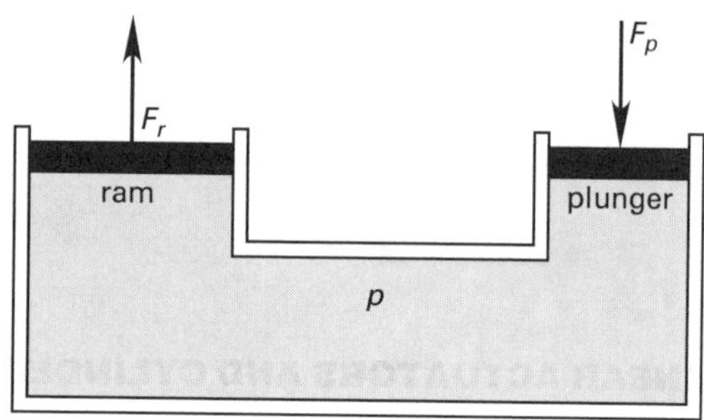

Since the pressure, p, is the same on both the plunger and ram, Eq. 19.24 can be solved for it.

$$F = pA = p\pi\left(\frac{D^2}{4}\right) \qquad 19.24$$

$$p_p = p_r \qquad 19.25$$

$$\frac{F_p}{A_p} = \frac{F_r}{A_r} \qquad 19.26$$

$$\frac{F_p}{D_p^2} = \frac{F_r}{D_r^2} \qquad 19.27$$

A small, manually actuated hydraulic ram will have a lever handle to increase the *mechanical advantage* of the ram from 1.0 to M, as illustrated in Figure 19.3. In most cases, the pivot and connection mechanism will not be frictionless, and some of the applied force will be used to overcome the friction. This friction loss is accounted for by a *lever efficiency* or *lever effectiveness*, η.

$$M = \frac{L_1}{L_2} \qquad 19.28$$

$$F_p = \eta M F_l \qquad 19.29$$

$$\frac{\eta M F_l}{A_p} = \frac{F_r}{A_r} \qquad 19.30$$

Figure 19.3 *Hydraulic Ram with Mechanical Advantage*

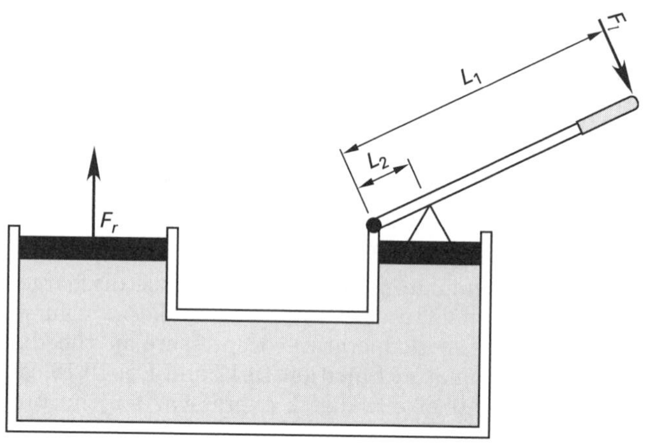

15. PNEUMATIC SYSTEMS

Pneumatic systems are fluid power systems that transfer energy using compressed gas (usually, air) as the working fluid.[21] Pneumatic systems require air compressors (instead of fluid pumps) and solenoid valves, but they are otherwise analogous to fluid power systems. Pneumatic systems require *filter-regulator-lubricator* (FRL) components to assure a clean, lubricated supply of air at constant pressure. Compressors and FRL components are rarely shown on pneumatic schematics. Pneumatic systems can be constructed to exhaust the gas (as to the atmosphere) directly without needing to have fluid return systems. Unlike hydraulic fluid applications that operate between 1000 psig and 10,000 psig (6.9 MPa to 69 MPa), most industrial pneumatic applications operate with gas pressures between 80 psig to 100 psig (550 kPa to 690 kPa). Due to the gas compressibility, pneumatic systems operate more slowly than liquid-based systems using liquid.

Kinetic energy (inertance, inductance) terms are negligible because gas masses are small. Gravimetric (i.e., mass) flow, not volumetric flow, is conserved, since the

[21]Oxygen-free nitrogen (OFN), a compressed gas supplied in bottled form, may be used for convenience or where flammability is an issue. Inert gases, primarily argon, may be used in aerospace applications where minimal chemical reactivity is needed.

working fluid is compressible. Gas flow can be laminar, but it is almost always turbulent.[22] Movement of gas within the system may be subsonic or supersonic.

Properties of gases at various points in pneumatic systems behave according to the ideal gas law, Eq. 19.31. Depending on geometry, insulation, and flow rates, processes within pneumatic systems are described by the *polytropic exponent* (*polytropic index*), n, and can be considered to be constant pressure (isobaric, $n = 0$), constant volume (isochoric, $n = \infty$), constant temperature (isothermal, $n = 1$), adiabatic/isentropic (very fast, $n = k$), or mechanically constrained polytropic. Table 19.4 summarizes the most important properties of atmospheric air at standard temperature and pressure (STP).

Ideal Gas

$$pV = mRT \quad \text{[ideal gas]} \qquad 19.31$$

Special Cases of Closed Systems (With No Change in Kinetic or Potential Energy)

$$pv^n = \text{constant for ideal gas} \quad \text{[polytropic process]} \qquad 19.32$$

Table 19.4 *Approximate Properties of STP Air**

property	customary U.S. value	SI value
specific gas constant, R	53.35 ft-lbf/lbm-°R	287.03 J/kg·K
specific heat at constant pressure, c_p	0.240 Btu/lbm-°R	1005 J/kg·K
specific heat at constant volume, c_v	0.171 Btu/lbm-°R	718 J/kg·K
ratio of specific heats, k	1.40	1.40
density, ρ	0.0807 lbm/ft^3	1.2885 kg/m^3
absolute (dynamic) viscosity, μ	3.599×10^{-7} lbf-sec/ft^2	1.723×10^{-5} Pa·s

(Multiply ft-lbf/lbm-°R by 5.3803 to obtain J/kg·K.)
(Multiply Btu/lbm-°R by 4186.8 to obtain J/kg·K.)
(Multiply lbm/ft^3 by 16.01 to obtain kg/m^3.)
(Multiply lbf-sec/ft^2 by 47.88 to obtain Pa·s.)

*STP = standard temperature and pressure, 492°R (273.15K), 14.696 lbf/in^2 (101.325 kPa)

16. MODELING HYDRAULIC AND PNEUMATIC SYSTEMS

In hydraulic and pneumatic systems, work and power are functions of pressure primarily, and as such, pressure is referred to as the *effort variable*. The product of pressure, p, and volumetric flow rate, Q, is fluid power, much like force × velocity in mechanical systems and voltage × current in electrical systems. The power, P, available at a point in a system where the pressure is p is

$$P = pA\text{v} = pQ \qquad 19.33$$

In addition to using Eq. 19.33 and continuity of mass and continuity of energy equations, performance of fluid and pneumatic power systems can be analyzed using lumped-parameter models. In a *lumped-parameter model*, a system is divided into small segments (i.e., lumps) that contain one or more components. Although the pressure and velocity can vary with time within the system, they are considered instantaneously fixed within the lump. The direction of flow is assumed to be one-dimensional. The concepts of fluid and pneumatic resistance, capacitance, and inductance are easily derived for the lump. Whether all three concepts are used to model a particular system depends on the judgment of the modeller.

In traditional fluid flow analysis of turbulent flow, the relationship between pressure drop and volumetric flow is nonlinear. Specifically, $h_f \propto \text{v}^2$. A relationship similar to $Q = K\sqrt{\Delta p}$ is usually expected. However, for simplicity in fluid power analyses, the relationships are often assumed to be linear over small variations in pressure. This assumption is valid in laminar flow (as in flow through capillary tubes), but it is a convenient simplification otherwise.

Since hydraulic fluid is essentially incompressible, fluid power system component and system models are based on the volumetric flow rate, Q.[23] Gases in pneumatic systems are compressible, so component and system models are based on mass flow rate, $\dot{m} = \rho Q$. As shown in Table 19.5, the characteristic equations are similar for hydraulic and pneumatic systems except for the use of basic variable.

Table 19.5 *Characteristic Equations of Fluid and Pneumatic Power Systems*

component	hydraulic fluid power	pneumatic fluid power
resistance, R	$Q = \dfrac{p_1 - p_2}{R_f}$	$\dot{m} = \dfrac{p_1 - p_2}{R_g}$
capacitance (compliance), C	$Q = C_f\dfrac{d}{dt}(p_1 - p_2)$	$Q = C_f\dfrac{d}{dt}(p_1 - p_2)$
inductance (inertance), I	$Q = \dfrac{1}{I_g}\int(p_1 - p_2)\,dt$	$\dot{m} = \dfrac{1}{I_g}\int(p_1 - p_2)\,dt$

[22]If the gas is incompressible and the flow is laminar, the Hagen-Poiseuille formula can be used. $\dot{V} = \pi D^4/128\mu L \times \Delta p \quad R_f = \Delta p/\dot{V} = 128\mu L/\pi D^4$

[23]Lowercase q is also encountered as the variable for volumetric flow rate in fluid power applications.

Fluids

17. FLUID RESISTANCE

Dissipation of energy in the form of heat occurs to some extent in all fluid systems. *Fluid resistance* (*hydraulic resistance*) represents the energy-dissipation aspect of a system, corresponding to friction and electrical resistance in mechanical and electrical systems.[24] Analogous to electrical systems $V_1 - V_2 = IR$, the driving force in hydraulic systems is a pressure difference, and flow, Q, is resisted by hydraulic resistance, R_f. The units of hydraulic resistance are lbf-sec/ft^5 (N·s/m^5).[25]

$$p_1 - p_2 = QR_f \tag{19.34}$$

Power dissipation in a hydraulic resistance is

$$P = Q\Delta p = Q^2 R_f = \frac{\Delta p^2}{R_f} \tag{19.35}$$

In pneumatic systems, pressure difference is correlated with mass flow rate. Pressure drops across components in subsonic pneumatic systems are usually small and fluctuate with time about a steady-state value. For turbulent flow, the pressure drop can be predicted by Eq. 19.36, where R_g is the *pneumatic resistance* to gas flow. The units of pneumatic resistance are lbf-sec/lbm-ft^2 (N·s/kg·m^2).

$$p_1 - p_2 = \dot{m}R_g \tag{19.36}$$

Power dissipation, P, in a pneumatic resistance is

$$P = Q\Delta p = \frac{\dot{m}\Delta p}{\rho} = \frac{\Delta p^2}{\rho R_g} \tag{19.37}$$

18. FLUID FLOW THROUGH AN ORIFICE

A valve or other flow restriction in a hydraulic system can be modelled as an orifice.[26] Flow through orifices is highly turbulent, and the mass flow rate depends on an experimentally determined *discharge coefficient*, C_d, to account for geometric and frictional effects. The product of the discharge coefficient and the orifice area, C_dA, is known as the *effective cross-sectional area*. The common formula for discharge of an incompressible fluid through an orifice is[27]

$$Q = C_dA\sqrt{\frac{2\Delta p}{\rho}} \quad \text{[SI]} \tag{19.38(a)}$$

$$Q = C_dA\sqrt{\frac{2g_c\Delta p}{\rho}} = C_dA\sqrt{\frac{2g\Delta p}{\gamma}} \quad \text{[U.S.]} \tag{19.38(b)}$$

For pneumatic systems, compressibility effects complicate the analysis of orifice flow. Figure 19.4 illustrates a region of pressurized gas communicating through an orifice with a region having a lower back pressure, p_b. The same assumptions are made as for compressible flow through a nozzle: the gas follows the ideal gas law, viscous and frictional effects are negligible, and flow is fast enough that heat loss does not occur (i.e., flow is isentropic). As with nozzle flow, there is a *critical back pressure* $p_b = p_{cr}$ that maximizes the flow rate (i.e., the *choked flow* condition). Flow will be subsonic for back pressures greater than p_{cr}. Flow will be sonic for back pressures equal to or less than the critical pressure. Due to the unfavorable geometry, supersonic flow is unlikely to occur. For air, $k = 1.4$, and Eq. 19.39 reduces to $p_{cr} = 0.528\,p_1$.

$$p_{cr} = p_1\left(\frac{2}{k+1}\right)^{k/(k-1)} \tag{19.39}$$

Figure 19.4 *Gas Flow Through an Orifice*

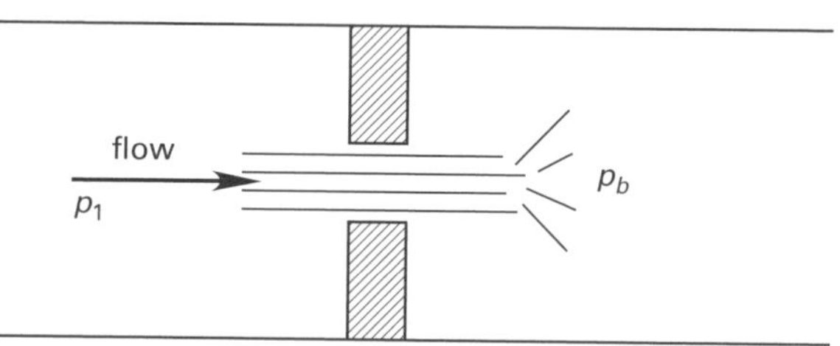

For subsonic flow through an orifice $p_b > p_{cr}$, the mass flow rate, $\dot{m}$, is

$$\dot{m} = C_dA\sqrt{\left(\frac{2}{RT_1}\right)p_b(p_1 - p_b)} \quad \text{[SI]} \tag{19.40(a)}$$

$$\dot{m} = C_dA\sqrt{\left(\frac{2g_c}{RT_1}\right)p_b(p_1 - p_b)} \quad \text{[U.S.]} \tag{19.40(b)}$$

[24] Although the term "fluid resistance" is used, both the component and the fluid contribute to resistance.
[25] It is also possible to define a model based on other flow rate units. If gallons per minute were used, the units of hydraulic resistance would be lbf-min/gal-ft^2.
[26] It will be necessary to determine the discharge coefficient and/or effective area, C_dA, of the valve experimentally.
[27] This is the common Torricelli equation derived from energy relationships, not from Eq. 19.34 or Eq. 19.36.

For sonic flow $p_b < p_{cr}$, the mass flow rate is

$$\dot{m} = C_d A p_1 \sqrt{\frac{1}{RT_1}} \sqrt{k\left(\frac{2}{k+1}\right)^{(k+1)/(k-1)}} \quad \text{[SI]} \qquad 19.41(a)$$

$$\dot{m} = C_d A p_1 \sqrt{\frac{g_c}{RT_1}} \sqrt{k\left(\frac{2}{k+1}\right)^{(k+1)/(k-1)}} \quad \text{[U.S.]} \qquad 19.41(b)$$

Example 19.3

0.12 kg/s of 293K air flows through a valve during a calibration test. The supply pressure is 160 kPa, and the air exhausts to the (standard) atmosphere. What is the effective area of the valve?

Solution

The ratio of back pressure to the supply pressure is

$$\frac{p_b}{p_1} = \frac{101.325 \text{ kPa}}{160 \text{ kPa}} = 0.633$$

Since this is greater than the critical back pressure ratio, 0.528, the flow is subsonic.

From Eq. 19.40, the effective area, $C_d A$, is

$$\begin{aligned} C_d A &= \frac{\dot{m}}{\sqrt{\left(\frac{2}{RT_1}\right) p_b (p_1 - p_b)}} \\ &= \frac{0.12 \ \frac{\text{kg}}{\text{s}}}{\sqrt{\left(\frac{(2)(101.325 \text{ kPa})}{\left(287.03 \ \frac{\text{J}}{\text{kg·K}}\right)(293\text{K})}\right) \times (160 \text{ kPa} - 101.325 \text{ kPa}) \times \left(1000 \ \frac{\text{Pa}}{\text{kPa}}\right)^2}} \\ &= 3.19 \times 10^{-4} \text{ m}^2 \end{aligned}$$

19. NATURAL FREQUENCY OF FLUID POWER SYSTEMS

Resonance in fluid power systems can occur if timing of pulsations from pumps and compressors corresponds to the natural frequency of the system. The *natural frequency*, ω, in units of rad/sec of a fluid line can be calculated from the *characteristic time* (i.e., the time that a pressure disturbance takes to travel to and return from the end of line, which is a function of speed of sound, c, in the line).[28]

$$T = \frac{2L}{c} \quad \text{[characteristic time]} \qquad 19.42$$

$$c = \sqrt{\frac{B}{\rho}} = \sqrt{\frac{1}{\beta\rho}} \quad \text{[wave velocity]} \qquad 19.43$$

$$\omega = 2\pi f = \frac{2\pi}{T} \quad \text{[natural frequency]} \qquad 19.44$$

Analogous to electrical systems, the natural frequency, ω, of a lossless (i.e., zero fluid resistance) single degree-of-freedom fluid system can also be calculated from its compliance and inertance.

$$\omega = \frac{1}{\sqrt{I_f C_f}} \quad \text{[hydraulic]} \qquad 19.45$$

$$\omega = \frac{1}{\sqrt{I_g C_g}} \quad \text{[pneumatic]} \qquad 19.46$$

20. MAXIMUM SURGE PRESSURE

Joukowsky's equation describes the pressure profile of a pressure fluctuation.

$$\frac{dp}{dt} = \rho c \frac{d\text{v}}{dt} \qquad 19.47$$

When a valve is closed suddenly (relative to the time for a pressure wave to travel the length of the pipe), dt approaches zero, and the overpressurization (i.e., the increase in pressure above the nominal pressure) in the pipe is given by Eq. 19.48. The second form of Eq. 19.48 is essentially Eq. 19.34 with the characteristic impedance substituted for the fluid resistance.

$$\Delta p = \rho c \Delta \text{v} = Q Z_o \qquad 19.48$$

[28] The speed of sound, c, is also known as the *characteristic wave celerity*.

Fluids

21. NOMENCLATURE

A	area	ft^2	m^2
B	bulk modulus	lbf/ft^2	Pa
c	specific heat	Btu/lbm-°R	J/kg·K
c	speed of sound	ft/sec	m/s
C	corrosion allowance	ft	m
C_d	discharge coefficient	–	–
C_f	hydraulic compliance	ft^5/lbf	m^5/N
C_g	pneumatic compliance	lbm-ft^2/lbf	kg·m^2/N
C_v	valve flow factor	–	–
d	inner diameter	ft	m
D	outer diameter	ft	m
f	frequency	Hz	Hz
f	friction factor	–	–
F	force	lbf	N
F_R	force to extend cylinder	lbf	N
g	gravitational acceleration, 32.17 (9.807)	ft/sec^2	m/s^2
g_c	gravitational constant, 32.17	lbm-ft/lbf-sec^2	n.a.
h_f	head loss due to friction	ft	m
I	effective (rms) line current	A	A
I_f	hydraulic inertance	lbf-sec^2/ft^5	N·s^2/m^5 (kg/m^4)
I_g	pneumatic inertance	lbf-sec^2/lbm-ft^2	N·s^2/kg·m^2
k	ratio of specific heats	–	–
K	constant	various	various
L	length	ft	m
m	mass	lbm	kg
$\dot{m}$	mass flow rate	lbm/sec	kg/s
M	mechanical advantage	–	–
n	polytropic exponent	–	–
N	rotational speed	rpm	rpm
p	pressure	lbf/ft^2	Pa
p_R	applied pressure	lbf/ft^2	Pa
pf	power factor	–	–
P	power	hp or ft-lbf/sec	W
Q	flow rate	ft^3/sec	m^3/s
R	electrical resistance	Ω	Ω
R	specific gas constant	ft-lbf/lbm-°R	J/kg·K
R_f	hydraulic resistance	lbf-sec/ft^5	N·s/m^5
R_g	pneumatic resistance	lbf-sec/lbm-ft^2	N·s/kg·m^2
Re	Reynolds number	–	–
S	maximum allowable tensile stress	lbf/ft^2	Pa
S	strength	lbf/ft^2	Pa
SG	specific gravity	–	–
t	nominal thickness	ft	m
t	time	sec	s
t'	design thickness	ft	m
T	temperature	°R	K
T	torque	ft-lbf	N·m
v	velocity	ft/sec	m/s
V	voltage	V	V
V	volume	ft^3	m^3
WHF	water hammer factor	–	–
y	temperature derating factor	–	–
Z	impedance	lbf-sec/ft^5	N·s/m^5

Symbols

β	compressibility	ft^2/lbf	1/Pa
η	efficiency	–	–
μ	absolute viscosity	lbf-sec/ft^2	Pa·s
ν	kinematic viscosity*	SSU	cS
ρ	density	lbm/ft^3	kg/m^3
ω	natural frequency	rad/sec	rad/s

*The following conversions may be used between Saybolt Seconds Universal (SSU) and centistokes (cS):

$$\nu_{\text{cS}} = 0.2253\nu_{\text{SSU}} - \frac{194.4}{\nu_{\text{SSU}}} \quad 32 < \text{SSU} < 100$$

$$\nu_{\text{cS}} = 0.2193\nu_{\text{SSU}} - \frac{134.6}{\nu_{\text{SSU}}} \quad 100 < \text{SSU} < 240$$

$$\nu_{\text{cS}} = \frac{\nu_{\text{SSU}}}{4.635} \quad \text{SSU} > 240$$

Subscripts

b	back
cr	critical
f	hydraulic (fluid)
g	pneumatic (gas)
l	lever
	max maximum
p	constant pressure or plunger
r	ram
ut	ultimate tensile
v	constant volume or valve

Topic III: Thermodynamics

20 Inorganic Chemistry

Content in blue refers to the *NCEES Handbook*.

NCEES EXAM SPECIFICATIONS AND RELATED CONTENT

MACHINE DESIGN AND MATERIALS EXAM

I.C.1. Material Properties: Physical
- 32. Galvanic Action
- 33. Corrosion

I.C.2. Material Properties: Chemical
- 32. Galvanic Action
- 33. Corrosion

I.C.3.a. Material Properties: Mechanical: Time-independent behavior
- 32. Galvanic Action
- 33. Corrosion

1. ATOMIC STRUCTURE

An *element* is a substance that cannot be decomposed into simpler substances during ordinary chemical reactions.[1] An *atom* is the smallest subdivision of an element that can take part in a chemical reaction. A *molecule* is the smallest subdivision of an element or compound that can exist in a natural state.

The atomic nucleus consists of neutrons and protons, known as *nucleons*. The masses of neutrons and protons are essentially the same—one *atomic mass unit*, amu. One amu is exactly $\frac{1}{12}$ of the mass of an atom of carbon-12, approximately equal to 1.66×10^{-27} kg.[2] The *relative atomic weight* or *atomic weight*, A, of an atom is approximately equal to the number of protons and neutrons in the nucleus.[3] (See App. 20.A.) The *atomic number*, Z, of an atom is equal to the number of protons in the nucleus.

The atomic number and atomic weight of an element E are written in symbolic form as ${}_Z\mathrm{E}^A$, E_Z^A, or ${}_Z^A\mathrm{E}$. For example, carbon is the sixth element; radioactive carbon has an atomic

[1]Atoms of an element can be decomposed into subatomic particles in nuclear reactions.

[2]Until 1961, the atomic mass unit was defined as $\frac{1}{16}$ of the mass of one atom of oxygen-16.

[3]The term *weight* is used even though all chemical calculations involve mass. The atomic weight of an atom includes the mass of the electrons. Published *chemical atomic weights* of elements are averages of all the atomic weights of stable isotopes, taking into consideration the relative abundances of the isotopes.

mass of 14. Therefore, the symbol for carbon-14 is C_6^{14}. Since the atomic number is superfluous if the chemical symbol is given, the atomic number can be omitted (e.g., C^{14}).

2. ISOTOPES

Although an element can have only a single atomic number, atoms of that element can have different atomic weights. Many elements possess *isotopes*. The nuclei of isotopes differ from one another only in the number of neutrons. Isotopes behave the same way chemically.[4] Therefore, isotope separation must be done physically (e.g., by centrifugation or gaseous diffusion) rather than chemically.

Hydrogen has three isotopes. H_1^1 is *normal hydrogen* with a single proton nucleus. H_1^2 is known as *deuterium* (*heavy hydrogen*), with a nucleus of a proton and neutron. (This nucleus is known as a *deuteron.*) Finally, H_1^3 (*tritium*) has two neutrons in the nucleus. While normal hydrogen and deuterium are stable, tritium is radioactive. Many elements have more than one stable isotope. Tin, for example, has 10.

The *relative abundance*, x_i, of an isotope, i, is equal to the fraction of that isotope in a naturally occurring sample of the element. The *chemical atomic weight* is the weighted average of the isotope weights.

$$A_{\text{average}} = x_1A_1 + x_2A_2 + \cdots \quad \text{20.1}$$

3. PERIODIC TABLE

The *periodic table*, as shown in App. 20.D, is organized around the *periodic law:* The properties of the elements depend on the atomic structure and vary with the atomic number in a systematic way. Elements are arranged in order of increasing atomic numbers from left to right. Adjacent elements in horizontal rows differ decidedly in both physical and chemical properties. However, elements in the same column have similar properties. Graduations in properties, both physical and chemical, are most pronounced in the *periods* (i.e., the horizontal rows). [**Periodic Table of the Elements**]

The vertical columns are known as *groups*, numbered in Roman numerals. Elements in a group are called *cogeners*. Each vertical group except 0 and VIII has A and B subgroups (*families*). The elements of a family resemble each other more than they resemble elements in the other family of the same group. Graduations in properties are definite but less pronounced in vertical families. The trend in any family is toward more *metallic properties* as the atomic weight increases.

Metals (elements at the left end of the periodic chart) have low electron affinities and electronegativities, are reducing agents, form positive ions, and have positive oxidation numbers. They have high electrical conductivities, luster, generally high melting points, ductility, and malleability.

Nonmetals (elements at the right end of the periodic chart) have high electron affinities and electronegativities, are oxidizing agents, form negative ions, and have negative oxidation numbers. They are poor electrical conductors, have little or no luster, and form brittle solids. Of the common nonmetals, fluorine has the highest electronic affinity and electronegativity, with oxygen having the next highest values.

The *metalloids* (e.g., boron, silicon, germanium, arsenic, antimony, tellurium, and polonium) have characteristics of both metals and nonmetals. Electrically, they are semiconductors.

The electron-attracting power of an atom is called its *electronegativity.* Metals have low electronegativities. Group VIIA elements (fluorine, chlorine, etc.) are most strongly electronegative. The alkali metals (Group IA) are the most weakly electronegative. Generally, the most *electronegative elements* are those at the right ends of the periods. Elements with low electronegativities are found at the beginning (i.e., left end) of the periods. Electronegativity decreases as you go down a group.

Elements in the periodic table are often categorized into the following groups.

- *actinides:* same as actinons
- *actinons:* elements 90–103[5]
- *alkali metals:* group IA
- *alkaline earth metals:* group IIA
- *halogens:* group VIIA
- *heavy metals:* metals near the center of the chart
- *inner transition elements:* same as transition metals
- *lanthanides:* same as lanthanons
- *lanthanons:* elements 58–71[6]
- *light metals:* elements in the first two groups
- *metals:* everything except the nonmetals
- *metalloids:* elements along the dark line in the chart separating metals and nonmetals

[4]There are slight differences, known as *isotope effects*, in the chemical behavior of isotopes. These effects usually influence only the rate of reaction, not the kind of reaction.

[5]The *actinons* resemble element 89, *actinium*. Therefore, element 89 is sometimes included as an actinon.

[6]The *lanthanons* resemble element 57, *lanthanum*. Therefore, element 57 is sometimes included as a lanthanon.

- *noble gases:* group 0
- *nonmetals:* elements 2, 5–10, 14–18, 33–36, 52–54, 85, and 86
- *rare earths:* same as lanthanons
- *transition elements:* same as transition metals
- *transition metals:* all B families and group VIII B[7]

4. OXIDATION NUMBER

The *oxidation number* (*oxidation state*) is an electrical charge assigned by a set of prescribed rules. It is actually the charge assuming all bonding is ionic.[8] The sum of the oxidation numbers equals the net charge. For monoatomic ions, the oxidation number is equal to the charge. The oxidation numbers of some common atoms and radicals are given in Table 20.1. **[Primary Bonds]**

In covalent compounds, all of the bonding electrons are assigned to the ion with the greater electronegativity. For example, nonmetals are more electronegative than metals. Carbon is more electronegative than hydrogen.

For atoms in a free-state molecule, the oxidation number is zero. Hydrogen gas is a diatomic molecule, H_2. Therefore, the oxidation number of the hydrogen molecule, H_2, is zero. The same is true for the atoms in O_2,N_2,Cl_2, and so on. Also, the sum of all the oxidation numbers of atoms in a neutral molecule is zero.

Fluorine is the most electronegative element, and it has an oxidation number of −1. Oxygen is second only to fluorine in electronegativity. Usually, the oxidation number of oxygen is −2, except in peroxides, where it is −1, and when combined with fluorine, where it is +2. Hydrogen is usually +1, except in hydrides, where it is −1.

For a charged *radical* (a group of atoms that combine as a single unit), the net oxidation number is equal to the charge on the radical, known as the *charge number.*

Example 20.1

What are the oxidation numbers of all the elements in the chlorate (ClO_3^{-1}) and permanganate (MnO_4^{-1}) ions?

Solution

For the chlorate ion, the oxygen is more electronegative than the chlorine. (Only fluorine is more electronegative than oxygen.) Therefore, the oxidation number of oxygen is −2. In order for the net oxidation number to be −1, the chlorine must have an oxidation number of +5.

Table 20.1 *Oxidation Numbers of Selected Atoms and Charge Numbers of Radicals*

name	symbol	oxidation or charge number
acetate	$C_2H_3O_2$	−1
aluminum	Al	+3
ammonium	NH_4	+1
barium	Ba	+2
borate	BO_3	−3
boron	B	+3
bromine	Br	−1
calcium	Ca	+2
carbon	C	+4, −4
carbonate	CO_3	−2
chlorate	ClO_3	−1
chlorine	Cl	−1
chlorite	ClO_2	−1
chromate	CrO_4	−2
chromium	Cr	+2, +3, +6
copper	Cu	+1, +2
cyanide	CN	−1
dichromate	Cr_2O_7	−2
fluorine	F	−1
gold	Au	+1, +3
hydrogen	H	+1 (−1 in hydrides)
hydroxide	OH	−1
hypochlorite	ClO	−1
iron	Fe	+2, +3
lead	Pb	+2, +4
lithium	Li	+1
magnesium	Mg	+2
mercury	Hg	+1, +2
nickel	Ni	+2, +3
nitrate	NO_3	−1
nitrite	NO_2	−1
nitrogen	N	−3, +1, +2, +3, +4, +5
oxygen	O	−2 (−1 in peroxides)
perchlorate	ClO_4	−1
permanganate	MnO_4	−1
phosphate	PO_4	−3
phosphorus	P	−3, +3, +5
potassium	K	+1
silicon	Si	+4, −4
silver	Ag	+1
sodium	Na	+1
sulfate	SO_4	−2
sulfite	SO_3	−2
sulfur	S	−2, +4, +6
tin	Sn	+2, +4
zinc	Zn	+2

[7]The *transition metals* are elements whose electrons occupy the d sublevel. They can have various oxidation numbers, including +2, +3, +4, +6, and +7.
[8]The three types of primary bonds are ionic (e.g., salts, metal oxides), covalent (e.g., within polymer molecules), and metallic (e.g., metals).

Thermodynamics

For the permanganate ion, the oxygen is more electronegative than the manganese. Therefore, the oxidation number of oxygen is −2. For the net oxidation number to be −1, the manganese must have an oxidation number of +7.

5. FORMATION OF COMPOUNDS

Compounds form according to the *law of definite (constant) proportions:* A pure compound is always composed of the same elements combined in a definite proportion by mass. For example, common table salt is always NaCl. It is not sometimes NaCl and other times Na_2Cl or $NaCl_3$ (which do not exist, in any case).

Furthermore, compounds form according to the *law of (simple) multiple proportions:* When two elements combine to form more than one compound, the masses of one element that combine with the same mass of the other are in the ratios of small integers.

In order to evaluate whether a compound formula is valid, it is necessary to know the *oxidation numbers* of the interacting atoms. Although some atoms have more than one possible oxidation number, most do not.

The sum of the oxidation numbers must be zero if a neutral compound is to form. For example, H_2O is a valid compound since the two hydrogen atoms have a total positive oxidation number of $2 \times 1 = +2$. The oxygen ion has an oxidation number of −2. These oxidation numbers sum to zero.

On the other hand, $NaCO_3$ is not a valid compound formula. The sodium (Na) ion has an oxidation number of +1. However, the carbonate radical has a charge number of −2. The correct sodium carbonate molecule is Na_2CO_3.

6. NAMING COMPOUNDS

Combinations of elements are known as *compounds. Binary compounds* contain two elements; *ternary (tertiary) compounds* contain three elements. A *chemical formula* is a representation of the relative numbers of each element in the compound. For example, the formula $CaCl_2$ shows that there are one calcium atom and two chlorine atoms in one molecule of calcium chloride.

Generally, the numbers of atoms are reduced to their lowest terms. However, there are exceptions. For example, acetylene is C_2H_2, and hydrogen peroxide is H_2O_2.

For binary compounds with a metallic element, the positive metallic element is listed first. The chemical name ends in the suffix "-ide." For example, NaCl is sodium chloride. If the metal has two oxidation states, the suffix "-ous" is used for the lower state, and "-ic" is used for the higher state. Alternatively, the element name can be used with the oxidation number written in Roman numerals. For example,

$FeCl_2$: ferrous chloride, or iron (II) chloride
$FeCl_3$: ferric chloride, or iron (III) chloride

For binary compounds formed between two nonmetals, the more positive element is listed first. The number of atoms of each element is specified by the prefixes "di-" (2), "tri-" (3), "tetra-" (4), and "penta-" (5), and so on. For example,

N_2O_5: dinitrogen pentoxide

Binary acids start with the prefix "hydro-," list the name of the nonmetallic element, and end with the suffix "-ic." For example,

HCl: hydrochloric acid

Ternary compounds generally consist of an element and a radical. The positive part is listed first in the formula. *Ternary acids* (also known as *oxyacids*) usually contain hydrogen, a nonmetal, and oxygen, and can be grouped into families with different numbers of oxygen atoms. The most common acid in a family (i.e., the root acid) has the name of the nonmetal and the suffix "-ic." The acid with one more oxygen atom than the root is given the prefix "per-" and the suffix "-ic." The acid containing one less oxygen atom than the root is given the ending "-ous." The acid containing two less oxygen atoms than the root is given the prefix "hypo-" and the suffix "-ous."

For example,

HClO: hypochlorous acid
$HClO_2$: chlorous acid
$HClO_3$: chloric acid (the root)
$HClO_4$: perchloric acid

A list of chemical names, along with their common names and formulas, is provided in App. 20.H.

7. MOLES AND AVOGADRO'S LAW

The *mole* is a measure of the quantity of an element or compound. Specifically, a mole of an element will have a mass equal to the element's atomic (or molecular) weight. The three main types of moles are based on mass being measured in grams, kilograms, and pounds. Obviously, a gram-based mole of carbon (12.0 grams) is not the same quantity as a pound-based mole of carbon (12.0 pounds). Although "mol" is understood in SI countries to mean a gram-mole, the term *mole* is ambiguous, and the units mol (gmol), kmol (kgmol), or lbmol must be specified.[9]

[9]There are also variations on the presentation of these units, such as g mol, gmole, g-mole, kmole, kg-mol, lb-mole, pound-mole, and p-mole. In most cases, the intent is clear.

One gram-mole of any substance has the same number of particles (atoms, molecules, ions, electrons, etc.), 6.022×10^{23}, *Avogadro's number*, N_A. A pound-mole contains approximately 454 times the number of particles in a gram-mole.

Avogadro's law (*hypothesis*) holds that equal volumes of all gases at the same temperature and pressure contain equal numbers of gas molecules. Specifically, at standard scientific conditions (1.0 atm and 0°C), one gram-mole of any gas contains 6.022×10^{23} molecules and occupies 22.4 L. A pound-mole occupies 454 times that volume, 359 ft^3.

"Molar" is used as an adjective when describing properties of a mole. For example, a *molar volume* is the volume of a mole.

Example 20.2

How many electrons are in 0.01 g of gold? ($A = 196.97$; $Z = 79$.)

Solution

The number of gram-moles of gold present is

$$n = \frac{m}{M} = \frac{0.01 \text{ g}}{196.97 \frac{\text{g}}{\text{mol}}} = 5.077 \times 10^{-5} \text{ mol}$$

The number of gold nuclei is

$$\begin{aligned} N = nN_A &= (5.077 \times 10^{-5} \text{ mol})\left(6.022 \times 10^{23} \frac{\text{nuclei}}{\text{mol}}\right) \\ &= 3.057 \times 10^{19} \text{ nuclei} \end{aligned}$$

Since the atomic number is 79, there are 79 protons and 79 electrons in each gold atom. The number of electrons is

$$N_{\text{electrons}} = (3.057 \times 10^{19})(79) = 2.42 \times 10^{21}$$

8. FORMULA AND MOLECULAR WEIGHTS

The *formula weight*, FW, of a molecule (compound) is the sum of the atomic weights of all elements in the molecule. The *molecular weight*, *M*, is generally the same as the formula weight. The units of molecular weight are g/mol, kg/kmol, or lbm/lbmol. However, units are sometimes omitted because weights are relative. For example,

$$CaCO_3\text{: FW} = M = 40.1 + 12 + 3 \times 16 = 100.1$$

An *ultimate analysis* (which determines how much of each element is present in a compound) will not necessarily determine the molecular formula. It will determine only the formula weight based on the relative proportions of each element. Therefore, except for hydrated molecules and other linked structures, the molecular weight will be an integer multiple of the formula weight.

For example, an ultimate analysis of hydrogen peroxide (H_2O_2) will show that the compound has one oxygen atom for each hydrogen atom. In this case, the formula would be assumed to be HO and the formula weight would be approximately 17, although the actual molecular weight is 34.

For *hydrated molecules* (e.g., $FeSO_4 \cdot 7H_2O$), the mass of the *water of hydration* (also known as the *water of crystallization*) is included in the formula and in the molecular weight.

9. EQUIVALENT WEIGHT

The *equivalent weight*, EW (i.e., an *equivalent*), is the amount of substance (in grams) that supplies one gram-mole (i.e., 6.022×10^{23}) of reacting units. For acid-base reactions, an acid equivalent supplies one gram-mole of H^+ ions. A base equivalent supplies one gram-mole of OH^- ions. In oxidation-reduction reactions, an equivalent of a substance gains or loses a gram-mole of electrons. Similarly, in electrolysis reactions an equivalent weight is the weight of substance that either receives or donates one gram-mole of electrons at an electrode.

The equivalent weight can be calculated as the molecular weight divided by the change in oxidation number experienced by a compound in a chemical reaction. A compound can have several equivalent weights.

$$\text{EW} = \frac{M}{\Delta \text{ oxidation number}} \qquad 20.2$$

Example 20.3

What are the equivalent weights of the following compounds?

(a) Al in the reaction

$$Al^{+++} + 3e^- \rightarrow Al$$

(b) H_2SO_4 in the reaction

$$H_2SO_4 + H_2O \rightarrow 2H^+ + SO_4^{-2} + H_2O$$

(c) NaOH in the reaction

$$NaOH + H_2O \rightarrow Na^+ + OH^- + H_2O$$

Solution

(a) The atomic weight of aluminum is approximately 27. Since the change in the oxidation number is 3, the equivalent weight is $27/3 = 9$.

Thermodynamics

(b) The molecular weight of sulfuric acid is approximately 98. Since the acid changes from a neutral molecule to ions with two charges each, the equivalent weight is $98/2 = 49$.

(c) Sodium hydroxide has a molecular weight of approximately 40. The originally neutral molecule goes to a singly charged state. Therefore, the equivalent weight is $40/1 = 40$.

10. GRAVIMETRIC FRACTION

The *gravimetric fraction*, x_i, of an element i in a compound is the fraction by weight m of that element in the compound. The gravimetric fraction is found from an *ultimate analysis* (also known as a *gravimetric analysis*) of the compound.

$$x_i = \frac{m_i}{m_1 + m_2 + \cdots + m_i + \cdots + m_n} = \frac{m_i}{m_t} \quad 20.3$$

The *percentage composition* is the gravimetric fraction converted to percentage.

$$\%\ \text{composition} = x_i \times 100\% \quad 20.4$$

If the gravimetric fractions are known for all elements in a compound, the *combining weights* of each element can be calculated. (The term *weight* is used even though mass is the traditional unit of measurement.)

$$m_i = x_i m_t \quad 20.5$$

11. EMPIRICAL FORMULA DEVELOPMENT

It is relatively simple to determine the *empirical formula* of a compound from the atomic and combining weights of elements in the compound. The empirical formula gives the relative number of atoms (i.e., the formula weight is calculated from the empirical formula).

step 1: Divide the gravimetric fractions (or percentage compositions) by the atomic weight of each respective element.

step 2: Determine the smallest ratio from step 1.

step 3: Divide all of the ratios from step 1 by the smallest ratio.

step 4: Write the chemical formula using the results from step 3 as the numbers of atoms. Multiply through as required to obtain all integer numbers of atoms.

Example 20.4

A clear liquid is analyzed, and the following gravimetric percentage compositions are recorded: carbon, 37.5%; hydrogen, 12.5%; oxygen, 50%. What is the chemical formula for the liquid?

Solution

step 1: Divide the percentage compositions by the atomic weights.

$$\text{C: } \frac{37.5}{12} = 3.125$$
$$\text{H: } \frac{12.5}{1} = 12.5$$
$$\text{O: } \frac{50}{16} = 3.125$$

step 2: The smallest ratio is 3.125.

step 3: Divide all ratios by 3.125.

$$\text{C: } \frac{3.125}{3.125} = 1$$
$$\text{H: } \frac{12.5}{3.125} = 4$$
$$\text{O: } \frac{3.125}{3.125} = 1$$

step 4: The empirical formula is CH_4O.

If it had been known that the liquid behaved as though it contained a hydroxyl (OH) radical, the formula would have been written as CH_3OH. This is recognized as methyl alcohol.

12. CHEMICAL REACTIONS

During chemical reactions, bonds between atoms are broken and new bonds are usually formed. The starting substances are known as *reactants*; the ending substances are known as *products*. In a chemical reaction, reactants are either converted to simpler products or synthesized into more complex compounds. There are four common types of reactions.

- *direct combination* (or *synthesis*): This is the simplest type of reaction where two elements or compounds combine directly to form a compound.

$$2H_2 + O_2 \rightarrow 2H_2O$$
$$SO_2 + H_2O \rightarrow H_2SO_3$$

- *decomposition* (or *analysis*): Bonds within a compound are disrupted by heat or other energy to produce simpler compounds or elements.

$$2HgO \rightarrow 2Hg + O_2$$
$$H_2CO_3 \rightarrow H_2O + CO_2$$

- *single displacement* (or *replacement*[10]): This type of reaction has one element and one compound as reactants.

$$2Na + 2H_2O \rightarrow 2NaOH + H_2$$
$$2KI + Cl_2 \rightarrow 2KCl + I_2$$

- *double displacement* (or *replacement*): These are reactions with two compounds as reactants and two compounds as products.

$$AgNO_3 + NaCl \rightarrow AgCl + NaNO_3$$
$$H_2SO_4 + ZnS \rightarrow H_2S + ZnSO_4$$

13. BALANCING CHEMICAL EQUATIONS

The coefficients in front of element and compound symbols in chemical reaction equations are the numbers of molecules or moles taking part in the reaction. (For gaseous reactants and products, the coefficients also represent the numbers of volumes. This is a direct result of Avogadro's hypothesis that equal numbers of molecules in the gas phase occupy equal volumes under the same conditions.)[11]

Since atoms cannot be changed in a normal chemical reaction (i.e., mass is conserved), the numbers of each element must match on both sides of the equation. When the numbers of each element match, the equation is said to be "balanced." The total atomic weights on both sides of the equation will be equal when the equation is balanced.

Balancing simple chemical equations is largely a matter of deductive trial and error. More complex reactions require use of oxidation numbers.

Example 20.5

Balance the following reaction equation.

$$Al + H_2SO_4 \rightarrow Al_2(SO_4)_3 + H_2$$

Solution

As written, the reaction is not balanced. For example, there is one aluminum on the left, but there are two on the right. The starting element in the balancing procedure is chosen somewhat arbitrarily.

step 1: Since there are two aluminums on the right, multiply Al by 2.

$$2Al + H_2SO_4 \rightarrow Al_2(SO_4)_3 + H_2$$

step 2: Since there are three sulfate radicals (SO_4) on the right, multiply H_2SO_4 by 3.

$$2Al + 3H_2SO_4 \rightarrow Al_2(SO_4)_3 + H_2$$

step 3: Now there are six hydrogens on the left, so multiply H_2 by 3 to balance the equation.

$$2Al + 3H_2SO_4 \rightarrow Al_2(SO_4)_3 + 3H_2$$

14. STOICHIOMETRIC REACTIONS

Stoichiometry is the study of the proportions in which elements and compounds react and are formed. A *stoichiometric reaction* (also known as a *perfect reaction* or an *ideal reaction*) is one in which just the right amounts of reactants are present. After the reaction stops, there are no unused reactants.

Stoichiometric problems are known as *weight and proportion problems* because their solutions use simple ratios to determine the masses of reactants required to produce given masses of products, or vice versa. The procedure for solving these problems is essentially the same regardless of the reaction.

step 1: Write and balance the chemical equation.

step 2: Determine the atomic (molecular) weight of each element (compound) in the equation.

step 3: Multiply the atomic (molecular) weights by their respective coefficients and write the products under the formulas.

step 4: Write the given mass data under the weights determined in step 3.

step 5: Fill in the missing information by calculating simple ratios.

[10]Another name for replacement is *metathesis*.

[11]When water is part of the reaction, the interpretation that the coefficients are volumes is valid only if the reaction takes place at a high enough temperature to vaporize the water.

Example 20.6

Caustic soda (NaOH) is made from sodium carbonate (Na_2CO_3) and slaked lime ($Ca(OH)_2$) according to the given reaction. How many kilograms of caustic soda can be made from 2000 kg of sodium carbonate?

Solution

	Na_2CO_3	$+ Ca(OH)_2$	$\rightarrow 2NaOH$	$+ CaCO_3$
molecular weights	106	74	2×40	100
given data	2000 kg		m kg	

The simple ratio used is

$$\frac{NaOH}{Na_2CO_3} = \frac{80}{106} = \frac{m}{2000 \text{ kg}}$$

Solving for the unknown mass, $m = 1509$ kg.

15. NONSTOICHIOMETRIC REACTIONS

In many cases, it is not realistic to assume a stoichiometric reaction because an excess of one or more reactants is necessary to assure that all of the remaining reactants take part in the reaction. Combustion is an example where the stoichiometric assumption is, more often than not, invalid. Excess air is generally needed to ensure that all of the fuel is burned.

With nonstoichiometric reactions, the reactant that is used up first is called the *limiting reactant*. The amount of product will be dependent on (limited by) the limiting reactant.

The *theoretical yield* or *ideal yield* of a product is the maximum mass of product per unit mass of limiting reactant that can be obtained from a given reaction if the reaction goes to completion. The *percentage yield* is a measure of the efficiency of the actual reaction.

$$\text{percentage yield} = \frac{\text{actual yield} \times 100\%}{\text{theoretical yield}} \qquad 20.6$$

16. SOLUTIONS OF GASES IN LIQUIDS

When a liquid is exposed to a gas, a small amount of the gas will dissolve in the liquid. Diffusion alone is sufficient for this to occur; bubbling or collecting the gas over a liquid is not necessary. Given enough time, at equilibrium, the concentration of the gas will reach a maximum known as the *saturation concentration*.

Due to the large amount of liquid compared to the small amount of dissolved gas, a liquid exposed to multiple gases will eventually become saturated by all of the gases; the presence of one gas does not affect the solubility of another gas.

The characteristics of a solution of one or more gases in a liquid is predicted by *Henry's law*. In the general formulation applicable to both liquids and vapor exposed to mixtures of gases, Henry's law states that, at equilibrium, the partial pressure, p_i, of a gas in a mixture will be proportional to the gas mole fraction, x_i for gas in the liquid or y_i for gas in the vapor, of that dissolved gas in solution.

In Eq. 20.7, Henry's law constant, h, has units of pressure, typically reported in the literature in atmospheres (same as atm/mole fraction). (See Table 20.2.)

Henry's Law at Constant Temperature

$$p_i = py_i = hx_i \qquad 20.7$$

Since, for mixtures of ideal gases, the mole fraction, x_i or y_i, volumetric fraction, B_i, and partial pressure fraction, p_i/p_{total}, all have the same numerical values, these measures can all be integrated into Henry's law.

Table 20.2 Approximate Values of Henry's Law Constant (solutions of gases in water)

	Henry's law constant, h (atm) (Multiply all values by 10^3.)	
gas	20°C	30°C
CO	53.6	62.0
CO_2	1.42	1.86
H_2S	48.3	60.9
N_2	80.4	92.4
NO	26.4	31.0
O_2	40.1	47.5
SO_2	0.014	0.016

Adapted from *Scrubber Systems Operational Review* (APTI Course SI:412C), Second Edition, *Lesson 11: Design Review of Absorbers Used for Gaseous Pollutants*, 1998, North Carolina State University for the U.S. Environmental Protection Agency.

As shown in Table 20.3, Henry's law is stated in several incompatible formulations, and the corresponding equations and Henry's law constants are compatible only with their own formulations. Also, a variety of variables are used for Henry's law constant, including h, k_h, K_h, and so on. The context and units of the Henry's law constant must be used to determine Henry's law.

The dimensionless form of the Henry's law constant is also known as the *absorption coefficient*, *coefficient of absorption*, and *solubility coefficient*. It represents the volume of a gas at a specific temperature and pressure that can be dissolved in a unit volume of liquid.

Typical units are L/L (dimensionless). Approximate values for gases in water at 1 atm and 20°C are: CO, 0.023; CO_2, 0.88; He, 0.009; H_2, 0.017; H_2S, 2.62; N_2, 0.015; NH_3, 710; O_2, 0.028.

Table 20.3 *Typical Units of Henry's Law Constant and Corresponding Equations*

equation	typical units of Henry's law constant	Henry's law statement (at equilibrium)
$p_i = py_i = hx_i$	h: pressure (atm)	Partial pressure is proportional to mole fraction.
$p_i = k_{h(p/C),i}C_i$	$k_{h(p/C)}$: pressure divided by concentration (atm·L/mol; atm·L/mg)	Partial pressure is proportional to concentration.*
$p_i = \dfrac{C_i}{k_{h(C/p),i}}$	$k_{h(C/p)}$: concentration divided by pressure (mol/atm·L; mg/atm·L)	
$C_{i,\text{gas}} = \alpha_i C_{i,\text{liquid}}$	α: dimensionless (L_{gas}/L_{liquid})	Concentration in the gas mixture is proportional to concentration in the liquid solution.

*The statements of Henry's law are the same. However, the values of Henry's law constants are inverses.

The amount of gas dissolved in a liquid varies with the temperature of the liquid and the concentration of dissolved salts in the liquid. Generally, the solubility of gases in liquids decreases with increasing temperature.

Appendix 20.B lists the saturation values of dissolved oxygen in water at various temperatures and for various amounts of chloride ion (also referred to as *salinity*).

Example 20.7

At 20°C and 1 atm, 1 L of water can absorb 0.043 g of oxygen and 0.017 g of nitrogen. Atmospheric air is 20.9% oxygen by volume, and the remainder is assumed to be nitrogen. What masses of oxygen and nitrogen will be absorbed by 1 L of water exposed to 20°C air at 1 atm?

Solution

Since partial pressure is volumetrically weighted,

$$\begin{aligned} m_{\text{oxygen}} &= (0.209)\left(0.043\ \frac{\text{g}}{\text{L}}\right) \\ &= 0.009\ \text{g/L} \\ m_{\text{nitrogen}} &= (1.000 - 0.209)\left(0.017\ \frac{\text{g}}{\text{L}}\right) \\ &= 0.0134\ \text{g/L} \end{aligned}$$

Example 20.8

At an elevation of 4000 ft, the barometric pressure is 660 mm Hg. What is the dissolved oxygen concentration of 18°C water with a 800 mg/L chloride concentration at that elevation?

Solution

From App. 20.B, oxygen's saturation concentration for 18°C water corrected for a 800 mg/L chloride concentration is

$$\begin{aligned} C_s &= 9.45\ \frac{\text{mg}}{\text{L}} - \left(\frac{800\ \frac{\text{mg}}{\text{L}}}{100\ \frac{\text{mg}}{\text{L}}}\right)\left(0.0083\ \frac{\text{mg}}{\text{L}}\right) \\ &= 9.384\ \text{mg/L} \end{aligned}$$

Use the App. 20.B footnote to correct for the barometric pressure.

$$\begin{aligned} C_s' &= \left(9.384\ \frac{\text{mg}}{\text{L}}\right)\left(\frac{660\ \text{mm} - 15.49\ \text{mm}}{760\ \text{mm} - 15.49\ \text{mm}}\right) \\ &= 8.12\ \text{mg/L} \end{aligned}$$

17. PROPERTIES OF SOLUTIONS

There are very few convenient ways of predicting the properties of nonreacting, nonvolatile organic and aqueous solutions (acids, brines, alcohol mixtures, coolants, etc.) from the individual properties of the components.

Volumes of two nonreacting organic liquids (e.g., acetone and chloroform) in a mixture are essentially additive. The volume change upon mixing will seldom be more than a few tenths of a percent. The volume change in aqueous solutions is often slightly greater, but is still limited to a few percent (e.g., 3% for some solutions of methanol and water). Therefore, the specific gravity (density, specific weight, etc.) can be considered to be a volumetric weighting of the individual specific gravities.

Most other fluid properties of aqueous solutions, such as viscosity, compressibility, surface tension, and vapor pressure, must be measured.

18. SOLUTIONS OF SOLIDS IN LIQUIDS

When a solid is added to a liquid, the solid is known as the *solute*, and the liquid is known as the *solvent*.[12] If the dispersion of the solute throughout the solvent is at the molecular level, the mixture is known as a *solution*. If the solute particles are larger than molecules, the mixture is known as a *suspension*.[13]

In some solutions, the solvent and solute molecules bond loosely together. This loose bonding is known as *solvation*. If water is the solvent, the bonding process is also known as *aquation* or *hydration*.

The solubility of most solids in liquid solvents usually increases with increasing temperature. Pressure has very little effect on the solubility of solids in liquids.

When the solvent has absorbed as much solute as it can, it is a *saturated solution*.[14] Adding more solute to an already saturated solution will cause the excess solute to settle to the bottom of the container, a process known as *precipitation*. Other changes (in temperature, concentration, etc.) can be made to cause precipitation from saturated and unsaturated solutions. Precipitation in a chemical reaction is indicated by a downward arrow (i.e., "↓"). For example, the precipitation of silver chloride from an aqueous solution of silver nitrate ($AgNO_3$) and potassium chloride (KCl) would be written as

$$\text{AgNO}_3(aq) + \text{KCl}(aq) \rightarrow \text{AgCl}(s)\downarrow + \text{KNO}_3(aq)$$

19. UNITS OF CONCENTRATION

Several units of concentration are commonly used to express solution strengths.

- F— *formality:* The number of gram formula weights (i.e., molecular weights in grams) per liter of solution.
- m— *molality:* The number of gram-moles of solute per 1000 grams of solvent. A "molal" solution contains 1 gram-mole per 1000 grams of solvent.
- M— *molarity:* The number of gram-moles of solute per liter of solution. A "molar" (i.e., 1 M) solution contains 1 gram-mole per liter of solution. Molarity is related to normality as shown in Eq. 20.8.

$$N = M \times \Delta\,\text{oxidation number} \tag{20.8}$$

- N— *normality:* The number of gram equivalent weights of solute per liter of solution. A solution is "normal" (i.e., 1 N) if there is exactly one gram equivalent weight per liter of solution.
- x— *mole fraction:* The number of moles of solute divided by the number of moles of solvent and all solutes.
- meq/L— *milligram equivalent weights of solute per liter of solution:* calculated by multiplying normality by 1000 or dividing concentration in mg/L by equivalent weight.
- mg/L— *milligrams per liter:* The number of milligrams of solute per liter of solution. Same as ppm for solutions of water.
- ppm— *parts per million:* The number of pounds (or grams) of solute per million pounds (or grams) of solution. Same as mg/L for solutions of water.
- ppb— *parts per billion:* The number of pounds (or grams) of solute per billion (10^9) pounds (or grams) of solution. Same as μg/L for solutions of water.

For compounds whose molecules do not dissociate in solution (e.g., table sugar), there is no difference between molarity and formality. There is a difference, however, for compounds that dissociate into ions (e.g., table salt). Consider a solution derived from 1 gmol of magnesium nitrate $Mg(NO_3)_2$ in enough water to bring the volume to 1 L. The formality is 1 F (i.e., the solution is 1 formal). However, 3 moles of ions will be produced: 1 mole of Mg^{++} ions and 2 moles of NO_3^- ions. Therefore, molarity is 1 M for the magnesium ion and 2 M for the nitrate ion.

The use of formality avoids the ambiguity in specifying concentrations for ionic solutions. Also, the use of formality avoids the problem of determining a molecular weight when there are no discernible molecules (e.g., as in a crystalline solid such as NaCl). Unfortunately, the distinction between molarity and formality is not always made, and molarity may be used as if it were formality.

20. pH AND pOH

A standard measure of the strength of an acid or base is the number of hydrogen or hydroxide ions in a liter of solution. Since these are very small numbers, a logarithmic scale is used.

$$\text{pH} = -\log_{10}[\text{H}^+] = \log_{10}\frac{1}{[\text{H}^+]} \tag{20.9}$$

[12] The term *solvent* is often associated with volatile liquids, but the term is more general than that. (A *volatile liquid* evaporates rapidly and readily at normal temperatures.) Water is the solvent in aqueous solutions.
[13] An *emulsion* is not a mixture of a solid in a liquid. It is a mixture of two immiscible liquids.
[14] Under certain circumstances, a *supersaturated solution* can exist for a limited amount of time.

$$\text{pOH} = -\log_{10}[\text{OH}^-] = \log_{10}\frac{1}{[\text{OH}^-]} \qquad 20.10$$

The quantities $[H^+]$ and $[OH^-]$ in square brackets are the *ionic concentrations* in moles of ions per liter. The number of moles can be calculated from Avogadro's law by dividing the actual number of ions per liter by 6.022×10^{23}. Alternatively, for a partially ionized compound in a solution of known molarity, M, the ionic concentration is

$$[\text{ion}] = XM \qquad 20.11$$

A *neutral solution* has a pH of 7. Solutions with a pH below 7 are acidic; the smaller the pH, the more acidic the solution. Solutions with a pH above 7 are basic.

The relationship between pH and pOH is

$$\text{pH} + \text{pOH} = 14 \qquad 20.12$$

21. BUFFERS

A *buffer solution* resists changes in acidity and maintains a relatively constant pH when a small amount of an acid or base is added to it. Buffers are usually combinations of weak acids and their salts. A buffer is most effective when the acid and salt concentrations are equal.

22. NEUTRALIZATION

Acids and bases neutralize each other to form water.

$$H^+ + OH^- \rightarrow H_2O$$

Assuming 100% ionization of the solute, the volumes, V, required for complete neutralization can be calculated from the normalities, N, or the molarities, M.

$$V_bN_b = V_aN_a \qquad 20.13$$

$$V_bM_b\Delta_{b,\text{charge}} = V_aM_a\Delta_{a,\text{charge}} \qquad 20.14$$

23. REVERSIBLE REACTIONS

Reversible reactions are capable of going in either direction and do so to varying degrees (depending on the concentrations and temperature) simultaneously. These reactions are characterized by the simultaneous presence of all reactants and all products. For example, the chemical equation for the exothermic formation of ammonia from nitrogen and hydrogen is

$$N_2 + 3H_2 \rightleftharpoons 2NH_3 \quad (\Delta H = -92.4\ \text{kJ})$$

At *chemical equilibrium*, reactants and products are both present. Concentrations of the reactants and products do not change after equilibrium is reached.

24. LE CHATELIER'S PRINCIPLE

Le Châtelier's principle predicts the direction in which a reversible reaction initially at equilibrium will go when some condition (e.g., temperature, pressure, concentration) is "stressed" (i.e., changed). The principle says that when an equilibrium state is stressed by a change, a new equilibrium is formed that reduces that stress.

Consider the formation of ammonia from nitrogen and hydrogen. (See Sec. 20.23.) When the reaction proceeds in the forward direction, energy in the form of heat is released and the temperature increases. If the reaction proceeds in the reverse direction, heat is absorbed and the temperature decreases. If the system is stressed by increasing the temperature, the reaction will proceed in the reverse direction because that direction absorbs heat and reduces the temperature.

For reactions that involve gases, the reaction equation coefficients can be interpreted as volumes. In the nitrogen-hydrogen reaction, four volumes combine to form two volumes. If the equilibrium system is stressed by increasing the pressure, then the forward reaction will occur because this direction reduces the volume and pressure.[15]

If the concentration of any participating substance is increased, the reaction proceeds in a direction away from the substance with the increase in concentration. (For example, an increase in the concentration of the reactants shifts the equilibrium to the right, increasing the amount of products formed.)

The *common ion effect* is a special case of Le Châtelier's principle. If a salt containing a common ion is added to a solution of a weak acid, almost all of the salt will dissociate, adding large quantities of the common ion to the solution. Ionization of the acid will be greatly suppressed, a consequence of the need to have an unchanged equilibrium constant.

Thermodynamics

[15]The exception to this rule is the addition of an inert or nonparticipating gas to a gaseous equilibrium system. Although there is an increase in total pressure, the position of the equilibrium is not affected.

25. IRREVERSIBLE REACTION KINETICS

The rate at which a compound is formed or used up in an irreversible (one-way) reaction is known as the *rate of reaction*, *speed of reaction*, *reaction velocity*, and so on. The rate, v, is the change in concentration per unit time, usually measured in mol/L·s.

$$\text{v} = \frac{\text{change in concentration}}{\text{time}} \qquad 20.15$$

According to the *law of mass action*, the rate of reaction varies with the concentrations of the reactants and products. Specifically, the rate is proportional to the molar concentrations (i.e., the molarities). The rate of the formation or conversion of substance A is represented in various forms, such as r_A, dA/dt, and $d[A]/dt$, where the variable A or [A] can represent either the mass or the concentration of substance A. Substance A can be either a pure element or a compound.

The rate of reaction is generally not affected by pressure, but it does depend on five other factors.

- *type of substances in the reaction:* Some substances are more reactive than others.
- *exposed surface area:* The rate of reaction is proportional to the amount of contact between the reactants.
- *concentrations:* The rate of reaction increases with increases in concentration.
- *temperature:* The rate of reaction approximately doubles with every 10°C increase in temperature.
- *catalysts:* If a catalyst is present, the rate of reaction increases. However, the equilibrium point is not changed. (A catalyst is a substance that increases the reaction rate without being consumed in the reaction.)

26. ORDER OF THE REACTION

The *order of the reaction* is the total number of reacting molecules in or before the slowest step in the process.[16] The order must be determined experimentally. However, for an irreversible elementary reaction, the order is usually assumed from the stoichiometric reaction equation as the sum of the combining coefficients for the reactants.[17,18] For example, for the reaction $mA + nB \rightarrow pC$, the overall order of the forward reaction is assumed to be $m + n$.

Many reactions (e.g., dissolving metals in acid or the evaporation of condensed materials) have *zero-order reaction rates.* These reactions do not depend on the concentrations or temperature at all, but rather, are affected by other factors such as the availability of reactive surfaces or the absorption of radiation. The formation (conversion) rate of a compound in a zero-order reaction is constant. That is, $dA/dt = -k_0$. k_0 is known as the *reaction rate constant.* (The subscript "0" refers to the zero-order.) Since the concentration (amount) of the substance decreases with time, dA/dt is negative. Since the negative sign is explicit in rate equations, the reaction rate constant is generally reported as a positive number.

Once a reaction rate equation is known, it can be integrated to obtain an expression for the concentration (mass) of the substance at various times. The time for half of the substance to be formed (or converted) is the *half-life*, $t_{1/2}$. Table 20.4 contains reaction rate equations and half-life equations for various types of low-order reactions.

27. REVERSIBLE REACTION KINETICS

Consider the following reversible reaction.

$$aA + bB \rightleftharpoons cC + dD \qquad 20.16$$

In Eq. 20.17 and Eq. 20.18, the *reaction rate constants* are k_{forward} and k_{reverse}. The order of the forward reaction is $a + b$; the order of the reverse reaction is $c + d$.

$$\text{v}_{\text{forward}} = k_{\text{forward}}[A]^a[B]^b \qquad 20.17$$

$$\text{v}_{\text{reverse}} = k_{\text{reverse}}[C]^c[D]^d \qquad 20.18$$

At equilibrium, the forward and reverse speeds of reaction are equal.

$$\text{v}_{\text{forward}} = \text{v}_{\text{reverse}}\big|_{\text{equilibrium}} \qquad 20.19$$

[16]This definition is valid for elementary reactions. For complex reactions, the order is an empirical number that need not be an integer.

[17]The overall order of the reaction is the sum of the orders with respect to the individual reactants. For example, in the reaction $2NO + O_2 \rightarrow 2NO_2$, the reaction is second order with respect to NO, first order with respect to O_2, and third order overall.

[18]In practice, the order of the reaction must be known, given, or determined experimentally. It is not always equal to the sum of the combining coefficients for the reactants. For example, in the reaction $H_2 + I_2 \rightarrow 2HI$, the overall order of the reaction is indeed 2, as expected. However, in the reaction $H_2 + Br_2 \rightarrow 2HBr$, the overall order is found experimentally to be 3/2, even though the two reactions have the same stoichiometry, and despite the similarities of iodine and bromine.

Table 20.4 Reaction Rates and Half-Life Equations

reaction	order	rate equation	integrated forms
$A \rightarrow B$	zero	$\frac{d[A]}{dt} = -k_0$	$[A] = [A]_0 - k_0 t$
			$t_{1/2} = \frac{[A]_0}{2k_0}$
$A \rightarrow B$	first	$\frac{d[A]}{dt} = -k_1[A]$	$\ln \frac{[A]}{[A]_0} = -k_1 t$
			$t_{1/2} = \frac{1}{k_1} \ln 2$
$A + A \rightarrow P$	second, type I	$\frac{d[A]}{dt} = -k_2[A]^2$	$\frac{1}{[A]} - \frac{1}{[A]_0} = k_2 t$
			$t_{1/2} = \frac{1}{k_2[A]_0}$
$aA + bB \rightarrow P$	second, type II	$\frac{d[A]}{dt} = -k_2[A][B]$	$\ln \frac{[A]_0 - [B]}{[B]_0 - \left(\frac{b}{a}\right)[X]} = \ln \frac{[A]}{[B]}$
			$= \left(\frac{b[A]_0 - a[B]_0}{a}\right) k_2 t + \ln \frac{[A]_0}{[B]_0}$
			$t_{1/2} = \left(\frac{a}{k_2\left(b[A]_0 - a[B]_0\right)}\right) \ln\left(\frac{a[B]_0}{2a[B]_0 - b[A]_0}\right)$

28. EQUILIBRIUM CONSTANT

For reversible reactions, the *equilibrium constant*, K, is proportional to the ratio of the reverse rate of reaction to the forward rate of reaction.[19] Except for catalysis, the equilibrium constant depends on the same factors affecting the reaction rate. For the complex reversible reaction given by Eq. 20.16, the equilibrium constant is given by the *law of mass action.*

$$K = \frac{[C]^c [D]^d}{[A]^a [B]^b} = \frac{k_{\text{forward}}}{k_{\text{reverse}}} \qquad 20.20$$

If any of the reactants or products are in pure solid or pure liquid phases, their concentrations are omitted from the calculation of the equilibrium constant. For example, in weak aqueous solutions, the concentration of water, H_2O, is very large and essentially constant; therefore, that concentration is omitted.

For gaseous reactants and products, the concentrations (i.e., the numbers of atoms) will be proportional to the partial pressures. Therefore, an equilibrium constant can be calculated directly from the partial pressures and is given the symbol K_p. For example, for the formation of ammonia gas from nitrogen and hydrogen, the equilibrium constant is

$$K_p = \frac{[p_{NH_3}]^2}{[p_{N_2}][p_{H_2}]^3} \qquad 20.21$$

K and K_p are not numerically the same, but they are related by Eq. 20.22. Δn is the number of moles of products minus the number of moles of reactants.

$$K_p = K(R^* T)^{\Delta n} \qquad 20.22$$

[19]The symbols K_c (in molarity units) and K_{eq} are occasionally used for the equilibrium constant.

Example 20.9

A particularly weak solution of acetic acid ($HC_2H_3O_2$) in water has the ionic concentrations (in mol/L) given. What is the equilibrium constant?

$$HC_2H_3O_2 + H_2O \rightleftharpoons H_3O^+ + C_2H_3O_2^-$$
$$[HC_2H_3O_2] = 0.09866$$
$$[H_2O] = 55.5555$$
$$[H_3O^+] = 0.00134$$
$$[C_2H_3O_2^-] = 0.00134$$

Solution

The concentration of the water molecules is not included in the calculation of the equilibrium or ionization constant. Therefore, the equilibrium constant is

$$K = K_a = \frac{[H_3O^+][C_2H_3O_2^-]}{[HC_2H_3O_2]} = \frac{(0.00134)(0.00134)}{0.09866} = 1.82 \times 10^{-5}$$

29. IONIZATION CONSTANT

The equilibrium constant for a weak solution is essentially constant and is known as the *ionization constant* (also known as a *dissociation constant*). (See Table 20.5 and App. 20.E.) For weak acids, the symbol K_a and name *acid constant* are used. (See App. 20.F.) For weak bases, the symbol K_b and the name *base constant* are used. (See App. 20.G.) For example, for the ionization of hydrocyanic acid,

$$HCN \rightleftharpoons H^+ + CN^-$$
$$K_a = \frac{[H^+][CN^-]}{[HCN]}$$

Pure water is itself a very weak electrolyte and ionizes only slightly.

$$2H_2O \rightleftharpoons H_3O^+ + OH^- \quad \text{20.23}$$

At equilibrium, the ionic concentrations are equal.

$$[H_3O^+] = 10^{-7}$$
$$[OH^-] = 10^{-7}$$

From Eq. 20.20, the ionization constant (*ion product*) for pure water is

$$K_w = K_{a,\text{water}} = [H_3O^+][OH^-] = (10^{-7})(10^{-7}) = 10^{-14} \quad \text{20.24}$$

If the molarity, M, and *fraction of ionization*, X, are known, the ionization constant can be calculated from Eq. 20.25.

$$K_{\text{ionization}} = \frac{MX^2}{1-X} \quad [K_a \text{ or } K_b] \quad \text{20.25}$$

The reciprocal of the ionization constant is the *stability constant* (*overall stability constant*), also known as the *formation constant*. Stability constants are used to describe complex ions that dissociate readily.

Example 20.10

A 0.1 molar (0.1 M) acetic acid solution is 1.34% ionized. Find the (a) hydrogen ion concentration, (b) acetate ion concentration, (c) un-ionized acid concentration, and (d) ionization constant.

Solution

(a) From Eq. 20.11, the hydrogen (hydronium) ion concentration is

$$[H_3O^+] = XM = (0.0134)(0.1) = 0.00134 \text{ mol/L}$$

(b) Since every hydronium ion has a corresponding acetate ion, the acetate and hydronium ion concentrations are the same.

$$[C_2H_3O_2^-] = [H_3O^+] = 0.00134 \text{ mol/L}$$

(c) The concentration of un-ionized acid can be derived from Eq. 20.11.

$$[HC_2H_3O_2] = (1-X)M = (1-0.0134)(0.1) = 0.09866 \text{ mol/L}$$

(d) The ionization constant is calculated from Eq. 20.25.

$$K_a = \frac{MX^2}{1-X} = \frac{(0.1)(0.0134)^2}{1-0.0134} = 1.82 \times 10^{-5}$$

Table 20.5 Approximate Ionization Constants of Common Water Supply Chemicals

substance	0°C	5°C	10°C	15°C	20°C	25°C
$Ca(OH)_2$						3.74×10^{-3}
$HClO$	2.0×10^{-8}	2.3×10^{-8}	2.6×10^{-8}	3.0×10^{-8}	3.3×10^{-8}	3.7×10^{-8}
$HC_2H_3O_2$	1.67×10^{-5}	1.70×10^{-5}	1.73×10^{-5}	1.75×10^{-5}	1.75×10^{-5}	1.75×10^{-5}
$HBrO$					$\approx 2 \times 10^{-9}$	
H_2CO_3 (K_1)	2.6×10^{-7}	3.04×10^{-7}	3.44×10^{-7}	3.81×10^{-7}	4.16×10^{-7}	4.45×10^{-7}
$HClO_2$					$\approx 1.1 \times 10^{-2}$	
NH_3	1.37×10^{-5}	1.48×10^{-5}	1.57×10^{-5}	1.65×10^{-5}	1.71×10^{-5}	1.77×10^{-5}
NH_4OH						1.79×10^{-5}
water*	14.9435	14.7338	14.5346	14.3463	14.1669	13.9965

*$-\log_{10}K$ given

30. IONIZATION CONSTANTS FOR POLYPROTIC ACIDS

A *polyprotic acid* has as many ionization constants as it has acidic hydrogen atoms. For oxyacids (see Sec. 20.6), each successive ionization constant is approximately 10^5 times smaller than the preceding one. For example, phosphoric acid (H_3PO_4) has three ionization constants.

$$K_1 = 7.1 \times 10^{-3} \quad (H_3PO_4)$$
$$K_2 = 6.3 \times 10^{-8} \quad (H_2PO_4^-)$$
$$K_3 = 4.4 \times 10^{-13} \quad (HPO_4^{-2})$$

31. SOLUBILITY PRODUCT

When an ionic solid is dissolved in a solvent, it dissociates. For example, consider the ionization of silver chloride in water.

$$AgCl(s) \rightleftharpoons Ag^+(aq) + Cl^-(aq)$$

If the equilibrium constant is calculated, the terms for pure solids and liquids (in this case, [AgCl] and [H_2O]) are omitted. Therefore, the *solubility product*, K_{sp}, consists only of the ionic concentrations. As with the general case of ionization constants, the solubility product for slightly soluble solutes is essentially constant at a standard value. (See App. 20.E.)

$$K_{sp} = [Ag^+][Cl^-] \qquad \textit{20.26}$$

When the product of terms exceeds the standard value of the solubility product, solute will precipitate out until the product of the remaining ion concentrations attains the standard value. If the product is less than the standard value, the solution is not saturated.

The solubility products of nonhydrolyzing compounds are relatively easy to calculate. (Example 20.11 demonstrates a method.) Such is the case for chromates (CrO_4^{-2}), halides (F^-, Cl^-, Br^-, I^-), sulfates (SO_4^{-2}), and iodates (IO_3^-). However, compounds that hydrolyze (i.e., combine with water molecules) must be treated differently. The method used in Ex. 20.11 cannot be used for hydrolyzing compounds. Appendix 20.E lists some common solubility products.

Example 20.11

At a particular temperature, it takes 0.038 grams of lead sulfate ($PbSO_4$, molecular weight = 303.25) per liter of water to prepare a saturated solution. What is the solubility product of lead sulfate if all of the lead sulfate ionizes?

Solution

Sulfates are not one of the hydrolyzing ions. Therefore, the solubility product can be calculated from the concentrations.

Since one liter of water has a mass of 1 kg, the number of moles of lead sulfate dissolved per saturated liter of solution is

$$n = \frac{m}{M} = \frac{0.038 \text{ g}}{303.25 \frac{\text{g}}{\text{mol}}} = 1.25 \times 10^{-4} \text{ mol}$$

Lead sulfate ionizes according to the following reaction.

$$PbSO_4(s) \rightleftharpoons Pb^{+2}(aq) + SO_4^{-2}(aq) \quad \text{[in water]}$$

Since all of the lead sulfate ionizes, the number of moles of each ion is the same as the number of moles of lead sulfate. Therefore,

$$K_{sp} = [Pb^{+2}][SO_4^{-2}] = (1.25 \times 10^{-4})(1.25 \times 10^{-4}) = 1.56 \times 10^{-8}$$

Thermodynamics

32. GALVANIC ACTION

Galvanic action (*galvanic corrosion* or *two-metal corrosion*) results from the difference in oxidation potentials of metallic ions. The greater the difference in oxidation potentials, the greater will be the galvanic action. If two metals with different oxidation potentials are placed in an electrolytic medium (e.g., seawater), a galvanic cell will be created. The metal with the higher potential (i.e., the more "active" metal) will act as an anode and will corrode. The metal with the lower potential (the more "noble" metal), being the cathode, will be unchanged. In one extreme type of intergranular corrosion known as *exfoliation*, open endgrains separate into layers.

Metals are often classified according to their positions in the *galvanic series* listed in Table 20.6. As would be expected, the metals in this series are in approximately the same order as their half-cell potentials. However, alloys and proprietary metals are also included in the series. [**Galvanic Series of Some Commercial Metals and Alloys in Seawater**]

Precautionary measures can be taken to inhibit or reduce galvanic action when use of dissimilar metals is unavoidable.

- Use dissimilar metals that are close neighbors in the galvanic series.
- Use *sacrificial anodes*. In marine saltwater applications, sacrificial zinc plates can be used.
- Use protective coatings, oxides, platings, or inert spacers to reduce the access of corrosive environments to the metals.[20]

33. CORROSION

Corrosion is an undesirable degradation of a material resulting from a chemical or physical reaction with the environment. Conditions within the crystalline structure can accentuate or retard corrosion. Corrosion rates are reported in units of mils per year (mpy) and micrometers per year (μm/y). (A *mil* is a thousandth of an inch.) [**Types of Corrosion**]

Uniform rusting of steel and oxidation of aluminum over entire exposed surfaces are examples of *uniform attack corrosion*. Uniform attack is usually prevented by the use of paint, plating, and other protective coatings.

Some metals are particularly sensitive to *intergranular corrosion*, IGC—selective or localized attack at metal-grain boundaries. For example, the Cr_2O_3 oxide film on stainless steel contains numerous imperfections at grain boundaries, and these boundaries can be attacked and enlarged by chlorides.

Intergranular corrosion may occur after a metal has been heated, in which case it may be known as *weld decay*. In the case of type 304 austenitic stainless steels, heating to 930–1300°F (500–700°C) in a welding process causes chromium carbides to precipitate out, reducing the corrosion resistance.[21] Reheating to 1830–2010°F (1000–1100°C) followed by rapid cooling will redissolve the chromium carbides and restore corrosion resistance.

Pitting is a localized perforation on the surface. It can occur even where there is little or no other visible damage. Chlorides and other halogens in the presence of water (e.g., HF and HCl) foster pitting in passive alloys, especially in stainless steels and aluminum alloys.

Concentration-cell corrosion (also known as *crevice corrosion* and *intergranular attack*, IGA) occurs when a metal is in contact with different electrolyte concentrations. It usually occurs in crevices, between two assembled parts, under riveted joints, or where there are scale and surface deposits that create stagnant areas in a corrosive medium. An electrode potentials table can be used as a point of reference to determine the most compatible material selections for minimizing corrosion in varying applications. [**Electrode Potentials (25°C, 1-Molar Solutions)**]

Erosion corrosion is the deterioration of metals buffeted by the entrained solids in a corrosive medium.

Selective leaching is the dealloying process in which one of the alloy ingredients is removed from the solid solution. This occurs because the lost ingredient has a lower corrosion resistance than the remaining ingredient. *Dezincification* is the classic case where zinc is selectively destroyed in brass. Other examples include the dealloying of nickel from copper-nickel alloys, iron from steel, and aluminum from copper-aluminum alloys.

Hydrogen damage occurs when hydrogen gas diffuses through and decarburizes steel (i.e., reacts with carbon to form methane). *Hydrogen embrittlement* (also known as *caustic embrittlement*) is hydrogen damage from hydrogen produced by caustic corrosion.

When subjected to sustained surface tensile stresses (including low residual stresses from manufacturing) in corrosive environments, certain metals exhibit catastrophic *stress corrosion cracking*, SCC. When the stresses are cyclic, *corrosion fatigue* failures can occur well below normal fatigue lives.

[20]While cadmium, nickel, chromium, and zinc are often used as protective deposits on steel, porosities in the surfaces can act as small galvanic cells, resulting in invisible subsurface corrosion.

[21]*Austenitic stainless steels* are the 300 series. They consist of chromium nickel alloys with up to 8% nickel. They are not hardenable by heat treatment, are nonmagnetic, and offer the greatest resistance to corrosion. *Martensitic stainless steels* are hardenable and magnetic. *Ferritic stainless steels* are magnetic and not hardenable.

Table 20.6 *Galvanic Series in Seawater (top to bottom: anodic (sacrificial, active) to cathodic (noble, passive))*

magnesium
zinc
Alclad 3S
cadmium
2024 aluminum alloy
low-carbon steel
cast iron
stainless steels (active)
  no. 410
  no. 430
  no. 404
  no. 316
Hastelloy A
lead
lead-tin alloys
tin
nickel
brass (copper-zinc)
copper
bronze (copper-tin)
90/10 copper-nickel
70/30 copper-nickel
Inconel
silver solder
silver
stainless steels (passive)
Monel metal
Hastelloy C
titanium
graphite
gold

Stress corrosion occurs because the more highly stressed grains (at the crack tip) are slightly more anodic than neighboring grains with lower stresses. Although intergranular cracking (at grain boundaries) is more common, corrosion cracking may be *intergranular* (between the grains), *transgranular* (through the grains), or a combination of the two, depending on the alloy. Cracks propagate, often with extensive branching, until failure occurs.

The precautionary measures that can be taken to inhibit or eliminate stress corrosion are as follows.

- Avoid using metals that are susceptible to stress corrosion in corrosive environments. These include austenitic stainless steels without heat treatment in seawater; certain tempers of the aluminum alloys 2124, 2219, 7049, and 7075 in seawater; and copper alloys exposed to ammonia.
- Protect open-grain surfaces from the environment. For example, press-fitted parts in drilled holes can be assembled with wet zinc chromate paste. Also, weldable aluminum can be "buttered" with pure aluminum rod.
- Stress-relieve by annealing heat treatment after welding or cold working.

Fretting corrosion occurs when two highly loaded members have a common surface (known as a *faying surface*) at which rubbing and sliding take place. The phenomenon is a combination of wear and chemical corrosion. Metals that depend on a film of surface oxide for protection, such as aluminum and stainless steel, are especially susceptible.

Fretting corrosion can be reduced by the following methods.

- Lubricate the faying surfaces.
- Seal the faying surfaces.
- Reduce vibration and movement.

Cavitation is the formation and sudden collapse of minute bubbles of vapor in liquids. It is caused by a combination of reduced pressure and increased velocity in the fluid. In effect, very small amounts of the fluid vaporize (i.e., boil) and almost immediately condense. The repeated collapse of the bubbles hammers and work-hardens the surface.

When the surface work-hardens, it becomes brittle. Small amounts of the surface flake away, and the surface becomes pitted. This is known as *cavitation corrosion*. Eventually, the entire piece may work-harden and become brittle, leading to structural failure.

34. WATER SUPPLY CHEMISTRY

Most municipal water supply composition data are not given in units of molarity, normality, molality, and so on. Rather, the most common measure of solution strength is the *$CaCO_3$ equivalent*. With this method, substances are reported in milligrams per liter (mg/L, same as parts per million, ppm) "as $CaCO_3$," even when $CaCO_3$ is unrelated to the substance or reaction that produced the substance.

Actual gravimetric amounts of a substance can be converted to amounts as $CaCO_3$ by use of the conversion factors in App. 20.C. These factors are easily derived from stoichiometric principles.

The reason for converting all substance quantities to amounts as $CaCO_3$ is to simplify calculations for nontechnical personnel. Equal $CaCO_3$ amounts constitute stoichiometric reaction quantities. For example, 100 mg/L as $CaCO_3$ of sodium ion (Na^+) will react with 100 mg/L as $CaCO_3$ of chloride ion (Cl^-) to produce 100 mg/L as $CaCO_3$ of salt (NaCl), even though the gravimetric quantities differ and $CaCO_3$ is not part of the reaction.

Example 20.12

Lime is added to water to remove carbon dioxide gas.

$$CO_2 + Ca(OH)_2 \rightarrow CaCO_3 \downarrow + H_2O$$

If water contains 5 mg/L of CO_2, how much lime is required for its removal?

Solution

From App. 20.C, the factor that converts CO_2 as substance to CO_2 as $CaCO_3$ is 2.27.

$$\begin{aligned} CO_2 \text{ as } CaCO_3 \text{ equivalent} &= (2.27)\left(5\ \frac{\text{mg}}{\text{L}}\right) \\ &= 11.35 \text{ mg/L as } CaCO_3 \end{aligned}$$

Therefore, the $CaCO_3$ equivalent of lime required will also be 11.35 mg/L.

From App. 20.C again, the factor that converts lime as $CaCO_3$ to lime as substance is (1/1.35).

$$\begin{aligned} Ca(OH)_2 \text{ substance} &= \frac{11.35\ \frac{\text{mg}}{\text{L}}}{1.35} \\ &= 8.41 \text{ mg/L as substance} \end{aligned}$$

This problem could also have been solved stoichiometrically.

35. ACIDITY AND ALKALINITY IN MUNICIPAL WATER SUPPLIES

Acidity is a measure of acids in solutions. Acidity in surface water (e.g., lakes and streams) is caused by formation of *carbonic acid* (H_2CO_3) from carbon dioxide in the air.[22] Acidity in water is typically given in terms of the $CaCO_3$ equivalent.

$$CO_2 + H_2O \rightarrow H_2CO_3 \qquad 20.27$$

$$H_2CO_3 + H_2O \rightarrow HCO_3^- + H_3O^+ \quad [\text{pH} > 4.5] \qquad 20.28$$

$$HCO_3^- + H_2O \rightarrow CO_3^{--} + H_3O^+ \quad [\text{pH} > 8.3] \qquad 20.29$$

Alkalinity is a measure of the amount of negative (basic) ions in the water. Specifically, OH^-, CO_3^{--}, and HCO_3^- all contribute to alkalinity.[23] The measure of alkalinity is the sum of concentrations of each of the substances measured as $CaCO_3$.

Alkalinity and acidity of a titrated sample are determined from color changes in indicators added to the titrant.

Example 20.13

Water from a city well is analyzed and is found to contain 20 mg/L as substance of HCO_3^- and 40 mg/L as substance of CO_3^{--}. What is the alkalinity of this water as $CaCO_3$?

Solution

From App. 20.C, the factors converting HCO_3^- and CO_3^{--} ions to $CaCO_3$ equivalents are 0.82 and 1.67, respectively.

$$\begin{aligned} \text{alkalinity} &= (0.82)\left(20\ \frac{\text{mg}}{\text{L}}\right) + (1.67)\left(40\ \frac{\text{mg}}{\text{L}}\right) \\ &= 83.2 \text{ mg/L as } CaCO_3 \end{aligned}$$

36. WATER HARDNESS

Water hardness is caused by multivalent (doubly charged, triply charged, etc., but not singly charged) positive metallic ions such as calcium, magnesium, iron, and manganese. (Iron and manganese are not as common, however.) Hardness reacts with soap to reduce its cleansing effectiveness and to form scum on the water surface and a ring around the bathtub.

Water containing bicarbonate (HCO_3^-) ions can be heated to precipitate carbonate molecules.[24] This hardness is known as *temporary hardness* or *carbonate hardness*.[25]

$$\begin{aligned} &Ca^{++} + 2HCO_3^- + \text{heat} \rightarrow \\ &\qquad CaCO_3\downarrow + CO_2 + H_2O \end{aligned} \qquad 20.30$$

$$\begin{aligned} &Mg^{++} + 2HCO_3^- + \text{heat} \rightarrow \\ &\qquad MgCO_3\downarrow + CO_2 + H_2O \end{aligned} \qquad 20.31$$

Remaining hardness due to sulfates, chlorides, and nitrates is known as *permanent hardness* or *noncarbonate hardness* because it cannot be removed by heating.

[22]Carbonic acid is very aggressive and must be neutralized to eliminate the cause of water pipe corrosion. If the pH of water is greater than 4.5, carbonic acid ionizes to form bicarbonate. (See Eq. 20.28.) If the pH is greater than 8.3, carbonate ions form that cause water hardness by combining with calcium. (See Eq. 20.29.)

[23]Other ions, such as NO_3^-, also contribute to alkalinity, but their presence is rare. If detected, they should be included in the calculation of alkalinity.

[24]Hard water forms scale when heated. This scale, if it forms in pipes, eventually restricts water flow. Even in small quantities, the scale insulates boiler tubes. Therefore, water used in steam-producing equipment must be essentially hardness-free.

[25]The hardness is known as *carbonate* hardness even though it is caused by *bicarbonate* ions, not carbonate ions.

The amount of permanent hardness can be determined numerically by causing precipitation, drying, and then weighing the precipitate.

$$Ca^{++} + SO_4^{--} + Na_2CO_3 \rightarrow 2Na^+ + SO_4^{--} + CaCO_3\downarrow \qquad 20.32$$

$$Mg^{++} + 2Cl^- + 2NaOH \rightarrow 2Na^+ + 2Cl^- + Mg(OH)_2\downarrow \qquad 20.33$$

Total hardness is the sum of temporary and permanent hardnesses, both expressed in mg/L as $CaCO_3$.

37. COMPARISON OF ALKALINITY AND HARDNESS

Hardness measures the presence of positive, multivalent ions in the water supply. Alkalinity measures the presence of negative (basic) ions such as hydrates, carbonates, and bicarbonates. Since positive and negative ions coexist, an alkaline water can also be hard.

If certain assumptions are made, it is possible to draw conclusions about the water composition from the hardness and alkalinity. For example, if the effects of Fe^{+2} and OH^- are neglected, the following rules apply. (All concentrations are measured as $CaCO_3$.)

- *hardness = alkalinity:* There is no noncarbonate hardness. There are no SO_4^-, Cl^-, or NO_3^- ions present.
- *hardness > alkalinity:* Noncarbonate hardness is present.
- *hardness < alkalinity:* All hardness is carbonate hardness. The extra HCO_3^- comes from other sources (e.g., $NaHCO_3$).

Titration with indicator solutions is used to determine the alkalinity. The *phenolphthalein alkalinity* (or "P reading" in mg/L as $CaCO_3$) measures hydrate alkalinity and half of the carbonate alkalinity. The *methyl orange alkalinity* (or "M reading" in mg/L as $CaCO_3$) measures the total alkalinity (including the phenolphthalein alkalinity). Table 20.7 can be used to interpret these tests.[26]

38. WATER SOFTENING WITH LIME

Water softening can be accomplished with lime and soda ash to precipitate calcium and magnesium ions from the solution. Lime treatment has the added benefits of disinfection, iron removal, and clarification. Practical limits of *precipitation softening* are 30 mg/L of $CaCO_3$ and 10 mg/L of $Mg(OH)_2$ (as $CaCO_3$) because of intrinsic solubilities. Water treated by this method usually leaves the softening apparatus with a hardness between 50 mg/L and 80 mg/L as $CaCO_3$.

Table 20.7 *Interpretation of Alkalinity Tests*

case	hydrate as $CaCO_3$	carbonate as $CaCO_3$	bicarbonate as $CaCO_3$
$P = 0$	0	0	M
$0 < P < \frac{M}{2}$	0	2P	M – 2P
$P = \frac{M}{2}$	0	2P	0
$\frac{M}{2} < P < M$	2P – M	2(M – P)	0
$P = M$	M	0	0

39. WATER SOFTENING BY ION EXCHANGE

In the *ion exchange process* (also known as *zeolite process* or *base exchange method*), water is passed through a filter bed of exchange material. This exchange material is known as *zeolite*. Ions in the insoluble exchange material are displaced by ions in the water.

The processed water will have a zero hardness. However, if there is no need for water with zero hardness (as in municipal water supply systems), some water can be bypassed around the unit.

There are three types of ion exchange materials. *Greensand (glauconite)* is a natural substance that is mined and treated with manganese dioxide. *Siliceous-gel zeolite* is an artificial solid used in small volume deionizer columns. *Polystyrene resins* are also synthetic and dominate the softening field.

The earliest synthetic zeolites were gelular *ion exchange resins* using a three-dimensional copolymer (e.g., styrene-divinylbenzene). Porosity through the continuous-phase gel was near zero, and dry contact surface areas of 500 ft^2/lbm (0.1 m^2/g) or less were common.

Macroreticular synthetic resins are discontinuous, three-dimensional copolymer beads in a rigid-sponge type formation. Each bead is made up of thousands of microspheres of the gel resin. Porosity is increased, and dry contact surface areas are approximately 270,000 ft^2/lbm to 320,000 ft^2/lbm (55 m^2/g to 65 m^2/g).

[26]The titration may be affected by the presence of silica and phosphates, which also contribute to alkalinity. The effect is small, but the titration may not be a completely accurate measure of carbonates and bicarbonates.

During operation, the calcium and magnesium ions are removed according to the following reaction in which R represents the zeolite anion. The resulting sodium compounds are soluble.

$$\begin{Bmatrix} \text{Ca} \\ \text{Mg} \end{Bmatrix} \begin{Bmatrix} (\text{HCO}_3)_2 \\ \text{SO}_4 \\ \text{Cl}_2 \end{Bmatrix} + \text{Na}_2\text{R} \rightarrow \text{Na}_2 \begin{Bmatrix} (\text{HCO}_3)_2 \\ \text{SO}_4 \\ \text{Cl}_2 \end{Bmatrix} + \begin{Bmatrix} \text{Ca} \\ \text{Mg} \end{Bmatrix} \text{R} \qquad 20.34$$

Typical saturation capacities of synthetic resins are 1.0 meq/mL to 1.5 meq/mL for anion exchange resins and 1.7 meq/mL to 1.9 meq/mL for cation exchange resins. However, working capacities are more realistic measures. Working capacities are approximately 10 kilograins/ft^3 to 15 kilograins/ft^3 (23 kg/m^3 to 35 kg/m^3) before regeneration.

Flow rates through the bed are typically 1 gpm/ft^3 to 6 gpm/ft^3 (2 L/s·m^3 to 13 L/s·m^3) of resin volume. The flow rate in terms of gpm/ft^2 (L/s·m^2) across the exposed surface will depend on the geometry of the bed, but values of 3 gpm/ft^2 to 15 gpm/ft^2 (2 L/s·m^2 to 10 L/s·m^2) are typical.[27]

Example 20.14

A municipal plant processes water with a total initial hardness of 200 mg/L. The designed discharge hardness is 50 mg/L. If an ion exchange unit is used, what is the bypass factor?

Solution

The water passing through the ion exchange unit is reduced to zero hardness. If x is the water fraction bypassed around the zeolite bed,

$$(1-x)\left(0\ \frac{\text{mg}}{\text{L}}\right) + x\left(200\ \frac{\text{mg}}{\text{L}}\right) = 50\ \text{mg/L}$$

$$x = 0.25$$

40. REGENERATION OF ION EXCHANGE RESINS

Ion exchange material has a finite capacity for ion removal. When the zeolite is saturated or has reached some other prescribed limit, it must be regenerated (rejuvenated).

Standard ion exchange units are regenerated when the alkalinity of their effluent increases to the *set point.* Most condensate polishing units that also collect crud are operated to a *pressure-drop endpoint.* The pressure drop through the ion exchange unit is primarily dependent on the amount of crud collected. When the pressure drop reaches a set point, the resin is regenerated.

Regeneration of synthetic ion exchange resins is accomplished by passing a *regenerating solution* over/through the resin. Although regeneration can occur in the ion exchange unit itself, external regeneration is common. This involves removing the bed contents hydraulically, backwashing to separate the components (for mixed beds), regenerating the bed components separately, washing, then recombining and transferring the bed components back into service.

Common regeneration compounds are NaCl (for water hardness removal units), H_2SO_4 (for cation exchange resins), and NaOH (for anion exchange resins). The amount of regeneration solution depends on the resin's degree of saturation. A rule of thumb is to expect to use 6 lbm to 10 lbm of regeneration compound per cubic foot of resin (100 kg to 160 kg per cubic meter). Alternatively, dosage of the regeneration compound may be specified in terms of hardness removed (e.g., 0.4 lbm of salt per 1000 grains of hardness removed). These rates are applicable to deionization plants for boiler make-up water. For condensate polishing, saturation levels of 10 lbm/ft^3 to 25 lbm/ft^3 (160 kg/m^3 to 400 kg/m^3) are used.

41. CHARACTERISTICS OF BOILER FEEDWATER

Water that is converted to steam in a boiler is known as *feedwater.* Water that is added to replace losses is known as *make-up water.* The purity of the feedwater returned to the boiler (after condensing) depends on the purity of the make-up water, since impurities continually build up. *Blowdown* is the intentional periodic release of some of the feedwater in order to remove chemicals whose concentrations have built up over time.

Water impurities cause *scaling* (which reduces fluid flow and heat transfer rates) and corrosion (which reduces strength). Deposits from calcium, magnesium, and silica compounds are particularly troublesome.[28] As feedwater impurity increases, deposits from copper, iron, and nickel oxides (corrosion products from pipe and equipment) become more problematic.

Water can also contain dissolved oxygen, nitrogen, and carbon dioxide. The nitrogen is inert and does not need to be considered. However, both the oxygen and carbon dioxide need to be removed. High-temperature dissolved oxygen readily attacks pipe and boiler metal. Most of the oxygen in make-up water is removed by heating the water. Although the solubility of oxygen in water decreases with temperature, water at high pressures can

[27]Much higher values, up to 15 gpm/ft^3 to 20 gpm/ft^3 or 40 gpm/ft^2 to 50 gpm/ft^2 (33 L/s·m^3 to 45 L/s·m^3 or 27 L/s·m^2 to 34 L/s·m^2), may occur in certain types of units and at certain times (e.g., start-up and leak conditions).

[28]Silica is most troubling because silicate deposits cannot be removed by chemical means. They must be removed mechanically.

hold large amounts of oxygen. Hydrazine (N_2H_4) and sodium sulfite (Na_2SO_3) are used for *oxygen scavenging.*[29] Because sodium sulfite forms sodium sulfate at high pressures, only hydrazine should be used above 1500 psi to 1800 psi (10.3 MPa to 12.4 MPa).

Carbon dioxide combines with water to form *carbonic acid.* In power plants, this is more likely found in condenser return lines than elsewhere. Acidity of the condensate is reduced by *neutralizing amines* such as morpholine, cyclohexylamine, ethanolamine (the tetrasodium salt of ethylenadiamine tetra-acetic acid, also known as Na_4EDTA), and diethylaminoethanol.

Most types of corrosion are greatly reduced when the pH is within the range of 9 to 10. Below this range, corrosion and deposits of sulfates, carbonates, and silicates become a major problem. For this reason, boilers are often shut down when the pH drops below 8.

Filming amines (*polar amines*) such as octadecylamine do not neutralize acidity or raise pH. They form a nonwettable layer which protects surfaces from corrosive compounds. Filming and neutralizing amines are often used in conjunction, though they are injected at different locations in the system.

Other purity requirements depend on the boiler equipment, and in particular, the steam pressure. Also, specific locations within the system can tolerate different concentrations.[30] The following guidelines apply to boilers operating in the 900 psig to 2500 psig (6.2 MPa to 17.3 MPa) range.[31]

- All water feedwater entering the boiler should be free from dissolved oxygen, carbon dioxide, suspended solids, and hardness.
- pH of water entering the boiler should be in the range of 8.5 to 9.0. pH of water in the boiler should be in the range of 10.8 to 11.5.
- Silica in boiler water should be limited to 5 ppm (as $CaCO_3$) at 900 psig (6.2 MPa), with a gradual reduction to 1 ppm (as $CaCO_3$) at 2500 psig (17.3 MPa).

42. MONITORING OF BOILER FEEDWATER QUALITY

Water chemistry of boiler feedwater is monitored through continuous *inline sampling* and periodic *grab sampling.* Water impurities are expressed in milligrams per liter (mg/L), parts per million (ppm), parts per billion (ppb), unit equivalents per million (epm), and milliequivalents per milliliter (meq/mL).[32,33] ppm and ppb can be "as $CaCO_3$" equivalents or "as substance." Electrical conductivity (a measure of total dissolved solids) is measured in microsiemens[34] (μS) or ppm.

Water quality is maintained by monitoring pH, electrical resistivity silica, and hardness (calcium, magnesium, and bicarbonates), the primary corrosion species (sulfates, chlorides, and sodium), dissolved gases (primarily oxygen), and the concentrations of buffering chemicals (e.g., phosphate, hydrazine, or AVT).

Continuous monitoring of ionic concentrations is complicated by a process known as *hideout,* in which a chemical species (e.g., *crevice salts*) disappears by precipitation or adsorption during low-flow and high heat transfer. Upon cooling (as during a shutdown), the hideout chemicals reappear.

43. PRODUCTION AND REGENERATION OF BOILER FEEDWATER

The processes used to treat raw water for use in power-generating plants depend on the incoming water quality. Depending on need, filters may be used to remove solids, softeners or exchange units may remove permanent (sulfate) hardness, and activated carbon may remove organics.

Bicarbonate hardness is converted to sludge when the water is boiled, is removed during blowdown, and does not need to be chemically removed. However, a strong *dispersant* (*sludge conditioner*) must be added to prevent scale formation.[35] Alternatively, the calcium salts can be converted to sludges of phosphate salts by adding phosphates. Prior to the 1970s, *caustic phosphate*

Thermodynamics

[29]8 ppm of sodium sulfite react with 1 ppm of oxygen. To achieve a complete reaction, an excess of 2 ppm to 3 ppm of sodium sulfite is required. Hydrazine reacts on a one-to-one basis and is commonly available as a 35% solution. Therefore, approximately 3 ppm of solution is required per ppm of oxygen.

[30]For example, silica in the boiler feedwater may be limited to 1 ppm to 5 ppm, but the silica content in steam should be limited to 0.01 ppm to 0.03 ppm.

[31]Tolerable concentrations at cold start-up can be as much as 5 to 100 times higher.

[32]Milligrams per liter (mg/L) is essentially the same as parts per million (ppm). The older units of grains per gallon are still occasionally encountered. 1 *grain* equals 1/7000th of a pound. Multiply grains per gallon (gpg) by 17.1 to get ppm. Unit equivalents per million are derived by dividing the concentration in ppm by the equivalent weight. Equivalent weight is the molecular weight divided by the valence or oxidation number.

[33]Use the following relationship to convert meq/mL to ppm.

$$\text{ppm} = (1000)\left(\frac{\text{meq}}{\text{mL}}\right)(\text{equivalent weight}) = \frac{(1000)\left(\frac{\text{meq}}{\text{mL}}\right)(\text{formula weight})}{\text{valence}}$$

[34]Microsiemens were previously known as "micromhos," where "ohms" is spelled backward to indicate its inverse.

[35]Typical sludge conditioner dispersants are natural organics (lignins, tannins, and starches) and synthetics (sodium polyacrylate, sodium polymethacrylate, sulfonated polystyrene, and maleic anhydride).

treatments using sodium phosphate, SP, in the form of Na_3PO_4, $NaHPO_4$, and NaH_2PO_4, were the most common method used to treat low-pressure boiler feedwater with high solids. Sodium phosphate converts impurities and buffers pH.

However, magnesium phosphate is a sticky sludge, and using phosphates may cause more problems than not using them. Also, with higher pressures and temperatures, pipe wall *wasting* becomes problematic.[36] Therefore, many modern plants now use an *all-volatile treatment*, AVT, also known as a *zero-solids treatment.* Early AVTs used ammonia (NH_3) for pH control, but ammonia causes *denting* in systems that operate without blowdown.[37]

Chelant AVTs do not add any solid chemicals to the boiler. They work by keeping calcium and magnesium in solution. The compounds are removed through continuous blowdown.[38] Chelant treatments can be used in boilers up to approximately 1500 psi (10.3 MPa) but require high-purity (low-solids) feedwater.[39]

Subsequent advances in corrosion protection have been based on use of stainless steel or titanium-tubed condensers, condensate polishing, and even better AVT formulations. Supplementing ammonia-AVTs are volatile amines, chelants, and boric acid treatments. Boric acid effectively combats denting, intergranular attack, and intergranular stress corrosion cracking.

Water that is processed through an ion-exchange unit is known as *deionized water.* Deionized water is "hungry" for minerals and picks up contamination as it passes through the power plant. When the operating pressure is 1500 psig (10.3 MPa) or higher, condensed steam cannot be reused without additional demineralization.[40] *Demineralization* is not generally needed for lower-pressure units, but some *condensate polishing* is still necessary.

In low-pressure plants, condensate polishing may consist of passing the water through a filter (of sand, anthracite, or pre-coat cellulose), a strong-acid cation unit, and a mixed-bed unit. The filter removes gross particles (referred to as *crud*), the strong-acid unit removes the dissolved iron, and the mixed-bed unit polishes the condensate.[41]

In modern high-pressure plants, the filtration step is often omitted, and crud removal occurs in the strong-acid unit. Furthermore, if the mixed-bed unit is made large enough or the cation component is increased, the strong-acid unit can also be omitted. A mixed-bed unit operating by itself is known as a *naked mixed-bed unit.* Demineralization of condensed steam is commonly accomplished with naked mixed beds.

44. NOMENCLATURE

A	atomic weight	lbm/lbmol	kg/kmol
B	volumetric fraction	–	–
C	concentration	n.a.	mg/L
C	molar heat capacity	Btu/°F-mol	J/K·mol
EW	equivalent weight	lbm/lbmol	kg/kmol
F	formality	n.a.	FW/L
FW	formula weight	lbm/lbmol	kg/kmol
H	molar enthalpy	Btu/lbmol	kcal/mol
h	Henry's law constant	atm	atm
i	isotope or element	–	–
k	reaction rate constant	1/min	1/min
K	equilibrium constant	–	–
m	mass	lbm	kg
m	molality	n.a.	mol/1000 g
M	molarity	n.a.	mol/L
M	molecular weight	lbm/lbmol	kg/kmol
n	number of moles	–	–
N	normality	n.a.	GEW/L
N_A	Avogadro's number, 6.022×10^{23}	n.a.	1/mol
p	pressure	lbf/ft^2	Pa
R^*	universal gas constant, 1545.35 (8314.47)	ft-lbf/lbmol-°R	J/kmol·K
t	time	min or sec	min or s
T	temperature	°R	K
v	rate of reaction	n.a.	mol/L·s
V	volume	ft^3	m^3
x	gravimetric fraction	–	–
x	mole fraction	–	–
x	relative abundance	–	–
X	fraction ionized	–	–
y	mole fraction	–	–
Z	atomic number	–	–

Subscripts

0	zero-order
a	acid
A	Avogadro
b	base
eq	equilibrium
f	formation
p	constant pressure
p	partial pressure
r	reaction
sp	solubility product
t	total
w	water

[36] *Wasting* is a term used in the power-generating industry to describe the process of general pipe wall thinning.
[37] *Denting* is the constricting of the intersection between tubes and support plates in boilers due to a buildup of corrosion.
[38] AVTs work in a completely different way than phosphates. It is not necessary to use phosphates and AVTs simultaneously.
[39] If the feedwater becomes contaminated, a supplementary phosphate treatment will be required.
[40] The term *deionization* refers to the process that produces makeup water. *Demineralization* refers to the process used to prepare condensed steam for reuse.
[41] *Crud* is primarily iron-corrosion products ranging from dissolved to particulate matter.

21 Fuels and Combustion

Content in blue refers to the *NCEES Handbook*.

NCEES EXAM SPECIFICATIONS AND RELATED CONTENT

THERMAL AND FLUID SYSTEMS EXAM

I.E.4. Thermodynamics: Combustion
- 5. Enthalpy of Reaction
- 27. Stoichiometric Air
- 31. Actual and Excess Air

1. HYDROCARBONS

With the exception of sulfur and related compounds, most fuels are hydrocarbons. A *hydrocarbon* is an organic compound containing only hydrogen and carbon. Hydrocarbons are further categorized into subfamilies including *alkynes* (C_nH_{2n-2}, such as acetylene C_2H_2), *alkenes* (C_nH_{2n}, such as ethylene C_2H_4), and *alkanes* (C_nH_{2n+2}, such as octane C_8H_{18}). *Aromatic hydrocarbons* (C_nH_{2n-6}, such as benzene C_6H_6) constitute another subfamily.

The alkanes are *saturated hydrocarbons*, composed entirely of single bonds and saturated with hydrogen. The alkanes are also known as the *paraffin series* and the *methane series*.

The alkenes and alkynes are *unsaturated hydrocarbons*. An alkene contains one or more carbon-carbon double bonds, and an alkyne contains one or more carbon-carbon triple bonds. The alkenes are subdivided into the chain-structured *olefin series* and the ring-structured *naphthalene series*.

2. FUEL ANALYSIS

Fuel analyses are reported as either percentages by weight (for liquid and solid fuels) or percentages by volume (for gaseous fuels). Percentages by weight are known as *gravimetric analyses*, while percentages by volume are known as *volumetric analyses*. An *ultimate analysis* is a type of gravimetric analysis in which the constituents are reported by atomic species rather than by compound. In an ultimate analysis, combined hydrogen from moisture in the fuel is added to hydrogen from the combustive compounds. (See Sec. 21.7.)

A *proximate analysis* (not "approximate") gives the gravimetric fraction of moisture, volatile matter, fixed carbon, and ash. Sulfur may be combined with the ash or may be specified separately.

Table 21.1 *Approximate Specific Heats (at Constant Pressure) of Gases (c_p in Btu/lbm-°R; at 1 atm)*

	temperature (°R)							
gas	500	1000	1500	2000	2500	3000	4000	5000
air	0.240	0.249	0.264	0.277	0.286	0.294	0.302	–
carbon dioxide	0.196	0.251	0.282	0.302	0.314	0.322	0.332	0.339
carbon monoxide	0.248	0.257	0.274	0.288	0.298	0.304	0.312	0.316
hydrogen	3.39	3.47	3.52	3.63	3.77	3.91	4.14	4.30
nitrogen	0.248	0.255	0.270	0.284	0.294	0.301	0.310	0.315
oxygen	0.218	0.236	0.253	0.264	0.271	0.276	0.286	0.294
sulfur dioxide	0.15	0.16	0.18	0.19	0.20	0.21	0.23	–
water vapor	0.444	0.475	0.519	0.566	0.609	0.645	0.696	0.729

(Multiply Btu/lbm-°R by 4.187 to obtain kJ/kg·K.)

A *combustible analysis* considers only the combustible components, disregarding moisture and ash.

Compositions of gaseous fuels are typically given as volumetric fractions of each component. For gas A in a mixture, its *volumetric fraction*, B_A, *mole fraction*, x_A, and *partial pressure ratio*, p_A/p_t, are all the same. This means the same value can be used with Dalton's and Henry's laws.

A volumetric fraction can be converted to a gravimetric fraction by multiplying by the molecular weight and then dividing by the sum of the products of all the volumetric fractions and molecular weights.

3. WEIGHTING OF THERMODYNAMIC PROPERTIES

Many gaseous fuels, and all gaseous combustion products, are mixtures of different compounds. Some thermodynamic properties of mixtures are gravimetrically weighted, while others are volumetrically weighted.[1]

Gravimetrically weighted properties include

- enthalpy
- entropy
- internal energy
- specific gas constant
- specific heat

For gases, volumetrically weighted properties include

- density
- molecular weight
- all molar properties, including molar specific heat, enthalpy per mole, and internal energy per mole

When a compound experiences a large temperature change, its thermodynamic properties should be evaluated at the average temperature. Table 21.1 can be used to find the specific heats of gases at various temperatures.

4. ENTHALPY OF FORMATION

Enthalpy, H, is the useful energy that a substance possesses by virtue of its temperature, pressure, and phase.[2] The *enthalpy of formation* (*heat of formation*), ΔH_f, of a compound is the energy absorbed during the formation of 1 gmol of the compound from the elements in their free, standard states (*e.g.*, O_2 as a gas, H_2 as a gas, C as a solid).[3] The enthalpy of formation is assigned a value of zero for elements in their free states at 25°C and 1 atm. This is the so-called *standard state* for enthalpies of formation. **[Heats of Reaction]**

Table 21.2 contains enthalpies of formation for some common elements and compounds. The enthalpy of formation depends on the temperature and phase of the compound. A standard temperature of 25°C is used in most tables of enthalpies of formation.[4] Compounds are solid (s) unless indicated to be gaseous (g) or liquid (l). Some aqueous (aq) values are also encountered.

5. ENTHALPY OF REACTION

The *enthalpy of reaction* (*heat of reaction*), ΔH_r, is the energy absorbed during a chemical reaction under constant volume conditions. It is found by summing the enthalpies of formation of all products and subtracting

[1]For gases, molar properties include molar specific heats, enthalpy per mole, and internal energy per mole.
[2]The older term *heat* is rarely encountered today.
[3]The symbol H_m is used to denote molar enthalpies. The symbol h is used for specific enthalpies (i.e., enthalpy per kilogram or per pound).
[4]It is possible to correct the enthalpies of formation to account for other reaction temperatures.

***Table 21.2** Standard Enthalpies of Formation (at 25°C)*

element/compound	ΔH_f (kcal/mol)
Al (*s*)	0.00
Al_2O_3 (*s*)	–399.09
C (graphite)	0.00
C (diamond)	0.45
C (*g*)	171.70
CO (*g*)	–26.42
CO_2 (*g*)	–94.05
CH_4 (*g*)	–17.90
C_2H_2 (*g*)	54.19
C_2H_4 (*g*)	12.50
C_2H_6 (*g*)	–20.24
CCl_4 (*g*)	–25.5
$CHCl_4$ (*g*)	–24
CH_2Cl_2 (*g*)	–21
CH_3Cl (*g*)	–19.6
CS_2 (*g*)	27.55
COS (*g*)	–32.80
$(CH_3)_2S$ (*g*)	–8.98
CH_3OH (*g*)	–48.08
C_2H_5OH (*g*)	–56.63
$(CH_3)_2O$ (*g*)	–44.3
C_3H_6 (*g*)	9.0
C_6H_{12} (*g*)	–29.98
C_6H_{10} (*g*)	–1.39
C_6H_6 (*g*)	19.82
Fe (*s*)	0.00
Fe (*g*)	99.5
Fe_2O_3 (*s*)	–196.8
Fe_3O_4 (*s*)	–267.8
H_2 (*g*)	0.00
H_2O (*g*)	–57.80
H_2O (*l*)	–68.32
H_2O_2 (*g*)	–31.83
H_2S (*g*)	–4.82
N_2 (*g*)	0.00
NO (*g*)	21.60
NO_2 (*g*)	8.09
NO_3 (*g*)	13
NH_3 (*g*)	–11.04
O_2 (*g*)	0.00
O_3 (*g*)	34.0
S (*g*)	0.00
SO_2 (*g*)	–70.96
SO_3 (*g*)	–94.45

(Multiply kcal/mol by 4.184 to obtain kJ/mol.)
(Multiply kcal/mol by 1800/*M* to obtain Btu/lbm.)

the sum of enthalpies of formation of all reactants. This is essentially a restatement of the energy conservation principle and is known as *Hess' law of energy summation.*

$$\Delta H_r = \sum_{\text{products}} \Delta H_f - \sum_{\text{reactants}} \Delta H_f \qquad 21.1$$

The associated energy for a chemical reaction can be defined in terms of heats of formation of the individual species ΔH_f° at the standard state. (The superscript circle indicates that all reactants and products are in their standard states.) The stoichiometric coefficient for species i is v_i.

Heats of Reaction

$$\Delta H_r^\circ = \sum_{\text{products}} v_i (\Delta H_f^\circ)_i - \sum_{\text{reactants}} v_i (\Delta H_f^\circ)_i \qquad 21.2$$

The heat of reaction will vary with the temperature.

Heats of Reaction

$$\Delta H_r^\circ(T) = \Delta H_r^\circ(T_{\text{ref}}) + \int_{T_{\text{ref}}}^{T} \Delta C_p \, dT \qquad 21.3$$

T_{ref} is a reference temperature (typically 25°C or 298K) and ΔC_p is the change in molar heat capacity for the reaction at constant pressure.

The heat of reaction for a combustion process that uses oxygen is also called the *heat of combustion.* The principal products are $CO_2(g)$ and $H_2O(l)$.

Reactions that give off energy (i.e., have negative enthalpies of reaction) are known as *exothermic reactions.* Many (but not all) exothermic reactions begin spontaneously. On the other hand, *endothermic reactions* absorb energy and require thermal or electrical energy to begin.

Example 21.1

Using enthalpies of formation, calculate the heat of stoichiometric combustion (standardized to 25°C) of gaseous methane (CH_4) and oxygen.

Solution

The balanced chemical equation for the stoichiometric combustion of methane is

$$CH_4 + 2O_2 \rightarrow 2H_2O + CO_2$$

Thermodynamics

The enthalpy of formation of oxygen gas (its free-state configuration) is zero. Using enthalpies of formation from Table 21.2 in Eq. 21.1, the enthalpy of reaction per mole of methane is

$$\begin{aligned}\Delta H_r &= 2\Delta H_{f,\mathrm{H_2O}} + \Delta H_{f,\mathrm{CO_2}} - \Delta H_{f,\mathrm{CH_4}} - 2\Delta H_{f,\mathrm{O_2}} \\ &= (2)\left(-57.80\ \frac{\mathrm{kcal}}{\mathrm{mol}}\right) + \left(-94.05\ \frac{\mathrm{kcal}}{\mathrm{mol}}\right) \\ &\quad - \left(-17.90\ \frac{\mathrm{kcal}}{\mathrm{mol}}\right) - (2)(0) \\ &= -191.75\ \mathrm{kcal/mol\ CH_4}\quad [\text{exothermic}]\end{aligned}$$

Using the footnote of Table 21.2, this value can be converted to Btu/lbm. The molecular weight of methane is

$$M_{\mathrm{CH_4}} = 12 + (4)(1) = 16$$

The heat of stoichiometric combustion is

$$\begin{aligned}\text{higher heating value} &= \frac{\left(191.75\ \frac{\mathrm{kcal}}{\mathrm{mol}}\right)\left(1800\ \frac{\mathrm{Btu\text{-}mol}}{\mathrm{lbm\text{-}kcal}}\right)}{16} \\ &= 21{,}572\ \mathrm{Btu/lbm}\end{aligned}$$

6. STANDARD CONDITIONS

Many gaseous fuel properties are specified per unit volume. The phrases "standard cubic foot" (SCF) and "normal cubic meters" (nm or $\mathrm{Nm^3}$) are incorporated into the designations of volumes and flow rates. These volumes are "standard" or "normal" because they are based on standard conditions.

There is not a single, universally agreed-on definition of *standard temperature and pressure* (*standard conditions, STP*). For nearly a century, the natural gas and oil industries in North America and the Organization of the Petroleum Exporting Countries (OPEC) defined standard conditions as 14.73 psia and 60°F (15.56°C; 288.71K). The *international standard metric conditions* for natural gas and similar fuels are 288.15K (59.00°F; 15.00°C) and 101.325 kPa. Some gas flows related to environmental engineering are based on standard conditions of either 15°C or 20°C and 101.325 kPa. These are different from the standard conditions used for scientific work, historically 32.00°F and 14.696 psia (0°C and 101.325 kPa) but more commonly now 0°C and 100.00 kPa, and the various temperatures, usually 20°C or 25°C, used to tabulate standard enthalpies of formation and reaction of components. Clearly, it is essential to know the "standard" conditions onto which any gaseous fuel flows are referred.

Some combustion equipment (e.g., particulate collectors) operates within a narrow range of temperatures and pressures. These conditions are referred to as *normal temperature and pressure* (NTP).

7. MOISTURE

If an ultimate analysis of a solid or liquid fuel is given, all the oxygen is assumed to be in the form of free water. The amount of hydrogen combined as free water is assumed to be one-eighth of the oxygen weight. All remaining hydrogen, known as the *available hydrogen*, is assumed to be combustible.

$$G_{\mathrm{H,combined}} = \frac{G_{\mathrm{O}}}{8} \qquad 21.4$$

$$G_{\mathrm{H,available}} = G_{\mathrm{H,total}} - \frac{G_{\mathrm{O}}}{8} \qquad 21.5$$

Moisture in fuel is undesirable because it increases fuel weight (transportation costs) and decreases available combustion heat. For coal, the *bed moisture level* is the moisture level when the coal is mined. The terms *dry* and *as fired* are often used in commercial coal specifications. The "as fired" condition corresponds to a specific moisture content when placed in the furnace. The "as fired" heating value (HV) should be used, because the moisture decreases the useful combustion energy. The approximate relationship between the two heating values is given by Eq. 21.6, where M is the fraction of moisture content from a proximate analysis.

$$\mathrm{HV_{as\ fired}} = \mathrm{HV_{dry}}(1 - M) \qquad 21.6$$

8. ASH AND MINERAL MATTER

Mineral matter is the noncombustible material in a fuel. *Ash* is the residue remaining after combustion. Ash may contain some combustible carbon as well as the original mineral matter. The two terms ("mineral matter" and "ash") are often used interchangeably when reporting fuel analyses.

Ash may also be categorized according to where it is recovered. Dry and wet *bottom ashes* are recovered from *ash pits*. However, as little as 10% of the total ash content may be recovered in the ash pit. *Fly ash* is carried out of the boiler by the flue gas. Fly ash can be deposited on walls and heat transfer surfaces. It will be discharged from the stack if not captured. *Economizer ash* and *air heater ash* are recovered from the devices the ashes are named after.

The finely powdered ash that covers combustion grates protects them from high temperatures.[5] If the ash has a low (i.e., below 2200°F; 1200°C) fusion temperature (melting point), it may form *clinkers* in the furnace and/or *slag* in other high-temperature areas. In extreme cases, it can adhere to the surfaces. Ashes with high melting (fusion) temperatures (i.e., above 2600°F; 1430°C) are known as *refractory ashes*. The T_{250} *temperature* is used as an index of slagging tendencies of an ash. This is the temperature at which the slag becomes molten with a viscosity of 250 poise. Slagging will be experienced when the T_{250} temperature is exceeded.

The actual melting point depends on the ash composition. Ash is primarily a mixture of silica (SiO_2), alumina (Al_2O_3), and ferric oxide (Fe_2O_3).[6] The relative proportions of each will determine the melting point, with lower melting points resulting from high amounts of ferric oxide and calcium oxide. The melting points of pure alumina and pure silica are in the 2700–2800°F (1480–1540°C) range.

Coal ash is either of a bituminous type or lignite type. Bituminous-type ash (from midwestern and eastern coals) contains more ferric oxide than lime and magnesia. Lignite-type ash (from western coals) contains more lime and magnesia than ferric oxide.

9. SULFUR

Several forms of sulfur are present in coal and fuel oils. *Pyritic sulfur* (FeS_2) is the primary form. *Organic sulfur* is combined with hydrogen and carbon in other compounds. *Sulfate sulfur* is iron sulfate and gypsum ($CaSO_4 \cdot 2H_2O$). Sulfur in elemental, organic, and pyritic forms oxidizes to sulfur dioxide. *Sulfur trioxide* can be formed under certain conditions. Sulfur trioxide combines with water to form sulfuric acid and is a major source of boiler/stack corrosion and acid rain.

$$SO_3 + H_2O \rightarrow H_2SO_4 \qquad 21.7$$

10. WASTE FUELS

Waste fuels are often used as fuels in industrial boilers and furnaces. Such fuels include digester and landfill gases, waste process gases, flammable waste liquids, and volatile organic compounds (VOCs) such as benzene, toluene, xylene, ethanol, and methane. Other waste fuels include oil shale, tar sands, green wood, seed and rice hulls, biomass refuse, peat, tire shreddings, and shingle/roofing waste.

The term *refuse-derived fuels* (RDF) is used to describe fuel produced from municipal waste. After separation (removal of glass, plastics, metals, corrugated cardboard, etc.), the waste is merely pushed into the combustion chamber. If the waste is to be burned elsewhere, it is compressed and baled.

The heating value of RDF depends on the moisture content and fraction of combustible material. For RDFs derived from typical municipal wastes, the heating value ranges from 3000 Btu/lbm to 6000 Btu/lbm (7 MJ/kg to 14 MJ/kg). Higher ranges (such as 7500–8500 Btu/lbm (17.5–19.8 MJ/kg)) can be obtained by careful selection of ingredients. Pelletized RDF (containing some coal and a limestone binder) with heating values around 8000 Btu/lbm (18.6 MJ/kg) can be used as a supplemental fuel in coal-fired units.

Scrap tires are an attractive fuel source due to their high heating values, which range from 12,000 Btu/lbm to 16,000 Btu/lbm (28 MJ/kg to 37 MJ/kg). To be compatible with existing coal-loading equipment, tires are chipped or shredded to 1 in (25 mm) size. Tires in this form are known as *tire-derived fuel* (TDF). Metal (from tire reinforcement) may or may not be present.

TDF has been shown to be capable of supplying up to 90% of a steam-generating plant's total heat input without any deterioration in particulate emissions, pollutants, and stack opacity, as long as the combustion device is properly designed, operated, and maintained. In fact, compared with some low-quality coals (e.g., lignite), TDF is far superior: about 2.5 times the heating value and less than half the sulfur.

11. INCINERATION

Many toxic wastes are incinerated rather than "burned." Incineration and combustion are not the same. *Incineration* is a disposal process that uses combustion to render wastes ineffective (nonharmful, nontoxic, etc.). Wastes and combustible fuel are combined in a furnace, and the heat of combustion destroys the waste. Wastes may themselves be combustible, though they may not be self-sustaining if the moisture content is too high.

Wastes destined for industrial incineration are usually classified by composition and heating value into seven broad categories. Type 0 is *trash* (highly combustible paper and wood, with 10% or less moisture); type 1 is *rubbish* (combustible waste with up to 25% moisture); type 2 is *refuse* (a mixture of rubbish and garbage, with up to 50% moisture); type 3 is *garbage* (residential waste with up to 70% moisture); type 4 is *animal solids and organic wastes* (85% moisture); type 5 is *gaseous, liquid, or semiliquid waste* from industrial processes;

[5]Some boiler manufacturers rely on the thermal protection the ash provides. For example, coal burned in cyclone boilers should have a minimum ash content of 7% to cover and protect the cyclone barrel tubes. Boiler wear and ash carryover will increase with lower ash contents.

[6]Calcium oxide (CaO), magnesium oxide ("magnesia," MgO), titanium oxide ("titania," TiO_2), ferrous oxide (FeO), and alkalies (Na_2O and K_2O) may be present in smaller amounts.

and type 6 is *semisolid and solid wastes* that require incineration in hearth, retort, or grate burning equipment.

12. COAL

Coal consists of volatile matter, fixed carbon, moisture, noncombustible mineral matter ("ash"), and sulfur. *Volatile matter* is driven off as a vapor when the coal is heated, and it is directly responsible for flame size. *Fixed carbon* is the combustible portion of the solid remaining after the volatile matter is driven off. Moisture is present in the coal as free water and (for some mineral compounds) as water of hydration. Sulfur, an undesirable component, contributes to heat content.

Coals are categorized into anthracitic, bituminous, and lignitic types. *Anthracite coal* is clean, dense, and hard. It is comparatively difficult to ignite but burns uniformly and smokelessly with a short flame. *Bituminous coal* varies in composition, but generally has a higher volatile content than anthracite, starts easily, and burns freely with a long flame. Smoke and soot are possible if bituminous coal is improperly fired. *Lignite coal* is a coal of woody structure, is very high in moisture, and has a low heating value. It normally ignites slowly due to its moisture, breaks apart when burning, and burns with little smoke or soot.

Coal burns efficiently in a particular furnace only if it is uniform in size, and different sizes of coal are best suited for different uses. Screen sizes are the most accurate way to grade coal, but descriptive terms are also widely used. Anthracite coal, for example, is available as *broken coal* (the largest size), *egg coal*, *nut coal*, and other sizes down to *barley coal* and *coal dust* (or *fines*). Companies differ, however, in the exact size limits they assign to these terms. *Run-of-mine* (ROM) is coal as mined.

13. LOW-SULFUR COAL

Switching to low-sulfur coal is a way to meet sulfur emission standards. Western and eastern low-sulfur coals have different properties.[7,8] Eastern low-sulfur coals are generally low-impact coals (that is, few changes need to be made to the power plant when switching to them). Western coals are generally high-impact coals. Properties of typical high- and low-sulfur fuels are shown in Table 21.3.

Lower sulfur content results in less boiler corrosion. However, of all the coal variables, the different ash characteristics are the most significant with regard to the steam generator components. The slagging and fouling tendencies are prime concerns.

Table 21.3 *Typical Properties of High- and Low-Sulfur Coals**

		low-sulfur	
property	high-sulfur	eastern	western
higher heating value,			
Btu/lbm	10,500	13,400	8000
(MJ/kg)	(24.4)	(31.2)	(18.6)
moisture content, %	11.7	6.9	30.4
ash content, %	11.8	4.5	6.4
sulfur content, %	3.2	0.7	0.5
slag melting			
temperature, °F	2400	2900	2900
(°C)	(1320)	(1590)	(1590)

(Multiply Btu/lbm by 2.326 to obtain kJ/kg.)

*All properties are "as received."

14. CLEAN COAL TECHNOLOGIES

A lot of effort has been put into developing technologies that will reduce acid rain, air toxics (also called air pollutants, primarily NOx and SO_2), and carbon dioxide emissions. These technologies are loosely labeled as *clean coal technologies* (CCTs). Whether or not these technologies can be retrofitted into an existing plant or designed into a new plant depends on the economics of the process.

With *coal cleaning*, coal is ground to ultrafine sizes to remove sulfur and ash-bearing minerals.[9] However, finely ground coal creates problems in handling, storage, and dust production. The risk of fire and explosion increases. Different approaches to reducing the problems associated with transporting and storing finely ground coal include the use of dust suppression chemicals, pelletizing, transportation of coal in liquid slurry form, and pelletizing followed by reslurrying. Some of these technologies may not be suitable for retrofit into existing installations.

With *coal upgrading*, moisture is thermally removed from low-rank coal (e.g., lignite or subbituminous coal). With some technologies, sulfur and ash are also removed when the coal is upgraded.

Reduction in sulfur dioxide emissions is the goal of *SO_2 control* technologies. These technologies include conventional use of lime and limestone in *flue gas desulfurization* (FGD) systems, *furnace sorbent-injection* (FSI), and *duct sorbent-injection*. *Advanced scrubbing* is included in FGD technologies.

[7] In the United States, low-sulfur coals predominately come from the western United States ("western subbituminous"), although some come from the east ("eastern bituminous").

[8] Some parameters dependent on coal type are coal preparation, firing rate, ash volume and handling, slagging, corrosion rates, dust collection and suppression, and fire and explosion prevention.

[9] 80% or more of *micronized coal* is 44 microns or less in size.

Redesigned burners and injectors and adjustment of the flame zone are typical types of *NOx control*. Use of secondary air, injection of ammonia or urea, and selective catalytic reduction (SCR) are also effective in NOx reduction.

Fluidized-bed combustion (FBC) reduces NOx emissions by reducing combustion temperatures to around 1500°F (815°C). FBC is also effective in removing up to 90% of the SO_2. *Atmospheric FBC* operates at atmospheric pressure, but higher thermal efficiencies are achieved in *pressurized FBC* units operating at pressures up to 10 atm.

Integrated gasification/combined cycle (IGCC) processes are able to remove 99% of all sulfur while reducing NOx to well below current emission standards. *Synthetic gas* (*syngas*) is derived from coal. Syngas has a lower heating value than natural gas, but it can be used to drive gas turbines in combined cycles or as a reactant in the production of other liquid fuels.

15. COKE

Coke, typically used in blast furnaces, is produced by heating coal in the absence of oxygen. The heavy hydrocarbons crack (i.e., the hydrogen is driven off), leaving only a carbonaceous residue containing ash and sulfur. Coke burns smokelessly. *Breeze* is coke smaller than 5/8 in (16 mm). It is not suitable for use in blast furnaces, but steam boilers can be adapted to use it. *Char* is produced from coal in a 900°F (500°C) carbonization process. The volatile matter is removed, but there is little cracking. The process is used to solidify tars, bitumens, and some gases.

16. LIQUID FUELS

Liquid fuels are lighter hydrocarbon products refined from crude petroleum oil. They include liquefied petroleum gas (LPG), gasoline, kerosene, jet fuel, diesel fuels, and heating oils. JP-4 ("jet propellant") is a 50-50 blend of kerosene and gasoline. JP-8 and Jet A are kerosene-like fuels for aircraft gas turbine engines. Important characteristics of a liquid fuel are its composition, ignition temperature, flash point,[10] viscosity, and heating value.

17. FUEL OILS

In the United States, fuel oils are categorized into grades 1 through 6 according to their viscosities.[11] Viscosity is the major factor in determining firing rate and the need for preheating for pumping or atomizing prior to burning. Grades 1 and 2 can be easily pumped at ambient temperatures. In the United States, the heaviest fuel oil used is grade 6, also known as *Bunker C oil*.[12,13] Fuel oils are also classified according to their viscosities as *distillate oils* (lighter) and *residual fuel oils* (heavier).

Like coal, fuel oils contain sulfur and ash that may cause pollution, slagging on the hot end of the boiler, and corrosion in the cold end. Table 21.4 lists typical properties of common commercial fuels, while Table 21.5 lists typical properties of fuel oils.

18. GASOLINE

Gasoline is not a pure compound. It is a mixture of various hydrocarbons blended to give a desired flammability, volatility, heating value, and octane rating. There is an infinite number of blends that can be used to produce gasoline.

Gasoline's heating value depends only slightly on composition. Within a variation of 1½%, the heating value can be taken as 20,200 Btu/lbm (47.0 MJ/kg) for regular gasoline and as 20,300 Btu/lbm (47.2 MJ/kg) for high-octane aviation fuel.

Since gasoline is a mixture of hydrocarbons, different fractions will evaporate at different temperatures. The *volatility* is the percentage of the fuel that evaporates by a given temperature. Typical volatility specifications call for 10% or greater at 167°F (75°C), 50% at 221°F (105°C), and 90% at 275°F (135°C). Low volatility causes difficulty in starting and poor engine performance at low temperatures.

The *octane number* (ON) is a measure of *knock resistance*. It is based on comparison, performed in a standardized one-cylinder engine, with the burning of isooctane and *n*-heptane. *n*-Heptane, C_7H_{16}, is rated zero and produces violent knocking. Isooctane, C_8H_{18}, is rated 100 and produces relatively knock-free operation. The percentage blend by volume of these fuels that matches the performance of the gasoline is the octane rating. The *research octane number* (RON) is a measure of the fuel's antiknock characteristics while idling; the *motor octane number* (MON) applies to high-speed, high-acceleration operations. The octane rating reported for commercial gasoline is an average of the two.

Gasolines with octanes greater than 100 (including aviation gasoline) are rated by their performance number. The *performance number* (PN) of gasoline containing antiknock compounds (e.g., tetraethyl lead, TEL, used in aviation gasoline) is related to the octane number.

$$\text{ON} = 100 + \frac{\text{PN} - 100}{3} \qquad 21.8$$

[10]This is different from the *flash point* that is the temperature at which fuel oils generate enough vapor to sustain ignition in the presence of spark or flame.

[11]Grade 3 became obsolete in 1948. Grade 5 is also subdivided into light and heavy categories.

[12]120°F (48°C) is the optimum temperature for pumping no. 6 fuel oil. At that temperature, no. 6 oil has a viscosity of approximately 3000 SSU. Further heating is necessary to lower the viscosity to 150–350 SSU for atomizing.

[13]To avoid *coking* of oil, heating coils in contact with oil should not be hotter than 240°F (116°C).

Table 21.4 Typical Properties of Common Commercial Fuels

	butane	no. 1 diesel	no. 2 diesel	ethanol	gasoline	JP-4	methanol	propane
chemical formula	C_4H_{10}	–	–	C_2H_5OH	–	–	CH_3OH	C_3H_8
molecular weight	58.12	≈ 170	≈ 184	46.07	≈ 126		32.04	44.09
heating value								
higher, Btu/lbm	21,240	19,240	19,110	12,800	20,260		9838	21,646
lower, Btu/lbm	19,620	18,250	18,000	11,500	18,900	18,400	8639	19,916
lower, Btu/gal	102,400	133,332	138,110	76,152	116,485	123,400	60,050	81,855
latent heat of vaporization, Btu/lbm		115	105	361	142		511 (20°C)	147
specific gravity*	2.01	0.876	0.920	0.794	0.68–0.74	0.8017	0.793	1.55

(Multiply Btu/lbm by 2.326 to obtain kJ/kg.)

(Multiply Btu/gal by 0.2786 to obtain MJ/m^3.)

*Specific gravities of propane and butane are with respect to air.

Table 21.5 Typical Properties of Fuel Oils[a]

grade	specific gravity	heating value (MBtu/gal)[b]	heating value (GJ/m^3)
1	0.805	134	37.3
2	0.850	139	38.6
4	0.903	145	40.4
5	0.933	148	41.2
6	0.965	151	41.9

(Multiply MBtu/gal by 0.2786 to obtain GJ/m^3.)

[a]Actual values will vary depending on composition.
[b]One MBtu equals one thousand Btus.

19. OXYGENATED GASOLINE

In parts of the United States, gasoline is "oxygenated" during the cold winter months. This has led to use of the term "winterized gasoline." The addition of *oxygenates* raises the combustion temperature, reducing carbon monoxide and unburned hydrocarbons.[14] Common oxygenates used in *reformulated gasoline* (RFG) include methyl tertiary-butyl ether (MTBE) and ethanol. Methanol, ethyl tertiary-butyl ether (ETBE), tertiary-amyl methyl ether (TAME), and tertiary-amyl ethyl ether (TAEE) may also be used. (See Table 21.6.) Oxygenates are added to bring the minimum oxygen level to 2–3% by weight.[15]

20. DIESEL FUEL

Properties and specifications for various grades of diesel fuel oil are similar to specifications for fuel oils. Grade 1-D ("D" for diesel) is a light distillate oil for high-speed engines in service requiring frequent speed and load changes. Grade 2-D is a distillate of lower volatility for engines in industrial and heavy mobile service. Grade 4-D is for use in medium speed engines under sustained loads.

Diesel oils are specified by a *cetane number*, which is a measure of the ignition quality (ignition delay) of a fuel. A cetane number of approximately 30 is required for satisfactory operation of low-speed diesel engines. High-speed engines, such as those used in cars, require a cetane number of 45 or more. Like the octane number for gasoline, the cetane number is determined by comparison with standard fuels. Cetane, $C_{16}H_{34}$, has a cetane number of 100. *n*-Methyl-naphthalene, $C_{11}H_{10}$, has a cetane number of zero. The cetane number can be increased by use of such additives as amyl nitrate, ethyl nitrate, and ether.

A diesel fuel's *pour point* number refers to its viscosity. A fuel with a pour point of 10°F (–12°C) will flow freely above that temperature. A fuel with a high pour point will thicken in cold temperatures.

The *cloud point* refers to the temperature at which wax crystals cloud the fuel at lower temperatures. The cloud point should be 20°F (–7°C) or higher. Below that temperature, the engine will not run well.

[14]Oxygenation may not be successful in reducing carbon dioxide. Since the heating value of the oxygenates is lower, fuel consumption of oxygenated fuels is higher. On a per-gallon (per-liter) basis, oxygenation reduces carbon dioxide. On a per-mile (per-kilometer) basis, however, oxygenation appears to increase carbon dioxide. In any case, claims of CO_2 reduction are highly controversial, as the CO_2 footprint required to plant, harvest, dispose of decaying roots, stalks, and leaves (i.e., silage), and refine alcohol is generally ignored.
[15]Other restrictions on gasoline during the winter months intended to reduce pollution may include maximum percentages of benzene and total aromatics and limits on Reid vapor pressure, as well as specifications covering volatile organic compounds, nitric oxide (NOx), and toxins.

Table 21.6 Typical Properties of Common Oxygenates

	ethanol	MTBE[a]	TAME	ETBE	TAEE
specific gravity	0.794	0.744	0.740	0.770	0.791
octane[b]	115	110	112	105	100
heating value, MBtu/gal[c]	76.2	93.6			
Reid vapor pressure, psig[d]	18	8	15–4	3–4	2
percent oxygen by weight	34.73	18.15	15.66	15.66	13.8
volumetric percent needed to achieve gasoline					
2.7% oxygen by weight	7.8	15.1	17.2	17.2	19.4
2.0% oxygen by weight	5.6	11.0	12.4	12.7	13.0

(Multiply MBtu/gal by 0.2786 to obtain MJ/m^3.)

[a]MTBE is water soluble and does not degrade. It imparts a foul taste to water and is a possible carcinogen. It has been legislatively banned in most states, including California.
[b]Octane is equal to (1/2)(MON + RON).
[c]One MBtu equals one thousand Btus.
[d]The Reid vapor pressure is the vapor pressure when heated to 100°F (38°C). This may also be referred to as the "blending vapor pressure."

Thermodynamics

21. ALCOHOL

Both methanol and ethanol can be used in internal combustion engines. *Methanol* (*methyl alcohol*) is produced from natural gas and coal, although it can also be produced from wood and organic debris. *Ethanol* (*ethyl alcohol, grain alcohol*) is distilled from grain, sugarcane, potatoes, and other agricultural products containing various amounts of sugars, starches, and cellulose.

Although methanol generally works as well as ethanol, only ethanol can be produced in large quantities from inexpensive agricultural products and by-products.

Alcohol is water soluble. The concentration of alcohol is measured by its *proof*, where 200 proof is pure alcohol. (180 proof is 90% alcohol and 10% water.)

Gasohol is a mixture of approximately 90% gasoline and 10% alcohol (generally ethanol).[16] Alcohol's heating value is less than gasoline's, so fuel consumption (per distance traveled) is higher with gasohol. Also, since alcohol absorbs moisture more readily than gasoline, corrosion of fuel tanks becomes problematic. In some engines, significantly higher percentages of alcohol may require such modifications as including larger carburetor jets, timing advances, heaters for preheating fuel in cold weather, tank lining to prevent rusting, and alcohol-resistant gaskets.

Mixtures of gasoline and alcohol can be designated by the first letter and the fraction of the alcohol. E10 is a mixture of 10% ethanol and 90% gasoline. M85 is a blend of 85% methanol and 15% gasoline.

Alcohol is a poor substitute for diesel fuel because alcohol's cetane number is low—from −20 to +8. Straight injection of alcohol results in poor performance and heavy knocking.

22. GASEOUS FUELS

Various gaseous fuels are used as energy sources, but most applications are limited to natural gas and *liquefied petroleum gases* (LPGs) (i.e., propane, butane, and mixtures of the two).[17,18] *Natural gas* is a mixture of methane (between 55% and 95%), higher hydrocarbons (primarily ethane), and other noncombustible gases. Typical heating values for natural gas range from 950 Btu/ft^3 to 1100 Btu/ft^3 (35 MJ/m^3 to 41 MJ/m^3).

The production of *synthetic gas* (*syngas*) through coal gasification may be applicable to large power generating plants. The cost of gasification, though justifiable to reduce sulfur and other pollutants, is too high for syngas to become a widespread substitute for natural gas.

23. IGNITION TEMPERATURE

The *ignition temperature* (*autoignition temperature*) is the minimum temperature at which combustion can be sustained. It is the temperature at which more heat is generated by the combustion reaction than is lost to the surroundings, after which combustion becomes self-sustaining. For coal, the minimum ignition temperature varies from around 800°F (425°C) for bituminous varieties to 900–1100°F (480–590°C)

[16]In fact, oxygenated gasoline may use more than 10% alcohol.
[17]A number of *manufactured gases* are of practical (and historical) interest in specific industries, including *coke-oven gas*, *blast-furnace gas*, *water gas*, *producer gas*, and *town gas*. However, these gases are not now in widespread use.
[18]At atmospheric pressure, propane boils at −44°F (−42°C), while butane boils at 31°F (−0.5°C).

for anthracite. For sulfur and charcoal, the ignition temperatures are approximately 470°F (240°C) and 650°F (340°C), respectively.

For gaseous fuels, the ignition temperature depends on the air-fuel ratio, temperature, pressure, and length of time the source of heat is applied. Ignition can be instantaneous or delayed, depending on the temperature. Generalizations can be made for any gas, but the generalized temperatures will be meaningless without specifying all of these factors.

24. ATMOSPHERIC AIR

It is important to make a distinction between "air" and "oxygen." Atmospheric air is a mixture of oxygen, nitrogen, and small amounts of carbon dioxide, water vapor, argon, and other inert ("rare") gases. For the purpose of combustion calculations, all constituents other than oxygen are grouped with nitrogen. (See Table 21.7.) It is necessary to supply 4.32 (i.e., 1/0.2315) masses of air to obtain one mass of oxygen. Similarly, it is necessary to supply 4.773 volumes of air to obtain one volume of oxygen. The average molecular weight of air is 28.97, and its specific gas constant is 53.35 ft-lbf/lbm-°R (287.03 J/kg·K).

Table 21.7 *Composition of Dry Air*[a]

component	percent by weight	percent by volume
oxygen	23.15	20.95
nitrogen/inerts	76.85	79.05
ratio of nitrogen to oxygen	3.320	3.773[b]
ratio of air to oxygen	4.320	4.773

[a]Inert gases and CO_2 are included as N_2.
[b]The value is also reported by various sources as 3.76, 3.78, and 3.784.

25. COMBUSTION REACTIONS

Combustion reactions contain a limited number of components. Carbon, hydrogen, sulfur, hydrocarbons, and oxygen are the reactants. Carbon dioxide and water vapor are the main products, with carbon monoxide, sulfur dioxide, and sulfur trioxide occurring in lesser amounts. Nitrogen and excess oxygen emerge hotter but unchanged from the stack.

Combustion reactions occur according to the normal chemical reaction principles. Balancing a combustion reaction is usually easiest if carbon is balanced first, followed by hydrogen and then by oxygen. When a gaseous fuel has several combustible gases, the volumetric fuel composition can be used as coefficients in the chemical equation.

Table 21.8 lists ideal combustion reactions. These reactions do not include any nitrogen or water vapor that are present in the combustion air.

Table 21.8 *Ideal Combustion Reactions*

fuel	formula	reaction equation (excluding nitrogen)
carbon (to CO)	C	$2C + O_2 \rightarrow 2CO$
carbon (to CO_2)	C	$C + O_2 \rightarrow CO_2$
sulfur (to SO_2)	S	$S + O_2 \rightarrow SO_2$
sulfur (to SO_3)	S	$2S + 3O_2 \rightarrow 2SO_3$
carbon monoxide	CO	$2CO + O_2 \rightarrow 2CO_2$
methane	CH_4	$CH_4 + 2O_2 \rightarrow CO_2 + 2H_2O$
acetylene	C_2H_2	$2C_2H_2 + 5O_2 \rightarrow 4CO_2 + 2H_2O$
ethylene	C_2H_4	$C_2H_4 + 3O_2 \rightarrow 2CO_2 + 2H_2O$
ethane	C_2H_6	$2C_2H_6 + 7O_2 \rightarrow 4CO_2 + 6H_2O$
hydrogen	H_2	$2H_2 + O_2 \rightarrow 2H_2O$
hydrogen sulfide	H_2S	$2H_2S + 3O_2 \rightarrow 2H_2O + 2SO_2$
propane	C_3H_8	$C_3H_8 + 5O_2 \rightarrow 3CO_2 + 4H_2O$
n-butane	C_4H_{10}	$2C_4H_{10} + 13O_2 \rightarrow 8CO_2 + 10H_2O$
octane	C_8H_{18}	$2C_8H_{18} + 25O_2 \rightarrow 16CO_2 + 18H_2O$
olefin series	C_nH_{2n}	$2C_nH_{2n} + 3nO_2 \rightarrow 2nCO_2 + 2nH_2O$
paraffin series	C_nH_{2n+2}	$2C_nH_{2n+2} + (3n+1)O_2 \rightarrow 2nCO_2 + (2n+2)H_2O$

(Multiply oxygen volumes by 3.773 to get nitrogen volumes.)

Example 21.2

A gaseous fuel is 20% hydrogen and 80% methane by volume. What stoichiometric volume of oxygen is needed to burn 120 volumes of fuel at the same conditions?

Solution

Write the unbalanced combustion reaction.

$$H_2 + CH_4 + O_2 \rightarrow CO_2 + H_2O$$

Use the volumetric analysis as coefficients of the fuel.

$$0.2H_2 + 0.8CH_4 + O_2 \rightarrow CO_2 + H_2O$$

Balance the carbons.

$$0.2H_2 + 0.8CH_4 + O_2 \rightarrow 0.8CO_2 + H_2O$$

Balance the hydrogens.

$$0.2H_2 + 0.8CH_4 + O_2 \rightarrow 0.8CO_2 + 1.8H_2O$$

Balance the oxygens.

$$0.2H_2 + 0.8CH_4 + 1.7O_2 \rightarrow 0.8CO_2 + 1.8H_2O$$

For gaseous components, the coefficients correspond to the volumes. Since one (0.2 + 0.8) volume of fuel requires 1.7 volumes of oxygen, the required oxygen is

$$(1.7)(120 \text{ volumes of fuel}) = 204 \text{ volumes of oxygen}$$

26. STOICHIOMETRIC REACTIONS

Stoichiometric quantities (*ideal quantities*) are the exact quantities of reactants that are needed to complete a combustion reaction without any reactants left over. Table 21.8 contains some of the more common chemical reactions. Stoichiometric volumes and masses can always be determined from the balanced chemical reaction equation. Table 21.9 can be used to quickly determine stoichiometric amounts for some fuels.

27. STOICHIOMETRIC AIR

Stoichiometric air (also called *theoretical air* or *ideal air*) is the air needed to provide the exact amount of oxygen for complete combustion of a fuel. Stoichiometric air includes atmospheric nitrogen. For each volume of oxygen, 3.773 volumes of nitrogen pass unchanged through the reaction.[19]

Stoichiometric air can be stated in units of mass (pounds or kilograms of air) for solid and liquid fuels, and in units of volume (cubic feet or cubic meters of air) for gaseous fuels. When stated in terms of mass, the stoichiometric ratio of air to fuel masses is known as the ideal *air-fuel ratio*, A/F.

Incomplete Combustion

$$A/F = \frac{\text{mass of air}}{\text{mass of fuel}} = \frac{m_{\text{air}}}{m_{\text{fuel}}} \qquad 21.9$$

The air-fuel ratio can also be stated in terms of moles. The *molar air-fuel ratio* is

Incomplete Combustion

$$\overline{A/F} = \frac{\text{no. of moles of air}}{\text{no. of moles of fuel}} = \frac{n_{\text{air}}}{n_{\text{fuel}}} \qquad 21.10$$

The air-fuel ratio and the molar air-fuel ratio are related by Eq. 21.11.

Incomplete Combustion

$$A/F = (\overline{A/F})\left(\frac{M_{\text{air}}}{M_{\text{fuel}}}\right) \qquad 21.11$$

The ideal air-fuel ratio can be determined from the combustion reaction equation. It can also be determined by adding the oxygen and nitrogen amounts listed in Table 21.9.

For fuels whose ultimate analysis is known, the approximate stoichiometric air (oxygen and nitrogen) requirement in pounds of air per pound of fuel (kilograms of air per kilogram of fuel) can be quickly calculated by using Eq. 21.12.[20] All oxygen in the fuel is assumed to be free moisture. All of the reported oxygen is assumed to be locked up in the form of water. Any free oxygen (i.e., oxygen dissolved in liquid fuels) is subtracted from the oxygen requirements.

$$A/F = 34.5\left(\frac{G_C}{3} + G_H - \frac{G_O}{8} + \frac{G_S}{8}\right) \quad \text{[solid and liquid fuels]} \qquad 21.12$$

For fuels consisting of a mixture of gases, Eq. 21.13 and the constant J from Table 21.10 can be used to quickly determine the stoichiometric air requirements.

$$A/F = \sum G_i J_i \quad \text{[gaseous fuels]} \qquad 21.13$$

For fuels consisting of a mixture of gases, the air-fuel ratio can also be expressed in volumes of air per volume of fuel using values of K from Table 21.10.

$$\text{volumetric air-fuel ratio} = \sum B_i K_i \quad \text{[gaseous fuels]} \qquad 21.14$$

[19]The only major change in the nitrogen gas is its increase in temperature. Dissociation of nitrogen and formation of nitrogen compounds can occur but are essentially insignificant.

[20]This is a "compromise" equation. Variations in the atomic weights will affect the coefficients slightly. The coefficient 34.5 is reported as 34.43 in some older books. 34.5 is the exact value needed for carbon and hydrogen, which constitute the bulk of the fuel. 34.43 is the correct value for sulfur, but the error is small and is disregarded in this equation.

Table 21.9 Consolidated Combustion Data[a,b,c,d]

fuel	for 1 mole of fuel					for 1 ft^3 of fuel[e]				for 1 lbm of fuel					units of fuel
	air		other products			air		other products		air		other products			
	O_2	N_2	CO_2	H_2O	SO_2	O_2	N_2	CO_2	H_2O	O_2	N_2	CO_2	H_2O	SO_2	
C carbon	1.0 379.5 32.0	3.773 1432 106	1.0 379.5 44.0							0.0833 31.63 2.667	0.3143 119.3 8.883	0.0833 31.63 3.667			moles ft^3 lbm
H_2 hydrogen	0.5 189.8 16.0	1.887 716.1 53.0		1.0 379.5 18.0		0.001317 0.5 0.04216	0.004969 1.887 0.1397		0.002635 1.0 0.04747	0.248 94.12 7.936	0.9357 355.1 26.29		0.496 188.25 8.936		moles ft^3 lbm
S sulfur	1.0 379.5 32.0	3.773 1432 106.0			1.0 379.5 64.06					0.03119 11.84 0.998	0.1177 44.67 3.306			0.03119 11.84 1.998	moles ft^3 lbm
CO carbon monoxide	0.5 189.8 16.0	1.887 716.1 53.0	1.0 379.5 44.01			0.001317 0.5 0.04216	0.004969 1.887 0.1397	0.002635 1.0 0.1160		0.01785 6.774 0.5712	0.06735 25.56 1.892	0.03570 13.55 1.572			moles ft^3 lbm
CH_4 methane	2.0 759 64.0	7.546 2864 212.0	1.0 379.5 44.01	2.0 758 36.03		0.00527 2.0 0.1686	0.01988 7.546 0.5586	0.002635 1.0 0.1160	0.00527 2.0 0.0949	0.1247 47.31 3.989	0.4705 178.5 13.21	0.06233 23.66 2.743	0.1247 47.31 2.246		moles ft^3 lbm
C_2H_2 acetylene	2.5 948.8 80.0	9.433 3580 265.0	2.0 758 88.02	1.0 379.5 18.02		0.006588 2.5 0.2108	0.02486 9.443 0.6983	0.00527 2.0 0.2319	0.002635 1.0 0.04747	0.09601 36.44 3.072	0.3622 137.5 10.18	0.07681 29.15 3.380	0.03841 14.57 0.6919		moles ft^3 lbm
C_2H_4 ethylene	3.0 1139 96.0	11.32 4297 318.0	2.0 758 88.02	2.0 758 36.03		0.007905 3.0 0.2530	0.02983 11.32 0.8380	0.00527 2.0 0.2319	0.00527 2.0 0.0949	0.1069 40.58 3.422	0.4033 153.1 11.34	0.07129 27.05 3.137	0.07129 27.05 1.284		moles ft^3 lbm
C_2H_6 ethane	3.5 1328 112.0	13.21 5010 371.0	2.0 758 88.02	3.0 1139 54.05		0.009223 3.5 0.2951	0.03480 13.21 0.9776	0.00527 2.0 0.2319	0.007905 3.0 0.1424	0.1164 44.17 3.724	0.4392 166.7 12.34	0.06651 25.24 2.927	0.09977 37.86 1.797		moles ft^3 lbm

(Multiply lbm/ft^3 by 0.06243 to obtain kg/m^3.)

[a]Rounding of molecular weights and air composition may introduce slight inconsistencies in the table values. This table is based on atomic weights with at least four significant digits, a ratio of 3.773 volumes of nitrogen per volume of oxygen, and 379.5 ft^3 per mole at 1 atm and 60°F.

[b]Volumes per unit mass are at 1 atm and 60°F (16°C). To obtain volumes at other temperatures, multiply by $(T_{°F} + 460°)/520°$ or $(T_{°C} + 273°)/289°$.

[c]The volume of water applies only when the combustion products are at such high temperatures that all of the water is in vapor form.

[d]This table can be used to directly determine some SI ratios. For kg/kg ratios, the values are the same as lbm/lbm. For L/L or m^3/m^3, use ft^3/ft^3. For mol/mol, use mole/mole. For mixed units (e.g., ft^3/lbm), conversions are required.

[e]Sulfur is not used in gaseous form.

*Table 21.10 Approximate Air-Fuel Ratio Coefficients for Components of Natural Gas**

fuel component	J (gravimetric)	K (volumetric)
acetylene, C_2H_2	13.25	11.945
butane, C_4H_{10}	15.43	31.06
carbon monoxide, CO	2.463	2.389
ethane, C_2H_6	16.06	16.723
ethylene, C_2H_4	14.76	14.33
hydrogen, H_2	34.23	2.389
hydrogen sulfide, H_2S	6.074	7.167
methane, CH_4	17.20	9.556
oxygen, O_2	–4.320	–4.773
propane, C_3H_8	15.65	23.89

*Rounding of molecular weights and air composition may introduce slight inconsistencies in the table values. This table is based on atomic weights with at least four significant digits and a ratio of 3.773 volumes of nitrogen per volume of oxygen.

Example 21.3

Use Table 21.9 to determine the theoretical volume of 90°F (32°C) air required to burn 1 volume of 60°F (16°C) carbon monoxide to carbon dioxide.

Solution

From Table 21.9, 0.5 volumes of oxygen are required to burn 1 volume of carbon monoxide to carbon dioxide. 1.887 volumes of nitrogen accompany the oxygen. The total amount of air at the temperature of the fuel is 0.5 + 1.887 = 2.387 volumes.

This volume will expand at the higher temperature. The volume at the higher temperature is

$$\begin{aligned} V_2 &= \frac{T_2 V_1}{T_1} \\ &= \frac{(90°\text{F} + 460°)(2.387 \text{ volumes})}{60°\text{F} + 460°} \\ &= 2.53 \text{ volumes} \end{aligned}$$

Example 21.4

How much air is required for the ideal combustion of (a) coal with an ultimate analysis of 93.5% carbon, 2.6% hydrogen, 2.3% oxygen, 0.9% nitrogen, and 0.7% sulfur; (b) fuel oil with a gravimetric analysis of 84% carbon, 15.3% hydrogen, 0.4% nitrogen, and 0.3% sulfur; and (c) natural gas with a volumetric analysis of 86.92% methane, 7.95% ethane, 2.81% nitrogen, 2.16% propane, and 0.16% butane?

Solution

(a) Use Eq. 21.12.

$$\begin{aligned} A/F &= 34.5\left(\frac{G_\text{C}}{3} + G_\text{H} - \frac{G_\text{O}}{8} + \frac{G_\text{S}}{8}\right) \\ &= (34.5)\left(\frac{0.935}{3} + 0.026 - \frac{0.023}{8} + \frac{0.007}{8}\right) \\ &= 11.58 \text{ lbm/lbm (kg/kg)} \end{aligned}$$

(b) Use Eq. 21.12.

$$\begin{aligned} A/F &= 34.5\left(\frac{G_\text{C}}{3} + G_\text{H} + \frac{G_\text{S}}{8}\right) \\ &= (34.5)\left(\frac{0.84}{3} + 0.153 + \frac{0.003}{8}\right) \\ &= 14.95 \text{ lbm/lbm (kg/kg)} \end{aligned}$$

(c) Use Eq. 21.14 and the coefficients from Table 21.10.

$$\begin{aligned} \begin{array}{c}\text{volumetric}\\ \text{air-fuel ratio}\end{array} &= \sum B_i K_i \\ &= (0.8692)\left(9.556 \ \frac{\text{ft}^3}{\text{ft}^3}\right) \\ &\quad +(0.0795)\left(16.723 \ \frac{\text{ft}^3}{\text{ft}^3}\right) \\ &\quad +(0.0216)\left(23.89 \ \frac{\text{ft}^3}{\text{ft}^3}\right) \\ &\quad +(0.0016)\left(31.06 \ \frac{\text{ft}^3}{\text{ft}^3}\right) \\ &= 10.20 \text{ ft}^3/\text{ft}^3 \quad (\text{m}^3/\text{m}^3) \end{aligned}$$

28. INCOMPLETE COMBUSTION

Incomplete combustion occurs when there is insufficient oxygen to burn all of the hydrogen, carbon, and sulfur in the fuel. Without enough available oxygen, carbon burns to carbon monoxide.[21] Carbon monoxide in the flue gas indicates incomplete and inefficient combustion. Incomplete combustion is caused by cold furnaces, low combustion temperatures, poor air supply, smothering from improperly vented stacks, and insufficient mixing of air and fuel.

[21]Toxic alcohols, ketones, and aldehydes may also be formed during incomplete combustion.

29. SMOKE

The amount of smoke can be used as an indicator of combustion completeness. Smoky combustion may indicate improper air-fuel ratio, insufficient draft, leaks, insufficient preheat, or misadjustment of the fuel system.

Smoke measurements are made in a variety of ways, with the standards depending on the equipment used. Photoelectric sensors in the stack are used to continuously monitor smoke. The *smoke spot number* (SSN) and ASTM smoke scale are used with continuous stack monitors. For coal-fired furnaces, the maximum desirable smoke number is SSN 4. For grade 2 fuel oil, the SSN should be less than 1; for grade 4, SSN 4; for grades 5L, 5H, and low-sulfur residual fuels, SSN 3; for grade 6, SSN 4.

The *Ringelmann scale* is a subjective method in which the smoke density is visually compared to five standardized white-black grids. Ringelmann chart no. 0 is solid white; chart no. 5 is solid black. Ringelmann chart no. 1, which is 20% black, is the preferred (and required) operating point for most power plants.

30. FLUE GAS ANALYSIS

Combustion products that pass through a furnace's exhaust system are known as *flue gases* (*stack gases*). Flue gases are almost all nitrogen.[22] Nitrogen oxides are not present in large enough amounts to be included separately in combustion reactions.

The actual composition of flue gases can be obtained in a number of ways, including by modern electronic detectors, less expensive "length-of-stain" detectors, and direct sampling with an Orsat-type apparatus.

The antiquated *Orsat apparatus* determines the volumetric percentages of CO_2, CO, O_2, and N_2 in a flue gas. The sampled flue gas passes through a series of chemical compounds. The first compound absorbs only CO_2, the next only O_2, and the third only CO. The unabsorbed gas is assumed to be N_2 and is found by subtracting the volumetric percentages of all other components from 100%. An Orsat analysis is a dry analysis; the percentage of water vapor is not usually determined. A wet volumetric analysis (needed to compute the dew-point temperature) can be derived if the volume of water vapor is added to the Orsat volumes.

The Orsat procedure is now rarely used, although the term "Orsat" may be generally used to refer to any flue gas analyzer. Modern electronic analyzers can determine free oxygen (and other gases) independently of the other gases.

31. ACTUAL AND EXCESS AIR

Complete combustion occurs when all of the fuel is burned. *Excess air* is usually needed to achieve complete combustion. The excess air is the difference between the theoretical air and the actual air supplied.

The actual air is usually expressed as a percentage of the theoretical air requirements; for example, if the actual air is 1.4 times the theoretical air, this can be expressed as 140% theoretical air. This can also be expressed as 40% excess air.

The *percent theoretical air* is

Incomplete Combustion

$$\begin{array}{c}\text{percent}\\ \text{theoretical air}\end{array} = \frac{(A/F)_{\text{actual}}}{(A/F)_{\text{stoichiometric}}} \times 100\% \quad 21.15$$

The *percent excess air* is

Incomplete Combustion

$$\begin{array}{c}\text{percent}\\ \text{excess air}\end{array} = \frac{(A/F)_{\text{actual}} - (A/F)_{\text{stoichiometric}}}{(A/F)_{\text{stoichiometric}}} \times 100\% \quad 21.16$$

Different fuel types burn more efficiently with different amounts of excess air. Coal-fired boilers need approximately 30–35% excess air, oil-based units need about 15%, and natural gas burners need about 10%.

The actual air-fuel ratio for dry, solid fuels with no unburned carbon can be estimated from the volumetric flue gas analysis and the gravimetric fractions of carbon and sulfur in the fuel.

$$A/F = \frac{m_{\text{air}}}{m_{\text{fuel}}} = \frac{3.04B_{N_2}\left(G_C + \frac{G_S}{1.833}\right)}{B_{CO_2} + B_{CO}} \quad 21.17$$

Too much free oxygen or too little carbon dioxide in the flue gas are indicative of excess air. Because the relationship between oxygen and excess air is relatively insensitive to fuel composition, oxygen measurements have replacedstandard carbon dioxide measurements in determining combustion efficiency.[23] The relationship between excess air and the volumetric fraction of oxygen in the flue gas is given in Table 21.11.

[22]This assumption is helpful in making quick determinations of the thermodynamic properties of flue gases.
[23]The relationship between excess air required and CO_2 is much more dependent on fuel type (e.g., liquid) and furnace design.

Table 21.11 Approximate Volumetric Percentage of Oxygen in Stack Gas

	excess air							
fuel*	0%	1%	5%	10%	20%	50%	100%	200%
fuel oils, no. 2–6	0	0.22	1.06	2.02	3.69	7.29	10.8	14.2
natural gas	0	0.25	1.18	2.23	4.04	7.83	11.4	14.7
propane	0	0.23	1.08	2.06	3.75	7.38	10.9	14.3

*Values for coal are only marginally lower than the values for fuel oils.

Reducing the air-fuel ratio will have several outcomes. (a) The furnace temperature will increase due to a reduction in cooling air. (b) The flue gas will decrease in quantity. (c) The heat loss will decrease. (d) The furnace efficiency will increase. (e) Pollutants will (usually) decrease.

With a properly adjusted furnace and good mixing, the flue gas will contain no carbon monoxide, and the amount of carbon dioxide will be maximized. The stoichiometric amount of carbon dioxide in the flue gas is known as the *ultimate CO_2* and is the theoretical maximum level of carbon dioxide. The air-fuel mixture should be adjusted until the maximum level of carbon dioxide is attained.

Example 21.5

Propane (C_3H_8) is burned completely with 20% excess air. What is the volumetric fraction of carbon dioxide in the flue gas?

Solution

From Table 21.8, the balanced chemical reaction equation is

$$C_3H_8 + 5O_2 \rightarrow 3CO_2 + 4H_2O$$

With 20% excess air, the oxygen volume is $(1.2)(5) = 6$.

$$C_3H_8 + 6O_2 \rightarrow 3CO_2 + 4H_2O + O_2$$

There are 3.773 volumes of nitrogen for every volume of oxygen.

$$(3.773)(6) = 22.6$$

$$C_3H_8 + 6O_2 + 22.6N_2 \rightarrow 3CO_2 + 4H_2O + O_2 + 22.6N_2$$

For gases, the coefficients can be interpreted as volumes. The volumetric fraction of carbon dioxide is

$$\begin{aligned} B_{CO_2} &= \frac{3}{3 + 4 + 1 + 22.6} \\ &= 0.0980 \quad (9.8\%) \end{aligned}$$

32. CALCULATIONS BASED ON FLUE GAS ANALYSIS

Equation 21.18 gives the approximate percentage (by volume) of actual excess air.

$$\begin{array}{c}\text{actual excess air} \\ \text{\% by volume}\end{array} = \frac{B_{O_2} - 0.5B_{CO}}{0.264B_{N_2} - B_{O_2} + 0.5B_{CO}} \times 100\% \qquad 21.18$$

The ultimate CO_2 (i.e., the maximum theoretical carbon dioxide) can be determined from Eq. 21.19.

$$\begin{array}{c}\text{ultimate } CO_2 \\ \text{\% by volume}\end{array} = \frac{B_{CO_2,\text{actual}}}{1 - 4.773B_{O_2,\text{actual}}} \times 100\% \qquad 21.19$$

The mass ratio of dry flue gases to solid fuel is given by Eq. 21.20.

$$\frac{\text{mass of flue gas}}{\text{mass of solid fuel}} = \frac{\begin{array}{r}(11B_{CO_2} + 8B_{O_2} \\ +7(B_{CO} + B_{N_2})) \\ \times\left(G_C + \dfrac{G_S}{1.833}\right)\end{array}}{3(B_{CO_2} + B_{CO})} \qquad 21.20$$

Example 21.6

A sulfur-free coal has a proximate analysis of 75% carbon. The volumetric analysis of the flue gas is 80.2% nitrogen, 12.6% carbon dioxide, 6.2% oxygen, and 1.0% carbon monoxide. Calculate the (a) actual air-fuel ratio, (b) percentage excess air, (c) ultimate carbon dioxide, and (d) mass of flue gas per mass of fuel.

Solution

(a) Use Eq. 21.17.

$$\begin{aligned} A/F &= \frac{3.04B_{N_2}G_C}{B_{CO_2} + B_{CO}} \\ &= \frac{(3.04)(0.802)(0.75)}{0.126 + 0.01} \\ &= 13.4 \text{ lbm air/lbm fuel} \\ &\qquad (\text{kg air/kg fuel}) \end{aligned}$$

Thermodynamics

(b) Use Eq. 21.18.

$$\begin{aligned}\text{actual excess air} &= \frac{B_{O_2} - 0.5B_{CO}}{0.264B_{N_2} - B_{O_2} + 0.5B_{CO}} \times 100\% \\ &= \frac{0.062 - (0.5)(0.01)}{(0.264)(0.802) - 0.062 + (0.5)(0.01)} \times 100\% \\ &= 36.8\% \text{ by volume}\end{aligned}$$

(c) Use Eq. 21.19.

$$\begin{aligned}\text{ultimate } CO_2 &= \frac{B_{CO_2,\text{actual}}}{1 - 4.773B_{O_2,\text{actual}}} \times 100\% \\ &= \frac{0.126}{1 - (4.773)(0.062)} \times 100\% \\ &= 17.9\% \text{ by volume}\end{aligned}$$

(d) Use Eq. 21.20.

$$\begin{aligned}\frac{\text{mass of flue gas}}{\text{mass of solid fuel}} &= \frac{\left(11B_{CO_2} + 8B_{O_2} + 7(B_{CO} + B_{N_2})\right) \times \left(G_C + \frac{G_S}{1.833}\right)}{3\left(B_{CO_2} + B_{CO}\right)} \\ &= \frac{\left((11)(0.126) + (8)(0.062) + (7)(0.01 + 0.802)\right)(0.75)}{(3)(0.126 + 0.01)} \\ &= 13.9 \text{ lbm flue gas/lbm fuel} \\ &\quad (\text{kg flue gas/kg fuel})\end{aligned}$$

33. TEMPERATURE OF FLUE GAS

The temperature of the gas at the furnace outlet—before the gas reaches any other equipment—should be approximately 550°F (300°C). Overly low temperatures mean there is too much excess air. Overly high temperatures—above 750°F (400°C)—mean that heat is being wasted to the atmosphere and indicate other problems (ineffective heat transfer surfaces, overfiring, defective combustion chamber, etc.).

The *net stack temperature* is the difference between the stack and local environment temperatures. The net stack temperature should be as low as possible without causing corrosion of the low end.

34. DEW POINT OF FLUE GAS MOISTURE

The *dew point* is the temperature at which the water vapor in the flue gas begins to condense in a constant pressure process. To avoid condensation and corrosion in the stack, the temperature of the flue gases must be above the dew point. When there is no sulfur in the fuel, the dew point is typically around 100°F (40°C). The presence of sulfur in virtually any quantity increases the actual dew point to approximately 300°F (150°C).[24]

Dalton's law predicts the dew point of moisture in the flue gas. The partial pressure of the water vapor depends on the mole fraction (i.e., the volumetric fraction) of water vapor. The higher the water vapor pressure, the higher the dew point. Air entering a furnace can also contribute to moisture in the flue gas. This moisture should be added to the water vapor from combustion when calculating the mole fraction.

$$\begin{array}{c}\text{water vapor} \\ \text{partial pressure}\end{array} = \begin{array}{c}(\text{water vapor mole fraction}) \\ \times (\text{flue gas pressure})\end{array} \qquad 21.21$$

Once the water vapor's partial pressure is known, the dew point can be found from steam tables as the saturation temperature corresponding to the partial pressure.

35. HEAT OF COMBUSTION

The *heating value* of a fuel can be determined experimentally in a *bomb calorimeter*, or it can be estimated from the fuel's chemical analysis. The *higher heating value* (HHV), or *gross heating value*, of a fuel includes the heat of vaporization (condensation) of the water vapor formed from the combustion of hydrogen in the fuel. The *lower heating value* (LHV), or *net heating value*, assumes that all the products of combustion remain gaseous. The LHV is generally the value to use in calculations of thermal energy generated, since the heat of vaporization is not recovered within the furnace.

Traditionally, heating values have been reported on an HHV basis for coal-fired systems but on an LHV basis for natural gas-fired combustion turbines. There is an 11% difference between HHV and LHV thermal efficiencies for gas-fired systems and a 4% difference for coal-fired systems, approximately.

The HHV can be calculated from the LHV if the enthalpy of vaporization, h_{fg}, is known at the pressure of the water vapor.[25] In Eq. 21.22, m_{water} is the mass of water produced per unit (lbm, mole, m^3, etc.) of fuel.

$$\text{HHV} = \text{LHV} + m_{\text{water}}h_{fg} \qquad 21.22$$

As presented in Sec. 21.7, only the hydrogen that is not locked up with oxygen in the form of water is combustible. This is known as the *available hydrogen*. The correct percentage of combustible hydrogen, $G_{\text{H,available}}$, is

[24]The theoretical dew point is even higher—up to 350–400°F (175–200°C). For complex reasons, the theoretical value is not attained.

[25]For the purpose of initial studies, the heat of vaporization is usually assumed to be 1040 Btu/lbm (2.42 MJ/kg). This corresponds to a partial pressure of approximately 1 psia (7 kPa) and a dew point of 100°F (40°C).

calculated from the hydrogen and oxygen fraction. Equation 21.23 (same as Eq. 21.5) assumes that all of the oxygen is present in the form of water.

$$G_{\text{H,available}} = G_{\text{H,total}} - \frac{G_{\text{O}}}{8} \qquad 21.23$$

Dulong's formula calculates the higher heating value of coals and coke with a 2–3% accuracy range for moisture contents below approximately 10%.[26] The gravimetric or volumetric analysis percentages for each combustible element (including sulfur) are multiplied by the heating value per unit (mass or volume) from a table of heating values and summed. [**Heating Values of Substances Occurring in Common Fuels**]

$$\text{HHV}_{\text{MJ/kg}} = 32.78G_{\text{C}} + 141.8\left(G_{\text{H}} - \frac{G_{\text{O}}}{8}\right) + 9.264G_{\text{S}} \quad \text{[SI]} \qquad 21.24(a)$$

$$\text{HHV}_{\text{Btu/lbm}} = 14{,}093G_{\text{C}} + 60{,}958 \times \left(G_{\text{H}} - \frac{G_{\text{O}}}{8}\right) + 3983G_{\text{S}} \quad \text{[U.S.]} \qquad 21.24(b)$$

The higher heating value of gasoline can be approximated from the Baumé specific gravity, °Be.

$$\text{HHV}_{\text{gasoline,MJ/kg}} = 42.61 + 0.093(°\text{Be} - 10) \quad \text{[SI]} \qquad 21.25(a)$$

$$\text{HHV}_{\text{gasoline,Btu/lbm}} = 18{,}320 + 40(°\text{Be} - 10) \quad \text{[U.S.]} \qquad 21.25(b)$$

The heating value of petroleum oils (including diesel fuel) can also be approximately determined from the oil's specific gravity. The values derived by using Eq. 21.26 may not exactly agree with values for specific oils because the equation does not account for refining methods and sulfur content. Equation 21.26 was originally intended for combustion at constant volume, as in a gasoline engine. However, variations in heating values for different oils are very small; therefore, Eq. 21.26 is widely used as an approximation for all types of combustion, including constant pressure combustion in industrial boilers.

$$\text{HHV}_{\text{fuel oil,MJ/kg}} = 51.92 - 8.792(\text{SG})^2 \quad \text{[SI]} \qquad 21.26(a)$$

$$\text{HHV}_{\text{fuel oil,Btu/lbm}} = 22{,}320 - 3780(\text{SG})^2 \quad \text{[U.S.]} \qquad 21.26(b)$$

36. MAXIMUM THEORETICAL COMBUSTION (FLAME) TEMPERATURE

It can be assumed that the maximum theoretical increase in flue gas temperature will occur if all the combustion energy is absorbed adiabatically by the smallest possible quantity of combustion products. This provides a method of estimating the *maximum theoretical combustion temperature*, also sometimes called the *maximum flame temperature* or *adiabatic flame temperature*.[27]

In Eq. 21.27, the mass of the products is the sum of the fuel, oxygen, and nitrogen masses for stoichiometric combustion. The mean specific heat is a gravimetrically weighted average of the values of c_p for all combustion gases. (Since nitrogen makes up the largest part of the combustion gases, the mixture's specific heat will be approximately that of nitrogen.) The heat of combustion can be found either from the lower heating value, LHV, or from a difference in air enthalpies across the furnace.

$$T_{\text{max}} = T_i + \frac{\text{LHV}}{m_{\text{products}}c_{p,\text{mean}}} \qquad 21.27$$

Due to thermal losses, incomplete combustion, and excess air, actual flame temperatures are always lower than the theoretical temperature. Most fuels produce flame temperatures in the range of 3350–3800°F (1850–2100°C).

37. COMBUSTION LOSSES

A portion of the combustion energy is lost in heating the dry flue gases, dfg.[28] This is known as *dry flue gas loss*. In Eq. 21.28, $m_{\text{flue gas}}$ is the mass of dry flue gas per unit mass of fuel. It can be estimated from Eq. 21.20. Although the full temperature difference is used, the specific heat should be evaluated at the average temperature of the flue gas. For quick estimates, the dry flue gas can be assumed to be pure nitrogen.

$$q_1 = m_{\text{flue gas}}c_p(T_{\text{flue gas}} - T_{\text{incoming air}}) \qquad 21.28$$

Heat is lost in vaporizing the water formed during the combustion of hydrogen. In Eq. 21.29, m_{vapor} is the mass of vapor per pound of fuel. G_{H} is the gravimetric fraction of hydrogen in the fuel. The coefficient 8.94 is essentially 8 + 1 = 9 and converts the gravimetric mass of hydrogen to gravimetric mass of water formed. h_g is the enthalpy of superheated steam at the flue gas

[26]The coefficients in Eq. 21.24 are slightly different from the coefficients originally proposed by Dulong. Equation 21.24 reflects currently accepted heating values that were unavailable when Dulong developed his formula. Equation 21.24 makes these assumptions: (1) None of the oxygen is in carbonate form. (2) There is no free oxygen. (3) The hydrogen and carbon are not combined as hydrocarbons. (4) Carbon is amorphous, not graphitic. (5) Sulfur is not in sulfate form. (6) Sulfur burns to sulfur dioxide.

[27]Flame temperature is limited by the dissociation of common reaction products (CO_2, N_2, etc.). At high enough temperatures (3400–3800°F; 1880–2090°C), the endothermic dissociation process reabsorbs combustion heat and the temperature stops increasing. The temperature at which this occurs is known as the *dissociation temperature* (*maximum flame temperature*). This definition of flame temperature is not a function of heating values and flow rates.

[28]The abbreviation "dfg" for dry flue gas is peculiar to the combustion industry. It may not be recognized outside of that field.

temperature and the partial pressure of the water vapor. h_f is the enthalpy of saturated liquid at the air's entrance temperature.

$$q_2 = m_{\text{vapor}}(h_g - h_f) = 8.94G_{\text{H}}(h_g - h_f) \quad 21.29$$

Heat is lost when it is absorbed by moisture originally in the combustion air (and by free moisture in the fuel, if any). In Eq. 21.30, $m_{\text{combustion air}}$ is the mass of combustion air per pound of fuel. W is the humidity ratio. h_g' is the enthalpy of superheated steam at the air's entrance temperature and partial pressure of the water vapor.

$$\begin{aligned} q_3 &= m_{\text{atmospheric water vapor}}(h_g - h_g') \\ &= Wm_{\text{combustion air}}(h_g - h_g') \end{aligned} \quad 21.30$$

When carbon monoxide appears in the flue gas, potential energy is lost in incomplete combustion. The higher heating value of carbon monoxide in combustion to CO_2 is 4347 Btu/lbm (9.72 MJ/kg).[29]

$$q_4 = \frac{2.334\text{HHV}_{\text{CO}}G_{\text{C}}B_{\text{CO}}}{B_{\text{CO}_2} + B_{\text{CO}}} \quad 21.31$$

For solid fuels, energy is lost in unburned carbon in the ash. (Some carbon may be carried away in the flue gas, as well.) This is known as *combustible loss* or *unburned fuel loss*. In Eq. 21.32, m_{ash} is the mass of ash produced per pound of fuel consumed. $G_{\text{C,ash}}$ is the gravimetric fraction of carbon in the ash. The heating value of carbon is 14,093 Btu/lbm (32.8 MJ/kg).

$$q_5 = \text{HHV}_{\text{C}}m_{\text{ash}}G_{\text{C,ash}} \quad 21.32$$

Energy is also lost through radiation from the exterior boiler surfaces. This can be calculated if enough information is known. The *radiation loss* is fairly insensitive to different firing rates, and once calculated it can be considered constant for different conditions.

Other conditions where energy can be lost include air leaks, poor pulverizer operation, excessive blowdown, steam leaks, missing or loose insulation, and excessive soot-blower operation. Losses due to these sources must be evaluated on a case-by-case basis. The term *manufacturer's margin* is used to describe an accumulation of various unaccounted-for losses, which can include incomplete combustion to CO, energy loss in ash, instrument errors, energy carried away by atomizing steam, sulfation and calcination reactions in fluidized bed combustion boilers, and loss due to periodic blowdown. It can be 0.25–1.5% of all energy inputs, depending on the type of combuster/boiler.

38. COMBUSTION EFFICIENCY

The *combustion efficiency* (also referred to as *boiler efficiency*, *furnace efficiency*, and *thermal efficiency*) is the overall thermal efficiency of the combustion reaction. Furnaces and boilers for all fuels (e.g., coal, oil, and gas) with air heaters and economizers have 75–85% efficiency ranges, with all modern installation trending to the higher end of this range.

In Eq. 21.33, m_{steam} is the mass of steam produced per pound of fuel burned. The useful heat may also be determined from the boiler rating. One *boiler horsepower* equals approximately 33,475 Btu/hr (9.8106 kW).[30]

$$\begin{aligned} \eta &= \frac{\text{useful heat extracted}}{\text{heating value}} \\ &= \frac{m_{\text{steam}}(h_{\text{steam}} - h_{\text{feedwater}})}{\text{HHV}} \end{aligned} \quad 21.33$$

Calculating the efficiency by subtracting all known losses is known as the *loss method*. Minor sources of thermal energy, such as the entering air and feedwater, are essentially disregarded.

$$\begin{aligned} \eta &= \frac{\text{HHV} - q_1 - q_2 - q_3 - q_4 - q_5 - \text{radiation}}{\text{HHV}} \\ &= \frac{\text{LHV} - q_1 - q_4 - q_5 - \text{radiation}}{\text{HHV}} \end{aligned} \quad 21.34$$

Combustion efficiency can be improved by decreasing either the temperature or the volume of the flue gas or both. Since the latent heat of moisture is a loss, and since the amount of moisture generated corresponds to the hydrogen content of the fuel, a minimum efficiency loss due to moisture formation cannot be eliminated. This minimum loss is approximately 13% for natural gas, 8% for oil, and 6% for coal.

39. DRAFT

The amount of air that flows through a furnace is determined by the furnace draft. *Draft* is the difference in static pressures that causes the flue gases to flow.[31] It is usually stated in inches of water (kPa). *Natural draft* (ND) furnaces rely on the *stack effect* (*chimney draft*) to

[29]2.334 is the ratio of the molecular weight of carbon monoxide (28.01) to carbon (12), which is necessary to convert the higher heating value of carbon monoxide from a mass of CO to a mass of C. The product $2.334\text{HHV}_{\text{CO}}$ is often stated as 10,160 Btu/lbm (23.63 MJ/kg), although the actual calculated value is somewhat less. Other values such as 10,150 Btu/lbm (23.61 MJ/kg) and 10,190 Btu/lbm (23.70 MJ/kg) are encountered.

[30]Boiler horsepower is sometimes equated with a gross heating rate of 44,633 Btu/hr (13.08 kW). However, this is the total incoming heating value assuming a standardized 75% combustion efficiency.

[31]The British term "draught" is synonymous with "draft."

draw off combustion gases. Air flows through the furnace and out the stack due to the pressure differential caused by reduced densities in the stack.[32]

Forced draft (FD) fans located before the furnace are used to supply air for burning. Combustion occurs under pressure, hence the descriptive term "pressure-fired unit" for such fans. FD fans are run at relatively high speeds (1200 rpm to 1800 rpm) with direct-drive motors. Two or more fans are used in parallel to provide for efficient operation at low furnace demand.

FD fans create a positive pressure (e.g., 2 in wg to 10 in wg; 0.5 kPa to 2.5 kPa). This pressure is reduced to a very small negative pressure after passing through the air heater, ducts, and windbox system. The negative pressure in the furnace serves to keep combustion gases from leaking out into the furnace and boiler areas. The pressure continues to drop as it passes through the boiler, economizer, air heater, and pollution-control equipment.

Whereas FD fans force air through the system, *induced draft* (ID) fans are used to draw combustion products through the furnace bed, stack, and pollution control system by injecting air into the stack after combustion. The term "suction units" is used with ID fans.

ID fans are located after dust collectors and precipitators (often at the base of the stack) in order to reduce the abrasive effects of fly ash. They are run at slower speeds than forced draft fans in order to reduce the abrasive effects even further. Unlike FD fans, ID fans are usually very large and powerful because they have to handle all the combustion gases, not just combustion air.

Modern welded stacks are essentially airtight and can operate at pressures above atmospheric, often eliminating the need for ID fans.

Pure FD and ID systems are rarely used. Modern stack systems operate in a condition of *balanced draft*. Balanced draft is the term used when the static pressure is equal to atmospheric pressure. This requires the use of both ID and FD fans. In order to keep combustion products inside the combustion chamber and stack system, balanced draft systems may actually operate with a slight negative pressure.

The *draft loss* is the static pressure drop due to friction through the boiler and stack. The *available draft* is the difference between the theoretical draft (stack effect) and the draft loss. The available draft is zero in a balanced system.

$$D_{\text{available}} = D_{\text{theoretical}} - D_{\text{friction}} \qquad 21.35$$

The *net rating* or *fan boost* of the fan is the total pressure (the difference of draft losses and draft gains) supplied by the fan at maximum operating conditions. The fan power for ID and FD fans is calculated, as with any fan application, from the net rating and the flow rate. It is customary in sizing ID and FD draft fans to include increases of approximately 15% for flow rate, 30% for pressure, and 20°F (10°C) for temperature.

40. STACK EFFECT

Stack effect (*chimney action* or *natural draft*) is a pressure difference caused by the difference in atmospheric air and flue gas densities. It can be relied on to draw combustion products at least partially up through the stack. The stack effect is determined with no flow of flue gas. With flow, some of the stack effect is converted to velocity head, and the remainder is used to overcome friction.

Generally, the higher the chimney, the greater the stack effect. However, the stack effect in modern plants is greatly reduced by the friction and cooling effects of economizers, air heaters, and precipitators. Fans supply the needed pressure differential. Therefore, the primary function of modern stacks is to carry the combustion products a sufficient distance upward to dilute the combustion products, not to generate draft. Modern stacks for coal-fired power plants are seldom built shorter than 200 ft (60 m) in order to meet their dispersion requirements, and most are higher than 500 ft (150 m).

The theoretical stack effect, $D_{\text{theoretical}}$, is calculated from the densities of the flue gas and atmospheric air. In Eq. 21.36, H_{stack} is the total height of the stack. The average temperature is used to determine the average density in Eq. 21.36.[33]

$$D_{\text{theoretical}} = H_{\text{stack}}(\gamma_{\text{air}} - \gamma_{\text{flue gas,ave}}) \qquad 21.36$$

If a few assumptions are made, the stack effect can be calculated from the average temperature of the flue gas (i.e., the temperature at the average stack elevation).[34] The average flue gas temperature is the temperature halfway up the stack. H_{stack} is the total height of the stack.

$$D_{\text{theoretical}} = \left(\frac{p_{\text{air}} H_{\text{stack}} g}{R_{\text{air}}}\right)\left(\frac{1}{T_{\text{air}}} - \frac{1}{T_{\text{flue gas,ave}}}\right) \quad \text{[SI]} \qquad 21.37(a)$$

$$D_{\text{theoretical}} = \left(\frac{p_{\text{air}} H_{\text{stack}} g}{R_{\text{air}} g_c}\right)\left(\frac{1}{T_{\text{air}}} - \frac{1}{T_{\text{flue gas,ave}}}\right) \quad \text{[U.S.]} \qquad 21.37(b)$$

[32]Chimneys that rely on natural draft are sometimes referred to as *gravity chimneys*.

[33]The actual stack effect will be less than the value calculated with Eq. 21.36. For realistic problems, the achievable stack effect probably should be considered to be 80% of the ideal.

[34]One assumption is that the pressure in the stack is equal to atmospheric pressure. Then, the density difference will be due to only the temperature difference. Also, the flue gas is assumed to be air. This makes it unnecessary to know the flue gas composition, molecular weight, and so on.

The *stack effect head*, h_{SE}, is used to calculate the stack velocity. The coefficient of velocity, C_v, is approximately 0.30 to 0.50.

$$h_{SE} = \frac{D_{theoretical}}{\gamma_{flue\,gas,ave}} \quad 21.38$$

$$v = C_v v_{ideal} = C_v\sqrt{2gh_{SE}} \quad 21.39$$

The required stack area is

$$A_{stack} = \frac{Q_{flue\,gas}}{v} \quad 21.40$$

Example 21.7

The air surrounding an 80 ft (24 m) tall vertical stack is at 70°F (21°C) and one standard atmospheric pressure. The temperature of the flue gas at the average stack elevation is 400°F (200°C). What is the theoretical natural draft?

SI Solution

The absolute temperatures are

$$T_{air} = 21°\text{C} + 273° = 294\text{K}$$
$$T_{flue\,gas} = 200°\text{C} + 273° = 473\text{K}$$

Calculate the stack effect from Eq. 21.37.

$$D_{theoretical} = \left(\frac{p_{air}H_{stack}g}{R_{air}}\right)\left(\frac{1}{T_{air}} - \frac{1}{T_{flue\,gas}}\right)$$
$$= \left(\frac{(101.3\text{ kPa})(24\text{ m})\left(9.81\ \frac{\text{m}}{\text{s}^2}\right)}{287\ \frac{\text{J}}{\text{kg·K}}}\right) \times \left(\frac{1}{294\text{K}} - \frac{1}{473\text{K}}\right)$$
$$= 0.107\text{ kPa}$$

Customary U.S. Solution

The absolute temperatures are

$$T_{air} = 70°\text{F} + 460° = 530°\text{R}$$
$$T_{flue\,gas} = 400°\text{F} + 460° = 860°\text{R}$$

Calculate the stack effect from Eq. 21.37.

$$D_{theoretical} = \left(\frac{p_{air}H_{stack}g}{R_{air}g_c}\right)\left(\frac{1}{T_{air}} - \frac{1}{T_{flue\,gas}}\right)$$
$$= \left(\frac{\left(14.7\ \frac{\text{lbf}}{\text{in}^2}\right)\left(12\ \frac{\text{in}}{\text{ft}}\right)^2(80\text{ ft})\left(32.2\ \frac{\text{ft}}{\text{sec}^2}\right)}{\left(53.3\ \frac{\text{ft-lbf}}{\text{lbm-°R}}\right)\left(32.2\ \frac{\text{lbm-ft}}{\text{lbf-sec}^2}\right)}\right) \times \left(\frac{1}{530°\text{R}} - \frac{1}{860°\text{R}}\right)$$
$$= 2.3\text{ lbf/ft}^2$$

Convert this to inches of water.

$$D_{theoretical} = \frac{p}{\gamma} = \frac{\left(2.3\ \frac{\text{lbf}}{\text{ft}^2}\right)\left(12\ \frac{\text{in}}{\text{ft}}\right)}{62.4\ \frac{\text{lbf}}{\text{ft}^3}}$$
$$= 0.44\text{ in wg}$$

41. STACK FRICTION

Flue gas velocities of 15 ft/sec to 40 ft/sec (4.5 m/s to 12 m/s) are typical. As with most turbulent fluids, the friction pressure drop, $D_{friction}$, through stack "piping" is proportional to the velocity head.[35]

$$D_{friction} = \frac{Kv^2}{2g} \quad 21.41$$

Values of the friction coefficient, K, for chimney fittings are similar to those used for ventilating ductwork. Values for specific pieces of equipment (burners, pressure regulators, vents, etc.) must be provided by the manufacturers.

The approximate friction loss coefficient for a circular section of ductwork or chimney of length L and diameter d is given by Eq. 21.42. Equation 21.42 assumes a Darcy friction factor of 0.0233. For other friction factors, the constant terms can be scaled up or down

[35]The term *piping* is used to mean the stack passages.

proportionately. For noncircular ducts, the diameter, d, can be replaced by four times the hydraulic radius (i.e., four times the duct area divided by the duct perimeter).

$$D_{\text{friction,kPa}} = \frac{0.119 L_{\text{m}} \rho_{\text{kg/m}^3} Q^2_{\text{L/s}}}{d^5_{\text{cm}}} \quad \text{[SI]} \quad 21.42(a)$$

$$D_{\text{friction,in wg}} = \frac{28 L_{\text{ft}} \gamma_{\text{lbf/ft}^3} Q^2_{\text{cfs}}}{d^5_{\text{in}}} \quad \text{[U.S.]} \quad 21.42(b)$$

42. NOMENCLATURE

A	area	ft^2	m^2
A/F	air-fuel ratio	lbm/lbm	kg/kg
B	volumetric fraction	–	–
°Be	Baumé specific gravity	degree	degree
c_p	specific heat capacity at constant pressure	Btu/lbm-°R	kJ/kg·K
C_p	molar heat capacity	Btu/lbmol-°R	kJ/kmol·K
C_{v}	coefficient of velocity	–	–
d	diameter	ft	m
D	draft	in wg	kPa
g	acceleration of gravity, 32.17 (9.807)	ft/sec^2	m/s^2
g_c	gravitational constant, 32.17	lbm-ft/lbf-sec^2	n.a.
G	gravimetric fraction	–	–
h	head	ft	m
h	specific enthalpy	Btu/lbm	kJ/kg
H	enthalpy	Btu	kJ
H	stack height	ft	m
HHV	higher heating value	Btu/lbm	kJ/kg
HV	heating value	Btu/lbm	kJ/kg
J	gravimetric air-fuel ratio	lbm/lbm	kg/kg
K	friction coefficient	–	–
K	volumetric air-fuel ratio	ft^3/ft^3	m^3/m^3
L	length	ft	m
LHV	lower heating value	Btu/lbm	kJ/kg
m	mass	lbm	kg
M	moisture fraction	–	–
n	number of moles	–	–
ON	octane number	–	–
p	pressure	lbf/ft^2	kPa
PN	performance number	–	–
q	heat loss	Btu/lbm	kJ/kg
Q	volumetric flow rate	ft^3/sec	m^3/s
SG	specific gravity	–	–
T	temperature	°R	K
v	velocity	ft/sec	m/s
V	volume	ft^3	m^3
W	humidity ratio	lbm/lbm	kg/kg

Symbols

γ	specific weight	lbf/ft^3	N/m^3
η	efficiency	–	–
ρ	density	lbm/ft^3	kg/m^3

Subscripts

A/F	air-fuel
C	carbon
f	liquid (fluid)
fg	vaporization (fluid to gas)
g	vapor (gas)
H	hydrogen
i	initial
max	maximum
O	oxygen
p	constant pressure
SE	stack effect
v	velocity
w	water

Thermodynamics

22 Energy, Work, and Power

Content in blue refers to the *NCEES Handbook*.

NCEES EXAM SPECIFICATIONS AND RELATED CONTENT

HVAC AND REFRIGERATION EXAM

I.F. Energy/Mass Balances
- 2. Law of Conservation of Energy
- 3. Work
- 4. Potential Energy of a Mass
- 5. Kinetic Energy of a Mass
- 6. Spring Energy

MACHINE DESIGN AND MATERIALS EXAM

I.B.3. Engineering Science and Mechanics: Dynamics
- 2. Law of Conservation of Energy
- 4. Potential Energy of a Mass
- 5. Kinetic Energy of a Mass
- 6. Spring Energy
- 10. Power
- 11. Efficiency

1. ENERGY OF A MASS

The *energy* of a mass represents the capacity of the mass to do work. Such energy can be stored and released. There are many forms that it can take, including mechanical, thermal, electrical, and magnetic energies. Energy is a positive, scalar quantity (although the change in energy can be either positive or negative).

The total energy of a body can be calculated from its mass, m, and its *specific energy*, u (i.e., the energy per unit mass).[1]

$$E = mu \qquad 22.1$$

Typical units of mechanical energy are foot-pounds and joules. (A joule is equivalent to the units of N·m and kg·m^2/s^2.) In traditional English-unit countries, the *British thermal unit* (Btu) is used for thermal energy, whereas the kilocalorie (kcal) is still used in some applications in SI countries. *Joule's constant*, or the *Joule equivalent* (778.17 ft-lbf/Btu, usually shortened to 778, three significant digits), is used to convert between English mechanical and thermal energy units.

$$\text{energy in Btu} = \frac{\text{energy in ft-lbf}}{J} \qquad 22.2$$

Two other units of large amounts of energy are the therm and the quad. A *therm* is 10^5 Btu (1.055×10^8 J). A *quad* is equal to a quadrillion (10^{15}) Btu. This is 1.055×10^{18} J, or roughly the energy contained in 200 million barrels of oil.

2. LAW OF CONSERVATION OF ENERGY

The *law of conservation of energy* says that energy cannot be created or destroyed. However, energy can be converted into different forms. Therefore, the sum of all energy forms is constant.

$$\sum E = \text{constant} \qquad 22.3$$

Expressed differently,

Conservation of Energy Law

$$\text{KE}_1 + \text{PE}_{1-2} = \text{KE}_2 \qquad 22.4$$

$$\text{KE}_1 + \text{PE}_1 = \text{KE}_2 + \text{PE}_2 + W \qquad 22.5$$

3. WORK

Work, W, is the act of changing the energy of a particle, body, or system. For a mechanical system, *external work* is work done by an external force, whereas *internal work* is

[1]The use of symbols E and u for energy is not consistent in the engineering field.

done by an internal force. Work is a signed, scalar quantity. Typical units are inch-pounds, foot-pounds, and joules. Mechanical work is seldom expressed in British thermal units or kilocalories.

For a mechanical system, work is positive when a force acts in the direction of motion and helps a body move from one location to another. Work is negative when a force acts to oppose motion. (Friction, for example, always opposes the direction of motion and can do only negative work.) The work done on a body by more than one force can be found by superposition.

From a thermodynamic standpoint, work is positive if a particle or body does work on its surroundings. Work is negative if the surroundings do work on the object. (For example, blowing up a balloon represents negative work to the balloon.) This is consistent with the conservation of energy, since the sum of negative work and the positive energy increase is zero (i.e., no net energy change in the system).[2]

The work and torque performed by a variable force can be found from Eq. 22.6 and Eq. 22.7.

Work

$$W_F = \int F \cos\theta \, ds \qquad 22.6$$

$$W_{\mathbf{T}} = \int \mathbf{T} \cdot d\theta \qquad 22.7$$

The work done by a force or torque of constant magnitude can be found from Eq. 22.8 and Eq. 22.9.

Work

$$W_F = (F_c \cos\theta)\,\Delta s \qquad 22.8$$

$$\begin{aligned} W_{\mathbf{T}} &= T\cdot\theta \\ &= Fr\theta\cos\phi \quad \text{[rotational systems]} \end{aligned} \qquad 22.9$$

The nonvector forms, Eq. 22.8 and Eq. 22.9, illustrate that only the component of force or torque in the direction of motion contributes to work. (See Fig. 22.1.)

Figure 22.1 Work of a Constant Force

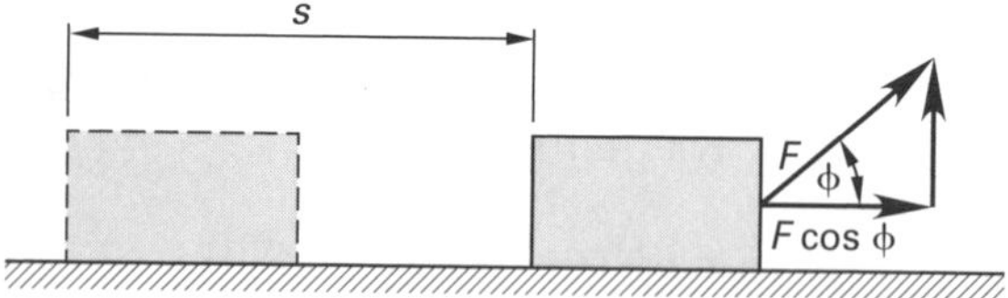

Common applications of the work done by a constant force are frictional work and gravitational work. The work to move an object a distance s against a frictional force of F_f is

$$W_{\text{friction}} = F_f s \qquad 22.10$$

The work done against gravity when a mass m changes in elevation from h_1 to h_2 is

$$W_{\text{gravity}} = mg(h_2 - h_1) \quad \text{[SI]} \qquad 22.11(a)$$

$$W_{\text{gravity}} = \frac{mg(h_2 - h_1)}{g_c} \quad \text{[U.S.]} \qquad 22.11(b)$$

The work done by or on a *linear spring* whose length or deflection changes from δ_1 to δ_2 is given by Eq. 22.12.[3] It does not make any difference whether the spring is a compression spring or an extension spring.

Work

$$W_{\text{spring}} = \tfrac{1}{2}k(s_2^2 - s_1^2) \qquad 22.12$$

Example 22.1

A lawnmower engine is started by pulling a cord wrapped around a sheave. The sheave radius is 8.0 cm. The cord is wrapped around the sheave two times. If a constant tension of 90 N is maintained in the cord during starting, what work is done?

Solution

The starting torque on the engine is

$$T = Fr = (90\ \text{N})\left(\frac{8\ \text{cm}}{100\ \frac{\text{cm}}{\text{m}}}\right) = 7.2\ \text{N·m}$$

The cord wraps around the sheave $(2)(2\pi\text{rad}) = 12.6$ rad. From Eq. 22.9, the work done by a constant torque is

$$\begin{aligned} W &= T\theta = (7.2\ \text{N·m})(12.6\ \text{rad}) \\ &= 90.7\ \text{J} \end{aligned}$$

Example 22.2

A 200 lbm crate is pushed 25 ft at constant velocity across a warehouse floor. There is a frictional force of 60 lbf between the crate and floor. What work is done by the frictional force on the crate?

Solution

From Eq. 22.10,

$$\begin{aligned} W_{\text{friction}} &= F_f s = (60\ \text{lbf})(25\ \text{ft}) \\ &= 1500\ \text{ft-lbf} \end{aligned}$$

4. POTENTIAL ENERGY OF A MASS

Potential energy (*gravitational energy*) is a form of mechanical energy possessed by a body due to its relative position in a gravitational field. Potential energy is

[2]This is just a partial statement of the *first law of thermodynamics*.
[3]A *linear spring* is one for which the linear relationship $F = kx$ is valid.

lost when the elevation of a body decreases. The lost potential energy usually is converted to kinetic energy or heat.

Potential Energy

$$\text{PE} = mgh \quad \text{[SI]} \qquad 22.13(a)$$

$$\text{PE} = \frac{mgh}{g_c} \quad \text{[U.S.]} \qquad 22.13(b)$$

In the absence of friction and other nonconservative forces, the change in potential energy of a body is equal to the work required to change the elevation of the body.

$$W = \Delta\text{PE} \qquad 22.14$$

5. KINETIC ENERGY OF A MASS

Kinetic energy is a form of mechanical energy associated with a moving or rotating body. The kinetic energy of a body moving with instantaneous linear velocity, v, is

Elements of Kinetic Energy

$$\text{KE} = \tfrac{1}{2}m\text{v}^2 \quad \text{[SI]} \qquad 22.15(a)$$

Units

$$\text{KE} = \frac{m\text{v}^2}{2g_c} \quad \text{[U.S.]} \qquad 22.15(b)$$

A body can also have rotational kinetic energy.

$$\text{KE} = \tfrac{1}{2}I\omega^2 \quad \text{[SI]} \qquad 22.16(a)$$

$$\text{KE} = \frac{I\omega^2}{2g_c} \quad \text{[U.S.]} \qquad 22.16(b)$$

Equation 22.15 and Eq. 22.16 can be combined to find the kinetic energy of a rotating body in motion.

Elements of Kinetic Energy

$$\text{KE} = \tfrac{1}{2}m\text{v}_c^2 + \tfrac{1}{2}I_c\omega^2 \qquad 22.17$$

According to the *work-energy principle* (see Sec. 22.9), the kinetic energy is equal to the work necessary to initially accelerate a stationary body or to bring a moving body to rest.

$$W = \Delta\text{KE} \qquad 22.18$$

Example 22.3

A solid disk flywheel ($I = 200$ kg·m^2) is rotating with a speed of 900 rpm. What is its rotational kinetic energy?

Solution

The angular rotational velocity is

$$\omega = \frac{\left(900\ \frac{\text{rev}}{\text{min}}\right)\left(2\pi\ \frac{\text{rad}}{\text{rev}}\right)}{60\ \frac{\text{s}}{\text{min}}} = 94.25\ \text{rad/s}$$

From Eq. 22.16, the rotational kinetic energy is

$$\begin{aligned} E &= \tfrac{1}{2}I\omega^2 = \left(\frac{1}{2}\right)(200\ \text{kg·m}^2)\left(94.25\ \frac{\text{rad}}{\text{s}}\right)^2 \\ &= 888 \times 10^3\ \text{J} \quad (888\ \text{kJ}) \end{aligned}$$

6. SPRING ENERGY

A spring is an energy storage device because the spring has the ability to perform work. In a perfect spring, the amount of energy stored is equal to the work required to compress the spring initially. The potential energy stored in the spring (*spring energy*) does not depend on the mass of the spring. Given a spring with spring constant (stiffness) k, the spring energy is

Spring Energy

$$U = k\left(\frac{x^2}{2}\right) \qquad 22.19$$

Example 22.4

A body of mass m falls from height h onto a massless, simply supported beam. The mass adheres to the beam. If the beam has a lateral stiffness k, what will be the deflection, δ, of the beam?

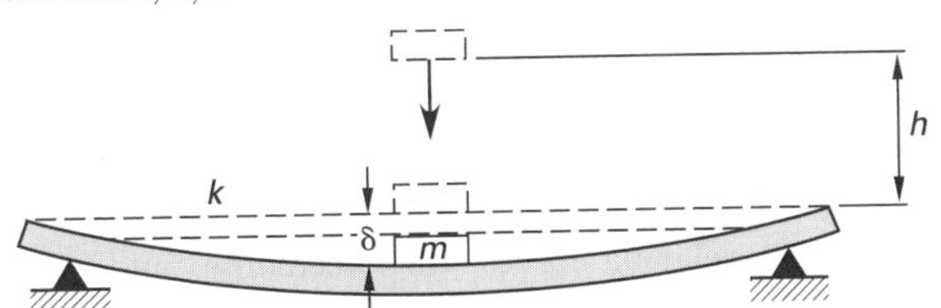

Solution

The initial energy of the system consists of only the potential energy of the body. Using consistent units, the change in potential energy is

$$E = mg(h + \delta) \quad \text{[consistent units]}$$

Thermodynamics

All of this energy is stored in the spring. Therefore, from Eq. 22.19,

Spring Energy

$$k\left(\frac{x^2}{2}\right) = mg(h+\delta)$$

Solving for the deflection,

$$\delta = \frac{mg \pm \sqrt{mg(2hk+mg)}}{k}$$

7. PRESSURE ENERGY OF A MASS

Since work is done in increasing the pressure of a system (e.g., work is done in blowing up a balloon), mechanical energy can be stored in pressure form. This is known as *pressure energy, static energy, flow energy, flow work,* and *p-V work (energy)*. For a system of pressurized mass m, the pressure energy is

$$E_{\text{flow}} = \frac{mp}{\rho} = mpv \quad [v = \text{specific volume}] \qquad 22.20$$

8. WORK-ENERGY PRINCIPLE

Since energy can neither be created nor destroyed, external work performed on a conservative system goes into changing the system's total energy. This is known as the *work-energy principle* (or *principle of work and energy*).

$$W = \Delta E = E_2 - E_1 \qquad 22.21$$

Generally, the term work-energy principle is limited to use with mechanical energy problems (i.e., conversion of work into kinetic or potential energies). When energy is limited to kinetic energy, the work-energy principle is a direct consequence of Newton's second law but is valid for only inertial reference systems.

By directly relating forces, displacements, and velocities, the work-energy principle introduces some simplifications into many mechanical problems.

- It is not necessary to calculate or know the acceleration of a body to calculate the work performed on it.
- Forces that do not contribute to work (e.g., are normal to the direction of motion) are irrelevant.
- Only scalar quantities are involved.
- It is not necessary to individually analyze the particles or component parts in a complex system.

Example 22.5

A 4000 kg elevator starts from rest, accelerates uniformly to a constant speed of 2.0 m/s, and then decelerates uniformly to a stop 20 m above its initial position. Neglecting friction and other losses, what work was done on the elevator?

Solution

By the work-energy principle, the work done on the elevator is equal to the change in the elevator's energy. Since the initial and final kinetic energies are zero, the only mechanical energy change is the potential energy change.

Taking the initial elevation of the elevator as the reference (i.e., $h_1 = 0$), and combining Eq. 22.11(a) and Eq. 22.21,

$$\begin{aligned} W &= E_{2,\text{potential}} - E_{1,\text{potential}} = mg(h_2 - h_1) \\ &= (4000 \text{ kg})\left(9.81 \ \frac{\text{m}}{\text{s}^2}\right)(20 \text{ m}) \\ &= 785 \times 10^3 \text{ J} \quad (785 \text{ kJ}) \end{aligned}$$

9. CONVERSION BETWEEN ENERGY FORMS

Conversion of one form of energy into another does not violate the conservation of energy law. However, most problems involving conversion of energy are really just special cases of the work-energy principle. For example, consider a falling body that is acted upon by a gravitational force. The conversion of potential energy into kinetic energy can be interpreted as equating the work done by the constant gravitational force to the change in kinetic energy.

In general terms, *Joule's law* states that one energy form can be converted without loss into another. There are two specific formulations of Joule's law. As related to electricity, $P = I^2R = V^2/R$ is the common formulation of Joule's law. As related to thermodynamics and ideal gases, Joule's law states that "the change in internal energy of an ideal gas is a function of the temperature change, not of the volume." This latter form can also be stated more formally as "at constant temperature, the internal energy of a gas approaches a finite value that is independent of the volume as the pressure goes to zero."

Example 22.6

A 2 lbm projectile is launched straight up with an initial velocity of 700 ft/sec. Neglecting air friction, calculate the (a) kinetic energy immediately after launch, (b) kinetic energy at maximum height, (c) potential energy at maximum height, (d) total energy at an elevation where the velocity has dropped to 300 ft/sec, and (e) maximum height attained.

Thermodynamics

Solution

(a) From Eq. 22.15, the kinetic energy is

Kinetic Energy

$$\text{KE} = \tfrac{1}{2}m\text{v}^2 = \left(\frac{1}{2}\right)\left(\frac{(2\text{ lbm})\left(700\ \frac{\text{ft}}{\text{sec}}\right)^2}{32.2\ \frac{\text{lbm-ft}}{\text{lbf-sec}^2}}\right)$$
$$= 15{,}217\text{ ft-lbf}$$

(b) The velocity is zero at the maximum height. Therefore, the kinetic energy is zero.

(c) At the maximum height, all of the kinetic energy has been converted into potential energy. Therefore, the potential energy is 15,217 ft-lbf.

(d) Although some of the kinetic energy has been transformed into potential energy, the total energy is still 15,217 ft-lbf.

(e) Since all of the kinetic energy has been converted into potential energy, the maximum height can be found from Eq. 22.13.

Potential Energy

$$\text{PE} = \frac{mgh}{g_c}$$
$$15{,}217\text{ ft-lbf} = \frac{(2\text{ lbm})\left(32.2\ \frac{\text{ft}}{\text{sec}^2}\right)h}{32.2\ \frac{\text{lbm-ft}}{\text{lbf-sec}^2}}$$
$$h = 7609\text{ ft}$$

Example 22.7

A 4500 kg ore car rolls down an incline and passes point A traveling at 1.2 m/s. The ore car is stopped by a spring bumper that compresses 0.6 m. A constant friction force of 220 N acts on the ore car at all times. What is the spring constant?

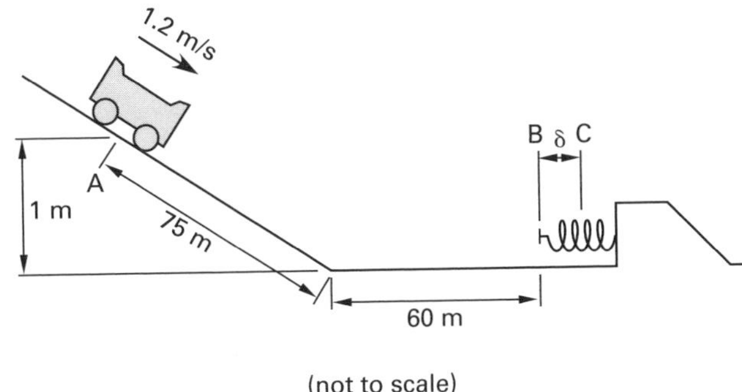

(not to scale)

Solution

The car's total energy at point A is the sum of the kinetic and potential energies.

Potential Energy

$$\text{PE} = mgh$$

Kinetic Energy

$$\text{KE} = \tfrac{1}{2}m\text{v}^2$$

$$\begin{aligned}E_{\text{total,A}} &= \text{KE} + \text{PE} = \tfrac{1}{2}m\text{v}^2 + mgh\\ &= \left(\frac{1}{2}\right)(4500\text{ kg})\left(1.2\ \frac{\text{m}}{\text{s}}\right)^2\\ &\quad + (4500\text{ kg})\left(9.81\ \frac{\text{m}}{\text{s}^2}\right)(1\text{ m})\\ &= 47\,385\text{ J}\end{aligned}$$

At point B, the potential energy has been converted into additional kinetic energy. However, except for friction, the total energy is the same as at point A. Since the frictional force does negative work, the total energy remaining at point B is

$$\begin{aligned}E_{\text{total,B}} &= E_{\text{total,A}} - W_{\text{friction}}\\ &= 47\,385\text{ J} - (220\text{ N})(75\text{ m} + 60\text{ m})\\ &= 17\,685\text{ J}\end{aligned}$$

At point C, the maximum compression point, the remaining energy has gone into compressing the spring a distance $x = 0.6$ m and performing a small amount of frictional work.

Spring Energy

$$\begin{aligned}E_{\text{total,B}} &= E_{\text{total,C}} = W_{\text{spring}} + W_{\text{friction}}\\ &= k\left(\frac{x^2}{2}\right) + F_f\delta\end{aligned}$$
$$17\,685 = \tfrac{1}{2}k(0.6\text{ m})^2 + (220\text{ N})(0.6\text{ m})$$

The spring constant can be determined directly.

$$k = 97\,520\text{ N/m}\quad (97.5\text{ kN/m})$$

10. POWER

Power is the amount of work done per unit time. It is a scalar quantity. (Although power is calculated from two vectors, the vector dot-product operation is seldom needed.) Typical basic units of power are ft-lbf/sec and watts (J/s), although *horsepower* is widely used. Table 22.1 shows factors used to convert units of power.

Power and Efficiency

$$P = \frac{dW}{dt} \qquad 22.22$$

Thermodynamics

For a body acted upon by a force or torque, the instantaneous power can be calculated from the velocity.

Power and Efficiency

$$P = \mathbf{F} \cdot \mathbf{v} \quad \text{[linear systems]} \qquad 22.23$$

$$P = T\omega \quad \text{[rotational systems]} \qquad 22.24$$

For a fluid flowing at a rate of $\dot{m}$, the unit of time is already incorporated into the flow rate (e.g., lbm/sec). If the fluid experiences a specific energy change of Δu, the power generated or dissipated will be

$$P = \dot{m}\Delta u \qquad 22.25$$

***Table 22.1** Useful Power Conversion Formulas*

1 hp	= 550 ft-lbf/sec
	= 33,000 ft-lbf/min
	= 0.7457 kW
	= 0.7068 Btu/sec
1 kW	= 737.6 ft-lbf/sec
	= 44,250 ft-lbf/min
	= 1.341 hp
	= 0.9483 Btu/sec
1 Btu/sec	= 778.17 ft-lbf/sec
	= 46,680 ft-lbf/min
	= 1.415 hp

Example 22.8

When traveling at 100 km/h, a car supplies a constant horizontal force of 50 N to the hitch of a trailer. What tractive power (in horsepower) is required for the trailer alone?

Solution

From Eq. 22.23, the power being generated is

Power and Efficiency

$$P = \mathbf{F} \cdot \mathbf{v} = \frac{(50\ \text{N})\left(100\ \frac{\text{km}}{\text{h}}\right)\left(1000\ \frac{\text{m}}{\text{km}}\right)}{\left(60\ \frac{\text{s}}{\text{min}}\right)\left(60\ \frac{\text{min}}{\text{h}}\right)\left(1000\ \frac{\text{W}}{\text{kW}}\right)}$$
$$= 1.389\ \text{kW}$$

Using a conversion from Table 22.1, the horsepower is

$$P = \left(1.341\ \frac{\text{hp}}{\text{kW}}\right)(1.389\ \text{kW}) = 1.86\ \text{hp}$$

11. EFFICIENCY

Energy-use efficiency, ϵ, of a system is the ratio of an output property to an input property for that system, as shown in Eq. 22.26. The property used is commonly work, power or, for thermodynamics problems, heat. When the rate of work is constant, either work or power can be used to calculate the efficiency. Otherwise, power should be used. Except in rare instances, the numerator and denominator of the ratio must have the same units.[4]

Power and Efficiency

$$\epsilon = \frac{P_{\text{out}}}{P_{\text{in}}} = \frac{W_{\text{out}}}{W_{\text{in}}} \qquad 22.26$$

Efficiency can be further discussed in terms of energy-using and energy-producing systems as a function of ideal and actual states. For energy-using systems (such as cars, electrical motors, elevators, etc.), the energy-use efficiency is the ratio of an ideal property to an actual property.

$$\epsilon = \frac{P_{\text{ideal}}}{P_{\text{actual}}} \quad [P_{\text{actual}} \geq P_{\text{ideal}}] \qquad 22.27$$

For energy-producing systems (such as electrical generators, prime movers, and hydroelectric plants), the energy-production efficiency is

$$\epsilon = \frac{P_{\text{actual}}}{P_{\text{ideal}}} \quad [P_{\text{ideal}} \geq P_{\text{actual}}] \qquad 22.28$$

The efficiency of an *ideal machine* is 1.0 (100%). However, all *real machines* have efficiencies of less than 1.0.

[4]The *energy-efficiency ratio* used to evaluate refrigerators, air conditioners, and heat pumps, for example, has units of Btu per watt-hour (Btu/W-hr).

12. NOMENCLATURE

E	energy	ft-lbf	J
F	force	lbf	N
g	gravitational acceleration, 32.2 (9.81)	ft/sec^2	m/s^2
g_c	gravitational constant, 32.2	lbm-ft/lbf-sec^2	n.a.
h	height	ft	m
I	mass moment of inertia	lbm-ft^2	kg·m^2
J	Joule's constant, 778.17	ft-lbf/Btu	n.a.
k	spring constant	lbf/ft	N/m
KE	kinetic energy	ft-lbf	J
m	mass	lbm	kg
p	pressure	lbf/ft^2	Pa
P	power	ft-lbf/sec	W
PE	potential energy	ft-lbf	J
r	radius	ft	m
s	deflection	Ft	m
s	distance	ft	m
t	time	sec	s
T	torque	ft-lbf	N·m
u	specific energy	ft-lbf/lbm	J/kg
U	energy	ft-lbf	J
v	velocity	ft/sec	m/s
W	work	ft-lbf	J
x	deflection	ft	m

Symbols

η	efficiency	–	–
θ	angular position	rad	rad
ρ	mass density	lbm/ft^3	kg/m^3
v	specific volume	ft^3/lbm	m^3/kg
ϕ	angle	deg	deg
ω	angular velocity	rad/sec	rad/s

Subscripts

f	frictional
p	constant pressure
T	torque
v	constant volume

23 Thermodynamic Properties of Substances

Content in blue refers to the *NCEES Handbook*.

NCEES EXAM SPECIFICATIONS AND RELATED CONTENT

HVAC AND REFRIGERATION EXAM

I.B.2. Thermodynamics: Properties

5. Temperature
7. Density
8. Specific Volume
9. Internal Energy
10. Enthalpy
11. Entropy
12. Heat Capacity
17. Quality
18. Gibbs Function
19. Helmholtz Function

THERMAL AND FLUID SYSTEMS EXAM

I.E.1 Thermodynamics: Thermodynamic properties

5. Temperature
7. Density
8. Specific Volume
9. Internal Energy
10. Enthalpy
11. Entropy
12. Heat Capacity
17. Quality
23. Using Saturation Tables
24. Using Superheat Tables
25. Using Gas Tables
29. Properties of Saturated Liquids
30. Properties of Liquid-Vapor Mixtures
31. Properties of Saturated Vapors
33. Equation of State for Ideal Gases
34. Properties of Ideal Gases
35. Specific Heats of Ideal Gases
36. Mass, Mole, and Volume Fractions
37. Partial Pressure and Partial Volume of Gas Mixtures

1. PHASES OF A PURE SUBSTANCE

Thermodynamics is the study of a substance's energy-related properties. The properties of a substance and the procedures used to determine those properties depend on the state and the phase of the substance.

The *thermodynamic state* of a substance is defined by two or more independent thermodynamic properties. For example, the temperature and pressure of a substance are two properties commonly used to define the state of a superheated vapor.

The common *phases* of a substance are solid, liquid, and gas. However, because substances behave according to different rules, it is convenient to categorize them into more than these three phases.[1]

solid: A solid does not take on the shape or volume of its container.

subcooled liquid: If a liquid is not saturated (i.e., the liquid is not at its boiling point), it is said to be subcooled. Water at 1 atm and room temperature is subcooled, as the addition of a small amount of heat will not cause vaporization.

saturated liquid: A saturated liquid has absorbed as much heat energy as it can without vaporizing. Liquid water at standard atmospheric pressure and 212°F (100°C) is an example of a saturated liquid.

liquid-vapor mixture: A liquid and vapor of the same substance can coexist at the same temperature and pressure. This is called a two-phase, liquid-vapor mixture.

saturated vapor: A vapor (e.g., steam at standard atmospheric pressure and 212°F (100°C)) that is on the verge of condensing is said to be saturated.

superheated vapor: A superheated vapor is one that has absorbed more heat than is needed merely to vaporize it. A superheated vapor will not condense when small amounts of heat are removed.

ideal gas: A gas that behaves according to the ideal gas law, $pV = \bar{R}T$, is an ideal gas.

real gas: A real gas does not behave according to the ideal gas law.

gas mixtures: Most gases mix together freely. Two or more pure gases together constitute a gas mixture.

vapor/gas mixtures: Atmospheric air is an example of a mixture of several gases and water vapor.

These phases and subphases can be illustrated with a pure substance in the piston/cylinder arrangement shown in Fig. 23.1. The pressure in this system is constant and is determined by the weight of the piston, which moves freely to permit volume changes.

In Fig. 23.1(a) the volume is minimum. This is usually the solid phase. (Water is an exception. Solid ice has a lower density than liquid water.) The temperature will rise as heat, Q, is added to the solid. This increase in temperature is accompanied by a small increase in volume. The temperature increases until the melting point is reached.

Figure 23.1 *Phase Changes at Constant Pressure*

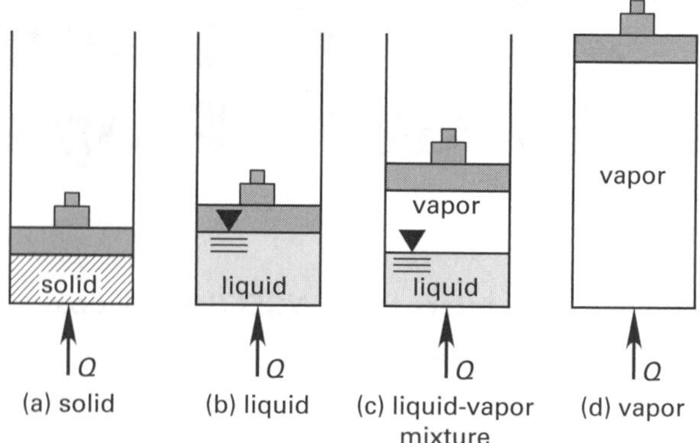

The solid will begin to melt as heat is added to it at the melting point. The temperature will not increase until all of the solid has been turned into liquid. Until the ice is completely melted, solid ice and liquid water will coexist. (The term *slush ice* describes a mixture of water and small chunks of solid ice at the freezing point.) The liquid phase, with its small increase in volume, is illustrated by Fig. 23.1(b).

If the subcooled liquid continues to receive heat, its temperature will rise. This temperature increase continues until evaporation is imminent. The liquid at this point is said to be saturated. Any increase in heat energy will cause a portion of the liquid to vaporize. This is shown in Fig. 23.1(c), in which a liquid-vapor mixture exists.

As with melting, evaporation occurs at constant temperature and pressure but with a very large increase in volume. The temperature cannot increase until the last drop of liquid has been evaporated, at which point the vapor is said to be saturated. This is shown in Fig. 23.1(d).

Additional heat will result in high-temperature superheated vapor. This vapor may or may not behave according to the ideal gas laws.

2. DETERMINING PHASE

It is possible to develop a three-dimensional surface that predicts the substance's phase based on the properties of pressure, temperature, and specific volume. Such an *equilibrium solid* is illustrated in Fig. 23.2.[2]

If one property is held constant during a process, a two-dimensional projection of the equilibrium solid can be used. This projection is known as an *equilibrium diagram* or a *phase diagram*, of which Fig. 23.3 is an example.

[1]Plasma, *cryogenic fluids* (*cryogens*) that boil at temperatures less than approximately 200°R (110K), and solids near absolute zero are not discussed in this chapter.

[2]Figure 23.2 is applicable for most substances, excluding water. In Fig. 23.2, the specific volume "steps in" (i.e., decreases) when the liquid turns to a solid as the temperature drops below freezing. However, water is unique in that it expands upon freezing. Thus, the pVT diagram for water shows a "step out" instead of a "step in" upon freezing. The remainder of the pVT diagram for water is the same as Fig. 23.2.

Figure 23.2 *Equilibrium Solid*

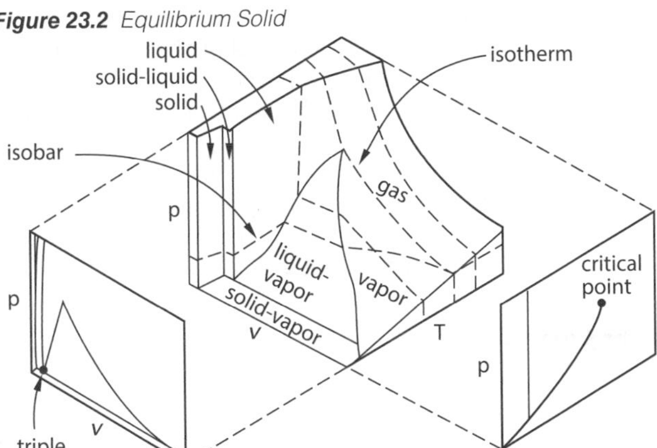

Figure 23.3 *Constant Temperature Phase Diagram*

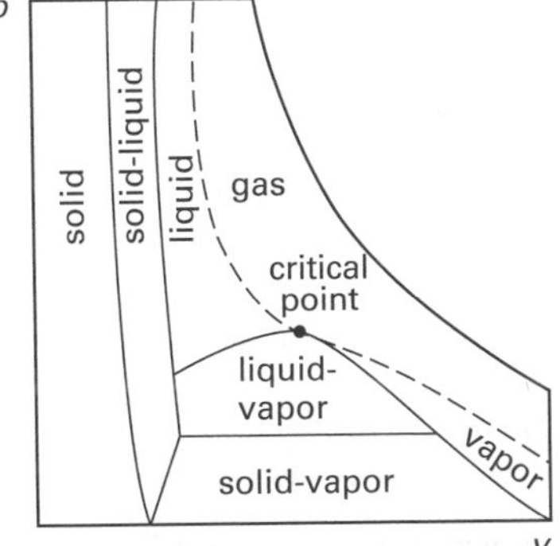

The most important part of a phase diagram is limited to the liquid-vapor region. A general phase diagram showing this region and the bell-shaped dividing line (known as the *vapor dome*) is shown in Fig. 23.4.

Figure 23.4 *Vapor Dome with Isobars*

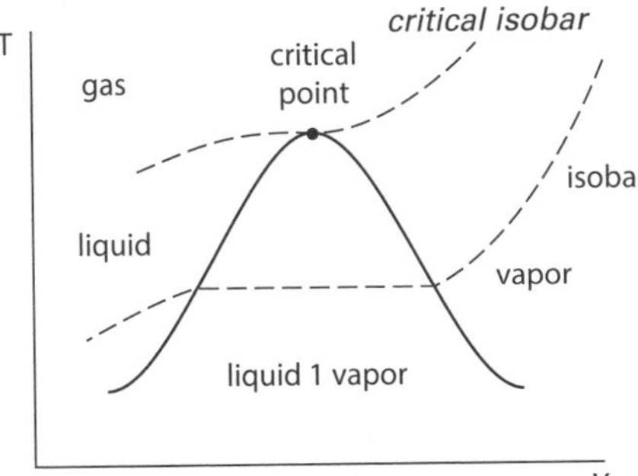

The vapor dome region can be drawn with many variables for the axes. For example, either temperature or pressure can be used for the vertical axis. Energy, specific volume, or entropy can be chosen for the horizontal axis. The principles presented here apply to all combinations.

The left-hand part of the vapor dome separates the liquid phase from the liquid-vapor phase. This part of the line is known as the *saturated liquid line*. Similarly, the right-hand part of the line separates the liquid-vapor phase from the vapor phase. This line is called the *saturated vapor line*.

Lines of constant pressure (*isobars*) can be superimposed on the vapor dome. Each isobar is horizontal as it passes through the two-phase region, indicating that both temperature and pressure remain unchanged as a liquid vaporizes.

Notice that there is no real dividing line between liquid and vapor at the top of the vapor dome. Far above the vapor dome, there is no distinction between liquids and gases, as their properties are identical. The phase is assumed to be a gas.

The implied dividing line between liquid and gas is the isobar that passes through the topmost part of the vapor dome. This is known as the *critical isobar*. The highest point of the vapor dome is known as the *critical point*. This critical isobar also provides a way to distinguish between a vapor and a gas. A substance below the critical isobar (but to the right of the vapor dome) is a vapor. Above the critical isobar, it is a gas.

Figure 23.5 illustrates a vapor dome for which pressure, p, has been chosen as the vertical axis, and enthalpy, h, has been chosen as the horizontal axis. The shape of the dome is essentially the same, but the lines of constant temperature (*isotherms*) have a different direction than isobars.

Figure 23.5 also illustrates the subscripting convention used to identify points on the saturation line. The subscript f (fluid) is used to indicate a saturated liquid. The subscript g (gas) is used to indicate a saturated vapor.[3] The subscript fg is used to indicate the difference in saturation properties and is used with vaporization properties.

Figure 23.5 *Vapor Dome with Isotherms*

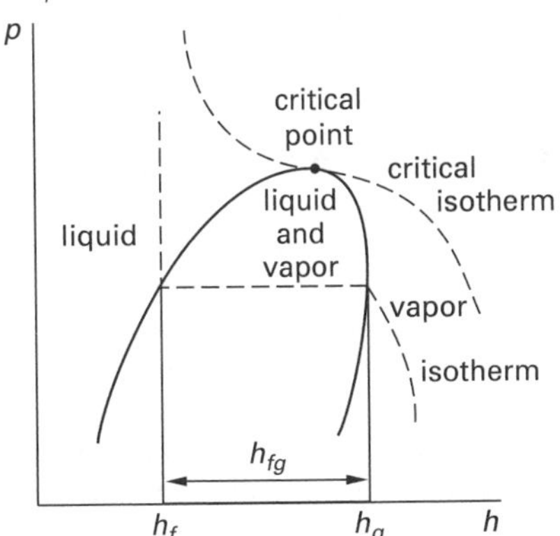

[3]Although this book makes it a rule never to call a vapor a gas, this convention is not adhered to in the field of thermodynamics. The subscript g is standard for a saturated vapor.

Thermodynamics

The vapor dome is a good tool for illustration, but it cannot be used to determine a substance's phase. Such a determination must be made based on the substance's pressure and temperature.

For example, consider water at a pressure of 1 atm. Its boiling temperature (the *saturation temperature*) at this pressure is 212°F (100°C). If the water has a lower temperature, for example, 85°F (29°C), the water will be liquid (i.e., will be subcooled). On the other hand, if the water's temperature (at 1 atm) is 270°F (132°C), the water must be in vapor form (i.e., must be superheated).

This example is valid only for water at atmospheric pressure. Water at other pressures will have other boiling temperatures. (The lower the pressure, the lower the boiling temperature.) However, the rules given here follow directly from the example.

Rule 1: A substance is a subcooled liquid if its temperature is less than the saturation temperature corresponding to its pressure.

Rule 2: A substance is in the liquid-vapor region if its temperature is equal to the saturation temperature corresponding to its pressure.

Rule 3: A substance is a superheated vapor if its temperature is greater than the saturation temperature corresponding to its pressure.

The rules that follow can be stated using pressure as the determining variable.

Rule 4: A substance is a subcooled liquid if its pressure is greater than the saturation pressure corresponding to its temperature.

Rule 5: A substance is in the liquid-vapor region if its pressure is equal to the saturation pressure corresponding to its temperature.

Rule 6: A substance is a superheated vapor if its pressure is less than the saturation pressure corresponding to its temperature.

3. PROPERTIES OF A SUBSTANCE

The thermodynamic *state* or condition of a substance is determined by its properties. *Intensive properties* are independent of the amount of substance present. Temperature, pressure, and stress are examples of intensive properties. *Extensive properties* are dependent on the amount of substance present. Examples of extensive properties are volume, strain, charge, and mass.

In this chapter, and in most books on thermodynamics, both lowercase and uppercase forms of the same characters are used to represent property variables. The two forms are used to distinguish between the units of mass. For example, lowercase h represents *specific enthalpy* (usually just called "enthalpy") in units of Btu/lbm or kJ/kg. Uppercase H is used to represent *total enthalpy* in units of Btu or kJ. H_m is used to represent the *molar enthalpy* in units of Btu/lbmol or kJ/kmol.

4. MASS: *m*

The mass of a substance is a measure of its quantity. Mass is independent of location and gravitational field strength. In thermodynamics, the customary U.S. and SI units of mass, m, are pounds-mass (lbm) and kilograms (kg), respectively.

5. TEMPERATURE: *T*

Temperature is a thermodynamic property of a substance that depends on internal energy content. Heat energy entering a substance will increase the temperature of that substance. Normally, heat energy will flow only from a hot object to a cold object. If two objects are in *thermal equilibrium* (i.e., are at the same temperature), no heat will flow between them.

If two systems are in thermal equilibrium, they must be at the same temperature. If both systems are in equilibrium with a third, then all three are at the same temperature. This concept is known as the *zeroth law of thermodynamics*.

The scales most commonly used for measuring temperature are the Fahrenheit and Celsius scales.[4] The relationship between these two scales is

Temperature Conversions

$$T_{°F} = 1.8T_{°C} + 32° \qquad 23.1$$

$$T_{°C} = \frac{T_{°F} - 32°}{1.8} \qquad 23.2$$

The *absolute temperature scale* defines temperature independently of the properties of any particular substance. This is unlike the Celsius and Fahrenheit scales, which are based on the freezing point of water. The absolute temperature scale should be used for all calculations.

In the customary U.S. system, the absolute scale is the *Rankine scale*.[5]

Temperature Conversions

$$T_{°R} = T_{°F} + 459.69° \qquad 23.3$$

[4]The term *centigrade* was replaced by the term *Celsius* in 1948.

[5]Normally, three significant temperature digits (i.e., 460°) are sufficient.

$$\Delta T_{°R} = \Delta T_{°F} \quad 23.4$$

The absolute temperature scale in the SI system is the *Kelvin scale*.[6]

Temperature Conversions

$$T_K = T_{°C} + 273.15° \quad 23.5$$

$$\Delta T_K = \Delta T_{°C} \quad 23.6$$

The relationships between temperature differences in the customary U.S. and SI systems are independent of the freezing point of water. (See Table 23.1.)

$$\Delta T_{°C} = \frac{\Delta T_{°F}}{1.8} \quad 23.7$$

$$\Delta T_K = \frac{\Delta T_{°R}}{1.8} \quad 23.8$$

Table 23.1 *Temperature Scales*

	Kelvin	Celsius	Rankine	Fahrenheit
normal boiling point of water	373.15K	100.00°C	671.67°R	212.00°F
triple point of water (see Sec. 14)	273.16K	0.01°C	491.69°R	32.02°F
ice point	273.15K	0.00°C	491.67°R	32.00°F
absolute zero	0K	−273.15°C	0°R	−459.67°F

6. PRESSURE: *p*

Pressure is the force exerted on one unit area of a surface. It is found by dividing the total force exerted on a surface by the surface's area.

$$p = \frac{F}{A} \quad 23.9$$

Customary U.S. pressure units are pounds per square inch (psi). Standard SI pressure units are kilopascals (kPa) or megapascals (MPa), although bars are also used in tabulations of thermodynamic data.

7. DENSITY: ρ

Density is the amount of mass contained in a unit volume of a substance. It is found by dividing the total mass of a substance by its total volume.

Density, Specific Weight, and Specific Gravity

$$\rho = \frac{m}{V} \quad 23.10$$

Customary U.S. density units in tabulations of thermodynamic data are pounds per cubic foot (lbm/ft^3). Standard SI density units are kilograms per cubic meter (kg/m^3). Density is the reciprocal of specific volume.

$$\rho = \frac{1}{v} \quad 23.11$$

8. SPECIFIC VOLUME: *v* AND *V*

Specific volume, v, is the volume occupied by one unit mass of a substance. It is found by dividing the total volume of a substance by its total mass.

Properties of Single-Component Systems: Definitions

$$v = \frac{V}{m} \quad 23.12$$

Customary U.S. units in tabulations of thermodynamic data are cubic feet per pound (ft^3/lbm). Standard SI specific volume units are cubic meters per kilogram (m^3/kg). Specific volume is the reciprocal of density.

$$v = \frac{1}{\rho} \quad 23.13$$

A related concept occasionally encountered is *molar volume*, V_m. Molar volume is the volume occupied by one mole of a substance, and it is found by dividing the substance's total volume by the number of moles of the substance. Units of molar volume are $ft^3/lbmol$ ($m^3/kmol$).

$$V_m = \frac{V}{N} \quad 23.14$$

9. INTERNAL ENERGY: *u* AND *U*

Internal energy, U, includes all of the potential and kinetic energies of the atoms or molecules in a substance. Energies in the translational, rotational, and vibrational modes are included. Since this movement increases as the temperature increases, internal energy is a function of temperature. It does not depend on the process or path taken to reach a particular temperature. In the United States, the *British thermal unit*, Btu, is used to measure all forms of thermodynamic energy. (One Btu is approximately the energy given off by burning one wooden match.)

[6]Normally, three significant temperature digits (i.e., 273°) are sufficient.

Specific internal energy, u, is the internal energy contained in one unit mass of a substance. Standard units of specific internal energy are Btu/lbm and kJ/kg.

Functions and Their Symbols and Units

$$u = \frac{U}{m} \tag{23.15}$$

10. ENTHALPY: *h* AND *H*

Enthalpy (also known at various times in history as *total heat* and *heat content*), H, represents the total useful energy of a substance. Useful energy consists of two parts—the internal energy, U, and the *flow energy* (also known as *flow work* and *p-V work*), pV. Therefore, enthalpy has the same units as internal energy.

$$H = U + pV \tag{23.16}$$

Specific enthalpy is enthalpy per unit mass.

Functions and Their Symbols and Units

$$h = u + pv = \frac{H}{m} \tag{23.17}$$

Enthalpy is defined as useful energy because, ideally, all of it can be used to perform useful tasks. It takes energy to increase the temperature of a substance. If that internal energy is recovered, it can be used to heat something else (e.g., to vaporize water in a boiler). Also, it takes energy to increase pressure and volume (as in blowing up a balloon). If pressure and volume are decreased, useful energy is given up.

The customary U.S. units of Eq. 23.18 and Eq. 23.19 are not consistent, since flow work, written as pV, has units of ft-lbf, not Btu. (There is also a consistency problem if pressure is defined in lbf/ft^2 and given in lbf/in^2.) Strictly, Eq. 23.16 should be written as

$$h = u + \frac{pv}{J} \quad \text{[U.S.]} \tag{23.18}$$

$$H = U + \frac{pV}{J} \quad \text{[U.S.]} \tag{23.19}$$

The conversion factor, J, in Eq. 23.18 is known as *Joule's constant* or the *mechanical equivalent of heat*. It has a value of 778.2 ft-lbf/Btu. (In SI units, Joule's constant has a value of 1.0 N·m/J and is unnecessary.) As in Eq. 23.16 and Eq. 23.17, Joule's constant is often omitted from the statement of generic thermodynamic equations, but it is always needed with customary U.S. units for dimensional consistency.

11. ENTROPY: *s* AND *S*

Absolute entropy, S, is a measure of the energy that is no longer available to perform useful work within the current environment. Other definitions used (the "disorder of the system," the "randomness of the system," etc.) are frequently quoted. Although these alternate definitions cannot be used in calculations, they are consistent with the *third law of thermodynamics* (also known as the *Nernst theorem*). This law states that the absolute entropy of a perfect crystalline solid in thermodynamic equilibrium is (approaches) zero when the temperature is (approaches) absolute zero.[7] Equation 23.20 expresses the third law mathematically.

$$\lim_{T \to 0\text{K}} S = 0 \tag{23.20}$$

The units of entropy are Btu/°R and kJ/K.

An increase in entropy is known as *entropy production*. The total absolute entropy in a system is equal to the summation of all absolute entropy productions that have occurred over the life of the system.

$$S = \sum \Delta S \tag{23.21}$$

For an isothermal heat transfer process taking place at a constant temperature T_o, the entropy production depends on the amount of energy transfer.

$$\Delta S = \frac{Q}{T_o} \tag{23.22}$$

For heat transfer processes that occur over a varying temperature, the entropy production must be found by integration.

$$\Delta S = \int dS = \int \frac{dQ}{T} \tag{23.23}$$

Specific absolute entropy, s, is absolute entropy per unit mass. Units of specific absolute entropy are Btu/lbm-°R and kJ/kg·K.

Functions and Their Symbols and Units

$$s = \frac{S}{m} \tag{23.24}$$

$$\Delta s = \frac{q}{T_o} \tag{23.25}$$

$$\Delta s = \int ds = \int \frac{dq}{T} \tag{23.26}$$

Unlike absolute entropy, *standardized entropy*, usually just called "entropy," is not referenced to absolute zero conditions but is measured with respect to some other

[7]A molecule with zero entropy exists in only one quantum state. The energy state is known precisely, without *uncertainty*.

convenient thermodynamic state. For water, the reference condition is the liquid phase at the triple point. (See Sec. 23.14.)

Example 23.1

Three planets of identical size and mass are oriented in space such that radiant energy transfers can occur. The average temperatures of planets A, B, and C are 530°R, 520°R, and 510°R (294K, 289K, and 283K), respectively. All three planets are massive enough that small energy losses or gains can be considered to be isothermal processes (i.e., they will not change the average temperature).

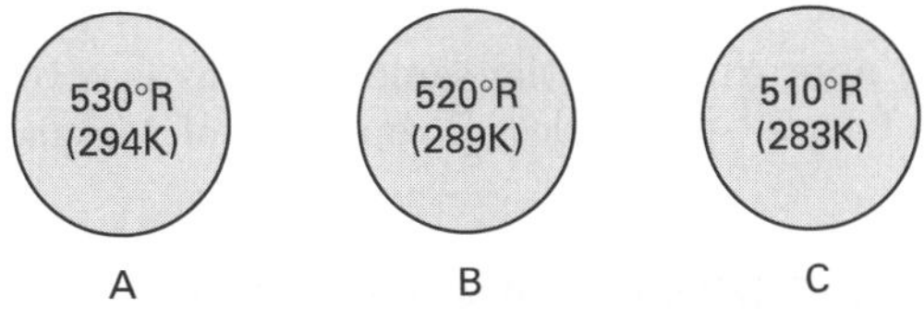

(a) Can a radiation energy transfer occur spontaneously from planet B to planet C? (b) What are the entropy productions for planets B and C if an energy transfer of 1000 Btu/lbm (2330 kJ/kg) occurs by radiation? (c) What is the overall entropy change as a result of the energy transfer in (b)? (d) Does entropy always increase? (e) Can planet B ever be returned to its original condition?

Solution

(a) A radiation transfer can occur spontaneously because planet B is hotter than planet C. Energy will flow spontaneously from a hot object to a cold object.

(b) In SI units, the entropy productions are

$$\Delta s_{\mathrm{B}} = \frac{q}{T_o} = \frac{-2330\ \frac{\mathrm{kJ}}{\mathrm{kg}}}{289\mathrm{K}} = -8.062\ \mathrm{kJ/kg{\cdot}K}$$

$$\Delta s_{\mathrm{C}} = \frac{2330\ \frac{\mathrm{kJ}}{\mathrm{kg}}}{283\mathrm{K}} = 8.233\ \mathrm{kJ/kg{\cdot}K}$$

In customary U.S. units, the entropy productions are

$$\Delta s_{\mathrm{B}} = \frac{q}{T_o} = \frac{-1000\ \frac{\mathrm{Btu}}{\mathrm{lbm}}}{520^\circ\mathrm{R}} = -1.923\ \mathrm{Btu/lbm\text{-}^\circ R}$$

$$\Delta s_{\mathrm{C}} = \frac{1000\ \frac{\mathrm{Btu}}{\mathrm{lbm}}}{510^\circ\mathrm{R}} = 1.961\ \mathrm{Btu/lbm\text{-}^\circ R}$$

(c) The entropy change is not the same for the two planets. Entropy is not conserved in an energy transfer process. The overall entropy production is

$$\begin{aligned}\Delta s &= \Delta s_{\mathrm{B}} + \Delta s_{\mathrm{C}} = -8.062\ \frac{\mathrm{kJ}}{\mathrm{kg{\cdot}K}} + 8.233\ \frac{\mathrm{kJ}}{\mathrm{kg{\cdot}K}} \\ &= 0.171\ \mathrm{kJ/kg{\cdot}K}\end{aligned}$$

In customary U.S. units,

$$\begin{aligned}\Delta s &= \Delta s_{\mathrm{B}} + \Delta s_{\mathrm{C}} = -1.923\ \frac{\mathrm{Btu}}{\mathrm{lbm\text{-}^\circ R}} + 1.961\ \frac{\mathrm{Btu}}{\mathrm{lbm\text{-}^\circ R}} \\ &= 0.038\ \mathrm{Btu/lbm\text{-}^\circ R}\end{aligned}$$

(d) Local entropy can decrease, as shown by planet B's negative entropy production. However, overall entropy always increases when the total universe is considered, as shown in part (c).

(e) Planet B can be brought back to its original condition if 1000 Btu/lbm (2330 kJ/kg) of energy is transferred from planet A to planet B. (Heat will not flow spontaneously from planet C to planet B, as heat will not flow spontaneously from a cold object to a hot object.[8])

12. HEAT CAPACITY: *c* AND *C*

To raise the temperature of a substance, the substance's internal energy must be increased. Different substances differ in how much heat must be added to achieve a given rise in temperature.

The *heat capacity*, C, of a given mass of material is defined as the amount of heat that must be added to raise its temperature by one unit. Common units of heat capacity are Btu/°F and kJ/K.

The *specific heat capacity* (more commonly called *specific heat*) of a substance is its heat capacity per unit mass. Common units of specific heat are Btu/lbm-°F and kJ/kg·K. For example, the specific heat of water at 68°F is 1 Btu/lbm-°F, so 1 Btu must be added to 1 lbm of 68°F water to raise it by 1°F. Because specific heats of solids and liquids are slightly temperature dependent, the mean specific heats are used when evaluating processes covering a large temperature range.

$$c = \frac{C}{m} \qquad 23.27$$

$$c = \frac{Q}{m\Delta T} \qquad 23.28$$

$$Q = mc\Delta T \qquad 23.29$$

The specific heat of a gas depends on the type of process during which the heat exchange occurs. In a constant-volume (or *isochoric*) process, the volume of gas is held

[8]This is one way of stating the *Second Law of Thermodynamics.*

Thermodynamics

constant while heat is added, so that all the added heat contributes to the rise in internal energy. The specific heat under these conditions is designated by c_v.

$$Q = mc_v \Delta T \quad \left[\begin{array}{c} \text{perfect gas} \\ \text{constant-volume process} \end{array} \right] \qquad 23.30$$

Properties of Single-Component Systems: Definitions

$$c_v = \left(\frac{\partial u}{\partial T} \right)_V \qquad 23.31$$

In a constant-pressure (or *isobaric*) process, the pressure of the gas is held constant while heat is added, so that the volume of the gas expands; some portion of the added heat contributes to the work done by the expansion rather than to the temperature rise. The specific heat under these conditions is designated by c_p. More heat must be added in a constant-pressure process than in a constant-volume process to effect the same rise in temperature, so for any gas, the value of c_p is always greater than the value of c_v. [**Thermal and Physical Properties of Ideal Gases (at Room Temperature)**]

$$Q = m\,c_p \Delta T \quad \left[\begin{array}{c} \text{perfect gas} \\ \text{constant-pressure process} \end{array} \right] \qquad 23.32$$

Properties of Single-Component Systems: Definitions

$$c_p = \left(\frac{\partial h}{\partial T} \right)_p \qquad 23.33$$

For a solid or liquid, there is essentially no difference between the values of c_p and c_v. However, the designation c_p rather than c is often encountered.

The *Dulong-Petit law* gives a way of predicting the approximate molar specific heat (in cal/mol·°C) at high temperatures as a function of the atomic weight.[9] This law is valid for solid elements having atomic weights greater than 40 and for most metallic elements. It is not valid at room temperature for carbon, silicon, phosphorus, and sulfur. 6.3 cal/mol·°C is known as the *Dulong-Petit value.*

$$c \approx \frac{6.3\ \frac{\text{cal}}{\text{mol}\cdot{}^\circ\text{C}}}{\text{AW}} \qquad 23.34$$

Example 23.2

Compare the value for the specific heat of pure iron calculated from Dulong and Petit's law with the value from a table of metal properties.

Solution

From a table of metal properties, the atomic weight of iron is 55.85. From Eq. 23.34,

$$\begin{aligned} c &\approx \frac{6.3\ \frac{\text{cal}}{\text{mol}\cdot{}^\circ\text{C}}}{\text{AW}} \\ &= \frac{6.3\ \frac{\text{cal}}{\text{mol}\cdot{}^\circ\text{C}}}{55.85} \\ &= 0.1128\ \text{cal/mol}\cdot{}^\circ\text{C} \end{aligned}$$

This is approximately the same value as is given in a table of metal properties. [**Properties of Metals - I-P Units**]

13. RATIO OF SPECIFIC HEATS: *k*

For gases, the *ratio of specific heats*, k, is defined by Eq. 23.35. For common gases other than hydrocarbons, the value of k ranges from about 1.29 to 1.67; for common hydrocarbons, k ranges from 1.09 to 1.30. For air, k = 1.4. [**Thermal and Physical Properties of Ideal Gases (at Room Temperature)**]

$$k = \frac{c_p}{c_v} \qquad 23.35$$

14. TRIPLE POINT PROPERTIES

The *triple point* of a substance is a unique state at which solid, liquid, and gaseous phases can coexist. Table 23.2 lists values of the triple point for several common substances.

Table 23.2 *Approximate Triple Points*

	pressure	temperature	
substance	atm	°R	K
ammonia	0.060	352	196
argon	0.676	151	84
carbon dioxide	5.10	390	217
helium	0.0508	4	2
hydrogen	0.0676	26	14
nitrogen	0.127	14	7.8
oxygen	0.00265	99	55
water	0.00592	492.02	273.34

[9]Dulong and Petit's law becomes valid at different temperatures for different substances, and a more specific definition of "high temperature" is impossible. For lead, the law is valid at 200K. For copper, it is not valid until above 400K.

15. CRITICAL PROPERTIES

If the temperature and pressure of a liquid are increased, a state will eventually be reached at which the liquid and gas phases are indistinguishable. This state is known as the *critical point*, and the properties (generally temperature, pressure, and specific volume) at that point are known as the *critical properties*. At the critical point, the heat of vaporization, h_{fg}, becomes zero. Above the critical temperature, the substance will be a gas no matter how high the pressure. [**Critical Properties**]

16. LATENT HEATS

The total energy (*total heat*, Q_t) entering a substance is the sum of the energy that changes the phase of the substance (*latent heat*, Q_l) and energy that changes the temperature of the substance (*sensible heat*, Q_s). During a phase change (solid to liquid, liquid to vapor, etc.), energy will be transferred to or from the substance without a change in temperature.[10]

$$Q_t = Q_s + Q_l \qquad 23.36$$

Examples of latent energies are the *latent heat of fusion* (i.e., change from solid to liquid), h_{sl}, *latent heat of vaporization*, h_{fg}, and *latent heat of sublimation* (i.e., direct change from solid to vapor without becoming liquid), h_{ig}.[11,12] The energy required for these latent changes to occur in water is given in Table 23.3.

Table 23.3 *Latent Heats for Water at One Atmosphere*

effect	Btu/lbm	kJ/kg	cal/g
fusion	143.4	333.5	79.7
vaporization	970.1	2256.5	539.0
sublimation	1220	2838	677.8

(Multiply Btu/lbm by 2.326 to obtain kJ/kg.)
(Multiply Btu/lbm by 5/9 to obtain cal/g.)

Example 23.3

How much energy is required to convert 1.0 lbm (0.45 kg) of water that is originally at 75°F (24°C) and 1 atm to vapor at 212°F (100°C) and 1 atm?

SI Solution

The sensible heat required to raise the temperature of the water from 24°C to 100°C is given by Eq. 23.32.

$$\begin{aligned} Q_s &= mc_p(T_2 - T_1) \\ &= (0.45\ \text{kg})\left(4.180\ \frac{\text{kJ}}{\text{kg}\cdot{}^\circ\text{C}}\right)(100^\circ\text{C} - 24^\circ\text{C}) \\ &= 143.0\ \text{kJ} \end{aligned}$$

From Table 23.3, the latent heat required to vaporize the water is

$$Q_l = mh_{fg} = (0.45\ \text{kg})\left(2256.5\ \frac{\text{kJ}}{\text{kg}}\right) = 1015.4\ \text{kJ}$$

The total heat required is

$$\begin{aligned} Q_t &= Q_s + Q_l = 143.0\ \text{kJ} + 1015.4\ \text{kJ} \\ &= 1158.4\ \text{kJ} \end{aligned}$$

Customary U.S. Solution

The sensible heat required to raise the temperature of the water from 75°F to 212°F is given by Eq. 23.32.

$$\begin{aligned} Q_s &= mc_p(T_2 - T_1) \\ &= (1\ \text{lbm})\left(1.0\ \frac{\text{Btu}}{\text{lbm-}^\circ\text{F}}\right)(212^\circ\text{F} - 75^\circ\text{F}) \\ &= 137.0\ \text{Btu} \end{aligned}$$

From Table 23.3, the latent heat required to vaporize the water is

$$Q_l = mh_{fg} = (1\ \text{lbm})\left(970.1\ \frac{\text{Btu}}{\text{lbm}}\right) = 970.1\ \text{Btu}$$

The total heat required is

$$\begin{aligned} Q_t &= Q_s + Q_l = 137.0\ \text{Btu} + 970.1\ \text{Btu} \\ &= 1107.1\ \text{Btu} \end{aligned}$$

17. QUALITY: *x*

Within the vapor dome, water is at its saturation pressure and temperature. When saturated, water can simultaneously exist in liquid and vapor phases in any proportion between 0 and 1. The *quality* is the fraction by weight of the total mass that is vapor.

[10]Changes in crystalline form are also latent changes.
[11]The subscript *s* (for "solid") is sometimes used in place of *i* (for "ice").
[12]*Sublimation* can only occur below the triple point, where it is too cold for the liquid phase to exist at all.

Properties for Two-Phase (Vapor-Liquid) Systems

$$x = \frac{m_g}{m_g + m_f} \quad 23.37$$

18. GIBBS FUNCTION: g AND G

The *Gibbs function* is defined for a pure substance by Eq. 23.38 through Eq. 23.40.

Functions and Their Symbols and Units

$$\begin{aligned} g &= h - Ts \\ &= u + pv - Ts \end{aligned} \quad 23.38$$

$$\begin{aligned} G &= H - TS \\ &= U + pV - TS \end{aligned} \quad 23.39$$

$$G = mg \quad 23.40$$

The Gibbs function is used in investigating latent changes and chemical reactions. For a constant-temperature, constant-pressure nonflow process approaching equilibrium, the Gibbs function approaches its minimum value.

$$(dG)_{T,p} < 0 \quad 23.41$$

Once the minimum value is obtained, the process will stop, and the Gibbs function will be constant.

$$(dG)_{T,p} = 0 \,|_{\text{equilibrium}} \quad 23.42$$

Like enthalpy of formation, the Gibbs function, $G°$, has been tabulated at the standard reference conditions of 25°C (77°F) and one atmosphere. A chemical reaction can occur spontaneously only if the change in Gibbs function is negative (i.e., the Gibbs function for the products is less than the Gibbs function for the reactants).

$$\sum_{\text{products}} nG° < \sum_{\text{reactants}} nG° \quad 23.43$$

19. HELMHOLTZ FUNCTION: a AND A

The *Helmholtz function* is defined for a pure substance by Eq. 23.44 through Eq. 23.46.[13]

Functions and Their Symbols and Units

$$\begin{aligned} a &= u - Ts \\ &= h - pv - Ts \end{aligned} \quad 23.44$$

$$\begin{aligned} A &= U - TS \\ &= H - pV - TS \end{aligned} \quad 23.45$$

$$A = ma \quad 23.46$$

Like the Gibbs function, the Helmholtz function is used in investigating equilibrium conditions. For a constant-temperature, constant-volume nonflow process approaching equilibrium, the Helmholtz function approaches its minimum value.

$$(dA)_{T,V} < 0 \quad 23.47$$

Once the minimum value is obtained, the process will stop, and the Helmholtz function will be constant.

$$(dA)_{T,V} = 0 \,|_{\text{equilibrium}} \quad 23.48$$

20. FREE ENERGY

The Helmholtz function has, in the past, also been known as the *free energy* of the system because its change in a reversible isothermal process equals the energy that can be "freed" and converted to mechanical work. Unfortunately, the same term has also been used for the Gibbs function under analogous conditions. For example, the difference in standard Gibbs functions of reactants and products has often been called the "free energy difference."

Since there is a great possibility for confusion, it is better to refer to the Gibbs and Helmholtz functions by their actual names.

21. COMPRESSIBILITY: β, κ, AND α

Three different compressibilities are distinguished. The *isobaric compressibility*, β, is defined as

$$\beta = \left(\frac{1}{v}\right)\left(\frac{\partial v}{\partial T}\right)_p \quad 23.49$$

The *isothermal compressibility*, κ, is

$$\kappa = \left(-\frac{1}{v}\right)\left(\frac{\partial v}{\partial p}\right)_T \quad 23.50$$

The *isentropic compressibility*, α, is

$$\alpha = \left(-\frac{1}{v}\right)\left(\frac{\partial v}{\partial p}\right)_s \quad 23.51$$

22. USING THE MOLLIER DIAGRAM

The *Mollier diagram* (*enthalpy-entropy diagram*) is a graph of enthalpy versus entropy for steam. (See App. 23.B.) It is particularly suitable for determining

[13]In older references, the symbol F was commonly used for the Helmholtz function.

property changes between the superheated vapor and the liquid-vapor regions. For this reason, the Mollier diagram covers only a limited region.

The Mollier diagram plots the enthalpy for a unit mass of steam as the ordinate and plots the entropy as the abscissa. Lines of constant pressure (isobars) slope upward from left to right. Below the saturation line, curves of *constant moisture content* (the complement of quality) slope down from left to right. Above the saturation line are lines of constant temperature and lines of constant superheat.

Example 23.4

Find the following properties using the Mollier diagram: (a) enthalpy and entropy of steam at 700 psia and 1000°F, (b) enthalpy of steam at 1 psia and 80% quality, and (c) final temperature of steam throttled from 700 psia and 1000°F to 450 psia.

Solution

(a) Reading directly from the Mollier diagram (see App. 23.B), $h = 1510$ Btu/lbm, and $s = 1.7$ Btu/lbm-°R.

(b) 80% quality is the same as 20% moisture. Reading at the intersection of 20% moisture and 1 psia, $h =$ 900 Btu/lbm.

(c) By definition, a throttling process does not change the enthalpy. This process is represented by a horizontal line to the right on the Mollier diagram. Starting at the intersection of 700 psia and 1000°F and moving horizontally to the right until 450 psia is reached defines the endpoint of the process. The final temperature is interpolated as approximately 990°F.

23. USING SATURATION TABLES

The information presented graphically on an enthalpy-entropy diagram can be obtained with greater accuracy from *saturation tables*, also known as *property tables* and (in the case of water) *steam tables*. These tables represent extensive tabulations of data for the liquid and vapor phases of a substance. **[Steam Tables]**

Saturation tables contain values of specific enthalpy, h, specific entropy, s, and specific volume, v. Within the vapor dome, these properties are functions of temperature.

However, as shown in Fig. 23.4, there is a unique pressure associated with each temperature (i.e., within the vapor dome there is only one horizontal isobar for each temperature). Since the pressure does not vary in even increments when temperature is changed, two sets of steam tables are generally provided, one organized by saturation temperature in even increments, and one organized by saturation pressure in even increments.

For each saturation temperature or saturation pressure, the table gives the specific volume of saturated liquid, v_f, and the specific volume of saturated vapor, v_g, at that temperature or pressure. For convenience, the table also gives the difference in specific volume between saturated liquid and saturated vapor, which is designated by the variable v_{fg}. The relationship between v_f, v_{fg}, and v_g is given by Eq. 23.52.

$$v_g = v_f + v_{fg} \qquad 23.52$$

The table lists the same data for enthalpy and entropy.

$$h_g = h_f + h_{fg} \qquad 23.53$$

$$s_g = s_f + s_{fg} \qquad 23.54$$

24. USING SUPERHEAT TABLES

In the superheated region, pressure and temperature are independent properties. Therefore, for each pressure, a large number of temperatures is possible. Superheated steam tables give the properties of specific volume, enthalpy, and entropy for various combinations of temperature and pressure. The *degrees of superheat* (e.g., "100°F of superheat") represent the difference between actual and saturation temperatures. **[Properties of Superheated Steam - I-P Units] [Properties of Superheated Steam - SI Units]**

25. USING GAS TABLES

Gas tables are essentially superheat tables for gases at "low pressure," which in this case means that pressure is less than several hundred psi. However, reasonably good results can be expected even if pressures are higher. **[Properties of Air at Low Pressure, per Pound]**

Gas tables are indexed by temperature. That is, implicit in their use is the assumption that properties are functions of temperature only. Gas tables are not arranged in the same way as other property tables.

The *relative pressure*, p_r, and the *relative volume*, v_r (also called the *pressure ratio* and *volume ratio*) are ratios that can make analysis of isentropic processes easier. They should not be confused with pressure and specific volume. Their use is illustrated in Ex. 23.6 and is based on Eq. 23.55 and Eq. 23.56.

$$\frac{v_{r,1}}{v_{r,2}} = \frac{V_1}{V_2} \quad [\Delta s = 0] \qquad 23.55$$

$$\frac{p_{r,1}}{p_{r,2}} = \frac{p_1}{p_2} \quad [\Delta s = 0] \qquad 23.56$$

The *entropy function*, ϕ, is not the same as specific entropy, although it has the same units. The entropy function can be used to calculate the change in entropy as the gas goes through a process. This entropy change is calculated with Eq. 23.57. Units of the gas constant must match those of s and ϕ.

$$s_2 - s_1 = \phi_2 - \phi_1 - R \ln \frac{p_2}{p_1} \tag{23.57}$$

Example 23.5

What is the specific enthalpy of air at 100°F and 50 psia?

SI Solution

Because the pressure is low (less than 300 psia), gas tables can be used. For a temperature of 100°F, the specific enthalpy is $h = 133.86$ Btu/lbm. [**Air at Low Pressure, per Pound**]

Example 23.6

Air is originally at 60°F and 14.7 psia. It is compressed isentropically to 86.5 psia. After compression, what are the temperature and enthalpy of the air?

Solution

For a table of air properties at low pressure, air at 60°F has a pressure ratio of 1.2147. [**Air at Low Pressure, per Pound**]

Use Eq. 23.56 to find the pressure ratio after compression.

$$\frac{p_{r,1}}{p_{r,2}} = \frac{p_1}{p_2}$$

$$p_{r,2} = \frac{p_{r,1}p_2}{p_1} = \frac{(1.2147)\left(86.5\ \frac{\text{lbf}}{\text{in}^2}\right)}{14.7\ \frac{\text{lbf}}{\text{in}^2}} = 7.148$$

Look in the p_r column of the gas table for a value of 7.148. This value most closely corresponds to a temperature of 400°F and an enthalpy of 206.46 Btu/lbm.

26. USING PRESSURE-ENTHALPY CHARTS

For convenience (and by tradition), properties of refrigerants are typically shown graphically in a *pressure-enthalpy diagram*, often called a *p-h diagram*. Figure 23.6 shows a skeleton *p-h* diagram. The vapor dome and saturation lines from Fig. 23.5 are recognizable. Lines of constant specific volume, constant entropy, and constant temperature are added.

Isotherms are horizontal within the vapor dome, corresponding to the constant saturation pressure. In the subcooled-liquid region to the left of the vapor dome,

Figure 23.6 *Pressure-Enthalpy Diagram*

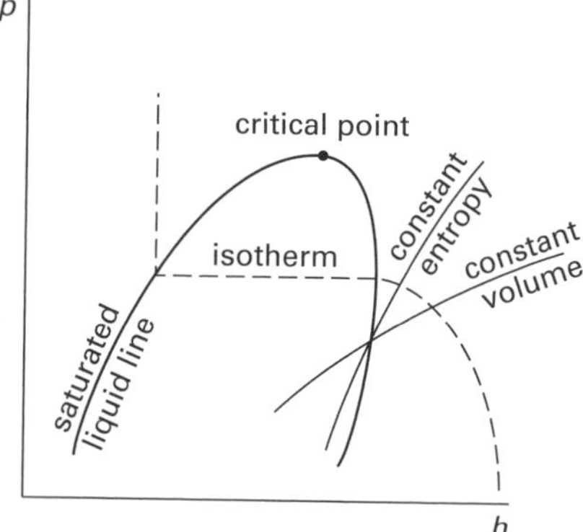

isotherms are essentially vertical, since the temperature (not the pressure) of a subcooled liquid determines enthalpy. In the superheated region, the isotherms gradually approach verticality as the gaseous refrigerant becomes more like an ideal gas.

The enthalpy of superheated refrigerant is easily determined from the pressure and temperature. Isentropic compression follows a line of constant entropy. Throttling follows a line of constant enthalpy (i.e., a vertical line).

Enthalpies (plotted on the horizontal axis) may be scaled to an arbitrary "zero point." Comparing charts in traditional units and SI units can be perplexing. For most calculations, including refrigeration calculations, only the difference between two values is important. The absolute value at any particular point is seldom a useful value.

27. PROPERTIES OF SOLIDS

There are few mathematical relationships that predict the thermodynamic properties of solids. Properties such as temperature, specific heat, and density, usually are known. If the properties are not known, they must be found from tables.

The reference point for properties of solids is usually absolute zero temperature. That is, properties such as enthalpy and entropy are defined to be zero at 0°R (0K). This is an arbitrary convention. The choice of reference point does not affect the *change* in properties between two temperatures.

28. PROPERTIES OF SUBCOOLED LIQUIDS

A liquid whose pressure is greater than the saturation pressure corresponding to its temperature is known as a *compressed liquid* or *subcooled liquid*. To put it another way, a compressed liquid's temperature is less than the saturation temperature corresponding to its pressure.

Liquids are only slightly compressible. For most thermodynamic problems, changes in properties for a liquid are negligible. If a table of property values for a compressed liquid is not available, specific volume, specific entropy, and specific internal energy can be approximated by using the values for a saturated liquid at the same temperature.

$$v_{T,p} \approx v_{f,T} \quad \text{23.58}$$

$$s_{T,p} \approx s_{f,T} \quad \text{23.59}$$

$$u_{T,p} \approx u_{f,T} \quad \text{23.60}$$

This is not true, however, for specific enthalpy. Because $h = u + pv$, the specific enthalpy for a compressed liquid is significantly dependent on its pressure as well as temperature. In this case, the approximation in Eq. 23.61, which adds a correction factor for the difference from saturation pressure, can be used.

$$h_{T,p} \approx h_{f,T} + (p - p_{\text{sat},T})v_{f,T} \quad \text{23.61}$$

29. PROPERTIES OF SATURATED LIQUIDS

Either the temperature or the pressure of a saturated liquid must be known in order to identify its thermodynamic state. One determines the other, because there is a one-to-one relationship between saturation pressure and saturation temperature. Steam tables can be used to determine saturation temperatures and pressures. [Steam Tables]

Steam tables are set up specifically for saturated liquid and vapor, so the properties of a saturated liquid can be read directly from the table. The subscript f refers to a saturated liquid (fluid), so specific volume is v_f, specific enthalpy is h_f, and so on. Density can be calculated as the reciprocal of the specific volume. The liquid's vapor pressure is the same as the saturation pressure listed in the table.

30. PROPERTIES OF LIQUID-VAPOR MIXTURES

When the thermodynamic state of a substance is within the vapor dome, there is a one-to-one correspondence between the saturation temperature and saturation pressure. One determines the other. The thermodynamic state is uniquely defined by any two independent properties (temperature and quality, pressure and enthalpy, entropy and quality, etc.).

If the quality, x, of a liquid-vapor mixture is known, it can be used to calculate all of the primary thermodynamic properties. If a thermodynamic property has a value between the saturated liquid and saturated vapor values (i.e., h is between h_f and h_g), any of Eq. 23.62 through Eq. 23.65 can be solved for the quality.

Properties for Two-Phase (Vapor-Liquid) Systems

$$\begin{aligned} v &= xv_g + (1-x)v_f \\ &= v_f + xv_{fg} \end{aligned} \quad \text{23.62}$$

$$\begin{aligned} u &= xu_g + (1-x)u_f \\ &= u_f + xu_{fg} \end{aligned} \quad \text{23.63}$$

$$\begin{aligned} h &= xh_g + (1-x)h_f \\ &= h_f + xh_{fg} \end{aligned} \quad \text{23.64}$$

$$\begin{aligned} s &= xs_g + (1-x)s_f \\ &= s_f + xs_{fg} \end{aligned} \quad \text{23.65}$$

Example 23.7

Superheated steam at 100 psia and 500°F (700 kPa and 250°C) is expanded isentropically (i.e., with no change in entropy) to 3 psia (20 kPa). What is the steam's specific enthalpy after expansion?

SI Solution

From superheated steam tables, the specific entropy of superheated steam at 700 kPa (0.70 MPa) and 250°C is 7.11 kJ/kg·K. As the expansion is isentropic, this is also the specific entropy after expansion. [Properties of Superheated Steam - SI Units]

From steam tables, at 20 kPa (0.02 MPa), the specific entropy of saturated liquid is $s_f = 0.8320$ kJ/kg·K, and the specific entropy of saturated vapor, s_g, is 7.9072 kJ/kg·K. [Properties of Saturated Water and Steam (Pressure) - SI Units]

Because its specific entropy is between s_f and s_g, the expanded steam is in the liquid-vapor region. Use Eq. 23.65, and solve for the quality after expansion.

Properties for Two-Phase (Vapor-Liquid) Systems

$$s = xs_g + (1-x)s_f$$

$$\begin{aligned} x &= \frac{s - s_f}{s_g - s_f} \\ &= \frac{7.11\ \frac{\text{kJ}}{\text{kg·K}} - 0.8320\ \frac{\text{kJ}}{\text{kg·K}}}{7.9072\ \frac{\text{kJ}}{\text{kg·K}} - 0.8320\ \frac{\text{kJ}}{\text{kg·K}}} \\ &= 0.8873 \end{aligned}$$

From steam tables, saturated liquid at 20 kPa has an enthalpy of $h_f = 251.4$ kJ/kg and a heat of vaporization of $h_{fg} = 2357.5$ kJ/kg. [Properties of Saturated Water and Steam (Pressure) - SI Units]

Use Eq. 23.65 to find the specific enthalpy of the steam after expansion.

Thermodynamics

Properties for Two-Phase (Vapor-Liquid) Systems

$$\begin{aligned} h &= h_f + x h_{fg} \\ &= 251.4\ \frac{\text{kJ}}{\text{kg}} + (0.8873)\left(2357.5\ \frac{\text{kJ}}{\text{kg}}\right) \\ &= 2343\ \text{kJ/kg} \end{aligned}$$

Customary U.S. Solution

From superheated steam tables, the specific entropy of superheated steam at 100 psia and 500°F is 1.712 Btu/lbm-°R. As the expansion is isentropic, this is also the specific entropy after expansion. [**Properties of Superheated Steam - I-P Units**]

From steam tables, at 3 psia the specific entropy of saturated liquid is $s_f = 0.2008$ Btu/lbm-°R, and the specific entropy of saturated vapor, s_g, is 1.8860 Btu/lbm-°R. [**Properties of Saturated Water and Steam (Pressure) — I-P Units**]

Because its specific entropy is between s_f and s_g, the expanded steam is in the liquid-vapor region. Use Eq. 23.65, and solve for the quality after expansion.

Properties for Two-Phase (Vapor-Liquid) Systems

$$s = x s_g + (1 - x) s_f$$

$$\begin{aligned} x &= \frac{s - s_f}{s_g - s_f} \\ &= \frac{1.712\ \frac{\text{Btu}}{\text{lbm-°R}} - 0.2008\ \frac{\text{Btu}}{\text{lbm-°R}}}{1.8860\ \frac{\text{Btu}}{\text{lbm-°R}} - 0.2008\ \frac{\text{Btu}}{\text{lbm-°R}}} \\ &= 0.8967 \end{aligned}$$

From steam tables, saturated liquid at 3 psi has an enthalpy of $h_f = 109.32$ Btu/lbm and a heat of vaporization of $h_{fg} = 1012.82$ Btu/lbm. [**Properties of Saturated Water and Steam (Pressure) — I-P Units**]

Use Eq. 23.64 to find the specific enthalpy of the steam after expansion.

Properties for Two-Phase (Vapor-Liquid) Systems

$$\begin{aligned} h &= h_f + x h_{fg} \\ &= 109.32\ \frac{\text{Btu}}{\text{lbm}} + (0.8967)\left(1012.82\ \frac{\text{Btu}}{\text{lbm}}\right) \\ &= 1018\ \text{Btu/lbm} \end{aligned}$$

Example 23.8

What is the enthalpy of 200°F (90°C, 363K) steam with a quality of 90%?

SI Solution

Use tables and Eq. 23.64. [**Properties of Saturated Water and Steam (Pressure) - SI Units**]

Properties for Two-Phase (Vapor-Liquid) Systems

$$\begin{aligned} h &= h_f + x h_{fg} \\ &= 377.04\ \frac{\text{kJ}}{\text{kg}} + (0.9)\left(2282.5\ \frac{\text{kJ}}{\text{kg}}\right) \\ &= 2431.3\ \text{kJ/kg} \end{aligned}$$

Customary U.S. Solution

Use tables and Eq. 23.64. [**Properties of Saturated Water and Steam (Pressure) — I-P Units**]

Properties for Two-Phase (Vapor-Liquid) Systems

$$\begin{aligned} h &= h_f + x h_{fg} \\ &= 168.13\ \frac{\text{Btu}}{\text{lbm}} + (0.9)\left(977.59\ \frac{\text{Btu}}{\text{lbm}}\right) \\ &= 1048\ \text{Btu/lbm} \end{aligned}$$

31. PROPERTIES OF SATURATED VAPORS

Properties of saturated vapors can be read directly from steam tables. The vapor's pressure or temperature can be used to define its thermodynamic state. The subscript g refers to a saturated vapor (gas), so specific volume is v_g, specific enthalpy is h_g, and so on. [**Steam Tables**]

Saturated vapors are sometimes redundantly described as "dry" saturated vapors (e.g., "dry saturated steam") to emphasize that the quality, x, is 1.0 and that none of the substance is in the liquid state.

32. PROPERTIES OF SUPERHEATED VAPORS

Unless a vapor is highly superheated, its properties should be found from a superheat table. Since the temperature and pressure are independent for a superheated vapor, both must be known in order to define the thermodynamic state.

If the vapor's temperature and pressure do not correspond to the superheat table entries, single or double interpolation will be required. Such interpolation can be avoided by using more complete tables, but where required, double linear interpolation is standard practice.

Example 23.9

What is the enthalpy of water at 300 psia (20 bars, 2.0 MPa) and 900°F (500°C, 773K)?

SI Solution

From steam tables, the saturation (boiling) temperature for water at 2.0 MPa is 212.38°C. The actual water temperature is higher than the saturation temperature, so the water is a superheated vapor. [**Properties of Saturated Water and Steam (Pressure) - SI Units**]

The enthalpy of water at 2.0 MPa and 500°C can be read directly from a superheat table as 3468.2 kJ/kg. [**Properties of Superheated Steam - SI Units**]

Customary U.S. Solution

From steam tables, the saturation (boiling) temperature for water at 300 psia is 417.33°F. The actual water termperature is higher than the saturation temperature, so the water is a superheated vapor. [**Properties of Saturated Water and Steam (Pressure) — I-P Units**]

The enthalpy of water at 300 psia and 900°F can be read directly from a superheat table as 1473.9 Btu/lbm. [**Properties of Superheated Steam - I-P Units**]

33. EQUATION OF STATE FOR IDEAL GASES

An *equation of state* is a relationship that predicts the state (a property such as pressure, temperature, volume, etc.) from a set of two other independent properties.

Avogadro's law states that equal volumes of different gases at the same temperature and pressure contain equal numbers of molecules. For one mole of any gas, Avogadro's law can be stated as the *equation of state for ideal gases*. (Temperature, T, in Eq. 23.66 must be absolute.)

$$\frac{pV}{T} = \overline{R} \qquad 23.66$$

In Eq. 23.66, $\bar{R}$ is known as the *universal gas constant*. It is "universal" (within a consistent system of units) because the same value can be used with any gas. Its value depends on the units used for pressure, temperature, and volume, as well as on the units of mass. (See Table 23.4.)

The ideal gas equation of state can be modified for more than one mole of gas. If there are n moles,

$$pV = n\overline{R}\,T \qquad 23.67$$

The number of moles can be calculated from the substance's mass and molecular weight.

$$n = \frac{m}{M} \qquad 23.68$$

Table 23.4 *Values of the Universal Gas Constant, $\bar{R}$*

units in SI and other metric systems
8.3143 kJ/kmol·K
8314.3 J/kmol·K
0.08206 atm·L/mol·K
1.986 cal/mol·K
8.314 J/mol·K
82.06 atm·cm^3/mol·K
0.08206 atm·m^3/kmol·K
8314.3 kg·m^2/s^2 kmol·K
8314.3 m^3·Pa/kmol·K
8.314 × 10^7 erg/mol·K
units in English systems
1545.33 ft-lbf/lbmol-°R
1.986 Btu/lbmol-°R
0.7302 atm-ft^3/lbmol-°R
10.73 ft^3-lbf/in^2-lbmol-°R

Equation 23.67 and Eq. 23.68 can be combined.

$$pV = \left(\frac{m}{M}\right)\overline{R}\,T = m\left(\frac{\overline{R}}{M}\right)T \qquad 23.69$$

Dividing the universal gas constant by the molecular weight of a particular gas gives R, the *specific gas constant*. It is "specific" because it is valid only for a gas with a molecular weight of M.

$$R = \frac{\overline{R}}{M} \qquad 23.70$$

Substituting Eq. 23.70 into Eq. 23.69 gives

Ideal Gas

$$pV = mRT \qquad 23.71$$

Dividing both sides of Eq. 23.71 by mass and solving for R gives

Ideal Gas

$$pv = RT \qquad 23.72$$

As R is constant for a given gas,

Ideal Gas

$$\frac{p_1 v_1}{T_1} = \frac{p_2 v_2}{T_2} \qquad 23.73$$

Example 23.10

What mass of nitrogen is contained in a 2000 ft^3 (57 m^3) tank at 1 atm and 70°F (21°C)?

SI Solution

First, convert to absolute temperature.

$$T = 21°\text{C} + 273° = 294\text{K}$$

From a table of ideal gas properties, the specific gas constant for nitrogen is 0.2968 kJ/kg·K. [**Thermal and Physical Properties of Ideal Gases (at Room Temperature)**]

Use the ideal gas law (Eq. 23.71), and solve for mass.

Ideal Gas

$$\begin{aligned} pV &= mRT \\ m &= \frac{pV}{RT} \\ &= \frac{(1\ \text{atm})\left(1.013\times 10^5\ \frac{\text{Pa}}{\text{atm}}\right)(57\ \text{m}^3)}{\left(0.2968\ \frac{\text{kJ}}{\text{kg·K}}\right)(294\text{K})} \\ &= 66.1\ \text{kg} \end{aligned}$$

Customary U.S. Solution

First, convert to absolute temperature.

$$T = 70°\text{F} + 460° = 530°\text{R}$$

From a table of ideal gas properties, the specific gas constant for nitrogen is 55.16 ft-lbf/lbm-°R. [**Thermal and Physical Properties of Ideal Gases (at Room Temperature)**]

Use the ideal gas law (Eq. 23.71), and solve for mass.

Ideal Gas

$$\begin{aligned} pV &= mRT \\ m &= \frac{pV}{RT} \\ &= \frac{(1\ \text{atm})\left(14.7\ \frac{\text{lbf}}{\text{in}^2\text{-atm}}\right)\left(12\ \frac{\text{in}}{\text{ft}}\right)^2 \times (2000\ \text{ft}^3)}{\left(55.16\ \frac{\text{ft-lbf}}{\text{lbm-°R}}\right)(530°\text{R})} \\ &= 144.8\ \text{lbm} \end{aligned}$$

Example 23.11

A 25 ft^3 (0.71 m^3) tank contains 10 lbm (4.5 kg) of an ideal gas. The gas has a molecular weight of 44 and is at 70°F (21°C). What is the pressure of the gas?

SI Solution

Use Eq. 23.70 to calculate the specific gas constant. 8314 J/kmol·K is the universal gas constant. [**Fundamental Constants**]

$$\begin{aligned} R &= \frac{\overline{R}}{M} = \frac{8314\ \frac{\text{J}}{\text{kmol·K}}}{44\ \frac{\text{kg}}{\text{kmol}}} \\ &= 189.0\ \text{J/kg·K} \end{aligned}$$

The absolute temperature is $T = 21°\text{C} + 273° = 294\text{K}$. Use the ideal gas law (Eq. 23.71), and solve for pressure.

Ideal Gas

$$\begin{aligned} pV &= mRT \\ p &= \frac{mRT}{V} \\ &= \frac{(4.5\ \text{kg})\left(189\ \frac{\text{J}}{\text{kg·K}}\right)(294\text{K})}{(0.71\ \text{m}^3)\left(1000\ \frac{\text{Pa}}{\text{kPa}}\right)} \\ &= 352.2\ \text{kPa} \end{aligned}$$

Customary U.S. Solution

Use Eq. 23.70 to calculate the specific gas constant. 1545 ft-lbf/lbm-°R is the universal gas constant. [**Fundamental Constants**]

$$\begin{aligned} R &= \frac{\overline{R}}{M} = \frac{1545\ \frac{\text{ft-lbf}}{\text{lbmol-°R}}}{44\ \frac{\text{lbm}}{\text{lbmol}}} \\ &= 35.11\ \text{ft-lbf/lbm-°R} \end{aligned}$$

The absolute temperature is $T = 70°\text{F} + 460° = 530°\text{R}$. Use the ideal gas law Eq. 23.71), and solve for pressure.

Ideal Gas

$$\begin{aligned} pV &= mRT \\ p &= \frac{mRT}{V} \\ &= \frac{(10\ \text{lbm})\left(35.11\ \frac{\text{ft-lbf}}{\text{lbm-°R}}\right)(530°\text{R})}{25\ \text{ft}^3} \\ &= 7443\ \text{lbf/ft}^2 \end{aligned}$$

34. PROPERTIES OF IDEAL GASES

A gas can be considered to behave ideally if its pressure is very low and the temperature is much higher than its critical temperature. (Otherwise, the substance is in vapor form.) Under these conditions, the molecule size is insignificant compared with the distance between molecules, and molecules do not come into contact. By definition, an *ideal gas* behaves according to the various ideal gas laws.

For an ideal gas at a given temperature, specific enthalpy does not vary with change in pressure, and specific internal energy does not vary with change in specific volume.

Ideal Gas

$$\left(\frac{\partial h}{\partial p}\right)_T = 0 \quad 23.74$$

$$\left(\frac{\partial u}{\partial v}\right)_T = 0 \quad 23.75$$

Values of h, u, and v for gases are usually read from gas tables. Because specific enthalpy contains a pv term, it can be related to the equation of state. Depending on the units chosen, a conversion factor may be needed.

Functions and Their Symbols and Units

$$\begin{aligned} h &= u + pv \\ &= u + RT \end{aligned} \quad 23.76$$

Furthermore, density is the reciprocal of specific volume, v. Therefore, the density of an ideal gas can be derived from the equation of state by setting $m = 1$ and solving for the reciprocal of specific volume.

$$\rho = \frac{1}{v} = \frac{p}{RT} \quad 23.77$$

35. SPECIFIC HEATS OF IDEAL GASES

The specific heats of an ideal gas can be calculated from its specific gas constant. Depending on the units chosen, a conversion factor may or may not be needed in Eq. 23.78 through Eq. 23.81. By definition, a *perfect gas* is an ideal gas whose specific heats are constant.

For an ideal gas,

Ideal Gas

$$c_p - c_v = R \quad 23.78$$

$$C_p - C_v = \overline{R} \quad 23.79$$

$$c_p = \frac{Rk}{k-1} \quad 23.80$$

$$C_p = \frac{\overline{R}k}{k-1} \quad 23.81$$

36. MASS, MOLE, AND VOLUME FRACTIONS

The *mass fraction*, y_i (also known as the *gravimetric fraction* or *weight fraction*), of a component i in a mixture of components is the ratio of mass of the component to the mass of the total mixture.

Ideal Gas Mixtures

$$y_i = \frac{m_i}{\sum m_i} = \frac{m_i}{m} \quad 23.82$$

y_i is a decimal fraction between zero and one. The sum of the mass fractions of all the components in a mixture is equal to one.

Ideal Gas Mixtures

$$\sum y_i = 1 \quad 23.83$$

The *mole fraction*, x_i, of a component i is the ratio of the number of moles of substance i to the total number of moles of all substances.

Ideal Gas Mixtures

$$x_i = \frac{N_i}{\sum N_i} = \frac{N_i}{N} \quad 23.84$$

x_i is a decimal fraction between zero and one. The sum of the mole fractions of all the components in a mixture is equal to one.

Ideal Gas Mixtures

$$\sum x_i = 1 \quad 23.85$$

It is possible to convert between mass and mole fractions.

Ideal Gas Mixtures

$$y_i = \frac{x_i M_i}{\sum x_i M_i} \quad 23.86$$

$$x_1 = \frac{\dfrac{y_i}{M_i}}{\sum \dfrac{y_i}{M_i}} \quad 23.87$$

The *volume fraction*, f_i, of a component i is the ratio of the partial volume of the component to the volume of the overall mixture.

$$f_i = \frac{V_i}{\sum V_i} = \frac{V_i}{V} \quad 23.88$$

f_i is a decimal fraction between zero and one. The sum of the volume fractions of all the components in a mixture is equal to one.

$$\sum f_i = 1 \quad 23.89$$

Thermodynamics

For nonreacting mixtures of ideal gases, the mole fraction and volumetric fraction (and partial pressure ratio) of a component are the same.

$$x_i = f_i\big|_{\text{ideal gas}} \qquad 23.90$$

37. PARTIAL PRESSURE AND PARTIAL VOLUME OF GAS MIXTURES

A *gas mixture* consists of an aggregation of molecules of each gas component, the molecules of any single component being distributed uniformly and moving as if they alone occupied the space. The *partial volume*, V_i, of a gas i in a mixture of nonreacting gases is the volume that gas i alone would occupy at the temperature and pressure of the mixture. (See Fig. 23.7.)

Figure 23.7 *Mixture of Ideal Gases*

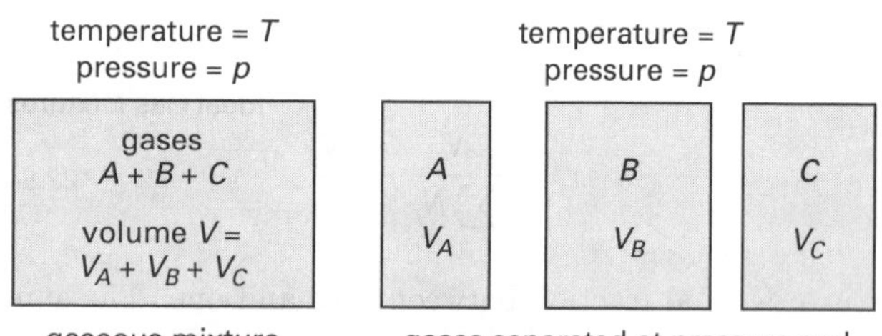

The partial volume of a gas i can be calculated from the volume fraction and total volume.

$$V_i = f_i V \qquad 23.91$$

From the ideal gas law, the partial volume of a gas i can also be calculated as in Eq. 23.92.

Ideal Gas Mixtures

$$V_i = \frac{m_i R_i T}{p} = \frac{N_i \bar{R} T}{p} \qquad 23.92$$

Amagat's law (also known as *Amagat-Leduc's rule*) states that the total volume of a mixture of nonreacting gases is equal to the sum of the partial volumes.

Ideal Gas Mixtures

$$V = \sum V_i \qquad 23.93$$

The *partial pressure*, p_i, of gas A in a mixture of nonreacting gases A, B, C, and so on, is the pressure gas A alone would exert in the total volume at the temperature of the mixture.

Ideal Gas Mixtures

$$p_i = \frac{m_i R_i T}{V} = \frac{N_i \bar{R} T}{V} \qquad 23.94$$

The partial pressure can also be calculated from the mole fraction and the total pressure. However, for ideal gases, the partial pressure ratio, mole fraction, and volumetric fraction are the same.

Ideal Gas Mixtures

$$x_i = \frac{p_i}{p} = \frac{V_i}{V} \quad \text{[ideal gas]} \qquad 23.95$$

If the average specific gas constant, $\bar{R}$, for the gas mixture is known, it can be used with the mass fraction to calculate the partial pressure.

$$p_i = \frac{y_i R_i p}{\bar{R}} \qquad 23.96$$

According to *Dalton's law of partial pressures*, the *total pressure* of a gas mixture is the sum of the partial pressures.

Ideal Gas Mixtures

$$p = \sum p_i \qquad 23.97$$

38. PROPERTIES OF NONREACTING IDEAL GAS MIXTURES

A mixture's average molecular weight and density are volumetrically weighted averages of its components' values.

$$M_{\text{ave}} = \sum f_i M_i \qquad 23.98$$

$$\rho_{\text{ave}} = \sum f_i \rho_i \qquad 23.99$$

On the other hand, a mixture's internal energy, enthalpy, specific heats, and specific gas constant are equal to the sum of the values of its individual components (i.e., the mixture average is weighted by mass). For example, a mixture's internal energy is

$$u = \frac{\sum m_i u_i}{m} \qquad 23.100$$

Therefore, for a mixture of ideal gases,

Ideal Gas Mixtures

$$u = \sum y_i u_i \qquad 23.101$$

Similarly,

Ideal Gas Mixtures

$$h = \sum y_i h_i \quad (23.102)$$

$$c_v = \sum y_i c_{v,i} \quad (23.103)$$

$$c_p = \sum y_i c_{p,i} \quad (23.104)$$

$$R = \sum y_i R_i \quad (23.105)$$

If the mixing is reversible and adiabatic, the entropy will also be equal to the sum of the individual entropies. However, each individual entropy, s_i, must be evaluated at the temperature and volume of the mixture and at the individual partial pressure, p_i.

Ideal Gas Mixtures

$$s = \sum y_i s_i \quad (23.106)$$

Equation 23.101, Eq. 23.102, and Eq. 23.106 are mathematical formulations of *Gibbs theorem* (also known as *Gibbs rule*). This theorem states that the total property (e.g., U, H, or S) of a mixture of nonreacting ideal gases is the sum of the properties that the individual gases would have if each occupied the total mixture volume alone at the same temperature.

Table 23.5 summarizes composite gas properties. Notice that the ratio of specific heats, k, is not a composite gas property. It is most expedient to find the composite ratio of specific heats from the mass-weighted specific heats.

***Table 23.5** Summary of Composite Ideal Gas Properties*

gravimetrically (mass) weighted	volumetrically (mole fraction) weighted
u	U
h	H
c_p	C_p
c_v	C_v
R	M
s	S
	ρ

Example 23.12

0.14 lbm (0.064 kg) of octane vapor (MW = 114) is mixed with 2.0 lbm (0.91 kg) of air (M = 29.0) in the manifold of an engine. The total pressure in the manifold is 12.5 psia (86.1 kPa), and the temperature is 520°R (290K). Assume octane behaves ideally. (a) What is the total volume of this mixture? (b) What is the partial pressure of the air in the mixture?

SI Solution

(a) The number of moles of octane and air are

$$N_{\text{octane}} = \frac{m_{\text{octane}}}{M_{\text{octane}}} = \frac{0.064\ \text{kg}}{114\ \frac{\text{kg}}{\text{kmol}}} = 5.61 \times 10^{-4}\ \text{kmol}$$

$$N_{\text{air}} = \frac{m_{\text{air}}}{M_{\text{air}}} = \frac{0.91\ \text{kg}}{29\ \frac{\text{kg}}{\text{kmol}}} = 0.0314\ \text{kmol}$$

From Eq. 23.92, the total volume of all gases is

$$V = \frac{N\bar{R}T}{p} = \frac{(5.61 \times 10^{-4}\ \text{kmol} + 0.0314\ \text{kmol}) \times \left(8314\ \frac{\text{J}}{\text{kmol·K}}\right)(290\text{K})}{86.1 \times 10^3\ \text{Pa}} = 0.895\ \text{m}^3$$

(b) The mole fraction of the air is

$$x_{\text{air}} = \frac{N_{\text{air}}}{N} = \frac{0.0314\ \text{kmol}}{5.61 \times 10^{-4}\ \text{kmol} + 0.0314\ \text{kmol}} = 0.982$$

The partial pressure of air is

$$p_{\text{air}} = x_{\text{air}} p = (0.982)(86.1\,\text{kPa}) = 84.6\,\text{kPa}$$

Customary U.S. Solution

(a) The number of moles of octane and air are

$$N_{\text{octane}} = \frac{m_{\text{octane}}}{M_{\text{octane}}} = \frac{0.14\ \text{lbm}}{114\ \frac{\text{lbm}}{\text{lbmol}}} = 0.001228\ \text{lbmol}$$

$$N_{\text{air}} = \frac{m_{\text{air}}}{M_{\text{air}}} = \frac{2.0\ \text{lbm}}{29.0\ \frac{\text{lbm}}{\text{lbmol}}} = 0.068966\ \text{lbmol}$$

Thermodynamics

From Eq. 23.92, the total volume of all gases is

$$V = \frac{N\bar{R}T}{p}$$

$$= \frac{(0.001228 \text{ lbmol} + 0.068966 \text{ lbmol}) \times \left(1545 \frac{\text{ft-lbf}}{\text{lbmol-°R}}\right)(520°\text{R})}{\left(12.5 \frac{\text{lbf}}{\text{in}^2}\right)\left(12 \frac{\text{in}}{\text{ft}}\right)^2}$$

$$= 31.34 \text{ ft}^3$$

(b) The mole fraction of the air is

$$x_{\text{air}} = \frac{N_{\text{air}}}{N}$$

$$= \frac{0.068966 \text{ lbmol}}{0.001228 \text{ lbmol} + 0.068966 \text{ lbmol}}$$

$$= 0.9825$$

The partial air pressure is

$$p_{\text{air}} = x_{\text{air}}p$$

$$= (0.9825)\left(12.5 \frac{\text{lbf}}{\text{in}^2}\right)$$

$$= 12.3 \text{ psia}$$

39. EQUATION OF STATE FOR REAL GASES

Real gases do not meet the basic assumptions defining an ideal gas. Specifically, the molecules of a real gas occupy a volume that is not negligible in comparison with the total volume of the gas. (This is especially true for gases at low temperatures.) Furthermore, real gases are subject to *van der Waals' forces*, which are attractive forces between gas molecules.

There are two methods of accounting for real gas behavior. The first method is to modify the ideal gas equation of state with various empirical correction factors. Since the modifications are empirical, the resulting equations of state are known as *correlations*. One well-known correlation is *van der Waals' equation of state*.

$$\left(p + \frac{a}{V^2}\right)(V - b) = N\bar{R}T \qquad 23.107$$

The van der Waals corrections usually need to be made only when a gas is below its critical temperature. For an ideal gas, the a and b terms are zero. When the spacing between molecules is close, as it would be at low temperatures, the molecules attract each other and reduce the pressure exerted by the gas. The pressure is then corrected by the a/V_2 term. b is a constant that accounts for the molecular volume in a dense state.

Other correlations of this type include the *Clausius, Bertholet, Dieterici,* and *Beattie-Bridgeman equations of state*. However, the most accurate empirical correlation is the *virial equation of state*, which has the form shown in Eq. 23.108 and Eq. 23.109. The constants B, C, D, and so on, are called the *virial coefficients*.

$$pV = N\bar{R}T\left(1 + \frac{B}{V} + \frac{C}{V^2} + \frac{D}{V^3} + \cdots\right) \qquad 23.108$$

$$pV = N\bar{R}T(1 + B'p + C'p^2 + D'p^3 + \cdots) \qquad 23.109$$

The *principle (law) of corresponding states* provides a second method of correcting for real gas behavior. This law states that the *reduced properties* of all gases are identical (i.e., all gases behave similarly when their reduced variables are used). Specifically, there is one property, the *compressibility factor*, Z, that has the same value for all gases when evaluated at the same values of the *reduced variables*.[14] (The reduced variables are not the same as the ratios defined in Sec. 23.25.)

$$Z = f(T_r, p_r, v_r) \qquad 23.110$$

$$T_r = \frac{T}{T_c} \qquad 23.111$$

$$p_r = \frac{p}{p_c} \qquad 23.112$$

$$v_r = \frac{v}{v_c} \qquad 23.113$$

Compressibility factors are almost always read from *generalized compressibility charts*. The compressibility factor can then be used to correct the ideal gas equation of state. **[Compressibility Factor and Charts]**

$$pv = ZRT \qquad 23.114$$

$$pV = mZRT \qquad 23.115$$

Example 23.13

What is the specific volume of carbon dioxide at 2680 psia (182 atm) and 300°F (150°C)?

SI Solution

The absolute temperature is

$$T = 150°\text{C} + 273° = 423\text{K}$$

[14]Some engineering disciplines (e.g., petroleum engineering) use the symbol z for the compressibility factor.

From Table 23.4, the critical temperature and pressure of carbon dioxide are 304.3K and 72.9 atm, respectively. The reduced variables are

$$T_r = \frac{T}{T_c} = \frac{423\text{K}}{304.3\text{K}} = 1.39$$

$$p_r = \frac{p}{p_c} = \frac{182 \text{ atm}}{72.9 \text{ atm}} = 2.5$$

From a generalized compressibility chart, Z is read as 0.75. [**Compressibility Factor and Charts**]

R is read from Table 23.6. Solving Eq. 23.114 for specific volume,

$$v = \frac{ZRT}{p} = \frac{(0.75)\left(189 \; \frac{\text{J}}{\text{kg·K}}\right)(423\text{K})}{(182 \text{ atm})\left(101.3 \times 10^3 \; \frac{\text{Pa}}{\text{atm}}\right)}$$

$$= 3.25 \times 10^{-3} \text{ m}^3/\text{kg}$$

Customary U.S. Solution

The absolute temperature is

$$T = 300°\text{F} + 460° = 760°\text{R}$$

From Table 23.4, the critical temperature and pressure of carbon dioxide are 547.8°R and 72.9 atm, respectively. The reduced variables are

$$T_r = \frac{T}{T_c} = \frac{760°\text{R}}{547.8°\text{R}} = 1.39$$

$$p_r = \frac{p}{p_c} = \frac{2680 \text{ psia}}{(72.9 \text{ atm})\left(14.7 \; \frac{\text{psia}}{\text{atm}}\right)} = 2.5$$

From a generalized compressibility chart, Z is read as 0.75. [**Compressibility Factor and Charts**]

R is read from Table 23.6. Solving Eq. 23.114 for specific volume,

$$v = \frac{ZRT}{p} = \frac{(0.75)\left(35.11 \; \frac{\text{ft-lbf}}{\text{lbm-°R}}\right)(760°\text{R})}{\left(2680 \; \frac{\text{lbf}}{\text{in}^2}\right)\left(12 \; \frac{\text{in}}{\text{ft}}\right)^2}$$

$$= 0.0519 \text{ ft}^3/\text{lbm}$$

40. SPECIFIC HEATS OF REAL GASES

By definition, the specific heat of an ideal (perfect) gas is independent of temperature. However, the specific heat of a real gas varies with temperature and (slightly) with pressure. There are several ways to find the specific heat of a gas at different temperatures: tables, graphs, and correlations.[15]

Equation 23.116 is a typical temperature correlation. The variation with pressure, being small at low pressures, is disregarded. The coefficients A, B, C, and D will depend on the units, the temperature range over which the correlation is to be used, and the desired accuracy. Typical coefficients are given in Table 23.6 at the end of this chapter.

$$c_p = A + BT + CT^2 + \frac{D}{\sqrt{T}} \qquad \textit{23.116}$$

$$c_v = c_p - R \qquad \text{[SI]} \qquad \textit{23.117(a)}$$

$$c_v = c_p - \frac{R}{J} \qquad \text{[U.S.]} \qquad \textit{23.117(b)}$$

41. NOMENCLATURE

a	Helmholtz free energy per unit mass	Btu/lbm	kJ/kg
a	van der Waals factor	atm-ft^6/lbmol	Pa·m^6/kmol
A	area	ft^2	m^2
A	Helmholtz free energy	Btu	kJ
AW	atomic weight	lbm/lbmol	kg/kmol
b	van der Waals factor	ft^3/lbmol	m^3/kmol
c	specific heat capacity	Btu/lbm-°F	kJ/kg·°C
C	heat capacity	Btu/°F	kJ/K
f	volume fraction	–	–
F	force	lbf	N
g	Gibbs free energy per unit mass	Btu/lbm	kJ/kg
G	Gibbs free energy	Btu	kJ
h	specific enthalpy	Btu/lbm	kJ/kg
H	enthalpy	Btu	kJ
H_m	molar enthalpy	Btu/lbmol	kJ/kmol
J	Joule's constant, 778.2	ft-lbf/Btu	n.a.
k	ratio of specific heats	–	–
m	mass	lbm	kg
M	molecular weight	lbm/lbmol	kg/kmol
N	number of moles	–	–

[15]It is also possible to calculate the specific heat over a small temperature range if the enthalpies are known (e.g., from an air table) from $c_p = \Delta h/\Delta T$. However, if the enthalpies are known, it is unlikely that the specific heat will be needed.

p	pressure	lbf/ft^2	Pa
q	heat per unit mass	Btu/lbm	kJ/kg
Q	total heat	Btu	kJ
R	specific gas constant	ft-lbf/lbm-°R	kJ/kg·K
$\overline{R}$	universal gas constant	ft-lbf/lbmol-°R	kJ/kmol·K
s	specific entropy	Btu/lbm-°R	kJ/kg·K
S	absolute entropy	Btu/°R	kJ/K
T	absolute temperature	°R	K
T	temperature	°F	°C
u	specific internal energy	Btu/lbm	kJ/kg
U	internal energy	Btu	kJ
v	specific volume	ft^3/lbm	m^3/kg
V	volume	ft^3	m^3
V_m	molar volume	ft^3/lbmol	m^3/kmol
x	mole fraction	–	–
x	quality	–	–
y	mass fraction	–	–
Z	compressibility factor	–	–

Symbols

α	isentropic compressibility	1/°R	1/K
β	isobaric compressibility	1/°R	1/K
κ	isothermal compressibility	1/°R	1/K
ρ	density	lbm/ft^3	kg/m^3
ϕ	entropy function	Btu/lbm-°R	kJ/kg·K

Subscripts

c	critical
f	saturated liquid (fluid)
fg	vaporization (liquid to vapor)
g	saturated vapor (gas)
i	ice
l	latent
m	molar
p	constant pressure
r	ratio, reduced, or relative
s	constant entropy, sensible, or solid
sat	saturated
t	total
T	constant temperature
v	constant volume
V	constant volume

***Table 23.6** Correlation Coefficients for Calculating Specific Heat (Btu/lbm-°R or Btu/lbm-°F)*

gas	temperature range °R	temperature range K	*A*	*B*	*C*	*D*
air	400 to 1200	220 to 670	0.2405	-1.186×10^{-5}	20.1×10^{-9}	0
	1200 to 4000	670 to 2220	0.2459	3.22×10^{-5}	-3.74×10^{-9}	−0.833
CH_4	400 to 1000	220 to 560	0.453	0.62×10^{-5}	268.8×10^{-9}	0
	1000 to 4000	560 to 2220	1.152	32.58×10^{-5}	-41.29×10^{-9}	−22.42
CO	400 to 1200	220 to 670	0.2534	-2.35×10^{-5}	26.88×10^{-9}	0
	1200 to 4000	670 to 2220	0.2763	3.04×10^{-5}	-3.89×10^{-9}	−1.5
CO_2	400 to 4000	220 to 2220	0.328	3.2×10^{-5}	-4.4×10^{-9}	−3.33
H_2	400 to 1000	220 to 560	2.853	145×10^{-5}	-883×10^{-9}	0
	1000 to 2500	560 to 1390	3.447	-4.7×10^{-5}	70.3×10^{-9}	0
	2500 to 4000	1390 to 2220	2.841	45×10^{-5}	-31.2×10^{-9}	0
H_2O	400 to 1800	220 to 1000	0.4267	2.425×10^{-5}	23.85×10^{-9}	0
	1800 to 4000	1000 to 2220	0.3275	14.67×10^{-5}	-13.59×10^{-9}	0
N_2	400 to 1200	220 to 670	0.2510	-1.63×10^{-5}	20.4×10^{-9}	0
	1200 to 4000	670 to 2220	0.2192	4.38×10^{-5}	-5.14×10^{-9}	−0.124
O_2	400 to 1200	220 to 670	0.213	0.188×10^{-5}	20.3×10^{-9}	0
	1200 to 4000	670 to 2220	0.340	-0.36×10^{-5}	0.616×10^{-9}	−3.19

(Multiply values in this table by 4.1868 to obtain coefficients for specific heats in kJ/kg·K or kJ/kg·°C.)

Adapted from a table published in *Engineering Thermodynamics*, C. O. Mackay, W. N. Barnard, and F. O. Ellenwood (John Wiley, New York, 1957).

24 Changes in Thermodynamic Properties

Content in blue refers to the *NCEES Handbook*.

NCEES EXAM SPECIFICATIONS AND RELATED CONTENT

HVAC AND REFRIGERATION EXAM

- I.B.1. Thermodynamics: Cycles
  - 12. Property Changes in Ideal Gases
- I.F. Energy/Mass Balances
  - 5. Reversible Processes

THERMAL AND FLUID SYSTEMS EXAM

- I.E.3. Thermodynamics: Energy balances
  - 5. Reversible Processes
  - 8. First Law of Thermodynamics for Closed Systems
  - 10. First Law of Thermodynamics for Open Systems
  - 12. Property Changes in Ideal Gases
  - 17. Second Law of Themodynamics

1. SYSTEMS

A *thermodynamic system* is defined as the matter enclosed within an arbitrary but precisely defined *control volume.* Everything external to the system is defined as the *surroundings, environment,* or *universe.* The environment and system are separated by the *system boundaries.* The surface of the control volume is known as the *control surface.* The control surface can be real (e.g., piston and cylinder walls) or imaginary.

If mass flows through the system across system boundaries, the system is an *open system.* Pumps, heat exchangers, and jet engines are examples of open systems. An important type of open system is the *steady-flow open system* in which matter enters and exits at the same rate. Pumps, turbines, heat exchangers, and boilers are all steady-flow open systems.

If no mass crosses the system boundaries, the system is said to be a *closed system.* The matter in a closed system may be referred to as a *control mass.* Closed systems can have variable volumes. The gas compressed by a piston in a cylinder is an example of a closed system with a variable control volume.

In most cases, energy in the form of heat, work, or electrical energy can enter or exit any open or closed system. Systems closed to both matter and energy transfer are known as *isolated systems.*

2. TYPES OF PROCESSES

Changes in thermodynamic properties of a system often depend on the type of process experienced. This is particularly true of gaseous systems. Several common types of processes are listed as follows, along with their heat, energy, and work relationships.[1]

- *adiabatic process*—a process in which no heat or other energy crosses the system boundary.[2] Adiabatic processes include isentropic and throttling processes.

$$Q = 0 \qquad 24.1$$

$$\Delta U = -W \qquad 24.2$$

[1]The heat, energy, and work relationships are easily derived from the *first law of thermodynamics.*

[2]An adiabatic process is not the same as a constant temperature process, however.

- *constant pressure process*—also known as an *isobaric process*

$$\Delta p = 0 \quad \text{24.3}$$

$$Q = \Delta H \quad \text{24.4}$$

- *constant temperature process*—also known as an *isothermal process*

$$\Delta T = 0 \quad \text{24.5}$$

$$Q = W \quad \text{24.6}$$

- *constant volume process*—also known as an *isochoric* or *isometric process*

$$\Delta V = 0 \quad \text{24.7}$$

$$Q = \Delta U \quad \text{24.8}$$

$$W = 0 \quad \text{24.9}$$

- *isenthalpic process*—a process in which the enthalpy of the system substance does not change

$$\Delta H = 0 \quad \text{24.10}$$

$$\Delta T = 0 \quad \text{[ideal gas]} \quad \text{24.11}$$

- *isentropic process*—an adiabatic process in which there is no change in system entropy (i.e., is reversible)

$$\Delta S = 0 \quad \text{24.12}$$

$$Q = 0 \quad \text{24.13}$$

- *throttling process*—an adiabatic process in which there is no change in system enthalpy (i.e., the process is isenthalpic), but for which there is a significant pressure drop[3]

$$\Delta H = 0 \quad \text{24.14}$$

$$p_2 < p_1 \quad \text{24.15}$$

A system that is in equilibrium at the start and finish of a process may or may not be in equilibrium during the process. A *quasistatic process* (*quasiequilibrium process*) is one that can be divided into a series of infinitesimal deviations (steps) from equilibrium. During each step, the property changes are small, and all intensive properties are uniform throughout the system. The interim equilibrium at each step is known as a *quasiequilibrium.*

Transport processes are generally more complex than can be analyzed by the simple thermodynamic relationships in this chapter. In a *transport process*, there is a transfer of some quantity. Drying, distillation, and evaporation are examples of heat transfer processes. Fluid flow, mixing, and sedimentation are examples of momentum transfer. Distillation, absorption, extraction, and leaching are examples of mass transfer processes.

3. POLYTROPIC PROCESSES

A *polytropic process* is one that obeys Eq. 24.16, the *polytropic equation of state.* Gases always constitute the system in polytropic processes.

$$p_1 V_1^n = p_2 V_2^n \quad \text{24.16}$$

n is the *polytropic exponent*, a property of the equipment, not of the gas. For efficient air compressors, n is typically between 1.25 and 1.30.

Depending on the value of the polytropic exponent, the polytropic equation of state can also be used for other processes.

$n = 0$	[constant pressure process]
$n = 1$	[constant temperature process]
$n = k$	[isentropic process]
$n = \infty$	[constant volume process]

The *polytropic specific heat capacity*, c_n, is defined as

$$c_n = \frac{n-k}{n-1} \times c_v \quad \text{24.17}$$

$$Q = mc_n \Delta T \quad \text{24.18}$$

4. THROTTLING PROCESSES

For ideal gases, throttling processes are constant temperature processes. Some real gases at normal temperatures, however, decrease in temperature when throttled.[4] Others increase in temperature. Whether or not an increase or a decrease in temperature occurs depends on the temperature and pressure at which the throttling occurs.

For any given set of initial conditions, there is one temperature at which no temperature change occurs when a real gas is throttled. This is called the *inversion temperature* or *inversion point.* The inversion temperature is dependent on the initial gas conditions. (See Table 24.1.)

[3]The classic throttling process is the expansion of a gas in a pipe through a porous plug that offers significant resistance to flow.
[4]This tendency can be used to liquefy gases and vapors (e.g., refrigerants) by passing them through an expansion valve.

Table 24.1 *Approximate Maximum Experimental Inversion Temperatures*

substance	°R	K
air	1085	603
argon	1301	723
carbon dioxide	≈ 2700	≈ 1500
helium	≈ 72	≈ 40
hydrogen	364	202
nitrogen	1118	621

(Multiply K by 1.8 to obtain °R.)

The *Joule-Thomson coefficient*, μ_J, (also known as the *Joule-Kelvin coefficient*), is defined as the ratio of the change in temperature to the change in pressure when a real gas is throttled at constant enthalpy H. The Joule-Thomson coefficient is zero for an ideal gas.

$$\mu_J = \left(\frac{\partial T}{\partial p}\right)_H \qquad 24.19$$

The Joule-Thomson coefficient can be calculated from the specific heat, c_p, and other properties.

$$\begin{aligned}\mu_J &= \left(\frac{1}{c_p}\right)\left(T\left(\frac{\partial V}{\partial T}\right)_p - v\right) = \left(\frac{v}{c_p}\right)(T\beta - 1) \\ &= \left(\frac{-1}{c_p}\right)\left(\frac{\partial h}{\partial p}\right)_T\end{aligned} \qquad 24.20$$

It is convenient to plot lines of constant enthalpy (*isenthalpic curves*) on a T-p diagram. The various curves in Fig. 24.1 represent different sets of starting conditions (p_1 and T_1). The slope of a curve represents the Joule-Thomson coefficient, which can be zero, positive, or negative at that point. The points where the slope is zero are known as *inversion points*. The locus of inversion points is known as the *inversion curve*.

States to the right of an inversion point will have negative Joule-Thomson coefficients. States to the left of an inversion point will have positive Joule-Thomson coefficients. Throttling totally within the region bounded by the temperature axis and the inversion curve will always result in a decrease in temperature. Similarly, throttling totally outside of the inversion curve will always produce an increase in temperature. Throttling across the inversion curve may produce a higher or lower final temperature, depending on initial and final conditions.

Figure 24.1 *Inversion Curve*

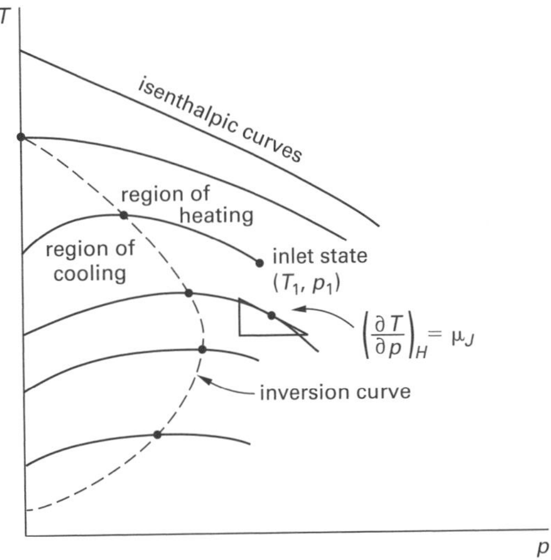

5. REVERSIBLE PROCESSES

Processes can be categorized as reversible and irreversible. A *reversible process* is one that is performed in such a way that at the conclusion of the process, *both* the system and the local surroundings can be restored to their initial states. A process that does not meet these requirements is an *irreversible process*. A reversible adiabatic process is implicitly isentropic.

All real-world processes result in an overall increase in entropy.

Entropy

$$\Delta S_{\text{total}} = \Delta S_{\text{system}} + \Delta S_{\text{surroundings}} \geq 0 \qquad 24.21$$

$$\begin{aligned}\Delta \dot{S}_{\text{total}} &= \sum \dot{m}_{\text{out}} s_{\text{out}} - \sum \dot{m}_{\text{in}} s_{\text{in}} \\ &\quad - \sum \frac{q_{\text{external}}}{T_{\text{external}}} \\ &\geq 0\end{aligned} \qquad 24.22$$

Although all real-world processes result in an overall increase in entropy, it is possible to conceptualize processes that have zero entropy change. For a reversible process,

$$\Delta s = 0\big|_{\text{reversible}} \qquad 24.23$$

Processes that contain friction are never reversible. Other processes that are irreversible are listed as follows.

- stirring a viscous fluid
- slowing down a moving fluid
- unrestrained expansion of gas
- throttling
- changes of phase (freezing, condensation, etc.)
- chemical reaction

Thermodynamics

- diffusion
- current flow through electrical resistance
- electrical polarization
- magnetization with hysteresis
- releasing a stretched spring
- inelastic deformation
- heat conduction

A system that has experienced an irreversible process can still be returned to its original state. An example of this is the water in a closed boiler-turbine installation. The entropy increases when the water is vaporized in the boiler (an irreversible process). The entropy decreases when the steam is condensed in the condenser. When the water returns to the boiler, its entropy is increased to its original value. However, the environment cannot be returned to its original condition; hence the cycle overall is irreversible.

6. FINDING WORK AND HEAT GRAPHICALLY

A process between two thermodynamic states can be represented graphically. The line representing the locus of quasiequilibrium states between the initial and final states is known as the *path* of the process.

It is sometimes convenient to see what happens to the pressure and volume of a system by plotting the path on a p-V diagram. In addition, the work done by or on the system can be determined from the graph. This is possible because the integral calculating p-V represents area under a curve in the p-V plane.

$$W = \int_{V_1}^{V_2} p \, dV \qquad 24.24$$

Similarly, the amount of heat absorbed or released from a system can be determined as the area under the path on the T-s diagram.

$$Q = \int_{s_1}^{s_2} T \, ds \qquad 24.25$$

The variables p, V, T, and s are *point functions* because their values are independent of the path taken to arrive at the thermodynamic state. Work and heat (W and Q), however, are *path functions* because they depend on the path taken. (See Fig. 24.2.)

Figure 24.2 *Process Work and Heat*

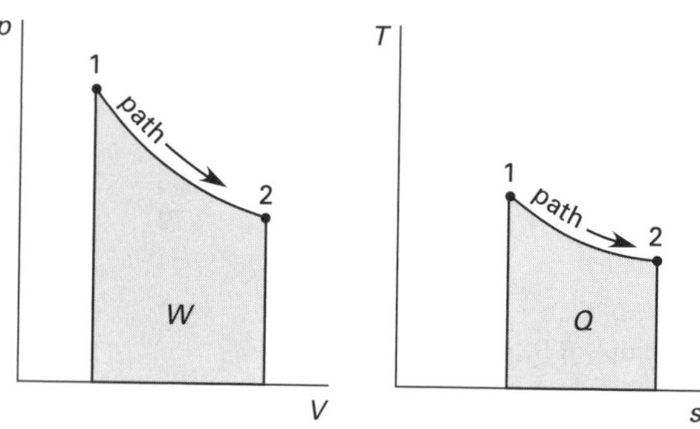

7. SIGN CONVENTION

A definite sign convention is used in calculating work, heat, and other property changes in systems.[5] This sign convention takes the system (not the environment) as the reference. (For example, a net heat gain would mean the system gained energy and the environment lost energy.)

- Heat, Q, is positive if heat flows into the system.
- Work, W, is positive if the system does work on the surroundings.
- Changes in enthalpy, entropy, and internal energy (ΔH, ΔS, and ΔU) are positive if these properties increase within the system.

8. FIRST LAW OF THERMODYNAMICS FOR CLOSED SYSTEMS

There is a basic principle that underlies all property changes as a system experiences a process: all energy must be accounted for. Energy that enters a system must either leave the system or be stored in some manner. Energy cannot be created or destroyed. These statements are the primary manifestations of the *first law of thermodynamics*: the work done in an adiabatic process depends solely on the system's endpoint conditions, not on the nature of the process.

For closed systems, the first law can be written in differential form as in Eq. 24.26. In most cases, kinetic energy, KE, and potential energy, PE, are negligible. Since Q and W are not properties, infinitesimal changes are designated as δ.

$$\delta Q - \delta W = dU + d\text{KE} + d\text{PE} \qquad 24.26$$

Most simple thermodynamic problems can be solved without resorting to differential calculus. The first law can then be written in finite terms.

[5]This sign convention is automatically followed if the formulas in this chapter are used.

Closed Thermodynamic Systems

$$Q - W = \Delta U + \Delta \text{KE} + \Delta \text{PE} \qquad 24.27$$

Again, KE and PE are usually negligible. Equation 24.27 says that the heat, Q, entering a closed system can either increase the temperature (i.e., increase U) or be used to perform work (increase W) on the surroundings. In a nonadiabatic closed system, heat energy entering the system also can leak to the surroundings. However, in Eq. 24.27, the Q term is understood to be the net heat entering the system, exclusive of the loss.

Q, U, and W must all have the same units. This is less of a problem with SI units than with English units. For example, if Q and U are in Btu and W is in ft-lbf, Joule's constant must be incorporated into the first law:

$$Q = \Delta U + \frac{W}{J} \quad \text{[U.S.]} \qquad 24.28$$

In accordance with the standard sign convention given in Sec. 24.7, Q will be negative if the net heat exchange to the system is a loss. ΔU will be negative if the internal energy of the system decreases. W will be negative if the surroundings do work on the system (e.g., a piston compressing gas in a cylinder).

9. THERMAL EQUILIBRIUM

A simple application of the first law is the calculation of a thermal equilibrium point for a nonreacting system. *Thermal equilibrium* is reached when all parts of the system are at the same temperature.

The drive toward thermal equilibrium occurs spontaneously, without the addition of external work, whenever masses (two liquids, a solid and a liquid, etc.) with different temperatures are combined. Therefore, the work term, W, in the first law is zero.

The energy comes from the heat given off by the cooling mass (i.e., mass A). Energy is stored by increasing the temperature of the warmed mass (mass B). These changes are equal, but opposite, since the net change is zero.

$$Q_{\text{net}} = Q_{\text{in},A} - Q_{\text{out},B} = 0 \qquad 24.29$$

The form of the equation for Q depends on the phases of the substances and the nature of the system. If the substances are either solid or liquid, Eq. 24.30 can be used. (For use with open systems, the m in Eq. 24.30 should be replaced with $\dot{m}$.)

$$m_A c_{p,A}(T_{1,A} - T_{2,A}) = m_B c_{p,B}(T_{2,B} - T_{1,B}) \qquad 24.30$$

Example 24.1

A block of steel is removed from a furnace and quenched in an insulated aluminum tank filled with water. The water and aluminum tank are initially in equilibrium at 75°F (24°C), and the final equilibrium temperature after quenching is 100°F (38°C). What is the initial temperature of the steel?

steel block mass:
m_{st} 2.0 lbm (0.9 kg)

steel specific heat:
$c_{p,\text{st}}$ 0.11 Btu/lbm-°F (0.460 kJ/kg·K)

aluminum tank mass:
m_{al} 5.0 lbm (2.25 kg)

aluminum specific heat:
$c_{p,\text{al}}$ 0.21 Btu/lbm-°F (0.880 kJ/kg·K)

water mass:
m_w 12.0 lbm (5.4 kg)

water specific heat:
$c_{p,w}$ 1.0 Btu/lbm-°F (4.190 kJ/kg·K)

SI Solution

The heat lost by the steel is equal to the heat gained by the tank and water.

$$m_{\text{st}}c_{p,\text{st}}(T_{1,\text{st}} - T_{\text{eq}}) = (m_{\text{al}}c_{p,\text{al}} + m_w c_{p,w})(T_{\text{eq}} - T_{1,w})$$

$$(0.9\ \text{kg})\left(0.460\ \frac{\text{kJ}}{\text{kg·K}}\right)(T_{1,\text{st}} - 38°\text{C}) = \left(\begin{array}{c}(2.25\ \text{kg})\left(0.880\ \frac{\text{kJ}}{\text{kg·K}}\right)\\ +(5.4\ \text{kg})\left(4.190\ \frac{\text{kJ}}{\text{kg·K}}\right)\end{array}\right)(38°\text{C} - 24°\text{C})$$

$$T_{1,\text{st}} = 870°\text{C}$$

Customary U.S. Solution

Proceeding as in the SI solution,

$$(2.0\ \text{lbm})\left(0.11\ \frac{\text{Btu}}{\text{lbm-°F}}\right)(T_{1,\text{st}} - 100°\text{F}) = \left(\begin{array}{c}(5.0\ \text{lbm})\left(0.21\ \frac{\text{Btu}}{\text{lbm-°F}}\right)\\ +(12.0\ \text{lbm})\left(1\ \frac{\text{Btu}}{\text{lbm-°F}}\right)\end{array}\right)(100°\text{F} - 75°\text{F})$$

$$T_{1,\text{st}} = 1583°\text{F}$$

10. FIRST LAW OF THERMODYNAMICS FOR OPEN SYSTEMS

The first law of thermodynamics can also be written for open systems, but more terms are required to account for the many energy forms. The first law formulation is essentially the Bernoulli energy conservation equation extended to nonadiabatic processes.

If the mass flow rate is constant, the system is a *steady-flow system*, and the first law is known as the *steady-flow energy equation*, SFEE, Eq. 24.31.

$$Q = \Delta U + \Delta\text{KE} + \Delta\text{PE} + W_{\text{flow}} + W_{\text{shaft}} \qquad 24.31$$

Some of the terms in Eq. 24.31 are illustrated in Fig. 24.3. Q is the net heat flow into or out of the system, inclusive of any losses. It can be supplied from furnace flame, electrical heating, nuclear reaction, or other sources. If the system is adiabatic, Q is zero.

ΔKE and ΔPE are the changes in the fluid's kinetic energy and potential energy. They are generally insignificant compared to the thermal energy transfers.

Figure 24.3 *Steady Flow Device*

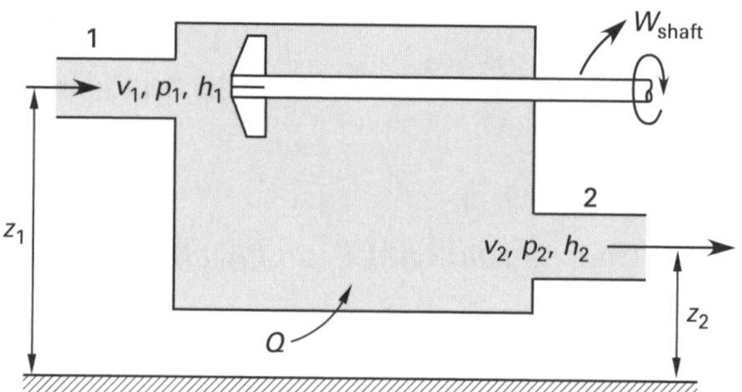

W_{shaft} is the *shaft work*—work that the steady-flow device does on the surroundings. The term is derived from the output shaft that serves to transmit energy out of the system. For example, turbines and internal combustion engines have output shafts. W_{shaft} can be negative, as in the case of a pump or compressor.

W_{flow} is the p-V (*flow energy*, *flow work*, etc.). There is a pressure, p_2, at the exit of the steady-flow device in Fig. 24.3. This exit pressure opposes the entrance of the fluid. Therefore, the flow work term represents the work required to cause the flow into the system against the exit pressure. The flow work can be calculated from Eq. 24.32. (Consistent units must be used.)

$$w_{\text{flow}} = p_2 v_2 - p_1 v_1 \qquad 24.32$$

ΔU is the change in the system's internal energy. Since the combination of internal energy and flow work constitutes enthalpy, it will seldom be necessary to work with either internal energy or flow work.

Functions and Their Symbols and Units

$$h = u + pv \qquad 24.33$$

$$\Delta h = \Delta u + w_{\text{flow}} \qquad 24.34$$

Equation 24.35 is a useful formulation of the SFEE. Energy flow rate can be obtained by multiplying both sides by $\dot{m}$.

$$q = h_2 - h_1 + \frac{\text{v}_2^2 - \text{v}_1^2}{2} + (z_2 - z_1)g + w_{\text{shaft}} \qquad \text{[SI]} \qquad 24.35(a)$$

$$q = h_2 - h_1 + \frac{\text{v}_2^2 - \text{v}_1^2}{2g_cJ} + \frac{(z_2 - z_1)g}{g_cJ} + \frac{w_{\text{shaft}}}{J} \qquad \text{[U.S.]} \qquad 24.35(b)$$

Example 24.2

4 lbm/sec (1.8 kg/s) of steam enter a turbine with a velocity of 65 ft/sec (20 m/s) and an enthalpy of 1350 Btu/lbm (3140 kJ/kg). After expansion in the turbine, the steam enters the condenser with an enthalpy of 1075 Btu/lbm (2500 kJ/kg) and at a velocity of 125 ft/sec (38 m/s). There is a total heat loss from the turbine casing of 50 Btu/sec (53 kJ/s). Potential energy changes are insignificant. What power is generated at the turbine shaft?

SI Solution

Equation 24.35 can be solved for the shaft work.

$$P_{\text{shaft}} = \dot{m}w_{\text{shaft}} = \dot{Q}_t + \dot{m}\left(h_1 - h_2 + \frac{\text{v}_1^2 - \text{v}_2^2}{2}\right)$$

$$= -53\,000\ \frac{\text{J}}{\text{s}} + \left(1.8\ \frac{\text{kg}}{\text{s}}\right)\left(\begin{array}{l} 3140 \times 10^3\ \frac{\text{J}}{\text{kg}} - 2500 \times 10^3\ \frac{\text{J}}{\text{kg}} \\ + \dfrac{\left(20\ \frac{\text{m}}{\text{s}}\right)^2 - \left(38\ \frac{\text{m}}{\text{s}}\right)^2}{2} \end{array}\right)$$

$$= 1.098 \times 10^6\ \text{W} \quad (1.1\ \text{MW})$$

Customary U.S. Solution

Proceeding as in the SI solution,

$$P_{\text{shaft}} = \dot{m}w_{\text{shaft}} = \dot{Q}_t + \dot{m}\left(h_1 - h_2 + \frac{\text{v}_1^2 - \text{v}_2^2}{2g_c J}\right)$$

$$= -50\ \frac{\text{Btu}}{\text{sec}} + \left(4.0\ \frac{\text{lbm}}{\text{sec}}\right)\left(\begin{array}{c}1350\ \frac{\text{Btu}}{\text{lbm}} - 1075\ \frac{\text{Btu}}{\text{lbm}} \\ + \dfrac{\left(65\ \frac{\text{ft}}{\text{sec}}\right)^2 - \left(125\ \frac{\text{ft}}{\text{sec}}\right)^2}{(2)\left(32.2\ \frac{\text{lbm-ft}}{\text{lbf-sec}^2}\right)\left(778\ \frac{\text{ft-lbf}}{\text{Btu}}\right)}\end{array}\right)$$

$$= 1049\ \text{Btu/sec}$$

11. BASIC THERMODYNAMIC RELATIONS

The following relations for a simple compressible substance (though not necessarily an ideal gas) can be derived from the first law of thermodynamics and other basic definitions.[6]

$$du = T\,ds - p\,dv \qquad 24.36$$

$$dh = T\,ds + v\,dp \qquad 24.37$$

12. PROPERTY CHANGES IN IDEAL GASES

For real solids, liquids, and vapors, changes in most properties can be determined only by subtracting the initial property value from the final property value. For simple compressible substances (i.e., ideal gases), however, many changes in properties can be found directly, without knowing the initial and final property values.

Some of the relations for determining property changes do not depend on the type of process. For example, a general relationship that applies to any ideal gas experiencing any process is easily derived from the equation of state.

Ideal Gas

$$\frac{p_1 v_1}{T_1} = \frac{p_2 v_2}{T_2} \qquad 24.38$$

When temperature is held constant, Eq. 24.38 reduces to *Boyle's law.*

Special Cases of Closed Systems (With No Change in Kinetic or Potential Energy)

$$pv = \text{constant} \quad [\text{ideal gas}]$$
$$p_1 v_1 = p_2 v_2 \qquad 24.39$$

When pressure is held constant, Eq. 24.38 reduces to *Charles' law.*

Special Cases of Closed Systems (With No Change in Kinetic or Potential Energy)

$$\frac{T}{v} = \text{constant} \quad [\text{ideal gas}] \qquad 24.40$$

When volume is held constant, Eq. 24.38 reduces to *Gay-Lussac's law.*

Special Cases of Closed Systems (With No Change in Kinetic or Potential Energy)

$$\frac{T}{p} = \text{constant} \quad [\text{ideal gas}] \qquad 24.41$$

Similarly, the changes in enthalpy, internal energy, and entropy are independent of the process. For perfect gases (i.e., ideal gases with constant specific heats),

$$\Delta h = c_p \Delta T \quad [\text{perfect gas}] \qquad 24.42$$

$$\Delta u = c_v \Delta T \quad [\text{perfect gas}] \qquad 24.43$$

$$\Delta s = c_p \ln\frac{T_2}{T_1} - R\ln\frac{p_2}{p_1} = c_v \ln\frac{T_2}{T_1} + R\ln\frac{v_2}{v_1} \quad [\text{perfect gas}] \qquad 24.44$$

Enthalpy and internal energy are *point functions*, so Eq. 24.42 and Eq. 24.43 are valid for all processes, even though the specific heats at constant pressure and volume are part of the calculation. These equations should not be confused with the heat transfer relations, which have a similar form. Heat, Q, is a path function and depends on the path taken.

$$q = c_p \Delta T\big|_p \quad [\text{perfect gas}] \qquad 24.45$$

$$q = c_v \Delta T\big|_v \quad [\text{perfect gas}] \qquad 24.46$$

Relationships between the properties follow for ideal gases experiencing specific processes. (The standard thermodynamic sign convention is automatically followed.) All relations can be written in slightly different forms, including per-unit mass and per-mole bases. For compactness, only the per-unit mass equations are listed. However, all equations can be converted to a

[6]It is impractical to list every relationship for property changes. Most formulas can be written in several forms. For example, all equations can be written either for a unit mass (i.e., per lbm or kg) or for a mole (i.e., per lbmol or kmol). For example, Eq. 24.36 on a molar basis would be written as $dU_m = T\,dS_m - p\,dV_m$.

molar basis by substituting H_m for h, V_m for v, $\bar{R}$ for R, $C_{p,m}$ for c_p, and so on. Other forms can be derived by substituting the equation of state where appropriate.[7]

Constant Pressure, Closed Systems (Charles's Law)

Special Cases of Closed Systems (With No Change in Kinetic or Potential Energy)

$$w = p\Delta v \tag{24.47}$$

$$w = R\Delta T \tag{24.48}$$

$$p_2 = p_1 \tag{24.49}$$

$$\frac{T}{v} = \text{constant} \quad \text{[ideal gas]} \tag{24.50}$$

$$\frac{T_1}{v_1} = \frac{T_2}{v_2} \tag{24.51}$$

$$v_2 = v_1\left(\frac{T_2}{T_1}\right) \tag{24.52}$$

$$q = h_2 - h_1 \tag{24.53}$$

$$q = c_p(T_2 - T_1) \tag{24.54}$$

$$q = c_v(T_2 - T_1) + p(v_2 - v_1) \tag{24.55}$$

$$u_2 - u_1 = c_v(T_2 - T_1) \tag{24.56}$$

$$u_2 - u_1 = \frac{c_v p(v_2 - v_1)}{R} \tag{24.57}$$

$$u_2 - u_1 = \frac{p(v_2 - v_1)}{k - 1} \tag{24.58}$$

$$s_2 - s_1 = c_p \ln \frac{T_2}{T_1} \tag{24.59}$$

$$s_2 - s_1 = c_p \ln \frac{v_2}{v_1} \tag{24.60}$$

$$h_2 - h_1 = q \tag{24.61}$$

$$h_2 - h_1 = c_p(T_2 - T_1) \tag{24.62}$$

$$h_2 - h_1 = \frac{kp(v_2 - v_1)}{k - 1} \tag{24.63}$$

Constant Volume, Closed Systems

Special Cases of Closed Systems (With No Change in Kinetic or Potential Energy)

$$w = 0 \tag{24.64}$$

$$\frac{T}{p} = \text{constant} \quad \text{[ideal gas]} \tag{24.65}$$

$$\frac{T_1}{p_1} = \frac{T_2}{p_2} \tag{24.66}$$

$$p_2 = p_1\left(\frac{T_2}{T_1}\right) \tag{24.67}$$

$$T_2 = T_1\left(\frac{p_2}{p_1}\right) \tag{24.68}$$

$$v_2 = v_1 \tag{24.69}$$

$$q = u_2 - u_1 \tag{24.70}$$

$$q = c_v(T_2 - T_1) \tag{24.71}$$

$$u_2 - u_1 = q \tag{24.72}$$

$$u_2 - u_1 = c_v(T_2 - T_1) \tag{24.73}$$

$$u_2 - u_1 = \frac{c_v v(p_2 - p_1)}{R} \tag{24.74}$$

$$u_2 - u_1 = \frac{v(p_2 - p_1)}{k - 1} \tag{24.75}$$

$$s_2 - s_1 = c_v \ln \frac{T_2}{T_1} \tag{24.76}$$

$$s_2 - s_1 = c_v \ln \frac{p_2}{p_1} \tag{24.77}$$

$$h_2 - h_1 = c_p(T_2 - T_1) \tag{24.78}$$

$$h_2 - h_1 = \frac{kv(p_2 - p_1)}{k - 1} \tag{24.79}$$

Constant Temperature, Closed Systems (Boyle's Law)

$$T_2 = T_1 \tag{24.80}$$

Special Cases of Closed Systems (With No Change in Kinetic or Potential Energy)

$$w = RT \ln \frac{v_2}{v_1} \tag{24.81}$$

$$w = RT \ln \frac{p_1}{p_2} \tag{24.82}$$

$$w = p_1 v_1 \ln \frac{v_2}{v_1} \tag{24.83}$$

[7]For example, RT can be substituted anywhere pv appears.

$$w = p_1 v_1 \ln \frac{p_1}{p_2} \qquad 24.84$$

$$pv = \text{constant} \quad \text{[ideal gas]} \qquad 24.85$$

$$p_1 v_1 = p_2 v_2 \qquad 24.86$$

$$p_2 = p_1 \left(\frac{v_1}{v_2} \right) \qquad 24.87$$

$$v_2 = v_1 \left(\frac{p_1}{p_2} \right) \qquad 24.88$$

$$q = w \qquad 24.89$$

$$q = T(s_2 - s_1) \qquad 24.90$$

$$q = p_1 v_1 \ln \frac{v_2}{v_1} \qquad 24.91$$

$$q = p_1 v_1 \ln \frac{p_1}{p_2} \qquad 24.92$$

$$q = RT \ln \frac{v_2}{v_1} \qquad 24.93$$

$$q = RT \ln \frac{p_1}{p_2} \qquad 24.94$$

$$u_2 - u_1 = 0 \qquad 24.95$$

$$s_2 - s_1 = \frac{q}{T} \qquad 24.96$$

$$s_2 - s_1 = R \ln \frac{v_2}{v_1} \qquad 24.97$$

$$s_2 - s_1 = R \ln \frac{p_1}{p_2} \qquad 24.98$$

$$h_2 - h_1 = 0 \qquad 24.99$$

Isentropic, Closed Systems (Reversible Adiabatic)

Special Cases of Closed Systems (With No Change in Kinetic or Potential Energy)

$$pv^k = \text{constant} \quad \text{[ideal gas]} \qquad 24.100$$

$$p_1 v_1^k = p_2 v_2^k \qquad 24.101$$

$$p_2 = p_1 \left(\frac{v_1}{v_2} \right)^k \qquad 24.102$$

$$v_2 = v_1 \left(\frac{p_1}{p_2} \right)^{1/k} \qquad 24.103$$

$$w = \frac{p_2 v_2 - p_1 v_1}{1 - k} \qquad 24.104$$

$$w = \frac{p_1 v_1}{k - 1} \left(1 - \left(\frac{p_2}{p_1} \right)^{(k-1)/k} \right) \qquad 24.105$$

$$w = \frac{R(T_2 - T_1)}{1 - k} \qquad 24.106$$

$$w = (c_p - R)(T_1 - T_2) = c_v(T_1 - T_2) \qquad 24.107$$

$$p_2 = p_1 \left(\frac{T_2}{T_1} \right)^{k/(k-1)} \qquad 24.108$$

$$T_2 = T_1 \left(\frac{v_1}{v_2} \right)^{k-1} \qquad 24.109$$

$$T_2 = T_1 \left(\frac{p_2}{p_1} \right)^{(k-1)/k} \qquad 24.110$$

$$v_2 = v_1 \left(\frac{T_1}{T_2} \right)^{1/(k-1)} \qquad 24.111$$

$$q = 0 \qquad 24.112$$

$$u_2 - u_1 = -w \qquad 24.113$$

$$u_2 - u_1 = c_v(T_2 - T_1) \qquad 24.114$$

$$u_2 - u_1 = \frac{c_v(p_2 v_2 - p_1 v_1)}{R} \qquad 24.115$$

$$u_2 - u_1 = \frac{p_2 v_2 - p_1 v_1}{k - 1} \qquad 24.116$$

$$w = u_1 - u_2 \qquad 24.117$$

$$s_2 - s_1 = 0 \qquad 24.118$$

$$h_2 - h_1 = c_p(T_2 - T_1) \qquad 24.119$$

$$h_2 - h_1 = \frac{k(p_2 v_2 - p_1 v_1)}{k - 1} \qquad 24.120$$

Polytropic, Closed Systems

For $n = 1$, use constant temperature equations. For $n = k$, use adiabatic equations. For $n = 0$, use constant pressure equations. For $n = \infty$, use constant volume equations.

Special Cases of Closed Systems (With No Change in Kinetic or Potential Energy)

$$pv^n = \text{constant} \quad \text{[ideal gas]} \qquad 24.121$$

Thermodynamics

$$p_1 v_1^n = p_2 v_2^n \quad 24.122$$

$$p_2 = p_1\left(\frac{v_1}{v_2}\right)^n \quad 24.123$$

$$v_2 = v_1\left(\frac{p_1}{p_2}\right)^{1/n} \quad 24.124$$

$$w = \frac{p_2 v_2 - p_1 v_1}{1-n} \quad [n \neq 1] \quad 24.125$$

$$w = \left(\frac{p_1 v_1}{n-1}\right)\left(1-\left(\frac{p_2}{p_1}\right)^{(n-1)/n}\right) \quad 24.126$$

$$w = \frac{R(T_1 - T_2)}{n-1} \quad 24.127$$

$$p_2 = p_1\left(\frac{T_2}{T_1}\right)^{n/(n-1)} \quad 24.128$$

$$T_2 = T_1\left(\frac{v_1}{v_2}\right)^{n-1} \quad 24.129$$

$$T_2 = T_1\left(\frac{p_2}{p_1}\right)^{(n-1)/n} \quad 24.130$$

$$v_2 = v_1\left(\frac{T_1}{T_2}\right)^{1/(n-1)} \quad 24.131$$

$$q = \frac{c_v(n-k)(T_2 - T_1)}{n-1} \quad 24.132$$

$$u_2 - u_1 = c_v(T_2 - T_1) \quad 24.133$$

$$u_2 - u_1 = q - w \quad 24.134$$

$$s_2 - s_1 = \frac{c_v(n-k)}{n-1}\ln\frac{T_2}{T_1} \quad 24.135$$

$$h_2 - h_1 = c_p(T_2 - T_1) \quad 24.136$$

$$h_2 - h_1 = \frac{n(p_2 v_2 - p_1 v_1)}{n-1} \quad 24.137$$

Constant Volume, Open Systems

Special Cases of Open Systems (With No Change in Kinetic or Potential Energy)

$$w = -v(p_2 - p_1) \quad 24.138$$

Constant Pressure, Open Systems

Special Cases of Open Systems (With No Change in Kinetic or Potential Energy)

$$w = 0 \quad 24.139$$

Constant Temperature, Open Systems

Equations for w are the same as for a closed system process. pv is constant, the same as for a closed system process.

Isentropic, Steady-Flow Systems

p_2, v_2, and T_2 are the same as for isentropic, closed systems.

Special Cases of Open Systems (With No Change in Kinetic or Potential Energy)

$$w = k\left(\frac{p_2 v_2 - p_1 v_1}{1-k}\right) = kR\left(\frac{T_2 - T_1}{1-k}\right) \quad 24.140$$

$$w = \left(\frac{k}{k-1}\right)RT_1\left(1-\left(\frac{p_2}{p_1}\right)^{(k-1)/k}\right) \quad 24.141$$

$$w = \left(\frac{k}{k-1}\right)p_1 v_1\left(1-\left(\frac{p_2}{p_1}\right)^{(k-1)/k}\right) \quad 24.142$$

$$w = c_p T_1\left(1-\left(\frac{p_2}{p_1}\right)^{(k-1)/k}\right) \quad 24.143$$

$$w = h_1 - h_2 \quad 24.144$$

$$q = 0 \quad 24.145$$

$$u_2 - u_1 = c_v(T_2 - T_1) \quad 24.146$$

$$h_2 - h_1 = -w \quad 24.147$$

$$h_2 - h_1 = c_p(T_2 - T_1) \quad 24.148$$

$$h_2 - h_1 = \frac{k(p_2 v_2 - p_1 v_1)}{k-1} \quad 24.149$$

$$s_2 - s_1 = 0 \quad 24.150$$

Polytropic, Steady-Flow Systems

p_2, v_2, and T_2 are the same as for polytropic, closed systems. The process is not adiabatic, so w is not equal to Δh.

$$w = h_1 - h_2 - |q_{\text{loss}}| \quad 24.151$$

Thermodynamics

$$h_2 - h_1 = -w + |q_{\text{loss}}| \quad 24.152$$

Special Cases of Open Systems (With No Change in Kinetic or Potential Energy)

$$w = \frac{n(p_2 v_2 - p_1 v_1)}{1 - n} \quad 24.153$$

$$w = \frac{nR(T_1 - T_2)}{n - 1} \quad 24.154$$

$$h_2 - h_1 = \frac{n(p_2 v_2 - p_1 v_1)}{n - 1} + |q_{\text{loss}}| \quad 24.155$$

$$q = \frac{c_v(n - k)(T_2 - T_1)}{n - 1} \quad 24.156$$

$$u_2 - u_1 = c_v(T_2 - T_1) \quad 24.157$$

$$h_2 - h_1 = c_p(T_2 - T_1) \quad 24.158$$

$$s_2 - s_1 = \frac{c_v(n - k)}{n - 1} \ln \frac{T_2}{T_1} \quad 24.159$$

Throttling, Steady-Flow Systems

$$p_1 v_1 = p_2 v_2 \quad 24.160$$

$$p_2 < p_1 \quad 24.161$$

$$v_2 > v_1 \quad 24.162$$

$$T_2 = T_1 \quad \text{[ideal gas]} \quad 24.163$$

$$q = 0 \quad 24.164$$

$$w = 0 \quad 24.165$$

$$u_2 - u_1 = 0 \quad 24.166$$

$$h_2 - h_1 = 0 \quad 24.167$$

$$s_2 - s_1 = R \ln \frac{p_1}{p_2} \quad 24.168$$

$$s_2 - s_1 = R \ln \frac{v_2}{v_1} \quad 24.169$$

Example 24.3

2.0 lbm (0.9 kg) of hydrogen are cooled from 760°F to 660°F (400°C to 350°C) in a constant volume process. The specific heat at constant volume, c_v, is 2.435 Btu/lbm-°R (10.2 kJ/kg·K). How much heat is removed?

SI Solution

Equation 24.71 is on a per unit mass basis. The total heat transfer for a mass m is

$$\begin{aligned} Q &= mc_v(T_2 - T_1) \\ &= (0.9 \text{ kg})\left(10.2 \ \frac{\text{kJ}}{\text{kg·K}}\right)(350\text{°C} - 400\text{°C}) \\ &= -459 \text{ kJ} \end{aligned}$$

The minus sign is consistent with the convention that a heat loss is negative.

Customary U.S. Solution

The total heat transfer for a mass m is

$$\begin{aligned} Q &= mc_v(T_2 - T_1) \\ &= (2.0 \text{ lbm})\left(2.435 \ \frac{\text{Btu}}{\text{lbm-°R}}\right)(660\text{°F} - 760\text{°F}) \\ &= -487 \text{ Btu} \end{aligned}$$

Example 24.4

4 lbmol (1.8 kmol) of air initially at 1 atm and 530°R (295K) are compressed isothermally to 8 atm. How much total heat is removed during the compression?

SI Solution

Equation 24.94 applies to constant temperature compression. $\bar{R}$ is used in place of R and Q_m instead of q to convert the calculations to a molar basis.

$$\begin{aligned} Q_m &= n\bar{R}T \ln \frac{p_1}{p_2} \\ &= (1.8 \text{ kmol})\left(8.314 \ \frac{\text{kJ}}{\text{kmol·K}}\right) \times (295\text{K}) \ln \frac{1 \text{ atm}}{8 \text{ atm}} \\ &= -9180 \text{ kJ} \end{aligned}$$

Customary U.S. Solution

Writing Eq. 24.94 on a molar basis,

$$\begin{aligned} Q_m &= n\bar{R}T \ln \frac{p_1}{p_2} \\ &= (4.0 \text{ lbmol})\left(1545 \ \frac{\text{ft-lbf}}{\text{lbmol-°R}}\right) \times (530\text{°R}) \ln \frac{1 \text{ atm}}{8 \text{ atm}} \\ &= -6.81 \times 10^6 \text{ ft-lbf} \end{aligned}$$

13. PROPERTY CHANGES IN INCOMPRESSIBLE FLUIDS AND SOLIDS

In order to simplify the solution to practical problems, enthalpy, entropy, internal energy, and specific volume in liquids and solids are often considered to be functions of temperature only. The effect of pressure is disregarded.

There are times, however, when it is necessary to evaluate the property changes in a solid or liquid system, no matter how small they may be. Changes for liquids can be evaluated by using compressed liquid tables. If certain assumptions are made, some property changes can also be calculated without knowing the initial and final values. The main assumptions are incompressibility and constant specific heats.

Entropy

$$ds = c\,\frac{dT}{T} = \frac{du}{T} \quad 24.170$$

$$c_p = c_v = c \quad 24.171$$

$$dv = 0 \quad 24.172$$

$$du = c\,dT \quad 24.173$$

$$dh = c\,dT + v\,dp \quad 24.174$$

If the specific heat is constant, Eq. 24.175 through Eq. 24.178 are valid.

$$v_2 - v_1 = 0 \quad 24.175$$

$$u_2 - u_1 = c(T_2 - T_1) \quad 24.176$$

$$s_2 - s_1 = \int c\,\frac{dT}{T} = c_{\text{mean}} \ln \frac{T_2}{T_1} \quad 24.177$$

$$h_2 - h_1 = c(T_2 - T_1) + v(p_2 - p_1) \quad 24.178$$

14. HEAT RESERVOIRS

It is convenient to show a source of energy as an infinite constant-temperature *reservoir*. Figure 24.4 illustrates a source of energy (known as a *high-temperature reservoir* or *source reservoir*). By convention, the reservoir temperature is designated T_H, and the heat transfer from it is Q_H. The energy derived from such a theoretical source might actually be supplied by combustion, electrical heating, or nuclear reaction.

Figure 24.4 *Energy Reservoirs*

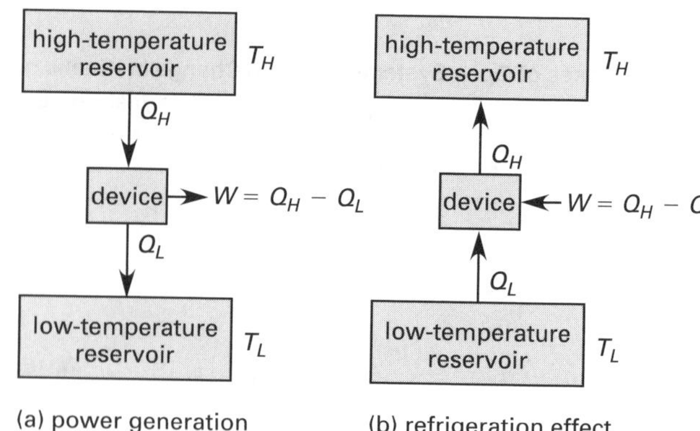

Similarly, energy is released to a *low-temperature reservoir* known as a *sink reservoir* or *energy sink*. The most common practical sink is the local environment. T_L and Q_L are used to represent the reservoir temperature and energy absorbed. It is common to refer to Q_L as the rejected energy or energy released to the environment.

15. CYCLES

Although heat can be extracted and work performed in a single process, a cycle is necessary to obtain work in a useful quantity and duration. A *cycle* is a series of processes that eventually brings the system back to its original condition. Most cycles are continually repeated.

A cycle is completely defined by the working substance, the high- and low-temperature reservoirs, the means of doing work on the system, and the means of removing energy from the system.[8] (See Fig. 24.5.)

A cycle will appear as a closed curve when plotted on p-V and T-s diagrams. The area within the p-V or T-s curve represents the net work or net heat, respectively.

Figure 24.5 *Net Work and Net Heat*

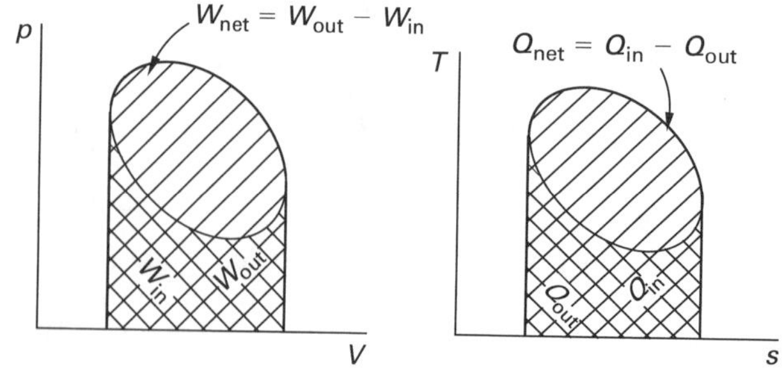

[8]The *Carnot cycle* depends only on the source and sink temperatures, not on the working fluid. However, most practical cycles depend on the working fluid.

Thermodynamics

16. THERMAL EFFICIENCY

The *thermal efficiency* of a *power cycle* is defined as the ratio of useful work output to the supplied input energy.[9] W in Eq. 24.179 is the *net work*, since some of the gross output work may be used to run certain parts of the cycle. For example, a small amount of turbine output power may run boiler feed pumps.

$$\begin{aligned}\eta_{\text{th}} &= \frac{\text{net work output}}{\text{energy input}} = \frac{W_{\text{net}}}{Q_{\text{in}}} \\ &= \frac{W_{\text{out}} - W_{\text{in}}}{Q_{\text{in}}}\end{aligned} \qquad 24.179$$

The first law can be written as

$$Q_{\text{in}} = Q_{\text{out}} + W_{\text{net}} \qquad 24.180$$

Equation 24.179 and Eq. 24.180 can be combined to define the thermal efficiency in terms of heat variables alone.

$$\eta_{\text{th}} = \frac{Q_{\text{in}} - Q_{\text{out}}}{Q_{\text{in}}} \qquad 24.181$$

Equation 24.181 shows that obtaining the maximum efficiency requires minimizing the Q_{out} term. The most efficient power cycle possible is the *Carnot cycle*.

17. SECOND LAW OF THERMODYNAMICS

The *second law of thermodynamics* can be stated in several ways. Equation 24.182 is the mathematical relation defining the second law. The equality holds for reversible processes; the inequality holds for irreversible processes.

$$\Delta s \geq \int_{T_1}^{T_2} \frac{dq}{T} \qquad 24.182$$

Equation 24.182 effectively states that net entropy must always increase in practical (irreversible) cyclical processes. Another way to state the second law mathematically is

Second Law of Thermodynamics

$$\Delta S_{\text{reservoir}} \geq \frac{Q}{T_{\text{reservoir}}} \qquad 24.183$$

A natural process that starts in one equilibrium state and ends in another will go in the direction that causes the entropy of the system and the environment to increase.

The *Kelvin-Planck statement* of the second law effectively says that it is impossible to build a cyclical engine that will have a thermal efficiency of 100%.

The *Clausius statement* of the second law effectively says that it is impossible for an engine operating in a cycle (i.e. refrigeration or heat pump) to have no other effect than to extract heat from a reservoir and turn it into an equivalent amount of work.

This formulation is not a contradiction of the first law of thermodynamics. The first law does not preclude the possibility of converting heat entirely into work—it only denies the possibility of creating or destroying energy. The second law says that if some heat is converted entirely into work, some other energy must be rejected to a low-temperature sink (i.e., lost to the surroundings).

Figure 24.6 illustrates violations of the second law.

Figure 24.6 *Second Law Violations*

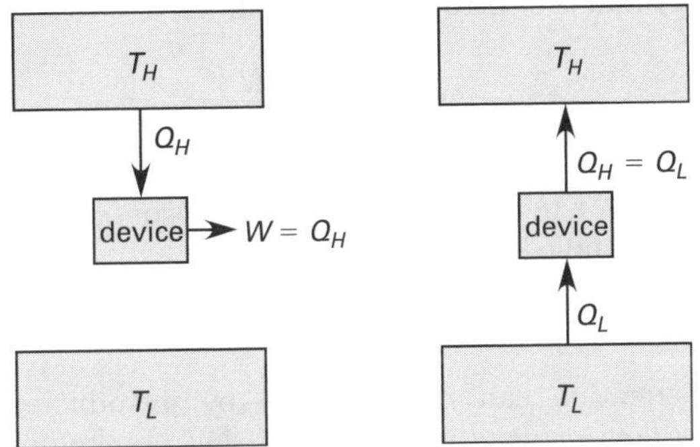

18. AVAILABILITY

The maximum possible work that can be obtained from a cycle is known as the *availability*. Availability is independent of the device but is dependent on the temperature of the local environment. Both the first and second law must be applied to determine availability.

In Fig. 24.7, the *heat engine* is surrounded by an environment at absolute temperature T_L. The net heat available for conversion to useful work is $Q = Q_H - Q_L$. Work, W, is performed at a constant rate on the environment by the engine. This is the traditional development of the availability concept. Actually, the symbols $\dot{Q}$, $\dot{m}$, and P (power) should be used to represent the time rate of heat, mass, and work. Assuming steady flow, and neglecting kinetic and potential energies, the first law can be written as

$$mh_1 + Q = mh_2 + W \qquad 24.184$$

[9]The effectiveness of refrigeration and compression cycles is measured by other parameters.

Figure 24.7 Typical Heat Engine

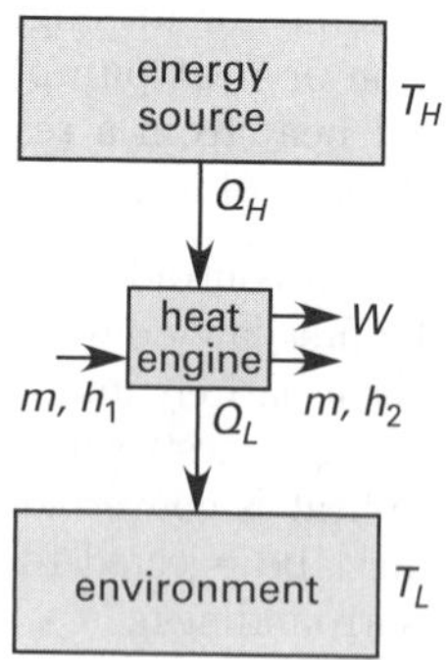

Entropy is not a part of the first law. The leaving entropy, however, can be calculated from the entering entropy and the entropy production.

$$ms_2 = ms_1 + \frac{Q}{T_L} \tag{24.185}$$

Since Eq. 24.184 and Eq. 24.185 both contain Q, they can be combined.

$$W = m(h_1 - T_L s_1 - h_2 + T_L s_2) \tag{24.186}$$

This equation can be simplified by introducing the *steady-flow availability function*, Φ. The maximum work output (availability) is

$$W_{\text{max}} = m(\Phi_1 - \Phi_2) \tag{24.187}$$

$$\Phi = h - T_L s \tag{24.188}$$

If the equality in Eq. 24.187 holds, both the process within the control volume and the energy transfers between the system and the environment must be reversible. Maximum work output, therefore, will be obtained in a reversible process. The difference between the maximum and the actual work output is known as the *process irreversibility*.

Example 24.5

What is the maximum useful work that can be produced per pound (per kilogram) of steam that enters a steady-flow system at 800 psia (5.0 MPa) saturated and leaves in equilibrium with the environment at 70°F and 14.7 psia (20°C and 0.1 MPa)?

SI Solution

From the saturated steam table, for saturated steam at 5.0 MPa, h_1 = 2794.2 kJ/kg, and s_1 = 5.9737 kJ/kg·K. [**Properties of Saturated Water and Steam (Temperature) - SI Units**]

The final properties are obtained from the saturated steam table for water at 20°C. h_2 = 83.91 kJ/kg, and s_2 = 0.2965 kJ/kg·K.

The availability is calculated from Eq. 24.186 using T_L = 20°C + 273° = 293K.

$$\begin{aligned} W_{\text{max}} &= m\big(h_1 - h_2 - T_L(s_1 - s_2)\big) \\ &= (1\ \text{kg})\left(\begin{array}{l} 2794.2\ \frac{\text{kJ}}{\text{kg}} - 83.91\ \frac{\text{kJ}}{\text{kg}} \\ -(293\text{K})\left(\begin{array}{l} 5.9737\ \frac{\text{kJ}}{\text{kg·K}} \\ -0.2965\ \frac{\text{kJ}}{\text{kg·K}} \end{array}\right) \end{array}\right) \\ &= 1047\ \text{kJ} \end{aligned}$$

Customary U.S. Solution

From the saturated steam table, h_1 = 1199.3 Btu/lbm, and s_1 = 1.4162 Btu/lbm-°R. [**Properties of Saturated Water and Steam (Temperature) - I-P Units**]

The final properties are obtained from the saturated steam table for water at 70°F. h_2 = 38.08 Btu/lbm, and s_2 = 0.07459 Btu/lbm-°R.

The availability is calculated from Eq. 24.186 using T_L = 70°F + 460° = 530°R.

$$\begin{aligned} W_{\text{max}} &= m\big(h_1 - h_2 - T_L(s_1 - s_2)\big) \\ &= (1\ \text{lbm})\left(\begin{array}{l} 1199.3\ \frac{\text{Btu}}{\text{lbm}} - 38.08\ \frac{\text{Btu}}{\text{lbm}} \\ -(530°\text{R})\left(\begin{array}{l} 1.4162\ \frac{\text{Btu}}{\text{lbm-°R}} \\ -0.07459\ \frac{\text{Btu}}{\text{lbm-°R}} \end{array}\right) \end{array}\right) \\ &= 450.2\ \text{Btu} \end{aligned}$$

19. NOMENCLATURE

c	specific heat capacity	Btu/lbm-°F	kJ/kg·K
g	acceleration of gravity, 32.17 (9.807)	ft/sec²	m/s²
g_c	gravitational constant, 32.17	lbm-ft/lbf-sec²	n.a.
h	specific enthalpy	Btu/lbm	kJ/kg
H	enthalpy	Btu	kJ
J	Joule's constant, 778.2	ft-lbf/Btu	n.a.
k	ratio of specific heats	–	–
KE	kinetic energy	Btu	kJ
$\dot{m}$	mass flow rate	lbm/sec	kg/s
m	mass	lbm	kg
n	number of moles	–	–
n	polytropic exponent	–	–
p	pressure	lbf/ft²	kPa

PE	potential energy	Btu	J
P	power	Btu/sec	kW
q	heat per unit mass	Btu/lbm	kJ/kg
q	heat rate	Btu/sec	kW
Q	heat	Btu	kJ
$\dot{Q}$	rate of heat transfer	Btu/hr	W
R	specific gas constant	ft-lbf/lbm-°R	kJ/kg·K
$\overline{R}$	universal gas constant	ft-lbf/lbmol-°R	kJ/kmol·K
s	specific entropy	Btu/lbm-°F	kJ/kg·K
S	entropy	Btu/°F	kJ/kmol·K
$\dot{S}$	rate of entropy	Btu/sec-°F	kJ/s·K
T	temperature	°F	°C
T	absolute temperature	°R	K
u	specific internal energy	Btu/lbm	kJ/kg
U	internal energy	Btu	kJ
v	specific volume	ft^3/lbm	m^3/kg
v	velocity	ft^3/sec	m/s
V	volume	ft^3	m^3
w	work per unit mass	Btu/lbm	kJ/kg
W	work	Btu	kJ
z	elevation	ft	m

Symbols

β	isobaric compressibility	1/°R	1/K
η	efficiency	–	–
μ_J	Joule-Thomson coefficient	°F-ft^2/lbf	K/Pa
Φ	availability function	Btu/lbm	kJ/kg

Subscripts

al	aluminum
eq	equilibrium
H	constant enthalpy or hot (high-temperature)
L	cold (low temperature)
n	polytropic
p	constant pressure
st	steel
th	thermal
v	constant volume
w	water

25 Compressible Fluid Dynamics

Content in blue refers to the *NCEES Handbook*.

NCEES EXAM SPECIFICATIONS AND RELATED CONTENT

THERMAL AND FLUID SYSTEMS EXAM

I.B.2. Fluid Mechanics: Compressible Flow
- 2. Steady-Flow Energy Equation
- 4. Isentropic Flow Factors
- 11. Nozzle Performance Characteristics
- 12. Shock Waves
- 14. Compressible Flow Through Orifices

1. INTRODUCTION

A *high-velocity gas* is defined as a gas moving with a velocity in excess of approximately 300 ft/sec (100 m/s). A high gas velocity is often achieved at the expense of internal energy. A drop in internal energy, u, is seen as a drop in enthalpy, h, since $h = u + pv$. Since the Bernoulli equation does not account for this conversion, it cannot be used to predict the thermodynamic properties of the gas. Furthermore, density changes and shock waves complicate the use of traditional evaluation tools such as energy and momentum conservation equations.

2. STEADY-FLOW ENERGY EQUATION

Consider a gas or vapor flowing from a high-pressure reservoir through a duct, as shown in Fig. 25.1. The properties in the source reservoir, designated by the subscript zero (0), are known alternatively as the *total properties*, *stagnation properties*, or *chamber properties*.

Figure 25.1 *High-Velocity Flow Locations*

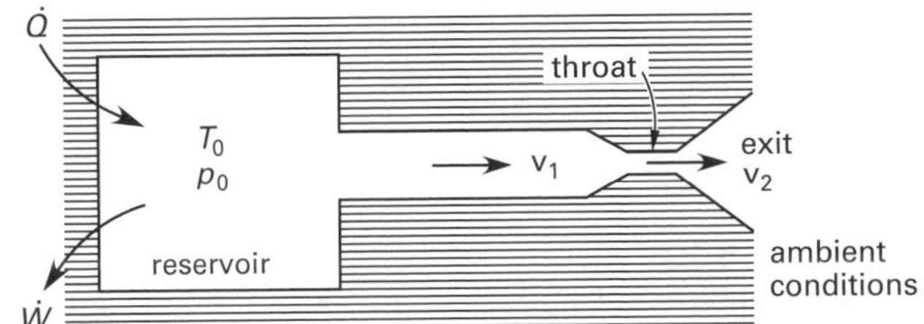

The flow can be assumed to be one dimensional as long as the flow cross section varies slowly along the path length. The gas possesses energy in static, kinetic, and internal (thermal) forms.

Steady-Flow Systems

$$\sum \dot{m}_i\left(h_i + \frac{\mathrm{v}_i^2}{2} + gz_i\right) - \sum \dot{m}_e\left(h_e + \frac{\mathrm{v}_e^2}{2} + gz_e\right) + \dot{Q}_{\text{in}} - \dot{W}_{\text{out}} = 0 \quad \text{25.1}$$

Since the flow is fast, there is no time for significant heat transfer to occur, so an assumption of adiabatic flow is appropriate. If the duct run is short, there will be little or no friction. Equation 25.2 is an energy balance of an adiabatic open-flow system for one unit mass of the gas. This is the *steady-flow energy equation*, SFEE, for an adiabatic system.[1]

$$h_i + \frac{\mathrm{v}_i^2}{2} + gz_i = h_e + \frac{\mathrm{v}_e^2}{2} + gz_e \quad \text{[SI]} \quad \text{25.2(a)}$$

$$Jh_i + \frac{\mathrm{v}_i^2}{2g_c} + z_i\left(\frac{g}{g_c}\right) = Jh_e + \frac{\mathrm{v}_e^2}{2g_c} + z_e\left(\frac{g}{g_c}\right) \quad \text{[U.S.]} \quad \text{25.2(b)}$$

[1]The *steady-flow energy equation*, SFEE, is also known as the *general flow equation*, GFE.

It is often necessary to write the velocity at a point in terms of the total properties. The velocity is zero in the source reservoir. Since the gas density is small, the potential energy terms are always disregarded. Equation 25.3 can be used to calculate the discharge velocity for high-pressure steam and other substances for which the ideal gas assumption would be inappropriate. The *theoretical maximum velocity* is achieved when all internal and pressure energies are converted to kinetic energy (i.e., when $h_e = 0$). This is never achieved in practice, however.

$$h_0 = h_e + \frac{v_e^2}{2} \quad \text{[SI]} \qquad 25.3(a)$$

$$Jh_0 = Jh_e + \frac{v_e^2}{2g_c} \quad \text{[U.S.]} \qquad 25.3(b)$$

Since h_0 is constant anywhere in the flow,

$$v_e = \sqrt{2(h_0 - h_e)} \quad \text{[SI]} \qquad 25.4(a)$$

$$v_e = \sqrt{2g_cJ(h_0 - h_e)} = 223.8\sqrt{h_0 - h_e} \quad \text{[U.S.]} \qquad 25.4(b)$$

Example 25.1

Steam at 200 psia (1.50 MPa) and 500°F (250°C) enters an insulated nozzle with negligible velocity and is expanded to 20 psia (0.15 MPa) and 98% quality. Find the steam's exit velocity.

SI Solution

Find the inlet enthalpy from superheated steam tables. **[Properties of Superheated Steam - SI Units]**

$$h_0 = 2923.9 \text{ kJ/kg}$$

Find the exit enthalpy at 0.15 MPa from steam tables. **[Properties of Saturated Water and Steam (Pressure) - SI Units]**

$$h_f = 467.1 \text{ kJ/kg}$$
$$h_{fg} = 2226.0 \text{ kJ/kg}$$

Properties for Two-Phase (Vapor-Liquid) Systems

$$\begin{aligned} h_e &= h_f + xh_{fg} \\ &= 467.1 \ \frac{\text{kJ}}{\text{kg}} + (0.98)\left(2226.0 \ \frac{\text{kJ}}{\text{kg}}\right) \\ &= 2648.6 \text{ kJ/kg} \end{aligned}$$

From Eq. 25.4(a), the exit velocity is

$$\begin{aligned} v_e &= \sqrt{2(h_0 - h_e)} \\ &= \sqrt{\begin{array}{l}(2)\left(2923.9 \ \frac{\text{kJ}}{\text{kg}} - 2648.6 \ \frac{\text{kJ}}{\text{kg}}\right) \\ \quad \times \left(1000 \ \frac{\text{J}}{\text{kJ}}\right)\end{array}} \\ &= 742.0 \text{ m/s} \end{aligned}$$

Customary U.S. Solution

Find the inlet enthalpy from superheated steam tables. **[Properties of Superheated Steam - I-P Units]**

$$h_0 = 1269.0 \text{ Btu/lbm}$$

Find the exit enthalpy at 20 psia from steam tables. **[Properties of Saturated Water and Steam (Pressure) — I-P Units]**

$$h_f = 196.25 \text{ Btu/lbm}$$
$$h_{fg} = 959.94 \text{ Btu/lbm}$$

Properties for Two-Phase (Vapor-Liquid) Systems

$$\begin{aligned} h_e &= h_f + xh_{fg} \\ &= 196.25 \ \frac{\text{Btu}}{\text{lbm}} + (0.98)\left(959.94 \ \frac{\text{Btu}}{\text{lbm}}\right) \\ &= 1137.0 \text{ Btu/lbm} \end{aligned}$$

From Eq. 25.4(b), the exit velocity is

$$\begin{aligned} v_e &= \sqrt{2g_cJ(h_0 - h_e)} \\ &= \sqrt{\begin{array}{l}(2)\left(32.2 \ \frac{\text{lbm-ft}}{\text{lbf-sec}^2}\right)\left(778 \ \frac{\text{ft-lbf}}{\text{Btu}}\right) \\ \quad \times \left(1269.0 \ \frac{\text{Btu}}{\text{lbm}} - 1137.0 \ \frac{\text{Btu}}{\text{lbm}}\right)\end{array}} \\ &= 2572 \text{ ft/sec} \end{aligned}$$

3. ISENTROPIC FLOW

If the gas flow is adiabatic and frictionless (that is, reversible), the entropy change is zero and the flow is known as *isentropic flow*. As a practical matter, completely isentropic flow does not exist. However, some high-velocity, steady-state flow processes proceed with little increase in entropy and are considered to be isentropic. The irreversible effects are accounted for by various correction factors, such as nozzle and discharge coefficients.

4. ISENTROPIC FLOW FACTORS

In isentropic flow, total pressure, total temperature, and total density remain constant, regardless of the flow area and velocity. The instantaneous properties, known as static properties, however, do change along the flow path.[2] Equation 25.6 through Eq. 25.8 predict these static properties as functions of the Mach number, Ma, for ideal gas flow.[3]

Mach Number

$$\mathrm{Ma} = \frac{\mathrm{v}}{c} = \frac{\mathrm{v}}{\sqrt{kRT}} = \frac{\mathrm{v}}{\sqrt{\frac{k\overline{R}T}{M}}} \quad \text{[SI]} \quad 25.5(a)$$

$$\mathrm{Ma} = \frac{\mathrm{v}}{c} = \frac{\mathrm{v}}{\sqrt{kg_cRT}} = \frac{\mathrm{v}}{\sqrt{\frac{kg_c\overline{R}T}{M}}} \quad \text{[U.S.]} \quad 25.5(b)$$

Isentropic Flow Relationships

$$\frac{T_0}{T} = 1 + \left(\frac{k-1}{2}\right)\mathrm{Ma}^2 \quad 25.6$$

$$\frac{p_0}{p} = \left(\frac{T_0}{T}\right)^{k/(k-1)} = \left(1 + \left(\frac{k-1}{2}\right)\mathrm{Ma}^2\right)^{k/(k-1)} \quad 25.7$$

$$\frac{\rho_0}{\rho} = \left(\frac{T_0}{T}\right)^{1/(k-1)} = \left(1 + \left(\frac{k-1}{2}\right)\mathrm{Ma}^2\right)^{1/(k-1)} \quad 25.8$$

The isentropic flow ratios given by Eq. 25.6 through Eq. 25.8 are functions only of the Mach number, Ma, and ratio of specific heats, k. Therefore, the ratios can be easily tabulated. The numbers in such tables are known as *isentropic flow factors*. **[One-Dimensional Isentropic Compressible-Flow Functions]**

Example 25.2

Air flows isentropically from a large tank at 530°R (294K) through a convergent-divergent nozzle and is expanded to supersonic velocities. The air has a ratio of specific heats of 1.4 and a molecular weight of 29.0 lbm/lbmol (29.0 kg/kmol). At a point where the Mach number is 2.5, what are the (a) gas temperature and (b) actual velocity?

SI Solution

The temperature isentropic flow factor T_0/T for a Mach number of 2.5 can be calculated from Eq. 25.6.

Isentropic Flow Relationships

$$\frac{T_0}{T} = 1 + \left(\frac{k-1}{2}\right)\mathrm{Ma}^2 = 1 + \left(\frac{1.4-1}{2}\right)(2.5)^2 = 2.25$$

Alternatively, the factor (or its inverse, $T/T_0 = 0.4444$) can be found in a table of isentropic flow factors using the Mach number of 2.5. **[One-Dimensional Isentropic Compressible-Flow Functions]**

(a) The gas temperature is

$$T = \frac{T_0}{\frac{T_0}{T}} = \frac{294\text{K}}{2.25} = 130.7\text{K}$$

(b) The Mach number is 2.5, so the gas velocity is 2.5 times the speed of sound. This speed of sound is calculated from the static temperature, not from the total temperature. The universal gas constant is 8314 J/kmol·K. **[Fundamental Constants]**

Mach Number

$$\mathrm{Ma} = \frac{\mathrm{v}}{c}$$

$$\mathrm{v} = (\mathrm{Ma})c = \mathrm{Ma}\sqrt{\frac{k\overline{R}T}{M}} = (2.5)\sqrt{\frac{(1.4)\left(8314\ \frac{\text{J}}{\text{kmol·K}}\right)(130.7\text{K})}{29.0\ \frac{\text{kg}}{\text{kmol}}}} = 572.6\ \text{m/s}$$

Customary U.S. Solution

(a) The gas temperature is

$$T = \frac{T_0}{\frac{T_0}{T}} = \frac{530°\text{R}}{2.25} = 235.6°\text{R}$$

Thermodynamics

[2]The *static properties* are not the same as the *stagnation properties*.
[3]For example, Eq. 25.6 can be derived from Eq. 25.2 by dividing both sides by the speed of sound.

(b) The Mach number is 2.5, so the gas velocity is 2.5 times the speed of sound. This speed of sound is calculated from the static temperature, not from the total temperature. The universal gas constant is 1545 ft-lbf/lbmol-°R. **[Fundamental Constants]**

Mach Number

$$\text{Ma} = \frac{\text{v}}{c}$$

$$\begin{aligned} \text{v} &= (\text{Ma})c = \text{Ma}\sqrt{\frac{kg_c\overline{R}T}{M}} \\ &= (2.5)\sqrt{\frac{(1.4)\left(32.2\ \frac{\text{lbm-ft}}{\text{lbf-sec}^2}\right)\left(1545\ \frac{\text{ft-lbf}}{\text{lbmol-°R}}\right)(235.6\text{°R})}{29.0\ \frac{\text{lbm}}{\text{lbmol}}}} \\ &= 1881\ \text{ft/sec} \end{aligned}$$

5. RATIO OF SPECIFIC HEATS

The ratio of specific heats, $k = c_p/c_v$, is remarkably similar for gases with similar structures. For monatomic gases (He, Ar, Ne, Kr, etc.), it is approximately 1.67. For diatomic gases (N_2, O_2, H_2, CO, NO, and air), it is approximately 1.4. For triatomic gases (H_2O and CO_2), it is approximately 1.3. The value of k is less than 1.3 for more complex gases.

In most problems, the ratio of specific heats is assumed to be constant over all temperatures and pressures encountered. However, when the gas experiences a large temperature change, the ratio of specific heats should be evaluated at the average temperature.

Rather than using mathematical relationships to calculate exact solutions, it is common to use factors from tables appropriate for that ratio of specific heat. Since tables are typically available only for k = 1.0, 1.1, 1.2, 1.3, 1.4, and 1.67, a solution will not be exact when the ratio of specific heats is some intermediate value.

6. CRITICAL CONSTANTS

The location where sonic velocity has been achieved (i.e., Ma = 1) is known as a *critical point*. Sonic properties are designated by an asterisk, such as p^* for critical pressure. The isentropic flow factors at that point are known as *critical ratios* or *critical constants*. For example, the *critical pressure ratio* can be calculated from Eq. 25.9.

$$R_{\text{cp}} = \frac{p^*}{p_0} = \left(\frac{2}{k+1}\right)^{k/(k-1)} \qquad 25.9$$

The *critical temperature ratio* and *critical density ratio* are also easily derived.

$$\frac{T^*}{T_0} = \frac{2}{k+1} \qquad 25.10$$

$$\frac{\rho^*}{\rho_0} = \left(\frac{2}{k+1}\right)^{1/(k-1)} \qquad 25.11$$

Example 25.3

What is the critical pressure ratio for air ($k = 1.4$)?

Solution

From Eq. 25.9,

$$R_{\text{cp}} = \left(\frac{2}{k+1}\right)^{k/(k-1)} = \left(\frac{2}{1.4+1}\right)^{1.4/(1.4-1)} = 0.5283$$

7. CHOKED FLOW

Equation 25.7 seems to indicate that the gas velocity (and hence, the mass flow rate) in a discharge duct can be increased by increasing the ratio of source reservoir to ambient pressures (or by decreasing the ratio of ambient to source reservoir pressures). This is a logical conclusion, but it is valid only up to a certain gas velocity in the duct—the *sonic velocity* (i.e., speed of sound).

Changes in ambient pressure required to change the gas velocity in the duct travel upstream at sonic velocity. Therefore, if the gas in the duct is already traveling at sonic velocity, the fact that the ambient pressure has been changed cannot be transmitted back to the source reservoir. Sonic velocity will occur when the ratio of ambient to source pressures drops to the *critical pressure ratio* (0.5283 for air). Once sonic velocity has been achieved in a duct or nozzle throat, the mass flow rate will be at its maximum and will remain constant for all subsequent decreases in ambient pressure.[4] This condition is called *choked flow*.

8. CHANGING DUCT AREA

Equation 25.12 defines the relationship between the change in flow area and the initial gas velocity for isentropic flow.

$$\frac{dA}{A} = \frac{dp}{\rho \text{v}^2}(1 - \text{Ma}^2) \qquad 25.12$$

Table 25.1 is based on Eq. 25.12 and summarizes the effects of changing flow area on the various thermodynamic properties for a gas in isentropic flow.

[4]When the flow is already choked and the back pressure is lowered, neither the throat velocity nor the mass flow rate will increase. When the flow is already choked and the source pressure is increased, the throat velocity will remain constant. However, the increased density in the source will cause the mass flow rate to increase.

Table 25.1 *Effect of Changing Duct Area on Isentropic Flow*

	initial Mach no.	
	Ma<1	Ma>1
area decreasing		
	total properties constant	total properties constant
	velocity increases	velocity decreases
	Mach no. increases	Mach no. decreases
	pressure decreases	pressure increases
	density decreases	density increases
	temperature decreases	temperature increases
	enthalpy decreases	enthalpy increases
	internal energy decreases	internal energy increases
	entropy is constant	entropy is constant
area increasing		
	total properties constant	total properties constant
	velocity decreases	velocity increases
	Mach no. decreases	Mach no. increases
	pressure increases	pressure decreases
	density increases	density decreases
	temperature increases	temperature decreases
	enthalpy increases	enthalpy decreases
	internal energy increases	internal energy decreases
	entropy is constant	entropy is constant

9. CONVERGENT-DIVERGENT NOZZLES

Sonic velocity from a reservoir can be achieved with almost any configuration of reservoir orifice. All that is necessary to achieve sonic velocity in an orifice is to lower the ambient pressure until the pressure ratio equals the critical pressure ratio. Achieving supersonic flow, however, will specifically require a *convergent-divergent nozzle* (also known as a *C-D nozzle* or *venturi nozzle*) in addition to the proper pressure conditions.

The properties at the point of flow constriction (i.e., the *throat*, subscript t) in a C-D nozzle are known as the *throat properties*. If sonic velocity is achieved in the throat, the properties are known as *critical properties*. Similarly, the *exit properties* (subscript e) represent the gas properties at the discharge point.

The rules for property changes given in Table 25.1, as well as the following statements, apply to a C-D nozzle attached to a source of (subsonic) pressurized gas or vapor.

- If sonic velocity occurs anywhere in the nozzle, it occurs in the throat.
- If sonic velocity occurs in the throat, the velocity may or may not be supersonic in the diverging section.
- If supersonic velocity occurs anywhere in the nozzle, it occurs in the diverging section.[5]
- If supersonic velocity occurs in the diverging section of the nozzle, the velocity is sonic in the throat.

From these four statements, it is clear that having sonic velocity in the throat is a necessary but not a sufficient condition for achieving supersonic velocity in the diverging section. In order to ensure supersonic velocity existing everywhere in the diverging section, there is an additional necessary condition. Specifically, the nozzle must be designed to expand the gases to the *design pressure ratio*, equal to p_{ambient}/p_0.

10. DESIGN OF SUPERSONIC NOZZLES

In order to design a nozzle capable of expanding a gas to some given velocity or Mach number, it is sufficient to have an expression for the flow area versus Mach number. Since the reservoir cross-sectional area is an unrelated variable, it is not possible to use the stagnation area as a reference area and to develop the ratio A_0/A as was done for temperature, pressure, and density. The usual choice for a reference area is the *critical area*—the area at which the gas velocity is (or could be) sonic. This area is designated as A^*.

$$\frac{A}{A^*} = \left(\frac{1}{\text{Ma}}\right)\left(\frac{1+\frac{1}{2}(k-1)\text{Ma}^2}{\frac{1}{2}(k+1)}\right)^{(k+1)/2(k-1)} \qquad 25.13$$

Figure 25.2 is a plot of Eq. 25.13 versus Mach number. As long as the Mach number is less than 1.0, the area must decrease in order for the velocity to increase. However, if the Mach number is greater than 1.0, the area must increase in order for the velocity to increase. This is consistent with Table 25.1.

[5]This is a valid statement only if the flow in the converging section is subsonic. If a C-D nozzle is placed in an existing supersonic flow, there will be supersonic flow in the converging section of the nozzle.

Figure 25.2 A/A* versus M

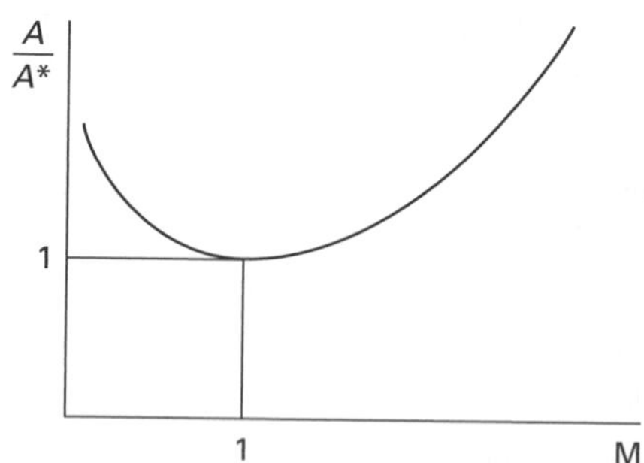

It is not possible to change the Mach number to any desired value over any arbitrary distance along the axis of a converging-diverging nozzle. If the rate of change, dA/dx, is too great, the assumptions of one-dimensional flow become invalid. Usually, the converging section has a steeper angle (known as the *convergent angle*) than the diverging section. If the diverging angle is too great, a normal shock wave may form in that part of the nozzle.

Example 25.4

The Mach number is 0.8 at a point in a nozzle where the area is 1.5 square units. Air ($k = 1.4$) is flowing. (a) What would the throat area have to be in order to achieve sonic velocity there? (b) What is the area at a point where the Mach number is 0.4?

Solution

(a) From the isentropic flow table for Ma = 0.8, $A/A^* = 1.0382$. **[One-Dimensional Isentropic Compressible-Flow Functions]**

$$A_t = \frac{A_{\text{Ma}=0.8}}{\left[\frac{A}{A^*}\right]_{\text{Ma}=0.8}} = \frac{1.5 \text{ units}^2}{1.0382} = 1.445 \text{ units}^2$$

(b) In this case, use the critical point as a reference point, even though sonic velocity may not actually be achieved. From an isentropic flow table for Ma = 0.4, $A/A^* = 1.5901$. **[One-Dimensional Isentropic Compressible-Flow Functions]**

$$\begin{aligned} A_{\text{Ma}=0.4} &= \left(\frac{A}{A^*}\right)_{\text{Ma}=0.4} A_{\text{Ma}=1} \\ &= \left(\frac{A}{A^*}\right)_{\text{Ma}=0.4} \left(\frac{A_{\text{Ma}=0.8}}{\left(\frac{A}{A^*}\right)_{\text{Ma}=0.8}}\right) \\ &= (1.5901)\left(\frac{1.5 \text{ units}^2}{1.0382}\right) \\ &= 2.30 \text{ units}^2 \end{aligned}$$

Example 25.5

A frictionless, adiabatic nozzle receives 10 lbm/sec (5.0 kg/s) of air from a reservoir whose stagnation properties are 200°F and 30 psia (100°C and 200 kPa). The atmospheric pressure is 14.7 psia (101.3 kPa). At the point in the nozzle where the cross-sectional area is 0.12 ft² (0.013 m²), a supersonic velocity of 1400 ft/sec (450 m/s) is attained. At this point, what are the (a) Mach number and (b) pressure?

SI Solution

The absolute total temperature is

$$T_0 = 100°\text{C} + 273° = 373\text{K}$$

Use the ideal gas law to find the total density in the reservoir. The specific gas constant for air is 0.287 kJ/kg·K. **[Fundamental Constants]**

Ideal Gas

$$\begin{aligned} pv &= RT \\ \rho_0 &= \frac{1}{v_0} = \frac{p_0}{RT_0} \\ &= \frac{(200 \text{ kPa})\left(1000 \frac{\text{Pa}}{\text{kPa}}\right)}{\left(\left(0.287 \frac{\text{kJ}}{\text{kg}\cdot\text{K}}\right)\left(1000 \frac{\text{J}}{\text{kJ}}\right)\right)(373\text{K})} \\ &= 1.868 \text{ kg/m}^3 \end{aligned}$$

From an isentropic flow table for Ma = 1, $T/T_0 = 0.8333$, and $[\rho/\rho_0] = 0.6339$. **[One-Dimensional Isentropic Compressible-Flow Functions]**

The sonic properties at the throat are

$$\begin{aligned} T^* &= \left[\frac{T}{T_0}\right] T_0 = (0.8333)(373\text{K}) \\ &= 310.8\text{K} \\ \rho^* &= \left[\frac{\rho}{\rho_0}\right]\rho_0 = (0.6339)\left(1.868 \frac{\text{kg}}{\text{m}^3}\right) \\ &= 1.184 \text{ kg/m}^3 \end{aligned}$$

Use the equation for the speed of sound to find the sonic velocity at the throat.

Mach Number

$$\begin{aligned} c &= \sqrt{kRT} \\ c^* &= \sqrt{kRT^*} \\ &= \sqrt{(1.4)\left(287 \frac{\text{J}}{\text{kg}\cdot\text{K}}\right)(310.8\text{K})} \\ &= 353.4 \text{ m/s} \end{aligned}$$

Use the continuity equation to find the throat area.

Continuity Equation

$$\dot{m} = \rho A\mathrm{v}$$

$$A^* = \frac{\dot{m}}{\rho^* c^*} = \frac{5.0\ \frac{\text{kg}}{\text{s}}}{\left(1.184\ \frac{\text{kg}}{\text{m}^3}\right)\left(353.4\ \frac{\text{m}}{\text{s}}\right)} = 0.01195\ \text{m}^2$$

The ratio of actual to sonic throat areas is

$$\frac{A}{A^*} = \frac{0.013\ \text{m}^2}{0.01195\ \text{m}^2} = 1.0879$$

From the isentropic flow table (interpolating as needed), a value for A/A^* of 1.0879 corresponds to a Mach number of 1.344. From the same table, the ratio of actual to stagnation pressures corresponding to this Mach number is $p/p_0 = 0.3404$. [**One-Dimensional Isentropic Compressible-Flow Functions**]

At the point where the velocity is 450 m/s, the static pressure is

$$p = \left(\frac{p}{p_0}\right)p_0 = (0.3404)(200\ \text{kPa}) = 68.08\ \text{kPa}$$

Customary U.S. Solution

The absolute total temperature is

$$T_0 = 200°\text{F} + 460° = 660°\text{R}$$

Use the ideal gas law to find the total density in the reservoir. The specific gas constant for air is 53.3 ft-lbf/lbm-°R. [**Fundamental Constants**]

Ideal Gas

$$pv = RT$$

$$\rho_0 = \frac{1}{v_0} = \frac{p_0}{RT_0} = \frac{\left(30\ \frac{\text{lbf}}{\text{in}^2}\right)\left(12\ \frac{\text{in}}{\text{ft}}\right)^2}{\left(53.3\ \frac{\text{ft-lbf}}{\text{lbm-°R}}\right)(660°\text{R})} = 0.1228\ \text{lbm/ft}^3$$

From an isentropic flow table for Ma = 1, $T/T_0 = 0.8333$, and $\rho/\rho_0 = 0.6339$. [**One-Dimensional Isentropic Compressible-Flow Functions**]

The sonic properties at the throat are

$$T^* = \left[\frac{T}{T_0}\right]T_0 = (0.8333)(660°\text{R}) = 550°\text{R}$$

$$\rho^* = \left[\frac{\rho}{\rho_0}\right]\rho_0 = (0.6339)\left(0.1228\ \frac{\text{lbm}}{\text{ft}^3}\right) = 0.07784\ \text{lbm/ft}^3$$

Use the equation for the speed of sound to find the sonic velocity at the throat.

Mach Number

$$c = \sqrt{kg_cRT}$$

$$c^* = \sqrt{kg_cRT^*} = \sqrt{(1.4)\left(32.2\ \frac{\text{lbm-ft}}{\text{lbf-sec}^2}\right)\times\left(53.3\ \frac{\text{ft-lbf}}{\text{lbm-°R}}\right)(550°\text{R})} = 1149.6\ \text{ft/sec}$$

Use the continuity equation to find the throat area.

Continuity Equation

$$\dot{m} = \rho A\mathrm{v}$$

$$A^* = \frac{\dot{m}}{\rho^* c^*} = \frac{10\ \frac{\text{lbm}}{\text{sec}}}{\left(0.07784\ \frac{\text{lbm}}{\text{ft}^3}\right)\left(1149.6\ \frac{\text{ft}}{\text{sec}}\right)} = 0.1118\ \text{ft}^2$$

The ratio of actual to sonic throat areas is

$$\frac{A}{A^*} = \frac{0.12\ \text{ft}^2}{0.1118\ \text{ft}^2} = 1.0733$$

From the isentropic flow table (interpolating as needed), a value for A/A^* of 1.0733 corresponds to a Mach number of 1.314. From the same table (again interpolating as needed), the ratio of actual to stagnation pressures corresponding to this Mach number is $p/p_0 = 0.3542$. [**One-Dimensional Isentropic Compressible-Flow Functions**]

At the point where the velocity is 450 m/s, the static pressure is

$$p = \left(\frac{p}{p_0}\right)p_0 = (0.3542)\left(30\ \frac{\text{lbf}}{\text{in}^2}\right) = 10.63\ \text{lbf/in}^2$$

Example 25.6

An attitude-adjustment jet in a satellite uses high-pressure gas with a ratio of specific heats of 1.4 and a molecular weight of 21 lbm/lbmol (kg/kmol). The chamber conditions are 450 psia and 4700°R (3.2 MPa and 2600K). The gas is expanded supersonically to 2.97 psia (20.5 kPa) at the nozzle exit. What are the (a) sonic velocity in the throat, (b) required exit-to-throat area ratio, and (c) exit velocity?

SI Solution

(a) The specific gas constant is

Mach Number

$$R = \frac{\overline{R}}{M} = \frac{8314\ \frac{\text{J}}{\text{kmol·K}}}{21\ \frac{\text{kg}}{\text{kmol}}} = 395.9\ \text{J/kg·K}$$

Use the chamber properties as the total properties. Since the exit velocity is supersonic, sonic flow is achieved in the throat. From the isentropic flow table, at Ma = 1, $T/T_0 = 0.8333$. [**One-Dimensional Isentropic Compressible-Flow Functions**]

The temperature at the throat is

$$T_t = \left[\frac{T}{T_0}\right]T_0 = (0.8333)(2600\text{K}) = 2167\text{K}$$

The velocity is sonic at the throat.

Mach Number

$$c = \sqrt{kRT}$$

$$c^* = \sqrt{kRT_t} = \sqrt{(1.4)\left(395.9\ \frac{\text{J}}{\text{kg·K}}\right) \times (2167\text{K})} = 1096\ \text{m/s}$$

(b) The static pressure ratio at the exit is

$$\frac{p}{p_0} = \frac{20.5\ \text{kPa}}{(3.2\ \text{MPa})\left(1000\ \frac{\text{kPa}}{\text{MPa}}\right)} = 0.006406$$

Locate this pressure ratio in the supersonic region of the isentropic flow table. The corresponding Mach number is approximately 4.0. Read $A/A^* = 10.7187$. [**One-Dimensional Isentropic Compressible-Flow Functions**]

(c) At Mach 4, $T/T_0 = 0.2381$. The exit temperature is

$$T_e = \left[\frac{T}{T_0}\right]T_0 = (0.2381)(2600\text{K}) = 619.1\text{K}$$

Find the exit velocity.

Mach Number

$$\text{Ma} = \frac{\text{v}}{c}$$

$$\text{v}_e = (\text{Ma})c_e = \text{Ma}\sqrt{kRT_e} = (4)\sqrt{(1.4)\left(395.9\ \frac{\text{J}}{\text{kg·K}}\right)(619.1\text{K})} = 2343\ \text{m/s}$$

Customary U.S. Solution

(a) Calculate the specific gas constant.

Mach Number

$$R = \frac{\overline{R}}{M} = \frac{1545\ \frac{\text{ft-lbf}}{\text{lbmol-°R}}}{21\ \frac{\text{lbm}}{\text{lbmol}}} = 73.57\ \text{ft-lbf/lbm-°R}$$

Use the chamber properties as the total properties. Since the exit velocity is supersonic, sonic flow is achieved in the throat. From the isentropic flow table, at Ma = 1, $T/T_0 = 0.8333$. [**One-Dimensional Isentropic Compressible-Flow Functions**]

The temperature at the throat is

$$T_t = \left[\frac{T}{T_0}\right]T_0 = (0.8333)(4700°\text{R}) = 3917°\text{R}$$

The velocity is sonic at the throat.

Thermodynamics

Mach Number

$$c = \sqrt{kg_cRT}$$

$$c^* = \sqrt{kg_cRT_t}$$

$$= \sqrt{(1.4)\left(32.2\ \frac{\text{lbm-ft}}{\text{lbf-sec}^2}\right) \times \left(73.57\ \frac{\text{ft-lbf}}{\text{lbm-°R}}\right)(3917\text{°R})}$$

$$= 3604\ \text{ft/sec}$$

(b) The static pressure ratio at the exit is

$$\frac{p}{p_0} = \frac{2.97\ \frac{\text{lbf}}{\text{in}^2}}{450\ \frac{\text{lbf}}{\text{in}^2}} = 0.0066$$

Locate this pressure ratio in the supersonic region of the isentropic flow table. The corresponding Mach number is approximately 4.0. Read $A/A^* = 10.7187$. [**One-Dimensional Isentropic Compressible-Flow Functions**]

(c) At Mach 4, $T/T_0 = 0.2381$. The exit temperature is

$$T_e = \left[\frac{T}{T_0}\right]T_0 = (0.2381)(4700\text{°R})$$

$$= 1119\text{°R}$$

Find the exit velocity.

Mach Number

$$\text{Ma} = \frac{\text{v}}{c}$$

$$\text{v}_e = (\text{Ma})c_e = \text{Ma}\sqrt{kRT_e}$$

$$= (4)\sqrt{(1.4)\left(32.2\ \frac{\text{lbm-ft}}{\text{lbf-sec}^2}\right) \times \left(73.57\ \frac{\text{ft-lbf}}{\text{lbm-°R}}\right)(1119\text{°R})}$$

$$= 7706\ \text{ft/sec}$$

11. NOZZLE PERFORMANCE CHARACTERISTICS

For a nozzle, there is no heat transfer, $\dot{Q}$, no work, $\dot{W}$, no change in mass flow rate, $\dot{m}$, and no significant change in elevation, z. The steady-flow equation, Eq. 25.14, reduces to

Special Cases of the Steady-Flow Energy Equation

$$h_i + \frac{\text{v}_i^2}{2} = h_e + \frac{\text{v}_e^2}{2} \qquad 25.14$$

The *nozzle efficiency*, η, is defined as the ratio of the actual to ideal (isentropic) energies extracted from the flowing gas. In 25.15(a), h_{es} is the isentropic enthalpy at the nozzle exit

Special Cases of the Steady-Flow Energy Equation

$$\eta = \frac{\text{v}_e^2 - \text{v}_i^2}{2(h_i - h_{es})} \qquad \text{[SI]} \qquad 25.15(a)$$

$$\eta = \frac{\text{v}_e^2 - \text{v}_i^2}{2g_c(h_i - h_{es})} \qquad \text{[U.S.]} \qquad 25.15(b)$$

The nozzle efficiency is also equal to square of the ratio of actual to ideal velocities.

$$\eta = \frac{\Delta h_{\text{actual}}}{\Delta h_{\text{ideal}}} = \left(\frac{\text{v}_{\text{actual}}}{\text{v}_{\text{ideal}}}\right)^2 \qquad 25.16$$

Example 25.7

What is the nozzle efficiency for the steam nozzle in Ex. 25.1?

SI Solution

Steam at 1.50 MPa and 250°C has a specific entropy of 6.71 kJ/kg·K. [**Properties of Superheated Steam - SI Units**]

Find the specific entropies of saturated liquid and saturated vapor at 0.15 MPa. [**Properties of Saturated Water and Steam (Pressure) - SI Units**]

$$s_f = 1.4337\ \text{kJ/kg·K}$$

$$s_g = 7.2230\ \text{kJ/kg·K}$$

If the steam were expanded to 0.15 MPa isentropically, its specific entropy would still be 6.71 Btu/lbm-°R; this is between s_f and s_g, so the steam would be a mix of saturated liquid and saturated vapor. Use the equation for specific entropy in a two-phase systems, and solve for the quality.

Properties for Two-Phase (Vapor-Liquid) Systems

$$s = xs_g + (1 - x)s_f$$

$$x = \frac{s - s_f}{s_g - s_f}$$

$$= \frac{6.71\ \frac{\text{kJ}}{\text{kg·K}} - 1.4337\ \frac{\text{kJ}}{\text{kg·K}}}{7.2230\ \frac{\text{kJ}}{\text{kg·K}} - 1.4337\ \frac{\text{kJ}}{\text{kg·K}}}$$

$$= 0.9114$$

From steam tables, find the specific enthalpy of saturated liquid and saturated vapor at 0.15 MPa. [**Properties of Saturated Water and Steam (Pressure) — I-P Units**]

$$h_f = 467.1 \text{ kJ/kg}$$
$$h_g = 2693.1 \text{ kJ/kg}$$

Find the specific enthalpy of the steam at 0.15 MPa and a quality of 0.9114. This is the exit enthalpy the steam would have if isentropically expanded.

Properties for Two-Phase (Vapor-Liquid) Systems

$$\begin{aligned} h_{es} &= xh_g + (1-x)h_f \\ &= (0.9114)\left(2693.1 \ \frac{\text{kJ}}{\text{kg}}\right) + (1-0.9114)\left(467.1 \ \frac{\text{kJ}}{\text{kg}}\right) \\ &= 2495.9 \text{ kJ/kg} \end{aligned}$$

The nozzle efficiency is defined by Eq. 25.15.

Special Cases of the Steady-Flow Energy Equation

$$\begin{aligned} \eta &= \frac{\mathrm{v}_e^2 - \mathrm{v}_i^2}{2(h_i - h_{es})} \\ &= \frac{\left(742.0 \ \frac{\text{m}}{\text{s}}\right)^2 - \left(0 \ \frac{\text{m}}{\text{s}}\right)^2}{(2)\left(2923.9 \ \frac{\text{kJ}}{\text{kg}} - 2495.9 \ \frac{\text{kJ}}{\text{kg}}\right)\left(1000 \ \frac{\text{J}}{\text{kJ}}\right)} \\ &= 0.6432 \quad (64.32\%) \end{aligned}$$

Customary U.S. Solution

Steam at 200 psia and 500°F has a specific entropy of 1.624 Btu/lbm-°R. [**Properties of Superheated Steam - I-P Units**]

Find the specific entropies of saturated liquid and saturated vapor at 20 psia. [**Properties of Saturated Water and Steam (Pressure) — I-P Units**]

$$s_f = 0.3358 \text{ Btu/lbm-°R}$$
$$s_g = 1.7319 \text{ Btu/lbm-°R}$$

If the steam were expanded to 20 psia isentropically, its specific entropy would still be 1.624 Btu/lbm-°R; this is between s_f and s_g, so the steam would be a mix of saturated liquid and saturated vapor. Use the equation for specific entropy in a two-phase systems, and solve for the quality.

Properties for Two-Phase (Vapor-Liquid) Systems

$$\begin{aligned} s &= xs_g + (1-x)s_f \\ x &= \frac{s - s_f}{s_g - s_f} \\ &= \frac{1.624 \ \frac{\text{Btu}}{\text{lbm-°R}} - 0.3358 \ \frac{\text{Btu}}{\text{lbm-°R}}}{1.7319 \ \frac{\text{Btu}}{\text{lbm-°R}} - 0.3358 \ \frac{\text{Btu}}{\text{lbm-°R}}} \\ &= 0.9227 \end{aligned}$$

From steam tables, find the specific enthalpy of saturated liquid and saturated vapor at 20 psia. [**Properties of Saturated Water and Steam (Pressure) — I-P Units**]

$$h_f = 196.25 \text{ Btu/lbm}$$
$$h_g = 1156.19 \text{ Btu/lbm}$$

Find the specific enthalpy of the steam at 20 psia and a quality of 0.9227. This is the exit enthalpy the steam would have if isentropically expanded.

Properties for Two-Phase (Vapor-Liquid) Systems

$$\begin{aligned} h_{es} &= xh_g + (1-x)h_f \\ &= (0.9227)\left(1156.19 \ \frac{\text{Btu}}{\text{lbm}}\right) + (1-0.9227)\left(196.25 \ \frac{\text{Btu}}{\text{lbm}}\right) \\ &= 1082.0 \text{ Btu/lbm} \end{aligned}$$

The nozzle efficiency is defined by Eq. 25.15. 778.2 ft-lbf/Btu is Joule's constant, used to convert from foot-pounds-force to British thermal units.

Special Cases of the Steady-Flow Energy Equation

$$\begin{aligned} \eta &= \frac{\mathrm{v}_e^2 - \mathrm{v}_i^2}{2g_c(h_i - h_{es})} \\ &= \frac{\left(2572 \ \frac{\text{ft}}{\text{sec}}\right)^2 - \left(0 \ \frac{\text{ft}}{\text{sec}}\right)^2}{\begin{array}{c}(2)\left(32.2 \ \frac{\text{lbm-ft}}{\text{lbf-sec}^2}\right)\left(778.2 \ \frac{\text{ft-lbf}}{\text{Btu}}\right) \\ \times\left(1269.0 \ \frac{\text{Btu}}{\text{lbm}} - 1082.0 \ \frac{\text{Btu}}{\text{lbm}}\right)\end{array}} \\ &= 0.7059 \quad (70.59\%) \end{aligned}$$

12. SHOCK WAVES

In Sec. 25.9, it was stated that a nozzle must be designed to the design pressure ratio in order to keep the flow supersonic in the diverging section of the nozzle. It is possible, though, to have supersonic velocity only in part of the diverging section. Once the flow is supersonic in a part of the diverging section, however, it cannot become subsonic by an isentropic process.

Therefore, the gas experiences a *shock wave* as the velocity drops from supersonic to subsonic. Shock waves are very thin (several molecules thick) and separate areas of radically different thermodynamic properties. Since the shock wave forms normal to the flow direction, it is known as a *normal shock wave*. The strength of a shock wave is measured by the change in Mach number across it.

The velocity always changes from supersonic to subsonic across a shock wave. Since there is no loss of heat energy, a shock wave is an adiabatic process, and total temperature is constant. However, the process is not isentropic, and total pressure decreases. Momentum is also conserved. Table 25.2 lists the property changes across a shock wave.

***Table 25.2** Property Changes Across a Normal Shock Wave*

property	change
total temperature	is constant
total pressure	decreases
total density	decreases
velocity	decreases
Mach number	decreases
pressure	increases
density	increases
temperature	increases
entropy	increases
internal energy	increases
enthalpy	is constant
momentum	is constant

The properties before a shock wave are given the subscript 1. The subscript 2 is used for the properties after a shock wave. These subscripts may be combined with the subscript 0 to represent a total (stagnation) property. For example, after a normal shock from supersonic to sonic conditions that occurs in a diverging nozzle, $p_{0,2} < p_{0,1}$.

The ratios of various properties before and after a normal shock wave can be calculated with Eq. 25.17 through Eq. 25.21. These values are dependent only on the ratio of specific heats, and the Mach number, Ma_1, before the shock wave. It is more convenient, however, to use normal shock factors from a normal shock table. [One-Dimensional Isentropic Compressible-Flow Functions]

Normal Shock Relationships

$$\text{Ma}_2 = \sqrt{\frac{(k-1)(\text{Ma}_1)^2+2}{2k(\text{Ma}_1)^2-(k-1)}} \qquad 25.17$$

$$\frac{T_2}{T_1} = \left(2+(k-1)(\text{Ma}_1)^2\right) \times \left(\frac{2k(\text{Ma}_1)^2-(k-1)}{(k+1)^2(\text{Ma}_1)^2}\right) \qquad 25.18$$

$$\frac{p_2}{p_1} = \left(\frac{1}{k+1}\right)\left(2k(\text{Ma}_1)^2-(k-1)\right) \qquad 25.19$$

$$\frac{\rho_2}{\rho_1} = \frac{v_1}{v_2} = \frac{(k+1)(\text{Ma}_1)^2}{(k-1)(\text{Ma}_1)^2+2} \qquad 25.20$$

$$T_{0,1} = T_{0,2} \qquad 25.21$$

Example 25.8

A shock wave in air ($k = 1.4$) occurs at a point where the Mach number is 2.4. If the static pressure before the shock wave is 5.0 atm, what are the (a) Mach number after the shock, (b) static pressure after the shock?

Solution

(a) The Mach number after the shock is

Normal Shock Relationships

$$\begin{aligned}\text{Ma}_2 &= \sqrt{\frac{(k-1)(\text{Ma}_1)^2+2}{2k(\text{Ma}_1)^2-(k-1)}} \\ &= \sqrt{\frac{(1.4-1)(2.4)^2+2}{(2)(1.4)(2.4)^2-(1.4-1)}} \\ &= 0.5231\end{aligned}$$

(b) The ratio of static pressures is

Normal Shock Relationships

$$\begin{aligned}\frac{p_2}{p_1} &= \left(\frac{1}{k+1}\right)(2k(\text{Ma}_1)^2-(k-1)) \\ &= \left(\frac{1}{1.4+1}\right)((2)(1.4)(2.4)^2-(1.4-1)) \\ &= 6.553\end{aligned}$$

The static pressure after the shock is

$$\begin{aligned}p_2 &= \left(\frac{p_2}{p_1}\right)p_1 = (6.553)(5.0\text{ atm}) \\ &= 32.8\text{ atm}\end{aligned}$$

Thermodynamics

13. CRITICAL (BACK) PRESSURE RATIO

Figure 25.3(a) illustrates a converging-diverging nozzle separating high- and low-pressure reservoirs. The total pressure in the high-pressure reservoir is constant. The pressure in the low-pressure reservoir (i.e., the *back pressure*) is variable. The nozzle geometry, particularly the throat and exit areas, A_t and A_e, is known. Figure 25.3 (b) illustrates the ratio of static pressure at the corresponding point to the total pressure (p/p_0) in the first reservoir. The following eight cases (*flow regimes*) represent decidedly different performances.

case A: If $p_2 = p_0$, there will be no flow.

case B: When p_2 is lowered to just slightly less than p_0, flow will be initiated. Refer to Fig. 25.3. As long as the pressure ratio p_t/p_0 is greater than the first critical pressure ratio, R_{cp} (0.5283 for air), flow in the divergent section will be subsonic and the exit pressure, p_e, will be equal to p_2. Flow will be isentropic, and the nozzle acts like a venturi. The conditions at any point in the nozzle where the area is known can be determined in the following manner.

Figure 25.3 *Flow Regimes in a C-D Nozzle*

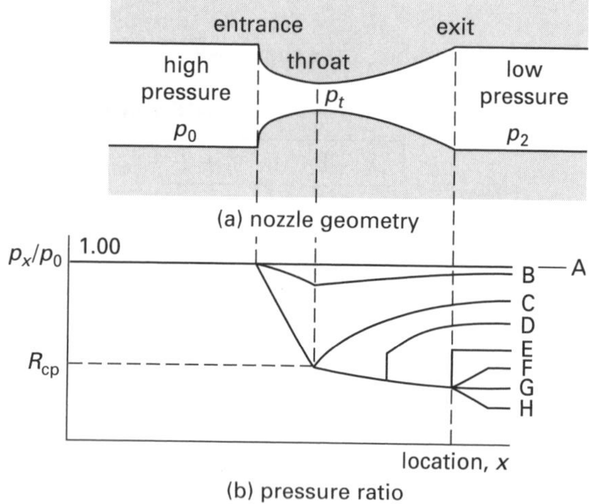

step 1: Calculate the exit pressure ratio. (For any other point, substitute the known pressure for p_2.)

$$\frac{p_e}{p_0} = \frac{p_2}{p_0} \qquad 25.22$$

step 2: Locate the pressure ratio calculated in step 1 in the isentropic flow table. Read the exit Mach number, Ma_e, and the critical area ratio $(A/A^*)_e$. (Since sonic velocity is not actually achieved, the theoretical critical area, A^*, does not exist in this nozzle.)

step 3: Calculate the ratio of actual throat area to the theoretical critical throat area.

$$\frac{A_t}{A^*} = \frac{A_t\left(\frac{A}{A^*}\right)_e}{A_e} \qquad 25.23$$

step 4: Locate the area ratio calculated in step 3 in the isentropic flow table. Read the throat Mach number, Ma_t.

case C: If p_2 is lowered so that p_t/p_0 equals the *first critical (back) pressure ratio* (0.5283 for air), sonic velocity will be achieved in the throat. Flow will be choked and isentropic. The gas properties in the nozzle can be found from the procedure given in case B. The exit velocity will be subsonic, and the exit pressure will equal the back pressure ($p_e = p_2$).

case D: If p_2 is lowered slightly below that of case C (but not below the pressure ratio corresponding to the area ratio ratio A_e/A^*), the mass flow rate will not increase above that of case C because the flow will be choked. However, since the back pressure p_2 is lower, the velocity will be higher. Supersonic velocity will be achieved in some parts of the nozzle. Flow will drop back to subsonic via a normal shock wave. The exit velocity will be subsonic, and the exit pressure will equal the back pressure ($p_e = p_2$). The location (i.e., the area) in the divergent section where the shock wave occurs can be found from the following trial-and-error procedure.

step 1: Assume a Mach number, Ma_1, at which the shock wave occurs.

step 2: For Ma_1, read the ratio of total pressures ($p_{0,2}/p_{0,1}$) from the normal shock table.

step 3: Calculate the ratio of static exit pressure to total pressure after the shock wave. $p_{0,1}$ is the total pressure in the source reservoir.

$$\frac{p_e}{p_{0,2}} = \frac{p_e}{p_{0,1}\left(\frac{p_{0,2}}{p_{0,1}}\right)} \qquad 25.24$$

step 4: Determine the exit Mach number, Ma_e, by locating the ratio calculated in step 3 in the p/p_0 column of the normal shock tables. Read the area ratio A/A^*.

step 5: Calculate the area ratio.

$$\frac{A_e}{A_t\left(\frac{p_{0,2}}{p_{0,1}}\right)} \qquad 25.25$$

step 6: Compare the area ratio from steps 4 and 5. If the values differ, repeat from step 1. (The area ratio from step 5 decreases when the assumed Mach number increases.) If the values are equal, the shock wave occurs at the assumed Mach number, M_x.

step 7: Calculate the area at which the shock wave occurs.

$$A = A_t\left(\frac{A}{A^*}\right) \qquad 25.26$$

case E: If the back pressure is equal to the *second critical (back) pressure ratio* (i.e., the pressure ratio corresponding to the area ratio (A_e/A^*), the shock wave will stand at the exit. The mass flow rate is the same as in case C.

case F: If the back pressure ratio is less than the second critical back pressure ratio but greater than the third, the condition is known as *over-expansion.* The exit area is too large, and gas will be discharged at a pressure less than the ambient conditions. The gas will "pop" back up to the ambient pressure through an *oblique compression shock wave* outside the nozzle, but this will not change the conditions inside the nozzle. The mass flow rate is the same as in case C.

case G: If the back pressure ratio is equal to the *third critical (back) pressure ratio*, there will be no shock wave. Pressure p_2 is known as the *design pressure* or the *isentropic pressure* for the nozzle. Flow will be isentropic. The mass flow rate is the same as in case C.

case H: If the pressure ratio is less than the third critical back pressure ratio, the condition is known as *underexpansion.* The exit area is too small for the back pressure. Expansion will be incomplete in the nozzle, and the gas will be discharged at a pressure above the local ambient pressure. Expansion continues outside the nozzle, and the pressure drops down to the ambient pressure by way of a *rarefaction shock wave* (also known as an *expansion wave* or an *oblique expansion shock wave*) outside the nozzle. The mass flow rate is the same as in case C.

Both over- and under-expansion reduce the efficiency of the nozzle. The effect of over-expansion is to reduce the gas exit velocity. In the case of rocket propulsion nozzles, the decreased exit velocity produces a proportional decrease in thrust. In the case of a fixed steam nozzle, the energy available to the steam turbine is reduced.

A given percentage of over-expansion (based on the ratio of actual-to-theoretical exit areas) can reduce the available energy as much as ten times the reduction for the same percentage of under-expansion. For that reason, steam nozzles feeding turbines are sometimes designed 10–20% too small to ensure under-expansion under light or partial loads.

Example 25.9

A nozzle is fed from a large reservoir containing 100 psia (0.7 MPa) air. The back pressure is adjustable. A shock wave stands at the exit of the nozzle. The Mach number just before the shock wave is 3.0. For which values of the back pressure will the shock wave at the exit disappear completely?

SI Solution

There are two ways the shock wave can disappear. If the back pressure is decreased to the design pressure (case G), the shock wave will move out of the nozzle and dissipate. At Mach 3, $p/p_0 = = 0.0272$. [**One-Dimensional Isentropic Compressible-Flow Functions**]

Therefore, the design pressure for this nozzle is

$$\begin{aligned} p_{\text{design}} &= \left(\frac{p}{p_0}\right)p_0 \\ &= (0.0272)(0.7\ \text{MPa})\left(1000\ \frac{\text{kPa}}{\text{MPa}}\right) \\ &= 19.04\ \text{kPa} \end{aligned}$$

If the back pressure is increased, the shock wave will move upstream and vanish at the throat (case C). From the supersonic portion of the isentropic flow tables at Mach 3, $A/A^* = 4.2346$. [**One-Dimensional Isentropic Compressible-Flow Functions**]

Find the same value in the subsonic part of the isentropic flow table. The Mach number corresponding to this value in the subsonic part of the table is about 0.14. At this Mach number, the shock wave will vanish and the flow will be subsonic everywhere in the nozzle. At Mach 0.14, $p/p_0 = 0.9864$. The back pressure is

$$\begin{aligned} p_{\text{back}} &= \left(\frac{p}{p_0}\right)p_0 = (0.9864)(0.7\ \text{MPa}) \\ &= 0.6905\ \text{MPa} \end{aligned}$$

Customary U.S. Solution

There are two ways the shock wave can disappear. If the back pressure is decreased to the design pressure (case G), the shock wave will move out of the nozzle and dissipate. At Mach 3, $p/p_0 = 0.0272$. [**One-Dimensional Isentropic Compressible-Flow Functions**]

Therefore, the design pressure for this nozzle is

$$\begin{aligned} p_{\text{design}} &= \left(\frac{p}{p_0}\right)p_0 \\ &= (0.0272)\left(100\ \frac{\text{lbf}}{\text{in}^2}\right) \\ &= 2.72\ \text{lbf/in}^2 \end{aligned}$$

Thermodynamics

If the back pressure is increased, the shock wave will move upstream and vanish at the throat (case C). From the supersonic portion of the isentropic flow tables at Mach 3, $A/A^* = 4.2346$. [**One-Dimensional Isentropic Compressible-Flow Functions**]

Find the same value in the subsonic part of the isentropic flow table. The Mach number corresponding to this value in the subsonic part of the table is about 0.14. At this Mach number, the shock wave will vanish and the flow will be subsonic everywhere in the nozzle. At Mach 0.14, $p/p_0 = 0.9864$. The back pressure is

$$\begin{aligned} p_{\text{back}} &= \left[\frac{p}{p_0}\right]p_0 = (0.9864)\left(100\ \frac{\text{lbf}}{\text{in}^2}\right) \\ &= 98.64\ \text{lbf/in}^2 \end{aligned}$$

14. COMPRESSIBLE FLOW THROUGH ORIFICES

Compressible flow through orifices (simple holes, perforations, reentrant tubes, etc.) is similar to flow through nozzles, cases A, B, and C. Compressible flow can be characterized by Eq. 25.27.

Orifices

$$Q = CA_0\sqrt{2g\left(\frac{p_1}{\gamma} + z_1 - \frac{p_2}{\gamma} - z_2\right)} \qquad 25.27$$

In Eq. 25.28, the *orifice coefficient*, C, is a function of the velocity and contraction coefficients and the ratio of orifice area, A_0, to area of flow, A_1.

Orifices

$$C = \frac{C_v C_c}{\sqrt{1 - C_c^2\left(\frac{A_0}{A_1}\right)^2}} \qquad 25.28$$

If the ratio of ambient-to-source pressures is less than the first critical pressure ratio, the velocity will be sonic in the orifice. However, since there is no diverging section, the flow can never become supersonic. Due to the high turbulence at the orifice, the coefficient of velocity and nozzle efficiency may be relatively low. This will be manifested in a lower-than-ideal mass flow rate.

Flow discharging to atmospheric pressure will be choked for most gases as long as the tank pressure is greater than 25–28 psia. Although the velocity through the orifice is always sonic, the mass flow rate is not constant because the pressure and the density of the gas in the tank and orifice are continually decreasing. The starting discharge rate, as well as any intermediate discharge rate, is an instantaneous value that cannot be used to determine time-to-empty. Since density is not constant, liquid discharge models cannot be applied.

As back pressure is reduced, any vena contracta that forms will increase in size and move upstream toward the orifice. The vena contracta will coincide with the orifice at choked flow. If the orifice is thin and frictionless, the orifice coefficient can be 1.0. If friction is present, the discharge velocity is reduced, resulting in an orifice coefficient less than 1.0. Orifice coefficients for choked flow can be correlated to the ratio of orifice plate thickness to orifice diameter, t/D. For $t/D = 0$, $C = 1.0$. For $0 < t/D < 1$, C varies linearly with t/D down to 0.81. For $1 < t/D < 7$, $C = 0.81$. For $7 < t/D < 7$, $C < 0.81$, and Fanno flow theory is applicable.

15. NOMENCLATURE

A	area	ft²	m²
c	specific heat	Btu/lbm-°F	kJ/kg·K
c	speed of sound	ft/sec	m/s
C	orifice coefficient	–	–
C_c	coefficient of contraction	–	–
C_v	coefficient of velocity	–	–
D	diameter	ft	m
g	acceleration of gravity, 32.17 (9.807)	ft/sec²	m/s²
g_c	gravitational constant, 32.17	lbm-ft/lbf-sec²	n.a.
h	specific enthalpy	Btu/lbm	kJ/kg
J	Joule's constant (mechanical equivalent of heat), 778.2	ft-lbf/Btu	n.a.
k	ratio of specific heats	–	–
$\dot{m}$	mass flow rate	lbm/sec	kg/s
Ma	Mach number	–	–
M	molecular weight	lbm/lbmol	kg/kmol
p	pressure	lbf/ft²	Pa
$\dot{Q}$	rate of heat transfer	Btu/hr	W
R	ratio	–	–
R	specific gas constant	ft-lbf/lbm-°R	kJ/kg·K
$\bar{R}$	universal gas constant	ft-lbf/lbmol-°R	kJ/kmol·K
R_{cp}	critical pressure ratio	–	–
s	specific entropy	Btu/lbm-°F	kJ/kg·K
T	absolute temperature	°R	K
u	specific internal energy	Btu/lbm	kJ/kg
v	specific volume	ft³/lbm	m³/kg
v	velocity	ft/sec	m/s
$\dot{W}$	rate of work	Btu/hr	W
x	distance	ft	m
x	quality	–	–
z	elevation	ft	m

Symbols

γ	specific weight	lbf/ft^3	N/m^3
η	efficiency	–	–
ρ	density	lbm/ft^3	kg/m^3

Subscripts

0	total (stagnation)
1	before the shock wave
2	after the shock wave
cp	critical pressure
e	exit
f	saturated liquid (fluid)
fg	vaporization (liquid to vapor)
g	saturated vapor (gas)
i	initial or inlet
p	constant pressure
s	isentropic
t	throat
v	constant volume

Topic IV: Power Cycles

Chapter

26 Vapor Power Equipment

Content in blue refers to the *NCEES Handbook*.

NCEES EXAM SPECIFICATIONS AND RELATED CONTENT

HVAC AND REFRIGERATION EXAM

I.F. Energy/Mass Balances
- 5. Pumps

II.B.2. Equipment and Components: Boilers and furnaces
- 18. Feedwater Heaters

II.B.4. Equipment and Components: Condensers/evaporators
- 18. Feedwater Heaters

THERMAL AND FLUID SYSTEMS EXAM

I.E.3. Thermodynamics: Energy balances
- 5. Pumps
- 9. Turbine Work, Power, and Efficiency

III.A.1. Energy/Power Equipment: Turbines
- 8. Turbines
- 9. Turbine Work, Power, and Efficiency
- 10. Two-Stage Expansion

III.A.2. Energy/Power Equipment: Boilers and steam generators
- 5. Pumps
- 6. Pump Efficiency
- 18. Feedwater Heaters

III.A.4. Energy/Power Equipment: Heat exchangers
- 18. Feedwater Heaters

1. TYPICAL SYSTEM INTEGRATION

This chapter takes a quick look at some of the equipment necessary to implement a typical utility power-generating plant. It is impractical to show all of the lines, valves, tanks, and redundant elements in even a simple power-generating system. There are numerous ways of combining the devices described in this chapter into a working power-generating system. However, Fig. 26.1(a) illustrates the main elements and interconnections for *unit operation* (i.e., in a typical drum boiler unit).

The integration of elements in high-pressure *once-through boilers* is somewhat different, as Fig. 26.1(b) illustrates. (The term "once-through" does not mean the steam is discarded after use. It means that the steam is heated along one long progressive transfer path.) Once-through boilers do not have steam drums. The steam flow rate is controlled by the boilerfeed pump. At start-up and at low load, the steam generator produces more steam than the turbine requires. At low loads, superheated steam is passed to a flash tank where it becomes available for feedwater heating, or it is passed directly to the condenser and returned to the boiler. This lets the flash tank act like a drum in a drum boiler unit. Valving and valve sequencing is more complex with once-through operation.

Power plant output may be qualified by the terms "thermal," "electrical," "gross," or "net." For example, a 1,000,000 Btu/hr *thermal* (i.e., "1 MBth" or "MBt") power plant indicates the energy transfer to the feedwater. A 1000 MW *electrical* (i.e., "1000 MWe") power plant refers to the generator output. Similarly, "MWt" refers to thermal power in megawatts.

Figure 26.1 Typical Integration of Power-Generating Elements

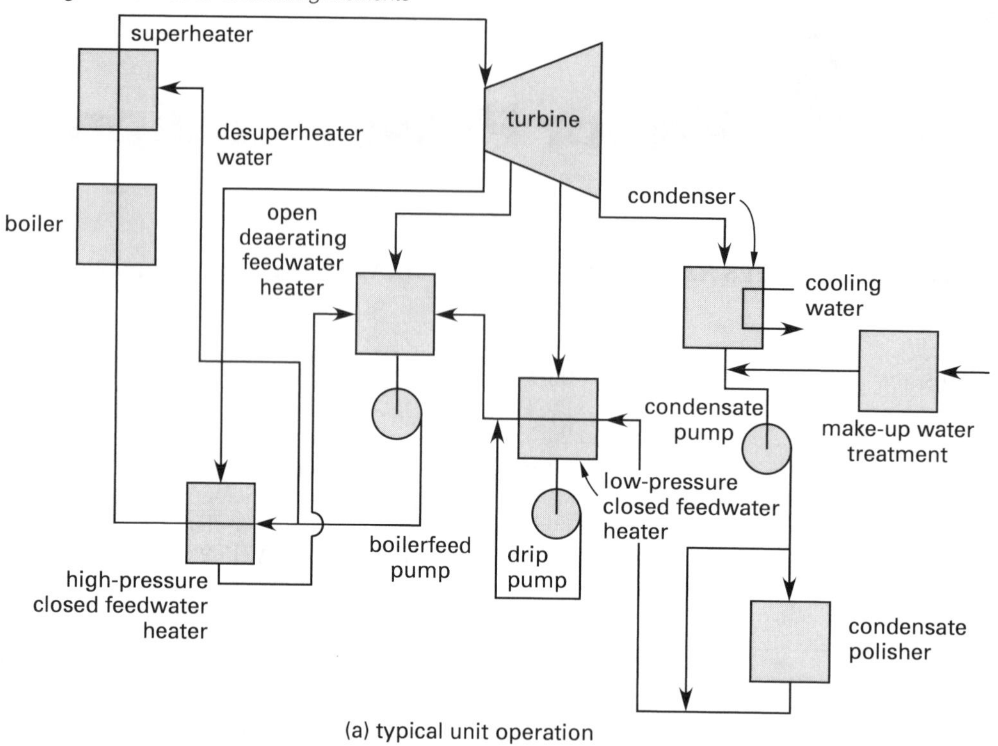

(a) typical unit operation

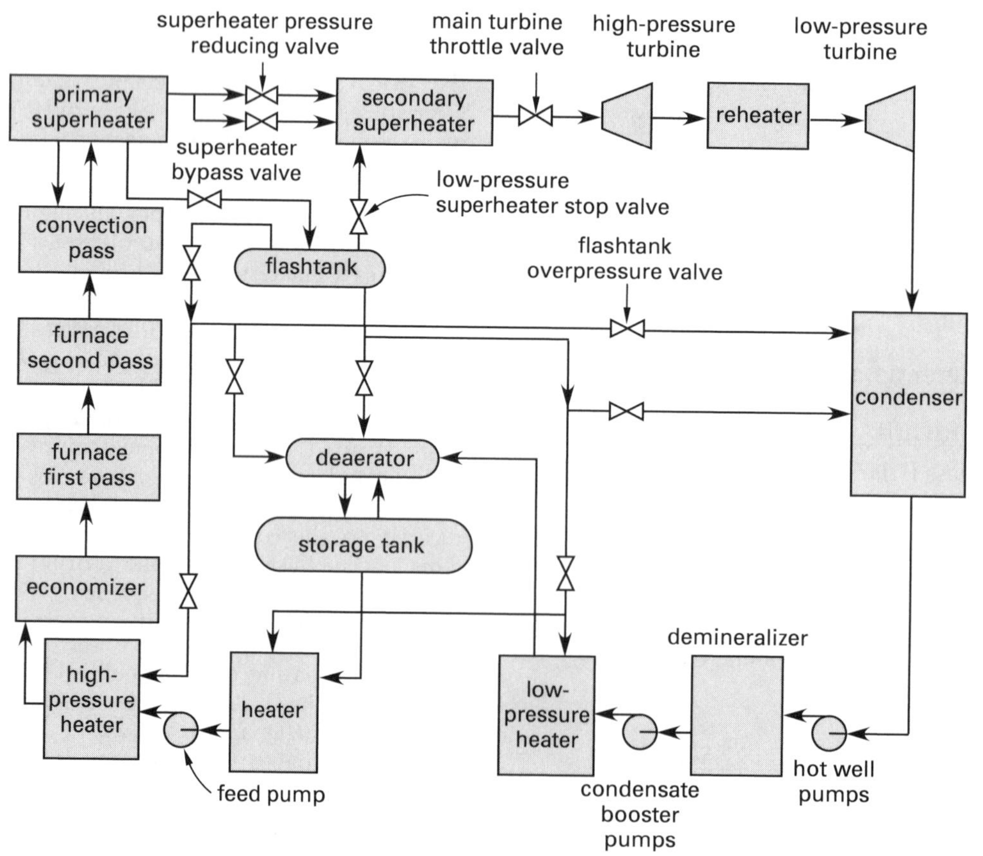

(b) typical once-through operation (bleeds not shown)

Some of the steam and some of the generated electrical power are used to drive auxiliary devices (motors, fans, soot-blowers, pumps, etc.). The *gross electrical output* is the power before the auxiliary loads have been removed; the *net electrical output* is after the auxiliary loads have been removed.

2. FURNACES

Most modern power-generating plants burn coal, some burn natural gas, and a few burn oil. In the past, solid fuel furnaces have been dominated by "pile burning" on grates, and more recently, by "suspension burning" of pulverized coal. Current fluidized-bed combustion offers additional benefits (particularly environmental) over both traditional methods.

The amount of air introduced into the combustion chamber affects the completeness of combustion, pollutant production, and furnace efficiency. For efficient combustion, air may be introduced at more than one point along the combustion path. *Primary air* is introduced first, followed by *secondary air* and *tertiary air*.

Coal in all modern power-generating plants is pulverized to some extent prior to use. In some cases, coal is ground into a fine or microfine powder.[1] The finely ground coal is suspended in a gaseous atmosphere while burning. There are two general ways of feeding *pulverized-coal furnaces*. The *bin system* (*central system*) stores dried and pulverized coal for later use.[2] Air transport, pressure pulse, and screw conveyor systems can be used to transfer pulverized coal to the furnace as needed.[3] *Direct-firing systems* (*unit systems*) pulverize coal on demand. Whether the bin or direct-firing system is used depends on the type of coal and the reliability of the pulverizing and feed equipment.[4]

High-sulfur, high-ash coals, "scrap" fuels, and toxic materials can be burned in *fluidized-bed furnaces*. Solid fuel is turned into a turbulent, fluid-like mass by mixing it with a bed material and blowing air through the mixture at a controlled rate. The bed material can be any solid substance that is not consumed. Sand and limestone are the most common options. The mixture of fuel and bed material becomes fluidized and assumes free-flowing properties when the air flow is at the *fluidizing velocity*.

Fluidized-bed combustion (FBC) uses limestone as the bed material, which considerably reduces some pollutants, thus reducing the need for expensive air-pollution control equipment. Approximately 90% of the sulfur is absorbed by the limestone, forming calcium sulfate. Combustion takes place rapidly at 1500°F to 1750°F (815°C to 950°C), about 400°F (220°C) lower than conventional boilers. This is below the point at which nitrogen oxides are formed.

3. STEAM GENERATORS

A *steam generator* is a combination of a furnace and a boiler. The *boiler* is constructed so that the combustion heat is transferred to feedwater. Modern boilers are *water-tube boilers* (i.e., water passes through tubes surrounded by combustion products).[5] Some of the tubes may be surface mounted or embedded in the walls of the furnace (known as the *setting* or "brickwork").

Other tubes may be placed across the path of the combustion gases. In the typical *cross-drum boiler*, the combustion gases flow perpendicular to the water tubes. Tubes may be straight or bent (i.e., *straight-tube boilers* or *bent-tube boilers*). Steam accumulates in one or more headers at the top of the boiler (known as the *steam drums*).[6] (See Fig. 26.2.)

Figure 26.2 *Cross-Drum, Straight-Tube, Water-Tube Steam Generator*

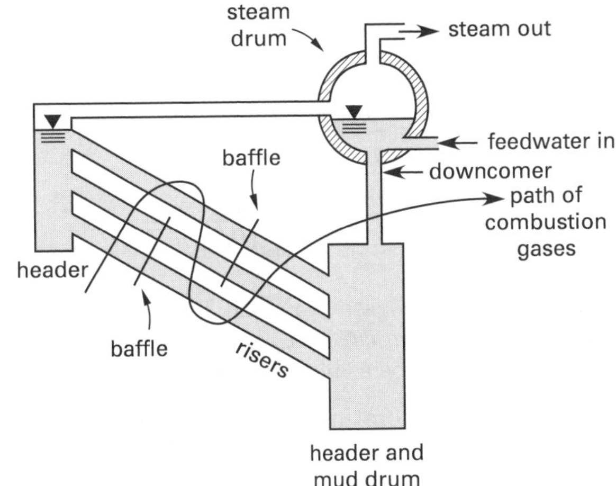

[1]Generally, approximately 65% to 70% of the pulverized coal will pass through a 200-mesh (74-micron) sieve, which has 200 openings per inch. The nominal aperture for 200 mesh is 0.0029 in (0.074 mm), which is like talcum powder. Anthracitic coals can be ground so that up to 90% passes through a 200 mesh sieve. *Micronized coal* is even finer: 80% to 90% will pass through a 325-mesh sieve (43 microns).

[2]An *eductor* is a special type of jet pump that uses air at 20 psig to 80 psig (140 kPa to 550 kPa) to withdraw pulverized coal from an unpressurized storage hopper, transport it in dense phase through a small-diameter feed line, and inject it into the burner. The coal flow rate is controlled by the air pressure. A control valve is not needed.

[3]Ground coal can also be carried for longer distances in slurry form.

[4]Reliability of pulverizers is not the problem it was when pulverizing was first introduced. Pulverizing on demand is now the norm.

[5]*Fire-tube boilers*, in which the combustion products pass through small-diameter tubes and transfer heat to a surrounding water jacket, are seldom used today because they are limited to low-pressure steam. The maximum operating pressure for fire-tube boilers is approximately 150 psi (1 MPa) for riveted construction and 400 psia (3 MPa) for welded construction.

[6]A drum for collecting sediment at the bottom of the boiler is called a *mud drum*.

Circulation of water through the boiler can occur in one of four ways. For pressures below approximately 2800 psi (19.3 MPa), *natural-circulation boilers* can be used. Density is the force driving circulation up the *riser* to the steam drum and back through the *downcomer* to the water-supply header or mud drum. With *controlled-circulation boilers*, common between 2400 psi and 2800 psi (16.6 MPa and 19.3 MPa), a pump keeps the water circulating between the steam drum and water-supply header. In once-through boilers (with moisture separators but no steam drums) and *supercritical-pressure boilers* (with neither steam drums nor separators) that operate above 3200 psi (22.1 MPa) and 1000°F (540°C), water circulates by use of the boilerfeed pump and must be ultrapure to prevent solids build-up.[7] Supercritical boilers are also known as *Benson boilers*. *Ultrasupercritical* (USC) *boilers* routinely operate at 1112°F (600°C) and 4420 psig (30.5 MPa), while demonstration units have operated as high as 1400°F (760°C) and 6090 psig (42 MPa).

Capacities of steam generators are given in mass of steam per hour. This is not an exact determination of the thermodynamic output, of course, unless the conditions of the entering water and leaving steam are also specified. The net energy input to a steam generator, known as the *heat absorption* , is given by Eq. 26.1.[8]

$$q_{\text{in,net}} = \dot{m}_{\text{steam}}(h_{\text{steam}} - h_{\text{feed}}) \qquad 26.1$$

Not all of the combustion energy released in a steam generator will be transferred to the water. The *boiler efficiency*, which ranges from 75% to 90%, is calculated from Eq. 26.2.

$$\eta_{\text{boiler}} = \frac{q_{\text{in,net}}}{q_{\text{in,gross}}} = \frac{\dot{m}_{\text{steam}}(h_{\text{steam}} - h_{\text{feed}})}{\dot{m}_{\text{fuel}}\text{HV}} \qquad 26.2$$

The steam mass flow rate in Eq. 26.2 will be as high as it can be at *maximum capacity*. At *normal capacity*, the boiler efficiency will be as high as it can be.

For an electrical generating system, the *heat rate* (*station rate*, *plant heat rate*, etc.) in Btu/kW-hr (kJ/kW·s) is defined as the total energy input to the steam generator divided by the electrical energy output. Typical full-throttle values in modern, large coal-fired plants are around 7500–8500 Btu/kW-hr (2.2–2.5 kJ/kW·s), with the lower values being more efficient. Values for partial throttle operation and values for older plants are higher. If the output is mechanical, the heat rate is the total energy input divided by the horsepower output.

Water passing through the boiler experiences a small pressure drop due to the friction and expansion. This pressure drop slightly increases the power to pump the water. Calculations are typically based on conditions at the boiler outlet. The error due to neglecting the pressure drop through the boiler is negligible.

Cross-coupling provides flexibility integrating redundant, back-up, excess-capacity, and auxiliary units. Cross-coupling permits any available source to be connected to any load, facilitating operation when units are idle or down for maintenance. In addition, when all loads are online, cross-coupling provides smoother transitions and faster responses to fluctuating demand. With a *two-two system* (i.e., two sources and two loads) in cross-coupled mode, both sources are connected to both loads during periods of high demand, but only one source is connected to either one load or the other during periods of low demand.

"BLEVE" is the acronym for *boiling liquid-expanding vapor explosion*, essentially the catastrophic failure of a container whose walls cannot withstand the pressure of vaporization. The term is most frequently used to describe explosions of tanks containing flammable liquids, but it is equally applicable to boilers. Common elements of BLEVE events are weakening of tank walls (usually by fire) and inadequate or nonfunctional pressure-relief valves.

4. BLOWDOWN

As water is evaporated in the boiler, dissolved and suspended solids are left behind. To prevent these solids from accumulating and causing fouling, constriction, and corrosion, some of the boiler water is bled off, a process known as *blowdown* (or *blowoff*). Blowdown may be intermittent or continuous. The blowdown rate is easily determined from the circulation rate (steam or water) and actual and permitted concentrations. With intermittent blowdown, the mass of water to be released after a period of accumulation, Δt, is given by Eq. 26.3. Concentrations of total solids, C, are typically given in ppm or mg/L.

$$m_{\text{blowdown}} = \left(\frac{C_{\text{feedwater}}}{C_{\text{maximum limit}} - C_{\text{feedwater}}}\right)\Delta t\dot{m}_{\text{steam}} \qquad 26.3$$

For continuous blowdown, the blowdown rate is

$$\dot{m}_{\text{blowdown}} = \left(\frac{C_{\text{feedwater}}}{C_{\text{maximum limit}} - C_{\text{feedwater}}}\right)\dot{m}_{\text{steam}} \qquad 26.4$$

[7]In the not-too-distant past, commercial operating temperatures were restricted to about 1050°F (570°C) by metallurgical considerations. Similarly, commercial pressures were limited to about 3500 psig (24.1 MPa). These limitations no longer restrict supercritical and ultracritical boilers.

[8]In the United States, various confusing and ambiguous abbreviations for units of heat absorption have been used, including kB, kBtu, kBH, kB/hr, and MBH (or MBh) (1000 Btus per hour); and mB, MB, mBtu, MBtu, MMBH, and MB/hr (1,000,000 Btus per hour). In some particularly unfortunate cases, "MB" is used to mean 1000 Btus (no hour rate), and "MMB" is used to mean 1,000,000 Btus (no hour rate). The actual meaning often has to be determined from the context.

Automatic (unattended) blowdown systems work in conjunction with total dissolved solids (TDS) monitors that continuously evaluate the ionic conductivity of the water. Sodium is completely soluble and is the most common ion in boiler water. Automatic blowdown systems operate by maintaining a preselected conductivity set-point in the range of 2400 μS to 2800 μS.

5. PUMPS

Centrifugal pumps are used in power-generation plants. Positive displacement pumps have become nearly obsolete for this application.[9]

The purpose of a pump is to increase the total energy content of the fluid flowing through it. Pumps can be considered adiabatic devices because the fluid gains (or loses) very little heat during the short time it passes through them. If the inlet and outlet are the same size and at the same elevation, the kinetic and potential energy terms can be neglected.[10] Then, the steady flow energy equation (SFEE) reduces to Eq. 26.5. (The second form of Eq. 26.5 is applicable to incompressible liquids only.)

$$W_{\text{pump}} = m(h_2 - h_1) = mv(p_2 - p_1) \qquad 26.5$$

$$P_{\text{pump}} = \dot{m}(h_2 - h_1) \qquad 26.6$$

Equation 26.5 assumes that the pump is capable of isentropic compression. However, due to inefficiencies, some of the input energy is converted to heat. This heat energy raises the temperature (and hence, the internal energy portion of enthalpy) without increasing the pressure. The actual exit enthalpy, h_2', takes the inefficiency into consideration.

$$h_2' = h_1 + \frac{h_2 - h_1}{\eta_s} \qquad 26.7$$

Combining Eq. 26.6 and Eq. 26.7, the actual input pump power is

$$P_{\text{pump}}' = \dot{m}(h_2' - h_1) = \frac{\dot{m}(h_2 - h_1)}{\eta_s \eta_m} \qquad 26.8$$

The ideal exit enthalpy, h_2, can be calculated from the specific volume and the exit pressure, p_2. The calculation is simplified if the fluid is assumed to be incompressible. (Equation 26.9 cannot be used with compression of gases and vapors.)

$$h_2 = h_1 + v(p_2 - p_1)\Big|_{\substack{\text{isentropic}\\ \text{incompressible}}} \qquad \text{[SI]} \qquad 26.9(a)$$

$$h_2 = h_1 + \frac{v(p_2 - p_1)}{J}\Big|_{\substack{\text{isentropic}\\ \text{incompressible}}} \qquad \text{[U.S.]} \qquad 26.9(b)$$

The SFEE can also be used to determine pump net work when written as an extended form of the Bernoulli equation in terms of energy per unit mass.

Mechanical Energy Equation in Terms of Energy Per Unit Mass

$$\frac{p_{\text{in}}}{\rho} + \frac{\text{v}_{\text{in}}^2}{2} + gh_{\text{in}} + w_{\text{shaft}} = \frac{p_{\text{out}}}{\rho} + \frac{\text{v}_{\text{out}}^2}{2} + gh_{\text{out}} + w_{\text{loss}} \qquad 26.10$$

The w_{loss} term includes the non-isentropic compression and pump bearing and mechanical friction losses. The w_{shaft} term represents net work (i.e., the remaining work available after adjusting for thermodynamic and mechanical losses).

Neglecting the kinetic and potential energy terms, the steady flow energy equation reduces to

$$w_{\text{pump}} = w_{\text{shaft}} - w_{\text{loss}} = \frac{p_{\text{out}}}{\rho} - \frac{p_{\text{in}}}{\rho} \qquad 26.11$$

6. PUMP EFFICIENCY

Line 1-to-2 in Fig. 26.3 illustrates the ideal condition line for a pump. Since the compression is isentropic, the line is directed vertically upward. The actual work, however, results in an increase in entropy. Line 1-to-2′ is the actual condition line for non-isentropic compression.

The ideal and actual work for a pump are

$$W_{\text{ideal}} = m(h_2 - h_1) = W_{\text{shaft}} - W_{\text{loss}} \qquad 26.12$$

$$W_{\text{actual}} = m(h_2' - h_1) \qquad 26.13$$

The definition of *isentropic efficiency*, η_s, for a pump is the inverse of what it is for a turbine.

$$\eta_s = \frac{W_{\text{ideal}}}{W_{\text{actual}}} = \frac{W_{\text{shaft}} - W_{\text{loss}}}{W_{\text{actual}}} = \frac{h_2 - h_1}{h_2' - h_1} \qquad 26.14$$

[9]Obsolescence is due to the difficulty in controlling reciprocating pumps during start-up and partial-load. With modern variable-frequency drives, however, this may no longer be the case.

[10]Even if the pump inlet and outlet are different sizes and at different elevations, the kinetic and potential energy changes are small compared to the pressure energy increase.

Power Cycles

Figure 26.3 *Compression by a Pump*

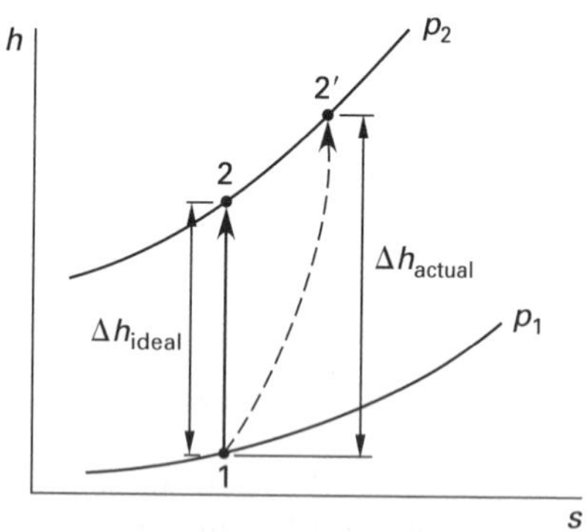

Equation 26.15 defines the pump efficiency. Since W_{ideal} is small—less than 5 Btu/lbm (10 kJ/kg)—and does not affect the thermal efficiency greatly, the pump mechanical efficiency is often ignored, and η_{pump} is taken as η_s.

$$\eta_{\text{pump}} = \eta_s \eta_m \tag{26.15}$$

Equation 26.16 provides a modified version of pump efficiency where W_{shaft} is the net shaft energy input and W_{loss} is the combined non-isentropic compression and pump bearing/mechanical friction losses.

Efficiency

$$\begin{aligned}\eta_{\text{pump}} = \eta_s \eta_m &= \frac{(W_{\text{shaft}} - W_{\text{loss}})}{W_{\text{shaft}}} \\ &= \frac{(w_{\text{shaft}} - w_{\text{loss}})}{w_{\text{shaft}}}\end{aligned} \tag{26.16}$$

Example 26.1

A pump in a steam power plant operates under an inlet state of 1.2 psi (8.3 kPa) and 108°F (42°C) with an overall pump efficiency of 85%. The pump work is approximately 2.6 Btu/lbm (6.0 kJ/kg). What is the pump outlet pressure?

Solution

From Eq. 26.16, neglecting mechanical losses,

Efficiency

$$\eta_{\text{pump}} = \eta_s = \frac{(w_{\text{shaft}} - w_{\text{loss}})}{w_{\text{shaft}}}$$

Remembering that w_{loss} includes the non-isentropic compression,

$$\begin{aligned}\frac{w_{\text{pump}}}{\eta_s} &= w_{\text{shaft}} - w_{\text{loss}} \\ &= \frac{p_{\text{out}}}{\rho} - \frac{p_{\text{in}}}{\rho} \\ &= v(p_{\text{out}} - p_{\text{in}}) \\ p_{\text{out}} &= \frac{w_{\text{pump}}}{v\eta_s} + p_{\text{in}}\end{aligned}$$

SI Solution

From a table of properties for saturated steam, the specific volume, v_f, for a pressure of 8.27 kPa is 0.0010 m³/kg. (Note: 1 kJ/m³ = 1 kPa) [**Properties of Saturated Water and Steam (Pressure) - SI Units**]

Solving for the outlet pressure gives

$$\begin{aligned}p_{\text{out}} &= \frac{w_{\text{pump}}}{v\eta_s} + p_{\text{in}} \\ &= \frac{6.0\ \frac{\text{kJ}}{\text{kg}}}{\left(0.0010\ \frac{\text{m}^3}{\text{kg}}\right)(0.85)} + 8.3\ \text{kPa} \\ &= 7067.1\ \text{kPa}\end{aligned}$$

U.S. Customary Solution

From a table of properties for saturated steam, the specific volume, v_f, for a pressure of 1.2 psi is 0.0162 ft³/lbm. [**Properties of Saturated Water and Steam (Pressure) - SI Units**]

Solving the equation for the outlet pressure gives

$$\begin{aligned}p_{\text{out}} &= \frac{w_{\text{pump}}}{v\eta_s} + p_{\text{in}} \\ &= \frac{\left(2.6\ \frac{\text{Btu}}{\text{lbm}}\right)\left(778.17\ \frac{\text{ft-lbf}}{\text{Btu}}\right)}{\left(0.0162\ \frac{\text{ft}^3}{\text{lbm}}\right)(0.85)\left(12\ \frac{\text{in}}{\text{ft}}\right)^2} + 1.5\ \frac{\text{lbf}}{\text{in}^2} \\ &= 1021.8\ \text{lbf/in}^2\end{aligned}$$

7. TEMPERATURE INCREASE IN PRESSURIZED LIQUIDS

The temperature of water (or any other liquid) increases when it is pressurized (i.e., passes through a pump). However, the increase in temperature is very small—far less than what would occur with a gas or vapor.

The increase in temperature can be thought of as being caused by two processes: an isentropic compression and an irreversible compression (if the pump is not 100% efficient). The increase due to isentropic compression is so small as to be negligible. For example, the increase in temperature of 100°F (38°C) water compressed isentropically from saturation to 1000 psi (6.9 MPa) is approximately 0.3°F (0.17°C).

The irreversible portion of the process incorporates the additional energy that must be put into the water to overcome friction and turbulence. These factors produce a profound heating effect (as compared to the isentropic portion). It is proper to consider only this irreversible work when calculating the increase in water temperature. This is because the change in enthalpy is a combination of changes in internal energy and flow work (i.e., $h = u + pV$), but only the internal energy change manifests as a temperature change.

Example 26.2

A boilerfeed pump increases the pressure of 90°F (30°C) water from 1 atm to 150 psia (1000 kPa). The pump's isentropic efficiency is 80%. The water's specific heat is 1.0 Btu/lbm-°F (4.187 kJ/kg·K). What is the temperature of the water leaving the pump?

SI Solution

Since the properties of a liquid are essentially independent of pressure, the properties of 30°C water can be read from a saturated steam table. [**Properties of Saturated Water and Steam (Temperature) - SI Units**]

$$h_{30°C} = 125.73 \text{ kJ/kg}$$
$$v_{30°C} = 1.0044 \text{ cm}^3/\text{g} = 0.0010044 \text{ m}^3/\text{kg}$$

The ideal enthalpy of the feedwater entering the boiler (point 2 on Fig. 26.3) is equal to the enthalpy at point 1 plus the energy put into the water by the pump. Assuming the water is incompressible, the specific volumes at points 1 and 2 are the same. From Eq. 26.9,

$$\begin{aligned} h_2 &= h_1 + v_1(p_2 - p_1) \\ &= 125.73 \text{ kJ/kg} \\ &\quad + \frac{\left(0.0010044 \frac{\text{m}^3}{\text{kg}}\right)(1000 \text{ kPa} - 101.3 \text{ kPa}) \times \left(1000 \frac{\text{Pa}}{\text{kPa}}\right)}{1000 \frac{\text{J}}{\text{kJ}}} \\ &= 126.63 \text{ kJ/kg} \end{aligned}$$

This calculation assumes that the pump is capable of isentropic compression. Because of the pump's inefficiency, not all of the 0.903 kJ/kg goes into raising the pressure. Some of the energy goes into raising the temperature. (Since $h = u + pv$, both energy contributions increase the enthalpy.) Therefore, to get to 1000 kPa, more than 0.903 kJ/kg must be added to the water. The actual enthalpy at point 2′ is

$$\begin{aligned} h_2' &= h_1 + \frac{h_2 - h_1}{\eta_s} = 125.73 \frac{\text{kJ}}{\text{kg}} + \frac{0.903 \frac{\text{kJ}}{\text{kg}}}{0.80} \\ &= 126.86 \text{ kJ/kg} \end{aligned}$$

The enthalpy was increased by 1.13 kJ/kg, but only the irreversible portion is considered when determining the temperature change.

$$\begin{aligned} h_{\text{irreversible}} &= 1.13 \frac{\text{kJ}}{\text{kg}} - 0.903 \frac{\text{kJ}}{\text{kg}} \\ &= 0.227 \text{ kJ/kg} \end{aligned}$$

Since the specific heat is 4.187 kJ/kg·K, the final temperature is

$$\begin{aligned} T_2' &= T_1 + \frac{\Delta h}{c_p} = 30°\text{C} + \frac{0.227 \frac{\text{kJ}}{\text{kg}}}{4.187 \frac{\text{kJ}}{\text{kg·K}}} \\ &= 30.05°\text{C} \end{aligned}$$

Customary U.S. Solution

Since the properties of a liquid are essentially independent of pressure, the properties of 90°F water can be read from a saturated steam table. [**Properties of Saturated Water and Steam (Temperature) - SI Units**]

$$h_{90°F} = 58.05 \text{ Btu/lbm}$$
$$v_{90°F} = 0.01610 \text{ ft}^3/\text{lbm}$$

The ideal enthalpy of the feedwater entering the boiler (point 2 on Fig. 26.3) is equal to the enthalpy at point 1 plus the energy put into the water by the pump. Assuming the water is incompressible, the specific volumes at points 1 and 2 are the same. From Eq. 26.9,

$$\begin{aligned} h_2 &= h_1 + \frac{v_1(p_2 - p_1)}{J} \\ &= 58.05 \text{ Btu/lbm} \\ &\quad + \frac{\left(0.01610 \frac{\text{ft}^3}{\text{lbm}}\right)\left(150 \frac{\text{lbf}}{\text{in}^2} - 14.7 \frac{\text{lbf}}{\text{in}^2}\right)\left(12 \frac{\text{in}}{\text{ft}}\right)^2}{778 \frac{\text{ft-lbf}}{\text{Btu}}} \\ &= 58.45 \text{ Btu/lbm} \end{aligned}$$

This calculation assumes that the pump is capable of isentropic compression. Because of the pump's inefficiency, not all of the 0.403 Btu/lbm goes into raising the pressure. Some of the energy goes into raising the temperature. (Since $h = u + pv$, both energy contributions

Power Cycles

increase the enthalpy.) Therefore, to get to 150 psia, more than 0.403 Btu/lbm must be added to the water. The actual enthalpy at point 2′ is

$$\begin{aligned} h_2' &= h_1 + \frac{h_2 - h_1}{\eta_s} = 58.05\ \frac{\text{Btu}}{\text{lbm}} + \frac{0.403\ \frac{\text{Btu}}{\text{lbm}}}{0.80} \\ &= 58.55\ \text{Btu/lbm} \end{aligned}$$

The enthalpy was increased by 0.5 Btu/lbm, but only the irreversible portion is considered when determining the temperature change.

$$\begin{aligned} \Delta h_{\text{irreversible}} &= 0.5\ \frac{\text{Btu}}{\text{lbm}} - 0.403\ \frac{\text{Btu}}{\text{lbm}} \\ &= 0.097\ \text{Btu/lbm} \end{aligned}$$

Since the specific heat is 1.0 Btu/lbm-°F, the corresponding temperature increase is 0.097°F. The final temperature is 90°F + 0.097°F ≈ 90.1°F.

8. TURBINES

There are two general categories of steam turbine operation. A *reaction turbine* consists of a rotating drum with small nozzles (reaction jets) located around the drum's periphery. Steam is discharged from the nozzles, and the drum turns in reaction to the steam's action. An *impulse turbine* is characterized by stationary jets discharging against vanes mounted on the periphery of a wheel. Modern steam turbines use both principles to extract energy from the steam. [**Turbines**]

There are several designations given to turbines. A *back pressure turbine* (*topping turbine* or *superposed turbine*) exhausts either to a high-pressure industrial process or to a second turbine operating in a lower-pressure range.[11] In a two-turbine installation, the first is designated as the *high-pressure* (HP) *turbine*, and the second is known as the *low-pressure* (LP) *turbine*. (If there are three turbines in series, the middle one is known as an *intermediate-pressure* (IP) *turbine*.) Turbines that operate at supercritical conditions are known as *supercritical-pressure* (SP) *turbines*.

High-, intermediate-, and low-pressure turbines that are integrated into a single cycle are known as *compound turbines*. *Tandem compound turbines* are high- and low-pressure turbines integrated into a single mounting (but having two separate casings for steam at different conditions) and sharing a single shaft. Partially expanded steam from the high-pressure turbine casing is routed through the *crossover pipe* to the low-pressure turbine casing. The single shaft is connected to a single load, such as a generator. *Cross-compound turbines* do not share a single shaft. Steam from the high-pressure or intermediate turbine is routed to a separate low-pressure turbine that has its own generator.

Some of the steam entering a turbine may be removed before it has expanded to the turbine exit pressure. This removal is known as a *bleed*, and the steam is known as *bleed steam*. The *extraction rate* or *bleed rate* is the rate, usually expressed in lbm/hr (kg/h) or in Btu/hr (kW), at which the partially expanded steam is bled off. A turbine with one or more bleeds is known as an *extraction turbine*.

In a *reheat turbine*, steam is removed from the turbine, returned to its original temperature in a reheater, and then returned to a lower-pressure portion of the same turbine.

9. TURBINE WORK, POWER, AND EFFICIENCY

Turbines can generally be thought of as pumps operating in reverse. A turbine extracts energy from the fluid which, in turn, decreases in temperature and pressure. The process is essentially adiabatic because the fluid loses very little heat during the short time it passes through the turbine.

Figure 26.4 diagrams the expansion of steam in a turbine from a high pressure, p_1, to a low pressure, p_2. The broken line is the *condition line*—the locus of all states of steam during the expansion process. The ideal condition line is isentropic, which is represented by a vertical line downward on the Mollier (h versus s) diagram.

Figure 26.4 *Single-Stage Turbine Expansion*

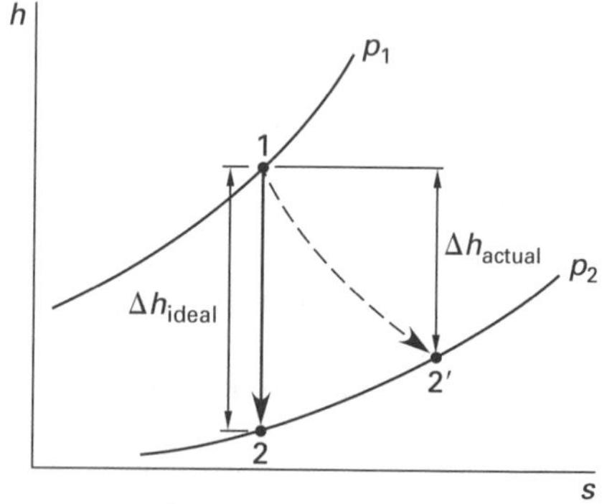

An expanded version of the steady flow energy equation can be used to determine the work output of the turbine.

[11] Topping turbines are added when older, low-pressure plants are "repowered." The furnace and steam generator are replaced so that high-pressure steam is produced, and a topping turbine is added. The topping turbine uses the high pressure steam. The original low-pressure turbine uses the exhaust steam from the topping turbine. Most of the remaining equipment in the cycle is unchanged.

Mechanical Energy Equation in Terms of Energy Per Unit Mass

$$\frac{p_{in}}{\rho} + \frac{v_{in}^2}{2} + gh_{in} = \frac{p_{out}}{\rho} + \frac{v_{out}^2}{2} + gh_{out} + w_{shaft} + w_{loss} \quad 26.17$$

w_{shaft} refers specifically to net shaft work (after adjusting for thermodynamic and mechanical losses). Equation 26.17 reduces to

$$w_{ideal} = w_{shaft} + w_{loss} = \frac{p_{in}}{\rho} - \frac{p_{out}}{\rho} \quad 26.18$$

Because changes in potential and kinetic energies are small and can be neglected, the ideal steady flow energy equation can also be reduced to Eq. 26.19 to predict the work output of the turbine. The turbine work is known as *shaft work* because it is transmitted from the turbine to a generator through a shaft.[12,13]

$$w_{ideal} = m(h_1 - h_2) \quad 26.19$$

Equation 26.19 assumes that the turbine is capable of isentropic expansion. However, not all of the available fluid energy can be extracted. Friction within the turbine increases the entropy without decreasing the pressure. The actual exit enthalpy, h_2', takes this inefficiency into consideration. If the expansion process is not isentropic, entropy will increase and the actual final enthalpy, h_2', will be higher than the ideal final enthalpy, h_2.

The *isentropic efficiency (adiabatic efficiency)*, η_s, of a turbine is the ratio of actual to ideal energy extractions.[14] Actual isentropic efficiencies vary from approximately 65% for 1 MW unit to over 90% for 100 MW and larger units.

$$\eta_s = \frac{w_{actual}}{w_{ideal}} = \frac{h_1 - h_2'}{h_1 - h_2} \quad 26.20$$

The actual energy extracted is

$$W_{actual} = m(h_1 - h_2') = m\eta_s(h_1 - h_2) \quad 26.21$$

$$h_2' = h_1 - \eta_s(h_1 - h_2) \quad 26.22$$

In addition to thermodynamic friction losses, there are additional friction losses in bearings and other moving mechanical parts.[15] The net turbine work is

$$W_{turbine} = W_{actual} - W_{friction} = \eta_m W_{actual} \quad 26.23$$

Most steam turbines have high mechanical efficiencies—in the order of 98%. Inasmuch as the friction losses are very small, the isentropic and turbine efficiencies are essentially identical. However, the *overall turbine efficiency* incorporates the mechanical friction losses.

$$\eta_{turbine} = \eta_s\eta_m \quad 26.24$$

Equation 26.25 provides a slightly modified version of turbine efficiency where W_{shaft} is the net shaft energy output and W_{loss} is equivalent to the sum of the turbine thermodynamic and mechanical friction losses.

Efficiency

$$\eta_{turbine} = \eta_s\eta_m = \frac{W_{shaft}}{W_{shaft} + W_{loss}} = \frac{w_{shaft}}{w_{shaft} + w_{loss}} \quad 26.25$$

The isentropic efficiency of a device affects both the flow rate and the thermodynamic properties of the substance flowing through the device. The mechanical efficiency affects the flow rate, but it does not affect the thermodynamic properties and should not be used to determine final enthalpies. Both the isentropic and mechanical efficiencies reduce the amount of useful energy that can be generated in a turbine, and they affect the amount of substance flowing through the turbine. The actual power generated in the turbine is

$$P_{turbine} = \dot{m}_{actual}\eta_{turbine}(h_1 - h_2) \quad 26.26$$

$$\dot{m}_{actual} = \frac{\dot{m}_{ideal}}{\eta_{turbine}} = \frac{\dot{m}_{ideal}}{\eta_s\eta_m} \quad 26.27$$

Example 26.3

A turbine in a steam power plant operates under an inlet state of 580 psi (4.0 MPa) and 725°F (385°C) and an outlet state of 1.2 psi (8.3 kPa). The actual work of the turbine is 410 Btu/lbm (950 kJ/kg). The turbine has an overall efficiency of 85% and a mechanical efficiency of 98%. What are the (a) friction loss, (b) the turbine isentropic efficiency, and (c) the turbine thermodynamic loss?

Power Cycles

[12] Other names for shaft work are *brake work*, *useful work*, and *net work*.
[13] Shaft work in this context is used in the most general sense.
[14] The terminology "adiabatic efficiency" is ambiguous because an "adiabatic system" does not require a 100% efficient turbine.
[15] W_{loss} includes both thermodynamic and mechanical friction losses, whereas $W_{friction}$ captures only the mechanical friction losses.

Solution

From Eq. 26.16, neglecting mechanical losses,

Efficiency

$$\eta_{\text{pump}} = \eta_s = \frac{(w_{\text{shaft}} - w_{\text{loss}})}{w_{\text{shaft}}}$$

Remembering that w_{loss} includes the non-isentropic compression,

$$\begin{aligned}\frac{w_{\text{pump}}}{\eta_s} &= w_{\text{shaft}} - w_{\text{loss}} \\ &= \frac{p_{\text{out}}}{\rho} - \frac{p_{\text{in}}}{\rho} \\ &= v(p_{\text{out}} - p_{\text{in}}) \\ p_{\text{out}} &= \frac{w_{\text{pump}}}{v\eta_s} + p_{\text{in}}\end{aligned}$$

SI Solution

(a) Use Eq. 26.23 to find the net work of the turbine.

$$\begin{aligned}W_{\text{turbine}} = W_{\text{shaft}} &= W_{\text{actual}} - W_{\text{friction}} \\ &= \eta_m W_{\text{actual}} \\ &= (0.98)\left(950.0\ \frac{\text{kJ}}{\text{kg}}\right) \\ &= 931.0\ \text{kJ/kg}\end{aligned}$$

Subtract the net work from the actual work to find the friction loss.

$$\begin{aligned}W_{\text{friction}} &= 950.0\ \frac{\text{kJ}}{\text{kg}} - 931.0\ \frac{\text{kJ}}{\text{kg}} \\ &= 19.0\ \text{kJ/kg}\end{aligned}$$

(b) From Eq. 26.25, for overall turbine efficiency,

Efficiency

$$\eta_{\text{turbine}} = \eta_s\eta_m$$

Solving for the isentropic efficiency gives,

$$\eta_s = \frac{\eta_{\text{turbine}}}{\eta_m} = \frac{0.85}{0.98} = 0.87$$

Use the equation for efficiency, and solve for the total losses, W_{loss}. Remember that W_{loss} includes both the non-isentropic component and mechanical friction losses.

Efficiency

$$\eta_{\text{turbine}} = \eta_s\eta_m = \frac{W_{\text{shaft}}}{(W_{\text{shaft}} + W_{\text{loss}})}$$

$$\begin{aligned}W_{\text{loss}} &= \frac{W_{\text{shaft}}}{\eta_{\text{turbine}}} - W_{\text{shaft}} \\ &= \frac{931.0\ \frac{\text{kJ}}{\text{kg}}}{0.85} - 931.0\ \frac{\text{kJ}}{\text{kg}} \\ &= 164.3\ \text{kJ/kg}\end{aligned}$$

(c) Total losses can be broken down into thermodynamic and friction losses.

$$W_{\text{loss}} = W_{\text{thermodynamic}} + W_{\text{friction}}$$

Solving for the thermodynamic loss gives

$$\begin{aligned}W_{\text{thermodynamic}} &= W_{\text{loss}} - W_{\text{friction}} \\ &= 164.3\ \frac{\text{kJ}}{\text{kg}} - 19.0\ \frac{\text{kJ}}{\text{kg}} \\ &= 145.3\ \text{kJ/kg}\end{aligned}$$

U.S. Customary Solution

(a) Use Eq. 26.23 to find the net work of the turbine.

$$\begin{aligned}W_{\text{turbine}} = W_{\text{shaft}} = W_{\text{actual}} - W_{\text{friction}} &= \eta_m W_{\text{actual}} \\ &= (0.98)\left(410\ \frac{\text{Btu}}{\text{lbm}}\right) \\ &= 401.8\ \text{Btu/lbm}\end{aligned}$$

Subtract the net work from the actual work to find the friction loss.

$$\begin{aligned}W_{\text{friction}} &= 410\ \frac{\text{Btu}}{\text{lbm}} - 401.8\ \frac{\text{Btu}}{\text{lbm}} \\ &= 8.2\ \text{Btu/lbm}\end{aligned}$$

(b) From Eq. 26.25, for overall turbine efficiency,

Efficiency

$$\eta_{\text{turbine}} = \eta_s\eta_m$$

Solving for the isentropic efficiency gives,

$$\eta_s = \frac{\eta_{\text{turbine}}}{\eta_m} = \frac{0.85}{0.98} = 0.87$$

Use the equation for efficiency, and solve for the total losses, W_{loss}. Remember that W_{loss} includes both the non-isentropic component and mechanical friction losses.

Efficiency

$$\eta_{turbine} = \eta_s\eta_m = \frac{W_{shaft}}{(W_{shaft} + W_{loss})}$$

$$\begin{aligned} W_{loss} &= \frac{W_{shaft}}{\eta_{turbine}} - W_{shaft} \\ &= \frac{401.8\ \frac{\text{Btu}}{\text{lbm}}}{0.85} - 401.8\ \frac{\text{Btu}}{\text{lbm}} \\ &= 70.9\ \text{Btu/lbm} \end{aligned}$$

(c) Total losses can be broken down into thermodynamic and friction losses.

$$W_{loss} = W_{thermodynamic} + W_{friction}$$

Solving for the thermodynamic loss gives

$$\begin{aligned} W_{thermodynamic} &= W_{loss} - W_{friction} \\ &= 70.9\ \frac{\text{Btu}}{\text{lbm}} - 8.2\ \frac{\text{Btu}}{\text{lbm}} \\ &= 62.7\ \text{Btu/lbm} \end{aligned}$$

10. TWO-STAGE EXPANSION

If expansion through the turbine is multiple stage or if the turbine has a bleed at an intermediate pressure, p_m, the expansion process for each stage will be as illustrated by Fig. 26.5. In this figure, a fraction y of the original steam mass is removed after expanding to pressure p_m. The remaining fraction $1 - y$ expands to pressure p_2.

Figure 26.5 *Two-Stage Turbine Expansion*

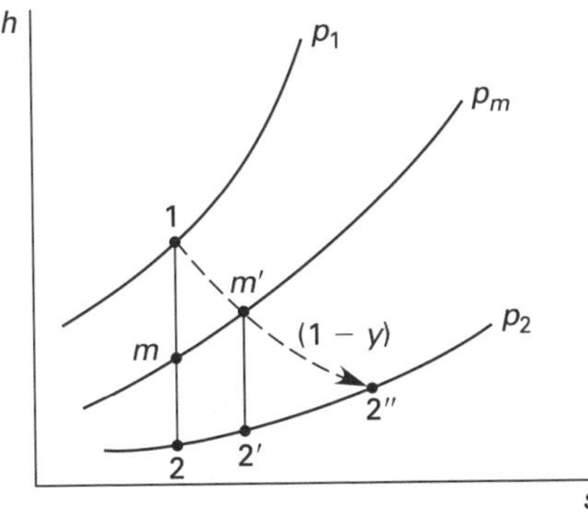

The ideal and actual work generated in the first stage are

$$W_{ideal,1} = m(h_1 - h_m) \qquad \text{26.28}$$

$$\begin{aligned} W_{actual,1} &= m(h_1 - h'_m) \\ &= \eta_{s,1}m(h_1 - h_m) \end{aligned} \qquad \text{26.29}$$

The ideal and actual work generated by the remaining fraction, $1 - y$, in the second stage are

$$W_{ideal,2} = (1-y)m(h'_m - h'_2) \qquad \text{26.30}$$

$$\begin{aligned} W_{actual,2} &= (1-y)m(h'_m - h''_2) \\ &= (1-y)m\eta_{s,2}(h'_m - h'_2) \end{aligned} \qquad \text{26.31}$$

The total work done per unit mass of the steam is

$$\begin{aligned} W_{turbine} &= W_{actual,1} + W_{actual,2} \\ &= m\big(h_1 - h'_m + (1-y)(h'_m - h''_2)\big) \\ &= m\big(\eta_{s,1}(h_1 - h_m) + (1-y)\eta_{s,2}(h'_m - h'_2)\big) \end{aligned} \qquad \text{26.32}$$

Example 26.4

Steam is expanded from 700°F (360°C) and 200 psia (1500 kPa) to 5 psia (50 kPa) in an 87% efficient turbine. What is the final enthalpy of the steam?

SI Solution

Refer to Fig. 26.4. Use superheated steam tables. [**Properties of Superheated Steam - SI Units**]

$$\begin{aligned} h_1 &= 3169.8\ \text{kJ/kg} \\ s_1 &= 7.14\ \text{kJ/kg·K} \end{aligned}$$

At this point, proceed as if the turbine is 100% efficient. Use the saturated steam tables for 50 kPa, to find the entropy and enthalpy values of the steam entering the turbine. [**Properties of Saturated Water and Steam (Pressure) - SI Units**]

$$\begin{aligned} s_f &= 1.0912\ \text{kJ/kg·K} \\ s_g &= 7.5930\ \text{kJ/kg·K} \\ h_f &= 340.5\ \text{kJ/kg} \\ h_{fg} &= 2304.7\ \text{kJ/kg} \end{aligned}$$

Since at this point it is assumed that the expansion is isentropic (i.e., 100% efficient), $s_2 = s_1$. The quality at point 2 can be found as

$$\begin{aligned} x &= \frac{s_2 - s_f}{s_g - s_f} = \frac{7.14\ \frac{\text{kJ}}{\text{kg·K}} - 1.0912\ \frac{\text{kJ}}{\text{kg·K}}}{7.5930\ \frac{\text{kJ}}{\text{kg·K}} - 1.0912\ \frac{\text{kJ}}{\text{kg·K}}} \\ &= 0.9303 \end{aligned}$$

Power Cycles

The ideal final enthalpy can be found from the quality.

$$\begin{aligned} h_2 &= h_f + x h_{fg} \\ &= 340.5\ \frac{\text{kJ}}{\text{kg}} + (0.9303)\left(2304.7\ \frac{\text{kJ}}{\text{kg}}\right) \\ &= 2484.6\ \text{kJ/kg} \end{aligned}$$

However, this value of h_2 assumes the expansion through the turbine is isentropic. Since the turbine is capable of extracting only 87% of the ideal energy, the actual final enthalpy is

$$\begin{aligned} h_2' &= h_1 - \eta_s(h_1 - h_2) \\ &= 3169.8\ \frac{\text{kJ}}{\text{kg}} - (0.87)\left(3169.8\ \frac{\text{kJ}}{\text{kg}} - 2484.6\ \frac{\text{kJ}}{\text{kg}}\right) \\ &= 2573.7\ \text{kJ/kg} \end{aligned}$$

Customary U.S. Solution

Refer to Fig. 26.4. Use superheated steam tables. [**Properties of Superheated Steam - I-P Units**]

$$\begin{aligned} h_1 &= 1374.1\ \text{Btu/lbm} \\ s_1 &= 1.724\ \text{Btu/lbm-°F} \end{aligned}$$

At this point, proceed as though the turbine is 100% efficient. Use saturated steam tables for 5 psia to find the entropy and enthalpy values for the the steam entering the turbine. [**Properties of Saturated Water and Steam (Pressure) — I-P Units**]

$$\begin{aligned} s_f &= 0.2348\ \text{Btu/lbm-°F} \\ s_{fg} &= 1.6092\ \text{Btu/lbm-°F} \\ h_f &= 130.13\ \text{Btu/lbm} \\ h_{fg} &= 1000.57\ \text{Btu/lbm} \end{aligned}$$

Since it is assumed that the expansion is isentropic (i.e., 100% efficient), $s_2 = s_1$. The quality at point 2 can be found as

$$\begin{aligned} x &= \frac{s_2 - s_f}{s_{fg}} = \frac{1.724\ \frac{\text{Btu}}{\text{lbm-°F}} - 0.2348\ \frac{\text{Btu}}{\text{lbm-°F}}}{1.6092\ \frac{\text{Btu}}{\text{lbm-°F}}} \\ &= 0.9254 \end{aligned}$$

The ideal final enthalpy can be found from the quality.

$$\begin{aligned} h_2 &= h_f + x h_{fg} \\ &= 130.13\ \frac{\text{Btu}}{\text{lbm}} + (0.9254)\left(1000.57\ \frac{\text{Btu}}{\text{lbm}}\right) \\ &= 1056.1\ \text{Btu/lbm} \end{aligned}$$

However, this value of h_2 assumes the expansion through the turbine is isentropic. Since the turbine is capable of extracting only 87% of the ideal energy, the actual final enthalpy is

$$\begin{aligned} h_2' &= h_1 - \eta_s(h_1 - h_2) \\ &= 1374.1\ \frac{\text{Btu}}{\text{lbm}} - (0.87)\left(1374.1\ \frac{\text{Btu}}{\text{lbm}} - 1056.1\ \frac{\text{Btu}}{\text{lbm}}\right) \\ &= 1097.4\ \text{Btu/lbm} \end{aligned}$$

Example 26.5

Repeat Ex. 26.4 using the Mollier (h versus s) diagram.

SI Solution

h_1 is read directly from the Mollier diagram at the intersection of 360°C and 1500 kPa. (Greater accuracy is possible with a large Mollier diagram.)

$$h_1 \approx 3169.8\ \text{kJ/kg}$$

The ideal final enthalpy, h_2, is found by dropping straight down to the 50 kPa line. h_2 is read as approximately 2485 kJ/kg. Since the turbine is capable of extracting only 87% of the ideal energy, the actual final enthalpy is calculated from Eq. 26.22. [**Properties of Superheated Steam - SI Units**]

$$\begin{aligned} h_2' &= h_1 - \eta_s(h_1 - h_2) \\ &= 3169.8\ \frac{\text{kJ}}{\text{kg}} - (0.87)\left(3169.8\ \frac{\text{kJ}}{\text{kg}} - 2485\ \frac{\text{kJ}}{\text{kg}}\right) \\ &= 2574\ \text{kJ/kg} \end{aligned}$$

Customary U.S. Solution

h_1 is read directly from the Mollier diagram at the intersection of 700°F and 200 psia. (Greater accuracy is possible with a large Mollier diagram.)

$$h_1 \approx 1374.1\ \text{Btu/lbm}$$

The ideal final enthalpy, h_2, is found by dropping straight down to the 5 psia line. h_2 is read as approximately 1055 Btu/lbm. Since the turbine is capable of extracting only 87% of the ideal energy, the actual final enthalpy is calculated from Eq. 26.22.

$$\begin{aligned} h_2' &= h_1 - \eta_s(h_1 - h_2) \\ &= 1374.1\ \frac{\text{Btu}}{\text{lbm}} - (0.87)\left(1374.1\ \frac{\text{Btu}}{\text{lbm}} - 1055\ \frac{\text{Btu}}{\text{lbm}}\right) \\ &= 1096\ \text{Btu/lbm} \end{aligned}$$

11. HEAT EXCHANGERS

A *heat exchanger* transfers energy from one fluid to another through a wall separating them. If the heat transfer is assumed to be adiabatic, the total energy of both input streams must be the same as the total energy of both output streams. No work is done within a heat exchanger, and the potential and kinetic energies of the fluids can be ignored.

$$\dot{m}_A h_{A,\text{in}} + \dot{m}_B h_{B,\text{in}} = \dot{m}_A h_{A,\text{out}} + \dot{m}_B h_{B,\text{out}} \qquad 26.33$$

12. CONDENSERS

Condensers are special-purpose heat exchangers that remove the heat of vaporization from steam.[16] Steam passes over tubes in which cooling water (or, occasionally, air) passes. Liquid water falls to the lower portion of the condenser known as the *hot well* or *condensate well*. Heat is transferred through the tubing walls to the cooling water and is then rejected to the environment. Since the condensation takes place on a cold surface, the term *surface condenser* is occasionally encountered.[17]

The heat flow, q, out of a condenser is referred to as the *heat load* and the *condenser duty*.

$$q = \dot{m}_{\text{steam}}(h_1 - h_2) \qquad 26.34$$

Equation 26.34 uses the steam properties to calculate the heat load. The heat load can also be calculated from the temperature increase in the cooling water. The *range* is the initial difference between the incoming cooling water temperature and the saturation temperature corresponding to the condenser pressure. The *terminal temperature difference* is the difference in temperatures of the cooling water and the condensed liquid leaving the condenser. It is seldom less than 5°F (2.7°C). A terminal temperature difference of 10°F (5.6°C) is usually assumed in initial performance estimates. The *rise* is the increase in the cooling water temperature as it passes through the condenser. The rise seldom exceeds 20°F (11°C).

$$q = \dot{m}_{\text{cooling water}} c_p (T_{\text{out}} - T_{\text{in}}) \qquad 26.35$$

The *hot well depression*, HWD, also known as the *condensate depression*, *condenser hot well subcooling*, and *degrees of freedom*, is the difference in temperature between the saturation temperature corresponding to the condenser pressure and the steam condensate in the condenser's hot well. It typically varies between 5°F and 10°F (2.7°C and 5.4°C). Large hot well depressions are undesirable because energy is wasted and because the solubility of gases in the condensate is higher at lower temperatures. To keep the temperature high, some steam is bypassed to the hot well.

$$\text{HWD} = T_{\text{sat},p} - T_{\text{condensate,out}} \qquad 26.36$$

The *cleanliness factor* is the ratio (expressed as a percentage) of the actual to the ideal (new, clean, etc.) heat transfer coefficient. For a given installation, it is the ratio of the actual to the ideal heat loads.

Condensation is assumed to occur at constant pressure. It is also assumed that the water will leave the condenser as a saturated liquid (corresponding to the condenser pressure).[18] In real power-generating systems, subcooling represents wastage of energy, as nothing is gained by subcooling the water.

Due to the volume decrease that occurs during condensation, the condenser pressure can be low and is usually far below atmospheric. The *back pressure* is the absolute pressure in the condenser, usually expressed in inches of mercury (bars). When the back pressure is lowered, more energy can be extracted from the steam.

Condenser flooding can be used to control condenser back pressure. Some of the cooling coils are submerged in liquid condensate in a *flooded condenser*. If condensate flow from the condenser is restricted, liquid will build up and flood the condenser. This reduces the heat exchange area, reduces the amount of vapor being condensed, and increases the pressure. If the condensate is allowed to leave, the liquid level will fall, increasing the heat exchange area and decreasing the pressure.

Utility condensers operate at a high vacuum. It is inevitable that some *noncondensing gases* (primarily oxygen and other air components) will enter the condenser. These gases, commonly referred to as "air" regardless of composition, enter dissolved in the steam and through air leaks.[19] At high temperatures, these gases are highly corrosive, particularly to the copper used in condenser tubing. In addition, the effect of these gases is to increase the back pressure. As Fig. 26.6 shows, increasing the back pressure reduces the energy available to the power cycle. Noncondensing gases are removed from

[16] *Dump condensers* are separate condensers used to condense steam when the turbine is not in service (i.e., are tripped).

[17] Besides surface condensers, *jet condensers* constitute the other major condenser category. While surface condensers both produce a vacuum and recover the condensate, jet condensers produce only a vacuum. The condensate escapes with the jet. Since the condensate is lost, jet condensers are not used in power-generating systems.

[18] The term "straight condensing" is sometimes used to mean converting saturated steam to saturated water at the same pressure and temperature. However, this may be confused with the *straight condensing cycle*. Therefore, "pure condensing" is probably a better choice. Besides pure condensing, other modes of condenser operation are desuperheating with condensing, condensing with subcooling, and desuperheating with condensing and subcooling.

[19] The most common *leak points* are cracks in expansion joints between the turbine and condenser, cracks in the condenser shell at condensate-return lines, low-pressure heater vents, and condensate pump seals.

condensers with steam-jet air ejectors. A general rule of thumb is that the air-removal rate should be less than 1 SCFM (0.5 L/s) per 100 MW thermal.

Figure 26.6 Effect of Noncondensing Gases on Power Cycle Performance

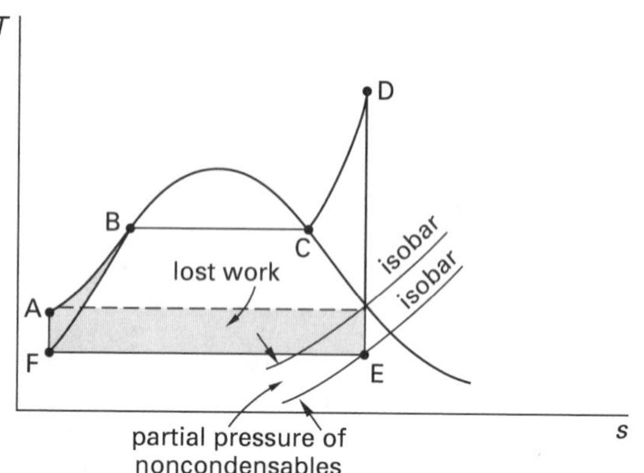

Turbine output cannot be increased indefinitely by decreasing the back pressure, however. As back pressure is lowered below what produces sonic velocity in the last turbine stage, condensate pressure is lowered without any increase in turbine output. This results in a loss similar to that of subcooling the condensate.

13. CONDENSER FOULING

As Fig. 26.6 illustrates, an increase in condenser pressure will decrease the energy extracted from the steam. The pressure increase is caused by an accumulation of noncondensing gases, but it can also occur when the condenser is unable to remove the heat fast enough, due to fouling. Within the vapor dome, reductions in enthalpy, temperature, and pressure go hand-in-hand. When the pressure is not reduced, the enthalpy is not reduced.

The most common causes of fouling on the water side of water-cooled, steam surface condensers are microbiological growth, scale formation, and tube pluggage by debris and aquatic organisms. These problems are more intense in once-through condensers that draw water from lakes, rivers, and oceans.

Since condensers are warm, they are prone to microbiological growth (slime and biofilm). This *microfouling* also provides a base for most *microbiologically influenced corrosion* (MIC). Microfouling is easily detected by a gradual increase in the terminal temperature difference. Growth of biological organisms is controlled along the coolant intake by continuous injection and *shock treatment* (sudden injection) of *biocides.*

Scaling is the formation of mineral compounds, most notably calcium carbonate and (to a lesser extent) calcium sulfate and phosphate, manganese compounds, and silicates, on the condenser surfaces. Sometimes adding sulfuric acid to the coolant will keep the pH low enough to prevent calcium carbonate deposits. Phosphates, phosphonates, and polymers can also be used. As with make-up water treatment, though, no generic treatment works for all installations. Once formed, scale must be mechanically removed or treated chemically. Calcium carbonate dissolves in acid, but other deposits require more exotic treatments.

Tube pluggage from leaves, vegetation, and anything else that can pass through intake screens will reduce coolant flow. This can be detected by weekly pressure-drop checks or sudden changes in terminal temperature difference. *Macrofouling* is a term referring to infestation by Asian clams, saltwater barnacles, marine blue mussels, zebra mussels, and similar marine life.

Zebra mussels, accidentally introduced in the United States in 1986, are particularly troublesome for several reasons. First, young mussels are microscopic and easily pass through intake screens. Second, they attach to anything, even other mussels, to produce thick colonies. Third, adult mussels quickly sense some biocides, most notably those that are halogen-based (including chlorine), and quickly close and remain closed for days or weeks. An ongoing biocide program aimed at pre-adult mussels, combined with slippery polymer-based surface coatings, is probably the best approach at prevention. Once a condenser is colonized, mechanical removal by scraping or water blasting is the only practical option.

Chlorine is a common treatment for once-through systems. However, chlorine reacts with organic precursors to produce trihalorganics, including trihalomethanes, which are carcinogenic compounds, and its use may be prohibited. Or, when chlorine is used, a *dehalogenation agent* (e.g., sodium bisulfite) may be needed downstream of the condenser. In addition to specialized *molluscicides,* other more expensive substitutes for chlorine include bromine, chlorine dioxide, ozone, nonoxidizing biocides, copper, cyanuric acid, hydrogen peroxide, polymers, potassium permanganate, and sodium hypochlorite.

If the condenser can be taken out of service for a few hours or more, heat treatment (*thermal backflushing*) is another alternative for zebra mussel prevention. Exposure to 95°F to 105°F (35°C to 41°C) water results in 100% mortality within approximately an hour. A temperature of 89°F (32°C) for six hours or more also appears to be effective. Water can be heated to these temperatures by recirculating condenser discharge or by steam sparging for more distant points.

14. NOZZLES, ORIFICES, AND VALVES

Since a flowing fluid is in contact with nozzle, orifice, and valve walls for only a very short period of time, flow through them can be considered adiabatic. No work is done on the fluid as it passes through. If the potential energy changes are neglected, the SFEE reduces to Eq. 26.37.

$$h_1 + \frac{v_1^2}{2} = h_2 + \frac{v_2^2}{2} \quad \text{[SI]} \quad 26.37(a)$$

$$h_1 + \frac{v_1^2}{2g_cJ} = h_2 + \frac{v_2^2}{2g_cJ} \quad \text{[U.S.]} \quad 26.37(b)$$

If the kinetic energy changes are neglected, $h_1 = h_2$. A constant enthalpy pressure drop is characteristic of a *throttling valve*. Pressure-reduction and superheater-bypass valves used to control turbine output in sliding-pressure operation are typically of the "tortuous-path" variety. Steam makes numerous abrupt direction changes, and pressure is reduced in a throttling process.

15. SUPERHEATERS

A *superheater* is essentially a heat exchanger used to increase the energy of the steam. A *radiant superheater* is exposed directly to combustion flame and receives energy by radiation. A *convection superheater* is typically exposed to the first pass of stack gases and is screened from direct view of the combustion flame by rows of boiler tubes. Alternatively, a superheater can be a separately fired, independent unit.

Since the overall heat transfer coefficient increases with increases in combustion product flow, the steam temperature increases with convection superheaters when the combustion rate increases. The opposite is true for radiant superheaters (since furnace temperature does not increase rapidly enough with increasing steam flow). In order to maintain a constant superheat, modern superheaters use both radiant and convection sections in series. Damper control and tiltable burners may also be used to maintain constant superheat.

A superheater does not increase steam pressure; it only increases temperature. In fact, steam pressure will decrease a slight amount (e.g., 2% to 5%) by virtue of the superheater's resistance to flow.

16. REHEATERS

The name *reheater* (*resuperheater*) is given to specific parts of a furnace or to separately fired superheaters that add energy to steam after it has already passed at least once through a turbine. This is done to reduce the moisture content of steam that, after partial expansion, may have already entered the vapor dome. Reheaters do not differ much from convection-type superheaters, except that the volume of steam handled is greater (since the pressure and density are less). In fact, reheaters are typically placed after the primary superheater tubing in the stack and before the secondary superheater tubing, if any are present. Radiant-type reheaters are also occasionally used.

The *reheat temperature* is usually the same as the original steam temperature (known as the *throttle temperature*), usually 1000°F to 1050°F (540°C to 570°C) in traditional drum boiler units and 1000°F to 1100°F (540°C to 590°C) or more in supercritical boilers. Optimum reheat pressure is approximately 25% of the throttle pressure. *Multiple reheat*, double or triple, is applicable only with supercritical throttle pressures (i.e., above 3206 psi (22.1 MPa)). Otherwise, multiple reheat will result in superheated exhaust.

17. DESUPERHEATERS

Some power plant auxiliary equipment is designed to use saturated (as opposed to superheated) steam. It is common to bleed steam from the superheater and run it through a *desuperheater* (*attemperator*) to decrease the steam enthalpy. This can be done by looping the steam back through cooler water in the steam drum or a feedwater heater. More commonly, it is done by injecting cooling water (i.e., a *fixed-orifice desuperheater*).[20] Since the water and steam leave at the same properties, the heat balance equation for water injection is given by Eq. 26.38. Kinetic energy is neglected, although it can be significant in some situations.

$$\begin{aligned} m_{\text{water}}h_{\text{water}} + m_{\text{steam}}h_{\text{steam}} \\ = (m_{\text{water}} + m_{\text{steam}})h_{\text{desuperheated steam}} \end{aligned} \quad 26.38$$

The *turndown ratio* (*capacity turndown*) is the ratio of maximum-to-minimum mass steam flow rates at which the temperature can be accurately controlled by the desuperheater. Minimum flow rates are influenced by the lowest velocity at which water droplets can be held in suspension.

Desuperheaters are usually needed to run auxiliary equipment (e.g., turbine-driven boilerfeed pumps); they should also be installed across high-pressure turbines. If a high-pressure turbine trips or is taken out of service, the high-pressure steam can be desuperheated and routed to the low-pressure turbine. Desuperheaters are also used to temper steam from intermediate- and low-pressure turbines on imbalanced or start-up conditions.

18. FEEDWATER HEATERS

A *feedwater heater* uses steam to increase the temperature of water entering the steam generator, thus improving the boiler efficiency. Each 10°F (5.6°C) increase in water temperature increases a corresponding improvement in boiler efficiency of approximately 1%. In a normal condensing cycle, the water being heated comes

[20]Other types of desuperheaters are surface-water absorption, steam atomizing, ejector-recycle, venturi, and variable orifice.

from the condenser. Steam for heating is bled off from an *extraction turbine.* Eight or more stages of feedwater heating, using both open and closed heaters, may be used in modern plants. The actual number of heaters is an economic decision, as the incremental savings from increased thermal efficiency eventually drops below the incremental installation cost. Equation 26.39 and Eq. 26.40 reflect the energy balance for open and closed feedwater heaters.

Special Cases of the Steady-Flow Energy Equation

$$\sum \dot{m}_i h_i = \sum \dot{m}_e h_e \qquad 26.39$$

$$\sum \dot{m}_i = \sum \dot{m}_e \qquad 26.40$$

Open heaters (also known as *direct-contact heaters* and *mixing heaters*) physically mix the steam and water, as shown in Fig. 26.7. Because they are essentially large boxes under internal steam pressure, open heaters are usually closest to the condenser (i.e., on the suction side of the boilerfeed pumps), where the pressure is low. Each open heater requires its own feed pump. Because open heaters are not highly pressurized, the temperature of the leaving water is seldom over 220°F (104°C) and is usually near 212°F (100°C).

Figure 26.7 *Open Feedwater Heater*

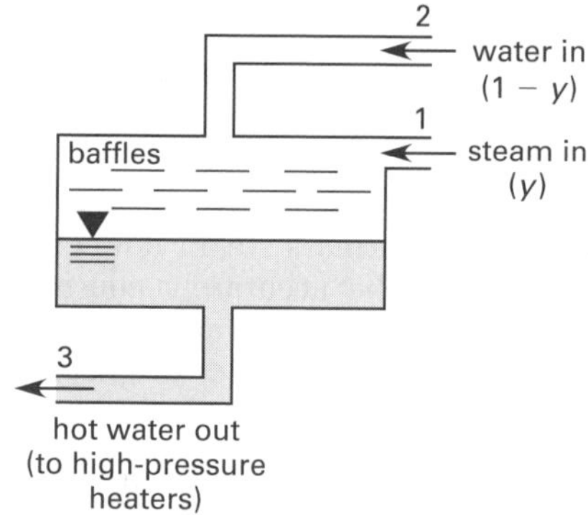

Open heaters may be simply vented to the atmosphere, or dissolved gases that come out of a solution may be extracted by a vent condenser or a vacuum pump.

When saturated or nearly saturated water is reduced in pressure, some of it may "flash" into steam. The reduction in pressure can occur in condensate return lines due to fluid friction or from pressure losses through valves and traps. When flashing occurs, the steam will impede liquid flow, a condition known as "binding."[21] To prevent boilerfeed pumps handling water near the boiling temperature from racing due to lack of feed liquid, open feedwater heaters are mounted quite high (i.e., 12 ft to 20 ft (3.6 m to 6 m) up or higher). This provides the necessary NPSHR for the pumps below.

The adiabatic energy balance around the open feedwater heater shown in Fig. 26.7 is

$$(1-y)h_2 + yh_1 = h_3 \qquad 26.41$$

If the three enthalpies are known, the *bleed fraction*, y, can be determined.

$$y = \frac{h_3 - h_2}{h_1 - h_2} \qquad 26.42$$

Equation 26.43 provides an alternate adiabatic energy balance around the open feedwater heater in terms of mass flow rate.

Open (Mixing) Feedwater Heater

$$\dot{m}_1 h_1 + \dot{m}_2 h_2 = h_3(\dot{m}_1 + \dot{m}_2) \qquad 26.43$$

A *closed feedwater heater* is a heat exchanger that can operate at either high or low pressures. There is no mixing of the water and steam in the feedwater heater. The cooled steam (known as the *drips*) leaves the feedwater heater in liquid form.

There are various methods of disposal of the drips. The condensate may be combined with the feedwater (as in Fig. 26.8) after passing through a *drip pump* to raise its pressure. (Since the drip pump work is small, it can be omitted for first approximations.) Alternatively, the drips can be returned to the condenser *hot well.* For the closed feedwater shown in Fig. 26.8 receiving bleed fraction y of steam, the adiabatic heat balance is

$$h_3 = yh_2 + (1-y)h_1 + W_{\text{drip pump per pound}} \qquad 26.44$$

Equation 26.45 provides an alternate adiabatic energy balance around the closed feedwater heater in terms of mass flow rate.

Closed (No Mixing) Feedwater Heater

$$\dot{m}_1 h_1 + \dot{m}_2 h_2 = \dot{m}_1 h_3 + \dot{m}_2 h_4 \qquad 26.45$$

For a feedwater heater, the *terminal temperature difference*, TTD (also known as the *approach*), is defined as the difference in saturation temperature corresponding to the steam pressure and the temperature of the leaving water. It typically varies between 5°F and 20°F (2.8°C and 11°C).

$$\text{TTD} = T_{\text{sat},p,2} - T_3 = T_4 - T_3 \qquad 26.46$$

---

[21]Flashing is intentional in a *flash evaporator* (*flash tank, flash chamber*, etc.). Since the temperature of the flash steam produced depends on the pressure maintained in the flash tank, precise temperature control can be maintained where needed. The enthalpy of the flash steam is the same as live steam at that pressure and temperature. The fraction of condensate that will flash when dropped from pressure p_1 to p_2 is

$$\text{flash fraction} = \frac{h_{f,p_1} - h_{f,p_2}}{h_{fg,p_2}}$$

Figure 26.8 *Closed Feedwater Heater*

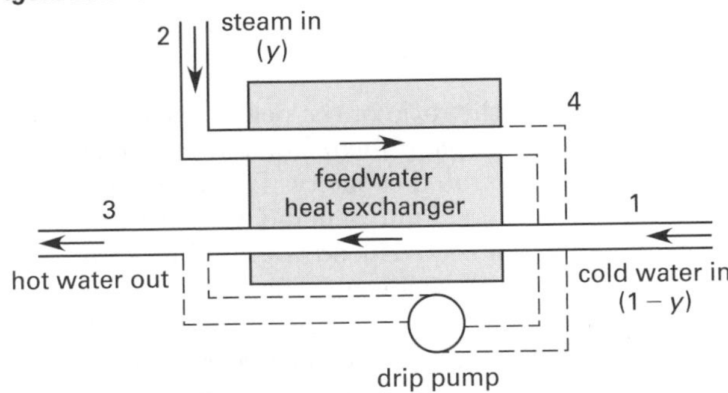

In the United States, *tubesheet feedwater heaters* have been used almost exclusively. Since tubesheet heaters are limited to temperature change rates of about 10°F/min (5.6°C/min), they are susceptible to thermal stresses when used with cycling plants. With *header-type feedwater heaters*, all the tubes are welded to headers. This type of design tolerates temperature change rates up to 30°F/min (17°C/min) and is more suitable for cycling loads.

19. EVAPORATORS

An *evaporator* is a closed shell-and-tube heat exchanger that is used to produce distilled boiler feedwater. Steam bled off from other points is the source of heating. Steam passes through the tubes. As the water is evaporated, the distilled water steam is routed to a holding tank, an open feedwater heater, or a special condenser.

Single-effect (*single-stage*) *evaporators* produce distilled water from the steam of a single evaporator. *Multiple-effect evaporators* are more common and use the steam produced in one evaporator as the heat source for the next evaporator. (See Fig. 26.9.) In that way, the heat of evaporization is used several times. Three and four effects in series are typical for power-generating plants.

Figure 26.9 *Double-Effect Evaporator*

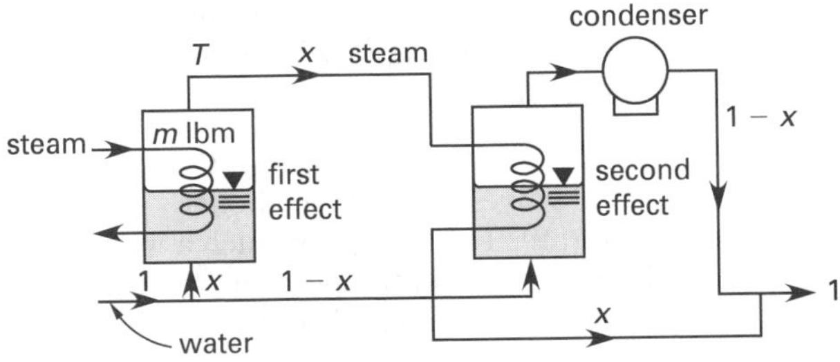

20. DEAERATORS

Deaerators (*deaerating heaters*, *deaerating tanks*) are a special category of open feedwater heaters. (Although the pressure is higher than in normal open feedwater heaters, deaerators are still direct-contact heaters.) They remove dissolved gases (e.g., oxygen and carbon dioxide) from feedwater. This is done in a baffled chamber in which the water is broken into droplets and heated by exposure to high-temperature steam. Gas solubility is essentially zero at the saturation temperature.[22] Gases are removed by a vent condenser.

21. ECONOMIZERS

An *economizer* is a water-tube heat exchanger consisting of tubes heated by the last pass of the combustion gases. It increases the temperature of boiler feedwater entering the steam generator and reduces the required combustion energy input by utilizing energy that would otherwise be wasted. The heat transfer takes place in the downstream of the boiler. Because an economizer can add significant resistance to the flow of stack gases, a forced draft fan is usually required.

The overall coefficient of heat transfer is in the range of approximately 2–3 Btu/hr-ft^2-°F (11–17 W/m^2·°C) for flue gases passing at 2000 lbm/hr-ft^2 (2.7 kg/s·m^2) to approximately 5–7 Btu/hr-ft^2-°F (28–40 W/m^2·°C) for flue gases passing at 6000 lbm/hr-ft^2 (8.1 kg/s·m^2).

External corrosion of the economizer heat transfer surface is avoided by keeping the temperature high enough (i.e., above the dew point of the stack gases) to prevent acid formation. This is generally 275°F to 350°F (135°C to 175°C), depending on the amount of sulfur dioxide in the stack gases. Internal corrosion is avoided by maintaining proper feedwater chemistry.[23]

22. AIR HEATERS

Increasing the temperature of combustion air increases combustion efficiency. While the temperature of incoming air for stoker furnaces is limited to approximately 250°F to 350°F (120°C to 175°C) by mechanical cooling considerations, at least some portion of combustion air used with pulverized coal must be heated to approximately 600°F (315°C) or more.

An *air heater* (*air preheater*) recovers energy from the stack gases and transfers it to incoming combustion air. *Convection preheaters* (also known as *recuperative heaters*) are conventional heat exchangers that transfer energy through tubes or flat plates. *Regenerative preheaters* use a slowly rotating drum with honeycomb-like passageways that are progressively exposed to incoming

[22]After heating, the residual oxygen concentration will be approximately 0.005 mg/L. Vacuum deaerators using steam-jet extractors or vacuum pumps without heating leave water with residual concentrations of approximately 0.2 mg/L of oxygen and 2 mg/L to 10 mg/L of carbon dioxide.
[23]Originally, economizer tubes were cast iron to protect them from corrosion. Now, with better furnace oxygen control, cast-iron tubes have been essentially replaced by steel tubes.

Power Cycles

air and outgoing stack gases. (See Fig. 26.10.) The overall heat transfer coefficient for both types is generally low, varying from approximately 1.5 Btu/hr-ft^2-°F (8.5 W/m^2·°C) for air passing at 2000 lbm/hr-ft^2 (2.7 kg/s·m^2) to approximately 3.0 Btu/hr-ft^2-°F (17 W/m^2·°C) for air passing at 6000 lbm/hr-ft^2 (8.1 kg/s·m^2).

Figure 26.10 *Regenerative Air Preheater*

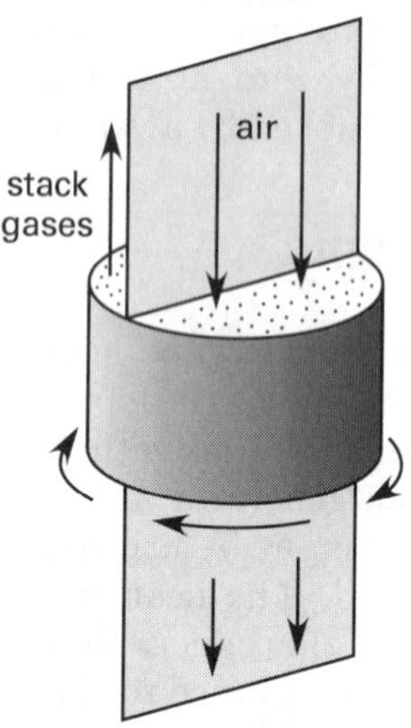

23. ELECTRICAL GENERATORS

Electrical generators convert energy extracted from the steam into electrical energy. Generator efficiencies are high—around 95% for 1 MW unit, 96% for 10 MW units, and 98% for 100 MW units and above. The *steam rate* (*water rate*) is defined as the steam mass flow rate divided by the generator output in kilowatts.[24] Typical units for steam rate are lbm/kW-hr (kg/kW·h). Values depend greatly on the throttle steam conditions and condenser pressure. For 1000°F (540°C) steam and condenser pressures of 2 in to 4 in of mercury, the steam rate is approximately 5.6 lbm/kW-hr to 5.9 lbm/kW-hr (7.1 g/kW·s to 7.4 g/kW·s), with the lower values corresponding to the lower condenser pressures.

24. LOAD AND CAPACITY

Electrical demand varies with time of day, month, and season. The *load curve* is a curve of fluctuating instantaneous load (electrical demand in MW or steam requirements) versus time of day for an average day in a particular season. The *load duration curve* is a curve of load (in MW) versus the number of hours (in a year) the load was experienced. Large power demand variations are met by taking entire power-generating plants on the electrical grid online or offline. Smaller variations are met at the plant level by taking individual boilers and turbines online or offline. Capacity at the plant and regional levels is categorized into *firm capacity* (always available, even in emergencies), *spinning reserve* (floating capacity on the electrical bus), *hot reserve* (in operation but not in service), and *cold reserve* (not in operation but operational).

The *load factor* is the ratio of the peak-to-average loads. The *capacity factor* (*plant factor*) of a unit is the ratio of the average load to rated capacity. The *demand factor* is the ratio of the maximum demand to the total maximum possible connected load (assuming everybody turned everything on at once). The *output factor* (*use factor*) accounts for downtime and is the ratio of actual energy output during a particular period to the output that would have occurred during that period if the plant had been operating at full load rating. The *operation factor* is a ratio of operational time to total time (including downtime).

A *base-load unit* is a unit or plant intended to satisfy a constant fixed demand by running continuously. A base-load unit will usually have a capacity factor of 50% or more. A *peaking unit* is intended to satisfy only peak loads.

25. CONTROL AND REGULATION OF TURBINES

Traditionally, turbines in the United States have been designed to operate at *constant throttle pressure*. Base-load units were not intended to drop below the minimum load they were designed for, typically 25% to 30% of capacity. However, daily cycling of base-load units (down to as low as 10%) has now become common.

Although actual steam powered generating plants can use variants or combinations of them, there are four basic boiler/turbine control strategies.

1. *conventional control* (*boiler-follow*): Most conventional drum-type steam units operate in the *boiler-follow mode.* Turbine steam control valves respond to generator loading and shaft speed changes; boiler controls sense steam flow and/or pressure variations. Within a reasonable range, the turbine responds rapidly by drawing on the stored energy in the boiler.

2. *turbine-follow:* With the *turbine-follow mode*, turbine control valves regulate the boiler's steam generation rate. This is done in such a way as to minimize pressure and temperature transients. Stored energy in the boiler is not used. Steam flow and turbine power are considerably slower than with conventional control.

3. *coordinated optimal:* The *coordinated optimal mode* combines conventional and turbine-follow modes. Response to load changes is relatively fast, steam quality is high, and boiler safety is enhanced due to fewer transients.

[24]For turbines acting as prime movers, the steam flow rate is divided by the horsepower output, with units of lbm/hp-hr.

4. *variable pressure:* In the *variable pressure mode* (*variable throttle pressure*, usually referred to as *sliding pressure operation*), steam generator pressure is varied according to turbine demand. Since saturation temperature and pressure are related, sliding pressure operation subjects the boiler, turbine, and other equipment to transient thermal stresses caused by frequent and repeated rapid temperature swings.

An alternative to using sliding pressure operation is using control valves to redirect the steam flow through different components.[25] The amount of steam entering the turbine is controlled by *boiler throttle valves* or *turbine (steam admission) throttle valves*. For example, at low loads, steam may be routed around the second superheater; at intermediate loads, superheated steam may pass through a pressure reduction valve; and at high loads, the turbine may receive fully superheated steam at full pressure. The process of implementing this control method in older plants is known as *cycle redesign*, *backfitting*, and *retrofitting*.

Different methods of governing can be used to maintain the turbine operating point. With *throttle control* (also known as *cut-off governing* and *full-arc admission*), the pressure and temperature of the steam admitted to the turbine are constant and all turbine nozzles are operational, but valves control the steam flow rate through each nozzle.[26] With *nozzle control*, also known as *partial-arc admission*, the steam flow rate to the turbine is controlled by "cutting out" some of the steam nozzles. With *bypass governing*, steam pressure and temperature are constant, but the stage at which steam is introduced to the turbine(s) is changed. With *blast governing*, the steam is not admitted to the turbine continuously. Rather, the steam flow is repeatedly turned on and off in a ratio that achieves the desired duty cycle.

At the steam level, variations in output are achieved by varying the firing (fueling) rate, by varying the input air and/or draft, and by using separately fired superheaters, desuperheaters, movable burner heads, and other techniques.

At the individual turbine, steam accumulators can supply short-term peak loads.[27] This method is commonly used in once-through boilers, but may also be used with sliding pressure operation to supply all low-load steam.

26. NOMENCLATURE

c_p	specific heat	Btu/lbm-°R	kJ/kg·K
C	concentration	ppm	ppm
g_c	gravitational constant, 32.2	lbm-ft/lbf-sec^2	n.a.
h	enthalpy	Btu/lbm	kJ/kg
HV	heating value	Btu/lbm	kJ/kg
HWD	hot well depression	°F	°C
J	Joule's constant, 778.17	ft-lbf/Btu	n.a.
m	mass	lbm	kg
$\dot{m}$	mass flow rate	lbm/sec	kg/s
p	pressure	lbf/ft^2	kPa
P	power	Btu/sec	kW
q	heat per unit mass	Btu/lbm	kJ/kg
$\dot{q}$	heat flow rate	Btu/sec	kW
Q	heat	Btu	kJ
t	time (duration)	sec	s
T	temperature	°R	K
TTD	terminal temperature difference	°F	°C
v	velocity	ft/sec	m/s
w	work per unit mass	Btu/lbm	kJ/kg
W	work	Btu	kJ
y	bleed fraction	–	–

Symbols

η	efficiency	–	–
υ	specific volume	ft^3/lbm	m^3/kg

Subscripts

e	exit
f	saturated fluid
fg	fluid-to-gas (vaporization)
g	gas
i	inlet
m	mechanical or intermediate
p	at pressure p or at constant pressure
s	isentropic
sat	saturated

[25] Actually, most modern plants use a combination of sliding pressure and backfitting, a system known as *hybrid throttle pressure operation*.

[26] The terminology is confusing. A *throttle valve* controls the mass flow rate, but a *throttling valve* changes the pressure. Actually, partly closing any valve will result in reductions in both the mass flow rate and the pressure. Phrases such as "...at 85% throttle..." usually refer to the capacity ratio, not the mass flow rate.

[27] A volume of saturated water at the system pressure is kept heated in the accumulator by boiler steam. When the load peaks and the system pressure drops (below the saturation pressure), some of the accumulator water flashes into steam, satisfying the majority of the increase in demand.

27 Vapor Power Cycles

Content in blue refers to the *NCEES Handbook*.

NCEES EXAM SPECIFICATIONS AND RELATED CONTENT

HVAC AND REFRIGERATION EXAM

I.B.1. Thermodynamics: Cycles
- 3. Thermal Efficiencies
- 5. Carnot Cycle
- 6. Basic Rankine Cycle
- 7. Rankine Cycle with Superheat

THERMAL AND FLUID SYSTEMS EXAM

I.E.2. Thermodynamics: Thermodynamic Cycles
- 6. Basic Rankine Cycle
- 7. Rankine Cycle with Superheat

1. GENERAL VAPOR POWER CYCLES

The most general form of a vapor (almost always steam) power cycle incorporates a vapor generator, turbine, condenser, and feed pump as illustrated in Fig. 27.1.

The following thermodynamic processes take place in a typical steam vapor power cycle.

A to B: Subcooled water is heated to the saturation temperature corresponding to the pressure in the steam generator.

B to C: Saturated water is vaporized in the steam generator, producing saturated vapor.

C to D: An optional superheating process increases the steam temperature and enthalpy.

***Figure 27.1** Simplified Vapor Power Cycle*

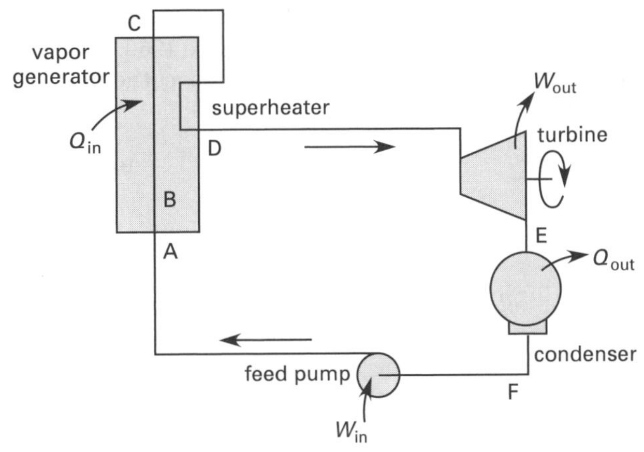

D to E: Steam expands in the turbine and does work as it decreases in temperature, pressure, and quality.

E to F: Vapor is returned to a liquid phase in the condenser.

F to A: The pressure of the liquid water is brought up to the steam generator pressure by the boiler feed pump.

After making a full cycle, the steam is brought back to the exact same thermodynamic conditions. While the entropy of the steam returns to its original value after each full cycle, the entropy of the environment increases due to the heat transfers (losses, etc.) from the steam generator and condenser. Thus, even if the pump compression and turbine expansion are isentropic processes, the cycle makes an irreversible thermal change to the environment.

2. SIGN CONVENTION

The sign convention normally adhered to in textbooks for changes in thermodynamic properties is not followed for pump work. Since the pump does work on the water, the pump work would be negative according to the traditional sign convention. However, since the power industry customarily refers to pump work as a positive quantity, that convention is followed in this chapter.[1]

[1]Nobody says "The pump work is negative four Btus per pound."

3. THERMAL EFFICIENCIES

The *thermal efficiency*, η_{th}, of a power cycle is the ratio of net work out (or net heat in) divided by the heat input.[2] The upper limit of thermal efficiency is the thermal efficiency of the theoretical Carnot cycle. (See Eq. 27.8.) Typical actual values of thermal efficiency depend to a great extent on the operating temperature and pressure and on the degree of cycle sophistication. Cycles with high-pressure superheat, reheat, and multiple stages of regeneration will be far more efficient in converting thermal energy to electricity than the simple low-pressure cycles that prevailed before 1960. However, the best single-cycle plants seldom have thermal efficiencies in excess of 40%.

Basic Cycles

$$\eta_{th} = \frac{W_{net}}{Q_H} = \frac{Q_H - Q_L}{Q_H} \qquad 27.1$$

The *heat rate*, HR (or *station rate*), is the ratio of the total energy input divided by the net electrical output. This is not merely the reciprocal of the thermal efficiency because of the units for heat rate. Heat rate has units of Btu/kW-hr (MJ/kW·h). It can be calculated from the thermal efficiency and the appropriate conversion factor.

$$\text{HR} = \frac{Q_{in}}{W_{out} - W_{in}} = \frac{3.6\ \frac{\text{MJ}}{\text{kW·h}}}{\eta_{th}} \qquad \text{[SI]} \quad 27.2(a)$$

$$\text{HR} = \frac{Q_{in}}{W_{out} - W_{in}} = \frac{3413\ \frac{\text{Btu}}{\text{kW-hr}}}{\eta_{th}} \qquad \text{[U.S.]} \quad 27.2(b)$$

For heat rates expressed in Btu/hp-hr, the relationship is

$$\text{HR} = \frac{2545\ \frac{\text{Btu}}{\text{hp-hr}}}{\eta_{th}} \qquad 27.3$$

4. CYCLE DESIGNATIONS

There are numerous descriptive terms used to describe categories of cycles. If a condenser is not part of the power-generating process, the steam cannot expand below atmospheric pressure. A cycle without a condenser is known as a *noncondensing cycle*. The steam may be used for process heating or feedwater heating, or may simply be discharged to the atmosphere. Steam may condense elsewhere (in a feedwater heater, for example). It is the absence of a condenser that identifies a noncondensing cycle, not the absence of condensation.

If there is a condenser, the cycle is known as a *condensing cycle*. If there are no bleeds (i.e., no extraction of steam prior to steam entering the condenser), the cycle is known as *straight condensing*. In an *extraction cycle*, steam is withdrawn from the turbine, usually for feedwater heating, although the bleed steam can also be used for other process heating.

5. CARNOT CYCLE

The Carnot cycle is an ideal power cycle that is impractical to implement. However, its theoretical work output sets the maximum attainable from any heat engine, as evidenced by the isentropic (reversible) processes between states (1 and 2) and (3 and 4) in Fig. 27.2. The working fluid in a Carnot cycle is irrelevant.

The processes involved are

1 to 2:	isentropic compression
2 to 3:	isothermal expansion of saturated liquid to saturated vapor
3 to 4:	isentropic expansion of vapor
4 to 1:	isothermal condensation of vapor

The properties at the various states can be found from the following solution methods. The letters *f* and *g* stand for saturated fluid and saturated gas, respectively. The Carnot cycle can be evaluated by working around, finding T, p, x, h, and s at each node.

at 2:	From the property table for T_H, read p_2, h_2, and s_2 for a saturated fluid.
at 3:	$T_3 = T_2$; $p_3 = p_2$; $x = 1$; h_3 is read from the table as h_g; s_3 is read as s_g.
at 4:	Either p_4 or T_4 must be known. Read p_4 from the T_4 line on the property table or vice versa; $x_4 = (s_3 - s_f)/s_{fg}$; $h_4 = h_f + x_4 h_{fg}$; $s_4 = s_3$.
at 1:	$T_1 = T_4$; $p_1 = p_4$; $x_1 = (s_2 - s_f)/s_{fg}$; $h_1 = h_f + x_1 h_{fg}$; $s_1 = s_2$.

The turbine and pump work terms per unit mass are

$$w_{turbine} = h_3 - h_4 \qquad 27.4$$

$$w_{pump} = h_2 - h_1 \qquad 27.5$$

[2] Only a *cycle* (i.e., an integrated system of devices) can have a *thermal* efficiency. A single device (e.g., a pump) can have mechanical and isentropic efficiencies, but it cannot have a thermal efficiency.

Figure 27.2 *Carnot Cycle*

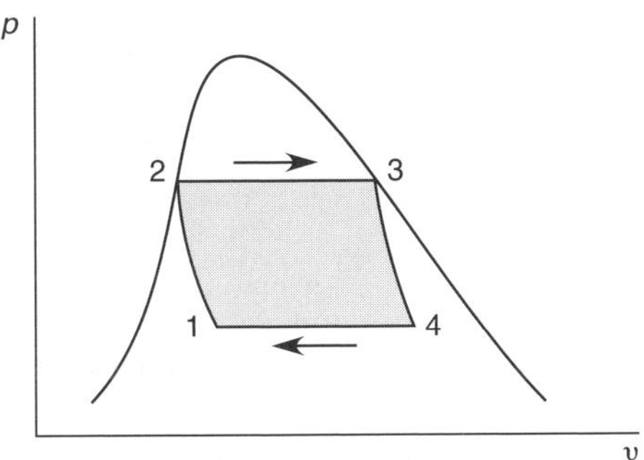

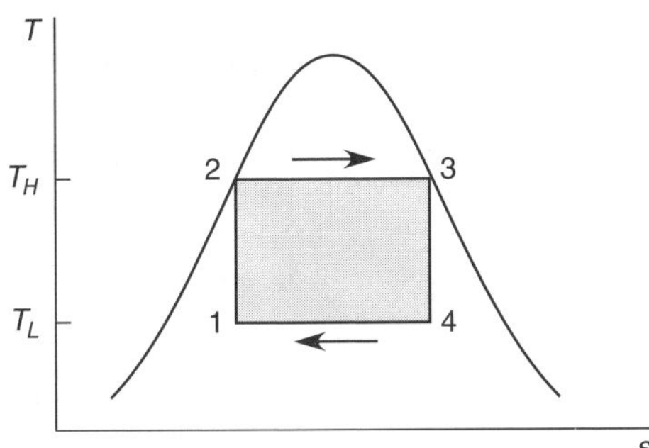

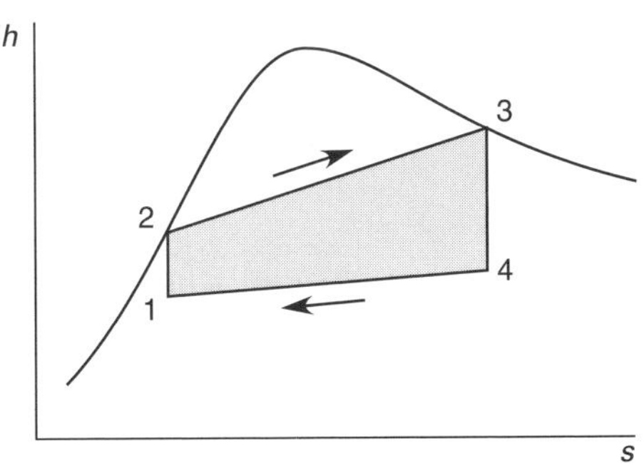

The heat flows per unit mass into and out of the system are

$$q_{\text{in}} = T_H(s_3 - s_2) = h_3 - h_2 \qquad 27.6$$

$$\begin{aligned} q_{\text{out}} &= T_L(s_4 - s_1) = T_L(s_3 - s_2) \\ &= h_4 - h_1 \end{aligned} \qquad 27.7$$

The thermal efficiency of the entire cycle is

Basic Cycles

$$\begin{aligned} \eta &= \frac{Q_H - Q_L}{Q_H} \\ &= \frac{q_H - q_L}{q_H} \\ &= \frac{w_{s,\text{turbine}} - w_{s,\text{pump}}}{q_H} \\ &= \frac{(h_3 - h_4) - (h_2 - h_1)}{h_3 - h_2} \end{aligned} \qquad 27.8$$

Equation 27.8 can also be expressed in terms of absolute temperatures as

Basic Cycles

$$\begin{aligned} \eta_c &= \frac{T_H - T_L}{T_H} \\ &= 1 - \frac{T_L}{T_H} \end{aligned} \qquad 27.9$$

If isentropic efficiencies for the pump and turbine are known, proceed as follows. Calculate all properties, assuming that the efficiencies are 100%. Then, modify h_4 and h_2 as shown by Eq. 27.10 and Eq. 27.11. Use the new values to find the actual thermal efficiency of the cycle.

$$h_4' = h_3 - \eta_{s,\text{turbine}}(h_3 - h_4) \qquad 27.10$$

$$h_2' = h_1 + \frac{h_2 - h_1}{\eta_{s,\text{pump}}} \qquad 27.11$$

$$w_{\text{turbine}}' = h_3 - h_4' \qquad 27.12$$

$$w_{\text{pump}}' = h_2' - h_1 \qquad 27.13$$

As the Carnot cycle is an ideal, reversible cycle, its processes can be reversed. Doing so reverses the direction of the cycle's path on the T-s diagram, and also reverses the heat and work flows, thus producing a refrigeration cycle. In a *reversed Carnot cycle*, the processes are

4 to 3:	isentropic compression (compressor)
3 to 2:	isothermal heat rejection from refrigerant (condenser)
2 to 1:	isentropic expansion (turbine)
1 to 4:	isothermal heat transfer to refrigerant (evaporator)

The thermal efficiency of a reversed Carnot cycle is

$$\eta_{c,\text{rev}} = \frac{T_H}{T_L - T_H}$$

6. BASIC RANKINE CYCLE

The basic Rankine cycle (see Fig. 27.3 and Fig. 27.4) is similar to the Carnot cycle except that the compression process occurs in the liquid region. The Rankine cycle is closely approximated in steam turbine plants. The efficiency of the Rankine cycle is lower than that of a Carnot cycle operating between the same temperature limits because the mean temperature at which heat is added to the system is lower than T_H.

Power Cycles

Figure 27.3 Basic Rankine Cycle

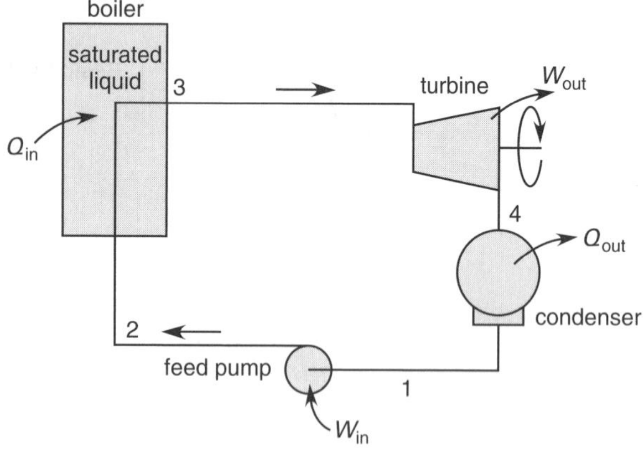

Figure 27.4 Basic Rankine Cycle

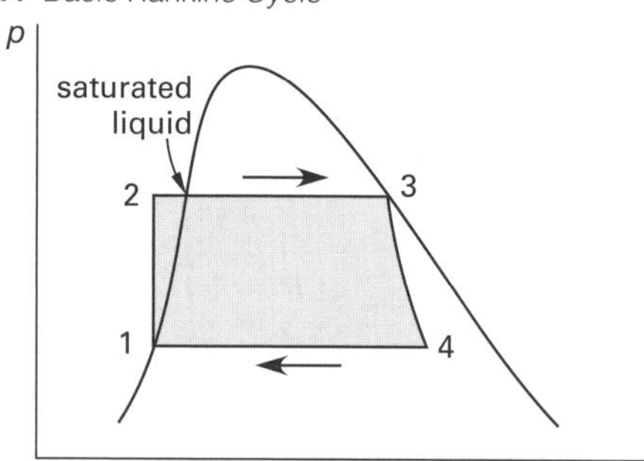

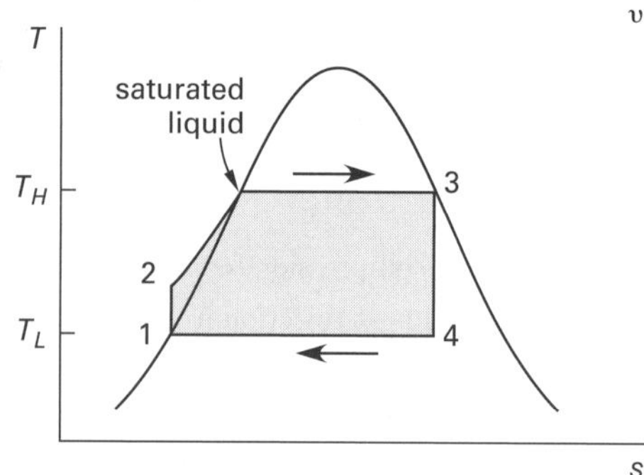

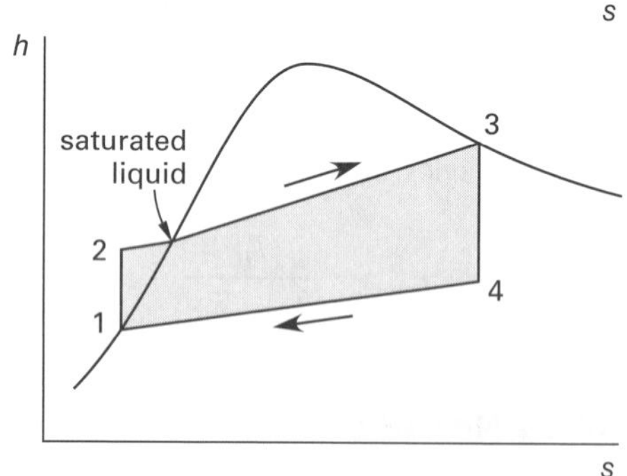

The processes used in the basic Rankine cycle are

1 to 2	adiabatic compression to boiler pressure
2 to 3:	vaporization in the boiler
3 to 4:	adiabatic expansion in the turbine
4 to 1:	condensation

The properties at each point can be found by working around the cycle, finding T, p, x, h, and s at each node.

boiler: From the property table for T_H, read p_{sat}, h_f, and s_f for a saturated liquid.

at 3: $T_3 = T_{\text{sat}}$; $p_3 = p_{\text{sat}}$; $x = 1$; h_3 is read from the table as h_g; s_3 is read as s_g.

at 4: Either p_4 or T_4 must be known. Read p_4 from the T_4 line on the property table or vice versa; $x_4 = (s_3 - s_f)/s_{fg}$; $h_4 = h_f + x_4 h_{fg}$; $s_4 = s_3$.

at 1: $T_1 = T_4$; $p_1 = p_4$; $x_1 = 0$; h_1 is read as h_f; s_1 is read as s_f; v_1 is read as v_f.

at 2: $p_2 = p_{\text{sat}}$; $s_2 = s_1$; $h_2 = h_1 + w_{\text{pump}} = h_1 + v_1(p_2 - p_1)$ (consistent units); T_2 is found as the saturation temperature for a liquid with enthalpy equal to h_2.

The per unit mass work and heat flow terms are

$$w_{\text{turbine}} = h_3 - h_4 \tag{27.14}$$

$$\begin{aligned} w_{\text{pump}} &= h_2 - h_1 \\ &\approx v_1(p_2 - p_1) \quad \text{[consistent units]} \end{aligned} \tag{27.15}$$

$$q_{\text{in}} = h_3 - h_2 \tag{27.16}$$

$$q_{\text{out}} = h_4 - h_1 \tag{27.17}$$

The thermal efficiency of the entire cycle is

Basic Cycles

$$\begin{aligned} \eta &= \frac{Q_H - Q_L}{Q_H} \\ &= \frac{w_{\text{turbine}} - w_{\text{pump}}}{q_H} \end{aligned} \tag{27.18}$$

Equation 27.18 can also be expressed in terms of enthalpies as

Internal Combustion Engines

$$\eta = \frac{(h_3 - h_4) - (h_2 - h_1)}{h_3 - h_2} \tag{27.19}$$

If isentropic efficiencies for the pump and the turbine are known, calculate all properties as if these efficiencies were 100%. Then, use the following relationships to modify h_4 and h_2. Use the new values to recalculate the thermal efficiency.

$$h_4' = h_3 - \eta_{s,\text{turbine}}(h_3 - h_4) \tag{27.20}$$

$$h_2' = h_1 + \frac{h_2 - h_1}{\eta_{s,\text{pump}}} \quad 27.21$$

$$w_{\text{turbine}}' = h_3 - h_4' \quad 27.22$$

$$w_{\text{pump}}' = h_2' - h_1 \quad 27.23$$

$$q_{\text{in}}' = h_3 - h_2' \quad 27.24$$

7. RANKINE CYCLE WITH SUPERHEAT

Superheating occurs when heat in excess of that required to produce saturated vapor is added to the water. Superheat is used to raise the vapor above the critical temperature, to raise the mean effective temperature at which heat is added, and to keep the expansion primarily in the vapor region to reduce wear on the turbine blades. A maximum practical metallurgical limit on superheat is approximately 1150°F (625°C), although some ultrasupercritical boilers operate above this temperature.

The processes in the Rankine cycle with superheat are similar to the basic Rankine cycle. (See Fig. 27.5 and Fig. 27.6.) [**Rankine Cycle**]

Figure 27.5 *Rankine Cycle with Superheat*

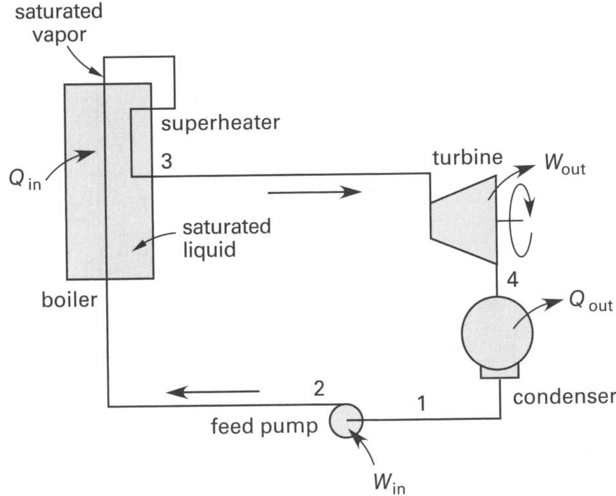

1 to 2 adiabatic compression to boiler pressure

2 to 3: vaporization of water in the boiler, followed by superheating steam in the superheater region of the boiler

3 to 4: adiabatic expansion in the turbine

4 to 1: condensation

The properties at each point can be found from the following procedure. The subscripts f and g refer to saturated fluid and saturated gas, respectively.

Figure 27.6 *Rankine Cycle with Superheat*

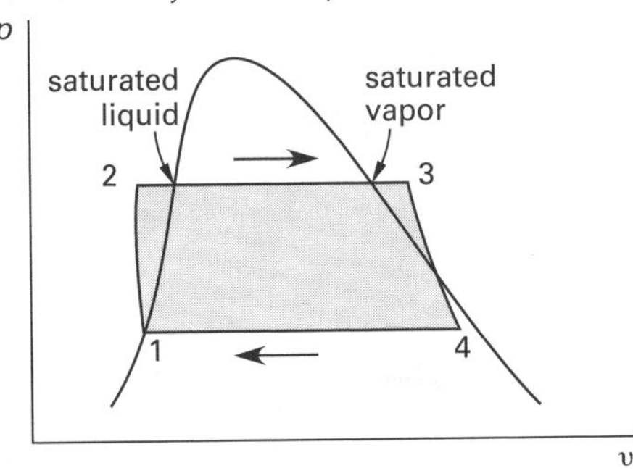

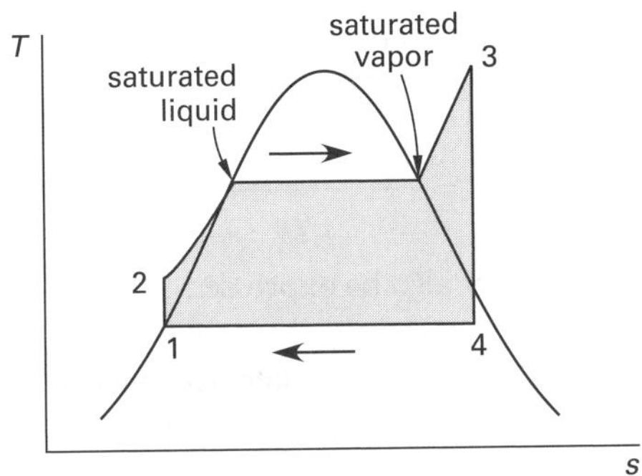

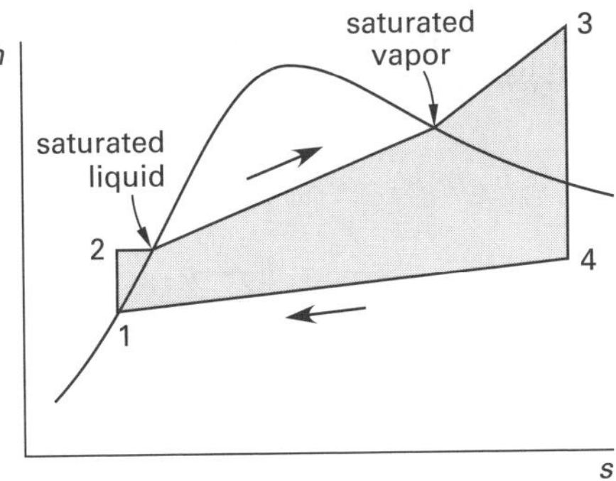

boiler: From the property table for T_{sat} or p_{sat}, read h_f and s_f for a saturated fluid.

superheater: Properties are T_{sat}, p_{sat}, h_g, and s_g. The quality, x, is 1.

at 3: T_3 is usually known; $p_3 = p_{\text{sat}}$; h_3 is read from superheat tables with T_3 and p_3 known. Same for s_3 and v_3.

at 4: T_4 is usually known; p_4 is read from the saturated table for T_4; $x_4 = (s_4 - s_f)/s_{fg}$; $h_4 = h_f + x_4 h_{fg}$; $s_4 = s_3$. If point 4 is in the superheated region, the Mollier diagram should be used to find the properties at point 4.

at 1: $T_1 = T_4$; $p_1 = p_4$; $x = 0$; h_1, s_1, and v_1 are read as h_f, s_f, and v_f from the saturated table.

at 2: $p_2 = p_{\text{sat}}$; $h_2 = h_1 + v_1(p_2 - p_1)$ (consistent units); $s_2 = s_1$. T_2 is equal to the saturation temperature for a liquid with enthalpy equal to h_2.

The per unit mass work and heat flow terms are

$$w_{\text{turbine}} = h_3 - h_4 \quad 27.25$$

$$\begin{aligned} w_{\text{pump}} &= h_2 - h_1 \\ &\approx v_1(p_2 - p_1) \quad \text{[consistent units]} \end{aligned} \quad 27.26$$

$$q_{\text{in}} = h_3 - h_2 \quad 27.27$$

$$q_{\text{out}} = h_4 - h_1 \quad 27.28$$

The thermal efficiency of the entire cycle is

Basic Cycles

$$\begin{aligned} \eta &= \frac{Q_H - Q_L}{Q_H} \\ &= \frac{w_{\text{turbine}} - w_{\text{pump}}}{q_H} \end{aligned} \quad 27.29$$

Equation 27.29 can also be expressed in terms of enthalpies as

Internal Combustion Engines

$$\eta = \frac{(h_3 - h_4) - (h_2 - h_1)}{h_3 - h_2} \quad 27.30$$

If pump and turbine isentropic efficiencies are known, calculate all quantities as if those efficiencies were 100%. Then, modify h_4 and h_2 before recalculating the thermal efficiency.

$$h_4' = h_3 - \eta_{s,\text{turbine}}(h_3 - h_4) \quad 27.31$$

$$h_2' = h_1 + \frac{h_2 - h_1}{\eta_{s,\text{pump}}} \quad 27.32$$

$$w_{\text{turbine}}' = h_3 - h_4' \quad 27.33$$

$$w_{\text{pump}}' = h_2' - h_1 \quad 27.34$$

$$q_{\text{in}}' = h_3 - h_2' \quad 27.35$$

8. RANKINE CYCLE WITH SUPERHEAT AND REHEAT

Reheat is used to increase the mean effective temperature at which heat is added without producing significant expansion in the liquid-vapor region. The analysis given assumes that $T_D = T_F$, as is usually the case. It is possible, however, that the two temperatures will be different. (See Fig. 27.7 and Fig. 27.8.)

Figure 27.7 *Reheat Cycle*

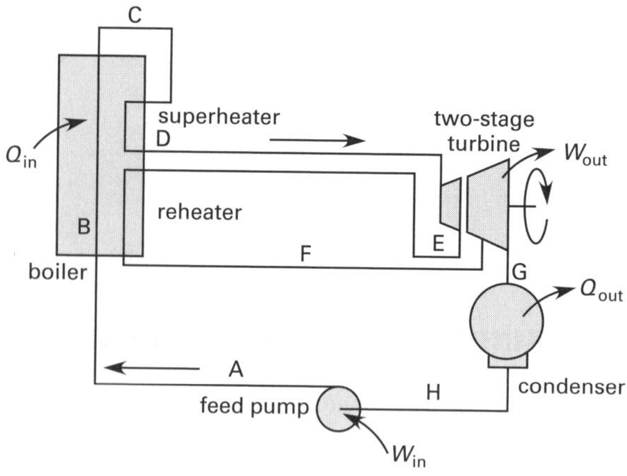

Figure 27.8 *Reheat Cycle*

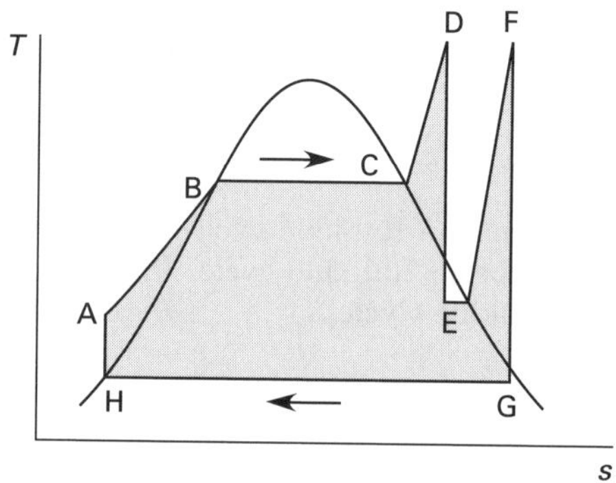

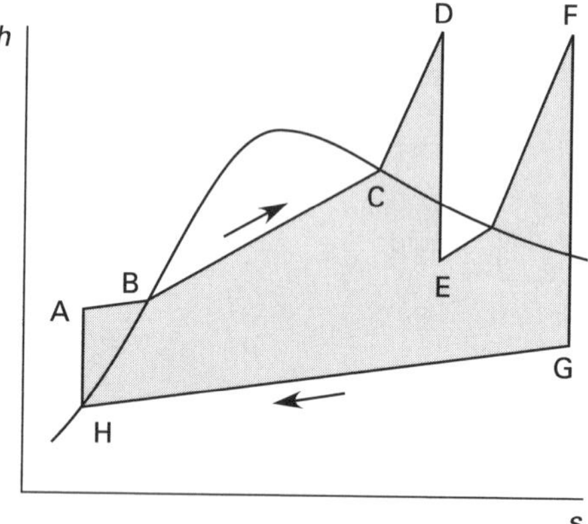

The properties at each point can be found from the following procedure.

at B: From the vapor table for T_B or p_B, read h_B and s_B for a saturated fluid.

at C: $T_C = T_B$; $p_C = p_B$; $x_C = 1$; h_C is read as h_g; s_C is read as s_g.

at D: T_D usually is known; $p_D = p_C$; h_D is read from superheat tables with T_D and p_D known. Same for s_D and v_D.

at E: p_E is usually known; T_E is read from the property table for p_E; $s_E = s_D$; $x_E = (s_E - s_f)/s_{fg}$; $h_E = h_f + x_E h_{fg}$. (Use the Mollier diagram if superheated.)

at F: T_F is usually known; $p_F = p_E$; h_F is read from superheat tables with T_F and p_F known. Same for s_F and v_F.

at G: T_G is usually known; p_G is read from property tables for T_G; $s_G = s_F$. $x_G = (s_G - s_f)/s_{fg}$; $h_G = h_f + x_G h_{fg}$. (Use the Mollier diagram if superheated.)

at H: $T_H = T_G$; $p_H = p_G$; $x_H = 0$; h_H, s_H, and v_H are read as h_f, s_f, and v_f.

at A: $p_A = p_B$; $h_A = h_H + v_H(p_A - p_H)$ (consistent units); $s_A = s_H$. T_A is equal to the saturation temperature for a liquid with enthalpy equal to h_A.

The per unit mass work and heat flow terms are

$$w_{\text{turbine}} = (h_D - h_E) + (h_F - h_G) \quad 27.36$$

$$w_{\text{pump}} = (h_A - h_H) = v_H(p_A - p_H) \quad 27.37$$

$$q_{\text{in}} = (h_D - h_A) + (h_F - h_E) \quad 27.38$$

$$q_{\text{out}} = h_G - h_H \quad 27.39$$

The thermal efficiency of the entire cycle is

Basic Cycles

$$\eta = \frac{Q_H - Q_L}{Q_H} = \frac{w_{\text{turbine}} - w_{\text{pump}}}{q_H} = \frac{(h_D - h_A) + (h_F - h_E) - (h_G - h_H)}{(h_D - h_A) + (h_F - h_E)} \quad 27.40$$

If pump and turbine isentropic efficiencies are known, calculate all quantities as if those efficiencies were 100%. Then, modify h_E, h_G, and h_A before recalculating the thermal efficiency.

$$h_E' = h_D - \eta_{s,\text{turbine}}(h_D - h_E) \quad 27.41$$

$$h_G' = h_F - \eta_{s,\text{turbine}}(h_F - h_G) \quad 27.42$$

$$h_A' = h_H + \frac{h_A - h_H}{\eta_{s,\text{pump}}} \quad 27.43$$

$$w_{\text{turbine}}' = (h_D - h_E') + (h_F - h_G') \quad 27.44$$

$$w_{\text{pump}}' = h_A' - h_H \quad 27.45$$

$$q_{\text{in}}' = (h_D - h_A') + (h_F - h_E') \quad 27.46$$

The *reheat factor*, RF, is the ratio of the actual turbine work (with all of the reheat stages) in a multistage expansion to the ideal turbine work assuming a one-stage, isentropic expansion from the same entering conditions to the same condenser pressure. The fractional improvement due to the reheat is RF − 1. The reheat factor depends on the steam properties and complexity of the cycle. Values are typically in the range of 1.05 to 1.10.

9. RANKINE CYCLE WITH REGENERATION (REGENERATIVE CYCLE)

If the mean effective temperature at which heat is added can be increased, the overall thermal efficiency of the cycle will be improved. This can be accomplished by raising the temperature at which the condensed fluid enters the boiler.

In the regenerative cycle, portions of the steam in the turbine are withdrawn at various points. Heat is transferred from this bleed stream to the feedwater coming from the condenser. Although only two bleeds are used in the following analysis, seven or more exchange locations can be used in a large installation. The regenerative cycle always involves superheating, although conceptually it does not need to. (See Fig. 27.9.)

In the following analysis, y_1 is the first bleed fraction, and y_2 is the second bleed fraction. (See Fig. 27.10.) [Rankine Cycle With Regeneration]

Figure 27.9 *Regenerative Cycle with Two Feedwater Heaters*

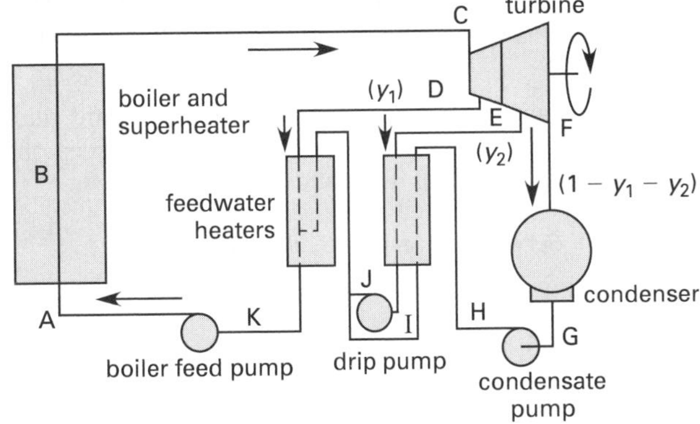

Figure 27.10 *Regenerative Cycle Property Plot*

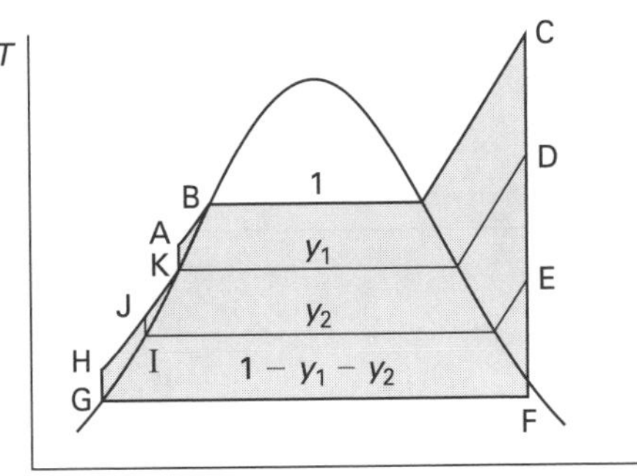

Power Cycles

The ideal per unit mass work and heat flow terms for the regenerative cycle are

$$w_{\text{turbine}} = (h_{\text{C}} - h_{\text{D}}) + (1 - y_1)(h_{\text{D}} - h_{\text{E}}) + (1 - y_1 - y_2)(h_{\text{E}} - h_{\text{F}}) \quad 27.47$$

$$w_{\text{pumps}} = (h_{\text{A}} - h_{\text{K}}) + y_2(h_{\text{J}} - h_{\text{I}}) + (1 - y_1 - y_2)(h_{\text{H}} - h_{\text{G}}) \quad 27.48$$

$$q_{\text{in}} = h_{\text{C}} - h_{\text{A}} \quad 27.49$$

$$q_{\text{out}} = (1 - y_1 - y_2)(h_{\text{F}} - h_{\text{G}}) \quad 27.50$$

The thermal efficiency of the entire cycle is

Basic Cycles

$$\eta = \frac{Q_H - Q_L}{Q_H} = \frac{(h_{\text{C}} - h_{\text{A}}) - (1 - y_1 - y_2)(h_{\text{F}} - h_{\text{G}})}{h_{\text{C}} - h_{\text{A}}} \quad 27.51$$

10. SUPERCRITICAL AND ULTRASUPERCRITICAL CYCLES

The thermal efficiency is increased by raising the average temperature at which heat is added. This is evident from Eq. 27.8, which shows that the ideal thermal efficiency depends on the highest temperature achieved in the cycle. Therefore, some modern plants are designed to achieve supercritical temperatures. The operating temperatures are limited only by metallurgical considerations.

Figure 27.11 shows that no heat is added at constant temperature. Reheating (resuperheating) is used to keep the steam quality high when it expands from point B.

Figure 27.11 *Supercritical Rankine Cycle Property Plot*

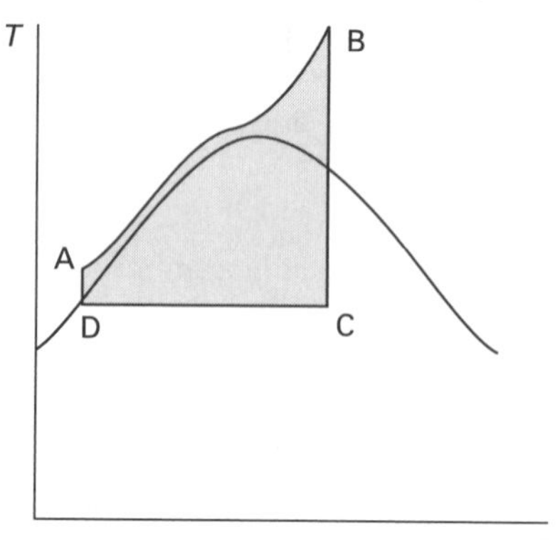

11. NOMENCLATURE

h	specific enthalpy	Btu/lbm	kJ/kg
HR	heat rate	Btu/kW-hr	MJ/kW·h
p	pressure	lbf/ft^2	Pa
q	heat per unit mass	Btu/lbm	kJ/kg
Q	heat	Btu	kJ
RF	reheat factor	–	–
s	specific entropy	Btu/lbm-°R	kJ/kg·K
T	absolute temperature	°R	K
v	specific volume	ft^3/lbm	m^3/kg
w	work per unit mass	Btu/lbm	kJ/kg
W	work	Btu	kJ
x	quality	–	–
y	bleed fraction	–	–

Symbols

η	efficiency	–	–

Subscripts

c	Carnot cycle
f	saturated fluid (liquid)
fg	vaporization (fluid to gas)
g	saturated vapor (gas)
H	high
L	low
rev	reversed
s	isentropic
th	thermal

28 Reciprocating Combustion Engine Cycles

Content in blue refers to the *NCEES Handbook*.

NCEES EXAM SPECIFICATIONS AND RELATED CONTENT

1. INTRODUCTION TO AIR-STANDARD CYCLES

Combustion power cycles differ from vapor power cycles in that the combustion products cannot be returned to their initial conditions for reuse. Combustion power cycles are often analyzed as air-standard cycles due to the computational difficulties of working with mixtures of fuel vapor, combustion products, and air.

An *air-standard cycle* is a closed system using a fixed amount of ideal air as the working fluid. In contrast to a combustion process, the heat of combustion is included in the calculations without consideration of the heat source or delivery mechanism. (That is, the combustion process is replaced by a process of instantaneous heat transfer from high-temperature surroundings.) Similarly, the cycle ends with an instantaneous transfer of waste heat to the surroundings. All processes are considered to be internally reversible. Because the air is ideal, it has a constant specific heat. The term *cold air-standard cycle* is sometimes used to describe a cycle where the specific heats are considered to be constant at their room-temperature values.

Actual engine efficiencies for internal combustion engine cycles may be up to 50% lower than the efficiencies calculated from air-standard analyses. Empirical corrections must be applied to theoretical calculations based on the characteristics of the engine. The large amount of excess air used in turbine combustion cycles results in better agreement between actual and ideal performance than for reciprocating engines.

2. ENGINE TERMINOLOGY

Internal combustion (IC) engines can be categorized into *spark ignition* (SI) and *compression ignition* (CI) categories. SI engines (i.e., typical gasoline engines) use a spark to ignite the air-fuel mixture, while the heat of compression ignites the air-fuel mixture in CI engines (i.e., typical diesel engines).

The diameter of the circular cylinder is the *bore*. The maximum distance traveled by the piston is the *stroke*. An engine with a bore of diameter D and stroke of length L is sometimes referred to as a $D \times L$ engine. At *top dead center* (TDC), the piston is furthest from the crankshaft centerline. At *bottom dead center* (BDC), the piston is closest to the crankshaft centerline.

The cylindrical *swept volume*, V_{swept}, is the product of the cylinder cross-sectional area and stroke.

$$V_{\text{swept}} = A_{\text{cylinder}} L = \frac{\pi D^2 L}{4} \qquad 28.1$$

The *clearance volume*, $V_{\text{clearance}}$, includes the combustion chamber volume plus the volume associated with gasket thickness, deck height, and piston dishing and valve relief cutouts, minus the volume of piston doming.

$$V_{\text{clearance}} = \frac{V_{\text{swept}}}{r_v - 1} \qquad 28.2$$

The *clearance*, c, is a fraction of the swept volume, usually expressed as a percentage.

$$c = \frac{V_{\text{clearance}}}{V_{\text{swept}}} \qquad 28.3$$

For reciprocating internal combustion engines, the *compression ratio*, r_v, is a ratio of volumes.[1]

$$\begin{aligned} r_v &= \frac{V_{\text{max}}}{V_{\text{min}}} = \frac{V_{\text{BDC}}}{V_{\text{TDC}}} = \frac{V_{\text{swept}} + V_{\text{clearance}}}{V_{\text{clearance}}} \\ &= \frac{1+c}{c} \end{aligned} \qquad 28.4$$

The expanding combustion gases act on one end of the piston in *single-acting engines* and on both ends of the piston in *double-acting engines*. Double-acting internal combustion engines are essentially nonexistent due to difficulties in sealing and power transmission. However, some double-acting reciprocating steam engines are used for stationary applications, and double-acting Stirling machines are common.[2] The total power generated in a double-acting engine is the sum of the power generated by the *head end* and *crank end*.

3. FOUR- AND TWO-STROKE ENGINES

Engines are categorized as either four-stroke or two-stroke. In a *four-stroke cycle*, four separate piston movements (*strokes*) and two complete crankshaft revolutions are required to accomplish all of the processes: *intake*, *compression*, *power*, and *exhaust strokes*.[3]

Two-stroke cycles are used in small gasoline engines (such as in lawn mowers and other garden equipment, outboards, and motorcycles) as well as in some diesel engines. In a two-stroke cycle, only two piston movements and one crankshaft revolution occur per power stroke. The air-fuel mixture is drawn in through the intake port, and exhaust gases expand out through the exhaust port, near the end of the power stroke. The intake, exhaust, and power strokes overlap. The term *scavenging* is used to describe the act of blowing the exhaust products out with the air-fuel mixture.

Because power strokes occur twice as often, a two-stroke engine produces more power (for a given engine weight and displacement) than a four-stroke engine. However, the increase in power is only approximately 70% to 90% (not 100%) due to various inefficiencies, including incomplete mixing and scavenging.

In the typical commercial two-stroke engine, some of the exhaust gases dilute the air-fuel mixture. Also, oil is mixed with the fuel to provide engine lubrication. Both of these practices lead to poorer fuel economy and higher emissions.

With the proper modifications (e.g., supercharging, indirect fuel injection, scavenging with pure air, lean-burning *stratified charge combustion*, and *exhaust gas dilution*), however, two-stroke gasoline engines may someday be an attractive alternative to traditional automobile engines. The primary benefits (as perceived by automobile engine manufacturers) of two-stroke engines are reduced engine vibration, more uniform torque and power, and the ability to achieve reduced air pollution emissions, particularly nitrogen oxides.[4] This theoretically eliminates the need for a nitrogen-reducing catalyst.[5]

4. CARBURETION AND FUEL INJECTION

In a normally aspirated engine, fuel is mixed with air outside of the cylinder in the *carburetor*. Mixing continues in the manifold. A butterfly valve in the carburetor throttles the flow of air. The air-fuel mixture enters through ports closed by flat-headed valves.[6]

In *fuel-injected* (FI) *engines*, pressurized fuel is injected directly into the cylinder at just the right moment in the cycle. The timing may be controlled electronically or mechanically (by distributor disk, camshaft lobes, etc.). Pressurization may be supplied by a fuel pump or by injector pistons (either spring-loaded or cam-actuated). With *direct injection*, fuel is injected directly into the combustion chamber, typically into the crown at the top of the piston. With *indirect injection*, fuel is injected into a *precombustion chamber* (*prechamber*) where it is mixed with air prior to moving into the combustion chamber. Diesel engines using indirect injection are quieter and less polluting, though they are slightly less fuel-efficient.

[1] This is different from reciprocating and turbine compressors and compression cycles where the compression ratio is a ratio of pressures.
[2] Stirling engines are not very common in the first place. But among Stirling engines, double action is common.
[3] The intake and exhaust strokes do not affect the thermodynamics of the cycle. These strokes do not appear on the cycle diagrams. The four strokes do not correspond to the four processes in the Otto and diesel cycles.
[4] Another benefit, that of fewer parts and similar design, is sacrificed in commercial vehicle designs because of the need for strict emissions control.
[5] *Two-way catalytic converters* (using platinum and palladium) are still required for hydrocarbon and carbon monoxide reduction. However, these catalysts are less expensive than the *three-way converters* that also use rhodium for nitrogen control.
[6] The term *poppet valve* (i.e., a valve that rises and returns to its seat) is almost never heard anymore.

5. OPERATING CHARACTERISTICS OF COMBUSTION ENGINES

The *specific fuel consumption* (SFC) is the fuel usage rate divided by the power generated. Typical units are lbm/hp-hr and kg/kW·h.

The *air-fuel ratio* (A/F) is the ratio of the air mass that enters the engine to each mass of fuel burned.

Incomplete Combustion

$$A/F = \frac{m_{\text{air}}}{m_{\text{fuel}}} \qquad 28.5$$

The *thermal efficiency*, η_{th}, of a combustion engine cycle is the ratio of net output power to input energy, both expressed in the same units.

Basic Cycles

$$\eta_{\text{th}} = \frac{W_{\text{net}}}{Q_H} = \frac{Q_H - Q_L}{Q_H} \qquad 28.6$$

The *mechanical efficiency*, η_m, of a combustion engine is

$$\eta_m = \frac{\text{bhp}}{\text{ihp}} = \frac{\text{actual power developed}}{\text{frictionless power developed}} \qquad 28.7$$

In reciprocating engines, the *volumetric efficiency*, η_v, is the ratio of the actual to ideal volumes of entering gases.

$$\eta_v = \frac{Q_{\text{actual}}}{Q_{\text{ideal}}} \qquad 28.8$$

The *relative efficiency*, η_r, is the ratio of the actual and ideal thermal efficiencies. The *ideal efficiency*, η_i, can usually be calculated from other cycle properties (e.g., the compression ratio).

$$\eta_r = \frac{\eta_{\text{th,actual}}}{\eta_{\text{th},i}} \qquad 28.9$$

6. BRAKE AND INDICATED PROPERTIES

The performance characteristics (e.g., horsepower) of internal combustion engines can be reported with or without the effect of power-reducing friction and other losses. A value of a property that includes the effect of friction is known as a *brake value*. If the effect of friction is removed, the property is known as an *indicated value*.[7]

Common brake properties are *brake horsepower* (bhp), *brake specific fuel consumption* (BSFC), and *brake mean effective pressure* (BMEP). Common indicated properties are *indicated horsepower* (ihp), *indicated specific fuel consumption* (ISFC), and *indicated mean effective pressure* (IMEP).

The brake and indicated horsepowers differ by the *friction horsepower* (fhp).

$$\text{fhp} = \text{ihp} - \text{bhp} \qquad 28.10$$

Except for Eq. 28.10, indicated and brake properties are usually not combined in calculations since they refer to two different operating conditions. For example, the calculation of the actual fuel mass flow rate requires two brake parameters.

$$\dot{m}_f = (\text{BSFC})(\text{bhp}) \qquad 28.11$$

7. ENGINE POWER AND TORQUE

The power and torque curves are graphs of maximum power and maximum torque that the engine can develop over its speed range. For any given speed, the maximum power that the engine can develop can be determined by loading the engine with progressively higher loads until stalling occurs. This is done on a *dynamometer* or with a *prony brake*, which applies a frictional resistance to a rotating drum attached to the engine's power takeoff, as shown in Fig. 28.1. Since the prony brake measures net available power (inclusive of frictional losses), the operating characteristics derived are brake values (i.e., brake torque and brake horsepower).

Figure 28.1 *Idealized Prony Brake*

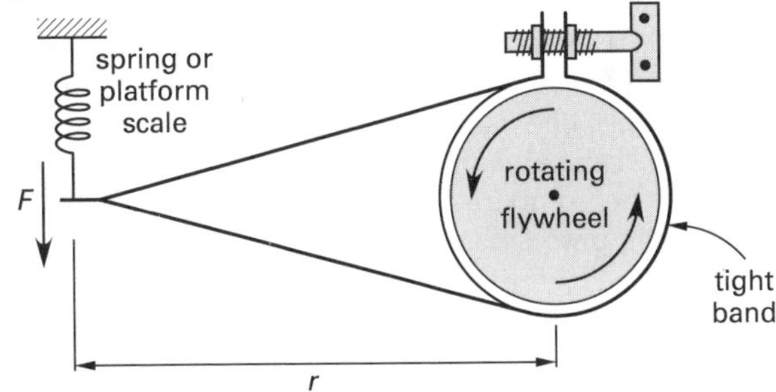

The *brake torque* developed is

$$T = rF \qquad 28.12$$

[7]It may be helpful to think of the i in "indicated" as meaning "ideal."

Power Cycles

The brake torque can also be calculated from the brake horsepower (brake kilowatts) and rotational speed, N, in rpm.

$$T_{\text{N}\cdot\text{m}} = \frac{60\,000(\text{BkW})}{2\pi N_{\text{rpm}}} = \frac{9549(\text{BkW})}{N_{\text{rpm}}} \quad \text{[SI]} \qquad 28.13(a)$$

$$T_{\text{ft-lbf}} = \frac{33{,}000(\text{bhp})}{2\pi N_{\text{rpm}}} = \frac{5252(\text{bhp})}{N_{\text{rpm}}} \quad \text{[U.S.]} \qquad 28.13(b)$$

The *brake horsepower* (*brake kilowatts*) can be calculated from the brake torque or directly from the results of the prony brake test. rF is the brake torque.

$$\text{BkW} = \frac{2\pi rFN}{60\,000} = \frac{rFN}{9549} \quad \text{[SI]} \qquad 28.14(a)$$

$$\text{bhp} = \frac{2\pi rFN}{33{,}000} = \frac{rFN}{5252} \quad \text{[U.S.]} \qquad 28.14(b)$$

Power ratings listed for most commercial engines correspond to the "standard conditions" of 500 ft (150 m) altitude, a dry barometric pressure of 29.00 in of mercury, water vapor pressure of 0.38 in of mercury, and temperature of 85°F (29°C).[8,9]

The *continuous duty rating* is the rated power that the manufacturer claims the engine is able to provide on a continuous (governed or steady rpm) basis without incurring damage. The *intermittent rating* represents the peak power that can be produced on an occasional basis.

Power Cycles

8. AIR-STANDARD CARNOT CYCLE

Theoretically, the *air-standard Carnot combustion cycle* can be implemented either as a reciprocating or steady-flow device. However, like the Carnot vapor cycle, the air-standard Carnot combustion cycle is not a practical engine cycle.[10] The value of the cycle is in establishing a maximum thermal efficiency against which all other cycles can be compared. The Carnot cycle consists of the following processes.

- 1 to 2: isentropic compression
- 2 to 3: isothermal expansion power stroke
- 3 to 4: isentropic expansion
- 4 to 1: isothermal compression

The T-s diagram shown in Fig. 28.2 is the same for any Carnot cycle. However, the isothermal expansion and compression do not occur within a vapor dome as they do in a vapor power cycle and, therefore, do not occur at constant pressure. Thus, the p-v diagram is different than for a vapor cycle.

Figure 28.2 *Air-Standard Carnot Cycle*

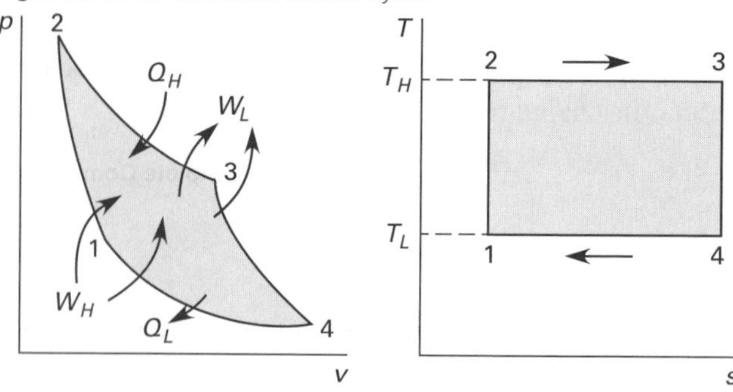

The *isentropic pressure ratio* is a ratio of pressures.

$$r_{p,s} = \frac{p_2}{p_1} = \frac{p_3}{p_4} = \left(\frac{T_4}{T_3}\right)^{k/(1-k)} \qquad 28.15$$

The *isentropic compression ratio* for reciprocating equipment is a ratio of volumes.

Internal Combustion Engines

$$r_{v,s} = \frac{V_1}{V_2} = \frac{V_4}{V_3} = \left(\frac{T_4}{T_3}\right)^{1/(1-k)} \qquad 28.16$$

The isentropic and isothermal relationships for ideal gases are as follows.

Internal Combustion Engines

$$\frac{p_2}{p_1} = \left(\frac{V_1}{V_2}\right)^k \qquad 28.17$$

$$\frac{p_3}{p_4} = \left(\frac{V_4}{V_3}\right)^k \qquad 28.18$$

$$\frac{T_2}{T_1} = \left(\frac{V_1}{V_2}\right)^{k-1} \qquad 28.19$$

$$\frac{T_3}{T_4} = \left(\frac{V_4}{V_3}\right)^{k-1} \qquad 28.20$$

[8] This "yet another" definition of standard conditions is specified in SAE Standard J816b.
[9] Some manufacturers rate their engines at sea level and 60°F (15.6°C). This increases the reported power by approximately 4%.
[10] In particular, it is not possible to design equipment that will transfer heat to a working fluid at a constant temperature in a reversible process over a reasonably finite time.

Formulas for pressure, temperature, and volume at any stage can be calculated from these relationships.

at 1:

$$p_1 = p_4\left(\frac{V_4}{V_1}\right) = p_2\left(\frac{V_2}{V_1}\right)^k = p_2\left(\frac{T_1}{T_2}\right)^{k/(k-1)} \quad 28.21$$

$$T_1 = T_4 = T_2\left(\frac{V_2}{V_1}\right)^{k-1} = T_2\left(\frac{p_1}{p_2}\right)^{(k-1)/k} \quad 28.22$$

$$V_1 = V_4\left(\frac{p_4}{p_1}\right) = V_2\left(\frac{p_2}{p_1}\right)^{1/k} = V_2\left(\frac{T_2}{T_1}\right)^{1/(k-1)} \quad 28.23$$

at 2:

$$p_2 = p_1\left(\frac{V_1}{V_2}\right)^k = p_1\left(\frac{T_2}{T_1}\right)^{k/(k-1)} - p_3\left(\frac{V_3}{V_2}\right) \quad 28.24$$

$$T_2 = T_1\left(\frac{V_1}{V_2}\right)^{k-1} = T_1\left(\frac{p_2}{p_1}\right)^{(k-1)/k} = T_3 \quad 28.25$$

$$V_2 = V_1\left(\frac{p_1}{p_2}\right)^{1/k} - V_1\left(\frac{T_1}{T_2}\right)^{1/(k-1)} = V_3\left(\frac{p_3}{p_2}\right) \quad 28.26$$

at 3:

$$p_3 = p_2\left(\frac{V_2}{V_3}\right) = p_4\left(\frac{V_4}{V_3}\right)^k = p_4\left(\frac{T_3}{T_4}\right)^{k/(k-1)} \quad 28.27$$

$$T_3 = T_2 = T_4\left(\frac{V_4}{V_3}\right)^{k-1} = T_4\left(\frac{p_3}{p_4}\right)^{(k-1)/k} \quad 28.28$$

$$V_3 = V_2\left(\frac{p_2}{p_3}\right) = V_4\left(\frac{p_4}{p_3}\right)^{1/k} = V_4\left(\frac{T_4}{T_3}\right)^{1/(k-1)} \quad 28.29$$

at 4:

$$p_4 = p_3\left(\frac{V_3}{V_4}\right)^k = p_3\left(\frac{T_4}{T_3}\right)^{k/(k-1)} = p_1\left(\frac{V_1}{V_4}\right) \quad 28.30$$

$$T_4 = T_3\left(\frac{V_3}{V_4}\right)^{k-1} = T_3\left(\frac{p_4}{p_3}\right)^{(k-1)/k} = T_1 \quad 28.31$$

$$V_4 = V_3\left(\frac{p_3}{p_4}\right)^{1/k} = V_3\left(\frac{T_3}{T_4}\right)^{1/(k-1)} = V_1\left(\frac{p_1}{p_4}\right) \quad 28.32$$

The work and heat flow terms per unit mass of working fluid are

$$W_L = c_v(T_3 - T_4) + T_2(s_3 - s_2) \quad 28.33$$

$$W_H = |c_v(T_1 - T_2)| + T_4(s_4 - s_1) \quad 28.34$$

$$Q_{L,4\text{-}1} = |T_L(s_1 - s_4)| = \left|p_4V_4 \ln \frac{V_1}{V_4}\right| \quad 28.35$$

$$Q_{H,2\text{-}3} = T_H(s_3 - s_2) = p_2V_2 \ln \frac{V_3}{V_2} \quad 28.36$$

The thermal efficiency of the air-standard Carnot cycle can be calculated from the heat and work terms, but it is easier to work with the two temperature extremes. The temperature forms of the thermal efficiency equation are easy to derive since $q = T\Delta s$.

Carnot Cycle

$$\eta_{\text{th}} = 1 - \frac{T_L}{T_H} = \frac{T_H - T_L}{T_H} \quad 28.37$$

In Eq. 28.37, $T_L = T_1 = T_4$, and $T_H = T_2 = T_3$.

The ideal thermal efficiency can also be calculated from the compression ratio.

$$\eta_{\text{th}} = 1 - r_{v,s}^{1-k} = 1 - r_{p,s}^{(1-k)/k} \quad 28.38$$

9. AIR-STANDARD OTTO CYCLE

The *air-standard Otto cycle* consists of the following processes and is illustrated in Fig. 28.3. The Otto cycle is a four-stroke cycle.

1 to 2: isentropic compression
2 to 3: constant volume heat addition
3 to 4: isentropic expansion
4 to 1: constant volume heat rejection

Figure 28.3 *Air-Standard Otto Cycle*

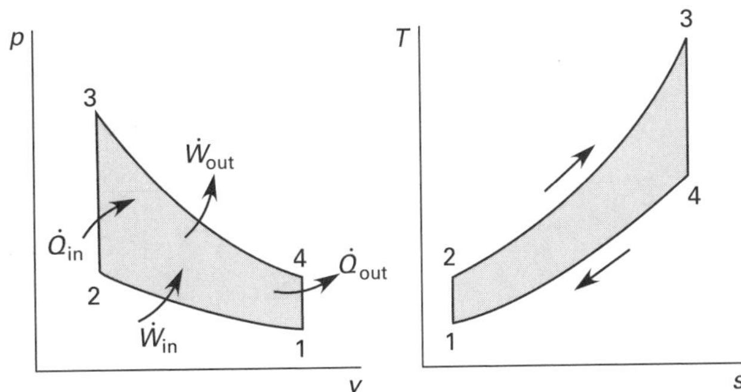

Power Cycles

Although it is possible to define a pressure ratio for the Otto cycle, usually only the *volumetric compression ratio*, r_v, is used.

Otto Cycle (Gasoline Engine)

$$r_v = \frac{v_1}{v_2} = \frac{v_4}{v_3} \quad 28.39$$

Depending on the octane, compression ratios are typically in the range of 5:1 to 10.5:1. Most modern automobile engines designed to run on unleaded gasoline have compression ratios of approximately 8:1 to 8.5:1.

The ratios of absolute temperatures are equal because the piston travels a fixed distance, regardless of which stroke the cycle is on.

$$\frac{T_4}{T_3} = \frac{T_1}{T_2} \quad 28.40$$

Pressure, volume, and temperature relationships for the isentropic and constant volume ideal gas processes can be evaluated using air tables or the ideal gas equations. Using the ideal gas equations makes the implicit assumption that the specific heats are constant during the four processes, while the specific heats actually vary greatly over the wide temperature extremes encountered. By using the air tables this source of error is eliminated, but because the working fluid is not actually pure air, the additional refinement is probably unwarranted.

Assuming an ideal gas, the work and heat flow terms are

$$q_{\text{in},2\text{-}3} = c_v(T_3 - T_2) \quad 28.41$$

$$q_{\text{out},4\text{-}1} = |c_v(T_1 - T_4)| \quad 28.42$$

$$w_{\text{in},1\text{-}2} = c_v(T_2 - T_1) = \frac{p_2v_2 - p_1v_1}{k-1} \quad 28.43$$

$$w_{\text{out},3\text{-}4} = |c_v(T_4 - T_3)| = \left|\frac{p_4v_4 - p_3v_3}{k-1}\right| \quad 28.44$$

The thermal efficiency for the Otto cycle can be calculated in a number of ways.

Otto Cycle (Gasoline Engine)

$$\eta_{\text{th}} = \frac{\dot{W}_{\text{net}}}{\dot{Q}_{\text{in}}} = \frac{\dot{W}_{\text{out}} - \dot{W}_{\text{in}}}{\dot{Q}_{\text{in}}} \quad 28.45$$

$$\eta_{\text{th}} = 1 - \frac{\dot{Q}_{\text{out}}}{\dot{Q}_{\text{in}}} = \frac{\dot{Q}_{\text{in}} - \dot{Q}_{\text{out}}}{\dot{Q}_{\text{in}}} \quad 28.46$$

$$\eta_{\text{th}} = \frac{T_3 - T_4}{T_3} = \frac{T_2 - T_1}{T_2} \quad 28.47$$

The ideal thermal efficiency can be calculated from the compression ratio.

Otto Cycle (Gasoline Engine)

$$\eta_{\text{th}} = 1 - \frac{1}{r_v^{k-1}} = 1 - r_v^{1-k} \quad 28.48$$

From Eq. 28.48, it can be inferred that the ideal thermal efficiency of an Otto cycle can be increased by increasing either the ratio of specific heats or the compression ratio or both. Increasing the ratio of specific heats can only be accomplished by substituting another gas (e.g., helium) for the atmospheric nitrogen. This is not a practical alternative. Increases in the compression ratio are limited to approximately 10:1 by fuel detonation.[11] (Cylinder sealing is a problem only for compression ratios above approximately 16:1.)

The air-standard Otto cycle is less efficient than the Carnot cycle operating between the same temperature limits. However, Eq. 28.38 and Eq. 28.48 show that equal efficiencies are obtained if the compression ratios are made equal.

Example 28.1

An internal combustion engine is to be evaluated on the basis of an air-standard Otto cycle. The pressure and temperature at the intake are 14.7 psia and 600°R (101.3 kPa and 330K), respectively. The maximum pressure and temperature in the cycle are 340 psia and 2610°R (2.3 MPa and 1450K), respectively. Consider air to be an ideal gas. What are the (a) pressure at the end of the compression stroke, (b) temperature at the end of the compression stroke, and (c) cycle efficiency?

[11] *Detonation* is a premature autoignition of the fuel. After the spark ignites one portion of the air-fuel mixture, the advancing flame front compresses the remainder of the mixture. If the compression ratio is too high, the compressed remainder may detonate.

SI Solution

Use the ideal gas law to find the specific volumes at states 1 and 2.

Ideal Gas

$$pv = RT$$

$$v_1 = \frac{RT_1}{p_1} = \frac{\left(0.2870\ \frac{\text{kJ}}{\text{kg}\cdot\text{K}}\right)\times\left(1000\ \frac{\text{J}}{\text{kJ}}\right)(330\text{K})}{(101.3\ \text{kPa})\left(1000\ \frac{\text{Pa}}{\text{kPa}}\right)} = 0.9349\ \text{m}^3/\text{kg}$$

$$v_2 = v_3 = \frac{RT_3}{p_3} = \frac{\left(0.2870\ \frac{\text{kJ}}{\text{kg}\cdot\text{K}}\right)\left(1000\ \frac{\text{J}}{\text{kJ}}\right)(1450\text{K})}{(2.3\ \text{MPa})\left(10^6\ \frac{\text{Pa}}{\text{MPa}}\right)} = 0.1809\ \text{m}^3/\text{kg}$$

Find the volumetric compression ratio.

Otto Cycle (Gasoline Engine)

$$r_v = \frac{v_1}{v_2} = \frac{0.9349\ \frac{\text{m}^3}{\text{kg}}}{0.1809\ \frac{\text{m}^3}{\text{kg}}} = 5.168$$

Use the equations for a constant entropy process to find the pressure and temperature at state 2.

Ideal Gas

$$\frac{p_2}{p_1} = \left(\frac{v_1}{v_2}\right)^k$$

$$p_2 = p_1\left(\frac{v_1}{v_2}\right)^k = (101.3\ \text{kPa})(5.168)^{1.4} = 1010\ \text{kPa}\quad(1.010\ \text{MPa})$$

$$\frac{T_2}{T_1} = \left(\frac{v_1}{v_2}\right)^{k-1}$$

$$T_2 = T_1\left(\frac{v_1}{v_2}\right)^{k-1} = (330\text{K})(5.168)^{1.4-1} = 636.6\text{K}$$

Use the volumetric compression ratio to find the thermal efficiency.

Otto Cycle (Gasoline Engine)

$$\eta_{\text{th}} = 1 - \frac{1}{r_v^{k-1}} = 1 - \frac{1}{(5.168)^{1.4-1}} = 0.4816\quad(48.16\%)$$

Customary U.S. Solution

Convert the pressures at states 1 and 3 to psf.

$$p_1 = \left(14.7\ \frac{\text{lbf}}{\text{in}^2}\right)\left(12\ \frac{\text{in}}{\text{ft}}\right)^2 = 2117\ \text{lbf/ft}^2$$

$$p_3 = \left(340\ \frac{\text{lbf}}{\text{in}^2}\right)\left(12\ \frac{\text{in}}{\text{ft}}\right)^2 = 48{,}960\ \text{lbf/ft}^2$$

Use the ideal gas law to find the specific volumes at states 1 and 2.

Ideal Gas

$$pv = RT$$

$$v_1 = \frac{RT_1}{p_1} = \frac{\left(53.35\ \frac{\text{ft-lbf}}{\text{lbm-°R}}\right)(600°\text{R})}{\left(14.7\ \frac{\text{lbf}}{\text{in}^2}\right)\left(12\ \frac{\text{in}}{\text{ft}}\right)^2} = 15.12\ \text{ft}^3/\text{lbm}$$

$$v_2 = v_3 = \frac{RT_3}{p_3} = \frac{\left(53.35\ \frac{\text{ft-lbf}}{\text{lbm-°R}}\right)(2610°\text{R})}{\left(340\ \frac{\text{lbf}}{\text{in}^2}\right)\left(12\ \frac{\text{in}}{\text{ft}}\right)^2} = 2.844\ \text{ft}^3/\text{lbm}$$

Find the volumetric compression ratio.

Otto Cycle (Gasoline Engine)

$$r_v = \frac{v_1}{v_2} = \frac{15.12 \ \frac{\text{ft}^3}{\text{lbm}}}{2.844 \ \frac{\text{ft}^3}{\text{lbm}}} = 5.316$$

Use the equations for a constant entropy process to find the pressure and temperature at state 2.

Ideal Gas

$$\frac{p_2}{p_1} = \left(\frac{v_1}{v_2}\right)^k$$

$$p_2 = p_1\left(\frac{v_1}{v_2}\right)^k = \left(14.7 \ \frac{\text{lbf}}{\text{in}^2}\right)(5.316)^{1.4} = 152.5 \ \text{lbf/in}^2$$

$$\frac{T_2}{T_1} = \left(\frac{v_1}{v_2}\right)^{k-1}$$

$$T_2 = T_1\left(\frac{v_1}{v_2}\right)^{k-1} = (600°\text{R})(5.316)^{1.4-1} = 1171°\text{R}$$

Use the volumetric compression ratio to find the thermal efficiency.

Otto Cycle (Gasoline Engine)

$$\eta_{\text{th}} = 1 - \frac{1}{r_v^{k-1}} = 1 - \frac{1}{(5.316)^{1.4-1}} = 0.4874 \quad (48.7\%)$$

Example 28.2

An Otto engine has 15% clearance. Air enters the engine at 14.0 psia and 580°R (96.6 kPa and 320K). The air-fuel ratio is 16.67. The heating value of gasoline is 19,500 Btu/lbm (45.5 MJ/kg), and the combustion efficiency is 76.9%. What are the temperatures and pressures at all points in the cycle?

SI Solution

(This solution takes a cold air-standard approach. The customary U.S. solution uses an air table.)

Refer to Fig. 28.3.

at 1:

$$V_1 = V_{\text{swept}} + V_{\text{clearance}} = (1+c)\,V_{\text{swept}} = (1+0.15)\,V_{\text{swept}} = 1.15\,V_{\text{swept}}$$

$$p_1 = 96.6 \text{ kPa} \quad \text{[given]}$$

$$T_1 = 320\text{K} \quad \text{[given]}$$

at 2:

$$V_2 = cV_{\text{swept}} = 0.15\,V_{\text{swept}}$$

The compression ratio is

$$r_v = \frac{V_1}{V_2} = \frac{1.15\,V_{\text{swept}}}{0.15\,V_{\text{swept}}} = 7.67$$

The compression from 1 to 2 is isentropic.

$$T_2 = T_1\left(\frac{V_1}{V_2}\right)^{k-1} = (320\text{K})(7.67)^{1.4-1} = 722.9\text{K}$$

$$p_2 = p_1\left(\frac{V_1}{V_2}\right)^k = (96.6 \text{ kPa})(7.67)^{1.4} = 1674 \text{ kPa} \quad (1.674 \text{ MPa})$$

at 3: The heat added by the fuel per kg of air is

$$q = \left(\frac{\text{LHV}}{A/F}\right)\eta = \left(\frac{45.5 \ \frac{\text{MJ}}{\text{kg fuel}}}{16.67 \ \frac{\text{kg air}}{\text{kg fuel}}}\right)(0.769) = 2.099 \text{ MJ/kg air}$$

The combustion heat increases the internal energy of the air. For any process, $\Delta u = c_v \Delta T$. The cold specific heat (0.718 kJ/kg·K) of air is used.

$$T_3 = T_2 + \frac{\Delta u}{c_v} = T_2 + \frac{q}{c_v} = 722.9\text{K} + \frac{\left(2.099 \ \frac{\text{MJ}}{\text{kg}}\right)\left(1000 \ \frac{\text{kJ}}{\text{MJ}}\right)}{0.718 \ \frac{\text{kJ}}{\text{kg}\cdot\text{K}}} = 3646\text{K}$$

$$p_3 = \frac{p_2 T_3}{T_2} = \frac{(1.674 \text{ MPa})(3646\text{K})}{722.9\text{K}} = 8.44 \text{ MPa}$$

at 4: The process from 3 to 4 is isentropic. Since the piston travels the same distance going from 1 to 2 as it does from 3 to 4, the 3-4 expansion ratio is the same as the 1-2 compression ratio.

$$T_4 = T_3\left(\frac{V_3}{V_4}\right)^{k-1} = (3646\text{K})\left(\frac{1}{7.67}\right)^{1.4-1}$$
$$= 1614\text{K}$$
$$p_4 = p_3\left(\frac{V_3}{V_4}\right)^{k} = (8.44\text{ MPa})\left(\frac{1}{7.67}\right)^{1.4}$$
$$= 0.487\text{ MPa}$$

Customary U.S. Solution

(This solution uses an air table. The SI solution takes a cold air-standard cycle approach.)

Refer to Fig. 28.3.

at 1:

$$V_1 = V_{\text{swept}} + V_{\text{clearance}} = (1+c)\,V_{\text{swept}}$$
$$= (1+0.15)\,V_{\text{swept}}$$
$$= 1.15\,V_{\text{swept}}$$
$$p_1 = 14.0\text{ psia} \quad \text{[given]}$$
$$T_1 = 580°\text{R} \quad \text{[given]}$$

From an air table, at 580°R, the relative volume is 120.70, and the relative pressure is 1.7800. [**Properties of Air at Low Pressure, per Pound**]

at 2:

$$V_2 = 0.15\,V_{\text{swept}}$$

The compression ratio is

$$r_v = \frac{V_1}{V_2} = \frac{1.15\,V_{\text{swept}}}{0.15\,V_{\text{swept}}} = 7.67$$

Consider the compression from 1 to 2 to be isentropic.

$$v_{r,2} = \frac{v_{r,2}}{r_v} = \frac{120.7}{7.67} = 15.74$$

The temperature corresponding to this volume ratio in the air table is 1273°R.

$$T_2 = 1273°\text{R}$$
$$u_2 = 222.72\text{ Btu/lbm}$$
$$p_{r,2} = 29.93$$
$$p_2 = \frac{p_1 p_{r,2}}{p_{r,1}} = \frac{(14.0\text{ psia})(29.93)}{1.78}$$
$$= 235.4\text{ psia}$$

at 3: The heat added by the fuel per pound of air is

$$q = \left(\frac{\text{HV}}{A/F}\right)\eta = \left(\frac{19{,}500\ \dfrac{\text{Btu}}{\text{lbm fuel}}}{16.67\ \dfrac{\text{lbm air}}{\text{lbm fuel}}}\right)(0.769)$$
$$= 899.55\text{ Btu/lbm air}$$

Enthalpy is the sum of internal energy and pV energy. Even though the enthalpy at point 2 could be found from the air table, the pressure at point 3 is not yet known. Therefore, this heat addition cannot merely be added to the enthalpy at point 2. However, internal energy does not include the pV term. The internal energy at the end of the heat addition is

$$u_3 = u_2 + q = 222.72\ \frac{\text{Btu}}{\text{lbm}} + 899.55\ \frac{\text{Btu}}{\text{lbm}}$$
$$= 1122.27\text{ Btu/lbm}$$

Locate this value of internal energy in the air table.

$$T_3 = 5300°\text{R}$$
$$v_{r,3} = 0.1710$$
$$p_{r,3} = 11{,}481$$

Use the ideal gas law to find the pressure at 3.

$$p_3 = \frac{p_2 T_3}{T_2} = \frac{(235.4\text{ psia})(5300°\text{R})}{1273°\text{R}}$$
$$= 980.1\text{ psia}$$

At 4: The process from 3 to 4 is isentropic.

$$v_{r,4} = v_{r,3} r_v = (0.1710)(7.67) = 1.3116$$

Power Cycles

Locate this volume ratio in the air table.

$$T_4 \approx 2900°\text{R}$$
$$p_{r,4} = 814.8$$
$$p_4 = \frac{p_3 p_{r,4}}{p_{r,3}} = \frac{(980.1 \text{ psia})(814.8)}{11{,}481} = 69.6 \text{ psia}$$

10. PLAN FORMULA

The performance of an internal combustion engine operating on the Otto cycle can be predicted from the PLAN formula, Eq. 28.43.[12] It is necessary to know the engine bore area for one cylinder (A in square inches or square meters), stroke (L in feet or meters), number of engine power strokes per minute (N), and mean effective pressure (MEP in psig or kilopascals). The power, P, will be in horsepower. The product LA is the *swept volume* of one cylinder. If the *engine displacement*, V, is known, it should be divided by the number of cylinders before being substituted for LA.

Internal Combustion Engines

$$P = (\text{MEP})\left(\frac{LAN}{K}\right) \qquad 28.49$$

It is essential to use the units as defined. The *brake horsepower* (bhp) will be calculated if the *brake mean effective pressure* (BMEP) is used. The *indicated horsepower* (ihp) will be calculated if the *indicated mean effective pressure* (IMEP) is used. In Eq. 28.49, the constant K is 33,000 is using U.S. units and 44.74 if using SI units. The PLAN formula with units specified, then, is

$$P_{\text{hp}} = (\text{MEP}_{\text{kPa}})\left(\frac{L_{\text{m}} A_{\text{m}^2} N}{44.74}\right) \qquad 28.50$$

$$P_{\text{hp}} = (\text{MEP}_{\text{psig}})\left(\frac{L_{\text{ft}} A_{\text{in}^2} N}{33{,}000}\right) \qquad 28.51$$

The variation of the PLAN formula can be used to give the power in kilowatts. In this case, the constant K is 60.

$$P_{\text{kW}} = (\text{MEP}_{\text{kPa}})\left(\frac{L_{\text{m}} A_{\text{m}^2} N}{60}\right) \qquad 28.52$$

Equation 28.53 gives the number of engine power strokes per minute for two- and four-stroke engines.

$$N = \frac{(2n)(\text{no. cylinders})}{\text{no. strokes per cycle}} \qquad 28.53$$

The indicated *mean effective pressure* (MEP) is determined graphically from an actual *p-v* plot (known as an *indicator drawing, indicator diagram*, or *indicator card*) of the actual cycle. This is done by taking the overall area of the indicator drawing and dividing it by the total width of the drawing. The resulting number is then multiplied by the indicator spring constant (scale) to obtain the net work per cycle, W_{net}.

The MEP is a theoretical average pressure that would produce the same amount of net work during one stroke as the varying pressure does in the entire cycle. MEP can be calculated if the cycle performance is known. MEP in nonsupercharged engines is typically limited to approximately 100 psi (700 kPa).[13]

$$\text{MEP} = \frac{W_{\text{net}}}{V_1 - V_2} \qquad 28.54$$

In some cases, the power generated by an engine is correlated with CND^2, where C is a constant. This is the same as Eq. 28.50 and Eq. 28.51, with the coefficient C incorporating all of the constant terms.

Figure 28.4 illustrates how the actual Otto cycle deviates from the ideal cycle (shown by the broken line) and the causes of the deviations. Due to heat transfers during the 1-to-2 and 3-to-4 processes, the lines are not true adiabats. Rather, these two processes are polytropic, with a polytropic exponent of approximately 1.3. The net effect of all of these deviations is to reduce the actual efficiency to approximately half of the ideal efficiency.

Figure 28.4 *Otto Cycle Indicator Drawing*

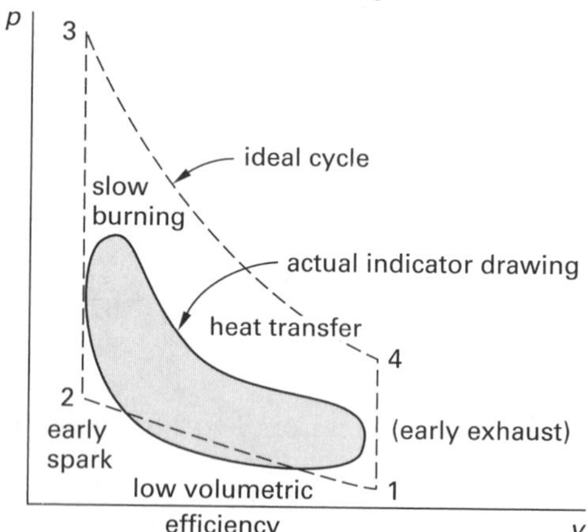

Example 28.3

A single-cylinder, four-stroke engine has a 10 in (25.4 cm) bore and an 18 in (45.7 cm) stroke. The engine is run at 200 rpm while being tested on a prony brake. The gross force exerted by the brake is 140 lbf (620 N), the tare is 25 lbf (110 N), and the arm length is 66 in (1.7 m). The indicator card shows an area of 1.2 in^2

[12] The PLAN formula is also applicable to reciprocating steam engines.

[13] With the advent of supercharging, *peak firing pressure*, also known as *peak cylinder pressure* (PCP), has become a benchmark for engine performance.

($7.7\ \text{cm}^2$) with an overall length of 3 in (7.6 cm). The spring scale used to draw the indicator card is 200 psi/in (540 kPa/cm). What are the (a) indicated main effective pressure, (b) indicated horsepower, (c) brake horsepower, and (d) mechanical efficiency?

SI Solution

(a) The indicated mean effective pressure is

$$\begin{aligned}\text{MEP} &= \frac{(\text{diagram area})(\text{pressure scale factor})}{\text{diagram length}}\\ &= \frac{(7.7\ \text{cm}^2)\left(540\ \frac{\text{kPa}}{\text{cm}}\right)}{7.6\ \text{cm}}\\ &= 547\ \text{kPa}\end{aligned}$$

(b) The stroke is

$$L = \frac{45.7\ \text{cm}}{100\ \frac{\text{cm}}{\text{m}}} = 0.457\ \text{m}$$

The cylinder area is

$$\begin{aligned}A &= \frac{\pi D^2}{4} = \frac{\pi(25.4\ \text{cm})^2}{4}\\ &= 506.7\ \text{cm}^2 \quad (0.05067\ \text{m}^2)\end{aligned}$$

The number of power strokes per minute is

$$\begin{aligned}N &= \frac{(2n)(\text{no. cylinders})}{\text{no. strokes per cycle}}\\ &= \frac{\left(2\ \frac{\text{strokes}}{\text{rev}}\right)\left(200\ \frac{\text{rev}}{\text{min}}\right)(1\ \text{cylinder})}{4\ \text{strokes per cycle}}\\ &= 100\ \text{power strokes/min}\end{aligned}$$

Equation 28.52 gives the indicated power.

$$\begin{aligned}P_{\text{kW}} &= (\text{MEP}_{\text{kPa}})\left(\frac{L_{\text{m}}A_{\text{m}^2}N}{60}\right)\\ &= \frac{(547\ \text{kPa})\left(1000\ \frac{\text{Pa}}{\text{kPa}}\right)(0.457\ \text{m}) \times(0.05067\ \text{m}^2)\left(100\ \frac{\text{power strokes}}{\text{min}}\right)}{\left(60\ \frac{\text{s}}{\text{min}}\right)\left(1000\ \frac{\text{W}}{\text{kW}}\right)}\\ &= 21.1\ \text{kW}\end{aligned}$$

(c) The brake power is found from the brake information. The brake arm length is

$$r = 1.7\ \text{m}$$

The force on the brake caused by the engine excludes the tare weight.

$$F = 620\ \text{N} - 110\ \text{N} = 510\ \text{N}$$

From Eq. 28.14, the brake power is

$$\begin{aligned}\text{BkW} &= \frac{2\pi rFN}{60\,000}\\ &= \frac{(2\pi)(1.7\ \text{m})(510\ \text{N})\left(200\ \frac{\text{rev}}{\text{min}}\right)}{\left(60\ \frac{\text{s}}{\text{min}}\right)\left(1000\ \frac{\text{W}}{\text{kW}}\right)}\\ &= 18.16\ \text{kW}\end{aligned}$$

(d) The mechanical efficiency is

$$\begin{aligned}\eta_m &= \frac{\text{brake power}}{\text{indicated power}} = \frac{18.16\ \text{kW}}{21.1\ \text{kW}}\\ &= 0.861 \quad (86.1\%)\end{aligned}$$

Customary U.S. Solution

(a) The indicated mean effective pressure is

$$\begin{aligned}\text{MEP} &= \frac{(\text{diagram area})(\text{pressure scale factor})}{\text{diagram length}}\\ &= \frac{(1.2\ \text{in}^2)\left(200\ \frac{\frac{\text{lbf}}{\text{in}^2}}{\text{in}}\right)}{3\ \text{in}}\\ &= 80\ \text{lbf/in}^2\end{aligned}$$

(b) The stroke is

$$L = \frac{18\ \text{in}}{12\ \frac{\text{in}}{\text{ft}}} = 1.5\ \text{ft}$$

The cylinder area is

$$A = \frac{\pi D^2}{4} = \frac{\pi(10\ \text{in})^2}{4} = 78.54\ \text{in}^2$$

Power Cycles

The number of power strokes per minute is

$$\begin{aligned} N &= \frac{(2n)(\text{no. cylinders})}{\text{no. strokes per cycle}} \\ &= \frac{\left(2\ \frac{\text{strokes}}{\text{rev}}\right)\left(200\ \frac{\text{rev}}{\text{min}}\right)(1\ \text{cylinder})}{4\ \text{strokes per cycle}} \\ &= 100\ \text{power strokes/min} \end{aligned}$$

From Eq. 28.51, the indicated horsepower is

$$\begin{aligned} P_{\text{hp}} &= (\text{MEP}_{\text{psig}})\left(\frac{L_{\text{ft}}A_{\text{in}^2}N}{33{,}000}\right) \\ &= \frac{\left(80\ \frac{\text{lbf}}{\text{in}^2}\right)(1.5\ \text{ft})(78.54\ \text{in}^2) \times\left(100\ \frac{\text{power strokes}}{\text{min}}\right)}{33{,}000\ \frac{\text{ft-lbf}}{\text{hp-min}}} \\ &= 28.56\ \text{hp} \end{aligned}$$

(c) The brake horsepower is found from the brake information. The brake arm length is

$$r = \frac{66\ \text{in}}{12\ \frac{\text{in}}{\text{ft}}} = 5.5\ \text{ft}$$

The force on the brake caused by the engine excludes the tare weight.

$$F = 140\ \text{lbf} - 25\ \text{lbf} = 115\ \text{lbf}$$

From Eq. 28.14, the brake horsepower is

$$\begin{aligned} \text{bhp} &= \frac{2\pi rFN}{33{,}000} \\ &= \frac{(2\pi)(5.5\ \text{ft})(115\ \text{lbf})\left(200\ \frac{\text{rev}}{\text{min}}\right)}{33{,}000\ \frac{\text{ft-lbf}}{\text{hp-min}}} \\ &= 24.09\ \text{hp} \end{aligned}$$

(d) The mechanical efficiency is

$$\begin{aligned} \eta_m &= \frac{\text{bhp}}{\text{ihp}} = \frac{24.09\ \text{hp}}{28.56\ \text{hp}} \\ &= 0.843 \quad (84.3\%) \end{aligned}$$

11. INTRODUCTION TO DIESEL ENGINES

A diesel engine is a compression-ignition, internal combustion engine. Spark plugs are not used for ignition.[14] Rather, high compression ratios produce auto-ignition of the air-fuel mixture. The higher the compression ratio, the higher the thermal efficiency. Compression ratios for diesel engines are ratios of volumes. The compression ratio varies from about 13.5:1 to 17.5:1, depending on whether or not the engine is turbocharged. Diesel engines can burn a variety of fuels, including No. 6 heavy fuel oil, natural gas, and light distillate fuel oils.

Four-stroke diesels offer many significant advantages over two-stroke diesels. (1) With two-stroke engines, a significant amount of energy is expended by a blower to supply air for scavenging. When the load on the engine drops, the blower continues to run at its peak load, reducing engine efficiency even more. (2) Two-cycle engines often require premium fuels, compared to No. 2 oil that most four-stroke engines can use. (3) Valve, piston, and ring burning are more common in two-stroke engines since there is less time for cooling during the cycle. Four-stroke engines can use aluminum pistons, while heavy cast-iron pistons are needed for heat dissipation in two-stroke engines. (4) In a two-stroke engine, piston rings pass over the intake ports on every stroke, leading to increased ring wear. (5) Since the valves and injectors operate every cycle with two-stroke engines, overhauls are required more frequently.

Supercharging compresses and increases the amount of air that enters the cylinder per stroke. (This increases the volumetric efficiency above 100%.) *Turbocharging* is a form of supercharging in which the exhaust gases drive the supercharger. With more air, more fuel can be burned, increasing the power per stroke. Supercharging also helps deliver air for combustion at higher altitudes. This results in better fuel economy than normally aspirated diesels. Supercharging also reduces smoke, particularly during lugging.

The temperature of compressed air from a turbocharger can be decreased by passing it through an aftercooler. An *aftercooler* is a closed heat exchanger that transfers heat from compressed air to cooler air. In vehicles, cool air flow is found in the manifold.

Emissions from diesel engines include nitrogen oxides, unburned hydrocarbons (*polycyclic aromatic hydrocarbons*—PAH), and carbon monoxide, much the same as from gasoline engines. Particulate emission (smoke or soot) and noise are also characteristic diesel pollutants.

[14]A *glow plug* may be used to improve cold-weather starting.

12. STATIONARY AND MARINE GAS DIESELS

Large stationary engines operating on the diesel cycle can be built to run on natural gas.[15] There are two types of gas-burning diesels: gas-diesel and lean-burn engines. Natural gas under pressure has poor ignition and combustion characteristics. Therefore, in the *gas-diesel engine*, a small amount (about 3%) of oil is burned in addition to the natural gas. The oil is injected into the regular chamber or a precombustion chamber. Compression ignites the oil, and the resulting flame "torch" ignites the natural gas.

Since most gas-diesel engines produce less than 10 MW of power, they can be used for small electrical generation plants, gas pipeline compression, refrigeration chillers, and so on.[16]

In the *lean-burn engine*, a spark ignites a very lean mixture in a precombustion chamber. The resulting flame jet issuing from the precombustion chamber ignites the main fuel mixture. Most commercial lean-burn engines produce power in the range of 400 kW to 3000 kW.

13. AIR-STANDARD DIESEL CYCLE

The processes in an air-standard diesel cycle are illustrated in Fig. 28.5.

1 to 2: isentropic compression
2 to 3: constant pressure heating
3 to 4: isentropic expansion
4 to 1: constant volume cooling

Figure 28.5 *Air-Standard Diesel Cycle*

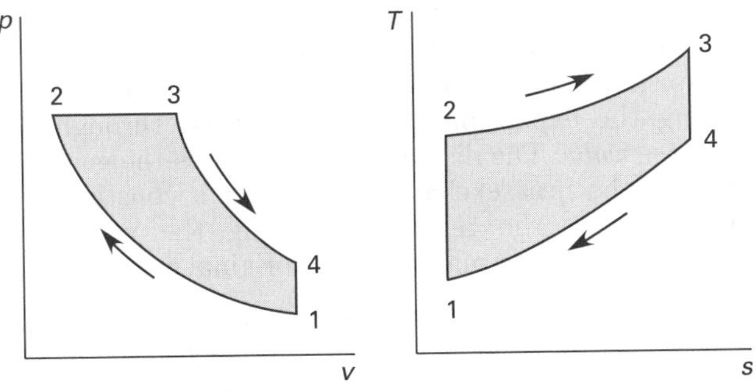

Diesel engines can be either four-stroke or two-stroke devices. The four-stroke and two-stroke engines are analyzed the same way, since the four processes on the p-v and T-s diagrams do not correspond to the four strokes. The main difference is the number of power strokes per revolution—one for the two-stroke engine and one-half for the four-stroke engine.

The diesel cycle *compression ratio*, r_v, is

$$r_v = \frac{V_1}{V_2} \qquad 28.55$$

The *cutoff ratio*, r_{cutoff}, is defined by Eq. 28.56. The volume V_3 is known as the *cutoff volume.*

$$r_{\text{cutoff}} = \frac{V_3}{V_2} = \frac{T_3}{T_2} \qquad 28.56$$

Pressure, volume, and temperature at each point in the cycle can be evaluated with ideal gas equations. The work and heat flow terms are evaluated with the following equations. (Equation 28.57 uses c_p because 2-to-3 is a constant pressure process.)

$$q_{\text{in}} = c_p(T_3 - T_2) \qquad 28.57$$

$$q_{\text{out}} = c_v(T_1 - T_4) \qquad 28.58$$

$$w_{\text{in}} = c_v(T_2 - T_1) \qquad 28.59$$

$$w_{\text{out}} = c_v(T_3 - T_4) \qquad 28.60$$

The air-standard diesel cycle is always less efficient than the air-standard Otto cycle for equal compression ratios. However, for a specific maximum cylinder pressure, the diesel cycle is more efficient than the Otto cycle. The thermal efficiency can be calculated in a number of ways.

Basic Cycles

$$\eta_{\text{th}} = \frac{W_{\text{net}}}{Q_{\text{in}}} = \frac{Q_{\text{in}} - Q_{\text{out}}}{Q_{\text{in}}} \qquad 28.61$$

Diesel Cycle

$$\eta_{\text{th,diesel}} = 1 - \frac{u_4 - u_1}{h_3 - h_2} = 1 - \frac{T_4 - T_1}{k(T_3 - T_2)} \qquad 28.62$$

In Eq. 28.62, $u_4 - u_1$ is the heat rejected, and $h_3 - h_2$ is the heat provided.

$$u_4 - u_1 = mc_v(T_4 - T_1) \qquad 28.63$$

$$h_3 - h_2 = mc_p(T_3 - T_2) \qquad 28.64$$

[15]Large diesels are those producing up to approximately 10 MW. This is large for diesels but small for gas turbines. In a *combined diesel* or *gas turbine system* (CODOG) for marine propulsion, diesels are used for low power and cruise operation; the turbine takes over when high speeds are needed.
[16]This is an awkwardly small size for gas turbines. Power turbines are not very practical in this size range.

In traditional diesels (i.e., those constructed through about 1980), thermal efficiency was not much higher than approximately 35%. Most manufacturers of modern stationary diesels report full-load thermal efficiencies of approximately 45%, and some as high as 50% for experimental test bed engines. Production diesel engines for commercial trucks routinely achieve efficiencies of 41–42%. These are the highest thermal efficiencies of any type of prime mover (including turbines). Significant technological advances in equipment, materials, and controls are required to achieve the highest efficiencies.

14. AIR-STANDARD DUAL CYCLE

The *air-standard dual cycle* shown in Fig. 28.6 is a combination of the diesel and Otto cycles. It more accurately predicts the performance of a spark-ignited internal combustion engine, since the combustion energy is added partly at constant volume and partly at constant pressure.

Figure 28.6 Air-Standard Dual Cycle

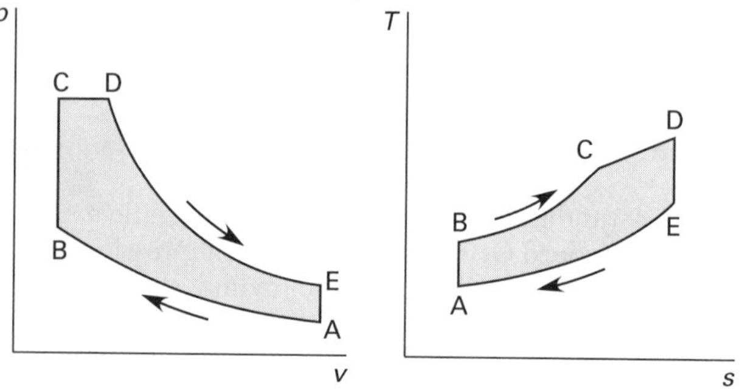

The processes in the dual cycle are

A to B: isentropic compression
B to C: constant volume heating
C to D: constant pressure heating
D to E: isentropic expansion
E to A: constant volume cooling

As with the Otto and diesel cycles, the *compression ratio*, r_v, is defined as a ratio of volumes.

$$r_v = \frac{V_A}{V_B} \quad 28.65$$

The *pressure ratio*, r_p, is

$$r_p = \frac{p_C}{p_B} \quad 28.66$$

The *cutoff ratio*, r_{cutoff}, is

$$r_{\text{cutoff}} = \frac{V_D}{V_C} \quad 28.67$$

Pressure, temperature, and volume at various points in the cycle can be evaluated with ideal gas relationships. The work and heat terms are defined as follows.

$$Q_{\text{in}} = c_v(T_C - T_B) + c_p(T_D - T_C) \quad 28.68$$

$$Q_{\text{out}} = |c_v(T_A - T_E)| \quad 28.69$$

$$W_{\text{in}} = c_v(T_B - T_A) \quad 28.70$$

$$\begin{aligned} W_{\text{out}} &= |p_C(v_D - v_C) + c_v(T_D - T_E)| \\ &= |(c_p - c_v)(T_D - T_C) + c_v(T_D - T_E)| \end{aligned} \quad 28.71$$

The efficiency of a dual cycle is between those of the Otto and diesel cycles.

Basic Cycles

$$\begin{aligned} \eta_{\text{th}} &= \frac{W_{\text{net}}}{Q_{\text{in}}} = \frac{Q_{\text{in}} - Q_{\text{out}}}{Q_{\text{in}}} \\ &= 1 - \frac{T_E - T_A}{(T_C - T_B) + k(T_D - T_C)} \end{aligned} \quad 28.72$$

15. STIRLING ENGINES

During a typical Stirling cycle, the gaseous working fluid (e.g., helium or hydrogen) is shuttled through a heat exchanger circuit consisting of a heat acceptor, a regenerator, and a heat rejector. A piston compresses the working fluid in a compression space near the *heat rejector*. A *displacer* then shuttles the gas through a heat exchanger circuit to an expansion space. Along the way, the gas absorbs heat stored in the *regenerator*.[17] The piston is then moved so that the gas is expanded. During the expansion, heat is absorbed through the *acceptor walls*. The displacer then shuttles the gas back through the heat exchanger circuit at a constant volume, heating the regenerator along the way. This returns the Stirling machine to its original condition.

Heating of the working fluid occurs externally. The heat source can be from a combustion, nuclear (typically radioisotope), or solar process. Cooling is usually by a water-glycol mixture. When combustion provides the energy, Stirling machines draw power from combustion in a separate *combustor*. Since the combustion process is external, Stirling engines are able to use a variety of fuels and burn them at lower temperatures, resulting in lower emissions.

[17]A *regenerator* is a device that captures heat energy and transfers it to the working fluid. In the case of the Stirling cycle, the heated gas passes through and heats a wire or ceramic mesh. When cool gas is passed back over the mesh, the heat transfer is reversed, and the gas is heated.

There are two methods of configuring Stirling machines. In a *kinematic Stirling engine*, double-acting pistons function separately as the compressor and displacer. Sealing of kinematic engines is a major challenge. *Free-piston engines* avoid leakage problems by eliminating the mechanical connection with the piston. Magnetic coupling (magnetic flux linkage) between the piston (and anything moving with it) and the surroundings limit the output of free-piston machines to electrical power generation.

Though Stirling engines are unlikely to be widely used in commercial vehicles, they are used to limited extents in small cryogenic applications. Since solar energy can be used as the heating source, Stirling engines are attractive for generating small amounts of electricity (i.e., 20 kW or less per high-efficiency engine/generator).

16. STIRLING CYCLE

The *Stirling cycle* has a thermal efficiency that can equal that of the Carnot cycle. The processes are shown in Fig. 28.7.

A to B: constant volume heating
B to C: isothermal heating and expansion
C to D: constant volume cooling
D to A: isothermal cooling and compression

Figure 28.7 *Stirling Cycle*

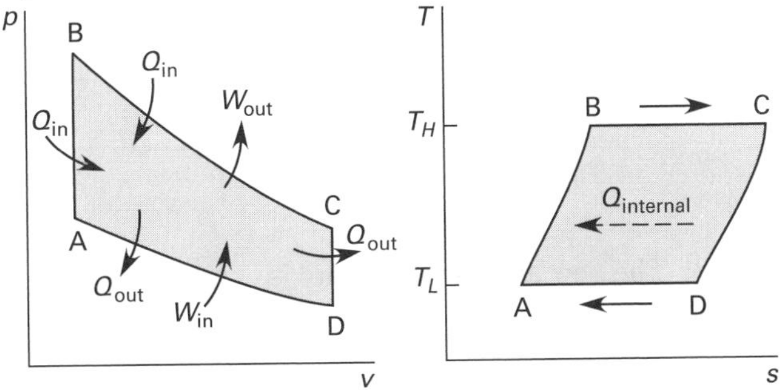

Pressure, temperature, and volume for the various points on the cycle can be evaluated from ideal gas relationships.

The work and heat terms are given by Eq. 28.73 through Eq. 28.76. There is also another heat transfer to the working fluid in the A-to-B process. However, this heat transfer takes place in a reversible regenerator, and the heat transferred is from the working fluid's C-to-D process. Since the heat remains within the system boundary, it is not included in the Q_{in} or Q_{out} terms.

$$Q_{\text{in,B-C}} = T_H(s_C - s_B) = RT_H \ln \frac{v_C}{v_B} = RT_H \ln \frac{p_B}{p_C} \qquad 28.73$$

$$Q_{\text{out,D-A}} = |T_L(s_A - s_D)| = \left| RT_L \ln \frac{v_A}{v_D} \right| = \left| RT_L \ln \frac{p_D}{p_A} \right| \qquad 28.74$$

$$W_{\text{in,D-A}} = Q_{\text{out,D-A}} \qquad 28.75$$

$$W_{\text{out,B-C}} = Q_{\text{in,B-C}} \qquad 28.76$$

The thermal efficiency of the Stirling cycle is equal to the Carnot cycle efficiency if the regenerator used to transfer heat from the C-to-D process to the A-to-B process is reversible. With a reversible regenerator, the thermal efficiency is

Basic Cycles

$$\eta_{th} = \frac{W_{net}}{Q_{in}} = \frac{Q_{in} - Q_{out}}{Q_{in}} = \frac{T_H - T_L}{T_H} \qquad 28.77$$

17. ERICSSON CYCLE

The processes of the Ericsson cycle are shown in Fig. 28.8. This cycle offers the best chance of achieving a thermal efficiency approaching that of the Carnot cycle.[18]

A to B: isothermal compression
B to C: constant pressure heating
C to D: isothermal expansion
D to A: constant pressure cooling

Pressure, temperature, and volume for the various points on the cycle can be evaluated from ideal gas relationships.

[18]The Ericsson cycle is approximated if the Brayton gas turbine cycle is modified to include regenerative heat exchange, intercooling, and reheating.

Figure 28.8 *Ericsson Cycle*

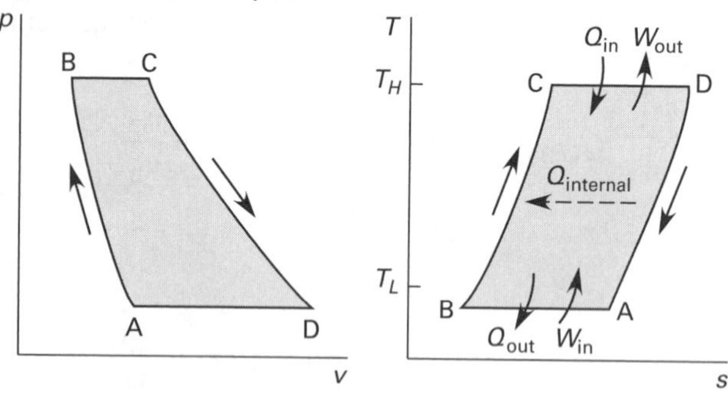

The work and heat terms are given by Eq. 28.78 through Eq. 28.81. There is also a heat transfer to the working fluid in the B-to-C process. However, if a reversible regenerator is used to capture heat from the D-to-A process, no external heat will be required for this.

$$\begin{aligned} Q_{\text{in,C-D}} &= T_H(s_\text{D} - s_\text{C}) = RT_H \ln \frac{v_\text{D}}{v_\text{C}} \\ &= RT_H \ln \frac{p_\text{C}}{p_\text{D}} \end{aligned} \quad \text{28.78}$$

$$\begin{aligned} Q_{\text{out,A-B}} &= |\, T_L(s_\text{B} - s_\text{A})| = \left| RT_L \ln \frac{v_\text{B}}{v_\text{A}} \right| \\ &= \left| RT_L \ln \frac{p_\text{A}}{p_\text{B}} \right| \end{aligned} \quad \text{28.79}$$

$$W_{\text{in,A-B}} = Q_{\text{out,A-B}} \quad \text{28.80}$$

$$W_{\text{out,C-D}} = Q_{\text{in,C-D}} \quad \text{28.81}$$

With a reversible regenerator, the thermal efficiency of the Ericsson cycle is equal to that of the Carnot cycle.

Basic Cycles

$$\begin{aligned} \eta_\text{th} &= \frac{W_\text{net}}{Q_\text{in}} = \frac{Q_\text{in} - Q_\text{out}}{Q_\text{in}} \\ &= \frac{T_H - T_L}{T_H} \end{aligned} \quad \text{28.82}$$

18. EFFECT OF ALTITUDE ON OUTPUT POWER

Since a lower atmospheric pressure decreases atmospheric density, the oxygen per intake stroke available to engines operating on the Otto, diesel, and dual cycles decreases with altitude. The following steps constitute a procedure for determining the variation in power when the altitude is changed. It is assumed that the engine speed is constant and that ideal gas behavior applies.

step 1: Let 1 and 2 be the lower and higher altitudes, respectively.

step 2: Calculate the frictionless power.

$$\text{ihp}_1 = \frac{\text{bhp}_1}{\eta_{m1}} \quad \text{28.83}$$

step 3: Calculate the friction power, which is assumed to be constant at constant speed.

$$\text{fhp} = \text{ihp}_1 - \text{bhp}_1 \quad \text{28.84}$$

step 4: Calculate the air densities ρ_{a1} and ρ_{a2} from App. 25.C.

step 5: Calculate the new frictionless power.

$$\text{ihp}_2 = \text{ihp}_1\left(\frac{\rho_{a2}}{\rho_{a1}}\right) \quad \text{28.85}$$

step 6: Calculate the new net power.

$$\text{bhp}_2 = \text{ihp}_2 - \text{fhp} \quad \text{28.86}$$

step 7: Calculate the new mechanical efficiency.

$$\eta_{m2} = \frac{\text{bhp}_2}{\text{ihp}_2} \quad \text{28.87}$$

step 8: The volumetric air flow rates are the same.

$$Q_{a2} = Q_{a1} \quad \text{28.88}$$

step 9: The original air and fuel rates are

$$\dot{m}_{f1} = (\text{BSFC}_1)(\text{bhp}_1) \quad \text{28.89}$$

$$\dot{m}_{a1} = (A/F)\dot{m}_{f1} \quad \text{28.90}$$

$$Q_{a1} = \frac{\dot{m}_{a1}}{\rho_{a1}} \quad \text{28.91}$$

step 10: The new air mass flow rate is

$$\dot{m}_{a2} = Q_{a2}\rho_{a2} \quad \text{[see step 8]} \quad \text{28.92}$$

step 11: For engines with metered injection and without air-fuel ratio (i.e., "wide range") sensors, $\dot{m}_{f2} = \dot{m}_{f1}$. For engines with carburetors, $\dot{m}_{f2} \approx \dot{m}_{f1}$. For engines with air-fuel ratio sensors,

$$\dot{m}_{f2} = \frac{\dot{m}_{a2}}{A/F} \quad \text{28.93}$$

step 12: The new fuel consumption is

$$\text{BSFC}_2 = \frac{\dot{m}_{f2}}{\text{bhp}_2} \quad \text{28.94}$$

Power Cycles

19. NOMENCLATURE

A	area	ft^2	m^2
A/F	air-fuel ratio	lbm/lbm	kg/kg
bhp	brake horsepower	hp	n.a.
BkW	brake kilowatts	n.a.	kW
BSFC	brake specific fuel consumption	lbm/hp-hr	kg/kW·h
c	clearance	%	%
c	specific heat capacity	Btu/lbm-°F	kJ/kg·K
D	diameter	ft	m
F	force	lbf	N
fhp	friction (horse) power	hp	kW
HHV	higher heating value	Btu/lbm	kJ/kg
ihp	indicated (horse) power	hp	kW
k	ratio of specific heats	–	–
L	stroke length	ft	m
m	mass	lbm	kg
$\dot{m}$	mass flow rate	lbm/sec	kg/s
MEP	mean effective pressure	lbf/ft^2	Pa
N	number of power strokes per minute	min^{-1}	min^{-1}
N	rotational speed	rev/min	rev/min
p	pressure	lbf/ft^2	Pa
q	heat rate	Btu/sec	W
q	heat per unit mass	Btu/lbm	kJ/kg
Q	heat	Btu	kJ
Q	volumetric flow rate	ft^3/sec	m^3/s
$\dot{Q}$	heat rate	Btu/sec	W
r	moment arm or radius	ft	m
r	ratio	–	–
R	specific gas constant	ft-lbf/lbm-°R	kJ/kg·K
s	specific entropy	Btu/lbm-°F	kJ/kg·K
T	temperature	°R	K
T	torque	ft-lbf	N·m
u	specific internal energy	Btu/lbm	kJ/kg
v	specific volume	ft^3/lbm	m^3/kg
V	volume	ft^3	m^3
w	work per unit mass	Btu/lbm	kJ/kg
W	work	Btu	kJ
$\dot{W}$	rate of work	Btu/sec	W

Symbols

η	efficiency	–	–
ρ	mass density	lbm/ft^3	kg/m^3

Subscripts

a	air
BDC	bottom dead center
f	fuel
H	high
i	ideal
L	low
m	mechanical
max	maximum
min	minimum
p	constant pressure or pressure
r	ratio or relative
s	isentropic
TDC	top dead center
th	thermal
v	constant volume or volumetric compression

Power Cycles

29 Combustion Turbine Cycles

Content in blue refers to the *NCEES Handbook*.

NCEES EXAM SPECIFICATIONS AND RELATED CONTENT

HVAC AND REFRIGERATION EXAM

I.B.1. Thermodynamics: Cycles
- 6. Brayton Gas Turbine Cycle
- 7. Air-Standard Brayton Cycle

THERMAL AND FLUID SYSTEMS EXAM

I.E.2. Thermodynamics: Thermodynamic Cycles
- 6. Brayton Gas Turbine Cycle
- 7. Air-Standard Brayton Cycle

1. GAS TURBINES

Combustion turbines (CTs), or "gas" turbines (GTs), are the preferred combustion engines in applications much above 10 MW. Large units regularly operate in the 100 MW to 200 MW range (up to approximately 340 MW).[1] Some smaller CTs—typically less than 40,000 hp (30 MW)—for such applications as marine propulsion and pipeline compression are rated in standard horsepower. Turbine size for the purpose of specifying environmental regulations is categorized by "MMBTU/hr," millions of British thermal units per hour.

There are two general CT categories: the traditional heavy-duty, industrial CT, and the smaller, lighter aeroderivative CT. *Heavy-duty turbines* typically have a single shaft. The rotor is supported on two bearings, and the thrust bearing is on the compressor end. The generator is direct-driven from the compressor (i.e., *cold-end drive*). Small individual combustors surround the hot end radially. The *axial exhaust* duct goes directly into the steam generator in cogeneration and combined cycle plants.

The *aeroderivative combustion turbine* is basically a jet engine that exhausts into a turbine generator. Output is less than 50 MW per unit, and most aeroderivative CTs produce less than 40 MW. *Split shafts* are common in this range. The power turbine and generator are usually mounted at the "hot end" of the gas generator. The generator is run from a gearbox, allowing the turbine compressor to run at higher, more efficient speeds.

The compression ratio (based on pressures) in the compression stage is typically 11:1 to 16:1, with most heavy-duty turbines in the 14:1 to 15:1 range. Aeroderivative turbines have higher compression ratios—typically 19:1 to 21:1, even as high as 30:1. Most heavy-duty combustion turbines have 16 to 18 compression stages. However, the Mach number at the tip of the of the first-stage rotor has risen from approximately 0.9 to approximately 1.4, reducing the number of compression stages to approximately 12 to 15 in modern transsonic turbines.

The temperature of the gas entering the expander section is typically 2200°F to 2350°F (1200°C to 1290°C).[2] The exhaust temperature is typically 1000°F to 1100°F (540°C to 590°C), which makes the exhaust an ideal heat source for combined cycles. Most combustion turbines have three to four expander stages. The exhaust flow rate in modern heavy-duty turbines per 100 MW is approximately 525 lbm/sec to 550 lbm/sec (240 kg/s to 250 kg/s).

Combustors vary widely in design. Large, single chambers (i.e., "cans") have large residence times and allow heavy fuels to be burned completely. Smaller, multiple chambers and annular burners perform better for gaseous and distillate fuels. Film cooling is no longer

[1]In the past, gas turbines were plagued by poor reliability, availability, and maintainability (RAM). The RAM record of modern turbines, particularly aeroderivative types, is excellent.

[2]New materials, coatings, and other devices in *advanced turbine systems* (ATS) have pushed the upper limit of temperatures in commercially available turbines to the 2600°F (1425°C) mark, while simultaneously dropping NOx emissions below 10 ppm.

adequate for wall cooling. Intensive convection cooling and thermal barrier coatings, including ceramic tiles, are needed.

2. TURBOJETS

Turbojets and turbofans ("jet" engines) are engines used for aircraft propulsion. In a *turbojet*, air flows more or less straight through from the cold end air inlet to the hot end, encountering compression, combustion, power extraction, and expansion nozzle sections. After the air is compressed, fuel is injected and ignited. Then, the heated air expands through the hot section to extract power for compression before leaving at high speed through the nozzle cone. Some of the air (*bleed air*) is usually extracted from the compressor section for cabin pressurization, deicing, and engine control. Ignition is maintained by a *flame holder*, but electric ignition can be used for starting. Since there is only one shaft, neglecting the work of accessories, the turbine work and compressor work are the same. The momentum of the exhaust stream is greater than the momentum of the incoming air, so thrust is generated. *Turbofans* are similar, although a fraction of the incoming air bypasses the compression stage, resulting in cooler temperatures and better fuel economy.

Neglecting the small effect of the exit pressure on forward thrust, the thrust developed by a subsonic turbojet engine depends on the air exit velocity and the *true airspeed* of the aircraft (relative to the surrounding air). The fuel mass is small and can be omitted for first approximations.

$$\begin{aligned} F &= (\dot{m}_{\text{air}} + \dot{m}_{\text{fuel}})\text{v}_{\text{exit}} - \dot{m}_{\text{air}}\text{v}_{\text{aircraft}} \\ &\approx \dot{m}_{\text{air}}(\text{v}_{\text{exit}} - \text{v}_{\text{aircraft}}) \end{aligned} \quad 29.1$$

Applicable to military jet aircraft, the terms *military power* and *full military power* refer to full thrust without afterburners. This power setting is generally time-limited. Lower "normal" power settings in decreasing magnitude include *maximum continuous power*, *climb power*, and *cruise power*. For an engine manufactured identically for both military and commercial (airline) use, the term "full military power" distinguishes the maximum power rating of the military engine (regardless of engine life) from the maximum derated power used by airlines in order to improve reliability and engine life.

3. STEAM INJECTION

Steam and water can be injected into the combustor to lower the exhaust temperature, inhibiting the formation of nitrogen oxides. An injection rate on the order of one-half pound of water per pound of fuel is sufficient to keep NOx emissions below older limits in the 75 ppm to 150 ppm range. More water can be used to reduce the emissions somewhat below those values. However, to reach the strictest limits (e.g., less than 10 ppm), selective catalytic reduction (SCR) is needed. NOx emissions can also be controlled to intermediate values (to approximately 25 ppm) without steam in "*dry combustors*" (*low-NOx burners*) based on *staged lean combustion*, also known as *sequential combustion*.

Steam is also injected into the expansion section (sometimes referred to as the *expander*) of CTs, a process known as *power-boosting*. Power-boost steam injection is particularly applicable with aeroderivative turbines where steam pressures are consistent with higher (20:1 or more) compression ratios. Steam injection is popular in cogeneration plants where process steam use is variable, and excess steam can be routed back to the turbine. NOx reduction occurs, but this is not the primary purpose. Steam, which absorbs heat better than air, is also widely used for cooling gas turbines.

4. TURBINE FUELS

Modern turbines can burn a wide range of gaseous and distillate fuels and can switch from one fuel to another over the entire load range. Natural gas is the most economical and, therefore, the most common fuel in combustion turbines used for electrical power generation. Propane, no. 2 oil, and kerosene are used as backup fuels.

Turbines can also be partially fueled by *synthetic gas* ("syngas") generated from coal. The *gasifier* may be a fixed-bed, fluid-bed, or entrained flow type. Both air-blown and oxygen-blown gasification processes can be used, although the oxygen-blown process requires a separate oxygen plant. The heating value of syngas is low, approximately 240 Btu/scf (8.9 MJ/m^3). So, combustion turbines cannot achieve full load operation on syngas alone.

In the *gasification process*, coal-water slurry is pumped into the gasifier. Oxygen or air is added, forming a hot, partially burned gas consisting of carbon monoxide, hydrogen sulfide, and carbonyl sulfide. Most of the non-carbon material in the coal melts and flows out of the gasifier as *slag*. Hot-gas clean-up equipment removes particulates, sulfur, and other impurities from the gas. Gasification is relatively insensitive to coal feedstock.

In future advances, finely ground coal may be used in *direct coal-fueled turbines* (DCFTs). In the past, attempts to inject coal directly into the turbine have been plagued by severe erosion of turbine blades by tiny (3–10 μm) ash particles, pluggage of the gas flow passages by ash deposits, and corrosion of high-temperature metallic surfaces by alkali compounds. This complication may be addressed in the future by coal-cleaning technologies, such as the thermal extraction/solid-liquid separation hyper-coal process. Other considerations are increased emissions and cost.

Traditional gas turbines burn fuel within the envelope of the engine. However, the in-line combustor does not provide adequate residence time to burn coal. Therefore, an "external combustor" must be used with DCFTs.[3] The shortcomings associated with burning ground coal may be overcome by some form of staged combustion, sometimes referred to as "rich-quench-lean" (RQL) firing.

With RQL external combustion, finely ground coal in powder or slurry form (known as *coal-water mixture* or CWM) is burned in a fuel-rich, oxygen-starved first-stage combustor at about 3000°F (1650°C). The low oxygen level inhibits NOx formation from fuel-bound nitrogen. The fuel-rich gas is quenched with water to approximately 2000°F (1100°C), inhibiting thermal NOx formation, and solidifying coal ash so that it can be removed in a cyclone separator, inertial slag separator, or ceramic filter.[4] Clean gas is then sent to the fuel-lean second-stage combustor where additional air is injected, and the temperature increases to approximately 2800°F (1540°C) as the carbon monoxide and hydrogen components burn. Sulfur emissions are controlled by injection of calcium-based sorbents in either the first- or second-stage combustors.

5. INDIRECT COAL-FIRED TURBINES

With *indirect coal-fired turbines* (ICFTs), the coal combustion products never enter the turbine expander. Rather, heat from combustion is transferred through a closed heat exchanger to the compressed air. Only clean air flows through the expander. Since the heat transfer occurs at over 2000°F (1100°C), well over the 1600°F (870°C) limit of traditional metallic heat exchangers, a special ceramic heat exchanger must be used. When used in a combined cycle, the term *externally fired combined cycle* (EFCC) is used to describe this process. (See Fig. 29.1.)

Theoretically, the thermal efficiency of the cycle can be increased by closing the process (i.e., making it a *closed turbine cycle*) where the exhaust from the expander is returned to the compressor for reuse.

6. BRAYTON GAS TURBINE CYCLE

Strictly speaking, the *Brayton gas turbine cycle* (also known as the *Joule cycle*) is an internal combustion cycle. It differs from the previous cycles in that each process is carried out in a different location, air flow and fuel injection are steady, and air-standard calculations are realistic since a large air-fuel ratio is used to keep combustion temperatures below metallurgical limits. [Brayton Cycle (Steady-Flow Cycle)]

Figure 29.1 Direct and Indirect Coal-Fired Turbines

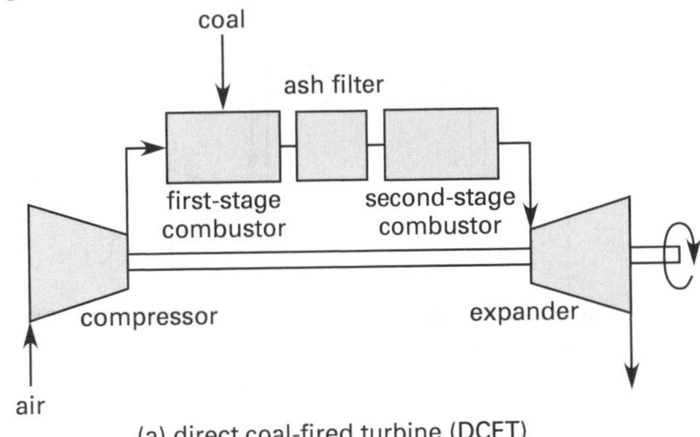

(a) direct coal-fired turbine (DCFT)

(b) indirect coal-fired turbine (ICFT)

Figure 29.2 illustrates the physical arrangement of components used to achieve the Brayton cycle. Almost all installations drive the compressor from the turbine. (Approximately 50% to 75% of the turbine power is required to drive the high-efficiency compressor.) The actual arrangement differs from the air-standard property plot shown in Fig. 29.3 in that the exhaust products exiting at point 4 are not cooled and returned to the compressor at point 1. The processes in the air-standard Brayton cycle follow.

1 to 2: isentropic compression in the compressor

2 to 3: constant pressure heat addition in the combustor

3 to 4: isentropic expansion in the turbine

4 to 1: constant pressure cooling to original conditions

Combustors are said to be *open* if the incoming air and combustion products flow to the turbine. A high percentage, 30% to 60%, of excess air is needed to cool the gases. Depending on the turbine construction details, the temperature of the air entering the turbine will be between 1200°F and 1800°F (650°C and 1000°C). Turbojet and turboprop engines typically use *open combustors*.

[3]The external combuster is located "off base" on a separate combustion "island."

[4]There is also promise in *pressurized slagging combustors* where slag is removed in liquid form.

Figure 29.2 *Gas Turbine*

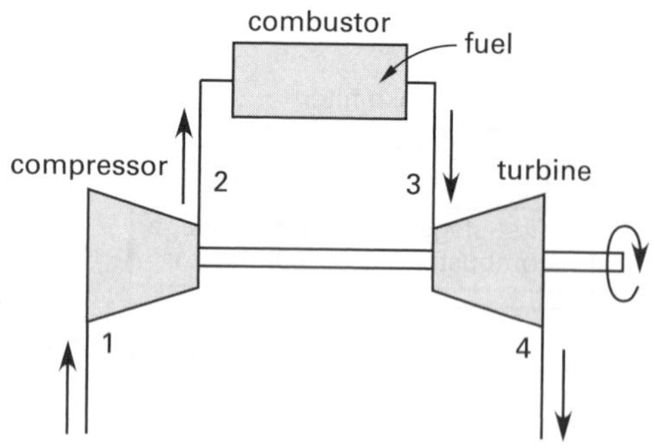

Figure 29.3 *Brayton Gas Turbine Cycle*

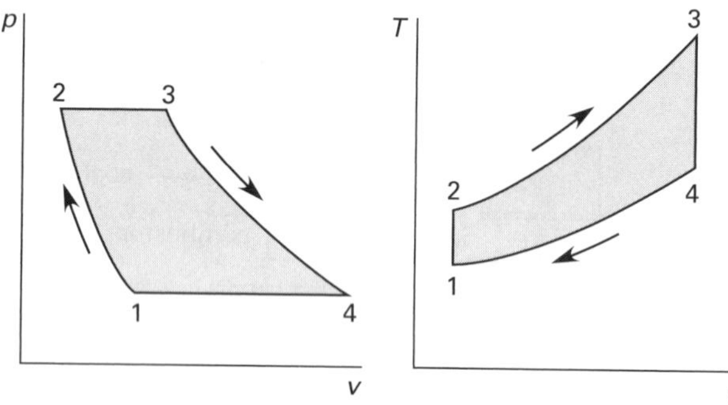

A *closed combustor* is a heat exchanger. Air flowing to the turbine is not combined with the combustion gases. This allows any type of fuel to be used, but the bulk of the heat exchanger limits closed combustors to stationary power plants and pipeline pumping stations.

The pressure, specific volume, and temperature at points 1, 2, 3, and 4 can be evaluated by using the ideal gas relationships for steady flow systems given in Eq. 29.2 through Eq. 29.4. Constant values of k, c_p, and c_v can be assumed if air is considered ideal.[5] An air table also can be used if the isentropic efficiencies are known. It is usually assumed that p_2 and p_3 are equal and that p_1 and p_4 are atmospheric pressure.

Ideal Gas

$$\frac{p_2}{p_1} = \left(\frac{v_1}{v_2}\right)^k \qquad 29.2$$

$$\frac{T_2}{T_1} = \left(\frac{p_2}{p_1}\right)^{(k-1)/k} \qquad 29.3$$

$$\frac{T_2}{T_1} = \left(\frac{v_1}{v_2}\right)^{k-1} \qquad 29.4$$

In Eq. 29.2 through Eq. 29.4, the ratio of specific heats, k, is

Ideal Gas

$$k = \frac{c_p}{c_v} \qquad 29.5$$

From these relationships, the values of p, v, and T can be evaluated as follows.

at 1:

$$\begin{aligned} p_1 &= p_2\left(\frac{v_2}{v_1}\right)^k = p_2\left(\frac{T_1}{T_2}\right)^{k/(k-1)} \\ &= p_4 \text{ [usually atmospheric]} \end{aligned} \qquad 29.6$$

$$\begin{aligned} T_1 &= T_4\left(\frac{v_1}{v_4}\right) = T_2\left(\frac{v_2}{v_1}\right)^{k-1} \\ &= T_2\left(\frac{p_1}{p_2}\right)^{(k-1)/k} \end{aligned} \qquad 29.7$$

$$\begin{aligned} v_1 &= v_4\left(\frac{T_1}{T_4}\right) = v_2\left(\frac{p_2}{p_1}\right)^{1/k} \\ &= v_2\left(\frac{T_2}{T_1}\right)^{1/(k-1)} \end{aligned} \qquad 29.8$$

at 2:

$$p_2 = p_3 = p_1\left(\frac{v_1}{v_2}\right)^k = p_1\left(\frac{T_2}{T_1}\right)^{k/(k-1)} \qquad 29.9$$

$$\begin{aligned} T_2 &= T_1\left(\frac{v_1}{v_2}\right)^{k-1} = T_1\left(\frac{p_2}{p_1}\right)^{(k-1)/1} \\ &= T_3\left(\frac{v_2}{v_3}\right) \end{aligned} \qquad 29.10$$

$$\begin{aligned} v_2 &= v_1\left(\frac{p_1}{p_2}\right)^{1/k} = v_1\left(\frac{T_1}{T_2}\right)^{1/(k-1)} \\ &= v_3\left(\frac{T_2}{T_3}\right) \end{aligned} \qquad 29.11$$

at 3:

$$p_3 = p_2 = p_4\left(\frac{v_4}{v_3}\right)^k = p_4\left(\frac{T_3}{T_4}\right)^{k/(k-1)} \qquad 29.12$$

$$\begin{aligned} T_3 &= T_2\left(\frac{v_3}{v_2}\right) = T_4\left(\frac{v_4}{v_3}\right)^{k-1} \\ &= T_4\left(\frac{p_3}{p_4}\right)^{(k-1)/k} \end{aligned} \qquad 29.13$$

[5]Another typical assumption, justified on the basis of large amounts of excess air, is that the gas is all nitrogen.

$$v_3 = v_2\left(\frac{T_3}{T_2}\right) = v_4\left(\frac{p_4}{p_3}\right)^{1/k} = v_4\left(\frac{T_4}{T_3}\right)^{1/(k-1)} \quad \text{29.14}$$

at 4:

$$p_4 = p_3\left(\frac{v_3}{v_4}\right)^k = p_3\left(\frac{T_4}{T_3}\right)^{k/(k-1)} \quad \text{29.15}$$

$$T_4 = T_3\left(\frac{v_3}{v_4}\right)^{k-1} = T_3\left(\frac{p_4}{p_3}\right)^{(k-1)/k} = T_1\left(\frac{v_4}{v_1}\right) \quad \text{29.16}$$

$$v_4 = v_3\left(\frac{p_3}{p_4}\right)^{1/k} = v_3\left(\frac{T_3}{T_4}\right)^{1/(k-1)} = v_1\left(\frac{T_4}{T_1}\right) \quad \text{29.17}$$

The work and heat flow terms are

Brayton Cycle (Steady-Flow Cycle)

$$\dot{W}_{12} = h_1 - h_2 = c_p(T_1 - T_2) \quad \text{[compressor]} \quad \text{29.18}$$

$$\dot{Q}_{23} = h_3 - h_2 = c_p(T_3 - T_2) \quad \text{[combustor]} \quad \text{29.19}$$

$$\dot{W}_{34} = h_3 - h_4 = c_p(T_3 - T_4) \quad \text{[turbine]} \quad \text{29.20}$$

$$\dot{Q}_{41} = h_1 - h_4 = c_p(T_1 - T_4) \quad \text{29.21}$$

$$\eta_{th} = \frac{\dot{W}_{net}}{\dot{Q}_{in}} = 1 - \frac{\dot{Q}_{out}}{\dot{Q}_{in}} = \frac{\dot{W}_{12} + \dot{W}_{34}}{\dot{Q}_{23}} = 1 - \frac{\dot{Q}_{41}}{\dot{Q}_{23}} \quad \text{29.22}$$

If the gas is ideal so that c_p is constant, then

$$\eta_{th} = \frac{(T_3 - T_2) - (T_4 - T_1)}{T_3 - T_2} \quad \text{29.23}$$

If the isentropic efficiency of the turbine, $\eta_{s,\text{turbine}}$, is less than 100%, the actual work produced by the turbine is less than the isentropic (or ideal) work.

Special Cases of the Steady-Flow Energy Equation

$$\eta_{s,\text{turbine}} = \frac{h_i - h_e}{h_i - h_{es}} = \frac{h_3 - h_{4,\text{actual}}}{h_3 - h_{4,\text{ideal}}} = \frac{w_{\text{turbine,actual}}}{w_{\text{turbine,ideal}}} \quad \text{29.24}$$

The actual enthalpy at the turbine exit can be calculated as

$$h_{4,\text{actual}} = h_3 - \eta_{s,\text{turbine}}(h_3 - h_{4,\text{ideal}}) \quad \text{29.25}$$

The actual temperature at the turbine exit can be calculated as

$$T_{4,\text{actual}} = T_3 - \eta_{s,\text{turbine}}(T_3 - T_{4,\text{ideal}}) \quad \text{29.26}$$

If the isentropic efficiency of the compressor, $\eta_{s,\text{compressor}}$, is less than 100%, the actual work consumed by the compressor is more than the isentropic (or ideal) work.

Special Cases of the Steady-Flow Energy Equation

$$\eta_{s,\text{compressor}} = \frac{h_{es} - h_i}{h_e - h_i} = \frac{h_{2,\text{ideal}} - h_1}{h_{2,\text{actual}} - h_1} = \frac{w_{\text{compressor,actual}}}{w_{\text{compressor,ideal}}} \quad \text{29.27}$$

The actual enthalpy at the compressor exit can be calculated as

$$h_{2,\text{actual}} = h_1 + \frac{h_{2,\text{ideal}} - h_1}{\eta_{s,\text{compressor}}} \quad \text{29.28}$$

The actual temperature at the compressor exit can be calculated as

$$T_{2,\text{actual}} = T_1 + \frac{T_{2,\text{ideal}} - T_1}{\eta_{s,\text{compressor}}} \quad \text{29.29}$$

Simple cycle (SC) operation means that the turbine is not part of a cogeneration or combined cycle system.[6] The full-load thermal efficiency of existing heavy-duty combustion turbines in simple cycles is approximately 34% to 36%, while new turbines on the cutting edge of technology (i.e., *advanced turbine systems*—ATS) are able to achieve 38% to 38.5%, with 40% being a reasonable future goal. Aeroderivative turbines commonly achieve efficiencies up to 41%. The heat rate can be found from the cycle efficiency.

[6]When integrated into combined cycle systems, overall thermal efficiencies are higher; however, these "combined efficiencies" are not the turbine thermal efficiencies.

Power Cycles

The *back work ratio* (BWR), which is approximately 50% to 75%, is the ratio of the compressor work to the turbine expansion work.

Example 29.1

Air enters the compressor of a gas turbine at 14.7 psia and 540°R (101.3 kPa and 300K). The pressure ratio is 4.5:1. The conditions at the turbine inlet are 64 psia and 2200°R (440 kPa and 1220K). The turbine's expansion pressure ratio is 1:4, and exhaust is to the atmosphere. The isentropic efficiency of the turbine is 85%. Compression is isentropic. What is the thermal efficiency of the cycle?

SI Solution

Assume an ideal gas. (The customary U.S. solution uses air tables to solve this example.)

Refer to Fig. 29.3.

at 1:

$$T_1 = 300\text{K} \quad \text{[given]}$$
$$p_1 = 101.3\ \text{kPa} \quad \text{[given]}$$

at 2:

Ideal Gas

$$\frac{T_2}{T_1} = \left(\frac{p_2}{p_1}\right)^{(k-1)/k}$$

$$T_2 = T_1\left(\frac{p_2}{p_1}\right)^{(k-1)/k} = (300\text{K})(4.5)^{(1.4-1)/1.4}$$
$$= 461\text{K}$$

$$p_2 = \left(\frac{p_2}{p_1}\right)p_1 = (4.5)(101.3\ \text{kPa})$$
$$= 456\ \text{kPa}$$

at 3:

$$T_3 = 1220\text{K} \quad \text{[given]}$$
$$p_3 = 440\ \text{kPa} \quad \text{[given]}$$

at 4:

$$p_4 = p_3\left(\frac{p_4}{p_3}\right) = (440\ \text{kPa})\left(\frac{1}{4}\right)$$
$$= 110\ \text{kPa}$$

Calculate the temperature if the expansion were isentropic.

Ideal Gas

$$\frac{T_{4,\text{ideal}}}{T_3} = \left(\frac{p_4}{p_3}\right)^{(k-1)/k}$$

$$T_{4,\text{ideal}} = T_3\left(\frac{p_4}{p_3}\right)^{(k-1)/k}$$
$$= (1220\text{K})\left(\frac{1}{4}\right)^{(1.4-1)/1.4}$$
$$= 821\text{K}$$

For ideal gases, the specific heats are constant. Therefore, the change in internal energy (and enthalpy, approximately) is proportional to the change in temperature. The actual temperature is

$$T_{4,\text{actual}} = T_3 - \eta_{s,\text{turbine}}(T_3 - T_{4,\text{ideal}})$$
$$= 1220\text{K} - (0.85)(1220\text{K} - 821\text{K})$$
$$= 881\text{K}$$

The thermal efficiency is given by Eq. 29.23.

$$\eta_{\text{th}} = \frac{(T_3 - T_2) - (T_{4,\text{actual}} - T_1)}{T_3 - T_2}$$
$$= \frac{(1220\text{K} - 461\text{K}) - (881\text{K} - 300\text{K})}{1220\text{K} - 461\text{K}}$$
$$= 0.235 \quad (23.5\%)$$

Customary U.S. Solution

Use an air table. (The SI solution assumes an ideal gas.) [Properties of Air at Low Pressure, per Pound]

at 1:

$$T_1 = 540°\text{R} \quad \text{[given]}$$
$$p_1 = 14.7\ \text{psia} \quad \text{[given]}$$
$$h_1 = 129.06\ \text{Btu/lbm}$$
$$p_{r,1} = 1.3860$$

at 2: The 1-to-2 process is isentropic.

$$p_{r,2} = p_{r,1}\left(\frac{p_2}{p_1}\right) = (1.3860)(4.5)$$
$$= 6.237$$

Locate this pressure ratio in the air table, interpolating as needed. [Properties of Air at Low Pressure, per Pound]

$$T_2 = 827.6°\text{R}$$
$$h_2 = 198.5\ \text{Btu/lbm}$$
$$p_2 = (4.5)(14.7\ \text{psia}) = 66.15\ \text{psia}$$

at 3: [**Properties of Air at Low Pressure, per Pound**]

$$T_3 = 2200°\text{R} \quad \text{[given]}$$
$$p_3 = 64 \text{ psia} \quad \text{[given]}$$
$$h_3 = 560.59 \text{ Btu/lbm}$$
$$p_{r,3} = 256.6$$

at 4:

$$p_{r,4} = p_{r,3}\left(\frac{p_4}{p_3}\right) = (256)\left(\frac{1}{4}\right) = 64.15$$

Locate this pressure ratio in the air table. [**Properties of Air at Low Pressure, per Pound**]

$$T_{4,\text{ideal}} = 1554.6°\text{R}$$
$$h_4 = 383.66 \text{ Btu/lbm}$$

Since the efficiency of the expansion process is 85%,

$$\begin{aligned} T_{4,\text{actual}} &= T_3 - \eta_{s,\text{turbine}}(T_3 - T_{4,\text{ideal}}) \\ &= 2200°\text{R} - (0.85)(2200°\text{R} - 1554.6°\text{R}) \\ &= 1651°\text{R} \end{aligned}$$

The thermal efficiency is given by Eq. 29.23.

$$\begin{aligned} \eta_{\text{th}} &= \frac{(T_3 - T_2) - (T_4 - T_1)}{T_3 - T_2} \\ &= \frac{(2200°\text{R} - 827.6°\text{R}) - (1651°\text{R} - 540°\text{R})}{2200°\text{R} - 827.6°\text{R}} \\ &= 0.1905 \quad (19\%) \end{aligned}$$

7. AIR-STANDARD BRAYTON CYCLE WITH REGENERATION

Regeneration is used to improve the efficiency of the Brayton cycle. Regeneration involves transferring some of the heat from the exhaust products to the air in the compressor. The transfer occurs in a *regenerator*, which is a crossflow heat exchanger. There is no effect on turbine work, compressor work, or net output. However, the cycle is more efficient since less heat is added. Of course, T_2 cannot be greater than T_6. Similarly, T_3 cannot be greater than T_5. (See Fig. 29.4 and Fig. 29.5.) [**Brayton Cycle with Regeneration**]

The processes are:

1 to 2:	isentropic compression
2 to 3:	constant pressure heat addition in regenerator
3 to 4:	constant pressure heat addition in combustor
4 to 5:	isentropic expansion
5 to 6:	constant pressure heat removal in the regenerator
6 to 1:	constant pressure heat removal in the sink

Figure 29.4 *Gas Turbine with Regeneration*

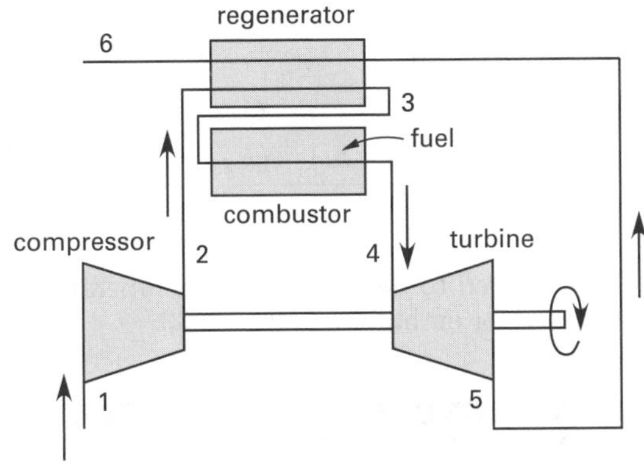

Figure 29.5 *Brayton Cycle with 100% Efficient Regeneration*

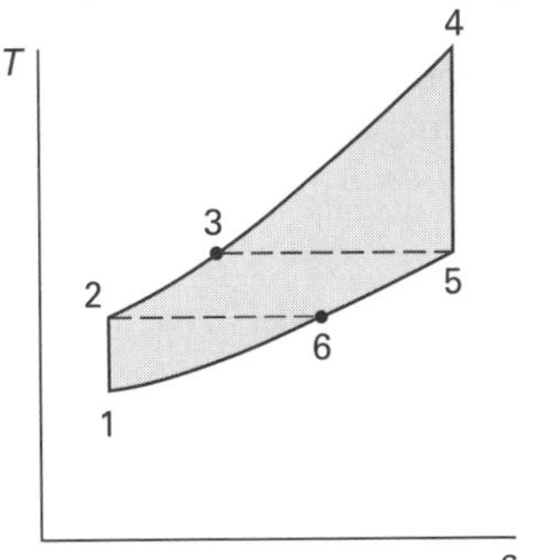

If T_3 is equal to T_5, the regenerator is said to be 100% efficient. Otherwise, the *regenerator efficiency*, also known as *regeneration effectiveness*, is calculated from Eq. 29.30. Actual regeneration efficiency rarely exceeds 75%.

Brayton Cycle with Regeneration

$$\eta_{\text{regen}} = \frac{q_{\text{actual}}}{q_{\text{max}}} = \frac{h_3 - h_2}{h_5 - h_2} \qquad 29.30$$

Using cold-air-standard assumptions (that the working fluid is room temperature air behaving as an ideal gas, that specific heats are constant, and that all processes in the cycle are reversible), this can be simplified further to

Brayton Cycle with Regeneration

$$\eta_{\text{regen}} \cong \frac{T_3 - T_2}{T_5 - T_2} \qquad 29.31$$

The thermal efficiency of the air-standard Brayton cycle with regeneration is

Brayton Cycle with Regeneration

$$\eta_{\text{th,regen}} = 1 - \left(\frac{T_1}{T_4}\right) r_p^{(k-1)/k}$$
$$= 1 - \left(\frac{T_1}{T_4}\right)\left(\frac{p_2}{p_1}\right)^{(k-1)/k} \quad 29.32$$

In terms of enthalpies, this is

$$\eta_{\text{th}} = \frac{W_{\text{net}}}{Q_{\text{in}}} = \frac{W_{\text{out}} - W_{\text{in}}}{Q_{\text{in}}}$$
$$= \frac{(h_4 - h_5) - (h_2 - h_1)}{h_4 - h_3} \quad 29.33$$

If air is considered to be an ideal gas, temperatures can be substituted for enthalpies in Eq. 29.33.

8. BRAYTON CYCLE WITH REGENERATION, INTERCOOLING, AND REHEATING

Multiple stages of compression and expansion can be used to improve the efficiency of the Brayton cycle. Physical limitations usually preclude more than two stages of intercooling and reheat. (This section assumes only one stage of each.) The physical arrangement is shown in Fig. 29.6.

The processes are:

A to B:	isentropic compression
B to C:	cooling at constant pressure (usually back to T_A)
C to D:	isentropic compression
D to F:	constant pressure heat addition
F to G:	isentropic expansion
G to H:	reheating at constant pressure in combustor or reheater (usually back to T_F)
H to I:	isentropic expansion
I to A:	constant pressure heat rejection

Figure 29.6 Augmented Gas Turbine

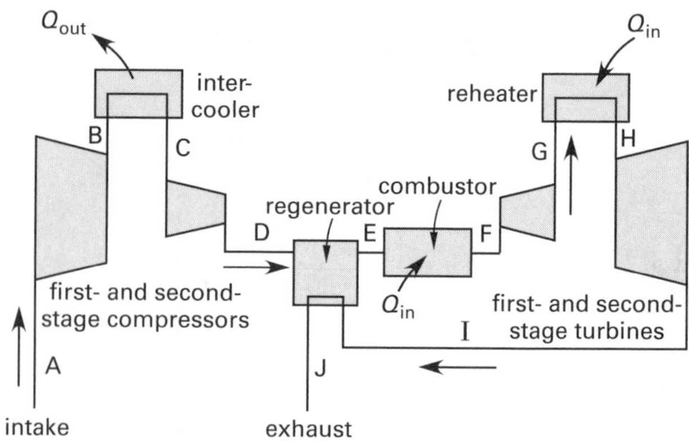

Calculation of the work and heat flow terms and of thermal efficiency is similar to that in the previous cycles, except that there are two W_{turbine}, $W_{\text{compressor}}$, and Q_{in} terms. If efficiencies for the compressor, turbine, and regenerator are known, the following relationships are required.

$$p'_{\text{B}} = p_{\text{B}} \quad 29.34$$

$$h'_{\text{B}} = h_{\text{A}} + \frac{h_{\text{B}} - h_{\text{A}}}{\eta_{s,\text{compressor}}} \quad 29.35$$

$$p'_{\text{D}} = p_{\text{D}} \quad 29.36$$

$$h'_{\text{D}} = h_{\text{C}} + \frac{h_{\text{D}} - h_{\text{C}}}{\eta_{s,\text{compressor}}} \quad 29.37$$

$$p'_{\text{E}} = p_{\text{E}} \quad 29.38$$

$$h'_{\text{E}} = h_{\text{D}} + \eta_{\text{regenerator}}(h_{\text{I}} - h_{\text{D}}) \quad 29.39$$

$$p'_{\text{G}} = p_{\text{G}} \quad 29.40$$

$$h'_{\text{G}} = h_{\text{F}} - \eta_{s,\text{turbine}}(h_{\text{F}} - h_{\text{G}}) \quad 29.41$$

$$h'_{\text{I}} = p_{\text{I}} \quad 29.42$$

$$h'_{\text{I}} = h_{\text{H}} - \eta_{s,\text{turbine}}(h_{\text{H}} - h_{\text{I}}) \quad 29.43$$

Optimum performance improvement with multiple staging is achieved when the pressure ratio across each turbine stage (either compression stages or expansion stages) is the same. Referring to Fig. 29.6 and Fig. 29.7, for the compressors, $p_{\text{B}} = p_{\text{C}} = \sqrt{p_{\text{A}} p_{\text{D}}}$. For the turbine, $p_{\text{G}} = p_{\text{H}} = \sqrt{p_{\text{F}} p_{\text{I}}}$.

9. HEAT RECOVERY STEAM GENERATORS

All simple power cycles end up discarding most of the incoming heat energy. For example, a thermal efficiency of 40% means that 60% of the heat energy is lost.

Figure 29.7 Augmented Brayton Cycle

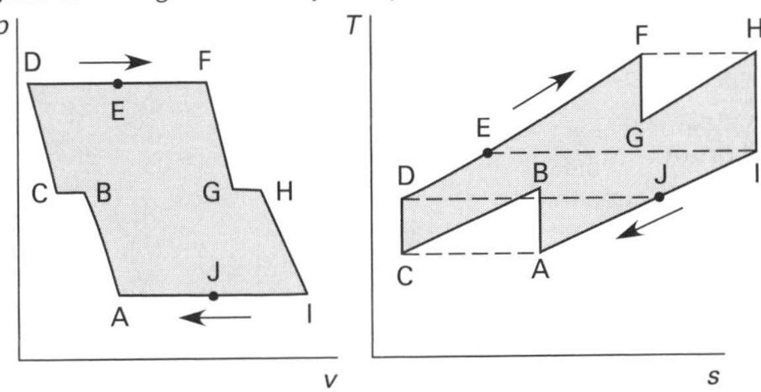

Combustion cycles (gas turbine cycles in particular) produce high-temperature, 800°F to 1100°F (430°C to 590°C) exhaust streams.[7] Some of this energy can be recovered.

All cogeneration and combined cycles make use of some type of *heat-recovery steam generator* (HRSG), also known as a *waste-heat boiler.* Although fire-tube designs are applicable to low gas flow rates—less than 100,000 lbm/hr (12.6 kg/s)—due to the large mass flow rates of exhaust gases typically encountered and the high-pressure 1000 psi to 2800 psi (7 MPa to 19 MPa) steam generated, HRSGs are typically of the water-tube design. When turbines are fueled by natural gas, the exhaust is clean and does not pose a corrosion problem for HRSGs.[8] (See Fig. 29.8.)

The exhaust gases pass sequentially through superheater, high-pressure evaporator, economizer, and low-pressure evaporator sections. (A selective catalytic conversion (SCR) section may also be used, though no heat transfer occurs in it.) To extract more heat energy, extended surfaces (i.e., fins) may be used on the tubes. The maximum fin temperature occurs at the tip, and fin material limits the operating temperature of the HRSG. Since the tube-side heat transfer coefficient is high, on the order of 1000 Btu/hr-ft^2-°F to 3000 Btu/hr-ft^2-°F (1.7 kW/m·K to 5.2 kW/m·K) in economizers and LP evaporators, the *fin density* is also high—on the order of 2 to 5 fins per inch. The fin density is lower in the superheater section.

As in traditional steam generators, steam drums in HRSGs can be either natural or forced circulation in design. The *circulation ratio* is the ratio of the mass of circulating steam-water (in the risers and downcomers) to the mass of generated steam. It varies from 10 to 40 for natural circulation and from 3 to 10 for forced circulation.

The temperature difference between the gas side and steam sides varies along the run of the HRSG and with variations in gas flow, inlet temperature, and extent of supplemental firing. The *pinch point* is the minimum temperature difference. The *approach point* is the difference between the saturation temperature and the temperature of the leaving water (in the economizer). Trial-and-error solutions are usually required to determine the entering and leaving temperatures at each section. For trial-and-error solutions, the pinch and approach points may initially be assumed to be typical values, 20°F and 15°F (11°C and 8°C), respectively.

HRSGs may be unfired, supplementary fired, or furnace fired. (Fired HRSGs are more common with combined cycles than with cogeneration.) *Unfired HRSGs* are usually one- or two-pass designs. Since the turbine exhaust contains approximately 16% oxygen by volume, additional fuel can be sprayed into the exhaust stream, usually after the superheater portion of the HRSG. This option is known as a *supplementary-fired HRSG.* Temperatures are limited to approximately 1700°F (930°C) by the liner material. Other than that, supplementary-fired units are similar to unfired versions.

Temperatures in excess of 1700°F (930°C) can be achieved with *furnace-fired HRSGs.* Temperatures up to about 2300°F (1260°C) can be achieved with a duct burner and water-cooled membrane walls; higher temperatures require special register burners with their own air chambers.

Referring to Fig. 29.9, the heat balance in the superheater and evaporator sections is

$$\begin{aligned}\dot{m}_{\text{gas}}c_{p,\text{gas}}(T_{\text{gas},1} - T_{\text{gas},3}) \\ = \dot{m}_{\text{steam}}(h_{\text{steam,out}} - h_{\text{water},2})\end{aligned} \quad 29.44$$

The heat balance for the complete HRSG is

$$\begin{aligned}\dot{m}_{\text{gas}}c_{p,\text{gas}}(T_{\text{gas},1} - T_{\text{gas},4}) \\ = \dot{m}_{\text{steam}}(h_{\text{steam,out}} - h_{\text{water},1})\end{aligned} \quad 29.45$$

The *transferred duty* (or just "duty") of the HRSG is the total heat transfer rate for all components (superheater, evaporator, economizer, etc.).

HRSGs often operate far from their design point. Equation 29.46 can be used to predict the performance of each section of the HRSG at one mass flow rate based on known performance at another mass flow rate.

$$\frac{U_1A_1}{U_2A_2} = \frac{\frac{\dot{Q}_1}{T_1}}{\frac{\dot{Q}_2}{T_2}} \approx \left(\frac{\dot{m}_1}{\dot{m}_2}\right)^{0.65} \quad 29.46$$

[7]The exhaust from municipal waste incinerators is approximately 1800°F (980°C), and some chemical waste incinerators produce 2000°F to 2400°F (1090°C to 1320°C) combustion gases.

[8]Combustion gas from municipal incinerators contains particulates that can cause slagging, and such gas is corrosive at both high and low temperatures. Above 800°F (430°C), hydrogen chloride (HCl) formed from the combustion of waste plastics, is very corrosive. Chlorine, formed when incinerating chlorinated wastes, is even more corrosive than HCl on carbon steel above 400°F to 450°F (200°C to 230°C). Therefore, HRSGs associated with the burning of such wastes must operate in a narrow temperature band below 400°F (200°C) and above the *acid-vapor dew point.*

Figure 29.8 Typical Heat Recovery Steam Generator

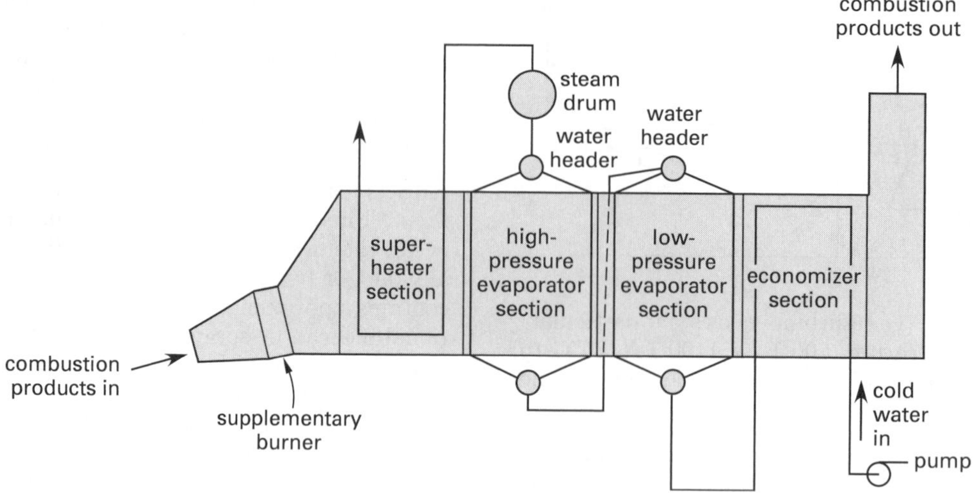

Figure 29.9 Pinch and Approach Points

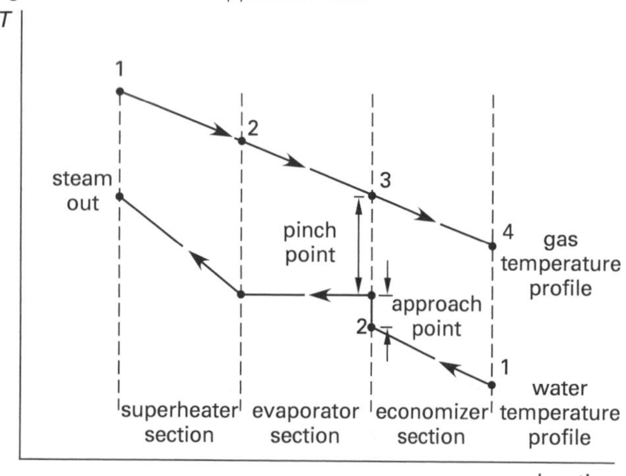

Figure 29.10 Simple Cogeneration Process

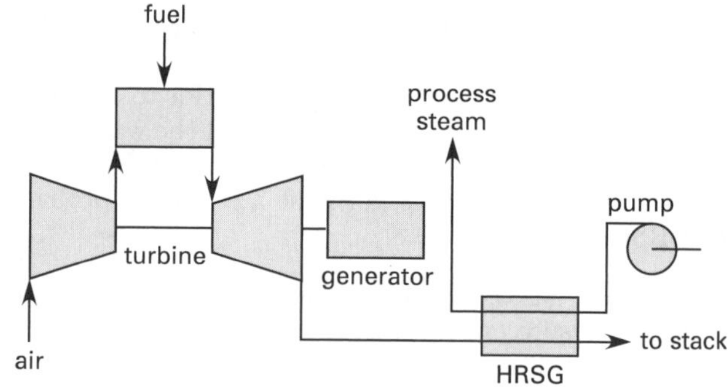

10. COGENERATION CYCLES

In *cogeneration* plants, HRSGs convert waste heat (generally from turbine exhaust) into low-pressure high-quality or saturated process steam with pressures of 10 psig to 300 psig (70 kPa to 210 kPa).[9] *Process steam* is steam that is used for some process other than electrical power generation,[10] such as *space heating* (also known as *district heating*). (See Fig. 29.10.)

Cogeneration systems can be configured as topping or bottoming cycles. In a *topping cycle*, the primary fuel produces electricity first. The turbine exhaust (either steam or gas) is captured to make steam. In a *bottoming cycle*, the waste heat from an industrial process is captured to produce steam first for the process and then to power a turbine. Bottoming cycles, being less efficient and requiring supplementary firing, are less common.

The *power-to-heat ratio* (PTR) is

$$\text{PTR} = \frac{P_{\text{turbine}}}{\dot{Q}_{\text{recovered}}} \qquad 29.47$$

The *fuel utilization* (FU) is not the same as thermal efficiency. Fuel utilizations for different units should be compared only for equal PTRs. Highly efficient cycles have fuel utilizations above 80%.

$$\text{FU} = \frac{P_{\text{turbine}} + \dot{Q}_{\text{recovered}}}{\dot{Q}_{\text{in}}} \qquad 29.48$$

11. COMBINED CYCLES

In a *combined cycle*, the heat recovered in an HRSG (i.e. waste heat boiler) is used to vaporize water for use in another Rankine steam cycle. Steam from the HRSG is high pressure, high temperature, exceeding 750 psi and 700°F (5 MPa and 370°C). Supplementary firing in the HRSG may be used. Combined cycle efficiencies are much higher than simple cycles, typically 45% to 55%,

[9]Not all combined cycles utilize exhaust from gas turbines. High-temperature diesel engine exhaust is rich in oxygen—ideal for use as preheated combustion air in smaller plants.

[10]Theoretically, the steam could be used for cooling in an absorption system. This probably is never done in practice, however.

Figure 29.11 Combined Cycle Process

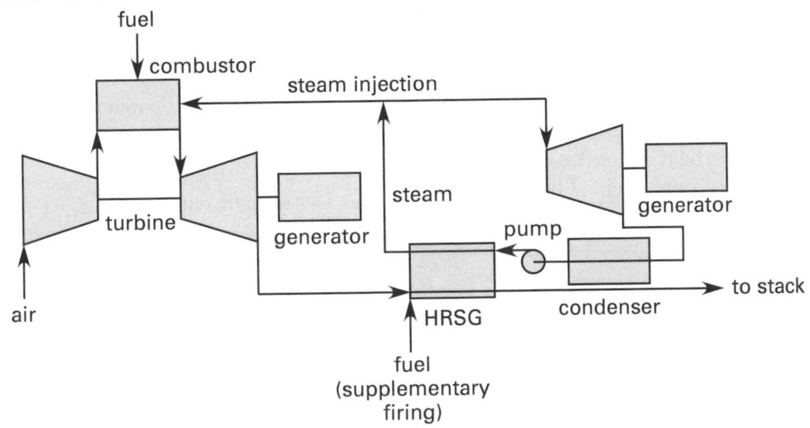

Figure 29.12 Integrated Gasification Combined Cycle

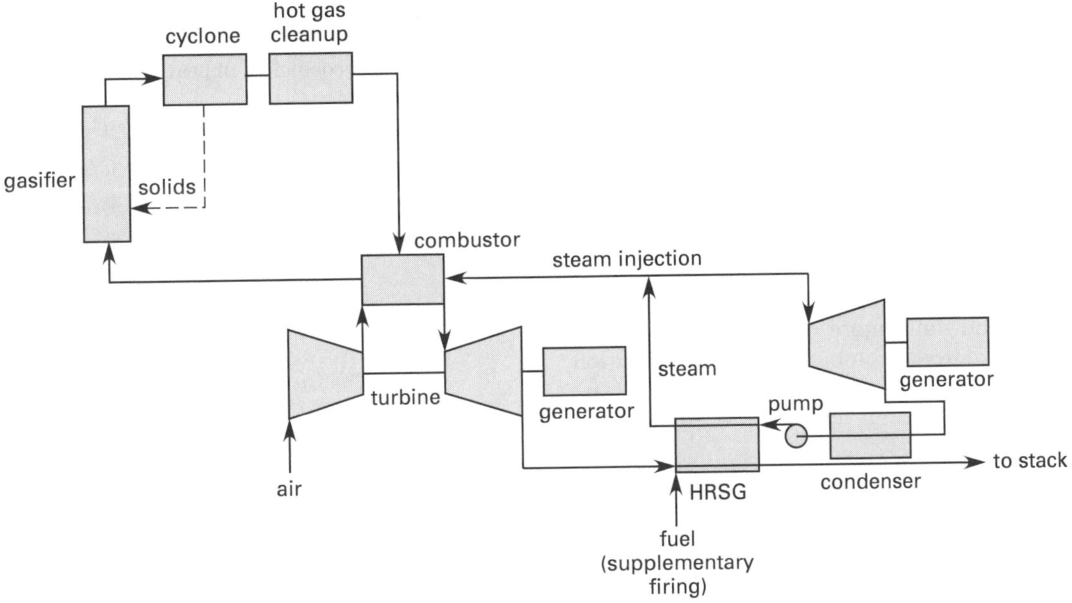

with plants providing only electrical power output achieving the highest efficiencies. Some state-of-the-art prototype ("proof of concept") installations are able to achieve combined efficiencies up to 59%. A combined efficiency of 60% appears to be achievable.[11] (See Fig. 29.11.) [**Combined Cycle**]

A plethora of confusing acronyms identify combined cycle (CC) variations, including GFCC (gas-fired turbine combined cycle), GTCC (gas turbine combined cycles), GCC (gasification combined cycle), CGCC (coal gasification combined cycle), DCCC (diesel coal combined cycle), EFCC (externally fired combined cycle), and IGCC (integrated gasification combined cycle). Some of these terms are synonymous.

Figure 29.12 illustrates an IGCC cycle. "Integrated" refers to the fact that the gasifier uses part of the compressor discharge rather than operating as an independent source. IGCC is an inherently low-emissions process because ash is removed in the gasification process and sulfur is removed from the fuel gas, not the flue gas. Overall efficiencies with IGCC are approximately 42% —lower than some combined cycles, but still much higher than the traditional coal-fired plant with emission controls.

12. HIGH-PERFORMANCE POWER SYSTEMS

It may be possible to combine all of the advanced technologies into a combined cycle and make coal combustion as efficient as natural gas.[12] In the prototypical *high-performance power system* (HIPPS), a *high-*

[11]The practical upper limit for combined cycle efficiencies is considered to be approximately 70%. The Carnot efficiency sets the ideal upper limit. This is 81% for cycles operating between 70°F (21°C) and 2350°F (1290°C).

[12]However, it may not be economical. Time will tell.

temperature advanced furnace (HITAF, which integrates the combustion of coal, heat transfer, and emissions control into a single unit) provides most of the heat for a Brayton topping cycle. Supplementary firing with clean fuel (such as natural gas) boosts the air temperature to achieve optimum gas turbine performance. Energy in the HITAF can also be used for superheating and reheating in the conventional Rankine bottoming cycle. Gas turbine exhaust is routed to the HITAF as combustion air.

13. REPOWERING AND LIFE EXTENSION

The average useful life of most coal-fired utility power plants is 30 to 40 years. However, because of increasing demand, utilities will assess upgrading their current infrastructure based on economic and environmental factors before retiring any significant portion of their old capacity.

Repowering is a popular term used by utilities to mean upgrading an existing plant. In a sense, repowering is a variation of *life extension*.[13] However, life extension also includes programs to increase reliability, availability, and maintainability (RAM) without changing equipment. For example, many programs concentrate on weak areas, such as boiler tube failure (BTF), the leading cause of forced outages.

Repowering concentrates on replacing the furnace and boiler with clean-burning, more efficient coal combustion (as in fluidized-bed combustion) or gasification while retaining the remainder of the existing plant (i.e., the coal handling, steam cycle, and generating equipment).

In the face of massive boiler repairs or replacement, a utility may choose to repower by replacing boilers with gas turbines and HRSGs, using the existing steam turbine in a topping cycle. This is a good option if the existing steam cycle is below that of the HRSG output, generally, less than 1450 psig (10 MPa). There are three general ways the gas turbine can be integrated into the existing plant. (1) In the *hot windbox system*, some or all of the turbine exhaust is routed to the furnace as combustion air. (2) Turbine exhaust can be used for feedwater heating. (3) Turbine exhaust can be used as the heat source in a *supplementary boiler* or separate superheater.

Though often overlooked, the fuel flexibility and efficiency of large diesel engines make diesel combined cycles (DCC) ideal candidates in repowering options in smaller power plants.

14. NOMENCLATURE

A	area	ft^2	m^2
c	specific heat capacity	Btu/lbm-°F	kJ/kg·K
F	force	lbf	N
FU	fuel utilization	–	–
h	specific enthalpy	Btu/lbm	kJ/kg
k	ratio of specific heats	–	–
$\dot{m}$	mass flow rate	lbm/sec	kg/s
p	pressure	lbf/ft^2	Pa
P	power	hp	kW
PTR	power-to-heat ratio	–	–
q	heat per unit mass	Btu/lbm	kJ/kg
$\dot{Q}$	heat flow rate	Btu/hr	kJ/h
r_p	pressure ratio	–	–
T	temperature	°R	K
U	overall coefficient of heat transfer	Btu/hr-ft^2-°F	W/m^2·K
v	specific volume	ft^3/lbm	m^3/kg
v	velocity	ft/sec	m/s
w	work per unit mass	Btu/lbm	kJ/kg
$\dot{W}$	work	Btu	kJ

Symbols

η	efficiency	–	–

Subscripts

e	exit
i	inlet
p	pressure or constant pressure
regen	regeneration
s	isentropic
th	thermal
v	constant volume

[13]Utilities in the United States are averse to using the term *life extension*. The reason lies in the 1970 Clean Air Act. Power plants built prior to 1971 were exempted from emissions control requirements. These "grandfathered" plants lose their exemption if they are significantly modified. "Refurbishment" and "repair," maybe; "life extension," no.

30 Advanced and Alternative Power-Generating Systems

Content in b lue refers to the *NCEES Handbook*.

1. INTRODUCTION

Many advanced energy technologies are maturing but in some cases uneconomical, are limited to specific applications, or are still in various stages of development. Renewables are limited by localization and capacity factor. (An installation's *capacity factor* is the actual power output over some period of time divided by the theoretical maximum output. A wind turbine's capacity factor, for example, is affected by the percentages of time the wind does not blow.) Not only are most renewables confined to specific locations, but even then, their capacity factors are often below 30%.

By comparison, new coal plants have capacity factors in excess of 85%.

The 1973 energy crisis and oil embargo illuminated the need for alternative energy sources. However, with the large fossil fuel reserves in the United States, wide geographic distribution of these reserves, and the high efficiencies being achieved in combined cycle plants, it is likely that coal and natural gas will continue to be the most economic source of baseload electricity well into the future. Of all the other renewable energy sources, wind energy comes closest in price. Even so, and even with tax credit incentives (i.e., "production credits"), wind power is still 25% to 100% more expensive than coal/gas power.

2. SOLAR THERMAL ENERGY

Solar thermal energy arrives at the outside of the earth's atmosphere at an average rate of 433 Btu/ft^2-hr (1.366 kW/m^2), a value known as the *solar constant.* 40% to 70% of this energy survives absorption in and reflection from the atmosphere and reaches the earth's surface. The actual incident energy, I, sometimes referred to as *insolation*, depends on many factors, including geographic location, tilt and orientation of the receiving surface, calendar day, time of day, and weather conditions. Average values for clear days are given in maps and tabulations.[1] Some of this energy can be captured in *active solar systems* and used for space heating, domestic hot water (DHW) generation, and cooling (using heat pumps and absorption chillers).[2]

Solar thermal energy is captured in *solar collectors.* The sun's energy enters the collector and, because of the *greenhouse effect*, is trapped inside.[3] Heat is absorbed by *heat transfer fluid* pumped through tubes mounted on the *absorber plate.*[4] The tubes and absorber plate can be left uncoated, painted black, or treated with a *selective surface* coating that absorbs more energy than it reradiates.[5] Water at 100°F to 150°F (38°C to 66°C) can easily be generated in this manner. Thermal energy in

[1]Some insolation maps use units of langleys/day. A *langley* is equal to 1 cal/cm^2, 3.69 Btu/ft^2, and 41 840 J/m^2.

[2]*Passive solar systems*, which include strategically oriented buildings, walls, and thermal collectors, and which rely on natural convection and conduction for storage and heat transfer, are not included in this chapter.

[3]As received, solar radiation has wavelengths of 0.2 μm to 3.0 μm. Thermal radiation reradiated from the collector plate has a wavelength of approximately 3 μm. Good covering materials have high transmittance (85% to 95%) of received radiation and significantly lower transmittance (less than 2%) of reradiated radiation. White crystal glass, low-iron tempered and sheet glasses, and tempered float glass satisfy these requirements. Polycarbonates, acrylics, and fiberglass also perform well but suffer from weathering and durability problems.

[4]Water can be used as the heat transfer medium, but it is subject to freezing, boiling, and chemical breakdown; and the system is subject to corrosion. To counteract these problems, ethylene glycol-water and glycerine-water mixtures are often used. Ethylene glycol, however, is toxic, and a heat exchanger must be used to keep the heat transfer fluid separate from domestic water. "State-of-the-art" fluids include silicones, hydrocarbon (aromatic and paraffinic) oils, and change-of-phase refrigerants (see Table 30.1).

[5]Common selective surface coatings with low-to-moderate costs include copper oxide, black nickel, black chrome, lead oxide, and aluminum conversion.

Table 30.1 Properties of Transfer Fluids

fluid	specific gravity	viscosity[a] cP	specific heat Btu/lbm-°F	specific heat kJ/kg·K	freezing point °F	freezing point °C
water	1.00	0.5–0.9	1.00	4.19	+32	0
50% water-50% ethylene glycol	1.05	1.2–4.4	0.83	3.5	–33	–36
50% water-50% propylene glycol	1.02	1.4–7.0	0.85	3.6	–28	–33
paraffinic oil	0.82	12–30	0.51	2.1	+15	–9
aromatic oil[b]	0.85	0.6–0.8	0.45	1.9	–100	–73
silicone oil[b]	0.94	10–20	0.38	1.6	–120	–84

[a]Viscosity in the range of 80°F to 140°F (27°C to 60°C).
[b]Typical values given.

the heat transfer fluid can be used directly (as in swimming pool heaters), or it can be transferred to and stored in a tank of water or rock pebble beds.

Most collectors in simple heating systems are *flat-plate collectors*. These are essentially wide, flat boxes with clear plastic or glass coverings known as the *glazing*. *Concentrating (focusing) collectors* use mirrors and/or lenses to focus the sun's energy on a small absorber area. Except for some parabolic mirror designs, focusing collectors use a controller and motor to track the sun across the sky. *Evacuated-tube collectors* are more complex, but their efficiencies are higher. A U-shaped tube carries the transfer fluid through an air-filled transparent cylinder, which itself is enclosed in a transparent vacuum cylinder. Evacuated collectors are useful when extremely hot transfer fluid is needed and are generally limited to commercial projects.

Ideally, collectors should point due south and be tilted (from the horizontal) an angle equal to the latitude. For winter use, the tilt should be somewhat higher (so that the maximum energy is received on the coldest days). The variations from south-pointing and latitude tilt may be ±10° to 15° without significant degradation in performance.

The heat absorbed by the solar collector can be calculated from the incident energy, the *absorptance*, α, of the absorber, and the *transmittance*, τ, of the cover plate. The *shading factor*, F_s, in Eq. 30.1 has a value of approximately 0.95 to 0.97 and accounts for dirt on the cover plates and shading from the glazing supports. It is neglected in most initial studies.

$$q_{\text{absorbed}} = F_s A \alpha \tau I \qquad 30.1$$

The heat absorbed by the transfer fluid is

$$q_{\text{fluid}} = \dot{m} c_p (T_{\text{out}} - T_{\text{in}}) \qquad 30.2$$

The difference between the heat absorbed by the collector and transfer fluid constitutes the conduction and convection losses.

$$\begin{aligned} q_{\text{loss}} &= q_{\text{absorbed}} - q_{\text{fluid}} \\ &= U_L A (T_{\text{plate}} - T_{\text{air}}) \end{aligned} \qquad 30.3$$

The average plate temperature, T_{plate}, in Eq. 30.3 is seldom known. However, the incoming fluid temperature is usually known and is used by convention in place of the plate temperature. The collector *heat removal efficiency factor*, F_R, is used to correct for the substitution in variables. For liquid collectors, F_R ranges from 0.8 to 0.95 and is usually specified by the collector manufacturer.

$$\begin{aligned} q_{\text{fluid}} &= q_{\text{absorbed}} - q_{\text{loss}} \\ &= F_R F_s A \alpha \tau I - F_R U_L A (T_{\text{in}} - T_{\text{air}}) \end{aligned} \qquad 30.4$$

The *collector efficiency* is the ratio of energy absorbed by the transfer fluid to the original incident energy striking the collector.

$$\eta = \frac{q_{\text{fluid}}}{IA} \qquad 30.5$$

Since convective and radiation losses increase at higher collector temperatures, the collector efficiency decreases as the difference between ambient air and average plate (or inlet) temperatures increases. When the efficiency is plotted against the ratio of inlet-ambient temperature difference to incident energy, $(T_{\text{in}} - T_{\text{ambient}})/I$, the falloff rate is approximately linear with a slope of $-F_R U_L$. (See Fig. 30.1.) Typical values of $F_R U_L$ for flat-plate collectors range from 0.6 Btu/hr-ft²-°F to 1.1 Btu/hr-ft²-°F (3.4 W/m²·K to 6.2 W/m²·K). When the temperature difference is zero, q_{loss} will also be zero, and the y-intercept will be the theoretical maximum efficiency, $F_R \tau \alpha$. Typical values of $F_R \tau \alpha$ range from 0.50 to 0.75.

Figure 30.1 *Solar Collector Efficiency Curve*

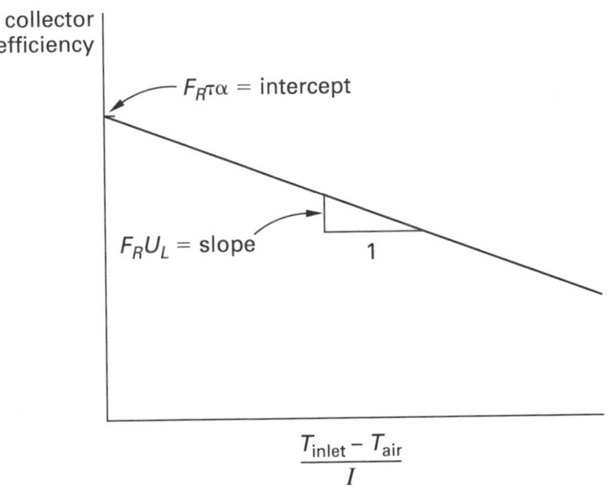

3. SOLAR POWER CYCLES

Solar energy generating systems (SEGS) use solar energy to generate electricity in traditional power cycles. Three main approaches are taken: *distributed collector systems* (DCS), also known as "trough-electric systems;" *central receiver systems* (CRS), also known as "power-tower systems;" and *dish/Stirling* (D/S) *systems*.[6] In a *trough-electric system*, parabolic tracking trough concentrators focus sunlight on evacuated glass tubes that run along the collectors' focal lines. Synthetic oil in tubes in the glass cylinders is heated to 575°F to 750°F (300°C to 400°C). Heat is transferred in a heat exchanger to steam used in a traditional Rankine cycle. Trough-electric technology is relatively mature, but due to the low temperatures, average annual thermal efficiencies are only 10% to 15%.

In a *power-tower system*, a field of *heliostats* (tracking mirrors) concentrates solar energy onto a receiver on a central tower. Since directly generated steam would be subject to variations in solar flux from passing clouds, heat generation and use are decoupled. Heat can be stored in molten eutectic salts such as sodium nitrate, which has a melting point of approximately 430°F (220°C), or mixtures of sodium and potassium nitrate. In one scenario, molten salt would be heated to approximately 1050°F (560°C) and cooled to approximately 550°F (290°C) in the heat exchanger. With these higher temperatures, typical thermal efficiencies of 15% to 20% are possible.

In a *dish-engine system*, the heat engine and electrical generator are located at the focus of a parabolic dish. Most installations use Stirling engines (kinematic and free-piston designs) since they are externally heated. Heat can be transferred to the Stirling cycle through direct irradiation, heat pipes, or pool boiling.[7] Due to size and wind loading, 25 kW to 50 kW is about the highest expected output per dish. However, since the units are modular, they can be grouped in a field to produce any desired output. With operating temperatures up to 1400°F (800°C), thermal efficiencies of 24% to 28% are already being realized, and over 30% has been achieved in some prototypes.

In addition to technical and economic problems, SEGS require large areas. Modern trough systems would require approximately 4500 ac (19 km^2) to generate 1000 MWe. Even with this space, practical and economic issues limit trough-electric systems to about 200 MWe and tower systems to approximately 100 MWe to 300 MWe.

4. PHOTOVOLTAIC ENERGY CONVERSION

A *photovoltaic cell* (*PV cell* or *solar cell*) generates a voltage from incident light, usually light in the visible region. Traditional PV cells are sliced from monolithic silicon crystals and (when cost is unimportant) from gallium arsenide (GaAs) in crystalline form. Thin vacuum-deposited semiconductor films (known as *amorphous thin films*), including silicon, copper indium diselenide (CIS), and cadmium telluride, are less costly than monolithic crystal slices but have much lower efficiencies. Spherical silicon particles on a metallic substrate constitute another emerging technology.

Each silicon cell produces the same current and a voltage of approximately 0.5 V. To obtain larger voltages or currents, cells are connected into *modules*. Large collections of modules are known as *arrays*.

Efficiency is measured by electrical energy output as a fraction of solar energy input. Although maximum efficiencies of single-crystal and thin-film cells are approximately 33% and 15%, respectively, module efficiencies are much less than single-cell efficiencies due to electrical and optical losses. Most commercial single-crystal and thin-film cells have efficiencies of approximately 20% and 10%, respectively.

Several approaches are taken to increase module efficiencies: (1) optical concentrating of solar energy onto existing technology cells, and (2) production of more efficient solar cells. Module efficiencies as high as 40% have been achieved with *triple junction solar cells* (cells with multiple layers, each responding to a different wavelength) and 40% to 43% using optical concentration. Commercial terrestrial concentration solar cells

[6]This discussion excludes the ultra-high concentrations and temperatures that are available with high-tech optical, two-stage collectors. These systems produce extremely high temperatures on extremely small targets. The discussion also omits the *solar chimney* concept in which collectors establish a 54°F to 72°F (30°C to 40°C) thermal gradient in the air within tall towers. The stack effect produces air movement, which drives fan-driven generators.

[7]One dish-engine design uses a *reflux pool-boiling receiver*. The receiver uses a pool of molten liquid metal (e.g., sodium, potassium, or a mixture thereof) to transfer heat from the exposed face of the receiver to the helium-filled heater tubes of the Stirling engine. As the liquid metal boils, it vaporizes. It gives up heat when it condenses on the heater tubes. Condensed liquid drips back to the pool by gravity (hence the term "reflux").

are available with 38% ratings. Since PV cells are wavelength sensitive, higher efficiencies can be achieved in layered *triple junction cells*.[8]

Existing utility PV "plants" (most of which are less than 1 MW) have capacity factors of approximately 18% to 22%. (Desert sites and two-axis tracking might increase capacity factors to 35%.) Availability is high, 90% to 97%. PV cell output is direct current. Output must be inverted (i.e., converted to alternating current) when connecting to the electric grid.

While PV power cannot compete economically with baseload power generation, it might someday be competitive with peak power alternatives. It is significant that peak PV output coincides with some peak power uses (e.g., air conditioning peaks on hot days). The main problems facing commercialization of PV technologies are high cost, limited lifetimes due to deterioration in performance, and storage of generated energy.

5. WIND POWER

Wind energy conversion systems (WECS), also known as "wind turbines," usually consist of a conventional induction generator driven through a drivetrain by a large rotor. Most rotors consist of a hub and two or three blades. The main shaft, gearbox, and generator are protected from the elements by an enclosure known as a *nacelle*. The WECS is mounted on a tower. Most WECS feeding the electrical grid turn at constant speed.

WECS are classified by the orientation of their rotational axes.[9] With *upwind machines* (*head-on*, *horizontal-axis rotors* or *wind-axis rotors*), the axis of rotation is parallel to the windstream and the blades face the direction from which the wind is coming. With *downwind machines*, the blades face away from the source of the wind. With *vertical-axis rotors* (*Darrieus rotors* or "eggbeater windmills"), the axis is perpendicular to both the surface of the earth and the windstream. (See Fig. 30.2.)

Most WECS are upwind machines, even though they require yaw drives to keep them facing into the wind. However, the propeller turbulence (*wash*) can put a strain on the tower. Downwind machines suffer from *tower shadow*. The tower eclipses some of the windstream, and the forces on the up- and down-blades are different.[10] The resulting vibration can be severe. However, downwind machines can yaw freely without a yaw drive, and blade stresses are less.

Vertical-axis rotors do not have to be turned to face the wind and require no support tower. They have low starting torques, but may not be self-starting. Their high tip/wind speed ratio produces relatively high power outputs. Since the rotor shape is the shape that a flexible cable would take if spun, the bending stresses are low, no matter how fast the rotor turns. The drivetrain and generator are located on the ground, providing for easy maintenance. However, most often they are not installed high enough above the ground to catch the highest velocity wind.

Figure 30.2 *Types of Rotors*

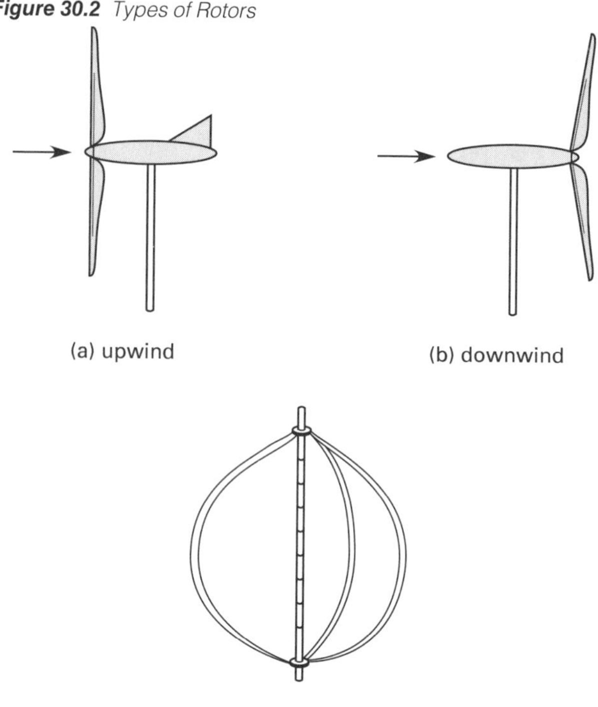

Because of frictional and inertial effects, WECS will not operate much below 5 mph (8 kph), and generally wind speeds of 10 mph (16 kph) or more are needed. 15 mph (24 kph) is generally considered an ideal average speed. Since power increases with the cube of the velocity, WECS are particularly efficient at high velocities. However, forces on WECS become unacceptably high much above 30 mph (48 kph). Various methods are used to limit speed. Most commercial designs are either *stall-limited* (stall-controlled) or *pitch-limited* so that the blades are automatically feathered at higher speeds. Other options include use of blade ailerons, yawing the rotor out of the wind, mechanical braking of the shaft, blade tip brakes, and electrical dynamic braking.

Most sites have a positive *wind shear*—the wind velocity increases exponentially (for a distance) with altitude. Because of this, WECS manufacturers use tower heights of 200 ft to 400 ft (60 m to 120 m) to generate the increased power available at higher altitudes.

[8]For example, blue and green light could be captured in amorphous silicon layers, while a silicon-germanium layer could intercept infrared light.
[9]*Cross-wind horizontal-axis rotors*, including *Davonius rotor* machines, where the axis is horizontal and perpendicular to the windstream, are very uncommon.
[10]This happens to a certain extent in all horizontal-axis machines because the wind speed is lower closer to the ground.

Graphs of the number of hours per year that the wind reaches each hourly mean velocity are known as *annual average velocity duration (AAVD) curves.* Curves showing the distribution of annual average wind power per unit subtended area as a function of wind speed are called *annual average power density distribution (AAPD) curves.* The annual average wind energy density distribution is equal to the annual average power density distribution multiplied by the number of hours per year the corresponding wind speeds occur.

The total ideal power available in a windstream is found by multiplying the kinetic energy per unit mass by the flow rate.

$$P_{\text{ideal}} = \dot{m}(\text{KE}) = \dot{Q}\rho(\text{KE}) = A\text{v}\rho\left(\frac{\text{v}^2}{2}\right)$$

$$= \frac{A\rho \text{v}^3}{2} \quad \text{[SI]} \quad 30.6(a)$$

$$= \frac{\pi (r_{\text{rotor}})^2 \rho \text{v}^3}{2}$$

$$P_{\text{ideal}} = \dot{m}(\text{KE}) = \dot{Q}\rho(\text{KE}) = A\text{v}\rho\left(\frac{\text{v}^2}{2g_c}\right)$$

$$= \frac{A\rho \text{v}^3}{2g_c} \quad \text{[U.S.]} \quad 30.6(b)$$

$$= \frac{\pi r_{\text{rotor}}^2 \rho \text{v}^3}{2g_c}$$

WECS are unable to extract the ideal power from the airstream. To do so would require decelerating the air flow to zero velocity (i.e., removing all of the kinetic energy). The actual power can be determined if the actual pressures and velocities before and after the WECS are known.

$$P_{\text{actual}} = A\text{v}\rho\left(\frac{p_2 - p_1}{\rho} + \frac{\text{v}_2^2 - \text{v}_1^2}{2}\right) \quad \text{[SI]} \quad 30.7(a)$$

$$P_{\text{actual}} = A\text{v}\rho\left(\frac{p_2 - p_1}{\rho} + \frac{\text{v}_2^2 - \text{v}_1^2}{2g_c}\right) \quad \text{[U.S.]} \quad 30.7(b)$$

The actual power generated will depend on the *power coefficient*, C_P. The power coefficient of an ideal wind machine rotor with a propeller rotor varies with the ratio of blade tip speed to free-flow windstream speed and approaches the maximum value of 0.593:1 (known as the *Betz coefficient*) when this ratio reaches a value of 5:1 or 6:1. The maximum power coefficients for the best two-blade rotors is approximately 0.47, and for Darrieus vertical-axis rotors, the maximum value is approximately 0.35. Figure 30.3 illustrates the variation in power coefficient for rotors.

***Figure 30.3** Power Coefficients*

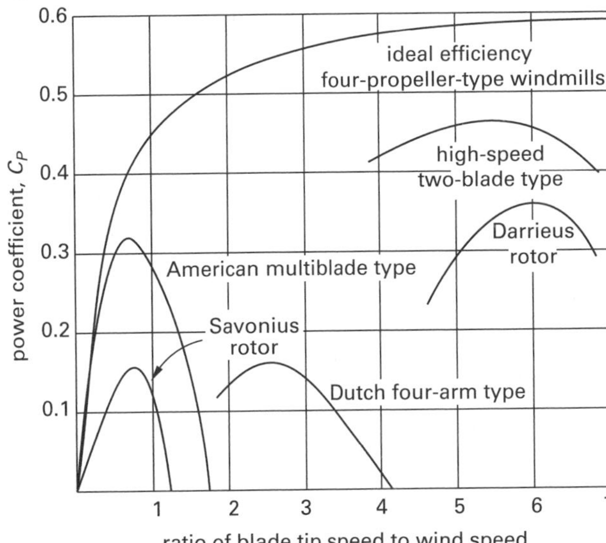

(source: National Technical Information Service)

$$P_{\text{actual}} = C_P P_{\text{ideal}} \quad 30.8$$

The velocity in the immediate wake of the rotor depends on the *interference factor*, a, and is given by $\text{v}_2 = \text{v}_1(1 - a)$. If some simplifying assumptions are made, the power coefficient can be calculated from the interference factor.[11]

$$C_P = 4a(1 - a)^2 \quad 30.9$$

Commercial WECS come in all sizes. Units for residential use are seldom larger than 2.5 kW. Modern units connected to utility grids are generally in the 1 MW to 2 MW range (as limited by economics and mechanical stresses), though some mammoth 7 MW to 8 MW units have been built over the years. Individual units can be combined into large "wind farms" to produce any amount of power needed.

Like solar energy, WECS produce little or no environmental impact.[12] However, wind power requires more space per watt generated than fossil fuel or nuclear power. Structural corrosion from air pollutants and fatigue failures from normal operation and random gust loads are prevalent. During normal operation, WECS are plagued by blade pitting and soiling by insects and air pollutants.

[11]The radial pressure gradient and the kinetic energy of the swirl velocity component in the rotor's wake are neglected.

[12]Many birds are killed by spinning rotors.

Teetering (*hinged* or *pivoted*) *two-blade machines* produce lower stresses. Numerous studies have been performed on how to improve laminar flow-related performance impacts, and blades have been redesigned with these studies in mind. Only laminar flow is sensitive to blade roughness due to pitting and soiling.

6. OCEAN THERMAL ENERGY CONVERSION

During the summer, the temperature at the surface of tropical oceans is almost constant at 77°F (25°C). 2000 ft to 3000 ft (600 m to 900 m) below, however, the temperature can be as low as 38°F to 40°F (4°C to 5°C). An *ocean thermal energy conversion* (OTEC) system uses differences in temperature at the surface of the ocean to drive a vapor cycle.

In a *closed-cycle plant*, warm water from the surface supplies heat to a boiler; cold water from the depths provides cooling in the condenser. A traditional low-pressure vapor cycle operating with ammonia, propylene, or other refrigerant drives a generator.

In an *open-cycle plant*, the warm seawater is injected into a near vacuum. The water vaporizes and the resulting steam expands through a low-pressure turbogenerator before being condensed with the cold seawater.[13]

The net-to-gross power ratio, typically around 0.20, is primarily a function of the power needed to pump water from the depths. In addition to low efficiencies due to the small temperature differential, other problems associated with OTEC systems include high component and material costs, corrosion, biofouling, and long-distance power transmission. Production of electrically intensive products (e.g., hydrogen from electrolysis of seawater or even desalinated fresh water) has been proposed as a solution to the latter problem. The major environmental effect would be a slight cooling at the surface of the ocean and a warming of the depths.

7. TIDAL AND WAVE POWER PLANTS

Electric power can be generated by capturing water brought into a lagoon or bay by tidal action, and then releasing the water through traditional hydroelectric generators when the tide recedes. The technology is neither new nor complex. However, only about a hundred sites in the world have large enough rises and falls for tidal power to be practical.

With *wave power plants*, advancing and receding ocean waves repeatedly compress and force air through air turbines. (The turbines spin in the same direction regardless of the wave direction.) The turbines drive electrical generators. Wave machines do not suffer from the power transmission problem of OTEC units, since they can be located close to shore.

8. GEOTHERMAL ENERGY

There are three general types of geothermal energy sources: vapor-dominated, liquid-dominated, and hot rock sources.[14] The 750 MW "Geysers" in northern California is an example of a *vapor-dominated reservoir* driving a *direct steam cycle*. Multiple wells provide steam at 100 psi to 120 psi (690 kPa to 820 kPa) and 400°F (205°C) which is collected, separated to remove liquid, filtered to remove abrasive particles, and passed through turbines. Evaporative air cooling towers are used to cool condensing water. Water lost in evaporation is replaced by condensed water, and approximately 80% of the steam is evaporated in this manner. The remaining 20%, containing minerals, is reinjected into the ground through deep injection wells. Substantial noncondensible gases (primarily hydrogen sulfide and ammonia, but also including carbon monoxide, hydrogen, methane, and nitrogen) constitute up to 5% of the steam volume and are removed by steam ejectors, vacuum pumps, or compressors and released through the cooling tower.[15] Thermal efficiency is approximately 22%, although 16% is more typical of similar plants worldwide.

Most geothermal sites are *liquid-dominated reservoirs* and produce mixtures of hot water (i.e., brine) and hot steam. In *hot water systems*, the hot water is discarded and only the steam is used. If the hot water temperature is approximately 330°F (165°C) or higher, a *flash steam cycle* can be used. Hot water and steam are pumped to an evacuated *flash tank*. The resulting high-pressure flash steam drives the turbine. In the *dual-flash cycle*, low-pressure hot water remaining in the first flash tank is routed to a second, low-pressure flash tank. Low-pressure steam is used in the LP turbine section. Liquid-dominated reservoirs typically do not produce large quantities of noncondensible gases.

If the temperature of the hot water is between approximately 250°F and 330°F (120°C and 165°C), a *binary cycle* using a separate heat transfer fluid, such as pentane or isobutane, is required. Hot water passes through a closed heat exchanger, vaporizing the heat transfer fluid. Cooled water is discarded in injection wells. The vapor drives a binary turbine generator in a traditional Rankine cycle. An air-cooled condenser maintains the turbine back pressure.

Fewer than ten naturally occurring vapor-dominated reservoirs are known to exist in the world, and only three or four of them are commercially exploited. With drilling, however, hot geological rock can be found

[13]The closed process was originally proposed in 1881 by French engineer Jacques-Arsene D'Arsonval. The open process is sometimes referred to as the *Claude cycle*, named after Georges Claude, a student of D'Arsonval, who demonstrated the practicality (at a net power loss) of the open process off the coast of Cuba in 1929.

[14]*Hydrothermal sources* (steam and water), *hot dry rock sources* (steam only), and *geopressured sources* (high-pressure liquid) are three slightly different categorizations of geothermal sources.

[15]Because of the high noncondensible component of steam at the Geysers, removal of noncondensibles is the largest single use of steam—up to 6%.

anywhere. The temperature of the earth's crust increases by 30°F (16.7°C) for every kilometer of depth. Anomalous temperatures of 570°F to 1300°F (300°C to 700°C) can be found within 20,000 ft (6000 m) of the surface at some locations. In *hot rock systems*, water is injected through injection wells into artificially made fractured rock beds 0.6 mi to 3.6 mi (1 km to 6 km) below the surface. Pressurized hot water at approximately 400°F (200°C) would be removed from an adjacent production well. Steam flashing from the hot water drives turbines located at the surface.

Since there are very few dry steam (i.e., vapor-dominated) sites, but many locations where very hot water is available, current geothermal energy plants exploit either binary cycle equipment, low-pressure flash operation, or both.

Difficulties associated with natural geothermal energy include the scarcity of natural sites, method of cooling, corrosion and fouling by mineral-laden water, and high moisture contents during steam expansion. Environmental problems include disposal of high-mineral waste water, pollution of the air by noncondensing noxious gases, drift, waste heat release, noise from steam venting, and ground subsidence.[16] Hot rock systems, still in their infancy, have many technical difficulties to overcome, including drilling and controlled fracturing techniques and loss of water.[17]

9. FUEL CELLS

A *fuel cell* converts chemical energy directly into electrical energy. With two electrodes separated by an electrolyte, a fuel cell is similar to a continuously fueled battery. (See Fig. 30.4.) Unlike a battery, however, the chemical process continues indefinitely as the spent reactants are replaced. Fuel (e.g., hydrogen or methane) is supplied to the anode (the negative terminal), and oxidizer (e.g., oxygen) is supplied to the cathode (the positive terminal). The anode and cathode are both porous to allow diffusion of the fuel and oxygen through them. Fuel reacts with the electrolyte at the cathode; oxygen reacts at the anode. Liquid water is produced, along with heat and electricity, in a process that is essentially the reverse of electrolysis. (Combustion does not occur in the fuel cell.) The voltage produced by a hydrogen-oxygen fuel cell is approximately 0.7 V.

When free hydrogen gas is not available, hydrogen-rich gas can be generated from fossil fuels in a *reformer* (*reformer reactor*). Reforming ("front-end reforming") can take place within the fuel cell if the temperature is high enough. This is referred to as the *direct fuel cycle* or *internal fuel reforming*. If the cell temperature is too low or if the fuel is not gaseous, a separate reformer is

Figure 30.4 Fuel Cell

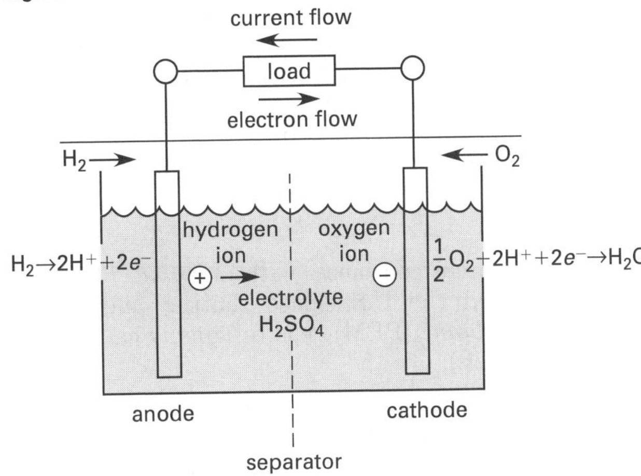

needed. Steam reforming of natural gas, adiabatic reforming of heavy distillates, and coal gasification are all applicable processes.

The major factor limiting the amount of power generated is the speed of the diffusion processes through the electrolyte between the electrodes. Because of this, fuel cells are named for the electrolyte used.

Commercial units combine individual cells, each of which is about 0.5 cm thick and generates approximately 700 mV, into *stacks*. Stacks containing hundreds of cells are used to generate up to 200 kW to 250 kW of power.[18] Multiple stacks can be used to produce any amount of power desired.

Since fuel cells are not heat engines, they are not limited by Carnot efficiencies. Theoretical fuel cell efficiencies are high, ideally on the order of 80% for hydrogen, 90% for methane, and up to 98% for more complex hydrocarbons.[19] However, not all of the heat generated can be recovered. This and other factors reduce the practical stack efficiency to 35% to 45% based on higher heating values (HHV).

Small hydrogen-oxygen fuel cells have been used in spacecraft for years. First-generation commercial utility-sized *phosphoric acid fuel cells* (PAFCs) operating at 400°F (200°C) are a mature technology. Efficiencies are typically around 40%.

Second-generation *molten carbonate fuel cells* (MCFCs), operating at 1200°F (650°C), offer higher efficiencies. By reforming coal, MCFCs could theoretically replace gas turbines in *integrated coal gasification combined cycle* (ICGCC) plants. HHV efficiencies of carbonate fuel cell stacks are projected to be approximately 50% to 55%.

[16]The salt content of geothermal waters varies from 2% to about 20%. By comparison, the salt content of seawater is approximately 3.3%. Hydrogen sulfide is soluble in water and escapes by evaporation during the cooling process.
[17]Nuclear blasts are no longer considered a viable method of producing a cavity deep within the earth.
[18]For example, a stack of 470 1 m^2 cells will generate approximately 750 kW.
[19]All fuel cell efficiencies are temperature-dependent.

Third-generation monolithic *solid oxide fuel cells* (MSOFC or SOFC) operating at 1800°F (1100°C) are being developed. A separate reformer reactor is not needed; the high temperature with or without a catalyst can reform the fuel gas (natural gas or gasified coal) internally. The high temperatures also make SOFCs ideal candidates for cogeneration and combined cycle plants, where overall efficiencies may eventually reach 50% to 70%.

Ongoing research continues with *alkaline fuel cells* (which powered the U.S. space shuttle) and *proton exchange membrane* (PEM) cells (originally called *solid polymer fuel cells*).

Fuel cells are attractive in the transportation sector because they produce clean energy from readily available reactants. Fuel can be supplied by hydrogen-rich liquids such as methanol, and oxygen is supplied by the air. Temperatures are low enough to keep NOx from being a problem, carbon dioxide is one-third of that from hydrocarbon combustion, and all other polluting emissions are substantially reduced. However, because of size, mass, poor start-up and transient responses, and expensive catalysts, fuel cells have been incorporated into only a few experimental and/or demonstration vehicles.

10. MAGNETOHYDRODYNAMICS

In a *magnetohydrodynamic* (MHD) *generator*, a high-temperature ionized plasma flows through a supersonic nozzle.[20] The high-velocity plasma in the expansion channel passes through a magnetic field generated by a toroidal coil or plates. A constant electric field is generated perpendicularly to the magnetic field (in the direction of the moving gas) by the moving ions.[21] Direct current is generated, so inverters are needed to produce grid-quality power.[22]

High magnetic fields (5 T) and high temperatures—4600°F (2800K) at the generator entrance and 4400°F (2700K) at the exit—are required to create the plasma. The high temperatures are achieved by combustion of gasified coal (with or without oxygen enrichment) in a high-temperature air heater (HTAH) prior to the additional combustion of fossil fuels (coal or natural gas) in a subsequent combustor.

The combustion products are doped (seeded) with salts of easily ionized elements, such as potassium or cesium, to obtain the necessary (1%) ionization. When supplied at approximately 150% of the stoichiometric rate, these alkali metal ions also combine with virtually all sulfur in the fuel. The resulting compounds (potassium sulfate) can be recovered in electrostatic precipitators for reuse as seed. With all sulfur removed, it may be possible to allow the stack gas to cool to as low as 395°F (200°C).

NOx generation was once thought to be problematic. Large amounts (10,000 ppm to 12,000 ppm) of NOx are produced at the high plasma temperatures. However, various modern control methods, including staged fuel-rich combustion, and decomposition in the radiant furnace apparently can reduce NOx to acceptable levels.

MHD generators are attractive because there are no highly stressed rotating parts and because the higher temperatures are compatible with bottoming cycles. Most of the heat remaining in the combustion gases can be recovered in a high-temperature air heater for use in cogeneration or combined bottoming cycles.

From a thermodynamic standpoint, the MHD cycle is the same as the Brayton gas turbine cycle, except that the work output during the expansion is electrical, not mechanical. Since the movement of ionized particles through a magnetic field results in a conversion of mechanical energy to electrical energy, the efficiency of that process is not limited to the Carnot maximum. The mechanical-electrical conversion efficiency, known as the *loading factor*, K, of this process is high: 0.80 to 0.90 (i.e., 80% to 90%). For an isentropic process, $K = 1$. For an isothermal process, $K = 0$.

$$K = \frac{E}{\mathrm{v}B} \qquad 30.10$$

The temperature change along the expansion channel is

$$\frac{T_2}{T_1} = \left(\frac{p_2}{p_1}\right)^{K(k-1)/k} \qquad 30.11$$

In a constant-velocity generator, the electrical energy removed decreases the enthalpy between the entrance and exit of the generator. Assuming a constant velocity generator and an ideal gas, the isentropic efficiency of the expansion process is

$$\eta_s = \frac{\Delta h_{\text{actual}}}{\Delta h_{\text{ideal}}} = \frac{\Delta T_{\text{actual}}}{\Delta T_{\text{ideal}}} \qquad 30.12$$

While the mechanical-electrical conversion efficiency is not limited to the Carnot efficiency, the thermal-electrical conversion efficiency is. Experimental installations have achieved a maximum 22% thermal efficiency, although 17% is more typical of coal-fired units. Since Rankine steam cycles can achieve 40% efficiency, MHD channels by themselves are not thermally attractive.

[20]MHD machines are not limited to plasmas. In MHD marine (submarine) propulsion systems, the cycle is reversed. Seawater flowing through the duct is acted upon by magnetic and electrical fields. The fields force the water through the duct, propelling the vessel forward.
[21]Generation of this electric field is known as the *Hall effect*.
[22]Momentum in MHD research has decreased significantly since the 1980s due to low efficiencies and pollution concerns.

When MHD generation is integrated with cogeneration and combined cycles, thermal efficiencies of up to 60% seem reasonable.

11. THERMOELECTRIC AND THERMIONIC GENERATORS

A *thermocouple* generates electric potential directly from heat and is an example of a *thermoelectric generator.* Depending on the temperature, certain semiconductors also exhibit thermoelectric effects: bismuth telluride and selenide at room temperature, lead telluride alloys at 390°F to 930°F (200°C to 500°C), and silicon germanium alloys at 750°F to 1830°F (400°C to 1000°C).

The *figure of merit*, Z, for a thermoelectric material is an indicator of the effectiveness of the thermoelectric conversion process. Typical values are $3 \times 10^{-3}\ \text{K}^{-1}$ for bismuth telluride alloys to $1 \times 10^{-3}\ \text{K}^{-1}$. In Eq. 30.13, S is the *Seebeck coefficient*, and k is the thermal conductivity.

$$Z = \frac{S^2}{\rho k} \qquad 30.13$$

Current *radioisotope thermoelectric generators* (RTGs) used in space probes contain radiatively coupled *unicouples* using plutonium dioxide as the heat source.[23] Typical output per unicouple is approximately 2.5 W at 3.5 V. Theoretically, any number of unicouples can be combined into a *multicouple*, and multicouples can be stacked to obtain power in any amount. Most RTG stacks for space missions produce less than 1 kWe.

Thermionic generators, also known as *thermionic energy converters* (TECs), consist of two closely spaced tungsten plates (an emitter cathode and a collector anode) known as *shoes.* The hot shoe is maintained at approximately 2600°F to 3150°F (1700K to 2000K), and the cold shoe is maintained at approximately 1350°F (1000K). Current is generated when electrons boil off the hot shoe by thermionic emission and flow to the cold shoe. Since the plates are separated by a low-pressure ionized gas (usually cesium vapor), the device is sometimes referred to as a *plasma diode.*

Thermoelectric and thermionic power generators are power cycles that convert heat into work. Therefore, their efficiencies are limited to the Carnot efficiency. Most RTGs have efficiencies of 10%, and TECs have efficiencies around 15%. For space missions, these low efficiencies are offset by other desirable operating characteristics.

12. METHODS OF ENERGY STORAGE

Thermal energy is stored in heated beds. In simple residential solar installations, pebble and rock beds or tanks filled with water and/or other high-density materials can be used. In power utility applications, large tanks or caverns containing oil and rock have been proposed. In high-tech applications, molten salt and liquid metals can be used. Molten salt is particularly attractive in two-stage solar electric plants.

Electrical energy can be stored electrically, mechanically, or chemically. Large energy storage capacitor banks and (theoretically) superconducting coils can store electrical energy in electric and magnetic field forms. Mechanical storage involves converting electrical energy into potential, gravitational, or kinetic energy. Tanks ("accumulators") or caverns of compressed air, gas, or liquid store energy in pressure.[24] Elevated water ("pumped hydro" storage) stores gravitational energy. Flywheels store rotational kinetic energy. Electrical energy can be used to charge batteries, or it can be used to produce hydrogen through electrolysis for later use in fuel cells.

13. BATTERIES

Batteries store and produce electrical energy electrochemically. The common flashlight cell and mercury-oxide "button" cells do not have enough energy or power for *electric vehicles* (EVs), however. In addition to traditional lead-acid batteries and the lithium-ion batteries widely used in commercial EVs, other technologies include nickel-cadmium, nickel-metal hydride, zinc-air, and molten sodium-sulfur. (See Table 30.2.) All other exotic batteries appear to be too expensive, toxic, or limited for commercial use.

Indicators of battery performance are their *specific power* (in W/kg); *specific energy*, also known as *energy density* (in W·h/kg); and *specific capacity* (in A·h/kg). Generally, specific energy decreases as specific power increases. Thus, batteries can have either high specific power or high specific energy. The best commercial EV batteries, NiMH and lithium-ion, operate with typical specific powers of 150–200 W/kg and 250–300 W/kg (and higher in the lab), respectively, and practical specific energies up to 70 W·h/kg and 120 W·h/kg, respectively.[25]

The specific capacity of lead-acid batteries, the most commercially mature technology, is theoretically around 120 A·h/kg; the typical energy density is around 30 W·h/kg. Drawbacks to using lead-acid batteries in EVs

[23]Some of the Pioneer and Voyager space probes used RTGs.
[24]CAES is the acronym used in the electrical power industry for *compressed-air energy storage.*
[25]For EVs, goals of 400 W/kg and 200 W·h/kg have been established.

Table 30.2 Battery Technology Comparison

battery system	negative electrode	positive electrode	electrolyte	nominal voltage (V)	theoretical specific energy (W·h/kg)	practical specific energy (W·h/kg)	practical energy density (W·h/L)	major issues
lead-acid	Pb	PbO_2	H_2SO_4	2.0	252	35	70	heavy, low cycle life, toxic materials
nickel iron	Fe	NiOOH	KOH	1.2	313	45	60	heavy, high maintenance
nickel cadmium	Cd	NiOOH	KOH	1.2	244	50	75	toxic materials, maintenance, cost
nickel hydrogen	H_2	NiOOH	KOH	1.2	434	55	60	cost, high pressure hydrogen, bulky
nickel metal hydride	H (as MH)	NiOOH	KOH	1.2	278–800 (depends on MH)	70	170	cost
nickel zinc	Zn	NiOOH	KOH	1.6	372	60	120	low cycle life
silver zinc	Zn	AgO	KOH	1.9	524	100	180	very expensive, limited life
zinc air	Zn	O_2	KOH	1.1	1320	110	80	low power, limited cycle life, bulky
zinc bromine	Zn	bromine complex	$ZnBr_2$	1.6	450	70	60	low power, hazardous components, bulky
lithium ion	Li	Li_xCoO_2	PC or DMC w/$LiPF_6$	4.0	766	120	200	safety issues, calendar life, cost
sodium sulfur	Na	S	beta alumina	2.0	792	100	>150	high temperature battery, safety, low power electrolyte
sodium nickel chloride	Na	$NiCl_2$	beta alumina	2.5	787	90	>150	high temperature operation, low power

are many: high mass, limited recharging cycles, low range, long recharge time, and life reduction if fully discharged.

Lithium-ion and nickel-metal hydride batteries appear to be the most advanced EV-capable batteries. They have long life spans and high power and energy densities. They recharge fully in approximately an hour.

Fiber-nickel-cadmium (FNC) batteries are already in limited use, particularly in military and civilian aircraft. FNC batteries can be recharged several thousand times, and they operate over a wide temperature range. Cadmium constitutes a disposal problem.

Sodium-sulfur batteries are attractive because they have three to four times the storage capacity of lead-acid batteries. Their major drawback is that they must be operated at elevated temperatures of 662°F to 716°F (350°C to 380°C) in order to keep the sulfur electrode molten. They also need a liquid cooling system.

Polymer-electrolyte technology, such as that used in flat cells for powering instant film packs and smart credit cards, holds promise for the future.

14. NOMENCLATURE

a	interference factor	–	–
A	area	ft^2	m^2
B	magnetic flux density	n.a.	T
c_p	specific heat	Btu/lbm-°F	kJ/kg·K
C_p	power factor	–	–
E	electric field	n.a.	V/m
F_L	loss factor	–	–
F_R	removal factor	–	–
F_s	shading factor	–	–
g	acceleration of gravity, 32.2 (9.81)	ft/sec^2	m/s^2
I	incident energy	Btu/ft^2-hr	W/m^2
k	ratio of specific heats	–	–
k	thermal conductivity	Btu-ft/hr-ft^2-°F	W/m·K
K	loading factor	–	–
KE	kinetic energy	ft-lbf/lbm	J/kg
$\dot{m}$	mass flow rate	lbm/sec	kg/s
p	pressure	lbf/ft^2	Pa
P	power	ft-lbf/sec	W
q	heat transfer	Btu/hr	W
$\dot{Q}$	volumetric flow rate	ft^3/sec	m^3/s
r	radius	ft	m
S	Seebeck coefficient	n.a.	V/K
T	temperature	°F	K
U	overall coefficient of heat transfer	Btu/hr-ft^2-°F	W/m^2·K
v	velocity	ft/sec	m/s
Z	figure of merit	1/°R	1/K

Symbols

α	absorptance	–	–
η	efficiency	–	–
ρ	density	lbm/ft^3	kg/m^3
ρ	electrical resistivity	Ω-ft	Ω·m
τ	transmittance	–	–

Subscripts

s	isentropic

Power Cycles

31 Gas Compression Cycles

Content in blue refers to the *NCEES Handbook*.

NCEES EXAM SPECIFICATIONS AND RELATED CONTENT

THERMAL AND FLUID SYSTEMS EXAM

II.A.2. Compressors
13. Dynamic Compressors

1. TYPES OF COMPRESSORS

There are two main types of gas compressors: reciprocating and centrifugal.[1] *Reciprocating compressors* are appropriate for high-pressure, low-volume applications, such as air conditioning systems. *Rotating compressors* (also known as *dynamic compressors* and *blowers*) are used in low- to moderate-pressure, high-volume applications such as gas turbines and turbojets.[2,3]

Since any compressor will be limited to operating at a finite *pressure ratio*, higher pressures are achieved by routing the output of one compression stage to the input of a subsequent compression stage. Compressors are categorized by the number of *compression stages* (e.g., a two-stage compressor). Subsequent compressions usually occur in different parts of a multistage compressor and are seldom carried out in separate compressors.

Like pumps and fans, dynamic compressors can be radial or axial in design.[4] In radial compressors (i.e., centrifugal compressors), air enters through the eye, is accelerated by the impeller blades, and leaves 90° from the original inlet direction. By mounting two or more impellers on a single shaft, centrifugal compressors can easily be constructed as multistage machines. In *axial-flow compressors*, air flows essentially straight through as it is compressed by blades in the rotor. Axial-flow compressors are explicitly multistage machines, since each row of blades represents a single stage.

In comparison, radial (centrifugal) machines compress small volumes through large pressure ratios, and axial machines compress large volumes through small pressure ratios. Combined *axial-centrifugal compressors* are used when large volumes need to be compressed through large pressure ratios. The axial stages first reduce the volume; the subsequent radial stages (on the same shaft) complete the pressure increase.

2. BOOSTERS, INTENSIFIERS, AND AMPLIFIERS

Most compressed air equipment (pneumatic tools, painting, instrumentation, etc.) operates in the range of 90 psig to 125 psig (620 kPa to 860 kPa). Some higher-pressure processes (e.g., plastic molding, metal-working, and pressure testing machines) may require up to 500 psig (3.5 MPa). This ultra-high pressure plant air can be obtained by dedicated multistage compressors, but a power savings can be achieved by starting with the lower-pressure plant air.

A *pressure booster* is essentially another electrically driven single-stage compressor (usually air-cooled) drawing on compressed plant air. Pressure ratios are typically 2:1 to 7:1. Flow rates up to 500 cfm (240 L/s) are typical.

[1]Other types of *rotary positive-displacement compressors*, such as screw, globe, water-ring, and sliding vane designs, are not covered in this chapter.
[2]There is much overlap in the operating characteristics of all types of compressors. The decision to use one type in favor of another will also be influenced by cost, speed, and available driving mechanisms.
[3]The terms *turbocompressor* and *turboblower* are also used to describe these dynamic machines. The "turbo" prefix does not require that the input power comes from engine exhaust or combustion products (i.e., a turbocompressor is not a *turbocharger*). Input power can be from any shaft drive.
[4]Traditional radial (centrifugal) compressors are used for compression of air and other gases (nitrogen, oxygen, carbon dioxide, etc.). Axial compressors are traditionally found in the steel industry to produce blast furnace air and in plants with large air-separation processes.

A *pressure amplifier* (*pressure intensifier*) is one method of producing high-pressure air from a lower-pressure source. A large piston is driven by plant air at low pressures. This piston, in turn, drives a smaller piston, producing a smaller quantity of higher-pressure air. Output is seldom higher than 25 cfm (12 L/s). The output pressure is determined by the input pressure and the ratio of piston areas. Amplifiers do not require any input electrical power. However, they exhaust substantial amounts of compressed plant air, and they require complex valving.

3. COMPRESSOR CONTROL

Reciprocating compressors used to supply plant air can either run continuously or intermittently. There is essentially no "control" when running continuously, although pressure will be limited to the relief valve setting, usually just above the highest working pressure required. For reciprocating compressors operating intermittently, the maximum pressure will be the cutoff pressure, usually 15 psi to 20 psi (100 kPa to 140 kPa) higher than the highest pressure needed.

There are three general ways of controlling dynamic compressors. The most effective method is *speed control*, controlling performance by varying rotational speed. This method is applicable to drivers (i.e., prime movers) such as steam or combustion turbines whose speeds can be varied.

When the driving speed is fixed, as it is with electrical motors, guide vane control and suction throttling can be used. With *guide vane control*, vanes in the compressor impart rotation to the inlet stream. By changing the guide vane settings, inlet velocity, flow rate, and pressure are changed. By imparting either a rotation or counter-rotation to the inlet stream, guide-vane control can either decrease or increase the flow and pressure, respectively. With the inefficient *suction throttling* control method, a butterfly valve in the suction line is partially closed. This decreases the entering suction pressure and flow rate.

4. STORAGE TANKS

Compressed-air storage tanks (*receiver tanks*) are used to reduce the duty cycle of plant compressors. The compressor starts when the tank pressure is reduced to the *cut-in pressure* and runs until the tank pressure reaches the *cut-out pressure*. Larger tanks result in longer off-periods for the compressor and fewer compressor on-cycles. However, they increase the running time per cycle.

5. COMPRESSOR CAPACITY

In the United States, the capacity of a compressor can be expressed in any of several ways.

- mass flow rate: $\dot{m}$
- volumetric flow rate referred to standard conditions (usually 14.7 psia and 60°F): SCFM (standard cubic feet per minute); SCFH (standard cubic feet per hour); MMSCFD (million standard cubic feet per 24-hour day)[5]
- volumetric flow rate referred to inlet conditions: ICFS (inlet cubic feet per second); ICFM (inlet cubic feet per minute); ICFH (inlet cubic feet per hour)

6. COMPRESSION PROCESSES

The thermodynamic behavior of a gas in a *reciprocating compressor* is controlled by the movement of the piston and heat transfer to compressor surfaces. Such compression and expansion are polytropic processes. The work done per cycle depends on the polytropic exponent, n, which is a function of the compressor. Efficient air compressors have polytropic exponents between 1.25 and 1.30, but values up to 1.35 are not uncommon. If the compression is known to be isentropic, the gas' ratio of specific heats, k, should be used in place of the polytropic exponent, n.

Ideal compression in an uncooled *dynamic compressor* is usually considered to be adiabatic (isentropic), with deviations from ideal performance accounted for by the *adiabatic (isentropic) compression efficiency*.[6] With intercooling (see Sec. 31.11), compression in dynamic compressors can also approach an isothermal process. Deviations from ideal isothermal performance are accounted for by the *isothermal compression efficiency*.

$$\eta_{\text{adiabatic}} = \frac{W_{\text{adiabatic}}}{W_{\text{actual}}} \quad 31.1$$

$$\eta_{\text{isothermal}} = \frac{W_{\text{isothermal}}}{W_{\text{actual}}} \quad 31.2$$

[5]The ASME Power Test Code defines "standard air" as air at a temperature of 68°F (20°C), a total pressure of 14.7 psia (101.3 kPa), and a relative humidity of 36%. The term *free air* is also used by manufacturers to describe the air demands of their equipment. Free air is air at atmospheric pressure and ambient temperature.

[6]Equation 31.1 is the definition of isentropic efficiency. For some reason, in the study of gas compression, the term *adiabatic* is used rather than the terms *reversible adiabatic* or *isentropic*.

Figure 31.1 shows that the work of an isothermal compression is less than the work of an adiabatic compression.[7] The work of a polytropic compression will be somewhere between the two.

Figure 31.1 *Comparison of Isothermal and Adiabatic Work (zero clearance)*

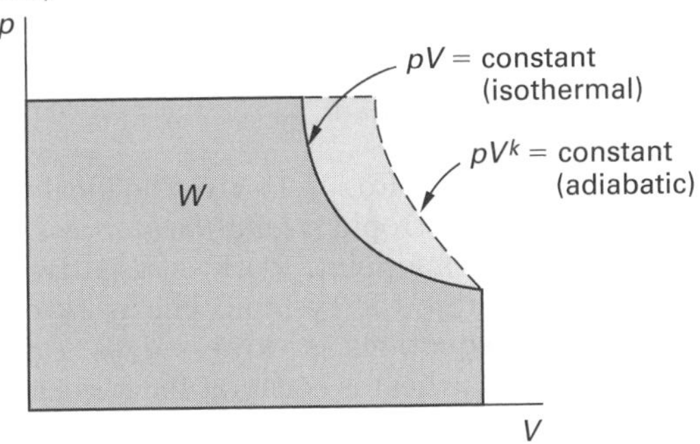

7. CLEARANCE AND CLEARANCE VOLUME

Reciprocating compressors are characterized by their *clearance volume*.[8] (See Fig. 31.2.) This is the volume (volume V_D in Fig. 31.3) at the head of the cylinder when the piston is at its most extended position in its stroke. (This position is known as *top dead center* or TDC.)[9] The gases remaining in the clearance volume after the discharge valve closes at top dead center are known as the *residual gases*.

Figure 31.2 *Clearance Volume*

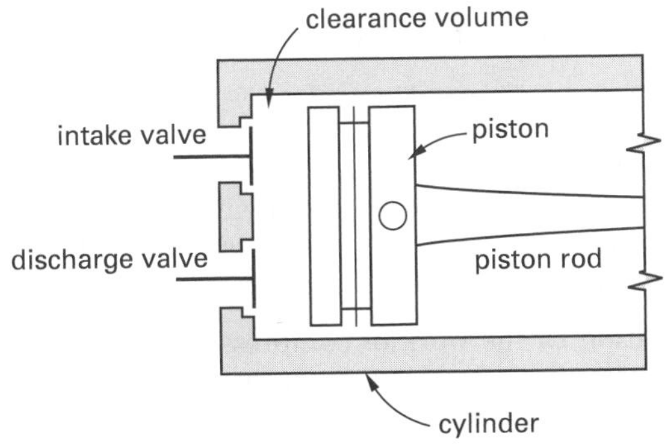

Figure 31.3 *Single-Stage Compression*

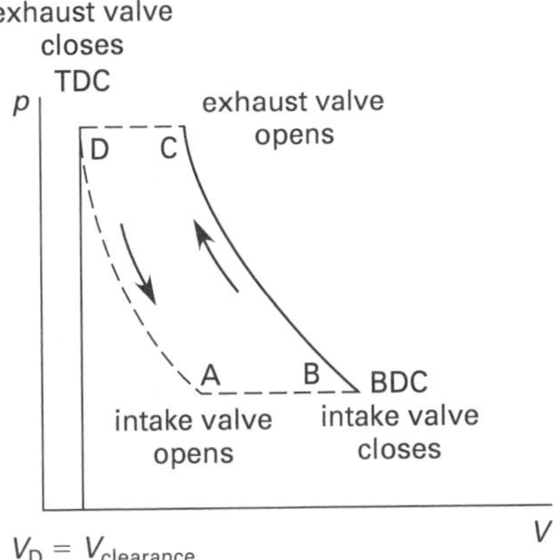

The *percent clearance* (also known as *clearance*), c, is calculated from the *swept volume* (*piston displacement*), $V_B - V_D$ in Eq. 31.3. Swept volume is not the total volume of the free gas.

$$\begin{aligned} c &= \frac{\text{clearance volume}}{\text{swept volume}} \times 100\% \\ &= \frac{V_D}{V_B - V_D} \times 100\% \end{aligned} \qquad 31.3$$

The residual gases expand along with the next intake of gas during the intake stroke to reduce the volumetric capacity per stroke. However, clearance affects only the volumetric efficiency; it does not affect the required input power. Two compressors with the same gas flow rate but with different clearances will require the same input power because the expanding residual gases give back their work of compression when they expand. The power requirement depends on only the mass of the gas passing through the compressor.

During steady-state compressor operation, the mass of gas entering the cylinder equals the mass of gas discharged, and clearance is not considered.

8. RECIPROCATING, SINGLE-STAGE COMPRESSION

The following processes describe a single stage of compression in a reciprocating compressor.

[7]Note this confusing point: The work done *in* an adiabatic (isentropic) *process* is less than the work done in an isothermal *process*, but (for the same pressure limits) the work done *by* an isentropic *compressor* is more than the work done by an isothermal *compressor*.
[8]Although there is clearance between the casing and blades of a dynamic compressor, the terms *clearance volume* and *clearance percent* always refer to reciprocating compressors.
[9]The piston is said to be at *bottom dead center* (BDC) when it is at its most retracted position in the stroke.

A to B: constant pressure suction (intake valve open, discharge valve closed)

B to C: polytropic compression (both valves closed)

C to D: constant pressure delivery (exhaust valve open, intake valve closed)

D to A: polytropic expansion (both valves closed)

The processes describing the compression do not constitute a true cycle because the gas mass in the B-to-C process is not the same as the gas mass in the C-to-D process. For that reason, the broken lines used in Fig. 31.3 show the cylinder volume, not the gas conditions.

The main parameters affecting performance are the polytropic exponent, n, compression ratio, r_p, and volumetric efficiency, η_v. For convenience, the polytropic exponents for the expansion and compression processes are assumed to be the same. (However, it is not difficult to consider different values.)

The *compression ratio* for reciprocating and rotating compressors, defined by Eq. 31.4, is the ratio of pressures (not the ratio of volumes as it is for internal combustion engines).[10]

$$r_p = \frac{p_D}{p_A} = \frac{p_C}{p_B} \quad \text{31.4}$$

The *volumetric efficiency*, is the ratio of the actual mass of gas compressed to the mass of gas in the swept volume. Volumetric efficiency and capacity are evaluated at the inlet conditions unless otherwise noted.

$$\eta_v = \left. \frac{\text{actual mass of gas compressed}}{\text{mass of gas in swept volume}} \right|_{\text{inlet conditions}} \quad \text{31.5}$$

$$\eta_v = \frac{V_B - V_A}{V_B - V_D} = 1 - \left(r_p^{1/n} - 1\right)\left(\frac{c}{100\%}\right) \quad \text{31.6}$$

The gas mass flow rate through the compressor can be determined from the mass per stroke and the number of strokes per minute. Assuming an ideal gas and evaluating the gas at the compressor's inlet conditions, the ideal and actual gas masses per stroke are

$$m_{\text{stroke,ideal}} = \frac{p_A(V_B - V_D)}{RT_A} \quad \text{31.7}$$

$$m_{\text{stroke,actual}} = \eta_v m_{\text{stroke,ideal}} \quad \text{31.8}$$

The mass flow rate is

$$\dot{m} = m_{\text{stroke,actual}} \times \text{rotational speed} \quad \text{31.9}$$

The relationships between the pressures and volumes are predicted by the ideal gas laws for polytropic processes.

$$\frac{V_A}{V_D} = \left(\frac{p_D}{p_A}\right)^{1/n} = r_p^{1/n} \quad \text{31.10}$$

$$\frac{V_B}{V_C} = \left(\frac{p_C}{p_B}\right)^{1/n} = r_p^{1/n} \quad \text{31.11}$$

Equation 31.12 through Eq. 31.14 give the ideal work of compression for a polytropic *steady-flow process*. (From a thermodynamic standpoint, work is negative if the surroundings compress the system. The work of compression in these equations is expressed as a positive number.) The specific heat is constant for an ideal gas.

$$\begin{aligned} w_{BC} &= h_C - h_B \\ &= c_p(T_C - T_B) \\ &= c_p(T_C - T_A) \end{aligned} \quad \text{31.12}$$

$$w_{BC} = \left(\frac{n}{1-n}\right)(p_C v_C - p_B v_B) \quad \text{31.13}$$

$$w_{BC} = \left(\frac{n p_B v_B}{1-n}\right)\left(\left(\frac{p_C}{p_B}\right)^{(n-1)/n} - 1\right) \quad \text{31.14}$$

During the D-to-A recovery stroke, the residual gases expand and do work on the piston. However, the gas mass has been greatly reduced, and the expansion work is disregarded. Therefore, the net work per cycle is

$$w_{\text{net}} = w_{BC} - w_{DA} \approx w_{BC} \quad \text{31.15}$$

"Work done by compressor" almost always means the change in energy imparted to the air. It does not mean the work supplied to the compressor (i.e., it does not include the compression efficiencies).

In addition to the work of compression, the compressor motor supplies power to overcome friction, discharge the gas against the receiver pressure, and move the piston during noncompression parts of the cycle. Equation 31.16 gives the *brake work*, which is expressed in brake horsepower (bhp) and brake kilowatts (BkW).

$$\text{brake work} = \frac{w_{\text{net}}}{\eta_m \eta_s} \quad \text{31.16}$$

Since brake work is expressed per unit mass, the power requirement is

$$P = \dot{m} \times \text{brake work} \quad \text{31.17}$$

[10]The term *ratio of compression*, r_c, is often used to distinguish the ratio of pressures from the *ratio of volumes*.

Typical single-stage compressors for common plant processes produce approximately 4 cfm to 5 cfm of 90 psig to 125 psig (2 L/s to 2.5 L/s of 620 kPa to 860 kPa) air per brake horsepower delivered. Single-stage pressure boosters starting with compressed plant air are highly efficient and can produce as much as 13 cfm (6.1 L/s) per brake horsepower.

Example 31.1

Air enters a reciprocating compressor at 14.7 psia and 70°F (101.3 kPa and 21°C). 100 ft³/min (50 L/s) of free air is compressed isentropically to 55 psia (379 kPa). The clearance is 6%. What is the volumetric efficiency?

SI Solution

Solve by using Eq. 31.6. (The customary U.S. solution takes a different approach.) Since the compression is isentropic, the polytropic exponent, n, is equal to the ratio of specific heats, k.

$$\begin{aligned}\eta_v &= 1-(r_p^{1/n}-1)\left(\frac{c}{100\%}\right)\\ &= 1-\left(\left(\frac{379\ \text{kPa}}{101.3\ \text{kPa}}\right)^{1/1.4}-1\right)\left(\frac{6\%}{100\%}\right)\\ &= 0.906\quad (90.6\%)\end{aligned}$$

Customary U.S. Solution

(The U.S. solution takes a different approach.) Refer to Fig. 31.3. Calculate the volumes per minute. Since the compression is isentropic, the polytropic exponent, n, is equal to the ratio of specific heats, k.

$$\begin{aligned}\frac{V_A}{V_D} &= \left(\frac{p_D}{p_A}\right)^{1/k} = \left(\frac{55\ \frac{\text{lbf}}{\text{in}^2}}{14.7\ \frac{\text{lbf}}{\text{in}^2}}\right)^{1/1.4}\\ &= 2.566\\ V_A &= 2.566\,V_D\end{aligned}$$

Since $p_C = p_D$ and $p_B = p_A$, $V_B/V_C = 2.566$ as well.

The actual volumetric flow rate entering the compressor each minute is 100 ft³ of free air.

$$\begin{aligned}V_B - V_A &= 100\ \text{ft}^3\\ V_B &= 100\ \text{ft}^3 + V_A\\ &= 100\ \text{ft}^3 + 2.566\,V_D\end{aligned}$$

The swept volume (piston displacement) is

$$\text{swept volume} = V_B - V_D$$

V_D is the clearance volume, which is 6% of the swept volume.

$$V_D = \left(\frac{c}{100\%}\right)(\text{swept volume}) = \left(\frac{6\%}{100\%}\right)(V_B - V_D)$$

Solving for V_B,

$$V_B = 17.667\,V_D$$

Combine the two equations for V_B.

$$\begin{aligned}17.667\,V_D &= 100\ \text{ft}^3 + 2.566\,V_D\\ V_D &= 6.622\ \text{ft}^3\\ V_B &= 17.667\,V_D = (17.667)(6.622\ \text{ft}^3)\\ &= 117.0\ \text{ft}^3\end{aligned}$$

The volumetric efficiency (calculated from volumes per minute) is

$$\begin{aligned}\eta_v &= \frac{V_B - V_A}{V_B - V_D}\\ &= \frac{100\ \text{ft}^3}{117.0\ \text{ft}^3 - 6.622\ \text{ft}^3}\\ &= 0.906\quad (90.6\%)\end{aligned}$$

9. NET WORK AND POLYTROPIC HEAD

The net work can be calculated per unit mass from Eq. 31.18. Z_{ave} is the compressibility factor averaged over the inlet and discharge conditions.

$$\begin{aligned}w_{net} = w_{ideal} &= \left(\frac{n}{n-1}\right)Z_{ave}R(T_1 - T_2)\\ &= \left(\frac{n}{n-1}\right)Z_{ave}RT_{inlet}\left(\left(\frac{p_2}{p_1}\right)^{(n-1)/n}-1\right)\\ &= \left(\frac{n}{n-1}\right)Z_{ave}RT_{inlet}\left(r_p^{(n-1)/n}-1\right)\end{aligned}\qquad 31.18$$

When the net work calculated in Eq. 31.15 or Eq. 31.18 is expressed in feet (numerically equivalent to ft-lbf/lbm) or meters, it may be referred to as the *polytropic head* or (if the process is adiabatic) as the *adiabatic head*. While the discharge pressure depends on the gas, the head does not.

10. POLYTROPIC EFFICIENCY

The *polytropic efficiency*, η_n, is the ratio of ideal polytropic compression work to actual compression work. (For isentropic processes, the polytropic efficiency is 100%.) H is the polytropic head added by the compressor.

$$\eta_n = \frac{w_{\text{ideal}}}{w_{\text{actual}}} = \frac{\dot{m}H}{P_{\text{actual}}} \quad \text{31.19}$$

The polytropic efficiency can also be calculated from the polytropic exponent, n.

$$\frac{n}{n-1} = \frac{\eta_n k}{k-1} \quad \text{31.20}$$

Example 31.2

A dynamic compressor receives 50,000 cfm (23.5 kL/s) of a gas at 15 psia (104 kPa) and 70°F (21°C). The pressure is increased by the compressor to 85 psia (590 kPa). The gas has a molecular weight of 31 and a ratio of specific heats of 1.17. The compressibility factor at the inlet is 0.99, and the average compressibility factor over the compression process is 0.98. The actual shaft power is 9225 hp (6880 kW). 183 hp (136 kW) are required to overcome friction and windage losses. What are the (a) theoretical compression power, (b) polytropic head, and (c) polytropic efficiency?

SI Solution

The absolute inlet temperature is

$$T_{\text{inlet}} = 21°\text{C} + 273° = 294\text{K}$$

The compression ratio is

$$r_p = \frac{p_{\text{discharge}}}{p_{\text{inlet}}} = \frac{590 \text{ kPa}}{104 \text{ kPa}} = 5.673$$

The specific gas constant is

$$R = \frac{\overline{R}}{M} = \frac{8314 \ \frac{\text{J}}{\text{kmol·K}}}{31 \ \frac{\text{kg}}{\text{kmol}}} = 268.2 \text{ J/kg·K}$$

The inlet density is

$$\rho_{\text{inlet}} = \frac{p_{\text{inlet}}}{ZRT_{\text{inlet}}} = \frac{(104 \text{ kPa})\left(1000 \ \frac{\text{Pa}}{\text{kPa}}\right)}{(0.99)\left(268.2 \ \frac{\text{J}}{\text{kg·K}}\right)(294\text{K})} = 1.332 \text{ kg/m}^3$$

The mass flow rate is

$$\dot{m} = Q\rho = \left(23.5 \ \frac{\text{kL}}{\text{s}}\right)\left(1 \ \frac{\text{kL}}{\text{m}^3}\right)\left(1.332 \ \frac{\text{kg}}{\text{m}^3}\right) = 31.3 \text{ kg/s}$$

This process is not known to be adiabatic or isentropic. The polytropic exponent is not known. The ideal polytropic work cannot be calculated directly from Eq. 31.18. An iterative process is needed. Assume the polytropic compression efficiency, η_n, is 80%. Use Eq. 31.18 and Eq. 31.20 to determine the theoretical polytropic head for this iteration using the assumed efficiency.

$$\frac{n}{n-1} = \frac{\eta_n k}{k-1} = \frac{(0.80)(1.17)}{1.17-1} = 5.506$$

$$\begin{aligned} w_{\text{net}} &= \left(\frac{n}{n-1}\right) Z_{\text{ave}} R T_{\text{inlet}} \left(r_p^{(n-1)/n} - 1\right) \\ &= (5.506)(0.98)\left(268.2 \ \frac{\text{J}}{\text{kg·K}}\right)(294\text{K}) \times \left((5.673)^{1/5.506} - 1\right) \\ &= 1.577 \times 10^5 \text{ J/kg} \end{aligned}$$

$$H = \frac{w_{\text{net}}}{g} = \frac{1.577 \times 10^5 \ \frac{\text{J}}{\text{kg}}}{9.81 \ \frac{\text{m}}{\text{s}^2}} = 16\,075 \text{ m}$$

The theoretical power for this iteration is

$$P_{\text{theoretical}} = \frac{P_{\text{ideal}}}{\eta_n} = \frac{\dot{m} w_{\text{net}}}{\eta_n} = \frac{\left(31.3 \ \frac{\text{kg}}{\text{s}}\right)\left(1.577 \times 10^5 \ \frac{\text{J}}{\text{kg}}\right)}{(0.80)\left(1000 \ \frac{\text{W}}{\text{kW}}\right)} = 6170 \text{ kW}$$

The actual compression power is known to be

$$P_{\text{actual}} = P_{\text{shaft}} - P_{\text{friction}} = 6880 \text{ kW} - 136 \text{ kW} = 6744 \text{ kW}$$

Since the actual power is higher, the assumed efficiency is too high.

For the next iteration, use an assumed efficiency of

$$\eta_{n,2} = \eta_{n,1}\left(\frac{P_{\text{theoretical}}}{P_{\text{actual}}}\right) = (0.80)\left(\frac{6170 \text{ kW}}{6744 \text{ kW}}\right)$$
$$= 0.732$$

Customary U.S. Solution

The absolute inlet temperature is

$$T_{\text{inlet}} = 70°\text{F} + 460° = 530°\text{R}$$

The compression ratio is

$$r_p = \frac{p_{\text{discharge}}}{p_{\text{inlet}}} = \frac{85 \text{ psia}}{15 \text{ psia}}$$
$$= 5.667$$

The specific gas constant is

$$R = \frac{\overline{R}}{M} = \frac{1545 \frac{\text{ft-lbf}}{\text{lbmol-°R}}}{31 \frac{\text{lbm}}{\text{lbmol}}}$$
$$= 49.84 \text{ ft-lbf/lbm-°R}$$

The inlet density is

$$\rho_{\text{inlet}} = \frac{p_{\text{inlet}}}{ZRT_{\text{inlet}}}$$
$$= \frac{\left(15 \frac{\text{lbf}}{\text{in}^2}\right)\left(12 \frac{\text{in}}{\text{ft}}\right)^2}{(0.99)\left(49.84 \frac{\text{ft-lbf}}{\text{lbm-°R}}\right)(530°\text{R})}$$
$$= 0.0826 \text{ lbm/ft}^3$$

The mass flow rate is

$$\dot{m} = Q\rho$$
$$= \left(50{,}000 \frac{\text{ft}^3}{\text{min}}\right)\left(0.0826 \frac{\text{lbm}}{\text{ft}^3}\right)$$
$$= 4130 \text{ lbm/min}$$

This process is not known to be adiabatic or isentropic. The polytropic exponent is not known. The ideal polytropic work cannot be calculated directly from Eq. 31.18. An iterative process is needed. Assume the polytropic compression efficiency, η_n, is 80%. Use Eq. 31.18 and Eq. 31.20 to determine the theoretical polytropic head for this iteration using the assumed efficiency.

$$\frac{n}{n-1} = \frac{\eta_n k}{k-1} = \frac{(0.80)(1.17)}{1.17-1}$$
$$= 5.506$$

$$w_{\text{net}} = \left(\frac{n}{n-1}\right)Z_{\text{ave}}RT_{\text{inlet}}(r_p^{(n-1)/n} - 1)$$
$$= (5.506)(0.98)\left(49.84 \frac{\text{ft-lbf}}{\text{lbm-°R}}\right)(530°\text{R}) \times \left((5.667)^{1/5.506} - 1\right)$$
$$= 52{,}784 \text{ ft-lbf/lbm}$$
$$H = 52{,}784 \text{ ft}$$

The theoretical horsepower for this iteration is

$$P_{\text{theoretical}} = \frac{P_{\text{ideal}}}{\eta_n} = \frac{\dot{m}w_{\text{net}}}{\eta_n}$$
$$= \frac{\left(4130 \frac{\text{lbm}}{\text{min}}\right)\left(52{,}784 \frac{\text{ft-lbf}}{\text{lbm}}\right)}{(0.80)\left(33{,}000 \frac{\text{ft-lbf}}{\text{hp-min}}\right)}$$
$$= 8257 \text{ hp}$$

The actual compression power is known to be

$$P_{\text{actual}} = P_{\text{shaft}} - P_{\text{friction}} = 9225 \text{ hp} - 183 \text{ hp}$$
$$= 9042 \text{ hp}$$

Since the actual power is higher, the assumed efficiency is too high.

For the next iteration, use an assumed efficiency of

$$\eta_{n,2} = \eta_{n,1}\left(\frac{P_{\text{theoretical}}}{P_{\text{actual}}}\right) = (0.80)\left(\frac{8257 \text{ hp}}{9042 \text{ hp}}\right)$$
$$= 0.731$$

After two additional iterations, the values are essentially stable.

$$P_{\text{theoretical}} = 9000 \text{ hp} \quad [9042 \text{ hp ideally}]$$
$$H = 53{,}460 \text{ ft}$$
$$\eta_n = 0.741$$

11. COOLING AND INTERCOOLING

For a given compression ratio, the work for isothermal compression is less than the work for adiabatic compression.[11] Furthermore, standard hydrocarbon lubricants used in air compressors are limited to approximately 365°F (185°C). Therefore, cooling is often used in compressors whose compression ratio exceeds 4.0. (A third benefit from cooling is a reduction in ring and valve loading.) For reciprocating compressors, the cooling is accomplished by surrounding the cylinder with a water jacket or by incorporating finned heat radiators in the jacket design.

Where a partially compressed gas is withdrawn, cooled, and compressed further, the term *intercooling* is used. Intercoolers are typically used with centrifugal compressors. The term *perfect intercooling* refers to the case where the gas is cooled to the original inlet temperature (i.e., $T_B = T_D$ in Fig. 31.4).[12]

If necessary, an *aftercooler* removes heat (and reduces the pressure) from the compressed gas after the compression process is complete.

Compressed air for plant use usually leaves the aftercooler at 25°F to 75°F (15°C to 45°C) above the ambient temperature.

Figure 31.4 *Two-Stage Compression with Intercooling*

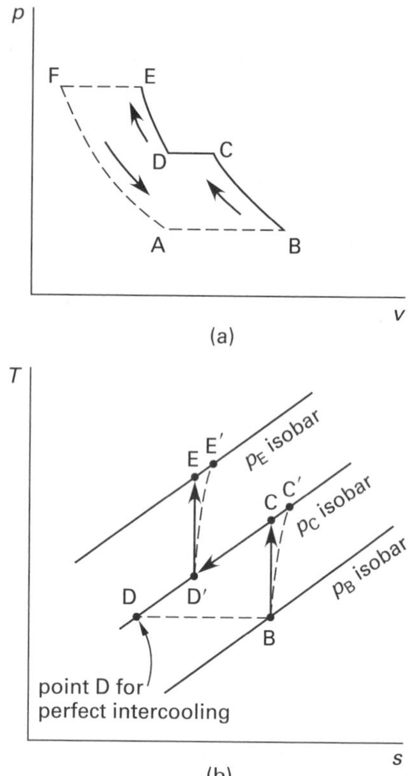

12. MULTISTAGE COMPRESSION

Figure 31.4 shows the path of a gas experiencing two-stage compression and intercooling. For a reciprocating compressor, the curved solid lines represent polytropic compressions, and the broken lines represent the cylinder volume. For a centrifugal compressor, the curved solid lines represent isentropic (or isothermal) compressions and the broken lines have no meaning.

Analysis of a multistage compressor is similar to that of a single-stage compressor. The minimum work occurs when the compression ratios of all stages are the same. This is known as *optimum staging*. In that case,

$$r_p = \frac{p_C}{p_B} = \frac{p_E}{p_D} \quad \text{31.21}$$

Since $p_C = p_D$,

$$p_C^2 = p_E p_B \quad \text{31.22}$$

Typical water-cooled two-stage compressors for common plant processes produce approximately 2 cfm to 3 cfm of 200 psig to 500 psig (1 to 1.5 L/s of 1.4 to 3.5 MPa) air per brake horsepower delivered.

13. DYNAMIC COMPRESSORS

The ideal gas relationships for adiabatic and isothermal processes can be used to analyze the performance of dynamic gas compressors. In fact, most of the equations in Sec. 31.8 can be used for isentropic processes if the ratio of specific heats, k, is used in place of the polytropic exponent, n.

For adiabatic compression, power can be calculated with Eq. 31.23.

Compressors

$$P_{\text{comp}} = \left(\frac{\dot{m}p_i k}{(k-1)\rho_i \eta_c}\right)\left(\left(\frac{p_e}{p_i}\right)^{1-(1/k)} - 1\right) \quad \text{31.23}$$

For isothermal compression, power can be calculated with Eq. 31.24.

Compressors

$$P_{\text{comp}} = \left(\frac{\overline{R}\,T_i}{M\eta_c}\ln\frac{p_e}{p_i}\right)\dot{m} \quad \text{31.24}$$

[11] In practice, the additional cost of the intercooler apparatus must be compared with the power savings.

[12] The discharge from an intercooler will usually be approximately 20°F (10°C) higher than the jacket water temperature.

Example 31.3

A turbojet is designed to operate at 35,000 ft and a Mach number of 0.75. At altitude, the compressor efficiency is 88%, the compression ratio is 8, and the inlet pressure and area are 100 psi and 2 ft^2, respectively. The density of air at 35,000 ft is 0.0237 lbm/ft^3, which increases to 0.6839 lbm/ft^3 at the compressor inlet. What is the horsepower generated by the turbojet's compressor?

Solution

Calculate the air temperature at 35,000 ft.

Temperature and Altitude Corrections for Air

$$\begin{aligned} T &= 59 - 0.00356620Z \\ &= 59°\text{F} - \left(0.00356620\ \frac{°\text{F}}{\text{ft}}\right)(35{,}000\ \text{ft}) \\ &= -65.82°\text{F} \end{aligned}$$

Convert to absolute temperature. [**Temperature Conversions**]

$$-65.82°\text{F} + 460° = 394.2°\text{R}$$

Find the speed of sound at 35,000 ft. For air, the ratio of specific heats, k, is 1.4, and the specific gas constant, R, is 53.35 ft-lbf/lbm-°R. [**Thermal and Physical Properties of Ideal Gases (at Room Temperature)**]

Mach Number

$$\begin{aligned} c &= \sqrt{kg_cRT} \\ &= \sqrt{(1.4)\left(32.17\ \frac{\text{lbm-ft}}{\text{lbf-sec}^2}\right)\left(53.35\ \frac{\text{ft-lbf}}{\text{lbm-°R}}\right)(394.2°\text{R})} \\ &= 973.2\ \text{ft/sec} \end{aligned}$$

Use the equation for Mach number to find the velocity of the air through the turbojet's compressor inlet section.

Mach Number

$$\text{Ma} = \frac{\text{v}}{c}$$

$$\begin{aligned} \text{v} = (\text{Ma})c &= (0.75)\left(973.2\ \frac{\text{ft}}{\text{sec}}\right) \\ &= 729.9\ \text{ft/sec} \end{aligned}$$

Use the continuity equation to find the mass flow rate of air into the turbojet's compressor inlet.

Continuity Equation

$$\begin{aligned} \dot{m} &= \rho A\text{v} \\ &= \left(0.0237\ \frac{\text{lbm}}{\text{ft}^3}\right)(2\ \text{ft}^2)\left(729.9\ \frac{\text{ft}}{\text{sec}}\right) \\ &= 34.60\ \text{lbm/sec} \end{aligned}$$

Use Eq. 31.23 to calculate the power generated by the turbojet's compressor. From the problem statement, the compressor efficiency is 88%, and the compression ratio, p_e/p_i, is 8.

Compressors

$$\begin{aligned} P_{\text{comp}} &= \left(\frac{\dot{m}p_ik}{(k-1)\rho_i\eta_c}\right)\left(\left(\frac{p_e}{p_i}\right)^{1-(1/k)} - 1\right) \\ &= \left(\frac{\left(34.60\ \frac{\text{lbm}}{\text{sec}}\right)\left(\left(100\ \frac{\text{lbf}}{\text{in}^2}\right)\left(12\ \frac{\text{in}}{\text{ft}}\right)^2\right)(1.4)}{(1.4-1)\left(0.6839\ \frac{\text{lbm}}{\text{ft}^3}\right)(0.88)}\right) \\ &\quad \times \left((8)^{1-(1/1.4)} - 1\right) \\ &= 2.35\times10^6\ \text{lbf/sec} \end{aligned}$$

Convert to horsepower. [**Measurement Relationships**]

$$\frac{2.35\times10^6\ \frac{\text{lbf}}{\text{sec}}}{550\ \frac{\text{ft-lbf}}{\text{hp-sec}}} = 4275\ \text{hp}$$

14. COMPRESSOR CHARACTERISTIC CURVES

Performance of compressors, like that of pumps and fans, is described by characteristic curves. Compressor manufacturers usually provide curves of discharge pressure versus inlet volumetric flow rate and brake power versus flow rate. Alternatively, head versus flow rate or pressure ratio versus flow rate may be substituted for the discharge pressure curve.

Characteristic curves are different for reciprocating and centrifugal compressors. Reciprocating compressors are essentially constant-volume, variable-pressure devices. Since their capacities are essentially fixed, the characteristic curve is quite steep. This is shown in Fig. 31.5.

Figure 31.5 illustrates the characteristic curve for a dynamic compressor. It is similar to curves for pumps and fans. The *surge limit* is the leftmost area on the curve. In surge, flow becomes unstable and the compressor pressure intermittently drops below the system pressure, resulting in backflow.[13] Figure 31.5 also illustrates that for any given speed, even with decreasing discharge pressure, the flow rate does not increase indefinitely. Flow approaches a maximum flow rate at the *choke* or *stone wall point*.

Figure 31.5 *Dynamic Compressor Head-Capacity Curves*

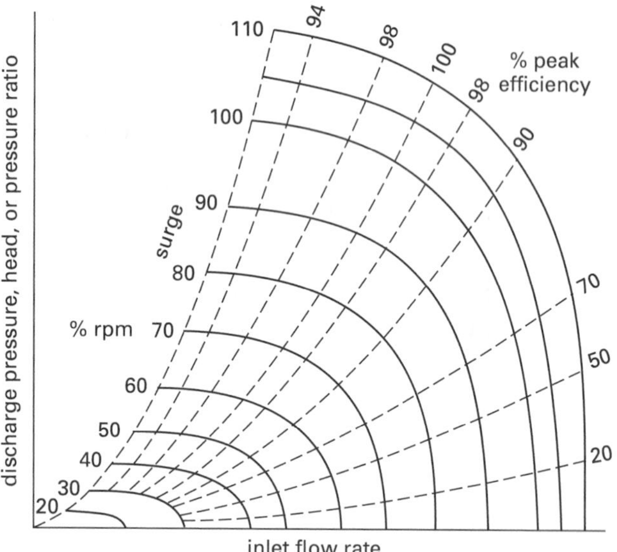

The system curve (not shown) will intersect the characteristic curve at the operating point, as it does with pumps and fans. Ideally, the *operating point* should coincide with the maximum efficiency line. Since each stage of compression will have its own characteristic curve, there will be as many operating points as there are stages. In a properly matched multistage compressor, all of the stages will be operating at high efficiencies. If the stages are not well matched or if the flow rate or inlet pressure change, one or more of the stages may operate "off" its design point. In the worst cases, a stage may be in a surge or choke condition.

15. OPERATION AT CHANGED INLET CONDITIONS

Discharge pressure and power curves can only be used for the gas and inlet conditions (typically, "standard" inlet conditions) intended. Those curves cannot be used if the gas or any of the inlet conditions are changed. In most cases, curves applicable for the new conditions are not available. In those cases, a curve for head versus inlet flow rate can be laboriously generated for the new conditions.[14]

Ideally, energy transfer (head) per unit mass is a function of only the velocity of the impeller. Head is the same for all gases, regardless of density. Therefore, while the gas properties affect the discharge pressure, they do not affect the head. Moving a compressor to another altitude will affect the discharge pressure, mass flow rate, and power requirement. It will not affect the ideal compression ratio or the volumetric flow rate.

In reality, compressors (particularly piston compressors) do not deliver as much flow at higher altitudes as they do at sea level. As the air viscosity decreases, leakage through compressor clearances increases. The effect is known as *droop*. (See Fig. 31.6.) The disparity between sea-level and high-altitude performance becomes more pronounced at higher discharge pressures.

Figure 31.6 *Reciprocating Compressor Droop*

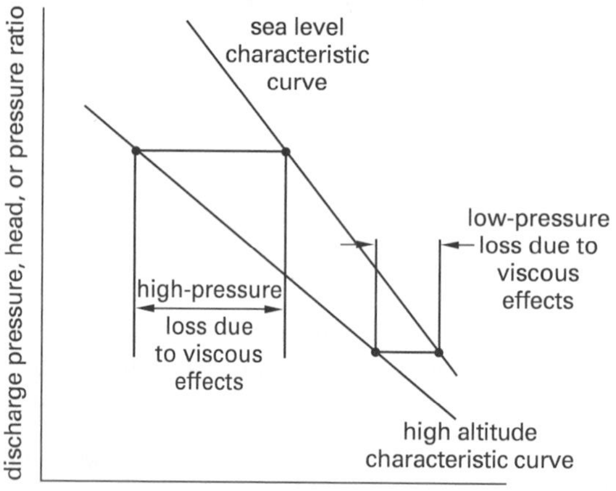

[13]Dynamic compressors should have antisurge controls to open a recirculation valve. These increase the flow sufficiently to keep operation out of surge.

[14]Generating the head-flow rate curve is not as easy as simply applying Eq. 31.18. The polytropic exponent, n, must simultaneously satisfy Eq. 31.18, Eq. 31.19, and Eq. 31.20. The actual compression power is taken from the existing power-flow rate curve provided by the manufacturer. Therefore, calculating even a single point on the head-flow rate curve is an iterative process. This is illustrated in Ex. 31.2.

16. NOMENCLATURE

A	area	ft^2	m^2
c	clearance	percent	percent
c	speed of sound	ft/sec	m/s
c_p	specific heat at constant pressure	Btu/lbm-°F	kJ/kg·K
g	gravitation acceleration, 32.17 (9.807)	ft/sec^2	m/s^2
g_c	gravitational constant, 32.17	lbm-ft/lbf-sec^2	n.a.
h	specific enthalpy	Btu/lbm	kJ/kg
H	head added	ft	m
k	ratio of specific heats	–	–
m	mass	lbm	kg
$\dot{m}$	mass flow rate	lbm/sec	kg/s
M	molecular weight	lbm/lbmol	kg/kmol
n	polytropic exponent	–	–
p	pressure	lbf/ft^2	Pa
P	power	Btu/lbm-sec	W/kg
Q	volumetric flow rate	ft^3/sec	m^3/s
r	ratio	–	–
R	specific gas constant	ft-lbf/lbm-°R	kJ/kg·K
$\bar{R}$	universal gas constant	ft-lbf/lbmol-°R	kJ/kmol·K
s	specific entropy	Btu/lbm-°F	kJ/kg·K
T	temperature	°R	K
v	specific volume	ft^3/lbm	m^3/kg
v	velocity	ft/sec	m/s
V	volume	ft^3	m^3
w	work per unit mass	Btu/lbm	kJ/kg
W	work	Btu	kJ
Z	altitude	ft	m
Z	compressibility factor	–	–

Symbols

η	efficiency	–	–
ρ	density	lbm/ft^3	kg/m^3

Subscripts

ave	average
c	compression
comp	compressor
e	exit
i	inlet
m	mechanical
n	polytropic
p	pressure
s	isentropic
v	volumetric

Power Cycles

32 Refrigeration Cycles

Content in blue refers to the *NCEES Handbook*.

NCEES EXAM SPECIFICATIONS AND RELATED CONTENT

HVAC AND REFRIGERATION EXAM

I.B.1. Thermodynamics: Cycles
- 4. Coefficient of Performance
- 5. Energy Efficiency Ratio
- 6. Refrigeration Capacity
- 7. Carnot Refrigeration Cycle
- 8. Vapor Compression Cycle
- 9. Air Refrigeration Cycle
- 11. Absorption Cycle

II.B.4. Equipment and Components: Condensers/evaporators
- 13. Low-Side Equipment

II.B.8. Equipment and Components: Refrigerants
- 2. Refrigerants
- 6. Refrigeration Capacity
- 14. Refrigerant Line Sizing

THERMAL AND FLUID SYSTEMS EXAM

I.E.2. Thermodynamics: Thermodynamic Cycles
- 4. Coefficient of Performance
- 5. Energy Efficiency Ratio
- 6. Refrigeration Capacity
- 7. Carnot Refrigeration Cycle
- 8. Vapor Compression Cycle
- 9. Air Refrigeration Cycle
- 11. Absorption Cycle

1. INTRODUCTION TO REFRIGERATION

Refrigeration is the process of transferring heat from a low-temperature area to a high-temperature area. Since heat flows spontaneously only from high- to low-temperature areas (according to the second law of thermodynamics), refrigeration needs an external energy source to force the heat transfer to occur. This energy source is a pump or compressor that does work in compressing the refrigerant. (See Fig. 32.1.) It is necessary to perform this work on the refrigerant in order to get it to discharge energy to the high-temperature area.

Figure 32.1 *Device Operating on a Refrigeration Cycle*

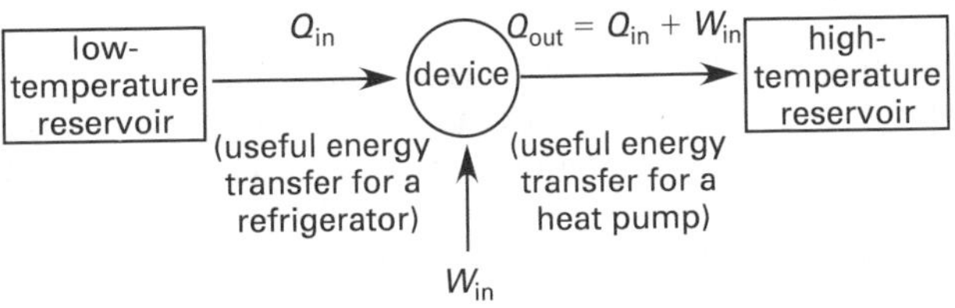

In a power cycle, heat from combustion is the input and work is the desired effect. Refrigeration cycles, though, are power cycles in reverse; work is the input and cooling is the desired effect. (For every power cycle, there is a corresponding refrigeration cycle.) In a refrigerator, the heat is absorbed from a low-temperature area and is rejected to a high-temperature area.[1] The pump work is also rejected to the high-temperature area.

General refrigeration devices consist of a coil (the *evaporator*) that absorbs heat, a *condenser* that rejects heat, a compressor, and a pressure-reduction device (the *expansion valve* or *throttling valve*).[2] (See Fig. 32.2.)

In operation, liquid refrigerant passes through the evaporator where it picks up heat from the low-temperature area and vaporizes, becoming slightly superheated. The vaporized refrigerant, that is, the "suction gas," is compressed by the compressor and, in so doing, increases even more in

[1]It is common thermodynamic jargon to refer to the low- and high-temperature areas as *environments* or *thermal reservoirs*. In particular, a low-temperature area is called a *source*, and a high-temperature area is referred to as a *sink*.

[2]In home refrigerators, the *expansion valve* takes the form of a long capillary tube. Other components include a motor to drive the compressor, a fan for the evaporator, a fan for the condenser, and the refrigerant itself.

Figure 32.2 *Refrigeration Device*

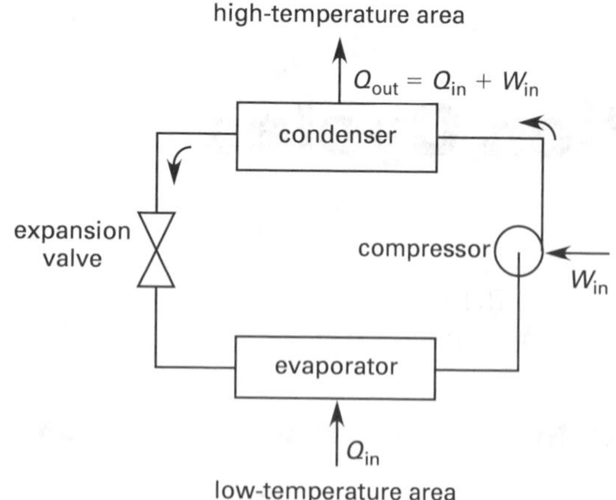

temperature. The high-pressure, high-temperature refrigerant passes through the condenser coils, and, being hotter than the high-temperature environment, loses energy. Finally, the pressure is reduced in the expansion valve, where some of the liquid refrigerant also flashes into a vapor.

If the low-temperature area from which the heat is being removed is air from occupied space (that is, air is being cooled), the device is known as an *air conditioner*. If the heat is being removed from water, the device is known as a *chiller*. An air conditioner produces cold air; a chiller produces cold water.

In small refrigeration systems, such as those used for in-window home air conditioning, the compressor, condenser, evaporator (cooling) coil, and fan are combined into a single housing. These are referred to as *unit* (or *unitary*) *air conditioners*. In large commercial systems, the compressor is separate from the evaporation coil.

2. REFRIGERANTS

There are more than three hundred commercial refrigerants commonly available. The refrigerant used depends on the temperatures of the low- and high-temperature areas, as well as the cooling load and the compressor power. *ASHRAE Handbook—Fundamentals* provides a table of comparative refrigerant performance and detailed safety information for some of the more common refrigerants. **[Comparative Refrigerant Performance per Ton of Refrigeration] [Refrigerant Data and Safety Classifications]**

Prior to the 1987 Montreal Protocol, chlorofluorocarbon (CFC) refrigerants R-11 and R-12, and hydrochlorofluorocarbon (HCFC) R-22, in pure form or in blends, were the economical and efficient choices for residential, automotive, appliance, and commercial applications.[3] All these formulations are now subject to production and use limitations or outright bans, with replacements coming from the hydrofluorocarbon (HFC) families, most notably HFC-134a (R-134a), HFC-407C, and HFC-410A. (HFC-141b is used as a blowing agent, not a refrigerant.) Production of all HCFCs will cease in 2030. Originally concerned only with ozone depletion potential (ODP), production and use limitations for CFCs and HCFCs have been extended even to HFCs due to an awareness of *global warming potential* (GWP). For example, due to its modest GWP, the sale of R-134a for refrigeration use has been banned in some states and countries. However, these restrictions are not specifically related to the Montreal Protocol. There are currently no restrictions on equipment or use of R-134a, R-407C, R410A, and R-417A.

R-11 was used in large (plant) centrifugal compressors. Worldwide production stopped in 1995.[4] Although there is no direct substitute (without incurring a performance loss), R-11 has been replaced by HFC-123 and, to a lesser extent, HFC-134a (R-134a).

R-12 was used in small- to medium-sized centrifugal compressors up to about 100 tons. It was commonly used in automotive air conditioning. Worldwide production stopped in 1995. It has been replaced for automotive use with R-134a. R-423A and R-437A are other replacements.

R-22 was the refrigerant "of choice" in unitary air conditioners (i.e., residential window units), chillers, freezers, residential air conditioning and heat pump units, and commercial and transport refrigeration and cooling. Under the Clean Air Act, R-22 may still be produced in the United States until 2020, but only for the purpose of servicing existing equipment; after the first day of 2020, it may not be produced or imported. R-410A is a replacement, as are R-407C, R-417A, R-422A, R-422D, R-500, and R-502.

R-114 was the most commonly used refrigerant for high-temperature coolers and heat pumps. As with R-11, because of the diversity of applications, there is no single replacement for R-114.

Ammonia was used in early residential refrigerators, but because it is toxic, it was replaced in the 1930s by Dupont's Freon R-12. However, ammonia is still used when extremely low temperatures are required and toxicity concerns are minimal, such as in industrial freezing and storage equipment, marine cargo ships, and ice rinks. It has potential, also, as a replacement for R-22.

The actual schedule for implementing production and use restrictions on these refrigerants is complicated by several factors. Developed countries have earlier compliance dates than developing countries. All HCFCs are grouped together in the phase-out schedule. Limitations

[3]Refrigerants often have a number of designations and trade names. R-22, for example, may be referred to as HCFC-22, Freon-22, and SUVA-22.

[4]R-11 and R-22 are still available for servicing equipment already in use, with existing inventory being reclaimed refrigerant recovered from "scrapped" equipment.

are based on mass production, so a country can phase out one compound (e.g., a blowing agent) more quickly in order to continue producing another compound (e.g., a refrigerant). Compounds remain available indefinitely for servicing older units, as long as stocks of refrigerant come from reclaimed equipment.

Depending on the operating conditions, backfilling older equipment with replacement refrigerants (e.g., substituting R-134a for R-12) will generally result in a 5% to 30% decrease in cooling capacities. If this reduction is unacceptable, performance can be restored by (1) substituting a more efficient impeller, (2) increasing the compressor's rotational speed, and (3) substituting a more efficient heat exchanger and tubing.

3. HEAT PUMPS

Heat pumps also operate on refrigeration cycles. Like standard refrigerators, they transfer heat from low-temperature areas to high-temperature areas. The device shown in Fig. 32.1 could represent either a heat pump or a refrigerator. There is no significant difference in the mechanisms or construction of heat pumps and refrigerators. The only difference is their purpose.

A refrigerator's main function is to cool the low-temperature area. The *useful energy transfer* for a refrigerator is the heat removed from the cold area. A heat pump's main function is to warm the high-temperature area. The useful energy transfer is the heat rejected to the high-temperature area. A heat pump is almost always used when *space heating* of an occupied area is needed. The attraction of *heat pump devices* is that they can provide both heating in the winter and cooling in the summer. During the summer, however, they run refrigeration cycles and are technically refrigerators. Strictly speaking, *heat pump cycles* are used only for space heating.

4. COEFFICIENT OF PERFORMANCE

The concept of thermal efficiency is not used with devices operating on refrigeration cycles. Rather, the *coefficient of performance* (COP) is defined as the ratio of useful energy transfer (see also Sec. 32.3) to the work input. The higher the coefficient of performance, the greater the effect will be for a given work input. Since the useful energy transfer is different for refrigerators and heat pumps, the coefficients of performance will also be different. [Basic Cycles]

In calculating coefficients of performance, the refrigerant is considered to be the system. Therefore, $\dot{Q}_L$ is the energy that enters the refrigerant. ($\dot{Q}_L$ is not the energy that enters the high-temperature area because that area is not the system.)

Basic Cycles

$$\begin{aligned}\text{COP}_{\text{refrigerator}} &= \frac{\dot{Q}_L}{\dot{W}}\\ &= \frac{\dot{Q}_L}{\dot{Q}_H - \dot{Q}_L}\\ &= \text{COP}_{\text{heat pump}} - 1\end{aligned} \tag{32.1}$$

The coefficient of performance for a heat pump includes the desired heating effect of the pump work input.

Basic Cycles

$$\begin{aligned}\text{COP}_{\text{heat pump}} &= \frac{\dot{Q}_H}{\dot{W}}\\ &= \frac{\dot{Q}_L + \dot{W}}{\dot{W}} = \frac{\dot{Q}_L + \dot{W}}{\dot{Q}_H - \dot{Q}_L}\\ &= \text{COP}_{\text{refrigerator}} + 1\end{aligned} \tag{32.2}$$

The upper limit of the coefficient of performance for both a refrigerator and a heat pump is based on the reversed Carnot cycle. T_L represents the absolute temperature of the low-temperature area, and T_H represents the absolute temperature of the high-temperature area.

Basic Cycles

$$\text{COP}_{c,\text{refrigerator}} = \frac{T_L}{T_H - T_L} \tag{32.3}$$

$$\text{COP}_{c,\text{heat pump}} = \frac{T_H}{T_H - T_L} \tag{32.4}$$

5. ENERGY EFFICIENCY RATIO

The *energy efficiency ratio* (EER) is defined as the useful energy transfer in Btu / hr divided by input power in watts. This is just the coefficient of performance expressed in mixed units.

$$\text{EER} = 3.412 \times \text{COP} \tag{32.5}$$

As with the coefficient of performance, the definition of useful energy transfer depends on whether the device is being used as a refrigerator or as a heat pump.

Efficiency

$$\begin{aligned}\text{EER}_{\text{refrigerator}} &= \frac{\text{output cooling energy (Btu/hr)}}{\text{input electrical energy (W)}}\\ &= \frac{\dot{Q}_{L,\text{Btu/hr}}}{\dot{W}_{\text{watts}}}\\ &= \text{EER}_{\text{heat pump}} - 3.412\end{aligned} \tag{32.6}$$

Power Cycles

$$\begin{aligned} \text{EER}_{\text{heat pump}} &= \frac{\text{output heating energy}}{\text{input electrical energy}} \\ &= \frac{(\dot{Q}_L + \dot{W})_{\text{Btu/hr}}}{\dot{W}_{\text{watts}}} \\ &= \text{EER}_{\text{refrigerator}} + 3.412 \end{aligned} \quad 32.7$$

6. REFRIGERATION CAPACITY

The rate of energy removal from the low-temperature area is known as the *refrigeration capacity* or *refrigeration effect*. While the kilowatt is the appropriate SI unit, in the United States capacity is measured in *refrigeration tons*, where one ton is equal to 200 Btu/min or 12,000 Btu/hr (3.517 kW) of heat removal. The ton is derived from the heat flow required to melt one ton of ice in 24 hours.

Since refrigeration capacity has the units of power (e.g., Btu/hr), it can be combined with the coefficient of performance to calculate the required pump horsepower.

$$\begin{aligned} \text{pump horsepower} &= \frac{4.715\dot{Q}_{\text{in,tons}}}{\text{COP}} \\ &= \frac{4.715\dot{Q}_{\text{in,Btu/hr}}}{12{,}000(\text{COP})} \end{aligned} \quad 32.8$$

The rate of energy removal can be used to calculate the refrigerant mass flow rate. *ASHRAE Handbook—Refrigeration* provides detailed tables for refrigerants R-22 and R-134a showing flow rates per ton of refrigerant and discharge and liquid line capacities. [**Flow Rate Per Ton of Refrigeration for Refrigerant 22**] [**Discharge and Liquid Line Capacities, in Tons, for Refrigerant 22**] [**Flow Rate Per Ton of Refrigeration for Refrigerant 134a**] [**Discharge and Liquid Line Capacities in Tons for Refrigerant 134a**]

$$\dot{m} = \frac{\dot{Q}_{\text{in}}}{\Delta h_{\text{evaporator}}} \quad 32.9$$

7. CARNOT REFRIGERATION CYCLE

The *Carnot refrigeration cycle* is a Carnot power cycle running in reverse. Because it is reversible, the Carnot refrigeration cycle has the highest coefficient of performance for any given temperature limits of all the refrigeration cycles. As shown in Fig. 32.3, the processes all occur within the vapor dome. [**Reversed Carnot Cycle**]

3 to 4: isentropic expansion
4 to 1: isothermal heating (vaporization)
1 to 2: isentropic compression
2 to 3: isothermal cooling (condensation)

Figure 32.3 *Carnot Refrigeration Cycle*

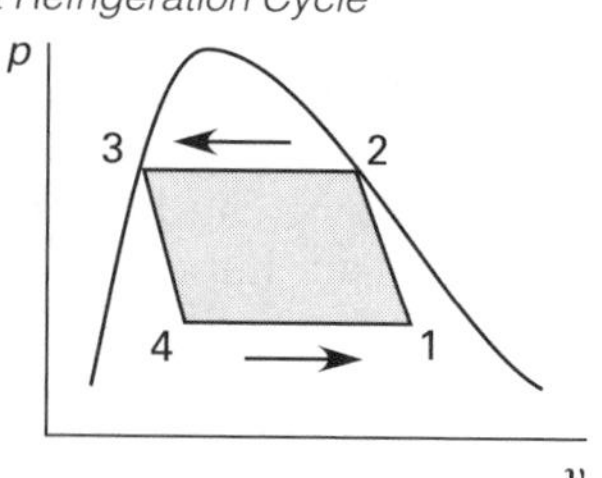

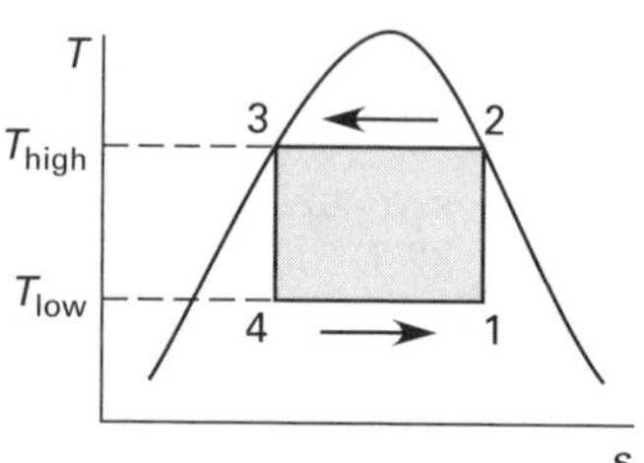

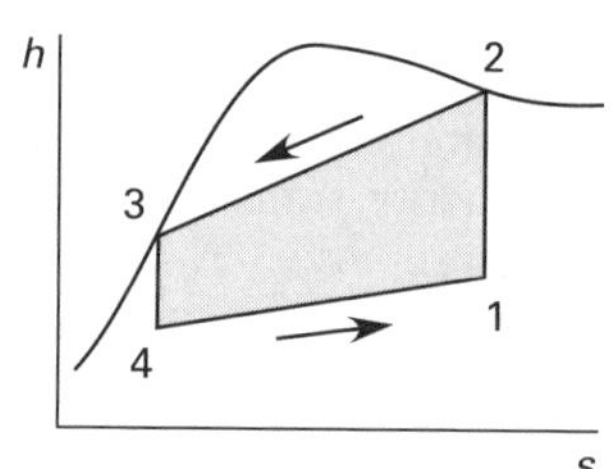

The analysis procedure for the Carnot refrigeration cycle is reversed from, but otherwise identical to, the procedure for analyzing the Carnot power cycle. The coefficients of performance are calculated using Eq. 32.1 and Eq. 32.2.

8. VAPOR COMPRESSION CYCLE

The *vapor compression cycle* is essentially a reversed Rankine vapor cycle.[5] It is the most common type of refrigeration cycle, finding application in household refrigerators, air conditioners for cars and houses, chillers, and so on. The processes are illustrated in Fig. 32.4.

3 to 4: isenthalpic expansion
4 to 1: constant pressure heating (vaporization)
1 to 2: isentropic compression
2 to 3: constant pressure cooling (condensation)

[5]An irreversible expansion through a throttling valve takes the place of the boiler.

Power Cycles

Figure 32.4 *Wet Vapor Compression Cycle*

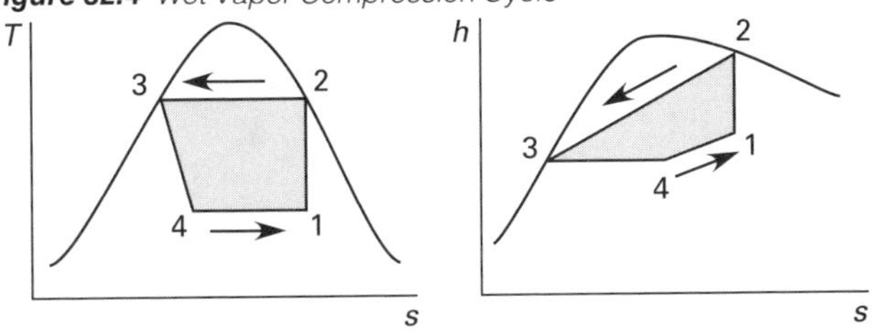

If the refrigerant leaves the evaporator (see Fig. 32.4, point 1) with a quality of less than 1.0, the cycle is known as a *wet vapor compression cycle. Wet compression* is undesirable because of compressor wear and performance problems. For that reason, refrigerators are designed so that the refrigerant leaves the evaporator either saturated or slightly superheated, as shown by point 1 in Fig. 32.5. Compression of saturated or superheated vapor is said to be *dry compression*, and the cycle is known as a *dry vapor compression cycle.*

Figure 32.5 *Dry Vapor Compression Cycle*

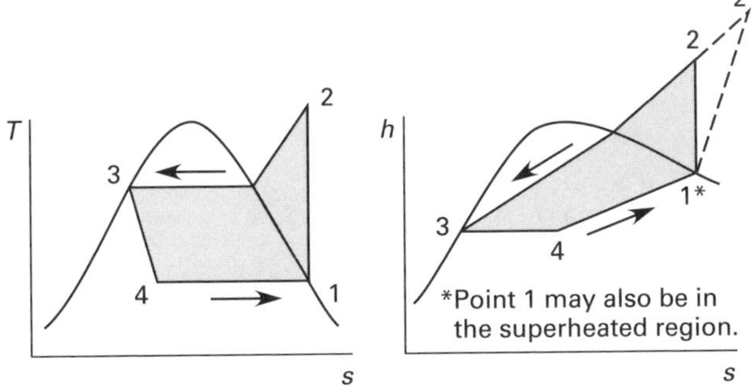

If the refrigerant is saturated when it leaves the evaporator, the following solution method can be used. (The subscripts f and g refer to saturated fluid and vapor properties read from a refrigerant table. They do not correspond to any point on the figures shown in this chapter.)

At 3: $x = 0$ (saturated liquid); either p_3 or T_3 must be known, or they can be found as the saturation temperature and pressure having an enthalpy of h_3. $h_3 = h_4$; $p_3 = p_2$.

At 4: $T_4 = T_1$; $h_4 = h_3$ (because 3-to-4 is isenthalpic).

At 1: $x = 1$ (saturated vapor); $T_1 = T_4$; $p_1 = p_{sat}$ for temperature T_1; h_1 and s_1 are read as h_g and s_g from the refrigerant table.

At 2: $p_2 = p_3$; if s_2 and either p_2 or T_2 are known, h_2 can be found by searching the refrigerant superheat table. $s_2 = s_1$ (because 1-to-2 is isentropic).

The refrigerant mass flow rate is

$$\dot{m} = \frac{\dot{Q}_{in}}{h_1 - h_4} \qquad 32.10$$

The compressor power is

$$\dot{W}_{in} = \dot{m}(h_2 - h_1) \qquad 32.11$$

The refrigeration effect is

$$\dot{Q}_{in} = \dot{m}(h_1 - h_4) \qquad 32.12$$

The coefficient of performance as a refrigerator is

Efficiency

$$\begin{aligned} \text{COP}_{\text{refrigerator}} &= \frac{\text{capacity in Btu/hr}}{\text{input energy}} \\ &= \frac{\dot{Q}_{in}}{\dot{W}_{in}} = \frac{h_1 - h_4}{h_2 - h_1} \end{aligned} \qquad 32.13$$

If the compression is not isentropic (i.e., an isentropic compressor efficiency is known), h_2 will be affected. This, in turn, will determine T_2. (p_2 is not affected.)

$$h_2' = h_1 + \frac{h_2 - h_1}{\eta_{s,\text{compressor}}} \qquad 32.14$$

Example 32.1

A refrigerator using R-134a has a cooling effect of 10,000 Btu/hr (2.9 kW). Refrigerant leaves the evaporator saturated at 0°F (−18°C). The pressure of the refrigerant entering the condenser is 120 psia (830 kPa; 8.3 bars). The refrigerant leaves the condenser as a saturated liquid. The compressor's isentropic efficiency is 80%. Find (a) the temperature of the refrigerant immediately after compression, (b) the refrigerant flow rate, (c) the compressor input shaft power, and (d) the refrigerator's coefficient of performance.

SI Solution

Refer to Fig. 32.5 and a pressure-enthalpy diagram for refrigerant R-134a. Values will vary somewhat with the precision to which the pressure-enthalpy chart can be read. [**Pressure Versus Enthalpy Curves for Refrigerant 134a**]

At point 1: Locate the intersection of the horizontal −18°C line and saturated vapor line.

$$\begin{aligned} T_1 &= -18^\circ\text{C} \quad \text{[given]} \\ p_1 &= 1.6 \text{ bars} \\ h_1 &= 385 \text{ kJ/kg} \end{aligned}$$

Power Cycles

At point 2: Follow a line of constant entropy upward and to the right to the horizontal 8.3 bars line.

$$T_2 = 40°\text{C}$$
$$p_2 = 8.3 \text{ bars} \quad \text{[given]}$$
$$h_2 = 425 \text{ kJ/kg}$$
$$h_2' = h_1 + \frac{h_2 - h_1}{\eta_{s,\text{compressor}}}$$
$$= 385 \frac{\text{kJ}}{\text{kg}} + \frac{425 \frac{\text{kJ}}{\text{kg}} - 385 \frac{\text{kJ}}{\text{kg}}}{0.80}$$
$$= 435 \text{ kJ/kg}$$

At point 3: Move horizontally to the saturated liquid line.

$$p_3 = p_2 = 8.3 \text{ bars}$$
$$T_3 = 32°\text{C}$$
$$h_3 = 244 \text{ kJ/kg}$$

At point 4: Throttling processes are constant-enthalpy processes. Follow a vertical line downward to the horizontal 1.6 bar line.

$$p_4 = p_1 = 1.6 \text{ bar}$$
$$h_4 = h_3 = 244 \text{ kJ/kg} \quad \text{[throttling process]}$$

(a) Use the chart to find the temperature corresponding to 8.3 bars and 435 kJ/kg. The temperature immediately after the compression process is $T_2' \approx 50°\text{C}$.

(b) The required refrigerant flow is found from the total heat load and the enthalpy change per pound.

$$\dot{m} = \frac{\dot{Q}_{\text{in}}}{h_1 - h_4}$$
$$= \frac{(2.9 \text{ kW})\left(1.0 \frac{\text{kJ}}{\text{kW·s}}\right)}{385 \frac{\text{kJ}}{\text{kg}} - 244 \frac{\text{kJ}}{\text{kg}}}$$
$$= 0.0206 \text{ kg/s}$$

(c) The compressor power is

$$\dot{W}_{\text{in}} = \dot{m}(h_2' - h_1)$$
$$= \left(0.0206 \frac{\text{kg}}{\text{s}}\right)\left(435 \frac{\text{kJ}}{\text{kg}} - 385 \frac{\text{kJ}}{\text{kg}}\right)$$
$$= 1.03 \text{ kW}$$

(d) The coefficient of performance of the refrigerator is

Efficiency

$$\text{COP}_{\text{refrigerator}} = \frac{\text{capacity in Btu/hr}}{\text{input energy}} = \frac{\dot{Q}_{\text{in}}}{\dot{W}_{\text{in}}}$$
$$= \frac{2.9 \text{ kW}}{1.03 \text{ kW}}$$
$$= 2.8$$

Customary U.S. Solution

Refer to Fig. 32.5 and a pressure-enthalpy diagram for refrigerant R-134a. Values will vary somewhat with the precision to which the pressure-enthalpy chart can be read. [**Pressure Versus Enthalpy Curves for Refrigerant 134a**]

At point 1: Locate the intersection of the horizontal 0°F line and saturated vapor line.

$$T_1 = 0°\text{F} \quad \text{[given]}$$
$$p_1 = 21 \text{ psia}$$
$$h_1 = 102 \text{ Btu/lbm}$$

At point 2: Follow a line of constant entropy upward and to the right to the horizontal 120 psia line.

$$T_2 = 103°\text{F}$$
$$p_2 = 120 \text{ psia} \quad \text{[given]}$$
$$h_2 = 119 \text{ Btu/lbm}$$
$$h_2' = h_1 + \frac{h_2 - h_1}{\eta_{s,\text{compressor}}}$$
$$= 102 \frac{\text{Btu}}{\text{lbm}} + \frac{119 \frac{\text{Btu}}{\text{lbm}} - 102 \frac{\text{Btu}}{\text{lbm}}}{0.80}$$
$$= 123 \text{ Btu/lbm}$$

(T_2' could be found if needed.)

At point 3: Move horizontally to the saturated liquid line.

$$p_3 = p_2$$
$$= 120 \text{ psia}$$
$$T_3 = 91°\text{F}$$
$$h_3 = 41 \text{ Btu/lbm}$$

At point 4: Throttling processes are constant-enthalpy processes. Follow a vertical line downward to the horizontal 21 psia line.

$$p_4 = p_1 = 21 \text{ psia}$$
$$h_4 = h_3 = 41 \text{ Btu/lbm} \quad \text{[throttling process]}$$

(a) Use the chart to find the temperature corresponding to 120 psia and 123 Btu/lbm. The temperature immediately after the compression process is $T_2' \approx 117°F$.

(b) The required refrigerant flow is found from the total heat load and the enthalpy change per pound.

$$\dot{m} = \frac{\dot{Q}_{in}}{h_1 - h_4} = \frac{10{,}000 \ \frac{\text{Btu}}{\text{hr}}}{102 \ \frac{\text{Btu}}{\text{lbm}} - 41 \ \frac{\text{Btu}}{\text{lbm}}} = 164 \text{ lbm/hr}$$

(c) The compressor power is

$$\dot{W}_{in} = \dot{m}(h_2' - h_1) = \left(164 \ \frac{\text{lbm}}{\text{hr}}\right)\left(123 \ \frac{\text{Btu}}{\text{lbm}} - 102 \ \frac{\text{Btu}}{\text{lbm}}\right) = 3444 \text{ Btu/hr}$$

(d) The coefficient of performance of the refrigerator is

Efficiency

$$\text{COP}_{\text{refrigerator}} = \frac{\text{capacity in Btu/hr}}{\text{input energy}} = \frac{\dot{Q}_{in}}{\dot{W}_{in}} = \frac{10{,}000 \ \frac{\text{Btu}}{\text{hr}}}{3444 \ \frac{\text{Btu}}{\text{hr}}} = 2.9$$

9. AIR REFRIGERATION CYCLE

Any gas will cool when expanded. This is the principle behind the air refrigeration cycle. The *air refrigeration cycle* (also known as a *Brayton cooling cycle*) is essentially a reversed Brayton turbine cycle. It is not common because of its high power consumption. However, air is nonflammable, readily available, and nontoxic. Therefore, the air refrigeration cycle is often used in aircraft air conditioning and gas liquefaction applications where a refrigerant with such characteristics is needed. [**Air Refrigeration Cycle**]

Compressed-air air conditioners, now essentially nonexistent, were common prior to the development of Freon and other chlorofluorocarbon refrigerants. These units consisted of a compressor, heat exchanger, and expansion turbine. The heat increase due to compression was dissipated by the heat exchanger. The expanded, cooler air was discharged directly into the room. (See Fig. 32.6.)

3 to 4: isentropic expansion
4 to 1: constant pressure heating
1 to 2: isentropic compression
2 to 3: constant pressure cooling

Figure 32.6 *Air Refrigeration Cycle*

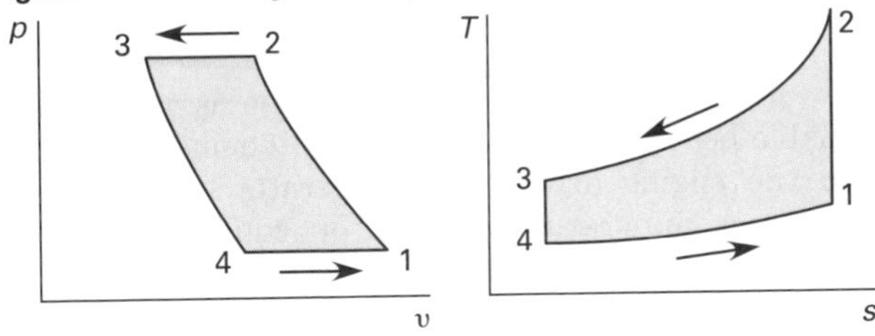

If air is considered to be an ideal gas, the ideal gas equations can be used to find the properties at each point.

$$\frac{T_3}{T_4} = \frac{T_2}{T_1} = \left(\frac{p_H}{p_L}\right)^{(k-1)/k} \qquad 32.15$$

Since $\Delta h = c_p \Delta T$ for an ideal gas, the coefficient of performance can be calculated from Eq. 32.16.

$$\text{COP}_{\text{refrigerator}} = \frac{T_1 - T_4}{(T_2 - T_3) - (T_1 - T_4)} \qquad 32.16$$

If air is not considered to be an ideal gas, an air table must be used to find the coefficient of performance.

Air Refrigeration Cycle

$$\text{COP}_{\text{refrigerator}} = \frac{h_1 - h_4}{(h_2 - h_3) - (h_1 - h_4)} \qquad 32.17$$

If operated as a heat pump cycle, the coefficient of performance is

Air Refrigeration Cycle

$$\text{COP}_{\text{heat pump}} = \frac{h_2 - h_3}{(h_2 - h_3) - (h_1 - h_4)} \qquad 32.18$$

The coefficient of performance can also be calculated from the pressure ratio, r_p. (This formula cannot be easily corrected for nonisentropic expansion or compression.)

$$\text{COP}_{\text{refrigerator}} = \frac{1}{r_p^{(k-1)/k} - 1} \qquad 32.19$$

$$r_p = \frac{p_{\text{high}}}{p_{\text{low}}} \qquad 32.20$$

Power Cycles

10. HEAT-DRIVEN REFRIGERATION CYCLES

A *heat-driven refrigeration cycle*, also known as a *heat-activated refrigeration cycle*, is practical when large quantities of waste or inexpensive heat energy are available. In addition to using waste heat, combustion of natural gas and LPG, solar collectors, geothermal sources, and waste steam can provide the energy to drive a refrigeration cycle.

One method of obtaining refrigeration is to use the available heat to drive a Stirling heat engine. The work from the engine drives the refrigerator's compressor. Thus, the apparatus consists of combined power-generating and refrigeration units, as shown in Fig. 32.7.

Figure 32.7 *Heat-Driven Cooling Cycle*

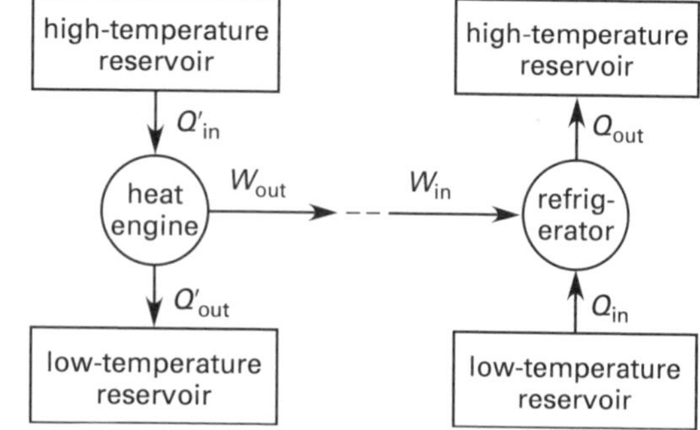

11. ABSORPTION CYCLE

The *absorption cycle* is similar to the vapor compression cycle, with one major change—there is no compressor, and no external work is used to compress the refrigerant.[6] Rather, a generator-absorber-recuperator apparatus (see Fig. 32.8) produces a solution of refrigerant in another liquid, and external heat superheats the refrigerant.

Two working fluids are required in the absorption cycle.[7] As in the vapor-compression cycle, the cycle starts with the refrigerant passing through the evaporator, removing heat from the low-temperature area. The refrigerant (in saturated vapor condition) then enters an *absorber* where it is absorbed by the liquid *absorbent.*

The mixture of liquid refrigerant and absorbent is next pumped into the higher pressure of the *generator.* (This operation requires very little work because the fluid is liquid.) Heat from an external source (e.g., solar collectors, geothermal generators, or combustion of natural gas) drives off the refrigerant in a superheated (though usually subatmospheric) condition. The absorbent left behind is returned to the absorber.

Figure 32.8 *Absorption Cycle Apparatus*

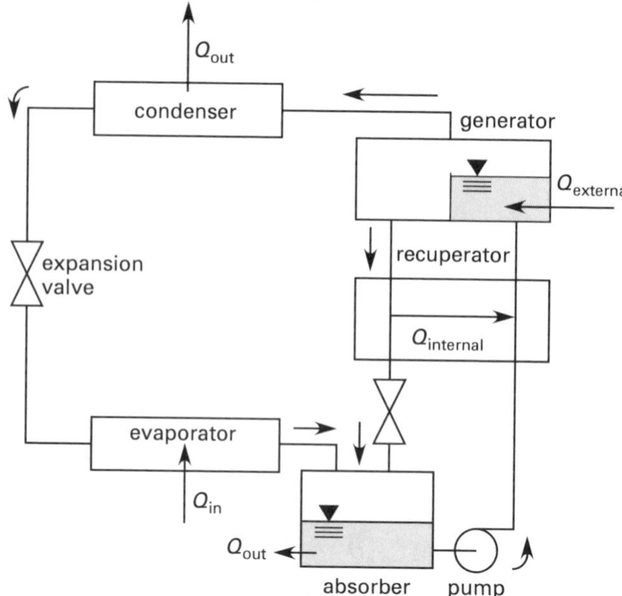

An optional heat exchanger (known as the *recuperator*) may be used to transfer heat from the returning absorbent to the mixture entering the generator. This heat transfer helps improve the system efficiency since the absorbent can carry more refrigerant at a lower temperature. If a recuperator is not used, it may be necessary to cool the liquid in the absorber by some other means (e.g., passing cooling water from another source through the absorber).

An optional *rectifier* (not shown) between the generator and condenser is needed in ammonia-water systems to remove any remaining traces of absorbent from the refrigerant.

The absorption cycle is not very efficient (e.g., coefficients of performance less than 1.0) and the initial equipment costs are higher than for the vapor compression cycle, but when cheap or free energy is available, the cycle can be economical. The cycle is also bulky and involves toxic fluids; hence it is unsuitable for home and auto cooling.

In Fig. 32.8, the coefficient of performance is

$$\mathrm{COP}_{\text{refrigerator}} = \frac{Q_{\text{in}}}{Q_{\text{external}}} = \frac{Q_{\text{in}}}{Q_{\text{out}} - Q_{\text{in}}} \qquad 32.21$$

The *generator-absorber-heat exchanger* (*GAX*) *cycle* is an advanced variation of the absorption cycle. With an ammonia-water absorption cycle, the lower range of the high-temperature absorption process overlaps the higher range of the low-temperature heat input process. With the proper equipment, a portion of the rejected absorber heat can be used to supply heat to the

[6]A very small amount of work may be used by the liquid return pump between the absorber and generator shown in Fig. 32.8. However, even this pump can be replaced by a *thermosiphon* (i.e., using an inert gas that circulates the liquid by expanding and contracting in response to heat).
[7]Common pairs are ammonia (the refrigerant) and water (the absorbent), and water (the refrigerant) and lithium bromide (the absorbent).

generator. *ASHRAE Handbook—Fundamentals* outlines three- and four-temperature thermal cycles and a single-effect absorption cycle.[8] **[Single-Effect Absorption Cycle] [Absorption Refrigeration Cycles]**

12. HIGH-SIDE EQUIPMENT

High-side equipment, primarily the compressor and condenser, is the equipment that operates at the high pressure in the refrigerator. Reciprocating piston compressors are used for small- and medium-sized refrigerators. For the smallest units, for example, those used in kitchen refrigerators and other unitary machines, hermetically sealed single-stage motor and compressor packages have traditionally been used. Sealed units eliminate the leakage through mechanical seals that is common in large commercial units. Compressor operators, power requirements, volumetric efficiency, and other thermodynamic considerations are similar to those for other compressors. Centrifugal compressors are used in the largest machines (usually 200 tons or larger).

Scroll compressors, also referred to as *scroll pumps* and *spiral compressors*, products of computer-aided manufacturing and precision machining, were introduced commercially in the late 1980s as replacements for reciprocating compressors in small residential air conditioners. They are approximately 10% more efficient than piston compressors, have higher reliabilities, and are suitable for use with modern refrigerants such as R-134a, R-407C, and R-410A.

Condensers used in small- and medium-sized refrigerators (i.e., up to approximately 100 tons) are typically air-cooled heat exchangers. For efficient operation, the condensing temperature should not be lower than 10°F (5°C) and not more than 30°F (17°C) above the initial air temperature. For larger capacities, water-cooled condensers are used.

Example 32.2

Refrigerant R-134a is used in a single-acting, two-cylinder compressor running at 300 rpm. The bore diameter is 6 in (152 mm), and the stroke length is 7 in (178 mm). Saturated vapor enters the compressor and is compressed isentropically to 150 psia (1.0 MPa; 10 bars). Saturated liquid enters the expansion valve. Evaporation occurs at 50 psia (350 kPa; 3.5 bars). The cooling effect is 17 tons (60.4 kW). What is the volumetric efficiency?

SI Solution

The swept volume per stroke is

$$\begin{aligned} V = AL &= \frac{\pi D^2 L}{4} \\ &= \frac{\pi(152 \text{ mm})^2\left(178 \ \frac{\text{mm}}{\text{stroke-cylinder}}\right)}{(4)\left(1000 \ \frac{\text{mm}}{\text{m}}\right)^3} \\ &= 0.00323 \text{ m}^3/\text{stroke-cylinder} \end{aligned}$$

Since the compressor is single-acting, each cylinder contributes one compression stroke each revolution. The ideal volumetric flow rate is

$$\begin{aligned} \dot{V}_{\text{ideal}} &= VN \\ &= \frac{\left(0.00323 \ \frac{\text{m}^3}{\text{stroke-cylinder}}\right) \times (300 \text{ rpm})(2 \text{ cylinders})}{60 \ \frac{\text{s}}{\text{min}}} \\ &= 0.0323 \text{ m}^3/\text{s} \end{aligned}$$

See Fig. 32.5. At point C, the pressure is 3.5 bars. From a pressure-enthalpy diagram for refrigerant R-134a, the enthalpy of saturated 3.5 bar vapor is approximately 400 kJ/kg. The specific volume is approximately 0.060 m³/kg. At point A, the pressure is 10 bars. The enthalpy of saturated 10 bar liquid is approximately 255 kJ/kg. This is also the enthalpy at point B. **[Pressure Versus Enthalpy Curves for Refrigerant 134a]**

The actual volumetric flow rate is

$$\begin{aligned} \dot{V}_{\text{actual}} = \dot{m}_C \upsilon_C &= \frac{\dot{Q}\upsilon_C}{h_C - h_B} \\ &= \frac{(60.4 \text{ kW})\left(0.060 \ \frac{\text{m}^3}{\text{kg}}\right)}{400 \ \frac{\text{kJ}}{\text{kg}} - 255 \ \frac{\text{kJ}}{\text{kg}}} \\ &= 0.0250 \text{ m}^3/\text{s} \end{aligned}$$

[8] *ASHRAE Handbook—Fundamentals* also covers forward absorption cycles, reverse absorption cycles, multistage absorption cycles, and double-effect absorption cycles.

The volumetric efficiency is

$$\eta_v = \frac{\dot{V}_{\text{actual}}}{\dot{V}_{\text{ideal}}} = \frac{0.0250\ \frac{\text{m}^3}{\text{s}}}{0.0323\ \frac{\text{m}^3}{\text{s}}} = 0.774 \quad (77\%)$$

Customary U.S. Solution

The swept volume per stroke is

$$\begin{aligned} V = AL &= \frac{\pi D^2 L}{4} \\ &= \frac{\pi(6\ \text{in})^2\left(7\ \frac{\text{in}}{\text{stroke-cylinder}}\right)}{(4)\left(12\ \frac{\text{in}}{\text{ft}}\right)^3} \\ &= 0.1145\ \text{ft}^3/\text{stroke-cylinder} \end{aligned}$$

Since the compressor is single-acting, each cylinder contributes one compression stroke each revolution. The ideal volumetric flow rate is

$$\begin{aligned} \dot{V}_{\text{ideal}} &= VN \\ &= \left(0.1145\ \frac{\text{ft}^3}{\text{stroke-cylinder}}\right)(300\ \text{rpm}) \\ &\quad \times (2\ \text{cylinders}) \\ &= 68.7\ \text{ft}^3/\text{min} \end{aligned}$$

See Fig. 32.5. At point C, the pressure is 50 psia. From a pressure-enthalpy diagram for refrigerant R-134a, the enthalpy of saturated 50 psia vapor is approximately 107 Btu/lbm. The specific volume is approximately 0.94 ft^3/lbm. At point A, the pressure is 150 psia. The enthalpy of saturated 150 psia liquid is approximately 46 Btu/lbm. This is also the enthalpy at point B. [**Pressure Versus Enthalpy Curves for Refrigerant 134a**]

The actual volumetric flow rate is

$$\begin{aligned} \dot{V}_{\text{actual}} = \dot{m}_C v_C &= \frac{\dot{Q} v_C}{h_C - h_B} \\ &= \frac{(17\ \text{tons})\left(200\ \frac{\text{Btu}}{\text{min-ton}}\right)\left(0.94\ \frac{\text{ft}^3}{\text{lbm}}\right)}{107\ \frac{\text{Btu}}{\text{lbm}} - 46\ \frac{\text{Btu}}{\text{lbm}}} \\ &= 52.4\ \text{ft}^3/\text{min} \end{aligned}$$

The volumetric efficiency is

$$\eta_v = \frac{\dot{V}_{\text{actual}}}{\dot{V}_{\text{ideal}}} = \frac{52.4\ \frac{\text{ft}^3}{\text{min}}}{68.7\ \frac{\text{ft}^3}{\text{min}}} = 0.763 \quad (76\%)$$

13. LOW-SIDE EQUIPMENT

Low-side equipment, primarily the evaporator, is the equipment that operates at the low pressure in the refrigerator. *Evaporators* for cooling air are usually externally finned and are known as *cooling coils*. Evaporators for cooling water and other liquids are mostly shell-and-coil and shell-and-tube heat exchangers known as *coolers* and *chillers*.[9] Tubes are generally smooth inside.[10]

Coolers and chillers for water generally operate with an average temperature difference of 6°F to 20°F (3°C to 11°C), and optimally with a 10°F to 14°F (5°C to 8°C) difference. To avoid freezing problems, the temperature of the entering refrigerant should be above 28°F (−2°C).

An evaporator can be either top-fed or bottom-fed with the entering refrigerant. Table 32.1 compares the advantages of both top-feed and bottom-feed evaporators.

Table 32.1 *Refrigeration Evaporators: Top-Feed Versus Bottom-Feed Evaporators*

advantages of top-feed evaporators	advantages of bottom-feed evaporators
allows for smaller low-pressure receiver	lower circulating rates of refrigerant due to less critical distribution considerations
better oil return	relative locations of evaporators and low-pressure receivers are less important
reduces or eliminates static pressure penalty	system design and layout are simpler
simpler defrost arrangement	
smaller refrigerant charge	

[9] In industrial settings, the shell-and-coil design is less desirable because the coil cannot be mechanically cleaned.
[10] *Microfins* inside the evaporator tube increase the heat transfer efficiency.

ASHRAE Handbook—Refrigeration provides additional details regarding the advantages and minimum refrigerant circulation rates for top- and bottom-feed evaporators. *ASHRAE Handbook—Refrigeration* also provides liquid overfeed recommendations for R-717 and R-22 refrigerants.[11] **[Refrigeration Evaporator: Top-Feed Versus Bottom-Feed] [Recommended Minimum Refrigerant Circulating Rate] [Liquid Overfeed Systems]**

14. REFRIGERANT LINE SIZING

The size of the refrigerant return line (from the condenser) is not a critical design parameter. The pressure drop in the throttling valve is far greater than the pressure drop from the fluid flow.

A small suction line, however, can greatly decrease compressor capacity. The suction line between the evaporator and the compressor is typically sized from tables based on the allowing pressure drop, refrigeration capacity, and length. Tables are presented in different formats depending on the source, and different tables are needed for each different refrigerant. Maximum allowable pressure drops of 3 psi (20 kPa) on high-temperature units and 1.5 psi (10 kPa) on low-temperature units are the established rules of thumb, and tables are given for this range of pressure drops. **[Flow Rate Per Ton of Refrigeration for Refrigerant 22] [Flow Rate Per Ton of Refrigeration for Refrigerant 134a] [Suction Line Capacities, in Tons, for Refrigerant 22 (Single- or High-Stage Applications)] [Suction Line Capacities in Tons for Refrigerant 134a (Single- or High-Stage Applications)]**

Suction lines should not be sized too large, as a reasonable velocity is needed to carry oil from the evaporator back to the compressor. For horizontal suction lines, 750 ft/min (3.8 m/s) is a recommended minimum velocity. For vertical suction lines, the velocity should be 1200 ft/min to 1400 ft/min (6.1 m/s to 7.1 m/s).

15. NOMENCLATURE

A	area	ft²	m²
c_p	specific heat at constant pressure	Btu/lbm-°F	kJ/kg·K
COP	coefficient of performance	–	–
D	diameter	ft	m
EER	energy efficiency ratio	Btu/W-hr	n.a.
h	enthalpy	Btu/lbm	kJ/kg
k	ratio of specific heats	–	–
L	length	ft	m
$\dot{m}$	mass flow rate	lbm/sec	kg/s
N	rotational speed	rpm	rpm
p	pressure	lbf/ft²	kPa
Q	heat	Btu	kJ
$\dot{Q}$	heat flow rate	Btu/hr	kW
r	ratio	–	–
s	entropy	Btu/lbm-°F	kJ/kg·K
T	temperature	°R	K
V	volume	ft³	m³
$\dot{V}$	volumetric flow rate	ft³/min	L/s
W	work	Btu/lbm	kJ/kg
$\dot{W}$	rate of work	Btu/hr	W
x	quality	–	–

Symbols

η	efficiency	–	–
v	specific volume	ft³/lbm	m³/kg

Subscripts

c	Carnot
f	saturated liquid
g	saturated vapor
H	high
L	low
p	pressure or constant pressure
s	isentropic
v	volumetric

[11]Liquid overfeed is the practice of providing liquid refrigerant to the evaporator at a rate above what physically evaporates.

Topic V: Heat Transfer

Chapter

33 Fundamental Heat Transfer

Content in blue refers to the *NCEES Handbook*.

NCEES EXAM SPECIFICATIONS AND RELATED CONTENT

HVAC AND REFRIGERATION EXAM

I.D. Heat Transfer
- 11. Biot Number
- 12. Thermal Diffusivity

II.C.4. Systems and Components: Energy recovery
- 21. Heat Transfer Through a Film

THERMAL AND FLUID SYSTEMS EXAM

I.C. Heat Transfer Principles
- 5. Thermal Resistance
- 11. Biot Number
- 12. Thermal Diffusivity
- 21. Heat Transfer Through a Film
- 22. Overall Coefficient of Heat Transfer
- 39. Lumped Parameter Electrical Analogy

1. INTRODUCTION TO CONDUCTIVE HEAT TRANSFER

Conduction is the flow of heat through solids or stationary fluids. Thermal conductance in metallic solids is due to molecular vibrations within the metallic crystalline lattice and movement of free valence electrons through the lattice. Insulating solids, which have fewer free electrons, conduct heat primarily by the agitation of adjacent atoms vibrating about their equilibrium positions. This vibrational mode of heat transfer is several orders of magnitude less efficient than conduction by free electrons.

In stationary liquids, heat is transmitted by longitudinal vibrations, similar to sound waves. The *net transport theory* explains heat transfer through gases. Hot molecules move faster than cold molecules. Hot molecules travel to cold areas with greater frequency than cold molecules travel to hot areas.

The other primary modes of heat transfer are convection (the transfer of heat in a moving fluid) and radiation (the transfer of energy between isolated bodies by electromagnetic waves through a vacuum or transparent gas or liquid).[1]

2. SIMPLIFYING ASSUMPTIONS

Determining heat transfer by conduction can be an easy task if sufficient simplifying assumptions are made. Major discrepancies can arise, however, when the simplifying assumptions are not met. The following assumptions are commonly made in simple problems.

- The heat transfer is steady-state.
- The heat path is one-dimensional. (Objects are infinite in one or more directions and do not have any end effects.)
- The heat path has a constant area.
- The heat path consists of a homogeneous material with constant conductivity.
- The heat path consists of an isotropic material.[2]
- There is no internal heat generation.

Many real heat transfer cases violate one or more of these assumptions. Unfortunately, problems with closed-form solutions (suitable for working by hand) are in the minority. More complex problems must be solved by appropriate iterative, graphical, or numerical methods.[3]

3. CONDUCTIVITY

The *thermal conductivity* (also known as the *thermal conductance*), k, is a measure of the rate at which a substance transfers thermal energy through a unit thickness.[4] Units of thermal conductivity are Btu-ft/hr-ft²-°F or Btu-in/hr-ft²-°F (W/m·K or W·cm/m²·K). The units of Btu-ft/hr-ft²-°F are the same as Btu/hr-ft-°F.[5] The conductivity of a substance should not be confused with the *overall conductivity*, U, of an object. (See Sec. 33.22.) Appendix 33.A, App. 33.B, and App. 33.C list representative thermal conductivities for commonly encountered substances, as does Table 33.1.

Table 33.1 *Typical Thermal Conductivities at 32°F (0°C)*

substance	k Btu/hr-ft-°F	k W/m·K
silver	242	419
copper	224	388
aluminum	117	202
brass	56	97
steel (1% C)	27	47
lead	20	35
ice	1.3	2.2
glass	0.63	1.1
concrete	0.50	0.87
water	0.32	0.55
fiberglass	0.030	0.052
cork	0.025	0.043
air	0.014	0.024

(Multiply Btu/hr-ft-°F by 1.7307 to obtain W/m·K.)

The thermal conductivity of foodstuffs is influenced by various factors, such as the water content, porosity, fiber orientation, and the state (i.e., above or below freezing). (See App. 33.D for tabulated values.) For foodstuffs above freezing, it can be estimated from the mass fraction of water, x_w, the mass fraction of the solids, $x_s = 1 - x_w$, the thermal conductivity of water, k_w, and the thermal conductivity of the solids portion of the material, k_s. k_s can be assumed to be 0.150 Btu/hr-ft-°F (0.259 W/m·°C) for first approximations.

$$k \approx k_w x_w + k_s(1 - x_w) \quad \text{[above freezing]} \qquad 33.1$$

For foodstuffs with $x_w > 0.5$, Eq. 33.2 may be used to estimate the thermal conductivity.

$$k_{\text{W/m}\cdot{}^\circ\text{C}} \approx 0.056 + 0.57x_w \quad [x_w > 0.5] \qquad 33.2$$

[1]There are two additional, but similar, methods of removing heat energy: change of phase and ablation. Heat transfer can result from the evaporation of liquids and the condensation of vapors. This is the principle used in heat pipes. (See Sec. 33.53.) In other cases, such as during high-speed reentry into the earth's atmosphere, the heat flux may be very high (e.g., up to 10,000 Btu/ft²-sec (114 MW/m²)) and the time of application very short (e.g., 1 or 2 min). It may be necessary to allow a portion of the material to melt or vaporize in order to remove the heat. This is known as *ablation*.

[2]Examples of *anisotropic materials*, materials whose heat transfer properties depend on the direction of heat flow, are crystals, plywood and other laminated sheets, and the core elements of some electrical transformers.

[3]Finite-difference methods are commonly used.

[4]Another (less-encountered) meaning for *conductivity* is the reciprocal of thermal resistance (i.e., conductivity $= kA/L$).

[5]Temperature units of °R can also be used in place of °F, since conductivity is always multiplied by ΔT, and $\Delta T_{°F} = \Delta T_{°R}$.

4. VARYING CONDUCTIVITY

Solids exhibit conductivities that vary with temperature.[6] (See Table 33.2.) Over limited ranges, thermal conductivity in common solids is assumed to vary linearly with temperature, as indicated in Eq. 33.3. k_{ref} is the conductivity at the reference temperature, usually 0°F (−18°C). Values of γ and γ' are not common, as graphs and tabulations of conductivity versus temperature are more readily available.

$$\begin{aligned} k_T &= k_{ref}(1+\gamma T) \\ &= k_{ref} + \gamma' T \end{aligned} \qquad 33.3$$

Table 33.2 Typical Ranges of Conductivity (see also App. 33.A, App. 33.B, and App. 33.C)

	conductivity range	
material	Btu-ft/hr-ft²-°F	W/m·K
gases (at 1 atm)	0.004–0.10	0.007–0.17
insulators	0.02–0.12	0.03–0.21
nonmetallic liquids	0.05–0.40	0.09–0.70
nonmetallic solids	0.02–1.5	0.03–2.6
liquid metals	5.0–45	8.7–78
metallic alloys	8.0–70	14–120
pure metals	30–240	52–420

(Multiply Btu-ft/hr-ft²-°F by 12 to obtain Btu-in/hr-ft²-°F.)
(Multiply Btu-ft/hr-ft²-°F by 1.7307 to obtain WMD.)
(Multiply Btu-ft/hr-ft²-°F by 4.1365×10^{-3} to obtain cal·cm/s·cm²·°C.)

Conductivity decreases with temperature for pure metals and increases with temperature for most alloys. For insulating materials, it increases with temperature. Conductivity in water and aqueous solutions increases with increases in temperature up to approximately 250°F (120°C) and then gradually decreases. Conductivity decreases with increases in concentrations of aqueous solutions, as it does with most other liquids. Conductivity increases with increases in pressure. Of the nonmetallic liquids, water is the best thermal conductor. Conductivity in gases increases almost linearly with temperature but is fairly independent of pressure in common ranges.

Thermal conductivity is often assumed to be constant over the entire length of the transmission path. In most calculations, either the average thermal conductivity or the conductivity at the arithmetic mean temperature is used.[7] When k varies linearly (or nearly so) with temperature, k is evaluated at the average temperature $\frac{1}{2}(T_1 + T_2)$.

In rare cases where k does not vary linearly with temperature, the solution must proceed iteratively. An initial estimate of the average conductivity yields a heat transfer which, in turn, is used to solve for the surface temperatures. These temperatures are used to determine a new average conductivity, and so on.

5. THERMAL RESISTANCE

Thermal resistance is a measure of a material's resistance to heat flow. The thermal resistance between points 1 and 2 is defined by Eq. 33.4. Typical units are °F-hr/Btu (K/W).

$$R_{th} = \frac{T_1 - T_2}{q} \qquad 33.4$$

Heat flow, then, can be calculated from the temperature difference and the total thermal resistance of all materials that the the heat flows through.

Thermal Resistance (*R*)

$$q = \frac{\Delta T}{R_{total}} \qquad 33.5$$

For a plate, the thermal resistance depends on the thickness (path length), L, and is

Thermal Resistance (*R*)

$$R = \frac{L}{kA} \qquad 33.6$$

For a film, the thermal resistance depends on the average film coefficient, h, and is

Thermal Resistance (*R*)

$$R = \frac{1}{hA} \qquad 33.7$$

For a curved layer (i.e., a layer of insulation on a pipe), the thermal resistance is

Thermal Resistance (*R*)

$$R = \frac{\ln \frac{r_2}{r_1}}{2\pi k L} \qquad 33.8$$

In Eq. 33.8, r_1 is the inner radius, and r_2 is the outer radius.

[6]Conductivity at absolute zero is zero because there is no atomic/molecular motion. For the first few degrees above absolute zero, conductivity increases with increases in temperature.
[7]The mean thermal conductivity is not the same as the thermal conductivity at the mean temperature. However, it is very nearly so, and this simplification is widely used.

Heat Transfer

6. *R*-VALUE

The *R-value* of a substance is the thermal resistance on a unit area basis. Typical units are °F-ft²-hr/Btu (K·m²/W). The *R*-value concept is usually encountered in the construction industry as a means of comparing insulating materials. It is not the same as the thermal resistance.

$$R\text{-value} = \frac{\Delta T}{\frac{Q}{A}} = RA \qquad 33.9$$

7. SPECIFIC HEAT CAPACITY

The *specific heat capacity* (often shortened to *specific heat*), c_p, is the energy needed to change the temperature of a unit mass of a body one degree. For solid bodies, $\Delta pV = 0$, so $Q = \Delta H = \Delta U$. Values of specific heat are given in App. 33.B and App. 33.C.

$$c_p = \frac{\Delta U}{m\Delta T} \qquad 33.10$$

The specific heats of foodstuffs can be estimated from the mass fraction of moisture, x_w. The specific heat of foodstuffs below freezing is approximately half the specific heat above freezing.

$$c_{p,\text{kJ/kg}\cdot°\text{C}} \approx 0.837 + 3.348x_w \quad \text{[above freezing]} \qquad 33.11$$

$$c_{p,\text{kJ/kg}\cdot°\text{C}} \approx 0.837 + 1.256x_w \quad \text{[below freezing]} \qquad 33.12$$

The specific heats of foodstuffs can also be calculated from correlations with either temperature or composition. Equation 33.13 correlates the specific heat above freezing with the mass fractions of moisture, protein, fat, carbohydrate, and ash.

$$c_{p,\text{kJ/kg}\cdot°\text{C}} \approx 4.180x_w + 1.711x_p + 1.928x_f + 1.547x_c + 0.908x_a \qquad 33.13 \quad \text{[above freezing]}$$

8. LATENT HEAT OF FUSION

The *latent heat of fusion* (*latent heat of freezing*), h_{fusion}, is the heat that is released or absorbed during a material's transition between solid and liquid phases without a change in the temperature of the material. This constitutes a significant cooling/refrigeration load in food processing when freezing is involved.

The latent heat of freezing for water is 143.4 Btu/lbm (333.5 kJ/kg). Most fruits and vegetables experience a freezing point depression of 1–4°F (0.5–2°C) due to their sugar content.

The latent heat of food products can be estimated based on the mass fraction of water, x_w, in the products. The units for latent heat are Btu/lbm (kJ/kg).

$$h_{\text{fusion,kJ/kg}} \approx \left(333.5\ \frac{\text{kJ}}{\text{kg}}\right)x_w \quad \text{[SI]} \qquad 33.14(a)$$

$$h_{\text{fusion,Btu/lbm}} \approx \left(143.4\ \frac{\text{Btu}}{\text{lbm}}\right)x_w \quad \text{[U.S.]} \qquad 33.14(b)$$

9. HEAT OF RESPIRATION

Recently harvested fruits and vegetables continue to respire, releasing water vapor and carbon dioxide at varying rates for days or even weeks after harvesting. The exothermic *heat of respiration*, $q_{\text{respiration}}$ (actually a heat-generation rate), adds to the cooling/refrigeration load when fruits and vegetables are cooled. The heat of respiration varies greatly with temperature, degree of ripeness, and condition (e.g., trimmed or topped). It decreases with time for most vegetables, but increases with time for fruits that ripen in storage (e.g., apples, peaches, and plums). Initial heat of respiration values should be used when calculating the cooling load of products for the first day or two; long-term equilibrium values should be used when calculating the cooling load for long-term cold storage.

10. CHARACTERISTIC DIMENSION

The *characteristic dimension* (*characteristic length*), L_c, of an object is the ratio of its volume to its surface area.

$$L_c = \frac{V}{A_s} \qquad 33.15$$

For a long cylinder of radius r, the characteristic dimension is

$$L_{c,\text{cylinder}} = \frac{V}{A_s} = \frac{\pi r^2 L}{2\pi r L} = \frac{r}{2} \qquad 33.16$$

For a sphere, the characteristic dimension is one-third of the radius (i.e., $L_{c,\text{sphere}} = r/3$). For an infinite slab, the characteristic dimension is one-half of the slab thickness. For an infinite square rod, the characteristic dimension is a quarter of the rod thickness.

11. BIOT NUMBER

The *Biot number*, Bi (also known as the *Biot modulus* and *transient modulus*), is a comparison of the internal thermal resistance to the external resistance of a body.[8] If the Biot number is small (i.e., less than 0.1), the internal thermal resistance will be small, and the body temperature will be essentially uniform throughout, during heating or cooling. The length used to calculate the Biot number is the characteristic length, not an external body dimension.

Transient Conduction Using the Lumped Capacitance Model

$$\mathrm{Bi} = \frac{hV}{kA_s} = \frac{hL_c}{k} \tag{33.17}$$

12. THERMAL DIFFUSIVITY

The *thermal diffusivity*, α, of a substance is a measure of the speed of propagation of a specific temperature into a solid. The higher the diffusivity, the faster a specific temperature will penetrate into the solid.

Thermal Diffusivity

$$\alpha = \frac{k}{\rho c_p} \tag{33.18}$$

See Table 33.3 and App. 33.D for compilations of representative thermal diffusivities.

The thermal diffusivity of foodstuffs can be calculated from Eq. 33.18. It can also be estimated from Eq. 33.19 using the mass fractions of the foodstuff's water, protein, fat, carbohydrate, and ash components. (See Table 33.4 for typical mass compositions.) Equation 33.19 gives the thermal diffusivity in units of $\mathrm{m^2/s}$.

$$\alpha \approx 0.146 \times 10^{-6} x_w + 0.100 \times 10^{-6} x_f + 0.075 \times 10^{-6} x_p + 0.082 \times 10^{-6} x_c \tag{33.19}$$

Table 33.3 *Representative Thermal Diffusivities (at 32°F (0°C) unless specified otherwise)*

material	ft²/hr	m²/s
aluminum, pure	3.33	8.59×10^{-5}
aluminum, 2024	1.76	4.54×10^{-5}
asbestos	0.010	2.58×10^{-7}
brass (70% Cu, 30% Zn, 68°F)	1.27	3.28×10^{-5}
brick, fire clay (400°F)	0.020	5.15×10^{-7}
copper	4.42	11.4×10^{-5}
cork	0.006	1.5×10^{-7}
glass, Pyrex™	0.023	5.93×10^{-7}
glass, window	0.013	3.35×10^{-7}
gold (68°F)	4.68	12.1×10^{-5}
ice	0.046	11.9×10^{-7}
iron, pure	0.70	1.81×10^{-5}
iron, cast (4% C, 68°F)	0.66	1.70×10^{-5}
lead (70°F)	0.95	2.45×10^{-5}
magnesium (60°F)	3.68	9.49×10^{-5}
mercury	0.172	44.4×10^{-7}
nickel	0.60	1.55×10^{-5}
rubber, soft	0.003	0.77×10^{-7}
steel (1% C)	0.48	1.24×10^{-5}
silver	6.60	17.0×10^{-5}
soil, dry	≈0.01	3×10^{-7}
soil, moist	≈0.03	8×10^{-7}
steel, stainless (68°F)	0.17	0.44×10^{-5}
steel (1% C, 68°F)	0.45	1.2×10^{-5}
tin	1.57	4.05×10^{-5}
tungsten	2.39	6.17×10^{-5}
water	0.005	1.3×10^{-7}
zinc	1.60	4.13×10^{-5}

(Multiply ft²/hr by 2.58×10^{-5} to obtain m²/s.)

(Multiply ft²/sec by 0.092903 to obtain m²/s.)

(Multiply ft²/sec by 3600 to obtain ft²/hr.)

13. FOURIER NUMBER

The *Fourier number*, Fo, also known as *relative time*, is the ratio of the rate of heat transferred by conduction to the rate of energy stored.[9] The Fourier number is useful in working transient heat transfer problems. The larger the Fourier number, the larger will be the amount of heat conducted through the solid as compared to the amount of heat stored. This is manifested as a deeper penetration of a specific temperature into the solid over a given period of time and a faster temperature change at a given depth in the object.

$$\mathrm{Fo} = \frac{kt}{\rho c_p L_c^2} = \frac{\alpha t}{L_c^2} \tag{33.20}$$

[8]Biot is pronounced "bee′-oe."

[9]The Fourier number is sometimes given the symbol τ.

Heat Transfer

Table 33.4 Compositions of Foodstuffs by Mass

item	water	protein	fat	carbohydrate	ash
apples, fresh	0.844	0.002	0.006	0.145	0.003
garlic	0.613	0.062	0.002	0.308	0.015
peaches	0.891	0.006	0.001	0.097	0.005
peas, raw	0.78	0.063	0.004	0.144	0.009
pineapple, raw	0.853	0.004	0.002	0.137	0.004
potatoes, raw	0.798	0.021	0.001	0.171	0.009
rice, white	0.120	0.067	0.004	0.804	0.005
spinach	0.907	0.032	0.003	0.043	0.015
tomatoes, ripe	0.935	0.011	0.002	0.047	0.005

14. MODIFIED FOURIER NUMBER

A modified Fourier number is used with graphical solutions to transient heat transfer problems. The modified Fourier number is similar in format to the Fourier number except that the length, L, is defined differently. For example, the length of a sphere used in calculating the modified Fourier number is the sphere's radius, r, not the characteristic length of $r/3$.

$$(\mathrm{Fo})_{\text{modified}} = \frac{\alpha t}{L^2} \qquad 33.21$$

15. TEMPERATURE

The *ambient temperature* is the same as *environment temperature, far-field temperature, local temperature*, or (occasionally) *air temperature*. However, due to the thermal resistance of a film, it is not the same as the *surface temperature*. The *bulk temperature* is the temperature of a thoroughly mixed liquid or gas. The term "bulk temperature" is occasionally used to represent the ambient temperature.

In most heat transfer calculations, the temperature variable shows up as a change in temperature. It is convenient to recognize that $\Delta T_{°F} = \Delta T_{°R}$ and $\Delta T_{°C} = \Delta T_K$.

16. TEMPERATURE PROFILE

A *temperature profile* (see Fig. 33.1) is a graph of the temperature versus location within an object. A profile will often be drawn on a representation of the cross section of the object through which the energy transfer occurs. In uniform materials, the profile will consist of straight lines, with steeper lines representing materials with lower thermal conductivities. The temperature scale is omitted and replaced with values of temperature at the interface points.

Figure 33.1 Typical Temperature Profile

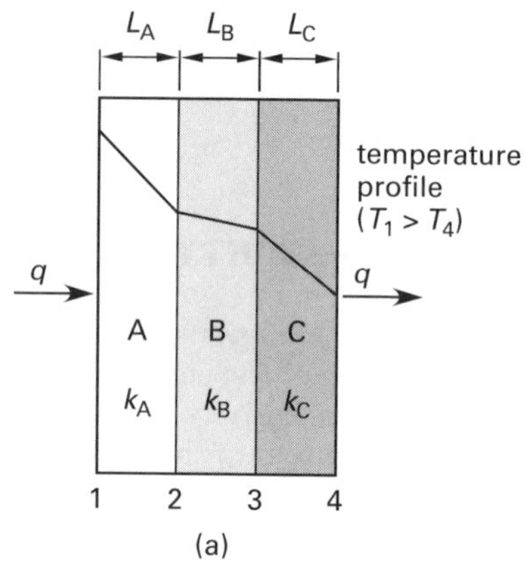

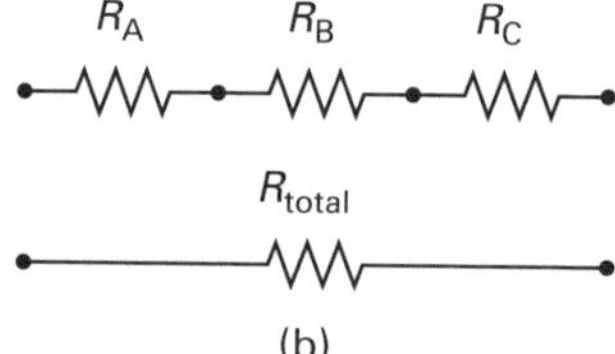

17. HEAT TRANSFER

The term *heat transfer* is used in three different contexts. The intended interpretation must be determined from the needs of the investigation.

The most common usage has units of energy per unit of time (e.g., Btu/hr).[10] Strictly speaking, this should be called a *heat transfer rate*. There is no general agreement on the variable used for this; it is sometimes represented by q, $\dot{Q}$, ΔQ, or even just Q, the same variable used for total heat. This book generally uses q or $\dot{Q}$, depending on the context.

[10]In the United States, the units of Btu/hr are often written as Btuh. Similarly, MBh generally means millions of Btus per hour, but can mean thousands of Btus per hour in some industries (e.g., HVAC). The unit of MMBtuh always means millions of Btus per hour.

A second usage, referred to in this book as *total heat transfer*, is the energy transfer taken over the entire body area or mass.

$$q = q''A \quad 33.22$$

The term "heat transfer" is also used when meaning the total energy change (in Btu) over some period of time. In this book, the symbol ΔU (change in internal energy) is used for this purpose.

$$\Delta U = mc_p \Delta T = Qt \quad 33.23$$

18. FOURIER'S LAW

If the assumptions listed in Sec. 33.2 are valid, *Fourier's law*, Eq. 33.24, is applicable. On its own, heat always flows from a higher temperature to a lower temperature. The heat transfer from high-temperature point 1 to lower-temperature point 2 through an infinite plane of thickness L and homogeneous conductivity, k, is

$$q''_{1-2} = \frac{-k(T_2 - T_1)}{L} = \frac{k(T_1 - T_2)}{L} \quad 33.24$$

$$q_{1-2} = -kA\frac{dT}{dx} = q''_{1-2}A = \frac{kA(T_1 - T_2)}{L} \quad 33.25$$

The temperature difference $T_2 - T_1$ is the *temperature gradient* or *thermal gradient*. Heat transfer is always positive. The minus sign in Eq. 33.24 indicates that the heat flow direction is opposite that of the thermal gradient. The direction of heat flow is obvious in most problems. Therefore, the minus sign is usually omitted.

19. ELECTRICAL ANALOGY

Ohm's law can be solved for current and written in terms of the electrical potential gradient, $E_1 - E_2$.

$$I = \frac{E_1 - E_2}{R} \quad 33.26$$

Drawing on the concept of thermal resistance, Eq. 33.26 and Eq. 33.27 are analogous, although R has different units in each.

$$q = \frac{T_1 - T_2}{R} \quad 33.27$$

The electrical analogy is particularly valuable in analyzing heat transfer through multiple layers (e.g., a layered wall or cylinder with several layers of different insulation). Each layer is considered a series resistance. The sum of the individual thermal resistances is the total thermal resistance. (See Fig. 33.1(b).)

Thermal Resistance (*R*)

$$R_{\text{total}} = \sum R \quad 33.28$$

20. SANDWICHED PLANES

Figure 33.1(a) illustrates the cross section of a *sandwiched plane* with three homogeneous, but different, materials. This type of construction is sometimes known as a *composite wall*. The heat flow due to conduction through a series ($i = 1$ to n) of plane surfaces is

$$q = \frac{T_1 - T_{n+1}}{R_{\text{total}}} = \frac{A(T_1 - T_{n+1})}{\sum_{i=1}^{n}\frac{L_i}{k_i}} \quad 33.29$$

$$R_{\text{total}} = \sum_{i=1}^{n}\frac{L_i}{k_i A} \quad 33.30$$

21. HEAT TRANSFER THROUGH A FILM

Heat conducted through solids is often removed by a physical transport process at an exposed fluid surface. For example, heat transmitted through a heat exchanger wall is removed by a moving coolant. Unless the fluid is extremely turbulent, the fluid molecules immediately adjacent to the exposed surface move much slower than molecules farther away. Molecules immediately adjacent to the wall may be stationary altogether. The fluid molecules that are affected by the exposed surface constitute a layer known as a film.[11] The film has a thermal resistance just like any other sandwiched plane.

Because the film thickness is not easily determined, the thermal resistance of a film is given by a *film coefficient* (*convective heat transfer coefficient* or *unit conductance*), h, with units of Btu/hr-ft^2-°F (W/m^2·K). If the film coefficient is not constant over the entire surface, the symbol for average film coefficient, $\bar{h}$ is used. Subscripts are used (e.g., h_o or h_i) to indicate whether the film is located on the outside or inside of the object.

[11]The film coefficient concept can also be used to quantify the thermal resistance of imperfect bonding between two planes (i.e., *contact resistance*), a dust layer, or scale buildup.

Heat Transfer

The heat flow through a film is

$$q = hA(T_1 - T_2) \quad 33.31$$

From the definition of thermal resistance (Eq. 33.4), then resistance of a film is

Thermal Resistance (*R*)

$$R = \frac{1}{hA} \quad 33.32$$

If the fluid is extremely turbulent, the thermal resistance will be small (i.e., h will be very large). In such cases, the wall temperature is essentially the same as the fluid temperature.

22. OVERALL COEFFICIENT OF HEAT TRANSFER

When conduction and convection (but not radiation) are the only modes of heat transfer, the *overall coefficient of heat transfer*, U, also known as the *overall conductivity*, *overall heat transmittance*, or simply *U-factor*, is defined by Eq. 33.35.

Composite Plane Wall

$$U = \frac{1}{AR_{\text{total}}} = \frac{1}{\sum_i \frac{L_i}{k_i} + \sum_j \frac{1}{h_j}} \quad 33.33$$

When there are two or more parallel paths of heat transfer through materials with different U-values, the overall U-value is calculated as the average of the individual U-values, each weighted by its surface area.

Composite Plane Wall

$$U_{\text{overall}} = \frac{U_aA_a + U_bA_b + U_cA_c + \cdots + U_nA_n}{A_o} \quad 33.34$$

In Eq. 33.34, A_o is the sum of all areas.

Composite Plane Wall

$$A_o = A_a + A_b + A_c + \cdots + A_n \quad 33.35$$

A common situation is heat transfer through a wall that contains doors and window. The overall U-value is calculated as

Composite Plane Wall

$$U_{\text{overall}} = \frac{U_{\text{wall}}A_{\text{wall}} + U_{\text{doors}}A_{\text{doors}} + U_{\text{windows}}A_{\text{windows}}}{A_o} \quad 33.36$$

The heat flow is calculated from

Composite Plane Wall

$$q = UA(T_{\text{inside}} - T_{\text{outside}}) \quad 33.37$$

The overall coefficient is usually used in conjunction with the outside (exposed) surface area (because it is easier to measure), but not always. Therefore, it will generally be written with a subscript (e.g., U_o or U_i) indicating which area it is based on. The heat transfer is

$$\begin{aligned} q &= U_oA_o(T_{\text{inside}} - T_{\text{outside}}) \\ &= U_iA_i(T_{\text{inside}} - T_{\text{outside}}) \end{aligned} \quad 33.38$$

When heat transfer occurs by two or three modes, the overall coefficient of heat transfer takes all active modes —conduction, convection, and radiation—into consideration. In that sense, the overall coefficient, U, is defined by Eq. 33.39, rather than being used in it.

$$U = \frac{q}{A(T_1 - T_2)} \quad 33.39$$

23. TEMPERATURE AT A POINT

The temperature at any point on or within a simple or complex wall can be found if the heat transfer, is known. The procedure is to calculate the thermal resistance up to the point of unknown temperature and then to solve Eq. 33.27 for the temperature difference. Since one of the temperatures is known, the unknown temperature is found from the temperature difference.

Another approach is to use the total thermal resistance and the known inside and outside temperatures. The temperature at a location x can be calculated from the following relationship.

Composite Plane Wall

$$\begin{aligned} \frac{R_x}{R_{\text{total}}} &= \frac{T_{\text{inside}} - T_x}{T_{\text{inside}} - T_{\text{outside}}} \\ T_x &= T_{\text{inside}} - \left(\frac{R_x}{R_{\text{total}}}\right)(T_{\text{inside}} - T_{\text{outside}}) \end{aligned} \quad 33.40$$

24. COMPLEX WALL

Figure 33.2 illustrates a complex wall. Heat transfer through such a wall is intrinsically two-dimensional. Since the temperature profile (see Sec. 33.16) will not be the same for each heat transfer path, there will be adjacent areas with different temperatures. Heat will naturally flow from the hotter to the cooler areas.

Heat Transfer

Figure 33.2 *Complex Wall Construction (top view)*

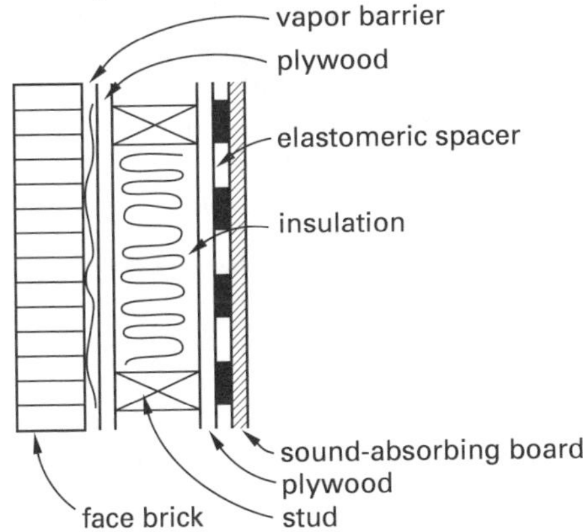

There is no easy way to analytically obtain an exact value of heat transfer through a complex wall. A lower bound on the heat transfer can be obtained by considering the wall as having multiple one-dimensional transmission paths (i.e., ignoring the interaction of adjacent paths). Similarly, an upper bound can be obtained by assuming the entire transmission path consists of a homogeneous material whose conductivity is obtained by weighting each material's conductivity by its volume within the wall.

Thermal conductivities and resistances for common wall constructions are widely available in heating, ventilating, and air conditioning (HVAC) textbooks. Values in HVAC tabulations usually include the inside and outside film coefficients and are based on some time of year and/or wind speed.

Example 33.1

A 100 ft² (9.3 m²) wall consists of 4 in (10 cm) of red brick (k = 0.38 Btu-ft/hr-ft²-°F (0.66 W/m·K)), 1 in (2.5 cm) of pine (k = 0.06 Btu-ft/hr-ft²-°F (0.10 W/m·K)), and ½ in (1.2 cm) of plasterboard (k = 0.30 Btu-ft/hr-ft²-°F (0.52 W/m·K)). The internal and external film coefficients are 1.65 Btu/hr-ft²-°F and 6.00 Btu/hr-ft²-°F (9.38 W/m²·K and 34.1 W/m²·K), respectively. The inside and outside air temperatures are 72°F (22°C) and 30°F (−1°C), respectively. Determine the heat transfer.

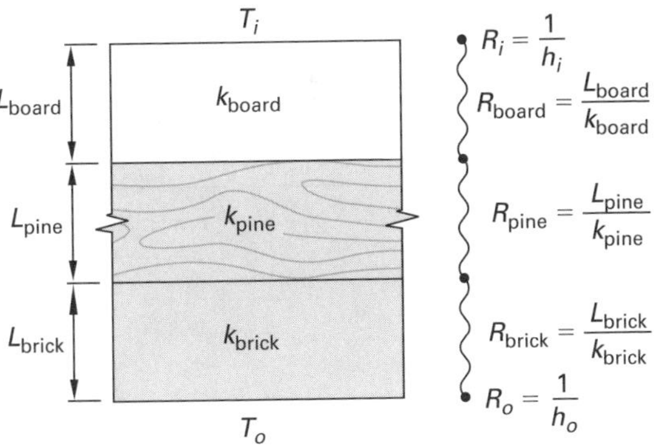

SI Solution

Convert the thicknesses from centimeters to meters.

$$L_{brick} = \frac{10 \text{ cm}}{100 \ \frac{\text{cm}}{\text{m}}} = 0.10 \text{ m}$$

$$L_{pine} = \frac{2.5 \text{ cm}}{100 \ \frac{\text{cm}}{\text{m}}} = 0.025 \text{ m}$$

$$L_{board} = \frac{1.2 \text{ cm}}{100 \ \frac{\text{cm}}{\text{m}}} = 0.012 \text{ m}$$

From Eq. 33.33, the overall coefficient of heat transfer is

$$U = \frac{1}{\sum_i \frac{L_i}{k_i} + \sum_j \frac{1}{h_j}}$$

$$\begin{aligned}\frac{1}{U} &= \frac{0.10 \text{ m}}{0.66 \ \frac{\text{W}}{\text{m·K}}} + \frac{0.025 \text{ m}}{0.10 \ \frac{\text{W}}{\text{m·K}}} + \frac{0.012 \text{ m}}{0.52 \ \frac{\text{W}}{\text{m·K}}} \\ &\quad + \frac{1}{9.38 \ \frac{\text{W}}{\text{m}^2\text{·K}}} + \frac{1}{34.1 \ \frac{\text{W}}{\text{m}^2\text{·K}}} \\ &= 0.561 \text{ m}^2\text{·K/W}\end{aligned}$$

$$\begin{aligned}U &= \frac{1}{0.561 \ \frac{\text{m}^2\text{·K}}{\text{W}}} \\ &= 1.78 \text{ W/m}^2\text{·K}\end{aligned}$$

From Eq. 33.37 (and recognizing that $\Delta T_{°C} = \Delta T_K$), the heat transfer is

$$\begin{aligned}q &= UA\Delta T \\ &= \left(1.78 \ \frac{\text{W}}{\text{m}^2\text{·K}}\right)(9.3 \text{ m}^2)\big(22°\text{C} - (-1°\text{C})\big) \\ &= 380.7 \text{ W}\end{aligned}$$

Customary U.S. Solution

Convert the thicknesses from inches to feet.

$$L_{brick} = \frac{4 \text{ in}}{12 \ \frac{\text{in}}{\text{ft}}} = 0.333 \text{ ft}$$

$$L_{pine} = \frac{1 \text{ in}}{12 \ \frac{\text{in}}{\text{ft}}} = 0.083 \text{ ft}$$

$$L_{board} = \frac{0.5 \text{ in}}{12 \ \frac{\text{in}}{\text{ft}}} = 0.042 \text{ ft}$$

Heat Transfer

From Eq. 33.33, the overall coefficient of heat transfer is

$$U = \frac{1}{\sum_i \frac{L_i}{k_i} + \sum_j \frac{1}{h_j}}$$

$$\begin{aligned}\frac{1}{U} &= \frac{0.333 \text{ ft}}{0.38 \frac{\text{Btu-ft}}{\text{hr-ft}^2\text{-°F}}} + \frac{0.083 \text{ ft}}{0.06 \frac{\text{Btu-ft}}{\text{hr-ft}^2\text{-°F}}} + \frac{0.042 \text{ ft}}{0.30 \frac{\text{Btu-ft}}{\text{hr-ft}^2\text{-°F}}} \\ &\quad + \frac{1}{1.65 \frac{\text{Btu}}{\text{hr-ft}^2\text{-°F}}} + \frac{1}{6.00 \frac{\text{Btu}}{\text{hr-ft}^2\text{-°F}}} \\ &= 3.17 \text{ hr-ft}^2\text{-°F/Btu}\end{aligned}$$

$$\begin{aligned}U &= \frac{1}{3.17 \frac{\text{hr-ft}^2\text{-°F}}{\text{Btu}}} \\ &= 0.315 \text{ Btu/hr-ft}^2\text{-°F}\end{aligned}$$

From Eq. 33.37, the heat transfer is

$$\begin{aligned}q &= UA\Delta T \\ &= \left(0.315 \frac{\text{Btu}}{\text{hr-ft}^2\text{-°F}}\right)(100 \text{ ft}^2)(72\text{°F} - 30\text{°F}) \\ &= 1323 \text{ Btu/hr}\end{aligned}$$

25. LOGARITHMIC MEAN AREA

Cylinders and spheres are examples of objects whose heat transfer paths increase in area from the inside to outside. In such instances, the *logarithmic mean area*, A_m, should be used.[12]

$$q = UA_m(T_1 - T_2) \qquad 33.41$$

$$A_m = \frac{A_o - A_i}{\ln \frac{A_o}{A_i}} \qquad 33.42$$

26. RADIAL CONDUCTION THROUGH A HOLLOW CYLINDER

The logarithmic mean area (excluding the ends) for a hollow cylinder of length L and wall thickness $r_2 - r_1$ is

$$A_m = \frac{2\pi L(r_2 - r_1)}{\ln \frac{r_2}{r_1}} \qquad 33.43$$

The overall radial heat transfer through an uninsulated hollow cylinder without films is given by Eq. 33.44. This equation disregards heat transfer from the ends and assumes that the length is sufficiently large so that the heat transfer is radial at all locations.

$$q = q''A_m = \frac{kA_m(T_1 - T_2)}{r_2 - r_1} \qquad 33.44$$

Substituting Eq. 33.43 into Eq. 33.44 or Eq. 33.8 into Eq. 33.27 gives

Conduction Through a Cylindrical Wall

$$q = \frac{2\pi kL(T_1 - T_2)}{\ln \frac{r_2}{r_1}} \qquad 33.45$$

The *ASHRAE Handbook* includes tables giving heat loss data for pipes of various materials and sizes. [**Heat Loss From Bare Steel Pipe in Still Air at 80°F**] [**Heat Loss from Bare Copper Tubing in Still Air at 80°F**] [**Approximate Heat Loss from Piping at 140°F Inlet, 70°F Ambient**]

The temperature at a point within a layer a distance r from the center is given by Eq. 33.46. T_i is the temperature at the inside of the layer. For a pipe wall, the inside wall temperature is not necessarily the same as the temperature of the pipe's contents.

$$T_r = T_i - (T_i - T_o)\left(\frac{\ln \frac{r}{r_i}}{\ln \frac{r_o}{r_i}}\right) \qquad 33.46$$

27. RADIAL CONDUCTION THROUGH A COMPOSITE CYLINDER

A *composite cylinder* consists of two or more concentric, cylindrical layers. (See Fig. 33.3.) This configuration is typically encountered as an *insulated pipe*. The logarithmic mean area and heat transfer are

$$q = q''A_m = \frac{2\pi L(T_1 - T_2)}{\frac{\ln \frac{r_b}{r_a}}{k_{\text{pipe}}} + \frac{\ln \frac{r_c}{r_b}}{k_{\text{insulation}}}} \qquad 33.47$$

[12]If $A_o/A_i \leq 2$, using the arithmetic mean $\frac{1}{2}(A_o + A_i)$ in place of the logarithmic mean area will result in a maximum error of 4%, with the heat transfer being too high.

Figure 33.3 Composite Cylinder (Insulated Pipe)

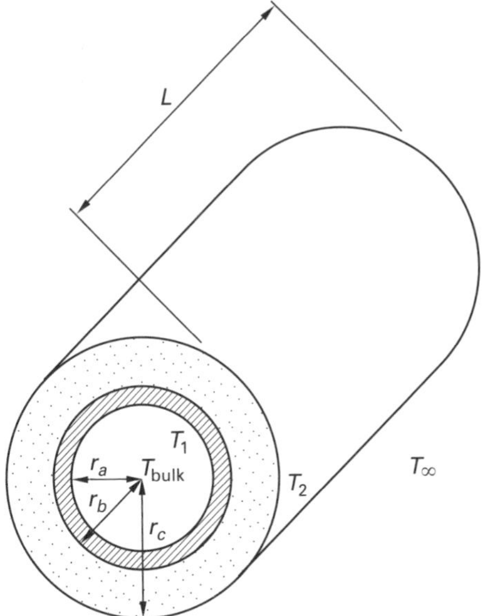

When films are present, Eq. 33.47 can be expanded. h_b is a film coefficient representing contact or interface resistance, if any.

$$q = \frac{2\pi L(T_{\text{bulk}} - T_\infty)}{\frac{1}{r_a h_a} + \frac{\ln \frac{r_b}{r_a}}{k_{\text{pipe}}} + \frac{1}{r_b h_b} + \frac{\ln \frac{r_c}{r_b}}{k_{\text{insulation}}} + \frac{1}{r_c h_c}} \quad \text{33.48}$$

Compared to the thermal resistance of the insulation, the thermal resistance of metal pipe walls is negligible. Little accuracy will be lost if the pipe wall thickness term is omitted from the denominator.[13] An equivalent assumption is that there is no temperature change across the thickness of the pipe wall.

Example 33.2

Liquid oxygen at −290°F (−180°C) is stored in a cylindrical tank with a thermal conductance of 28.0 Btu-ft/hr-ft²-°F (48.0 W/m·K), an inside diameter of 5 ft (1.5 m), and a length of 20 ft (6.0 m). The wall thickness is ⅜ in (1 cm). The tank is insulated with 1.0 ft (30 cm) of powdered diatomaceous silica with an average thermal conductivity of 0.022 Btu-ft/hr-ft²-°F (0.038 W/m·K). The surrounding air temperature is 70°F (21°C), and the outside film coefficient is 6.0 Btu/hr-ft²-°F (34 W/m²·K). The inside film coefficient is assumed to be infinite. Disregard heat transfer through the tank ends. Calculate the radial heat gain to the liquid oxygen.

SI Solution

The wall thickness is

$$t = \frac{1 \text{ cm}}{100 \frac{\text{cm}}{\text{m}}} = 0.01 \text{ m}$$

In Fig. 33.3, the corresponding radii are

$$\begin{aligned} r_a &= \frac{1.5 \text{ m}}{2} \\ &= 0.75 \text{ m} \\ r_b &= r_a + t = 0.75 \text{ m} + 0.01 \text{ m} \\ &= 0.76 \text{ m} \\ r_c &= r_b + t_{\text{insulation}} = 0.76 \text{ m} + 0.30 \text{ m} \\ &= 1.06 \text{ m} \end{aligned}$$

The heat transfer is calculated directly from Eq. 33.48.

$$\begin{aligned} q &= \frac{2\pi L(T_{\text{bulk}} - T_\infty)}{\frac{1}{r_a h_a} + \frac{\ln \frac{r_b}{r_a}}{k_{\text{tank}}} + \frac{\ln \frac{r_c}{r_b}}{k_{\text{insulation}}} + \frac{1}{r_c h_c}} \\ &= \frac{(2\pi)(6 \text{ m})\big(21°\text{C} - (-180°\text{C})\big)}{\frac{1}{(0.75 \text{ m})(\infty)} + \frac{\ln \frac{0.76 \text{ m}}{0.75 \text{ m}}}{48.0 \frac{\text{W}}{\text{m·K}}} + \frac{\ln \frac{1.06 \text{ m}}{0.76 \text{ m}}}{0.038 \frac{\text{W}}{\text{m·K}}} + \frac{1}{(1.06 \text{ m})\left(34 \frac{\text{W}}{\text{m}^2\text{·K}}\right)}} \\ &= 862.7 \text{ W} \end{aligned}$$

Customary U.S. Solution

The wall thickness is

$$\begin{aligned} t &= \frac{\frac{3}{8} \text{ in}}{12 \frac{\text{in}}{\text{ft}}} \\ &= 0.031 \text{ ft} \end{aligned}$$

[13]Since the thermal conductivity values may be in error by as much as 20%, it makes little sense to strive for perfection by including the thermal resistance of a thin-walled metal pipe.

Heat Transfer

In Fig. 33.3, the corresponding radii are

$$\begin{aligned} r_a &= \frac{5 \text{ ft}}{2} \\ &= 2.5 \text{ ft} \\ r_b &= r_a + t \\ &= 2.5 \text{ ft} + 0.031 \text{ ft} \\ &= 2.53 \text{ ft} \\ r_c &= r_b + t_{\text{insulation}} \\ &= 2.53 \text{ ft} + 1 \text{ ft} \\ &= 3.53 \text{ ft} \end{aligned}$$

The heat transfer is calculated directly from Eq. 33.48.

$$\begin{aligned} q &= \frac{2\pi L(T_{\text{bulk}} - T_\infty)}{\dfrac{1}{r_a h_a} + \dfrac{\ln \dfrac{r_b}{r_a}}{k_{\text{tank}}} + \dfrac{\ln \dfrac{r_c}{r_b}}{k_{\text{insulation}}} + \dfrac{1}{r_c h_c}} \\ &= \frac{(2\pi)(20 \text{ ft})\big(70°\text{F} - (-290°\text{F})\big)}{\begin{array}{l} \dfrac{1}{(2.5 \text{ ft})(\infty)} + \dfrac{\ln \dfrac{2.53 \text{ ft}}{2.5 \text{ ft}}}{28.0 \dfrac{\text{Btu-ft}}{\text{hr-ft}^2\text{-°F}}} \\ \quad + \dfrac{\ln \dfrac{3.53 \text{ ft}}{2.53 \text{ ft}}}{0.022 \dfrac{\text{Btu-ft}}{\text{hr-ft}^2\text{-°F}}} \\ \quad + \dfrac{1}{(3.53 \text{ ft})\left(6.0 \dfrac{\text{Btu-ft}}{\text{hr-ft}^2\text{-°F}}\right)} \end{array}} \\ &= 2979 \text{ Btu/hr} \end{aligned}$$

28. PIPE INSULATION

Commercial pipe insulation is usually fiberglass or calcium silicate.[14] Fiberglass has a high insulation value, is low in cost and low in weight, and is easy to install. However, conventional fiberglass is limited to uses below approximately 850°F (450°C).[15] Mineral wool (mineral-fiber), calcium silicate, and composite materials are used for high-temperature (i.e., above approximately 1000°F (540°C)) installations. Traditional calcium silicate, known as *cal sil*, can withstand more mechanical abuse than other insulating materials. However, it requires a saw for cutting, while fiberglass and mineral wool can easily be cut with a knife. An outer jacket (cover) of plastic (white Kraft all-service jacket (ASJ)), aluminum, or stainless steel will protect insulation from dirt and moisture.[16,17]

Insulation products are available as pipe wrap, blankets, and boards. Integral protective jacketing of pipe wrap can be provided, with sealing tape closing the hinged wrap around the pipe. Insulating boards are designed for large flat surfaces, or, when manufactured with multiple strip hinges, for large curved surfaces such as cylindrical tanks. Compressible blankets are highly flexible and are used for irregular surfaces. (See Table 33.5.)

29. CRITICAL INSULATION THICKNESS

The addition of insulation to a bare pipe or wire increases the surface area. (See Fig. 33.3.) Adding insulation to a small-diameter pipe may actually increase the heat loss above bare-pipe levels. Adding insulation up to the *critical thickness* is dominated by the increase in surface area. Only adding insulation past the critical thickness will decrease heat loss.[18] The *critical radius* is usually very small (e.g., a few millimeters), and it is most relevant in the case of insulating thin wires. The critical radius, measured from the center of the pipe or wire, is

$$r_{\text{critical,cylinder}} = \frac{k_{\text{insulation}}}{h_\infty} \qquad 33.49$$

The critical radius of insulation for a spherical tank is

$$r_{\text{critical,sphere}} = \frac{2k_{\text{insulation}}}{h_\infty} \qquad 33.50$$

A typical value for h_∞ is 1.2 Btu/hr-ft^2-°F (6.8 W/m^2·°C) assuming natural convection.

30. ECONOMIC INSULATION THICKNESS

Optimizing an insulation installation usually requires choosing an insulating material and then selecting its thickness. Fiberglass and mineral wool insulations are

[14]Asbestos insulation, commonly used in the past, has fallen from favor. This includes 85% *magnesia* insulation consisting of 85% magnesium carbonate and 15% asbestos fiber.

[15]High-temperature resin binders can increase the useful range of fiberglass insulation to approximately 1000°F (540°C).

[16]Metallic jacketing may represent a safety hazard for plant personnel since the surface is generally conductive and reflective. Jacket temperature should be less than 130°F (54°C) in areas where contact by personnel is possible.

[17]An exterior jacket should always be removed when adding a second layer of insulation. Retaining the original jacket may produce a condensation site or damage the jacket material or adhesive.

[18]There is another, less commonly used, meaning for the term *critical thickness*: the thickest required insulation. In situations where the required insulation thickness is different for energy conservation, condensation control, personnel protection, and process temperature control, the "critical" thickness is the thickness that controls the design.

Table 33.5 *Representative Conductivities of Pipe Insulation (Btu-ft/hr-ft²-°F)*

	insulation temperature					
	100°F (38°C)	200°F (93°C)	300°F (149°C)	400°F (204°C)	500°F (260°C)	600°F (316°C)
calcium silicate	0.033	0.037	0.041	0.046	0.057	0.060
cellular glass	0.039	0.047	0.055	0.064	0.074	0.085
fiberglass	0.026	0.030	0.034			
magnesia, 85%	0.034	0.037	0.041	0.044		
polyurethane	0.016	0.016	0.016			

(Multiply Btu-ft/hr-ft^2-°F by 12 to obtain Btu in/hr-ft^2-°F.)

(Multiply Btu-ft/hr-ft^2-°F by 1.7307 to obtain W/m·K.)

(Multiply Btu-ft/hr-ft^2-°F by 4.1365×10^{-3} to obtain cal·cm/s·cm^2·°C.)

commonly less expensive than traditional calcium silicate. However, other factors are included in the economic analysis, including cost of installation labor, thermal efficiency, current fuel cost, and useful life.[19]

Adding more insulation conserves more energy, but the costs of material and installation are also greater. Insulation thickness is optimized by balancing the heat losses against the cost of insulating. The *economic insulation thickness* is the thickness that minimizes the annual cost of ownership and operation. The economic thickness will vary with pipe diameter.

The *ASHRAE Handbook—Fundamentals* contains tables that give minimum pipe insulation thicknesses for varying pipe sizes and operating temperature ranges, and minimum supply and return duct insulation thicknesses for varying installation locations and climate zones. **[Minimum Pipe Insulation Thickness][Minimum Duct Insulation R-Value of Cooling-Only and Heating-Only Supply Ducts and Return Ducts]**

31. INSULATION THICKNESS TO PREVENT FREEZING OF WATER PIPE

Freezing of water flowing in a pipe will be prevented if the exit water temperature is kept from dropping below 32°F. The maximum allowable heat loss per unit mass from the water is

$$\frac{q_{\text{max}}}{\dot{m}} = c_p(T_{\text{entrance,water}} - 32°\text{F}) \qquad 33.51$$

If there is no water flow, a pipe exposed for long periods to subfreezing temperatures cannot be prevented from freezing.

The *ASHRAE Handbook—Fundamentals* contains a table that gives the time needed to freeze water in a pipe (assuming no flow) for various pipe sizes and insulation thicknesses. **[Time Needed to Freeze Water, in Hours]**

32. INSULATION THICKNESS TO PREVENT SWEATING

Sweating is the condensation of moisture from the surrounding air on the surface of the pipe insulation. Sweating is prevented by keeping the surface temperature above the air's dew-point temperature.[20]

33. RADIAL CONDUCTION THROUGH A SPHERICAL SHELL

For a spherical shell, the logarithmic mean area and heat transfer are

$$A_m = \sqrt{A_o A_i} \qquad 33.52$$

$$\begin{aligned} q'' &= qA_m \\ &= k\sqrt{A_o A_i}\left(\frac{T_i - T_o}{r_o - r_i}\right) \\ &= \frac{4\pi k r_o r_i (T_i - T_o)}{r_o - r_i} \end{aligned} \qquad 33.53$$

The thermal resistance of each shell layer is

$$R = \left(\frac{1}{4\pi k}\right)\left(\frac{r_o - r_i}{r_o r_i}\right) \qquad 33.54$$

[19]The North American Insulation Manufacturers Association (NAIMA) can provide manual and computerized methods of evaluating the economics of different insulations.

[20]The minimum thickness is found by equating the heat transfer through the insulation by conduction to the heat transfer from the surface by convection and radiation. As a first approximation, a total heat transfer coefficient, h_{total}, for convection and radiation of 0.65 Btu/hr-ft^2-°F (3.7 W/m^2·K) can be used.

Spheres have the smallest surface area-to-volume ratio, so spherical tanks are used where heat transfer is to be minimized. Unlike for plane surfaces and cylindrical tanks, steady-state heat transfer from spheres cannot be reduced indefinitely by the addition of increasing amounts of insulation. For a sphere, steady-state conduction can never be less than the minimum value given by Eq. 33.55, no matter how thick the wall is. If r_o is infinite, Eq. 33.53 becomes

$$q_{\text{min}} = 4\pi k r_i (T_i - T_o) \qquad 33.55$$

The radial heat transfer through a series of concentric spherical layers and an outside film, h_c, at radius r_c, is given by Eq. 33.56. (See Fig. 33.4.)

$$q = \frac{4\pi(T_1 - T_\infty)}{\dfrac{\dfrac{1}{r_a} - \dfrac{1}{r_b}}{k_{\text{inner layer}}} + \dfrac{\dfrac{1}{r_b} - \dfrac{1}{r_c}}{k_{\text{outer layer}}} + \dfrac{1}{r_c^2 h_c}} \qquad 33.56$$

Figure 33.4 Radial Heat Transfer Through Concentric Spherical Layers and Outside Film

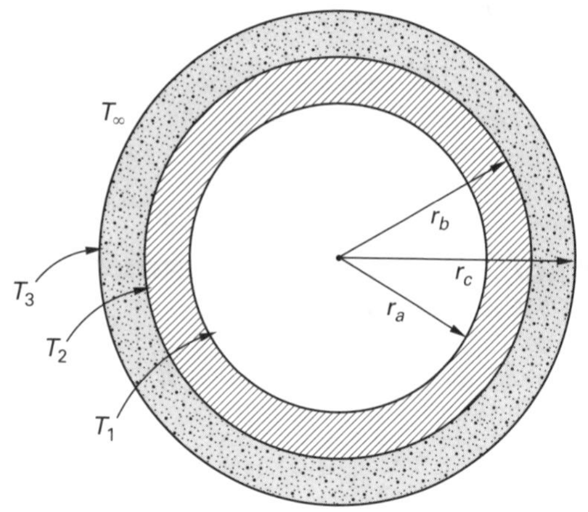

34. CONDUCTION THROUGH A CUBE

Conduction through a roughly cubic shell consisting of a small cavity surrounded by thick walls can be calculated from Eq. 33.57. The constant 0.725 is a semi-empirical correction factor.[21]

$$q \approx 0.725k\sqrt{A_o A_i}\left(\frac{T_i - T_o}{t_{\text{wall}}}\right) \quad \left[\frac{A_o}{A_i} \geq 2\right] \qquad 33.57$$

35. TRANSIENT CONDUCTION

Matter changes temperature when it gains or loses thermal energy. A change in the temperature gradient over time invalidates the use of Eq. 33.22 except for the calculation of an instantaneous heat transfer. As the temperature gradient decreases, the heat transfer also decreases. *Transient heat transfer* (also known as *unsteady-state heat transfer*) problems with constant-temperature surroundings can be solved by a variety of methods, including Newton's method (for simplified problems), the lumped parameter method, and graphical methods.

36. NEWTON'S METHOD

If the temperature difference between a body and the surroundings is not too large, the rate of heat transfer by all modes is approximately proportional to the temperature difference. This empirical relationship is valid for all modes of heat transfer and is known as *Newton's law of cooling* (or *heating*) and the *Newtonian method*.[22,23] The temperature at time t (starting at zero) is

$$T_t = T_\infty + (T_0 - T_\infty)e^{-rt} \qquad 33.58$$

For each configuration, r is a rate constant that is usually found from other known information. Either the rate constant or the time for a change in temperature from T_0 to T_t can be found from Eq. 33.59.

$$rt = -\ln\frac{T_t - T_\infty}{T_0 - T_\infty} \qquad 33.59$$

Example 33.3

Water in a 5 in (12.7 cm) diameter, 10 ft (3.0 m) long tank cools from 200°F (93°C) to 190°F (88°C) in 1.25 hr. The ambient temperature is 75°F (24°C). Approximately how long will it take for the water to cool from 200°F to 125°F (93°C to 52°C)?

SI Solution

From Eq. 33.59,

$$\begin{aligned} r &= -\frac{1}{t}\ln\frac{T_t - T_\infty}{T_0 - T_\infty} \\ &= -\frac{1}{1.25 \text{ h}}\ln\frac{88°\text{C} - 24°\text{C}}{93°\text{C} - 24°\text{C}} \\ &= 0.0602 \text{ 1/h} \end{aligned}$$

[21] Studies describing more accurate methods are available.

[22] *Newton's law of cooling* applies to the combined heat transfer mechanisms of conduction, convection, and radiation.

[23] Newton's law of cooling is essentially the same as the lumped parameter method (see Sec. 33.37). The choice of the two methods depends on what information is known. If information about the temperature history is known, Newton's law of cooling is applicable. If material properties are known, the lumped parameter method should be used.

Solve Eq. 33.59 for the unknown time.

$$\begin{aligned} t &= -\frac{1}{r}\ln\frac{T_t - T_\infty}{T_0 - T_\infty} \\ &= -\frac{1}{0.0602\ \frac{1}{\text{h}}}\ln\frac{52^\circ\text{C} - 24^\circ\text{C}}{93^\circ\text{C} - 24^\circ\text{C}} \\ &= 14.98\ \text{h} \end{aligned}$$

Customary U.S. Solution

From Eq. 33.59,

$$\begin{aligned} r &= -\frac{1}{t}\ln\frac{T_t - T_\infty}{T_0 - T_\infty} \\ &= -\frac{1}{1.25\ \text{hr}}\ln\frac{190^\circ\text{F} - 75^\circ\text{F}}{200^\circ\text{F} - 75^\circ\text{F}} \\ &= 0.0667\ 1/\text{hr} \end{aligned}$$

Solve Eq. 33.59 for the unknown time.

$$\begin{aligned} t &= -\frac{1}{r}\ln\frac{T_t - T_\infty}{T_0 - T_\infty} \\ &= -\frac{1}{0.0667\ \frac{1}{\text{hr}}}\ln\frac{125^\circ\text{F} - 75^\circ\text{F}}{200^\circ\text{F} - 75^\circ\text{F}} \\ &= 13.74\ \text{hr} \end{aligned}$$

37. LUMPED PARAMETER METHOD

If the Biot number, $\text{Bi} = hV/kA_s$, is much less than one (say, approximately 0.10 or less), the internal thermal resistance of a body will be smaller in comparison to the external thermal resistance. In that case, the *lumped parameter method* can be used to approximate the transient (time-dependent) heat flow.[24,25] The temperature at time t (starting at zero) is

$$T_t = T_\infty + (T_0 - T_\infty)e^{-\text{BiFo}} \qquad 33.60$$

The instantaneous heat transfer at time t is

$$q_t = hA_s(T_t - T_\infty) = hA_s(T_0 - T_\infty)e^{-\text{BiFo}} \qquad 33.61$$

38. TOTAL AMOUNT OF HEAT TRANSFERRED (COOLING)

Since the heat transfer at any time t is proportional to the temperature difference, the cumulative thermal energy, U_{total}, transferred up to time, t, is proportional to the original energy content and the remaining temperature gradient.

$$\begin{aligned} U_{\text{total},t} &= \int Q_t\,dt = \left(\frac{T_0 - T_t}{T_0 - T_\infty}\right)U_{\text{total},0} \\ &= \left(1 - \frac{T_t - T_\infty}{T_0 - T_\infty}\right)U_{\text{total},0} \end{aligned} \qquad 33.62$$

$$U_{\text{total},0} = mc_p(T_0 - T_\infty) = V\rho c_p(T_0 - T_\infty) \qquad 33.63$$

39. LUMPED PARAMETER ELECTRICAL ANALOGY

The analogy of a capacitor discharging through a resistor is often used to illustrate lumped parameter performance.[26] This analogy has given rise to the concepts of thermal capacitance and thermal resistance. The equivalent *thermal capacitance* (also known as the *capacity* of the system) and equivalent *thermal resistance* are

$$C_{\text{eq}} = c_p\rho V \qquad 33.64$$

$$R_{\text{eq}} = \frac{1}{hA_s} \qquad 33.65$$

These equivalent variables can be used to calculate the performance of a transient system.

Constant Fluid Temperature

$$\begin{aligned} T - T_\infty &= (T_0 - T_\infty)e^{-\beta t} \\ &= (T_0 - T_\infty)e^{-t/R_{\text{eq}}C_{\text{eq}}} \end{aligned} \qquad 33.66$$

In Eq. 33.66,

Constant Fluid Temperature

$$\beta = \frac{hA_s}{\rho V c_p} \qquad 33.67$$

[24]Because the assumptions are the same, this method is also referred to as *Newton's method.*
[25]The maximum error will be less than 7%.
[26]The analogy is often made, but there is no significant advantage to solving problems in this manner.

The instantaneous heat transfer at time t is

$$\begin{aligned} Q_t &= -\rho V c_p \left(\frac{dT}{dt}\right) \\ &= hA_s(T_t - T_\infty) \\ &= hA_s(T_0 - T_\infty)e^{-t/R_{eq}C_{eq}} \\ &= hA_s(T_0 - T_\infty)e^{-\beta t} \end{aligned} \qquad 33.68$$

The *thermal time constant*, τ, so named because it has units of time, is a measure of the rate of decay of the thermal gradient. The heat content (i.e., temperature gradient, $T_t - T_\infty$) of the source will be reduced by 63.2% in one time constant (i.e., will be reduced to 36.8% of its original value, $T_0 - T_\infty$).

$$\tau = \frac{c_p \rho L_c}{h} = \frac{c_p \rho V}{hA_s} \qquad 33.69$$

The thermal time constant is thus the inverse of β in Eq. 33.67.

Constant Fluid Temperature

$$\beta = \frac{1}{\tau} \qquad 33.70$$

Example 33.4

A 0.03125 in (0.8 mm) diameter copper wire is heated by a short circuit to 300°F (150°C) before a slow-blow fuse burns out and all heating ceases. The ambient temperature is 100°F (38°C), and the film coefficient on the wire is 1.65 Btu/hr-ft²-°F (9.37 W/m²·K). Use the following copper properties to determine how long it will take for the wire temperature to drop to 120°F (49°C).

conductivity, k	224 Btu-ft/hr-ft²-°F	(388 W/m·K)
specific heat, c_p	0.091 Btu/lbm-°F	(380 J/kg·K)
density, ρ	558 lbm/ft³	(8940 kg/m³)

SI Solution

The characteristic length of the wire is half of the radius.

$$\begin{aligned} L_c &= \frac{r}{2} \\ &= \frac{D}{4} \\ &= \frac{0.0008 \text{ m}}{4} \\ &= 0.0002 \text{ m} \end{aligned}$$

The Biot number is

$$\begin{aligned} \text{Bi} &= \frac{hL_c}{k} = \frac{\left(9.37 \ \frac{\text{W}}{\text{m}^2\text{·K}}\right)(0.0002 \text{ m})}{388 \ \frac{\text{W}}{\text{m·K}}} \\ &= 4.83 \times 10^{-6} \quad [<0.1] \end{aligned}$$

The Fourier number is

$$\begin{aligned} \text{Fo} &= \frac{kt}{\rho c_p L_c^2} \\ &= \frac{\left(388 \ \frac{\text{W}}{\text{m·K}}\right)t}{\left(8940 \ \frac{\text{kg}}{\text{m}^3}\right)\left(380 \ \frac{\text{J}}{\text{kg·K}}\right)(0.0002 \text{ m})^2} \\ &= 2855t \quad [t \text{ in seconds}] \end{aligned}$$

From Eq. 33.60,

$$\begin{aligned} T_t &= T_\infty + (T_0 - T_\infty)e^{-\text{BiFo}} \\ 49°\text{C} &= 38°\text{C} + (150°\text{C} - 38°\text{C})e^{-(4.83\times10^{-6})(2855t)} \\ 0.0982 &= e^{-0.0138t} \\ t &= \frac{-1}{0.0138}\ln 0.0982 = 168 \text{ s} \quad (0.047 \text{ h}) \end{aligned}$$

Customary U.S. Solution

From Sec. 33.10, the characteristic length of the wire is half of the radius.

$$\begin{aligned} L_c &= \frac{r}{2} = \frac{D}{4} = \frac{0.03125 \text{ in}}{(4)\left(12 \ \frac{\text{in}}{\text{ft}}\right)} \\ &= 0.00065 \text{ ft} \end{aligned}$$

The Biot number is

$$\begin{aligned} \text{Bi} &= \frac{hL_c}{k} = \frac{\left(1.65 \ \frac{\text{Btu}}{\text{hr-ft}^2\text{-°F}}\right)(0.00065 \text{ ft})}{224 \ \frac{\text{Btu}}{\text{hr-ft}^2\text{-°F}}} \\ &= 4.79 \times 10^{-6} \quad [<0.1] \end{aligned}$$

The Fourier number is

$$\begin{aligned} \text{Fo} &= \frac{kt}{\rho c_p L_c^2} \\ &= \frac{\left(224\ \frac{\text{Btu-ft}}{\text{hr-ft}^2\text{-°F}}\right)t}{\left(558\ \frac{\text{lbm}}{\text{ft}^3}\right)\left(0.091\ \frac{\text{Btu}}{\text{lbm-°F}}\right)(0.00065\ \text{ft})^2} \\ &= 1.044\times 10^7 t \quad [t \text{ in hours}] \end{aligned}$$

From Eq. 33.60,

$$\begin{aligned} T_t &= T_\infty + (T_0 - T_\infty)e^{-\text{BiFo}} \\ 120°\text{F} &= 100°\text{F} \\ &\quad + (300°\text{F} - 100°\text{F})e^{-(4.79\times 10^{-6})(1.044\times 10^7 t)} \\ 0.10 &= e^{-50t} \\ t &= \left(\frac{-1}{50}\right)\ln 0.1 = 0.046\ \text{hr} \quad (166\ \text{sec}) \end{aligned}$$

40. GRAPHICAL SOLUTIONS TO TRANSIENT HEAT TRANSFER

If the Biot number is greater than approximately 0.1, the internal resistance of the object cannot be disregarded. If Fo $\geq$ 0.2 and the geometry is sufficiently simple, a graphical approach can be used. Graphs for transient heat flow problems are available for simple shapes, including *semi-infinite solids* (i.e., an object that is infinite in one direction only), spheres, large (infinite) slabs, and long (infinite) cylinders.

Graphs for solving transient heat transfer problems are known as *temperature-time charts*. (See App. 33.E, App. 33.F, and App. 33.G.) *Heisler charts* exhibit similar information for a particular point in the solid, but use a logarithmic scale to cover a greater range of values (in particular, longer time periods).[27] (See App. 33.H, App. 33.I, and App. 33.J.)

Values in temperature-time charts are plotted as functions of the modified Biot and modified Fourier numbers. (See Sec. 33.14.) Temperature-time charts may be general enough to determine the temperature history at any point in the solid, or they may be unique to some point within the shape (e.g., the center of a sphere or the centerline of a slab).[28] Example 33.5 illustrates the use of temperature-time charts.

Example 33.5

Whole peaches at 80°F (27°C) are to be cooled in 14.7 psia, 10°F (101 kPa, −12°C) air by natural convection prior to being completely frozen. The average film coefficient on the peaches is 5.8 Btu/hr-ft²-°F (33 W/m²·K). The peaches are placed on trays and are spaced far enough apart so that they do not affect one another. The peaches are considered to be homogeneous spheres with 3 in (7.6 cm) diameters. Use the properties listed to determine how long it will take for the peach centers to cool to 40°F (4°C).

conductivity, k	0.4 Btu-ft/hr-ft²-°F	(0.7 W/m·K)
specific heat, c_p	1.0 Btu/lbm-°F	(4.2 kJ/kg·K)
thermal diffusivity, α	0.04 ft²/hr	(1.0×10^{-6} m²/s)

SI Solution

The characteristic length of a sphere is

$$\begin{aligned} L_{c,\text{sphere}} &= \frac{r}{3} = \frac{D}{6} = \frac{0.076\ \text{m}}{6} \\ &= 0.0127\ \text{m} \end{aligned}$$

The Biot number is

$$\begin{aligned} \text{Bi} &= \frac{hL_c}{k} \\ &= \frac{\left(33\ \frac{\text{W}}{\text{m}^2\cdot\text{K}}\right)(0.0127\ \text{m})}{0.7\ \frac{\text{W}}{\text{m}\cdot\text{K}}} \\ &= 0.599 \end{aligned}$$

Since Bi > 0.1, the lumped parameter method cannot be used. Calculate the parameters needed to use App. 33.E.

$$\begin{aligned} \frac{T_t - T_\infty}{T_0 - T_\infty} &= \frac{4°\text{C} - (-12°\text{C})}{27°\text{C} - (-12°\text{C})} = 0.410 \\ r_o &= \frac{D}{2} = \frac{0.076\ \text{m}}{2} = 0.038\ \text{m} \\ \frac{k}{hr_o} &= \frac{0.7\ \frac{\text{W}}{\text{m}\cdot\text{K}}}{\left(33\ \frac{\text{W}}{\text{m}^2\cdot\text{K}}\right)(0.038\ \text{m})} \\ &= 0.56 \end{aligned}$$

[27]The charts are named after H. P. Heisler, who published his original graphical methods in the *ASME Transactions* in 1947.

[28]Auxiliary *position-correction charts* have also been developed by researchers and are available to correct the center-line or surface temperature to other locations within the body. These are available in most heat transfer textbooks.

Since the desired temperature corresponds to the center of the peach, use App. 33.E. The value of $\alpha t/r_o^2$ is found from the chart to be approximately 0.3.

$$\frac{\alpha t}{r_o^2} = 0.3$$

$$t = \frac{0.3r_o^2}{\alpha} = \frac{(0.3)(0.038\ \text{m})^2}{1.0\times 10^{-6}\ \frac{\text{m}^2}{\text{s}}}$$

$$= 433\ \text{s} \quad (0.120\ \text{h})$$

Customary U.S. Solution

The characteristic length of a sphere is

$$L_{c,\text{sphere}} = \frac{r}{3} = \frac{D}{6} = \frac{3\ \text{in}}{(6)\left(12\ \frac{\text{in}}{\text{ft}}\right)}$$

$$= 0.0417\ \text{ft}$$

The Biot number is

$$\text{Bi} = \frac{hL_c}{k}$$

$$= \frac{\left(5.8\ \frac{\text{Btu}}{\text{hr-ft}^2\text{-°F}}\right)(0.0417\ \text{ft})}{0.4\ \frac{\text{Btu-ft}}{\text{hr-ft}^2\text{-°F}}}$$

$$= 0.605$$

Since Bi > 0.1, the lumped parameter method cannot be used. Calculate the parameters needed to use App. 33.E.

$$\frac{T_t - T_\infty}{T_0 - T_\infty} = \frac{40\text{°F} - 10\text{°F}}{80\text{°F} - 10\text{°F}} = 0.43$$

$$r_o = \frac{D}{2} = \frac{3\ \text{in}}{(2)\left(12\ \frac{\text{in}}{\text{ft}}\right)} = 0.125\ \text{ft}$$

$$\frac{k}{hr_o} = \frac{0.4\ \frac{\text{Btu-ft}}{\text{hr-ft}^2\text{-°F}}}{\left(5.8\ \frac{\text{Btu}}{\text{hr-ft}^2\text{-°F}}\right)(0.125\ \text{ft})}$$

$$= 0.55$$

Since the desired temperature corresponds to the center of the peach, use App. 33.E. The value of $\alpha t/r_o^2$ is found from the chart to be approximately 0.3.

$$\frac{\alpha t}{r_o^2} = 0.3$$

$$t = \frac{0.3r_o^2}{\alpha} = \frac{(0.3)(0.125\ \text{ft})^2}{0.04\ \frac{\text{ft}^2}{\text{hr}}}$$

$$= 0.117\ \text{hr} \quad (422\ \text{sec})$$

41. INTERNAL HEAT GENERATION

Some objects generate their own heat internally, for example, electrical heating elements, nuclear fuel rods, chemical reaction beds, beds of fine coal on furnace grates, and composting heaps. The relationship between the heat generation rate, G (in Btu/hr-ft^3 or W/m^3), and distribution of temperature within the object can be evaluated for simple steady-state cases involving homogeneous material and uniform heat generation.

42. FLAT PLATE WITH HEAT GENERATION

An infinite flat plate with thickness $t = 2L$ will dissipate internally generated heat equally and uniformly from its two surfaces.[29] The temperature a distance x from the closest surface is given by Eq. 33.71. (x is measured from the surface, not from the center.)

$$T_x = T_s + \frac{GLx}{k} - \frac{Gx^2}{2k} \quad [x \le L] \qquad 33.71$$

$$T_s = T_{\text{center}} - \frac{GL^2}{2k} \qquad 33.72$$

43. CYLINDER WITH HEAT GENERATION

An infinitely long cylinder with outside radius r_o will dissipate internally generated heat uniformly along its entire length. The temperature at distance r from the center is given by Eq. 33.73.

$$T_r = T_s + \left(\frac{Gr_o^2}{4k}\right)\left(1 - \left(\frac{r}{r_o}\right)^2\right) = T_{\text{center}} - \frac{Gr^2}{4k} \qquad 33.73$$

$$T_s = T_{\text{center}} - \frac{Gr_o^2}{4k} \qquad 33.74$$

[29] Assuming the plate to be infinite in two out of its three dimensions eliminates the need to determine the heat transfer from its ends.

Example 33.6

Both surfaces of a large 0.25 in (6.4 mm) thick copper plate are kept at 80°F (27°C) by a circulating coolant. The thermal conductivity of the copper is 224 Btu-ft/hr-ft^2-°F (388 W/m·K). The electrical resistivity of the copper is 0.68×10^{-6} Ω-in (1.7×10^{-6} Ω·cm). A DC electrical potential of 0.01 V is applied across the two surfaces. What is the temperature at the geometric midpoint of the plate?

SI Solution

The thickness of the copper plate is

$$\begin{aligned} t &= 2L \\ &= 0.0064 \text{ m} \end{aligned}$$

The heat generation rate, G, is per unit volume (i.e., per cubic meter) of the copper plate. Therefore, it is convenient to work with a square section of the plate having a volume of 1 m^3. The length of the sides of a square of the copper plate is

$$\begin{aligned} V &= \text{thickness} \times w^2 \\ 1 \text{ m}^3 &= (0.0064 \text{ m})w^2 \\ w &= 12.5 \text{ m} \end{aligned}$$

From the definition of resistivity, the electrical resistance of the plate is

$$\begin{aligned} R_{\text{electrical}} &= \frac{\rho(\text{thickness})}{A} \\ &= \frac{(1.7 \times 10^{-6} \text{ }\Omega\cdot\text{cm})(0.0064 \text{ m})}{(12.5 \text{ m})^2\left(100 \frac{\text{cm}}{\text{m}}\right)} \\ &= 6.96 \times 10^{-13} \text{ }\Omega \end{aligned}$$

The thermal power generation, G, in the section is the same as the electrical power dissipated. For a DC voltage, $P = IE = E^2/R = I^2R$.

$$\begin{aligned} G &= \frac{E^2}{R_{\text{electrical}}} \\ &= \frac{(0.01 \text{ V})^2}{6.96 \times 10^{-13} \text{ }\Omega} \\ &= 1.44 \times 10^8 \text{ W/m}^3 \end{aligned}$$

Equation 33.72 is used to calculate the center temperature. By definition, L is half of the plate thickness.

$$L = \frac{t}{2} = \frac{0.0064 \text{ m}}{2} = 0.0032 \text{ m}$$

$$\begin{aligned} T_{\text{center}} &= T_s + \frac{GL^2}{2k} \\ &= 27°\text{C} + \frac{\left(1.44 \times 10^8 \frac{\text{W}}{\text{m}^3}\right)(0.0032 \text{ m})^2}{(2)\left(388 \frac{\text{W}}{\text{m}\cdot\text{K}}\right)} \\ &= 28.9°\text{C} \end{aligned}$$

Customary U.S. Solution

The thickness of the copper plate is

$$t = 2L = \frac{0.25 \text{ in}}{12 \frac{\text{in}}{\text{ft}}} = 0.02083 \text{ ft}$$

The heat generation rate, G, is per unit volume (i.e., per cubic foot) of the copper plate. Therefore, it is convenient to work with a square section of the plate having a volume of 1 ft^3. The length of the sides of a square of the copper plate is

$$\begin{aligned} V &= \text{thickness} \times w^2 \\ 1 \text{ ft}^3 &= (0.02083 \text{ ft})w^2 \\ w &= 6.93 \text{ ft} \end{aligned}$$

From the definition of resistivity, the electrical resistance of the plate is

$$\begin{aligned} R_{\text{electrical}} &= \frac{\rho(\text{thickness})}{A} \\ &= \frac{(0.68 \times 10^{-6} \text{ }\Omega\text{-in})(0.02083 \text{ ft})}{(6.93 \text{ ft})^2\left(12 \frac{\text{in}}{\text{ft}}\right)} \\ &= 2.46 \times 10^{-11} \text{ }\Omega \end{aligned}$$

The thermal power generation, G, in the section is the same as the electrical power dissipated. For a DC voltage, $P = IE = E^2/R = I^2R$.

$$\begin{aligned} G &= \frac{E^2}{R_{\text{electrical}}} = \frac{(0.01 \text{ V})^2}{2.46 \times 10^{-11} \text{ }\Omega} \\ &= 4.07 \times 10^6 \text{ W/ft}^3 \end{aligned}$$

Convert from watts to Btu/hr.

$$G = \left(4.07 \times 10^6 \ \frac{\text{W}}{\text{ft}^3}\right)\left(3.412 \ \frac{\text{Btu}}{\text{W-hr}}\right)$$
$$= 1.39 \times 10^7 \ \text{Btu/hr-ft}^3$$

Equation 33.72 is used to calculate the center temperature. By definition, L is half of the plate thickness.

$$L = \frac{t}{2} = \frac{0.25 \ \text{in}}{(2)\left(12 \ \frac{\text{in}}{\text{ft}}\right)} = 0.01042 \ \text{ft}$$

$$T_{\text{center}} = T_s + \frac{GL^2}{2k}$$
$$= 80°\text{F} + \frac{\left(1.39 \times 10^7 \ \frac{\text{Btu}}{\text{hr-ft}^3}\right)(0.01042 \ \text{ft})^2}{(2)\left(224 \ \frac{\text{Btu-ft}}{\text{hr-ft}^2\text{-°F}}\right)}$$
$$= 83.4°\text{F}$$

44. FINS

Fins (*extended surfaces*) are objects that receive and move thermal energy by conduction along their length and width prior to (in most cases) convective and radiative heat removal. Radiators include simple fins, fin tubes, finned channels, and heat pipes. Some simple configurations can be considered and evaluated as fins even though that is not their intended function.[30]

External fins are attached at their base to a source of thermal energy at temperature T_b. It is assumed that the temperature is constant across the face of the fin at any point along its extent.

The *radiator efficiency* (*fin efficiency*), η, is the ratio of the actual to ideal heat transfers assuming the entire fin is at the base temperature, T_b. (See Fig. 33.5.)

$$\eta_f = \frac{q_{\text{actual}}}{q_{\text{ideal}}} = \frac{q_{\text{actual}}}{hA_f(T_b - T_\infty)} = \frac{q_{\text{actual}}}{h\big(2Lw + t(2L + w)\big)(T_b - T_\infty)} \qquad 33.75$$

The fin efficiency can also be calculated from Eq. 33.76.[31,32] P is the perimeter length of the exposed face, A_c.

$$\eta_f = \frac{\tanh mL}{mL} \qquad 33.76$$

$$m = \sqrt{\frac{hP}{kA_c}} = \sqrt{\frac{h\big(2(w+t)\big)}{ktw}} \qquad 33.77$$

(The function tanh is the hyperbolic tangent, not the tangent of the film coefficient.) The area, A_c, in Eq. 33.77 is the cross-sectional area of the fin. In rectangular and cylindrical fins, this area is uniform along the fin length. For a pipe, it is the solid annular area. (See Fig. 33.6.)

Not all heat radiators increase the heat transfer. In some cases, the fin can act as insulation. The functionality of a radiator depends on its size (specifically, its face perimeter, P). If the quantity hA/Pk is less than 1.0, the perimeter will be large enough and the radiator will increase the heat transfer.[33] If hA/Pk is greater than 1.0, the radiator will insulate. This may be the case where the film coefficient, h, is large, as it is with boiling or high-velocity (turbulent) liquids.

The *radiator effectiveness* (*fin effectiveness*) is the ratio of the actual heat transfer with a fin to the heat transfer that would have occurred without a fin. This is not the same as the radiator efficiency. Fin efficiencies are less than 100%; effectiveness is generally greater than 100%.

$$E = \frac{q_{\text{actual}}}{q_{\text{no fin}}} = \frac{q_{\text{actual}}}{hA_c(T_b - T_\infty)} \qquad 33.78$$

The film coefficient, h, should be evaluated at the average of the fin surface and environmental temperatures. Since the fin surface temperature varies along its length, the midpoint surface temperature is used in the calculation.

$$T_h = \tfrac{1}{2}\left(\tfrac{1}{2}(T_b + T_{x=L}) + T_\infty\right) \qquad 33.79$$

Radiation may be significant along all or a part of the fin. A *combined* (*total, overall,* etc.) *film coefficient*, h_{total}, that accounts for both convective and radiative heat transfers should be used.

$$h_{\text{total}} = h + h_r \qquad 33.80$$

Unless stated otherwise, the formulas for temperature of and heat transfer from a fin do not consider the heat transfer from the exposed end. For that reason, the fin is said to possess an *adiabatic tip* or *insulated tip*. If the tip area is significant or if it is desired to include the heat transfer from the tip for other reasons, the film

[30] As an example, a wire being soldered at one end can be considered as a rod-shaped fin. Two pipes being butt-welded together end to end can be considered as two infinite fins, each dissipating half of the welding energy.

[31] The arguments of hyperbolic functions are in radians.

[32] Graphs of fin efficiency are available for rectangular, triangular, parabolic, and annual fins.

[33] In practice, the extra material, installation, and maintenance expense associated with fins is rarely justified unless $hA/Pk < 0.25$.

Figure 33.5 *Fin Efficiency (refer to Fig. 33.6 for dimensions of fins)*

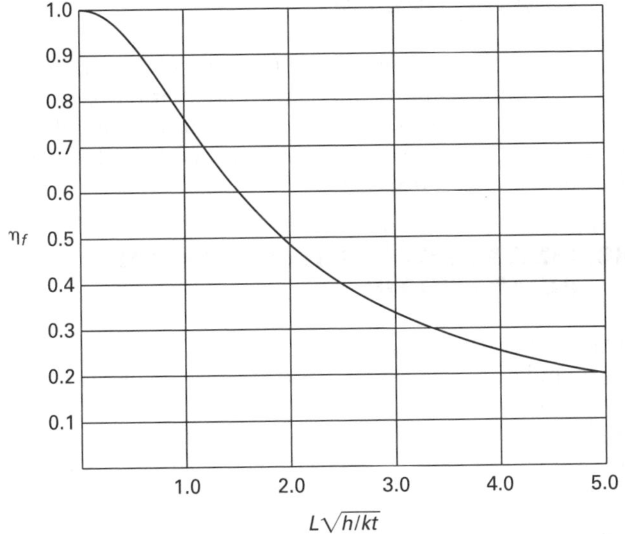

(a) rectangular (straight) fins

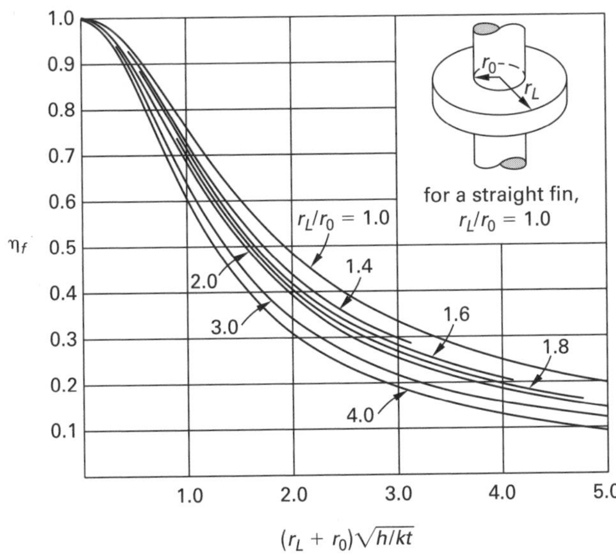

(b) circular (transverse) fins

Adapted from *Engineering Heat Transfer*, by James R. Welty, John Wiley & Sons, copyright © 1974.

Figure 33.6 *Finned Heat Radiators*

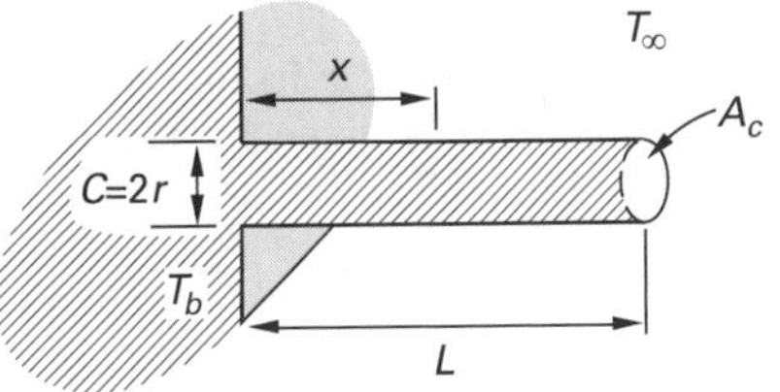

(a) finite cylindrical fin (pin fin)

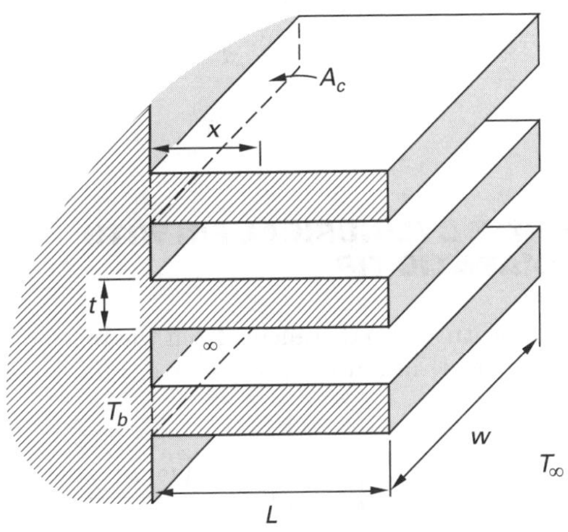

(b) finite rectangular fin (straight fin)

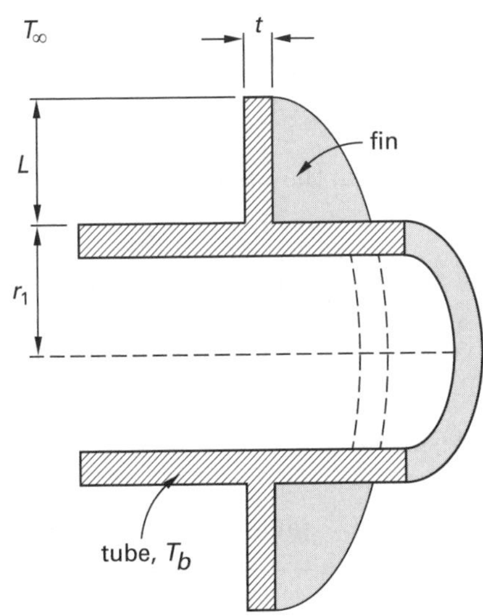

(c) annular (circular) fin (transverse fin)

coefficient for the fin can be used as an approximation.[34] Closed form solutions for most common nonadiabatic cases are available in heat transfer textbooks.

Fins should not be used inside fluid-carrying pipes. Aside from the difficulty of manufacturing, spiral or transverse fins within a pipe would trap fluid and substantially decrease the internal film coefficient. Longitudinal fins would "laminarize" the flow, also decreasing the internal film coefficient.

[34]Even this is an approximation, as the additional heat transfer will change the temperature distribution along the fin length. However, the effect is probably minimal.

45. INFINITE CYLINDRICAL FIN

The temperature at a distance x from the base of an infinite cylindrical (rod) fin is given by Eq. 33.81. The variable m is given by Eq. 33.77.

$$T_x = T_\infty + (T_b - T_\infty)e^{-mx} \qquad 33.81$$

The heat transfer from the barrel of the cylinder is given by Eq. 33.82. The quantity under the square root is not the same as the variable defined by Eq. 33.77.

$$q = \sqrt{hPkA_c}\,(T_b - T_\infty) \qquad 33.82$$

$$A_c = \pi r^2 \qquad 33.83$$

$$P = \pi D = 2\pi r \qquad 33.84$$

46. FINITE CYLINDRICAL FIN WITH ADIABATIC TIP

The temperature at a distance x from the base of a finite cylindrical (rod) fin, also known as a *pin fin*, is given by Eq. 33.85.

$$T_x = T_\infty + (T_b - T_\infty)\left(\frac{\cosh\big(m(L-x)\big)}{\cosh mL}\right) \qquad 33.85$$

$$P = \pi D = 2\pi r \qquad 33.86$$

$$A_c = \pi r^2 \qquad 33.87$$

$$m = \sqrt{\frac{hP}{kA_c}} = \sqrt{\frac{2h}{kr}} \qquad 33.88$$

The heat transfer from the barrel of the cylinder is

Fins

$$q = \sqrt{hPkA_c}\,(T_b - T_\infty)\tanh mL_c \qquad 33.89$$

The corrected fin length is not needed for a fin with an adiabatic tip, so L_c in Eq. 33.89 is equal to L.

The fin efficiency is

$$\eta_f = \frac{\tanh\sqrt{\frac{2hL^2}{kr}}}{\sqrt{\frac{2hL^2}{kr}}} \qquad 33.90$$

47. FINITE CYLINDRICAL FIN WITH CONVECTIVE TIP

Within a small margin of error (less than 8%), the equations for finite cylindrical fins with adiabatic tips can be used when the tip contributes to heat transfer if the length, L, is replaced with the corrected length $L_c = L + A_c/P$. [**Fins**]

48. FINITE RECTANGULAR FIN WITH ADIABATIC TIP

The temperature at a distance x from the base of a finite rectangular fin (also known as a *straight fin* or *longitudinal fin*) of length L is given by Eq. 33.91. The variable m is given by Eq. 33.77. The film coefficient, h, is an average value that can be used with the entire exposed surface of the fin (excluding the adiabatic tip).

The product Lt is the *profile area* of the fin.

$$T_x = T_\infty + (T_b - T_\infty)\left(\frac{\cosh\big(m(L-x)\big)}{\cosh mL}\right) \qquad 33.91$$

The heat transfer from the body of the fin is

Fins

$$q = \sqrt{hPkA_c}\,(T_b - T_\infty)\tanh mL_c \qquad 33.92$$

$$P = 2(w+t) \qquad 33.93$$

$$A_c = wt \qquad 33.94$$

$$mL = L\sqrt{\frac{hP}{kA}} \approx L^{1.5}\sqrt{\frac{2h}{kLt}} \quad [w \gg t] \qquad 33.95$$

$$\eta_f = \frac{\tanh mL}{mL} \qquad 33.96$$

49. FINITE RECTANGULAR FIN WITH CONVECTIVE TIP

Most fins are not insulated on their tips, and they experience heat loss from the tip by convection.

$$T_x = T_\infty + (T_b - T_\infty) \times \left(\frac{\cosh\big(m(L-x)\big) + \frac{h}{mk}\sinh\big(m(L-x)\big)}{\cosh mL + \frac{h}{mk}\sinh mL}\right) \qquad 33.97$$

Heat Transfer

$$P = 2(w + t) \qquad 33.98$$

$$A_c = wt \qquad 33.99$$

The heat transfer from the body of the fin is

$$q = \sqrt{hPkA_c}\,(T_b - T_\infty) \times \left(\frac{\sinh mL + \frac{h}{mk} \cosh mL}{\cosh mL + \frac{h}{mk} \sinh mL} \right) \qquad 33.100$$

The approximate efficiency of the fin can be calculated with a margin of error of less than 8% if the length in Eq. 33.96 is replaced with the *corrected fin length*, $L_c = L + t/2$.

$$\eta_f = \frac{\tanh\left(m\left(L + \frac{t}{2}\right)\right)}{m\left(L + \frac{t}{2}\right)} \qquad 33.101$$

50. CIRCULAR FIN

Figure 33.6(c) shows a *circular fin*, also known as a *transverse fin* and *annular fin*, installed on the exterior of a pipe. The exterior transverse area (corresponding to the "tip" area on straight fins) is substantial and is probably never insulated. Therefore, the corrected length should be used in the fin equations.

$$L_c = L + \frac{t}{2} \qquad 33.102$$

The heat transfer from the two surfaces is

$$q = \eta_f h A_f (T_b - T_\infty) \qquad 33.103$$

$$A_f = 2\pi\left((L_c + r_1)^2 - r_1^2\right) \qquad 33.104$$

The fin efficiency can be read from Fig. 33.5(b).

Example 33.7

The base of a 0.5 in × 0.5 in × 10 in (1.2 cm × 1.2 cm × 25 cm) long rectangular rod is maintained at 300°F (150°C) by an electrical heating element. The conductivity of the rod is 80 Btu-ft/hr-ft²-°F (140 W/m·K). The ambient air temperature is 80°F (27°C), and the average film coefficient is 1.65 Btu/hr-ft²-°F (9.4 W/m²·K). (a) What is the temperature of the rod 3 in (7.6 cm) from the base? (b) What energy input is required to maintain the base temperature?

SI Solution

(a) Since the tip area is approximately 1% of the total surface, the rod is assumed to have an adiabatic tip.

The perimeter length is

$$P = 2(w + t) = (2)(0.012 \text{ m} + 0.012 \text{ m}) = 0.048 \text{ m}$$

The cross-sectional area of the fin at its base is

$$A_c = wt = (0.012 \text{ m})(0.012 \text{ m}) = 0.000144 \text{ m}^2$$

From Eq. 33.77,

$$m = \sqrt{\frac{hP}{kA_c}} = \sqrt{\frac{\left(9.4 \ \frac{\text{W}}{\text{m}^2\cdot\text{K}}\right)(0.048 \text{ m})}{\left(140 \ \frac{\text{W}}{\text{m}\cdot\text{K}}\right)(0.000144 \text{ m}^2)}} = 4.73 \text{ 1/m}$$

From Eq. 33.91, the temperature 7.6 cm from the base is

$$\begin{aligned} T_x &= T_\infty + (T_b - T_\infty)\left(\frac{\cosh\left(m(L - x)\right)}{\cosh mL}\right) \\ &= 27^\circ\text{C} + (150^\circ\text{C} - 27^\circ\text{C}) \times \left(\frac{\cosh\left(\left(4.73 \ \frac{1}{\text{m}}\right)(0.25 \text{ m} - 0.076 \text{ m})\right)}{\cosh\left(\left(4.73 \ \frac{1}{\text{m}}\right)(0.25 \text{ m})\right)} \right) \\ &= 120.6^\circ\text{C} \end{aligned}$$

(b) At steady state, the energy input is equal to the energy loss. From Eq. 33.89, the total heat loss is

$$\begin{aligned} q &= \sqrt{hPkA_c}\,(T_b - T_\infty)\tanh mL_c \\ &= \sqrt{\left(9.4 \ \frac{\text{W}}{\text{m}^2\cdot\text{K}}\right)(0.048 \text{ m}) \times \left(140 \ \frac{\text{W}}{\text{m}\cdot\text{K}}\right)(0.000144 \text{ m}^2)} \\ &\quad \times (150^\circ\text{C} - 27^\circ\text{C})\tanh\left(\left(4.73 \ \frac{1}{\text{m}}\right)(0.25 \text{ m})\right) \\ &= 9.72 \text{ W} \end{aligned}$$

Customary U.S. Solution

(a) Since the tip area is approximately 1% of the total surface, the rod is assumed to have an adiabatic tip.

The perimeter length is

$$P = 2(w+t) = \frac{(2)(0.5\text{ in} + 0.5\text{ in})}{12\ \frac{\text{in}}{\text{ft}}} = 0.167\text{ ft}$$

The cross-sectional area of the fin at its base is

$$A_c = wt = \frac{(0.5\text{ in})(0.5\text{ in})}{\left(12\ \frac{\text{in}}{\text{ft}}\right)^2} = 0.00174\text{ ft}^2$$

From Eq. 33.77,

$$m = \sqrt{\frac{hP}{kA_c}} = \sqrt{\frac{\left(1.65\ \frac{\text{Btu}}{\text{hr-ft}^2\text{-°F}}\right)(0.167\text{ ft})}{\left(80\ \frac{\text{Btu-ft}}{\text{hr-ft}^2\text{-°F}}\right)(0.00174\text{ ft}^2)}} = 1.41\ 1/\text{ft}$$

From Eq. 33.91, the temperature 3 in from the base is

$$\begin{aligned} T_x &= T_\infty + (T_b - T_\infty)\left(\frac{\cosh\big(m(L-x)\big)}{\cosh mL}\right) \\ &= 80°\text{F} + (300°\text{F} - 80°\text{F}) \\ &\quad\times\left(\frac{\cosh\frac{\left(1.41\ \frac{1}{\text{ft}}\right)(10\text{ in} - 3\text{ in})}{12\ \frac{\text{in}}{\text{ft}}}}{\cosh\frac{\left(1.41\ \frac{1}{\text{ft}}\right)(10\text{ in})}{12\ \frac{\text{in}}{\text{ft}}}}\right) \\ &= 248.4°\text{F} \end{aligned}$$

(b) At steady state, the energy input is equal to the energy loss. From Eq. 33.89, the total heat loss is

$$\begin{aligned} q &= \sqrt{hPkA_c}\,(T_b - T_\infty)\tanh mL_c \\ &= \sqrt{\begin{array}{l}\left(1.65\ \frac{\text{Btu}}{\text{hr-ft}^2\text{-°F}}\right)(0.167\text{ ft}) \\ \quad\times\left(80\ \frac{\text{Btu-ft}}{\text{hr-ft}^2\text{-°F}}\right)(0.00174\text{ ft}^2)\end{array}} \\ &\quad\times(300°\text{F} - 80°\text{F})\tanh\frac{\left(1.41\ \frac{1}{\text{ft}}\right)(10\text{ in})}{12\ \frac{\text{in}}{\text{ft}}} \\ &= 35.6\text{ Btu/hr} \end{aligned}$$

51. SHAPE FACTORS

A horizontal pipe buried in the ground is a case of an object experiencing multidimensional heat transfer. Heat transfer will be by conduction, but the heat flow path will not be radial to the pipe everywhere. The complexity of this and similar problems is handled by use of a *shape factor*, S. Shape factors for several common configurations are given in Table 33.6.[35]

$$q = Sk(T_1 - T_2) \qquad 33.105$$

In Eq. 33.105, k is the thermal conductivity of the earth, which varies approximately from 0.4 Btu-ft/hr-ft²-°F (0.7 W/m·K) for dry sand to approximately 1.5 Btu-ft/hr-ft²-°F (2.6 W/m·K) for saturated sand.

52. STEAM AND ELECTRIC HEAT TRACING

Viscous fluids become difficult to pump if their temperatures drop significantly. This is a concern in lines that cannot be flushed, blown down, or drained while a process is on-line. Pipes exposed to low temperatures can be heated to keep their contents flowing by wrapping with *heat tracing*.

Electric heat tracing, also called *heat tape*, consists of long, flexible heating elements. There are two basic types: self-regulating and constant-watt products. *Self-regulating heat tape* is more common. The lower the ambient temperature, the greater the current.[36] *Constant-watt heat tape* puts out a constant heat (e.g., 8 W/ft, 12 W/ft, or 15 W/ft) at all times. The current drawn from a constant-watt heat tape can be calculated from Eq. 33.106, where P_L is the rated power per unit length.

$$I = \frac{P_L L}{E} \qquad 33.106$$

[35] The case of two parallel cylinders is covered in most heat transfer textbooks.

[36] At very low temperatures, this can result in a high initial current and can trip circuit breakers.

Table 33.6 Shape Factors of Buried Objects

configuration	illustration	shape factor, S
sphere, buried ($D=2r$)		$S=\dfrac{2\pi D}{1-\dfrac{D}{4z}}$
finite horizontal isothermal cylinder, buried ($z \ll L$) ($D \ll L$) ($D=2r$)		$S=\dfrac{2\pi L}{\operatorname{arccosh}\dfrac{2z}{D}}$
infinite horizontal isothermal cylinder, buried (per unit length) ($D=2r$)		$S/L=\dfrac{2\pi}{\operatorname{arccosh}\dfrac{2z}{D}}$
finite vertical isothermal cylinder, top at surface, buried ($D \ll L$)		$S=\dfrac{2\pi L}{\ln\dfrac{4L}{D}}$
intersecting planes of thickness t ($a>t/5$) ($b>t/5$)		$S=\dfrac{aL}{t}+\dfrac{bL}{t}+0.54L$

Design and installation of steam tracing is more of an art than a science. *Steam tracing* typically consists of 3/8 in or 1/2 in (9.5 mm or 12.7 mm) types K and L copper or stainless steel tubing through which low-pressure, 15 psig to 125 psig (100 kPa to 860 kPa) steam runs.[37] (See Fig. 33.7.) A *steam header* (*manifold*) will feed up to 20 or so tracings. Each tracing should have its own isolation/control valve. Each tracing runs approximately 75 ft to 150 ft (20 m to 45 m) between its supply valve and its terminating steam trap. Temperature-sensing (thermostatic or thermodynamic) *steam traps* should be spaced regularly, including at the ends of each tracing run and the bottom of vertical runs, to remove condensate.[38] Vacuum breakers and strainers are included in most designs.

Tracing should be located inside the pipe insulation, along the bottom of the pipe.[39] Tracing is attached by bands or graphite-based heat transfer cement. However, a union should be inserted at pipe flange points, located outside of the pipe insulation. It is also common to install an expansion loop wherever a union is installed.

[37]Low-pressure steam is not only cheaper than high-pressure steam, but also it liberates more heat during condensation.

[38]If the pressure drops to zero at the end of the tracing, temperature-triggered steam traps are required.

[39]Two other methods (used when the heating load is high) of using steam to heat process piping are running the steam line inside the process pipe or jacketing the entire process pipe with steam. These two methods essentially create shell-in-tube heat exchangers.

Figure 33.7 *Steam Tracing*

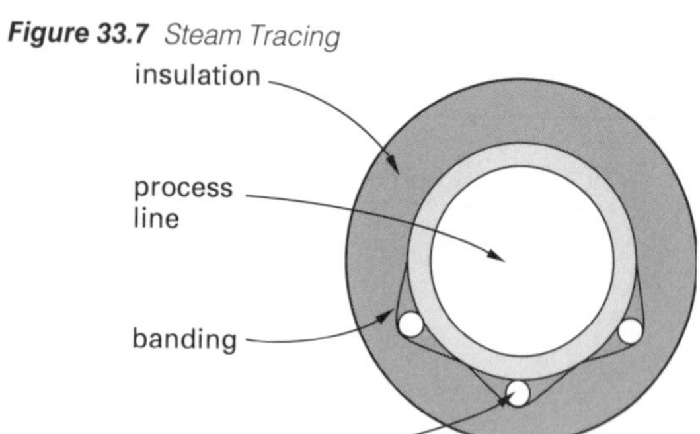

53. HEAT PIPES

A *heat pipe* is a cylindrical device with a high thermal conductance used to transfer large amounts of thermal energy on a continuous basis without moving parts or additional energy input. Heat pipes can be used on a small scale (as in satellites to remove heat from electronics) or on a large scale (as in a steam condenser or furnace economizer).

Operation of a heat pipe is based on the latent heat of vaporization of a working fluid. Heat is absorbed at the evaporator end of the heat pipe. The working fluid vaporizes and moves to the condenser end. The vapor condenses at the cooler end, releasing the heat of vaporization. The condensed working fluid returns to the evaporator end where the cycle is repeated on a continuous basis. The heat transfer occurs without any external power other than the thermal energy transferred.

There are two types of heat pipes. In the *thermo-siphon heat pipe*, gravity, buoyancy, and vapor pressure are the forces that move the different phases of the working fluid. While the thermo-siphon heat pipe does not have to be vertical, the heat source (i.e., the evaporator) must be below the heat sink (i.e., the condenser).[40] The vapor generated travels upward because of its buoyancy; condensed fluid moves downward under the influence of gravity. (See Fig. 33.8.)

A *capillary-action heat pipe* contains a wick constructed of gauze, wire mesh, or other material throughout its length. The condensate is returned to the evaporator end through this wick. There are no restrictions on orientation, and because gravity is not required, capillary action heat pipes can be used in zero gravity.

Figure 33.8 *Thermo-Siphon Heat Pipe*

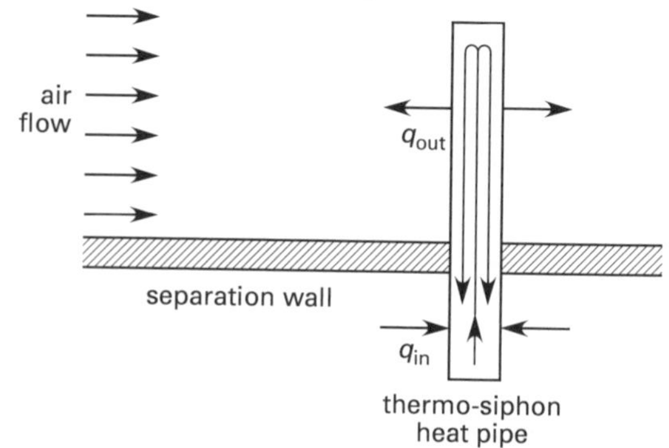

54. NOMENCLATURE

A	area	ft^2	m^2
Bi	Biot number	–	–
c_p	specific heat capacity	Btu/lbm-°F	J/kg·K
C	thermal capacitance	Btu/°F	J/K
D	diameter	ft	m
E	effectiveness	–	–
E	voltage	V	V
Fo	Fourier number	–	–
G	heat generation rate	Btu/hr-ft^3	W/m^3
h	specific enthalpy	Btu/lbm	J/kg
h	film coefficient	Btu/hr-ft^2-°F	W/m^2·K
H	enthalpy	Btu	J
I	current	A	A
k	thermal conductivity	Btu-ft/hr-ft^2-°F	W/m·K
L	length	ft	m
m	mass	lbm	kg
m	$\sqrt{hP/kA}$	1/ft	1/m
P	face perimeter	ft	m
P	power	W	W
q''	heat transfer per unit area	Btu/hr-ft^2	W/m^2
q	rate of heat transfer	Btu/hr	W
q	specific heat of respiration	Btu/hr-lbm	W/kg
Q	heat	Btu	J
r	empirical rate constant	1/hr	1/s
r	radius	ft	m
R	electrical resistance	Ω	Ω
R	thermal resistance	hr-°F/Btu	K/W
S	shape factor	ft	m
t	thickness	ft	m
t	time	hr	s
T	temperature	°F	K
U	internal energy	Btu	J
U	overall heat transfer coefficient	Btu/hr-ft^2-°F	W/m^2·K

[40]Depending on the heat pipe, an orientation of as little as 10° from the horizontal may be satisfactory for common thermo-siphon heat pipes. With a separate condensate-collection tube returning condensate directly to the evaporator end, completely horizontal operation is possible.

V	volume	ft^3	m^3
w	width	ft	m
x	distance	ft	m
x	mass fraction	–	–

Symbols

α	thermal diffusivity	ft^2/hr	m^2/s
β	reciprocal of thermal time constant	hr^{-1}	s^{-1}
γ	empirical constant	1/°F	1/K
η	efficiency	–	–
ρ	electrical resistivity	Ω-in	Ω·cm
ρ	density	lbm/ft^3	kg/m^3
τ	thermal time constant	hr	s

Subscripts

0	initial
a	ash
b	base
c	carbohydrate, characteristic corrected, or cross-sectional
eq	equivalent
f	fat or fin
h	film
i	inner or ith layer
j	jth film
L	per unit length
m	logarithmic mean
min	minimum
o	outer or overall
p	protein
r	radiative or at radius r
ref	reference
s	solids or surface
t	at time t
th	thermal
T	at temperature T
w	water
x	at distance x
∞	at infinity

34 Natural Convection, Evaporation, and Condensation

Content in blue refers to the *NCEES Handbook*.

NCEES EXAM SPECIFICATIONS AND RELATED CONTENT

HVAC AND REFRIGERATION EXAM

I.D. Heat Transfer
- 4. Nusselt Number

THERMAL AND FLUID SYSTEMS EXAM

I.C. Heat Transfer Principles
- 4. Nusselt Number
- 5. Prandtl Number
- 11. Nusselt Equation

1. INTRODUCTION

Natural convection (also known as *free convection*) is the removal of heat from a surface by a fluid that moves vertically under the influence of a density gradient. As a fluid warms, it becomes lighter and rises from the heating surface. The fluid is acted on by buoyant and gravitational forces. The fluid does not have a component of motion parallel to the surface.[1]

Natural convection is attractive from an engineering design standpoint because no motors, fans, pumps, or other equipment with moving parts are required. However, the transfer surface must be much larger than it would be with forced convection.[2]

Heat Transfer

2. HEAT TRANSFER BY NATURAL CONVECTION

Equation 34.1 is the basic equation used to calculate the steady-state heat transfer by natural convection in both heating and cooling configurations. The *film coefficient* (*heat transfer coefficient*), h, is seldom known to great accuracy.[3] The average film coefficient, $\bar{h}$, is used where there are variations over the heat transfer surface.[4]

Newton's Law of Cooling

$$q = hA(T_w - T_\infty) \qquad 34.1$$

3. FILM COEFFICIENTS

Typical values of film coefficients for natural convection are listed in Table 34.1.

[1]Rotating spheres and cylinders and vertical plane walls are special categories of convective heat transfer where the fluid has a component of relative motion parallel to the heat transfer surface.
[2]Natural convection requires approximately 2 to 10 times more surface area than does forced convection.
[3]An error of up to 25% can be expected.
[4]Though $\bar{h}$ has traditionally been used in books on the subject of heat transfer, most modern books and this book use the symbol h. The fact that the film coefficient is an inaccurate, average value is implicit.

*Table 34.1 Typical Film Coefficients for Natural Convection**

	Btu/hr-ft²-°F	W/m²·K
no change in phase:		
still air	0.8–4.4	5.0–25.0
condensing:		
steam		
horizontal surface	1700–4300	9600–24 400
vertical	700–2000	4000–11 300
organic solvents	150–500	850–2800
ammonia	500–1000	2800–5700
evaporating:		
water	800–2000	4500–11 300
organic solvents	100–300	550–1700
ammonia	200–400	1100–2300

(Multiply Btu/hr-ft²-°F by 5.6783 to obtain W/m²·K.)

*Values outside these ranges have been observed. However, these ranges are typical of those encountered in industrial processes.

4. NUSSELT NUMBER

The *Nusselt number*, Nu, is defined by Eq. 34.2. The Nusselt number is sometimes written with a subscript (e.g., Nu_h or Nu_f) to indicate that the fluid properties are evaluated at the film temperature. (See Eq. 34.11.)

$$\text{Nu} = \frac{hD}{k} \quad \text{34.2}$$

5. PRANDTL NUMBER

The dimensionless *Prandtl number*, Pr, is defined by Eq. 34.3. It represents the ratio of momentum diffusion to thermal diffusion. The values used are for the fluid, not for the surface material. For gases, the values used in calculating the Prandtl number do not vary significantly with temperature, and hence neither does the Prandtl number itself.

Convection: Terms

$$\text{Pr} = \frac{c_p\mu}{k} = \frac{c_p\nu\rho}{k} = \frac{\nu}{\alpha} \quad \text{34.3}$$

6. GRASHOF NUMBER

The dimensionless *Grashof number*, Gr, is the ratio of buoyant to viscous forces. Dynamic similarity in free convection problems is assured by equating the Grashof numbers.

The *characteristic length*, L, is defined in Table 34.2 for various configurations.[5] The coefficient of volumetric expansion, β, for ideal gases is the reciprocal of the absolute film temperature. Gravitational acceleration, g, and viscosity, μ, must have the same unit of time in order to make Gr dimensionless. The quantity $g\beta\rho^2/\mu^2$ is tabulated in App. 34.A through App. 34.F, so the component values generally do not need to be evaluated individually.

$$\text{Gr} = \frac{L^3 g\beta\rho^2(T_s - T_\infty)}{\mu^2} = \frac{L^3 g\beta(T_s - T_\infty)}{\nu^2} \quad \text{34.4}$$

The Grashof number may be written with a subscript indicating which dimension is to be used as the characteristic length. For example, the symbol Gr_D could be used to represent the Grashof number in which diameter is the characteristic length.

For air, the critical Grashof number for laminar flow is approximately 10^9. Below 10^9, the air flow will be laminar; above 10^9, it will be turbulent.

7. RAYLEIGH NUMBER

The *Rayleigh number*, Ra, is the product of the Grashof and Prandtl numbers.

$$\text{Ra} = \text{GrPr} = \frac{L^3 g\beta\rho^2(T_s - T_\infty)c_p}{k\mu} \quad \text{34.5}$$

The quantity $g\beta\rho^2 c_p/k\mu$ is tabulated in some books and given the symbol a.

$$\text{Ra} = aL^3(T_s - T_\infty) \quad \text{34.6}$$

$$a = \frac{g\beta\rho^2 c_p}{k\mu} \quad \text{34.7}$$

[5] The length of the side of a square, the mean length of a rectangle, and 90% of the diameter of a circle have historically been used as the *characteristic length*. However, the ratio of surface area to perimeter gives better agreement with experimental data.

Table 34.2 *Parameters for the Nusselt Equation (any substance; isothermal surfaces, U.S. or SI units)*

configuration	L	GrPr	C	n
vertical plate or vertical cylinder[a]	height[b]	$< 10^4$	1.36	0.20
		10^4 to 10^9	0.59	$\frac{1}{4}$
		10^9 to 10^{12}	0.10	$\frac{1}{3}$
inclined plate (θ measured from horizontal)	Use vertical plate constants, substituting $\sin\theta$ Gr for Gr.			
horizontal cylinder[c]	outside	10^4 to 10^7	0.48	$\frac{1}{4}$
	diameter	10^7 to 10^{12}	0.13	$\frac{1}{3}$
thin horizontal wire	diameter	$< 10^{-5}$	0.49	0
		10^{-5} to 10^{-3}	0.71	0.04
		10^{-3} to 1	1.09	0.10
		1 to 10^4	1.09	0.20
		10^4 to 10^9		
	(See also Eq. 34.20 and Fig. 34.1.)			
horizontal plate:[d]				
hot surface facing up	$\frac{1}{2}(s_1 + s_2)$	10^5 to 2×10^7	0.54	$\frac{1}{4}$
or cold surface facing down	or $0.9d$	2×10^7 to 3×10^{10}	0.14	$\frac{1}{3}$
hot surface facing down	$\frac{1}{2}(s_1 + s_2)$	3×10^5 to 3×10^{10}	0.27	$\frac{1}{4}$
or cold surface facing up	or $0.9d$			
sphere[e]	radius	10^3 to 10^9	0.53	$\frac{1}{4}$
		$> 10^9$	0.15	$\frac{1}{3}$

[a]A vertical cylinder can be considered a vertical plate as long as $d/L \geq 35/(\mathrm{Gr}_L)^{1/4}$.
[b]For short vertical plates, the characteristic length is approximately (height × width)/(height + width).
[c]The values for the laminar range can also be used for heat transfer to liquid metals.
[d]For a circular flat disc, the characteristic length is 90% of the disc diameter.
[e]Ranges and values reported by different researchers show significant variation. Some correlations use diameter as the characteristic length of the sphere. Values can also be used for short cylinders and blocks with a characteristic length of (height × width)/(height + width).

8. REYNOLDS NUMBER

The free-stream velocity is always zero with natural convection, so the traditional Reynolds number is also always zero. The Grashof and Rayleigh numbers take the place of determining whether flow is laminar or turbulent.[6] The film Reynolds number (see Sec. 34.21) is used to determine whether condensation is turbulent.

9. CORRELATIONS

Equation 34.1 is simple to use. The main difficulty is finding the film coefficient, h. Various theoretical, empirical, and semi-empirical correlations have been developed using dimensional analysis and experimentation. These correlations are of several forms.

Theoretical correlations are developed completely from dimensional analysis and theoretical considerations.

Empirical correlations are determined by fitting a curve through observed data points. The film coefficient has traditionally been correlated with the *heat flux*, $q'' = q/A$ (in Btu/hr-ft² or kW/m²), or with the difference in temperature between the heated surface and the fluid. For example, the convective film coefficient for boiling water can be predicted approximately by Eq. 34.8 and Eq. 34.9.[7]

$$\log(h_b) \approx -2.05 + 2.5\log(T_s - T_{\mathrm{bulk}}) + 0.014\,T_{\mathrm{sat}} \quad \text{[U.S. only]} \qquad 34.8$$

$$h_b \approx 190 + 0.43q'' \quad \text{[U.S. only]} \qquad 34.9$$

[6]Rising air nevertheless has a velocity. The critical Reynolds number for laminar flow of air is approximately 550 (corresponding to a Grashof number of 10^9).
[7]Equation 34.8 and Eq. 34.9 actually yield some "pretty good" initial estimates.

Semi-empirical correlations, derived from dimension analysis with exponents and constants determined from experimentation, are the form of Eq. 34.10.[8] Exponents m and n are often sufficiently close so that a common value can be used.[9]

$$\begin{aligned} \text{Nu} &= C \times (\text{Pr}^m \text{Gr}^n) \\ &\approx C \times (\text{PrGr})^n \\ &= C \times (\text{Ra})^n \end{aligned} \qquad 34.10$$

Each correlation can only be used in particular configurations (i.e., a correlation for horizontal cylinders cannot generally be used for vertical cylinders), and even then, the correlation will be valid only within a particular range of parameters (e.g., Prandtl or Grashof numbers).

The usefulness of a correlation depends on how well it predicts actual performance. Though correlations should always be accompanied by parameter ranges, the percentage accuracy of the correlation is generally not stated. Considering that correlations are often accurate to only ±20%, a value derived from a correlation near the end of its applicable range should be considered "ballpark."

Often, two or more parameter ranges will be given for a particular configuration. (See the GrPr ranges in Table 34.2.) The lower range of parameters corresponds to laminar air flow, while the higher parameter ranges correspond to turbulent air flow. Correlations are less reliable near the transition region between the two regimes.

10. FILM TEMPERATURE

Film properties are evaluated at the average of the surface temperature, T_s, and the *bulk temperature*, T_∞. When there is a variation of the surface temperature, as there could be along the length of a long tube used for heat transfer, the surface temperature is assumed to be the temperature at midlength along the tube.[10]

$$T_h = \tfrac{1}{2}(T_s + T_\infty) \qquad 34.11$$

11. NUSSELT EQUATION

The *Nusselt equation* and equations of its form are often used to find the film coefficient for natural convective heating and cooling. The thermal conductivity, k, in Eq. 34.12 is for the transfer fluid, not for the surface wall, and it is evaluated at the film temperature, T_h.

$$\frac{hL}{k} = C(\text{GrPr})^n = C(\text{Ra})^n \qquad 34.12$$

For a flat plate, the film coefficient is

Vertical Flat Plate in Large Body of Stationary Fluid

$$\bar{h} = C\left(\frac{k}{L}\right)\text{Ra}_L^n \qquad 34.13$$

For a horizontal cylinder, the film coefficient is

Long Horizontal Cylinder in Large Body of Stationary Fluid

$$\bar{h} = C\left(\frac{k}{D}\right)\text{Ra}_D^n \qquad 34.14$$

For laminar convection ($1000 < \text{GrPr} < 10^9$), n has a value of approximately $1/4$. For turbulent convection ($\text{GrPr} > 10^9$), n is approximately $1/3$. For sublaminar convection ($\text{GrPr} < 1000$), n is less than $1/4$ (typically taken as $1/5$), and graphical solutions are commonly used. [**Vertical Flat Plate in Large Body of Stationary Fluid**]

The values of the dimensionless empirical constants C and n in Eq. 34.12 and given in Table 34.2 can be used with all fluids and any consistent systems of units. Table 34.2 is limited in application to single heat transfer surfaces (i.e., a single tube or a single plate).

Horizontal pipe diameters greater than approximately 8 in (20 cm) and plate heights greater than approximately 2 ft (0.6 m) have little effect on film coefficients.[11] Therefore, characteristic lengths for tall plates and large-diameter pipes should be limited to 2 ft (0.6 m) and 0.67 ft (0.2 m), respectively, when calculating the film coefficients for free convection.[12]

12. FILM COEFFICIENTS FOR AIR

Table 34.3 contains simplified equations for calculating film coefficients for air at standard atmospheric pressure. These equations are derived from the Nusselt equation using simplifying assumptions.[13]

[8] Equation 34.10 is known as a *Nusselt-type correlation*.
[9] The implication of exponents m and n being identical is that the temperature differences and fluid velocities are small.
[10] Variations in the surface temperature with time, however, cannot be so easily handled.
[11] Some researchers report 3 ft (0.9 m) as the limiting value.
[12] These limiting values for characteristic length are taken into consideration in the simplified equations given in Table 34.3. The characteristic length, L, does not appear in the turbulent flow equations.
[13] The main assumption is the film temperature.

Table 34.3 *Natural Convection Film Coefficients: Simplified Equations for Air (isothermal surfaces, 1 atm)*

configuration	GrPr	simplified equation[a]
vertical plate or vertical cylinder[b]	10^4 to 10^9	$h = 1.37\left(\frac{T_s - T_\infty}{L}\right)^{1/4}$ [SI]
		$h = 0.29\left(\frac{T_s - T_\infty}{L}\right)^{1/4}$ [U.S.]
	10^9 to 10^{12}	$h = 1.24(T_s - T_\infty)^{1/3}$ [SI]
		$h = 0.19(T_s - T_\infty)^{1/3}$ [U.S.]
horizontal cylinder	10^3 to 10^9	$h = 1.32\left(\frac{T_s - T_\infty}{D}\right)^{1/4}$ [SI]
		$h = 0.27\left(\frac{T_s - T_\infty}{D}\right)^{1/4}$ [U.S.]
	10^9 to 10^{12}	$h = 1.24(T_s - T_\infty)^{1/3}$ [SI]
		$h = 0.18(T_s - T_\infty)^{1/3}$ [U.S.]
horizontal plate (square):[c]		
hot surface facing up or cold surface facing down	10^5 to 2×10^7	$h = 1.32\left(\frac{T_s - T_\infty}{D}\right)^{1/4}$ [SI]
		$h = 0.27\left(\frac{T_s - T_\infty}{L}\right)^{1/4}$ [U.S.]
	2×10^7 to 3×10^{10}	$h = 1.52(T_s - T_\infty)^{1/3}$ [SI]
		$h = 0.22(T_s - T_\infty)^{1/3}$ [U.S.]
cold surface facing up or hot surface facing down	3×10^5 to 3×10^{10}	$h = 0.59\left(\frac{T_s - T_\infty}{D}\right)^{1/4}$ [SI]
		$h = 0.12\left(\frac{T_s - T_\infty}{L}\right)^{1/4}$ [U.S.]

[a]Units defined as in nomenclature.
[b]A vertical cylinder can be considered a vertical plate as long as $d/L \geq 35/(\mathrm{Gr}_L)^{1/4}$.
[c]For horizontal circular disc, use $L = 0.9 \times$ disc diameter.

13. FILM COEFFICIENTS FOR AIR AT OTHER PRESSURES

The film coefficients derived from Table 34.3 can be multiplied by Eq. 34.15 or Eq. 34.16 to correct for pressures other than atmospheric.

$$\text{correction factor} = \left(\frac{p_{\text{actual}}}{p_{\text{std. atmosphere}}}\right)^{1/2} \quad \text{[laminar]} \qquad 34.15$$

$$\text{correction factor} = \left(\frac{p_{\text{actual}}}{p_{\text{std. atmosphere}}}\right)^{2/3} \quad \text{[turbulent]} \qquad 34.16$$

14. FILM COEFFICIENTS FOR WATER

Equation 34.17 is a simplified equation for calculating the film coefficient for water on vertical planes and cylinders at room temperature (i.e., 70°F (21°C)) and in the range $10^4 < \mathrm{GrPr} < 10^9$.

$$h = 127\left(\frac{T_s - T_\infty}{L}\right)^{1/4} \quad \text{[SI]} \qquad 34.17(a)$$

$$h = 26\left(\frac{T_s - T_\infty}{L}\right)^{1/4} \quad \text{[U.S.]} \qquad 34.17(b)$$

Heat Transfer

Example 34.1

A horizontal 4.0 in (10 cm) diameter (actual) pipe carries steam through a 20 ft (6 m) long room. The temperature of the exterior of the pipe surface is 300°F (150°C). The temperature of the air in the room is 100°F (36°C). What is the natural convective heat loss from the exterior of the pipe?

SI Solution

The characteristic length is the pipe outside diameter.

$$L = 0.10 \text{ m}$$

The temperature gradient is

$$T_s - T_\infty = 150°\text{C} - 36°\text{C} = 114°\text{C}$$

The air properties are evaluated at the film temperature.

$$\begin{aligned} T_h &= \tfrac{1}{2}(T_s + T_\infty) \\ &= \left(\frac{1}{2}\right)(150°\text{C} + 36°\text{C}) \\ &= 93°\text{C} \end{aligned}$$

The air properties at 93°C are found from App. 34.D.

$$\begin{aligned} k &= 0.03115 \text{ W/m·K} \\ \rho &= 0.964 \text{ kg/m}^3 \\ \mu &= 2.15 \times 10^{-5} \text{ kg/s·m} \\ \beta &= 2.74 \times 10^{-3} \text{ 1/K} \\ \text{Pr} &= 0.694 \end{aligned}$$

The Grashof number is

$$\begin{aligned} \text{Gr} &= \frac{L^3 g\beta\rho^2(T_s - T_\infty)}{\mu^2} \\ &= \frac{(0.10 \text{ m})^3\left(9.81 \frac{\text{m}}{\text{s}^2}\right)\left(2.74 \times 10^{-3} \frac{1}{\text{K}}\right) \times \left(0.964 \frac{\text{kg}}{\text{m}^3}\right)^2 (114\text{K})}{\left(2.15 \times 10^{-5} \frac{\text{kg}}{\text{s·m}}\right)^2} \\ &= 6.16 \times 10^6 \end{aligned}$$

$$\text{GrPr} = (6.16 \times 10^6)(0.694) = 4.28 \times 10^6$$

Using Eq. 34.12 and values from Table 34.2,

$$\begin{aligned} h &= \frac{kC(\text{GrPr})^n}{L} \\ &= \frac{\left(0.03115 \frac{\text{W}}{\text{m·K}}\right)(0.53)(4.28 \times 10^6)^{1/4}}{0.10 \text{ m}} \\ &= 7.51 \text{ W/m}^2\text{·K} \end{aligned}$$

The heat transfer from the pipe is

$$\begin{aligned} q'' = qA &= \pi D L h(T_s - T_\infty) \\ &= \pi(0.10 \text{ m})(6 \text{ m})\left(7.51 \frac{\text{W}}{\text{m}^2\text{·K}}\right)(150°\text{C} - 36°\text{C}) \\ &= 1614 \text{ W} \end{aligned}$$

Customary U.S. Solution

The characteristic length is the pipe outside diameter.

$$L = \frac{4 \text{ in}}{12 \frac{\text{in}}{\text{ft}}} = 0.333 \text{ ft}$$

The temperature gradient is

$$T_s - T_\infty = 300°\text{F} - 100°\text{F} = 200°\text{F}$$

The air properties are evaluated at the film temperature.

$$\begin{aligned} T_h &= \tfrac{1}{2}(T_s + T_\infty) \\ &= \left(\frac{1}{2}\right)(300°\text{F} + 100°\text{F}) \\ &= 200°\text{F} \end{aligned}$$

The air properties at 200°F are found from App. 34.C.

$$\begin{aligned} k &= 0.0174 \text{ Btu-ft/hr-ft}^2\text{-°F} \\ \rho &= 0.060 \text{ lbm/ft}^3 \\ \mu &= 1.44 \times 10^{-5} \text{ lbm/ft-sec} \\ \beta &= 1.52 \times 10^{-3} \text{ 1/°F} \\ \text{Pr} &= 0.72 \end{aligned}$$

The Grashof number is

$$\mathrm{Gr} = \frac{L^3 g\beta\rho^2(T_s - T_\infty)}{\mu^2}$$

$$= \frac{(0.333\ \text{ft})^3\left(32.2\ \frac{\text{ft}}{\text{sec}^2}\right)\left(1.52\times10^{-3}\ \frac{1}{^\circ\text{F}}\right)\times\left(0.060\ \frac{\text{lbm}}{\text{ft}^3}\right)^2(200^\circ\text{F})}{\left(1.44\times10^{-5}\ \frac{\text{lbm}}{\text{ft-sec}}\right)^2}$$

$$= 6.28\times10^6$$

Using Eq. 34.12 and values from Table 34.2,

$$h = \frac{kC(\mathrm{GrPr})^n}{L}$$

$$= \frac{\left(0.0174\ \frac{\text{Btu-ft}}{\text{hr-ft}^2\text{-}^\circ\text{F}}\right)(0.53)(4.52\times10^6)^{1/4}}{0.333\ \text{ft}}$$

$$= 1.28\ \text{Btu/hr-ft}^2\text{-}^\circ\text{F}$$

The heat transfer from the pipe is

$$q'' = qA = \pi DLh(T_s - T_\infty)$$

$$= \pi(0.333\ \text{ft})(20\ \text{ft})\left(1.28\ \frac{\text{Btu}}{\text{hr-ft}^2\text{-}^\circ\text{F}}\right)\times(300^\circ\text{F} - 100^\circ\text{F})$$

$$= 5356\ \text{Btu/hr}$$

15. FILM COEFFICIENTS FOR AIR ON HEATED FLAT PLATES

If the film coefficient for air heated on a flat plate is known for either the vertical or horizontal configuration, an approximate film coefficient for the corresponding configuration can be determined from Eq. 34.18 and Eq. 34.19.

$$h_{\text{horizontal, facing upward}} \approx 1.27h_{\text{vertical}} \qquad 34.18$$

$$h_{\text{horizontal, facing downward}} \approx 0.67h_{\text{vertical}} \qquad 34.19$$

16. FILM COEFFICIENTS FOR THE OUTSIDE OF A COIL

For air, the film coefficient on the outside of a coil of tubing is approximately the same as the film coefficient for a single horizontal tube.

17. FILM COEFFICIENTS FOR THIN WIRES

Since the boundary layer (i.e., the film) thickness is not small compared to the diameter, correlations derived from the traditional Nusselt equation may not be adequate. Equation 34.20 is a theoretical relationship for calculating the film coefficient for horizontal thin wires with laminar convection.[14] Table 34.2 and Fig. 34.1 can also be used.

$$h = \frac{2k}{D\times\ln\left(1+\dfrac{2}{1+\dfrac{2}{0.4\mathrm{Gr}^{1/4}}}\right)} \qquad [\mathrm{GrPr} < 10^3] \qquad 34.20$$

Figure 34.1 *Correlation for Horizontal Small Tubes and Thin Wires*

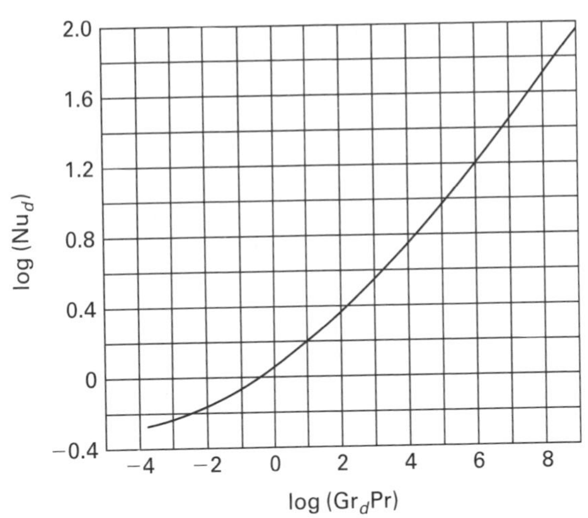

18. FILM COEFFICIENT WITHIN ENCLOSED AIR SPACES

A double-paned window is an example of a configuration where heat is transferred across an enclosed air space. The heat flow mechanism is largely conduction across the air layer at low Grashof numbers. At higher Grashof numbers, convection becomes more of a factor.

Equation 34.21 through Eq. 34.23 can be used to calculate the film coefficient for gases enclosed between two vertical plates (or inside a vertical annulus) and for which the ratio of height to separation distance, L/B, is

[14]The assumptions used to derive this theoretical equation include(1) the film thickness is much smaller than the diameter, and (2) the diameter is 2.5 times the height of an equivalent vertical surface.

greater than 3. The temperature difference used in calculating the Grashof number is the difference in surface temperatures for the two plates. The characteristic length used in calculating the Nusselt and Grashof numbers is the separation distance (clear spacing), B, between the two plates.

$$\mathrm{Nu}_B = 1.0 \quad [\mathrm{Gr}_B\mathrm{Pr} < 2 \times 10^3] \qquad 34.21$$

$$\mathrm{Nu}_B = \frac{0.20(\mathrm{Gr}_B\mathrm{Pr})^{1/4}}{\left(\frac{L}{B}\right)^{1/9}} \quad \left[\begin{array}{l} 6 \times 10^3 < \mathrm{Gr}_B\mathrm{Pr} < 2 \times 10^5 \\ 0.5 < \mathrm{Pr} < 2.0 \\ 11 < \frac{L}{B} < 42 \end{array}\right] \qquad 34.22$$

$$\mathrm{Nu}_B = \frac{0.073(\mathrm{Gr}_B\mathrm{Pr})^{1/3}}{\left(\frac{L}{B}\right)^{1/9}} \quad \left[\begin{array}{l} 2 \times 10^5 < \mathrm{Gr}_B\mathrm{Pr} < 2 \times 10^7 \\ 0.5 < \mathrm{Pr} < 2.0 \\ 11 < \frac{L}{B} < 42 \end{array}\right] \qquad 34.23$$

Equation 34.24 and Eq. 34.25 can be used for liquids between two vertical plates (or inside a vertical annulus).

$$\mathrm{Nu}_B = 1.0 \quad [\mathrm{Gr}_B\mathrm{Pr} < 1 \times 10^3] \qquad 34.24$$

$$\mathrm{Nu}_B = \frac{0.028(\mathrm{Gr}_B\mathrm{Pr})^{1/4}}{\left(\frac{L}{B}\right)^{1/4}} \quad [1 \times 10^3 < \mathrm{Gr}_B\mathrm{Pr} < 1 \times 10^7] \qquad 34.25$$

Equation 34.26 and Eq. 34.27 can be used for gases between horizontal plates with the lower plate hotter than the upper.

$$\mathrm{Nu}_B = 0.21(\mathrm{Gr}_B\mathrm{Pr})^{1/4} \quad [7 \times 10^3 < \mathrm{Gr}_B\mathrm{Pr} < 3 \times 10^5] \qquad 34.26$$

$$\mathrm{Nu}_B = 0.061(\mathrm{Gr}_B\mathrm{Pr})^{1/3} \quad [\mathrm{Gr}_B\mathrm{Pr} > 3 \times 10^5] \qquad 34.27$$

Equation 34.28 can be used for liquids between horizontal plates with the lower plate hotter than the upper.

$$\mathrm{Nu}_B = 0.069(\mathrm{Gr}_B\mathrm{Pr})^{1/3}\mathrm{Pr}^{0.074} \quad [1.5 \times 10^5 < \mathrm{Gr}_B\mathrm{Pr} < 1 \times 10^9] \qquad 34.28$$

19. CONDENSING VAPOR

When a vapor condenses on a cooler surface, the condensate forms a thin layer on the surface. This layer insulates the surface and creates a thermal resistance. However, if the condensate falls or flows from the surface (as it would from a horizontal tube), the condensate also removes thermal energy from the surface. Film coefficients are relatively high (e.g., in the order of 2000 Btu/hr-ft²-°F to 4000 Btu/hr-ft²-°F (11.5 kW/m²·K to 22.7 kW/m²·K)).[15]

Filmwise condensation occurs when the condensing surface is smooth and free from impurities.[16] A continuous film of condensate covers the entire surface. The film flows smoothly down over the surface under the action of gravity and eventually falls off. However, if the surface contains impurities or irregularities that prevent complete wetting, the film will be discontinuous, a condition known as *dropwise condensation*.[17]

Equation 34.29, based on Nusselt's theoretical work, predicts film coefficients for the laminar filmwise condensation of a pure saturated vapor on the outside of a horizontal tube with a diameter between 1 in and 3 in (2.5 cm and 7.6 cm).[18] Equation 34.29 is in fair agreement with experimental data, with calculated values generally being low.

The actual surface temperature is often unknown in initial studies. However, for steam, condensation frequently occurs with a temperature difference of $T_{\text{sat},v} - T_s$ between 5°F and 40°F (3°C and 22°C). For first approximations, the vapor density, ρ_v, can be taken as zero. Proper units must be observed in order to keep the argument of exponentiation unitless.[19]

$$h_c = 0.725\left(\frac{\rho_l(\rho_l - \rho_v)gh'_{fg}k_l^3}{D\mu_l(T_{\text{sat},v} - T_s)}\right)^{1/4} \quad \text{[SI and U.S.]} \qquad 34.29$$

The fluid (subscript l) properties are evaluated at the film temperature.

$$T_h = \tfrac{1}{2}(T_{\text{sat},v} + T_s) \qquad 34.30$$

The latent heat of condensation, h_{fg}, corresponding to the steam pressure (not the film temperature), is used for the effective heat of condensation, h'_{fg}, unless there is significant subcooling of the condensate as it flows over

[15]A film coefficient of 2000 Btu/hr-ft²-°F (11.5 kW/m²·K) is routinely assumed as a first estimate for condensation of steam on the outside of tubes.
[16]Filmwise condensation can always be expected with clean steel and aluminum tubes under ordinary conditions, as well as with heavily contaminated tubes. Dropwise condensation generally requires smooth surfaces with minute amounts of contamination, rather than rough surfaces. Since dropwise condensation can be expected only under carefully controlled conditions, the assumption of filmwise condensation is generally warranted.
[17]Film coefficients for dropwise condensation can be 4 to 8 times larger than for filmwise condensation because the film is thinner and the thermal resistance is smaller.
[18]When a noncondensing gas is present simultaneously with the condensing vapor, the gas adds to the thermal resistance. Even small amounts of air (e.g., less than 5% by volume) can reduce the film coefficient significantly (e.g., reductions up to 80%). Noncondensable gases should be avoided if the highest rates of condensation are required.
[19]In particular, μ should be in units of lbm/ft-hr (kg/s·m) and g should be 4.17×10^8 ft/hr² (9.81 m/s²).

the tube (i.e., the tube is much colder than the saturation temperature). In that case, the effective heat of condensation is given by Eq. 34.31.[20]

$$h'_{fg} = h_{fg} + 0.68c_p(T_{\text{sat},v} - T_s) \quad 34.31$$

Superheated vapor has essentially the same film coefficient as saturated vapor. If the vapor is condensing from a superheated condition, or if the quality of the vapor surrounding the cooling surface is less than 100%, Eq. 34.29 is still used with the vapor's saturation temperature corresponding to the system pressure. The vapor's superheated temperature is not used and the effect of superheat is disregarded.

The temperature difference used to calculate the heat transfer is the difference between the vapor's saturation temperature and the surface temperature.

$$q = h_cA(T_{\text{sat},v} - T_s) \quad 34.32$$

In the case of a tube bank with N layers of horizontal tubes arranged vertically over one another, the condensate will drop from one layer to another. A conservative estimate of the film coefficient is given by Eq. 34.33, which assumes that the increased thickness of the film on the lower tubes due to the accumulation of condensate will be partially offset by the increase in liquid agitation (i.e., turbulence).

$$h_c = 0.725\left(\frac{\rho_l(\rho_l - \rho_v)gh'_{fg}k_l^3}{ND\mu_l(T_{\text{sat},v} - T_s)}\right)^{1/4} \quad \text{[SI and U.S.]} \quad 34.33$$

20. CONDENSATION ON VERTICAL AND INCLINED SURFACES

Filmwise condensation of pure saturated vapors on vertical surfaces (including the insides and outsides of tubes) or on flat surfaces inclined at an angle θ from the horizontal is predicted by Eq. 34.34.[21,22,23,24] As with condensation on horizontal surfaces, the latent heat of condensation is evaluated at the vapor temperature, while the remaining fluid properties are evaluated at the film temperature. The characteristic length, L, in Eq. 34.34 is the surface length.

$$h_c = 0.943\left(\frac{\rho_l(\rho_l - \rho_v)gh'_{fg}k_l^3 \sin\theta}{L\mu_l(T_{\text{sat},v} - T_s)}\right)^{1/4} \quad \text{[SI and U.S.]} \quad 34.34$$

Nusselt showed that the film coefficient for condensation on a vertical flat surface of height L is the same as for a tube of diameter $L/2.86$. This explains the similarity between Eq. 34.29 and Eq. 34.34.

21. TURBULENT CONDENSATION ON VERTICAL PLATES AND TUBES

On a tall vertical surface or on a surface with a large amount of condensate, the condensate on the lower portion of the surface may flow turbulently. This will increase the heat transfer rate.

The condensation will remain laminar as long as the *film Reynolds number* (also known as the *condensation Reynolds number*), Re_h, is less than approximately 1800 for a vertical tube.[25] The variable P in Eq. 34.35 is the wetted perimeter, equal to the width, w, for a vertical plate with condensation on one side, and equal to πD for a vertical tube with diameter D and with condensation all around. The maximum value of Re_h occurs at the lower edge of the condensing surface.

$$\text{Re}_h = \frac{4\dot{m}}{P\mu_h} \quad 34.35$$

The condensation Reynolds number is sometimes expressed in terms of mass flow per unit width of plate, Γ, so that[26]

$$\text{Re}_h = \frac{4\Gamma}{\mu_h} \quad 34.36$$

$$\Gamma = \frac{\dot{m}}{P} \quad 34.37$$

[20]Equation 34.31 was derived by Nusselt with a constant of 3/8. For Pr > 0.5 and $c_p(T_{\text{sat},v} - T_s) < h_{fg}$, the constant 0.68 yields values that are in better agreement with experimental data.

[21]Equation 34.34 cannot be used for condensation on inclined tubes. The film flow is not parallel with the longitudinal axis of an inclined tube, resulting in an effective inclination angle that varies with location along the tube.

[22]Equation 34.34 was derived by Nusselt with a coefficient of 0.943. However, ripples in the laminar film appear at condensation Reynolds numbers (see Sec. 34.21) as low as 30 or 40. Experimental data show actual film coefficients are approximately 20% higher than the theoretical. A coefficient of 1.13 in place of 0.943 reflects this increase. Retaining the 0.943 value, however, yields a conservative value.

[23]Some researchers measure their angles from the vertical, in which case the cosine function will be used. It should be noted that $\cos\theta_{\text{vertical}} = \sin\theta_{\text{horizontal}}$

[24]Equation 34.34 can be used to find the condensing film coefficient on a vertical tube when $\dot{m} < 1020$, where $\dot{m}$ is the total condensation in pounds per hour.

[25]The critical film Reynolds number for horizontal tubes is 3600 because the film flows down two sides. Turbulent condensing flow inside a tube requires a large-diameter tube, and rarely occurs otherwise.

[26]Some references define Re_f as Γ/μ, in which case the critical Reynolds number would be 450 instead of 1800.

From a practical standpoint, the condensate generation rate, $\dot{m}$, cannot be calculated until h is known, resulting in an iterative solution procedure.

$$\dot{m} = \rho_l A\text{v} = \frac{q}{h_{fg}} = \frac{hA(T_{\text{sat},v} - T_s)}{h_{fg}} \qquad 34.38$$

The approximate average heat film coefficient for turbulent condensation with $\text{Re}_h > 1800$ is given by Eq. 34.39.

$$h_c = 0.0076\text{Re}_h^{2/5}\left(\frac{\rho_l(\rho_l - \rho_v)gk_l^3}{\mu_l^2}\right)^{1/3} \quad \text{[SI and U.S.]} \qquad 34.39$$

Generally, the assumption of a laminar film should be checked with Eq. 34.39 for every condensation problem.

22. CONDENSATION ON THE OUTSIDE OF A SPHERE

Equation 34.40 can be used to calculate the film coefficient on the exterior of an isothermal sphere.

$$h_c = 0.815\left(\frac{\rho_l(\rho_l - \rho_v)gh'_{fg}k_l^3}{D\mu_l(T_{\text{sat},v} - T_s)}\right)^{1/4} \quad \text{[SI and U.S.]} \qquad 34.40$$

23. CONDENSATION INSIDE TUBES

Equation 34.34 (or Eq. 34.39 in the turbulent case) can be used for condensation inside vertical tubes. Condensation inside horizontal tubes is complicated by the issue of condensate removal. Since the condensate specific volume is much less than the specific volume of the vapor, the accumulation will not occlude much of the wall surface. With a large enough diameter pipe and/or the assumption of vapor traps to remove the condensate, use of Eq. 34.34 is justified.[27]

If condensate is forced rapidly through the tube by a pump, then the film coefficient will be essentially the same as for forced convection. For low vapor velocities inside tubes, Eq. 34.34 can be used with a leading coefficient of 0.612 for condensing steam (the *Kern correlation*) and 0.555 for condensing refrigerants.

Example 34.2

A horizontal brass tube with 1 in (2.54 cm) outside and ⅞ in (2.22 cm) inside diameters is surrounded by steam at 7.5 psia (50 kPa). The wall temperature is 60°F (15°C). Disregard the subcooling of the condensate. What is the film coefficient on the outside of the tube?

SI Solution

The saturation temperature for 50 kPa steam is 81°C.

$$T_{\text{sat},v} = 81^\circ\text{C}$$

The film properties are evaluated at the average of the wall and saturation temperatures.

$$\begin{aligned} T_h &= \tfrac{1}{2}(T_{\text{sat},v} + T_s) \\ &= \left(\frac{1}{2}\right)(81^\circ\text{C} + 15^\circ\text{C}) \\ &= 48^\circ\text{C} \quad [\text{say } 50^\circ\text{C}] \end{aligned}$$

Film properties for 50°C water are obtained by interpolation from App. 34.B.

$$\begin{aligned} k_{50^\circ\text{C}} &= 0.6435 \text{ W/m·K} \\ \mu_{50^\circ\text{C}} &= 5.72 \times 10^{-4} \text{ kg/s·m} \\ \rho_{l,50^\circ\text{C}} &= 988.0 \text{ kg/m}^3 \\ \rho_{v,50^\circ\text{C}} &\approx 0 \\ h_{fg,50\,\text{kPa}} &= 2304.7 \text{ kJ/kg} \quad (2.3047 \times 10^6 \text{ J/kg}) \end{aligned}$$

The diameter is

$$D = \frac{2.54 \text{ cm}}{100 \ \frac{\text{cm}}{\text{m}}} = 0.0254 \text{ m}$$

From Eq. 34.29, the film coefficient is

$$h_c = 0.725\left(\frac{\rho_l(\rho_l - \rho_v)gh'_{fg}k_l^3}{D\mu_l(T_{\text{sat},v} - T_s)}\right)^{1/4}$$

$$= (0.725)\left(\frac{\left(988.0 \ \frac{\text{kg}}{\text{m}^3}\right)\left(988.0 \ \frac{\text{kg}}{\text{m}^3} - 0 \ \frac{\text{kg}}{\text{m}^3}\right) \times \left(9.81 \ \frac{\text{m}}{\text{s}^2}\right) \times \left(2.3047 \times 10^6 \ \frac{\text{J}}{\text{kg}}\right) \times \left(0.6435 \ \frac{\text{W}}{\text{m·K}}\right)^3}{(0.0254 \text{ m})\left(5.72 \times 10^{-4} \ \frac{\text{kg}}{\text{s·m}}\right) \times (81^\circ\text{C} - 15^\circ\text{C})}\right)^{1/4}$$

$$= 6416 \text{ W/m}^2\text{·K}$$

[27] A value of 1200 Btu/hr-ft²-°F (6800 W/m²·K) is routinely used for condensation of steam inside radiators and fan coils.

Customary U.S. Solution

The saturation temperature for 7.5 psia steam is 180°F.

$$T_{\text{sat},v} = 180°\text{F}$$

The film properties are evaluated at the average of the wall and saturation temperatures.

$$T_h = \tfrac{1}{2}(T_{\text{sat},v} + T_s) = \left(\frac{1}{2}\right)(180°\text{F} + 60°\text{F}) = 120°\text{F}$$

Film properties for 120°F water are obtained by interpolation from App. 34.A.

$$k_{120°\text{F}} = 0.372 \text{ Btu/hr-ft-°F}$$

$$\begin{aligned}\mu_{120°\text{F}} &= \left(0.392 \times 10^{-3} \frac{\text{lbm}}{\text{sec-ft}}\right)\left(3600 \frac{\text{sec}}{\text{hr}}\right)\\ &= 1.41 \text{ lbm/hr-ft}\end{aligned}$$

Use steam tables to obtain the liquid density, interpolating if needed. [**Properties of Saturated Water and Steam (Temperature) - I-P Units**]

$$\rho_{l,120°\text{F}} = \frac{1}{v_{l,120°\text{F}}} = \frac{1}{0.01621 \frac{\text{ft}^3}{\text{lbm}}} = 61.69 \text{ lbm/ft}^3$$

$$\begin{aligned}\rho_{v,120°\text{F}} &= \frac{1}{v_{v,120°\text{F}}} = \frac{1}{203.0 \frac{\text{ft}^3}{\text{lbm}}}\\ &\approx 0\end{aligned}$$

$$h_{fg,7.5\text{psia}} = 990.0 \text{ Btu/lbm}$$

The diameter is

$$D = \frac{1 \text{ in}}{12 \frac{\text{in}}{\text{ft}}} = 0.0833 \text{ ft}$$

Gravitational acceleration is

$$\begin{aligned}g &= \left(32.2 \frac{\text{ft}}{\text{sec}^2}\right)\left(3600 \frac{\text{sec}}{\text{hr}}\right)^2\\ &= 4.17 \times 10^8 \text{ ft/hr}^2\end{aligned}$$

From Eq. 34.29, the film coefficient is

$$h_c = 0.725\left(\frac{\rho_l(\rho_l - \rho_v)gh'_{fg}k_l^3}{D\mu_l(T_{\text{sat},v} - T_s)}\right)^{1/4}$$

$$= (0.725)\left(\frac{\left(61.69 \frac{\text{lbm}}{\text{ft}^3}\right)\left(61.69 \frac{\text{lbm}}{\text{ft}^3} - 0 \frac{\text{lbm}}{\text{ft}^3}\right) \times\left(4.17 \times 10^8 \frac{\text{ft}}{\text{hr}^2}\right) \times\left(990.0 \frac{\text{Btu}}{\text{lbm}}\right) \times\left(0.372 \frac{\text{Btu}}{\text{hr-ft-°F}}\right)^3}{(0.0833 \text{ ft})\left(1.41 \frac{\text{lbm}}{\text{hr-ft}}\right) \times(180°\text{F} - 60°\text{F})}\right)^{1/4}$$

$$= 1122 \text{ Btu/hr-ft}^2\text{-°F}$$

24. EVAPORATION FROM HORIZONTAL TUBES

If a liquid is vaporizing from a heated surface, the change of phase can occur in three distinctly different ways: pool, nucleate, and film boiling. *Pool boiling* occurs when the temperature of the heating element is near the vaporization (boiling or saturation) temperature. There will be little or no bubble formation and liquid agitation (rolling). Heat transfer will be essentially convective in nature.

As the temperature of the heating element exceeds the vaporization temperature, vapor bubbles begin to form, a mechanism known as *nucleate boiling*. Equations for determining the film coefficient for nucleate boiling are complex, and correlations typically depend on the heat flux, $q'' = q/A$.

When the surface temperature is much greater (i.e., approximately 200°F (93°C) or more) than the vaporization temperature, a film of vapor will cover the surface of the heating element. This is known as *film boiling*.

The *Bromley equation*, Eq. 34.41, can be used for laminar film boiling around horizontal tubes with diameters up to approximately ½ in (1.2 cm).[28] This makes the equation useful for liquids that are heated by commercial electrical heating rods. No forced movement of the fluid across the heating element is permitted. Agitation is solely the result of boiling. All properties should be evaluated at the saturation temperature corresponding to the vapor pressure.

$$h_b = 0.62\left(\frac{\rho_v(\rho_l - \rho_v)g(h_{fg} + 0.4c_{p,v}) \times (T_s - T_{\text{sat},v})k_v^3}{D\mu_v(T_s - T_s - T_{\text{sat},v})}\right)^{1/4} \quad \text{34.41}$$

[SI and U.S.]

25. EVAPORATION FROM FLAT PLATES

Most methods for calculating the film coefficient for evaporation from flat plates are unwieldy, and graphical methods are often used. (See Fig. 34.2.)

Figure 34.2 *Film Coefficient for Evaporating Water on Horizontal and Vertical Plates (atmospheric pressure)*

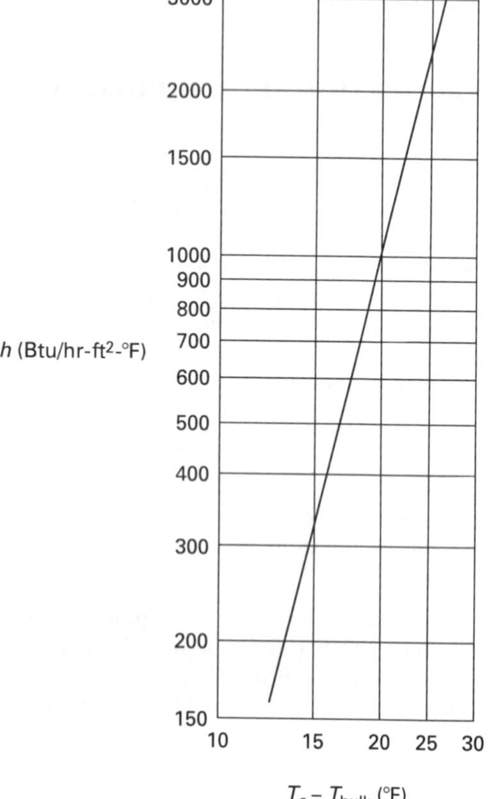

26. SIMPLIFIED VAPORIZATION EQUATIONS FOR WATER

Many researchers have developed empirical equations for calculating the film coefficient for boiling water. These are of the form of Eq. 34.42 and are correlated with heat flux, $q = Q/A$. Values of the coefficient C and exponent n for water boiling at atmospheric pressure are listed in Table 34.4. Equation 34.43 is used to correct the film coefficients derived from Eq. 34.42 for other pressures.

$$h_0 = C(T_s - T_{\text{sat},v})^n \quad \text{34.42}$$

$$h_p = h_0\left(\frac{p_{\text{actual}}}{p_{\text{std. atmosphere}}}\right)^{2/5} \quad \text{34.43}$$

Table 34.4 *Simplified Evaporation Constants for Water (atmospheric pressure; SI units only)*

orientation	Q/A (kW/m²)	C	n
horizontal surface	<16	1042	⅓
(in wide vessel)	16 to 240	5.56	3
vertical surface	<3	537	⅐
(in wide vessel)	3 to 63	7.96	3

(Multiply kW/m² by 317 to obtain Btu/hr-ft².)

27. EFFECTS OF RADIATION ON EVAPORATION FILM COEFFICIENTS

Since the temperature of a heating element is likely to be high, the effects of radiation must be considered. The combined heat transfer is more than the sum of the convective and radiative components. Radiation increases the film thickness, reducing the values of the film coefficient and convective heat transfer. If the radiation film coefficient, h_r, is known, the total film coefficient can be found by solving Eq. 34.44 iteratively.

$$h_{\text{total}} = h_b\left(\frac{h_b}{h_{\text{total}}}\right)^{1/3} + h_r \quad \text{34.44}$$

The total heat transfer in film boiling is

$$q'' = qA = h_{\text{total}}A(T_s - T_{\text{sat},v}) \quad \text{34.45}$$

[28]The flow is assumed to be laminar when the flow path around the tube is short.

28. NOMENCLATURE

A	area	ft^2	m^2
B	separation distance	ft	m
c_p	specific heat capacity at constant pressure	Btu/lbm-°F	J/kg·K
C	constant	–	–
D	diameter	ft	m
g	gravitational acceleration, 32.17 (9.807)	ft/sec^2	m/s^2
Gr	Grashof number	–	–
h	film coefficient	Btu/hr-ft²-°F	W/m²·K
h_{fg}	heat of vaporization	Btu/lbm	J/kg
k	thermal conductivity	Btu-ft/hr-ft²-°F	W/m·K
L	characteristic length	ft	m
m	exponent	–	–
$\dot{m}$	mass flow rate	lbm/hr	kg/s
n	exponent	–	–
N	number of tube layers	–	–
Nu	Nusselt number	–	–
p	pressure	lbf/ft^2	kPa
P	wetted perimeter	ft	m
Pr	Prandtl number	–	–
q''	heat transfer per unit area	Btu/hr-ft²	W/m^2
q	heat transfer rate	Btu/hr	W
Ra	Rayleigh number	–	–
Re	Reynolds number	–	–
s	side length	ft	m
T	temperature	°F	K
v	velocity	ft/hr	m/s

Symbols

α	thermal diffusivity	ft^2/sec	m^2/s
β	volumetric coefficient of expansion	1/°R	1/K
Γ	mass flow rate per unit width	lbm/hr-ft	kg/s·m
θ	angle	deg	deg
μ	viscosity[a,b,c]	lbm/hr-ft	kg/s·m
ν	kinematic viscosity	ft^2/sec	m^2/s
ρ	density	lbm/ft^3	kg/m^3

[a]The use of mass units in viscosity values is typical in the subject of convective heat transfer.
[b]Most data compilations give fluid viscosity in units of seconds. In the United States, heat transfer is traditionally given on a per hour basis. Therefore, a conversion factor of 3600 is needed when calculating dimensionless numbers from table data.
[c]The combination of units kg/s·m is the same as a N·s/m² or Pa·s.

Subscripts

0	initial, or zero gage pressure
b	boiling
B	separation distance
c	condensation
D	diameter
h	film
l	liquid
p	pressure
r	radiation
s	surface
sat	saturated
std	standard
v	vapor
w	wall surface
∞	at infinity

35 Forced Convection and Heat Exchangers

Content in blue refers to the *NCEES Handbook*.

NCEES EXAM SPECIFICATIONS AND RELATED CONTENT

HVAC AND REFRIGERATION EXAM

I.D. Heat Transfer
- 2. Heat Transfer by Forced Convection
- 3. Dimensionless Numbers
- 6. Film Temperature
- 8. Laminar Flow over Isothermal Flat Plates
- 13. Turbulent Flow Inside Straight Tubes
- 14. Turbulent Flow Inside Coiled Tubes
- 20. Crossflow over a Single Cylinder
- 23. Flow over Spheres

II.B.3. Equipment and Components: Heat exchangers
- 32. Logarithmic Temperature Difference
- 34. Heat Transfer in Heat Exchangers

THERMAL AND FLUID SYSTEMS EXAM

I.C. Heat Transfer Principles
- 2. Heat Transfer by Forced Convection
- 3. Dimensionless Numbers
- 6. Film Temperature
- 7. Flow over Flat Plates
- 8. Laminar Flow over Isothermal Flat Plates
- 10. Turbulent Flow over Isothermal Flat Plates
- 11. Laminar Flow Inside Tubes with Constant Wall Temperature
- 12. Laminar Flow Inside Tubes with Constant Heat Flux
- 13. Turbulent Flow Inside Straight Tubes
- 14. Turbulent Flow Inside Coiled Tubes

III.A.4. Energy/Power Equipment: Heat exchangers
- 32. Logarithmic Temperature Difference
- 34. Heat Transfer in Heat Exchangers
- 37. Fouling
- 39. NTU Method

1. INTRODUCTION

As with natural convection, *forced convection* depends on the movement of a fluid to remove heat from a surface. With forced convection, a fan, a pump, or relative motion causes the fluid motion. If the flow is over a flat surface, the fluid particles near the surface will flow more slowly due to friction with the surface. The *boundary layer* of slow-moving particles comprises the major thermal resistance. The thermal resistance of the tube and other heat exchanger components is often disregarded.

2. HEAT TRANSFER BY FORCED CONVECTION

Newton's law of convection, Eq. 35.1, gives the heat transfer for Newtonian fluids in forced convection over exterior surfaces.[1,2] The film coefficient, h, is also known as the *coefficient of forced convection.* T_∞ is the *free-stream temperature.*

Newton's Law of Cooling

$$\begin{aligned}\dot{Q} &= qA \\ &= hA(T_s - T_\infty)\end{aligned} \qquad 35.1$$

For flow within a tube, the more easily determined *bulk temperature* (see Sec. 35.5) is used in place of the free-stream temperature.

$$\begin{aligned}\dot{Q} &= qA \\ &= hA(T_s - T_b)\end{aligned} \qquad 35.2$$

3. DIMENSIONLESS NUMBERS

The dimensionless Nusselt number, Nu, Prandtl number, Pr, and Reynolds number, Re, are

$$\text{Nu} = \frac{hD}{k} \qquad 35.3$$

Convection: Terms

$$\text{Pr} = \frac{c_p\mu}{k} = \frac{\nu}{\alpha} \qquad 35.4$$

Reynolds Number

$$\text{Re} = \frac{\text{v}D}{\nu} = \frac{\text{DG}}{\mu} \qquad 35.5$$

$$G = \text{v}_\infty \rho_\infty \qquad 35.6$$

The viscosity, μ, used in Eq. 35.4 must have the same units of time as the conductivity, k.

The density and velocity used in Eq. 35.6 must correspond to the same point. In the most common case, both are free-stream values. It is incorrect to use the free-stream velocity with the density evaluated at the film temperature.

The *Peclet number*, Pe, is the product of the Reynolds and Prandtl numbers.

$$\text{Pe} = \text{Re}\,\text{Pr} = \frac{D\text{v}\rho c_p}{k} \qquad 35.7$$

The *Graetz number*, Gz, is used in the reporting of empirical data for laminar flow in tubes.

$$\text{Gz} = \text{Re}_d\,\text{Pr}\left(\frac{D}{L}\right) \qquad 35.8$$

The *Stanton number*, St, is encountered in correlations of fluid friction and heat transfer.

$$\text{St} = \frac{\text{Nu}}{\text{Re}\,\text{Pr}} = \frac{h}{\text{v}\rho c_p} \qquad 35.9$$

4. DIMENSIONLESS NUMBER RANGES

All heat transfer correlations have associated ranges of the Prandtl and Reynolds numbers, whether stated or not. The endpoints of these ranges are indistinct and depend on the fluid properties. For example, Eq. 35.34 can be used with a Reynolds number as low as 2100 as long as the Prandtl number is less than 10 (i.e., it cannot be used at that lower limit for fluids with viscosities more than twice that of water). Therefore, the lower limit is established as 10,000 instead of 2100, and the equation is deemed applicable to all fluids. This explains why researchers report different ranges for the same correlation.

5. BULK TEMPERATURE

The *bulk temperature*, T_b, also known as the *mixing cup temperature*, is the energy-average fluid temperature. The bulk temperature concept is usually encountered with tube flow where there is no free-stream temperature. The centerline temperature is a candidate for theoretical considerations, but it cannot be easily measured. Therefore, the bulk temperature is used to calculate the heat transfer for flow in tubes.

The bulk temperature used to calculate local properties is evaluated over the tube cross-sectional area at the point along the tube length being considered. The bulk temperature used in the calculation of an average film coefficient over the entire length of a tube or heat exchanger is evaluated as the average of the entrance and exit temperatures. For this reason, it is often referred to as the *mean bulk temperature.*

$$T_b = \tfrac{1}{2}(T_{\text{in}} + T_{\text{out}}) \qquad 35.10$$

[1]Newton's law of convection is the same for natural and forced convection. Only the methods used to evaluate the film coefficient are different.
[2]The results of this chapter do not generally apply to non-Newtonian fluids.

The bulk temperature should be used to calculate the film coefficient of a fluid flowing in a tube or heat exchanger unless the properties change a lot (i.e., as they do with high-viscosity fluids).[3] It may be necessary to solve heat transfer equations iteratively in order to determine the bulk temperature, since the film coefficient (based on the bulk temperature) is needed in order to determine the outlet temperature.

The mass flow rate in a tube is constant everywhere. Where there are large variations in temperature along the length of a tube, the density, velocity, and temperature must all be consistent. It is incorrect to use a density evaluated at the midpoint temperature with an entrance velocity.

6. FILM TEMPERATURE

As with natural convection, the free-field temperature, T_∞, is used to calculate the film temperature in external flow configurations.

Film Temperature of a Tube

$$T_{\text{film}} = \frac{(T_s + T_\infty)}{2} \qquad 35.11$$

Film properties inside tubes are evaluated at the average of the surface (e.g., wall) temperature, T_s, and the bulk temperature, T_b. When there is a variation of the surface temperature, as there could be along the length of a tube used for heat transfer, the surface temperature is assumed to be the temperature at midlength along the tube.

$$T_{\text{film}} = \frac{(T_s + T_b)}{2} \qquad 35.12$$

7. FLOW OVER FLAT PLATES

The boundary layer of a fluid flowing over a flat plate is assumed to have a parabolic velocity distribution.[4] (See Fig. 35.1.) The layer has three distinct regions: laminar, transition, and turbulent. From the leading edge, the layer is laminar and the thickness increases gradually until the transition region where the thickness increases dramatically. Thereafter, the boundary layer is turbulent. The laminar region is always present, though its length decreases as velocity increases. Turbulent flow may not develop at all with short plates.

Figure 35.1 Flow Over a Flat Plate

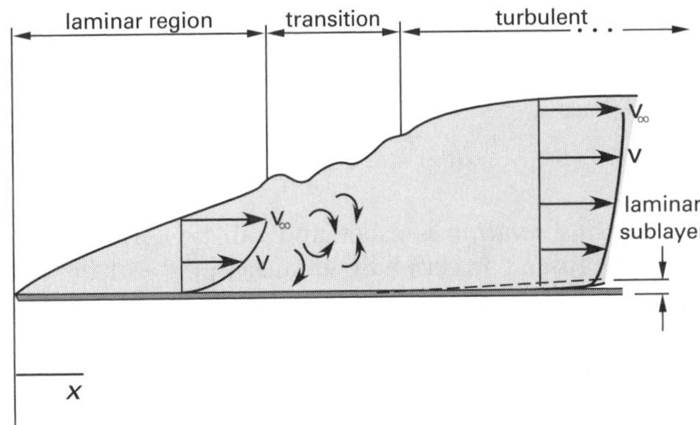

The Reynolds number is used to determine which of the three flow regimes is applicable. Laminar flow on smooth flat plates occurs for Reynolds numbers up to approximately 2×10^5; turbulent flow exists for Reynolds numbers greater than approximately 3×10^6.[5,6] Transition flow is in between. The distance from the leading edge at which turbulent flow is initially experienced is determined from the *critical Reynolds number*, commonly taken as $\text{Re} = 5 \times 10^5$ for smooth flat plates, though the actual value is highly dependent on surface roughness. Distance, x, is measured from the leading edge.

$$\text{Re}_x = \frac{\text{v}_\infty x \rho}{\mu} = \frac{\text{v}_\infty x}{\nu} \qquad 35.13$$

For a flat plate with a length L in parallel flow,

External Flow

$$\text{Re}_L = \frac{\rho \text{v}_\infty L}{\mu} \qquad 35.14$$

The heat transfer from a flat plate is

$$\dot{Q} = \bar{h} A (T_s - T_\infty) \qquad 35.15$$

8. LAMINAR FLOW OVER ISOTHERMAL FLAT PLATES

The *Pohlhausen solution* is theoretically exact and can be used to determine the local thermal film coefficient

[3] In that case, evaluate the film coefficient at the inlet and outlet and take the logarithmic mean $((h_2 - h_1)/(\ln(h_2/h_1))$ of the two. This requires calculating the film coefficients twice.
[4] The velocity distribution does not have to be parabolic. In *Couette flow*, there are two closely spaced parallel surfaces, one which is stationary and the other moving with constant velocity. The velocity gradient is assumed to be linear between the plates.
[5] Turbulent flow can begin at Reynolds numbers less than 3×10^5 if the plate is rough. This discussion assumes the plate is smooth.
[6] *The NCEES PE Mechanical Handbook* only considers average temperature across the entire plate length for modified laminar and turbulent flow regimes. Sec. 35.7–Sec. 35.10 cover localized and average methodologies across more accurately bounded laminar and turbulent flow regimes.

at a distance x from the leading edge of an isothermal flat plate in laminar flow.[7] Fluid properties in Eq. 35.16[8] are evaluated at the film temperature.

$$\mathrm{Nu}_{x,\mathrm{local}} = 0.332(\mathrm{Re}_x)^{1/2}\mathrm{Pr}^{1/3} \quad [0.6 < \mathrm{Pr} < 1;\ \mathrm{Re}_x < 2\times 10^5] \qquad 35.16$$

The *Blasius solution* is exact and can be derived from the Pohlhausen correction by setting $(\mathrm{Pr})^{1/3} = 1.0$.[9] The Blasius solution is useful with water and gases.[10] The fluid properties in Eq. 35.17 are evaluated at the film temperature.

$$\mathrm{Nu}_{x,\mathrm{local}} = 0.332(\mathrm{Re}_x)^{1/2} \quad [0.6 < \mathrm{Pr} < 50;\ \mathrm{Re}_x < 2\times 10^5] \qquad 35.17$$

The local *skin-friction coefficient*, $C_{f,x}$, for laminar flow a distance x from the leading edge is[11,12].

$$C_{f,x} = \frac{0.664}{(\mathrm{Re}_x)^{1/2}} \quad [\mathrm{Re}_x < 2\times 10^5] \qquad 35.18$$

The average film coefficient over a distance L (in the direction of flow) from the leading edge can be calculated from Eq. 35.19. Fluid properties in Eq. 35.19[13,14] are evaluated at the film temperature.

External Flow

$$\overline{\mathrm{Nu}}_L = \frac{\bar{h}L}{k} = 0.6640\mathrm{Re}_L^{1/2}\mathrm{Pr}^{1/3} \quad [\mathrm{Pr} > 0.6;\ \mathrm{Re}_L < 2\times 10^5] \qquad 35.19$$

$$\mathrm{Re}_L = \frac{\mathrm{v}_\infty L}{\nu} \qquad 35.20$$

The average skin-friction coefficient is

$$\overline{C}_f = \frac{1.328}{(\mathrm{Re}_L)^{1/2}} \qquad 35.21$$

9. LAMINAR FLOW OVER FLAT PLATES WITH CONSTANT HEAT FLUX

Equation 35.22 can be used to determine the local thermal film coefficient at a distance x from the leading edge for a fluid in laminar flow over a flat plate with constant heat flux (i.e., constant heat transfer). Fluid properties are evaluated at the film temperature.

$$\mathrm{Nu}_{x,\mathrm{local}} = 0.453(\mathrm{Re}_x)^{1/2}\mathrm{Pr}^{1/3} \quad [0.6 < \mathrm{Pr} < 50;\ \mathrm{Re}_x < 5\times 10^5] \qquad 35.22$$

10. TURBULENT FLOW OVER ISOTHERMAL FLAT PLATES

For completely turbulent flow ($\mathrm{Pr} > 0.7; 3\times 10^6 < \mathrm{Re} < 10^8$), the local and average film coefficients are given by Eq. 35.23 and Eq. 35.24, respectively. All of the fluid properties in Eq. 35.23[15] are evaluated at the film temperature.

$$\mathrm{Nu}_{x,\mathrm{local}} = 0.0288(\mathrm{Re}_x)^{0.8}\mathrm{Pr}^{1/3} \quad [0.6 < \mathrm{Pr} < 60;\ 3\times 10^6 < \mathrm{Re}_x < 10^8] \qquad 35.23$$

Equation 35.24[16,17] is applicable for the entire plate and includes the heat transfer contribution of the leading laminar portion of flow.[18] This is sometimes called a *laminar-turbulent average*.

Heat Transfer

[7]The x in the subscripts does not mean "at point x." Rather, it means that the value of x is used as the characteristic length in the dimensionless number.

[8]According to some authorities: $[\mathrm{Re} < 5\times 10^5]$. In the *NCEES PE Mechanical Handbook*, $\mathrm{Re}_L < 10^5$.

[9]The Blasius solution can also be derived from dimensional analysis.

[10]When $\mathrm{Pr} = 1$, the velocity profile (i.e., the hydrodynamic boundary layer) of the boundary layer for laminar fluid flow is the same as the temperature profile (i.e., the thermal boundary layer) for laminar convective heat transfer. This is essentially true for water, and approximately true for most gases that have Prandtl numbers in the 0.6 to 1.0 range. (Most gases have Prandtl numbers in the 0.65 to 0.84 range. These numbers raised to fractional powers result in a value very close to 1.0.) For liquids, however, the Prandtl number ranges from very small (liquid metals) to very large (viscous oils).

[11]The *skin-friction coefficient* is also known as the *Fanning friction factor*. The *Darcy friction factor*, more commonly used in fluid flow problems, is $f_{\mathrm{Darcy}} = 4f_{\mathrm{Fanning}}$

[12]The *drag coefficient* for submerged bodies is the sum of the *skin-friction coefficient* and the *profile drag coefficient*. For flat plates, there is no profile drag. Therefore, the drag coefficient is the same as the skin-friction coefficient, and the drag per unit area is

$$\tau_x = \frac{\mathrm{drag}}{A} = \frac{C_{f,x}\rho \mathrm{v}_\infty^2}{2g_c}$$

$$\mathrm{drag}_{\mathrm{ave}} = A_{\mathrm{surface}}\tau_{x,\mathrm{ave}}$$

[13]According to some authorities: $[\mathrm{Re}_L < 3\times 10^5]$. In the *NCEES PE Mechanical Handbook*, $\mathrm{Re}_L < 10^5$.

[14]According to some authorities: $[\mathrm{Pr} > 0.7]$.

[15]Some authorities report an empirical coefficient of 0.0296 instead of the theoretical 0.0288.

[16]Some authorities report the coefficient as 0.036 or 0.037.

[17]Some authorities report the 871 correction as 850 or 23,200, or, if there is no laminar flow at all on the plate, they omit it altogether.

[18]*The NCEES PE Mechanical Handbook* does not include the boundary for the Re range. The adjustment (−871) is provided in Eq. 35.24.

External Flow

$$\overline{\mathrm{Nu}}_L = \frac{\bar{h}L}{k} = 0.0366\left(\mathrm{Re}_L^{0.8} - 871\right)\mathrm{Pr}^{1/3} \qquad 35.24$$

$$[0.6 < \mathrm{Pr} < 60;\ 5\times 10^5 < \mathrm{Re}_x < 10^7;\ \mathrm{Re}_L < 10^8]$$

The local and average skin-friction coefficients for turbulent flow are

$$C_{f,x} = \frac{0.0576}{(\mathrm{Re}_x)^{1/5}} \qquad 35.25$$

$$\overline{C}_f = \frac{0.072}{(\mathrm{Re}_L)^{1/5}} \qquad 35.26$$

Equation 35.25 and Eq. 35.26 do not include the skin friction for the laminar section. The total friction from the leading edge to point x is obtained by subtracting the turbulent drag for the critical length and adding the laminar drag for the entire section (including the laminar section).

$$C_{f,\mathrm{total}} = \frac{0.072}{(\mathrm{Re}_L)^{1/5}} - \frac{0.072\left(\frac{x_{\mathrm{critical}}}{L}\right)}{(\mathrm{Re}_{x,\mathrm{critical}})^{1/5}} + \frac{1.328\left(\frac{x_{\mathrm{critical}}}{L}\right)}{(\mathrm{Re}_{x,\mathrm{critical}})^{1/2}} \qquad 35.27$$

For a critical Reynolds number of 5×10^5, this simplifies to

$$C_{f,\mathrm{total}} = \frac{0.072}{(\mathrm{Re}_L)^{1/5}} - 0.00334\left(\frac{x_{\mathrm{critical}}}{L}\right) \qquad 35.28$$

11. LAMINAR FLOW INSIDE TUBES WITH CONSTANT WALL TEMPERATURE

Laminar flow in smooth tubes occurs at Reynolds numbers less than 2000. As with flow over a flat plate, the velocity distribution is parabolic, but the extent of the parabola is limited to the tube radius. In the *entrance region*, the parabola does not extend to the centerline. Further on, a point is reached where the parabolic distribution is complete, and the flow is said to be *fully developed* laminar flow.[19] At that point, the average velocity is one-half of the maximum (centerline) velocity.

For laminar flow inside a tube with constant wall temperature, the film coefficient decreases with distance from the entrance and approaches the fully developed laminar value given by Eq. 35.29. (All fluid properties in Eq. 35.29[20] are evaluated at the bulk temperature.)

Laminar Flow in Circular Tubes

$$\mathrm{Nu}_D = 3.66 \quad [\mathrm{Re} < 2000;\ \mathrm{Pr} > 0.6] \qquad 35.29$$

The tube length required to establish fully developed laminar film is known as the *thermal entry length*.

$$x_{\mathrm{thermal\ entry}} = 0.05 D_H \mathrm{Re}_{D_H}\mathrm{Pr} \qquad 35.30$$

The *Sieder-Tate* (also known as *Sieder-Tate correlation*) predicts the average film coefficient along the entire length of laminar flow. All of the fluid properties are evaluated at the bulk temperature except μ_s, which is evaluated at the surface (wall) temperature.

$$\overline{\mathrm{Nu}}_D = 1.86\left(\frac{\mathrm{Re}_D\mathrm{Pr}}{\frac{L}{D}}\right)^{1/3}\left(\frac{\mu_b}{\mu_s}\right)^{0.14} \qquad 35.31$$

$$\left[\mathrm{Pr} > 0.48;\mathrm{Re}_D < 2000;\ 0.0044 < \frac{\mu_b}{\mu_s} < 9.75;\ \left(\mathrm{Re}_D\mathrm{Pr}\left(\frac{D}{L}\right)\right)^{1/3}\left(\frac{\mu_b}{\mu_s}\right)^{0.14} > 2\right]$$

12. LAMINAR FLOW INSIDE TUBES WITH CONSTANT HEAT FLUX

Some tubes experience a constant *heat flux* (i.e., a constant heat transfer) per unit length (area).[21] The film coefficient decreases with distance from the tube entrance. The laminar flow is *fully developed* when the difference between the surface (wall) and mean fluid temperature is constant.[22] This essentially occurs when $(x/r_i)/\mathrm{Re}_D\mathrm{Pr} > 0.100$. As Table 35.1 indicates, the fully developed laminar Nusselt number is 4.364.

Laminar Flow in Circular Tubes

$$\mathrm{Nu}_D = 4.36 \quad [\mathrm{Re}_D < 2000;\ \mathrm{Pr} > 0.6] \qquad 35.32$$

[19] The term *fully developed* is also used when referring to full turbulence. In this section it is understood that the flow is laminar.

[20] Some authorities report the value as 3.656.

[21] An example of a tube with constant heat flux is a pipe that has been uniformly wrapped with an electric heat strip along the pipe length.

[22] Standard Re values in the industry for laminar flow are typically less than 2000. *The NCEES PE Mechanical Handbook* references laminar flow at less than 2300, which is not typical.

Table 35.1 Nusselt Numbers for Laminar Flow Inside Tubes with Constant Heat Flux

$\dfrac{x/r}{\text{Re}_D\text{Pr}}$	Nu_D
0	∞
0.002	12.00
0.004	9.93
0.010	7.49
0.020	6.14
0.040	5.19
0.100	4.51
∞	4.364

13. TURBULENT FLOW INSIDE STRAIGHT TUBES

The theoretical *Nusselt equation* can be used to find the inside film coefficient for turbulent flow inside round, horizontal tubes.[23] In Eq. 35.33[24], all fluid properties except specific heat are evaluated at the film temperature. Specific heat is evaluated at the bulk temperature.

$$\text{Nu} = 0.023\,\text{Re}_\text{D}^{0.8}\text{Pr}^{1/3}$$

$$\left[0.6 < \text{Pr} < 160; \text{Re} > 10^4; \frac{L}{D} > 60\right] \qquad 35.33$$

Equation 35.33 is difficult to use in design work because the film temperature is an inconvenient concept with tube flow. With tube flow, it is more common to base all fluid properties on the bulk temperature. Equation 35.34[25] is the *Dittus-Boelter equation,* as it was modified by W. H. McAdams. It evaluates all fluid properties at the bulk temperature.

$$\text{Nu} = 0.023\,\text{Re}^{0.8}\text{Pr}^n$$

$$\left[0.7 < \text{Pr} < 120; \text{Re} > 10^4; \frac{L}{D} > 60\right] \qquad 35.34$$

The exponent, n, in Eq. 35.34 has a value of 0.3 when the surface (wall) temperature is less than the bulk fluid temperature, and n is 0.4 when the surface (wall) temperature is greater than the bulk fluid temperature.

Within the normal range of most gases, $\text{Pr}^n \approx 1.0$, resulting in Eq. 35.35.

$$\text{Nu} = 0.023\,\text{Re}^{0.8} \qquad 35.35$$

If there is a large change in viscosity during the heat transfer process, as there would be with oils and other viscous fluids heated in a long tube, Eq. 35.35 is modified into the *Sieder-Tate equation* for turbulent flow. All fluid properties in Eq. 35.36[26] are evaluated at the bulk temperature except for μ_s, which is evaluated at the surface temperature.

Turbulent Flow in Circular Tubes

$$\text{Nu}_D = 0.023\,\text{Re}_D^{0.8}\text{Pr}^{1/3}\left(\frac{\mu_b}{\mu_s}\right)^{0.14}$$

$$\left[0.7 < \text{Pr} < 160; \text{Re} > 10^4; \frac{L}{D} > 60\right] \qquad 35.36$$

For L/D ratios less than 60 occurring in pipes with sharp leading edges, the right-hand side of Eq. 35.36 can be multiplied by either Eq. 35.37 or Eq. 35.38.

$$1 + \left(\frac{D}{L}\right)^{0.7} \quad \left[2 < \frac{L}{D} < 20\right] \qquad 35.37$$

$$1 + \left(\frac{6D}{L}\right) \quad \left[20 < \frac{L}{D} < 60\right] \qquad 35.38$$

Example 35.1

Water flows at 10 ft/sec (3 m/s) through the inside of a 2.00 in (51 mm) inside diameter, 2.125 in (54 mm) outside diameter tube. The tube wall temperature is 170°F (75°C) along its entire length. The water enters at 70°F (20°C) and is heated to 130°F (56°C). What is the inside film coefficient?

SI Solution

The average bulk temperature of water is

$$T_b = \tfrac{1}{2}(T_\text{in} + T_\text{out}) = \left(\frac{1}{2}\right)(20°\text{C} + 56°\text{C}) = 38°\text{C}$$

From App. 34.B, the fluid properties evaluated at the average bulk temperature are

$$\rho_{38°\text{C}} = 994.7\ \text{kg/m}^3$$
$$c_{p,38°\text{C}} = 4.183\ \text{kJ/kg·K}$$
$$\mu_{38°\text{C}} = 0.682 \times 10^{-3}\ \text{kg/m·s}$$
$$k_{38°\text{C}} = 0.6283\ \text{W/m·K}$$
$$\text{Pr} = 4.51$$

[23]Equation 35.33 can also be used to obtain conservative values for turbulent flow in vertical tubes.
[24]According to some authorities: $[0.5 < \text{Pr} < 100]$. According to some authorities: $[L/d > 10]$ when used with Eq. 35.37 and Eq. 35.38.
[25]According to some authorities: $[\text{Pr} < 100]$. According to some authorities: $[L/d > 10]$ when used with Eq. 35.37 and Eq. 35.38.
[26]Some authorities report the coefficient as 0.027 instead of 0.023. According to some authorities: $[\text{Pr} > 0.6]$. The upper Prandtl number limit is also reported as 700, 16,700, and 17,000. According to some authorities: $[L/d > 10]$ when used with Eq. 35.37 and Eq. 35.38.

Heat Transfer

The Reynolds number is

Reynolds Number

$$\text{Re} = \frac{\text{v}D}{\nu} = \frac{\text{v}D\rho}{\mu} = \frac{\left(3\ \frac{\text{m}}{\text{s}}\right)(0.051\ \text{m})\left(994.7\ \frac{\text{kg}}{\text{m}^3}\right)}{0.682\times10^{-3}\ \frac{\text{kg}}{\text{m}\cdot\text{s}}}$$
$$= 2.23\times10^5 \quad [\text{turbulent}]$$

From Eq. 35.3 and Eq. 35.34, the film coefficient is

$$h = 0.023(\text{Re}^{0.8}\text{Pr}^n)\left(\frac{k}{D}\right)$$
$$= \frac{(0.023)(2.23\times10^5)^{0.8}(4.51)^{0.4}\left(0.6283\ \frac{\text{W}}{\text{m}\cdot\text{K}}\right)}{0.051\ \text{m}}$$
$$= 9832\ \text{W/m}^2\cdot\text{K}$$

Customary U.S. Solution

The average bulk temperature of water is

$$T_b = \tfrac{1}{2}(T_{\text{in}} + T_{\text{out}}) = \left(\frac{1}{2}\right)(70°\text{F} + 130°\text{F})$$
$$= 100°\text{F}$$

From App. 34.A, the fluid properties evaluated at the average bulk temperature are

$$\rho_{100°\text{F}} = 62.0\ \text{lbm/ft}^3$$
$$c_{p,100°\text{F}} = 0.998\ \text{Btu/lbm-°F}$$
$$\nu_{100°\text{F}} = 0.74\times10^{-5}\ \text{ft}^2\text{/sec}$$
$$k_{100°\text{F}} = 0.364\ \text{Btu-ft/hr-ft}^2\text{-°F}$$
$$\text{Pr} = 4.52$$

The Reynolds number is

Reynolds Number

$$\text{Re} = \frac{\text{v}D}{\nu} = \frac{\left(10\ \frac{\text{ft}}{\text{sec}}\right)(2.00\ \text{in})}{\left(0.74\times10^{-5}\ \frac{\text{ft}^2}{\text{sec}}\right)\left(12\ \frac{\text{in}}{\text{ft}}\right)}$$
$$= 2.25\times10^5 \quad [\text{turbulent}]$$

From Eq. 35.3 and Eq. 35.34, the film coefficient is

$$h = 0.023(\text{Re}^{0.8}\text{Pr}^n)\left(\frac{k}{D}\right)$$
$$= \frac{(0.023)(2.25\times10^5)^{0.8}(4.52)^{0.4}\left(0.364\ \frac{\text{Btu-ft}}{\text{hr-ft}^2\text{-°F}}\right)}{\frac{2\ \text{in}}{12\ \frac{\text{in}}{\text{ft}}}}$$
$$= 1757\ \text{Btu/hr-ft}^2\text{-°F}$$

14. TURBULENT FLOW INSIDE COILED TUBES

For flow inside helically coiled tubes and Reynolds numbers above 10^4, the film coefficient derived for a straight pipe should be multiplied by

$$1 + \frac{3.5D_{\text{tube}}}{D_{\text{coil}}} \quad [\text{Re} > 10^4] \qquad 35.39$$

15. TURBULENT AIR FLOW IN TUBES

A reasonably accurate approximation for turbulent air at one atmospheric pressure and between 0°F and 240°F (−17°C and 116°C) flowing in a circular tube is given by Eq. 35.40.

$$h \approx (0.00351 + 0.000001583\,T_{°\text{F}}) \times \left(\frac{G^{0.8}_{\text{lbm/hr-ft}^2}}{D^{0.2}_{\text{ft}}}\right) \quad [\text{U.S}] \qquad 35.40$$

Equation 35.41(a) and Eq. 35.41(b) are approximations that do not depend on the air temperature.

$$h \approx \frac{3.52\text{v}^{0.8}_{\text{m/s}}}{D^{0.2}_{\text{m}}} \quad [\text{SI}] \qquad 35.41(a)$$

$$h \approx \frac{0.5\text{v}^{0.8}_{\text{ft/sec}}}{D^{0.2}_{\text{in}}} \quad [\text{U.S.}] \qquad 35.41(b)$$

16. TURBULENT WATER FLOW IN TUBES

A reasonably accurate approximation for turbulent water between 40°F and 220°F (4°C and 105°C) flowing in a round tube is

$$h \approx \frac{1429(1 + 0.0146\,T_{b,°\text{C}})\text{v}^{0.8}_{\text{m/s}}}{D^{0.2}_{\text{m}}} \quad [\text{SI}] \qquad 35.42(a)$$

$$h \approx \frac{150(1 + 0.011\,T_{b,°\text{F}})\text{v}^{0.8}_{\text{ft/sec}}}{D^{0.2}_{\text{in}}} \quad [\text{U.S.}] \qquad 35.42(b)$$

17. TURBULENT OIL FLOW IN TUBES

A reasonably accurate approximation for oil heated in pipes is

$$h \approx \frac{0.034\text{v}_{\text{ft/sec}}}{\mu_b^{0.63}} \quad [\text{U.S.}] \qquad 35.43$$

18. TURBULENT LIQUID METAL FLOW IN TUBES

When the surface (wall) temperature is constant, Eq. 35.44 can be used to calculate the average film coefficient for liquid metals (e.g., mercury, sodium, and lead-bismuth alloys) experiencing fully developed turbulent flow inside tubes.[27] Fluid properties are evaluated at the mean bulk temperature.

$$\mathrm{Nu}_D = 5.0 + 0.025(\mathrm{Re}_D\mathrm{Pr})^{0.8} \quad \left[\mathrm{Re}_D\mathrm{Pr} > 100;\ \frac{L}{D} > 60\right] \qquad 35.44$$

With a constant heat flux, Eq. 35.45 through Eq. 35.47 can be used to calculate the average film coefficient for liquid metals with fully developed turbulent flow inside tubes. Equation 35.47 is known as the *Lubarsky-Kaufman correlation.*

$$\mathrm{Nu}_D = 4.82 + 0.0185(\mathrm{Re}_D\mathrm{Pr})^{0.827} \quad \left[\begin{array}{c} 3.6 \times 10^3 < \mathrm{Re}_D < 9.05 \times 10^5; \\ 100 < \mathrm{Re}_D\mathrm{Pr} < 10{,}000 \end{array}\right] \qquad 35.45$$

$$\mathrm{Nu}_D = 7.0 + 0.025(\mathrm{Re}_D\mathrm{Pr})^{0.8} \quad \left[\mathrm{Re}_D\mathrm{Pr} > 100;\ \frac{L}{D} > 60\right] \qquad 35.46$$

$$\mathrm{Nu}_D = 0.625(\mathrm{Re}_D\mathrm{Pr})^{0.4} \quad \left[100 < \mathrm{Re}_D\mathrm{Pr} > 10{,}000;\ \frac{L}{D} > 60\right] \qquad 35.47$$

19. FLOW THROUGH NONCIRCULAR DUCTS

Dimensional analysis shows that a *characteristic length* is required in the Nusselt number, but it does not identify the length to be used. It has been common practice to correlate empirical pressure drop and heat transfer data with the *hydraulic diameter*, D_H, of noncircular (e.g., rectangular, square, elliptical, polygonal) ducts.[28,29] The Nusselt number for laminar and turbulent flow through noncircular ducts is given by Eq. 35.48.

$$\mathrm{Nu} = \frac{hD_H}{k} \qquad 35.48$$

$$D_H = 4\left(\frac{\text{area in flow}}{\text{wetted perimeter}}\right) \qquad 35.49$$

For rectangular ducts,

$$D_H = \frac{1.3(\text{short side} + \text{long side})^{5/8}}{(\text{short side} + \text{long side})^{1/4}} \qquad 35.50$$

Annular flow is the flow of fluid through an annulus. Fluid flow is annular in simple tube-in-tube heat exchangers.[30] For an annulus, Eq. 35.51 gives the hydraulic diameter.[31]

$$D_H = D_{i,\text{shell}} - D_{o,\text{tube}} \qquad 35.51$$

The film coefficient for fully developed laminar flow through noncircular ducts can be calculated from the Nusselt numbers in Table 35.2.

Table 35.2 *Nusselt Numbers for Fully Developed Laminar Flow Ducts*

	Nusselt number	
configuration	constant wall temperature	constant heat flux
circular	3.658	4.364
square	2.98	3.63
rectangular, aspect ratio		
$1{:}\sqrt{2}$	–	3.78
1:2	3.39	4.11
1:3	–	4.77
1:4	4.44	5.35
1:8	5.95	6.60
parallel plates	7.54	8.235
triangular, isosceles	2.35	3.00

[27] The term *fully developed* is also used when referring to full laminar flow. In this section it is understood that the flow is turbulent.

[28] A *duct* is any closed channel through which a fluid flows. Tubes and pipes are examples of round ducts. "Ducts" are not limited to air conditioning ducts.

[29] The use of the hydraulic diameter as the characteristic length is convenient and logical. Though an approximation, empirical data supports using the hydraulic diameter in most cases. Notable exceptions are flow through ducts with narrow angles (e.g., an equilateral triangle with a narrow angle) and flow *parallel* to banks of tubes.

[30] Flow is not annular through more complex shell-and-tube heat exchangers.

[31] Though the hydraulic diameter is widely used as the characteristic length in calculating the Nusselt number for annuli, it is not a universal choice. Some researchers recommend using an equivalent diameter defined as

$$D_{\text{eq}} = \frac{4(\text{area in flow})}{\text{heated perimeter}} = \frac{D_{i,\text{shell}}^2 - D_{o,\text{tube}}^2}{D_{o,\text{tube}}}$$

Heat Transfer

20. CROSSFLOW OVER A SINGLE CYLINDER

Figure 35.2 illustrates a cylinder (e.g., a tube or wire) in crossflow. Equation 35.52 can be used with any fluid to calculate the film coefficient.[32] The fluid properties are evaluated at the film temperature. The entire surface area of the tube is used when calculating the heat transfer. Flow is laminar up to a Reynolds number of approximately 5×10^5. (See Fig. 35.3 and Table 35.3.)

Cylinder of Diameter D in Cross Flow

$$\overline{\text{Nu}}_D = \frac{\bar{h}D}{k} = C_1\text{Re}_D{}^n\text{Pr}^{1/3} \quad [\text{Pr} \geq 0.7] \qquad 35.52$$

$$\text{Re}_D = \frac{\rho v_\infty D}{\mu} \qquad 35.53$$

Figure 35.2 *Single Cylinder in Crossflow*

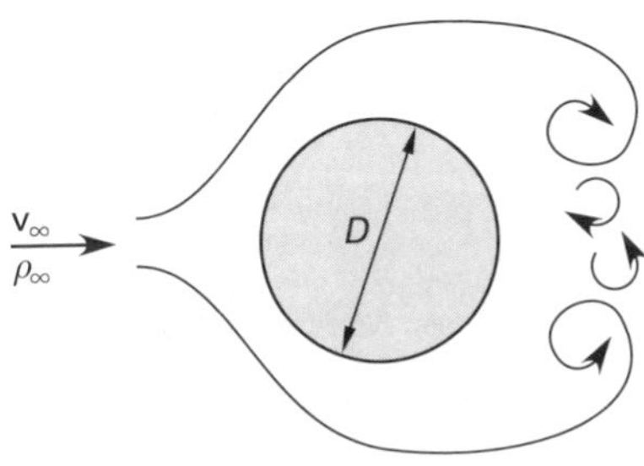

Figure 35.3 *Average Nusselt Number for Cylinder in Crossflow (air)*

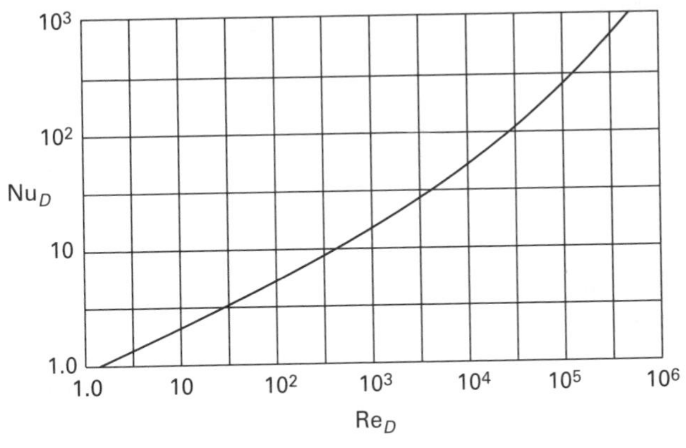

$$\text{Nu} = C_2(\text{Re}_D)^n \qquad 35.54$$

$$C_2 \approx 0.9C_1 \qquad 35.55$$

Table 35.3 *Constants for Tubes in Crossflow (air and other gases)*

Re_D	C_1	C_2	n
0.4[a]–4	0.989	0.891	0.330
4–40	0.911	0.821	0.385
40–4000	0.683	0.615	0.466
4000–40,000	0.193	0.174	0.618
40,000–400,000[b]	0.0266	0.0239	0.805

[a]Some sources give the lower limit as 1.0.
[b]Some sources give the upper limit as 250,000.

Equation 35.52 can be simplified for air since $\text{Pr}^{1/3} \approx 0.9$. Equation 35.54 is sometimes referred to as the *Hilbert-Morgan equation.*

21. CROSSFLOW OVER TUBE BUNDLES

Industrial heat exchangers contain a large number of tubes arranged in layers to increase the heat transfer surface. Two or more layers are considered a tube bundle.[33] The tubes can be arranged in several ways, including a square (i.e., in-line) or rotated square (i.e., staggered), as shown in Fig. 35.4(a) and (b). Figure 35.4 also defines the *longitudinal pitch* (spacing), s_l, and *transverse pitch* (spacing), s_t. The number of transverse rows, M, is the number of tubes in the first layer encountered by the flow. N is the number of layers.

Figure 35.4 *Tube Bundles in Crossflow*

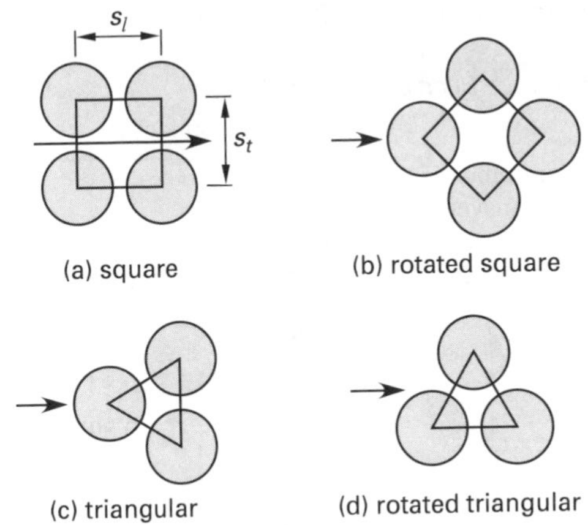

For the square and rotated square configurations, the maximum fluid velocity based on the minimum area is used to calculate the Reynolds number.[34] L is the length of each tube.

$$v_{max} = \frac{Q}{A_{min}} \qquad 35.56$$

[32]There are more sophisticated correlations. Two popular ones are the Žukauskas correlation and the Churchill and Bernstein correlation.
[33]The tubes in a single layer perform as single tubes in crossflow. See Sec. 35.20.
[34]Equation 35.57 is an example of where common sense needs to be used. If there are M tubes in the layer, there are only $M - 1$ openings. However, there is also space on the outside of the last tubes in the layer. The actual configuration will determine the true nature of the calculation.

$$A_{min} = LM(s_t - D_o) \quad \text{[in-line]} \qquad 35.57$$

$$A_{min} = LM \times \text{minimum}\begin{Bmatrix} 2s_t - D_o \\ \sqrt{s_t^2 + s_l^2} - D_o \end{Bmatrix} \qquad 35.58$$

[staggered]

The *Colburn equation*, Eq. 35.59, is used to determine the film coefficient for turbulent flow in heat exchangers with 10 or more transverse rows (i.e., $M \geq 10$) in the square or rotated square configurations.[35,36,37] Fluid properties in Eq. 35.59[38] and Eq. 35.60[39] are evaluated at the film and bulk temperatures, according to their subscripts.

$$\text{Nu}_D = \frac{hD_o}{k_{film}} = 0.26(\text{Re}_{max})^{0.6}(\text{Pr}_{film})^{0.3} \qquad 35.59$$

[in-line; Re > 5000]

$$\text{Nu}_D = \frac{hD_o}{k_{film}} = 0.33(\text{Re}_{max})^{0.6}(\text{Pr}_{film})^{0.3} \qquad 35.60$$

[staggered; Re > 5000]

$$\text{Re}_{max} = \frac{G_{max}D_o}{\mu_{film}} = \frac{\text{v}_{max}\rho D_o}{\mu_{film}} \qquad 35.61$$

Most commercial heat exchangers have 10 or more transverse rows. However, if there are fewer than 10 transverse rows, the film coefficient calculated from Eq. 35.59 and Eq. 35.60 should be multiplied by the correction factor from Table 35.4.

Table 35.4 *Tube Bundle Correction Factor*

number of transverse rows	correction factor: in-line	correction factor: staggered
1	0.64	0.68
2	0.80	0.75
3	0.87	0.83
4	0.90	0.89
5	0.92	0.92
6	0.94	0.95
7	0.96	0.97
8	0.98	0.98
9	0.99	0.99
10	1.00	1.00

22. PRESSURE DROP ACROSS TUBE BUNDLES

Empirical relationships for the pressure drop for flow across tube bundles are available in most heat transfer books. Though these relationships are of theoretical interest, the shellside pressure drop through a commercial heat exchanger with baffles is much more difficult to calculate. Though some correlations exist for specific configurations, most predictions are derived from the performance of similar units.

23. FLOW OVER SPHERES

The general Nusselt correlation for fluid flow over a sphere (see Fig. 35.5) is given by Eq. 35.62. Flow is laminar up to a Reynolds number of approximately 3×10^5. Fluid properties should be evaluated at the film temperature.

$$\text{Nu}_D = (0.97 + 0.68\sqrt{\text{Re}_D})\text{Pr}^{1/3} \qquad 35.62$$

$[1 < \text{Re}_D < 2000]$

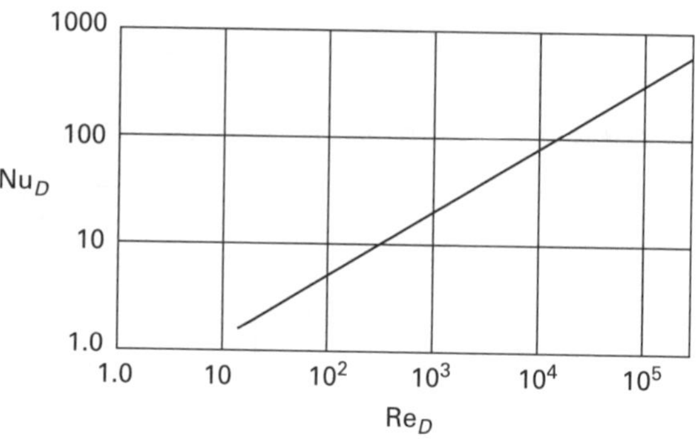

Figure 35.5 *Average Nusselt Number for Flow Over a Sphere*

An empirical correlation with a higher Reynolds number limit is

Flow Over a Sphere of Diameter D

$$\overline{\text{Nu}}_D = \frac{\bar{h}D}{k} = 2.0 + 0.6\left(\text{Re}_D\right)^{1/2}\text{Pr}^{1/3} \qquad 35.63$$

$[1 < \text{Re}_D < 70{,}000]$, $[0.6 < \text{Pr} < 400]$

For gases, Eq. 35.64[40] can be used.

$$\text{Nu}_D = 0.37(\text{Re}_D)^{0.6}(\text{Pr})^{1/3} \qquad 35.64$$

$[25 < \text{Re}_D < 150{,}000]$

[35] A theoretical derivation shows that the exponent is 1/3, not 0.3. However, the 0.3 is consistent with experimental data.
[36] Each tube in the first transverse row essentially has the same heat transfer as a single tube.
[37] Other researchers omit the Prandtl number term and report slightly different coefficients. Such a formulation makes the assumption that $(\text{Pr})^{0.3} \approx 1$.
[38] According to some authorities: [Re > 6000].
[39] See Ftn. 45.
[40] According to some authorities: $[25 < \text{Re}_D < 100{,}000]$ and $[17 < \text{Re}_D < 70{,}000]$.

For air, Eq. 35.65 can be used.

$$\mathrm{Nu}_D = 0.33(\mathrm{Re}_D)^{0.6} \quad [20 < \mathrm{Re}_D < 150{,}000] \qquad 35.65$$

24. CHANGE OF PHASE IN TUBES WITH FORCED CONVECTION

The processes of vaporization and condensation inside tubes experiencing forced convection are complex. Limited empirical relationships have been developed, but all are very specific in their derivations and limited in application. The ability to extrapolate these results to different conditions is questionable.

25. HEAT TRANSFER IN PACKED BEDS

Catalytic reactors, pebble-bed heat exchangers, and fluidized-bed furnaces are examples of *packed beds*. The heat transfer is based on the bed volume, not on surface area. The film coefficient per unit volume of packed bed, h_V, is a function of the solid particle surface area and the empty cross-sectional area of the bed per unit volume.

$$\dot{Q} = h_V V_{\mathrm{bed}}(T_{s,\mathrm{bed\,particle}} - T_{b,\mathrm{gas}}) \qquad 35.66$$

26. HEAT EXCHANGERS

Two fluids flow through or over a heat exchanger.[41] Heat from the hot fluid passes through the exchanger walls to the cold fluid.[42] The heat transfer mechanism is essentially completely forced convection.

Heat exchangers are categorized into simple *tube-in-tube heat exchangers* (also known as *jacketed pipe heat exchangers*), single-pass shell-and-tube heat exchangers, multiple-pass shell-and-tube heat exchangers, and crossflow heat exchangers.[43] *Shell-and-tube heat exchangers*, also known as *sathes* and *S & T heat exchangers*, consist of a large housing, the *shell*, with many smaller tubes running through it. The *tube fluid* passes through the tubes, while the *shell fluid* passes through the shell and around tubes.[44]

In a *single-pass heat exchanger*, each fluid is exposed to the other fluid only once. (See Fig. 35.6.) Operation is known as *parallel flow* (same as *cocurrent flow*) if both fluids flow in the same direction along the longitudinal axis of the exchanger and *counterflow* (same as *counter current flow*) if the fluids flow in opposite directions.[45,46] Counterflow is more efficient, and the heat transfer area required is less than that with parallel flow since the temperature gradient is more constant.

Figure 35.6 *Simple Heat Exchanger (single-pass, counterflow, tube-in-tube)*

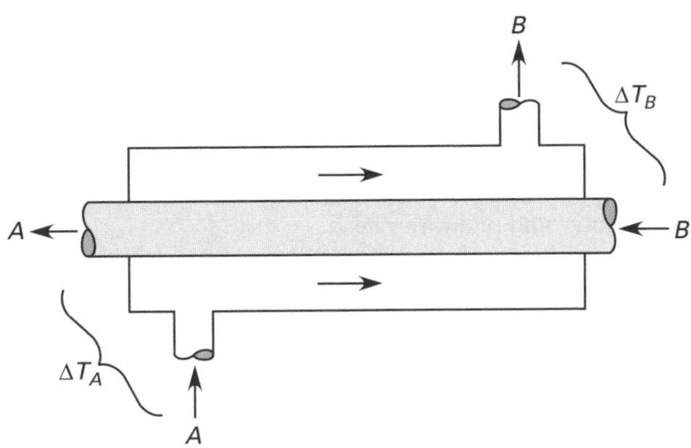

For increased efficiency, most exchangers are *multiple-pass heat exchangers*. The tubes pass through the shell more than once, and the shell fluid is routed around baffles.

In a *crossflow heat exchanger*, one fluid flow is normal to the other.[47] Crossflow exchangers can operate with both fluids unmixed (typical when fluids are constrained to move through tubes and passageways), or one or both fluids may be mixed within the heat exchanger by forcing the fluids around tubes, baffles, or passages. If the fluid is mixed, its temperature is essentially uniform across the outlet. In TEMA X shells (see Sec. 35.27) experiencing pure crossflow and air-cooled exchangers, the fluids are generally unmixed.

One of the fluids in a crossflow heat exchanger can have multiple passes through the other fluid. Since the flow cannot be parallel in a crossflow heat exchanger, the designations used are *counter-crossflow* and *cocurrent-crossflow*. The distinction between mixed and unmixed fluids is further complicated by whether the fluids are mixed or unmixed between passes.

[41]These fluids do not have to be liquids. *Air-cooled exchangers* reduce water consumption in traditional cooling applications.

[42]A *recuperative heat exchanger*, typified by the traditional shell-and-tube exchanger, maintains separate flow channels for each of the fluids. A *regenerative heat exchanger* has only one flow path, to which the two fluids are exposed on an alternating basis.

[43]Fin coil heat exchangers are a special case of crossflow heat exchangers.

[44]Tubular heat exchangers are also known as *shell-and-tube heat exchangers*.

[45]Flow through shell-and-tube heat exchangers is neither purely parallel nor purely counterflow. Thus, these exchangers are sometimes designated as *parallel counterflow exchangers*.

[46]The designation *cocurrent* is not an abbreviation for *counter current*.

[47]An automobile radiator is an example of a crossflow exchanger.

27. HEAT EXCHANGER DESIGNATIONS

A heat exchanger with X shell passes and Y tube passes is designated as an *X-Y heat exchanger*. In addition, most manufacturers follow the TEMA standards for design, fabrication, and material selection.[48,49,50] Heat exchanger types can be described by a three-character TEMA designation (see Table 35.5). For example, a one-two TEMA E shell and tube heat exchanger would be a shell-and-tube heat exchanger with one shell pass and two tube passes.

***Table 35.5** TEMA Heat Exchanger Designations*

front-end head type[a]	
A	channel and removable head
B	bonnet (integral, removable head)
C	channel (integral with tubesheet; removable plate cover)
D	special, high-pressure closure[b]
shell type	
E	one-pass shell
F	two-pass shell with longitudinal baffle[b]
F	split flow, one-pass tube[b]
G	split flow, two-pass tube[b]
H	double split flow[b]
J	divided flow, one-pass tube[b]
K	kettle type reboiler[b]
rear-end head type	
L	fixed tubesheet (like "A" head)
M	fixed tubesheet (like "B" head)
N	fixed tubesheet (like "C" head)
P	outside packed floating head
S	floating head with backing device (including clamp ring)
T	pull-through floating head
U	U-tube bundle
W	packed floating tubesheet with lantern ring
X	crossflow heat exchanger

[a]The term *head* is synonymous with *cover*.
[b]Specialty exchangers such as reboilers, steam heaters, vapor condensers, and feedwater heaters.

28. COMMERCIAL HEAT EXCHANGERS

Fixed tubesheet and U-tube bundles are the two most common commercial heat exchanger designs.[51] Fixed tubesheet heat exchangers (e.g., TEMA types BEM, AEM, and CEN) use straight tubes and offer the lowest cost heat transfer surface. A series of straight tubes is sealed between flat, perforated metal tubesheets.

The straight tubes are replaceable and easily cleaned on the inside. Since the tube bundle cannot be removed, the shellside of the tubes can only be cleaned chemically. Therefore, the shellside fluid must be nonfouling. There are no gaskets on the shellside, and the two fluids cannot accidentally mix. The no-gasket, closed-shell design is applicable to high vacuum and high pressure work as well as to potentially toxic fluids.

Removable-bundle heat exchangers differ in the types of removable heads. *Floating-head exchangers* (TEMA types AEW and BEW) have straight-through tubes with one tubesheet that is fixed to the shell and another that is fixed only to the tubes and "floats" within the shell. Floating-head exchangers are able to compensate for thermal expansion and contraction. Since they are separated by only a gasket, both fluids must be nonvolatile and nontoxic, and operation must be below approximately 300°F (150°C) and 300 psig (2.1 MPa).

With *outside-packed, floating-head exchangers* (TEMA types BEP and AEP), only the shellside fluid is exposed to the packing (i.e., the gasket). Corrosive gases, vapors, and liquids can be circulated in the tubes.

Internal clampring, floating-head exchangers (TEMA types AES and BES) are useful for applications with high-fouling fluids that require frequent inspection and cleaning or with high temperature differentials between the two fluids. Multipass arrangements are possible.

Pull-through floating-head exchangers (TEMA types AET and BET) have a floating head that is bolted directly to the floating tubesheet. The bundle can be removed without removing the shell or floating head covers. The design accommodates fewer tubes for a given shell diameter and offers less surface area than other removable-bundle exchangers. Although expensive, this design permits frequent cleaning.

U-tube heat exchangers (TEMA types AEU and BEU) have a bundle of tubes, each bent in a series of concentrically tighter U-shapes. They are lower in initial cost since there is only one tubesheet. The tube bundle can be removed for inspection and cleaning. The individual tubes automatically compensate for thermal expansion and contraction. These exchangers are ideal for intermittent service or where thermal shock is expected. However, the insides of the tubes cannot be mechanically cleaned along their entire length, and the tubes cannot be replaced.[52] The design cannot be single-pass

[48]The Tubular Exchanger Manufacturers Association (TEMA) publishes the definitive standards for shell-and-tube heat exchanger construction and performance.
[49]Other applicable standards are published by the American Society of Mechanical Engineers (ASME) and the American Petroleum Institute (API).
[50]Similar to ASME's Pressure Vessel Code, TEMA standards B, C, and R are applicable to shell-and-tube heat exchangers with shell diameters not exceeding 60 in (152 cm), pressures not exceeding 3000 psi (20.67 MPa), and product of shell diameter and pressure not exceeding 60,000 lbf/in (10,500 N/mm).
[51]Other variations and commercial designs include *packed floating head*, *bayonet*, *thimble*, *jacketed pipe*, and *platecoil* (*flat*) *heat exchangers*.
[52]A few of the outer bends can be replaced. However, a complete retubing is usually necessary.

on the tube side, and true countercurrent flow is not possible. *Plate-type heat exchangers* (often called *panel coil exchangers*, *welded-plate exchangers*, or simply *plate exchangers*) are used as immersion heaters in tanks for the heating or cooling of solutions and as evaporator surfaces where liquids cascade down their sides.[53] They are constructed from two sheets welded and embossed or expanded to form a series of passes through which one of the fluids flows. (See Fig. 35.7.)

Plate-and-frame heat exchangers combine several thin-gauge plates with embossed flow paths that are separated by an elastomeric or asbestos fiber gasket.[54] Several two-plate units are clamped together into a compact unit that is suitable for large temperature crosses or small temperature approaches. (See Sec. 35.30.)

Figure 35.7 Cross Section of Plate-Type Heat Exchanger

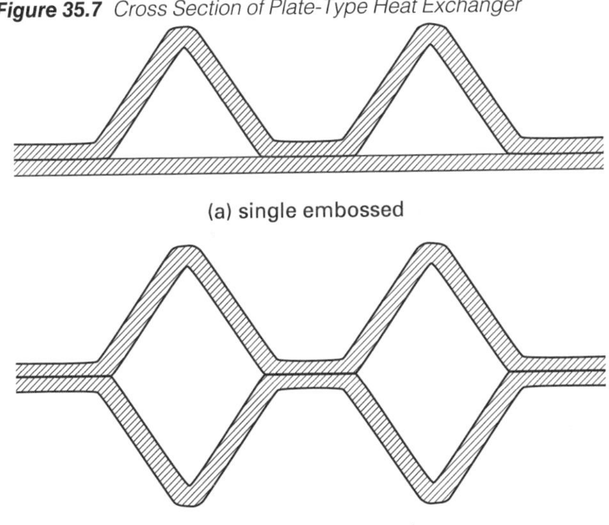

In addition to a high heat transfer rate due to their thin plate material, other advantages of plate and plate-and-frame heat exchangers include compactness, accessibility to both sides of each plate for maintenance and cleaning, and flexibility of design (since the number of plates in the frame can be varied).

For a given heat transfer coefficient, plate heat exchangers have a lower pressure drop. However, the narrow passageways produce high pressure drops in high-volume, low-pressure gas applications. These exchangers are limited to approximately 300 psig (2.1 MPa) and compatible gasket materials.

Air-cooled exchangers include a motor and fan assembly that forces air over a series of (typically coiled) tubes. To increase the heat transfer, the tubes are usually finned, hence the names *fin-tube exchanger* and *heavy-duty coil*.

29. TRANSVERSE BAFFLES IN HEAT EXCHANGERS

Several types of *baffles* are used inside the shells of commercial heat exchangers to increase the velocity (and, accordingly, the film coefficient) on the shellside. The most common type is the *segmental baffle*, also known as the *crossflow baffle*. (See Fig. 35.8.)

Figure 35.8 Types of Baffles

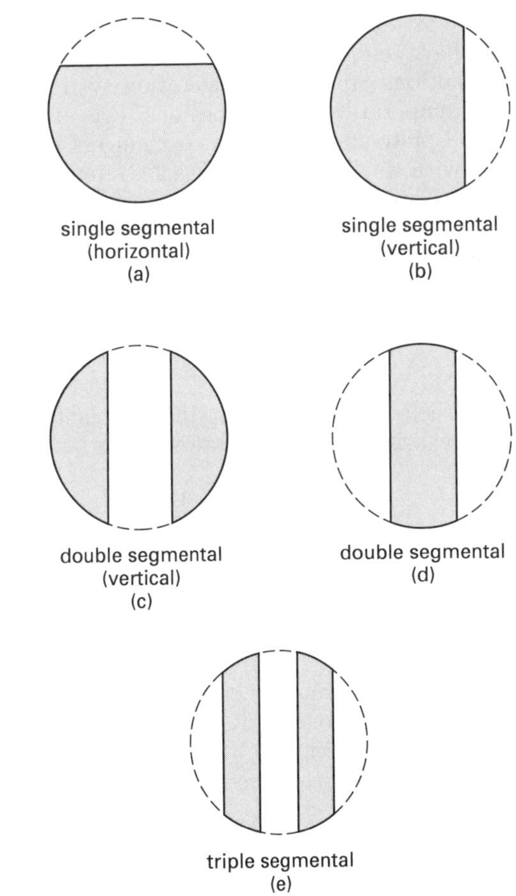

[53]"Platecoil" and "Temp-Plate" are trade names for the plate-type heat exchanger.

[54]The choice of gasket material depends on the temperatures, pressures, and fluids encountered. The temperature limits of popular gasket materials are 230°F to 275°F (110°C to 135°C) for nitrile, 300°F (150°C) for resin-cured butyl rubber, 320°F to 350°F (160°C to 175°C) for ethylene-propylene diene monomer (EPDM), and 212°F to 350°F (100°C to 175°C) for fluorocarbon rubber base. Nitrile is a general service material suitable for water applications. EPDM is suitable for steam and higher temperature aqueous solutions. All of the foregoing materials are limited to approximately 400 psig (2.8 MPa). Compressed asbestos gaskets can be used up to approximately 600°F (320°C), but they are limited to approximately 250 psig (1.7 MPa).

30. TEMPERATURE DIFFERENCE TERMINOLOGY

There are several specialized temperature difference terms used in the analysis of heat transfer.

The difference in hot and cold fluid temperatures is seldom constant in a heat exchanger. The *approach* (*temperature approach* or *approach temperature*) is the smallest difference in temperature between the two fluids anywhere along the heat exchange path. For a double-pipe, single-pass counterflow heat exchanger, the temperature approach is defined by Eq. 35.67.

$$\Delta T_{\text{approach}} = T_{\text{hot,in}} - T_{\text{cold,out}} \quad \left[\begin{array}{c}\text{single pass,}\\ \text{counterflow}\end{array}\right]$$
$$\Delta T_{\text{approach}} = T_{\text{hot,out}} - T_{\text{cold,out}} \quad \left[\begin{array}{c}\text{single pass,}\\ \text{parallel flow}\end{array}\right] \qquad 35.67$$

Traditional cost-effective shell-and-tube heat exchangers seldom have temperature approaches less than 10°F (6°C). Exceptions are some refrigeration systems that work with temperature approaches of 5°F to 9°F (3°C to 5°C), and plate-and-frame heat exchangers that can work well with as little as a 2°F (1°C) temperature approach. In combustion air preheaters using flue gas as the heating source, the temperature approach should be approximately 36°F (20°C).

The *extreme temperature difference* for heat exchangers is defined as

$$\Delta T_{\text{extreme}} = T_{\text{hot,in}} - T_{\text{cold,in}} \qquad 35.68$$

The ratio of the cold fluid change to the extreme temperature difference, a form of "temperature efficiency," is[55]

$$\eta_T = \frac{T_{\text{cold,out}} - T_{\text{cold,in}}}{T_{\text{hot,in}} - T_{\text{cold,in}}} \qquad 35.69$$

31. TEMPERATURE CROSS

A *temperature cross* occurs when the exit temperature of the cold fluid is above the exit temperature of the hot fluid. This occurs predominantly with counterflow heat exchangers, although it can also occur with a shell-and-tube exchanger with one shell pass and multiple tube passes. A temperature cross indicates that there is a relatively small temperature difference between the two fluids. This requires either a large heat transfer area or a relative high fluid velocity (to increase the overall heat transfer coefficient).

32. LOGARITHMIC TEMPERATURE DIFFERENCE

The temperature difference between two fluids is not constant in a heat exchanger. When calculating the heat transfer for a tube whose temperature difference changes along its length, the *logarithmic mean temperature difference*, ΔT_{lm} or LMTD, is used.[56,57,58] In Eq. 35.70, ΔT_A and ΔT_B are the temperature differences at ends A and B, respectively, regardless of whether the fluid flow is parallel or counterflow, as shown in Fig. 35.9.[59,60]

$$\Delta T_{\text{lm}} = \frac{\Delta T_A - \Delta T_B}{\ln \dfrac{\Delta T_A}{\Delta T_B}} = \frac{\Delta T_A - \Delta T_B}{2.3\log_{10} \dfrac{\Delta T_A}{\Delta T_B}} \qquad 35.70$$

Figure 35.9 *Temperature Profile for LMTD Calculation (single-pass, counterflow exchanger, no change of phase)*

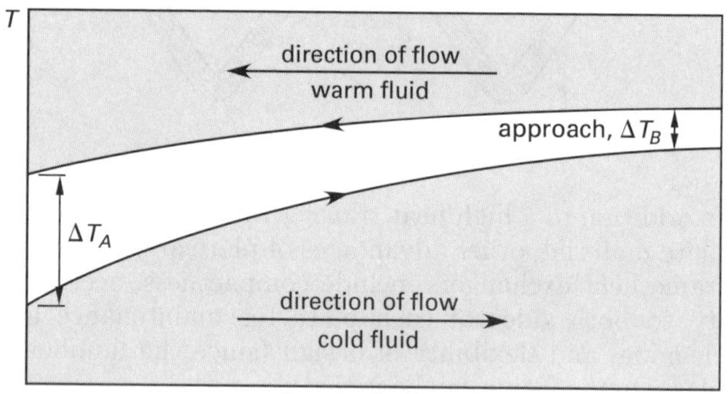

[55]The symbol S is also used for this quantity by some authors.

[56]An exception occurs in HVAC calculations where ΔT at midlength has traditionally been used to calculate the heat transfer in air conditioning ducts. Considering the imprecise nature of HVAC calculations, the added sophistication of using the logarithmic mean temperature difference is probably unwarranted.

[57]The symbol ΔT_m is also widely used for the log-mean temperature difference. However, this can also be interpreted as the arithmetic mean temperature.

[58]The logarithmic temperature difference is used even with change of phase (e.g., boiling liquid or condensing vapor) and the temperature is constant in one tube.

[59]It doesn't make any difference which end is A and which is B. If the numerator in Eq. 35.70 is negative, the denominator will also be negative.

[60]Using Eq. 35.70 presents many difficulties, particularly with computer-based heat transfer analysis. As ΔT_A and ΔT_B become equal, Eq. 35.70 becomes indeterminate, even though the correct relationship is $\Delta T_{\text{lm}} = \Delta T_A = \Delta T_B$. Also, the first derivative of Eq. 35.70, used in some calculations, is undefined when ΔT_A and ΔT_B are equal, even though the correct value is 0.5. A replacement expression (Underwood, 1933) that avoids these difficulties with (generally) less than a 0.3 % error is

$$\Delta T_{\text{lm}} \approx \left(\frac{\Delta T_A^{1/3} + \Delta T_B^{1/3}}{2}\right)^3$$

Equation 35.70 can be combined with Eq. 35.67 to determine the log mean temperature difference for counterflow and parallel flow heat exchangers as a function of their inlet and outlet conditions.

Log Mean Temperature Difference (LMTD)

$$\Delta T_{\text{lm}} = \frac{(T_{\text{hot,out}} - T_{\text{cold,in}}) - (T_{\text{hot,in}} - T_{\text{cold,out}})}{2.3\log_{10}\left(\frac{T_{\text{hot,out}} - T_{\text{cold,in}}}{T_{\text{hot,in}} - T_{\text{cold,out}}}\right)} \quad 35.71$$

[counterflow heat exchanger]

$$\Delta T_{\text{lm}} = \frac{(T_{\text{hot,out}} - T_{\text{cold,out}}) - (T_{\text{hot,in}} - T_{\text{cold,in}})}{2.3\log_{10}\left(\frac{T_{\text{hot,out}} - T_{\text{cold,out}}}{T_{\text{hot,in}} - T_{\text{cold,in}}}\right)} \quad 35.72$$

[parallel flow heat exchanger]

For multiple-pass and crossflow heat exchangers, a multiplicative correction factor, F_c, is required for ΔT_{lm}. The correction factor is 1.00 for single-pass parallel and counterflow tube-in-tube heat exchangers. When one of the fluids does not change temperature, as in a feedwater heater or other condensation/evaporation environment, the correction factor is also 1.00 for parallel, counter-, and crossflow.[61] The procedure in all other cases is to calculate ΔT_{lm} as if the fluids were in counterflow. The correction factor, F_c, depends on the type of heat exchanger and is almost always given graphically.[62] (See App. 35.A and App. 35.B.)

33. RULES OF THUMB FOR COOLING WATER

The following rules of thumb will result in more efficient heat transfer operations when water is used for cooling.

- For shell-and-tube heat exchangers, the water temperature should not increase by more than 20°F (10°C) when $\Delta T_{\text{lm}} < 70°\text{F}$ (40°C) and should not increase by more than 35°F (20°C) when $\Delta T_{\text{lm}} > 70°\text{F}$ (40°C).
- The water's exit temperature should not exceed 122°F (50°C) due to the increased potential for solids precipitation and fouling.
- The water's entrance temperature should be at least 10°F (5°C) above the freezing temperature of the liquid being cooled.

34. HEAT TRANSFER IN HEAT EXCHANGERS

Equation 35.73 calculates the steady-state heat transfer (also known as the *heat duty* and *heat load*) in a heat exchanger or feedwater heater.[63,64,65] The *overall heat transfer coefficient*, U, also known as the *overall conductance* and the *overall coefficient of heat transfer*, can be specified for use with either the outside or inside tube areas.[66] The heat transfer is independent of whether the outside or inside area is used.

Rate of Heat Transfer

$$\dot{Q} = UAF_c\Delta T_{\text{lm}} = U_oA_oF_c\Delta T_{\text{lm}} = U_iA_iF_c\Delta T_{\text{lm}} \quad 35.73$$

35. OVERALL HEAT TRANSFER COEFFICIENT

The overall heat transfer coefficient, U, is calculated from the film coefficients and the tube material conductivities. (See Table 35.6 and Table 35.7.) The overall heat transfer coefficient, based on outside and inside areas and exclusive of a fouling factor, is

$$\frac{1}{U_o} = \frac{1}{h_o} + \left(\frac{r_o}{k_{\text{tube}}}\right)\ln\left(\frac{r_o}{r_i}\right) + \frac{r_o}{r_ih_i} \quad 35.74$$

$$\frac{1}{U_i} = \frac{1}{h_i} + \left(\frac{r_i}{k_{\text{tube}}}\right)\ln\left(\frac{r_o}{r_i}\right) + \frac{r_i}{r_oh_o} \quad 35.75$$

The second term in Eq. 35.74 and Eq. 35.75 is sometimes approximated by the term t/k when the tube diameters are "large." However, the thermal resistance of the tube is very small and is often omitted entirely. If the tube resistance term is kept at all, it is not very difficult to use Eq. 35.74 and Eq. 35.75 as written.

The overall heat transfer coefficient, U, can also be determined using area and length.

$$\frac{1}{U_o} = \frac{A_o}{A_i}\frac{1}{h_i} + \frac{A_o\ln\left(\frac{r_o}{r_i}\right)}{2\pi k_{\text{tube}}L} + \frac{1}{h_o} \quad 35.76$$

$$\frac{1}{U_i} = \frac{1}{h_i} + \frac{A_i\ln\left(\frac{r_o}{r_i}\right)}{2\pi k_{\text{tube}}L} + \frac{A_i}{A_o}\frac{1}{h_o} \quad 35.77$$

[61] A general rule for good designs of boiling/evaporative systems is that the logarithmic mean temperature difference should be kept below 110°F (60°C).
[62] Calculating the F_c factor is preferred over reading it from a chart. The error of the F_c factor read from a chart can be as great as 5%.
[63] *Closed feedwater heaters* are heat exchangers whose purpose is to heat water with condensing steam.
[64] There are three heat loads referred to in heat exchanger specifications: the *specific heat load*, which is the design heat transfer; the heat released by the hot fluid; and the heat absorbed by the cold fluid. All three would be the same if operation was adiabatic, but due to practical losses, they are not. If they differ by more than 10%, the cause of the discrepancy should be evaluated.
[65] Transient (i.e., start-up) performance of heat exchangers is poorly covered in most textbooks. An excellent article on this subject is Chester A. Plant's "Evaluate Heat-Exchanger Performance," *Chemical Engineering Magazine*, p. 104, July 1992.
[66] It is more common (and preferred) to use the outside tube area because the tube outside diameter is more easily measured.

Table 35.6 Typical Values of Overall Coefficient of Heat Transfer (U-values)* (Btu/hr-ft²-°F)

		clean surface	with normal fouling
heating applications			
hot side	*cold side*		
steam	aqueous solution	300–550	150–275
steam	light oils	110–140	60–110
steam	medium lube oils	110–130	50–100
steam	Bunker C or no. 6 oil	70–90	60–80
steam	air or gases	5–10	4–8
hot water	aqueous solution	200–250	110–160
cooling applications			
cold side	*hot side*		
water	aqueous solution	200–250	105–155
water	medium lube oil	20–30	10–20
water	air or gases	5–10	4–8
freon/ ammonia	aqueous solution	60–90	40–60
calcium or sodium brine	aqueous solution	175–200	80–125

(Multiply Btu/hr-ft²-°F by 5.6783 to obtain W/m²·K.)

*Overall heat transfer coefficient values are strongly dependent on the type of heat exchanger, as well as on the hot- and cold-side fluids.

Since the heat transfer is independent of whether the inside or outside area is used, Eq. 35.76 and Eq. 35.77 can be simplified.

$$\frac{1}{UA} = \frac{1}{h_iA_i} + \frac{\ln\left(\frac{r_o}{r_i}\right)}{2\pi k_{\text{tube}}L} + \frac{1}{h_oA_o} \qquad 35.78$$

If the tube thermal resistance term is omitted entirely and if the tube is thin-walled, Eq. 35.79 and Eq. 35.80 can be used as approximations.

$$\frac{1}{U_o} \approx \frac{1}{h_o} + \frac{r_o}{r_ih_i} \approx \frac{1}{h_o} + \frac{1}{h_i} \qquad 35.79$$

$$\frac{1}{U_i} \approx \frac{1}{h_i} + \frac{r_i}{r_oh_o} \approx \frac{1}{h_i} + \frac{1}{h_o} \qquad 35.80$$

In reality, it is very difficult to predict the heat transfer coefficient for most types of commercial heat exchangers. Values can be predicted by comparison with similar units, or "tried-and-true" rules of thumb can be used. One such rule of thumb for baffled shell-and-tube heat exchangers predicts the clean, average heat transfer coefficient as 60% of the value for the same arrangement of tubes in pure crossflow.

36. TUBE LENGTH REQUIRED

For a simple, tube-in-tube counterflow heat exchanger as shown in Fig. 35.6, the length of tube required for a fluid to change temperature from T_{in} to T_{out} is given by Eq. 35.81. All units must be dimensionally consistent with regard to distance and time. The normal maximum length as limited by practical assembly and maintenance requirements for straight tubes is 20 ft (6 m). Multiple-pass heat exchangers are needed for longer lengths.

$$L = \frac{\rho \text{v} D_i^2 c_p (T_{\text{out}} - T_{\text{in}})}{4U_oD_o\Delta T_{\text{lm}}} \qquad 35.81$$

Example 35.2

Water flows at 10 ft/sec (3 m/s) through the inside of a 2.00 in (51 mm) inside diameter, 2.125 in (54 mm) outside diameter tube. The tube wall temperature is 170°F (75°C) along its entire length. The water enters at 70°F (20°C) and is heated to 130°F (56°C). The inside film coefficient is 1757 Btu/hr-ft²-°F (9832 W/m²·K).

All thermal resistance other than the internal film for the tube can be disregarded. What tube length is required?

SI Solution

The logarithmic mean temperature difference is

Log Mean Temperature Difference (LMTD)

$$\Delta T_{\text{lm}} = \frac{(T_{\text{hot,out}} - T_{\text{cold,in}}) - (T_{\text{hot,in}} - T_{\text{cold,out}})}{2.3\log_{10}\left(\frac{T_{\text{hot,out}} - T_{\text{cold,in}}}{T_{\text{hot,in}} - T_{\text{cold,out}}}\right)}$$

$$= \frac{(75^\circ\text{C} - 20^\circ\text{C}) - (75^\circ\text{C} - 56^\circ\text{C})}{2.3\log_{10}\left(\frac{75^\circ\text{C} - 20^\circ\text{C}}{75^\circ\text{C} - 56^\circ\text{C}}\right)}$$

$$= 33.9^\circ\text{C}$$

Since $U_oA_o = U_iA_i$, $U_i = h_i$, and $A = \pi DL$,

$$U_o = \frac{h_iA_i}{A_o} = h_i\left(\frac{D_i}{D_o}\right)$$

$$= \frac{\left(9832\ \frac{\text{W}}{\text{m}^2\cdot\text{K}}\right)(51\ \text{mm})}{54\ \text{mm}}$$

$$= 9286\ \text{W/m}^2\cdot\text{K}$$

Heat Transfer

Table 35.7 *Approximate Thermal Conductivity of Common Heat Exchanger Materials (Btu-ft/hr-ft²-°F)*

	temperature, °F													
	200	300	400	500	600	700	800	900	1000	1100	1200	1300	1400	1500
aluminum (annealed)														
type 1100-0	126	124	123	122	121	120	118							
type 3003-0	111	111	111	111	111	111	111							
type 3004-0	97	98	99	100	102	103	104							
type 6061-0	102	103	104	105	106	106	106							
aluminum (tempered)														
type 1100 (all tempers)	123	122	121	120	118	118	118							
type 3003 (all tempers)	96	97	98	99	100	102	104							
type 3004 (all tempers)	97	98	99	100	102	103	104							
type 6061 (T4 & T6)	95	96	97	98	99	100	102							
type 6063 (T5 & T6)	116	116	116	116	116	115	114							
type 6063 (T42)	111	111	111	111	111	111	111							
cast iron	31	31	30	29	28	27	26	25						
carbon steel	29.2	28.4	27.6	26.6	25.6	24.6	23.5	22.5	21.4	20.2	19.0	17.6	16.2	15.6
carbon moly steel (½% C)	25.2	25.1	24.8	24.3	23.7	23.0	22.2	21.4	20.4	19.5	18.4	16.7	15.3	15.0
chrome moly steels														
1% Cr, ½% Mo	21.9	22.0	21.9	21.7	21.3	20.8	20.2	19.7	19.1	18.5	17.7	16.5	15.0	14.8
2¼% Cr, 1% Mo	21.3	21.5	21.5	21.4	21.1	20.7	20.2	19.7	19.1	18.5	18.0	17.2	15.6	15.3
5% Cr, ½% Mo	18.1	18.7	19.1	19.2	19.2	19.0	18.7	18.4	18.0	17.6	17.1	16.6	16.0	15.8
12% Cr	14	15	15	15	16	16	16	16	17	17	17	18		
austenitic stainless steels														
18% Cr, 8% Ni	9.3	9.8	10	11	11	12	12	13	13	14	14	14	15	15
25% Cr, 20% Ni	7.8	8.4	8.9	9.5	10	11	11	12	12	13	14	14	15	15
admiralty brass	70	75	79	84	89									
naval brass	71	74	77	80	83									
copper	225	225	224	224	223									
copper and nickel alloys														
90% Cu, 10% Ni	30	31	34	37	42	47	49	51	53					
80% Cu, 20% Ni	22	23	25	27	29	31	34	37	40					
70% Cu, 30% Ni	18	19	21	23	25	27	30	33	37					
30% Cu, 70% Ni Alloy 400	15	15	16	16	17	18	18	19	20	20				

(Multiply Btu-ft/hr-ft²-°F by 1.731 to obtain W/m·K.)

From *Standards of the Tubular Exchanger Manufacturers Association*, copyright © 2007, by Tubular Exchanger Manufacturers Association. Reproduced with permission.

The tube length required is

$$L = \frac{\rho v D_i^2 c_p (T_{out} - T_{in})}{4 U_o D_o \Delta T_{lm}}$$

$$= \left(994.7\ \frac{kg}{m^3}\right)\left(3\ \frac{m}{s}\right)(0.051\ m)^2 \times \left(\frac{\left(4.183\ \frac{kJ}{kg{\cdot}K}\right)\left(1000\ \frac{J}{kJ}\right)(56°C - 20°C)}{(4)\left(9286\ \frac{W}{m^2{\cdot}K}\right)(0.054\ m)(33.9°C)}\right)$$

$$= 17.2\ m$$

Customary U.S. Solution

The logarithmic mean temperature difference is

Log Mean Temperature Difference (LMTD)

$$\Delta T_{lm} = \frac{(T_{hot,out} - T_{cold,in}) - (T_{hot,in} - T_{cold,out})}{2.3\log_{10}\left(\frac{T_{hot,out} - T_{cold,in}}{T_{hot,in} - T_{cold,out}}\right)}$$

$$= \frac{(170°F - 70°F) - (170°F - 130°F)}{2.3\log_{10}\left(\frac{170°F - 70°F}{170°F - 130°F}\right)}$$

$$= 65.5°F$$

Since $U_oA_o = U_iA_i$, $U_i = h_i$, and $A = \pi DL$,

$$U_o = \frac{h_i A_i}{A_o} = h_i\left(\frac{D_i}{D_o}\right)$$

$$= \frac{\left(1757\ \frac{Btu}{hr\text{-}ft^2\text{-}°F}\right)(2.00\ in)}{2.125\ in}$$

$$= 1654\ Btu/hr\text{-}ft^2\text{-}°F$$

The tube length required is

$$L = \frac{\rho v D_i^2 c_p (T_{out} - T_{in})}{4 U_o D_o \Delta T_{lm}}$$

$$= \left(62.0\ \frac{lbm}{ft^3}\right)\left(10\ \frac{ft}{sec}\right)\left(3600\ \frac{sec}{hr}\right)\left(\frac{2.00\ in}{12\ \frac{in}{ft}}\right)^2 \times \left(\frac{\left(0.998\ \frac{Btu}{lbm\text{-}°F}\right)(130°F - 70°F)}{(4)\left(1654\ \frac{Btu}{hr\text{-}ft^2\text{-}°F}\right)\left(\frac{2.125\ in}{12\ \frac{in}{ft}}\right)(65.5°F)}\right)$$

$$= 48.4\ ft$$

37. FOULING

Fouling is corrosion, precipitation of compounds in solution (i.e., *scaling*), settling of suspended particulate solids, and biological activity (i.e., growth of algae and other life forms) that adhere to a heat transfer surface. Fouling of heat transfer surfaces decreases the heat transfer and increases pumping power.[67,68]

Unless the tube walls are known to be "new and clean," a *fouling factor*, R_f, should be incorporated into the expression for the overall heat transfer coefficient.[69,70] On a macroscopic basis, where inside and outside coefficients and fouling factors are combined, Eq. 35.82 can be used.

$$R_f = \frac{1}{U_{fouled}} - \frac{1}{U_{clean}} \qquad 35.82$$

The tubeside and shellside fouling factors can be individually combined to determine the fouled overall heat transfer coefficient. Values of the fouling factor for normal operation are given in Table 35.8.

[67] Fouling is known in the heat exchanger industry as the "silent thief."

[68] Condenser tube fouling accounts for up to 50% of the total thermal resistance in steam power plants. For refrigeration condensers, each 0.0001 hr-ft²-°F/Btu (1.75×10^{-5} K·m²/W) increase in the fouling factor increases the centrifugal refrigeration compressor power by approximately 1.1%.

[69] In a shell-and-tube heat exchanger, it is preferable to run a fouling liquid through straight tubes, since the interior of straight tubes is accessible for inspection and cleaning. This is not the case for the exteriors of tubes and U-tube bundles.

[70] Fouled tubes can be cleaned chemically or mechanically to return them to almost "new" performance. Mechanical cleaning that occurs during normal operation is known as *online cleaning*, while *offline* mechanical cleaning requires the unit to be taken out of service. The two most common online mechanical cleaning systems are *brush-and-basket* and *sponge rubber ball* types. Brushes, rubber, scrapers, cutters, vibrators, and water lances are common offline cleaning systems.

Table 35.8 *Typical Fouling Factors for Shell-and-Tube Heat Exchangers *(hr-ft²-°F/Btu)*

fluid	below 125°F (50°C)	above 125°F (50°C)
air	0.002	0.002
water		
city/well	0.001	0.002
brine, salt, or seawater	0.0005	0.001
hard water	0.003	0.005
distilled	0.0005	0.0005
treated boiler feedwater	0.001	0.001
untreated cooling tower	0.002	0.002
steam		
clean	0.0005	0.0005
oil-bearing	0.001	0.001
diesel exhaust	–	0.01

(Multiply hr-ft²-°F/Btu by 0.17611 to obtain m²·K/W.)

*Fouling factors depend on the heat exchanger type. Lower values apply to heat exchangers with lower fouling tendencies.

$$\frac{1}{U_{o,\text{fouled}}} = \frac{1}{h_o} + \left(\frac{r_o}{k_{\text{tube}}}\right)\ln\left(\frac{r_o}{r_i}\right) + R_{f,o} + \left(\frac{r_o}{r_i}\right)\left(R_{f,i} + \frac{1}{h_i}\right) \approx \frac{1}{h_o} + \frac{1}{h_i} + R_{f,o} + R_{f,i} \qquad 35.83$$

In terms of area and length,

Overall Heat-Transfer Coefficient for Concentric Tube and Shell-and-Tube Heat Exchangers

$$\frac{1}{UA} = \frac{1}{h_iA_i} + \frac{R_{f,i}}{A_i} + \frac{\ln\left(\frac{D_o}{D_i}\right)}{2\pi k_{\text{tube}}L} + \frac{R_{f,o}}{A_o} + \frac{1}{h_oA_o} \qquad 35.84$$

$$\frac{1}{UA} = \frac{1}{h_iA_i} + \frac{R_{f,i}}{A_i} + \frac{\ln\left(\frac{r_o}{r_i}\right)}{2\pi k_{\text{tube}}L} + \frac{R_{f,o}}{A_o} + \frac{1}{h_oA_o} \qquad 35.85$$

Fouling is intrinsically a changing characteristic. Equation 35.83 calculates the fouled overall heat transfer coefficient but does not predict when that value will be reached. Fouling as a function of time can be modeled as *linear*, *falling-rate*, or *asymptotic* (approaching $R_{f,\infty}$). The values of coefficients a, b, $R_{f,\infty}$, and t' must be known or determined from empirical data or other knowledge.

$$R_{f,t} = at \quad \text{[linear]} \qquad 35.86$$

$$R_{f,t}^2 = R_{f,0}^2 + bt \quad \text{[falling-rate]} \qquad 35.87$$

$$R_{f,t} = R_{f,\infty} + (1 - e^{-t/t'}) \quad \text{[asymptotic]} \qquad 35.88$$

Example 35.3

A single-pass shell-in-tube heat exchanger is constructed of 1¼ in schedule-40 pipe for the tube and 2½ in schedule-40 pipe for the shell. The thermal conductivity of the pipe material is 35 Btu-ft/hr-ft²-°F (61 W/m·K). Flow is counterflow.

The heat exchanger cools 10 gal/min (0.6 L/s) of municipal city water flowing in the shell from 55°F (13°C) to 45°F (7°C). The water film coefficient is 246 Btu/hr-ft²-°F (1400 W/m²·K). The cooling solution is 10 gal/min (0.6 L/s) of brine initially at 10°F (−12°C). The brine has a specific heat of 0.68 Btu/lbm-°F (2.85 kJ/kg·K) and a density of 77.0 lbm/ft³ (1233 kg/m³). The brine film coefficient is 265 Btu/hr-ft²-°F (1500 W/m²·K).

How long should the heat exchanger be in a typical fouled condition?

SI Solution

The dimensions of the schedule-40 tube and the shell pipes are

	tube	shell
inside diameter	35.05 mm	62.71 mm
outside diameter	42.16 mm	73.03 mm

The water's bulk temperature is

$$T_{b,\text{water}} = \left(\frac{1}{2}\right)(7°\text{C} + 13°\text{C}) = 10°\text{C}$$

The fluid properties at 10°C are interpolated from App. 34.B, or can also be obtained directly from the steam tables. **[Properties of Water (SI Units)]**

$$\rho_{10°\text{C}} = 999 \text{ kg/m}^3$$
$$c_{p,10°\text{C}} = 4.20 \text{ kJ/kg·K}$$

The mass flow rates are

$$\dot{m}_{\text{water}} = Q\rho = \frac{\left(0.6 \frac{\text{L}}{\text{s}}\right)\left(999 \frac{\text{kg}}{\text{m}^3}\right)}{1000 \frac{\text{L}}{\text{m}^3}} = 0.599 \text{ kg/s}$$

$$\dot{m}_{\text{brine}} = Q\rho = \frac{\left(0.6 \frac{\text{L}}{\text{s}}\right)\left(1233 \frac{\text{kg}}{\text{m}^3}\right)}{1000 \frac{\text{L}}{\text{m}^3}} = 0.740 \text{ kg/s}$$

The heat transfer is found from the temperature gain of the water. The logarithmic mean temperature difference is not used here, even though this is a heat exchanger.

$$\begin{aligned}\dot{Q} &= \dot{m}c_p\Delta T\\ &= \left(0.599\ \frac{\text{kg}}{\text{s}}\right)\left(4.20\ \frac{\text{kJ}}{\text{kg·K}}\right)\left(1000\ \frac{\text{J}}{\text{kJ}}\right)\\ &\quad\times(13°\text{C}-7°\text{C})\\ &= 1.509\times10^4\ \text{W}\end{aligned}$$

The brine's final temperature is

$$\begin{aligned}T_{\text{brine,out}} &= T_{\text{brine,in}}+\frac{\dot{Q}}{\dot{m}_{\text{brine}}c_p}\\ &= -12°\text{C}+\frac{1.509\times10^4\ \text{W}}{\left(0.740\ \frac{\text{kg}}{\text{s}}\right)\left(2.85\ \frac{\text{kJ}}{\text{kg·K}}\right)\left(1000\ \frac{\text{J}}{\text{kJ}}\right)}\\ &= -4.84°\text{C}\end{aligned}$$

The logarithmic mean temperature difference is

Log Mean Temperature Difference (LMTD)

$$\begin{aligned}\Delta T_{\text{lm}} &= \frac{(T_{\text{hot,out}}-T_{\text{cold,out}})-(T_{\text{hot,in}}-T_{\text{cold,in}})}{2.3\log_{10}\left(\frac{T_{\text{hot,out}}-T_{\text{cold,out}}}{T_{\text{hot,in}}-T_{\text{cold,in}}}\right)}\\ &= \frac{(13°\text{C}-(-4.84°\text{C}))-(7°\text{C}-(-12°\text{C}))}{2.3\log_{10}\left(\frac{13°\text{C}-(-4.84°\text{C})}{7°\text{C}-(-12°\text{C})}\right)}\\ &= 18.41°\text{C}\end{aligned}$$

Fouling factors are converted from the values given in Table 35.8.

$$\begin{aligned}R_{f,o} &= \left(0.001\ \frac{\text{hr-ft}^2\text{-°F}}{\text{Btu}}\right)\left(0.17611\ \frac{\text{m}^2\text{·K·Btu}}{\text{W-hr-ft}^2\text{-°F}}\right)\\ &= 1.8\times10^{-4}\ \text{m}^2\text{·K/W}\quad[\text{water below }125°\text{F}]\end{aligned}$$

$$\begin{aligned}R_{f,i} &= \left(0.0005\ \frac{\text{hr-ft}^2\text{-°F}}{\text{Btu}}\right)\left(0.17611\ \frac{\text{m}^2\text{·K·Btu}}{\text{W-hr-ft}^2\text{-°F}}\right)\\ &= 8.8\times10^{-5}\ \text{m}^2\text{·K/W}\quad[\text{brine}]\end{aligned}$$

The overall heat transfer coefficient can be calculated from Eq. 35.83. (The ratios of dimensions can remain in millimeters.)

$$\begin{aligned}\frac{1}{U_o} &= \frac{1}{h_o}+\left(\frac{r_o}{k_{\text{tube}}}\right)\ln\left(\frac{r_o}{r_i}\right)+R_{f,o}+\left(\frac{r_o}{r_i}\right)\left(R_{f,i}+\frac{1}{h_i}\right)\\ &= \frac{1}{1400\ \frac{\text{W}}{\text{m}^2\text{·K}}}+\left(\frac{\frac{0.04216\ \text{m}}{2}}{61\ \frac{\text{W}}{\text{m·K}}}\right)\ln\left(\frac{42.16\ \text{mm}}{35.05\ \text{mm}}\right)\\ &\quad+\left(1.8\times10^{-4}\ \frac{\text{m}^2\text{·K}}{\text{W}}\right)+\left(\frac{42.16\ \text{mm}}{35.05\ \text{mm}}\right)\\ &\quad\times\left(8.8\times10^{-5}\ \frac{\text{m}^2\text{·K}}{\text{W}}+\frac{1}{1500\ \frac{\text{W}}{\text{m}^2\text{·K}}}\right)\\ &= 1.866\times10^{-3}\ \text{m}^2\text{·K/W}\\ U_o &= \frac{1}{1.866\times10^{-3}\ \frac{\text{m}^2\text{·K}}{\text{W}}}\\ &= 536\ \text{W/m}^2\text{·K}\end{aligned}$$

The heat transfer is known. Therefore, the length can be calculated from Eq. 35.73.

Rate of Heat Transfer

$$\dot{Q} = U_oA_oF_c\Delta T_{\text{lm}} = U_o\pi D_oLF_c\Delta T_{\text{lm}}$$

$$\begin{aligned}L &= \frac{\dot{Q}}{U_o\pi D_oF_c\Delta T_{\text{lm}}}\\ &= \frac{1.509\times10^4\ \text{W}}{\left(536\ \frac{\text{W}}{\text{m}^2\text{·K}}\right)\pi(0.04216\ \text{m})(1)(18.41°\text{C})}\\ &= 11.55\ \text{m}\end{aligned}$$

Customary U.S. Solution

The dimensions of the schedule-40 tube and shell pipes are

	tube	shell
inside diameter	1.38 in	2.47 in
outside diameter	1.66 in	2.88 in

The water's bulk temperature is

$$T_{b,\text{water}} = \left(\frac{1}{2}\right)(45°\text{F}+55°\text{F}) = 50°\text{F}$$

The fluid properties at 50°F are obtained from App. 34.A, or can also be obtained directly from the steam tables. [**Properties of Water (I-P Units)**]

$$\rho_{50°\text{F}} = 62.41 \text{ lbm/ft}^3$$
$$c_{p,50°\text{F}} = 1.00 \text{ Btu/lbm-°F}$$

The mass flow rates are

$$\dot{m}_{\text{water}} = Q\rho = \frac{\left(10 \ \frac{\text{gal}}{\text{min}}\right)\left(60 \ \frac{\text{min}}{\text{hr}}\right)\left(62.41 \ \frac{\text{lbm}}{\text{ft}^3}\right)}{7.48 \ \frac{\text{gal}}{\text{ft}^3}} = 5005 \text{ lbm/hr}$$

$$\dot{m}_{\text{brine}} = Q\rho = \frac{\left(10 \ \frac{\text{gal}}{\text{min}}\right)\left(60 \ \frac{\text{min}}{\text{hr}}\right)\left(77.0 \ \frac{\text{lbm}}{\text{ft}^3}\right)}{7.48 \ \frac{\text{gal}}{\text{ft}^3}} = 6176 \text{ lbm/hr}$$

The heat transfer is found from the temperature gain of the water. The logarithmic mean temperature difference is not used here, even though this is a heat exchanger.

$$\dot{Q} = \dot{m}c_p\Delta T = \left(5005 \ \frac{\text{lbm}}{\text{hr}}\right)\left(1.0 \ \frac{\text{Btu}}{\text{lbm-°F}}\right)(55°\text{F} - 45°\text{F}) = 50{,}050 \text{ Btu/hr}$$

The brine's final temperature is

$$T_{\text{brine,out}} = T_{\text{brine,in}} + \frac{\dot{Q}}{\dot{m}_{\text{brine}}c_p} = 10°\text{F} + \frac{50{,}050 \ \frac{\text{Btu}}{\text{hr}}}{\left(6176 \ \frac{\text{lbm}}{\text{hr}}\right)\left(0.68 \ \frac{\text{Btu}}{\text{lbm-°F}}\right)} = 21.9°\text{F}$$

The logarithmic mean temperature difference is

Log Mean Temperature Difference (LMTD)

$$\Delta T_{\text{lm}} = \frac{(T_{\text{hot,out}} - T_{\text{cold,out}}) - (T_{\text{hot,in}} - T_{\text{cold,in}})}{2.3\log_{10}\left(\frac{T_{\text{hot,out}} - T_{\text{cold,out}}}{T_{\text{hot,in}} - T_{\text{cold,in}}}\right)} = \frac{(45°\text{F} - 10°\text{F}) - (55°\text{F} - (21.9°\text{F}))}{2.3\log_{10}\left(\frac{45°\text{F} - 10°\text{F}}{55°\text{F} - (21.9°\text{F})}\right)} = 34.0°\text{F}$$

Fouling factors are given in Table 35.8.

$$R_{f,o} = 0.001 \text{ hr-ft}^2\text{-°F/Btu} \quad \text{[water below 125°F]}$$
$$R_{f,i} = 0.0005 \text{ hr-ft}^2\text{-°F/Btu} \quad \text{[brine]}$$

The overall heat transfer coefficient can be calculated from Eq. 35.83. (The ratios of dimensions can remain in inches.)

$$\begin{aligned}
\frac{1}{U_o} &= \frac{1}{h_o} + \left(\frac{r_o}{k_{\text{tube}}}\right)\ln\left(\frac{r_o}{r_i}\right) + R_{f,o} + \left(\frac{r_o}{r_i}\right)\left(R_{f,i} + \frac{1}{h_i}\right) \\
&= \frac{1}{246 \ \frac{\text{Btu}}{\text{hr-ft}^2\text{-°F}}} + \left(\frac{\frac{1.66 \text{ in}}{2}}{\left(12 \ \frac{\text{in}}{\text{ft}}\right)\left(35 \ \frac{\text{Btu-ft}}{\text{hr-ft}^2\text{-°F}}\right)}\right) \\
&\quad \times\ln\left(\frac{1.66 \text{ in}}{1.38 \text{ in}}\right) + 0.001 \ \frac{\text{hr-ft}^2\text{-°F}}{\text{Btu}} + \left(\frac{1.66 \text{ in}}{1.38 \text{ in}}\right) \\
&\quad \times\left(0.0005 \ \frac{\text{hr-ft}^2\text{-°F}}{\text{Btu}} + \frac{1}{265 \ \frac{\text{Btu}}{\text{hr-ft}^2\text{-°F}}}\right) \\
&= 0.01057 \text{ hr-ft}^2\text{-°F/Btu}
\end{aligned}$$

$$U_o = \frac{1}{0.01057 \ \frac{\text{hr-ft}^2\text{-°F}}{\text{Btu}}} = 94.6 \text{ Btu/hr-ft}^2\text{-°F}$$

The heat transfer is known. Therefore, the length can be calculated from Eq. 35.73.

Rate of Heat Transfer

$$\dot{Q} = U_oA_oF_c\Delta T_{\text{lm}} = U_o\pi D_oLF_c\Delta T_{\text{lm}}$$

$$L = \frac{\dot{Q}}{U_o\pi D_oF_c\Delta T_{\text{lm}}} = \frac{\left(50{,}050 \ \frac{\text{Btu}}{\text{hr}}\right)\left(12 \ \frac{\text{in}}{\text{ft}}\right)}{\left(94.6 \ \frac{\text{Btu}}{\text{hr-ft}^2\text{-°F}}\right)\pi(1.66 \text{ in})(1)(34.0°\text{F})} = 35.8 \text{ ft}$$

38. COOLING SUPERHEATED STEAM

It is commonly believed that highly superheated steam renders heat transfer cooling surfaces inefficient due to the steam's low heat-transfer properties. However, the controlling factor is the cooling surface temperature, not the degree of superheat. If the cooling surface temperature is maintained below the cooling fluid condensation temperature, performance will be the same for superheated steam as it is for saturated steam. If the superheated steam flow rate and/or the superheat is sufficiently high, the steam will heat the cooling surface to above the condensation temperature, and condensation will slow or cease. In the latter case, a desuperheater will be required.

39. NTU METHOD

Use of Eq. 35.73 is known as the *F-method* or the *LMTD method.* The F-method is suitable for design (i.e., the calculation of the required heat transfer area).

Some heat exchanger analysis problems, such as where both outlet temperatures are unknown, appear to be unsolvable or solvable only by trial and error using the F-method.[71] However, the *number of transfer units* (NTU) *method* (also known as the *efficiency method* and *effectiveness method*) can be used to handle these problems more easily.[72] The steps in the NTU method depend on whether or not both exit temperatures are known.

The first step in the NTU method is to calculate the *thermal capacity rates, C,* for the two fluids. It is possible for the two capacity rates to be equal, but usually they are not. The smaller capacity rate is designated C_{min}. The larger is designated C_{max}. The fluid with the smaller capacity rate, C_{min}, will experience the larger temperature change.

$$C = \dot{m}c_p \tag{35.89}$$

The *heat exchanger effectiveness,* ϵ, is defined as the ratio of the actual heat transfer to the maximum possible heat transfer.[73,74] This ratio is generally not known in advance.

$$\epsilon = \frac{\dot{Q}_{\text{actual}}}{\dot{Q}_{\text{ideal}}} \tag{35.90}$$

If the hot fluid has the minimum capacity rate (i.e., $C_{\text{min}} = C_{\text{hot}}$), the effectiveness is given by Eq. 35.91.

Heat Exchanger Effectiveness, ϵ

$$\epsilon = \frac{C_{\text{cold}}(T_{\text{cold,out}} - T_{\text{cold,in}})}{C_{\text{min}}(T_{\text{hot,in}} - T_{\text{cold,in}})} \tag{35.91}$$

If the cold fluid has the minimum capacity rate (i.e., $C_{\text{min}} = C_{\text{cold}}$), the effectiveness is given by Eq. 35.92.

Heat Exchanger Effectiveness, ϵ

$$\epsilon = \frac{C_{\text{hot}}(T_{\text{hot,in}} - T_{\text{hot,out}})}{C_{\text{min}}(T_{\text{hot,in}} - T_{\text{cold,in}})} \tag{35.92}$$

If the effectiveness is known, the heat transfer can be calculated from Eq. 35.93. Notice that the temperature difference is the difference of two entering temperatures, which are generally both known.

$$\dot{Q} = \epsilon C_{\text{min}}(T_{\text{hot,in}} - T_{\text{cold,in}}) \tag{35.93}$$

The outlet temperatures are generally not known, and Eq. 35.91 and Eq. 35.92 cannot be used. In this case, the NTU method starts by calculating the ratio $C_{\text{min}}/C_{\text{max}}$ and then finding the number of transfer units, NTU, from Eq. 35.94.

Number of Exchanger Transfer Units (NTU)

$$\text{NTU} = \frac{AU_{\text{avg}}}{C_{\text{min}}} = \frac{U_oA_o}{C_{\text{min}}} = \frac{U_iA_i}{C_{\text{min}}} \tag{35.94}$$

Next, the effectiveness is determined from Eq. 35.95 or Eq. 35.99 or from traditional charts such as those in App. 35.E. The heat capacity ratio, C_r, is $C_{\text{min}}/C_{\text{max}}$. For a single-pass heat exchanger operating in counterflow, the effectiveness is

Effectiveness-NTU Relations

$$\begin{aligned}\epsilon &= \frac{1-\exp(-\text{NTU}(1-C_r))}{1-C_r\exp(-\text{NTU}(1-C_r))} \\ &= \frac{1-\exp\left(-\text{NTU}\left(1-\dfrac{C_{\text{min}}}{C_{\text{max}}}\right)\right)}{1-\left(\dfrac{C_{\text{min}}}{C_{\text{max}}}\right)\exp\left(-\text{NTU}\left(1-\dfrac{C_{\text{min}}}{C_{\text{max}}}\right)\right)} \quad [\text{for } C_r < 1]\end{aligned} \tag{35.95}$$

$$\epsilon = \frac{\text{NTU}}{1+\text{NTU}} \quad [\text{for } C_r = 1] \tag{35.96}$$

Solving for the number of transfer units, NTU, gives

Effectiveness-NTU Relations

$$\text{NTU} = \frac{1}{C_r - 1}\ln\left(\frac{\epsilon - 1}{\epsilon C_r - 1}\right) \quad [\text{for } C_r < 1] \tag{35.97}$$

$$\text{NTU} = \frac{\epsilon}{1-\epsilon} \quad [\text{for } C_r = 1] \tag{35.98}$$

---

[71] Actually, any shell-and-tube heat exchanger with an even number of tube passes has a closed-form analytical solution for the outlet temperature. However, the mathematics are laborious and the form of the solution varies with the type of flow and heat exchanger design.

[72] The names *number of thermal units* (NTU), *heat transfer units* (HTU), and *temperature ratio* (TR) are synonymous with *number of transfer units* (NTU).

[73] The maximum possible transfer can occur only if the heat exchanger has an infinite length.

[74] Other names used in the literature to define the effectiveness are *efficiency, thermodynamic efficiency, temperature efficiency,* and *performance parameter.* The symbol P is also used in place of ϵ.

For a single-pass heat exchanger operating in parallel flow operation, the effectiveness is

$$\epsilon = \frac{1 - \exp(-\mathrm{NTU}(1+C_r))}{1+C_r} = \frac{1 - \exp\left(-\mathrm{NTU}\left(1+\frac{C_{\min}}{C_{\max}}\right)\right)}{1+\frac{C_{\min}}{C_{\max}}} \quad \text{35.99}$$

Equation 35.95 and Eq. 35.99 and similar (though more complex) relationships for the remaining exchanger configurations, given in heat transfer textbooks, were used to develop the charts in App. 35.E.

Example 35.4

40,000 lbm/hr (5.0 kg/s) of lubrication oil are cooled in a single-pass countercurrent heat exchanger by 10,000 lbm/hr (1.3 kg/s) of water. The oil enters at 225°F (105°C). The water enters at 95°F (35°C). The oil's specific heat is 0.45 Btu/lbm-°F (1.9 kJ/kg·K). The overall coefficient of heat transfer for the heat exchanger is 50 Btu/hr-ft²-°F (280 W/m²·K) based on an effective heat transfer area of 150 ft² (14 m²). (a) What is the heat transfer? (b) What is the water's exit temperature?

SI Solution

(a) Assume an initial water exit temperature of 93°C to obtain the water's specific heat directly from App. 34.B.

$$c_{p,93^\circ\mathrm{C,water}} = 4.229\ \mathrm{kJ/kg{\cdot}K}$$

The thermal capacity rates are

$$C_{\mathrm{water}} = \dot{m}c_p = \left(1.3\ \frac{\mathrm{kg}}{\mathrm{s}}\right)\left(4.229\ \frac{\mathrm{kJ}}{\mathrm{kg{\cdot}K}}\right) = 5.5\ \mathrm{kW/K}$$

$$C_{\mathrm{oil}} = \dot{m}c_p = \left(5.0\ \frac{\mathrm{kg}}{\mathrm{s}}\right)\left(1.9\ \frac{\mathrm{kJ}}{\mathrm{kg{\cdot}K}}\right) = 9.5\ \mathrm{kW/K}$$

$$C_{\min} = C_{\mathrm{water}} = 5.5\ \mathrm{kW/K}$$

$$C_{\max} = C_{\mathrm{oil}} = 9.5\ \mathrm{kW/K}$$

$$\frac{C_{\min}}{C_{\max}} = \frac{5.5\ \frac{\mathrm{kW}}{\mathrm{K}}}{9.5\ \frac{\mathrm{kW}}{\mathrm{K}}} = 0.579$$

From Eq. 34.94, the number of transfer units is

Number of Exchanger Transfer Units (NTU)

$$\mathrm{NTU} = \frac{AU_{\mathrm{avg}}}{C_{\min}} = \frac{\left(280\ \frac{\mathrm{W}}{\mathrm{m^2{\cdot}K}}\right)(14\ \mathrm{m^2})}{\left(5.5\ \frac{\mathrm{kW}}{\mathrm{K}}\right)\left(1000\ \frac{\mathrm{J}}{\mathrm{kJ}}\right)} = 0.71$$

From App. 35.E, (a) for a single-pass counterflow heat exchanger with NTU = 0.71 and $C_{\min}/C_{\max} = 0.579$, the effectiveness is $\epsilon \approx 0.45$. (The actual value calculated from Eq. 35.95 is 0.453.)

The heat transfer is

$$\dot{Q} = \epsilon C_{\min}(T_{\mathrm{hot,in}} - T_{\mathrm{cold,in}}) = (0.45)\left(5.5\ \frac{\mathrm{kW}}{\mathrm{K}}\right)\left(1000\ \frac{\mathrm{W}}{\mathrm{kW}}\right)(105^\circ\mathrm{C} - 35^\circ\mathrm{C}) = 1.73\times 10^5\ \mathrm{W}$$

(b) The water's exit temperature is

$$T_{\mathrm{cold,out}} = T_{\mathrm{cold,in}} + \frac{\dot{Q}}{\dot{m}c_p} = 35^\circ\mathrm{C} + \frac{1.73\times 10^5\ \mathrm{W}}{\left(1.3\ \frac{\mathrm{kg}}{\mathrm{s}}\right)\left(4.229\ \frac{\mathrm{kJ}}{\mathrm{kg{\cdot}K}}\right)\left(1000\ \frac{\mathrm{J}}{\mathrm{kJ}}\right)} = 66.5^\circ\mathrm{C}$$

Customary U.S. Solution

(a) Assume an initial water exit temperature of approximately 200°F in order to obtain the water's specific heat. (The value is not sensitive to temperature.) From App. 34.A,

$$c_{p,200^\circ\mathrm{F,water}} = 1.00\ \mathrm{Btu/lbm\text{-}^\circ F}$$

Heat Transfer

The thermal capacity rates are

$$C_{\text{water}} = \dot{m}c_p = \left(10{,}000\ \frac{\text{lbm}}{\text{hr}}\right)\left(1.00\ \frac{\text{Btu}}{\text{lbm-°F}}\right)$$
$$= 10{,}000\ \text{Btu/hr-°F}$$

$$C_{\text{oil}} = \dot{m}c_p = \left(40{,}000\ \frac{\text{lbm}}{\text{hr}}\right)\left(0.45\ \frac{\text{Btu}}{\text{lbm-°F}}\right)$$
$$= 18{,}000\ \text{Btu/hr-°F}$$

$$C_{\text{min}} = C_{\text{water}} = 10{,}000\ \text{Btu/hr-°F}$$

$$C_{\text{max}} = C_{\text{oil}} = 18{,}000\ \text{Btu/hr-°F}$$

$$\frac{C_{\text{min}}}{C_{\text{max}}} = \frac{10{,}000\ \frac{\text{Btu}}{\text{hr-°F}}}{18{,}000\ \frac{\text{Btu}}{\text{hr-°F}}} = 0.556$$

From Eq. 34.94, the number of transfer units is

Number of Exchanger Transfer Units (NTU)

$$\text{NTU} = \frac{AU_{\text{avg}}}{C_{\text{min}}} = \frac{\left(50\ \frac{\text{Btu}}{\text{hr-ft}^2\text{-°F}}\right)(150\ \text{ft}^2)}{10{,}000\ \frac{\text{Btu}}{\text{hr-°F}}} = 0.75$$

From App. 35.E, (a) for a single-pass counterflow heat exchanger with $\text{NTU} = 0.75$ and $C_{\text{min}}/C_{\text{max}} = 0.556$, the effectiveness is $\epsilon \approx 0.47$. (The actual value calculated from Eq. 35.95 is 0.471.)

The heat transfer is

$$\dot{Q} = \epsilon C_{\text{min}}(T_{\text{hot,in}} - T_{\text{cold,in}})$$
$$= (0.47)\left(10{,}000\ \frac{\text{Btu}}{\text{hr-°F}}\right)(225\text{°F} - 95\text{°F})$$
$$= 6.11 \times 10^5\ \text{Btu/hr}$$

(b) The water's exit temperature is

$$T_{\text{cold,out}} = T_{\text{cold,in}} + \frac{\dot{Q}}{\dot{m}c_p}$$
$$= 95\text{°F} + \frac{6.11 \times 10^5\ \frac{\text{Btu}}{\text{hr}}}{\left(10{,}000\ \frac{\text{lbm}}{\text{hr}}\right)\left(1.00\ \frac{\text{Btu}}{\text{lbm-°F}}\right)}$$
$$= 156\text{°F}$$

40. HEAT TRANSFER FLUIDS

Indirect heating (*cooling*) occurs when a process fluid is heated (cooled) by an intermediate fluid. Indirect heat transfer is used when the process fluid cannot be directly exposed to sources of heat or cooling, as when the process fluid is sensitive to hot spots or requires uniform heating.

A *heat transfer fluid* is any substance that is used for indirect heat transfer. Water, brines, mineral oils and other inorganic liquids, gases, molten inorganic salts, and liquid metals (sodium and mercury) are all used in indirect heat transfer.[75] The term "heat transfer fluid," however, is generally used to refer to commercial inorganic liquids.

Water is often the best heat transfer fluid, though it is limited in unpressurized applications to approximately 35°F to 190°F (2°C to 90°C). Adding propylene glycol or the toxic and environmentally unfriendly ethylene glycol extends the range to approximately −50°F to 250°F (−45°C to 120°C), but heat transfer properties are impaired.

Commercial heat transfer fluids occupy specific low- and high-niches: indirect heat transfer operations below 32°F (0°C) and above (approximately) 350°F (175°C) up to 600°F to 800°F (315°C to 425°C). Heat transfer fluids may be used in liquid-phase heat transfer or vapor-phase heat transfer.

A heat transfer fluid can be selected in one of three ways, depending on whether the process is being designed from the ground up, the fluid is being retrofitted into an existing installation, or the pumping energy is to be minimized. From Eq. 35.34, the film coefficient for turbulent flow in a tube is

$$h = 0.023\left(\frac{k}{D}\right)\left(\frac{D\mathrm{v}\rho}{\mu}\right)^{0.8}\left(\frac{C_p\mu}{k}\right)^n \qquad 35.100$$

[75]Liquids used for commerical heat transfer fluids include alkylated benzenes, alkylated biphenyls, alkylated napthalenes, hydrogenated and unhydrogenated polyphenyls, benzylated aromatics, diphenyl/diphenyl oxide eutectics, aromatic-ether-based fluids, polyalkylene glycols, dicarboloxylic acid esters, polymethyl siloxanes, mineral oils, and inorganic nitrate salts.

For a constant mass flow rate per unit area, G, with turbulent flow,

$$h = 0.023G^{0.8}D^{-0.2}K_G \quad 35.101$$

$$K_G = c_p^n k^{1-n}\mu^{n-0.8} \quad 35.102$$

For a constant volumetric flow rate per unit area, Q, with turbulent flow,

$$h = 0.023Q^{0.8}D^{-0.2}K_V \quad 35.103$$

$$K_V = \rho^{0.8}c_p^n k^{1-n}\mu^{n-0.8} \quad 35.104$$

The two *heat transfer factors*, K_G and K_V, contain only physical properties of the fluid. The design engineer keeps the mass flow rate per unit area constant and compares K_G values; the plant engineer keeps the volumetric flow rate per unit area constant and compares K_V values.[76]

The *Fried heat transfer efficiency factor*, HTEF, is a third factor, usually intermediate between K_G and K_V, whose values can be compared when selecting heat transfer fluids. It represents the ratio of the film coefficient to the frictional energy expended in pumping the fluid.

$$\text{HTEF} = \frac{19.75\rho^{0.57}c_p^n k^{1-n}}{\mu^{0.52}} \quad 35.105$$

Choice of heat transfer fluid also depends on cost, thermal (heat transfer) properties, fire safety, environmental and toxicological issues, containment issues, chemical stability over the required temperature range, and compatibility with packing, gasket, and wall materials.

Film coefficients for heat transfer fluids are calculated in the same manner as for other fluids (i.e., with Eq. 35.36 for tubeside flow). Fluid properties are generally provided by the manufacturer.[77] Equation 35.106 is a very simplified relationship that can be used for initial estimates with organic liquids.

$$h \approx 423\frac{\text{v}_{\text{m/s}}^{0.8}}{D_{\text{m}}^{0.2}} \quad \text{[SI]} \quad 35.106(a)$$

$$h \approx 60\frac{\text{v}_{\text{ft/sec}}^{0.8}}{D_{\text{in}}^{0.2}} \quad \text{[U.S.]} \quad 35.106(b)$$

41. CONTROL OF PROCESS HEAT EXCHANGE OPERATIONS

Correct operation of many chemical and other processes requires consistent outlet temperatures, regardless of flow rates. Traditional systems control the inlet flow rate of one fluid by monitoring the outlet temperature of the other. This is generally satisfactory.

Control of steam-heated exchangers at high downturn is more problematic. (See Fig. 35.10.) (*Downturn* is operation with one fluid's flow rate substantially reduced.) Control of steam-heated exchangers has traditionally been with simple steam-inlet control and condensate removal. However, condensate-flow is more successful at high downturn.[78]

42. COOLING ELECTRONIC ENCLOSURES

Electronic equipment generates significant amounts of heat. In some cases, the components generate enough heat to damage themselves. The heat can be calculated from Eq. 35.107, where the current and voltage are effective values, also known as rms (root-mean-square) values, measured at the power cord.

$$P = \text{current} \times \text{voltage} \quad 35.107$$

Equipment cooling is accomplished with forced air (fans or blowers), heat exchangers, and air conditioners.[79] Component cooling is usually accomplished by mounting a heat radiator (i.e., fins) on the component.

Fans occupy minimal space and can move large volumes of air against low static heads. Blowers operate (and are most efficient when operating) against higher static heads.[80] Cabinets should always be pressurized by introducing filtered air, rather than drawing the air in. (A vacuum inside the cabinet will attract unfiltered air through the cracks around the panels and door.)

When the cabinet is sealed, or when the ambient temperature near the enclosure is near the maximum permissible operating temperature of the equipment, a heat exchanger (air-to-air or air-to-water) or air conditioner is needed. Air conditioners are the only alternative when the ambient temperature is above the maximum permissible operating temperature.[81]

Heat Transfer

[76]Some commercial literature uses the variables "Const. G" and "Const. V" in place of K_G and K_V.
[77]Commercial heat transfer fluids often have the word "therm" in their names, for example, "Dowtherm" (Dow Chemical), "Therminol" (Monsanto), "Mobiltherm" (Mobil Oil), and so on.
[78]There are problems with condensate-return control, as well. Water hammer and repeated thermal expansion and contraction due to alternating exposure to steam and condensate shorten the exchanger's life. Condensate leaving the exchanger may be well below the saturation temperature, and the condensate will readily absorb uncondensable gases, becoming aggressively corrosive.
[79]Air travels parallel to the rotational axis of a *fan* (i.e., radial flow); air travels normal to the rotational axis of a (centrifugal) *blower*.
[80]Multistage blowers can reach pressures up to 100 in of water.
[81]Environmental heat gain must be added to the equipment power dissipation. It is common to include the surface area of all four sides but to omit half the area of the top and all the area of the base when calculating the heat gain. Insulation should be used liberally to limit environmental heat gain.

Figure 35.10 *Control Methods for Steam-Heated Exchangers*

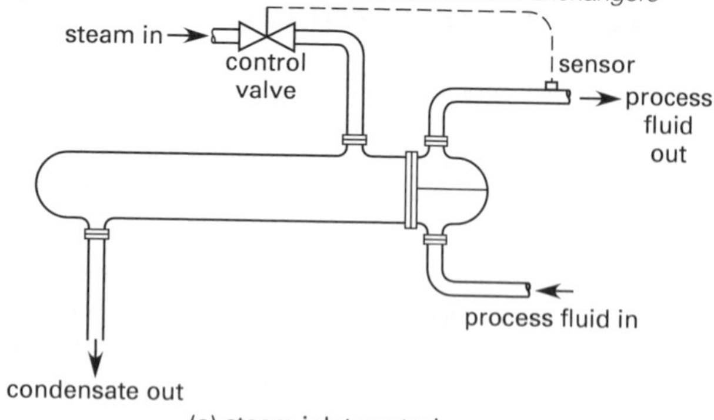

(a) steam inlet control

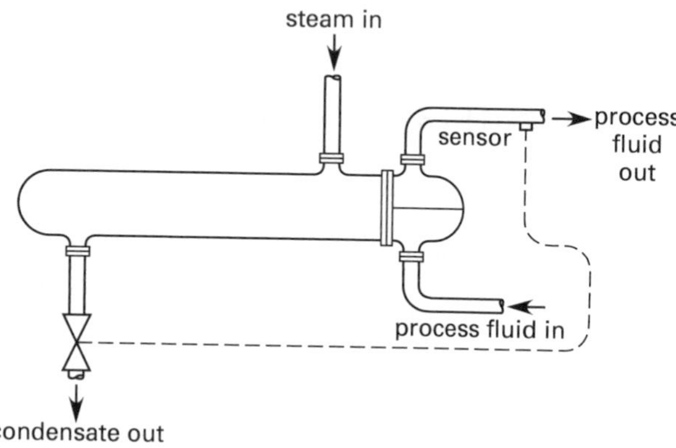

(b) condensate flow control

The following are some rules of thumb regarding forced air cooling.

- The effective area of an air intake will be approximately 65% of the grill area.
- To prevent choking, the flow area inside the cabinet should equal the effective intake area.
- The flow path should not be short circuited. The exhaust area should be downstream, beyond all of the heat-producing components.
- A single excessively hot component can be effectively protected with a baffle (to direct a small amount of high-velocity air over it) or with a dedicated fan.
- Using a booster fan in a two-stage configuration can eliminate the need for a larger, noisier fan.
- Static pressure drop through an enclosure is usually 0.25–0.50 in wg (inches of water gage), and is seldom greater than 1.0 in wg.
- Depending on the type of unit and mounting arrangement, the fan/blower motor power may or may not need to be added to the cooled equipment's power.

The airflow required to dissipate equipment power P with an airflow temperature rise of ΔT is given by Eq. 35.108 and Eq. 35.109, which assume a constant specific heat of 0.24 Btu/lbm-°F (1010 J/kg·K) and an air density of 0.075 lbm/ft³ (1.2 kg/m³). A 25% increase in air flow is usually added to the calculated air flow as a safety factor.

$$Q_{\text{cfm}} = \frac{3.16 P_{\text{watts}}}{\Delta T_{^\circ\text{F}}} = \frac{1.76 P_{\text{watts}}}{\Delta T_{^\circ\text{C}}} \qquad 35.108$$

$$Q_{\text{L/s}} = \frac{0.825 P_{\text{watts}}}{\Delta T_{^\circ\text{C}}} \qquad 35.109$$

Maximum temperature occurs at the inside top of a cabinet. This is a relevant factor only in critical components. Usually, the average cabinet temperature is used in designing the cooling system.

43. NOMENCLATURE

A	area	ft²	m²
c_p	specific heat	Btu/lbm-°F	J/kg·K
C	constant or coefficient	–	–
C	thermal capacity rate	Btu/hr-°F	W/K
D	diameter	ft	m
f	friction factor.	–	–
F	factor	–	–
G	mass flow rate per unit area	lbm/hr-ft²	kg/s·m²
Gz	Graetz number	–	–
h	film coefficient	Btu/hr-ft²-°F	W/m²·K
HTEF	heat transfer efficiency factor	–	–
k	thermal conductivity	Btu-ft/hr-ft²-°F	W/m·K
L	length	ft	m
m	exponent	–	–
m	mass	lbm	kg
$\dot{m}$	mass flow rate	lbm/hr	kg/s
M	number of tube rows per layer	–	–
n	exponent	–	–
N	number of tube layers	–	–
Nu	Nusselt number	–	–
p	pressure	lbf/ft²	kPa
P	power	hp	kW
Pe	Peclet number	–	–
Pr	Prandtl number	–	–
q	heat transfer per unit area	Btu/hr-ft²	W/m²
$\dot{Q}$	heat transfer rate	Btu/hr	W
Q	volumetric flow rate	ft³/hr	m³/s
r	radius	ft	m
R	thermal resistance	hr-ft²-°F/Btu	m²·K/W
Ra	Rayleigh number	–	–

Re	Reynolds number	–	–
s	spacing or side length	ft	m
St	Stanton number	–	–
T	temperature[a,b]	°F	K
U	overall coefficient of heat transfer	Btu/hr-ft²-°F	W/m²·K
v	velocity	ft/hr	m/s
x	distance x	ft	m

[a]The symbol θ is used for temperature in some books.
[b]It is common in heat exchanger literature to use lowercase t as the cold side temperature. This eliminates the requirement for "cold" and "hot" designations.

Symbols

α	thermal diffusivity	ft²/sec	m²/s
ϵ	effectiveness	–	–
η	efficiency	–	–
μ	viscosity[a,b,c]	lbm/hr-ft	kg/s·m
ν	kinematic viscosity	ft²/sec	m²/s
ρ	mass density	lbm/ft³	kg/m³

[a]The use of mass units in viscosity values is typical in the subject of convective heat transfer.
[b]Most data compilations give fluid viscosity in units of seconds. In the United States, heat transfer is traditionally stated on a per hour basis. Therefore, a conversion factor of 3600 sec/hr is needed when calculating dimensionless numbers from table data.
[c]The combination of units kg/s·m is the same as a Pa·s and N·s/m².

Subscripts

0	initial
a	atmospheric
A	at end A
b	bulk
B	at end B
c	correction
D	based on diameter
D	drag
f	fouling or friction
G	constant mass flow rate
H	hydraulic
i	inside
l	liquid or longitudinal
L	over length L
lm	log mean
m	mean
max	maximum
min	minimum
o	outside
Q	constant volumetric flow rate
s	surface
t	transverse or at time t
T	temperature
x	at point x
∞	free-stream (far field) or at time $= \infty$

Heat Transfer

36 Radiation and Combined Heat Transfer

Content in blue refers to the *NCEES Handbook*.

NCEES EXAM SPECIFICATIONS AND RELATED CONTENT

THERMAL AND FLUID SYSTEMS EXAM

I.C. Heat Transfer Principles
1. Thermal Radiation
2. Black, Real, and Gray Bodies
6. Net Radiation Heat Transfer
7. Reciprocity
8. Radiation with Reflection/Reradiation

1. THERMAL RADIATION

Thermal radiation is electromagnetic radiation with wavelengths in the 0.1 to 100 μm range. All bodies, even "cold" ones, radiate thermal radiation.

Thermal radiation incident to a body can be absorbed, reflected, or transmitted. A body's *absorptivity*, α, reflectivity, ρ, and transmissivity, τ, are the fractions of incident energy that the body absorbs, reflects, and transmits, respectively.

The *radiation conservation law* is[1]

Types of Bodies

$$\alpha + \rho + \tau = 1 \qquad 36.1$$

2. BLACK, REAL, AND GRAY BODIES

The rate of thermal radiation emitted per unit area of a body is the *emissive power*, *E*.

$$E = \frac{q_{\text{radiation}}}{A} \qquad 36.2$$

Since absorptivity, α, cannot exceed 1.0, Kirchhoff's law places an upper limit on emissive power. Bodies that radiate at this upper limit (i.e., $\alpha = 1$) are known as *black bodies* or *ideal radiators*. A black body emits the maximum possible radiation for its temperature and absorbs all incident energy.[2]

Real bodies do not radiate at the ideal level and are typically modeled as gray bodies. The ratio of actual to ideal emissive powers is the *emissivity*, ϵ.

$$\epsilon = \frac{E_{\text{actual}}}{E_{\text{black}}} \qquad 36.3$$

Emissivity generally has the following characteristics.

- Emissivity varies widely with the surface condition of a material.
- Emissivity is low with highly polished metals.
- Emissivity is high with most nonmetals.
- Emissivity increases with increases in temperature.

The emissivity (and, therefore, the emissive power) usually depends on the temperature of the body.[3] A body that emits at constant emissivity, regardless of wavelength, is known as a *gray body*.

Kirchhoff's radiation law states that for a body, the emissivity, ϵ, and absorptivity, α, are equal. At a given temperature, the ratios of emissive power to absorptivity for all bodies are equal. (Bodies at the same temperature are said to be in *thermal equilibrium*.)

$$\frac{\epsilon_1}{\alpha_1} = \frac{\epsilon_2}{\alpha_2} = \epsilon_{\text{black}}\Big|_T \qquad 36.4$$

[1]Notice that emissivity, ϵ, does not appear in the conservation law.
[2]Black body performance can be approximated but not achieved in practice.
[3]This is equivalent to saying the emissivity depends on the wavelength of the radiation.

For a black body, both absorptivity and emissivity are 1.0.

Types of Bodies

$$\alpha = \epsilon = 1 \qquad 36.5$$

However, emissivity also equals absorptivity for any body in thermal equilibrium.[4,5]

$$\epsilon = \alpha \quad \text{[thermal equilibrium]} \qquad 36.6$$

For a gray body, reflectivity is constant and

Types of Bodies

$$\epsilon + \rho = 1 \qquad 36.7$$

For an opaque body, transmissivity is zero, and

Types of Bodies

$$\alpha + \rho = 1 \qquad 36.8$$

3. RADIATION FROM A BODY

The *Stefan-Boltzmann law*, also known as the *fourth-power law*, gives the total emissive power, E, from a black body. The temperature, T, is expressed in degrees Rankine or in kelvins. σ is the *Stefan-Boltzmann constant*.

$$E_{\text{black}} = \sigma T^4 \qquad 36.9$$

Fundamental Constants

$$\sigma = 0.1713 \times 10^{-8}\ \text{Btu/hr-ft}^2\text{-}^\circ\text{R}^4 \qquad 36.10$$

$$\sigma = 5.67 \times 10^{-8}\ \text{W/m}^2\text{·K}^4 \qquad 36.11$$

The radiation from a gray body follows directly from the definition of emissivity.

$$E_{\text{gray}} = \epsilon E_{\text{black}} = \epsilon \sigma T^4 \qquad 36.12$$

4. BLACK BODY SHAPE FACTOR

For two black bodies, body 1 and body 2, that are radiating to each other, a *shape factor*, F_{12}, accounts for the bodies' spatial arrangement. (The smaller body is usually designated as body 1.) The shape factor for two black bodies is the fraction of the total radiation leaving body 1 that will travel directly to body 2. For this reason, the shape factor is also often referred to as the *black body shape factor* or *view factor*. (Less common terms include *arrangement factor*, *geometric shape factor*, *angle factor*, *interaction factor*, and *configuration factor*.)

The shape factor is easily seen to be 1.0 for two infinite, parallel planes, for two infinite coaxial cylinders, or for two concentric spheres, because all emitted radiation is absorbed. It is more difficult to evaluate the shape factor for more complex arrangements of bodies and surfaces. However, many of the simpler cases have been solved, and their graphical solutions are available. (See Fig. 36.1 and Fig. 36.2.)

Figure 36.1 *Black Body Shape Factor Adjacent Perpendicular Rectangles*

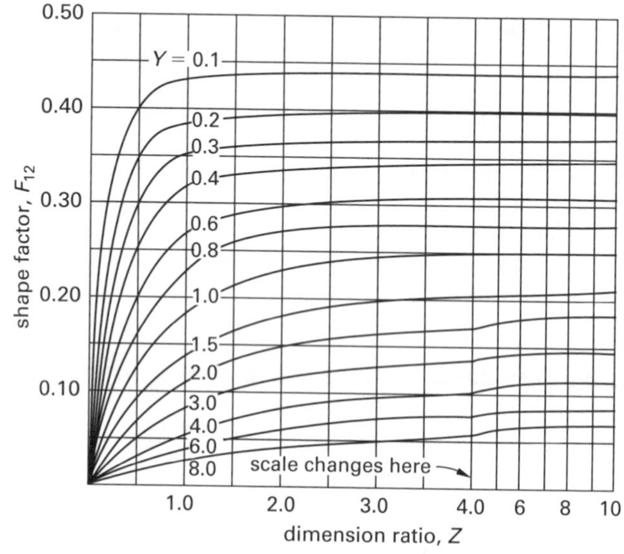

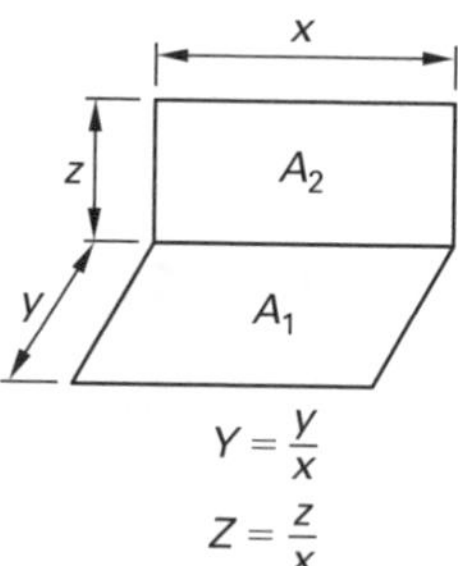

Shape factors for perpendicular rectangles with a common side. From H. C. Hottel, "Radiant Heat Transmission," *Mechanical Engineering* magazine, Volume 52 (1930). By permission of The American Society of Mechanical Engineers.

[4]"Steady-state operation" would be a better term here since the term "equilibrium" implies temperature equality with another body. In fact, the phrase "the bodies are in thermal equilibrium" means that the body temperatures are equal. In this case, the term "equilibrium" means that the body's temperature is constant.

[5]Equation 36.6 follows directly from Kirchhoff's law (see Eq. 36.4).

Figure 36.2 *Black Body Shape Factor (for directly opposed, parallel, finite surfaces)*

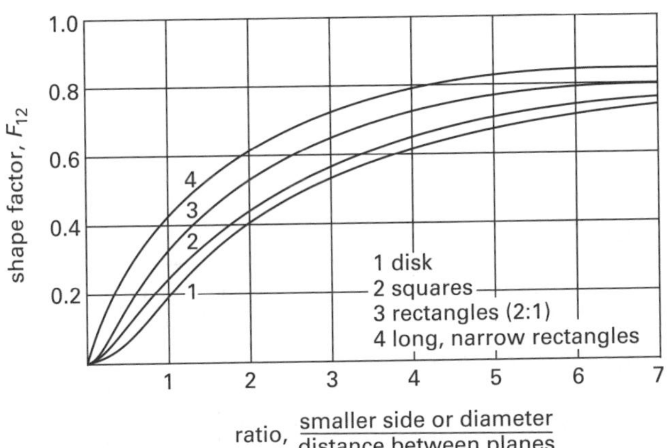

Shape factors for parallel squares, rectangles, and disks. From H. C. Hottel, "Radiant Heat Transmission," *Mechanical Engineering* magazine, Volume 52 (1930). By permission of The American Society of Mechanical Engineers.

5. GRAY BODY SHAPE FACTOR

Real bodies deviate from black body behavior. To account for the effect of less than ideal emissivities, the black body shape factor, F_{12}, is replaced by the gray body shape factor, $\mathcal{F}_{12}$. The *gray body shape factor* accounts for the spatial arrangements of the bodies and their emissivities. The lower limit for the gray body shape factor is $\epsilon_1\epsilon_2$; the upper limit is 1.0.

The gray body shape factor can be written as the product of the black body shape factor and the *emissivity factor*, F_e. The emissivity factor accounts for the departure of the surface from black-body conditions. (See Table 36.1.)

$$\mathcal{F}_{12} = F_{12}F_e \qquad 36.13$$

For two gray bodies that radiate to each other (and to no others), the gray body shape factor can be calculated from the black body shape factor.[6] A_1 in Eq. 36.14 is the smaller area.

$$A_1\mathcal{F}_{12} = \frac{1}{\dfrac{1-\epsilon_1}{\epsilon_1 A_1} + \dfrac{1}{A_1 F_{12}} + \dfrac{1-\epsilon_2}{\epsilon_2 A_2}} \qquad 36.14$$

A special case is two gray bodies with uniform thermal radiation and $F_{12} = 1$. Examples of this case include infinite parallel gray plates, infinite length concentric gray cylinders, and concentric gray spheres. In such cases, Eq. 36.15 can be used. A_1 is the smaller area.

$$A_1\mathcal{F}_{12} = \frac{1}{\dfrac{1-\epsilon_1}{\epsilon_1 A_1} + \dfrac{1}{A_1} + \dfrac{1-\epsilon_2}{\epsilon_2 A_2}} \qquad 36.15$$

With two infinite parallel plates, $A_1 = A_2$. Then, the gray body shape factor is

$$\mathcal{F}_{12} = \frac{1}{\dfrac{1}{\epsilon_1} + \dfrac{1}{\epsilon_2} - 1} \qquad 36.16$$

For a small gray body enclosed by a black body,

$$\mathcal{F}_{12} = \epsilon_1 \qquad 36.17$$

6. NET RADIATION HEAT TRANSFER

The net heat transfer due to radiation between two gray bodies at different temperatures is given by Eq. 36.18. The area of body 1 must be used with $\mathcal{F}_{12}$, and the area of body 2 must be used with $\mathcal{F}_{21}$. Whether $\mathcal{F}_{12}$ or $\mathcal{F}_{21}$ is used depends on which is easier to evaluate.

Net Energy Exchange by Radiation Between Two Bodies

$$\begin{aligned} q_{12} &= \epsilon\sigma A(T_1^4 - T_2^4) \\ &= A_1 E_{\text{net},12} \end{aligned} \qquad 36.18$$

$$\begin{aligned} E_{\text{net},12} &= \sigma\mathcal{F}_{12}(T_1^4 - T_2^4) \\ &= \sigma F_{12}F_e(T_1^4 - T_2^4) \end{aligned} \qquad 36.19$$

For the special case of a low-emissivity shield (see Fig. 36.3),

One-Dimensional Geometry With Thin, Low-Emissivity Shield Inserted Between Two Parallel Plates

$$q_{12} = \frac{\sigma(T_1^4 - T_2^4)}{\dfrac{1-\epsilon_1}{\epsilon_1 A_1} + \dfrac{1}{A_1 F_{13}} + \dfrac{1-\epsilon_{3,1}}{\epsilon_{3,1}A_3} + \dfrac{1-\epsilon_{3,2}}{\epsilon_{3,2}A_3} + \dfrac{1}{A_3 F_{32}} + \dfrac{1-\epsilon_2}{\epsilon_2 A_2}} \qquad 36.20$$

[6]Although Eq. 36.14 is limited to two bodies that exchange heat with each other and with no other bodies, not all of each body's radiation has to reach the other body. This is evident in the presence of the black body shape factor, F_{12}.

Table 36.1 *Arrangement and Emissivity Factors*

arrangement	area	F_{12}	F_e
infinite parallel planes	A_1 or A_2	1	$\dfrac{1}{\dfrac{1}{\epsilon_1}+\dfrac{1}{\epsilon_2}-1}$
completely enclosed body; small compared with enclosure[a]	A_1	1	ϵ_1
completely enclosed body; large compared with enclosure[a]	A_1	1	$\dfrac{1}{\dfrac{1}{\epsilon_1}+\dfrac{1}{\epsilon_2}-1}$
concentric spheres or infinite cylinders with diffuse radiation[a]	A_1	1	$\dfrac{1}{\dfrac{1}{\epsilon_1}+\left(\dfrac{A_1}{A_2}\right)\left(\dfrac{1}{\epsilon_2}-1\right)}$
concentric spheres or infinite cylinders with specular (mirror-like) radiation[a]	A_1	1	$\dfrac{1}{\dfrac{1}{\epsilon_1}+\dfrac{1}{\epsilon_2}-1}$
two perpendicular rectangles with a common edge	A_1 or A_2	(See Fig. 36.1.)[b]	$\epsilon_1\epsilon_2$
directly opposed, parallel disks, squares, or rectangles of equal size	A_1 or A_2	(See Fig. 36.2.)[b]	$\epsilon_1\epsilon_2$
directly opposed, parallel disks, squares, or rectangles of equal size, connected by nonconducting, reradiating walls	A_1 or A_2	(See Fig. 36.4.)[b]	$\epsilon_1\epsilon_2$

[a]Object 1 is the smaller, enclosed body.
[b]Arrangement factors for these configurations are presented graphically in most heat transfer books.

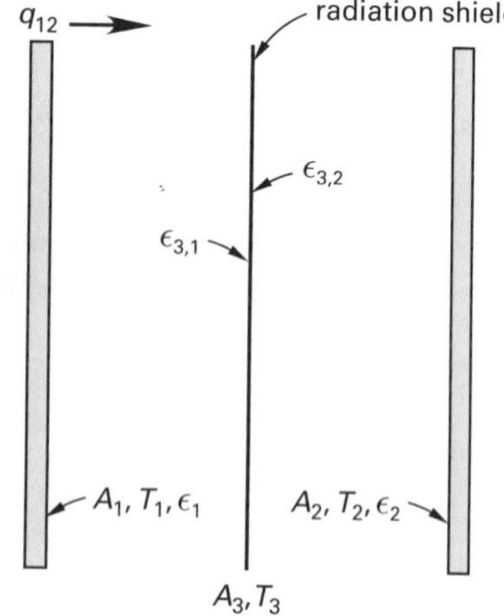

Figure 36.3 *Low-Emissivity Shield*

7. RECIPROCITY THEOREM

From Eq. 36.21 and Eq. 36.22,

$$\begin{aligned} q_{\text{net}} &= \sigma A_1\mathcal{F}_{12}(T_1^4 - T_2^4) \\ &= \sigma A_2\mathcal{F}_{21}(T_1^4 - T_2^4) \end{aligned} \qquad 36.21$$

This can be reduced to

Reciprocity

$$A_1\mathcal{F}_{12} = A_2\mathcal{F}_{21} \qquad 36.22$$

Equation 36.22 is known as the *reciprocity theorem for radiation.*

The product of the area and the shape factor is known as the *geometric flux, G.*

$$G_{12} = A_1\mathcal{F}_{12} \qquad 36.23$$

The reciprocity theorem can be written in terms of the geometric flux.

$$G_{12} = G_{21} \qquad 36.24$$

Energy is conserved, so all energy radiated from a body in an enclosure will end up striking the surface of one or another body in the enclosure. For example, if bodies 1, 2, and 3 are in a hollow within body 4, then all radiation that leaves body 1 must strike the surface of body 1 (if body 1 is concave), body 2, body 3, or body 4. From this it follows that the sum of the shape factors from body 1 must be one. This is known as the *summation rule* and is expressed mathematically in Eq. 36.22.

Summation Rule for N Surfaces

$$\sum_{j=1}^{N} F_{ij} = 1$$

8. RADIATION WITH REFLECTION/ RERADIATION

Surfaces that reradiate absorbed thermal radiation are known as *refractory materials* or *refractories.* A furnace wall that reradiates almost all the thermal energy it receives from combustion flames back to boiler tubes is an example of a refractory. The shape factor for cases with reradiation is traditionally given the symbol $\bar{F}$. Equation 36.25 (similar to Eq. 36.14) can used to calculate the gray body shape factor when there are two communicating refractory bodies. (See Fig. 36.4.)

$$A_1\mathcal{F}_{12} = \frac{1}{\dfrac{1-\epsilon}{\epsilon_1 A_1} + \dfrac{1}{A_1\bar{F}_{12}} + \dfrac{1-\epsilon^2}{\epsilon_2 A_2}} \qquad 36.25$$

The rate of heat flow from body 1 to body 2 is

Reradiating Surfaces

$$q_{12} = \frac{\sigma(T_1^4 - T_2^4)}{\dfrac{1-\epsilon}{\epsilon_1 A_1} + \dfrac{1}{A_1F_{12} + \left(\dfrac{1}{A_1F_{1R}} + \dfrac{1}{A_2F_{2R}}\right)^{-1}} + \dfrac{1-\epsilon^2}{\epsilon_2 A_2}}$$

Figure 36.4 *Black Body Shape Factor (for parallel finite squares, rectangles, and disks connected by nonconducting, reradiating wall)*

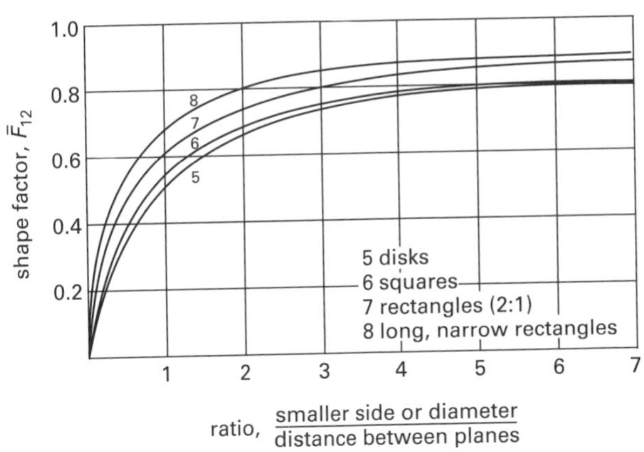

9. COMBINED HEAT TRANSFER

When heat is transferred by both radiation and convection, it is convenient to define the *radiant heat transfer coefficient*, $h_{\text{radiation}}$. T_∞ is the free-stream (far-field) temperature for convective heat transfer.[7] T_1 and T_2 should both be expressed as absolute temperatures.

$$\begin{aligned} h_{\text{radiation}} &= \frac{q_{\text{net}}}{A_1(T_1 - T_\infty)} \\ &= \frac{E_{\text{net}}}{T_1 - T_\infty} \\ &= \frac{\sigma F_{12}F_e(T_1^4 - T_2^4)}{T_1 - T_\infty} \\ &= \frac{\sigma \mathcal{F}_{12}(T_1^4 - T_2^4)}{T_1 - T_\infty} \end{aligned} \qquad 36.26$$

The *combined heat transfer coefficient* is[8]

$$h_{\text{total}} = h_{\text{radiation}} + h_{\text{convective}} \qquad 36.27$$

The combined radiation and convective heat transfer is

$$\begin{aligned} q &= q_{\text{radiation}} + q_{\text{convection}} \\ &= h_{\text{total}}A_1(T_1 - T_\infty) \end{aligned} \qquad 36.28$$

Example 36.1

A 1.0 ft (30 cm) diameter uninsulated horizontal duct carries hot air through a basement. The duct surface temperature is 200°F (95°C); the duct has an emissivity of 0.8. Air surrounding the duct in the basement is at 40°F (5°C); the basement walls are at 0°F (−20°C). The convective film coefficient on the exterior of the duct is 0.96 Btu/hr-ft²-°F (5.5 W/m²·K). What is the heat loss per unit area of duct?

[7] T_∞ can be the same as either T_1 or T_2. In that case, the solutions to common types of problems will be greatly simplified since the heat transfer can be calculated separately. If the temperatures are different, trial and error will be necessary to determine the surface temperature.

[8] It is understood that radiation and convection are the combination of heat transfer mechanisms. Conductive heat transfer does not use the film coefficient concept.

SI Solution

Since the convective film coefficient for the duct is known, the heat losses from convection and radiation can be calculated independently.

The absolute temperatures are

$$T_{\text{duct}} = 95°\text{C} + 273° = 368\text{K}$$
$$T_{\infty} = 5°\text{C} + 273° = 278\text{K}$$
$$T_{\text{walls}} = -20°\text{C} + 273° = 253\text{K}$$

The convective heat loss is

$$q''_{\text{convection}} = \frac{q}{A} = h(T_{\text{duct}} - T_{\infty})$$
$$= \left(5.5\ \frac{\text{W}}{\text{m}^2\text{·K}}\right)(368\text{K} - 278\text{K})$$
$$= 495\ \text{W/m}^2$$

Since the duct is entirely enclosed by the basement, the shape factor is $F_{12} = 1.0$. The emissivity factor is $F_e = \epsilon_{\text{duct}}$. The radiation heat transfer is

$$E_{\text{net}} = \sigma F_{12} F_e (T_{\text{duct}}^4 - T_{\text{walls}}^4)$$
$$= \left(5.67 \times 10^{-8}\ \frac{\text{W}}{\text{m}^2 \cdot \text{K}^4}\right)(1.0) \times (0.8)\left((368\text{K})^4 - (253\text{K})^4\right)$$
$$= 646\ \text{W/m}^2$$

The total heat loss is

$$q''_{\text{total}} = q''_{\text{convection}} + E_{\text{net}} = 495\ \frac{\text{W}}{\text{m}^2} + 646\ \frac{\text{W}}{\text{m}^2}$$
$$= 1141\ \text{W/m}^2$$

Customary U.S. Solution

Since the convective film coefficient for the duct is known, the heat losses from convection and radiation can be calculated independently.

The absolute temperatures are

$$T_{\text{duct}} = 200°\text{F} + 460° = 660°\text{R}$$
$$T_{\infty} = 40°\text{F} + 460° = 500°\text{R}$$
$$T_{\text{walls}} = 0°\text{F} + 460° = 460°\text{R}$$

The convective heat loss is

$$q''_{\text{convection}} = \frac{q}{A} = h(T_{\text{duct}} - T_{\infty})$$
$$= \left(0.96\ \frac{\text{Btu}}{\text{hr-ft}^2\text{-°F}}\right)(660°\text{R} - 500°\text{R})$$
$$= 153.6\ \text{Btu/hr-ft}^2$$

Since the duct is entirely enclosed by the basement, the arrangement factor is $F_a = 1.0$. The emissivity factor is $F_e = \epsilon_{\text{duct}}$. The radiation heat transfer is

$$E_{\text{net}} = \sigma F_{12} F_e (T_{\text{duct}}^4 - T_{\text{walls}}^4)$$
$$= \left(0.1713 \times 10^{-8}\ \frac{\text{Btu}}{\text{hr-ft}^2\text{-°R}^4}\right)(1.0) \times (0.8)\left((660°\text{R})^4 - (460°\text{R})^4\right)$$
$$= 198.7\ \text{Btu/hr-ft}^2$$

The total heat loss is

$$q''_{\text{total}} = q''_{\text{convection}} + E_{\text{net}}$$
$$= 153.6\ \frac{\text{Btu}}{\text{hr-ft}^2} + 198.7\ \frac{\text{Btu}}{\text{hr-ft}^2}$$
$$= 352.3\ \text{Btu/hr-ft}^2$$

Example 36.2

The air temperature in the duct is increased, so that the duct in Ex. 36.1 loses heat at the rate of 500 Btu/hr-ft^2 (1600 W/m^2). The duct temperature and film coefficient are unknown. All other values are the same. What is the duct surface temperature?

SI Solution

The unknown surface temperature cannot be extracted directly from the combined heat transfer equation. It is more convenient to solve this problem by trial and error.

Assume laminar natural convective heat transfer and a duct temperature of 393K (120°C). From Table 36.1, the convective film coefficient on the outside of the duct is approximately

$$h_{\text{convective}} = 1.32\left(\frac{T_{\text{duct}} - T_{\infty}}{D}\right)^{0.25}$$
$$= (1.32)\left(\frac{393\text{K} - 278\text{K}}{0.3\ \text{m}}\right)^{0.25}$$
$$= 5.84\ \text{W/m}^2\text{·K}$$

The convective heat loss is

$$q''_{\text{convection}} = \frac{q}{A} = h(T_{\text{duct}} - T_{\infty})$$
$$= \left(5.84\ \frac{\text{W}}{\text{m}^2\text{·K}}\right)(393\text{K} - 278\text{K})$$
$$= 672\ \text{W/m}^2$$

The radiation loss is

$$\begin{aligned} E_{\text{net}} &= \sigma F_{12} F_e (T_{\text{duct}}^4 - T_{\text{walls}}^4) \\ &= \left(5.67 \times 10^{-8} \ \frac{\text{W}}{\text{m}^2\text{·K}^4}\right)(1.0) \\ &\quad \times (0.8)\left((393\text{K})^4 - (253\text{K})^4\right) \\ &= 896 \ \text{W/m}^2 \end{aligned}$$

The total heat loss for this iteration is

$$\begin{aligned} q''_{\text{total}} &= q''_{\text{convection}} + E_{\text{net}} = 672 \ \frac{\text{W}}{\text{m}^2} + 896 \ \frac{\text{W}}{\text{m}^2} \\ &= 1568 \ \text{W/m}^2 \end{aligned}$$

Since the calculated heat loss agrees with the known heat loss, the assumed surface temperature is correct. (Usually, several trial and error iterations would be needed to converge on the solution.)

Customary U.S. Solution

The unknown surface temperature cannot be extracted directly from the combined heat transfer equation. It is more convenient to solve this problem by trial and error.

Assume laminar convective heat transfer and a duct temperature of 710°R (250°F). From Table 36.1, the convective film coefficient on the outside of the duct is approximately

$$\begin{aligned} h_{\text{convective}} &= 0.27\left(\frac{T_{\text{duct}} - T_\infty}{D}\right)^{0.25} \\ &= (0.27)\left(\frac{710°\text{R} - 500°\text{R}}{1 \ \text{ft}}\right)^{0.25} \\ &= 1.03 \ \text{Btu/hr-ft}^2\text{-°F} \end{aligned}$$

The convective heat loss is

$$\begin{aligned} q''_{\text{convection}} &= \frac{q}{A} = h(T_{\text{duct}} - T_\infty) \\ &= \left(1.03 \ \frac{\text{Btu}}{\text{hr-ft}^2\text{-°F}}\right)(710°\text{R} - 500°\text{R}) \\ &= 216.3 \ \text{Btu/hr-ft}^2 \end{aligned}$$

The radiation loss is

$$\begin{aligned} E_{\text{net}} &= \sigma F_{12} F_e (T_{\text{duct}}^4 - T_{\text{walls}}^4) \\ &= \left(0.1713 \times 10^{-8} \ \frac{\text{Btu}}{\text{hr-ft}^2\text{-°R}^4}\right)(1.0) \\ &\quad \times (0.8)\left((710°\text{R})^4 - (460°\text{R})^4\right) \\ &= 286.9 \ \text{Btu/hr-ft}^2 \end{aligned}$$

The total heat loss for this iteration is

$$\begin{aligned} q''_{\text{total}} &= q''_{\text{convection}} + E_{\text{net}} \\ &= 216.3 \ \frac{\text{Btu}}{\text{hr-ft}^2} + 286.9 \ \frac{\text{Btu}}{\text{hr-ft}^2} \\ &= 503.2 \ \text{Btu/hr-ft}^2 \end{aligned}$$

Since the calculated heat loss agrees with the known heat loss, the assumed surface temperature is correct. (Usually, several trial-and-error iterations would be required to converge on the solution.)

10. EQUILIBRIUM CONDITION WITH COMBINED HEAT TRANSFER

A single body that remains at a constant temperature is said to be in an equilibrium condition.[9] To be in equilibrium, the body must continually lose all of the energy gained. Depending on the situation, a body might radiate all of the energy gained by convection, or it may lose by convection all of the energy gained by radiation.

Example 36.3

A small temperature probe with an emissivity of 0.8 measures the temperature of a gas flowing in a large pipe as 850°F (450°C). The pipe walls are 350°F (180°C). The convective film coefficient for the probe in the gas flow is 27 Btu/hr-ft²-°F (150 W/m²·K). The probe's surface temperature is constant. There is no heat transfer to the probe by conduction. What is the actual gas temperature?

SI Solution

The absolute temperatures are

$$\begin{aligned} T_{\text{probe}} &= 450°\text{C} + 273° = 723\text{K} \\ T_{\text{pipe}} &= 180°\text{C} + 273° = 453\text{K} \end{aligned}$$

[9] See Ftn. 4.

Heat Transfer

Since the probe is entirely enclosed by the large pipe, the shape factor is $F_{12} = 1.0$, and the emissivity factor is $F_e = 0.8$.

Since the probe's surface temperature is constant, the heat gained by the probe through convection from the gas stream is being lost through radiation to the cooler pipe walls.

$$q''_{\text{gain,convection}} = E_{\text{loss}}$$
$$h(T_{\text{gas}} - T_{\text{probe}}) = \sigma F_{12} F_e (T^4_{\text{probe}} - T^4_{\text{pipe}})$$

$$\begin{aligned} T_{\text{gas}} &= T_{\text{probe}} + \frac{\sigma F_{12} F_e (T^4_{\text{probe}} - T^4_{\text{pipe}})}{h} \\ &= 723\text{K} + \frac{\left(5.67 \times 10^{-8} \frac{\text{W}}{\text{m}^2 \cdot \text{K}^4}\right) \times (1.0)(0.8) \times \left((723\text{K})^4 - (453\text{K})^4\right)}{150 \frac{\text{W}}{\text{m}^2 \cdot \text{K}}} \\ &= 793\text{K} \end{aligned}$$

Customary U.S. Solution

The absolute temperatures are

$$T_{\text{probe}} = 850°\text{F} + 460° = 1310°\text{R}$$
$$T_{\text{pipe}} = 350°\text{F} + 460° = 810°\text{R}$$

Since the probe is entirely enclosed by the large pipe, the arrangement factor is $F_{12} = 1.0$. and the emissivity factor is $F_e = 0.8$.

Since the probe's surface temperature is constant, the heat gained by the probe through convection from the gas stream is being lost through radiation to the cooler pipe walls.

$$q''_{\text{gain,convection}} = E_{\text{loss}}$$
$$h(T_{\text{gas}} - T_{\text{probe}}) = \sigma F_{12} F_e (T^4_{\text{probe}} - T^4_{\text{pipe}})$$

$$\begin{aligned} T_{\text{gas}} &= T_{\text{probe}} + \frac{\sigma F_{12} F_e (T^4_{\text{probe}} - T^4_{\text{pipe}})}{h} \\ &= 1310°\text{R} + \frac{\left(0.1713 \times 10^{-8} \frac{\text{Btu}}{\text{hr-ft}^2\text{-}°\text{R}^4}\right) \times (1.0)(0.8) \times \left((1310°\text{R})^4 - (810°\text{R})^4\right)}{27 \frac{\text{Btu}}{\text{hr-ft}^2\text{-}°\text{F}}} \\ &= 1438°\text{R} \end{aligned}$$

11. SOLAR RADIATION

The average solar energy hitting the outer edge of the earth's atmosphere is approximately 433 Btu/hr-ft² (1.366 kW/m²) and is known as the *solar constant.* The actual instantaneous value reaching the surface depends on the altitude, latitude, time of year, time of day, sky conditions, and orientation angle of the receiving body.[10,11]

12. NOCTURNAL RADIATION

Measurements of the night sky show that the effective temperature of the sky for purposes of radiation is approximately 410°R (210K) on cold clear nights. Since this temperature is below the freezing point of water, when the air is calm it is possible for standing water and tree fruits to freeze even when the air temperature is above the freezing point of water.

Example 36.4

On a cold, clear night, the surface of a pond has a convective film coefficient of 5 Btu/hr-ft²-°R (28 W/m²·K). The emissivity of the pond surface is 0.96. The air temperature is 60°F (15°C). The pond temperature remains the same all night long. Evaporative losses are negligible. What is the water temperature?

SI Solution

With an effective sky temperature of 210K, the pond will lose heat through radiation. Since the pond's temperature is constant, it gains heat from the air through convection. Therefore, the pond temperature is less than 15°C.

[10]The value of the solar constant given can be used with modification at altitudes higher than approximately 50,000 ft (15 000 m).
[11]Most heat transfer textbooks detail the method of calculating the exact instantaneous solar heat gain. Books on the subject of HVAC cover this subject particularly well.

The absolute temperatures, when the air is calm, are

$$T_{sky} = 210\text{K}$$
$$T_{air} = 15°\text{C} + 273° = 288\text{K}$$

The equilibrium equation is

$$q''_{gain,convection} = E_{loss}$$
$$h(T_{air} - T_{pond}) = \sigma F_{12} F_e (T^4_{pond} - T^4_{sky})$$

$$\left(28 \ \frac{\text{W}}{\text{m}^2\text{·K}}\right)(288\text{K} - T_{pond}) = \left(5.67 \times 10^{-8} \ \frac{\text{W}}{\text{m}^2\text{·K}^4}\right) \times (1.0)(0.96) \times \left(T^4_{pond} - (210\text{K})^4\right)$$

Solving by trial and error,

$$T_{pond} \approx 280\text{K} \quad (7°\text{C})$$

Customary U.S. Solution

With an effective sky temperature of 410°R, the pond will lose heat through radiation. Since the pond's temperature is constant, it gains heat from the air through convection. Therefore, the pond temperature is less than 60°F.

The absolute temperatures when the air is calm are

$$T_{sky} = 410°\text{R}$$
$$T_{air} = 60°\text{F} + 460° = 520°\text{R}$$

The equilibrium equation is

$$q''_{gain,convection} = E_{loss}$$
$$h(T_{air} - T_{pond}) = \sigma F_{12} F_e (T^4_{pond} - T^4_{sky})$$

$$\left(5 \ \frac{\text{Btu}}{\text{hr-ft}^2\text{-°R}}\right)(520°\text{R} - T_{pond}) = \left(0.1713 \times 10^{-8} \ \frac{\text{Btu}}{\text{hr-ft}^2\text{-°R}^4}\right)(1.0)(0.96) \times \left(T^4_{pond} - (410°\text{R})^4\right)$$

Solving by trial and error,

$$T_{pond} \approx 508°\text{R} \quad (48°\text{F})$$

13. NOMENCLATURE

A	area	ft^2	m^2
D	diameter	ft	m
E	emissive power	Btu/hr-ft^2	W/m^2
F	factor	–	–
F	black body shape factor	–	–
$\mathcal{F}$	gray body shape factor	–	–
G	geometric flux	ft^2	m^2
h	film coefficient	Btu/hr-ft^2-°F	W/m^2·K
q	heat transfer rate	Btu/hr	W
q''	heat transfer per unit area	Btu/hr-ft^2	W/m^2
T	temperature	°R	K

Symbols

α	absorptivity	–	–
ϵ	emissivity	–	–
ρ	reflectivity	–	–
σ	Stefan-Boltzmann constant, 0.1713×10^{-8} (5.67×10^{-8})	Btu/hr-ft^2-°R^4	W/m^2·K^4
τ	transmissivity	–	–

Subscripts

12	from body 1 to body 2
21	from body 2 to body 1
e	emissivity

Heat Transfer

Topic VI: HVAC

Chapter

37 Psychrometrics

Content in blue refers to the *NCEES Handbook*.

NCEES EXAM SPECIFICATIONS AND RELATED CONTENT

HVAC AND REFRIGERATION EXAM

I.C.1. Psychrometrics: Heating/cooling processes
- 2. Pressure and Temperature Corrections for Air
- 3. Properties of Atmospheric Air
- 6. The Psychrometric Chart
- 10. Adiabatic Mixing of Two Air Streams

I.C.2. Psychrometrics: Humidification/dehumidification processes
- 12. Sensible Heat Ratio
- 16. Cooling with Coil Dehumidification
- 19. Cooling with Humidification
- 20. Cooling with Spray Dehumidification
- 21. Heating with Humidification
- 22. Heating and Dehumidification

II.C.4. Systems and Components: Energy recovery
- 23. Energy-Recovery Ventilators (ERV)/Heat-Recovery Ventilators (HRV)

THERMAL AND FLUID SYSTEMS EXAM

I.F.3. Supportive Knowledge: Psychrometrics
- 2. Pressure and Temperature Corrections for Air
- 3. Properties of Atmospheric Air
- 6. The Psychrometric Chart
- 10. Adiabatic Mixing of Two Air Streams
- 12. Sensible Heat Ratio

I.F.4. Supportive Knowledge: Codes and standards
- 16. Cooling with Coil Dehumidification
- 19. Cooling with Humidification
- 20. Cooling with Spray Dehumidification
- 21. Heating with Humidification
- 22. Heating and Dehumidification

III.C. Energy Recovery
- 23. Energy-Recovery Ventilators (ERV)/Heat-Recovery Ventilators (HRV)

1. INTRODUCTION TO PSYCHROMETRICS

Atmospheric air contains small amounts of moisture and can be considered to be a mixture of two ideal gases—dry air and water vapor. All of the thermodynamic rules relating to the behavior of nonreacting gas mixtures apply to atmospheric air. From Dalton's law, for example, the total atmospheric pressure is the sum of the dry air partial pressure and the water vapor pressure.[1]

$$p_t = p_a + p_w \quad (37.1)$$

The study of the properties and behavior of atmospheric air is known as *psychrometrics*. Properties of atmospheric air are seldom evaluated, however, from theoretical thermodynamic principles. Rather, specialized techniques and charts have been developed for that purpose.

2. PRESSURE AND TEMPERATURE CORRECTIONS FOR AIR

As air pressure and temperature change with altitude, corrections are sometimes required depending on the application.

[1]Equation 37.1 points out a problem in semantics. The term *air* means *dry air*. The term *atmosphere* refers to the combination of dry air and water vapor. It is common to refer to the atmosphere as *moist air*.

HVAC

Pressure in pounds per square inch as a function of altitude, Z, is

Temperature and Altitude Corrections for Air

$$p = \left(14.696\ \frac{\text{lbm}}{\text{in}^2}\right)(1 - 6.8754 \times 10^{-6} Z)^{5.2559} \quad 37.2$$

Temperature in Fahrenheit as a function of altitude, Z, is

Temperature and Altitude Corrections for Air

$$T = 59 - 0.00356620Z \quad 37.3$$

3. PROPERTIES OF ATMOSPHERIC AIR

At first, psychrometrics seems complicated by three different definitions of temperature. These three terms are not interchangeable.

- *dry-bulb temperature*, T_{db}: This is the equilibrium temperature that a regular thermometer measures if exposed to atmospheric air.
- *wet-bulb temperature*, T_{wb}: This is the temperature of air that has gone through an adiabatic saturation process. (See Sec. 37.17.)
- *dew-point temperature*, T_{dp}: This is the dry-bulb temperature at which water starts to condense out when moist air is cooled in a constant pressure process.

For every temperature, there is a unique vapor pressure, p_{sat}, which represents the maximum pressure the water vapor can exert. The actual vapor pressure, p_w, can be less than or equal to, but not greater than, the saturation value. The saturation pressure is found from steam tables as the pressure corresponding to the dry-bulb temperature of the atmospheric air.

$$p_w \leq p_{\text{sat}} \quad 37.4$$

If the vapor pressure equals the saturation pressure, the air is said to be saturated.[2] *Saturated air* is a mixture of dry air and saturated water vapor. When the air is saturated, all three temperatures are equal.

$$T_{\text{db}} = T_{\text{wb}} = T_{\text{dp}}\Big|_{\text{sat}} \quad 37.5$$

Unsaturated air is a mixture of dry air and superheated water vapor.[3] When the air is unsaturated, the dew-point temperature will be less than the wet-bulb temperature. The *wet-bulb depression* is the difference between the dry-bulb and wet-bulb temperatures.

$$T_{\text{dp}} < T_{\text{wb}} < T_{\text{db}}\Big|_{\text{unsat}} \quad 37.6$$

The amount of water vapor in atmospheric air is specified by three different parameters. The *humidity ratio*, W, is the mass ratio of water vapor to dry air. If both masses are expressed in pounds (kilograms), the units of humidity ratio are lbm/lbm (kg/kg). However, since there is so little water vapor, the water vapor mass is often reported in *grains* of water. (There are 7000 grains per pound.) Accordingly, the humidity ratio will have the units of grains per pound.

Psychrometric Properties

$$W = \frac{m_w}{m_a} \quad 37.7$$

Since $m = \rho V$, and since $V_w = V_a$, the humidity ratio can be written as

$$W = \frac{\rho_w}{\rho_a} \quad 37.8$$

The density of a typical air mixture is the ratio of total mass to total volume.

Psychrometric Properties

$$\rho = \frac{m_a + m_w}{V} = \left(\frac{1}{v}\right)(1 + W) \quad 37.9$$

The specific volume, v, of moist air as a function of dry air mass can be determined by Eq. 37.10 .

Psychrometric Properties

$$v = \frac{V}{m_a} = \frac{V}{28.97n_a} \quad 37.10$$

From the equation of state for an ideal gas, $m = pV/RT$, the gas relationship for both dry and moist air can be expressed as

Psychrometric Properties

$$p_a V = n_a R T \quad 37.11$$

$$p_w V = n_w R T \quad 37.12$$

Combine Eq. 37.11 and Eq. 37.12 to express the gas relationship of a mixture containing both dry air and water vapor.

$$(p_a + p_w)V = (n_a + n_w)RT \quad 37.13$$

Since $V_w = V_a$ and $T_w = T_a$, the humidity ratio can be written in one additional form.

$$W = \frac{R_a p_w}{R_w p_a} = \frac{53.35 p_w}{85.78 p_t - p_w} = 0.622\left(\frac{p_w}{p_t - p_w}\right) \quad 37.14$$

The humidity ratio is also expressed as the product of the mole fraction ratio (x_w/x_a) and the ratio of the molecular masses of water to dry air.

[2] Actually, the water vapor is saturated, not the air. However, this particular inconsistency in terms is characteristic of psychrometrics.

[3] As strange as it sounds, atmospheric water vapor is almost always superheated. This can be shown by drawing an isotherm passing through the vapor dome on a p-V diagram. The only place where the water vapor pressure is less than the saturation pressure is in the superheated region.

HVAC

Psychrometric Properties

$$\frac{18.01528 \ \frac{\text{g}}{\text{mol}}}{28.9845 \ \frac{\text{g}}{\text{mol}}} = 0.62198 \quad (0.622) \qquad 37.15$$

$$W = 0.622\frac{x_w}{x_a} \qquad 37.16$$

Specific humidity, γ, is the ratio of the water vapor mass to the combined water vapor and dry air masses of the air being sampled.

Psychrometric Properties

$$\gamma = \frac{m_w}{m_w + m_a} \qquad 37.17$$

Specific humidity in terms of humidity ratio is

Psychrometric Properties

$$\gamma = \frac{W}{1 + W} \qquad 37.18$$

Absolute humidity of water vapor, d_v, also known as water vapor density, is the ratio of water vapor mass to the total selected volume of the air being sampled.

Psychrometric Properties

$$d_v = \frac{m_w}{V} \qquad 37.19$$

Saturation humidity, W_{sat}, is the humidity ratio of saturated moist air to water at the same temperature and pressure conditions.

Psychrometric Properties

$$W_{\text{sat}} = 0.622 p_{w,\text{sat}}(p - p_{w,\text{sat}}) \qquad 37.20$$

The *degree of saturation*, μ (also known as the *saturation ratio* and the *percentage humidity*), is the ratio of the actual humidity ratio to the saturated humidity ratio at the same temperature and pressure.

$$\mu = \frac{W}{W_{\text{sat}}} \qquad 37.21$$

A third index of moisture content is the *relative humidity*—the partial pressure of the water vapor divided by the saturation pressure.

$$\phi = \frac{p_w}{p_{\text{sat}}} \qquad 37.22$$

Relative humidity can also be expressed as the ratio of the mole fraction of water vapor, x_w, to the mole fraction of water vapor saturated at the same temperature and pressure, $x_{w,\text{sat}}$.

Psychrometric Properties

$$\phi = \left.\frac{x_w}{x_{w,\text{sat}}}\right|_{t,p} \qquad 37.23$$

From the equation of state for an ideal gas, $\rho = p/RT$, so the relative humidity can be written as

$$\phi = \frac{\rho_w}{\rho_{\text{sat}}} \qquad 37.24$$

Combining the definitions of specific and relative humidities,

$$\phi = 1.608 W\left(\frac{p_a}{p_{\text{sat}}}\right) \qquad 37.25$$

The degree of saturation, μ, and relative humidity, ϕ, are interdependent.

Psychrometric Properties

$$\mu = \frac{\phi}{\left(\frac{1 + (1 - \phi) W_{\text{sat}}}{0.622}\right)} \qquad 37.26$$

$$\phi = \frac{\mu}{1 - (1 - \mu)\left(\frac{p_{w,\text{sat}}}{p}\right)} \qquad 37.27$$

4. VAPOR PRESSURE

There are at least six ways of determining the partial pressure, p_w, of the water vapor in the air. The first method, derived from Eq. 37.22, is to multiply the relative humidity, ϕ, by the water's saturation pressure. The saturation pressure, in turn, is obtained from steam tables as the pressure corresponding to the air's dry-bulb temperature.

$$p_w = \phi p_{\text{sat,db}} \qquad 37.28$$

A more direct method is to read the saturation pressure (from the steam tables) corresponding to the air's dew-point temperature.

$$p_w = p_{\text{sat,dp}} \qquad 37.29$$

The third method can be used if water's mole (volumetric) fraction is known.

$$p_w = x_w p_t = B_w p_t \qquad 37.30$$

The fourth method is to calculate the actual vapor pressure from the empirical *Carrier equation*, valid for customary U.S. units only.[4]

$$p_w = p_{\text{sat,wb}} - \frac{(p_t - p_{\text{sat,wb}})(T_{\text{db}} - T_{\text{wb}})}{2830 - 1.44T_{\text{wb}}} \quad \text{[U.S. only]} \qquad 37.31$$

The fifth method is based on the humidity ratio.

$$p_w = \frac{p_t W}{0.622 + W} \qquad 37.32$$

The sixth (and easiest) method is to read the water vapor pressure from a psychrometric chart. Some, but not all, psychrometric charts have water vapor scales.

Example 37.1

Use the methods described in the previous section to determine the partial pressure of water vapor in standard atmospheric air at 60°F (16°C) dry-bulb temperature and 50% relative humidity.

SI Solution

method 1: From the steam tables, the saturation pressure corresponding to 16°C is 0.01819 bars. The partial pressure of the vapor is

$$\begin{aligned} p_w &= \phi p_{\text{sat}} \\ &= (0.50)(0.01819\ \text{bars})\left(100\ \frac{\text{kPa}}{\text{bar}}\right) \\ &= 0.910\ \text{kPa} \end{aligned}$$

method 2: The dew-point temperature (reading straight across on the psychrometric chart) is approximately 5°C. The saturation pressure from the steam table corresponding to 5°C is approximately 0.0087 bars (0.87 kPa).

method 3: The humidity ratio is 0.0056 kg/kg. From Eq. 37.32,

$$\begin{aligned} p_w &= \frac{p_t W}{0.622 + W} \\ &= \frac{(101.3\ \text{kPa})\left(0.0056\ \frac{\text{kg}}{\text{kg}}\right)}{0.622 + 0.0056\ \frac{\text{kg}}{\text{kg}}} \\ &= 0.904\ \text{kPa} \end{aligned}$$

Customary U.S. Solution

method 1: From the steam tables, the saturation pressure corresponding to 60°F is 0.2564 lbf/in^2. The partial pressure of the vapor is

$$\begin{aligned} p_w = \phi p_{\text{sat}} &= (0.50)\left(0.2564\ \frac{\text{lbf}}{\text{in}^2}\right) \\ &= 0.128\ \text{lbf/in}^2 \end{aligned}$$

method 2: The dew-point temperature (reading straight across the psychrometric chart) is approximately 41°F. The saturation pressure from the steam table corresponding to 41°F is approximately 0.127 lbf/in^2.

method 3: Use the Carrier equation. The wet-bulb temperature of the air is approximately 50°F. From the steam tables, the saturation pressure corresponding to that temperature is 0.1780 lbf/in^2.

$$\begin{aligned} p_w &= p_{\text{sat,wb}} - \frac{(p_t - p_{\text{sat,wb}})(T_{\text{db}} - T_{\text{wb}})}{2830 - 1.44T_{\text{wb}}} \\ &= 0.1780\ \frac{\text{lbf}}{\text{in}^2} - \frac{\left(14.7\ \frac{\text{lbf}}{\text{in}^2} - 0.1780\ \frac{\text{lbf}}{\text{in}^2}\right) \times (60°\text{F} - 50°\text{F})}{2830 - (1.44)(50°\text{F})} \\ &= 0.125\ \text{lbf/in}^2 \end{aligned}$$

5. ENERGY CONTENT OF AIR

Since moist air is a mixture of dry air and water vapor, its total enthalpy, h (i.e., energy content), takes both components into consideration. Total enthalpy is conveniently shown on the diagonal scales of the psychrometric chart, but it can also be calculated. As Eq. 37.34 indicates, the reference temperature (i.e., the temperature that corresponds to a zero enthalpy) for the enthalpy of dry air is 0°F (0°C). Steam properties correspond to a low-pressure superheated vapor at room temperature.

$$h_t = h_a + Wh_w \qquad 37.33$$

$$h_a = c_{p,\text{air}}T \approx \left(1.005\ \frac{\text{kJ}}{\text{kg}\cdot°\text{C}}\right)T_{°\text{C}} \quad \text{[SI]} \qquad 37.34(a)$$

$$h_a = c_{p,\text{air}}T \approx \left(0.240\ \frac{\text{Btu}}{\text{lbm-}°\text{F}}\right)T_{°\text{F}} \quad \text{[U.S.]} \qquad 37.34(b)$$

[4]Equation 37.31 uses updated constants and is more accurate than the equation originally published by Carrier.

HVAC

$$h_w = c_{p,\text{water vapor}}T + h_{fg} \approx \left(1.805\ \frac{\text{kJ}}{\text{kg·°C}}\right)T_{°C} + 2501\ \frac{\text{kJ}}{\text{kg}} \quad \text{[SI]} \qquad 37.35(a)$$

$$h_w = c_{p,\text{water vapor}}T + h_{fg} \approx \left(0.444\ \frac{\text{Btu}}{\text{lbm-°F}}\right)T_{°F} + 1061\ \frac{\text{Btu}}{\text{lbm}} \quad \text{[U.S.]} \qquad 37.35(b)$$

6. THE PSYCHROMETRIC CHART

It is possible to develop mathematical relationships for enthalpy and specific volume (the two most useful thermodynamic properties) for atmospheric air. However, these relationships are almost never used. Rather, psychrometric properties can be read directly from *psychrometric charts* ("psych charts," as they are usually referred to). There are different psychrometric charts for low, medium, and high temperature ranges, as well as charts for different atmospheric pressures (i.e., elevations). These various ASHRAE psychrometric charts are shown in App. 37.A to App. 37.D. **[ASHRAE Psychrometric Chart No. 1—Sea Level] [ASHRAE Psychrometric Chart No. 3 - High Temperature] [ASHRAE Psychrometric Chart No. 4 - Normal Temperature]**

The usage of several scales varies somewhat from chart to chart. In particular, the use of the enthalpy scale depends on the chart used. Furthermore, not all psychrometric charts contain all scales.

A psychrometric chart is easy to use, despite the multiplicity of scales. The thermodynamic state (i.e., the position on the chart) is defined by specifying the values of any two parameters on intersecting scales (e.g., dry-bulb and wet-bulb temperature, or dry-bulb temperature and relative humidity). Once the state point has been located on the chart, all other properties can be read directly.

7. ENTHALPY CORRECTIONS

Some psychrometric charts have separate lines or scales for wet-bulb temperature and enthalpy. However, the deviation between lines of constant wet-bulb temperature and lines of constant enthalpy is small. Therefore, other psychrometric charts use only one set of diagonal lines for both scales. The error introduced is small—seldom greater than 0.1 – 0.2 Btu/lbm (0.23 – 0.46 kJ/kg). When extreme precision is needed, correction factors from the psychrometric chart can be used.

Example 37.2

Air at 50°F (10°C) dry bulb has a humidity ratio of 0.006 lbm/lbm (0.006 kg/kg). (a) Use the psychrometric chart to determine the enthalpy of the air. (b) Calculate the enthalpy of the air directly. (c) How much heat is needed to heat one unit mass of the air from 50°F to 140°F (10°C to 60°C) without changing the moisture content?

SI Solution

(a) Use the moisture content and dry-bulb temperature scales to locate the point corresponding to the original conditions. From the psychrometric chart, the enthalpy is approximately 25 kJ/kg.

(b) Use Eq. 37.34 and Eq. 37.35.

$$\begin{aligned} h_a &= c_{p,\text{air}}T \approx \left(1.005\ \frac{\text{kJ}}{\text{kg·°C}}\right)T_{°C} \\ &= \left(1.005\ \frac{\text{kJ}}{\text{kg·°C}}\right)(10°\text{C}) \\ &= 10\ \text{kJ/kg} \end{aligned}$$

$$\begin{aligned} h_w &= c_{p,\text{water vapor}}T + h_{fg} \\ &\approx \left(1.805\ \frac{\text{kJ}}{\text{kg·°C}}\right)T_{°C} + 2501\ \frac{\text{kJ}}{\text{kg}} \\ &= \left(1.805\ \frac{\text{kJ}}{\text{kg·°C}}\right)(10°\text{C}) + 2501\ \frac{\text{kJ}}{\text{kg}} \\ &= 2519\ \text{kJ/kg} \end{aligned}$$

$$\begin{aligned} h_t &= h_a + Wh_w \\ &= 10\ \frac{\text{kJ}}{\text{kg}} + \left(0.006\ \frac{\text{kg}}{\text{kg}}\right)\left(2519\ \frac{\text{kJ}}{\text{kg}}\right) \\ &= 25.1\ \text{kJ/kg} \end{aligned}$$

(c) The psychrometric chart does not go up to 60°C. Therefore, the energy difference must be calculated mathematically. Although the initial enthalpy could be subtracted from the calculated final enthalpy, it is equivalent merely to calculate the difference based on the variable terms.

$$\begin{aligned} q &= h_{t,2} - h_{t,1} = (c_{p,\text{air}} + Wc_{p,\text{water vapor}})(T_2 - T_1) \\ &= \left(\left(1.005\ \frac{\text{kJ}}{\text{kg·°C}}\right) + \left(0.006\ \frac{\text{kg}}{\text{kg}}\right)\left(1.805\ \frac{\text{kJ}}{\text{kg·°C}}\right)\right) \\ &\quad \times (60°\text{C} - 10°\text{C}) \\ &= 50.8\ \text{kJ/kg} \end{aligned}$$

Customary U.S. Solution

(a) Use the moisture content and dry-bulb temperature scales to locate the point corresponding to the original conditions. From the psychrometric chart, the enthalpy is approximately 18.5 Btu/lbm.

HVAC

(b) Use Eq. 37.34 and Eq. 37.35.

$$\begin{aligned} h_a &= c_{p,\text{air}}T \approx \left(0.240\ \frac{\text{Btu}}{\text{lbm-°F}}\right)T_{\text{°F}} \\ &= \left(0.240\ \frac{\text{Btu}}{\text{lbm-°F}}\right)(50\text{°F}) \\ &= 12\ \text{Btu/lbm} \end{aligned}$$

$$\begin{aligned} h_w &= c_{p,\text{water vapor}}T + h_{fg} \\ &\approx \left(0.444\ \frac{\text{Btu}}{\text{lbm-°F}}\right)T_{\text{°F}} + 1065\ \frac{\text{Btu}}{\text{lbm}} \\ &= \left(0.444\ \frac{\text{Btu}}{\text{lbm-°F}}\right)(50\text{°F}) + 1065\ \frac{\text{Btu}}{\text{lbm}} \\ &= 1087.2\ \text{Btu/lbm} \end{aligned}$$

$$\begin{aligned} h_t &= h_a + Wh_w \\ &= 12\ \frac{\text{Btu}}{\text{lbm}} + \left(0.006\ \frac{\text{lbm}}{\text{lbm}}\right)\left(1087.2\ \frac{\text{Btu}}{\text{lbm}}\right) \\ &= 18.5\ \text{Btu/lbm} \end{aligned}$$

(c) Psychrometric charts for room temperature do not go up to 140°F. (App. 37.D could be used.) Therefore, the energy difference must be calculated mathematically. Although the initial enthalpy could be subtracted from the calculated final enthalpy, it is equivalent merely to calculate the difference based on the variable terms.

$$\begin{aligned} q &= h_{t,2} - h_{t,1} \\ &= (c_{p,\text{air}} + Wc_{p,\text{water vapor}})(T_2 - T_1) \\ &= \left(\begin{array}{c} 0.240\ \frac{\text{Btu}}{\text{lbm-°F}} + \left(0.006\ \frac{\text{lbm}}{\text{lbm}}\right) \\ \times\left(0.444\ \frac{\text{Btu}}{\text{lbm-°F}}\right) \end{array}\right) \\ &\quad \times(140\text{°F} - 50\text{°F}) \\ &= 21.84\ \text{Btu/lbm} \end{aligned}$$

8. BASIS OF PROPERTIES

Several of the properties read from the psychrometric chart (specific volume, enthalpy, etc.) are given "per pound of dry air." This basis does not mean that the water vapor's contribution is absent. For example, if the enthalpy of atmospheric air is 28.0 Btu per pound of dry air, the energy content of the water vapor has been included. However, to get the energy of a mass of moist air, the enthalpy of 28 Btu/lbm would be multiplied by the mass of the dry air (m_a) only, not by the combined air and water masses.

$$h_t = m_a h_{\text{chart}} \qquad 37.36$$

Example 37.3

During the summer, air in a room reaches 75°F and 50% relative humidity. Find the air's (a) wet-bulb temperature, (b) humidity ratio, (c) enthalpy, (d) specific volume, (e) dew-point temperature, (f) actual vapor pressure, and (g) degree of saturation.

Solution

Locate the point where the 75°F vertical line intersects the curved 50% humidity line. Read all other values directly from the chart.

(a) Follow the diagonal line up to the left until it intersects the wet-bulb temperature scale. Read $T_{\text{wb}} = 62.6$°F.

(b) Follow the horizontal line to the right until it intersects the humidity ratio scale. Read $W = 64.8$ gr (0.0093 lbm) of moisture per pound of dry air.

(c) Finding the enthalpy is different on different charts. Some charts use the same diagonal lines for wet-bulb temperature and humidity. Corrections are required in such cases. Other charts employ two alignment scales to use in conjunction with a straightedge. Read 28.1 Btu per pound of dry air.

(d) Interpolate between diagonal specific volume lines. Read $v = 13.68$ cubic feet per pound of dry air.

(e) Follow the horizontal line to the left until it intersects the dew-point scale. Read $T_{\text{dp}} = 55.1$°F.

(f) From the steam tables, the saturation pressure corresponding to a dry-bulb temperature of 75°F is approximately 0.43 psia. From Eq. 37.22, the water vapor pressure is

$$p_w = \phi p_{\text{sat}} = (0.50)\left(0.43\ \frac{\text{lbf}}{\text{in}^2}\right) = 0.215\ \text{lbf/in}^2$$

(g) The humidity ratio at 75°F saturated is 131.5 gr (0.0188 lbm) per pound. From Eq. 37.21, the degree of saturation is

$$\mu = \frac{W}{W_{\text{sat}}} = \frac{64.8\ \frac{\text{gr}}{\text{lbm}}}{131.5\ \frac{\text{gr}}{\text{lbm}}} = 0.49$$

9. LEVER RULE

With few exceptions (e.g., relative humidity and enthalpy correction), the scales on a psychrometric chart are linear. Because they are linear, any one property can be used as the basis for interpolation or extrapolation for another property on an intersecting linear scale. This applies regardless of orientation of the scales. The scales do not have to be orthogonal.

Furthermore, since psychrometric properties are extensive properties (i.e., they depend on the quantity of air present), the mass of air can be used as the basis for interpolation or extrapolation. This principle, known as the *lever rule* or *inverse lever rule*, is used when determining the properties of a mixture. (See Sec. 37.10.)

10. ADIABATIC MIXING OF TWO AIR STREAMS

Figure 37.1 shows the mixing of two moist air streams. The state of the mixture can be determined if the flow rates and psychrometric properties of the two component streams are known.

Figure 37.1 *Mixing of Two Air Streams*

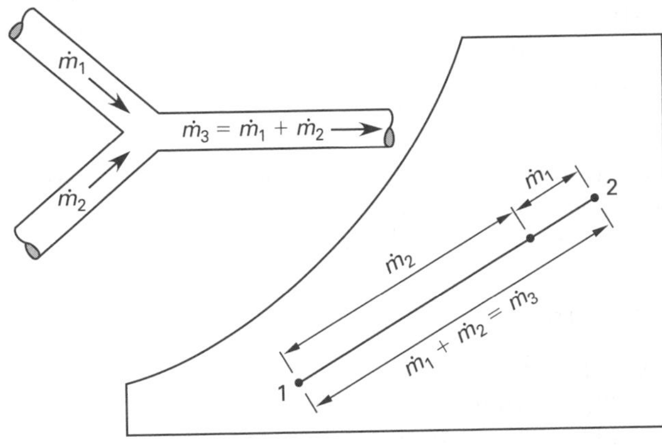

The two input states are located on the psychrometric chart and a straight line is drawn between them. The state of the mixture air will be on the straight line. The lever rule based on air masses is used to locate the mixture point. (Since the water vapor adds little to the mixture mass, the ratio of moist air masses can be approximated by the ratio of dry air masses, which, in turn, can be approximated by the ratio of air flow volumes.)

The lever rule can be used to find the mixture properties algebraically. Density changes can generally be disregarded, allowing volumetric flow rates to be used in place of mass flow rates. For the dry-bulb temperature (or any other property with a linear scale), the mixture temperature is

$$T_{\text{mixture}} = T_1 + \left(\frac{\dot{m}_2}{\dot{m}_1 + \dot{m}_2}\right)(T_2 - T_1) \approx T_1 + \left(\frac{Q_2}{Q_1 + Q_2}\right)(T_2 - T_1) \quad 37.37$$

Adiabatic mixing of two moist airstreams can also be expressed in terms of enthalpies and humidity ratios.

Adiabatic Mixing of Two Moist Airstreams

$$\frac{h_2 - h_3}{h_3 - h_1} = \frac{W_2 - W_3}{W_3 - W_1} = \frac{\dot{m}_{a,1}}{\dot{m}_{a,2}} \quad 37.38$$

Example 37.4

5000 ft^3/min (2.36 m^3/s) of air at 40°F (4°C) dry-bulb temperature and 35°F (2°C) wet-bulb temperature are mixed with 15,000 ft^3/min (7.08 m^3/s) of air at 75°F (24°C) dry-bulb and 50% relative humidity. Find the mixture dry-bulb temperature.

SI Solution

An approximate mixture temperature can be found by disregarding the change in density and taking a volumetrically weighted average. (The psychrometric chart can be used to determine a more precise value mixture temperature. The more precise approach is used in the customary U.S. solution.)

$$T_{\text{mixture}} \approx \frac{Q_1T_1 + Q_2T_2}{Q_1 + Q_2} = \frac{\left(2.36\ \frac{\text{m}^3}{\text{s}}\right)(4°\text{C}) + \left(7.08\ \frac{\text{m}^3}{\text{s}}\right)(24°\text{C})}{2.36\ \frac{\text{m}^3}{\text{s}} + 7.08\ \frac{\text{m}^3}{\text{s}}} = 19°\text{C}$$

Customary U.S. Solution

Locate the two points on the psychrometric chart, and draw a line between them. Estimate the specific volumes.

$$v_1 = 12.65\ \text{ft}^3/\text{lbm}$$
$$v_2 = 13.68\ \text{ft}^3/\text{lbm}$$

HVAC

Calculate the dry air masses.

$$\dot{m}_1 = \frac{Q_1}{v_1} = \frac{5000\ \frac{\text{ft}^3}{\text{min}}}{12.65\ \frac{\text{ft}^3}{\text{lbm}}}$$
$$= 395\ \text{lbm/min}$$

$$\dot{m}_2 = \frac{Q_2}{v_2} = \frac{15{,}000\ \frac{\text{ft}^3}{\text{min}}}{13.68\ \frac{\text{ft}^3}{\text{lbm}}}$$
$$= 1096\ \text{lbm/min}$$

Use Eq. 37.37.

$$T_{\text{mixture}} = T_1 + \left(\frac{\dot{m}_2}{\dot{m}_1 + \dot{m}_2}\right)(T_2 - T_1)$$
$$= 40°\text{F} + \left(\frac{1096\ \frac{\text{lbm}}{\text{min}}}{1096\ \frac{\text{lbm}}{\text{min}} + 395\ \frac{\text{lbm}}{\text{min}}}\right) \times (75°\text{F} - 40°\text{F})$$
$$= 65.7°\text{F}$$

11. AIR CONDITIONING PROCESSES

The psychrometric chart is particularly useful in analyzing air conditioning processes because the paths of many processes are straight lines. Sensible heating and cooling processes, for example, follow horizontal straight lines. Adiabatic saturation processes follow lines of constant enthalpy (essentially parallel to lines of constant wet-bulb temperature). The paths of pure humidification and dehumidification follow vertical paths. Figure 37.2 summarizes the directions of these paths.

Figure 37.2 *Common Psychrometric Processes*

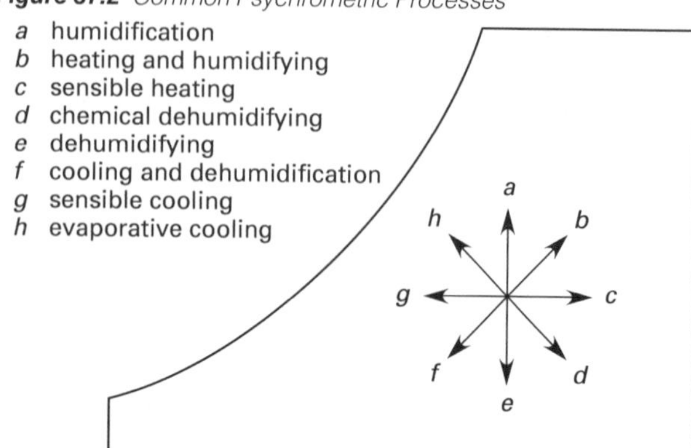

12. SENSIBLE HEAT RATIO

In general, the slope of any process line on the psychrometric chart is determined from the *sensible heat ratio*, also known as the *sensible heat factor*, SHF, and *sensible-total ratio*, S/T, scale on the chart. In an air conditioning process, the sensible heat ratio, SHR, is the ratio of sensible heat added (or removed) to total heat added (or removed). (The use of such scales varies from chart to chart. The process slope is determined from the sensible heat factor protractor and then translated (i.e., moved) to the appropriate point on the chart.)

$$\text{SHR} = \frac{q_s}{q_t} = \frac{q_s}{q_s + q_l} \qquad 37.39$$

The sensible heat ratio is always the slope of the line representing the change from the beginning point to the ending point on the psychrometric chart. Different designations are given to the sensible heat ratio, however, depending on where the changes occur.

If the sensible and latent energies change as the air passes through an occupied room, the term *room sensible heat ratio*, RSHR, is used. If the changes occur as the air passes through an air conditioning coil (apparatus), the term *coil* (or *apparatus*) *sensible heat ratio* is used, CSHR. Since the air conditioning apparatus usually removes heat and moisture from both the conditioned room and from outside makeup air, the term *grand sensible heat ratio*, GSHR, can be used in place of the coil sensible heat ratio. The *effective sensible heat ratio*, ESHR, is the slope of the line between the apparatus dew point on the saturation line and the design conditions of the conditioned space.

The sensible heat ratio is a psychrometric slope; it is not a geometric slope.

Example 37.5

During the summer, air from a conditioner enters an occupied space at 55°F (13°C) dry-bulb and 30% relative humidity. The ratio of sensible to total loads in the space is 0.45:1. The humidity ratio of the air leaving the room is 60 gr/lbm (8.6 g/kg). What is the dry-bulb temperature of the leaving air?

SI Solution

The sensible heat ratio is 0.45. Use the psychrometric chart (see App. 37.B) to determine the slope corresponding to this ratio. Draw a temporary line from the center of the protractor to the 0.45 mark on the sensible heat factor (inside) scale.

Locate 13°C dry-bulb and 30% relative humidity on the psychrometric chart. Draw a line through this point parallel to the temporary line, which is drawn with a slope of 0.45. The intersection of this line and the horizontal line corresponding to 8.6 g/kg determines the condition of the leaving air. The dry-bulb temperature is approximately 25.2°C.

HVAC

Customary U.S. Solution

The sensible heat ratio is 0.45. Use the psychrometric chart (see App. 37.A) to determine the slope corresponding to this ratio. Draw a temporary line from the center of the protractor to 0.45 on the sensible heat factor (inside) scale.

Locate 55°F dry-bulb and 30% relative humidity on the psychrometric chart. Draw a line through this point parallel to the temporary line, which is drawn with a slope of 0.45. The intersection of this line and the horizontal line corresponding to 60 gr/lbm determines the condition of the leaving air. The dry-bulb temperature is approximately 76°F. **[ASHRAE Psychrometric Chart No. 1—Sea Level]**

13. STRAIGHT HUMIDIFICATION

Straight (pure) *humidification* increases the water content of the air without changing the dry-bulb temperature. This is represented by a vertical condition line on the psychrometric chart. The *humidification load* is the mass of water added to the air per unit time (usually per hour).

14. BYPASS FACTOR AND COIL EFFICIENCY

Conditioning of air is accomplished by passing it through cooling or heating coils. Ideally, all of the air will come into contact with the coil for a long enough time and will leave at the coil temperature. In reality, this does not occur, and the air does not reach the coil temperature. The *bypass factor* can be thought of as the percentage of the air that is not cooled (or heated) by the coil. Under this interpretation, the remaining air (which is cooled or heated by the coil) is assumed to reach the coil temperature. The bypass factor expressed in decimal form is

$$\text{BF} = \frac{T_{\text{db,out}} - T_{\text{coil}}}{T_{\text{db,in}} - T_{\text{coil}}} \tag{37.40}$$

Bypass factors depend largely on the type of coil used. Bypass factors for large commercial units (such as those used in department stores) are small—around 10%. For small residential units, they are approximately 35%.

The *coil efficiency* is the complement of the bypass factor.

$$\eta_{\text{coil}} = 1.0 - \text{BF} \tag{37.41}$$

15. SENSIBLE COOLING AND HEATING

There is no change in the dew point or moisture content of the air with *sensible heating* and *cooling*. Since the moisture content is constant, these processes are represented by horizontal condition lines on the psychrometric chart (moving right for heating and left for cooling). (See Fig. 37.3.)

The energy change during the process can be calculated from enthalpies read directly from the psychrometric chart or approximated from the dry-bulb temperatures. In Eq. 37.42, $c_{p,\text{air}}$ is usually taken as 0.240 Btu/lbm-°F (1.005 kJ/kg·°C), and $c_{p,\text{moisture}}$ is taken as approximately 0.444 Btu/lbm-°F (1.805 kJ/kg·°C).

$$\begin{aligned} q &= m_a(h_2 - h_1) \\ &= m_a(c_{p,\text{air}} + \omega c_{p,\text{moisture}})(T_2 - T_1) \end{aligned} \tag{37.42}$$

Figure 37.3 *Sensible Cooling*

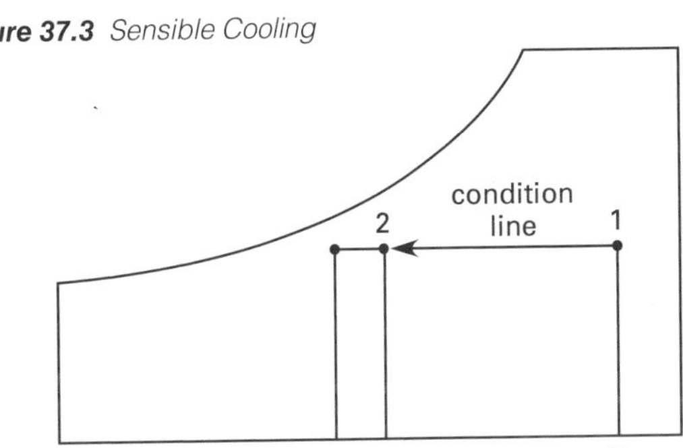

16. COOLING WITH COIL DEHUMIDIFICATION

If the cooling coil's temperature is below the air's dew point (as is usually the case), moisture will condense on the coil. The *effective coil temperature* in this instance is referred to as the *apparatus dew point*, ADP (also known as the *coil apparatus dew point*), and is determined from the intersection of the condition line (i.e., *coil load line*) and the curved saturation line on the psychrometric chart. The apparatus dew point is the temperature to which the air would be cooled if 100% of it contacted the coil. (The term "apparatus dew point" is generally only used with cooling-dehumidification processes.)

The mass of condensing water will be

Moist-Air Cooling and Dehumidification

$$\dot{m}_w = \dot{m}_a(W_1 - W_2) \tag{37.43}$$

The total energy removed from the air includes both sensible and latent components. The latent heat is calculated from the heat of vaporization evaluated at the pressure of the water vapor.

Moist-Air Cooling and Dehumidification

$$q_t = q_s + q_l = \dot{m}_a\left((h_1 - h_2) - (W_1 - W_2)h_w\right) \tag{37.44}$$

$$\dot{q}_l = \dot{m}_a(W_1 - W_2)h_{fg} \tag{37.45}$$

Referring to Fig. 37.4, it is convenient to think of air experiencing sensible cooling from point 1 to point 3, after which the air follows the saturation line down from point 3 to point 4 (the apparatus dew point). Water condenses out between points 3 and 4. For convenience, the condition line is drawn as a straight line between points 1 and 4. The slope of the ADP-2-1 line corresponds to the sensible heat ratio. (See Sec. 37.12.) Since some of the air does not contact the coil at all, the final condition of the air will actually be at point 2 on the condition line.

Figure 37.4 *Cooling and Dehumidification*

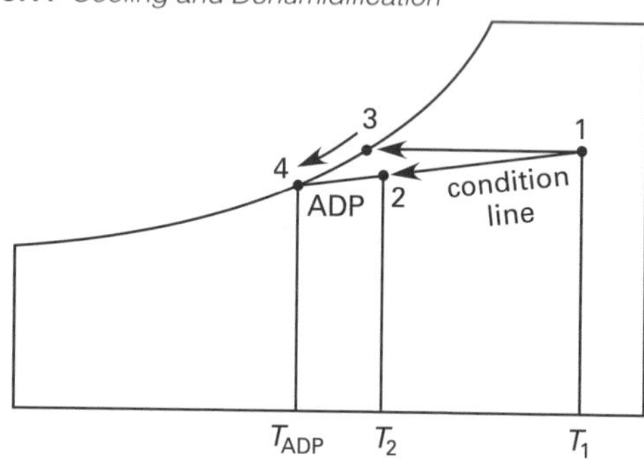

In practice, point 1 is usually known and either point 2 or point 4 are unknown. If point 2 is known, point 4 (the apparatus dew point) can be found graphically by extending the condition line over to the saturation line. (In some cases, the sensible heat ratio must be used to locate the apparatus dew point.) If point 4 is known, point 2 can be found from the bypass factor. The *contact factor*, CF, is essentially a dehumidification efficiency, calculated as the complement of the bypass factor.

$$\text{CF} = 1 - \text{BF} = 1 - \frac{T_{2,\text{db}} - \text{ADP}}{T_{1,\text{db}} - \text{ADP}} \qquad 37.46$$

Water condenses out over the entire temperature range from point 3 to point 4. The temperature of the water being removed is assumed to be the dew-point temperature at point 2.

Example 37.6

A coil has a bypass factor of 20% and an apparatus dew point of 55°F (13°C). Air enters the coil at 85°F (29°C) dry-bulb and 69°F (21°C) wet-bulb. What are the (a) latent heat loss, (b) sensible heat loss, and (c) sensible heat ratio?

SI Solution

(a) Locate the point corresponding to the entering air on the psychrometric chart. [**ASHRAE Psychrometric Chart No. 1—Sea Level**]

The enthalpy and humidity ratio are approximately

$$h_1 = 60.4 \text{ kJ/kg}$$
$$W_1 = 0.0123 \text{ kg/kg}$$

Use Eq. 37.46 to calculate the dry-bulb temperature of the air leaving the coil.

$$\begin{aligned} T_{2,\text{db}} &= \text{ADP} + \text{BF}(T_{1,\text{db}} - \text{ADP}) \\ &= 13°\text{C} + (0.20)(29°\text{C} - 13°\text{C}) \\ &= 16.2°\text{C} \end{aligned}$$

Draw a condition line between the entering air and the apparatus dew point on the psychrometric chart. Locate the point corresponding to 16.2°C dry-bulb on the condition line. The leaving enthalpy and humidity ratio are approximately

$$h_2 = 40.8 \text{ kJ/kg}$$
$$W_2 = 0.0100 \text{ kg/kg}$$

The total energy loss per kilogram is

$$\begin{aligned} q_t = h_1 - h_2 &= 60.4 \ \frac{\text{kJ}}{\text{kg}} - 40.8 \ \frac{\text{kJ}}{\text{kg}} \\ &= 19.6 \text{ kJ/kg of dry air} \end{aligned}$$

Since the partial pressure of the water vapor is unknown, estimate $h_{fg} \approx 2501$ kJ/kg.

From Eq. 37.45, on a kilogram basis,

$$\begin{aligned} q_l &= (W_1 - W_2)h_{fg} \\ &= \left(0.0123 \ \frac{\text{kg}}{\text{kg}} - 0.0100 \ \frac{\text{kg}}{\text{kg}}\right)\left(2501 \ \frac{\text{kJ}}{\text{kg}}\right) \\ &= 5.75 \text{ kJ/kg of dry air} \end{aligned}$$

(b) From Eq. 37.44, the sensible heat loss is

Moist-Air Cooling and Dehumidification

$$q_t = q_s + q_l$$

$$\begin{aligned} q_s = q_t - q_l &= 19.6 \ \frac{\text{kJ}}{\text{kg}} - 5.75 \ \frac{\text{kJ}}{\text{kg}} \\ &= 13.9 \text{ kJ/kg of dry air} \end{aligned}$$

(c) From Eq. 37.39, the sensible heat ratio is

$$\begin{aligned} \text{SHR} = \frac{q_s}{q_t} &= \frac{13.9 \ \frac{\text{kJ}}{\text{kg}}}{19.6 \ \frac{\text{kJ}}{\text{kg}}} \\ &= 0.71 \end{aligned}$$

Customary U.S. Solution

(a) Locate the point corresponding to the entering air on the psychrometric chart. [**ASHRAE Psychrometric Chart No. 1—Sea Level**]

The enthalpy and humidity ratio are approximately

$$h_1 = 33.1 \text{ Btu/lbm}$$
$$W_1 = 0.0116 \text{ lbm/lbm}$$

Use Eq. 37.46 to calculate the dry-bulb temperature of the air leaving the coil.

$$\begin{aligned} T_{2,\text{db}} &= \text{ADP} + \text{BF}(T_{1,\text{db}} - \text{ADP}) \\ &= 55°\text{F} + (0.20)(85°\text{F} - 55°\text{F}) \\ &= 61°\text{F} \end{aligned}$$

Draw a condition line between the entering air and apparatus dew point on the psychrometric chart. Locate the point corresponding to 61°F dry-bulb on the condition line. The leaving enthalpy and humidity ratio are approximately

$$h_2 = 25.1 \text{ Btu/lbm}$$
$$W_2 = 0.0097 \text{ lbm/lbm}$$

The total energy loss per pound is

$$\begin{aligned} q_t = h_1 - h_2 &= 33.1 \ \frac{\text{Btu}}{\text{lbm}} - 25.1 \ \frac{\text{Btu}}{\text{lbm}} \\ &= 8.0 \text{ Btu/lbm of dry air} \end{aligned}$$

Since the partial pressure of the water vapor is not known, estimate $h_{fg} \approx 1060$ Btu/lbm.

From Eq. 37.45, on a pound basis,

$$\begin{aligned} q_l &= (W_1 - W_2)h_{fg} \\ &= \left(0.0116 \ \frac{\text{lbm}}{\text{lbm}} - 0.0097 \ \frac{\text{lbm}}{\text{lbm}}\right)\left(1060 \ \frac{\text{Btu}}{\text{lbm}}\right) \\ &= 2.01 \text{ Btu/lbm of dry air} \end{aligned}$$

(b) From Eq. 37.44, the sensible heat loss is

Moist-Air Cooling and Dehumidification

$$q_t = q_s + q_l$$

$$\begin{aligned} q_s = q_t - q_l &= 8.0 \ \frac{\text{Btu}}{\text{lbm}} - 2.0 \ \frac{\text{Btu}}{\text{lbm}} \\ &= 6.0 \text{ Btu/lbm of dry air} \end{aligned}$$

(c) From Eq. 37.39, the sensible heat ratio is

$$\begin{aligned} \text{SHR} &= \frac{q_s}{q_t} = \frac{6.0 \ \frac{\text{Btu}}{\text{lbm}}}{8.0 \ \frac{\text{Btu}}{\text{lbm}}} \\ &= 0.75 \end{aligned}$$

17. ADIABATIC SATURATION PROCESSES

To measure the wet-bulb temperature, air must experience an *adiabatic saturation process*, also known as *evaporative cooling*. Adiabatic saturation processes occur in cooling towers, air washers, and evaporative coolers ("swamp coolers"). To become saturated, the air must pick up the maximum amount of moisture it can hold at that temperature. This moisture comes from the vaporization of liquid water. For the process to be adiabatic, there can be no external source of energy to vaporize the liquid water needed to saturate the air.

At first analysis, the terms adiabatic and saturation seem contradictory. Adiabatic saturation is possible, however, if the latent heat of vaporization comes from the air itself. If the air gives up sensible heat, that energy can be used to vaporize liquid water. Of course, the air temperature decreases when sensible heat is given up. That is the reason that the wet-bulb temperature is generally less than the dry-bulb temperature. Only when the air is saturated will the two temperatures be equal.

An adiabatic saturation process can be produced with a *sling psychrometer*, which is essentially a regular thermometer with its bulb wrapped in wet cotton or gauze. Rapidly twirling the thermometer through the air at the end of a cord will cause the water in the gauze to evaporate. The latent heat needed to vaporize the water will come from the sensible heat of the air, and the thermometer will measure the wet-bulb temperature.

Since the increase in the water vapor's latent heat content equals the decrease in the air's sensible heat, the total enthalpies before and after adiabatic saturation are the same. Therefore, an adiabatic saturation process follows a line of constant enthalpy on the psychrometric chart. These lines are, for approximation purposes, parallel to lines of constant wet-bulb temperature.

The bypass factor concept is not used with adiabatic saturation processes. Instead, the *saturation efficiency* (*humidification efficiency*) is used. The saturation efficiency of large commercial air washers is typically 90% to 95%. The wet bulb temperature does not change during the saturation process.

$$\eta_{\text{sat}} = \frac{T_{\text{db,air,in}} - T_{\text{db,air,out}}}{T_{\text{db,air,in}} - T_w} \qquad 37.47$$

HVAC

18. AIR WASHERS

An *air washer* is a device that passes air through a dense spray of recirculating water. The water is used to change the properties of the air. Air washers are used in air purifying and cleaning processes (i.e., removal of solids, liquids, gases, vapors, and odors), as well as for evaporative cooling and dew-point control.[5]

The difference between a spray humidifier and spray dehumidifier is the temperature of the spray water. In an *adiabatic air washer*, the spray water is recirculated without being heated or cooled. After equilibrium is reached, the water temperature will be equal to the air's entering wet-bulb temperature. The air will be cooled and humidified, leaving partially or completely saturated at its entering wet-bulb temperature. However, if the spray water is chilled, the air will be cooled and dehumidified. And, if the spray water is heated, the air will be humidified and (possibly) heated.

An air washer's *saturation efficiency*, typically 90% to 95%, is measured by the drop in dry-bulb temperature relative to the entering wet-bulb depression.

$$\eta_{\text{sat}} = \frac{T_{\text{in,db}} - T_{\text{out,db}}}{T_{\text{in,db}} - T_{\text{in,wb}}} \quad \text{37.48}$$

Air velocity through washers is approximately 500 ft/min (2.6 m/s). Velocities outside the range of 300 ft/min to 750 ft/min (1.5 m/s to 3.8 m/s) are probably faulty. The water pressure is typically 20 psig to 40 psig (140 kPa to 280 kPa). The spray quantity per bank of nozzles is in the range of 1.5 gal/min to 5 gal/min per 1000 ft^3 (3.3 L/s to 11 L/s per 1000 m^3) of air. Screens, louvers, and mist eliminator plates will generate a static pressure drop of approximately 0.2 in wg to 0.5 in wg (50 kPa to 125 kPa) at 500 ft/min (2.6 m/s). Other operating parameters used to describe air washer performance include air mass flow rate per unit area (lbm/hr-ft^2 or kg/m^2·s), air and liquid heat transfer coefficients per volume of chamber (Btu/hr-°F-ft^3 or kW/°C·m^3), and the *spray ratio* (the mass of water sprayed to the mass of air passing through the washer per unit time).

19. COOLING WITH HUMIDIFICATION

When air passes through a water spray (as in an *air washer*), an *adiabatic saturation process* known as *evaporative cooling* occurs.[6] (See Sec. 37.17.) The air leaves with a lower temperature and a higher moisture content. This is graphically represented on the psychrometric chart by a condition line parallel to the lines of constant enthalpy (essentially constant wet-bulb temperature), and is mathematically represented using the ratio of the change in enthalpy to the change in the humidity ratio.

Adiabatic Mixing of Water Injected Into Moist Air (Evaporative Cooling)

$$\frac{h_2 - h_1}{W_2 - W_1} = \frac{\Delta h}{\Delta W} = h_w \quad \text{37.49}$$

Adiabatic saturation is a constant-enthalpy process, since any evaporation of the water requires heat to be drawn from the air. Since the removed heat goes into the remaining water, the water temperature increases. When the water spray is continuously recirculated, the water temperature gradually increases to the wet-bulb temperature of the incoming air. The minimum leaving air temperature will be the water temperature (i.e., the wet-bulb temperature of the incoming air).

The direct saturation efficiency, η_e, is the extent to which the temperature of the air leaving an evaporative cooler approaches the wet-bulb temperature of the air entering the cooler.

Direct Evaporative Air Coolers

$$\epsilon_e = 100\frac{T_1 - T_2}{T_1 - T_{\text{wb,in}}} \quad \text{37.50}$$

During steady-state operation, the temperature of the water spray will normally be stable at the air's wet-bulb temperature. However, the water temperature can also be artificially maintained by refrigeration at less than the wet-bulb temperature (but more than the dew-point temperature). Line 1–3 in Fig. 37.5 illustrates such a process.

Figure 37.5 Cooling with Humidification (adiabatic saturation)

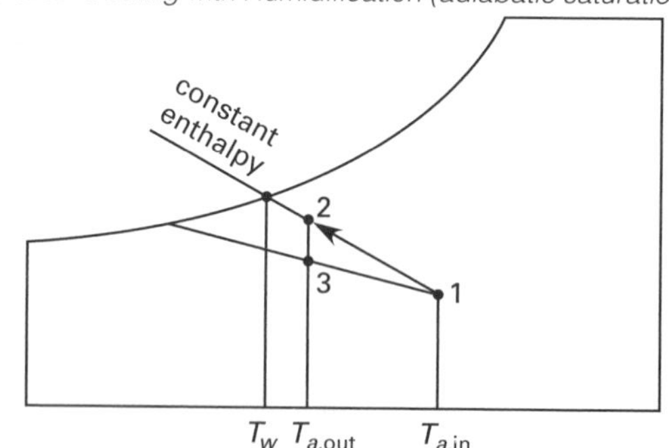

To prevent ice buildup, the cooled air temperature should be kept from dropping below the freezing point of water. The entering wet-bulb temperature should be kept above 35°F (1.7°C).

[5]Air washers are generally not used for removing carbonaceous or greasy particles.

[6]An *air washer* is basically a *spray chamber* through which air passes. When supplied with chilled water from a refrigeration source, the air washer can cool, dehumidify, or humidify the air. Air washers can be used without refrigeration to cool and humidify the air through an evaporative cooling process.

Example 37.7

Air at 90°F (32°C) dry-bulb and 65°F (18°C) wet-bulb enters an evaporative cooler. The air leaves at 90% relative humidity. The continuously recirculated spray water is stable at 65°F (18°C). What is the dry-bulb temperature of the leaving air?

Solution

Since the spray water is the same temperature as the wet-bulb temperature of the entering air, the cooler has reached its steady-state operating conditions. Locate the entering point on the psychrometric chart and draw a line of constant enthalpy (or constant 65°F (18°C) wet-bulb temperature) up to the 90% relative humidity curve. Read the dry-bulb temperature as approximately 67°F (19°C). **[ASHRAE Psychrometric Chart No. 1—Sea Level]**

20. COOLING WITH SPRAY DEHUMIDIFICATION

If air passes through a water spray whose temperature is less than the entering air's wet-bulb temperature, both the dry-bulb and wet-bulb temperatures will decrease.[7] If the leaving water temperature is below the entering air's dew point, dehumidification will occur. As with any evaporative cooling, the air will give up thermal energy to the water. The final water temperature will depend on the thermal energy pickup and water flow rate. All air temperatures decrease, and some moisture condenses. The *performance factor* is defined as

$$F_p = 1 - \frac{T_{\text{air,wb,out}} - T_{w,\text{out}}}{T_{\text{air,wb,in}} - T_{w,\text{in}}} \qquad 37.51$$

The performance factor in terms of enthalpy is defined as

Evaporative Dehumidifiers

$$F_p = \frac{h_{\text{air,wb,in}} - h_{\text{air,wb,out act}}}{h_{\text{air,wb,in}} - h_{\text{air,wb,out dh}}} \qquad 37.52$$

21. HEATING WITH HUMIDIFICATION

If air is humidified by injecting steam (*steam humidification*) or by passing the air through a hot water spray, the dry-bulb temperature and enthalpy of the air will increase.[8] The final air enthalpy and/or the required steam enthalpy can be determined from a conservation of energy equation. In Eq. 37.53, the mass of the air used is the dry air mass, which does not change. h_a, though expressed per pound of dry air, includes the energy of all vaporized water.

Adiabatic Mixing of Water Injected Into Moist Air (Evaporative Cooling)

$$\dot{m}_a h_{a,1} + \dot{m}_w h_w = \dot{m}_a h_{a,2} \qquad 37.53$$

This equation can be rearranged into a commonly used form.

$$\dot{m}_w h_w = \dot{m}_a (h_2 - h_1) \qquad 37.54$$

From a conservation of mass for the water,

Adiabatic Mixing of Water Injected Into Moist Air (Evaporative Cooling)

$$\dot{m}_a W_1 + \dot{m}_w = \dot{m}_a W_2 \qquad 37.55$$

Combining Eq. 37.54 and Eq. 37.55 gives

Adiabatic Mixing of Water Injected Into Moist Air (Evaporative Cooling)

$$\frac{h_2 - h_1}{W_2 - W_1} = \frac{\Delta h}{\Delta W} = h_w \qquad 37.56$$

The humidification load, H, can be determined by Eq. 37.57.

Humidifiers

$$H = \rho V R (W_1 - W_2) - S + L \qquad 37.57$$

For a fixed outdoor air supply,

Humidifiers

$$H = 60 \rho Q_{\text{outdoor}} (W_{\text{indoor}} - W_{\text{outdoor}}) - S + L \qquad 37.58$$

For a variable outdoor air supply,

Humidifiers

$$H = 60 \rho Q_{\text{total}} (W_{\text{indoor}} - W_{\text{outdoor}}) \times \left(\frac{T_{\text{indoor}} - T_{\text{mix}}}{T_{\text{indoor}} - T_{\text{outdoor}}} \right) - S + L \qquad 37.59$$

Figure 37.6 illustrates that the condition line will be above the line of constant enthalpy that radiates from the point corresponding to the incoming air. However, even though heat is added to the water, the air temperature can either decrease (as in the 1–2 process shown) or increase (as in the 1–2′ process shown).

HVAC

[7]This can unintentionally occur during the start-up of an air washer used for humidification, or the water can be kept intentionally chilled.

[8]When a spray of hot water is used, the water must be continually heated. Unlike a cold water spray, a natural equilibrium water temperature is not achieved.

Figure 37.6 *Heating with Steam Humidification*

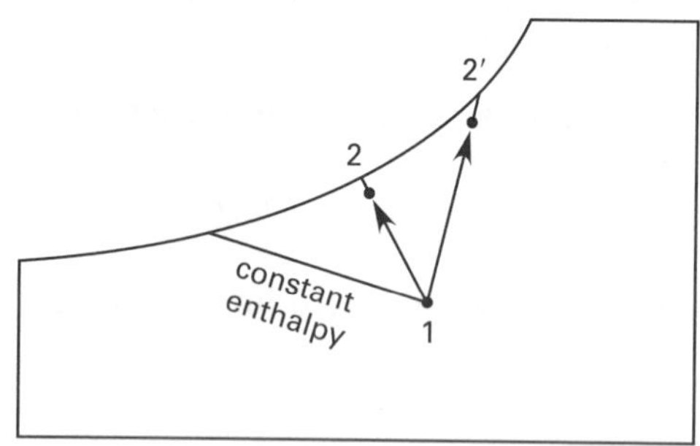

22. HEATING AND DEHUMIDIFICATION

Air passing through a solid or liquid *adsorbent bed*, such as silica gel or activated alumina, will decrease in humidity at constant wet-bulb temperature. This is sometimes referred to as *chemical dehumidification*, *chemical dehydration*, or *"absorbent" dehumidification*.[9] If only latent heat was involved, this process would be the reverse of an adiabatic saturation process. However, as moisture is removed, exothermic chemical energy is generated in addition to the heat of vaporization liberated. Since thermal energy is generated, this is not an adiabatic process. (See Fig. 37.7.) [**Desiccant Dehumidification**]

Figure 37.7 *Heating and Dehumidification*

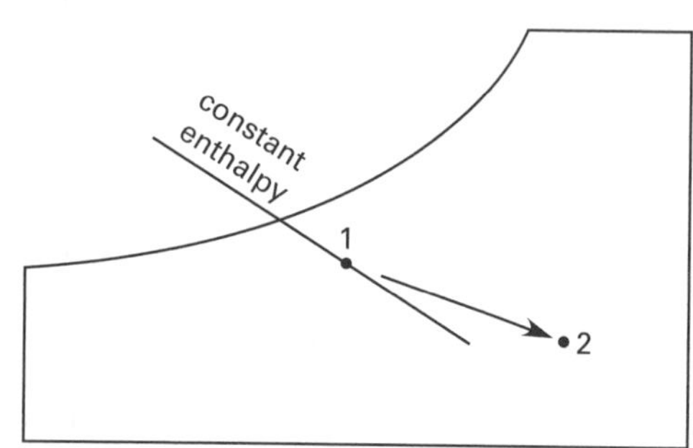

23. ENERGY-RECOVERY VENTILATORS (ERV)/HEAT-RECOVERY VENTILATORS (HRV)

Heat- and energy-recovery ventilator devices reduce overall heating and cooling equipment sizing and associated operating costs by transferring the sensible (HRV) or the latent and sensible (ERV) heat from the residual exhaust air to the supply air as shown in Fig. 37.8.

Figure 37.8 *Energy-Recovery Ventilator/Heat-Recovery Ventilator*

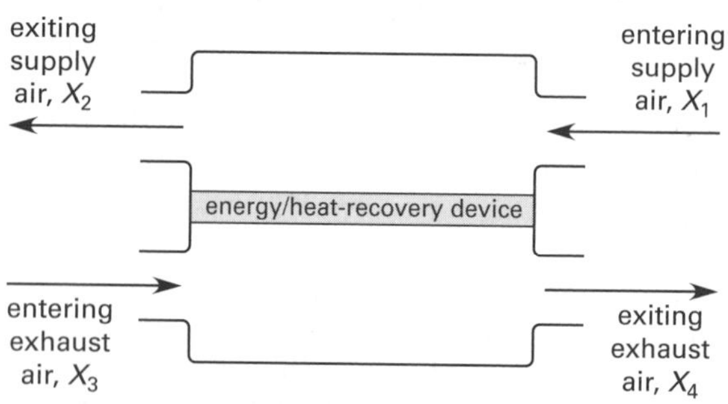

Energy-Recovery Ventilators (ERV)

The latent effectiveness, ϵ_l, of an energy-recovery ventilator (ERV) is

Energy-Recovery Ventilator (ERV)

$$\begin{aligned}\epsilon_l &= \frac{q_l}{q_{l,\max}} \\ &= \frac{\dot{m}_{\text{supply}} h_{fg}(W_1 - W_2)}{\dot{m}_{\min} h_{fg}(W_1 - W_3)} \\ &= \frac{\dot{m}_{\text{exhaust}} h_{fg}(W_4 - W_3)}{\dot{m}_{\min} h_{fg}(W_1 - W_3)}\end{aligned} \quad \text{37.60}$$

The moisture effectiveness, ϵ_m, and total effectiveness, ϵ_t, of an ERV is

Energy-Recovery Ventilator (ERV)

$$\begin{aligned}\epsilon_m &= \frac{\dot{m}_w}{\dot{m}_{w,\max}} \\ &= \frac{\dot{m}_{\text{supply}}(W_1 - W_2)}{\dot{m}_{\min}(W_1 - W_3)} \\ &= \frac{\dot{m}_{\text{exhaust}}(W_4 - W_3)}{\dot{m}_{\min}(W_1 - W_3)}\end{aligned} \quad \text{37.61}$$

$$\begin{aligned}\epsilon_t &= \frac{q_t}{q_{t,\max}} \\ &= \frac{\dot{m}_{\text{supply}}(h_2 - h_1)}{\dot{m}_{\min}(h_3 - h_1)} \\ &= \frac{\dot{m}_{\text{exhaust}}(h_3 - h_4)}{\dot{m}_{\min}(h_3 - h_1)}\end{aligned} \quad \text{37.62}$$

[9]The correct term for a substance that collects water on its surface is an *adsorbent*. By virtue of their great porosities, adsorbent particles have large surface areas. The attractive forces on the surfaces of these solids cause a thin layer of condensed water to form. Adsorbents are reactivated by heating.

The total energy transfer between the exhaust and supply streams, q_t, is

Energy-Recovery Ventilator (ERV)

$$\begin{aligned} q_t &= q_s + q_l \\ &= 60\dot{m}_{\text{supply}}(h_{1,\text{supply}} - h_{2,\text{supply}}) \\ &= 60Q_{\text{supply}}\rho_{\text{supply}}(h_{1,\text{supply}} - h_{2,\text{supply}}) \\ &= 60\left[\begin{array}{l} \dot{m}_{\text{supply}}c_{p,\text{supply}}(T_1 - T_2) \\ \quad + \dot{m}_{\text{supply}}h_{fg}(W_1 - W_2) \end{array}\right] \end{aligned} \qquad \text{37.63}$$

$$\begin{aligned} q_t &= q_s + q_l \\ &= 60\dot{m}_{\text{exhaust}}(h_{4,\text{exhaust}} - h_{3,\text{exhaust}}) \\ &= 60Q_{\text{exhaust}}\rho_{\text{exhaust}}(h_{4,\text{exhaust}} - h_{3,\text{exhaust}}) \\ &= 60\left[\begin{array}{l} \dot{m}_{\text{exhaust}}c_{p,\text{exhaust}}(T_4 - T_3) \\ \quad + \dot{m}_{\text{exhaust}}h_{fg}(W_4 - W_3) \end{array}\right] \end{aligned} \qquad \text{37.64}$$

$$q_t = 60\epsilon_t\dot{m}_{\text{min}}(h_{1,\text{supply}} - h_{3,\text{exhaust}}) \qquad \text{37.65}$$

The supply and exhaust air fan power can be estimated using Eq. 37.66 and Eq. 37.67.

Energy-Recovery Ventilator (ERV)

$$P_{\text{supply}} = \left(\frac{Q_{\text{supply}}\Delta p_{\text{supply}}}{6356}\right)\eta_{\text{fan}} \qquad \text{37.66}$$

$$P_{\text{exhaust}} = \left(\frac{Q_{\text{exhaust}}\Delta p_{\text{exhaust}}}{6356}\right)\eta_{\text{fan}} \qquad \text{37.67}$$

Heat-Recovery Ventilators (HRV)

The sensible effectiveness, ϵ_s, of a heat-recovery ventilator (HRV) is

Heat-Recovery Ventilator (HRV)—Sensible Energy Recovery

$$\begin{aligned} \epsilon_s &= \frac{q_s}{q_{s,\text{max}}} \\ &= \frac{\dot{m}_{\text{supply}}c_{p,\text{supply}}(T_2 - T_1)}{C_{\text{min}}(T_3 - T_1)} \\ &= \frac{\dot{m}_{\text{exhaust}}c_{p,\text{exhaust}}(T_3 - T_4)}{C_{\text{min}}(T_3 - T_1)} \end{aligned} \qquad \text{37.68}$$

24. NOMENCLATURE

ADP	apparatus dew point	°F	°C
B	volumetric fraction	–	–
BF	bypass factor	–	–
C	cycles of concentration	ppm	mg/L
CF	contact factor	–	–
c_p	specific heat	Btu/lbm-°F	kJ/kg·°C
F_p	performance factor	–	–
G	gravimetric fraction	–	–
h	enthalpy	Btu/lbm	kJ/kg
H	humidification load	lbm of water/hr	-
m	mass	lbm	kg
n	number of moles	–	–
p	pressure	lbf/ft^2	kPa
P	power	Btu/sec	W
q	heat	Btu/lbm	J/kg
Q	volumetric flow rate	gal/min	L/s
R	specific gas constant	ft-lbf/lbm-°R	kJ/kg·K
RF	rating factor	–	–
SHR	sensible heat ratio	–	–
T	temperature	°F	°C
V	volume	ft^3	m^3
x	mole fraction	–	–
W	humidity ratio	lbm/lbm	kg/kg
Z	altitude	ft/ft	m/m

Symbols

ϵ	effectiveness	–	–
η	efficiency	–	–
μ	degree of saturation	–	–
ρ	mass density	lbm/ft^3	kg/m^3
v	specific volume	ft^3/lbm	m^3/kg
ϕ	relative humidity	–	–

Subscripts

a	dry air
db	dry-bulb
dp	dew-point
e	evaporative
fg	vaporization
l	latent
p	pressure
s	sensible
sat	saturation
t	total
unsat	unsaturated
v	vapor
w	water
wb	wet-bulb

38 Cooling Towers, Cooling Ponds, and Water-Cooled Condensers

Content in blue refers to the *NCEES Handbook*.

NCEES EXAM SPECIFICATIONS AND RELATED CONTENT

HVAC AND REFRIGERATION EXAM

II.B.1. Equipment and Components: Cooling towers and fluid coolers
- 2. Cooling Ponds
- 3. Cooling Tower Blowdown
- 5. Water-Cooled Condensers

THERMAL AND FLUID SYSTEMS EXAM

III.A.5. Energy/Power Equipment: Cooling towers
- 2. Cooling Ponds
- 3. Cooling Tower Blowdown
- 5. Water-Cooled Condensers

1. WET COOLING TOWERS

Conventional *wet cooling towers* cool warm water by exposing it to colder air.[1] They are usually used to provide cold water to power plant and large refrigeration condensers. The air is used to change the properties of the water, which leaves cooler. As it leaves, the saturated (or nearly saturated) warmed air takes sensible and latent heat from the water.

Cooling towers are generally counterflow, crossflow, or a combination. Though natural-draft and atmospheric towers exist, limited space usually requires that cooling towers operate with mechanical draft. Fans are located at the base of *forced draft towers* and blow air into the water cascading down. With *induced mechanical draft*, fans are located at the top of the tower, drawing air upward. Some portion of the exhaust air might reenter the cooling tower. This is known as *recycle air* (*recirculation air*). Recycle air decreases the efficiency of the tower.

During countercurrent operation, warm water is introduced at the top of the tower and is distributed by troughs or spray nozzles. The water passes over staggered slats or interior *fill* (also known as *packing*).[2] Air flows upward, contacting the water on its downward path. A portion of the water evaporates, cooling the remainder of the water. The water temperature cannot decrease below the wet-bulb temperature. The actual final water temperature depends on a number of factors, including the state of the incoming air, the heat load, and the design (and efficiency) of the cooling tower. (See Fig. 38.1.)

There are several environmental issues associated with wet cooling towers. Makeup water, though relatively little is needed, may be difficult to obtain. Moist plume discharges cause shadowing of adjacent areas and fogging and icing on nearby highways. Disposal of blowdown wastewater is also problematic.

Equation 38.1 is a per-unit energy balance that can be used to evaluate cooling tower performance. Each term, including the circulating water flow rate, is per unit mass (e.g., pound or kilogram) of dry air. Since the energy contribution of the makeup water is small, that term can be omitted for a first approximation.

$$\begin{aligned} m_{w,\text{in}}h_{w,\text{in}} + h_{a,\text{in}} + (W_{a,\text{out}} - W_{a,\text{in}})h_{\text{makeup}} &= m_{w,\text{out}}h_{w,\text{out}} + h_{a,\text{out}} \\ &= (m_{w,\text{in}} + W_{a,\text{in}} - W_{a,\text{out}})h_{w,\text{out}} + h_{a,\text{out}} \end{aligned} \quad \text{38.1}$$

If operation is at standard pressure, a psychrometric chart can be used to obtain the air enthalpies. For operation at different altitudes (i.e., different atmospheric pressures), the mathematical psychrometric relationships in Sec. 37.4 can be used to calculate the enthalpy. From Sec. 37.16, the humidity ratio is

$$W = \frac{0.622p_{\text{water vapor}}}{p_{\text{total}} - p_{\text{water vapor}}} \quad \text{38.2}$$

[1]Though larger in size, a cooling tower is similar in operation to an air washer. In fact, an air washer can be used to cool water. Since air washer operation is not countercurrent, however, larger air flows are required to obtain the same cooling effect.

[2]Modern *filled towers* use corrugated *cellular fill* to maximize the air-water contact area. Standard polyvinyl chloride (PVC) fill is useful up to about 125°F (52°C). From 125°F to 140°F (52°C to 60°C), chlorinated PVC fill is recommended. Polypropylene fill should be used above 140°F (60°C). "Fill-less" towers, where the sprayed water merely falls through oncoming air, are used in some industries (food, steel, and paper processes) where a high-product carryover can lead to coating or buildup on the fill material.

Figure 38.1 *Counterflow Wet Cooling Tower*

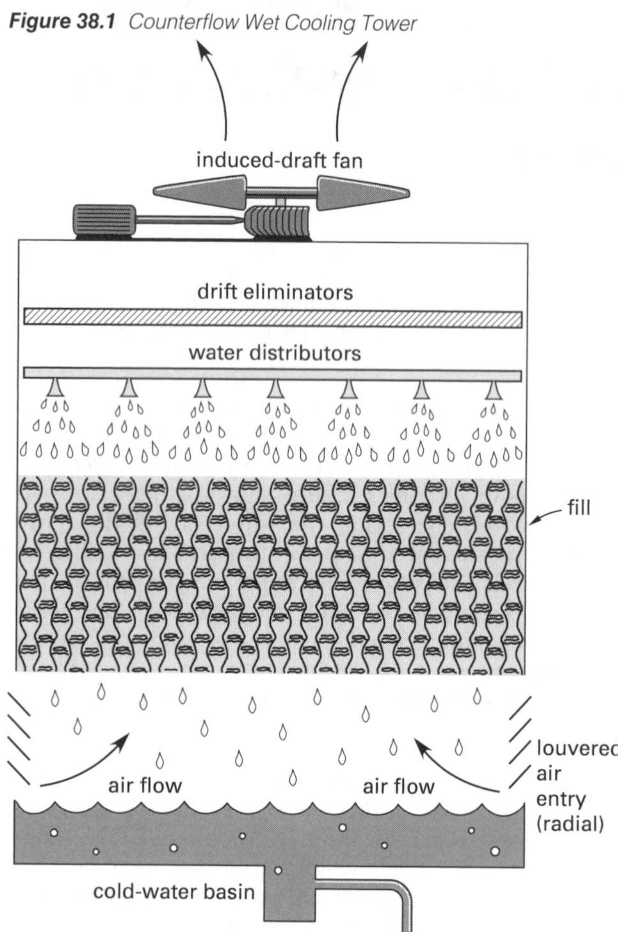

When a cooling tower is used to provide cold water for the condenser of a refrigeration system, the water circulation will be approximately 3 gal/min per ton (0.19 L/s) of refrigeration. Approximately 2 gal/min to 4 gal/min of water are distributed per square foot (1.4 L/s to 2.7 L/s per square meter) of tower, and the air velocity should be approximately 700 ft/min (3.6 m/s) through the net free area. Coolants for condensers in reciprocating refrigeration systems usually call for an 85°F to 90°F (29°C to 32°C) water temperature. (This corresponds to a condensing temperature of approximately 100°F to 110°F (38°C to 43°C).) Various valves, mixing, louvers, and dampers are used to maintain a constant output water temperature.

The lowest temperature to which water can be cooled by purely evaporative means is the wet-bulb temperature of the entering air. The *cooling efficiency*, η_w, is based on the water temperature. The *water range* (*cooling range* or *range*) is defined as the actual difference between the entering and leaving water temperatures. (For water-cooled refrigeration condensers, this is equal to the water's temperature increase in the condenser.) The *approach* is defined as the difference between the leaving water temperature and the entering air wet-bulb temperatures.[3] Cooling efficiency is typically 50% to 70%.[4] Natural draft towers can cool the water to within 10°F to 12°F (5.5°C to 6.7°C) of the wet-bulb temperature. Forced draft towers can cool the water to within 5°F to 6°F (2.8°C to 3.3°C).

$$\eta_w = \frac{\text{range}}{\text{approach} + \text{range}} = \frac{T_{w,\text{in}} - T_{w,\text{out}}}{T_{w,\text{in}} - T_{\text{air,wb,in}}} \qquad 38.3$$

As Eq. 38.3 indicates, the actual wet-bulb temperature of the cooling air is particularly important in determining cooling tower performance. The higher the wet-bulb temperature, the lower the efficiency. (This is because when the denominator in Eq. 38.3 decreases, the numerator decreases even more.) In rating their cooling towers, most manufacturers have adopted the practice of using wet-bulb temperatures that will be exceeded only 2.5% of the time or less.

Performance of a cooling tower also depends on the relative humidity of the air. High relative humidities decrease the water evaporation rate, decreasing the efficiency.

The *heat rejection rate* (*tower load* or *cooling duty*) is calculated from the range and the water mass flow rate.

$$q_o = \dot{m}_w c_p (T_{w,\text{in}} - T_{w,\text{out}}) = \dot{m}_w (h_{w,\text{in}} - h_{w,\text{out}}) \qquad 38.4$$

Cooling towers are sometimes rated in tower units, which are essentially proportional to the tower cost. The number of *tower units*, TU, is equal to a rating factor multiplied by the flow rate. *Rating factors* define the relative difficulty in cooling, essentially the relative amount of contact area or fill volume required. Manufacturers provide charts showing the relationship between rating factor, approach, range, and wet-bulb temperature.

$$\text{TU} = \text{RF} \times Q_{\text{gpm}} \qquad 38.5$$

2. COOLING PONDS

Cooling ponds are used to reject heat using a once-through cooling process in which thermal energy is dissipated through evaporation. The *evaporation rate*, w_p, for a cooling pond is

Cooling Ponds

$$w_p = \frac{A(95 + 0.425\text{v})}{h_{fg}} (p_w - p_a) \qquad \text{[U.S.]} \qquad 38.6$$

Example 38.1

A cooling pond with an area of 8 ac is used to reject waste heat from an industrial complex. The average daily wind velocity is 5 mph from the south. The

[3]Thus, approach for a cooling tower is analogous to the terminal temperature difference in the surface condenser.
[4]The term "thermal efficiency" is sometimes used here inappropriately.

temperature of the water surface averages 70°F with a daily ambient air temperature of 80°F and 60% relative humidity. What is the evaporation rate of the water in the cooling pond?

Customary U.S. Solution

The evaporation rate can be determined using Eq. 38.6.

Convert the area of the cooling pond from acres to square feet.

$$A = \left(43{,}560\ \frac{\text{ft}^2}{\text{ac}}\right)(8\ \text{ac}) = 348{,}480\ \text{ft}^2$$

Convert the wind velocity to feet per minute.

$$\text{v} = \frac{\left(5\ \frac{\text{mi}}{\text{hr}}\right)\left(5280\ \frac{\text{ft}}{\text{mi}}\right)}{60\ \frac{\text{min}}{\text{hr}}} = 440\ \text{ft/min}$$

From a psychrometric chart for 80°F and 60% relative humidity, the dew-point temperature of ambient air is 64.9°F. [**ASHRAE Psychrometric Chart No. 1—Sea Level**]

From a table of properties for saturated steam by temperature, the pressure of the air, p_a, at 64.9°F is 0.30 psia, and the pressure of the water, p_w, at 70°F is 0.36 psia. Use the absolute pressure column, which is the saturation pressure for the referenced temperature. [**Properties of Saturated Water and Steam (Temperature) - I-P Units**]

Convert the air and water pressure values from pounds per square inch to inches mercury.

$$\left(0.30\ \frac{\text{lbm}}{\text{in}^2}\right)\left(2.036\ \frac{\text{in Hg}}{\frac{\text{lbm}}{\text{in}^2}}\right) = 0.61\ \text{in Hg}$$

$$\left(0.36\ \frac{\text{lbm}}{\text{in}^2}\right)\left(2.036\ \frac{\text{in Hg}}{\frac{\text{lbm}}{\text{in}^2}}\right) = 0.74\ \text{in Hg}$$

From a table of properties for saturated steam by temperature, h_{fg} at 70°F is 1053.7 Btu/lbm. [**Properties of Saturated Water and Steam (Pressure) — I-P Units**]

Substituting these values into Eq. 38.6 gives

Cooling Ponds

$$\begin{aligned} w_p &= \frac{A(95 + 0.425\text{v})}{h_{fg}}(p_w - p_a) \\ &= \left(\frac{(348{,}480\ \text{ft}^2)\left(95 + (0.425)\left(440\ \frac{\text{ft}}{\text{min}}\right)\right)}{1053.7\ \frac{\text{Btu}}{\text{lbm}}}\right) \\ &\quad \times (0.74\ \text{in Hg} - 0.61\ \text{in Hg}) \\ &= 12{,}124\ \text{lbm/hr} \end{aligned}$$

3. COOLING TOWER BLOWDOWN

Water losses occur from evaporation, windage, and blowdown. *Evaporation loss* can be calculated from the humidity ratio increase and is approximately 0.1% per °F (0.18% per °C) decrease in water temperature.[5] *Windage loss*, also known as *drift*, is water lost in small droplets and carried away by the air flow. Windage loss is typically in the 0.1% to 0.3% range for mechanical draft towers. Since windage droplets are a mechanical mixture (not a thermodynamic solution of two gases), they are not adequately accounted for by the humidity ratio.

Makeup water must be provided to replace all water losses. The makeup water needed to replace water lost to evaporation is

Cooling Tower Evaporation

$$w_p = \frac{q}{h_{fg}} \qquad 38.7$$

As more and more water enters the system, *total dissolved solids*, TDS (e.g., chlorides), will build up over time. Water can be treated to prevent deposit, and a portion of the water can be periodically or continuously bled off. *Cycles of concentration* (*ratio of concentration*), *C*, is the ratio of total dissolved solids in the recirculating water to the total dissolved solids in the makeup water.[6]

$$C = \frac{(\text{TDS})_{\text{recirculating}}}{(\text{TDS})_{\text{makeup}}} = \frac{m_{\text{evaporation}} + m_{\text{blowdown}} + m_{\text{windage}}}{m_{\text{blowdown}} + m_{\text{windage}}} \qquad 38.8$$

[5]This value is approximate and is reported in various ways. Some authorities state "0.1% per degree Fahrenheit"; others say "1% per 10°F"; and yet others, "1% per 10°F to 13°F."
[6]Multiply grains/gallon (gr/gal) by 17.1 to obtain parts per million (ppm) or milligrams per liter (mg/L).

HVAC

Though windage removes some of the solids, most must be removed by bleeding some of the water off. This is known as *blowdown* or *bleed-off.* If the maximum cycles of concentration are known, the blowdown is

$$m_{\text{blowdown}} = \frac{m_{\text{evaporation}} + (1 - C_{\text{max}})m_{\text{windage}}}{C_{\text{max}} - 1} \quad 38.9$$

Additives should be used to prevent specific problems encountered, such as scale buildup, corrosion, biological growth, foaming, and discoloration.

4. DRY COOLING TOWERS

Dry cooling is used when environmental protection and water conservation are issues. It is used primarily by nonutility generators (e.g., waste-to-energy and cogeneration plants).

There are two types of dry cooling towers. Both use finned-tube heat exchangers. In a *direct-condensing tower*, steam travels through large-diameter "trunks" to a crossflow heat exchanger where it is condensed and cooled by the cooler air.[7] In an *indirect-condensing dry cooling tower*, steam is condensed by cold water jets (surface or jet condenser) and is subsequently cooled by air. The hot condensate is then pumped to crossflow heat exchangers where it is sensibly cooled (no condensation) by the air. Air flow may be mechanical or natural draft. Most U.S. installations are direct-condensing. Worldwide, natural-draft indirect systems are more predominant, particularly for power plants with capacities in excess of 100 MW. (See Fig. 38.2.)

***Figure 38.2** Dry Cooling Towers*

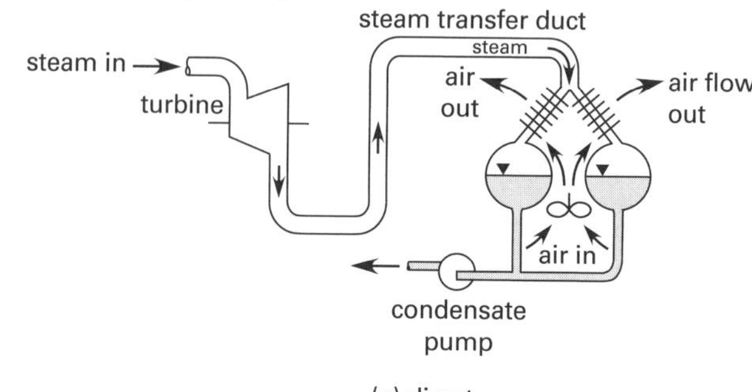

(a) direct

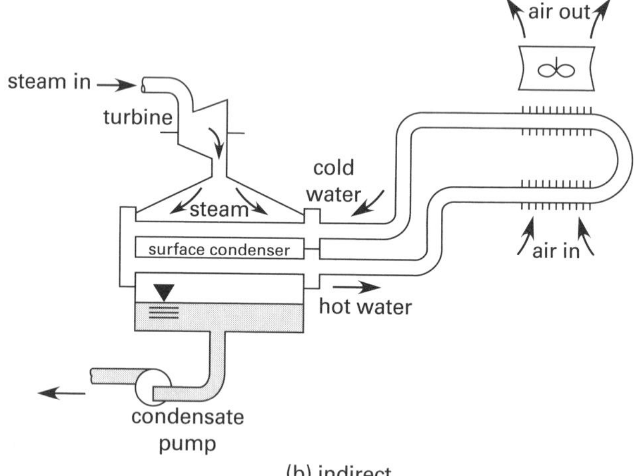

(b) indirect

5. WATER-COOLED CONDENSERS

A water-cooled condenser is a device that rejects refrigeration heat through a water-cooled heat exchanger (e.g. double tube, shell-and-coil, or shell-and-tube). Water-cooled condensers are typically used in systems with refrigerant capacities greater than 100 tons. The volumetric flow rate (in cubic feet per hour) needed to achieve a specified heat rejection rate based on entering and leaving temperatures is

Water-Cooled Condensers

$$Q = \frac{q_o}{\rho_w c_p (T_{w,\text{out}} - T_{w,\text{in}})} \quad 38.10$$

To calculate the volumetric flow rate, Q, in cubic feet per hour (ft^3/hr), the heat rejection rate, q_o, must be in British thermal units per hour (Btu/hr); the temperature, T, must be in degrees Fahrenheit; and the density, ρ, must be in pounds per cubic feet (lbm/ft^3). To calculate the volumetric flow rate in gallons per minute, multiply this result by 0.125.

6. NOMENCLATURE

A	pond surface area	ft^2	m^2
c_p	specific heat	Btu/lbm-°F	kJ/kg·°C
C	cycles of concentration	ppm	mg/L
h	enthalpy	Btu/lbm	kJ/kg
m	mass	lbm	kg
$\dot{m}$	mass flow rate	lbm/hr	kg/hr
p	pressure	lbf/ft^2	kPa
q	heat rate	Btu/hr	J/s
q_o	heat rejection rate	Btu/hr	J/s
Q	volumetric flow rate	gal/min	L/s
RF	rating factor	–	–
T	temperature	°F	°C
TDS	total dissolved solids	ppm	mg/L
TU	tower units	–	–
v	air velocity	ft/min	m/s
w_p	water evaporation rate	lbm/hr	kg/h
W	humidity ratio	lbm/lbm	kg/kg

[7]The term "direct contact" does not mean that the air and steam are combined in a single vessel.

Symbols

η	efficiency	–	–
ρ	density	lbm/ft^3	kg/m^3

Subscripts

a	air
fg	vaporization (fluid to gas)
w	water
wb	wet-bulb

39 Ventilation

Content in blue refers to the *NCEES Handbook*.

NCEES EXAM SPECIFICATIONS AND RELATED CONTENT

HVAC AND REFRIGERATION EXAM

II.D.2. Supportive Knowledge: Air quality and ventilation

- 2. Ventilation Standards
- 3. ASHRAE Ventilation Rate: Single Zone
- 5. Special ASHRAE Ventilation Requirements
- 10. Oxygen Needs
- 13. Sensible and Latent Heat

1. VENTILATION

Ventilation primarily refers to air that is necessary to satisfy the needs of occupants.[1] The term may mean the air that is introduced into an occupied space, or it may refer to the new air that is deliberately drawn in from the outside and mixed with return air. Ventilation, however, does not normally include unintentional infiltration through cracks and openings.

Ventilation air is provided to the occupied space primarily to remove heat and moisture generated in the space. Heat and moisture can both be generated metabolically as well as by equipment and processes. To a lesser extent, ventilation is also used to remove odors, provide oxygen, prevent carbon dioxide buildup, and remove noxious fumes. Generally, however, all of the other needs will be met if removal of body heat is accomplished.

2. VENTILATION STANDARDS

Few ventilation codes have the force of law, but there are several recommended standards. The U.S. Occupational Safety and Health Act (OSHA) contains a handful of mandatory standards, including 29 CFR 1910.146, dealing with minimum oxygen percentage in confined spaces, and 29 CFR 1910.94, dealing primarily with local exhaust systems. Almost all cities, counties, states, and municipalities have building codes, a few of which have their own ventilation requirements. Most building codes now incorporate provisions of the *International Mechanical Code* (IMC), published by the International Code Council (ICC). The American National Standards Institute (ANSI) has joined forces with the American Society of Heating, Refrigerating and Air-Conditioning Engineers (ASHRAE) in publishing indoor air quality standards, but it has several specialized standards of its own, primarily ANSI Z9. ANSI Z9 is published in conjunction with the American Industrial Hygiene Association (AIHA) and covers, for example, ventilation of laboratories, paint spray booths, and grinding stations. The U.S. National Institute for Occupational Safety and Health (NIOSH) has a few standards covering foundry ventilation, recirculation, and push-pull hoods. The National Fire Protection Association (NFPA) has standards that cover some specialized ventilation requirements such as NFPA 45, which covers lab fume hoods. The ASTM International ASTM D6245 describes how carbon dioxide can be used as an index of ventilation rate and objectionable body odor. The Air Movement and Control Association (AMCA) and the Sheet Metal and Air Conditioning Contractors' National Association (SMACNA) cover ventilation peripherally in their fan and duct publications, such as AMCA 201, but they do not recommend ventilation rates.

[1]The term *process air* is the most common designation given to "ventilation" needed for manufacturing processes.

Perhaps the most authoritative and comprehensive ventilation requirements are published by American Conference of Governmental Industrial Hygienists (ACGIH) and ASHRAE. The ACGIH Industrial Ventilation Committee publishes *Industrial Ventilation: A Manual of Recommended Practice for Design*, which is used throughout the world. ASHRAE publishes, in conjunction with ANSI, the most complete guidance, in particular, *Ventilation for Acceptable Indoor Air Quality* (ASHRAE Standard 62.1) and *Ventilation and Acceptable Indoor Air Quality in Low-Rise Residential Buildings* (ASHRAE Standard 62.2). ASHRAE Standard 62.1 sets voluntary minimum standards for new and substantially renovated commercial buildings. ASHRAE Standard 62.1 is a consensus standard that has been incorporated into NFPA 5000 and LEED green building qualifications. ASHRAE Standard 62.2 specifies voluntary minimum standards for single-family houses and multifamily dwellings three stories or less in height. The *ASHRAE Handbook: Fundamentals* volume also provides useful design guidance.

Compliance with *Energy Standard for Buildings Except Low-Rise Residential Buildings* (ASHRAE Standard 90.1), along with state and federal regulations intended to minimize environmental impact and energy loss (e.g., California's *Energy Efficiency Standards for Residential and Nonresidential Buildings*, California Energy Code, Title 24, Part 6 of the *California Code of Regulations*), may also be required.

Minimum ventilation requirements are given by local building codes, local ordinances, health regulations, and construction specifications. A common minimum design standard (as specified by ASHRAE Standard 62.1) is 20 ft^3/min (0.57 m^3/min; 9.4 L/s) of new, outside air per person. In interior areas that permit heavy smoking (e.g., casinos and smoking lounges), 30 ft^3/min to 60 ft^3/min (0.84 m^3/min to 1.68 m^3/min; 14.1 L/s to 28.2 L/s) per person is required. Some nonsmoking areas may require more than 20 ft^3/min (0.57 m^3/min; 9.4 L/s) anyway to avoid the "sick building syndrome" (as when formaldehyde-emitting furniture and building materials are present). **[Minimum Ventilation Rates in the Breathing Zone, Based on ANSI/ASHRAE Standard 62.1-2007]**

Some ventilation requirements are specified by the number of air changes (i.e., room volumes or "cubical contents" without allowance for room contents) required per hour, ACH. This is known as the *air change method.* A typical minimum value for toilet rooms, for example, is four air changes per hour. For other uses (automotive, boiler rooms, engine rooms, etc.) the number of air changes can be significantly higher (e.g., 25 to 100 per hour).

HVAC

3. ASHRAE VENTILATION RATE: SINGLE ZONE

ASHRAE Standard 62.1 prescribes two methods for determining the amount of outdoor ventilation air: a *ventilation rate procedure* (VRP) and an *indoor air quality procedure* (IAQP). Since it is not practical to monitor all air contaminants in all locations, and since some contaminants (e.g., mold and fungi spores) cannot be monitored in real time, using the IAQP is associated with significant risk. Due to a multiplicity of air contaminants that (a) are not monitored, (b) are not detected, and (c) do not even have definite limits, and since the straightforward VRP is so much simpler, the rate-based methodology is preferred.

Modern rate-based ventilation standards, including ASHRAE Standard 62.1, specify the required *breathing zone outdoor air* (i.e., the outdoor ventilation air in the breathing zone), Q_{bz}, as a function of both zone occupancy, P_z, and zone floor area, A_z. The first term accounts for contaminants produced by occupants, while the second term accounts for contaminants produced by the building. The maximum number of occupants expected in the zone during typical usage is normally used rather than a value based on building code classification occupancy densities. However, different short-term time-averaging methods prescribed by ASHRAE Standard 62.1 may also be used if the zone population fluctuates.[2] ASHRAE Standard 62.1 requires that the rate specified by Eq. 39.1 be maintained during operation under all load conditions.

$$Q_{bz} = R_pP_z + R_aA_z \qquad 39.1$$

Table 39.1 contains representative values of R_p and R_a, although local codes, federal regulations, and contract requirements take precedence. The default columns are used only if the actual occupant density is unknown. A more detailed list of *minimum ventilation rates* in the breathing zone is available in ASHRAE Standard 62.1. **[Minimum Ventilation Rates in the Breathing Zone, Based on ANSI/ASHRAE Standard 62.1-2007]**

The outdoor ventilation rate specified by Eq. 39.1 is affected by the *distribution effectiveness*, E_z, as specified in Table 39.2. The *zone outdoor airflow*, Q_{oz}, at the diffusers is given by Eq. 39.2. For single-zone systems, this is also the system total outdoor air requirement, Q_{ot}, as shown in Eq. 39.3.

$$Q_{oz} = \frac{Q_{bz}}{E_z} \qquad 39.2$$

$$Q_{oz} = Q_{ot} \quad \text{[single-zone system]} \qquad 39.3$$

[2]Outdoor airflow rates can be reduced dynamically in the critical zones that have variable occupancy. Changes in outdoor air demand (i.e., changes in occupancy) can be detected several ways, including measurement of carbon dioxide, CO_2.

Table 39.1 Representative Minimum Ventilation Rates in Breathing Zone[a,b]

	people outdoor air rate, R_p		area outdoor air rate, R_a		default values			
					occupant density, people per 1000 ft^2 (100 m^2)	combined outdoor air rate		
occupancy category	cfm/person	L/s·person	cfm/ft^2	L/s·m^2		cfm/person	L/s·person	cfm/ft^2
educational:								
classrooms (age 9 and up)	10	5	0.12	0.6	35	13	6.7	0.46
science laboratories	10	5	0.18	0.9	25	17	8.6	0.43
general:								
conference and meeting rooms	5	2.5	0.06	0.3	50	6	3.1	0.30
hotel:								
bedrooms and living rooms	5	2.5	0.06	0.3	10	11	5.5	0.11
office building:								
main lobbies	5	2.5	0.06	0.3	10	11	5.5	0.11
offices	5	2.5	0.06	0.3	5	17	8.5	0.09
miscellaneous:								
bank vaults	5	2.5	0.06	0.3	5	17	8.5	0.09
public assembly:								
auditorium seating areas	5	2.5	0.06	0.3	150	5	2.7	0.75
retail:								
sales areas	7.5	3.8	0.12	0.6	15	16	7.8	0.24
sports and entertainment:								
spectator areas	7.5	3.8	0.06	0.3	150	8	4.0	1.2

(Multiply cfm by 0.02832 to obtain m^3/min.)
(Multiply cfm by 0.4719 to obtain L/s.)
(Multiply cfm/ft^2 by 5.08 to obtain L/s·m^2.)
(Multiply people/1000 ft^2 by 0.929 to obtain people/100 m^2.)
(Multiply cfm/ft^2 by 0.3048 to obtain m^3/min·m^2.)

[a]This table applies to environmental tobacco smoke (ETS)-free areas only. Refer to ASHRAE Standard 62.1 Sec. 5.17 for requirements for buildings containing ETS areas and ETS-free areas.

[b]Rates are based on an air density of 0.075 lbm/ft^3 (1.2 kg/m^3).

HVAC

Table 39.2 Zone Air Distribution Effectiveness

air distribution configuration	effectiveness, E_z
ceiling supply of cool air	1.0
ceiling supply of warm air with floor return	1.0
ceiling supply of warm air 15°F (8°C) or more above space temperature, with ceiling return	0.8
ceiling supply of warm air less than 15°F (8°C) above space temperature, with ceiling return, provided that the 150 fpm (0.8 m/s) supply air jet reaches to within 4.5 ft (1.4 m) of floor level	1.0
ceiling supply of warm air less than 15°F (8°C) above space temperature, with ceiling return, with supply jet air velocity less than 150 fpm (0.8 m/s)	0.8
floor supply of cool air with ceiling return, provided that the 150 fpm (0.8 m/s) supply air jet reaches 4.5 ft (1.4 m) or more above the floor level*	1.0
floor supply of cool air with ceiling return, provided the low-velocity displacement ventilation achieves unidirectional flow and thermal stratification	1.2
floor supply of warm air with floor return	1.0
floor supply of warm air with ceiling return	0.7

*This describes most underfloor air distribution systems.

Example 39.1

A single-zone high school classroom has a floor area of 1600 ft^2 and seats 35 students. Cool air is supplied from ceiling diffusers. How much outdoor ventilation air is required per ASHRAE Standard 62.1?

Solution

From Table 39.1 (educational: classroom (age 9 and up)), $R_p = 10$ cfm/person, and $R_a = 0.12$ cfm/ft^2. From Eq. 39.1,

$$\begin{aligned} Q_{\text{bz}} &= R_pP_z + R_aA_z \\ &= \left(10\ \frac{\text{ft}^3}{\text{min-person}}\right)(35 \text{ people}) \\ &\quad + \left(0.12\ \frac{\text{ft}^3}{\text{min-ft}^2}\right)(1600 \text{ ft}^2) \\ &= 542 \text{ ft}^3/\text{min} \end{aligned}$$

Since cool air is supplied from the ceiling, from Table 39.2, $E_z = 1.0$. The total outdoor air to the zone is

$$Q_{\text{oz}} = \frac{Q_{\text{bz}}}{E_z} = \frac{542\ \frac{\text{ft}^3}{\text{min}}}{1.0} = 542 \text{ ft}^3/\text{min}$$

4. ASHRAE VENTILATION RATE: MULTIZONE

There are two types of multiple-zone (multizone) systems drawing outside air: 100% outside air (OA) systems and recirculating air systems. For 100% outside air systems, the system total outside air requirement is the sum of all of the zonal requirements.

$$Q_{\text{ot}} = \sum Q_{\text{oz}} \quad \text{[multiple 100\% OA zones]} \qquad 39.4$$

For recirculating systems with outside air intakes, a correction is made for *occupancy diversity* (also known as *occupant diversity* or *population diversity*), D, which is the ratio of the *system population* (the maximum number of simultaneous occupants in the space served by the system), P_s, to the sum of the zonal peak occupancies.

$$D = \frac{P_s}{\sum P_z} \qquad 39.5$$

Only the population outdoor air component is affected by the diversity term. The uncorrected system outdoor air requirement is corrected for diversity but not for distribution effectiveness.

$$Q_{\text{ou}} = D\sum R_pP_z + \sum R_aA_z \qquad 39.6$$

Since multizone recirculating systems are not as efficient as 100% OA systems, the system outdoor rate is determined from the *system ventilation efficiency*, E_v. The system ventilation efficiency, in turn, depends on the maximum primary outdoor air fraction. ASHRAE Standard 62.1 gives two methods for determining this efficiency: the default method using Table 39.3 (ASHRAE Table 6-3), and the more accurate and more involved calculated method using ASHRAE App. A. These methods produce significantly different results, but either may be used.

The system ventilation efficiency depends on the maximum *primary outdoor air fraction* evaluated over all of the zones. A zone's primary outdoor air fraction, Z_p, is the fraction of total air (including the outdoor and recirculated airflows) from the air handler, known as the *zonal primary airflow*, Q_{pz}, that is outdoor air.

$$Z_p = \frac{Q_{\text{oz}}}{Q_{\text{pz}}} \qquad 39.7$$

Table 39.3 can be used with the maximum value of Z_p to determine the system ventilation efficiency, E_v.

$$Q_{\text{ot}} = \frac{Q_{\text{ou}}}{E_v} \quad \left[\begin{array}{c}\text{multiple zones}\\ \text{with recirculation}\end{array}\right] \qquad 39.8$$

Table 39.3 *System Ventilation Efficiency**

maximum Z_p	E_v
≤ 0.15	1.0
≤ 0.25	0.9
≤ 0.35	0.8
≤ 0.45	0.7
≤ 0.55	0.6
>0.55	Use ASHRAE Standard 62.1 App. A method.

*Interpolation may be used between tabulated values.

From ASHRAE Standard 62.1, Table 6-3, copyright © 2010, by American Society of Heating, Refrigerating and Air-Conditioning Engineers, Inc. Reproduced with permission.

5. SPECIAL ASHRAE VENTILATION REQUIREMENTS

Some conditions pertaining to the quality of the air in the immediate and surrounding areas can trigger special ventilation requirements. These specific conditions are highlighted below.

- Air intake separation distance requirements are provided in ASHRAE Standard 62.1. [**Air Intake Minimum Separation Distance, Based on ANSI/ASHRAE Standard 62.1-2007**]
- Natural ventilation may be relied on when certain requirements are met.
- Outdoor air drawn from National Ambient Air Quality Standards (NAAQS) *nonattainment areas* must be treated to reduce particulate matter, ozone, carbon dioxide, sulfur oxides, nitrogen dioxide, and/or lead to specified levels. Specifically, coils and other devices with wetted surfaces must have MERV 6 filters upstream if the outdoor air does not meet NAAQS for PM-10 particulates. An ozone air cleaner (minimum 40% efficiency) is generally required if the average ozone concentration exceeds 0.107 ppm.
- Ventilation in areas exposed to *environmental tobacco smoke* (ETS) (i.e., in smoking areas) requires the use of methods in ASHRAE Standard 62.1 Sec. 5.17. Smoke-free areas must be maintained at higher static pressures relative to adjacent ETS areas.
- Variable air volume (VAV) systems with fixed outside air dampers must comply at the minimum supply airflow.
- Residential spaces in buildings over three stories have special requirements.

ASHRAE Standard 90.1 Sec. 6.4.3.9 (*Ventilation Controls for High-Occupancy Areas*) specifies that *demand controlled ventilation* (DCV) is required for spaces larger than 500 ft^2 (47 m^2) and with a design occupancy for ventilation of greater than 40 people per 1000 ft^2 (100 m^2) of floor area and served by systems with one or more of the following: (a) an air-side economizer, (b) automatic modulating control of the outdoor air damper, or (c) a design outdoor airflow greater than 3000 cfm.

6. INFILTRATION

Infiltration (also known as *accidental infiltration*) refers to the air that unintentionally enters an occupied space through cracks around doors and windows and through openings in a building. Accidental infiltration may be as high as 0.4 to 1.0 ACH.

With the *crack length method*, the amount of infiltration, Q, is determined from the *crack coefficient*, B, and the crack length, L. Values of the crack coefficient vary greatly and depend on the type of window or door, wind velocity, orientation, and degree of closure.[3] Alternatively, the infiltration may be found from the plane area. (This method is more common when determining infiltration through entire walls.) As with the crack length method, the crack area coefficient B' depends on many factors.

$$Q = BL = B'A \qquad 39.9$$

More sophisticated correlations recognize the dependence on the difference in outside and inside pressures. Values of B'' and n must be known or assumed and must be consistent with the units of pressure.

$$Q = B''A(\Delta p)^n \qquad 39.10$$

Since air entering through cracks on the windward side must leave through cracks on the leeward side, only half of the total crack length is used when all four sides of a building are exposed to wind. However, the amount of crack length used also depends on the building orientation. When only one wall is exposed to wind, that wall's total crack length is used. With two exposed walls, the wall with the larger crack length is used. When three walls are exposed, only two walls contribute to crack length. The crack length used should never be less than half of the total crack length.

The air change method can also be used for infiltration. Infiltration into modern (tight) residential construction may be as low as 0.2 air changes per hour, while older residences in good condition may experience ten times as much. In the past, a rule of thumb used (to size

[3]Typical values of the crack length coefficient are given in most HVAC books. Manufacturers' literature should be used for specific name-brand windows and doors.

furnaces) in the absence of other information was that infiltration into residences with windows on one, two, or three sides would be one, one and one-half, or two air changes per hour, respectively. Experience is needed to modify these values for use with modern, energy-efficient construction.

7. INFILTRATION IN TALL BUILDINGS

For tall buildings (i.e., those over 100 ft (30 m) in height), the pressure difference, Δp, in Eq. 39.11 is a combination of the wind velocity pressure and the *stack effect* (*chimney effect*). The stack effect is particularly important during the winter. The combined infiltration due to static pressure (including wind velocity pressure) and the stack effect is proportional to the square root of the sum of the heads acting on the building.

$$\Delta p = \sqrt{p_w^2 + \Delta p_{\text{stack effect}}^2} \quad \text{39.11}$$

If the infiltrations due to wind alone and stack effect alone are known, Eq. 39.12 is equivalent to taking the square root of the sum of the heads.

$$Q = \sqrt{Q_w^2 + Q_{\text{stack effect}}^2} \quad \text{39.12}$$

In addition to traditional HVAC methods based on crack length, infiltration from wind may be calculated from extensions of traditional fluid principles. Specifically, if the opening area or effective leakage area, A, and *entrance* or *discharge coefficient*, C_d, are known, the infiltration is

$$Q_w = C_d A \sqrt{\frac{2p_w}{\rho}} \quad \text{[SI]} \quad \text{39.13(a)}$$

$$Q_w = C_d A \sqrt{\frac{2g_c p_w}{\rho}} \quad \text{[U.S.]} \quad \text{39.13(b)}$$

The *wind pressure*, p_w, is based on the theoretical *velocity pressure* (*stagnation pressure*) modified by a *wind surface pressure coefficient*, C_p, which is a function of wind direction, building orientation, and vertical location. Ideally, $C_p = 1.0$ for wind perpendicular to a surface, but rarely does C_p exceed 0.9 in practice, and values between 0.5 and 0.9 are typical. In cases where the wind flow is affected by adjacent structures or vegetation, a *shelter factor* (*sheltering coefficient*), s, may be incorporated into the calculation of wind pressure. For an unsheltered building projecting vertically above level surroundings, $s = 1$.

$$p_w = \frac{C_p s^2 \rho v^2}{2} \quad \text{[SI]} \quad \text{39.14(a)}$$

$$p_w = \frac{C_p s^2 \rho v^2}{2g_c} \quad \text{[U.S.]} \quad \text{39.14(b)}$$

Several different methods (theoretical, heuristic, and code-based) can be used to determine the stack effect. The theoretical value is given by Eq. 39.15, where h_{NPL} is the elevation of the *neutral pressure level* and all temperatures are absolute.

$$\Delta p_{\text{stack effect}} = (\rho_{\text{outside}} - \rho_{\text{inside}})g(h - h_{\text{NPL}}) = \frac{\rho_{\text{inside}} g(h - h_{\text{NPL}}) \times (T_{\text{abs,outside}} - T_{\text{abs,inside}})}{T_{\text{abs,inside}}} \quad \text{[SI]} \quad \text{39.15(a)}$$

$$\Delta p_{\text{stack effect}} = \frac{(\rho_{\text{outside}} - \rho_{\text{inside}})g(h - h_{\text{NPL}})}{g_c} = \frac{\rho_{\text{inside}} g(h - h_{\text{NPL}}) \times (T_{\text{abs,outside}} - T_{\text{abs,inside}})}{g_c T_{\text{abs,inside}}} \quad \text{[U.S.]} \quad \text{39.15(b)}$$

Some methods determine infiltration by using an *effective wind velocity* (*equivalent wind velocity*), $v_{\text{effective}}$, to combine the effects of wind and stack effect. For example, a table of crack coefficients usually requires knowing the wind velocity. For short buildings, the actual wind velocity, v_o, at the opening (window, door, opening, crack, etc.) elevation is used. For tall buildings, v_o at the opening elevation is modified for stack effect. For rough estimates of infiltration made by such methods, Eq. 39.16 gives an effective wind velocity at a location y above or below a building's midheight, the assumed location of the neutral pressure level. v_o is the wind velocity that would be used if the stack effect was neglected. y is positive above the midheight and negative below the midheight. This correctly reflects infiltration into the building due to both wind and stack effect below midheight, but infiltration due to wind balanced against exfiltration due to stack effect above midheight.

$$v_{\text{effective,mph}} = \sqrt{v_{o,\text{mph}}^2 - 1.75 y_{\text{ft}}} \quad \text{[U.S. only]} \quad \text{39.16}$$

8. INDOOR DESIGN CONDITION

The *indoor design condition* refers to the thermodynamic state of the air that is removed from an occupied space. The indoor design temperature, T_{id}, represents the maximum dry-bulb air temperature—a not-to-be-exceeded limit—that the space will reach.

HVAC books contain tables of recommended inside design conditions and charts of comfort ranges that can be used to select suitable combinations of temperature and humidity. Within a *comfort range*, choice of the actual inside design condition is subjective and requires modification based on experience for the needs of the particular industries, the season, and the levels of physical exertion.

Most people feel comfortable when the dry-bulb temperature is kept between 74°F and 77°F (23.3°C and 25°C) and the relative humidity is 30% to 35% (in the winter) or 45% to 50% (in the summer). 75°F (23.9°C) dry-bulb and 50%

HVAC

relative humidity is often selected as an inside design condition for initial studies. This temperature is for the breathing line, 3 ft to 5 ft (0.9 m to 1.5 m) above the floor.[4]

Attention also needs to be given to the *temperature swing*, the difference between the thermostat's on and off settings. For commercial applications, the swing during the summer should be approximately 2°F to 4°F (1.1°C to 2.2°C) above the indoor design (i.e., "off") setting, and the swing during the winter should be approximately 4°F (2.2°C) below the indoor design (i.e., "off") setting.

9. HUMIDIFICATION

Ventilation provides humidification to the occupied space, particularly during the winter. Air should not be completely dry when it enters an occupied space. Air that is too dry will cause discomfort and susceptibility to respiratory ailments. Also, some pathogenic bacteria that survive in low- and high-humidity air will die very quickly in air with midrange humidities.[5]

Some manufacturing and materials handling processes require specific humidity for efficient operation. *Hygroscopic materials*, such as wood, paper, textiles, leather, and many food and chemical products, readily absorb moisture. A constant humidity level is required to obtain consistent manufacturing conditions with such products.

Dry air prevents static electricity from dissipating (into the air). Therefore, dry air can cause intermittent electrical/electronic failures; affect the handling of static-prone materials such as paper, films, and plastics; and ignite potentially explosive atmospheres of dust and gases.

Humidification can be provided by evaporating water in the occupied space (the *evaporative pan method*) or by injecting water (the *water spray method*) or steam into the duct flow. Most commercial humidification is accomplished by placing one or more steam manifolds in the air distribution duct. *Booster humidification* (*spot humidification*) from a separate, independent source is required when a higher humidity is needed in a limited area within a larger controlled space. Steam flow is controlled by humidistats placed downstream of the steam manifold.[6]

10. OXYGEN NEEDS

Providing oxygen is not an issue in traditional buildings. Infiltration alone provides the oxygen needed. However, in closed and confined environments such as mines, tunnels, manholes, and closed tanks, forced ventilation and/or oxygen masks are necessary.

Air is approximately 20.9% oxygen by volume, independent of altitude. For confined spaces, OSHA (29 CFR 1910.134) specifies a minimum oxygen content of 19.5%, which is adequate for elevations below 3000 ft (914 m). NIOSH and ACGIH specifications differ slightly. Concentrations less than 19.5% are known as *oxygen-deficient atmospheres*. Reaction to oxygen deficiency varies with individuals, but in general, significant impairments to work rate, perception, concentration, and judgment can be expected with lower values. Some individuals may experience coronary, pulmonary, and circulatory problems. Concentrations below 12% pose immediate danger to life. OSHA defines an *oxygen-enriched atmosphere* as one with an oxygen concentration greater than 23.5% (29 CFR 1910.146(b)). (OSHA 1915.12(a)(2) pertaining to shipyard operations specifies concentrations above 22% as being enriched.) Oxygen-enriched environments pose extreme fire and explosion hazards, especially if combustible material is present. ASHRAE Standard 62.1-2007, App. C provides a table of heart rate and oxygen consumption as a function of level of exertion. [**Heart Rate and Oxygen Consumption at Different Activity Levels**]

11. CARBON DIOXIDE BUILDUP

Diluting exhaled carbon dioxide, like providing oxygen, is only an issue in completely closed environments. Infiltration alone provides the dilution needed. Healthy individuals can usually tolerate a concentration of 0.5% (by volume), though the air will be noticeably stale.[7] Equation 39.17 is used for finding the approximate time (in hours) for a 3% buildup of carbon dioxide in a closed area.[8] The carbon dioxide concentration should not exceed 5% under any circumstances.

$$t_{\text{h}} = \frac{1.4V_{\text{room,m}^3}}{\text{no. of occupants}} \quad \text{[SI]} \qquad 39.17(a)$$

$$t_{\text{hr}} = \frac{0.04V_{\text{room,ft}^3}}{\text{no. of occupants}} \quad \text{[U.S.]} \qquad 39.17(b)$$

12. ODOR REMOVAL

The airflow required through a room to remove body odors depends on the room size and level of activity. Body odors become more pronounced when the relative humidity is above 55%. Except for very cramped areas, common ventilation standards are normally sufficient for odor removal. Good practice requires approximately one-third of the air to be new air.

HVAC

[4]The temperature variation with height above the floor is approximately 0.75°F/ft (1.4°C/m).

[5]In particular, airborne type 1 pneumococcus, group C staphylococcus, and staphylococcus are quickly killed in relative humidities of 45% to 55%. Other viruses, including measles, influenza, and encephalomyelitis, survive longer in very dry air than in midrange relative humidities.

[6]Steam flow is turned off when the humidity reaches the high-limit humidistat setting, typically 90% relative humidity. This prevents oversaturation of the air when there is a failure in the air conditioning system or controlling humidistat.

[7]Some submarines have operated at 1% by volume carbon dioxide.

[8]This equation assumes an initial (atmospheric) carbon dioxide content of 0.03% and carbon dioxide production of 0.011 ft^3/min (0.00031 m^3/min; 5.2 mL/s) per person.

13. SENSIBLE AND LATENT HEAT

Metabolic heat contains both sensible and latent components. *Sensible heat* is pure thermal energy that increases the air's dry-bulb temperature. *Latent heat* is moisture that increases air's humidity ratio. Table 39.4 gives the approximate amounts of metabolic heat in a 75°F (23.9°C) environment. The "adjusted" column refers to a normal mix of men, women, and children. For design purposes, the heat gain for an adult female is approximately 85% of the adult male rate; the heat gain for a child is 75% of the adult male rate. Additionally, *ASHRAE Handbook — Fundamentals* provides a table of rates at which heat and moisture are given off during different states of activity. [**Representative Rates at Which Heat and Moisture Are Given Off by People in Different States of Activity**]

***Table 39.4** Approximate Heat Generation by Occupants (Btu/hr)*

activity	total adult males	total adjusted	sensible* adjusted	latent* adjusted
seated, at rest, theater, classroom	390	350	245	105
moderately active office work	475	450	250	200
standing, light work, slowly walking	550	450	250	200
moderate dancing	900	850	305	545
walking 3 mph, moderately heavy work	1000	1000	375	625
heavy work	1500	1450	580	870

(Multiply Btu/hr by 0.293 to obtain watts.)

*The sensible-latent splits given are for a 75°F (23.9°C) environment. For an 80°F (26.7°C) environment, total heat remains the same, but sensible heat decreases approximately 20% and latent heat increases accordingly.

14. VENTILATION FOR HEAT REMOVAL

Ventilation requirements can be calculated from sensible heat and/or moisture (i.e., latent heat) generation rates. In Eq. 39.18, T_{in} is the dry-bulb temperature of the air entering the room.

$$\begin{aligned} q_s &= \dot{m}c_p(T_{\text{id}} - T_{\text{in}}) \\ &= Q\rho c_p(T_{\text{id}} - T_{\text{in}}) \end{aligned} \quad 39.18$$

Equation 39.18 can be written in terms of the number of air changes per hour, ACH, and the temperature of the ventilation air, T_{out}.

$$q = \frac{\rho c_p V_{\text{room}}(\text{ACH})(T_{\text{in}} - T_{\text{out}})}{60\ \frac{\text{min}}{\text{hr}}} \quad 39.19$$

In ventilation work, volumetric flow rates are traditionally given in ft³/min (cfm), m³/min, or L/s. The constant 1.08 (0.02) in Eq. 39.20 is the product of an air density of 0.075 lbm/ft³ (1.2 kg/m³), a specific heat of 0.24 Btu/lbm-°F (1.0 kJ/kg·°C), and the factor 60 min/hr (60 s/min).[9]

$$Q_{\text{m}^3/\text{min}} = \frac{q_{s,\text{kW}}}{\left(0.02\ \frac{\text{kJ·min}}{\text{m}^3\text{·s·°C}}\right)(T_{\text{id,°C}} - T_{\text{in,°C}})} \quad \text{[SI]} \quad 39.20(a)$$

$$Q_{\text{ft}^3/\text{min}} = \frac{q_{s,\text{Btu/hr}}}{\left(1.08\ \frac{\text{Btu-min}}{\text{ft}^3\text{-hr-°F}}\right)(T_{\text{id,°F}} - T_{\text{in,°F}})} \quad \text{[U.S.]} \quad 39.20(b)$$

$$Q_{\text{L/s}} = \frac{q_{s,\text{W}}}{\left(1.20\ \frac{\text{W·s}}{\text{L·°C}}\right)(T_{\text{id,°C}} - T_{\text{in,°C}})} \quad \text{[SI]}$$

$$Q_{\text{L/s}} = \frac{q_{s,\text{Btu/hr}}}{\left(2.28\ \frac{\text{Btu-sec}}{\text{L-hr-°F}}\right)(T_{\text{id,°F}} - T_{\text{in,°F}})} \quad \text{[mixed units]}$$

The sensible heat loads will usually be more significant than the latent load, and ventilation will be determined solely on that basis. When large moisture sources are present, however, the latent loads may control.

$$\begin{aligned} q_l &= \dot{m}_{\text{water}}h_{fg} = \dot{m}_{\text{air}}\Delta W h_{fg} \\ &= Q\rho\Delta W h_{fg} \end{aligned} \quad 39.21$$

The constant 4775 (49.36) in Eq. 39.22 is the product of the air density of 0.075 lbm/ft³ (1.2 kg/m³), a latent heat of vaporization at the approximate partial pressure of the water vapor in the air of 1061 Btu/lbm (2468 kJ/kg), and the factor 60 min/hr (60 s/min).[10]

$$Q_{\text{m}^3/\text{min}} = \frac{q_{l,\text{kW}}}{\left(49.36\ \frac{\text{kJ·min}}{\text{m}^3\text{·s}}\right)\Delta W_{\text{kg/kg}}} \quad \text{[SI]} \quad 39.22(a)$$

$$Q_{\text{ft}^3/\text{min}} = \frac{q_{l,\text{Btu/hr}}}{\left(4775\ \frac{\text{Btu-min}}{\text{ft}^3\text{-hr}}\right)\Delta W_{\text{lbm/lbm}}} \quad \text{[U.S.]} \quad 39.22(b)$$

[9] c_p = 0.24 Btu/lbm-°F (1.0 kJ/kg·°C) is applicable to dry air. For air with normal amounts of water vapor, the specific heat is closer to c_p = 0.244 Btu/lbm-°F (1.02 kJ/kg·°C).

[10] There is some variation in these constants depending on what heat of vaporization is used. For example, some sources use 1076 Btu/lbm (2503 kJ/kg), in which case the constant is 4840 Btu-min/ft³-hr (50.06 kJ·min/m³·s).

15. VENTILATION FOR MOLD CONTROL

Humidity as low as 70%, even without condensing infiltration, can provide sufficient moisture for mold and fungi growth in as little as six hours. Moisture management should be specifically considered in ventilation design, and it should take precedence over energy management. Best practice requires that (a) infiltration of unfiltered and unconditioned humid air be prevented; (b) negative interior pressures be avoided and net positive pressure with respect to outdoors be maintained (in the absence of wind and stack effects) while dehumidification is occurring; (c) building and system design, operation, and maintenance provide for dehumidification (drying) of surfaces and materials prone to moisture accumulation under normal operating conditions; and (d) the HVAC should specifically monitor and control humidity.

With only a few exceptions, ASHRAE Standard 62.1 Sec 5.9.1 limits maximum humidity in occupied spaces to 65% during periods of peak outdoor dew point.[11] Without dehumidification equipment, meeting this limit may be difficult with high outdoor air latent loading and low inside space sensible heat ratio. Meeting acceptable humidity ratios is facilitated in buildings maintained at small positive pressures.

Even if the interior relative humidity is maintained below 65%, as prescribed by ASHRAE Standard 62.1, local areas with higher humidity can exist. Areas with higher spot humidity include carpet over concrete, window sills, under sinks, under outside sliding doors, near defective roof installations, and, ironically, near dehumidification equipment. Although not required by ASHRAE, relative humidities less than 35% may be required to avoid condensation on cold surfaces during the winter. Alternatively, relative humidity should be kept above 30% to prevent generation of static electricity.

16. FUME EXHAUST HOODS

Air hoods (also known as *fume hoods* or *exhaust hoods*) are used to provide localized protection from and removal of hazardous vapors, dusts, and biological materials. Table 39.5 lists representative *control velocities* needed for air to capture and entrain materials generated by various processes. When there is sufficient air in the work room to create the face velocity, a separate blower bringing air from the outside may be used, in which case, the term *auxiliary-air hood* is used. Hoods may be of the nonenclosure or closure varieties. *Nonenclosure air hoods* for nontoxic materials may be simple canopies over open tanks (for vapors that rise) or periphery slots (for contents that do not rise). Flanges around the exterior of canopies extending over the edges of the tank increase the collection efficiency and reduce the required airflow, but nonenclosure hoods are particularly inefficient at best.

Table 39.5 *Minimum Control Velocities for Enclosure Hoods*

process	minimum control velocity ft/min	minimum control velocity m/min
evaporation from open tanks	50–100	15–30
paint spraying, welding, plating	100–200	36–60
stone cutting, mixing, conveying	200–500	60–150
grinding, crushing	500–2000	150–600

(Multiply ft/min by 0.3048 to obtain m/min.)

With a conventional *enclosure air hood*, air is drawn through the front opening into the *fume chamber* and across the work surface, entraining the captured material. In *nonbypass air hoods*, the resulting high-velocity air jet sweeping over the work surface and noise is often disconcerting, leading to a reluctance by users to close the sash fully. *Bypass air hoods* address this issue. (See Fig. 39.1.) In air hoods with bypass, air enters the fume chamber through the bypass opening when the sash is closed. When the sash is open, air enters the fume chamber through the sash opening. The bypass area is typically 20–30% of the all-open sash area, and the resulting terminal face velocities with a fully closed sash are in the 300–500 ft/min (90–150 m/s) range, significantly higher than required for most processes. Therefore, most users prefer a combination of bypass and partially open sash.

Figure 39.1 *Conventional Cabinet Bypass Air Hood*

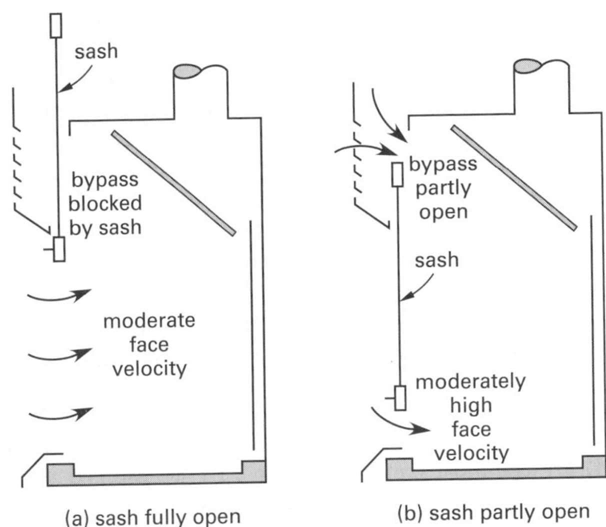

With *auxiliary-air hood units* (also known as *makeup air fume hoods*), outside air is drawn in by a supply blower and discharges downward through a face plenum along the top width of the bypass hood. Typically, 50–70% of the discharged air can be supplied by the auxiliary blower. Unfortunately, even when adjusted properly, auxiliary air hoods provide relatively poor containment

[11]Exceptions where humidity may exceed 65% include kitchens, hot tub rooms, refrigerated areas including ice rinks, shower rooms, spas, and pools.

and result in significantly higher worker exposure compared to conventional (non-auxiliary air) hoods. The air curtain created may even pull vapors out of the hood interior. Therefore, most authorities, including *Laboratory Ventilation* (ANSI/AIHA Standard Z9.5), recommend against auxiliary air hoods.

Some contaminants are released with almost no velocity of their own. This is the case with products of natural evaporation. Other contaminants (paint booth overspray, dust from grinding wheels, etc.) are released at high velocity. With distance, the velocity dissipates and reaches zero at the *null point*. Although capture is easiest at the null point, it is difficult in most situations to determine the distance to the null point. Even when the distance to the null point is known, the directionality may vary. Therefore, a high air intake velocity is needed to capture the moving contaminants near the point of generation.

Air velocity decreases with increasing distance from the source, varying almost inversely with the square of the distance. Therefore, the hood opening should be as close as possible to the contaminant source.

Since heated air rises, different design principles are needed for high-temperature (e.g., molten metal) processes. The heated air mixes with the surrounding air, and a larger volume of diluted air must be captured.

17. DILUTION VENTILATION

Dilution ventilation (*toxicity dilution*) refers to the dilution of contaminated air in order to reduce its health or explosion hazard. Dilution is less effective than outright removal of hazards by exhaust ventilation. Dilution is generally applicable to organic liquids and solvents whose toxicities are low, when the workers are not too close to the source, and when vapor generation is fairly uniform.[12] The required volume of dilution air must not be so great as to make air velocities unreasonable.

Dilution is achieved by providing enough air to reduce a vapor's concentration to an acceptable level. Various designations are given to acceptable levels, although the *threshold limit value* (TLV) in parts per million (ppm) by volume (mg/m^3 in SI) and *lower explosive limit* (LEL) in parts per hundred (pph) by volume (mg/m^3 in SI) are the most common.[13] The TLV is assumed to be the concentration that workers may be continuously exposed to during a certain period. TLVs are subject to ever-changing legislation and ongoing research.[14]

Three types of TLVs are used. The *time-weighted average* (TLV-TWA) is the time-weighted average concentration that workers may be exposed to for eight hours per day, day after day, without experiencing adverse effects. The *short-term exposure limit* (TLV-STEL) is the time-weighted average concentration that workers may be exposed to for fifteen minutes, up to four times per eight-hour period.[15] The *ceiling value* (TLV-C) is the concentration that should not be exceeded, even instantaneously. Depending on the substance, one, two, or all three of these limits may be applicable.[16]

Assuming that a contaminant is uniformly distributed throughout the plant air, at equilibrium the contaminant generation rate, R, is equal to the ventilated removal rate.

$$R = CQ \qquad 39.23$$

To account for irregular vapor evolution, inefficient ventilation, and toxicity, an empirical multiplicative "effectiveness of mixing factor" (the K-factor) between 3 and 10 is used.

$$KR = CQ \qquad 39.24$$

The maximum concentration, C_{max}, is usually the threshold limit value or the *permissible exposure limit* (PEL), both with a safety factor. The required airflow is

$$Q_{L/s} = \frac{(4.02\times 10^8)KR_{kg/min}}{(M)(C_{max})} = \frac{(4.02\times 10^8)K(SG)R_{L/min}}{(M)(C_{max})} \quad \text{[SI]} \qquad 39.25(a)$$

$$Q_{ft^3/min} = \frac{(3.86\times 10^8)KR_{lbm/min}}{(M)(C_{max})} = \frac{(4.03\times 10^8)K(SG)R_{pints/min}}{(M)(C_{max})} \quad \text{[U.S.]} \qquad 39.25(b)$$

When two or more hazardous substances that have similar toxicologic effects are simultaneously present (i.e., act on the same organ or metabolic process), the combined effect should be considered. If the air is to be

[12] Dilution ventilation is not recommended for carbon tetrachloride, chloroform, and gasoline, among others. Dusts are seldom removed successfully by dilution.

[13] Another unit used for dust concentrations in respirable air is *millions of particles per cubic foot* (mppcf) determined by *midget impinger* techniques. The conversion between mppcf and other units is not exact, depending primarily on the particle size and density. However, equivalences of 5.6 mppcf and 6.4 mppcf to 1.0 mg/m^3 are quoted. In the absence of any other information, an average value of 6 mppcf is recommended.

[14] In the United States, TLVs are updated annually in *Industrial Ventilation: A Manual of Recommended Practice*, published by the American Conference of Governmental Industrial Hygienists.

[15] Other restrictions may apply to TLV-STEL. For example, there may be a sixty-minute waiting period between successive exposures at this level.

[16] For example, irritant gases may be controlled only by the TLV-C value.

HVAC

breathed, the airflow rates for each substance must be calculated and the separate flow rates summed. The mixture threshold value is exceeded when

$$\left(\frac{C}{\text{TLV}}\right)_1 + \left(\frac{C}{\text{TLV}}\right)_2 + \cdots > 1 \qquad 39.26$$

The additive nature implied by Eq. 39.26 is assumed unless the two substances are known to act independently instead of additively. In that case, the threshold limit is exceeded only when the ratio C/TLV for at least one component in the mixture exceeds unity. The highest ventilation rate calculated for each component independently is the design ventilation rate.

18. RECIRCULATION OF CLEANED AIR

The volume of ventilation air required will be reduced if some of the contaminated air can be cleaned and returned. Dust and particulate matter in air can be removed by two types of air cleaners. *Air filters* are applicable when the concentration is between 0.5 grains and 50 grains per 1000 cubic feet (1.1 milligrams per cubic meter to 110 milligrams per cubic meter). *Dust collectors* are used at the higher concentrations normally found in manufacturing processes. Equilibrium will be achieved when the contaminant generation rate equals the rate at which the air filter removes particles from the air.

$$R = \eta_{\text{filter}} CQ \qquad 39.27$$

19. CLEAN ROOMS

Clean rooms are defined by the number of particles (pollen, skin flakes, etc.) above a given size (usually 0.5 microns) in a cubic foot of air. Most semiconductor clean rooms are Class-100 or better, meaning that there will be no more than one hundred 0.5 micron-sized particles per cubic foot.

Clean room technology generally relies on high-efficiency, prefilters, either *high-efficiency particulate arresting* (HEPAs) or *ultra-low penetration air* (ULPA) filters in the supply, positive room pressure, fast air movement, and floor grates (i.e., downflow air movement). A positive pressure of approximately 0.1 in of water (25 Pa) is typical. Twenty air changes per hour is a typical minimum, while airflow velocities of 75 ft/min to 100 ft/min (0.38 m/s to 0.5 m/s) are used in the best clean rooms.

High-efficiency (60% to 90%) prefilters reduce the load on HEPAs. Usually, HEPAs (99.97% efficiency at the 0.3 micron level) are suitable for Class-100 clean rooms, while ULPAs are needed for Class-10 or better clean rooms. Large centralized equipment may be used, or modular *air handling units* (AHUs) drawing air from the main general supply may be used for individual clean rooms. Adjustable-frequency drives can be used to change the airflow in order to reduce energy usage or change the cleanliness. Stainless steel is the preferred material for ducts and hoods, as it does not have the flaking problem associated with galvanized metals.

Ventilation requirements are similar to regular designs. Makeup air should be 25% of the total airflow and not less than 20 ft^3/min (0.57 m^3/min; 9.4 L/s) per person. Clean rooms are normally maintained with a positive pressure relative to the surroundings.

20. VENTILATION AND PRESSURIZATION IN LABORATORIES AND CLEAN ROOMS

Clean rooms should be maintained at positive pressure with respect to the surrounding areas. However, OSHA, NFPA, and ANSI require most laboratories to be maintained at negative pressure. Recirculation of air from laboratories is strongly discouraged, if not prohibited.

21. CLOSED RECIRCULATING ATMOSPHERES

The closed recirculating atmosphere in submarines and spacecraft presents unique challenges. There are four primary requirements for closed recirculation of atmosphere within closed environments such as submarines and spacecraft: oxygen replacement, carbon dioxide removal, moisture removal, and, in some cases, heat removal. Water vapor is removed in a dehumidification process. Replacement oxygen is added from tanks, electrolysis of water, or oxygen generators. Oxygen can be released continuously by a monitoring system that senses the percentage of oxygen in the air, or it can be released periodically in bulk.

Exhaled air is 4–5% by volume carbon dioxide. The carbon dioxide content is chemically reduced to a normal atmospheric concentration, approximately 0.04% by volume, in a *scrubber*.[17,18] Scrubbers use chemical aqueous absorbents (e.g., soda lime, consisting of mostly calcium hydroxide ($Ca(OH)_2$) with small amounts of sodium hydroxide (NaOH) and/or potassium hydroxide (KOH)) to remove carbon dioxide. The absorbent can be rejuvenated by heating.

HVAC

[17]Similar but separate scrubbing operations are required to remove other contaminants, such as carbon monoxide, hydrogen, and refrigerants in the closed system.

[18]Atmospheric air is 0.038% (380 ppm) carbon dioxide by volume.

22. NOMENCLATURE

A	area	ft^2	m^2
ACH	number of air changes per hour	1/hr	1/h
B	crack coefficient	ft^3/ft-min	m^3/m·min
c_p	specific heat	Btu/lbm-°F	kJ/kg·°C
C	concentration	lbm/ft^3	kg/m^3
C_d	coefficient of discharge	–	–
C_p	coefficient of pressure	–	–
D	occupancy diversity	–	–
E	efficiency	–	–
g	gravitational acceleration, 32.2 (9.81)	ft/sec^2	m/s^2
g_c	gravitational constant, 32.2	lbm-ft/lbf-sec^2	n.a.
h	elevation (height)	ft	m
h	enthalpy	Btu/lbm	kJ/kg
K	mixing factor	–	–
L	length	ft	m
$\dot{m}$	mass flow rate	lbm/min	kg/min
M	molecular weight	lbm/lbmol	kg/kmol
n	exponent	–	–
p	pressure	lbf/in^2	Pa
P	population or occupants	persons	persons
q	heat transfer rate	Btu/min	W
Q	flow rate	ft^3/min	m^3/min
R	contaminant generation rate	lbm/min	kg/min
R_a	rate per unit area	ft^3/min-ft^2	L/s·m^2
R_p	rate per person	ft^3/min-person	L/s·person
s	sheltering coefficient	–	–
SG	specific gravity	–	–
t	time	hr	h
T	temperature	°F or °R	°C or K
TLV	threshold limit value	various	various
v	velocity	mi/hr	km/h
V	volume	ft^3	m^3
W	humidity ratio	lbm/lbm	kg/kg
y	elevation	ft	m

Symbols

η	efficiency	–	–
ρ	density	lbm/ft^3	kg/m^3

Subscripts

a	area
abs	absolute
ADP	apparatus dew point
bz	breathing zone
fg	vaporization
id	indoor design
in	entering the room
l	latent
NPL	neutral pressure level
o	opening
ot	outdoor total
ou	outdoor uncorrected
oz	outdoor zone
p	people or primary
pz	primary zonal
s	sensible or system
v	ventilation
w	wind
z	zone

40 HVAC System Components

Content in blue refers to the *NCEES Handbook*.

NCEES EXAM SPECIFICATIONS AND RELATED CONTENT

HVAC AND REFRIGERATION EXAM

II.B.7. Equipment and Components: Control systems components
- 38. Dampers

II.C.1. Systems and Components: Air distribution
- 3. Velocity Pressure
- 5. Air Handlers
- 6. Variable Air Volume Boxes
- 7. Axial Fans
- 8. Centrifugal Fans
- 10. Fan Power
- 13. Fan Performance Curves
- 19. Fan Similarity
- 26. Friction Losses in Round Ducts
- 28. Rectangular Ducts

MACHINE DESIGN AND MATERIALS EXAM

II.A.5. Mechanical Components: Dampers
- 38. Dampers

THERMAL AND FLUID SYSTEMS EXAM

II.A.2. Hydraulic and Fluid Equipment: Compressors
- 3. Velocity Pressure
- 5. Air Handlers
- 6. Variable Air Volume Boxes
- 7. Axial Fans
- 8. Centrifugal Fans
- 10. Fan Power
- 13. Fan Performance Curves
- 19. Fan Similarity
- 26. Friction Losses in Round Ducts
- 28. Rectangular Ducts

1. STANDARD AND ACTUAL FLOW RATES

Airflow through fans is typically measured in units of cubic feet per minute, ft^3/min (L/s). When the flow is at the *standard conditions* of 70°F (21°C) and 14.7 psia (101 kPa), the airflow is designated as SCFM (*standard cubic feet per minute*). As described in Sec. 40.3, the density, ρ, of standard air is 0.075 lbm/ft^3 (1.2 kg/m^3). The specific volume, v, of standard air is the reciprocal of the density of standard air, or 13.33 ft^3/lbm

(0.8333 m^3/kg). Airflow at any other condition is designated as ACFM (*actual cubic feet per minute*). The two quantities are related by the density factor, K_d.[1] In Eq. 40.2, absolute temperature must be used.[2] Table 40.1 gives the ratio of p_{actual} to p_{std}.

$$\text{SCFM} = \frac{\text{ACFM}}{K_d} \quad 40.1$$

$$K_d = \frac{\rho_{std}}{\rho_{actual}} = \left(\frac{p_{std}}{p_{actual}}\right)\left(\frac{T_{actual}}{T_{std}}\right) \quad 40.2$$

Standard air is implicitly dry air. A correction for water vapor can be made if the relative humidity, ϕ, is known. The saturation pressure, p_{sat}, is read from a saturated steam table for the dry bulb temperature of the air.

$$\begin{aligned} K_d &= \left(\frac{p_{std}}{p_{actual} - p_{vapor}}\right)\left(\frac{T_{actual}}{T_{std}}\right) \\ &= \left(\frac{p_{std}}{p_{actual} - \phi p_{sat}}\right)\left(\frac{T_{actual}}{T_{std}}\right) \end{aligned} \quad 40.3$$

Table 40.1 *Pressure at Altitudes*

altitude (ft (m))	ratio of p_{actual}/p_{std}
sea level (0)	1.00
1000 (305)	0.965
2000 (610)	0.930
3000 (915)	0.896
4000 (1220)	0.864
5000 (1525)	0.832
6000 (1830)	0.801
7000 (2135)	0.772

(Multiply ft by 0.3048 to obtain m.)

Other airflow designations are ICFM (inlet cubic feet per minute), SDCFM (standard dry cubic feet per minute, the time rate of DSCF, dry standard cubic feet), MSCFD (thousand standard cubic feet per day), and MMSCFD (million standard cubic feet per day). The term ICFM is not normally used in duct design. It is used by compressor manufacturers and suppliers to specify conditions before and after filters, boosters, and other equipment. If the conditions before and after the equipment are the same, then ICFM and ACFM will be identical. Otherwise, Eq. 40.4 can be used.

$$\text{ACFM} = \text{ICFM}\left(\frac{p_{before}}{p_{after}}\right)\left(\frac{T_{after}}{T_{before}}\right) \quad 40.4$$

Example 40.1

A manufacturing application in Denver, Colorado (p_{actual} = 12.2 psia, T_{actual} = 60°F, relative humidity of 75%) requires 100 SCFM of compressed air at 125 psig. What is the ICFM in Denver?

Solution

Use Eq. 40.3 to calculate the density factor, K_d. From App. 23.A, the saturation pressure at 60°F is 0.2564 lbf/in^2.

$$\begin{aligned} K_d &= \left(\frac{p_{std}}{p_{actual} - \phi p_{sat}}\right)\left(\frac{T_{actual}}{T_{std}}\right) \\ &= \left(\frac{14.7\ \frac{\text{lbf}}{\text{in}^2}}{12.2\ \frac{\text{lbf}}{\text{in}^2} - (0.75)\left(0.2564\ \frac{\text{lbf}}{\text{in}^2}\right)}\right)\left(\frac{60°\text{F} + 460°}{70°\text{F} + 460°}\right) \\ &= 1.20 \end{aligned}$$

From Eq. 40.1, the inlet flow rate is

$$\begin{aligned} \text{ICFM} = \text{ACFM} = K_d(\text{SCFM}) &= (1.20)\left(100\ \frac{\text{ft}^3}{\text{min}}\right) \\ &= 120\ \text{ft}^3/\text{min} \end{aligned}$$

2. STATIC PRESSURE

The force of moving or stationary air perpendicular to a duct wall is known as the *static pressure*, p_s. Static pressure can be measured in the field by a static tube or static tap. It is usually reported in inches of water, abbreviated "in wg" (or "in. w.g.") for "inches of water gage," or in pascals. The pressure, height of a fluid column, and specific weight of the fluid are related by Eq. 40.5. The specific weight, γ, of water is approximately 0.0361 lbf/in^3 (9810 N/m^3).

$$h = \frac{p}{\gamma} \quad 40.5$$

[1] Some sources use an *air density ratio* that is the reciprocal of the density factor, K_d, defined by Eq. 40.2. In some confusing cases, the same name (i.e., density factor) is used with the reciprocal value.

[2] The temperature correction should be based on the temperature and pressure of the air through the duct system. Though atmospheric pressure and temperature both decrease with higher altitudes, air entering any occupied space will generally be heated to normal temperatures. Therefore, the temperature correction will not generally be used unless the duct system carries air for process heating or cooling.

$$p_{s,\text{in wg}} = \frac{p_{\text{psig}}}{0.0361 \frac{\text{lbf}}{\text{in}^3}} \quad \text{40.6}$$

3. VELOCITY PRESSURE

The *velocity pressure*, p_v, is the kinetic energy of the air expressed in inches of water. Velocity pressure is measured with a pitot tube, hand-held velometer, hot wire or rotating vane anemometer, or calibrated orifice or nozzle. The velocity head of a mass of moving air is

Minor Losses in Pipe Fittings, Contractions, and Expansions

$$h_v = \frac{v^2}{2g} \quad \text{40.7}$$

Air velocities are typically measured in ft/min (fpm) in the United States and in m/s in countries that use SI units. (The designation "LFM," for "linear feet per minute," is occasionally encountered.) As expressed, Eq. 40.7 calculates the kinetic energy as the height of an air column, not a height of a water column. The specific weights of air and water are approximately 0.075 lbf/ft^3 (density of 1.2 kg/m^3) and 62.4 lbf/ft^3 (density of 1000 kg/m^3), respectively. Assuming these conditions represent the flow being evaluated, the velocity pressure in inches of water may be estimated as[3]

Duct Design: Bernoulli Equation

$$\begin{aligned} p_{v,\text{in wg}} &= \rho\left(\frac{v_{\text{ft/min}}}{1097}\right)^2 \\ &\approx \left(\frac{v_{\text{ft/min}}}{4005}\right)^2 \quad \text{[standard conditions]} \\ &= \left(\frac{\left(\dfrac{v}{60 \frac{\text{sec}}{\text{min}}}\right)^2 \left(12 \frac{\text{in}}{\text{ft}}\right)}{(2)\left(32.2 \frac{\text{ft}}{\text{sec}^2}\right)}\right)\left(\frac{0.075 \frac{\text{lbf}}{\text{ft}^3}}{62.4 \frac{\text{lbf}}{\text{ft}^3}}\right) \end{aligned} \quad \text{40.8}$$

In SI units, with pressure in pascals, the velocity pressure is

$$\begin{aligned} p_{v,\text{Pa}} &= \frac{\rho v_{\text{m/s}}^2}{2} = \frac{\left(1.2 \frac{\text{kg}}{\text{m}^3}\right) v_{\text{m/s}}^2}{2} \\ &= 0.6 v_{\text{m/s}}^2 \quad \text{[standard conditions]} \end{aligned} \quad \text{40.9}$$

4. TOTAL PRESSURE

The *total pressure*, p_t, is the sum of the velocity and static pressures. Contributions from potential energy are insignificant in virtually all duct design problems.

The total pressure decreases in the direction of flow. (However, the static pressure can increase with increased diameter and decrease with increased velocity.)

$$p_t = p_s + p_v = p_s + \rho\left(\frac{v_{\text{ft/min}}}{1097}\right)^2 \quad \text{40.10}$$

The change in total pressure is the algebraic sum of the changes in static and velocity pressures. In straight ducts with no branches or diameter changes, the change in total pressure is the same as the friction loss. Δp_t is positive when total pressure decreases.

$$\begin{aligned} \Delta p_t &= p_{t,1} - p_{t,2} = \Delta p_s + \Delta p_v \\ &= p_{s,1} - p_{s,2} + p_{v,1} - p_{v,2} \end{aligned} \quad \text{40.11}$$

5. AIR HANDLERS

Air handling unit (*AHU* or *air handler*) is the term used to describe a unit that combines a fan with other process equipment, such as heating coils, cooling/dehumidification coils, and filters, as well as various dampers and bypass paths. An air handler that conditions outside air only, without receiving any recirculated air, is known as a *make-up air unit* (MAU). An air handler designed for exterior use is known as a *packaged unit* (PU) or *rooftop unit* (RTU). Smaller units containing an air filter, coil, and blower are known as *terminal units*, *blower-coil units*, or *fan-coil units*. (See *ASHRAE Handbook—HVAC Applications* for a diagram of a typical single-zone air handling unit control arrangement.) **[HVAC System Components] [Typical Single Zone Air Handling Unit] [Single-Zone VAV Fan System Diagram]**

For air handling units with mixed-air plenums in which the temperature difference between the outdoor air and return air is more than 20°F, the mixed air temperature is

Air-Handling Unit Mixed-Air Plenums

$$\begin{aligned} T_m &= (\text{fraction of outdoor air})\,T_o \\ &\quad + (\text{fraction of return air})\,T_r \\ &= \frac{Q_o}{Q_t}T_o + \frac{Q_r}{Q_t}T_r \end{aligned} \quad \text{40.12}$$

Q_o, Q_r, and Q_t are the volumetric flow rates for outdoor air, return air, and combined outdoor air and return air, respectively.

[3]The constant 4005 is reported as 4004 and 4004.4 in some references. 4005 is the most common.

HVAC

6. VARIABLE AIR VOLUME BOXES

Historically, most HVAC systems were constant air volume (CAV) systems that varied the temperature of the delivered air in order to maintain space conditions. In its simplest form, a *variable air volume* (VAV) *terminal box* is a unit that varies the amount of air entering a zone while keeping the temperature of the delivered air constant. A VAV box typically contains a motorized damper controlled by a controller that receives signals from sensors. Zone air temperature is the primary input variable affecting flow quantity, although velocity in the supply duct may also be measured. The controller may be pneumatic, single loop digital, or microprocessor. In addition to increased reliability and accuracy, microprocessor controllers accommodate daily schedules, automatic adjustment of hot and cold set points, multiple after-hours (unoccupied) setbacks, outdoor air ventilation control, and on-demand operation. The controller's *deadband* (*deadzone* or *neutral zone*) is the temperature range over which the controller does not generate any control signal. For a VAV with both heating and cooling, a deadband separates the ranges of temperatures over which heating and cooling coils are turned on. [**Single-Duct, Constant Volume Reheat**]

There are many types of VAV systems, including single- and multiple-zone systems; variable and constant volume systems (with and without constant- and variable-speed fans); and systems with and without reheat units. Many hybrid heating and cooling systems make use of VAV systems in combination with other approaches for maintaining space conditions. *ASHRAE Standard* 55 lists the following commonly used types of VAV systems:

- bypass unit
- changeover-bypass system
- cooling-only unit
- dual duct system
- fan-powered reheat unit
- induction unit
- reheat unit

Single-zone, VAV-with-reheat (VRH) and parallel fan-powered VAV (FPV) units are most common. Selection (sizing and design) of VAV systems is complex, as input and output duct quantity and sizes, pressure drop across the box, noise generation, heating and cooling capacity, and installation size are all related. Selection is further complicated by consideration of life-cycle costs that include energy usage. *ASHRAE Handbook—HVAC Applications* contains schematics for many types of VAV units. [**Single-Duct, Variable Air Volume (VAV)**] [**Variable Air Volume, Dual-Maximum**] [**Series Fan-Powered VAV Terminal Unit**] [**Parallel Fan-Powered VAV Terminal Unit**]

In a true VRH installation, the central air handler's heating and cooling coils remove sensible and latent heat from all zones served, but the air supplied to the VAV box is heated only to 55°F. Any additional heating required by the zone is done by the reheat coil in the VAV terminal box. In a cooling mode, the VAV damper will be all or mostly open, supplying as much 55°F air as is required to cool the zone. Since VAV boxes vary the volume of supplied air instead of the supply air temperature, operation in cooling mode is energy efficient. In the heating mode, or in the cooling mode after the zone has been adequately cooled, electric or hydronic heating elements turn on, and the VAV damper closes substantially, supplying only enough air to meet outside air ventilation requirements. Air typically enters the VAV terminal box at 55°F (13°C) and leaves (enters the zone) after its temperature is increased approximately 20°F (11°C), to 75°F (24°C), by the reheat coil. The reheat coil should be sized to satisfy the conditioned zone's heating load as well as to heat the supply air from 55°F (13°C) to the required temperature to satisfy the room setpoint temperature. For safe operation, electric heaters should include redundant high-limit sensors and a minimum airflow switch to address low airflow conditions caused by duct blockages (e.g., fire damper closure). [**Electric Heaters**]

VAV controllers have minimum and maximum volumetric setpoints. Regardless of the type of VAV, the maximum is typically the airflow required to provide the design cooling. For a VRH unit, the minimum airflow is the largest of (1) outside air ventilation requirements, (2) the airflow that meets the design heating load at a reasonable temperature (e.g., below 90°F (32°C)), and (3) the airflow required to prevent dumping and poor distribution. For an FPV unit, typically only the ventilation requirement is important because the parallel fan operation ensures high supply rates. Both California Title 24 and ASHRAE Standard 90.1 limit the minimum airflow to the largest of (1) 30% of the maximum airflow, (2) the minimum required for ventilation, (3) 0.4 cfm/ft^2 (2 L/s per m^2), and (4) 300 cfm (142 L/s).

VAV boxes in the duct system's index run will affect the system pressure. Equipment in the index run will also affect the design pressure. The index run is generally the longest run in the system. Since the inlet velocity will normally be greater than the outlet velocity, there will be static regain, so the static pressure drop across a VAV box will normally be less than the total pressure drop (the sum of static and velocity pressure changes). Total pressure drop is the true measure of fan power required. Drops in total pressure should be used to evaluate and select VAV boxes, since the fan has to supply both static and velocity pressures.

7. AXIAL FANS

Axial-flow fans are essentially propellers mounted with small tip clearances in ducts. They develop static pressure by changing the airflow velocity. Axial flow fans are usually used when it is necessary to move large quantities of air (i.e., greater than 500,000 ft^3/min; 235 000 L/s) against low static pressures (i.e., less than 12 in of water, 3 kPa), although the pressures and flow rates are much lower at most installations. An axial-flow fan may be followed by a *diffuser* (i.e., an *evase* or *volute*) to convert some of the kinetic energy to static pressure.

Compared with centrifugal fans, axial flow fans are more compact and less expensive. However, they run faster than centrifugals, draw more power, are less efficient, and are noisier. Axial flow fans are capable of higher velocities than centrifugal fans. In addition, overloading is less likely due to the flatter power curve. Fan noise is lowest at maximum efficiencies.

Axial fans can be further categorized into propeller, tubeaxial, and vaneaxial varieties. *Propeller fans* (such as the popular ceiling-mounted fans) are usually used only for exhaust and make-up air duty, where the system static pressure is not more than ½ in wg (125 Pa). Since they generally don't have housings, they are not capable of generating static pressures in excess of about 1 in wg (250 Pa). Though they are light and inexpensive, they are the least efficient (efficiency of about 50%) and the most noisy axial fans. *ASHRAE Handbook—HVAC Systems and Equipment* provides additional axial fan impeller and housing design details. [**Types of Fans**]

Tubeaxial fans, also known as *duct fans*, generally move air against less than 3 in of water (750 Pa). They have four to eight blades, and the clearance between the blade tips and surrounding duct is small. Fan efficiency is approximately 75–80%. Tubeaxials can be recognized by their hub diameters, which are less than 50% of the tip-to-tip diameters.

Vaneaxial fans can be distinguished from tubeaxial fans by their hub diameters, which are greater than 50% of the tip-to-tip diameters. Furthermore, the fan assembly will usually have vanes downstream from the fan to straighten the airflow and recover the rotational kinetic energy that would otherwise be lost. Vaneaxials typically have as many as 24 blades, and the blades may have cross sections similar to airfoils. Because they recover the rotational energy, vaneaxials are capable of moving air against pressures of up to 12 in of water (3.8 kPa). Their efficiencies are typically 85–90%.

8. CENTRIFUGAL FANS

Centrifugal fans are used in installations moving less than 1×10^6 ft^3/min (470 000 L/s) and pressures less than 60 in of water (15 kPa). Like centrifugal pumps, they develop static pressure by imparting a centrifugal force on the rotating air. Depending on the blade curvature, kinetic energy can be made greater (forward-curved blades) or less (backward-curved blades) than the tangential velocity of the impeller blades. [**Types of Fans**]

Forward-curved centrifugals (also called *squirrel cage fans*) are the most widely used centrifugals for general ventilation and packaged units. They operate at relatively low speeds, about half that of backward-curved fans. This makes them useful in high-temperature applications where stress due to rotation is a factor. Compared with backward-curved centrifugals, forward-curved blade fans have a greater capacity (due to their higher velocities) but require larger scrolls. However, since the fan blades are "cupped," they cannot be used when the air contains particles or contaminants. Efficiencies are the lowest of all centrifugals—70–75%. [**Types of Fans**]

Motors driving centrifugal fans with forward-curved blades can be overloaded if the duct losses are not calculated correctly. The power drawn increases rapidly with increases in the delivery rate. The motors are usually sized with some safety factor to compensate for the possibility that the actual system pressure will be less than the design pressure. For forward-curved blades, the maximum efficiency occurs near the point of maximum static pressure. Since their tip speeds are low, these fans are quiet. The fan noise is lowest at maximum pressure.

Radial fans (also called *straight-blade fans*, *paddle wheel fans*, and *shaving wheel fans*) have blades that are neither forward- nor backward-inclined. Radial fans are the workhorses of most industrial exhaust applications and can be used in material-handling and conveying systems where large amounts of bulk material pass through them. Such fans are low-volume, high-pressure (up to 60 in wg; 15 kPa), high-noise, high-temperature, and low-efficiency (65–70%) units. *Radial tip fans* constitute a subcategory of radial fans. Their performance characteristics are between those of forward-curved and conventional radial fans. [**Types of Fans**]

Backward-curved centrifugals are quiet, medium- to high-volume and pressure, and high-efficiency units. They can be used in most applications with clean air below 1000°F (540°C) and up to about 40 in wg (10 kPa). They are available in three styles: flat, curved, and airfoil. Airfoil fans have the highest efficiency (up to 90%), while the other types have efficiencies between 80% and 90%. Because of these high efficiencies, power savings easily compensate for higher installation or replacement costs. [**Types of Fans**]

Motor overloading is less likely with backward-curved blades than with forward-curved blades, and, therefore, may be referred to as *non-overloading fans*. These fans are normally equipped with motors sized to the peak power requirement so that the motors will not overload at any other operating condition. [**Fan Performance**]

Such fans operate over a great range of flows without encountering unstable air. Though their efficiencies are greater, they are noisier than forward-curved fans. The fan noise is lowest at the highest efficiencies. For the same operating speed, backward-curved blade fans develop more pressure than forward-curved fans.

9. FAN SPECIFIC SPEED

The fan specific speed is calculated from Eq. 40.13. Specific speed ranges will be 10,000 to 20,000 (110 to 220) for radial centrifugals; 12,000 to 50,000 (130 to 550) for centrifugals; 40,000 to 170,000 (440 to 1900) for vaneaxials; 100,000 to 200,000 (1100 to 2200) for tubeaxials; and 120,000 to 300,000 (1300 to 3300) for propeller fans.

$$N_s = \frac{N_{\text{rpm}}\sqrt{Q_{\text{L/s}}}}{p_{s,\text{Pa}}^{0.75}} \quad \text{[SI]} \qquad 40.13(a)$$

$$N_s = \frac{N_{\text{rpm}}\sqrt{Q_{\text{ft}^3\text{/min}}}}{p_{s,\text{in wg}}^{0.75}} \quad \text{[U.S.]} \qquad 40.13(b)$$

10. FAN POWER

The *air horsepower* (*blower horsepower*), ahp, or *air kilowatts*, akW, is the power required to move the air.

$$\text{akW} = \frac{Q_{\text{L/s}}(p_{t,\text{Pa}})}{10^6} \quad \text{[SI]} \qquad 40.14(a)$$

Fan Power Requirements

$$\text{ahp} = 0.000157 Q_{\text{ft}^3\text{/min}}(p_{t,\text{in wg}}) \quad \text{[U.S.]} \qquad 40.14(b)$$

The actual power delivered to a fan from its motor is the *brake horsepower*, bhp, or *brake kilowatts*, bkW. Centrifugal fan efficiencies are in the range of 70–75%, although values as high as 90% are possible. The mechanical efficiency, $\eta_{\text{mechanical}}$, of a fan is normally read from the *total efficiency* fan curves once the operating point is known, but can be calculated from Eq. 40.15.

$$\eta_{\text{mechanical}} = \frac{\text{ahp}}{\text{bhp}} = \frac{\text{akW}}{\text{bkW}} \qquad 40.15$$

The electrical power delivered to the motor will include the friction, windage, and other electrical losses in the motor, which are factored in as $\eta_{\text{mechanical}}$, plus drive losses, DL, or drive belt efficiency losses, η_{drive}. The electrical power is

Fan Power Requirements

$$P_{\text{motor}} = \frac{(1+\text{DL})\text{bhp}}{\eta_{\text{mechanical}}\eta_{\text{drive}}} \qquad 40.16$$

The primary advantage of using a variable speed fan motor is the ability to accommodate minor changes in flow rates and resistance (e.g., changes in the system friction, as when filter resistance increases over time), which results in considerable fan-related energy savings. Variable speed fan drives are often called variable frequency drives (VFD).

Example 40.2

A fan moves 27,000 ft^3/min (12 700 L/s) of air at 1800 ft/min (9.2 m/s) against a static pressure of 2 in wg (500 Pa). The electrical motor driving the fan delivers 13.12 hp (9.77 kW) to it. What is the mechanical efficiency?

SI Solution

From Eq. 40.9, the velocity pressure is

$$p_{\text{v,Pa}} = 0.6\text{v}_{\text{m/s}}^2 = (0.6)\left(9.2\ \frac{\text{m}}{\text{s}}\right)^2 = 51\ \text{Pa}$$

From Eq. 40.10, the total pressure is

$$p_t = p_s + p_\text{v} = 500\ \text{Pa} + 51\ \text{Pa} = 551\ \text{Pa}$$

From Eq. 40.14, the air kilowatts are

$$\begin{aligned}\text{akW} &= \frac{Q_{\text{L/s}}(p_{t,\text{Pa}})}{10^6} = \frac{\left(12{,}700\ \frac{\text{L}}{\text{s}}\right)(551\ \text{Pa})}{10^6} \\ &= 7\ \text{kW}\end{aligned}$$

From Eq. 40.15, the fan efficiency is

$$\eta_{\text{mechanical}} = \frac{\text{ahp}}{\text{bhp}} = \frac{7\ \text{kW}}{9.77\ \text{kW}} = 0.72 \quad (72\%)$$

Customary U.S. Solution

From Eq. 40.8, the velocity pressure is

Duct Design: Bernoulli Equation

$$p_{\text{v,in wg}} = \left(\frac{\text{v}_{\text{ft/min}}}{4005}\right)^2 = \left(\frac{1800\ \frac{\text{ft}}{\text{min}}}{4005}\right)^2 = 0.2\ \text{in wg}$$

From Eq. 40.10, the total pressure is

$$\begin{aligned}p_t &= p_s + p_\text{v} = 2.0\ \text{in wg} + 0.2\ \text{in wg} \\ &= 2.2\ \text{in wg}\end{aligned}$$

From Eq. 40.14, the air horsepower is

Fan Power Requirements

$$\begin{aligned}\text{ahp} &= 0.000157 Q_{\text{ft}^3/\text{min}} p_{t,\text{in wg}} \\ &= (0.000157)\left(27{,}000\ \frac{\text{ft}^3}{\text{min}}\right)(2.2\ \text{in wg}) \\ &= 9.33\ \text{hp}\end{aligned}$$

From Eq. 40.15, the fan efficiency is

$$\eta_{\text{mechanical}} = \frac{\text{ahp}}{\text{bhp}} = \frac{9.33\ \text{hp}}{13.12\ \text{hp}} = 0.71 \quad (71\%)$$

11. VARIABLE FLOW RATES

Ventilation and air conditioning rates usually vary with time. It is essential in modern, large systems to be able to vary the flow rate, as large amounts of energy are saved. *Variable flow rates* (i.e., *capacity control, flow rate modulation*) can be achieved through use of system dampers, fan speed control, variable blade pitch, inlet vanes, and variable frequency drives.

System dampers downstream of the fan are rarely used for capacity control. They increase friction loss, are noisy, and are nonlinear in their response. Speed control of the fan through fluid or magnetic coupling has low noise levels, but a high initial cost. For that reason, it also is seldom used.

With *blade pitch control* (*controllable pitch*), all blades are linked and simultaneously controlled while the fan is operating. Changing the blade angle of attack is efficient, quiet, and linear in response.

Inlet vanes are commonly used with centrifugal and inline fans. Inlet vanes pre-spin and throttle the air prior to its entry into the wheel. Inlet vanes are relatively inefficient, noisy, and nonlinear in response.

Though a direct drive with flexible coupling can be used when flow is steady, most fans are run by v-belts.[4] In some applications requiring variable volume, either *variable-pitch pulleys* or variable (multispeed) motors can be used. However, most modern designs use pulse width modulation (PWM) (also known as *pulse duration modulation*) with variable frequency drives to achieve energy-saving and efficient control of supply and exhaust fans, pumps, and other HVAC system components.

12. TEMPERATURE INCREASE ACROSS THE FAN

The difference between the brake horsepower and air horsepower represents the power lost in the fan. This *friction horsepower*, fhp, heats the air passing through the fan. If the fan motor is also in the airstream, it will also contribute to the heating effect to the extent that the motor's efficiency is not 100%.[5]

$$\text{fhp} = \text{bhp} - \text{ahp} = \text{bhp}(1 - \eta_{\text{mechanical}}) \qquad 40.17$$

The temperature increase across the fan is given by Eq. 40.18. Consistent units must be used.

$$\Delta T = \frac{\text{fhp}}{\dot{m}c_p} \quad \text{(friction horsepower)}$$

or

$$\Delta T = \frac{0.0278\Delta p}{\rho\eta} \quad \text{(pressure drop (in wg))} \qquad 40.18$$

If traditional units are used and the air is essentially at standard conditions, Eq. 40.19 and Eq. 40.20 give the relationship between temperature change and the sensible heating or cooling effect.

$$\begin{aligned}\Delta T_{^\circ\text{F}} &= \frac{3160(\text{heating or cooling effect in kW})}{Q_{\text{ft}^3/\text{min}}} \\ &= \frac{0.926\left(\text{heating or cooling effect in }\dfrac{\text{Btu}}{\text{hr}}\right)}{Q_{\text{ft}^3/\text{min}}} \\ &= \frac{2356(\text{heating or cooling effect in hp})}{Q_{\text{ft}^3/\text{min}}}\end{aligned} \qquad 40.19$$

$$\begin{aligned}\Delta T_{^\circ\text{C}} &= \frac{829(\text{heating or cooling effect in kW})}{Q_{\text{L/s}}} \\ &= \frac{1760(\text{heating or cooling effect in kW})}{Q_{\text{ft}^3/\text{min}}}\end{aligned} \qquad 40.20$$

13. FAN PERFORMANCE CURVES

The operational parameters of fans are usually presented graphically by fan manufacturers. Total pressure, power, and efficiency are typically plotted on *fan characteristic curves*. Figure 40.1(a)–(c) contains typical curves for the three main types of fans: axial flow, forward-curved centrifugal, and backward-curved centrifugal. The dip in total pressure for axial flow and forward-curved centrifugal fans is characteristic. **[Fan Performance]**

[4]When selecting v-belts for fans, use a load factor of 1.4 (i.e., a *power rating* of 1.4 times the motor power).

[5]If the electrical input power and air horsepower are known, it isn't necessary to determine where the friction losses occur. The bearings, pulleys, and belts may all contribute to friction. However, the heating depends only on the difference between the input power and the power contributing to pressure and velocity.

HVAC

Figure 40.1 Typical Fan Curves

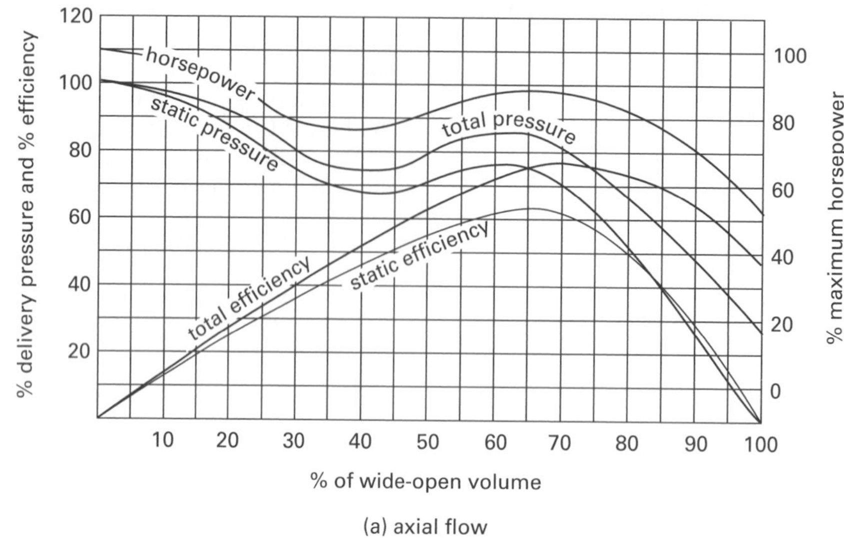

(a) axial flow

(b) forward-curved centrifugal

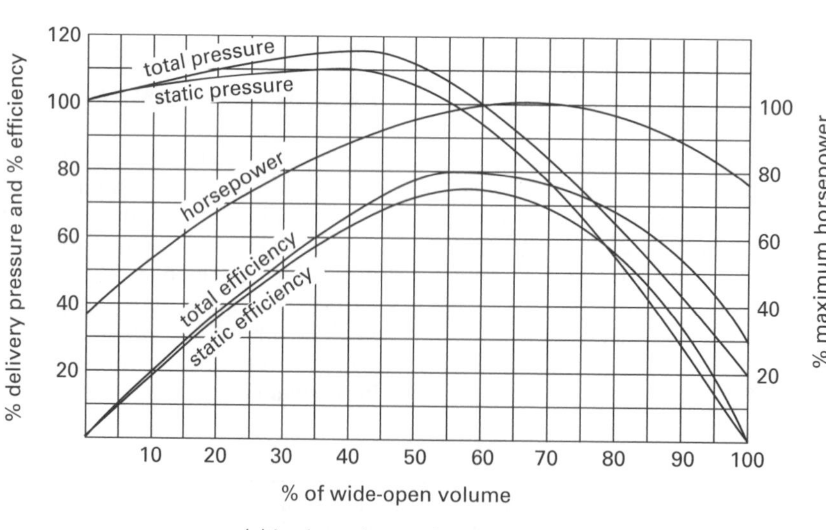

(c) backward-curved centrifugal

14. MULTIRATING TABLES

Some manufacturers provide *fan rating tables* similar to Table 40.2. These tables, known as *multirating tables*, give the fan curve data in tabular, rather than in graphical, format. The highest mechanical efficiency for each pressure range will be in the middle third of the flow rate, Q, range. Manufacturers often indicate (by underlining or shading) points of operation that are within 2% of the peak efficiency. If the peak efficiency point is not indicated, the actual efficiency can be calculated for each point using Eq. 40.15. Otherwise, selections can be limited to the middle third of the column.

Table 40.2 *Typical Fan Rating Table (portion)*

	$p_s = 1$ in wg		$p_s = 2$ in wg	
Q (ft^3/min)	n (rev/min)	P (bhp)	n (rev/min)	P (bhp)
5000	440	1.20	617	2.67
10,000	492	2.18	626	4.20
15,000	600	4.06	706	6.45
20,000	816	9.59	830	10.83

15. SYSTEM CURVE

The ductwork's *index run* (or, *critical path*) has the highest overall pressure drop and determines the total pressure, and therefore, the fan power required. *Resistance pressure*, *friction pressure*, and *external static pressure* are terms that are used to designate the minimum total system pressure that the fan must provide.

As with liquid flow in pipes, the pressure loss due to friction of air flowing in ducts varies with the square of the velocity. And, since $Q = Av$, the pressure loss varies with the square of Q. The graph of the friction loss versus the flow rate is the *system characteristic curve* (*system curve*), as shown in Fig. 40.2.

If one point on the curve is known, the remainder of the curve can be found or generated from Eq. 40.21. Static head is assumed to be insignificant, and due to the low density of air, this is almost always true.

$$\frac{p_2}{p_1} = \left(\frac{Q_2}{Q_1}\right)^2 \qquad 40.21$$

Figure 40.2 *Typical System Curve*

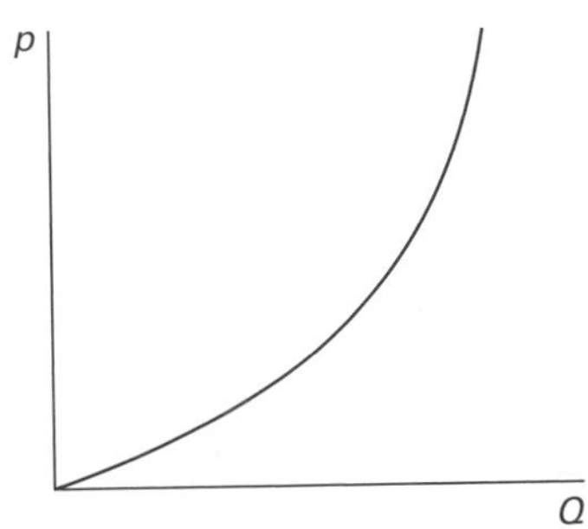

16. SYSTEM EFFECT

Almost all fans are rated under ideal laboratory conditions. Not only is the air at standard conditions, but also many of the physical features that would normally cause turbulence (bearings, diffusers, plenums, duct corners, etc.) are not present when the fan is tested. For that reason, rated performance is rarely achieved in practice.

Most fans are tested without being attached to a duct system.[6] Merely connecting a duct system to a fan will produce a degradation in fan performance from rated values. This degradation, known as the *system effect*, is in addition to the duct friction and other losses. The *system effect factor* is the additional pressure (in inches of water) that must be added to the calculated duct friction. The system effect factor depends on the flow rate (velocity) through the fan and the type of fan. For that reason, it must be based on information provided by the fan manufacturer.

17. OPERATING POINT

The intersection of the fan and system curves defines the *operating point* (*point of operation*).[7] If a fan is to be chosen by plotting the system curve on various fan curves, the following guidelines should be observed.

- To minimize the required motor power, the operating point should be as close as possible to the peak efficiency.
- For fans whose pressure characteristics have a dip (e.g., forward-curved centrifugals and axial flow fans), the operating point should be to the right of the peak fan pressure. This will avoid the noise and uneven motor loading that accompany pressure and volume fluctuations. (See Fig. 40.3.) The unstable region is shown in Fig. 40.3 as being to the left of the peak. Technically, the region of instability includes a horizontal band containing all system pressures for which there are two or

[6]The Air Movement and Control Association provides specifications for test setups. Discharge ductwork is specified, including straighteners and length (i.e., the number of equivalent duct diameters).

[7]The *rating point* (*point of rating*) is the one single point on the fan curve that corresponds to the stated (often the rated) performance. The *duty point* (*point of duty*) is one single point on the system curve where a fan is to operate.

HVAC

three different airflows. Depending on transient factors, the airflow could fluctuate between the corresponding rates while satisfying the requirement that duct friction equals the supply pressure. This is known as airflow *surging*. The curves of axial fans and those with forward curved airfoils and backward curved blades have regions of instability and are more prone to surging. Surging is more difficult with steep fan curves.

- A fan with a steep pressure curve should be chosen to avoid large variations in flow rate with changes in system pressure.

Figure 40.3 *Operating Point and Unstable Region*

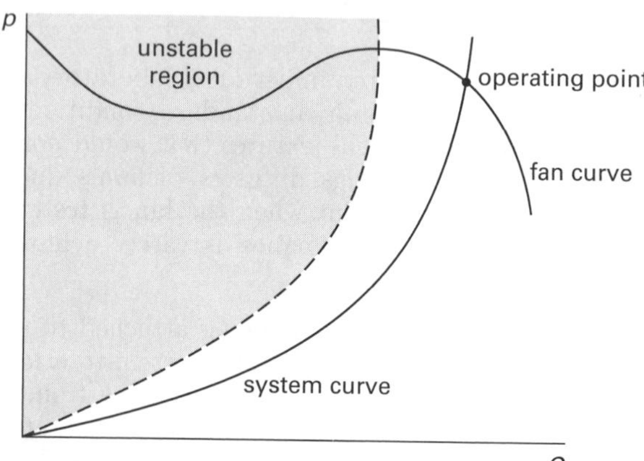

System configurations (length of runs, bends, equipment, etc.) are usually less flexible than fans when it comes to changing performance. If the operating point for a specific fan does not provide the required airflow (efficiency, power, etc.), there are several different steps that can be taken.

- Use a different fan.
- Change the fan speed.
- Change the fan size.
- Use two fans in parallel. The combined flow at a particular pressure will be the sum of the individual fan flows corresponding to that pressure.
- Use two fans in series. The combined pressure at a particular flow rate will be the sum of the individual fan pressures corresponding to that flow rate.

Example 40.3

The pressure loss due to friction in a system is 1.5 in wg (375 Pa) when the flow rate is 3500 ft³/min (1650 L/s). Velocity head and outlet pressure are negligible. What will be the flow rate if a fan with the characteristics shown is used?

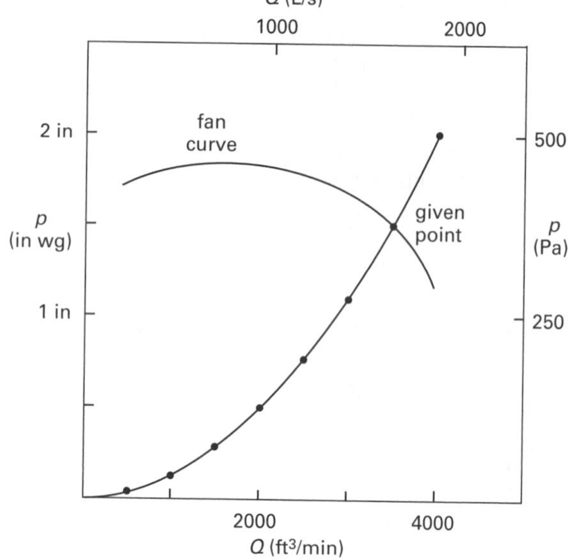

SI Solution

The fan curve is given. One point on the system curve is known. Use Eq. 40.21 to derive the remainder of the system curve. For 1500 L/s, the pressure drop would be

$$p_2 = p_1\left(\frac{Q_2}{Q_1}\right)^2 = (375\text{ Pa})\left(\frac{1500\ \frac{\text{L}}{\text{s}}}{1650\ \frac{\text{L}}{\text{s}}}\right)^2 = 310\text{ Pa}$$

The remainder of the system curve can be determined in the same manner. The fan and system curves intersect at approximately 1550 L/s.

Q (L/s)	p (Pa)
500	35
750	80
1000	140
1250	215
1500	310
1650	375
1750	420

Customary U.S. Solution

The fan curve is given. One point on the system curve is known. Use Eq. 40.21 to derive the remainder of the system curve. For 3000 ft^3/min, the pressure drop would be

$$p_2 = p_1\left(\frac{Q_2}{Q_1}\right)^2 = (1.5\text{ in wg})\left(\frac{3000\ \frac{\text{ft}^3}{\text{min}}}{3500\ \frac{\text{ft}^3}{\text{min}}}\right)^2$$
$$= 1.1\text{ in wg}$$

The remainder of the system curve can be determined in the same manner. The fan and system curves intersect at approximately 3300 ft^3/min.

Q (ft^3/min)	p (in wg)
500	0.03
1000	0.12
1500	0.28
2000	0.49
2500	0.77
3000	1.1
3500	1.5
4000	2.0

18. AFFINITY LAWS

Within reasonable limits, the speed of v-belt driven fans can be easily changed by changing pulleys. The following *affinity laws* (*fan laws*) can be used to predict performance of a particular fan at different speeds. These fan laws assume the fan size, fan efficiency, and air density are the same.[8,9]

$$\frac{Q_2}{Q_1} = \frac{N_2}{N_1} \qquad 40.22$$

$$\frac{p_2}{p_1} = \left(\frac{N_2}{N_1}\right)^2 \quad \left[\text{static, velocity, or total pressure}\right] \qquad 40.23$$

$$\frac{\text{ahp}_2}{\text{ahp}_1} = \left(\frac{N_2}{N_1}\right)^3 \qquad 40.24$$

Since the efficiency at the two speeds is assumed to be the same, Eq. 40.24 can be rewritten in terms of power drawn.

$$\frac{\text{bhp}_2}{\text{bhp}_1} = \left(\frac{N_2}{N_1}\right)^3 \qquad 40.25$$

A change in the fan speed cannot be used to put the operating point into a stable region. Changing the fan speed does not change the relative position of the operating point with respect to the dip present in some fan curves. The locus of peak points follows the Q^2 rule also. So, if an operating point is to the left of the peak point (i.e., is in an unstable region) at one fan speed, the new operating point will be to the left of the peak point on the new fan curve corresponding to the new speed.

19. FAN SIMILARITY

The performance of one fan can be used to predict the performance of a *dynamically similar* (*homologous*) fan. This can be done by using Eq. 40.26 through Eq. 40.34. Pressures may be static, velocity, or total.

Equation 40.26 to Eq. 40.28 are used to determine capacity, pressure, and power as a function of changes in size, speed, or density.

Fan Laws

$$Q_1 = Q_2\left(\frac{D_1}{D_2}\right)^3\left(\frac{N_1}{N_2}\right) = Q_2\left(\frac{D_1}{D_2}\right)^2\sqrt{\frac{p_1}{p_2}}\sqrt{\frac{\gamma_2}{\gamma_1}} \qquad 40.26$$

$$p_1 = p_2\left(\frac{D_1}{D_2}\right)^2\left(\frac{N_1}{N_2}\right)^2\left(\frac{\rho_1}{\rho_2}\right) = p_2\left(\frac{D_1}{D_2}\right)^2\left(\frac{N_1}{N_2}\right)^2\left(\frac{\gamma_1}{\gamma_2}\right) \qquad 40.27$$

$$P_1 = P_2\left(\frac{D_1}{D_2}\right)^5\left(\frac{N_1}{N_2}\right)^3\left(\frac{\rho_1}{\rho_2}\right) = P_2\left(\frac{D_1}{D_2}\right)^2\left(\frac{p_1}{p_2}\right)^{3/2}\sqrt{\frac{\gamma_2}{\gamma_1}} = \left(\frac{Q_1}{Q_2}\right)^2\left(\frac{p_1}{p_2}\right) \qquad 40.28$$

HVAC

[8]These fan laws are simplifications of the similarity laws presented in the next section. The similarity laws must be used if the density changes.

[9]For any given efficiency, the locus of equal-efficiency points on the pressure-capacity (p-Q) diagram is a parabola starting at the origin and crossing the different fan curves corresponding to different speeds. The intersection points of the fan curves and the parabolic equal-efficiency curve are known as *corresponding points*. Theoretically, the fan laws can only be used at these points.

Equation 40.29 to Eq. 40.31 are used to determine capacity, speed, and power as a function of changes in size, pressure, or density.

Fan Laws

$$Q_1 = Q_2\left(\frac{D_1}{D_2}\right)^2\left(\frac{p_1}{p_2}\right)^{1/2}\left(\frac{\rho_2}{\rho_1}\right)^{1/2} \quad 40.29$$

$$N_1 = N_2\left(\frac{D_2}{D_1}\right)\left(\frac{p_1}{p_2}\right)^{1/2}\left(\frac{\rho_2}{\rho_1}\right)^{1/2} = N_2\sqrt{\frac{Q_2}{Q_1}}\left(\frac{p_1}{p_2}\right)^{3/4}\left(\frac{\gamma_2}{\gamma_1}\right)^{3/4} \quad 40.30$$

$$P_1 = P_2\left(\frac{D_2}{D_1}\right)^2\left(\frac{p_1}{p_2}\right)^{3/2}\left(\frac{\rho_2}{\rho_1}\right)^{1/2} \quad 40.31$$

Equation 40.32 to Eq. 40.34 are used to determine speed, pressure, and power as a function of changes in size, capacity, or density.

Fan Laws

$$N_1 = N_2\left(\frac{D_2}{D_1}\right)^3\left(\frac{Q_1}{Q_2}\right) \quad 40.32$$

$$p_1 = p_2\left(\frac{D_2}{D_1}\right)^4\left(\frac{Q_1}{Q_2}\right)^2\left(\frac{\rho_1}{\rho_2}\right) \quad 40.33$$

$$P_1 = P_2\left(\frac{D_2}{D_1}\right)^4\left(\frac{Q_1}{Q_2}\right)^3\left(\frac{\rho_1}{\rho_2}\right) \quad 40.34$$

Similarity laws may be used to predict the performance of a larger fan from a smaller fan's performance, since the efficiency of the larger fan can be expected to exceed that of the smaller fan. Larger fans should not be used to predict the performance of smaller fans. Even extrapolations to larger fans should be viewed cautiously when there is a significant decrease in air density or when the ratio of the larger-to-smaller fan diameters, the ratio of speed, or the product of the diameter and speed ratios exceed 3.0.

HVAC

Example 40.4

A fan turning at 800 rev/min develops 6.2 hp (4.6 kW) against a static pressure of 2.25 in wg (560 Pa). (a) If the fan is driven at 1400 rev/min, what will be the power developed? (b) If duct length is increased such that the static pressure loss is 2.85 in wg (713 Pa) and the fan efficiency remains the same, what speed will be required?

SI Solution

(a) Use Eq. 40.28. Since there is no change to the impeller diameter or fluid density, those terms can be ignored.

Fan Laws

$$\begin{aligned} P_1 &= P_2\left(\frac{D_1}{D_2}\right)^5\left(\frac{N_1}{N_2}\right)^3\left(\frac{\rho_1}{\rho_2}\right) \\ &= P_2\left(\frac{N_1}{N_2}\right)^3 \\ &= (4.6\ \text{kW})\left(\frac{1400\ \frac{\text{rev}}{\text{min}}}{800\ \frac{\text{rev}}{\text{min}}}\right)^3 \\ &= 24.7\ \text{kW} \end{aligned}$$

(It is unlikely that the same motor will be able to provide this increased power.)

(b) Use Eq. 40.30. Since there is no change to the impeller diameter or fluid density, those terms can be ignored.

Fan Laws

$$\begin{aligned} N_1 &= N_2\left(\frac{D_2}{D_1}\right)\left(\frac{p_1}{p_2}\right)^{1/2}\left(\frac{\rho_2}{\rho_1}\right)^{1/2} \\ &= N_2\left(\frac{p_1}{p_2}\right)^{1/2} \\ &= \left(800\ \frac{\text{rev}}{\text{min}}\right)\left(\frac{713\ \text{Pa}}{560\ \text{Pa}}\right)^{1/2} \\ &= 903\ \text{rev/min} \end{aligned}$$

Customary U.S. Solution

Use Eq. 40.28 since there is no change to the impeller diameter or fluid density.

Fan Laws

$$\begin{aligned} P_1 &= P_2\left(\frac{D_1}{D_2}\right)^5\left(\frac{N_1}{N_2}\right)^3\left(\frac{\rho_1}{\rho_2}\right) \\ &= P_2\left(\frac{N_1}{N_2}\right)^3 \\ &= (6.2\ \text{hp})\left(\frac{1400\ \frac{\text{rev}}{\text{min}}}{800\ \frac{\text{rev}}{\text{min}}}\right)^3 \\ &= 33.2\ \text{hp} \end{aligned}$$

(It is unlikely that the same motor will be able to provide this increased power.)

(b) Use Eq. 40.30. Since there is no change to the impeller diameter or fluid density, those terms can be ignored.

Fan Laws

$$\begin{aligned} N_1 &= N_2\left(\frac{D_2}{D_1}\right)\left(\frac{p_1}{p_2}\right)^{1/2}\left(\frac{\rho_2}{\rho_1}\right)^{1/2} \\ &= N_2\left(\frac{p_1}{p_2}\right)^{1/2} \\ &= \left(800\ \frac{\text{rev}}{\text{min}}\right)\left(\frac{2.85 \text{ in wg}}{2.25 \text{ in wg}}\right)^{1/2} \\ &= 900 \text{ rev/min} \end{aligned}$$

20. OPERATION AT NONSTANDARD CONDITIONS

Fan tests used to develop curves and rating tables are based on air at standard conditions—70°F and 14.7 psia (21°C and 101 kPa). Small variations in density due to normal temperature and humidity fluctuations can be disregarded. However, if the system operates at extremely elevated temperatures or reduced atmospheric pressures, corrections will be necessary.

Fans are constant-volume devices. They deliver the same volume of air (at the same fan speed) regardless of temperature, pressure, and humidity ratio. Therefore, the actual flow rate, ACFM, should be used to select a fan from rating tables. The speed can be read directly from the fan table. The standard (i.e., table) values of power and pressure (static, velocity, and total) should be modified by the *density factor*, K_d (see Eq. 40.2), with the density.[10]

$$\begin{aligned} K_d &= \frac{p_{s,\text{std}}}{p_{s,\text{actual}}} = \frac{p_{\text{v,std}}}{p_{\text{v,actual}}} = \frac{p_{t,\text{std}}}{p_{t,\text{actual}}} \\ &= \frac{p_{s,\text{std}}}{p_{s,\text{actual}}} = \frac{\text{bhp}_{\text{std}}}{\text{bhp}_{\text{actual}}} = \frac{\text{bkW}_{\text{std}}}{\text{bkW}_{\text{actual}}} \\ &= \frac{\text{ahp}_{\text{std}}}{\text{ahp}_{\text{actual}}} = \frac{\text{akW}_{\text{std}}}{\text{akW}_{\text{actual}}} \end{aligned} \qquad 40.35$$

Example 40.5

A fan is chosen to move 18,000 SCFM (8500 L/s) of air against a static pressure of 0.85 in wg (210 Pa). The fan draws 4.2 hp (3.1 kW) when moving standard air in that configuration. The fan is used in a nonstandard environment to provide 150°F (66°C) air for drying. What will be the (a) required power, and (b) friction loss?

SI Solution

(a) Equation 40.2 gives the density factor. The pressure is unchanged.

$$K_d = \frac{T_{\text{actual}}}{T_{\text{std}}} = \frac{66°\text{C} + 273°}{21°\text{C} + 273°} = 1.15$$

From Eq. 40.35,

$$\text{akW}_{\text{actual}} = \frac{\text{akW}_{\text{std}}}{K_d} = \frac{3.1 \text{ kW}}{1.15} = 2.7 \text{ kW}$$

(b) The friction loss is

$$p_{f,\text{actual}} = \frac{p_{f,\text{std}}}{K_d} = \frac{210 \text{ Pa}}{1.15} = 183 \text{ Pa}$$

Customary U.S. Solution

(a) Equation 40.2 gives the density factor. The pressure is unchanged.

$$K_d = \frac{T_{\text{actual}}}{T_{\text{std}}} = \frac{150°\text{F} + 460°}{70°\text{F} + 460°} = 1.15$$

From Eq. 40.35,

$$\text{bhp}_{\text{actual}} = \frac{\text{bhp}_{\text{std}}}{K_d} = \frac{4.2 \text{ hp}}{1.15} = 3.7 \text{ hp}$$

(b) The friction loss is

$$p_{f,\text{actual}} = \frac{p_{f,\text{std}}}{K_d} = \frac{0.85 \text{ in wg}}{1.15} = 0.74 \text{ in wg}$$

Example 40.6

70°F (21°C) air in a duct located at an altitude of 5000 ft (1525 m) moves at 1500 ft/min (7.7 m/s). The actual flow rate is 39,000 ft³/min (18 300 L/s). The duct resistance at that altitude is 1.5 in wg (375 Pa). (a) What duct resistance should be used with fan rating tables? (b) At standard conditions, what input power to the fan is required if the fan efficiency is 75%?

SI Solution

(a) From Table 40.1, the atmospheric pressure ratio at 1525 m altitude is 0.832. The pressure is

$$p_{\text{actual}} = (0.832)(101.3 \text{ kPa}) = 84.3 \text{ kPa}$$

From Eq. 40.2, the density factor is

$$K_d = \frac{p_{\text{std}}}{p_{\text{actual}}} = \frac{101.3 \text{ kPa}}{84.3 \text{ kPa}} = 1.2$$

From Eq. 40.35, the standard friction pressure loss is

$$p_{f,\text{std}} = K_d(p_{f,\text{actual}}) = (1.2)(375 \text{ Pa}) = 450 \text{ Pa}$$

[10]There is also a correction for viscosity. However, the viscosity change is so insignificant that it is disregarded.

HVAC

(b) Since the volumetric flow rate does not change, the duct speed is also unchanged. The fan supplies both velocity and static pressure. From Eq. 40.9, the velocity pressure is

$$\begin{aligned} p_{\text{v,Pa}} &= \frac{\rho \text{v}^2_{\text{m/s}}}{2} \\ &= (0.832)\left(\frac{1.2\ \frac{\text{kg}}{\text{m}^3}}{2}\right)\left(7.7\ \frac{\text{m}}{\text{s}}\right)^2 \\ &= 30\ \text{Pa} \end{aligned}$$

From Eq. 40.10, the total pressure energy supplied by the fan is

$$p_t = p_s + p_\text{v} = 375\ \text{Pa} + 30\ \text{Pa} = 405\ \text{Pa}$$

From Eq. 40.14, the original power drawn is

$$\begin{aligned} \text{akW} &= \frac{Q_{\text{L/s}}(p_{t,\text{Pa}})}{10^6} = \frac{\left(18\,300\ \frac{\text{L}}{\text{s}}\right)(405\ \text{Pa})}{10^6} \\ &= 7.41\ \text{kW} \end{aligned}$$

From Eq. 40.15 and Eq. 40.35, the standardized brake kilowatts are

$$\begin{aligned} \text{bkW}_{\text{std}} &= \frac{K_d(\text{akW}_{\text{actual}})}{\eta_{\text{mechanical}}} = \frac{(1.2)(7.41\ \text{kW})}{0.75} \\ &= 11.9\ \text{kW} \end{aligned}$$

Customary U.S. Solution

(a) From Table 40.1, the atmospheric pressure ratio at 5000 ft altitude is 0.832. The pressure is

$$p_{\text{actual}} = (0.832)(14.7\ \text{psia}) = 12.2\ \text{psia}$$

From Eq. 40.2, the density factor is

$$K_d = \frac{p_{\text{std}}}{p_{\text{actual}}} = \frac{14.7\ \text{psia}}{12.2\ \text{psia}} = 1.2$$

From Eq. 40.35, the standard friction pressure loss is

$$\begin{aligned} p_{f,\text{std}} &= K_d(p_{f,\text{actual}}) = (1.2)(1.5\ \text{in wg}) \\ &= 1.8\ \text{in wg} \end{aligned}$$

(b) Since the volumetric flow rate does not change, the duct speed is also unchanged. The fan supplies both velocity and static pressure. From Eq. 40.8, the velocity pressure is

$$\begin{aligned} p_\text{v} &= (0.832)\left(\frac{\text{v}}{4005}\right)^2 = (0.832)\left(\frac{1500\ \frac{\text{ft}}{\text{min}}}{4005}\right)^2 \\ &= 0.12\ \text{in wg} \end{aligned}$$

From Eq. 40.10, the total pressure energy supplied by the fan is

$$\begin{aligned} p_t &= p_s + p_\text{v} = 1.5\ \text{in wg} + 0.12\ \text{in wg} \\ &= 1.62\ \text{in wg} \end{aligned}$$

From Eq. 40.14, the original power drawn is

Fan Power Requirements

$$\begin{aligned} \text{ahp} &= 0.000157 Q_{\text{ft}^3/\text{min}} p_t \\ &= (0.000157)\left(39{,}000\ \frac{\text{ft}^3}{\text{min}}\right)(1.62\ \text{in wg}) \\ &= 9.94\ \text{hp} \end{aligned}$$

From Eq. 40.15 and Eq. 40.35, the standardized brake horsepower is

$$\begin{aligned} \text{bhp}_{\text{std}} &= \frac{K_d(\text{ahp}_{\text{actual}})}{\eta_{\text{mechanical}}} = \frac{(1.2)(9.94\ \text{hp})}{0.75} \\ &= 15.9\ \text{hp} \end{aligned}$$

21. DUCT SYSTEMS

There are two primary types of air ducting designs: trunk and radial. The most common duct design is the *trunk system*, also known as an *extended plenum system*. A large main supply duct (trunk duct) connects to and extends the air handler plenum. Smaller branch ducts, known as *runout ducts*, deliver air from the trunk to the individual outlets. Particularly in residential applications, the trunk is usually rectangular, while the branch ducts are usually round. In a *reducing trunk system*, the trunk is proportionately reduced once the cumulative pressure drop across branches allows for a significant reduction in trunk dimension. Because of additional design and construction costs, reducing trunk systems are generally used only in commercial and high-end residential applications.

Radial duct systems are used less often than trunk systems and are typically used where the air handling equipment may be centrally located and where it is not necessary to conceal ductwork. All of the radial branch ducts connect directly to the equipment plenum. *Gravity duct systems* are essentially radial systems that circulate heated air through ductwork by natural convection, without fan assistance. They are suitable only for small residences, and they are typically associated with coal- and wood-burning furnaces.

In *dual duct systems*, heated and cooled air flowing from hot and cold "decks," respectively, are both available to the zone. The mixture of the two flows is determined by sensors in the conditioned zone, and the two airflows are combined in a *mixing plenum* through use of a *mixing damper*. The entire apparatus is usually combined into a single *dual duct VAV box*. Numerous variations exist. The system can be constant- or variable-volume; a single fan can be used, or each deck can have its own fan; the mixing plenum can have its own heating and cooling capabilities.

22. SHEET METAL DUCT

Most commercial and residential ductwork, whether rectangular or round, is manufactured from plain and galvanized sheet steel, stainless steel, and aluminum with folded seams. Ducts may subsequently be painted or powder coated or wrapped with fiberglass insulation. Table 40.3 lists the sheet metal gauges used in ducts.

Table 40.3 *Thickness of Sheet Metal Used in Ducts (in (mm))*

gauge	plain mild steel	galvanized steel	stainless steel	aluminum
16	0.0598 (1.52)	0.0635 (1.62)	0.0625 (1.59)	0.0508 (1.29)
18	0.0478 (1.21)	0.0516 (1.31)	0.0500 (1.27)	0.0403 (1.02)
20	0.0359 (0.91)	0.0396 (1.01)	0.0375 (0.953)	0.0319 (0.810)
22	0.0299 (0.76)	0.0336 (0.85)	0.0312 (0.792)	0.0253 (0.643)
24	0.0239 (0.61)	0.0276 (0.70)	0.0250 (0.635)	0.0201 (0.511)
26	0.0179 (0.45)	0.0217 (0.55)	0.0187 (0.475)	0.0159 (0.404)

(Multiply in by 25.4 to obtain mm.)

23. SPIRAL DUCT

Spiral duct (with a spiral seam, as distinguished from traditional round duct with a longitudinal seam) is preferred for high velocity systems, when visual aesthetics are important, and where space is available. It can be manufactured from any sheet metal. Double-wall and oval varieties are available. Spiral duct can be painted, powder coated, and/or insulated just like rectangular ducting. *Polyvinyl coated spiral duct* (PCD) is available for underground ducts and fume exhaust systems. PCD combines the strength of steel and the chemical inertness of plastic; it is lightweight, weather resistant, and corrosion resistant. Although seams for spiral duct can be welded, a folded lock seam is adequate for normal ductwork. The seam is external to the duct, presenting a smooth surface to the airflow. Standard spiral duct is uncorrugated; corrugations add mechanical strength and are used with underground ducts. Purchase and installation costs for spiral duct are comparable or lower than for traditional rectangular sheet metal ducts. Flow resistance (friction) is generally lower. Spiral duct generally has less leakage, reduced noise, greater mechanical strength, greater bursting (seam failure) resistance to positive pressures, and greater collapse resistance to negative pressures.

There is no significant difference between spiral and traditional ducts in duct layout and design methodology, although accurate determination of system friction may require use of manufacturer's charts and tables. Traditional duct has a specific roughness of 0.0005 ft (0.15 mm). Spiral duct is usually put into the ASHRAE category of "medium smooth" and has a specific roughness of 0.0003 ft (0.09 mm). Corrugated spiral duct is categorized as "medium rough" and has a specific roughness of 0.0024 ft (0.74 mm). Corrugations increase the duct resistance by 10–30%. Spiral ducts have less friction pressure loss, FP, than traditional ducts, so standard friction charts predict pressure losses for spiral ducts that are slightly (i.e., 5–10%) higher than actual. The error is conservative and well within the acceptable range considering all other inaccuracies and assumptions.

Burst and collapse pressures of ductwork depend somewhat on manufacturing methods, and so, are based on a combination of theoretical methods and testing. Pressures are typically correlated to the wall thickness-diameter ratio, t/D, as illustrated in Fig. 40.4.

24. FIBERGLASS DUCT

Fiberglass (also known as *fiber-reinforced plastic*, *fiber-reinforced polymer* (FRP) and *fibrous*) *duct*, of rigid and flexible varieties, is generally more expensive than standard metal, but may be easier to install without sheet metal training. Rectangular ducts can be manufactured using FRP *ductboard*. FRP duct has comparable friction but is lighter and stronger than light gauge steel. It generally has better acoustical qualities and is corrosion resistant. FRP duct may be the best material in corrosive environments and where the duct must resist positive and negative pressure extremes. Since FRP duct cannot be grounded, it can accumulate static electricity

Figure 40.4 *Typical Collapse Pressure of Steel Spiral Ducts*

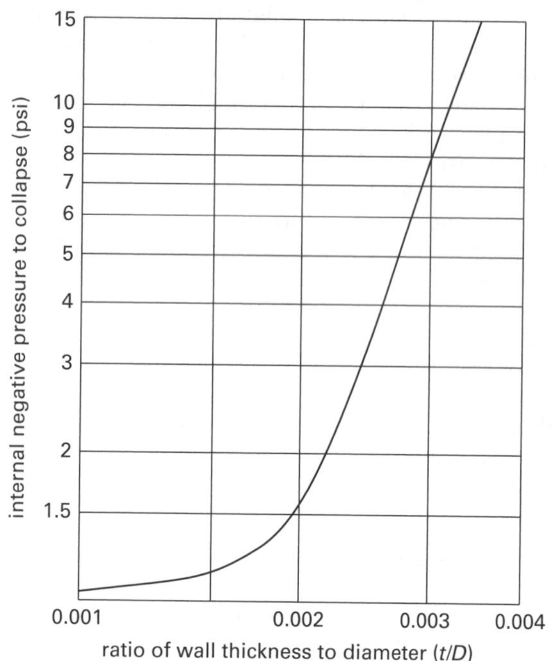

From *A Complete Line of High Pressure Ductwork*, Vol. 5, copyright © 2006, by Spiral Manufacturing Co., Inc. Reproduced with permission.

and become an ignition source in dusty environments. Therefore, FRP duct should not be used in dust collection systems.

25. FLEXIBLE DUCT

Flexible duct (*flex*) is typically manufactured by wrapping a plastic sheet over a metal wire coil. Flex is commonly used for connecting supply/return grills to trunk and branch lines. Due to the significantly increased friction (e.g., three times as much as smooth metal ductwork), runs of flex line are kept as short as possible, generally less than 15 ft (5 m). Since flex does not tolerate large negative pressures, its use in return air systems is not favored.[11]

26. FRICTION LOSSES IN ROUND DUCTS

Friction loss (*friction pressure*, p_f) can be calculated from the standard Moody equation. Equation 40.36 expresses the Moody equation in typical air-moving units for standard air (0.075 lbm/ft^3) through average, clean, round galvanized duct (specific roughness of 0.0005 ft) having a typical number of connections, joints, and slip couplings.[12,13]

$$p_{f\,\text{in wg},100\,\text{ft}} = \frac{2.74}{D_{\text{in}}^{1.22}}\left(\frac{\text{v}_{\text{fpm}}}{1000}\right)^{1.9} = \frac{0.109\,Q_{\text{cfm}}^{1.9}}{D_{\text{in}}^{5.02}} \quad 40.36$$

Another approach to calculating friction loss is to use a modified Darcy-Weisbach equation.

Duct Design: Bernoulli Equation

$$\Delta p_{f,\text{in wg}} = \left(\frac{12fL}{D_h} + \sum C\right)\rho\left(\frac{\text{v}}{1097}\right)^2 \quad 40.37$$

$\sum C$ is the sum of the duct section local losses. The dimensionless, local loss coefficient, C, can be determined by

$$C = \frac{\Delta p_t}{\rho\left(\frac{\text{v}}{1097}\right)^2} = \frac{\Delta p_t}{\Delta p_\text{v}} \quad 40.38$$

However, Eq. 40.36 and Eq. 40.37 are almost never used in the HVAC industry. Rather, friction losses are typically determined from graphs. Figure 40.5 and Fig. 40.6 assume clean, commercial-quality round ducts with a normal number of joints and conditions close to standard air. For smooth ducts with no joints, the friction loss is 60–95% of the value determined from Fig. 40.5 and Fig. 40.6.[14]

Duct flow areas are calculated from their nominal dimensions. Any size duct can be manufactured. However, there are standard sizes of premanufactured round duct, and these sizes should be chosen to minimize cost. Generally, commercial duct manufacturers produce every whole-inch size up to at least 20 in (510 mm) in diameter. After that, ducts are available in 2 in (50 mm) increments. Odd-number sizes may be available with premium pricing.

Example 40.7

2000 ft^3/min (1000 L/s) of air flows in a 13 in (315 mm) diameter duct. What are the (a) velocity and (b) friction loss per 100 ft (per meter) of duct?

SI Solution

(a) Use Fig. 40.6. Locate the intersection of the 1000 L/s and 315 mm lines. The velocity is approximately 13 m/s.

[11] UL181 testing specification requires functionality at a negative pressure of 0.03 lbf/in^2 (200 Pa), so flexible duct can tolerate moderate negative pressures.
[12] Specific roughness is the reciprocal of the number of duct diameters required to cause a static pressure loss of one velocity pressure. Typical values for galvanized duct are 0.0003 ft to 0.0005 ft (0.09 mm to 0.15 mm).
[13] Some authorities, such as Carrier, report the 1.9 exponent as 1.82.
[14] For corrugated ductwork, the friction loss is approximately twice that shown in Fig. 40.5 and Fig. 40.6. However, the analysis is not precise enough to make most corrections, including those for a different number of joints or duct material. Corrections should be limited to extreme cases when the ductwork deviates significantly from commercial (e.g., airflow through brickwork or corrugated duct).

Figure 40.5 *Standard Friction Loss in Standard Duct* (inches of water per 100 ft of duct; ±5% for temperatures of 40°F to 100°F, elevations to 1500 ft, and duct pressures of −20 in wg to +20 in wg) (Recommended operating points shown as shaded region.)*

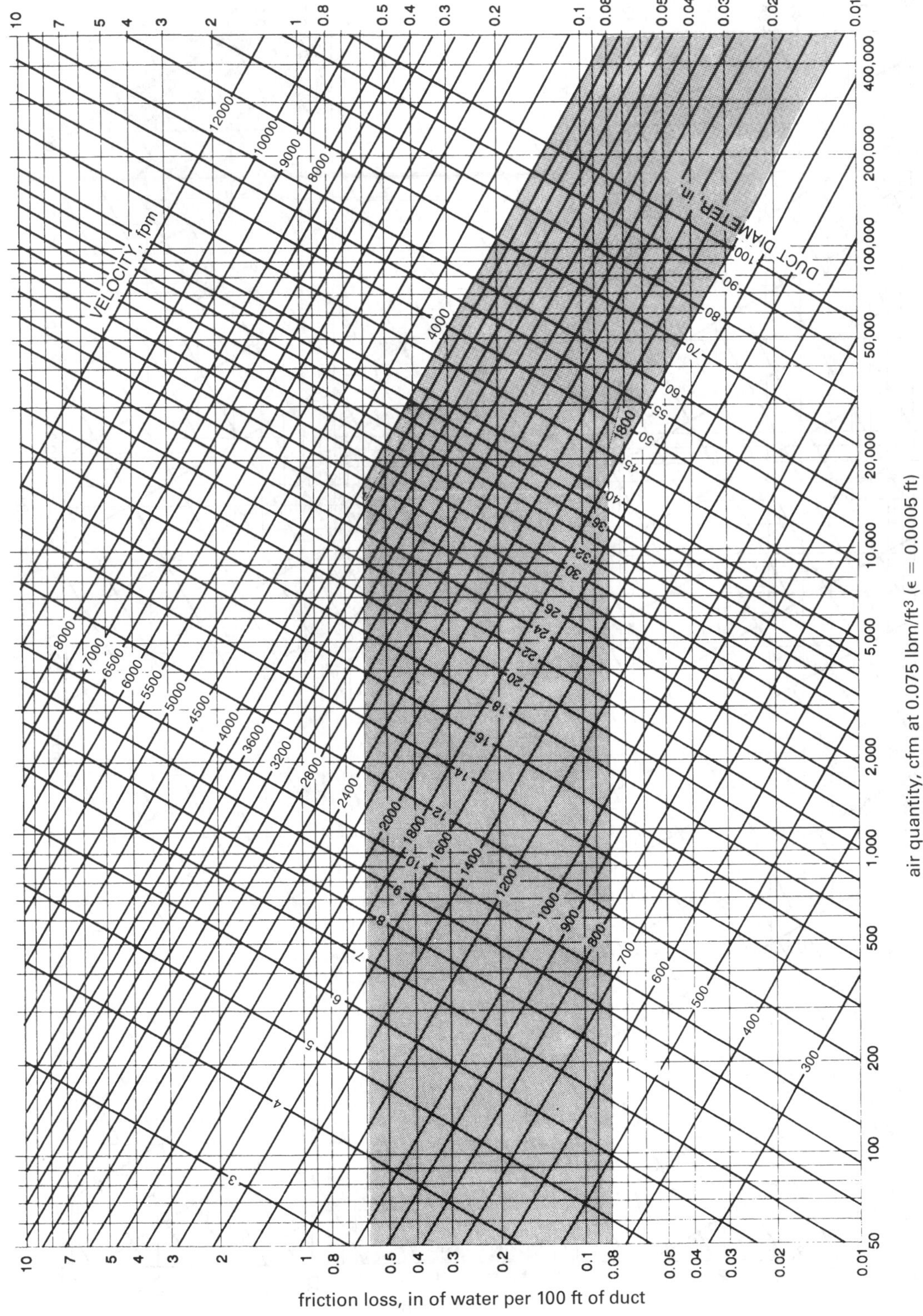

*Clean, round galvanized duct with a specific roughness of 0.0005 ft (0.15 mm) and approximately 25 beaded slip-couplings (joints) per 100 ft (30 m). Can also be used for smooth commercial spiral duct with about 10 joints per 100 ft (30 m).

Figure 40.6 *Standard Friction Loss in Standard Duct* (pascals per meter of duct; ±5% for temperatures of 5°C to 35°C, elevations to 500 m, and duct pressures of −5 kPa to +5 kPa) (Recommended operating points shown as shaded region.)*

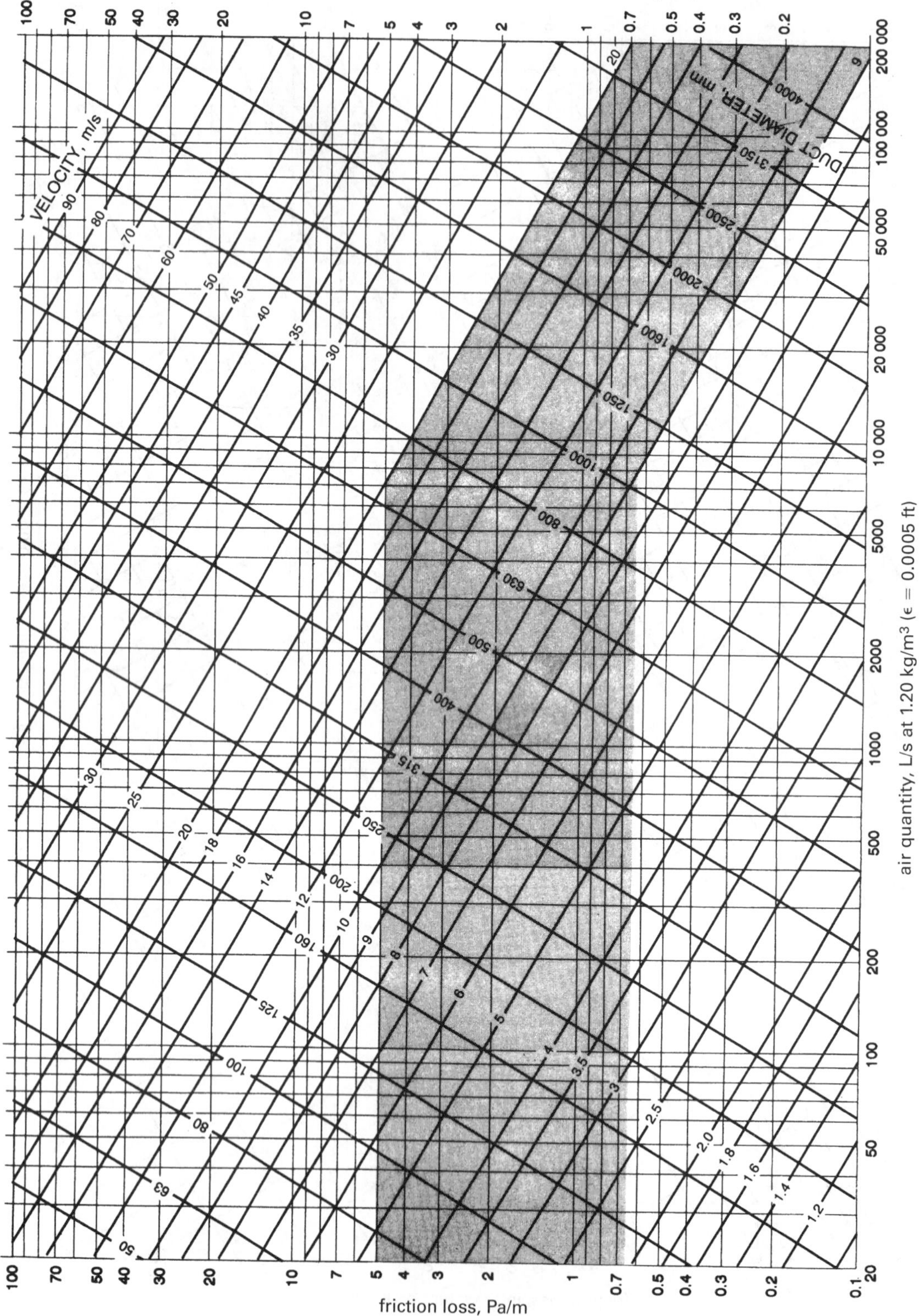

*Clean, round galvanized duct with a specific roughness of 0.0005 ft (0.15 mm) and approximately 1 beaded slip-coupling (joint) per meter. Can also be used for smooth commercial spiral duct with about 1 joint per 3 m.

(b) Move horizontally to the left and read from the vertical scale. The friction loss is approximately 6 Pa/m of duct.

Customary U.S. Solution

(a) Use Fig. 40.5. Locate the intersection of the 2000 ft^3/min and 13 in lines. The velocity is approximately 2200 ft/min. (This answer could also be calculated using the continuity equation, $v = Q/A$.) [Continuity Equation]

(b) Drop straight down to the horizontal scale. The friction loss is approximately 0.5 in wg per 100 ft of duct.

Example 40.8

2000 ft^3/min (1000 L/s) of standard air moves through a duct with a velocity of 1600 ft/min (8 m/s). (a) What size duct is required? (b) What is the friction loss?

SI Solution

(a) Use Fig. 40.6. Locate the intersection of 1000 L/s and 8 m/s. The required duct diameter is 400 mm.

(b) Move horizontally to the left and read from the vertical scale. The friction loss is approximately 1.7 Pa/m.

Customary U.S. Solution

(a) Use Fig. 40.5. Locate the intersection of 2000 ft^3/min and 1600 ft/min. The required duct diameter is approximately 15 in.

(b) Drop straight down to the horizontal scale. The friction loss is approximately 0.23 in wg per 100 ft of duct.

27. PRESSURE DROP FOR NONSTANDARD CONDITIONS

Common pressure drop equations and friction charts assume standard air with a density of 0.075 lbm/ft^3, which corresponds to air at sea level at about 65°F and 40% relative humidity, or air at 70°F and 0% relative humidity. The density of air depends on the temperature, T, and elevation above sea level, z. For nonstandard temperatures and elevations, the pressure drop is given by Eq. 40.39.

$$p_{f,\text{actual}} = K_{\text{elevation}} K_{\text{temperature}} p_{f,\text{std}} \tag{40.39}$$

$$K_{\text{elevation}} = (1 - 6.8754 \times 10^{-6} z_{\text{ft}})^{4.73} \tag{40.40}$$

$$K_{\text{temperature}} = \left(\frac{530°\text{R}}{T_{°\text{F}} + 460°} \right)^{0.825} \tag{40.41}$$

28. RECTANGULAR DUCTS

Duct systems are initially designed for round ducts. Then, conversions to rectangular ducts are made as required. A round duct with diameter D can be converted to a rectangular duct with equal friction per unit length if the desired aspect ratio is known. The *aspect ratio*, R, of a rectangular duct should be kept below 8 for ease of manufacture.

$$\text{short side} = \frac{D(1+R)^{1/4}}{1.3R^{5/8}} \tag{40.42}$$

$$R = \frac{\text{long side}}{\text{short side}} \tag{40.43}$$

The *equivalent diameter* of a rectangular duct with an aspect ratio less than 8 is given by the *Huebscher equation*, Eq. 40.44. The round duct will have the same friction and capacity as the rectangular duct. Figure 40.5 and Fig. 40.6 can be used with D_{eq} and the actual flow rate to find the friction loss.[15] The velocity indicated by the chart will be incorrect but can be calculated as Q/A.

$$D_{\text{eq}} = \frac{1.30(\text{short side} \times \text{long side})^{5/8}}{(\text{short side} + \text{long side})^{1/4}} \tag{40.44}$$

Equation 40.45 shows an alternate form of the *Huebscher equation* in which a is the length of one side of the duct, and b is the length of the adjacent side of the duct.

Rectangular Ducts

$$D_{\text{eq}} = \frac{1.30(ab)^{0.625}}{(a+b)^{0.250}} \tag{40.45}$$

29. FRICTION LOSSES IN FITTINGS

Friction losses (*dynamic losses*) due to bends, fittings, enlargements, contractions, and obstructions are calculated in the same ways as for liquid friction losses—either by loss coefficient or equivalent length methods.[16,17]

With the *loss coefficient method*, the friction loss is calculated as a multiple of the velocity pressure. Though reported values vary widely, typical values of the loss coefficient, K, for common features are given in Table 40.4.[18] Loss coefficients are usually based on the upstream velocity pressure. However, there are some cases (i.e., where plenum air with negligible velocity enters a duct) where the downstream velocity is used by

[15]This is the formula used by ASHRAE and most other authorities for the *equivalent* diameter. Some sources merely equate formulas for round and rectangular areas and use the *hydraulic* diameter, which is $\sqrt{(4 \times \text{short side} \times \text{long side})/\pi}$. When rounded to the nearest whole duct size, the difference is often insignificant. Although the two ducts may have the same cross-sectional area, they will not have the same capacity or friction.

[16]Unlike losses for liquid flows, however, fitting losses for duct systems are significant. They are not "minor" losses.

[17]Any fitting or feature that causes a static pressure loss of 0.75 in wg (200 Pa) or higher is a potential source of unwanted noise.

[18]Static pressure losses due to equipment (e.g., filters, coils, and heat exchangers) are determined from manufacturers' literature.

Table 40.4 *Typical Fitting Loss Coefficients*[a,b,c]

feature			K
abrupt expansion	$v_{down}/v_{up} = A_{up}/A_{down} =$	0	1.00
from A_{up} to A_{down}		0.20	0.64
(referred to v_{up})		0.4	0.36
$K_{up} = (1-(A_{up}/A_{down}))^2$		0.6	0.16
		0.8	0.04
abrupt contraction	$v_{up}/v_{down} = A_{down}/A_{up} =$	0.20	0.32
from A_{up} to A_{down}		0.25	0.30
(referred to v_{down})		0.40	0.25
		0.50	0.20
		0.60	0.16
		0.75	0.10
		0.80	0.06
round pipe of diameter	$D_1/D_2 =$	0.10	0.20
D_1 across (through)		0.25	0.55
duct of diameter D_2		0.50	2.00
(referred to v_{up} or v_{down})			
tapered reducing section[d]	taper angle[e] =	20°	0.012
(referred to v_{down})		30°	0.020
		40°	0.032
		45°	0.040
		60°	0.070
bell-mouthed entrance[f]			0.040
(referred to v_{down})			
90° round elbows[g] continuous	$r/D =$	0.00	1.20
die-stamped (bend radius, r)		(miter)	
(referred to v_{up} or v_{down})		0.50	0.83
		0.75	0.46
		1.00	0.31
		1.25	0.27
		1.50	0.22
		1.75	0.20
		2.00	0.19
		2.25	0.18
		2.50	0.17
		2.75	0.16
		3.00	0.15

30°, 45°, 60° continuous, die-stamped elbows

multiply 90° loss coefficients by 0.33 (30°), 0.50 (45°), and 0.67 (60°)

90° mitered and gored elbows (round)

straight miter: 1.2

straight miter with turning vanes: 0.5

	$r/D = 0.75$	$r/D = 1.0$	$r/D = 1.5$	$r/D = 2.0$
3 piece	0.54	0.42	0.34	0.33
4 piece	0.50	0.37	0.27	0.24
5 piece	0.46	0.33	0.24	0.19

30°, 45°, 60° gored elbows

multiply 90° loss coefficients by 0.45 (30°), 0.60 (45°), and 0.78 (60°)

[a]Subscripts "up" and "down" refer to upstream and downstream, respectively.

[b]In multiples of velocity pressure.

[c]Specific roughness is $\epsilon = 0.0005$ ft; $f \approx 0.0185$.

[d]The total energy loss is small for all taper angles. The advantage of a very long taper is insignificant.

[e]The *taper angle* is the angle one side makes with the straight wall. The *included angle* refers to twice the taper angle.

[f]When stationary air is drawn into a bell-mouthed opening, the fan must supply the velocity pressure (1.0) as well as overcome the entrance friction (0.04). Because of that, some sources report this value as 1.04.

[g]Also, see Table 40.5.

convention. The coefficient should always be used with the velocity at the point corresponding to the coefficient's subscript.

$$p_{f,\text{Pa}} = Kp_{v,\text{Pa}} = 0.6K\text{v}^2_{\text{m/s}} \quad \text{[SI]} \quad 40.46(a)$$

$$p_{f,\text{in wg}} = Kp_{v,\text{in wg}} = K\left(\frac{\text{v}_{\text{ft/min}}}{4005}\right)^2 \quad \text{[U.S.]} \quad 40.46(b)$$

In a straight duct of constant diameter, the velocity pressure is unchanged. The change in static pressure is the friction loss. Since there is no change in the velocity pressure, the friction loss will produce an equivalent decrease in total pressure. In other words, the friction loss is the change in total pressure. The same loss coefficient is used for calculating the change in static pressure and the change in total pressure.

$$\Delta p_{t,1} = p_f = p_{t,1} - p_{t,2} = p_{s,1} - p_{s,2}$$
$$= Kp_{v,\text{up}} \quad \begin{bmatrix}\text{constant area duct} \\ \text{with no branches}\end{bmatrix} \quad 40.47$$

Loss coefficients are *zero-length losses*. That is, they only include the dynamic effects. If a large fitting has a specific length, that length must be included in the run-of-duct length when calculating the duct friction.

Another approach, called the equivalent length method, is to equate the fitting losses with an equivalent length of ducts. These lengths are given in multiples of duct diameter in Table 40.5.

Table 40.5 *Typical Equivalent Lengths*[a,b]

			L_e
90° continuous, round	$r/D =$	0.00 (miter)	65D
elbows of bend radius		0.50	
r and diameter D		0.75	45D
		1.00	23D
		1.25	17D
		1.50	15D
		1.75	12D
		2.00	11D
		2.25	10D
		2.50	9.7D
		2.75	9.2D
		3.00	8.6D
			8.1D

30°, 45°, 60° continuous elbows multiply 90° equivalent lengths by 0.33 (30°), 0.50 (45°), and 0.67 (60°)

90° mitered and gored elbows (round)

straight miter: 65D

straight miter with turning vanes: 27D

	$r/D = 0.75$	$r/D = 1.0$	$r/D = 1.5$	$r/D = 2.0$
3 piece	29D	23D	18D	18D
4 piece	27D	20D	15D	13D
5 piece	25D	18D	13D	10D

30°, 45°, 60° gored elbows

multiply 90° equivalent lengths by 0.45 (30°), 0.60 (45°), and 0.78 (60°)

[a]In terms of inside duct diameter, D.

[b]Specific roughness is $\epsilon = 0.0005$ ft; $f \approx 0.0185$.

Duct elbows can be die-formed (i.e., stamped) or gored. Gore elbows typically have three or five gores per 90° of bend (e.g., a "3-gore, 90° elbow"), although 2- and 4-piece elbows are available. A *radius-to-diameter ratio* (i.e., the ratio of *centerline radius* (CLR) and diameter) of 1.5 is typical, although values of 2.0 and 2.5 are also used. Data on friction losses in bends are usually dependent on the radius-diameter ratio. The *throat radius* (i.e., the radius to the inside of the bend) is not commonly used to categorize friction losses.

The equivalent length of a smooth-radius rectangular duct elbow depends on aspect ratio and radius. For a duct with a width w, height h, and bend radius r, the equivalent length of a 90° smooth rectangular elbow can be estimated from Eq. 40.48.

$$L_{\text{eq}} = w\left(0.33\frac{r}{w}\right)^{-2.13(h/w)^{0.126}} \quad 40.48$$

30. COEFFICIENT OF ENTRY

The *coefficient of entry*, C_e, is the ratio of the actual to ideal velocities as stationary air is drawn into an inlet.[19] The coefficient of entry is not the same as the loss coefficient, K_e, which is the entrance friction expressed as a multiple of the velocity pressure. Typical values of the coefficient of entry are given in Table 40.6. Equation 40.51 is the relationship between the coefficient of entry and the loss coefficient.

$$C_e = \sqrt{\frac{p_{\text{v,duct}}}{p_{s,\text{duct}}}} \quad 40.49$$

$$\begin{aligned} Q &= \text{v}_{\text{actual}}A = C_e\text{v}_{\text{ideal}}A \\ &= 4005\,C_eA\sqrt{p_{s,\text{duct}}} \end{aligned} \quad 40.50$$

$$K_e = \frac{1 - C_e^2}{C_e^2} \quad 40.51$$

Table 40.6 *Typical Coefficients of Entry*

entrance	C_e
plain opening (round, rectangular, square)	0.72
flanged openings (round, rectangular, square)	0.82
bell-mouthed	0.98
tapered square*	0.93
conical*	0.96

*For included taper angle of 30° to 60°. Values of other angles are close.

31. STATIC REGAIN

Disregarding friction loss, energy is constant along the run of a duct. If the velocity pressure decreases due to an increase in duct area or a branch takeoff, the static pressure will increase. This increase is known as the *static regain*, SR.[20]

Ideally, the regain would be exactly equal to the decrease in velocity pressure. Actually, 10% to 25% of the energy is lost due to friction, turbulence, and other factors. The portion of the theoretical regain that is realized is given by the *static regain coefficient*, R. R has typical values of 0.75 to 0.90 for well-designed ducts without reducing sections. (When $\text{v}_{\text{down}} > \text{v}_{\text{up}}$, the regain will be a static pressure loss. Use $R = 1.1$ in that case.)

$$R = \frac{\text{SR}_{\text{actual}}}{\text{SR}_{\text{ideal}}} = \frac{\text{SR}_{\text{actual,Pa}}}{0.6(\text{v}_{\text{up}}^2 - \text{v}_{\text{down}}^2)} \quad \text{[SI]} \quad 40.52(a)$$

$$R = \frac{\text{SR}_{\text{actual}}}{\text{SR}_{\text{ideal}}} = \frac{\text{SR}_{\text{actual,in wg}}}{\frac{\text{v}_{\text{up}}^2 - \text{v}_{\text{down}}^2}{(4005)^2}} \quad \text{[U.S.]} \quad 40.52(b)$$

32. DIVIDED-FLOW FITTINGS

Figure 40.7 illustrates typical commercial *divided-flow* (i.e., *branch takeoff*) *fittings* and the terminology that describes them. Air enters upstream, from the left. After the air reduction at the branch takeoff, the downstream velocity will (generally) be less than the upstream velocity. The change in static pressure due to the change in velocity is

$$\begin{aligned} p_{s,\text{down}} - p_{s,\text{up}} &= (p_{t,\text{down}} - p_{t,\text{up}}) - (p_{\text{v,down}} - p_{\text{v,up}}) \\ &= R(p_{\text{v,down}} - p_{\text{v,up}}) \end{aligned} \quad 40.53$$

The total pressure change from upstream to downstream is

$$\begin{aligned} p_{t,\text{down}} - p_{t,\text{up}} &= (p_{s,\text{down}} - p_{s,\text{up}}) + (p_{\text{v,down}} - p_{\text{v,up}}) \\ &= (R+1)(p_{\text{v,down}} - p_{\text{v,up}}) \end{aligned} \quad 40.54$$

The total pressure change from upstream through the branch is

$$\begin{aligned} p_{t,\text{br}} - p_{t,\text{up}} &= (p_{s,\text{br}} - p_{s,\text{up}}) + (p_{\text{v,br}} - p_{\text{v,up}}) \\ &= -K_{\text{br}}(p_{\text{v,up}}) \end{aligned} \quad 40.55$$

[19]This is analogous to the coefficient of velocity, C_v, used in liquid flow measurement devices.

[20]If a downstream or branch velocity is low enough, it is possible for the regain to actually exceed the dynamic losses due to fitting turbulence. This may hide the true inefficiency of the fitting.

The change in static pressure from upstream through the branch is

$$\begin{aligned} p_{s,\mathrm{br}} - p_{s,\mathrm{up}} &= (p_{t,\mathrm{br}} - p_{t,\mathrm{up}}) - (p_{\mathrm{v,br}} - p_{\mathrm{v,up}}) \\ &= -K_{\mathrm{br}}(p_{\mathrm{v,up}}) - (p_{\mathrm{v,br}} - p_{\mathrm{v,up}}) \\ &= (1 - K_{\mathrm{br}})p_{\mathrm{v,up}} - p_{\mathrm{v,br}} \end{aligned} \qquad 40.56$$

Manufacturers of commercial fittings provide graphs of the loss coefficient as a function of the ratio of branch-to-upstream velocities.[21] Typical values of the *branch loss coefficient*, K_{br}, are given in Table 40.7.

***Table 40.7** Typical Branch Loss Coefficient (K_{br}) Values**

ratio of	angle of takeoff		
$\mathrm{v_{br}/v_{up}}$	90°	60°	45°
0.5	1.1	0.8	0.5
1.0	1.5	0.8	0.5
1.5	2.2	1.1	0.9
2.0	3.0	2.9	2.8
2.5	4.3	3.3	3.2
3.0	5.6	5.2	4.9

*Round ducts only.

Example 40.9

The velocity in a main duct before a branch is 3200 ft/min (16 m/s). The velocity in the branch duct is 2560 ft/min (12.8 m/s). What is the change in static pressure from the main duct through the branch?

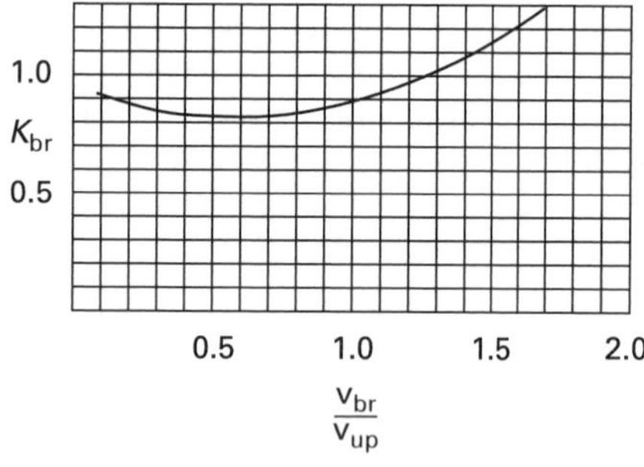

SI Solution

The ratio of the branch to upstream velocity is

$$\frac{\mathrm{v_{br}}}{\mathrm{v_{up}}} = \frac{12.8\ \frac{\mathrm{m}}{\mathrm{s}}}{16\ \frac{\mathrm{m}}{\mathrm{s}}} = 0.8$$

From the graph, the loss coefficient is $K_{\mathrm{br}} = 0.85$.

From Eq. 40.9, the upstream and branch velocity pressures are

$$p_{\mathrm{v,up}} = 0.6\mathrm{v}^2_{\mathrm{m/s}} = (0.6)\left(16\ \frac{\mathrm{m}}{\mathrm{s}}\right)^2 = 154\ \mathrm{Pa}$$

$$p_{\mathrm{v,br}} = (0.6)\left(12.8\ \frac{\mathrm{m}}{\mathrm{s}}\right)^2 = 98\ \mathrm{Pa}$$

From Eq. 40.56, the change in static pressure is

$$\begin{aligned} p_{s,\mathrm{br}} - p_{s,\mathrm{up}} &= (1 - K_{\mathrm{br}})p_{\mathrm{v,up}} - p_{\mathrm{v,br}} \\ &= (1 - 0.85)(154\ \mathrm{Pa}) - 98\ \mathrm{Pa} \\ &= -75\ \mathrm{Pa} \end{aligned}$$

Customary U.S. Solution

The ratio of the branch to upstream velocity is

$$\frac{\mathrm{v_{br}}}{\mathrm{v_{up}}} = \frac{2560\ \frac{\mathrm{ft}}{\mathrm{min}}}{3200\ \frac{\mathrm{ft}}{\mathrm{min}}} = 0.8$$

From the graph, the loss coefficient is $K_{\mathrm{br}} = 0.85$. From Eq. 40.8, the upstream and branch velocity pressures are

$$p_{\mathrm{v,up}} = \left(\frac{\mathrm{v_{up}}}{4005}\right)^2 = \left(\frac{3200\ \frac{\mathrm{ft}}{\mathrm{min}}}{4005}\right)^2 = 0.64\ \mathrm{in\ wg}$$

$$p_{\mathrm{v,br}} = \left(\frac{2560\ \frac{\mathrm{ft}}{\mathrm{min}}}{4005}\right)^2 = 0.41\ \mathrm{in\ wg}$$

From Eq. 40.56, the change in static pressure is

$$\begin{aligned} p_{s,\mathrm{br}} - p_{s,\mathrm{up}} &= (1 - K_{\mathrm{br}})p_{\mathrm{v,up}} - p_{\mathrm{v,br}} \\ &= (1 - 0.85)(0.64\ \mathrm{in\ wg}) - 0.41\ \mathrm{in\ wg} \\ &= -0.31\ \mathrm{in\ wg} \end{aligned}$$

33. DUCT DESIGN PRINCIPLES

Duct systems are categorized as low-velocity (up to 2500 ft/min or to 12.8 m/s) and high-velocity (above 2500 ft/min or 12.8 m/s). An important consideration in system design is fan outlet static pressure recovery conditions (i.e., velocity pressure to static pressure shift) which are based on equivalent (or effective) duct length, L_{eq}, and fan outlet duct area, A_{o}. *Low-velocity systems*, also known as *conventional systems*, are usually

[21]Some manufacturers also provide direct-reading charts that give the friction loss directly in terms of in wg.

HVAC

Figure 40.7 *Types of Commercial Divided-Flow Fittings*

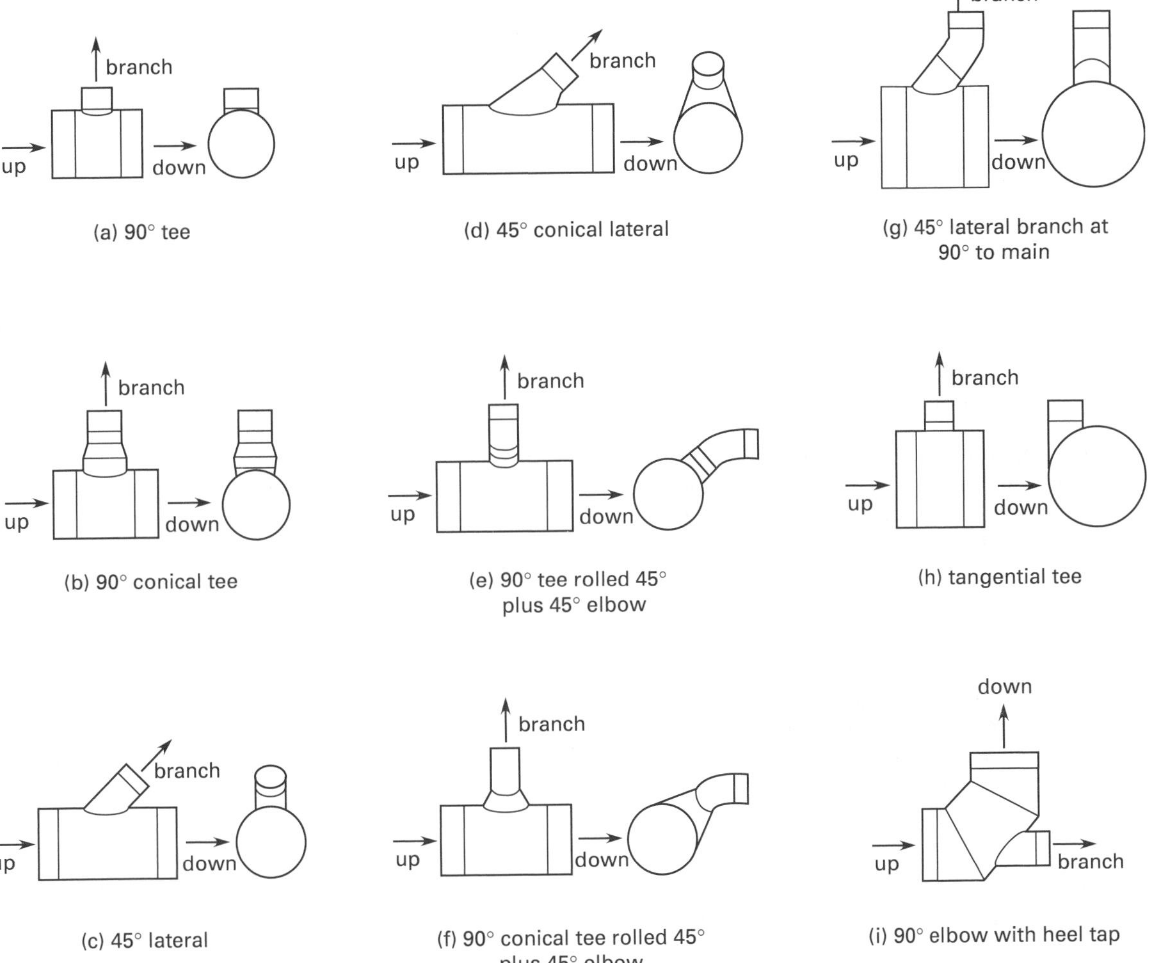

designed with the velocity-reduction and equal-friction methods. For 100% recovery at the fan outlet in low-velocity systems,

$$L_{\text{eq}} = \frac{\sqrt{A_{\text{o}}}}{4.3} \tag{40.57}$$

High-velocity systems are able to take advantage of benefits associated with the static regain method. For 100% recovery in high-velocity systems,

$$L_{\text{eq}} = \frac{\text{v}_o\sqrt{A_{\text{o}}}}{10{,}600} \tag{40.58}$$

Duct systems are categorized according to the static pressure at the fan: low-pressure (0–2 in wg; 0–500 Pa), medium-pressure (2–6 in wg; 500–1500 Pa), and high-pressure (6–10 in wg; 1500–2500 Pa). Residential and commercial ducts are typically designed such that the pressure drop at the fan is between 0.08 in wg and 0.15 in wg.

Supply duct systems take air from the fan and bring it to the ventilated space. *Exhaust duct systems* (*return air systems*) carry air from the ventilated space back to the fan.

Compared with conventional systems, high-pressure and high-velocity systems require less space, cost less for ductwork, and provide better control of the conditioned space. However, they are noisier, require a more precise duct design, require larger (more expensive) fans, and use more energy over the course of the year, which will likely cause greater peak kW demand charges from the electric utility. Because the economic cost benefit of long-term energy savings tends to outweigh any initial construction savings, the overall design trend continues toward low-pressure systems.

The following general recommendations apply to all duct designs and design methods.

- Make duct routes as direct as possible.
- Avoid sudden changes in direction and diameter.
- Use radius-to-diameter ratios of 1.5 or higher.

- Eliminate obstructions in and through the ducts.
- Use radiused elbows whenever possible, and when not, use turning vanes.
- Make rectangular ducts as square as possible. Avoid aspect ratios greater than 8:1, and use 4:1 or less whenever space permits.
- Use smooth metal construction whenever possible.
- Maintain an incremental size difference of at least 2 in in adjacent duct sections.
- Include a small volume allowance above the sum of all the outlet volumes to account for leakage.[22]
- Size the fan with excess capacity to compensate for inaccuracies in the design.
- Install balancing dampers in all branches, even when the static regain method is used for the design. In order to minimize noise, install dampers as close as possible to the main duct.
- Use the lowest possible duct velocities in order to minimize fan power and noise.
- In practice, to prevent undersizing supply ducts in residential applications, supply-side designs should be based on no greater than 0.1 in wg per 100 ft; and, a value of 0.06 in wg is more appropriate. For return lines, values in the range of 0.04–0.05 in wg are appropriate.
- Conventional rule-of-thumb "wisdom" specifies the gross area of return grilles as 1 ft^2 (144 in^2) per ton of refrigeration. However, this generally results in the average system return being undersized by 30% or more. A better rule of thumb is to have 1 in^2 of gross grille area for every 2 ft^3/min of airflow.

34. ECONOMICAL DUCT DESIGN

All other factors being equal, economical duct design is achieved by using standard, factory-manufactured round duct, keeping runs straight, minimizing the aspect ratio of rectangular duct, minimizing the total amount of sheet metal (i.e., minimizing the total mass) used, and by maintaining trunk size until a reduction of 2 in (51 mm) or more is warranted (this is known as the *2-inch rule*).

35. LEAKAGE

Ducts are not intended to be leak-free. However, the volumetric leakage should be less than 1% for well-sealed ducts and 2–5% for unsealed ducts.[23] Ducts are not pressure vessels and are not intended to be tested by sealing and pressurization. (The term "airtight" should be avoided.) Leakage should be tested volumetrically with the air in motion.

Leakage can be classified and quantified by a duct leakage class. The *leakage class*, C_L, is defined as actual leakage in cubic feet per minute per 100 square feet of duct surface area, a quantity known as the *leakage factor*, F, when the gage pressure, p, within the duct is 1 in wg. An exponent, N, is used to correlate the leakage class and leakage factor. N depends on turbulence within the duct, but it has a reliable average value of 0.65. Duct leakage is essentially independent of duct velocity. Equation 40.59 is valid for both positive and negative pressures.

$$Q_{\text{leakage}} = C_L p^N \qquad 40.59$$

For convenience, the leakage class of ducts constructed by skilled, trained technicians can be predicted by the SMACNA *seal class*.[24] Seal class A, applicable to ducts with pressurizations 4 in wg and higher, requires all transverse joints, longitudinal seams, and duct wall penetrations to be sealed. For seal class A, the leakage class, C_L, can be estimated as 3 cfm/100 ft^2 for round metal ducts and as 6 cfm/100 ft^2 for rectangular metal ducts. Seal class B, applicable to pressurizations of 3–4 in wg, requires sealing of transverse joints and longitudinal seams. The approximate leakage class is 6 cfm/100 ft^2 for round metal ducts and 12 cfm/100 ft^2 for rectangular metal ducts. Seal class C, applicable to pressurizations of 2 in wg and less, requires sealing only of transverse joints. The approximate leakage class is 12 cfm/100 ft^2 for round metal ducts and 24 cfm/100 ft^2 for rectangular metal ducts. Unsealed ductwork can be expected to exhibit a leakage class of 24 cfm/100 ft^2 for round metal ducts and 48 cfm/100 ft^2 for rectangular metal ducts.

Actual construction and sealing can be used to predict the seal class when ducts are manufactured customarily. Since Eq. 40.59 was developed from measurements of ducts constructed with normal and customary quality, it should only be used to predict leakage from ducts whose construction is customary for the intended pressurization range. It should not be used when construction is inconsistent with intended use. For example, a duct expected to operate at a pressure less than 2 in wg

[22]Some sources say to include up to 10% excess air to account for leaks. While this may sound nominal, the fan laws show that increasing the fan speed 10% to obtain the extra flow will increase the horsepower 30%. It is unlikely that a motor would be able to provide 30% more power. Therefore, more reliance should be placed on tight ductwork than on excess air.
[23]ASHRAE and SMACNA recommendations.
[24]*Duct Construction Standards—Metal and Flexible*, SMACNA, 1985 ed.

would not normally be constructed with a class A seal class, and Eq. 40.59 cannot be expected to predict leakage accurately in that instance.

Ductwork carries flows of 2–5 cfm/ft^2 (cfm per square foot of duct surface area). Systems with a lot of ductwork and small airflows are nearer to the lower end, while systems with minimal ductwork and large airflows are nearer to the upper end. The leakage as a percentage of the supplied airflow is

$$\begin{aligned} \text{leakage}_{\%\,\text{of supply}} &= \frac{Q_{\text{leakage,cfm}}}{Q_{\text{supply,cfm}}} \times 100\% \\ &= \frac{Q_{\text{leakage,cfm}/100\,\text{ft}^2}}{Q_{\text{supply,cfm}/\text{ft}^2}} \end{aligned} \qquad 40.60$$

Example 40.10

Air flows through a rectangular duct (seal class C) at the rate of 3 cfm/ft^2. The gage pressure in the duct is 1.7 in wg. What is the leakage as a percentage of the supply rate?

Solution

Rectangular duct with seal class C can be expected to have a leakage class, C_L, of 24 cfm/100 ft^2.

From Eq. 40.59,

$$\begin{aligned} Q_{\text{leakage}} &= C_L p^N = \left(24\ \frac{\text{cfm}}{100\ \text{ft}^2}\right)(1.7\ \text{in wg})^{0.65} \\ &= 33.88\ \text{cfm}/100\ \text{ft}^2 \end{aligned}$$

From Eq. 40.60, combining the area and flow rate terms,

$$\begin{aligned} \text{leakage}_{\%\,\text{of supply}} &= \frac{Q_{\text{leakage,cfm}/100\,\text{ft}^2}}{Q_{\text{supply,cfm}/\text{ft}^2}} = \frac{33.88\ \frac{\text{cfm}}{100\ \text{ft}^2}}{3\ \frac{\text{cfm}}{\text{ft}^2}} \\ &= 11.29\% \end{aligned}$$

36. COLLAPSE OF DUCTS

Under certain conditions, ducts may collapse inward. This may happen in medium- and high-velocity systems when a fire damper or blast gate suddenly closes, but can also occur in long, large-diameter air return systems. The negative pressure created between a closed damper and the retreating mass of air may collapse the duct. The negative pressure required to collapse a duct depends on the duct construction and must be specified by the duct manufacturer.

To prevent collapse in air return systems, increased metal gage and/or angle rings may be used. To prevent collapse due to sudden closures, a negative-pressure relief valve can be installed immediately downstream of each fire damper.

37. FILTERS

Duct filters are used to remove dust, pollen, spores, bacteria, and other particles. In residential and light commercial applications, traditional pleated furnace-type filters are used, along with fiberglass and foam media filters. Most commercial/industrial filter units contain two or more stages of successively finer filtration, starting with a pleated pre-filter. *Hogs hair* filters (made from latex-coated organic fibers) are washable and reusable and can be used as pre-filters. Commercial and industrial environments may require box and bag cloth filters. In more demanding environments, such as clean rooms and hospitals, multi-stage, high-efficiency particulate air (HEPA) and electrostatic filters can be used. Filters with layers of granulated activated carbon (GAC) are somewhat useful in removing gases, VOCs, and odors.

Filtration efficiency is measured by the percentage of particles removed. For most filters, the removal efficiency varies nonlinearly with the particle size. The *arrestance* is the percentage of macroscopic particles (lint, hair, dust, etc.) removed. Arrestance for most filter types is usually well above 80%. However, most filters have lower *filtration efficiencies* with smaller particles. The 0.3 micron particle size is an industry standard comparison point. For example, when new, a typical, high-quality pleated furnace filter for residential use has a removal efficiency of approximately 80% at 10 microns, 40% at 1 micron, and less than 10% at 0.3 micron. Electrostatic filters have efficiencies approaching 95%. HEPA filters have efficiencies of 99.97% at the 0.3 micron level, and some manufacturers claim efficiencies better than "5 nines" (i.e., 99.999%) for multi-stage units. Although filters with traditional filter elements (e.g., foam and pleated) increase in efficiency over time (albeit with an increase in pressure drop), efficiencies of electrostatic filters decrease as the filter is used and the collection surface becomes coated.

Filters are rated by their *minimum efficiency reporting value* (MERV), a standard used to categorize the overall efficiency of the filter. MERV ratings range from 1 to 16, with the more efficient filters receiving the higher values. Typical residential pleated filters have poor performance below 10 microns and have MERV ratings of 1 to 4. High-quality filters with MERV ratios 5 to 8 can remove particles as small as 3 microns. Filters with 9 to 12 MERV ratings are used in commercial and industrial applications and will stop particles in the 1 to 3 micron range. The most efficient filters have MERV ratings of 13 to 16 and will stop particles as small as 0.3 microns. These filters are used in hospitals and clean rooms.

Pressure drops in most pleated filters are less than 0.5 in wg, and for residential and light industrial applications are generally 0.2–0.3 in wg. Manufacturer's data must be used for accurate assessments. Filters with MERV ratings greater than 13 generally have high pressure drops. Because of this, they may be installed in parallel (not inline) with the return duct and filter only a portion of the air at a time.

38. DAMPERS

Since friction loss is proportional to the distance from the fan to the outlet, duct runs to outlets near the fan will have lower friction losses than the main duct run. If not constrained, a disproportionate amount of air will be distributed to the lower-friction runs. Limited pressure balancing can be achieved with *jumper ducts*, also known as *crossover ducts*, which are ducts (without equipment) that run between zones and terminate at simple grilles in order to equalize pressure in the zones. Jumper ducts are generally only used in residential construction where duct runs are short.

Duct systems are not self-equalizing or self-balancing. Even when designed well, installed ducts rarely perform as designed, and each run must be adjusted individually after installation. Balancing adjusts the flow rate in each duct to match the design value for the corresponding zone. Dampers are used to balance airflow in ducts and to regulate the quantity of outside make-up air drawn in. Dampers can be motorized. Motorized dampers typically have a default position when there is no control voltage applied. This is considered the failsafe position, which can be fail open, fail closed, or fail-in-last position. Most dampers are usually operated manually. A significant characteristic (i.e., disadvantage) of manual dampers is that they are generally left forever in their originally installed position. Pressure loss through dampers can be substantial, even when fully open. A typical pressure-loss coefficient, C, for a fully open damper is 0.52. [**Damper Types**]

Dampers are included in all runs to keep the pressure drop the same in all duct runs, even those that are short. *Balancing* is the act of closing down the dampers to equalize the friction losses. It is a good idea to install dampers even when sophisticated design methods are used. Dampers can be manually operated (for balancing or other occasional use), motorized (for zone control and variable volume), or gravity operated.

Figure 40.8 illustrates three general types of dampers. *Volume dampers* should be used only in branch ducts when a splitter damper cannot be used.[25] *Splitter dampers* should be used at the junction of the main and branch ducts. *Automatic dampers* are usually chosen with parallel blades for applications with two distinct positions. Dampers with opposed blades are chosen when airflow is to be controlled over a wide range. *Gravity dampers* are self-closing and are intended to prevent backflow. *Fire dampers* and *smoke dampers* close automatically to prevent the spread of smoke throughout the system.

Balancing/volume-adjusting dampers should be installed close to the main supply, as far away as possible from the outlets. However, outlets can be designed to act as dampers. Such decorative *grilles* may be *fixed pattern dampers* (i.e., *perforated plate* or *fixed-bar grilles*) or *adjustable bar grilles* with a manual control lever. The disadvantage of combining terminal air distribution with dampering is that the noise created by the friction is projected directly into the room. In order to limit noise, the maximum flow velocities and flow rates are specified with standard commercial grilles by the manufacturers.

Damper authority is the ratio of the fully open control damper pressure drop to the total branch pressure drop with the fully opened control damper at full-flow conditions.

Damper Authority

$$\text{damper authority } \% = \frac{\text{open damper resistance}}{\text{total system resistance}} \times 100 \quad 40.61$$

The intent is to design systems with high damper authority to ensure good controllability. *ASHRAE Handbook—Fundamentals* provides diagrams of characteristic curves (see Chap. 7, "Fundamentals of Control"). [**Characteristic Curves of Installed Dampers With Fully Ducted Arrangement**]

Figure 40.8 *Types of Dampers*

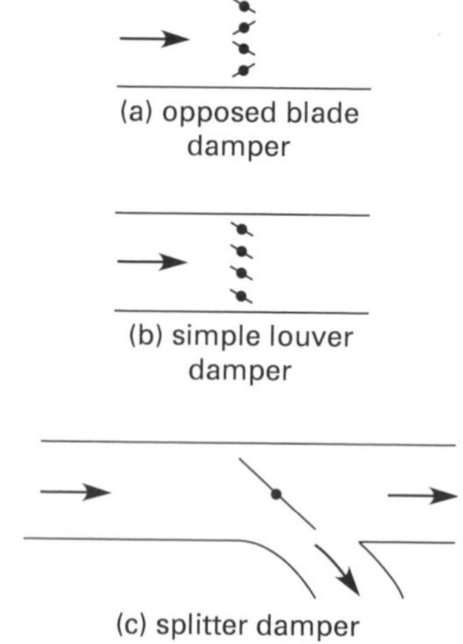

39. EQUAL-FRICTION METHOD

The *equal-friction method* is applicable to simple low-velocity systems. The method gets its name from the procedure that arbitrarily keeps the friction loss per unit length the same in all duct runs. Velocity pressure

[25] In the ventilation industry, dampers used only to adjust volume are also known as *blast gates*.

and regain are disregarded. The system will require extensive dampering, as no attempt is made to equalize pressure drops in the branches.

A common assumption in preliminary design studies is a friction loss of 0.08 in wg per 100 ft (0.65 Pa/m) for virtually all situations except offices (0.10 in wg per 100 ft; 0.82 Pa/m) and industrial uses (0.15 in wg per 100 ft; 1.2 Pa/m). *ASHRAE Handbook—Fundamentals* provides guidance for selecting appropriate friction factors for low-pressure systems, especially for systems downstream of terminal boxes (see *ASHRAE Handbook—Fundamentals* Chap. 21, "Duct Design"). This value is referred to as the *design pressure drop* (dpd). These values ensure duct velocities are low enough to avoid excessive noise, and they represent a good compromise between duct and fan installation and operating costs.

step 1: Select the main duct velocity from Table 40.8, contract specifications, or judgment.

Table 40.8 *Typical Maximum Duct Velocities (in ft/min)*

application	large supply ducts	small supply ducts	return ducts
residences	800	600	600
apartments/hotel bedrooms	1500	1100	1000
theaters	1600	1200	1200
deluxe offices		1100	800
average offices		1300	1000
general offices	2200	1400	1200
restaurants	1800	1400	1200
small shops		1500	1200
department stores			
lower floors	2100	1600	1200
upper floors	1800	1400	1200

(Multiply ft/min by 0.00508 to obtain m/s.)

step 2: From the velocity and flow rate in the main duct, find the friction loss per unit length from Fig. 40.5 or Fig. 40.6.

step 3: After each branch, reduce the main duct flow rate by the branch flow. Find the new velocity and duct size to keep the same friction loss per unit length. (This means that all points will be along a vertical line on Fig. 40.5 or Fig. 40.6.)

step 4: Determine the static pressure drop in the highest-resistance duct. The fan must supply this static pressure drop plus the desired outlet pressure.

step 5: Compare the actual system pressure with the design pressure, if known. If they are significantly different, repeat all the steps with a different main duct velocity.

step 6: Size branch runs the same way—keeping the same friction loss per unit length. Use dampers to equalize the pressure drops.

Example 40.11

A theater duct system is shown. All bends have a radius-to-diameter ratio of 1.75. The branch takeoff between sections A and B has a branch loss coefficient of $K_{br} = 1.5$. The design pressure at each outlet is 0.15 in wg (38 Pa). Disregard the divided flow fitting loss. Use the equal-friction method to size the system.

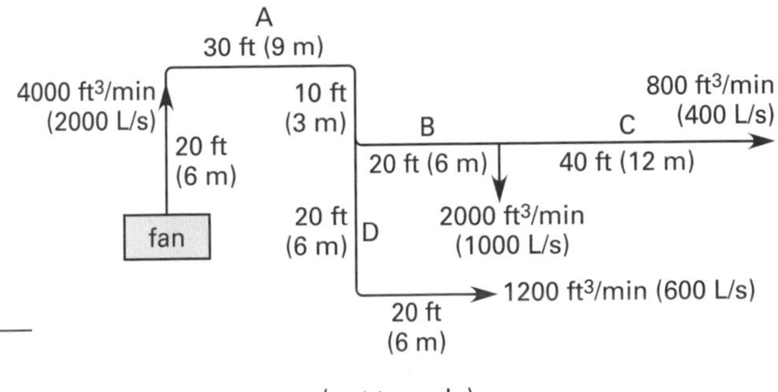

SI Solution

step 1: From Table 40.8, choose the main duct velocity (section A) as 1600 ft/min. From the table footnote, the SI velocity is

$$v_{main} = \left(1600 \ \frac{\text{ft}}{\text{min}}\right)\left(0.00508 \ \frac{\text{m·min}}{\text{s·ft}}\right) = 8.1 \text{ m/s}$$

step 2: The total airflow from the fan is 2000 L/s. From Fig. 40.6, the main duct diameter (section A) is approximately 560 mm. (This may not correspond to a standard duct size.) The friction loss is 1.2 Pa/m.

step 3: After the first takeoff, the flow rate in section B is

$$2000 \ \frac{\text{L}}{\text{s}} - 600 \ \frac{\text{L}}{\text{s}} = 1400 \text{ L/s}$$

From Fig. 40.6 for 1400 L/s and 1.2 Pa/m, the diameter is 490 mm, and the velocity is 7.4 m/s.

step 4: By inspection, the longest run is ABC. From Table 40.5, the equivalent length of each bend is $12D$.

$$L_{eq,bend} = 12D = \frac{(12)(560 \text{ mm})}{1000 \ \frac{\text{mm}}{\text{m}}} = 6.72 \text{ m} \quad (7 \text{ m})$$

The equivalent length of the entire run is

HVAC

	6 m	(from fan to first bend)
	7 m	(equivalent length of first bend)
	9 m	(first bend to second bend)
	7 m	(equivalent length of second bend)
	3 m	(jog between sections A and B)
	6 m	(section B)
	12 m	(section C)
total:	50 m	

The straight-through friction loss in the longest run is

$$(50\ \text{m})\left(1.2\ \frac{\text{Pa}}{\text{m}}\right) = 60\ \text{Pa}$$

Use Eq. 40.55 to find the friction loss in the branch take-off between sections A and B.

$$\begin{aligned} p_{t,\text{A}} - p_{t,\text{B}} &= K_{\text{br}}(p_{\text{v,up}}) = K_{\text{br}}(0.6)\text{v}_{\text{up}}^2 \\ &= (1.5)(0.6)\left(8.1\ \frac{\text{m}}{\text{s}}\right)^2 \\ &= 59\ \text{Pa} \end{aligned}$$

The fan must be able to supply a static pressure of

$$p_{s,\text{fan}} = 60\ \text{Pa} + 59\ \text{Pa} + 38\ \text{Pa} = 157\ \text{Pa}$$

The total pressure supplied by the fan is

$$\begin{aligned} p_{t,\text{fan}} &= p_{s,\text{fan}} + 0.6\text{v}_{\text{m/s}}^2 \\ &= 157\ \text{Pa} + (0.6)(8.1\ \text{m/s})^2 \\ &= 196\ \text{Pa} \end{aligned}$$

Customary U.S. Solution

step 1: From Table 40.8, choose the main duct velocity (section A) as 1600 ft/min.

step 2: The total airflow from the fan is 4000 ft^3/min. From Fig. 40.5, the main duct diameter (section A) is 21 in. The friction loss is 0.15 in wg per 100 ft.

step 3: After the first takeoff, the flow rate in section B is

$$4000\ \frac{\text{ft}^3}{\text{min}} - 1200\ \frac{\text{ft}^3}{\text{min}} = 2800\ \text{ft}^3/\text{min}$$

From Fig. 40.5 for 2800 ft^3/min and 0.15 in wg per 100 ft, the diameter is 18 in, and the velocity is 1500 ft/min. Similarly, the diameters at sections C and D are 11.5 in (say 12 in) and 13 in, respectively. The velocity at section C is 1000 ft/min.

step 4: By inspection, the longest run is ABC. From Table 40.5, the equivalent length of each bend is $12D$.

$$L_{\text{eq,bend}} = 12D = \frac{(12)(21\ \text{in})}{12\ \frac{\text{in}}{\text{ft}}} = 21\ \text{ft}$$

The equivalent length of the entire run is

	20 ft	(from fan to first bend)
	21 ft	(equivalent length of first bend)
	30 ft	(first bend to second bend)
	21 ft	(equivalent length of second bend)
	10 ft	(jog between sections A and B)
	20 ft	(section B)
	40 ft	(section C)
total:	162 ft	

The straight-through friction loss in the longest run is

$$\left(\frac{162\ \text{ft}}{100\ \text{ft}}\right)(0.15\ \text{in wg per 100 ft}) = 0.24\ \text{in wg}$$

Use Eq. 40.55 to find the friction loss in the branch take-off between sections A and B.

$$\begin{aligned} p_{t,\text{A}} - p_{t,\text{B}} &= K_{\text{br}}(p_{\text{v,up}}) \\ &= (1.5)\left(\frac{1600\ \frac{\text{ft}}{\text{min}}}{4005}\right)^2 \\ &= 0.24\ \text{in wg} \end{aligned}$$

The fan must be able to supply a static pressure of

$$\begin{aligned} p_{s,\text{fan}} &= 0.24\ \text{in wg} + 0.24\ \text{in wg} + 0.15\ \text{in wg} \\ &= 0.63\ \text{in wg} \end{aligned}$$

The total pressure supplied by the fan is

$$\begin{aligned} p_{t,\text{fan}} &= p_{s,\text{fan}} + p_{\text{v,up}} \\ &= 0.63\ \text{in wg} + \left(\frac{1600\ \frac{\text{ft}}{\text{min}}}{4005}\right)^2 \\ &= 0.79\ \text{in wg} \end{aligned}$$

40. COMBINATION METHOD

A *combination method* is sometimes used. The main duct is sized by the equal-friction method. The branch runs are sized so as to dissipate the remaining friction (compared with the main duct run) and to (theoretically) eliminate the need for dampers.

The desired outlet pressure is subtracted from the pressure at the main duct branch takeoff to get the pressure that must be dissipated in the branch run. This pressure is divided by the estimated equivalent length to find the pressure drop per unit length. Figure 40.5 and Fig. 40.6 can be used to find the duct size and velocity.[26]

Example 40.12

An air supply system consists of a long run and two branches. The total friction loss in the long run is 0.15 in wg (38 Pa). The longest duct was sized with the equal-friction method using a pressure drop of 0.2 in wg per 100 ft (1.6 Pa/m). The equivalent length of the branch takeoff at A is 12 ft (3.6 m). The equivalent length of the elbow in duct A is 18 ft (5.4 m). Rather than use a damper in duct A to equal the pressure drop, duct A will be sized small enough to equalize the losses through increased velocity. Use the combination method to size duct A.

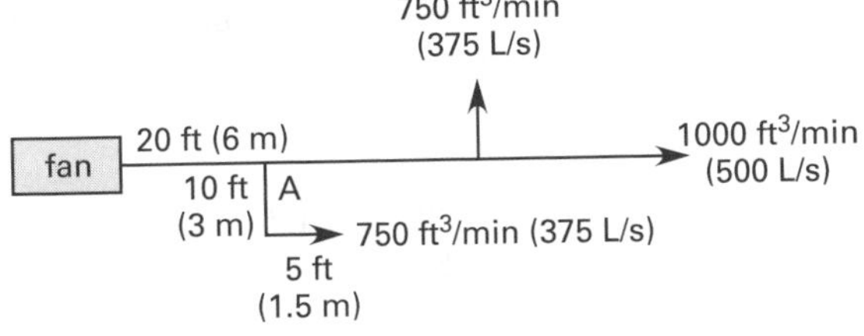

SI Solution

Subtracting the pressure drop from the fan to the branch takeoff, the pressure left to be dissipated in duct A is

$$38 \text{ Pa} - (6 \text{ m})\left(1.6 \ \frac{\text{Pa}}{\text{m}}\right) = 28 \text{ Pa}$$

The required loss per meter in duct A is

$$\frac{28 \text{ Pa}}{3 \text{ m} + 1.5 \text{ m} + 3.6 \text{ m} + 5.4 \text{ m}} = 2.07 \text{ Pa/m}$$

Use Fig. 40.6. With 375 L/s and 2.07 Pa/m, the velocity is approximately 6.9 m/s, and the diameter is approximately 260 mm.

Customary U.S. Solution

Subtracting the pressure drop from the fan to the branch takeoff, the pressure left to be dissipated in duct A is

$$\begin{aligned} 0.15 \text{ in wg} - \left(\frac{20 \text{ ft}}{100 \text{ ft}}\right)(0.2 \text{ in wg per } 100 \text{ ft}) \\ = 0.11 \text{ in wg} \end{aligned}$$

The required loss per 100 ft in duct A is

$$\frac{(0.11 \text{ in wg})\left(100 \ \frac{\text{ft}}{100 \text{ ft}}\right)}{10 \text{ ft} + 5 \text{ ft} + 12 \text{ ft} + 18 \text{ ft}} = 0.24 \text{ in wg per } 100 \text{ ft}$$

Use Fig. 40.5. With 750 ft^3/min and 0.24 in wg, the velocity is 1240 ft/min, and the diameter is 10 in.

41. STATIC REGAIN METHOD

In the *static regain method*, the diameter of each successive branch is reduced in order to increase the static pressure at the branch entrance back to the fan discharge pressure. The reduction is such that the static regain offsets the friction loss in the succeeding section. (This method can also be used to size branch ducts as long as the duct sizes are reasonable.) In Eq. 40.62, point A is before the branch takeoff, point B is immediately after the branch takeoff, and point C is just prior to the next branch takeoff. (See Fig. 40.9.)

$$p_{s,\text{A}} - p_{s,\text{B}} = p_{f,\text{B-C}} \qquad 40.62$$

Figure 40.9 *Duct Runs for Static Regain*

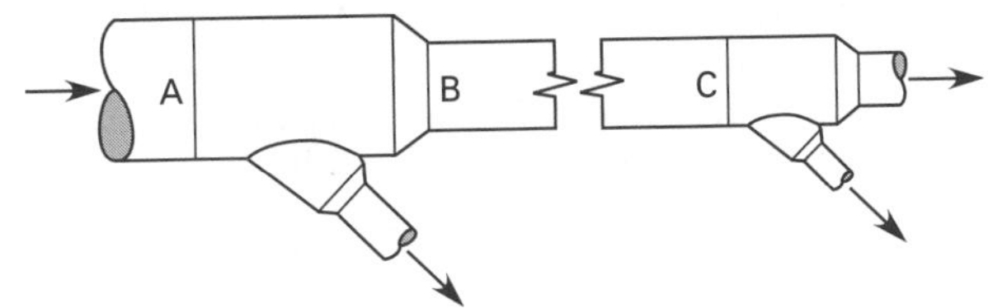

The analytical relationship between the velocities in the sections is given by Eq. 40.63. C is equal to 0.0832 for a static regain coefficient, R, of 0.75. Values of C for other values of R can be calculated from Eq. 40.64.[27]

$$\text{v}_\text{A}^2 - \text{v}_\text{B}^2 = C\text{v}_\text{B}^{2.43}\left(\frac{L_\text{BC}}{Q_\text{B}^{0.61}}\right) \quad \text{[U.S. only]} \qquad 40.63$$

$$C = \frac{6.256 \times 10^{-2}}{R} \quad \text{[U.S. only]} \qquad 40.64$$

[26]To limit noise, very high velocities should be avoided.
[27]Equation 40.63 is based on a friction factor of 0.0270.

The velocity v_B appears on both sides, making Eq. 40.63 difficult to use. For that reason, duct diameters are often chosen by trial and error. However, graphical aids (e.g., Fig. 40.10) can be used to determine the unknown velocity without extensive trial and error iterations. To use Fig. 40.10, the quantity $L_{AB}/Q_B^{0.61}$ is calculated. The intersection of the L/Q line and the v_A line defines v_B. In most cases, it is assumed that the regain will equal the friction loss in the following section. In that case, v_B is read directly from the horizontal scale. However, Fig. 40.10 can also be used to determine a velocity that will increase or decrease the static pressure by some given amount. If a loss in static pressure is required, move to the right of the intersection point until the vertical separation between the two curves equals the desired loss. (Use the vertical scale on the right edge to determine the vertical separation.) Then, drop down and read the velocity from the horizontal scale. A gain is handled similarly—by moving to the left.

The following steps constitute a simplified static regain method. In practice, the prediction of the regain coefficient, R, is quite difficult, rendering this method generally unusable.

step 1: Use Table 40.8 to choose a velocity in the main run.

step 2: Size the main run using the continuity equation, $A = Q/v$. [**Continuity Equation**]

step 3: Find the equivalent length of the main duct from the fan to the first branch takeoff. Assume any unknown bend radii.

step 4: Use Fig. 40.5 or Fig. 40.6 to find the friction loss, $p_{f,\text{main}}$ in the main run up to the branch takeoff.

step 5: Determine the fan pressure. Assuming that all subsequent friction after the first branch takeoff will be recovered with static regain, the static pressure supplied by the fan will be

$$p_{s,\text{fan}} = p_{f,\text{main}} + \text{grille discharge pressure} \quad 40.65$$

Equation 40.65 assumes that the fan discharge velocity and velocity in the main duct run are the same. If the velocities are different, Eq. 40.52 is used to determine a static regain that reduces the pressure supplied by the fan.

step 6: Calculate the flow rate in the duct after the branch takeoff.

$$Q = Q_{\text{main}} - \text{branch flow} \quad 40.66$$

step 7: Knowing the flow rate and length of the next section, determine the velocity in that section from Fig. 40.10. For other values of R, or when a regain chart is not available, the duct size must be found by trial and error. The duct size is varied until the friction loss equals the regain.

step 8: Solve for the duct size using the continuity equation, $A = Q/v$. [**Continuity Equation**]

Example 40.13

The fan in the duct system shown moves a total of 1500 ft³/min. The velocity of the air in the fan is 1700 ft/min. Bends have equivalent lengths of 15 ft. The required outlet grille pressure is 0.25 in wg. Use the static regain method with a static regain coefficient of 0.75 to size the main duct run fan-A-F.

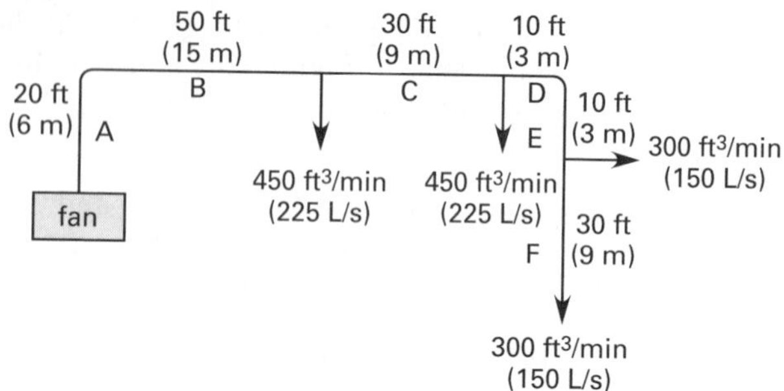

Solution

step 1: Choose 1500 ft/min as the main duct velocity.

step 2: The area and diameter of the main duct are

Continuity Equation

$$Q = Av$$

$$A = \frac{Q}{v} = \frac{1500 \ \frac{\text{ft}^3}{\text{min}}}{1500 \ \frac{\text{ft}}{\text{min}}} = 1 \ \text{ft}^2$$

$$D = \sqrt{\frac{4A}{\pi}} = \sqrt{\frac{(4)(1 \ \text{ft}^2)}{\pi}}\left(12 \ \frac{\text{in}}{\text{ft}}\right) = 13.5 \ \text{in} \quad (14 \ \text{in})$$

step 3: The equivalent length of the main duct from the fan to the first takeoff and bend is

$$L = 20 \ \text{ft} + 15 \ \text{ft} + 50 \ \text{ft} = 85 \ \text{ft}$$

step 4: From Fig. 40.5, the friction loss in the main run up to the branch takeoff is approximately 0.20 in wg per 100 ft. The actual friction loss is

$$p_{f,\text{main}} = (0.20 \ \text{in wg per } 100 \ \text{ft})\left(\frac{85 \ \text{ft}}{100 \ \text{ft}}\right) = 0.17 \ \text{in wg}$$

step 5: Since the main duct velocity is lower than the fan discharge velocity, there will be static regain from the fan. From Eq. 40.52,

Figure 40.10 Static Regain Chart ($R = 0.75$)

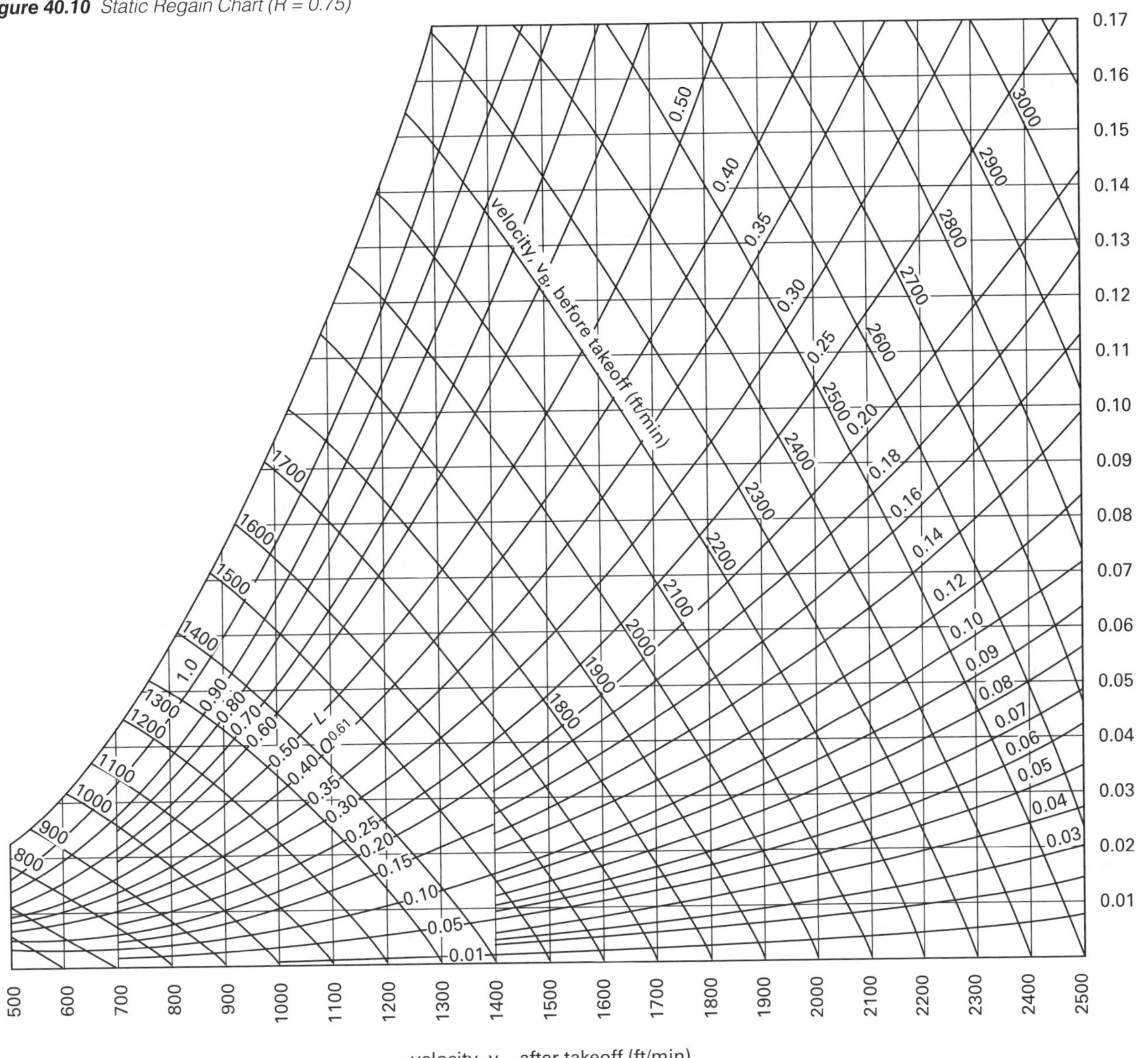

$$
\begin{aligned}
SR_{fan} &= \frac{R(v_{fan}^2 - v_{main}^2)}{(4005)^2} \\
&= (0.75) \\
&\quad \times \left(\frac{\left(1700\ \frac{ft}{min}\right)^2 - \left(1500\ \frac{ft}{min}\right)^2}{(4005)^2} \right) \\
&= 0.03 \text{ in wg}
\end{aligned}
$$

$$
\begin{aligned}
p_{s,fan} &= p_{f,main} + \text{grille pressure} - SR_{fan} \\
&= 0.17 \text{ in wg} + 0.25 \text{ in wg} - 0.03 \text{ in wg} \\
&= 0.39 \text{ in wg}
\end{aligned}
$$

step 6: The flow rates, equivalent lengths, and $L/Q^{0.61}$ ratios for each section are

section	L	Q	$\frac{L}{Q^{0.61}}$
C	30	1050	0.43
D to E	$10 + 15 + 10 = 35$	600	0.71
F	30	300	0.92

step 7: From Fig. 40.10, the velocities are

section	v (ft/min)
C	1130
D to E	800
F	560

step 8: Solve for the duct size using the continuity equation.

$$D = \sqrt{\frac{4A}{\pi}} = \sqrt{\frac{4Q}{\pi \text{v}}}$$

$$D_C = \sqrt{\frac{(4)\left(1050\ \frac{\text{ft}^3}{\text{min}}\right)}{\pi\left(1130\ \frac{\text{ft}}{\text{min}}\right)}}\left(12\ \frac{\text{in}}{\text{ft}}\right)$$

$$= 13.1 \text{ in} \quad (13 \text{ in})$$

$$D_{D/E} = \sqrt{\frac{(4)\left(600\ \frac{\text{ft}^3}{\text{min}}\right)}{\pi\left(800\ \frac{\text{ft}}{\text{min}}\right)}}\left(12\ \frac{\text{in}}{\text{ft}}\right)$$

$$= 11.7 \text{ in} \quad (12 \text{ in})$$

$$D_F = \sqrt{\frac{(4)\left(300\ \frac{\text{ft}^3}{\text{min}}\right)}{\pi\left(560\ \frac{\text{ft}}{\text{min}}\right)}}\left(12\ \frac{\text{in}}{\text{ft}}\right)$$

$$= 9.91 \text{ in} \quad (10 \text{ in})$$

42. TOTAL PRESSURE DESIGN METHOD

The previous duct design methods focus on static pressure. The static regain method actually compensates for the inefficiency of a fitting by increasing subsequent duct sizes. None of the duct design methods mentioned attempts to minimize the friction loss. Nevertheless, energy is lost due to friction and turbulence at fittings.

The true measure of the loss in a fitting is represented by the change in total pressure it causes. Unlike static pressure, total pressure along a duct run will always decrease, never increase. Features (i.e., fittings) where the total pressure drops by a significant amount represent inefficient features (i.e., the "wrong" fitting for that location). These should be replaced with fittings with lower loss coefficients. This is the basic premise of the *total pressure design method.*

43. AIR DISTRIBUTION

An *outlet* is a supply opening through which air enters the ventilated space. An *inlet* is a return opening through which air is removed from the ventilated space. In residential construction, outlets are usually placed in floors under windows, and inlets are placed in the ceiling or on walls near the ceilings. In commercial construction, locations are determined by numerous factors and can be anywhere. *ASHRAE Handbook—HVAC Applications* also identifies recommended return inlet face velocities as a function of inlet location. The terms *grille*, *register*, and *diffuser* are used to describe coverings for the openings, and the terms are used somewhat interchangeably. A *grille* is a decorative covering for an opening. For example, perforated plate grilles are used to cover inlets to return air ducts. A *filter grille* accommodates a furnace filter behind its face. A *diffuser* is a grille with fixed or moveable louvers that guides the supply or return air. The number of louvers is given in bars per unit length. In residential applications, units with 2 bars to the inch are used for heating, while units with 3 bars to the inch are for mixed use. Diffusers can be 1-, 2-, 3-, and 4-way, referring to the number of orthogonal directions the air is directed by louvers. A *register* is a grille with an internal damper. Registers may have louvers. [**Recommended Return Inlet Face Velocity**]

Incorrect location of registers results in drafts, hot and cold spots, and noise. Locating a register requires knowledge of register performance regarding throw, spread, drop, and terminal velocity for the given airflow and velocity. The *Coanda effect* (*ceiling effect*) causes air to adhere to the ceiling after discharge from a wall register at the ceiling level or a supply diffuser located in the ceiling, which is most common in commercial HVAC applications. The suction effect is proportional to the square of the discharge velocity. With improper designs, the terminal velocity is too low, eventually decreasing to a point (about 4.5 ft/sec (1.5 m/s)) where the ceiling effect suction is inadequate, and the discharged air drops downward, a characteristic known as *dumping.* Occupants find dumping to be uncomfortable, as it places them in drafts. Once dumping begins, the discharge velocity must be increased to 30–40% above the original velocity in order to reattach the airflow to the ceiling. [**Recommended Return Inlet Face Velocity**]

ASHRAE has two suggestions: For systems in the cooling mode, diffuser selection should be based on the ratio of the diffuser's throw to the length of the zone being supplied. The magnitude of the zone length (characteristic length) often depends on the arrangement of the diffusers as well as the dimensions of the space. For systems in the heating mode, the diffuser to room temperature difference (DT) should not exceed 15°F to avoid excessive temperature stratification.

For systems in the cooling mode, ASHRAE has developed the *air diffusion performance index* (ADPI) to categorize occupant thermal comfort in sedentary environments with ceilings of at least 8 ft. The ADPI is a single-digit index derived from temperatures and velocities at specific locations (prescribed by the ASHRAE test method) around the outlet diffuser. The ADPI essentially represents the percentage of occupants that would feel comfortable. In general, velocities experienced by occupants should be less than 70 fpm (50 fpm ideally), and the ADPI should predict greater than 70%, and preferably 80–90%, of occupant acceptance.

ASHRAE suggests combinations of velocity and temperatures that accomplish these goals.[28] Using manufacturer's data, diffusers should be selected that satisfy the suggested T_{50}/L *throw ratios*. T_{50} is the manufacturer's reported throw to 50 fpm. L is the *space characteristic length*. This is usually the distance from the outlet to the wall or mid-plane between outlets. The desired throw value can be determined by multiplying the desired throw ratio by the characteristic length. The throw ratio is based on a 9 ft ceiling height. The throw can be increased or decreased by the same amount that the ceiling height exceeds or is less than 9 ft. [**Air Diffusion Performance Index (ADPI) Selection Guide**] [**Characteristic Room Length for Several Diffusers**]

For an outlet (grille, register, etc.) to work properly, the air must have a minimum static pressure (typically 0.1–0.3 in wg; 25–75 Pa) at the grille outlet. This *grille pressure* (*terminal pressure*) is added to the static and velocity pressures of the air when the fan is sized.

The *gross area* or *core area* of the grille is its total cross-sectional area. The net opening left when the gross area is reduced by the area of the louvers or dividers is the *free area* or *daylight area*, also known as the *effective area*. The outlet velocity can be found from the core velocity and outlet's coefficient of discharge, C_d, which is typically between 0.7 and 0.9.

$$\text{v}_{\text{outlet}} = \frac{\text{v}_{\text{core}} A_{\text{core}}}{C_d A_{\text{free}}} \qquad 40.67$$

The *throw* (also known as the *blow*) is the distance from the outlet to the *distribution point*. (See Fig. 40.11.) When an outlet has been properly selected, the average *terminal velocity* at the distribution point should be approximately 50 ft/min (0.25 m/s) for sedentary occupants up to 75 ft/min (0.38 m/s) for slightly active occupants.[29,30] Therefore, the throw is roughly the distance from the outlet to the location where the average air velocity is 50 ft/min (0.25 m/s). As it emerges from the outlet, duct air will entrain room air. The increase in airflow width is known as the *rise*, and the absolute width of the airflow is the *spread*. (Even straight outlets have airflows that diverge with a total included angle of up to 20°.)

Figure 40.11 *Air Distribution Terminology*

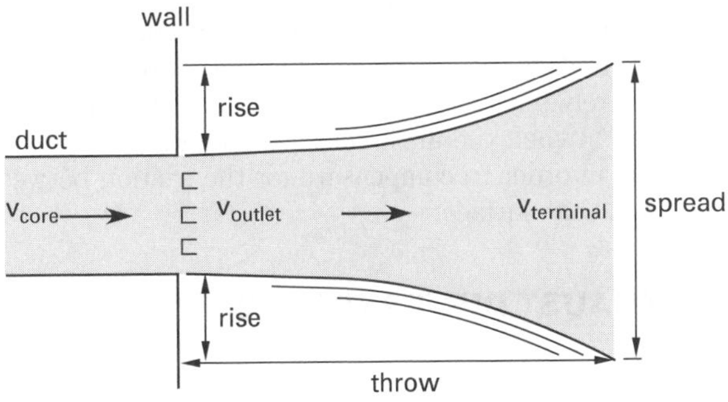

The conservation of momentum law predicts the amount of entrained air. The velocity of the room air is initially zero. The *induction ratio*, IR, is the ratio of combined to outlet air masses.

$$(m_{\text{outlet}})(\text{v}_{\text{outlet}}) = (m_{\text{outlet}} + m_{\text{entrained}})\text{v}_{\text{combined}} \qquad 40.68$$

$$\text{IR} = \frac{\text{v}_{\text{outlet}}}{\text{v}_{\text{combined}}} = \frac{m_{\text{outlet}} + m_{\text{entrained}}}{m_{\text{outlet}}} \qquad 40.69$$

The centerline velocity can be predicted for a distance x from the outlet. In Eq. 40.70, K is the *outlet constant* supplied by the outlet manufacturer, and R_{fa} is the ratio of free area to core (gross) area. The outlet velocity should be 300–500 ft/min (1.5–2.5 m/s) for ultra-quiet areas, 500–750 ft/min (2.5–3.8 m/s) for residences, theaters, and libraries, and 600–1000 ft/min (3.0–5.1 m/s) for offices and service areas. In noisy industrial areas, velocities as high as 2000 ft/min (10.2 m/s) may be tolerable.

$$\text{v}_{\text{centerline at distance } x} = \frac{KQ_{\text{outlet}}}{x\sqrt{C_d A_{\text{core}} R_{\text{fa}}}} = \frac{KQ_{\text{outlet}}}{x\sqrt{C_d A_{\text{free}}}} \qquad 40.70$$

The centerline velocity of the air emerging from a duct is approximately twice that of the average velocity across the duct face. Since the throw is roughly the distance at which the average distribution velocity is 50 ft/min (0.25 m/s), the throw is

$$\text{throw} = \frac{KQ_{\text{outlet}}}{\left(100\ \frac{\text{ft}}{\text{min}}\right)\sqrt{C_d A_{\text{free}}}} \quad \left[\text{consistent units}\right] \qquad 40.71$$

Generally, the throw should be 75% of the distance from the outlet face to the opposing normal surface. However this is completely dependent upon the type of diffuser/grill and the geometry of the installation. (Chapter

[28] *ASHRAE Handbook: Fundamentals, Inch-Pound Edition*, American Society of Heating, Refrigerating and Air-Conditioning Engineers, Inc., Atlanta, GA, 2009.
[29] For industrial work, the velocity may be as high as 300 ft/min (1.5 m/s).
[30] Some sources say the minimum air movement should be 20 ft/min (0.1 m/s) or above.

HVAC

20, "Room Air Distribution Equipment" in *ASHRAE Handbook—Fundamentals* provides descriptions of various arrangements.) For example, for a ceiling-mounted outlet and a 12 ft (3.6 m) ceiling height, the throw would be approximately 9 ft (2.7 m). The throw should be increased 25% to 50% when the air is released along a wall or near the ceiling in order to compensate for the friction between the air and that surface.

44. EXHAUST DUCT SYSTEMS

Exhaust (return air) duct systems are designed somewhat differently from supply systems.[31] Low-velocity designs often use the equal-friction method. Since the duct operates under a negative pressure, collapse is always a consideration. The negative suction rating of the fan should also not be exceeded.

With a single-fan system, in order for air to enter the return ducts through return grilles and then exhaust through relief dampers, the return air must enter the return air duct above atmospheric pressure. This requires the building to be continuously over-pressurized, causing doors to blow open and other problems. In order to avoid these over-pressurization problems, a *return air fan* after the return grilles and before the relief dampers is used. Air enters the return ductwork near atmospheric pressure, and the return air fan raises the static pressure in the duct to above atmospheric as required for the relief dampers. Since no additional friction sources (as compared to the single-fan system) are added, this does not increase the total fan power required by the system, although it does increase the total initial installation cost.

Converging-flow fittings behave differently than do divided-flow fittings. Figure 40.12 shows a typical converging-flow fitting. The branch loss coefficient (branch-to-downstream) for the fitting is defined by Eq. 40.72.

$$K_{\text{br}} = \frac{p_{t,\text{br}} - p_{t,\text{down}}}{p_{\text{v,down}}} \quad \text{40.72}$$

The *main loss coefficient* (upstream to downstream) is

$$K_{\text{main}} = \frac{p_{t,\text{up}} - p_{t,\text{down}}}{\text{V}p_{\text{v,down}}} \quad \text{40.73}$$

The change in static pressure from the branch to downstream is

$$p_{s,\text{br}} - p_{s,\text{down}} = (1 + K_{\text{br}})p_{\text{v,down}} - p_{\text{v,br}} \quad \text{40.74}$$

The change in static pressure from upstream to downstream is

$$p_{s,\text{up}} - p_{s,\text{down}} = (1 + K_{\text{main}})p_{\text{v,down}} - p_{\text{v,up}} \quad \text{40.75}$$

Figure 40.12 Converging-Flow Fitting

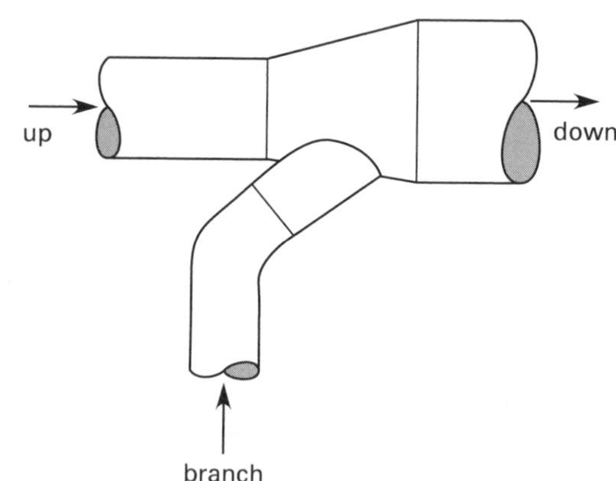

45. DUCT SYSTEM NOISE

Fan noise and *duct noise* have many sources.

1. *Vortex shedding* describes air separation from the blade surface and trailing edge. The resulting noise is broadband (i.e., containing a wide range of frequencies). Noise from vortex shedding is minimized by good blade profile design, use of proper pitch angle, and the presence of notched or serrated trailing blade edges.

2. Broadband noise can also be caused by turbulence in the air stream caused by inlet and outlet disturbances, sharp edges, and bends. High-impedance systems (i.e., those with high backpressure) are noisier than low-impedance systems.

3. Fan speed is a major factor in fan noise. The variation in sound level, L (measured in decibels), with rotational speed, N, is predicted by Eq. 40.76.[32]

$$L_2 = L_1 - 50\log_{10}\frac{N_1}{N_2} \quad \text{40.76}$$

4. A fan is generally quieter when operated near its peak efficiency. Noise will vary as the system load varies and the operating point shifts.

5. Substantial noise can be generated by structural vibration related to unbalance, bearings, rotor to stator eccentricity, and motor mounting.

46. DUST COLLECTION SYSTEMS

Design of dust collection systems draws on concepts similar to the design of duct systems for conditioned air, with notable differences. Dust collection systems are

[31]Some buildings (e.g., those with high-velocity supply systems and where space is limited) don't have return air systems. Even when there is a return air system, some rooms (e.g., those generating odors) may not have return air inlets.

[32]A 3 dB change is barely noticeable; a 5 dB change is clearly noticeable; a 10 dB change is twice (or half) as loud.

primarily return-air systems, and the fan is often incorporated into a cyclone type *dust collector*.[33] *Two-stage dust collectors* combine a cyclone with a fine-dust filter. Each dust type should be collected separately. To minimize explosion hazards, wood dust, metal dust, and fumes should be separated. Solvent fumes must be collected using non-ferrous ducts (e.g., aluminum) and explosion-proof blowers.

Spiral duct is preferred for dust collection because of its smoothness. 22 gauge metal duct is the most common; 18 gauge is used for heavy duty systems where high collapsing strength and abrasion resistance are required. 24 gauge and 20 gauge can also be used. Branch ducts serve individual dust-generating equipment; branch ducts join a common main duct feeding the dust collector. Branches should enter the main duct horizontally to prevent dust in the mains from falling back into the branches. Ducts should be equipped with access doors for duct cleaning and blockage clearing.

Dust-generating equipment attached to the system is categorized as primary or secondary. All primary machines operate simultaneously and are served by direct branch connections to the main duct. Secondary machines operate sporadically and are isolated from the main duct by sliding-blade blast gates. Equipment requiring the highest airflow should be placed closest to the dust collector. Branch diameters can be determined from the sizes of factory-installed equipment collars or from the continuity equation if the airflows are known. Main duct diameters increase as new primary branches enter the main duct in order to accommodate the airflow of all branches while maintaining the main duct velocity. Duct velocities are high in order to entrain dust in the airflow. Duct noise on the suction side can be minimized by using heavier gauge duct, proper hanging, and exterior insulation. Duct silencers (i.e., mufflers) should only be placed in the discharge side. Table 40.9 lists typical duct speeds.

Table 40.9 *Typical Duct Velocities for Dust Collection Systems (fpm (m/s))*

type of dust	branch duct velocity	main duct velocity
metalworking	4500 (23)	4000 (20)
woodworking	4000 (20)	3500 (18)
other light dust	4000 (20)	3500 (18)

(Multiply ft/min by 0.00508 to obtain m/s.)

The system friction pressure loss is determined using standard methods (e.g., loss coefficients, equivalent lengths, and standard duct friction loss charts). Friction in the ducts, silencers, and up-blast stack caps on the discharge side of the dust collection should be added to the friction from the suction index run ducts, along with losses from entrances, capture nozzles, fittings, filters, and flexible duct. Rules of thumb and assumptions may be needed in the absence of manufacturer's data. Convenient assumptions include (1) 1 in wg entrance loss, (2) 2 in wg filter loss, and (3) flex hose loss equal to three times smooth duct loss.

47. NOMENCLATURE

A	area	ft^2	m^2
ACFM	actual flow rate	ft^3/min	L/s
ahp	air horsepower	hp	n.a.
akW	air kilowatts	n.a.	kW
bhp	brake horsepower	hp	n.a.
bkW	brake kilowatts	n.a.	kW
c_p	specific heat at constant pressure	Btu/lbm-°F	kJ/kg·K
C	coefficient	–	–
C_L	leakage class	cfm/100 ft^2	n.a.
D	diameter	ft	m
DL	drive losses	–	–
E	energy	ft-lbf	J
EHP	electrical power	hp	n.a.
F	leakage factor	cfm/100 ft^2	n.a.
fhp	friction horsepower	hp	n.a.
g	acceleration of gravity, 32.2 (9.81)	ft/sec^2	m/s^2
h	head or height	ft	m
H	duct height	ft	mm
ICFM	inlet flow rate	cfm	L/s
IR	induction ratio	–	–
K	factor	–	–
L	length	ft	m
L	sound pressure level	dB	dB
L	space characteristic length	ft	mm
$\dot{m}$	mass flow rate	lbm/sec	kg/s
N	fan speed	rev/min	rev/min
N	exponent	–	–
p	pressure	in wg	Pa
P	power	hp	kW
Q	volumetric flow rate	ft^3/min	L/s
r	radius	ft	m
R	aspect ratio	–	–
R	regain coefficient	–	–
S	perimeter	ft	mm
SCFM	standard flow rate	ft^3/min	L/s
SE	static efficiency	–	–
SR	static regain	in wg	Pa
t	thickness	in	mm
T	absolute temperature	°R	K
T_{50}	throw to 50 fpm	ft	mm

[33]Cyclone design is covered in Chap. 70.

HVAC

v	velocity	ft/min	m/s
w	duct width	ft	mm
z	elevation above sea level	ft	m

There is only marginal consistency in the symbols used by this industry. For example, the symbol for total pressure can be p_t, p_T, TP, T_p, h_t, and many other variations. The symbol for fitting loss coefficient is almost universally K in the industry; ASHRAE uses C.

Symbols

γ	specific weight	lbf/ft^3	N/m^3
η	efficiency	–	–
ρ	density	lbm/ft^3	kg/m^3
υ	specific volume	ft^3/lbm	m^3/kg
ϕ	relative humidity	%	%

Subscripts

br	branch
d	density or discharge
down	downstream
e	entry
eq	equivalent
f	friction
k	kinetic
m	mixed
m	motor
o	outdoor
r	return
s	specific
s	static
sat	saturation
std	standard
t	total
up	upstream
v	velocity

41 Heating Load

Content in blue refers to the *NCEES Handbook*.

NCEES EXAM SPECIFICATIONS AND RELATED CONTENT

HVAC AND REFRIGERATION EXAM

- I.D. Heat Transfer
  - 5. Walls and Ceilings
- II.A. Heating/Cooling Loads
  - 5. Walls and Ceilings
  - 6. Air Spaces
  - 7. Ground Slabs
  - 9. Internal Heat Sources

THERMAL AND FLUID SYSTEMS EXAM

- III.B. Cooling/Heating
  - 5. Walls and Ceilings
  - 6. Air Spaces
  - 7. Ground Slabs
  - 9. Internal Heat Sources

1. INTRODUCTION

A building's *heating load* is the maximum heat loss (typically expressed in Btu/hr or kW) during the heating season.[1] The *maximum heating load* occurs when the outside temperature is the lowest. The maximum heating load corresponds to the minimum furnace size, even though the lowest temperature occurs only a few times each year. The *average heating load* can be derived from the maximum heating load and is used to determine the annual fuel requirements.

Heating load consists of heat to make up for transmission and infiltration losses. Determining transmission losses is essentially a heat transfer problem. *Transmission loss* is heat lost through the walls, roof, and floor. *Infiltration loss* is heat required to warm ventilation and infiltration air. Though no "credit" for solar heat gain is taken in heating load calculations, reliable sources of internal heating are considered.[2] Modifications for thermal inertia due to high-mass walls and ceilings are generally not made in calculations of heating load. When thermal inertia is considered, the approach taken is simplistic. (See Sec. 41.10.)

Calculation of the heating load is greatly simplified by having access to tabulations of data.[3] Data on climatological conditions are essential, and data on building materials and construction will greatly simplify the task of calculating heat transfer coefficients. Data of this nature is available in a variety of formats. Heat transmission data are available for specific materials as well as for composite walls of specific construction. Both types of data are useful.

Calculations of heating load are based on many assumptions. Because of the intrinsic unreliability of some of the data, an *exposure allowance* of up to 15% may be added to the calculated ideal heating load. This helps to account for unexpected heat losses and severe climatic conditions.

2. INSIDE DESIGN CONDITIONS

For the purposes of initial heating load calculations for residences and office spaces, the *inside design temperature* is generally taken as 70°F to 72°F (21.1°C to 22.2°C).

For industrial spaces, such as factories and warehouses, the inside design temperature is lower: 60°F to 65°F (15.6°C to 18.3°C). Humidity is typically 30% to 35%

[1] A *therm* per hour is 100,000 Btu/hr.
[2] The sky is assumed to be overcast during the heating season, so solar heat gain is minimal. Reliable sources of internal heating include permanently mounted equipment and lights.
[3] It is essential that engineers working in this area obtain their own compilations of this type of data.

relative humidity, but is generally not considered except in certain manufacturing industries (e.g., textiles and printing) where moisture content is critical.

3. OUTSIDE DESIGN CONDITIONS

The outside temperature and average wind speed (for infiltration) are needed to determine heating load. For estimates of annual heating costs, information on the winter degree days is needed. (See Sec. 41.12.) These values are almost always obtained from tables of climatological data.

Design conditions are probabilistic in nature. There is always some probability that a temperature will be exceeded. Depending on the nature of the facility, it may be desirable to select an outside design temperature that will be exceeded (for example) 5 days out of 100 days. Many tables give design temperatures for 1%, 2.5%, and 5% exceedance probabilities.

4. ADJACENT SPACE CONDITIONS

Since the conductive heat transfer through shared walls depends on the temperatures on both faces, determining the heating load for single rooms and separately heated offices also requires knowing the temperatures in adjacent spaces. For residential calculations, this may require knowing the temperature in attics, large closets, and basements. For attics ventilated by large open louvers, the approximate attic temperature is the average of the inside and outside design temperatures.

5. WALLS AND CEILINGS

Each material used in constructing a wall, ceiling, and so on, contributes resistance to heat flow. This resistance can be specified in a variety of ways. *Total resistance*, R (with units of hr-ft²-°F/Btu or m²·°C/W), is the total resistance to heat flow through all of the material. *Unit resistance* (with units of hr-ft²-°F/Btu-ft or m²·°C/W·cm) is the resistance per unit thickness of the material.[4] The total resistance is the product of the resistivity and the material thickness. *Conductance*, C, and *conductivity*, k, are the reciprocals of total resistance and unit resistance, respectively. *ASHRAE Handbook — Fundamentals* includes conductance, conductivitity, and the total and unit resistance values for various materials. [**Thermal Resistance of Building Materials**]

$$R = \frac{L}{k} = \frac{1}{C} \tag{41.1}$$

The heat transfer through walls, doors, windows, and ceilings is calculated from the traditional heat transfer equation, Eq. 41.2. The *overall coefficient of heat transfer*, U, can be calculated for each transmission path from the conductivities and resistances of the individual components in that path, or it can be obtained from tabulations of typical wall/ceiling construction. Typical values are found in *ASHRAE Handbook — Fundamentals*. [**U-Factors for Various Fenestration Products**] [**Design U-Factors of Swinging Doors**]

$$q = UA(T_i - T_o) \tag{41.2}$$

$$U = \frac{1}{\sum R_i} = \frac{1}{\frac{1}{h_i} + \sum\frac{L}{k} + \sum\frac{1}{C} + \frac{1}{h_o}} \tag{41.3}$$

Unlike most heat transfer problems, little effort is expended in calculating surface heat transfer (film) coefficients from theoretical correlations. Tables of typical values are used. (See *ASHRAE Handbook–Fundamentals*.) When tabulations of overall coefficients of heat transfer are used, it is important to know if an outside film coefficient has been included. If it has, the table's assumption about outside wind speed must be known and compared with actual wind conditions. Multiplicative corrections for other wind speeds generally accompany such tables. [**Surface Film Coefficients/Resistances for Air**]

6. AIR SPACES

The thermal conductance of an air space, C, used in Eq. 41.3 depends on the space thickness, emissivities of both sides, orientation, mean temperature, and temperature differential. Typical values for a 50°F (27.8°C) temperature differential are given in *ASHRAE Handbook–Fundamentals*. The effective space emissivity, E, used in the table is a function of the surface emissivity, ϵ, and is defined by Eq. 41.4. Surface emissivities range from approximately 0.05 for bright aluminum foil through 0.25 for bright galvanized steel to 0.90 for typical building materials (wood, sheetrock, masonry, etc.). [**Thermal Resistances of Plane Air Spaces**]

$$E = \frac{1}{\frac{1}{\epsilon_1} + \frac{1}{\epsilon_2} - 1} \tag{41.4}$$

[4] Unit resistance and conductivity are combined with a length when calculating thermal resistance. There are several sets of units in use for conductivity, depending on how the length is measured. For lengths in feet, Btu/hr-ft-°F and Btu-ft/hr-ft²-°F are the same. For lengths in inches (centimeters), the units Btu-in/hr-ft²-°F (W·cm/h·m²·°C) must be used.

HVAC

7. GROUND SLABS

Heat is lost from ground slabs both through the face and from the exposed edges. Since the ground temperature is usually higher than the winter air temperature, a lower temperature difference (typically 5°F (2.8°C)) should be used to find the heat loss through the face.

This loss is generally small, and an overall coefficient of 0.05 Btu/hr-ft^2-°F (0.3 W/m^2·°C) is typical. The loss is essentially constant throughout the year since the soil temperature under a building does not vary appreciably. The radial loss from the edges can be found from the *slab edge coefficient*, F, and from the perimeter of the exposed edge. Thickness of the slab is disregarded. Coefficients for concrete slabs on grade depend on the construction method, the amount of insulation, and weather conditions. The coefficient varies from approximately 0.5 Btu/hr-ft-°F (0.9 W/m·°C) for slabs with insulated edges in warm climates to approximately 2.7 Btu/hr-ft-°F (4.9 W/m·°C) for slabs with no insulation located in cold climates.[5] **[Typical Apparent Thermal Conductivity Values for Soils]**

$$q = PF(T_i - T_o) \qquad 41.5$$

8. VENTILATION AND INFILTRATION AIR

Outside ventilation air must be warmed before being introduced into the occupied space. All infiltrated air (from window sashes, door jams, and other cracks) will also go through the air conditioner, so its sensible heat load is combined with the ventilation air deliberately drawn in. Outside air entering at very low temperatures may need to be humidified.

Based on dry or low-humidity air, the sensible heating load for infiltration and ventilation air whose temperature is increased from T_1 to T_2 is[6]

$$q_{\text{kW}} = \left(72\ \frac{\text{kW·min}}{\text{m}^3\text{·°C}}\right) Q_{\text{m}^3/\text{min}}(T_{2,°\text{C}} - T_{1,°\text{C}}) \qquad \text{[SI]} \quad 41.6(a)$$

$$= \left(1.2\ \frac{\text{kW·s}}{\text{m}^3\text{·°C}}\right) Q_{\text{m}^3/\text{s}}(T_{2,°\text{C}} - T_{1,°\text{C}})$$

$$q_{\text{W}} = \left(1.2\ \frac{\text{W·s}}{\text{L·°C}}\right) Q_{\text{L/s}}(T_{2,°\text{C}} - T_{1,°\text{C}})$$

$$= \left(1200\ \frac{\text{W·s}}{\text{m}^3\text{·°C}}\right) Q_{\text{m}^3/\text{s}}(T_{2,°\text{C}} - T_{1,°\text{C}})$$

$$q_{\text{Btu/hr}} = \left(1.08\ \frac{\text{Btu-min}}{\text{hr-ft}^3\text{-°F}}\right) Q_{\text{cfm}}(T_{2,°\text{F}} - T_{1,°\text{F}}) \qquad \text{[U.S.]} \quad 41.6(b)$$

$$= \left(0.018\ \frac{\text{Btu}}{\text{ft}^3\text{-°F}}\right) Q_{\text{cfh}}(T_{2,°\text{F}} - T_{1,°\text{F}})$$

If the outside design temperature is 32°F (0°C) or cooler, there will be little or no moisture in the incoming air. However, for air with higher temperatures, heat is required to warm any moisture that enters with outside air.

Although the sensible heating of the dry air and accompanying moisture can be calculated separately and added, it is more expedient to use enthalpy values read from the psychrometric chart.

$$q_{\text{kW}} = \dot{m}_{a,\text{kg/s}}(h_i - h_o)_{\text{kJ/kg}}$$

$$= Q_{\text{m}^3/\text{s}}\rho_{\text{kg/m}^3}(h_i - h_o)_{\text{kJ/kg}}$$

$$\approx \left(1.2\ \frac{\text{kg}}{\text{m}^3}\right) Q_{\text{m}^3/\text{s}}(h_i - h_o)_{\text{kJ/kg}} \qquad \text{[SI]} \quad 41.7(a)$$

$$\approx \left(0.0012\ \frac{\text{kg}}{\text{L}}\right) Q_{\text{L/s}}(h_i - h_o)_{\text{kJ/kg}}$$

$$q_{\text{Btu/hr}} = \dot{m}_{a,\text{lbm/hr}}(h_i - h_o)_{\text{Btu/lbm}}$$

$$= \left(60\ \frac{\text{min}}{\text{hr}}\right) Q_{\text{ft}^3/\text{min}}\rho_{\text{lbm/ft}^3} (h_i - h_o)_{\text{Btu/lbm}} \qquad \text{[U.S.]} \quad 41.7(b)$$

$$\approx \left(4.5\ \frac{\text{lbm-min}}{\text{ft}^3\text{-hr}}\right) Q_{\text{ft}^3/\text{min}} (h_i - h_o)_{\text{Btu/lbm}}$$

[5]The slab edge coefficient may be given in Btu/hr-°F (W/°C) per unit distance.
[6]The constant 1.08 in Eq. 41.6(c) is reported by other authorities as 1.085 and 1.1 depending on the degree of precision intended.

Special accounting for infiltrated air is necessary when the conditioned space supplies combustion air. This normally is the case for residential fireplaces and space heaters. Each volume of gaseous fuel gas burned uses approximately nine volumes of heated air. Unless the face of the fireplace is blocked off, heated air continues to be lost up the flue, even after combustion stops.

9. INTERNAL HEAT SOURCES

Internal heat sources are heat sources within the conditioned space.[7] Heat sources may introduce sensible and latent loads. Two general methods are used to estimate internal sources. The load can be based on the equipment nameplate rating. Alternatively, tables of typical values can be used. Such tables are helpful in estimating the fractions of each load that are sensible and latent.[8] The sensible heat supplied by permanent machinery and lighting should be subtracted from the heating load. *ASHRAE Handbook—Fundamentals* provides recommended rates of *radiant* and *convective heat gain*. **[Representative Rates at Which Heat and Moisture Are Given Off by People in Different States of Activity] [Average Efficiencies and Related Data Representative of Typical Electric Motors] [Recommended Rates of Radiant and Convective Heat Gain: Unhooded Appliances During Idle (Ready-to-Cook) Conditions] [Recommended Rates of Radiant Heat Gain: Hooded Appliances During Idle (Ready-to-Cook) Conditions]**

Some equipment is rated in horsepower; other equipment is rated by wattage.[9] The *service factor*, SF, used in Eq. 41.8 is the fraction of the rated power being developed. Service factors greater than 1.00 are possible in overload conditions.

$$q_{\text{kW}} = \left(0.7457\ \frac{\text{kW}}{\text{hp}}\right)(\text{SF})\left(\frac{P_{\text{hp}}}{\eta}\right) = (\text{SF})\left(\frac{P_{\text{kW}}}{\eta}\right) \quad \text{[SI]} \quad 41.8(a)$$

$$q_{\text{Btu/hr}} = \left(2545\ \frac{\text{Btu}}{\text{hp-hr}}\right)(\text{SF})\left(\frac{P_{\text{hp}}}{\eta}\right) = \left(3413\ \frac{\text{Btu}}{\text{kW-hr}}\right)(\text{SF})\left(\frac{P_{\text{kW}}}{\eta}\right) \quad \text{[U.S.]} \quad 41.8(b)$$

The equipment efficiency, η, is included in the denominator in Eq. 41.8. This is appropriate when the motor and equipment being driven are both in the conditioned space. If the motor is not actually in the space but the driven equipment is, then the efficiency should be omitted.[10] (An example of this configuration would be an in-duct fan that is driven by a motor located outside of the duct.) Typical motor efficiencies are 80% for 1 hp motors and 90% for 10 hp and higher motors.

Sensible heat gain for typical appliances can be calculated using Eq. 41.9 or Eq. 41.10. F_u is the *usage factor*, and F_r is the *radiation factor*. Eq. 41.9 is valid for hooded appliances typically found in commercial kitchens, where only the radiant heat is considered part of the overall indoor heat load (latent and sensible convective heat are ducted externally and do not enter the space).

Heat Gain for Generic Appliances

$$q_s = q_{\text{input}} F_u F_r \quad 41.9$$

For unhooded appliances, all latent and sensible heat enters the space. Adjusting the previous equation accordingly, $F_u = F_l$, where F_l is the ratio of sensible heat gain to the manufacturer's rated energy input, q_{input}.

Heat Gain for Generic Appliances

$$q_s = q_{\text{input}} F_l \quad 41.10$$

Sensible heat gain from electric lighting can be calculated using Eq. 41.11. F_{ul} is the lighting use factor expressed as a percentage of use, and F_{sa} is the special allowance factor, or the ratio of the total lighting power consumption including ballasts to the nominal rated power of the lighting fixture.

Electric Lighting

$$q_{s,\text{electric}} = 3.412 q_{\text{watts}} F_{\text{ul}} F_{\text{sa}} \quad 41.11$$

As a rule of thumb, for fluorescent lights, the rated wattage should be increased by 20% to 25% to account for wound (transformer type) ballast heating. Therefore, the *special allowance factor*, F_{sa}, is conservatively 1.25. (For incandescent lights or fluorescent lights with power-saving electronic ballasts, the rated wattage is not increased, so the F_{sa} is 1.0.) However, it may be inappropriate to assume that all heat generated enters the conditioned space. If the air space above the lights in a dropped ceiling is not directly conditioned, then only some fraction (e.g., 60%) of the heat enters the occupied space.

[7]Residential heating sources such as toasters and coffee brewers are sometimes referred to as *domestic heat sources*.

[8]In the absence of tabular data, simple, logical rules of thumb can be used. For example, when cooking under a hood, all moisture is captured, the latent load is zero, and the sensible load transferred due to radiation is taken as some assumed fraction of the nameplate rating. If not cooking under a hood, two-thirds of the load is sensible and one-third is latent. However, tables are usually essential when determining the heat gain for esoteric sources such as doughnut machines, deep-fry kettles, and waffle irons for ice cream sandwiches.

[9]The *watt density* (strictly, the *linear watt density*) for baseboard heaters, heat tracing, and similar devices is the amount of heating generated per foot (meter). It is not based on area.

[10]There is a third, less likely alternative. The conditioned space may contain only the motor, with the driven machinery being someplace else. In that case, only the energy lost in the energy conversion appears in the load. The nameplate power is multiplied by $(1-\eta)/\eta$.

Unless the room is reasonably occupied on a permanent basis, the heating load should not be reduced by the metabolic heat of the occupants.

Some buildings have internal sources generating large amounts of energy. Under no circumstance, however, should the theoretical heating load be reduced by these internal heat gains to a point where the inside temperature would be 40°F (4.4°C) or lower in the absence of these internal sources.

Example 41.1

A window shaker unit is to be installed to heat a 14 ft × 17 ft tiny house. The design heat load for the space is 14,000 Btu/hr. The space contains a coffee brewer, a small freezer, a hot plate, and a refrigerator. What is the capacity of the window shaker unit needed to meet the heat load for the space?

Solution

In residential applications where appliances are unhooded, q_s must be adjusted accordingly. Calculate the radiant and total heat gain contribution from each appliance using the ASHRAE fundamental energy rating, usage factor, and radiation factor. [**Recommended Rates of Radiant and Convective Heat Gain: Unhooded Appliances During Idle (Ready-to-Cook) Conditions**]

The needed capacity of the window shaker unit is determined by the freezer and refrigerator heat load; the coffee brewer and hot plate are not continuously operated.

Use Eq. 41.9 to calculate the radiant heat gain for each appliance.

For the small freezer,

Heat Gain for Generic Appliances

$$\begin{aligned} q_{s,r} &= q_{\text{input}} F_u F \\ &= \left(2700\ \frac{\text{Btu}}{\text{hr}}\right)(0.41)(0.45) \\ &= 498.15\ \text{Btu/hr} \quad (498\ \text{Btu/hr}) \end{aligned}$$

Results for all appliances are shown in the table.

Use Eq. 41.10 to calculate the total heat gain (usage) for each appliance. Since the residential load has no externally ducted hoods, $F_u = F_l$.

For the small freezer,

Heat Gain for Generic Appliances

$$\begin{aligned} q_{s,u} &= q_{\text{input}} F_l = q_{\text{input}} F_u \\ &= \left(2700\ \frac{\text{Btu}}{\text{hr}}\right)(0.41) \\ &= 1107\ \text{Btu/hr} \end{aligned}$$

Results for all appliances are shown in the table.

		heat gain (Btu/hr)		usage factor	radiation factor
appliance	energy rating (Btu/hr)	sensible radiant	total	F_u	F_r
coffee brewer	13,000	199	1170	0.09	0.17
small freezer	2700	498	1107	0.41	0.45
hot plate	3800	900	3002	0.79	0.30
refrigerator	4800	300	1200	0.25	0.25

The overall heat gain for the freezer and refrigerator is 2307 Btu/hr. The window shaker unit will need to provide 14,000 Btu/hr − 2307 Btu/hr = 11,693 Btu/hr.

Therefore, a one-ton (12,000 Btu/hr) window shaker unit is enough to meet the heat load for the space.

10. THERMAL INERTIA

Buildings with massive masonry (including concrete) walls are thermally more stable than those with thin walls. However, the effect of *thermal lag* (*thermal inertia* or *thermal flywheel effect*) is difficult to incorporate in most studies. A simplified approach known as the *M-factor method* has been developed. This method correlates a factor, M, with the degree-days and mass per unit wall area (lbm/ft² or kg/m²). (See Fig. 41.1.) The M-factor modifies the overall coefficient of heat transfer, U. The actual heat transfer is

$$q = MUA(T_i - T_o) \qquad 41.12$$

Figure 41.1 *M-Factor*

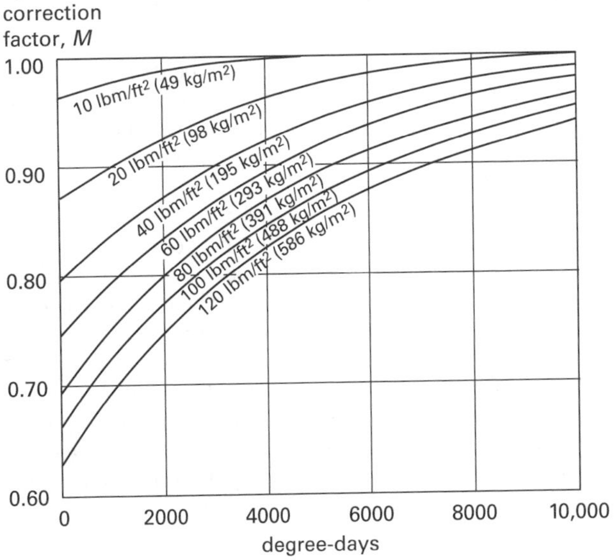

Source: "Mass Masonry Energy," Masonry Industry Committee, ca. 1979.

As written, Eq. 41.12 is appropriate for analyzing the heat loss through a wall. For design, particularly when wall construction with a maximum heat loss coefficient, U, is specified by the building code, a wall may be designed to have an instantaneous heat loss coefficient of U/M and still be up to code.

11. FURNACE SIZING

A furnace must be capable of keeping a building warm on the coldest days of the heating season. Therefore, a furnace should be sized based on the coldest temperature reasonably expected (i.e., on the outside design temperature). However, not all days in the heating season will have temperatures that cold, and the entire capacity will rarely be utilized.

Pickup load is the furnace capacity needed to bring a cold building up to the inside design temperature in a reasonable time. Since the outside design temperature provides excess capacity most of the year, a pickup allowance may not be necessary. However, churches, auditoriums, and office buildings needing to be "ready" at specific times need special attention.[11] If a building is heated during the day only, a 10% increase in the rated furnace size is sometimes added as the pickup load to allow for starting up. If a building is left unheated for extended periods, the increase should be 25% or higher.

Some furnaces have nameplate ratings for both fuel input and output heating. The furnace efficiency is incorporated in a rating for "output Btus." Output ratings apply to newly installed furnaces. Output can be expected to decrease with time.

12. DEGREE-DAYS AND KELVIN-DAYS

The *degree-day* (*kelvin-day*) concept can be used to estimate heating costs during the heating season.[12] The assumption is made that the heating system operates when degree-days are being accumulated, and there is an indoor *base temperature*, T_b, that triggers the heating system to turn on. In the past, the base temperature was typically 65°F in the United States (19°C in Europe), although other temperatures such as 50°F and 15.5°C are used. The base temperature affects the estimate materially, so it should not be selected casually. The base temperature depends on the temperature that a building is heated to and the sources of internal heat gain. For example, if a building was going to be heated to 19°C, and if the average internal heat gain from people, lights, and equipment in a building was estimated to increase the temperature by 3.5°C, it would be appropriate to use 19°C − 3.5°C = 15.5°C as the base temperature.

The heating season is assumed to last for N days, different for each geographical location, and determined from meteorological data. Each day whose 24-hour average temperature, $\overline{T}$, is less than the base temperature, T_b, will accumulate $T_b - \overline{T}$ degree-days (kelvin-days), DD. The sum of these degree-day terms over the entire heating season of n days is the total *heating degree-days* (*heating kelvin-days*), HDD, also known as *winter degree-days*, for that location.[13]

$$\text{HDD} = \sum_{i=1}^{n}(T_b - \overline{T}_i) \qquad 41.13$$

Ultimately, use of the degree-day concept comes down to the meteorological data that are available. Degree-day information is not always available, although average monthly temperature usually is available from meteorological data. *Hitchin's formula* (1983) can be used to estimate degree-days from the average weekly or monthly temperature. The average number of degree-days per day during the heating period is calculated from Eq. 41.14. The empirical exponent k has a value of 0.39 1/°F (0.71 1/°C).

$$\frac{\text{HDD}}{n} = \frac{T_b - \overline{T}}{1 - e^{-k(T_b - \overline{T})}} \qquad 41.14$$

The degree-day concept is intended to help in estimating heating cost during the heating season. There may be times during some days when the heating system does not operate, and for that reason, the degree-day accumulation does not represent a complete record of the exterior temperatures encountered. Only the cold parts of the heating season are represented. Degree-days cannot be used to determine the average temperature during the entire heating season. However, Hitchin's formula can be extended to the entire heating season and used to calculate the average outside temperature. For regions where heating is continuous throughout the heating season (as represented by large differences between the base and average temperatures), this is equivalent to Eq. 41.15.

$$\frac{\text{HDD}}{n} = T_b - \overline{T} \qquad 41.15$$

[11] There is not much information on this subject. The increase for pickup load is largely a matter of judgment.

[12] Although the degree-day concept continues to appear in ASHRAE publications, ASHRAE stopped promoting the method in the 1980s in favor of more accurate surface-by-surface heat loss calculations.

[13] A similar concept for *cooling degree-days* (*summer degree-days*), CDD, exists for use in estimating average cooling costs during the cooling season.

13. FUEL CONSUMPTION

The design heating load used to size the furnace is not the average heating load over the heating season. However, when combined with degree-days, the design heating load can be used with smaller, simple structures to calculate the total fuel consumption and heating costs during the entire heating season. (This method assumes that there will be no heating when the outside temperature is equal to the base temperature or higher. This assumption is appropriate, particularly in residential buildings, even though the interior temperature is maintained at 68°F to 72°F (20°C to 22.2°C) because of the existence of internal heat sources.) The units of fuel consumption in Eq. 41.16 depend on the units of the heating value. Equation 41.16 does not include factors for operating pumps, fans, stokers, and other devices, nor are costs of maintenance, tank insurance, and so on, included.[14] The minimum efficiency, η, of gas- and oil-fired residential furnaces is approximately 70% to 75%.

$$\text{fuel consumption (in units/heating season)} = \frac{\left(86\,400\ \frac{\text{s}}{\text{d}}\right)q_{\text{kW}}(\text{HDD})}{(T_i - T_o)(\text{HV}_{\text{kJ/unit}})\eta_{\text{furnace}}} \quad \text{[SI]} \qquad 41.16(a)$$

$$\text{fuel consumption (in units/heating season)} = \frac{\left(24\ \frac{\text{hr}}{\text{day}}\right)q_{\text{Btu/hr}}(\text{HDD})}{(T_i - T_o)(\text{HV}_{\text{Btu/unit}})\eta_{\text{furnace}}} \quad \text{[U.S.]} \qquad 41.16(b)$$

Equation 41.16 traditionally has been used to estimate fuel consumption. In recent years, various improvements to the "model" have been made. Specifically, Eq. 41.16 is multiplied by an empirical correction, C_D, to correct for the difference between calculated values and actual performance.[15] Values of C_D range from about 0.60 to 0.87 and are correlated with the number of degree-days. Also, the furnace efficiency, η, is replaced with an efficiency factor, k, that includes the effects of rated full-load efficiency, part-load performance, over-sizing, and energy conservation devices. A value of 1.0 should be used for k for electric heating. Values of 0.55 and 0.65 are appropriate for older and energy-efficient houses, respectively.

14. CONSERVATION THROUGH THERMOSTAT SETBACK

A variety of *conservation methods* are used to reduce energy consumption during the heating season. These include installing insulation, weather stripping, and reducing the thermostat setting for all or part of the day. An extreme case of thermostat setback occurs when the heating system is turned off entirely at night.

If a building is to be maintained at two different temperatures during different parts of the day, the average heating can be found from the duration-weighted average of the two inside temperatures. Alternatively, separate calculations can be made for the periods of different temperatures.

15. FREEZE-UP OF HEATING AND PREHEATING COILS

Coil freeze-up (*freezing of the coils*) has two different causes. In the winter, cold make-up (ventilation) air from outside may cause water to freeze inside the heating coils. In the summer, coils with effective surface temperatures of 32°F (0°C) or less can cause moisture (humidity) in the air to freeze on the outside of the cooling (evaporator) coils.

Steam is usually the source of heating in coils because the heat of vaporization is so much larger than the sensible energy available from hot water alone. Steam in the coil condenses as it releases the heat of vaporization. To prevent freezing inside the coil, the condensate must drain from the coil rapidly, before the cold outside air can reduce the water's temperature to freezing. A rapid-drain plumbing design includes a vacuum breaker, large trap, long drip line, drain line below the supply line, adequate drain line slope, and clean strainers, among other features. Regular maintenance is required to ensure everything works as designed.

In the winter, when the outside air temperature is below freezing, coil freeze-up can be avoided by ensuring cold make-up air is thoroughly mixed with warm return air before the mixture enters the coils.[16] In the event that the mixture is still too cold, freeze-resistant heating coils with adequate piping should be used. In rare cases, some form of electric air preheating can be provided prior to the steam coils.

16. NOMENCLATURE

A	area	ft^2	m^2
C	thermal conductance	Btu/hr-ft^2-°F	W/m^2·°C
CDD	cooling degree-days	°F-day	°C·d
DD	degree-days	°F-day	°C·d
E	effective emissivity	–	–
F	slab edge coefficient	Btu/hr-ft-°F	W/m·°C
F	factor	–	–

[14]Equation 41.16 appears to imply that the lower T_o is, the lower the fuel consumption will be. This is obviously untrue. The temperature difference in the denominator actually cancels the same temperature difference used to calculate the heat transfer terms in q. Thus, q is put on a per-degree basis. The average temperature difference used in the calculation of degree days converts the per-degree heat loss to an average heat loss.
[15]It has been shown that Eq. 41.16 overestimates the fuel requirement in most cases. C_D simply reduces the estimate.
[16]Mixing louvers can fail. So, proper coil protection should always be provided.

HVAC

F_l	ratio of sensible heat gain to the manufacturer's rated energy input	–	–
h	enthalpy	Btu/lbm	kJ/kg
h	surface heat transfer coefficient	Btu/hr-ft^2-°F	W/m^2·°C
HDD	heating degree-days	°F-day	°C·d
HV	heating value	various	various
k	efficiency factor	–	–
k	empirical Hitchin's exponent	1/°F	1/°C
k	thermal conductivity	Btu-ft/hr-ft^2-°F	W·m/m^2·°C
L	length	ft	m
$\dot{m}$	mass flow rate	lbm/hr	kg/h
M	masonry M factor	–	–
n	number of days in heating season	–	–
P	perimeter length	ft	m
P	power	hp	W
q	heat transfer rate	Btu/hr	W
Q	volumetric flow rate	ft^3/min	m^3/s
R	total thermal resistance	hr-ft^2-°F/Btu	m^2·°C/W
SF	service factor	–	–
T	temperature	°F	°C
U	overall coefficient of heat transfer	Btu/hr-ft^2-°F	W/m^2·°C

Symbols

ϵ	emissivity	–	–
η	efficiency	–	–
ρ	density	lbm/ft^3	kg/m^3

Subscripts

a	dry air
b	base
fg	vaporization
i	inside design
o	outside design
r	radiation
s	sensible
sa	lighting special allowance
u	usage
ul	lighting use

42 Cooling Load

Content in blue refers to the *NCEES Handbook*.

NCEES EXAM SPECIFICATIONS AND RELATED CONTENT

HVAC AND REFRIGERATION EXAM

- II.A. Heating/Cooling Loads
  - 3. Refrigeration System Size and Ratings
  - 6. Instantaneous Cooling Load from Walls and Roofs
  - 7. Instantaneous Cooling Load from Windows
  - 9. Ventilation and Infiltration
  - 10. Heat Gain to Air Conditioning Ducts
  - 13. Latent and Sensible Loads
  - 16. Changeover Temperature and Reheating Cooled Air
- II.C.5. Systems and Components: Basic control concepts
  - 2. Source of Cooling

THERMAL AND FLUID SYSTEMS EXAM

- III.B. Cooling/Heating
  - 9. Ventilation and Infiltration

1. INTRODUCTION

The procedure for finding the *cooling load* (also referred to as the *air conditioning load*) is similar in some respects to the procedure for finding the heating load.[1] The aspects of determining inside and outside design conditions, heat transfer from adjacent spaces, ventilation air requirements, and internal heat gains are the same as for heating load calculations and are not covered in this chapter.

However, the calculation of cooling load is complicated considerably by the thermal lag of the exterior surfaces (i.e., walls and roof). Depending on construction, the solar energy absorbed by exterior surfaces can take hours to appear as an interior cooling load.[2] Further complicating the determination of cooling load are the direct transmission of solar energy through windows and the facts that the delay is different for each surface, the solar energy absorbed changes with time of day, and instantaneous heat gain into the room contributes to instantaneous and delayed cooling loads.

It is important to distinguish between three terms. The *instantaneous heat absorption* is the solar energy that is absorbed at a particular moment. The *instantaneous heat gain* is the energy that enters the conditioned space at that moment. Due to solar lag, the heat gain is a complex combination of heat absorptions from previous hours. The *instantaneous cooling load* is a portion (i.e., is essentially the convective portion) of the instantaneous heat gain.

2. SOURCE OF COOLING

Once the cooling load is determined, the source and size of the cooling unit must be considered. Cooling normally comes from liquefied refrigerant passing through a cooling coil. Some or all of the airflow passes across the cooling coil. The refrigerant is continuously vaporized in the coil as part of a complete refrigeration cycle. Alternatively, cold water may be used in the coil to cool the airflow. In such cases, the water is cooled in a *chiller* running its own refrigeration cycle.

An *economizer* is an electromechanical system that changes a portion of the cooling process in order to decrease cost, usually by taking advantage of cold ambient air. A *water-side economizer* substitutes natural

[1] The *refrigeration load* or *coil loading* is the cooling load expressed in appropriate units (e.g., tons of refrigeration).
[2] That is, the *instantaneous heat gain* is the heat that enters the conditioned space. Due to thermal lag, it is not the same as the *instantaneous heat absorption* by the building at that same moment. This terminology is not rigidly adhered to.

HVAC

cooling from a cooling tower for the chiller's more expensive refrigeration cycle when the ambient air temperature drops below the desired coil temperature. An *air-side economizer* increases the amount of outside air that is brought into a space when the ambient air characteristics (temperature, humidity, or enthalpy) are better than the return airflow. The outside air may still be conditioned by passing through coils, but less change will be required. Use of cold ambient air by either type of economizer is known as *free cooling*. [**Air-Side Economizer Cycle**] [**Economizer High-Limit Controls**]

Chilled beam systems are a relatively new source of cooling introduced in the industry. These systems use modes of radiation and convective heat transfer to cool interior spaces with beams that act as heat exchangers for large-scale facilities. These systems are either passive or active. Passive systems use convective currents to provide cooling to the space at roughly 400 Btu/hr-ft, while active systems use induction nozzles to mix room air and primary air and to provide cooling at roughly 800 Btu/hr-ft. Design considerations include ensuring the dew-point temperature of the air in the room is low enough to avoid condensation on the beam surfaces. [**Chilled Beam Systems**] [**Passive and Active Chilled-Beam Operation**]

3. REFRIGERATION SYSTEM SIZE AND RATINGS

Refrigeration equipment for HVAC use is typically rated in tons of cooling, where a *ton of cooling* is equal to 12,000 Btu/hr (3517 W). Prior to the emphasis on energy conservation and detailed energy analyses, a common rule of thumb, particularly in residential and small commercial installations, was to size refrigeration systems at one ton for every 400 ft^2 (37 m^2) of conditioned space. For modern homes meeting minimum insulation, window, and sealing code requirements, the rule of thumb has increased to one ton for every 600 ft^2, 800 ft^2, or 1000 ft^2 (56 m^2, 74 m^2, or 93 m^2), although oversizing is still likely without a detailed analysis.

The theoretical power dissipated by a refrigeration cycle's single-phase compressor motor is calculated from the motor's actual measured voltage and current draw or from the cooling load. Various efficiencies must be considered.

$$\begin{aligned} P_{\text{compressor,watts}} &= I_{\text{amps}} V_{\text{volts}} \\ &= \frac{q_{c,\text{Btu/hr}}}{3.412\eta_{\text{electrical}}\eta_{\text{compressor}}} \end{aligned} \qquad 42.1$$

Similarly, the theoretical motor horsepower is calculated as

$$P_{\text{compressor,hp}} = \frac{q_{c,\text{Btu/hr}}}{2544\eta_{\text{electrical}}\eta_{\text{compressor}}} \qquad 42.2$$

In reality, the efficiencies depend on the refrigerant and its condition. Compressor manufacturers provide charts showing horsepower versus cooling load at various suction pressures for various refrigerants. A general rule for air conditioning is 1 hp per 10,000 Btu/hr (2.9 kW), although for colder temperatures (e.g., as required by freezers), 1 hp for each 3000–5000 Btu/hr (0.9–1.5 kW) may be required. Power requirements are also affected by ambient conditions and by the amount of free cooling used.

Because of consumer difficulties in evaluating the various efficiencies and other details affecting operating costs of unitary systems, the *energy-efficiency ratio*, EER, is used as a simple, comparable measure to describe the efficiency of cooling systems (see Eq. 42.3).[3] The total input power, P_{total}, includes the power required to run the compressor as well as fans, controls, and all other parts of the air conditioning system. EER is typically evaluated at 50% relative humidity with a 95°F (35°C) outside temperature and an 80°F (27°C) inside temperature (return air). EER is equivalent to 3.412 times the coefficient of performance, COP, as shown.

Efficiency

$$\begin{aligned} \text{EER} &= \frac{\text{output cooling energy (Btu/hr)}}{\text{input electrical energy (W)}} \\ &= \frac{q_{c,\text{Btu/hr}}}{P_{\text{total,watts}}} \end{aligned} \qquad 42.3$$

$$\text{EER} = 3.412(\text{COP}) \qquad 42.4$$

Efficiency

$$\begin{aligned} \text{COP} &= \frac{\text{capacity in Btu/hr}}{\text{input in watts} \times 3.412} \\ &= \frac{q}{3.412P} \end{aligned} \qquad 42.5$$

The *seasonal energy-efficiency ratio*, SEER, of the Air-Conditioning, Heating, and Refrigeration Institute (AHRI) has the same definition as the EER, but is averaged over a range of outside temperatures using a standardized cooling season. This is unlike the EER, which is evaluated at a specific operating point.[4] Although subject to local and future legislation as well as exceptions by type of unit, by U.S. law, all newly manufactured and installed air conditioning equipment must be at least 13 SEER. "High efficiency" units are 14 SEER and higher. Commercially available units up to

[3] "Other details" affecting energy usage include the local ambient air conditions, ducting and sealing, type of unit (unitary or split), and the differences between air-cooled and water-cooled units.
[4] ANSI/AHRI 210/240: *Standard for Performance Rating of Unitary Air-Conditioning & Air-Source Heat Pump Equipment*, AHRI.

approximately 20 SEER are available, but most newly installed equipment is 16 SEER or lower. Equation 42.6 gives an approximate relationship between EER and SEER.

$$\text{EER} = 1.12(\text{SEER}) - 0.02(\text{SEER})^2 \qquad 42.6$$

4. ENERGY STAR CONSTRUCTION

Energy Star is a joint program of the U.S. Environmental Protection Agency (EPA) and the U.S. Department of Energy (DOE) that promotes and rewards energy-efficient construction and remodeling, primarily for residential houses. For construction to be Energy Star certified, it must have energy-efficient features that enable the structure to reduce total energy consumption by 15–20% (depending on climate) compared to a structure built according to the local energy code. Such features typically include high-performance windows, increased insulation levels, high-efficiency heating, cooling and water heating equipment, fluorescent lighting, Energy Star appliances, duct sealing, and air sealing of the building envelope.

Certification is performed by a Home Energy Rating System (HERS) rater. The rater determines the structure's characteristics via computer modeling, inspects the insulation and sealing of the structure, and tests the structure's duct and building envelope tightness.

5. INSIDE AND OUTSIDE DESIGN CONDITIONS

It is impossible to maintain the air passing through a conditioned space at a particular temperature. Cool air enters a space, and warm air leaves. The *inside design temperature*, T_i, is understood to be the temperature of the air removed from the conditioned space. Indoor design temperatures for summer use are generally a few degrees warmer than winter design temperatures—approximately 75°F (23.9°C).

6. INSTANTANEOUS COOLING LOAD FROM WALLS AND ROOFS

Three general methods are used to determine instantaneous cooling load: (1) total equivalent temperature difference method, (2) transfer function method, and (3) cooling load factor and temperature difference methods. Each method is derived from the same basic heat transfer equation, where T_b is the average outdoor air temperature and T_i is the conditioned space temperature.

Heat Gain Through Interior Surfaces

$$q = UA(T_b - T_i) \qquad 42.7$$

The total equivalent temperature difference (TETD) method determines the instantaneous heat gain. The *total equivalent temperature difference*, ΔT_{te}, depends on the type of construction, geographical location, time of day, and wall orientation. It is read from extensive tabulations. The instantaneous heat gain consists of stored radiant and convective portions. In the TETD/TA (time averaging) method, weighting factors are used to average the radiant portions from current and previous hours. The sum of the convective portions and the weighted average of the series of radiant portions are taken as the cooling load. Computer analysis and considerable judgment are required to use this method.

$$q_{\text{heat gain}} = UA\Delta T_{\text{te}} \qquad 42.8$$

The *transfer function method* is similar to the total equivalent temperature difference method. A series of weighting factors, known as *room transfer functions*, is applied to cooling load values from the current and previous hours. The transfer functions are related to spatial geometry, configuration, mass, and other characteristics.

The only modern method of calculating the cooling load suitable for quick (manual) analysis is the *cooling load temperature difference method*, CLTD, using the related *solar cooling load factor*, SCL, and *cooling load factor*, CLF, covered in subsequent sections. For exterior surfaces, the cooling load is calculated by Eq. 42.9. Tables of CLTD are needed. Values depend on time of year, location and orientation, type, configuration, and orientation of the surface, as well as other factors. Using the CLTD/SCL/CLF method, the instantaneous cooling load for conduction through opaque walls and roofs is

$$q_c = UA(\text{CLTD}_{\text{corrected}}) \qquad 42.9$$

The base conditions used to calculate the values of CLTD are

- a clear sky on July 21
- exposed, sunlit, flat roofs
- walls at 40°N latitude based on roof and wall construction and orientation
- an indoor temperature of 78°F
- an outdoor maximum temperature of 95°F with a mean temperature of 85°F
- a daily temperature range of 21°F

These base conditions generally don't coincide precisely with actual conditions during the study period, so CLTD is corrected according to Eq. 42.10. T_i is the indoor design temperature, and T_m is the mean outdoor temperature.

$$\begin{aligned}\text{CLTD}_{\text{corrected}} &= \text{CLTD}_{\text{table}} + (78°\text{F} - T_i) \\ &\quad + (T_m - 85°\text{F})\end{aligned} \qquad 42.10$$

$$T_m = T_{\text{outdoor,max}} - \tfrac{1}{2}(\text{daily range}) \qquad 42.11$$

HVAC

7. INSTANTANEOUS COOLING LOAD FROM WINDOWS

The steady-state energy equation through a window is found using Eq. 42.12. U is the heat transfer coefficient in units of Btu/hr-ft^2-°F, and A_p is the total projected area in ft^2 (i.e., rough opening less installation clearances). SHGC is the solar heat gain coefficient, E_t is the total irradiance, C is a dimensional conversion factor equal to 60 min/hr, and AL is air leakage in ft^3/ft^2.

Fenestration

$$q = UA_p(T_{\text{out}} - T_{\text{in}}) + (\text{SHGC})A_pE_t + C(\text{AL})A_p\rho c_p(T_{\text{out}} - T_{\text{in}}) \quad 42.12$$

The heat transfer coefficient can be further developed through weighted U-factors as shown in Eq. 42.13.

Fenestration

$$U = \frac{U_{\text{cg}}A_{\text{cg}} + U_{\text{eg}}A_{\text{eg}} + U_fA_f}{A_p} \quad 42.13$$

Using the CLTD/SCL/CLF method, the cooling load due to solar energy received through windows is calculated in two parts.[5] The first is an immediate conductive part; the second is a radiant part. Appropriate tables are needed to evaluate the *shading coefficient*, SC, and the SCL for the radiant portions.

$$\begin{aligned} q_c &= q_{\text{conductive}} + q_{\text{radiant}} \\ &= UA(\text{CLTD}_{\text{corrected}}) + A(\text{SC})(\text{SCL}) \end{aligned} \quad 42.14$$

Design considerations for limiting condensation employ the condensation resistance factor, CRF.[6] The condensation resistance factor is the ratio of the surface temperature difference to the ambient temperature difference. Minimum condensation resistance requirements vary by region. **[Minimum Condensation Resistance Requirements (t_h = 68°F)]**

Condensation Resistance Factor (CRF) or Temperature Index (I)

$$\text{CRF} = \frac{T - T_c}{T_h - T_c} \quad 42.15$$

8. COOLING LOAD FROM INTERNAL HEAT SOURCES

Latent loads (including metabolic latent loads) are considered instantaneous cooling loads. Only a portion, given by the cooling load factor, CLF, of the sensible heat sources show up as instantaneous cooling load. CLF is a function of time and depends on zone type, occupancy period, interior and exterior shading, and other factors. Although tables are usually necessary to evaluate CLF, there are some cases where CLF is assumed to be 1.0. These include when the cooling system is shut down during the night, when there is a high occupant density (as in theaters and auditoriums), and when lights and other sources are operated for 24 hours a day.

$$q_{c,\text{internal sources}} = q_l + q_s(\text{CLF}) \quad 42.16$$

9. VENTILATION AND INFILTRATION

Since all air passes through the air conditioner, sensible and latent loads from ventilation and infiltration air are instantaneous cooling loads. *Building air leakage*, Q, is a function of the pressure differential, Δp, between the inside and outside of a building. The latent and sensible heat gain from ventilation and infiltration, q_t, can be determined by Eq. 42.17 and Eq. 42.18. The total heat factor, C_t, is 4.5 at sea level conditions or 3.74 at an altitude of 5000 ft.

Heat Gain Calculations Using Standard Air Values

$$q_t = 60 \times 0.075Q_s\Delta h = 4.5Q_s\Delta h \quad 42.17$$

$$q_t = C_tQ_s\Delta h \quad 42.18$$

Breaking q_t down into its sensible components gives

Heat Gain Calculations Using Standard Air Values

$$q_s = 60 \times 0.075(0.24 + 0.45W)Q_s\Delta T \quad 42.19$$

Equation 42.19 can be simplified, where the sensible heat factor, C_s, is 1.1 at sea level conditions or 0.92 at an altitude of 5000 ft.

Heat Gain Calculations Using Standard Air Values

$$q_s = C_sQ_s\Delta T \quad 42.20$$

Breaking q_t down into its latent components gives

Heat Gain Calculations Using Standard Air Values

$$q_l = 60 \times 0.075 \times 1076Q_s\Delta W \quad 42.21$$

Equation 42.21 can be simplified, where C_t is 4840 at sea level conditions or 4027 at an altitude of 5000 ft.

Heat Gain Calculations Using Standard Air Values

$$q_l = C_tQ_s\Delta W \quad 42.22$$

Other elevations can be determined by Eq. 42.23, where the heat factor, $C_{x,\text{sea level}}$, can be any of the sea-level C values.

[5] The term *fenestration* refers to windows or other openings transparent to solar radiation.
[6] The temperature index, I, is a variable that is often used in place of the condensation resistance factor, CRF.

HVAC

Elevation Corrections for Total, Sensible, and Latent Heat Equations

$$C_{x,\text{altitude}} = \frac{C_{x,\text{sea level}}p}{p_{\text{sea level}}} \qquad 42.23$$

$$\frac{p}{p_{\text{sea level}}} = \left(1 - (\text{elev in ft} \times 6.875 \times 10^{-6})\right)^{5256} \qquad 42.24$$

The *building leakage curve* is defined by a coefficient, C, and an exponent, n, both obtained from a curve fit correlation of at least 12 test points between 15–20 Pa and 60–75 Pa. In U.S. models, the values of C and n are used to determine leakage in ft^3/min even though Δp is in pascals.

$$Q_{\text{ft}^3/\text{min}} = C\Delta p_{\text{Pa}}^{n} \qquad 42.25$$

Air leaks can be identified and leakage can be quantified and minimized by pressure testing.[7] It is common to base leakage testing on ACH50 (also known as ACH-50, ACH_{50}, or n50), a fan-induced pressurization at 50 Pa ($0.00725\ \text{lbf/in}^2$). ACH50 is used to determine the *air change rate*, the number of air changes per hour. The pressurization is accomplished by temporarily inserting a *blower door*, a frame with a built-in fan, pressure and airflow rate measurements, and other instrumentation, into an exterior door opening. The instrumentation in the blower door gives a direct reading of CFM50, the leakage in cubic feet per minute, which can also be calculated from ACH50 as shown in Eq. 42.26. Once the pressurization level is reached, leaks are subsequently located by a smoke puffer or an infrared camera.

$$\text{CFM50} = \frac{(\text{ACH50})\,V_{\text{structure,ft}^3}}{60\ \frac{\text{min}}{\text{hr}}} \qquad 42.26$$

CFM50 can be used to determine the *effective leakage area*, ELA (or EfLA). ELA was defined by the Lawrence Berkeley Laboratory in its infiltration model as the area of a nozzle-shaped orifice (with rounded edges, similar to the inlet of a blower door fan) that leaks the same amount of air that the building leaks at a pressure difference of 4 Pa.[8] The coefficient of discharge, C_d, for a smooth, rounded orifice is 0.97–0.98, essentially 1.00. The coefficient C and exponent n are the same as in Eq. 42.25. The approximate density, ρ, of dry air is $0.075\ \text{lbm/ft}^3$ ($1.2\ \text{kg/m}^3$).

$$\text{ELA}_{\text{in}^2} = \frac{Q\sqrt{\frac{\rho}{2g_c\Delta p}}}{C_d} \quad \text{[consistent units]}$$

$$= \frac{C\Delta p_{\text{Pa}}^{n}\left(12\ \frac{\text{in}}{\text{ft}}\right)^2 \times \sqrt{\frac{\rho_{\text{lbm/ft}^3}}{(2)\left(32.2\ \frac{\text{lbm-ft}}{\text{lbf-sec}^2}\right) \times \Delta p_{\text{Pa}}\left(0.02089\ \frac{\frac{\text{lbf}}{\text{ft}^2}}{\text{Pa}}\right)}}}{C_d\left(60\ \frac{\text{sec}}{\text{min}}\right)}$$

$$\approx \frac{\text{CFM50}}{18} \qquad 42.27$$

Specific leakage is the ELA reported per unit floor area or per unit building envelope area.[9] The *air permeability* is the leakage rate per unit building envelope area. In Eq. 42.28, the above-grade surface area, S, includes floor, ceiling, wall, and window areas. Various units are used to report air permeability. When reported in units of $\text{m}^3/\text{m}^2\cdot\text{h}$, the designation "Q50" is often used. When reported in $\text{ft}^3/\text{ft}^2\text{-min}$, the designation "MLR," the *Minneapolis leakage ratio*, may be used.

$$\text{Q50} = \frac{(\text{ACH50})\,V_{\text{structure}}}{S} \qquad 42.28$$

In order to compare buildings with different floor areas and heights, ASHRAE defines the *normalized leakage*, NL, as a measure of the tightness of a building envelope relative to the building size and number of stories.[10] The "normal" reference condition is a new, 1100 ft² (100 m²), single-story, non-energy efficient, slab-on-grade, non-low income house. The *reference height*, h_{ref}, in Eq. 42.29 is 98 in (2.5 m).

$$\text{NL} = 1000\left(\frac{\text{ELA}_{\text{in}^2}}{A_{\text{floor,in}^2}}\right)\left(\frac{h_{\text{building,in}}}{h_{\text{ref,in}}}\right)^{0.3} \qquad 42.29$$

HVAC

[7]The greatest accuracy is achieved when results of pressurization and depressurization (i.e., negative pressure) tests are averaged.

[8]The *equivalent leakage area*, EqLA, is defined by the Canadian National Research Council as the area of a sharp-edged orifice (a sharp, round hole cut into a thin plate with $C_d = 0.61$) that would leak the same amount of air as the building does at a pressurization of 10 Pa. The 10 Pa EqLA is approximately two times the 4 Pa ELA.

[9]The building envelope is also referred to as the *building fabric*.

[10]ASHRAE Standard 119: *Air Leakage Performance for Detached Single-Family Residential Buildings*, ASHRAE.

Pressurization to 50 Pa corresponds to a 20 mph (32 km/h) wind impacting all sides of a structure. Since infiltration from all exterior surfaces would not occur naturally, an estimate of the natural infiltration rate, ACHnat (also known as ACHn or ENIR), is estimated from the *LBL factor* (also known as the *N factor* or the *energy climate factor*), which is dependent on the climate region, the number of stories of the structure, and sheltering from wind.[11] LBL factors range from 4 to 40, with typical values ranging from 10 to 20. Energy Star has established a natural infiltration target threshold of 0.35 air changes per hour, but well-sealed structures can achieve ACHnat values much lower than this. As calculated, ACHnat is only a rough estimate, and true values can range from 50% lower to 100% higher. Table 42.1 uses ACHnat and other factors to categorize the air-tightness of moderately sized houses.

$$\text{ACHnat} = \frac{\text{ACH50}}{\text{LBL}} \qquad 42.30$$

***Table 42.1** Approximate Air-Tightness of Moderately Sized Houses*

	tight	moderate	leaky
CFM50	<1500	1500–4000	>4000
ACH50	<5	5–10	>10
ACHnat	<0.35	0.35–1	>1

The annual cost of the additional cooling load due to air leakage with a SEER-rated appliance can be estimated from Eq. 42.31.

$$\text{annual cooling cost} = \frac{q_{c,\text{Btu/hr}}t_{\text{hr/yr}}(\text{cost})_{\$/\text{kW-hr}}}{\left(1000\ \frac{\text{W}}{\text{kW}}\right)(\text{SEER})} \approx \frac{0.26(\text{CDD})(\text{CFM50}) \times (\text{cost})_{\$/\text{kW-hr}}}{(\text{LBL})(\text{SEER})} \qquad 42.31$$

10. HEAT GAIN TO AIR CONDITIONING DUCTS

The calculation of heat absorbed by air conditioning ducts that pass through unconditioned spaces can be estimated using Eq. 42.32 and Eq. 42.33. The duct exit temperature is

Duct Heat Gain or Loss

$$T_{\text{exit}} = T_{\text{enter}} - T_{\text{drop}} \quad \text{[warm air duct]}$$
$$T_{\text{exit}} = T_{\text{enter}} - T_{\text{gain}} \quad \text{[cold air duct]} \qquad 42.32$$

$$T_{\text{drop}} = 0.2\left(\frac{q''PL}{\text{v}c_p\rho A}\right)$$
$$T_{\text{gain}} = 0.2\left(\frac{q''PL}{\text{v}c_p\rho A}\right) \qquad 42.33$$

The heat transfer is generally estimated from tables or figures of standard configurations (e.g., the heat loss per fixed length of duct per 10 degrees of temperature difference). Extrapolation is used for other duct lengths and temperature differences.

The logarithmic mean temperature difference (generally used when the temperature difference varies along the length) is seldom used in the HVAC industry.[12] Rather, the heat transfer is based on the temperature difference between the environment and the midlength temperature of the duct. If the temperature of the duct at its midlength is not known, one or more iterations will be needed to calculate the temperature drop and heat transfer.

11. DEGREE-DAYS

Some sources present tables of *cooling degree-days*, CDD (*summer degree-days*).[13] Data in these tables are usually related to a base temperature of 65°F (18.3°C), but they may be related to a 70°F (21.1°C) base. Cooling is considered to occur only when the temperature is higher than 65°F (18.3°C).

12. SEASONAL COOLING ENERGY

The approximate total energy used during the cooling season can be calculated from the cooling degree days. SEER is the *seasonal energy-efficiency ratio*, which incorporates various equipment and process efficiencies as well as the conversion from Btus to kilowatt-hours. SEER for electrically driven refrigeration is typically in

[11]The LBL factor is named after Lawrence Berkeley Labs, where the correlations between a structure's characteristics and leakiness were first evaluated in the 1980s.
[12]This is probably because the accuracy of other data does not warrant a high level of sophistication.
[13]Tables of cooling degree-days are far less common than tables of heating degree-days.

the range of 10 Btu/W-hr to 12 Btu/W-hr (2.9 W/W to 3.5 W/W). The seasonal cooling cost is determined from the cost per kilowatt-hour.

$$\text{energy}_{\text{kW·h/season}} = \frac{\left(24\ \frac{\text{h}}{\text{d}}\right) q_{\text{design cooling,W}}(\text{CDD})}{\left(1000\ \frac{\text{W}}{\text{kW}}\right)(T_o - T_i)(\text{SEER}_{\text{W/W}})} \quad \text{[SI]} \quad 42.34(a)$$

$$\text{energy}_{\text{kW-hr/season}} = \frac{\left(24\ \frac{\text{hr}}{\text{day}}\right) q_{\text{design cooling,Btu/hr}}(\text{CDD})}{\left(1000\ \frac{\text{W}}{\text{kW}}\right)(T_o - T_i)(\text{SEER}_{\text{Btu/W-hr}})} \quad \text{[U.S.]} \quad 42.34(b)$$

13. LATENT AND SENSIBLE LOADS

Latent loads increase the cooling load. The main sources of residential latent loads are infiltration, perspiration and exhalation by occupants, cooking, laundry, showering, and bathing. The often quoted rule of thumb is that residential latent load is 30% of the total load, although the actual latent load varies widely depending on infiltration rate, climate, and occupancy. **[Representative Rates at Which Heat and Moisture Are Given Off by People in Different States of Activity] [Recommended Rates of Radiant and Convective Heat Gain: Unhooded Appliances During Idle (Ready-to-Cook) Conditions]**

The *sensible heat ratio*, SHR (also known as the *sensible heat factor*, SHF), is the sensible load divided by the total load (including the latent load). The *latent factor*, LF, is the reciprocal of the SHR. Most air conditioning equipment is designed to operate at a sensible heat ratio in the range of 0.70–0.75. According to ASHRAE, a latent factor of 1.3 or a sensible heat ratio of 0.77 matches the performance of typical residential vapor compression cooling systems.

$$\text{SHR} = \frac{1}{\text{LF}} = \frac{q_{\text{sensible}}}{q_{\text{sensible}} + q_{\text{latent}}} = \frac{c_p \Delta T}{\Delta h} \quad 42.35$$

14. RECIRCULATING AIR BYPASS

Some of the return air may be bypassed around the air conditioner through a bypass channel in the air handling unit. Figure 42.1 illustrates such a *recirculating air bypass* configuration.[14]

Figure 42.1 *Recirculating Air Bypass*

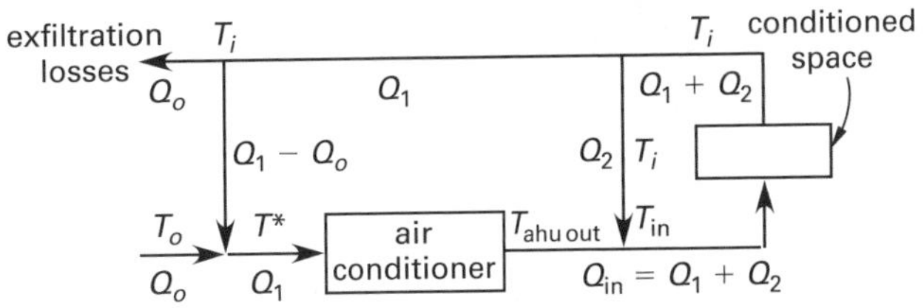

The inside design temperature, T_i, and the sensible and latent loads are generally known, as are the outside design conditions. The temperature, T_{in}, and flow rate of the air entering the conditioned space are generally not known. The following procedure can be used to determine these unknowns.[15] This procedure can also be used when there is no bypass (i.e., straight recirculation) by setting $Q_2 = 0$. The procedure is presented in terms of volumetric flows, making use of flow rates that are typically known at each branch. Using volumetric flow rates is sufficiently accurate for situations that do not involve large temperature differences (i.e., $T_i - T_o$ is not too large). In situations where this assumption is inappropriate, Eq. 42.36, Eq. 42.38, and others should be reformulated in terms of mass flow rates.

step 1: Locate the indoor, i, and outdoor, o, design conditions on the psychrometric chart. Read h_i, h_o, and v_o.

step 2: Draw a line between the indoor and outdoor points. This line represents all possible ratios of mixing indoor and outdoor air. The ratio of outdoor ventilation air, Q_o, to the fraction of conditioned recirculating air, Q_1, determines the actual mixture point, *. For an initial estimate, assume that the densities of the two air streams are the same. Then, the air masses are proportional to the air volumes. Calculate the temperature T^* from Eq. 42.36.[16] (See Fig. 42.2.)

$$\frac{T^* - T_i}{T_o - T_i} = \frac{Q_o}{Q_1} \quad 42.36$$

[14]The separate bypass duct shown in Fig. 42.1 does not actually exist. Bypassed air actually flows unchanged through a separate channel in the air conditioner. There is a variation of this configuration in which the outdoor air is mixed with the return air before the bypass takeoff. The primary difference between the variations is when the Q_o term is used.

[15]Slight modifications of the procedure may be necessary, depending on what is known. Care must be taken in distinguishing between subscripts "i," "in," and "1."

[16]Though dry-bulb temperatures are commonly used in Eq. 42.36, they need not be. Since all of the temperature scales are linear on the psychrometric chart, wet-bulb and dew-point temperatures could be used.

Figure 42.2 *Adiabatic Mixing of Inside and Outside Air*

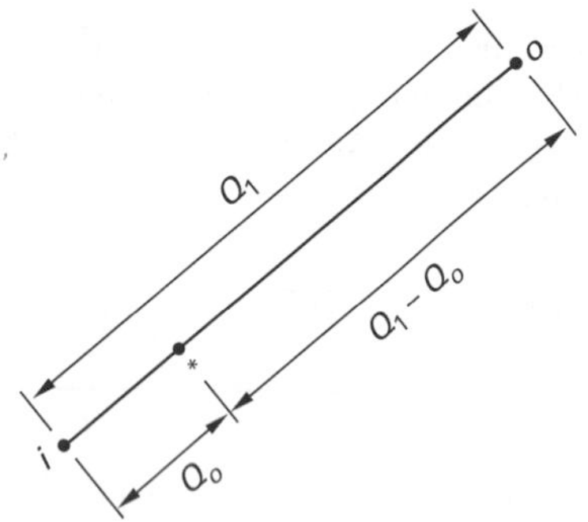

step 3: If the sensible and latent loads in the conditioned space are known, calculate the *room sensible heat ratio*, RSHR. The latent and sensible loads from outside ventilation air are not included.

$$\text{RSHR} = \frac{q_s}{q_s + q_l} = \frac{c_p \Delta T}{\Delta h} \quad \text{42.37}$$

step 4: Draw a line with the slope RSHR (based on the psychrometric chart's sensible heat ratio scale) through point "*i*." Since the air conditioner must bring the air through a process that takes non-bypassed air from condition "*i*" to condition "ahu output," the "ahu out" point must be somewhere along this line.

Alternatively, if the condition of the air leaving the air conditioner is known, draw a line from that point ("ahu out" for conditioner output) to point "*i*." (See Fig. 42.3.)

Figure 42.3 *Adiabatic Mixing of Conditioned and Bypass Air*

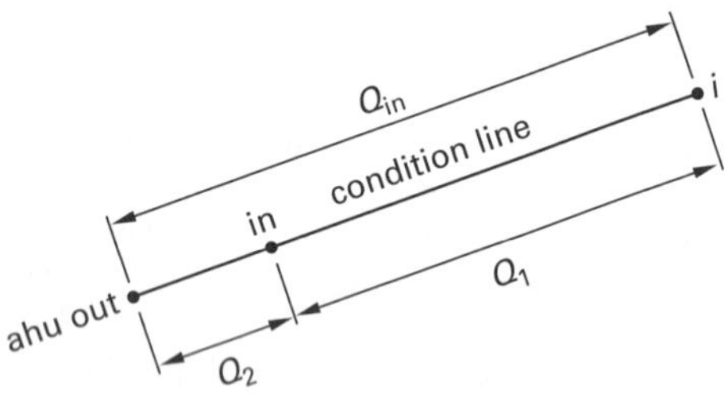

step 5: The condition of the air entering the conditioned space lies along the line drawn in step 4. The line represents all possible ratios of mixing conditioned and bypassed air. The amounts of conditioned air, Q_1, and bypassed air, Q_2, determine the mixture point "in." For an initial estimate, assume that the densities of the two air streams are the same. Calculate the temperature T_{in}. The ratio of air flows determines the *system bypass factor.*

$$\begin{aligned} \text{BF}_{\text{system}} &= \frac{T_{\text{in}} - T_{\text{ahu out}}}{T_i - T_{\text{ahu out}}} \\ &= \frac{Q_2}{Q_1 + Q_2} \end{aligned} \quad \text{42.38}$$

step 6: If neither T_{in} nor Q_{in} is known, T_{in} should be chosen such that it is 15°F to 20°F (8°C to 11°C) less than T_i. The temperature of the air entering the conditioned space and the flow rate through the space are related; one determines the other. The larger the temperature difference $T_i - T_{\text{in}}$ (representing the temperature rise as the air flows through the conditioned space), the lower the flow rate. However, very large temperature differences require extremely efficient mixing within the space, and very low T_{in} temperatures are uncomfortable for occupants near the discharge registers. Therefore, the temperature difference should not exceed 15°F to 20°F (8°C to 11°C).

step 7: If Q_{in} is known, calculate T_{in} from the sensible heating relationship, Eq. 42.39. (Also, see Eq. 42.36.)

$$\begin{aligned} Q_{\text{in,L/s}} &= Q_1 + Q_2 \\ &= \frac{q_{s,\text{W}}}{\left(1.20 \ \frac{\text{J}}{\text{L}\cdot{}^\circ\text{C}}\right)(T_i - T_{\text{in}})} \end{aligned} \quad \text{[SI]} \quad \text{42.39(a)}$$

$$\begin{aligned} Q_{\text{in,ft}^3/\text{min}} &= Q_1 + Q_2 \\ &= \frac{q_{s,\text{Btu/hr}}}{\left(1.08 \ \frac{\text{Btu-min}}{\text{hr-ft}^3\text{-}{}^\circ\text{F}}\right)(T_i - T_{\text{in}})} \end{aligned} \quad \text{[U.S.]} \quad \text{42.39(b)}$$

step 8: Locate point "in" corresponding to T_{in} on the RSHR condition line from step 4.

step 9: Knowing T_{in} establishes the ratio of Q_1 and Q_2 in Eq. 42.38. Calculate Q_1 and Q_2.

$$\begin{aligned} Q_2 &= (\text{BF}_{\text{system}})Q_{\text{in}} \\ &= (\text{BF}_{\text{system}})(Q_1 + Q_2) \end{aligned} \quad \text{42.40}$$

$$Q_1 = Q_{\text{in}} - Q_2 \quad \text{42.41}$$

step 10: Draw a line through the mixture point * and the conditioner output point "ahu out." This line represents the process occurring in the air conditioner. Heat from outside air and from within the space are both removed by the air conditioner. Therefore, the slope of this line is the *grand sensible heat ratio*, GSHR, also known as the *coil sensible heat ratio* and *grand sensible heat factor*. If this slope is known in advance, it can be used (with the sensible heat ratio scale on the psychrometric chart) to draw a line through either * or "ahu out," thereby establishing point "ahu out" or *, respectively. If it is not known in advance, it can be determined from the sensible heat ratio scale.

$$\text{GSHR} = \frac{q_{s,\text{room}} + q_{s,\text{outside air}}}{q_{t,\text{room}} + q_{t,\text{outside air}}} \qquad 42.42$$

The air will be cooled and dehumidified as it passes through the coil. (See Fig. 42.4.) The coil *apparatus dew point* (ADP) is determined by extending the line containing points "ahu out" and * to the saturation line. The ADP should be greater than 32°F (0°C) so that moisture does not freeze on the coil.

Figure 42.4 *Total Heat Removal Process*

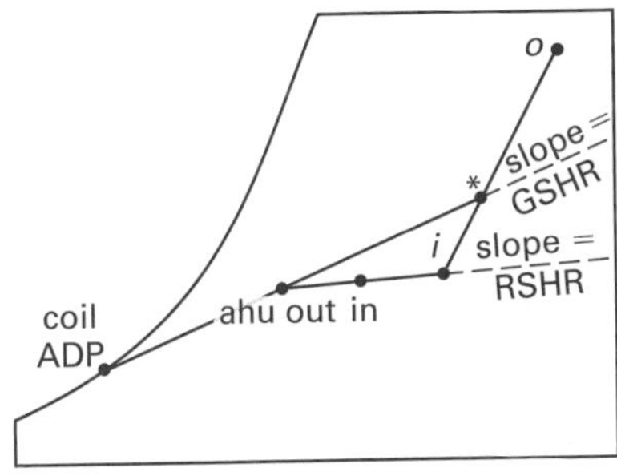

step 11: The required air conditioning capacity is

$$\begin{aligned} q_t &= q_{t,\text{room}} + q_{t,\text{outside air}} \\ &= q_{s,\text{room}} + q_{l,\text{room}} \\ &\quad + \frac{(h_o - h_i) Q_o}{\upsilon_o} \end{aligned} \qquad 42.43$$

Equation 42.44 expresses the air conditioning capacity in traditional HVAC units. The constant 4.5 lbm-min/ft³-hr is the product of air density (0.075 lbm/ft³) and 60 min/hr. The constant 1.2 kg/L is the product of air density (1.2 kg/m³) and conversions 1000 W/kW and 0.001 m³/L.

$$\begin{aligned} q_{t,\text{W}} &= q_{t,\text{room}} + q_{l,\text{room}} \\ &\quad + \left(1.2 \ \frac{\text{kg}}{\text{L}}\right)(h_o - h_i) Q_{o,\text{L/s}} \end{aligned} \qquad \text{[SI]} \quad 42.44(a)$$

$$\begin{aligned} q_{t,\text{Btu/hr}} &= q_{s,\text{room}} + q_{l,\text{room}} \\ &\quad + \left(4.5 \ \frac{\text{lbm-min}}{\text{ft}^3\text{-hr}}\right) \\ &\quad \times (h_o - h_i) Q_{o,\text{ft}^3/\text{min}} \end{aligned} \qquad \text{[U.S.]} \quad 42.44(b)$$

Example 42.1

The inside design condition for a conditioned space with partial recirculation is 75°F dry-bulb (23.9°C) and 62.5°F (16.9°C) wet-bulb. The outside air is at 94°F dry-bulb (34.4°C) and 78°F (25.6°C) wet-bulb. The sensible space load is 160,320 Btu/hr (46.5 kW). The latent load from occupants and infiltration, but excluding intentional ventilation, is 19,210 Btu/hr (5.6 kW). A total of 1275 ft³/min (600 L/s) of outside air is required. The air temperature increases 19°F (10.6°C) as it passes through the conditioned space. Air leaves the conditioner saturated.

Find the (a) coil apparatus dew point, (b) volume of air passing through the space, (c) wet-bulb temperature of the air entering the space, (d) system bypass ratio, and (e) dry-bulb temperature of the air entering the conditioner.

SI Solution

(a) The room sensible heat ratio, RSHR, is

$$\begin{aligned} \text{RSHR} &= \frac{q_s}{q_s + q_l} \\ &= \frac{46.5 \text{ kW}}{46.5 \text{ kW} + 5.6 \text{ kW}} \\ &= 0.89 \end{aligned}$$

Locate the indoor, i, and outdoor, o, design conditions on the psychrometric chart. Draw a line with the slope 0.89 through point "i." Extend the line to the left to 11.9°C on the saturation line. This is T_{co}. Since the air leaves the conditioner saturated, the coil apparatus dew point coincides with the conditioner output.

(b) The dry-bulb temperature of the air as it enters the conditioned space is

$$T_{\text{in}} = T_i - 10.6°\text{C} = 23.9°\text{C} - 10.6°\text{C} = 13.3°\text{C}$$

Calculate the air flow through the space.

$$\begin{aligned} Q_{\text{in,L/s}} &= \frac{q_{s,\text{W}}}{\left(1.20 \ \frac{\text{J}}{\text{L}\cdot°\text{C}}\right)(T_i - T_{\text{in}})} \\ &= \frac{(46.5 \text{ kW})\left(1000 \ \frac{\text{W}}{\text{kW}}\right)}{\left(1.20 \ \frac{\text{J}}{\text{L}\cdot°\text{C}}\right)(23.9°\text{C} - 13.3°\text{C})} \\ &= 3656 \text{ L/s} \end{aligned}$$

(c) Locate point "in" corresponding to T_{in} on the RSHR condition line from step 4. Read the wet-bulb temperature at this point as 12.5°C.

HVAC

(d) Calculate the system bypass ratio from Eq. 42.38.

$$\begin{aligned}\mathrm{BF}_{\text{system}} &= \frac{T_{\text{in}} - T_{\text{ahu out}}}{T_i - T_{\text{ahu out}}} = \frac{13.3^\circ\text{C} - 11.9^\circ\text{C}}{23.9^\circ\text{C} - 11.9^\circ\text{C}} \\ &= 0.117\end{aligned}$$

(e) The flow rates are

$$Q_2 = (\mathrm{BF}_{\text{system}})\,Q_{\text{in}} = (0.117)\left(3656\ \frac{\text{L}}{\text{s}}\right) = 428\ \text{L/s}$$

$$Q_1 = 3656\ \frac{\text{L}}{\text{s}} - 428\ \frac{\text{L}}{\text{s}} = 3228\ \text{L/s}$$

Use Eq. 42.36 to locate the point corresponding to the air entering the conditioner.

$$\frac{Q_o}{Q_1} = \frac{600\ \frac{\text{L}}{\text{s}}}{3228\ \frac{\text{L}}{\text{s}}} = 0.186$$

$$\frac{T^* - T_i}{T_o - T_i} = 0.186$$

$$\begin{aligned}T^* &= T_i + (0.186)(T_o - T_i) \\ &= 23.9^\circ\text{C} + (0.186)(34.4^\circ\text{C} - 23.9^\circ\text{C}) \\ &= 25.9^\circ\text{C}\end{aligned}$$

Customary U.S. Solution

(a) The room sensible heat ratio, RSHR, is

$$\begin{aligned}\mathrm{RSHR} &= \frac{q_s}{q_s + q_l} = \frac{160{,}320\ \frac{\text{Btu}}{\text{hr}}}{160{,}320\ \frac{\text{Btu}}{\text{hr}} + 19{,}210\ \frac{\text{Btu}}{\text{hr}}} \\ &= 0.89\end{aligned}$$

Locate the indoor, i, and outdoor, o, design conditions on the psychrometric chart. Draw a line with the slope 0.89 through point "i." Extend the line to the left to 53.4°F on the saturation line. This is $T_{\text{ahu out}}$. Since the air leaves the conditioner saturated, the coil apparatus dew point coincides with the conditioner output.

(b) The dry-bulb temperature of the air as it enters the conditioned space is

$$\begin{aligned}T_{\text{in}} &= T_i - 19^\circ\text{F} = 75^\circ\text{F} - 19^\circ\text{F} \\ &= 56^\circ\text{F}\end{aligned}$$

Calculate the air flow through the space.

$$\begin{aligned}Q_{\text{in,ft}^3/\text{min}} &= \frac{q_{s,\text{Btu/hr}}}{\left(1.08\ \frac{\text{Btu-min}}{\text{hr-ft}^3\text{-}^\circ\text{F}}\right)(T_i - T_{\text{in}})} \\ &= \frac{160{,}320\ \frac{\text{Btu}}{\text{hr}}}{\left(1.08\ \frac{\text{Btu-min}}{\text{hr-ft}^3\text{-}^\circ\text{F}}\right)(75^\circ\text{F} - 56^\circ\text{F})} \\ &= 7813\ \text{ft}^3/\text{min}\end{aligned}$$

(c) Locate point "in" corresponding to T_{in} on the RSHR condition line from step 4. Read the wet-bulb temperature at this point as 54.6°F.

(d) Calculate the system bypass ratio. From Eq. 42.38,

$$\begin{aligned}\mathrm{BF}_{\text{system}} &= \frac{T_{\text{in}} - T_{\text{ahu out}}}{T_i - T_{\text{ahu out}}} = \frac{56^\circ\text{F} - 53.4^\circ\text{F}}{75^\circ\text{F} - 53.4^\circ\text{F}} \\ &= 0.120\end{aligned}$$

(e) The flow rates are

$$\begin{aligned}Q_2 &= (\mathrm{BF}_{\text{system}})\,Q_{\text{in}} = (0.120)\left(7813\ \frac{\text{ft}^3}{\text{min}}\right) \\ &= 938\ \text{ft}^3/\text{min}\end{aligned}$$

$$\begin{aligned}Q_1 &= 7813\ \frac{\text{ft}^3}{\text{min}} - 938\ \frac{\text{ft}^3}{\text{min}} \\ &= 6875\ \text{ft}^3/\text{min}\end{aligned}$$

Use Eq. 42.36 to locate the point corresponding to the air entering the conditioner.

$$\frac{Q_o}{Q_1} = \frac{1275\ \frac{\text{ft}^3}{\text{min}}}{6875\ \frac{\text{ft}^3}{\text{min}}} = 0.185$$

$$\frac{T^* - T_i}{T_o - T_i} = 0.185$$

$$\begin{aligned}T^* &= T_i + (0.185)(T_o - T_i) \\ &= 75^\circ\text{F} + (0.185)(94^\circ\text{F} - 75^\circ\text{F}) \\ &= 78.5^\circ\text{F}\end{aligned}$$

15. FREEZE-UP OF COOLING COILS

Coil freeze-up (*freezing of the coils*) has two different causes. In the winter, cold make-up (ventilation) air from outside may cause water to freeze inside the heating coils. In the summer, coils with effective surface temperatures of 32°F (0°C) or less can cause moisture (humidity) in the air to freeze on the outside of the cooling (evaporator) coils.

In the summer, water (humidity) condensing on the outside of a cooling coil is normally removed by a built-in pan and drain. However, frost (rather than liquid condensate) will form when the dew point temperature is below freezing. Coil freeze-up can be prevented by designing the system so that the coil temperature is above freezing. If the extension of the line containing points "ahu out" and * does not intersect the saturation line above 32°F (0°C) (or, if the line does not intersect the saturation line at all), heating may be required.

Having an above-freezing coil apparatus dew point may not be economical or practical, and with preexisting systems, it may not be possible at all. Freeze-up can be functionally prevented by ensuring that (a) the cooling load is high, and (b) the refrigerant flow rate is full. Low cooling load is usually attributable to low airflow or low entering temperature. Low airflow can be caused by a plugged coil, fouled air filter, low blower speed, broken fan belt, failed blower motor, closed distribution register, or undersized ductwork. Low entering air temperatures can be caused by setting the thermostat too low during the noncooling part of the daily schedule or by cold outside (make-up) air. Low refrigerant flow rates are usually attributable to refrigerant leaks and can be detected by pressure gauges, but low flow rates can also be caused by restricted liquid or suction filter-driers and faulty metering devices.

16. CHANGEOVER TEMPERATURE AND REHEATING COOLED AIR

The changeover temperature is the outdoor temperature at which heat gain to the interior can be offset using cold primary air to balance the transmission loss.

In-Room Terminal Systems

$$T_{co} = T_r - \frac{q_{is} + q_{es} - 1.1Q_p(T_r - T_p)}{\Delta q_{td}} \qquad 42.45$$

Use the air-to-transmission ratio, A/T, to determine Δq_{td}. The air-to-transmission ratio is the ratio of the primary airflow to a given space, A, divided by the transmission per degree of that space, T. The transmission per degree is the difference between the space temperature and the outdoor temperature, assuming steady-state heat transfer.

Transmission of Heat in a Space

$$\frac{A}{T} = \frac{\text{primary airflow}}{\text{transmission per degree}} \qquad 42.46$$

If the sensible heat load from the occupied region decreases, or if the air leaving the air conditioner ("ahu out") is too cold for any reason (not necessarily because of below-freezing temperatures), the air can be reheated in a *reheat coil*. Reheating can occur within the air conditioner, in an intermediate distribution box, or in an air terminal device. As the moisture content does not change, sensible reheating is represented on the psychrometric chart by a horizontal line from the point representing the conditioner output ("ahu out") toward the right. If the air leaves the conditioner saturated, the line will start at the equipment's apparatus dew point on the psychrometric chart's saturation curve. Since the energy of heating must be removed by the coil when the air returns through it, the cooling load (refrigeration load) will be increased by the reheat, as shown in Fig. 42.5. The additional cooling load can be calculated from either the enthalpy difference or (since the specific heat and humidity ratio are known) the dry-bulb temperature difference.

Figure 42.5 Reheat

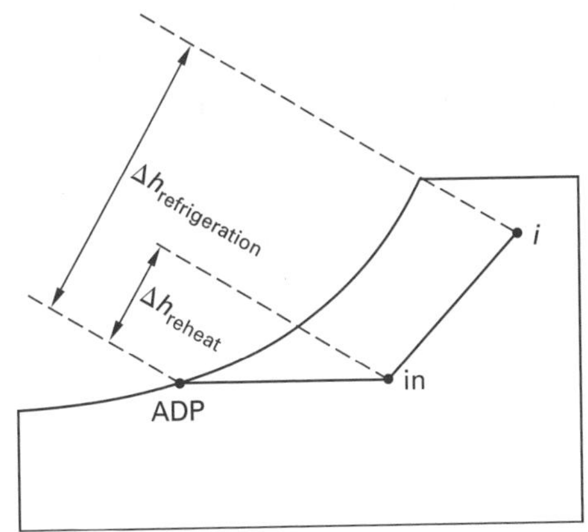

17. NOMENCLATURE

A	area	ft^2	m^2
A	duct area	in^2	cm^2
ACH50	air changes per hour at 50 Pa	1/hr	1/h
ACH-nat	air changes per hour at 4 Pa	1/hr	1/h
AL	air leakage	ft^3/ft^2	
BF	bypass factor	–	–
c_p	specific heat	Btu/lbm-°F	kJ/kg·°C
C	leakage curve coefficient	–	–
C_d	coefficient of discharge	–	–
CDD	cooling degree days	°F-day	°C·d
CFM50	airflow rate at 50 Pa	ft^3/min	n.a.

HVAC

CLF	cooling load factor	–	–
CLTD	cooling load temperature difference	°F	°C
COP	coefficient of performance	–	–
CRF	condensation resistance factor	–	–
E_t	total irradiance	Btu/hr-ft^2	–
EER	energy-efficiency ratio	Btu/W-hr	n.a.
ELA	effective leakage area	in^2	m^2
g_c	gravitational constant, 32.2	lbm-ft/lbf-sec^2	n.a.
GSHR	grand sensible heat ratio	–	–
h	enthalpy	Btu/lbm	kJ/kg
h	height	in	m
I	current	A	A
L	duct length	ft	m
LBL	energy climate factor	–	–
LF	latent factor	–	–
n	leakage curve exponent	–	–
NL	normalized leakage	1/hr	1/h
Δp	pressurization	Pa	Pa
P	duct perimeter	in	cm
P	power	hp	W
q	heat transfer rate	Btu/hr	W
q''	heat loss	Btu/hr-ft^2	W/m^2
Q	volumetric flow rate	ft^3/min	m^3/h, L/s
Q50	air permeability	ft^3/ft^2-min	m^3/m^2·h
RSHR	room sensible heat ratio	–	–
S	building envelope surface area	ft^2	m^2
SC	shading coefficient	–	–
SCL	solar cooling load factor	Btu/hr-ft^2	W/m^2
SEER	seasonal energy-efficiency ratio	Btu/W-hr	n.a.
SHGC	solar heat gain coefficient	–	–
SHR	sensible heat ratio	–	–
t	time (duration)	hr	h
T	temperature	°F	°C
T^*	mixture temperature	°F	°C
U	overall coefficient of heat transfer	Btu/hr-ft^2-°F	W/m^2·°C
v	duct velocity	ft/min	m/s
V	building volume	ft^3	m^3
V	voltage	V	V

Symbols

η	efficiency	–	–
ρ	density	lbm/ft^3	kg/m^3
υ	specific volume	ft^3/lbm	m^3/kg

Subscripts

ahu	air handling unit
b	adjacent or outdoor
c	cooling, cold
cg	center of glass
co	changeover
eg	edge of glass
es	external sensible
f	frame
h	hot
i	indoor design
i	conditioned or indoor
in	in (entering the space)
is	internal sensible
l	latent
m	mean
o	outdoor design
p	primary
p	projected
ref	reference
s	sensible
t	total
te	total equivalent
td	temperature difference

HVAC

43 Air Conditioning Systems and Controls

Content in blue refers to the *NCEES Handbook*.

NCEES EXAM SPECIFICATIONS AND RELATED CONTENT

HVAC AND REFRIGERATION EXAM

II.B.7. Equipment and Components: Control systems components
- 2. Control Equipment
- 3. Hvac Process Control
- 4. Typical Controls Integration
- 6. Feedback and Control
- 7. Digital Control
- 8. Relay Logic Diagrams

1. TYPES OF AIR CONDITIONING SYSTEMS

Depending on the medium delivered to the conditioned space, air conditioning systems are categorized as all-air, air-and-water, all-water, and unitary.

All-air systems maintain the temperature by distributing only air, and some less efficient systems rely on internal loads for heating, sending only cold air to the space. Most central units are *single-duct*, which means that the cooling and heating coils are in series. In *dual-duct* units, the heating and cooling coils are in parallel ducts.

In *air-and-water systems*, which are common in commercial applications, air and water are both distributed to the conditioned space. In *all-water systems*, the cooling and heating effects are provided solely by cooled and/or heated water pumped to the conditioned space. With *unitary equipment*, the fan, condenser, compressor, and cooling and heating coils are combined in a stand-alone unit for window and through-the-wall installation. There are also *split systems* that consist of separate fan/coil and condenser/compressor assemblies.

2. CONTROL EQUIPMENT

Control equipment in HVAC systems consists of sensors, actuators, motors, relays, and controllers. *Sensors* (*transducers*) are used to monitor temperature (*thermostats* or just "*stats*"), enthalpy, and humidity (*humidistats* or *hygrostats*). In complex delivery systems, pressure may also be monitored (i.e., by *pressurestats*). Thermostats are designated as "room," "insertion," or "immersion" according to their placement (i.e., in the occupied space, in a duct, or in a water/steam manifold, respectively). [**Temperature Controls: Terminology**]

The signal from a sensor is received by the *controller*, which energizes or deenergizes the appropriate equipment. Since most control systems operate at low voltages (e.g., 24 V AC), a *relay* must be used when the equipment operates at a higher voltage. Relays allow low-voltage thermostats and controllers to control high-voltage, high-drain devices. For example, a room thermostat might energize a relay that, in turn, would provide power to a line-voltage fan motor. Alternatively, the thermostat could activate a pneumatic relay (*electropneumatic switch*). [**Control System Types**]

Actuators (also referred to as *operators* and *motors*) provide the force to open and close valves and dampers. The control signal can be electrical, electronic, or pneumatic. Actuators are designated as *normally open* (NO) or *normally closed* (NC) depending on their position when deenergized. Most actuators act relatively slowly. *Solenoids*, however, act quickly in response to signals.

A *pneumatic actuator* is essentially a piston/cylinder arrangement or diaphragm/bellows. With pneumatic actuators, a separate compressed-air system is required to supply the force for changing damper settings. With a typical 18 psig (124 kPa) source, the pneumatic signal will be approximately 3 psig to 15 psig (21 kPa to 100 kPa). Pneumatic actuator performance is essentially linear. The actuator position is proportional to the air pressure.

3. HVAC PROCESS CONTROL

Control of basic commercial HVAC systems has traditionally meant controlling either the amount of bypass air or the amount of reheat. These two constant volume methods are known as face-and-bypass damper control and reheat control, respectively. *Face-and-bypass damper control* is normally used to control only the dry-bulb temperature. Because of the possibility of bringing in too much moisture (an uncontrolled variable), this method is generally not used with a high percentage of outside air unless the outside air can be dehumidified. *Reheat control* is needed when both room temperature and humidity control are needed. Once the proper humidity level is achieved, reheat ensures the proper room temperature. [**Control System Types**]

A third method, *air volume control*, relies on variations in flow rate through the conditioned space. Only one parameter (i.e., dry-bulb temperature) can be adequately controlled in this manner. Volume control is more applicable in the largest systems where the additional cost and complexity can be economically justified. However, the reduction in costs of variable speed drives has begun to change the economics of these systems. Advances in noise control, monitoring of other comfort parameters, and the ability to provide sufficient outside ventilation when volume is low may overcome the criticisms this method has received.

4. TYPICAL CONTROLS INTEGRATION

Control systems are typically classified as pneumatic, electric, electronic, or direct digital. There are numerous variations in equipment layout, mixing sequence, and control methodology. Figure 43.1 schematically illustrates a central station *air handling unit* (AHU) in a multizone system with reheat control. Return air enters from the top; ventilation air enters from the left. Conditioned air is discharged into the distribution ductwork. The cooling effect may be either provided by the evaporator coils of a vapor-compression refrigeration cycle or from liquid chiller cooling coils. The heating effect (within the individual zone ducts) may be provided by hot water or steam coils or from the condenser section of a vapor-compression heat pump. The air flow through various components is controlled by remotely actuated dampers. [**Control System Types**]

Until the fan motor starts, all the dampers are in their deenergized positions. The bypass damper is normally open. The outside air damper is normally closed. (Interlocking the outside air damper to the fan motor prevents induction of cold air by the stack effect and potential coil freeze-up whenever the fan is not running.)

The control sequence begins when the fan motor starts. The fan voltage energizes a relay and/or electric-pneumatic valve (EP), which provides air to the controllers. When the fan starts, damper motor (or damper actuator) DM1 opens the outside damper to a predetermined minimum position, permitting outside air to enter.

Damper motor DM2 is controlled by two sensors: the return air humidistat (humidity controller, HC) and the supply air temperature controller (TC), also known as a *mixing thermostat*. The duplex pressure selector (DS) selects the higher of the two pressure signals from either the HC or TC sensors and positions damper motor DM2 appropriately.[1]

The cooling coil both cools and dehumidifies the air stream. If the latent loads are low, the air is merely cooled to TC's set point temperature. If latent loads are high, more air is passed through the cooling coil to remove the moisture. Reheating is used to prevent overcooling of the space. Reheating is controlled by room thermostats (TR1, TR2, and TR3). When a TR set point is reached, the corresponding steam or hot water valve (V) is opened.

Temperature controller TC also acts as a high-limit controller, preventing the supply air control from increasing above what is required for adequate zone cooling.

When the space has low latent loads, humidification is required. Figure 43.1 does not show the humidification system and controlling humidistats. Small humidification increases can be obtained by increasing the amount of bypass air. Steam or hot water injection is needed for large humidity changes.

Not shown is an outside enthalpy controller. This controller compares the heat content of the return air with that of the outside air. When the refrigeration load can be reduced, the enthalpy controller overrides the temperature controller and increases the outdoor air damper opening. In smaller installations, an *outside air thermostat* can work almost as well.

5. FREEZE PROTECTION

When the outdoor air is at subfreezing temperature, water in preheat, reheat, and chilled water coils can freeze whether or not the coils are in operation. Freezing is caused by direct contact with or incomplete mixing (*stratification*) of the cold outside air with return air, although reduced warm air flows due to clogged filters can also be a contributing factor. In some systems, freeze-up can be prevented by using antifreeze or by draining the coils when they are not in use.

It is appropriate to protect the coils with thermostats (*freeze stats*). For example, the face dampers upstream of the coils can be closed down when the plenum temperature drops to approximately 35°F (2°C) or when the temperature of the incoming heated water in the heating coils (as determined by an *immersion thermostat*) drops below 120°F to 150°F (50°C to 65°C). Furthermore, when the fan is not running, the outside air dampers should be closed and minimum heat should continue to be provided to the heating coils.

[1]Motor DM2 controls two dampers to vary the air passing through the bypass and coil. The bypass damper is normally open; the coil damper is normally closed. In some systems, the coil damper is a *face damper*. The face damper is installed immediately before the face of the cooling coil, hence its name.

HVAC

Figure 43.1 Bypass Air Handling Unit (reheat control configuration)

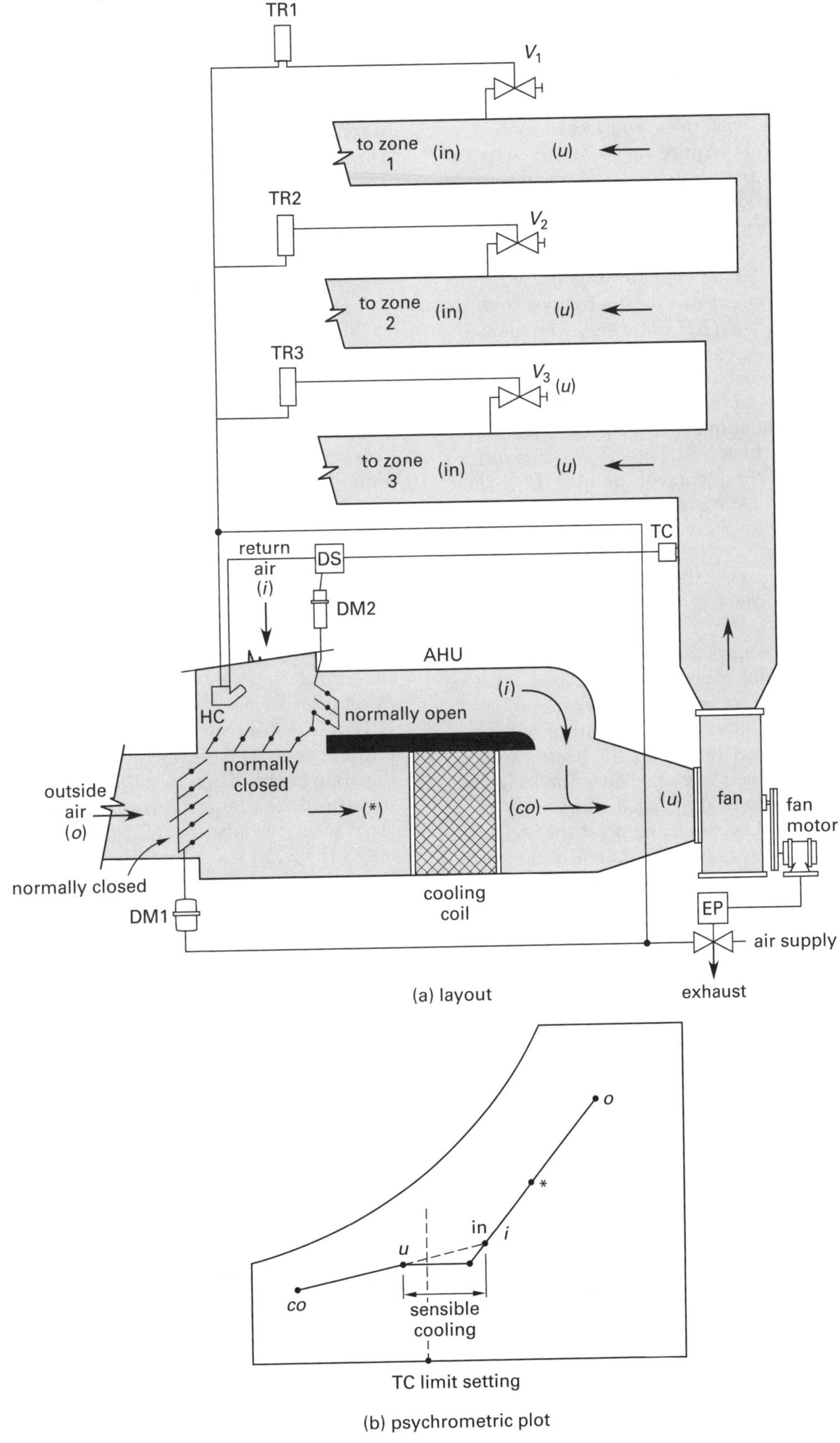

(a) layout

(b) psychrometric plot

6. FEEDBACK AND CONTROL

Sensors and their controllers have traditionally constituted an analog feedback loop and control system, but these analog systems are becoming less prevalent as the industry transitions to digital systems. (See Fig. 43.2.) HVAC applications use both open and closed loop controls. Open loop controls require no feedback between the controlled variable (e.g. temperature) and the controller. This configuration is less desirable because it results in off-nominal heating and cooling conditions. In closed loop controls, the controlled variable provides an input to the controller. This configuration is typically used in commercial applications as it provides greater system stability. (See *ASHRAE Handbook—Fundamentals* for a schematic of an example of feedback control.) Using temperature as the controlled variable, the *set point*, $T_{\text{set point}}$, is the temperature that the conditioned space would like to maintain. The *control point* is the actual temperature in the room. The *offset* is the difference between the set and control points. The time required for the room temperature to become established at the set point is known as the *settling time.* **[Control System Types]**

$$T_{\text{offset}} = T_{\text{set point}} - T_{\text{control point}} \qquad 43.1$$

There are several basic control methods. The most simple is *two-position control.* The controlled device (e.g., a valve or damper) is either fully on or fully off. Because of thermal inertia and other delays, the temperature will continue to increase for a short time after the heat is turned off. This is known as *temperature overshoot.* Similarly, the temperature will continue to drop for a short time after the heat is turned on. The room temperature oscillates around the set point. The range of temperatures experienced by the room is the *operating differential,* while the difference between the on and off set point temperatures is the *control differential.*

For *timed two-position control* (*anticipation control*), a small heater is built into the thermostat. While the room is being heated, the thermostat is also being heated, and this turns the thermostat off sooner than otherwise. The overshoot is reduced considerably.

With *floating action control,* the controlled device has three positions: fully on, fully off, and a fixed intermediate position. The controller has three corresponding signals: fully on, fully off, and a *neutral position* signal that is generated while the temperature is within a *dead band* range. Within the dead band, there is an intermediate amount of air heating (or cooling). If the amount of heating (cooling) corresponds to the heat loss (gain), the temperature remains within the dead band. Otherwise, the controller will generate a fully on or fully off signal.

With *proportional action,* the position (e.g., percentage opening) of the damper or valve is proportional to the offset. The *throttling range,* R_{throttle}, is the temperature range over which the damper or valve changes from fully closed to fully open. The throttling range should coincide with the normal range of temperatures encountered. (When the temperature is outside of the throttling range, the system is "out of control.") Since the damper or valve should be 50% open at the set point, within the limits of 0 to 1.00, the fraction open is calculated using Eq. 43.2. Valve gain (% change in control signal/% change in control variable) can be used to ensure fine-tuned controllability.

$$\text{fraction open} = 0.50 + \frac{T_{\text{offset}}}{R_{\text{throttle}}} \qquad 43.2$$

Proportional action does not provide extra heating (cooling) to compensate for changes; it tends to maintain the existing control point. With proportional action, the settling time is very long.

Proportional action with automatic reset, also known as *proportional plus integral control* (PI), attempts to return the room temperature back to the set point. In effect, the controller "overreacts" and the signal is more than proportional to the offset.

With *proportional plus integral plus derivative control* (PID), the control action responds to three different parameters: (1) the magnitude of the offset, (2) the duration of the offset, and (3) the rate at which the temperature is changing. These three aspects correspond to the terms "proportional," "integral," and "derivative," respectively, in the name. Because of the complexity of the algorithm, and since an accurate time base is needed, PID control is implemented through digital control.

7. DIGITAL CONTROL

Direct digital control (DDC) is an alternative to traditional analog control. Devices in analog and digital control systems are analogous. (See Fig. 43.3.) Digital sensors replace analog sensors one for one and are located in the same locations. Digital controllers replace analog controllers. Digital actuators replace analog (electric and pneumatic) actuators. Digital devices are connected by simple "twisted pair" control wiring. Each device has its own address, and the digital signal generated by the controller includes the device address. **[Control System Types]**

HVAC

Figure 43.2 Response of Temperature Control Methods

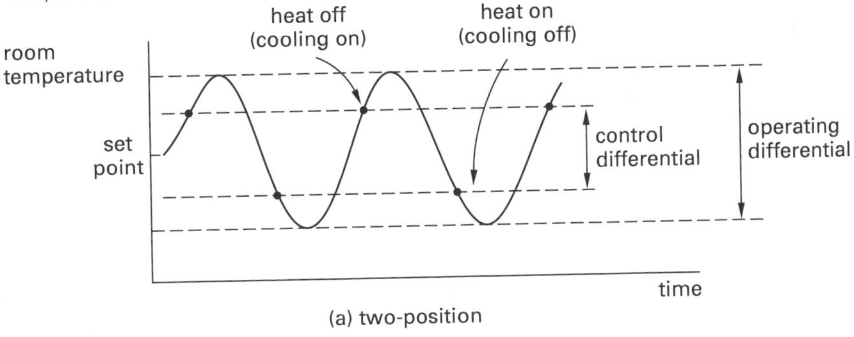

(a) two-position

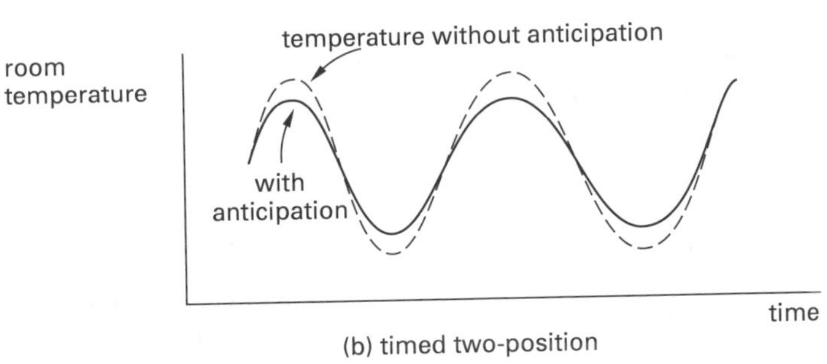

(b) timed two-position

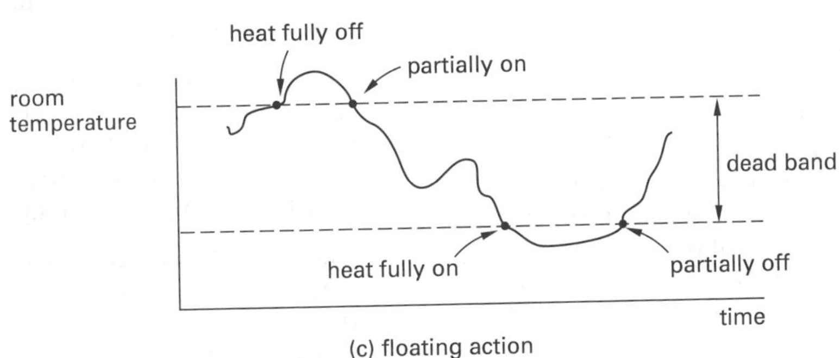

(c) floating action

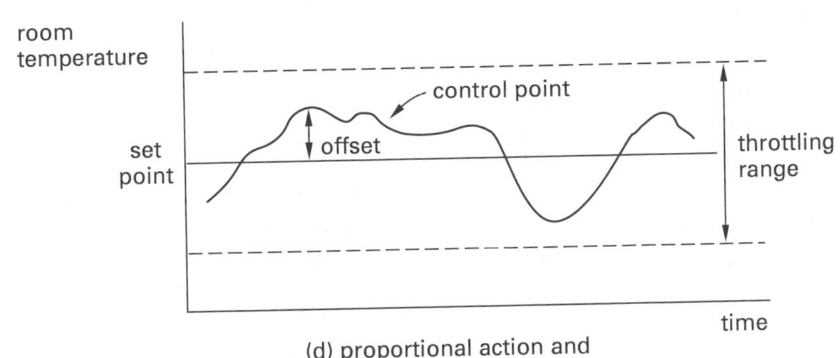

(d) proportional action and
proportion with reset

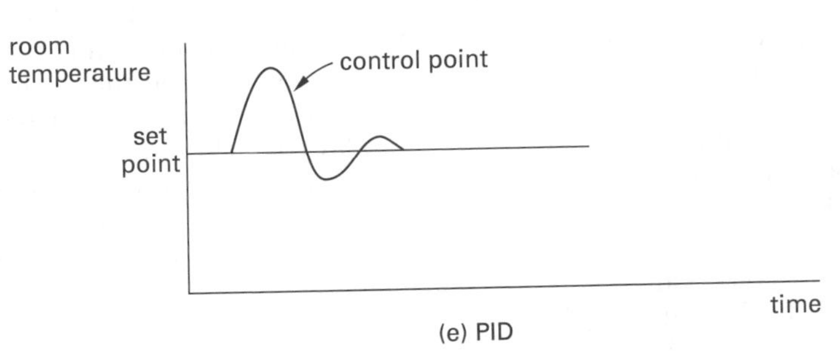

(e) PID

HVAC

Figure 43.3 *Basic Hook-Up Diagrams*

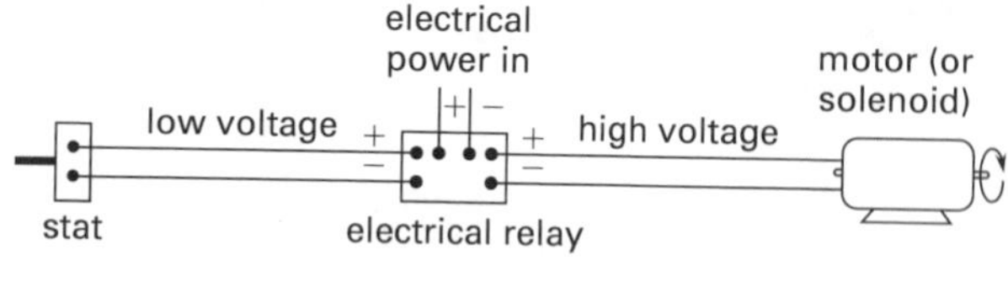

(a) analog all-electric

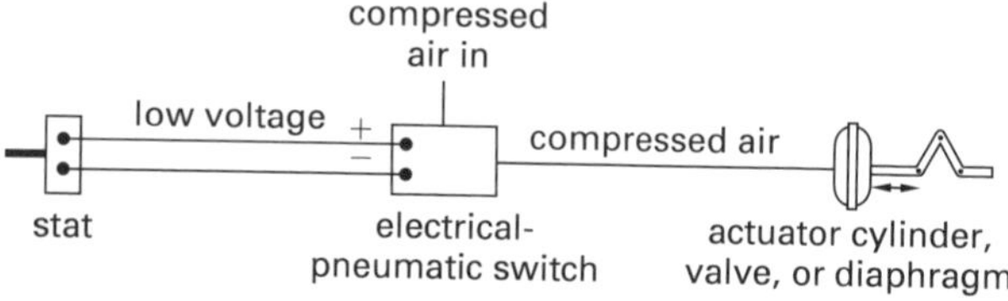

(b) analog pneumatic

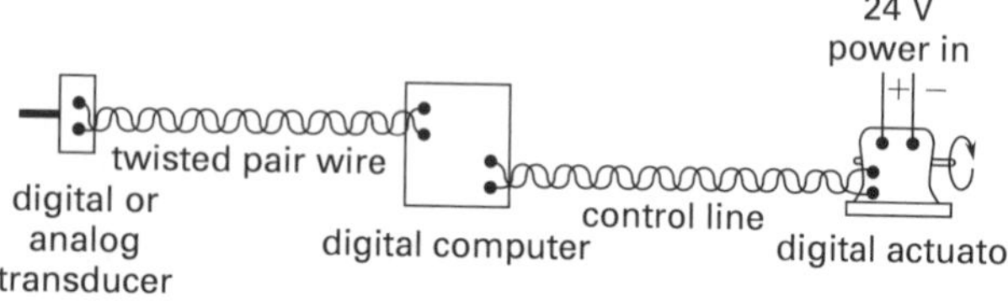

(c) digital

Digital controllers are essentially *local control computers* (LCC) running algorithms preprogrammed by the manufacturer that can be modified through remote software changes over a network. Digital controllers can be operated independently, or they can be integrated into a building management system (BMS). PI and (in some cases) PID control algorithms are easily implemented. Changes to the dead band, proportional band, set points, low- and high-limits, lockouts, and so on, can be programmed for all the controlled devices after installation. [**Digital Controllers**]

When installing a digital controller in a system already equipped with analog (pneumatic or electric) control devices, a *digital-to-analog* (*digital-to-proportional*) *staging module* is needed. The staging module is essentially an electropneumatic switch that translates digital signals into signals compatible with the pre-existing devices.

8. RELAY LOGIC DIAGRAMS

Controllers used for HVAC systems receive analog and digital inputs from temperature, humidity, and pressure sensors, as well as from other devices. Controller outputs can be analog (on/off for relays or infinitely variable for dampers, valves, and actuators) or digital to control and communicate with digital devices. Analog and digital relays can, themselves, be used to implement simple control sequences. Contacts in a relay can be normally open (NO) or normally closed (NC), and relays may have several sets of contacts in any combination of NO and NC. Since relay contacts are either off or on (open or closed), *relay logic* is binary or *Boolean*. For example, two basic relays in parallel constitute a simple OR gate, while two relays in series constitute a simple AND gate. Figure 43.4 contains the symbols of devices that primarily behave in a Boolean manner or appear in HVAC circuit diagrams. [**Control System Types**] [**Digital Controllers**]

The symbols in Fig. 43.4 are combined into *relay logic diagrams*, also known as *ladder logic diagrams*, *line diagrams*, and *elementary diagrams*. Each row in the diagram is known as a *rung*. In such a diagram, the voltage sources (120 V, 24 V, etc.) are shown as a *supply rail* (*supply bus*), typically labeled "L1," and a *ground rail* (*ground bus*) labeled "L2." Relay logic diagrams are drawn according to the following rules.

1. A rung is numbered on its left-hand side. Optional comments and descriptions of a rung's function appear on the right-hand side.
2. Control devices (inputs, such as switches) appear to the left of a rung, while controlled devices (outputs, such as motors) appear to the right of a rung. Controlled devices cannot appear between control devices.
3. Control devices can be connected in series and are shown on the same rung, while control devices connected in parallel are shown on a rung and its sub-rungs. The same control device (i.e., a switch) may appear on multiple rungs.
4. Output devices cannot be placed in series. Each output device is shown on its own rung or sub-rung. For example, a switch that controls a motor and a pilot light will be represented by either a rung with a sub-rung, or by two rungs.
5. Rungs can be electrically connected (with a *tie line* or *tie bus*) vertically, but control and controlled devices cannot be connected between rungs in the tie line.
6. When analyzing functionality, the control sequence (or, alternatively, the current) always moves from left-to-right (or, occasionally, vertically). The control sequence never moves right-to-left, not even for a single sub-rung or component.
7. A relay's coil and its contacts are shown on separate rungs. (For example, a switch may energize a relay coil that controls a fan. The input switch and output coil are shown on a single rung. The relay contacts and the output fan motor would be shown on a separate rung.) The numbers of the rungs containing a relay's contacts are listed as a comment for the rung that contains the relay's coil.

Figure 43.4 *NEMA Boolean Control Element Symbols**

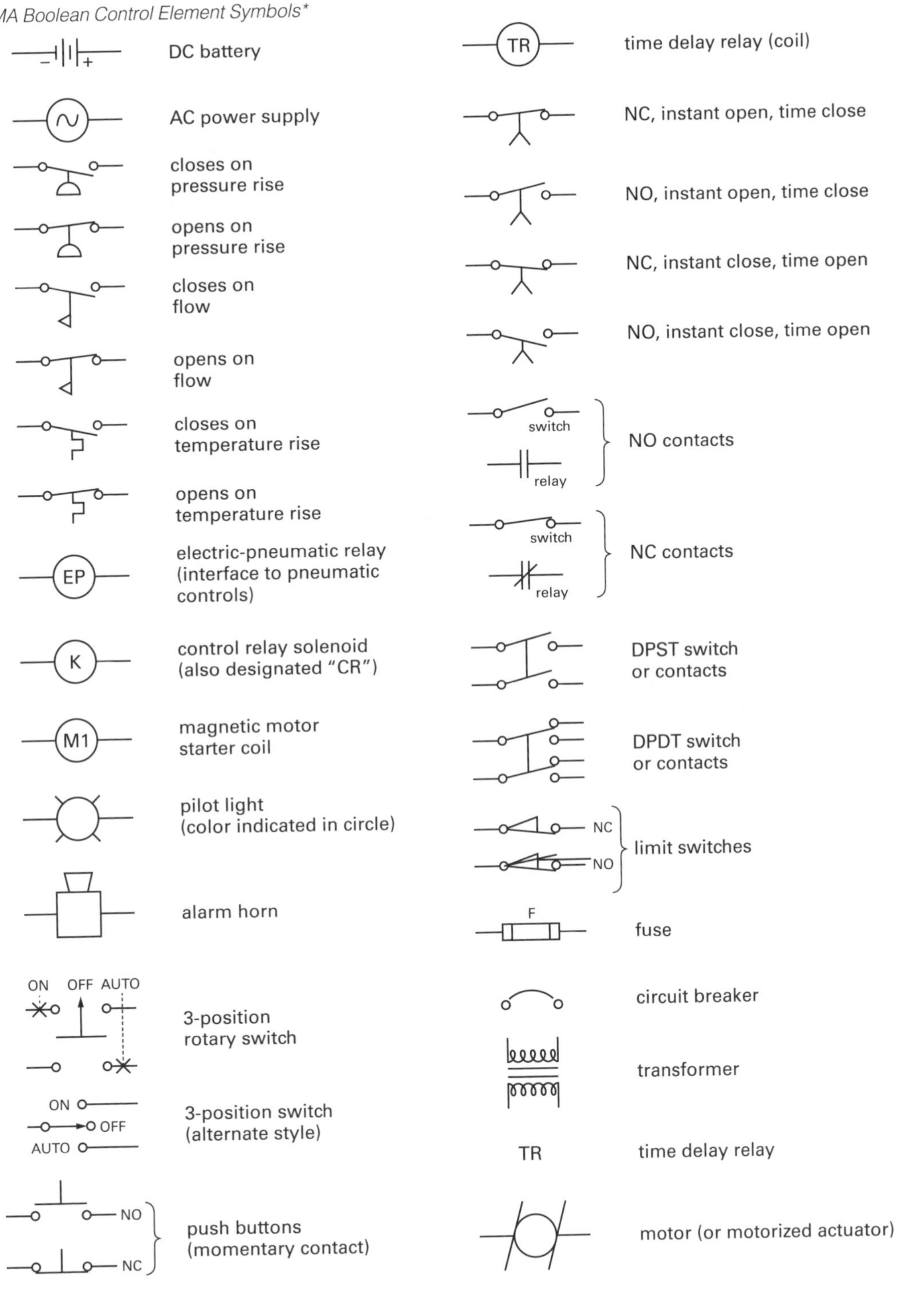

	DC battery
	AC power supply
	closes on pressure rise
	opens on pressure rise
	closes on flow
	opens on flow
	closes on temperature rise
	opens on temperature rise
EP	electric-pneumatic relay (interface to pneumatic controls)
K	control relay solenoid (also designated "CR")
M1	magnetic motor starter coil
	pilot light (color indicated in circle)
	alarm horn
ON OFF AUTO	3-position rotary switch
ON, OFF, AUTO	3-position switch (alternate style)
NO, NC	push buttons (momentary contact)
TR	time delay relay (coil)
	NC, instant open, time close
	NO, instant open, time close
	NC, instant close, time open
	NO, instant close, time open
switch, relay	NO contacts
switch, relay	NC contacts
	DPST switch or contacts
	DPDT switch or contacts
NC, NO	limit switches
F	fuse
	circuit breaker
	transformer
TR	time delay relay
	motor (or motorized actuator)

*Slight differences exist between NEMA and IEC symbols.

Example 43.1

Describe the function of the relay logic diagram shown.

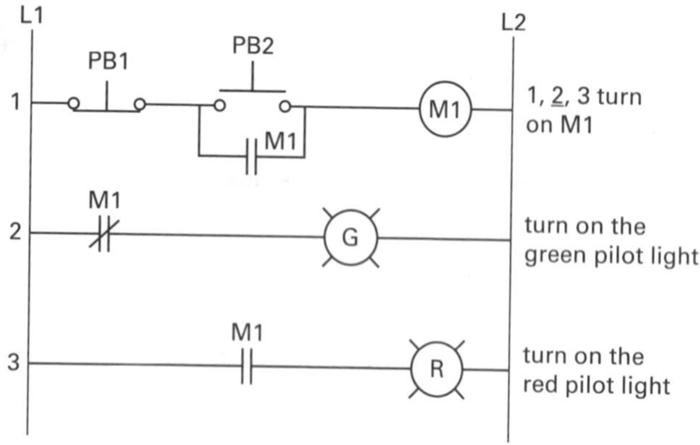

Solution

The appropriate voltage is applied across rails L1 and L2. There are three rungs, labeled 1, 2, and 3. The controlled device on rung 1 is a relay coil. The controlled device on rung 2 is a green pilot light. The controlled device on rung 3 is a red pilot light. Both pilot lights are controlled by relay M1 contacts. The physical circuit contains two momentary push-buttons, PB1 and PB2, a relay M1, and green and red pilot lights. The relay has two NO (normally open) contacts and one NC (normally closed) contact. The numbers to the right of the first rung show that coil M1 controls contacts on rungs 1, 2, and 3. PB1 is normally closed, and pushing it disconnects the rung from power; PB1 probably acts as an "off" or interrupt switch. Pushing PB2 energizes the (circled) M1 coil. The energized coil closes the M1 contacts on sub-rung 1, maintaining continuous power to relay coil (circled) M1. When relay coil M1 is energized, the NC M1 contacts on rung 2 open, extinguishing the green (safe) light. When relay coil M1 is energized, the NO contacts on rung 3 close, illuminating the red (running) light. **[Digital Controllers]**

9. NOMENCLATURE

R	range	°F	°C
T	temperature	°F	°C

HVAC

Topic VII: Statics

Chapter

44 Determinate Statics

Content in blue refers to the *NCEES Handbook*.

NCEES EXAM SPECIFICATIONS AND RELATED CONTENT

MACHINE DESIGN AND MATERIALS EXAM

I.B.1. Engineering Science and Mechanics: Statics
- 4. Concentrated Forces
- 6. Moment of a Force about a Point
- 7. Varignon's Theorem
- 9. Components of a Moment
- 17. Conditions of Equilibrium

1. INTRODUCTION TO STATICS

Statics is a part of the subject known as *engineering mechanics*.[1] It is the study of rigid bodies that are stationary. To be stationary, a rigid body must be in static equilibrium. In the language of statics, a stationary rigid body has no *unbalanced forces* or moments acting on it.

2. INTERNAL AND EXTERNAL FORCES

An *external force* is a force on a rigid body caused by other bodies. The applied force can be due to physical contact (i.e., pushing) or close proximity (e.g., gravitational, magnetic, or electrostatic forces). If unbalanced, an external force will cause motion of the body.

An *internal force* is one that holds parts of the rigid body together. Internal forces are the tensile and compressive forces within parts of the body as found from the product of stress and area. Although internal forces can cause deformation of a body, motion is never caused by internal forces.

3. UNIT VECTORS

A *unit vector* is a vector of unit length directed along a coordinate axis.[2] In the rectangular coordinate system, there are three unit vectors, **i**, **j**, and **k**, corresponding to

[1]Engineering mechanics also includes the subject of dynamics. Interestingly, the subject of mechanics of materials (i.e., strength of materials) is not part of engineering mechanics.
[2]Although polar, cylindrical, and spherical coordinate systems can have unit vectors also, this chapter is concerned only with the rectangular coordinate system.

the three coordinate axes, x, y, and z, respectively. (There are other methods of representing vectors, in addition to bold letters. For example, the unit vector **i** is represented as $\bar{i}$ or $\hat{i}$ in other sources.) Unit vectors are used in vector equations to indicate direction without affecting magnitude. For example, the vector representation of a 97 N force in the negative x-direction would be written as $\mathbf{F} = -97\mathbf{i}$.

4. CONCENTRATED FORCES

A *force* is a push or pull that one body exerts on another. A *concentrated force*, also known as a *point force*, is a vector having magnitude, direction, and location (i.e., point of application) in three-dimensional space. (See Fig. 44.1.) In this chapter, the symbols **F** and F will be used to represent the vector and its magnitude, respectively. (As with the unit vectors, the symbols **F**, $\bar{F}$, and $\hat{F}$ are used in other sources to represent the same vector.)

The vector representation of a three-dimensional force is given by Eq. 44.1, and the vector representation of a plane force is given by Eq. 44.2. Of course, vector addition is required.

Force

$$\mathbf{F} = F_x\mathbf{i} + F_y\mathbf{j} + F_z\mathbf{k} \qquad 44.1$$

$$F = F_x i + F_y j \qquad 44.2$$

Figure 44.1 Components and Direction Angles of a Force

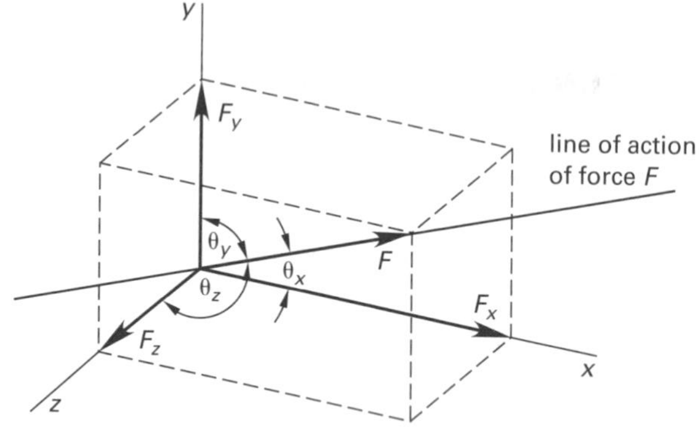

If **u** is a *unit vector* in the direction of the force, the force can be represented as

$$\mathbf{F} = F\mathbf{u} \qquad 44.3$$

The components of the force can be found from the *direction cosines*, the cosines of the true angles made by the force vector with the x-, y-, and z-axes.

Resolution of a Force

$$F_x = F\cos\theta_x \qquad 44.4$$

$$F_y = F\cos\theta_y \qquad 44.5$$

$$F_z = F\cos\theta_z \qquad 44.6$$

$$F = \sqrt{F_x^2 + F_y^2 + F_z^2} \qquad 44.7$$

The *line of action* of a force is the line in the direction of the force extended forward and backward. The force, **F**, and its unit vector, **u**, are along the line of action.

5. MOMENTS

Moment is the name given to the tendency of a force to rotate, turn, or twist a rigid body about an actual or assumed pivot point. (Another name for moment is *torque*, although torque is used mainly with shafts and other power-transmitting machines.) When acted upon by a moment, unrestrained bodies rotate. However, rotation is not required for the moment to exist. When a restrained body is acted upon by a moment, there is no rotation.

An object experiences a moment whenever a force is applied to it.[3] Only when the line of action of the force passes through the center of rotation (i.e., the actual or assumed pivot point) will the moment be zero.

Moments have primary dimensions of length × force. Typical units include foot-pounds, inch-pounds, and newton-meters.[4]

6. MOMENT OF A FORCE ABOUT A POINT

Moments are vectors. The moment vector, $\mathbf{M}_O$, for a force about point O is the *cross product* of the force, **F**, and the vector from point O to the point of application of the force, known as the *position vector*, **r**. The scalar product $|\mathbf{r}|\sin\theta$ is known as the *moment arm, d.*

Moments (Couples)

$$\mathbf{M}_O = \mathbf{r} \times \mathbf{F} \qquad 44.8$$

$$M_O = |\mathbf{M}_O| = |\mathbf{r}|\,|\mathbf{F}|\sin\theta = d|\mathbf{F}| \quad [\theta \le 180°] \qquad 44.9$$

The line of action of the moment vector is normal to the plane containing the force vector and the position vector. The sense (i.e., the direction) of the moment is determined from the *right-hand rule.*

Right-hand rule: Place the position and force vectors tail to tail. Close your right hand and position it over the pivot point. Rotate the position vector into the force

[3]The moment may be zero, as when the moment arm length is zero, but there is a (trivial) moment nevertheless.
[4]Units of kilogram-force-meter have also been used in metricated countries. Foot-pounds and newton-meters are also the units of energy. To distinguish between moment and energy, some authors reverse the order of the units, so pound-feet and meter-newtons become the units of moment. This convention is not universal and is unnecessary since the context is adequate to distinguish between the two.

vector, and position your hand such that your fingers curl in the same direction as the position vector rotates. Your extended thumb will coincide with the direction of the moment.[5] (See Fig. 44.2.)

Figure 44.2 *Right-Hand Rule*

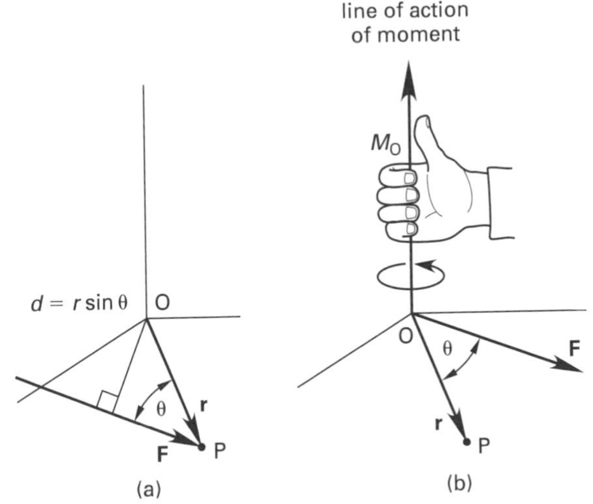

7. VARIGNON'S THEOREM

Varignon's theorem is a statement of how the total moment is derived from a number of forces acting simultaneously at a point. It states that the sum of individual moments about a point caused by multiple concurrent forces is equal to the moment of the resultant force about the same point.

$$(\mathbf{r}\times\mathbf{F}_1)+(\mathbf{r}\times\mathbf{F}_2)+\cdots=\mathbf{r}\times(\mathbf{F}_1+\mathbf{F}_2+\cdots) \quad 44.10$$

Systems of n Forces

$$\mathbf{M}=\sum M_n=\sum r_n\times F_n \quad 44.11$$

8. MOMENT OF A FORCE ABOUT A LINE

Most rotating machines (motors, pumps, flywheels, etc.) have a fixed rotational axis. That is, the machines turn around a line, not around a point. The moment of a force about the rotational axis is not the same as the moment of the force about a point. In particular, the moment about a line is a scalar.[6] (See Fig. 44.3.)

Figure 44.3 *Moment of a Force About a Line*

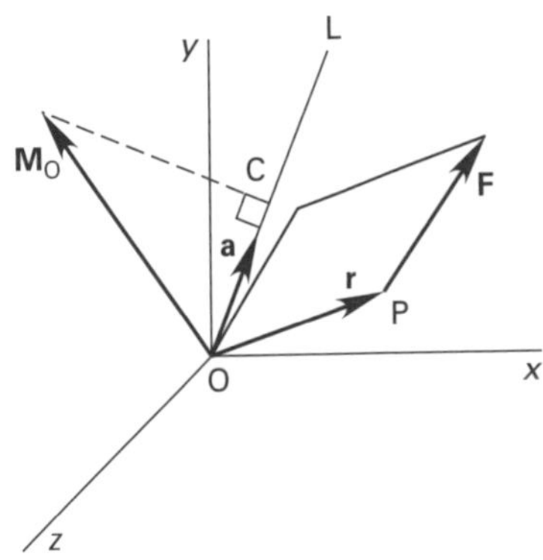

Moment M_{OL} of force **F** about line OL is the projection, OC, of moment $\mathbf{M}_O$ onto the line. Equation 44.12 gives the moment of a force about a line. **a** is the unit vector directed along the line, and a_x, a_y, and a_z are the direction cosines of the axis OL. Notice that Eq. 44.12 is a dot product (i.e., a scalar).

$$M_{OL}=\mathbf{a}\cdot\mathbf{M}_O=\mathbf{a}\cdot(\mathbf{r}\times\mathbf{F}) = \begin{vmatrix} a_x & a_y & a_z \\ x_P-x_O & y_P-y_O & z_P-z_O \\ F_x & F_y & F_z \end{vmatrix} \quad 44.12$$

If point O is the origin, then Eq. 44.12 will reduce to Eq. 44.13.

$$M_{OL}=\begin{vmatrix} a_x & a_y & a_z \\ x & y & z \\ F_x & F_y & F_z \end{vmatrix} \quad 44.13$$

9. COMPONENTS OF A MOMENT

The direction cosines of a force (vector) can be used to determine the components of the moment about the coordinate axes.

$$M_x = M\cos\theta_x \quad 44.14$$

$$M_y = M\cos\theta_y \quad 44.15$$

$$M_z = M\cos\theta_z \quad 44.16$$

Alternatively, the following three equations can be used to determine the components of the moment from a force applied at point (x, y, z) referenced to an origin at $(0, 0, 0)$.

[5]The direction of a moment also corresponds to the direction a right-hand screw would progress if it was turned in the direction that rotates **r** into **F**.
[6]Some sources say that the moment of a force about a line can be interpreted as a moment directed along the line. However, this interpretation does not follow from vector operations.

Moments (Couples)

$$M_x = yF_z - zF_y \quad (44.17)$$

$$M_y = zF_x - xF_z \quad (44.18)$$

$$M_z = xF_y - yF_x \quad (44.19)$$

The resultant moment magnitude can be reconstituted from its components.

$$M = \sqrt{M_x^2 + M_y^2 + M_z^2} \quad (44.20)$$

10. COUPLES

Any pair of equal, opposite, and parallel forces constitutes a *couple*. A couple is equivalent to a single moment vector. Since the two forces are opposite in sign, the x-, y-, and z-components of the forces cancel out. Therefore, a body is induced to rotate without translation. A couple can be counteracted only by another couple. A couple can be moved to any location within the plane without affecting the equilibrium requirements.

In Fig. 44.4, the equal but opposite forces produce a moment vector, $\mathbf{M}_O$, of magnitude Fd. The two forces can be replaced by this moment vector, which can be moved to any location on a body. (Such a moment is known as a *free moment*, *moment of a couple*, or *coupling moment*.)

$$M_O = 2rF \sin\theta = Fd \quad (44.21)$$

Figure 44.4 *Couple*

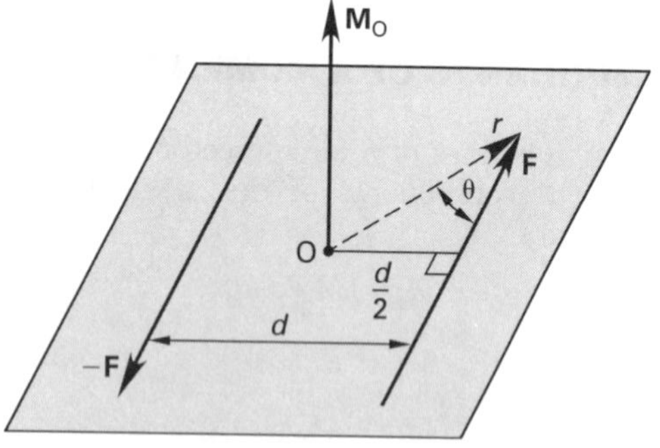

11. EQUIVALENCE OF FORCES AND FORCE-COUPLE SYSTEMS

If a force, F, is moved a distance, d, from the original point of application, a couple, M, equal to Fd must be added to counteract the induced couple. The combination of the moved force and the couple is known as a *force-couple system*. Alternatively, a force-couple system can be replaced by a single force located a distance $d = M/F$ away.

12. RESULTANT FORCE-COUPLE SYSTEMS

The equivalence described in the previous section can be extended to three dimensions and multiple forces. Any collection of forces and moments in three-dimensional space is statically equivalent to a single resultant force vector plus a single resultant moment vector. (Either or both of these resultants can be zero.)

The x-, y-, and z-components of the resultant force are the sums of the x-, y-, and z-components of the individual forces, respectively.

$$F_{R,x} = \sum_i (F \cos\theta_x)_i \quad (44.22)$$

$$F_{R,y} = \sum_i (F \cos\theta_y)_i \quad (44.23)$$

$$F_{R,z} = \sum_i (F \cos\theta_z)_i \quad (44.24)$$

The resultant moment vector is more complex. It includes the moments of all system forces around the reference axes plus the components of all system moments.

$$M_{R,x} = \sum_i (yF_z - zF_y)_i + \sum_i (M \cos\theta_x)_i \quad (44.25)$$

$$M_{R,y} = \sum_i (zF_x - xF_z)_i + \sum_i (M \cos\theta_y)_i \quad (44.26)$$

$$M_{R,z} = \sum_i (xF_y - yF_x)_i + \sum_i (M \cos\theta_z)_i \quad (44.27)$$

13. LINEAR FORCE SYSTEMS

A *linear force system* is one in which all forces are parallel and applied along a straight line. (See Fig. 44.5.) A straight beam loaded by several concentrated forces is an example of a linear force system.

Figure 44.5 *Linear Force System*

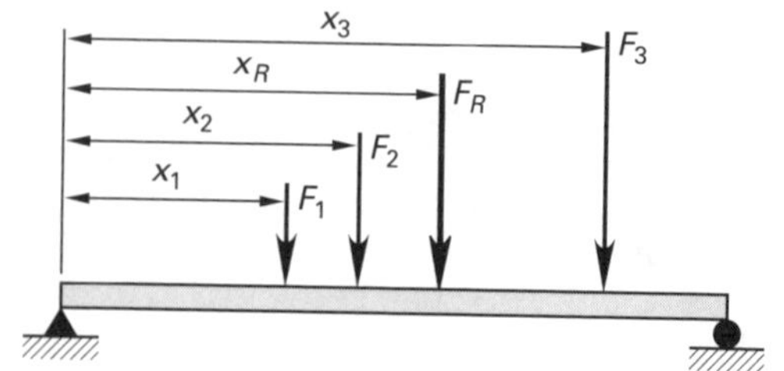

Statics

For the purposes of statics, all of the forces in a linear force system can be replaced by an *equivalent resultant force*, F_R, equal to the sum of the individual forces. The location of the equivalent force coincides with the location of the centroid of the force group.

$$F_R = \sum_i F_i \qquad 44.28$$

$$x_R = \frac{\sum_i F_i x_i}{\sum_i F_i} \qquad 44.29$$

14. DISTRIBUTED LOADS

If an object is continuously loaded over a portion of its length, it is subject to a *distributed load.* Distributed loads result from *dead load* (i.e., self-weight), hydrostatic pressure, and materials distributed over the object.

If the load per unit length at some point x is $w(x)$, the statically equivalent concentrated load, F_R, can be found from Eq. 44.30. The equivalent load is the area under the loading curve. (See Fig. 44.6.)

$$F_R = \int_{x=0}^{x=L} w(x)\,dx \qquad 44.30$$

Figure 44.6 *Distributed Loads on a Beam*

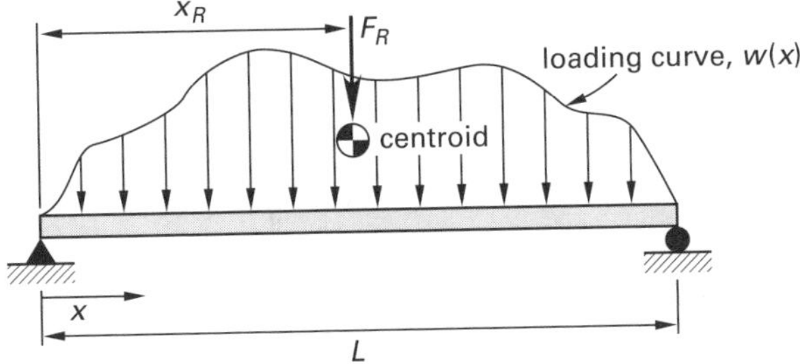

The location, x_R, of the equivalent load is calculated from Eq. 44.31. The location coincides with the centroid of the area under the loading curve and is referred to in some problems as the *center of pressure.*

$$x_R = \frac{\int_{x=0}^{x=L} x\,w(x)\,dx}{F_R} \qquad 44.31$$

For a straight beam of length L under a uniform transverse loading of w pounds per foot (newtons per meter),

$$F_R = wL \qquad 44.32$$

$$x_R = \frac{L}{2} \qquad 44.33$$

Equation 44.34 and Eq. 44.35 can be used for a straight beam of length L under a triangular distribution that increases from zero (at $x = 0$) to w (at $x = L$), as shown in Fig. 44.7.

$$F_R = \frac{wL}{2} \qquad 44.34$$

$$x_R = \frac{2L}{3} \qquad 44.35$$

Figure 44.7 *Special Cases of Distributed Loading*

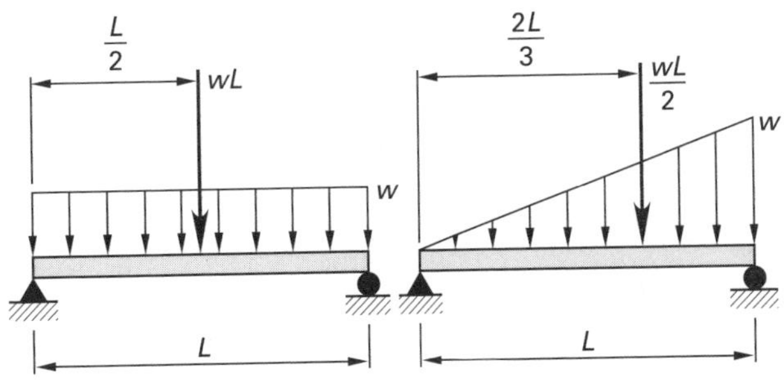

Example 44.1

Find the magnitude and location of the two equivalent forces on the two spans of the beam.

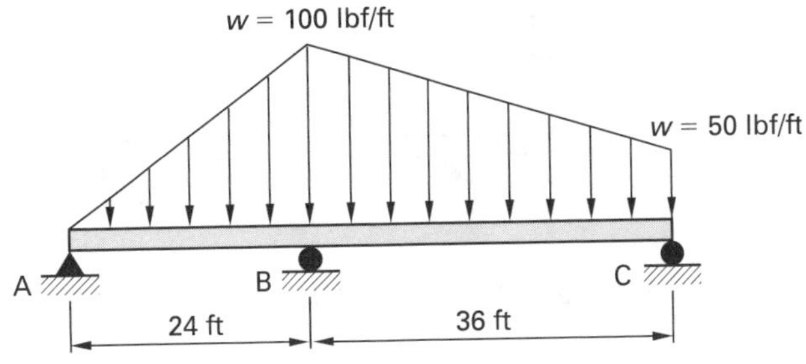

Solution

Span A–B

The area under the triangular loading curve is the total load on the span.

$$\begin{aligned} A &= \tfrac{1}{2}bh = \left(\frac{1}{2}\right)(24\text{ ft})\left(100\ \frac{\text{lbf}}{\text{ft}}\right) \\ &= 1200\text{ lbf} \end{aligned}$$

The centroid of the loading triangle is located at

Properties of Various Shapes

$$\begin{aligned} x_c &= \frac{2b}{3} \\ &= \left(\frac{2}{3}\right)(24\text{ ft}) \\ &= 16\text{ ft} \end{aligned}$$

Statics

Span B–C

The area under the loading curve consists of a uniform load of 50 lbf/ft over the entire span B–C, plus a triangular load that starts at zero at point C and increases to 50 lbf/ft at point B. The area under the loading curve is

$$A = wL + \tfrac{1}{2}bh = \left(50\ \frac{\text{lbf}}{\text{ft}}\right)(36\text{ ft}) + \left(\frac{1}{2}\right)(36\text{ ft})\left(50\ \frac{\text{lbf}}{\text{ft}}\right)$$
$$= 2700\text{ lbf}$$

Reversing the axes to obtain a simpler expression, the centroid of the *trapezoidal loading curve* is located at

Properties of Various Shapes

$$y_c = \frac{h(2a+b)}{3(a+b)}$$
$$= x_{R,B-C} = \left(\frac{36\text{ ft}}{3}\right)\left(\frac{2\left(50\ \frac{\text{lbf}}{\text{ft}}\right) + 100\ \frac{\text{lbf}}{\text{ft}}}{50\ \frac{\text{lbf}}{\text{ft}} + 100\ \frac{\text{lbf}}{\text{ft}}}\right)$$
$$= 16\text{ ft}$$

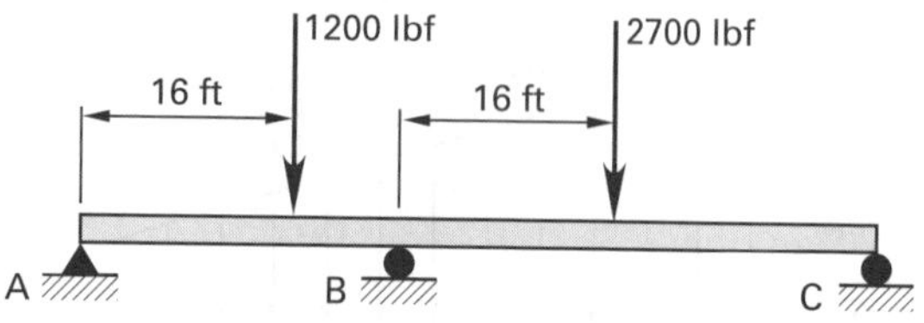

15. MOMENT FROM A DISTRIBUTED LOAD

The total force from a uniformly distributed load, w, over a distance, x, is wx. For the purposes of statics, the uniform load can be replaced by a concentrated force of wx located at the centroid of the distributed load, that is, at the midpoint, $x/2$, of the load. The moment taken about one end of the distributed load is

$$M_{\text{distributed load}} = \text{force} \times \text{distance} = wx\left(\frac{x}{2}\right) = \tfrac{1}{2}wx^2 \quad 44.36$$

In general, the moment of a distributed load, uniform or otherwise, is the product of the total force and the distance to the centroid of the distributed load.

16. TYPES OF FORCE SYSTEMS

The complexity of methods used to analyze a statics problem depends on the configuration and orientation of the forces. Force systems can be divided into the following categories.

- *concurrent force system:* All of the forces act at the same point.
- *collinear force system:* All of the forces share the same line of action.
- *parallel force system:* All of the forces are parallel (though not necessarily in the same direction).
- *coplanar force system:* All of the forces are in a plane.
- *general three-dimensional system:* This category includes all other combinations of nonconcurrent, nonparallel, and noncoplanar forces.

17. CONDITIONS OF EQUILIBRIUM

An object is static when it is stationary. To be stationary, all of the forces on the object must be in equilibrium.[7] For an object to be in equilibrium, the resultant force and moment vectors must both be zero.

Systems of n Forces

$$\mathbf{F} = \sum \mathbf{F}_n \quad 44.37$$

$$\sum \mathbf{F}_n = 0 \quad 44.38$$

$$F_R = \sqrt{F_{R,x}^2 + F_{R,y}^2 + F_{R,z}^2} = 0 \quad 44.39$$

$$\mathbf{M}_R = \sum \mathbf{M} = 0 \quad 44.40$$

$$\mathbf{M} = \sum M_n = \sum r_n F_n \quad 44.41$$

$$\sum \mathbf{M}_n = 0 \quad 44.42$$

$$M_R = \sqrt{M_{R,x}^2 + M_{R,y}^2 + M_{R,z}^2} = 0 \quad 44.43$$

Since the square of any nonzero quantity is positive, Eq. 44.44 through Eq. 44.49 follow directly from Eq. 44.39 and Eq. 44.43.

$$F_{R,x} = 0 \quad 44.44$$

$$F_{R,y} = 0 \quad 44.45$$

$$F_{R,z} = 0 \quad 44.46$$

[7]The term *static equilibrium*, though widely used, is redundant.

$$M_{R,x} = 0 \quad 44.47$$

$$M_{R,y} = 0 \quad 44.48$$

$$M_{R,z} = 0 \quad 44.49$$

Equation 44.44 through Eq. 44.49 seem to imply that six simultaneous equations must be solved in order to determine whether a system is in equilibrium. While this is true for general three-dimensional systems, fewer equations are necessary with most problems. Table 44.1 can be used as a guide to determine which equations are most helpful in solving different categories of problems.

Table 44.1 *Number of Equilibrium Conditions Required to Solve Different Force Systems*

type of force system	two-dimensional	three-dimensional
general	3	6
coplanar	3	3
concurrent	2	3
parallel	2	3
coplanar, parallel	2	2
coplanar, concurrent	2	2
collinear	1	1

18. TWO- AND THREE-FORCE MEMBERS

Members limited to loading by two or three forces are special cases of equilibrium. A *two-force member* can be in equilibrium only if the two forces have the same line of action (i.e., are collinear) and are equal but opposite. In most cases, two-force members are loaded axially, and the line of action coincides with the member's longitudinal axis. By choosing the coordinate system so that one axis coincides with the line of action, only one equilibrium equation is needed.

A *three-force member* can be in equilibrium only if the three forces are concurrent or parallel. Stated another way, the force polygon of a three-force member in equilibrium must close on itself.

19. REACTIONS

The first step in solving most statics problems is to determine the reaction forces (i.e., the *reactions*) supporting the body. The manner in which a body is supported determines the type, location, and direction of the reactions. Conventional symbols are often used to define the type of support (such as pinned, roller, etc.). Examples of the symbols are shown in Table 44.2.

Table 44.2 *Types of Two-Dimensional Supports*

type of support	reactions and moments	number of unknowns*
simple, roller, rocker, ball, or frictionless surface	reaction normal to surface, no moment	1
cable in tension, or link	reaction in line with cable or link, no moment	1
frictionless guide or collar	reaction normal to rail, no moment	1
built-in, fixed support	two reaction components, one moment	3
frictionless hinge, pin connection, or rough surface	reaction in any direction, no moment	2

*The number of unknowns is valid for two-dimensional problems only.

For beams, the two most common types of supports are the roller support and the pinned support. The *roller support*, shown as a cylinder supporting the beam, supports vertical forces only. Rather than support a horizontal force, a roller support simply rolls into a new equilibrium position. Only one equilibrium equation (i.e., the sum of vertical forces) is needed at a roller support. Generally, the terms *simple support* and *simply supported* refer to a roller support.

The *pinned support*, shown as a pin and clevis, supports both vertical and horizontal forces. Two equilibrium equations are needed.

Generally, there will be vertical and horizontal components of a reaction when one body touches another. However, when a body is in contact with a *frictionless surface*, there is no frictional force component parallel to the surface. The reaction is normal to the contact surfaces. The assumption of frictionless contact is particularly useful when dealing with systems of spheres and cylinders in contact with rigid supports. Frictionless contact is also assumed for roller and rocker supports.[8]

20. DETERMINACY

When the equations of equilibrium are independent, a rigid body force system is said to be *statically determinate*. A statically determinate system can be solved for all unknowns, which are usually reactions supporting the body.

When the body has more supports than are necessary for equilibrium, the force system is said to be *statically indeterminate*. In a statically indeterminate system, one or more of the supports or members can be removed or reduced in restraint without affecting the equilibrium position.[9] Those supports and members are known as *redundant members*. The number of redundant members is known as the *degree of indeterminacy*. Figure 44.8 illustrates several common indeterminate structures.

A statically indeterminate body requires additional equations to supplement the equilibrium equations. The additional equations typically involve deflections and depend on mechanical properties of the body.

21. TYPES OF DETERMINATE BEAMS

Figure 44.9 illustrates the terms used to describe determinate beam types.

Figure 44.8 *Examples of Indeterminate Systems*

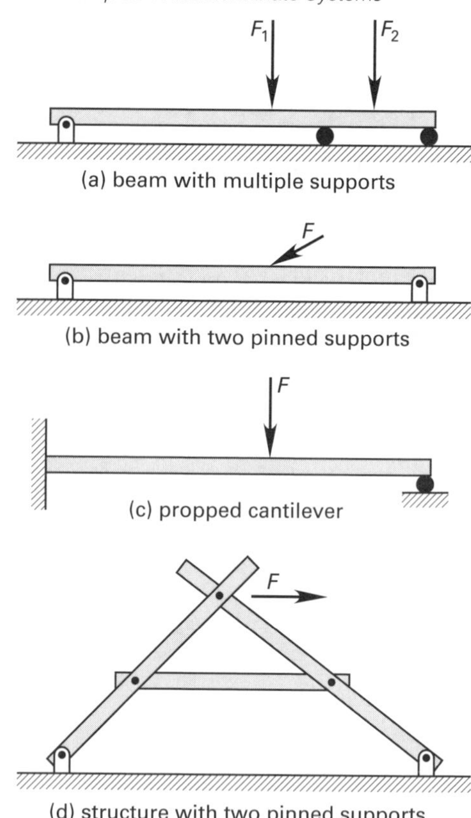

Figure 44.9 *Types of Determinate Beams*

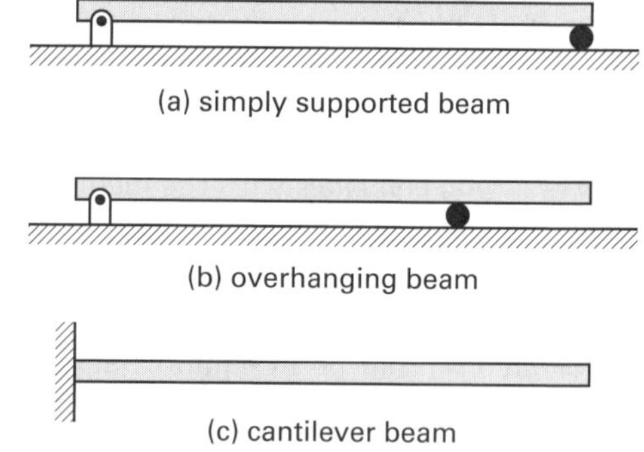

22. FREE-BODY DIAGRAMS

A *free-body diagram* is a representation of a body in equilibrium. It shows all applied forces, moments, and reactions. Free-body diagrams do not consider the internal structure or construction of the body, as Fig. 44.10 illustrates.

[8]Frictionless surface contact, which requires only one equilibrium equation, should not be confused with a frictionless pin connection, which requires two equilibrium equations. A pin connection with friction introduces a moment at the connection, increasing the number of required equilibrium equations to three.
[9]An example of a support reduced in restraint is a pinned joint replaced by a roller joint. The pinned joint restrains the body vertically and horizontally, requiring two equations of equilibrium. The roller joint restrains the body vertically only and requires one equilibrium equation.

Figure 44.10 *Bodies and Free Bodies*

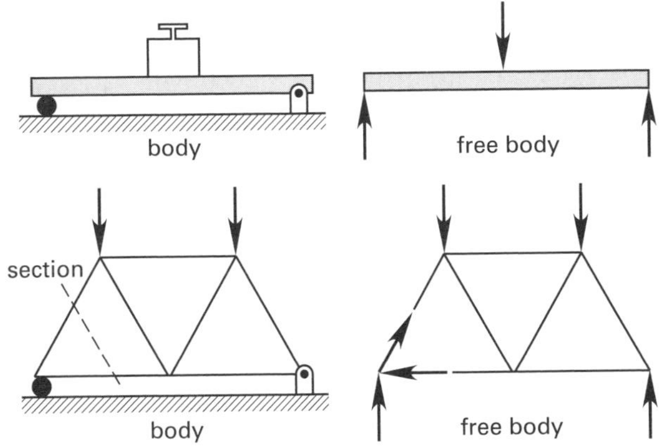

Since the body is in equilibrium, the resultants of all forces and moments on the free body are zero. In order to maintain equilibrium, any portions of the body that are removed must be replaced by the forces and moments those portions impart to the body. Typically, the body is isolated from its physical supports in order to help evaluate the reaction forces. In other cases, the body may be sectioned (i.e., cut) in order to determine the forces at the section.

23. FINDING REACTIONS IN TWO DIMENSIONS

The procedure for finding determinate reactions in two-dimensional problems is straightforward. Determinate structures will have either a roller support and a pinned support or two roller supports.

step 1: Establish a convenient set of coordinate axes. (To simplify the analysis, one of the coordinate directions should coincide with the direction of the forces and reactions.)

step 2: Draw the free-body diagram.

step 3: Resolve the reaction at the pinned support (if any) into components normal and parallel to the coordinate axes.

step 4: Establish a positive direction of rotation (e.g., clockwise) for the purpose of taking moments.

step 5: Write the equilibrium equation for moments about the pinned connection. (By choosing the pinned connection as the point about which to take moments, the pinned connection reactions do not enter into the equation.) This will usually determine the vertical reaction at the roller support.

step 6: Write the equilibrium equation for the forces in the vertical direction. Usually, this equation will have two unknown vertical reactions.

step 7: Substitute the known vertical reaction from step 5 into the equilibrium equation from step 6. This will determine the second vertical reaction.

step 8: Write the equilibrium equation for the forces in the horizontal direction. Since there is a maximum of one unknown reaction component in the horizontal direction, this step will determine that component.

step 9: If necessary, combine the vertical and horizontal force components at the pinned connection into a resultant reaction.

Example 44.2

Determine the reactions, R_1 and R_2, on the following beam.

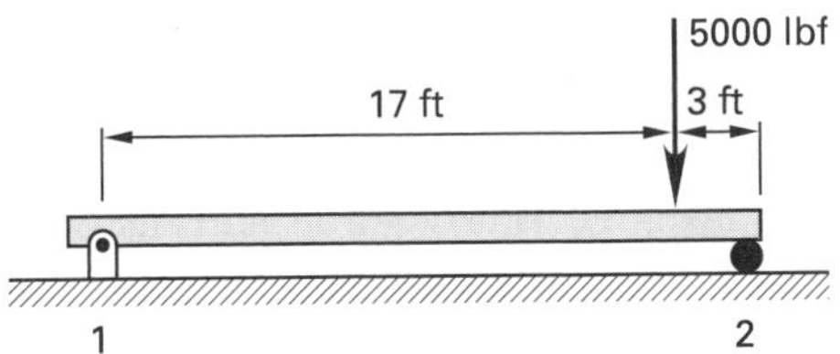

Solution

step 1: The x- and y-axes are established parallel and perpendicular to the beam.

step 2: The free-body diagram is

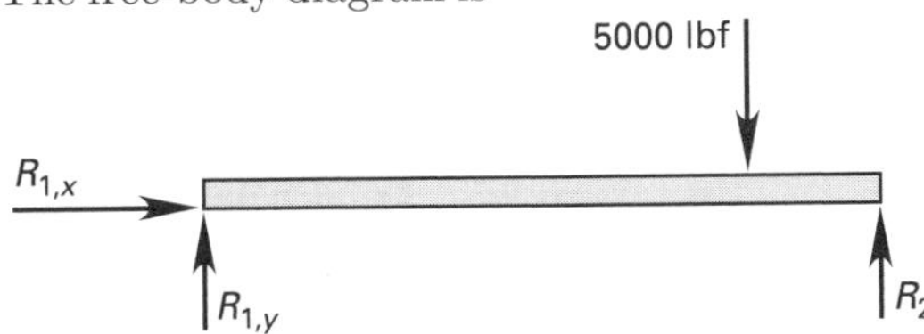

step 3: R_1 is a pinned support, so it has two components, $R_{1,x}$ and $R_{1,y}$.

step 4: Assume clockwise moments are positive.

step 5: Take moments about the left end and set them equal to zero. Use Eq. 44.40.

$$\sum M_{\text{left end}} = (5000 \text{ lbf})(17 \text{ ft}) - R_2(20 \text{ ft}) = 0$$
$$R_2 = 4250 \text{ lbf}$$

step 6: The equilibrium equation for the vertical direction is given by Eq. 44.38.

$$\sum F_y = R_{1,y} + R_2 - 5000 \text{ lbf} = 0$$

step 7: Substituting R_2 into the vertical equilibrium equation,

$$R_{1,y} + 4250 \text{ lbf} - 5000 \text{ lbf} = 0$$
$$R_{1,y} = 750 \text{ lbf}$$

Statics

step 8: There are no applied forces in the horizontal direction. Therefore, the equilibrium equation is given by Eq. 44.38.

$$\sum F_x = R_{1,x} + 0 = 0$$
$$R_{1,x} = 0$$

24. COUPLES AND FREE MOMENTS

Once a couple on a body is known, the derivation and source of the couple are irrelevant. When the moment on a body is 80 N·m, it makes no difference whether the force is 40 N with a lever arm of 2 m, or 20 N with a lever arm of 4 m, and so on. Therefore, the point of application of a couple is disregarded when writing the moment equilibrium equation. For this reason, the term *free moment* is used synonymously with *couple*.

Figure 44.11 illustrates two diagrammatic methods of indicating the application of a free moment.

Figure 44.11 *Free Moments*

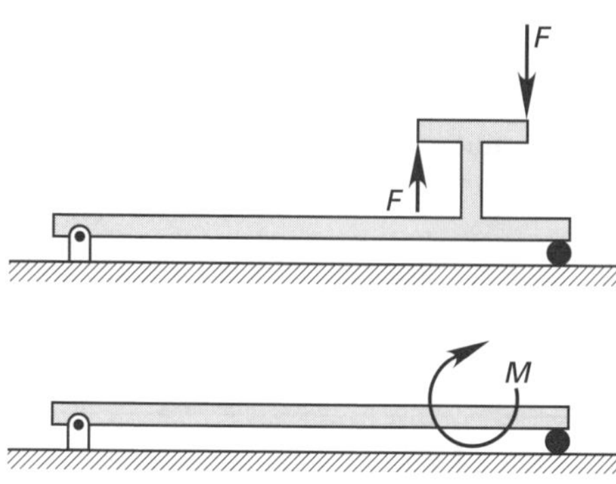

Example 44.3

What is the reaction, R_2, for the beam shown?

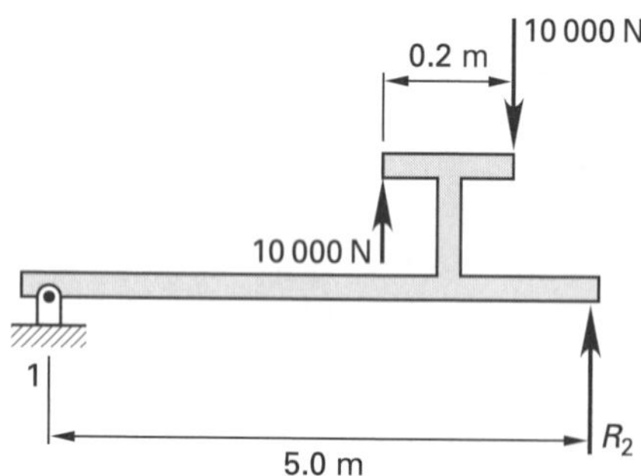

Solution

The two couple forces are equal and cancel each other as they come down the stem of the tee bracket. There are no applied vertical forces.

The couple has a value given by Eq. 44.8.

Moments (Couples)

$$\mathbf{M} = \mathbf{r} \times \mathbf{F}$$

$$M = rF = (0.2 \text{ m})(10\,000 \text{ N}) = 2000 \text{ N·m} \quad \text{[clockwise]}$$

Choose clockwise as the direction for positive moments. Taking moments about the pinned connection and using Eq. 44.42,

Systems of n Forces

$$\sum \mathbf{M}_n = 0$$

$$\sum M = 2000 \text{ N·m} - R_2(5 \text{ m}) = 0$$
$$R_2 = 400 \text{ N}$$

25. INFLUENCE LINES FOR REACTIONS

An *influence line* (also known as an *influence graph* and *influence diagram*) is a graph of the magnitude of a reaction as a function of the load placement.[10] The x-axis of the graph corresponds to the location on the body (along the length of a beam). The y-axis corresponds to the magnitude of the reaction.

By convention (and to generalize the graph for use with any load), the load is taken as one force unit. Therefore, for an actual load of F units, the actual reaction, R, is the product of the actual load and the influence line ordinate.

$$R = F \times \text{influence line ordinate} \qquad 44.50$$

Example 44.4

Draw the influence line for the left reaction for the beam shown.

Solution

If the unit load is at the left end, the left reaction will be 1.0. If the unit load is at the right end, it will be supported entirely by the right reaction, so the left reaction will be zero. The influence line for the left reaction varies linearly for intermediate load placement.

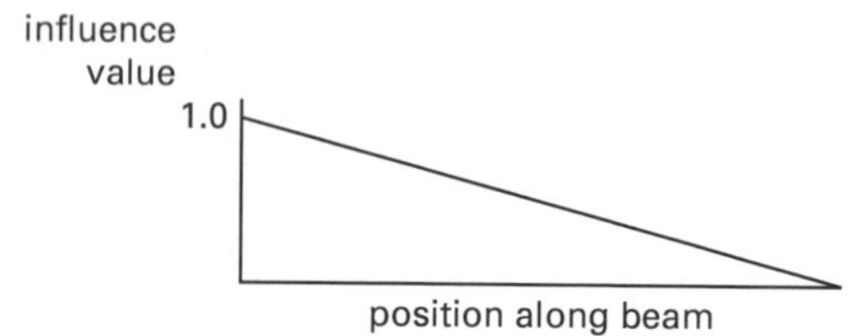

[10]Influence diagrams can also be drawn for moments, shears, and deflections.

Statics

26. HINGES

Hinges are added to structures to prevent translation while permitting rotation. A frictionless hinge can support a force, but it cannot transmit a moment. Since the moment is zero at a hinge, a structure can be sectioned at the hinge and the remainder of the structure can be replaced by only a force.

Example 44.5

Calculate the reaction, R_3, and the hinge force on the two-span beam shown.

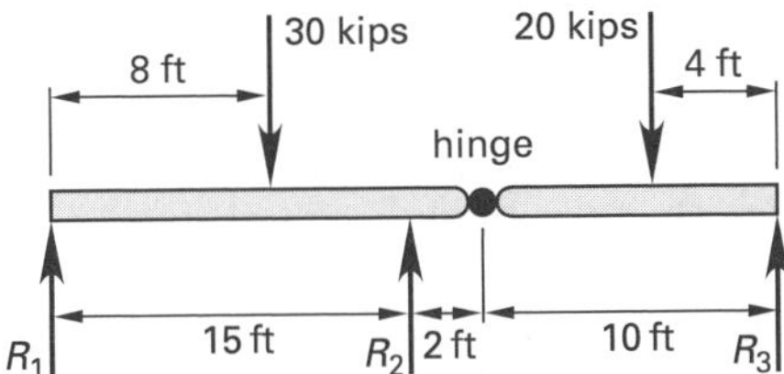

Solution

At first, this beam may appear to be statically indeterminate since it has three supports. However, the moment is known to be zero at the hinge, so the hinged portion of the span can be isolated.

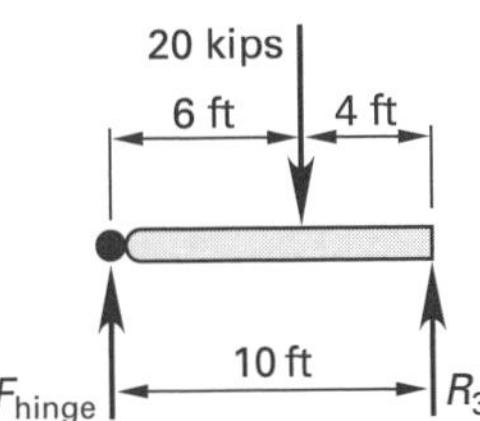

Reaction R_3 is found by taking moments about the hinge. Assume clockwise moments are positive.

$$\sum M_{\text{hinge}} = (20{,}000 \text{ lbf})(6 \text{ ft}) - R_3(10 \text{ ft}) = 0$$
$$R_3 = 12{,}000 \text{ lbf}$$

The hinge force is found by summing vertical forces on the isolated section.

$$\sum F_y = F_{\text{hinge}} + 12{,}000 \text{ lbf} - 20{,}000 \text{ lbf} = 0$$
$$F_{\text{hinge}} = 8000 \text{ lbf}$$

27. LEVERS

A *lever* is a simple mechanical machine able to increase an applied force. The ratio of the load-bearing force to applied force (i.e., the *effort*) is known as the *mechanical advantage* or *force amplification*. As Fig. 44.12 shows, the mechanical advantage is equal to the ratio of lever arms.

$$\begin{aligned} \text{mechanical advantage} &= \frac{F_{\text{load}}}{F_{\text{applied}}} \\ &= \frac{\text{applied force lever arm}}{\text{load lever arm}} \\ &= \frac{\text{distance moved by applied force}}{\text{distance moved by load}} \end{aligned} \quad 44.51$$

Figure 44.12 *Lever*

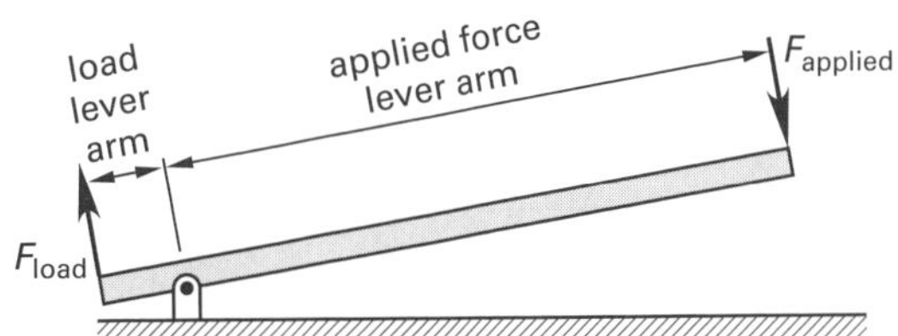

28. PULLEYS

A *pulley* (also known as a *sheave*) is used to change the direction of an applied tensile force. A series of pulleys working together (known as a *block and tackle*) can also provide *pulley advantage* (i.e., *mechanical advantage*). A *hoist* is any device used to raise or lower an object. A hoist may contain one or more pulleys.

If the pulley is attached by a bracket or cable to a fixed location, it is said to be a *fixed pulley*. If the pulley is attached to a load, or if the pulley is free to move, it is known as a *free pulley*.

Most simple problems disregard friction and assume that all ropes are parallel.[11] In such cases, the pulley advantage is equal to the number of ropes coming to and going from the load-carrying pulley. The diameters of the pulleys are not factors in calculating the pulley advantage. (See Table 44.3.)

In other cases, a *loss factor*, ϵ, is used to account for rope rigidity. For most wire ropes and chains with 180° contact, the loss factor at low speeds varies between 1.03 and 1.06. The loss factor is the reciprocal of the *pulley efficiency*, η.

$$\epsilon = \frac{\text{applied force}}{\text{load}} = \frac{1}{\eta} \quad 44.52$$

[11]Although the term *rope* is used here, the principles apply equally well to wire rope, cables, chains, belts, and so on.

Table 44.3 Mechanical Advantages of Rope-Operated Machines

	fixed sheave	free sheave	ordinary pulley block (*n* sheaves)	differential pulley block
F_{ideal}	W	$\frac{W}{2}$	$\frac{W}{n}$	$\left(\frac{W}{2}\right)\left(1-\frac{d}{D}\right)$
F to raise load	ϵW	$\frac{\epsilon W}{1+\epsilon}$	$\frac{\epsilon^n(\epsilon-1)W}{\epsilon^n-1}$	$\frac{\left(\epsilon^2-\frac{d}{D}\right)W}{1+\epsilon}$
F to lower load	$\frac{W}{\epsilon}$	$\frac{W}{1+\epsilon}$	$\left(\frac{\frac{1}{\epsilon}-1}{1-\epsilon^n}\right)W$	$\left(\frac{\epsilon W}{1+\epsilon}\right)\left(\frac{1}{\epsilon^2}-\frac{d}{D}\right)$
ratio of distance of force to distance of load	1	2	n	$\frac{2D}{D-d}$

29. AXIAL MEMBERS

An *axial member* is capable of supporting axial forces only and is loaded only at its joints (i.e., ends). This type of performance can be achieved through the use of frictionless bearings or smooth pins at the ends. Since the ends are assumed to be pinned (i.e., rotation-free), an axial member cannot support moments. The weight of the member is disregarded or is included in the joint loading.

An axial member can be in either tension or compression. It is common practice to label forces in axial members as (T) or (C) for tension or compression, respectively. Alternatively, tensile forces can be written as positive numbers, while compressive forces are written as negative numbers.

The members in simple trusses are assumed to be axial members. Each member is identified by its endpoints, and the force in a member is designated by the symbol for the two endpoints. For example, the axial force in a member connecting points C and D will be written as **CD**. Similarly, $\mathbf{EF}_y$ is the y-component of the force in the member connecting points E and F.

For equilibrium, the resultant forces at the two joints must be equal, opposite, and collinear. This applies to the total (resultant) force as well as to the x- and y-components at those joints.

30. FORCES IN AXIAL MEMBERS

The line of action of a force in an axial member coincides with the longitudinal axis of the member. Depending on the orientation of the coordinate axis system, the direction of the longitudinal axis will have both x- and y-components. Therefore, the force in an axial member will generally have both x- and y-components.

The following four general principles are helpful in determining the force in an axial member.

- A horizontal member carries only horizontal loads. It cannot carry vertical loads.
- A vertical member carries only vertical loads. It cannot carry horizontal loads.
- The vertical component of an axial member's force is equal to the vertical component of the load applied to the member.
- The total and component forces in an inclined member are proportional to the sides of the triangle outlined by the member and the coordinate axes.[12]

Example 44.6

Member BC is an inclined axial member pinned at B and sliding frictionless at C, and oriented as shown. A vertical 1000 N force is applied to the top end. What are the x- and y-components of the force in member BC? What is the total force in the member?

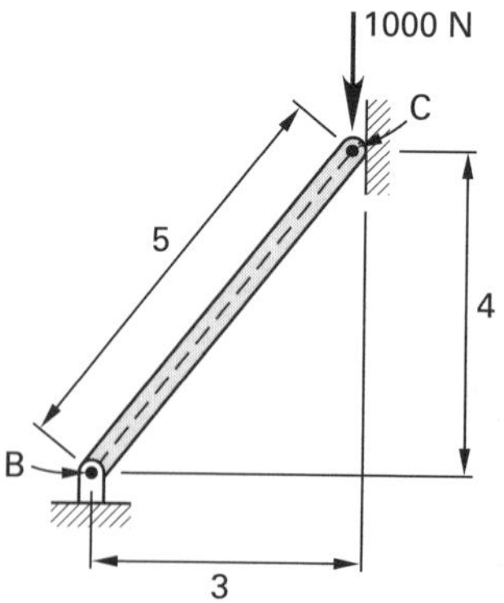

[12]This is an application of the principle of similar triangles.

Solution

From the third principle,

$$\mathbf{BC}_y = 1000 \text{ N}$$

From the fourth principle,

$$\mathbf{BC}_x = \tfrac{3}{4}\mathbf{BC}_y = \left(\frac{3}{4}\right)(1000 \text{ N}) = 750 \text{ N}$$

The resultant force in member BC can be calculated from the Pythagorean theorem. However, it is easier to use the fourth principle.

$$\mathbf{BC} = \tfrac{5}{4}\mathbf{BC}_y = \left(\frac{5}{4}\right)(1000 \text{ N}) = 1250 \text{ N}$$

31. TRUSSES

A *truss* or *frame* is a set of *pin-connected axial members* (i.e., *two-force members*). The connection points are known as *joints*. Member weights are disregarded, and truss loads are applied only at joints. A *structural cell* consists of all members in a closed loop of members. For the truss to be stable (i.e., to be a *rigid truss*), all of the structural cells must be triangles. Figure 44.13 identifies *chords*, *end posts*, *panels*, and other elements of a typical *bridge truss*.

Figure 44.13 *Parts of a Bridge Truss*

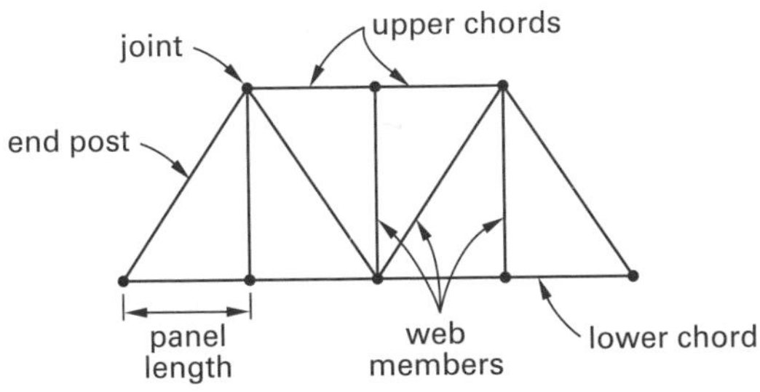

A *trestle* is a braced structure spanning a ravine, gorge, or other land depression in order to support a road or rail line. Trestles usually are indeterminate, have multiple earth contact points, have redundant members, and are more difficult to evaluate than simple trusses.

Several types of trusses have been given specific names. Some of the more common types of named trusses are shown in Fig. 44.14.

Figure 44.14 *Special Types of Trusses*

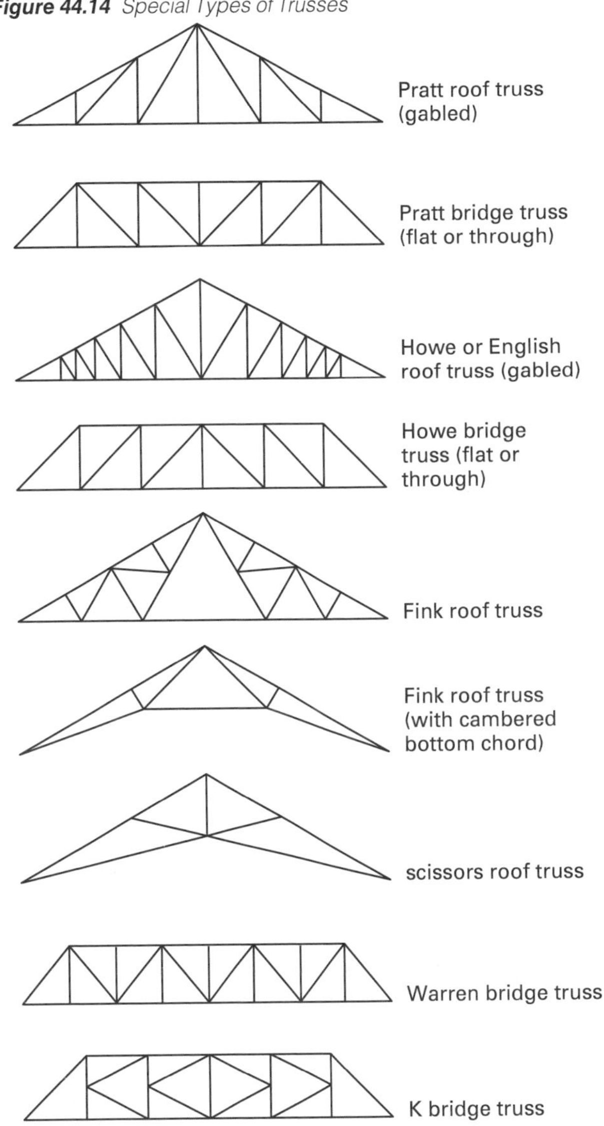

Truss loads are considered to act only in the plane of a truss. Trusses are analyzed as two-dimensional structures. Forces in truss members hold the various truss parts together and are known as *internal forces*. The internal forces are found by applying equations of equilibrium to appropriate free-body diagrams.

Although free-body diagrams of truss members can be drawn, this is not usually done. Instead, free-body diagrams of the pins (i.e., the joints) are drawn. A pin in compression will be shown with force arrows pointing toward the pin, away from the member. (Similarly, a pin in tension will be shown with force arrows pointing away from the pin, toward the member.)[13]

Statics

[13] The method of showing tension and compression on a truss drawing may appear incorrect. This is because the arrows show the forces on the pins, not on the members.

With typical bridge trusses supported at the ends and loaded downward at the joints, the upper chords are almost always in compression, and the end panels and lower chords are almost always in tension.

32. DETERMINATE TRUSSES

A truss will be statically determinate if Eq. 44.53 holds.

$$\text{no. of members} = 2(\text{no. of joints}) - 3 \qquad 44.53$$

If the left-hand side is greater than the right-hand side (i.e., there are *redundant members*), the truss is statically indeterminate. If the left-hand side is less than the right-hand side, the truss is unstable and will collapse under certain types of loading.

Equation 44.53 is a special case of the following general criterion.

$$\begin{aligned} &\text{no. of members} \\ &\quad + \text{no. of reactions} \\ &\qquad - 2(\text{no. of joints}) = 0 \quad [\text{determinate}] \\ &\qquad\qquad > 0 \quad [\text{indeterminate}] \\ &\qquad\qquad < 0 \quad [\text{unstable}] \end{aligned} \qquad 44.54$$

Furthermore, Eq. 44.53 is a necessary, but not sufficient, condition for truss stability. It is possible to arrange the members in such a manner as to not contribute to truss stability. This will seldom be the case in actual practice, however.

33. ZERO-FORCE MEMBERS

Forces in truss members can sometimes be determined by inspection. One of these cases is where there are *zero-force members*. A third member framing into a joint already connecting two collinear members carries no internal force unless there is a load applied at that joint. Similarly, both members forming an apex of the truss are zero-force members unless there is a load applied at the apex. (See Fig. 44.15.)

Figure 44.15 *Zero-Force Members*

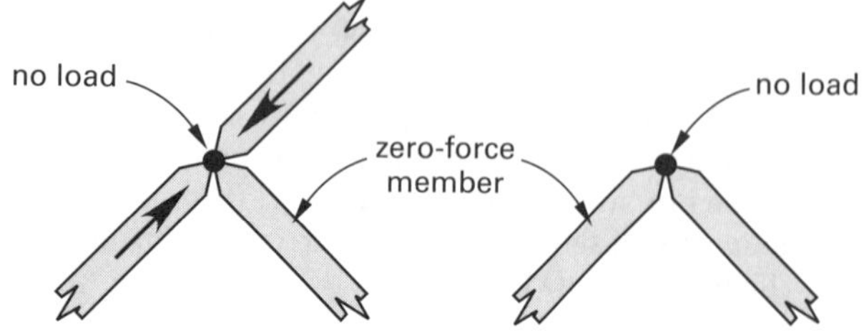

34. METHOD OF JOINTS

The *method of joints* is one of three methods that can be used to find the internal forces in each truss member. This method is useful when most or all of the truss member forces are to be calculated. Because this method advances from joint to adjacent joint, it is inconvenient when a single isolated member force is to be calculated.

The method of joints is a direct application of the equations of equilibrium in the x- and y-directions. Traditionally, the method starts by finding the reactions supporting the truss. Next, the joint at one of the reactions is evaluated, which determines all the member forces framing into the joint. Then, knowing one or more of the member forces from the previous step, an adjacent joint is analyzed. The process is repeated until all the unknown quantities are determined.

At a joint, there may be up to two unknown member forces, each of which can have dependent x- and y-components.[14] Since there are two equilibrium equations, the two unknown forces can be determined. Even though determinate, however, the sense of a force will often be unknown. If the sense cannot be determined by logic, an arbitrary decision can be made. If the incorrect direction is chosen, the calculated force will be negative.

Example 44.7

Use the method of joints to calculate the force **BD** in the truss shown.

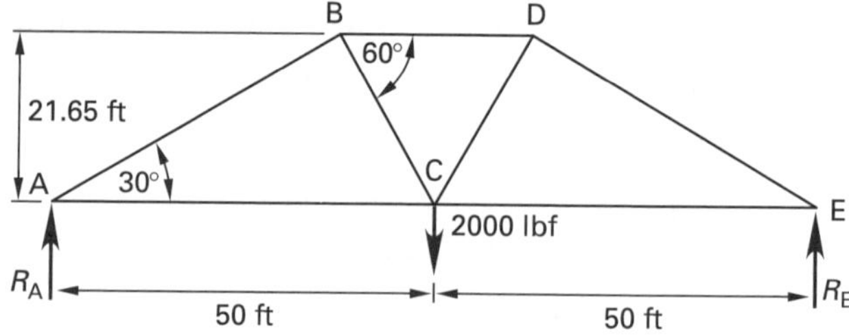

Solution

First, find the reactions. Assume clockwise is positive and take moments about point A.

$$\begin{aligned} \sum M_{\text{A}} &= (2000 \text{ lbf})(50 \text{ ft}) - R_{\text{E}}(50 \text{ ft} + 50 \text{ ft}) = 0 \\ R_{\text{E}} &= 1000 \text{ lbf} \end{aligned}$$

Since the sum of forces in the y-direction is also zero,

$$\begin{aligned} \sum F_y &= R_{\text{A}} + 1000 \text{ lbf} - 2000 \text{ lbf} = 0 \\ R_{\text{A}} &= 1000 \text{ lbf} \end{aligned}$$

There are three unknowns at joint B (and also at D). The analysis must start at joint A (or E) where there are only two unknowns (forces **AB** and **AC**).

[14]Occasionally, there will be three unknown member forces. In that case, an additional equation must be derived from an adjacent joint.

The free-body diagram of pin A is shown. The direction of R_A is known to be upward. The directions of forces **AB** and **AC** can be assumed, but logic can be used to determine them. Only the vertical component of **AB** can oppose R_A. Therefore, **AB** is directed downward. (This means that member AB is in compression.) Similarly, **AC** must oppose the horizontal component of **AB**, so **AC** is directed to the right. (This means that member AC is in tension.)

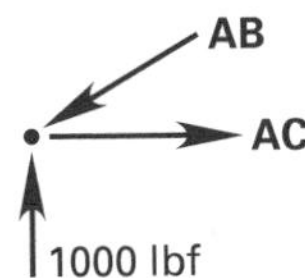

Resolve force **AB** into horizontal and vertical components using trigonometry, direction cosines, or similar triangles. (R_A and **AC** are already parallel to an axis.) Then, use the equilibrium equations to determine the forces.

By inspection, $AB_y = 1000$ lbf.

$$\begin{aligned} \mathbf{AB}_y &= \mathbf{AB}\sin 30° \\ 1000 \text{ lbf} &= \mathbf{AB}(0.5) \\ \mathbf{AB} &= 2000 \text{ lbf} \quad \text{(C)} \\ \mathbf{AB}_x &= \mathbf{AB}\cos 30° \\ &= (2000 \text{ lbf})(0.866) \\ &= 1732 \text{ lbf} \end{aligned}$$

Now, draw the free-body diagram of pin B. (Notice that the direction of force **AB** is toward the pin, just as it was for pin A.) Although the true directions of the forces are unknown, they can be determined logically. The direction of force **BC** is chosen to counteract the vertical component of force **AB**. The direction of force **BD** is chosen to counteract the horizontal components of forces **AB** and **BC**.

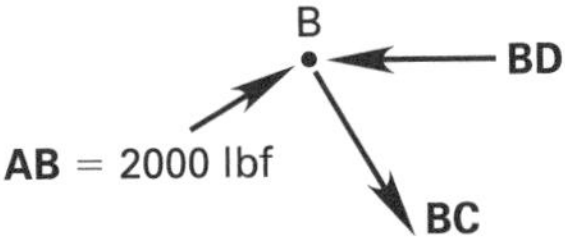

$\mathbf{AB}_x$ and $\mathbf{AB}_y$ are already known. Resolve the force **BC** into horizontal and vertical components.

$$\begin{aligned} \mathbf{BC}_x &= \mathbf{BC}\sin 30° = \mathbf{BC}(0.5) \\ \mathbf{BC}_y &= \mathbf{BC}\cos 30° = \mathbf{BC}(0.866) \end{aligned}$$

Now, write the equations of equilibrium for point B.

$$\begin{aligned} \sum F_x &= 1732 \text{ lbf} + 0.5\mathbf{BC} - \mathbf{BD} = 0 \\ \sum F_y &= 1000 \text{ lbf} - 0.866\mathbf{BC} = 0 \end{aligned}$$

From the second equation, **BC** = 1155 lbf. Substituting this into the first equation,

$$\begin{aligned} 1732 \text{ lbf} + (0.5)(1155 \text{ lbf}) - \mathbf{BD} &= 0 \\ \mathbf{BD} &= 2310 \text{ lbf} \quad \text{(C)} \end{aligned}$$

Since **BD** turned out to be positive, its direction was chosen correctly.

The direction of the arrow indicates that the member is compressing the pin. Consequently, the pin is compressing the member. Member BD is in compression.

If the process is continued, all forces can be determined. However, the truss is symmetrical, and it is not necessary to evaluate every joint to calculate all forces.

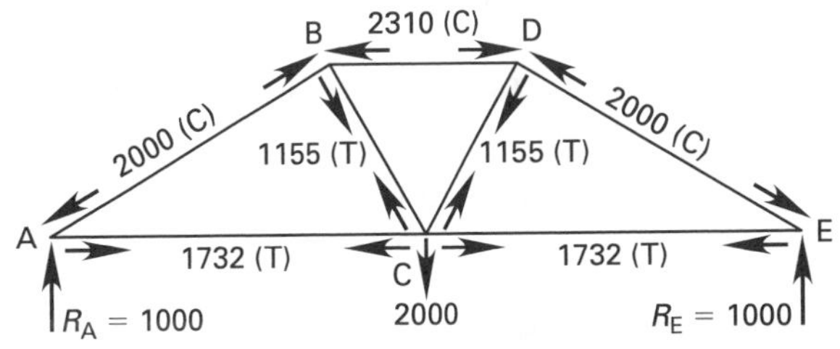

35. CUT-AND-SUM METHOD

The *cut-and-sum method* can be used to find forces in inclined members. This method is strictly an application of the vertical equilibrium condition ($\sum F_y = 0$).

The method starts by finding all of the support reactions on a truss. Then, a cut is made through the truss in such a way as to pass through one inclined or vertical member only. (At this point, it should be clear that the vertical component of the inclined member must balance all of the external vertical forces.) The equation for vertical equilibrium is written for the free body of the remaining truss portion.

Example 44.8

Find the force in member BC for the truss in Ex. 44.7.

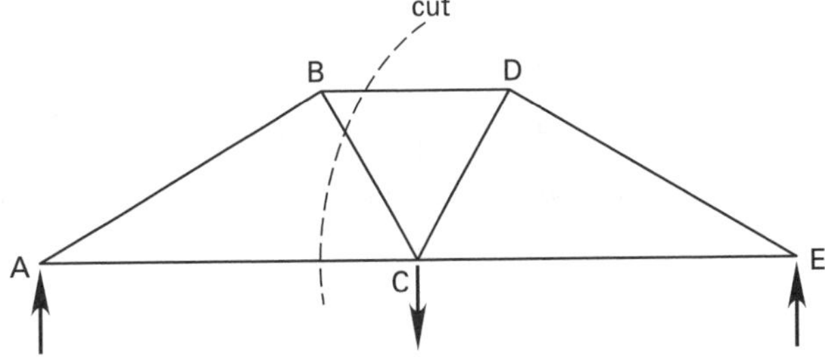

Statics

Solution

The reactions were determined in Ex. 44.7. The truss is cut in such a way as to pass through member BC but through no other inclined member. The free body of the remaining portion of the truss is

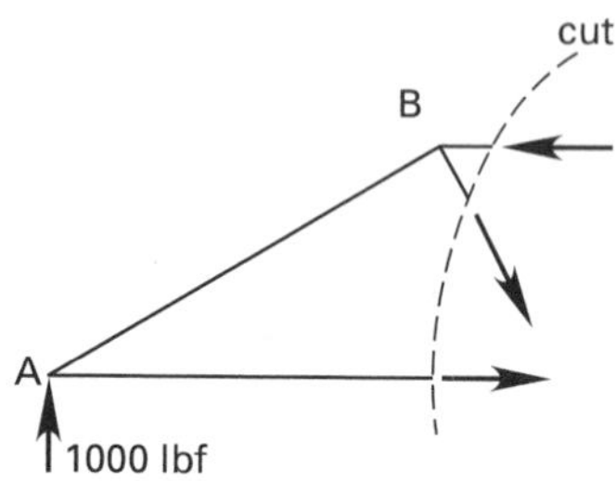

The vertical equilibrium equation is

$$\sum F_y = R_A - \mathbf{BC}_y = 1000 \text{ lbf} - 0.866\mathbf{BC} = 0$$

$$\mathbf{BC} = 1155 \text{ lbf} \quad \text{(T)}$$

36. METHOD OF SECTIONS

The *method of sections* is a direct approach to finding forces in any truss member. This method is convenient when only a few truss member forces are unknown.

As with the previous two methods, the first step is to find the support reactions. Then, a cut is made through the truss, passing through the unknown member.[15] Finally, all three conditions of equilibrium are applied as needed to the remaining truss portion. (Since there are three equilibrium equations, the cut cannot pass through more than three members in which the forces are unknown.)

Example 44.9

Find the forces in members CD and CE. The support reactions have already been determined.

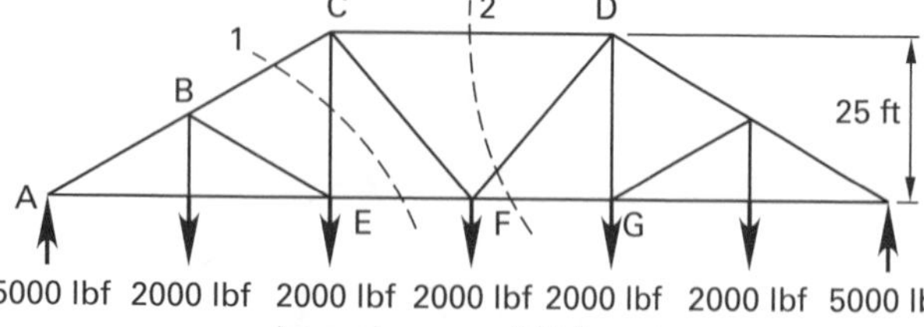

Solution

To find the force **CE**, the truss is cut at section 1.

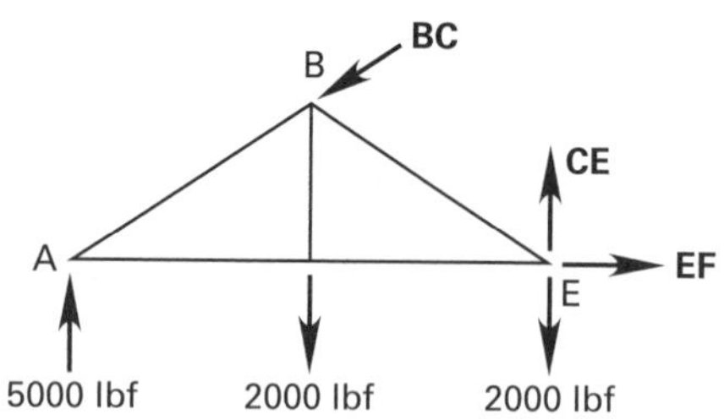

Taking moments about point A will eliminate all of the unknown forces except **CE**. Assume clockwise moments are positive.

$$\sum M_A = (2000 \text{ lbf})(20 \text{ ft}) + (2000 \text{ lbf})(40 \text{ ft}) - (40 \text{ ft})\mathbf{CE} = 0$$

$$\mathbf{CE} = 3000 \text{ lbf} \quad \text{(T)}$$

To find the force **CD**, the truss is cut at section 2.

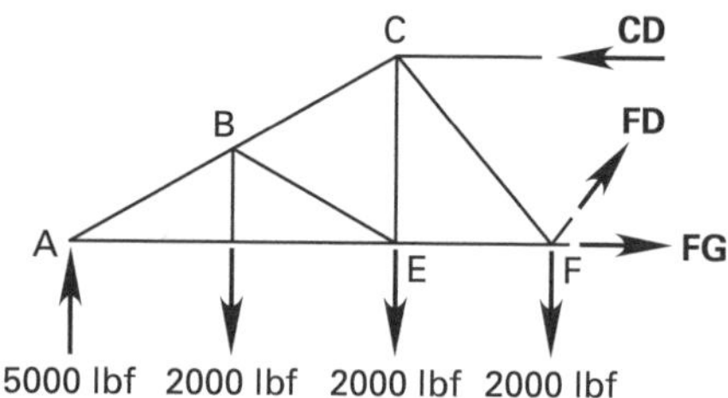

Taking moments about point F will eliminate all unknowns except **CD**. Assume clockwise moments are positive.

$$\sum M_F = (5000 \text{ lbf})(60 \text{ ft}) - (25 \text{ ft})\mathbf{CD} - (2000 \text{ lbf})(20 \text{ ft}) - (2000 \text{ lbf})(40 \text{ ft}) = 0$$

$$\mathbf{CD} = 7200 \text{ lbf} \quad \text{(C)}$$

37. SUPERPOSITION OF LOADS

Superposition is a term used to describe the process of determining member forces by considering loads one at a time. Suppose, for example, that the force in member FG is unknown and that the truss carries three loads. If the method of superposition is used, the force in member FG (call it $\mathbf{FG}_1$) is determined with only the first load acting on the truss. $\mathbf{FG}_2$ and $\mathbf{FG}_3$ are similarly found. The true member force **FG** is found by adding $\mathbf{FG}_1$, $\mathbf{FG}_2$, and $\mathbf{FG}_3$.

[15] Knowing where to cut the truss is the key part of this method. Such knowledge is developed only by practice.

Superposition should be used with discretion since trusses can change shape under load. If a truss deflects such that the load application points are significantly different from those in the undeflected truss, superposition cannot be used for that truss.

In simple truss analysis, change of shape under load is neglected. Superposition, therefore, can be assumed to apply.

38. TRANSVERSE TRUSS MEMBER LOADS

Truss members are usually designed as axial members, not as beams. Trusses are traditionally considered to be loaded at joints only. Figure 44.16, however, illustrates cases of nontraditional *transverse loading* that can actually occur. For example, a truss member's own weight would contribute to a uniform load, as would a severe ice buildup.

Figure 44.16 *Transverse Truss Member Loads*

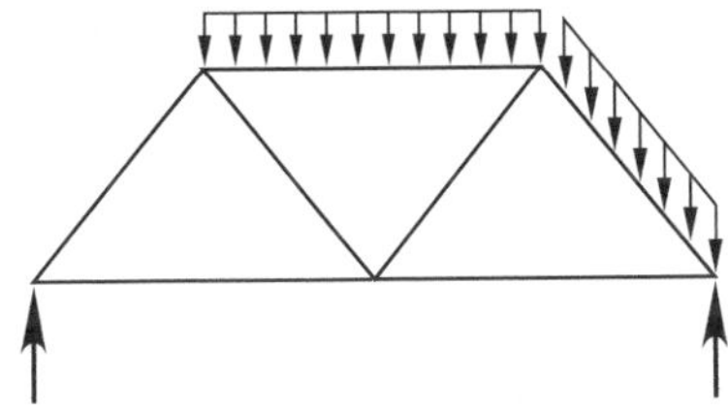

Transverse loads add two solution steps to a truss problem. First, the truss member must be individually considered as a beam simply supported at its pinned connections, and the reactions needed to support the transverse loading must be found. These reactions become additional loads applied to the truss joints, and the truss can then be evaluated in the normal manner.

The second step is to check the structural adequacy (deflection, bending stress, shear stress, buckling, etc.) of the truss member under transverse loading.

39. CABLES CARRYING CONCENTRATED LOADS

An *ideal cable* is assumed to be completely flexible, massless, and incapable of elongation. It acts as an axial two-force tension member between points of concentrated loading. In fact, the term *tension* or *tensile force* is commonly used in place of "member force" when dealing with cables. (See Fig. 44.17.)

Figure 44.17 *Cable with Concentrated Load*

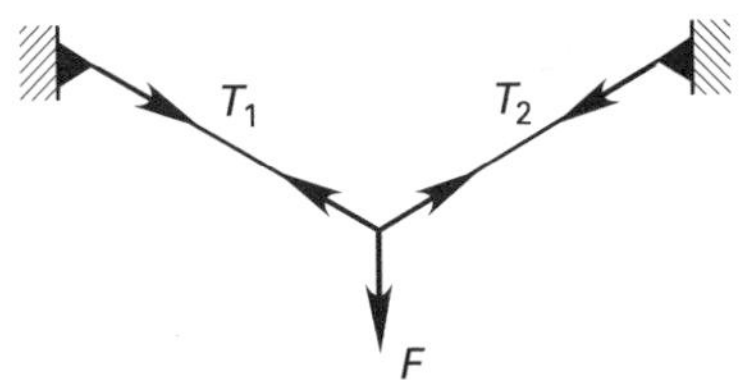

The methods of joints and sections used in truss analysis can be used to determine the tensions in cables carrying concentrated loads. After separating the reactions into x- and y-components, it is particularly useful to sum moments about one of the reaction points. All cables will be found to be in tension, and (with vertical loads only) the horizontal tension component will be the same in all cable segments. Unlike the case of a rope passing over a series of pulleys, however, the total tension in the cable will not be the same in every cable segment.

Example 44.10

What are the tensions **AB**, **BC**, and **CD**?

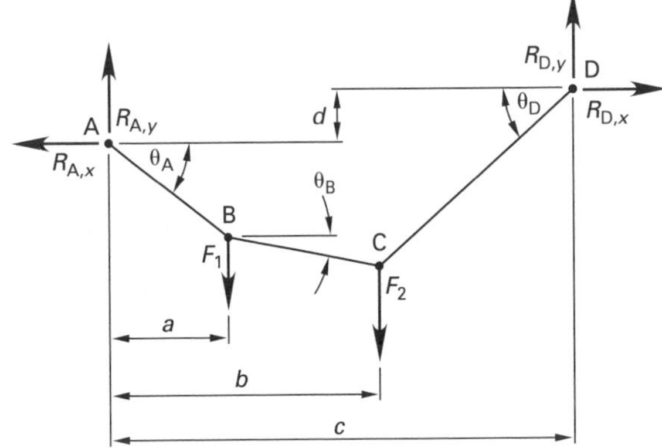

Solution

Separate the two reactions into x- and y-components. (The total reactions R_A and R_D are also the tensions **AB** and **CD**, respectively.)

$$R_{A,x} = -\mathbf{AB}\cos\theta_A$$
$$R_{A,y} = -\mathbf{AB}\sin\theta_A$$
$$R_{D,x} = -\mathbf{CD}\cos\theta_D$$
$$R_{D,y} = -\mathbf{CD}\sin\theta_D$$

Next, take moments about point A to find **CD**. Assume clockwise to be positive.

$$\sum M_A = aF_1 + bF_2 - d\mathbf{CD}_x + c\mathbf{CD}_y = 0$$

None of the applied loads are in the x-direction, so the only horizontal loads are the x-components of the reactions. To find tension **AB**, take the entire cable as a free body. Then, sum the external forces in the x-direction.

$$\sum F_x = R_{D,x} - R_{A,x} = 0$$

$$\mathbf{CD}\cos\theta_D = \mathbf{AB}\cos\theta_A$$

The x-component of force is the same in all cable segments. To find **BC**, sum the x-direction forces at point B.

$$\sum F_x = \mathbf{BC}_x - \mathbf{AB}_x = 0$$

$$\mathbf{BC}\cos\theta_B = \mathbf{AB}\cos\theta_A$$

Statics

40. PARABOLIC CABLES

If the distributed load per unit length, w, on a cable is constant with respect to the horizontal axis (as is the load from a bridge floor), the cable will be parabolic in shape.[16] This is illustrated in Fig. 44.18.

Figure 44.18 Parabolic Cable

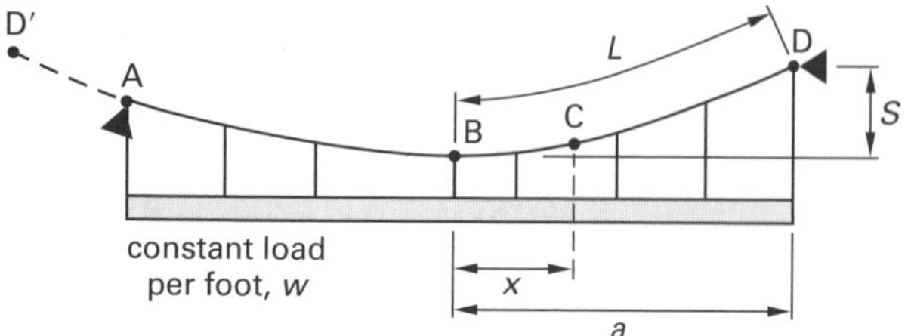

The maximum sag is designated as S. If the location of the maximum sag (i.e., the lowest cable point) is known, the horizontal component of tension, H, can be found by taking moments about a reaction point. If the cable is cut at the maximum sag point, B, the cable tension on the free body will be horizontal since there is no vertical component to the cable. Cutting the cable in Fig. 44.18 at point B and taking moments about point D will determine the minimum cable tension, H.

$$\sum M_\text{D} = wa\left(\frac{a}{2}\right) - HS = 0 \quad 44.55$$

$$H = \frac{wa^2}{2S} \quad 44.56$$

$$w = mg \quad \text{[SI]} \quad 44.57(a)$$

$$w = \frac{mg}{g_c} \quad \text{[U.S.]} \quad 44.57(b)$$

Since the load is vertical everywhere, the horizontal component of tension is constant everywhere in the cable. The tension, T_C, at any point C can be found by applying the equilibrium conditions to the cable segment BC.

$$T_{\text{C},x} = H = \frac{wa^2}{2S} \quad 44.58$$

$$T_{\text{C},y} = wx \quad 44.59$$

$$T_\text{C} = \sqrt{T_{\text{C},x}^2 + T_{\text{C},y}^2} = w\sqrt{\left(\frac{a^2}{2S}\right)^2 + x^2} \quad 44.60$$

The angle of the cable at any point is

$$\tan\theta = \frac{wx}{H} \quad 44.61$$

The tension and angle are maximum at the supports.

If the lowest sag point, point B, is used as the origin, the shape of the cable is

$$y(x) = \frac{wx^2}{2H} \quad 44.62$$

The approximate length of the cable from the lowest point to the support (i.e., length BD) is

$$L \approx a\left(1 + \frac{2}{3}\left(\frac{S}{a}\right)^2 - \frac{2}{5}\left(\frac{S}{a}\right)^4\right) \quad 44.63$$

Example 44.11

A pedestrian foot bridge has two suspension cables and a flexible floor weighing 28 lbf/ft. The span of the bridge is 100 ft. When the bridge is empty, the tension at point C is 1500 lbf. Assuming a parabolic shape, what is the maximum cable sag, S?

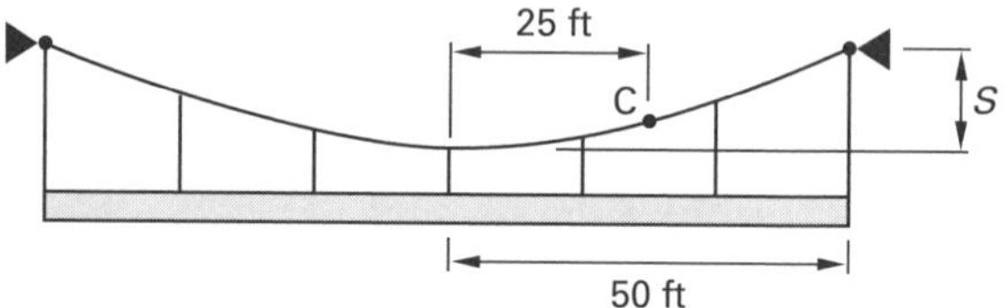

Solution

Since there are two cables, the floor weight per suspension cable is

$$w = \frac{28\ \frac{\text{lbf}}{\text{ft}}}{2} = 14\ \text{lbf/ft}$$

From Eq. 44.60,

$$T_\text{C} = w\sqrt{\left(\frac{a^2}{2S}\right)^2 + x^2}$$

$$1500\ \text{lbf} = 14\ \frac{\text{lbf}}{\text{ft}}\sqrt{\left(\frac{(50\ \text{ft})^2}{2S}\right)^2 + (25\ \text{ft})^2}$$

$$S = 12\ \text{ft}$$

[16]The parabolic case can also be assumed with cables loaded only by their own weight (e.g., telephone and trolley wires) if both ends are at the same elevations and if the sag is no more than 10% of the distance between supports.

41. CABLES CARRYING DISTRIBUTED LOADS

An idealized tension cable with a distributed load is similar to a linkage made up of a very large number of axial members. The cable is an axial member in the sense that the internal tension acts tangentially to the cable everywhere.

Since the load is vertical everywhere, the horizontal component of cable tension is constant along the cable. The cable is horizontal at the point of lowest sag. There is no vertical tension component, and the cable tension is minimum. By similar reasoning, the cable tension is maximum at the supports.

Figure 44.19 illustrates a general cable with a distributed load. The shape of the cable will depend on the relative distribution of the load. A free-body diagram of segment BC is also shown.

Figure 44.19 *Cable with Distributed Load*

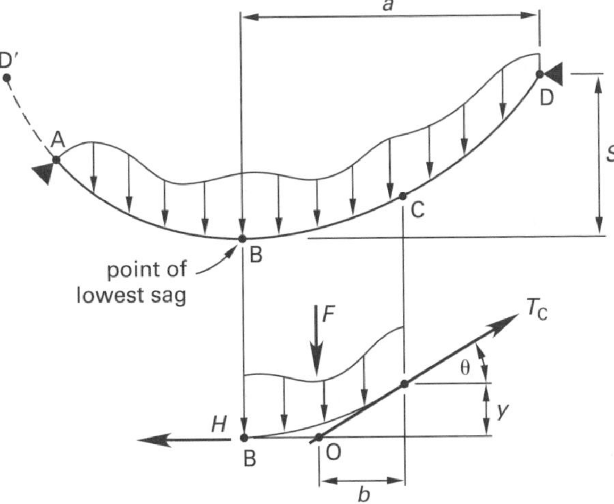

F is the resultant of the distributed load on segment BC, T is the cable tension at point C, and H is the tension at the point of lowest sag (i.e., the point of minimum tension). Since segment BC taken as a free body is a three-force member, the three forces (H, F, and T) must be concurrent to be in equilibrium. The horizontal component of tension can be found by taking moments about point C.

$$\sum M_C = Fb - Hy = 0 \qquad 44.64$$

$$H = \frac{Fb}{y} \qquad 44.65$$

Also, $\tan\theta = y/b$. So,

$$H = \frac{F}{\tan\theta} \qquad 44.66$$

The basic equilibrium conditions can be applied to the free-body cable segment BC to determine the tension in the cable at point C.

$$\sum F_x = T_C\cos\theta - H = 0 \qquad 44.67$$

$$\sum F_y = T_C\sin\theta - F = 0 \qquad 44.68$$

The resultant tension at point C is

$$T_C = \sqrt{H^2 + F^2} \qquad 44.69$$

42. CATENARY CABLES

If the distributed load is constant along the length of the cable, as it is with a loose cable loaded by its own weight, the cable will have the shape of a *catenary*. A vertical axis catenary's shape is determined by Eq. 44.70, where c is a constant and cosh is the *hyperbolic cosine*. The quantity x/c is in radians.[17]

$$y(x) = c\cosh\frac{x}{c} \qquad 44.70$$

Referring to Fig. 44.20, the vertical distance, y, to any point C on the catenary is measured from a reference plane located a distance c below the point of greatest sag, point B. The distance c is known as the *parameter of the catenary*. Although the value of c establishes the location of the x-axis, the value of c does not correspond to any physical distance, nor is the reference plane the ground level.

Figure 44.20 *Catenary Cable*

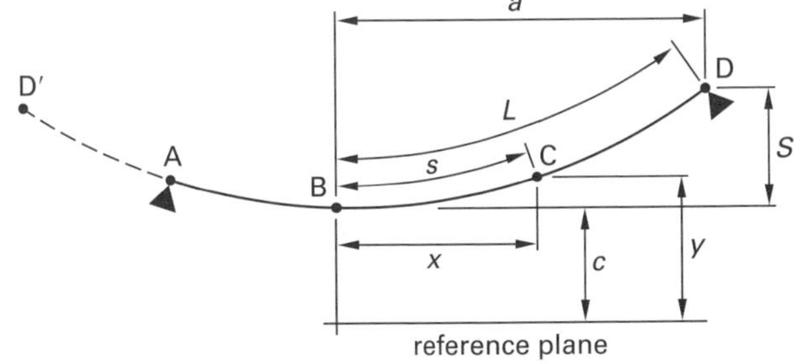

In order to define the cable shape and determine cable tensions, it is necessary to have enough information to calculate c. For example, if a and S are known, Eq. 44.73

[17]In order to use Eq. 44.70 through Eq. 44.73, you must reset your calculator from degrees to radians.

Statics

can be solved by trial and error for c.[18] Once c is known, the cable geometry and forces are determined by the remaining equations.

For any point C, the equations most useful in determining the shape of the catenary are

$$y = \sqrt{s^2 + c^2} = c\cosh\frac{x}{c} \tag{44.71}$$

$$s = c\sinh\frac{x}{c} \tag{44.72}$$

$$\text{sag} = S = y_\text{D} - c = c\left(\cosh\frac{a}{c} - 1\right) \tag{44.73}$$

$$\tan\theta = \frac{s}{c} \tag{44.74}$$

The equations most useful in determining the cable tensions are

$$H = wc \tag{44.75}$$

$$F = ws \tag{44.76}$$

$$T = wy \tag{44.77}$$

$$\tan\theta = \frac{ws}{H} \tag{44.78}$$

$$\cos\theta = \frac{H}{T} \tag{44.79}$$

Example 44.12

A cable 100 m long is loaded by its own weight. The maximum sag is 25 m, and the supports are on the same level. What is the distance between the supports?

Solution

Since the two supports are on the same level, the cable length, L, between the point of maximum sag and support D is half of the total length.

$$L = \frac{100\text{ m}}{2} = 50\text{ m}$$

Combining Eq. 44.71 and Eq. 44.73 for point D (with $S = 25$ m),

$$y_\text{D} = c + S = \sqrt{L^2 + c^2}$$

$$c + 25\text{ m} = \sqrt{(50\text{ m})^2 + c^2}$$

$$c = 37.5\text{ m}$$

Substituting a for x and $L = 50$ for s in Eq. 44.72,

$$s = c\sinh\frac{x}{c}$$

$$50\text{ m} = (37.5\text{ m})\sinh\frac{a}{37.5}$$

$$a = 41.2\text{ m}$$

The distance between supports is

$$2a = (2)(41.2\text{ m}) = 82.4\text{ m}$$

43. CABLES WITH ENDS AT DIFFERENT ELEVATIONS

A cable will be asymmetrical if its ends are at different elevations. In some cases, as shown in Fig. 44.21, the cable segment will not include the lowest point, B. However, if the location of the theoretical lowest point can be derived, the positions and elevations of the cable supports will not affect the analysis. The same procedure is used in proceeding from theoretical point B to either support. In fact, once the theoretical shape of a cable has been determined, the supports can be relocated anywhere along the cable line without affecting the equilibrium of the supported segment.

Figure 44.21 *Asymmetrical Segment of Symmetrical Cable*

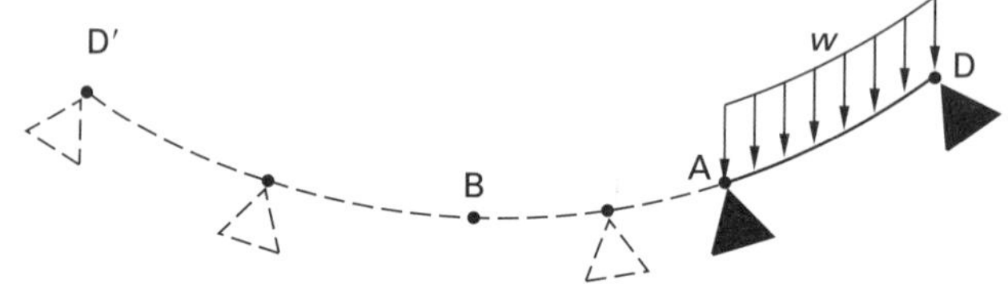

44. TWO-DIMENSIONAL MECHANISMS

A two-dimensional *mechanism* (*machine*) is a nonrigid structure. Although parts of the mechanism move, the relationships between forces in the mechanism can be determined by statics. In order to determine an unknown force, one or more of the mechanism components must be considered as a free body. All input forces and reactions must be included on this free body. In general, the resultant force on such a free body will not be in the direction of the member.

Several free bodies may be needed for complicated mechanisms. Sign conventions of acting and reacting forces must be strictly adhered to when determining the effect of one component on another.

[18]Because obtaining the solution may require trial and error, it will be advantageous to assume a parabolic shape if the cable is taut. (See Ftn. 16.) The error will generally be small.

Example 44.13

A 70 N·m couple is applied to the mechanism shown. All connections are frictionless hinges. What are the x- and y-components of the reactions at B?

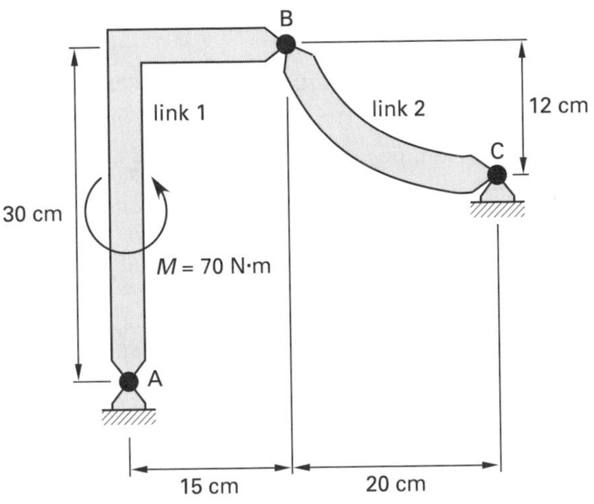

Solution

Isolate links 1 and 2 and draw their free bodies.

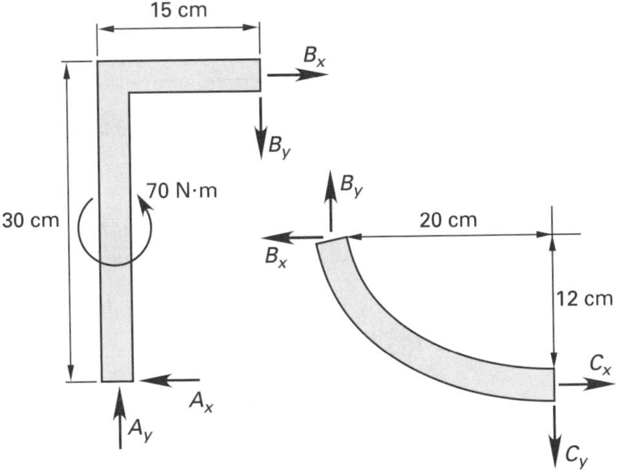

Assume clockwise moments are positive. Take moments about point A on link 1.

$$\sum M_A = B_x(0.3\ \text{m}) + B_y(0.15\ \text{m}) - 70\ \text{N·m} = 0$$

Assume clockwise moments are positive. Take moments about point C on link 2.

$$\sum M_C = B_y(0.20\ \text{m}) - B_x(0.12\ \text{m}) = 0$$

Solving these two equations simultaneously determines the force at joint B.

$$B_x = 179\ \text{N}$$

$$B_y = 108\ \text{N}$$

45. EQUILIBRIUM IN THREE DIMENSIONS

The basic equilibrium equations can be used with vector algebra to solve a three-dimensional statics problem. When a manual calculation is required, however, it is often more convenient to write the equilibrium equations for one orthogonal direction at a time, thereby avoiding the use of vector notation and reducing the problem to two dimensions. The following method can be used to analyze a three-dimensional structure.

step 1: Establish the $(0, 0, 0)$ origin for the structure.

step 2: Determine the (x, y, z) coordinates of all load and reaction points.

step 3: Determine the x-, y-, and z-components of all loads and reactions. This is accomplished by using direction cosines calculated from the (x, y, z) coordinates.

step 4: Draw a *coordinate free-body diagram* of the structure for each of the three coordinate axes. Include only forces, reactions, and moments that affect the coordinate free body.

step 5: Apply the basic two-dimensional equilibrium equations.

Example 44.14

Beam AC is supported at point A by a frictionless ball joint and at points B and C by cables. A 100 lbf load is applied vertically to point C, and a 180 lbf load is applied horizontally to point B. What are the cable tensions, T_1 and T_2?

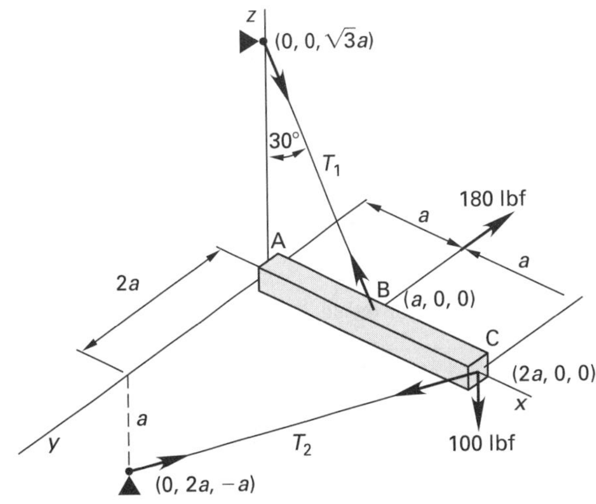

Solution

step 1: Point A has already been established as the origin.

step 2: The locations of all support and load points are shown on the illustration.

Statics

step 3: By inspection, for the 180 lbf horizontal load at point B,

$$F_x = 0$$

$$F_y = -180 \text{ lbf}$$

$$F_z = 0$$

By inspection, for the 100 lbf vertical load at point C,

$$F_x = 0$$

$$F_y = 0$$

$$F_z = -100 \text{ lbf}$$

The length of cable 1 is

$$\begin{aligned} L_1 &= \sqrt{x^2 + y^2 + z^2} \\ &= \sqrt{(a-0)^2 + (0-0)^2 + (0-\sqrt{3}\,a)^2} \\ &= 2a \end{aligned}$$

The direction cosines of the force from cable 1 at point B are

$$\cos\theta_x = \frac{d_x}{L_1} = \frac{0-a}{2a} = -0.5$$

$$\cos\theta_y = \frac{d_y}{L_1} = \frac{0-0}{2a} = 0$$

$$\cos\theta_z = \frac{d_z}{L_1} = \frac{\sqrt{3}\,a-0}{2a} = 0.866$$

Therefore, the components of the tension in cable 1 are

$$T_{1,x} = -0.5\,T_1$$

$$T_{1,y} = 0$$

$$T_{1,z} = 0.866\,T_1$$

Similarly, for cable 2,

$$\begin{aligned} L_2 &= \sqrt{x^2 + y^2 + z^2} \\ &= \sqrt{(2a-0)^2 + (0-2a)^2 + \big(0-(-a)\big)^2} \\ &= 3a \end{aligned}$$

The direction cosines for the force from cable 2 at point C are

$$\cos\theta_x = \frac{d_x}{L_2} = \frac{0-2a}{3a} = -0.667$$

$$\cos\theta_y = \frac{d_y}{L_2} = \frac{2a-0}{3a} = 0.667$$

$$\cos\theta_z = \frac{d_z}{L_2} = \frac{-a-0}{3a} = -0.333$$

Therefore, the components of the tension in cable 2 are

$$T_{2,x} = -0.667\,T_2$$

$$T_{2,y} = 0.667\,T_2$$

$$T_{2,z} = -0.333\,T_2$$

step 4: The three coordinate free-body diagrams are

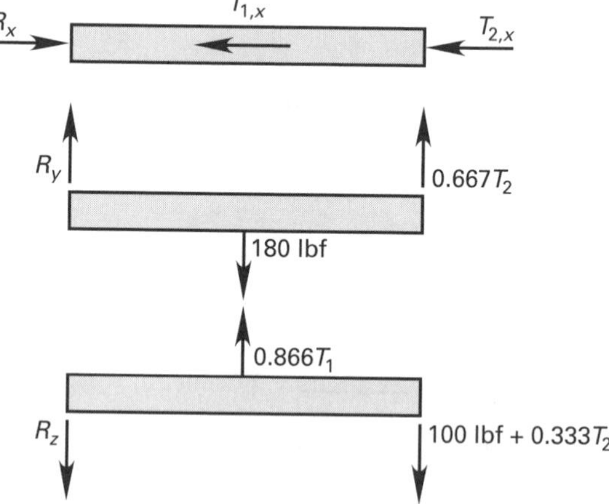

step 5: Tension T_2 can be found by taking moments about point A on the *y*-coordinate free body.

$$\begin{aligned} \sum M_A &= (0.667\,T_2)2a - (180 \text{ lbf})a = 0 \\ T_2 &= 135 \text{ lbf} \end{aligned}$$

Tension T_1 can be found by taking moments about point A on the *z*-coordinate free body.

$$\begin{aligned} \sum M_A &= (0.866\,T_1)a - (0.333)(135 \text{ lbf})2a \\ &\quad - (100 \text{ lbf})2a = 0 \\ T_1 &= 335 \text{ lbf} \end{aligned}$$

46. TRIPODS

A *tripod* is a simple three-dimensional truss (frame) that consists of three axial members. (See Fig. 44.22.) One end of each member is connected at the *apex* of the tripod, while the other ends are attached to the supports. All connections are assumed to allow free rotation in all directions.

Statics

Figure 44.22 Tripod

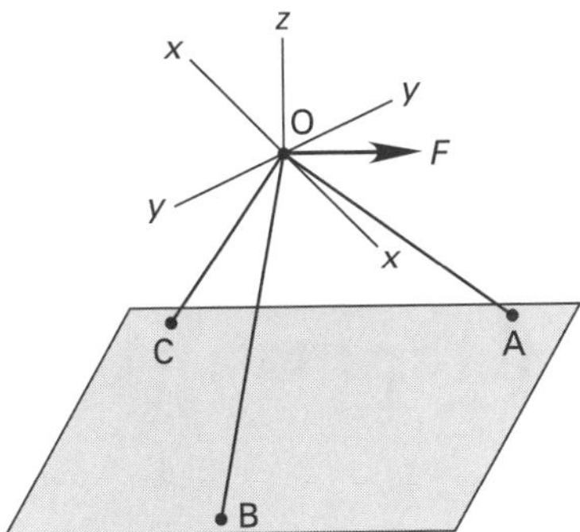

The general solution procedure given in the preceding section can be made more specific for tripods.

step 1: Establish the apex as the origin.

step 2: Determine the x-, y-, and z-components of the force applied to the apex.

step 3: Determine the (x, y, z) coordinates of points A, B, and C—the three support points.

step 4: Determine the length of each tripod leg from the coordinates of the support points.

$$L = \sqrt{x^2 + y^2 + z^2} \qquad 44.80$$

step 5: Determine the direction cosines for the leg forces at the apex. For leg A, for example,

$$\cos\theta_{A,x} = \frac{x_A}{L} \qquad 44.81$$

$$\cos\theta_{A,y} = \frac{y_A}{L} \qquad 44.82$$

$$\cos\theta_{A,z} = \frac{z_A}{L} \qquad 44.83$$

step 6: Write the x-, y-, and z-components of each leg force in terms of the direction cosines. For leg A, for example,

$$F_{A,x} = F_A \cos\theta_{A,x} \qquad 44.84$$

$$F_{A,y} = F_A \cos\theta_{A,y} \qquad 44.85$$

$$F_{A,z} = F_A \cos\theta_{A,z} \qquad 44.86$$

step 7: Write the three sum-of-forces equilibrium equations for the apex.

$$F_{A,x} + F_{B,x} + F_{C,x} + F_x = 0 \qquad 44.87$$

$$F_{A,y} + F_{B,y} + F_{C,y} + F_y = 0 \qquad 44.88$$

$$F_{A,z} + F_{B,z} + F_{C,z} + F_z = 0 \qquad 44.89$$

Example 44.15

Determine the force in each leg of the tripod.

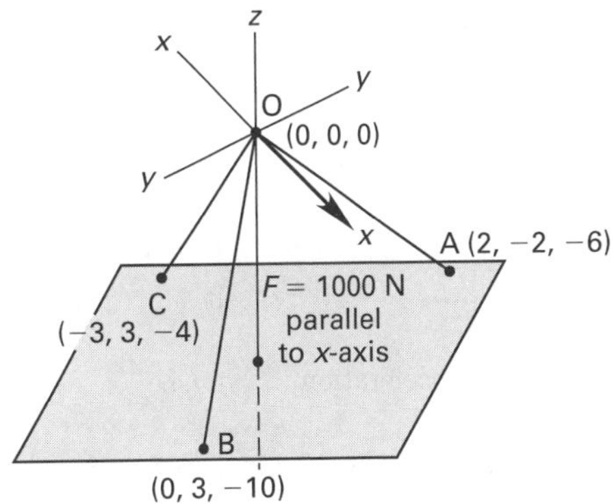

Solution

step 1: The origin is at the apex.

step 2: By inspection, $F_x = +1000$ N. All other components are zero.

step 3: The coordinates of the tripod support are given.

step 4: The lengths and direction cosines for the tripod legs have been calculated and are presented in the following table.

member	x^2	y^2	z^2	L^2	L	$\cos\theta_x$	$\cos\theta_y$	$\cos\theta_z$
OA	4	4	36	44	6.63	0.3015	−0.3015	−0.9046
OB	0	9	100	109	10.44	0.0	0.2874	−0.9579
OC	9	9	16	34	5.83	−0.5146	0.5146	−0.6861

step 5: See step 4.

step 6: The equilibrium equations are

$$\begin{array}{rcrcrcrl} 0.3015F_A & + & 0F_B & - & 0.5146F_C & + & 1000 & = 0 \\ -0.3015F_A & + & 0.2874F_B & + & 0.5146F_C & & & = 0 \\ -0.9046F_A & - & 0.9579F_B & - & 0.6861F_C & & & = 0 \end{array}$$

The solution to these simultaneous equations is

$$F_A = +1531 \text{ N} \quad \text{(T)}$$

$$F_B = -3480 \text{ N} \quad \text{(C)}$$

$$F_C = +2841 \text{ N} \quad \text{(T)}$$

step 7: See step 6.

47. NOMENCLATURE

a	distance to lowest cable point	ft	m
A	area	ft^2	m^2
b	base	ft	m
c	parameter of the catenary	ft	m
d	distance or diameter	ft	m
d	couple separation distance	ft	m
D	diameter	ft	m
F	force	lbf	N
g	gravitational acceleration, 32.2 (9.81)	ft/sec^2	m/s^2
g_c	gravitational constant, 32.2	$lbm\text{-}ft/lbf\text{-}sec^2$	n.a.
h	height	ft	m
H	horizontal cable force	lbf	N
L	length	ft	m
M	moment	ft-lbf	N·m
n	number of sheaves	–	–
r	position vector or radius	ft	m
R	reaction force	lbf	N
s	distance along cable	ft	m
S	sag	ft	m
T	tension	lbf	N
w	load per unit length	lbf/ft	N/m
W	weight	lbf	n.a.
x	horizontal distance or position	ft	m
y	vertical distance or position	ft	m
z	distance or position along z-axis	ft	m

Symbols

ϵ	pulley loss factor	–	–
η	pulley efficiency	–	–
θ	angle	deg	deg

Subscripts

O	origin
P	point P
R	resultant

45 Indeterminate Statics

Content in blue refers to the *NCEES Handbook*.

1. INTRODUCTION TO INDETERMINATE STATICS

A structure that is *statically indeterminate* is one for which the equations of statics are not sufficient to determine all reactions, moments, and internal forces. Additional formulas involving deflection are required to completely determine these variables.

Although there are many configurations of statically indeterminate structures, this chapter is primarily concerned with beams on more than two supports, trusses with more members than are required for rigidity, and miscellaneous composite structures.

2. DEGREE OF INDETERMINACY

The *degree of indeterminacy* (*degree of redundancy*) is equal to the number of reactions or members that would have to be removed in order to make the structure statically determinate. For example, a two-span beam on three simple supports is indeterminate (redundant) to the first degree. The degree of indeterminacy of a pin-connected, two-dimensional truss is given by Eq. 45.1.

$$I = r + m - 2j \qquad 45.1$$

The degree of indeterminacy of a pin-connected, three-dimensional truss is

$$I = r + m - 3j \qquad 45.2$$

Rigid frames have joints that transmit moments. The degree of indeterminacy of two-dimensional rigid plane frames is more complex. In Eq. 45.3, s is the number of *special conditions* (also known as the number of *equations of conditions*). s is 1 for each internal hinge or a shear release, 2 for each internal roller, and 0 if neither hinges nor rollers are present.

$$I = r + 3m - 3j - s \qquad 45.3$$

3. INDETERMINATE BEAMS

Figure 45.1 Types of Indeterminate Beams

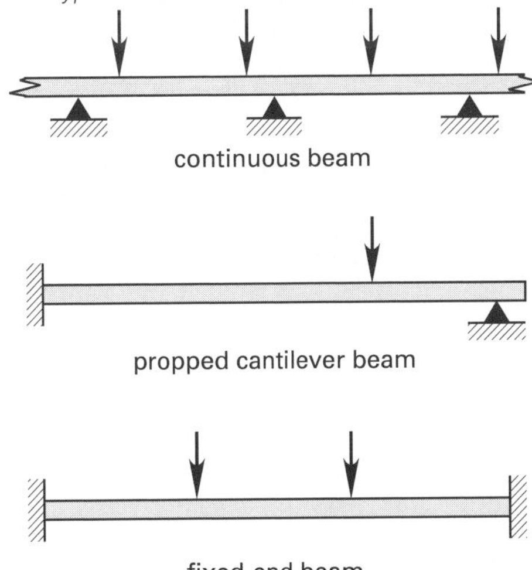

Three common configurations of beams can easily be recognized as being statically indeterminate. These are the *continuous beam*, *propped cantilever beam*, and *fixed-end beam*, as illustrated in Fig. 45.1.

4. REVIEW OF ELASTIC DEFORMATION

When an axial force, F, acts on an object with length L, cross-sectional area A, and modulus of elasticity E, the deformation[1] is

$$\delta = \frac{FL}{AE} \qquad 45.4$$

[1]The terms *deformation* and *elongation* are often used interchangeably in this context.

When an object with initial length L_o and coefficient of thermal expansion α experiences a temperature change of ΔT degrees, the deformation is

$$\delta = \alpha L_o \Delta T \qquad 45.5$$

5. CONSISTENT DEFORMATION METHOD

The *consistent deformation method*, also known as the *compatibility method*, is one of the methods of solving indeterminate problems. This method is simple to learn and to apply. First, geometry is used to develop a relationship between the deflections of two different members (or for one member at two locations) in the structure. Then, the deflection equations for the two different members at a common point are written and equated, since the deformations must be the same at a common point. This method is illustrated by the following examples.

Example 45.1

A pile carrying an axial compressive load is constructed of concrete with a steel jacket. The end caps are rigid, and the steel-concrete bond is perfect. What are the forces in the steel and concrete if a load F is applied?

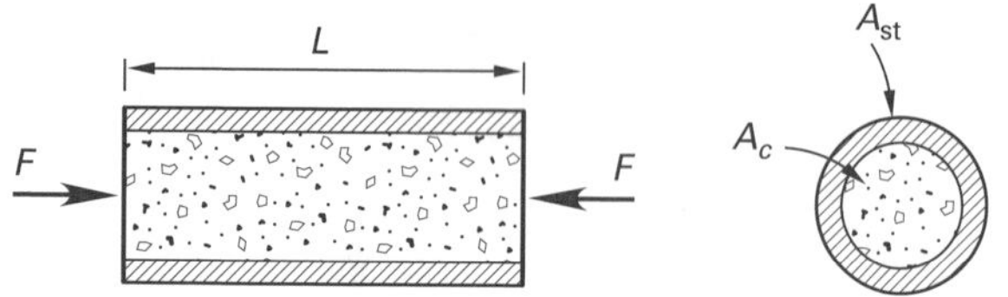

Solution

Let F_c and F_{st} be the loads carried by the concrete and steel, respectively. Then,

$$F_c + F_{st} = F$$

The deformation of the steel is given by Eq. 45.4.

$$\delta_{st} = \frac{F_{st}L}{A_{st}E_{st}}$$

Similarly, the deflection of the concrete is

$$\delta_c = \frac{F_c L}{A_c E_c}$$

But, $\delta_c = \delta_{st}$ since the bonding is perfect. Therefore,

$$\frac{F_c L}{A_c E_c} - \frac{F_{st}L}{A_{st}E_{st}} = 0$$

The first and last equations are solved simultaneously for F_c and F_{st}.

$$F_c = \frac{F}{1 + \frac{A_{st}E_{st}}{A_c E_c}}$$

$$F_{st} = \frac{F}{1 + \frac{A_c E_c}{A_{st}E_{st}}}$$

Example 45.2

A uniform bar is clamped at both ends and the axial load applied near one of the supports. What are the reactions?

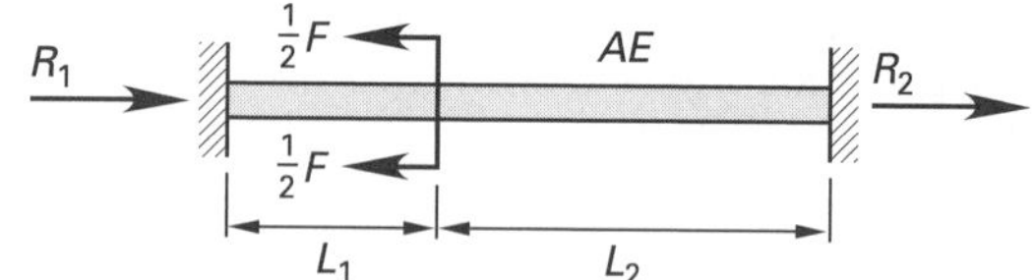

Solution

The first required equation is

$$R_1 + R_2 = F$$

The shortening of section 1 due to the reaction R_1 is

$$\delta_1 = \frac{-R_1 L_1}{AE}$$

The elongation of section 2 due to the reaction R_2 is

$$\delta_2 = \frac{R_2 L_2}{AE}$$

However, the bar is continuous, so $\delta_1 = -\delta_2$. Therefore,

$$R_1 L_1 = R_2 L_2$$

The first and last equations are solved simultaneously to find R_1 and R_2.

$$R_1 = \frac{F}{1 + \frac{L_1}{L_2}}$$

$$R_2 = \frac{F}{1 + \frac{L_2}{L_1}}$$

Example 45.3
The nonuniform bar shown is clamped at both ends and constrained from changing length. What are the reactions if a temperature change of ΔT is experienced?

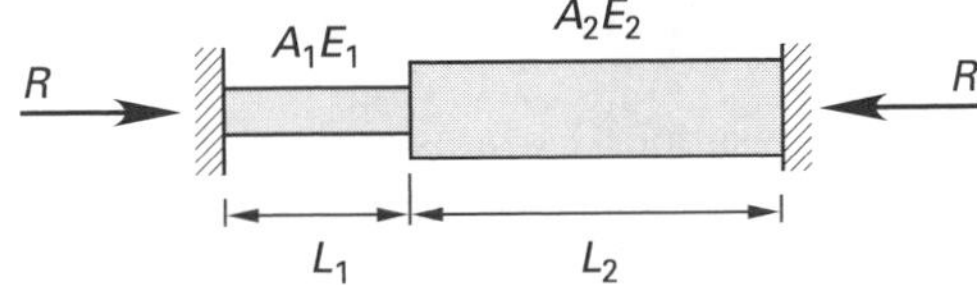

Solution

The thermal deformations of sections 1 and 2 can be calculated directly. Use Eq. 45.5.

$$\delta_1 = \alpha_1 L_1 \Delta T$$

$$\delta_2 = \alpha_2 L_2 \Delta T$$

The total deformation is $\delta = \delta_1 + \delta_2$. However, the deformation can also be calculated from the principles of mechanics.

$$\delta = \frac{RL_1}{A_1E_1} + \frac{RL_2}{A_2E_2}$$

Combine these equations and solve directly for R.

$$R = \frac{(\alpha_1 L_1 + \alpha_2 L_2)\Delta T}{\frac{L_1}{A_1E_1} + \frac{L_2}{A_2E_2}}$$

Example 45.4
The beam shown is supported by dissimilar members. The bar is rigid and remains horizontal.[2] The beam's mass is insignificant. What are the forces in the members?

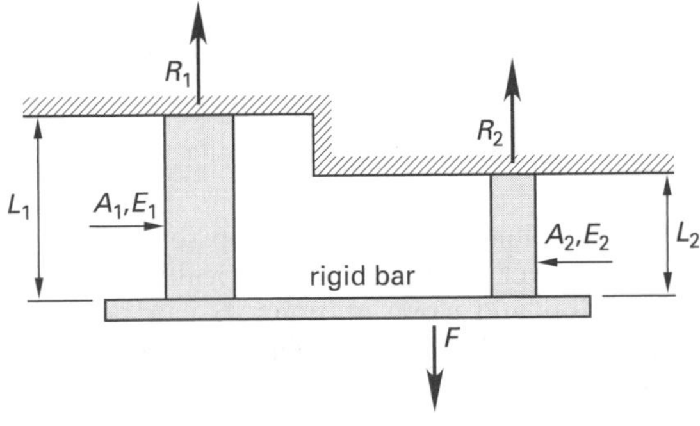

Solution

The required equilibrium condition is

$$R_1 + R_2 = F$$

The elongations of the two tension members are

$$\delta_1 = \frac{R_1L_1}{A_1E_1}$$

$$\delta_2 = \frac{R_2L_2}{A_2E_2}$$

Since the horizontal bar remains horizontal, $\delta_1 = \delta_2$.

$$\frac{R_1L_1}{A_1E_1} = \frac{R_2L_2}{A_2E_2}$$

The first and last equations are solved simultaneously to find R_1 and R_2.

$$R_1 = \frac{F}{1 + \frac{L_1A_2E_2}{L_2A_1E_1}}$$

$$R_2 = \frac{F}{1 + \frac{L_2A_1E_1}{L_1A_2E_2}}$$

Example 45.5
The beam shown is supported by dissimilar members. The bar is rigid but is not constrained to remain horizontal. The beam's mass is insignificant. Develop the simultaneous equations needed to determine the reactions in the vertical members.

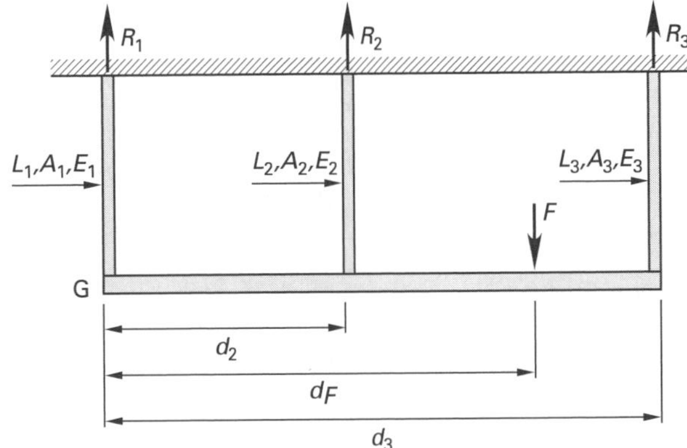

Solution

The forces in the supports are R_1, R_2, and R_3. Any of these may be tensile (positive) or compressive (negative).

$$R_1 + R_2 + R_3 = F$$

[2]This example is easily solved by summing moments about a point on the horizontal beam.

Statics

The changes in length are given by Eq. 45.4.

$$\delta_1 = \frac{R_1 L_1}{A_1 E_1}$$

$$\delta_2 = \frac{R_2 L_2}{A_2 E_2}$$

$$\delta_3 = \frac{R_3 L_3}{A_3 E_3}$$

Since the bar is rigid, the deflections will be proportional to the distance from point G.

$$\delta_2 = \delta_1 + \left(\frac{d_2}{d_3}\right)(\delta_3 - \delta_1)$$

Moments can be summed about point G to give a third equation.

$$\begin{aligned} M_G &= R_3 d_3 + R_2 d_2 - F d_F \\ &= 0 \end{aligned}$$

Example 45.6

A load is supported by three tension members. Develop the simultaneous equations needed to find the forces in the three members.

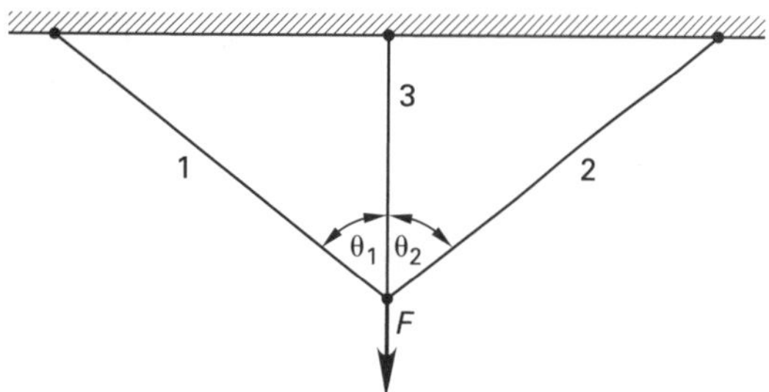

Solution

The equilibrium requirement is

$$F_{1y} + F_3 + F_{2y} = F$$

$$F_1 \cos\theta_1 + F_3 + F_2 \cos\theta_2 = F$$

Assuming the elongations are small compared to the member lengths, the angles θ_1 and θ_2 are unchanged. Then, the vertical deflections are the same for all three members.

$$\frac{F_1 L_1 \cos\theta_1}{A_1 E_1} = \frac{F_3 L_3}{A_3 E_3} = \frac{F_2 L_2 \cos\theta_2}{A_2 E_2}$$

These equations can be solved simultaneously to find F_1, F_2, and F_3. (It may be necessary to work with the x-components of the deflections in order to find a third equation.)

6. SUPERPOSITION METHOD

Two-span (three-support) beams and propped cantilevers are indeterminate to the first degree. Their reactions can be determined from a variation of the consistent deformation procedure known as the *superposition method*.[3] This method requires finding the deflection with one or more supports removed and then satisfying the known conditions.

step 1: Remove enough redundant supports to reduce the structure to a statically determinate condition.

step 2: Calculate the deflections at the locations of the removed redundant supports. Use consistent sign conventions.

step 3: Apply each redundant support as an isolated load, and find the deflections at the redundant support points as functions of the redundant support forces.

step 4: Use superposition to combine (i.e., add) the deflections due to the actual loads and the redundant support loads. The total deflections must agree with the known deflections (usually zero) at the redundant support points.

Example 45.7

A propped cantilever is loaded by a concentrated force at midspan. Determine the reaction, S, at the prop.

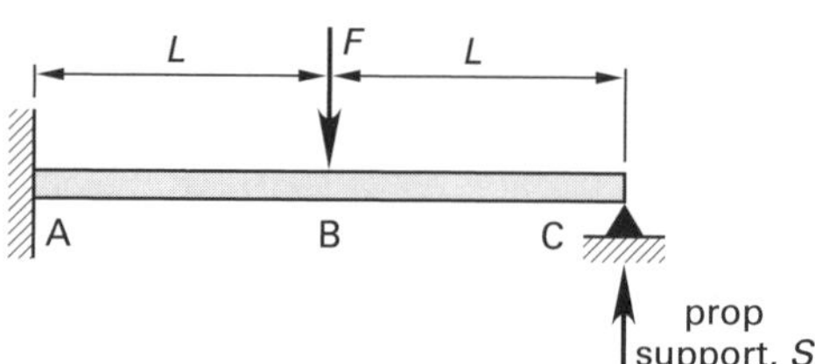

Solution

Start by removing the unknown prop reaction at point C. The cantilever beam is then statically determinate. The deflection and slope at point B can be found or derived from the elastic beam deflection equations. For a cantilever with end load, the deflection and slope are calculated as follows. (The deflection at point B is the same as for a tip-loaded cantilever of length L.)

$$\text{deflection at point B: } \delta_B = \frac{-FL^3}{3EI}$$

[3] Superposition can also be used with higher-order indeterminate problems. However, the simultaneous equations that must be solved may make superposition unattractive for manual calculations.

Statics

$$\text{slope at point B: } y'_B = \frac{-FL^2}{2EI}$$

The slope remains constant to the right of point B. The deflection at point C due to the load at point B is

$$\delta_{C,F} = \delta_B + y'_B L = \frac{-5FL^3}{6EI}$$

The upward deflection at the cantilever tip due to the prop support, S, alone is given by

$$\delta_{C,S} = \frac{S(2L)^3}{3EI} = \frac{8SL^3}{3EI}$$

It is known that the actual deflection at point C is zero (the boundary condition). Therefore, the prop support, S, can be determined as a function of the applied load.

$$\delta_{C,S} + \delta_{C,F} = 0$$

$$\frac{8SL^3}{3EI} - \frac{5FL^3}{6EI} = 0$$

$$S = \frac{5F}{16}$$

7. THREE-MOMENT EQUATION

A *continuous beam* has two or more spans (i.e., three or more supports) and is statically indeterminate. (See Fig. 45.2.) The *three-moment equation* is a method of determining the reactions on continuous beams. It relates the moments at any three adjacent supports. The three-moment method can be used with a two-span beam to directly find all three reactions.

Figure 45.2 *Portion of a Continuous Beam*

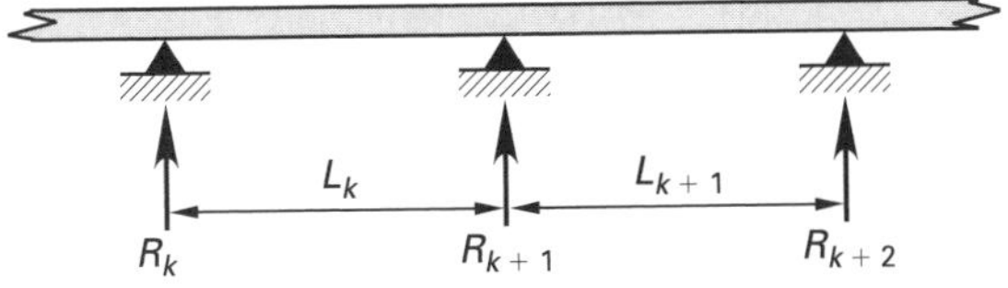

When a beam has more than two spans, the equation must be used with three adjacent supports at a time, starting with a support whose moment is known. (The moment is known to be zero at a simply supported end. For a cantilever end, the moment depends only on the loads on the cantilever portion.)

In its most general form, the three-moment equation is applicable to beams with nonuniform cross sections. In Eq. 45.6, I_k is the moment of inertia of span k.

$$\frac{M_k L_k}{I_k} + 2M_{k+1}\left(\frac{L_k}{I_k} + \frac{L_{k+1}}{I_{k+1}}\right) + \frac{M_{k+2}L_{k+1}}{I_{k+1}} = -6\left(\frac{A_k a}{I_k L_k} + \frac{A_{k+1}b}{I_{k+1}L_{k+1}}\right) \quad 45.6$$

Equation 45.6 uses the following special nomenclature.

- a distance from the left support to the centroid of the moment diagram on the left span
- b distance from the right support to the centroid of the moment diagram on the right span
- I_k the moment of inertia of the open span between supports k and $k+1$
- L_k length of the span between supports k and $k+1$
- M_k bending moment at support k
- A_k area of moment diagram between supports k and $k+1$, assuming that the span is simply and independently supported

The products Aa and Ab are known as *first moments of the areas*. It is convenient to derive simplified expressions for Aa and Ab for commonly encountered configurations. Several are presented in Fig. 45.3.

For beams with uniform cross sections, the moment of inertia terms can be eliminated.

$$M_k L_k + 2M_{k+1}(L_k + L_{k+1}) + M_{k+2}L_{k+1} = -6\left(\frac{A_k a}{L_k} + \frac{A_{k+1}b}{L_{k+1}}\right) \quad 45.7$$

Figure 45.3 *Simplified Three-Moment Equation Terms*

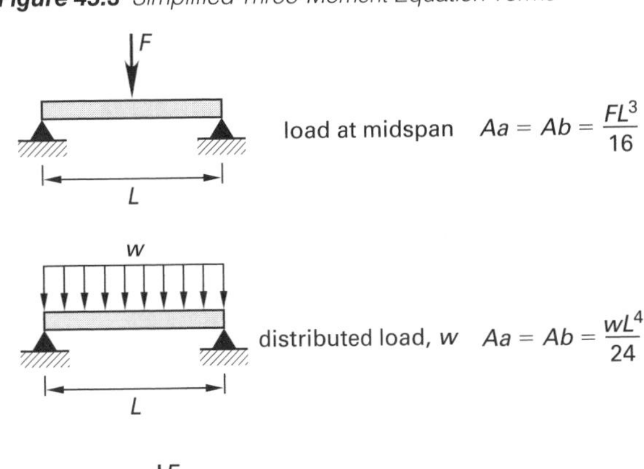

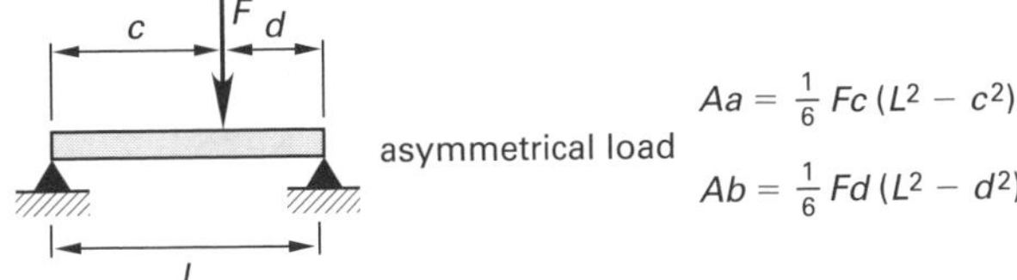

Statics

Example 45.8

Find the four reactions supporting the beam. EI is constant.

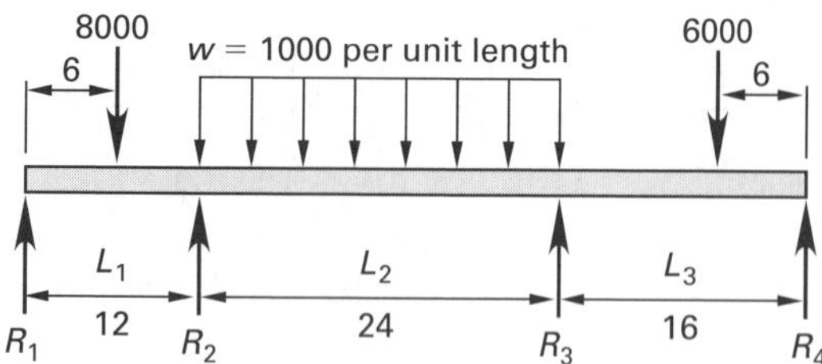

Solution

Spans 1 and 2:

Since the three-moment method can be applied to only two spans at a time, work first with the left and middle spans (spans 1 and 2).

From Fig. 45.3, the quantities A_1a and A_2b are

$$A_1a = \frac{FL^3}{16} = \frac{(8000)(12)^3}{16} = 864{,}000$$

$$A_2b = \frac{wL^4}{24} = \frac{(1000)(24)^4}{24} = 13{,}824{,}000$$

Since the left end of the beam is simply supported, M_1 is zero. From Eq. 45.7, the three-moment equation becomes

$$2M_2(L_1 + L_2) + M_3L_2 = -6\left(\frac{A_1a}{L_1} + \frac{A_2b}{L_2}\right)$$

$$2M_2(12 + 24) + M_3(24) = (-6)\left(\frac{864{,}000}{12} + \frac{13{,}824{,}000}{24}\right)$$

After simplification,

$$3M_2 + M_3 = -162{,}000$$

Spans 2 and 3:

From the previous calculations,

$$A_2a = A_2b = 13{,}824{,}000$$

From Fig. 45.3 for the third span,

$$\begin{aligned} A_3b &= \tfrac{1}{6}Fd(L^2 - d^2) \\ &= \left(\frac{1}{6}\right)(6000)(6)\left((16)^2 - (6)^2\right) \\ &= 1{,}320{,}000 \end{aligned}$$

Since the right end is simply supported, $M_4 = 0$ and the three-moment equation is

$$M_2L_2 + 2M_3(L_2 + L_3) = -6\left(\frac{A_2a}{L_2} + \frac{A_3b}{L_3}\right)$$

$$M_2(24) + 2M_3(24 + 16) = (-6)\left(\frac{13{,}824{,}000}{24} + \frac{1{,}320{,}000}{16}\right)$$

After simplifying,

$$0.3M_2 + M_3 = -49{,}388$$

There are two equations in two unknowns (M_2 and M_3). A simultaneous solution yields

$$M_2 = -41{,}708$$

$$M_3 = -36{,}875$$

Finding reactions:

M_2 can be written in terms of the loads and reactions to the left of support 2. Assuming clockwise moments are positive,

$$\begin{aligned} M_2 &= 12R_1 - (6)(8000) \\ &= -41{,}708 \end{aligned}$$

$$R_1 = 524.3$$

Once R_1 is known, moments can be taken from support 3 to the left.

$$\begin{aligned} M_3 &= (36)(524.3) + 24R_2 - (30)(8000) - (12)(24{,}000) \\ &= -36{,}875 \end{aligned}$$

$$R_2 = 19{,}677.1$$

Similarly, R_4 and R_3 can be determined by working from the right end to the left. Assuming counterclockwise moments are positive,

$$\begin{aligned} M_3 &= 16R_4 - (10)(6000) \\ &= -36{,}875 \end{aligned}$$

$$R_4 = 1445.3$$

$$\begin{aligned} M_2 &= (40)(1445) + 24R_3 - (34)(6000) - (12)(24{,}000) \\ &= -41{,}708 \end{aligned}$$

$$R_3 = 16{,}353.3$$

Check:

Check for equilibrium in the vertical direction.

$$\sum \text{loads} = 8000 + 24{,}000 + 6000 = 38{,}000$$

$$\sum \text{reactions} = 524.3 + 19{,}677.1 + 1445.3 + 16{,}353.3 = 38{,}000$$

8. FIXED-END MOMENTS

When the end of a beam is constrained against rotation, it is said to be a *fixed end* (also known as a *built-in end*). The ends of fixed-end beams are constrained to remain horizontal. Cantilever beams have a single fixed end. Some beams, as illustrated in Fig. 45.1, have two fixed ends and are known as *fixed-end beams*.[4]

Fixed-end beams are inherently indeterminate. To reduce the work required to find end moments and reactions, tables of fixed-end moments are often used.

9. INDETERMINATE TRUSSES

It is possible to manually calculate the forces in all members of an indeterminate truss. However, due to the time required, it is preferable to limit such manual calculations to trusses that are indeterminate to the first degree. The following *dummy unit load method* can be used to solve trusses with a single redundant member.

step 1: Draw the truss twice. Omit the redundant member on both trusses. (There may be a choice of redundant members.)

step 2: Load the first truss (which is now determinate) with the actual loads.

step 3: Calculate the force, S, in each of the members. Assign a positive sign to tensile forces.

step 4: Load the second truss with two unit forces acting collinearly toward each other along the line of the redundant member.

step 5: Calculate the force, u, in each of the members.

step 6: Calculate the force in the redundant member from Eq. 45.8.

$$S_{\text{redundant}} = \frac{-\sum \frac{SuL}{AE}}{\sum \frac{u^2L}{AE}} \qquad 45.8$$

If AE is the same for all members,

$$S_{\text{redundant}} = \frac{-\sum SuL}{\sum u^2L} \qquad 45.9$$

The true force in member j of the truss is

$$F_{j,\text{true}} = S_j + S_{\text{redundant}}u_j \qquad 45.10$$

Example 45.9

Find the force in members BC and BD. $AE = 1$ for all members except for CB, which is 2, and AD, which is 1.5.

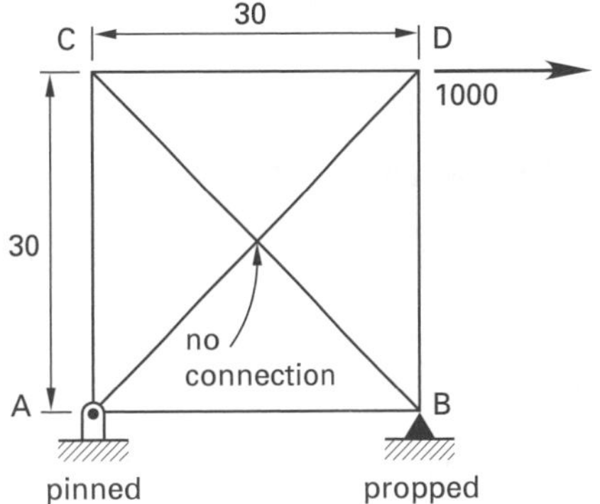

Solution

The two trusses are shown appropriately loaded.

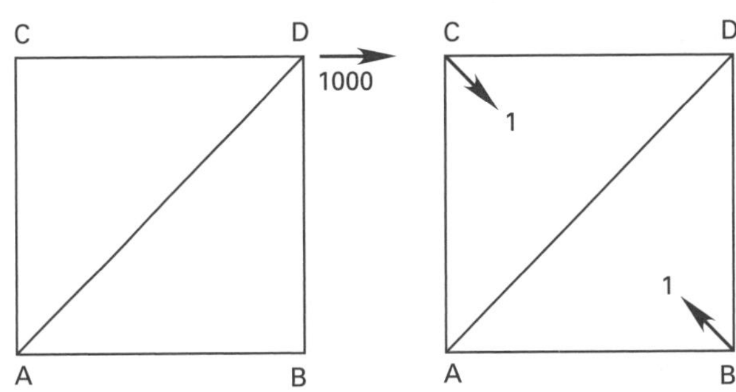

member	L	AE	S	u	$\frac{SuL}{AE}$	$\frac{u^2L}{AE}$
AB	30	1	0	−0.707 (C)	0	15
BD	30	1	−1000 (C)	−0.707	21,210	15
DC	30	1	0	−0.707	0	15
CA	30	1	0	−0.707	0	15
CB	42.43	2	0	1.0	0	21.22
AD	42.43	1.5	1414 (T)	1.0	39,997	28.29
					61,207	109.51

[4]The definition is loose. The term *fixed-end beam* can also be used to mean any indeterminate beam with at least one built-in end (e.g., a propped cantilever).

From Eq. 45.8,

$$S_{BC} = \frac{-\sum \frac{SuL}{AE}}{\sum \frac{u^2L}{AE}} = \frac{-61{,}207}{109.51} = -558.9 \text{ (C)}$$

From Eq. 45.10,

$$\begin{aligned} F_{BD,true} &= S_{BD} + S_{redundant}u_{BD} \\ &= -1000 + (-558.9)(-0.707) \\ &= -604.9 \text{ (C)} \end{aligned}$$

10. INFLUENCE DIAGRAMS

Shear, moment, and reaction influence diagrams (influence lines) can be drawn for any point on a beam or truss. This is a necessary first step in the evaluation of stresses induced by moving loads. It is important to realize, however, that the influence diagram applies only to one point on the beam or truss.

Influence Diagrams for Beam Reactions

In a typical problem, the load is fixed in position and the reactions do not change. If a load is allowed to move across a beam, the reactions will vary. An influence diagram can be used to investigate the value of a chosen reaction as the load position varies.

To make the influence diagram as general in application as possible, a unit load is used. As an example, consider a 20 ft, simply supported beam and determine the effect on the left reaction of moving a 1 lbf load across the beam.

If the load is directly over the right reaction ($x = 0$), the left reaction will not carry any load. Therefore, the ordinate of the influence diagram is zero at that point. (Even though the right reaction supports 1 lbf, this influence diagram is being drawn for one point only—the left reaction.) Similarly, if the load is directly over the left reaction ($x = L$), the ordinate of the influence diagram will be 1. Basic statics can be used to complete the rest of the diagram, as shown in Fig. 45.4.

Use this rudimentary example of an influence diagram to calculate the left reaction for any placement of any load by multiplying the actual load by the ordinate of the influence diagram.

Figure 45.4 *Influence Diagram for Reaction of Simple Beam*

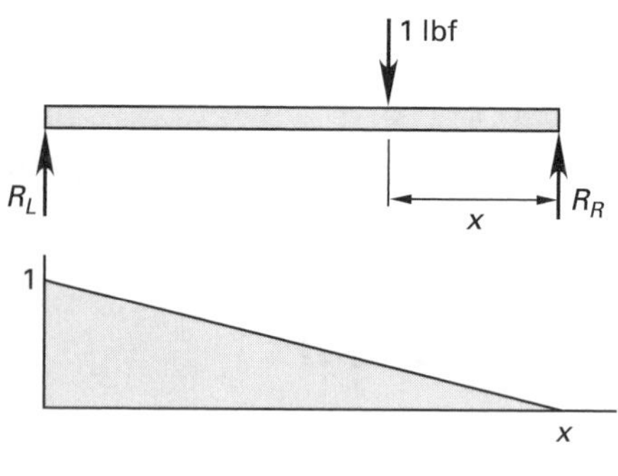

$$R_L = P \times \text{ordinate} \qquad 45.11$$

Even though the influence diagram was drawn for a point load, it can still be used when the beam carries a uniformly distributed load. In the case of a uniform load of w distributed over the beam from x_1 to x_2, the left reaction can be calculated from Eq. 45.12.

$$\begin{aligned} R_L &= \int_{x_1}^{x_2} (w \times \text{ordinate})\, dx \\ &= w \times \text{area under curve} \end{aligned} \qquad 45.12$$

Example 45.10

A 500 lbf load is placed 15 ft from the right end of a 20 ft, simply supported beam. Use the influence diagram to determine the left reaction.

Solution

Since the influence line increases linearly from 0 to 1, the ordinate is the ratio of position to length. That is, the ordinate is $15/20 = 0.75$. From Eq. 45.11, the left reaction is

$$\begin{aligned} R_L &= (0.75)(500 \text{ lbf}) \\ &= 375 \text{ lbf} \end{aligned}$$

Example 45.11

A uniform load of 15 lbf/ft is distributed between $x = 4$ ft and $x = 10$ ft along a 20 ft, simply supported beam. What is the left reaction?

Statics

Solution

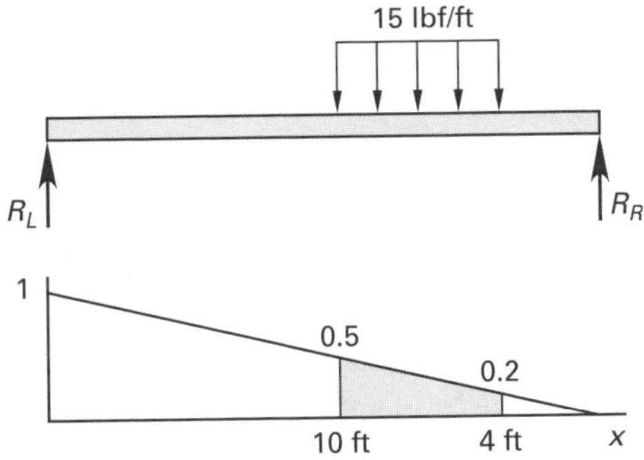

From Eq. 45.12, the left reaction can be calculated from the area under the influence diagram between the limits of loading.

$$\begin{aligned} A &= \left(\frac{1}{2}\right)(10\text{ ft})(0.5) - \left(\frac{1}{2}\right)(4\text{ ft})(0.2) \\ &= 2.1\text{ ft} \end{aligned}$$

The left reaction is

$$\begin{aligned} R_L &= \left(15\ \frac{\text{lbf}}{\text{ft}}\right)(2.1\text{ ft}) \\ &= 31.5\text{ lbf} \end{aligned}$$

Finding Reaction Influence Diagrams Graphically

Since the reaction will always have a value of 1 when the unit load is directly over the reaction and since the reaction is always directly proportional to the distance x, the reaction influence diagram can be easily determined from the following steps.

step 1: Remove the support being investigated.

step 2: Displace (lift) the beam upward a distance of one unit at the support point. The resulting beam shape will be the shape of the reaction influence diagram.

Example 45.12

What is the approximate shape of the reaction influence diagram for reaction 2?

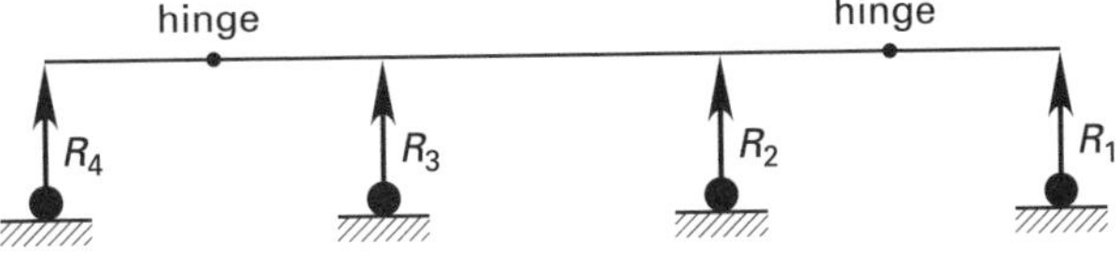

Solution

Pushing up at reaction 2 such that the deflection is one unit results in the shown shape.

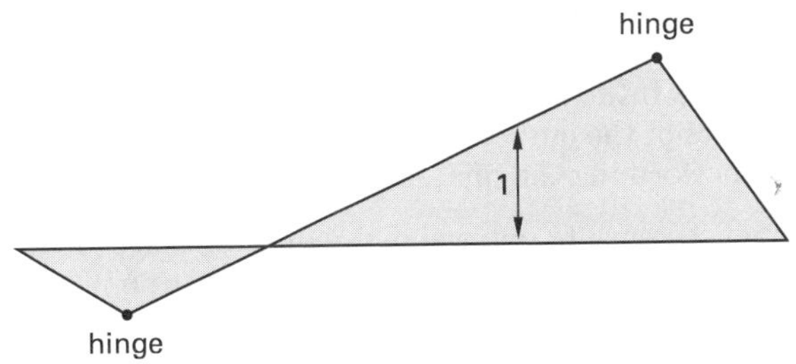

Influence Diagrams for Beam Shears

A shear influence diagram (not the same as a shear diagram) illustrates the effect on the shear at a particular point in the beam of moving a load along the beam's length. As an illustration, consider point A along the simply supported beam of length 20.

In all cases, principles of statics can be used to calculate the shear at point A as the sum of loads and reactions on the beam from point A to the left end. (With the appropriate sign convention, summation to the right end could be used as well.) If the unit load is placed between the right end ($x = 0$) and point A, the shear at point A will consist only of the left reaction, since there are no other loads between point A and the left end. From the reaction influence diagram, the left reaction varies linearly. At $x = 12$ ft, the location of point A, the shear is $V = R_L = 12/20 = 0.6$.

When the unit load is between point A and the left end, the shear at point A is the sum of the left reaction (upward and positive) and the unit load itself (downward and negative). Therefore, $V = R_L - 1$. At $x = 12$ ft, the shear is $V = 0.6\text{ lbf} - 1\text{ lbf} = -0.4\text{ lbf}$.

Figure 45.5 is the shear influence diagram. In the diagram, the shear goes through a reversal of 1, and the slopes of the two inclined sections are the same.

Figure 45.5 *Shear Influence Diagram for Simple Beam*

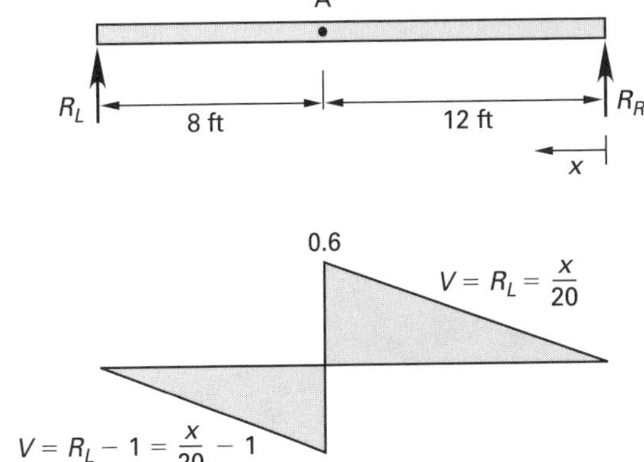

Statics

Shear influence diagrams are used in the same manner as reaction influence diagrams. The shear at point A for any position of the load can be calculated by multiplying the ordinate of the diagram by the actual load. Distributed loads are found by multiplying the uniform load by the area under the diagram between the limits of loading. If the loading extends over positive and negative parts of the curve, the sign of the area is considered when performing the final summation.

If it is necessary to determine the distribution of loading that will produce the maximum shear at a point whose influence diagram is available, the load should be positioned in order to maximize the area under the diagram.[5] This can be done by "covering" either all of the positive area or all of the negative area.[6]

Shear Influence Diagrams by Virtual Displacement

A difficulty in drawing shear influence diagrams for continuous beams on more than two supports is finding the reactions. The method of *virtual displacement* or *virtual work* can be used to find the influence diagram without going through that step.

step 1: Replace the point being investigated (i.e., point A) with an imaginary link with unit length. (It may be necessary to think of the link as having a length of 1 ft, but the link does not add to or subtract from any length of the beam.) If the point being investigated is a reaction, place a hinge at that point and lift the hinge upward a unit distance.

step 2: Push the two ends of the beam (with the link somewhere in between) toward each other a very small amount until the linkage is vertical. The distance between supports does not change, but the linkage allows the beam sections to assume a slope. The sections to the left and right of the linkage displace δ_1 and δ_2, respectively, from their equilibrium positions. The slope of both sections is the same. Points of support remain in contact with the beam.

step 3: Determine the ratio of δ_1 and δ_2. Since the slope on the two sections is the same, the longer section will have the larger deflection. If $L = a + b$ is the length of the beam, the relationships between the deflections can be determined from Eq. 45.14 through Eq. 45.16.

$$\delta_1 + \delta_2 = 1 \qquad 45.13$$

$$\frac{\delta_1}{\delta_2} = \frac{a}{b} \qquad 45.14$$

$$\delta_1 = \left(\frac{a}{L}\right)\delta \qquad 45.15$$

$$\delta_2 = \left(\frac{b}{L}\right)\delta \qquad 45.16$$

Since $\delta = \delta_1 + \delta_2$ was chosen as 1, Eq. 45.15 and Eq. 45.16 really give the relative proportions of the unit link that extend below and above the reference line in Fig. 45.6.

Figure 45.6 *Virtual Beam Displacements*

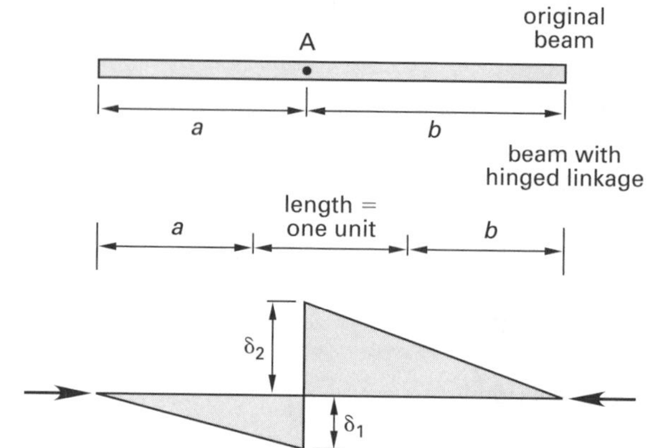

Knowing that the total shear reversal through point A is one unit and that the slopes are the same, the relative proportions of the reversal below and above the line will determine the shape of the displaced beam. The shape of the influence diagram is the shape taken on by the beam.

step 4: As required, use equations of straight lines to obtain the shear influence ordinate as a function of position along the beam.

Example 45.13

For the simply supported beam shown, draw the shear influence diagram for a point 10 ft from the right end.

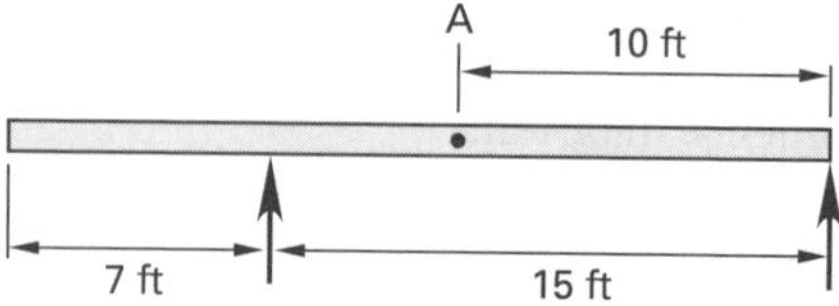

[5]If the *minimum shear* is requested, the maximum negative shear is implied. The minimum shear is not zero in most cases.
[6]Usually, the dead load is assumed to extend over the entire length of the beam. The uniform live loads are distributed in any way that will cause the maximum shear.

Solution

If a unit link is placed at point A and the beam ends are pushed together, the following shape will result. The beam must remain in contact with the points of support, and the two slopes are the same.

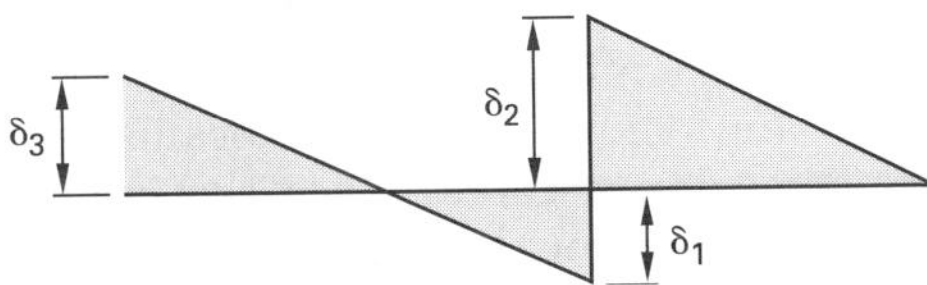

The overhanging 7 ft of beam do not change the shape of the shear influence diagram between the supports. The deflections can be evaluated assuming a 15 ft long beam using Eq. 45.15 and Eq. 45.16.

$$\delta_1 = \frac{a}{L} = \frac{5 \text{ ft}}{15 \text{ ft}} = 0.333$$

$$\delta_2 = \frac{b}{L} = \frac{10 \text{ ft}}{15 \text{ ft}} = 0.667$$

The slope in both sections of the beam is the same. This slope can be used to calculate δ_3.

$$m = \frac{\delta_1}{a} = \frac{0.333}{5 \text{ ft}} = 0.0667 \text{ 1/ft}$$

$$\delta_3 = (7 \text{ ft})\left(0.0667 \ \frac{1}{\text{ft}}\right) = 0.467$$

Example 45.14

Where should a uniformly distributed load be placed on the following beam to maximize the shear at section A?

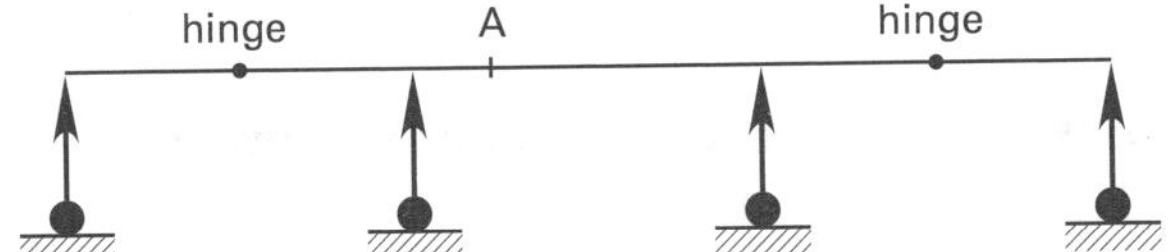

Solution

Using the principle of virtual displacement, the following shear influence diagram results by inspection. (It is not necessary to calculate the relative displacements to answer this question. It is only necessary to identify the positive and negative parts of the influence diagram.)

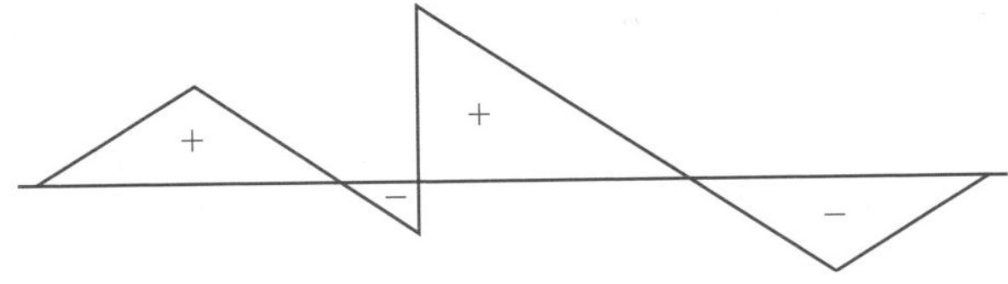

To maximize the shear, the uniform load should be distributed either over all positive or all negative sections of the influence diagram.

Moment Influence Diagrams by Virtual Displacement

A moment influence diagram (not the same as a moment diagram) gives the moment at a particular point for any location of a unit load. The method of virtual displacement can be used in this situation to simplify finding the moment influence diagram. (See Fig. 45.7.)

step 1: Replace the point being investigated (i.e., point A) with an imaginary hinge.

step 2: Rotate the beam one unit rotation by applying equal but opposite moments to each of the two beam sections. Except where the point being investigated is at a support, this unit rotation can be achieved simply by "pushing up" on the beam at the hinge point.

Figure 45.7 *Moment Influence Diagram by Virtual Displacement*

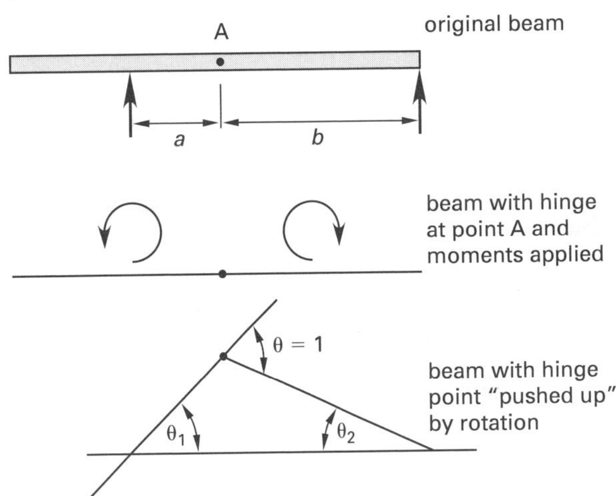

step 3: The angles made by the sections on either side of the hinge will be proportional to the lengths of the opposite sections. (Since the angle is small for a virtual displacement, the angle and its tangent, or slope, are the same.)

$$\theta_1 = \frac{b}{L} \qquad 45.17$$

$$\theta_2 = \frac{a}{L} \qquad 45.18$$

$$L = a + b \qquad 45.19$$

Example 45.15

What are the approximate shapes of the moment influence diagrams for points A and B on the beam shown?

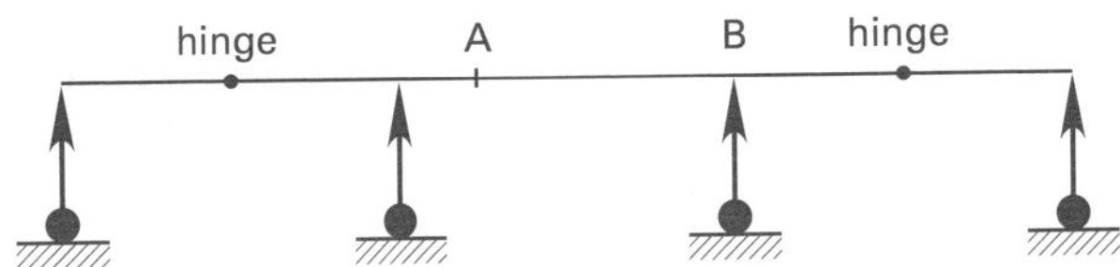

Statics

Solution

By placing an imaginary hinge at point A and rotating the two adjacent sections of the beam, the following shape results.

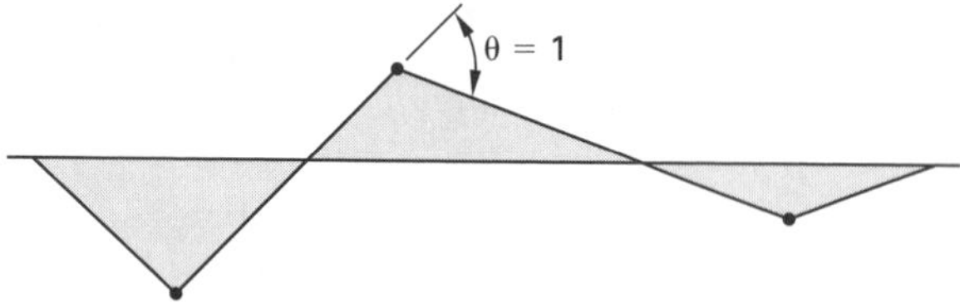

The moment influence diagram for point B is found by placing an imaginary hinge at point B and applying a rotating moment. Since the beam must remain in contact with all supports, and since there is no hinge between the two middle supports, the moment influence diagram must be horizontal in that region.

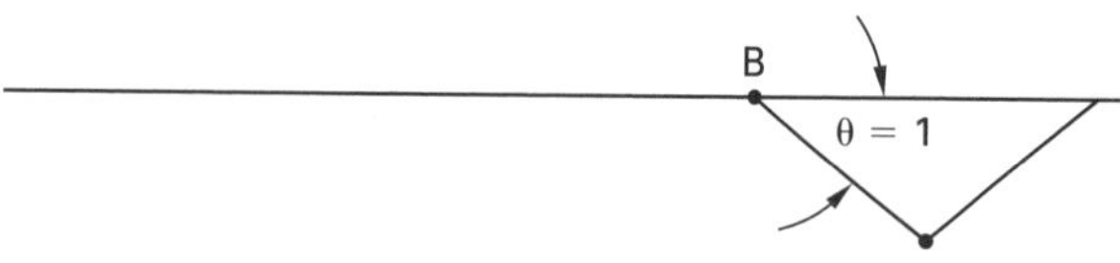

Shear Influence Diagrams on Cross-Beam Decks

When girder-type construction is used to construct a road or bridge deck, the traffic loads will not be applied directly to the girder. Rather, the loads will be transmitted to the girder at panel points from cross beams (floor beams). Figure 45.8 shows a typical construction detail involving girders and cross beams.

Figure 45.8 *Cross-Beam Decking*

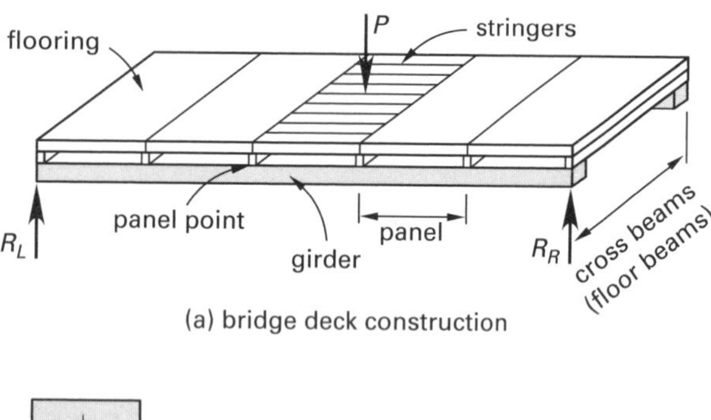

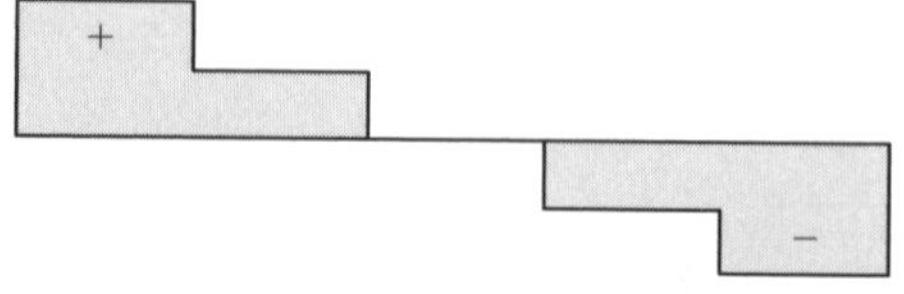

(b) shear diagram for girder

A load applied to the deck stringers will be transmitted to the girder only at the panel points. Because the girder experiences a series of concentrated loads, the shear between panel points is horizontal. Since the shear is always constant between panel points, we speak of *panel shear* rather than shear at a point. Accordingly, shear influence diagrams are drawn for a panel, not for a point. Moment influence diagrams are similarly drawn for a panel.

Influence Diagrams on Cross-Beam Decks

Shear and moment influence diagrams for girders with cross beams are identical to simple beams, except for the panel being investigated. Once the influence diagram has been drawn for the simple beam, the influence diagram ordinates at the ends of the panel being investigated are connected to obtain the influence diagram for the girder. This is illustrated in Fig. 45.9.

Figure 45.9 *Comparison of Influence Diagrams for Simple Beams and Girders (panel bc)*

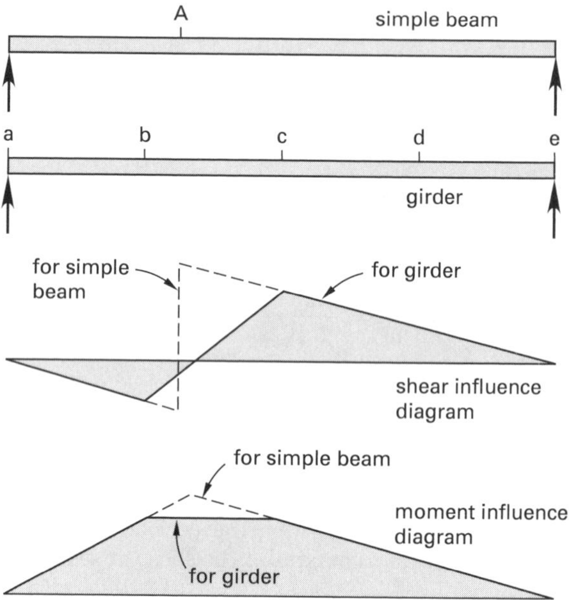

Influence Diagrams for Truss Members

Since members in trusses are assumed to be axial members, they cannot carry shears or moments. Therefore, shear and moment influence diagrams do not exist for truss members. However, it is possible to obtain an influence diagram showing the variation in axial force in a given truss member as the load varies in position.

There are two general cases for finding forces in truss members. The force in a horizontal truss member is proportional to the moment across the member's panel. The force in an inclined truss member is proportional to the shear across that member's panel.

So, even though only the axial load in a truss member may be wanted, it is still necessary to construct the shear and moment influence diagrams for the entire truss in order to determine the applications of loading on the truss that produce the maximum shear and moment across the member's panel.

Example 45.16

(a) Draw the influence diagram for vertical shear in panel DF of the through truss shown. (b) What is the maximum force in member DG if a 1000 lbf load moves across the truss?

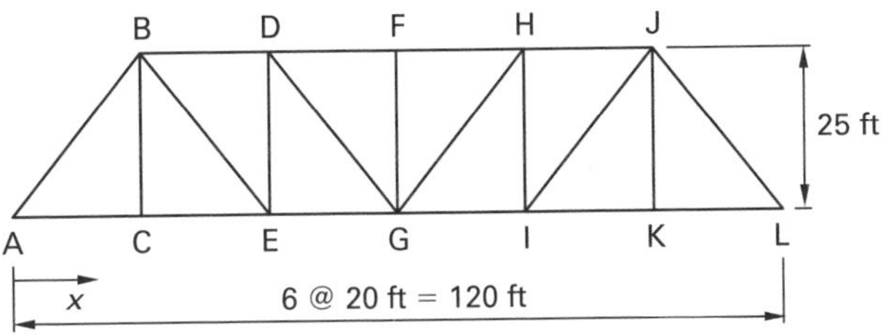

Solution

(a) Allow a unit load to move from joint L to joint G along the lower chords. If the unit vertical load is at a distance x from point L, the right reaction will be $+(1-(x/120))$. The unit load itself has a value of -1, so the shear at distance x is just $-x/120$.

Allow a unit load to move from joint A to joint E along the lower chords. If the unit load is a distance x from point L, the left reaction will be $x/120$, and the shear at distance x will be $(x/120)-1$.

These two lines can be graphed.

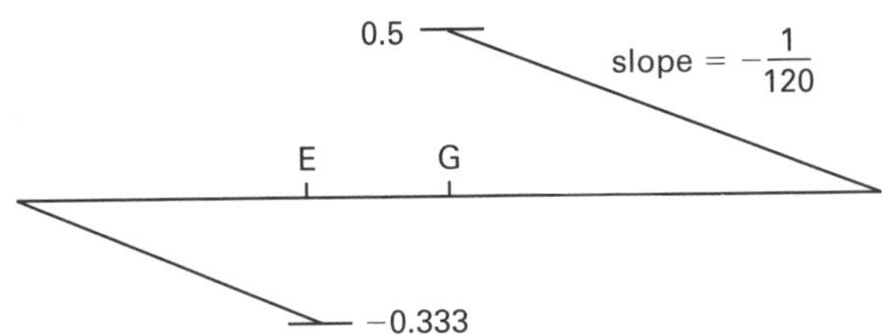

The influence line is completed by connecting the two lines as shown. Therefore, the maximum shear in panel DF will occur when a load is at point G on the truss.

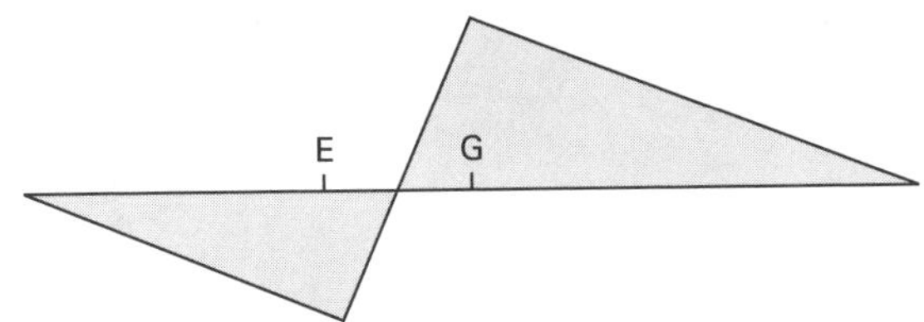

(b) If the 1000 lbf load is at point G, the two reactions at points A and L will each be 500 lbf. The cut-and-sum method can be used to calculate the force in member DG simply by evaluating the vertical forces on the free body to the left of point G.

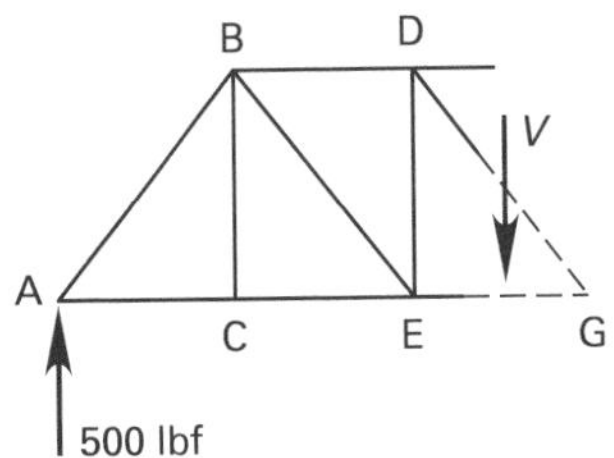

For equilibrium to occur, V must be 500 lbf. This vertical shear is entirely carried by member DG. The length of member DG is

$$\sqrt{(20\text{ ft})^2+(25\text{ ft})^2}=32\text{ ft}$$

The force in member DG is

$$\left(\frac{32\text{ ft}}{25\text{ ft}}\right)(500\text{ lbf})=640\text{ lbf}$$

Example 45.17

(a) Draw the moment influence diagram for panel DF on the truss shown in Ex. 45.16. (b) What is the maximum force in member DF if a 1000 lbf load moves across the truss?

Solution

(a) The left reaction is $x/120$, where x is the distance from the unit load to the right end. If the unit load is to the right of point G, the moment can be found by summing moments from point G to the left. The moment is $(x/120)(60)=0.5x$.

If the unit load is to the left of point E, the moment will again be found by summing moments about point G. The distance between the unit load and point G is $x-60$.

$$\left(\frac{x}{120}\right)(60)-(1)(x-60)=60-0.5x$$

These two lines can be graphed. The moment for a unit load between points E and G is obtained by connecting the two end points of the lines derived above. Therefore, the maximum moment in panel DF will occur when the load is at point G on the truss.

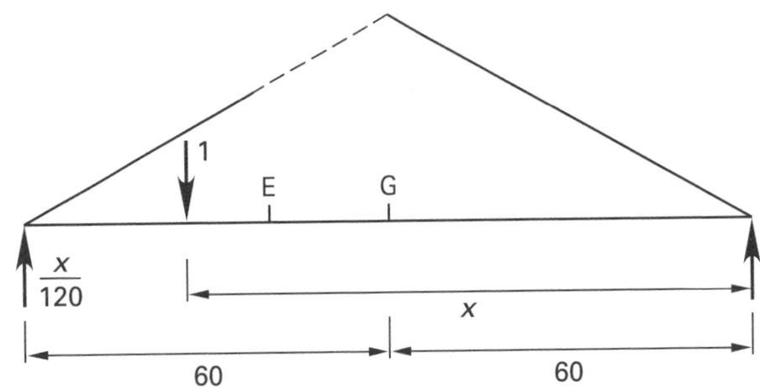

(b) If the 1000 lbf load is at point G, the two reactions at points A and L will each be 500 lbf. The method of sections can be used to calculate the force in member DF by taking moments about joint G.

Statics

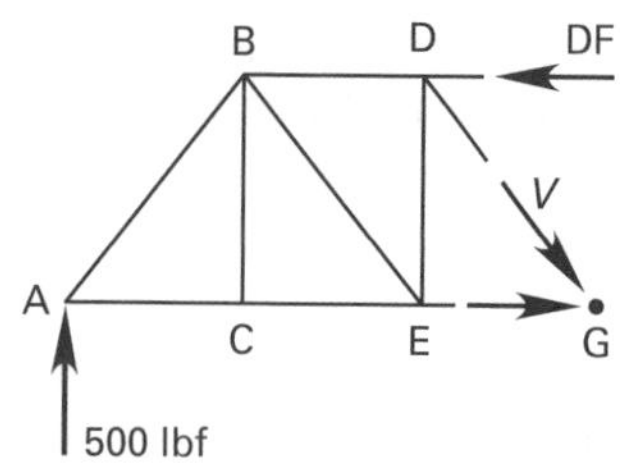

$$\sum M_G = (500 \text{ lbf})(60 \text{ ft}) - (\text{DF})(25 \text{ ft}) = 0$$
$$\text{DF} = 1200 \text{ lbf}$$

11. MOVING LOADS ON BEAMS

Global Maximum Moment and Shear Stresses Anywhere on Beam

If a beam supports a single moving load, the maximum bending and shearing stresses at any point can be found by drawing the moment and shear influence diagrams for that point. Once the positions of maximum moment and maximum shear are known, the stresses at the point in question can be found from Mc/I and QV/Ib.

If a simply supported beam carries a set of moving loads (which remain equidistant as they travel across the beam), the following procedure can be used to find the *dominant load*. (The dominant load is the one that occurs directly over the point of maximum moment.)

step 1: Calculate and locate the resultant of the load group.

step 2: Assume that one of the loads is dominant. Place the group on the beam such that the distance from one support to the assumed dominant load is equal to the distance from the other support to the resultant of the load group.

step 3: Check to see that all loads are on the span and that the shear changes sign under the assumed dominant load. If the shear does not change sign under the assumed dominant load, the maximum moment may occur when only some of the load group is on the beam. If it does change sign, calculate the bending moment under the assumed dominant load.

step 4: Repeat steps 2 and 3, assuming that the other loads are dominant.

step 5: Find the maximum shear by placing the load group such that the resultant is a minimum distance from a support.

Placement of Load Group to Maximize Local Moment

In the design of specific members or connections, it is necessary to place the load group in a position that will maximize the load on those members or connections. The procedure for finding these positions of local maximum loadings is different from the global maximum procedures.

The solution to the problem of local maximization is somewhat trial-and-error oriented. It is aided by use of the influence diagram. In general, the variable being evaluated (reaction, shear, or moment) is maximum when one of the wheels is at the location or section of interest.

When there are only two or three wheels in the load group, the various alternatives can be simply evaluated by using the influence diagram for the variable being evaluated. When there are many loads in the load group (e.g., a train loading), it may be advantageous to use heuristic rules for predicting the dominant wheel.

12. NOMENCLATURE

A	area	ft^2	m^2
b	width of beam	ft	m
c	distance from neutral axis to extreme fiber	ft	m
d	distance	ft	m
E	modulus of elasticity	lbf/ft^2	Pa
F	force	lbf	N
I	area moment of inertia	ft^4	m^4
I	degree of indeterminacy	–	–
j	number of joints	–	–
L	length	ft	m
m	number of members (bars)	–	–
M	moment	ft-lbf	N·m
P	load	lbf	N
Q	statical moment	ft^3	m^3
r	number of reactions	–	–
R	reaction	lbf	N
s	number of special conditions	–	–
S	force	lbf	N
T	degree of indeterminacy	–	–
T	temperature	°F	°C
u	force	lbf	N
V	shear	lbf	N
w	distributed load	lbf/ft	N/m
y'	slope	ft/ft	m/m

Symbols

α	coefficient of thermal expansion	1/°F	1/°C
δ	deformation	ft	m
θ	angle	deg	deg

Subscripts

c	concrete
F	force
L	left
o	original
R	right
st	steel

Statics

Topic VIII: Materials

Chapter

46 Engineering Materials

Content in blue refers to the *NCEES Handbook*.

NCEES EXAM SPECIFICATIONS AND RELATED CONTENT

MACHINE DESIGN AND MATERIALS EXAM

I.D.1. Strength of Materials: Stress/strain
28. Modern Composite Materials

1. CHARACTERISTICS OF METALS

Metals are the most frequently used materials in engineering design. Steel is the most prevalent engineering metal because of the abundance of iron ore, simplicity of production, low cost, and predictable performance. However, other metals play equally important parts in specific products.

Most metals are characterized by the properties in Table 46.1.

***Table 46.1** Properties of Most Metals and Alloys*

high thermal conductivity (low thermal resistance)
high electrical conductivity (low electrical resistance)
high chemical reactivity[a]
high strength
high ductility[b]
high density
high radiation resistance
highly magnetic (ferrous alloys)
optically opaque
electromagnetically opaque

[a]Some alloys, such as stainless steel, are more resistant to chemical attack than pure metals.
[b]Brittle metals, such as some cast irons, are not ductile.

Metallurgy is the subject that encompasses the procurement and production of metals. *Extractive metallurgy* is the subject that covers the refinement of pure metals from their ores.

2. UNIFIED NUMBERING SYSTEM

The Unified Numbering System (UNS) was introduced in the mid-1970s to provide a consistent identification of metals and alloys for use throughout the world. The UNS designation consists of one of seventeen single uppercase letter prefixes followed by five digits. Many of the letters are suggestive of the family of metals, as Table 46.2 indicates.

For each UNS designation, there is a specific percentage range of critical alloying elements. However, the UNS designation is a description, not a specification. Specifications are administered by the American Society of Testing and Materials (ASTM) and similar organizations. The UNS designation refers only to the major alloying and residual elements. It is not an exact specification. One manufacturer may produce an alloy in the middle of the UNS ranges, while another manufacturer may operate at the low or high end of the ranges.[1] The presence of small amounts of

[1]In addition to economically lowering the carbon content, *argon-oxygen-decarburization* (AOD) during refining has made it possible to control nitrogen and other alloying ingredients precisely. The percentage of expensive alloying ingredients (e.g., molybdenum in stainless steels) will generally be at the low ends of the allowable ranges.

Table 46.2 UNS Alloy Prefixes

A	aluminum
C	copper
E	rare-earth metals
F	cast irons
G	AISI and SAE carbon and alloy steels
H	AISI and SAE H-steels
J	cast steels (except tool steels)
K	miscellaneous steels and ferrous alloys
L	low-melting metals
M	miscellaneous nonferrous metals
N	nickel
P	precious metals
R	reactive and refractory metals
S	heat- and corrosion-resistant steels (stainless and valve steels and superalloys)
T	tool steels (wrought and cast)
W	welding filler metals
Z	zinc

residual elements, directionality due to manufacturing processes (e.g., rolling), and heat treatments are also not part of the specification. Furthermore, the UNS designations are not always sufficiently unique to differentiate between two existing products. Additional specifications or trade names are still required in those instances.

A cross-reference index is generally needed to convert older designations to the UNS designation.[2] However, for many stainless steels, the first three digits are the same as the AISI numbering system. Straight AISI type 304 is written simply as UNS S30400. The last two digits may be used to designate some differentiating characteristic. For example, stainless 304L, with a maximum of 0.03% carbon, is designated as S30403.[3] The UNS designations for other families (e.g., aluminum, copper, and nickel) also incorporate some or all of the common designations in use prior to the UNS.

3. BLAST FURNACE IRON PRODUCTION

Iron (chemical symbol Fe) is obtained from its oxides Fe_2O_3 (*hematite*, 69.9% iron) and, to a lesser extent, Fe_3O_4 (*magnetite*, 72.4% iron).[4] Only about 50% of iron ore consists of iron oxides, the remainder being the gangue. *Gangue* is the earth and stone mixed with the iron oxides.

The process used to reduce iron oxides to pure iron takes place in a *blast furnace*. The furnace is charged with alternate layers of iron ore, coke, and limestone in the approximate ratio of 4:2:1, respectively.[5] The limestone serves as a flux for the gangue and the coke ash, enabling the molten gangue and impurities to be drawn off as *slag*.

A traditional blast furnace is shown in Fig. 46.1. The top of the furnace is provided with a pair of conical bells for loading the charge (when open) and limiting the escape of gases (when closed). High-temperature air is injected through nozzles around the periphery of the lower portion of the furnace. These openings are known as *tuyères*.[6]

Figure 46.1 Blast Furnace

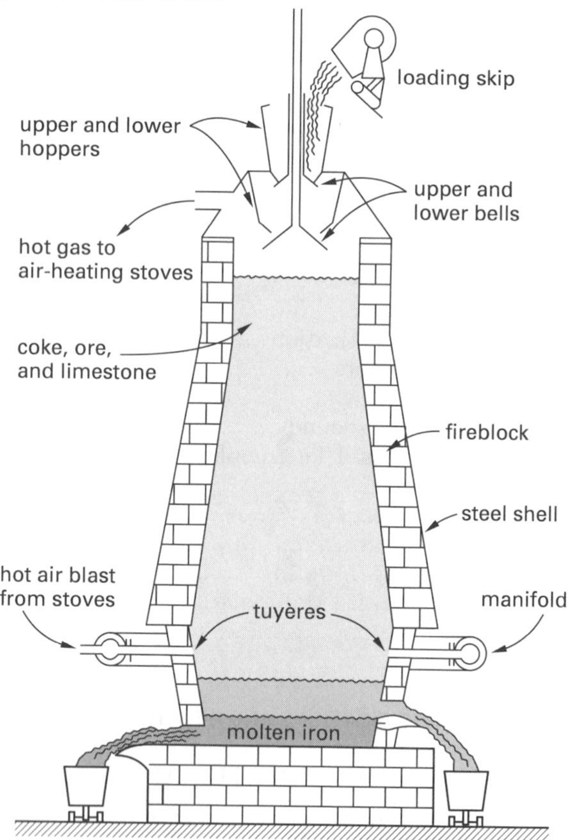

The hot combustion air is produced in preheaters (stoves) that adjoin the blast furnace. Generally, four stoves are provided for each furnace. Each stove is heated in rotation by burning the carbon monoxide-rich

[2]SAE and ASTM jointly publish *Metals and Alloys in the Unified Numbering System*, which includes a cross-reference index.
[3]Since straight type 304 has a maximum of 0.08% carbon, some engineers prefer to write S30408 for uniformity.
[4]Since an additional roasting process is needed to remove the sulfur, iron pyrite, FeS_2, is not used in iron production, despite its abundance. Other lower-grade ores, such as $FeCO_3$ (siderite, 48.3% iron) and $Fe_2O_3 \cdot n[H_2O]$ (limonite, 60 to 65% iron), are used only in the absence of better ores.
[5]*Coke* is coal that has been previously burned in an oxygen-poor environment. The remaining carbonaceous material has a high-combustion energy content. (Clean-air legislation has had a substantial impact on coke making processes.)
[6]*Tuyère* is pronounced twee-yer and too-ur.

furnace gases. Cold air enters one stove while the remaining stoves are being heated. The air is heated to 1000°F to 1300°F (550°C to 700°C) before being injected into the blast furnace.

The injected air oxidizes the coke, producing heat and large amounts of carbon monoxide. The carbon monoxide rises to the top of the furnace and, at a temperature of approximately 600°F (300°C), reduces the iron oxide to FeO. The following chemical reactions describe the production of FeO.

$$\begin{aligned} C + O_2 &\rightarrow CO_2 \\ CO_2 + C &\rightarrow 2CO \\ 2C + O_2 &\rightarrow 2CO \\ 3Fe_2O_3 + CO &\rightarrow 2Fe_3O_4 + CO_2 \\ Fe_3O_4 + CO &\rightarrow 3FeO + CO_2 \end{aligned}$$

As the reduction process continues, the FeO temperature drops down to 1300°F to 1500°F (700°C to 800°C). The FeO is reduced to a spongy mass of pure iron by the carbon monoxide.

$$FeO + CO \rightarrow Fe + CO_2$$

The molten iron then drops into a region where the temperature is 1500°F to 2500°F (800°C to 1400°C). The iron becomes saturated with carbides and free carbon. The absorbed carbon lowers the melting point of the iron from approximately 2800°F (1550°C) to approximately 2100°F (1150°C) so that it runs as a liquid to the bottom of the furnace.

The slag melts at approximately the same temperature as the iron but, being less dense, floats on the liquid iron. This allows the slag and iron to be drawn off separately. The iron usually goes in a liquid state to a subsequent refinement process, but may be allowed to cool in molds (forming blocks of iron known as *pigs*). The slag is discarded in a *slag heap*.

Since liquid iron is an excellent solvent, pig iron contains all of the minerals that are not fluxed away by the liquid limestone. The approximate composition of pig iron is 3% to 4% carbon, 1% to 3% silicon, 0.1% to 2% phosphorus, 0.5% to 2% manganese, and 0.01% to 0.1% sulfur. The actual composition will depend on the gangue elements. Calcium, magnesium, and aluminum oxides are fluxed out by the molten limestone and appear in the slag.

The subsequent process that the pig iron undergoes depends on the desired end product (i.e., the desired carbon content). This is shown in Fig. 46.2. The oxygen, open hearth, and electric furnace processes are used to produce steel.

Figure 46.2 *Methods of Refining Pig Iron*

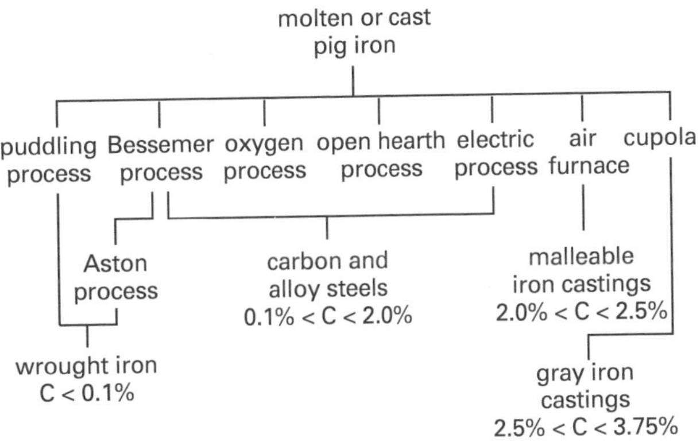

4. OXYGEN PROCESSES

Pig iron is saturated with carbon and contains other impurities. The *oxygen processes* (also known as the *dissolved oxygen process*, the L-D process, and the *Linz-Donawitz process*) are used to reduce the carbon content and purify the iron.[7]

The chemical refinement takes place in a pear-shaped steel crucible (a *converter*) lined with refractory material. The crucible is filled with a *bath* of molten pig iron at approximately 2200°F (1200°C), steel scrap, and lime. In the oxygen process, a water-cooled oxygen *lance* is lowered to within several feet (approximately a meter) of the bath surface. (See Fig. 46.3.)

High-pressure oxygen flows through the lance at high velocity, pushing aside the molten slag and exposing the molten iron. The silicon and manganese impurities are oxidized first, causing a temperature rise to approximately 3500°F (1900°C). Carbon is oxidized at the higher temperatures. Since the reaction is violent, the bath churns and circulates naturally.

The impurities are completely oxidized in approximately 25 minutes, although loading and unloading extends the cycle time to approximately one hour. Since the refinement also eliminates beneficial elements, measured amounts of carbon, manganese, and other alloying ingredients are subsequently added at the end of the refinement process to obtain the desired steel grade.

5. ELECTRIC ARC FURNACE

Electric furnaces common to "minimills" utilizing electric arc or induction heating are used to produce tool and special alloy steels.[8] It is possible to produce a high-

[7]Linz and Donawitz are the two Austrian towns in which the oxygen process was perfected.
[8]The term "minimill" has become a misnomer. Electric arc furnaces can produce up to 130 tons/hr, not much less than the 150 tons/hr to 300 tons/hr capacity of a standard basic oxygen furnace.

Figure 46.3 *Oxygen Process Crucible*

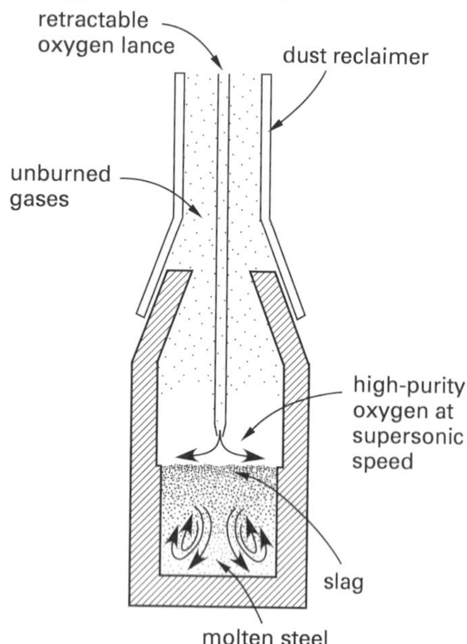

quality steel because air and gaseous fuels are not required, and the impurities they introduce are eliminated.

To further increase the quality and to reduce the expensive refining time, the charge is usually select scrap or *direct-reduced iron* rather than molten pig iron. As with the open hearth process, iron oxide is added as an oxidizing agent, and the composition is modified following refinement.

In an *electric arc furnace*, heat is generated by electrical arcs from three electrodes (for three-phase current) extending through the furnace wall down into the charge space. Although the electrode voltage is low (approximately 40 V), the current is high (approximately 12,000 A). Coils surround an *induction furnace*, and the heating is created from eddy current flowing within the melt.

The two major problems associated with electric arc furnaces are (1) steel contamination from trace metals present in the charging scrap, and (2) the formation of ionized nitrogen in the arc, a cause of undesirable hardening.

6. ADVANCED STEEL-MAKING PROCESSES

Air-quality legislation has had a substantial impact on the processes used to produce coke. New steel-making technologies are being used to reduce or eliminate the need for coke entirely (as in *direct iron-making processes*). Existing blast furnaces can be retrofitted to use pulverized coal (*coal injection*). In *reduced-coke processes*, coal, iron pellets and fines, and limestone are added to an already molten iron bath. Carbon from the coal combines with oxygen from the ore to produce carbon monoxide and molten iron. Oxygen is injected to burn some of the gas before it leaves the vessel.

7. STEEL AND ALLOY STEEL GRADES

The properties of steel can be adjusted by the addition of alloying ingredients. Some steels are basically mixtures of iron and carbon. Other steels are produced with a variety of ingredients.

The simplest and most common grades of steel belong to the group of *carbon steels*. Carbon is the primary noniron element, although sulfur, phosphorus, and manganese can also be present. Carbon steel can be subcategorized into *plain carbon steel* (*nonsulfurized carbon steel*), *free-machining steel* (*resulfurized carbon steel*), and *resulfurized and rephosphorized carbon steel*. Plain carbon steel is subcategorized into *low-carbon steel* (less than 0.30% carbon), *medium-carbon steel* (0.30% to 0.70% carbon), and *high-carbon steel* (0.70% to 1.40% carbon).

Low-carbon steels are used for wire, structural shapes, and screw machine parts. Medium-carbon steels are used for axles, gears, and similar parts requiring medium to high hardness and high strength. High-carbon steels are used for drills, cutting tools, and knives.

Low-alloy steels (containing less than 8.0% total alloying ingredients) include the majority of steel alloys but exclude the high-chromium content *corrosion-resistant* (*stainless*) *steels*. Generally, low-alloy steels will have higher strength (e.g., double the yield strength) of plain carbon steel. *Structural steel*, *high-strength steel*, and *ultrahigh-strength steel* are general types of low-alloy steel.[9]

High-alloy steels contain more than 8.0% total alloying ingredients.

Table 46.3 lists typical alloying ingredients and their effects on steel properties. The percentages represent typical values, not maximum solubilities.

Since steel properties are dependent on composition, steels are designated by composition. Table 46.4 shows the AISI-SAE four-digit designations for typical steels and alloys.[10] The first two digits designate the type of steel; the last two digits designate the percentage of carbon in hundredths of a percent. (For example, AISI steel 1035 is a plain carbon steel with 0.35% carbon. This is

[9]The *ultrahigh-strength steels*, also known as *maraging steels*, are very low-carbon (less than 0.03%) steels with 15% to 25% nickel and small amounts of cobalt, molybdenum, titanium, and aluminum. With precipitation hardening, ultimate tensile strengths up to 400,000 lbf/in^2 (2.8 GPa), yield strengths up to 250,000 lbf/in^2 (1.7 GPa), and elongations in excess of 10% are achieved. Maraging steels are used for rocket motor cases, aircraft and missile turbine housings, aircraft landing gear, and other applications requiring high strength, low weight, and toughness.
[10]The abbreviations stand for the American Iron and Steel Institute and the Society of Automotive Engineers.

also referred to as "35-point carbon" steel.) A number following an alloying ingredient is the nominal percentage of that ingredient.

Also, an optional capital letter may be added as a prefix to designate the manufacturing process (A, acid Bessemer; B, basic Bessemer; C, basic open hearth; CB, either B or C at steel mill option; O, basic oxygen).[11]

Table 46.3 *Steel Alloying Ingredients*

ingredient	range (%)	purpose
aluminum	–	deoxidation
boron	0.001–0.003	increase hardness
carbon	0.1–4.0	increase hardness and strength
chromium	0.5–2	increase hardness and strength
	4–18	increase corrosion resistance
copper	0.1–0.4	increase atmospheric corrosion resistance
iron sulfide	–	increase brittleness
manganese	0.23–0.4	reduce brittleness, combine with sulfur
	> 1.0	increase hardness
manganese sulfide	0.8–0.15	increase machinability
molybdenum	0.2–5	increase dynamic and high-temperature strength and hardness
nickel	2–5	increase toughness, increase hardness
	12–20	increase corrosion resistance
	> 30	reduce thermal expansion
phosphorus	0.04–0.15	increase hardness and corrosion resistance
silicon	0.2–0.7	increase strength
	2	increase spring steel strength
	1–5	improve magnetic properties
sulfur	–	(see *iron sulfide* and *manganese sulfide*)
titanium	–	fix carbon in inert particles; reduce martensitic hardness
tungsten	–	increase high-temperature hardness
vanadium	0.15	increase strength

Table 46.4 *AISI-SAE Steel Designations*

carbon steels	
10XX	nonsulfurized carbon steel (plain-carbon)
11XX	resulfurized carbon steel (free-machining)
12XX	resulfurized and rephosphorized carbon steel
low-alloy steels	
13XX	manganese 1.75
23XX	nickel 3.50
25XX	nickel 1.25, chromium 0.65
31XX	nickel 3.50, chromium 1.55
33XX	nickel 3.50, chromium 1.55
40XX	molybdenum 0.25
41XX	chromium 0.50 or 0.95, molybdenum 0.12 or 0.20
43XX	nickel 1.80, chromium 0.50 or 0.80, molybdenum 0.25
46XX	nickel 1.55 or 1.80, molybdenum 0.20 or 0.25
47XX	nickel 1.05, chromium 0.45, molybdenum 0.20
48XX	nickel 3.50, molybdenum 0.25
50XX	chromium 0.38 or 0.40
51XX	chromium 0.80, 0.90, 0.95, 1.00, or 1.05
5XXXX	chromium 0.50, 1.00, or 1.45, carbon 1.00
61XX	chromium 0.60, vanadium 0.10–0.15; or chromium 0.95, vanadium 0.15
86XX	nickel 0.55, chromium 0.50 or 0.65, molybdenum 0.20
87XX	nickel 0.55, chromium 0.50, molybdenum 0.25
92XX	manganese 0.85, silicon 2.00
93XX	nickel 3.25, chromium 1.20, molybdenum 0.12
98XX	nickel 1.00, chromium 0.80, molybdenum 0.25
heat-and corrosion-resistant steels	
2XX	chromium-nickel-manganese (nonhardenable, austenitic, nonmagnetic)
3XX	chromium-nickel (nonhardenable, austenitic, nonmagnetic)
4XX	chromium (hardenable, martensitic, magnetic)
4XX	chromium (generally not hardenable, ferritic, magnetic)
5XX	chromium (low-chromium, heat-resisting)

8. TOOL STEEL

Each grade of tool steel is designed for a specific purpose, and as such, there are few generalizations that can be made about tool steel. Each tool steel exhibits its own blend of the three main performance criteria: toughness, wear resistance, and *hot hardness*.[12]

Some of the few generalizations possible are listed as follows.

[11]The electric furnace process may be designated by a C, D, or E prefix.

[12]The ability of a steel to resist softening at high temperatures is known as *hot hardness* and *red hardness*.

- An increase in carbon content increases wear resistance and reduces toughness.
- An increase in wear resistance reduces toughness.
- Hot hardness is independent of toughness.
- Hot hardness is independent of carbon content.

Group A steels are air-hardened, medium-alloy cold-work tool steels. Air-hardening allows the tool to develop a homogeneous hardness throughout, without distortion. This hardness is achieved by large amounts of alloying elements and comes at the expense of wear resistance.

Group D steels are high-carbon, high-chromium tool steels suitable for cold-working applications. These steels are high in abrasion resistance but low in machinability and ductility. Some steels in this group are air hardened, while others are oil quenched. Typical uses are blanking and cold-forming punches.

Group H steels are hot-work tool steels, capable of being used in the 1100°F to 2000°F (600°C to 1100°C) range. They possess good wear resistance, hot hardness, shock resistance, and resistance to surface cracking. Carbon content is low, between 0.35% and 0.65%. This group is subdivided according to the three primary alloying ingredients: chromium, tungsten, or molybdenum. For example, a particular steel might be designated as a "chromium hot-work tool steel."

Group M steels are molybdenum high-speed steels. Properties are very similar to the group T steels, but group M steels are less expensive since one part molybdenum can replace two parts tungsten. For that reason, most high-speed steel in common use is produced from the M group. Cobalt is added in large percentages (5% to 12%) to increase high-temperature cutting efficiency in heavy-cutting (high-pressure cutting) applications.

Group O steels are oil-hardened, cold-work tool steels. These high-carbon steels use alloying elements to permit oil quenching of large tools and are sometimes referred to as *nondeforming steels*. Chromium, tungsten, and silicon are typical alloying elements.

Group S steels are shock-resistant tool steels. Toughness (not hardness) is the main characteristic, and either water or oil may be used for quenching. Group S steels contain chromium and tungsten as alloying ingredients. Typical uses are hot header dies, shear blades, and chipping chisels.

Group T steels are tungsten high-speed tool steels that maintain a sharp hard cutting edge at temperatures in excess of 1000°F (550°C). The ubiquitous 18-4-1 grade T1 (named after the percentages of tungsten, chromium, and vanadium, respectively) is part of this group. Increases in hot hardness are achieved by simultaneous increases in carbon and vanadium (the key ingredient in these tool steels) and special, multiple-step heat treatments.[13]

Group W steels are water-hardened tool steels. These are plain high-carbon steels (modified with small amounts of vanadium or chromium, resulting in high surface hardness but low hardenability). The combination of high surface hardness and ductile core makes group W steels ideal for rock drills, pneumatic tools, and cold header dies. The limitation on this tool steel group is the loss of hardness that begins at temperatures above 300°F (150°C) and is complete at 600°F (300°C).

Example 46.1

The composition of a group M tool steel is being formulated to replace the 18-4-1 group T steel. What are the percentages of alloying ingredients if two-thirds of the tungsten are to be replaced with molybdenum?

Solution

Since one part molybdenum replaces two parts tungsten, the alloy would be designated 6-6-4-1, representing 6% molybdenum, 6% tungsten, 4% chromium, and 1% vanadium.

9. STAINLESS STEEL

Adding chromium improves steel's corrosion resistance. Moderate corrosion resistance is obtained by adding 4% to 6% chromium to low-carbon steel. (Other elements, specifically less than 1% each of silicon and molybdenum, are also usually added.)

For superior corrosion resistance, larger amounts of chromium are needed. At a minimum level of 12% chromium, steel is *passivated* (i.e., an inert film of chromic oxide forms over the metal and inhibits further oxidation). The formation of this protective coating is the basis of the corrosion resistance of *stainless steel*.[14]

Passivity is enhanced by oxidizers and aeration but is reduced by abrasion that wears off the protective oxide coating. An increase in temperature may increase or decrease the passivity, depending on the abundance of oxygen.

Stainless steels are generally categorized into ferritic, martensitic (heat-treatable), austenitic, duplex, and high-alloy stainless steels.[15,16] Table 46.5 categorizes some of the more popular AISI grades of stainless steel.

[13]For example, the 18-4-1 grade is heated to approximately 1050°F (550°C) for two hours, air cooled, and then heated again to the same temperature. The term *double-tempered steel* is used in reference to this process. Most heat treatments are more complex.

[14]Stainless steels are corrosion resistant in oxidizing environments. In reducing environments (such as with exposure to hydrochloric and other halide acids and salts), the steel will corrode.

[15]There is a fifth category, that of *precipitation-hardened stainless steels*, widely used in the aircraft industry. (Precipitation hardening is also known as *age hardening*.) These steels have been given the AISI designation 630. UNS designations for precipitation-hardened stainless steels include S13800, S15500, S17400, and S17700.

[16]The *sigma phase* structure that appears at very high chromium levels (e.g., 24% to 50%) is usually undesirable in stainless steels because it reduces corrosion resistance and impact strength. A notable exception is in the manufacture of automobile engine valves.

Table 46.5 *Characteristics of Common Stainless Steels*

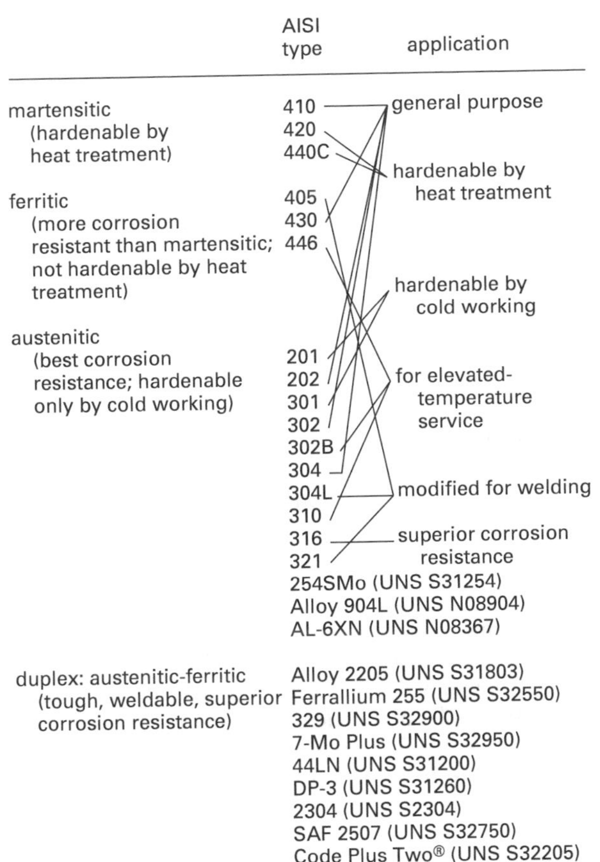

Ferritic stainless steels, grouped with the AISI 400 series, contain more than 10% to 27% chromium. The body-centered cubic ferrite structure is stable (i.e., does not transform to austenite, a face-centered cubic structure) at all temperatures. For this reason, ferritic steels cannot be hardened significantly. Since ferritic stainless steels contain no nickel, they are less expensive than austenitic steels. Turbine blades are typical of the heat-resisting products manufactured from ferritic stainless steels.

The so-called *superferritics*, such as Alloy 2904C (S44735), Sea-Cure (UNS S44660), and Alloy 2903, are highly resistant to chloride pitting and crevice corrosion. Superferritics have been incorporated into marine tubing and heat exchangers for power plant condensers. Like all ferritics, however, superferritics experience embrittlement above 885°F (475°C).

The *martensitic* (*heat-treatable*) *stainless steels* (also part of the AISI 400 series) contain no nickel and differ from ferritic stainless steels primarily in higher carbon contents. Cutlery and surgical instruments are typical applications requiring both corrosion resistance and hardness.

The *austenitic stainless steels* are commonly used for general corrosive applications. The stability of the austenite (a face-centered cubic structure) depends primarily on 4% to 22% nickel as an alloying ingredient. The basic composition is approximately 18% chromium and 8% nickel, hence the term "18 to 8 type." (See Table 46.6.)

Table 46.6 *Typical Compositions of Stainless Steels*

element	ferritic	martensitic	austenitic
carbon	0.08–0.20%	0.15–1.2%	0.03–0.25%
manganese	1–1.5%	1%	2%
silicon	1%	1%	1–2%
chromium	11–27%	11.5–18%	16–26%
nickel	–	–	3.5–22%
phosphorus and sulfur	–	–	normal
molybdenum	–	–	some cases
titanium	–	–	some cases

The so-called *superaustenitics*, such as the 317 and 316L series, achieve superior corrosion resistance by adding more molybdenum and nitrogen, respectively. AISI type 317 (UNS S31700) contains 3% to 4% molybdenum. Variants of type 317L (UNS S317XX) contain up to 5% molybdenum. Variants of type 316L achieve superior corrosion resistance by adding 10% to 14% nitrogen. "6-Mo" superaustenitics with approximately 6% molybdenum, 20% chromium, and 0.10% nitrogen are well established, particularly in the chemical process industry. "7-Mo" alloys probably represent the ultimate in corrosion resistance while still remaining commercially viable.

Austenitic stainless steels are the most weldable of the stainless steels but are nevertheless susceptible to sensitization. They are nonmagnetic and are hardenable only by cold working. They can be polished to a mirror finish, which makes them useful in food-industry applications. Because of the nickel, they are more expensive than ferritic stainless steels.

Welding stainless steels is possible when proper welding rod alloys are used, but is difficult for several reasons.

- Stainless steels, particularly austenitic types, possess relatively low thermal conductivities and high coefficients of thermal expansion. The maintenance of a high temperature gradient (because the heat is not readily dissipated) and a high expansion increases the possibility of *weld bead cracking* (i.e., longitudinal cracking along the weld). Weld bead cracking can be minimized by welding at as low a temperature as possible.
- High temperature sensitizes the steel adjacent to welds, producing local chromium deficits. This phenomenon is known as *sensitization*, and the resulting corrosion is known as *weld decay*.
- Substantial grain growth occurs in ferritic steels, since there is no gamma-alpha transformation to

keep grains small. Growth of grains is substantial at normal welding temperatures.

- When stainless steels cool from welding temperatures, martensite forms unless cooling is slowed down. The martensite makes the metal brittle and reduces its ductility.

Because of their nominal costs, austenitic AISI 304 and 316 are commonly used corrosion-resistant steels. However, their low strengths make them unsuitable for high-pressure applications. They are also susceptible to wear and galling, and they have limited resistance to localized corrosion and *stress corrosion cracking* (SCC), particularly *chloride stress corrosion cracking* (CSCC) above 130°F (54°C).[17] Alternatives include duplex and high-alloy austenitic stainless steels.

Second-generation *duplex stainless steels* are austenitic-ferritic stainless steels that have the toughness and weld-ability of austenitic stainless steels and yield strengths and corrosion/wear resistances greater than the 300 series.[18] Alloy 2205 is the most widely used. Types 2304 and 2507 (UNS S32304 and S32507, respectively) are third-generation duplex steels designed to reduce mill costs. With the highest percentages of nitrogen, molybdenum, and nickel of any duplex stainless steel, 2507 has been dubbed a "super-duplex" stainless steel.

The *high-alloy austenitic stainless steels* containing 22% to 28% chromium, 24% to 32% nickel, and 4% to 6% molybdenum provide superior corrosion resistance at lower cost than the nickel- and titanium-based alloys they replace.

10. CAST IRON

Cast iron is a general name given to a wide range of alloys containing iron, carbon, and silicon, and to a lesser extent, manganese, phosphorus, and sulfur. Generally, the carbon content will exceed 2%. The properties of cast iron depend on the amount of carbon present, as well as the form (i.e., graphite or carbide) of the carbon.

Carbon in the form of carbide is stable only at low temperatures. (The carbide is said to be a *metastable structure.*) At high temperatures, *graphitization* takes place according to the following reaction.

$$Fe_3C \rightarrow 3Fe + C \text{ (graphite)}$$

The most common type of cast iron is *gray cast iron.* The carbon in gray cast iron is in the form of graphite flakes. Graphite flakes are very soft and constitute points of weakness in the metal, which simultaneously improve machinability and decrease ductility. Gray cast iron is categorized into classes according to its tensile strength, as shown in Table 46.7. Compressive strength is three to five times the tensile strength.

Table 46.7 *Classes of Gray Cast Iron**

	minimum tensile strength		tensile modulus of elasticity	
class	lbf/in^2	MPa	lbf/in^2	GPa
20	20,000	138	10–14 × 10^6	69–97
25	25,000	172	12–15 × 10^6	83–104
30	30,000	207	13–16.5 × 10^6	90–114
35	35,000	242	14.5–17 × 10^6	100–117
40	40,000	276	16–20 × 10^6	110–138
45	45,000	310	–	–
50	50,000	345	18.8–23 × 10^6	130–159
60	60,000	414	20.4–23.5 × 10^6	141–162

(Multiply lbf/in^2 by 0.006895 to obtain MPa.)

*ASTM specification A48

Magnesium and cerium can be added to improve the ductility of gray cast iron. The resulting *nodular cast iron* (also known as *ductile cast iron*) has the best tensile and yield strengths of all the cast irons. It also has good ductility (typically 5%) and machinability. Because of these properties, it is often used for automobile crankshafts. (See Table 46.8.)

Table 46.8 *Common Grades of Ductile Iron[a]*

	minimum tensile strength		minimum yield strength		elongation
class/grade	ksi	MPa	ksi	MPa	%
60-40-18[b]	60	410	40	280	18
65-45-12[c]	65	450	45	310	12
80-55-06[d]	80	550	55	380	6
100-70-03[e]	100	690	70	480	3
120-90-02[f]	120	830	90	620	2

(Multiply ksi by 6.895 to obtain MPa.)

[a]ASTM A-536-70
[b]May be annealed after casting.
[c]An as-cast grade.
[d]An as-cast grade with higher manganese content.
[e]Usually obtained by a normalizing heat treatment.
[f]Oil quenched and tempered to specified hardness.

White cast iron has been cooled quickly from a molten state. No graphite is produced from the cementite, and the carbon remains in the form of a carbide, Fe_3C.[19] The carbide is hard and is the reason that white cast iron is difficult to machine. White cast iron is used primarily in the production of malleable cast iron.

[17]Chloride stress corrosion cracking is an important issue in heat exchangers for use with ocean and inland water.
[18]The first generation duplex stainless steels developed in the 1930s, such as AISI type 329 (UNS S32900), lost much of their corrosion resistance after welding unless they were given a post-weld heat treatment. Second-generation stainless steels contain less carbon and 0.15% to 0.30% nitrogen and, when combined with proper welding technique, offer the same level of corrosion resistance as mill-annealed material.
[19]White and gray cast irons get their names from the coloration at a fracture.

Malleable cast iron is produced by reheating white cast iron to between 1500°F and 1850°F (800°C and 1000°C) for several days, followed by slow cooling. During this treatment, the carbide is partially converted to nodules of graphitic carbon known as *temper carbon*. The tensile strength is increased to approximately 55,000 lbf/in^2 (380 MPa), and the elongation at fracture increases to approximately 18%.

Mottled cast iron contains both cementite and graphite and is between white and gray cast irons in composition and performance.

Compacted graphitic iron (CGI) is a unique form of cast iron with worm-shaped graphite particles. The shape of the graphite particles gives CGI the best properties of both gray and ductile cast iron: twice the strength of gray cast iron and half the cost of aluminum. The higher strength permits thinner sections. (Some engine blocks are 25% lighter than gray iron castings.) Using computer-controlled refining, volume production of CGI with the consistency needed for commercial applications is possible.

Silicon is the most important element affecting graphitization. The effects of various elements in cast iron are listed in Table 46.9. Most of the elements that increase hardness do so by promoting the formation of iron carbide.

Table 46.9 *Effects of Elements in Cast Iron*

element	effect
aluminum	deoxidizes molten cast iron
carbon	depending on form, affects machinability, ductility, and shrinkage
manganese	below 0.5%, reduces hardness by combining with sulfur; above 0.5%, increases hardness
phosphorus	increases fluidity and lowers melting temperature
silicon	below 3.25%, softens iron and increases ductility; above 3.25%, hardens iron; above 13%, increases acid and corrosion resistance
sulfur	increases hardness, sulfur is removed by addition of manganese

11. WROUGHT IRON

Wrought iron is low-carbon (less than 0.1%) iron with small amounts (approximately 3%) of slag and gangue in the form of fibrous inclusions. It has good ductility and corrosion resistance. Prior to the use of steel, wrought iron was the most important structural metal.

In the ancient *puddling process*, wrought iron is produced in a *reverberatory furnace* similar to the open hearth furnace.[20] The molten iron floats on a layer of iron oxide that provides the oxygen for removal of almost all of the carbon, sulfur, and manganese. The limestone flux combines with silicon and phosphorus to form slag.

As the iron becomes purer, its melting temperature increases to above the furnace temperature. Spongy masses of congealing iron and slag are collected on the ends of rods inserted into the pool of molten metal. These masses are then removed and forged or hammered to squeeze out most of the slag. The remaining product consists of slag-coated iron particles welded together by the forging processes. The deformed slag particles contribute to the fatigue resistance of wrought iron.

In the modern *Aston process* (*Byers-Aston process*), pig metal is melted in a cupola and is then highly purified in a Bessemer converter.[21] Simultaneously and separately, molten slag is prepared in an open hearth furnace and transferred to a mixing ladle. The molten iron is poured into the cooler slag in the mixing ladle (a process known as *shotting*), where the iron rapidly solidifies and releases dissolved gases. The gases fracture the solidifying iron, and molten slag enters the fissures. The excess molten slag is poured off, and the metal mass is pressed and rolled into blooms, billets, and slabs to remove most of the interior slag.[22] The slabs are hot-rolled together to form larger pieces of wrought iron.

12. PRODUCTION OF ALUMINUM

Aluminum is produced from *bauxite ore*, a mixture of hydroxides of aluminum ($Al_2O_3 \cdot n[H_2O]$) and oxides of iron, silicon, and titanium. Most of the nonrecycled aluminum produced today is produced electrochemically.

In the *Bayer process*,[23] the ore is crushed and ground into a fine powder. It is then treated with a hot solution of sodium hydroxide, producing a solution of sodium aluminate. The solution is drawn off into a separate tank, leaving the remaining ore constituents (known as *red mud* because of the iron coloration) as a solid deposit to be discarded.

$$Al(OH)_3 + NaOH \rightarrow NaAl(OH)_4$$

[20] A *reverberatory furnace* is a cavernous, brick-lined chamber. The metal and ore are melted by flames that play over the top of the melt.

[21] A *cupola* is a tall, open-top vertical stack lined with furnace brick. An air blast is introduced at the base. Heat is produced from the combustion of coke mixed with the ore.

[22] In the rolling mill industry, a *bloom* is a long piece having a cross section greater than approximately 6 in (15 cm) square. A *billet* is smaller than a bloom, with a cross section greater than approximately 1.5 in (4 cm). A *slab* has a minimum thickness of 1.5 in (4 cm) and width between 10 and 15 in (25 and 40 cm). The width of a slab is always at least three times the thickness.

[23] Other metals that are refined in a similar electrochemical process are magnesium, copper, zinc, and (to lesser extents) gold and silver.

As the solution cools, aluminum hydroxide precipitates, leaving a sodium hydroxide solution. The aluminum hydroxide is collected in solid, crystalline form and baked to form *alumina* (aluminum oxide, Al_2O_3). Because of its high melting temperature (3720°F, 2050°C), alumina cannot be economically reduced in a furnace.

Final reduction (*smelting*) is accomplished through the electrolytic *Hall-Heroult process* using molten *cryolite* (Na_3AlF_6) as the electrolyte. Large carbon blocks act as anodes, and the carbon-lined steel tank acts as the cathode.

The aluminum oxide dissolves in the cryolite and is separated into molten aluminum and oxygen gas by the electric current. The aluminum collects in the bottom of the tank. Carbon dioxide is released at the anodes. The cryolite recomposes after decomposition and can be reused. (The following reactions disregard the cryolite.)

$$Al_2O_3 \rightarrow 2Al^{+++} + 3O^{--}$$

$$Al^{+++} + 3e^- \rightarrow Al$$

$$C + 2O^{--} \rightarrow CO_2 + 4e^-$$

13. PROPERTIES OF ALUMINUM

Aluminum satisfies applications requiring low weight, corrosion resistance, and good electrical and thermal conductivities. Its corrosion resistance derives from the oxide film that forms over the raw metal, inhibiting further oxidation. The primary disadvantages of aluminum are its cost and low strength.

In pure form, aluminum is soft, ductile, and not very strong. Copper, manganese, magnesium, and silicon can be added to increase its strength, at the expense of other properties, primarily corrosion resistance.[24] Aluminum is hardened by the *precipitation hardening* (*age hardening*) process.

The oxide coating that forms readily (particularly at high temperatures) on aluminum complicates welding. However, special processes that perform the welding under a blanket of inert gas (e.g., helium or argon) overcome this complication.[25]

14. ALUMINUM ALLOYS

Except for use in electrical work, most aluminum is alloyed with other elements, primarily copper, magnesium, and silicon.[26] Aluminum alloys are identified by a four-digit number and a letter suffix (e.g., 2014-T4). The number indicates the major alloying ingredient and chemical composition of the alloy, as determined from Table 46.10. The suffix indicates the condition of the alloy, as determined from Table 46.11 and Table 46.12.

Table 46.10 *Aluminum Designations*

designation	major alloying ingredient
1XXX	commercially pure (99+%)
2XXX	copper
3XXX	manganese
4XXX	silicon
5XXX	magnesium
6XXX	magnesium and silicon
7XXX	zinc
8XXX	other

Table 46.11 *Conditions of Aluminum Alloys*

letter suffix	meaning
F	as fabricated
O	soft (after annealing)
H	strain hardened (cold worked) temper
T	heat treated

Table 46.12 *Aluminum Treatment Conditions*

suffix	meaning
H1	strain hardened by working to desired dimensions
H2	strain hardened by cold working, followed by partial annealing
H3	strain hardened and stabilized
T2	annealed (castings only)
T3	solution heat treated, followed by cold working (strain hardening)
T4	solution heat treated, followed by natural aging at room temperature
T5	artificial aging only
T6	solution heat treated, followed by artificial aging
T7	solution heat treated, followed by stabilizing by overaging heat treating
T8	solution heat treated, followed by cold working and subsequent artificial aging
T9	solution heat treated, followed by artificial aging and subsequent cold working

Silicon occurs as a normal impurity in aluminum, and in natural amounts (less than 0.4%), it has little effect on properties. If moderate quantities (above 3%) of silicon are added, the molten aluminum will have high fluidity,

[24]One ingenious method of having both corrosion resistance and strength is to produce a composite material. *Alclad* is the name given to aluminum alloy that has a layer of pure aluminum bonded to the surface. The alloy provides the strength, and the pure aluminum provides the corrosion resistance.

[25]TIG (tungsten-inert gas) and MIG (metal-inert gas) processes are commonly used to weld aluminum.

[26]*EC (electrical-conductor) grade aluminum* consists of approximately 99.45% aluminum.

making it ideal for castings. Above 12%, silicon improves the hardness and wear resistance of the alloy. When combined with copper and magnesium (as Mg_2Si and AlCuMgSi) in the alloy, silicon improves age hardenability. Silicon has negligible effect on the corrosion resistance of aluminum.

Copper improves the age hardenability of aluminum, particularly in conjunction with silicon and magnesium. Thus, copper is a primary element in achieving high mechanical strength in aluminum alloys at elevated temperatures. Copper also increases the conductivity of aluminum, but decreases its corrosion resistance.

Magnesium is highly soluble in aluminum and is used to increase strength by improving age hardenability. Magnesium improves corrosion resistance and may be added when exposure to saltwater is anticipated.

Some aluminum alloys can be work-hardened (e.g., 1100, 3003, 5052). The ductility of these alloys decreases as strength is increased through working. Most aluminum alloys (e.g., 2014, 2017, 2024, 6061), however, must be precipitation-hardened.[27] The decrease in ductility with increased strength through heat treatment is small or nonexistent.

The letter suffixes H and T are followed by numbers that provide additional detail about the type of hardening process used to achieve the material properties. Table 46.12 lists the types of treatments associated with the H and T suffix letters.

15. PRODUCTION OF COPPER

Copper occurs in the free (metallic) state as well as in ores containing its oxides, sulfides, and carbonates. *Native copper* is recovered by the simple process of heating highly crushed ore. Molten copper flows to the bottom of the furnace.

Oxides and carbonates of copper are reduced in a blast or reverberatory furnace.

Sulfides of copper are heated in air, and the sulfur is replaced by oxygen. The product, which contains both copper and iron oxides, is further reduced in a reverberatory furnace. This step removes the oxygen but leaves some sulfur and iron. The final sulfur-removal process takes place in a furnace similar to a Bessemer converter in which air is injected into the molten copper. The iron oxide combines with a silica furnace liner.

Very low-grade ores are leached with sulfuric acid to recover the copper.

Regardless of the primary recovery method, *electrolysis* (*electrodeposition*) is generally required to remove the remaining impurities. Thick sheets of impure copper and thin sheets of pure copper are immersed together in an electrolyte of copper sulfate. The pure copper acts as the cathode. A direct current causes copper from the impure sheets to migrate to the pure sheets. Impurities drop to the bottom of the tank as they are released.

16. ALLOYS OF COPPER

Zinc is the most common alloying ingredient in copper. It constitutes a significant part (up to 40% zinc) in brass.[28] (Brazing rod contains even more, approximately 45% to 50%, zinc.) Zinc increases copper's hardness and tensile strength. Up to approximately 30%, it increases the percent elongation at fracture. It decreases electrical conductivity considerably. *Dezincification*, a loss of zinc in the presence of certain corrosive media or at high temperatures, is a special problem that occurs in brasses containing more than 15% zinc.

Tin constitutes a major (up to 20%) component in most bronzes. Tin increases fluidity, which improves casting performance. In moderate amounts, corrosion resistance in saltwater is improved. (*Admiralty metal* has approximately 1%; *government bronze* and *phosphorus bronze* have approximately 10% tin.) In moderate amounts (less than 10%), tin increases the alloy's strength without sacrificing ductility. Above 15%, however, the alloy becomes brittle. For this reason, most bronzes contain less than 12% tin. Tin is more expensive than zinc as an alloying ingredient.

Lead is practically insoluble in solid copper. When present in small to moderate amounts, it forms minute soft particles that greatly improve machinability (2% to 3% lead) and wearing (bearing) properties (10% lead).

Silicon increases the mechanical properties of copper by a considerable amount. On a per unit basis, silicon is the most effective alloying ingredient in increasing hardness. *Silicon bronze* (96% copper, 3% silicon, 1% zinc) is used where high strength combined with corrosion resistance is needed (e.g., in boilers).

If aluminum is added in amounts of 9% to 10%, copper becomes extremely hard. Thus, *aluminum bronze* (as an example) trades an increase in brittleness for increased wearing qualities. Aluminum in solution with the copper makes it possible to precipitation harden the alloy.

Beryllium in small amounts (less than 2%) improves the strength and fatigue properties of copper. These properties make precipitation-hardened *copper-beryllium* (*beryllium-copper*, *beryllium bronze*, etc.) ideal for small springs. These alloys are also used for producing nonsparking tools.

[27]These alloys are all known as *duralumin*.

[28]*Brass* is an alloy of copper and zinc. *Bronze* is an alloy of copper and tin. Unfortunately, brasses are often named for the color of the alloys, leading to some very misleading names. For example, *nickel silver*, *commercial bronze*, and *manganese bronze* are all brasses.

17. NICKEL AND ITS ALLOYS

Like aluminum, nickel is largely hardened by precipitation hardening. Nickel is similar to iron in many of its properties, except that it has higher corrosion resistance and a higher cost. Also, nickel alloys have special electrical and magnetic properties.

Copper and iron are completely miscible with nickel. Copper increases formability. Iron improves electrical and magnetic properties markedly.

Some of the better-known nickel alloys are *monel metal* (30% copper, used hot-rolled where saltwater corrosion resistance is needed), *K-monel metal* (29% copper, 3% aluminum, precipitation-hardened for use in valve stems), *inconel* (14% chromium, 6% iron, used hot-rolled in gas turbine parts), and *inconel-X* (15% chromium, 7% iron, 2.5% titanium, aged after hot rolling for springs and bolts subjected to corrosion). *Hastelloy* (22% chromium) is another well-known nickel alloy.[29]

Nichrome (15% to 20% chromium) has high electrical resistance, high corrosion resistance, and high strength at red heat temperatures, making it useful in resistance heating. *Constantan* (40% to 60% copper, the rest nickel) also has high electrical resistance and is used in thermocouples.

Alnico (14% nickel, 8% aluminum, 24% cobalt, 3% copper, the rest iron) and *cunife* (20% nickel, 60% copper, the rest iron) are two well-known nickel alloys with magnetic properties ideal for permanent magnets. Other magnetic nickel alloys are *permalloy* and *permivar*.

Invar, *Nilvar*, and *Elinvar* are nickel alloys with low or zero thermal expansion and are used in thermostats, instruments, and surveyors' measuring tapes.

For decades, C-family Alloy C-276 (Alloy 276) was the nickel-chromium-molybdenum workhorse for piping and reaction vessels in chemical process industries. Newer alloys from the C-family (e.g., Alloy C-22, also designated as Alloy 22, 622, and 5621 hMoW) with more chromium and less tungsten are extremely corrosion resistant, even in extremely aggressive, mixed-acid environments. Alloy 59 (UNS N06059), also referred to as Alloy 5923 hMo, maintains its corrosion resistance in strongly oxidizing environments, including the most severe corrosion conditions in modern pollution-control equipment.

18. REFRACTORY METALS

Reactive and *refractory metals* include alloys based on titanium, tantalum, zirconium, molybdenum, niobium (also known as columbium), and tungsten. These metals are used when superior properties (i.e., corrosion resistance) are needed. They are most often used where high-strength acids are used or manufactured.

19. NATURAL POLYMERS

A *polymer* is a large molecule in the form of a long chain of repeating units. The basic repeating unit is called a *monomer* or just *mer*. (A large molecule with two alternating mers is known as a *copolymer* or *interpolymer*. Vinyl chloride and vinyl acetate form one important family of copolymer plastics.)

Many of the natural organic materials (e.g., rubber and asphalt) are polymers. (Polymers with elastic properties similar to rubber are known as *elastomers*.) Natural rubber is a polymer of the *isoprene latex* mer (formula $[C_5H_8]_n$, repeating unit of $CH_2{=}CCH_3{-}CH{=}CH_2$, systematic name of 2-methyl-1,3-butadiene). The strength of natural polymers can be increased by causing the polymer chains to cross-link, restricting the motion of the polymers within the solid.

Cross-linking of natural rubber is known as *vulcanization*. Vulcanization is accomplished by heating raw rubber with small amounts of sulfur. The process raises the tensile strength of the material from approximately 300 lbf/in^2 (2.1 MPa) to approximately 3000 lbf/in^2 (21 MPa). The addition of carbon black as a reinforcing *filler* raises this value to approximately 4500 lbf/in^2 (31 MPa) and provides tear resistance and toughness.

The amount of cross-linking between the mers determines the properties of the solid. Figure 46.4 shows how sulfur joins two adjacent isoprene (natural rubber) mers in *complete cross-linking*.[30] If sulfur does not replace both of the double carbon bonds, *partial cross-linking* is said to have occurred.

Example 46.2

What is the approximate fraction (by mass) of sulfur in a completely cross-linked natural rubber polymer?

Solution

Figure 46.4 shows that, for complete cross-linking, each mer of the natural rubber mer requires one sulfur atom.

The atomic weight of sulfur is approximately 32 g/mol. The molecular weight of the rubber mer is

$$(5)\left(12\ \frac{\text{g}}{\text{mol}}\right) + (8)\left(1\ \frac{\text{g}}{\text{mol}}\right) = 68$$

[29] K-monel is one of four special forms of monel metal. There are also H-monel, S-monel, and R-monel forms.

[30] A tire tread may contain 3% to 4% sulfur. Hard rubber products, which do not require flexibility, may contain as much as 40% to 50% sulfur.

Figure 46.4 Vulcanization of Natural Rubber

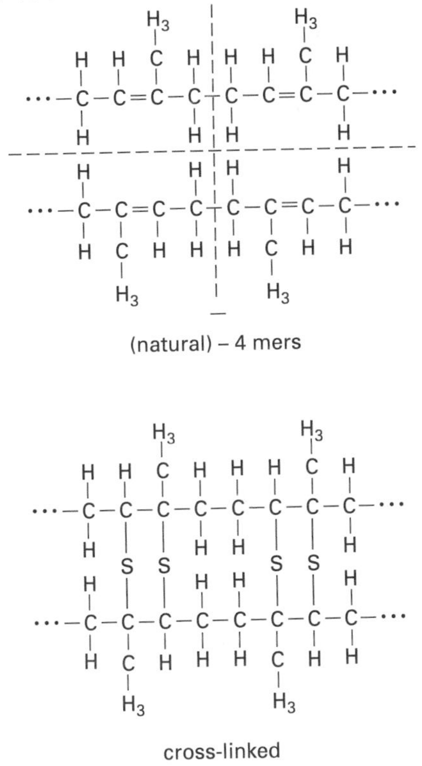

(natural) – 4 mers

cross-linked

The fraction of sulfur is

$$\frac{m_{\text{sulfur}}}{m_{\text{sulfur}} + m_{\text{mer}}} = \frac{32 \ \frac{\text{g}}{\text{mol}}}{32 \ \frac{\text{g}}{\text{mol}} + 68 \ \frac{\text{g}}{\text{mol}}} = 0.32 \quad (32\%)$$

20. DEGREE OF POLYMERIZATION

The *degree of polymerization*, DP, is the average number of mers in the molecule, typically several hundred to several thousand.[31] (In general, compounds with degrees of less than ten are called *telenomers* or *oligomers*.) The degree of polymerization can be calculated from the mer and polymer molecular weights.

$$\text{DP} = \frac{M_{\text{polymer}}}{M_{\text{mer}}} \qquad \textit{46.1}$$

A polymer batch usually will contain molecules with different length chains. Therefore, the degree of polymerization will vary from molecule to molecule, and an average degree of polymerization is reported.

The stiffness and hardness of polymers vary with their degrees. Polymers with low degrees are liquids or oils. With increasing degree, they go through waxy to hard resin stages. High-degree polymers have hardness and strength qualities that make them useful for engineering applications. Tensile strength and melting (softening) point also increase with increasing degree of polymerization.

Example 46.3

A polyvinyl chloride molecule (mer molecular weight of 62.5) is found to contain 860 carbon atoms, 1290 hydrogen atoms, and 430 chlorine atoms. What is the degree of polymerization?

Solution

The approximate atomic weights of carbon, hydrogen, and chlorine are 12 g/mol, 1 g/mol, and 35.5 g/mol, respectively. The molecular weight of the molecule is

$$\begin{aligned} M_{\text{polymer}} &= (860)\left(12 \ \frac{\text{g}}{\text{mol}}\right) + (1290)\left(1 \ \frac{\text{g}}{\text{mol}}\right) \\ &\quad + (430)\left(35.5 \ \frac{\text{g}}{\text{mol}}\right) \\ &= 26{,}875 \text{ g/mol} \end{aligned}$$

The degree of polymerization is

$$\text{DP} = \frac{M_{\text{polymer}}}{M_{\text{mer}}} = \frac{26{,}875 \ \frac{\text{g}}{\text{mol}}}{62.5 \ \frac{\text{g}}{\text{mol}}} = 430$$

Example 46.4

What is the degree of polymerization of a polyvinyl acetate (mer of $C_4H_6O_2$, molecular weight of 86 g/mol) sample having the following analysis of molecular weights?

range of molecular weights (g/mol)	mole fraction
5001–15,000	0.30
15,001–25,000	0.47
25,001–35,000	0.23

Solution

Using the midpoint of each range, the average polymer molecular weight is

$$\begin{aligned} M_{\text{polymer}} &= (0.30)\left(10{,}000 \ \frac{\text{g}}{\text{mol}}\right) \\ &\quad + (0.47)\left(20{,}000 \ \frac{\text{g}}{\text{mol}}\right) \\ &\quad + (0.23)\left(30{,}000 \ \frac{\text{g}}{\text{mol}}\right) \\ &= 19{,}300 \text{ g/mol} \end{aligned}$$

[31]Degrees of polymerization for commercial plastics are usually less than 1000.

The degree of polymerization is

$$\mathrm{DP} = \frac{M_{\text{polymer}}}{M_{\text{mer}}} = \frac{19{,}300 \ \frac{\text{g}}{\text{mol}}}{86 \ \frac{\text{g}}{\text{mol}}} = 224$$

21. SYNTHETIC POLYMERS

Table 46.13 lists some of the common mers. Polymers are named by adding the prefix "poly" to the name of the basic mer. For example, C_2H_4 is the chemical formula for ethylene. Chains of C_2H_4 are called polyethylene.

***Table 46.13** Names of Common Mers*

name	repeating unit	combined formula
ethylene	CH_2CH_2	C_2H_4
propylene	$CH_2(HCCH_3)$	C_3H_6
styrene	$CH_2CH(C_6H_5)$	C_8H_8
vinyl acetate	$CH_2CH(C_2H_3O_2)$	$C_4H_6O_2$
vinyl chloride	CH_2CHCl	C_2H_3Cl
isobutylene	$CH_2C(CH_3)_2$	C_4H_8
methyl methacrylate	$CH_2C(CH_3)(COOCH_3)$	$C_5H_8O_2$
acrylonitrile	CH_2CHCN	C_3H_3N
epoxide (ethoxylene)	CH_2CH_2O	C_2H_4O
amide (nylon)	$CONH_2$ or CONH	$CONH_2$ or CONH

Polymers are able to form when double (covalent) bonds break and produce reaction sites. The number of bonds in the mer that can be broken open for attachment to other mers is known as the *functionality* of the mer. The ethylene mer in Fig. 46.5 is *bifunctional* since the C=C bond can be broken to form two reaction sites (e.g., –C–C–). Other mers are *trifunctional* or *tetrafunctional.* When combining into chains, bifunctional mers form *linear polymers*, whereas trifunctional and tetrafunctional mers form *network polymers.*

***Figure 46.5** Steps in Forming Polyethylene*

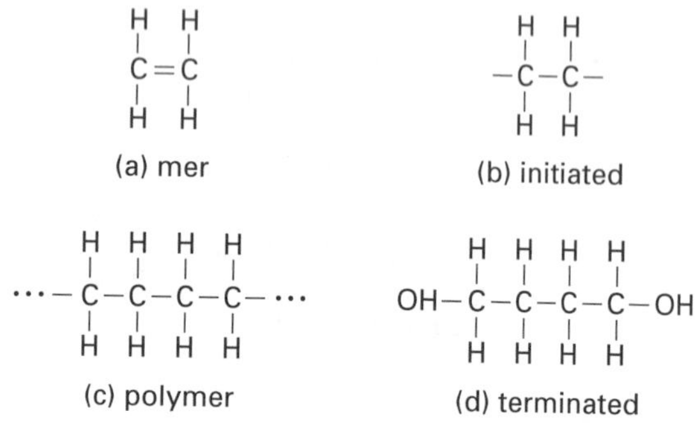

```
H H
| |
C=C
| |
H H
```

(a) mer

```
  H H
  | |
 -C-C-
  | |
  H H
```

(b) initiated

```
      H H H H
      | | | |
 ···- C-C-C-C- ···
      | | | |
      H H H H
```

(c) polymer

```
    H H H H
    | | | |
 OH-C-C-C-C-OH
    | | | |
    H H H H
```

(d) terminated

Two processes produce reaction sites: addition polymerization and condensation polymerization. With *addition polymerization*, mers simply combine sequentially into chains by breaking double (covalent) bonds. No other products are produced. For example, the formation of polyethylene is given by the following reaction and Fig. 46.5.

$$2CH_2 = CH_2 \rightarrow -CH_2 - CH_2 - CH_2 - CH_2 -$$

Substances called *initiators* are used to start addition polymerization.[32] Initiators break down under heat, light, or other energy and provide free radicals. A free radical disrupts the double bond in a mer and attaches to one side of the mer, opening up the mer's double bond and releasing enough energy to open another mer's double bond.

The process continues (*propagation*) until the initiator is used up, until the mers are used up, or until another free radical terminates the chain. The latter occurrence (i.e., when there is no possibility for additional joining) is known as *saturation.* Since the same radicals that initiated a polymer chain can terminate the chain, initiators are also known as *terminators.*

Condensation polymerization (*step reaction polymerization*) also requires opening bonds in two molecules to form a larger molecule. This formation is often accompanied by the release of small molecules such as H_2O, CO_2, or N_2. The repeating units derived from the condensation process are not the same as the monomers from which they are formed since portions of the original monomer form the small molecules that are lost.[33] The formation of *phenolic plastics* from formaldehyde and ammonia illustrates condensation polymerization.

$$6HCHO + 4NH_3 \rightarrow (CH_2)_6N_4 + 6H_2O$$

22. FLUOROPOLYMERS

Fluoropolymers (*fluoroplastics*) are a class of paraffinic, thermoplastic polymers in which some or all of the hydrogens have been replaced by fluorine.[34] There are seven major types of fluoropolymers, with overlapping characteristics and applications. They include the fully fluorinated fluorocarbon polymers of Teflon® PTFE (polytetrafluoroethylene), FEP (fluorinated ethylene propylene), and PFA (perfluoroalkoxy), as well as the partially fluorinated polymers of PCTFE (polychlorotrifluoroethylene), ETFE (ethylene tetrafluoroethylene), ECTFE (ethylene chlorotrifluoroethylene), and PVDF (polyvinylidene fluoride). (See Table 46.14.)

[32] *Initiators* are usually *organic peroxides* such as *hydrogen peroxide* (H_2O_2) and *benzoyl peroxide* (C_6H_5COO).
[33] The formation of nylon and the famous *Bakelite material* from phenol (C_6H_5OH) are examples of condensation polymerization.
[34] *Fluoroelastomers* are uniquely different from fluoropolymers. They have their own areas of application.

Fluoropolymers compete with metals, glass, and other polymers in providing corrosion resistance. Choosing the right fluoropolymer depends on the operating environment including temperature, chemical exposure, and mechanical stress.

Table 46.14 Characteristics of Fluoropolymers

fluoro-polymer	specific gravity	tensile strength (ksi)	flexural modulus (ksi)	upper service temperature[a] (°C)
PTFE[b]	2.13–2.22	2.0–6.5	70–110	287
FEP	2.12–2.17	2.7–3.1	90	204
PFA	2.12–2.17	4.0–4.5	100	260
PCTFE	2.08–2.2	4.0–6.0	150–260	199
ETFE	1.70	6.5	200	149–182
ECTFE	1.68	6.6–7.8	240	149–179
PVDF[c]	1.76–1.78	3.5–6.2	70–320	150

(Multiply ksi by 6.895 to obtain MPa.)

[a]Upper service temperature is also known as *temperature of continuous heat resistance*.

[b]Properties of PTFE are highly variable, depending on type of resin and method of processing.

[c]Properties of PVDF include those of its copolymers, hence the range of values.

PTFE, the first available fluoropolymer, is probably the most inert compound known. It has been used extensively for pipe and tank linings, fittings, gaskets, valves, and pump parts. It has the highest operating temperature—approximately 500°F (260°C). Unlike the other fluoropolymers, however, it is not a melt-processed polymer. Like a powdered metallurgy product, PTFE is processed by compression and isostatic molding, followed by sintering. PTFE is also the weakest of all the fluoropolymers.

23. ELASTOMERIC COMPOUNDS

Flexible parts, such as gaskets and O-rings, are manufactured from *elastomeric compounds.* Common elastomerics include natural rubber, butyl rubber, buna-N (nitrile rubber), neoprene, ethylene-propylene-diene monomer (EPDM) rubber, chlorosulfonated polyethylene, and various fluoroelastomers. Chemical resistance, temperature, and pressure are the primary factors considered in choosing sealing compounds.

24. THERMOSETTING AND THERMOPLASTIC POLYMERS

Most polymers can be softened and formed by applying heat and pressure. These are known by various terms including *thermoplastics*, *thermoplastic resins*, and *thermoplastic polymers*. Polymers that are resistant to heat (and that actually harden or "kick over" through the formation of permanent cross-linking upon heating) are known as *thermosetting plastics*. Table 46.15 lists the common polymers in each category. Thermoplastic polymers retain their chain structures and do not experience any chemical change (i.e., bonding) upon repeated heating and subsequent cooling. Thermoplastics can be formed in a cavity mold, but the mold must be cooled before the product is removed. Thermoplastics are particularly suitable for injection molding. The mold is kept relatively cool, and the polymer solidifies almost instantly.

Thermosetting polymers form complex, three-dimensional networks. Thus, the complexity of the polymer increases dramatically, and a product manufactured from a thermosetting polymer may be thought of as one big molecule. Thermosetting plastics are rarely used with injection molding processes.

Table 46.15 Thermosetting and Thermoplastic Polymers

thermosetting
- epoxy
- melamine
- natural rubber (polyisoprene)
- phenolic (phenol formaldehyde, Bakelite®)
- polyester (DAP)
- silicone
- urea formaldehyde

thermoplastic
- acetal
- acrylic
- acrylonitrile-butadiene-styrene (ABS)
- cellulosics (e.g., cellophane)
- polyamide (nylon)
- polyarylate
- polycarbonate
- polyester (PBT and PET)
- polyethylene
- polymethyl-methacrylate (Plexiglas®, Lucite®)
- polypropylene
- polystyrene
- polytetrafluoroethylene (Teflon®)
- polyurethane
- polyvinyl chloride (PVC)
- synthetic rubber (Neoprene®)
- vinyl

Bakelite® is a trademark of Union Carbide.
Plexiglass® is a trademark of Atohass.
Lucite® is a trademark of ICI Acrylics.
Teflon® is a trademark of Du Pont Company.

25. GLASS

Glass is a term used to designate any material that has a volumetric expansion characteristic similar to Fig. 46.6. Glasses are sometimes considered to be *supercooled liquids* because their crystalline structures solidify in random orientation when cooled below their melting points. It is a direct result of the high liquid viscosities of oxides, silicates, borates, and phosphates that the molecules cannot move sufficiently to form large crystals with cleavage planes.

As a liquid glass is cooled, its atoms develop more efficient packing arrangements. This leads to a rapid decrease in volume (i.e., a steep slope on the temperature-

Figure 46.6 Behavior of a Glass

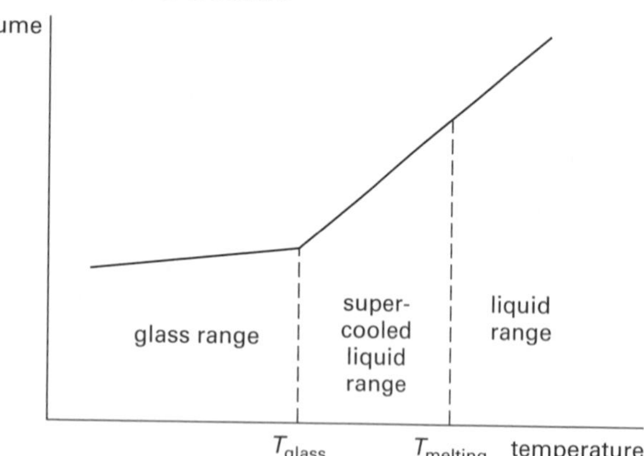

volume curve). Since no crystallization occurs, the liquid glass simply solidifies without molecular change when cooled below the melting point. (This is known as *vitrification.*) The more efficient packing continues past the point of solidification.

At the *glass transition temperature* (*fictive temperature*), the glass viscosity increases suddenly by several orders of magnitude. Since the molecules are more restrained in movement, efficient atomic rearrangement is curtailed, and the volume-temperature curve changes slope. This temperature also divides the region into flexible and brittle regions. At the glass transition temperature, there is a 100-fold to 1000-fold increase in stiffness (modulus of elasticity).

Both organic and inorganic compounds may behave as glasses. *Common glasses* are mixtures of SiO_2, B_2O_3, and various other compounds to modify performance, as shown in Table 46.16.[35]

26. CERAMICS

Ceramics are compounds of metallic and nonmetallic elements. Ceramics form crystalline structures but have no free valence electrons. All electrons are shared ionically or in covalent bonds. Common examples include brick, portland cement, refractories, and abrasives. (Glass is also considered a ceramic even though it does not crystallize.) Typical ceramic properties are listed in Table 46.17.

Although perfect ceramic crystals have extremely high tensile strengths (e.g., some glass fibers have ultimate strengths of 100,000 lbf/in^2 (700 MPa)), the multiplicity of cracks and other defects in natural crystals reduces their tensile strengths to near-zero levels.

Due to the absence of free electrons, ceramics are typically poor conductors of electrical current, although some (e.g., magnetite, Fe_3O_4) possess semiconductor properties. Other ceramics, such as $BaTiO_3$, SiO_2, and $PbZrO_3$, have *piezoelectric* (*ferroelectric*) *qualities* (i.e., generate a voltage when compressed).

Table 46.17 Properties of Typical Ceramics

high melting point
high hardness
high compressive strength
high tensile strength (perfect crystals)
low ductility (brittleness)
high shear resistance (low slip)
low electrical conductivity
low thermal conductivity
high corrosion (acid) resistance
low coefficient of thermal expansion

Ceramics with similar structures behave similarly. Table 46.18 lists several structure designations (e.g., sodium chloride structure) as used in the study of ceramics.

Polymorphs are compounds that have the same chemical formula but have different physical structures. Some ceramics, of which *silica* (SiO_2) is a common example, exhibit *polymorphism.* At room temperature, silica is in the form of *quartz.* At 1607°F (875°C), the structure changes to *tridymite.* A change to a third structure, that of *cristobalite*, occurs at 2678°F (1470°C).

Table 46.18 Structural Designations (A and B are metals; X is a nonmetal)

compound	formula	basic form	other examples
sodium chloride	NaCl	AX	FeO, MgO, CaO
cesium chloride	CsCl	AX	
calcium fluoride	CaF_2	AX_2	GeO_2, MgF_2,TO_2
silica	SiO_2	AX_2	
corundum	Al_2O_3	A_2X_3	Cr_2O_3,Fe_2O_3
spinels (ferrites)	$MgAl_2O_4$	AB_2X_4	
perovskite	$CaTiO_3$	ABX_3	$BaTiO_3$, $PbZrO_3$
zircon	$ZrSiO_4$	ABX_4	

Ferrimagnetic materials (*ferrites*, *spinels*, or *ferrispinels*) are ceramics with valuable magnetic qualities. Advances in near-room-temperature superconductivity have been based on *lanthanum barium copper oxide* ($La_{2-x}Ba_xCuO_4$), a ceramic oxide, as well as compounds based on yttrium (Y-Ba-Cu-O), bismuth, thallium, and others.

Common ceramics are listed in Table 46.19.

[35]This excludes lead-alkali glasses that contain 30% to 60% PbO.

Materials

Table 46.16 Analyses and Properties of Representative Glasses

type of glass	analysis, percent by weight: SiO_2	modifiers[a,b]	Al_2O_3	B_2O_3	PbO	softening temperature,[a] °C	coefficient of expansion, $°C^{-1}$	characteristics	use
fused silica	99.9	–	–	–	–	1667	5.5×10^{-7}	thermal shock resistant	laboratory equipment
96% silica	96.0	–	–	4.0	–	1500	8.0×10^{-7}	thermal shock resistant	laboratory equipment
borosilicate (Pyrex®)[c]	80.5	4.2	2.2	12.9	–	820	32.0×10^{-7}	thermal shock resistant, easy to form	cooking utensils
lead-alkali	54.0	11.0	–	–	35	630	89.0×10^{-7}	high index of refraction	cut glass
aluminasilicate	57.7	9.5	25.3	7.4	–	915	42.0×10^{-7}	thermal shock resistant	thermometers
soda-lime silica	73.6	25.4	1.0	–	–	696	92.0×10^{-7}	easy to form	plate, bulbs
lead-alkali	35.0	7.0	–	–	58	580	91.0×10^{-7}	dielectric	capacitors

(Multiply $°F^{-1}$ by 9/5 to obtain $°C^{-1}$.)

[a]The temperature at which glass will sag appreciably under its own weight.
[b]Sum of Na_2O, K_2O, Ni_2O, CaO, MgO, and BaO.
[c]Pyrex® is a trademark of Corning.
Data derived with permission from *Metals, Ceramics and Polymers*, by Oliver H. Wyatt and David Dew-Hughes, Cambridge University Press, © 1974. Reprinted with permission of Cambridge University Press.

Table 46.19 Common Ceramics

compound	mineral name	use[a]
Al_2O_3	corundum, alumina	abrasives, firebrick
$Al_2Si_2O_5(OH)_4$	kaolinite clay	porcelain paste
$BaTiO_3$	barium titanate	piezoelectricity
BN	boron nitride	refractory
CaF_2	fluorite	flux
CaO		refractory
Fe_2O_3		refractory
Fe_3O_4 or $FeFe_2O_4$	magnetite	thermistors
MgO	periclase	refractory
Mg_2SiO_4	forsterite	refractory
$MnFe_2O_4$		ferrimagnetism
$MgCr_2O_4$	magnesium chromate	piezoelectricity
$MgFe_2O_4$		antiferromagnetism
$NiFe_2O_4$		ferrimagnetism
NaCl	table salt	food, chemicals
$PbZrO_3$		piezoelectricity
SiC	silicon carbide	refractory
SiO_2	quartz[b]	refractory
TiC	titanium carbide	refractory
TiO_2	titanium dioxide	refractory
UO_2	uranium dioxide	nuclear fuel
ZrN	zirconium nitride	refractory
$ZnFe_2O_4$		ferrimagnetism

[a]The term *refractory* means the ceramic is used in firebrick, stoneware, and other containers intended for use at high temperatures.
[b]SiO_2 has several temperature-dependent polymorphs, including coesite, cristobalite, and tridymite.

27. ABRASIVES

An *abrasive* is a hard material that can cut other materials. Abrasives are typically ceramic compounds embedded in stiff binders. *Natural abrasives* include *emery* (50% to 60% Al_2O_3, rest iron oxide), corundum, quartz, garnets, and diamonds. *Artificial abrasives* include cemented carbides (e.g., SiC) and artificially made aluminum oxide (Al_2O_3).

Binders for rigid wheel abrasives are kiln-fired vitreous materials derived from clays or feldspars. Other grinding wheels use rubber or synthetic elastomers as the binders. Grains of abrasive are mixed thoroughly with a binder and molded into final form. Diamond wheels are often in the form of thin metal discs with the diamond grains bonded only to the periphery.

Carbides (*cemented carbides, sintered carbides*) have extreme hardness, wear resistance, and thermal stability.[36] These properties make them useful for high-speed metal cutting. Silicon carbide (SiC, sold under the name

[36]*Sintering* is the process where a physical mixture of carbide and powdered metal is heated in order to solidify the powder into a single piece. When the metal melts, it acts as the binder for the carbide.

of *carborundum*) is probably the best known. Carbides of tungsten, molybdenum, titanium, vanadium, tantalum, and zirconium are also widely used.

28. MODERN COMPOSITE MATERIALS

There are many types of modern composite material systems, including dispersion-strengthened, particle-strengthened, and fiber-strengthened materials. (Steel-reinforced concrete and steel-reinforced wood systems are also composite systems.)

In *dispersion-strengthened systems* (e.g., aluminum-aluminum oxide systems known as *SAP alloys*, *toughened ceramics*, *metal-metal systems* in which tungsten whiskers are blended in a copper alloy matrix, and NiO_2ThO_2 mixtures known as *TD-nickel*), the matrix is an actual load-bearing element.[37] The discrete particles (*dispersoids*) distributed throughout the matrix occupy less than 15% of the volume, and particle sizes are in the 0.01 to 0.1 μm range.

In *particle-strengthened systems* (e.g., tungsten carbide, known as WC, in a cobalt matrix, and *cermets* produced by sintering), the dispersoid particles are larger than 1.0 μm in size, and they occupy up to 25% of the material volume. The matrix is not the major load-carrying element, but it does contribute to strength.

In *fiber-strengthened systems* (e.g., glass-reinforced epoxies), reinforcing materials vary widely in size, and sizes may range up to several mils.[38] The matrix transmits the loads to the fibers and protects the fibers from chemical attack.

Initial attempts at fiber-reinforced polymers involved the impregnation of natural fibers (cotton, wood, etc.) with Bakelite and phenolic resins. The introduction of various types of glass (e.g., rovings, windings, and woven cloth) as reinforcement was the next development step.[39] Currently, higher-strength epoxy resins have been used as matrices with graphite, boron, beryllium, steel, titanium, aluminum, or magnesium fibers.

Many fiber-reinforced composites, particularly those involving cloth, are highly *anisotropic*. Properties vary with the orientation of lay-ups as well as with weave orientation in the reinforcing materials. In directions transverse to fiber orientation, tensile and compressive strengths are a function of the matrix material. Loads parallel to the fibers are carried by the reinforcement, while flexural strength is limited by the shear bond between the filaments and the matrix material. Equation 46.2, Eq. 46.3, and Eq. 46.4 can be used to find the density, heat capacity, and strength parallel to the fiber direction of fiber-reinforced composites.

Composite Materials

$$\rho_c = \sum f_i \rho_i \qquad 46.2$$

$$C_c = \sum f_i c_i \qquad 46.3$$

$$\sigma_c = \sum f_i \sigma_i \qquad 46.4$$

The modulus of elasticity for a fiber-reinforced composite is a function of the modulus of elasticity of each material in the composite and the volumetric fraction of that material in the composite, as shown in Eq. 46.5.

Composite Materials

$$\left(\sum \frac{f_i}{E_i}\right)^{-1} \le E_c \le \sum f_i E_i \qquad 46.5$$

The strains of the components are equal for axially oriented, long, fiber-reinforced composites.

Composite Materials

$$\left(\frac{\Delta L}{L}\right)_1 = \left(\frac{\Delta L}{L}\right)_2 \qquad 46.6$$

Typical reinforcing materials for fiberglass include E- and S-glass.[40] *S-glass* is a silica-alumina-magnesia compound with improved tensile properties. It is used mainly in nonwoven, monodirectional, and wound configurations. *E-glass* is a lime-alumina-borosilicate ($CaOAl_2O_3$-SiO_2) compound used primarily in woven fabrics. (See Table 46.20.)

Graphite (i.e., carbon) fibers are used where high stiffnesses and low coefficients of thermal expansion are needed.[41] These advantages are balanced by the disadvantages of brittleness and high cost. The ultimate tensile strength for graphite varies inversely with modulus of elasticity. Graphite fibers range in strength from 180 ksi (1.2 GPa) for yarn

[37] Ceramics typically fail catastrophically, without warning. To counteract this tendency, *toughened ceramics* incorporate discrete solids such as SiC or TiC whiskers throughout the ceramic matrix. Cracks that start are arrested by the dispersoids. Dispersoids can increase ceramic toughness by as much as 40%.

[38] A *whisker* is a single crystal grown by vapor deposition. Although lengths of several millimeters are typical, diameters are only a few micrometers.

[39] A *roving* consists of a number of parallel strands of fiber. The strands are side by side (not interwoven or twisted together), forming a flat ribbon.

[40] *A-glass* is common soda-lime glass used for windows, bottles, and jars. Other types of glass used for reinforcement include *C-glass* (developed for greater chemical and corrosion resistance), *D-glass* (glass possessing a low dielectric constant), and *M-glass* (glass containing BeO to increase the elastic modulus).

[41] It is interesting that graphite in *fiber* or *whisker* form is used to provide strength, while graphite in *powder* form is used as a solid lubricant. The *laminar* structure of graphite permits particles to easily slide over one another. This laminar structure does not easily break down, making graphite particularly valuable as a lubricant at high temperatures and pressures—up to at least 3600°F (2000°C). In fact, graphite's coefficient of friction decreases with temperature. Another excellent solid lubricant, *molybdenum disulfide* (MoS_2), also has a laminar structure, but its friction coefficient increases sharply above 1600°F (900°C).

Table 46.20 Typical Properties of E- and S-Glass Fibers at Room Temperature

property	E-glass	S-glass
specific gravity	2.54	2.48
density (lbm/in^3)	0.092	0.090
ultimate tensile strength (lbf/in^2)		
• monofilament	5.0×10^5	6.6×10^5
• 12-end roving	3.7×10^5	5.5×10^5
modulus of elasticity (lbf/in^2)	10.5×10^6	12.5×10^6
coefficient of thermal expansion (1/°F)	2.8×10^{-6}	$1.6\text{–}2.2 \times 10^{-6}$
specific heat (Btu/lbm-°F)	0.192	0.176

(Multiply lbm/in^2 by 27,680 to obtain kg/m^3.)
(Multiply lbf/in^2 by 0.006895 to obtain MPa.)

configurations to 350 ksi (2.4 GPa) for tow, while the modulus of elasticity varies from 60×10^6 lbf/in^2 to 20×10^6 lbf/in^2 (410 GPa to 140 GPa).[42]

There are three general categories of carbon fibers: standard, high-modulus, and high-strength. *Kevlar aramid* fibers provide strength and stiffness that are essentially in between that of glass and carbon fibers.

Resistance to chemical corrosion, rather than strength-to-weight ratio, is the primary factor in selecting *fiber-reinforced plastics* (also referred to as *fiber-reinforced polymers*, both abbreviated FRP) for process tanks, reaction vessels, and pipes.[43] The main resins used for this purpose are vinyl esters, epoxies, polyesters, furans, and phenolics. Vinyl esters, which can handle both acidic and basic fluids as well as strong oxidizers such as chlorine, are the most widely used. Epoxies have better thermal and mechanical properties, but epoxies cannot be used in highly acidic environments (i.e., pH less than 3). Polyesters, on the other hand, are acid resistant, but are not alkali resistant (i.e., pH more than 9). Furans and phenolics are relatively weak and are used in special applications.

A typical FRP tank or pipe consists of three layers: the veil, liner, and structural laminate (from the inside out). The *veil* consists of a thin layer (e.g., 0.25 mm) of glass or polyester fibers saturated with approximately 90% by weight of the resin. The 2.5 mm thick *liner* consists of glass in resin in a 3:1 ratio, respectively. The liner provides chemical resistance. The outermost fiberglass *structural laminate* bears all of the pressure, stresses, and mechanical forces imposed on the tank.[44] Although the structural laminate can be laid up by hand, filament-wound tanks are more popular because manufacturing costs are lower. (See Table 46.21.)

Table 46.21 Typical Properties of Laid-Up Composites (linear lay-up)

composite	fiber content (%)	specific gravity	modulus of elasticity (ksi) axial	trans-verse	shear
graphite-epoxy					
high strength	65	1.58	20,000	1000	650
high modulus	65	1.61	29,000	1000	700
ultrahigh modulus	65	1.69	44,000	1000	950
Kevlar™ 49-epoxy	65	1.39	12,500	800	300
E-glass-epoxy	65	1.99	6000	1500	300
chopped glass-polyester					
and sheet molding	30	1.88	2500	2500	100
compound (SMC)	65	1.99	3500	3500	150

(Multiply ksi by 6.895 to obtain MPa.)

29. NOMENCLATURE

C	specific heat capacity	Btu/lbm-°F	J/kmol-°C
D	diameter	in	m
DP	degree of polymerization	–	–
E	modulus of elasticity	lbf/in^2	Pa
f	fraction	–	–
f'_c	compressive strength	lbf/in^2	Pa
I	moment of inertia	in^4	m^4
L	length	in	m
m	mass	lbm	kg
M	moment	in-lbf	N·m
MC	moisture content	%	%
M	molecular weight	–	–
p	pressure	lbf/in^2	Pa
P	force	lbf	N
S_H	circumferential stress	lbf/in^2	Pa
SG	specific gravity	–	–
V	volume	ft^3	m^3
w_c	specific weight of concrete (ACI 318 nomenclature)	lbf/ft^3	kg/m^3
W	weight	lbf	n.a.

[42] *Tow* consists of loose, untwisted fibers.
[43] Other characteristics for which FRP may be used include cost, weight, and ease of manufacture of complex shapes.
[44] ASTM standards D3299 and D4097 both specify the common $pD/2S_H$ formula for wall thickness, where S_H is the allowable circumferential (hoop) stress that produces a maximum strain of 0.0010.

Symbols

γ	specific weight	lbf/ft^3	N/m^3
ρ	density	lbm/ft^3	kg/m^3
σ	strength	lbf/in^2	MPa

Subscripts

c	concrete
ct	cylinder tensile (splitting)
i	
r	rupture
SSD	saturated surface dry

47 Material Properties and Testing

Content in blue refers to the *NCEES Handbook*.

NCEES EXAM SPECIFICATIONS AND RELATED CONTENT

MACHINE DESIGN AND MATERIALS EXAM

I.C.3.a. Material Properties: Mechanical: Time-independent behavior
- 10. Fracture Toughness

I.C.3.b. Material Properties: Mechanical: Time-dependent behavior
- 23. Creep Test

I.D.1. Strength of Materials: Stress/strain
- 2. Tensile Test
- 8. Ductility
- 9. True Stress and Strain
- 11. Strain Energy

1. CLASSIFICATION OF MATERIALS

When used to describe engineering materials, the terms "strong" and "tough" are not synonymous. Similarly, "weak," "soft," and "brittle" have different engineering meanings. A *strong material* has a high ultimate strength, whereas a *weak material* has a low ultimate strength. A *tough material* will yield greatly before breaking, whereas a *brittle material* will not. (A brittle material is one whose strain at fracture is less than approximately 0.05 (5%).) A *hard material* has a high modulus of elasticity, whereas a *soft material* does not. Figure 47.1 illustrates some of the possible combinations of these classifications.

Figure 47.1 *Types of Engineering Materials*

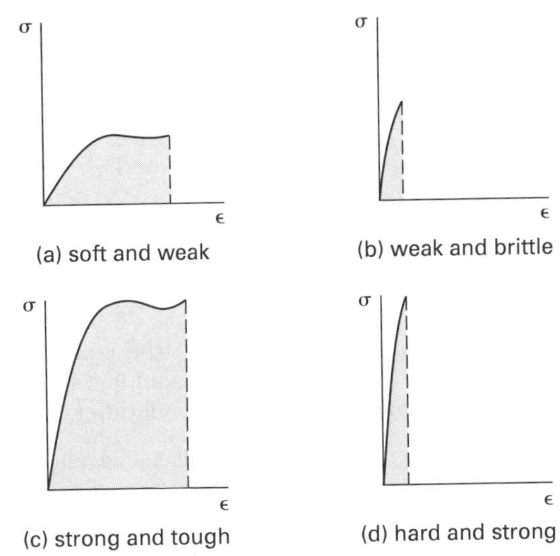

2. TENSILE TEST

Many useful material properties are derived from the results of a standard *tensile test*. In this test, a prepared material sample (i.e., a *specimen*) is axially loaded in tension, and the resulting elongation, ΔL or δ, is measured as the load, P, increases. A *load-elongation curve* of tensile test data for a ductile ferrous material (e.g., low-carbon steel or other BCC transition metal) is shown in Fig. 47.2.

Materials

When elongation is plotted against the applied load, the graph is applicable only to an object with the same length and area as the test specimen. To generalize the test results, the data are converted to stresses and strains by the use of Eq. 47.1 and Eq. 47.3.[1]

Figure 47.2 *Typical Tensile Test Results for a Ductile Material*

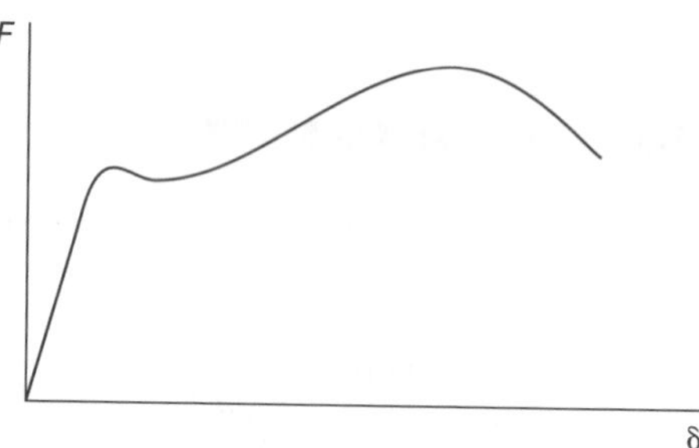

Engineering stress, σ (usually called *stress*), is the load per unit original area. Typical engineering stress units are lbf/in^2 and MPa. *Engineering strain*, ϵ (usually called *strain*), is the elongation of the test specimen expressed as a percentage or decimal fraction of the original length. Units of in/in and m/m are used for strain.

Uniaxial Loading and Deformation

$$\sigma = \frac{P}{A} \quad 47.1$$

Engineering Strain

$$\epsilon = \frac{\Delta L}{L_o} \quad 47.2$$

$$e = \frac{\delta}{L_o} \quad 47.3$$

If the stress-strain data are plotted, the shape of the resulting line will be essentially the same as the force-elongation curve, although the scales will differ.

Segment O-A in Fig. 47.3 is a straight line. The relationship between the stress and strain in this linear region is given by *Hooke's law*, Eq. 47.4. The slope of line segment O-A is the *modulus of elasticity*, *E*, also known as *Young's modulus*. Table 47.1 lists approximate values of the modulus of elasticity for materials at room temperature. The modulus of elasticity will be lower at higher temperatures. For steel at higher temperatures, the modulus of elasticity is reduced approximately as shown in Table 47.2.

Hooke's Law

$$\sigma = E\epsilon \quad 47.4$$

Figure 47.3 *Typical Stress-Strain Curve for Steel*

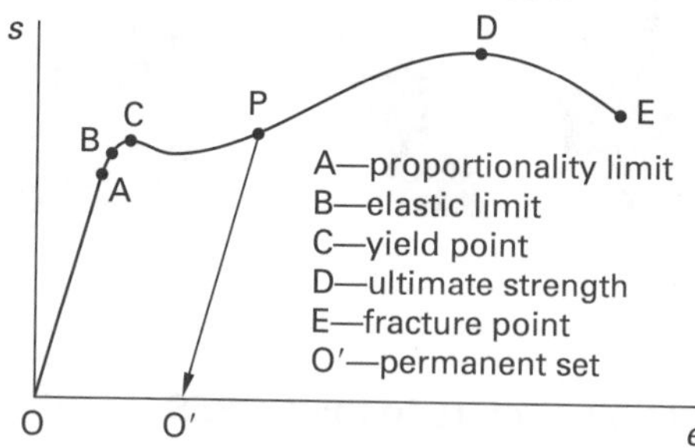

Table 47.1 *Approximate Modulus of Elasticity of Representative Materials at Room Temperature*

material	lbf/in^2	GPa
aluminum alloys	10–11 × 10^6	70–80
brass	15–16 × 10^6	100–110
cast iron	15–22 × 10^6	100–150
cast iron, ductile	22–25 × 10^6	150–170
cast iron, malleable	26–27 × 10^6	180–190
copper alloys	17–18 × 10^6	110–112
glass	7–12 × 10^6	50–80
magnesium alloys	6.5 × 10^6	45
molybdenum	47 × 10^6	320
nickel alloys	26–30 × 10^6	180–210
steel, hard*	30 × 10^6	210
steel, soft*	29 × 10^6	200
steel, stainless	28–30 × 10^6	190–210
titanium	15–17 × 10^6	100–110

(Multiply lbf/in^2 by 6.895 × 10^{-6} to obtain GPa.)

*Common values are given.

Table 47.2 *Approximate Reduction of Steel's Modulus of Elasticity at Higher Temperatures*

temperature		
°F	°C	% of original value
70	20	100%
400	200	90%
800	425	75%
1000	540	65%
1200	650	60%

The stress at point A in Fig. 47.3 is known as the *proportionality limit* (i.e., the maximum stress for which the linear relationship is valid). Strain in the *proportional region* is called *proportional strain*.

[1]The most common *test specimen* in the United States has a length of 2.00 in and a diameter of 0.505 in. Since the cross-sectional area of this *0.505 bar* is 0.2 in^2, the stress in lbf/in^2 is calculated by multiplying the force in pounds by five.

The *elastic limit*, point B in Fig. 47.3, is slightly higher than the proportionality limit. As long as the stress is kept below the elastic limit, there will be no *permanent set* (permanent deformation) when the stress is removed. Strain that disappears when the stress is removed is known as *elastic strain*, and the stress is said to be in the *elastic region*. When the applied stress is removed, the *recovery* is 100%, and the material follows the original curve back to the origin.

If the applied stress exceeds the elastic limit, the recovery will be along a line parallel to the straight-line portion of the curve, as shown in line segment P-O′ in Fig. 47.3. The strain that results (line O-O′) is *permanent set* (i.e., a permanent deformation). The terms *plastic strain* and *inelastic strain* are used to distinguish this behavior from elastic strain.

For steel, the *yield point*, point C, is very close to the elastic limit. For all practical purposes, the *yield strength* or *yield stress*, S_y (or S_{yt} to indicate yield in tension), can be taken as the stress that accompanies the beginning of plastic strain. Yield strengths are reported in lbf/in^2, ksi, and MPa.[2] Table 47.3 presents comparative yield strengths for common metal alloys.

Figure 47.3 does not show the full complexity of the stress-strain curve near the yield point. Rather than being smooth, the curve is ragged near the yield point. At the upper yield strength, there is a pronounced drop (i.e., "drop of beam") in load-carrying ability to a plateau yield strength after the initial yielding occurs. The plateau value is known as the *lower yield strength* and is commonly reported as the yield strength. Figure 47.4 shows upper and lower yield strengths.

The *ultimate strength* or *tensile strength*, S_u (or S_{ut} to indicate an ultimate tensile strength), point D in Fig. 47.3, is the maximum stress the material can support without failure.

The *breaking strength* or *fracture strength*, S_f, is the stress at which the material actually fails (point E in Fig. 47.3). For ductile materials, the breaking strength is less than the ultimate strength due to the necking down in the cross-sectional area that accompanies high plastic strains.

Figure 47.4 *Upper and Lower Yield Strengths*

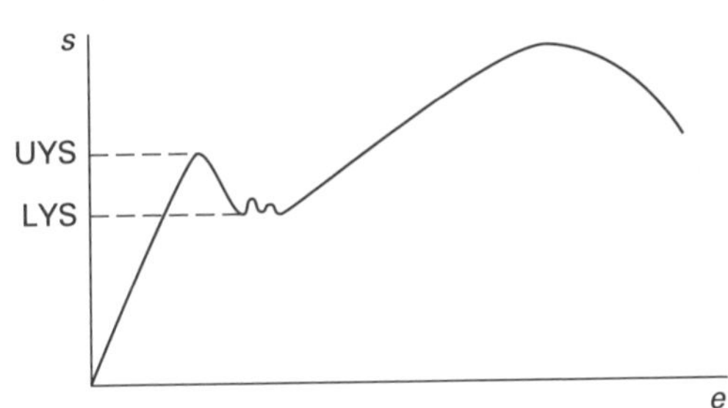

Table 47.3 *Approximate Yield Strengths of Representative Materials*

material	yield strength lbf/in^2	yield strength MPa
iron and steel		
1020	43,000	300
A36	36,000	250
A992/A572 grade 50	50,000	345
stainless (304)	43,000	300
pure	24,000	160
copper		
beryllium	130,000	900
brass	11,000	75
pure	10,000	70
aluminum		
2024	50,000	345
6061	21,000	145
pure	5000	35
titanium		
alloy 6% Al, 4% V	160,000	1100
pure	20,000	140
nickel		
hastelloy	55,000	380
inconel	40,000	280
monel	35,000	240
pure	20,000	140

(Multiply lbf/in^2 by 6.895×10^{-3} to obtain MPa.)

3. STRESS-STRAIN CHARACTERISTICS OF NONFERROUS METALS

Most nonferrous materials, such as aluminum, magnesium, copper, and other FCC and HCP metals, do not have well-defined yield points. The stress-strain curve starts to bend at low stresses, as illustrated by Fig. 47.5. In such cases, the yield strength is commonly defined as the stress that will cause a 0.2% *parallel offset* (i.e., a plastic strain of 0.002) for metals and 2% parallel offset for plastics.[3] However, the yield strength can also be defined by other offset values (e.g., 0.1% for metals and 1.0% for plastics).

The yield strength is found by extending a line from the offset strain value parallel to the linear portion of the curve until it intersects the curve.

With nonferrous metals, the difference between parallel offset and total strain characteristics is important. Sometimes, the yield point will be defined as the stress

[2]A *kip* is a thousand pounds. *ksi* is the abbreviation for kips per square inch (thousands of lbf/in^2).
[3]The 0.2% parallel offset strength is also known as the *proof stress*.

Figure 47.5 *Typical Stress-Strain Curve for a Nonferrous Metal*

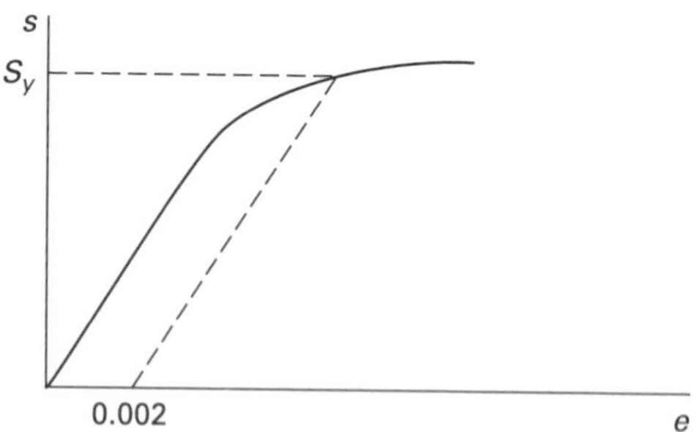

accompanying *0.5% total strain* (i.e., a strain of 0.005) determined by extending a line from the stress-strain curve vertically downward to the strain axis.

4. STRESS-STRAIN CHARACTERISTICS OF BRITTLE MATERIALS

Brittle materials, such as glass, cast iron, and ceramics, can support only small strains before they fail catastrophically (i.e., without warning). As the stress is increased, the elongation is linear and Hooke's law (given in Eq. 47.4) can be used to predict the strain. Failure occurs within the linear region, and there is very little, if any, necking down. Since the failure occurs at a low strain, brittle materials are not ductile. Figure 47.6 is typical of the stress-strain curve of a brittle material.

Figure 47.6 *Stress-Strain Curve of a Brittle Material*

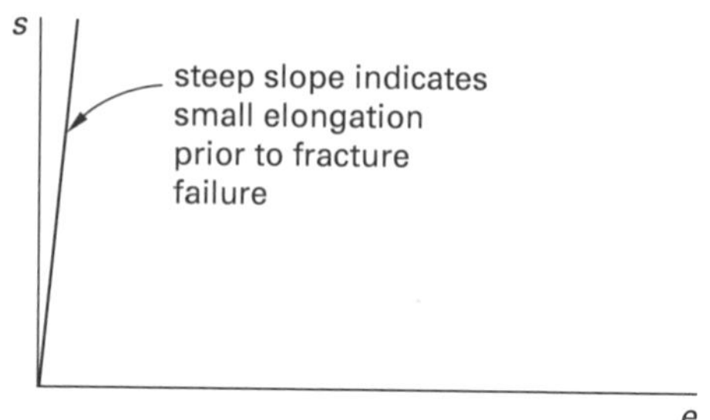

5. SECANT MODULUS

The modulus of elasticity, E, is usually determined from the steepest portion of the stress-strain curve. (This avoids the difficulty of locating the starting part of the curve.) For materials with variable modulus of elasticity, or for linear materials operating in the nonlinear region, the *secant modulus* gives the average ratio of stress to strain. The secant modulus is the slope of the straight line connecting the origin and the point of operation. Some designs using elastomers, concrete, and prestressing wire may be based on the secant modulus. (See Fig. 47.7.)

Figure 47.7 *Secant Modulus*

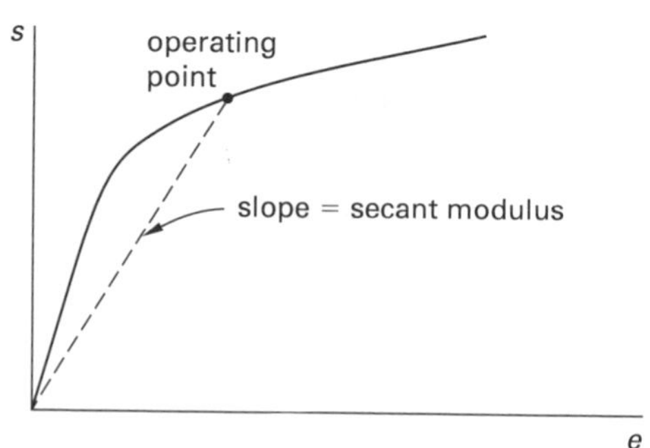

6. POISSON'S RATIO

As a specimen elongates axially during a tensile test, it will also decrease slightly in diameter or breadth. For any specific material, the percentage decrease in diameter, known as the *lateral strain*, will be a fraction of the *axial strain*. The ratio of the lateral strain to the axial strain is known as *Poisson's ratio*, ν, which is approximately 0.3 for most metals. (See Table 47.4.)

$$\nu = \frac{\epsilon_{\text{lateral}}}{\epsilon_{\text{axial}}} = \frac{\frac{\Delta D}{D_o}}{\frac{\Delta L}{L_o}} \quad 47.5$$

Poisson's ratio applies only to elastic strain. When the stress is removed, the lateral strain disappears along with the axial strain.

Table 47.4 *Approximate Values of Poisson's Ratio*

material	Poisson's ratio
rubber	0.49
thermosetting plastics	0.40–0.45
aluminum	0.32–0.34 (0.33)*
magnesium	0.35
copper	0.33–0.36 (0.33)*
titanium	0.34
brass	0.33–0.36
stainless steel	0.30
steel	0.26–0.30 (0.30)*
nickel	0.30
beryllium	0.27
cast iron	0.21–0.33 (0.27)*
glass (SiO_2)	0.21–0.27 (0.23)*
diamond	0.20

*commonly used for design

Materials

7. STRAIN HARDENING AND NECKING DOWN

When the applied stress exceeds the yield strength, the specimen will experience plastic deformation and will strain harden. (Plastic deformation is primarily due to the shear stress-induced movement of dislocations.) Since the specimen volume is constant (i.e., $A_oL_o = AL$), the cross-sectional area decreases. Initially, the strain hardening more than compensates for the decrease in area, so the material's strength increases and the engineering stress increases with larger strains.

Eventually, a point is reached when the available strain hardening and increase in strength cannot keep up with the decrease in cross-sectional area. The specimen then begins to neck down (at some local weak point), and all subsequent plastic deformation is concentrated at the neck. The cross-sectional area decreases even more rapidly thereafter since only a small portion of the specimen volume is strain hardening. The engineering stress decreases to failure.

Figure 47.8 shows necking down in two different specimens tested to failure. Very ductile materials pull out to a point, while most moderately ductile materials exhibit a *cup-and-cone failure*. (Failed brittle materials, not shown, do not exhibit any significant reduction in area.)

Figure 47.8 *Types of Tensile Ductile Failure*

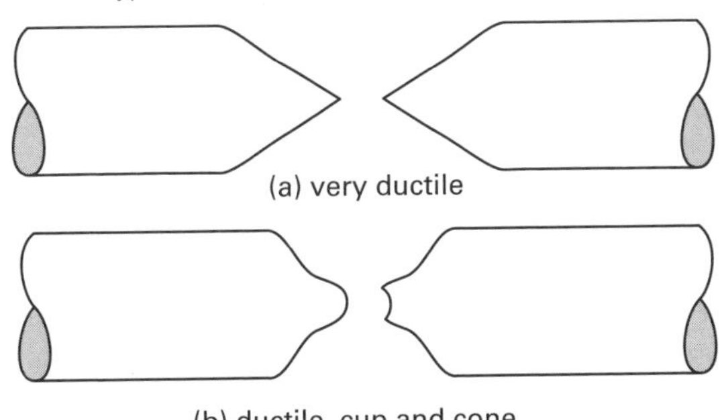

8. DUCTILITY

A material that deforms and elongates a great deal before failure is said to be a *ductile material*.[4] (Steel, for example, is a ductile material.) The *percent elongation*, short for *percent elongation at failure*, is the total plastic strain at failure. (See Fig. 47.9.) (Percent elongation does not include the elastic strain, because even at ultimate failure the material snaps back an amount equal to the elastic strain.)

Percent Elongation

$$\%\,\text{elongation} = \frac{\Delta L}{L_o} \times 100\% \qquad 47.6$$

Figure 47.9 *Percent Elongation*

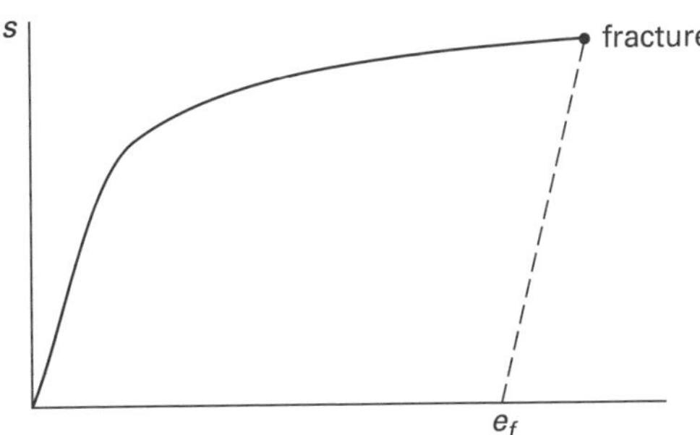

The value of the final strain to be used in Eq. 47.6 is found by extending a line from the failure point downward to the strain axis, parallel to the linear portion of the curve. This is equivalent to putting the two broken specimen pieces together and measuring the total length.

Highly ductile materials exhibit large percent elongations. However, percent elongation is not the same as *ductility*.

$$\text{ductility} = \frac{\text{ultimate failure strain}}{\text{yielding strain}} \qquad 47.7$$

The *reduction in area* (at the point of failure), %RA, expressed as a percentage or decimal fraction, is a third measure of a material's ductility. The reduction in area due to necking down will be 50% or greater for ductile materials and less than 10% for brittle materials.[5]

Percent Reduction in Area (RA)

$$\%\text{RA} = \frac{A_o - A_f}{A_o} \times 100\% \qquad 47.8$$

9. TRUE STRESS AND STRAIN

Engineering stress, given by Eq. 47.1, is calculated for all stress levels from the original cross-sectional area. However, during a tensile test, the area of a specimen decreases as the stress increases. The decrease is only slight in the elastic region but is much more significant after plastic deformation begins.

If the instantaneous area is used to calculate the stress, the stress is known as *true stress* or *physical stress*, σ. Equation 47.9 assumes a homogeneous strain distribution along the gage length and that there is no change in total volume with strain, which are valid up to the point

[4]The words "brittle" and "ductile" are antonyms.
[5]*Notch-brittle materials* have reductions in area that are moderate (e.g., 25–35%) when tested in the usual manner, but close to zero when the test specimen is given a small notch or crack.

Materials

of necking. ($q = (A_o - A)/A_o$, the fractional *reduction in area*, used in Eq. 47.9, is expressed as a number between 0 and 1.)

$$\sigma = \frac{P}{A} = \frac{P}{\left(1 - \frac{A_o - A}{A_o}\right)A_o} = \frac{P}{(1-q)A_o} = \frac{\sigma}{1-q} = \sigma(1+\epsilon) \quad \text{[prior to necking, circular specimen]} \qquad 47.9$$

For circular or square specimens prior to necking, Eq. 47.9 may be written as Eq. 47.10.

$$\sigma = \frac{\sigma}{(1-\nu\epsilon)^2} \qquad 47.10$$

Engineering strain, given by Eq. 47.3, is calculated from the original length, although the actual length increases during the tensile test. The *true strain*, *physical strain*, or *log strain*, ϵ_T, is found from Eq. 47.11.

True Strain

$$\epsilon_T = \ln\frac{L}{L_o} = \ln\frac{A}{A_o}\epsilon = \ln(1+\epsilon) \qquad 47.11$$

Since the plastic deformation occurs through a shearing process, there is essentially no volume decrease during elongation.

$$A_oL_o = AL \qquad 47.12$$

Therefore, true strain can be calculated from the cross-sectional areas and, for a circular specimen, from diameters. If necking down has occurred, true strain must be calculated from the areas or diameters, not the lengths.

$$\epsilon_T = \ln\frac{A_o}{A} = \ln\left(\frac{D_o}{D}\right)^2 = 2\ln\frac{D_o}{D} \qquad 47.13$$

A graph of true stress and true strain is known as a *flow curve*. Log σ_T can also be plotted against log ϵ_T, resulting in a straight-line relationship.

The flow curve of many metals in the plastic region can be expressed by Eq. 47.14, known as a *power curve*. K is known as the *strength coefficient*, and n is the *strain-hardening exponent*. Values of both vary greatly with material, composition, and heat treatment. n can vary from 0 (for a perfectly inelastic solid) to 1.0 (for an elastic solid). Typical values are between 0.1 and 0.5. For annealed steel with 0.05% carbon, for example, $K \approx$ 77,000 psi (530 MPa) and $n = 0.26$.

$$\sigma = K\epsilon^n \qquad 47.14$$

Figure 47.10 compares engineering and true stresses and strains for a ferrous alloy. As can be seen, the two curves coincide throughout the elastic region. Although true stress and strain are more accurate, almost all engineering work is based on engineering stress and strain, which is justifiable for two reasons: (a) design using ductile materials is limited to the elastic region where engineering and true values differ little, and (b) the reduction in area of most parts at their service stresses is not known; only the original area is known.

***Figure 47.10** True and Engineering Stresses and Strains for a Ferrous Alloy*

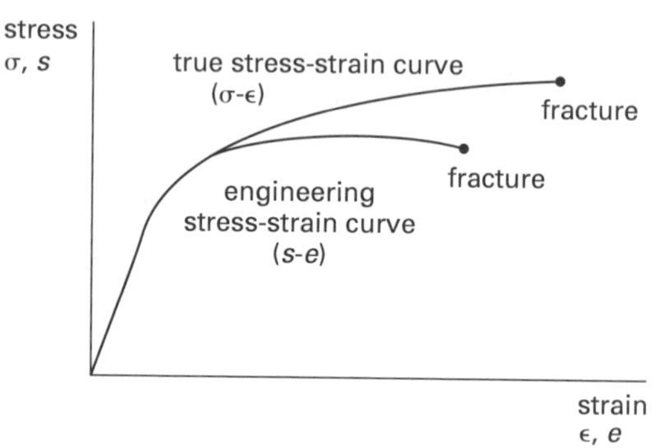

Example 47.1

The engineering stress in a solid tension member was 47,000 lbf/in^2 at failure. The reduction in area was 80%. What were the true stress and strain at failure?

Solution

Since engineering stress is F/A_o, from Eq. 47.9 the true stress is

$$\sigma_T = \frac{\sigma}{1-q} = \frac{47{,}000\ \frac{\text{lbf}}{\text{in}^2}}{1-0.80} = 235{,}000\ \text{lbf/in}^2$$

From Eq. 47.8, the ratio of the initial area to the final area is

Percent Reduction in Area (RA)

$$\%\text{RA} = \frac{A_o - A_f}{A_o} \times 100\%$$

$$\frac{A_i}{A_f} = \frac{A_o}{A} = \frac{1}{1 - \frac{\%\text{RA}}{100\%}} = \frac{1}{1-0.80} = 5$$

From Eq. 47.13, the true strain is

$$\epsilon_T = \ln\frac{A_o}{A} = \ln(5) = 1.61 \quad (161\%)$$

Materials

10. FRACTURE TOUGHNESS

If a material contains a crack, stress is concentrated at the tip or tips of the crack. A crack in the surface of the material will have one tip (i.e., a stress concentration point); an internal crack in the material will have two tips. This increase in stress can cause the crack to propagate (grow) and can significantly reduce the material's ability to bear loads. Other things being equal, a crack in the surface of a material has a more damaging effect.

There are three modes of crack propagation, as illustrated by Fig. 47.11,

- *opening or tensile:* forces act perpendicular to the crack, which pulls the crack open, as shown in Fig. 47.11(a). This is known as mode I.
- *in-plane shear or sliding:* forces act parallel to the crack, which causes the crack to slide along itself, as shown in Fig. 47.11(b). This is known as mode II.
- *out-of-plane shear or pushing (pulling):* forces act perpendicular to the crack, tearing the crack apart, as shown in Fig. 47.11(c). This is known as mode III.

Figure 47.11 *Crack Propagation Modes*

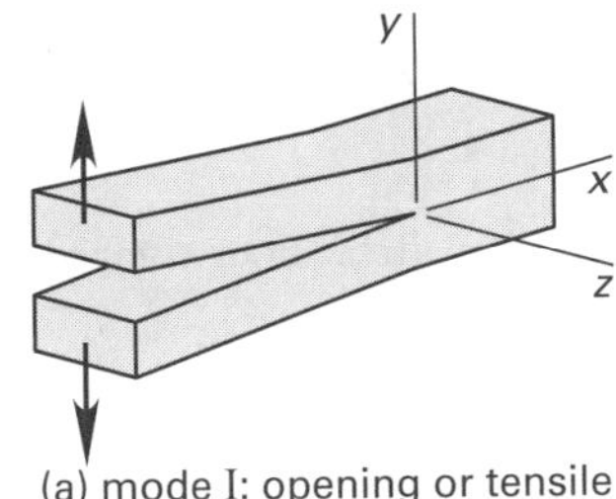

(a) mode I: opening or tensile

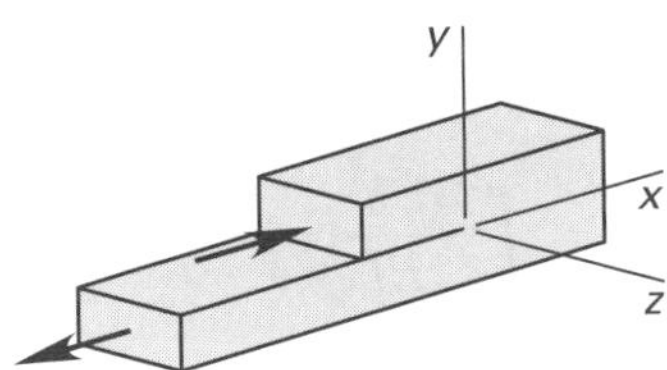

(b) mode II: in-plane shear or sliding

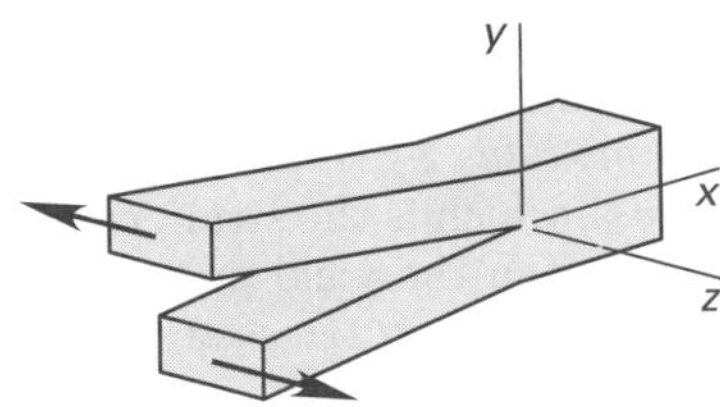

(c) mode III: out-of-plane shear or pushing (pulling)

Fracture toughness is the amount of energy required to propagate a preexisting flaw. Fracture toughness is quantified by a stress intensity factor, K. For a mode I crack (see Fig. 47.11(a)), the stress intensity factor for a crack is designated K_{LC}. The stress intensity factor can be used to predict whether an existing crack will propagate through the material. When K_{LC} reaches a critical value, fast fracture occurs. The crack suddenly begins to propagate through the material at the speed of sound, leading to catastrophic failure. This critical value at which fast fracture occurs is called *fracture toughness*, and it is a property of the material.

The stress intensity factor is calculated from Eq. 47.15, where σ is the applied engineering stress and a is the crack length. For a surface crack, a is measured from the crack tip to the surface of the material, as shown in Fig. 47.11(a). For an internal crack, a is half the distance from one tip to the other, as shown in Fig. 47.11 (b). Y is a dimensionless geometrical factor that is dependent on the location of the crack as shown in Fig. 47.12.

Mechanical Properties

$$K_{LC} = Y\sigma\sqrt{\pi a} \qquad 47.15$$

Figure 47.12 *Crack Length and Geometrical Factors*

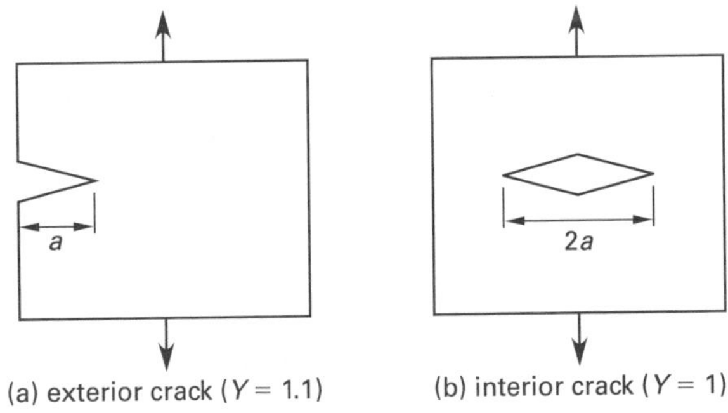

When a is measured in meters and σ is measured in MPa, typical units for both the stress intensity factor and fracture toughness are MPa·$\sqrt{\text{m}}$. Typical values of fracture toughness for various materials are given in Table 47.5.

Table 47.5 *Representative Values of Fracture Toughness*

material	K_{LC} MPa·$\sqrt{\text{m}}$	K_{LC} $\sqrt{\text{in}}$
Al 2014-T651	24.2	22
Al 2024-T3	44	40
52100 steel	14.3	13
4340 steel	46	42
alumina	4.5	4.1
silicon carbide	3.5	3.2

Materials

Example 47.2

An aluminum alloy plate containing a 2 cm long crack is 10 cm wide and 0.5 cm thick. The plate is pulled with a uniform tensile force of 10 000 N. What is most nearly the stress intensity factor at the end of the crack?

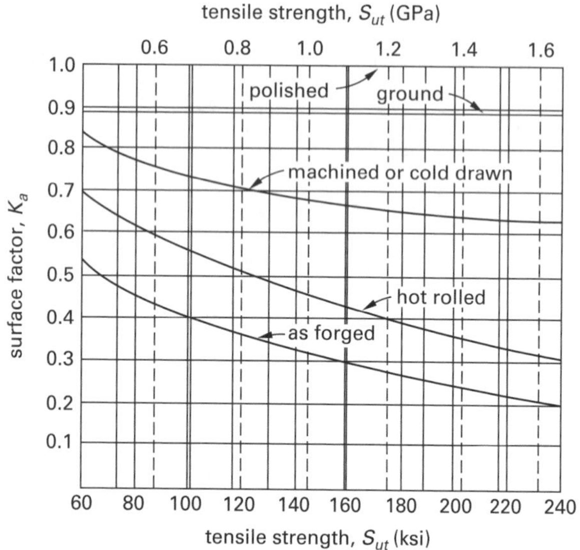

Reprinted with permission from *Mechanical Engineering Design*, 3rd ed. by Joseph Edward Shigley, © 1977, The McGraw-Hill Companies, Inc.

Solution

From Eq. 47.1, the nominal engineering stress is

Uniaxial Loading and Deformation

$$\sigma = \frac{P}{A_o} = \frac{(10\,000 \text{ N})\left(100 \ \frac{\text{cm}}{\text{m}}\right)^2}{(10 \text{ cm})(0.5 \text{ cm})}$$
$$= 20 \times 10^6 \text{ N/m}^2 \quad (20 \text{ MPa})$$

Since this is an exterior crack, $Y = 1.1$. Using Eq. 47.15, the stress intensity factor is

Mechanical Properties

$$K_{\text{LC}} = Y\sigma\sqrt{\pi a}$$
$$= (1.1)(20 \text{ MPa})\sqrt{\pi\left(\frac{2 \text{ cm}}{100 \ \frac{\text{cm}}{\text{m}}}\right)}$$
$$= 5.51 \text{ MPa}\cdot\sqrt{\text{m}} \quad (5.5 \text{ MPa}\cdot\sqrt{\text{m}})$$

11. STRAIN ENERGY

Strain energy, also known as *internal work*, is the energy per unit volume stored in a deformed material. The strain energy is equivalent to the work done by the applied tensile force. Simple work is calculated as the product of a force moving through a distance.

$$\text{work} = \text{force} \times \text{distance} = \int F\,dL \qquad 47.16$$

$$\text{work per unit volume} = \int \frac{F\,dL}{AL} = \int_0^{\epsilon_{\text{final}}} \sigma d\epsilon \qquad 47.17$$

This work per unit volume corresponds to the area under the true stress-strain curve. Units are in-lbf/in^3 (i.e., inch-pounds (a unit of energy) per cubic inch (a unit of volume)), usually shortened to lbf/in^2, or J/m^3 (i.e., joules per cubic meter), less frequently shown as Pa. (Equation 47.17 cannot be simplified further because stress is not proportional to strain for the entire curve; however, for the elastic region, the stress-strain curve is essentially a straight line, and the area is triangular. Therefore, if a body of length L is deformed under force F or torque T, the resulting strain energy be expressed in terms of Eq. 47.18 through Eq. 47.22.

Strain Energy

$$U = \tfrac{1}{2}F\delta \quad \text{[strain energy]} \qquad 47.18$$

$$U = \frac{F^2L}{2AE} \quad \text{[tension or compression]} \qquad 47.19$$

$$U = \frac{T^2L}{2GJ} \quad \text{[torsion]} \qquad 47.20$$

$$U = \frac{F^2L}{2AG} \quad \text{[shear]} \qquad 47.21$$

$$U = \int \frac{M^2\,dx}{2EI} \quad \text{[bending]} \qquad 47.22$$

12. RESILIENCE

A *resilient material* is able to absorb and release *strain energy* without permanent deformation. *Resilience* is measured by the *modulus of resilience*, also known as the *elastic toughness*, which is the strain energy per unit volume required to reach the yield point. This is represented by the area under the stress-strain curve up to

Materials

the yield point. Since the stress-strain curve is essentially a straight line up to that point, the area is triangular.

$$U_R = \int_0^{\epsilon_y} \sigma \, d\epsilon = E\int_0^{\epsilon_y} \epsilon \, d\epsilon = \frac{E\epsilon_y^2}{2} = \frac{S_y\epsilon_y}{2} \quad 47.23$$

The modulus of resilience (shown in Fig. 47.13) varies greatly for steel. It can be more than ten times higher for high-carbon spring steel (U_R = 320 in-lbf/in^3 and 2.2 MJ/m^3) than for low-carbon steel (U_R = 20 in-lbf/in^3 and 0.14 MJ/m^3).

Figure 47.13 *Modulus of Resilience*

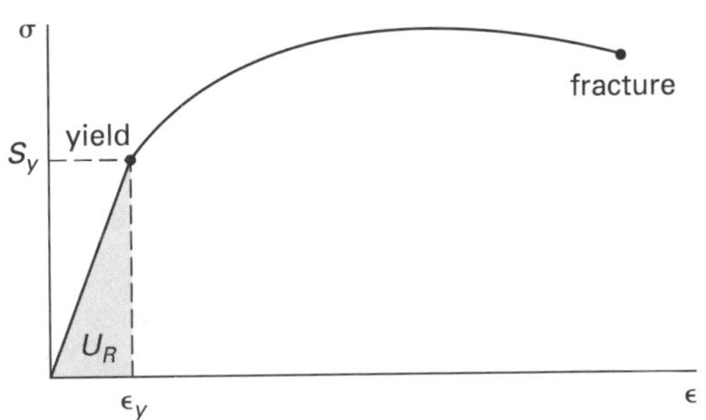

13. TOUGHNESS

A *tough material* will be able to withstand occasional high stresses without fracturing. Products subjected to sudden loading, such as chains, crane hooks, railroad couplings, and so on, should be tough. One measure of a material's *toughness* is the *modulus of toughness* (i.e, the strain energy or work per unit volume required to cause fracture). (See Fig. 47.14.) This is the total area under the stress-strain curve, given the symbol U_T. Since the area is irregular, the modulus of toughness cannot be exactly calculated by a simple formula. However, the modulus of toughness of ductile materials (with large strains at failure) can be approximately calculated from either Eq. 47.24 or Eq. 47.25.

$$U_T \approx S_u\epsilon_u \quad \text{[ductile]} \quad 47.24$$

$$U_T \approx \left(\frac{S_y + S_u}{2}\right)\epsilon_u \quad \text{[ductile]} \quad 47.25$$

Figure 47.14 *Modulus of Toughness*

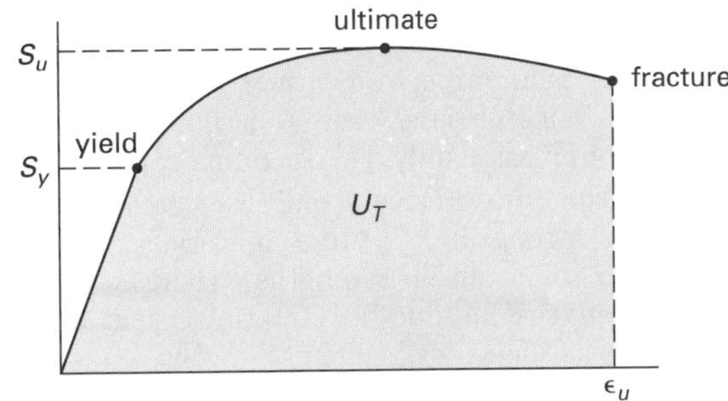

For brittle materials, the stress-strain curve may be either linear or parabolic. If the curve is parabolic, Eq. 47.26 approximates the modulus of toughness.

$$U_T \approx \tfrac{2}{3}S_u\epsilon_u \quad \text{[brittle]} \quad 47.26$$

14. UNLOADING AND RELOADING

If the load is removed after a specimen is stressed elastically, the material will return to its original state. If the load is removed after a specimen is stressed into the plastic region, the *unloading curve* will follow a sloped path back to zero stress. The slope of the unloading curve will be equal to the original modulus of elasticity, E, illustrated by Fig. 47.15.

If this same material is subsequently reloaded, the *reloading curve* will follow the previous unloading curve up to the continuation of the original stress-strain curve. Therefore, the *apparent yield stress* of the reloaded specimen will be higher. This extra strength is the result of the strain hardening that has occurred.[6] Although the material will have a higher strength, its ductility and toughness will have been reduced.

Figure 47.15 *Unloading and Reloading Curves*

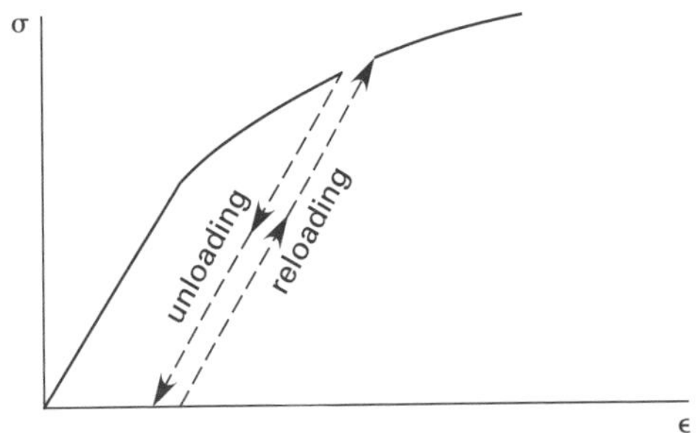

[6]The additional strength is lost if the material is subsequently annealed.

15. COMPRESSIVE STRENGTH

Compressive strength, S_{uc} (i.e., ultimate strength in compression), is an important property for brittle materials such as concrete and cast iron that are primarily loaded in compression only. (f'_c is commonly used as the symbol for the compressive strength of concrete.) The compressive strengths of these materials are much greater than their tensile strengths, whereas the compressive strengths for ductile materials, such as steel, are the same as their tensile yield strengths.

Within the linear (elastic) region, Hooke's law is valid for compression of both brittle and ductile materials.

The failure mechanism for ductile materials is plastic deformation alone. Such materials do not rupture in compression. Therefore, a ductile material can support a load long after the material is distorted beyond a useful shape.

The failure mechanism for brittle materials is shear along an inclined plane. The characteristic plane and hourglass failures for brittle materials are shown in Fig. 47.16.[7] Theoretically, only *cohesion* contributes to compressive strength, and the *angle of rupture* (i.e., the incline angle), θ, should be 45°. In real materials, however, internal friction also contributes strength. The angles of rupture for cast iron, concrete, brick, and so on vary roughly between 50° and 60°. If the *angle of internal friction*, ϕ, is known for the material, the angle of rupture can be calculated exactly from *Mohr's theory of rupture*.

$$\theta = 45^\circ + \frac{\phi}{2} \qquad 47.27$$

Figure 47.16 *Compressive Failures*

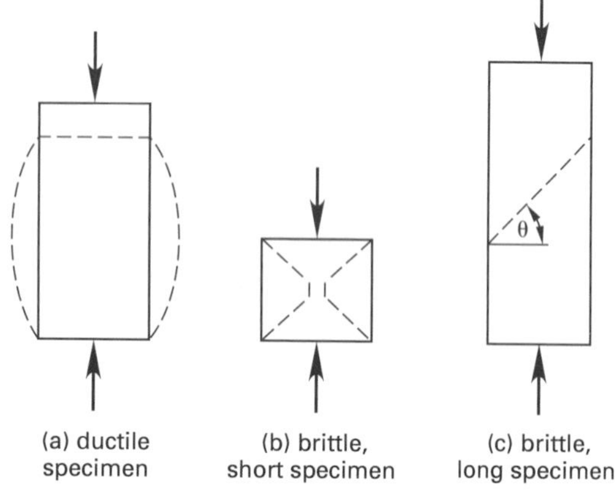

16. TORSION TEST

Figure 47.17 illustrates a simple cube loaded by a shear stress, τ. The volume of the cube does not decrease when loaded, but the shape changes. The *shear strain* is the angle, θ, expressed in radians. The shear strain is proportional to the shear stress, analogous to Hooke's law for tensile loading. G is the *shear modulus*, also known as the *modulus of shear*, *modulus of elasticity in shear*, and *modulus of rigidity*.

$$\tau = G\theta \qquad 47.28$$

Figure 47.17 *Cube Loaded in Shear*

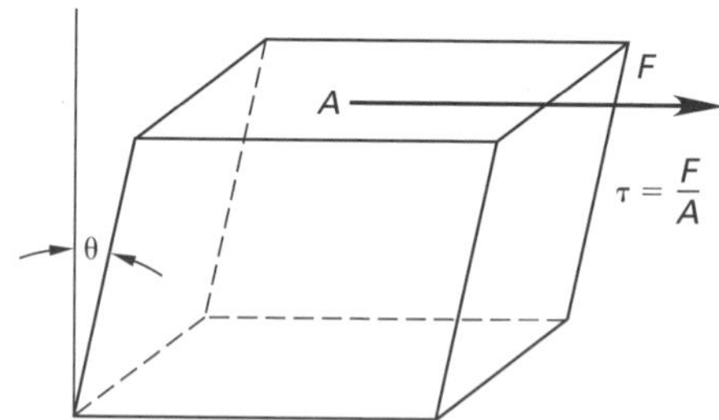

The shear modulus can be calculated from the modulus of elasticity and Poisson's ratio and can be derived from the results of a tensile test. (See Table 47.6.)

Shear Stress-Strain

$$G = \frac{E}{2(1+\nu)} \qquad 47.29$$

The shear stress can also be calculated from a torsion test, as illustrated by Fig. 47.18. Equation 47.30 relates the angle of twist (in radians) to the shear modulus.

Table 47.6 *Approximate Values of Shear Modulus*

material	lbf/in²	GPa
aluminum	3.8×10^6	26
brass	5.5×10^6	38
copper	6.2×10^6	43
cast iron	8.0×10^6	55
magnesium	2.4×10^6	17
steel	11.5×10^6	79
stainless steel	10.6×10^6	73
titanium	6.0×10^6	41
glass	4.2×10^6	29

(Multiply lbf/in² by 6.895×10^{-6} to obtain GPa.)

[7]The *hourglass failure* appears when the material is too short for a complete failure surface to develop.

Figure 47.18 *Uniform Bar in Torsion*

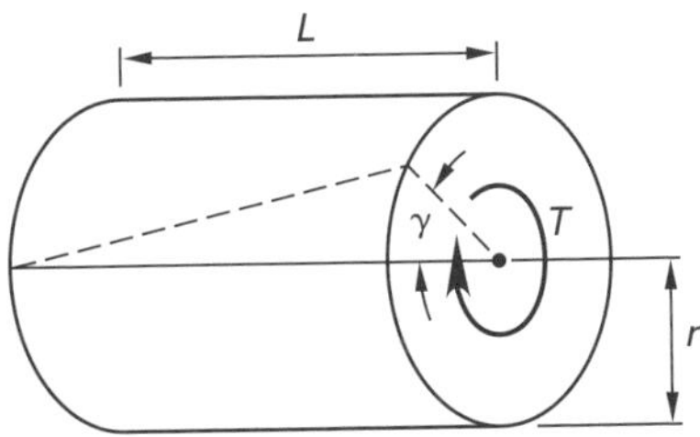

Shear Stress-Strain

$$\gamma = \frac{TL}{JG} = \frac{\tau L}{rG} \quad \text{[radians]} \qquad 47.30$$

The *shear strength*, S_s or S_{ys}, of a material is the maximum shear stress that the material can support without yielding in shear. (The ultimate shear strength, S_{us}, is rarely encountered.) For ductile materials, *maximum shear stress theory* predicts the shear strength as one-half of the tensile yield strength. A more accurate relationship is derived from the *distortion energy theory* (also known as *von Mises theory*). [**Variable Loading Failure Theories**]

$$S_{ys} = \frac{S_{yt}}{2} \quad \text{[maximum shear stress theory]} \qquad 47.31$$

$$S_{ys} = \frac{S_{yt}}{\sqrt{3}} = 0.577S_{yt} \qquad 47.32$$

[distortion energy theory]

17. RELATIONSHIP BETWEEN THE ELASTIC CONSTANTS

The elastic constants (modulus of elasticity, shear modulus, bulk modulus, and Poisson's ratio) are related in elastic materials. Table 47.7 lists the common relationships.

18. FATIGUE TESTING

A material can fail after repeated stress loadings even if the stress level never exceeds the ultimate strength, a condition known as *fatigue failure*. [**Variable Loading Failure Theories**]

The behavior of a material under repeated loadings is evaluated by a fatigue test. A specimen is loaded repeatedly to a specific stress amplitude, S, and the number of applications of that stress required to cause failure, N, is counted. Rotating beam tests that load the specimen in bending are more common than alternating deflection and push-pull tests but are limited to round specimens.[8] (See Fig. 47.19.)

Figure 47.19 *Rotating Beam Test*

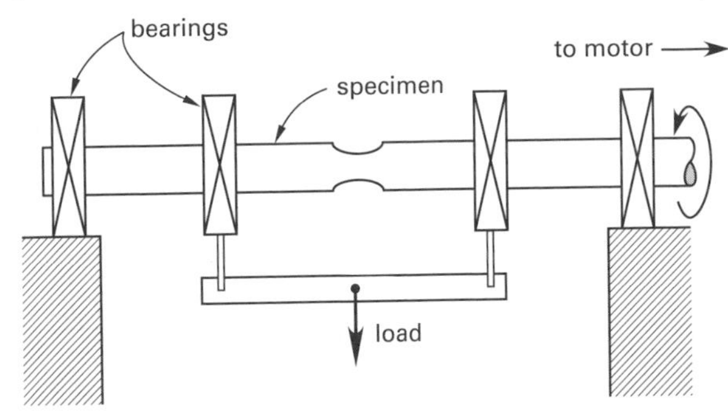

This procedure is repeated for different stresses, using eight to fifteen specimens. The results of these tests are graphed, resulting in an *S-N curve* (i.e., stress-number of cycles), also known as a *Wöhler curve*, which is shown in Fig. 47.20.

Figure 47.20 *Typical S-N Curve for Steel*

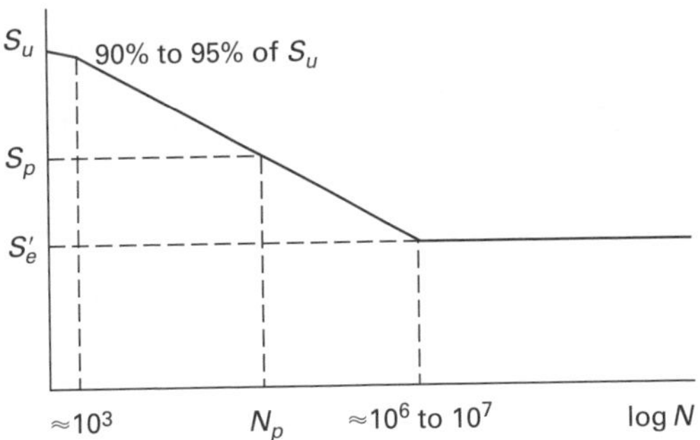

For an alternating stress test, the stress plotted on the *S-N* curve can be the maximum, minimum, or mean value. The choice depends on the method of testing as well as the intended application. The maximum stress should be used in rotating beam tests, since the mean stress is zero. For cyclic, one-dimensional bending, the maximum and mean stresses are commonly used.

For a particular stress level, say S_p in Fig. 47.20, the number of cycles required to cause failure, N_p, is the *fatigue life*. S_p is the *fatigue strength* corresponding to N_p.

For steel in bending subjected to fewer than approximately 10^3 loadings, the fatigue strength starts at the ultimate strength and drops to 90–95% of the ultimate strength at 10^3 cycles.[9] (Although *low-cycle fatigue* theory has its own peculiarities, a part experiencing a small number of cycles can usually be designed or analyzed for

[8]In the design of ductile steel buildings, the static case is assumed up to 20,000 cycles. However, in critical applications (such as nuclear steam vessels, turbines, and so on) that experience temperature swings, fatigue failure can occur with a smaller number of cycles due to *cyclic strain*, not due to *cyclic stress*.

[9]Steel in tension has a lower fatigue strength at 10^3 cycles, approximately 72–75% of S_u.

Table 47.7 *Relationships Between Elastic Constants*

elastic constants	in terms of E, ν	E, G	B, ν	B, G	E, B
E	–	–	$3(1-2\nu)B$	$\frac{9BG}{3B+G}$	–
ν	–	$\frac{E}{2G}-1$	–	$\frac{3B-2G}{2(3B+G)}$	$\frac{3B-E}{6B}$
G	$\frac{E}{2(1+\nu)}$	–	$\frac{3(1-2\nu)B}{2(1+\nu)}$	–	$\frac{3EB}{9B-E}$
B	$\frac{E}{3(1-2\nu)}$	$\frac{GE}{3(3G-E)}$	–	–	–

static loading.) The curve is linear between 10^3 and 10^6 cycles if a logarithmic N-scale is used. Beyond 10^6 to 10^7 cycles, there is no further decrease in strength.

Therefore, below a certain stress level, called the *endurance limit* or *endurance strength*, S_e', the material will withstand an almost infinite number of loadings without experiencing failure.[10] This is characteristic of steel and titanium. Therefore, if a dynamically loaded part is to have an infinite life, the stress must be kept below the endurance limit. Ratio S_e'/S_u is known as the *endurance ratio* or *fatigue ratio.* For carbon steel, the endurance ratio is approximately 0.4 for pearlitic, 0.60 for ferritic, and 0.25 for martensitic microstructures. For martensitic alloy steels, it is approximately 0.35.

For steel whose microstructure is unknown, the endurance strength is given approximately by Eq. 47.33.[11] **[Variable Loading Failure Theories]**

$$S'_{e,\text{steel}} \begin{cases} = 0.5S_u & [S_u < 200{,}000 \text{ lbf/in}^2] \\ & [S_u < 1.4 \text{ GPa}] \\ = 100{,}000 \text{ lbf/in}^2 & [S_u > 200{,}000 \text{ lbf/in}^2] \\ \quad (700 \text{ MPa}) & [S_u > 1.4 \text{ GPa}] \end{cases} \qquad 47.33$$

For cast iron, the endurance ratio is lower.

$$S'_{e,\text{cast iron}} = 0.4S_u \qquad 47.34$$

Steel and titanium are the most important engineering materials that have well-defined endurance limits. Many nonferrous metals and alloys, such as aluminum, magnesium, and copper alloys, do not have well-defined endurance limits. (See Fig. 47.21.) The strength continues to decrease with cyclic loading and never levels off. In such cases, the endurance limit is taken as the stress that causes failure at 10^7 loadings (less typically, at 5×10^7, 10^8, or 5×10^8 loadings). Alternatively, the endurance strength is approximated by Eq. 47.35.

$$S'_{e,\text{aluminum}} = \begin{cases} 0.3S_u & [\text{cast}] \\ 0.4S_u & [\text{wrought}] \end{cases} \qquad 47.35$$

Figure 47.21 *Typical S-N Curve for Aluminum*

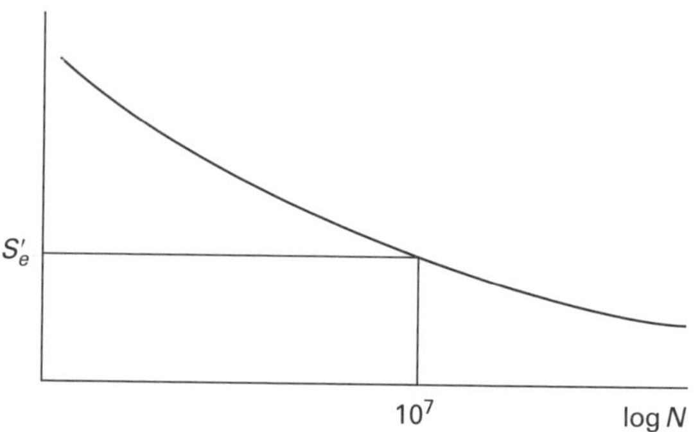

The yield strength is an irrelevant factor in cyclic loading. Fatigue failures are fracture failures; they are not yielding failures. They start with microscopic cracks at the material surface. Some of the cracks are present initially; others form when repeated cold working reduces the ductility in strain-hardened areas. These cracks grow minutely with each loading. Since cracks start at the location of surface defects, the endurance limit is increased by proper treatment of the surface. Such treatments include polishing, surface hardening, shot peening, and filleting joints.

The endurance limit is not a true property of the material since the other significant influences, particularly surface finish, are never eliminated. However,

[10] Most endurance tests use some form of sinusoidal loading. However, the fatigue and endurance strengths do not depend much on the shape of the loading curve. Only the maximum amplitude of the stress is relevant. Therefore, the endurance limit can be used with other types of loading (sawtooth, square wave, random, etc.).

[11] The coefficient in Eq. 47.33 actually varies between 0.25 and 0.6. However, 0.5 is commonly quoted.

representative values of S_e' obtained from ground and polished specimens provide a baseline to which other factors can be applied to account for the effects of surface finish, temperature, stress concentration, notch sensitivity, size, environment, and desired reliability. These other influences are accounted for by fatigue strength reduction (derating) factors, k_i, which are used to calculate a working endurance strength, S_e, for the material. (See Fig. 47.22.) [**Variable Loading Failure Theories**]

$$S_e = \prod k_i S_e' \qquad 47.36$$

Figure 47.22 *Surface Finish Reduction Factors for Endurance Strength*

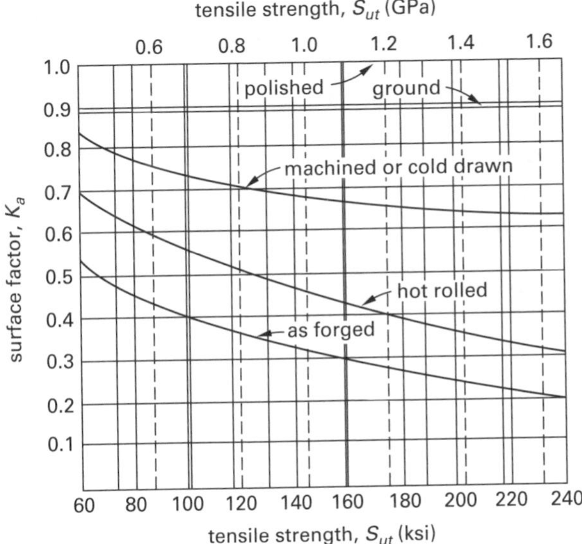

Reprinted with permission from *Mechanical Engineering Design*, 3rd ed. by Joseph Edward Shigley, © 1977, The McGraw-Hill Companies, Inc.

Since a rough surface significantly decreases the endurance strength of a specimen, it is not surprising that notches (and other features that produce stress concentration) do so as well. In some cases, the theoretical tensile *stress concentration factor*, K_t, due to notches and other features can be determined theoretically or experimentally. The ratio of the fatigue strength of a polished specimen to the fatigue strength of a notched specimen at the same number of cycles is known as the *fatigue notch factor*, K_f, also known as the *fatigue stress concentration factor*. The *fatigue notch sensitivity*, q, is a measure of the degree of agreement between the stress concentration factor and the fatigue notch factor.

$$q = \frac{K_f - 1}{K_t - 1} \quad [K_f > 1] \qquad 47.37$$

19. TESTING OF PLASTICS

With reasonable variations, mechanical properties of plastics are evaluated using the same methods as for metals.[12] Although temperature is an important factor in the testing of plastics, tests for tensile strength, endurance, hardness, toughness, and creep rate are similar or the same as for metals. Figure 47.23 illustrates typical tensile test results.[13]

Figure 47.23 *Typical Tensile Test Performance for Plastics (loaded at 2 in/min)*

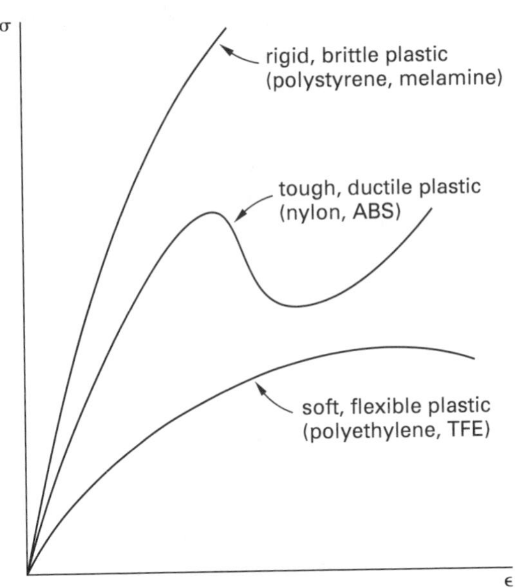

Unlike metals, which follow Hooke's law, plastics are non-Hookean. (They may be Hookean for a short-duration loading.) The modulus of elasticity, for example, changes with stress level, temperature, time, and chemical environment. A plastic that appears to be satisfactory under one set of conditions can fail quickly under slightly different conditions. Therefore, properties of plastics determined from testing (and from tables) should be used only to compare similar materials, not to predict long-term behavior. Plastic tests are used to determine material specifications, not performance specifications.

[12]For example, plastic specimens for tensile testing can be produced by injection molding as well as by machining from compression-molded plaques, rather than by machining from bar stock.

[13]Plastics are sensitive to the rate of loading. Figure 47.23 illustrates tensile performance based on a 2 in/min loading rate. However, for a fast loading rate (such as 2 in/sec), most plastics would exhibit brittle performance. On the other hand, given a slow loading rate (such as 2 in/mo), most would behave as a soft and flexible plastic. Therefore, with different rates of loading, all three types of stress-strain performance shown in Fig. 47.23 can be obtained from the same plastic.

On a short-term basis, plastics behave elastically. They distort when loaded and spring back when unloaded. Under prolonged loading, however, creep (cold flow) becomes significant. When loading is removed, there is some instantaneous recovery, some delayed recovery, and some permanent deformation. The recovery might be complete if the load is removed within 10 hours, but if the loading is longer (e.g., 100 hours), recovery may be only partial. Because of this behavior, plastics are subjected to various other tests.

Additional tests used to determine the mechanical properties of plastics include deflection temperature, long-term (e.g., 3000 hours) tensile creep, creep rupture, and *stress-relaxation* (long-duration, constant-strain tensile testing at elevated temperatures).[14] Because some plastics deteriorate when exposed to light, plasma, or chemicals, performance under these conditions can be evaluated, as can be the insulating and dielectric properties.

The *creep modulus* (also known as *apparent modulus*), determined from tensile creep testing, is the instantaneous ratio of stress to creep strain. The creep modulus decreases with time. The *deflection temperature* test indicates the dimensional stability of a plastic at high temperatures. A plastic bar is loaded laterally (as a beam) to a known stress level, and the temperature of the bar is gradually increased. The temperature at which the deflection reaches 0.010 in (0.254 mm) is taken as the *thermal deflection temperature* (TDT). The *Vicat softening point*, primarily used with polyethylenes, is the temperature at which a loaded standard needle penetrates 1 mm when the temperature is uniformly increased at a standard rate.

20. NONDESTRUCTIVE TESTING

Nondestructive testing (NDT) or *nondestructive evaluation* (NDE) is used when it is impractical or uneconomical to perform destructive sampling on manufactured products and their parts. Typical applications of NDT are inspection of helicopter blades, cast aluminum wheels, and welds in nuclear pressure vessels. Some procedures are particularly useful in providing quality monitoring on a continuous, real-time basis. In addition to visual processes, the main types of nondestructive testing are magnetic particle, eddy current, liquid penetrant, ultrasonic imaging, acoustic emission, and infrared testing, as well as radiography. [**Nondestructive Test Methods**]

The *visual-optical* process differs from normal visual inspection in the use of optical scanning systems, borescopes, magnifiers, and holographic equipment. Flaws are identified as changes in light intensity (reflected, transmitted, or refracted), color changes, polarization changes, or phase changes. This method is limited to the identification of surface flaws or interior flaws in transparent materials.

Liquid penetrant testing is based on a fluorescent dye being drawn by capillary action into surface defects. A developer substance is commonly used to aid in visual inspection. This method can be used with any nonporous material, including metals, plastics, and glazed ceramics. It is capable of finding cracks, porosities, pits, seams, and laps.

Liquid penetrant tests are simple, can be used with complex shapes, and can be performed on site. Workpieces must be clean and nonporous. However, only small surface defects are detectable.

Magnetic particle testing takes advantage of the attraction of ferromagnetic powders (e.g., the *Magnaflux*™ *process*) and fluorescent particles (e.g., the *Magnaglow*™ *process*) to leakage flux at surface flaws in magnetic materials. The particles accumulate and become visible at such flaws when an intense magnetic field is set up in a workpiece.

This method can locate most surface flaws (such as cracks, laps, and seams) and, in some special cases, subsurface flaws. The procedure is fast and simple to interpret. However, workpieces must be ferromagnetic and clean. Following the test, demagnetization may be required. A high-current power source is required.

Eddy current testing uses alternating current from a test coil to induce eddy currents in electrically conducting, metallic objects. Flaws and other material properties affect the current flow. The change in current is monitored by a detection circuit or on a meter or screen. This method can be used to locate defects of many types, including cracks, voids, inclusions, and weld defects, as well as to find changes in composition, structure, hardness, and porosity. Intimate contact between the material and the test coil is not required. Operation can be continuous, automatic, and monitored electronically. Sensitivity is easily adjusted. Therefore, this method is ideal for unattended continuous processing. Many variables, however, can affect the current flow, and only electrically conducting materials can be tested with this method.

With *infrared testing*, infrared radiation emitted from objects can be detected and correlated with quality. Any discontinuities that interrupt heat flow, such as flaws, voids, and inclusions, can be detected.

Infrared testing requires access to only one side and is highly sensitive. It is applicable to complex shapes and assemblies of dissimilar components but is relatively slow. The detection can be performed electronically. Results are affected by variations in material size, coatings, and colors, and hot spots can be hidden by cool surface layers.

[14]Plastic pipes have their own special tests (e.g., ASTM D1598, D1785, and D2444).

In *ultrasound imaging testing* (*ultrasonics*), mechanical vibrations in the 0.1–50 MHz range are induced by pressing a piezoelectric transducer against a workpiece.[15] The transmitted waves are normally reflected back, but the waves are scattered by interior defects. The results are interpreted by reading a screen or meter. The method can be used for metals, plastics, glass, rubber, graphite, and concrete. It is excellent for detecting internal defects such as inclusions, cracks, porosities, laminations, and changes in material structure.

Ultrasound testing is extremely flexible. It can be automated and is very fast. Results can be recorded or interpreted electronically. Penetration through thick steel layers is possible. Direct contact (or immersion in a fluid) is required, but only one surface needs to be accessible. Rough surfaces and complex shapes may cause difficulties, however. A related method, *acoustic emission monitoring*, is used to test pressurized systems.

Radiography (i.e., *nuclear sensing*) uses neutron, X-ray, gamma-ray (e.g., Ce-137), and isotope (e.g., Co-60) sources. (When neutrons are used, the method is known as *neutron radiography* or *neutron gaging*.) The intensity of emitted radiation is changed when the rays pass through defects, and the intensity changes are monitored on a fluoroscope or recorded on film. This method can be used to detect internal defects, changes in material structure, thickness, and the absence of internal workpieces. It is also used to check liquid levels in filled containers.

Up to 30 in (0.75 m) of steel can be penetrated by X-ray sources. Gamma sources, which are more portable and lower in cost than X-ray sources, can be used with steel up to 10 in (0.25 m).

Radiography requires access to both sides of the workpiece. Radiography involves some health risk, and there may be government standards associated with its use. Electrical power and cooling water may be required in large installations. Shielding and possibly film processing are also required, making this the most expensive form of nondestructive testing.

There are two types of *holographic NDT methods*. *Acoustic holography* is a form of ultrasonic testing that passes an ultrasonic beam through the workpiece (or through a medium such as water surrounding the workpiece) and measures the displacement of the workpiece (or medium). With suitable processing, a three-dimensional hologram is formed that can be visually inspected.

In one form of *optical holography*, a hologram of the unloaded workpiece is imposed on the actual workpiece. If the workpiece is then loaded (stressed), the observed changes (e.g., deflections) from the holographic image will be nonuniform when discontinuities and defects are present.

21. HARDNESS TESTING

Hardness tests measure the capacity of a surface to resist deformation. (See Table 47.8.) The main use of hardness testing is to verify heat treatments, an important factor in product service life. Through empirical correlations, it is also possible to predict the ultimate strength and toughness of some materials. [**Material Hardness**]

The *scratch hardness test*, also known as the *Mohs test*, compares the hardness of the material to that of minerals. Minerals of increasing hardness are used to scratch the sample. The resulting *Mohs scale* hardness can be used or correlated to other hardness scales, as shown in Fig. 47.24.

Figure 47.24 *Mohs Hardness Scale*

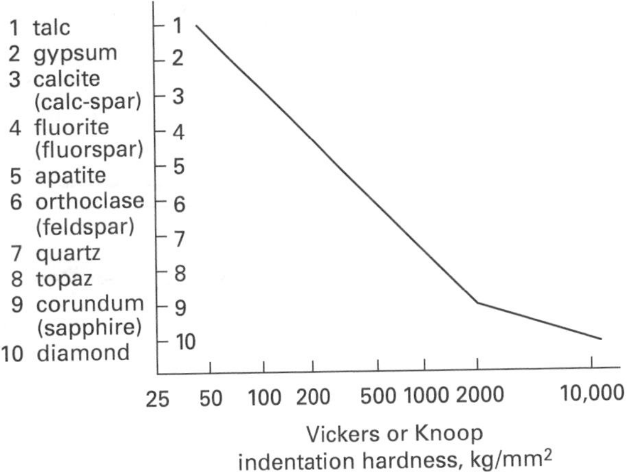

The *file hardness* test is a combination of the cutting and scratch tests. Files of known hardness are drawn across the sample. The file ceases to cut the material when the material and file hardnesses are the same.

The *Brinell hardness test* is used primarily with iron and steel castings, although it can be used with softer materials. (See Table 47.9.) The *Brinell hardness number*, BHN (or HB or H_B), is determined by pressing a hardened steel ball into the surface of a specimen. The diameter of the resulting depression is correlated to the hardness. The standard ball is 10 mm in diameter and loads are 500 kg and 3000 kg for soft and hard materials, respectively.

The Brinell hardness number is the load per unit contact area. If a load, P (in kilograms), is applied through a steel ball of diameter, D (in millimeters), and produces a depression of diameter, d (in millimeters), and depth, t (in millimeters), the Brinell hardness number can be calculated from Eq. 47.38.

[15]Theoretically, any frequency can be used. However, as frequency goes up, the detail available increases, while the penetration decreases. Biomedical applications operate below 8 MHz, and industrial NDT uses 2–10 MHz waves.

Materials

Table 47.8 *Hardness Penetration Tests*

test	penetrator	diagram	measured dimension	hardness[a]
Brinell	sphere	(a)	diameter, d	$\text{BHN} = \dfrac{2P}{\pi D(D - \sqrt{D^2 - d^2})}$
Rockwell[b]	sphere or penetrator	(b)	depth, t	$R = C_1 - C_2 t$
Vickers	square pyramid	(b)	mean diagonal, d_1	$\text{VHN} = \dfrac{1.854P}{d_1^2}$
Meyer	sphere	(a)	diameter, d	$\text{MHN} = \dfrac{4P}{\pi d^2}$
Meyer-Vickers	square pyramid	(b)	mean diagonal, d_1	$M_V = \dfrac{2P}{d_1^2}$
Knoop	asymmetrical pyramid	(c)	long diagonal, L	$K = \dfrac{14.2P}{L^2}$

[a]All forces, P, in kgf. All length and depth measurements in mm.
[b]C_1 and C_2 are constants that depend on the scale.

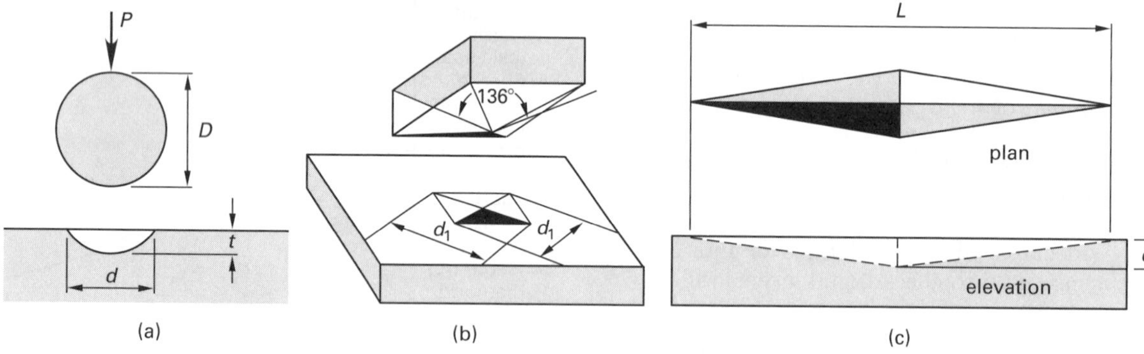

$$\begin{aligned}\text{BHN} &= \frac{P}{A_{\text{contact}}} = \frac{P}{\pi D t} \\ &= \frac{2P}{\pi D(D - \sqrt{D^2 - d^2})}\end{aligned} \qquad 47.38$$

For heat-treated plain-carbon steels, the ultimate tensile strength can be approximately calculated from the steel's Brinell hardness number. **[Relationship Between Hardness and Tensile Strength]**

$$S_{u,\text{MPa}} \approx 515(\text{BHN}) \quad [\text{BHN} \leq 175] \qquad 47.39$$

$$S_{u,\text{lbf/in}^2} \approx 490(\text{BHN}) \quad [\text{BHN} > 175] \qquad 47.40$$

The *Rockwell hardness test* is similar to the Brinell test. A steel ball or diamond spheroconical penetrator (known as a *brale indenter*) is pressed into the material. The machine applies an initial load (60 kgf, 100 kgf, or 150 kgf) that sets the penetrator below surface imperfections.[16] Then, a significant load is applied. The Rockwell hardness, R (or HR or H_R), is determined from the depth of penetration and is read directly from a dial.

Although a number of Rockwell scales (A through G) exist, the B and C scales are commonly used for steel. The *Rockwell B scale* is used with a steel ball for mild steel and high-strength aluminum. The *Rockwell C scale* is used with the brale indenter for hard steels having ultimate tensile strengths up to 300 ksi (2 GPa). The *Rockwell A scale* has a wide range and can be used with both soft materials (such as annealed brass) and hard materials (such as cemented carbides).

Other penetration hardness tests include the *Meyer*, *Vickers*, *Meyer-Vickers*, and *Knoop* tests, as described in Table 47.8.

Cutting hardness is a measure of the force per unit area to cut a chip at low speed. (See Fig. 47.25.)

$$\text{cutting hardness} = \frac{F}{bt} \qquad 47.41$$

[16]Other Rockwell tests use 15 kgf, 30 kgf, and 45 kgf. The use of kgf units is traditional, and even modern test equipment is calibrated in kgf. Multiply kgf by 9.80665 to get newtons.

Figure 47.25 *Cutting Hardness*

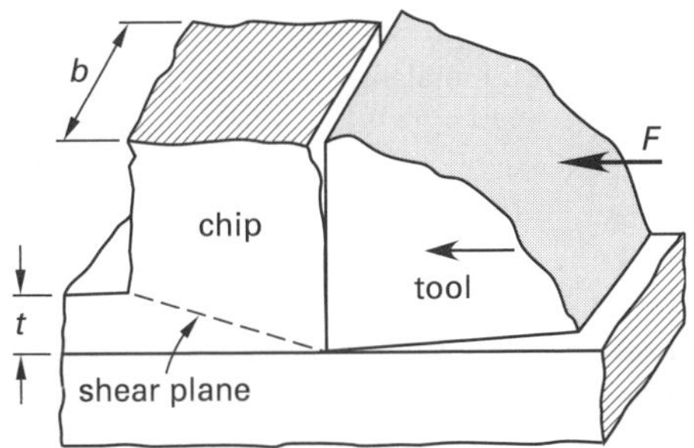

All of the preceding hardness tests are *destructive tests* because they mar the material surface. However, *ultrasonic tests* and various *rebound tests* (e.g., the *Shore hardness test* and the *scleroscopic hardness test*) are *nondestructive tests*. In a rebound test, a standard object, usually a diamond-tipped hammer, is dropped from a standard height onto the sample. The height of the rebound is measured and correlated to other hardness scales.

The various hardness tests do not measure identical properties of the material, so correlations between the various scales are not exact. For steel, the Brinell and Vickers hardness numbers are approximately the same below values of 320 Brinell. Also, the Brinell hardness is approximately ten times the Rockwell C hardness, R_C, for $R_C > 20$. Table 47.9 is an accepted correlation between several of the scales for steel. The table should not be used for other materials.

Table 47.9 *Correlations Between Hardness Scales for Steel*

Brinell number	Vickers number	Rockwell numbers C	Rockwell numbers B	scleroscope number
780	1150	70	...	106
712	960	66	...	95
653	820	62	...	87
601	717	58	...	81
555	633	55	120	75
514	567	52	119	70
477	515	49	117	65
429	454	45	115	59
401	420	42	113	55
363	375	38	110	51
321	327	34	108	45
293	296	31	106	42
277	279	29	104	39
248	248	24	102	36
235	235	22	99	34
223	223	20	97	32
207	207	16	95	30
197	197	13	93	29
183	183	9	90	27
166	166	4	86	25
153	153	...	82	23
140	140	...	78	21
131	131	...	74	20
121	121	...	70	...
112	112	...	66	...
105	105	...	62	...
99	99	...	59	...
95	95	...	56	...

22. TOUGHNESS TESTING

During World War II, the United States experienced spectacular failures in approximately 25% of its Liberty ships and T-2 tankers. The mild steel plates of these ships were connected by welds that lost their ductility and became brittle in winter temperatures. Some of the ships actually broke into two sections. Such *brittle failures* are most likely to occur when three conditions are met: (a) triaxial stress, (b) low temperature, and (c) rapid loading.

Toughness is a measure of the material's ability to yield and absorb highly localized and rapidly applied stresses. *Notch toughness* is evaluated by measuring the *impact energy* that causes a notched sample to fail.[17] **[Impact Test]**

In the *Charpy test* (see Fig. 47.26), popular in the United States, a standardized beam specimen is given a 45° notch. The specimen is then centered on simple supports with the notch down. A falling pendulum striker hits the center of the specimen. This test is performed several times with different heights and different specimens until a sample fractures.

Figure 47.26 *Charpy Test*

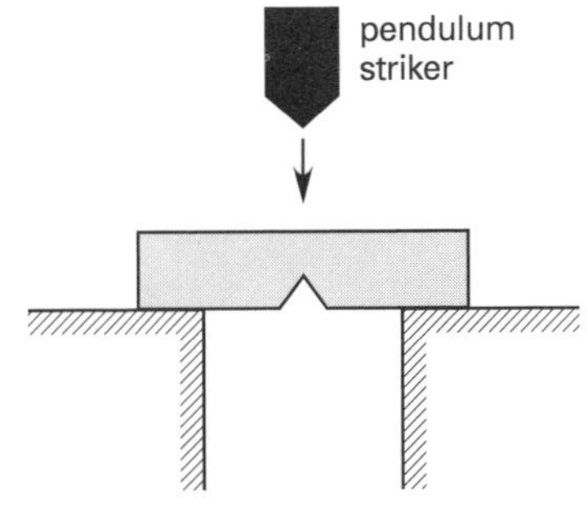

[17]Without a notch, the specimen would experience uniaxial stress (tension and compression) at impact. The notch allows triaxial stresses to develop. Most materials become more brittle under triaxial stresses than under uniaxial stresses.

Materials

The kinetic energy expended at impact, equal to the initial potential energy less the rebound or follow-through height of the pendulum striker, is calculated from measured heights. It is designated C_V and is expressed in either foot-pounds (ft-lbf) or joules (J).[18] The energy required to cause failure is a measure of toughness.

At 70°F (21°C), the energy required to cause failure ranges from 45 ft-lbf (60 J) for carbon steels to approximately 110 ft-lbf (150 J) for chromium-manganese steels. As temperature is reduced, however, the toughness decreases. In BCC metals, such as steel, at a low enough temperature the toughness decreases sharply. The transition from high-energy ductile failures to low-energy brittle failures begins at the *fracture transition plastic* (FTP) temperature.

Since the transition occurs over a wide temperature range, the *transition temperature* (also known as the *ductile-brittle transition temperature*, DBTT) is taken as the temperature at which an impact of 15 ft-lbf (20.4 J) will cause failure. (15 ft-lbf is used for low-carbon ship steels. Other values may be used with other materials.) This occurs at approximately 30°F (−1°C) for low-carbon steel. Table 47.10 gives ductile transition temperatures for some forms of steel.

The appearance of the fractured surface is also used to evaluate the transition temperature. The fracture can be fibrous (from shear fracture) or granular (from cleavage fracture), or a mixture of both. The fracture planes are studied and the percentages of ductile failure plotted against temperature. The temperature at which the failure is 50% fibrous and 50% granular is known as the *fracture appearance transition temperature*, FATT.

Table 47.10 *Approximate Ductile Transition Temperatures*

type of steel	ductile transition temperature (°F (°C))
carbon steel	30° (−1°)
high-strength, low-alloy steel	0° to 30° (−16° to −1°)
heat-treated, high-strength, carbon steel	−25° (−32°)
heat-treated, construction alloy steel	−40° to −80° (−40° to −62°)

Not all materials have a ductile-brittle transition. Aluminum, copper, other FCC metals, and most HCP metals do not lose their toughness abruptly. Figure 47.27 illustrates the failure energy curves for several different types of materials.

Figure 47.27 *Failure Energy versus Temperature*

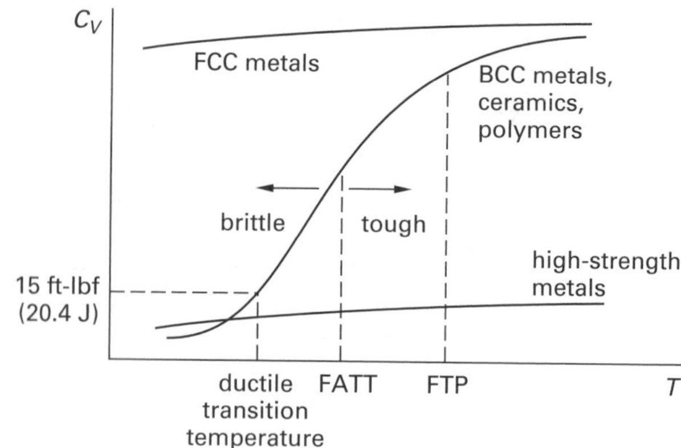

Another toughness test is the *Izod test.* This is illustrated in Fig. 47.28 and is similar to the Charpy test in its use of a notched specimen. The height to which a swinging pendulum follows through after causing the specimen to fail determines the energy of failure.

Figure 47.28 *Izod Test*

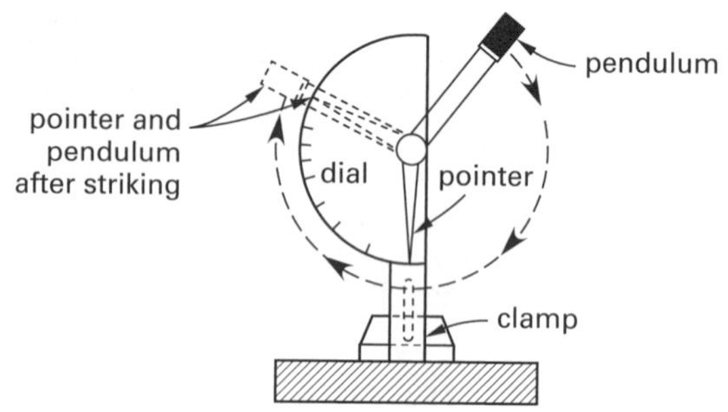

23. CREEP TEST

Creep or *creep strain* is the continuous yielding of a material under constant stress. For metals, creep is negligible at low temperatures (i.e., less than half of the absolute melting temperature), although the usefulness of nonreinforced plastics as structural materials is seriously limited by creep at room temperature.

During a *creep test*, a low tensile load of constant magnitude is applied to a specimen, and the strain is measured as a function of time. The *creep strength* is the stress that results in a specific creep rate, usually 0.001% or 0.0001% per hour. The *rupture strength*, determined from a *stress-rupture test*, is the stress that results in a failure after a given amount of time, usually 100, 1000, or 10,000 hours.

[18]The energy may also be expressed per unit cross section of specimen area.

If strain is plotted as a function of time, three different curvatures will be apparent following the initial elastic extension.[19] (See Fig. 47.29.) During the first stage, the *creep rate* ($d\epsilon/dt$) decreases since strain hardening (dislocation generation and interaction with grain boundaries and other barriers) is occurring at a greater rate than annealing (annihilation of dislocations, climb, cross-slip, and some recrystallization). This is known as *primary creep*.

During the second stage, the creep rate is constant, with strain hardening and annealing occurring at the same rate. This is known as *secondary creep* or *cold flow*. During the third stage, the specimen begins to neck down, and rupture eventually occurs. This region is known as *tertiary creep*.

Figure 47.29 *Stages of Creep*

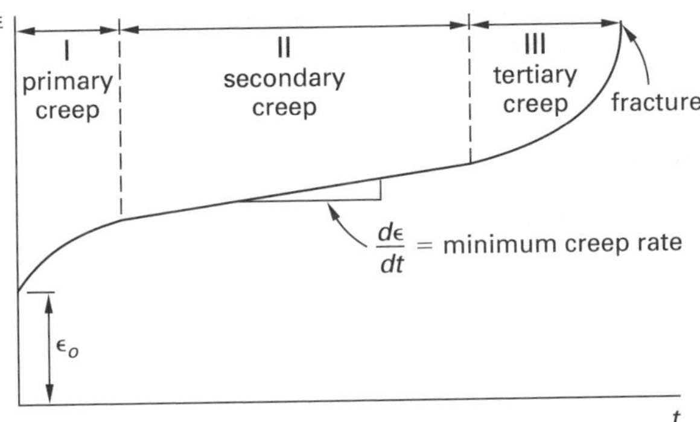

The secondary creep rate is lower than the primary and tertiary creep rates. The secondary creep rate, represented by the slope (on a log-log scale) of the line during the second stage, is temperature and stress dependent. This slope increases at higher temperatures and stresses. The steady-state creep rate can be represented by Eq. 47.42, where Q is the activation energy for creep, A is the pre-exponential constant, and n is the stress sensitivity.

Mechanical Properties

$$\frac{d\epsilon}{dt} = A\sigma^n e^{-Q/RT} \qquad 47.42$$

Dislocation climb (glide and creep) is the primary creep mechanism, although diffusion creep and grain boundary sliding also contribute to creep on a microscopic level. On a larger scale, the mechanisms of creep involve slip, subgrain formation, and grain-boundary sliding. (See Fig. 47.30.)

Figure 47.30 *Effect of Stress on Creep Rates*

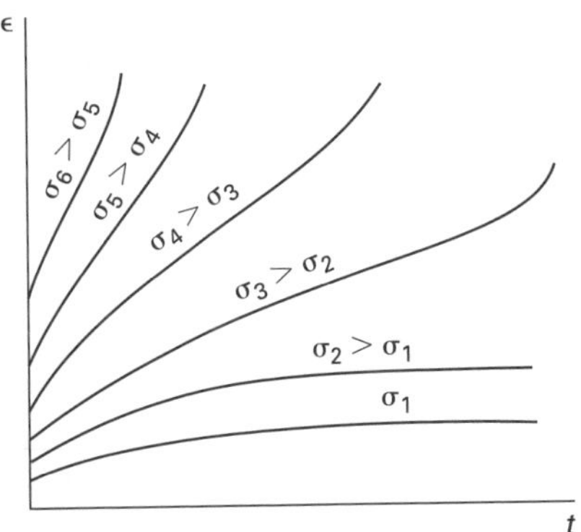

24. EFFECTS OF IMPURITIES AND STRAIN ON MECHANICAL PROPERTIES

Anything that restricts the movement of dislocations will increase the strength of metals and reduce ductility. Alloying materials, impurity atoms, imperfections, and other dislocations produce stronger materials. This is illustrated in Fig. 47.31.

Figure 47.31 *Effect of Impurities on Mechanical Properties*

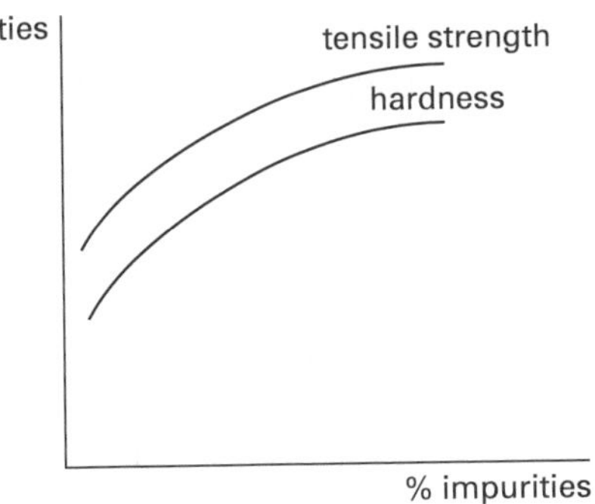

Additional dislocations are generated by the plastic deformation (i.e., cold working) of metals, and these dislocations can strain-harden the metal. Figure 47.32 shows the effect of strain-hardening on mechanical properties.

[19]In Great Britain, the initial elastic elongation, ϵ_o, is considered the first stage. Therefore, creep has four stages in British nomenclature.

Figure 47.32 *Effect of Strain-Hardening on Mechanical Properties*

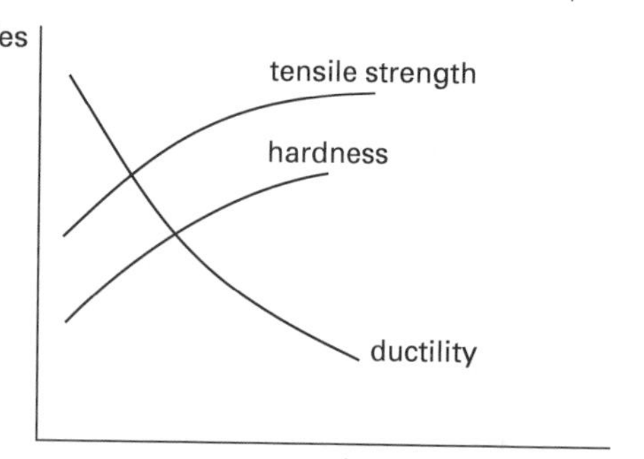

25. NOMENCLATURE

a	length	ft	m
A	area	in^2	m^2
b	width	in	m
B	bulk modulus	lbf/in^2	MPa
C	constant	–	–
C_V	impact energy	ft-lbf	J
d	diameter of impression	in	mm
D	diameter	in	m
e	engineering strain	in/in	m/m
E	modulus of elasticity	lbf/in^2	MPa
F	force	lbf	N
G	shear modulus	lbf/in^2	MPa
J	polar moment of inertia	in^4	m^4
k	exponent	1/hr	1/h
k	strength derating factor	–	–
K	stress concentration factor	–	–
K	strength coefficient	lbf/in^2	MPa
L	length	in	m
LYS	lower yield strength	lbf/in^2	MPa
n	exponent	–	–
N	number of cycles	–	–
P	force of impression or load	n.a.	kgf
q	fatigue notch sensitivity factor	–	–
q	reduction in area	–	–
Q	activation energy	ft-lbf	J
r	radius	in	m
R	ideal gas law constant	ft-lbf/lbm-°R	J/kg·K
RA	reduction in area	–	–
S	stress	lbf/in^2	MPa
S	strength	lbf/in^2	MPa
t	depth (thickness)	in	mm
t	time	hr	h
T	temperature	°F	°C
T	torque	in-lbf	N·m
U	strain energy	lbf/in^2	MPa
U_R	modulus of resilience	in-lbf/in^3	MJ/m^3
U_T	modulus of toughness	lbf/in^2	MPa
UYS	upper yield strength	lbf/in^2	MPa
Y	geometrical factor	–	–

Symbols

γ	angle of twist	rad	rad
δ	elongation	in	m
ϵ	true strain or creep	in/in	m/m
θ	angle of rupture	deg	deg
θ	shear strain	rad	rad
ν	Poisson's ratio	–	–
σ	true stress	lbf/in^2	MPa
τ	shear stress	lbf/in^2	MPa
ϕ	angle of internal friction	deg	deg

Subscripts

c	compressive
e	endurance
f	fatigue or fracture
o	original
p	particular
s	shear
t	tensile
T	toughness or true
u	ultimate
y	yield

48 Thermal Treatment of Metals

Content in blue refers to the *NCEES Handbook*.

NCEES EXAM SPECIFICATIONS AND RELATED CONTENT

MACHINE DESIGN AND MATERIALS EXAM

I.C.1. Material Properties: Physical
- 1. Soluble Alloy Equilibrium Diagrams (freezing and melting points)
- 11. Properties Versus Grain Size

I.C.2. Material Properties: Chemical
- 3. The Lever Rule
- 4. Gibbs Phase Rule

I.C.3.b. Material Properties: Mechanical: Time-dependent behavior
- 7. Quenching and Rates of Cooling
- 8. TTT and CCT Curves

1. SOLUBLE ALLOY EQUILIBRIUM DIAGRAMS

Most engineering materials are not pure elements but are alloys of two or more elements. Alloys of two elements are known as *binary alloys*. Steel, for example, is an alloy of primarily iron and carbon. Usually, one of the elements is present in a much smaller amount, and this element is known as the *alloying ingredient*. The primary ingredient is known as the *host ingredient*, *base metal*, or *parent ingredient*.

Sometimes, such as with alloys of copper and nickel, the alloying ingredient is 100% soluble in the parent ingredient. Nickel-copper alloy is said to be a *completely miscible alloy* or a *solid-solution alloy*.

The presence of the alloying ingredient changes the thermodynamic properties, notably the freezing (or melting) temperatures of both elements.[1] Usually the freezing temperatures decrease as the percentage of alloying ingredient is increased. Since the freezing points of the two elements are not the same, one of them will start to solidify at a higher temperature than the other. Thus, for any given composition, the alloy might consist of all liquid, all solid, or a combination of solid and liquid, depending on the temperature.

A *phase* of a material at a specific temperature will have a specific composition and crystalline structure and distinct physical, electrical, and thermodynamic properties. (In metallurgy, the word *phase* refers to more than just solid, liquid, and gas phases.)

The regions of an *equilibrium diagram*, also known as a *phase diagram*, illustrate the various alloy phases. The phases are plotted against temperature and composition. (The composition is usually a gravimetric fraction of the alloying ingredient. Only one ingredient's gravimetric fraction needs to be plotted for a binary alloy.) Sometimes, the amount of alloying ingredient is specified in *atomic fraction* or *atomic percent*. The conversions between gravimetric fractions, G_A and G_B, and atomic fractions, A_A and A_B, depend on the ratio of the molecular weights, M_A and M_B.

$$A_A = \frac{M_B G_A}{M_B G_A + M_A G_B} \quad \text{48.1}$$

$$A_B = \frac{M_A G_B}{M_B G_A + M_A G_B} \quad \text{48.2}$$

$$G_A = \frac{M_A A_A}{M_A A_A + M_B A_B} \quad \text{48.3}$$

[1]The term *freezing point* or *melting point* is used depending on whether heat is being removed or added, respectively.

$$G_B = \frac{M_B A_B}{M_A A_A + M_B A_B} \quad \text{48.4}$$

The equilibrium conditions do not occur instantaneously, and an equilibrium diagram is applicable only to the case of slow cooling.

Figure 48.1 is an equilibrium diagram for copper-nickel alloy. (Most equilibrium diagrams are much more complex.) The *liquidus line* is the boundary above which no solid can exist. The *solidus line* is the boundary below which no liquid can exist. The area between these two lines represents a mixture of solid and liquid phase materials.

Figure 48.1 *Copper-Nickel Phase Diagram*

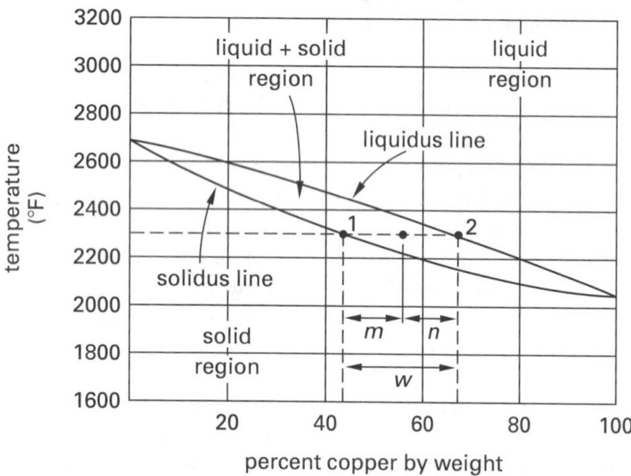

Curve (a) in Fig. 48.2 is a *time-temperature* or *temperature-time curve* for a pure metal cooling from liquid to solid state. At a particular point, the temperature remains constant (i.e., there is a *thermal arrest*). This temperature is the *freezing point* of the liquid, indicated on the graph by a horizontal line known as a *shelf* or *plateau*. The metal continues to lose heat energy—its *heat of fusion*—as the phase change from liquid to solid occurs.

With an alloy of two elements, it is logical to expect two plateaus, since the two elements solidify at different temperatures. However, there is a range of temperatures over which the solidification occurs, and the transformation curve is smooth with an inflection point. This is illustrated by curve (b) in Fig. 48.2.

If the time-temperature curve is plotted for various compositions, the transition temperature will vary with proportions of the constituents. The locus of the curve's inflection points coincides with the liquidus and solidus lines in the equilibrium diagram (see Fig. 48.3).

Figure 48.2 *Time-Temperature Curves*

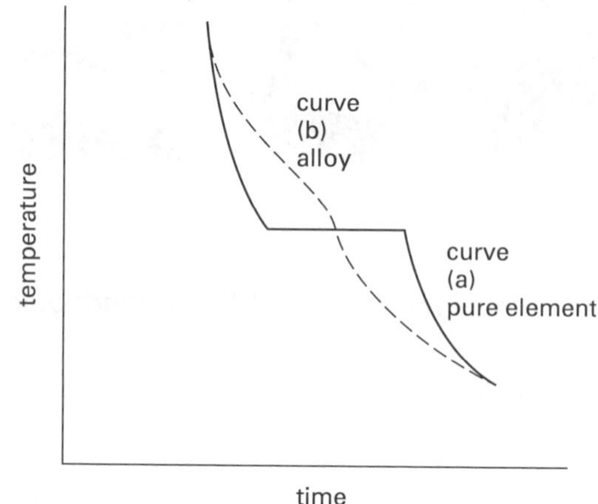

Figure 48.3 *Time-Temperature Curves for a Binary Alloy*

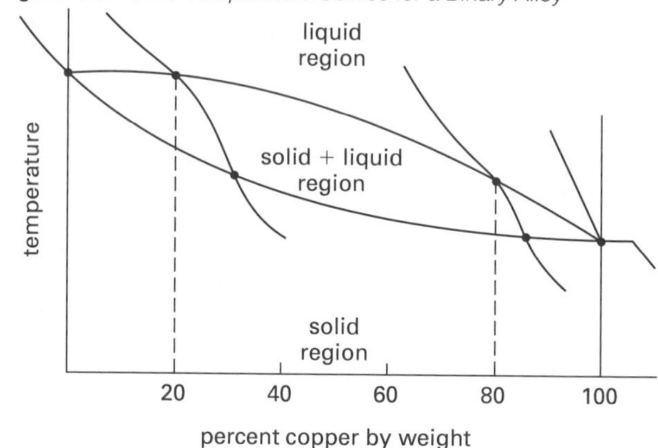

2. EUTECTIC ALLOY EQUILIBRIUM DIAGRAMS: PARTIAL SOLUBILITY

Just as only a limited amount of salt can be absorbed by water, there are many instances where a limited amount of the alloying ingredient can be absorbed by the solid mixture. The elements of a binary alloy may be completely soluble in the liquid state but only partially soluble in the solid state.

When the alloying ingredient is present in amounts above the maximum solubility percentage, the alloying ingredient precipitates out. In aqueous solutions, the precipitate falls to the bottom of the container. In metallic alloys, the precipitate remains suspended as pure crystals dispersed throughout the primary metal.

Figure 48.4 is typical of an equilibrium diagram for ingredients displaying a limited solubility.

In chemistry, a *mixture* is different from a *solution*. Salt in water forms a solution. Sugar crystals mixed with salt crystals form a mixture. An alloy consisting of a mixture of two solid ingredients with a melting point lower than the melting point of either ingredient is known as a *eutectic alloy*.

Figure 48.4 *Equilibrium Diagram of a Limited Solid Solubility Alloy*

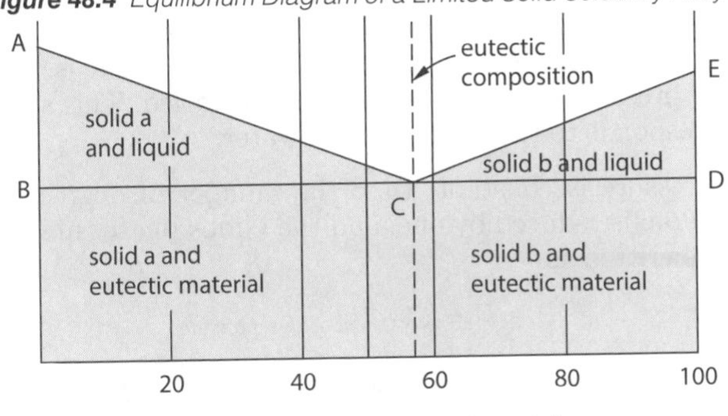

In Fig. 48.4, the components α and β are perfectly miscible at point C only. This point is known as the *eutectic composition.* The material in the region ABC consists of a mixture of solid component α crystals in a liquid of components α and β. This liquid is known as the *eutectic material,* and it will not solidify until the line B–D (the *eutectic line, eutectic point,* or *eutectic temperature*) is reached, the lowest point at which the eutectic material can exist in liquid form.[2]

Since the two ingredients do not mix, reducing the temperature below the eutectic line results in crystals (layers or plates) of both pure ingredients forming. This is the microstructure of a solid eutectic alloy: alternating pure crystals of the two ingredients. Since two solid substances are produced from a single liquid substance, the process could be written in chemical reaction format as: liquid $\rightarrow \alpha + \beta$. (Alternatively, upon heating the reaction would be: $\alpha + \beta \rightarrow$ liquid) For this reason, the phase change is called a *eutectic reaction.*

There are similar reactions involving other phases and states. Table 48.1 and Table 48.2 illustrate these. [**Binary Phase Diagrams**]

Table 48.1 *Types of Equilibrium Reactions*

reaction name	type of reaction upon cooling
eutectic	liquid $\rightarrow$ solid α + solid β
peritectic	liquid + solid α $\rightarrow$ solid β
monotectic	liquid α $\rightarrow$ liquid β + solid α
eutectoid	solid γ $\rightarrow$ solid α + solid β
peritectoid	solid α + solid γ $\rightarrow$ solid β

3. THE LEVER RULE

Within the liquid-solid region, the percentage of solid and liquid phases is a function of temperature and composition. Near the liquidus line, there is very little solid phase. Near the solidus line, there is very little liquid phase. The *lever rule* is used to find the relative amounts of solid and liquid phase at any composition. These percentages are given in fraction (or percent) by weight.

Table 48.2 *Typical Appearance of Equilibrium Diagram at Reaction Points*

reaction name	phase reaction	phase diagram
eutectic	$L \rightarrow \alpha(s) + \beta(s)$ cooling	α, α + L, L, β + L, β, α + β
peritectic	$\alpha(s) + L \rightarrow \beta(s)$ cooling	α, α + L, L, α + β, β, β + L
eutectoid	$\gamma(s) \rightarrow \alpha(s) + \beta(s)$ cooling	α, α + γ, γ, β + γ, β, α + β
peritectoid	$\alpha(s) + \gamma(s) \rightarrow \beta(s)$ cooling	α, α + γ, γ, α + β, β, β + γ

Figure 48.1 shows an alloy with an average composition of 55% copper at 2300°F. (A horizontal line representing different conditions at a single temperature is known as a *tie line.*) The liquid composition is defined by point 2, and the solid composition is defined by point 1.

The fractions of solid and liquid phases depend on the distances m, n, and w (equal to $m+n$), which are measured using any convenient scale. (Although the distances can be measured in millimeters or tenths of an inch, it is more convenient to use the percentage alloying ingredient scale. This is illustrated in Ex. 48.1.) Then, the fractions of solid and liquid can be calculated from Eq. 48.5 and Eq. 48.6.

$$\text{fraction solid} = \frac{n}{w} = 1 - \text{fraction liquid} \qquad 48.5$$

$$\text{fraction liquid} = \frac{m}{w} = 1 - \text{fraction solid} \qquad 48.6$$

The lever rule and method of determining the composition of the two components of a two-phase system are applicable to any solution or mixture, liquid or solid, in which two phases are present. The lever rule states that a fraction of any given phase is the opposite lever arm divided by the distance of the tie line. For a material with a composition of x at a specified temperature T, the weight percent composition of each phase (in this case designated α and β) can be determined using the lever rule equations, Eq. 48.7 and Eq. 48.8. x, x_α, and x_β are generally found from a lever rule diagram. [**Lever Rule Diagram**]

[2]The term *point* usually can be interpreted as temperature. Thus, the eutectic point really refers to the eutectic temperature.

Lever Rule

$$\text{wt\%}\alpha = \frac{x_\beta - x}{x_\beta - x_\alpha} \times 100\% \qquad 48.7$$

Lever Rule

$$\text{wt\%}\beta = \frac{x - x_\alpha}{x_\beta - x_\alpha} \times 100\% \qquad 48.8$$

Example 48.1

A mixture of 55% copper and 45% nickel exists at 2300°F. What are the fractions of solid and liquid phases and the compositions of each?

Solution

Referring back to Fig. 48.1, the solid portion of the mixture will have the composition at point 1 (44% copper), while the liquid will be at composition 2 (68% copper).

If the solid phase is phase α and the liquid phase is phase β, the phase fractions are

Lever Rule

$$\begin{aligned}\text{wt\%}\alpha &= \frac{x_\beta - x}{x_\beta - x_\alpha} \times 100\% \\ &= \frac{0.68 - 0.55}{0.68 - 0.44} \times 100\% \\ &= 54\%\end{aligned}$$

Lever Rule

$$\begin{aligned}\text{wt\%}\beta &= \frac{x - x_\alpha}{x_\beta - x_\alpha} \times 100\% \\ &= \frac{0.55 - 0.44}{0.68 - 0.44} \times 100\% \\ &= 46\%\end{aligned}$$

4. GIBBS PHASE RULE

The *Gibbs phase rule* defines the relationship between the number of phases and elements in an equilibrium mixture. For such an equilibrium mixture to exist, the alloy must have been slowly cooled, and thermodynamic equilibrium must have been achieved along the way. At equilibrium, and considering both temperature and pressure to be independent variables, the Gibbs phase rule is

Phase Relations

$$P + F = C + 2 \qquad 48.9$$

P is the number of phases existing simultaneously; F is the number of independent variables, known as *degrees of freedom*; and C is the number of elements in the alloy. Composition, temperature, and pressure are examples of degrees of freedom that can be varied.

For example, if water is to be stored in a condition where three phases (solid, liquid, gas) are present simultaneously, then $P = 3$, $C = 1$, and $F = 0$. That is, neither pressure nor temperature can be varied. This state corresponds to the *triple point* of water.

If pressure is constant, then the number of degrees of freedom is reduced by one, and the Gibbs phase rule can be rewritten as

$$P + F = C + 1\big|_{\text{constant pressure}} \qquad 48.10$$

If the Gibbs rule predicts $F = 0$, then an alloy can exist in only one composition.

5. ALLOTROPIC CHANGES IN STEEL

Allotropes have the same compositions but different atomic structures (microstructures), volumes, electrical resistances, and magnetic properties. In the case of iron, *allotropic changes* are reversible changes that occur at the *critical points* (i.e., *critical temperatures*). (See Table 48.3.)

Table 48.3 Allotropic Points for Pure Iron (upon heating)

1674°F (912°C): alpha (BCC) to gamma (FCC) transition
2541°F (1394°C): gamma (FCC) to delta (BCC) transition
2800°F (1538°C): delta (BCC) to liquid transition

Iron exists in three allotropic forms: alpha-iron, delta-iron, and gamma-iron. The changes are brought about by varying the temperature of the iron. As shown in Fig. 48.5, heating pure iron from room temperature changes its structure from body-centered cubic (BCC) to face-centered cubic (FCC) and then back to body-centered cubic.

Figure 48.5 Allotropic Changes of Iron

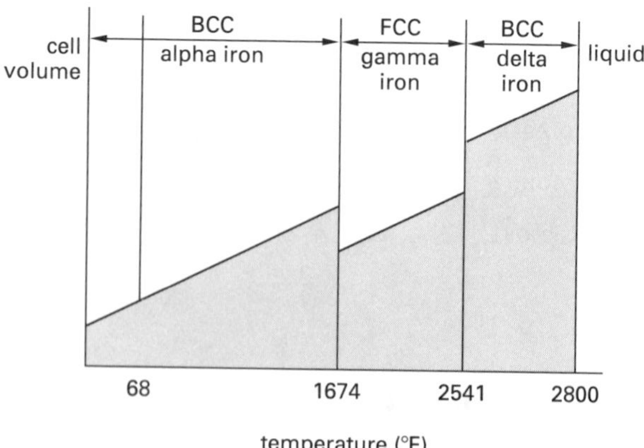

Alpha-iron, also known as *ferrite*, is a BCC structure that exists only below the A_3 line (defined as follows). The maximum carbon solubility is 0.03%, the lowest of all three allotropic forms. Alpha-iron is stable from

Figure 48.6 *Iron-Iron Carbide Diagram*

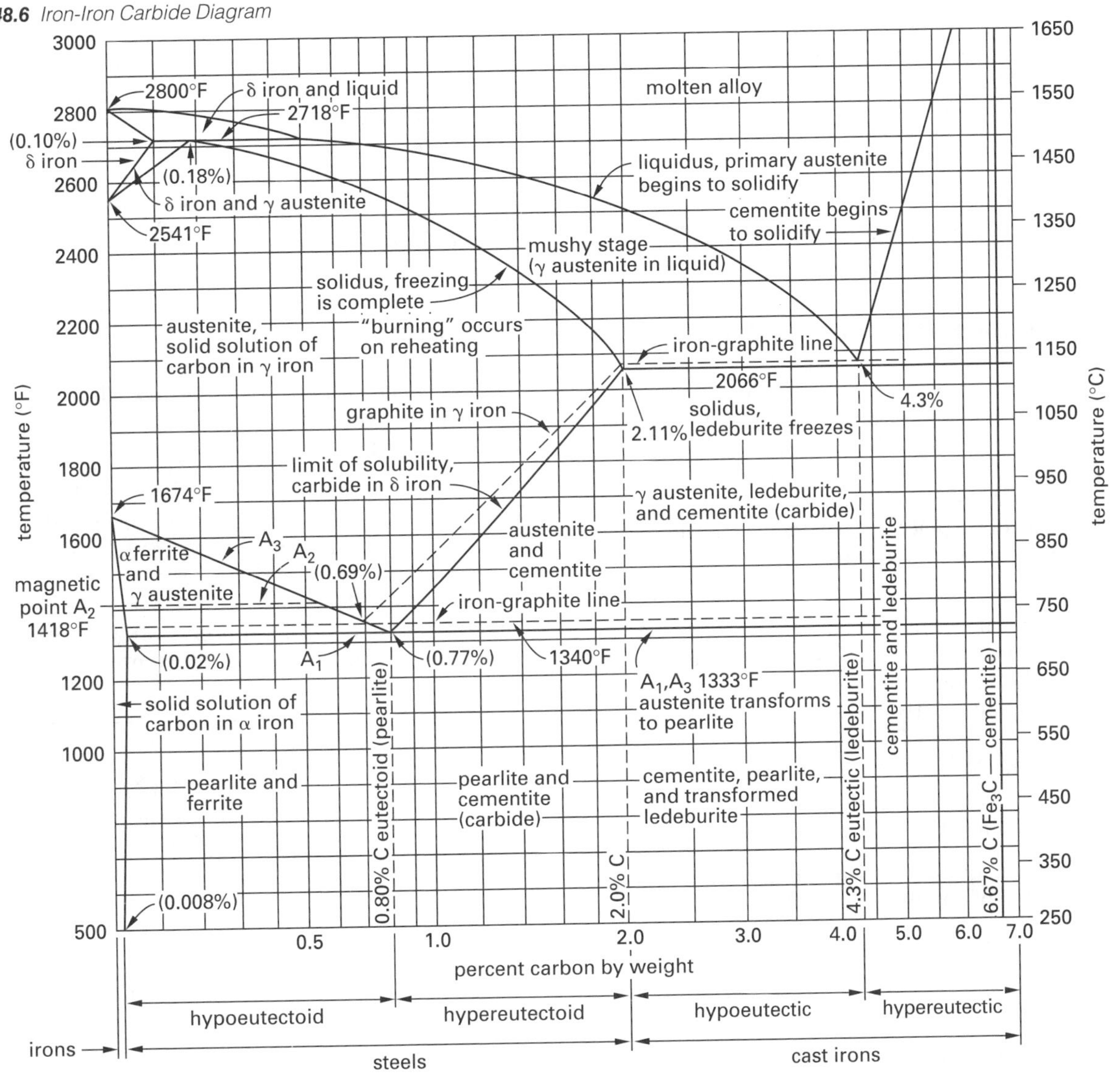

−460°F to 1674°F (−273°C to 912°C), soft, and strongly magnetic up to approximately 1414°F (768°C). (*Beta-iron* is a nonmagnetic form of BCC alpha-iron that exists between 1418°F and 1674°F (770°C and 912°C). The distinction between alpha- and beta-iron is not usually made.)

Gamma-iron is an FCC arrangement of iron atoms, stable between 1674°F and 2541°F (912°C and 1394°C), and nonmagnetic. The maximum carbon solubility for solid iron is 2.11% (also reported as 2.08%, 2.03%, and 1.7%).

Delta-iron is a BCC form of iron existing above 2541°F (1394°C).

Since allotropic changes occur at differing temperatures in an iron-carbon mixture dependent on composition, there are *critical lines* but no critical points. Depending on the authority, these critical lines may be labeled A_c, A_r, or just A.[3] Refer to Fig. 48.6.

- A_0: the critical line, about 410°F (210°C), above which cementite becomes nonmagnetic
- A_1: the so-called *lower critical point*, or *eutectoid temperature*, 1333°F (723°C) and 0.8% carbon, a line above which the austenite-to-ferrite and -cementite transformation occurs
- A_2: the critical line, about 1418°F (770°C), below which the alloy becomes magnetic[4]

[3]Labeling the critical lines as A_{c1}, A_{c2}, and so on is a reference to the French word *chauffage* ("heating"). Such critical temperatures are encountered if iron is slowly heated from room temperature. Critical lines labels of A_{r1}, A_{r2}, and so on refer to the French word *refroidissement* ("cooling"), as such critical temperatures are observed upon cooling the iron back to room temperature. As the temperatures (A_{c1} and A_{r1}, etc.) are approximately the same, the distinction is not always made, as it is not in this book.

[4]The A_2 temperature 1418°F (770°C) is also known as the *Curie point*.

- A_3: the so-called *upper critical point*, or critical line forming a division between austenite (above) and ferrite (below), with the actual temperature being dependent on composition. (See Fig. 48.6.)
- A_4: the critical point, 2541°F (1394°C), at which the gamma-delta transformation occurs

In some parts of the iron-carbon diagram, two critical lines coincide. For example, the tie line (1333°F or 723°C) for more than 0.8% carbon is labeled $A_{1,3}$ or A_{13}.

6. THE IRON-IRON-CARBIDE DIAGRAM

The *iron-iron carbide phase diagram* is much more complex than idealized equilibrium diagrams due to the existence of many different phases. Each of these phases has a different microstructure and, therefore, different mechanical properties. By treating the steel in a manner that forces the occurrence of particular phases, steel with desired wear and endurance properties can be produced.

Iron-carbon mixtures are categorized into *steel* (less than 2% carbon) and *cast iron* (more than 2% carbon) according to the amounts of carbon in the mixtures. Iron-carbon alloys are further classified as follows.

- *steel:* iron alloy with less than 2.0% carbon

  hypoeutectoid steel: iron alloy with less than 0.8% carbon, consisting of ferrite and pearlite

  eutectoid steel: equilibrium iron alloy with 0.8% carbon, consisting of ferrite and pearlite

  hypereutectoid steel: iron alloy with 0.8% to 2.0% carbon, consisting of cementite and pearlite

- *cast iron:* iron alloy with more than 2% carbon

  hypoeutectic cast iron: iron alloy with 2.0% to 4.3% carbon

  eutectic cast iron: iron alloy with 4.3% carbon

  hypereutectic cast iron: iron alloy with more than 4.3% carbon

The most important eutectic reaction in the iron-carbon system is the formation of a solid mixture of austenite and cementite at approximately 2065°F (1129°C). *Austenite* is a solid solution of carbon in gamma-iron. It is nonmagnetic, decomposes on slow cooling, and does not normally exist below 1333°F (723°C), though it can be partially preserved by extremely rapid cooling.

Cementite (Fe_3C), also known as *carbide* or *iron carbide*, has approximately 6.67% carbon. Cementite is the hardest of all forms of iron, has low tensile strength, and is quite brittle. Cementite ceases to be magnetic above the A_0 line.

The most important eutectoid reaction in the iron-carbon system is the formation of *pearlite* from the decomposition of austenite at approximately 1333°F (723°C). Pearlite is actually a mixture of two solid components, ferrite and cementite, with the common *lamellar (layered) appearance.* (The name pearlite is derived from similarity in appearance to mother-of-pearl.)

Ferrite is essentially pure iron (less than 0.025% carbon) in BCC alpha-iron structure. It is magnetic and has properties complementary to cementite, since it has low hardness, high tensile strength, and high ductility.

Example 48.2

An iron alloy at 1500°F contains 2.0% carbon by weight. (a) How much carbon is present in each of the phases? (b) What are the percentages of austenite and cementite in the mixture?

Solution

(a) Referring to Fig. 48.6, the iron-iron carbide diagram, compositions of the phases are read from the carbon contents at the intersection of the 1500°F tie line and the phase boundary lines. The austenite will have the composition at point 1 (approximately 1.08% carbon), while the cementite will be at composition 2 (6.67% carbon).

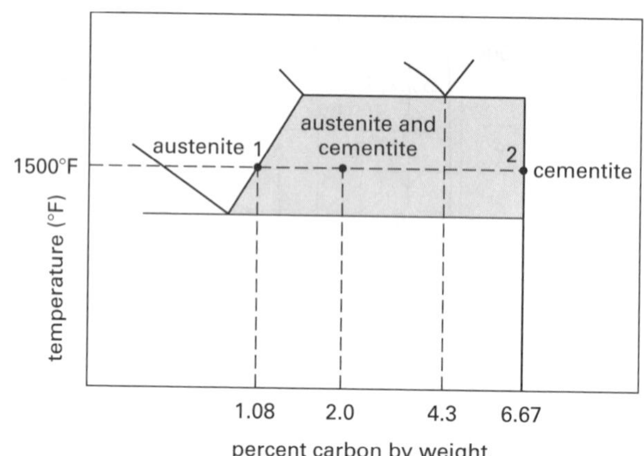

(b) Treating austenite as phase α and cementite as phase β, the phase fractions are

Lever Rule

$$\begin{aligned}\text{wt\%}\alpha &= \frac{x_\beta - x}{x_\beta - x_\alpha} \times 100\% \\ &= \frac{0.0667 - 0.0200}{0.0667 - 0.0108} \times 100\% \\ &= 83.5\%\end{aligned}$$

Lever Rule

$$\begin{aligned}\text{wt\%}\beta &= \frac{x - x_\alpha}{x_\beta - x_\alpha} \times 100\% \\ &= \frac{0.0200 - 0.0108}{0.0667 - 0.0108} \times 100\% \\ &= 16.5\%\end{aligned}$$

7. QUENCHING AND RATES OF COOLING

As steel is heated, the grain sizes remain the same until the A_1 line is reached. Between the A_1 and A_2 lines, the average grain size of austentite in solution decreases.

This characteristic is used in heat treatments wherein steel is heated and then quenched. *Quenching* is rapid cooling from elevated temperature, preventing the formation of equilibrium phases from occurring in the finished product. The *quenching* can be performed with gases (usually air), oil, water, or brine. Agitation or spraying of these fluids during the quenching process increases the severity of the quenching.

The *rate of cooling* determines the hardness and ductility. Rapid cooling in water or brine is necessary to quench low- and medium-carbon steels, since steels with small amounts of pearlite are difficult to harden. Oil is used to quench high-carbon and alloy steel or parts with nonuniform cross sections (to prevent warping). Figure 48.7 illustrates the relative rates of cooling for different quenching media.

Figure 48.7 *Relative Cooling Rates for Different Quenching Media*

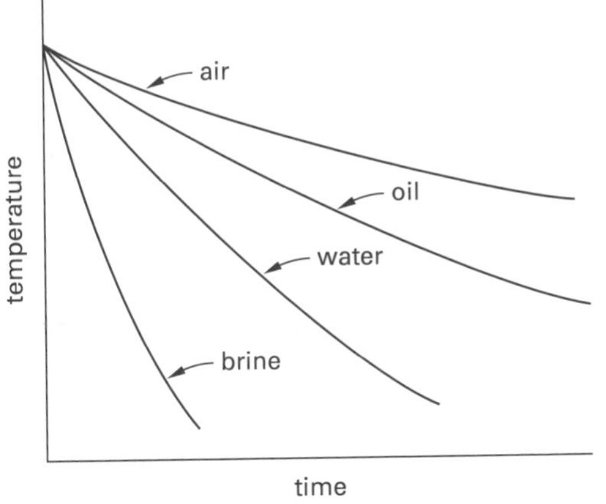

8. TTT AND CCT CURVES

Controlled-cooling-transformation (CCT) curves and *time-temperature-transformation (TTT) curves* are used to determine how fast an alloy should be cooled to obtain a desired microstructure. Although these curves show different phases, they are not equilibrium diagrams. On the contrary, they show the microstructures that are produced with controlled temperatures or when quenching interrupts the equilibrium process.

TTT curves are determined under ideal, isothermal conditions. For that reason, they are also known as *isothermal transformation diagrams.* CCT curves are experimentally determined under conditions of continuous cooling. Therefore, CCT curves are better suited for designing cooling processes. However, TTT curves are more readily available than CCT curves and are used in lieu of them. Both curves are similar in shape, although the CCT curves are displaced downward and to the right from TTT curves.

Figure 48.8 shows a TTT diagram for a high-carbon (0.80% carbon or more) steel. Curve 1 represents extremely rapid quenching. The transformation begins at 420°F (216°C) and continues for 8–30 seconds, changing all of the austenite to martensite. Such a material is seldom used because martensite has almost no ductility.

Figure 48.8 *TTT Diagram for High-Carbon Steel*

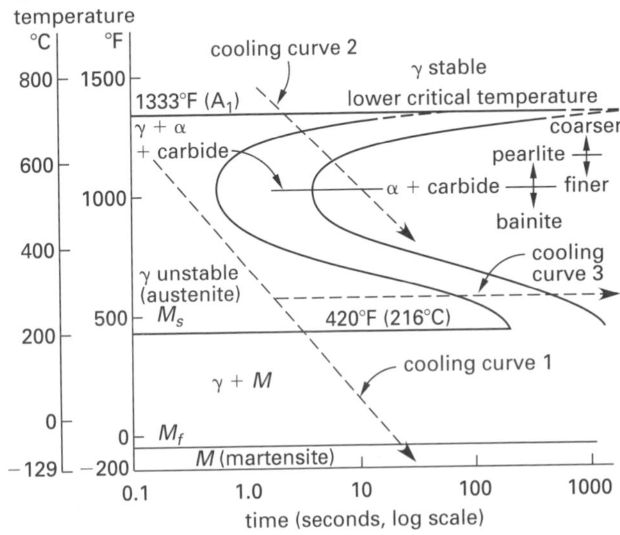

Curve 2 is a slower quench that converts all of the austenite to fine pearlite. This corresponds to a *normalizing process.*

A horizontal line below the critical temperature is a *tempering process.* If the temperature is decreased rapidly along curve 1 to 520°F (270°C) and is then held constant along cooling curve 3, *bainite* is produced. This is the principle of *austempering.* Performing the same procedure at 350°F to 400°F (180°C to 200°C) is *martempering*, which produces *tempered martensite*, a soft and tough steel.

The austenite-martensite transformation is extremely important, and the temperatures at which the transition starts and finishes are sometimes referred to as *critical temperatures.*

- M_s: the temperature at which austenite first begins to transform into martensite
- M_f: the temperature at which austenite is fully transformed into martensite

Materials

9. STEEL HARDENING PROCESSES

Hard steel resists plastic deformation. Steel is hard if it has a homogeneous, austenitic structure with coarse grains. Some steels (e.g., those that have little carbon) are difficult to harden. The *hardenability* of a steel specimen can be determined in a standard *Jominy end-quench test.*

The basic hardening processes consist of heating to approximately 100°F (50°C) above the A_3 critical line, allowing austenite to form, and then quenching rapidly. Hardened steel consists primarily of martensite or bainite. *Martensite* is a supersaturated solution of carbon in alpha-iron.[5] *Bainite* is not as hard as martensite, but it does have good impact strength and fairly high hardness. Neither martensite nor bainite are equilibrium substances—they are not found on the iron-carbon equilibrium diagram but are formed during the quenching operation.

The maximum hardness obtained depends on the carbon content. (See Fig. 48.9.) An upper limit of R_C 66–67 is reached with approximately 0.5% carbon. No further increase in hardness is achieved by increasing the carbon content. Since hardening is accompanied by a decrease in toughness, it is usually followed by tempering.

Figure 48.9 *Maximum Hardness of Steel*

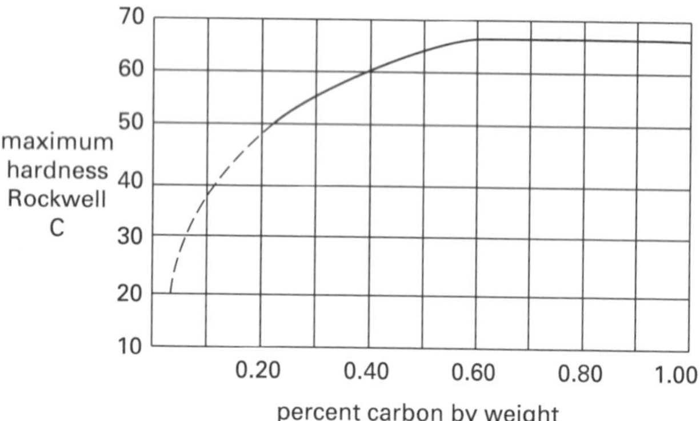

There are many steel-hardening processes with special names, listed alphabetically as follows.

- *austempering:* an interrupted quenching process resulting in an austenite-to-bainite transition. Steel is quenched to below approximately 800°F (430°C) but above 400°F (200°C), and is allowed to reach equilibrium. No martensite is formed, and further tempering is not required.
- *austenitizing:* quenching after heating above the A_3 line (for steel with up to 0.8% carbon) or above the A_1 line (for steel with more than 0.8% carbon).
- *martempering:* an interrupted quenching process resulting in an austenite-to-(tempered) martensite transition. Steel is quenched to below 400°F (200°C) and allowed to reach equilibrium. Further tempering is not required. **[Material Hardness]**

10. HEAT TREATING CAST IRON

Hardness in low-carbon steels results from the presence of martensite or bainite—supersaturated solutions of carbon in alpha-iron that begin as austenite at higher temperatures and are "frozen" in place by rapid cooling. Cast iron starts as iron carbide (cementite), not austenite, and cast iron is not normally hardened by heating and quenching. Hardness in cast irons is primarily obtained by including alloying ingredients that promote the formation and retention of iron carbide (Fe_3C). Although it is hard, iron carbide is also very brittle.

Heating can actually reduce hardness in cast iron. Iron carbide dissociates into iron and graphite at high temperatures. Graphite in flake form (*gray cast iron*) and in spheroidal form (*nodular cast iron*) greatly decreases hardness. The presence of silicon in cast iron greatly affects hardness, since silicon promotes the formation of graphite. Cast irons typically contain 1½% or more silicon, and less than 1% is needed to ensure that the carbon in iron carbide dissociates into iron and graphite. Cast iron with no graphite is known as *white cast iron.* Cast iron that has been heated and slowly cooled to permit graphite to form is known as *malleable cast iron.*

Small castings (e.g., with dimensions less than 3 in or 4 in) or white cast iron contain less silicon and can be hardened by rapid quenching. This occurs because rapid cooling prevents the dissociation of iron carbide into iron and graphite. However, the interiors of larger castings cannot be cooled fast enough to prevent the formation of weakening graphite.

11. PROPERTIES VERSUS GRAIN SIZE

Many properties are related to grain size, which initially depends on composition but can be changed by heat treatment. Coarse-grained structures have less toughness and ductility but have greater machineability and case hardenability.

As low-carbon steels are heated from room temperature, the grain size remains constant up to the A_1 line. Above the A_1 line, ferrite and pearlite are transformed into austenite, and the grain size decreases. The grain size is minimum at the upper critical line, A_3, and then increases again as the steel is heated above the A_3 line.

Aluminum in small quantities is an important alloying ingredient in steel. As a deoxidizer, it raises the temperature at which rapid grain growth takes place. In steels that have been deoxidized with aluminum (e.g., medium-carbon and alloy steels), no grain growth occurs until the *coarsening temperature* is reached, which is well above the critical temperature.

[5]The carbon in martensite distorts the BCC structure of iron. The distorted BCC lattice is known as a *body-centered tetragonal* (BCT) structure.

12. RECRYSTALLIZATION

Recrystallization can be used with all metals to relieve stresses induced during cold working. It involves heating the material in a furnace to a specific temperature (the *recrystallization temperature*) and holding it there for a long time. This induces the formation and growth of strain-free grains within the grains already formed. The resulting microstructure is essentially the microstructure that existed before any cold working but is softer and more ductile than the cold-worked microstructure.

The recrystallization process is more sensitive to temperature than it is to exposure time. Recrystallization will occur naturally over a wide range of temperatures; however, the reaction rates increase at higher temperatures. Table 48.4 lists approximate temperatures that will produce complete recrystallization in one hour. [Thermal and Mechanical Processing]

Table 48.4 Approximate Recrystallization Temperatures

material	recrystallization temperature °F	°C
copper (99.999% pure)	250	120
(5% zinc)	600	315
(5% aluminum)	550	290
(2% beryllium)	700	370
aluminum (99.999% pure)	175	80
(99.0+% pure)	550	290
(alloys)	600	315
nickel (99.99% pure)	700	370
(99.4% pure)	1100	590
(monel metal)	1100	590
iron (pure)	750	400
(low-carbon steel)	1000	540
magnesium (99.99% pure)	150	65
(alloys)	450	230
zinc	50	10
tin	25	–4
lead	25	–4

13. STRESS RELIEF PROCESSES FOR STEEL

The annealing, normalizing, and tempering processes are used to relieve the internal stresses, refine the grain size, and soften the material (to improve machineability). The high temperatures used in these processes allow some of the carbon to migrate out of the martensite, thereby relieving stresses in the crystalline structure. (See Fig. 48.10.)

Figure 48.10 Products of Cooling and Reheating Iron-Carbon Alloys

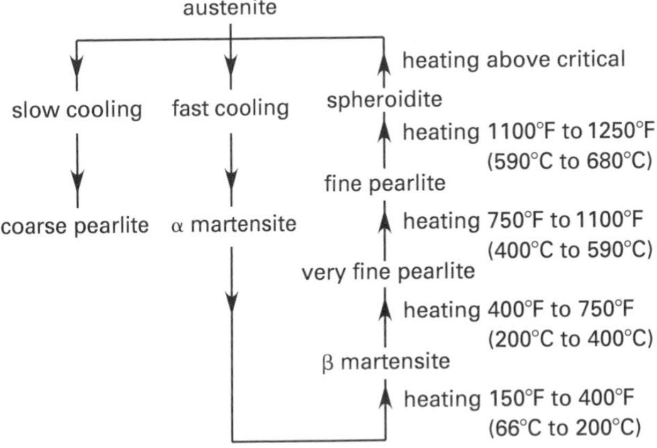

The basic *full annealing* process involves heating to approximately 100°F (50°C) above the critical A_3 point, allowing austenite to form fully, and then cooling slowly in a furnace to produce coarse pearlite. Three separate stages constitute annealing. First, the material is stress-relieved by heating in the *recovery stage*. Next, during the *recrystallization stage*, new crystals form within the existing distorted structure. Finally, during the *grain growth stage*, some of the crystals grow in size (by eliminating smaller grains).

The *partial annealing process*, also known as *process annealing*, *spheroidize-annealing*, or just *spheroidizing*, also softens the material and relieves stresses, but heating is to below the A_3 point. The term *spheroidizing* gets its name from the spherical cementite particles that appear in the ferrite matrix, as opposed to the lamellar structure of cementite and ferrite in pearlite.[6]

Normalizing is similar to annealing but is more rapid than furnace cooling because it uses air cooling. Heating is to approximately 200°F (100°C) higher than the critical point, generally about 1650°F (900°C). The typical air cooling rate is 100°F/min (50°C/min). Normalizing produces a harder and stronger steel than full annealing.

Tempering, also known as *drawing* or *toughening*, is used with hypoeutectoid steels to change martensite into pearlite. It is used after hardening to produce softer

[6]Although the cementite concentrations have changed in shape, the phase is the same: cementite and ferrite.

and tougher steel. The steel is heated to below its critical temperature. However, the higher the temperature, the more soft and ductile the steel becomes.

To avoid *blue embrittlement* (named after the resulting blue surface finish) in steel, tempering should not be done between 450°F and 700°F (230°C and 370°C). Within this range, the *notch toughness* of the steel (as determined from an impact test), is lowered considerably. The reason for this effect has been traced to free nitrogen in the steel. [**Thermal and Mechanical Processing**]

14. COLD WORKING VERSUS HOT WORKING

As a material is worked and the dislocations move, plastic strain builds up. If the strain occurs at a high enough temperature (i.e., above the recrystallization temperature), there will be sufficient thermal energy to anneal out the lattice distortions. Forming operations above the recrystallization temperature are known as *hot working*, since annealing occurs simultaneously with the plastic forming. The material remains ductile.

If the plastic forming takes place at a low temperature (i.e., below the recrystallization temperature), there will be insufficient thermal energy to anneal out the dislocations. The material will become progressively stronger, harder, and more brittle until it eventually fails. This is known as *cold working*. [**Thermal and Mechanical Processing**]

15. HARDENING OF NONALLOTROPIC ALLOYS

The properties of nonferrous substances that do not readily form allotropes cannot be changed by rapid heating followed by controlled cooling. Such substances are known as *nonallotropic alloys* and include aluminum, copper, and magnesium alloys as well as stainless steels containing nickel.

The primary method of hardening nonallotropic alloys is *solution heat treatment*, which consists of two or three steps: precipitation, quenching, and (optionally) artificial aging. Because of these steps, solution heat treating is also known as *precipitation hardening* and *age hardening*.

Precipitation involves the formation of a new crystalline structure through the application of controlled quenching and tempering. Precipitation disperses hard particles throughout the existing more ductile material. These particles disrupt the long dislocation planes of the material, restricting the movement of dislocations and increasing the strength and stiffness of the alloy. The ultimate strength is raised to the rupture strength of either the particles or the surrounding matrix.

Solution heat treatment culminates in rapid quenching. Quenching speeds must be consistent with the size of the object. Massive specimens may require slower processes that use oil or boiling water.

The final step is to hold the material at a specific temperature for a given amount of time. This is known as *aging* or *artificial aging*. Post-treatment cooling for precipitation hardening is relatively unimportant.

It is important not to over-age aluminum. If the precipitation process goes on too long, the precipitates will not be effective in strengthening the material. Precipitation hardening is optimum at the point where the particles are just starting to form.

Table 48.5 lists some of the more common *temper designations* for 2XXX-, 6XXX-, and 7XXX-series aluminum alloys that can be precipitation hardened. For example, 2024-T4 is a widely used alloy having strength and toughness when hardened and aged.

Table 48.5 *Aluminum Tempers*

temper	description
T2	annealed (castings only)
T3	solution heat-treated, followed by cold working
T4	solution heat-treated, followed by natural aging
T5	artificial aging only
T6	solution heat-treated, followed by artificial aging
T7	solution heat-treated, followed by stabilizing by overaging heat treating
T8	solution heat treated, followed by cold working and subsequent artificial aging

16. SURFACE HARDENING

Often, it is desirable to have a hard (wear-resistant) outer surface with a ductile interior. This combination is needed when the product is subjected to fatigue. There are several processes used to *surface harden* (also known as *case harden* and *differential harden*) steel.

- *boron diffusion:* exposure to boron (a powerful hardening ingredient) at low temperatures; slow but distortion-free; suitable for high carbon, spring, and tool steels as well as bonded steel carbides and some age-hardenable alloys.
- *carburizing:* heating for up to 24 hours at approximately 1650°F (900°C) in contact with a carbonaceous material (usually carbon monoxide, CO, gas), followed by rapid cooling. Carburizing is used for steels with less than 0.2% carbon. Carburizing is also known as *cementation*.
- *cyaniding:* heating at 1700°F (925°C) in a cyanide-rich atmosphere or immersion in a cyanide salt bath.

- *flame hardening:* supplying flame heat at the surface in quantities and rates higher than can be conducted into the material's interior, followed by drastic spray quenching. Typically used with steels containing more than 0.4% carbon.
- *induction hardening:* using high-frequency electric currents to heat the metal surface, followed by normal quenching. Typically used with steels containing more than 0.4% carbon.
- *nitriding:* heating at 1000°F (540°C) for up to 100 hours (but usually less than 70 hours) in an ammonia atmosphere, followed by slow cooling (no quenching required).

17. SHOT-PEENING

Shot-peening is the "bombardment" of a metal surface by high-velocity particles (e.g., hard steel shot). As each particle strikes, the target's surface stretches and deforms plastically, creating residual compressive stresses at the surface. The induced compressive stress from shot-peening removes tensile stresses left over from manufacturing operations and offsets the effects of applied tensile operating loads. In gears, the compressive layer improves load-carrying capacity by increasing the bending fatigue strength of the teeth.[7] A 20% improvement in both strength (i.e., endurance limit) and wear is typical for shot-peened parts.

18. NOMENCLATURE

A	atomic fraction	–	–
C	number of elements	–	–
F	degrees of freedom	–	–
G	gravimetric fraction	–	–
M	martensite transformation temperature	°F	°C
M	molecular weight	lbm/lbmol	kg/kmol
P	number of phases	–	–
R_C	Rockwell C hardness	–	–
x	average composition	–	–

Subscripts

f	finish
s	start
x	average composition

[7]Fatigue failure never starts in an area under compressive stress.

Materials

49 Properties of Areas

Content in blue refers to the *NCEES Handbook*.

NCEES EXAM SPECIFICATIONS AND RELATED CONTENT

MACHINE DESIGN AND MATERIALS EXAM

I.B.1. Engineering Science and Mechanics: Statics
1. Centroid of an Area
2. First Moment of the Area
4. Moment of Inertia of an Area
5. Parallel Axis Theorem
7. Radius of Gyration
8. Product of Inertia

1. CENTROID OF AN AREA

The *centroid of an area* is analogous to the center of gravity of a homogeneous body.[1] The centroid is often described as the point at which a thin homogeneous plate would balance. This definition, however, combines the definitions of centroid and center of gravity and implies that gravity is required to identify the centroid, which is not true.

The location of the centroid of an area bounded by the x- and y-axes and the mathematical function $y = f(x)$ can be found by the *integration method* Another approach is to use Eq. 49.1 and Eq. 49.2. The centroidal location depends only on the geometry of the area and is identified by the coordinates (x_c, y_c). Some references place a bar over the coordinates of the centroid to indicate an average point, such as $(\overline{x}, \overline{y})$.

Centroids of Masses, Areas, Lengths, and Volumes

$$x_{ac} = \frac{M_{ay}}{A} = \sum x_n\left(\frac{a_n}{A}\right) \qquad 49.1$$

$$y_{ac} = \frac{M_{ax}}{A} = \sum y_n\left(\frac{a_n}{A}\right) \qquad 49.2$$

$$A = \int f(x)\,dx \qquad 49.3$$

$$dA = f(x)\,dx = g(y)\,dy \qquad 49.4$$

The locations of the centroids of *basic shapes*, such as triangles and rectangles, are well known. The most common basic shapes are tabulated in multiple sources, and there should be no need to derive centroidal locations for these shapes by the integration method. [**Properties of Various Shapes**]

The centroid of a complex area can be found from Eq. 49.5 and Eq. 49.6 if the area can be divided into the basic shapes. This process is simplified when all or most of the subareas adjoin the reference axis.

$$x_{ac} = \frac{\sum_n A_n x_{ac,n}}{\sum_n A_n} \qquad 49.5$$

$$y_{ac} = \frac{\sum_n A_n y_{ac,n}}{\sum_n A_n} \qquad 49.6$$

Example 49.1

Find the y-coordinate of the centroid of the area shown.

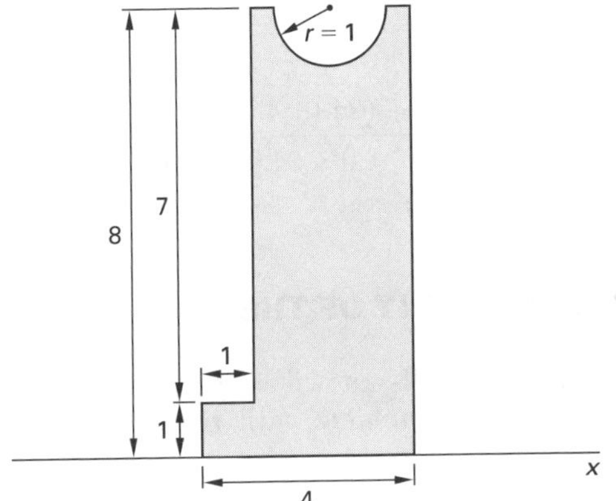

[1]The analogy has been simplified. A three-dimensional body also has a centroid. The centroid and center of gravity will coincide when the body is homogeneous.

Solution

The x-axis is the reference axis. The area is divided into basic shapes of a 1×1 square, a 3×8 rectangle, and a half-circle of radius 1. (The area could also be divided into 1×4 and 3×7 rectangles and the half-circle, but then the 3×7 rectangle would not adjoin the x-axis.)

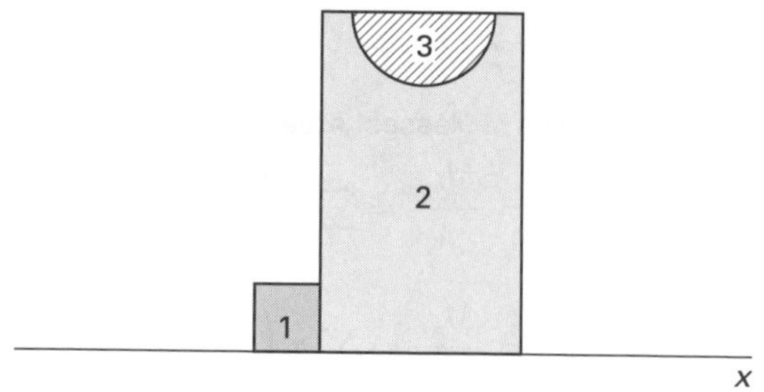

First, calculate the areas of the basic shapes. Notice that the half-circle area is negative since it represents a cutout.

$$A_1 = (1.0)(1.0) = 1.0 \text{ units}^2$$
$$A_2 = (3.0)(8.0) = 24.0 \text{ units}^2$$
$$A_3 = -\tfrac{1}{2}\pi r^2 = -\frac{1}{2}\pi(1.0)^2 = -1.57 \text{ units}^2$$

Next, find the y-components of the centroids of the basic shapes. Most are found by inspection, but the equation for the half-circle can be found in various engineering references. The centroidal location for the half-circle is positive. [**Properties of Various Shapes**]

Properties of Various Shapes

$$y_{c,1} = 0.5 \text{ units}$$
$$y_{c,2} = 4.0 \text{ units}$$
$$y_{c,3} = \frac{4a}{3}\pi = 8.0 - 0.424 = 7.576 \text{ units}$$

Finally, use Eq. 49.6.

$$\begin{aligned} y_c &= \frac{\sum A_i y_{c,i}}{\sum A_i} \\ &= \frac{(1.0)(0.5) + (24.0)(4.0) + (-1.57)(7.576)}{1.0 + 24.0 - 1.57} \\ &= 3.61 \text{ units} \end{aligned}$$

2. FIRST MOMENT OF THE AREA

The quantity $\int x\,dA$ is known as the *first moment of the area* or *first area moment* with respect to the y-axis. Similarly, $\int y\,dA$ is known as the first moment of the area with respect to the x-axis. These expressions can be simplified for basic shapes, as shown in Eq. 49.8 and Eq. 49.10.

$$M_{ay} = \int x\,dA \qquad 49.7$$

Centroids of Masses, Areas, Lengths, and Volumes

$$M_{ay} = \sum x_n a_n \qquad 49.8$$

$$M_{ax} = \int y\,dA \qquad 49.9$$

Centroids of Masses, Areas, Lengths, and Volumes

$$M_{ax} = \sum y_n a_n \qquad 49.10$$

The centroid of the area for basic shapes is found using Eq. 49.11 and Eq. 49.12.

$$A = \frac{M_{ay}}{x_{ac}} \qquad 49.11$$

$$A = \frac{M_{ax}}{y_{ac}} \qquad 49.12$$

In basic engineering, the two primary applications of the first moment concept are to determine centroidal locations and shear stress distributions. In the latter application, the first moment of the area is known as the *statical moment.*

3. CENTROID OF A LINE

The location of the *centroid of a line* can be defined by Eq. 49.13 and Eq. 49.14, which are analogous to the equations used for centroids of areas.

$$x_c = \frac{\int x\,dL}{L} \qquad 49.13$$

$$y_c = \frac{\int y\,dL}{L} \qquad 49.14$$

Since equations of lines are typically in the form $y = f(x)$, dL must be expressed in terms of x or y.

$$dL = \left(\sqrt{\left(\frac{dy}{dx}\right)^2 + 1}\right)dx \qquad 49.15$$

$$dL = \left(\sqrt{\left(\frac{dx}{dy}\right)^2 + 1}\right)dy \qquad 49.16$$

4. MOMENT OF INERTIA OF AN AREA

The *moment of inertia*, I, of an area is needed in mechanics of materials problems. It is convenient to think of the moment of inertia of a beam's cross-sectional area as a measure of the beam's ability to resist bending. Given equal loads, a beam with a small moment of inertia will bend more than a beam with a large moment of inertia.

Since the moment of inertia represents a resistance to bending, it is always positive. Since a beam can be unsymmetrical (e.g., a rectangular beam) and can be stronger in one direction than another, the moment of inertia depends on orientation. Therefore, a reference axis or direction must be specified.

The moment of inertia taken with respect to one of the axes in the rectangular coordinate system is sometimes referred to as the *rectangular moment of inertia.*

The symbol I_x is used to represent a moment of inertia with respect to the x-axis. Similarly, I_y is the moment of inertia with respect to the y-axis. I_x and I_y do not normally combine and are not components of some resultant moment of inertia.

Any axis can be chosen as the reference axis, and the value of the moment of inertia will depend on the reference selected. The moment of inertia taken with respect to an axis passing through the area's centroid is known as the *centroidal moment of inertia*, I_{cx} or I_{cy}. The centroidal moment of inertia is the smallest possible moment of inertia for the shape.

The *integration method* can be used to calculate the moment of inertia of a function that is bounded by the x- and y-axes and a curve $y = f(x)$. From Eq. 49.17 and Eq. 49.18, it is apparent why the moment of inertia is also known as the *second moment of the area* or *second area moment.*

Moment of Inertia

$$I_x = \int y^2 dA \qquad 49.17$$

$$I_y = \int x^2 dA \qquad 49.18$$

$$dA = f(x)\,dx = g(y)\,dy \qquad 49.19$$

Example 49.2
What is the centroidal moment of inertia with respect to the x-axis of a rectangle 5.0 units wide and 8.0 units tall?

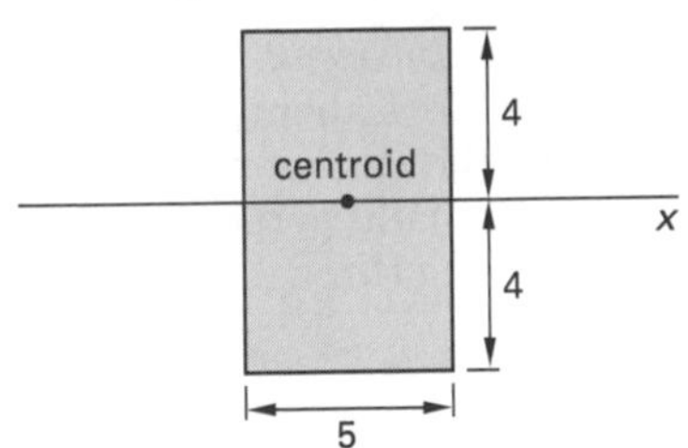

Solution

Since the centroidal moment of inertia is needed, the reference line passes through the centroid. The centroidal moment of inertia is

Properties of Various Shapes

$$I_{cx} = \frac{bh^3}{12} = \frac{(5)(8)^3}{12} = 213.3 \text{ units}^4$$

Example 49.3
What is the moment of inertia with respect to the y-axis of the area bounded by the y-axis, the line $y = 8.0$, and the parabola $y^2 = 8x$?

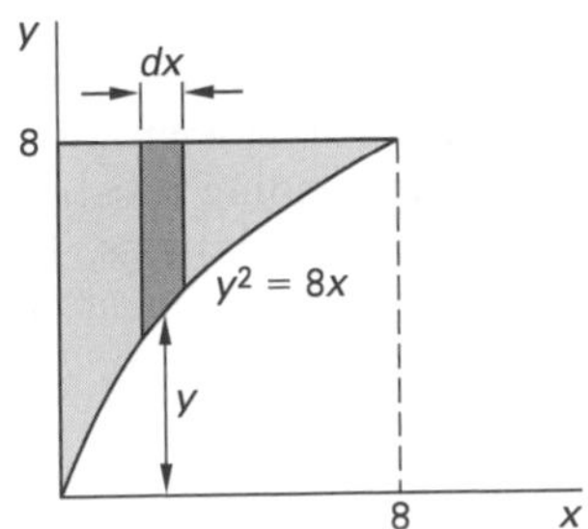

Solution

This problem is more complex than it first appears, since the area is above the curve, bounded not by $y = 0$ but by $y = 8$. In particular, dA must be determined correctly.

$$y = \sqrt{8x}$$

Use Eq. 49.4.

$$dA = \big(8 - f(x)\big)\,dx = (8 - y)\,dx$$
$$= (8 - \sqrt{8x})\,dx$$

Equation 49.18 is used to calculate the moment of inertia with respect to the y-axis.

Moment of Inertia

$$I_y = \int x^2 dA = \int_0^8 x^2(8 - \sqrt{8x})\,dx$$
$$= \tfrac{8}{3}x^3 - \left(\frac{4\sqrt{2}}{7}\right)x^{7/2}\Bigg|_0^8$$
$$= 195.0 \text{ units}^4$$

5. PARALLEL AXIS THEOREM

If the moment of inertia is known with respect to one axis, the moment of inertia with respect to another parallel axis can be calculated from the *parallel axis theorem*, also known as the *transfer axis theorem.* In

Materials

Eq. 49.20, d is the distance between the centroidal axis and the second, parallel axis, and I_c is the moment of inertia of the second, parallel axis. This can be further broken down into x and y components as shown in Eq. 49.21and Eq. 49.22.

Moment of Inertia Parallel Axis Theorem

$$I' = I_c + d^2A \qquad 49.20$$

$$I'_x = I_{x_c} + d_y^2A \qquad 49.21$$

$$I'_y = I_{y_c} + d_x^2A \qquad 49.22$$

The second term in Eq. 49.20 is often much larger than the first term. Areas close to the centroidal axis do not affect the moment of inertia significantly. This principle is exploited by structural steel shapes (such as is shown in Fig. 49.1) that derive bending resistance from *flanges* located away from the centroidal axis. The *web* does not contribute significantly to the moment of inertia.

Figure 49.1 *Structural Steel W-Shape*

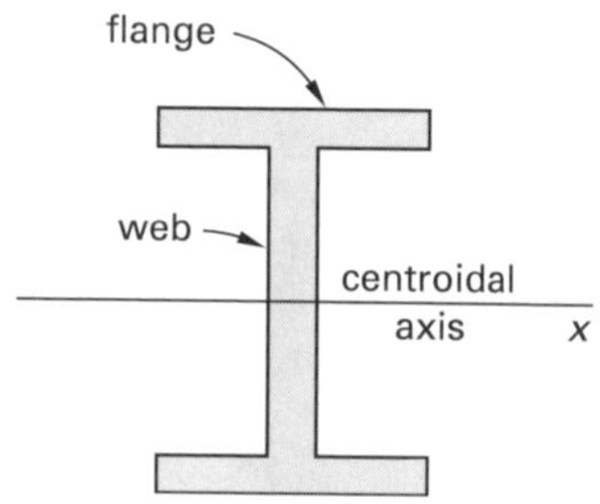

Example 49.4

Find the moment of inertia about the x-axis for the inverted-T area shown.

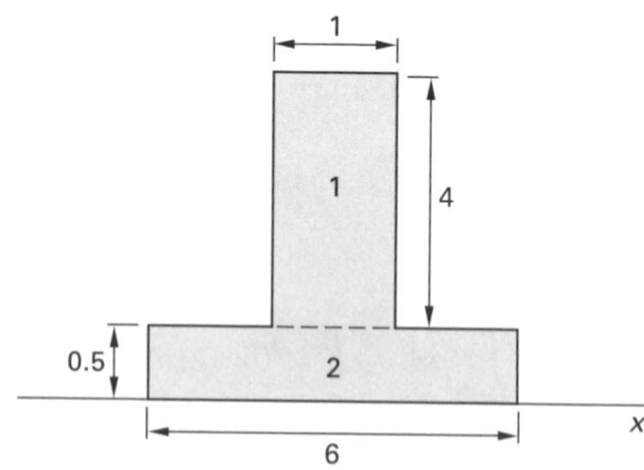

Solution

The area is divided into two basic shapes: 1 and 2. The moment of inertia of basic shape 2 with respect to the x-axis is

Properties of Various Shapes

$$I_{x,2} = \frac{bh^3}{3} = \frac{(6.0)(0.5)^3}{3} = 0.25 \text{ units}^4$$

The moment of inertia of basic shape 1 about its own centroid is

Properties of Various Shapes

$$I_{cx,1} = \frac{bh^3}{12} = \frac{(1)(4)^3}{12} = 5.33 \text{ units}^4$$

The x-axis is located 2.5 units from the centroid of basic shape 1. From the parallel axis theorem, Eq. 49.20, the moment of inertia of basic shape 1 about the x-axis is

Moment of Inertia Parallel Axis Theorem

$$I_{x,1} = I_{xc,1} + d_1^2A_1 = 5.33 + (4)(2.5)^2$$
$$= 30.33 \text{ units}^4$$

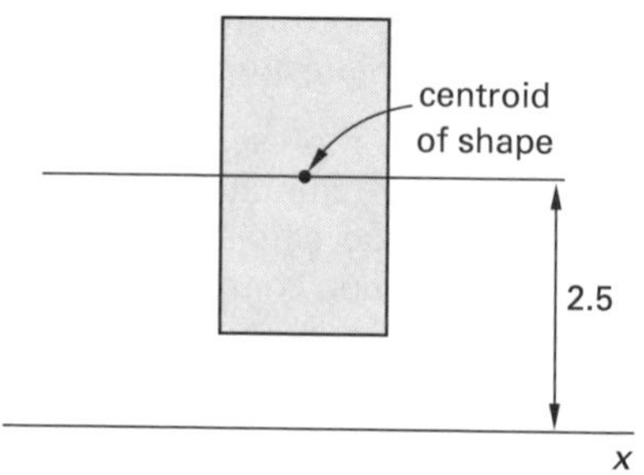

The total moment of inertia of the T-area is

$$I_x = I_{x,1} + I_{x,2} = 30.33 \text{ units}^4 + 0.25 \text{ units}^4$$
$$= 30.58 \text{ units}^4$$

6. POLAR MOMENT OF INERTIA

The *polar moment of inertia*, J, is required in torsional shear stress calculations. It can be thought of as a measure of an area's resistance to torsion (twisting). The definition of a polar moment of inertia of a two-dimensional area requires three dimensions because the reference axis for a polar moment of inertia of a plane area is perpendicular to the plane area.

The polar moment of inertia is derived from Eq. 49.23.

$$J = \int (x^2 + y^2)dA \qquad 49.23$$

It is often easier to use the *perpendicular axis theorem* to quickly calculate the polar moment of inertia.

- *perpendicular axis theorem:* The polar moment of inertia of a plane area about an axis normal to the plane is equal to the sum of the moments of inertia about any two mutually perpendicular axes lying in the plane and passing through the given axis.

$$J_c = I_x + I_y \qquad 49.24$$

Since the two perpendicular axes can be chosen arbitrarily, it is most convenient to use the centroidal moments of inertia.

$$J_c = I_{cx} + I_{cy} \qquad 49.25$$

Example 49.5

What is the centroidal polar moment of inertia of a circular area of radius r?

Solution

The centroidal moment of inertia of a circle with respect to the x-axis is

Properties of Various Shapes

$$I_{cx} = \frac{\pi a^4}{4}$$

Since the area is symmetrical, I_{cy} and I_{cx} are the same. From Eq. 49.25,

$$J_c = I_{cx} + I_{cy} = \frac{\pi a^4}{4} + \frac{\pi a^4}{4} = \frac{\pi a^4}{2}$$

7. RADIUS OF GYRATION

Every nontrivial area has a centroidal moment of inertia. Usually, some portions of the area are close to the centroidal axis and other portions are farther away. The *transverse radius of gyration*, or just *radius of gyration*, r, is an imaginary distance from the centroidal axis at which the entire area can be assumed to exist without affecting the moment of inertia. Despite the name "radius," the radius of gyration is not limited to circular shapes or to polar axes. This concept is illustrated in Fig. 49.2.

Figure 49.2 *Radius of Gyration*

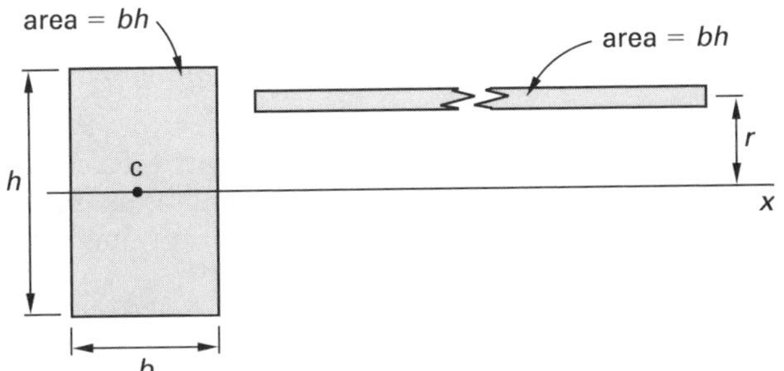

The method of calculating the radius of gyration is based on the parallel axis theorem. If all of the area is located a distance, r, from the original centroidal axis, there will be no I_c term in Eq. 49.20. Only the Ad^2 term will contribute to the moment of inertia.

$$I = r^2 A \qquad 49.26$$

Radius of Gyration

$$r_x = \sqrt{\frac{I_x}{A}} \qquad 49.27$$

$$r_y = \sqrt{\frac{I_y}{A}} \qquad 49.28$$

The concept of *least radius of gyration* comes up frequently in column design problems. (The column will tend to buckle about an axis that produces the smallest radius of gyration.) Usually, finding the least radius of gyration for symmetrical section will mean solving Eq. 49.27 twice: once with I_x to find r_x and once with I_y to find r_y. The smallest value of r is the least radius of gyration.

The analogous quantity in the polar system is

Radius of Gyration

$$r_p = \sqrt{\frac{J}{A}} \qquad 49.29$$

Just as the polar moment of inertia, J, can be calculated from the two rectangular moments of inertia, the polar radius of gyration can be calculated from the two rectangular radii of gyration.

$$r^2 = r_x^2 + r_y^2 \qquad 49.30$$

Example 49.6

What is the radius of gyration of the rectangular shape in Ex. 49.2?

Solution

The area of the rectangle is

$$A = bh = (5)(8) = 40 \text{ units}^2$$

From Eq. 49.27, the radius of gyration is

Radius of Gyration

$$r_x = \sqrt{\frac{I_x}{A}} = \sqrt{\frac{213.3 \text{ units}^4}{40 \text{ units}^2}} = 2.31 \text{ units}$$

2.31 units is the distance from the centroidal x-axis that an infinitely long strip with an area of 40 square units would have to be located in order to have a moment of inertia of 213.3 units^4.

8. PRODUCT OF INERTIA

The *product of inertia*, I_{xy}, of a two-dimensional area is useful when relating properties of areas evaluated with respect to different axes. It is found by multiplying each differential element of area by its x- and y-coordinate and then summing over the entire area. (See Fig. 49.3.)

Product of Inertia

$$I_{xy} = \int xy\,dA \qquad 49.31$$

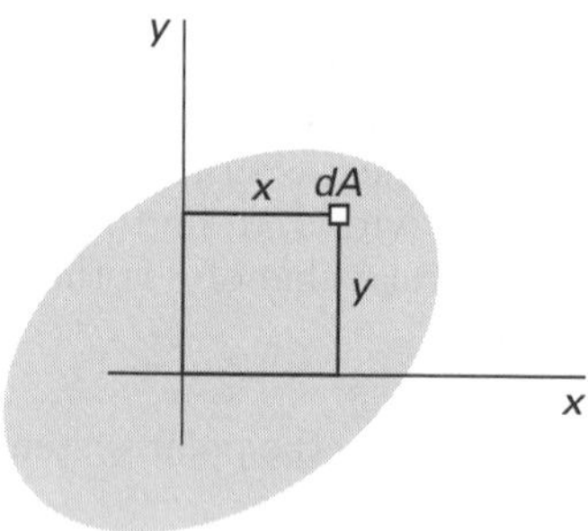

Figure 49.3 Calculating the Product of Inertia

The product of inertia is zero when either axis is an axis of symmetry. Since the axes can be chosen arbitrarily, the area may be in one of the negative quadrants, and the product of inertia may be negative.

The parallel axis theorem for products of inertia, as shown in Fig. 49.4, is given by Eq. 49.32. (Both axes are allowed to move to new positions.) x_c' and y_c' are the coordinates of the centroid in the new coordinate system.

Product of Inertia

$$I_{xy} = I_{xc\,yc} + d_x d_y A \qquad 49.32$$

Figure 49.4 Parallel Axis Theorem for Products of Inertia

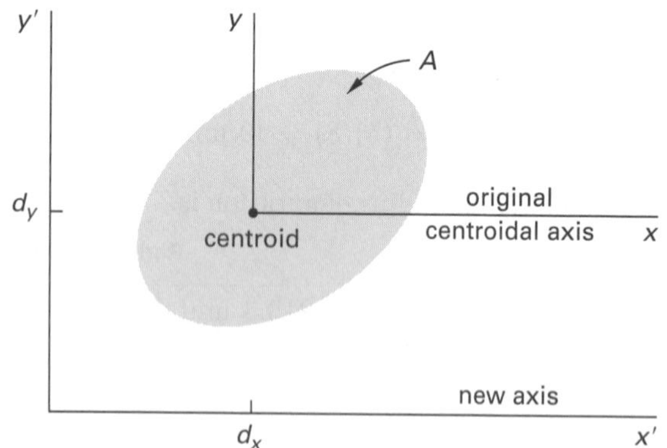

9. SECTION MODULUS

In the analysis of beams, the outer compressive (or tensile) surface is known as the *extreme fiber*. The distance, c, from the centroidal axis of the beam cross section to the extreme fiber is the distance to the extreme fiber. The *section modulus*, S, combines the centroidal moment of inertia and the distance to the extreme fiber.

$$S = \frac{I_c}{c} \qquad 49.33$$

10. ROTATION OF AXES

Figure 49.5 shows rotation of the x-y axes through an angle, θ, into a new set of u-v axes, without rotating the area. If the moments and product of inertia of the area are known with respect to the old x-y axes, the new properties can be calculated from Eq. 49.34 through Eq. 49.36.

Figure 49.5 Rotation of Axes

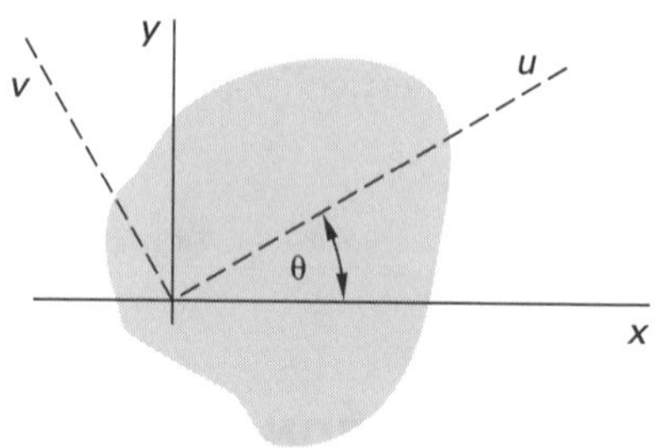

$$\begin{aligned} I_u &= I_x\cos^2\theta - 2I_{xy}\sin\theta\cos\theta + I_y\sin^2\theta \\ &= \tfrac{1}{2}(I_x + I_y) + \tfrac{1}{2}(I_x - I_y)\cos 2\theta \\ &\quad - I_{xy}\sin 2\theta \end{aligned} \qquad 49.34$$

$$\begin{aligned} I_v &= I_x\sin^2\theta + 2I_{xy}\sin\theta\cos\theta + I_y\cos^2\theta \\ &= \tfrac{1}{2}(I_x + I_y) - \tfrac{1}{2}(I_x - I_y)\cos 2\theta \\ &\quad + I_{xy}\sin 2\theta \end{aligned} \qquad 49.35$$

$$\begin{aligned} I_{uv} &= I_x\sin\theta\cos\theta + I_{xy}(\cos^2\theta - \sin^2\theta) \\ &\quad - I_y\sin\theta\cos\theta \\ &= \tfrac{1}{2}(I_x - I_y)\sin 2\theta + I_{xy}\cos 2\theta \end{aligned} \qquad 49.36$$

Since the polar moment of inertia about a fixed axis perpendicular to any two orthogonal axes in the plane is constant, the polar moment of inertia is unchanged by the rotation.

$$\begin{aligned} J_{xy} &= I_x + I_y \\ &= I_u + I_v = J_{uv} \end{aligned} \qquad 49.37$$

11. PRINCIPAL AXES FOR AREA PROPERTIES

Referring to Fig. 49.5, there is one angle, θ, that will maximize the moment of inertia, I_u. This angle can be found from calculus by setting $dI_u/d\theta = 0$. The resulting equation defines two angles, one that maximizes I_u and one that minimizes I_u.

$$\tan 2\theta = \frac{-2I_{xy}}{I_x - I_y} \qquad 49.38$$

The two angles that satisfy Eq. 49.38 are 90° apart. The set of u-v axes defined by Eq. 49.38 are known as *principal axes*. The moments of inertia about the principal axes are defined by Eq. 49.39 and are known as the *principal moments of inertia.*

$$I_{\text{max,min}} = \tfrac{1}{2}(I_x + I_y) \pm \sqrt{\tfrac{1}{4}(I_x - I_y)^2 + I_{xy}^2} \qquad 49.39$$

12. MOHR'S CIRCLE FOR AREA PROPERTIES

Once I_x, I_y, and I_{xy} are known, *Mohr's circle* can be drawn to graphically determine the moments of inertia about the principal axes. The procedure for drawing Mohr's circle is given as follows.

step 1: Determine I_x, I_y, and I_{xy} for the existing set of axes.

step 2: Draw a set of I-I_{xy} axes.

step 3: Plot the center of the circle, point **c**, by calculating distance c along the I-axis. (See Fig. 49.6.)

$$c = \tfrac{1}{2}(I_x + I_y) \qquad 49.40$$

Figure 49.6 *Mohr's Circle*

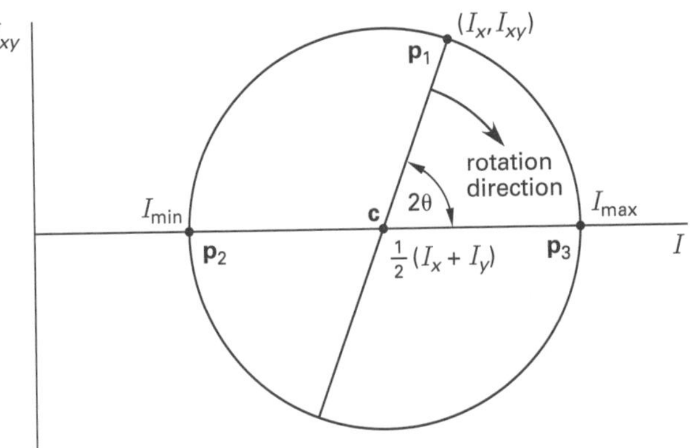

step 4: Plot the point $\mathbf{p}_1 = (I_x, I_{xy})$.

step 5: Draw a line from point $\mathbf{p}_1$ through center **c** and extend it an equal distance below the I-axis. This is the diameter of the circle.

step 6: Using the center **c** and point $\mathbf{p}_1$, draw the circle. An alternate method of constructing the circle is to draw a circle of radius r.

$$r = \sqrt{\tfrac{1}{4}(I_x - I_y)^2 + I_{xy}^2} \qquad 49.41$$

step 7: Point $\mathbf{p}_2$ defines I_{min}. Point $\mathbf{p}_3$ defines I_{max}.

step 8: Determine the angle θ as half of the angle 2θ on the circle. This angle corresponds to I_{max}. (The axis giving the minimum moment of inertia is perpendicular to the maximum axis.) The sense of this angle and the sense of the rotation are the same. That is, the direction that the diameter would have to be turned in order to coincide with the I_{max}-axis has the same sense as the rotation of the x-y axes needed to form the principal u-v axes. (See Fig. 49.7.)

Figure 49.7 *Principal Axes from Mohr's Circle*

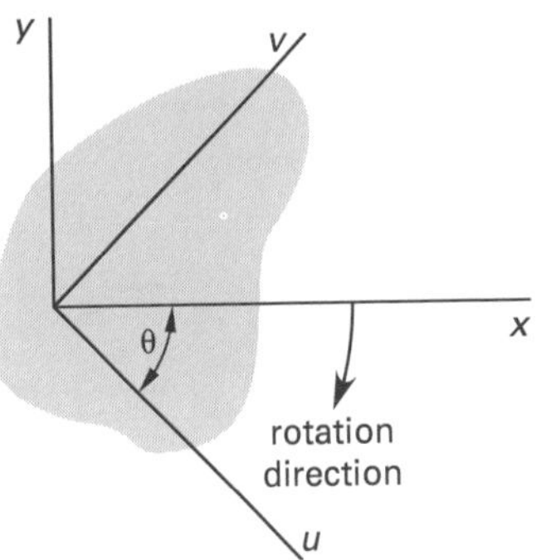

Materials

13. NOMENCLATURE

a	definition radius	units
A	area	units^2
b	base distance	units
c	distance to extreme fiber	units
d	separation distance	units
h	height distance	units
I	moment of inertia	units^4
J	polar moment of inertia	units^4
L	length	units
P	product of inertia	units^4
M	moment of area	units^3
r	radius	units
r	radius of gyration	units
S	section modulus	units^3
u	distance in the u-direction	units
v	distance in the v-direction	units
V	volume	units^3
x	distance in the x-direction	units
y	distance in the y-direction	units

Symbols

θ	angle	deg

Subscripts

a	area
c	centroidal

50 Strength of Materials

Content in blue refers to the *NCEES Handbook*.

NCEES EXAM SPECIFICATIONS AND RELATED CONTENT

MACHINE DESIGN AND MATERIALS EXAM

I.D.1. Strength of Materials: Stress/strain
- 2. Hooke's Law
- 4. Total Strain Energy
- 6. Thermal Deformation
- 8. Combined Stresses (Biaxial Loading)
- 9. Mohr's Circle for Stress
- 10. General Strain (Three-Dimensional Strain)

I.D.2. Strength of Materials: Shear
- 14. Shear Stress in Beams
- 15. Shear Flow

I.D.3. Strength of Materials: Bending
- 13. Shear and Bending Moment Diagrams
- 16. Bending Stress in Beams

1. BASIC CONCEPTS

Strength of materials (known also as *mechanics of materials*) deals with the elastic behavior of loaded engineering materials.[1]

Stress is force per unit area, P/A. Typical units of stress are lbf/in^2, ksi (thousands of pounds per square inch), and MPa. Although there are many names given to stress, there are only two primary types, differing in the orientation of the loaded area. With *normal stress*, σ, the area is normal to the force carried. With *shear stress*, τ, the area is parallel to the force. (See Fig. 50.1.)

Figure 50.1 *Normal and Shear Stress*

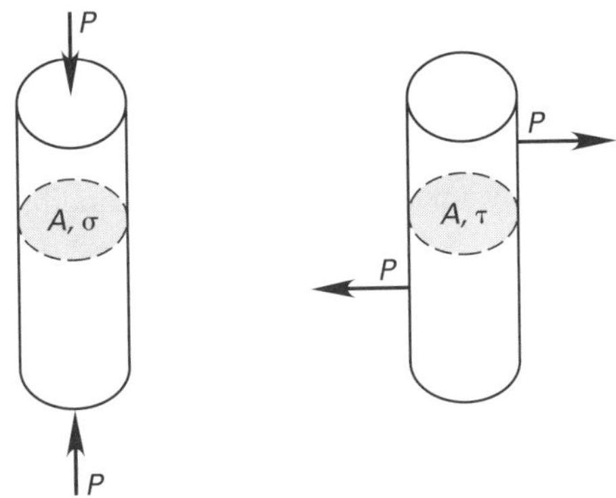

Strain, ϵ, is elongation expressed on a fractional or percentage basis. It may be listed as having units of in/in, mm/mm, percent, or no units at all. A strain in one direction will be accompanied by strains in orthogonal directions in accordance with Poisson's ratio. *Dilation* is the sum of the strains in the three coordinate directions.

$$\text{dilation} = \epsilon_x + \epsilon_y + \epsilon_z \qquad 50.1$$

[1]Plastic behavior and ultimate strength design are not covered in this chapter.

2. HOOKE'S LAW

Hooke's law is a simple mathematical statement of the relationship between elastic stress and strain: Stress is proportional to strain. (See Fig. 50.2.) For normal stress, the constant of proportionality is the *modulus of elasticity*, *E*.

Hooke's Law

$$\sigma = E\epsilon \qquad 50.2$$

Figure 50.2 *Application of Hooke's Law*

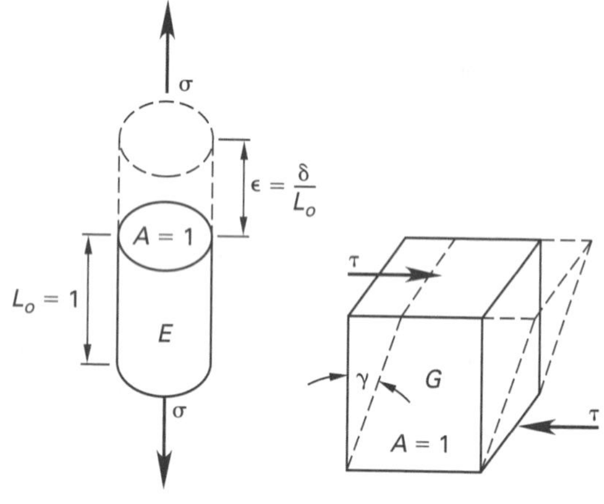

(a) normal stress on unit cylinder

(b) shear stress on a unit cube

Hooke's law applies also to a plane element in pure shear. For such an element, the shear stress is linearly related to the shear strain, γ, by the shear modulus, *G*.

Hooke's Law

$$\gamma_{xy} = \frac{\tau_{xy}}{G} \qquad 50.3$$

For an elastic, isotropic material, the modulus of elasticity, shear modulus, and Poisson's ratio are related by Eq. 50.4.

Shear Stress-Strain

$$G = \frac{E}{2(1+\nu)} \qquad 50.4$$

3. ELASTIC DEFORMATION

Since stress is P/A and strain is δ/L, Hooke's law can be rearranged to give the elongation of an axially loaded member with a uniform cross section experiencing normal stress. Tension loading is considered positive; compressive loading is negative.

$$\delta = L_o\epsilon = \frac{L_o\sigma}{E} = \frac{L_oF}{EA} \qquad 50.5$$

The actual length of a member under loading is given by Eq. 50.6. The algebraic sign of the deformation must be observed.

$$L = L_o + \delta \qquad 50.6$$

4. TOTAL STRAIN ENERGY

The energy stored in a loaded member is equal to the work required to deform the member. Below the proportionality limit, the total *strain energy* for a member loaded in tension or compression is given by Eq. 50.7. Equation 50.8 shows an alternate way to find the strain energy.

Strain Energy

$$U = \tfrac{1}{2}F\delta \qquad 50.7$$

$$U = \frac{F^2L_o}{2AE} = \frac{\sigma^2L_oA}{2E} \qquad 50.8$$

5. STIFFNESS AND RIGIDITY

Stiffness is the amount of force required to cause a unit of deformation (displacement) and is often referred to as a *spring constant*. Typical units are pounds per inch and newtons per meter. The stiffness of a spring or other structure can be calculated from the deformation equation by solving for F/δ. Equation 50.9 is valid for tensile and compressive normal stresses. For torsion and bending, the stiffness equation will depend on how the deflection is calculated.

$$k = \frac{F}{\delta} \quad \text{[general form]} \qquad \text{[SI]} \quad 50.9(a)$$

$$k = \frac{AE}{L_o} \quad \text{[normal stress form]} \qquad \text{[U.S.]} \quad 50.9(b)$$

When more than one spring or resisting member share the load, the relative stiffnesses are known as *rigidities*. Rigidities have no units, and the individual rigidity values have no significance. (See Fig. 50.3.) A ratio of two rigidities, however, indicates how much stiffer one member is compared to another. Equation 50.10 is one method of calculating rigidity of member *j* in a multimember structure. (Since rigidities are relative numbers, they can be multiplied by the least common denominator to obtain integer values.)

$$R_j = \frac{k_j}{\sum_i k_i} \qquad 50.10$$

Figure 50.3 *Stiffness and Rigidity*

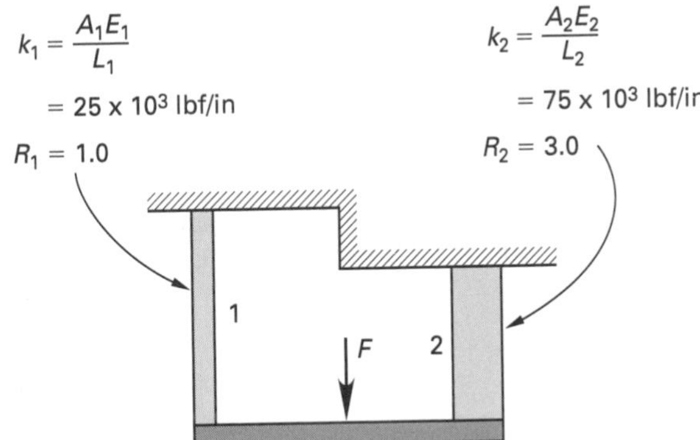

Rigidity is proportional to the reciprocal of deflection. (See Table 50.1.) *Flexural rigidity* is the reciprocal of deflection in members that are acted upon by a moment (i.e., are in bending), although that term may also be used to refer to the product, *EI*, of the modulus of elasticity and the moment of inertia.

6. THERMAL DEFORMATION

If the temperature of an object is changed, the object will experience changes in length, area, and volume. The magnitude of these changes will depend on the *coefficient of linear expansion*, α, which is widely tabulated for solids. (See Table 50.2.) The *coefficient of volumetric expansion*, β, can be calculated from the coefficient of linear expansion. The change in length caused by a change in temperature is called the *thermal deformation* and is designated as δ_{th}.

Thermal Deformations

$$\delta_{\text{th}} = \alpha L(T - T_o) \qquad 50.11$$

$$\Delta A = \gamma A_o(T_2 - T_1) \qquad 50.12$$

$$\gamma \approx 2\alpha \qquad 50.13$$

$$\Delta V = \beta V_o(T_2 - T_1) \qquad 50.14$$

$$\beta \approx 3\alpha \qquad 50.15$$

It is a common misconception that a hole in a plate will decrease in size when the plate is heated (because the surrounding material "squeezes in" on the hole). However, changes in temperature affect all dimensions the same way.

Table 50.1 *Deflection and Stiffness for Various Systems (due to bending moment alone)*

system	maximum deflection, x	stiffness, k
	$\frac{Fh}{AE}$	$\frac{AE}{h}$
	$\frac{Fh^3}{3EI}$	$\frac{3EI}{h^3}$
	$\frac{Fh^3}{12EI}$	$\frac{12EI}{h^3}$
(w is load per unit length)	$\frac{wL^4}{8EI}$	$\frac{8EI}{L^3}$
	$\frac{Fh^3}{12E(I_1+I_2)}$	$\frac{12E(I_1+I_2)}{h^3}$
	$\frac{FL^3}{48EI}$	$\frac{48EI}{L^3}$
(w is load per unit length)	$\frac{5wL^4}{384EI}$	$\frac{384EI}{5L^3}$
	$\frac{FL^3}{192EI}$	$\frac{192EI}{L^3}$
(w is load per unit length)	$\frac{wL^4}{384EI}$	$\frac{384EI}{L^3}$

In this case, the circumference of the hole is a linear dimension that follows Eq. 50.11. As the circumference increases, the hole area also increases. (See Fig. 50.4.)

Table 50.2 *Average Coefficients of Linear Thermal Expansion (multiply all values by 10^{-6})*

substance	1/°F	1/°C
aluminum alloy	12.8	23.0
brass	10.0	18.0
cast iron	5.6	10.1
chromium	3.8	6.8
concrete	6.7	12.0
copper	8.9	16.0
glass (plate)	4.9	8.9
glass (Pyrex™)	1.8	3.2
invar	0.39	0.7
lead	15.6	28.0
magnesium alloy	14.5	26.1
marble	6.5	11.7
platinum	5.0	9.0
quartz, fused	0.2	0.4
steel	6.5	11.7
tin	14.9	26.9
titanium alloy	4.9	8.8
tungsten	2.4	4.4
zinc	14.6	26.3

(Multiply 1/°F by 9/5 to obtain 1/°C.)
(Multiply 1/°C by 5/9 to obtain 1/°F.)

Figure 50.4 *Thermal Expansion of an Area*

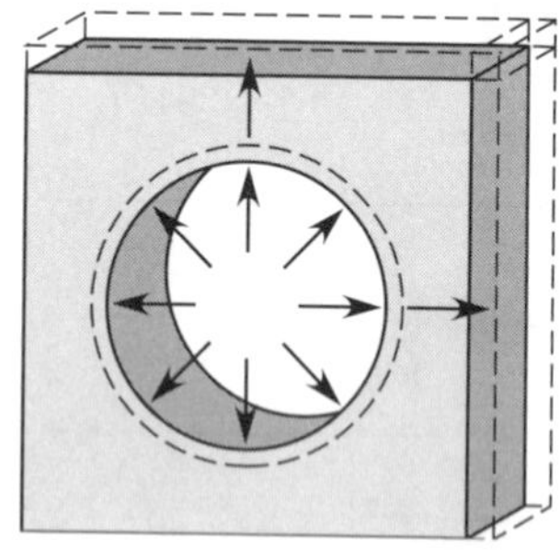

If Eq. 50.11 is rearranged, an expression for the *thermal strain* is obtained.

$$\epsilon_{\text{th}} = \frac{\delta_{\text{th}}}{L_o} = \alpha(T_2 - T_1) \qquad 50.16$$

Thermal strain is handled in the same manner as strain due to an applied load. For example, if a bar is heated but is not allowed to expand, the stress can be calculated from the thermal strain and Hooke's law.

$$\sigma_{\text{th}} = E\epsilon_{\text{th}} \qquad 50.17$$

Low values of the coefficient of expansion, such as with Pyrex™ glassware, result in low thermally induced stresses and high insensitivity to temperature extremes. Differences in the coefficients of expansion of two materials are used in *bimetallic elements*, such as thermostatic springs and strips.

Example 50.1

A replacement steel railroad rail ($L = 20.0$ m, $A = 60 \times 10^{-4}$ m^2) was installed when its temperature was 5°C. The rail was installed tightly in the line, without an allowance for expansion. If the rail ends are constrained by adjacent rails and if the spikes prevent buckling, what is the compressive force in the rail at 25°C?

Solution

From Table 50.2, the coefficient of linear expansion for steel is 11.7×10^{-6} 1/°C. From Eq. 50.16, the thermal strain is

$$\begin{aligned}\epsilon_{\text{th}} &= \alpha(T_2 - T_1) = \left(11.7 \times 10^{-6}\ \frac{1}{^\circ\text{C}}\right)(25^\circ\text{C} - 5^\circ\text{C}) \\ &= 2.34 \times 10^{-4}\ \text{m/m}\end{aligned}$$

The modulus of elasticity of steel is 20×10^{10} N/m^2 (20×10^4 MPa). The compressive stress is given by Hooke's law. Use Eq. 50.17.

$$\begin{aligned}\sigma_{\text{th}} &= E\epsilon_{\text{th}} = \left(20 \times 10^{10}\ \frac{\text{N}}{\text{m}^2}\right)\left(2.34 \times 10^{-4}\ \frac{\text{m}}{\text{m}}\right) \\ &= 4.68 \times 10^7\ \text{N/m}^2\end{aligned}$$

The compressive force is

$$\begin{aligned}F &= \sigma_{\text{th}}A = \left(4.68 \times 10^7\ \frac{\text{N}}{\text{m}^2}\right)(60 \times 10^{-4}\ \text{m}^2) \\ &= 281\,000\ \text{N}\end{aligned}$$

7. STRESS CONCENTRATIONS

A *geometric stress concentration* occurs whenever there is a discontinuity or nonuniformity in an object. Examples of nonuniform shapes are stepped shafts, plates with holes and notches, and shafts with keyways. It is convenient to think of stress as lines of force following streamlines within an object. (See Fig. 50.5.) There will be a stress concentration wherever local geometry forces the streamlines closer together.

Figure 50.5 *Streamline Analogy to Stress Concentrations*

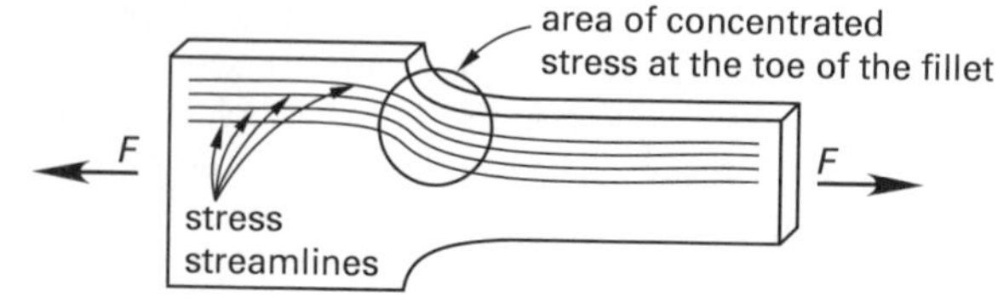

Materials

Stress values determined by simplistic F/A, Mc/I, or Tr/J calculations will be greatly understated. *Stress concentration factors* (*stress risers*) are correction factors used to account for the nonuniform stress distributions. The symbol K is often used, but this is not universal. The actual stress is determined as the product of the stress concentration factor, K, and the *nominal stress*, σ_0. Values of the stress concentration factor are almost always greater than 1.0 and can run as high as 3.0 and above. The exact value for a given application must be determined from extensive experimentation or from published tabulations of standard configurations.

$$\sigma' = K\sigma_0 \tag{50.18}$$

Stress concentration factors are normally not applied to members with multiple redundancy, for static loading of ductile materials, or where local yielding around the discontinuity reduces the stress. For example, there will be many locations of stress concentration in a lap rivet connection. However, the stresses are kept low by design, and stress concentrations are disregarded.

Stress concentration factors are not applicable to every point on an object; they apply only to the point of maximum stress. For example, with filleted shafts, the maximum stress occurs at the toe of the fillet. Therefore, the stress concentration factor should be applied to the stress calculated from the smaller section's properties. For objects with holes or notches, it is important to know if the nominal stress to which the factor is applied is calculated from an area that includes or excludes the holes or notches.

In addition to geometric stress concentrations, there are also *fatigue stress concentrations*. The *fatigue stress concentration factor* is the ratio of the fatigue strength without a stress concentration to the fatigue stress with a stress concentration. Fatigue stress concentration factors depend on the material, material strength, and geometry of the stress concentration (i.e., radius of the notch). Fatigue stress concentration factors can be less than the geometric factors from which they are computed.

8. COMBINED STRESSES (BIAXIAL LOADING)

Loading is rarely confined to a single direction. Many practical cases have different normal and shear stresses on two or more perpendicular planes. Sometimes, one of the stresses may be small enough to be disregarded, reducing the analysis to one dimension. In other cases, however, the shear and normal stresses must be combined to determine the maximum stresses acting on the material.

For any point in a loaded specimen, a plane can be found where the shear stress is zero. The normal stresses associated with this plane are known as the *principal stresses*, which are the maximum and minimum stresses acting at that point in any direction.

For two-dimensional (biaxial) loading (i.e., two normal stresses combined with a shearing stress), the normal and shear stresses on a plane whose normal line is inclined an angle θ from the horizontal can be found from Eq. 50.19 and Eq. 50.20. Proper sign convention must be adhered to when using the combined stress equations. The positive senses of shear and normal stresses are shown in Fig. 50.6. As is usually the case, tensile normal stresses are positive; compressive normal stresses are negative. In two dimensions, shear stresses are designated as clockwise (positive) or counterclockwise (negative).

$$\sigma_\theta = \tfrac{1}{2}(\sigma_x + \sigma_y) + \tfrac{1}{2}(\sigma_x - \sigma_y)\cos 2\theta + \tau \sin 2\theta \tag{50.19}$$

$$\tau_\theta = -\tfrac{1}{2}(\sigma_x - \sigma_y)\sin 2\theta + \tau \cos 2\theta \tag{50.20}$$

At first glance, the orientation of the shear stresses may seem confusing. However, the arrangement of stresses shown produces equilibrium in the x- and y-directions without causing rotation. Other than a mirror image or a trivial rotation of the arrangement shown in Fig. 50.6, no other arrangement of shear stresses will produce equilibrium.

Figure 50.6 *Sign Convention for Combined Stress*

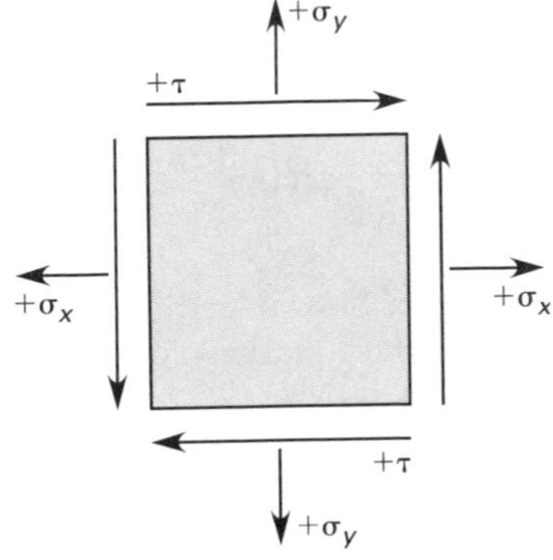

The maximum and minimum values (as θ is varied) of the normal stress, σ_θ, are the *principal stresses*, which can be found by differentiating Eq. 50.19 with respect to θ, setting the derivative equal to zero, and substituting θ back into Eq. 50.19. Equation 50.21 is derived in this manner. A similar procedure is used to derive the *extreme shear stresses* (i.e., maximum and minimum shear stresses) in Eq. 50.22 from Eq. 50.20. (The term *principal stress* implies a normal stress, never a shear stress.)

$$\sigma_1, \sigma_2 = \tfrac{1}{2}(\sigma_x + \sigma_y) \pm \tau_1 \tag{50.21}$$

Materials

$$\tau_1,\tau_2 = \pm\tfrac{1}{2}\sqrt{(\sigma_x - \sigma_y)^2 + (2\tau)^2} \qquad 50.22$$

The angles of the planes on which the normal stresses are minimum and maximum are given by Eq. 50.23. (See Fig. 50.7.) θ is measured from the x-axis, clockwise if negative and counterclockwise if positive. Equation 50.23 will yield two angles, 90° apart. These angles can be substituted back into Eq. 50.19 and Eq. 50.20 to determine which angle corresponds to the minimum normal stress and which angle corresponds to the maximum normal stress.[2]

$$\theta_{\sigma_1,\sigma_2} = \tfrac{1}{2}\arctan\left(\frac{2\tau}{\sigma_x - \sigma_y}\right) \qquad 50.23$$

Figure 50.7 *Stresses on an Inclined Plane*

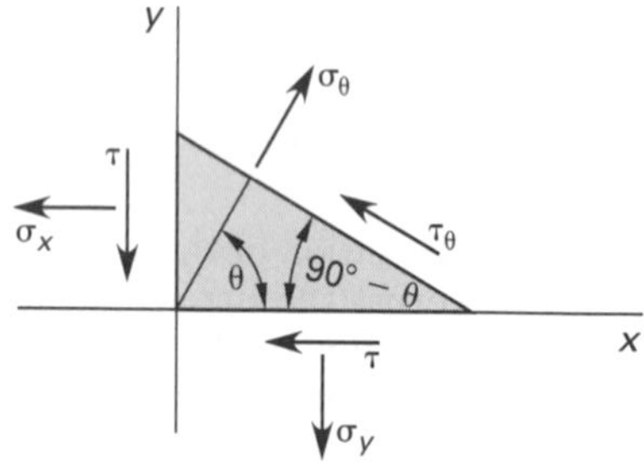

The angles of the planes on which the shear stresses are minimum and maximum are given by Eq. 50.24. These planes will be 90° apart and will be rotated 45° from the planes of principal normal stresses. As with Eq. 50.23, θ is measured from the x-axis, clockwise if negative and counterclockwise if positive. Generally, the sign of a shear stress on an inclined plane will be unimportant.

$$\theta_{\tau_1,\tau_2} = \tfrac{1}{2}\arctan\left(\frac{\sigma_x - \sigma_y}{-2\tau}\right) \qquad 50.24$$

For the special case of biaxial loading (i.e., a two-dimensional stress state), the equations for principal stress are given by equation Eq. 50.25.

Principal Stresses

$$\sigma_1,\sigma_2 = \frac{\sigma_x + \sigma_y}{2} \pm \sqrt{\left(\frac{\sigma_x - \sigma_y}{2}\right)^2 + \tau_{xy}^2} \qquad 50.25$$

Example 50.2

(a) Find the maximum normal and shear stresses on the object shown. (b) Determine the angle of the plane of principal normal stresses.

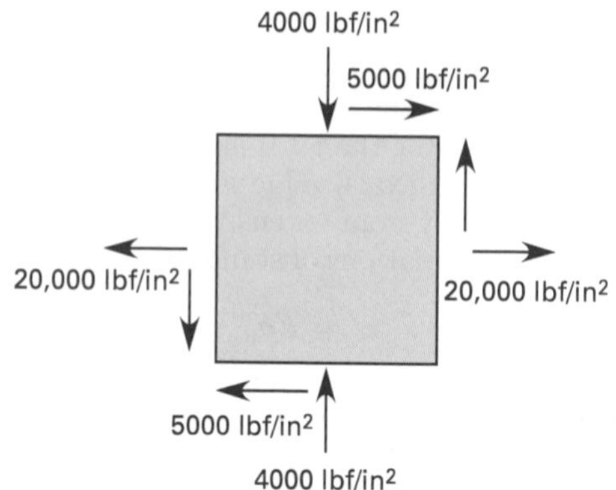

Solution

(a) Find the principal shear stresses first. The applied 4000 lbf/in^2 compressive stress is negative. From Eq. 50.21 and Eq. 50.25,

$$\sigma_1,\sigma_2 = \tfrac{1}{2}(\sigma_x + \sigma_y) \pm \tau_1$$

Principal Stresses

$$\sigma_1,\sigma_2 = \frac{\sigma_x + \sigma_y}{2} \pm \sqrt{\left(\frac{\sigma_x - \sigma_y}{2}\right)^2 + \tau_{xy}^2}$$

$$\sigma_1,\sigma_2 = \frac{\sigma_x + \sigma_y}{2} \pm \tau_1 = \frac{\sigma_x + \sigma_y}{2} \pm \sqrt{\left(\frac{\sigma_x - \sigma_y}{2}\right)^2 + \tau_{xy}^2}$$

Rearranging, the principal shear stress is

$$\begin{aligned}\tau_1 &= \sqrt{\left(\frac{\sigma_x - \sigma_y}{2}\right)^2 + \tau_{xy}^2}\\ &= \sqrt{\left(\frac{20{,}000\ \frac{\text{lbf}}{\text{in}^2} - \left(-4000\ \frac{\text{lbf}}{\text{in}^2}\right)}{2}\right)^2 + \left(5000\ \frac{\text{lbf}}{\text{in}^2}\right)^2}\\ &= 13{,}000\ \text{lbf/in}^2\end{aligned}$$

Materials

[2]Alternatively, the following procedure can be used to determine the direction of the principal planes. Let σ_x be the algebraically larger of the two given normal stresses. The angle between the direction of σ_x and the direction of σ_1, the algebraically larger principal stress, will always be less than 45°.

From Eq. 50.25, the maximum normal stress is

Principal Stresses

$$\sigma_1 = \frac{\sigma_x + \sigma_y}{2} \pm \sqrt{\left(\frac{\sigma_x - \sigma_y}{2}\right)^2 + \tau_{xy}^2}$$

$$= \frac{20{,}000\ \frac{\text{lbf}}{\text{in}^2} + \left(-4000\ \frac{\text{lbf}}{\text{in}^2}\right)}{2} \pm \sqrt{\left(\frac{20{,}000\ \frac{\text{lbf}}{\text{in}^2} - \left(-4000\ \frac{\text{lbf}}{\text{in}^2}\right)}{2}\right)^2 + \left(5000\ \frac{\text{lbf}}{\text{in}^2}\right)^2}$$

$$= 21{,}000\ \text{lbf/in}^2 \quad \text{[tension]}$$

(b) The angle of the principal normal stresses is given by Eq. 50.23.

$$\theta = \tfrac{1}{2}\arctan\left(\frac{2\tau}{\sigma_x - \sigma_y}\right)$$

$$= \tfrac{1}{2}\arctan\left(\frac{(2)\left(5000\ \frac{\text{lbf}}{\text{in}^2}\right)}{20{,}000\ \frac{\text{lbf}}{\text{in}^2} - \left(-4000\ \frac{\text{lbf}}{\text{in}^2}\right)}\right)$$

$$= \left(\frac{1}{2}\right)(22.6°, 202.6°)$$

$$= 11.3°, 101.3°$$

It is not obvious which angle produces which normal stress. One of the angles can be substituted back into Eq. 50.19 for σ_θ.

$$\sigma_{11.3°} = \tfrac{1}{2}(\sigma_x + \sigma_y) + \tfrac{1}{2}(\sigma_x - \sigma_y)\cos 2\theta + \tau \sin 2\theta$$

$$= \left(\frac{1}{2}\right)\left(20{,}000\ \frac{\text{lbf}}{\text{in}^2} + \left(-4000\ \frac{\text{lbf}}{\text{in}^2}\right)\right) + \left(\frac{1}{2}\right)\left(20{,}000\ \frac{\text{lbf}}{\text{in}^2} - \left(-4000\ \frac{\text{lbf}}{\text{in}^2}\right)\right) \times \cos\big((2)(11.3°)\big) + \left(5000\ \frac{\text{lbf}}{\text{in}^2}\right)\sin\big((2)(11.3°)\big)$$

$$= 21{,}000\ \text{lbf/in}^2$$

The 11.3° angle corresponds to the maximum normal stress of 21,000 lbf/in².

9. MOHR'S CIRCLE FOR STRESS

Mohr's circle can be constructed to graphically determine the principal stresses. (See Fig. 50.8.)

Figure 50.8 *Mohr's Circle for Stress*

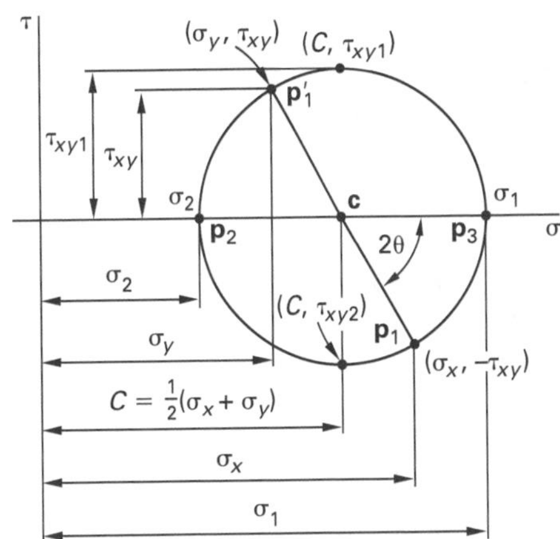

step 1: Determine the applied stresses: σ_x, σ_y, and τ. (Tensile normal stresses are positive; compressive normal stresses are negative. Clockwise shear stresses are positive; counterclockwise shear stresses are negative.)

step 2: Draw a set of σ-τ axes.

step 3: Plot the center of the circle, C

Mohr's Circle—Stress, 2D

$$C = \frac{\sigma_x + \sigma_y}{2} \qquad 50.26$$

step 4: Plot the point $\mathbf{p}_1 = (\sigma_x, -\tau_{xy})$. (Alternatively, plot $\mathbf{p}_1'$ at $(\sigma_y, +\tau_{xy})$.)

step 5: Draw a line from point $\mathbf{p}_1$ through center C and extend it an equal distance beyond the σ-axis. This is the diameter of the circle.

step 6: Using the center C and point $\mathbf{p}_1$, draw the circle. An alternative method is to draw a circle of radius R about point C.

Mohr's Circle—Stress, 2D

$$R = \sqrt{\left(\frac{\sigma_x - \sigma_y}{2}\right)^2 + \tau_{xy}^2} \qquad 50.27$$

step 7: Point $\mathbf{p}_2$ defines the smaller principal stress, σ_2. Point $\mathbf{p}_3$ defines the larger principal stress, σ_1.

Mohr's Circle—Stress, 2D

$$\sigma_1 = C + R \qquad 50.28$$

$$\sigma_2 = C - R \qquad 50.29$$

step 8: Determine the angle θ as half of the angle 2θ on the circle. This angle corresponds to the larger principal stress, σ_1. On Mohr's circle, angle 2θ is measured counterclockwise from the $\mathbf{p}_1 - \mathbf{p}_1'$ line to the horizontal axis.

Example 50.3

Construct Mohr's circle for Ex. 50.2.

Solution

Mohr's Circle—Stress, 2D

$$C = \frac{\sigma_x + \sigma_y}{2} = \frac{20{,}000\ \frac{\text{lbf}}{\text{in}^2} + \left(-4000\ \frac{\text{lbf}}{\text{in}^2}\right)}{2} = 8000\ \text{lbf/in}^2$$

Mohr's Circle—Stress, 2D

$$R = \sqrt{\left(\frac{\sigma_x - \sigma_y}{2}\right)^2 + \tau_{xy}^2} = \sqrt{\left(\frac{20{,}000\ \frac{\text{lbf}}{\text{in}^2} - \left(-4000\ \frac{\text{lbf}}{\text{in}^2}\right)}{2}\right)^2 + \left(5000\ \frac{\text{lbf}}{\text{in}^2}\right)^2} = 13{,}000\ \text{lbf/in}^2$$

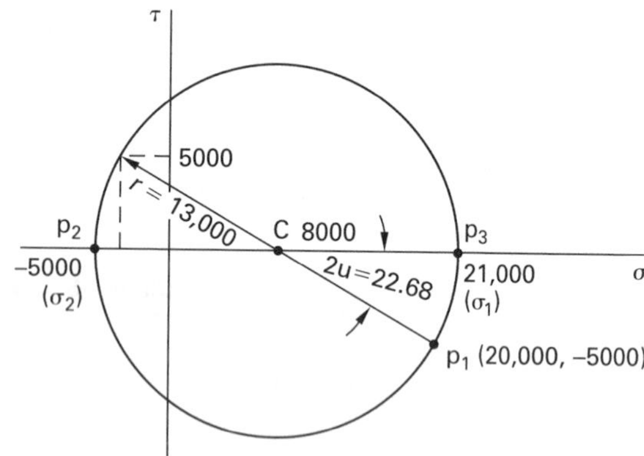

10. GENERAL STRAIN (THREE-DIMENSIONAL STRAIN)

Hooke's law, previously defined for axial loads and for pure shear, can be extended to three-dimensional stress-strain relationships and written in terms of the three elastic constants, E, G, and v. Equation 50.30 through Eq. 50.39 can be used to find the stresses and strains on the differential element in Figure 50.9.

Figure 50.9 *Sign Conventions for Positive Stresses in Three Dimensions*

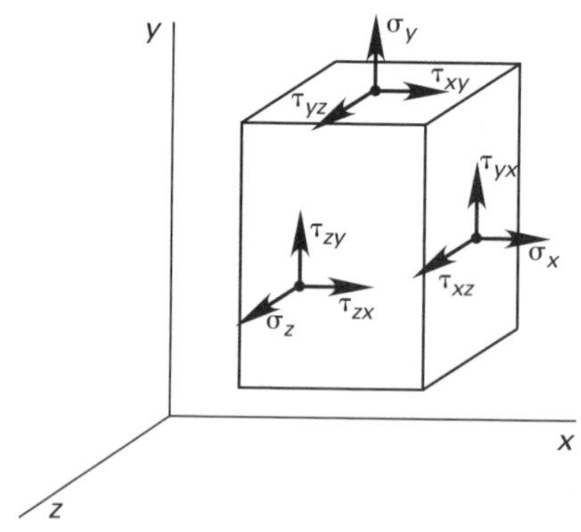

For the three-dimensional case (i.e., triaxial stress-strain),

Hooke's Law

$$\epsilon_x = \left(\frac{1}{E}\right)\left(\sigma_x - \nu(\sigma_y + \sigma_z)\right) \qquad 50.30$$

$$\epsilon_y = \left(\frac{1}{E}\right)\left(\sigma_y - \nu(\sigma_z + \sigma_x)\right) \qquad 50.31$$

$$\epsilon_z = \left(\frac{1}{E}\right)\left(\sigma_z - \nu(\sigma_x + \sigma_y)\right) \qquad 50.32$$

$$\gamma_{xy} = \frac{\tau_{xy}}{G} \qquad 50.33$$

$$\gamma_{yz} = \frac{\tau_{yz}}{G} \qquad 50.34$$

$$\gamma_{zx} = \frac{\tau_{zx}}{G} \qquad 50.35$$

For the plane stress case (i.e., $\sigma_z = 0$),

Hooke's Law

$$\epsilon_x = \left(\frac{1}{E}\right)(\sigma_x - \nu\sigma_y) \qquad 50.36$$

$$\epsilon_y = \left(\frac{1}{E}\right)(\sigma_y - \nu\sigma_x) \qquad 50.37$$

$$\epsilon_z = -\left(\frac{1}{E}\right)(\nu\sigma_x + \nu\sigma_y) \qquad 50.38$$

$$\begin{bmatrix}\sigma_x\\ \sigma_y\\ \tau_{xy}\end{bmatrix} = \frac{E}{1-\nu^2}\begin{bmatrix}1 & \nu & 0\\ \nu & 1 & 0\\ 0 & 0 & \frac{1-\nu}{2}\end{bmatrix}\begin{bmatrix}\epsilon_x\\ \epsilon_y\\ \gamma_{xy}\end{bmatrix} \qquad 50.39$$

11. IMPACT LOADING

If a load is applied to a structure suddenly, the structure's response will be composed of two parts: a transient response (which decays to zero) and a steady-state response. (These two parts are also known as the *dynamic* and *static responses*, respectively.) It is not unusual for the transient loading to be larger than the steady-state response.

Although a *dynamic analysis* of the structure is preferred, the procedure is lengthy and complex. Therefore, arbitrary multiplicative factors may be applied to the steady-state stress to determine the maximum transient. For example, if a load is applied quickly as compared to the natural period of vibration of the structure (e.g., the classic definition of an *impact load*), a dynamic factor of 2.0 might be used. Actual dynamic factors should be determined or validated by testing.

The energy-conservation method (i.e., the work-energy principle) can be used to determine the maximum stress due to a falling mass. The total change in potential energy of the mass from the change in elevation and the deflection, δ, is equated to the appropriate expression for total strain energy. (See Sec. 50.4.)

12. SHEAR AND MOMENT

Shear at a point is the sum of all vertical forces acting on an object. It has units of pounds, kips, tons, newtons, and so on. Shear is not the same as shear stress, since the area of the object is not considered.

A typical application is shear at a point on a beam, V, defined as the sum of all vertical forces between the point and one of the ends.[3] The direction (i.e., to the left or right of the point) in which the summation proceeds is not important. Since the values of shear will differ only in sign for summations to the left and right ends, the direction that results in the fewest calculations should be selected.

$$V = \sum_{\substack{\text{point to}\\\text{one end}}} F_i \qquad 50.40$$

Shear is taken as positive when there is a net upward force to the left of a point and negative when there is a net downward force between the point and the left end.

Moment at a point is the total bending moment acting on an object. In the case of a beam, the moment, M, will be the algebraic sum of all moments and couples located between the investigation point and one of the beam ends. As with shear, the number of calculations required to calculate the moment can be minimized by careful choice of the beam end.[4]

$$M = \sum_{\substack{\text{point to}\\\text{one end}}} F_i d_i + \sum_{\substack{\text{point to}\\\text{one end}}} C_i \qquad 50.41$$

Moment is taken as positive when the upper surface of the beam is in compression and the lower surface is in tension. Since the beam ends will usually be higher than the midpoint, it is commonly said that "a positive moment will make the beam smile."

13. SHEAR AND BENDING MOMENT DIAGRAMS

The value of the shear and moment, V and M, will depend on the location along the beam. Both shear and moment can be described mathematically for simple loadings, but the formulas are likely to become discontinuous as the loadings become more complex. It is much more convenient to describe the shear and moment functions graphically. Graphs of shear and moment as functions of position along the beam are known as *shear* and *moment diagrams*. Drawing these diagrams does not require knowing the shape or area of the beam.

The following guidelines and conventions should be observed when constructing a *shear diagram*.

- The shear at any point is equal to the sum of the loads and reactions from the point to the left end.
- The magnitude of the shear at any point is equal to the slope of the moment line at that point.

Beams

$$V = \frac{dM(x)}{dx} \qquad 50.42$$

- Loads and reactions acting upward are positive.
- The shear diagram is straight and sloping over uniformly distributed loads.
- The shear diagram is straight and horizontal between concentrated loads.
- The shear is a vertical line and is undefined at points of concentrated loads.

The following guidelines and conventions should be observed when constructing a *bending moment diagram*. By convention, the moment diagram is drawn on the compression side of the beam.

[3]The conditions of equilibrium require that the sum of all vertical forces on a beam be zero. However, the *shear* can be nonzero because only a portion of the beam is included in the analysis. Since that portion extends to the beam end in one direction only, the shear is sometimes called *resisting shear* or *one-way shear*.

[4]The conditions of equilibrium require that the sum of all moments on a beam be zero. However, the *moment* can be nonzero because only a portion of the beam is included in the analysis. Since that portion extends to the beam end in one direction only, the moment is sometimes called *bending moment*, *flexural moment*, *resisting moment*, or *one-way moment*.

- The moment at any point is equal to the sum of the moments and couples from the point to the left end.[5]
- Clockwise moments about the point are positive.
- The magnitude of the moment at any point is equal to the area under the shear line up to that point. This is equivalent to the integral of the shear function.

Beams

$$M_2 - M_1 = \int_{x_1}^{x_2} V(x)dx \qquad 50.43$$

- The *maximum moment* occurs where the shear is zero.
- The moment diagram is straight and sloping between concentrated loads.
- The moment diagram is curved (parabolic upward) over uniformly distributed loads.

These principles are illustrated in Fig. 50.10.

Figure 50.10 *Drawing Shear and Moment Diagrams*

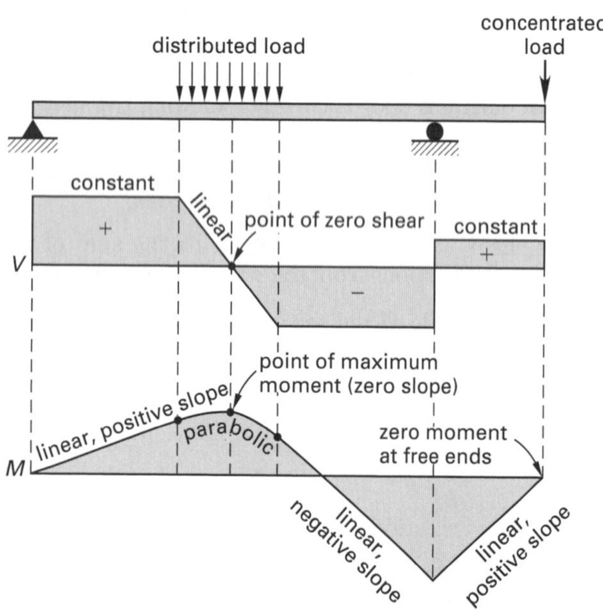

Example 50.4

Draw the shear and bending moment diagrams for the following beam.

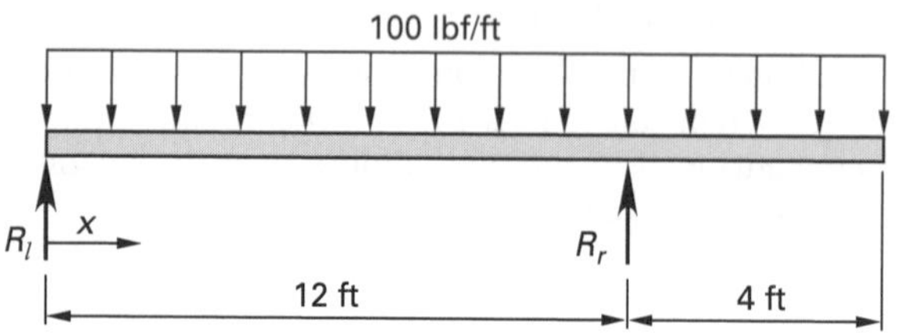

Solution

First, determine the reactions. Take moments about the left end. The uniform load of $100x$ can be assumed to be concentrated at $L/2$.

$$R_r = \frac{\frac{1}{2}wL^2}{x} = \frac{\left(\frac{1}{2}\right)(16\text{ ft})(16\text{ ft})\left(100\ \frac{\text{lbf}}{\text{ft}}\right)}{12\text{ ft}} = 1066.7\text{ lbf}$$

$$R_l = (16\text{ ft})\left(100\ \frac{\text{lbf}}{\text{ft}}\right) - R_r = 533.3\text{ lbf}$$

The shear diagram starts at +533.3 at the left reaction but decreases linearly at the rate of 100 lbf/ft between the two reactions. Measuring x from the left end, the shear line goes through zero at

$$x = \frac{533.3\text{ lbf}}{100\ \frac{\text{lbf}}{\text{ft}}} = 5.333\text{ ft}$$

The shear just to the left of the right reaction is

$$533.3\text{ lbf} - (12\text{ ft})\left(100\ \frac{\text{lbf}}{\text{ft}}\right) = -666.7\text{ lbf}$$

The shear just to the right of the right reaction is

$$-666.7\text{ lbf} + R_r = +400\text{ lbf}$$

To the right of the right reaction, the shear diagram decreases to zero at the same constant rate: 100 lbf/ft. This is sufficient information to draw the shear diagram.

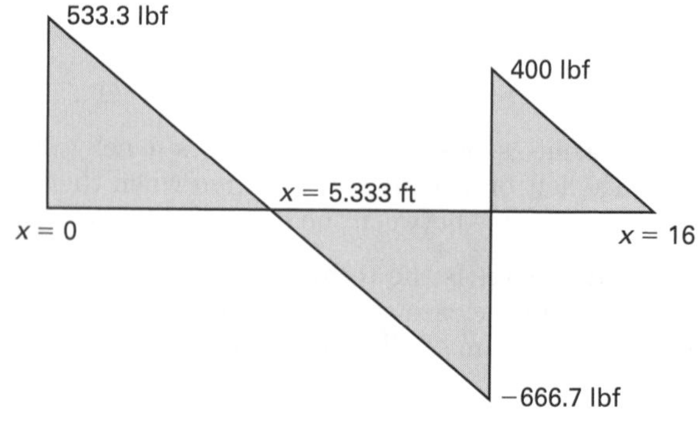

[5]If the beam is cantilevered with its built-in end at the left, the fixed-end moment will be unknown. In that case, the moment must be calculated to the right end of the beam.

The bending moment at a distance x to the right of the left end has two parts. The left reaction of 533.3 lbf acts with moment arm x. The moment between the two reactions is

$$M_x = 533.3x - 100x\left(\frac{x}{2}\right)$$

This equation describes a parabolic section (curved upward) with a peak at $x = 5.333$ ft, where the shear is zero. The maximum moment is

$$\begin{aligned} M_{x=5.333\,\text{ft}} &= (533.3\ \text{lbf})(5.333\ \text{ft}) - \left(50\ \frac{\text{lbf}}{\text{ft}}\right)(5.333\ \text{ft})^2 \\ &= 1422.0\ \text{ft-lbf} \end{aligned}$$

The moment at the right reaction (where $x = 12$ ft) is

$$\begin{aligned} M_{x=12\,\text{ft}} &= (533.3\ \text{lbf})(12\ \text{ft}) - \left(50\ \frac{\text{lbf}}{\text{ft}}\right)(12\ \text{ft})^2 \\ &= -800\ \text{ft-lbf} \end{aligned}$$

The right end is a free end, so the moment is zero. The moment between the right reaction and the right end could be calculated by summing moments to the left end, but it is more convenient to sum moments to the right end. Measuring x from the right end, the moment is derived only from the uniform load.

$$M = 100x\left(\frac{x}{2}\right) = 50x^2$$

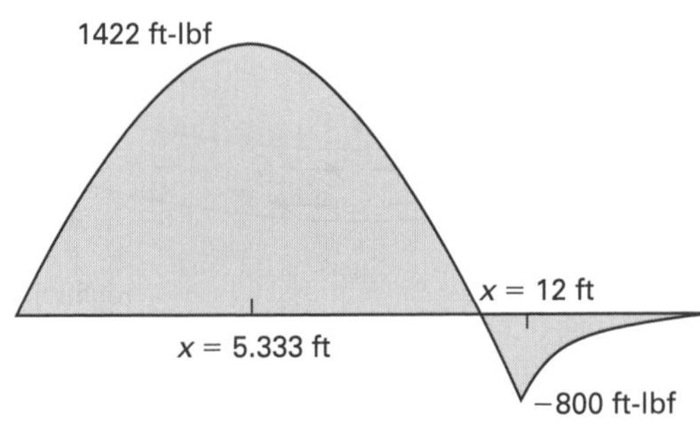

This is sufficient information to draw the moment diagram. Once the maximum moment is located, no attempt is made to determine the exact curvature. The point where $M = 0$ is of limited interest, and no attempt is made to determine the exact location.

Notice that the cross-sectional area of the beam was not needed in this example.

14. SHEAR STRESS IN BEAMS

Shear stress is generally not the limiting factor in most designs. However, it can control the design (or be limited by code) in wood, masonry, and concrete beams and in thin tubes.

The average shear stress experienced at a point along the length of a beam depends on the shear, V, at that point and the area, A, of the beam. The shear can be found from the shear diagram.

$$\tau = \frac{V}{A} \qquad 50.44$$

In most cases, the entire area, A, of the beam is used in calculating the average shear stress. However, in flanged beam calculations it is assumed that only the web carries the average shear stress.[6] (See Fig. 50.11.) The flanges are not included in shear stress calculations.

$$\tau = \frac{V}{A_{\text{web}}} = \frac{V}{t_w d} \qquad 50.45$$

Figure 50.11 *Web of a Flanged Beam*

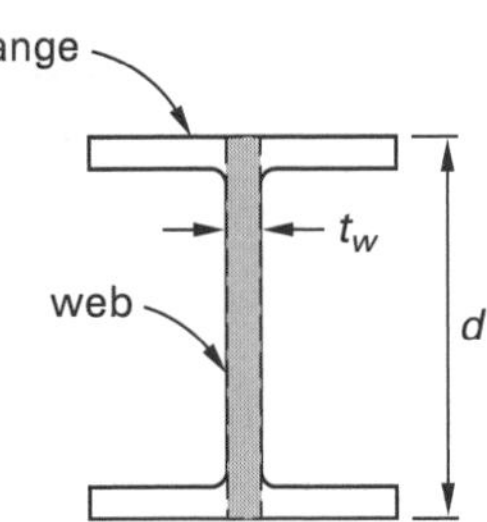

Shear stress is also induced in a beam due to flexure (i.e., bending). Figure 50.6 shows that for biaxial loading, identical shear stresses exist simultaneously in all four directions. One set of parallel shears (a couple) counteracts the rotational moment from the other set of parallel shears. The horizontal shear exists even when the loading is vertical (e.g., when a horizontal beam is loaded by a vertical force). For that reason, the term *horizontal shear* is sometimes used to distinguish it from the applied shear load.

The exact value of the horizontal shear stress is dependent on the location, y_1, within the depth of the beam. The shear stress distribution is given by Eq. 50.46. The shear stress is zero at the top and bottom surfaces of the beam and is usually maximum at the neutral axis (i.e., the center).

Stresses in Beams

$$\tau_{xy} = \frac{VQ}{Ib} \qquad 50.46$$

[6]This is more than an assumption; it is a fact. There are several reasons the flanges do not contribute to shear resistance, including a non-uniform shear stress distribution in the flanges. This nonuniformity is too complex to be analyzed by elementary methods.

In Eq. 50.46, V is the vertical shear at the point along the length of the beam where the shear stress is wanted. I is the beam's centroidal moment of inertia, and b is the width of the beam at the depth y_1 within the beam where the shear stress is wanted. Q is the *statical moment* of the area, as defined by Eq. 50.47.

$$Q = \int_{y_1}^{c} y\,dA \qquad 50.47$$

For rectangular beams, $dA = b\,dy$. Then, the statical moment of the area A' above layer y_1 is equal to the product of the area and the distance from the centroidal axis to the centroid of the area. (See Fig. 50.12.)

$$Q = A'y' \qquad 50.48$$

Figure 50.12 *Shear Stress Distribution Within a Rectangular Beam*

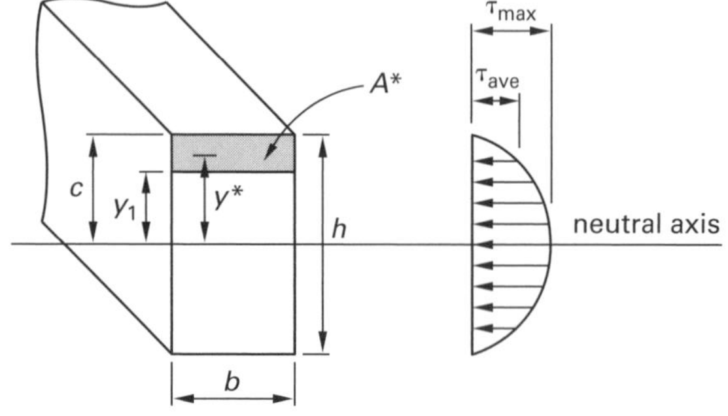

Equation 50.49 calculates the maximum shear stress in a rectangular beam. It is 50% higher than the average shear stress.

Formulas for Maximum Shear Stress Due to Bending

$$\tau_{\text{max,rectangular}} = \frac{3V}{2A} \qquad 50.49$$

For a beam with a circular cross section, the maximum shear stress is

Formulas for Maximum Shear Stress Due to Bending

$$\tau_{\text{max,circular}} = \frac{4V}{3A} \qquad 50.50$$

For a hollow cylinder used as a beam, the maximum shear stress occurs at the plane of the neutral axis and is

Formulas for Maximum Shear Stress Due to Bending

$$\tau_{\text{max,hollow cylinder}} = \frac{2V}{A} \qquad 50.51$$

15. SHEAR FLOW

The shear flow, q, is the shear per unit length, as found from Eq. 50.52. The vertical shear, V, is a function of location, x, along the beam, generally designated as V_x. This shear is resisted by the entire cross section, although the shear stress depends on the distance from the neutral axis. The shear stress is usually considered to be vertical (i.e., in the y-direction), in line with the shearing force, but the same shear stress that acts in the y-z plane also acts on the x-z plane.

Stresses in Beams

$$q = \frac{VQ}{I} \qquad 50.52$$

16. BENDING STRESS IN BEAMS

Normal stress occurs in a bending beam, as shown in Fig. 50.13, where the beam is acted upon by a *transverse force*. Although it is a normal stress, the term *bending stress* or *flexural stress* is used to indicate the cause of the stress. The lower surface of the beam experiences tensile stress (which causes lengthening). The upper surface of the beam experiences compressive stress (which causes shortening). There is no normal stress along a horizontal plane passing through the centroid of the cross section, a plane known as the *neutral plane* or the *neutral axis*.

Figure 50.13 *Normal Stress Due to Bending*

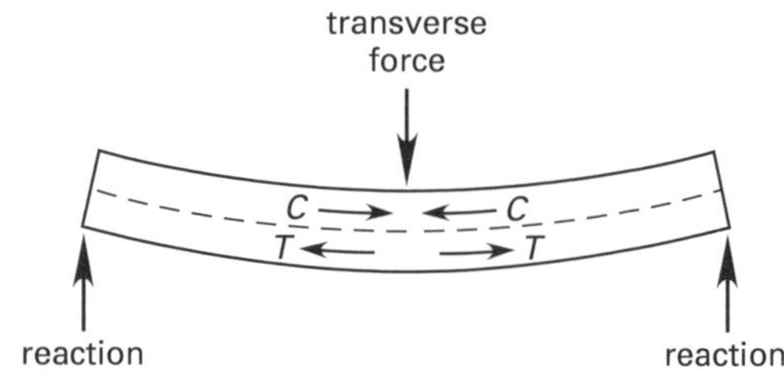

Bending stress varies with location (depth) within the beam. It is zero at the neutral axis and increases linearly with distance from the neutral axis, as predicted by Eq. 50.53. (See Fig. 50.14.)

Stresses in Beams

$$\sigma_x = -\frac{My}{I_c} \qquad 50.53$$

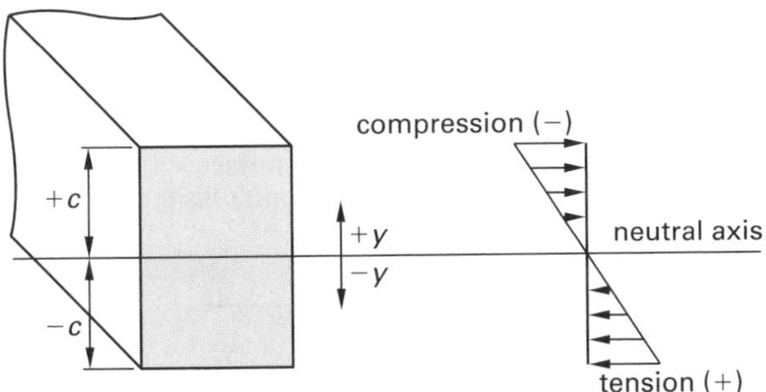

Figure 50.14 *Bending Stress Distribution in a Beam*

In Eq. 50.53, M is the *bending moment.* I_c is the centroidal moment of inertia of the beam's cross section. The negative sign in Eq. 50.53, required by the convention that compression is negative, is commonly omitted.

Since the maximum stress will govern the design, y can be set equal to c to obtain the *extreme fiber stress.* c is the distance from the neutral axis to the *extreme fiber* (i.e., the top or bottom surface most distant from the neutral axis).

Stresses in Beams

$$\sigma_x = \pm\frac{Mc}{I} \qquad 50.54$$

Equation 50.54 shows that the maximum bending stress will occur where the moment along the length of the beam is maximum. The region immediately adjacent to the point of maximum bending moment is called the *dangerous section* of the beam. The dangerous section can be found from a bending moment or shear diagram.

For any given beam cross section, I_c and c are fixed. Therefore, these two terms can be combined into the *section modulus, s.*[7]

Stresses in Beams

$$\sigma_x = -\frac{M}{S} \qquad 50.55$$

$$s = \frac{I_c}{c} \qquad 50.56$$

Since $c = h/2$, the section modulus of a rectangular $b \times h$ section ($I = bh^3/12$) is

$$s_{\text{rectangular}} = \frac{bh^2}{6} \qquad 50.57$$

Example 50.5

The beam in Ex. 50.4 has a 6 in × 8 in cross section. What are the maximum shear and bending stresses in the beam?

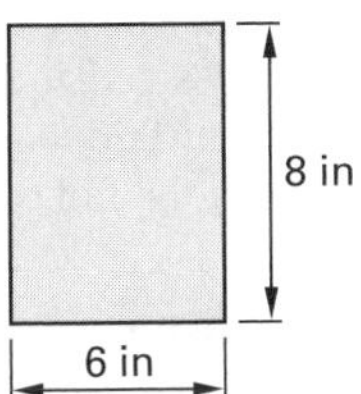

Solution

The maximum shear (taken from the shear diagram) is 666.7 lbf. (The negative sign can be disregarded.) From Eq. 50.49, the maximum shear stress in a rectangular beam is

Stresses in Beams

$$\tau_{\text{max}} = \frac{3V}{2A} = \frac{(3)(666.7\ \text{lbf})}{(2)(6\ \text{in})(8\ \text{in})} = 20.8\ \text{lbf/in}^2$$

The centroidal moment of inertia is

$$I_c = \frac{bh^3}{12} = \frac{(6\ \text{in})(8\ \text{in})^3}{12} = 256\ \text{in}^4$$

The maximum bending moment (from the bending moment diagram) is 1422 ft-lbf. From Eq. 50.54, the maximum bending stress is

Stresses in Beams

$$\begin{aligned}\sigma_{x,\text{max}} &= \frac{Mc}{I_c} = \frac{(1422\ \text{ft-lbf})\left(12\ \frac{\text{in}}{\text{ft}}\right)(4\ \text{in})}{256\ \text{in}^4} \\ &= 266.6\ \text{lbf/in}^2\end{aligned}$$

17. STRAIN ENERGY DUE TO BENDING MOMENT

The elastic strain energy due to a bending moment stored in a beam is

$$U = \frac{1}{2EI}\int M^2(x)\,dx \qquad 50.58$$

The use of Eq. 50.58 is illustrated by Ex. 50.10.

[7]The symbol Z is also commonly used for the section modulus, particularly when referring to the plastic section modulus.

18. ECCENTRIC LOADING OF AXIAL MEMBERS

If a load is applied through the centroid of a tension or compression member's cross section, the loading is said to be *axial loading* or *concentric loading*. *Eccentric loading* occurs when the load is not applied through the centroid.

If an axial member is loaded eccentrically, it will bend and experience bending stress in the same manner as a beam. Since the member experiences both axial stress and bending stress, it is known as a *beam-column*. In Fig. 50.15, e is known as the *eccentricity*.

Figure 50.15 *Eccentric Loading of an Axial Member*

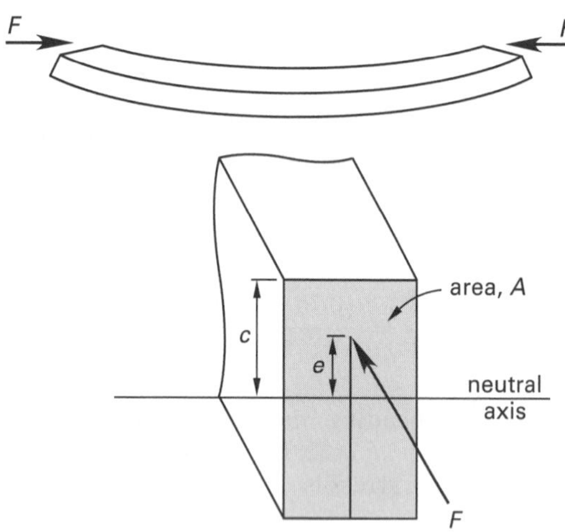

Both the axial stress and bending stress are normal stresses oriented in the same direction; therefore, simple addition can be used to combine them. Combined stress theory is not applicable. By convention, F is negative if the force compresses the member, which is shown in Fig. 50.15.

$$\sigma_{\text{max,min}} = \frac{F}{A} \pm \frac{Mc}{I_c} \qquad 50.59$$

$$\sigma_{\text{max,min}} = \frac{F}{A} \pm \frac{Fec}{I_c} \qquad 50.60$$

If a pier or column (primarily designed as a compression member) is loaded with an eccentric compressive load, part of the section can still be placed in tension. (See Fig. 50.16.) Tension will exist when the Mc/I_c term in Eq. 50.59 is larger than the F/A term. It is particularly important to eliminate or severely limit tensile stresses in unreinforced concrete and masonry piers, since these materials cannot support tension.

Figure 50.16 *Tension in a Pier*

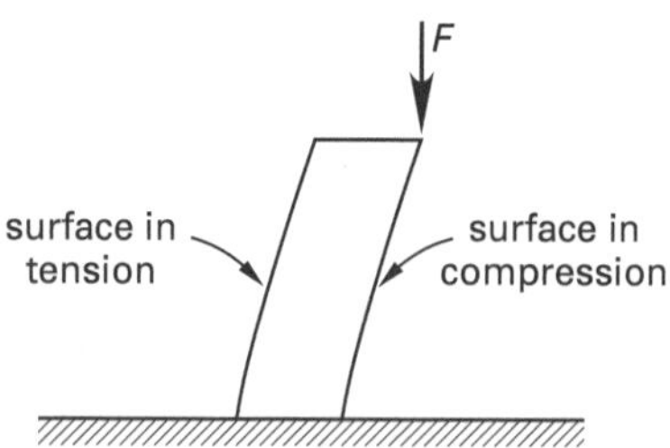

Regardless of the magnitude of the load, there will be no tension as long as the eccentricity is low. In a rectangular member, the load must be kept within a rhombus-shaped area formed from the middle thirds of the centroidal axes. This area is known as the *core*, *kern*, or *kernel*. Figure 50.17 illustrates the kerns for other cross sections.

Example 50.6

A built-in hook with a cross section of 1 in × 1 in carries a load of 500 lbf, but the load is not in line with the centroidal axis of the hook's neck. What are the minimum and maximum stresses in the neck? Is the neck in tension everywhere?

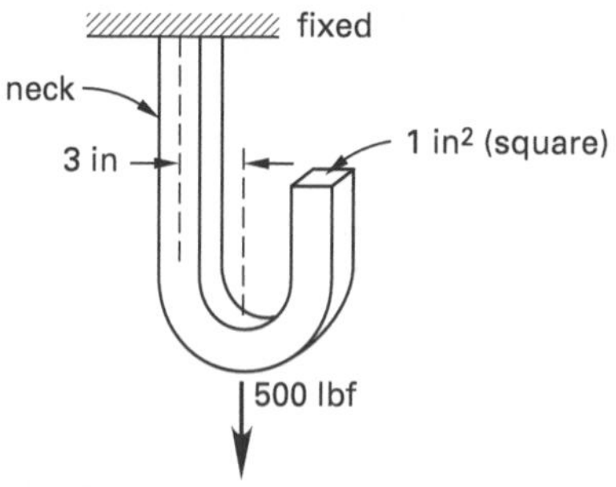

Solution

The centroidal moment of inertia of a 1 in × 1 in section is

$$I_c = \frac{bh^3}{12} = \frac{(1 \text{ in})(1 \text{ in})^3}{12} = 0.0833 \text{ in}^4$$

The hook is eccentrically loaded with an eccentricity of 3 in. From Eq. 50.60, the total stress is the sum of the direct axial tension and the bending stress. As the hook bends to reduce the eccentricity, the inner face of the neck will experience a tensile bending stress. The outer face of the neck will experience a compressive bending stress.

$$\begin{aligned}\sigma_{\text{max,min}} &= \frac{F}{A} \pm \frac{Fec}{I_c} \\ &= \frac{500 \text{ lbf}}{1 \text{ in}^2} \pm \frac{(500 \text{ lbf})(3 \text{ in})(0.5 \text{ in})}{0.0833 \text{ in}^4} \\ &= 500 \; \frac{\text{lbf}}{\text{in}^2} \pm 9000 \; \frac{\text{lbf}}{\text{in}^2} \\ &= +9500 \text{ lbf/in}^2, \; -8500 \text{ lbf/in}^2\end{aligned}$$

Figure 50.17 Kerns of Common Cross Sections

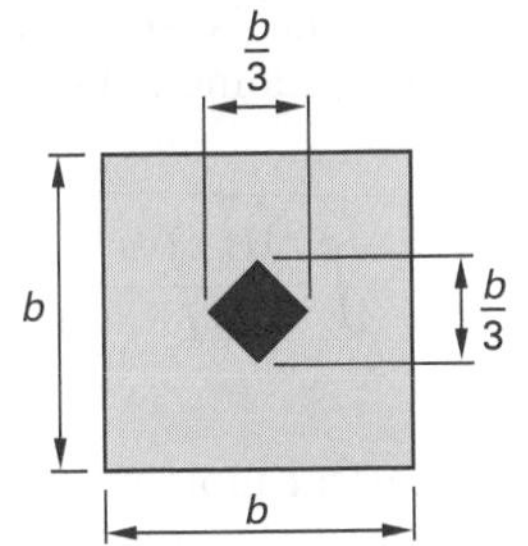

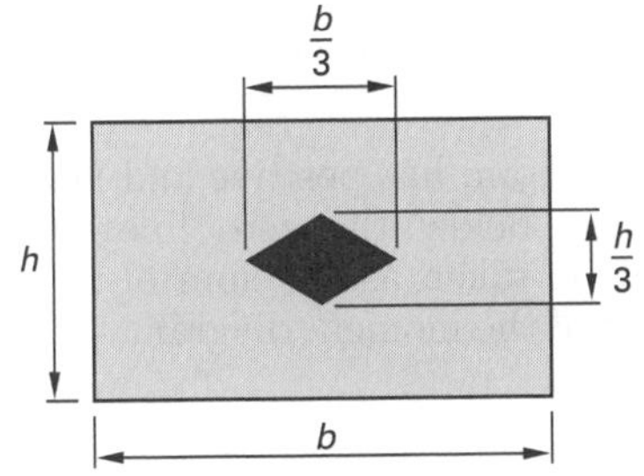

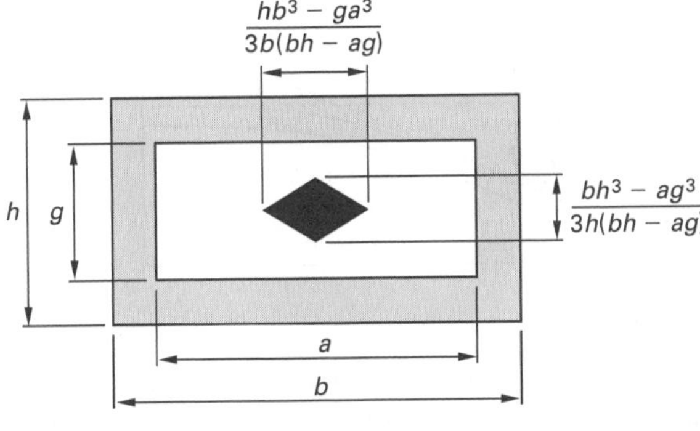

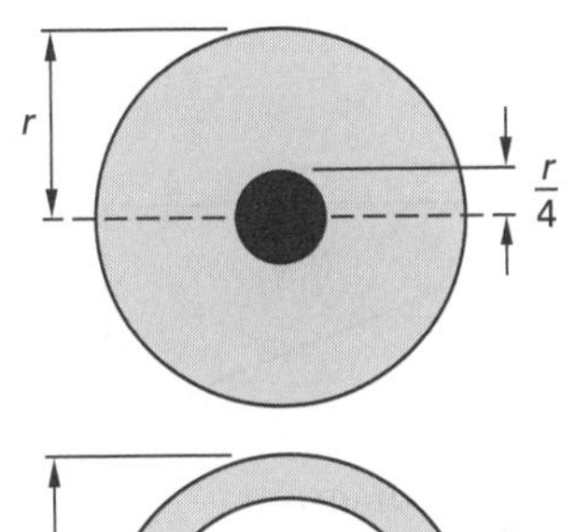

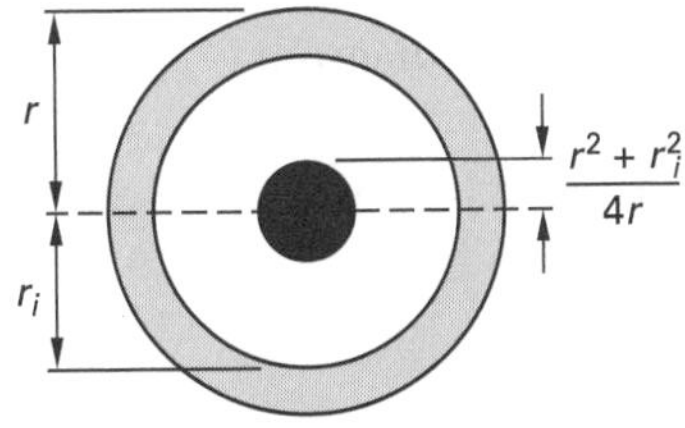

The 500 lbf/in^2 direct stress is tensile, and the inner face experiences a total tensile stress of 9500 lbf/in^2. However, the compressive bending stress of 9000 lbf/in^2 counteracts the direct tensile stress, resulting in an 8500 lbf/in^2 compressive stress at the outer face of the neck.

19. BEAM DEFLECTION: DOUBLE INTEGRATION METHOD

The deflection and the slope of a loaded beam are related to the moment and shear by Eq. 50.61 through Eq. 50.65.

$$y = \text{deflection} \qquad 50.61$$

$$y' = \frac{dy}{dx} = \text{slope} \qquad 50.62$$

$$y'' = \frac{d^2y}{dx^2} = \frac{M(x)}{EI} \qquad 50.63$$

$$y''' = \frac{d^3y}{dx^3} = \frac{V(x)}{EI} \qquad 50.64$$

If the *moment function*, $M(x)$, is known for a section of the beam, the deflection at any point can be found from Eq. 50.65.

$$y = \frac{1}{EI}\int\left(\int M(x)\,dx\right)dx \qquad 50.65$$

In order to find the deflection, constants must be introduced during the integration process. For some simple configurations, these constants can be found from Table 50.3.

Table 50.3 *Beam Boundary Conditions*

end condition	y	y'	y''	V	M
simple support	0				0
built-in support	0	0			
free end			0	0	0
hinge					0

Example 50.7

Find the tip deflection of the beam shown. EI is 5×10^{10} lbf-in^2 everywhere.

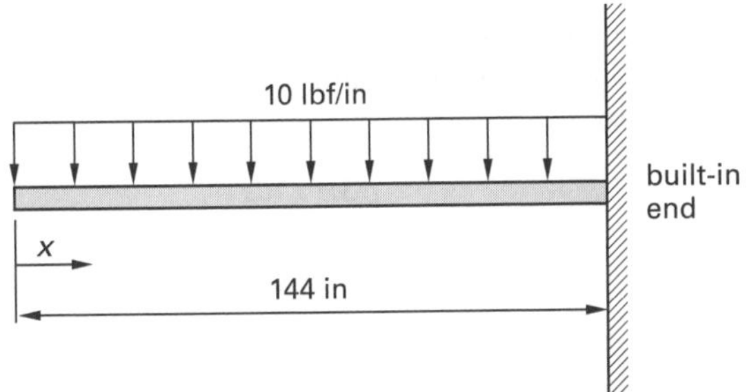

Solution

The moment at any point x from the left end of the beam is

$$M(x) = -10x\left(\tfrac{1}{2}x\right) = -5x^2$$

This is negative by the left-hand rule convention. From Eq. 50.63,

$$y'' = \frac{M(x)}{EI}$$
$$EIy'' = M(x) = -5x^2$$
$$EIy' = \int -5x^2 dx = -\tfrac{5}{3}x^3 + C_1$$

Since $y' = 0$ at a built-in support (from Table 50.3) and $x = 144$ in at the built-in support,

$$0 = \left(-\frac{5}{3}\right)(144)^3 + C_1$$
$$C_1 = 4.98 \times 10^6$$
$$EIy = \int (-\tfrac{5}{3}x^3 + 4.98 \times 10^6)\,dx$$
$$= -\tfrac{5}{12}x^4 + (4.98 \times 10^6)x + C_2$$

Again, $y = 0$ at $x = 144$ in, so $C_2 = -5.38 \times 10^8$ lbf-in³. Therefore, the deflection as a function of x is

$$y = \frac{1}{EI}\left(-\tfrac{5}{12}x^4 + (4.98 \times 10^6)x - 5.38 \times 10^8\right)$$

At the tip $x = 0$, so the deflection is

$$y_{\text{tip}} = \frac{-5.38 \times 10^8 \text{ lbf-in}^3}{5 \times 10^{10} \text{ lbf-in}^2} = -0.0108 \text{ in}$$

20. BEAM DEFLECTION: MOMENT AREA METHOD

The moment area method is a semigraphical technique that is applicable whenever slopes of deflection beams are not too great. This method is based on the following two theorems.

- *Theorem I:* The angle between tangents at any two points on the *elastic line* of a beam is equal to the area of the moment diagram between the two points divided by *EI*.

$$\phi = \int \frac{M(x)\,dx}{EI} \qquad 50.66$$

- *Theorem II:* One point's deflection away from the tangent of another point is equal to the *statical moment* of the bending moment between those two points divided by *EI*.

$$y = \int \frac{xM(x)\,dx}{EI} \qquad 50.67$$

If *EI* is constant, the statical moment $\int xM(x)\,dx$ can be calculated as the product of the total moment diagram area times the horizontal distance from the point whose deflection is wanted to the centroid of the moment diagram.

If the moment diagram has positive and negative parts (areas above and below the zero line), the statical moment should be taken as the sum of two products, one for each part of the moment diagram.

Example 50.8

Find the deflection, y, and the angle, ϕ, at the free end of the cantilever beam shown. Neglect the beam weight.

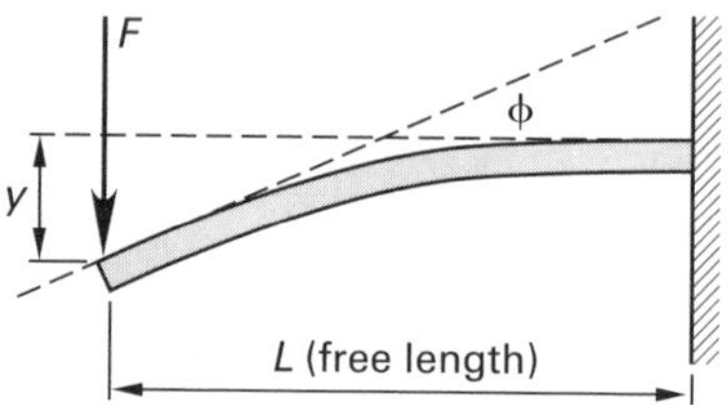

Solution

The deflection angle, ϕ, is the angle between the tangents at the free and built-in ends (Theorem I). The moment diagram is

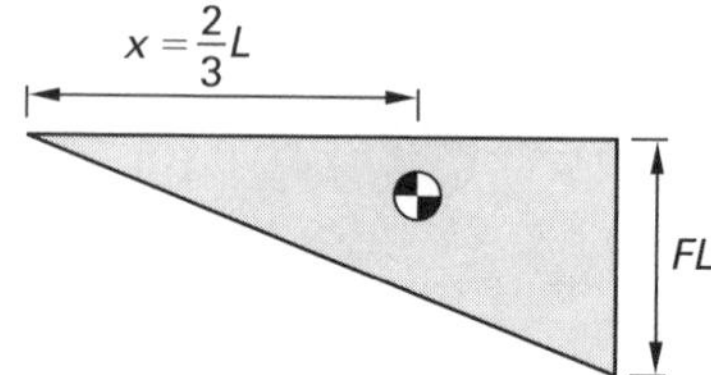

The area of the moment diagram is

$$\int M(x)\,dx = \tfrac{1}{2}(FL)(L) = \tfrac{1}{2}FL^2$$

From Eq. 50.66,

$$\phi = \int \frac{M(x)\,dx}{EI} = \frac{FL^2}{2EI}$$

The centroid of the moment diagram is located at $x = \frac{2}{3}L$. From Eq. 50.67,

$$y = \int \frac{xM(x)\,dx}{EI} = \left(\frac{FL^2}{2EI}\right)\left(\tfrac{2}{3}L\right) = \frac{FL^3}{3EI}$$

Example 50.9

Find the deflection of the free end of the cantilever beam shown. Neglect the beam weight.

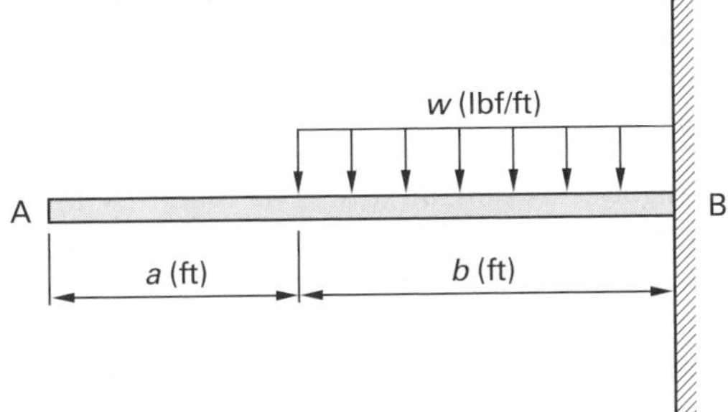

Solution

The distance from point A (where the deflection is wanted) to the centroid is $a + 0.75b$. The area of the moment diagram is $wb^3/6$. From Eq. 50.67,

$$y = \int \frac{xM(x)\,dx}{EI} = \left(\frac{wb^3}{6EI}\right)(a + 0.75b)$$

21. BEAM DEFLECTION: STRAIN ENERGY METHOD

The deflection at a point of load application can be found by the strain energy method. This method uses the work-energy principle and equates the external work to the total internal strain energy. Since work is a force moving through a distance (which in this case is the deflection), Eq. 50.68 holds true.

$$\tfrac{1}{2}Fy = \sum U \qquad 50.68$$

Example 50.10

Find the deflection at the tip of the stepped beam shown.

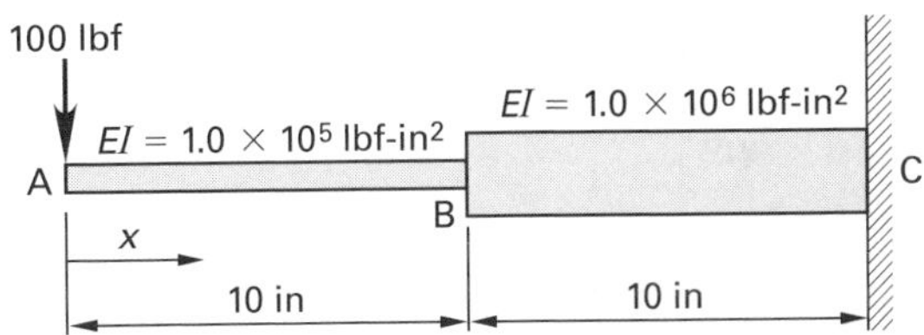

Solution

In section AB, $M(x) = 100x$ in-lbf.

From Eq. 50.58,

$$\begin{aligned} U &= \frac{1}{2EI}\int M^2(x)\,dx \\ &= \frac{1}{(2)(1\times 10^5\ \text{lbf-in}^2)}\int_{0\ \text{in}}^{10\ \text{in}} (100x)^2 \\ &= 16.67\ \text{in-lbf} \end{aligned}$$

In section BC, $M = 100x$.

$$U = \frac{1}{(2)(1\times 10^6\ \text{lbf-in}^2)}\int_{10\ \text{in}}^{20\ \text{in}} (100x)^2 = 11.67\ \text{in-lbf}$$

Equating the internal work, U, and the external work,

$$\begin{aligned} \sum U &= W \\ 16.67\ \text{in-lbf} + 11.67\ \text{in-lbf} &= \left(\frac{1}{2}\right)(100\ \text{lbf})y \\ y &= 0.567\ \text{in} \end{aligned}$$

22. BEAM DEFLECTION: CONJUGATE BEAM METHOD

The *conjugate beam method* changes a deflection problem into one of drawing moment diagrams. The method has the advantage of being able to handle beams of varying cross sections (e.g., stepped beams) and materials. It has the disadvantage of not easily being able to handle beams with two built-in ends. The following steps constitute the conjugate beam method.

step 1: Draw the moment diagram for the beam as it is actually loaded.

step 2: Construct the M/EI diagram by dividing the value of M at every point along the beam by EI at that point. If the beam has a constant cross section, EI will be constant, and the M/EI diagram will have the same shape as the moment diagram. However, if the beam cross section varies with x, I will change. In that case, the

Materials

M/EI diagram will not look the same as the moment diagram.

step 3: Draw a conjugate beam of the same length as the original beam. The material and the cross-sectional area of this conjugate beam are not relevant.

(a) If the actual beam is simply supported at its ends, the conjugate beam will be simply supported at its ends.

(b) If the actual beam is simply supported away from its ends, the conjugate beam has hinges at the support points.

(c) If the actual beam has free ends, the conjugate beam has built-in ends.

(d) If the actual beam has built-in ends, the conjugate beam has free ends.

step 4: Load the conjugate beam with the M/EI diagram. Find the conjugate reactions by methods of statics. Use the superscript * to indicate conjugate parameters.

step 5: Find the conjugate moment at the point where the deflection is wanted. The deflection is numerically equal to the moment as calculated from the conjugate beam forces.

Example 50.11

Find the deflections at the two load points. EI has a constant value of 2.356×10^7 lbf-in^2.

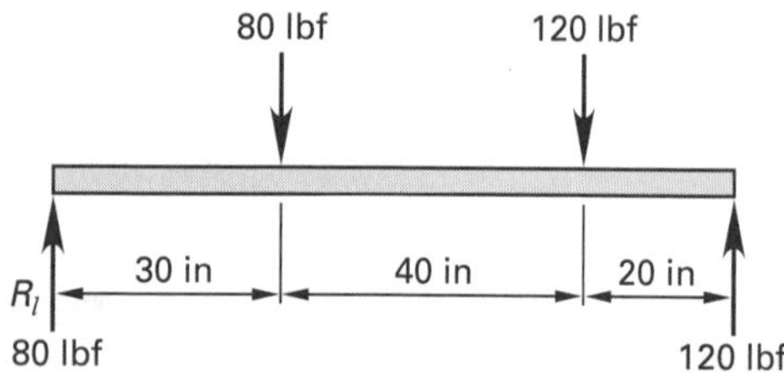

Solution

Applying step 1, the moment diagram for the actual beam is

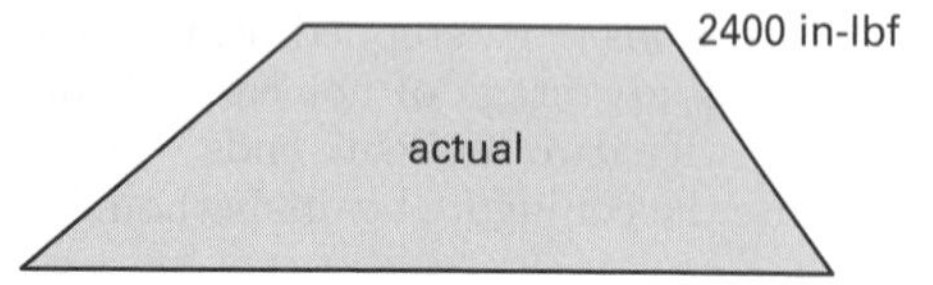

Applying steps 2, 3, and 4, since the beam cross section is constant, the conjugate load has the same shape as the original moment diagram. The peak load on the conjugate beam is

$$\frac{M}{EI} = \frac{2400 \text{ in-lbf}}{2.356 \times 10^7 \text{ lbf-in}^2} = 1.019 \times 10^{-4} \text{ 1/in}$$

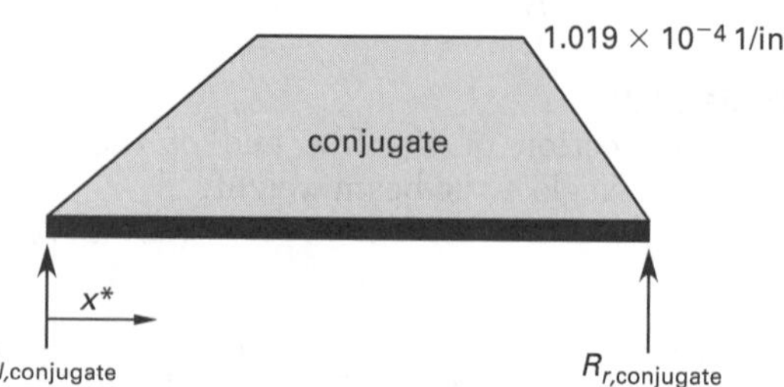

The conjugate reaction, $R_{l,\text{conjugate}}$, is found by the following method. The loading diagram is assumed to be made up of a rectangular load and two negative triangular loads. The area of the rectangular load (which has a centroid at $x_{\text{conjugate}} = 45$ in) is taken as $(90 \text{ in})(1.019 \times 10^{-4} \text{ 1/in}) = 9.171 \times 10^{-3}$.

Similarly, the area of the left triangle (which has a centroid at $x_{\text{conjugate}} = 10$ in) is $(0.5)(30 \text{ in})(1.019 \times 10^{-4} \text{ 1/in}) = 1.529 \times 10^{-3}$. The area of the right triangle (which has a centroid at $x_{\text{conjugate}} = 83.33$ in) is taken as $(0.5)(20 \text{ in})(1.019 \times 10^{-4} \text{ 1/in}) = 1.019 \times 10^{-3}$.

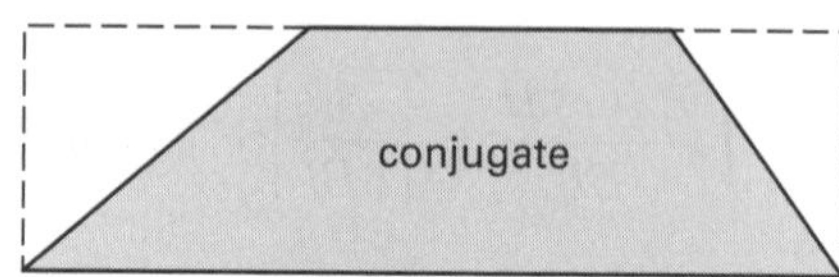

$$\begin{aligned} \sum M_l^* &= (90 \text{ in}) R_{r,\text{conjugate}} \\ &\quad + (1.019 \times 10^{-3})(83.3 \text{ in}) \\ &\quad + (1.529 \times 10^{-3})(10 \text{ in}) \\ &\quad - (9.171 \times 10^{-3})(45 \text{ in}) \\ &= 0 \\ R_{r,\text{conjugate}} &= 3.472 \times 10^{-3} \end{aligned}$$

Then,

$$\begin{aligned} R_{l,\text{conjugate}} &= (9.171 - 1.019 - 1.529 - 3.472) \times 10^{-3} \\ &= 3.151 \times 10^{-3} \end{aligned}$$

Materials

In step 5, the conjugate moment at $x_{\text{conjugate}} = 30$ in is the deflection of the actual beam at that point.

$$\begin{aligned} M_{\text{conjugate}} &= (3.151 \times 10^{-3})(30 \text{ in}) \\ &\quad + (1.529 \times 10^{-3})(30 \text{ in} - 10 \text{ in}) \\ &\quad - (9.171 \times 10^{-3})\left(\frac{30 \text{ in}}{90 \text{ in}}\right)(15 \text{ in}) \\ &= 7.926 \times 10^{-2} \text{ in} \end{aligned}$$

The conjugate moment (the deflection) at the rightmost load is

$$\begin{aligned} M_{\text{conjugate}} &= (3.472 \times 10^{-3})(20 \text{ in}) \\ &\quad + (1.019 \times 10^{-3})(13.3 \text{ in}) \\ &\quad - (9.171 \times 10^{-3})\left(\frac{20 \text{ in}}{90 \text{ in}}\right)(10 \text{ in}) \\ &= 6.261 \times 10^{-2} \text{ in} \end{aligned}$$

23. BEAM DEFLECTION: SUPERPOSITION

When multiple loads act simultaneously on a beam, all of the loads contribute to deflection. The principle of *superposition* permits the deflections at a point to be calculated as the sum of the deflections from each individual load acting singly.[8] This principle is valid as long as none of the deflections is excessive and all stresses are kept less than the yield point of the beam material.

24. BEAM DEFLECTION: TABLE LOOKUP METHOD

Most commonly used beam deflection formulas can be found from tables. [**Bending Moment, Vertical Shear, and Deflection of Beams of Uniform Cross Section, Under Various Conditions of Loading**]

The actual deflection of very *wide beams* (i.e., those whose widths are larger than approximately 8 or 10 times the thickness) is less than that predicted by typical equations for elastic behavior. (This is particularly true for leaf springs.) [**Bending Moment, Vertical Shear, and Deflection of Beams of Uniform Cross Section, Under Various Conditions of Loading**]

The large width prevents lateral expansion and contraction of the beam material, reducing the deflection. For wide beams, the calculated deflection should be reduced by multiplying by $(1 - \nu^2)$.

25. INFLECTION POINTS

The *inflection point* (also known as a *point of contraflexure*) on a horizontal beam in elastic bending occurs where the curvature changes from concave up to concave down, or vice versa. There are three ways of determining the inflection point.

1. If the elastic deflection equation, $y(x)$, is known, the inflection point can be found from differentiation (i.e., by determining the value of x for which $y'(x) = M(x) = 0$).

2. From Eq. 50.63, $y''(x) = M(x)/EI$. $y''(x)$ is also the reciprocal of the *radius of curvature*, ρ, of the beam.

$$\begin{aligned} y''(x) &= \frac{1}{\rho(x)} \\ &= \frac{M(x)}{EI} \end{aligned} \qquad 50.69$$

   Since the flexural rigidity, EI, is always positive, the radius of curvature, $\rho(x)$, changes sign when the moment equation, $M(x)$, changes sign.

3. If a shear diagram is known, the inflection point can sometimes be found by noting the point at which the positive and negative shear areas on either side of the point balance.

26. TRUSS DEFLECTION: STRAIN ENERGY METHOD

The deflection of a truss at the point of a single load application can be found by the *strain energy method* if all member forces are known. This method is illustrated by Ex. 50.12.

Example 50.12

Find the vertical deflection of point A under the external load of 707 lbf. $AE = 10^6$ lbf for all members. The internal forces have already been determined.

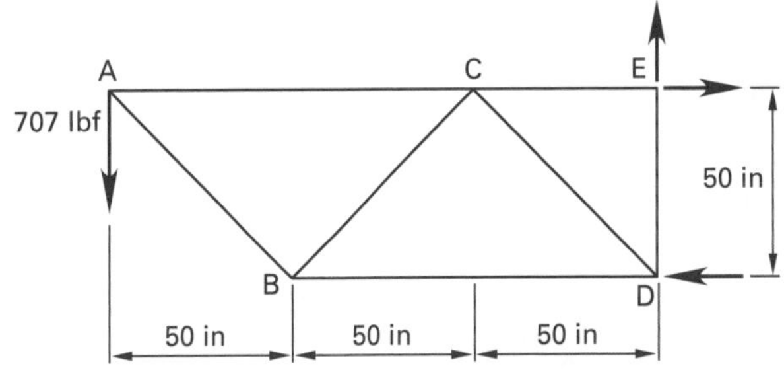

[8]The principle of superposition is not limited to deflections. It can also be used to calculate the shear and moment at a point and to draw the shear and moment diagrams.

Materials

Solution

The length of member AB is $\sqrt{(50\text{ in})^2 + (50\text{ in})^2} = 70.7$ in. From Eq. 50.8, the internal strain energy in member AB is

$$U = \frac{F^2 L_o}{2AE} = \frac{(-1000\text{ lbf})^2 (70.7\text{ in})}{(2)(10^6\text{ lbf})} = 35.4\text{ in-lbf}$$

Similarly, the energy in all members can be determined.

member	L (in)	F (lbf)	U (in-lbf)
AB	70.7	−1000	+35.4
BC	70.7	+1000	+35.4
AC	100	+707	+25.0
BD	100	−1414	+100.0
CD	70.7	−1000	+35.4
CE	50	+2121	+112.5
DE	50	+707	+12.5
			356.2

The work done by a constant force, F, moving through a distance, y, is Fy. In this case, the force increases with y. The average force is $\frac{1}{2}F$. The external work is $W_{\text{ext}} = \left(\frac{1}{2}\right)(707\text{ lbf})y$, so

$$\left(\frac{1}{2}\right)(707\text{ lbf})y = 356.2\text{ in-lbf}$$

$$y = 1\text{ in}$$

27. TRUSS DEFLECTION: VIRTUAL WORK METHOD

The *virtual work method* (also known as the *unit load method*) is an extension of the strain energy method. It can be used to determine the deflection of any point on a truss.

step 1: Draw the truss twice.

step 2: On the first truss, place all the actual loads.

step 3: Find the forces, S, due to the actual applied loads in all the members.

step 4: On the second truss, place a dummy one-unit load, f, in the direction of the desired displacement.

step 5: Find the forces, u, due to the one-unit dummy load in all members.

step 6: Find the desired displacement from Eq. 50.70. The summation is over all truss members that have nonzero forces in *both* trusses.

$$f\delta = \sum \frac{SuL}{AE} \quad [f = 1] \qquad 50.70$$

Example 50.13

What is the horizontal deflection of joint F on the truss shown? Use $E = 3 \times 10^7$ lbf/in^2. Joint A is restrained horizontally. Member lengths and areas are listed in the accompanying table.

Solution

Applying steps 1 and 2, use the truss as drawn.

Applying step 3, the forces in all the truss members are summarized in step 5.

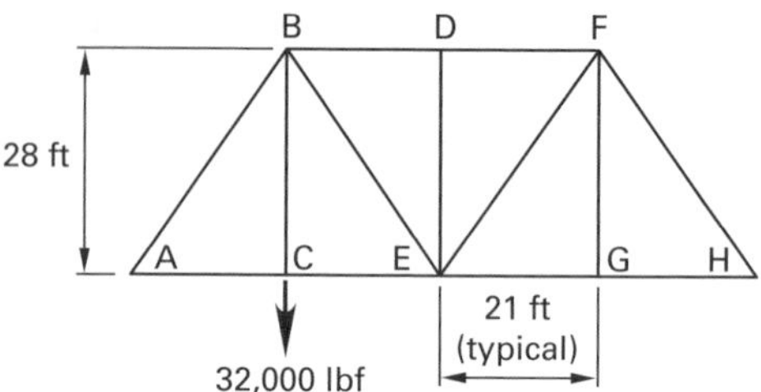

Applying step 4, draw the truss and load it with a unit horizontal force at point F.

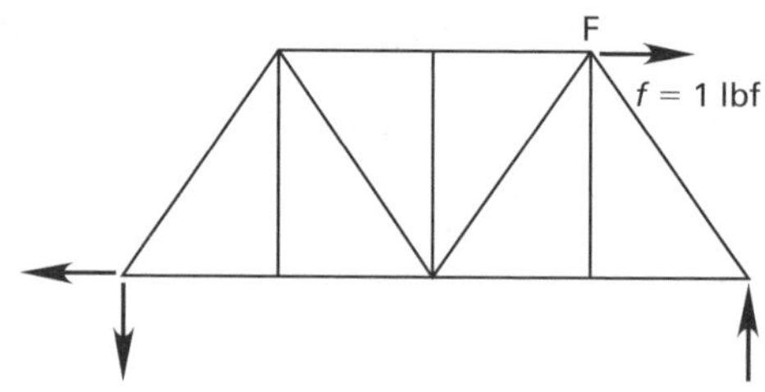

Applying step 5, find the forces, u, in all members of the second truss. These are summarized in the following table. Notice the sign convention: + for tension and − for compression.

member	S (lbf)	u (lbf)	L (ft)	A (in^2)	$\frac{SuL}{AE}$ (ft-lbf)
AB	−30,000	5/12	35	17.5	-8.33×10^{-4}
CB	32,000	0	28	14	0
EB	−10,000	−5/12	35	17.5	2.78×10^{-4}
ED	0	0	28	14	0
EF	10,000	5/12	35	17.5	2.78×10^{-4}
GF	0	0	28	14	0
HF	−10,000	−5/12	35	17.5	2.78×10^{-4}
BD	−12,000	1/2	21	10.5	-4.00×10^{-4}
DF	−12,000	1/2	21	10.5	-4.00×10^{-4}
AC	18,000	3/4	21	10.5	9.00×10^{-4}
CE	18,000	3/4	21	10.5	9.00×10^{-4}
EG	6000	1/4	21	10.5	1.00×10^{-4}
GH	6000	1/4	21	10.5	1.00×10^{-4}
					12.01×10^{-4}

Materials

Applying step 6, the deflection is conveniently calculated by summing the last column in the table. Since 12.01×10^{-4} is positive, the deflection is in the direction of the dummy unit load. In this case, the deflection is to the right.

28. MODES OF BEAM FAILURE

Beams can fail in different ways, including excessive deflection, local buckling, lateral buckling, and rotation.

Excessive deflection occurs when a beam bends more than a permitted amount.[9] The deflection is elastic and no yielding occurs. For this reason, the failure mechanism is sometimes called *elastic failure*. Although the beam does not yield, the excessive deflection may cause cracks in plaster and sheetrock, misalignment of doors and windows, and occupant concern and reduction of confidence in the structure.

Local buckling is an overload condition that occurs near large concentrated loads. Such locations include where a column frames into a supporting girder or a reaction point. *Vertical buckling* and *web crippling*, two types of local buckling, can be eliminated by use of *stiffeners*. Such stiffeners can be referred to as *intermediate stiffeners*, *bearing stiffeners*, *web stiffeners*, or *flange stiffeners*, depending on the location and technique of stiffening. (See Fig. 50.18.)

Lateral buckling, such as illustrated in Fig. 50.19, occurs when a long, unsupported member rolls out of its normal plane. To prevent lateral buckling, either the beam's compression flange must be supported continuously or at frequent intervals along its length, or the beam must be restrained against twisting about its longitudinal axis.

Figure 50.18 *Local Buckling and Stiffeners*

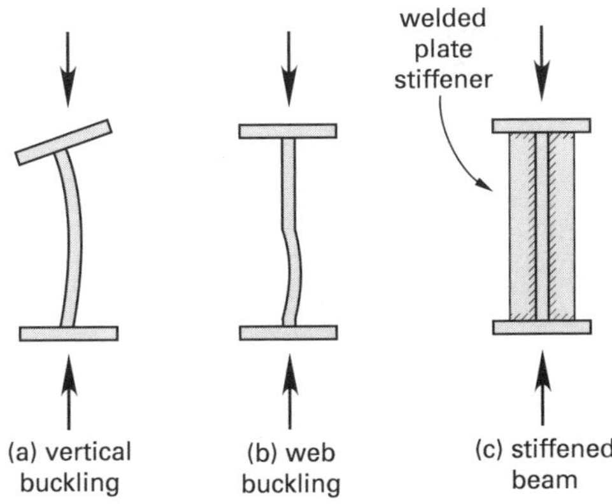

Figure 50.19 *Lateral Buckling and Flange Support*

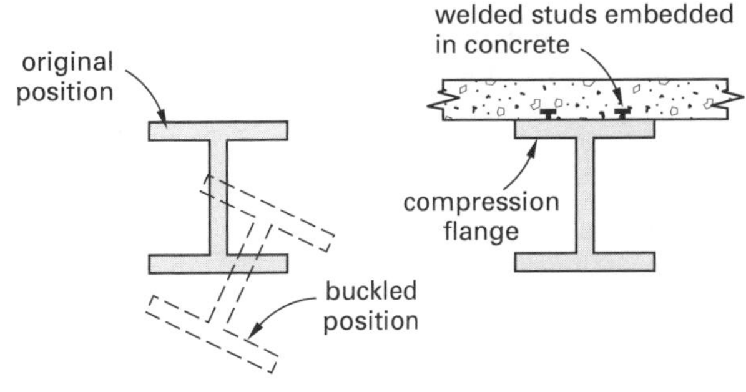

Figure 50.20 *Beam Failure by Rotation*

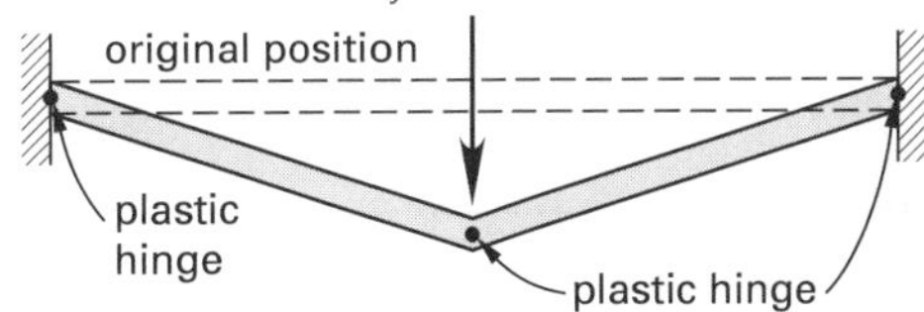

Rotation is an inelastic (plastic) failure of the beam. When the bending stress at a point exceeds the strength of the beam material, the material yields, as shown in Fig. 50.20. As the beam yields, its slope changes. Since the beam appears to be rotating at a hinge at the yield point, the term *plastic hinge* is used to describe the failure mechanism.

29. CURVED BEAMS

Many members (e.g., hooks, chain links, clamps, and machine frames) have curved main axes. The distribution of bending stress in a curved beam is nonlinear. Compared to a straight beam, the stress at the inner radius is higher because the inner radius fibers are shorter. Conversely, the stress at the outer radius is lower because the outer radius fibers are longer. Also, the neutral axis is shifted from the center inward toward the center of curvature.

Since the process of finding the neutral axis and calculating the stress amplification is complex, tables and graphs are used for quick estimates and manual computations. The forms of these computational aids vary, but the straight-beam stress is generally multiplied by factors, K, to obtain the stresses at the extreme faces. The factor values depend on the beam cross section and radius of curvature.

$$\sigma_{\text{curved}} = K\sigma_{\text{straight}} = \frac{KMc}{I} \quad 50.71$$

[9]Building codes specify maximum permitted deflections in terms of beam length.

Table 50.4 is typical of compilations for round and rectangular beams. Factors K_A and K_B are the multipliers for the inner (high stress) and outer (low stress) faces, respectively. The ratio h/r is the fractional distance that the neutral axis shifts inward toward the radius of curvature.

***Table 50.4** Curved Beam Correction Factors*

solid rectangular section	r/c	K_A	K_B	h/r
	1.2	2.89	0.57	0.305
	1.4	2.13	0.63	0.204
	1.6	1.79	0.67	0.149
	1.8	1.63	0.70	0.112
	2.0	1.52	0.73	0.090
	3.0	1.30	0.81	0.041
	4.0	1.20	0.85	0.021
	6.0	1.12	0.90	0.0093
	8.0	1.09	0.92	0.0052
	10.0	1.07	0.94	0.0033

solid circular section	r/c	K_A	K_B	h/r
	1.2	3.41	0.54	0.224
	1.4	2.40	0.60	0.151
	1.6	1.96	0.65	0.108
	1.8	1.75	0.68	0.084
	2.0	1.62	0.71	0.069
	3.0	1.33	0.79	0.03
	4.0	1.23	0.84	0.016
	6.0	1.14	0.89	0.007
	8.0	1.10	0.91	0.0039
	10.0	1.08	0.93	0.0025

30. COMPOSITE STRUCTURES

A *composite structure* is one in which two or more different materials are used. Each material carries part of an applied load. Examples of composite structures include steel-reinforced concrete and steel-plated timber beams.

Most simple composite structures can be analyzed using the *method of consistent deformations*, also known as the *area transformation method*. This method assumes that the strains are the same in both materials at the interface between them. Although the strains are the same, the stresses in the two adjacent materials are not equal, since stresses are proportional to the moduli of elasticity.

The following steps comprise an analysis method based on area transformation.

step 1: Determine the modulus of elasticity for each of the materials used in the structure.

step 2: For each of the materials used, calculate the *modular ratio*, n.

$$n = \frac{E}{E_{\text{weakest}}} \qquad 50.72$$

E_{weakest} is the smallest modulus of elasticity of any of the materials used in the composite structure. For two materials that experience the same strains (i.e., are perfectly bonded), n is also the ratio of stresses.

step 3: For all of the materials except the weakest, multiply the actual material stress area by n. Consider this expanded (*transformed*) area to have the same composition as the weakest material.

step 4: If the structure is a tension or compression member, the distribution or placement of the transformed area is not important. Just assume that the transformed areas carry the axial load. For beams in bending, the transformed area can add to the width of the beam, but it cannot change the depth of the beam or the thickness of the reinforcement.

step 5: For compression or tension members, calculate the stresses in the weakest and stronger materials.

$$\sigma_{\text{weakest}} = \frac{F}{A_t} \qquad 50.73$$

$$\sigma_{\text{stronger}} = \frac{nF}{A_t} \qquad 50.74$$

step 6: For flexural (bending) members, find the centroid of the transformed beam, and complete steps 7 through 9.

step 7: Find the centroidal moment of inertia of the transformed beam $I_{c,t}$.

step 8: Find V_{max} and M_{max} by inspection or from the shear and moment diagrams.

step 9: Calculate the stresses in the weakest and stronger materials.

$$\sigma_{\text{weakest}} = \frac{Mc_{\text{weakest}}}{I_{c,t}} \qquad 50.75$$

$$\sigma_{\text{stronger}} = \frac{nMc_{\text{stronger}}}{I_{c,t}} \qquad 50.76$$

Example 50.14

A short circular steel core is surrounded by a copper tube. The assemblage supports an axial compressive load of 100,000 lbf. The core and tube are well bonded, and the load is applied uniformly. Find the compressive stress in the inner steel core and the outer copper tube.

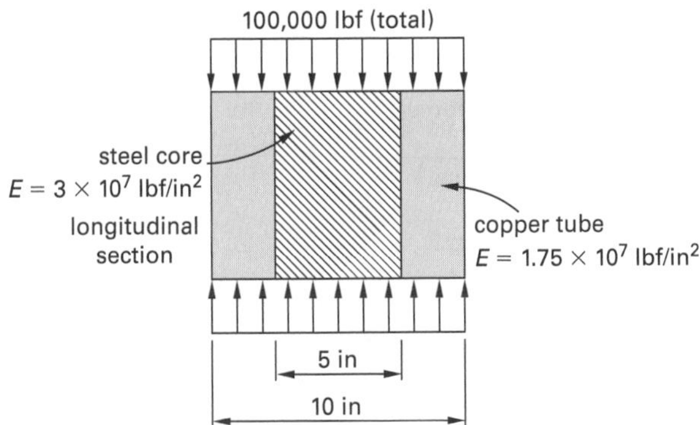

Solution

The moduli of elasticity are given in the illustration. From step 2, the modular ratio is

$$n = \frac{E_{\text{steel}}}{E_{\text{copper}}} = \frac{3\times 10^7\ \frac{\text{lbf}}{\text{in}^2}}{1.75\times 10^7\ \frac{\text{lbf}}{\text{in}^2}} = 1.714$$

The actual cross-sectional area of the steel is

$$A_{\text{steel}} = \tfrac{\pi}{4}D^2 = \left(\frac{\pi}{4}\right)(5\ \text{in})^2 = 19.63\ \text{in}^2$$

The actual cross-sectional area of the copper is

$$\begin{aligned} A_{\text{copper}} &= \tfrac{\pi}{4}(D_o^2 - D_i^2) = \left(\frac{\pi}{4}\right)\left((10\ \text{in})^2 - (5\ \text{in})^2\right) \\ &= 58.90\ \text{in}^2 \end{aligned}$$

The steel is the stronger material. Its area must be expanded to an equivalent area of copper. From step 3, the total transformed area is

$$\begin{aligned} A_t &= A_{\text{copper}} + nA_{\text{steel}} = 58.90\ \text{in}^2 + (1.714)(19.63\ \text{in}^2) \\ &= 92.55\ \text{in}^2 \end{aligned}$$

Since the two pieces are well bonded and the load is applied uniformly, both pieces experience identical strains. From step 5, the compressive stresses are

$$\begin{aligned} \sigma_{\text{copper}} &= \frac{F}{A_t} = \frac{-100{,}000\ \text{lbf}}{92.55\ \text{in}^2} \\ &= -1080\ \text{lbf/in}^2\quad \text{[compression]} \end{aligned}$$

$$\begin{aligned} \sigma_{\text{steel}} &= \frac{nF}{A_t} = n\sigma_{\text{copper}} = (1.714)\left(-1080\ \frac{\text{lbf}}{\text{in}^2}\right) \\ &= -1851\ \text{lbf/in}^2 \end{aligned}$$

Example 50.15

At a particular point along the length of a steel-reinforced wood beam, the moment is 40,000 ft-lbf. Assume the steel reinforcement is lag-bolted to the wood at regular intervals along the beam. What are the maximum bending stresses in the wood and steel?

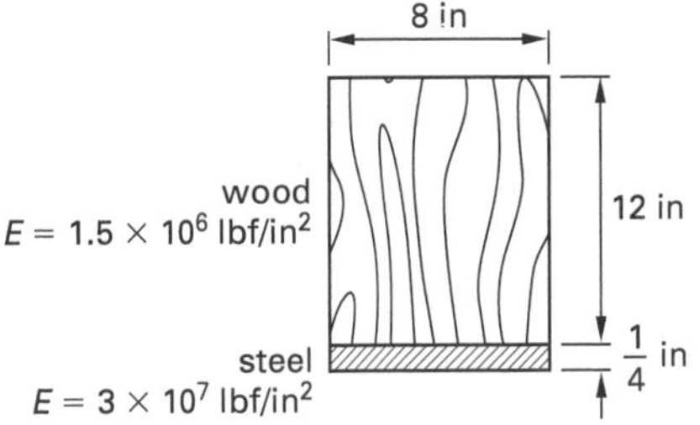

Solution

The moduli of elasticity are given in the illustration. From step 2, the modular ratio is

$$n = \frac{E_{\text{steel}}}{E_{\text{wood}}} = \frac{3\times 10^7\ \frac{\text{lbf}}{\text{in}^2}}{1.5\times 10^6\ \frac{\text{lbf}}{\text{in}^2}} = 20$$

The actual cross-sectional area of the steel is

$$A_{\text{steel}} = (0.25\ \text{in})(8\ \text{in}) = 2\ \text{in}^2$$

The steel is the stronger material. Its area must be expanded to an equivalent area of wood. Since the depth of the beam and reinforcement cannot be increased (step 4), the width must increase. The width of the transformed steel plate is

$$b' = nb = (20)(8\ \text{in}) = 160\ \text{in}$$

The centroid of the transformed section is 4.45 in from the horizontal axis. The centroidal moment of inertia of the transformed section is $I_{c,t} = 2211.5\ \text{in}^4$. (The calculations for centroidal location and moment of inertia are not presented here.)

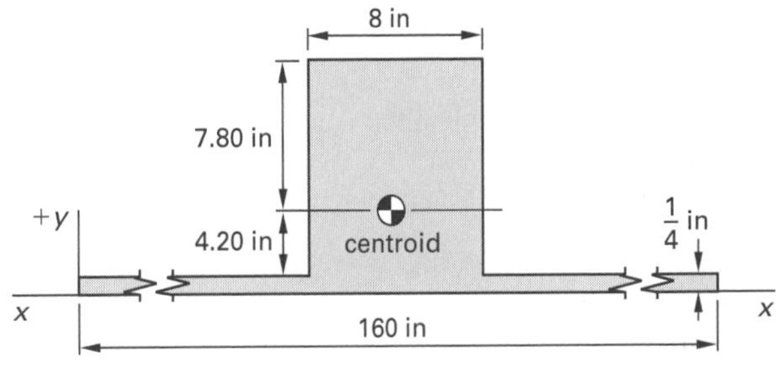

Since the steel plate is bolted to the wood at regular intervals, both pieces experience the same strain. From step 9, the stresses in the wood and steel are

$$\sigma_{\text{max,wood}} = \frac{Mc_{\text{wood}}}{I_{c,t}} = \frac{(40{,}000 \text{ ft-lbf})\left(12 \ \frac{\text{in}}{\text{ft}}\right)(7.8 \text{ in})}{2211.5 \text{ in}^4} = 1693 \text{ lbf/in}^2$$

$$\sigma_{\text{max,steel}} = \frac{nMc_{\text{steel}}}{I_{c,t}} = \frac{(20)(40{,}000 \text{ ft-lbf})\left(12 \ \frac{\text{in}}{\text{ft}}\right)(4.45 \text{ in})}{2211.5 \text{ in}^4} = 19{,}317 \text{ lbf/in}^2$$

31. NOMENCLATURE

a	width	in	m
A	area	in^2	m^2
b	width	in	m
c	distance from neutral axis to extreme fiber	in	m
C	center	lbf/in^2	MPa
C	compressive force	lbf	N
C	constant	–	–
C	couple	in-lbf	N·m
d	depth or distance	in	m
D	diameter	in	m
e	eccentricity	in	m
E	modulus of elasticity	lbf/in^2	MPa
f	unit force	lbf	N
F	force	lbf	N
g	width	in	m
G	shear modulus	lbf/in^2	MPa
h	height	in	m
I	moment of inertia	in^4	m^4
J	polar moment of inertia	in^4	m^4
k	spring constant	lbf/in	N/m
K	stress concentration factor	–	–
L	length	in	m
M	moment	in-lbf	N·m
n	modular ratio	–	–
P	force	lbf	N
Q	statical moment	in^3	m^3
r	radius	in	m
R	radius	ft	m
R	reaction	lbf	N
R	rigidity	–	–
s	section modulus	in^3	m^3
S	force	lbf	N
t	thickness	in	m
T	temperature	°F	°C
T	tensile force	lbf	N
T	torque	in-lbf	N·m
u	load due to unit force	lbf	N
U	energy	in-lbf	N·m
V	vertical shear force	lbf	N
V	volume	in^3	m^3
w	load per unit length	lbf/in	N/m
W	work	in-lbf	N·m
x	location	in	m
x	maximum deflection	in	m
y	deflection	in	m
y	distance	in	m
y	location	in	m

Symbols

α	coefficient of linear thermal expansion	1/°F	1/°C
β	coefficient of volumetric thermal expansion	1/°F	1/°C
γ	coefficient of area thermal expansion	1/°F	1/°C
δ	deformation	in	m
ϵ	strain	–	–
θ	angle	deg	deg
ν	Poisson's ratio	–	–
ρ	radius of curvature	in	m
σ	normal stress	lbf/in^2	MPa
τ	shear stress	lbf/in^2	MPa
ϕ	angle	rad	rad

Subscripts

0	nominal
A	inner face (high stress)
b	bending
B	outer face (low stress)
c	centroidal
ext	external
i	inside
j	jth member
l	left
o	original
r	range or right
t	thermal or transformed
th	thermal
w	web

51 Failure Theories

Content in blue refers to the *NCEES Handbook*.

NCEES EXAM SPECIFICATIONS AND RELATED CONTENT

MACHINE DESIGN AND MATERIALS EXAM

- I.D.6. Strength of Materials: Fatigue
  - 9. Alternating Stress: Soderberg Line
  - 10. Alternating Stress: Goodman Line
  - 12. Cumulative Fatigue
- I.D.7. Strength of Materials: Failure theories
  - 2. Static Loading of Brittle Materials: Maximum Normal Stress Theory
  - 6. Static Loading of Ductile Materials: Maximum Shear Stress Theory
  - 8. Static Loading of Ductile Materials: Distortion Energy Theory
  - 11. Endurance Limit Modifying Factors

1. STATIC LOADING OF BRITTLE MATERIALS: UNIAXIAL LOADING

Brittle materials such as gray cast iron fail by sudden fracturing, not by yielding. The basic method of designing with brittle materials is to keep the maximum stress below the ultimate strength. Stress concentration factors must be included. The failure criterion is

$$\sigma > S_u \quad \text{[failure criterion]} \qquad 51.1$$

2. STATIC LOADING OF BRITTLE MATERIALS: MAXIMUM NORMAL STRESS THEORY

The *maximum normal stress theory* predicts failure stress reasonably well for brittle materials under static loading.[1] If $\sigma_1 \geq \sigma_2 \geq \sigma_3$, failure is assumed to occur either if the largest tensile principal stress, σ_1, is greater than the ultimate tensile strength, or if the magnitude of the largest compressive principal stress, σ_3, is greater than the magnitude of the ultimate compressive strength. Brittle materials generally have much higher compressive strengths than tensile strengths, so both tensile and compressive stresses must be checked.

Figure 51.1 illustrates a standard graphical method of describing the safe operating region. Notice that the axes represent the principal stresses, not the stresses in the x- and y-directions.

Stress concentration factors are applicable to brittle materials under static loading. The *factor of safety*, FS, is the ultimate strength, S_u, divided by the actual stress, σ. (The *margin of safety* is MS = FS − 1.) Where a factor of safety is known in advance, the *allowable stress*, S_a, can be calculated by dividing the ultimate strength by it. The allowable operating region can be constructed from the allowable stresses rather than from the ultimate stresses.

$$\text{FS} = \frac{S_u}{\sigma} \qquad 51.2$$

$$S_a = \frac{S_u}{\text{FS}} \qquad 51.3$$

[1] As described in subsequent sections, the maximum normal stress theory is limited to cases where both principal stresses have the same sign (i.e., both are compressive or both are tensile).

The failure criteria are given by Eq. 51.4 and Eq. 51.5.

Brittle Materials

$$\sigma_1 \geq S_{ut} \qquad 51.4$$

$$\sigma_3 \leq -S_{uc} \qquad 51.5$$

Figure 51.1 Maximum Normal Stress Theory

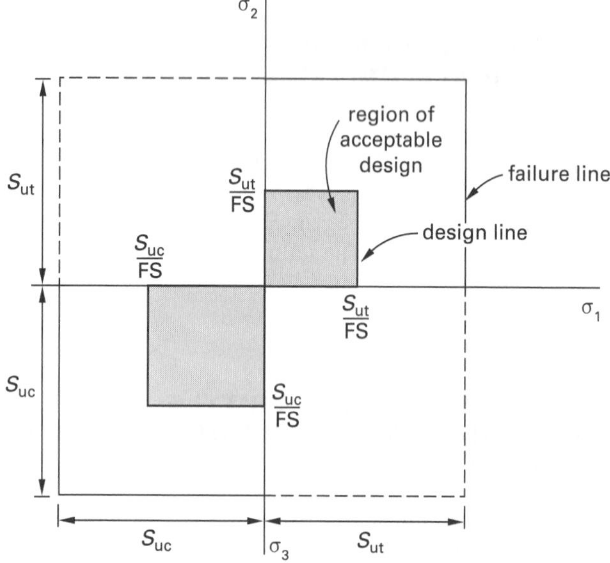

Example 51.1

A lathe bed is made of gray cast iron. The ultimate strengths for the cast iron are 30,000 lbf/in² (tension) and 110,000 lbf/in² (compression). The bed is subjected to maximum stresses of 17,150 lbf/in² in tension and 42,800 lbf/in² in compression. What are the factors of safety?

Solution

The factor of safety in tension is

$$\text{FS} = \frac{S_{ut}}{\sigma_t} = \frac{30{,}000 \ \frac{\text{lbf}}{\text{in}^2}}{17{,}150 \ \frac{\text{lbf}}{\text{in}^2}} = 1.75$$

The factor of safety in compression is

$$\text{FS} = \frac{S_{uc}}{\sigma_c} = \frac{110{,}000 \ \frac{\text{lbf}}{\text{in}^2}}{42{,}800 \ \frac{\text{lbf}}{\text{in}^2}} = 2.57$$

Tensile failure is the limiting case.

3. STATIC LOADING OF BRITTLE MATERIALS: COULOMB-MOHR THEORY

The maximum normal stress theory is somewhat in conflict with experimental evidence. Reliable operation in the second and fourth quadrants (i.e., when the two principal stresses have opposite signs) has not been observed, even though the stresses are less than the ultimate strengths. The *Coulomb-Mohr theory* is a conservative theory that reduces the acceptable operating region in the second and fourth quadrants. As with the maximum normal stress theory, the factor of safety is calculated from Eq. 51.2.

The failure criterion is

$$\frac{\sigma_1}{S_{ut}} + \frac{\sigma_2}{S_{uc}} > 1 \quad \text{[failure criterion]} \qquad 51.6$$

4. STATIC LOADING OF BRITTLE MATERIALS: MODIFIED MOHR THEORY

The Coulomb-Mohr theory is considered conservative because failures typically "miss" the diagonal line in Fig. 51.2 by a considerable margin. The *modified Mohr theory* more closely predicts the envelope of observed failures. (See Fig. 51.3.) Though the failure criterion can be based on deriving equations of the straight lines in the second and fourth quadrants, graphical solutions are more common.

Figure 51.2 Coulomb-Mohr Theory

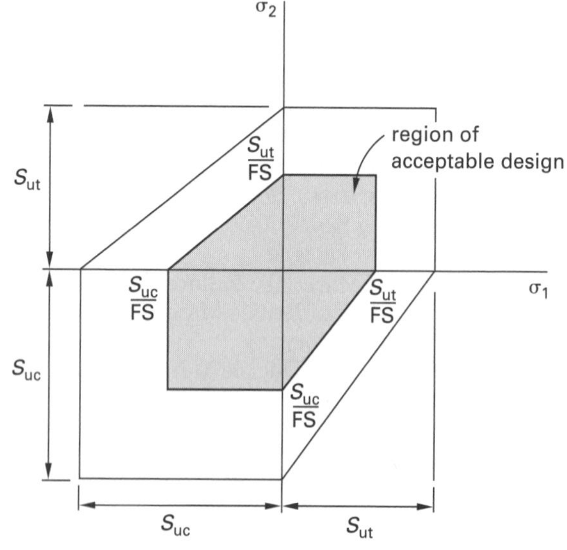

Figure 51.3 Modified Mohr Theory

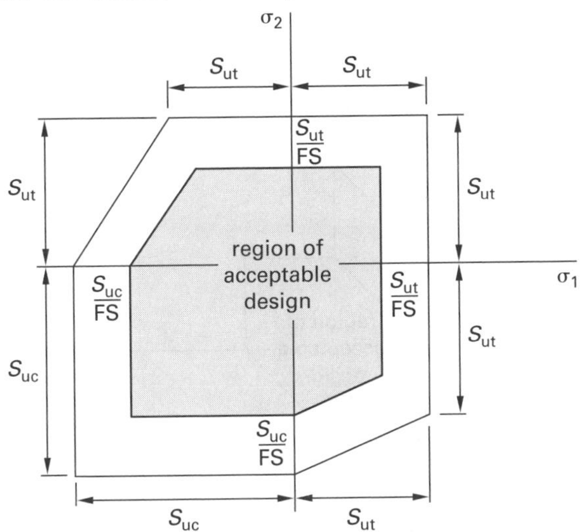

Example 51.2

A 0.5 in diameter dowel is made from cast iron (ultimate tensile strength of 40,000 lbf/in²; ultimate compressive strength of 135,000 lbf/in²). The dowel supports a compressive load of 15,000 lbf and is subjected simultaneously to an unknown torsional shear stress. Use the modified Mohr theory to determine the torsional shear stress that will cause failure.

Solution

The compressive normal stress in the dowel is

$$\sigma_c = \frac{F}{A} = \frac{-15{,}000 \text{ lbf}}{\left(\frac{\pi}{4}\right)(0.5 \text{ in})^2}$$
$$= -76{,}394 \text{ lbf/in}^2 \quad \text{[negative because compressive]}$$

To use the modified Mohr theory, the principal stresses must be known. However, the shear stress in this problem is the unknown, so the principal stress cannot be calculated directly. Use a trial-and-error approach.

Assume the shear stress is 10,000 lbf/in². With $\tau =$ 10,000 lbf/in², $\sigma_x = -76{,}400$ lbf/in², and $\sigma_y = 0$, the principal stresses are found (from standard methods) to be 1300 lbf/in² and −77,700 lbf/in². Draw the modified Mohr diagram and plot these principal values. The point is not close to the failure line, so the assumed shear stress of 10,000 lbf/in² is too low.

Repeat the process with shear stresses of 30,000 lbf/in² and 40,000 lbf/in².

τ	σ_1	σ_2
10,000	1300	−77,700
30,000	10,400	−86,800
40,000	17,100	−93,500

The third point (corresponding to a shear stress of 40,000 lbf/in²) is essentially on the failure line, so this is the maximum allowable shear stress. (It is a coincidence that this is also the ultimate tensile strength.)

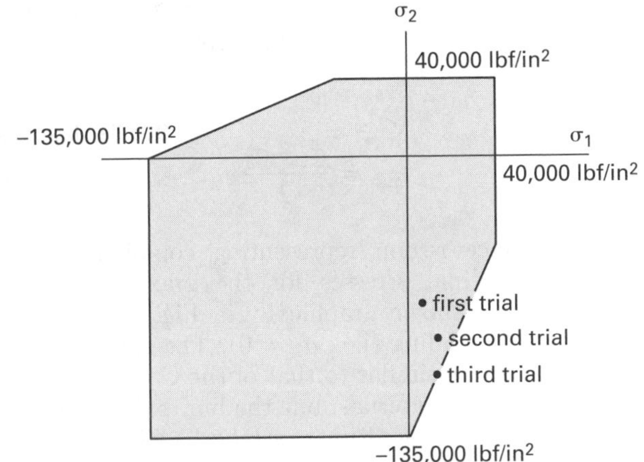

5. STATIC LOADING OF DUCTILE MATERIALS: UNIAXIAL LOADING

Ductile materials fail by yielding, not by fracture. The basic method of designing with ductile materials in uniaxial loading is to keep the maximum stress below the yield strength. This is known as the *maximum stress theory.* Stress concentration factors must be included. The failure criterion is

$$\sigma > S_y \quad \text{[failure criterion]} \qquad 51.7$$

The failure criterion for the equivalent *maximum strain theory* is

$$\epsilon > \frac{S_y}{E} \quad \text{[failure criterion]} \qquad 51.8$$

6. STATIC LOADING OF DUCTILE MATERIALS: MAXIMUM SHEAR STRESS THEORY

With the conservative *maximum shear stress theory,* shear stress is used to indicate yielding (i.e., failure). Loading is not limited to shear and torsion. Loading can include normal stresses as well as shear stresses.

According to the maximum shear stress theory, yielding occurs when the maximum shear stress exceeds the yield strength in shear.[2] It is implicit in this theory that the yield strength in shear is half of the tensile yield strength.[3]

The failure criterion for ductile materials is

Ductile Materials

$$\tau_{\text{max}} \geq \frac{S_y}{2} \quad 51.9$$

From the combined stress theory, if $\sigma_1 \geq \sigma_2 \geq \sigma_3$, the maximum shear stress, τ_{max}, for *triaxial loading* is given by Eq. 51.10.

Ductile Materials

$$\tau_{\text{max}} = \frac{\sigma_1 - \sigma_3}{2} \quad 51.10$$

The acceptance region representing combinations of allowable principal stresses for the maximum shear stress theory is shown graphically in Fig. 51.4 for the case of biaxial loading (i.e., $\sigma_3 = 0$). The shape of this failure envelope is similar to that of the Coulomb-Mohr theory for brittle materials, but the limits are based on yield strengths, not on ultimate strengths. Since the tensile and compressive yield strengths are assumed equal for ductile materials, the failure envelope is symmetrical. In Fig. 51.4, the limits are not divided by 2 as might be suspected from Eq. 51.9 and Eq. 51.10. This is because the failure envelope is used with the principal stress, not with the shear stress.

The factor of safety with the maximum shear stress theory is

$$\text{FS} = \frac{S_y}{2\tau_{\text{max}}} \quad 51.11$$

Example 51.3

Strain gauges attached to a bearing support show the three principal stresses to be 10,600 lbf/in^2, 2400 lbf/in^2, and −9200 lbf/in^2. The support is cast aluminum with a tensile yield strength of 24,000 lbf/in^2. Use the maximum shear stress theory to determine the factor of safety.

Solution

Calculate the combined shear stresses.

Figure 51.4 *Maximum Shear Stress Theory*

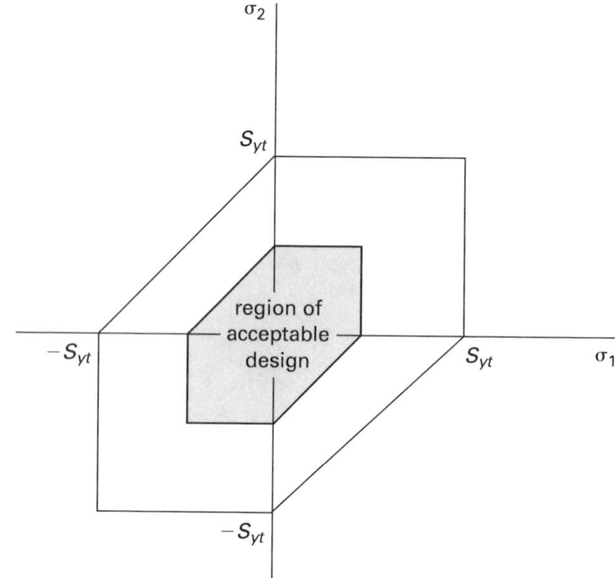

Ductile Materials

$$\tau_{\text{max}} = \frac{\sigma_1 - \sigma_3}{2} = \frac{10{,}600\ \frac{\text{lbf}}{\text{in}^2} - \left(-9200\ \frac{\text{lbf}}{\text{in}^2}\right)}{2} = 9900\ \text{lbf/in}^2$$

The maximum shear stress is 9900 lbf/in^2.

From Eq. 51.11, the factor of safety is

$$\text{FS} = \frac{S_y}{2\tau_{\text{max}}} = \frac{24{,}000\ \frac{\text{lbf}}{\text{in}^2}}{(2)\left(9900\ \frac{\text{lbf}}{\text{in}^2}\right)} = 1.21$$

7. STATIC LOADING OF DUCTILE MATERIALS: STRAIN ENERGY THEORY

The *strain energy theory* calculates the energy per unit volume. This energy is compared to the energy that causes yielding, which is known as the *modulus of resilience*, MR. Since the stress-strain curve is essentially a straight line, as shown by Fig. 51.5, the strain energy (i.e., the area under the curve) for any uniaxial loading less than the yield strength is

$$U = \frac{\sigma\epsilon}{2} \quad 51.12$$

[2]The application of this theory, as well as those that follow, depends on being able to find the maximum shear stress, τ_{max}. Even in some cases of uniaxial loading, finding the maximum shear stress can be tricky. Two cylindrical rollers or two spheres in contact, for example, represent cases of uniaxial compressive loading, which have well-documented but nonobvious maximum shear stresses.

[3]If the symmetrical shape of the failure envelope is accepted, it is easy to justify the assumption that the yield strength in shear is one-half of the yield strength in tension. For pure shear loading, the two principal stresses will each be equal and opposite (with magnitudes equal to the applied shear stress). Plotting the locus of points with $\sigma_1 = -\sigma_2$, the failure envelope is encountered at $S_{yt}/2$.

For biaxial loading on the principal planes, the strain energy is calculated from superposition.

$$U = \frac{\sigma_1\epsilon_1 + \sigma_2\epsilon_2}{2} \qquad 51.13$$

Figure 51.5 *Strain Energy as Area Under the Stress-Strain Curve*

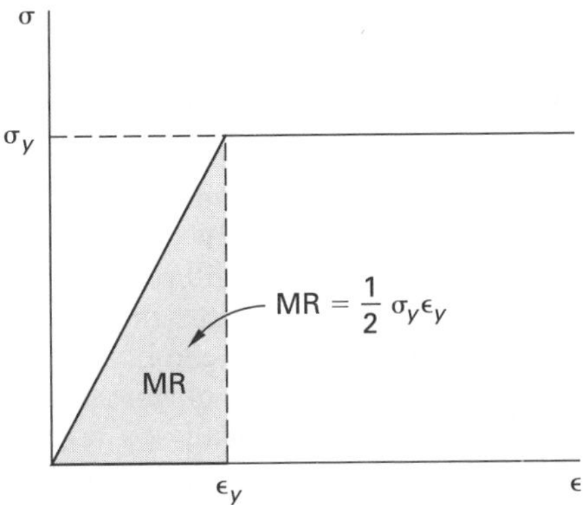

Strain is generally not measured directly, but it can be calculated from the stress and material properties.

$$\epsilon_1 = \frac{1}{E}(\sigma_1 - \nu\sigma_2) \qquad 51.14$$

$$\epsilon_2 = \frac{1}{E}(\sigma_2 - \nu\sigma_1) \qquad 51.15$$

$$U = \frac{1}{2E}(\sigma_1^2 + \sigma_2^2 - 2\nu\sigma_1\sigma_2) \qquad 51.16$$

Failure is assumed to occur when the strain energy, U, exceeds the modulus of resilience. The modulus of resilience can be determined from a simple tensile test and is calculated as the area under the stress-strain curve up to the yield point.[4] At failure in a simple tensile test, $\sigma_1 = S_{yt}$ and $\sigma_2 = 0$. Therefore, from Eq. 51.16, the modulus of resilience is

$$\text{MR} = \frac{1}{2E}\left(S_{yt}^2 + (0)^2 - (2)(0)\right) = \frac{S_{yt}^2}{2E} \qquad 51.17$$

Eliminating the $1/2E$ terms, the failure criterion is

$$\sigma_1^2 + \sigma_2^2 - 2\nu\sigma_1\sigma_2 > S_{yt}^2 \quad \text{[failure criterion]} \qquad 51.18$$

The factor of safety is

$$\text{FS} = \frac{S_{yt}}{\sqrt{\sigma_1^2 + \sigma_2^2 - 2\nu\sigma_1\sigma_2}} \qquad 51.19$$

8. STATIC LOADING OF DUCTILE MATERIALS: DISTORTION ENERGY THEORY

The *distortion energy theory* (also known as the *theory of constant energy of distortion, von Mises theory, von Mises-Hencky theory,* and *octahedral shear-stress theory*) is similar in development to the strain energy method but is more strict. It is commonly used to predict tensile and shear failure in steel parts. The *von Mises stress* (also known as the *effective stress*), σ′, is calculated from the principal stresses. For *biaxial loading,* either Eq. 51.20 or Eq. 51.21 can be used to find the effective stress.

Ductile Materials

$$\sigma' = \left(\sigma_A^2 - \sigma_A\sigma_B + \sigma_B^2\right)^{1/2} \qquad 51.20$$

$$\sigma' = \left(\sigma_x^2 - \sigma_x\sigma_y + \sigma_y^2 + 3\tau_{xy}^2\right)^{1/2} \qquad 51.21$$

For *triaxial loading* (i.e., where there are three orthogonal normal stresses), the *von Mises stress* is

Ductile Materials

$$\left(\frac{(\sigma_1-\sigma_2)^2 + (\sigma_2-\sigma_3)^2 + (\sigma_1-\sigma_3)^2}{2}\right)^{1/2} \geq S_y \qquad 51.22$$

The failure criterion is

$$\sigma' > S_y \quad \text{[failure criterion]} \qquad 51.23$$

The factor of safety is

$$\text{FS} = \frac{S_y}{\sigma'} \qquad 51.24$$

If the loading is pure torsion at failure, then $\sigma_1 = -\sigma_2 = \tau_{max}$, and $\sigma_3 = 0$. If τ_{max} is substituted for σ in Eq. 51.22 (with $\sigma_3 = 0$), an expression for the yield strength in shear is derived. Equation 51.25 predicts a larger yield strength in shear than does the maximum shear stress theory ($0.5S_{yt}$).

$$\tau_{\text{max,failure}} = \frac{S_y}{\sqrt{3}} = 0.577S_y \qquad 51.25$$

Example 51.4

The steel used in a shaft has a tensile yield strength of 110,000 lbf/in^2 and a Poisson's ratio of 0.3. The shaft is simultaneously acted upon by a longitudinal compressive stress of 60,000 lbf/in^2 and by a torsional stress of 40,000 lbf/in^2. Compare the factor of safety using (a) maximum shear stress, (b) strain energy, and (c) distortion energy.

[4]The strain energy theory is seldom used to predict failures. However, it is similar in development to the distortion energy theory, which has supplanted it, so the strain energy theory is presented here as part of the logical progression of failure theories.

Solution

From the combined stress theory and using $\sigma_y = -60{,}000$ lbf/in^2, the maximum shear stress is

$$\tau_{max} = \frac{1}{2}\sqrt{(\sigma_x - \sigma_y)^2 + (2\tau)^2}$$

$$= \left(\frac{1}{2}\right)\sqrt{\left(-60{,}000\ \frac{\text{lbf}}{\text{in}^2}\right)^2 + \left((2)\left(40{,}000\ \frac{\text{lbf}}{\text{in}^2}\right)\right)^2}$$

$$= 50{,}000\ \text{lbf/in}^2$$

The principal stresses are

$$\sigma_1, \sigma_2 = \tfrac{1}{2}(\sigma_x + \sigma_y) \pm \tau_{max}$$

$$= \left(\frac{1}{2}\right)\left(-60{,}000\ \frac{\text{lbf}}{\text{in}^2}\right) \pm 50{,}000\ \frac{\text{lbf}}{\text{in}^2}$$

$$= 20{,}000\ \text{lbf/in}^2,\ -80{,}000\ \text{lbf/in}^2$$

(a) From Eq. 51.11, the factor of safety for the maximum shear stress theory is

$$\text{FS} = \frac{S_{yt}}{2\tau_{max}} = \frac{110{,}000\ \frac{\text{lbf}}{\text{in}^2}}{(2)\left(50{,}000\ \frac{\text{lbf}}{\text{in}^2}\right)} = 1.1$$

(b) From Eq. 51.19, the factor of safety for the strain energy theory is

$$\text{FS} = \frac{S_{yt}}{\sqrt{\sigma_1^2 + \sigma_2^2 - 2\nu\sigma_1\sigma_2}}$$

$$= \frac{110{,}000\ \frac{\text{lbf}}{\text{in}^2}}{\sqrt{\begin{array}{l}\left(-80{,}000\ \frac{\text{lbf}}{\text{in}^2}\right)^2 + \left(20{,}000\ \frac{\text{lbf}}{\text{in}^2}\right)^2 \\ \quad -(2)(0.3)\left(-80{,}000\ \frac{\text{lbf}}{\text{in}^2}\right)\left(20{,}000\ \frac{\text{lbf}}{\text{in}^2}\right)\end{array}}}$$

$$= 1.25$$

(c) From Eq. 51.20, the von Mises stress is

Ductile Materials

$$\sigma' = \left(\sigma_A^2 - \sigma_A\sigma_B + \sigma_B^2\right)^{1/2}$$

$$= \sqrt{\begin{array}{l}\left(-80{,}000\ \frac{\text{lbf}}{\text{in}^2}\right)^2 - \left(-80{,}000\ \frac{\text{lbf}}{\text{in}^2}\right) \\ \quad \times \left(20{,}000\ \frac{\text{lbf}}{\text{in}^2}\right) + \left(20{,}000\ \frac{\text{lbf}}{\text{in}^2}\right)^2\end{array}}$$

$$= 91{,}652\ \text{lbf/in}^2$$

From Eq. 51.24, the factor of safety for the distortion energy theory is

$$\text{FS} = \frac{S_{yt}}{\sigma'} = \frac{110{,}000\ \frac{\text{lbf}}{\text{in}^2}}{91{,}652\ \frac{\text{lbf}}{\text{in}^2}} = 1.20$$

9. ALTERNATING STRESS: SODERBERG LINE

Many parts are subjected to a combination of static and reversed loadings, as illustrated in Fig. 51.6 for sinusoidal loadings. For these parts, failure cannot be determined solely by comparing stresses with the yield strength or endurance limit. The combined effects of the average stress and the amplitude of the reversal must be considered. This is done graphically on a diagram that plots the mean stress versus the alternating stresses.

Figure 51.6 *Sinusoidal Fluctuating Stress*

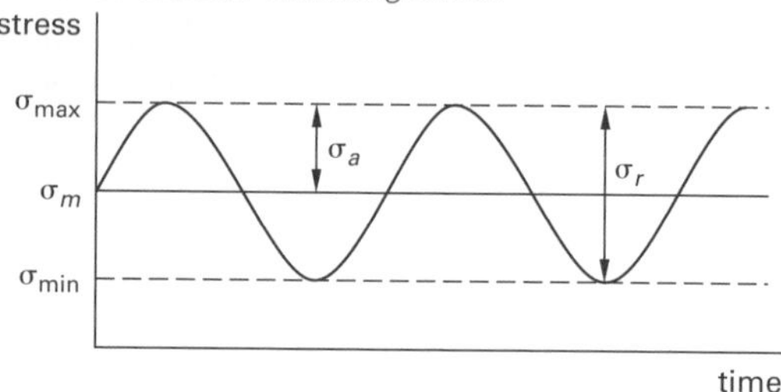

The *mean stress* is

$$\sigma_m = \frac{\sigma_{max} + \sigma_{min}}{2} \qquad 51.26$$

The *alternating stress* is half of the *range stress.*

$$\sigma_r = \sigma_{max} - \sigma_{min} \qquad 51.27$$

$$\sigma_a = \tfrac{1}{2}\sigma_r = \tfrac{1}{2}(\sigma_{max} - \sigma_{min}) \qquad 51.28$$

A criterion for acceptable design (or for failure) is established by graphically relating the yield strength and the endurance limit. One method of relating this information is a Soderberg line, which is particularly suited for normal stresses in ductile materials. However, it is the most conservative of the fluctuating stress theories.

Figure 51.7 illustrates how an area of acceptable design is developed by drawing a straight line (the *Soderberg line* or the *failure line*) from the endurance limit, S_e, to the yield strength, S_{yt}. Both of these values should be divided by a suitable factor of safety to define the allowable design area. If the point (σ_m, σ_a) falls below the allowable stress line, the design is acceptable. The fatigue failure criteria is provided by Eq. 51.29.

Variable Loading Failure Theories

$$\frac{\sigma_a}{S_e} + \frac{\sigma_m}{S_{yt}} \geq 1 \qquad 51.29$$

Stress concentration factors, such as K_f, are applied to the alternating stress only. This is justified with ductile materials, such as steel, which yield around discontinuities, reducing constant mean stress. Increasing the alternating stress is equivalent to reducing S_e by K_f.

Figure 51.7 illustrates how the Soderberg *equivalent stress*, σ_{eq}, is determined from the operating point. The factor of safety is

$$\text{FS} = \frac{S_e}{\sigma_{eq}} = \frac{S_e}{\sigma_a + \left(\frac{S_e}{S_{yt}}\right)\sigma_m} \qquad 51.30$$

Figure 51.7 *Soderberg Line and Equivalent Stress*

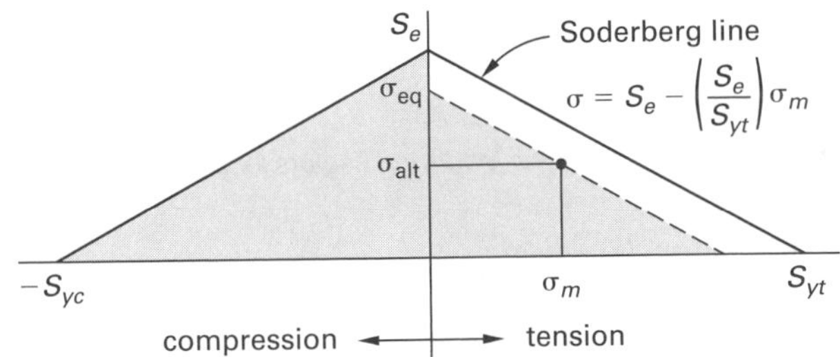

10. ALTERNATING STRESS: GOODMAN LINE

The Soderberg line is a conservative criterion, and it is not often used. Also, the envelope of failures is not truly linear, but follows more of a parabolic line, named the *Gerber line* or *Gerber parabolic relationship*, extending above the Soderberg line from the endurance limit to the ultimate tensile strength.

The *Goodman line* (also known as the *modified Goodman line*) is less conservative than the Soderberg line and is more easily constructed than the Gerber line.[5] It is applicable for steel, aluminum, titanium, and some magnesium alloys.[6] Figure 51.8 illustrates the Goodman line, as well as the method for determining the Goodman equivalent stress.[7]

Variable Loading Failure Theories

$$\sigma_{eq} = \sigma_a + \left(\frac{S_e}{S_{ut}}\right)\sigma_m \qquad 51.31$$

The Goodman factor of safety is

$$\text{FS} = \frac{S_e}{\sigma_{eq}} = \frac{S_e}{\sigma_a + \left(\frac{S_e}{S_{ut}}\right)\sigma_m} \qquad 51.32$$

As with the Soderberg criterion, only the alternating stress should be increased by K_f.

Figure 51.8 *Goodman Line and Equivalent Stress*

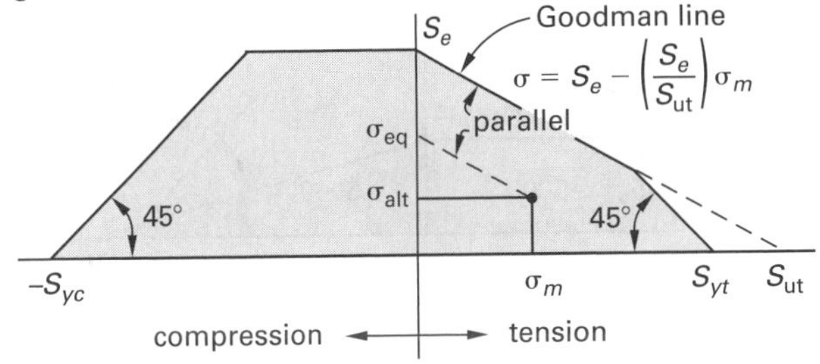

Example 51.5

An aircraft bell crank is made from aluminum with a yield strength of 40,000 lbf/in^2, an ultimate strength of 45,000 lbf/in^2, and an endurance limit of 13,500 lbf/in^2, reduced to 8500 lbf/in^2 by various derating factors. The bell crank is subjected to minimum and maximum tensile stresses of 6500 lbf/in^2 and 9500 lbf/in^2, respectively. The bell crank is to have an indefinite life, and a factor of safety in excess of 3 is required. Is the design acceptable?

Solution

The mean and alternating stresses are

$$\begin{aligned}\sigma_m &= \tfrac{1}{2}(\sigma_{max} + \sigma_{min}) = \left(\frac{1}{2}\right)\left(9500\ \frac{\text{lbf}}{\text{in}^2} + 6500\ \frac{\text{lbf}}{\text{in}^2}\right)\\ &= 8000\ \text{lbf/in}^2\\ \sigma_a &= \tfrac{1}{2}(\sigma_{max} - \sigma_{min}) = \left(\frac{1}{2}\right)\left(9500\ \frac{\text{lbf}}{\text{in}^2} - 6500\ \frac{\text{lbf}}{\text{in}^2}\right)\\ &= 1500\ \text{lbf/in}^2\end{aligned}$$

[5]There is no significant distinction between a *Goodman line* and a *modified Goodman line*. The modifications were made early in the theory's development, and both names are now used to represent a straight line drawn on the fatigue diagram between the endurance and ultimate tensile strengths.
[6]The Goodman line should not be used with gray cast iron and some types of magnesium. Failures for these materials do not follow the theory well. The envelope of failure for these materials is bounded by the parabolic *Smith line*, which is similar to the Gerber line, except that it runs under the straight line connecting the endurance and ultimate strengths. However, brittle materials such as gray cast iron are not usually considered satisfactory for stresses that fluctuate significantly unless very large factors of safety are included.
[7]The Soderberg, Goodman, Gerber, and Smith lines are drawn on *fatigue diagrams*. Strictly speaking, they are Soderberg *lines*, not Soderberg *diagrams*. The distinction is more critical for Goodman lines, because there is a Goodman diagram for fatigue loading, but it is constructed differently. However, the correct names are not consistently applied, and the terms "Soderberg diagram" and "Goodman diagram" are often heard.

The material properties are divided by the safety factor.

$$\frac{S_{yt}}{3} = \frac{40{,}000\ \frac{\text{lbf}}{\text{in}^2}}{3} = 13{,}333\ \text{lbf/in}^2$$

$$\frac{S_{ut}}{3} = \frac{45{,}000\ \frac{\text{lbf}}{\text{in}^2}}{3} = 15{,}000\ \text{lbf/in}^2$$

$$\frac{S_e}{3} = \frac{8500\ \frac{\text{lbf}}{\text{in}^2}}{3} = 2833\ \text{lbf/in}^2$$

Since the operating point lies above the failure line, the design is not acceptable.

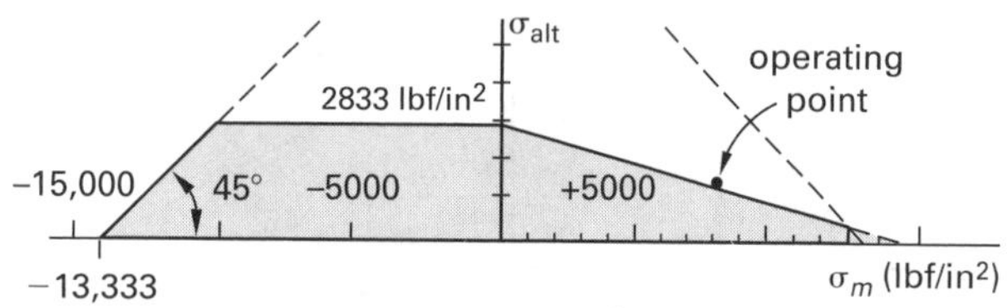

11. ENDURANCE LIMIT MODIFYING FACTORS

The endurance limit, S_e', is not a true property of the material, since the other significant influences, particularly surface finish, are never eliminated. However, representative values of S_e' obtained from ground and polished specimens provide a baseline to which other factors can be applied to account for the effects of surface finish, temperature, stress concentration, notch sensitivity, size, environment, and desired reliability. These other influences are accounted for by endurance limit modifying factors that are used to calculate a working endurance strength, S_e, for the material, as shown in Eq. 51.33.

Variable Loading Failure Theories

$$S_e = k_a k_b k_c k_d k_e S_e' \qquad 51.33$$

In the absence of test data. the endurance limit for steels is commonly taken as provided in Eq. 51.34.

Variable Loading Failure Theories

$$S_e' = \begin{Bmatrix} 0.5S_{ut},\ S_{ut} \leq 1400\ \text{MPa} \\ 700\ \text{MPa},\ S_{ut} > 1400\ \text{MPa} \end{Bmatrix} \qquad 51.34$$

The surface factor, k_a, is calculated from Eq. 51.35 using values of the factors from Table 51.1.

Variable Loading Failure Theories

$$k_a = aS_{ut}^b \qquad 51.35$$

Table 51.1 *Factors for Calculating k_a*

surface finish	a (kpsi)	a (MPa)	b
ground	1.34	1.58	−0.085
machined or cold-drawn (CD)	2.70	4.51	−0.265
hot rolled	14.4	57.7	−0.718
as forged	39.9	272.0	−0.995

The size factor, k_b, and load factor, k_c, are determined for axial loadings from Table 51.2 and for bending and torsion from Table 51.3.

Table 51.2 *Endurance Limit Modifying Factors for Axial Loading*

size factor, k_b	1
load factor, k_c	
$S_{ut} \leq 1520$ MPa	0.923
$S_{ut} > 1520$ MPa	1

Table 51.3 *Endurance Limit Modifying Factors for Bending and Torsion*

size factor, k_b	
$d \leq 8$mm	1
8 mm $\leq d \leq$ 250 mm	$k_b = 1.189d_{\text{eff}}^{-0.097}$
$d > 250$ mm	between 0.6 and 0.75
load factor, k_c	
bending	1
torsion	0.577

As the size of the material gets larger, the endurance limit decreases, due to the increased number of defects in a larger volume. Since the endurance strength is derived from a circular specimen with a diameter of 7.6 mm, the size modification factor is 1.0 for bars of that size. The effective dimension, d_{eff}, takes on different values based on the cross section and on how the cross section is being fatigued. Simplistically, for noncircular cross sections, the smallest cross sectional dimension should be used, and for a solid circular specimen in rotating bending, $d_{\text{eff}} = d$. For a nonrotating or noncircular cross section, d_{eff} is obtained by equating the area of material stressed above 95% of the maximum stress to the same area in the rotating-beam specimen of the same length. That area is designated $A_{0.95}\sigma$. For a nonrotating solid rectangular section with width w and thickness t, the effective dimension is

$$d_{\text{eff}} = 0.808\sqrt{wt}$$

Values of the temperature factor, k_d, and the miscellaneous effects factor, k_e, are found from Table 51.4. The miscellaneous effects factor is used to account for various factors that reduce strength, such as corrosion, plating, and residual stress.

Table 51.4 *Additional Endurance Limit Modifying Factors*

temperature factor, k_d	1	$[T \leq 450°C]$
miscellaneous effects factor, k_e	1,	unless otherwise specified

12. CUMULATIVE FATIGUE

If a part is subjected to $\sigma_{max,1}$ for n_1 cycles, $\sigma_{max,2}$ for n_2 cycles, and so on, the part will accumulate varying amounts of fatigue damage during each series of cycles. *Miner's rule* (also known as the *Palmgren-Miner cycle ratio summation formula, fatigue interaction formula, and cumulative usage factor rule*) can be used to evaluate cumulative damage.[8] In Eq. 51.36, the N_i are the fatigue lives for the corresponding stress levels.

Variable Loading Failure Theories

$$\sum \frac{n_i}{N_i} > C \quad \text{[failure criterion]} \qquad 51.36$$

A value of 1.0 is commonly used for C. However, the exact value should be determined from experimentation appropriate to the material and application. Values between approximately 0.7 and 2.2 have been reported.

13. ALTERNATING COMBINED STRESSES

If there are variations in biaxial or triaxial stresses, a conservative approach is to use a combination of the distortion energy and Goodman line.

step 1: Calculate the principal mean stresses from the combined stress theory.

step 2: Calculate the principal alternating stresses.

step 3: Calculate the mean and alternating von Mises stresses.

$$\sigma'_m = \sqrt{\sigma^2_{m,1} + \sigma^2_{m,2} - \sigma_{m,1}\sigma_{m,2}} \qquad 51.37$$

$$\sigma'_{alt} = \sqrt{\sigma^2_{alt,1} + \sigma^2_{alt,2} - \sigma_{alt,1}\sigma_{alt,2}} \qquad 51.38$$

step 4: Plot the mean and alternating von Mises stresses in relationship to a standard Goodman line.

14. NOMENCLATURE

A	area	in^2	m^2
C	constant	–	–
d	dimension	ft	m
E	modulus of elasticity	lbf/in^2	Pa
F	force	lbf	N
FS	factor of safety	–	–
k	definition factor	–	–
K_f	stress concentration factor	–	–
MR	modulus of resilience	in-lbf/in^3	J/m^3
n	number of cycles	–	–
N	endurance life	–	–
S	strength or stress	lbf/in^2	Pa
U	strain energy	in-lbf/in^3	J/m^3

Symbols

ϵ	strain	in/in	m/m
ν	Poisson's ratio	–	–
σ	normal stress	lbf/in^2	Pa
σ_1	maximum principal stress	lbf/in^2	Pa
σ_2	intermediate principal stress	lbf/in^2	Pa
σ_3	minimum principal stress	lbf/in^2	Pa
τ	shear stress	lbf/in^2	Pa

Subscripts

a	allowable
alt	alternating
c	compressive
e	endurance
eq	equivalent
m	mean
r	range
s	shear
t	tensile
u	ultimate
uc	ultimate compressive
ut	ultimate tensile
y	yield

Materials

[8]Miner's rule does not take into account the increase in the endurance limit that results when virgin material is understressed.

Topic IX: Machine Design

52 Basic Machine Design

Content in blue refers to the *NCEES Handbook*.

NCEES EXAM SPECIFICATIONS AND RELATED CONTENT

MACHINE DESIGN AND MATERIALS EXAM

- I.D.4. Strength of Materials: Buckling
  - 3. Slender Columns
  - 4. Intermediate Columns
- I.D.5. Strength of Materials: Torsion
  - 12. Circular Shaft Design
- II.A.1. Mechanical Components: Pressure vessels and piping
  - 6. Interference Fits
- II.A.9. Mechanical Components: Shafts and keys
  - 12. Circular Shaft Design
  - 13. Torsion in Thin-Walled Shells
- II.B.2. Joints and Fasteners: Bolts, screws, rivets
  - 9. Bolt Preload
  - 10. Bolt Torque to Obtain Preload
  - 16. Eccentrically Loaded Bolted Connections

1. ALLOWABLE STRESS DESIGN

Once an actual stress has been determined, it can be compared to the *allowable stress*. In engineering design, the term "allowable" always means that a factor of safety has been applied to the governing material strength.

$$\text{allowable stress} = \frac{\text{material strength}}{\text{factor of safety}} \qquad 52.1$$

For ductile materials, the material strength used is the yield strength. For steel, the factor of safety ranges from 1.5 to 2.5, depending on the type of steel and the application. Higher factors of safety are seldom necessary in normal, noncritical applications, due to steel's predictable and reliable performance.

$$\sigma_a = \frac{S_y}{\text{FS}} \quad \text{[ductile]} \qquad 52.2$$

For brittle materials, the material strength used is the ultimate strength. Since brittle failure is sudden and unpredictable, the factor of safety is high (e.g., in the 6 to 10 range).

$$\sigma_a = \frac{S_u}{\text{FS}} \quad \text{[brittle]} \qquad 52.3$$

If an actual stress is less than the allowable stress, the design is considered acceptable. This is the principle of the *allowable stress design method*, also known as the *working stress design method*.

$$\sigma_{\text{actual}} \leq \sigma_a \qquad 52.4$$

2. ULTIMATE STRENGTH DESIGN

The allowable stress method has been replaced in most structural work by the *ultimate strength design method*, also known as the *load factor design method*, *plastic design method*, or just *strength design method*. This design method does not use allowable stresses at all. Rather, the member is designed so that its actual *nominal strength* exceeds the required ultimate strength.[1]

[1]It is a characteristic of the ultimate strength design method that the term "strength" actually means load, shear, or moment. Strength seldom, if ever, refers to stress. Thus, the nominal strength of a member might be the load (in pounds, kips, newtons, etc.) or moment (in ft-lbf, ft-kips, or N·m) that the member supports at plastic failure.

The *ultimate strength* (i.e., the required strength) of a member is calculated from the actual *service loads* and multiplicative factors known as *overload factors* or *load factors*. Usually, a distinction is made between dead loads and live loads.[2] For example, the required ultimate moment-carrying capacity in a concrete beam designed according to ACI 318 would be[3]

$$M_u = 1.2M_{\text{dead load}} + 1.6M_{\text{live load}} \quad 52.5$$

The *nominal strength* (i.e., the actual ultimate strength) of a member is calculated from the dimensions and materials. A *capacity reduction factor*, ϕ, of 0.70 to 0.90 is included in the calculation to account for typical workmanship and increase required strength. The moment criteria for an acceptable design is

$$M_n \geq \frac{M_u}{\phi} \quad 52.6$$

3. SLENDER COLUMNS

Very short compression members are known as *piers*. Long compression members are known as *columns*. Failure in piers occurs when the applied stress exceeds the yield strength of the material. However, very long columns fail by sideways *buckling* long before the compressive stress reaches the yield strength. Buckling failure is sudden, often without significant initial sideways bending. The load at which a column fails is known as the *critical load* or *Euler load*.

The critical buckling load is the theoretical maximum load that an initially straight column can support without buckling. For columns with frictionless or pinned ends, this load is given by Eq. 52.7. r is the *radius of gyration*. K is the *end-restraint coefficient* (also known as *effective-length factor*), which varies from 0.5 to 2.0. For most real columns, the design values of K should be used since infinite stiffness of the supporting structure is not achievable.

Long Columns

$$P_{\text{cr}} = \frac{\pi^2 EI}{(KL)^2} = \frac{\pi^2 EA}{\left(\frac{KL}{r}\right)^2} \quad 52.7$$

The corresponding critical buckling stress is given by Eq. 52.8. In order to use Euler's theory, this stress cannot exceed half of the compressive yield strength of the column material.

Long Columns

$$\sigma_{\text{cr}} = \frac{P_{\text{cr}}}{A} = \frac{\pi^2 E}{\left(\frac{KL}{r}\right)^2} \quad [\sigma_{\text{cr}} < \tfrac{1}{2}S_y] \quad 52.8$$

The quantity L/r is known as the *slenderness ratio*. Long columns have high slenderness ratios. The smallest slenderness ratio for which Eq. 52.8 is valid, found by setting the critical buckling stress equal to half the compressive yield strength, is the *critical slenderness ratio*. Typical critical slenderness ratios range from 80 to 120. The critical slenderness ratio becomes smaller as the compressive yield strength increases.

L is the longest unbraced column length. If a column is braced against buckling at some point between its two ends, the column is known as a *braced column*, and L will be less than the full column height. Columns with rectangular cross sections have two radii of gyration, r_x and r_y, and so will have two slenderness ratios. The largest slenderness ratio will govern the design.

Euler's curve for columns, line BCD in Fig. 52.1, is generated by plotting the Euler stress versus the slenderness ratio. (See Eq. 52.8.) Since the material's compressive yield strength cannot be exceeded, a horizontal line, AC, is added to limit applications to the region below. Theoretically, members with slenderness ratios less than $(S_r)_C$ could be treated as pure compression members. However, this is not done in practice.

Figure 52.1 *Euler's Curve*

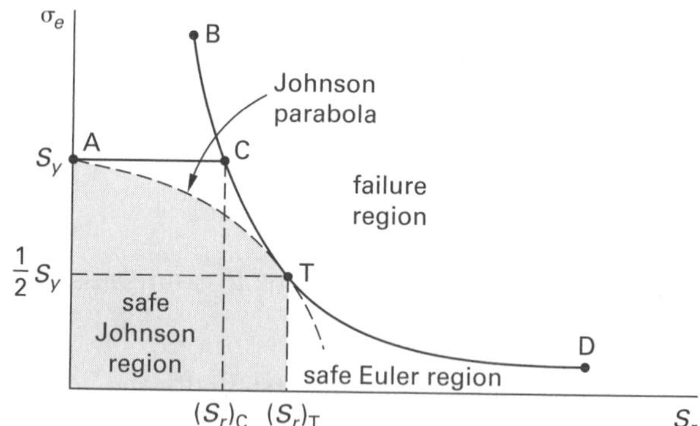

Defects in materials, errors in manufacturing, inabilities to achieve theoretical end conditions, and eccentricities frequently combine to cause column failures in the region around point C. This region is excluded by designers.

[2] *Dead load* is an inert, inactive load, primarily due to the structure's own weight. *Live load* is the weight of all nonpermanent objects, including people and furniture, in the structure.

[3] ACI 318 has been adopted as the source of concrete design rules in the United States.

The empirical *Johnson procedure* used to exclude the failure area is to draw a parabolic curve from point A through a tangent point T on the Euler curve at a stress of $(1/2)S_y$. The corresponding value of the slenderness ratio is

Intermediate Columns

$$(S_r)_T = \sqrt{\frac{2\pi^2 E}{KS_y}} \qquad 52.9$$

Example 52.1

A steel member is used as an 8.5 ft long column. The ends are pinned. The data for the column is shown.

$$E = 2.9 \times 10^7 \text{ lbf/in}^2$$
$$S_y = 36{,}000 \text{ lbf/in}^2$$
$$r = 0.569 \text{ in}$$

What is the maximum allowable compressive stress in order to have a factor of safety of 3.0?

Solution

First, check the slenderness ratio to see if this is a long column. $K = 1$ since the ends are pinned. [**Approximate Values of Effective Length Factor, K**]

From Eq. 52.9,

Intermediate Columns

$$(S_r)_T = \sqrt{\frac{2\pi^2 E}{KS_y}} = \sqrt{\frac{2\pi^2\left(2.9 \times 10^7 \ \frac{\text{lbf}}{\text{in}^2}\right)}{36{,}000 \ \frac{\text{lbf}}{\text{in}^2}}}$$
$$= 126.1$$

Find the slenderness ratio to be sure it is acceptable.

Long Columns

$$\frac{KL}{r} = \frac{(1)(8.5 \text{ ft})\left(12 \ \frac{\text{in}}{\text{ft}}\right)}{0.569 \text{ in}} = 179.3 \quad [>126.1, \text{ so OK}]$$

The Euler stress is found from Eq. 52.8.

Long Columns

$$\sigma_{cr} = \frac{\pi^2 E}{\left(\frac{KL}{r}\right)^2} = \frac{\pi^2\left(2.9 \times 10^7 \ \frac{\text{lbf}}{\text{in}^2}\right)}{(179.3)^2}$$
$$= 8903 \text{ lbf/in}^2$$

Since 8903 lbf/in^2 is less than half of the yield strength of 36,000 lbf/in^2, the Euler formula is valid. From Eq. 52.2, the allowable working stress is

$$\sigma_a = \frac{S_y}{\text{FS}} = \frac{\sigma_{cr}}{\text{FS}}$$
$$= \frac{8903 \ \frac{\text{lbf}}{\text{in}^2}}{3.0}$$
$$= 2968 \text{ lbf/in}^2$$

4. INTERMEDIATE COLUMNS

Columns with slenderness ratios less than or equal to the column stress determination factor ($S_r < (S_r)_D$), but that are too long to be short piers, are known as *intermediate columns.*

Intermediate Columns

$$(S_r)_D = \sqrt{\frac{2\pi^2 E}{K^2 S_y}} \qquad 52.10$$

The critical axial load (the buckling load) is given by Eq. 52.11.

Intermediate Columns

$$P_{cr} = A\left(S_y - \frac{K^2}{E}\left(\frac{S_y S_r}{2\pi}\right)^2\right) \qquad 52.11$$

5. THICK-WALLED CYLINDERS

A thick-walled cylinder has a wall thickness-to-radius ratio greater than 0.2 (i.e., a wall thickness-to-diameter ratio greater than 0.1). Figure 52.2 illustrates a thick-walled tank under either internal or external pressure.

In thick-walled tanks, radial stress is significant and cannot be disregarded. In *Lamé's solution,* a thick-walled cylinder is assumed to be made up of thin laminar rings. This method shows that the radial and circumferential stresses vary with location within the tank wall. (The term *circumferential stress* is preferred over *hoop stress* when dealing with thick-walled cylinders.) Compressive stresses are negative.

$$\sigma_c = \frac{r_i^2 p_i - r_o^2 p_o + \frac{(p_i - p_o) r_i^2 r_o^2}{r^2}}{r_o^2 - r_i^2} \qquad 52.12$$

$$\sigma_r = \frac{r_i^2 p_i - r_o^2 p_o - \frac{(p_i - p_o) r_i^2 r_o^2}{r^2}}{r_o^2 - r_i^2} \qquad 52.13$$

$$\sigma_l = \frac{p_i r_i^2}{r_o^2 - r_i^2} \quad \left[\begin{array}{c} p_o \text{ does not act} \\ \text{longitudinally on the ends} \end{array}\right] \qquad 52.14$$

At every point in the cylinder, the circumferential, radial, and long stresses are the principal stresses. Unless an external torsional shear stress is added, it is not necessary to use the combined stress equations. Failure theories can be applied directly.

Figure 52.2 *Thick-Walled Cylinder*

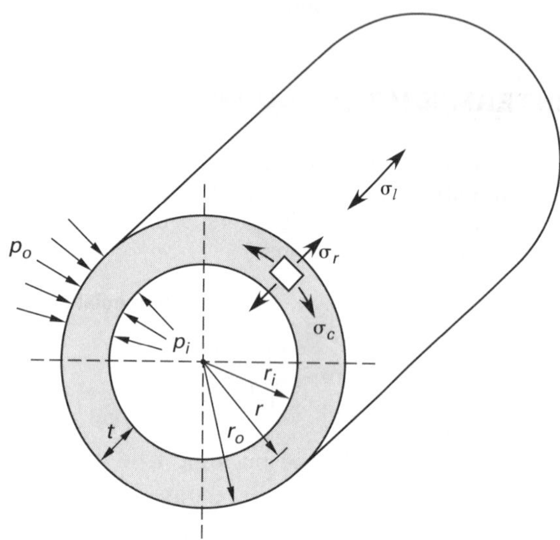

The cases of main interest are those of internal or external pressure only. The stress equations for these cases are summarized in Table 52.1. The maximum shear and normal stresses occur at the inner surface for both internal and external pressure.

Table 52.1 *Stresses in Thick-Walled Cylinders**

stress	external pressure, p	internal pressure, p
$\sigma_{c,o}$	$\dfrac{-(r_o^2 + r_i^2)p_o}{r_o^2 - r_i^2}$	$\dfrac{2r_i^2 p_i}{r_o^2 - r_i^2}$
$\sigma_{r,o}$	$-p_o$	0
$\sigma_{c,i}$	$\dfrac{-2r_o^2 p_o}{r_o^2 - r_i^2}$	$\dfrac{(r_o^2 + r_i^2)p_i}{r_o^2 - r_i^2}$
$\sigma_{r,i}$	0	$-p_i$
$\tau_{\max}$	$\frac{1}{2}\sigma_{c,i}$	$\frac{1}{2}(\sigma_{c,i} + p_i)$

*This table can be used with thin-walled cylinders. However, in most cases it will not be necessary to do so.

The *diametral strain* (which is the same as the *circumferential* and *radial strains*) is given by Eq. 52.15. Radial stresses are always compressive (and negative), and algebraic signs must be observed with Eq. 52.15. Since the circumferential and radial stresses depend on location within the wall thickness, the strain can be evaluated at inner, outer, and any intermediate locations within the wall.

$$\begin{aligned} \epsilon &= \frac{\Delta D}{D} = \frac{\Delta C}{C} = \frac{\Delta r}{r} \\ &= \frac{\sigma_c - \nu(\sigma_r + \sigma_l)}{E} \end{aligned} \qquad 52.15$$

6. INTERFERENCE FITS

When assembling two pieces, interference fitting is often more economical than pinning, keying, or splining. The assembly operation can be performed in a hydraulic press, either with both pieces at room temperature or after heating the outer piece and cooling the inner piece. The former case is known as a *press fit* or *interference fit*; the latter is known as a *shrink fit*.

If two cylinders are pressed together, the pressure acting between them will expand the outer cylinder (placing it into tension) and will compress the inner cylinder. The radial *interference*, δ, is the difference in dimensions between the two cylinders.[4]

Interference-Fit Stresses

$$\delta = |\delta_o| + |\delta_i| \qquad 52.16$$

$$\begin{aligned} \delta &= \delta_o - \delta_i \\ &= I_{\text{radial}} = \frac{I_{\text{diametral}}}{2} \\ &= r_{o,\text{inner}} - r_{i,\text{outer}} \\ &= |\Delta r_{o,\text{inner}}| + |\Delta r_{i,\text{outer}}| \\ &= |\delta_o| + |\delta_i| \end{aligned} \qquad 52.17$$

If the two cylinders have the same length, the thick-wall cylinder equations can be used. The materials used for the two cylinders do not need to be the same. Since there is no longitudinal stress from an interference fit and since the radial stress is negative, the strain Eq. 52.15 is

$$\begin{aligned} \epsilon &= \frac{\Delta D}{D} = \frac{\Delta C}{C} = \frac{\Delta r}{r} \\ &= \frac{\sigma_c - \nu\sigma_r}{E} \end{aligned} \qquad 52.18$$

[4]The equations for interference fits as given in the *NCEES PE Mechanical Reference Handbook* (*NCEES Handbook*) use different designations for the member radii than are used in common engineering practice. a represents the inner radius of the shaft, b represents the outer radius of the shaft, and c represents the outer radius of the hub. The *NCEES Handbook* equations are included here, along with variations which use industry-standard variables and subscripts. The PE exam will require the use of the *NCEES Handbook* equations, but the variations may make these equations easier to understand and apply.

Equation 52.19 applies to the general case where both cylinders are hollow and have different moduli of elasticity and Poisson's ratios. The outer cylinder is designated as the *hub*; the inner cylinder is designated as the *shaft*. If the shaft is solid, use $r_{i,\text{shaft}} = 0$ in Eq. 52.19.

Contact pressure is defined as

Interference-Fit Stresses

$$p = \frac{\delta}{bA} = \frac{\delta}{r_{o,\text{shaft}}A} \qquad 52.19$$

$$\begin{aligned} A &= \left(\frac{1}{E_i}\right)\left(\frac{b^2+a^2}{b^2-a^2} - \nu_i\right) + \left(\frac{1}{E_o}\right)\left(\frac{c^2+b^2}{c^2-b^2} + \nu_o\right) \\ &= \frac{1}{E_i}\left(\frac{r_{o,\text{shaft}}^2 + r_{i,\text{shaft}}^2}{r_{o,\text{shaft}}^2 - r_{i,\text{shaft}}^2} - \nu_i\right) \\ &\quad + \frac{1}{E_o}\left(\frac{r_{o,\text{hub}}^2 + r_{o,\text{shaft}}^2}{r_{o,\text{hub}}^2 - r_{o,\text{shaft}}^2} - \nu_o\right) \end{aligned} \qquad 52.20$$

$$A = \left(\frac{1}{E_i}\right)\left(\frac{b^2+a^2}{b^2-a^2} - \nu_i\right) + \left(\frac{1}{E_o}\right)\left(\frac{c^2+b^2}{c^2-b^2} + \nu_o\right) \qquad 52.21$$

$$\delta_o = \left(\frac{Pb}{E_o}\right)\left(\frac{c^2+b^2}{c^2-b^2} + \nu_o\right) \qquad 52.22$$

$$\delta_i = \left(\frac{Pb}{E_i}\right)\left(\frac{b^2+a^2}{b^2-a^2} - \nu_i\right) \qquad 52.23$$

$$\begin{aligned} \delta_o - \delta_i &= \left(\frac{pr_{o,\text{shaft}}}{E_\text{hub}}\right)\left(\frac{r_{o,\text{hub}}^2 + r_{o,\text{shaft}}^2}{r_{o,\text{hub}}^2 - r_{o,\text{shaft}}^2} + \nu_\text{hub}\right) \\ &\quad + \left(\frac{pr_{o,\text{shaft}}}{E_\text{shaft}}\right)\left(\frac{r_{o,\text{shaft}}^2 + r_{i,\text{shaft}}^2}{r_{o,\text{shaft}}^2 - r_{i,\text{shaft}}^2} - \nu_\text{shaft}\right) \end{aligned} \qquad 52.24$$

In the special case where the shaft is solid and is made from the same material as the hub (i.e. moduli and Poisson's ratio are identical), the radial interference is given by Eq. 52.25.

$$\delta = \left(\frac{2pr_\text{shaft}}{E}\right)\left(\frac{1}{1 - \left(\frac{r_\text{shaft}}{r_{o,\text{hub}}}\right)^2}\right) \qquad 52.25$$

The resulting contact pressure is given by Eq. 52.26.

Interference-Fit Stresses

$$\begin{aligned} p &= \left(\frac{E\delta}{2bc^2}\right)(c^2 - b^2) \\ &= \left(\frac{E\delta}{2r_{o,\text{shaft}}r_{o,\text{hub}}^2}\right)(r_{o,\text{hub}}^2 - r_{o,\text{shaft}}^2) \end{aligned} \qquad 52.26$$

If the shaft is not solid but the moduli and Poisson's ratio are still identical, the resulting contact pressure is given by Eq. 52.27.

Interference-Fit Stresses

$$\begin{aligned} p &= \left(\frac{E\delta}{b}\right)\left(\frac{(c^2-b^2)(b^2-a^2)}{2b^2(c^2-a^2)}\right) \\ &= \left(\frac{E\delta}{r_{o,\text{shaft}}}\right)\frac{(r_{o,\text{hub}}^2 - r_{o,\text{shaft}}^2)(r_{o,\text{shaft}}^2 - r_{i,\text{shaft}}^2)}{2r_{o,\text{shaft}}^2(r_{o,\text{hub}}^2 - r_{i,\text{shaft}}^2)} \end{aligned} \qquad 52.27$$

The maximum assembly force required to overcome friction during a press-fitting operation is given by Eq. 52.28. This relationship is approximate because the coefficient of friction is not known with certainty, and the assembly force affects the pressure, p, through Poisson's ratio. The coefficient of friction is highly variable. Values in the range of 0.03–0.33 have been reported. In the absence of experimental data, it is reasonable to use 0.12 for lightly oiled connections and 0.15 for dry assemblies.

$$F_\text{max} = \mu N = 2\pi\mu p r_{o,\text{shaft}} L_\text{engagement} \qquad 52.28$$

The maximum torque that the press-fitted hub can withstand or transmit is given by Eq. 52.29. This can be greater or less than the shaft's torsional shear capacity. Both values should be calculated.

Interference-Fit Stresses

$$T = 2\pi b^2 \mu p L = 2\pi\mu p r_{o,\text{shaft}}^2 L_\text{engagement} \qquad 52.29$$

Most interference fits are designed to keep the contact pressure or the stress below a given value. Designs of interference fits limited by strength generally use the distortion energy failure criterion. That is, the maximum shear stress is compared with the shear strength determined from the failure theory.

Example 52.2

A steel cylinder has inner and outer diameters of 1.0 in and 2.0 in, respectively. The cylinder is pressurized internally to 10,000 lbf/in^2. The modulus of elasticity is 2.9×10^7 lbf/in^2, and Poisson's ratio is 0.3. What is the radial strain at the inside face?

Solution

The longitudinal stress is

$$\sigma_l = \frac{F}{A} = \frac{p_i \pi r_i^2}{\pi(r_o^2 - r_i^2)} = \frac{p_i r_i^2}{r_o^2 - r_i^2}$$
$$= \frac{\left(10{,}000\ \frac{\text{lbf}}{\text{in}^2}\right)(0.5\ \text{in})^2}{(1.0\ \text{in})^2 - (0.5\ \text{in})^2}$$
$$= 3333\ \text{lbf/in}^2$$

The stresses at the inner face are found from Table 52.1.

$$\sigma_{c,i} = \frac{(r_o^2 + r_i^2)p}{r_o^2 - r_i^2}$$
$$= \frac{\left((1.0\ \text{in})^2 + (0.5\ \text{in})^2\right)\left(10{,}000\ \frac{\text{lbf}}{\text{in}^2}\right)}{(1.0\ \text{in})^2 - (0.5\ \text{in})^2}$$
$$= 16{,}667\ \text{lbf/in}^2$$
$$\sigma_{r,i} = -p = -10{,}000\ \text{lbf/in}^2$$

The circumferential and radial stresses increase the radial strain; the longitudinal stress decreases the radial strain. The radial strain is

$$\frac{\Delta r}{r} = \frac{\sigma_{c,i} - \nu(\sigma_{r,i} + \sigma_l)}{E}$$
$$= \frac{16{,}667\ \frac{\text{lbf}}{\text{in}^2} - (0.3)\left(-10{,}000\ \frac{\text{lbf}}{\text{in}^2} + 3333\ \frac{\text{lbf}}{\text{in}^2}\right)}{2.9 \times 10^7\ \frac{\text{lbf}}{\text{in}^2}}$$
$$= 6.44 \times 10^{-4}$$

Example 52.3

A hollow aluminum cylinder is pressed over a hollow brass cylinder as shown. Both cylinders are 2 in long. The interference is 0.004 in. The average coefficient of friction during assembly is 0.25. (a) What is the maximum shear stress in the brass? (b) What initial disassembly force is required to separate the two cylinders?

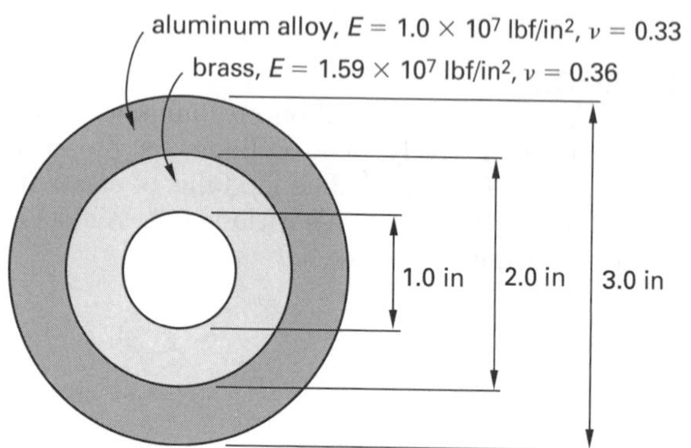

Solution

(a) Work with the aluminum outer cylinder, which is under internal pressure.

$$\sigma_{c,i} = \frac{(r_o^2 + r_i^2)p}{r_o^2 - r_i^2} = \frac{\left((1.5\ \text{in})^2 + (1.0\ \text{in})^2\right)p}{(1.5\ \text{in})^2 - (1.0\ \text{in})^2}$$
$$= 2.6p$$
$$\sigma_{r,i} = -p$$

From Eq. 52.15, the diametral strain is

$$\epsilon = \frac{\sigma_{c,i} - \nu(\sigma_{r,i} + \sigma_l)}{E} = \frac{2.6p - 0.33(-p)}{1.0 \times 10^7\ \frac{\text{lbf}}{\text{in}^2}}$$
$$= 2.93 \times 10^{-7}p$$
$$\Delta D = \epsilon D = (2.93 \times 10^{-7}p)(2.0\ \text{in})$$
$$= 5.86 \times 10^{-7}p$$

Now work with the brass inner cylinder, which is under external pressure. Use Table 52.1.

$$\sigma_{c,o} = \frac{-(r_o^2 + r_i^2)p}{r_o^2 - r_i^2} = \frac{-\left((1.0\ \text{in})^2 + (0.5\ \text{in})^2\right)p}{(1.0\ \text{in})^2 - (0.5\ \text{in})^2}$$
$$= -1.667p$$
$$\sigma_{r,o} = -p$$

From Eq. 52.15, the diametral strain is

$$\epsilon = \frac{\sigma_{c,o} - \nu(\sigma_{r,o} + \sigma_l)}{E}$$
$$= \frac{-1.667p - 0.36(-p)}{1.59 \times 10^7\ \frac{\text{lbf}}{\text{in}^2}}$$
$$= -0.822 \times 10^{-7}p$$
$$\Delta D = \epsilon D = (-0.822 \times 10^{-7}p)(2.0\ \text{in})$$
$$= -1.644 \times 10^{-7}p$$

The diametral interference is known to be 0.004 in. From Eq. 52.16,

$$I_{\text{diametral}} = |\Delta D_{o,\text{inner}}| + |\Delta D_{i,\text{outer}}|$$
$$0.004\ \text{in} = |5.86 \times 10^{-7}p| + |-1.644 \times 10^{-7}p|$$
$$p = 5330\ \text{lbf/in}^2$$

(Equation 52.19 could have been used to find p directly.)

From Table 52.1, the circumferential stress at the inner face of the brass (under external pressure) is

$$\sigma_{c,i} = \frac{-2r_o^2 p}{r_o^2 - r_i^2} = \frac{(-2)(1.0\ \text{in})^2\left(5330\ \frac{\text{lbf}}{\text{in}^2}\right)}{(1.0\ \text{in})^2 - (0.5\ \text{in})^2}$$
$$= -14{,}213\ \text{lbf/in}^2$$

Also from Table 52.1, the maximum shear stress is

$$\tau_{\max} = \tfrac{1}{2}\sigma_{c,i} = \left(\frac{1}{2}\right)\left(-14{,}213\ \frac{\text{lbf}}{\text{in}^2}\right)$$
$$= -7107\ \text{lbf/in}^2$$

(b) The initial force necessary to disassemble the two cylinders is the same as the maximum assembly force. Use Eq. 52.28.

$$F_{\max} = \mu N = 2\pi\mu p r_{o,\text{shaft}} L_{\text{engagement}}$$
$$= (2\pi)(0.25)\left(5330\ \frac{\text{lbf}}{\text{in}^2}\right)(1\ \text{in})(2\ \text{in})$$
$$= 16{,}745\ \text{lbf}$$

7. BOLTS

There are three leading specifications for bolt thread families: American National Standards Institute (ANSI), International Organization for Standardization (ISO) metric, and Deutsches Institut für Normung (DIN) metric.[5] ANSI is widely used in the United States. DIN fasteners are widely available and broadly accepted.[6] ISO metric fasteners are used in large volume by U.S. car manufacturers. The European Committee for Standardization (CEN) standards promulgated by the European Community (EC) have essentially adopted the ISO standards.

An American National (Unified) thread is specified by the sequence of parameters S(×L)-N-F-A-(H-E), where S is the thread outside diameter (nominal size), L is the optional shank length, N is the number of threads per inch, F is the thread pitch family, A is the class (allowance), and H and E are the optional hand and engagement length designations. The letter R can be added to the thread pitch family to indicate that the thread roots are radiused (for better fatigue resistance). For example, a 3/8 × 1-16UNC-2A bolt has a 3/8 in diameter, a 1 in length, and 16 Unified Coarse threads per inch rolled with a class 2A accuracy.[7] A UNRC bolt would be identical except for radiused roots. Table 52.2 lists some (but not all) values for these parameters.

Table 52.2 *Representative American National (Unified) Bolt Thread Designations*[a]

S	Size
	1–12
	1/4–9/16 in in 1/16 in increments
	5/8–1 1/2 in in 1/8 in increments
	1 3/4–4 in in 1/4 in increments
F	Thread Family
	UNC and NC—Unified Coarse[b]
	UNF and NF—Unified Fine[b]
	UNEF and NEF—Unified Extra Fine[c]
	8N—8 threads per inch
	12UN and 12N—12 threads per inch
	16UN and 16N—16 threads per inch
	UN, UNS, and NS—special series
A	Allowance (A—external threads, B—internal threads)[d]
	1A and 1B—liberal allowance for each of assembly with dirty or damaged threads
	2A and 2B—normal production allowance (sufficient for plating)
	3A and 3B—close tolerance work with no allowance
H	Hand
	blank—right-hand thread
	LH—left-hand thread

[a]In addition to fastener thread families, there are other special-use threads, such as Acme, stub, square, buttress, and worm series.
[b]Previously known as United States Standard or American Standard.
[c]The UNEF series is the same as the Society of Automotive Engineers (SAE) fine series.
[d]Allowance classes 2 and 3 (without the A and B designation) were used prior to industry transition to the Unified classes.

The *grade* of a bolt indicates the fastener material and is marked on the bolt cap.[8] In this regard, the marking depends on whether an SAE grade or ASTM designation is used. The minimum *proof load* (i.e., the maximum stress the bolt can support without acquiring a permanent set) increases with the grade. (The term *proof strength* is less common.) If a bolt is manufactured in the United States, its cap must also show the logo or mark of the manufacturer.

A metric thread is specified by an M or MJ and a diameter and a pitch in millimeters, in that order. For example, M10 × 1.5 is a thread having a nominal major diameter of 10 mm and a pitch of 1.5 mm. The MJ series have rounded root fillets and larger minor diameters.

[5]Other fastener families include the Italian UNI, Swiss VSM, Japanese JIS, and United Kingdom's BS series.
[6]To add to the confusion, many DIN standards are identical to ISO standards, with only slight differences in the tolerance ranges. However, the standards are not interchangeable in every case.
[7]Threads are generally rolled, not cut, into a bolt.
[8]The *type* of a structural bolt should not be confused with the *grade* of a structural rivet.

Head markings on metric bolts indicate their *property class* and correspond to the approximate tensile strength in MPa divided by 100. For example, a bolt marked 8.8 would correspond to a medium carbon, quenched and tempered bolt with an approximate tensile strength of 880 MPa. (The minimum of the tensile strength range for property class 8.8 is 830 MPa.)

8. RIVET AND BOLT CONNECTIONS

Figure 52.3 illustrates a tension *lap joint* connection using rivet or bolt connectors.[9] Unless the plate material is very thick, the effects of eccentricity are disregarded. A connection of this type can fail in shear, tension, or bearing. A common design procedure is to determine the number of connectors based on shear stress and then to check the bearing and tensile stresses.

Figure 52.3 *Tension Lap Joint*

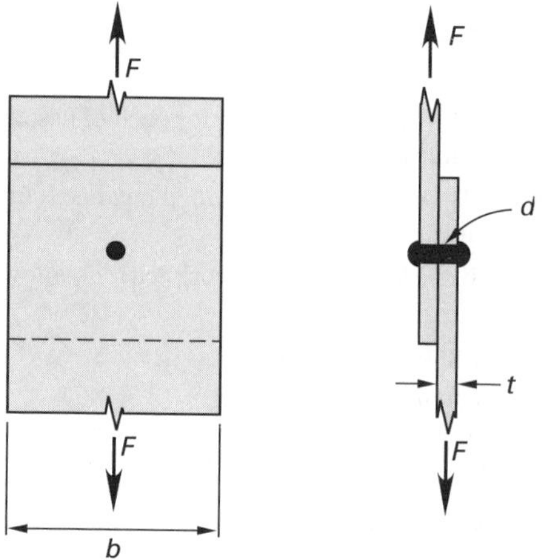

One of the failure modes is shearing of the connectors. In the case of *single shear*, each connector supports its proportionate share of the load. In *double shear*, each connector has two shear planes, and the stress per connector is halved.[10] (See Fig. 52.4.) The shear stress in a cylindrical connector is

$$\tau = \frac{F}{A} = \frac{F}{\frac{\pi D^2}{4}} \qquad 52.30$$

The number of required connectors, as determined by shear, is

$$n = \frac{\tau}{\text{allowable shear stress}} \qquad 52.31$$

Figure 52.4 *Single and Double Shear*

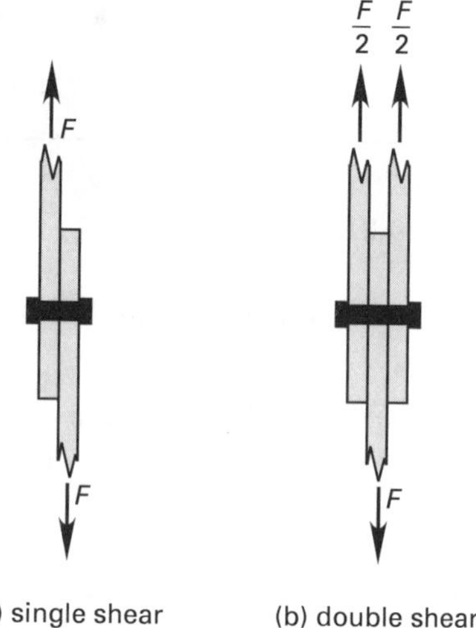

The plate can fail in tension. Although the design of structural steel building members in tension is codified, theoretically, if there are n connector holes of diameter D in a line across the width, b, of the plate, the cross-sectional area in the plate remaining to resist the tension is

$$A_t = t(b - nD) \qquad 52.32$$

The number of connectors across the plate width must be chosen to keep the tensile stress less than the allowable stress. The maximum tensile stress in the plate will be

$$\sigma_t = \frac{F}{A_t} \qquad 52.33$$

The plate can also fail by *bearing* (i.e., crushing). The number of connectors must be chosen to keep the actual *bearing stress* below the allowable bearing stress. For one connector, the bearing stress in the plate is

$$\sigma_p = \frac{F}{Dt} \qquad 52.34$$

$$n = \frac{\sigma_p}{\sigma_{a,\text{bearing}}} \qquad 52.35$$

The plate can also fail by shear tear-out, as illustrated in Fig. 52.5. The shear stress is

$$\tau = \frac{F}{2A_s} = \frac{F}{2t\left(L - \frac{D}{2}\right)} \qquad 52.36$$

[9]Rivets are no longer used in building construction, but they are still extensively used in manufacturing.

[10]"Double shear" is not the same as "double rivet" or "double butt." *Double shear* means that there are two shear planes in one rivet. *Double rivet* means that there are two rivets along the force path. *Double butt* refers to the use of two backing plates (i.e., "scabs") used on either side to make a tension connection between two plates. Similarly, *single butt* refers to the use of a single backing plate to make a tension connection between two plates.

Figure 52.5 Shear Tear-Out

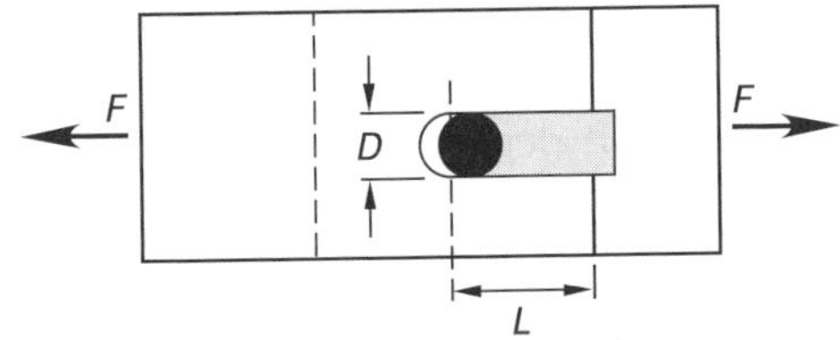

The *joint efficiency* is the ratio of the strength of the joint divided by the strength of a solid (i.e., unpunched or undrilled) plate.

9. BOLT PRELOAD

Consider the ungasketed connection shown in Fig. 52.6. The load varies from F_{min} to F_{max}. If the bolt is initially snug but without initial tension, the force in the bolt also will vary from F_{min} to F_{max}. If the bolt is tightened so that there is an initial *preload force*, F_i, greater than F_{max} in addition to the applied load, the bolt will be placed in tension and the parts held together will be in compression.[11] When a load is applied, the bolt tension will increase even more, but the compression in the members will decrease.

Figure 52.6 Bolted Tension Joint with Varying Load

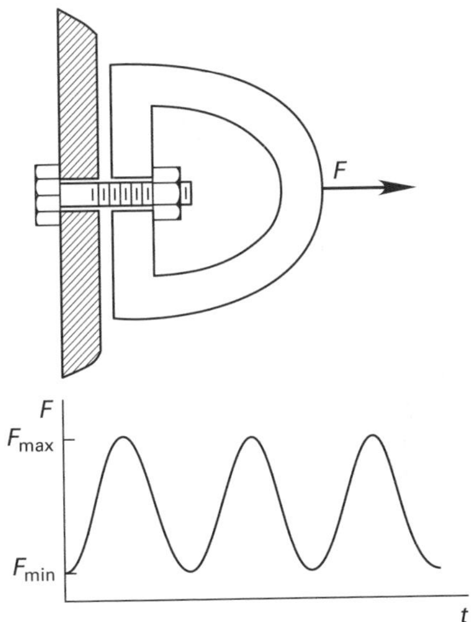

The amount of compression in the parts will vary as the applied load varies. The clamped members will carry some of the applied load, since this varying load has to "uncompress" the clamped part as well as lengthen the bolt. The net result is the reduction of the variation of the force in the bolt. The initial tension produces a larger mean stress, but the overall result is the reduction of the alternating stress. Preloading is an effective method of reducing the alternating stress in bolted tension connections.

It is convenient to define the *spring constant*, k, of the bolt. The *grip*, L, is the thickness of the members being connected by the bolt (not the bolt length). It is common to use the nominal diameter of the bolt, disregarding the reduction due to threading.

$$k_b = \frac{F}{\Delta L} = \frac{A_b E_b}{L} \qquad 52.37$$

The actual spring constant for a bolted member, k_m, is difficult to determine if the clamped area is not small and well defined. The only accurate way to determine the stiffness of a member in a bolted joint is through experimentation. If the clamped members are flat plates, various theories can be used to calculate the effective load-bearing areas of the flanges, but doing so is a laborious process.

One simple rule of thumb is that the bolt force spreads out to three times the bolt-hole diameter. Of course, the hole diameter needs to be considered (i.e., needs to be subtracted) in calculating the effective force area. If the modulus of elasticity is the same for the bolt and the clamped members, using this rule of thumb, the larger area results in the members being eight times stiffer than the bolts.

$$k_m = \frac{A_{e,m} E_m}{L} \qquad 52.38$$

If the clamped members have different moduli of elasticity, or if a gasket constitutes one of the layers compressed by the bolt, the composite spring constant can be found from Eq. 52.39.[12]

$$\frac{1}{k_{m,\text{composite}}} = \frac{1}{k_1} + \frac{1}{k_2} + \frac{1}{k_3} + \cdots \qquad 52.39$$

The corresponding stiffness constant of the joint is

Joining Methods

$$C = \frac{k_b}{k_b + k_m} \qquad 52.40$$

The bolt and the clamped members all carry parts of the applied load, P. F_i is the initial preload force.

Joining Methods

$$F_b = CP + F_i = \left(\frac{k_b}{k_b + k_m}\right)P + F_i \qquad 52.41$$

[11]If the initial preload force is less than F_{max}, the bolt may still carry a portion of the applied load. Equation 52.42 can be solved for the value of F that will result in a loss of compression ($F_{parts} = 0$) and cause the bolt to carry the entire applied load.

[12]If a soft washer or gasket is used, its spring constant can control Eq. 52.39.

$$F_m = (1 - C)P - F_i = \left(1 - \frac{k_b}{k_b + k_m}\right)P - F_i \quad \text{52.42}$$

O-ring (metal and elastomeric) seals permit metal-to-metal contact and affect the effective spring constant of the members very little. However, the seal force tends to separate bolted members and must be added to the applied force. The seal force can be obtained from the seal deflection and seal stiffness or from manufacturer's literature.

For static loading, recommended amounts of preloading often are specified as a percentage of the *proof load* (or *proof strength*) in psi or MPa.[13] For bolts, the proof load is slightly less than the yield strength. Traditionally, preload has been specified conservatively as 75% of proof for reusable connectors and 90% of proof for one-use connectors.[14] Connectors with some ductility can safely be used beyond the yield point, and 100% is now in widespread use.[15] When understood, advantages of preloading to 100% of proof load often outweigh the disadvantages.[16]

If the applied load varies, the forces in the bolt and members will also vary. In that case, the preload must be determined from an analysis of the Goodman line.

Tightening of a tension bolt will induce a torsional stress in the bolt.[17] Where the bolt is to be locked in place, the torsional stress can be removed without greatly affecting the preload by slightly backing off the bolt. If the bolt is subject to cyclic loading, the bolt will probably slip back by itself, and it is reasonable to neglect the effects of torsion in the bolt altogether. (This is the reason that well-designed connections allow for a loss of 5–10% of the initial preload during routine use.)

Stress concentrations at the beginning of the threaded section are significant in cyclic loading.[18] To avoid a reduction in fatigue life, the alternating stress used in the Goodman line should be multiplied by an appropriate stress concentration factor, K. For fasteners with rolled threads, an average factor of 2.2 for SAE grades 0–2 (metric grades 3.6–5.8) is appropriate. For SAE grades 4–8 (metric grades 6.6–10.9), an average factor of 3.0 is appropriate. Stress concentration factors for the fillet under the bolt head are different, but lower than these values. Stress concentration factors for cut threads are much higher.

The stress in a bolt depends on its load-carrying area. This area is typically obtained from a table of bolt properties. In practice, except for loading near the bolt's failure load, working stresses are low, and the effects of threads are usually ignored, so the area is based on the major (nominal) diameter.

$$\sigma_{\text{bolt}} = \frac{KF}{A} \quad \text{52.43}$$

10. BOLT TORQUE TO OBTAIN PRELOAD

During assembly, the preload tension is not monitored directly. Rather, the torque required to tighten the bolt is used to determine when the proper preload has been reached. Methods of obtaining the required preload include the standard torque wrench, the *run-of-the-nut method* (e.g., turning the bolt some specific angle past snugging torque), *direct-tension indicating* (DTI) washers, and computerized automatic assembly.

The standard manual torque wrench does not provide precise, reliable preloads, since the fraction of the torque going into bolt tension is variable.[19] Torque-, angle-, and time-monitoring equipment, usually part of an automated assembly operation, is essential to obtaining precise preloads on a consistent basis. It automatically applies the snugging torque and specified rotation, then checks the results with torque and rotation sensors. The computer warns of out-of-spec conditions.

The *Maney formula* is a simple relationship between the initial bolt tension, F_i, and the installation torque, T. The *torque coefficient*, K_T (also known as the *bolt torque factor* and the *nut factor*) found from Eq. 52.45 depends mainly on the coefficient of friction, μ, and is not the same as the coefficient of friction, as shown. The torque coefficient for lubricated bolts generally varies from 0.15 to 0.20, and a value of 0.2 is commonly used.[20] With antiseize lubrication, it can drop as low as 0.12.

[13]This is referred to as a "rule of thumb" specification, because a mathematical analysis is not performed to determine the best preload.

[14]Some U.S. military specifications call for 80% of proof load in tension fasteners and only 30% for shear fasteners. The object of keeping the stresses below yielding is to be able to reuse the bolts.

[15]Even under normal elastic loading of a bolt, local plastic deformation occurs in the bolt-head fillet and thread roots. Since the stress-strain curve is nearly flat at the yield point, a small amount of elongation into the plastic region does not increase the stress or tension in the bolt.

[16]The disadvantages are (a) field maintenance probably will not be possible, as manually running up bolts to 100% proof will result in many broken bolts; (b) bolts should not be reused, as some will have yielded; and (c) the highest-strength bolts do not exhibit much plastic elongation and ordinarily should not be run up to 100% proof load.

[17]An argument for the conservative 75% of proof load preload limit is that the residual torsional stress will increase the bolt stress to 90% or higher anyway, and the additional 10% needed to bring the preload up to 100% probably will not improve economic performance much.

[18]Stress concentrations are frequently neglected for static loading.

[19]Even with good lubrication, about 50% of the torque goes into overcoming friction between the head and collar/flange, another 40% is lost in thread friction, and only the remaining 10% goes into tensioning the connector.

[20]With a coefficient of friction of 0.15, the torque coefficient is approximately 0.20 for most bolt sizes, regardless of whether the threads are coarse or fine.

Torque Requirements

$$T = KF_iD$$

$$\left[F_i = \begin{array}{l} 0.75F_p \text{ for preload connections} \\ 0.90F_p \text{ for permanent connections} \end{array}\right] \quad 52.44$$

$$K = \frac{\mu_c r_c}{D} + \left(\frac{r_t}{D}\right)\left(\frac{\tan\theta + \mu_t \sec\alpha}{1 - \mu_t \tan\theta \sec\alpha}\right) \quad 52.45$$

μ_c is the coefficient of friction at the collar (fastener bearing face). r_c is the mean collar radius (i.e., the effective radius of action of the friction forces on the bearing face). r_t is the effective radius of action of the frictional forces on the thread surfaces. f_t is the coefficient of friction between the thread contact surfaces. θ is the *lead angle*, also known as the *helix thread angle*, found from Eq. 52.46. α is the *thread half angle* (30° for UNF threads), and D is the mean thread diameter.

$$\tan\theta = \frac{\text{lead per revolution}}{2\pi r_t} \quad 52.46$$

The roof load is obtained from Eq. 52.47. A_t is the tensile stress area of the threaded portion, and S_p is the proof stress.

Torque Requirements

$$F_p = A_t S_p \quad 52.47$$

The *NCEES Handbook* provides additional guidance for determining specifications for steel bolts and other materials, as well as torque coefficient factors for different bolt conditions.

11. WELD STRESS

The common *fillet weld* is shown in Fig. 52.7. Such welds are used to connect one plate to another. The applied load, P, is assumed to be carried in shear by the *effective weld throat*. The *effective throat size*, t_e, is related to the weld size, h, by Eq. 52.48.

$$t_e = \frac{\sqrt{2}}{2}h = 0.707h \quad 52.48$$

Neglecting any increased stresses due to eccentricity, the shear stress in a fillet lap weld depends on the *effective throat thickness*, t_e, and is

$$\sigma = \frac{P}{t_eL} = \frac{1.414P}{hL} \quad 52.49$$

For plates of equal thicknesses and held by two welds of the same size (see Fig. 52.8), each weld carries half the load. Therefore the stress identified in Eq. 52.50 is also reduced in half for each weld in such a configuration.

Figure 52.7 *Fillet Lap Weld and Symbol*

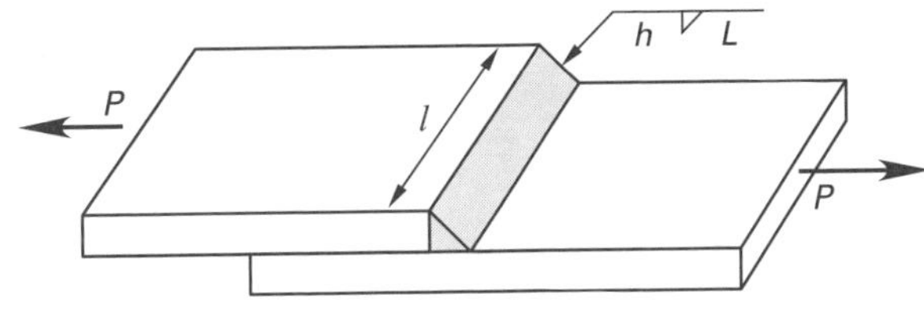

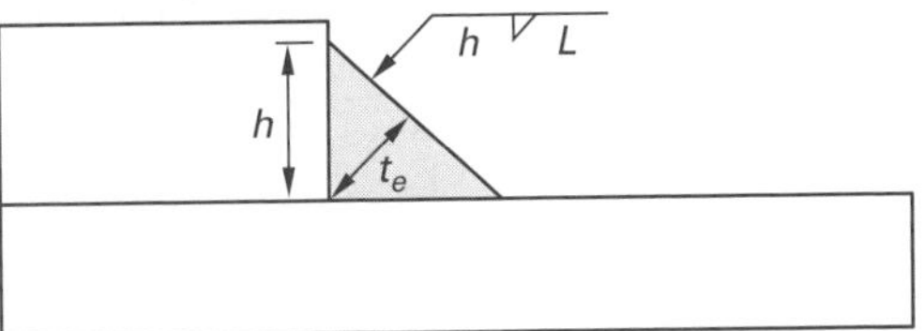

Figure 52.8 *Double Filet Lap Weld*

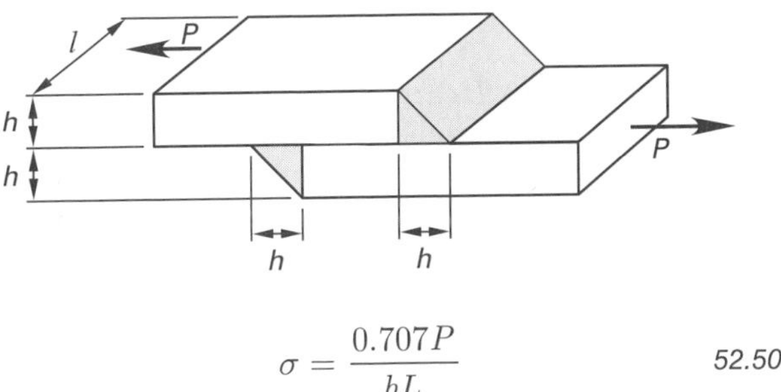

$$\sigma = \frac{0.707P}{hL} \quad 52.50$$

The American Welding Association's *Welding Handbook* provides equations to calculate the stress for a variety of weld configurations. [**Types of Welds**]

Weld (filler) metal should have a strength equal to or greater than the base material. Properties of filler metals are readily available from their manufacturers and, for standard rated welding rods, from engineering handbooks. [**Minimum Weld-Metal Properties**]

Various codes often reduce the allowable stress in weld metals versus base metals to ensure a higher factor of safety is present in the weld. The American Institute of Steel Construction (AISC) codes specify allowable stresses and corresponding corresponding factors of safety, n, based on the distortion-energy theory. [**Stresses Permitted by the AISC Code for Weld Material**]

12. CIRCULAR SHAFT DESIGN

Shear stress occurs when a shaft is placed in torsion. The shear stress at the outer surface of a bar of radius r, which is torsionally loaded by a torque, T, is

Torsion

$$\tau = \frac{Tr}{J} \quad 52.51$$

The total strain energy due to torsion is

Strain Energy

$$U = \frac{T^2 L}{2GJ} \quad 52.52$$

J is the shaft's polar moment of inertia. For a solid round shaft,

Properties of Various Shapes

$$J = \frac{\pi a^4}{2} \quad 52.53$$

For a hollow round shaft,

Properties of Various Shapes

$$J = \frac{\pi(a^4 - b^4)}{2} \quad 52.54$$

If a shaft of length L carries a torque T, as in Fig. 52.9, the angle of twist (in radians) will be

Torsional Strain

$$\phi = \frac{TL}{GJ} \quad 52.55$$

$$\phi = \frac{L\theta}{r} \quad 52.56$$

Figure 52.9 *Torsional Deflection of a Circular Shaft*

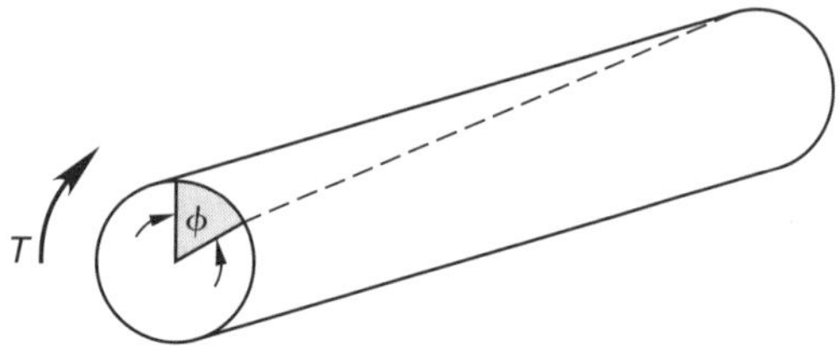

G is the *shear modulus*. For steel, it is approximately 11.5×10^6 lbf/in^2 (8.0×10^4 MPa). The shear modulus also can be calculated from the modulus of elasticity.

Shear Stress-Strain

$$G = \frac{E}{2(1+\nu)} \quad 52.57$$

The torque, T, carried by a shaft spinning at n revolutions per minute is related to the transmitted power.

$$T_{\text{N·m}} = \frac{9549 P_{\text{kW}}}{n_{\text{rpm}}} \quad \text{[SI]} \quad 52.58(a)$$

$$T_{\text{in-lbf}} = \frac{63{,}025 P_{\text{horsepower}}}{n_{\text{rpm}}} \quad \text{[U.S.]} \quad 52.58(b)$$

If a statically loaded shaft without axial loading experiences a bending stress, $\sigma_x = My/I$ (i.e., is loaded in flexure), in addition to torsional shear stress, $\tau = Tr/J$, the maximum shear stress from the combined stress theory is

$$\tau_{\max} = \sqrt{\left(\frac{\sigma_x}{2}\right)^2 + \tau^2} \quad 52.59$$

$$\tau_{\max} = \frac{16}{\pi D^3}\sqrt{M^2 + T^2} \quad 52.60$$

The equivalent normal stress from the distortion energy theory is

$$\sigma' = \frac{16}{\pi D^3}\sqrt{4M^2 + 3T^2} \quad 52.61$$

The diameter can be determined by setting the shear and normal stresses equal to the maximum allowable shear (as calculated from the maximum shear stress theory, $S_y/2(\text{FS})$, or from the distortion energy theory, $\sqrt{3}\,S_y/2(\text{FS})$, and normal stresses, respectively).

Equation 52.59 and Eq. 52.60 should not be used with dynamically loaded shafts (i.e., those that are turning). Fatigue design of shafts should be designed according to a specific code (e.g., ANSI or ASME) or should use a fatigue analysis (e.g., Goodman, Soderberg, or Gerber).

Example 52.4

The press-fitted, aluminum alloy-brass cylinder described in Ex. 52.3 is used as a shaft. The press fit is adequate to maintain nonslipping contact between the two materials. The shaft carries a steady torque of 24,000 in-lbf. There is no bending stress. What is the maximum torsional shear stress in the (a) aluminum and (b) brass?

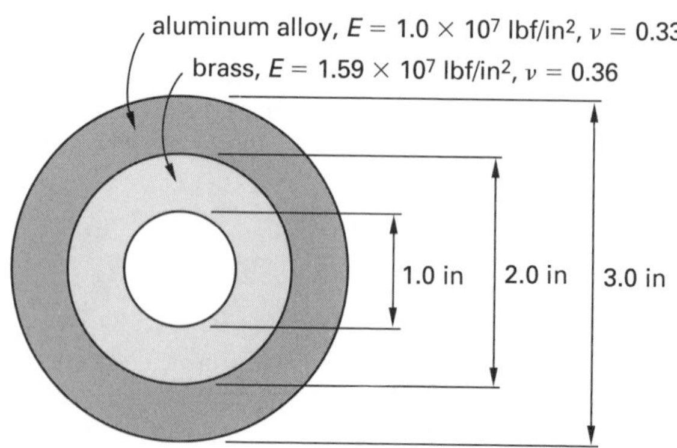

Solution

The stronger material (as determined from the shear modulus, G) should be converted to an equivalent area of the weaker material.

From Eq. 52.57, for the aluminum,

Shear Stress-Strain

$$G_{\text{aluminum}} = \frac{E}{2(1+\nu)} = \frac{1.0 \times 10^7 \ \frac{\text{lbf}}{\text{in}^2}}{(2)(1+0.33)}$$
$$= 3.76 \times 10^6 \ \text{lbf/in}^2$$

From Eq. 52.57, for the brass,

Shear Stress-Strain

$$G_{\text{brass}} = \frac{E}{2(1+\nu)} = \frac{1.59 \times 10^7 \ \frac{\text{lbf}}{\text{in}^2}}{(2)(1+0.36)}$$
$$= 5.85 \times 10^6 \ \text{lbf/in}^2$$

Brass has the higher shear modulus. The modular shear ratio is

$$n = \frac{G_{\text{brass}}}{G_{\text{aluminum}}} = \frac{5.85 \times 10^6 \ \frac{\text{lbf}}{\text{in}^2}}{3.76 \times 10^6 \ \frac{\text{lbf}}{\text{in}^2}}$$
$$= 1.56$$

The polar moment of inertia of the aluminum is

Properties of Various Shapes

$$J_{\text{aluminum}} = \frac{\pi(a^4 - b^4)}{2}$$
$$= \left(\frac{\pi}{2}\right)\left((1.5 \ \text{in})^4 - (1.0 \ \text{in})^4\right)$$
$$= 6.38 \ \text{in}^4$$

The equivalent polar moment of inertia of the brass is

$$J_{\text{brass}} = n\left(\frac{\pi}{2}\right)(a^4 - b^4)$$
$$= (1.56)\left(\frac{\pi}{2}\right)\left((1.0 \ \text{in})^4 - (0.5 \ \text{in})^4\right)$$
$$= 2.30 \ \text{in}^4$$

The total equivalent polar moment of inertia is

$$J_{\text{total}} = J_{\text{aluminum}} + J_{\text{brass}} = 6.38 \ \text{in}^4 + 2.30 \ \text{in}^4$$
$$= 8.68 \ \text{in}^4$$

(a) The maximum torsional shear stress in the aluminum occurs at the outer edge. Use Eq. 52.51.

Torsion

$$\tau = \frac{Tr}{J} = \frac{(24{,}000 \ \text{in-lbf})(1.5 \ \text{in})}{8.68 \ \text{in}^4}$$
$$= 4147 \ \text{lbf/in}^2$$

(b) Using the composite structures analysis methodology, the maximum torsional shear stress in the brass is

$$\tau = \frac{nTr}{J} = \frac{(1.56)(24{,}000 \ \text{in-lbf})(1.0 \ \text{in})}{8.68 \ \text{in}^4}$$
$$= 4313 \ \text{lbf/in}^2$$

13. TORSION IN THIN-WALLED SHELLS

Shear stress due to torsion in a thin-walled shell (also known as a *closed box*) acts around the perimeter of the shell, as shown in Fig. 52.10. The shear stress, τ, is given by Eq. 52.62. A_m is the mean area enclosed by the shaft as measured to the midpoint of the wall.

Hollow, Thin-Walled Shafts

$$\tau = \frac{T}{2A_m t} \qquad 52.62$$

$$A_m = \pi r_m^2 \qquad 52.63$$

Figure 52.10 *Torsion in Thin-Walled Shells*

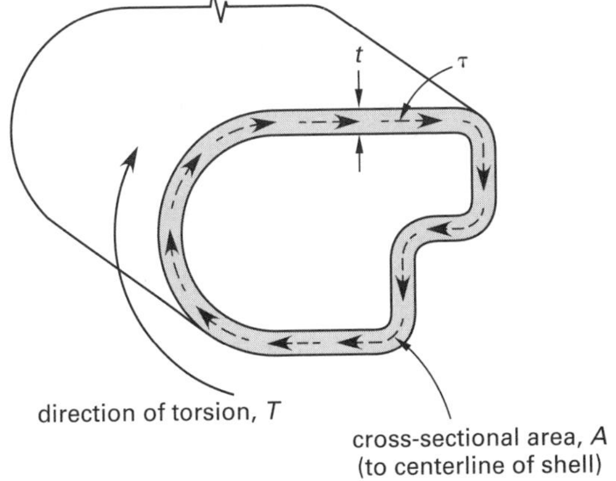

For non-circular thin walled shells, shear stress at any point is not proportional to the distance from the centroid of the cross section. Rather, the *shear flow*, q, around the shell is constant, regardless of whether the wall thickness is constant or variable.[21] The shear flow is the shear per-unit length of the centerline path.[22] At any point where the shell thickness is t,

$$q = \tau t = \frac{T}{2A_m} \quad \text{[constant]} \qquad 52.64$$

When the wall thickness, t, is constant, the angular twist depends on the perimeter, p, of the shell as measured along the centerline of the shell wall.

$$\gamma = \frac{TLp}{4A_m^2tG} \qquad 52.65$$

14. TORSION IN SOLID, NONCIRCULAR MEMBERS

When a noncircular solid member is placed in torsion, the shear stress is not proportional to the distance from the centroid of the cross section. The maximum shear usually occurs close to the point on the surface that is nearest the centroid.

Shear stress, τ, and angular deflection (angle of twist), ϕ, due to torsion are functions of the cross-sectional shape. They cannot be specified by simple formulas that apply to all sections. Table 52.3 lists the governing equations for several basic cross sections. These formulas have been derived by dividing the member into several concentric, thin-walled, closed shells and summing the torsional strength provided by each shell.

15. SHEAR CENTER FOR BEAMS

A beam with a symmetrical cross section supporting a transverse force that is offset from the longitudinal centroidal axis will be acted upon by a torsional moment, and the beam will tend to "roll" about a longitudinal axis known as the *bending axis* or *torsional axis*. For solid, symmetrical cross sections, this bending axis passes through the centroid of the cross section. However, for an asymmetrical beam (e.g., a channel beam on its side), the bending axis passes through the *shear center* (*flexural center*, *torsional center*, or *center of twist*), not the centroid. (See Fig. 52.11.) The shear center is a point that does not experience rotation (i.e., is a point about which all other points rotate) when the beam is in torsion.

Table 52.3 *Torsion in Solid, Noncircular Shapes*

cross section	K in formula $\gamma = TL/KG$	τ_{max}
ellipse (axes $2a$, $2b$)	$\frac{\pi a^3b^3}{a^2+b^2}$	$\frac{2T}{\pi ab^2}$ (maximum at ends of minor axis)
square (side a)	$0.1406a^4$	$\frac{T}{0.208a^3}$ (maximum at midpoint of each side)
rectangle ($2a$ by $2b$)	*	$\frac{T(3a+1.8b)}{8a^2b^2}$ (maximum at midpoint of each longer side)
equilateral triangle (side a)	$\frac{a^4\sqrt{3}}{80}$	$\frac{20T}{a^3}$ (maximum at midpoint of each side)
slotted tube (radius r, thickness t)	$\frac{2\pi rt^3}{3}$	$\frac{T(6\pi r+1.8t)}{4\pi^2r^2t^2}$ (maximum along both edges remote from ends)
I-beam (width b, height h, t_f, t_w)	$\frac{2bt_f^3+ht_w^3}{3}$	$\frac{3Tt_f}{2bt_f^3+ht_w^3}$ $[t_w < t_f]$

$$*ab^3\left(\frac{16}{3} - \left(\frac{3.36b}{a}\right)\left(1 - \frac{b^4}{12a^4}\right)\right)$$

[21]The concept of shear flow can also be applied to a regular beam in bending, although there is little to be gained by doing so. Removing the dimension b in the general beam shear stress equation, $q = VQ/I$.

[22]Shear flow is not analogous to magnetic flux or other similar quantities because the shear flow path does not need to be complete (i.e., does not need to return to its starting point).

Figure 52.11 *Channel Beam in Pure Bending (shear resultant, V, directed through shear center, O)*

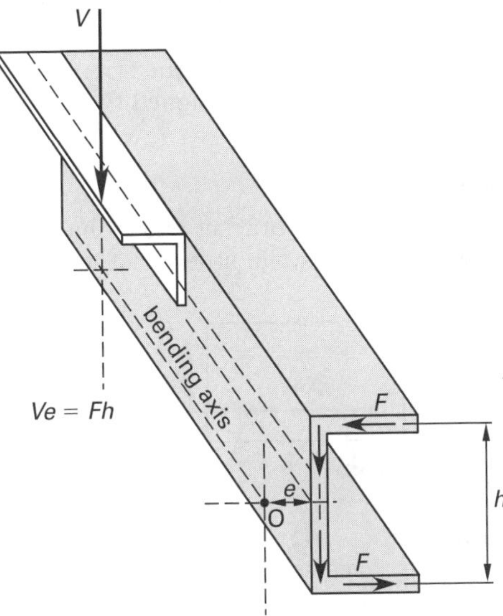

For beams with transverse loading, simple bending without torsion can only occur if the transverse load (*shear resultant*, *V*, or *shear force of action*) is directed through the shear center. Otherwise, a torsional moment calculated as the product of the shear resultant and the torsional eccentricity will cause the beam to twist. The *torsional eccentricity* is the distance between the line of action of the shear resultant and the shear center.

The shear center is always located on an axis of symmetry. The location of the shear center for any particular beam geometry is determined by setting the torsional moment equal to the shear resisting moment, as calculated from the total shear flow and the appropriate moment arm. In practice, however, shear centers for common shapes are located in tables similar to Fig. 52.12.

Figure 52.12 *Shear Centers of Selected Thin-Walled Open Sections**

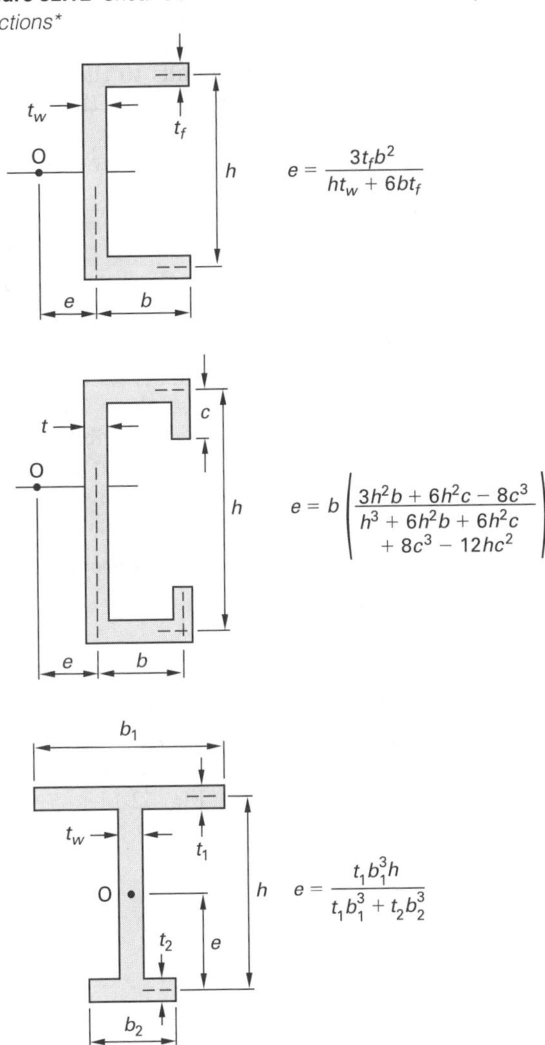

*Distance *e* from shear center, O, measured from the reference point shown. Distance *e* is not the torsional eccentricity, which is the separation of the shear center and the line of action of the shear force.

16. ECCENTRICALLY LOADED BOLTED CONNECTIONS

An eccentrically loaded connection is illustrated in Fig. 52.13. The bracket's natural tendency is to rotate about the centroid of the connector group. The shear stress in the connectors includes both the direct vertical shear and the torsional shear stress. The sum of these shear stresses is limited by the shear strength of the critical connector, which in turn determines the capacity of the connection, as limited by bolt shear strength.[23]

Figure 52.13 *Eccentrically Loaded Connection*

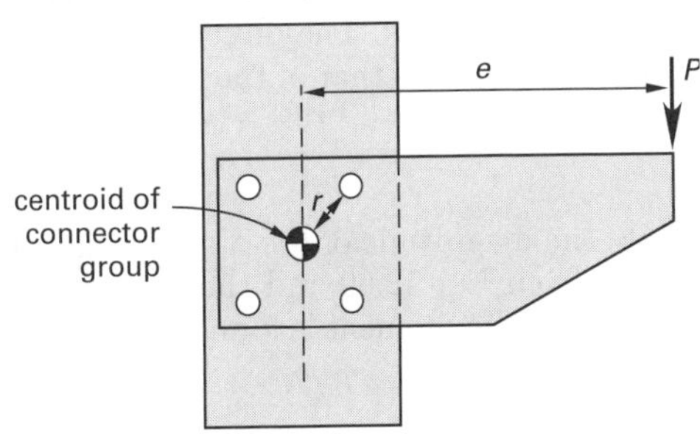

[23]This type of analysis is known as an *elastic analysis* of the connection. Although it is traditional, it tends to greatly understate the capacity of the connection.

Analysis of an eccentric connection is similar to the analysis of a shaft under torsion. The shaft torque, T, is analogous to the moment due to the load, P, on the connection. The shaft's radius corresponds to the distance from the centroid of the fastener group to the *critical fastener*. The critical fastener is the one for which the vector sum of the vertical and torsional shear stresses is the greatest.

Torsion

$$\tau = \frac{Tr}{J} \qquad 52.66$$

The polar moment of inertia, J, is calculated from the parallel axis theorem. Since bolts and rivets have little resistance to twisting in their holes, their individual polar moments of inertia are omitted.[24] Only the r^2A terms in the parallel axis theorem are used. r_i is the distance from the fastener group centroid to the centroid (i.e., center) of the ith fastener, which has an area of A_i.

Moment of Inertia

$$\begin{aligned} I_z = J = I_y + I_x &= \int (x^2 + y^2)\,dA \\ &= r_p^2 A \end{aligned} \qquad 52.67$$

The torsional shear stress is directed perpendicularly to a line between each fastener and the connector group centroid. The direction of the shear stress is the same as the rotation of the connection. (See Fig. 52.14.)

Figure 52.14 *Direction of Torsional Shear Stress*

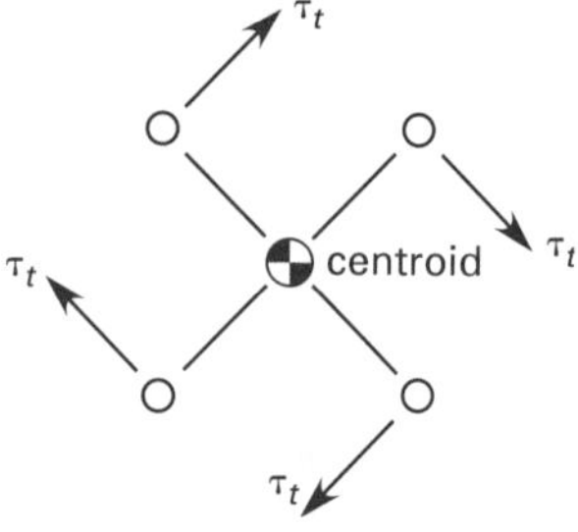

Once the torsional shear stress has been determined in the critical fastener, it is added in a vector sum to the direct vertical shear stress. The direction of the vertical shear stress is the same as that of the applied force.

$$\tau_v = \frac{P}{nA} \qquad 52.68$$

In Eq. 52.68, the magnitude of the direct shear force due to P is found from Eq. 52.69 and the magnitude of the shear force due to the moment is found from Eq. 52.70.

$$|F| = P/N \qquad 52.69$$

$$|F| = \frac{Mr_i}{\sum\limits_{i=1}^{n} r_i^2} \qquad 52.70$$

Typical connections gain great strength from the frictional slip resistance between the two surfaces. By preloading the connection bolts, the normal force between the plates is greatly increased. The connection strength from friction will rival or exceed the strength from bolt shear in connections that are designed to take advantage of preload.

Example 52.5

All fasteners used in the bracket shown have a nominal ½ in diameter. What is the stress in the most critical fastener?

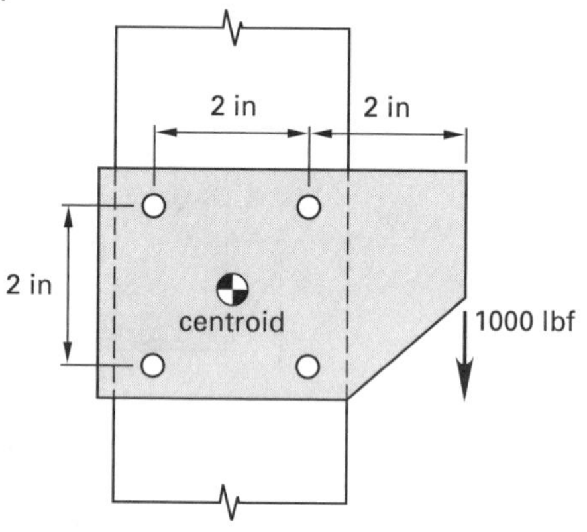

Solution

Since the fastener group is symmetrical, the centroid is centered within the four fasteners. This makes the eccentricity of the load equal to 3 in. Each fastener is located a distance r from the centroid, where

$$r = \sqrt{x^2 + y^2} = \sqrt{(1\text{ in})^2 + (1\text{ in})^2} = 1.414\text{ in}$$

The area of each fastener is

$$A_i = \frac{\pi D^2}{4} = \left(\frac{\pi(0.5\text{ in})^2}{4}\right) = 0.1963\text{ in}^2$$

Using the parallel axis theorem for polar moments of inertia and disregarding the individual torsional resistances of the fasteners,

$$\begin{aligned} J &= \sum_i r_i^2 A_i = (4)(1.414\text{ in})^2(0.1963\text{ in}^2) \\ &= 1.570\text{ in}^4 \end{aligned}$$

The torsional shear stress in each fastener is

$$\begin{aligned} \tau_t &= \frac{Fer}{J} = \frac{(1000\text{ lbf})(3\text{ in})(1.414\text{ in})}{1.570\text{ in}^4} \\ &= 2702\text{ lbf/in}^2 \end{aligned}$$

[24] In spot-welded and welded stud connections, the torsional resistance of each connector can be considered.

Each torsional shear stress can be resolved into a horizontal shear stress, τ_{tx}, and a vertical shear stress, τ_{ty}. Both of these components are equal to

$$\tau_{tx} = \tau_{ty} = \frac{(\sqrt{2})\left(2702\ \frac{\text{lbf}}{\text{in}^2}\right)}{2}$$
$$= 1911\ \text{lbf/in}^2$$

The direct vertical shear downward is

$$\tau_v = \frac{F}{nA} = \frac{1000\ \text{lbf}}{(4)(0.1963\ \text{in}^2)}$$
$$= 1274\ \text{lbf/in}^2$$

The two right fasteners have vertical downward components of torsional shear stress. The direct vertical shear is also downward. These downward components add, making both right fasteners critical.

The total stress in each of these fasteners is

$$\tau = \sqrt{\tau_{tx}^2 + (\tau_{ty} + \tau_v)^2}$$
$$= \sqrt{\left(1911\ \frac{\text{lbf}}{\text{in}^2}\right)^2 + \left(1911\ \frac{\text{lbf}}{\text{in}^2} + 1274\ \frac{\text{lbf}}{\text{in}^2}\right)^2}$$
$$= 3714\ \text{lbf/in}^2$$

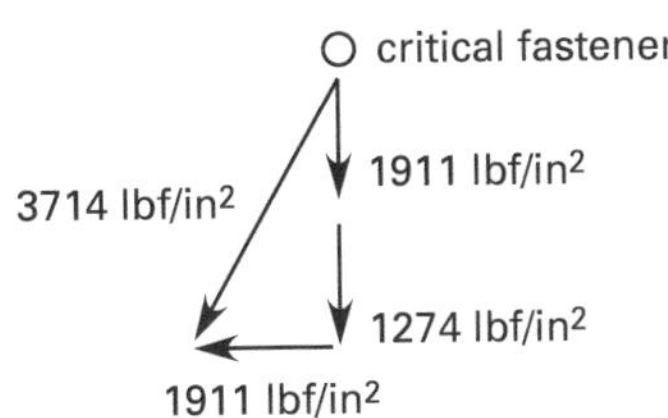

17. ECCENTRICALLY LOADED WELDED CONNECTIONS

The traditional elastic analysis of an eccentrically loaded welded connection is virtually the same as for a bolted connection, with the additional complication of having to determine the polar moment of inertia of the welds.[25] This can be done either by taking the welds as lines or by assuming each weld has an arbitrary thickness, t. After finding the centroid of the weld group, the rectangular moments of inertia of the individual welds are taken about that centroid using the parallel axis theorem. These rectangular moments of inertia are added to determine the polar moment of inertia. This laborious process can be shortened by use of equations for the bending and torsional properties of fillet welds, which can be found in tables in various engineering references. [Bending and Torsional Properties of Fillet Welds]

The torsional shear stress, calculated from Mr/J (where r is the distance from the centroid of the weld group to the most distant weld point), is added vectorially to the direct shear to determine the maximum shear stress at the critical weld point.

18. FLAT PLATES

Flat plates under uniform pressure are separated into two edge-support conditions: simply supported and built-in edges.[26] Commonly accepted working equations are summarized in Table 52.4. It is assumed that (a) the plates are of "medium" thickness (meaning that the thickness is equal to or less than one-fourth of the minimum dimension of the plate), (b) the pressure is no more than will produce a maximum deflection of one-half of the thickness, (c) the plates are constructed of isotropic, elastic material, and (d) the stress does not exceed the yield strength.

Example 52.6

A steel pipe with an inside diameter of 10 in (254 mm) is capped by welding round mild steel plates on its ends. The allowable stress is 11,100 lbf/in^2 (77 MPa). The internal gage pressure in the pipe is maintained at 500 lbf/in^2 (3.5 MPa). What plate thickness is required?

SI Solution

A fixed edge approximates the welded edges of the plate. From Table 52.4, the maximum bending stress is

$$\sigma_{\text{max}} = \frac{3pr^2}{4t^2}$$

$$t = \sqrt{\frac{3pr^2}{4\sigma_{\text{max}}}} = \sqrt{\frac{(3)(3.5\ \text{MPa})\left(\frac{254\ \text{mm}}{2}\right)^2}{(4)(77\ \text{MPa})}}$$
$$= 23.4\ \text{mm}$$

Customary U.S. Solution

A fixed edge approximates the welded edges of the plate. From Table 52.4, the maximum bending stress is

$$\sigma_{\text{max}} = \frac{3pr^2}{4t^2}$$

$$t = \sqrt{\frac{3pr^2}{4\sigma_{\text{max}}}} = \sqrt{\frac{(3)\left(500\ \frac{\text{lbf}}{\text{in}^2}\right)\left(\frac{10\ \text{in}}{2}\right)^2}{(4)\left(11{,}100\ \frac{\text{lbf}}{\text{in}^2}\right)}}$$
$$= 0.919\ \text{in}$$

[25] Steel building design does not use an elastic analysis to design eccentric brackets, either bolted or welded. The design methodology is highly proceduralized and codified.

[26] Fixed-edge conditions are theoretical and are seldom achieved in practice. Considering this fact and other simplifying assumptions that are made to justify the use of Table 52.4, a value of $\nu = 0.3$ can be used without loss of generality.

Machine Design

Table 52.4 *Flat Plates Under Uniform Pressure*

shape	edge condition	maximum stress	deflection at center
circular	simply supported	$\frac{\frac{3}{8}pr^2(3+\nu)}{t^2}$ (at center)	$\frac{\frac{3}{16}pr^4(1-\nu)(5+\nu)}{Et^3}$
	built-in	$\frac{\frac{3}{4}pr^2}{t^2}$ (at edge)	$\frac{\frac{3}{16}pr^4(1-\nu^2)}{Et^3}$
rectangular	simply supported	$\frac{C_1pb^2}{t^2}$ (at center)	$\frac{C_2pb^4}{Et^3}$
	built-in	$\frac{C_3pb^2}{t^2}$ (at centers of long edges)	$\frac{C_4pb^4}{Et^3}$

$\frac{a}{b}$	1.0	1.2	1.4	1.6	1.8	2	3	4	5	∞
C_1	0.287	0.376	0.453	0.517	0.569	0.610	0.713	0.741	0.748	0.750
C_2	0.044	0.062	0.077	0.091	0.102	0.111	0.134	0.140	0.142	0.142
C_3	0.308	0.383	0.436	0.487	0.497	0.500	0.500	0.500	0.500	0.500
C_4	0.0138	0.0188	0.023	0.025	0.027	0.028	0.028	0.028	0.028	0.028

19. NOMENCLATURE

a	curve-fit constant	–	–
a	dimension	ft	m
a	radius	ft	m
A	area	ft²	m²
b	curve-fit constant	–	–
b	dimension	ft	m
b	radius	ft	m
b	width	ft	m
c	dimension	ft	m
c	distance from neutral axis to extreme fiber	ft	m
C	circumference	ft	m
C	stiffness constant	–	–
D	diameter	ft	m
e	distance from shear center	ft	m
e	eccentricity	ft	m
E	energy	ft-lbf	J
E	Young's modulus	lbf/ft²	Pa
f	coefficient of friction	–	–
F	force	lbf	N
FS	factor of safety	–	–
g	acceleration of gravity, 32.2 (9.81)	ft/sec²	m/s²
g_c	gravitational constant, 32.2	lbm-ft/lbf-sec²	n.a.
G	shear modulus	lbf/ft²	Pa
h	height	ft	m
I	interference	ft	m
I	moment of inertia	ft⁴	m⁴
J	polar moment of inertia	ft⁴	m⁴
k	spring constant	lbf/ft	N/m
K	end-restraint coefficient	–	–
K	torque coefficient	–	–
K	stress concentration factor	–	–
L	grip	in	mm
L	length	ft	m
L'	effective length	ft	m
m	mass	lbm	kg
M	moment	ft-lbf	N·m
n	modular ratio	–	–
n	number of connectors	–	–
n	rotational speed	rev/min	rev/min
N	normal force	lbf	N
p	perimeter	ft	m
p	pressure	lbf/ft²	Pa
P	load	lbf	N
P	power	hp	kW
q	shear flow	lbf/ft	N/m
r	radius	ft	m
r	radius of gyration	ft	m
S	strength	lbf/ft²	Pa
S_r	slenderness ratio	–	–
t	thickness	ft	m

T	torque	ft-lbf	N·m
U	energy	ft-lbf	J
V	shear resultant	lbf	N
w	load per unit length	lbf/ft	N/m
y	weld size	ft	m

Symbols

α	thread half angle	deg	deg
ϕ	angle of twist	rad	rad
δ	deflection	ft	m
δ	radial interference	ft	m
ϵ	strain	ft/ft	m/m
θ	lead angle	deg	deg
θ	shear strain	rad	rad
μ	coefficient of friction	–	rad
ν	Poisson's ratio	–	–
σ	normal stress	lbf/ft^2	Pa
τ	shear stress	lbf/ft^2	Pa
ϕ	angle	rad	rad
ϕ	load factor	–	–

Subscripts

a	allowable
ave	average
b	bending or bolt
c	centroidal, circumferential, or collar
cr	critical
e	Euler or effective
eq	equivalent
f	flange
h	hoop
i	inside or initial
l	longitudinal
m	mean or member
max	maximum
min	minimum
n	nominal
o	outside
p	bearing or potential
r	radial or rope
sh	sheave
t	tension, thread, or transformed
tm	thread mean
T	torque
ut	ultimate tensile
u	ultimate
v	vertical
w	wire, web, or width
y	yield

53 Advanced Machine Design

Content in blue refers to the *NCEES Handbook*.

NCEES EXAM SPECIFICATIONS AND RELATED CONTENT

MACHINE DESIGN AND MATERIALS EXAM

II.A.2. Mechanical Components: Bearings
- 39. Roller Bearing Capacity
- 43. Journal Bearing Frictional Losses

II.A.3. Mechanical Components: Gears
- 13. Velocity, Power, and Torque in Rotating Members
- 14. Spur Gear Terminology
- 18. Force Analysis of Spur Gears
- 22. Spur Gear Strength: Lewis Beam Strength
- 26. Bevel Gears
- 27. Worm Gear Sets

II.A.4. Mechanical Components: Springs
- 1. Springs
- 3. Allowable Spring Stresses: Static Loading
- 5. Helical Compression Springs: Static Loading
- 11. Helical Torsion Springs

II.A.6. Mechanical Components: Belt, pulley and chain drives
- 28. Flat Belt Drives

II.A.7. Mechanical Components: Clutches and brakes
- 35. Disk and Plate Clutches

1. SPRINGS

An *ideal spring* is assumed to be perfectly elastic within its working range. The deflection is assumed to follow Hooke's law.[1] The force in a linear elastic spring can be found from Eq. 53.1.

Spring Energy

$$F_s = kx \qquad 53.1$$

[1]A spring can be perfectly elastic even though it does not follow Hooke's law. The deviation from proportionality, if any, occurs at very high loads. The difference in theoretical and actual spring forces is known as the *straight line error*.

The *spring constant*, k, is also known as the *stiffness*, *spring rate*, *scale*, and *k-value*.[2] The spring constant can be calculated as the ratio of the difference in forces applied to a spring over the difference in deflection of the spring.

$$k = \frac{F_1 - F_2}{x_1 - x_2} \quad 53.2$$

A spring stores energy when it is compressed or extended. By the *work-energy principle*, the energy storage is equal to the work required to displace the spring. The potential energy of a spring whose ends have been displaced a total distance x is

Spring Energy

$$U = k\frac{x^2}{2} \quad 53.3$$

If a mass, m, is dropped from a height h onto and is captured by a spring, the compression, x, can be found by equating the change in potential energy to the energy storage. Potential energy is equal to mgh, which can also be written as $mg(h+x)$. **[Potential Energy]**

$$mg(h+x) = \frac{1}{2}kx^2 \quad \text{[SI]} \quad 53.4(a)$$

$$m\left(\frac{g}{g_c}\right)(h+x) = \frac{1}{2}kx^2 \quad \text{[U.S.]} \quad 53.4(b)$$

Within the elastic region, this energy can be recovered by restoring the spring to its original unstressed condition. It is assumed that there is no permanent set, and no energy is lost through external friction or *hysteresis* (internal friction) when the spring returns to its original length.[3]

The entire applied load is felt by each spring in a series of springs linked end-to-end. The *equivalent* (*composite*) *spring constant* for springs in series is

$$\frac{1}{k_{eq}} = \frac{1}{k_1} + \frac{1}{k_2} + \frac{1}{k_3} + \cdots \quad \left[\begin{array}{c}\text{series}\\ \text{springs}\end{array}\right] \quad 53.5$$

Springs in parallel (e.g., concentric springs) share the applied load. The equivalent spring constant for springs in parallel is

$$k_{eq} = k_1 + k_2 + k_3 + \cdots \quad \left[\begin{array}{c}\text{parallel}\\ \text{springs}\end{array}\right] \quad 53.6$$

2. SPRING MATERIALS

A wide variety of materials are used for springs, including high-carbon steel, stainless steel and various alloys, nickel-based alloys (e.g., inconel), and copper-based alloys (e.g., phosphor-bronze and silicon-bronze). (See Table 53.1.) "Super-alloys" are used for high-temperature and highly corrosive environments. Spring rate, fatigue strength, temperature range, corrosion resistance, magnetic properties, and cost are all considerations.

Springs manufactured from prehardened materials are generally stress-relieved in a low-temperature process by heating to between 400°F and 800°F (200°C and 430°C) after forming. Springs with intricate shapes must be manufactured from annealed materials and be subsequently strengthened in high-temperature processes. They are first quenched to full hardness and then tempered. Age-hardenable materials (e.g., beryllium copper) can be strengthened simply by heating after forming.

Most springs are cold-wound. Springs with wire diameters much in excess of 1/2 in or 5/8 in (12 mm or 16 mm) are wound while red hot. Although the design methods are essentially the same for hot-wound and cold-wound springs, the allowable stresses are reduced approximately 20%, and the modulus is reduced slightly (approximately 9% for the shear modulus and approximately 5% for the elastic modulus). There are other unique issues and special needs associated with the manufacturing of hot-wound springs, as well.

Materials suitable for fatigue service include music wire (ASTM A228), carbon and alloy valve spring wire (ASTM A230), chrome-vanadium (ASTM A232), beryllium copper (ASTM B197), phosphor bronze (ASTM B159), and, to a lesser degree, type-302 stainless steel (ASTM A313). *Shotpeening* (*stresspeening*) is one of the best methods for increasing a spring's fatigue life.

3. ALLOWABLE SPRING STRESSES: STATIC LOADING

Helical compression and extension springs experience torsional shear stresses. The yield strength in shear is the theoretical maximum stress. There are three common ways of choosing the maximum allowable shear stress for static service.[4]

1. Selecting the allowable stress based on some percentage of the ultimate tensile strength is the most common method. For ferrous materials except for austenitic stainless, the percentage is approximately 45% to 65%. For nonferrous and austenitic stainless, the percentage is

[2]Another unfortunate name for the spring constant, k, that is occasionally encountered is the *spring index*. This is not the same as the spring index, C, used in helical coil spring design. The units will determine which meaning is intended.

[3]There is essentially no hysteresis in properly formed compression, extension, or open-wound helical torsion springs.

[4]These methods apply to helical compression and extension springs. Recommended percentages are different for other types of springs.

Table 53.1 Properties of Typical Spring Materials (room temperature[a])

material	modulus of elasticity (lbf/in^2)	shear modulus (lbf/in^2)	density (lbm/in^3)
high-carbon wire			
music wire ASTM A228	30×10^6	11.5×10^6	0.284
hard-drawn ASTM A227	30×10^6	11.5×10^6	0.284
oil-tempered ASTM A229	30×10^6	11.5×10^6	0.284
valve spring ASTM A230	30×10^6	11.5×10^6	0.284
alloy-steel wire			
chrome-vanadium SAE 6150, AISI 6150, ASTM A232	30×10^6	11.5×10^6	0.284
chrome-silicon AISI 9254, ASTM A401	30×10^6	11.5×10^6	0.284
silicon manganese AISI 9260	30×10^6	11.5×10^6	0.284
stainless steel wire			
AISI 302, ASTM A313	28×10^6	10.0×10^6	0.280
AISI 410, 420	28×10^6	11.0×10^6	0.286
17-7 PH[b]	29.5×10^6	11.0×10^6	0.286
18-2	29×10^6	9.8×10^6	0.272
nickel-chrome A286	29×10^6	10.4×10^6	0.290
copper alloys			
phosphor bronze ASTM B159	15×10^6	6.3×10^6	0.320
silicon bronze ASTM B99(A)	15×10^6	5.6×10^6	0.308
silicon bronze ASTM B99(A)	17×10^6	6.4×10^6	0.316
beryllium-copper ASTM B197	18.5×10^6	7.0×10^6	0.297
nickel alloys			
inconel 600	31×10^6	11.0×10^6	0.307
inconel X750	31.5×10^6	11.5×10^6	0.298
Ni Span C902	27×10^6	9.7×10^6	0.294

(Multiply lbf/in^2 by 6.89×10^{-6} to obtain GPa.)
(Multiply lbm/in^3 by 27.7×10^3 to obtain kg/m^3.)

[a]Properties vary with temperature. Compiled from various sources.
[b]PH—precipitation hardened.
Adapted from *Design Handbook*, 1987 Edition, Barnes Group (Associated Spring), Bristol, CT.

approximately 35% to 55%. The lower limit should be used for unconstrained designs (i.e., the spring diameter can be as large as necessary to keep the stresses low). The higher limit is used when space is limited and higher stresses are unavoidable.[5] The ultimate strength used to calculate the allowable stress can be found either through tabular information, such as can be found in Table 53.2, or can be calculated using Eq. 53.7. m is a constant based on the type of wire used for the spring: 0.163 for music wire, 0.193 for oil-tempered wire, 0.201 for hard-drawn wire, 0.155 for chrome vanadium wire, or 0.091 for chrome silicon wire. [**Mechanical Springs**]

Mechanical Springs

$$S_{ut} = \frac{A}{D_{\text{wire}}^m} \qquad \text{53.7}$$

2. Selecting the allowable stress based on the yield strength in shear is probably the most theoretically rigorous method. The yield strength in shear can be calculated from the tensile yield strength using either the maximum shear stress or the distortion energy theory.[6] If called for, a factor of safety of approximately 1.5 is appropriate for ferrous springs.

3. Some specifications limit the torsional shear stress to a percentage of the tensile yield strength.

Some springs (e.g., flat leaf springs and helical torsion springs) experience a bending stress. Such springs are limited by the tensile yield strength of the spring material.

4. SAFE SPRING STRESSES: FATIGUE LOADING

Two methods can be used to design or analyze springs that are repeatedly stressed. The more rigorous method is to use a modified Goodman diagram. (The Wahl factor is applied to both the mean and alternating stress.) A simpler method is to design for static loading using a reduced maximum shear stress. An approximation of the endurance strength in shear can be calculated from the ultimate tensile strength (see Table 53.2) and a factor from Table 53.3. Since the table values are conservative, the lives of most springs designed to them will exceed the numbers of cycles listed. However, a factor of safety may still be used or required.

$$\tau_{\max} = \frac{S_e}{\text{FS}} = \frac{(\text{factor})S_{ut}}{\text{FS}} \qquad \text{53.8}$$

[5]For highly precise springs with minimum hysteresis, creep, and drift, the percentages quoted in this section should be reduced.
[6]The tensile yield strength can be estimated for ferrous spring materials as 75% of the ultimate tensile strength.

Table 53.2 Approximate Minimum Ultimate Tensile Strengths[a] of Typical Spring Materials (ksi, at room temperature[b])

	wire size								
	in: 0.02	0.04	0.06	0.08	0.10	0.15	0.20	0.30	0.40
material	mm: 0.5	1.0	1.5	2.0	2.5	3.8	5.1	7.6	10
high-carbon wire									
music wire ASTM A228	350	315	296	282	271	253	240		
hard-drawn ASTM A227	285	255	235	225	215	205	190	175	16
oil-tempered ASTM A229	295	265	250	235	230	215	195	180	17
valve spring ASTM A230					210	205	195		
alloy-steel wire									
chrome-vanadium SAE 6150, AISI 6150, ASTM A232		280	265	260	245	230	220	205	
stainless steel wire									
AISI 302, ASTM A313	300	275	260	245	235	215	190	160	
17-7 PH[c]	275	255	242	235	223	211			
18-2	296	270	265	255	253				
copper alloys									
phosphor bronze ASTM B159	150	140	138	135	132	130	128	122	

(Multiply in by 25.4 to obtain mm.)

(Multiply ksi by 6.895 to obtain MPa.)

[a]ASTM specifications provide a range of ultimate strengths. Table values approximately correspond to the minimum of the range.
[b]Properties vary with temperature.
[c]Precipitation hardened.
Adapted from *Design Handbook*, 1987 Edition, Barnes Group (Associated Spring), Bristol, CT.

Table 53.3 Fatigue Correction Factor (fraction of ultimate tensile strength)

	shear stress		bending stress	
fatigue life (cycles)	not shot-peened spring	shot-peened spring	not shot-peened spring	shot-peened spring
10,000	0.45*	0.45	0.80	0.80
100,000	0.35	0.42	0.53	0.62
1,000,000	0.33	0.40	0.50	0.60
10,000,000	0.30	0.36	0.48	0.58

*35% for phosphor bronze and type-302 stainless steel.
Adapted from *Design Handbook*, 1987 Edition, Barnes Group (Associated Spring), Bristol, CT.

5. HELICAL COMPRESSION SPRINGS: STATIC LOADING

The *spring index*, C, is the ratio of the mean coil diameter to the wire diameter.[7] (See Fig. 53.1.) It is difficult to wind springs with small (i.e., less than 4) spring indexes, and the operating stresses will be high. The wire cannot be easily bent to the desired small radius. On the other hand, springs with large indexes (i.e., greater than 12) are flimsy and tend to buckle. Most springs have indexes between 8 and 10, although 5 is a typical value for clutch springs.

Mechanical Springs

$$C = \frac{D_{\text{mean coil}}}{D_{\text{wire}}} \quad 53.9$$

The mean coil diameter can be calculated as the average of the outside and inside diameters of the spring.

$$D_{\text{mean coil}} = \frac{D_i + D_o}{2} \quad 53.10$$

The spring deflection can be written in terms of the mean coil diameter or the spring index.

$$x = \frac{F}{k} = \frac{8FD^3_{\text{mean coil}}N}{GD^4_{\text{wire}}} = \frac{8FC^3N}{GD_{\text{wire}}} \quad 53.11$$

[7]This section is only for helical compression springs manufactured from round wire. Springs can also be manufactured from wire with a square or rectangular cross section. The design equations are different in that case.

Figure 53.1 *Helical Compression Spring*

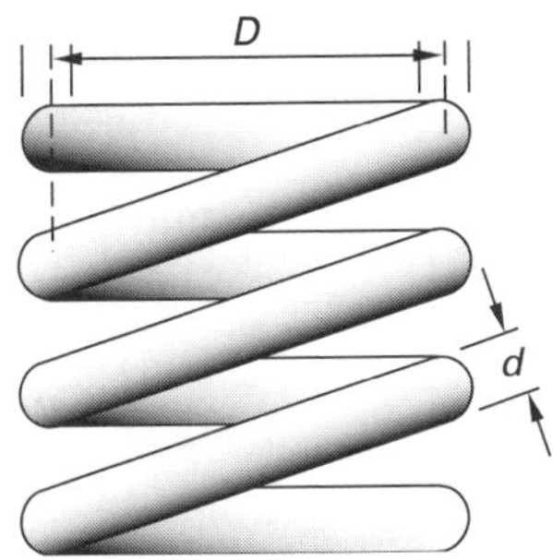

The number of *active coils* (i.e., "turns") in a helical spring is less than or equal to the total number of coils, depending on the method of finishing the ends. When a helical compression spring has *plain ends* (i.e., neither squared nor ground), all of the coils contribute to spring force. In that case, the total number of coils, N_t, is equal to the number of active coils, N. However, most designs call for squaring and/or grinding the ends in order to obtain better seating for the spring. The number of end coils, N_e, is given in Table 53.4 for various end treatments. Normally, there should be at least two active coils.

$$N = N_t - N_e \quad \text{53.12}$$

The *spring constant* is given by the *load-deflection equation*, also called the *spring rate equation*.

Mechanical Springs

$$k = \frac{D_{\text{wire}}^4 G}{8D_{\text{mean coil}}^3 N} \quad \text{53.13}$$

The load-deflection equation can also be written in terms of the force and deflection, or in terms of the spring index, as shown in Eq. 53.11.

$$k = \frac{F}{x} = \frac{D_{\text{wire}} G}{8C^3 N} \quad \text{53.14}$$

Helical spring stress equations are derived from a superposition of torsional and direct shear stresses.[8] The *stress correction factor*, K_s, corrects for the curvature.[9] It is not a true stress concentration factor, but rather a factor that corrects the average stress to the maximum stress. The maximum shear stress occurs at the inner face of the wire coil where the torsional and direct shear stresses are additive.

Mechanical Springs

$$K_s = \frac{4C+2}{4C-3} \quad \text{53.15}$$

$$\tau = K_s \frac{8FD_{\text{mean coil}}}{\pi D_{\text{wire}}^3} \quad \text{53.16}$$

The end treatment shown in Fig. 53.2 affects a spring's solid length and pitch. The exact effects are not always obvious. Table 53.4 contains relationships that are accepted for design use.

Figure 53.2 *End Treatment of Helical Compression Springs*

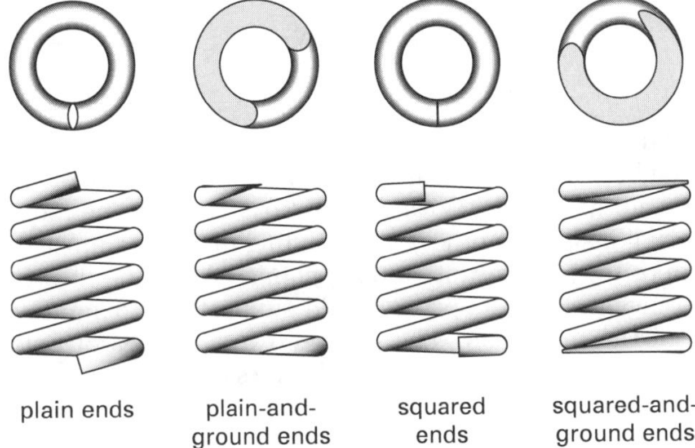

plain ends | plain-and-ground ends | squared ends | squared-and-ground ends

Table 53.4 *Effect of End Treatment on Helical Compression Springs*

	end treatment			
variable	plain	plain and ground	squared only	squared and ground
end coils, N_e	0	1	2	2
total coils, N_t	N	$N+1$	$N+2$	$N+2$
free length, L_0	$pN+d$	$p(N+1)$	$pN+3d$	$pN+2d$
solid length*, L_s	$d(N_t+1)$	dN_t	$d(N_t+1)$	dN_t
pitch, p	$\frac{L_0-d}{N}$	$\frac{L_0}{N+1}$	$\frac{L_0-3d}{N}$	$\frac{L_0-2d}{N}$

*Use the effective wire diameter when calculating the solid height. This includes the nominal wire diameter, the wire tolerance, and the paint or plating thickness.

The solid deflection is calculated from the free and solid lengths.

$$x_s = L_0 - L_s \quad \text{53.17}$$

The *spring pitch* (*coil pitch*), p, is the mean coil separation. The *solid length* (*compressed length*) is given in

[8]The *direct shear stress* is simply the force supported by the spring divided by the wire cross-sectional area, $\pi D^2/4$.
[9]Equation 53.15 is approximate, but the error is less than 2%.

Table 53.4. The *clash allowance*, A_c, is the percentage difference in solid and working deflections. It should be approximately 20%.

$$A_c = \frac{x_s - x_w}{x_w} \qquad 53.18$$

The total spring *wire length* is the length of wire in the coils. The *active wire length* is calculated from the number of active coils.

$$L_a = \pi DN \qquad 53.19$$

The *helix direction* for single springs can be either right hand or left hand. If the spring works over a threaded member, the winding direction should be opposite of the thread direction. With two *nested springs* (i.e., one spring inside the other), the winding directions must be opposite to prevent intermeshing. Also for nested springs, the outer spring should support approximately ⅔ of the total load. The solid and free lengths of both springs should be approximately the same.

6. HELICAL COMPRESSION SPRINGS: DESIGN

Conventional spring design is an iterative procedure. One or more parameters are varied until the requirements are satisfied. Often one or more parameters are unknown and must be assumed to complete the design. For example, when the wire diameter is unknown, the allowable stress and Wahl factor can both only be estimated. (For the initial iteration, it is common to assume a Wahl factor of 1.1.) The outside diameter of the spring may also be a limited factor, as when the spring must fit in a hole.

An important decision is whether the allowable stress is comparable to the maximum working stress or the stress when the spring is compressed solid. Since most helical compression springs will be compressed solid sometime in their lives, it seems logical to use the solid height stress. In the absence of guidance, either interpretation would apply.

Using the standard commercial wire size (or sheet metal gauge for flat springs) will result in the most economical design.[10] However, this usually cannot be done without slightly altering one of the other parameters (coil diameter, spring rate, etc.).

Example 53.1

A steel spring is manufactured from 0.2253 in diameter wire with a spring index of 6. The load on the spring fluctuates between 140.2 lbf and 219.8 lbf (620 N and 970 N). The spring material has a yield strength in shear of 120,000 lbf/in^2 (830 MPa) and an endurance strength in shear of 100,000 lbf/in^2 (690 MPa). The spring will seldom be compressed solid. A factor of safety of 1.5 is required. Is the spring satisfactory?

SI Solution

Convert the wire diameter to millimeters and calculate the mean spring diameter.

$$D_{\text{wire}} = (0.2253\text{ in})\left(25.4\ \frac{\text{mm}}{\text{in}}\right) = 5.723\text{ mm}$$

Mechanical Springs

$$C = \frac{D_{\text{mean coil}}}{D_{\text{wire}}}$$

$$D_{\text{mean coil}} = CD_{\text{wire}} = (6)(5.723\text{ mm}) = 34.34\text{ mm}$$

The stress correction factor is

Mechanical Springs

$$K_s = \frac{4C+2}{4C-3} = \frac{(4)(6)+2}{(4)(6)-3} = 1.2381$$

From Eq. 53.16, the maximum and minimum stresses are

Mechanical Springs

$$\tau_{\max} = K_s\frac{8F_{\max}D_{\text{mean coil}}}{\pi D_{\text{wire}}^3} = (1.2381)\left(\frac{(8)(970\text{ N})\left(\frac{34.34\text{ mm}}{1000\ \frac{\text{mm}}{\text{m}}}\right)}{\pi\left(\frac{5.723\text{ mm}}{1000\ \frac{\text{mm}}{\text{m}}}\right)^3}\right)$$

$$= 560{,}239{,}000\text{ Pa}\quad(560.24\text{ MPa})$$

Mechanical Springs

$$\tau_{\min} = K_s\frac{8F_{\min}D_{\text{mean coil}}}{\pi D_{\text{wire}}^3} = (1.2381)\left(\frac{(8)(620\text{ N})\left(\frac{34.34\text{ mm}}{1000\ \frac{\text{mm}}{\text{m}}}\right)}{\pi\left(\frac{5.723\text{ mm}}{1000\ \frac{\text{mm}}{\text{m}}}\right)^3}\right)$$

$$= 358{,}110{,}000\text{ Pa}\quad(358.11\text{ MPa})$$

The mean and alternating stresses are

$$\tau_m = \tfrac{1}{2}(\tau_{\max}+\tau_{\min}) = \left(\frac{1}{2}\right)(560.24\text{ MPa}+358.11\text{ MPa})$$
$$= 459.18\text{ MPa}$$

$$\tau_{\text{alt}} = \tfrac{1}{2}(\tau_{\max}-\tau_{\min}) = \left(\frac{1}{2}\right)(560.24\text{ MPa}-358.11\text{ MPa})$$
$$= 101.07\text{ MPa}$$

[10] To avoid misunderstanding, it is generally better to call out wire (flat stock) by diameter (thickness) rather than gauge number.

The allowable region is defined by the material properties reduced by the factor of safety.

$$\frac{S_{es}}{\text{FS}} = \frac{690 \text{ MPa}}{1.5} = 460 \text{ MPa}$$

$$\frac{S_{ys}}{\text{FS}} = \frac{830 \text{ MPa}}{1.5} = 553.3 \text{ MPa} \quad (553 \text{ MPa})$$

The point corresponding to the actual stresses is just outside the allowable region. The spring is not satisfactory.

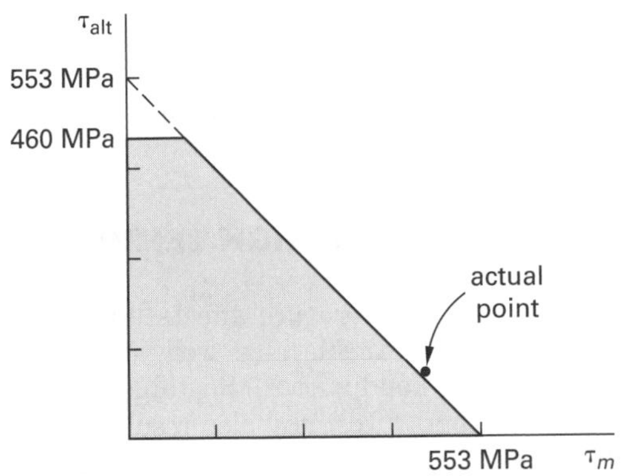

Customary U.S. Solution

The mean spring diameter and correction factor is

Mechanical Springs

$$C = \frac{D_{\text{mean coil}}}{D_{\text{wire}}}$$

$$D_{\text{mean coil}} = CD_{\text{wire}} = (6)(0.2253 \text{ in}) = 1.3518 \text{ in}$$

$$K_s = \frac{4C+2}{4C-3} = \frac{(4)(6)+2}{(4)(6)-3} = 1.2381$$

From Eq. 53.16, the maximum and minimum stresses are

Mechanical Springs

$$\begin{aligned}\tau_{\max} &= K_s \frac{8F_{\max}D_{\text{mean coil}}}{\pi D_{\text{wire}}^3} = (1.2381)\left(\frac{(8)(219.8 \text{ lbf})(1.3518 \text{ in})}{\pi(0.2253 \text{ in})^3}\right) \\ &= 81{,}913 \text{ lbf/in}^2\end{aligned}$$

$$\begin{aligned}\tau_{\min} &= K_s \frac{8F_{\min}D_{\text{mean coil}}}{\pi D_{\text{wire}}^3} = (1.2381)\left(\frac{(8)(140.2 \text{ lbf})(1.3518 \text{ in})}{\pi(0.2253 \text{ in})^3}\right) \\ &= 52{,}248 \text{ lbf/in}^2\end{aligned}$$

The mean and alternating stresses are

$$\begin{aligned}\tau_m &= \tfrac{1}{2}(\tau_{\max} + \tau_{\min}) = \left(\frac{1}{2}\right)\left(81{,}913 \ \frac{\text{lbf}}{\text{in}^2} + 52{,}248 \ \frac{\text{lbf}}{\text{in}^2}\right) \\ &= 67{,}081 \text{ lbf/in}^2\end{aligned}$$

$$\begin{aligned}\tau_{\text{alt}} &= \tfrac{1}{2}(\tau_{\max} - \tau_{\min}) = \left(\frac{1}{2}\right)\left(81{,}913 \ \frac{\text{lbf}}{\text{in}^2} - 52{,}248 \ \frac{\text{lbf}}{\text{in}^2}\right) \\ &= 14{,}833 \text{ lbf/in}^2\end{aligned}$$

The allowable region is defined by the material properties reduced by the factor of safety.

$$\frac{S_{es}}{\text{FS}} = \frac{100{,}000 \ \frac{\text{lbf}}{\text{in}^2}}{1.5} = 66{,}667 \text{ lbf/in}^2$$

$$\frac{S_{ys}}{\text{FS}} = \frac{120{,}000 \ \frac{\text{lbf}}{\text{in}^2}}{1.5} = 80{,}000 \text{ lbf/in}^2$$

The point corresponding to the actual stresses is just outside the allowable region. The spring is not satisfactory.

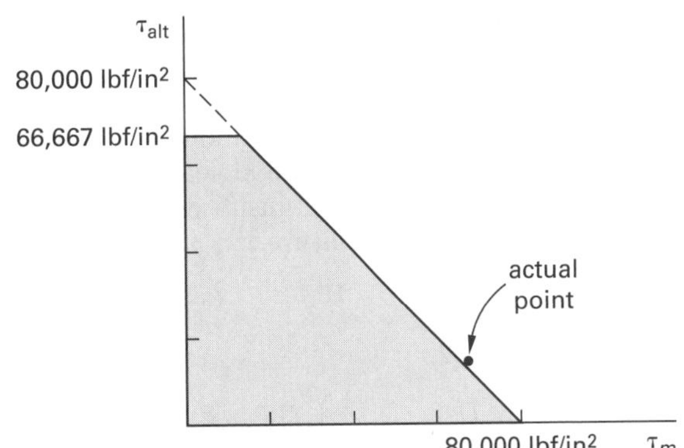

7. BUCKLING OF HELICAL COMPRESSION SPRINGS

Tall springs (i.e., with free lengths more than 4 to 5 times their mean coil diameters) and weak springs (i.e., with ratios of mean coil to wire diameters less than 5) can buckle when heavily loaded. A simple method of checking for buckling is to compare the ratio of working deflection to free length against the limiting values given in Table 53.5. Springs that tend to buckle can be guided by placing them over a tube or rod.

Table 53.5 *Approximate Maximum δ_w/h_t to Prevent Spring Buckling*

L_0/D	both ends pivoting ball	one end pivoting ball, other squared and ground	both ends squared and ground
1	0.72	–	0.72
2	0.63	–	0.71
3	0.38	–	0.71
4	0.20	0.46	0.63
5	0.11	0.25	0.53
6	0.07	0.18	0.38
7	0.05	0.14	0.26
8	0.04	0.10	0.19
9	–	0.08	0.16
10	–	0.07	0.14

Adapted from *Design Handbook*, 1987 Edition, Barnes Group (Associated Spring), Bristol, CT.

8. HELICAL COMPRESSION SPRINGS: DYNAMIC RESPONSE

At certain speeds, resonance between the frequency of the applied force and one of the spring's natural frequencies may occur. The spring may experience excessive vibrations (*surging*) and fail prematurely.

The *fundamental frequency* (*fundamental harmonic frequency*), f_0, in Hz, of a helical compression spring with both ends fixed can be calculated from Eq. 53.20. To avoid resonance with any of the harmonics, the spring's fundamental frequency should be at least 13 times the exciting frequency.[11] (Units of unsubscripted variables are as defined in the nomenclature.)

$$f_0 = \frac{D_{\text{wire}}}{2\pi R^2 n_a}\sqrt{\frac{G}{32\rho}} = (112)\left(\frac{D_{\text{wire,mm}}}{D^2_{\text{mean coil,mm}} n_a}\right)\sqrt{\frac{G}{\rho}} \quad \text{[SI]} \quad 53.20(a)$$

$$f_0 = \frac{D_{\text{wire}}}{2\pi R^2 n_a}\sqrt{\frac{Gg_c}{32\rho}} = \left(\frac{D_{\text{wire,in}}}{8.89 D^2_{\text{mean coil,in}} n_a}\right)\sqrt{\frac{Gg_c}{\rho}} \quad \text{[U.S.]} \quad 53.20(b)$$

Table 53.1 lists values of the shear modulus, G, for various spring materials. The density, ρ, of virtually all iron-based spring materials is 0.284 lbm/in³ (7870 kg/m³), and $G = 11.5 \times 10^6$ lbf/in² (79 GPa). Equation 53.21 can be used directly for steel springs.

$$f_0 \approx (3.5 \times 10^5)\left(\frac{D_{\text{wire,mm}}}{D^2_{\text{mean coil,mm}} n_a}\right) \quad \text{[SI]} \quad 53.21(a)$$

$$f_0 \approx (14{,}000)\left(\frac{D_{\text{wire,in}}}{D^2_{\text{mean coil,in}} n_a}\right) \quad \text{[U.S.]} \quad 53.21(b)$$

Other methods of preventing resonance include winding the spring with a variable pitch (which changes the natural frequency), using a rubbing friction device, and using combinations of multiple springs to achieve the desired spring rate.[12]

9. HELICAL SPRINGS: HIGH TEMPERATURE

Operation at high temperature affects both the elastic modulus (shear and tensile), as well as the strength (yield, ultimate, and endurance). Sustained operation at high enough temperatures may also result in creep, set, and relaxation. Evaluating the effects of high temperatures requires knowing the high-temperature material properties.

Recommended maximum spring service temperatures are 250°F to 300°F (90°C to 150°C) for high-carbon steel, 350°F (175°C) for oil-tempered carbon steel, 425°F to 475°F (220°C to 245°C) for alloy steel, 500°F to 900°F (260°C to 480°C) for stainless steel, and 1000°F to 1100°F (540°C to 590°C) for nickel alloys.

10. HELICAL EXTENSION SPRINGS

Helical *extension springs* can be wound with loose or tight coils. With tight-wound coils, there is an *initial tension*, F_i, that must be applied in order to separate the coils. (See Fig. 53.3.)

Figure 53.3 *Deflection Curves for Extension Springs*

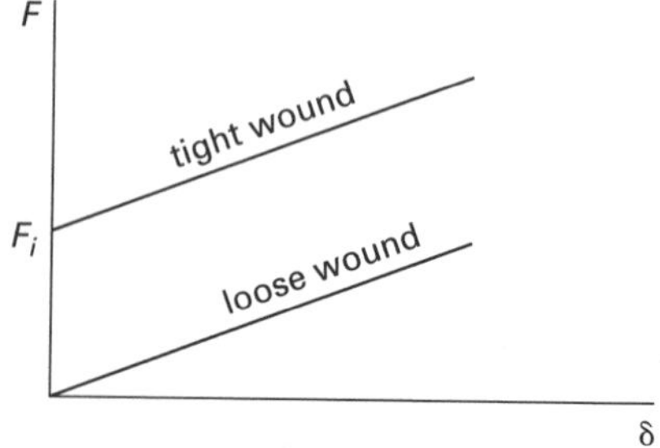

[11] Some authorities recommend 15 to 20 times the fundamental frequency.
[12] In extreme cases, a special spring can be manufactured from stranded wire.

The initial stress is given by Eq. 53.22.

$$\tau_i = \frac{8WDF_i}{\pi d^3} \quad 53.22$$

After the applied load has increased to F_i, the spring behavior is predicted by compression spring formulas.

$$\tau = \tau_i + \frac{8FDW}{\pi d^3} \quad 53.23$$

$$\delta = (8n_a D^3)\left(\frac{F - F_i}{Gd^4}\right) \quad 53.24$$

Extension springs often fail at their hooks from stress concentrations due to hook curvature. Two critical points are shown in Fig. 53.4. Point A is highly stressed in bending. Point B is highly stressed in shear. Equation 53.25 and Eq. 53.26 are simplified expressions for the maximum bending and shear stresses in the hook. In effect, the ratio of the radii constitute the stress concentration factors. It is recommended that r_4 be greater than twice the wire diameter.

$$\sigma_A = \left(\frac{16FD}{\pi d^3}\right)\left(\frac{r_1}{r_3}\right) \quad 53.25$$

$$\tau_B = \left(\frac{8FD}{\pi d^3}\right)\left(\frac{r_2}{r_4}\right) \quad 53.26$$

Figure 53.4 *Points of Stress Concentration in Extension Springs*

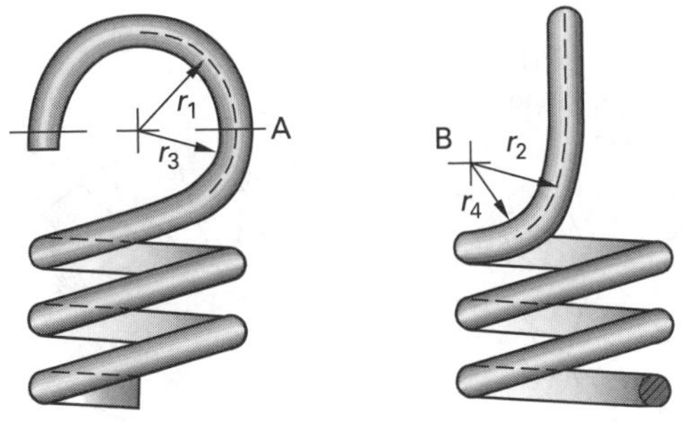

11. HELICAL TORSION SPRINGS

Helical torsion springs manufactured from round wire are essentially round cantilever beams.[13] Loading produces a bending stress. Most torsion springs operate over an arbor. A clearance of about 10% between the arbor and spring is generally adequate to prevent binding. The bending stress is largest at the inner radius of the spring. (See Fig. 53.5.) Equation 53.27 finds the stress on the inner fiber of the spring, and Eq. 53.28 can be used to find either the inner or outer fiber.

Mechanical Springs

$$\sigma = K_i \frac{32Fr}{\pi D^3_{\text{wire}}} \quad 53.27$$

$$\sigma = K_b \frac{32M}{\pi D^3_{\text{wire}}} \quad 53.28$$

$$Fr = k\theta \quad 53.29$$

The Wahl stress correction factors for bending are calculated using Eq. 53.30 for either the inner or outer faces.

Mechanical Springs

$$K_i = \frac{4C^2 - C - 1}{4C(C-1)} \quad 53.30$$

In the absence of friction, the spring constant, k, for helical torsion springs will be the same for any angular deflection. The angular spring rate is given first in units of torque/rad and then in units of torque/rev in Eq. 53.31, but does not include the contribution of the long spring ends, which can be treated as springs as well.[14]

$$k = \frac{Fr}{\theta} = \frac{d^4E}{64DN} \approx \frac{d^4E}{10.8DN} \quad 53.31$$

If the long ends are flexible and are of the same wire diameter, their contribution to the active coils is one-third of their moment arm (measured from the center of the coil).

$$n_a = n_{\text{body}} + n_{\text{ends}} = n_{\text{body}} + \frac{L_1 + L_2}{3\pi D} \quad 53.32$$

Figure 53.5 *Helical Torsion Spring*

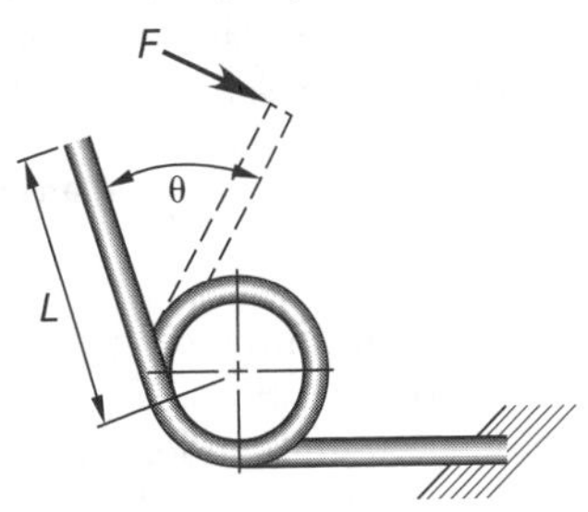

[13]There are two basic types of torsion springs: the *flat coil spring* (also known as a *power spring* or *clock spring*) and the helical torsion spring. Flat coil springs are not covered in this section.

[14]The factor 10.8 is a practical replacement for the theoretical conversion from radians to revolutions of $64/2\pi = 10.18$ that accounts for friction between adjacent spring coils and between the spring and the arbor.

12. FLAT AND LEAF SPRINGS

Flat springs are constructed as simply supported or cantilever beams. They can be flat, curved, or nested (as in a leaf spring). The traditional beam deflection tables can be used with simple flat springs when the deflections are small.[15]

Since beam bending stress is proportional to the thickness of the beam, stress can be kept low by using several low-thickness springs instead of a single thick spring. *Leaf springs* (*leaf set*), as commonly used in cars, consist of several flat springs, each atop another. The capacity of a leaf set is the sum of the capacities of all springs. The springs slide longitudinally over one another. The effect of this sliding and the resultant friction is difficult to evaluate.

In vehicles, vibrations and impacts produce leaf spring deflections so that potential energy is as strain energy. Increasing the energy storage capability of a leaf spring ensures a more compliant suspension system. The amount of elastic energy that can be stored by a leaf spring per unit volume is given by Eq. 53.33. Materials with large strengths and small moduli of elasticity are the most suitable materials for leaf springs.

$$U_V = \frac{\sigma^2}{2E} \qquad 53.33$$

A modification of the elastic modulus of the basic beam equations is required if the spring is very wide (i.e., those with w/t ratios greater than 8). Different methods (typically graphical) are required if the deflection is large (i.e., 30% or more of the spring length).[16]

13. VELOCITY, POWER, AND TORQUE IN ROTATING MEMBERS

Equation 53.34 gives the tangential velocity for a rotating circular member with radius r. If the pitch circle diameter is used with gears, the tangential velocity is referred to as the *pitch circle velocity*.

Plane Circular Motion

$$\text{v}_t = r\omega \qquad 53.34$$

This equation can also be expressed in other forms if the velocity is found in specific units. Equation 53.35 finds the tangential velocity in units of length per second, and Eq. 53.36 finds the tangential velocity in units of ft/min.

$$\text{v}_t = \frac{\pi D n_{\text{rpm}}}{60} \quad \text{[length/sec]} \qquad 53.35$$

$$\begin{aligned}\text{v}_t &= r\omega = rn_{\text{rpm}}\left(\frac{2\pi}{\text{rev}}\right)\left(\frac{1\text{ min}}{60\text{ sec}}\right)\\ &= \frac{\pi r n_{\text{rpm}}}{30} = \frac{\pi D n_{\text{rpm}}}{60} \quad \text{[units of length/sec]} \qquad 53.36\\ \text{v}_t &= \frac{\pi D n_{\text{rpm}}}{12} \quad \text{[ft/min]}\end{aligned}$$

The relationship between transmitted horsepower, torque, and rotational speed is

$$P_{\text{kW}} = \frac{T_{\text{N·m}} n_{\text{rpm}}}{9549} \quad \text{[SI]} \qquad 53.37(a)$$

Shaft-Horsepower Relationship and Force-Horsepower Relationship

$$\text{HP} = \frac{T_{\text{in-lbf}} n_{\text{rpm}}}{63{,}025} \quad \text{[U.S.]} \qquad 53.37(b)$$

14. SPUR GEAR TERMINOLOGY

Spur gears have the simplest type of teeth. The teeth faces of a spur gear are parallel to the axis of rotation. Figure 53.6 illustrates a typical spur gear and some of the terminology used to describe gear and tooth geometry.

Figure 53.6 Spur Gear Terminology

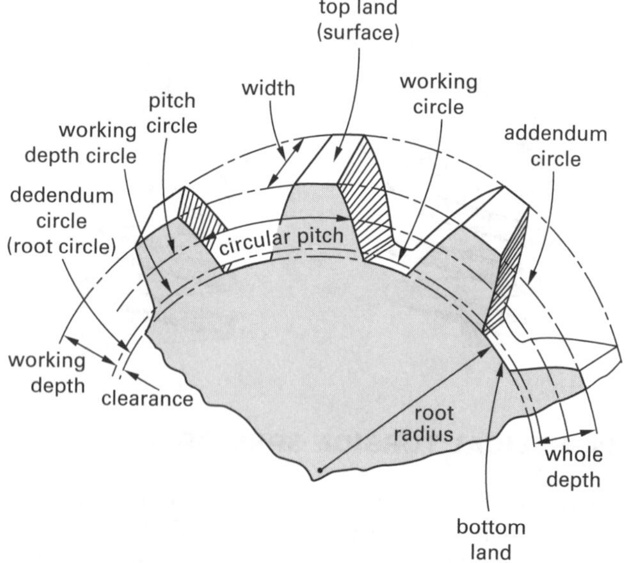

[15]The common beam equations assume that the deflection is small (i.e., less than a few percent of the length) and that the load remains perpendicular to the beam at all times.
[16]When the beam is bent extensively, the maximum stress in a cantilever beam may not even occur at the fixed end. Complexities such as this call for a different analysis method.

The *pitch circle* is an imaginary circle on which the gear lever arm is based. The *pitch point* is an imaginary point of tangency between the pitch circles of two meshing gears.

The velocity of a pitch point is the *pitch-line velocity* (*pitch velocity*). The pitch velocity can be calculated directly from the pitch diameter, D_p, or indirectly from the torque and power using Eq. 53.37.

The *addendum* is the radial distance from the pitch circle to the top of the tooth. For full-depth gears, it is equal to the reciprocal of the diametral pitch (i.e., $1/p_d$). The *base circle* is the circle that is tangent to the line of action. The *clearance* is the separation between the dedendum circle of gear 1 and the addendum circle of gear 2 when both gears are in mesh.

Equations for minimum clearance and other characteristics for U.S. standard full-depth involute spur gear system are given in Table 53.6.

***Table 53.6** Characteristics of Full-Depth Involute Gear Systems*

characteristic	based on diametral pitch, p_d
addendum	$1/p_d$
minimum dedendum	$1.157/p_d$
working depth	$2/p_d$
minimum total depth	$2.157/p_d$
basic tooth thickness on pitch line	$1.5708/p_d$
minimum clearance	$0.157/p_d$

The *clearance circle* is the circle that is tangent to the addendum of the meshing gear. The *dedendum* is the radial distance from the pitch circle to the root circle. For full-depth gears, it is equal to either $1.25/p_d$ or $1.35/p_d$, depending on the clearance wanted. The *tooth face* is the tooth area between the pitch circle and the addendum circle. The *face width* is the axial width of the tooth. The *flank* is the tooth area between the pitch circle and the dedendum. The *land* is the flat surface at the top of each tooth.

Whole depth is the distance from the addendum circle to the dedendum circle. It is equal to the working depth plus the clearance. The *working depth* is the distance that a tooth from a meshing gear extends into the space between two teeth.

Three meanings for the term "pitch" are used for spur gears. The *diametral pitch*, p_d, is the number of teeth per inch of pitch circle diameter. The diametral pitch is the same for all meshing teeth. The *circular pitch*, p_c, is the distance between corresponding tooth points along the pitch circle. It is equal to the tooth thickness plus the curved separation distance between teeth. The circular pitch is the same for all meshing teeth.

The *normal pitch* (also known as the *base pitch*) is the distance from a point on one gear to the corresponding point measured along the base circle. It is also the distance from a point to the same corresponding point on the meshing gear tooth.

The *module*, m, of a gear is the ratio of the pitch diameter to the number of teeth. Thus, the module is the reciprocal of the diametral pitch. The module (with units of mm per tooth) is the common SI index of tooth size.

Involute Gear Tooth Nomenclature

$$m = \frac{D_p}{N} \quad \text{[module]} \qquad 53.38$$

$$p_d = \frac{N}{D_p} \quad \text{[diametral pitch]} \qquad 53.39$$

$$p_c = \frac{\pi D_p}{N} = \pi m \quad \text{[circular pitch]} \qquad 53.40$$

$$D_p = \frac{N}{p_d} = mN \quad \text{[pitch diameter]} \qquad 53.41$$

Not every diametral pitch is available. To be economical, designs should make use of the standard diametral pitches. Common "coarse" series diametral pitches include 1, 1¼, 1½, 1¾, 2, 2¼, 2½, 3, 4, 6, 8, 10, 12, and 16 teeth/in. Common "fine" series diametral pitches include 20, 24, 32, 40, 48, 80, 96, 120, 150, and 200 teeth/in. SI modules are divided into series. Although smaller, intermediate, and fractional modules exist in other (less desirable) series, the preferred standard series 1 modules are 1, 1.25, 1.5, 2, 2.5, 3, 4, 5, 6, 8, 10, 12, 16, 20, 25, 32, 40, and 50 mm (standards ISO 53 and JIS B 1701).

Most gears in use are *involute gears* (i.e., have involute-cut teeth). An involute of a circle is the curve traced by the end of a taut string that is unwound from the circumference. The base circle is defined as the circle from which the involute is generated.

The *line of action* (also known as the *pressure line* and *generating line*) is a line passing through the pitch point that is tangent to both base circles. It is the distance between the intersections of the pressure line and the addendum circles. In Fig. 53.7, line A-A is tangent to the base circles and is the line of action. Angle ϕ, determined by the points of tangency, is the *pressure angle* (also known as the *angle of obliquity*). This involute is called a "ϕ° involute." In the United States, 20°, 22½°, and 25° pressure angles are in common use. The once-used older 14½° pressure angle has essentially become obsolete as it produces larger gears.

The pressure angle is proportional to the center-to-center distance of the gears, but a small deviation (i.e., error) in center-to-center distance will change the pressure angle slightly. However, changes in center-to-center spacing and backlash don't change the velocity ratio or

Figure 53.7 *Meshing Gear Terminology*

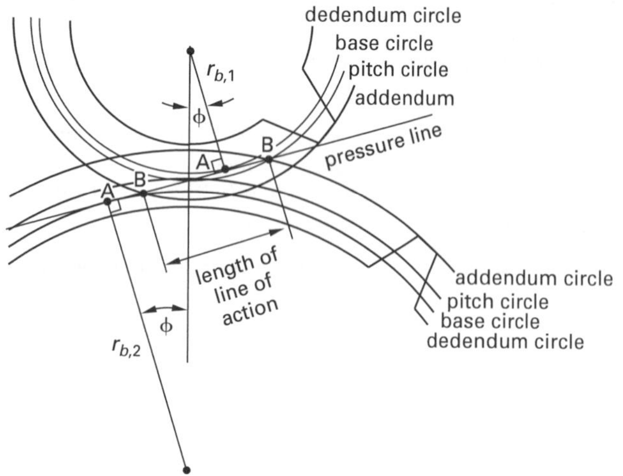

the general performance of gear sets with involute gears. This is the main reason that involute gearing is widely used.

Figure 53.7 also illustrates the length of the line of contact between teeth on meshing gears. Line B-B is the section of the line of action between the points where it crosses the two addenda circles. This is sometimes called the *length of the line of action*. The length of the line of action is designated L_{ab}.

15. TOOTH THICKNESS

Ideally, the tooth thickness will be one-half of the circular pitch. However, to avoid binding from gears whose teeth are inadvertently manufactured with thicker teeth, a clearance must be included in the design. The clearance is known as the *backlash*. The amount of backlash is the distance along the pitch circle, but it is generally measured with feeler gauges. Backlash generally varies between $0.03/p_d$ and $0.05/p_d$. Backlash has no effect on gear tooth action for involute-cut gears.

16. GEAR SETS AND GEAR DRIVES

In a simple set of two external gears in contact (referred to as a *mesh* or a *gear set*), one gear drives the other. Often, the smaller gear drives the larger, in which case the smaller gear is referred to as the driving *pinion*. The larger gear in the set is referred to as the *driven gear*. The *center distance* is the distance between the centers of the pinion and gear.

The force between two meshing teeth will be the same. Since the same power ideally will be transmitted by each gear, the product of torque and rotational speed are the same.

$$\mathrm{v}_{\mathrm{pinion}} = \mathrm{v}_{\mathrm{gear}} \qquad 53.42$$

$$F_{\mathrm{pinion}} = F_{\mathrm{gear}} \qquad 53.43$$

The angular *velocity ratio*, VR, for a pair of gears can be calculated in a number of ways.

$$\mathrm{VR} = \frac{n_{\mathrm{pinion}}}{n_{\mathrm{gear}}} = \frac{\omega_{\mathrm{pinion}}}{\omega_{\mathrm{gear}}} = \frac{r_{\mathrm{gear}}}{r_{\mathrm{pinion}}} = \frac{N_{\mathrm{gear}}}{N_{\mathrm{pinion}}} \qquad 53.44$$

At times, one pair of teeth will carry all of the force and will transmit all of the power. At other times, two (or more) pairs may be in contact. The average number of tooth pairs in contact is the *contact ratio*, CR. The contact ratio is usually between 1.2:1 and 1.6:1. (1.2:1 means that one pair of teeth is in contact at all times, and a second pair is in contact 20% of the time.) For good design, the contact ratio should be approximately 1.5:1.

$$\mathrm{CR} = \frac{L_{\mathrm{ab}}}{p\cos\phi} = \frac{L_{\mathrm{ab}}P}{\pi\cos\phi} \qquad 53.45$$

The length of the line of action is calculated from Eq. 53.46. r_1 and r_2 are the respective pitch radii of the gears. For full-depth teeth, $m = 1$, and for stub teeth, $m = 0.8$.

$$L_{\mathrm{ab}} = \sqrt{\left(r_1 + \frac{m_1}{P}\right)^2 - (r_1\cos\phi)^2} - r_1\sin\phi + \sqrt{\left(r_2 + \frac{m_2}{P}\right)^2 - (r_2\cos\phi)^2} - r_2\sin\phi \qquad 53.46$$

For two gears whose centers are separated by a distance, C, and with outside diameters D_o and d_o, and with base diameters D_b and d_b, the contact ratio can also be calculated as

$$\mathrm{CR} = \frac{\sqrt{D_o^2 - D_b^2} + \sqrt{d_o^2 - d_b^2} - 2C\sin\phi}{2p\cos\phi} \qquad 53.47$$

A gear set is *prime* when the numbers of teeth on each meshing gear have no common factor except 1. This is a desirable condition, as all teeth tend to wear evenly.

The *service factor* is a single measure that combines the external load dynamics of an application with a gear drive's reliability and operating life. The service factor is applied to the motor's rated performance. When selecting components, either the actual load must be multiplied by the service factor, or all catalog ratings (horsepower, torque, and overhung load) must be divided by the service factor. Acceptable service factor values are determined from experience or are specified by various authorities. The American Gear Manufacturer's Association (AGMA) publishes service factors that depend on the application service class number and type of prime mover. Values range from 0.8 to 2.25.

17. MESH EFFICIENCY

Ideally, the input power is passed through each gear to the next in line.

$$P_{\text{gear}} = P_{\text{pinion}} \qquad 53.48$$

$$T_{\text{gear}} n_{\text{gear}} = T_{\text{pinion}} n_{\text{pinion}} \qquad 53.49$$

In reality, each gear set will dissipate some of the input power. This is accounted for by the *efficiency of the gear train* (i.e., the *mesh efficiency*), η_{mesh}.

$$\eta_{\text{mesh}} = \frac{P_{\text{output}}}{P_{\text{input}}} \qquad 53.50$$

In the case of a pinion driving a gear,

$$P_{\text{gear}} = \eta_{\text{mesh}} P_{\text{pinion}} \qquad 53.51$$

$$T_{\text{gear}} n_{\text{gear}} = \eta_{\text{mesh}} T_{\text{pinion}} n_{\text{pinion}} \qquad 53.52$$

18. FORCE ANALYSIS OF SPUR GEARS

The *tangential force* (also known as the *transmitted load*) at the pitch circle is the useful force, and it is the only component that transmits power. It can be found from the transmitted torque. The radial force can be determined from the tangential component. There is no axial force on a spur gear.

Spur Gears

$$W_t = \frac{60{,}000 H_{\text{kW}}}{\pi D_{\text{gear}} n} \qquad \text{[SI]} \qquad 53.53(a)$$

$$W_t = \frac{\left(33{,}000 \ \frac{\text{ft-lbf}}{\text{hp-min}}\right) H_{\text{hp}}}{\text{v}_{\text{ft/min}}} \qquad \text{[U.S.]} \qquad 53.53(b)$$

The radial force on a spur gear tooth is

$$F_r = F_t \tan\phi \qquad 53.54$$

The total force on a spur gear tooth is

$$F_{\text{total}} = \sqrt{F_t^2 + F_r^2} = \frac{F_t}{\cos\phi} = \frac{F_r}{\sin\phi} \qquad 53.55$$

19. HELICAL GEARS

Figure 53.8 illustrates the three angles associated with a *helical gear*. In addition to the pressure angle (*normal pressure angle*), ϕ_n, there is a *tangential pressure angle*, ϕ_t, and a *helix angle*, ψ. The 20° normal pressure angle is the most common. Helix angles commonly range from 0° to 30° and seldom approach 45°.

$$\cos\psi = \frac{\tan\phi_n}{\tan\phi_t} \qquad 53.56$$

Figure 53.8 *Helical Gear*

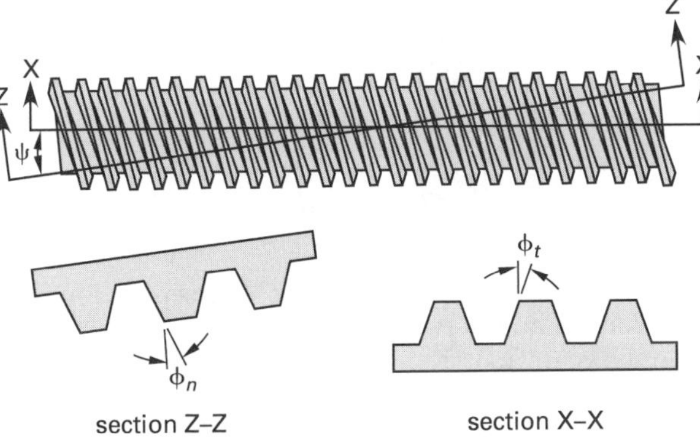

The relationship between the *transverse circular pitch* (usually called *circular pitch*), p_t, and the *normal circular pitch*, p_n, is

$$p_t = \frac{p_n}{\cos\psi} \qquad 53.57$$

The *axial pitch* is

$$p_a = \frac{p_t}{\tan\psi} \qquad 53.58$$

The *normal diametral pitch* is

$$P_n = \frac{P_t}{\cos\psi} \qquad 53.59$$

20. FORCE ANALYSIS OF HELICAL GEARS

There are three force components on a helical gear: radial, axial, and tangential. (See Fig. 53.9.) As with spur gears, the tangential force is the useful force, and it is the only component that transmits power. It is calculated from Eq. 53.53. The radial and axial components are calculated from the tangential force.

Depending on the rotation, the axial component of force (also known as the *thrust force*) compresses or stretches the shaft, while the radial component works to separate the meshing gears. For that reason, the radial force is

Figure 53.9 Forces on a Helical Gear

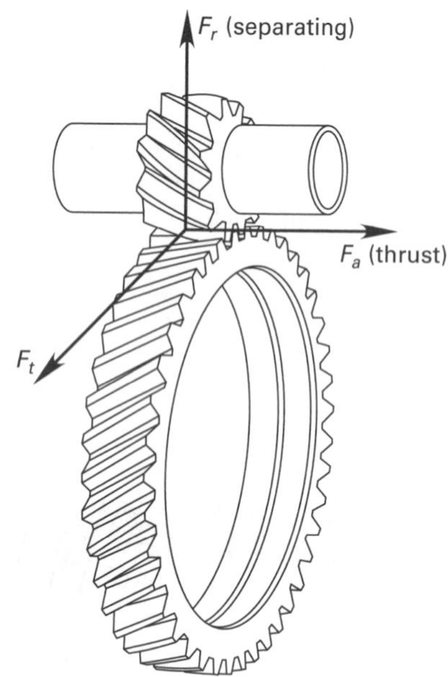

also known as the *separation force*. The separation force acts along a line between the gear centers and tries to drive the gears apart.

$$F_r = F_t \tan\phi_t = \frac{F_t \tan\phi_n}{\cos\psi} \qquad \text{53.60}$$

$$F_a = F_t \tan\psi \qquad \text{53.61}$$

The three force components combine into the total force, F, on the gear. Although total force is of academic interest, the transmitted tangential force, F_t, is used to design the gear.

$$F = \sqrt{F_r^2 + F_a^2 + F_t^2} \qquad \text{53.62}$$

The radial, axial, and tangential force components in terms of the total force are

$$F_r = F \sin\phi_n \qquad \text{53.63}$$

$$F_a = F \cos\phi_n \sin\psi \qquad \text{53.64}$$

$$F_t = F \cos\phi_n \cos\psi \qquad \text{53.65}$$

21. ALLOWABLE STRESSES FOR GEAR DESIGN

The simplest method to calculate an allowable stress for gear tooth design is to divide the tensile yield strength by a substantial factor (e.g., 3 to 5). This method requires judgment and experience.

Another method is to calculate the allowable stress as some percentage (e.g., 75%) of the endurance strength in reversed bending. Since most gear stresses vary repeatedly only from zero to a maximum value (i.e., are not completely reversed), this recommendation builds in a considerable factor of safety.

The AGMA *bending strength* should only be used with the AGMA procedures. It is not an allowable stress in the general sense.

22. SPUR GEAR STRENGTH: LEWIS BEAM STRENGTH

The *Lewis beam strength theory* has been largely superseded by AGMA procedures, but it is of historical interest.[17] The theory assumes that one tooth carries the entire tangential load as a cantilever beam.[18]The force is applied at the tip of the tooth, parallel to the top land. The radial component of force is disregarded. The maximum stress occurs at the tooth root. The maximum allowable tangential force is given by Eq. 53.66. The *form factor*, Y, varies approximately between 0.2 and 0.5 and has been widely tabulated in gear and machine-design textbooks.[19]

$$W_t = \frac{\sigma F Y}{P_d} \qquad \text{53.66}$$

The form factor depends on the pressure angle and the type of gear (e.g., spur, helical, etc.). Figure 53.10 illustrates the variation in form factor for spur gears.

Alternatively, the actual bending stress can be determined.

Spur Gears

$$\sigma = \frac{W_t P_d}{F Y} \qquad \text{53.67}$$

The original Lewis beam strength theory assumed a static loading of the tooth. This is appropriate for low-speed operation only. To account for speed and dynamic effects, the allowable Lewis beam strength force is reduced by a *speed factor*. In the *Barth speed factor*,

[17]The Lewis method can be used for quick estimates.

[18]The Lewis beam strength theory was first proposed in 1892.

[19]There are two factors named "form factor." The *Lewis form factor* is typically written with a lower case y. The *form factor* is written with an upper case Y. Since $Y = \pi y$, one can be derived from the other.

Figure 53.10 *Lewis Form Factor, Y, for Spur Gears*

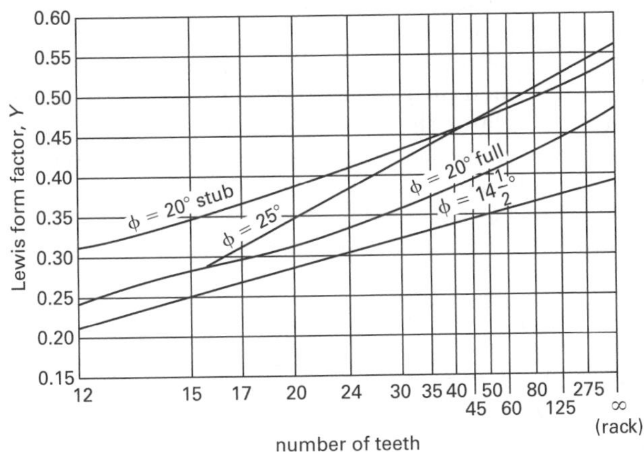

given by Eq. 53.69, the constant a is 600 for ordinary industrial gears and gears with cast teeth and 1200 for accurately cut gears running as high as 6000 ft/min.[20]

$$W_t = \frac{k_d \sigma F Y}{P_d} \qquad 53.68$$

$$k_d = \frac{a}{a + \mathrm{v}_{t,\mathrm{ft/min}}} \qquad 53.69$$

Additional extensions have been made to the Lewis theory to account for the stress concentration at the root of the tooth.

23. DESIGN GUIDELINES FOR GEARS

The following general rules can be used to simplify designs of gears.

- The face width, w, of spur gears should be 3 to 5 times the circular pitch, π/P.
- Increasing the face width will decrease wear and fatigue, but the dynamic load will increase.
- A coarser tooth (smaller diametral pitch) will improve the fatigue performance but not the wear performance. For a given tooth design, only a harder material will improve the wear.
- Larger pitch diameters have lower tangential forces and lower dynamic loadings.
- Decreasing the tooth error decreases the dynamic loading.

24. AGMA GEAR DESIGN METHOD

The AGMA formula for bending stress in the tooth (referred to as a *bending stress number*, s_b) is given by Eq. 53.70. K_a is an application factor, K_v is the dynamic (velocity) factor, K_s is the size factor, K_m is the load distribution factor, and J is the geometry factor. P is the diametral pitch in teeth per inch, and m is the module in mm per tooth.

$$s_{b,\mathrm{MPa}} = \left(\frac{F_t K_a}{K_\mathrm{v}}\right)\left(\frac{1}{mw}\right)\left(\frac{K_s K_m}{J}\right) \qquad \text{[SI]} \qquad 53.70(a)$$

$$s_{b,\mathrm{lbf/in^2}} = \left(\frac{F_t K_a}{K_\mathrm{v}}\right)\left(\frac{P}{w}\right)\left(\frac{K_s K_m}{J}\right) \qquad \text{[U.S.]} \qquad 53.70(b)$$

The allowable design stress (referred to as the *allowable bending stress number*) is calculated from the AGMA *allowable tensile bending strength number*, S_t, obtained from AGMA tables.[21] K_L is the life factor, K_T is the temperature factor, and K_R is the reliability factor.

$$\sigma_a = \frac{S_t K_L}{K_T K_R} \qquad 53.71$$

AGMA publishes a similar procedure for comparing the surface stress to the allowable contact stress. This is sometimes referred to as the AGMA *pitting resistance* or AGMA *surface durability.*

All AGMA methods are highly dependent on having the AGMA charts and tables.[22] Since different charts and tables are needed for different tooth depths and pressure angles, these charts and tables are extensive. Some are included in engineering handbooks and machine design textbooks, but the material is too extensive to include here.

25. LIMIT WEAR DESIGN

Gear teeth can fail by compressive surface fatigue (excessive gear tooth contact stress) as well as by tooth breakage. Either factor can be the controlling factor in determining tooth width. *Limit wear design* is the

[20]Over the years, more sophisticated methods for incorporating the speed effects have been proposed. Since this section covers a methodology that has been essentially superseded by the AGMA procedures, only the historical Barth speed factor is presented.

[21]Some authorities have suggested that some percentage (e.g., 75%) of the endurance strength in reversed bending can be used. However, the pure AGMA procedure requires that the specific strengths from its tables be used. Conversely, the AGMA table values should not be used for any other purpose than gear tooth design.

[22]AGMA standard numbers are subject to change, and the standards themselves are extremely dissociated. For example, bending geometry factors and pitting geometry factors may both be needed, but will come from different standards. Different gear types (spur, helical, bevel, spiral, etc.) have their own standards. Various standards (203, 218, 908, 2000, 2001, 2002, 2003, and 2101) may be useful. AGMA offers two methods to rate spur gears. Both are contained in *Fundamental Rating Factors and Calculation Methods for Involute Spur and Helical Gear Teeth* (ANSI/AGMA 2001). One method calculates the allowable transmitted horsepower on the pitting resistance of gear tooth contact surfaces, while the other calculates transmitted horsepower based on gear tooth bending strength. A similar standard, ANSI/AGMA 2003-C10, is used for bevel and spiral gears.

procedure that accounts for this aspect of gear design. The general procedure is to keep the limit wear load (capacity) greater than the dynamic load. The dynamic load is calculated from the transmitted tangential force, F_t, with corrections for *errors in action* (e.g., inaccuracies in tooth form, cutting, and spacing, mounting misalignments, and rotational inertia).[23]

26. BEVEL GEARS

Bevel gear sets are used with intersecting shafts. Gears can be straight or spiral.[24] Figure 53.11 illustrates a straight bevel gear set. The most common shaft angle is 90°, although any angle can be accommodated. The circular pitch and pitch diameter are as defined for spur gears. The *pitch angles*, γ and Γ, are the angles of the forward *pitch cones* projected back to the apex.

$$\tan\gamma = \frac{N_{\text{pinion}}}{N_{\text{gear}}} \quad 53.72$$

$$\tan\Gamma = \frac{N_{\text{gear}}}{N_{\text{pinion}}} \quad 53.73$$

Figure 53.11 *Straight Bevel Gear Set*

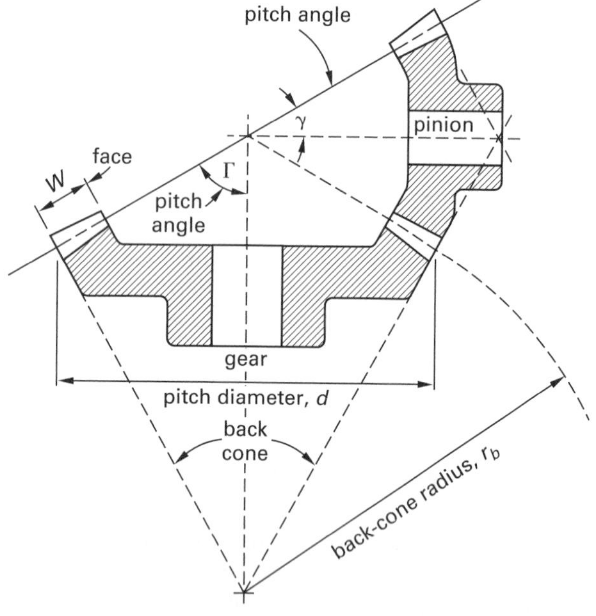

Bevel gears experience a *separation force* (*thrust force*) away from the apex. Installation (choice of bearings) must be designed to ensure that gears are not thrown out of alignment as they are loaded. A force analysis of bevel gears depends on the pressure angle, φ, and the pinion pitch cone angle, γ.

The forces on a bevel gear set are shown in Fig. 53.12 and can be calculated from the following equations, which assume the pinion, *P*, drives the gear, *G*. *T* is the torque on the pinion shaft, W_s is the separating force, W_a is the pinion thrust, and W_r is the gear thrust. Equation 53.53 relates the torque to power transfer and can be used calculate the tangential force on the pinion, W_t. The pitch diameter, D_p, is twice the mean radius, r_{av}.

Figure 53.12 *Forces on Bevel Gear Set*

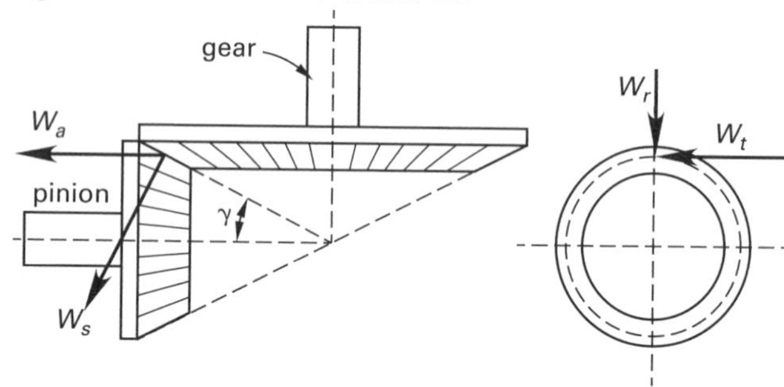

$$T = r_{av}W_t = \frac{D_pW_t}{2} \quad 53.74$$

$$W_s = W_t\tan\phi \quad 53.75$$

Bevel Gears

$$W_a = W_t\tan\phi\sin\gamma \quad 53.76$$

Bevel Gears

$$W_r = W_t\tan\phi\cos\gamma \quad 53.77$$

27. WORM GEAR SETS

Worm drives, consisting of a *worm* and a *worm gear*, are used to turn noncoplanar shafts oriented at right angles with high speed ratios (e.g., 2:1 or 3:1). (See Fig. 53.13.) Worm drives are ordinarily irreversible: A worm will turn a gear, but a gear will not turn a worm. Therefore, worm sets are normally self-locking, although this characteristic should not be depended on for safe operation.

Worms can have a single thread, but they usually have more (as many as six) threads. The number of threads (teeth) is known as the *lead ratio*, LR (N_W). A double-threaded worm has a lead ratio of 2:1 and so on. When a double-threaded (triple-threaded, etc.) worm is rotated through a complete revolution, two (three, etc.) worm pitches pass by a fixed point. When in mesh, one revolution of the worm moves two (three, etc.) pitches of the

[23]The *Buckingham equation* was widely used for many years to calculate dynamic effects until the AGMA equations were developed.
[24]*Hypoid gears* are essentially spiral bevel gears for shafts that do not intersect.

gear. If a worm gear has 60 teeth, it would take 30 revolutions of a double-threaded worm to turn the gear one complete revolution, and the speed ratio would be 30:1. The worm always turns faster than the worm gear. The axial distance that one worm thread moves for one worm revolution is the *lead*, L. For a single-threaded worm, the lead is equivalent to the circular pitch. For a double-threaded (triple-threaded) worm, the lead is double (triple) the circular pitch. The angle of the thread is known as the *lead angle*, λ. The *helix angle*, ψ_W, is the complement of the lead angle ($\theta + \psi_W = 90°$).[25] The relationship between the lead angle, lead, and worm diameter is

Worm Gears

$$L = p_x N_W \quad 53.78$$

$$\tan\lambda = \frac{L}{\pi D_W} \quad 53.79$$

$$\tan\lambda = \frac{p_x(LR)}{\pi D_W} = \frac{p_x N_W}{\pi D_W} \quad 53.80$$

It is customary to calculate the velocity ratio, VR, using the number of threads on the worm and number of teeth on the gear.

Worm Gears

$$\text{VR} = \frac{N_G}{N_W} = \frac{D_G}{D_W \tan\lambda} \quad 53.81$$

The power and torque transmission by a worm-gear train is the same as for any two gears in mesh.

$$T_{\text{gear}} n_{\text{gear}} = T_{\text{worm}} n_{\text{worm}} \quad 53.82$$

28. FLAT BELT DRIVES

There are two types of *belt drives*, as shown in Fig. 53.14. With the *open belt drive*, the shafts turn in the same direction. With the *crossed belt drive*, they turn in opposite directions.

Analysis of flat belts is essentially an application of pulley friction. The relationship between the tight and slack side tensions, T_1 and T_2, respectively, is given by Eq. 53.83 and Eq. 53.84 for both open and crossed belts. The angle of contact, θ, must be in radians. Although leather belts were used extensively in the past, they have been largely replaced by polyamide and urethane belts. The coefficient of friction, μ, for belts is approximately 0.4 for leather, 0.8 for urethane, and 0.5 to 0.8 for polyamide. The coefficient of friction for dry (unlubricated) steel bands on cast iron is 0.18.

Figure 53.13 *Worm and Worm Gear Terminology*

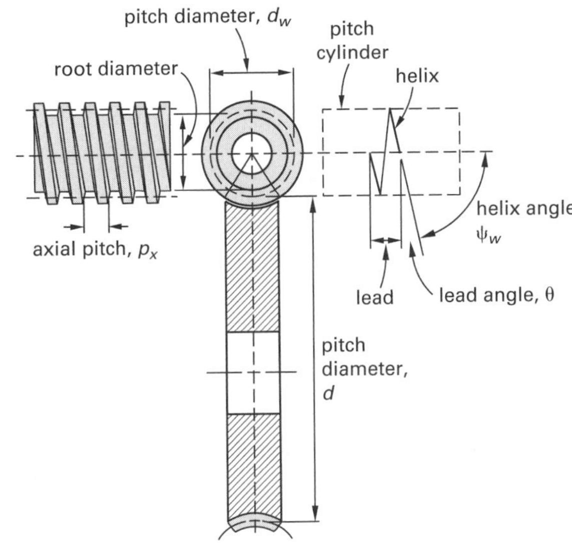

Figure 53.14 *Open and Crossed Belt Drives*

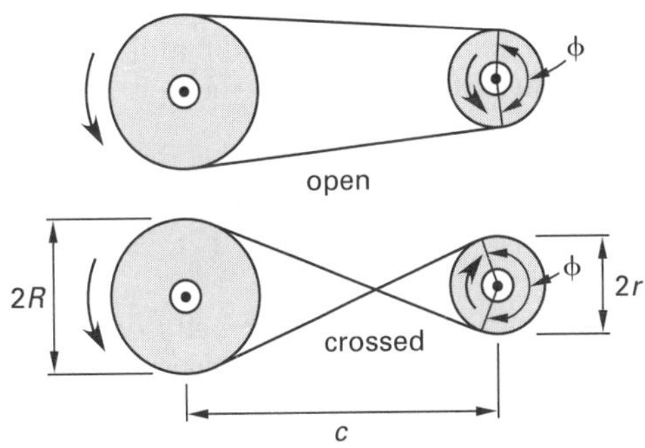

Tensions in Belts and Bands

$$\frac{T_1}{T_2} = e^{\mu\theta} \quad \left[\begin{array}{c} T_1 > T_2; \\ \text{slippage impending} \end{array}\right] \quad 53.83$$

Centrifugal Force (Belt)

$$\frac{F_1 - F_c}{F_2 - F_c} = e^{\mu\theta} \quad [\text{slippage impending}] \quad 53.84$$

Equation 53.85 gives the centrifugal loading per unit length due to the belt. The density of leather belting is approximately 0.035 lbm/in^3 to 0.045 lbm/in^3 (970 kg/m^3 to 1250 kg/m^3), and the density of modern polyamide and urethane belting is approximately 0.038 lbm/in^3 to 0.045 lbm/in^3 (1050 kg/m^3 to 1250 kg/m^3). m is the mass of the belt per unit length.

[25]Some authorities state that the helix angle is the same as the lead angle. The intended meaning can be determined from the values, as worm lead angles are seldom greater than 30°. Above that value, the efficiency gain is minimal, the friction angle decreases, and the mesh loses its self-locking characteristic.

Centrifugal Force (Belt)

$$F_c = m\mathrm{v}^2 \quad \text{[SI]} \quad 53.85(a)$$

$$F_c = \frac{m\mathrm{v}^2}{g_c} \quad \text{[U.S.]} \quad 53.85(b)$$

The theoretical power-torque relationships from Eq. 53.37 apply. The *net tension* responsible for the transmission of horsepower is $F_1 - F_2$.

$$P_{\mathrm{kW}} = \frac{F\mathrm{v}_{\mathrm{m/s}}}{1000\,\frac{\mathrm{W}}{\mathrm{kW}}} \quad \text{[SI]} \quad 53.86(a)$$

Shaft-Horsepower Relationship and Force-Horsepower Relationship

$$\mathrm{HP} = \frac{F\mathrm{v}}{33{,}000} \quad \text{[U.S.]} \quad 53.86(b)$$

Equation 53.87 relates the maximum and minimum tensions to the initial tension, F_i, in the belt at the time of assembly.

$$F_i = \frac{F_1 + F_2}{2} \quad 53.87$$

The angle of contact, θ, for open belts depends on the radii of the two pulleys and the center-to-center distance, C.

Open and Crossed Belts

$$\theta_{D_o} = \pi + 2\sin^{-1}\left(\frac{D_o - D_i}{2C}\right) \quad \text{[large pulley]} \quad 53.88$$

$$\theta_{D_i} = \pi - 2\sin^{-1}\left(\frac{D_o - D_i}{2C}\right) \quad \text{[small pulley]} \quad 53.89$$

Equation 53.90 and Eq. 53.91 approximate the belt lengths.

Open and Crossed Belts

$$L_{\text{open drive}} = \sqrt{4C^2 - (D_o - D_i)^2} + \tfrac{1}{2}(D_o\theta_{D_o} + D_i\theta_{D_i}) \quad 53.90$$

$$L_{\text{crossed drive}} = \sqrt{4C^2 - (D_o - D_i)^2} + \tfrac{1}{2}(D_o + D_i)\theta \quad 53.91$$

Ultimately, a belt is limited by the maximum tension it can support. From Eq. 53.86, with $F_2 = 0$ and $F_1 = 2F_i$, the maximum power that can be transmitted is approximately

$$P_{\mathrm{kW}} = \frac{F_i\mathrm{v}_{\mathrm{m/s}}}{500} \quad \text{[SI]} \quad 53.92(a)$$

$$\mathrm{HP} = \frac{F\mathrm{v}}{33{,}000} = \frac{2F_i\mathrm{v}_{\mathrm{ft/min}}}{33{,}000} = \frac{F_i\mathrm{v}_{\mathrm{ft/min}}}{16{,}500} \quad \text{[U.S.]} \quad 53.92(b)$$

Manufacturers rate belt materials in terms of the horsepower (per unit width) that can be transmitted at various speeds. The rated power is derated by various factors. In Eq. 53.93, C_p is the pulley correction factor, C_v is the velocity correction factor, and K is the service factor. F_a is the allowable belt tension as specified by the belt manufacturer. Tables of allowable belt tension derating factors are included in most engineering handbooks.

$$P_{\mathrm{kW}} = \frac{C_pC_\mathrm{v}F_a\mathrm{v}_{\mathrm{m/s}}}{500K_s} \quad \text{[SI]} \quad 53.93(a)$$

$$\mathrm{HP} = \frac{C_pC_\mathrm{v}F_a\mathrm{v}_{\mathrm{ft/min}}}{16{,}500K_s} \quad \text{[U.S.]} \quad 53.93(b)$$

29. V-BELTS

Design of V-belt drives is largely dependent on use of manufacturers' literature and tables from engineering handbooks. Sheave diameters are understood to be the diameter of the pitch diameters, as needed in determining the velocity ratios. V-belt cross sections have been standardized and are designated as A, B, C, D, or E and the inside diameter. Thus, a B100 V-belt would have a standard B cross section and an inside length of 100 in.

A cross section of a V-belt and a sheave groove is shown in Fig. 53.15.

Figure 53.15 *Cross Section of V-Belt and Sheave Groove*

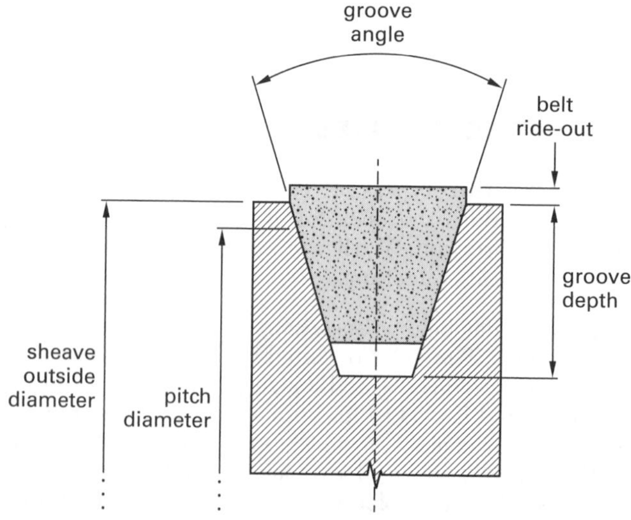

As with flat belts, the power that can be transmitted is derated by various factors, including a *service factor* that accounts for the type of prime mover and the nature of the load. In some cases, the arc of contact may be considered in derating the belt.

The number of belts is calculated from the transmitted power, the rated power per belt, and a contact arc factor (needed when the contact angle is less than 180°).

The center-to-center distance should be between 70% and 200% of the sum of the two sheave diameters. Belt length (referred to as *pitch length* and *effective length*) can be calculated in the same manner as for open flat belts. The length of a known belt can also be calculated by adding a small correction to the inside diameter of the belt. Depending on the belt type, those corrections are: A, 1.3 in; B, 1.8 in; C, 2.9 in; D, 3.3 in; and E, 4.5 in.

30. TIMING BELTS

The function of a *timing belt* (also referred to as a *synchronous belt*, *serpentine belt*, *power transmission belt*, or *timing chain*) includes maintaining the phase (synchronization, angle relationship, alignment, etc.) between the connected shafts as well as transmitting power. (See Fig. 53.16.) It is common to use timing belts in modern vehicle engines to drive overhead camshafts, usually at half the speed of the crankshaft. Belts can also be used to run water pumps and other accessories, and the flat back surface may be used for applications that do not require exact timing. Timing belts are also used in linear positioning applications as well as in motorcycle drives.

Figure 53.16 *Typical V6 Timing Belt Installation*

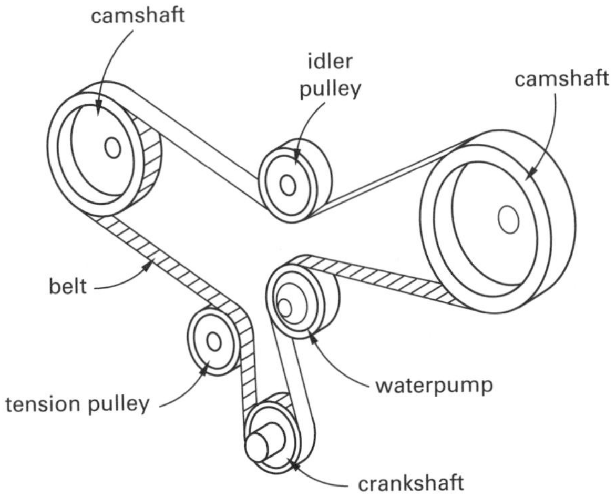

Pulleys (also referred to as *sprockets* and *gears*) are manufactured from aluminum, steel, nylon, and acetal, or a combination of two. *Idler pulleys* are used to guide belts into proper orientation, while tensioner pulleys maintain belt tension.

Timing belts are manufactured from rubber, neoprene, or (poly)urethane reinforced with high tensile strength fiberglass, Twaron, or Kevlar. Neoprene has the best solvent resistance, which is the reason it is used in automotive applications. Table 53.7 lists properties of these matrix materials.

Timing belts almost always have teeth (*cogs*). The shape and pitch (spacing) of the teeth constitute the *profile*. Trapezoidal, curvilinear, and European tooth profiles are used. Profile designations include MXL, XL, L, HTD3, HTD5, HTD8, T2.5, T5, T10, AT5, and AT10. For example, common profiles and pitches are XL 1/5 in), L (3/8 in), and HTD (3 mm, 5 mm, and 8 mm). *Precision belts* are also available with very small pitches. Table 53.8 lists some common profiles.

When the belt is in contact with the pulley, the tensile (reinforcing) cord axis will coincide with the pitch circle. The pitch circle diameter, D_p (larger than the pulley diameter), is a function of the diametral pitch, p_d, and the number of teeth, N, and is given by Eq. 53.94. The bottom on the belt tooth will coincide with the outer diameter of the pulley. The distance, U, between the tensile cord axis and the bottom of the belt tooth will determine the outside pulley diameter. Slight corrections from Eq. 53.95 are needed with small-diameter pulleys and certain profiles (e.g., HTD8). These corrections are given in belt catalogs.

$$D_p = \frac{p_d N}{\pi} \qquad 53.94$$

$$\text{pulley outer diameter} = D_p - 2U \qquad 53.95$$

Belt wear is minimized by ensuring that belt teeth come into contact with different parts of the pulleys on subsequent cycles. This is accomplished by making the number of teeth on the driving and driven pulleys *coprime* (i.e., the numbers have no common factors).

Since timing belts are difficult to inspect thoroughly and can fail without warning, they are replaced after a certain period of time or vehicle distance traveled (e.g., 50,000 mi to 100,000 mi) as part of preventative maintenance.

Engines that can tolerate a broken timing belt without suffering catastrophic collisions between pistons and valves are known as *noninterference engines*. Engines that leave valves cracked open in piston space are known as *interference engines*.

31. BRAKES AND BRAKE MATERIALS

A *brake* is a device for bringing motion to a halt by converting kinetic energy into heat. Brakes can be designed to be self-energizing, self-locking, or neither.[26] Normally, a brake should be self-energizing but not self-locking. With a *self-energizing brake*, the braking drag force augments (increases) the application force of the brake.

[26] *Hydraulic braking* or *hydraulically actuated braking* refers to the transmission of the applied braking force (e.g., from a foot pedal) to the brakes via a hydraulic brake line. The actuating braking force is reconstituted in the brake cylinder.

Table 53.7 Properties of Timing Belt Materials

	tensile strength range		operating temperature				
material	psi	MPa	°F	°C	oil resistance	solvent resistance	ultraviolet resistance
polyisoprene (natural rubber)	500 to 3500	3.45 to 24.1	−60 to 175	−51 to 79	poor	poor	poor
neoprene	500 to 3000	3.45 to 20.7	−50 to 185	−46 to 85	fair	fair	good
urethane (polyurethane)	500 to 6000	3.45 to 41.3	−30 to 170	−35 to 77	good	poor	excellent

Table 53.8 Timing Belt Profiles

common description	distance from pitch line to belt tooth bottom, *U*
minipitch 0.080 in MXL	0.010 in
40 D.P.	0.007 in
1/5 in XL	0.010 in
3/8 in L	0.015 in
3 mm HTD	0.015 in
5 mm HTD	0.0225 in
8 mm HTD	0.027 in
2 mm GT	0.010 in
3 mm GT	0.015 in
5 mm GT	0.0225 in
T2.5 (2.5 mm)	0.3 mm
T5 (5 mm)	0.5 mm
T10 (10 mm)	1.0 mm

Basically, the rotating surface drags the brake shoe onto the moving surface, reducing the application effort required to achieve braking action. A self-energizing brake will require some finite application force, F. A *self-locking brake* does not require any external application force. The frictional force is so large that braking action occurs when $F \leq 0$. That is, no application force is needed at all. A self-locking brake is seldom desirable, as a negative force is required to disengage it. Brakes are normally self-energizing and/or self-locking in one direction of rotation only.

Chrysotile asbestos-based friction brake products (*pads* and *linings*) have been highly refined since the material was introduced in the early 1900s. Four classes of non-asbestos products are used in response to environmental and health concerns about asbestos: non-asbestos organic (NAO), resin-bonded metallic (semimetallic), sintered metallic, and carbon-carbon. Only the semimetallic and NAO varieties are useful for common automotive brake linings. Semimetallic linings are not readily applicable to most drum brake applications. NAO linings contain primarily synthetic aramid (aromatic polyamide) fibers (such as Kevlar and Twaron), which have high strength and are resistant to heat.

32. ENERGY DISSIPATION IN BRAKES AND CLUTCHES

Energy dissipated in brakes and clutches can be calculated as the change in kinetic (rotational or translational) energy of the braking mass. For example, the change in rotational kinetic energy of a spinning mass with rotational mass moment of inertia, I, is

$$\Delta E_k = \frac{I(\omega_1^2 - \omega_2^2)}{2} \quad \text{[SI]} \qquad 53.96(a)$$

$$\Delta E_k = \frac{I(\omega_1^2 - \omega_2^2)}{2g_c} \quad \text{[U.S.]} \qquad 53.96(b)$$

The maximum theoretical temperature increase occurs if all of the kinetic energy is absorbed by the brake or clutch material. This is an extreme assumption, as some of the energy will be removed, particularly for braking devices that are run "wet."[27] For steel brake and clutch parts the specific heat, c_p is approximately 0.12 Btu/lbm-°F (0.5 kJ/kg·K).

$$\Delta T = \frac{\Delta E_k}{m_{\text{brake material}} c_p} \qquad 53.97$$

Various limits are imposed on clutches and brakes, including maximum energy dissipation rate per unit area, maximum temperature, and maximum contact pressure. Table 53.9 lists general limits based on application. Table 53.10 lists limits based on brake material.

[27] Frictional heat can be removed by an oil spray or bath.

Table 53.9 *Typical Clutch and Brake Design Parameters**

	coefficient of friction		maximum temperature		maximum pressure	
application	oily	dry	°F	°C	psi	kPa
CI/CI	0.05	0.15–0.20	600	320	150–250	1000–1750
PM/CI	0.05–0.1	0.1–0.4	1000	540	150	1000
PM/HS	0.05–0.1	0.1–0.3	1000	540	300	2100
WA/CI-ST	0.1–0.2	0.3–0.6	350–500	175–260	50–100	350–700
MA/CI-ST	0.08–0.12	0.2–0.5	500	260	50–150	350–1000
IA/CI-ST	0.12	0.32	500–750	260–400	150	1000

(Multiply psi by 6.9 to obtain kPa.)

*CI—cast iron; PM—powdered metal; HS—hardened steel; CI-ST—cast iron or steel; WA—woven asbestos; MA—molded asbestos; IA—impregnated asbestos.
Compiled from various sources.

Table 53.10 *General Limits on Energy Dissipation Rates for Clutches and Brakes*

	maximum energy dissipation rate	
condition	hp/in^2	kW/m^2
continuous load, poor heat removal	0.85	980
intermittent load, poor heat removal, long recovery period	1.7	2000
continuous load, good heat removal (e.g., oil bath)	2.5	2900
vehicular brakes	5.0	5800

(Multiply hp/in^2 by 1156 to obtain kW/m^2.)

Compiled from various sources.

The energy dissipation rate (i.e., power) can be calculated from Eq. 53.98.

$$P_{\text{braking}} = \left(\frac{F_f}{A}\right)\text{v} = \left(\frac{Ff}{A}\right)\text{v} = fp_{\max}\text{v} \qquad 53.98$$

During braking, the number of revolutions (or distance traveled) required to bring a spinning (moving) mass to a standstill has little relationship to the original speed. With a uniform deceleration, the average speed will be one-half of the original speed. Then, the braking time to come to a complete standstill can be calculated from the original kinetic energy and the average rate of frictional energy dissipated.

$$t_{\text{braking}} = \frac{E_{k,\text{original}}}{F_f \text{v}_{\text{ave}}} \qquad 53.99$$

33. BLOCK BRAKES

A *block brake* (i.e., an *external shoe brake*) is illustrated in Fig. 53.17. The pressure will be distributed approximately uniformly if the block *contact angle*, ϕ, is less than 60°. The coefficient of sliding friction, f, is used. The frictional retarding force on the shoe is calculated from the normal force.

$$F_f = fN \qquad 53.100$$

Regardless of the location of the pivot point, the normal force can be calculated from the torque capacity, T, of the brake.

$$N = \frac{F_f}{f} = \frac{T}{rf} \qquad 53.101$$

Figure 53.17 *Block Brake*

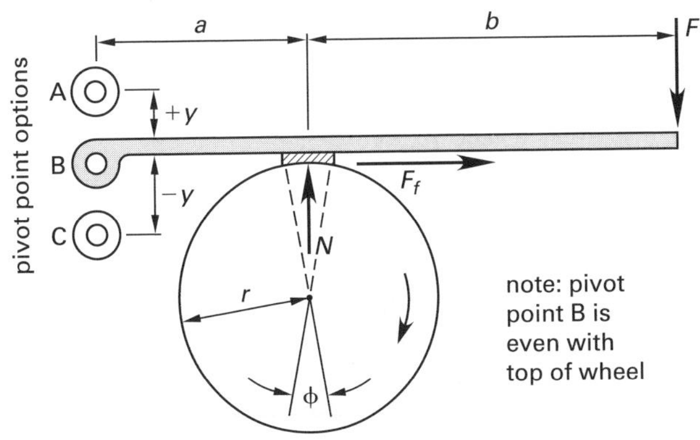

If the contact angle is greater than 60°, an equivalent

coefficient of friction, f', should be used in place of the normal coefficient of sliding friction. In Eq. 53.102, ϕ must be in radians.

$$f' = \frac{f\left(4\sin\left(\frac{\phi}{2}\right)\right)}{\phi + \sin\phi} \quad \text{53.102}$$

The required actuating force, F, at the end of the brake arm can be determined by taking moments about the pivot point. Depending on the pivot point, moments will result from the normal force, N, and the frictional force, F_f, as well as from the actuating force, F.

If the wheel turns clockwise, pivot point C results in a self-energizing brake. Similarly, pivot point A is self-energizing for counterclockwise rotation. If $a = fy$, the brake will be completely self-energizing, and the actuating force will be zero. If $a < fy$, the brake will be self-locking.

34. BAND BRAKES

The band brake shown in Fig. 53.18 is evaluated from belt friction equations. The *braking torque* (*braking moment*) is given by Eq. 53.103, in which F_1 is the tight-side tension. The angle ϕ is in radians.

$$T = r(F_1 - F_2) \quad \text{53.103}$$

$$F_1 - F_2 = F_2(e^{f\phi} - 1) \quad \text{53.104}$$

$$\frac{F_1}{F_2} = e^{f\phi} \quad \text{53.105}$$

Figure 53.18 Band Brake

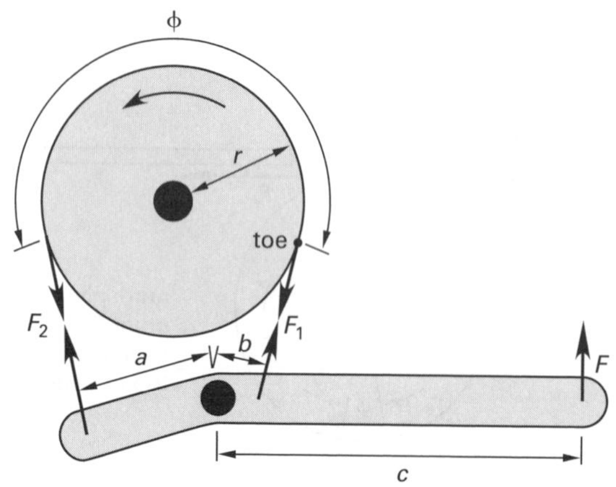

The applied force, F, is found by taking moments about the pivot point.

$$F = \frac{aF_2 - bF_1}{c} = \left(\frac{T}{cr}\right)\left(\frac{a - be^{f\phi}}{e^{f\phi} - 1}\right) \quad \text{53.106}$$

If $be^{f\phi} > a$, the applied force F will be negative and the brake will be self-locking.

The contact pressure is maximum at the toe of the brake. (The *toe* is where the anchored end of the band first contacts the drum.)

$$p = \frac{F_1}{A} = \frac{F_1}{rw} \quad \text{53.107}$$

The required contact surface width, w, is calculated from the allowable contact pressure of the band material.

35. DISK AND PLATE CLUTCHES

A clutch is a device for the connection of an initially stationary shaft to a rotating shaft. (See Fig. 53.19.) The torque and power transmitted through a clutch are related by Eq. 53.37.

The torque capacity of a clutch is proportional to the number of friction planes. In the case of a single two-sided automobile clutch disk between two plates, there are two friction planes. The maximum transmitted torque will be twice the torque value per surface calculated from Eq. 53.37.

If a clutch assembly is considered to be rigid, the initial wear will be in the outer areas, where the frictional work is greater than in the inner areas. After a period of time (break-in), the pressure distribution will change and the wear will become uniform. *Uniform wear* is equivalent to assuming that all work occurs at the average frictional radius. Uniform wear is the conservative assumption in terms of allowable torque and transmitted horsepower.

With uniform wear, the maximum pressure, p_{max}, will occur at the inner radius. The pressure at any other radius, r, is

$$p_r = \frac{p_{max} r_i}{r} = \frac{p_{max} D_i}{D_o} \quad \text{[uniform wear]} \quad \text{53.108}$$

The axial application force is

Uniform Wear and Pressure on Clutches and Brakes

$$F = \frac{\pi p_{max} D_i}{2}(D_o - D_i) \quad \text{[uniform wear]} \quad \text{53.109}$$

The torque per contact surface is carried at the mean radius.

Uniform Wear and Pressure on Clutches and Brakes

$$T = \frac{Ff}{4}(D_o + D_i) \quad \text{[uniform wear]} \qquad 53.110$$

If the clutch assembly is known to be semiflexible, a *uniform pressure* distribution can be assumed. The axial application force, F, is also the normal force. Since the pressure is uniform, the maximum pressure, p_{max}, occurs everywhere on the clutch.

Uniform Wear and Pressure on Clutches and Brakes

$$F = \frac{\pi p_{max}}{4}(D_o^2 - D_i^2) \quad \text{[uniform pressure]} \qquad 53.111$$

The frictional radius, r_f, is the mean radius for frictional purposes. It can be used to determine the limiting torque per contact plane.

$$r_f = \frac{2}{3}\left(\frac{r_o^3 - r_i^3}{r_o^2 - r_i^2}\right) \quad \text{[uniform pressure]} \qquad 53.112$$

Uniform Wear and Pressure on Clutches and Brakes

$$T = \frac{Ff}{3}\left(\frac{D_o^3 - D_i^3}{D_o^2 - D_i^2}\right) \quad \text{[uniform pressure]} \qquad 53.113$$

Figure 53.19 *Generalized Plate Clutch*

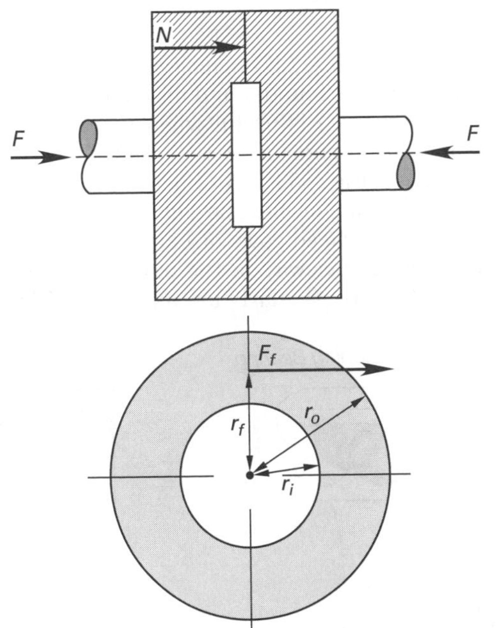

36. CONE CLUTCHES

Figure 53.20 illustrates a *cone clutch*, with its *cone angle*, α. Cone clutches, like disk clutches, can be evaluated for uniform wear or uniform pressure. For uniform wear, the axial force, F, necessary to hold the cone in the cup and generate a force, N, normal to the contact surface is

$$F = N\sin\alpha = \left(\frac{\pi p_{max} d}{2}\right)(D - d) \qquad 53.114$$

The frictional force is

$$F_f = fN = \frac{fF}{\sin\alpha} \qquad 53.115$$

The frictional torque is

$$\begin{aligned} T_f &= \frac{F_f d_m}{2} = \frac{fF_{axial} d_m}{2\sin\alpha} \\ &= \left(\frac{\pi f p_{max} d}{8\sin\alpha}\right)(D^2 - d^2) \\ &= \left(\frac{Ff}{4\sin\alpha}\right)(D + d) \end{aligned} \qquad 53.116$$

For uniform pressure,

$$F = \left(\frac{\pi p}{4}\right)(D^2 - d^2) \qquad 53.117$$

$$\begin{aligned} T &= \left(\frac{\pi f p}{12\sin\alpha}\right)(D^3 - d^3) \\ &= \left(\frac{Ff}{3\sin\alpha}\right)\left(\frac{D^3 - d^3}{D^2 - d^2}\right) \end{aligned} \qquad 53.118$$

Figure 53.20 *Cone Clutch*

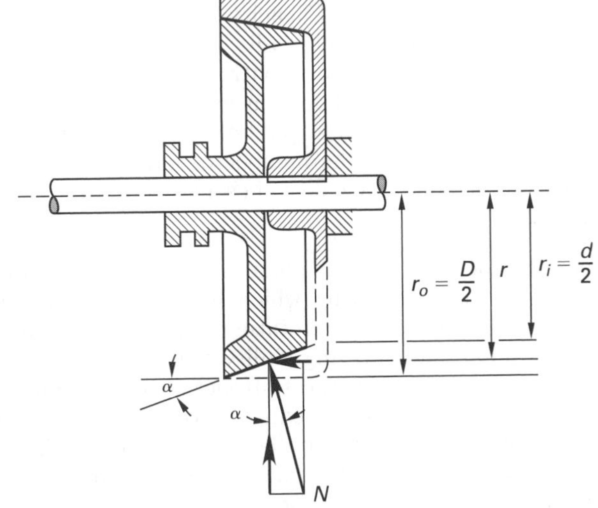

37. BUSHINGS

A *bushing* (*mechanical bushing*, *friction bearing*, or *plain bearing*) is a removable or permanent cylindrical sleeve-shaped lining. A bushing is inserted into a hole in a parent material in order to reduce friction between a rotating shaft and the parent material or to constrain the movement of the shaft. Bushings may or may not be lubricated, although the term *journal bearing* is generally used with lubrication.

Low-load, low-speed applications may be able to use bushings without lubrication. This option is particularly attractive with low-temperature applications where keeping cold oil thin enough to circulate would be difficult. Bushings can be constructed from plastic (e.g., nylon), brass, carbon graphite, or oil-impregnated powdered metal. Bushings are generally limited by lateral load and heat. So, the bushing design and selection procedure is to make checks of the contact pressure (based on the projected bearing area), the maximum tangential velocity, and the product of pressure and velocity against upper limits established for the material.

38. BALL, ROLLER, AND NEEDLE BEARINGS

Ball, roller, and needle bearings (i.e., *anti-friction bearings*) and their variations use rolling-element bearings such as balls or cylinders constrained within *bearing races* to reduce friction. The races, in turn, support and resist the applied loadings. One race is usually fixed, while the other race moves with the rotating element (e.g., a shaft). The balls or rollers themselves rotate as the races rotate relative to each other. *Ball bearings* can support radial and thrust loads, but they are limited in capacity by the compressive contact pressure between them and the races. Therefore, they are used where loads are relatively low. Capacity is a function of the number of balls in play. The balls may be loose or their positions may be constrained with a *cage* (*separator* or *retainer*), in which case the term *caged bearing* or *Conrad bearing* may be used. Caged bearings generally have lower capacities since the number of balls is reduced. Examples of rolling element bearings are shown in Fig. 53.21.

Since the locus of contact points represents a line (as opposed to a ball bearing which has a small contact point), *roller bearings* use rotating cylinders and have greater radial load carrying ability but require greater alignment between the races. Roller bearings cannot normally support thrust loads. *Needle bearings* use small diameter cylinders (needles or pins) and can fit in less space. Since there is no space between the rollers, no cage is required. Although they require less space, needle bearings have greater friction and must run slower than ball and standard roller bearings.

Figure 53.21 *Rolling Element Bearings*

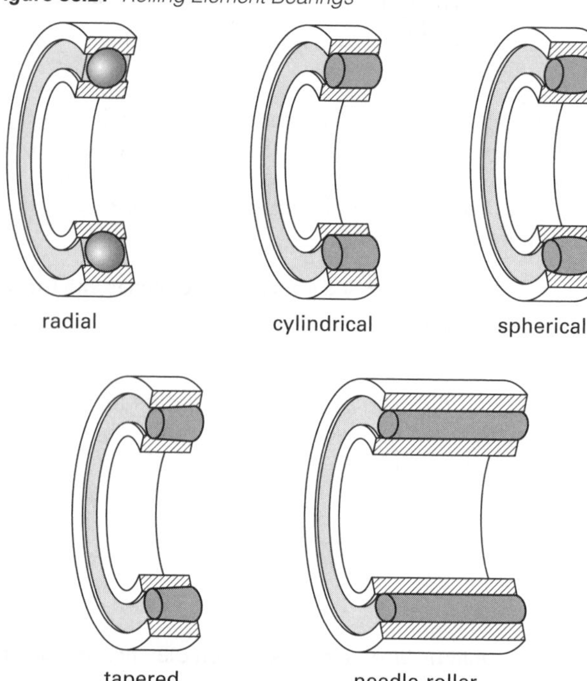

Ball and roller thrust bearings are used with primarily axial loading. Where there is both radial and axial loading, tapered roller bearings are used, often in opposite-facing pairs to support axial loading from both directions.

A *self-aligning bearing* (*Wingquist bearing* or *SKF bearing*, named after the founder and his company that manufactured them first) is useful when there is deflection and angular misalignment of the shaft relative to the housing. (See Fig. 53.22.) Such misalignment is expected with long or flexible shafts. Such a bearing typically has two caged rows of balls that run within a concave spherical raceway in the outer ring. Self-aligning bearings have particularly low axial capacities.

Figure 53.22 *Self-Aligning Bearing*

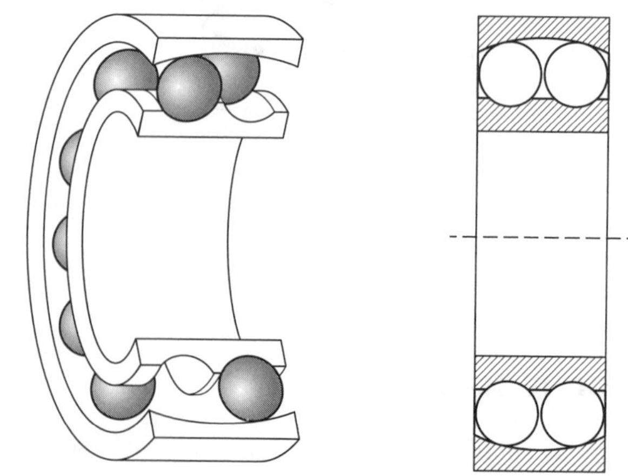

Thrust bearings are used with the largest axial loads, such as propeller shafts and those having substantial vertical weights. Thrust bearings may be simple flat faying pads, be hydrodynamic, or contain ball or roller bearings.

39. ROLLER BEARING CAPACITY

Bearings are specified by both their static and dynamic load capacities. Bearings are selected (i.e., as from a manufacturer's catalog) so that the equivalent bearing capacity, C, is greater than the applied or design load (i.e., the cataloged capacity), P. Static capacity can be used if rotational speed is slow, intermittent, and/or subject to shocks. Dynamic capacity is used if the rotational speed is smooth and relatively constant.

The *basic static load rating*, C_o, is determined by the manufacturer. Similarly, for a rotating bearing, the *basic dynamic load rating*, C_r, is determined by the manufacturer. When a bearing is subjected to both a radial loading, F_r, and an axial load, F_a, an *equivalent radial load*, P_{eq}, must be calculated in order to select bearings or to calculate bearing life. (The term *equivalent* implies that radial and axial loads have been combined into a single radial parameter.) For a single-row ball bearing, the equivalent radial load can be calculated as

Ball/Roller Bearing Selection

$$P_{\text{eq}} = XVF_r + YF_a \qquad 53.119$$

V, X, and Y are factors that are best supplied by the manufacturer, but can be determined from accepted methods. $V = 1$ when the outer ring (raceway) is stationary and the inner ring rotates; $V = 1.2$ when the inner ring is stationary and the outer ring rotates.

X and Y depend on the relative ratio of the axial and radial forces. For low axial loading as defined by Eq. 53.120, $X = 1$, and $Y = 0$. C_0 is the basic static load rating from the manufacturer's catalog.

Ball/Roller Bearing Selection

$$\frac{F_a}{VF_r} \leq e \quad \text{[low axial loading]} \qquad 53.120$$

$$e = 0.513\left(\frac{F_a}{C_0}\right)^{0.236} \qquad 53.121$$

For high axial loading as defined by Eq. 53.122, $X = 0.56$, and Y is calculated from Eq. 53.123.

Ball/Roller Bearing Selection

$$\frac{F_a}{VF_r} > e \quad \text{[high axial loading]} \qquad 53.122$$

$$Y = 0.840\left(\frac{F_a}{C_0}\right)^{-0.247} \qquad 53.123$$

40. ROLLING ELEMENT BEARING LIFE

Lubrication of bearings is provided for four the common reasons: (a) reduce friction by separating the surfaces in contact, (b) dissipate heat, (c) prevent corrosion, and (d) remove dirt and contamination. Bearing life is affected by proper lubrication. *Bearing life* (*bearing fatigue life*) is a measure of how long a bearing can be expected to last under standard operating conditions. Bearing life depends primarily on the loading, although care, cleaning, and lubrication are significant external influences on bearing life. The bearing is assumed serviceable as long as 90% of the rolling elements are functional, and the parameter that predicts when 10% of the rolling elements have failed is designated L_{10} and referred to as "L-ten".[28] Interestingly, the mean time before failure, MTBF, of each individual rolling element is approximately five times the lifetime of the complete bearing. Therefore, at MTBF, 50% of the rolling elements can be expected to have failed. L_{10} is known as the *basic life rating*, *basic rating life*, and *rating life*.

The *basic dynamic load rating*, also known as the *dynamic basic load rating*, C_r, for rotating bearings, is measured in lbf or kN. The *equivalent dynamic bearing load*, also known as the *equivalent design radial load*, P_r, is also measured in lbf or kN. The equivalent dynamic bearing load is the hypothetical load, constant in magnitude and direction, that would have the same influence on bearing life as the actual loads to which the bearing is subjected.

The basic life rating, measured in millions of revolutions, depends on the minimum required basic dynamic load rating, C, and the equivalent dynamic radial load, P. The equation for the minimum required basic dynamic load rating is

Ball/Roller Bearing Selection

$$C = PL^{1/a} \qquad 53.124$$

By rearranging Eq. 53.124 into Eq. 53.125, the life can be found. The L-ten is found from Eq. 53.126.

$$L = \left(\frac{C}{P}\right)^a \qquad 53.125$$

$$L_{10} = \left(\frac{C}{P}\right)^{10/3} \quad \left[\begin{array}{l}\text{millions of revolutions;} \\ \text{straight roller bearings}\end{array}\right] \qquad 53.126$$

[28]In other countries, the designation B_{10} may be used in place of L_{10}.

The basic life rating, measured in hours, depends on the rotational speed, n.

$$L_{10} = \left(\frac{10^6}{n_{\text{rpm}}\left(60\ \frac{\text{min}}{\text{hr}}\right)} \right) \left(\frac{C}{P}\right)^3 \quad \left[\begin{array}{c}\text{hours; single-row}\\ \text{ball bearings}\end{array}\right] \tag{53.127}$$

$$L_{10} = \left(\frac{10^6}{n_{\text{rpm}}\left(60\ \frac{\text{min}}{\text{hr}}\right)} \right) \left(\frac{C}{P}\right)^{10/3} \quad \left[\begin{array}{c}\text{hours; straight}\\ \text{roller bearings}\end{array}\right] \tag{53.128}$$

Equation 53.129 is a specific case of the more generic bearing life equation. In Eq. 53.129, a_1 is a life adjustment factor for reliability, a_2 is a life adjustment factor for ball or roller material, a_3 is a life adjustment factor for operating conditions, and f_B is a dynamic load rating adjustment factor to account for the number of adjacently mounted bearing elements. Most of these factors have to be obtained from the manufacturer.

$$L_n = a_1 a_2 a_3 \left(\frac{10^6}{n_{\text{rpm}}\left(60\ \frac{\text{min}}{\text{hr}}\right)} \right) \left(\frac{f_B C}{P}\right)^p \quad [\text{hours}] \tag{53.129}$$

Friction in rolling element bearings generates heat directly related to the rotational speed. Equation 53.130 is a common rule of thumb used to determine the allowable speed of ball and straight roller bearings. d is the bore (inside) diameter, and D is the outside diameter. (See Fig. 53.23.)

$$n_{\text{rpm}} < \frac{10^6}{d_{\text{mm}} + D_{\text{mm}}} \tag{53.130}$$

41. LUBRICANTS AND LUBRICATION

In the United States, *crankcase (engine) oils* and *gear oils* are typically called out by *SAE grades* (10–70), although the AGMA also has a grading method. In countries that use the SI (metric) system, the ISO *viscosity grade*, VG, is used.[29] Viscosity of a lubricating oil should be evaluated at its bulk temperature.

Figure 53.23 *Ball Bearing Dimensions*

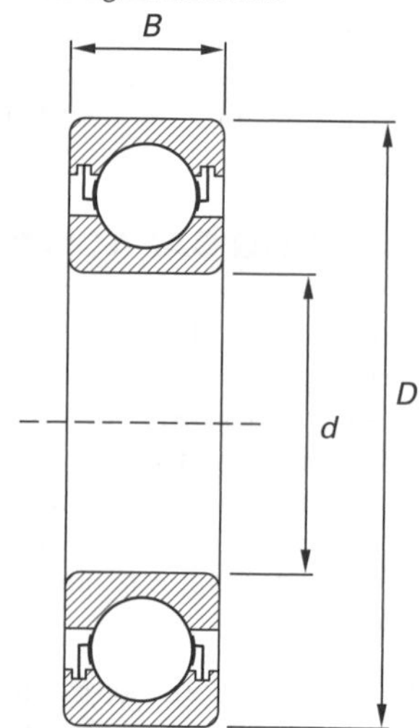

In Fig. 53.24, the viscosity of lubricants is expressed in *reyns*, which is equivalent to lbf-sec/in^2. Viscosities in centipoise (cP), centistokes (cSt), and Saybolt seconds (SSU) (also known as *Saybolt universal viscosity*, SUV) are also widely encountered.

$$\mu_{\text{reyns}} = \frac{\mu_{\text{cP}}}{6.89 \times 10^6} \tag{53.131}$$

$$\mu_{\text{Pa·s}} = \frac{\mu_{\text{cP}}}{1000} \tag{53.132}$$

$$\mu_{\text{Pa·s}} = \rho_{\text{kg/m}^3} \nu_{\text{m}^2/\text{s}} \tag{53.133}$$

$$\nu_{\text{m}^2/\text{s}} = \nu_{\text{cSt}} \times 10^{-6} \tag{53.134}$$

$$\nu_{\text{cSt}} = 0.22\ \text{SSU} - \frac{180}{\text{SSU}} \tag{53.135}$$

The *pour point* is the lowest temperature at which the oil will flow when chilled under standard test procedures.[30] Actual values vary from manufacturer to manufacturer. For example, the average pour point of modern formulations of SAE 10W-30 is approximately 30°F (−1°C); actual values for all grades range from 10°F (−12°C) to 70°F (21°C), depending on the manufacturer. Additives (i.e., *pour point depressants*) can be used to depress the pour points (i.e., make them usable at lower temperatures). Since naphthenic petroleum oils and synthetics contain no wax, they can be cooled below the pour points of most oils. Pour points are given in intervals of 5°F. Actual pour points of 10.5°F and 14.5°F would both be reported as 15°F.

Although the pour point is of academic interest, the viscosity limit is a more practical temperature cutoff. The *viscosity limit* is the temperature below which the oil simply will not flow fast enough to return to the pan or pass through suction and drain pipes, pumps, and

[29] The VG rating is the nominal kinematic viscosity in centistokes (cSt) at 40°C (104°F).
[30] ASTM Method D97.

Figure 53.24 *Viscosity of Common Lubricating Oils*

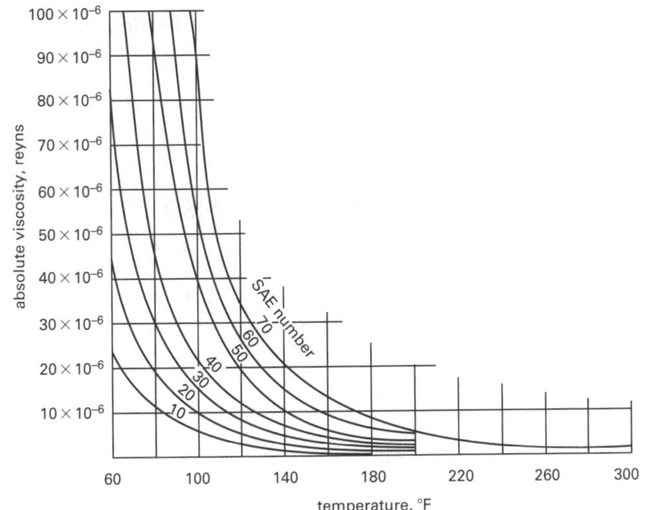

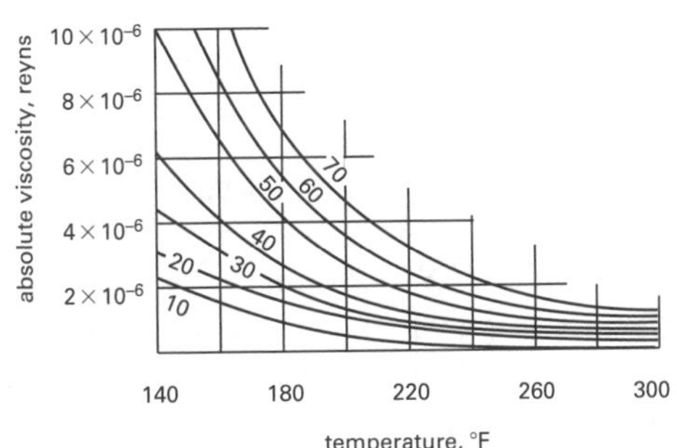

filters. The most restrictive viscosity requirements are found in large circulating oil systems, such as those used in industrial turbomachinery. For these systems, the viscosity limit may be close to room temperature.

42. JOURNAL BEARINGS

Figure 53.25 illustrates a *journal bearing.* (See Table 53.11 for typical parameters.) The *eccentricity ratio*, ϵ (also known as the *attitude*), of the shaft in the bearing will vary with loading. The *diametral clearance*, c_d, is twice the *radial clearance*, c_r. However, the clearances c_d/d and c_r/r are the same.[31]

$$\epsilon = \frac{2e}{c_d} \qquad 53.136$$

$$c_d = D - d \qquad 53.137$$

$$c_d = 2c_r \qquad 53.138$$

Figure 53.25 *Journal Bearing Nomenclature*

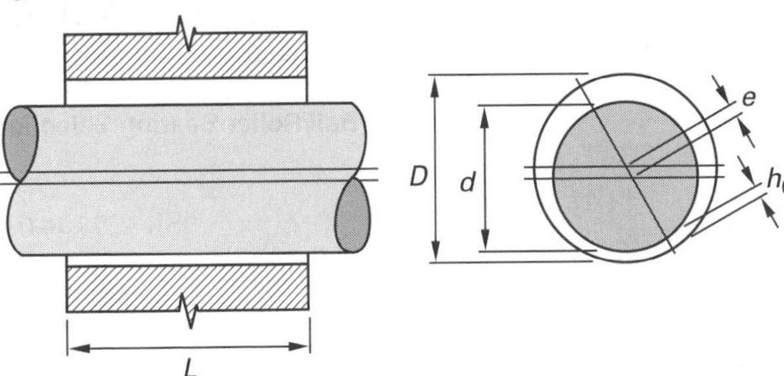

The bearing "pressure" is based on the projected area of the shaft.

$$p = \frac{\text{lateral shaft load}}{\text{projected area}} = \frac{\text{lateral shaft load}}{LD} \qquad 53.139$$

The relationship between eccentricity, diametral clearance, and minimum film thickness, h_0, is

$$h_0 = \tfrac{1}{2}c_d(1 - \epsilon) \qquad 53.140$$

Table 53.11 *Typical Journal Bearing Parameters*

application	L/D	average pressure* psi	average pressure* kPa
steam turbines	–	2–10	14–70
pumps	2	80–100	550–700
machine tools	2–4	80–100	550–700
electric motors	2	100–200	700–1400
automotive mains	0.5–0.8	500–600	3500–4100
automotive wristpins	–	2000–3000	14 000–21 000

(Multiply psi by 6.9 to obtain kPa.)

*Maximum pressure can be twice as high as the average pressure.
Compiled from various sources.

43. JOURNAL BEARING FRICTIONAL LOSSES

The viscosity of journal bearings will cause a frictional torque that opposes rotation. If the bearing is unloaded (i.e., the eccentricity is zero), or if the eccentricity is low

[31]When the ratio is written as "c/r," it is understood that c is the radial clearance. A radial clearance would not be combined with a diameter.

and the rotational speed is high, the *Petroff equation* can be used to find the frictional torque. In Eq. 53.141, r is the bearing radius and L is its length.

Ball/Roller Bearing Selection

$$T_{f,\text{N·m}} = \frac{4\pi^2 r^3 L\mu_{\text{Pa·s}} N_{\text{rps}}}{c} \quad \text{[SI]} \quad 53.141(a)$$

$$T_{f,\text{in-lbf}} = \frac{4\pi^2 r^3 L\mu_{\text{reyns}} N_{\text{rps}}}{c} \quad \text{[U.S.]} \quad 53.141(b)$$

The effective coefficient of friction is

$$f = 2\pi^2 \frac{\mu N}{P}\frac{r}{c} = \frac{2\pi^2 \mu_{\text{reyns}} N_{\text{rps}} D}{p c_d} \quad 53.142$$

The frictional horsepower, $P_{f\text{,hp}}$, can be found from frictional torque by using Eq. 53.37. Frictional heating varies with the square of the rotational speed. The usual range of operating temperatures is 140°F to 160°F (60°C to 70°C). Most lubricants start to deteriorate above 200°F (90°C). Frictional heating, q_f, is dissipated in an oil cooler or through contact with a cooler surface. The specific heat for petroleum oils is approximately 0.42 Btu/lbm-°F to 0.49 Btu/lbm-°F (1.8 kJ/kg·K to 2.0 kJ/kg·K).

$$q_f = P_{f,\text{hp}} = \dot{m} c_p \Delta T \quad 53.143$$

The ratio of loaded to unloaded friction torque can be correlated to the load number. The *load number* is calculated from Eq. 53.144.

$$N_L = \left(\frac{p}{\mu_{\text{reyns}} n_{\text{rps}}}\right)\left(\frac{c_d}{L}\right)^2 \quad 53.144$$

Example 53.2

SAE 20 oil at 150°F (65°C) bulk temperature is used in a lightly loaded journal bearing. The bearing has a length of 5 in (12.7 cm) and a radius of 1.5 in (3.81 cm). The shaft turns at 1800 rpm. The radial clearance is 0.005 in (0.127 mm). What is the frictional power?

SI Solution

From Fig. 53.24, the viscosity of SAE 20 oil at 65°C is approximately 2.5×10^{-6} reyns. Combining Eq. 53.131 and Eq. 53.132,

$$\begin{aligned}\mu_{\text{Pa·s}} &= \frac{(6.89 \times 10^6)\mu_{\text{reyns}}}{1000} \\ &= \frac{(6.89 \times 10^6)(2.5 \times 10^{-6}\ \text{reyns})}{1000} \\ &= 0.017225\ \text{Pa·s}\end{aligned}$$

From Eq. 53.141, the frictional torque is

Ball/Roller Bearing Selection

$$\begin{aligned}T_{\text{N·m}} &= \frac{4\pi^2 r^3 L\mu_{\text{Pa·s}} N_{\text{rps}}}{c} \\ &= \frac{4\pi^2 (0.0381\ \text{m})^3 (0.127\ \text{m}) \times (0.017225\ \text{Pa·s})\left(1800\ \frac{\text{rev}}{\text{min}}\right)}{\left(60\ \frac{\text{sec}}{\text{min}}\right)\left(\frac{0.127\ \text{mm}}{1000\ \frac{\text{mm}}{\text{m}}}\right)} \\ &= 1.1283\ \text{N·m}\end{aligned}$$

Equation 53.37 gives the frictional power.

$$\begin{aligned}P_{\text{kW}} &= \frac{T_{\text{N·m}} n_{\text{rpm}}}{9549} \\ &= \frac{(1.1283\ \text{N·m})\left(1800\ \frac{\text{rev}}{\text{min}}\right)}{9549} \\ &= 0.2127\ \text{kW}\end{aligned}$$

Customary U.S. Solution

From Fig. 53.24, the viscosity of SAE 20 oil at 150°F is approximately 2.5×10^{-6} reyns. From Eq. 53.141, the frictional torque is

Ball/Roller Bearing Selection

$$\begin{aligned}T_{f,\text{in-lbf}} &= \frac{4\pi^2 r^3 L\mu_{\text{reyns}} N_{\text{rps}}}{c} \\ &= \frac{4\pi^2 (1.5\ \text{in})^3 (5\ \text{in})\left(2.5 \times 10^{-6}\ \frac{\text{lbf-sec}}{\text{in}^2}\right)\left(1800\ \frac{\text{rev}}{\text{min}}\right)}{\left(60\ \frac{\text{sec}}{\text{min}}\right)(0.005\ \text{in})} \\ &= 9.99\ \text{in-lbf}\end{aligned}$$

Equation 53.37 gives the frictional power.

Shaft-Horsepower Relationship and Force-Horsepower Relationship

$$\begin{aligned}\text{HP} &= \frac{T_{\text{in-lbf}} n_{\text{rpm}}}{63{,}025} = \frac{(9.99\ \text{in-lbf})\left(1800\ \frac{\text{rev}}{\text{min}}\right)}{63{,}025} \\ &= 0.285\ \text{hp}\end{aligned}$$

44. JOURNAL BEARINGS: RAIMONDI AND BOYD METHOD

With the *Raimondi and Boyd method*, journal bearing performance is correlated with the *bearing characteristic number*, S, also known as the *Sommerfeld number*.[32]

$$S = \left(\frac{D}{c_D}\right)^2 \left(\frac{\mu n_{\text{rps}}}{p}\right) \quad 53.145$$

The coefficient of friction, f, is needed when calculating the frictional torque and power. It is calculated from Eq. 53.146.

$$f = \left(\frac{c_D}{D}\right)(\text{coefficient of friction variable}) \quad 53.146$$

The minimum film thickness, h_0, is calculated similarly from Eq. 53.147.

$$h_0 = r\left(\frac{c_D}{D}\right)\left(\begin{array}{c}\text{minimum film}\\ \text{thickness variable}\end{array}\right) \quad 53.147$$

45. NOMENCLATURE

a	brake dimension	in	m
a	dynamic speed constant	ft/min	n.a.
a	life adjustment factor	–	–
A	allowance	–	–
A	area	in^2	m^2
b	belt width	in	m
b	brake dimension	in	m
c	brake dimension	–	–
c	center-to-center distance	in	m
c	clearance	in	m
c	specific heat	Btu/lbm-°F	kJ/kg·°C
C	basic bearing load rating	lbf	N
C	belt correction factor	–	–
C	spring index	–	–
C_0	basic bearing static load rating	lbf	N
C_r	basic bearing dynamic load rating	lbf	N
CR	contact ratio	–	–
d	diameter	in	mm
D	diameter	in	mm
e	eccentricity	in	mm
e	rolling element bearing factor	–	–
E	energy	in-lbf	J
E	modulus of elasticity	lbf/in^2	Pa
f	coefficient of friction	–	–
f	frequency	Hz	Hz
f	normal force	Lbf	N
f_B	dynamic load rating adjustment factor	–	–
F	force	lbf	N
FS	factor of safety	–	–
g	gravitational acceleration, 32.2 (9.81)	ft/sec^2	m/s^2
g_c	gravitational constant, 32.2	lbm-ft/lbf-sec^2	n.a.
G	shear modulus	lbf/in^2	Pa
h	film thickness	in	mm
h	height	in	m
H	power	hp	kW
HP	horsepower	hp	-
I	rotational mass moment of inertia	lbm-in^2	kg·m^2
J	AGMA geometry factor	–	–
k	gear speed factor	–	–
k	spring constant	lbf/in	N/m
k	torsional spring constant	in-lbf/rev	N·m/rev
K	factor	–	–
KE	kinetic energy	in-lbf	J
L	lead	in	m
L	length	in	m
L	lifetime	various	various
L_{ab}	length of line of action	in	m
LR	lead ratio	–	–
m	belt mass per unit length	lbm/in	kg/m
m	mass	lbm	kg
m	module	in	mm
M	moment	in-lbf	N·m
n	number of coils	–	–
n	rotational speed	rpm	rpm
N	normal force	lbf	N
N	number	–	–
N	number of teeth on gear	–	–
p	circular pitch	in	mm
p	pressure	lbf/in^2	Pa
p	spring pitch	in	m
P	diametral pitch	1/in	1/m
P	force or load	lbf	N
P	power	hp	kW
q	generated heat	Btu/min	kW
r	radius	in	m
R	radius	in	m
s	AGMA stress number	lbf/in^2	Pa
S	Sommerfeld number	–	–
S	strength	lbf/in^2	Pa
t	thickness	in	mm
t	time	sec	s
T	torque	in-lbf	N·m
U	energy	in-lbf	J
v	velocity	ft/sec	m/s
V	rolling element bearing factor	–	–
VR	velocity ratio	–	–
w	width	in	mm
W	load	lbf	N
W	Wahl correction factor	–	–
x	change in length	ft	m
X	rolling element bearing factor	–	–
Y	Lewis form factor	–	–
Y	rolling element bearing factor	–	–

[32]Correlations have also been developed for various lubricant flows, maximum film pressure, and location of the minimum film thickness in addition to the correlations for the coefficient of friction and minimum film thickness.

Symbols

α	cone angle	deg	deg
γ	pitch angle	deg	deg
Γ	pitch angle	deg	deg
δ	deflection	in	m
ϵ	eccentricity ratio	–	–
η	efficiency	–	–
θ	angle	deg	deg
μ	absolute viscosity	lbf-sec/in^2	Pa·s
ν	kinematic viscosity	SSU	m^2/s
ρ	density	lbm/in^3	kg/m^3
σ	normal stress	lbf/in^2	Pa
τ	shear stress	lbf/in^2	Pa
ϕ	angle	rad	rad
ϕ	pressure angle	deg	deg
ψ	helix angle	deg	deg
ω	rotational speed	rad/sec	rad/s

Subscripts

0	minimum
a	active, allowable, application, or axial
alt	alternating
b	bending
c	centrifugal or clash
d	diametral or dynamic
e	endurance
eq	equivalent
f	free or frictional
G	gear
i	initial or inner
k	kinetic
L	life or load
m	mean, mechanical, or load distribution
n	normal
o	outer
p	at pitch circle, pitch, potential, or pulley
P	pinion
r	radial, at radius r, rated, or rating
R	reliability
s	separation, service, shear, size, or solid
t	tangential, tensile, or total
T	temperature
u	ultimate
v	velocity
V	per unit volume
w	working
W	worm
y	yield

54 Pressure Vessels

Content in blue refers to the *NCEES Handbook*.

NCEES EXAM SPECIFICATIONS AND RELATED CONTENT

THERMAL AND FLUID SYSTEMS EXAM

II.A.3. Hydraulic and Fluid Equipment: Pressure vessels

18. Actual Stresses
25. Wall Thickness

1. INTRODUCTION

The American Society of Mechanical Engineers (ASME) established its Boiler and Pressure Vessel Committee in 1911. The Committee establishes rules governing the design, fabrication, inspection, and repair of boilers and pressure vessels and interprets these rules when questions arise. These rules constitute the *ASME Boiler and Pressure Vessel Code* (BPVC and "Code"), which consists of the 12 sections listed in Table 54.1.

This chapter covers only pressure vessels with curved shells that are under internal pressure and are designed in accordance with Sec. VIII, "Pressure Vessels," Div. 1.[1,2,3] Division 1 of Sec. VIII covers pressure vessels operating between 15 psig and 3000 psig (103 kPa and 20.7 MPa).[4,5] Section VIII covers only nonnuclear applications. (Pressure vessels intended for use in commercial nuclear power plants are covered in Sec. III of the BPVC.)

BPVC jurisdiction over pipes and fittings extends from the pressure vessel only up to the first connection (i.e., pipe joint) to piping and added equipment, regardless of whether the connection is a welded, bolted, or threaded flange or a proprietary sealing pipe joint.[6,7]

[1]The BPVC also covers pressure vessels under external pressure.

[2]The alternate rules in BPVC Sec. VIII, Div. 2 place greater constraints on the materials and on the design, manufacturing, and testing processes. However, Div. 2 permits the pressure vessel to be designed to a lower factor of safety (i.e., a factor of safety of 3 compared to the theoretical values of 3.5 used in Div. 1). The shell and heads of a pressure vessel designed in accordance with Div. 2 will be thinner than a comparable pressure vessel designed to Div. 1. The end result is a reduction in material cost. Most of the rules in Div. 2 pertain only to stationary vessels installed in fixed locations.

[3]Even so, this chapter is only an introduction to the full BPVC.

[4]Division 2, being analysis-oriented, has no pressure limits.

[5]Large low-pressure vessels operating above 0.5 psig (3.4 kPa) but below 15 psig (103 kPa) are designed and built in accordance with API standards. Atmospheric vessels containing flammable and combustible liquids are designed and built in accordance with codes developed by Underwriters' Laboratories (UL) and the American Petroleum Institute (API) and with other acceptable good standards.

[6]Except for threaded plug closures used for inspection, BPVC jurisdiction ends at the first thread joint.

[7]The term *joint* is used to refer to a *welded seam*. The words "joint," "seam," and (sometimes) "weld" are synonymous.

Table 54.1 *Sections in the ASME Boiler and Pressure Vessel Code*

section	title
I	Power Boilers
II	Materials
III	Rules for Construction of Nuclear Facility Components
IV	Heating Boilers
V	Nondestructive Examination
VI	Care and Operation of Heating Boilers
VII	Care of Power Boilers
VIII	Pressure Vessels
IX	Welding, Brazing, and Fusing Qualifications
X	Fiber-Reinforced Plastic Pressure Vessels
XI	Rules for Inservice Inspection of Nuclear Power Plant Components
XII	Transport Tanks

2. RELATED CODES AND STANDARDS

Other codes and standards are frequently mentioned in conjunction with the ASME BPVC.

API 510: *Pressure Vessel Inspection Code: In-Service Inspection, Rating, Repair, and Alteration* covers the maintenance, inspection, repair, alteration, and rerating procedures for pressure vessels used by the petroleum and chemical process industries.

API 570: *Piping Inspection Code: Inspection, Repair, Alteration, and Rerating of In-Service Piping Systems* covers metallic piping systems that have been in service. Pressure vessels are excluded. Minimum thickness requirement calculations are not included and are referred to ASME B31.3.

API 620: *Design and Construction of Large, Welded, Low-Pressure Storage Tanks* covers the design and construction of large, welded, field-erected low-pressure carbon steel above-ground storage tanks (including flat-bottom tanks) with a single vertical axis of revolution that are designed for metal temperatures not greater than 250°F (121°C) and with pressures no greater than 15 psig (103 kPa). This standard also covers the design of cryogenic, double-wall storage tanks.

API 650: *Welded Tanks for Oil Storage* covers design, material, fabrication, erection, and testing of vertical, cylindrical, above-ground, closed- and open-top, welded carbon, stainless steel storage, and aluminum tanks. API 650 applies only to tanks whose entire bottoms are uniformly supported, in non-refrigerated service, whose maximum design temperature is 200°F (93°C) or less, and whose pressure at the liquid surface (not including the pressure from roof plates) is essentially atmospheric. Welded aluminum tanks are covered in API 650 App. AL.

ASME B31.3: *Process Piping* covers the design of piping for all fluids (with some exceptions for fluid service and high-pressure applications, but which can include hazardous material piping). Piping systems with internal pressures below 15 psig (103 kPa) are excluded.

ASME B31.9: *Building Services Piping* covers piping in industrial, commercial, and public buildings not requiring the sizes, pressures, and temperatures covered in ASME's *Power Piping* (ASME B31.1).

3. COMPLIANCE

Compliance with the BPVC by manufacturers is voluntary. However, most customers prefer knowing that the pressure vessels they order are designed and built to a known standard, and most states require use of ASME certificated pressure vessels as part of their state law. Manufacturers who build unfired pressure vessels in compliance with the code are able to mark the name plate of their products with the U stamp (mark) shown in Fig. 54.1.[8]

Figure 54.1 *Pressure Vessel Stamp*

4. EXEMPTIONS

Some pressure vessels fall outside the jurisdiction of the BPVC. These include vessels of any size operating below 15 psig (103 kPa); hot water supply storage tanks heated by indirect means and with a nominal capacity not exceeding 120 gal, water temperature not exceeding 210°F (99°C), and heat input not exceeding 200,000 Btu/hr (58.6 kW); and vessels less than 6 in (152 mm) in diameter (without limitation to pressure or length). Any vessel exempt from inspection, but nevertheless built in accordance with the BPVC, may be stamped with the U symbol.

[8]The mark shown in Fig. 54.1 is one of many specified by ASME: A, field assembly of power boilers; E, electric boilers; H, heating boilers (steel plate or cast-iron sectional); HLW, lined potable water heaters; M, miniature boilers; N, nuclear components; NPT, nuclear component partials; NA, nuclear installation/assembly; NV, nuclear safety valves; PP, pressure piping; RP, reinforced plastic pressure vessels; U, U2, pressure vessels; UM, miniature pressure vessels; UV, pressure vessel safety valves; and V, boiler safety valves.

5. STAMPING REQUIREMENTS

Pressure vessels must be permanently marked with information about their construction and type of service. This information may be stamped on the vessel in a conspicuous location (e.g., near an opening or manway) or may be on a permanently attached nameplate as shown in Fig. 54.2.

Nameplates describing a pressure vessel are attached directly to the shell.[9] The nameplate of a pressure vessel designed and constructed in accordance with the BPVC will contain the official "U" stamp and all of the following: manufacturer's name (listed after the words "certified by"), vessel serial number, year built, maximum allowable working pressure and corresponding temperature, and minimum design metal temperature and corresponding pressure.[10] For pressure vessels that are intended for service below −20°F (−29°C), the minimum allowable temperature is also listed.

Figure 54.2 *Typical Nameplate*

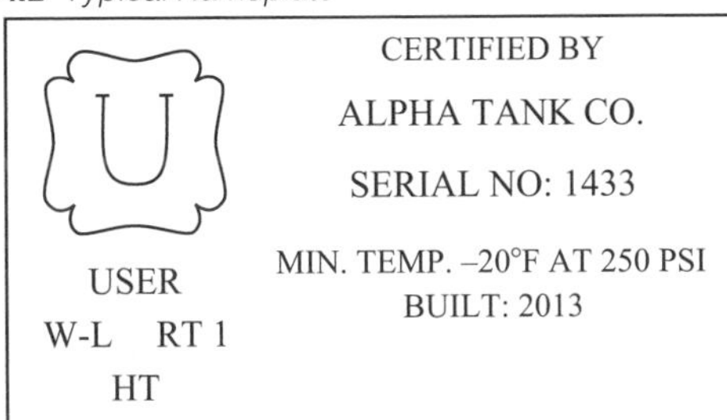

The method of construction and type of service must also be listed. One or more of the following abbreviations will be used: W, arc or gas welded; RES, resistance welded; B, brazed; L, lethal service; UB, unfired steam boiler; DF, direct firing; RT-1, fully radiographed; RT-2, some joints partially radiographed; RT-3, spot radiographed; RT-4, radiographed but other categories not applicable; HT, postweld heat treated; or PHT, parts of vessel heat treated.

6. DESIGN ELEMENTS

Figure 54.3 illustrates the various parts of a pressure vessel. The pressure vessel can be divided into shell-type and plate-type elements. A *shell-type element* resists internal pressure through tension (i.e., "membrane action"). A *plate-type element* resists internal pressure through bending. Shell-type elements can be cylindrical, spherical, ellipsoidal, torispherical, or toriconical.

The main body of a pressure vessel is known as the *shell*. A shell can be seamless or seamed. External pipes and equipment are connected to a pressure vessel at *nozzles*. Seamless pipe used for a nozzle is an example of a *seamless shell*.

7. SERVICE APPLICATION

Special restrictions are placed on vessels that contain lethal substances, operate below −20°F (−29°C), are used for steam generation, or are subject to direct firing.

The minimum wall thickness of shells and heads is $\frac{1}{16}$ in (1.6 mm), exclusive of any corrosion allowance. However, for pressure vessels intended for compressed air, steam, and water service, the minimum wall thickness of shells and heads is $\frac{3}{32}$ in (2.4 mm). Pressure vessels intended for compressed air service and those subject to internal corrosion or erosion must be designed with a suitable inspection opening.

A *lethal substance* is a poisonous gas or liquid that is toxic or lethal in even small quantities. Vessels intended for use with lethal substances may not be constructed from SA-283 carbon steel or cast iron. Expanded connections may not be used. All butt-welded joints must be fully radiographed. Postweld heat treatment is required when vessels carrying lethal substances are constructed of carbon or low-alloy steel.[11,12,13]

The BPVC does not specify material selection based on vessel content. However, hydrogen and caustic embrittlement are important noncode issues that must be dealt with during the design process. Pressure vessels exposed to hydrogen at high temperatures and high pressures are subject to *hydrogen embrittlement*. The vessel can be considered to be subject to hydrogen damage if the hydrogen partial pressure exceeds 87 psia (600 kPa) and the vessel temperature is 300°F (150°C) or more, or if the hydrogen partial pressure exceeds 145 psia (1000 kPa absolute) regardless of temperature. A *Nelson diagram* is used to determine materials that are appropriate for various combinations of temperature and pressure.[14]

Caustic soda (sodium hydroxide) is a common alkaline material. Carbon steel is subject to *caustic cracking* (sometimes called *caustic embrittlement*) and rapid corrosion at elevated temperatures.

[9]Duplicate nameplates on supports or at other locations must be marked "Duplicate."

[10]It has been common in some metric countries to specify pressure in either bars or kilograms per square cm (kg/cm^2). Multiply lbf/in^2 by 0.06895 to obtain bars. Multiply lbf/in^2 by 0.07031 to obtain kg/cm^2.

[11]A corrosion allowance is not required for seamless vessel parts designed with 0.85 joint efficiency or for vessels that are not radiographed.

[12]A good rule of thumb for determining the minimum thickness of a pressure vessel is $t_{\text{min,in}} = (d_{\text{in}} + 100)/1000$. This thickness will result in a vessel that is stiff enough for handling during fabrication and erection.

[13]Several lethal substances regularly encountered in the chemical processing industry include hydrocyanic acid, carbonyl chloride, cyanogen, mustard gas, and xylyl bromide.

[14]A *Nelson diagram* is a *p-T* diagram indicating the pressure-temperature combinations below which failure of a particular alloy have not been reported. Since the Nelson diagram is based on a historical record, it is possible that failures will still occur in the regions previously indicated as eventless. Refer to API RP 941.

Figure 54.3 *Parts of a Pressure Vessel*

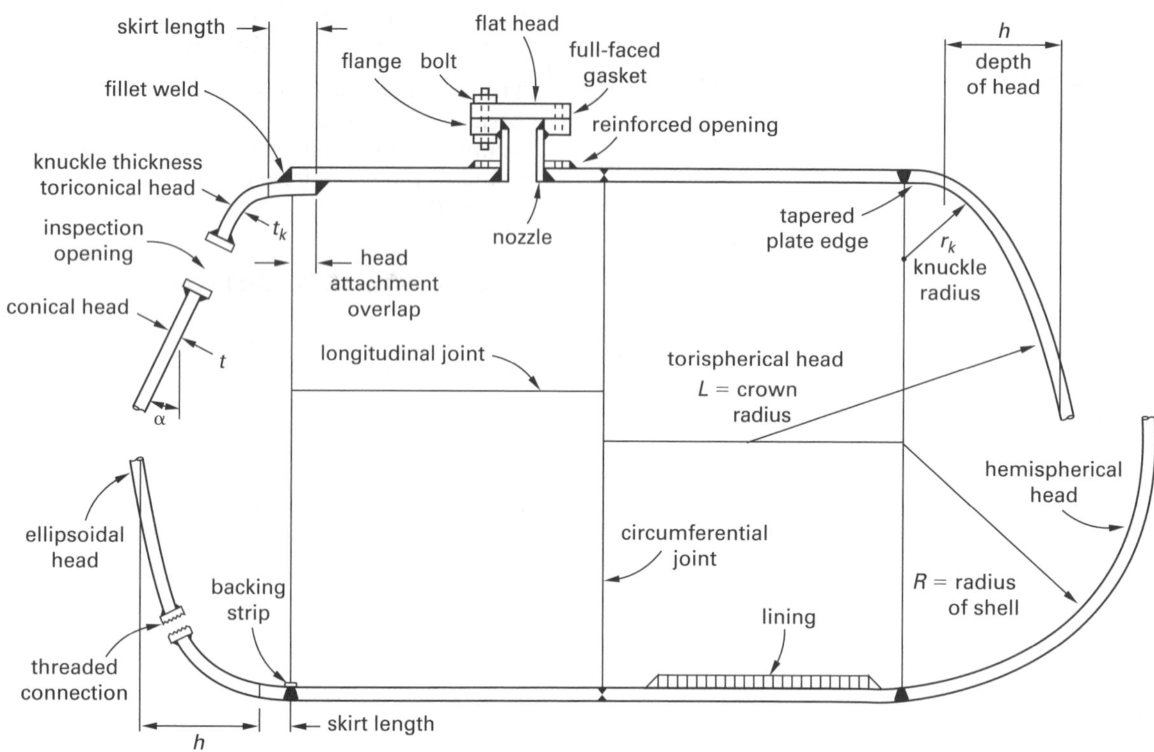

Reproduced from R. Chuse and S. Eber, *Pressure Vessels*, sixth ed., copyright © 1984, with permission of the publisher, McGraw-Hill.

8. MATERIALS

Pressure vessels can be constructed from various materials including carbon steel, low-alloy steel, high-alloy steel, nonferrous metal, cast iron, and integrally clad-plate. The BPVC specifies which materials and at what temperatures these materials can be used. To ensure compatibility in welding, material categories are designated by P-numbers, as shown in Table 54.2.

Carbon steels are the most common material chosen for noncorrosive environments between −20°F and 800°F (−29°C and 426°C). However, carbon steel weakens with long exposure to temperatures higher than 785°F (418°C) through a process known as *graphitization*. For service above 800°F (426°C), materials must be selected carefully.

Table 54.2 *Material Designations*

designation	material
P-1	carbon steels with tensile strengths between 40,000 lbf/in^2 and 75,000 lbf/in^2 (276 MPa and 517 MPa)
P-3	alloy steel with up to $^3/_4$% chromium; alloy steels with up to 2% total alloying ingredients
P-4	alloy steel with chromium between $^3/_4$% and 2%; alloy steels with more than 2% total alloying ingredients
P-5	alloy steels with up to 10% total alloying ingredients
P-8	austenitic stainless steels
P-9	nickel alloy steels
P-10	other steel alloys

9. HEADS

Heads may be spherical, ellipsoidal, torispherical, conical, or flat, as shown in Fig. 54.4. Ellipsoidal and hemispherical heads are common, while torispherical heads appear more frequently in thin vessels. BPVC formulas for calculating the minimum thickness for each head type are given in Table 54.10.

Torispherical (i.e., *flanged and dished*) *heads* are specified by their inside diameter, D, crown (dish) radius, L, knuckle radius, r_k, and head thickness, t_h. These variables are shown in Fig. 54.4. An *ASME flanged and dished head* is a torispherical head for which the knuckle radius is 6% of the inside crown radius. *2:1 ellipsoidal heads* have a radius that is twice the height (projection) and are more economical than deeper hemispherical heads.

Figure 54.4 Head Shapes

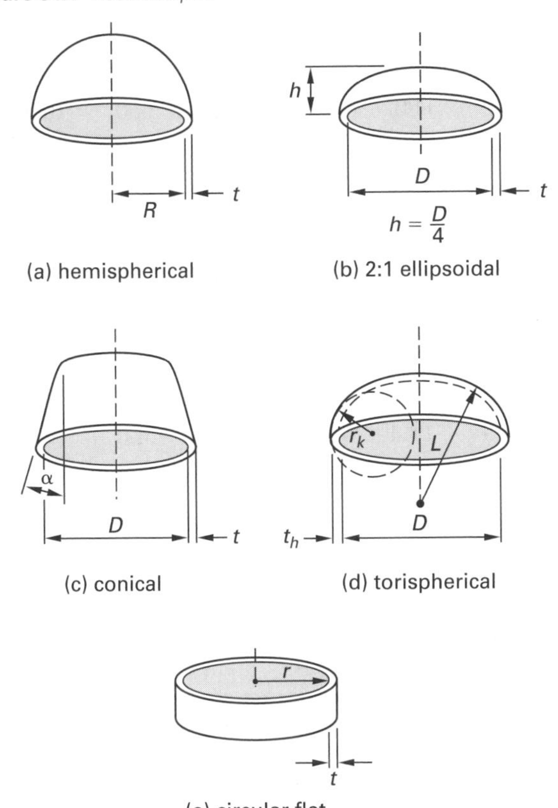

When a head is no thicker than its shell, the head does not need a flange and may be butt-welded to the shell. In practice, however, most nonhemispherical heads have *straight flanges* (i.e., straight longitudinal necks).

10. LOADING TYPE

In addition to loads from internal pressurization, pressure vessels must be designed for the weight of the vessel and its contents, static reactions from attachments, wind pressure, seismic forces, temperature gradients and thermal expansion, and dynamic forces from cyclic variations in pressure or temperature.

11. ALLOWABLE STRESSES

The base allowable stress in tension, S, is specified in the BPVC (see Table 54.3) and depends on the material and the temperature. (The allowable stress for parts in compression also depends on the slenderness ratio.) The BPVC specifies that the *maximum allowable stress* used for design purposes is S for tensile stresses and for general primary membrane stress induced by any combination of loading. For most internal-pressure configurations, allowable stress in compression is not the controlling factor.[15]

The maximum allowable stress is $1.5S$ for the sum of primary membrane stress and bending stress due to any combination of loadings except those for wind and earthquake.

Forces and stresses from wind, snow, earthquakes, and other transient loads are calculated in ways consistent with the *International Building Code* (IBC), using load factors instead of increasing the allowable stresses beyond $1.0S$.[16] Wind and earthquake forces should be calculated from ASCE 7. The various load combinations given in BPVC Sec. VIII, Div. 2, Table 5-3 should be used. Seismic loads have a load factor of 0.7, and wind and seismic loads do not act simultaneously.

Table 54.3 lists the maximum allowable stress for common carbon and low-alloy steels. Values in Table 54.3 may be interpolated.

12. DISCONTINUITY STRESSES

Pressure vessels inevitably contain locations with abrupt changes in geometry, material, and loading. These locations are known as *discontinuity areas*, and the BPVC calls the associated stresses *discontinuity stresses*. A *gross structural discontinuity* is a geometrical, structural, or material discontinuity that affects the stress and strain distribution across the entire wall thickness over a region of significant size. Examples are nozzle-to-cylinder and shell-to-head junctions (particularly when the shell and head centerlines are not coincident). *Local structural discontinuities* affect the stress and strain locally across part of the wall thickness. Common examples are weld toes in butt-welds, notches, and areas of undercutting.

Each of the parts of a discontinuity area behaves differently to pressure and thermal loads. Modeled as disconnected free bodies, the parts have different displacements. As connected parts, the displacements are the same. The difference between the free and actual displacements is a *forced displacement*, which produces the discontinuity stresses.

The BPVC defines two classes of stress: primary and secondary. A *primary stress* (*load-controlled stress*), P_m, is the direct result of mechanical loading and can be calculated from the laws of equilibrium. Examples are stresses from pressure and self-weight. In a region away from any discontinuities, all stresses are primary stresses. A primary stress increases in direct proportion to the applied load, irrespective of the shape of the stress-strain curve, until failure. In a thin section

[15]The *critical height* of a vertical vessel under internal pressure and subject to wind loading is $pR/16t$. Above this height, compressive stress may govern.
[16]The $\frac{1}{3}$ stress increase (i.e., a multiplicative factor of $\frac{4}{3}$ on allowable stress) for wind and transient loading is no longer permitted by any building code.

Table 54.3 *Typical Maximum Allowable Stress for Carbon and Low-Alloy Steels (thousands of lbf/in²)*

		temperature (°F)										
specification	type/ grade	−20 to 400	500	600	650	700	750	800	850	900	950	1000
SA-283	C	15.7	15.7	15.3	14.8	–	–	–	–	–	–	–
SA-285	C	15.7	15.7	15.3	14.8	14.3	13.0	10.8	8.7	5.9	–	–
SA-515	60	17.1	17.1	16.4	15.8	15.3	13.0	10.8	8.7	5.9	4.0	2.5
SA-515	65	18.6	18.6	17.9	17.3	16.7	13.9	11.4	8.7	5.9	4.0	2.5
SA-515	70	20.0	20.0	19.4	18.8	18.1	14.8	12.0	9.3	6.7	4.0	2.5
SA-516	55	15.7	15.7	15.3	14.8	14.3	13.0	10.8	8.7	5.9	4.0	2.5
SA-516	60	17.1	17.1	16.4	15.8	15.3	13.0	10.8	8.7	5.9	4.0	2.5
SA-516	65	18.6	18.6	17.9	17.3	16.7	13.9	11.4	8.7	5.9	4.0	2.5
SA-516	70	20.0	20.0	19.4	18.8	18.1	14.8	12.0	9.3	6.7	4.0	2.5
SA-105	–	20.0	19.6	18.4	17.8	17.2	14.8	12.0	9.3	6.7	4.0	2.5
SA-181	–	20.0	19.6	18.4	17.8	17.2	14.8	12.0	9.3	6.7	4.0	2.5
SA-350	LF1	17.1	16.3	15.3	14.8	14.3	13.0	10.8	8.7	5.9	4.0	2.5
SA-350	LF2	20.0	19.6	18.4	17.8	17.2	14.8	12.0	9.3	6.7	4.0	2.5
SA-53	E/B	17.1	17.1	17.1	17.1	15.6	13.0	10.8	8.7	5.9	–	–
SA-106	B	17.1	17.1	17.1	17.1	15.6	13.0	10.8	8.7	5.9	4.0	2.5
SA-193	B7 $\leq$ 2.5 in	25.0	25.0	25.0	25.0	25.0	23.6	21.0	17.0	12.5	8.5	4.5

(Multiply lbf/in² by 6.895 to obtain kPa.)

Printed with permission from Eugene F. Megyesy, *Pressure Vessel Handbook*, copyright © 1989, by Pressure Vessel Handbook Publishing, Inc.

membrane such as a pressure vessel, *general primary membrane stresses* are categorized into (1) uniform (single value) tensile and compressive stresses, and (2) linearly varying, bending stresses, P_b. A *bending stress* varies linearly over the thickness of a thin component. *Local primary membrane stresses*, P_L, consider discontinuities but not stress concentrations.

A *secondary stress*, Q, is the result of a geometric discontinuity, stress concentration, or temperature difference. Both primary and secondary stresses increase in proportion to the loading until the yield point is reached. However, while primary stresses continue to increase with increased loading, secondary stresses are considered *self-limiting stresses*. Once the yield point has been reached at the stress concentration, local yielding and some distortion ameliorate the conditions causing stress (i.e., the direct relationship between load and stress is annulled), and the secondary stress remains essentially constant. Failure direct from a single application of a secondary stress is, therefore, not expected. However, secondary stresses become particularly important in fatigue analyses of cyclic loading. A *peak stress* is an increment of stress over and above the primary and secondary stresses, caused by discontinuities or local thermal conditions. Although fatigue cracks are possible, there is no gross deformation due to peak stress.

Different allowable stresses are prescribed for primary membrane, primary bending, and secondary stresses produced by mechanical loads.[17] (See Table 54.4.) The limiting value of a *primary membrane stress*, usually tensile in nature, will be reached over the full vessel cross section simultaneously. To prevent a catastrophic plastic failure (i.e., a burst under pressure), BPVC Sec. VIII, Div. 2 limits primary membrane stresses to S (essentially $^2/_3$ of the yield strength). Since maximum bending stress occurs only at a localized point most distant from the neutral axis, the local primary membrane stress, P_L (i.e., primary membrane stress plus primary localized bending stress), is limited to $1.5S$ (i.e., essentially the yield strength). Secondary stresses can comfortably exceed the yield strength, so primary plus self-relieving secondary stresses are limited to $3S$ (i.e., twice the yield strength).

[17]BPVC Sec. VIII, Div. 2, Table 5.6 gives guidance in categorizing stresses as primary, secondary, localized, and so on, according to vessel component and location.

Table 54.4 *Allowable Stress Combinations, Normal Operations*[a,b]

stress combination	allowable stress[c]
P_m [general primary membrane stress]	S
P_b [general primary bending stress]	$1.5S$
$P_L = P_m + Q_{ms}$ [local primary membrane stress]	$1.5S$
Q_m [secondary membrane stress]	$1.5S$
Q_b [secondary bending stress]	$3.0S < 2.0F_y <$ UTS
F_p [peak stress]	$2.0S_n$
$P_L + P_b = P_m + Q_{ms} + P_b$	$1.5S$
$P_m + P_b + Q_m^* + Q_b$	$3.0S < 2.0F_y <$ UTS
$P_m + P_b + Q_m^* + Q_b + F_p$	S_n

[a]P, Q, and F do not represent single quantities, but represent sets of six stress components: σ_t, σ_r, σ_l, τ_{tl}, τ_{lr}, and τ_{rt}.
[b]Different stress limitations will hold for stresses induced by emergency operations, earthquakes, and hydraulic tests.
[c]S is the allowable stress, per BPVC Sec. VIII, Div. 1, at the design temperature. Values of yield strength as a function of temperature are provided in BPVC Sec. II, Part D, Table Y-I. Values of ultimate strength as a function of temperature are provided in BPVC Sec. II, Part D, Table U.
Source: BPVC Sec. VIII, Div. 2, Fig. 5.1

13. SHAKEDOWN AND RATCHETING

Pressure vessels often experience repeated cycles of pressurization/depressurization and heating/cooling. A vessel's response can be categorized as shakedown or ratcheting, as defined in BPVC Sec. VIII, Div. 2, Part 5.12. (See Fig. 54.5.) Paraphrasing, *shakedown* is caused by cyclic loads or cyclic temperature distributions that produce plastic deformations at a point when the loading or temperature distribution is applied, but upon removal of the loading or temperature distribution, only elastic primary and secondary stresses remain at that point, except in small areas associated with local stress (strain) concentrations. These small areas exhibit a stable hysteresis loop, with no indication of progressive advancement. Further loading and unloading, or applications and removals of the temperature distribution, produce only elastic primary and secondary stresses. Shakedown is not a failure mode, although it does justify using fatigue curve data.

Paraphrasing, *ratcheting* is a progressive, incremental, inelastic deformation or strain that occurs at a point subjected to cycles of thermal stress or cycles of mechanical stress superimposed on a mean stress, or both. (*Thermal stress ratcheting* is partly or wholly caused by thermal stress.) Ratcheting is produced by a sustained load acting over the full cross section at that point, in combination with a strain-controlled cyclic load or temperature distribution that is alternately applied and removed. Ratcheting causes cyclic straining of the material, which can result in failure by fatigue and, at the same time, produces cyclic incremental growth of a failure mechanism, which can ultimately lead to collapse. Ratcheting is a failure mode as it refers to incremental growth in gross dimensions.

Figure 54.5 *Shakedown and Ratcheting*

stress, strain

(a) shakedown

stress, strain

(b) ratcheting

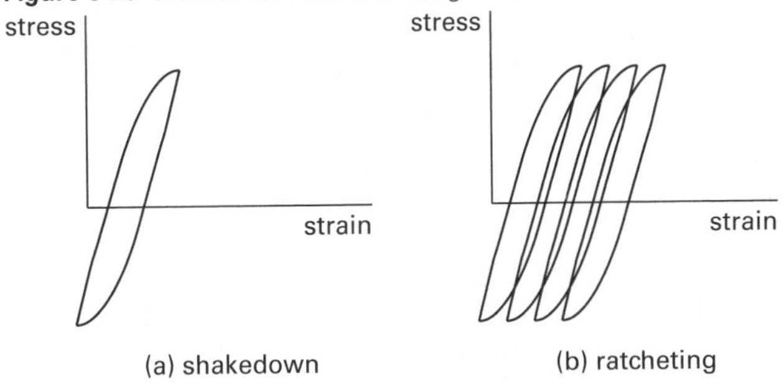

14. TEMPERATURE EFFECTS

Table 54.5 lists the modulus of elasticity for ferrous materials as a function of temperature.

Table 54.5 *Modulus of Elasticity of Ferrous Materials (millions of lbf/in²)*

temperature		carbon steels		austenitic stainless steel
(°F)	(°C)	$C \leq 0.3\%$	$C > 0.3\%$	
70	21	29.5	29.3	28.3
200	93	28.8	28.6	27.6
300	149	28.3	28.1	27.0
400	204	27.7	27.5	26.5
500	260	27.3	27.1	25.8
600	316	26.7	26.5	25.3
700	371	25.5	25.3	24.8
800	427	24.2	24.0	24.1
900	482	22.4	22.3	23.5
1000	538	20.4	20.2	22.8
1100	593	18.0	17.9	22.1

(Multiply lbf/in² by 6.895 to obtain kPa.)

15. IMPACT TESTING

Brittle fracture is a concern when a pressure vessel is designed to handle cryogenic fluids or is exposed to severe winter temperatures. Only *notch-tough materials* should be used when the design temperature is very low. Notch toughness is determined by *impact testing* the pressure vessel material. The cutoff temperature below which impact testing is required depends on the material and the shell thickness. In general, most ferrous metals require special attention whenever the design temperature is less than −20°F (−29°C).

Because there is no marked decrease in ductility at low temperature, vessels manufactured from wrought aluminum do not require impact tests until the design temperature is below −452°F (−269°C). Copper and copper alloys, nickel and nickel alloys, and cast aluminum alloys can be used down to −325°F (−254°C), and titanium or zirconium and their alloys can be used down to −75°F (−59°C) before impact testing is required.

Impact testing is not required for grade 1 or 2 carbon steels provided that the following five conditions are met: (1) the wall thickness is not greater than ½ in (12.7 mm) or 1 in (25.4 mm) depending on the material, (2) the completed vessel is hydrostatically tested, (3) the design temperature is between −20°F and 650°F (−29°C and 343°C), (4) thermal or mechanical shock loadings are not a controlling design element, and (5) cyclic loading is not a controlling design requirement.

16. MAXIMUM ALLOWABLE WORKING PRESSURE

The *maximum allowable working pressure* (MAWP) is specified by the manufacturer. It is the maximum pressure permissible at the top of the vessel in its normal operating position and temperature, in corroded condition, and while under the effects of other expected loadings (e.g., wind and external pressure).[18] MAWP is calculated for different parts of the pressure vessel based on BPVC equations adjusted for static head. The overall MAWP is the smallest of the adjusted values. The maximum allowable working pressure is the design pressure in the thickness equations.

The term *maximum allowable pressure new and cold*, MAWP N&C, is specified by the manufacturer. It is the maximum pressure for the vessel when new (not corroded) and at room temperature.

17. DESIGN PRESSURE AND TEMPERATURE

The BPVC requires that pressure vessels be designed for the most severe combination of pressure and temperature that will be encountered during normal operation, regardless of whether the combination is short term or infrequent.

The maximum design temperature must not be less than the mean metal temperature under expected operating conditions. Design temperatures in excess of those listed in Table 54.3 are not allowed. The *minimum design metal temperature* (MDMT) is the lowest expected service temperature, except where a lower value is allowed by the BPVC. The MDMT marked on the nameplate will correspond to an associated MAWP. As temperature can vary with location and through the thickness, the mean temperature is specified on the nameplate.

The *operating pressure*, p, is the pressure on the top of the vessel at which the vessel normally operates. The maximum difference in pressures between the inside and outside of a vessel or between any two chambers of a combination vessel should also be evaluated. The *design pressure* is the operating pressure plus a reasonable safety margin.[19] Frequently, the design pressure is equal to the MAWP. The design pressure and MDMT are used to determine the minimum allowable thickness of the vessel.

18. ACTUAL STRESSES

Cylindrical pressure vessels are classified as either thin walled or thick walled depending on the vessel's wall thickness, t, and diameter, D. In a *thick-walled pressure vessel*, the wall thickness is at least one-twentieth of the diameter (that is, $D/t \le 20$). In a *thin-walled pressure vessel*, the wall thickness is less than one-twentieth of the diameter ($D/t > 20$).

A cylindrical vessel with end caps is subject to three principal stresses: tangential stress, radial stress, and axial stress. *Tangential stress* (also called *hoop stress* or *circumferential stress*), σ_t, is stress that acts in a circumferential direction. *Radial stress*, σ_r, is stress that acts toward or away from the longitudinal axis. *Axial stress* (also called *longitudinal stress*), σ_a, is stress that acts in the direction of the length of the cylinder.

These stresses vary as a function of the internal and external pressures, p_i and p_o, respectively, as well as of the inner and outer radii of the vessel wall, r_i and r_o, respectively. Within the vessel wall, tangential and radial stress (but not axial stress) also vary with the distance, r, from the center of the vessel. Figure 54.6 shows a cylindrical vessel subjected to both internal and external pressures.

Stresses are calculated differently for thick- and thin-walled vessels. For a thick-walled vessel, the tangential stress at a distance r from the center of the vessel is given by Eq. 54.1.

Cylindrical Pressure Vessel

$$\sigma_t = \frac{p_i r_i^2 - p_o r_o^2 - \dfrac{r_i^2 r_o^2 (p_o - p_i)}{r^2}}{r_o^2 - r_i^2} \qquad 54.1$$

For a thick-walled vessel, the radial stress at a distance r from the center of the vessel is given by Eq. 54.2.

[18] A pressure vessel can have more than one operating temperature and hence more than one MAWP.
[19] A reasonable safety margin is 10% or 25 lbf/in^2 (170 kPa), whichever is greater.

Figure 54.6 *Cylindrical Vessel Subjected to Internal and External Pressures*

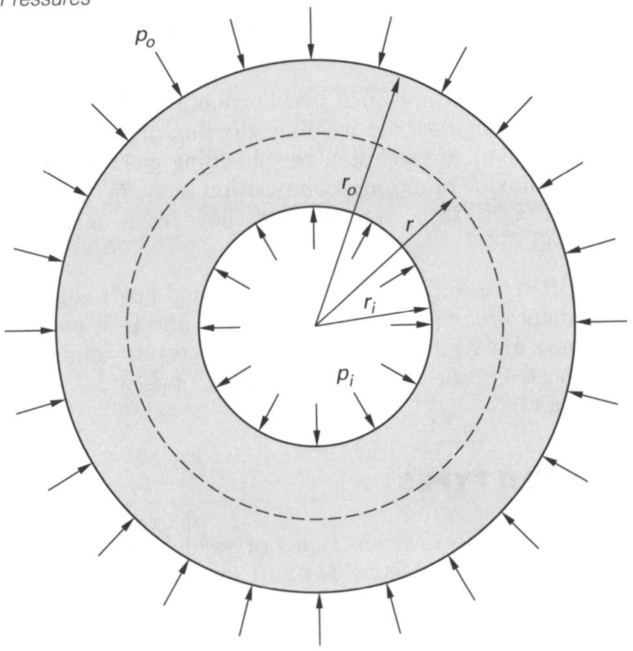

Cylindrical Pressure Vessel

$$\sigma_r = \frac{p_i r_i^2 - p_o r_o^2 + \dfrac{r_i^2 r_o^2 (p_o - p_i)}{r^2}}{r_o^2 - r_i^2} \quad 54.2$$

For a thick-walled vessel, the axial stress is given by Eq. 54.3. Axial stress is constant within the wall; it does not vary with the distance from the center of the vessel.

Cylindrical Pressure Vessel

$$\sigma_a = \frac{p_i r_i^2 - p_o r_o^2}{r_o^2 - r_i^2} \quad 54.3$$

Equation 54.1 through Eq. 54.3 can often be simplified for specific conditions. When calculating stress on the inner surface, r_i is equal to r; for stress on the outer surface, r_o is equal to r. If the stress is based only on internal pressure, p_o is zero; if the stress is based only on external pressure, p_i is zero. For example, in calculating the tangential stress on the inner surface of a cylindrical vessel ($r_i = r$) due to internal pressure ($p_o = 0$), Eq. 54.1 can be simplified to Eq. 54.4.

Cylindrical Pressure Vessel

$$\sigma_t = p_i \left(\frac{r_o^2 + r_i^2}{r_o^2 - r_i^2} \right) \quad 54.4$$

For a thin-walled cylindrical vessel, tangential and axial stresses vary only with the vessel's diameter, D, and the internal pressure, p_i. Tangential stress can be found from Eq. 54.5, and axial stress can be found from Eq. 54.6. Tangential stress is twice axial stress, so it is tangential stress that drives pressure vessel design. In a thin-walled vessel, radial stress is significantly less than both the tangential and axial stresses and is typically neglected.

Cylindrical Pressure Vessel

$$\sigma_t = \frac{p_i D}{2t} \quad 54.5$$

$$\sigma_l = \frac{p_i D}{4t} \quad 54.6$$

Equation 54.5 and Eq. 54.6 can also be used to calculate the stresses in circumferential and longitudinal joints[20], but cannot be used to calculate wall thickness. Wall thickness calculations are addressed in Sec. 54.25.

For a thin-walled pipe with a maximum allowable tangential stress of S, the maximum working pressure can be found from the Barlow formula.

Cylindrical Pressure Vessel

$$p = \frac{2St}{D} \quad 54.7$$

19. CORROSION ALLOWANCE

An optional *corrosion allowance* compensates for any wall thinning expected over the lifetime of the vessel. The BPVC does not provide guidance in determining the allowance. Values will be unique to each situation and must be determined by the user considering vessel duty and corrosiveness of the contents. Generic total corrosion allowances of one-sixth of the required thickness, 1/16 in (1.6 mm), and 1/8 in (3.2 mm) are common, as is a corrosion rate of 0.005 in/yr (0.13 mm/yr), but these have no specific basis other than tradition. A corrosion allowance of zero is common for stainless steel and other nonferrous materials not subject to corrosion. An allowance of 1 mm is typical for air receivers where moisture condensation is inevitable. Usually, the maximum corrosion allowance for carbon steel is 6 mm.

Every dimension used in a formula should be a corroded dimension.[21] When calculating the thickness from a known pressure, if a formula (such as from Table 54.9 or Table 54.10) used to calculate the theoretical thickness includes a nominal radius term, r, the corrosion allowance should be added to the radius before calculating the theoretical thickness. This will account for the slight

[20]The term girth seam is sometimes used when referring to a circumferential joint.

[21]As illustrated in BPVC Sec. VIII, Div. 1, App. L (e.g., Ex. L.9.2.1), this admonition includes adding the corrosion allowance to the nominal radius as well as to the resulting required thickness. The effect of increasing the radius is small, and many authorities simply add the corrosion allowance to the thickness as a final step.

increase in radius (and subsequent increase in stress) when the wall corrodes. The required starting wall thickness is obtained by also adding the corrosion allowance to the calculated theoretical thickness. The corrosion allowance is subtracted from an actual known ending radius to obtain the starting radius. Similarly, a vessel with starting inner diameter D will have a diameter of $D+2c$ at the end of its useful life.

When calculating an allowable pressure from a known starting thickness, the ending (corroded) thickness, $t_{\text{starting}} - c$, should be used.

$$t_{\text{starting}} = t_{\text{calculated}} + c = t_{\text{ending}} + c \quad \text{54.8}$$

$$r_{\text{starting}} = r_{\text{ending}} - c \quad \text{54.9}$$

20. WELDING SPECIFICATIONS

Specifications for all of the variables in the welding process are contained in a *welding procedure specification* (WPS). The WPS details such items as the welding method, type of joint, filler metal, preheating, and postweld heat treatment. The validity of the WPS is supported by a *procedure qualification record* (PQR), which includes test results on weldability and other factors. Only *welders* and *welder operators* qualified under Sec. IX of the BPVC may construct or repair a pressure vessel.[22]

Welding may be automatic, semiautomatic, or manual, though not all methods may be useful in all cases.[23] Electrodes used in inert gas metal-arc welding may be consumable or nonconsumable. Single or multiple electrodes may be used. Necessary information to completely specify the joint type includes the number of weld passes, the type of pass (i.e., string or weaving bead), the size of weld wire or electrode, and (for automatic welding) the rate of travel and wire feed.

Backing strips may be used to ensure that butt welding from one side penetrates the entire plate thickness. The backing strip becomes fused to the weld.

Fillet (F) and groove (G) welds are commonly used. When specifying welds, the position needs to be identified, as not all welders are certified in all positions. For plates, categories are 1F/1G (flat), 2F/2G (horizontal), 3F/3G (vertical), and 4F/4G (overhead). For pipes, categories are 1F/1G (flat), 2F/2G (horizontal), 4F (overhead), 5F/5G (multiple position, pipe horizontal), and 6G (multiple position, pipe inclined). While it is best to restrict welding to a particular position, welding in multiple positions is permitted. If vertical welding (position 3G) is used, the direction should be specified as uphill or downhill.

Filler metals are specified by filler metal group number (F). For submerged arc welding, the flux must be specified. For inert arc welding, the shielding gas type (e.g., carbon dioxide or argon), composition (e.g., 75% carbon dioxide and 25% argon), and flow rate must be specified.

The BPVC specifies when preheating and postweld heat treatment are required. Some (e.g., P-4 and P-5) materials must always be preheated. Above certain minimum thicknesses, the BPVC requires postweld heat treatment.[24]

21. WELD TYPES

The BPVC specified six types of weld joints. Type 1 weld joints are double-welded butt joints. The quality of weld is the same inside and outside of the vessel with double-welded butt joints. Backing strips, if used, are removed after welding. After the weld is made on one side, the second side of the joint is cleaned and rewelded. The weld quality is identical on both sides of the joint. Type 2 welds are single-welded butt joints with backing strips that remain in place after welding. Type 3 welds are single-welded butt joints without backing strips. Type 4 joints are double full-fillet lap joints. Type 5 joints are single full-fillet lap joints with plug welds. Type 6 joints are single full-fillet lap joints without plug welds. The weld types are shown in Fig. 54.7 with their typical welding symbols and in Fig. 54.8 with their typical locations.[25]

22. WELD APPLICATIONS

Certain weld types are more appropriate for certain locations on the pressure vessel. Table 54.6 lists the types of welds by pressure vessel location permitted by the BPVC for gas and arc welding methods.

Type 1 joints have no limitations and may be used for all applications. Type 2 joints can also be used anywhere, except that a joint to a plate with a preformed offset can only be used for circumferential joints. Other restrictions are listed in the footnotes of Table 54.6.

[22] A *welder* is qualified in manual and semiautomatic welding. A *welding operator* is qualified to use automatic welding equipment.

[23] In addition to BPVC restrictions, the circumstances of welding limit certain operations. For example, in large vessels without manways, the closing joint must be welded from the outside. Also, small vessels less than approximately 18 in to 24 in (460 mm to 610 mm) cannot be welded from the inside.

[24] For P-1 carbon steel vessels welded without preheating, postweld heat treatment is required for thicknesses greater than $1\frac{1}{4}$ in (31.8 mm). The limit increases to $1\frac{1}{2}$ in (38.1 mm) thickness if the piece has been preheated to 200°F (93°C). Vessels intended for lethal service must be postweld heat treated.

[25] Standard American Welding Society (AWS) symbols are used.

Figure 54.7 *Types of Welds and Symbols*

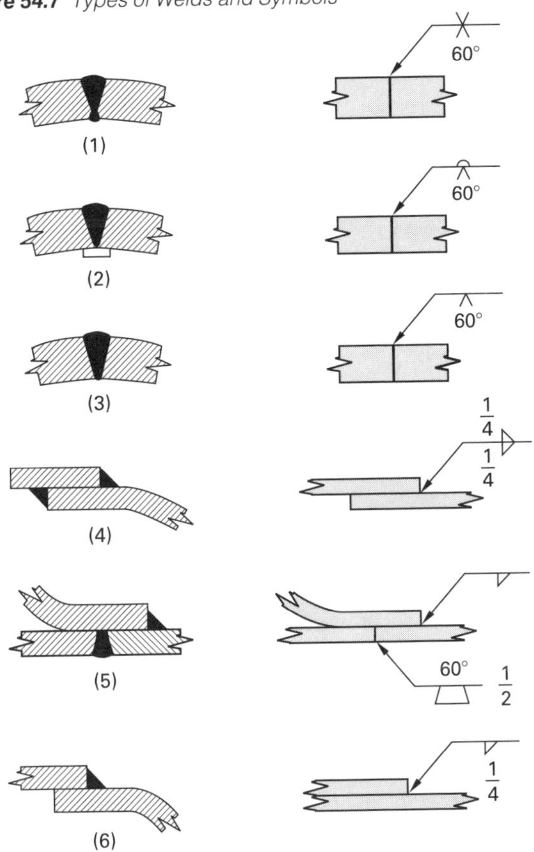

Designations: 1, double-weld butt joint; 2, single-weld butt joint with integral backing strip; 3, single-weld butt joint without backing strip; 4, double-full fillet lap joint; 5, single-full fillet lap joint with plug welds; 6, single-full fillet lap joint without plug welds.

Table 54.6 *Permitted Weld Types by Location*

	general joint description			
joint type[a]	longitudinal[b] (A)[c]	circumferential (B)[c]	flange-plate attachment (See Fig. 54.8.) (C)	openings and nozzle attachment (See Fig. 54.8.) (D)
1	yes	yes	yes	yes
2	yes	yes	yes	yes
3	yes[c]	yes[c]	yes[c]	no
4	yes[d]	yes[e]	yes[e]	no
5	no	yes[f]	yes[f]	no
6	yes[f]	yes[f]	no	no

[a]Designations: 1, double-weld butt joint; 2, single-weld butt joint with integral backing strip; 3, single-weld butt joint without backing strip; 4, double-full fillet lap joint; 5, single-full fillet lap joint with plug welds; 6, single-full fillet lap joint without plug welds.

[b]Including welds in and on hemispherical heads, and any weld in a sphere.

[c]Circumferential butt joints not over $^5/_8$ in (16 mm) and outside diameter not over 24 in (610 mm) thick.

[d]Longitudinal joints not over $^3/_8$ in (9.5 mm) thick.

[e]Longitudinal joints not over $^5/_8$ in (16 mm) thick.

[f]Restrictions apply.

Category A welds are longitudinal welds that include (1) longitudinal joints in the main shell, connecting chambers, diameter transitions, and nozzles; (2) any joint in sphere, formed or flat head, or the side plates of a flat-sided vessel; and (3) circumferential joints connecting hemispherical heads to main shells, to transitions in diameter, to nozzles, or to connecting chambers.

Category B welds are circumferential welds that include (1) circumferential welded joints in the main shell, connecting chambers, nozzles, and diameter transitions; and (2) circumferential welds connecting formed heads (other than hemispherical) to main shells, diameter transitions, nozzles, and connecting chambers.

Category C welds are flange and nozzle attachments that include (1) welded joints connecting flanges, tube sheets or flat heads to main shells, formed heads, diameter transitions, nozzles, and connecting chambers; and (2) any welded joint connecting one side plate to another side plate of a flat-sided vessel.

Category D welds are flange and nozzle attachments connecting (1) chambers or nozzles to main shells, spheres, diameter transitions, heads, and flat sided vessels, and (2) nozzles to communicating chambers.

23. JOINT EFFICIENCY

Welded joints are common in the construction of pressure vessels. These joints can be of several different types and subjected to different degrees of radiographic inspection. The type and quality of the weld will affect its ability to reach the strength of the parent material. Consequently, the joint efficiency, E, is used in many calculations as a derating factor. For welded joints subjected to tension, efficiency values depend on the type of weld and inspection (typically radiography) performed.[26] The strongest joints are double-welded butt joints (joint type 1). Joint efficiencies for various types of joints in shells and heads are shown in Table 54.7 and Table 54.8. For seamless shells, fully and spot radiographed shells have an efficiency of 1.0. Shells that do not meet spot radiography requirements have an efficiency of 0.85.

Table 54.7 *Efficiencies of Welded Joints in Shells (categories A, B, C, and D)*

	radiography		
joint type*	full	spot	none
1	1.00	0.85	0.70
2	0.90	0.80	0.65
3	–	–	0.60
4	–	–	0.55
5	–	–	0.50
6	–	–	0.45

*Designations: 1, double-weld butt joint; 2, single-weld butt joint with integral backing strip; 3, single-weld butt joint without backing strip (rarely used); 4, double-full fillet lap joint (rarely used); 5, single-full fillet lap joint with plug welds (rarely used); 6, single-full fillet lap joint without plug welds (rarely used).

[26]The efficiency of a butt weld joint in compression is 100%.

Figure 54.8 Four Types of Welds by Location (category)

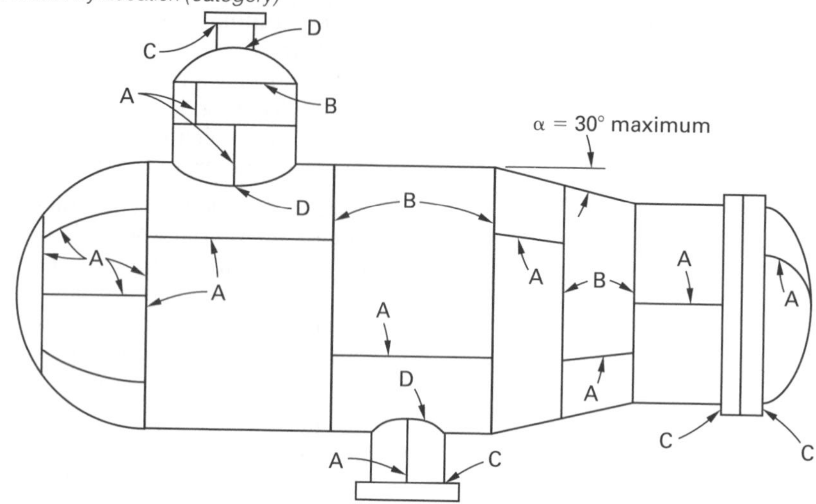

Reproduced from R. Chuse and S. Eber, *Pressure Vessels*, sixth ed., copyright © 1984, with permission of the publisher, McGraw-Hill.

Table 54.8 Efficiencies of Welded Joints in Heads

		radiography		
head type	joint type*	full	spot	none
hemispherical	1	1.00	0.85	0.70
	2	0.90	0.80	0.65
others	any	1.00	1.00	0.85

*1, double-weld butt joint; 2, single-weld butt joint with integral backing strip.

24. WELD EXAMINATION

Common methods of *nondestructive examination* (NDE) used on welded joints of pressure vessels are radiography and ultrasonic examination.[27,28] Full radiographic inspection of joints is mandatory for (1) all longitudinal welds and welds at openings when the joint efficiency is taken as 1.0 or 0.9; (2) all butt welds joined by electrogas welding with any single pass greater than 1½ in and all electrogas welds; (3) with some exceptions, all butt welds where the material thickness exceeds 1½ in (38 mm); (4) all butt welds in the head and shells of unfired steam boilers with design pressures exceeding 50 psig (345 kPa); and (5) all butt welds in vessels containing lethal substances.[29]

Spot radiographic inspection can be used to obtain an economical spot check of welding quality.

Radiographic inspection is optional for butt-welded joints that are not required to be fully radiographed, and is not required when the vessel is designed for external pressure or when the joint has been designed (i.e., the joint efficiency value has been chosen) for no inspection.

Full ultrasonic inspection is required for electroslag and electrogas welds in ferritic material.

The *RT marking system* is used to indicate the extent of radiographic examination. "RT-1" means 100% of all longitudinal and circumferential seams were radiographed. It also indicates that 100% of nozzle welds over 10 in (25 mm) diameter were radiographed. This level is considered "full radiography" and results in a 1.0 joint efficiency on all welds. RT-1 is mandatory for head/shell thicknesses greater than 1.25 in (32 mm). (Vessels designed to BPVC Sec. I and certain nozzles require 100% RT on circumferential and longitudinal seams.) "RT-2" means 100% of longitudinal weld seams were radiographed, and spot RT was done on circumferential seams. This level is also considered "full radiography" and results in a 1.0 joint efficiency for thickness calculations. No radiographic testing is done on nozzle welds for this level. "RT-3" means, with only a few exceptions, spot radiographic inspection was performed on all longitudinal and circumferential seams. RT-3 results in a 0.85 joint efficiency. No nozzle connection welds were radiographed. "RT-4" means some radiographic examination took place, but the amount can't be described with the RT numbering system. RT-4 results in a 0.70 joint efficiency.

[27]Most other methods of NDE, including eddy current, acoustic emission, liquid penetrant, and magnetic particle testing, are also used with pressure vessels.

[28]Radiography is sometimes referred to as "X-raying." However, radiography can use either X-rays generated from high electrical voltages or gamma rays generated from a radioactive isotope capsule.

[29]Ultrasonic examination may be substituted for radiography for the final closure seam of a pressure vessel if the construction of the vessel does not permit interpretable radiographs in accordance with the BPVC.

25. WALL THICKNESS

For a thin-walled cylindrical vessel, the tangential stress is greater than the axial and radial stresses and usually governs wall thickness. The minimum wall thickness of such a vessel (exclusive of any corrosion allowance) is

Cylindrical Pressure Vessel

$$t = \frac{p_i r_i}{Se - 0.6p_i} \quad \begin{bmatrix} p_i < 0.385Se \\ t \leq 0.5r_i \end{bmatrix} \qquad 54.10$$

In Eq. 54.10, Se is the *effective allowable stress*, which is the product of the BPVC's tabulated allowable stress value, S, and the joint efficiency, e.

For a thin-walled spherical shell or hemispherical head with radius r, the minimum thickness (exclusive of any corrosion allowance) is calculated from Eq. 54.11. For a head without a flange, use the efficiency of the head-to-shell joint if it is less than the efficiency of the joint in the head.

$$t = \frac{p_i r_i}{2Se - 0.2p_i} \quad \begin{bmatrix} p_i < 0.665Se \\ t \leq 0.356r_i \end{bmatrix} \qquad 54.11$$

Table 54.9 contains the code formulas for designing thin-walled cylindrical shells.[30]

Table 54.9 *Required Wall Thickness of Thin Cylindrical Shells*[a]

member	minimum thickness,[a] t	maximum pressure, p	limitation
(in terms of inside radius) longitudinal joint (circumferential stress)	$\frac{pr}{Se-0.6p}$	$\frac{Set}{r+0.6t}$	$p \leq 0.385Se$[b] $t \leq 0.5r$
(in terms of inside radius) circumferential joint (longitudinal stress)	$\frac{pr}{2Se+0.4p}$	$\frac{2Set}{r-0.4t}$	$p \leq 1.25Se$ $t \leq 0.5r$
(in terms of outside radius, r_o) longitudinal joint (circumferential stress)	$\frac{pr_o}{Se+0.4p}$	$\frac{Set}{r_o-0.4t}$	$p \leq 0.385Se$ $t \leq 0.5r$

[a]The minimum thickness of shells and heads used for compressed air, water, or steam is 3/32 in (2.4 mm). The minimum thickness of shells for plates of welded construction is 1/16 in (1.6 mm).
[b]Equations based on stress across a circumferential joint govern only if the efficiency of the circumferential joint is less than one-half of the longitudinal joint's efficiency.

The minimum thickness of a head depends on its shape, as indicated in Table 54.10.

Although plates of any thickness can be produced upon order and heads can be "spun" to any thickness, maximum economy will be achieved when standard-thickness plates are used. Table 54.11 lists the thicknesses of plates commonly available in the United States.

Example 54.1

A cylindrical pressure vessel is being designed to operate at a 125 psig (862 kPa) design pressure and a 700°F (371°C) design temperature. When new, the vessel will have a 90 in (2286 mm) inside diameter. It will be constructed of SA-515 grade-70 plate with a spot-radiographed, longitudinal, double-welded butt joint. A 0.125 in (3.175 mm) corrosion allowance is to be included.

(a) What shell wall thickness is required?

(b) What standard wall thickness plate should be used?

SI Solution

(a) The joint efficiency of a spot-radiographed double-welded butt joint is 0.85. From Table 54.3, the allowable stress for SA-515-70 plate at 700°F is 18,100 lbf/in^2.

$$S = \left(18{,}100 \ \frac{\text{lbf}}{\text{in}^2}\right)\left(6.895 \ \frac{\text{kPa·in}^2}{\text{lbf}}\right) = 124\,800 \text{ kPa}$$

In its corroded condition, the vessel's inside radius will be

$$r_i = \frac{2286 \text{ mm}}{2} + 3.175 \text{ mm} = 1146.175 \text{ mm}$$

Check the first of the two conditions for using Eq. 54.11.

$$p < 0.385Se$$
$$862 \text{ kPa} < (0.385)(124\,800 \text{ kPa})(0.85) < 40\,841 \text{ kPa} \quad [\text{OK}]$$

From Eq. 54.11, the shell wall thickness without corrosion allowance is

Cylindrical Pressure Vessel

$$t = \frac{p_i r_i}{Se - 0.6p_i} = \frac{(862 \text{ kPa})(1146.175 \text{ mm})}{(124\,800 \text{ kPa})(0.85) - (0.6)(862 \text{ kPa})} = 9.359 \text{ mm}$$

[30]In the absence of additional loads from wind, earthquake, or attachments, the stress in a circumferential joint will govern only when the efficiency of the circumferential joint is less than one-half of the longitudinal joint efficiency.

Table 54.10 Required Thickness of Heads

head type	minimum thickness, t	maximum internal pressure, p	limitation
thin hemispherical			
(inside radius)	$\frac{pr}{2Se-0.2p}$	$\frac{2Set}{r+0.2t}$	$p \leq 0.665Se$
(outside radius)	$\frac{pr_o}{2Se+0.8p}$	$\frac{2Set}{r_o-0.8t}$	$t \leq 0.356r$
ellipsoidal*			
(inside diameter)	$\frac{pDK}{2Se-0.2p}$	$\frac{2Set}{KD+0.2t}$	$K=\left(\frac{1}{6}\right)\left[2+\left(\frac{D}{2h}\right)^2\right]$
(outside diameter)	$\frac{pD_oK}{2Se+2p(K-0.1)}$	$\frac{2Set}{KD_o-2t(K-0.1)}$	$K=1$ for 2 : 1 ellipsoidal heads
torispherical			
(inside length)	$\frac{pLM}{2Se-0.2p}$	$\frac{2Set}{LM+0.2t}$	$\frac{L}{r_k} \leq 16\frac{2}{3}$
(outside length)	$\frac{pL_oM}{2Se+p(M-0.2)}$	$\frac{2Set}{ML_o-t(M-0.2)}$	$M=\left(\frac{1}{4}\right)\left(3+\sqrt{\frac{L}{r_k}}\right)$
conical			
(inside diameter)	$\frac{pD}{2\cos\alpha(Se+0.6p)}$	$\frac{2Set\cos\alpha}{D+1.2t\cos\alpha}$	$\alpha \leq 30°$
(outside diameter)	$\frac{pD_o}{2\cos\alpha(Se+0.4p)}$	$\frac{Set\cos\alpha}{D_o+0.8t\cos\alpha}$	
circular flat unstayed			
(not bolted)	$t=d\sqrt{\frac{Cp}{Se}}$	$p=\frac{St^2e}{d^2C}$	d and C as per BPVC Fig. UG-34
(bolted)	$t=d\sqrt{\frac{Cp}{Se}+\frac{1.9Wh_G}{Sed^3}}$	$p=\frac{Se}{C}\left(\frac{t^2}{d^2}-\frac{1.9Wh_G}{Sed^3}\right)$	p and W as per BPVC App. 2

*Ellipsoidal shells in which $K>1.0$ and all torispherical heads made of materials having a specified minimum tensile strength exceeding 80,000 lbf/in² must be designed using a value of S equal to 20,000 lbf/in² at room temperature and reduced in proportion to the reduction in maximum allowable stress values at temperature for the material.

The shell wall thickness with corrosion allowance included is

$$\begin{aligned} t_c &= t+c = 9.359 \text{ mm} + 3.175 \text{ mm} \\ &= 12.53 \text{ mm} \end{aligned}$$

Check the second condition for using Eq. 54.11.

$$\begin{aligned} t_c &\leq 0.5r_i \\ 12.53 \text{ mm} &\leq (0.5)(1146.175 \text{ mm}) \\ &= 573 \text{ mm} \quad [\text{OK}] \end{aligned}$$

(b) From Table 54.11, select a ½ in plate with a thickness of 12.7 mm.

Customary U.S. Solution

(a) The joint efficiency of a spot-radiographed, double-welded butt joint is 0.85. From Table 54.3, the allowable stress for SA-515-70 plate at 700°F is 18,100 lbf/in². In its corroded condition, the vessel's inside radius will be

$$r_i = \frac{90 \text{ in}}{2} + 0.125 \text{ in} = 45.125 \text{ in}$$

Table 54.11 *Standard Plate Thicknesses*

in		mm
1/16	(0.062)	1.59
5/64	(0.078)	1.98
3/32	(0.090)	2.38
1/8	(0.125)	3.18
3/16	(0.188)	4.76
1/4	(0.250)	6.35
5/16	(0.313)	7.94
3/8	(0.375)	9.53
7/16	(0.437)	11.1
1/2	(0.500)	12.7
every 1/16 to 1 in		
every 1/8 to 2 in		

(Multiply in by 25.4 to obtain mm.)

Check the first of the two conditions for using Eq. 54.11.

$$p < 0.385Se$$

$$125\ \text{lbf/in}^2 < (0.385)\left(18{,}100\ \frac{\text{lbf}}{\text{in}^2}\right)(0.85)$$
$$< 5923\ \text{lbf/in}^2 \quad \text{[OK]}$$

From Eq. 54.11, the shell wall thickness without corrosion allowance is

Cylindrical Pressure Vessel

$$t = \frac{p_i r_i}{Se - 0.6p_i} = \frac{\left(125\ \frac{\text{lbf}}{\text{in}^2}\right)(45.125\ \text{in})}{\left(18{,}100\ \frac{\text{lbf}}{\text{in}^2}\right)(0.85) - (0.6)\left(125\ \frac{\text{lbf}}{\text{in}^2}\right)} = 0.3684\ \text{in}$$

The shell wall thickness with corrosion allowance included is

$$t_c = t + c = 0.3684\ \text{in} + 0.125\ \text{in} = 0.4934\ \text{in}$$

Check the second condition for using Eq. 54.11.

$$t_c \le 0.5r_i$$
$$0.4934\ \text{in} \le (0.5)(45.125\ \text{in}) = 22.56\ \text{in} \quad \text{[OK]}$$

(b) From Table 54.11, select a ½ in plate with a thickness of 0.500 in.

26. NOZZLE NECKS

A *nozzle* is an opening in a pressure vessel. The nozzle may connect the vessel to other parts of the piping network, or it may be normally closed off. *Handholes* and larger *manways* are nozzles that are opened for inspection, cleaning, and repair.[31] A nozzle typically ends at a bolted flange plate. The cylindrical section between a pressure vessel and the flange is the *nozzle neck*.

The procedure for determining the minimum wall thickness of nozzle necks, $t_{\text{UG-45}}$, is specified in BPVC Sec. VIII, UG-45. For necks of openings used only for inspection, $t_{\text{UG-45}}$ is simply t_a, the minimum neck thickness required for internal pressure plus any corrosion allowance for the neck, c_{neck}.

$$t_{\text{UG-45}} = t_a \quad \left[\begin{array}{c}\text{inspection openings; internal}\\ \text{pressure; } t_a \text{ includes } c_{\text{neck}}\end{array}\right] \qquad 54.12$$

Most nozzles are not used for inspection, and for them, the minimum neck thickness is given by Eq. 54.13. t_{b1} is the thickness required for the shell or head at the location where the neck is attached, calculated using $e = 1$, plus the shell/head corrosion allowance, c_{shell}, but not less than $\frac{1}{16}$ in.[32] t_{b3} is the minimum wall thickness (specified in Table UG-45) for the nearest (larger outside diameter) standard (i.e., schedule-40) national pipe size (NPS), plus the neck corrosion allowance, c_{neck}. Allowing for pipe manufacturing tolerance, the minimum standard NPS wall thickness is 87.5% of the nominal thickness.

$$t_{\text{UG-45}} = \max\left\{\begin{array}{l} t_a \\ t_b = \min\left\{\begin{array}{c} t_{b1} \ge \frac{1}{16}\ \text{in} \\ t_{b3}\end{array}\right\}\end{array}\right\} \qquad 54.13$$

$$\left[\begin{array}{c}\text{non-inspection openings;}\\ \text{internal pressure}\end{array}\right]$$

[31] A *davit* (*davit arm*) is a swinging support constructed as part of the vessel, supporting the manway cover when it is unbolted and moved aside.
[32] UG-45 says, "...but in no case less than the minimum thickness specified for the material in UG-16(b)." 1/16 in (plus the corrosion allowance) is the minimum thickness for most shells and heads, but there are special cases in UG-16(b). For example, for carbon steel, the minimum thickness is 3/32 in plus the corrosion allowance.

27. FLAT UNSTAYED HEADS

Flat surfaces appear extensively in pressure vessels. Circular surfaces are most common, although other shapes may be used. A flat surface used as the end closure or head of a pressure vessel may be an integral part of the vessel (when formed with the cylindrical shell or welded to it), or it may be a removable plate attached by bolts through a gasket to a flange.[33] Because they are the weakest of head configurations, they may be reinforced with rods, spars, ribs, braces, stiffeners, and so on, known as *stays*, in order to prevent excessive deflection. Flat plates and heads connected around their peripheries but without any other reinforcement are known as *flat unstayed heads*. Unreinforced flat plates are generally two to five times thicker than the surrounding shell. Figure 54.9 illustrates two of the many acceptable ways of attaching flat heads.[34]

Figure 54.9 *Flat Unstayed Heads*

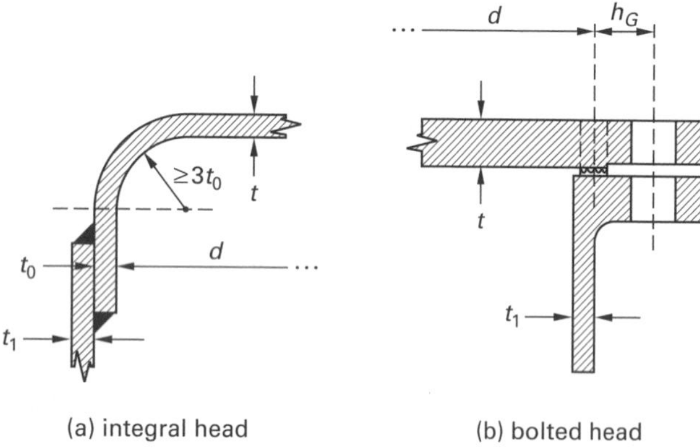

From BPVC Sec. VIII-1, UG-34 c(2), the minimum required thickness of flat, unstayed circular covers and heads is given by Eq. 54.14. C is the *flange attachment factor* (from BPVC Sec. VIII, Fig. UG-34(j)) that depends on the methods of attachment and gasketing and that accounts for the degree of edge fixity. D may be the inside, outside, or bolt line diameter of the shell, depending on the design. The efficiency, e, is 1.0 for all configurations except for butt welded heads, and in that case, efficiency is determined by the extent of radiography. S is the maximum allowable stress in tension at temperature.

$$t = D\sqrt{\frac{Cp}{Se}} + c \quad \text{[circular welded plates]} \qquad 54.14$$

The gasket and bolt loading of some flat covers (e.g., Fig. 54.9(b)) adds an edge moment. The thickness of the flat cover depends on the flange *design bolt load*, W, and the *gasket moment arm*, h_G, both calculated per BPVC Sec. VIII, Div. III, App. 2. W is the sum of the bolt loads required to resist the end pressure load and to maintain tightness of the gasket. Equation 54.15 is to be used twice—once with the design conditions, and again with the gasket seating conditions (i.e., $p = 0$, S for room temperature, and W as required for gasket seating per Eq. 4 of BPVC App. 2-5(e))—and the greater of the two values used.

$$t = D\sqrt{\frac{Cp}{Se} + \frac{1.9 W h_G}{SeD^3}} + c \quad \left[\begin{array}{c}\text{circular bolted}\\ \text{plates}\end{array}\right] \qquad 54.15$$

28. REINFORCEMENT OF OPENINGS

Depending on the shell wall thickness and the size of the opening, reinforcement at the shell-neck connection point may be required.[35] The basic rule is that the cross-sectional area of the reinforcement must equal the cross-sectional area removed (for the opening) less any "excess" area not needed to resist the internal pressure.

Reinforcement is most easily obtained by adding a *pad*, an annular flat plate, exterior to and surrounding the nozzle. Some manufacturers replace all of the removed area with pad area as a general practice. This requires little or no calculation, but results in over-reinforced openings. By considering the shell and nozzle thicknessess and nozzle diameter, a minimum required area can be computed for vessels under internal pressure.

$$A_r = Dt_r + 2t_n t_r(1 - f_{r1}) \qquad 54.16$$

In Eq. 54.16, t_r is the theoretical shell or head thickness as calculated from the BPVC equations,[36,37] D is the opening inside diameter, t_n is the specified nozzle thickness, and f_r is the allowable stress ratio. As these are nominal conditions, corrosion will eventually reduce the available area. Corrosion is accounted for by subtracting the corrosion allowance, c, from the actual shell and nozzle thicknesses, and by adding twice the corrosion allowance to the opening diameter.

Openings can be reinforced or unreinforced. Figure 54.10 shows the general case of a reinforced opening. When designing a vessel, it is important to determine that sufficient area is provided. The area criteria for vessels with

[33]A toroidal knuckle with a flat head is an example of an integral flat plate head.
[34]BPVC Sec. VIII, Div. 3, Fig. UG-34 shows all of the acceptable standard designs.
[35]Reinforcement of pressure vessels not subject to rapid pressure fluctuations is not required when the vessel thickness is $^3\!/_8$ in (9.5 mm) or less and the opening diameter is 3.5 in (88.9 mm) or less, or when the vessel thickness is more than $^3\!/_8$ in (9.5 mm) and the opening diameter is 2.375 in (60.3 mm) or less.
[36]Reinforcement of extraordinarily large openings requires a different procedure. For pressure vessels with diameters of 60 in (1.52 m) or less, large openings are those that are more than one-half the vessel diameter or larger than 20 in (508 mm). For pressure vessels with diameters of more than 60 in (1524 mm), large openings are those that are more than one-third the vessel diameter or larger than 40 in (1016 mm).
[37]Refer to the BPVC for calculating the required thickness for reinforcement purposed for openings in conical and ellipsoidal heads.

Figure 54.10 *Pad Reinforcement of an Opening (simplified for non-integral reinforcement)*

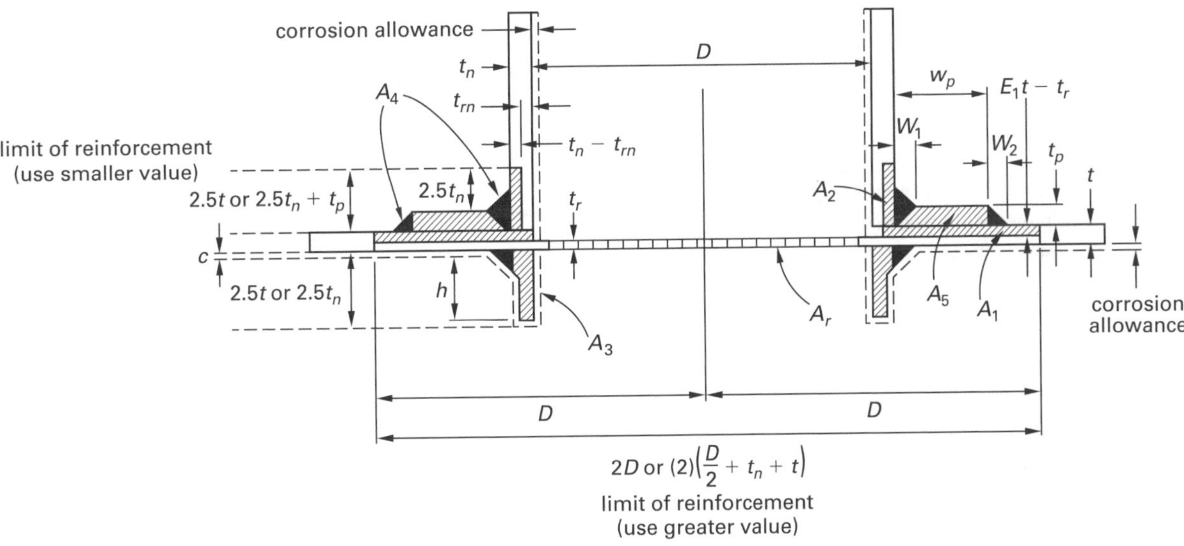

and without reinforcement are given in the following equations, where the terms are defined in Fig. 54.10 and Table 54.12.

$$A_1 + A_2 + A_3 + A_4 \geq A_r \quad \text{[without reinforcement]} \qquad 54.17$$

$$A_1 + A_2 + A_3 + A_4 + A_5 \geq A_r \quad \text{[with reinforcement]} \qquad 54.18$$

If the sum of the available areas meets the requirements (i.e., the inequalities are satisfied), there is no need for (additional) reinforcing plates. However, should the sum of the available areas be less than that required, a pad must be used to provide that area (or the available reinforcement must be increased). The additional area required is

$$A_{p,r} = A_r - (A_1 + A_2 + A_3 + A_4 + A_5)$$

If used, reinforcement pads must be welded to both the shell and nozzle wall as shown in Fig. 54.10. The welds at the outer edge of the pad and inner edge of the pad, which does not abut the nozzle neck, must be fillet welds with a minimum throat dimension of $\frac{1}{2}t_{\text{min}}$, where t_{min} is the smaller of $\frac{3}{4}$ in or the thickness of the thinner of the parts joined by the weld. Welds attaching a pad to a nozzle neck abutting the vessel wall must have a fillet weld with minimum throat dimension t_w not less than $0.7t_{\text{min}}$.

Not all material around an opening is effective reinforcement. To be considered effective, the material must be located within the *limits of reinforcement.* Material located around a $2D$-diameter (approximately) circle concentric with the opening axis is considered effective, as is material located up to (approximately) $2.5t$ or $2.5t_n$ above and below the vessel surface. This assumption is built into the calculation of A_2. (Refer to Fig. 54.10 for exact limits of reinforcement.)

Some openings are designed to be integrally reinforced. This type of reinforcement is provided in the form of extended or thickened necks, built-up shell plates, or weld buildup that is an integral part of the shell or vessel wall. The installed pad shown in Fig. 54.10 is not considered integral reinforcement. When used, integral reinforcement can require the use of a correction factor, F, in the calculation of A_r and A_1 in the previous equations to compensate for the variation of the stress due to internal pressure on different planes with respect to the vessel axis.

Example 54.2

A pressure vessel has an inside diameter of 46 in. Vessel walls are 0.5625 in an SA-516-60 plate. The vessel is intended for service at 350 psig and 200°F. An 8 in (nominal) diameter nozzle with 0.593 in wall thickness is constructed of SA-53-B material. The nozzle extends 1.50 in into the pressure vessel and does not interrupt any of the main vessel seams. $\frac{3}{8}$ in fillet welds are used throughout. All seams are fully radiographed. Corrosion is not a factor in the design. Determine whether reinforcement is required.

Solution

Determine the required thicknesses of the shell and nozzle. From Table 54.3, the maximum allowable stress is 17,100 lbf/in^2 for both the shell and nozzle materials. The efficiency of fully radiographed welds is $e = 1.00$.

The shell inside radius is

$$r_i = \frac{46 \text{ in}}{2} = 23 \text{ in}$$

*Table 54.12 Areas for Unreinforced and Reinforced Openings**

	without reinforcement	with reinforcement
A_1 (area available in shell, use large value)	$D(et-t_r)-2t_n(et-t_r)(1-f_{r1})$ or $2(t+t_n)(et-t_r)-2t_n(et-t_r)(1-f_{r1})$	
A_2 (area available in nozzle projecting outward, use smaller value)	$5t(t_n-t_{rn})f_{r2}$ or $5t_n(t_n-t_{rn})f_{r2}$	$5t(t_n-t_{rn})f_{r2}$ or $2(t_n-t_{rn})(2.5t_n+t_p)f_{r2}$
A_3 (area available from nozzle projecting inward, use smallest value)	$5tt_if_{r2}$ or $5t_it_if_{r2}$ or $2ht_if_{r2}$; t_i is the thickness of the inward-projecting nozzle material, less the corrosion allowance	
A_{41} (outward weld area)	W^2f_{r2}	W^2f_{r3}
A_{42}(outer element weld area)	–	W^2f_{r4}
A_{43} (inward weld area)	–	W^2f_{r2}
A_5 (added pad)		$(D_P-D-2t_n)t_pf_{r4}$

*f_{r1}, f_{r2} = allowable stress in nozzle wall/allowable stress in vessel (shell) wall ≤ 1.0; f_{r3} = lesser of f_{r1} and $f_{r4} \leq 1.0$; f_{r4} = allowable stress in reinforcing pad element/allowable stress in vessel (shell) wall ≤ 1.0. $f_{r1} < f_{r2}$. $f_{r1} = 1.0$ for nozzle walls abutting a vessel wall and for some other cases. (See BPVC Sec. VIII, Div. 1.) These factors compensate for any strength reduction if the material used for the nozzle and reinforcing element is not as strong as the vessel material. No reduction is required if the reinforcing element, nozzle, or weld material is stronger than the shell material. However, no excess can be claimed if the nozzle, weld, or reinforcing element material is stronger than that of the shell.

From Eq. 54.11, the required shell wall thickness is

Cylindrical Pressure Vessel

$$t_r = \frac{p_i r_i}{Se-0.6p_i} = \frac{\left(350\ \frac{\text{lbf}}{\text{in}^2}\right)(23\ \text{in})}{\left(17{,}100\ \frac{\text{lbf}}{\text{in}^2}\right)(1.00)-(0.6)\left(350\ \frac{\text{lbf}}{\text{in}^2}\right)} = 0.477\ \text{in}$$

The 0.593 in wall thickness and 8 in nominal diameter indicate that the nozzle is constructed from schedule-100 pipe. The actual inside diameter of schedule-100 pipe is 7.439 in. The nozzle inside radius is

$$r_i = \frac{7.439\ \text{in}}{2} = 3.7195\ \text{in}$$

From Eq. 54.11, the required nozzle wall thickness is

Cylindrical Pressure Vessel

$$t_{rn} = \frac{p_i r_i}{Se-0.6p_i} = \frac{\left(350\ \frac{\text{lbf}}{\text{in}^2}\right)(3.7195\ \text{in})}{\left(17{,}100\ \frac{\text{lbf}}{\text{in}^2}\right)(1.00)-(0.6)\left(350\ \frac{\text{lbf}}{\text{in}^2}\right)} = 0.077\ \text{in}$$

$$f_{r1} = f_{r2} = \frac{17{,}100\ \frac{\text{lbf}}{\text{in}^2}}{17{,}100\ \frac{\text{lbf}}{\text{in}^2}} = 1$$

From Eq. 54.16, the total reinforcement required (with the factor $1-f_{r1}$ equal to zero) is

$$\begin{aligned} A_r &= Dt_r + 2t_nt_r(1-f_{r1}) = Dt_r \\ &= (7.439\ \text{in})(0.477\ \text{in}) + 0 \\ &= 3.548\ \text{in}^2 \end{aligned}$$

The reinforcement available is calculated in four parts. (Areas are rounded down to understate the available reinforcement.)

$$\begin{aligned} D(et - t_r) - 2t_n(et - t_r)(1 - f_{r1}) &= (7.439\ \text{in})\big((1.00)(0.5625) - 0.477\ \text{in}\big) - 0 \\ &= 0.636\ \text{in}^2 \end{aligned}$$

$$\begin{aligned} 2(t + t_n)(et - t_r) - 2t_n(et - t_r)(1 - f_{r1}) &= (2)(0.5625\ \text{in} + 0.593\ \text{in}) \\ &\quad \times\big((1.00)(0.5625\ \text{in}) - 0.477\ \text{in}\big) - 0 \\ &= 0.1.98\ \text{in}^2 \end{aligned}$$

$$\begin{aligned} A_1 &= \text{greater of} \begin{cases} 0.636\ \text{in}^2 \\ 0.198\ \text{in}^2 \end{cases} \\ &= 0.636\ \text{in}^2 \end{aligned}$$

$$\begin{aligned} 5t(t_n - t_{rn})f_{r2} &= (5)(0.5625\ \text{in}) \\ &\quad \times(0.593\ \text{in} - 0.077\ \text{in}) \\ &\quad \times(1) \\ &= 1.451\ \text{in}^2 \end{aligned}$$

$$\begin{aligned} 5t_n(t_n - t_{rn})f_{r2} &= (5)(0.593\ \text{in}) \\ &\quad \times(0.593\ \text{in} - 0.077\ \text{in}) \\ &\quad \times(1) \\ &= 1.530\ \text{in}^2 \end{aligned}$$

$$\begin{aligned} A_2 &= \text{smaller of} \begin{cases} 1.451\ \text{in}^2 \\ 1.530\ \text{in}^2 \end{cases} \\ &= 1.451\ \text{in}^2 \end{aligned}$$

The nozzle projects 1.50 in into the vessel. Corrosion is not a factor in the design, so there is no corrosion allowance and $t_i = t_n - 0.593$ in. (See Table 54.13.) The limit of reinforcement parallel to the nozzle wall is the smallest of

$$\begin{cases} 5tt_if_{r2} = (5)(0.5625\ \text{in})(0.593\ \text{in})(1) = 1.668\ \text{in}^2 \\ 5t_it_if_{r2} = (5)(0.593\ \text{in})(0.593\ \text{in})(1) = 1.758\ \text{in}^2 \\ 2ht_if_{r2} = (2)(1.5\ \text{in})(0.593\ \text{in})(1) = 1.779\ \text{in}^2 \end{cases}$$

$$A_3 = 1.668\ \text{in}^2$$

Reinforcement is provided by the nozzle-to-shell weld. (If a reinforcement pad is used, reinforcement is also provided by the pad-to-shell weld if that weld is within the limits of reinforcement. The pad-to-shell area is disregarded.)

$$\begin{aligned} A_4 = W^2 &= (0.375\ \text{in})^2 \\ &= 0.140\ \text{in}^2 \end{aligned}$$

The total reinforcement available is

$$\begin{aligned} A_1 + A_2 + A_3 + A_4 &= 0.636\ \text{in}^2 + 1.451\ \text{in}^2 \\ &\quad + 1.668\ \text{in}^2 + 0.140\ \text{in}^2 \\ &= 3.895\ \text{in}^2 \end{aligned}$$

Since $A_{\text{available}} > A_r$, no reinforcement is needed.

29. FLANGED JOINTS

Flanged joints (with flat cover plates) are needed to disassemble, inspect, and clean pressure vessels. Joints may be bolted or boltless pressure-actuated. *Bolted joints*, where sealing gaskets are compressed by bolt forces, are more common. In boltless joints (which may be of the *axially locked joint* and *pressure-actuated joint* varieties), the internal pressure compresses and seals the gasket. Because of the relative size advantages, a boltless joint may be superior at pressures over 2000 psig (14 MPa) and when the shell diameter is roughly 20 in (510 mm) or when the flange thickness exceeds 1½ in (38 mm).

The three main types of bolted flanges are the ring flange, the tapered hub (also known as *welding neck*) flange, and the lap-joint flange, shown in Fig. 54.11. The *lap-joint flange* is used for low-pressure, low-cost pressure vessels. Joints may be hubbed or hubless. Advantages of this flange type are low cost and ease of bolt hole alignment. The backing ring can be constructed of a different material from the shell and lap ring, an important consideration when expensive alloys are used.

The *ring flange* is suitable for low and moderate pressure. It consists of an annular plate welded to the end of a cylindrical nozzle (shell). Bolts are spaced equidistantly around the bolt circle. The number of bolts is commonly a multiple of four. An unconfined gasket is used between the annular plate and the closure plate. The gasket usually extends to the inner edge of the bolt line so that the bolts can help center the gasket. A full-face gasket may cover the entire flange area and extend beyond the bolt circle, but is typically used for pressure less than 100 psig (700 kPa).[38]

[38]In full-face gaskets, the material outside of the bolt ring is not effective in sealing.

Figure 54.11 Types of Flanges

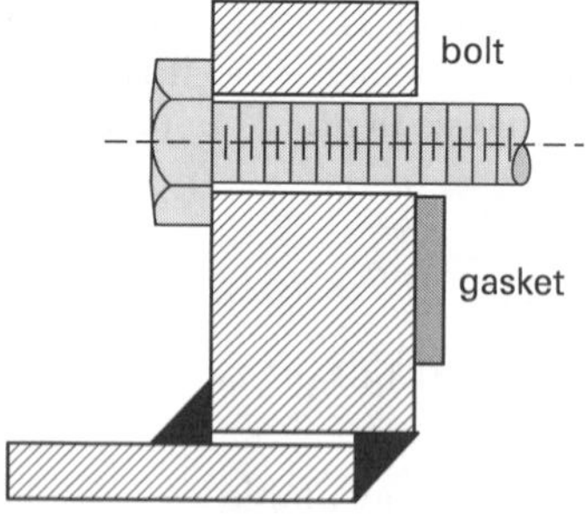

(a) ring flange with flat face

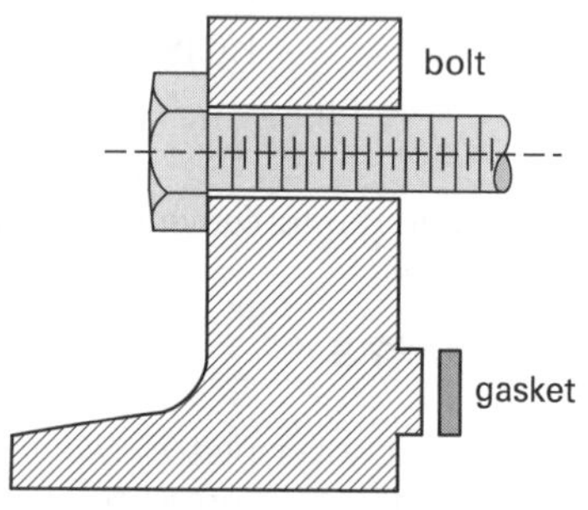

(b) welding neck flange with tongue facing

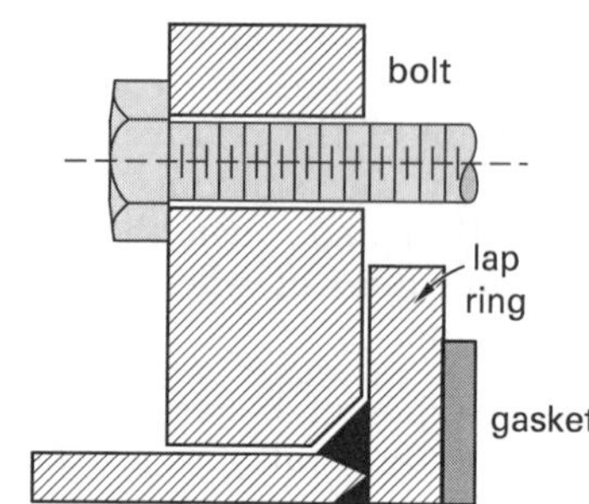

(c) lap joint with raised-face flange

For reliable and safe operation up to a pressure of approximately 5000 psig (35 MPa), the *tapered-hub flange* can be used. This flange is (roughly) L-shaped and is butt-welded to the shell opening.

Depending on the design, the mating area of the flange surfaces and/or cover plate may or may not compress the gasket. An unconfined and prestressed gasket is commonly used with flat-faced ring flanges. Such a gasket can expand inward and outward when tightened. Since the gasket is unconfined, there is no protection against *gasket blowout.*

Semiconfined gaskets are confined in single-step male-female types of joints. Gaskets are completely confined in tongue-and-groove, ring, double-step male-female joints. Fully confined gaskets are appropriately chosen when there are significant fluctuations in pressure and temperature.

The major concern in regard to flange and gasket choice is flange leakage. Gaskets chosen for operation under internal pressure are much less effective under vacuum. Sheet gaskets may be "sucked in," though this is countered by specifying a spiral-wound gasket.

30. BOLT STRENGTHS

Bolt materials and strength parameters (yield and ultimate) are specified in BPVC Sec. VIII, Div. 1, Annex 3. A as functions of operating temperature. The allowable tensile load per bolt is the product of the allowable temperature stress and the bolt area. Table 54.13 lists approximate maximum tensile forces for standard- and 8-thread series bolts at room temperature.

31. SEAMLESS PIPE VESSELS AND NECKS

The required thickness and maximum allowable pressure of pipes under internal pressure are given by Eq. 54.19 and Eq. 54.20, which are consistent with the cylindrical shell equations in Table 54.9. The allowable stress, S, for common pipe materials A53B and A106B steel pipe is 17,100 lbf/in^2 (118 MPa) within the temperature range of −20°F to 650°F (−29°C to 343°C).[39] The joint efficiency, e, of seamless pipe is 1.0. The diameter, D_i, and radius, r_i, are inside values.

$$t = \frac{p_i r_i}{Se - 0.6p_i} + c \quad \left[\begin{array}{c}\text{when calculating required}\\ \text{thickness for given } p_i\end{array}\right] \qquad 54.19$$

$$p_i = \frac{Se(t-c)}{r_i + 0.6(t-c)} \quad \left[\begin{array}{c}\text{when calculating allowable}\\ \text{pressure for given } t\end{array}\right] \qquad 54.20$$

When selecting pipe using nominal dimensions, a manufacturing tolerance of ±12.5% must be considered. That is, it must be assumed that the pipe selected will have an actual wall thickness of 0.875 times the nominal thickness. This reduction is in addition to the corrosion allowance. Additional thickness is required if the pipe is to be threaded.

Example 54.3

An 8 in (nominal) diameter, schedule-80 seamless pipe with welded flanges is constructed of A106B steel. The pipe carries 700°F steam. The corrosion allowance is 0.050 in. What is the maximum allowable pressure for this configuration?

[39]A53B pipe may not be used above 900°F (482°C). A106B pipe may not be used above 1000°F (538°C). Both A53B and A106B pipe are subject to graphitization above 800°F (427°C).

Solution

For a schedule-80, 8 in pipe, the wall thickness is 0.500 in and the outside diameter is 8.625 in.

Taking the manufacturing tolerance of ±12.5% into account, the minimum wall thickness is

$$t = (0.875)(0.500\ \text{in}) = 0.4375\ \text{in}$$

Subtracting the corrosion allowance, the design wall thickness is

$$t - c = 0.4375\ \text{in} - 0.05\ \text{in} = 0.3875\ \text{in}$$

The inside radius is

$$\begin{aligned} r_i &= \frac{D_i}{2} = \frac{D_o - 2(t-c)}{2} \\ &= \frac{8.625\ \text{in} - (2)(0.3875\ \text{in})}{2} \\ &= 3.925\ \text{in} \end{aligned}$$

From Table 54.3 for 700°F operation, the maximum allowable stress is 15,600 lbf/in^2.

From Eq. 54.20, the maximum allowable pressure is

$$\begin{aligned} p_i &= \frac{Se(t-c)}{r_i + 0.6(t-c)} = \\ &= \frac{\left(15{,}600\ \frac{\text{lbf}}{\text{in}^2}\right)(1.00)(0.3875\ \text{in})}{3.925\ \text{in} + (0.6)(0.3875\ \text{in})} \\ &= 1454\ \text{lbf/in}^2 \end{aligned}$$

32. PRESSURE TESTING

Pressure vessels under internal pressure are normally tested hydrostatically with water.[40] However, vessels that cannot safely be filled with water, that cannot be dried, or that cannot tolerate traces of the test liquid can be tested pneumatically with air.

Most testing is not carried out at the operating temperature. Since material strengths decrease at higher temperatures, the test pressure is increased according to the ratio of the allowable stress at the test temperature to the allowable stress at the design temperature.

According to the ASME Code Sec. VIII, Div. 1, the hydrostatic test pressure is 130% of the MAWP multiplied by the ratio of the allowable stress at the test temperature to the allowable stress at the design temperature. When hydrotesting, it is recommended that the metal temperature is at least 30°F (17°C) above the MDMT, but not greater than 120°F (48°C), to minimize the risk of brittle fracture. For pneumatic tests, the test pressure is 110% of the MAWP multiplied by the ratio of the allowable stress at the test temperature to the allowable stress at the design temperature. The metal temperature during a pneumatic test must be at least 30°F (17°C) above the MDMT to minimize the risk of brittle fracture.

For cast-iron pressure vessels, the test pressure is 200% of the MAWP unless the design working pressure is less than 30 psig (207 kPa), in which case, the test pressure is 60 psig (414 kPa) or 250% of the design working pressure, whichever is less. A corrosion allowance is included when calculating all test pressures.

Following the application of hydrostatic and pneumatic pressures, all joints and connections must be visually inspected. Leakage is not allowed, except for openings intended for welded connections and at temporary test closures. Additionally, for pneumatically tested vessels, the full length of all welds around openings and attachment welds having a throat greater than 1/4 in (6 mm) must be examined.

Special rules apply to pressure vessels whose operating pressure is limited by flange strength or that have multiple chambers, are subject to external pressure, or operate at below-atmospheric pressures.

33. PRESSURE RELIEF DEVICES

BPVC Sec. VIII, Div. 1, UG-125 through UG-140 require all pressure vessels to be equipped with *overpressure protection devices* such as *rupture disks*, RDs (*burst disks*), *pressure relief valves* (*safety relief valves*), PRVs, or combinations thereof to prevent catastrophic failure during abnormal conditions. Generally, a PRV is a normally closed, spring-actuated device that automatically opens to relieve pressure. When the overpressure situation abates, the PRV closes, preventing further loss of contents. Two PRVs can be mounted on a *three-way valve* fitting such that either of the PRVs can be removed for maintenance. Only one PRV should be active at any given moment, however. The three-way valve should normally be back-seated to reduce the possibility of leakage through the stem packing.

The set-point pressure, opening pressure, tolerance, overpressure, and accumulation pressure are related concepts. The marked *set-point pressure* (*set pressure*, *setting pressure*) is the value of increasing pressure at which a pressure relief device is intended to (begin to) open. Per UG-134(a), the protection device set point pressure must be at or below the MAWP. The actual *opening pressure* (*popping pressure*, *start-to-leak pressure*, *burst pressure*, or *breaking pressure*) may be slightly different from the set-point pressure due to intrinsic manufacturing batch *tolerances*. *Overpressure* is the pressure above the set pressure, expressed either as an absolute value or as a percentage of the set pressure. "Overpressure" is associated with the device. *Accumulation* is

[40]Even pressure vessels that are normally under internal pressure may sometimes draw a vacuum, as during a steamout. Other less predictable instances of vacuum failures occur when the contents of a vessel are being drained while the vent line is closed or blocked, or when a filter element is clogged or under-sized.

Table 54.13 Approximate Maximum Allowable Tensile Force per Bolt (room temperature)

			allowable material stress (lbf/in^2) (See Table 54.3 for specific materials.)			
bolt diameter	number of threads	area at bottom of thread	7000[a] carbon steel	16,250 alloy steel	18,750 alloy steel	20,000 alloy steel
(in)	per inch	(in^2)	tensile load per bolt[b] (lbf)			
standard (UNC) thread[c]						
1/2	13	0.126	882	2047	2362	2520
5/8	11	0.202	1414	3282	3787	4040
3/4	10	0.302	2114	4907	5662	6040
7/8	9	0.419	2933	6808	7856	8380
1	8	0.551	3857	8953	10,331	11,020
1 1/8	7	0.693	4851	11,261	12,993	13,860
1 1/4	7	0.890	6230	14,462	16,687	17,800
1 3/8	6	1.054	7378	17,127	19,762	21,080
1 1/2	6	1.294	9058	21,027	24,262	25,880
1 5/8	5 1/2	1.515	10,605	24,618	28,406	30,300
1 3/4	5	1.744	12,208	28,340	32,700	34,880
1 7/8	5	2.049	14,343	33,296	38,418	40,980
2	4 1/2	2.300	16,100	37,375	43,125	46,000
2 1/4	4 1/2	3.020	21,140	49,075	56,625	60,400
2 1/2	4	3.715	26,005	60,368	69,656	74,300
2 3/4	4	4.618	32,326	75,042	86,587	92,360
3	4	5.620	39,340	91,325	105,375	112,400
8-thread (8 UN) series						
1 1/8	8	0.728	5096	11,830	13,650	14,560
1 1/4	8	0.929	6503	15,096	17,418	18,580
1 3/8	8	1.155	8085	18,768	21,656	23,100
1 1/2	8	1.405	9835	22,831	26,343	28,100
1 5/8	8	1.680	11,760	27,300	31,500	33,600
1 3/4	8	1.980	13,860	32,175	37,125	39,600
1 7/8	8	2.304	16,128	37,440	43,200	46,080
2	8	2.652	18,564	43,095	49,725	53,040
2 1/4	8	3.423	23,961	55,623	64,181	68,460
2 1/2	8	4.292	30,044	69,745	80,475	85,840
2 3/4	8	5.259	36,813	85,458	98,606	105,180
3	8	6.324	44,268	102,765	118,575	126,480

(Multiply in by 25.4 to obtain mm.)
(Multiply in^2 by 645.16 to obtain mm^2.)
(Multiply lbf/in^2 by 6.895 to obtain kPa.)
(Multiply lbf by 4.448 to obtain N.)

[a]Maximum temperature, 450°F.
[b]Calculated as the product of allowable stress of specific materials.
[c]Typically used up to 1 in diameter, after which 8 UN bolts are used.

Reproduced from R. Chuse and S. Eber, *Pressure Vessels*, sixth ed., copyright © 1984, with permission of the publisher, McGraw-Hill.

pressure above the maximum MAWP of the vessel, expressed either as an absolute value or a percentage of the MAWP. "Accumulation" is associated with the vessel. (See Fig. 54.12.)

Figure 54.12 *Overpressure and Accumulation*

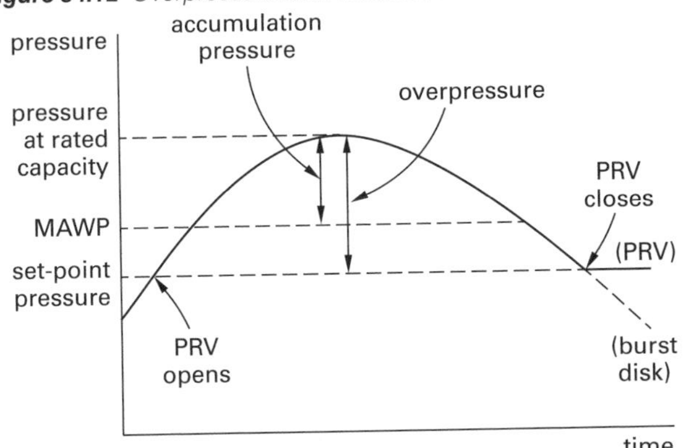

While the pressure set point must not exceed the MAWP, the maximum allowed accumulation pressure is 3 psi (20 kPa) or 110% of MAWP, whichever is greater. When multiple pressure relief devices are used, the maximum accumulation pressure is 4 psi (30 kPa) or 116% of MAWP. In a fire contingency, the maximum accumulation pressure is 21% (i.e., the vessel pressure is permitted to reach 121% of MAWP) [UG-125(c)].

Manufacturing tolerances are not generally considered when choosing devices based on set pressures. However, the BPVC requires that the tolerance of a PRV cannot exceed ±2 psi (±15 kPa) for pressures up to and including 70 psi (500 kPa), and ±3% for pressures above 70 psi (500 kPa) [UG-126(d)]. Similarly, the burst pressure tolerance of an RD cannot exceed ±2 psi (±15 kPa) for marked burst pressures up to and including 40 psi (300 kPa), and ±3% for burst pressures above 40 psi (300 kPa) [UG-127(a)(1)].

When used in a RD/PRV combination, the RD should be installed in the relief line closest to the pressure vessel so that it can protect the PRV from corrosive pressure vessel contents and prevent fugitive emissions. The RD and PRV should essentially have the same nominal set pressures, disregarding tolerances [UG-127 ftn. 52]. The marked burst pressure must be between 90% and 100% of the valve's set pressure [UG-132(a)(4)(a)]. A pressure gauge or other sensing device (try cock, free vent, telltale indicator, or other sensor) must be placed between the RD and the PRV to continuously indicate the status of the RD.

The *relieving capacity* of a commercial pressure relief device will be provided by the manufacturer, commonly determined at the accumulation pressure, not the set pressure. Therefore, capacities are rated at 110% of the set pressure. Since the RD in an RD-PRV assembly will open at 90% of the PRV's set pressure, the rated capacity of the PRV should be reduced by 10% [UG-127(a)(3)(b)(2)]. Capacities of nonreclosing pressure relief devices can be calculated from theoretical formulas appropriate for the contents, using a discharge coefficient of 0.62.

Flow resistance of a rupture disk device is usually specified by the manufacturer as a multiple of the velocity head, K_R. The flow resistance of an entire pressure relief system can be determined from traditional fluid flow/piping methods. The total system flow resistance should consider the rupture disk, piping, valves and fittings, and exit losses. Only 90% of the calculated capacity should be used [UG-127(a)(2)(b)].

34. NOMENCLATURE

A	area	in^2	mm^2
c	corrosion allowance	in	mm
C	flange attachment factor	–	–
D	insert diameter	in	mm
e	weld joint efficiency	–	–
f_r	allowable stress ratio	–	–
F	correction factor	–	–
F_p	peak stress	lbf/in^2	Pa
F_y	yield strength	lbf/in^2	Pa
h	depth of head	in	mm
h_G	gasket moment arm	in	mm
K	factor	–	–
L	crown radius	in	mm
M	factor	–	–
p	pressure[a,b]	lbf/in^2	Pa
P	primary stress	lbf/in^2	Pa
Q	secondary stress	lbf/in^2	Pa
Q_m^*	secondary membrane stress from relenting, self-limiting loads	lbf/in^2	Pa
Q_{ms}^*	secondary membrane stress from sustained loads	lbf/in^2	Pa
r	radius	in	mm
S	stress	lbf/in^2	Pa
S_n	allowable fatigue stress for n cycles	lbf/in^2	Pa
t	thickness	in	mm
UTS	ultimate tensile stress	lbf/in^2	Pa
W	total flange bolt load	lbf	N
W	weld size	in	mm

[a]The variable for pressure in the *ASME Boiler and Pressure Vessel Code* is the uppercase P. A lowercase p is used in this chapter for consistency with the rest of this book.
[b]All pressures expressed in this chapter are gage pressures.

Symbols

α	one-sided taper angle (half of apex angle)	deg	deg
σ	stress	lbf/ft^2	Pa

Subscripts

a	allowable or axial
b	bending
c	corroded
h	head
i	inside
k	knuckle
L	local
m	mechanical
n	nozzle
o	outside
p	pad
r	radial, ratio, or required
t	tangential

Topic X: Dynamics and Vibrations

Chapter

55 Properties of Solid Bodies

Content in blue refers to the *NCEES Handbook*.

1. SOLID BODIES

Understanding the properties of solid bodies is an integral part of many engineering applications. It is important to determine the behavior of solid bodies when subjected to various forces and types of loading. General equations for determining essential properties (center of gravity, centroid, radius of gyration, moment of inertia, and product of inertia) of various solid bodies are presented in Sec. 55.2 through Sec. 55.8. Appendix 55.A provides equations for determining the mass moment of inertia for many common solid bodies, and App. 7.B provides general volume and area equations for several common shapes. [**Properties of Various Solids**] [**Mass and Mass Moments of Inertia of Geometric Shapes**]

2. CENTER OF GRAVITY

A solid body will have both a center of gravity and a centroid, but the locations of these two points will not necessarily coincide. The earth's attractive force, called *weight*, can be assumed to act through the *center of gravity* (also known as the *center of mass*). Only when the body is homogeneous will the *centroid of the volume* coincide with the center of gravity.[1,2]

For simple objects and regular polyhedrons, the location of the center of gravity can be determined by inspection. It will always be located on an axis of symmetry. The location of the center of gravity can also be determined mathematically if the object can be described mathematically.

$$x_c = \frac{\int x \, dm}{m} \qquad 55.1$$

$$y_c = \frac{\int y \, dm}{m} \qquad 55.2$$

$$z_c = \frac{\int z \, dm}{m} \qquad 55.3$$

If the object can be divided into several smaller constituent objects, the location of the composite center of gravity can be calculated from the centers of gravity of each of the constituent objects.

$$x_c = \frac{\sum m_i x_{ci}}{\sum m_i} \qquad 55.4$$

$$y_c = \frac{\sum m_i y_{ci}}{\sum m_i} \qquad 55.5$$

$$z_c = \frac{\sum m_i z_{ci}}{\sum m_i} \qquad 55.6$$

3. MASS AND WEIGHT

The *mass*, m, of a homogeneous solid object is calculated from its mass density and volume. Mass is independent of the strength of the gravitational field.

$$m = \rho V \qquad 55.7$$

The *weight*, w, of an object depends on the strength of the gravitational field, g.

$$w = mg \qquad \text{[SI]} \qquad 55.8(a)$$

$$w = \frac{mg}{g_c} \qquad \text{[U.S.]} \qquad 55.8(b)$$

[1]The study of nonhomogeneous bodies is beyond the scope of this book. Homogeneity is assumed for all solid objects.

[2]Section 49.1provides the equations for determining the centroid.

4. INERTIA

For solid bodies, inertia (the *inertial force* or *inertia vector*), $m\mathbf{a}$, is the resistance the object offers to attempt to accelerate it (i.e., change its velocity) in a linear direction. Although the mass, m, is a scalar quantity, the acceleration, $\mathbf{a}$, is a vector.[3,4]

5. MASS MOMENT OF INERTIA

The *mass moment of inertia* measures a solid object's resistance to changes in rotational speed about a specific axis. I_x, I_y, and I_z are the mass moments of inertia with respect to the x-, y-, and z-axes. They are not components of a resultant value.[5]

The *centroidal mass moment of inertia*, I_c, is obtained when the origin of the axes coincides with the object's center of gravity. Although it can be found mathematically from Eq. 55.9 through Eq. 55.11, it is easier to use a table of properties of solid bodies, such as App. 55.A, for simple objects. [**Mass and Mass Moments of Inertia of Geometric Shapes**]

Mass Moment of Inertia

$$I_x = \int (y^2 + z^2)\, dm \qquad 55.9$$

$$I_y = \int (x^2 + z^2)\, dm \qquad 55.10$$

$$I_z = \int (x^2 + y^2)\, dm \qquad 55.11$$

6. PARALLEL AXIS THEOREM

Once the centroidal mass moment of inertia is known, the *parallel axis theorem* is used to find the mass moment of inertia about any parallel axis.[6]

Parallel-Axis Theorem

$$I_{\text{new}} = I_c + md^2 \qquad 55.12$$

For a composite object, the parallel axis theorem must be applied for each of the constituent objects.

$$I = I_{c,1} + m_1 d_1^2 + I_{c,2} + m_2 d_2^2 + L \qquad 55.13$$

7. RADIUS OF GYRATION

The *mass radius of gyration*, also called the *radius of gyration* of a solid object, r_m, represents the distance from the rotational axis at which the object's entire mass could be located without changing the mass moment of inertia.[7,8]

Mass Radius of Gyration

$$r_m = \sqrt{\frac{I}{m}} \qquad 55.14$$

$$I = r_m^2 m \qquad 55.15$$

Example 55.1

A solid cylindrical rod is 4 ft (4 m) long and has a diameter of 0.1 ft (0.1 m) and a density of 400 lbm/ft^3 (7000 kg/m^3). With respect to a centroidal axis perpendicular to the rod's length, what are (a) the mass moment of inertia (b) radius of gyration?

SI Solution

(a) The cross-sectional area of the rod is

$$A = \frac{\pi D^2}{4} = \frac{\pi (0.1\ \text{m})^2}{4} = 0.007\,853\ \text{m}^2$$

The mass of the rod is

Mass and Mass Moments of Inertia of Geometric Shapes

$$\begin{aligned} m &= \rho L A \\ &= \left(7000\ \frac{\text{kg}}{\text{m}^3}\right)(4\ \text{m})(0.007\,853\ \text{m}^2) \\ &= 219.9\ \text{kg} \end{aligned}$$

Use the formula for the mass moment of inertia of a cylindrical rod.

Mass and Mass Moments of Inertia of Geometric Shapes

$$\begin{aligned} I_{y,c} &= \frac{mL^2}{12} \\ &= \frac{(219.9\ \text{kg})(4\ \text{m})^2}{12} \\ &= 293.2\ \text{kg·m}^2 \end{aligned}$$

---

[3] Section 49.4 details the moment of inertia of an area.
[4] Section 49.6 details the polar moment of inertia of an area.
[5] At first, it may be confusing to use the same symbol, I, for area and mass moments of inertia. However, the problem types are distinctly dissimilar, and both moments of inertia are seldom used simultaneously.
[6] Section 49.5 provides the parallel axis theorem for area.
[7] The symbol for radius of gyration can be either k or r. k is preferred when the radius of gyration is that of a solid body. r is the more common symbol for the radius of gyration of an area (not of a solid body), although that radius of gyration cannot be determined from Eq. 55.14. The symbol r can also be used (e.g., shaft radii) with shapes, objects, and mechanisms that are inherently noncircular, such as I-beam cross sections and pendulums, that do not have a circular (radial) dimension.
[8] Section 49.7 details the radius of gyration, r, for areas.

(b) Use Eq. 55.14 to find the mass radius of gyration.

Mass Radius of Gyration

$$r_m = \sqrt{\frac{I}{m}} = \sqrt{\frac{293.2\ \text{kg·m}^2}{219.9\ \text{kg}}} = 1.155\ \text{m}$$

Alternatively, use the formula for the mass radius of gyration of a cylindrical rod.

Mass and Mass Moments of Inertia of Geometric Shapes

$$r_{y,c}^2 = \frac{L^2}{12}$$

$$r_{y,c} = \sqrt{\frac{L^2}{12}} = \sqrt{\frac{(4\ \text{m})^2}{12}} = 1.155\ \text{m}$$

U.S. Customary Solution

(a) The cross-sectional area of the rod is

$$A = \frac{\pi D^2}{4} = \frac{\pi (0.1\ \text{ft})^2}{4} = 0.007853\ \text{ft}^2$$

The mass of the rod is

Mass and Mass Moments of Inertia of Geometric Shapes

$$m = \rho LA = \left(400\ \frac{\text{lbm}}{\text{ft}^3}\right)(4\ \text{ft})(0.007853\ \text{ft}^2) = 12.565\ \text{lbm}$$

Use the formula for the mass moment of inertia of a cylindrical rod.

Mass and Mass Moments of Inertia of Geometric Shapes

$$I_{y,c} = \frac{mL^2}{12} = \frac{(12.565\ \text{lbm})(4\ \text{ft})^2}{12} = 16.753\ \text{lbm-ft}^2$$

(b) Use Eq. 55.14 to find the mass radius of gyration.

Mass Radius of Gyration

$$r_m = \sqrt{\frac{I}{m}} = \sqrt{\frac{16.753\ \text{lbm-ft}^2}{12.565\ \text{lbm}}} = 1.155\ \text{ft}$$

Alternatively, use the formula for the mass radius of gyration of a cylindrical rod.

Mass and Mass Moments of Inertia of Geometric Shapes

$$r_{y,c}^2 = \frac{L^2}{12}$$

$$r_{y,c} = \sqrt{\frac{L^2}{12}} = \sqrt{\frac{(4\ \text{ft})^2}{12}} = 1.155\ \text{ft}$$

8. PRINCIPAL AXES

An object's mass moment of inertia depends on the orientation of axes chosen. The *principal axes* are the axes for which the *products of inertia* are zero.[9] Equipment rotating about a principal axis will draw minimum power during speed changes. [**Product of Inertia**]

Finding the principal axes through calculation is too difficult and time consuming to be used with most rotating equipment. Furthermore, the rotating axis is generally fixed. *Balancing operations* are used to change the distribution of mass about the rotational axis. A device, such as a rotating shaft, flywheel, or crank, is said to be *statically balanced* if its center of mass lies on the axis of rotation. It is said to be *dynamically balanced* if the center of mass lies on the axis of rotation and the products of inertia are zero.

9. NOMENCLATURE

a	acceleration	ft/sec^2	m/s^2
A	area	ft^2	m^2
d	distance	ft	m
D	diameter	ft	m
g	gravitational acceleration, 32.2 (9.81)	ft/sec^2	m/s^2
g_c	gravitational constant, 32.2	lbm-ft/lbf-sec^2	n.a.
h	height	ft	m
I	mass moment of inertia	lbm-ft^2	kg·m^2
L	length	ft	m
m	mass	lbm	kg
r	radius	ft	m
r	radius of gyration	ft	m
r_p	radius of gyration with respect to the polar moment of inertia	ft	m
r_x	radius of gyration with respect to the x-axis	ft	m
r_y	radius of gyration with respect to the y-axis	ft	m
V	volume	ft^3	m^3
w	weight	lbf	N

[9]Section 49.8 defines the product of inertia used to assess a system's asymmetry.

Symbols

ρ	density	lbm/ft^3	kg/m^3

Subscripts

c	centroidal
i	inner
m	mass
o	outer

56 Kinematics

Content in blue refers to the *NCEES Handbook*.

NCEES EXAM SPECIFICATIONS AND RELATED CONTENT

MACHINE DESIGN AND MATERIALS EXAM

I.B.2. Engineering Science and Mechanics: Kinematics

- 8. Uniform Acceleration
- 10. Projectile Motion
- 11. Rotational Particle Motion
- 12. Relationship Between Linear and Rotational Variables
- 13. Normal Acceleration
- 16. Relative Motion

1. INTRODUCTION TO KINEMATICS

Dynamics is the study of moving objects. The subject is divided into kinematics and kinetics. *Kinematics* is the study of a body's motion independent of the forces on the body. It is a study of the geometry of motion without consideration of the causes of motion. Kinematics deals only with relationships among position, velocity, acceleration, and time.

2. PARTICLES AND RIGID BODIES

Bodies in motion can be considered *particles* if rotation is absent or insignificant. Particles do not possess rotational kinetic energy. All parts of a particle have the same instantaneous displacement, velocity, and acceleration.

A *rigid body* does not deform when loaded and can be considered a combination of two or more particles that remain at a fixed, finite distance from each other. At any given instant, the parts (particles) of a rigid body can have different displacements, velocities, and accelerations.

3. COORDINATE SYSTEMS

The position of a particle is specified with reference to a *coordinate system*. The description takes the form of an ordered sequence (q_1, q_2, q_3, ...) of numbers called *coordinates*. A coordinate can represent a position along an axis, as in the rectangular coordinate system, or it can represent an angle, as in the polar, cylindrical, and spherical coordinate systems.

In general, the number of *degrees of freedom* is equal to the number of coordinates required to completely specify the state of an object. If each of the coordinates is independent of the others, the coordinates are known as *holonomic coordinates*.

The state of a particle is completely determined by the particle's location. In three-dimensional space, the locations of particles in a system of m particles must be specified by $3m$ coordinates. However, the number of required coordinates can be reduced in certain cases. The position of each particle constrained to motion on a surface (i.e., on a two-dimensional system) can be specified by only two coordinates. A particle constrained to moving on a curved path requires only one coordinate.[1]

The state of a rigid body is a function of orientation as well as position. Six coordinates are required to specify the state: three for orientation and three for location.

4. CONVENTIONS OF REPRESENTATION

Consider the particle shown in Fig. 56.1. Its position (as well as its velocity and acceleration) can be specified in three primary forms: vector form, rectangular coordinate form, and unit vector form.

Figure 56.1 Position of a Particle

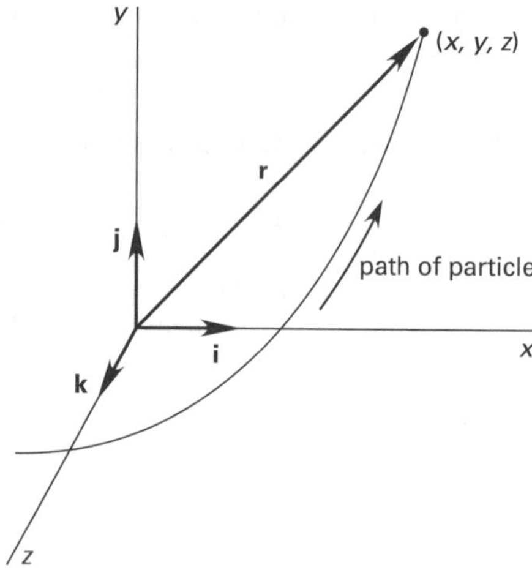

The vector form of the particle's position is **r**, where the vector **r** has both magnitude and direction. The rectangular coordinate form is (x, y, z). The unit vector form is given by Eq. 56.1.

$$\mathbf{r} = x\mathbf{i} + y\mathbf{j} + z\mathbf{k} \qquad 56.1$$

5. LINEAR PARTICLE MOTION

A *linear system* is one in which particles move only in straight lines. (It is also known as a *rectilinear system.*) The relationships among position, velocity, and acceleration for a linear system are given by Eq. 56.2 through Eq. 56.4. When values of t are substituted into these equations, the position, velocity, and acceleration are known as *instantaneous values.*

$$s(t) = \int \mathrm{v}(t)\,dt = \int\left(\int a(t)\,dt\right)dt \qquad 56.2$$

$$\mathrm{v}(t) = \frac{ds(t)}{dt} = \int a(t)\,dt \qquad 56.3$$

$$a(t) = \frac{d\mathrm{v}(t)}{dt} = \frac{d^2s(t)}{dt^2} \qquad 56.4$$

The average velocity and acceleration over a period from t_1 to t_2 are

$$\mathrm{v}_{\text{ave}} = \frac{\int_1^2 \mathrm{v}(t)\,dt}{t_2 - t_1} = \frac{s_2 - s_1}{t_2 - t_1} \qquad 56.5$$

$$a_{\text{ave}} = \frac{\int_1^2 a(t)\,dt}{t_2 - t_1} = \frac{\mathrm{v}_2 - \mathrm{v}_1}{t_2 - t_1} \qquad 56.6$$

Example 56.1

A particle is constrained to move along a straight line. The velocity and location are both zero at $t = 0$. The particle's velocity as a function of time is

$$\mathrm{v}(t) = 8t - 6t^2$$

(a) What are the acceleration and position functions?

(b) What is the instantaneous velocity at $t = 5$?

Solution

(a) From Eq. 56.4,

$$\begin{aligned} a(t) &= \frac{d\mathrm{v}(t)}{dt} = \frac{d(8t - 6t^2)}{dt} \\ &= 8 - 12t \end{aligned}$$

From Eq. 56.2,

$$\begin{aligned} s(t) &= \int \mathrm{v}(t)\,dt = \int (8t - 6t^2)\,dt \\ &= 4t^2 - 2t^3 \text{ when } s(t=0) = 0 \end{aligned}$$

(b) Substituting $t = 5$ into the $\mathrm{v}(t)$ function given in the problem statement,

$$\begin{aligned} \mathrm{v}(5) &= (8)(5) - (6)(5)^2 \\ &= -110 \quad [\text{backward}] \end{aligned}$$

6. DISTANCE AND SPEED

The terms "displacement" and "distance" have different meanings in kinematics. *Displacement* (or *linear displacement*) is the net change in a particle's position as determined from the position function, $s(t)$. *Distance traveled* is the accumulated length of the path traveled during all direction reversals, and it can be found by

[1]The curve can be a straight line, as in the case of a mass hanging on a spring and oscillating up and down. In this case, the coordinate will be a linear coordinate.

adding the path lengths covered during periods in which the velocity sign does not change. Therefore, distance is always greater than or equal to displacement.

$$\text{displacement} = s(t_2) - s(t_1) \qquad 56.7$$

Similarly, "velocity" and "speed" have different meanings: *velocity* is a vector, having both magnitude and direction; *speed* is a scalar quantity, equal to the magnitude of velocity. When specifying speed, direction is not considered.

Example 56.2

What distance is traveled during the period $t = 0$ to $t = 6$ by the particle described in Ex. 56.1?

Solution

Start by determining when, if ever, the velocity becomes negative. (This can be done by inspection, graphically, or algebraically.) Solving for the roots of the velocity equation, $v(t) = 8t - 6t^2$, the velocity changes from positive to negative at

$$t = \frac{4}{3}$$

The initial displacement is zero. From the position function found in Ex. 56.1(a), the position at $t = 4/3$ is

$$s\left(\frac{4}{3}\right) = (4)\left(\frac{4}{3}\right)^2 - (2)\left(\frac{4}{3}\right)^3 = 2.37$$

The displacement while the velocity is positive is

$$\Delta s = s\left(\frac{4}{3}\right) - s(0) = 2.37 - 0 = 2.37$$

The position at $t = 6$ is

$$s(6) = (4)(6)^2 - (2)(6)^3 = -288$$

The displacement while the velocity is negative is

$$\Delta s = s(6) - s\left(\frac{4}{3}\right) = -288 - 2.37 = -290.37$$

The total distance traveled is

$$2.37 + 290.37 = 292.74$$

7. UNIFORM MOTION

The term *uniform motion* means uniform velocity. The velocity is constant and the acceleration is zero. For a constant velocity system, the position function varies linearly with time. (See Fig. 56.2.)

$$s(t) = s_0 + vt \qquad 56.8$$

$$v(t) = v \qquad 56.9$$

$$a(t) = 0 \qquad 56.10$$

Figure 56.2 *Constant Velocity*

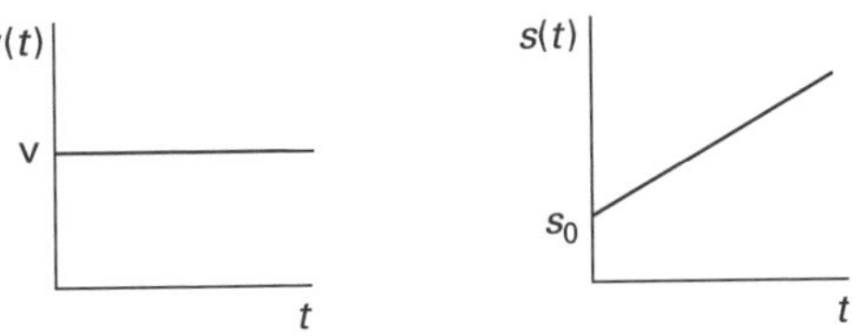

8. UNIFORM ACCELERATION

The acceleration is constant in many cases, as shown in Fig. 56.3. (Gravitational acceleration, where $a = g$, is a notable example.) If the acceleration is constant, the a term can be taken out of the integrals in Eq. 56.2 and Eq. 56.3.

Constant Acceleration

$$a(t) = a_0 \qquad 56.11$$

$$v(t) = a_0(t - t_0) + v_0 \qquad 56.12$$

$$s(t) = \frac{a_0(t - t_0)^2}{2} + v_0(t - t_0) + s_0 \qquad 56.13$$

If the initial time is assumed to be zero, these equations can instead be expressed as Eq. 56.14 through Eq. 56.16.

$$a(t) = a \qquad 56.14$$

$$v(t) = a\int dt = v_0 + at \qquad 56.15$$

$$s(t) = a\iint dt^2 = s_0 + v_0 t + \tfrac{1}{2}at^2 \qquad 56.16$$

Figure 56.3 *Uniform Acceleration*

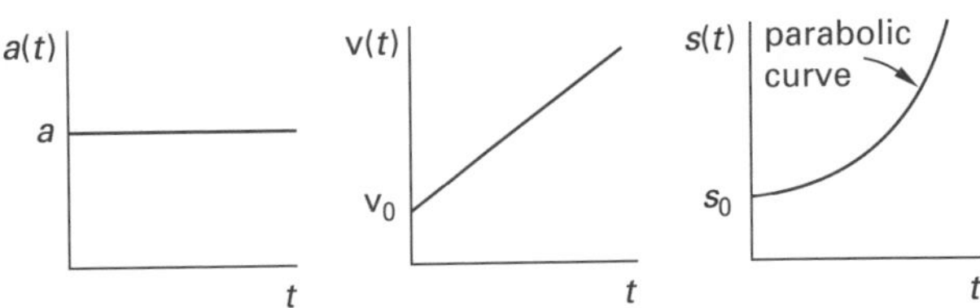

Table 56.1 summarizes the equations required to solve most uniform acceleration problems.

*Table 56.1 Uniform Acceleration Formulas**

to find	given these	use this equation
a	t, v_0, v	$a=\frac{v-v_0}{t}$
a	t, v_0, s	$a=\frac{2s-2v_0t}{t^2}$
a	v_0, v, s	$a=\frac{v^2-v_0^2}{2s}$
s	t, a, v_0	$s=v_0t+\frac{1}{2}at^2$
s	a, v_0, v	$s=\frac{v^2-v_0^2}{2a}$
s	t, v_0, v	$s=\frac{1}{2}t(v_0+v)$
t	a, v_0, v	$t=\frac{v-v_0}{a}$
t	a, v_0, s	$t=\frac{\sqrt{v_0^2+2as}-v_0}{a}$
t	v_0, v, s	$t=\frac{2s}{v_0+v}$
v_0	t, a, v	$v_0=v-at$
v_0	t, a, s	$v_0=\frac{s}{t}-\frac{1}{2}at$
v_0	a, v, s	$v_0=\sqrt{v^2-2as}$
v	t, a, v_0	$v=v_0+at$
v	a, v_0, s	$v=\sqrt{v_0^2+2as}$

*The table can be used for rotational problems by substituting α, ω, and θ for a, v, and s, respectively.

Example 56.3

A locomotive traveling at 80 km/h locks its wheels and skids 95 m before coming to a complete stop. If the deceleration is constant, how many seconds will it take for the locomotive to come to a standstill?

Solution

First, convert the 80 km/h to meters per second.

$$v_0=\frac{\left(80\ \frac{\text{km}}{\text{h}}\right)\left(1000\ \frac{\text{m}}{\text{km}}\right)}{3600\ \frac{\text{s}}{\text{h}}}=22.22\ \text{m/s}$$

In this problem, t is the unknown. Using Eq. 56.12 and solving for the acceleration,

Constant Acceleration

$$v(t)=a_0(t-t_0)+v_0=at+v_0$$

$$a_0=\frac{(v(t)-v_0)}{t}=\frac{-v_0}{t}$$

Substituting a_0 into the position equation,

$$\begin{aligned} s(t) &= \frac{a_0(t-t_0)^2}{2}+v_0(t-t_0)+s_0 \\ &= \frac{\frac{-v_0}{t}(t-0)^2}{2}+v_0(t-0)+0 \\ &= \frac{-v_0t}{2}+v_0t \\ s(t) &= \frac{v_0t}{2} \end{aligned}$$

Solving for t,

$$t=\frac{2s(t)}{v_0+v}=\frac{(2)(95\ \text{m})}{22.22\ \frac{\text{m}}{\text{s}}+0\ \frac{\text{m}}{\text{s}}}$$

$$=8.55\ \text{s}$$

9. LINEAR ACCELERATION

Linear acceleration means that the acceleration increases uniformly with time. Figure 56.4 shows how the velocity and position vary with time.[2]

Figure 56.4 Linear Acceleration

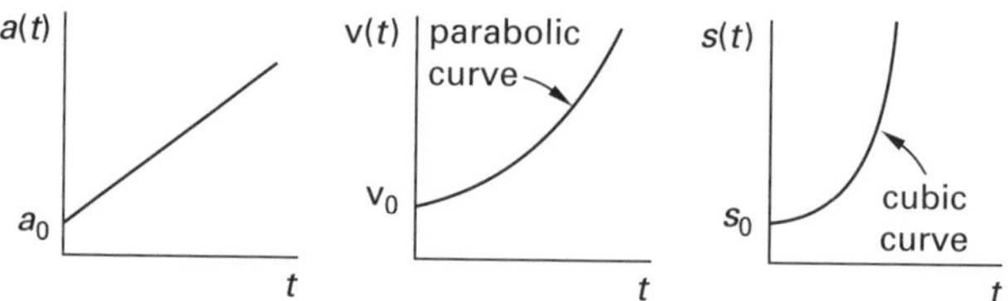

10. PROJECTILE MOTION

A *projectile* is placed into motion by an initial impulse. (Kinematics deals only with dynamics during the flight. Projectile motion is a special case of motion under constant acceleration. The force acting on the projectile during the launch phase is covered in kinetics.) Neglecting air drag, once the projectile is in motion, it is acted upon only by the downward gravitational acceleration (i.e., its own weight).

[2]Because of the successive integrations, if the acceleration function is a polynomial of degree n, the velocity function will be a polynomial of degree $n+1$. Similarly, the position function will be a polynomial of degree $n+2$.

Consider a general projectile set into motion at an angle of θ (from the horizontal plane) and initial velocity v_0. Its range is R, the maximum altitude attained is H, and the total flight time is T. In the absence of air drag, the following rules apply to the case of a level target.[3]

- The trajectory is parabolic.
- The impact velocity is equal to the initial velocity, v_0.
- The impact angle is equal to the initial launch angle, θ.
- The range is maximum when $\theta = 45°$.
- The time for the projectile to travel from the launch point to the apex is equal to the time to travel from apex to impact point.
- The time for the projectile to travel from the apex of its flight path to impact is the same time an initially stationary object would take to fall a distance H.

The equations for projectile motion can be obtained from the equations for constant acceleration.

Projectile Motion

$$a_x = 0 \quad 56.17$$

$$v_x = v_0 \cos\theta \quad 56.18$$

$$x = (v_0 \cos\theta)t + x_0 \quad 56.19$$

$$a_y = -g \quad 56.20$$

$$v_y = -gt + v_0 \sin\theta \quad 56.21$$

$$y = -\frac{gt^2}{2} + (v_0 \sin\theta)t + y_0 \quad 56.22$$

$$v_y^2 = (v_{y,0})^2 - 2g(y - y_0) \quad 56.23$$

Table 56.2 contains more solutions to common projectile problems. These equations are derived from the laws of uniform acceleration and conservation of energy.

Example 56.4

A projectile is launched at 600 ft/sec (180 m/s) with a 30° inclination from the horizontal. The launch point is on a plateau 500 ft (150 m) above the plane of impact. Neglecting friction, find the maximum altitude above the plane of impact, the total flight time, and the range.

SI Solution

The maximum altitude above the impact plane includes the height of the plateau and the elevation achieved by the projectile.

First find the initial velocity in the y-direction at $t = 0$.

Projectile Motion

$$v_y = -gt + v_0 \sin\theta = \left(180\ \frac{\text{m}}{\text{s}}\right)\sin 30° = 90\ \text{m/s}$$

The max height the projectile reaches from above the plane of impact occurs when $v_y = 0$. Rearrange and solve for y to obtain the maximum altitude above the impact plane.

Projectile Motion

$$\begin{aligned} v_y^2 &= (v_{y,0})^2 - 2g(y - y_0) \\ y &= \frac{v_y^2 - (v_{y,0})^2}{-2g} + y_0 \\ &= \frac{0 - \left(90\ \frac{\text{m}}{\text{s}}\right)^2}{(-2)\left(9.81\ \frac{\text{m}}{\text{s}^2}\right)} + 150\ \text{m} \\ &= 562.8\ \text{m} \end{aligned}$$

The total flight time includes the time to reach the maximum altitude and the time to fall from the maximum altitude to the impact plane below.

To determine the time to rise to the apex, let, $v_y = 0$ and rearrange the equation for v_y to solve for t.

Projectile Motion

$$\begin{aligned} v_y &= -gt_{\text{rise}} + v_0 \sin\theta \\ t_{\text{rise}} &= \frac{v_0 \sin\theta - v_y}{g} \\ &= \frac{90\ \frac{\text{m}}{\text{s}} - 0}{9.81\ \frac{\text{m}}{\text{s}^2}} \\ &= 9.174\ \text{s} \end{aligned}$$

To determine the time to fall from the apex, rearrange the equation for y.

Projectile Motion

$$\begin{aligned} y &= -\frac{gt_{\text{fall}}^2}{2} + (v_0 \sin\theta)t_{\text{fall}} + y_0 = \frac{-gt_{\text{fall}}^2}{2} + y_0 \\ t_{\text{fall}} &= \sqrt{\frac{2(y - y_0)}{-g}} \\ &= \sqrt{\frac{(2)(0 - 562.8\ \text{m})}{-9.81\ \frac{\text{m}}{\text{s}^2}}} \\ &= 10.712\ \text{s} \end{aligned}$$

[3]The case of projectile motion with air friction cannot be handled in kinematics, since a retarding force acts continuously on the projectile. In kinetics, various assumptions (e.g., friction varies linearly with the velocity or with the square of the velocity) can be made to include the effect of air friction.

Table 56.2 Projectile Motion Equations (θ may be negative for projection downward)

	level target	target above	target below	horizontal projection ($v_0 = v_x$, $\theta = 0°$)
$x(t)$	$(v_0\cos\theta)t$			$v_0 t$
$y(t)$	$(v_0\sin\theta)t - \frac{1}{2}gt^2$			$H - \frac{1}{2}gt^2$
$v_x(t)$[a]	$v_0\cos\theta$			v_0
$v_y(t)$[b]	$v_0\sin\theta - gt$			$-gt$
$v(t)$[c]	$\sqrt{v_0^2 - 2gy} = \sqrt{v_0^2 - 2gtv_0\sin\theta + g^2t^2}$			$\sqrt{v_0^2 + g^2t^2}$
$v(y)$[d]	$\sqrt{v_0^2 - 2gy}$			$\sqrt{v_0^2 + 2g(H-y)}$
H	$\frac{v_0^2\sin^2\theta}{2g}$	$\frac{v_0^2\sin^2\theta}{2g}$	$z + \frac{v_0^2\sin^2\theta}{2g}$	$\frac{1}{2}gt^2$
R	$v_0 T\cos\theta$			$v_0 t$
	$\frac{v_0^2\sin 2\theta}{g}$	$\left(\frac{v_0\cos\theta}{g}\right)\left(v_0\sin\theta + \sqrt{v_0^2\sin^2\theta - 2gz}\right)$	$\left(\frac{v_0\cos\theta}{g}\right)\left(v_0\sin\theta + \sqrt{2gz + v_0^2\sin^2\theta}\right)$	
T	$\frac{R}{v_0\cos\theta}$			$\sqrt{\frac{2H}{g}}$
	$\frac{2v_0\sin\theta}{g}$	$\frac{v_0\sin\theta}{g} + \sqrt{\frac{2(H-z)}{g}}$	$\frac{v_0\sin\theta}{g} + \sqrt{\frac{2H}{g}}$	
t_H	$\frac{v_0\sin\theta}{g} = \frac{T}{2}$	$\frac{v_0\sin\theta}{g}$		

[a]horizontal velocity component
[b]vertical velocity component
[c]resultant velocity as a function of time
[d]resultant velocity as a function of vertical elevation above the launch point

The total time is

$$t_{\text{total}} = t_{\text{rise}} + t_{\text{fall}} = 9.174\ \text{s} + 10.712\ \text{s} = 19.89\ \text{s}$$

The max range is

Projectile Motion

$$x = (v_0\cos\theta)t + x_0 = \left(180\ \frac{\text{m}}{\text{s}}\right)\cos 30\ (19.89\ \text{s}) + 0$$
$$= 3101\ \text{m}$$

Customary U.S. Solution

The maximum altitude above the impact plane includes the height of the plateau and the elevation achieved by the projectile.

First find the initial velocity in the y-direction at $t = 0$.

Projectile Motion

$$v_y = -gt + v_0\sin\theta = \left(600\ \frac{\text{ft}}{\text{sec}}\right)\sin 30° = 300\ \text{ft/sec}$$

The max height the projectile reaches from above the plane of impact occurs when $v_y = 0$. Rearrange and solve for y to obtain the maximum altitude above the impact plane.

Projectile Motion

$$v_y^2 = (v_{y,0})^2 - 2g(y - y_0)$$

$$y = \frac{v_y^2 - (v_{y,0})^2}{-2g} + y_0$$

$$= \frac{0 - \left(300\ \frac{\text{ft}}{\text{sec}}\right)^2}{(-2)\left(32.2\ \frac{\text{ft}}{\text{sec}^2}\right)} + 500\ \text{ft}$$

$$= 1897.5\ \text{ft}$$

The total flight time includes the time to reach the maximum altitude and the time to fall from the maximum altitude to the impact plane below.

To determine the time to rise to the apex, let $v_y = 0$ and rearrange the equation for v_y to solve for t.

Projectile Motion

$$v_y = -gt_{\text{rise}} + v_0 \sin\theta$$

$$t_{\text{rise}} = \frac{v_0 \sin\theta - v_y}{g}$$

$$= \frac{300\ \frac{\text{ft}}{\text{sec}} - 0}{32.2\ \frac{\text{ft}}{\text{sec}^2}}$$

$$= 9.317\ \text{sec}$$

To determine the time to fall from the apex, rearrange the equation for y.

Projectile Motion

$$y = -\frac{gt_{\text{fall}}^2}{2} + (v_0\sin\theta)t_{\text{fall}} + y_0 = \frac{-gt_{\text{fall}}^2}{2} + y_0$$

$$t_{\text{fall}} = \sqrt{\frac{(2)(y - y_0)}{-g}}$$

$$= \sqrt{\frac{(2)(0 - 1897.5\ \text{ft})}{-32.2\ \frac{\text{ft}}{\text{sec}^2}}}$$

$$= 10.856\ \text{sec}$$

The total time is

$$t_{\text{total}} = t_{\text{rise}} + t_{\text{fall}} = 9.317\ \text{sec} + 10.856\ \text{sec} = 20.17\ \text{sec}$$

The max range is

Projectile Motion

$$x = (v_0\cos\theta)t + x_0$$

$$= \left(600\ \frac{\text{ft}}{\text{sec}}\right)\cos 30(20.17\ \text{sec}) + 0$$

$$= 10{,}481\ \text{ft} \quad (1.98\ \text{mi})$$

11. ROTATIONAL PARTICLE MOTION

Rotational particle motion (also known as *angular motion* and *circular motion*) is motion of a particle around a circular path. (See Fig. 56.5.) The particle travels through 2π radians per complete revolution. [**Plane Circular Motion**]

Figure 56.5 *Rotational Particle Motion*

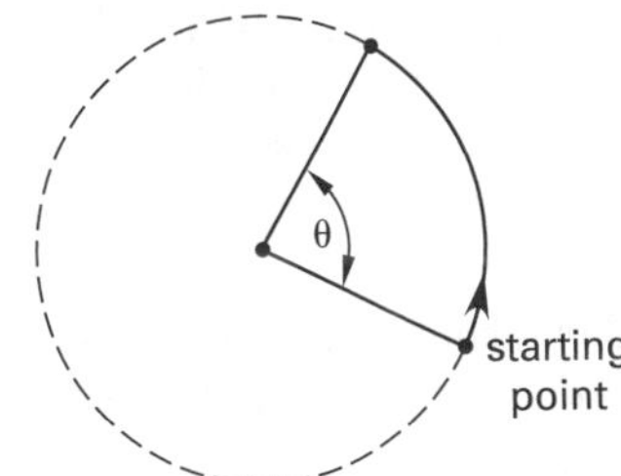

The behavior of a rotating particle is defined by its *angular position*, θ, *angular velocity*, ω, and *angular acceleration*, α, functions. These variables are analogous to the $s(t)$, $v(t)$, and $a(t)$ functions for linear systems. Angular variables can be substituted one-for-one in place of linear variables in most equations.

The relationships among angular position, velocity, and acceleration for a rotational system are given by Eq. 56.24 through Eq. 56.26. When values of t are substituted into these equations, the position, velocity, and acceleration are known as *instantaneous values*.

$$\theta(t) = \int \omega(t)\,dt = \iint \alpha(t)\,dt^2 \qquad 56.24$$

$$\omega(t) = \frac{d\theta(t)}{dt} = \int \alpha(t)\,dt \qquad 56.25$$

$$\alpha(t) = \frac{d\omega(t)}{dt} = \frac{d^2\theta(t)}{dt^2} \qquad 56.26$$

The average velocity and acceleration are

$$\omega_{\text{ave}} = \frac{\int_1^2 \omega(t)\,dt}{t_2 - t_1} = \frac{\theta_2 - \theta_1}{t_2 - t_1} \qquad 56.27$$

$$\alpha_{\text{ave}} = \frac{\int_1^2 \alpha(t)\,dt}{t_2 - t_1} = \frac{\omega_2 - \omega_1}{t_2 - t_1} \qquad 56.28$$

Dynamics and Vibrations

For constant angular acceleration, the equations for angular velocity and displacement are provided by Eq. 56.29 to Eq. 56.31.

Normal and Tangential Components

$$\alpha(t) = \alpha_0 \qquad 56.29$$

$$\omega(t) = \alpha_0(t - t_0) + \omega_0 \qquad 56.30$$

$$\theta(t) = \alpha_0 \frac{(t - t_0)^2}{2} + \omega_0(t - t_0) + \theta_0 \qquad 56.31$$

Equation 56.32 directly relates angular velocity as a function of angular position.

Normal and Tangential Components

$$\omega^2 = \omega_0^2 + 2\alpha_0(\theta - \theta_0) \qquad 56.32$$

Example 56.5

A turntable starts from rest and accelerates uniformly at 1.5 rad/sec². How many revolutions will it take before a rotational speed of 33⅓ rpm is attained?

Solution

First, convert 33⅓ rpm into radians per second. Since there are 2π radians per complete revolution,

$$\omega = \frac{\left(33\frac{1}{3}\ \frac{\text{rev}}{\text{min}}\right)\left(2\pi\ \frac{\text{rad}}{\text{rev}}\right)}{60\ \frac{\text{sec}}{\text{min}}} = 3.49\ \text{rad/sec}$$

Use the equation for angular velocity as a function of angular position. Let $\theta_0 = 0$ and $\omega_0 = 0$.

Normal and Tangential Components

$$\omega^2 = \omega_0^2 + 2\alpha_0(\theta - \theta_0)$$

Rearrange and solve for θ:

$$\theta = \frac{\omega^2}{2\alpha_0} = \frac{\left(3.49\ \frac{\text{rad}}{\text{sec}}\right)^2}{(2)\left(1.5\ \frac{\text{rad}}{\text{sec}^2}\right)} = 4.06\ \text{rad}$$

Converting from radians to revolutions,

$$n = \frac{4.06\ \text{rad}}{2\pi\ \frac{\text{rad}}{\text{rev}}} = 0.646\ \text{rev}$$

Example 56.6

A flywheel is brought to a standstill from 400 rpm in 8 sec. (a) What was its average angular acceleration in rad/sec² during that period? (b) How far (in radians) did the flywheel travel?

Solution

(a) The initial rotational speed must be expressed in radians per second. Since there are 2π radians per revolution,

$$\omega_0 = \frac{\left(400\ \frac{\text{rev}}{\text{min}}\right)\left(2\pi\ \frac{\text{rad}}{\text{rev}}\right)}{60\ \frac{\text{sec}}{\text{min}}} = 41.89\ \text{rad/sec}$$

t, t_0, ω_0, and ω are known, and α_0 is unknown.

Rearrange the equation for angular velocity and solve for α_0.

Normal and Tangential Components

$$\begin{aligned}\omega(t) &= \alpha_0(t - t_0) + \omega_0 \\ \alpha_0 &= \frac{(\omega - \omega_0)}{(t - t_0)} \\ &= \frac{\left(0\ \frac{\text{rad}}{\text{sec}} - 41.89\ \frac{\text{rad}}{\text{sec}}\right)}{(8\ \text{sec} - 0\ \text{sec})} \\ &= -5.236\ \text{rad/sec}^2\end{aligned}$$

(b) t, ω_0, and ω are known, and θ is unknown.

Normal and Tangential Components

$$\begin{aligned}\theta(t)_0 &= \alpha_0 \frac{(t - t_0)^2}{2} + \omega_0(t - t_0) + \theta_0 \\ &= \left(-5.236\ \frac{\text{rad}}{\text{sec}^2}\right)\left(\frac{(8\ \text{sec} - 0\ \text{sec})^2}{2}\right) \\ &\quad + \left(41.89\ \frac{\text{rad}}{\text{sec}}\right)(8\ \text{sec} - 0\ \text{sec}) + 0\ \text{rad} \\ &= 167.6\ \text{rad} \quad (26.67\ \text{rev})\end{aligned}$$

12. RELATIONSHIP BETWEEN LINEAR AND ROTATIONAL VARIABLES

A particle moving in a curvilinear path will also have instantaneous linear velocity and linear acceleration. These linear variables will be directed tangentially to the path and, therefore, are known as *tangential velocity* and *tangential acceleration*, respectively. (See Fig. 56.6.) In general, the linear variables can be obtained by multiplying the rotational variables by the path radius, r.

Plane Circular Motion

$$\text{v}_t = r\omega \qquad 56.33$$

$$\text{v}_{t,x} = \text{v}_t \cos\phi = \omega r \cos\phi \qquad 56.34$$

$$\text{v}_{t,y} = \text{v}_t \sin\phi = \omega r \sin\phi \qquad 56.35$$

Plane Circular Motion

$$a_t = r\alpha \quad \text{56.36}$$

If the path radius is constant, as it would be in rotational motion, the linear distance (i.e., the *arc length*) traveled is

Plane Circular Motion

$$s = r\theta \quad \text{56.37}$$

13. NORMAL ACCELERATION

A moving particle will continue tangentially to its path unless constrained otherwise. For example, a rock twirled on a string will move in a circular path only as long as there is tension in the string. When the string is released, the rock will move off tangentially.

Figure 56.6 *Tangential Variables*

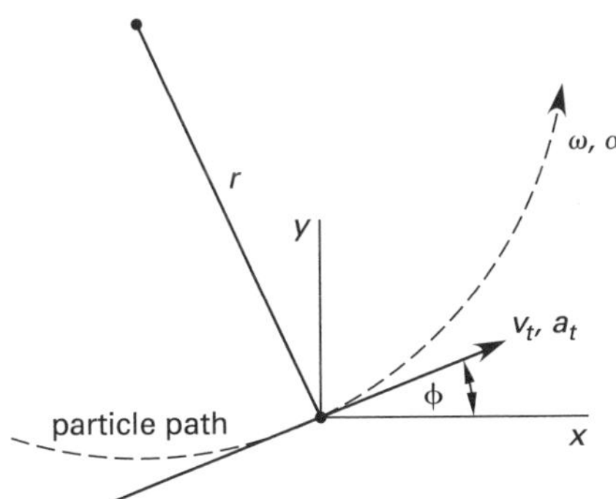

The twirled rock is acted upon by the tension in the string. In general, a restraining force will be directed toward the center of rotation. Whenever a mass experiences a force, an acceleration is acting.[4] The acceleration has the same sense as the applied force (i.e., is directed toward the center of rotation). Since the inward acceleration is perpendicular to the tangential velocity and acceleration, it is known as *normal acceleration*, a_n. (See Fig. 56.7.)

Plane Circular Motion

$$a_n = -r\omega^2 \quad \text{56.38}$$

$$a_n = \frac{v_t^2}{r} = v_t\omega \quad \text{56.39}$$

The *resultant acceleration*, *a*, is the vector sum of the tangential and normal accelerations. The magnitude of the resultant acceleration is

$$a = \sqrt{a_t^2 + a_n^2} \quad \text{56.40}$$

[4]This is a direct result of Newton's second law of motion.

The *x*- and *y*-components of the resultant acceleration are

$$a_x = a_n \sin\phi \pm a_t \cos\phi \quad \text{56.41}$$

$$a_y = a_n \cos\phi \mp a_t \sin\phi \quad \text{56.42}$$

The normal and tangential accelerations can be expressed in terms of the *x*- and *y*-components of the resultant acceleration (not shown in Fig. 56.7).

$$a_n = a_x \sin\phi \pm a_y \cos\phi \quad \text{56.43}$$

$$a_t = a_x \cos\phi \mp a_y \sin\phi \quad \text{56.44}$$

Figure 56.7 *Normal Acceleration*

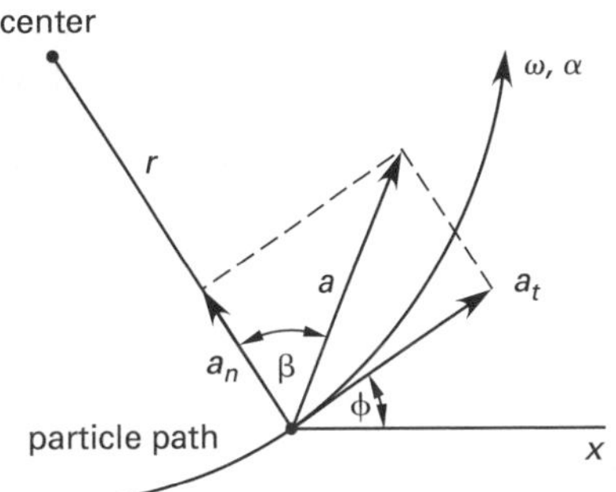

14. CORIOLIS ACCELERATION

Consider a particle moving with linear radial velocity v_r away from the center of a flat disk rotating with constant velocity ω. Since $v_t = \omega r$, the particle's tangential velocity will increase as it moves away from the center of rotation. This increase is understood to be produced by the tangential *Coriolis acceleration*, a_c. (See Fig. 56.8.)

$$a_c = 2v_r\omega \quad \text{56.45}$$

Figure 56.8 *Coriolis Acceleration on a Rotating Disk*

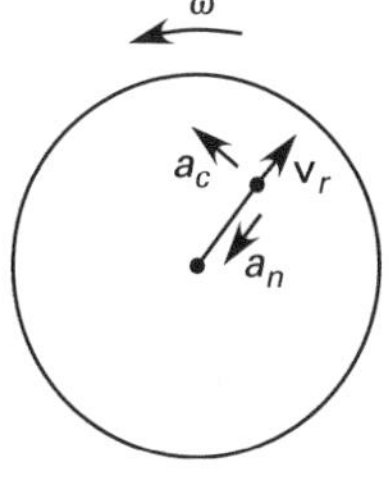

Coriolis acceleration also acts on particles moving on rotating spheres. Consider an aircraft flying with constant air speed v from the equator to the north pole while the earth (a sphere of radius *R*) rotates below it. Three accelerations act on the aircraft: normal, radial, and Coriolis accelerations, shown in Fig. 56.9. The

Coriolis acceleration depends on the latitude, ϕ, because the earth's tangential velocity is less near the poles than at the equator.

$$a_n = r\omega^2 = R\omega^2 \cos\phi \tag{56.46}$$

$$a_r = \frac{v^2}{R} \tag{56.47}$$

$$a_c = 2\omega v_x = 2\omega v \sin\phi \tag{56.48}$$

Figure 56.9 *Coriolis Acceleration on a Rotating Sphere*

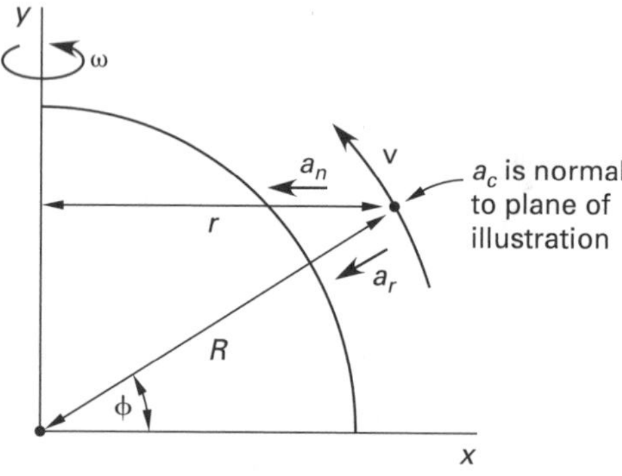

Example 56.7

A slider moves with a constant velocity of 20 ft/sec along a rod rotating at 5 rad/sec. What is the magnitude of the slider's total acceleration when the slider is 4 ft from the center of rotation?

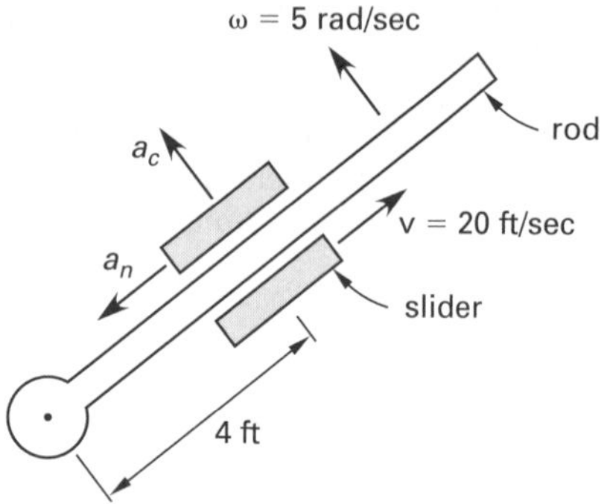

Solution

The normal acceleration is given by Eq. 56.46.

$$\begin{aligned} a_n &= r\omega^2 = (4\ \text{ft})\left(5\ \frac{\text{rad}}{\text{sec}}\right)^2 \\ &= 100\ \text{ft/sec}^2 \end{aligned}$$

The Coriolis (tangential) acceleration is given by Eq. 56.48.

$$\begin{aligned} a_c &= 2\omega v = (2)\left(5\ \frac{\text{rad}}{\text{sec}}\right)\left(20\ \frac{\text{ft}}{\text{sec}}\right) \\ &= 200\ \text{ft/sec}^2 \end{aligned}$$

The total acceleration is given by Eq. 56.40.

$$\begin{aligned} a &= \sqrt{a_c^2 + a_n^2} = \sqrt{\left(200\ \frac{\text{ft}}{\text{sec}^2}\right)^2 + \left(100\ \frac{\text{ft}}{\text{sec}^2}\right)^2} \\ &= 223.6\ \text{ft/sec}^2 \end{aligned}$$

15. PARTICLE MOTION IN POLAR COORDINATES

In polar coordinates, the path of a particle is described by a radius vector, **r**, and an angle, θ. Since the velocity of a particle is not usually directed radially out from the center of the coordinate system, it can be divided into two perpendicular components. The terms *normal* and *tangential* are not used with polar coordinates. Rather, the terms *radial* and *transverse* are used. Figure 56.10 illustrates the *radial* and *transverse components* of velocity in a polar coordinate system.

Figure 56.10 also illustrates the unit radial and unit transverse vectors, $\mathbf{e}_r$ and $\mathbf{e}_\theta$, used in the vector forms of the motion equations.

$$\text{position: } \mathbf{r} = r\mathbf{e}_r \tag{56.49}$$

$$\text{velocity: } \mathbf{v} = v_r\mathbf{e}_r + v_\theta\mathbf{e}_\theta = \frac{dr}{dt}\mathbf{e}_r + r\frac{d\theta}{dt}\mathbf{e}_\theta \tag{56.50}$$

$$\begin{aligned} \text{acceleration: } \mathbf{a} &= a_r\mathbf{e}_r + a_\theta\mathbf{e}_\theta \\ &= \left(\frac{d^2r}{dt^2} - r\left(\frac{d\theta}{dt}\right)^2\right)\mathbf{e}_r \\ &\quad + \left(r\frac{d^2\theta}{dt^2} + 2\frac{dr}{dt}\frac{d\theta}{dt}\right)\mathbf{e}_\theta \end{aligned} \tag{56.51}$$

Figure 56.10 *Radial and Transverse Components*

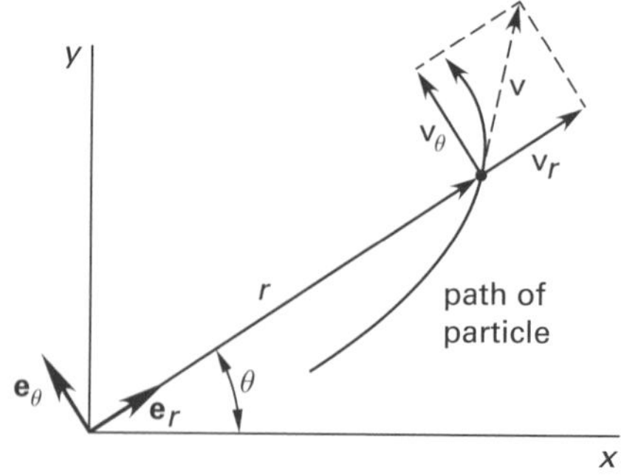

The magnitudes of the radial and transverse components of velocity and acceleration are given by Eq. 56.52 through Eq. 56.55.

$$v_r = \frac{dr}{dt} \tag{56.52}$$

$$v_\theta = r\frac{d\theta}{dt} \quad (56.53)$$

$$a_r = \frac{d^2r}{dt^2} - r\left(\frac{d\theta}{dt}\right)^2 \quad (56.54)$$

$$a_\theta = r\frac{d^2\theta}{dt^2} + 2\frac{dr}{dt}\frac{d\theta}{dt} \quad (56.55)$$

If the radial and transverse components of acceleration and velocity are known, they can be used to calculate the tangential and normal accelerations in a rectangular coordinate system.

$$a_t = \frac{a_r v_r + a_\theta v_\theta}{v_t} \quad (56.56)$$

$$a_n = \frac{a_\theta v_r - a_r v_\theta}{v_t} \quad (56.57)$$

16. RELATIVE MOTION

The term *relative motion* is used when motion of a particle is described with respect to something else in motion. The particle's position, velocity, and acceleration may be specified with respect to another moving particle or with respect to a moving frame of reference, known as a *Newtonian* or *inertial frame of reference*.

In Fig. 56.11 the relative position, $\mathbf{r}_A$, velocity, $\mathbf{v}_A$, and acceleration, $\mathbf{a}_A$, with respect to a translating axis can be calculated from Eq. 56.58, Eq. 56.59, and Eq. 56.60, respectively. The angular velocity, ω, and angular acceleration, α, are the magnitudes of the relative position vector, $\mathbf{r}_{A/B}$.

Figure 56.11 *Translating Axis*

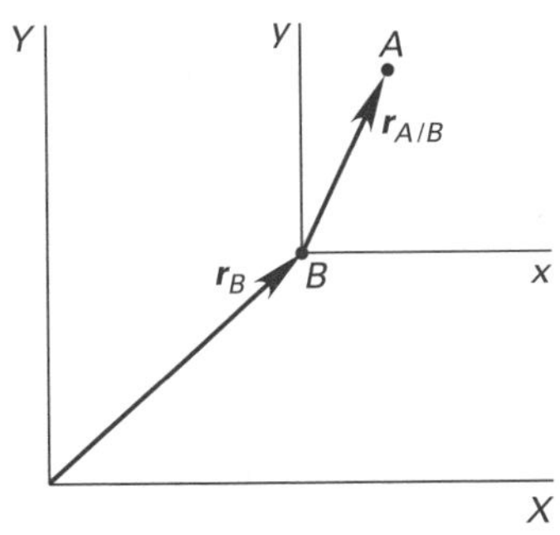

Relative Motion

$$\mathbf{r}_A = \mathbf{r}_B + \mathbf{r}_{A/B} \quad (56.58)$$

$$\mathbf{v}_A = \mathbf{v}_B + (\omega \times \mathbf{r}_{A/B}) = \mathbf{v}_B + \mathbf{v}_{A/B} \quad (56.59)$$

$$\mathbf{a}_A = \mathbf{a}_B + (\alpha \times \mathbf{r}_{A/B}) + \omega \times (\omega \times \mathbf{r}_{A/B}) = \mathbf{a}_B + \mathbf{a}_{A/B} \quad (56.60)$$

Since vector subtraction and addition operations can be performed graphically, many relative motion problems can be solved by a simplified graphical process.

In Fig. 56.12 the relative position, $\mathbf{r}_A$, velocity, $\mathbf{v}_A$, and acceleration, $\mathbf{a}_A$, with respect to a translating and rotating axis can be calculated from Eq. 56.58, Eq. 56.61, and Eq. 56.62 respectively.

Figure 56.12 *Translating and Rotating Axis*

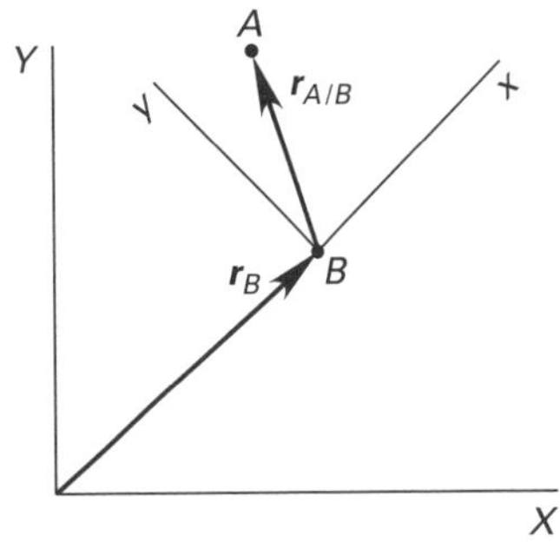

Relative Motion

$$\mathbf{v}_A = \mathbf{v}_B + \left(\mathbf{\Omega} \times \mathbf{r}_{A/B}\right) + \mathbf{v}_{A/B} \quad (56.61)$$

$$\mathbf{a}_A = \mathbf{a}_B + \left(\dot{\mathbf{\Omega}} \times \mathbf{r}_{A/B}\right) + \mathbf{\Omega} \times \left(\mathbf{\Omega} \times \mathbf{r}_{A/B}\right) + 2\mathbf{\Omega} \times \mathbf{v}_{A/B} + \mathbf{a}_{A/B} \quad (56.62)$$

Example 56.8

A stream flows at 5 km/h. At what upstream angle, ϕ, should a 10 km/h boat be piloted in order to reach the shore directly opposite the initial point?

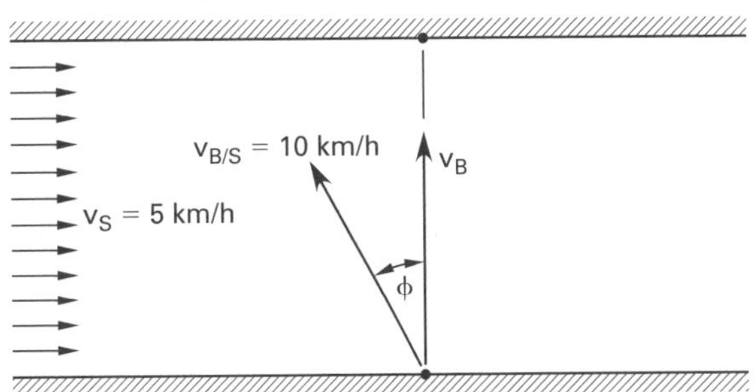

Solution

From Eq. 56.59, the absolute velocity of the boat, v_B, with respect to the shore is equal to the vector sum of the absolute velocity of the stream, v_S, and the relative velocity of the boat with respect to the stream, $v_{B/S}$. The magnitudes of these two velocities are known.

$$v_B = v_S + v_{B/S}$$

Since vector addition is accomplished graphically by placing the two vectors head to tail, the angle can be determined from trigonometry.

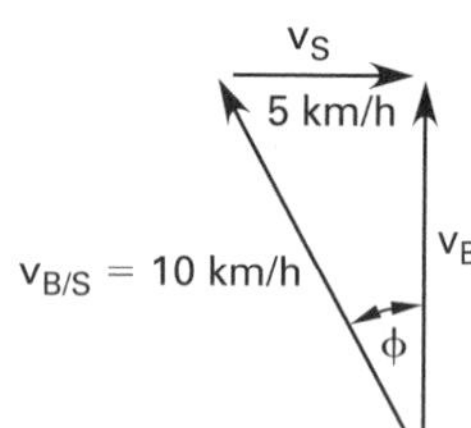

$$\sin\phi = \frac{v_S}{v_{B/S}} = \frac{5\ \frac{\text{km}}{\text{h}}}{10\ \frac{\text{km}}{\text{h}}} = 0.5$$

$$\phi = \arcsin 0.5 = 30°$$

Example 56.9

A stationary member of a marching band tosses a 2.0 ft long balanced baton straight up into the air and then begins walking forward at 4 mi/hr. At a particular moment, the baton is 20 ft in the air and is falling back toward the earth with a velocity of 30 ft/sec. The tip of the baton is rotating at 140 rpm in the orientation shown.

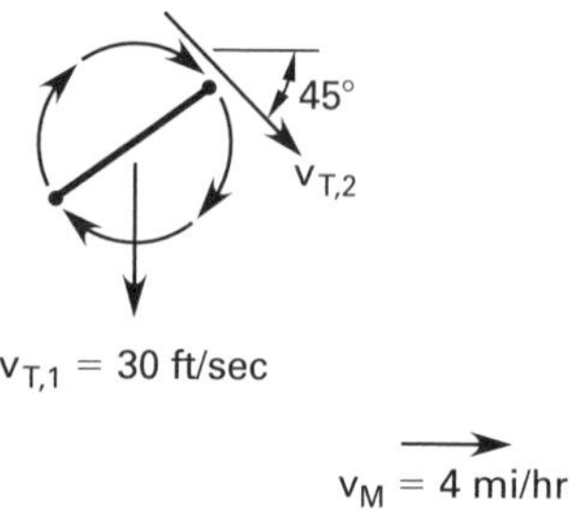

(a) What is the speed of the baton tip with respect to the ground? (b) What is the speed of the baton tip with respect to the band member?

Solution

(a) The baton tip has two absolute velocity components. The first, with a magnitude of $v_{T,1} = 30$ ft/sec, is directed vertically downward. The second, with a magnitude of $v_{T,2}$, is directed as shown in the illustration. The baton's radius, r, is 1 ft, center of rotation to tip. From Eq. 56.33,

Plane Circular Motion

$$v_{T,2} = r\omega = \frac{(1\ \text{ft})\left(140\ \frac{\text{rev}}{\text{min}}\right)\left(2\pi\ \frac{\text{rad}}{\text{rev}}\right)}{60\ \frac{\text{sec}}{\text{min}}}$$

$$= 14.7\ \text{ft/sec}$$

The vector sum of these two absolute velocities is the velocity of the tip, $\mathbf{v}_T$, with respect to the earth.

$$\mathbf{v}_T = \mathbf{v}_{T,1} + \mathbf{v}_{T,2}$$

$v_{T,2}$ = 14.7 ft/sec
ϕ = 135°
v_T
$v_{T,1}$ = 30 ft/sec
α

The velocity of the tip is found from the law of cosines.

$$v_T = \sqrt{v_{T,1}^2 + v_{T,2}^2 - 2v_{T,1}v_{T,2}\cos\phi}$$

$$= \sqrt{\begin{array}{l}\left(30\ \frac{\text{ft}}{\text{sec}}\right)^2 + \left(14.7\ \frac{\text{ft}}{\text{sec}}\right)^2 \\ \quad -(2)\left(30\ \frac{\text{ft}}{\text{sec}}\right)\left(14.7\ \frac{\text{ft}}{\text{sec}}\right)(\cos 135°)\end{array}}$$

$$= 41.7\ \text{ft/sec}$$

(b) From Eq. 56.59, the velocity of the tip with respect to the band member is

$$\mathbf{v}_{T/M} = \mathbf{v}_T - \mathbf{v}_M$$

Subtracting a vector is equivalent to adding its negative. The velocity triangle is as shown. The law of cosines is used again to determine the relative velocity.

The band member's absolute velocity, v_M, is

$$v_M = \frac{\left(4\ \frac{\text{mi}}{\text{hr}}\right)\left(5280\ \frac{\text{ft}}{\text{mi}}\right)}{3600\ \frac{\text{sec}}{\text{hr}}} = 5.87\ \text{ft/sec}$$

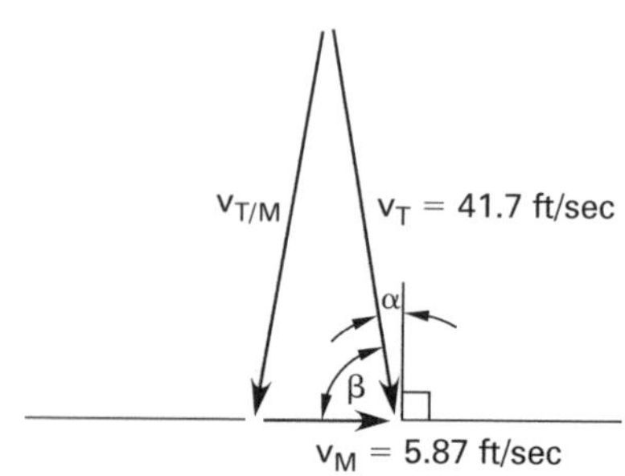

The angle α is found from the law of sines.

$$\frac{\sin\alpha}{14.7\ \frac{\text{ft}}{\text{sec}}} = \frac{\sin 135°}{41.7\ \frac{\text{ft}}{\text{sec}}}$$

$$\alpha = 14.4°$$

$$\beta = 90° - \alpha = 90° - 14.4° = 75.6°$$

$$\begin{aligned} v_{T/M} &= \sqrt{v_T^2 + v_M^2 - 2v_T v_M \cos\beta} \\ &= \sqrt{\begin{array}{l}\left(41.7\ \frac{\text{ft}}{\text{sec}}\right)^2 + \left(5.87\ \frac{\text{ft}}{\text{sec}}\right)^2 \\ -(2)\left(41.7\ \frac{\text{ft}}{\text{sec}}\right)\left(5.87\ \frac{\text{ft}}{\text{sec}}\right)(\cos 75.6°)\end{array}} \\ &= 40.6\ \text{ft/sec} \quad (27.7\ \text{mi/hr}) \end{aligned}$$

17. DEPENDENT MOTION

When the position of one particle in a multiple-particle system depends on the position of one or more other particles, the motions are said to be "dependent." A block-and-pulley system with one fixed rope end, as illustrated by Fig. 56.13, is a *dependent system.*

Figure 56.13 *Dependent System*

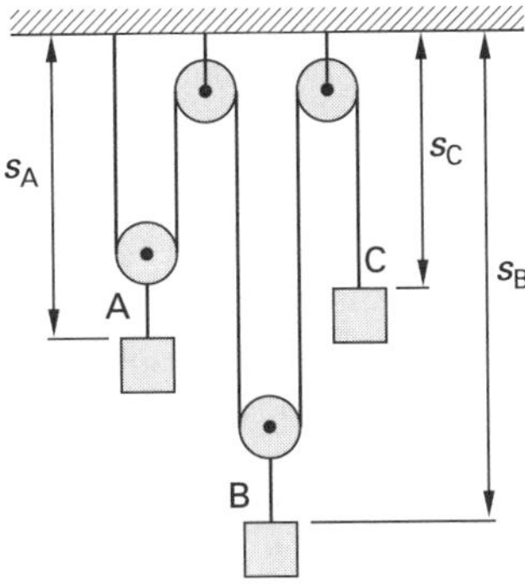

The following statements define the behavior of a dependent block-and-pulley system.

- Since the length of the rope is constant, the sum of the rope segments representing distances between the blocks and pulleys is constant. By convention, the distances are measured from the top of the block to the support point.[5] Since in Fig. 56.13 there are two ropes supporting block A, two ropes supporting block B, and one rope supporting block C,

$$2s_A + 2s_B + s_C = \text{constant} \qquad 56.63$$

- Since the position of the nth block in an n-block system is determined when the remaining $n - 1$ positions are known, the number of *degrees of freedom* is one less than the number of blocks.

- The movement, velocity, and acceleration of a block supported by two ropes are half the same quantities of a block supported by one rope.

- The relative relationships between the blocks' velocities or accelerations are the same as the relationships between the blocks' positions. For Fig. 56.13,

$$2v_A + 2v_B + v_C = 0 \qquad 56.64$$

$$2a_A + 2a_B + a_C = 0 \qquad 56.65$$

18. GENERAL PLANE MOTION

Rigid body *plane motion* can be described in two dimensions. Examples include rolling wheels, gear sets, and linkages. Plane motion can be considered as the sum of a translational component and a rotation about a fixed axis, as illustrated by Fig. 56.14.

Figure 56.14 *Components of Plane Motion*

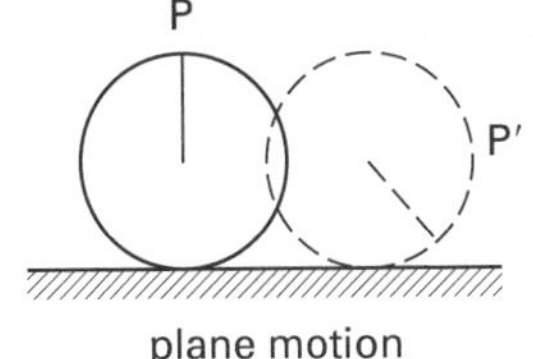

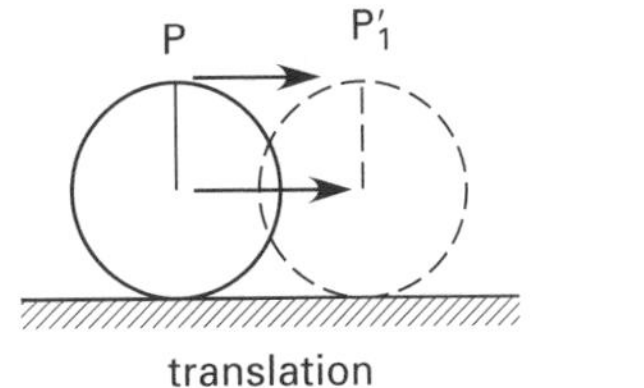

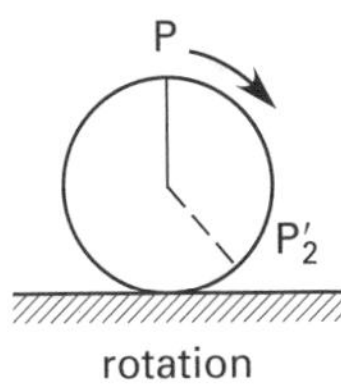

19. ROTATION ABOUT A FIXED AXIS

Analysis of the rotational component of a rigid body's plane motion can sometimes be simplified if the location of the body's instantaneous center is known. Using the instantaneous center reduces many relative motion problems to simple geometry. The *instantaneous center* (also known as the *instant center* and IC) is a point at which the body could be fixed (pinned) without changing the instantaneous angular velocities of any point on the body. With the angular velocities, the body seems to rotate about a fixed instantaneous center. [**Instantaneous Center of Rotation (Instant Centers)**]

[5] In measuring distances, the finite diameters of the pulleys and the lengths of rope wrapped around the pulleys are disregarded.

The instantaneous center is located by finding two points for which the absolute velocity directions are known. Lines drawn perpendicular to these two velocities will intersect at the instantaneous center. (This graphical procedure is slightly different if the two velocities are parallel, as Fig. 56.15 shows. In that case, use is made of the fact that the tangential velocity is proportional to the distance from the instantaneous center.) For a rolling wheel, the instantaneous center is the point of contact with the supporting surface.

Figure 56.15 *Graphical Method of Finding the Instantaneous Center*

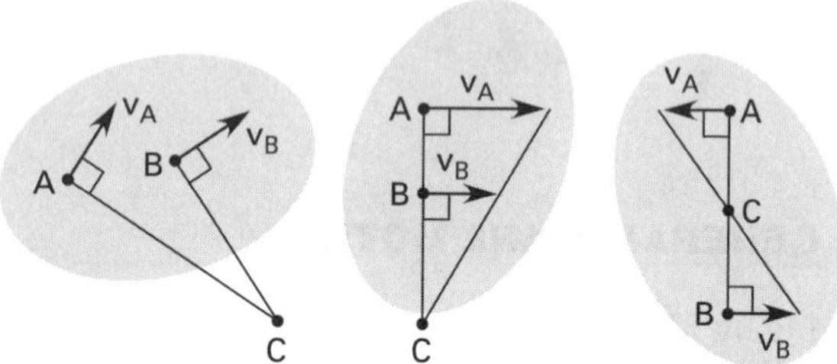

The absolute velocity of any point, P, on a wheel rolling with translational velocity, v_O, can be found by geometry. (See Fig. 56.16.) Assume that the wheel is pinned at C and rotates with its actual angular velocity, $\omega = v_O/r$. The direction of the point's velocity will be perpendicular to the line of length L between the instantaneous center and the point.

$$v = L\omega = \frac{L v_O}{r} \qquad 56.66$$

Figure 56.16 *Instantaneous Center of a Rolling Wheel*

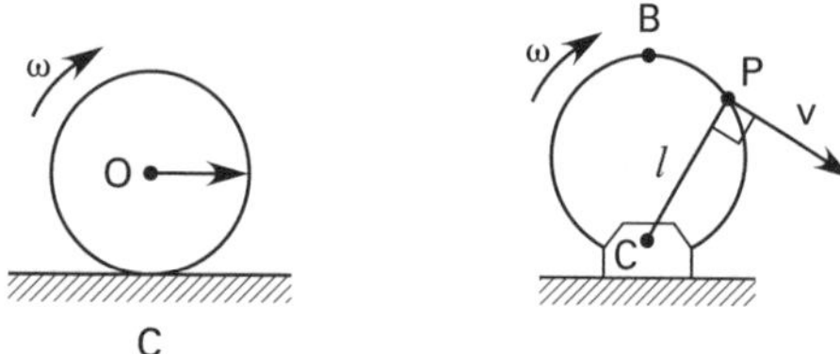

Equation 56.66 is valid only for a velocity referenced to the instantaneous center, point C. Table 56.3 can be used to find the velocities with respect to other points.

Table 56.3 *Relative Velocities of a Rolling Wheel*

	reference point		
point	O	C	B
v_O	0	$v_O\rightarrow$	$\leftarrow v_O$
v_C	$\leftarrow v_O$	0	$\leftarrow 2v_O$
v_B	$v_O\rightarrow$	$2v_O\rightarrow$	0

Example 56.10

A truck with 35 in diameter tires travels at a constant 35 mi/hr. What is the absolute velocity of point P on the circumference of the tire?

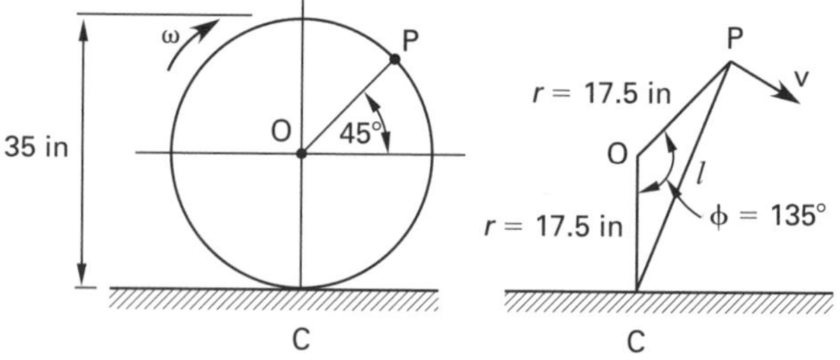

Solution

The translational velocity of the center of the wheel is

$$v_O = \frac{\left(35\ \frac{\text{mi}}{\text{hr}}\right)\left(5280\ \frac{\text{ft}}{\text{mi}}\right)}{3600\ \frac{\text{sec}}{\text{hr}}} = 51.33\ \text{ft/sec}$$

The wheel radius is

$$r = \frac{35\ \text{in}}{(2)\left(12\ \frac{\text{in}}{\text{ft}}\right)} = 1.458\ \text{ft}$$

The angular velocity of the wheel is

$$\omega = \frac{v_O}{r} = \frac{51.33\ \frac{\text{ft}}{\text{sec}}}{1.458\ \text{ft}} = 35.21\ \text{rad/sec}$$

The instantaneous center is the contact point, C. The law of cosines is used to find the distance L.

$$L^2 = r^2 + r^2 - 2r^2\cos\phi = 2r^2(1-\cos\phi)$$

$$L = \sqrt{(2)(1.458\ \text{ft})^2(1-\cos 135°)} = 2.694\ \text{ft}$$

From Eq. 56.66, the absolute velocity of point P is

$$v_P = L\omega = (2.694\ \text{ft})\left(35.21\ \frac{\text{rad}}{\text{sec}}\right) = 94.9\ \text{ft/sec}$$

20. INSTANTANEOUS CENTER OF ACCELERATION

The *instantaneous center of acceleration* is used to compute the absolute acceleration of a point as if a body were in pure rotation about that point. It is the same as the instantaneous center of rotation, only for a body starting from rest and accelerating uniformly with angular acceleration, α. The absolute acceleration, a, determined from Eq. 56.67 is the same as the *resultant acceleration* in Fig. 56.7.

$$a = L\alpha = \frac{La_O}{r} \qquad 56.67$$

In general, the instantaneous center of acceleration, C_a, will be deflected at angle β from the absolute acceleration vectors, as shown in Fig. 56.17. The relationship among the angle, β, the instantaneous acceleration, α, and the instantaneous velocity, ω, is

$$\tan\beta = \frac{\alpha}{\omega^2} \qquad 56.68$$

Figure 56.17 *Instantaneous Center of Acceleration*

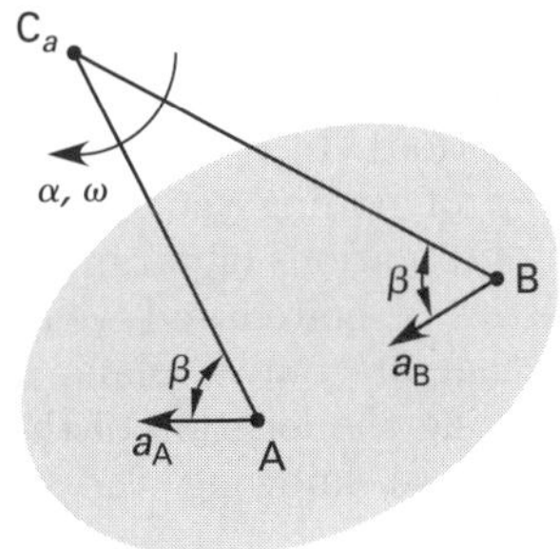

21. SLIDER RODS

The absolute velocity of any point, P, on a slider rod assembly can be found from the instantaneous center concept. (See Fig. 56.18.) The instantaneous center, C, is located by extending perpendiculars from the velocity vectors.

Figure 56.18 *Instantaneous Center of Slider Rod Assembly*

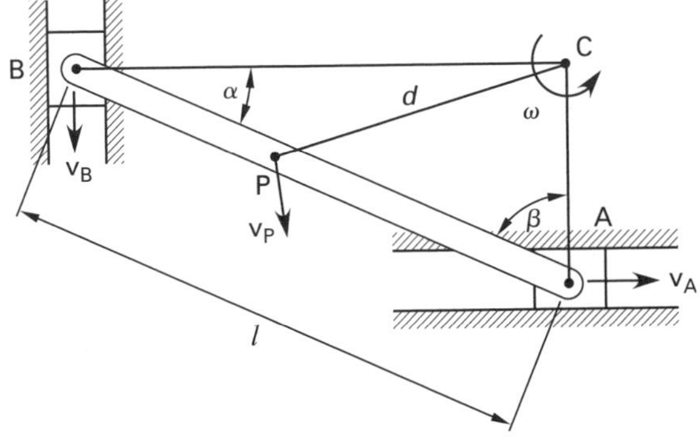

If the velocity with respect to point C of one end of the slider is known, for example v_A, then v_B can be found from geometry. Since the slider can be assumed to rotate about point C with angular velocity ω,

$$\omega = \frac{v_A}{AC} = \frac{v_A}{L\cos\beta} = \frac{v_B}{BC} = \frac{v_B}{L\cos\alpha} \qquad 56.69$$

Since $\cos\alpha = \sin\beta$,

$$v_B = v_A\tan\beta \qquad 56.70$$

If the velocity with respect to point C of any other point P is required, it can be found from

$$v_P = d\omega \qquad 56.71$$

22. SLIDER-CRANK ASSEMBLIES

Figure 56.19 illustrates a slider-crank assembly for which points A and D are in the same plane and at the same elevation. The instantaneous velocity of any point, P, on the rod can be found if the distance to the instantaneous center is known.

$$v_P = d\omega_1 \qquad 56.72$$

Figure 56.19 *Slider-Crank Assembly*

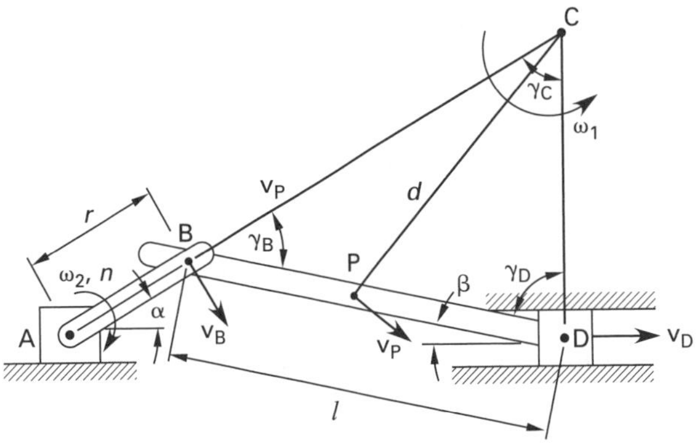

The tangential velocity of point B on the crank with respect to point A is perpendicular to the end of the crank. Slider D moves with a horizontal velocity. The intersection of lines drawn perpendicular to these velocity vectors locates the instantaneous center, point C. At any given instant, the connecting rod seems to rotate about point C with instantaneous angular velocity ω_1.

The velocity of point D, v_D, is

$$v_D = (CD)\omega_1 = v_B\left(\frac{CD}{BC}\right) = \omega_2\left(\frac{AB \times CD}{BC}\right) \qquad 56.73$$

Similarly, the velocity of point B, v_B, is

$$v_B = (AB)\omega_2 = (BC)\omega_1 = v_D\left(\frac{BC}{CD}\right) \qquad 56.74$$

$$\omega_2 = \frac{2\pi n}{60} \qquad 56.75$$

The following geometric relationships exist between the various angles.

$$\frac{\omega_1}{\omega_2} = \frac{\text{AB}}{\text{BC}} \qquad 56.76$$

$$\frac{\sin\alpha}{L} = \frac{\sin\beta}{r} \qquad 56.77$$

$$\frac{\sin\gamma_{\text{D}}}{\text{BC}} = \frac{\sin\gamma_{\text{B}}}{\text{CD}} = \frac{\sin\gamma_{\text{C}}}{L} \qquad 56.78$$

$$\gamma_{\text{B}} = \alpha + \beta \qquad 56.79$$

$$\gamma_{\text{D}} = 90 - \beta \qquad 56.80$$

$$\gamma_{\text{C}} = 90 - \alpha \qquad 56.81$$

23. INSTANTANEOUS CENTER OF CONNECTOR GROUPS

Parts that are bolted or welded do not generally experience significant differential rotation, but even so, a point around which rotation would occur can be imagined. With traditional elastic analysis, that point is the centroid (i.e., center of gravity) of the connector group. However, rotation is not always about the centroid, and stresses in the connectors do not always develop in proportion to distance from the centroid, particularly when the connected parts simultaneously translate and rotate. The actual instantaneous center (IC) will generally not coincide with the centroid of the connector group, as shown in Fig. 56.20. Also, the connection's centroid may shift as loading increases and the parts slip.

Figure 56.20 Connector Group Instantaneous Center

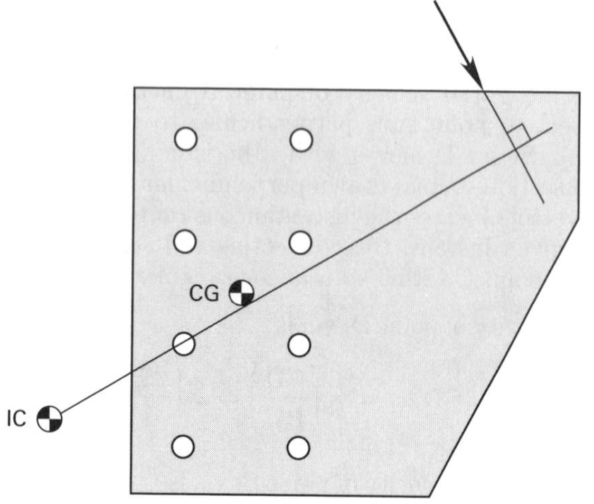

IC methods for structural connector analysis and design assume that translational and rotational events happen simultaneously about an instantaneous center of rotation that is located near a line that is perpendicular to the applied force and that passes near the original center of gravity of the connector group. The challenge is to find the location of the IC along this line.

For the special cases of an applied load that is either vertical or horizontal, the IC lies on the line that is perpendicular to the line of force and passes through the center of gravity of the bolt group, although the location on the line must still be found. In other cases, finding both the line and IC is done by trial and error using a combination of theory and empirical assumptions to establish a relationship between the applied force and the connector forces.

In the case of bolted connections, the relationship depends on whether the bolted pieces are allowed to slip and bring the bolts into bearing against the sides of their holes (i.e., a bearing connection) or are strong (tight) enough to hold the pieces in alignment (i.e., a slip critical connection). In practical structural design, however, the step of finding the location of the IC is skipped by using tabulations of parameters for the most common geometries of bolted and welded connections, including those that are out of plane with the line of eccentricity. Knowing the location of the IC is inconsequential compared to knowing the capacity of the connection.

24. NOMENCLATURE

a	acceleration	ft/sec^2	m/s^2
d	distance	ft	m
g	gravitational acceleration, 32.2 (9.81)	ft/sec^2	m/s^2
H	height	ft	m
L	length	ft	m
n	rotational speed	rev/min	rev/min
r	radius	ft	m
R	earth's radius	ft	m
R	range	ft	m
s	position	ft	m
t	time	sec	s
T	total duration	sec	s
v	velocity	ft/sec	m/s
X	dimension	ft	m
Y	dimension	ft	m
z	elevation	ft	m

Symbols

α	angle	deg	deg
α	angular acceleration	rad/sec^2	rad/s^2
β	angle	deg	deg
γ	angle	deg	deg
θ	angular position	rad	rad
ϕ	angle or latitude	deg	deg
ω	angular velocity	rad/sec	rad/s

Subscripts

ϕ	transverse
0	initial
a	acceleration
ave	average
c	Coriolis
H	to maximum altitude
n	normal
O	at center
r	radial
t	tangential
x	x-component
y	y-component

57 Kinetics

Content in blue refers to the *NCEES Handbook*.

NCEES EXAM SPECIFICATIONS AND RELATED CONTENT

MACHINE DESIGN AND MATERIALS EXAM

I.B.3. Engineering Science and Mechanics: Dynamics
- 5. Linear Momentum
- 7. Angular Momentum
- 9. Newton's Second Law of Motion
- 10. Centripetal Force
- 13. Flat Friction
- 18. Motion of Rigid Bodies
- 25. Coefficient of Restitution

1. INTRODUCTION TO KINETICS

Kinetics is the study of motion and the forces that cause motion. Kinetics includes an analysis of the relationship between the force and mass for translational motion and between torque and moment of inertia for rotational motion. Newton's laws form the basis of the governing theory in the subject of kinetics.

2. RIGID BODY MOTION

The most general type of motion is *rigid body motion*. There are five types.

- *pure translation:* The orientation of the object is unchanged as its position changes. (Motion can be in straight or curved paths.)
- *rotation about a fixed axis:* All particles within the body move in concentric circles around the *axis of rotation*.
- *general plane motion:* The motion can be represented in two dimensions (i.e., in the *plane of motion*).
- *motion about a fixed point:* This describes any three-dimensional motion with one fixed point, such as a spinning top or a truck-mounted crane. The distance from a fixed point to any particle in the body is constant.
- *general motion:* This is any motion not falling into one of the other four categories.

Figure 57.1 illustrates the terms yaw, pitch, and roll as they relate to general motion. *Yaw* is a left or right swinging motion of the leading edge. *Pitch* is an up or down swinging motion of the leading edge. *Roll* is rotation about the leading edge's longitudinal axis.

Figure 57.1 *Yaw, Pitch, and Roll*

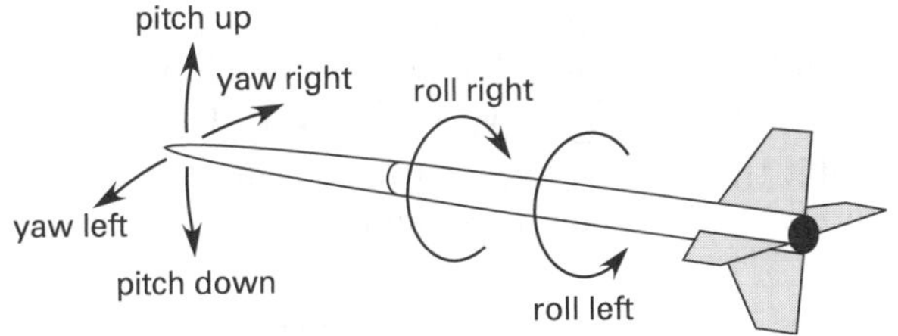

3. STABILITY OF EQUILIBRIUM POSITIONS

Stability is defined in terms of a body's relationship with an equilibrium position. *Neutral equilibrium* exists if a body, when displaced from its equilibrium position, remains in its displaced state. *Stable equilibrium* exists if the body returns to the original equilibrium position after experiencing a displacement. *Unstable equilibrium* exists if the body moves away from the equilibrium position. These terms are illustrated by Fig. 57.2.

Figure 57.2 *Types of Equilibrium*

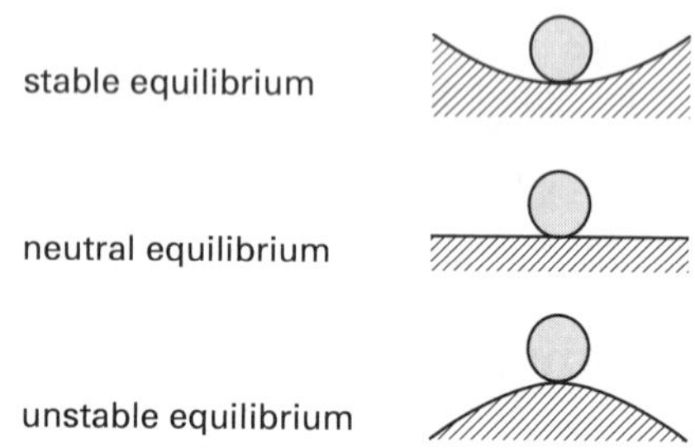

4. CONSTANT FORCES

Force is a push or a pull that one body exerts on another, including gravitational, electrostatic, magnetic, and contact influences. Forces that do not vary with time are *constant forces.*

Actions of other bodies on a rigid body are known as *external forces*. External forces are responsible for external motion of a body. *Internal forces* hold together parts of a rigid body.

5. LINEAR MOMENTUM

The vector *linear momentum* (usually just *momentum*) is defined by Eq. 57.1.[1] It has the same direction as the velocity vector. Momentum has units of force × time (e.g., lbf-sec or N·s).

$$\mathbf{p} = m\mathbf{v} \quad \text{[SI]} \qquad 57.1(a)$$

$$\mathbf{p} = \frac{m\mathbf{v}}{g_c} \quad \text{[U.S.]} \qquad 57.1(b)$$

Momentum is conserved when no external forces act on a particle. If no forces act on the particle, the velocity and direction of the particle are unchanged. The *law of conservation of momentum* states that the linear momentum is unchanged if no unbalanced forces act on the particle (i.e. $\sum m_1\mathbf{v}_1 = \sum m_2\mathbf{v}_2$). This does not prohibit the mass and velocity from changing, however. Only the product of mass and velocity is constant.

Depending on the nature of the problem, momentum can be conserved in any or all of the three coordinate directions. Equation 57.2 reflects momentum changes when external forces act on a particle.

Linear Momentum

$$\sum m_i(\mathbf{v}_i)_{t_2} = \sum m_i(\mathbf{v}_i)_{t_1} + \sum \int_{t_1}^{t_2} F_i\,dt \qquad 57.2$$

6. BALLISTIC PENDULUM

Figure 57.3 illustrates a *ballistic pendulum*. A projectile of known mass but unknown velocity is fired into a hanging target (the *pendulum*). The projectile is captured by the pendulum, which moves forward and upward. Kinetic energy is not conserved during impact because some of the projectile's kinetic energy is transformed into heat. However, momentum is conserved during impact, and the movement of the pendulum can be used to calculate the impact velocity of the projectile.[2]

Figure 57.3 *Ballistic Pendulum*

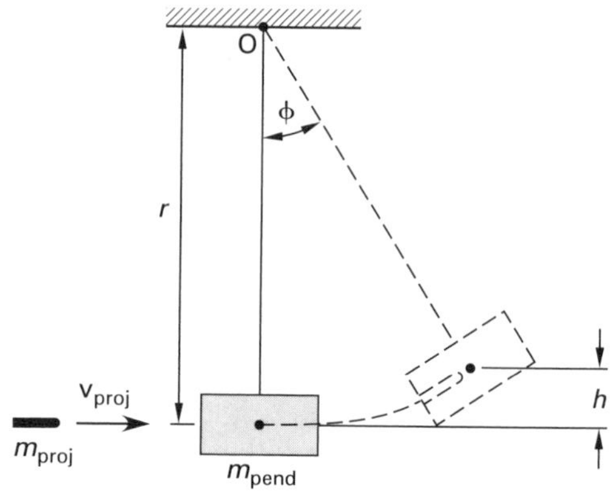

Since no external forces act on the block during impact, the momentum of the system is conserved.

$$\mathbf{p}_{\text{before impact}} = \mathbf{p}_{\text{after impact}} \qquad 57.3$$

$$m_{\text{proj}}\mathbf{v}_{\text{proj}} = (m_{\text{proj}} + m_{\text{pend}})\mathbf{v}_{\text{pend}} \qquad 57.4$$

Although kinetic energy before impact is not conserved, the total remaining energy after impact is conserved. That is, once the projectile has been captured by the

[1]The symbols **P**, **mom**, mv, and others are also used for momentum. Some authorities assign no symbol and just use the word momentum.

[2]In this type of problem, it is important to be specific about when the energy and momentum are evaluated. During impact, kinetic energy is not conserved, but momentum is conserved. After impact, as the pendulum swings, energy is conserved but momentum is not conserved because gravity (an external force) acts on the pendulum during its swing.

pendulum, the kinetic energy of the pendulum-projectile combination is converted totally to potential energy as the pendulum swings upward.

$$\frac{1}{2}(m_{\text{proj}} + m_{\text{pend}})\text{v}_{\text{pend}}^2 = (m_{\text{proj}} + m_{\text{pend}})gh \quad 57.5$$

$$\text{v}_{\text{pend}} = \sqrt{2gh} \quad 57.6$$

The relationship between the rise of the pendulum, h, and the swing angle, ϕ, is

$$h = r(1 - \cos\phi) \quad 57.7$$

Since the time during which the force acts is not well defined, there is no single equivalent force that can be assumed to initiate the motion. Any force that produces the same impulse over a given contact time will be applicable.

7. ANGULAR MOMENTUM

The vector *angular momentum* (also known as *moment of momentum*) taken about a point O is the moment of the linear momentum vector. Angular momentum has units of distance × force × time (e.g., ft-lbf-sec or N·m·s). It has the same direction as the rotation vector and can be determined by use of the right-hand rule. (That is, it acts in a direction perpendicular to the plane containing the position and linear momentum vectors.) (See Fig. 57.4.)

Figure 57.4 *Angular Momentum*

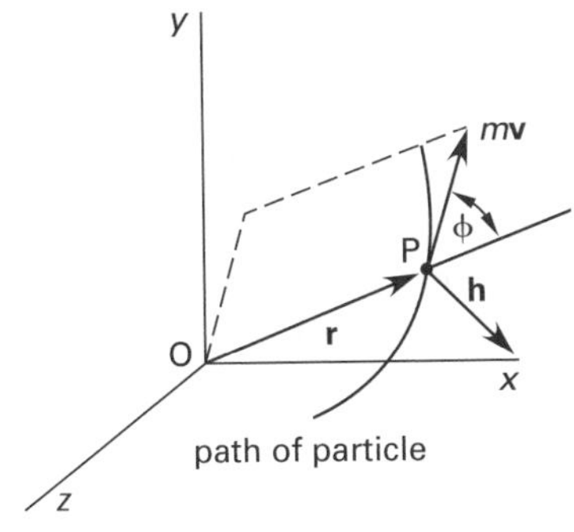

Angular Momentum

$$\mathbf{H}_{\text{O}} = \mathbf{r} \times m\mathbf{v} \quad \text{[SI]} \quad 57.8(a)$$

$$\mathbf{H}_{\text{O}} = \frac{\mathbf{r} \times m\mathbf{v}}{g_c} \quad \text{[U.S.]} \quad 57.8(b)$$

Any of the methods normally used to evaluate cross-products can be used with angular momentum. The scalar form of Eq. 57.8 is

$$h_{\text{O}} = rm\text{v}\sin\phi \quad \text{[SI]} \quad 57.9(a)$$

$$h_{\text{O}} = \frac{rm\text{v}\sin\phi}{g_c} \quad \text{[U.S.]} \quad 57.9(b)$$

For a rigid body rotating about an axis passing through its center of gravity located at point O, the angular momentum is given by Eq. 57.10.

Angular Momentum

$$\mathbf{H}_{\text{O}} = I_{\text{O}}\omega \quad \text{[SI]} \quad 57.10(a)$$

$$\mathbf{H}_{\text{O}} = \frac{I_{\text{O}}\omega}{g_{\text{O}}} \quad \text{[U.S.]} \quad 57.10(b)$$

8. NEWTON'S FIRST LAW OF MOTION

Much of this chapter is based on Newton's laws of motion. *Newton's first law of motion* can be stated in several forms.

- *common form:* A particle will remain in a state of rest or will continue to move with constant velocity unless an unbalanced external force acts on it.
- *law of conservation of momentum form:* If the resultant external force acting on a particle is zero, then the linear momentum of the particle is constant.

9. NEWTON'S SECOND LAW OF MOTION

Newton's second law of motion is stated as follows.

- *second law:* The acceleration of a particle is directly proportional to the force acting on it and inversely proportional to the particle mass. The direction of acceleration is the same as the direction of the force.

This law can be stated in terms of the force vector required to cause a change in momentum. The resultant force is equal to the rate of change of linear momentum.

$$\sum\mathbf{F} = \frac{d(m\mathbf{v})}{dt} \quad 57.11$$

If the mass is constant with respect to time,

Newton's Second Law (Equations of Motion)

$$\sum \mathbf{F} = m\frac{d\mathbf{v}}{dt} = m\mathbf{a} \qquad \text{[SI]} \quad 57.12(a)$$

$$\sum \mathbf{F} = \frac{m}{g_c}\frac{d\mathbf{v}}{dt} = \frac{m\mathbf{a}}{g_c} \qquad \text{[U.S.]} \quad 57.12(b)$$

Equation 57.12 can be written in rectangular coordinates form (i.e., in terms of x- and y-component forces), in polar coordinates form (i.e., tangential and normal components), and in cylindrical coordinates form (i.e., radial and transverse components).

Although Newton's laws do not specifically deal with rotation, there is an analogous relationship between torque and change in angular momentum. For a rotating body, the torque, **T**, required to change the angular momentum is

$$\sum T = \frac{dH_O}{dt} = \dot{H}_O = \frac{d(I_O\omega)}{dt}$$

$$\text{where } \sum(H_{Oi})_{t_2} = \sum(H_{Oi})_{t_1} + \sum\int_{t_1}^{t_2} T_{Oi}dt \qquad 57.13$$

If the moment of inertia is constant, the scalar form of Eq. 57.13 is given by Eq. 57.14.

$$\sum T = I\left(\frac{d\omega}{dt}\right) = I\alpha \qquad \text{[SI]} \quad 57.14(a)$$

$$\sum T = \left(\frac{I}{g_c}\right)\left(\frac{d\omega}{dt}\right) = \frac{I\alpha}{g_c} \qquad \text{[U.S.]} \quad 57.14(b)$$

Example 57.1

The acceleration in m/s² of a 40 kg body is specified by the equation

$$a(t) = 8 - 12t$$

What is the instantaneous force acting on the body at $t = 6$ s?

Solution

The acceleration is

$$a(6) = 8\ \frac{\text{m}}{\text{s}^2} - \left(12\ \frac{\text{m}}{\text{s}^3}\right)(6\ \text{s}) = -64\ \text{m/s}^2$$

From Newton's second law, the instantaneous force is

Newton's Second Law (Equations of Motion)

$$\begin{aligned}\sum F &= ma\\ f &= ma\\ &= (40\ \text{kg})\left(-64\ \frac{\text{m}}{\text{s}^2}\right)\\ &= -2560\ \text{N}\end{aligned}$$

10. CENTRIPETAL FORCE

Newton's second law says there is a force for every acceleration a body experiences. For a body moving around a curved path, the total acceleration can be separated into tangential and normal components. Acceleration associated with the normal component is defined as centripetal acceleration, where

Centripetal Acceleration

$$a_n = \frac{d\text{v}}{dt} = \frac{\text{v}_t^2}{r} \qquad 57.15$$

By Newton's second law, there are corresponding forces in the tangential and normal directions. The force associated with the normal acceleration is known as the *centripetal force*.[3]

$$F_c = ma_n = \frac{m\text{v}_t^2}{r} \qquad \text{[SI]} \quad 57.16(a)$$

$$F_c = \frac{ma_n}{g_c} = \frac{m\text{v}_t^2}{g_c r} \qquad \text{[U.S.]} \quad 57.16(b)$$

The centripetal force is a real force on the body toward the center of rotation. The so-called *centrifugal force* is an apparent force on the body directed away from the center of rotation. The centripetal and centrifugal forces are equal in magnitude but opposite in sign, as shown in Fig. 57.5.

Figure 57.5 *Centripetal Force*

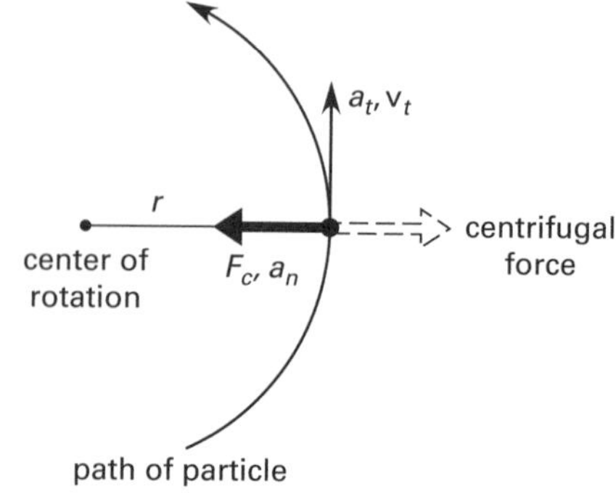

[3]The term *normal force* is reserved for the plane reaction in friction calculations.

An unbalanced rotating body (vehicle wheel, clutch disk, rotor of an electrical motor, etc.) will experience a dynamic *unbalanced force*. Though the force is essentially centripetal in nature and is given by Eq. 57.16, it is generally difficult to assign a value to the radius. For that reason, the force is often determined directly on the rotating body or from the deflection of its supports.

Since the body is rotating, the force will be experienced in all directions perpendicular to the axis of rotation. If the supports are flexible, the force will cause the body to vibrate, and the frequency of vibration will essentially be the rotational speed. If the supports are rigid, the bearings will carry the unbalanced force and transmit it to other parts of the frame.

Plane circular motion equations provide the basis for constant radius rotation, as shown.

Plane Circular Motion

$$\mathbf{r} = r\mathbf{e}_\theta \qquad 57.17$$

$$\mathbf{v} = r\omega\mathbf{r}_t \qquad 57.18$$

$$\mathbf{a} = (-r\omega^2)\mathbf{e}_r + r\alpha\mathbf{e}_\theta \qquad 57.19$$

The normal acceleration is given by Eq. 57.20

Plane Circular Motion

$$a_n = -r\omega^2 \qquad 57.20$$

Example 57.2

A 4500 lbm (2000 kg) car travels at 40 mph (65 kph) around a curve with a radius of 200 ft (60 m). What is the centripetal force?

SI Solution

The tangential velocity is

$$\mathrm{v}_t = \frac{\left(65\ \frac{\text{km}}{\text{h}}\right)\left(1000\ \frac{\text{m}}{\text{km}}\right)}{3600\ \frac{\text{s}}{\text{h}}} = 18.06\ \text{m/s}$$

From Eq. 57.16(a), the centripetal force is

$$F_c = \frac{m\mathrm{v}_t^2}{r} = \frac{(2000\ \text{kg})\left(18.06\ \frac{\text{m}}{\text{s}}\right)^2}{60\ \text{m}} = 10\,872\ \text{N}$$

Customary U.S. Solution

The tangential velocity is

$$\mathrm{v}_t = \frac{\left(40\ \frac{\text{mi}}{\text{hr}}\right)\left(5280\ \frac{\text{ft}}{\text{mi}}\right)}{3600\ \frac{\text{sec}}{\text{hr}}} = 58.67\ \text{ft/sec}$$

From Eq. 57.16(b), the centripetal force is

$$F_c = \frac{m\mathrm{v}_t^2}{g_c r} = \frac{(4500\ \text{lbm})\left(58.67\ \frac{\text{ft}}{\text{sec}}\right)^2}{\left(32.2\ \frac{\text{lbm-ft}}{\text{lbf-sec}^2}\right)(200\ \text{ft})} = 2405\ \text{lbf}$$

11. NEWTON'S THIRD LAW OF MOTION

Newton's third law of motion is as follows.

- *third law:* For every acting force between two bodies, there is an equal but opposite reacting force on the same line of action.

$$\mathbf{F}_{\text{reacting}} = -\mathbf{F}_{\text{acting}} \qquad 57.21$$

12. DYNAMIC EQUILIBRIUM

An accelerating body is not in static equilibrium. Accordingly, the familiar equations of statics ($\Sigma F = 0$ and $\Sigma M = 0$) do not apply. However, if the *inertial force*, $m\mathbf{a}$, is included in the static equilibrium equation, the body is said to be in *dynamic equilibrium*.[4,5] This is known as *D'Alembert's principle*. Since the inertial force acts to oppose changes in motion, it is negative in the summation.

$$\sum\mathbf{F} - m\mathbf{a} = 0 \qquad \text{[SI]} \qquad 57.22(a)$$

$$\sum\mathbf{F} - \frac{m\mathbf{a}}{g_c} = 0 \qquad \text{[U.S.]} \qquad 57.22(b)$$

It should be clear that D'Alembert's principle is just a different form of Newton's second law, with the ma term transposed to the left-hand side.

[4]Other names for the inertial force are *inertia vector* (when written as $m\mathbf{a}$), *dynamic reaction*, and *reversed effective force*. The term $\sum\mathbf{F}$ is known as the *effective force*.

[5]*Dynamic* and *equilibrium* are contradictory terms. A better term is *simulated equilibrium*, but this form has not caught on.

The analogous rotational form of the dynamic equilibrium principle is

$$\sum \mathbf{T} - I\alpha = 0 \quad \text{[SI]} \quad 57.23(a)$$

$$\sum \mathbf{T} - \frac{I\alpha}{g_c} = 0 \quad \text{[U.S.]} \quad 57.23(b)$$

13. FLAT FRICTION

Friction is a force that always resists motion or impending motion. It always acts parallel to the contacting surfaces. The frictional force, F_f, exerted on a stationary body is known as *static friction*, *Coulomb friction*, and *fluid friction*. If the body is moving, the friction is known as *dynamic friction* and is less than the static friction.

The actual magnitude of the frictional force depends on the *normal force*, N, and the *coefficient of friction*, μ, between the body and the surface. For a body resting on a horizontal surface, the normal force is the weight of the body. (See Fig. 57.6.)

$$N = mg \quad \text{[SI]} \quad 57.24(a)$$

$$N = \frac{mg}{g_c} \quad \text{[U.S.]} \quad 57.24(b)$$

If the body rests on an inclined surface, the normal force depends on the incline angle.

$$N = mg\cos\phi \quad \text{[SI]} \quad 57.25(a)$$

$$N = \frac{mg\cos\phi}{g_c} \quad \text{[U.S.]} \quad 57.25(b)$$

Figure 57.6 *Frictional and Normal Forces*

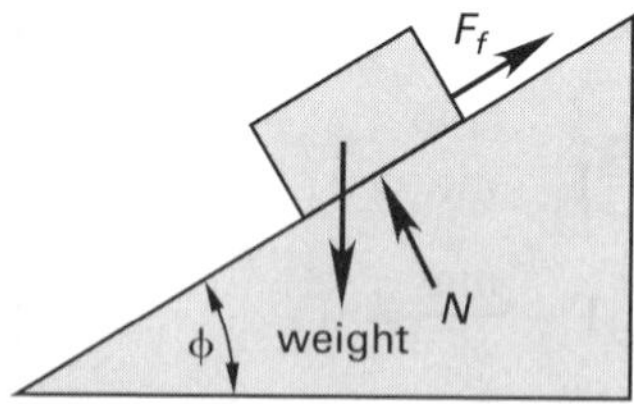

The maximum static frictional force (limiting friction), F_{max}, is the product of the coefficient of friction, μ, and the normal force, N. (The subscripts s and k are used to distinguish between the static and dynamic (kinetic) coefficients of friction.)

Friction

$$F_{f,max} \le \mu N \quad 57.26$$

The frictional force acts only in response to a disturbing force. If a small disturbing force (i.e., a force less than $F_{f,max}$) acts on a body, then the frictional force will equal the disturbing force, and the maximum frictional force will not develop. This occurs during the *equilibrium phase*. The *motion impending phase* is when the disturbing force equals the maximum frictional force, $F_{f,max}$. Once motion begins, however, the coefficient of friction drops slightly, and a lower frictional force opposes movement. These cases are illustrated in Fig. 57.7.

Figure 57.7 *Frictional Force Versus Disturbing Force*

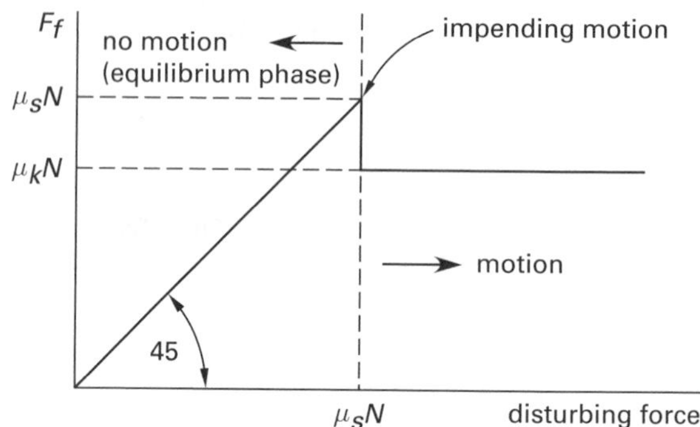

A body on an inclined plane will not begin to slip down the plane until the component of weight parallel to the plane exceeds the frictional force. If the plane's inclination angle can be varied, the body will not slip until the angle reaches a critical angle known as the *angle of repose* or *angle of static friction*, ϕ. Equation 57.27 relates this angle to the coefficient of static friction.

$$\tan\phi = \mu_s \quad 57.27$$

Tabulations of coefficients of friction distinguish between types of surfaces and between static and dynamic cases. They might also list values for dry conditions and oiled conditions. The term *dry* is synonymous with *nonlubricated*. The ambiguous term *wet*, although a natural antonym for *dry*, is sometimes used to mean *oily*. However, it usually means wet with water, as in tires on a wet roadway after a rain. Typical values of the coefficient of friction are given in Table 57.1.[6]

[6]Experimental and reported values of the coefficient of friction vary greatly from researcher to researcher and experiment to experiment. The values in Table 57.1 are more for use in solving practice problems than serving as the last word in available data.

Table 57.1 Typical Coefficients of Friction

materials	condition	dynamic	static
cast iron on cast iron	dry	0.15	1.00
plastic on steel	dry	0.35	0.45
grooved rubber on pavement	dry	0.40	0.55
bronze on steel	oiled	0.07	0.09
steel on graphite	dry	0.16	0.21
steel on steel	dry	0.42	0.78
steel on steel	oiled	0.08	0.10
steel on asbestos-faced steel	dry	0.11	0.15
steel on asbestos-faced steel	oiled	0.09	0.12
press fits (shaft in hole)	oiled	–	0.10–0.15

A special case of the angle of repose is the *angle of internal friction*, ϕ, of soil, grain, or other granular material. (See Fig. 57.8.) The angle made by a pile of granular material depends on how much friction there is between the granular particles. Liquids have angles of internal friction of zero, because they do not form piles.

Figure 57.8 *Angle of Internal Friction of a Pile*

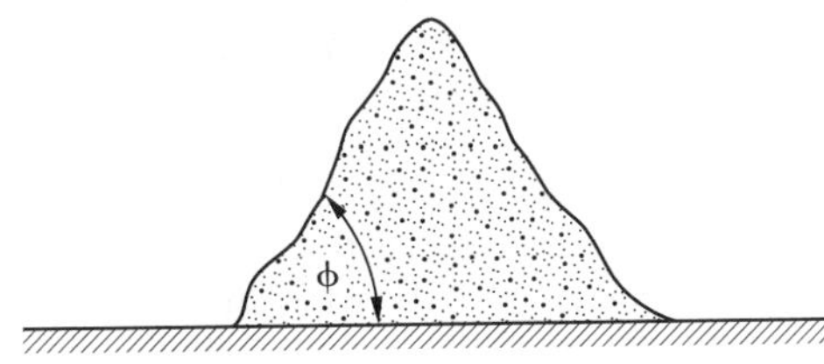

14. WEDGES

Wedges are machines that are able to raise heavy loads. The wedge angles are chosen so that friction will keep the wedge in place once it is driven between the load and support. As with any situation where friction is present, the frictional force is parallel to the contacting surfaces. (See Fig. 57.9.)

Figure 57.9 *Using a Wedge to Raise a Load*

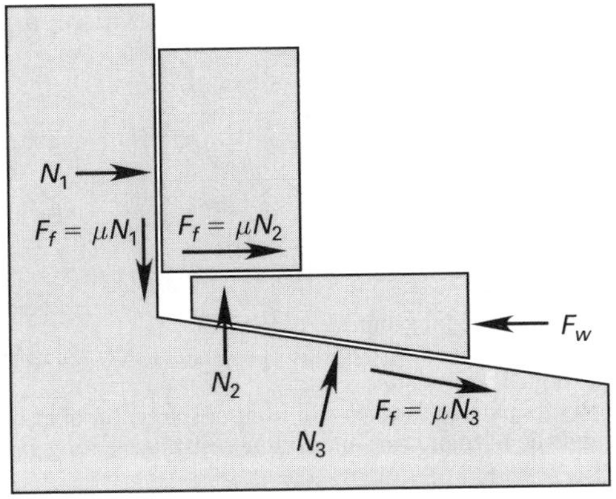

15. BELT FRICTION

Friction from a flat belt, rope, or band wrapped around a pulley or sheave is responsible for the transfer of torque. Except at start-up, one side of the belt (the tight side) will have a higher tension than the other (the slack side). The basic relationship between these belt tensions and the coefficient of friction neglects centrifugal effects and is given by Eq. 57.28.[7] (The angle of wrap, ϕ, must be expressed in radians.) (See Fig. 57.10.)

$$\frac{F_{\text{max}}}{F_{\text{min}}} = e^{\mu\phi} \qquad 57.28$$

Figure 57.10 *Flat Belt Friction*

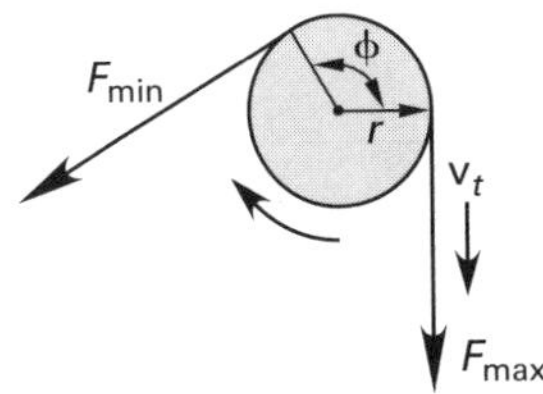

The net transmitted torque is

$$T = (F_{\text{max}} - F_{\text{min}})r \qquad 57.29$$

The power transmitted by the belt running at tangential velocity v_t is given by Eq. 57.30.[8]

$$P = (F_{\text{max}} - F_{\text{min}})v_t \qquad 57.30$$

The change in belt tension caused by centrifugal force should be considered when the velocity or belt mass is very large. Equation 57.31 can be used, where m is the mass per unit length of belt.

$$\frac{F_{\text{max}} - m v_t^2}{F_{\text{min}} - m v_t^2} = e^{\mu\phi} \qquad \text{[SI]} \qquad 57.31(a)$$

$$\frac{F_{\text{max}} - \dfrac{m v_t^2}{g_c}}{F_{\text{min}} - \dfrac{m v_t^2}{g_c}} = e^{\mu\phi} \qquad \text{[U.S.]} \qquad 57.31(b)$$

Example 57.3

During start-up, a 4.0 ft diameter pulley with centroidal moment of inertia of 1610 lbm-ft^2 is subjected to tight-side and loose-side belt tensions of 200 lbf and 100 lbf, respectively. A frictional torque of 15 ft-lbf is acting to resist pulley rotation. (a) What is the angular acceleration? (b) How long will it take the pulley to reach a speed of 120 rpm?

Solution

(a) From Eq. 57.29, the net torque is

$$\begin{aligned} T &= rF_{\text{net}} = (2 \text{ ft})(200 \text{ lbf} - 100 \text{ lbf}) - 15 \text{ ft-lbf} \\ &= 185 \text{ ft-lbf} \end{aligned}$$

From Eq. 57.14, the angular acceleration is

$$\begin{aligned} \alpha &= \frac{g_c T}{I} = \frac{\left(32.2 \, \frac{\text{lbm-ft}}{\text{lbf-sec}^2}\right)(185 \text{ ft-lbf})}{1610 \text{ lbm-ft}^2} \\ &= 3.7 \text{ rad/sec}^2 \end{aligned}$$

(b) The angular velocity is

$$\omega = \frac{\left(120 \, \frac{\text{rev}}{\text{min}}\right)\left(2\pi \, \frac{\text{rad}}{\text{rev}}\right)}{60 \, \frac{\text{sec}}{\text{min}}} = 12.6 \text{ rad/sec}$$

This is a case of constant angular acceleration starting from rest.

$$\begin{aligned} t &= \frac{\omega}{\alpha} = \frac{12.6 \, \frac{\text{rad}}{\text{sec}}}{3.7 \, \frac{\text{rad}}{\text{sec}^2}} \\ &= 3.4 \text{ sec} \end{aligned}$$

16. ROLLING RESISTANCE

Rolling resistance is a force that opposes motion, but it is not friction. Rather, it is caused by the deformation of the rolling body and the supporting surface. Rolling resistance is characterized by a *coefficient of rolling resistance*, a, which has units of length.[9] (See Fig. 57.11.) Since this deformation is very small, the rolling resistance in the direction of motion is

$$F_r = \frac{mga}{r} \qquad \text{[SI]} \qquad 57.32(a)$$

$$F_r = \frac{mga}{rg_c} = \frac{wa}{r} \qquad \text{[U.S.]} \qquad 57.32(b)$$

Figure 57.11 *Wheel Rolling Resistance*

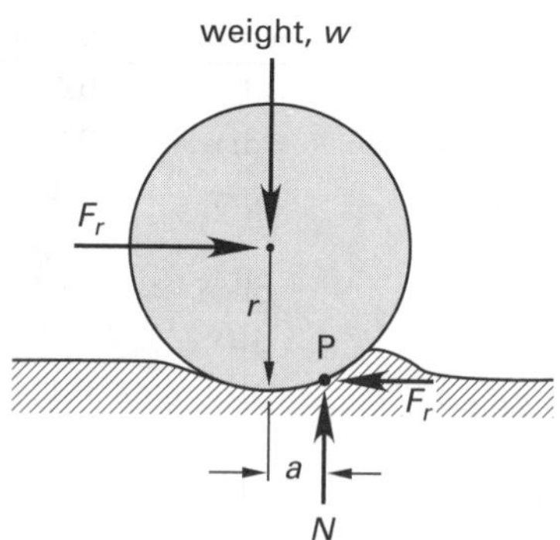

The term *coefficient of rolling friction*, μ_r, is occasionally encountered, although friction is not the cause of rolling resistance.

$$\mu_r = \frac{F_r}{w} = \frac{a}{r} \qquad 57.33$$

17. ROADWAY/CONVEYOR BANKING

If an object (e.g., a vehicle) travels in a circular path with instantaneous radius r and tangential velocity v_t, it will experience an apparent centrifugal force. The centrifugal force is resisted by a combination of surface (e.g., roadway) banking (*superelevation rate*) and *sideways friction*.[10] If the surface is banked so that friction is not required to resist the centrifugal force, the superelevation angle, ϕ, can be calculated from Eq. 57.34.[11]

$$\tan\phi = \frac{v_t^2}{gr} \qquad 57.34$$

Equation 57.34 can be solved for the *normal speed* corresponding to the geometry of a curve.

$$v_t = \sqrt{gr \tan\phi} \qquad 57.35$$

When friction is used to counteract some of the centrifugal force, the *side friction factor*, μ, between the object and surface (e.g., the tires and the roadway) is incorporated into the calculation of the superelevation angle.

$$e = \tan\phi = \frac{v_t^2 - \mu gr}{gr + \mu v_t^2} \qquad 57.36$$

[7]This equation does not apply to V-belts. V-belt design and analysis is dependent on the cross-sectional geometry of the belt.

[8]When designing a belt system, the horsepower to be transmitted should be multiplied by a *service factor* to obtain the *design power*. Service factors range from 1.0 to 1.5 and depend on the nature of the power source, the load, and the starting characteristics.

[9]Rolling resistance is traditionally derived by assuming the roller encounters a small step in its path a distance a in front of the center of gravity. The forces acting on the roller are the weight and driving force acting through the centroid and the normal force and rolling resistance acting at the contact point. Equation 57.32 is derived by taking moments about the contact point, P.

[10]The *superelevation rate* is the slope (in ft/ft or m/m) in the transverse direction (i.e., across the roadway).

[11]Generally it is not desirable to rely on roadway banking alone, since a particular superelevation angle would correspond to a single speed only.

If the banking angle, ϕ, is set to zero, Eq. 57.36 can be used to calculate the maximum velocity of an object in a turn when there is no banking.

For highway design, *superelevation rate*, e, is the amount of rise or fall of the cross slope per unit amount of horizontal width (i.e., the tangent of the slope angle above or below horizontal). Customary U.S. units are expressed in feet per foot, such as 0.06 ft/ft, or inch fractions per foot, such as 3/4 in/ft. SI units are millimeters per meter, such as 60 mm/m. The slope can also be expressed as a percent cross slope, such as 6% cross slope; or as a ratio, 1:17. The *superelevation*, y, is the difference in heights of the inside and outside edges of the curve.

When the speed, superelevation, and radius are such that no friction is required to resist sliding, the curve is said to be "balanced." There is no tendency for a vehicle to slide up or down the slope at the *balanced speed.* At any speed other than the balanced speed, some friction is needed to hold the vehicle on the slope. Given a friction factor, μ, the required value of the superelevation rate, e, can be calculated. (See Fig. 57.12.)

Figure 57.12 *Roadway Banking*

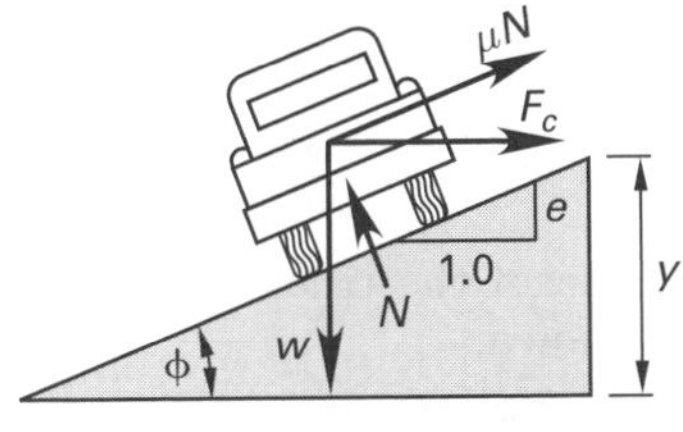

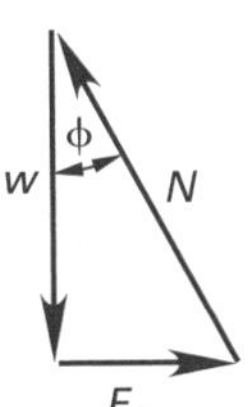

Equation 57.36 can be solved for the curve radius. For small banking angles (i.e., $\phi \leq 8°$), this simplifies to

$$r \approx \frac{v_t^2}{g(e+\mu)} \qquad 57.37$$

$$r_{\text{m}} = \frac{v_{\text{km/h}}^2}{127(e_{\text{m/m}}+\mu)} \qquad \text{[SI]} \qquad 57.38(a)$$

$$r_{\text{ft}} = \frac{v_{\text{mph}}^2}{15(e_{\text{ft/ft}}+\mu)} \qquad \text{[U.S.]} \qquad 57.38(b)$$

Example 57.4

A vehicle is traveling at 70 mph when it enters a circular curve of a test track. The curve radius is 240 ft. The sideways sliding coefficient of friction between the tires and the roadway is 0.57. (a) At what minimum angle from the horizontal must the curve be banked in order to prevent the vehicle from sliding off the top of the curve? (b) If the roadway is banked at 20° from the horizontal, what is the maximum vehicle speed such that no sliding occurs?

Solution

(a) The speed of the vehicle is

$$\begin{aligned} v_t &= \frac{\left(70\ \frac{\text{mi}}{\text{hr}}\right)\left(5280\ \frac{\text{ft}}{\text{mi}}\right)}{3600\ \frac{\text{sec}}{\text{hr}}} \\ &= 102.7\ \text{ft/sec} \end{aligned}$$

From Eq. 57.36, the required banking angle is

$$\begin{aligned} \phi &= \arctan\left(\frac{v_t^2 - \mu g r}{g r + \mu v_t^2}\right) \\ &= \arctan\left(\frac{\left(102.7\ \frac{\text{ft}}{\text{sec}}\right)^2 - (0.57)\left(32.2\ \frac{\text{ft}}{\text{sec}^2}\right)(240\ \text{ft})}{\left(32.2\ \frac{\text{ft}}{\text{sec}^2}\right)(240\ \text{ft}) + (0.57)\left(102.7\ \frac{\text{ft}}{\text{sec}}\right)^2}\right) \\ &= 24.1° \end{aligned}$$

(b) Solve Eq. 57.36 for the velocity.

$$\begin{aligned} v_t &= \sqrt{\frac{rg(\tan\phi+\mu)}{1-\mu\tan\phi}} \\ &= \sqrt{\frac{(240\ \text{ft})\left(32.2\ \frac{\text{ft}}{\text{sec}^2}\right)(\tan 20° + 0.57)}{1-0.57\tan 20°}} \\ &= 95.4\ \text{ft/sec} \quad (65.1\ \text{mi/hr}) \end{aligned}$$

18. MOTION OF RIGID BODIES

When a rigid body experiences pure translation, its position changes without any change in orientation. At any instant, all points on the body have the same displacement, velocity, and acceleration. The behavior of a rigid body in translation is generally given by Eq. 57.39 through Eq. 57.46. All equations are written for the center of mass. (These equations represent Newton's second law written in component form.)

Motion of a Rigid Body

$$a_x = \frac{F_x}{m} \qquad 57.39$$

$$\sum F_x = m a_x \qquad \text{[consistent units]} \qquad 57.40$$

If F is time dependent,

Motion of a Rigid Body

$$a_x(t) = \frac{F_x(t)}{m} \quad 57.41$$

$$\mathrm{v}_x(t) = \int_{t_0}^{t} a_x(\tau)\,d\tau + \mathrm{v}_{xt_0} \quad 57.42$$

$$x(t) = \int_{t_0}^{t} \mathrm{v}_x(\tau)\,d\tau + x_{t_0} \quad 57.43$$

If the force is constant,

Motion of a Rigid Body

$$a_x = \frac{F_x}{m} \quad 57.44$$

$$\mathrm{v}_x = a_x(t - t_0) + \mathrm{v}_{xt_0} \quad 57.45$$

$$x = a_x\frac{(t-t_0)^2}{2} + \mathrm{v}_{xt_0}(t - t_0) + x_{t_0} \quad 57.46$$

When a torque acts on a rigid body, the rotation will be about the center of gravity unless the body is constrained otherwise. In the case of rotation, the torque and angular acceleration are related by Eq. 57.47.

$$T = I\alpha \quad \text{[SI]} \quad 57.47(a)$$

$$T = \frac{I\alpha}{g_c} \quad \text{[U.S.]} \quad 57.47(b)$$

Euler's equations of motion are used to analyze the motion of a rigid body about a fixed point, O. This class of problem is particularly difficult because the mass moments of products of inertia change with time if a fixed set of axes is used. Therefore, it is more convenient to define the x-, y-, and z-axes with respect to the body. Such an action is acceptable because the angular momentum about the origin, $\mathbf{h}_\mathrm{O}$, corresponding to a given angular velocity, ω, is independent of the choice of coordinate axes. For fixed body rotation, Eq. 57.48 through Eq. 57.50 apply.

Motion of a Rigid Body

$$\omega = \frac{d\theta}{dt} \quad 57.48$$

$$\alpha = \frac{d\omega}{dt} \quad 57.49$$

$$\alpha\,d\theta = \omega\,d\omega \quad 57.50$$

An infinite number of axes can be chosen. (A general relationship between moments and angular momentum is given in most dynamics textbooks.) However, if the origin is at the mass center and the x-, y-, and z-axes coincide with the principal axes of inertia of the body (such that the product of inertia is zero), the angular momentum of the body about the origin (i.e., point O at (0, 0, 0)) is given by the simplified relationship

$$\mathbf{h}_\mathrm{O} = I_x\omega_x\mathbf{i} + I_y\omega_y\mathbf{j} + I_z\omega_z\mathbf{k} \quad 57.51$$

The three scalar Euler equations of motion can be derived from this simplified relationship.

$$\sum M_x = I_x\alpha_x - (I_y - I_z)\omega_y\omega_z \quad 57.52$$

$$\sum M_y = I_y\alpha_y - (I_z - I_x)\omega_z\omega_x \quad 57.53$$

$$\sum M_z = I_z\alpha_z - (I_x - I_y)\omega_x\omega_y \quad 57.54$$

Example 57.5

A 5000 lbm truck skids with a deceleration of 15 ft/sec^2. (a) What is the coefficient of sliding friction? (b) What are the frictional forces and normal reactions (per axle) at the tires?

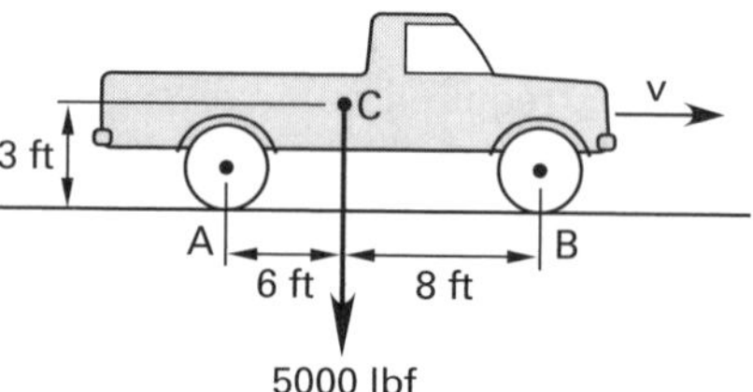

Solution

(a) The free-body diagram of the truck in equilibrium with the inertial force is shown.

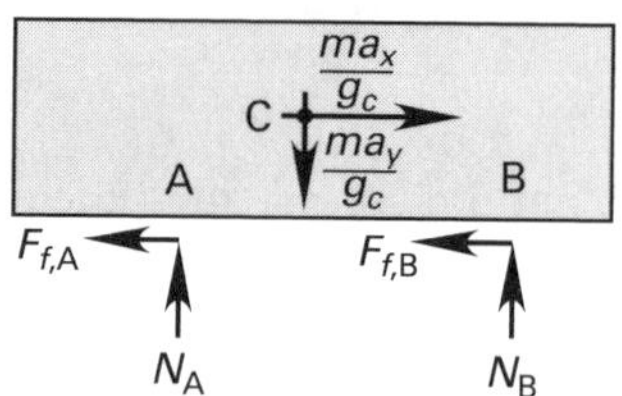

From Eq. 57.44, dynamic equilibrium in the horizontal direction is

Motion of a Rigid Body

$$a_x = \frac{F_x}{m}$$

$$\sum F_x = \frac{ma_x}{g_c} - F_{f,\mathrm{A}} - F_{f,\mathrm{B}} = 0$$

$$= \frac{ma_x}{g_c} - (N_\mathrm{A} + N_\mathrm{B})\mu = 0$$

$$\frac{(5000\ \text{lbm})\left(15\ \frac{\text{ft}}{\text{sec}^2}\right)}{32.2\ \frac{\text{lbm-ft}}{\text{lbf-sec}^2}} - (5000\ \text{lbf})\mu = 0$$

The coefficient of friction is

$$\mu = 0.466$$

(b) The vertical reactions at the tires can be found by taking moments about one of the contact points.

$$\sum M_{\text{A}}\text{: } 14N_{\text{B}} - (6 \text{ ft})(5000 \text{ lbf}) - (3 \text{ ft})\left(\frac{5000 \text{ lbm}}{32.2 \ \frac{\text{lbm-ft}}{\text{lbf-sec}^2}}\right)\left(15 \ \frac{\text{ft}}{\text{sec}^2}\right) = 0$$

$$N_{\text{B}} = 2642 \text{ lbf}$$

The remaining vertical reaction is found by summing vertical forces.

$$\sum F_y\text{: } N_{\text{A}} + N_{\text{B}} - \frac{mg}{g_c} = 0$$

$$N_{\text{A}} + 2642 \text{ lbf} - 5000 \text{ lbf} = 0$$

$$N_{\text{A}} = 2358 \text{ lbf}$$

The horizontal frictional forces at the front and rear axles are

$$F_{\mu,\text{A}} = (0.466)(2358 \text{ lbf}) = 1099 \text{ lbf}$$

$$F_{\mu,\text{B}} = (0.466)(2642 \text{ lbf}) = 1231 \text{ lbf}$$

19. CONSTRAINED MOTION

Figure 57.13 shows a cylinder (or sphere) on an inclined plane. If there is no friction, there will be no torque to start the cylinder rolling. Regardless of the angle, the cylinder will slide down the incline in *unconstrained motion.*

Figure 57.13 *Unconstrained Motion*

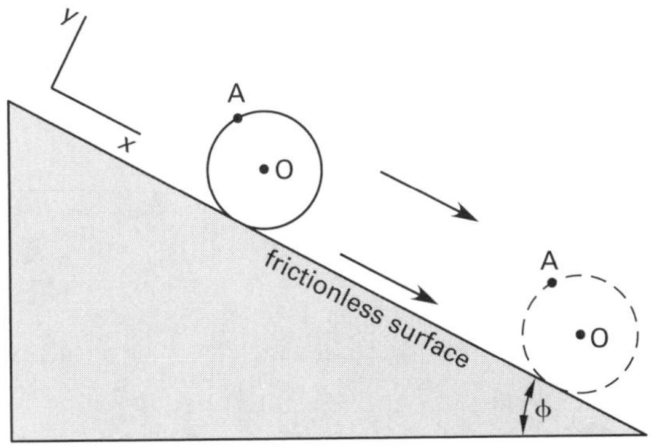

The acceleration sliding down the incline can be calculated by writing Newton's second law for an axis parallel to the plane.[12] Once the acceleration is known, the velocity can be found from the constant-acceleration equations.

$$ma_{\text{O},x} = mg \sin \phi \qquad 57.55$$

If friction is sufficiently large, or if the inclination is sufficiently small, there will be no slipping. This condition occurs if

$$\phi < \arctan \mu_s \qquad 57.56$$

The frictional force acting at the cylinder's radius, r, supplies a torque that starts and keeps the cylinder rolling. The frictional force is

$$F_f = \mu N \qquad 57.57$$

$$F_f = \mu mg \cos \phi \qquad \text{[SI]} \qquad 57.58(a)$$

$$F_f = \frac{\mu mg \cos \phi}{g_c} \qquad \text{[U.S.]} \qquad 57.58(b)$$

With no slipping, the cylinder has two degrees of freedom (the x-directional and angle of rotation), and motion of the center of mass must simultaneously satisfy (i.e., is constrained by) two equations. (This excludes motion perpendicular to the plane.) This is called *constrained motion.* (See Fig. 57.14.)

$$mg \sin \phi - F_\mu = ma_{\text{O},x} \qquad \text{[consistent units]} \qquad 57.59$$

$$F_\mu r = I_{\text{O}}\alpha \qquad \text{[consistent units]} \qquad 57.60$$

Figure 57.14 *Constrained Motion*

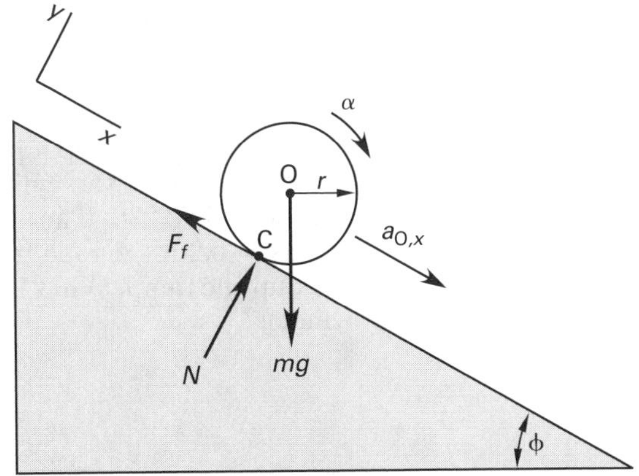

The mass moment of inertia used in calculating angular acceleration can be either the centroidal moment of inertia, I_{O}, or the moment of inertia taken about the contact point, I_{C}, depending on whether torques (moments) are evaluated with respect to point O or point C, respectively.

[12] Most inclined plane problems are conveniently solved by resolving all forces into components parallel and perpendicular to the plane.

If moments are evaluated with respect to point O, the coefficient of friction, μ, must be known, and the centroidal moment of inertia can be used. If moments are evaluated with respect to the contact point, the frictional and normal forces drop out of the torque summation. The cylinder instantaneously rotates as though it were pinned at point C. (See Fig. 57.15.) If the centroidal moment of inertia, I_O, is known, the parallel axis theorem can be used to find the required moment of inertia.

$$I_C = I_O + mr^2 \quad \text{57.61}$$

Figure 57.15 *Instantaneous Center of a Constrained Cylinder*

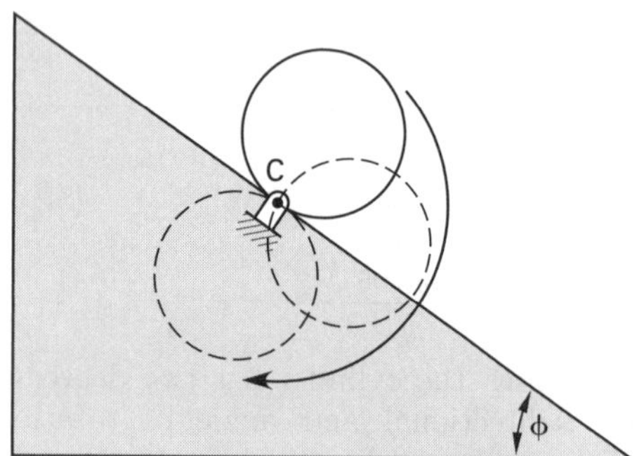

When there is no slipping, the cylinder will roll with constant linear and angular accelerations. The distance traveled by the center of mass can be calculated from the angle of rotation.

$$s_O = r\theta \quad \text{57.62}$$

If $\phi \geq \arctan \mu_s$, the cylinder will simultaneously roll and slide down the incline. The analysis is similar to the no-sliding case, except that the coefficient of sliding friction is used. Once sliding has started, the inclination angle can be reduced to $\arctan \mu_k$, and rolling with sliding will continue.

Example 57.6

A 150 kg cylinder with radius 0.3 m is pulled up a plane inclined at 30° as fast as possible without the cylinder slipping. The coefficient of friction is 0.236. There is a groove in the cylinder at radius = 0.2 m. A rope in the groove applies a force of 500 N up the ramp. What is the linear acceleration of the cylinder?

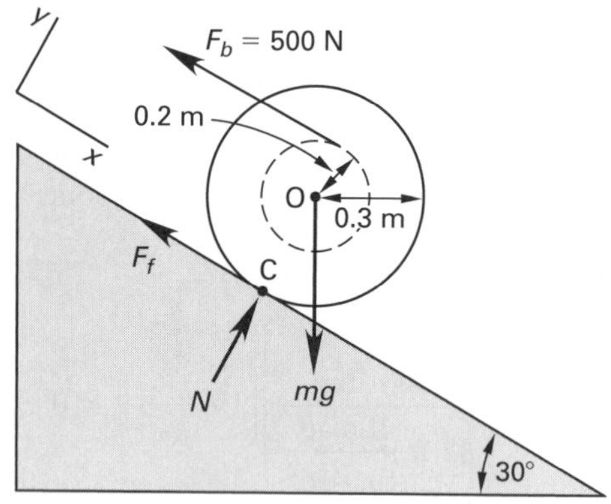

Solution

To solve by summing forces in the x-direction:

The normal force is

$$\begin{aligned} N &= mg\cos\phi = (150 \text{ kg})\left(9.81 \ \frac{\text{m}}{\text{s}^2}\right)(\cos 30°) \\ &= 1274.4 \text{ N} \end{aligned}$$

The frictional maximum (friction impending) force is

$$\begin{aligned} F_f &= \mu N = (0.236)(1274.4 \text{ N}) \\ &= 300.8 \text{ N} \end{aligned}$$

The summation of forces in the x-direction is

$$ma_{O,x} = mg\sin\phi - F_\mu - F_b$$

$$\begin{aligned} a_{O,x} &= \frac{(150 \text{ kg})\left(9.81 \ \frac{\text{m}}{\text{s}^2}\right)(\sin 30°) - 300.8 \text{ N} - 500 \text{ N}}{150 \text{ kg}} \\ &= -0.434 \text{ m/s}^2 \quad \text{[up the incline]} \end{aligned}$$

To solve by taking moment about the contact point:[13]

From App. 55.A, the centroidal mass moment of inertia of the cylinder is

$$\begin{aligned} I_O &= \tfrac{1}{2}mr^2 = (0.5)(150 \text{ kg})(0.3 \text{ m})^2 \\ &= 6.75 \text{ kg·m}^2 \end{aligned}$$

[13]This example can also be solved by summing moments about the center. If this is done, the governing equations are

$$\begin{aligned} I_O &= \tfrac{1}{2}mr^2 \\ M_O &= I_O\alpha = F_b r' - F_f r \\ \tfrac{1}{2}mr^2\alpha &= F_b r' - \mu mgr\cos\phi \\ a_{O,x} &= r\alpha \end{aligned}$$

The mass moment of inertia with respect to the contact point, C, is given by the parallel axis theorem.

$$I_C = I_O + mr^2 = \tfrac{1}{2}mr^2 + mr^2 = \tfrac{3}{2}mr^2$$
$$= \left(\frac{3}{2}\right)(150\ \text{kg})(0.3\ \text{m})^2$$
$$= 20.25\ \text{kg·m}^2$$

The x-component of the weight acts through the center of gravity. (This term dropped out when moments were taken with respect to the center of gravity.)

$$(mg)_x = mg\sin\phi = (150\ \text{kg})\left(9.81\ \frac{\text{m}}{\text{s}^2}\right)(\sin 30°)$$
$$= 735.8\ \text{N}$$

The summation of torques about point C gives the angular acceleration with respect to point C.

$$(735.8\ \text{N})(0.3\ \text{m}) - (500\ \text{N})(0.3\ \text{m} + 0.2\ \text{m}) = (20.25\ \text{kg·m}^2)\alpha$$
$$\alpha = -1.445\ \text{rad/s}^2$$

The linear acceleration can be calculated from the angular acceleration and the distance between points C and O.

$$a_{O,x} = r\alpha = (0.3\ \text{m})\left(-1.445\ \frac{\text{rad}}{\text{s}^2}\right)$$
$$= -0.433\ \text{m/s}^2 \quad \text{[up the incline]}$$

20. CABLE TENSION FROM AN ACCELERATING SUSPENDED MASS

When a mass hangs motionless from a cable, or when the mass is moving with a uniform velocity, the cable tension will equal the weight of the mass. However, when the mass is accelerating downward, the weight must be reduced by the inertial force. If the mass experiences a downward acceleration equal to the gravitational acceleration, there is no tension in the cable. Therefore, the two cases shown in Fig. 57.16 are not the same.

Figure 57.16 *Cable Tension from a Suspended Mass*

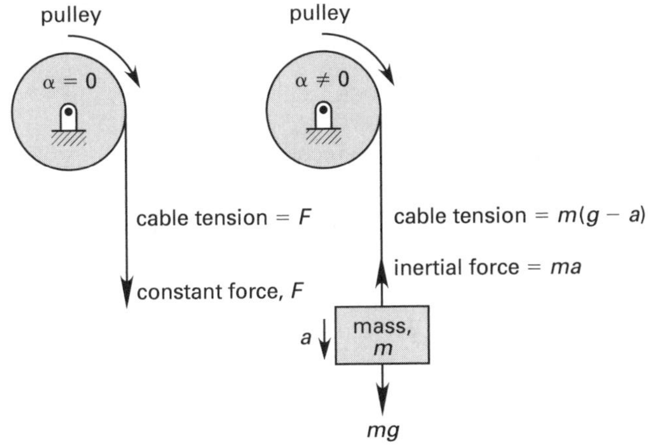

Example 57.7

A 300 lbm cylinder (I_O = 710 lbm-ft^2) has a narrow groove cut in it as shown. One end of the cable is wrapped around the cylinder in the groove, while the other end supports a 200 lbm mass. The pulley is massless and frictionless, and there is no slipping. Starting from a standstill, what are the linear accelerations of the 200 lbm mass and the cylinder?

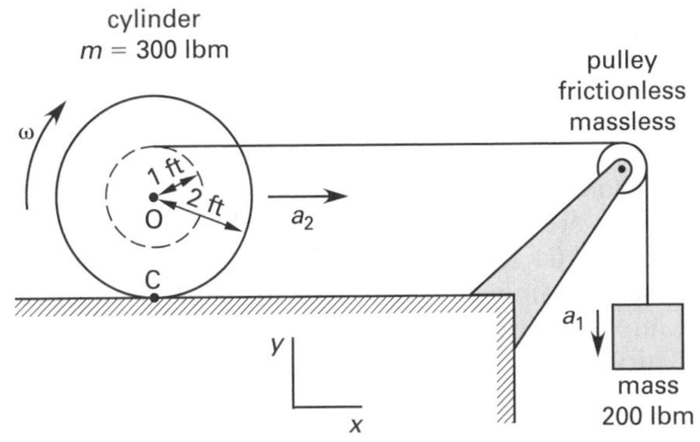

Solution

Since there is no slipping, there is friction between the cylinder and the plane. However, the coefficient of friction is not given. Therefore, moments must be taken about the contact point (the instantaneous center). The moment of inertia about the contact point is

$$I_C = I_O + mr^2$$
$$= 710\ \text{lbm-ft}^2 + (300\ \text{lbm})(2\ \text{ft})^2$$
$$= 1910\ \text{lbm-ft}^2$$

The first equation is a summation of forces on the mass.

$$\sum F_y\!: \ T + \frac{ma_1}{g_c} - \frac{mg}{g_c} = 0$$

The second equation is a summation of moments about the instantaneous center. The frictional force passes through the instantaneous center and is disregarded.

$$\sum M_C\!: \ T(2\ \text{ft} + 1\ \text{ft}) = \frac{I_C\alpha}{g_c}$$

Since there are three unknowns, a third equation is needed. This is the relationship between the linear and angular accelerations. a_2 is the acceleration of point O, located 2 ft from point C. a_1 is the acceleration of the cable, whose groove is located 3 ft from point C.

$$\alpha = \frac{a}{r} = \frac{a_2}{2\ \text{ft}} = \frac{a_1}{3\ \text{ft}}$$
$$a_1 = \tfrac{3}{2}a_2$$

Dynamics and Vibrations

Solving the three equations simultaneously yields

$$\text{cylinder: } a_2 = 10.4 \text{ ft/sec}^2$$
$$\text{mass: } a_1 = 15.6 \text{ ft/sec}^2$$
$$T = 103 \text{ lbf}$$
$$\alpha = 5.2 \text{ rad/sec}^2$$

21. IMPULSE

Impulse, **Imp**, is a vector quantity equal to the change in momentum.[14] Units of linear impulse are the same as for linear momentum: lbf-sec and N·s. Units of lbf-ft-sec and N·m·s are used for angular impulse. Equation 57.63 and Eq. 57.64 define the scalar magnitudes of *linear impulse* and *angular impulse*. Figure 57.17 illustrates that impulse is represented by the area under the F-t (or T-t) curve.

$$\text{Imp} = \int_{t_1}^{t_2} F\,dt \quad \text{[linear]} \qquad 57.63$$

$$\text{Imp} = \int_{t_1}^{t_2} T\,dt \quad \text{[angular]} \qquad 57.64$$

Figure 57.17 *Impulse*

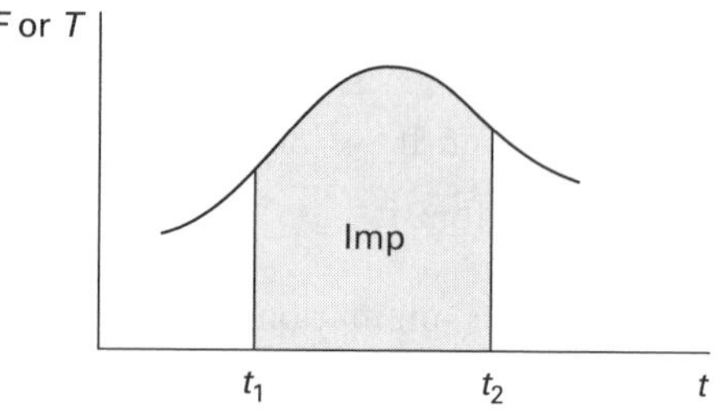

If the applied force or torque is constant, impulse is easily calculated. A large force acting for a very short period of time is known as an *impulsive force*.

$$\text{Imp} = F(t_2 - t_1) \quad \text{[linear]} \qquad 57.65$$

$$\text{Imp} = T(t_2 - t_1) \quad \text{[angular]} \qquad 57.66$$

If the impulse is known, the average force acting over the duration of the impulse is

$$F_{\text{ave}} = \frac{\text{Imp}}{\Delta t} \qquad 57.67$$

22. IMPULSE-MOMENTUM PRINCIPLE

The change in momentum is equal to the applied *impulse*. This is known as the *impulse-momentum* principle. For a linear system with constant force and mass, the scalar magnitude form of this principle is

$$\text{Imp} = \Delta p \qquad 57.68$$

$$F(t_2 - t_1) = m(\text{v}_2 - \text{v}_1) \quad \text{[SI]} \qquad 57.69(a)$$

$$F(t_2 - t_1) = \frac{m(\text{v}_2 - \text{v}_1)}{g_c} \quad \text{[U.S.]} \qquad 57.69(b)$$

For an angular system with constant torque and moment of inertia, the analogous equations are

$$T(t_2 - t_1) = I(\omega_2 - \omega_1) \quad \text{[SI]} \qquad 57.70(a)$$

$$T(t_2 - t_1) = \frac{I(\omega_2 - \omega_1)}{g_c} \quad \text{[U.S.]} \qquad 57.70(b)$$

Example 57.8

A 2000 kg cannon fires a 10 kg projectile horizontally at 600 m/s. It takes 0.007 s for the projectile to pass through the barrel and 0.01 s for the cannon to recoil. The cannon has a spring mechanism to absorb the recoil. (a) What is the cannon's initial recoil velocity? (b) What force is exerted on the recoil spring?

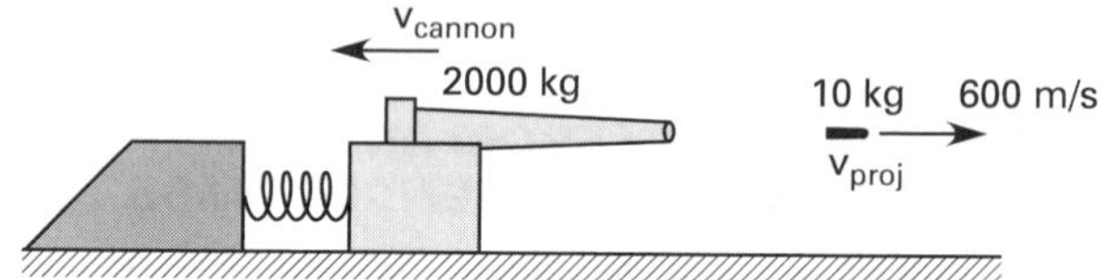

Solution

(a) The accelerating force is applied to the projectile quickly, and external forces such as gravity and friction are not significant factors. Therefore, momentum is conserved.

$$\sum p\text{: } m_{\text{proj}}\Delta\text{v}_{\text{proj}} = m_{\text{cannon}}\Delta\text{v}_{\text{cannon}}$$
$$(10 \text{ kg})\left(600 \ \frac{\text{m}}{\text{s}}\right) = (2000 \text{ kg})\text{v}_{\text{cannon}}$$
$$\text{v}_{\text{cannon}} = 3 \text{ m/s}$$

(b) From Eq. 57.69(a), the recoil force is

$$F = \frac{m\Delta\text{v}}{\Delta t} = \frac{(2000 \text{ kg})\left(3 \ \frac{\text{m}}{\text{s}}\right)}{0.01 \text{ s}} = 6 \times 10^5 \text{ N}$$

[14]Although **Imp** is the most common notation, engineers have no universal symbol for impulse. Some authors use I, **I**, and i, but these symbols can be mistaken for moment of inertia. Other authors merely use the word *impulse* in their equations.

23. IMPULSE-MOMENTUM PRINCIPLE IN OPEN SYSTEMS

The impulse-momentum principle can be used to determine the forces acting on flowing fluids (i.e., in open systems). This is the method used to calculate forces in jet engines and on pipe bends, and forces due to other changes in flow geometry. Equation 57.71 is rearranged in terms of a mass flow rate.

$$F = \frac{m\Delta \text{v}}{\Delta t} = \dot{m}\Delta \text{v} \quad \text{[SI]} \qquad 57.71(a)$$

$$F = \frac{m\Delta \text{v}}{g_c\Delta t} = \frac{\dot{m}\Delta \text{v}}{g_c} \quad \text{[U.S.]} \qquad 57.71(b)$$

Example 57.9

Air enters a jet engine at 1500 ft/sec (450 m/s) and leaves at 3000 ft/sec (900 m/s). The thrust produced is 10,000 lbf (44 500 N). Disregarding the small amount of fuel added during combustion, what is the mass flow rate?

SI Solution

From Eq. 57.71(a),

$$\dot{m} = \frac{F}{\Delta \text{v}} = \frac{44\,500 \text{ N}}{900 \frac{\text{m}}{\text{s}} - 450 \frac{\text{m}}{\text{s}}} = 98.9 \text{ kg/s}$$

Customary U.S. Solution

From Eq. 57.71(b),

$$\dot{m} = \frac{Fg_c}{\Delta \text{v}} = \frac{(10{,}000 \text{ lbf})\left(32.2 \frac{\text{lbm-ft}}{\text{lbf-sec}^2}\right)}{3000 \frac{\text{ft}}{\text{sec}} - 1500 \frac{\text{ft}}{\text{sec}}}$$
$$= 215 \text{ lbm/sec}$$

Example 57.10

20 kg of sand fall continuously each second on a conveyor belt moving horizontally at 0.6 m/s. What power is required to keep the belt moving?

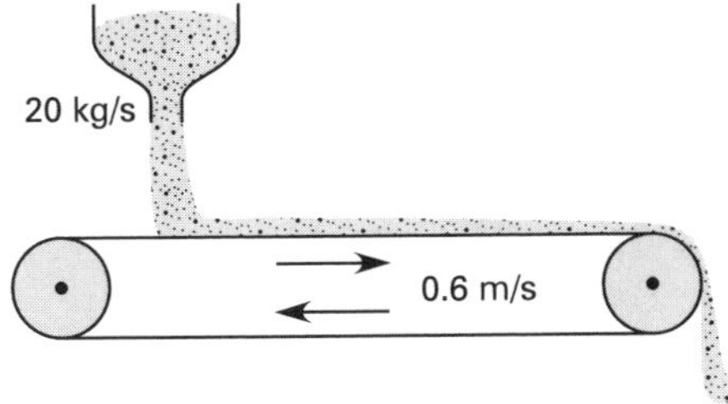

Solution

From Eq. 57.71(a), the force on the sand is

$$F = \dot{m}\Delta \text{ v} = \left(20 \frac{\text{kg}}{\text{s}}\right)\left(0.6 \frac{\text{m}}{\text{s}}\right) = 12 \text{ N}$$

The power required is

$$P = F\text{v} = (12 \text{ N})\left(0.6 \frac{\text{m}}{\text{s}}\right)$$
$$= 7.2 \text{ W}$$

Example 57.11

A 6 × 9, ⅝ in diameter hoisting cable (area of 0.158 in^2, modulus of elasticity of 12 × 10^6 lbf/in^2) carries a 1000 lbm load at its end. The load is being lowered vertically at the rate of 4 ft/sec. When 200 ft of cable have been reeled out, the take-up reel suddenly locks. Neglect the cable mass. What are the (a) cable stretch, (b) maximum dynamic force in the cable, (c) maximum dynamic stress in the cable, and (d) approximate time for the load to come to a stop vertically?

Solution

(a) The stiffness of the cable is

$$k = \frac{F}{x} = \frac{AE}{L} = \frac{(0.158 \text{ in}^2)\left(12 \times 10^6 \frac{\text{lbf}}{\text{in}^2}\right)}{(200 \text{ ft})\left(12 \frac{\text{in}}{\text{ft}}\right)}$$
$$= 790 \text{ lbf/in}$$

Neglecting the cable mass, the kinetic energy of the moving load is

$$E_k = \frac{m\text{v}^2}{2g_c} = \frac{(1000 \text{ lbm})\left(4 \frac{\text{ft}}{\text{sec}}\right)^2\left(12 \frac{\text{in}}{\text{ft}}\right)}{(2)\left(32.2 \frac{\text{lbm-ft}}{\text{lbf-sec}^2}\right)}$$
$$= 2981 \text{ in-lbf}$$

By the work-energy principle, the decrease in kinetic energy is equal to the work of lengthening the cable (i.e., the energy stored in the spring).

$$\Delta E_k = \tfrac{1}{2}k\delta^2$$
$$2981 \text{ in-lbf} = \left(\frac{1}{2}\right)\left(790 \frac{\text{lbf}}{\text{in}}\right)\delta^2$$
$$\delta = 2.75 \text{ in}$$

(b) The maximum dynamic force in the cable is

$$F = k\delta = \left(790 \ \frac{\text{lbf}}{\text{in}}\right)(2.75 \text{ in}) = 2173 \text{ lbf}$$

(c) The maximum dynamic tensile stress in the cable is

$$\sigma = \frac{F}{A} = \frac{2173 \text{ lbf}}{0.158 \text{ in}^2} = 13{,}753 \text{ lbf/in}^2$$

(d) Since the tensile force in the cable increases from zero to the maximum while the load decelerates, the average decelerating force is half of the maximum force. From the impulse momentum principle, Eq. 57.69,

$$F\Delta t = \frac{m\Delta \text{v}}{g_c}$$

$$\left(\frac{1}{2}\right)(2173 \text{ lbf})\Delta t = \frac{(1000 \text{ lbm})\left(4 \ \frac{\text{ft}}{\text{sec}}\right)}{32.2 \ \frac{\text{lbm-ft}}{\text{lbf-sec}^2}}$$

$$\Delta t = 0.114 \text{ sec}$$

24. IMPACTS

According to Newton's second law, momentum is conserved unless a body is acted upon by an external force such as gravity or friction from another object. In an *impact* or *collision*, contact is very brief and the effect of external forces is insignificant. Therefore, momentum is conserved, even though energy may be lost through heat generation and deformation of the bodies.

Consider two particles, initially moving with velocities v_1 and v_2 on a collision path, as shown in Fig. 57.18. The conservation of momentum equation can be used to find the velocities after impact, v_1' and v_2'. (Observe algebraic signs with velocities.)

$$m_1\text{v}_1 + m_2\text{v}_2 = m_1\text{v}_1' + m_2\text{v}_2' \qquad 57.72$$

Figure 57.18 *Direct Central Impact*

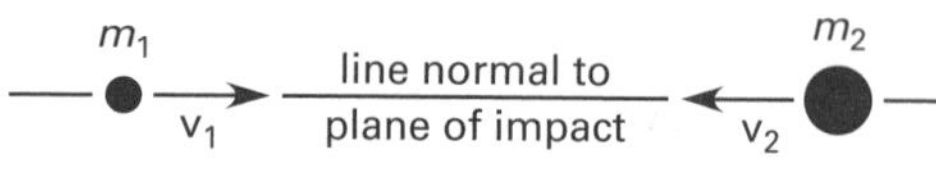

The impact is said to be an *inelastic impact* if kinetic energy is lost. (Other names for an inelastic impact are *plastic impact* and *endoergic impact*.[15]) The impact is said to be *perfectly inelastic* or *perfectly plastic* if the two particles stick together and move on with the same final velocity.[16] The impact is said to be *elastic* only if kinetic energy is conserved.

$$m_1\text{v}_1^2 + m_2\text{v}_2^2 = m_1\text{v}_1'^2 + m_2\text{v}_2'^2\Big|_{\text{elastic impact}} \qquad 57.73$$

25. COEFFICIENT OF RESTITUTION

A simple way to determine whether the impact is elastic or inelastic is by calculating the *coefficient of restitution*, e. The collision is inelastic if $e < 1.0$, perfectly inelastic if $e = 0$, and elastic if $e = 1.0$. The coefficient of restitution is the ratio of relative velocity differences along a mutual straight line where velocity components are normal to the plane of impact. (When both impact velocities are not directed along the same straight line, the coefficient of restitution should be calculated separately for each velocity component.)

$$e = \frac{\text{relative separation velocity}}{\text{relative approach velocity}} = \frac{(\text{v}_2')_n - (\text{v}_1')_n}{(\text{v}_1)_n - (\text{v}_2)_n} \qquad 57.74$$

Using e, post-impact velocities can be found from Eq. 57.75 and Eq. 57.76.

Coefficient of Restitution

$$(\text{v}_1')_n = \frac{m_2(\text{v}_2)_n(1+e) + (m_1 - em_2)(\text{v}_1)_n}{m_1 + m_2} \qquad 57.75$$

$$(\text{v}_2')_n = \frac{m_1(\text{v}_1)_n(1+e) + (em_1 - m_2)(\text{v}_2)_n}{m_1 + m_2} \qquad 57.76$$

26. REBOUND FROM STATIONARY PLANES

Figure 57.19 illustrates the case of an object rebounding from a massive, stationary plane.[17] This is an impact where $m_2 = \infty$ and $\text{v}_2 = 0$. The impact force acts perpendicular to the plane, regardless of whether the impact is elastic or inelastic. Therefore, the x-component of velocity is unchanged. Only the y-component of velocity is affected, and even then, only if the impact is inelastic.

$$\text{v}_x = \text{v}_x' \qquad 57.77$$

[15]Theoretically, there is also an *exoergic impact* (i.e., one in which kinetic energy is gained during the impact). However, this can occur only in special cases, such as in nuclear reactions.

[16]In traditional textbook problems, clay balls should be considered perfectly inelastic.

[17]The particle path is shown as a straight line in Fig. 57.19 for convenience. The path will be a straight line only when the particle is dropped straight down, or essentially a straight line when velocities are high, time in flight is short, and distances are short. Otherwise, the path will be parabolic.

Figure 57.19 *Rebound from a Stationary Plane*

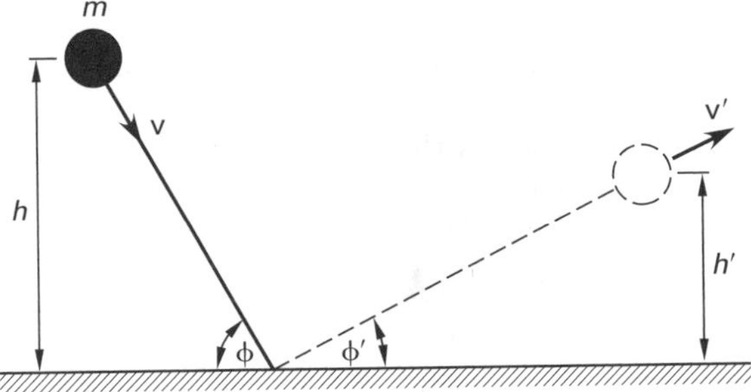

The coefficient of restitution can be used to calculate the *rebound angle*, *rebound height*, and *rebound velocity*.

$$e = \frac{\tan\phi'}{\tan\phi} = \sqrt{\frac{h'}{h}} = \frac{-v'_y}{v_y} \quad \text{57.78}$$

Example 57.12

A golf ball dropped vertically from a height of 8.0 ft (2.4 m) onto a hard surface rebounds to a height of 6.0 ft (1.8 m). What are the (a) impact velocity, (b) rebound velocity, and (c) coefficient of restitution?

SI Solution

The impact velocity of the golf ball is found by equating the decrease in potential energy to the increase in kinetic energy.

$$-\tfrac{1}{2}mv^2 = mgh$$

$$\begin{aligned} v &= -\sqrt{2gh} = -\sqrt{(2)\left(9.81\ \frac{\text{m}}{\text{s}^2}\right)(2.4\ \text{m})} \\ &= -6.86\ \text{m/s} \quad \text{[negative because down]} \end{aligned}$$

The rebound velocity can be found from the rebound height.

$$\begin{aligned} v' &= \sqrt{2gh'} = \sqrt{(2)\left(9.81\ \frac{\text{m}}{\text{s}^2}\right)(1.8\ \text{m})} \\ &= 5.94\ \text{m/s} \end{aligned}$$

From Eq. 57.74 with $v_2 = v_2' = 0$ m/s,

$$\begin{aligned} e &= \frac{v_2' - v_1'}{v_1 - v_2} = \frac{0\ \frac{\text{m}}{\text{s}} - \left(5.94\ \frac{\text{m}}{\text{s}}\right)}{-6.86\ \frac{\text{m}}{\text{s}} - 0\ \frac{\text{m}}{\text{s}}} \\ &= 0.866 \end{aligned}$$

Customary U.S. Solution

The impact velocity is

$$\begin{aligned} v &= -\sqrt{2gh} = -\sqrt{(2)\left(32.2\ \frac{\text{ft}}{\text{sec}^2}\right)(8.0\ \text{ft})} \\ &= -22.7\ \text{ft/sec} \quad \text{[negative because down]} \end{aligned}$$

Similarly, the rebound velocity is

$$\begin{aligned} v' &= \sqrt{2gh'} = \sqrt{(2)\left(32.2\ \frac{\text{ft}}{\text{sec}^2}\right)(6.0\ \text{ft})} \\ &= 19.7\ \text{ft/sec} \end{aligned}$$

From Eq. 57.74 with $v_2 = v_2' = 0$ ft/sec,

$$\begin{aligned} e &= \frac{v_2' - v_1'}{v_1 - v_2} = \frac{0\ \frac{\text{ft}}{\text{sec}} - 19.7\ \frac{\text{ft}}{\text{sec}}}{-22.7\ \frac{\text{ft}}{\text{sec}} - 0\ \frac{\text{ft}}{\text{sec}}} \\ &= 0.868 \end{aligned}$$

27. COMPLEX IMPACTS

The simplest type of impact is the direct central impact, shown in Fig. 57.18. An impact is said to be a *direct impact* when the velocities of the two bodies are perpendicular to the contacting surfaces. *Central impact* occurs when the force of the impact is along the line of connecting centers of gravity. Round bodies (i.e., spheres) always experience central impact, whether or not the impact is direct.

When the velocities of the bodies are not along the same line, the impact is said to be an *oblique impact*, as illustrated in Fig. 57.20. The coefficient of restitution can be used to find the x-components of the resultant velocities. Since impact is central, the y-components of velocities will be unaffected by the collision.

$$v_{1y} = v_{1y}' \quad \text{57.79}$$

$$v_{2y} = v_{2y}' \quad \text{57.80}$$

$$e = \frac{v_{1x}' - v_{2x}'}{v_{2x} - v_{1x}} \quad \text{57.81}$$

$$m_1 v_{1x} + m_2 v_{2x} = m_1 v_{1x}' + m_2 v_{2x}' \quad \text{57.82}$$

Figure 57.20 *Central Oblique Impact*

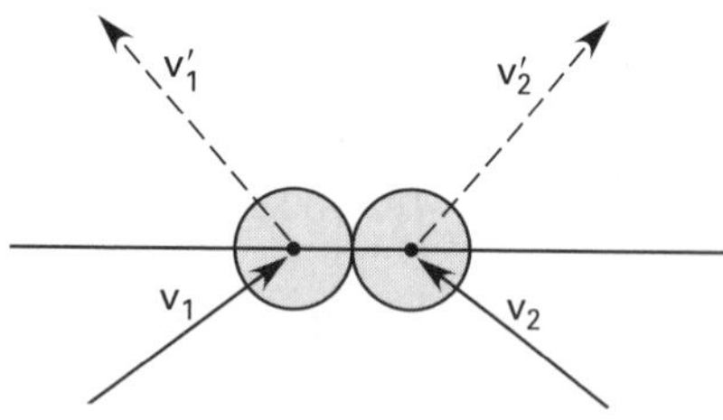

Eccentric impacts are neither direct nor central. The coefficient of restitution can be used to calculate the linear velocities immediately after impact along a line normal to the contact surfaces. Since the impact is not central, the bodies will rotate. Other methods must be used to calculate the rate of rotation. (See Fig. 57.21.)

$$e = \frac{v'_{1n} - v'_{2n}}{v_{2n} - v_{1n}} \qquad 57.83$$

Figure 57.21 *Eccentric Impact*

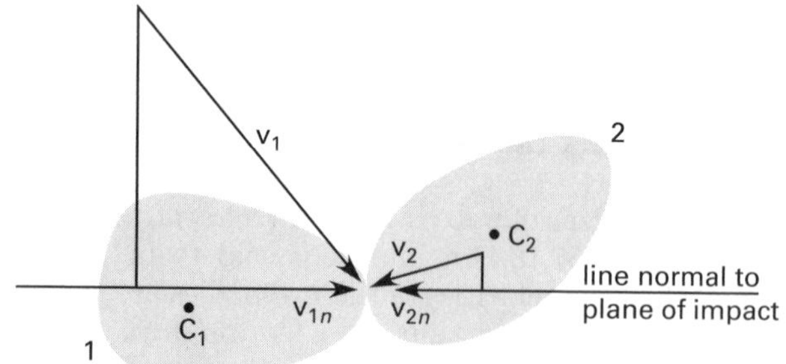

28. VELOCITY-DEPENDENT FORCE

A force that is a function of velocity is known as a *velocity-dependent force.* A common example of a velocity-dependent force is the *viscous drag* a particle experiences when falling through a fluid. There are two main cases of viscous drag: linear and quadratic.

A *linear velocity-dependent force* is proportional to the first power of the velocity. A linear relationship is typical of a particle falling slowly through a fluid (i.e., viscous drag in laminar flow). In Eq. 57.84, C is a constant of proportionality known as the *viscous coefficient* or *coefficient of viscous damping.*

$$F_b = C\mathrm{v} \qquad 57.84$$

In the case of a particle falling slowly through a viscous liquid, the differential equation of motion and its solution are derived from Newton's second law.

$$mg - C\mathrm{v} = ma \qquad \text{[SI]} \qquad 57.85(a)$$

$$\frac{mg}{g_c} - C\mathrm{v} = \frac{ma}{g_c} \qquad \text{[U.S.]} \qquad 57.85(b)$$

$$\mathrm{v}(t) = \mathrm{v}_t(1 - e^{-Ct/m}) \qquad \text{[SI]} \qquad 57.86(a)$$

$$\mathrm{v}(t) = \mathrm{v}_t(1 - e^{-Cg_ct/m}) \qquad \text{[U.S.]} \qquad 57.86(b)$$

Equation 57.86 shows that the velocity asymptotically approaches a final value known as the *terminal velocity*, v_t. For laminar flow, the terminal velocity is

$$\mathrm{v}_t = \frac{mg}{C} \qquad \text{[SI]} \qquad 57.87(a)$$

$$\mathrm{v}_t = \frac{mg}{Cg_c} \qquad \text{[U.S.]} \qquad 57.87(b)$$

A *quadratic velocity-dependent force* is proportional to the second power of the velocity. A quadratic relationship is typical of a particle falling quickly through a fluid (i.e., turbulent flow).

$$F_b = C\mathrm{v}^2 \qquad 57.88$$

In the case of a particle falling quickly through a liquid under the influence of gravity, the differential equation of motion is

$$mg - C\mathrm{v}^2 = ma \qquad \text{[SI]} \qquad 57.89(a)$$

$$\frac{mg}{g_c} - C\mathrm{v}^2 = \frac{ma}{g_c} \qquad \text{[U.S.]} \qquad 57.89(b)$$

For turbulent flow, the terminal velocity is

$$\mathrm{v}_t = \sqrt{\frac{mg}{C}} \qquad \text{[SI]} \qquad 57.90(a)$$

$$\mathrm{v}_t = \sqrt{\frac{mg}{Cg_c}} \qquad \text{[U.S.]} \qquad 57.90(b)$$

If a skydiver falls far enough, a turbulent terminal velocity of approximately 125 mph (200 kph) will be achieved. With a parachute (but still in turbulent flow), the terminal velocity is reduced to approximately 25 mph (40 kph).

29. VARYING MASS

Integral momentum equations must be used when the mass of an object varies with time. Most varying mass problems are complex, but the simplified case of an ideal rocket can be evaluated. This discussion assumes constant gravitational force, constant fuel usage, and constant exhaust velocity. (For brevity of presentation, all of the following equations are presented in consistent form only.)

The forces acting on the rocket are its thrust, F, and gravity. Newton's second law is

$$F(t) - F_{\text{gravity}} = \frac{d}{dt}\big(m(t)\mathrm{v}(t)\big) \qquad 57.91$$

If $\dot{m}$ is the constant fuel usage, the thrust and gravitational forces are

$$F(t) = \dot{m}\mathrm{v}_{\text{exhaust,absolute}} = \dot{m}\big(\mathrm{v}_{\text{exhaust}} - \mathrm{v}(t)\big) \quad \text{57.92}$$

$$F_{\text{gravity}} = m(t)g \quad \text{57.93}$$

The velocity as a function of time is found by solving the following differential equation.

$$\dot{m}\mathrm{v}_{\text{exhaust,absolute}} - m(t)g = \dot{m}\mathrm{v}(t) + m(t)\frac{d\mathrm{v}(t)}{dt} \quad \text{57.94}$$

$$\mathrm{v}(t) = \mathrm{v}_0 - gt + \mathrm{v}_{\text{exhaust}} \ln \frac{m_0}{m_0 - \dot{m}t} \quad \text{57.95}$$

The final burnout *velocity*, v_f, depends on the initial mass, m_0, and the final mass, m_f.

$$\mathrm{v}_f = \mathrm{v}_0 - g\left(\frac{m_0}{\dot{m}}\right)\left(1 - \frac{m_f}{m_0}\right) + \mathrm{v}_{\text{exhaust}} \ln \frac{m_0}{m_f} \quad \text{57.96}$$

A simple relationship exists for a rocket starting from standstill in a gravity-free environment (i.e., $\mathrm{v}_0 = 0$ and $g = 0$).

$$\frac{m_f}{m_0} = e^{-\mathrm{v}_f/\mathrm{v}_{\text{exhaust}}} \quad \text{57.97}$$

30. NEWTON'S LAW OF GRAVITATION

For a particle far enough away from a large body, gravity can be considered to be a central force field. *Newton's law of gravitation*, also known as *Newton's law of universal gravitation*, describes the force of attraction between the two masses. The law states that the attractive gravitational force between the two masses is directly proportional to the product of masses, is inversely proportional to the square of the distance between their centers of mass, and is directed along a line passing through the centers of gravity of both masses.

$$F = \frac{Gm_1m_2}{r^2} \quad \text{57.98}$$

G is *Newton's gravitational constant* (*Newton's universal constant*). Approximate values of G for the earth are given in Table 57.2 for different sets of units. For an earth-particle combination, the product Gm_{earth} has the value of 4.39×10^{14} lbf-ft^2/lbm (4.00×10^{14} N·m^2/kg).

Table 57.2 *Approximate Values of Newton's Gravitational Constant, G*

value	units
6.674×10^{-11}	N·m^2/kg^2 (m^3/kg·s^2)
6.674×10^{-8}	cm^3/g·s^2
3.440×10^{-8}	lbf-ft^2/slug^2
3.323×10^{-11}	lbf-ft^2/lbm^2
3.440×10^{-8}	ft^4/lbf-sec^4

31. NOMENCLATURE

symbol	description	U.S.	SI
a	acceleration	ft/sec^2	m/s^2
a	coefficient of rolling resistance	ft	m
a	semimajor axis length	ft	m
A	area	ft^2	m^2
b	semiminor axis length	ft	m
C	coefficient of viscous damping (linear)	lbf-sec/ft	N·s/m
C	coefficient of viscous damping (quadratic)	lbf-sec^2/ft^2	N·s^2/m^2
C	constant used in space mechanics	1/ft	1/m
D	diameter	ft	m
e	coefficient of restitution	–	–
e	superelevation	ft/ft	m/m
E	energy	ft-lbf	J
E	modulus of elasticity	lbf/ft^2	Pa
f	coefficient of friction	–	–
F	force	lbf	N
g	gravitational acceleration, 32.2 (9.81)	ft/sec^2	m/s^2
g_c	gravitational constant, 32.2	lbm-ft/lbf-sec^2	n.a.
G	universal gravitational constant, 3.320×10^{-11} (6.673×10^{-11})	lbf-ft^2/lbm^2	N·m^2/kg^2
H	angular momentum	ft^2-lbm/sec	m^2·kg/s
h	height	ft	m
h	specific angular momentum	ft^2/sec	m^2/s
I	mass moment of inertia	lbm-ft^2	kg·m^2
Imp	angular impulse	lbf-ft-sec	N·m·s
Imp	linear impulse	lbf-sec	N·s
k	spring constant	lbf/ft	N/m
m	mass	lbm	kg
$\dot{m}$	mass flow rate	lbm/sec	kg/s
M	mass of the earth	lbm	kg
M	moment	ft-lbf	N·m
n	normal direction	–	–
N	normal force	lbf	N
p	momentum	lbf-sec	N·s
P	power	ft-lbf/sec	W

Dynamics and Vibrations

r	radius	ft	m
s	distance	ft	m
t	time	sec	s
T	torque	ft-lbf	N·m
v	velocity	ft/sec	m/s
w	weight	lbf	N
W	work	ft-lbf	J
y	superelevation	ft	m

Symbols

α	angular acceleration	rad/sec^2	rad/s^2
δ	deflection	ft	m
ϵ	eccentricity	–	–
θ	angular position	rad	rad
μ	coefficient of friction	–	–
σ	stress	lbf/ft^2	Pa
ϕ	angle	deg or rad	deg or rad
ω	angular velocity	rad/sec	rad/s

Subscripts

0	initial
b	braking
c	centripetal
C	instant center
f	final or frictional
i	initial
k	kinetic (dynamic)
n	normal
O	center or centroidal
p	periodic
r	rolling
s	static
t	tangential or terminal
w	wedge

58 Mechanisms and Power Transmission Systems

Content in blue refers to the *NCEES Handbook*.

NCEES EXAM SPECIFICATIONS AND RELATED CONTENT

MACHINE DESIGN AND MATERIALS EXAM

I.D.1. Strength of Materials: Stress/strain
 4. Rotating Rings
II.A.6. Mechanical Components: Belt, pulley and chain drives
 18. Chain Drives
II.A.8. Mechanical Components: Power screws
 1. Power Screws and Screw Jacks

1. POWER SCREWS AND SCREW JACKS

A *power screw* changes angular position into linear position (i.e., changes rotary motion into traversing motion). The linear positioning can be horizontal (as in vices and lathes) or vertical, as in jacks. Cross sections of square and Acme threads, both commonly used in power screws, are shown in Fig. 58.1.[1] For square threads, the *thread angle*, ϕ, is zero.

Figure 58.1 *Power Screw Threads*

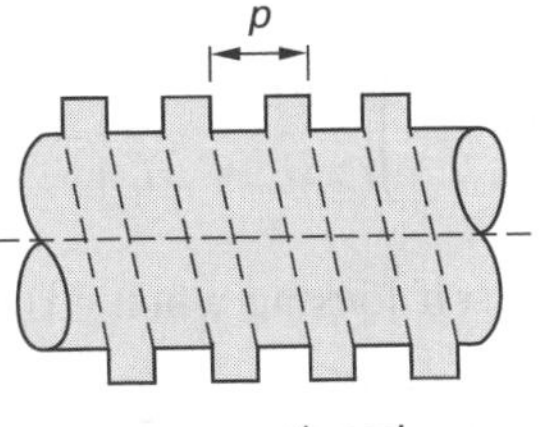

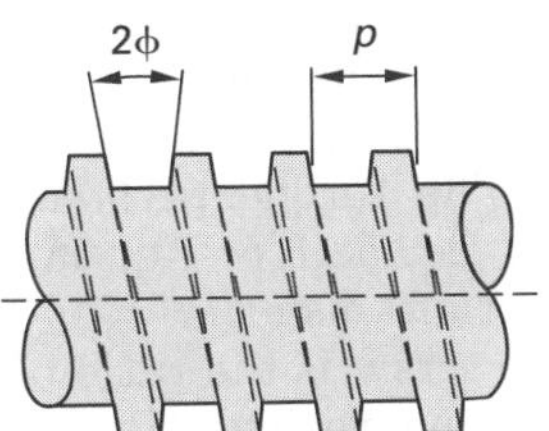

Square power screws are designated by the mean thread diameter, d_m, pitch, p, and *pitch angle*, α. The *pitch*, p, is the distance between corresponding points on a thread. The *lead*, L, is the distance the screw advances each revolution. Often, double- and triple-threaded screws are used, increasing the number of starts, N, to two or three, respectively. The lead is one, two, or three times the pitch for single-, double-, and triple-threaded screws, respectively.

Power Screws

$$L = Np \quad (58.1)$$

For a square thread screw-jack, the external moment, M, applied to the axis of the screw is found from Eq. 58.2. r is the mean thread radius, and ϕ is the coefficient of friction, which can also be represented by μ. The value of the coefficient of friction is positive for screw tightening, and negative for screw loosening.

Power Screws

$$M = Pr\,\tan(\alpha \pm \phi) \quad (58.2)$$

The torque required to turn a square screw in motion against an axial force F (i.e., "raise" the load) is found from Eq. 58.3. The torque must be sufficient to overcome the load through the threads, the friction in the threads, and (in the absence of an antifriction ring) the friction in the collar.[2]

[1]The 10° modified thread is essentially equivalent to the square thread, but it is more economical to manufacture than the square thread.
[2]The relationship is different for Acme and other threads.

Power Screws

$$T_R = \left(\frac{FD_m}{2}\right)\left(\frac{L + \pi\mu D_m}{\pi D_m - \mu L}\right) + \frac{F\mu_c D_c}{2} \tag{58.3}$$

The torque required to turn the screw in motion in the direction of the applied axial force (i.e., "lower" the load) is given by Eq. 58.4. If the torque is zero or negative (as it would be if the lead was large or friction was low), then the screw is not self-locking and the load will lower by itself, causing the screw to spin (i.e., will "overhaul").

Power Screws

$$T_L = \left(\frac{FD_m}{2}\right)\left(\frac{\pi\mu D_m - L}{\pi D_m + \mu L}\right) + \frac{F\mu_c D_c}{2} \tag{58.4}$$

The screw will be self-locking when either $\pi\mu D_m > L$ or $\mu > \tan\alpha$.

The torque calculated in Eq. 58.3 and Eq. 58.4 is required to overcome thread friction and to raise the load (i.e., axially compress the screw). Typically, only 10% to 15% of the torque goes into axial compression of the screw. The remainder is used to overcome friction. The torque without friction can be calculated from Eq. 58.3 or Eq. 58.4 (depending on the travel direction) using $\mu = 0$. The mechanical efficiency of the screw is the ratio of torque without friction to the torque with friction.

Power Screws

$$\eta = \frac{FL}{2\pi T_R} \tag{58.5}$$

Figure 58.2 *Screw with Collar*

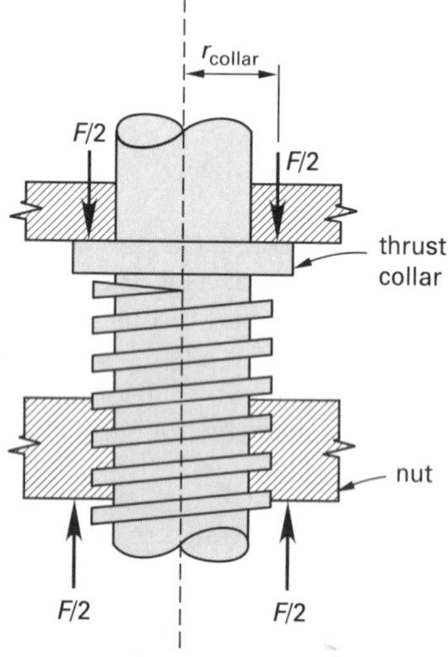

2. ENERGY STORAGE IN FLYWHEELS

A *flywheel* is a device that stores energy, reduces speed fluctuations in machinery, and provides large amounts of energy for external work. The rotational kinetic energy of a flywheel is given by Eq. 58.6.

$$E_{k,r} = \frac{I\omega^2}{2} = \frac{I(2\pi n_{\text{rps}})^2}{2} \quad \text{[SI]} \tag{58.6(a)}$$

$$E_{k,r} = \frac{I\omega^2}{2g_c} = \frac{I(2\pi n_{\text{rps}})^2}{2g_c} \quad \text{[U.S.]} \tag{58.6(b)}$$

The mass moment of inertia of a flywheel can be calculated from the total flywheel mass and the radius of gyration, k. The quantity mk^2 is sometimes referred to as the *flywheel effect*.

$$I = mk^2 \tag{58.7}$$

Speed fluctuations occur when work is performed at the expense of flywheel energy. By the work-energy principle, the change in kinetic energy is equal to the work performed. For example, the work performed by a punch press in punching N holes of diameter D through a sheet of thickness t is

$$W = F_{\text{ave}}t = \frac{F_{\text{max}}t}{2} = \frac{\pi NDt^2S_{us}}{2} \tag{58.8}$$

The change in rotational kinetic energy is

$$\Delta E_{k,r} = \frac{I(\omega_2^2 - \omega_1^2)}{2} = C_f I\omega_m^2 \quad \text{[SI]} \tag{58.9(a)}$$

$$\Delta E_{k,r} = \frac{I(\omega_2^2 - \omega_1^2)}{2g_c} = \frac{C_f I\omega_m^2}{g_c} \quad \text{[U.S.]} \tag{58.9(b)}$$

The *coefficient of fluctuation*, C_f, can be as low as 0.02 to 0.05 for geared drives, generators, and alternators. For punching, shearing, and pressing operations, it is approximately 0.05 to 0.10. For stamping mills and crushing operations, it can be as high as 0.20.

$$C_f = \left|\frac{\omega_2 - \omega_1}{\omega_m}\right| \tag{58.10}$$

The required mass to keep the fluctuation within the desired limits is given by Eq. 58.11. v_c is the tangential velocity at the centroid of the rim cross-sectional area.

$$m = \frac{\Delta E_k}{C_f v_c^2} \quad \text{[SI]} \qquad 58.11(a)$$

$$m = \frac{g_c \Delta E_k}{C_f v_c^2} \quad \text{[U.S.]} \qquad 58.11(b)$$

The relationship between the horsepower and torque of the prime mover that recharges the flywheel is

$$P_{\text{kW}} = \frac{T_{\text{N}\cdot\text{m}} n_{\text{rpm}}}{9549 \ \frac{\text{N}\cdot\text{m}}{\text{kW}\cdot\text{min}}} \quad \text{[SI]} \qquad 58.12(a)$$

$$P_{\text{hp}} = \frac{T_{\text{in-lbf}} n_{\text{rpm}}}{63{,}025 \ \frac{\text{in-lbf}}{\text{hp-min}}} \quad \text{[U.S.]} \qquad 58.12(b)$$

Given a constant (or average) torque, the time to accelerate the flywheel from ω_1 to ω_2 is given by

$$T = I\alpha = \frac{I(\omega_2 - \omega_1)}{t} \quad \text{[SI]} \qquad 58.13(a)$$

$$T = \frac{I\alpha}{g_c} = \frac{I(\omega_2 - \omega_1)}{g_c t} \quad \text{[U.S.]} \qquad 58.13(b)$$

Flywheels are characterized by their *specific power* (i.e., the power output divided by the mass of the flywheel system) and their *specific energy* (the delivered energy divided by the total mass). These two parameters are relatively independent because the specific power is governed by the discharging device, while the specific energy is governed by the flywheel design (geometry and materials).

3. ROTATING SOLID DISKS

For a solid disk (i.e., with no hole for a shaft, as shown in Fig. 58.3) whose thickness is small compared with its outer radius, r_o, the radial and tangential stresses at a radius r are given by Eq. 58.14 and Eq. 58.15. Both stresses are maximum and equal at the center, where $r = 0$.

$$\sigma_r = \left(\frac{3+\nu}{8}\right)\rho\omega^2(r_o^2 - r^2) \quad \text{[SI]} \qquad 58.14(a)$$

$$\sigma_r = \left(\frac{3+\nu}{8}\right)\left(\frac{\rho\omega^2}{g_c}\right)(r_o^2 - r^2) \quad \text{[U.S.]} \qquad 58.14(b)$$

$$\sigma_t = \rho\omega^2\left(\left(\frac{3+\nu}{8}\right)r_o^2 - \left(\frac{1+3\nu}{8}\right)r^2\right) \quad \text{[SI]} \qquad 58.15(a)$$

$$\sigma_t = \left(\frac{\rho\omega^2}{g_c}\right)\left(\left(\frac{3+\nu}{8}\right)r_o^2 - \left(\frac{1+3\nu}{8}\right)r^2\right) \quad \text{[U.S.]} \qquad 58.15(b)$$

Figure 58.3 Rotating Bodies

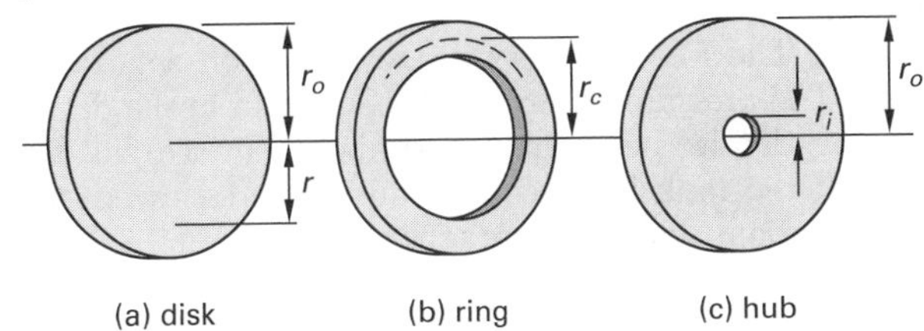

4. ROTATING RINGS

A rotating annular disk, also known as a *ring*, is a solid disk with a circular hole in the center. (See Fig. 58.3.) For a ring with constant thickness, the tangential and radial stresses at a radius r are

Rotating Rings

$$\sigma_t = \rho\omega^2\left(\frac{3+\nu}{8}\right)\left(r_i^2 + r_o^2 + \frac{r_i^2 r_o^2}{r^2} - \left(\frac{1+3\nu}{3+\nu}\right)r^2\right) \qquad 58.16$$

$$\sigma_r = \rho\omega^2\left(\frac{3+\nu}{8}\right)\left(r_i^2 + r_o^2 - \frac{r_i^2 r_o^2}{r^2} - r^2\right) \qquad 58.17$$

The tangential stress will be maximum at the inner boundary.[3] The maximum tangential stress will usually be larger than the maximum radial stress. However, the maximum radial stress should also be checked. For a disk with a central hole, the maximum radial stress occurs at the *geometric mean radius*.

$$\sqrt{r_o r_i} \qquad 58.18$$

The change in the inner radius (i.e., the change in radial interference), Δr_i, at the inside radius of the hub due to rotation is found from the tangential stress.

$$\sigma_{t,\max} = E\epsilon_i = E\left(\frac{\Delta r_i}{r_i}\right) \qquad 58.19$$

5. ROTATING FLUID MASSES

When a full cylindrical tank containing liquid is spun around a vertical axis, the centripetal force of the liquid against the tank wall causes the same stresses (e.g., hoop stresses) as would an equivalent internal pressurization. Equation 58.20 gives the hydrostatic pressure in the fluid at any radius r from the axis of rotation. The pressure is the same in all directions.

$$p_r = \frac{\rho\omega^2 r^2}{2} \quad \text{[SI]} \qquad 58.20(a)$$

$$p_r = \frac{\rho\omega^2 r^2}{2g_c} \quad \text{[U.S.]} \qquad 58.20(b)$$

[3]It is interesting that the maximum tangential stress approaches a value that is twice as large as that of a solid disk. This is an example of a *geometric stress concentration factor* caused by a hole.

Equation 58.20 is for a tank that is completely full. It is more difficult to determine pressure when the tank is partially full. The free liquid surface will take on a parabolic shape. The pressure at any point is found from the density and the depth (beneath the parabolic surface) at that point. If the tank is closed and initially pressurized, that pressurization will increase the pressure at a depth accordingly.

6. RIM FLYWHEELS

A *rim flywheel* consists of a heavy rotating ring connected by spokes to a hub. (See Fig. 58.4.) The contributions of the hub and spokes to the moment of inertia are generally small (i.e., less than 10%).

Assuming that the arms do not restrain the rim from expanding, and the rim thickness is small compared to the mean radius, the tangential (hoop) stress in a rim flywheel is given by Eq. 58.21. v_c is the tangential velocity at the centroid of the rim cross-sectional area.[4] The *burst velocity* can be calculated by setting the tangential stress equal to the ultimate tensile strength with no factor of safety. A large factor of safety (e.g., 10 or larger) has traditionally been used with the cast-iron flywheels that were common in the past.

$$\sigma_t = \rho v_c^2 = \rho(\omega r_c)^2 = \rho(2\pi r_c n_{\text{rps}})^2 \quad \text{[SI]} \quad 58.21(a)$$

$$\sigma_t = \frac{\rho v_c^2}{g_c} = \frac{\rho(\omega r_c)^2}{g_c} = \frac{\rho(2\pi r_c n_{\text{rps}})^2}{g_c} \quad \text{[U.S.]} \quad 58.21(b)$$

Figure 58.4 *Rim Flywheel*

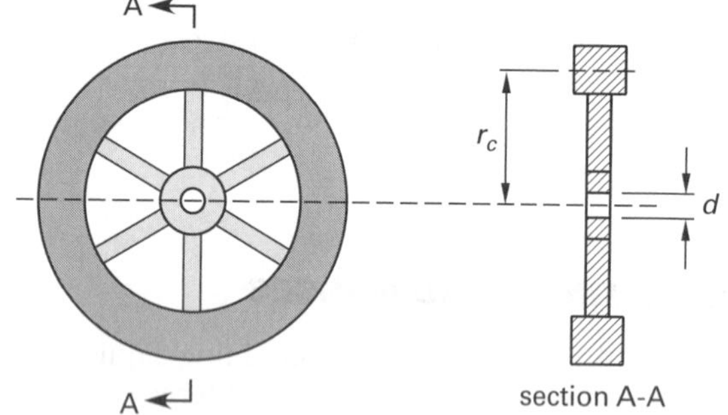

7. ADVANCED FLYWHEEL SYSTEMS

Advanced flywheel systems (high-performance *flywheel rotors*) are used in innovative energy conservation, transportation, and space applications, but particularly in low-emission electric and hybrid vehicles. They consist of a high-strength flywheel, low-loss bearings, and a charging/discharging system. Advanced flywheels are constructed of high-strength composite materials to enable them to resist the high speeds (35,000 rpm to 50,000 rpm and above), high tangential velocities (1200 m/s and above), and high temperatures needed for efficient operation.[5] Table 58.1 lists typical characteristics of materials used in advanced flywheels.

Table 58.1 *Typical Properties of Materials Used in Advanced Flywheels*

material	ultimate strength[a] (GPa)	density (kg/m³)	specific energy[b] (kJ/kg)
high-strength steel	1.5–2.1	7700–8000	250
E-glass/epoxy	1.4	1900	700
S-glass/epoxy	2.1	1900	1100
Kevlar™/epoxy	1.9	1400	1400
graphite/epoxy	1.6–3.2	1500	1100[c]

(Multiply GPa by 1.45×10^5 to obtain lbf/in².)
(Multiply kg/m³ by 0.0624 to obtain lbm/in³.)
(Multiply kJ/kg by 334 to obtain ft-lbf/lbm.)

[a]Strength of the matrix. Individual fibers will have much higher strengths.
[b]Values given merely indicate the approximate range. Actual values would depend on the design.
[c]Kevlar™ is a trademark of the DuPont Company.

Rotor suspension systems have historically used ball bearings. However, modern designs focus on noncontact magnetic bearings.

Charging and discharging systems can be mechanical or electrical. (For maximum efficiency in transportation applications, a *continuously variable transmission* (*CVT*) is needed in mechanically coupled systems.) Charging does not have to be by the same methods as discharging. For example, with a *tandem flywheel system*, one end of the rotor shaft is used for charging the system through mechanical or electrical means, while the discharging equipment is mounted on the other end. In outerspace, where volumetric efficiency is more critical, *concentric flywheel systems* combine electrical charging and discharging methods. Magnetic energy transfer occurs between the rotating rotor and the stationary stator in the frame.

8. GEAR TRAIN TERMINOLOGY

An *external gear* is any gear whose teeth "point" away from the axis of rotation, compared to an *internal gear*, whose teeth point in toward the axis of rotation. Figure 58.5 shows both an external and an internal gear.

[4]A general rule of thumb for cast-iron rim flywheels of average size is that the mean velocity should not exceed 6000 ft/min.
[5]Since rotation is in a vacuum, rotors cannot be convectively cooled.

Figure 58.5 *External and Internal Gears*

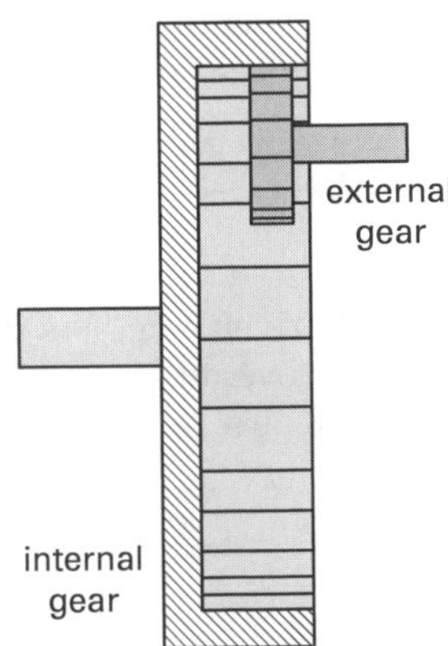

A *simple gear set* (*gear train*) consists of two gears with fixed centers in mesh. The gears turn in opposite directions if both are external gears. If one of the gears is an internal gear, the gears turn in the same direction. Each pair of gears constitutes a *stage*. For multiple gears in mesh, each pair of gears in succession constitute a stage, so that one gear can be part of two stages. An *idler gear*, also known as an *intermediate gear*, is a gear that rotates freely on its bearings and changes the transmission path without changing the gear ratio. (See Fig. 58.6.)

Figure 58.6 *Idler Gear*

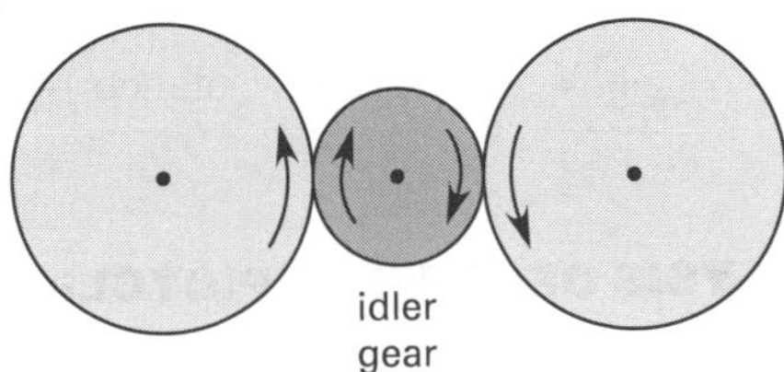

A *compound gear* consists of two gears mounted on a single (usually short) shaft. *Reverted gear sets* are those whose input and output shafts are in-line, usually using one or more compound gears. (See Fig. 58.7.) A reverted gear set always has an even number of stages. Only the gears in mesh need to have the same diametral pitches.

Figure 58.7 *Reverted Gear Set and Compound Gears*

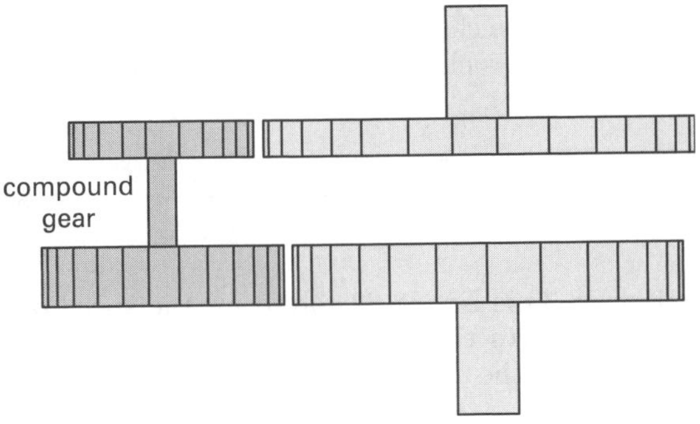

The *velocity ratio* (*mesh ratio* or *gear ratio*), VR, is the ratio of the input to output speeds. In Eq. 58.22, D is the pitch circle diameter.

$$\text{VR} = \frac{n_{\text{in}}}{n_{\text{out}}} = \frac{N_{\text{out}}}{N_{\text{in}}} = \frac{D_{p,\text{out}}}{D_{p,\text{in}}} = \frac{1}{\text{TV}} \qquad 58.22$$

If all of the gears have fixed centers (i.e., fixed axes of rotation), the velocity ratio is the product of all of the numbers of teeth on the driven gears divided by the product of all of the numbers of teeth on the driving gears.

The *train value*, TV, is the reciprocal of the velocity ratio. (Figure 58.10 can be used to determine the train values for common configurations.)

$$\text{TV} = \frac{n_{\text{out}}}{n_{\text{in}}} = \frac{N_{\text{in}}}{N_{\text{out}}} = \frac{1}{\text{VR}} \qquad 58.23$$

9. DESIGN OF GEAR TRAINS

Finding the number of teeth that each gear should have in order to achieve a particular train ratio is time consuming, as a trial-and-error solution may be needed. A particularly desired gear ratio may not exactly be achievable, since each gear must contain an integral number of teeth. Other constraints may be placed on the design, including the minimum and maximum numbers of teeth and the maximum number of stages.

All gears in mesh must have the same diametral pitch. Therefore, for any two gears in mesh, the relationship between the diametral pitch, pitch circle diameters, and number of teeth is

$$P_1 = P_2 \qquad 58.24$$

$$\frac{N_1}{D_1} = \frac{N_2}{D_2} \qquad 58.25$$

The center-to-center distance, C, is the sum of the two pitch radii.

$$C = r_1 + r_2 = \frac{D_1 + D_2}{2} = \frac{\frac{N_1}{P} + \frac{N_2}{P}}{2} \qquad 58.26$$

Equation 58.25 and Eq. 58.26 can be written for all pairs of meshing gears and solved simultaneously for unknown quantities. Knowledge of various gear ratios between the meshing gears can be used to simplify the simultaneous equations.

For reverted gear trains, the total train ratio should ideally be shared equally between all stages. Since this ratio may not be feasible, the total train ratio should be factored into stage ratios that numerically are not too far apart.

10. SIMPLE EPICYCLIC GEAR SETS

Epicyclic gear sets (also known as *planetary gear sets*) are characterized by one or more gears that do not have fixed axes of rotation. They have two inputs (one of which may be fixed or stationary) and one output. Compared with gear sets where all gears have fixed centers, epicyclic gear sets can have much higher gear ratios, are more compact, have lower tooth loadings and pitch-line velocities, offer in-line input and output shafts, may be easier to lubricate, and are generally less expensive.

The simplest type of epicyclic gear set is shown in Fig. 58.8. It consists of a *sun gear*, *ring gear* (also known as an *annulus gear*), and one or more *planet gears* (also referred to as *planets*, *planet pinions*, and *spider gears*). The rotating bent yoke that connects the planets to their shaft is known as the *planet carrier*, *arm*, and *spider*. **[Planetary Gear Terms and Ratios]**

Figure 58.8 *Simple Epicyclic Gear Set*

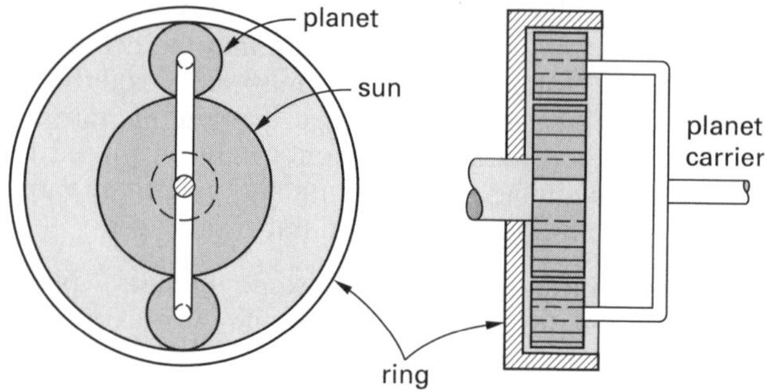

The planets rotate about their own axes and revolve around the sun gear. During rotation, a point on a planet gear traces out epicyclic curves, hence the name. There are generally one to four planets. The number of planets does not affect the output speed, but the maximum power transmission is essentially proportional to the number of planets. The number of planets is limited by space so that the planets do not "overlap." Either the ring gear or the carrier can be fixed. If one is fixed, then the other must rotate.

When the carrier is fixed and the ring gear is the output, the arrangement is known as a *star gear set*. When the sun gear is fixed and the input is the ring gear, the system is called a *solar gear set*.

Simple epicyclic gear trains operate in one of three different modes.

1. The driving and driven members turn in the same direction, and the overall gear ratio is between 2 and ∞. The overall gear ratio for this mode is

$$\text{VR} = \frac{N_{\text{sun}} + N_{\text{ring}}}{N_{\text{sun}}} \qquad 58.27$$

2. The driving and driven members turn in the same direction, and the overall gear ratio is between 1 and 2. The overall gear ratio for this mode is

$$\text{VR} = \frac{N_{\text{sun}} + N_{\text{ring}}}{N_{\text{ring}}} \qquad 58.28$$

3. The driving and driven members turn in opposite directions, and the overall gear ratio is between 1 and ∞.

$$\text{VR} = \frac{N_{\text{ring}}}{N_{\text{sun}}} \qquad 58.29$$

When the planet gears are equally spaced around the sun gear, the fundamental relationships between the number of teeth on the gears are derived from the fact that all gears are in mesh and must have the same diametral pitch. In Eq. 58.31, M is some whole number.

$$N_{\text{sun}} + 2N_{\text{planet}} = N_{\text{ring}} \qquad 58.30$$

$$N_{\text{sun}} + N_{\text{ring}} = M(\text{no. of planets}) \qquad 58.31$$

11. ANALYSIS OF SIMPLE EPICYCLIC GEAR SETS

Analysis of a simple epicyclic gear set starts with calculating the *train value*, TV. This value is the same regardless of whether the gear set is reducing or augmenting. However, it is negative if the sun and ring gears rotate in different directions when the arm is locked. (TV is always negative when the ring gear is an internal gear.) Although the symbol ω is used for convenience, the relationship is valid for rotational speed in rad/sec, rev/min, and rev/sec. A consistent sign convention should be chosen for directions of rotation (e.g., positive for clockwise and negative for counterclockwise).

$$\text{TV} = \frac{N_{\text{ring}}}{N_{\text{sun}}} = \frac{\omega_{\text{sun}} - \omega_{\text{carrier}}}{\omega_{\text{ring}} - \omega_{\text{carrier}}} \qquad 58.32$$

Equation 58.33 is Eq. 58.32 solved for the speed of the sun gear. It gives the relationship between the known train value and the unknown rotational speeds.

$$\omega_{\text{sun}} = (\text{TV})\omega_{\text{ring}} + (1 - \text{TV})\omega_{\text{carrier}} \qquad 58.33$$

Planet speed is not part of Eq. 58.33. The rotational speed (with respect to their own axis) and direction of the planets can be found from Eq. 58.34. The quantity is negative because the planets and sun gear turn in opposite directions. A consistent sign convention must be used.

$$\frac{N_{\text{planets}}}{N_{\text{sun}}} = \frac{-(\omega_{\text{sun}} - \omega_{\text{carrier}})}{\omega_{\text{planets}} - \omega_{\text{carrier}}} \qquad 58.34$$

Example 58.1

A simple epicyclic gear set with two planets is similar to Fig. 58.8. The sun gear has 32 teeth, the planets have 16 teeth, and the ring has 64 teeth. The sun gear turns clockwise at 100 rev/min. The ring is fixed. Find the (a) speed and (b) direction of the carrier.

Solution

(a) If the arm was locked and the ring was free to rotate, the sun and ring gears would rotate in different directions. Therefore, the train value is negative.[6]

$$\text{TV} = \frac{-64 \text{ teeth}}{32 \text{ teeth}} = -2$$

Since the ring gear is fixed, $\omega_{\text{ring}} = 0$. For this example, decide that clockwise rotation is positive. Use Eq. 58.33 with the speeds given in rev/min.

$$\begin{aligned} 100 \; \frac{\text{rev}}{\text{min}} &= (0)(-2) + \omega_{\text{carrier}}(1-(-2)) \\ \omega_{\text{carrier}} &= 33.3 \text{ rev/min} \end{aligned}$$

(b) Since ω_{carrier} is positive, the carrier rotation is clockwise.

12. DESIGN OF SIMPLE EPICYCLIC GEAR TRAINS

As with fixed-center gear trains, analysis of epicyclic gear trains is easier than the design process. However, the following procedure can be used.

step 1: Determine the mode of operation (see Sec. 58.10) and gear ratio.

step 2: Use Table 58.2 to determine which gear in the set must be the smallest.

step 3: Depending on which gear is the smallest, use either Table 58.3 or Table 58.4 to determine the number of teeth on the other gears in terms of the gear ratio and the number of teeth on the smallest gear.[7]

Table 58.2 *Smallest Epicyclic Gear*

mode	gear ratio	smallest gear
1	$2 < \text{VR} \le 4$	planet
1	$4 \le \text{VR} < \infty$	sun
2	$1 < \text{VR} \le 4/3$	sun
2	$4/3 \le \text{VR} < 2$	planet
3	$1 < \text{VR} \le 3$	planet
3	$3 \le \text{VR} < \infty$	sun

Table 58.3 *Number of Teeth When the Sun Gear Is the Smallest (N_{sun} and VR known)*

no. of teeth	mode 1	mode 2	mode 3
N_{planet}	$\frac{(\text{VR}-2)N_{\text{sun}}}{2}$	$\frac{(2-\text{VR})N_{\text{sun}}}{2\text{VR}-2}$	$\frac{(\text{VR}-1)N_{\text{sun}}}{2}$
N_{ring}	$(\text{VR}-1)N_{\text{sun}}$	$\frac{N_{\text{sun}}}{\text{VR}-1}$	$\text{VR}N_{\text{sun}}$

Table 58.4 *Number of Teeth When the Planet Gear Is the Smallest (N_{planet} and VR known)*

no. of teeth	mode 1	mode 2	mode 3
N_{sun}	$\frac{2N_{\text{planet}}}{\text{VR}-2}$	$\frac{(2\text{VR}-2)N_{\text{planet}}}{2-\text{VR}}$	$\frac{2N_{\text{planet}}}{\text{VR}-1}$
N_{ring}	$\frac{(2\text{VR}-2)N_{\text{planet}}}{\text{VR}-2}$	$\frac{2N_{\text{planet}}}{2-\text{VR}}$	$\frac{2\text{VR}N_{\text{planet}}}{\text{VR}-1}$

Example 58.2

A simple epicyclic gear drive has a speed ratio of 1.7:1. The driving and driven members turn in the same direction. Each planet is to have 18 teeth. (a) How many teeth are on each gear? (b) How many planets should there be?

Solution

(a) From Sec. 58.10, the gear set operates in mode 2. From Table 58.2, the planets are the smallest gears. There are two relationships from Table 58.4.

$$\begin{aligned} N_{\text{sun}} &= \frac{(2\text{VR}-2)N_{\text{planet}}}{2-\text{VR}} = \frac{\big((2)(1.7)-2\big)(18)}{2-1.7} \\ &= 84 \text{ teeth} \\ N_{\text{ring}} &= \frac{2N_{\text{planet}}}{2-\text{VR}} = \frac{(2)(18)}{2-1.7} \\ &= 120 \end{aligned}$$

[6]It may be easier to remember that the train value is always negative whenever the ring gear is in an internal gear.
[7]Generally, no gear should have fewer than 15 to 18 teeth for proper operation.

Figure 58.9 *Graphic Representation of Epicyclic Gear Sets*

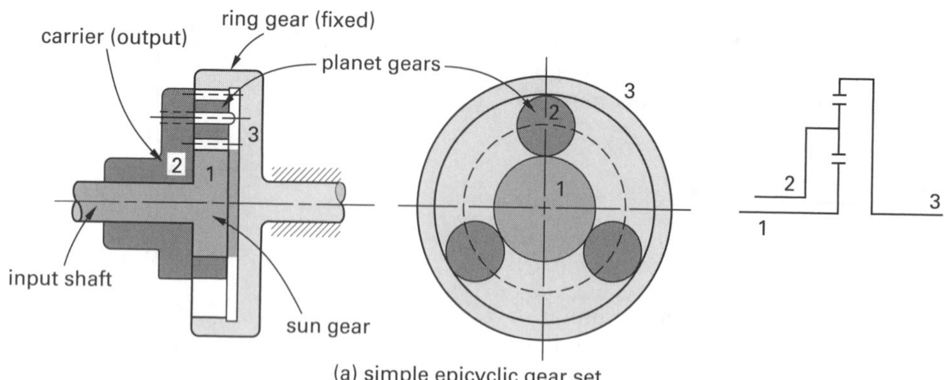

(a) simple epicyclic gear set

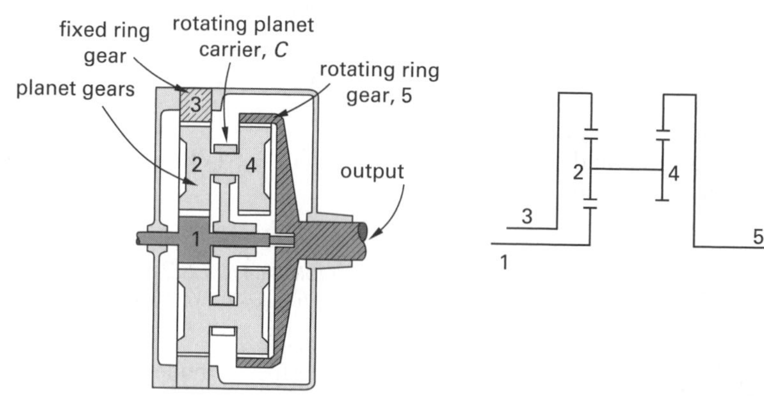

(b) complex epicyclic gear set

(b) Use Eq. 58.31.

$$N_{sun} + N_{ring} = M(\text{no. of planets})$$
$$84 + 120 = 204 = M(\text{no. of planets})$$

Both M and the number of planets are whole numbers, and both must be factors of 204—that is, 2, 3, 4, or 6. There could be 2, 3, 4, or 6 planets.

13. GRAPHIC REPRESENTATION OF EPICYCLIC GEAR SETS

A standard graphic method (with variations) is used for showing more complex epicyclic gear trains in two dimensions. Figure 58.9 illustrates how a simple gear set with three planets would be diagrammed. The gears are shown as vertical lines, and the shafts are shown as horizontal lines. In both Fig. 58.9(a) and Fig. 58.9(b), notice that only one planet is shown in the two-dimensional representation. Appendix 58.A lists the relationships between rotational speeds and numbers of gear teeth for some less common gear arrangements.

14. ANALYSIS OF COMPLEX EPICYCLIC GEAR TRAINS

There are several ways of analyzing more complex epicyclic gear trains. Some methods are intuitive, some are procedural, some are graphical, and some are tabular. The procedure presented in this section is tabular and does not require any significant visualization or intuitive reasoning.

step 1: Identify all of the gears that have the same center of rotation as the arm (i.e., have the same axis as the carrier axis). Include stationary gears (such as fixed ring gears), but do not include the arm, any gear that is attached to it, or any fixed gear that is not centered on the planetary axis. Prepare a three-row matrix with a column for each of these gears.

step 2: Write "$\omega_{carrier}$" in the first row for each gear in the table. (Items in quotation marks are to be written exactly as indicated.) Do not substitute numerical values until step 7.[8]

[8]The method presented is well founded in rigorous theory. However, it is somewhat "magical" in its manner of deriving the answer from innocuous steps and is sometimes referred to as a "hocus-pocus" solution method.

step 3: The speed of one of the gears in the table may be unknown and desired. Refer to this gear as gear z. If all of the gear speeds except the carrier speed are known, choose gear z arbitrarily. For that gear, put "ω_z" into the third row for column z.

step 4: Put "$\omega_z - \omega_{\text{carrier}}$" into the second row for column z.

step 5: Use Fig. 58.10 to determine the velocity relationships, ω/ω_z, between gear z and all other gears in the table. Use a consistent sign convention and consider the directions of rotation if other configurations are used in the gear set.

step 6: Multiply the velocity relationships from step 5 by "$\omega_z - \omega_{\text{carrier}}$" and put the product into row 2.

step 7: Substitute all known values of ω into the third row for all empty columns and into ω_{carrier} wherever it appears. If the velocity of any gear i is unknown, insert "ω_i."

step 8: For each gear whose speed is known, write the *characteristic equation.*

$$\text{row 1} + \text{row 2} = \text{row 3} \qquad 58.35$$

Example 58.3

A complex epicyclic gear train is constructed as shown. The ring gear is fixed. The planet carrier is the output and turns at 20 rev/min. Gear A has 16 teeth, gear B has 32 teeth, gear C has 24 teeth, and gear D has 72 teeth. What are the (a) speed and (b) direction of the sun input gear A?

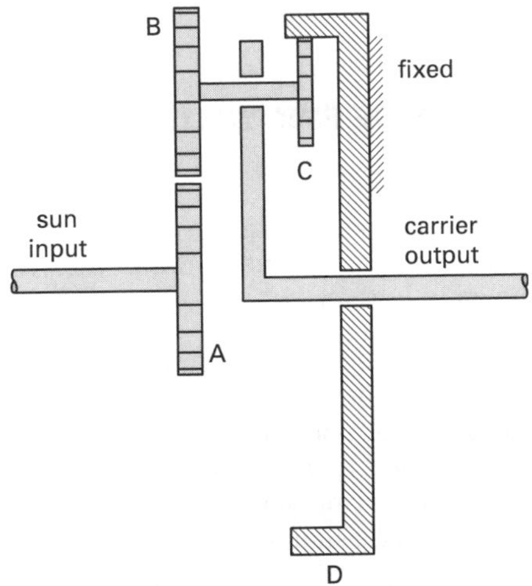

Figure 58.10 *Speed Relationships Between Gears with Fixed Centers (gear A is always the input)*

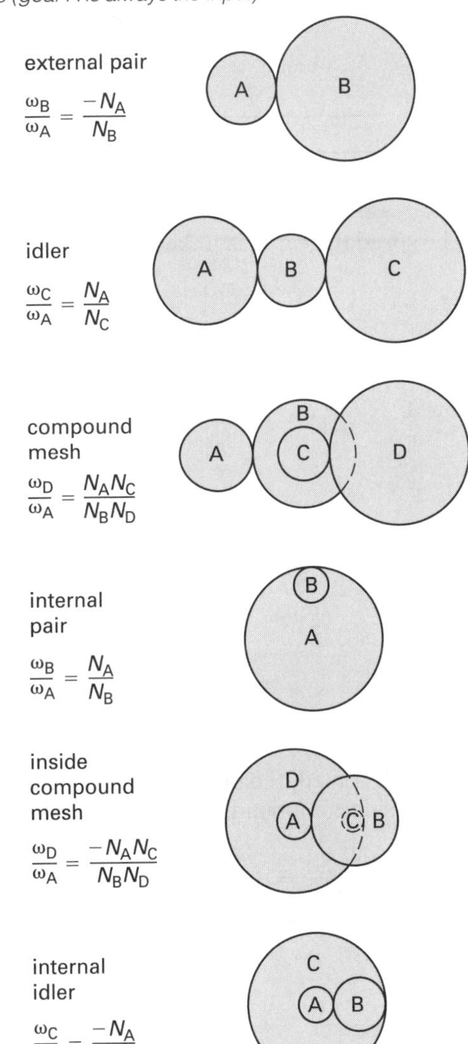

Solution

step 1: Gears A and D are concentric with the arm.

	A	D
row 1		
row 2		
row 3		

step 2:

	A	D
row 1	ω_{carrier}	ω_{carrier}
row 2		
row 3		

step 3: The speed of gear A is unknown.

	A	D
row 1	ω_{carrier}	ω_{carrier}
row 2		
row 3	ω_A	

step 4:

	A	D
row 1	ω_{carrier}	ω_{carrier}
row 2	$\omega_A - \omega_{\text{carrier}}$	
row 3	ω_A	

step 5: From Fig. 58.10, the path from gear A to gear D is an inside compound mesh. The speed ratio is

$$\frac{\omega_D}{\omega_A} = \frac{-N_A N_C}{N_B N_D} = \frac{-(16)(24)}{(32)(72)} = -\frac{1}{6}$$

step 6:

	A	D
row 1	ω_{carrier}	ω_{carrier}
row 2	$\omega_A - \omega_{\text{carrier}}$	$(-1/6)(\omega_A - 20)$
row 3	ω_A	

step 7: Choose the direction of rotation of the carrier as the positive direction. Since gear D is fixed, $\omega_D = 0$. Also, $\omega_{\text{carrier}} = +20$.

	A	D
row 1	20	20
row 2	$\omega_A - 20$	$(-1/6)(\omega_A - 20)$
row 3	ω_A	0

step 8: The characteristic equation for column D is

$$\text{row 1} + \text{row 2} = \text{row 3}$$

$$20\ \frac{\text{rev}}{\text{min}} + \left(-\tfrac{1}{6}\right)\left(\omega_A - 20\ \frac{\text{rev}}{\text{min}}\right) = 0$$

$$\omega_A = 140\ \text{rev/min}$$

[positive, so in the same direction as the carrier]

Example 58.4

Solve Ex. 58.3 using App. 58.A.

Solution

This corresponds to case B in App. 58.A. Rotations are as viewed looking at the sun input (from left to right) with clockwise motion designated as positive.

gear	gear in case B	N	n
A	2	16	
B	4	32	
C	5	24	
D	6	72	0
carrier	3	–	+20

$$n_3\left(1 + \frac{N_2 N_5}{N_4 N_6}\right) - n_2\left(\frac{N_2 N_5}{N_4 N_6}\right) = n_6$$

$$(20)\left(1 + \frac{(16)(24)}{(32)(72)}\right) - n_2\left(\frac{(16)(24)}{(32)(72)}\right) = 0$$

$$(20)\left(\frac{7}{6}\right) - n_2\left(\frac{1}{6}\right) = 0$$

$$n_2 = 140$$

15. ANALYSIS OF EPICYCLIC BEVEL GEAR SETS

Many epicyclic gear sets make use of bevel gears. This is the standard design for traditional automotive differentials, as shown in Fig. 58.11.

In the simple epicyclic gear set shown in Fig. 58.8, the power path is from the planet carrier, through the planets, and on to the sun, or the reverse. With a bevel gear set shown in Fig. 58.11, the power path is from the drive shaft to the rotating ring gear, through the bevel gears, and on to the two axle shafts.

The simple formula used in Sec. 58.11 can be used with bevel gear sets if an analogy is made to the traditional (simple) epicyclic gear set. The two identical bevel gears are essentially idlers and are analogous to the planets. The two identical gears driving the axle shafts are analogous to the ring and sun gears that are normally

Figure 58.11 Automative Differential

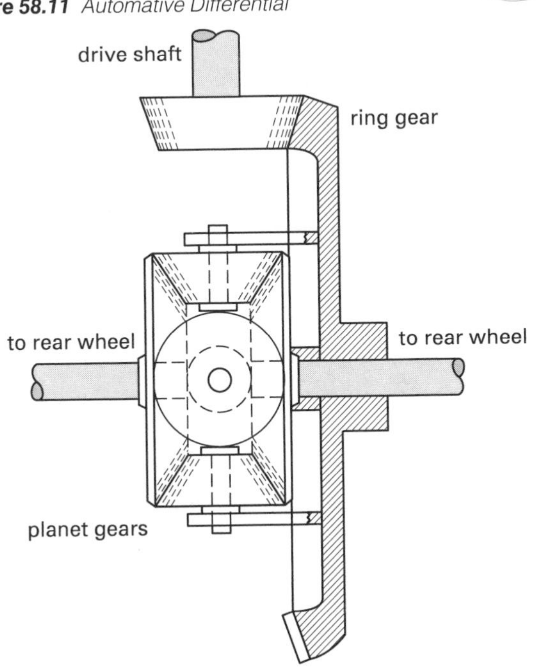

separated by planets. The actual automotive ring gear turns with the bevel gears (i.e., the "planets") and is analogous to the carrier.

Since the analogous sun-planet-ring combination is "rectangular," and since the analogous sun and ring gears would turn at the same speeds but in opposite directions if the carrier was locked, the train value, TV, is always −1.

Equation 58.33 is written in terms of the actual automotive gear names using a train value of −1.[9] The speed of the ring gear is found from the number of teeth on it and the drive shaft gear.

$$\omega_{\text{left rear wheel}} + \omega_{\text{right rear wheel}} = 2\omega_{\text{ring gear}} \quad \textit{58.36}$$

When a vehicle is traveling in a straight line, both wheels turn at the same speed, equal to the rotational speed of the ring gear. When the vehicle travels around a bend, the inner wheel turns more slowly than the outer wheel.

Example 58.5

A car is jacked up on one side such that the left wheel is stationary on the road surface while the right wheel is free to turn in the air. The engine is run such that the drive shaft turns at 1000 rev/min. An automotive differential similar to Fig. 58.11 is used. The drive shaft gear has 17 teeth, and the ring gear has 54 teeth. What is the speed of the right wheel?

Solution

The speed of the ring gear is

$$\begin{aligned}\omega_{\text{ring gear}} &= \left(\frac{N_{\text{drive shaft}}}{N_{\text{ring gear}}}\right)\omega_{\text{drive shaft}} = \left(\frac{17}{54}\right)\left(1000\ \frac{\text{rev}}{\text{min}}\right)\\ &= 314.8\ \text{rev/min}\end{aligned}$$

Use Eq. 58.36.

$$\begin{aligned}\omega_{\text{left rear wheel}} &= 2\omega_{\text{ring gear}} - \omega_{\text{right rear wheel}}\\ 0 &= (2)\left(314.8\ \frac{\text{rev}}{\text{min}}\right) - \omega_{\text{right rear wheel}}\\ \omega_{\text{right rear wheel}} &= 629.6\ \text{rev/min}\end{aligned}$$

16. CAMS

Cams are devices that convert regular rotating motion into irregular translating motion. Cams drive another element, known as a *follower*, *lifter*, or *tappet*. There are various types of cams. This section is concerned only with *plate cams*, the type traditionally used in internal combustion engines to open and close valves. There are various types of followers for plate cams, although the *flat-faced follower* and *roller lifter* are the most common.[10] Where cost is not a factor, the roller lifter is preferred because of its lower friction.

Cam lobes can be shaped in a variety of ways, depending on what type of displacement is desired. Theoretically, any curve can be used to design the cam profile, including circular arcs, parabolas, cubics, quartics, and sinusoids.[11] Common standard-motion lobe shapes (i.e., *profiles*) are designated as uniform, parabolic (constant acceleration), harmonic, and cycloidal.[12] Figure 58.12 illustrates the displacement, velocity, and acceleration curves for some of these lobe designs over the rotation necessary to cause full rise.

Some of the curves have advantages over the others. *Jerk* (*geometric jerk*) is a term used to describe the *third derivative of motion*, sometimes called the *second acceleration*. The parabolic cam has three infinite jerks per rise. It is the worst of the three cam types for high-speed use. The harmonic cam also has one infinite jerk. The cycloidal cam has a higher acceleration than the other two cam profiles, but the jerk is finite. Therefore, the cycloidal cam is among the best of the three for high-speed operation.

Figure 58.13 illustrates a generalized cam with a roller follower on its base circle. (The lobe is not shown.) The *base circle* is the smallest circle that can be drawn

[9]The ring gear referenced in Eq. 58.36 is the actual ring gear shown in Fig. 58.11.
[10]Pointed and spherical-faced followers can be used for special applications.
[11]A *trapezoidal profile*, for example, is an alternating sequence of cubic and parabolic sections.
[12]Though they are easy to manufacture, cam lobes constructed from circular arcs are not covered here.

tangent to the cam surface. x is the *follower offset* (which is zero for *radial cams*), and the follower's movement is radial with respect to the cam profile. r is the separation of the cam and follower centers. R is the vertical follower rise (displacement) at a particular angle of cam rotation (not necessarily the maximum rise or *lift*, L). The *dwell* is the duration, usually measured in cam angle degrees, for which the rise is constant (i.e., the lifter position is unchanged at the top or bottom position). The subscript b refers to when the follower is on the base circle (i.e., is at its lowest point). From the Pythagorean theorem,

Figure 58.12 *Cam Performance vs. Lobe Profile*

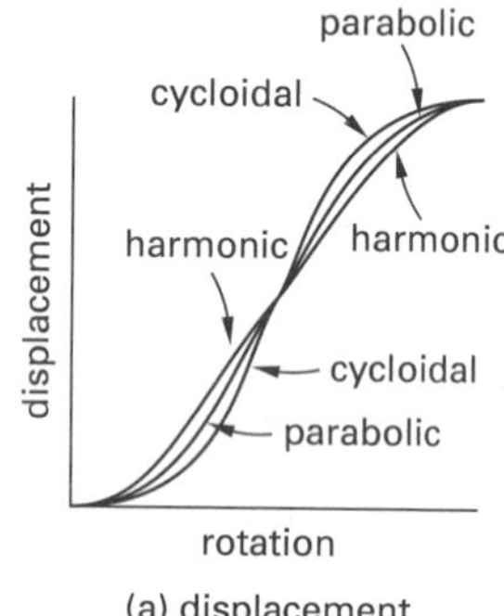

(a) displacement

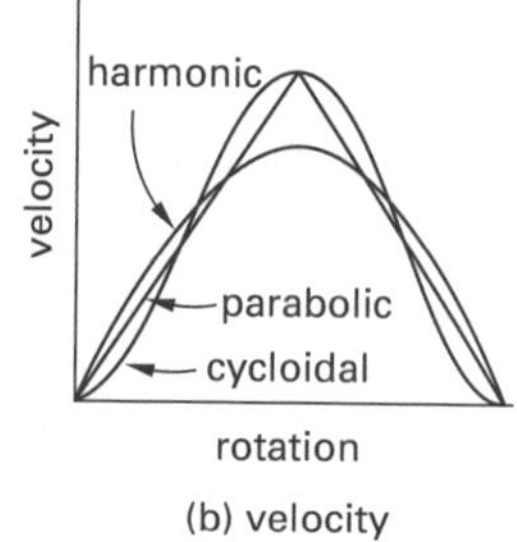

(b) velocity

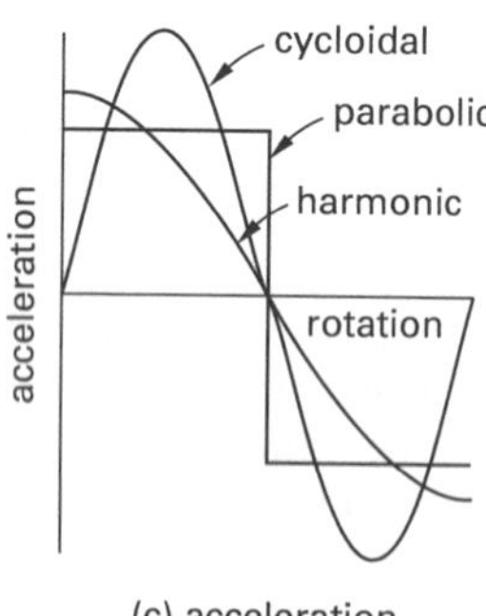

(c) acceleration

$$\begin{aligned} r^2 &= y^2 + x^2 = (y_b + R)^2 + x^2 \\ &= y_b^2 + 2y_bR + R^2 + x^2 \\ &= r_b^2 + 2y_bR + R^2 \end{aligned} \qquad 58.37$$

For flat-faced lifters, the force on a cam and follower is in line with the vertical axis of the follower. For roller lifters, the force may be offset by a *pressure angle*, ϕ, as shown in Fig. 58.13.

Figure 58.13 *Cam and Follower*

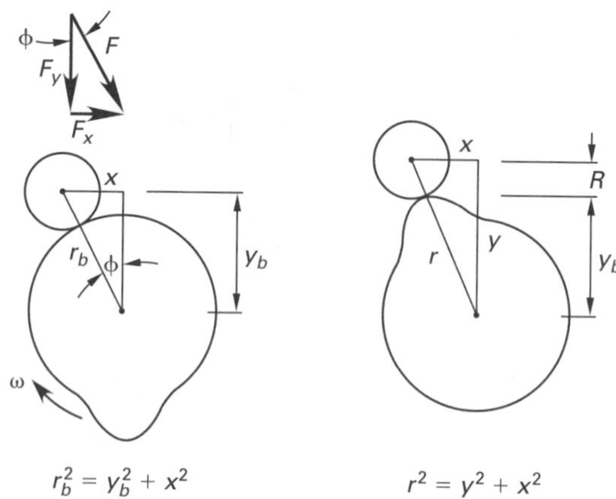

The pressure angle is usually kept below 30° to 35° to prevent the lifter from binding or locking the cam. The pressure angle is lower for larger base-circle diameters. Referring to Eq. 58.38, it can be seen that the larger the offset, the smaller the pressure angle. For this reason, pressure angle considerations are less important when an offset follower is used.

$$\tan\phi = \frac{\frac{v}{\omega} - x}{r_b + R} \qquad 58.38$$

For a radial cam, $x = 0$ and $r_b = y_b$.

$$\tan\phi = \frac{v}{\omega(y_b + R)} \quad \text{[radial cams]} \qquad 58.39$$

The resisting torque on the camshaft due to the follower force depends on the vertical force component.

$$T = \frac{F_y v}{\omega} \qquad 58.40$$

In high-speed operation, a cam follower may bounce out of contact with the cam. This is referred to as *jump*. One theory proposes that jump will not occur if at least two full cycles of vibration can occur in the follower train during the positive acceleration time interval of motion. (Jump may not occur even with fewer than two cycles, but this must be investigated.) In Eq. 58.41, θ_1 is the angle through which the cam rotates during the positive acceleration period. ω_n is the natural frequency of the spring system, and ω is the cam's speed of rotation.

$$\left(\frac{\theta_1}{360°}\right)\left(\frac{\omega_n}{\omega}\right) \ge 2 \qquad 58.41$$

Representing the stiffness of the spring used to hold the follower against the cam as k_1 and the stiffness of the follower train as k_2, the natural frequency of the follower system is given by Eq. 58.42. m is the lumped follower mass.

$$\omega_n = \sqrt{\frac{k_1 + k_2}{m}} \qquad 58.42$$

17. ANALYSIS OF CAM PERFORMANCE

Table 58.5 contains theoretical expressions for the displacement, velocity, and acceleration for three different lobe profiles. In some cases, simplified expressions may be available. For example, the maximum velocity, acceleration, and displacement experienced by a lifter moving a maximum distance L during cam rotation θ_0 are given by Eq. 58.43 and Eq. 58.44. Values of velocity and acceleration coefficients are given in Table 58.6. Maximum deceleration is the same as maximum acceleration. To use Eq. 58.43 and Eq. 58.44, ω and θ_0 must both use the same units (e.g., radians).[13]

$$\mathrm{v}_{\max} = \frac{C_\mathrm{v}\omega L}{\theta_0} \qquad 58.43$$

$$a_{\max} = \frac{C_a\omega^2 L}{\theta_0^2} \qquad 58.44$$

Example 58.6

A cycloidal cam produces a full lift of 1.5 in during an acceleration stroke occurring over 50° of cam rotation. Determine the (a) maximum velocity and (b) maximum acceleration.

Solution

The rotational speed of the cam is unknown. The velocity must be determined in units per degree rather than in units per second.

(a) Use Eq. 58.43 and Table 58.6. The maximum velocity is

$$\mathrm{v}_{\max} = \frac{C_\mathrm{v}L}{\theta_0} = \frac{(2)(1.5\ \text{in})}{50°} = 0.06\ \text{in/deg}$$

(b) Use Eq. 58.44 and Table 58.6. The maximum acceleration is

$$a_{\max} = \frac{C_a L}{\theta_0^2} = \frac{(2\pi)(1.5\ \text{in})}{(50°)^2} = 0.00377\ \text{in/deg}^2$$

Example 58.7

A radial cam with cycloidal lobes turns at 1200 rpm. The rise stroke occurs over 120° of cam rotation. The total lift is 25 mm. The follower is held against the cam by a spring with a spring constant of 5 N/mm. The spring compression is 5 mm when the follower is on its base circle. The lumped mass of the follower is 0.8 kg. Stiffness of the follower train is to be disregarded. When the cam turns 50° from the start of the lifter movement, what are the (a) angular velocity, (b) rise, (c) inertial force on the lifter, (d) total force on the spring, (e) net force on the cam, and (f) shaft torque on the cam due to one lifter?

Solution

(a) The angular velocity is

$$\omega = 2\pi n_{\text{rps}} = \frac{(2\pi)\left(1200\ \frac{\text{rev}}{\text{min}}\right)}{60\ \frac{\text{s}}{\text{min}}} = 125.66\ \text{rad/s}$$

(b) The angular rotation for rise is

$$\theta_0 = \frac{(120°)(2\pi)}{360°} = 2.094\ \text{rad}$$

A cam rotation of 50° is

$$\theta = \frac{(50°)(2\pi)}{360°} = 0.8727\ \text{rad}$$

From Table 58.5, and evaluating the argument of the sine term in radians, the lifter displacement is

$$\begin{aligned} R &= \left(\frac{L}{\pi}\right)\left(\frac{\pi\theta}{\theta_0} - \tfrac{1}{2}\sin\frac{2\pi\theta}{\theta_0}\right) \\ &= \left(\frac{25\ \text{mm}}{\pi}\right)\left(\begin{array}{l}\pi\left(\dfrac{0.8727\ \text{rad}}{2.094\ \text{rad}}\right) \\ \quad -\dfrac{1}{2}\sin\left(\dfrac{(2\pi)(0.8727\ \text{rad})}{2.094\ \text{rad}}\right)\end{array}\right) \\ &= 8.43\ \text{mm} \end{aligned}$$

Table 58.6 Maximum Cam Performance Coefficients

lobe profile	C_v	C_a	C_{jerk}
parabolic	2	4	∞
cycloidal	2	2π	$4\pi^2$
simple harmonic	$\frac{\pi}{2}$	$\frac{\pi^2}{2}$	$\frac{\pi^3}{2}$

[13]As written, Eq. 58.43 and Eq. 58.44 have units of distance/sec and distance/sec², respectively. In some problems, units of distance/deg and distance/deg² are needed. Results in these forms can be calculated by omitting ω from the equations.

Table 58.5 Cam Lobe Dynamics *(θ and θ_0 are in radians; arguments of sin and cos terms are in radians)

type of lobe	displacement (rise, R) maximum at $\theta/\theta_0 = 1$	velocity maximum at $\theta/\theta_0 = 0.5$	acceleration
parabolic	accelerating, $R = 2L\left(\frac{\theta}{\theta_0}\right)^2$ $\left(\frac{\theta}{\theta_0} \le 0.5\right)$	$\frac{dR}{dt} = \frac{4L\omega\theta}{\theta_0^2}$	$\frac{d^2R}{dt^2} = \frac{4L\omega^2}{\theta_0^2}$ (constant)
	decelerating, $R = L\left(1 - 2\left(1 - \frac{\theta}{\theta_0}\right)^2\right)$ $\left(\frac{\theta}{\theta_0} \ge 0.5\right)$	$\frac{dR}{dt} = \left(\frac{4L\omega}{\theta_0}\right)\left(1 - \frac{\theta}{\theta_0}\right)$	$\frac{d^2R}{dt^2} = -\frac{4L\omega^2}{\theta_0^2}$ (constant)
harmonic	$R = \left(\frac{L}{2}\right)\left(1 - \cos\frac{\pi\theta}{\theta_0}\right)$	$\frac{dR}{dt} = \left(\frac{\pi L\omega}{2\theta_0}\right)\sin\frac{\pi\theta}{\theta_0}$	$\frac{d^2R}{dt^2} = \left(\frac{\pi^2 L\omega^2}{2\theta_0^2}\right)\cos\frac{\pi\theta}{\theta_0}$ (maximum at $\theta/\theta_0 = 0$)
cycloidal	$R = \left(\frac{L}{\pi}\right)\left(\frac{\pi\theta}{\theta_0} - \frac{1}{2}\sin\frac{2\pi\theta}{\theta_0}\right)$	$\frac{dR}{dt} = \left(\frac{L\omega}{\theta_0}\right)\left(1 - \cos\frac{2\pi\theta}{\theta_0}\right)$	$\frac{d^2R}{dt^2} = \left(\frac{2\pi L\omega^2}{\theta_0^2}\right)\sin\frac{2\pi\theta}{\theta_0}$ (maximum at $\theta/\theta_0 = 0.25$)

*θ_0 is the cam rotation for full lifter rise, L. This table can be used for nonsymmetrical lobes if the analysis is performed in two phases, with two values of θ_0. In each phase, θ_0 is taken as twice the angular range for that phase.

(c) From Table 58.5, the lifter acceleration is

$$a = \left(\frac{2\pi L\omega^2}{\theta_0^2}\right)\sin\frac{2\pi\theta}{\theta_0}$$

$$= \left(\frac{(2\pi)(25\text{ mm})\left(125.66\ \frac{\text{rad}}{\text{s}}\right)^2}{(2.094\text{ rad})^2}\right) \times\sin\left(\frac{(2\pi)(0.8727\text{ rad})}{2.094\text{ rad}}\right)$$

$$= 2.825\times 10^5\ \text{mm/s}^2$$

The inertial force is

$$F = ma = \frac{(0.8\text{ kg})\left(2.825\times 10^5\ \frac{\text{mm}}{\text{s}^2}\right)}{1000\ \frac{\text{mm}}{\text{m}}} = 226.0\text{ N}$$

(d) The force on the spring is

$$F = k\delta = \left(5\ \frac{\text{N}}{\text{mm}}\right)(5\text{ mm} + 8.43\text{ mm}) = 67.2\text{ N}$$

(e) The inertial and spring forces both oppose lift since $50° < \frac{1}{2}\theta_0$.

$$F_y = 226.0\text{ N} + 67.2\text{ N} = 293.2\text{ N}$$

(f) From Table 58.5, the lifter velocity is

$$\text{v} = \left(\frac{L\omega}{\theta_0}\right)\left(1 - \cos\frac{2\pi\theta}{\theta_0}\right)$$
$$= \left(\frac{(25\text{ mm})\left(125.66\ \frac{\text{rad}}{\text{s}}\right)}{2.094\text{ rad}}\right) \times \left(1 - \cos\left(\frac{(2\pi)(0.8727\text{ rad})}{2.094\text{ rad}}\right)\right)$$
$$= 2800\text{ mm/s}$$

The shaft torque on the cam due to one lifter

$$T = \frac{F_y\text{v}}{\omega} = \frac{(293.2\text{ N})\left(2800\ \frac{\text{mm}}{\text{s}}\right)}{\left(125.66\ \frac{\text{rad}}{\text{s}}\right)\left(1000\ \frac{\text{mm}}{\text{m}}\right)} = 6.53\text{ N·m}$$

18. CHAIN DRIVES

A *chain drive* comprises a closed loop of chain that meshes with *sprockets* (*sprocket gears*) having teeth shaped to mesh with the chain openings. *Roller chains* (also known as *drive chains*) and *engineering steel link chains* (also known as *offset link chains*, *offset sidebar chains*, *closed joint conveyor chains*, and *Ewart link chains*) are both widely used for power transmission.

Most roller chain is made from plain carbon or alloy steel. Each link has a plate with two pins, two bushings, and two rollers. The links are closed with pin link plates held by cotter pins. 304 stainless steel and occasionally brass and nylon are used in food processing machinery and other applications where it is desirable to avoid using lubricating oils that might contaminate products. **[Belts, Pulleys, and Chain Drives]**

Roller chains are commonly used in bicycles and motorcycles. Applications ordinarily use a *master link* (also known as a *connecting link*) which has a pin held by a C clip rather than friction fit. *Half links* (also known as *offsets*) are used to increase the chain length by a single roller. Typically, chain speed will be less than 800 ft/min (4 m/s), although high-speed roller chains can achieve 3000 ft/min (15 m/s). Figure 58.14 shows a roller chain.

Figure 58.14 *Roller Chain*

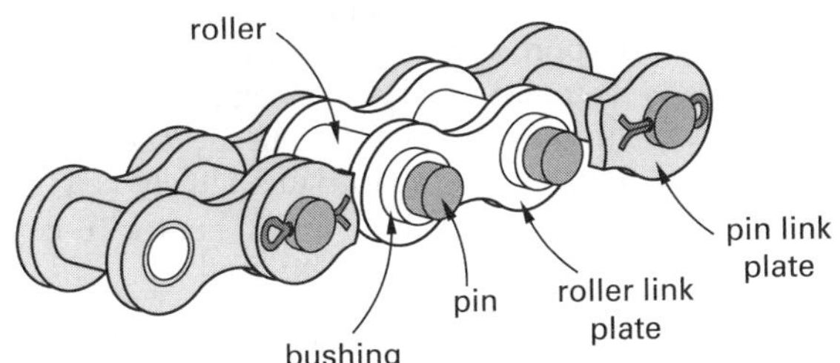

Standard roller chain size designations (per ASME/ANSI Standard B29.1) are 25, 35, 41, 40, 50, 60, 80, 100, 120, 140, 160, 180, 200, and 240. (See Table 58.7.) The last digit represents the chain variety—0 for standard chain, 1 for lightweight chain, and 5 for bushed chain with no rollers. The first digits indicate the pitch of the chain in eighths of an inch. A typical bicycle chain with a half-inch pitch would have a no. 40 designation. Metric pitches are expressed in sixteenths of an inch. A metric no. 8 chain designated "08B-1" is equivalent to size no. 40.

Table 58.7 *Properties of Standard Roller Chain*

designation	pitch (in)	roller diameter (in)	ultimate strength (lbf)	working load (lbf)
25	0.250	0.130	780	140
35	0.375	0.200	1760	480
41	0.500	0.306	1500	500
40	0.500	0.312	3125	810
50	0.625	0.400	4880	1430
60	0.750	0.469	7030	1980
80	1.000	0.625	12,500	3300
100	1.250	0.750	19,530	5070
120	1.500	0.875	28,125	6830
140	1.750	1.000	38,280	9040
160	2.000	1.125	50,000	11,900
180	2.250	1.460	63,280	13,670
200	2.500	1.562	78,125	16,090
240	3.000	1.1875	112,500	22,270

(Multiply in by 25.4 to obtain mm.)
(Multiply lbf by 4.448 to obtain N.)

Normal operating tension (known as *allowable pull*) is the ultimate tensile strength (UTS) of the chain reduced by a factor of 6 to 9. The minimum ultimate tensile strength, P_u, of an ASME/ANSI B29.1 standard steel chain is calculated from the pitch, p, in inches (millimeters). The width of a single chain varies and does not affect the load capacity. The strengths of *siamesed chains* will be proportional to the number of link plates attached to each pin.

$$P_u = 12{,}500p^2 \qquad \textit{58.45}$$

The *engineering steel chain* is distinguished by its offset/cranked link sidebar. (See Fig. 58.15.) Engineering steel chains can have much larger pitches (and, accordingly, higher strengths) than standard roller chains. They are often used in elevating and conveying equipment.

Figure 58.15 Engineering Steel Link Chain

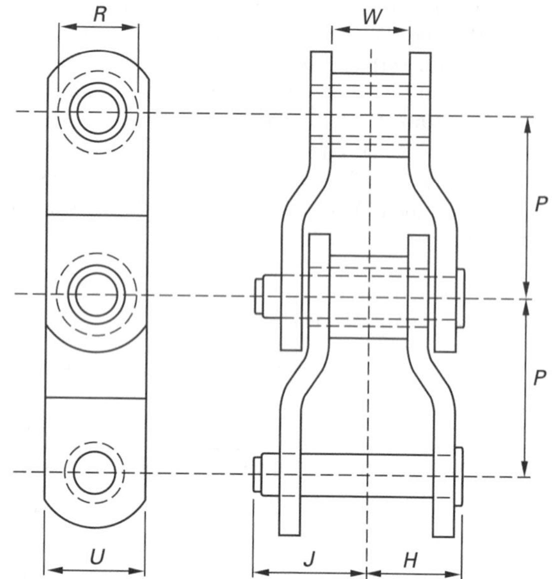

H = pin head to centerline of chain
J = pin end to centerline of chain
P = chain pitch
R = roller diameter
U = maximum chain height
W = width between sidebars at roller end

Drive ratios should normally not be greater than 6:1. When larger reductions are required, multiple stages are preferable. Properties of steel chains are given in Table 58.8.

Table 58.8 Properties of Standard Steel Link Chain

designation	pitch (in)	roller diameter (in)	distance between sidebars (in)	working load (lbf)
2010	2.500	1.250	1.500	4650
2512	3.067	1.625	1.562	6000
2814	3.500	1.750	1.500	7600
3315	4.073	1.781	1.938	10,000
3618	4.500	2.250	2.062	12,000
4020	5.000	2.500	2.750	17,500
4824	6.000	3.000	3.000	23,600
5628	7.000	3.500	3.250	30,500

(Multiply in by 25.4 to obtain mm.)
(Multiply lbf by 4.448 to obtain N.)

19. DESIGN/ANALYSIS OF ROLLER CHAIN DRIVES

Roller chain drives are useful for transmitting large torques. Equation 58.46 gives the relationship between applied torque, transmitted power, and rotational speed.

$$P_{\text{kW,transmitted}} = \frac{T_{\text{N·m}} n_{\text{rpm}}}{9549 \ \frac{\text{N·m}}{\text{kW·min}}} \qquad \text{[SI]} \qquad 58.46(a)$$

$$P_{\text{hp,transmitted}} = \frac{T_{\text{in-lbf}} n_{\text{rpm}}}{63{,}025 \ \frac{\text{in-lbf}}{\text{hp-min}}} = \frac{T_{\text{ft-lbf}} n_{\text{rpm}}}{5252 \ \frac{\text{ft-lbf}}{\text{hp-min}}} \qquad \text{[U.S.]} \qquad 58.46(b)$$

The *chain velocity* depends on the number of teeth on the sprocket, Z, and the *chain pitch*, p. Designations of standard American (ANSI) chain pitches are given in Table 58.7. Commercial chain pitches are 6 mm, 1/4 in, 8 mm, 3/8 in, 1/2 in, 5/8 in, 3/4 in, 1 in, 1 1/4 in, 1 1/2 in, 1 3/4 in, 2 in, 2 1/2 in, and 3 in. Roller chains used for fractional horsepower motor applications are almost always 1/4 in pitch, no. 25 single strand.

$$\text{v}_{\text{m/s}} = \frac{Zp_{\text{mm}} n_{\text{rpm}}}{60\,000 \ \frac{\text{mm·s}}{\text{m·min}}} \qquad \text{[SI]} \qquad 58.47(a)$$

$$\text{v}_{\text{ft/sec}} = \frac{Zp_{\text{in}} n_{\text{rpm}}}{720 \ \frac{\text{in-sec}}{\text{ft-min}}} \qquad \text{[U.S.]} \qquad 58.47(b)$$

The driving *sprocket* (*sprocket wheel*) is also known as the *pinion*, *chainwheel*, or *chainring*. Generally, traditional chain drives are limited to use with horizontal shafts. The line between sprocket centers should not be inclined more than 60° from the horizontal (specifically including parallel, vertically displaced shafts) without an automatic tensioning device. The driven shaft can be above or below the driving shaft, but it is preferred to have the upper strand be the tight-side strand. For the best chain life, the distance between sprocket/shaft centers should be 30–50 chain pitches, though satisfactory operation up to approximately 80 pitches is possible if chain tension is maintained by slack-side idlers/tensioners, drive support guides are used, the chain is lightly loaded, and/or the chain runs slowly (e.g., less than 3 ft/sec (1 m/s)). For large center-to-center distances, two-stage drives should be used.

The ratio of high-to-low rotational speeds is known as the *drive ratio*, *gear ratio*, *speed ratio*, and *velocity ratio*. When determining the drive ratio and/or the rotational speeds of sprockets, the chain is ignored. The number of sprocket teeth is used to determine rotational speeds, as with a meshing gear system. Although drive ratios up to 5:1 are effective, efficiency is maximized with ratios of 3:1 or less. Two-stage drives with idler shafts are useful when drive ratios greater than 5:1 are needed.

The number of teeth, Z, on the sprockets is determined from the drive ratio, DR. It is not usually possible to satisfy a drive ratio based on rotational speed exactly. As a rule of thumb, the sum of the numbers of teeth on the driving and driven sprockets should not be less than 50. Generally, the minimum number of teeth is 17, unless shaft speeds are very low, although 15 tooth sprockets are common. For high-speed and sprockets experiencing anything but smooth operation, sprockets should have at least 25 teeth. For normal service life, sprockets should not have more than 67 teeth. To ensure accurate meshing, chainwheels should not have more than 114 teeth.

$$\text{DR} = \frac{n_{\text{high speed}}}{n_{\text{low speed}}} = \frac{Z_{\text{large}}}{Z_{\text{small}}} \qquad 58.48$$

A slight velocity variation known as the *chordal velocity variation*, Δv, occurs each time the chain engages and disengages a tooth. The variation, which is generally less than 5%, is minimal for sprockets with numbers of teeth greater than 25 and is given by Eq. 58.49.

$$\Delta\text{v} = \frac{\text{v}_{\text{max}} - \text{v}_{\text{min}}}{\text{v}_{\text{max}}} = \left(1 - \cos\frac{180°}{Z}\right) \times 100\% \qquad 58.49$$

Since most drives use chains with an even number of pitches, uniform wear distribution in the chains and sprockets is facilitated by selecting sprockets with odd numbers of teeth. An exception is made for 1:1 drives where even-toothed sprockets are preferred in order to minimize the chordal variation on the drive. To prevent errors in mesh between chains and sprockets, chains should be replaced when their elongation exceeds the larger of 2% or $200\%/Z$.

Power transmission can be increased by using a multi-stranded chain. Each *strand* is essentially a single chain. A single-strand chain is known as a *simplex chain*. Figure 58.16 illustrates *double-strand* (*duplex*), *triple-strand* (*triplex*), and *quadruple-strand* (*quadplex*) chains. The power transmission capacities for duplex, triplex, and quadplex chains are calculated by multiplying the *multiple strand factors*—approximately 1.7, 2.5, and 3.3, respectively—by the capacity of the basic simplex chain.

Chain design and selection depends heavily on the manufacturer's literature and knowledge of the operating environment. The usual procedure for designing a chain drive is to (1) determine the *required horsepower rating*, from the amount of horsepower to be transmitted by the motor and the service factor, (2) determine the number of teeth on the smaller sprocket, and then (3) use the chain manufacturer's literature (i.e., a graph or table) to select the chain pitch and type. Other factors may be included if known.

Figure 58.16 *Double-, Triple-, and Quadruple-Strand Chains*

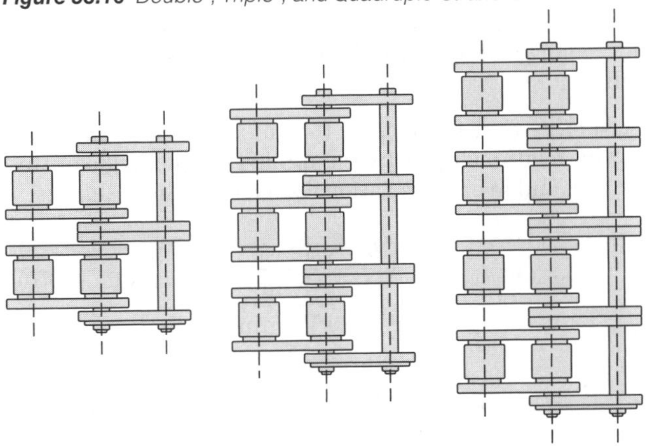

(a) double strand

(b) triple strand

(c) quadruple strand

Horsepower Ratings for Roller Chain-1986

$$\text{required hp rating} = \frac{\text{hp to be transmitted} \times \text{service factor}}{\text{multiple strand factor}} \qquad 58.50$$

The *service factor*, depends on the smoothness characteristics of both the driver and the driven machine, and are typically taken from tables such as those in *Machinery's Handbook*. [**Horsepower Ratings for Roller Chain-1986**]

Driven machines are also categorized by smoothness. For example, centrifugal and rotary compressors run smoothly, while reciprocating compressors run with heavy shock. Belt conveyors run smoothly, while bucket conveyors run with heavy shock. Centrifugal, gear, and rotary pumps run with moderate shock.

The multiple strand factor, is 1.7 for two strands, 2.5 for three strands, and 3.3 for four strands.

20. DRIVE SHAFT UNIVERSAL JOINTS

Drive shafts (also known as *shaft drives* or *propeller shafts*) are used to transmit power and torque from a prime mover to a nearby application. *Universal joints* (also known as *U-joints* or *Cardan joints*) are used to accommodate minor angular misalignments, while *splined couplings* (also known as *splined joints*) are used to accommodate minor length changes. Drive shafts with single U-joints are known as *torque tubes*. Drive shafts with U-joints at both ends are known as *Hotchkiss drives*.

Torque tubes are typically limited to approximately 15° of angular misalignment, while Hotchkiss drives can be used with up to approximately 30° of misalignment. Shafts are limited by imbalance but are otherwise limited only by their own shear strength, which translates into a maximum torque rating. U-joints themselves are similarly limited by their own torque ratings.

Efficiencies of new, oiled U-joints that are used for simple alignment (as opposed to being used specifically for change of direction) are seldom less than 99.5% efficient.[14]

Assuming that there are n U-joints, each with efficiency η, the relationship between applied torque, transmitted power, and average rotational speed in a drive shaft is

$$P_{\text{kW,transmitted}} = \frac{\eta^n T_{\text{N·m}} n_{\text{rpm}}}{9549 \frac{\text{N·m}}{\text{kW·min}}} \quad \text{[SI]} \quad 58.51(a)$$

$$P_{\text{hp,transmitted}} = \frac{\eta^n T_{\text{in-lbf}} n_{\text{rpm}}}{63{,}025 \frac{\text{in-lbf}}{\text{hp-min}}} = \frac{\eta^n T_{\text{ft-lbf}} n_{\text{rpm}}}{5252 \frac{\text{ft-lbf}}{\text{hp-min}}} \quad \text{[U.S.]} \quad 58.51(b)$$

Constant-velocity joints (also known as *CV joints*) are similar to U-joints in functionality but are different in construction. A CV joint's driving members are steel balls constrained in curved grooves between the forks of the joint.[15] A CV joint may operate with an angular deviation up to 80° without significant power loss or vibration, although joint life decreases as angular deviation increases. When two CV joints are used, the angle of the angular deviation can be even larger.

21. NOMENCLATURE

a	acceleration	ft/sec^2	m/s^2
C	center-to-center distance	ft	m
C	coefficient	–	–
D	diameter	ft	m
D	pitch circle diameter	ft	m
DR	drive ratio	–	–
E	energy	ft-lbf	J
E	modulus of elasticity	lbf/ft^2	Pa
f	chain adjustment factor	–	–
f	coefficient of friction	–	–
F	force	lbf	N
g_c	gravitational constant, 32.2	lbm-ft/lbf-sec^2	n.a.
I	mass moment of inertia	lbm-ft^2	kg·m^2
k	radius of gyration	ft	m
k	stiffness	lbf/ft	N/m
L	full lifter rise	ft	m
L	lead	in	mm
m	mass	lbm	kg
M	moment	lbf	kg
M	whole number	–	–
n	rotational speed	rev/sec	rev/s
N	number (quantity)	–	–
p	pitch	in	mm
p	pressure	lbf/ft^2	Pa
P	diametral pitch	1/ft	1/m
P	load	lbf	kg
P	power	hp	kW
r	radial distance	ft	m
r	radius	ft	m
R	instantaneous lifter rise	ft	m
S	strength	lbf/ft^2	Pa
t	thickness	ft	m
t	time	sec	s
T	torque	ft-lbf	N·m
TV	train value	–	–
v	velocity	ft/sec	m/s
VR	velocity ratio	–	–
W	work	ft-lbf	J
x	offset distance	ft	m
x	vertical lifter rise	ft	m
y	distance	ft	m
Z	number of teeth	–	–

Symbols

α	angle	deg	deg
α	rotational acceleration	rad/sec^2	rad/s^2
ϵ	strain	–	–
η	efficiency	–	–
μ	coefficient of friction	–	–
θ	angle of rotation	deg	deg
θ	lead angle	deg	deg
ν	Poisson's ratio	–	–
ρ	density	lbm/ft^3	kg/m^3
σ	normal stress	lbf/ft^2	Pa
ϕ	coefficient of friction	–	–
ϕ	pressure angle	deg	deg
ϕ	thread angle	deg	deg
ω	rotational velocity	rad/sec	rad/s

Subscripts

a	acceleration
ave	average
b	base circle
c	centroidal or collar
d	design
f	fluctuation

[14]Interestingly, efficiency isn't necessarily improved when alignment is improved. U-joint drives actually require a certain amount of misalignment in order to function reliably. A minimum of ½° is recommended by most U-joint manufacturers for lubrication and sealing, without which the needle bearings in U-joints will fail.

[15]A single U-joint is not a constant-velocity joint (i.e., does not maintain an instantaneous output rotational speed equal to the input rotational speed). A double U-joint arrangement is a constant-velocity joint, although the term *double Cardan joint* is preferred over "constant-velocity joint."

i	inner
k	kinetic
L	lower
m	mean
m	mechanical or motor
max	maximum
n	natural
o	outer
p	pitch
r	at radius r, radial, or rotational
R	raise
s	shear
t	tangential
u	ultimate
v	velocity
y	y- (vertical) component

59 Vibrating Systems

Content in blue refers to the *NCEES Handbook*.

NCEES EXAM SPECIFICATIONS AND RELATED CONTENT

HVAC AND REFRIGERATION EXAM

II.D.3. Supportive Knowledge: Vibration control
- 4. Free Vibration
- 13. Damped Free Vibrations
- 16. Damped Forced Vibrations
- 17. Vibration Isolation and Control

MACHINE DESIGN AND MATERIALS EXAM

I.E.1. Vibration: Natural frequencies and acoustics
- 4. Free Vibration
- 5. Initial Conditions
- 8. Free Rotation

I.E.2. Vibration: Damping
- 13. Damped Free Vibrations
- 16. Damped Forced Vibrations

I.E.3. Vibration: Forced vibrations
- 15. Magnification Factor
- 17. Vibration Isolation and Control
- 18. Isolation from Active Base

1. TYPES OF VIBRATIONS

Vibration is an oscillatory motion about an equilibrium point.[1] If the motion is the result of a disturbing force that is applied once and then removed, the motion is known as *natural* (or *free*) *vibration*. If a force of impulse is applied repeatedly to a system, the motion is known as *forced vibration*.

Within both of the categories of natural and forced vibrations are the subcategories of damped and undamped vibrations. If there is no damping (i.e., no friction), a system will experience free vibrations indefinitely. This is known as *free vibration* and *simple harmonic motion*. (See Fig. 59.1.)

Figure 59.1 *Types of Vibrations*

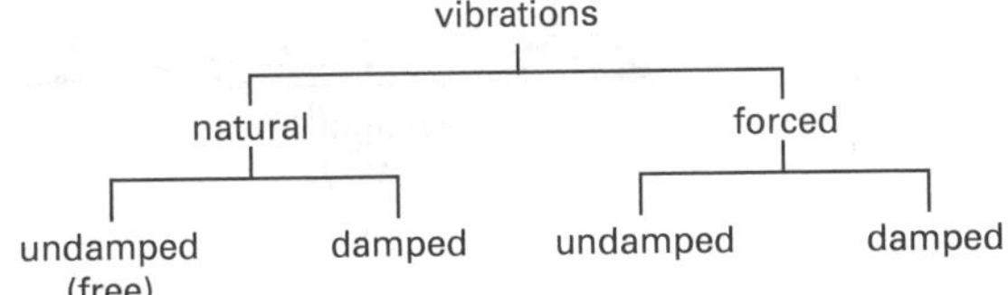

The performance (behavior) of some mechanical systems can be defined by a single variable. Such systems are referred to as *single degree of freedom* (*SDOF*) *systems*. For example, the position of a mass hanging from a spring is defined by the one variable $x(t)$.[2] Systems requiring two or more variables to define the positions of all parts are known as *multiple degree of freedom* (*MDOF*) *systems*. (See Fig. 59.2.)

[1]Although this chapter is presented in terms of mechanical vibrations, the concepts are equally applicable to electrical, fluid, and other types of systems.

[2]Although the convention is by no means universal, the variable x is commonly used as the position variable in oscillatory systems, even when the motion is in the vertical (y) direction.

Figure 59.2 *Single and Multiple Degree of Freedom Systems*

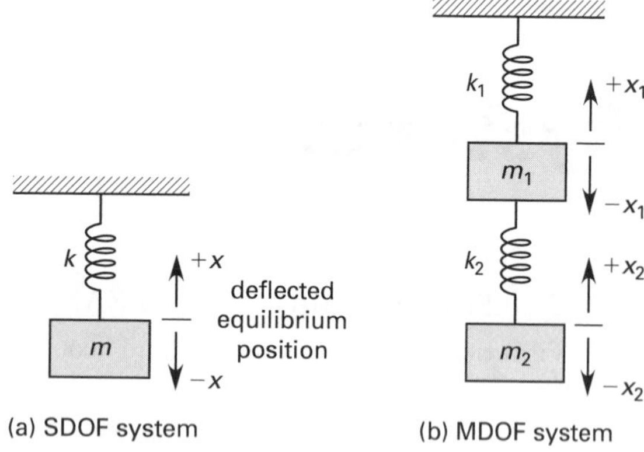

2. IDEAL COMPONENTS

When used to describe components in a vibrating system, the adjectives *perfect* and *ideal* generally imply *linearity* and the absence of friction and damping. The behavior of a *linear component* can be described by a linear equation. For example, the linear equation $F = kx$ describes a linear spring; however, the quadratic equation $F = cv^2$ describes a nonlinear dashpot. Similarly, $F = ma$ and $F = cv$ are linear inertial and viscous forces, respectively.

3. STATIC DEFLECTION

An important concept used in calculating the behavior of a vibrating system is the *static deflection*, δ_{st}. This is the deflection of a mechanical system due to gravitational force alone.[3] (The disturbing force is not considered.) In calculating the static deflection, it is extremely important to distinguish between mass and weight. Figure 59.3 illustrates several cases of static deflection.

Figure 59.3 *Examples of Static Deflection*

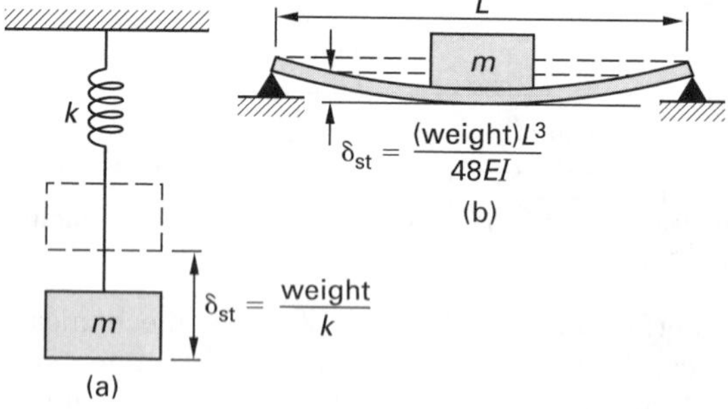

4. FREE VIBRATION

The simple mass and ideal spring illustrated in Fig. 59.3 is an example of a system that can experience free vibration. After the mass is displaced and released, it will oscillate up and down. Since there is no friction (i.e., the vibration is undamped), the oscillations will continue forever. (See Fig. 59.4.)

Figure 59.4 *Free Vibration*

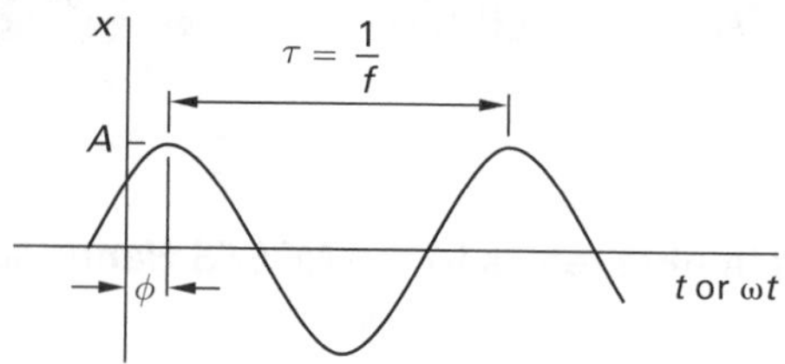

The system is initially at rest. The mass is hanging on the spring, and the *equilibrium position* is at the position of static deflection.

Free Vibration

$$mg = k\delta_{st} \qquad 59.1$$

The system is disturbed by a downward force (i.e., the mass is pulled downward from its static deflection and released).

After the initial disturbing force is removed, the object will be acted upon by the restoring force ($-kx$) and the inertial force ($-ma$). Both of these forces are proportional to the displacement from the equilibrium point, and they are opposite in sign from the displacement. From D'Alembert's principle,

Free Vibration

$$m\ddot{x} + kx = 0 \qquad \text{[SI]} \qquad 59.2(a)$$

$$\frac{m\ddot{x}}{g_c} + kx = 0 \qquad \text{[U.S.]} \qquad 59.2(b)$$

$$\sum F = 0{:}\ -ma - kx = 0$$

$$m\frac{d^2x}{dt^2} = -kx \qquad \text{[SI]} \qquad 59.3(a)$$

$$\frac{m}{g_c}\frac{d^2x}{dt^2} = -kx \qquad \text{[U.S.]} \qquad 59.3(b)$$

The solution to this second-order differential equation is easily derived. x_0 and v_0 are the initial displacement and initial velocity of the object, respectively.

Free Vibration

$$x(t) = x_0\cos\omega_n t + \left(\frac{v_0}{\omega_n}\right)\sin\omega_n t \qquad 59.4$$

ω_n is known as the *angular natural frequency of vibration*. It has units of radians per second. It is not the same as the *linear natural frequency*, *f*, which has units

[3]The term *deformation* is used synonymously with *deflection*.

of hertz (formerly known as cycles per second). The undamped *natural period of vibration*, τ_n, is the reciprocal of the linear frequency.

Free Vibration

$$\omega_n = \sqrt{\frac{k}{m}} = \sqrt{\frac{g}{\delta_{st}}} \quad \text{[SI]} \qquad 59.5(a)$$

$$\omega_n = \sqrt{\frac{kg_c}{m}} = \sqrt{\frac{g}{\delta_{st}}} \quad \text{[U.S.]} \qquad 59.5(b)$$

$$f = \frac{\omega_n}{2\pi} = \frac{1}{\tau_n} \qquad 59.6$$

$$\tau_n = \frac{2\pi}{\omega_n} = \frac{2\pi}{\sqrt{\frac{k}{m}}} = \frac{2\pi}{\sqrt{\frac{g}{\delta_{st}}}} = \frac{1}{f} \qquad 59.7$$

Equation 59.5 can be written in terms of the weight of the object suspended from the spring or in terms of the static deflection.

$$\text{weight} = mg \quad \text{[SI]} \qquad 59.8(a)$$

$$\text{weight} = \frac{mg}{g_c} \quad \text{[U.S.]} \qquad 59.8(b)$$

Free Vibration

$$\omega_n = \sqrt{\frac{g}{\delta_{st}}} = \sqrt{\frac{kg}{\text{weight}}} \qquad 59.9$$

Equation 59.9 is extremely useful. It is not limited to the simple mass-on-a-spring arrangement that is shown in Fig. 59.3(a). It can be used with a variety of systems, including those involving beams, shafts, and plates. Example 59.2 illustrates how this is done.

Equation 59.10 is an alternate form of the solution to Eq. 59.3. A is the *amplitude* and ϕ is the phase angle.

$$x(t) = A\cos(\omega_n t - \phi) \qquad 59.10$$

$$A = \sqrt{x_0^2 + \left(\frac{\text{v}_0}{\omega_n}\right)^2} \qquad 59.11$$

$$\phi = \arctan\left(\frac{\text{v}_0}{\omega_n x_0}\right) \qquad 59.12$$

The position, velocity, and acceleration are all sinusoidal with time. The maximum values are

$$x_{\text{max}} = A \qquad 59.13$$

$$\text{v}_{\text{max}} = A\omega_n \qquad 59.14$$

$$a_{\text{max}} = A\omega_n^2 \qquad 59.15$$

5. INITIAL CONDITIONS

With natural, undamped vibrations, the initial conditions (i.e., initial position and velocity) do not affect the natural period of oscillation. However, the amplitude of the oscillations will be affected, as indicated by Eq. 59.11.

Example 59.1

A 120 lbm (54 kg) mass is supported by three springs as shown. The initial displacement is 2.0 in (5.0 cm) downward from the static equilibrium position. No external forces act on the mass after it is released. What are the maximum velocity and acceleration?

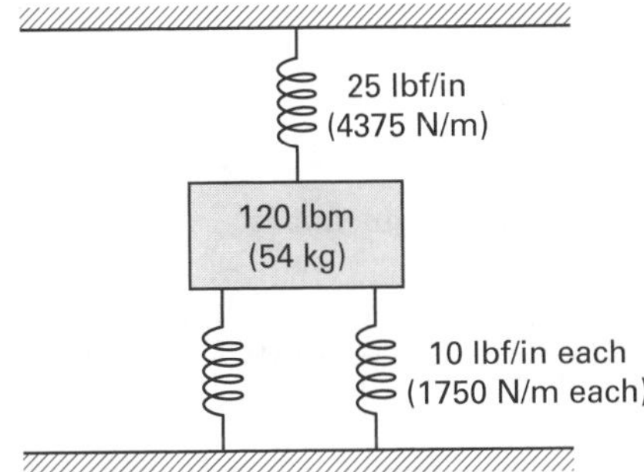

SI Solution

Since the springs are in parallel, they all share the applied load. The equivalent spring constant is

$$k_{eq} = k_1 + k_2 + k_3 = 4375\ \frac{\text{N}}{\text{m}} + 1750\ \frac{\text{N}}{\text{m}} + 1750\ \frac{\text{N}}{\text{m}}$$
$$= 7875\ \text{N/m}$$

The static deflection is

$$\delta_{st} = \frac{\text{weight}}{k} = \frac{mg}{k} = \frac{(54\ \text{kg})\left(9.81\ \frac{\text{m}}{\text{s}^2}\right)}{7875\ \frac{\text{N}}{\text{m}}}$$
$$= 0.0673\ \text{m}$$

The *natural frequency*, ω_n, is given by Eq. 59.9. (Compare this to the value calculated from Eq. 59.5.)

Free Vibration

$$\omega_n = \sqrt{\frac{g}{\delta_{st}}} = \sqrt{\frac{9.81\ \frac{\text{m}}{\text{s}^2}}{0.0673\ \text{m}}}$$
$$= 12.07\ \text{rad/s}$$

Since the mass is pulled down and released, the initial conditions are

$$\text{v}_0 = 0$$
$$x_0 = 5.0\ \text{cm} \quad (0.05\ \text{m})$$

From Eq. 59.11, the amplitude of oscillation is $A = 0.05$ m. From Eq. 59.14 and Eq. 59.15, the maximum velocity and acceleration are

$$\begin{aligned} v_{max} &= A\omega_n = (0.05\ \text{m})\left(12.07\ \frac{\text{rad}}{\text{s}}\right) \\ &= 0.604\ \text{m/s} \\ a_{max} &= A\omega_n^2 = (0.05\ \text{m})\left(12.07\ \frac{\text{rad}}{\text{s}}\right)^2 \\ &= 7.28\ \text{m/s}^2 \end{aligned}$$

(Radians are dimensionless.)

Customary U.S. Solution

The equivalent spring constant is

$$k_{eq} = 25\ \frac{\text{lbf}}{\text{in}} + 10\ \frac{\text{lbf}}{\text{in}} + 10\ \frac{\text{lbf}}{\text{in}} = 45\ \text{lbf/in}$$

Referring to Fig. 59.3 and Eq. 59.8, the static deflection is

$$\begin{aligned} \delta_{st} &= \frac{\text{weight}}{k} = \frac{mg}{kg_c} = \frac{(120\ \text{lbm})\left(32.2\ \frac{\text{ft}}{\text{sec}^2}\right)}{\left(45\ \frac{\text{lbf}}{\text{in}}\right)\left(32.2\ \frac{\text{lbm-ft}}{\text{lbf-sec}^2}\right)} \\ &= 2.67\ \text{in} \end{aligned}$$

The natural frequency is given by Eq. 59.9.

Free Vibration

$$\begin{aligned} \omega_n &= \sqrt{\frac{g}{\delta_{st}}} = \sqrt{\frac{\left(32.2\ \frac{\text{ft}}{\text{sec}^2}\right)\left(12\ \frac{\text{in}}{\text{ft}}\right)}{2.67\ \text{in}}} \\ &= 12.03\ \text{rad/sec} \end{aligned}$$

Since the mass is pulled down and released, the initial conditions are

$$\begin{aligned} v_0 &= 0 \\ x_0 &= \frac{2\ \text{in}}{12\ \frac{\text{in}}{\text{ft}}} = 0.167\ \text{ft} \end{aligned}$$

From Eq. 59.11, the amplitude of oscillation is $A = 0.167$ ft. From Eq. 59.14 and Eq. 59.15, the maximum velocity and acceleration are

$$\begin{aligned} v_{max} &= A\omega_n = (0.167\ \text{ft})\left(12.03\ \frac{\text{rad}}{\text{sec}}\right) \\ &= 2.0\ \text{ft/sec} \\ a_{max} &= A\omega_n^2 = (0.167\ \text{ft})\left(12.03\ \frac{\text{rad}}{\text{sec}}\right)^2 \\ &= 24.2\ \text{ft/sec}^2 \end{aligned}$$

(Radians are dimensionless.)

Example 59.2

A diving board is supported by a frictionless pivot at one end and by an unyielding, frictionless fulcrum, as indicated. A diver of mass m stands at the free end and bounces up and down. What is the frequency of oscillation?

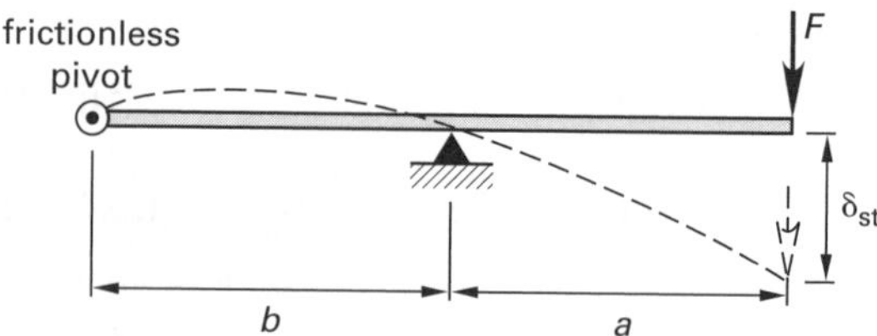

Solution

The deflection curve of the beam is shown by the dotted line. The tip force is

$$F = mg \quad \text{[SI]}$$
$$F = \frac{mg}{g_c} \quad \text{[U.S.]}$$

Use standard beam tables to determine the deflection. If the diver were to stand perfectly still, the static deflection at the tip would be

Simple Beam with Overhung Load

$$\delta_{st} = \frac{Fa^2(a+b)}{3EI}$$

From Eq. 59.6 and Eq. 59.9, the linear natural frequency is

$$f = \frac{1}{2\pi}\sqrt{\frac{g}{\delta_{st}}} = \frac{1}{2\pi}\sqrt{\frac{3EIg}{Fa^2(a+b)}}$$

6. VERTICAL VERSUS HORIZONTAL OSCILLATION

As long as friction is absent, the two cases of oscillation shown in Fig. 59.5 are equivalent (i.e., will have the same frequency and amplitude). Although it may seem that there is an extra gravitational force with vertical motion, the weight of the body is completely canceled by the opposite spring force when the system is in equilibrium. Therefore, vertical oscillations about an equilibrium point are equivalent to horizontal oscillations about the unstressed point.

Figure 59.5 *Vertical and Horizontal Oscillations*

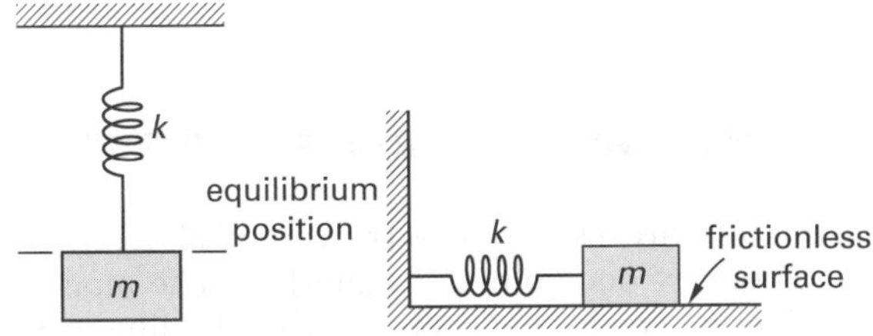

7. CONSERVATION OF ENERGY

The conservation of energy in vibrating systems requires the kinetic energy at the static equilibrium position to equal the stored elastic energy at the position of maximum displacement. For the mass-spring system shown in Fig. 59.5, the energy conservation equation is

$$\tfrac{1}{2}kx_{\text{max}}^2 = \tfrac{1}{2}m\text{v}_{\text{max}}^2 \quad \text{[SI]} \quad 59.16(a)$$

$$\tfrac{1}{2}kx_{\text{max}}^2 = \frac{m\text{v}_{\text{max}}^2}{2g_c} \quad \text{[U.S.]} \quad 59.16(b)$$

The velocity function is derived by taking the derivative of the position function.

$$x(t) = x_{\text{max}} \sin \omega_n t \quad 59.17$$

$$\text{v}(t) = \frac{dx(t)}{dt} = \omega_n x_{\text{max}} \cos \omega_n t \quad 59.18$$

Equation 59.18 shows that $\text{v}_{\text{max}} = \omega_n x_{\text{max}}$. Substituting this into Eq. 59.16 derives the natural frequency of vibration.

$$\omega_n^2 = \frac{k}{m} \quad \text{[SI]} \quad 59.19(a)$$

$$\omega_n^2 = \frac{kg_c}{m} \quad \text{[U.S.]} \quad 59.19(b)$$

8. FREE ROTATION

The so-called *torsional pendulum* in Fig. 59.6 can be analyzed in a manner similar to the spring-mass combination. Ignoring the mass and moment of inertia of the shaft, the differential equation is

Torsional Vibration

$$\ddot{\theta} + \left(\frac{k_t}{I}\right)\theta = 0 \quad \text{[SI]} \quad 59.20(a)$$

$$\ddot{\theta} + \left(\frac{k_t g_c}{I}\right)\theta = 0 \quad \text{[U.S.]} \quad 59.20(b)$$

Figure 59.6 *Torsional Pendulum*

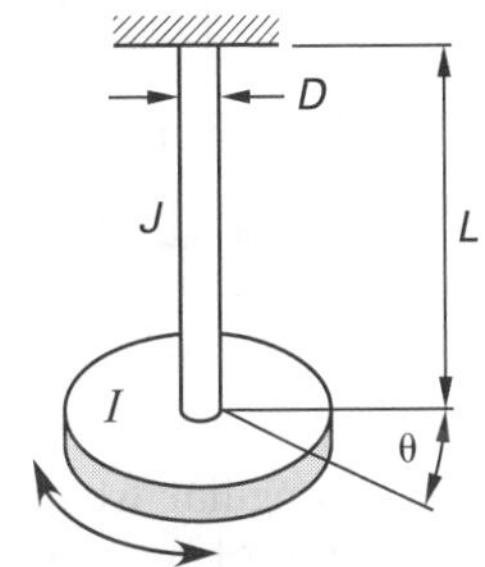

The *torsional spring constant*, k_t, used in Eq. 59.20 is

Torsional Vibration

$$k_t = \frac{GJ}{L} = \frac{\pi D^4 G}{32L} \quad 59.21$$

If the shaft is connected to the support through a torsional spring, or if the shaft consists of several sections of different diameters (i.e., a *stepped shaft*), the equivalent torsional spring constant can be calculated in the same manner as for springs in series.

Equivalent Springs

$$\frac{1}{k_{\text{eq}}} = \frac{1}{k_{t1}} + \frac{1}{k_{t2}} + \cdots + \frac{1}{k_n} \quad 59.22$$

The solution to Eq. 59.20 is directly analogous to the solution for the spring-mass system. Equation 59.23 through Eq. 59.29 summarize the governing equations.

$$\begin{aligned} \phi(t) &= \phi_0 \cos \omega_n t + \left(\frac{\omega_0}{\omega_n}\right)\sin \omega_n t \\ &= A\cos(\omega_n t - \phi) \end{aligned} \quad 59.23$$

Alternate forms of Eq. 59.23 use θ_0 for the initial angular velocity.[4]

[4]For consistency with the rest of this manual, the variable ω_0 will be used for the initial angular velocity.

Torsional Vibration

$$\theta(t) = \theta_0\cos(\omega_n t) + \left(\frac{\dot{\theta}_0}{\omega_n}\right)\sin(\omega_n t)$$

$$\omega_n = \sqrt{\frac{k_t}{I}} \quad \text{[SI]} \qquad 59.24(a)$$

$$\omega_n = \sqrt{\frac{k_t g_c}{I}} \quad \text{[U.S.]} \qquad 59.24(b)$$

$$A = \sqrt{\theta_0^2 + \left(\frac{\omega_0}{\omega_n}\right)^2} \qquad 59.25$$

$$\theta = \arctan\left(\frac{\omega_0}{\omega_n\theta_0}\right) \qquad 59.26$$

$$\phi_{\text{max}} = A \qquad 59.27$$

$$\omega_{\text{max}} = A\omega_n \qquad 59.28$$

$$\alpha_{\text{max}} = A\omega_n^2 \qquad 59.29$$

The undamped, natural angular frequency for a system with a solid round supporting rod is provided by Eq. 59.30. The undamped natural period of vibration is provided by Eq. 59.31.

Torsional Vibration

$$\omega_n = \sqrt{\frac{GJ}{IL}} \qquad 59.30$$

$$\tau_n = \frac{2\pi}{\omega_n} = \frac{2\pi}{\sqrt{\frac{k_t}{I}}} = \frac{2\pi}{\sqrt{\frac{GJ}{IL}}} \qquad 59.31$$

9. SUMMARY OF FREE VIBRATION PERFORMANCE EQUATIONS

Most equations of motion can be easily derived for free vibrations without damping. Table 59.1 provides a convenient summary of several common cases. [**Free Vibration**]

10. RAYLEIGH'S METHOD

Usually, the mass of the spring (beam, bar, shaft, etc.) is disregarded when calculating the frequency or period of vibration of a simple system. This is done to simplify the solution, although the mass of the spring element actually does affect the frequency. The exact solution is generally complex, but *Rayleigh's method* can be used to derive answers that will usually be less than 5% in error.

Rayleigh's method is to increase the oscillating object's mass by a fraction of the spring mass. [**Equivalent Masses, Springs, and Dampers**]

- For spring-mass systems, add 1/3 of the spring mass to the oscillating object mass.
- For simply supported beams loaded at the center, add 17/35 of the beam mass to the carried mass.
- For cantilever beams loaded at the free end, add 33/140 of the beam mass to the carried mass.
- For circular shafts in torsion, add 1/3 of the shaft mass moment of inertia to the mass moment of inertia of the rotating load.

11. TRANSFORMED LINEAR STIFFNESS

Transformers are used in electrical and electronic circuits to convert one voltage to another. The analogous mechanical device is the lever. Figure 59.7 illustrates a simple lever-spring system. The pivot point is frictionless; the springs are ideal; the lever is infinitely stiff and has negligible mass.

The *lever ratio* is

$$\frac{L_1}{L_2} = \frac{\delta_1}{\delta_2} \qquad 59.32$$

In ideal electrical circuits, the energies per unit time (i.e., the power) transferred across the two transformer windings are equal. In ideal mechanical systems, energy is similarly conserved. Specifically, the change in a mass' gravitational potential energy is equal to the change in stored spring energy. Referring to Fig. 59.7(a), a unit deflection of the mass ($\delta_2 = 1$) will result in stored spring energies such that the total energy change is zero. Equation 59.33 disregards internal energy.

$$\frac{\Delta E}{\delta_2} = \tfrac{1}{2}k_2\left(\frac{a}{L_2}\right)^2 + \tfrac{1}{2}k_1\left(\frac{L_1}{L_2}\right)^2 - mg = 0 \quad \text{[SI]} \qquad 59.33(a)$$

$$\frac{\Delta E}{\delta_2} = \tfrac{1}{2}k_2\left(\frac{a}{L_2}\right)^2 + \tfrac{1}{2}k_1\left(\frac{L_1}{L_2}\right)^2 - \frac{mg}{g_c} = 0 \quad \text{[U.S.]} \qquad 59.33(b)$$

An equivalent system is shown in Fig. 59.7(b). All of the spring force is concentrated at the position of spring 2. The angular frequency of the mass is unchanged. The equivalent spring constant, k', referred to spring 2, is

$$k' = k_2 + \left(\frac{L_1}{a}\right)^2 k_1 \qquad 59.34$$

Table 59.1 *Performance of Simple Oscillatory Systems (small deflections; consistent units)**

mechanism	natural frequency (ω_n)	linear frequency (f)	period (τ_n)
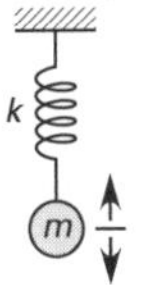 *k*, *m*; mass and spring	$\sqrt{\frac{k}{m}}$	$\frac{1}{2\pi}\sqrt{\frac{k}{m}}$	$2\pi\sqrt{\frac{m}{k}}$
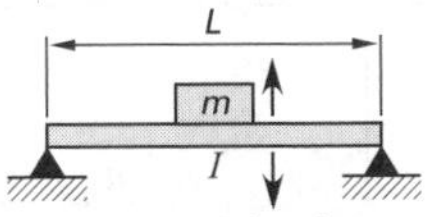 *L*, *m*, *I*; mass on massless beam (I = area moment of inertia of cross section)	$\sqrt{\frac{48EI}{mL^3}}$	$\frac{1}{2\pi}\sqrt{\frac{48EI}{mL^3}}$	$2\pi\sqrt{\frac{mL^3}{48EI}}$
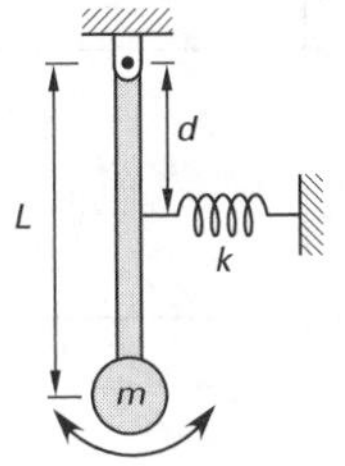 *d*, *L*, *k*, *m*; constrained compound pendulum (massless bar, frictionless pivot)	$\sqrt{\frac{mgL+kd^2}{mL^2}}$	$\frac{1}{2\pi}\sqrt{\frac{mgL+kd^2}{mL^2}}$	$2\pi\sqrt{\frac{mL^2}{mgL+kd^2}}$
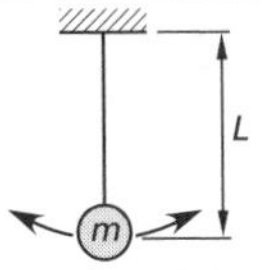 *L*, *m*; simple pendulum	$\sqrt{\frac{g}{L}}$	$\frac{1}{2\pi}\sqrt{\frac{g}{L}}$	$2\pi\sqrt{\frac{L}{g}}$
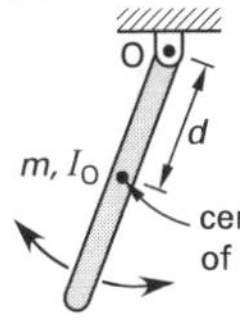 O, *d*, *m*, I_O, center of mass; compound pendulum	$\sqrt{\frac{mgd}{I_O}}$	$\frac{1}{2\pi}\sqrt{\frac{mgd}{I_O}}$	$2\pi\sqrt{\frac{I_O}{mgd}}$
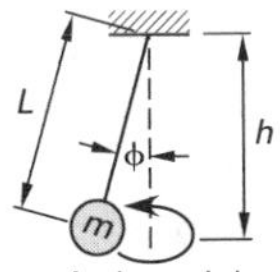 *L*, ϕ, $h = L\cos\phi$, *m*; conical pendulum	$\sqrt{\frac{g}{h}}$	$\frac{1}{2\pi}\sqrt{\frac{g}{h}}$	$2\pi\sqrt{\frac{h}{g}}$
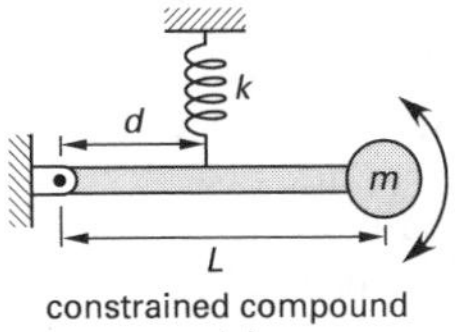 *k*, *d*, *m*, *L*; constrained compound pendulum	$\sqrt{\frac{kd^2}{mL^2}}$	$\frac{1}{2\pi}\sqrt{\frac{kd^2}{mL^2}}$	$2\pi\sqrt{\frac{mL^2}{kd^2}}$
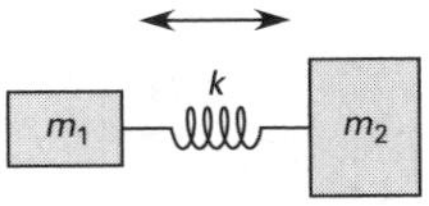 m_1, *k*, m_2; two masses and spring	$\sqrt{\frac{k(m_1+m_2)}{m_1m_2}}$	$\frac{1}{2\pi}\sqrt{\frac{k(m_1+m_2)}{m_1m_2}}$	$2\pi\sqrt{\frac{m_1m_2}{k(m_1+m_2)}}$
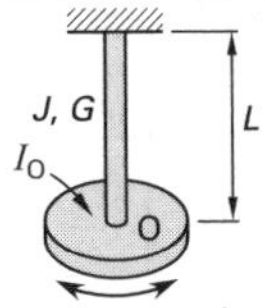 *J*, *G*, I_O, *L*, O; torsional mass and spring	$\sqrt{\frac{JG}{I_OL}}$	$\frac{1}{2\pi}\sqrt{\frac{JG}{I_OL}}$	$2\pi\sqrt{\frac{I_OL}{JG}}$

*Replace m with m/g_c for customary U.S. units. Replace I_O with I_O/g_c for customary U.S. units.

*Table 59.1 Performance of Simple Oscillatory Systems (small deflections; consistent units) (continued)**

mechanism	natural frequency (ω_n)	linear frequency (f)	period (τ_n)
two torsional masses	$\sqrt{\frac{JG(I_1+I_2)}{I_1I_2L}}$	$\frac{1}{2\pi}\sqrt{\frac{JG(I_1+I_2)}{I_1I_2L}}$	$2\pi\sqrt{\frac{I_1I_2L}{JG(I_1+I_2)}}$
oscillating pulley	$\sqrt{\frac{k}{m_1+\frac{m_2}{2}}}$	$\frac{1}{2\pi}\sqrt{\frac{k}{m_1+\frac{m_2}{2}}}$	$2\pi\sqrt{\frac{m_1+\frac{m_2}{2}}{k}}$
floating block	$\sqrt{\frac{g(\mathrm{SG}_l)}{L(\mathrm{SG}_b)}}$	$\frac{1}{2\pi}\sqrt{\frac{g(\mathrm{SG}_l)}{L(\mathrm{SG}_b)}}$	$2\pi\sqrt{\frac{L(\mathrm{SG}_b)}{g(\mathrm{SG}_l)}}$
U-tube and liquid	$\sqrt{\frac{2g}{L}}$	$\frac{1}{2\pi}\sqrt{\frac{2g}{L}}$	$2\pi\sqrt{\frac{L}{2g}}$
cantilever	$k_i^2\sqrt{\frac{EI}{mL^3}}$ $k_1 = 1.8751$ $k_2 = 4.6941$ $k_3 = 7.8532$	$\frac{k_i^2}{2\pi}\sqrt{\frac{EI}{mL^3}}$	$\frac{2\pi}{k_i^2}\sqrt{\frac{mL^3}{EI}}$
cantilever spring	$\sqrt{\frac{3EI}{mL^3}}$	$\frac{1}{2\pi}\sqrt{\frac{3EI}{mL^3}}$	$2\pi\sqrt{\frac{mL^3}{3EI}}$

*Replace m with m/g_c for customary U.S. units. Replace I_O with I_O/g_c for customary U.S. units.

Another transformation (not shown) is not as immediately useful. An equivalent spring constant, referred to spring 1, could be derived. This would have the effect of concentrating all of the spring force at the position of spring 1.

$$k'' = k_1 + \left(\frac{a}{L_1}\right)^2 k_2 \tag{59.35}$$

12. TRANSFORMED ANGULAR STIFFNESS

In a meshing gear set, the transmitted force and power are the same for both meshing gears. However, a gear set changes the torque. Therefore, a meshing gear set is a torsional analog to an electrical transformer.

Consider the loaded gear set shown in Fig. 59.8. The masses of shafts 1 and 2 are insignificant compared with the torsional loads, I_1 and I_2. Also, the tooth stiffness is typically disregarded in analyzing this type of system. Power transfer across the gear set is without loss.

The original system can be converted to the equivalent simple, torsional system shown in Fig. 59.8(b). The equivalent load I_2' of shaft 2 (referred to shaft 1) is

$$I_2' = \left(\frac{\omega_2}{\omega_1}\right)^2 I_2 = \left(\frac{N_1}{N_2}\right)^2 I_2 = \left(\frac{D_1}{D_2}\right)^2 I_2 \tag{59.36}$$

The equivalent torsional stiffness of shaft 2 is

$$k_2' = \left(\frac{N_1}{N_2}\right)^2 k_2 = \left(\frac{D_1}{D_2}\right)^2 k_2 \tag{59.37}$$

Figure 59.7 *Lever-Coupled Linear System*

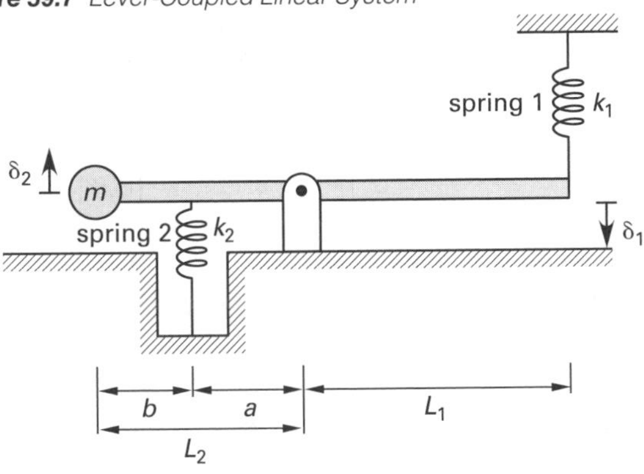

(a) original system

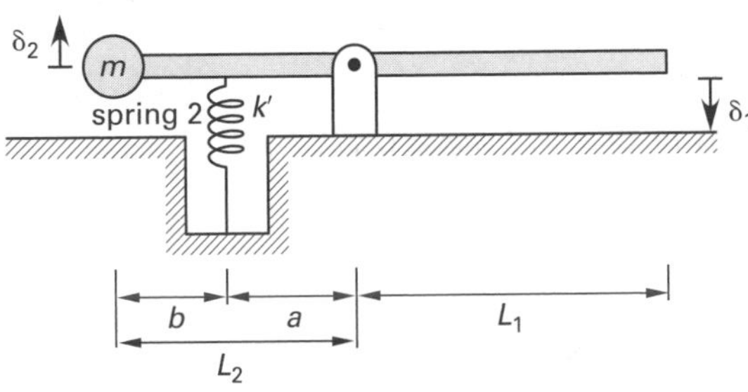

(b) equivalent system

Figure 59.8 *Gear-Coupled Torsional System*

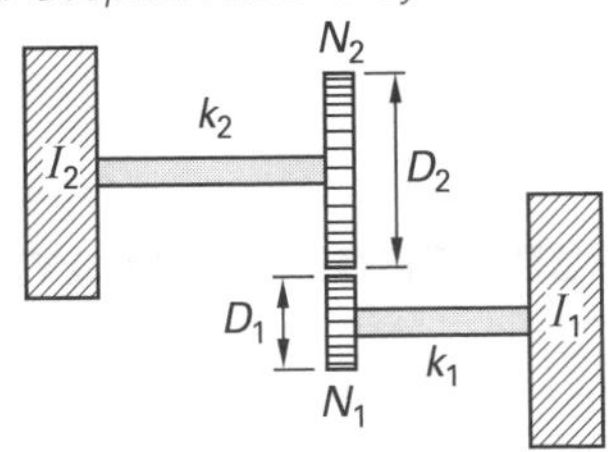

(a) original system

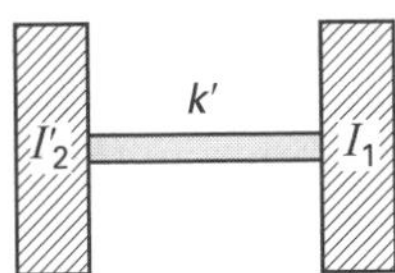

(b) equivalent system

The total equivalent torsional stiffness, k', is

$$\frac{1}{k'} = \frac{1}{k_1} + \frac{1}{k_2'} \quad 59.38$$

The natural frequency is

$$\omega_n = \sqrt{k'\left(\frac{I_1 + I_2'}{I_1 I_2'}\right)} \quad 59.39$$

13. DAMPED FREE VIBRATIONS

When friction resists the oscillatory motion, the system is said to be damped. Friction can occur internally, between two surfaces, or due to motion through a liquid. (Air friction can be disregarded in most problems.) The third type of friction is known as *viscous damping.* Figure 59.9 illustrates the *dashpot* symbol used to represent a source of viscous damping.

Figure 59.9 *Spring-Mass System with Dashpot*

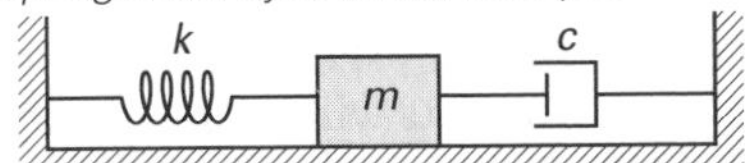

The viscous damping force can be a function of v or v^2. If velocity is high through the liquid, the viscous damping force will be a function of v^2. Only low-velocity, *linear damping* is covered in this chapter. With linear damping, the damping force is proportional to velocity. The constant of proportionality, c, is also known as the *coefficient of viscous damping.*

Viscous Damping

$$F = c\dot{x} = c\text{v} \quad 59.40$$

$$F = c\frac{dx}{dt} = c\text{v} \quad 59.41$$

The differential equation of motion is

$$m\ddot{x} + c\dot{x} + kx = 0 \quad 59.42$$

$$m\frac{d^2x}{dt^2} = -kx - c\frac{dx}{dt} \quad 59.43$$

The general solution is given by Eq. 59.44. The constants A and B must be determined from initial conditions.

$$x(t) = Ae^{r_1 t} + Be^{r_2 t} \quad 59.44$$

The roots of Eq. 59.44 are

$$r_1, r_2 = \frac{-c}{2m} \pm \sqrt{\left(\frac{c}{2m}\right)^2 - \frac{k}{m}} \quad \text{[SI]} \quad 59.45(a)$$

$$r_1, r_2 = g_c\left(\frac{-c}{2m}\sqrt{\left(\frac{c}{2m}\right)^2 - \frac{k}{mg_c}}\right) \quad \text{[U.S.]} \quad 59.45(b)$$

The *damping ratio* (*damping factor*), ζ, is defined as

Vibration Transmissibility, Base Motion

$$\zeta = \frac{c}{c_{\text{critical}}} = \frac{c}{2m\omega_n} = \frac{c}{2\sqrt{mk}} \quad \text{[SI]} \qquad 59.46(a)$$

$$\zeta = \frac{c}{c_{\text{critical}}} = \frac{cg_c}{2m\omega_n} = \frac{c}{2\sqrt{\frac{mk}{g_c}}} \quad \text{[U.S.]} \qquad 59.46(b)$$

If $\zeta < 1.0$ (i.e., $c < c_{\text{critical}}$), the radical in Eq. 59.45 will be negative, and both roots will be imaginary. The oscillations are said to be *underdamped*. This case is also known as *light damping*. Motion will be oscillatory with diminishing magnitude, as illustrated in Fig. 59.10.

Figure 59.10 *Underdamped Oscillation*

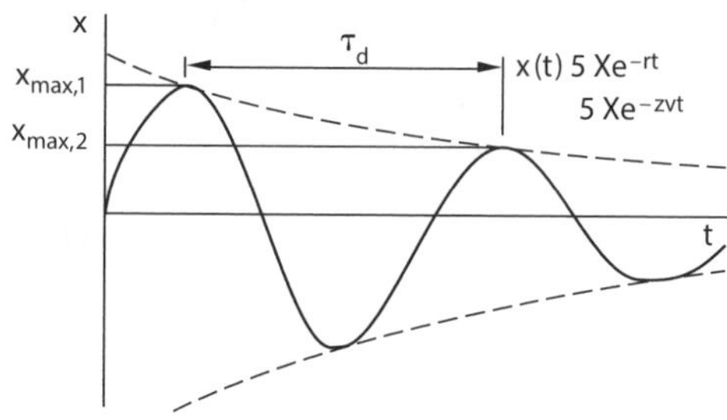

ω_d is the *damped frequency*. It is not the same as the natural frequency, ω_n, which is calculated assuming that $c = 0$.

Torsional Vibration

$$\omega_d = \omega_n\sqrt{1 - \zeta^2} \qquad 59.47$$

The *logarithmic decrement*, δ, is the natural logarithm of the ratio of two successive amplitudes.

Torsional Vibration

$$\delta = \ln\frac{x_n}{x_{n+1}} = \frac{2\pi\zeta}{\sqrt{1-\zeta^2}} = \zeta\omega_n\tau_d \qquad 59.48$$

If $\zeta > 1.0$ (i.e., $c > c_{\text{critical}}$), the radical is positive, and both roots are real. The motion is said to be *overdamped*. This case is also known as *heavy damping*. There will be a gradual return to the equilibrium, but no oscillation. (See Fig. 59.11.)

If $\zeta = 1.0$ (i.e., $c = c_{\text{critical}}$), the radical is zero, and the motion is said to be *critically damped*. (See Fig. 59.12.) Such motion is also known as *dead-beat motion*. There is no overshoot, and the return is the fastest of the three types of damped motion. The *critical damping coefficient* is

Torsional Vibration

$$c_{\text{critical}} = 2m\omega_n = 2\sqrt{km} = \frac{c}{\zeta} \quad \text{[SI]} \qquad 59.49(a)$$

$$c_{\text{critical}} = \frac{2m\omega_n}{g_c} = 2\sqrt{\frac{km}{g_c}} = \frac{c}{\zeta} \quad \text{[U.S.]} \qquad 59.49(b)$$

Figure 59.11 *Overdamped Oscillation*

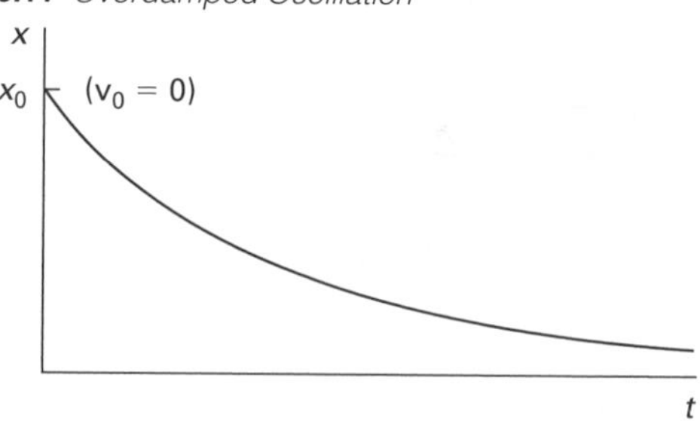

Figure 59.12 *Critically Damped Oscillation*

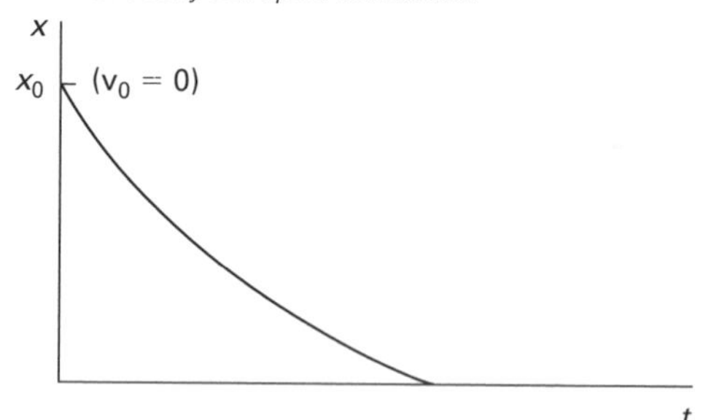

14. UNDAMPED FORCED VIBRATIONS

When an external disturbing force, $F(t)$, acts on the system, the system is said to be forced. Although the *forcing function* is usually considered to be periodic, it need not be (as in the case of impulse, step, and random functions).[5] However, an initial disturbance (i.e., when a mass is displaced and released to oscillate freely) is not an example of forced vibration. (See Fig. 59.13.)

Figure 59.13 *Forced Vibrations*

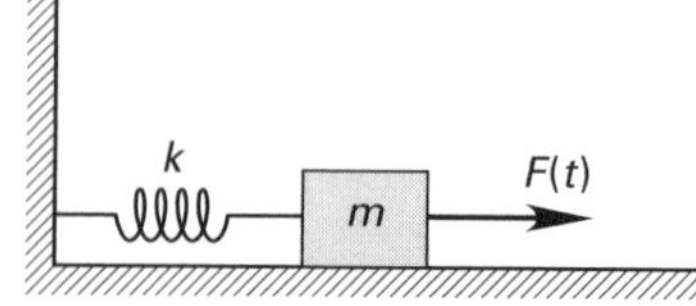

Consider a sinusoidal periodic force with a *forcing frequency* of ω_f and maximum value of F_0.

$$F(t) = F_0 \cos \omega_f t \qquad 59.50$$

[5]The sinusoidal case is important, since Fourier transforms can be used to model any forcing function in terms of sinusoids.

The differential equation of motion is

$$m\ddot{x} = -kx + F_0 \cos \omega_f t \qquad 59.51$$

$$m\frac{d^2x}{dt^2} = -kx + F_0 \cos \omega_f t \qquad 59.52$$

The solution to Eq. 59.52 consists of the sum of two parts: a complementary solution and a particular solution. The *complementary solution* is obtained by setting $F_0 = 0$ (i.e., solving the homogeneous differential equation). The solution is

$$x_c(t) = A \cos \omega_n t + B \sin \omega_n t \qquad 59.53$$

The *particular solution* is found by assuming its form and substituting that function into Eq. 59.54.

$$x_p(t) = D \cos \omega_f t \qquad 59.54$$

$$D = \frac{F_0}{m(\omega_n^2 - \omega_f^2)} \quad \text{[SI]} \qquad 59.55(a)$$

$$D = \frac{F_0 g_c}{m(\omega_n^2 - \omega_f^2)} \quad \text{[U.S.]} \qquad 59.55(b)$$

The solution of Eq. 59.52 is

$$x(t) = A \cos \omega_n t + B \sin \omega_n t + \left(\frac{F_0}{m(\omega_n^2 - \omega_f^2)} \right) \cos \omega_f t \quad \text{[SI]} \qquad 59.56(a)$$

$$x(t) = A \cos \omega_n t + B \sin \omega_n t + \left(\frac{F_0 g_c}{m(\omega_n^2 - \omega_f^2)} \right) \cos \omega_f t \quad \text{[U.S.]} \qquad 59.56(b)$$

15. MAGNIFICATION FACTOR

The *magnification factor*, β (also known as the *amplitude ratio* and *amplification factor*), is defined as the ratio of the steady-state vibration amplitude, D, and the *pseudo-static deflection*, F_0/k.

$$\beta = \frac{D}{\frac{F_0}{k}} = \left| \frac{1}{1 - \left(\frac{\omega_f}{\omega_n}\right)^2} \right| \qquad 59.57$$

Figure 59.14 illustrates the magnification factor for various values of ω_f/ω_n. *Resonance* occurs when ω_f equals or nearly equals ω_n. Oscillations are theoretically infinite.[6] Resonance leads to rapid failure in structures and mechanical equipment. **[Vibration Transmissibility, Base Motion]**

Figure 59.14 *Magnification Factor (no damping)*

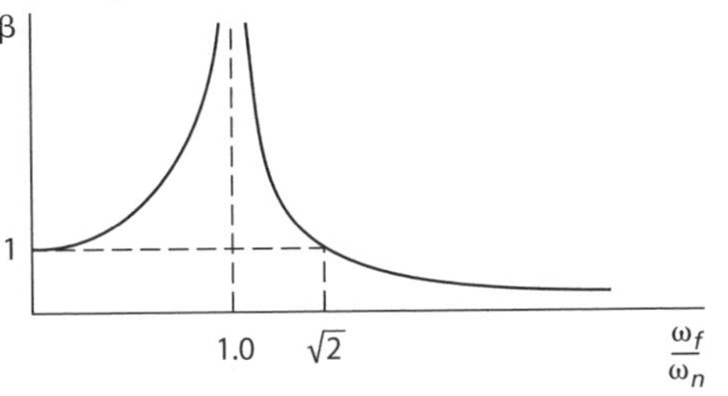

If the ratio $\omega_f/\omega_n = \sqrt{2}$, the magnification factor is

$$\beta = \left| \frac{1}{1 - (\sqrt{2})^2} \right| = 1 \qquad 59.58$$

Therefore, ω_f/ω_n must be greater than $\sqrt{2}$ for the system to have an oscillation magnitude smaller than the static deflection alone.

When ω_f is significantly greater than ω, the magnification factor is close to zero, and the system will be nearly stationary.

Example 59.3

A 250 lbm (113.6 kg) motor turns at a rate of 1000 rpm (16.66 rps). It is mounted on a resilient pad having a stiffness of 3000 lbf/in (525 kN/m). Due to an unbalanced condition, a periodic force of 20 lbf (89 N) is applied in the vertical direction once each revolution. If the motor is constrained to move vertically and damping is negligible, what is the amplitude of vibration?

SI Solution

The natural frequency of the system is

Free Vibration

$$\omega_n = \sqrt{\frac{k}{m}} = \sqrt{\frac{\left(525 \frac{\text{kN}}{\text{m}}\right)\left(1000 \frac{\text{N}}{\text{kN}}\right)}{113.6 \text{ kg}}} = 67.98 \text{ rad/s}$$

The forcing frequency is

$$\omega_f = \left(16.66 \frac{\text{rev}}{\text{s}}\right)\left(2\pi \frac{\text{rad}}{\text{rev}}\right) = 104.7 \text{ rad/s}$$

[6]Damping is always present in real systems and keeps the excursions finite.

The pseudo-static deflection is

$$\frac{F_0}{k} = \frac{89 \text{ N}}{525\,000 \ \frac{\text{N}}{\text{m}}} = 1.70 \times 10^{-4} \text{ m}$$

The magnification factor is

$$\beta = \left| \frac{1}{1 - \left(\frac{\omega_f}{\omega_n}\right)^2} \right| = \left| \frac{1}{1 - \left(\frac{104.7 \ \frac{\text{rad}}{\text{s}}}{67.98 \ \frac{\text{rad}}{\text{s}}}\right)^2} \right|$$
$$= 0.729$$

The amplitude of oscillation is calculated from Eq. 59.57.

$$D = \beta\left(\frac{F_0}{k}\right) = (0.729)(1.70 \times 10^{-4} \text{ m})$$
$$= 1.24 \times 10^{-4} \text{ m}$$

Customary U.S. Solution

The natural frequency of the system is

$$\omega_n = \sqrt{\frac{kg_c}{m}}$$
$$= \sqrt{\frac{\left(3000 \ \frac{\text{lbf}}{\text{in}}\right)\left(32.2 \ \frac{\text{lbm-ft}}{\text{lbf-sec}^2}\right)\left(12 \ \frac{\text{in}}{\text{ft}}\right)}{250 \text{ lbm}}}$$
$$= 68.09 \text{ rad/sec}$$

The forcing frequency is

$$\omega_f = \frac{\left(1000 \ \frac{\text{rev}}{\text{min}}\right)\left(2\pi \ \frac{\text{rad}}{\text{rev}}\right)}{60 \ \frac{\text{sec}}{\text{min}}} = 104.7 \text{ rad/sec}$$

The pseudo-static deflection is

$$\frac{F_0}{k} = \frac{20 \text{ lbf}}{3000 \ \frac{\text{lbf}}{\text{in}}} = 0.00667 \text{ in}$$

The magnification factor is

$$\beta = \left| \frac{1}{1 - \left(\frac{\omega_f}{\omega_n}\right)^2} \right| = \left| \frac{1}{1 - \left(\frac{104.7 \ \frac{\text{rad}}{\text{sec}}}{68.09 \ \frac{\text{rad}}{\text{sec}}}\right)^2} \right|$$
$$= 0.733$$

The amplitude of oscillation is calculated from Eq. 59.57.

$$D = \beta\left(\frac{F_0}{k}\right) = (0.733)(0.00667 \text{ in})$$
$$= 0.00489 \text{ in}$$

16. DAMPED FORCED VIBRATIONS

If a viscous damping force, $cv = c(dx/dt)$, is added to a sinusoidally forced system, as in Fig. 59.15, the differential equation of motion is

$$m\frac{d^2x}{dt^2} = -kx - c\frac{dx}{dt} + F_0 \cos \omega_f t \quad \text{[SI]} \quad 59.59(a)$$

$$\frac{m}{g_c}\frac{d^2x}{dt^2} = -kx - c\frac{dx}{dt} + F_0 \cos \omega_f t \quad \text{[U.S.]} \quad 59.59(b)$$

$$m\ddot{x} = -kx - c\dot{x} + F_0 \cos\omega_f t \quad \text{[SI]} \quad 59.60(a)$$

$$\frac{m}{g_c}\ddot{x} = -kx - c\dot{x} + F_0 \cos\omega_f t \quad \text{[U.S.]} \quad 59.60(b)$$

Figure 59.15 *Damped Forced Oscillations*

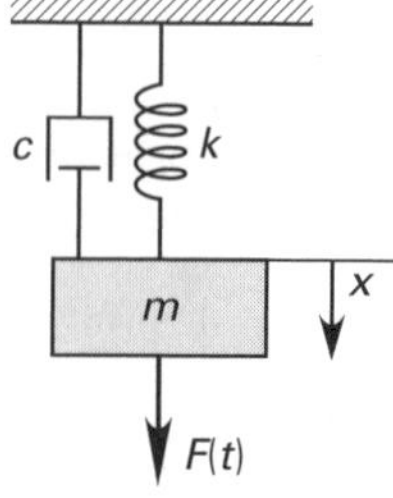

The solution to Eq. 59.59 has several terms. As a result of the damping force, the complementary solution has decaying exponentials. Therefore, the complementary solution is also known as the *transient component* because its contribution to the system performance decreases rapidly. However, the transient terms do contribute to the initial performance. For this reason, initial cycles may experience displacements greater than the steady-state values. The particular solution is known as the *steady-state component*.

Equation 59.61 defines the damped magnification factor, β_d, for steady-state damped forced vibrations. (See Fig. 59.16.) The magnification factor for the undamped case (see Eq. 59.57) can be derived by setting $\zeta = 0$.

$$\begin{aligned}\beta_d &= \left|\frac{D}{\frac{F_0}{k}}\right| \\ &= \frac{1}{\sqrt{\left(1-\left(\frac{\omega_f}{\omega_n}\right)^2\right)^2+\left(\frac{c\omega_f}{m\omega_n^2}\right)^2}} \quad \text{[SI only]} \\ &= \frac{1}{\sqrt{\left(1-\left(\frac{\omega_f}{\omega_n}\right)^2\right)^2+\left(\frac{2c\omega_f}{c_{\text{critical}}\omega_n}\right)^2}} \quad \text{[U.S. and SI]} \\ &= \frac{1}{\sqrt{(1-r^2)^2+(2\zeta r)^2}} \quad \text{[U.S. and SI]}\end{aligned} \qquad 59.61$$

Vibration Transmissibility, Base Motion

$$r = \frac{\omega_f}{\omega_n} \qquad 59.62$$

Figure 59.16 *Damped Magnification Factor*

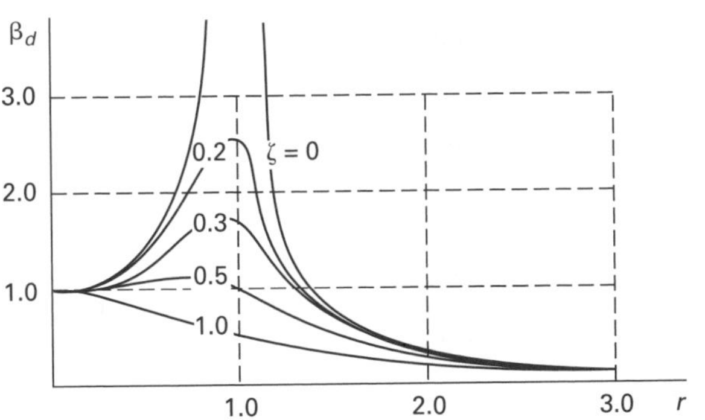

17. VIBRATION ISOLATION AND CONTROL

It is often desired to isolate a rotating machine from its surroundings, to limit the vibrations that are transmitted to the supports, and to reduce the amplitude of the machine's vibrations.

The *transmissibility* (i.e., *linear transmissibility*) is the ratio of the transmitted force (i.e., the force transmitted to the supports) to the applied force (i.e., the force from the imbalance). Equation 59.63 is illustrated in Fig. 59.17.

$$\begin{aligned}\text{TR} &= \left|\frac{F_{\text{transmitted}}}{F_{\text{applied}}}\right| \\ &= \frac{F}{F_f} \\ &= \beta_d\sqrt{1+(2r\zeta)^2}\end{aligned} \qquad 59.63$$

$$\beta_d = \frac{1}{\sqrt{(1-r^2)^2+(2\zeta r)^2}} \qquad 59.64$$

Equation 59.65 shows another form of the equation for vibration transmissibility. F_{st} is the static deflection of the system.

Vibration Transmissibility, Base Motion

$$\frac{x}{y} = \frac{F_t}{F_{\text{st}}} = \frac{\sqrt{1+(2r\zeta)^2}}{\sqrt{(1-r^2)^2+(2\zeta r)^2}} \qquad 59.65$$

In some cases, the transmissibility may be reported in units of *decibels*. Unlike linear transmissibility, which is always positive, logarithmic transmissibility can be negative.

$$\text{TR}_{\text{dB}} = 20\log(\text{TR}_{\text{linear}}) \qquad 59.66$$

Figure 59.17 *Linear Transmissibility*

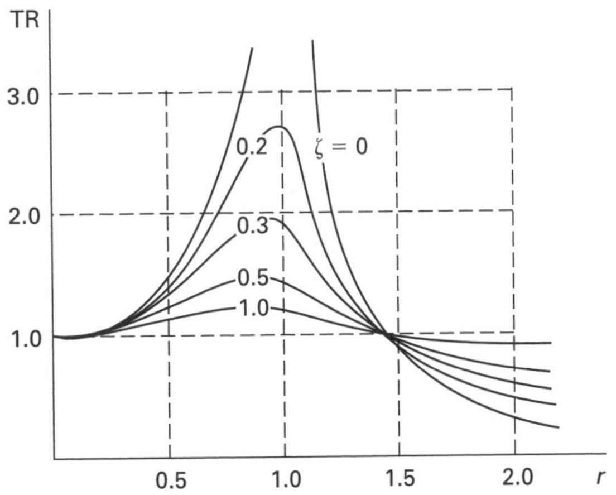

The magnitude of oscillations in vibrating equipment can be reduced and the equipment isolated from the surroundings by mounting on resilient pads or springs. The isolated system must have a natural frequency less than $1/\sqrt{2} = 0.707$ times the disturbing (forcing) frequency. That is, the transmissibility will be reduced below 1.0 only if $\omega_f/\omega_n > \sqrt{2}$. Otherwise, the attempted isolation will actually increase the transmitted force.

The natural frequency is found from Eq. 59.9, where the static deflection is calculated from the mass and stiffness of the pad or spring.

The amount of isolation is characterized by the *isolation efficiency*, also known as the *percent of isolation* and *degree of isolation*. Suggested isolation efficiencies for satisfactory operation are listed in Table 59.2.

$$\eta = 1 - \left| \frac{1}{\left(\frac{f_f}{f}\right)^2 - 1} \right| \qquad 59.67$$

Table 59.2 *Typical Isolation Efficiencies*

equipment	isolation efficiency
centrifugal compressor	0.98
reciprocating compressor	
0 hp to 15 hp	0.85
15 hp to 150 hp	0.90
centrifugal fans	
800 rpm and higher	0.90 to 0.95
centrifugal pumps	0.95
pipe mounts	0.95

Isolation materials and isolator devices have specific deflection characteristics. If the isolation efficiency is known, it can be used to determine the type of isolator or isolation device used based on the static deflection. Table 59.3 lists typical ranges of isolation materials and devices. Table 59.4 lists typical damping factors. *ASHRAE Handbook—HVAC Applications* contains a selection guide with recommendations for isolating vibrations from several common types of equipment and installation locations. [**Selection Guide for Vibration Isolation**]

A *tuned system* is one for which the natural frequency of the vibration absorber is equal to the frequency that is to be eliminated (i.e., the forcing frequency). In theory, this is easy to accomplish: the mass and spring constant of the absorber are varied until the desired natural frequency is achieved. This is known as "tuning" the system.

Table 59.3 *Typical Deflection of Isolation Materials and Devices*

approximate deflection		
in	mm	materials
0–1/16	0–2	cork, natural rubber, felt, lead/asbestos, fiberglass
1/16–1/4	2–6	neoprene pads, neoprene mounts, multiple layers of felt or cork
1/4–1 1/2	6–40	steel coil springs, multiple layers of natural rubber or neoprene pads
1 1/2–15	40–380	steel coil or leaf springs

(Multiply in by 25.4 to obtain mm.)

Table 59.4 *Typical Damping Factors of Isolators*

material	ζ
steel spring	0.005
natural rubber	0.05
neoprene	0.05
cork	0.06
felt	0.06
metal mesh	0.12
air damper	0.17
friction-damped spring	0.33

Example 59.4

A 4000 lbm (1800 kg) machine is rotating at 1000 rpm. The rotating component has an imbalance of 100 lbm (45 kg) acting with an eccentricity of 2 in (50 mm). The machine is already supported by isolation mounts having a combined stiffness of 30,000 lbf/in (5.3 MN/m), but vibration is still excessive. In order to reduce the amplitude of vibration, a viscous damper is connected between the machine and a rigid support. The damping ratio is 0.2. What are the (a) amplitude of oscillation and (b) transmitted force?

SI Solution

(a) The static deflection of the mounts is

Free Vibration

$$mg = k\delta_{\text{st}}$$

$$\delta_{\text{st}} = \frac{mg}{k} = \frac{(1800\ \text{kg})\left(9.81\ \frac{\text{m}}{\text{s}^2}\right)}{\left(5.3\ \frac{\text{MN}}{\text{m}}\right)\left(10^6\ \frac{\text{N}}{\text{MN}}\right)}$$
$$= 0.00333\ \text{m}$$

The undamped natural frequency is

$$f = \frac{1}{2\pi}\sqrt{\frac{g}{\delta_{st}}} = \frac{1}{2\pi}\sqrt{\frac{9.81\ \frac{\text{m}}{\text{s}^2}}{0.00333\ \text{m}}} = 8.638\ \text{Hz}$$

The forcing frequency is

$$f_f = \frac{1000\ \frac{\text{rev}}{\text{min}}}{60\ \frac{\text{s}}{\text{min}}} = 16.67\ \text{Hz}$$

The angular forcing frequency is

$$\omega_f = 2\pi f_f = 2\pi(16.67\ \text{Hz}) = 104.7\ \text{rad/sec}$$

The out-of-balance force caused by the rotating eccentric mass with a radial eccentricity of e is

$$F_f = m\omega_f^2 e = \frac{(45\ \text{kg})\left(104.7\ \frac{\text{rad}}{\text{s}}\right)^2(50\ \text{mm})}{1000\ \frac{\text{mm}}{\text{m}}} = 24\,665\ \text{N}$$

From Eq. 59.62, the ratio of frequencies is

Vibration Transmissibility, Base Motion

$$r = \frac{\omega_f}{\omega_n} = \frac{f_f}{f} = \frac{16.67\ \text{Hz}}{8.638\ \text{Hz}} = 1.93$$

From Eq. 59.61, the magnification factor is

$$\beta = \left|\frac{1}{\sqrt{(1-r^2)^2 + (2\zeta r)^2}}\right| = \frac{1}{\sqrt{\left(1-(1.93)^2\right)^2 + \left((2)(0.2)(1.93)\right)^2}} = 0.353$$

The amplitude of oscillation is

$$A = \beta\left(\frac{F_f}{k}\right) = (0.353)\left(\frac{24\,665\ \text{N}}{\left(5.3\ \frac{\text{MN}}{\text{m}}\right)\left(10^6\ \frac{\text{N}}{\text{MN}}\right)}\right) = 0.00164\ \text{m}$$

(b) From Eq. 59.63, the transmissibility is

$$\text{TR} = \beta\sqrt{1+(2r\zeta)^2} = 0.353\sqrt{1+\left((2)(1.93)(0.2)\right)^2} = 0.446$$

The transmitted force is

$$F = (\text{TR})F_f = (0.446)(24\,665\ \text{N}) = 11\,000\ \text{N}$$

Customary U.S. Solution

(a) The static deflection of the mounts is

$$\delta_{st} = \frac{\text{weight}}{k} = \left(\frac{m}{k}\right)\left(\frac{g}{g_c}\right) = \left(\frac{4000\ \text{lbm}}{30{,}000\ \frac{\text{lbf}}{\text{in}}}\right)\left(\frac{32.2\ \frac{\text{ft}}{\text{sec}^2}}{32.2\ \frac{\text{lbm-ft}}{\text{lbf-sec}^2}}\right) = 0.1333\ \text{in}$$

The undamped natural frequency is

$$f = \frac{1}{2\pi}\sqrt{\frac{g}{\delta_{st}}} = \frac{1}{2\pi}\sqrt{\frac{\left(32.2\ \frac{\text{ft}}{\text{sec}^2}\right)\left(12\ \frac{\text{in}}{\text{ft}}\right)}{0.1333\ \text{in}}} = 8.569\ \text{Hz}$$

The forcing frequency is

$$f_f = \frac{1000\ \frac{\text{rev}}{\text{min}}}{60\ \frac{\text{s}}{\text{min}}} = 16.67\ \text{Hz}$$

The angular forcing frequency is

$$\omega_f = 2\pi f_f = 2\pi(16.67\ \text{Hz}) = 104.7\ \text{rad/sec}$$

The out-of-balance force caused by the rotating eccentric mass with a radial eccentricity of e is

$$F_f = \frac{m\omega_f^2 e}{g_c} = \frac{(100\ \text{lbm})\left(104.7\ \frac{\text{rad}}{\text{sec}}\right)^2(2\ \text{in})}{\left(32.2\ \frac{\text{lbm-ft}}{\text{lbf-sec}^2}\right)\left(12\ \frac{\text{in}}{\text{ft}}\right)} = 5674\ \text{lbf}$$

Dynamics and Vibrations

From Eq. 59.62, the ratio of frequencies is

Vibration Transmissibility, Base Motion

$$r = \frac{\omega_f}{\omega_n} = \frac{f_f}{f} = \frac{16.67 \text{ Hz}}{8.569 \text{ Hz}} = 1.945$$

From Eq. 59.61, the magnification factor is

$$\beta = \left| \frac{1}{\sqrt{(1-r^2)^2 + (2\zeta r)^2}} \right| = \frac{1}{\sqrt{(1-(1.945)^2)^2 + ((2)(0.2)(1.945))^2}} = 0.346$$

The amplitude of oscillation is

$$A = \beta\left(\frac{F_f}{k}\right) = (0.346)\left(\frac{5674 \text{ lbf}}{30{,}000 \ \frac{\text{lbf}}{\text{in}}}\right) = 0.0654 \text{ in}$$

(b) From Eq. 59.63, the transmissibility is

$$\text{TR} = \beta\sqrt{1+(2r\zeta)^2} = 0.346\sqrt{1+((2)(1.945)(0.2))^2} = 0.438$$

The transmitted force is

$$F = (\text{TR})F_f = (0.438)(5674 \text{ lbf}) = 2485 \text{ lbf}$$

18. ISOLATION FROM ACTIVE BASE

In some cases, a machine is to be isolated from an active base. The base (floor, supports, etc.) vibrates, and the magnitude of the vibration seen by the machine is to be limited or reduced. This case is not fundamentally different from the case of a vibrating machine being isolated from a stationary base.

The concept of transmissibility is replaced by the *amplitude ratio* (*magnification factor* or *amplification factor*). This is the ratio of the transmitted displacement (deflection, excursion, motion, etc.) to the applied displacement. That is, it is the ratio of the maximum mass motion to the maximum base motion. The amplification ratio is numerically identical to the transmissibility calculated in Eq. 59.63.

$$\text{AR} = \frac{\delta_{\text{dynamic}}}{\delta_{\text{static}}} = \frac{\delta_{\text{dynamic}}}{\frac{F}{k}} = \text{TR} \qquad 59.68$$

Example 59.5

The suspension system of a car is modeled as a perfect spring and dashpot connecting a massless wheel and a supported mass, m, as shown. The car enters a bumpy area with forward velocity, v, and initial mass position, y_0. The profile of the road surface is modeled as a perfect sinusoid with a peak-to-peak distance of L. The sinusoid is described mathematically by the equation $y(x) = y_{\text{max}}\sin 2\pi(x - x_0)/L$.

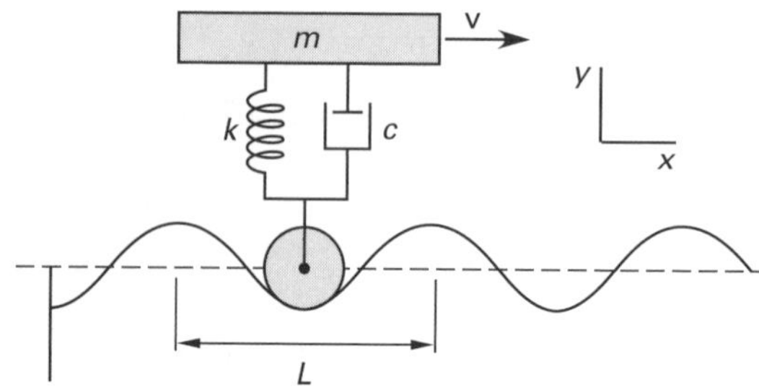

What are the (a) forcing frequency, (b) undamped natural frequency of the system, and (c) horizontal velocity at resonance? (d) For a particular speed, how would the steady-state amplitude of the mass be determined?

Solution

(a) Since the horizontal car velocity is v, the frequency of the forcing function will be

$$f_f = \frac{\text{v}}{L}$$

(b) The undamped natural frequency is found from Eq. 59.5 and Eq. 59.6 (Use consistent units.)

$$f = \frac{1}{2\pi}\sqrt{\frac{k}{m}}$$

(c) Resonance occurs when the forcing frequency equals the natural frequency. (Use consistent units.)

$$f_f = f$$

$$\frac{\text{v}}{L} = \frac{1}{2\pi}\sqrt{\frac{k}{m}}$$

$$\text{v} = \frac{L}{2\pi}\sqrt{\frac{k}{m}}$$

(d) From Eq. 59.46, the damping ratio is

Vibration Transmissibility, Base Motion

$$\zeta = \frac{c}{c_{\text{critical}}}$$

The amplitude ratio can be found from either Fig. 59.15 or Eq. 59.68. The steady-state amplitude will be $(\text{AR})y_{\text{max}}$.

19. VIBRATIONS IN SHAFTS

A shaft's natural frequency of vibration is referred to as the *critical speed*. This is the rotational speed in revolutions per second that just equals the lateral natural frequency of vibration. Therefore, vibration in shafts is basically an extension of lateral vibrations (e.g., whipping "up and down") in beams. Rotation is disregarded, and the shaft is considered only from the standpoint of lateral vibrations.

The shaft will have multiple modes of vibration. General practice is to keep the operating speed well below the first critical speed. For shafts with distributed or multiple loadings, it may be important to know the second critical speed. However, higher critical speeds are usually well out of the range of operation.

For shafts with constant cross-sectional areas and simple loading configurations, the static deflection due to pulleys, gears, and self-weight can be found from beam formulas. Shafts with single antifriction (i.e., ball and roller) bearings at each end can be considered to be simply supported, while shafts with sleeve bearings or two side-by-side antifriction bearings at each shaft end can be considered to have fixed built-in supports.[7]

A shaft carrying no load other than its own weight can be considered as a uniformly loaded beam. The maximum deflection at midspan can be found from beam tables.

Once the deflection is known, Eq. 59.9 can be used to find the critical speed.

The classical analysis of a shaft carrying single or multiple inertial loads (flywheel, pulley, etc.) assumes that the shaft itself is weightless.[8]

Equation 59.69 gives the critical speed (i.e., the fundamental frequency of vibration) of a shaft carrying multiple masses in terms of the static deflections at the masses.[9] In theory, all that is necessary is to stop the rotation and measure the deflection (from the horizontal) at each mass. The assumption that the static and rotating deflection curves are identical is not exactly true. However, the speed error is less than approximately 5%, generally on the high side.

$$f = \frac{1}{2\pi}\sqrt{\frac{g\sum m_i\delta_{\text{st},i}}{\sum m_i\delta_{\text{st},i}^2}} \qquad 59.69$$

When there are only two rotating masses on the shaft, the total deflections can be calculated by superposition. However, Eq. 59.69 is laborious to use when there are more than two masses, as the number of deflection calculations is the square of the number of masses. In that case, the *Dunkerley approximation* is used.[10] In Eq. 59.70, the f_i are the natural frequencies of vibration when mass m_i alone is on the shaft. The Dunkerley equation generally underestimates the critical speed.

$$\left(\frac{1}{f}\right)^2 = \sum\left(\frac{1}{f_i}\right)^2 \qquad 59.70$$

If a shaft's critical speed is unsuitable, Eq. 59.71 can be used to calculate the approximate diameter of a shaft that will be acceptable.

$$D_{\text{new}} = D_{\text{old}}\sqrt{\frac{f_{\text{new}}}{f_{\text{old}}}} \qquad 59.71$$

Typical values used in the analysis of steel shafts are: density, ρ, 0.28 lbm/in^3 (7750 kg/m^3); modulus of elasticity, E, 2.9×10^7 lbf/in2 (200 GPa); and gravitational constant, g_c, 386.4 in-lbm/lbf-sec^2.

Example 59.6

A 30 lbm (14 kg) flywheel is supported on an overhanging steel shaft as shown. The shaft diameter is 1.0 in (25 mm), and the shaft mass is negligible. What is the critical speed of the shaft?

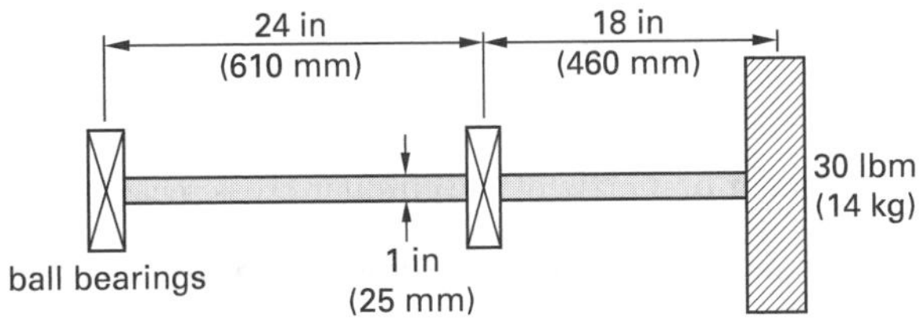

SI Solution

The radius of the shaft is

$$r = \frac{D}{2} = \frac{25 \text{ mm}}{2} = 12.5 \text{ mm}$$

The moment of inertia of the circular cross section is

Properties of Various Shapes

$$I = \frac{\pi r^4}{4} = \frac{\pi\left(\frac{12.5 \text{ mm}}{1000 \frac{\text{mm}}{\text{m}}}\right)^4}{4} = 1.917 \times 10^{-8} \text{ m}^4$$

[7]Sleeve bearings are assumed to be fixed supports, not because they have the mechanical strength to prevent binding, but because sleeve bearings cannot operate and would not be operating with an angled shaft.
[8]The mass of the shaft can be included with Rayleigh's method and Dunkerley's approximation.
[9]This equation is sometimes known as the *Rayleigh equation* or the *Rayleigh-Ritz equation*, as it is based on the Rayleigh method of equating the maximum kinetic energy to the maximum potential energy.
[10]The spelling "Dunkerly" is also found in the literature.

The ball bearings prevent vertical but not angular deflection. Consider the shaft to be simply supported. The deflection at the flywheel is

Simple Beam with Overhung Load

$$\delta_{st} = \frac{Fa^2(a+b)}{3EI} = \frac{mga^2(a+b)}{3EI}$$

$$= \frac{(14\ \text{kg})\left(9.81\ \frac{\text{m}}{\text{s}^2}\right)(0.46\ \text{m})^2(0.46\ \text{m} + 0.61\ \text{m})}{(3)(200\ \text{GPa})\left(10^9\ \frac{\text{Pa}}{\text{GPa}}\right)(1.917\times 10^{-8}\ \text{m}^4)}$$

$$= 2.703\times 10^{-3}\ \text{m} \quad (2.703\ \text{mm})$$

From Eq. 59.6, the natural frequency is

$$f = \frac{1}{2\pi}\sqrt{\frac{g}{\delta_{st}}} = \frac{1}{2\pi}\sqrt{\frac{9.81\ \frac{\text{m}}{\text{s}^2}}{2.703\times 10^{-3}\ \text{m}}}$$
$$= 9.59\ \text{Hz}$$

Customary U.S. Solution

The radius of the shaft is

$$r = \frac{D}{2} = \frac{1\ \text{in}}{2} = 0.5\ \text{in}$$

The moment of inertia of the circular cross section is

Properties of Various Shapes

$$I = \frac{\pi r^4}{4} = \frac{\pi(0.5\ \text{in})^4}{4}$$
$$= 0.0491\ \text{in}^4$$

The ball bearings prevent vertical but not angular deflection. Consider the shaft to be simply supported. The deflection at the flywheel is

$$\delta_{st} = \frac{Fa^2(a+b)}{3EI} = \left(\frac{ma^2(a+b)}{3EI}\right)\left(\frac{g}{g_c}\right)$$

$$= \left(\frac{(30\ \text{lbm})(18\ \text{in})^2(18\ \text{in} + 24\ \text{in})}{(3)\left(2.9\times 10^7\ \frac{\text{lbf}}{\text{in}^2}\right)(0.0491\ \text{in}^4)}\right) \times \left(\frac{32.2\ \frac{\text{ft}}{\text{sec}^2}}{32.2\ \frac{\text{lbm-ft}}{\text{lbf-sec}^2}}\right)$$

$$= 0.0956\ \text{in}$$

From Eq. 59.6, the natural frequency is

$$f = \frac{1}{2\pi}\sqrt{\frac{g}{\delta_{st}}} = \frac{1}{2\pi}\sqrt{\frac{\left(32.2\ \frac{\text{ft}}{\text{sec}^2}\right)\left(12\ \frac{\text{in}}{\text{ft}}\right)}{0.0956\ \text{in}}}$$
$$= 10.1\ \text{Hz}$$

20. SECOND CRITICAL SHAFT SPEED

The second critical speed for a simply supported shaft uniformly loaded along its length is four times the fundamental critical speed.

The second critical speed for a massless shaft carrying two concentrated masses, m_1 and m_2, can be derived from Eq. 59.72. The positive roots are $1/\omega_1$ and $1/\omega_2$, where ω_1 and ω_2 in this bi-quadratic equation are the first and second natural frequencies (critical speeds), respectively. The a_{ij} influence constants are the deflections at the location of mass i due to a unit force at the location of mass j. Only three deflection calculations are needed because $a_{12} = a_{21}$.

$$\frac{1}{\omega^4} - \left(\frac{1}{\omega^2}\right)(a_{11}m_1 + a_{22}m_2) + (a_{11}a_{22} - a_{12}a_{21})m_1m_2 = 0 \qquad 59.72$$

21. CRITICAL SPEED OF STEPPED SHAFTS

The Dunkerley equation is one of the few simple methods for evaluating the critical speed of a stepped shaft. The mass of each shaft section is assumed to be concentrated at the midpoints of their respective lengths. The "shaft" itself is considered to be massless. This rather crude approximation provides surprisingly good results. As a further simplification, the shaft section with the smallest diameter can be disregarded.

22. VIBRATIONS IN THIN PLATES

The natural frequency of thin plates and diaphragms depends on the shape (e.g., circular, square, or rectangular) and the method of mounting (e.g., fixed or free edges). Completely fixed edges are difficult to achieve in practice, so the formulas used to calculate the natural frequencies are derived from a blend of heuristic and theoretical methods.

Equation 59.73 is an approximate equation based on a modification of Eq. 59.6 and Eq. 59.9. As an approximation, this equation can be used with any uniformly loaded round, square, rectangular, elliptical, or triangular plates with any edge conditions.[11] The maximum error is small, generally less than 3%. δ_{st} is the maximum

[11]The deflection calculated must correspond to the actual edge conditions.

static deflection produced by the self-mass of the plate and any uniformly distributed mass attached to the plate and vibrating with it.

$$f = \frac{1.277}{2\pi}\sqrt{\frac{g}{\delta_{st}}} \quad 59.73$$

If certain assumptions are made, the fundamental natural frequency of a plate can be derived. These assumptions are that the plate material is elastic, homogeneous, and of uniform thickness. Also, the plate thickness and deflections are assumed to be small in comparison to the size of the plate. w is the uniform load per unit area, including the self-weight per unit area (calculated as the specific weight times the thickness). In Eq. 59.75, a is the shorter edge distance. Values of K are given in Table 59.5.

$$f = \frac{K}{2\pi}\sqrt{\frac{Dg}{wr^4}} \quad \text{[circular plates]} \quad 59.74$$

$$f = \frac{K}{2\pi}\sqrt{\frac{Dg}{wa^4}} \quad \text{[rectangular plates]} \quad 59.75$$

$$D = \frac{Et^3}{12(1-\nu^2)} \quad 59.76$$

Table 59.5 *Vibration Coefficients for Plates (fundamental mode)*

case		K					
circular plate							
fixed edges		10.2					
simply supported edges		4.99					
free edges*		5.25					
rectangular plate							
$a \times b$ sides	$a/b =$	1.0	0.8	0.6	0.4	0.2	0.0
fixed edges		36.0	29.9	25.9	23.6	22.6	22.4
simply supported edges		19.7	16.2	13.4	11.5	10.3	9.87

*Supported at inner surface; edges free to flutter.

23. BALANCING

To be "balanced," a rotating component must be statically and dynamically in equilibrium. For two masses, m_i, rotating at radii, r_i, and mounted 180° out of phase, the requirement for *static balance* is given by Eq. 59.77. Static balancing is usually sufficient for rotating disks, thin wheels, and gears.

$$\sum m_i r_i = 0 \quad \text{[static balance]} \quad 59.77$$

Dynamic balance requires that there also be no unbalanced couples. When all of the masses are in the same plane of rotation, static balance and dynamic balance will be achieved simultaneously. When rotation is not in the same plane, all of the couples must also balance. In general, two balancing masses are required to provide static and dynamic balance where there is an unbalanced couple. (See Fig. 59.18.)

$$\sum m_i r_i x_i = 0 \quad \text{[dynamic balance]} \quad 59.78$$

Figure 59.18 *Static and Dynamic Balance*

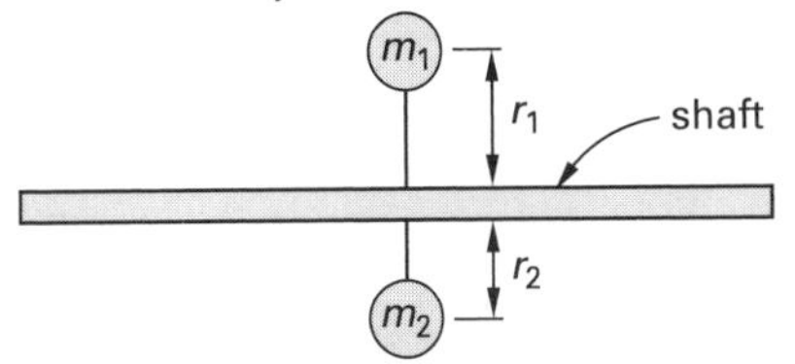

(a) in plane static balance

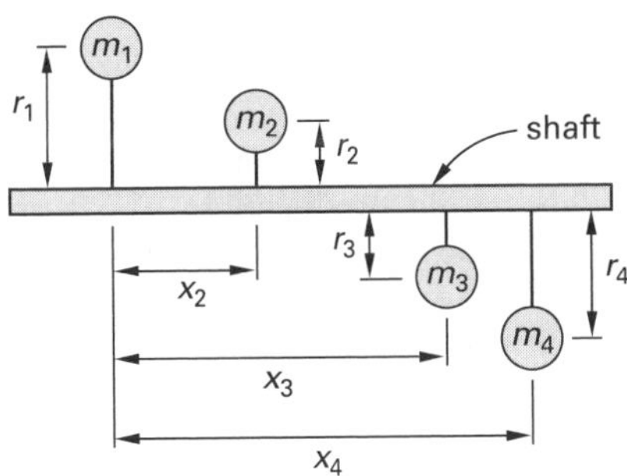

(b) out of plane dynamic balance

Example 59.7

A 2 ft (610 mm) diameter steel disk is mounted with a 0.33 ft (100 mm) eccentricity on a shaft. The mass of the disk is 570 lbm (260 kg). The assembly is balanced by two counterweights located 1.5 ft (460 mm) and 3.0 ft (910 mm) from the disk, respectively, each mounted at a radius of 1.0 ft (300 mm) from the shaft's centerline. What are the masses of the two counterweights?

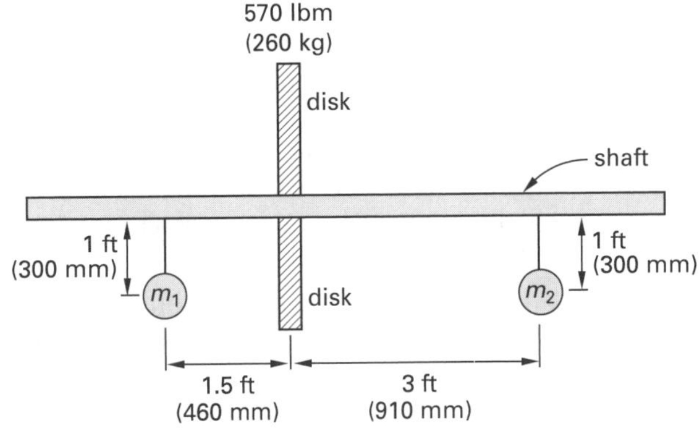

SI Solution

Take moments about the position of the left-hand counterweight. The couple produced by the eccentric disk mass must be balanced by the right counterweight. Use Eq. 59.78.

$$m_{\text{disk}}x_{\text{disk}}e = m_{\text{right}}x_{\text{right}}r_{\text{right}}$$

$$\begin{aligned}(260 \text{ kg})(460 \text{ mm}) \\ \times(100 \text{ mm})\end{aligned} = \begin{aligned}m_{\text{right}}(460 \text{ mm} + 910 \text{ mm}) \\ \times(300 \text{ mm})\end{aligned}$$

$$m_{\text{right}} = 29.1 \text{ kg}$$

Use Eq. 59.77 to satisfy the criterion for static balance.

$$\begin{aligned}m_{\text{left}}r_{\text{left}} + m_{\text{right}}r_{\text{right}} - m_{\text{disk}}e &= 0 \\ m_{\text{left}}(300 \text{ mm}) + (29.1 \text{ kg})(300 \text{ mm}) & \\ - (260 \text{ kg})(100 \text{ mm}) &= 0 \\ m_{\text{left}} &= 57.6 \text{ kg}\end{aligned}$$

Customary U.S. Solution

Take moments about the position of the left-hand counterweight. The couple produced by the eccentric disk mass must be balanced by the right counterweight. Use Eq. 59.78.

$$\begin{aligned}m_{\text{disk}}x_{\text{disk}}e &= m_{\text{right}}x_{\text{right}}r_{\text{right}} \\ (570 \text{ lbm})(1.5 \text{ ft})(0.33 \text{ ft}) &= m_{\text{right}}(1.5 \text{ ft} + 3 \text{ ft})(1 \text{ ft}) \\ m_{\text{right}} &= 62.7 \text{ lbm}\end{aligned}$$

Use Eq. 59.77 to satisfy the criterion for static balance.

$$\begin{aligned}m_{\text{left}}r_{\text{left}} + m_{\text{right}}r_{\text{right}} - m_{\text{disk}}e &= 0 \\ m_{\text{left}}(1 \text{ ft}) + (62.7 \text{ lbm})(1 \text{ ft}) & \\ - (570 \text{ lbm})(0.33 \text{ ft}) &= 0 \\ m_{\text{left}} &= 125.4 \text{ lbm}\end{aligned}$$

24. MODAL VIBRATIONS

In addition to the fundamental (first) frequency emphasized up to this point, there can be higher-order frequencies (*harmonics* or higher *modes*). Generally, if a system is protected against resonance at its fundamental frequency, it will be protected against resonance at the even higher harmonic frequencies.

When a beam, shaft, or plate vibrates laterally at its fundamental frequency, all of it will be on one side of the equilibrium position at any given moment. This is not true with higher modes. There will be one or more positions (i.e., *nodes*) where the deflection curve will pass through the equilibrium position. The location of the nodes and the *modal shape* are given in handbooks.

The nth harmonic frequency for beams and shafts with uniformly distributed masses on simple supports is n^2 times the fundamental frequency. If both ends are fixed, then the higher modal frequencies are $\frac{1}{9}(2n+1)^2$ times the fundamental frequency. Cantilever beams with distributed masses are not handled so easily.

Equation 59.79 calculates the nth natural frequency for a cantilever beam with uniformly distributed load, w, (including its own weight) per unit length over its entire length, L. Table 59.6 gives the values of the vibration constant, K_n, and the locations of the nodes.

$$f_n = \frac{K_n}{2\pi}\sqrt{\frac{EIg}{wL^4}} \qquad 59.79$$

Table 59.6 *Vibration Constants for Cantilever Beams**

mode	K_n	position of nodes (x/L, from fixed end)				
1	3.52	0.0				
2	22.0	0.0	0.783			
3	61.7	0.0	0.504	0.868		
4	121	0.0	0.358	0.644	0.905	
5	200	0.0	0.279	0.500	0.723	0.926

*Any uniform cross section; uniformly distributed load w (including self-weight) per unit length; length, L.

25. MULTIPLE DEGREE OF FREEDOM SYSTEMS

A system is designed as a *multiple degree of freedom (MDOF) system* when it takes two or more independent variables to define the position of independent parts of the system.

MDOF systems may oscillate at a single natural frequency (e.g., when all of the masses are moving in the same direction), or different parts of the system may oscillate independently at their own frequencies. The *amplitude ratio* is the ratio of the maximum deflection (excursion, movement, etc.) of one part of the system to the maximum deflection of another.

Closed-form solutions for various simple systems can be calculated, but real-world situations are more difficult. For example, consider a circuit board mounted on a bulkhead (plate or diaphragm). The circuit board has its own vibration characteristics; the bulkhead also has its own. When they are assembled together, a third vibration characteristic is produced. The circuit board and the bulkhead are said to be "coupled" because each affects the other. (See Fig. 59.19.)

Even when it is difficult to determine the actual performance characteristics of an MDOF system, damage from resonance can still be prevented by proper design.

Figure 59.19 *System with Three Degrees of Freedom*

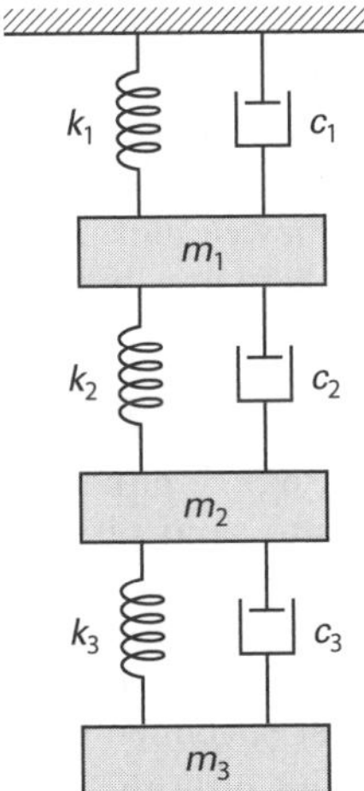

The modes are "uncoupled" as much as possible. The *octave rule* is a simple rule of thumb for uncoupling the modes. The octave rule requires the uncoupled natural frequency of each added component to be at least twice the natural frequency of the element to which it is attached. In other words, the natural frequency should be doubled every time an additional degree of freedom is added to the system.

When space, weight, or other constraints make it impossible to design modal interaction out of the system, testing and other (usually heuristic or graphical) methods must be used to determine the maximum dynamic forces on the assembly.

26. SHOCK

Vibration is a steady-state, regular phenomenon. *Shock* is a transient phenomenon. Shock results in a sharp, nearly sudden change in velocity. A *shock pulse* (*shock impulse*) is a disturbing force characterized by a rise and subsequent decay of acceleration in a very short period of time. A shock pulse is described by its peak amplitude (usually in gravities), its duration (in milliseconds), and an overall shape (triangular, rectangular, half-sine, etc.). Shock pulses, particularly those that are complex, are often depicted graphically as an acceleration-time curve.

The change in velocity, $\Delta \mathrm{v}$, can be found as area under an acceleration-time curve or as the area under a force-time curve divided by the mass. For a complex curve, the area can be found by integration or by breaking it into simpler sections. Equation 59.80 through Eq. 59.85 give the change in velocity for several regular shocks of maximum acceleration amplitude, A_0, and duration, t_0.

$$\Delta \mathrm{v} = \sqrt{2gh} \quad \text{[inelastic drop from height } h] \qquad 59.80$$

$$\Delta \mathrm{v} = 2\sqrt{2gh} \quad \text{[elastic drop from height } h] \qquad 59.81$$

$$\Delta \mathrm{v} = \frac{2A_0 g t_0}{\pi} \quad \text{[half-sine acceleration]} \qquad 59.82$$

$$\Delta \mathrm{v} = A_0 g t_0 \quad \text{[rectangular acceleration]} \qquad 59.83$$

$$\Delta \mathrm{v} = \frac{A_0 g t_0}{2} \quad \text{[triangular acceleration]} \qquad 59.84$$

$$\Delta \mathrm{v} = \frac{A_0 g t_0}{2} \quad \text{[versed sine acceleration]} \qquad 59.85$$

The *shock transmission* (*transmitted shock*), G, is an acceleration parameter (with units of gravities) that depends on the natural frequency, f, and the change in velocity, $\Delta \mathrm{v}$, due to the shock.

$$G = \frac{2\pi f \Delta \mathrm{v}}{g} = \frac{\omega_n \Delta \mathrm{v}}{g} \qquad 59.86$$

The *dynamic deflection* of a linear isolator that experiences a shock pulse is

$$\delta = \frac{\Delta \mathrm{v}}{2\pi f} = \frac{\Delta \mathrm{v}}{\omega_n} \qquad 59.87$$

Isolating a system against shocks is very different than isolating the system against vibration. Isolators must be capable of absorbing shock energy instantly. The energy may be dissipated in an inelastic isolator (e.g., crush insulation), or the energy may be released at the damped natural frequency of the system later in the cycle.

Example 59.8

A sensitive piece of electronic equipment is mounted on an isolation system with a natural frequency of 15 Hz. The equipment and mount are subjected to a standard 15 g, 11 msec, half-sine shock test. What are the (a) maximum shock transmission and (b) isolation deflection?

SI Solution

(a) From Eq. 59.82, the change in velocity is

$$\Delta \mathrm{v} = \frac{2A_0 g t_0}{\pi} = \frac{(2)(15\ \mathrm{g})\left(9.81\ \frac{\mathrm{m}}{\mathrm{s}^2 \cdot \mathrm{g}}\right)(11\ \mathrm{ms})}{\pi\left(1000\ \frac{\mathrm{ms}}{\mathrm{s}}\right)}$$

$$= 1.03\ \mathrm{m/s}$$

From Eq. 59.86, the shock transmission is

$$G = \frac{2\pi f \Delta \mathrm{v}}{g} = \frac{2\pi(15\ \mathrm{Hz})\left(1.03\ \frac{\mathrm{m}}{\mathrm{s}}\right)}{9.81\ \frac{\mathrm{m}}{\mathrm{s}^2 \cdot \mathrm{g}}}$$

$$= 9.9\ \text{g's} \quad \text{(9.9 gravities)}$$

(b) Use Eq. 59.87 to find the dynamic linear deflection.

$$\delta = \frac{\Delta \text{v}}{2\pi f} = \frac{1.03\ \frac{\text{m}}{\text{s}}}{2\pi(15\ \text{Hz})} = 0.0109\ \text{m}$$

Customary U.S. Solution

From Eq. 59.82, the change in velocity is

$$\Delta \text{v} = \frac{2A_0 g t_0}{\pi} = \frac{(2)(15\ \text{g})\left(32.2\ \frac{\text{ft}}{\text{sec}^2\text{-g}}\right)\left(12\ \frac{\text{in}}{\text{ft}}\right)(11\ \text{msec})}{\pi\left(1000\ \frac{\text{msec}}{\text{sec}}\right)} = 40.59\ \text{in/sec}$$

From Eq. 59.86, the shock transmission is

$$G = \frac{2\pi f \Delta \text{v}}{g} = \frac{2\pi(15\ \text{Hz})\left(40.59\ \frac{\text{in}}{\text{sec}}\right)}{\left(32.2\ \frac{\text{ft}}{\text{sec}^2\text{-g}}\right)\left(12\ \frac{\text{in}}{\text{ft}}\right)} = 9.9\ \text{g's}\quad (9.9\ \text{gravities})$$

(b) Use Eq. 59.87 to find the dynamic linear deflection.

$$\delta = \frac{\Delta \text{v}}{2\pi f} = \frac{40.59\ \frac{\text{in}}{\text{sec}}}{2\pi(15\ \text{Hz})} = 0.431\ \text{in}$$

27. VIBRATION AND SHOCK TESTING

There are two basic types of vibration testing: constant and random-frequency tests. Constant-frequency *sinusoidal tests* (also known as *harmonic tests*) were the earliest types of tests used. They detect, one at a time, the resonant frequencies of an item. Random tests excite all of the resonant frequencies simultaneously and duplicate the buffeting that equipment will typically experience.

Most vibration tests are performed on a *shaker table* (*exciter*). Tables can be electrodynamic, hydraulic, or mechanical. Electrodynamic shakers produce the highest frequencies, but they are limited to shaking smaller pieces of equipment with low forces and displacements. Hydraulic shakers (also known as *hydrashakers*) use hydraulic fluid to drive a piston. Test frequencies are limited to approximately 500 Hz. Forces can be quite large. Mechanical shakers use eccentric cams to produce the traditional test. They cannot produce random vibrations. Mechanical shakers are limited to approximately 55 Hz.

Since there are many types of shock, there are many types of shock tests used. Hammers, spring-loaded rigs, air guns, and drops into sand pits are in use. Shipping containers are often subjected to a crude *drop test*, where the containerized equipment is mounted on a table and then dropped onto a concrete floor. The *swing test* is similar—the test specimen swings into a concrete wall. A common test for aircraft-mounted equipment is a *machine drop* test producing a half-sine shock of 15 g's lasting 11 ms. (The nature of the shock is controlled by the resilience of the contact surface.) Missile components subjected to explosive impulses (e.g., explosive separation of stages) may be tested with a 6 ms, 100 g shock.

28. VIBRATION INSTRUMENTATION

There are two main types of *vibration sensors*: proximity and casing devices. *Proximity displacement transducers* (*proximity probes*) do not touch the vibrating item. They sense vibrations by establishing an electric/magnetic field between the probe tip and the object. Changes in the field produced by the minute movement of the conductive surface are detected by the instrument.

Casing transducers that touch or are mounted on vibrating equipment come in two varieties: accelerometers and velocity transducers. The output of an *accelerometer* is proportional to acceleration. Accelerometers generally use a piezoelectric crystal with an internally mounted reference mass. The voltage produced by the crystal varies as the attached mass is vibrated. Accelerometers can also be constructed as strain gauges mounted on small flexible members. A *velocity transducer* is an electromagnetic device with a coil and core (or magnet), one of which is stationary and the other that is moving. The voltage produced is proportional to the relative velocity of the core through the coil. The terms *vibration pick-up* and *vibrometer* refer to devices that produce velocity-dependent voltages.

29. NOMENCLATURE

a	acceleration	ft/sec^2	m/s^2
a	plate short side dimension	ft	m
A	amplitude	ft	m
A	amplitude	rad	rad
A	maximum acceleration amplitude	g's	g's
AR	amplitude ratio	–	–
b	plate long side dimension	ft	m
c	coefficient of viscous damping (linear)	lbf-sec/ft	N·s/m
D	diameter or distance	ft	m
D	displacement	ft	m
D	vibration parameter	ft-lbf	N·m
e	eccentricity	ft	m
E	energy	ft-lbf	J
E	modulus of elasticity	lbf/ft^2	Pa
f	frequency	Hz	Hz
F	force	lbf	N
g	gravitational acceleration, 32.2 (9.81)	ft/sec^2	m/s^2
g_c	gravitational constant, 32.2	lbm-ft/lbf-sec^2	n.a.
G	shear modulus	lbf/ft^2	Pa
G	shock transmission	g's	g's
h	height	ft	m
I	area moment of inertia	–	–
I	mass moment of inertia	lbm-ft^2	kg·m^2
J	polar area moment of inertia	ft^4	m^4
k	spring constant	lbf/ft	N/m
k_r	torsional spring constant	ft-lbf/rad	N·m/rad
K	vibration constant	–	–
L	length	ft	m
m	mass	lbm	kg
r	radius	ft	m
r	ratio of forcing to natural frequency	–	–
r	root	sec^{-1}	s^{-1}
SG	specific gravity	–	–
t	thickness of thin plate	ft	m
t	time	sec	s
TR	transmissibility	–	–
v	velocity	ft/sec	m/s
w	load per unit area	lbf/ft^2	N/m^2
w	load per unit length	lbf/ft	N/m
x	position	ft	m

Symbols

α	angular acceleration	rad/sec^2	rad/s^2
β	magnification factor	–	–
δ	deflection	ft	m
δ	logarithmic decrement	–	–
ζ	damping ratio	–	–
η	isolation efficiency	–	–
ν	Poisson's ratio	–	–
ρ	density	lbm/ft^3	kg/m^3
τ	period	sec	s
ϕ	angle	rad	rad
ω	frequency	rad/sec	rad/s

Subscripts

0	initial
b	block
c	complementary
d	damped
eq	equivalent
f	forced
l	liquid
n	natural
n	nth mode
p	particular
r	rotational
st	static
t	torsional or transmitted

Topic XI: Control Systems

Chapter

60 Modeling of Engineering Systems

Content in blue refers to the *NCEES Handbook*.

1. INTRODUCTION TO ENGINEERING SYSTEMS MODELING

The ultimate benefit derived from a model is the ability to predict the behavior (known as the *response*) of a real-world system. Some models (scale models, working models, mock-ups, etc.) are physical, but others (such as the ones in this chapter) are mathematical. The goal of modeling is to develop a differential equation or other mathematical function, known as the *response function*, that predicts how the system will behave (i.e., what position, acceleration, or velocity it will have). The response function is commonly transformed into the Laplace s-variable domain.

It is rarely possible to write the response function by observation, and several methods of developing the response function are available. This chapter takes the *two-port black box* (see Fig. 60.1) approach—"two-port" because there is one input pair and one output pair to the model, and "black box" because the inner workings of the model are irrelevant once the response function is known.[1]

Figure 60.1 *Two-Port Black Box Model*

Many types of real-world systems (e.g., long-term weather prediction) are too complex to be modeled mathematically. Others lend themselves to special types of modeling theory. This chapter is limited to systems that can be modeled by *idealized* (*linear*) *elements* (components, devices, etc.).[2] (See Sec. 60.2.) In some cases, *nonlinear elements* can be considered linear in limited operation ranges. Elements such as mechanical springs and electrical resistors that absorb and dissipate energy are known as *passive elements*. Energy sources are *active elements*.

The response function for a model is derived from a *system equation*, which is usually a differential equation. The *order of the system* (*model*) is the highest order derivative in the system equation.

It is not necessary to work separate problems to find position, velocity, and acceleration response functions for a system. If the position function $x(t)$ is known, it can be differentiated to give $\mathrm{v}(t)$ and $a(t)$. If any one of the three response functions is known, the other two can be derived.

A *single degree of freedom* (*SDOF*) *system* can be completely defined by one response variable. A single mass on a spring, a swinging pendulum, and a rotating pulley are examples of SDOF systems. In each case, one variable (x or θ) defines the position of the major system element.

Systems in which multiple components have their own values of the dependent variable are known as *multiple degree of freedom* (*MDOF*) *systems*. The degree of freedom of the system is the number of unrelated dependent variables needed to specify the behavior of all major system elements.

2. ELEMENTS

Each physical device in the system is an *element*. All of the ideal elements are considered to be two-port devices. Springs and dashpots have two ends, each of which can have a different value of the response variable. For example, the velocity of both ends of a shock absorber need not be the same. Even though masses do not have ends in the traditional sense, they are considered to be two-port devices as well.

[1]It is important to recognize that the mathematical response function is the model. Drawing system diagrams and taking other steps to derive the response function are merely developmental aids.

[2]Almost any linear flow process can be modeled in this manner. Fluid flow and heat transfer systems are other common applications.

A dependent variable that describes the performance of an element is a *response variable* (also known as a *state variable*). Each element in the model will have its own response variable, such as position, velocity, and acceleration in mechanical systems and voltage in electrical systems. Time is the independent variable.

3. ENERGY SOURCES AND FORCING FUNCTIONS

Some systems start with and gradually use up stored energy; other systems receive energy on a one-time basis. Still others receive energy on a continuous basis. The equation describing the amount of energy introduced as a function of time is the *forcing function*. Figure 60.2 illustrates common forcing function profiles. Although a wide variety of forcing functions is possible, engineering systems easily become too complicated for manual solutions unless limited to simple types.

Figure 60.2 *Common Forcing Function Profiles*

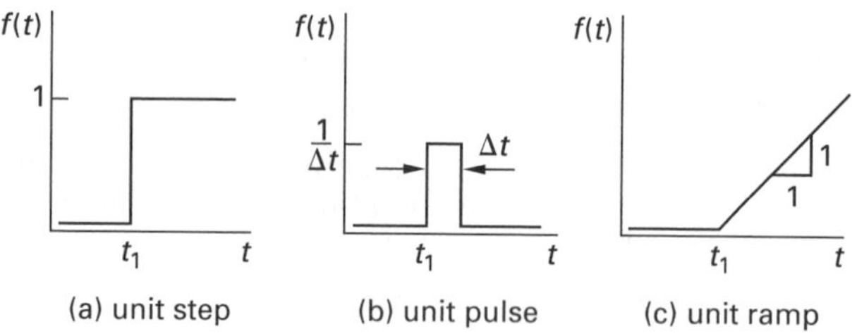

An energy source need not be an actual physical component such as a wound spring, battery, or fuel cell. Anything that produces motion in the system, including potential energy or an applied force, can be an energy source.

In modeling, the source or method of energy generation may not be known. For example, a velocity source may produce a specific velocity regardless of the system mass, or a current source may produce a specific current regardless of circuit elements. How this is accomplished need not be explicitly known.

The *homogeneous forcing function* is the zero function (i.e., no energy at all).[3] The homogeneous case does not preclude an initial disturbance or a previous amount of potential energy. However, the forcing function is homogeneous if it ceases to act as soon as the system begins to move. For example, a spring-mass system that is displaced, released, and allowed to oscillate freely experiences a zero force after the release.

A *unit step* has zero magnitude up to a particular instant (t_1) and a magnitude of 1 thereafter. It can be multiplied by a scalar if the magnitude of the actual forcing function has any other value.

$$f(t) = \begin{cases} 0 & t < t_1 \\ 1 & t \geq t_1 \end{cases} \qquad 60.1$$

Hitting a bell with a hammer is an example of an impulse. A *unit impulse* (also known as a *pulse*) is a limited-duration force whose total impulse (that is, $f\Delta t$) is 1. The unit pulse can be multiplied by a scalar if the actual pulse has an impulse different than 1. Usually the time, Δt, is extremely short.

$$f(t) = \begin{cases} 0 & t < t_1 \\ \dfrac{1}{\Delta t} & t_1 \leq t < t_1 + \Delta t \\ 0 & t \geq t + \Delta t \end{cases} \qquad 60.2$$

Some forces increase with time. The slope of a *unit ramp* forcing function is 1. If a forcing function changes at any other rate, the unit ramp can be multiplied by a scalar.

The unit step, ramp, impulse, parabola, and so on, are known as *singularity functions* because of the singularity that exists at the point of discontinuity in each function. The effects of integrating and differentiating these functions are listed in Table 60.1.

Table 60.1 *Operations on Forcing Functions*

function	function when differentiated	function when integrated
unit impulse	–	unit step
unit step	unit impulse	unit ramp
unit ramp	unit step	unit parabola
unit parabola	unit ramp	(third degree)
unit exponential	unit exponential	unit exponential
unit sinusoid	unit sinusoid	unit sinusoid

The most common forcing functions used in the analysis of engineering systems are sinusoids. When combined with Fourier series analysis, sinusoids can be used to approximate all other forcing functions.

4. THROUGH- AND ACROSS-VARIABLES

Through-variables have different values at the two ends of an element; *across-variables* have the same value. For example, force is a through-variable since it is passed through objects. Consider a mass hanging on a spring. The gravitational force on the mass is passed through the spring to the support.

[3]This is consistent with homogeneous differential equations defined in Sec. 60.1.

The velocity of a mass, however, is measured with respect to an inertial (stationary) frame of reference. Since the "ground" is connected to a mass (see Rule 1, Sec. 60.5), velocity is the across-variable.

5. SYSTEM DIAGRAMS

The system elements are interconnected to produce a system diagram with the following procedure. The diagram is then used to derive the system equation. System diagrams for mechanical systems do not always resemble the topology of the systems they represent. Example 60.1(a) illustrates a system of two elements (a mass and dashpot) connected in series that actually has a parallel system diagram.

step 1: Decide on a dependent response variable. Position (x or θ) is preferred in mechanical systems, but velocity (v or ω) and acceleration (a or α) can be used if desired.

step 2: Identify all parts of the system that have different values of the response variable. It is not necessary to know the actual values, only to recognize where the variable changes. One of the values will always be zero (corresponding to zero velocity) and is known as the *ground level.*

step 3: Start the system diagram by drawing a horizontal line for each different value of the dependent variable identified in step 2.

step 4: Insert and connect the passive elements (masses, springs, etc.) to the appropriate horizontal lines.

Rule 1: One end of a mass always connects to the lowest (ground) level.

step 5: Insert and connect the active (energy) sources to the appropriate horizontal lines.

Rule 2: One end of an energy source always connects to the lowest (ground) level.

A *line diagram* is a variation of the system diagram in which the levels associated with different values of the response variable (the horizontal lines from step 3) are replaced with nodes, and the element symbols are replaced with their values.

6. SYSTEM EQUATIONS

The *system equation* is derived from a system or line diagram. The system equation is a differential equation containing the dependent variable, but it does not explicitly give the response variable. Traditional methods and Laplace transforms can be used to derive the response variable.

The principle used to obtain a system equation from a system diagram is a conservation law analogous to Kirchhoff's current law that says "...what goes in must come out..."[4] For mechanical systems, force is the conserved quantity. Specifically, the total force supplied by the source equals the forces leaving through all parallel branches (known as *legs*) in the system diagram. Use of the following two rules is illustrated in Ex. 60.1.

Rule 3: The force passing through a leg consisting of elements in series can be determined from the conditions across any of the elements in that leg.

Corollary to Rule 3: The force passing through a leg consisting of elements in series can be determined from the conditions across the first element in that leg.

Rule 4: When writing a difference in response variable values (e.g., v_2-v_1 or x_2-x_1), the first subscript is the same as the node number for which the equation is being written.

7. MECHANICAL SYSTEMS

A mechanical system consists of interconnected lumped masses, linear springs, linear dashpots, and energy sources. The response variable for a mechanical system is usually the position of one of the masses, although velocity and acceleration can also be chosen.

A *lumped mass* is a rigid body that acts like a particle. All parts of the mass experience identical displacements, velocities, and accelerations. An *ideal lumped mass* is one for which Newtonian (i.e., nonrelativistic) physics applies. The governing equation for a lumped mass is Newton's second law.[5]

$$F = ma = mx'' \quad \text{[ideal lumped mass]} \quad \text{[SI]} \qquad 60.3(a)$$

$$F = \frac{ma}{g_c} = \frac{mx''}{g_c} \quad \text{[ideal lumped mass]} \quad \text{[U.S.]} \qquad 60.3(b)$$

An ideal spring is massless, has a constant stiffness, k, and is immune to set, creep, and fatigue. Hooke's law predicts its performance but is rewritten to explicitly include the positions of both spring ends. (If both ends are displaced the same distance, there will be no change in extension, compression, or force.)

$$F = k(x_2 - x_1) \quad \text{[ideal spring]} \qquad 60.4$$

A *dashpot* (*damper*), of which an automobile shock absorber is an example, is a device that opposes motion and dissipates energy. An ideal linear dashpot is

[4]Electrical current is a through-variable. Kirchhoff's current law applies to any through-variable.
[5]It is understood that force, F, and acceleration, a, are functions of time.

massless and applies a force opposing motion that is proportional to velocity. Since both ends of the dashpot move, the governing equation is

$$\begin{aligned} F &= C(v_2 - v_1) \\ &= C(x_2' - x_1') \quad \text{[ideal dashpot]} \end{aligned} \qquad 60.5$$

Example 60.1

For the three systems shown, draw the system diagram, draw the line diagram, and write the system equation. Use consistent units.

(a) A mass is connected through a damper to a solid wall as shown. The mass slides without friction on its support and is acted upon by a force.

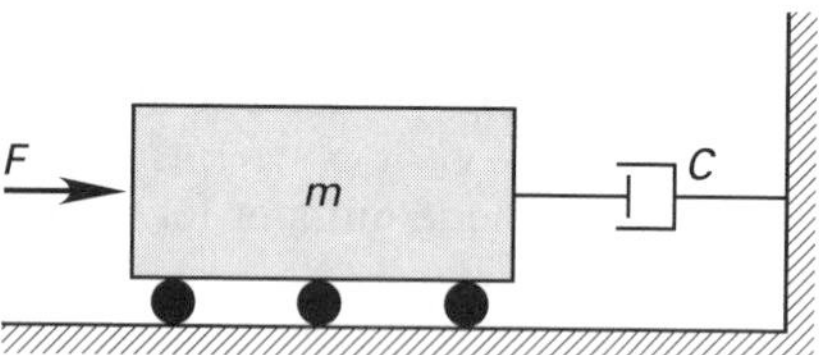

(b) A force is applied through a damper to a mass. The mass rolls on frictionless bearings.

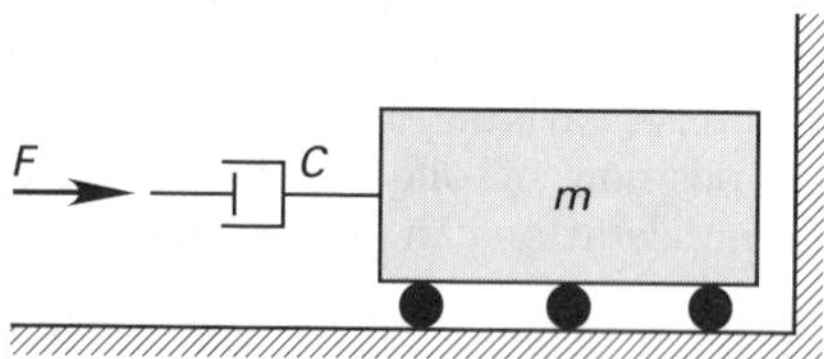

(c) A shock absorber with an integral coil spring is acted upon by a force applied to one end.

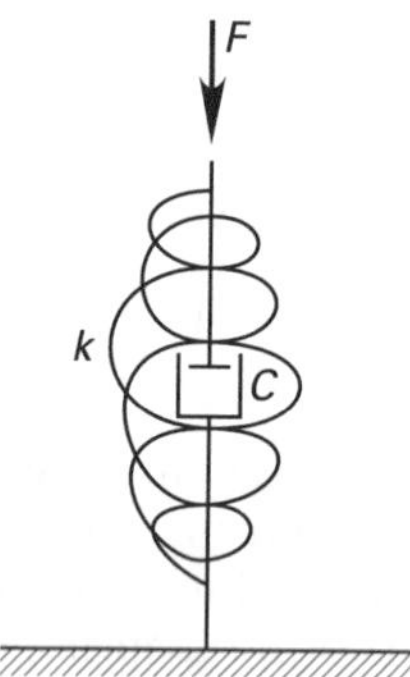

Solution

(a) Choose velocity as the response variable. All parts of the mass move with the same velocity: Call it v_1. The plunger also moves with velocity v_1 since it is attached to the mass. The body of the damper is attached to the stationary wall. Call this velocity $v = 0$.

Two horizontal lines are drawn: the top line for v_1 and the bottom line for $v = 0$. One end of the damper travels at v_1; the other end travels at $v = 0$. Therefore, connect the dashpot to these lines. The mass moves at v_1, so one of its lines connects to v_1. By Rule 1 (see Sec. 60.5), the other end connects to $v = 0$. The force contacts the mass, so one end of the force connects to v_1. By Rule 2, the other end connects to $v = 0$.

The system and line diagrams are

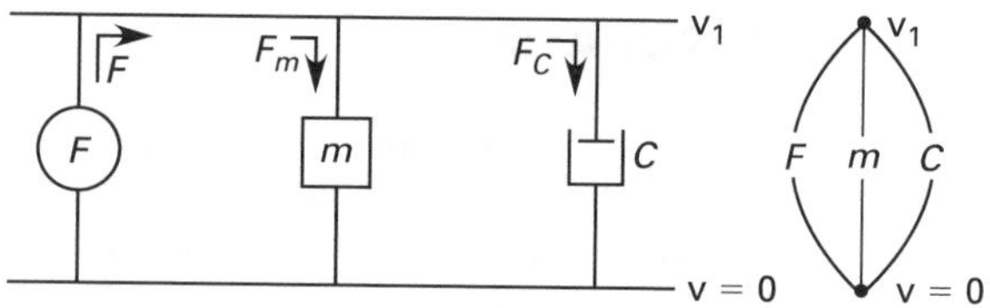

The force leaving the "source" splits—some of it going through the mass and some of it going through the dashpot. The conservation law is written to conserve the force in the v_1 line.

$$F = F_m + F_C$$

Expanding Rule 4 (see Sec. 60.6) with Eq. 60.3 and Eq. 60.5,

$$F = ma_1 + C(v_1 - 0)$$

The differential equation is

$$F = mx_1'' + Cx_1'$$

(b) There are three velocities in this example: the velocity of the plunger (call this v_1), the velocity of the dashpot body and mass (call this v_2), and $v = 0$. The system line diagrams are

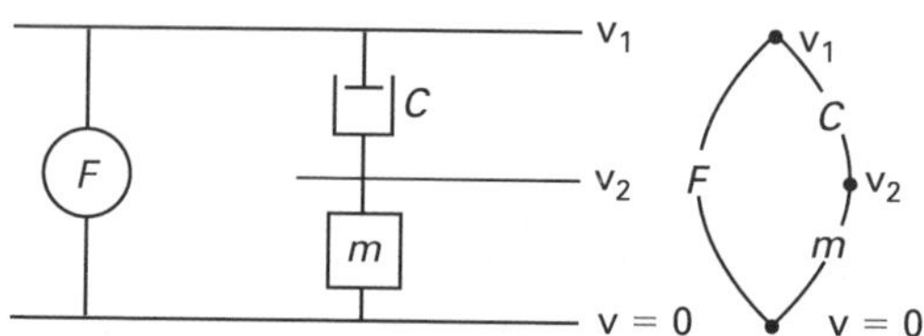

The system equation is written to conserve force in the v_1 line. The force passing through the dashpot is the same force experienced by the mass and is not additive. (This is the essence of Rule 3.) Rule 4 is used to write the dashpot force as $C(v_1 - v_2)$.

$$\begin{aligned} F &= F_C \\ F &= C(v_1 - v_2) = C(x_1' - x_2') \end{aligned}$$

Since the same force is experienced by both the dashpot and mass, the system equation could be written in terms of the governing equation of the mass. Whether or not this is a better choice will depend on the initial conditions and other information that is available.

$$F = F_m = ma_2 = mx_2''$$

(c) There are only two velocities here. The system and line diagrams are

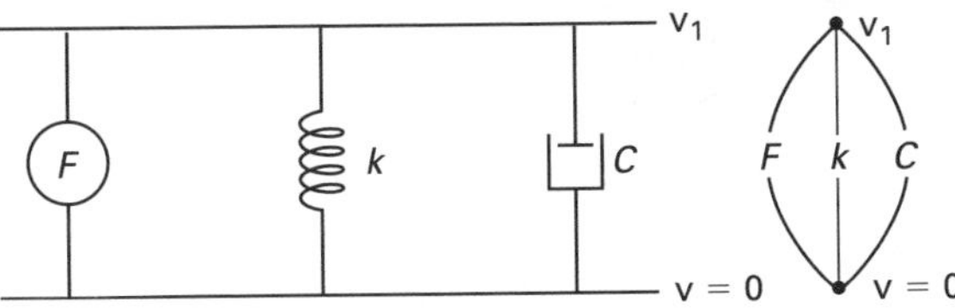

Force in the v_1 line is conserved. The system equation is

$$F = F_C + F_k = C v_1 + k x_1$$
$$= C x_1' + k x_1$$

8. MECHANICAL ENERGY TRANSFORMATIONS

Levers transform one force into another at the expense of the distance the ends travel. The ratio of transformation depends on the lengths of the lever on both sides of the fulcrum. Whether the ratio is less than or greater than one is a matter of preference as long as the transformed force and velocity are correct. The ratio will be negative because the direction of the force and velocity is reversed by a lever. The transformation is represented in system diagrams by the symbol for an electrical transformer.

Example 60.2

What are the system equations for the system shown?

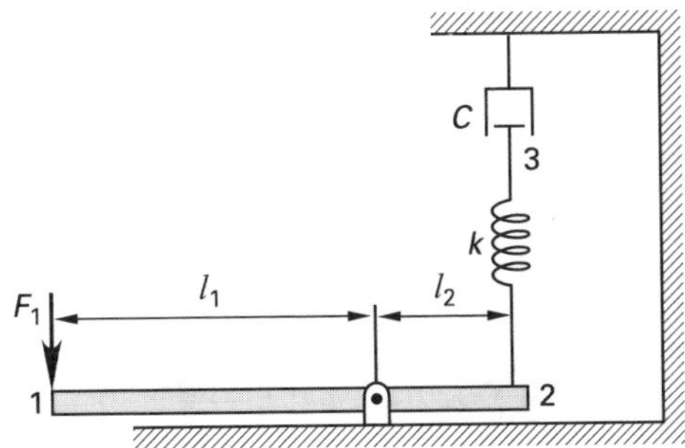

Solution

The lever transforms the force and displacement at point 1 into force and displacement at point 2. The system diagram is

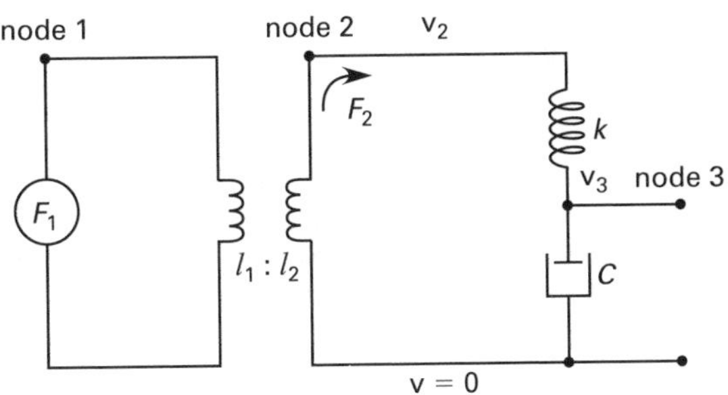

One of the system equations is based on node 2.

$$F_2 = k(x_2 - x_3)$$

The following additional equations are based on the lever transformation ratio. The minus signs indicate that the displacement and force at opposite ends of the lever are in opposite directions.

$$x_2 = \left(-\frac{l_2}{l_1}\right)x_1$$

$$F_2 = \left(-\frac{l_1}{l_2}\right)F_1$$

9. ROTATIONAL SYSTEMS

Rotational systems are directly analogous to mechanical systems. Flywheels have rotational mass moments of inertia, I, torsion springs have torsional stiffness, k_r, and fluid couplings have rotational viscous damping, C_r. The governing equations for these elements are

$$T = I\alpha \quad \text{[ideal flywheel]} \quad \text{[SI]} \qquad 60.6(a)$$

$$T = \frac{I\alpha}{g_c} \quad \text{[ideal flywheel]} \quad \text{[U.S.]} \qquad 60.6(b)$$

$$T = k_r(\theta_2 - \theta_1) \quad \text{[ideal torsion spring]} \qquad 60.7$$

$$T = C_r(\omega_2 - \omega_1) \quad \text{[ideal fluid coupling]} \qquad 60.8$$

Angular velocity, ω (in rad/s), is usually chosen as the response variable, although angular position, θ (in radians), or acceleration, α (in rad/s^2), can be used if desired. As with the translational mechanical systems, any one of these three response variables can be used to write the system equation, after which the others can be found by integration or differentiation.

Energy can be provided to rotational systems by constant-torque or constant-velocity sources. Gear sets transform torque and velocity. The ratio of transformation, a, is the ratio of numbers of teeth or diameters. The ratio is negative because each set of gears reverses the direction of rotation. Whether or not the ratio n_1/n_2 or n_2/n_1 is used is not important as long as the transformed dependent variable is correct. When a gear set increases the rotational speed, the torque decreases, and vice versa. (See Fig. 60.3.)

$$a = -\frac{T_1}{T_2} = -\frac{\omega_2}{\omega_1} = \frac{n_1}{n_2} \qquad 60.9$$

Figure 60.3 Rotational Transformer

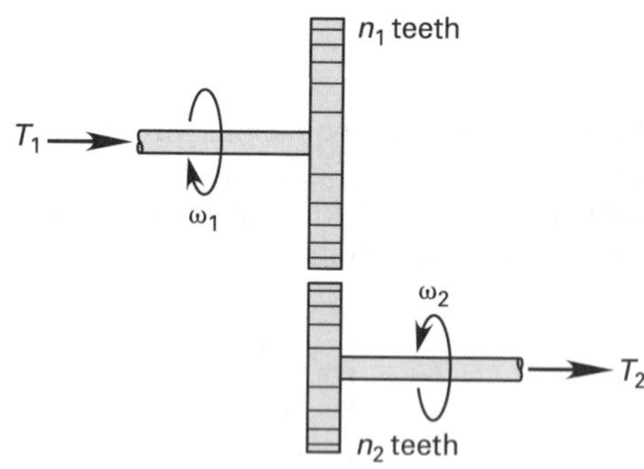

Example 60.3

Two flywheels are connected by a flexible shaft. The second flywheel is acted upon by a linear viscous force. Draw the system diagram and write the system equation.

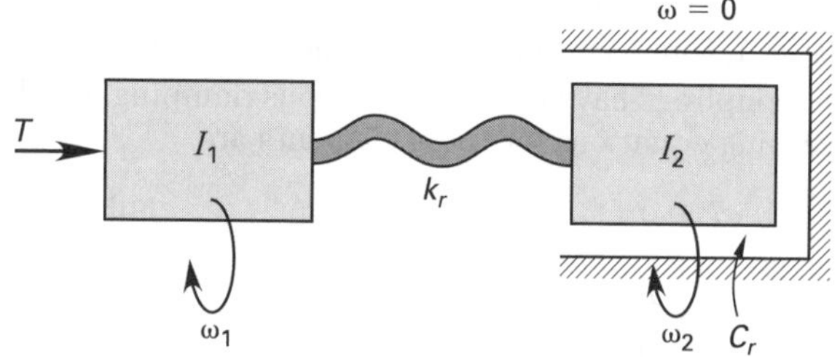

Solution

Choose angular velocity as the response variable. There are three different angular velocities—two for the flywheels and one for the stationary reference. The system diagram is

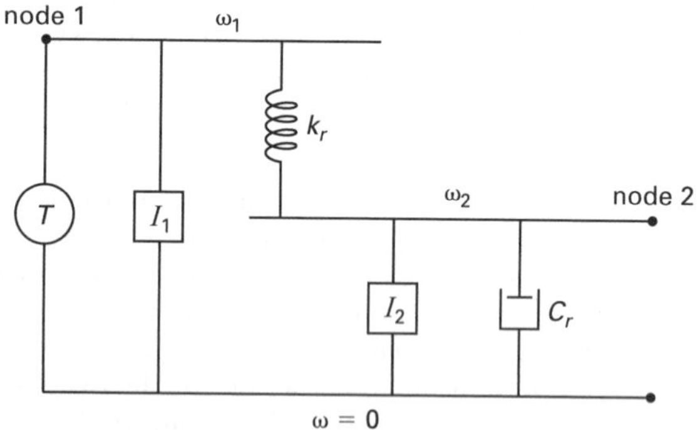

The system equations are

$$\begin{aligned} \textit{node 1:}\quad T &= I_1\alpha_1 + k_r(\theta_1 - \theta_2) \\ &= I_1\theta_1'' + k_r(\theta_1 - \theta_2) \\ \textit{node 2:}\quad 0 &= C_r\omega_2 + I_2\alpha_2 + k_r(\theta_2 - \theta_1) \\ &= C_r\theta_2' + I_2\theta_2'' + k_r(\theta_2 - \theta_1) \end{aligned}$$

Example 60.4

A motor drives a flywheel through a set of reduction gears. The flywheel is connected to the driven gear by a flexible shaft. All other gears and shafts have infinite stiffness. What is the system equation?

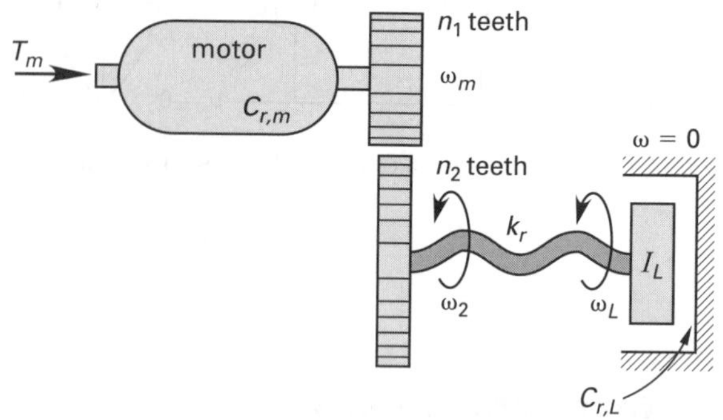

Solution

Angular velocity, ω, is chosen as the response variable. The gear set is represented by the symbol for an electrical transformer. The ratio of transformation is less than 1. This means that the motor's torque will be decreased while the rotational speed is increased. The system equations are

$$\begin{aligned} \textit{node m:}\quad T_m &= I_m\alpha_m + C_{r,m}\omega_m + aT_2 \\ \textit{node 2:}\quad T_2 &= k_r(\theta_2 - \theta_L) \\ \textit{node L:}\quad 0 &= C_{r,L}\omega_L + I_L\alpha_L + k_r(\theta_L - \theta_2) \end{aligned}$$

Since ω_m, ω_2, ω_L, and T_2 are all unknown, a fourth equation is needed.

$$a\omega_m = -\omega_2$$

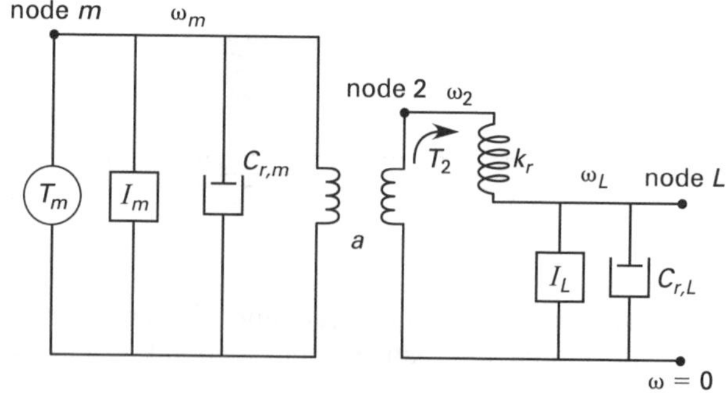

10. FLUID SYSTEMS

Figure 60.4 illustrates an open fluid *reservoir* with a constant vertical cross section. The flow rate out of the reservoir can be calculated from the cross-sectional area and the rate of change in the surface elevation. In

Eq. 60.10, C_f is known as the *fluid capacitance*. In system diagrams, open reservoirs always connect to the lowest (ground) level.

$$Q = A\left(\frac{dh}{dt}\right) = \frac{A\left(\frac{dp}{dt}\right)}{\gamma} = C_f\left(\frac{dp}{dt}\right) \qquad 60.10$$

$$C_f = \frac{A}{\gamma} \qquad 60.11$$

Figure 60.4 *Fluid Reservoir*

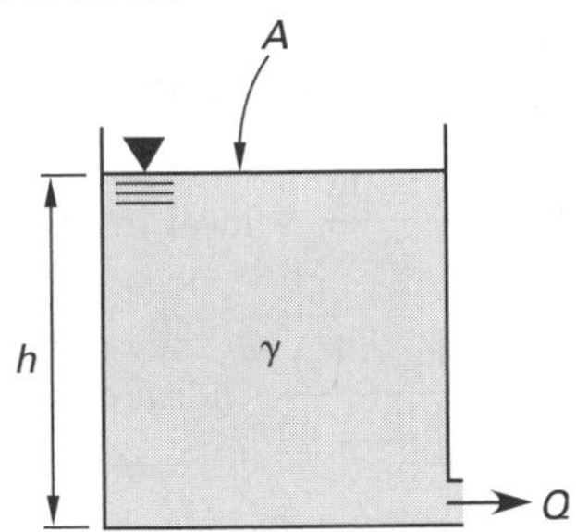

The Darcy equation indicates that *flow resistance* is proportional to the square of the flow quantity. As a simplification over a narrow range of flows, the flow resistance in simple systems analysis problems is assumed to be proportional to the flow quantity. R_f is the *fluid resistance coefficient* and is often found by experimentation.

$$p_2 - p_1 = R_f Q \qquad 60.12$$

Newton's second law (or the impulse-momentum principle) is the basis for defining the *fluid inertance* (*fluid inductance*), I, which accounts for the inertia of the fluid flow.

$$F = m\frac{d\text{v}}{dt} \qquad 60.13$$

$$A(p_2 - p_1) = \gamma A l \frac{\frac{dQ}{dt}}{A} \qquad 60.14$$

$$p_2 - p_1 = \frac{\gamma l}{A}\frac{dQ}{dt} = I\frac{dQ}{dt} \qquad 60.15$$

$$I = \frac{\gamma l}{A} \qquad 60.16$$

Equation 60.15 is integrated to obtain an expression for the flow quantity.

$$Q = \frac{1}{I}\int (p_2 - p_1)\,dt \qquad 60.17$$

Example 60.5

A pump is used to keep liquid flowing through a filter pack. Pipe friction and potential head are insignificant. What is the system equation?

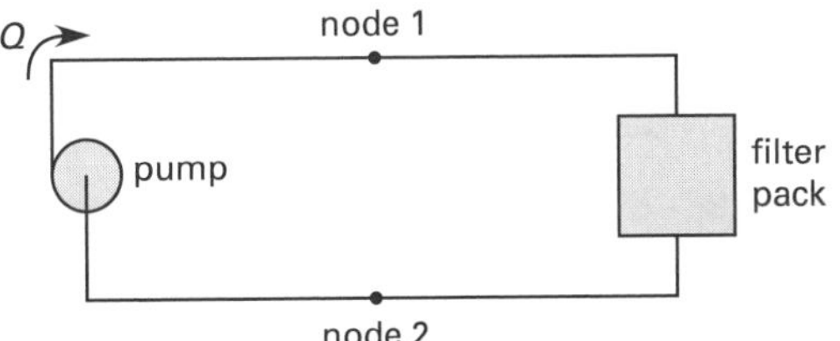

Solution

Pressure varies along the flow path and is the response variable. There are two different pressures: before and after the filter pack. This is analogous to a series electrical circuit of a battery and resistor. The system diagram is

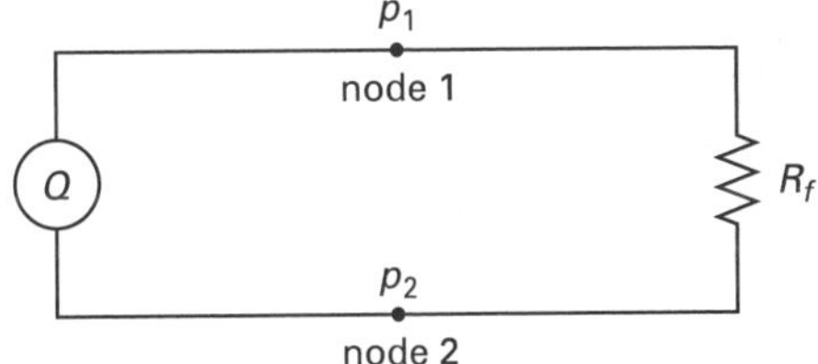

The system equation is

$$Q = \frac{1}{R_f}(p_2 - p_1)$$

Example 60.6

A reservoir discharges through a long pipe and is not refilled. What is the system equation that defines the pressure at the tank bottom as a function of time?

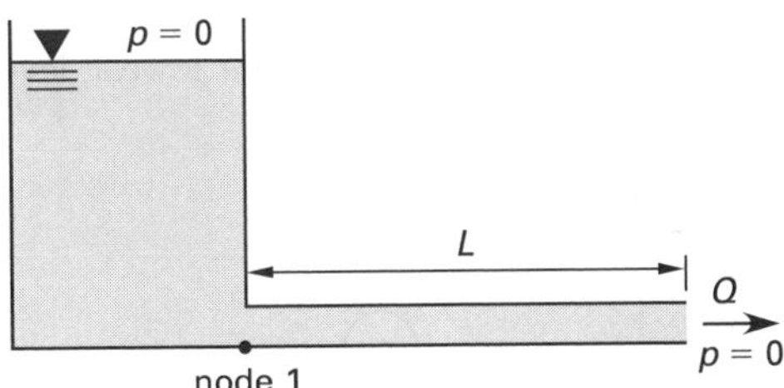

Solution

Although there is flow, there is no external energy source (i.e., there is no pump) in this system. This is analogous to an electrical capacitor discharging through a resistor. The system diagram is

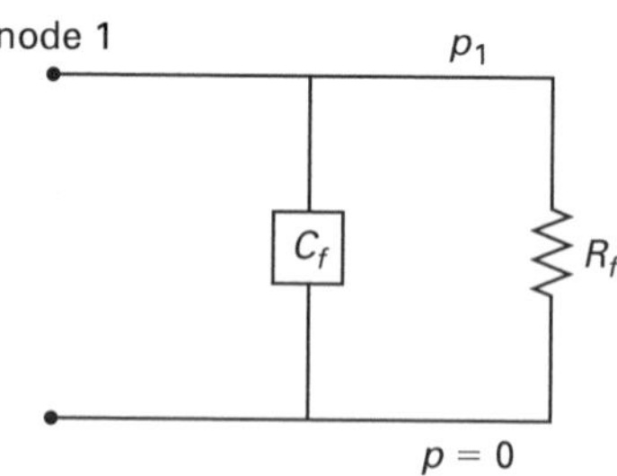

The system equation is

$$0 = \frac{p_1}{R_f} + C_f \frac{dp_1}{dt}$$

Example 60.7

A pump transfers liquid from one reservoir to another. Pipe friction is insignificant. What are the system equations that define the pressures at the bottoms of the reservoirs?

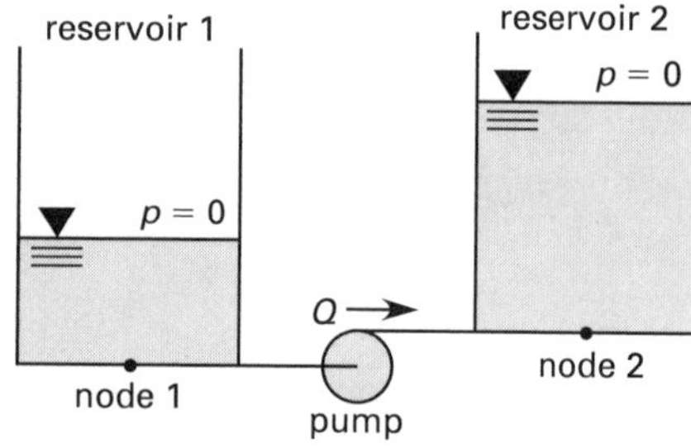

Solution

Both open reservoirs connect to the lowest level. The system diagram is

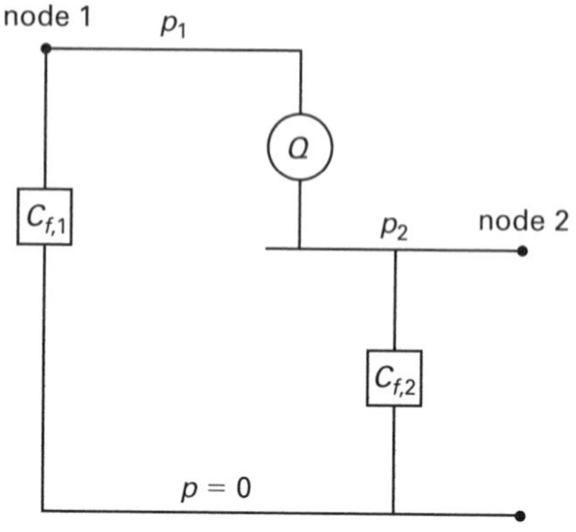

The system equations are

$$\textit{node 1:} \quad Q = C_{f,1} \frac{dp_1}{dt}$$

$$\textit{node 2:} \quad Q = C_{f,2} \frac{dp_2}{dt}$$

11. ELECTRICAL SYSTEMS

Resistors, capacitors, and inductors are passive elements in an electrical circuit. Voltage (an across-variable) is chosen as a convenient dependent variable, and the governing equations are written in terms of the conserved quantity, current (a through-variable).

$$I = \frac{V_1 - V_2}{R} \quad \text{[ideal resistor]} \qquad 60.18$$

$$I = C \frac{d(V_1 - V_2)}{dt} \quad \text{[ideal capacitor]} \qquad 60.19$$

$$I = \frac{1}{L} \int (V_1 - V_2)\,dt \quad \text{[ideal inductor]} \qquad 60.20$$

The system diagram is the same as the electrical circuit. It is not necessary to draw a different diagram.

Example 60.8

What are the system equations for the electrical circuits shown?

(a)

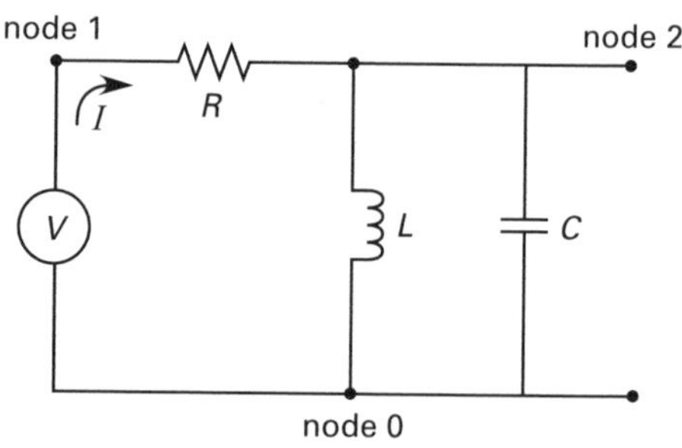

(b)

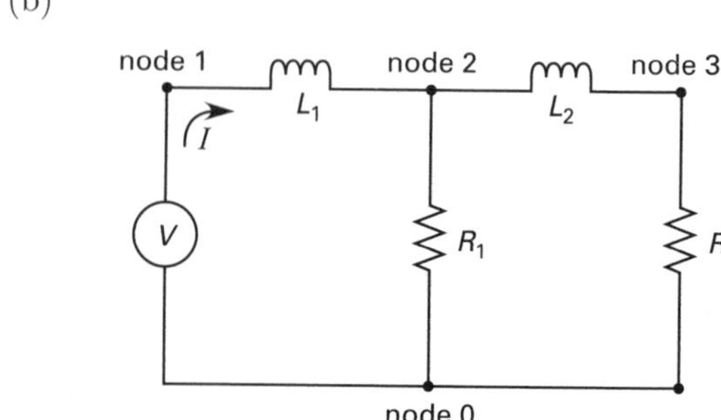

Solution

(a) Nodes 1 and 2 are at different potentials (voltages).

node 1: $I = \frac{V_1 - V_2}{R}$

node 2: $0 = C\frac{dV_2}{dt} + \frac{1}{L}\int V_2 dt + \frac{V_2 - V_1}{R}$

(b) Nodes 1, 2, and 3 have different potentials.

node 1: $I = \frac{1}{L_1}\int (V_1 - V_2)\,dt$

node 2:

$$0 = \frac{V_2}{R_1} + \frac{1}{L_2}\int (V_2 - V_3)\,dt + \frac{1}{L_1}\int (V_2 - V_1)\,dt$$

node 3: $0 = \frac{1}{L_2}\int (V_3 - V_2)\,dt + \frac{V_3}{R_2}$

12. NOMENCLATURE

a	acceleration	ft/sec^2	m/s^2
a	ratio of transformation	–	–
A	area	ft^2	m^2
C	capacitance	F	F
C	capacitance (fluid)	ft^5/lbf	m^5/N
C	damping coefficient	lbf-sec/ft	N·s/m
C_r	rotational damping coefficient	ft-lbf-sec/rad	N·m·s/rad
f	forcing function	various	various
F	force	lbf	N
g_c	gravitational constant, 32.2	ft-lbm/lbf-sec^2	n.a.
h	depth	ft	m
I	current	A	A
I	inertance (fluid)	lbf/ft^4	N/m^4
I	mass moment of inertia	lbm-ft^2	kg·m^2
k	spring constant	lbf/ft	N/m
k_r	rotational spring constant	ft-lbf/rad	N·m/rad
l	length	ft	m
L	inductance	H	H
m	mass	lbm	kg
n	number of teeth	–	–
p	pressure	lbf/ft^2	N/m^2
Q	flow rate	ft^3/sec	m^3/s
r	response function	various	various
R	resistance	Ω	Ω
t	time	sec	s
T	torque	ft-lbf	N·m
v	velocity	ft/sec	m/s
V	voltage	V	V
x	position	ft	m

Symbols

α	angular acceleration	rad/sec^2	rad/s^2
γ	specific weight	lbf/ft^3	N/m^3
θ	angular position	rad	rad
ω	angular velocity	rad/sec	rad/s

Subscripts

C	dashpot
f	fluid
L	liquid or load
m	mass or motor
r	rotational

Control Systems

61 Analysis of Engineering Systems

Content in blue refers to the *NCEES Handbook*.

1. TYPES OF RESPONSE

Natural response (also known as *initial condition response*, *homogeneous response*, and *unforced response*) is how a system behaves when energy is applied and is subsequently removed. The system is left alone and is allowed to do what it would naturally, without the application of further disturbing forces. In the absence of friction, natural response is characterized by a sinusoidal response function. With friction, the response function will contain exponentially decaying sinusoids.

Forced response is the behavior of a system that is acted on by a force that is applied periodically. Forced response in the absence of friction is characterized by sinusoidal terms having the same frequency as the forcing function.

Natural and forced responses are present simultaneously in forced systems. The sum of the two responses is the *total response*.[1] This is the reason that differential equations are solved by adding a particular solution to the homogeneous solution. The homogeneous solution corresponds to the natural response; the particular solution corresponds to the forced response.

$$\begin{array}{c}\text{total}\\ \text{response}\end{array} = \begin{array}{c}\text{natural}\\ \text{response}\end{array} + \begin{array}{c}\text{forced}\\ \text{response}\end{array} \qquad \textit{61.1}$$

Since the influence of decaying functions disappears after a few cycles, natural response is sometimes referred to as *transient response*. Once the transient response effects have died out, the total response will consist entirely of forced terms. This is the *steady-state response*.

2. GRAPHICAL SOLUTION

Graphical solutions are available in limited cases, particularly those with homogeneous, step, and sinusoidal inputs. When the system equation is a homogeneous second-order linear differential equation with constant coefficients in the form of Eq. 61.2, the natural time response can be determined from Fig. 61.1. ω is the natural frequency, and ζ is the damping ratio.

$$x'' + 2\zeta\omega x' + \omega^2 x = 0 \qquad \textit{61.2}$$

When the system equation is a second-order linear differential equation with constant coefficients, and the forcing function is a step of height h (as in Eq. 61.3), the time response can be determined from Fig. 61.2.

$$x'' + 2\zeta\omega x' + \omega^2 x = \omega^2 h \qquad \textit{61.3}$$

[1]This response is a function of time, and, therefore, is referred to as *time response* to distinguish it from *frequency response*. (See Sec. 61.13.)

Figure 61.1 *Natural Response*

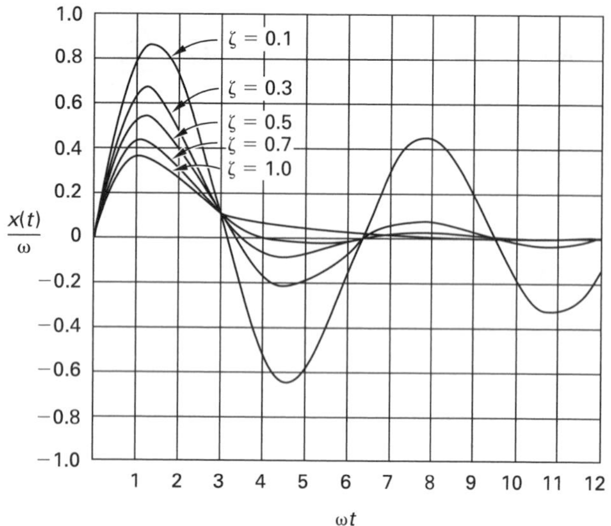

Figure 61.2 *Response to a Unit Step*

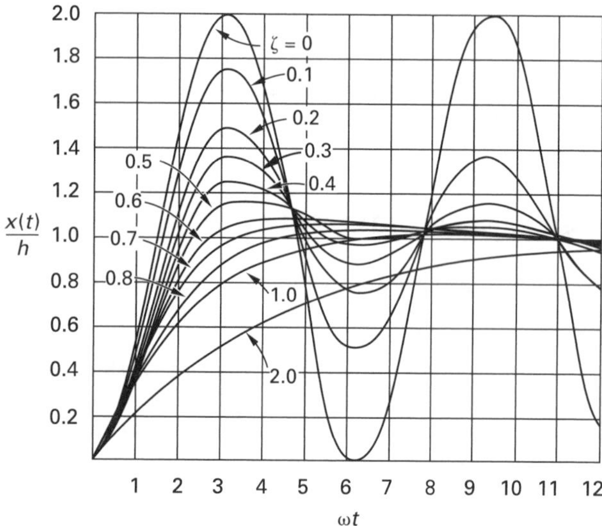

Figure 61.2 illustrates that a system responding to a step will eventually settle to the steady-state position of the step, but that when damping is low ($\zeta < 1$), there will be *overshoot.* Figure 61.3 illustrates this and other parameters of second-order response to a step input. The settling time, t_s, depends on the *tolerance* (i.e., the separation of the actual and steady-state responses). The *time delay*, t_d, in Figure 61.3 is the time to reach 50% of the steady-state value. The *time constant*, τ, is the time to reach approximately 63% of the steady-state value.

$$\omega_d = \text{damped frequency} = \omega\sqrt{1-\zeta^2} \quad \text{61.4}$$

$$t_r = \text{rise time} = \frac{\pi - \arccos\zeta}{\omega_d} \quad \text{61.5}$$

$$t_p = \text{peak time} = \frac{\pi}{\omega_d} \quad \text{61.6}$$

$$M_p = \text{peak gain} \quad [\text{fraction overshoot}]$$

$$= \exp\left(\frac{-\pi\zeta}{\sqrt{1-\zeta^2}}\right) \quad \text{61.7}$$

$$t_s = \text{settling time} = \frac{3.91}{\zeta\omega} \quad [2\% \text{ criterion}] \quad \text{61.8}$$

$$t_s = \frac{3.00}{\zeta\omega} \quad [5\% \text{ criterion}] \quad \text{61.9}$$

$$\tau = \text{time constant} = \frac{1}{\zeta\omega} \quad \text{61.10}$$

Figure 61.3 *Second-Order Step Time Response Parameters*

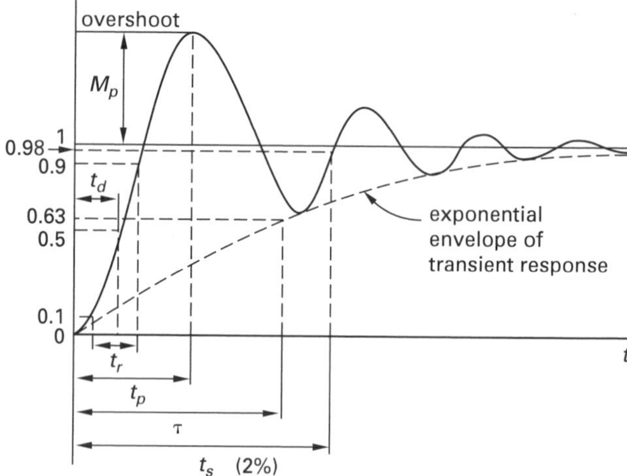

3. CLASSICAL SOLUTION METHOD

A system model can be thought of as a block where the *input signal* is the forcing function, $f(t)$, and the *output signal* is the response function, $r(t)$. The standard analytical method of determining the response of a system from its system equation uses Laplace transforms. Accordingly, the classical solution approach does not determine the output signal or response function directly. Rather, it derives the *transfer function*, $T(s)$.[2] The transfer function is also known as the *rational function.*

$$T(s) = \mathcal{L}\left(\frac{r(t)}{f(t)}\right) \quad \text{61.11}$$

Transfer functions in the s-domain are Laplace transformations of the corresponding time-domain functions. Transforming $r(t)/f(t)$ into $T(s)$ is more than a simple change of variables. The s symbol can be thought of as a

[2]Strictly speaking, $r(t)/f(t)$ is the *transfer function* and $(\mathcal{L}(r(t)/f(t)))$ is the *transform of the transfer function.* However, the distinction is seldom made.

derivative operator; similarly, the integration operator is represented as $1/s$. By convention, Laplace transforms are represented by uppercase letters, while operand functions are represented by lowercase letters.

Example 61.1

A system equation in differential equation form has been derived for the output force from a mechanical network. Convert the system equation to the s-domain.

$$f(t) = k\int (\mathrm{v}_1 - \mathrm{v}_2)\,dt + m\,\frac{d(\mathrm{v}_2 - \mathrm{v}_1)}{dt}$$

Solution

Replace all derivative operators by s; replace all integration operators by $1/s$. Replace time velocity, $\mathrm{v}(t)$, with s-domain velocity, $\mathrm{v}(s)$.

$$F(s) = \frac{k\mathrm{v}_1}{s} - \frac{k\mathrm{v}_2}{s} + sm\mathrm{v}_2 - sm\mathrm{v}_1$$

Example 61.2

Determine the transfer function for the mechanical system shown and draw its black box representation.

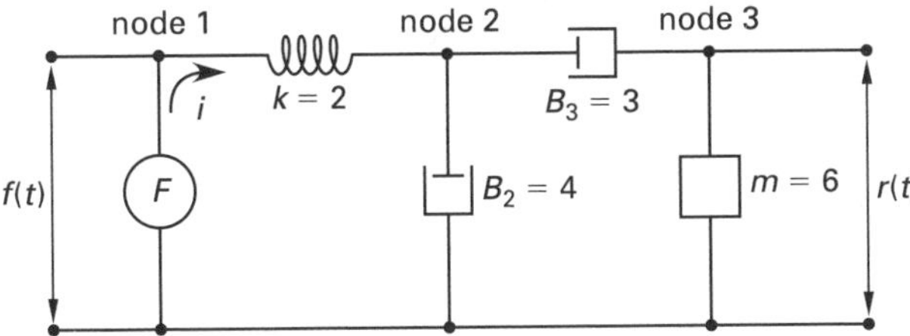

Solution

First, write the system equations. There are two loop "currents," so two simultaneous equations will be needed. One system equation can be written for each of the three nodes, so there will be one redundant system equation. (At this point, it is not obvious which of the two system equations can be used to derive the transfer function.)

node 1:

$$f(t) = k\int (\mathrm{v}_1 - \mathrm{v}_2)\,dt = 2\int (\mathrm{v}_1 - \mathrm{v}_2)\,dt$$

node 2:

$$\begin{aligned} 0 &= k\int (\mathrm{v}_2 - \mathrm{v}_1)\,dt + B_2\mathrm{v}_2 + B_3(\mathrm{v}_2 - \mathrm{v}_3) \\ &= 2\int (\mathrm{v}_2 - \mathrm{v}_1)\,dt + 4\mathrm{v}_2 + 3(\mathrm{v}_2 - \mathrm{v}_3) \end{aligned}$$

node 3:

$$0 = m\,\frac{d\mathrm{v}_3}{dt} + B_3(\mathrm{v}_3 - \mathrm{v}_2) = 6\,\frac{d\mathrm{v}_3}{dt} + 3(\mathrm{v}_3 - \mathrm{v}_2)$$

Next, convert the system equations to the s-domain by substituting s for derivative operations and $1/s$ for integration operations.

node 1: $F(s) = \dfrac{2(\mathrm{v}_1 - \mathrm{v}_2)}{s}$

node 2: $0 = \dfrac{2(\mathrm{v}_2 - \mathrm{v}_1)}{s} + 4\mathrm{v}_2 + 3(\mathrm{v}_2 - \mathrm{v}_3)$

node 3: $0 = 6s\mathrm{v}_3 + 3(\mathrm{v}_3 - \mathrm{v}_2)$

The transfer function is the ratio of the output to input velocities, $\mathrm{v}_3/\mathrm{v}_1$. It does not depend on v_2. The equation for node 1 cannot be used unless $i(t)$ is known, which it (generally) is not. v_2 is eliminated from the equations for nodes 2 and 3. From the second and third nodes,

$$\begin{aligned} \mathrm{v}_2 &= \frac{2\mathrm{v}_1 + 3s\mathrm{v}_3}{2 + 7s} \\ &= \mathrm{v}_3(1 + 2s) \end{aligned}$$

The transfer function, $T(s)$, is found by equating these two expressions and solving for $\mathrm{v}_3/\mathrm{v}_1$.

$$T(s) = \frac{\mathrm{v}_3}{\mathrm{v}_1} = \frac{1}{7s^2 + 4s + 1}$$

The black box representation of this system is

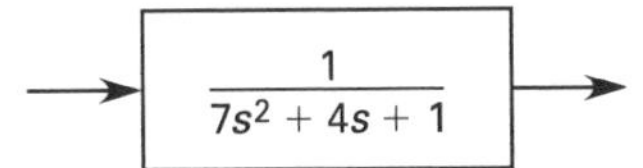

4. FEEDBACK THEORY

The output signal is returned as input in a feedback loop (feedback system). As illustrated in Fig. 61.4, a basic feedback system consists of two black box units (a *dynamic unit* and a *feedback unit*), a *pick-off point* (*take-off point*), and a *summing point* (*comparator* or *summer*). The summing point is assumed to perform positive addition unless a minus sign is present. The incoming signal, V_i, is combined with the feedback signal, V_f, to give the *error* (*error signal*), E. Whether addition or subtraction is used in Eq. 61.12 depends on whether the summing point is additive (i.e., a positive feedback system) or subtractive (i.e., a negative feedback system), respectively. $E(s)$ is the *error transfer function* (*error gain*).

$$\begin{aligned} E(s) &= \mathcal{L}(e(t)) = V_i(s) \pm V_f(s) \\ &= V_i(s) \pm H(s)V_o(s) \end{aligned} \qquad 61.12$$

Control Systems

Figure 61.4 Feedback System

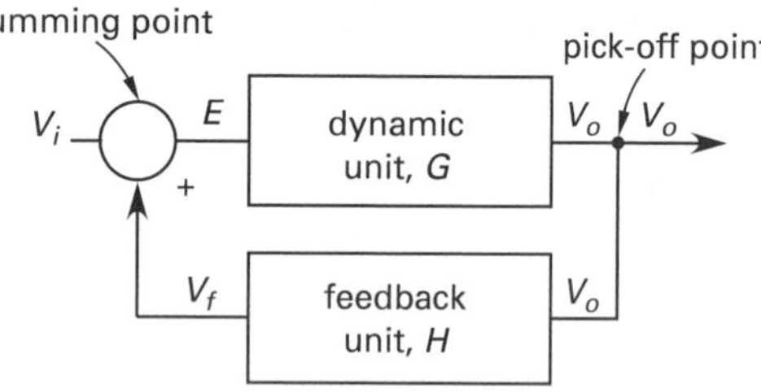

The ratio $E(s)/V_i(s)$ is the *error ratio* (*actuating signal ratio*).

$$\frac{E(s)}{V_i(s)} = \frac{1}{1 + G(s)H(s)} \quad \text{[negative feedback]} \qquad 61.13$$

$$\frac{E(s)}{V_i(s)} = \frac{1}{1 - G(s)H(s)} \quad \text{[positive feedback]} \qquad 61.14$$

Since the dynamic and feedback units are black boxes, each has an associated transfer function. The transfer function of the dynamic unit is known as the *forward transfer function* (*direct transfer function*), $G(s)$. In most feedback systems—amplifier circuits in particular—the magnitude of the forward transfer function is known as the *forward gain* or *direct gain*. $G(s)$ can be a scalar if the dynamic unit merely scales the error. However, $G(s)$ is normally a complex operator that changes both the magnitude and the phase of the error.

$$V_o(s) = G(s)E(s) \qquad 61.15$$

The pick-off point transmits the output signal, V_o, from the dynamic unit back to the feedback element. The output of the dynamic unit is not reduced by the pick-off point. The transfer function of the feedback unit is the *reverse transfer function* (*feedback transfer function*, *feedback gain*, etc.), $H(s)$, which can be a simple magnitude-changing scalar or a phase-shifting function.

$$V_f(s) = H(s)V_o(s) \qquad 61.16$$

The ratio $V_f(s)/V_i(s)$ is the *feedback ratio* (*primary feedback ratio*).

$$\frac{V_f(s)}{V_i(s)} = \frac{G(s)H(s)}{1 + G(s)H(s)} \quad \text{[negative feedback]} \qquad 61.17$$

$$\frac{V_f(s)}{V_i(s)} = \frac{G(s)H(s)}{1 - G(s)H(s)} \quad \text{[positive feedback]} \qquad 61.18$$

The *loop transfer function* (*loop gain*, *open-loop gain*, or *open-loop transfer function*) is the gain after going around the loop one time, $\pm G(s)H(s)$.

The *overall transfer function* (*closed-loop transfer function*, *control ratio*, *system function*, *closed-loop gain*, etc.), $G_{\text{loop}}(s)$, is the overall transfer function of the feedback system. The quantity $1 + G(s)H(s) = 0$ is the *characteristic equation*. The *order of the system* is the largest exponent of s in the characteristic equation. (This corresponds to the highest-order derivative in the system equation.)

$$G_{\text{loop}}(s) = \frac{V_o(s)}{V_i(s)} = \frac{G(s)}{1 + G(s)H(s)} \quad \text{[negative feedback]} \qquad 61.19$$

$$G_{\text{loop}}(s) = \frac{G(s)}{1 - G(s)H(s)} \quad \text{[positive feedback]} \qquad 61.20$$

With positive feedback and $G(s)H(s)$ less than 1.0, G_{loop} will be larger than $G(s)$. This increase in gain is a characteristic of positive feedback systems. As $G(s)H(s)$ approaches 1.0, the closed-loop transfer function increases without bound, usually an undesirable effect.

In a negative feedback system, the denominator of Eq. 61.19 will be greater than 1.0. Although the closed-loop transfer function will be less than $G(s)$, there may be other desirable effects. Generally, a system with negative feedback will be less sensitive to variations in temperature, circuit component values, input signal frequency, and signal noise. Other benefits include distortion reduction, increased stability, and impedance matching.[3]

Example 61.3

A high-gain, noninverting operational amplifier has a gain of 10^6. Feedback is provided by a resistor voltage divider. What is the closed-loop gain?

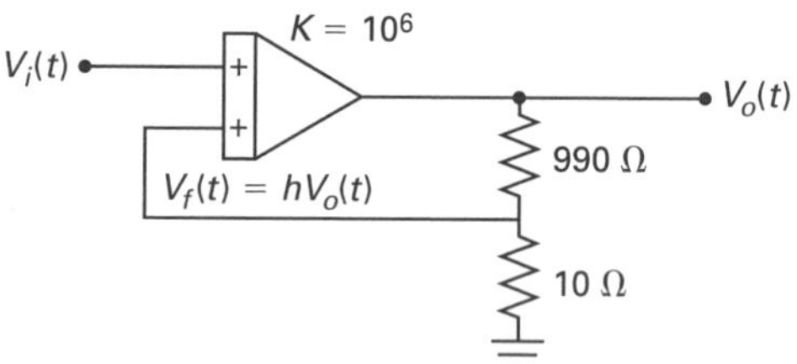

Solution

The fraction of the output signal appearing at the summing point depends on the resistances in the divider circuit.

$$h = \frac{10\ \Omega}{10\ \Omega + 990\ \Omega} = 0.01$$

[3]For circuits to be directly connected in series without affecting their performance, all input impedances must be infinite, and all output impedances must be zero.

Since the feedback path only scales the feedback signal, the feedback will be positive. From Eq. 61.19, the closed-loop gain is

$$K_{\text{loop}} = \frac{K}{1 - Kh} = \frac{10^6}{1 - (10^6)(0.01)} \approx -100$$

5. SENSITIVITY

For large loop gains ($G(s)H(s) \gg 1$) in negative feedback systems, the overall gain will be approximately $1/H(s)$. Thus, the forward gain will not be a factor, and by choice of $H(s)$, the output can be made insensitive to variations in $G(s)$.

In general, the *sensitivity* of any variable, A, with respect to changes in another parameter, B, is

$$S_B^A = \frac{d \ln A}{d \ln B} = \frac{\frac{dA}{A}}{\frac{dB}{B}} \approx \left(\frac{\Delta A}{\Delta B}\right)\left(\frac{B}{A}\right) \qquad 61.21$$

The sensitivity of the loop transfer function with respect to the forward transfer function is

$$S_{G(s)}^{G_{\text{loop}}(s)} = \left(\frac{\Delta G_{\text{loop}}(s)}{\Delta G(s)}\right)\left(\frac{G(s)}{G_{\text{loop}}(s)}\right) = \frac{1}{1 + G(s)H(s)} \quad \text{[negative feedback]} \qquad 61.22$$

$$S_{G(s)}^{G_{\text{loop}}(s)} = \frac{1}{1 - G(s)H(s)} \quad \text{[positive feedback]} \qquad 61.23$$

Example 61.4

A closed-loop gain of -100 is required from a circuit, and the output signal must not vary by more than $\pm 1\%$. An amplifier is available, but its output varies by $\pm 20\%$. How can this amplifier be used?

Solution

A closed-loop sensitivity of 0.01 is required. This means

$$\frac{\Delta V_o}{V_o} = \frac{\Delta G_{\text{loop}} V_i}{G_{\text{loop}} V_i} = \frac{\Delta G_{\text{loop}}}{G_{\text{loop}}} = 0.01$$

Similarly, the existing amplifier has a sensitivity of 20%. This means

$$\frac{\Delta G}{G} = 0.20$$

The ratio of these variations corresponds to the definition of sensitivity (see Eq. 61.21). For positive feedback,

$$\frac{\Delta G_{\text{loop}} G}{\Delta G G_{\text{loop}}} = \frac{0.01}{0.20} = \frac{1}{1 - GH}$$

Solving, $GH = -19$.

Solving for G from Eq. 61.20,

$$G_{\text{loop}} = -100 = \frac{G}{1 - GH} = \frac{G}{1 - (-19)}$$

Solving, $G = -2000$. Finally, solve for H.

$$H = \frac{GH}{G} = \frac{-19}{-2000} = 0.0095$$

6. BLOCK DIAGRAM ALGEBRA

The functions represented by several interconnected black boxes (*cascaded blocks*) can be simplified into a single block operation. Some of the most important simplification rules of block diagram algebra are shown in Fig. 61.5. Case 3 represents the standard feedback model.

Example 61.5

A complex block system is constructed from five blocks and two summing points. What is the overall transfer function?

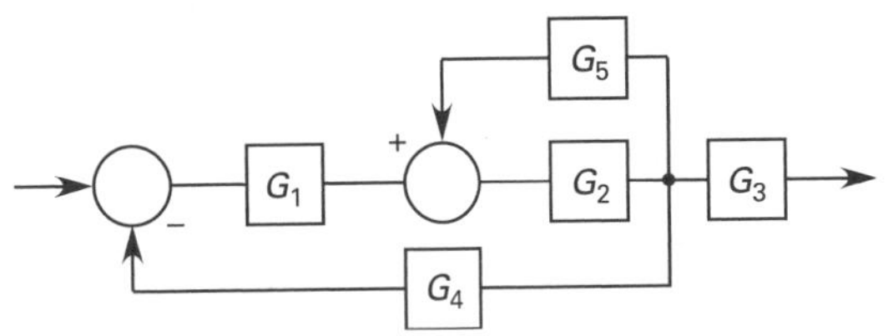

Solution

Use case 5 to move the second summing point back to the first summing point.

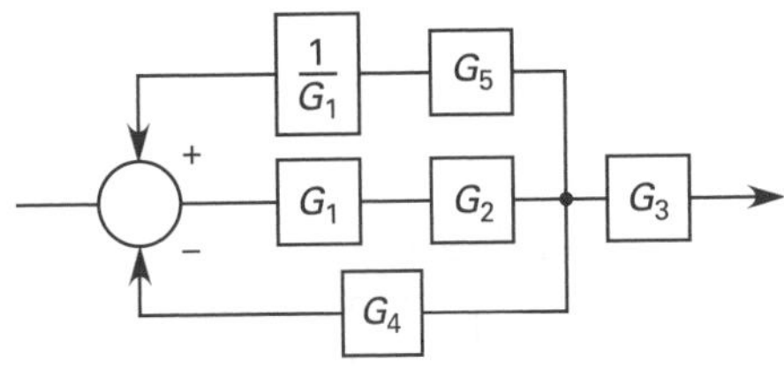

Control Systems

Figure 61.5 *Rules of Simplifying Block Diagrams*

case	original structure	equivalent structure
1		
2		
3		
4		
5		
6		
7		
8		

Use case 1 to combine boxes in series.

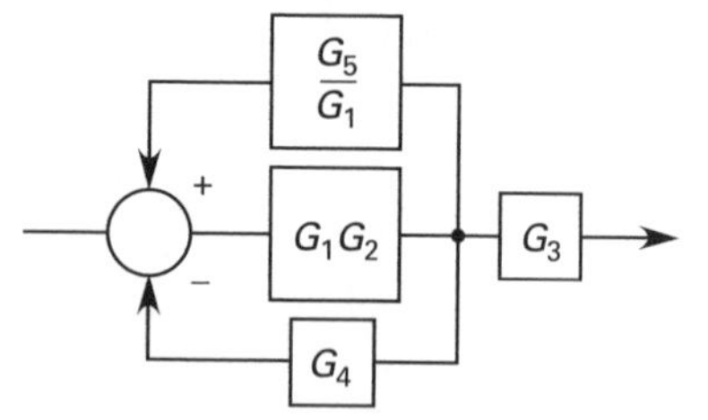

Use case 2 to combine the two feedback loops.

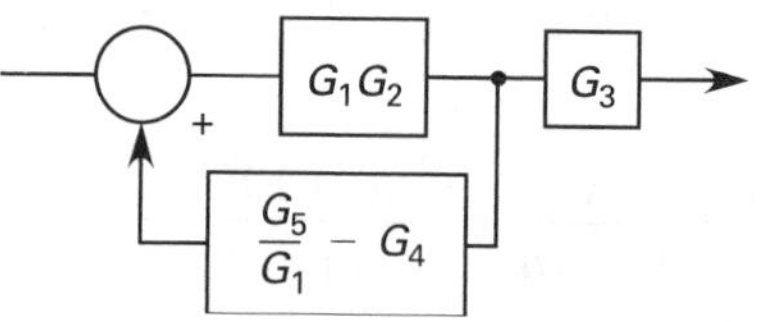

Use case 8 to move the pick-off point outside the G_3 box.

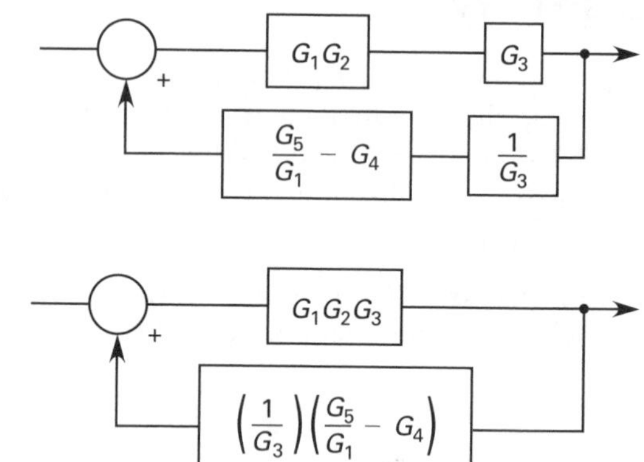

Use case 3 to determine the system gain.

$$G_{\text{loop}} = \frac{G_1G_2G_3}{1 - G_1G_2G_3\left(\frac{1}{G_3}\right)\left(\frac{G_5}{G_1} - G_4\right)}$$

$$= \frac{G_1G_2G_3}{1 - G_2G_5 + G_1G_2G_4}$$

7. PREDICTING SYSTEM TIME RESPONSE

The transfer function is derived without knowledge of the input and is insufficient to predict the time response of the system. The system time response will depend on the form of the input function. Since the transfer function is expressed in the s-domain, the forcing and response functions must be also.

$$R(s) = T(s)F(s) \qquad 61.24$$

The time-based response function, $r(t)$, is found by performing the inverse Laplace transform.

$$r(t) = \mathcal{L}^{-1}(R(s)) \qquad 61.25$$

Example 61.6

A mechanical system is acted upon by a constant force of eight units starting at $t = 0$. Determine the time-based response function, $r(t)$, if the transfer function is

$$T(s) = \frac{6}{(s+2)(s+4)}$$

Solution

The forcing function is a step of height 8 at $t = 0$. The Laplace transform of a unit step is $1/s$. Therefore, $F(s) = 8/s$.

From Eq. 61.24,

$$R(s) = T(s)F(s) = \left(\frac{6}{(s+2)(s+4)}\right)\left(\frac{8}{s}\right) = \frac{48}{s(s+2)(s+4)}$$

The response function is found from Eq. 61.25 and a table of Laplace transforms. (A product of linear terms in the denominator of $R(s)$ is equivalent to a sum of terms in $r(t)$.)

$$r(t)\mathcal{L}^{-1} = \frac{48}{s(s+2)(s+4)} = 6 - 12e^{-2t} + 6e^{-4t}$$

The last two terms are decaying exponentials, which represent the transient natural response. The first term does not vary with time; it is the steady-state response.

8. PREDICTING TIME RESPONSE FROM A RELATED RESPONSE

In some cases, it may be possible to use a known response to one input to determine the response to another input. For example, the impulse function is the derivative of the step function, so the response to an impulse is the derivative of the response to a step function.

9. INITIAL AND FINAL VALUES

The initial and final (steady-state) values of any function, $P(s)$, can be found from the *initial* and *final value theorems*, respectively, providing the limits exist. Equation 61.26 and Eq. 61.27 are particularly valuable in determining the steady-state response (substitute $R(s)$ for $P(s)$) and the steady-state error (substitute $E(s)$ for $P(s)$).

$$\lim_{t\to 0^+} p(t) = \lim_{s\to\infty} \big(sP(s)\big) \quad \text{[initial value]} \qquad 61.26$$

$$\lim_{t\to\infty} p(t) = \lim_{s\to 0} \big(sP(s)\big) \quad \text{[final value]} \qquad 61.27$$

Example 61.7

Determine the final value of the response function $r(t)$ if

$$R(s) = \frac{1}{s(s+1)}$$

Solution

From Eq. 61.27, $R(s)$ is multiplied by s and the limit taken as s tends to zero.

$$R(\infty) = \lim_{s\to 0}\left(\frac{s}{s(s+1)}\right) = \lim_{s\to 0}\left(\frac{1}{s+1}\right) = 1$$

10. SPECIAL CASES OF STEADY-STATE RESPONSE

In addition to determining the steady-state response from the final value theorem (see Sec. 61.9), the steady-state response to a specific input can be easily derived from the transfer function, $T(s)$, in a few specialized cases. For example, the steady-state response function for a system acted upon by an impulse is simply the transfer function. That is, a pulse has no long-term effect on a system.

$$R(\infty) = T(s) \quad \text{[pulse input]} \qquad 61.28$$

The steady-state response for a *step input* (often referred to as a *DC input*) is obtained by substituting 0 for s everywhere in the transfer function. (If the step has magnitude h, the steady-state response is multiplied by h.)

$$R(\infty) = T(0) \quad \text{[unit step input]} \qquad 61.29$$

The steady-state response for a sinusoidal input is obtained by substituting $j\omega_j$ for s everywhere in the transfer function, $T(s)$. The output will have the same frequency as the input. It is particularly convenient to perform sinusoidal calculations using phasor notation (as illustrated in Ex. 61.9).

$$R(\infty) = T(j\omega_f) \qquad 61.30$$

Example 61.8

What is the steady-state response of the system in Ex. 61.6 when acted upon by a step of height 8?

Solution

Substitute 0 for s in $T(s)$ and multiply by 8.

$$R(\infty) = (8)\big(T(0)\big) = (8)\left(\frac{6}{(0+2)(0+4)}\right) = 6$$

Control Systems

Example 61.9

Determine the steady-state response when a sinusoidal forcing function of 4 sin $(2t + 45°)$ is applied to a system whose transfer function is

$$T(s) = \frac{-1}{7s^2 + 7s + 1}$$

Solution

The angular frequency of the forcing function is $\omega_f = 2$ rad/s. Substitute $j2$ for s in $T(s)$, and simplify the expression by recognizing that $j^2 = -1$.

$$\begin{aligned} T(j2) &= \frac{-1}{7(j2)^2 + 7(j2) + 1} = \frac{-1}{-28 + 14j + 1} \\ &= \frac{-1}{-27 + 14j} \end{aligned}$$

Next, convert $T(j2)$ to phasor (polar) form. The magnitude and angle of the denominator are

$$\text{magnitude} = \sqrt{(14)^2 + (-27)^2} = 30.41$$

$$\text{angle} = 180° - \arctan \frac{14}{27} = 152.6°$$

However, this is the negative reciprocal of $T(j2)$.

$$T(j2) = \frac{-1}{30.41\angle 152.6°} = -0.0329\angle -152.6°$$

The forcing function expressed in phasor form is $4\angle 45°$. From Eq. 61.11, the steady-state response is

$$\begin{aligned} v(t) &= T(t)f(t) = (-0.0329\angle -152.6°)(4\angle 45°) \\ &= -0.1316\angle -107.6° \end{aligned}$$

11. POLES AND ZEROS

A *pole* is a value of s that makes a function, $P(s)$, infinite. Specifically, a pole makes the denominator of $P(s)$ zero.[4] A *zero* of the function makes the numerator of $P(s)$ (and hence $P(s)$ itself) zero. Poles and zeros need not be real or unique; they can be imaginary and repeated within a function.

A *pole-zero diagram* is a plot of poles and zeros in the *s-plane*—a rectangular coordinate system with real and imaginary axes. Zeros are represented by ○s; poles are represented as ×s. Poles off the real axis always occur in conjugate pairs known as *pole pairs*.

Sometimes it is necessary to derive the function $P(s)$ from its pole-zero diagram. This will be only partially successful since repeating identical poles and zeros are not usually indicated on the diagram. Also, scale factors (scalar constants) are not shown.

Example 61.10

Draw the pole-zero diagram for the following transfer function.

$$T(s) = \frac{5(s + 3)}{(s + 2)(s^2 + 2s + 2)}$$

Solution

The numerator is zero when $s = -3$. This is the only zero of the transfer function.

The denominator is zero when $s = -2$ and $s = -1 \pm j$. These three values are the poles of the transfer function.

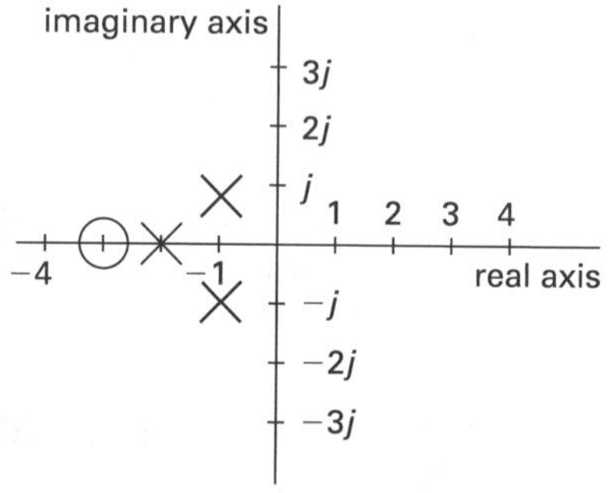

Example 61.11

A pole-zero diagram for a transfer function $T(s)$ has a single pole at $s = -2$ and a single zero at $s = -7$. What is the corresponding function?

Solution

$$T(s) = \frac{K(s + 7)}{s + 2}$$

The scale factor K must be determined by some other means.

12. PREDICTING SYSTEM TIME RESPONSE FROM RESPONSE POLE-ZERO DIAGRAMS

A response pole-zero diagram (see Fig. 61.6) based on $R(s)$ can be used to predict how the system responds to a specific input. (Note that this pole-zero diagram must be based on the product $T(s)F(s)$ since that is how $R(s)$ is calculated. Plotting the product $T(s)F(s)$ is equivalent to plotting $T(s)$ and $F(s)$ separately on the same diagram.)

[4]Pole values are the system *eigenvalues*.

Figure 61.6 *Types of Response Determined by Pole Location*

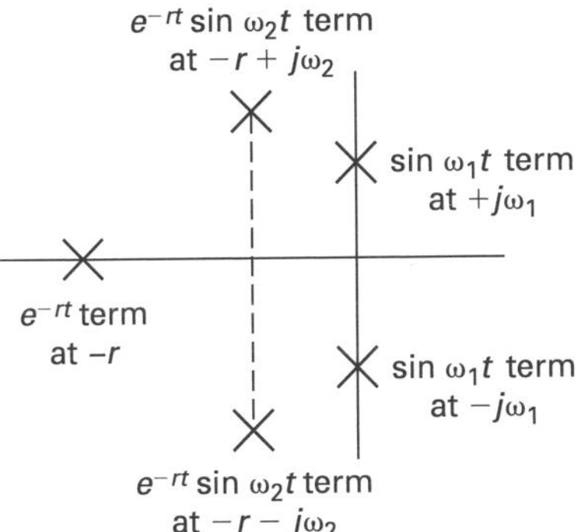

The system will experience an *exponential decay* when a single pole falls on the real axis. A pole with a value of $-r$, corresponding to the linear term $(s + r)$, will decay at the rate of e^{-rt}. The quantity $1/r$ is the decay *time constant*, the time for the response to achieve approximately 63% of its steady-state value. Thus, the farther left the point is located from the vertical imaginary axis, the faster the motion will die out.

Undamped sinusoidal oscillation will occur if a pole pair falls on the imaginary axis. A conjugate pole pair with the value of $\pm j\omega$ indicates oscillation with a natural frequency of ω rad/s.

Pole pairs to the left of the imaginary axis represent *decaying sinusoidal* response. The closer the poles are to the real (horizontal) axis, the slower will be the oscillations. The closer the poles are to the imaginary (vertical) axis, the slower will be the decay. The *natural frequency*, ω, of undamped oscillation can be determined from a *conjugate pole pair* having values of $r \pm \omega_j$.

$$\omega = \sqrt{r^2 + \omega_f^2} \qquad 61.31$$

The magnitude and phase shift can be determined for any input frequency from the pole-zero diagram (see Fig. 61.7) with the following procedure: Locate the angular frequency, ω_f, on the imaginary axis. Draw a line from each pole (i.e., a pole-line) and from each zero (i.e., a zero-line) of $T(s)$ to this point. The angle of each of these lines is the angle between it and the horizontal real axis. The overall magnitude is the product of the lengths of the zero-lines divided by the product of the lengths of the pole-lines. (The scale factor must also be included because it is not shown on the pole-zero diagram.) The phase is the sum of the pole-angles less the sum of the zero-angles.

$$|R| = \frac{K\prod_z |L_z|}{\prod_p |L_p|} \qquad 61.32$$

$$\angle R = \sum_p \alpha - \sum_z \beta \qquad 61.33$$

Figure 61.7 *Calculating Magnitude and Phase from a Pole-Zero Diagram*

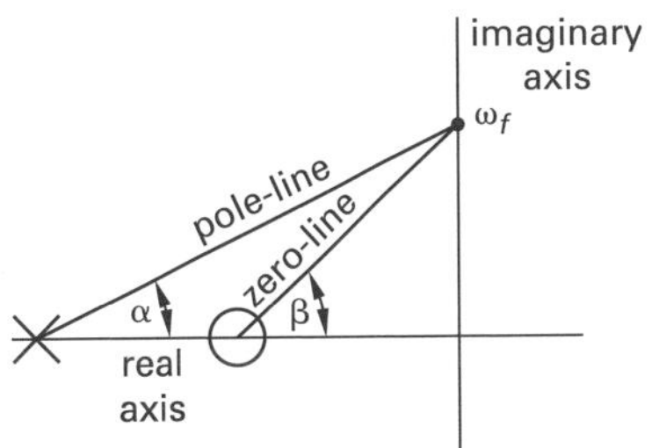

Example 61.12

What is the response of a system with the response pole-zero diagram representing $R(s) = T(s)F(s)$ shown?

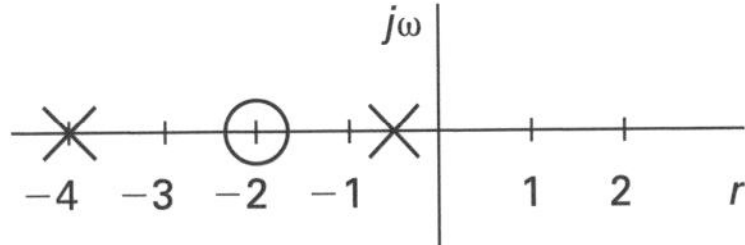

Solution

The poles are at $r = -\frac{1}{2}$ and $r = -4$. The response is

$$r(t) = C_1 e^{-\frac{1}{2}t} + C_2 e^{-4t}$$

Constants C_1 and C_2 must be found from other data.

Example 61.13

What is the system response if $T(s) = (s + 2)/(s + 3)$ and the input is a unit step?

Solution

The transform of a unit step is $1/s$. The response is

$$R(s) = T(s)F(s) = \frac{s+2}{s(s+3)}$$

The response pole-zero diagram is

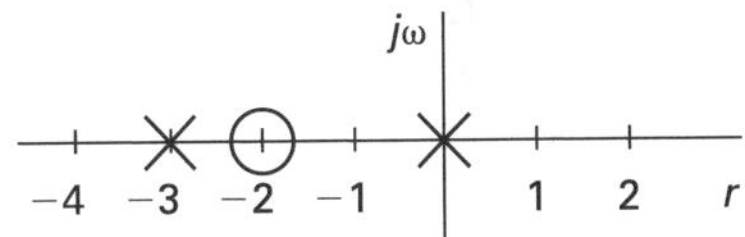

The pole at $r = 0$ contributes the exponential $C_1 e^{-0t}$ (or just C_1) to the total response. The pole at $r = -3$ contributes the term $C_2 e^{-3t}$. The total response is

$$r(t) = C_1 + C_2 e^{-3t}$$

Constants C_1 and C_2 must be found from other data.

Control Systems

13. FREQUENCY RESPONSE

The gain and phase angle frequency response of a system will change as the forcing frequency is varied. (The dependence of $r(t)$ on ω_f was illustrated in Ex. 61.9.) The *frequency response* is the variation in these parameters, always with a sinusoidal input. *Gain* and *phase characteristics* are plots of the steady-state gain and phase angle responses with a sinusoidal input versus frequency. While a linear frequency scale can be used, frequency response is almost always presented with a logarithmic frequency scale.

The steady-state gain response is expressed in decibels, while the steady-state phase angle response is expressed in degrees. The gain is calculated from Eq. 61.34,where $|T(s)|$ is the absolute value of the steady-state response.

$$\text{gain} = 20\log|T(j\omega)| \quad \text{[in dB]} \qquad 61.34$$

A doubling of $|T(j\omega)|$ is referred to as an *octave* and corresponds to a 6.02 dB increase. A tenfold increase in $|T(j\omega)|$ is a *decade* and corresponds to a 20 dB increase.

$$\begin{aligned}\text{number of octaves} &= \frac{\text{gain}_2 - \text{gain}_1}{6.02} \quad \text{[in dB]} \\ &= 3.32 \times \text{number of decades}\end{aligned} \qquad 61.35$$

$$\begin{aligned}\text{number of decades} &= \frac{\text{gain}_2 - \text{gain}_1}{20} \quad \text{[in dB]} \\ &= 0.301 \times \text{number of octaves}\end{aligned} \qquad 61.36$$

14. GAIN CHARACTERISTIC

The *gain characteristic* (M-curve for magnitude) is a plot of the gain as ω_f is varied. It is possible to make a rough sketch of the gain characteristic by calculating the gain at a few points (pole frequencies, $\omega = 0$, $\omega = \infty$, etc.). The curve will usually be asymptotic to several lines. The frequencies at which these asymptotes intersect are *corner frequencies*. The peak gain, M_p, occurs at the natural (resonant) frequency of the system.[5] Large peak gains indicate lowered stability and large overshoots. The *gain crossover point*, if any, is the frequency at which log(gain) = 0.

The *half-power points* (*cut-off frequencies*) are the frequencies for which the gain is 0.707 (i.e., $\sqrt{2}/2$) times the peak value. (This is equivalent to saying the gain is 3 dB less than the peak gain.) The *cut-off rate* is the slope of the gain characteristic in dB/octave at a half-power point. The frequency difference between the half-power points is the *bandwidth*, BW. (See Fig. 61.8.) The *closed-loop bandwidth* is the frequency range over which the closed-loop gain falls 3 dB below its value at $\omega = 0$. (The term "bandwidth" often means closed-loop bandwidth.) The *quality factor*, Q, is

$$Q = \frac{\omega_n}{\text{BW}} \qquad 61.37$$

Figure 61.8 Bandwidth

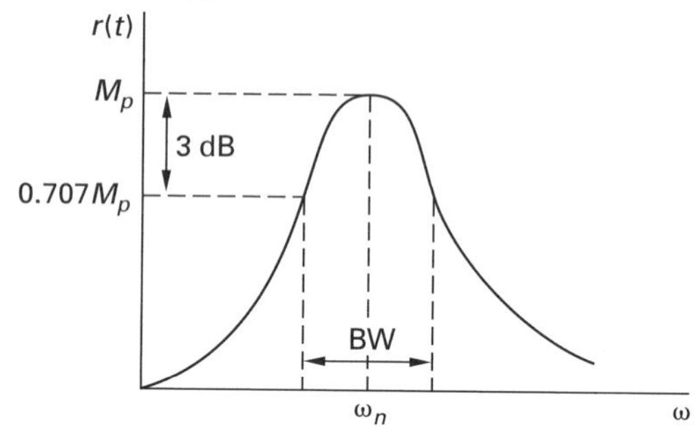

Since a low or negative gain (compared to higher parts of the curve) effectively represents attenuation, the gain characteristic can be used to distinguish between low- and high-pass filters. A low-pass filter will have a large gain at low frequencies and a small gain at high frequencies. Conversely, a high-pass filter will have a high gain at high frequencies and a low gain at low frequencies.

It may be possible to determine certain parameters (e.g., the natural frequency and bandwidth) from the transfer function directly. For example, when the denominator of $T(s)$ is a single linear term of the form $s + r$, the bandwidth will be equal to r. Thus, the bandwidth and time constant are reciprocals.

Another important case is when $T(s)$ has the form of Eq. 61.38. (Compare the form of $T(s)$ here to its form in Eq. 61.11 in Sec. 61.3.) The coefficient of the s^2 term must be 1, and ω_n must be much larger than BW so that the pole is close to the imaginary axis. The zero defined by constants a and b in the numerator is not significant.

$$T(s) = \frac{as + b}{s^2 + (\text{BW})s + \omega_n^2} \qquad 61.38$$

Example 61.14

Determine the maximum gain, bandwidth, upper half-power frequency, and half-power gain of a system whose transfer function is

$$T(s) = \frac{1}{s + 5}$$

[5]The gain characteristic peaks when the forcing frequency equals the natural frequency. It is also said that this peak corresponds to the resonant frequency. Strictly speaking, this is true, although the gain may not actually be resonant (i.e., be infinite).

Solution

The steady-state response to sinusoidal input is determined by substituting $j\omega$ for s in $T(s)$. When $\omega = 0$, $T(s)$ will have a value of $1/5 = 0.2$, and $T(s)$ decreases thereafter. The maximum gain is 0.2. The bandwidth is 5 rad/s. Since the lower half-power frequency is implicitly 0 rad/s, the upper half-power frequency is 5 rad/s, at which point the gain will be $(0.707)(0.2) = 0.141$.

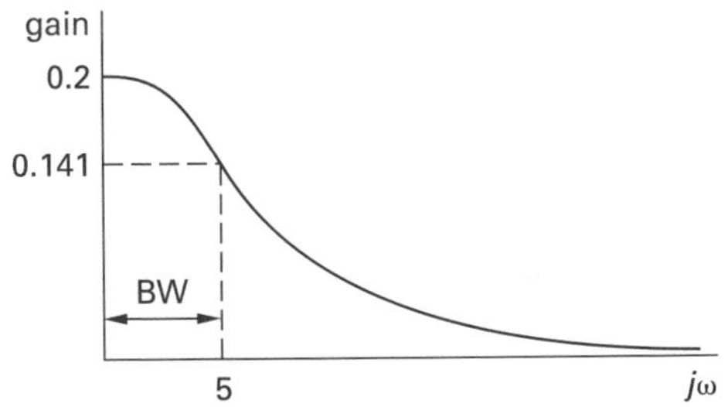

Example 61.15

Predict the natural frequency and bandwidth for the following transfer function.

$$T(s) = \frac{s + 19}{s^2 + 7s + 1000}$$

Solution

The form of this equation is the same as Eq. 61.38. The bandwidth is $\text{BW} = 7$, and the natural frequency is $\sqrt{1000} = 31.62$ rad/s.

15. PHASE CHARACTERISTIC

The phase angle response will also change as the forcing frequency is varied. The *phase characteristic* (α *curve*) is a plot of the phase angle as ω_f is varied.

16. STABILITY

A stable system will remain at rest unless disturbed by external influence, and it will return to a rest position once the disturbance is removed. A pole with a value of $-r$ on the real axis corresponds to an exponential response of e^{-rt}. Since e^{-rt} is a decaying signal, the system is stable. Similarly, a pole of $+r$ on the real axis corresponds to an exponential response of e^{rt}. Since e^{rt} increases without limit, the system is unstable.

Since any pole to the right of the imaginary axis corresponds to a positive exponential, a *stable system* will have poles only in the left half of the s-plane. If there is an isolated pole on the imaginary axis, the response is stable. However, a conjugate pole pair on the imaginary axis corresponds to a sinusoid that does not decay with time. Such a system is considered to be unstable.

Passive systems (i.e., the homogeneous case) are not acted upon by a forcing function and are always stable. In the absence of an energy source, exponential growth cannot occur. *Active systems* contain one or more energy sources and may be stable or unstable.

There are several *frequency response* (*domain*) *analysis* techniques for determining the stability of a system, including Bode plot, root-locus diagram, Routh stability criterion, Hurwitz test, and Nichols chart. The term *frequency response* almost always means the steady-state response to a sinusoidal input.

The value of the denominator of $T(s)$ is the primary factor affecting stability. When the denominator approaches zero, the system increases without bound. In the typical feedback loop, the denominator is $1 \pm GH$, which can be zero only if $|GH| = 1$. It is logical, then, that most of the methods for investigating stability (e.g., Bode plots, root-locus, Nyquist analysis, and the Nichols chart) investigate the value of the open-loop transfer function, GH. Since $\log(1) = 0$, the requirement for stability is that $\log(GH)$ must not equal 0 dB.

A negative feedback system will also become unstable if it changes to a positive feedback system, which can occur when the feedback signal is changed in phase more than 180°. Therefore, another requirement for stability is that the phase angle change must not exceed 180°.

17. BODE PLOTS

Bode plots are gain and phase characteristics for the open-loop $G(s)H(s)$ transfer function used to determine the *relative stability* of a system. (See Fig. 61.9.) The gain characteristic is a plot of $20 \log(|G(s)H(s)|)$ versus ω for a sinusoidal input. (It is important to recognize that the Bode plots, though similar in appearance to the gain and phase frequency response charts, are used to evaluate stability and do not describe the closed-loop system response.)

The *gain margin* is the number of decibels that the open-loop transfer function, $G(s)H(s)$, is below 0 dB at the *phase crossover frequency* (i.e., where the phase angle is $-180°$). (If the gain happens to be plotted on a linear scale, the gain margin is the reciprocal of the gain at the phase crossover point.) The gain margin must be positive for a stable system, and the larger it is, the more stable the system will be.

The *phase margin* is the number of degrees the phase angle is above $-180°$ at the *gain crossover point* (i.e., where the logarithmic gain is 0 dB or the actual gain is 1).

In most cases, large positive gain and phase margins will ensure a stable system. However, the margins could have been measured at other than the crossover frequencies. Therefore, a Nyquist stability plot is needed to verify the absolute stability of a system.

Figure 61.9 *Gain and Phase Margin Bode Plots*

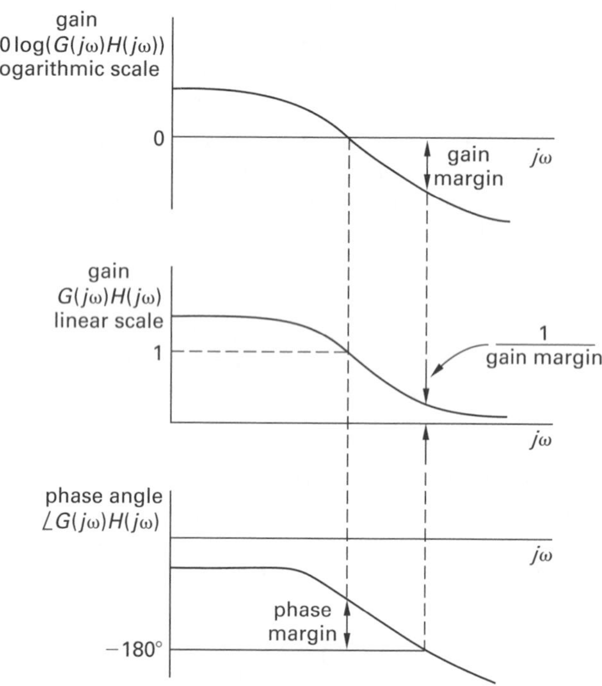

18. ROOT-LOCUS DIAGRAMS

A *root-locus diagram* is a pole-zero diagram showing how the poles of $G(s)H(s)$ move when one of the system parameters (e.g., the gain factor) in the transfer function is varied. The diagram gets its name from the need to find the roots of the denominator (i.e., the poles). The locus of points defined by the various poles is a line or curve that can be used to predict *points of instability* or other critical operating points. A point of instability is reached when the line crosses the imaginary axis into the right-hand side of the pole-zero diagram.

A root-locus curve may not be contiguous, and multiple curves will exist for different sets of roots. Sometimes the curve splits into two branches. In other cases, the curve leaves the real axis at *breakaway points* and continues on with constant or varying slopes approaching asymptotes. One branch of the curve will start at each open-loop pole and end at an open-loop zero.

Example 61.16

Draw the root-locus diagram for a feedback system with open-loop transfer function $G(s)H(s)$. K is a scalar constant that can be varied.

$$G(s)H(s) = \frac{Ks(s+1)(s+2)}{s(s+2)+K(s+1)}$$

Solution

The poles are the zeros of the denominator.

$$s_1, s_2 = -\tfrac{1}{2}(2+K) \pm \sqrt{1+\tfrac{1}{4}K^2}$$

Since the second term can be either added or subtracted, there are two roots for each value of K. Allowing K to vary from zero to infinity produces a root-locus diagram with two distinct branches. The first branch extends from the pole at the origin to the zero at $s = -1$. The second branch extends from the pole at $s = -2$ to $-\infty$. All poles and zeros are not shown. Since neither branch crosses into the right half, the system is stable for all values of K.

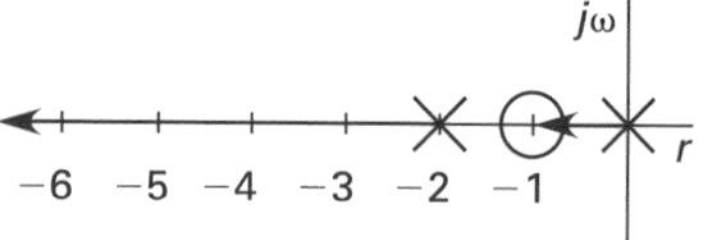

19. HURWITZ TEST

A stable system has poles only in the left half of the s-plane. These poles correspond to roots of the characteristic equation (see Sec. 61.4). The characteristic equation can be expanded into a polynomial of the form

$$a_0 s^n + a_1 s^{n-1} + \cdots + a_{n-1}s + a_n = 0 \qquad 61.39$$

The *Hurwitz stability criterion* requires that all coefficients be present and have the same sign (which is equivalent to requiring all coefficients to be positive). If the coefficients differ in sign, the system is unstable. If the coefficients are all alike in sign, the system may or may not be stable. The Routh criterion (test) should be used in that case.

20. ROUTH CRITERION

The *Routh criterion*, like the Hurwitz test, uses the coefficients of the polynomial characteristic equation. A table (the *Routh table*) of these coefficients is formed. The Routh-Hurwitz criterion states that the number of sign changes in the first column of the table equals the number of positive (unstable) roots. Therefore, the system will be stable if all entries in the first column have the same sign.

The table is organized in the following manner.

a_0	a_2	a_4	a_6	...
a_1	a_3	a_5	a_7	...
b_1	b_2	b_3	b_4	...
c_1	c_2	c_3	c_4	...
⋮	⋮	⋮	⋮	

The remaining coefficients are calculated in the following pattern until all values are zero.

$$b_1 = \frac{a_1 a_2 - a_0 a_3}{a_1} \qquad 61.40$$

$$b_2 = \frac{a_1 a_4 - a_0 a_5}{a_1} \quad 61.41$$

$$b_3 = \frac{a_1 a_6 - a_0 a_7}{a_1} \quad 61.42$$

$$c_1 = \frac{b_1 a_3 - a_1 b_2}{b_1} \quad 61.43$$

$$c_2 = \frac{b_1 a_5 - a_1 b_3}{b_1} \quad 61.44$$

Special methods are used if there is a zero in the first column but nowhere else in that row. One of the methods is to substitute a small number, represented by ϵ or δ, for the zero and calculate the remaining coefficients as usual.

Example 61.17

Evaluate the stability of a system that has a characteristic equation of

$$s^3 + 5s^2 + 6s + C = 0$$

Solution

All of the polynomial terms are present (Hurwitz criterion), so the Routh table is

s^3	1	6
s^2	5	C
s^1	$\frac{30 - C}{5}$	0
s^0	C	0

To be stable, all of the entries in the first column must be positive, which requires $0 < C < 30$.

21. NYQUIST ANALYSIS

The Nyquist analysis is a particularly useful graphical method when time delays are present in a system or when frequency response data are available. *Nyquist's stability criterion* is $N = P - Z$, where P is the number of poles in the right half of the s-plane, Z is the number of zeros in the right half of the s-plane, and N is the number of encirclements (revolutions) of $1 + G(s)H(s)$ around the critical point. N may be positive, negative, or zero. Essentially, if P is zero, then N must be zero. Otherwise, N and P must be equal for the system to be stable, which is another way of saying the number of zeros in the right half of the pole-zero diagram must be zero.

22. APPLICATION TO CONTROL SYSTEMS

A control system monitors a process and makes adjustments to maintain performance within certain acceptable limits. Feedback is implicitly a part of all control systems.[6] The *controller* (*control element*) is the part of the control system that establishes the acceptable limits of performance, usually by setting its own reference inputs. The controller transfer function for a proportional controller is a constant: $G_1(s) = K$.[7] The *plant* (*controlled system*) is the part of the system that responds to the controller. Both of these are in the forward loop. The input signal, $R(s)$, in Fig. 61.10 is known in a control system as the *command* or *reference value*. Figure 61.10 is known as a *control logic diagram* or *control logic block diagram*.

Figure 61.10 *Typical Feedback Control System*

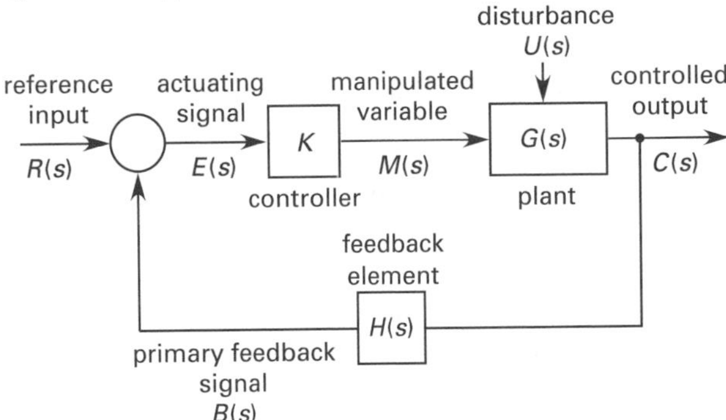

A *servomechanism* is a special type of control system in which the controlled variable is mechanical position, velocity, or acceleration. In many servomechanisms, $H(s) = 1$ (i.e., unity feedback), and it is desired to keep the output equal to the reference input (i.e., maintain a zero error function). If the input, $R(s)$, is constant, the descriptive terms *regulator* and *regulating system* are used.

23. APPLICATION TO TACHOMETER CONTROL

Tachometers are used to measure rotational speeds. Each rotation of the tachometer shaft produces one or more inductive or photoelectric pulses. The rotational speed is determined by counting and scaling the number of pulses per period.

[6]Not all controlled systems are feedback systems. The positions of many precision devices (e.g., print heads in dot-matrix printers, or cutting heads on some numerically controlled machines) are controlled by precision *stepper motors*. However, unless the device has feedback (e.g., a position sensor), it will have no way of knowing if it gets out of control.

[7]Sometimes the notation K_n is used for K, where n is the type of the system. In a *type 0 system*, a constant error signal results in a constant value of the output signal. In a *type 1 system*, a constant error signal results in a constant rate of change of the output signal. In a *type 2 system*, a constant error signal produces a constant second derivative of the output variable.

To maintain a particular rotational speed, the output of the tachometer is fed back to a control circuit. The control logic block diagram of a *constant-speed control system* is shown in Fig. 61.11. Control circuits may contain various electrical and electronic devices.[8]

The desired speed is usually set by adjusting a voltage level. This is illustrated as an input potentiometer, which feeds the desired speed into a *summing point*. The tachometer feeds the actual speed into the summing point as well, which actually computes the difference (known as the *error*, e) between the two signals.

The error is amplified from e to Ke, and the amplified signal is fed into an appropriate *control mechanism*. There are many methods, both analog and digital, by which the signal can be translated by the control mechanism into a change in motor input. Figure 61.11 only requires that the transfer function (i.e., from Ke to m) be known. The changed motor speed, n, is converted to a tachometer signal, b, which is fed back to the summing point to complete the loop.

24. STATE MODEL REPRESENTATION

While the classical methods of designing and analyzing control systems are adequate for most situations, state model representations are preferred for more complex cases, particularly those with multiple inputs and outputs or when behavior is nonlinear or varies with time.

The state variables completely define the dynamic state (position, voltage, pressure, etc.), $x_i(t)$, of the system at time t. (In simple problems, the number of state variables corresponds to the number of *degrees of freedom*, n, of the system.) The n state variables are written in matrix form as a state vector, **X**.

$$\mathbf{X} = \begin{pmatrix} x_1 \\ x_2 \\ x_3 \\ \vdots \\ x_n \end{pmatrix} \qquad 61.45$$

It is a characteristic of state models that the state vector is acted upon by a first-degree derivative operator, d/dt, to produce a differential term, $\mathbf{X}'$, of order 1.

$$\mathbf{X}' = \frac{d\mathbf{X}}{dt} \qquad 61.46$$

Equation 61.47 and Eq. 61.48 show the general form of a state model representation: **U** is an r-dimensional (i.e., an $r \times 1$ matrix) *control vector*; **Y** is an m-dimensional (i.e., an $m \times 1$ matrix) *output vector*; **A** is an $n \times n$ *system matrix*; **B** is an $n \times r$ *control matrix*; and **C** is an $m \times n$ *output matrix*. The actual unknowns are the x_i variables. The y_i variables, which may not be needed in all problems, are only linear combinations of the x_i variables. (For example, the x variables might represent spring end positions; the y variables might represent stresses in the spring. Then, $y = k\Delta x$.) Eq. 61.47 is the *state equation*, and Eq. 61.48 is the *response equation*.

$$\mathbf{X}' = \mathbf{AX} + \mathbf{BU} \qquad 61.47$$

$$\mathbf{Y} = \mathbf{CX} \qquad 61.48$$

A conventional block diagram can be modified to show the multiplicity of signals in a state model, as shown in Fig. 61.12.[9] The actual physical system does not need to be a feedback system. The form of Eq. 61.47 and Eq. 61.48 is the sole reason that a feedback diagram is appropriate.

A state variable model permits only first-degree derivatives, so additional x_i state variables are used for higher-order terms (e.g., acceleration).

System controllability exists if all of the system states can be controlled by the inputs, **U**. In state model language, system controllability means that an arbitrary initial state can be steered to an arbitrary target state in a finite amount of time. *System observability* exists if the initial system states can be predicted from knowing the inputs, **U**, and observing the outputs, **Y**.[10]

Example 61.18

Write the state variable formulation of a mechanical system's transfer function $T(s)$.

$$T(s) = \frac{7s^2 + 3s + 1}{s^3 + 4s^2 + 6s + 2}$$

Solution

Recognize the transfer function as the quotient of two terms, and multiply $T(s)$ by the dimensionless quantity x/x.

$$T(s) = \frac{Y(s)}{U(s)} = \left(\frac{7s^2 + 3s + 1}{s^3 + 4s^2 + 6s + 2}\right)\left(\frac{x}{x}\right)$$

$$Y(s) = 7s^2x + 3sx + x$$

$$U(s) = s^3x + 4s^2x + 6sx + 2x$$

$Y(s)$ and $U(s)$ represent the following differential equations.

$$y(t) = 7x''(t) + 3x'(t) + x(t)$$

$$u(t) = x'''(t) + 4x''(t) + 6x'(t) + 2x(t)$$

[8]The electronic circuit may contain low-pass filters, differential amplifiers, booster amplifiers, an adjustable voltage source, and so on. The actual integration of these devices generally requires specific product knowledge.

[9]The block $\mathbf{I}/s$ is a diagonal identity matrix with elements of $1/s$. This effectively is an integration operator.

[10]*Kalman's theorem*, based on matrix rank, is used to determine system controllability and observability.

Figure 61.11 Block Diagram for Constant-Speed Motor Control

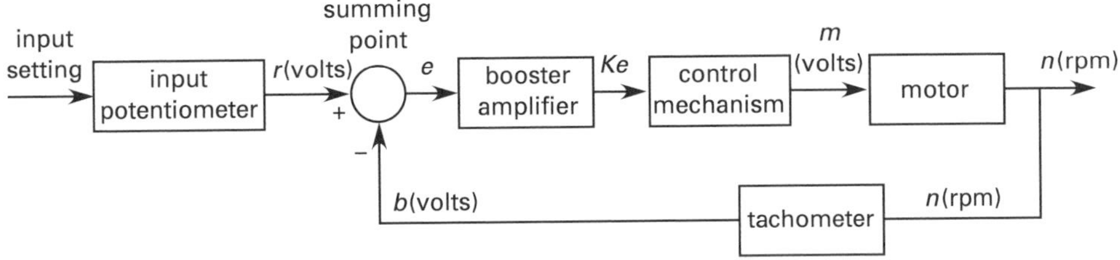

Figure 61.12 State Variable Diagram

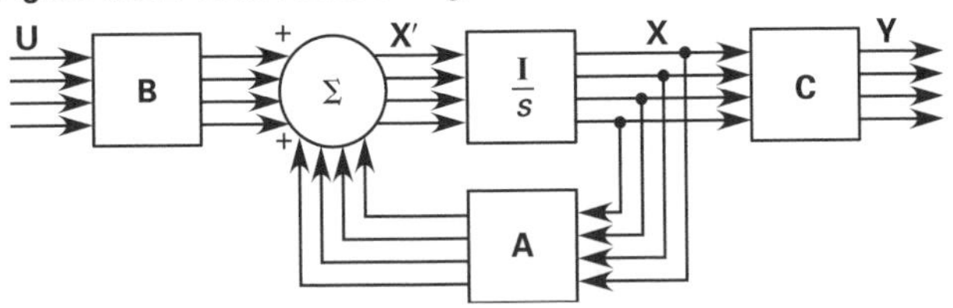

Make the following substitutions. ($x_4(t)$ is not needed because one level of differentiation is built into the state model.)

$$x_1(t) = x(t)$$
$$x_2(t) = x'(t) = x_1'(t)$$
$$x_3(t) = x''(t) = x_2'(t)$$

Write the first derivative variables in terms of the $x_i(t)$ to get the **A** matrix entries.

$$x_1' = 0x_1(t) + 1x_2(t) + 0x_3(t) + 0$$
$$x_2' = 0x_1(t) + 0x_2(t) + 1x_3(t) + 0$$
$$x_3' = -2x_1(t) - 6x_2(t) - 4x_3(t) + u(t)$$

Determine the coefficients of the **C** matrix by rewriting $y(t)$ in the same variable order.

$$y(t) = 1x_1(t) + 3x_2(t) + 7x_3(t)$$

From Eq. 61.47 and Eq. 61.48, the state variable representation of $T(s)$ is

$$\begin{bmatrix} x_1' \\ x_2' \\ x_3' \end{bmatrix} = \begin{bmatrix} 0 & 1 & 0 \\ 0 & 0 & 1 \\ -2 & -6 & -4 \end{bmatrix} \begin{bmatrix} x_1 \\ x_2 \\ x_3 \end{bmatrix} + \begin{bmatrix} 0 \\ 0 \\ 1 \end{bmatrix} [r(t)]$$

$$[y(t)] = [1 \; 3 \; 7] \begin{bmatrix} x_1 \\ x_2 \\ x_3 \end{bmatrix}$$

25. NOMENCLATURE

A	steady-state response
B	damping coefficient
BW	bandwidth
C	capacitance or constant
$e(t)$	error
$E(s)$	error, $\mathcal{L}(e(t))$
$f(t)$	forcing function
$F(s)$	forcing function, $\mathcal{L}(f(t))$
$G(s)$	forward transfer function
h	step height
$H(s)$	reverse transfer function
$i(t)$	current
$I(s)$	current, $\mathcal{L}(i(t))$
j	$\sqrt{-1}$
k	spring stiffness
K	gain
L	inductance
L	length of line
m	mass
M	fraction overshoot or magnitude
n	system type
N	Nyquist's number
$p(t)$	arbitrary function
P	number of poles
$P(s)$	arbitrary function, $\mathcal{L}(p(t))$
Q	quality factor
r	real value (root)
$r(t)$	time response
$R(s)$	response function, $\mathcal{L}(r(t))$
S	sensitivity
t	time
$T(s)$	transfer function, $\mathcal{L}(T(t))$
$T(t)$	transfer function
u	input variable
V	voltage
x	position
x	state variable
y	output variable
Z	number of zeros

Symbols

α	pole angle
β	zero angle
δ	a small number
ϵ	a small number
ζ	damping ratio
τ	time constant
ω	natural frequency

Subscripts

d	damped or delay
f	feedback or forced
i	in
n	natural
o	out
p	peak or pole
r	rise
s	settling
t	total
z	zero

Topic XII: Plant Engineering

Chapter

62 Management Science

Content in blue refers to the *NCEES Handbook*.

NCEES EXAM SPECIFICATIONS AND RELATED CONTENT

MACHINE DESIGN AND MATERIALS EXAM

I.A.3. Basic Engineering Practice: Quality assurance/quality control

12. Quality Control Charts: X-Bar Charts and R-Charts

13. Quality Control Charts

1. INTRODUCTION

Management science, also known as *quantitative business analysis*, *operations research*, and *management systems modeling*, is used to develop mathematical models of real-world situations. This chapter presents various quantitative business analysis techniques used to model and analyze manufacturing and industrial environments. Accordingly, this chapter is more concerned with solutions to problems than with explaining why the problems need to be solved or with listing advantages and disadvantages of solutions. Though they may seem to be obscure, all of the techniques presented in this chapter are commonly taught in operations research (OR), industrial engineering (IE), and MBA curricula.[1]

A *deterministic model* is a mathematical model that is built around a set of fixed rules such that any given input always results in a specific output. If an input can produce a variety of outputs determined by rules of probability, the model is known as a *probabilistic* or *stochastic model.*

A common aspect of most management science techniques is the goal of arriving at an optimum solution (regardless of whether the goal is actually realized in practice). The process of optimizing is unique to each type of problem. Calculus is not generally used in optimizing.[2] Optimizing real-world problems always requires a computer, though optimization by hand is possible with simple problems.[3]

Some management science methods attempt to optimize a specific mathematical function known as the *objective function*, Z. If Z is a profit function, it is optimized by maximization; if Z is a cost or time function, it is optimized by minimization.[4] Some management science techniques can maximize only, so in cases requiring minimization, the negative of the objective function is maximized.

[1] *Operations research* developed as a field of its own during World War II when optimizing modeling techniques were used to determine the best way for a submarine to patrol a specific region.

[2] One obvious exception is how the economic order quantity is calculated. (See Ex. 62.1.) The EOQ formula is derived by taking the derivative of the total cost function.

[3] Some management science techniques, though interesting, are too obtuse, time consuming, or complex for solving by hand. Subjects that have been omitted from or given only a mere mention in this book include nonlinear programming, dynamic programming, and integer programming. Furthermore, most management science subjects have many complicated variations that are omitted from this chapter. Simple forecasting and the economic order quantity (EOQ) model are also traditional management science subjects.

[4] There is an important difference between *cost* and *price*. Both represent an amount paid, but the distinction depends on who makes the payment and when the payment is made. To one party, the cost of materials incorporated into a manufactured item is the price paid by that party for those materials. That is, there is no difference. However, the cost to one party to acquire or produce an item is much lower than the price at which the item is later sold to a second party.

To restrict objective functions from increasing without bound, *constraints* are placed on one or more of the function's variables. These constraints are typically mathematical representations of how resources are limited or combined. Non-negativity constraints are common in mathematical programming problems.

If the objective function and its constraints are linear combinations of the independent variables, the model is said to be a *linear model*. Otherwise, the model is nonlinear.

Not all manufacturing management problems need to be solved by complex or obscure procedures. Some problems (e.g., facilities layout) do not have a general solution procedure and must be solved by exhaustive enumeration. Many problems, such as Ex. 62.1 and Ex. 62.2, can be solved simply by using common sense and logical thinking to minimize the total cost.

Example 62.1

A particular part is used by a company at a uniform rate of 120 units per month. The part is obtained from the supplier at a cost of $20 per unit. The company's cost of stocking the product is $0.06 per unit per month. The prorated cost of placing an order and putting shipments into inventory is $0.07 per unit over all normal order quantities. The company's effective monthly interest rate on borrowed money is 0.5% per month. The inventory is initially full. Orders arrive instantaneously when needed, and shortages do not occur. What is the optimum stocking quantity of the product?

Solution

This is essentially an economic order quantity (EOQ) problem. However, the ordering cost is expressed per unit ordered, and therefore, the total ordering cost is initially unknown. An iterative approach is necessary.

The interest expense on money tied up in a unit of inventory is

$$\left(\frac{0.005}{\text{month}}\right)\left(\frac{\$20}{\text{unit}}\right) = \frac{\$0.10}{\text{unit-month}}$$

Let Q be the quantity ordered. The total cost per month is

$$\begin{aligned} C_t &= \text{cost of ordering} + \text{cost of stocking} \\ &= \left(\frac{\text{no. of orders}}{\text{month}}\right)\left(\frac{\text{cost}}{\text{order}}\right) \\ &\quad + (\text{average monthly inventory}) \\ &\quad \times (\text{stocking} + \text{interest costs}) \end{aligned}$$

Initially assume that the order quantity, Q, is 100. Then, the cost of placing an order will be

$$\left(\frac{100 \text{ units}}{\text{order}}\right)\left(\frac{\$0.07}{\text{unit}}\right) = \$7.00/\text{order}$$

The inventory drops linearly from Q to zero over time, so the average inventory at any moment is $Q/2$.

$$\begin{aligned} C_t &= \left(\frac{\frac{120}{\text{month}}}{Q}\right)(\$7.00) + \left(\frac{Q}{2}\right)(\$0.06 + \$0.10) \\ &= \frac{\$840}{Q} + \$0.08Q \end{aligned}$$

This is minimized by setting the derivative of the cost function equal to zero.

$$\frac{dC_t}{dQ} = \frac{-\$840}{Q^2} + 0.08 = 0$$

$$Q = 102.5 \text{ units}$$

This is close to the initial estimate of Q. (Further iterations refine the value to exactly 105 units.)

Example 62.2

Three different semi-automatic machines are being evaluated as a replacement to a completely manual operation. Each machine is mutually exclusive, and only one machine can be selected. Each machine has a different level of automation, cost of operation, and fraction of generated defects. Which machine should be selected in each production quantity range?

machine	set-up cost	per unit material cost	per unit labor cost	fraction defective
A	$200	$0.47	$0.56	0.06
B	$700	$0.52	$0.35	0.03
C	$1200	$0.54	$0.27	0.02

Solution

When a process has a *scrap rate* greater than zero or a *yield* less than 100% (i.e., a fraction of the items produced are defective), the total cost per saleable item produced should be minimized.

$$\frac{\text{cost}}{\text{saleable item}} = \frac{\frac{\text{cost}}{\text{aggregate item}}}{1 - \text{fraction defective}}$$

Solving this example requires determining each machine's cost per saleable item as a function of production quantity. The setup cost is a *fixed cost*, allocated over all units produced. The material and labor costs are *variable costs*. The fraction defective is a scale factor that increases the cost of all saleable items produced.

Plant Engineering

Let x represent the total number of all items (saleable and nonsaleable) produced. The unit cost per saleable item is

$$C = \frac{\text{material cost} + \text{labor cost} + \frac{\text{set-up cost}}{x}}{1 - \text{fraction defective}}$$

$$C_A = \frac{\$0.47 + \$0.56 + \frac{\$200}{x}}{1 - 0.06} = \$1.10 + \frac{\$212.77}{x}$$

$$C_B = \frac{\$0.52 + \$0.35 + \frac{\$700}{x}}{1 - 0.03} = \$0.90 + \frac{\$721.65}{x}$$

$$C_C = \frac{\$0.54 + \$0.27 + \frac{\$1200}{x}}{1 - 0.02} = \$0.83 + \frac{\$1224.49}{x}$$

The three cost equations are not linear functions, and straight lines cannot be drawn between two points on the curves. Graphical or algebraic operations show that C_A is minimum for $0 < x < 2544$, C_B is minimum for $2544 < x < 7183$, and C_C is minimum for $x > 7183$.

2. CRITICAL PATH TECHNIQUES

Definitions

activity: any subdivision of a project whose execution requires time and other resources.

critical path: a path connecting all activities that have minimum or zero slack times. The critical path is the longest path through the network.

duration: the time required to perform an activity. All durations are *normal durations* unless otherwise referred to as *crash durations.*

event: the beginning or completion of an activity.

event time: actual time at which an event occurs.

float: same as slack time.

slack time: the minimum time that an activity can be delayed without causing the project to fall behind schedule. Slack time is always minimum or zero along the critical path.

Introduction

Critical path techniques are used to graphically represent the multiple relationships between stages in a complicated project. The graphical network shows the *precedence relationships* between the various activities. The graphical network can be used to control and monitor the progress, cost, and resources of a project. A critical path technique will also identify the most critical activities in the project.

Critical path techniques use *directed graphs* to represent a project. These graphs are made up of *arcs* (arrows) and *nodes* (junctions). The placement of the arcs and nodes completely specifies the precedences of the project. Durations and precedences are usually given in a *precedence table* (*precedence matrix*).

Activity-on-Node Networks: The Critical Path Method

The *critical path method*, CPM, is a deterministic method applicable when all activity durations are known in advance. CPM is usually represented as an *activity-on-node* model since arcs are used to specify precedence and the nodes represent the activities. Two *dummy nodes* taking zero time may be used to specify the start and finish of the project. Other starting and finishing events are not represented on the graph, other than as the heads and tails of the arcs.

Solving a CPM Problem

The solution to a critical path method problem reveals the earliest and latest times that an activity can be started and finished. It also identifies the *critical path* and generates the *slack times* for each activity.

The following procedure may be used to solve a CPM problem. To facilitate the solution, each node should be replaced by a square that has been quartered. The compartments have the meanings indicated by the key.

ES	EF
LS	LF

key
ES Earliest Start
EF Earliest Finish
LS Latest Start
LF Latest Finish

step 1: Place the project start time or date in the **ES** and **EF** positions of the start activity. The start time is zero for relative calculations.

step 2: Consider any unmarked activity, all of whose predecessors have been marked in the **EF** and **ES** positions. (Go to step 4 if there are no such activities.) Mark in its **ES** position the largest number marked in the **EF** position of those predecessors.

step 3: Add the activity time to the **ES** time and write this in the **EF** box. Go to step 2.

step 4: Place the value of the latest finish date in the **LS** and **LF** boxes of the finish mode.

Plant Engineering

step 5: Consider unmarked predecessors whose successors have all been marked. Their **LF** is the smallest **LS** of the successors. Go to step 7 if there are no unmarked predecessors.

step 6: The **LS** for the new node is **LF** minus its activity time. Go to step 5.

step 7: The slack time for each node is **LS** – **ES** or **LF** – **EF**.

step 8: The critical path encompasses nodes for which the slack time equals **LS** – **ES** from the start node. There may be more than one critical path.

Example 62.3

Using the precedence table given, construct the precedence matrix and draw an activity-on-node network.

activity	duration (days)	predecessors
A, start	0	–
B	7	A
C	6	A
D	3	B
E	9	B, C
F	1	D, E
G	4	C
H, finish	0	F, G

Solution

The precedence matrix is

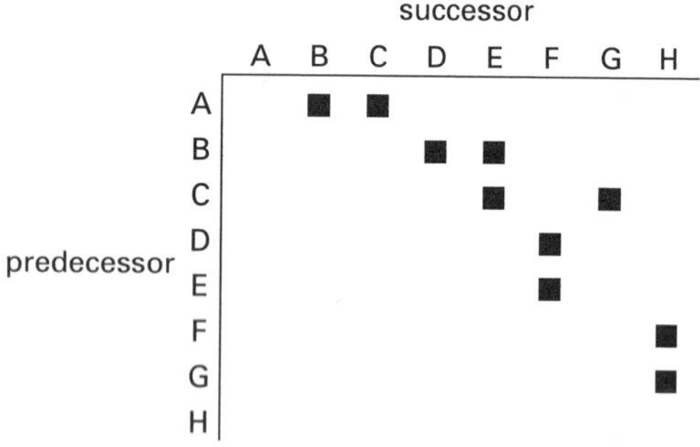

The activity-on-node network is

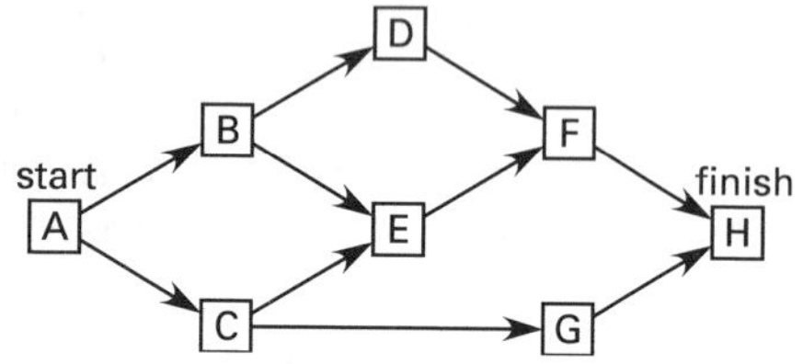

Example 62.4

Complete the network for the previous example and find the critical path. Assume the desired completion date is in 19 days.

Solution

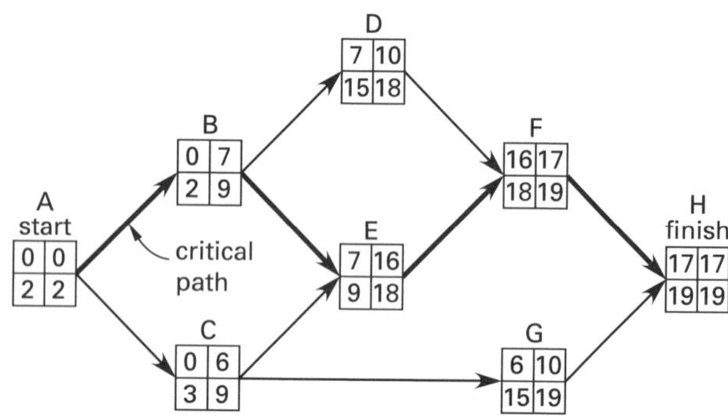

Activity-on-Branch Networks

In an *activity-on-branch* network, the arcs represent the activities, which are labeled with letters of the alphabet, and the nodes represent events, which are numbered. The activity durations may appear in parentheses near the activity letter.

The activity-on-branch method is complicated by the frequent requirement for *dummy activities* and *dummy nodes* to maintain precedence. Consider the following part of a precedence table.

activity	predecessors
L	–
M	–
N	L, M
P	M

Note that activity P depends on the completion of only activity M. Figure 62.1(a) is an activity-on-branch representation of this precedence. However, N depends on the completion of both L and M. It would be incorrect to draw the network as Fig. 62.1(b) since the activity N appears twice. To represent the project, the dummy activity X must be used, as shown in Fig. 62.1(c).

If two activities have the same starting and ending events, a dummy node is required to give one activity a uniquely identifiable completion event. This is illustrated in Fig. 62.2(b).

The solution method for an activity-on-branch problem is essentially the same as for the activity-on-node problem, requiring forward and reverse passes to determine earliest and latest dates.

Figure 62.1 Activity-on-Branch Networks

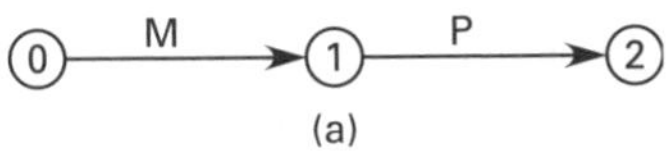

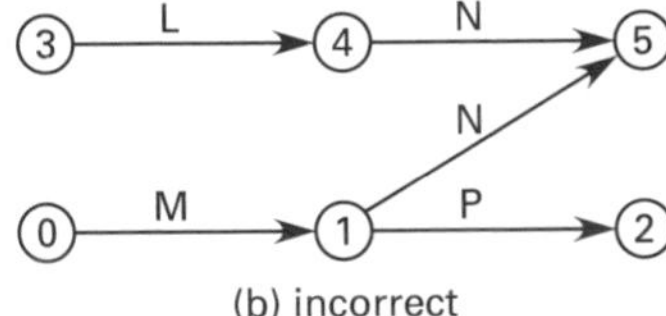

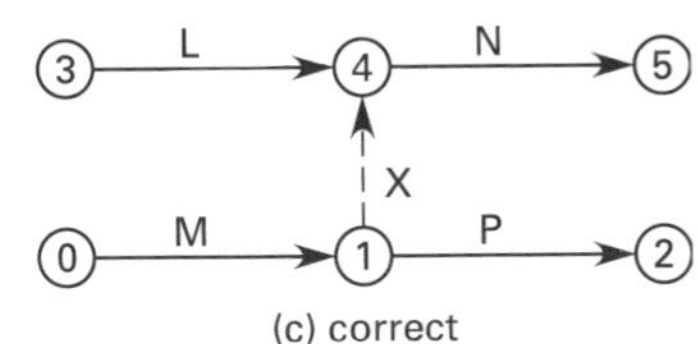

Figure 62.2 Use of a Dummy Node

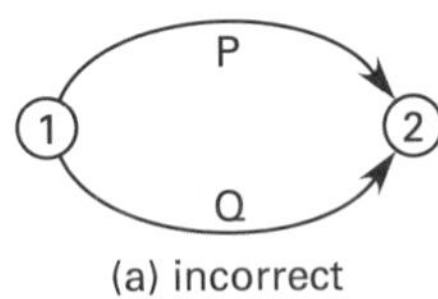

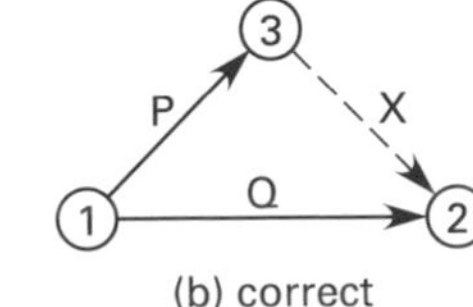

Example 62.5

Represent the project in Ex. 62.3 as an activity-on-branch network.

Solution

event	event description
0	start project
1	finish B, start D
2	finish C, start G
3	finish B and C, start E
4	finish D and E, start F
5	finish F and G

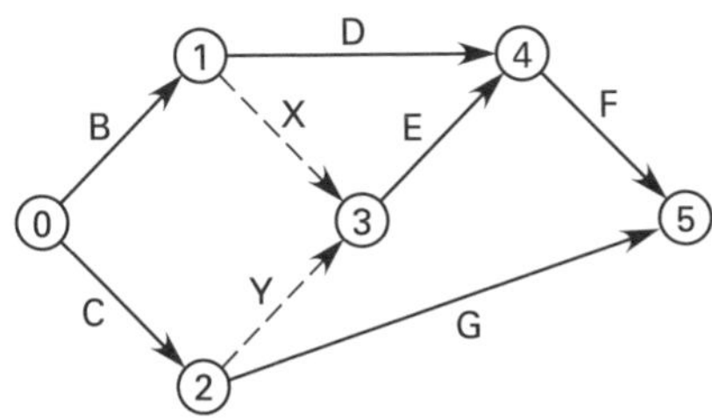

Stochastic Critical Path Models

Stochastic models differ from deterministic models only in the way in which the activity durations are found. Whereas durations are known explicitly for the deterministic model, the time for a stochastic activity is distributed as a random variable.

This stochastic nature complicates the problem greatly since the actual distribution is often unknown. Such problems are solved as a deterministic model using the mean of an assumed duration distribution as the activity duration.

The most common stochastic critical path model is PERT, which stands for *program evaluation and review technique*. In PERT, all duration variables are assumed to come from a *beta distribution*, with mean and standard deviation given by Eq. 62.1 and Eq. 62.2, respectively.

$$\mu = \tfrac{1}{6}(t_{\text{min}} + 4t_{\text{most likely}} + t_{\text{max}}) \quad \text{62.1}$$

$$\sigma = \tfrac{1}{6}(t_{\text{max}} - t_{\text{min}}) \quad \text{62.2}$$

The project *completion time* for large projects is assumed to be normally distributed with mean equal to the critical path length and with overall variance equal to the sum of the variances along the critical path.

The probability that a project duration will exceed some length, D, can be found from Eq. 62.3. z is the standard normal variable.

$$p\{\text{duration} > D\} = p\{t > z\} \quad \text{62.3}$$

$$z = \frac{D - \mu_{\text{critical path}}}{\sigma_{\text{critical path}}} \quad \text{62.4}$$

Example 62.6

The mean times and variances for activities along a PERT critical path are given. What is the probability that the project's completion time will be (a) less than 14 days, (b) more than 14 days, (c) more than 23 days, and (d) between 14 and 23 days?

activity mean time (days)	activity standard deviation (days)
9	1.3
4	0.5
7	2.6

Solution

The most likely completion time is the sum of the mean activity times.

$$\begin{aligned}\mu_{\text{critical path}} &= 9 \text{ days} + 4 \text{ days} + 7 \text{ days}\\ &= 20 \text{ days}\end{aligned}$$

The variance of the project's completion times is the sum of the variances along the critical path. Variance, σ^2, is the square of the standard deviation, σ.

$$\begin{aligned}(\sigma_{\text{critical path}})^2 &= (1.3\text{ days})^2 + (0.5\text{ days})^2 \\ &\quad +(2.6\text{ days})^2 \\ &= 8.7\text{ days}^2 \\ \sigma_{\text{critical path}} &= \sqrt{8.7\text{ days}^2} \\ &= 2.95\text{ days} \quad [\text{use 3 days}]\end{aligned}$$

The standard normal variable corresponding to 14 days is

$$\begin{aligned}z &= \frac{D-\mu_{\text{critical path}}}{\sigma_{\text{critical path}}} = \frac{14\text{ days} - 20\text{ days}}{3\text{ days}} \\ &= -2.0\end{aligned}$$

The area under the standard normal curve is 0.4772 for $0.0 < z < -2.0$. Since the normal curve is symmetrical, the negative sign is irrelevant in determining the area.

The standard normal variable corresponding to 23 days is

$$\begin{aligned}z &= \frac{D-\mu_{\text{critical path}}}{\sigma_{\text{critical path}}} = \frac{23\text{ days} - 20\text{ days}}{3\text{ days}} \\ &= 1.0\end{aligned}$$

The area under the standard normal curve is 0.3413 for $0.0 < z < 1.0$.

(a) $$\begin{aligned}p\{\text{duration} < 14\} &= p\{z < -2.0\} = 0.5 - 0.4772 \\ &= 0.0228 \quad (2.28\%)\end{aligned}$$

(b) $$\begin{aligned}p\{\text{duration} > 14\} &= p\{z > -2.0\} = 0.4772 + 0.5 \\ &= 0.9772 \quad (97.72\%)\end{aligned}$$

(c) $$\begin{aligned}p\{\text{duration} > 23\} &= p\{z > 1.0\} = 0.5 - 0.3413 \\ &= 0.1587 \quad (15.87\%)\end{aligned}$$

(d) $$\begin{aligned}p\{14 < \text{duration} < 23\} &= p\{-2.0 < z < 1.0\} \\ &= 0.4772 + 0.3413 \\ &= 0.8185 \quad (81.85\%)\end{aligned}$$

3. QUEUING THEORY

Introduction

A *queue* is a waiting line. *Queuing theory* can predict the length of a waiting line, the average time a customer can expect to spend in the queue, and the probability that n customers will be in the queue.

Different queuing models have been developed to account for differences in the distribution of arrivals, distribution of service times, number of servers, size of the calling population, and order in which the customers are served.

Most of the models are too complicated to be presented here. However, two models are given after a brief listing of the general relationships. The relationships are for *steady-state operation.* This means that the service facility has been open and in operation long enough for the queue to stabilize.

General Relationships

$$L = \lambda W \qquad 62.5$$

$$L_q = \lambda W_q \qquad 62.6$$

$$W = W_q + \frac{1}{\mu} \qquad 62.7$$

$$\lambda < \mu s \qquad 62.8$$

$$\text{average service time} = 1/\mu \qquad 62.9$$

$$\text{average time between arrivals} = 1/\lambda \qquad 62.10$$

The *M/M/1* System

The following points are assumed in the $M/M/1$ model.[5]

- There is only one server ($s = 1$).
- The calling population is infinite.
- The service times are exponentially distributed with mean service rate μ. That is, the probability of a customer's remaining service time exceeding h (after already spending time with the server) is

$$p\{t > h\} = e^{-\mu h} \qquad 62.11$$

Notice that Eq. 62.11 is independent of the time already spent with the server.

- The arrival rate is distributed as Poisson with mean λ. The probability of x customers arriving in the next period (the time period used to express μ and λ) is

$$p\{x\} = \frac{e^{-\lambda}\lambda^x}{x!} \qquad 62.12$$

The following relationships are valid for the $M/M/1$ system.

$$p\{0\} = 1 - \rho = \frac{\mu - \lambda}{\mu} \qquad 62.13$$

$$p\{n\} = p\{0\}\rho^n \qquad 62.14$$

[5]In the nomenclature of queuing theory, an exponentially distributed interarrival or service time is given the symbol M (for Markovian). Other options include D (deterministic, constant, regular, etc.), E or K (for Erlangian or gamma distribution), and G (general distribution).

$$W = \frac{1}{\mu - \lambda} = W_q + \frac{1}{\mu} = \frac{L}{\lambda} \qquad 62.15$$

$$W_q = \frac{\rho}{\mu - \lambda} = \frac{L_q}{\lambda} \qquad 62.16$$

$$L = \frac{\lambda}{\mu - \lambda} = L_q + \rho \qquad 62.17$$

$$L_q = \frac{\rho\lambda}{\mu - \lambda} \qquad 62.18$$

Example 62.7

Given an $M/M/1$ system with an average service rate of $\mu = 20$ customers per hr and an average arrival rate of $\lambda = 12$ per hr, find the steady-state values of W, W_q, L, and L_q. What is the probability that there will be five customers in the system?

Solution

$$\rho = \frac{\lambda}{\mu} = \frac{12}{20} = 0.6$$

$$W = \frac{1}{\mu - \lambda} = \frac{1}{20 - 12} = 0.125 \text{ hr}$$

$$W_q = \frac{\rho}{\mu - \lambda} = \frac{0.6}{20 - 12} = 0.075 \text{ hr}$$

$$L = \frac{\lambda}{\mu - \lambda} = \frac{12}{20 - 12} = 1.5 \text{ customers}$$

$$L_q = \frac{\rho\lambda}{\mu - \lambda} = \frac{(0.6)(12)}{20 - 12} = 0.9 \text{ customers}$$

$$p\{0\} = 1 - \rho = 1 - 0.6 = 0.4$$

$$p\{5\} = p\{0\}\rho^n = (0.4)(0.6)^5 = 0.031$$

The *M/M/s* System

The same assumptions are used for the $M/M/s$ system as were used for the $M/M/1$ system, except that there are s servers instead of only one. Each server has a mean service rate μ. All servers draw from a single line so that the first person in line goes to the first available server. Each server does not have its own line. However, if customers are allowed to change the lines they are in so that they may go to any available server, this model may also be used to predict the performance of a multiple server system where each server has its own line.

$$\rho = \frac{\lambda}{s\mu} \qquad 62.19$$

$$W = W_q + \frac{1}{\mu} \qquad 62.20$$

$$W_q = \frac{L_q}{\lambda} \qquad 62.21$$

$$L_q = \frac{p\{0\}\left(\frac{\lambda}{\mu}\right)^s \rho}{s!\,(1-\rho)^2} \qquad 62.22$$

$$L = L_q + \frac{\lambda}{\mu} \qquad 62.23$$

$$p\{0\} = \frac{1}{\dfrac{\left(\frac{\lambda}{\mu}\right)^s}{s!\,(1-\rho)} + \displaystyle\sum_{j=0}^{s-1} \frac{\left(\frac{\lambda}{\mu}\right)^j}{j!}} \qquad 62.24$$

$$p\{n\} = \frac{p\{0\}\left(\frac{\lambda}{\mu}\right)^n}{n!} \quad [n \le s] \qquad 62.25$$

$$p\{n\} = \frac{p\{0\}\left(\frac{\lambda}{\mu}\right)^n}{s!\, s^{n-s}} \quad [n > s] \qquad 62.26$$

Figure 62.3 is a graphical solution to the $M/M/s$ multiple server model.

Figure 62.3 *Mean Number in System for M/M/s System*

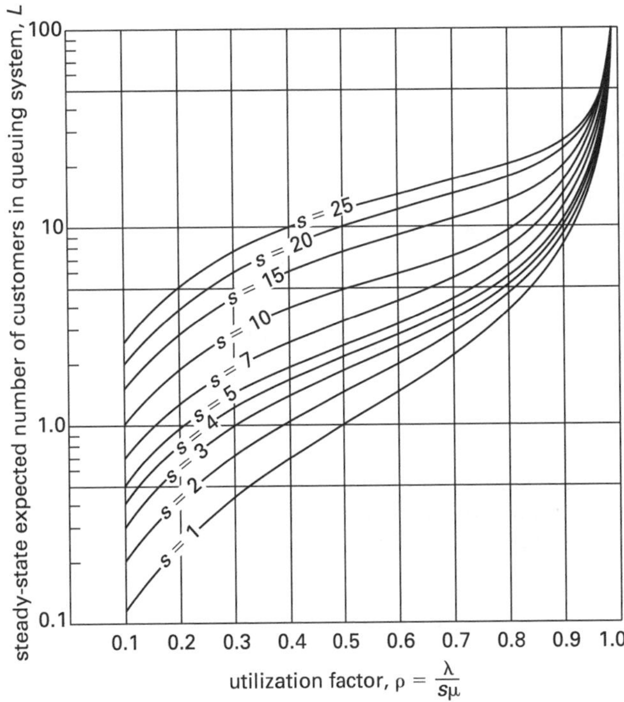

Reprinted from *Operations Research*, 6th Ed., by Frederick S. Hillier and Gerald J. Lieberman, Holden-Day, Inc., with permission from the McGraw-Hill Companies, © 1974.

Plant Engineering

Example 62.8

A company has several identical machines operating in parallel. The average breakdown rate is 0.7 machines per week. There is one repair station for the entire company. It takes a maintenance worker one entire week to repair a machine, although the time is reduced in proportion to the number of maintenance workers assigned to the repair. Each maintenance worker is paid \$400 per week. Machine downtime is valued at \$800 per week. Other costs (additional tools, etc.) are to be disregarded. What is the optimum number of maintenance workers at the repair station?

Solution

Assume the number of breakdowns per week can be represented by a Poisson distribution. Then, this example can be solved with queuing theory. Using two or more maintenance workers only decreases the repair time, so this is a single-server model, even with multiple maintenance workers.

The average number of machines breaking down each week is the mean arrival rate, $\lambda = 0.7$ per week. With one worker, the repair rate, μ, is 1.0 per week.

The average time a machine is out of service (waiting for its turn to be repaired and during the repair) is W, the "time in the system."

$$\begin{aligned} W &= \frac{1}{\mu - \lambda} = \frac{1}{1 - 0.7} \\ &= 3.33 \text{ weeks} \end{aligned}$$

With one maintenance worker, the average downtime cost in a week is

$$\begin{aligned} &\left(\frac{\text{downtime cost}}{\text{machine-week}}\right)(\text{no. of machines})(\text{no. of weeks}) \\ &\quad = (\text{downtime cost})\lambda W \\ &\quad = \left(\frac{\$800}{\text{machine-week}}\right)(0.7 \text{ machines})(3.33 \text{ weeks}) \\ &\quad = \$1865 \end{aligned}$$

However, the product of λ and W is the same as the average number of machines in the system, L. The average number of machines in the system, L (i.e., being repaired or waiting for repair), is

$$\begin{aligned} L_1 &= \frac{\lambda}{\mu - \lambda} = \frac{0.7}{1 - 0.7} \\ &= 2.33 \end{aligned}$$

The total average weekly cost with one worker is the sum of the costs of the worker and the downtime.

$$C_{t,1} = (1 \text{ worker})\left(\frac{\$400}{\text{week}}\right) + (2.33)(\$800) = \$2264$$

With two workers, the values are

$$\mu = (2 \text{ workers})\left(\frac{1}{\text{worker-week}}\right) = 2/\text{week}$$

$$L_2 = \frac{0.7}{2 - 0.7} = 0.538$$

$$C_{t,2} = (2)\left(\frac{\$400}{\text{week}}\right) + (0.538)(\$800) = \$1230$$

With three workers, the values are

$$\mu = (3 \text{ workers})\left(\frac{1}{\text{worker-week}}\right) = 3/\text{week}$$

$$L_3 = \frac{0.7}{3 - 0.7} = 0.304$$

$$C_{t,3} = (3)\left(\frac{\$400}{\text{week}}\right) + (0.304)(\$800) = \$1443$$

Adding workers will increase the costs above C_3. Two maintenance workers should staff the maintenance station as this number minimizes the weekly cost.

Example 62.9

Twenty identical machines are in operation. The hourly reliability for any one machine is 90%. (That is, the probability of a machine breaking down in any given hour is 10%.) The cost of downtime is \$5 per hour. Each broken machine requires one technician for repair, and the average repair time is one hour. If all technicians are busy, broken machines wait idle. Each technician costs \$2.50 per hour. How many separate technicians should be used?

Solution

This is a multiple-server model. It is assumed that the $M/M/s$ assumptions are satisfied. The mean arrival rate, λ, is $(0.10)(20) = 2$ per hour. The repair rate, μ, is 1 per hour. Clearly, one technician cannot handle the workload, nor can two technicians. The average number of machines in the system, L (i.e., being repaired or waiting for repair), is calculated for two, three, four, and five servers. The total cost per hour is calculated as

$$C = 2.50s + 5L$$

s	p_0	L_q	L	cost per hour (\$)
2	–	–	–	infinite
3	0.11	0.91	2.91	22.00
4	0.13	0.17	2.17	20.90
5	0.13	0.04	2.04	22.70

The minimum hourly cost is achieved with four technicians.

4. RELIABILITY

A *fault* in a machine or other system is a known cause of breakdown. An *error* is an undesired state within the machine that might lead to improper operation. A *failure* occurs when the machine fails to operate as expected or intended. A *fault-tolerant system* contains provisions to avoid failures after faults occur.

In the most common case, units fail permanently and are neither repaired nor replaced. Reliability of a single item (machine, unit, piece of equipment, etc.) is characterized by its *mean time to failure*, MTTF. The term *mean time between forced outages*, MTBFO, is used with redundant systems in place of MTTF. *Coverage* is the probability of the system reconfiguring itself when a fault occurs. *Redundancy* is the primary tool used to increase reliability and coverage. Systems with two units in parallel are known as *duplex systems*. Systems with three units are known as *triple modular redundancy*, TMR, systems.[6]

The exponential distribution is most frequently used in reliability calculations.[7] The *failure rate*, λ, is the expected number of failures per unit time. The *reliability* is

$$R_t = e^{-\lambda t} \qquad 62.27$$

The MTTF is found by integrating the reliability function. Therefore, the MTTF is the reciprocal of the failure (arrival) rate.

$$\begin{aligned} \text{MTTF} &= \int_0^\infty R_t\,dt = \int_0^\infty e^{-\lambda t}dt \\ &= \frac{1}{\lambda} \end{aligned} \qquad 62.28$$

The probability of exactly c failures in time t is given by the Poisson distribution. A cumulative probability chart can be used to find the probability of c or fewer failures in time t.

$$p\{c\} = \frac{\left(\frac{t}{\text{MTTF}}\right)^c e^{-t/\text{MTTF}}}{c!} \qquad 62.29$$

For a 1-out-of-n *fully redundant system* (a parallel system with n identical redundant units, only one of which needs to be operational for the system to operate), the reliability is

$$R_{\text{1-out-of-}n\text{ system}} = 1 - (1 - R_t)^n \qquad 62.30$$

For a 1-out-of-2 redundant system (also known as a *2-1-0 system*),

$$R_{\text{1-out-of-2}} = 2e^{-\lambda t} - e^{-2\lambda t} \qquad 62.31$$

$$\begin{aligned} \text{MTTF}_{\text{1-out-of-2}} &= \int_0^\infty R_t\,dt \\ &= \int_0^\infty 2e^{-\lambda t} - e^{-2\lambda t} \\ &= \frac{1.5}{\lambda} \end{aligned} \qquad 62.32$$

For a 1-out-of-3 fully redundant system (also known as a *3-2-1-0 system*),

$$R_{\text{1-out-of-3}} = 3e^{-\lambda t} - 3e^{-2\lambda t} + e^{-3\lambda t} \qquad 62.33$$

$$\text{MTTF}_{\text{1-out-of-3}} = \frac{11}{6\lambda} \qquad 62.34$$

In some cases, failed units are repaired online. The average repair time is the *mean time to repair*, MTTR, and is the reciprocal of the repair rate, μ. The *mean time to failure*, MTTF, is the average time a unit operates before failing. The *mean time between failures*, MTBF, is the length of time between when the original and repaired units start.

$$\text{MTBF} = \text{MTTF} + \text{MTTR} \qquad 62.35$$

$$\text{MTTR} = \frac{1}{\mu} \qquad 62.36$$

Availability is the probability that a system will be operating at any given time. The system *uptime* is calculated by multiplying the availability by the theoretically maximum number of operational hours (e.g., 8760 hours per year).[8] *Unavailability* and *downtime* are similarly calculated.

$$\text{availability} = \frac{\text{MTTF}}{\text{MTBF}} \qquad 62.37$$

$$\text{unavailability} = 1 - \text{availability} \qquad 62.38$$

$$\text{uptime} = (\text{availability})\left(8760\ \frac{\text{hr}}{\text{yr}}\right) \qquad 62.39$$

$$\text{downtime} = (\text{unavailability})\left(8760\ \frac{\text{hr}}{\text{yr}}\right) \qquad 62.40$$

Plant Engineering

[6]With TMR systems, only one unit is required for successful operation. When one unit fails, the system becomes a duplex system until the failed unit is repaired. With logic, software, electronic, and computer systems, failure can be determined by comparing the output of each of the three units. In effect, the two good units "vote" to determine which unit is faulty and should be shut down.

[7]The three-parameter *Weibull distribution* is more descriptive, flexible, and powerful than the negative exponential distribution. It has gained acceptance primarily in the aerospace industry because of its ability to model the failure distribution more exactly. Its complexity, however, makes application to noncritical applications cumbersome.

[8]If the machine does not operate 24 hours per day or 365 days per year, the number of hours will be accordingly reduced.

The MTBF values for fully redundant systems can be calculated from a Markov model of the system. (See Sec. 62.22.)

$$\text{MTBF}_{\text{1-out-of-2}} = \frac{\mu}{2\lambda^2} \quad \text{62.41}$$

$$\text{MTBF}_{\text{1-out-of-3}} = \frac{\mu^2}{3\lambda^3} \quad \text{62.42}$$

Repairing a machine as soon as it breaks down is always the preferred course of action. In some cases, however, a faulty machine can be repaired only at regular intervals. This is particularly true for unattended equipment that is inspected only at periodic intervals, often called the *proof test interval*, PTI. A complete failure will occur if the system's redundancy is not adequate to sustain multiple faults during the PTI.

It is not unexpected that the system's MTBF is a function of the PTI. A small decrease in the PTI can greatly increase the MTBF. The MTBFs, availabilities, and *average downtime*, ADT, for periodically repaired equipment operating with full redundancy is

$$\text{ADT}_{\text{1-out-of-2}} = \frac{\text{PTI}}{2} \quad \text{62.43}$$

$$\text{MTBF}_{\text{1-out-of-2}} = \frac{1}{\lambda^2(\text{PTI})} \quad \text{62.44}$$

$$\text{availability}_{\text{1-out-of-2}} = \frac{1-\lambda^2(\text{PTI})^2}{3} \quad \text{62.45}$$

$$\text{ADT}_{\text{1-out-of-3}} = \frac{\text{PTI}}{\sqrt{3}} \quad \text{62.46}$$

$$\text{MTBF}_{\text{1-out-of-3}} = \frac{1}{\lambda^3(\text{PTI})^2} \quad \text{62.47}$$

$$\text{availability}_{\text{1-out-of-3}} = \frac{1-\lambda^3(\text{PTI})^3}{4} \quad \text{62.48}$$

The *mean down time*, MDT, for a k-out-of-n system with periodic repair is

$$\text{MDT} = \frac{\text{PTI}}{n-k+2} \quad \text{62.49}$$

Example 62.10

One hundred items are tested to failure. Two failed at $t=1$, 5 at $t=2$, 7 at $t=3$, 20 at $t=4$, 35 at $t=5$, and 31 at $t=6$. Find the probability of failure in any period, the conditional probability of failure, and the mean time to failure.

Solution

elapsed time t	failures $F(t)$	survivors $S(t)$	probability of failure $0.01F(t)$	conditional probability of failure $F(t)/S(t-1)$
0	0	100	0	0
1	2	98	0.02	0.02
2	5	93	0.05	0.051
3	7	86	0.07	0.075
4	20	66	0.20	0.233
5	35	31	0.35	0.530
6	31	0	0.31	1.00

The mean time to failure is

$$\begin{aligned}\text{MTTF} &= \frac{\begin{array}{c}(2)(1)+(5)(2)+(7)(3)\\ +(20)(4)+(35)(5)+(31)(6)\end{array}}{100}\\ &= 4.74\end{aligned}$$

Example 62.11

The overall reliability of a fully redundant system must be at least 0.99 over a year's time. The system will be designed to operate as long as any one of multiple identical, parallel units is operational. The MTTF of each unit is 0.8 years. What level of redundancy is required?

Solution

For each unit,

$$\lambda = \frac{1}{\text{MTTF}} = \frac{1}{0.8\ \text{yr}} = 1.25\ 1/\text{yr}$$

$$\begin{aligned}R_{1\ \text{yr}} &= e^{-\lambda t} = e^{(-1.25\ 1/\text{yr})(1\ \text{yr})}\\ &= 0.2865\end{aligned}$$

From Eq. 62.30,

$$\begin{aligned}R_{\text{1-out-of-}n\ \text{system}} &= 1-(1-R_t)^n\\ 0.99 &= 1-(1-0.2865)^n\\ n &= 13.6\quad [\text{use } 14]\end{aligned}$$

5. PREVENTATIVE MAINTENANCE

The value of *preventative maintenance*, PM, to prevent breakdowns is undisputed.[9] However, it is not as easy to decide on the frequency and timing of PM, the size of maintenance facilities and number of staff, location and

[9]Maintenance to correct disrepair is known as *remedial maintenance*.

centralization issues, and the quantity of spares to be carried. Quantitative business analysis techniques can be used to formulate some of these PM policies.[10]

The general goal in optimizing PM policies is to minimize the total cost of operation, taking into consideration the costs of preventative maintenance, downtime, and repair. Sometimes the costs are fixed, as when specific penalties must be paid when output is not achieved. At other times, the costs are related to hourly rates and the duration of downtime. The time to failure of a machine and the times for both repair and preventative maintenance are generally not fixed, and they are not always normally distributed either. However, unless simulation is used, it is almost always necessary to work with the average times (e.g., *mean time to failure*, MTTF).

The following guidelines should be considered when establishing PM policies, particularly for single machines. When there are several identical machines operating in parallel, the problem more closely resembles a waiting-line (queuing) problem. Breakdowns are comparable to arrivals in the line, and repair stations (repair crews) are the stations. The optimum solution takes into consideration the costs of idle maintenance crews.

- PM is more applicable when the time-to-breakdown distribution has low variability because the time before a breakdown can be more accurately predicted.
- PM is only useful when its cost is less than the cost of the breakdown. In the absence of cost information, PM is useful when the average PM time is less than the average repair time.
- PM is more applicable when there is little or no inventory of the item produced by the broken machine.

6. REPLACEMENT

Introduction

Replacement and *renewal models* determine the most economical time to replace existing equipment. Replacement processes fall into two categories, depending on the life pattern of the equipment, which either deteriorates gradually (becomes obsolete or less efficient) or fails suddenly.

In the case of gradual deterioration, the solution consists of balancing the cost of new equipment against the cost of maintenance or decreased efficiency of the old equipment. Several models are available for cases with specialized assumptions, but no general solution methods exist.

In the case of sudden failure (e.g., light bulbs), the solution method consists of finding a replacement frequency that minimizes the costs of the required new items, the labor for replacement, and the expected cost of failure. The solution is made difficult by the probabilistic nature of the life spans.

Deterioration Models

The replacement decision criterion with deterioration models is the present worth of all future costs associated with each policy. Solution is by trial and error, calculating the present worth of each policy and incrementing the replacement period by one time period for each iteration.

Example 62.12

Item A is currently in use. Its maintenance cost is \$400 this year and is increasing each year by \$30. Item A can be replaced by item B at a current cost of \$3500. However, the purchase cost of B is increasing by \$50 each year. Item B has no maintenance costs. Disregarding income taxes, find the optimum replacement year. Use 10% as the interest rate.

Solution

Calculate the present worth, P, of the various policies.

policy 1: Replacement at $t=5$ (starting the 6th year)

$$\begin{aligned} P(A) &= (-\$400)(P/A,10\%,5) - (\$30)(P/G,10\%,5) \\ &= (-\$400)(3.7908) - (\$30)(6.8618) \\ &= -\$1722 \end{aligned}$$

$$\begin{aligned} P(B) &= -\big(\$3500 + (5)(\$50)\big)(P/F,10\%,5) \\ &= -\big(\$3500 + (5)(\$50)\big)(0.6209) \\ &= -\$2328 \end{aligned}$$

policy 2: Replacement at $t=6$

$$\begin{aligned} P(A) &= (-\$400)(P/A,10\%,6) - (\$30)(P/G,10\%,6) \\ &= (-\$400)(4.353) - (\$30)(9.6842) \\ &= -\$2032 \end{aligned}$$

$$\begin{aligned} P(B) &= -\big(\$3500 + (6)(\$50)\big)(P/F,10\%,6) \\ &= -\big(\$3500 + (6)(\$50)\big)(0.5645) \\ &= -\$2145 \end{aligned}$$

[10]Queuing theories are covered in Sec. 62.3. Replacement policies are covered in Sec. 62.6. Even when there are no specific quantitative techniques for solving a particular problem type, simulation can be used to determine the outcome of most preventative maintenance policies.

policy 3: Replacement at $t = 7$

$$\begin{aligned} P(A) &= (-\$400)(P/A, 10\%, 7) - (\$30)(P/G, 10\%, 7) \\ &= (-\$400)(4.8684) - (\$30)(12.7631) \\ &= -\$2330 \end{aligned}$$

$$\begin{aligned} P(B) &= -\big(\$3500 + (7)(\$50)\big)(P/F, 10\%, 7) \\ &= -\big(\$3500 + (7)(\$50)\big)(0.5132) \\ &= -\$1976 \end{aligned}$$

The present worth of A drops below the present worth of B sometime between $t = 6$ and $t = 7$. Replacement should take place at that time.

Sudden Failure Models

The time between installation and failure is not constant for members in the general equipment population. Therefore, in order to solve a sudden failure model, it is necessary to have the distribution of individual item lives (*mortality curve*). The conditional probability of failure in a small time interval, say from t to $t + \delta t$, is calculated from the mortality curve. This probability is *conditional* since it is conditioned on nonfailure up to time t.

The conditional probability of failure may decrease with time (as with *infant mortality*), remain constant (as with an exponential reliability distribution and failure from random causes), or increase with time (as with items that deteriorate with use). If the conditional probability of failure decreases or remains constant over time, operating items should never be replaced prior to failure.

It is usually assumed that all failures occur at the end of a period. The problem is to find the period that minimizes the total cost. The expression for the number of units failing in time t is

$$F(t) = n\left(\begin{array}{l} p\{t\} + \sum\limits_{i=1}^{t-1} p\{i\}p\{t-i\} + \sum\limits_{j=2}^{t-1} \\ \quad \times\left(\sum\limits_{i=1}^{j-1} p\{i\}p\{j-1\}p\{t-j\} + \cdots\right) \end{array}\right) \qquad 62.50$$

The term $np\{t\}$ gives the number of failures in time t from the original group.

The term $n\sum p\{i\}p\{t-i\}$ gives the number of failures in time t from the set of items that replaced the original items.

The third probability term times n gives the number of failures in time t from the set of items that replaced the first replacement set.

It can be shown that $F(t)$ with replacement will converge to a steady state limiting rate of

$$\overline{F(t)} = \frac{n}{\text{MTBF}} \qquad 62.51$$

The optimum policy is to replace all items in the group, including items just recently installed, when the total cost per period is minimized. That is, try to find T such that $K(T)/T$ is minimized.

$$K(T) = nC_1 + C_2\sum_{t=0}^{T-1} F(t) \qquad 62.52$$

Discounting (i.e., taking into consideration the time value of money) is usually not included in the total cost formula since the time periods are considered short. If the equipment has an unusually long life, discounting would be required.

There are some cases where group replacement is always less expensive than replacing the failures as they occur. Group replacement will be the most economical policy if Eq. 62.53 is valid.

$$C_2\big(\overline{F(t)}\big) > \left.\frac{K(T)}{T}\right|_{\min} \qquad 62.53$$

If the opposite inequality holds, group replacement may still be the optimum policy. Further analysis is then required.

7. FACILITIES LAYOUT

Facilities layout (*plant layout*) problems are numerous in variety and complexity. Laying out facilities involves locating departments and/or operations with respect to one another. In traditional *process layout*, machines with the same function are grouped together. In *product layout* (product-oriented layout), the layout depends on the sequencing of production operations. If the same equipment is used at two different times, it is duplicated in a product layout.

Some computerized methods exist for exhaustively evaluating alternatives. Manual layout techniques are even more limited. Often, paper-cutting is combined with intuition to come up with a layout. Departments are sized to a particular scale and are cut out of paper. The pieces of paper are slid around until a layout "works."

Except for the artificial case of a small number of equally sized, equally shaped departments or operations whose locations are limited to a rectangular grid, it is unlikely that all possible layouts will be considered.[11] The "optimum" layout may actually be merely the best that could be found given the amount of time available.

[11]The number of layout variations, including mirror images, with n equally sized square departments is $n!$.

An alternate manual method is to construct a graph whose "vertices" (nodes) are the departments or operations. The "edges" (line segments) are drawn between two vertices if adjacent associated departments are desired. The edges may be weighted to indicate the level of traffic between the departments. The goal is to rearrange the vertices so that no edges cross. If this can be done, then the layout can be planar. If the departments are somewhat flexible in terms of size and shape, it is possible to have the desired adjacencies.

Certain simplifying assumptions are usually made with both computerized and manual methods. For example, all layouts may be required to be two dimensional. Departments may be assumed to be square or rectangular. When the locations of specific pieces of equipment within the department are unknown, it is assumed that all movement into and out of the department originates and terminates at the centroid of the departmental area. Also, only highly repetitive movements between departments are considered. Once-in-a-while travel is excluded from the analysis.

Almost all facility layout procedures—manual and computerized, exact, trial-and-error, and heuristic—attempt to minimize the transportation cost, sometimes referred to as *movement*.[12] In simple cases, this may mean minimizing the product of trips between departments and the distances between their centroids. In more complex cases, the product of trips and distances may also be multiplied by volumes, weights, and labor rates.

Nonquantitative factors also need to be considered. Sometimes, as when equipment, records, or personnel are shared, it is absolutely necessary that departments be located next to each other. In other cases, as when safety is compromised, it may be absolutely essential to separate departments. In most cases, the *nearness priorities* characterize the adjacency requirements between being absolutely necessary and being absolutely undesirable.[13] The ways that nonquantitative factors are presented and incorporated into the solution vary from case to case.

8. WORK MEASUREMENT

Work measurement determines how long it takes to complete an operation. The *standard time* (also known as the *product standard* or *production standard*) for an operation is the time required by a qualified and trained worker working at a normal pace. Standard times can be determined in a variety of ways, including by statistical analysis of actual measurements (*time studies*), by analysis of historical data, or by analysis of the physical motions associated with the process (i.e., *predetermined systems*). *Learning curves* can be incorporated where appropriate.

Standard times are used to develop production schedules (*factory loading*), estimate costs and output, determine worker effectiveness, and establish the basis for incentive labor plans. Standard times are combined with actual labor rates to develop *standard costs* of products.

Time standards are usually not developed directly for large operations (e.g., building a television). For various reasons (including reduction of variance), standards are developed for each significant *element* (also known as a *task*) of the operation (e.g., tightening a cover screw or soldering a resistor to the printed circuit board). Such task data can then be used to determine standard times for other products with identical tasks.

In determining the tasks requiring time standards, it is normal to separate fixed and variable tasks. For example, setting up the workplace at the beginning of each shift is a fixed task that would be studied separately. Applying adhesive strips to the product on an assembly line is a variable task. It is also common to separate handling and material transport tasks from processing tasks.

Actual time measurements during normal operation can be taken by hand with a stopwatch or derived from the analysis of video recordings.[14] Operations can be continuously observed, or measurements can be taken on a random basis.[15] To establish a truly representative average, measurements should be taken from as many individuals as possible.

The raw average time for an operation is not used as a production standard for two reasons. First, different people work at different levels of effort, and workers do not always work at a normal pace when being observed. This is taken into consideration by the observer who applies a subjective *performance factor* (*performance rating*) to the raw data.[16] Expressing the performance factor as a decimal, the *normal time* is

$$t_{\text{normal}} = t_{\text{ave}}(\text{performance factor}) \quad 62.54$$

Second, *time allowances* must be added for personal time, normal delays, defective material, and fatigue.[17] Such allowances, usually expressed as percentages, depend greatly on the nature of the operation.

[12]A *trial-and-error method* depends on insight, intuition, and ingenuity to come up with a solution. A *heuristic method* follows a procedure and/or uses rules of thumb to derive an answer. Neither is an optimizing technique.

[13]The *Muther nearness priorities* (developed by Richard Muther in the 1950s) are (1) absolutely necessary, (2) very important, (3) important, (4) OK (ordinary importance), (5) unimportant, and (6) undesirable.

[14]The stopwatch is used in two different ways. In *snapback recording*, the stopwatch is zeroed out after each measurement. In *continuous recording*, the elapsed times are recorded. The interval times are determined later by subtraction.

[15]Though the sampling is random, the random schedule is established in advance with the aid of random numbers. The investigator follows a predetermined observation schedule that is unknown to the worker being observed.

[16]In theory, either the standard time or the performance rating must be known. However, the investigator traditionally observes the time and rates the performance simultaneously.

[17]A minimum of 5% should be added for personal time. Allowances for normal delays are determined by observation. Allowances for fatigue are highly subjective and depend on the nature of the operation.

Allowances for hard manual labor are much larger than allowances for sitting workers. Expressing the performance factor and allowance as decimals, the standard time is[18]

$$t_{std} = \frac{t_{ave}(\text{performance factor})}{1 - \text{allowance}} = \frac{t_{normal}}{1 - \text{allowance}} \quad 62.55$$

The variability of time study data is commonly expressed in terms of a *coefficient of variation*. This is simply the percentage variation calculated from Eq. 62.56. Notice that the sample standard deviation, s_t, of the time measurements is used.

$$\text{coefficient of variation} = \left(\frac{s_t}{t_{ave}}\right) \times 100\% \quad 62.56$$

$$s_t = \sqrt{\frac{\sum t^2 - \frac{\left(\sum t\right)^2}{N}}{N-1}} \quad 62.57$$

The number of observations is determined from the desired precision of the final answer and the observed variability of the times measured. The more variable the times and the greater the required *precision* ($\pm P$), the more observations will be required to achieve a desired *confidence level* (C).[19,20] The most common confidence level is 95%, although 90% and 99% are also used.[21] A 95% confidence level (i.e., a 0.95 *confidence coefficient*) means that the probability is 95% that the desired precision will be met.

Equation 62.58 can be used for estimating the sample size required. If N is already greater than n, no further sampling is required.

$$N = \left(\frac{t_C s_t}{P t_{ave}}\right)^2 \quad 62.58$$

The actual maximum inaccuracy (error), ϵ, is calculated from Eq. 62.59, which uses a two-tail distribution factor, t_C, from Table 62.1 or App. 11.C.[22]

$$\epsilon = P t_{ave} = t_C s_t \quad 62.59$$

$$P = \frac{\epsilon}{t_{ave}} = \frac{t_C s_t}{t_{ave}} \quad 62.60$$

***Table 62.1** Two-Tail Distribution Factors**

confidence level, C	z_C or t_C (∞)
90%	1.645
95%	1.960
97.5%	2.240
98%	2.326
99%	2.476
99.9%	3.270

*Standard normal distribution is equivalent to a Student's t-distribution with infinite degrees of freedom.

Time standards can also be developed from systems of *standard data*, also known as *universal standard data* and *synthetic standard data*.[23] Standard data systems are particularly valuable when there is no operation to observe, as when the product is still at the blueprint stage. An analyst (who is familiar with the steps and equipment necessary to complete an operation) subdivides the operation into tasks and the tasks into even finer micromotions.

For example, the task of moving a small object from a conveyor belt to a workstation might be subdivided into the micromotions of eye focus, reach, pickup and grasp, move, position, and release. Times for these micromotions are taken from extensive tables of standard data. Such tables may list time in minutes or in TMUs (*time measurement units*) equal to 0.0006 minutes.

Example 62.13

A highly experienced worker was repeatedly observed in a time study. The average time taken by the worker to complete a task was 0.80 min. The observing analyst gave the worker a performance rating of 120%. The standard allowance is 10%. What is the standard time for the task?

[18]It is also common to simply add the allowance percentage, though the two methods are not equivalent. The method used is usually a matter of company policy.

$t_{std} = t_{normal}(1 + \text{allowance})$

[19]The *confidence interval* and *confidence level* are different. The confidence interval is twice the precision, expressed in absolute terms. If a ±2% precision is required for a 1.0 minute operation, the confidence interval will be 0.04 minute.

[20]The precision is sometimes called the *required accuracy*, even though a required accuracy of ±5% is obviously inappropriate. "Required maximum inaccuracy" is more descriptive of the actual intent of the term.

[21]The actual value used will be a matter of company policy.

[22]When the standard deviation of the underlying population is not known and must be estimated from the sample, as is usually the case, *Student's t-distribution* should be used in place of the normal distribution. This level of sophistication is particularly necessary when the number of observations is small. Values of t_C are determined from App. 11.C. When the *actual* error is being calculated from n actual observations, t_C should be determined with $n-1$ degrees of freedom. When the required number of observations, N, is being calculated, and the degrees of freedom is unknown (i.e., $N-1$ is unknown) or expected to be large, the degrees of freedom is assumed to be infinite. Values from Student's t-distribution for infinite degrees of freedom are identical to z-values from the normal table. (See Table 62.1.)

[23]*Methods-Time-Measurement* (MTM), *Work Factor*, and MODAPTS are commercial systems of predetermined standard data systems.

Solution

From Eq. 62.55, the standard time is

$$t_{\text{std}} = \frac{t_{\text{ave}}(\text{performance factor})}{1 - \text{allowance}} = \frac{(0.80\ \text{min})(1.2)}{1 - 0.10}$$
$$= 1.07\ \text{min}$$

Example 62.14

The estimate of an operation's time is to be obtained from 10 measurements. (a) Given a required confidence level of 90%, what is the precision of the estimate? (b) Given a required precision of 5%, what is the confidence level? (c) How many additional observations are required to achieve the 5% precision with a 90% confidence level?

times: 0.32, 0.35, 0.34, 0.36, 0.38, 0.40, 0.40, 0.36, 0.31, 0.38

Solution

$$\sum t = 0.32 + 0.35 + 0.34 + 0.36 + 0.38 + 0.40 + 0.40 + 0.36 + 0.31 + 0.38$$
$$= 3.6$$

$$\sum t^2 = (0.32)^2 + (0.35)^2 + (0.34)^2 + (0.36)^2 + (0.38)^2 + (0.40)^2 + (0.40)^2 + (0.36)^2 + (0.31)^2 + (0.38)^2$$
$$= 1.3046$$

(a) The average time and sample standard deviation are

$$t_{\text{ave}} = \frac{\sum t}{N} = \frac{3.60}{10} = 0.36$$

From Eq. 62.57, the sample standard deviation is

$$s_t = \sqrt{\frac{\sum t^2 - \frac{\left(\sum t\right)^2}{N}}{N-1}} = \sqrt{\frac{1.3046 - \frac{(3.6)^2}{10}}{10 - 1}}$$
$$= 0.03091$$

Since the number of samples is less than 20 (or 25 or 30), the normal curve should not be used. From App. 11.C, the t_C factor for 90% two-tailed confidence limits with $n-1 = 10-1 = 9$ degrees of freedom is 1.83. The maximum error (with 90% confidence) is

$$\epsilon = t_C s_t = (1.83)(0.03091)$$
$$= 0.0566$$

The precision is

$$P = \frac{\epsilon}{t_{\text{ave}}} = \frac{0.0566}{0.36}$$
$$= 0.157 \quad (\pm 15.7\%)$$

(b) With a precision of $\pm P$, the maximum error is

$$\epsilon = Pt_{\text{ave}} = (0.05)(0.36) = 0.018$$

The value of t_C is

$$t_C = \frac{\epsilon}{s_t} = \frac{0.018}{0.03091}$$
$$= 0.582$$

From App. 11.C with $n - 1 = 9$ degrees of freedom, the value of 0.582 can be interpolated between the 40% and 50% (two-tail) columns. The confidence level is between 40% and 50%.

(c) From Eq. 62.58, the total number of observations required is

$$N = \left(\frac{t_C s_t}{Pt_{\text{ave}}}\right)^2$$
$$= \left(\frac{(1.83)(0.03091)}{(0.05)(0.36)}\right)^2$$
$$= 9.9 \quad \text{[use 10 observations]}$$

In this case, no additional observations are required.

9. WORK SAMPLING

Work sampling is a method of achieving the results of a stopwatch study without using a stopwatch.[24] It directly determines the fraction of time spent in a particular operation. Work sampling can be used to indirectly establish time standards and to rate performance.

The basic procedure is to observe a worker at random times over a specific time period (e.g., 8 or 40 hours) and to simply note whether or not the worker was performing a given task. The fraction of observations is

[24]Work sampling is particularly useful in determining the fraction of time that a worker or group of workers is idle.

identical to the fraction of time the worker spends on the task. This fraction is combined with the time period to determine the average time for the task.

$$t_{\text{std}} = \frac{(\text{fixed time period})(\text{fraction observed}) \times (\text{performance factor})}{(1 - \text{allowance})(\text{no. of pieces produced})} \qquad 62.61$$

The accuracy of the average depends on the total number of observations, N. It is a fundamental principle in work sampling studies that the total required number of observations is proportional to the amount of time spent on the given task. Since the time is initially unknown, an estimate is made and the number of observations is iteratively refined as observations are taken. The observed fraction of observations of the task is

$$\overline{p} = \frac{n}{N} = \frac{\text{no. of observations of task}}{\text{total no. of observations}} \qquad 62.62$$

The sample standard deviation of the observed fraction is the standard deviation of the binomial distribution.

$$s_p = \sqrt{\frac{\overline{p}(1-\overline{p})}{N}} \qquad 62.63$$

It is important not to confuse the maximum fractional error, commonly referred to as the *precision*, with the maximum absolute error. Since $\overline{p}$ varies from 0 to 1, the fractional and absolute errors are both less than 1.00. The maximum actual inaccuracy is

$$\epsilon = P\,\overline{p} = t_C s_p \qquad 62.64$$

The total number of observations required to achieve the desired precision and confidence is

$$N = \left(\frac{t_C \overline{p}(1-\overline{p})}{\epsilon}\right)^2 = t_C^2\left(\frac{1-\overline{p}}{P^2\overline{p}}\right) \qquad 62.65$$

Example 62.15

Work sampling is to be used to determine the fraction of time that a worker is idle. The initial estimate of idleness is 12%. The required precision is ±10%. The required confidence level is 95%. (a) Determine the number of observations that should be initially scheduled. (b) Assuming the fraction of time idle is 12%, with the given confidence level, what is the range of values of the fraction of idleness?

Solution

(a) The initial estimate of the fraction is $\overline{p} = 0.12$, and the required precision, P, is 0.10. Assuming a large number of samples, the two-tail normal curve parameter for a 95% confidence level is found in Table 62.1 to be 1.960.

From Eq. 62.65, the number of observations is

$$\begin{aligned} N &= \frac{t_C^2(1-\overline{p})}{P^2\overline{p}} \\ &= \frac{(1.960)^2(1.00-0.12)}{(0.10)^2(0.12)} \\ &= 2817 \end{aligned}$$

(b) With a precision (i.e., fractional error) of 10%, the absolute error is

$$\epsilon = P\overline{p} = (0.10)(0.12) = 0.012$$

The range of values will be

$$\begin{aligned} \overline{p} \pm \epsilon &= 0.12 \pm 0.012 \\ &= 0.132,\ 0.108 \end{aligned}$$

10. ASSEMBLY LINE BALANCING

Line balancing determines which tasks will be performed progressively at multiple assembly stations. Some tasks must precede others; some tasks can be performed at any point in the assembly; some tasks (e.g., installation of fasteners and final tightening) can be split between stations. Line balancing can also determine how many stations are needed and which tasks will be performed in parallel to increase throughput.

The cumulative durations of all the tasks at a particular station cannot exceed the cycle time, which is the reciprocal of the *production rate*. Generally, the *cycle time* is known.[25]

$$\text{cycle time} = \frac{1}{\text{production rate}} \qquad 62.66$$

Line balancing establishes the capacity of the line. Work cannot pass down the line any faster than it can pass through the *bottleneck* of the line (i.e., the station with the longest set of tasks). Therefore, the longest set of tasks establishes the minimum cycle time. The *target cycle time* is the ideal case and can be calculated if the number of stations is known.

$$\text{target cycle time} = \frac{\sum(\text{all task times})}{\text{no. of stations}} \qquad 62.67$$

[25]The problem is much more difficult when the cycle time is not known. Not only do all arrangements of tasks need to be evaluated, but also the arrangements need to be evaluated over all reasonable cycle times.

As a management science problem, the simultaneous goals are to maximize the throughput of the line, minimize the cumulative idle time of all the stations, minimize the total labor cost, and minimize the initial investment in station equipment. In most initial studies, only the goal of minimizing idle time is considered. The total idle time is

$$\text{idle time} = (\text{no. of stations})(\text{cycle time}) - \sum(\text{all task times}) \qquad 62.68$$

The percentage *delay* (also known as *balance delay*) is defined by Eq. 62.69. The best arrangement will have the lowest balance delay. However, an optimal solution does not require a zero balance delay. In fact, a zero balance delay may not be possible, as when the sum of task times is not an integer multiple of the cycle time. Optimality is still assured if the number of stations is the smallest integer greater than the sum of task times divided by the cycle time.

$$\text{balance delay} = \frac{\text{idle time} \times 100\%}{(\text{no. of stations})(\text{cycle time})} \qquad 62.69$$

Trial-and-error methods of balancing lines by hand are of the shuffle-and-reshuffle variety. Heuristic methods also require considerable personal ingenuity. The number of possibilities is often very large, and therefore, modern linear and dynamic programming models, as well as techniques based on exhaustive enumeration, are solved with computers. Regardless of the method used, solutions are often not unique.

The following three different heuristic methods are used to balance lines.

- The task with the largest time (that will fit in the station's available time) is assigned to that station.
- The task with the most successors (that will fit in the station's available time) is assigned to that station.
- The task with the greatest sum of task times of its successor tasks (that will fit in the station's available time) is assigned to that station.

Standard times (see Sec. 62.8) to perform the various station tasks are used in the balancing process. However, except for robot operators and machine-controlled lines, task times are actually random. Sometimes tasks take longer; sometimes they take less time. This implies that almost every line will be unavoidably unbalanced. If the work is rigidly paced to the cycle time (as it would be if the work were permanently attached to the conveyance system), some work pieces might be left unfinished if the previous part required more time than usual. This scenario introduces what is probably one of the most important requirements for maximizing the line throughput: station inventory.

Line throughput will be maximized if each station has a backlog of unfinished work.[26] In this case, the conveyance system is used merely to bring work to stations rather than to pace the stations. If a station is busy when a new piece of work arrives, the station operator merely places that piece into his or her inventory of unfinished work. If all tasks are finished early, before a new piece of work arrives, the operator begins on a piece from the inventory. The station is never idle.

Example 62.16

(a) Design a heuristically balanced line for a cycle time of 10. (b) Calculate the idle time. (c) Calculate the balance delay. (d) What would be the effect of decreasing the cycle time to 9?

task	predecessors	time
start	none	0
A	start	7
B	start	4
C	B	5
D	A, C	2
E	C	8
F	D, E	4
finish	F	0

Solution

(a) First, draw a simple precedence diagram to visualize the precedences.

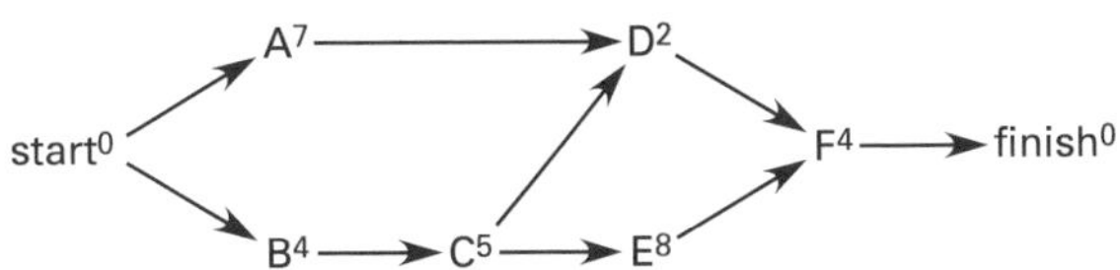

Station 1: The task with the largest time without predecessors will be assigned. (The "start" task, used merely for consistency in drawing directed graphs, is disregarded.) Task A is the largest and is assigned to station 1. That takes up 7 of the cycle time, which leaves 10 − 7 = 3 still available. B is too long, and C's predecessor is incomplete. D's predecessor, C, is incomplete. No tasks can be added to fill the remaining time.

[26]Measurable increases in line output have been reported by selectively unbalancing the line and ensuring that each station has a backlog of work.

Plant Engineering

Station 2: Task A is taken; tasks B and D are the only possibilities. Task B, with the larger time of 4, is assigned to station 2. Continuing down the list, tasks C and D can be performed. Task C has a larger task time than D. Assigning task C to station 2 increases the station time to a total of 9. No further tasks can be added.

Station 3: Tasks A, B, and C have been taken. Tasks D and E are available and both are assigned to station 3.

Station 4: Tasks A, B, C, D, and E have been taken. Task F is assigned to station 4.

The heuristic solution is

Station 1:	A	(busy 7; idle 3)
Station 2:	B, C	(busy 9; idle 1)
Station 3:	D, E	(busy 10; idle 0)
Station 4:	F	(busy 4; idle 6)

(b) The sum of task times is

$$7+4+5+2+8+4=30$$

The idle time is

$$\begin{aligned}\text{idle time} &= (\text{no. of stations})(\text{cycle time}) \\ &\quad -\sum(\text{all task times}) \\ &= (4\ \text{stations})(10)-30=10\end{aligned}$$

(c) The balance delay is

$$\begin{aligned}\text{balance delay} &= \frac{\text{idle time}\times 100\%}{(\text{no. of stations})(\text{cycle time})} \\ &= \frac{(10)(100\%)}{(4)(10)} \\ &= 25\%\end{aligned}$$

(d) Since station 3 has no idle time, decreasing the cycle time would require rebalancing the line. Task D could be shifted to station 4, leaving station 3 only with task E. The assignments to stations 1 and 2 would not change. The production rate would be increased, and the percentage balance delay would be decreased.

$$\text{balance delay}=\frac{(36-30)(100\%)}{(4)(9)}=16.7\%$$

11. WAGE INCENTIVE PLANS

The goal of *wage incentive plans* is to motivate workers to produce more by allowing them to earn more.[27] *Individual incentive plans* are tied to each worker's output; *group incentive plans* are tied to the output of a department or other group. Individual plans can usually be modified for use as group plans.

The simplest wage incentive plan is *piecework* or *piece-rate*, in which workers are paid in proportion to the number of pieces completed. Piecework payment can be very effective, but it is difficult to administer under minimum-wage laws.

Standard-hour plans are similar to piecework, but they guarantee workers a *base wage* regardless of output. Above the *performance standard* (i.e., the output corresponding to the standard payment), workers are paid in proportion to the output. The difference between the base and actual wages is the *bonus*. *Productivity* (also called *efficiency*) is the percentage of the standard output achieved. Output may be expressed in units or standard hours.

The bonus can be 100% of the increase in output above the standard or it can be less (though usually no less than 50% of the increase).[28] Plans with a 100% bonus are known as *one-for-one plans* or *100% premium plans*, while plans with bonuses less than 100% are known as *gain-sharing plans*. The percentage of the earned bonus kept by the worker is known as *labor's participation* in the plan.

Effective plans allow workers to earn bonuses of at least 25% to 30% of their base wages (i.e., the *relative earnings* are 125% to 130%). However, depending on the standard, some workers may be unable to consistently achieve even the standard output. Rather than encouraging higher output, such plans might discourage workers, resulting in even lower output than without the incentive program. For this reason, many incentive programs begin paying a bonus at a *threshold* output less than 100% of standard. Such *reduced-standard plans* can be one-for-one or gain sharing.

Wage incentive plans can be identified in one of three ways: by name, by specifying the participation and productivity at which the bonus begins, or by specifying the standard bonus. The *standard bonus* is the bonus earned at a productivity of 100%. Obviously, the standard bonus is zero for any 100% standard plan. For any linear plan, the standard bonus depends on the *bonus factor*, BF, defined as the difference between 100% and the threshold productivity. For one-for-one plans, BF = 0.

$$\text{standard bonus}=(\text{BF})(\text{participation}) \quad \text{[linear plans]} \qquad 62.70$$

[27]As a very general rule, the output before wage incentives are implemented will be 60% to 90% of the output based on standard times. With wage incentives, productivity will increase to 130% or more. Although workers may not work significantly faster with wage incentives, they reduce their idle time.

[28]The lower bonus may be justified on the basis that the remainder is needed to pay for the administrative costs of running the incentive program.

The *relative earnings*, RE, are

$$\text{RE} = 1 + (\text{participation})(\text{productivity} + \text{BF} - 1) \quad \left[\begin{array}{c}\text{linear plans,}\\ \text{productivity} \geq 1 - \text{BF}\end{array}\right] \qquad 62.71$$

The *relative bonus*, RB, is

$$\text{RB} = \text{RE} - 1 \qquad 62.72$$

Figure 62.4 illustrates four types of *linear incentive plans*. *Curvilinear plans*, also known as *self-regulating plans*, are also used, particularly when there is the possibility of runaway production or cheating. (See Fig. 62.5.) Payment under curvilinear plans asymptotically approaches a maximum "ceiling" wage, which is often 1 + labor's participation. The maximum relative earnings is one plus the fractional participation. The nature of the curve is specified by the productivity level at which the bonus begins (i.e., the threshold productivity) and at least one other point on the curve.

Figure 62.4 *Four Types of Linear Incentive Plans*

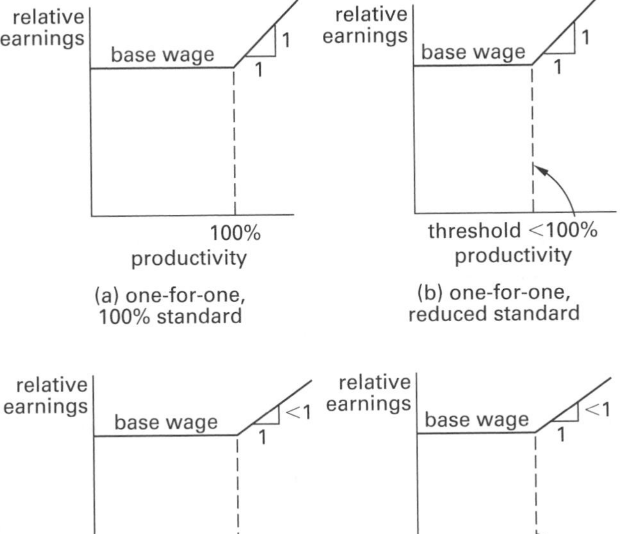

For curvilinear plans, the standard bonus and relative earnings are

$$\text{standard bonus} = \frac{(\text{BF})(\text{participation})}{1 + \text{BF}} \quad [\text{curvilinear plans}] \qquad 62.73$$

Figure 62.5 *Curvilinear Plan*

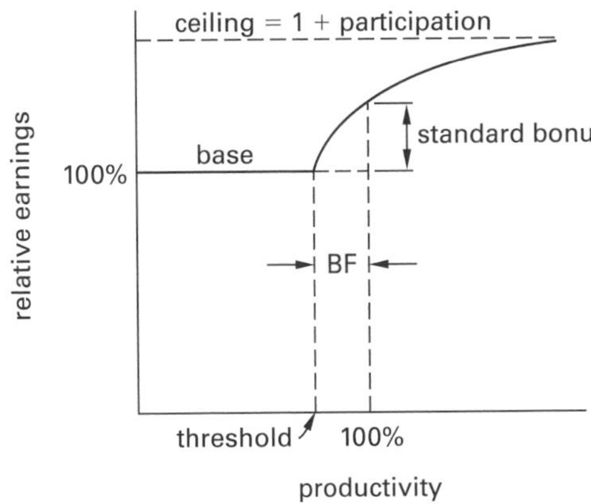

$$\text{RE} = 1 + \frac{(\text{participation})\left(\begin{array}{c}\text{productivity}\\ +(\text{productivity})\\ \times(\text{BF}) - 1\end{array}\right)}{\text{productivity} + (\text{productivity})(\text{BF})} \quad [\text{curvilinear plans}] \qquad 62.74$$

For both linear and curvilinear reduced-standard plans, the threshold productivity at which the bonus begins can be found from Eq. 62.75.

$$\begin{array}{c}\text{threshold}\\ \text{productivity}\end{array} = 1 - \text{BF} \quad \left[\begin{array}{c}\text{linear and}\\ \text{curvilinear plans}\end{array}\right] \qquad 62.75$$

Example 62.17

A linear wage incentive program provides for 50% participation and a 25% standard bonus. A worker produces 1900 units in 8 hr. The standard time for each unit is 0.004 hr. What are (a) the bonus factor, (b) the worker's relative earnings, and (c) the worker's relative bonus?

Solution

(a) From Eq. 62.70, the bonus factor is

$$\begin{aligned}\text{BF} &= \frac{\text{standard bonus}}{\text{participation}} = \frac{0.25}{0.50}\\ &= 0.50 \quad (50\%)\end{aligned}$$

(b) The worker's productivity is

$$\begin{aligned}\text{productivity} &= \frac{\text{standard time}}{\text{actual time}}\\ &= \frac{(1900 \text{ units})\left(0.004 \ \frac{\text{hr}}{\text{unit}}\right)}{8 \text{ hr}}\\ &= 0.95 \quad (95\%)\end{aligned}$$

From Eq. 62.71, the relative earnings are

$$\begin{aligned}\text{RE} &= 1 + (\text{participation})(\text{productivity} + \text{BF} - 1)\\ &= 1 + (0.50)(0.95 + 0.50 - 1)\\ &= 1.225 \quad (122.5\%)\end{aligned}$$

(c) From Eq. 62.72, relative bonus is

$$\begin{aligned}\text{RB} &= \text{RE} - 1 = 1.225 - 1\\ &= 0.225 \quad (22.5\%)\end{aligned}$$

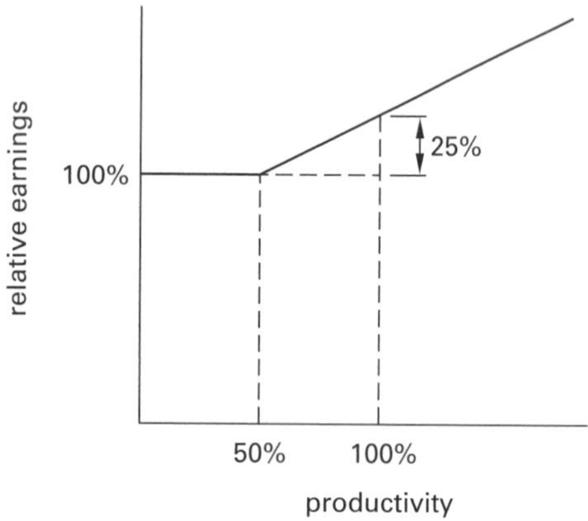

Example 62.18

A curvilinear wage incentive program provides for 75% participation and a 37.5% standard bonus. A worker produces 1900 units in 8 hr. The standard time for each unit is 0.004 hr. What are the worker's relative earnings?

Solution

As in Ex. 62.17, the productivity is 95%. The bonus factor is calculated from Eq. 62.73.

$$\begin{aligned}\text{standard bonus} &= \frac{(\text{BF})(\text{participation})}{1 + \text{BF}}\\ 0.375 &= \frac{(\text{BF})(0.75)}{1 + \text{BF}}\\ \text{BF} &= 1.00\end{aligned}$$

From Eq. 62.74, the relative earnings are

$$\begin{aligned}\text{RE} &= 1 + \frac{(\text{participation})\begin{pmatrix}\text{productivity}\\ +(\text{productivity})\\ \times(\text{BF}) - 1\end{pmatrix}}{\text{productivity} + (\text{productivity})(\text{BF})}\\ &= 1 + (0.75)\left(\frac{0.95 + (0.95)(1.0) - 1.0}{0.95 + (0.95)(1.00)}\right)\\ &= 1.355 \quad (135.5\%)\end{aligned}$$

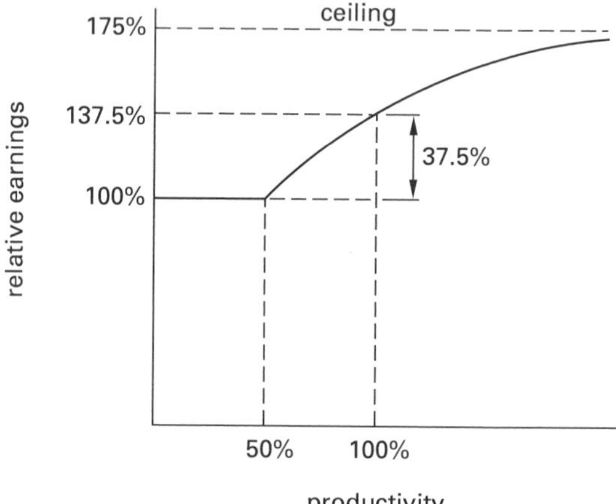

12. QUALITY CONTROL CHARTS: X-CHARTS AND R-CHARTS

Statistical quality control (SQC), also known as *statistical process control* (SPC), uses several techniques to ensure that a minimum quality level is consistently obtained from production processes. Typical SQC tasks include routine monitoring of process output, sampling incoming raw materials, and testing finished work.

Monitoring process output and charting the results are often the most visible aspects of SQC. Small samples of work are tested at regular or random intervals, and the results are shown graphically.[29] The graphs are known as *control charts or Shewhart control charts* because they show, in addition to the measured values, the *control limits* (i.e., the limits of acceptable values).[30] (See Fig. 62.6.)

[29]The graphs are often conspicuously posted at the entrances of departments.

[30]Control limits have nothing to do with *specification limits*. Specification limits determine if the product is acceptable to the customer. Control limits determine if the process is statistically in control.

Figure 62.6 *Interpretation of SPC Charts*

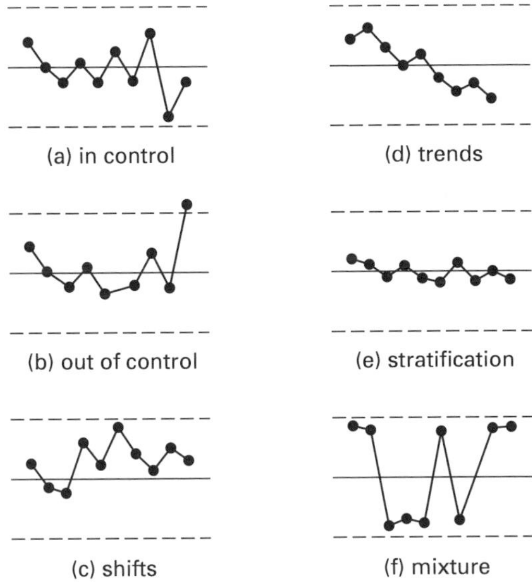

Control charts can be prepared for the average value of some process variable (the *x-bar chart*) and for the range (*R-chart*), and other types of charts discussed in Sec. 62.13.

It is a basic assumption that random effects are present in every process. Variation within certain limits is inevitable, and if the magnitudes of the limits or the variation are unacceptable, they must be reduced by changes in manufacturing or product design. If the magnitudes are acceptable, then corrections are required only when the results exceed the magnitude expected on the basis of random effects.

For a process in control, the variable is assumed to be normally distributed with population mean σ and population standard deviation σ' (based on combined sample set data used to determine the sample set mean, μ, and sample set standard deviation, σ). When the process changes by more than $\pm <2\sigma'$ (the warning limits), a change might be occurring, and the process should be watched. When the process changes by more than $\pm\ 3\sigma'$ (the action limits), the process is considered to have gone out of control.[31] In that case, the process is evaluated and, if necessary, changed.[32]

The *average run length*, ARL, is the average number of samples from an in-control process that will be tested before a point outside the control limits will be encountered and is used to determine control chart performance.

$$\text{ARL} = \frac{1}{1-C} \qquad 62.76$$

Typical parameters for 3σ control limits state that a process is out of control if 1 point falls outside control limits; 2 out of 3 consecutive points fall on the same side of the centerline and more than 2σ from the centerline; 4 out of 5 consecutive points fall on the same side of and more than σ from the centerline; 8 consecutive points fall on the same side of the centerline; 7 consecutive points trend up or trend down; 8 consecutive points fall on either side of the centerline and more than σ from the centerline; 14 consecutive points alternate sides of the centerline; or 15 consecutive points are within σ of the centerline. **[Tests for Out of Control, for Three-Sigma Control Limits]**

In a typical control chart application, a sample of n objects is taken and the quality parameter, x, is measured for each of the n items. The average of the sample, $\overline{X}$, is calculated. This process is repeated K times (e.g., over K days, etc.).[33] The weighted average, $\overline{X}_W$, and (in theory) the standard deviation, $\sigma_{\overline{x}}$, of the process averages are calculated from the K values of $\overline{x}$.[34]

The unbiased estimators of the population average, μ^*, and standard deviation, σ^*, of the $n \times K$ individual values are $\overline{X}_W$ and $\sqrt{n}\sigma^*$ where $A = 3/\sqrt{n}$ when $n \geq 25$ samples, or can be taken from tables of factors for control-chart limits. **[Factors for Control-Chart Limits]**

Equation 62.77 applies only to x-bar charts, typically for existing processes where μ and σ are known (i.e., standards given). To calculate values for other types of quality control charts, see Sec. 62.13.

As long as the process is in control and the population parameters are known, the sample averages will be within the *upper control limit*, UCL, and *lower control limit*, LCL, as shown in Eq. 62.77 and Eq. 62.78.

Control-Limit Calculations

$$\text{UCL}_{\overline{X}_W,\text{STD}} = \mu^* + A\sigma^* \qquad 62.77$$

$$\text{LCL}_{\overline{X}_W,\text{STD}} = \eta^* - A\sigma^* \qquad 62.78$$

[31]With 3σ control limits, no more than 0.27% (i.e., 27 times out of 1000) of the points are expected to be outside of the control limits.

[32]Although three standard deviations is the usual value, any multiple of standard deviations can be chosen for the action limits.

[33]For the assumption of a normal distribution to be valid, K must be greater than approximately 20.

[34]Though s^2 is an unbiased estimator of the population variance, σ'^2, s is a biased estimator of σ' and is not used. However, if s is calculated for each sample, it can be used if corrected according to

$$\sigma' = \frac{1}{c_2}\sqrt{\left(\frac{n-1}{n}\right)}\overline{s}$$

When the population parameters are not known, the lower and upper control limits for the x-bar chart can be written as

Control-Limit Calculations

$$\text{UCL}_{\overline{X}_W,\text{NO STD}} = \overline{X}_W + A_2\overline{R} \quad 62.79$$

$$\text{LCL}_{\overline{X}_W,\text{NO STD}} = \overline{X}_W - A_2\overline{R} \quad 62.80$$

For various reasons, including timely response and job enrichment, the steps and calculations necessary to maintain x-bar (and other) charts are often performed by workers who are not trained in higher mathematics or statistical analysis. Therefore, a simplified method based on the *range* of values (i.e., the maximum measurement less the minimum measurement) can be used to calculate the standard deviation.[35]

$$r = \text{maximum sample value} - \text{minimum sample value} \quad 62.81$$

The range is calculated for each of the K samples. Then, the average range, $\overline{R}$, is calculated and converted to the population standard deviation, σ^*. A similar calculation can be done if the individual standard deviation, σ, is calculated for each of the K samples. The factors A_2 and d used in Eq. 62.82 depend on the sample size, n, and can be found from tables of control-chart limit factors. **[Factors for Control-Chart Limits]**

$$\sigma^* = \frac{\overline{R}}{d} \text{ and } A_2 = \frac{3}{d\sqrt{n}} \quad 62.82$$

The variability of the process can also be tracked by charting the range values in an *R-chart*.[36] The average range is assumed to be zero, and Eq. 62.83 and Eq. 62.84 give the upper and lower control limits, respectively. The values depend on the confidence level (i.e., the percentage of points that are expected to fall outside the limits when the process is in control). Since control limits on the R-chart are not symmetrical, values of D_3 and D_4 are generally taken from tables. **[Factors for Control-Chart Limits]**.[37]

Control-Limit Calculations

$$\text{UCL}_{\overline{R},\text{NO STD}} = D_4\overline{R} \quad 62.83$$

$$\text{LCL}_{\overline{R}\text{ NO STD}} = D_3\overline{R} \quad 62.84$$

The upper and lower control limits can also be written as Eq. 62.85 and Eq. 62.86, if the population parameters are known.

Control-Limit Calculations

$$\text{UCL}_{\overline{R}\text{ STD}} = D_2\sigma^* \quad 62.85$$

$$\text{LCL}_{\overline{R}\text{ STD}} = D_1\sigma^* \quad 62.86$$

There is an important difference between nonconformances and nonconformities. A *nonconformance* is a *defective*—an item that does not meet specification. A *nonconformity* is a *defect*. For example, an acceptance specification may say that a painted object may not have more than three paint defects. An object with two bubbles in its paint will have two nonconformities but will not be a nonconformance.

There are three common variations with time of control charts: (1) When the sample size changes from time to time, the horizontal control limits will move inward or outward with the changes. (2) The average may show a steady, shifting trend represented by an inclined average line, as when there is uniform tool wear. The parallel control limits will be similarly inclined. (3) In continuous processes, such as liquid product manufacturing or refining, an argument can be made for using a moving average (i.e., recalculating the average over the most recent M samples). The smoothing effect of a moving average accurately represents the effect of product blending.

Example 62.19

A factory dispenses candy into individual bags. Three observations of the process are made every 2 hr until 10 overall samples have been taken. The results of those observations are as shown.

sample	observations			$\overline{X}_i$	r_i
1	20	19	21	20.00	2
2	20	20	20	20.00	0
3	21	20	19	20.00	2
4	19	22	20	20.33	3
5	21	20	19	20.00	2
6	20	20	21	20.33	1
7	19	20	21	20.00	2
8	20	22	19	20.33	3
9	19	20	22	20.33	3
10	20	21	20	20.33	1

Determine the X-chart and R-chart upper control limit (UCL) and lower control limit (LCL) assuming no standards given.

[35]Most quality monitoring today, even on the "factory floor," uses the computer. The R-chart is still an option on most quality software, but there is little to justify its use. The s-chart and σ-chart are more powerful and detect changes in variability faster.

[36]σ charts are also used. For more information on σ charts, refer to a textbook on statistical quality control.

[37]Though the factors are not symmetrical, the probabilities are. For example, with the 95% factors, 2.5% will be outside of each control limit. However, asymmetrical probabilities could be chosen if desired.

Solution

The overall average is

Statistical Quality Control

$$\overline{X}_W = \sum_{i=1}^{K} \frac{\overline{X}_i}{K} = \frac{\begin{array}{c}20.00 + 20.00 + 20.00 + 20.33 + 20.00 \\ +20.33 + 20.00 + 20.33 + 20.33 + 20.33\end{array}}{10} = 20.17$$

The average range is

Statistical Quality Control

$$\overline{R} = \sum_{i=1}^{K} \frac{r_i}{K} = \frac{\begin{array}{c}2 + 0 + 2 + 3 + 2 \\ +1 + 2 + 3 + 3 + 1\end{array}}{10} = 1.90$$

For three observations, the control-chart limit factors are $A_2 = 1.02$, $D_3 = 0$, and $D_4 = 2.57$. [**Factors for Control-Chart Limits**]

The X-chart UCL is

Statistical Quality Control

$$\text{UCL}_{\overline{X}_W,\text{NO STD}} = \overline{X}_W + A_2\overline{R} = 20.17 + (1.02)(1.90) = 22.11$$

The X-chart LCL is

$$\text{LCL}_{\overline{X}_W,\text{NO STD}} = \overline{X}_W - A_2\overline{R} = 20.17 - (1.02)(1.90) = 18.23$$

The R-chart UCL is

$$\text{UCL}_{\overline{R}\ \text{NO STD}} = D_4\overline{R} = (2.57)(1.90) = 4.88$$

The R-chart LCL is

$$\text{LCL}_{\overline{R},\text{NO STD}} = D_3\overline{R} = (0)(1.90) = 0$$

Since no observations are out of the observation and range control limits, the process can be considered stable.

Example 62.20

A quality analysis is performed on the thickness of sheets produced by a rolling operation. A sample of five sheets is taken from each of four shifts, resulting in the data shown.

shift, i	sample thickness (cm)					$\overline{X}_W$ (cm)	R_i (cm)
1	2.3	2.7	2.6	2.7	2.4	2.54	0.4
2	2.2	2.6	2.0	2.3	2.4	2.30	0.6
3	3.0	2.8	3.1	3.0	3.5	3.08	0.7
4	2.6	1.9	3.3	1.6	3.3	2.54	1.7

Is the process out of control?

Solution

The overall range is

Dispersion, Mean, Median, and Mode Values

$$\overline{X} = \sum_{i=1}^{K} \frac{\overline{X}_i}{K} = \sum_{i=1}^{K} \frac{\overline{X}_W}{K} = \frac{2.54 \text{ cm} + 2.30 \text{ cm} + 3.08 \text{ cm} + 2.54 \text{ cm}}{4} = 2.615 \text{ cm}$$

The *sample range*, r, is the largest sample value minus the smallest sample value.

$$r_i = X_{i,\max} - X_{i,\min}$$

The average range is

Dispersion, Mean, Median, and Mode Values

$$\overline{R} = \sum_{i=1}^{K} \frac{r_i}{K} = \frac{0.4 \text{ cm} + 0.6 \text{ cm} + 0.7 \text{ cm} + 1.7 \text{ cm}}{4} = 0.85 \text{ cm}$$

Find the control limits for the X-chart. For 5 samples, $A_2 = 0.58$. [**Factors for Control-Chart Limits**]

Statistical Quality Control

$$\text{UCL} = \overline{X} + A_2\overline{R} = 2.615 \text{ cm} + (0.58)(0.85 \text{ cm}) = 3.108 \text{ cm}$$

$$\text{LCL} = \overline{X} - A_2\overline{R} = 2.615 \text{ cm} - (0.58)(0.85 \text{ cm}) = 2.122 \text{ cm}$$

Find the control limits for the R-chart. For 5 samples, $D_4 = 2.11$ and $D_3 = 0$. [**Factors for Control-Chart Limits**]

Statistical Quality Control

$$\text{UCL} = D_4\overline{R} = (2.11)(0.85 \text{ cm}) = 1.7935 \text{ cm}$$

$$\text{LCL} = D_3\overline{R} = (0)(0.85 \text{ cm}) = 0 \text{ cm}$$

There are 3 points outside the upper control limit of 3.108 cm (3.5 cm, 3.3 cm, 3.3 cm), and 3 points outside the lower control limit of 2.122 cm (2.0 cm, 1.9 cm, 1.6 cm), so the first test criterion (a single point falls outside the control limits) is fulfilled, and the process is out of control. [**Tests for Out of Control, for Three-Sigma Control Limits**]

Plant Engineering

13. QUALITY CONTROL CHARTS

In addition to the x-bar chart and R-chart, control charts can also be prepared for other measures of dispersion (*s-chart* or *σ-chart*), for the fraction defective (*p-chart*), or for the number of defects per unit (*c-chart*), or any combination thereof. The p-chart and c-chart are examples of *attribute charts*, where only the condition of an item needs to be determined.

The *p-chart* is particularly useful in monitoring the fraction of nonquantitative rejects (i.e., *nonconformances*—the quantity of units that don't work or "just aren't right" for any reason). The *np-chart* monitors the number of rejects in a standard sample size. The variable p is the fraction (or percentage) defective in each sample of N items. $\overline{p}$ is the average fraction defective taken over M samples, which is equal to the total number of defects found in M samples divided by the total number of samples, $N \times M$.

The average and control limits both depend on $\overline{p}$, which is determined historically. The lower and upper control limits for the p-chart are given by Eq. 62.87. N_i is the number of samples in period i, usually a constant. The second term in Eq. 62.87 is three times the standard deviation of a binomial distribution.[38]

$$\text{LCL}_p,\ \text{UCL}_p = \overline{p} \pm 3\sqrt{\frac{\overline{p}(1-\overline{p})}{n_i}} \qquad 62.87$$

The *c-chart* tracks the number of defects (i.e., *nonconformities*) per sample of constant size. (The sample size may be one complete assembly or single unit.) The defects may be of any variety, anywhere in the assembly. Over a period of M samples, the average number of defects, $\overline{c}$, is determined. The probability of any number of defects per sample is assumed to be a Poisson distribution, and since the mean and variance of a Poisson distribution are the same, the upper and lower control limits are given by Eq. 62.88.[39]

$$\text{LCL}_c,\ \text{UCL}_c = \overline{c} \pm 3\sqrt{\overline{c}} \qquad 62.88$$

For most other charts, the average and limits can be allowed to change and shift over time. However, for an in-control process monitored by a c-chart, the average and limits are fixed. If quality deterioration occurs and $\overline{c}$ increases, changing the average and limits would imply that the process is still in control when it is not.

14. QUALITY ACCEPTANCE SAMPLING

Acceptance sampling is the testing of samples taken from a *lot* (batch or process) in order to determine if the entire lot should be accepted or rejected. Acceptance sampling is appropriate when testing is destructive or when 100% testing would be too expensive. To design an *acceptance plan* (also known as a *Dodge-Romig plan*), the number in the sample and the *acceptance number* (i.e., the maximum allowable number of defects in the sample) must be specified.

In *single acceptance plans*, a sample of size n out of a total lot size of N is tested. If the number of defects is equal to or less than c, the lot is accepted. The plan can be described graphically by an *operating characteristic* (OC) *curve* (see Fig. 62.7), which plots the *probability of acceptance* (also known as the *producer's acceptance risk*) versus the *lot quality* (i.e., the true fraction or percentage defective).[40] Points on the OC curve are determined from the binomial or, more preferably, from the Poisson approximation to the binomial. In practice, however, acceptance plans are generally designed by referring to tables of predetermined plans.

Figure 62.7 *Acceptance Plan Operation Characteristic Curve*

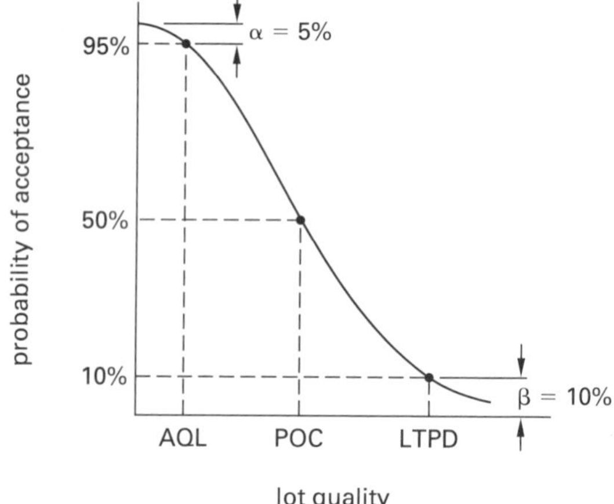

The *producer's risk*, α, also known as the *supplier's risk*, is the probability of a *Type I error* (i.e., rejecting a good lot—that is, of finding trouble when none exists). The *consumer's risk* is the probability of a *Type II error* (i.e., accepting a bad lot—that is, of not finding trouble when it exists). Once a sampling plan has been determined (i. e., the OC curve is established) and the actual incoming fraction defective is known, Eq. 62.89 can be applied.

$$\alpha + \beta = 100\% \qquad 62.89$$

[38] The p-chart and np-chart use the normal approximation to the binomial distribution. For this approximation to be reliable, the sample size should be greater than approximately 25, $\overline{np}$ should be greater than or equal to approximately 4, and np should be greater than or equal to approximately 1.

[39] Unlike the normal distribution, the Poisson distribution is not symmetrical. Since the control limits in the c-chart are symmetrical, they do not represent equal probabilities of exceedance.

[40] Operating characteristic curves are actually a series of discontinuous points since lot items are finite and discrete. However, they are never drawn in that manner.

The only way to decrease the consumer's risk without increasing the producer's risk is to increase the sample size (i.e., to change the sampling plan). Sampling plans based on large samples have steeper OC curves. (The ideal curve, corresponding to 100% sampling, is a vertical line at the lot quality.) Plans with smaller samples have more gradually inclined curves. When the sample history shows unsatisfactory quality, a *tightened plan* with more samples can be implemented. When the part or supplier has a history of extremely high quality, a *reduced plan* with fewer samples can be used.

Three points, usually those corresponding to 95%, 50%, and 10% probabilities, are given special consideration on the OC curve. The lot quality corresponding to a 95% probability of acceptance (i.e., a 95% consumer's risk) is known as the *acceptable quality level*, AQL.[41] The AQL is sometimes written as the "$p_{95\%}$ quality" or the "$p_{0.95}$ point." The AQL is the percentage defective that is "satisfactory" (i.e., a lot with percentage defective equal to $p_{95\%}$ or less is a good lot). The probability will be $1 - \alpha = 5\%$ that the lot will be rejected if the fraction defective is $p_{95\%}$ or less.

The lot quality corresponding to a 50% probability of acceptance (i.e., producer's and consumer's risks both equal to 50%) is known as the "$p_{50\%}$ quality" or "$p_{0.50}$ point," sometimes referred to as the "*point of control*" (POC) or "*indifference quality level*."

The lot quality corresponding to a 10% probability of acceptance (i.e., a 10% consumer's risk) is known as the *lot tolerance percent (fraction) defective*, LTPD, also known as the "$p_{10\%}$ quality" or, less frequently, the *rejectable quality level*, RQL.

It is generally easier to design a sampling plan that has specific values of AQL and LTPD than to find the AQL and LTPD for a known plan. There are few published tables for the latter case. The traditional method is to plot the OC curve and read the values of fraction defective corresponding to probabilities of acceptance of 95% and 10%. (See Ex. 62.22.)

A Poisson distribution is used to describe the probability of any item being defective. The average number of defects in a sample is

$$\lambda = (\text{fraction defective})(\text{sample size}) \qquad 62.90$$

$$p\{m \text{ defects}\} = \frac{\lambda^m e^{-\lambda}}{m!} \qquad 62.91$$

The probability that a lot will be accepted is

$$p\{\text{acceptance}\} = \sum_{m=0}^{c} p\{m \text{ defects}\} \qquad 62.92$$

In practice, most acceptance plans are "designed" by using published tables or, at the very least, by using *cumulative summation* ("*cumsum*") *tables* for a particular distribution. However, if the $p_{50\%}$ quality is known, a single-sampling plan can be easily designed. Once the number of defects, c, has been (arbitrarily) chosen, Eq. 62.93 is used to calculate the sample size. There will be a family of plans (with different values of c) that satisfies Eq. 62.93. Plans with large values of N and c are better than plans with small values, as the slope of the OC curve is steeper near the $p_{50\%}$ point. In Eq. 62.93, $p_{50\%}$ is expressed in decimal form.

$$N = \frac{c + 0.67}{p_{50\%}} \qquad 62.93$$

If all items in rejected lots are subsequently screened (i.e., 100% tested) with only good items being passed, the screened lots will contain no defectives.[42] When the various lots are combined, some lots will have been acceptance-sample passed and some will have been 100% screened. The *average outgoing quality limit*, AOQL, is the maximum fraction (percentage) defective of accepted items in the combined lots. In the long run, regardless of incoming quality, item quality will be better than or equal to the AOQL.[43]

The AOQL cannot be read directly from the OC chart. Given a specific OC curve, the AOQL is the maximum value of the product of the lot quality and the probability of acceptance (i.e., the abscissa and ordinate of points on the OC curve). The AOQL can be rapidly found from the curve by trial and error.

With a *double-acceptance plan*, a sample of n_1 is taken out of the original lot size of N. The lot is accepted if the number of defects is c_1 or fewer and rejected if greater than c_2. Otherwise, a second sample of n_2 is taken. The lot is accepted if the total number of defectives from both samples is c_2 or fewer.

Example 62.21

A sample of 100 items is taken from a large lot known to have a 3% defect rate. The acceptance number is 2. What is the probability of acceptance?

Plant Engineering

[41]Any probability could be used. However, the 95% consumer's risk (5% producer's risk) is traditional.

[42]Theoretically, each defective item should be replaced with a good item. However, if the lot size is very large, it doesn't really make any difference if a defective item is merely discarded.

[43]Exceptions are possible in the short run.

Solution

The lot will be accepted if 0, 1, or 2 defects are found. The Poisson probability distributed is assumed. The average number of defects in the sample is

$$\begin{aligned}\lambda &= (\text{fraction defective})(\text{sample size})\\ &= (0.03)(100)\\ &= 3\end{aligned}$$

$$p\{m \text{ defects}\} = \frac{\lambda^m e^{-\lambda}}{m!}$$

$$\begin{aligned}p\{\text{acceptance}\} &= p\{0 \text{ defects}\} + p\{1 \text{ defect}\}\\ &\quad + p\{2 \text{ defects}\}\\ &= \frac{3^0e^{-3}}{0!} + \frac{3^1e^{-3}}{1!} + \frac{3^2e^{-3}}{2!}\\ &= 0.0498 + 0.1494 + 0.2240\\ &= 0.4232 \quad (42.32\%)\end{aligned}$$

(This is the same value that is obtained from a cumulative probability chart with $\lambda = 3$ and $c = 2$.)

Example 62.22

A sampling plan is designed to accept the entire lot if there are 2 or fewer defective in a sample of 100 items. The actual fraction defective of the lot is unknown. (a) Draw the operating characteristic curve. (b) What is the approximate AQL for a producer's risk of 5%? (c) What is the approximate LTPD for a consumer's risk of 10? (d) What is the approximate AOQL?

Solution

(a) Points on the curve can be generated by repeating Ex. 62.21 for different values of the fraction defective, p.

fraction defective, p	Np	p (acceptance)
0.00	0.00	1.000
0.01	1.00	0.920
0.02	2.00	0.677
0.03	3.00	0.423
0.04	4.00	0.238
0.05	5.00	0.125
0.06	6.00	0.062

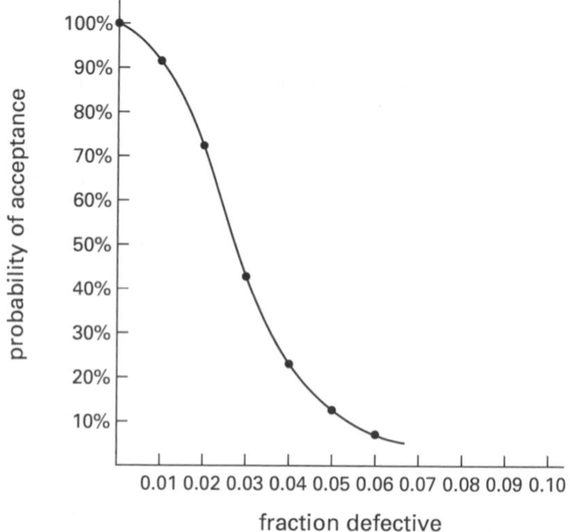

(b) The producer's risk is 5%, so the probability of acceptance is 95%. The AQL is between 0% and 1%. (The actual value is 0.82%.)

(c) The probability of acceptance is the same as the consumer's risk: 10%. The LTPD is between 5% and 6%. (The actual value is 5.3%.)

(d) The AOQL is the maximum product of the fraction defective and the probability of acceptance. This is approximately $(0.02)(0.677) = 0.01354$ (1.35%).

15. RISK ASSESSMENT

Risk assessment is an analytical process that determines the probability that some mishap will occur. It is applied primarily to human health risks and safety hazards or accidents from industrial practices.

Health risk assessment is based on toxicological studies. Data are usually gathered from exposure of laboratory animals to chemicals or processes, although human exposure data can be used if available.[44] Tests are run to determine if a chemical is a *carcinogen* (i.e., causes cancer), a *mutagen* (i.e., causes mutations), a *teratogen* (i.e., causes birth defects), a *nephrotoxin* (i.e., harms the kidneys), a *neurotoxin* (i.e., damages the brain or nervous system), or a *genotoxin* (i.e., harms the genes).

Safety risk assessment of potentially unsafe equipment or practices is traditionally called *hazard analysis.* Attention has traditionally focused on processes and equipment. However, hazard analysis is also applicable to material storage, shipping, transportation, and waste disposal. An analysis of each point in the process is performed using a fault-tree or event-tree.

Fault-tree analysis, FTA, is a deductive logic modeling method. Unwanted events (i.e., failures or accidents) are first assumed. Then, the conditions are identified that could bring about the events. All possible contributors

[44]The best-known exposure test is the *Ames test,* whereby microbes, cells, or test animals are exposed to a chemical in order to determine its carcinogenicity.

to an unwanted event are considered. The event is shown graphically at the top of a network. It is linked through various event statements, logic gates, and probabilities to more basic fault events located laterally and below, producing a graphical tree having branches of sequences and system conditions.

Failure modes and effect analysis, FMEA, is essentially the reverse of fault-tree analysis. It starts with the components of the system and, by focusing on the weaknesses and failure susceptibilities, evaluates how the components can contribute to the unwanted event. The basis for FMEA is, essentially, a parts list showing how each assembly is broken down into subassemblies and basic components. The appearance of an FMEA analysis is tabular, with the columns being used for failure modes, causes, symptoms, redundancies, consequences of failure, frequencies, probabilities, and so on.

Life-cycle analysis is used to prevent failures of items that have limited lives or that accumulate damage. Though the probability of failure may ideally be low, the history of the equipment may also be unknown. Life cycle analysis commonly focuses on time, history, and condition to determine the *degree of damage* (i.e., any reduction in useful life) present in a unit. One approach is to use reporting systems (i.e., operating logs and historical data). Another approach uses ongoing *material conditions monitoring*, MCM, and testing of the actual equipment.

16. WORK METHODS

The goal of *work methods* (*methods engineering*) is the reduction of fabrication and assembly time, worker effort, and manufacturing cost. This is accomplished in a variety of ways, including initial *design for manufacture and assembly* (DFMA) selection of methods to be used, human factors engineering, work measurement, plant layout, assembly line balancing, and administration of the manufacturing process. Work methods, which is more worker- and workplace-oriented, goes beyond traditional *value engineering*, which is product-design oriented.

17. HUMAN FACTORS ENGINEERING

Unlike behavioral science that deals with the psychological effects of work on the individual, *human factors engineering*, also known as *human engineering* and *ergonomics*, is concerned with the physical effects of work. Human factors attempts to increase or optimize the efficiency and reliability of the *machine-worker system* (also known as the *man-machine system*). The "optimization" process used is not mathematical but is based on thoughtful design technique.[45]

The three most common goals of human factors engineering are protecting the operator, minimizing the required conscious thought and effort, and assuring that the machine-worker system will operate successfully. These goals are usually accomplished by foolproofing the *machine-worker interface* (i.e., the controls and physical devices used) and designing the work environment for long-term health and safety.

Typical foolproofing techniques include locating controls in visible and accessible locations, varying the control handle and knob shapes for different functions, sizing and locating levers and wheels so that adequate operating force can be applied, mounting switches with uniform on-off directions, and arranging gauges so that their dials point in a uniform direction during proper operation.

Anthropometric data may be used to correctly size the equipment and work environment. Such items as range of motion, sitting height, leg room, and speed of motion are extensively tabulated by age, gender, nationality, and population percentile. Since it is impossible to design the machine-worker interface for every individual, the reach (as one example) required to operate a control should be no greater than the shortest reach of all workers expected to operate it. Designs frequently intentionally exclude the upper and lower 5% of the population sizes.

Static anthropometry deals with human dimensions such as height, length of forearms, foot size, and so on. *Dynamic anthropometry* deals with limits of motion, such as how far a person can reach in front of them. Both types of data are used in workplace design. Table 62.2 lists typical dimensions developed from anthropometric data for video display terminals, VDTs.[46]

Attention is also given to the work environment, including temperature, humidity, illumination, and noise level. The "comfort range" for temperature and humidity depend on the nature of the work, dress, and the ventilation. Special consideration is required when the working temperature is not in the range of 65°F to 80°F (26.4°C to 38.4°C). Ear protection should be provided, used, and required when the noise level is above an action level of 85 dB.[47] The level of illumination provided depends on the nature of the work. Typically, storage areas require 5 fc to 20 fc (54 lux to 229 lux), office work

Plant Engineering

[45]It is sometimes difficult to decide when improving the system involves human factors engineering and when it does not. Human factors engineering encompasses the subtopics of safety, industrial hygiene, design of training, and management and supervision, as well as system design.

[46]Local laws, regulations, and other standards may apply.

[47]The U.S. Occupational Safety and Health Act (OSHA) of 1970 established 90 dB as the maximum noise level for an 8 hour exposure. Higher levels, up to a maximum of 115 dB, are permitted for shorter intervals, as short as 15 minutes.

Table 62.2 Recommended Dimensions for VDT Workstations

	inches	centimeters
workstation surface		
minimum height	29	74
width	48–64	122–163
depth	32	81
legroom		
minimum height	26	66
minimum width	26	66
eye-to-screen distance	20–30	51–76
minimum height of home-row key from floor	29	74
keyboard thickness	1–2	2.5–5

(Multiply in by 2.54 to obtain cm.)

requires 100 fc to 200 fc (1100 lux to 2200 lux), and fine assembly requires 500 to 1000 fc (5400 lux to 1100 lux).[48]

18. MACHINE SAFEGUARDING

Most machines must be safeguarded to prevent injury to workers. For example, for a machine that is belt-driven, the belt must be enclosed by a safeguard (i.e., a *guard* or *shield*) to prevent a worker's clothing, hair, body, tools, or workpieces from being drawn into the drive mechanism. In general, safeguards should (1) prevent contact, (2) be secure, (3) protect from entering objects, (4) create no new hazards, (5) create no interference with work, and (6) allow safe repair and maintenance.

A worker should not be able to easily remove or tamper with a safeguard. A safeguard that can be made ineffective accidentally or in order to speed up operation is little better than no safeguard. Safeguards and safety devices should be made of durable materials that will withstand the conditions of normal use. They must be firmly secured to the machine.

The safeguard should ensure that no objects can fall into moving parts. A small tool that is dropped into a rotating machine can become a dangerous projectile.

A safeguard defeats its own purpose if it creates a hazard of its own. Safeguards should not contain jagged holes or unfinished, sharp edges. Edges should be rolled or bolted in such a way that sharp edges are eliminated.

A safeguard that impedes a worker from performing the job quickly and comfortably might be overridden or disregarded. Proper safeguarding should enhance efficiency as it relieves worker apprehension about injury.

It should be possible to lubricate, maintain, and repair equipment without removing any safeguards. Locating an oil reservoir outside the guard with a supply line leading to the lubrication point will eliminate the need for a worker or maintenance worker to enter hazardous areas.

There are many ways to safeguard machinery. The type of operation, nature of the raw material, method of handling, physical layout of the work area (see Fig. 62.8), and the production requirements or limitations all have to be taken into consideration. In general, power transmission apparatus is best protected by fixed guards that enclose the danger area. For hazards at the point of operation, where moving equipment performs work on stock, Table 62.3 lists options for safeguarding. Miscellaneous methods, such as signs and awareness barriers, should also be considered.

Figure 62.8 Typical Design of Computer Workstations

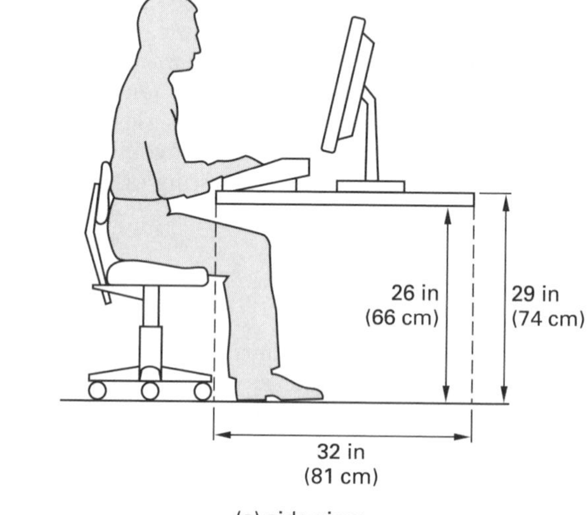

(a) side view

32 in
(81 cm)

48 in
(122 cm)

64 in
(163 cm)

(b) top view

[48]The *foot-candle* is equal to a lumen per square foot. The unit of illumination in SI units is the *lux*, equal to a lumen per square meter.

19. SIMULATION

Introduction

Simulation is a technique of performing sampling experiments on a model of the system. Simulation is appropriate when experimenting with the real system would be unsafe, inconvenient, expensive, or time-consuming, or when analytical techniques are not available.

Time Control

Time control can either be *fixed-interval incrementing* (also known as *uniform time flow*) or next-event incrementing (also known as *variable time flow*). With fixed-interval incrementing, the model's master clock is advanced by one unit and the system model is updated by determining what has happened during the elapsed time.

With next-event incrementing, the master clock is incremented by a variable amount of time. In this case, the computer actually proceeds by keeping track of when the simulated events occur and jumping ahead to the next of these events.

Generating Random Numbers

A truly random process cannot be repeated. However, since all *random number* sequences in a simulation test can be duplicated, such sequences are called *pseudo-random number* sequences. A mathematical procedure for creating this sequence is known as a *random number generator.* This generator should have the following properties.

- The numbers generated should come from a uniform distribution.
- The generation should be fast.
- The sequence should not repeat.
- The sequence should not deteriorate to a single value.

Historically, the *midsquare*, *midproduct*, and *Fibonacci methods* have been used to generate random numbers, but these do not always yield satisfactory results. The *linear congruential method* is frequently used in modern simulation programs.

A number D is said to be congruent with N with modulus M if $(D - N)/M$ is an integer.

The *mixed congruential method* calculates the next random variate, r_{i+1}, from the current variate, r_i, by using Eq. 62.94.

$$r_{i+1} = (ar_i + c)(\textbf{mod}\ T) \qquad 62.94$$

mod is the remaindering modulus function. This means that $(ar_i + c)$ is to be divided by T and r_{i+1} set equal to the remainder. The starting value, r_0, is known as the *seed* and must be supplied by the user. If $c = 0$, the method is known as a *multiplicative congruential method.* If $a = 1$, it is an *additive congruential method.*

The constants, a, c, and T are usually chosen according to established rules to ensure the desired properties of the generator.

The *Mersenne twister algorithm* generates random number sequences that are higher in quality than linear congruential methods.

Variance Reduction Techniques

Methods of increasing the accuracy of the sample estimation without increasing computer time are called *variance reduction techniques.* Two such techniques, *stratified sampling* and *complementary random numbers,* are complicated. They are used when extreme accuracy is required.

Table 62.3 *Machine Safeguarding Options*

guard types
- fixed
- interlocked
- adjustable
- self-adjusting

presence-sensing devices
- electromechanical (pressure-sensitive body bar, floor mat, etc.)
- foot switch
- photoelectric (optical)
- infrared
- capacitive (radio-frequency)

actuation control
- proper presence
- proper sequence
- two-switch (two-hand) actuation
- remote actuation

feeding and ejection options
- automatic
- semiautomatic
- remote (robotic)

20. MONTE CARLO SIMULATION

Introduction

Monte Carlo simulation evaluates the interactions of random variables whose distributions are known but whose combined effects are too complex to be specified mathematically. *Crude Monte Carlo* simulation is essentially random sampling from distributions to obtain values of each interacting variable. The following steps are similar in all Monte Carlo problems.

step 1: Establish the probability distribution for each variable in the study. This does not need to be a mathematical formula—a histogram is sufficient.

step 2: Form the cumulative distribution function for each variable.

step 3: Multiply the cumulative variable by 100 to obtain a cumulative axis that runs from 0 to 100 instead of 0 to 1.

step 4: Generate random numbers between 0 and 100. Either a random number table or a pseudo-random number generator may be used.

step 5: For each random number generated, locate the corresponding variable value. If the original distribution is continuous, take the midpoint of the range as the variable value.

A typical application of the Monte Carlo technique is finding the average line size, average waiting time, or percent idleness of servers in queuing problems.

Example 62.23

A small bank has only one teller. The distribution of service times is

time	probability
1/4 minute	0.55
1/2 minute	0.40
1 minute	0.05

Use the crude Monte Carlo method to simulate the operation of the teller's waiting line. What is the average time the teller spends with a customer?

Solution

step 1: Given.

step 2: The cumulative and converted distributions are

time (minutes)	cumulative probability	(100) (cumulative probability)	range
1/4	0.55	55	1–55
1/2	0.95	95	56–95
1	1.00	100	96–100

step 3: See step 2.

step 4: Select ten random numbers. For this problem, use the sequence 01, 90, 25, 29, 09, 37, 67, 07, 15, 38.

step 5: The first random number is 01. This corresponds to a service time of 1/4 minute. The random number 90 corresponds to a service time of 1/2 minute, etc. The times of the first 10 customers are found similarly: 1/4, 1/2, 1/4, 1/4, 1/4, 1/4, 1/2, 1/4, 1/4, 1/4.

The average service time of the first ten customers is

$$\begin{aligned}\bar{t} &= \left(\frac{1}{10}\right)\left(\frac{1}{4}+\frac{1}{2}+\frac{1}{4}+\frac{1}{4}+\frac{1}{4}+\frac{1}{4}+\frac{1}{2}+\frac{1}{4}+\frac{1}{4}+\frac{1}{4}\right)\\ &= 0.3 \text{ minutes}\end{aligned}$$

(The actual average service time could have been found as the expected value. However, more iterations would be required to obtain this value by the simulation method.)

$$\begin{aligned}\bar{t} &= (0.55)(0.25 \text{ min}) + (0.4)(0.5) + (0.05)(1)\\ &= 0.388 \text{ min}\end{aligned}$$

Variance Reduction

In Ex. 62.23, many of the random numbers were located nearer to 0 than to 100. Since the number formed by subtracting a random number from 100 is also a random number, it can be used to improve the accuracy of the estimation. This is known as the *complementary Monte Carlo technique* or *complementary random number technique*.

Example 62.24

Repeat Ex. 62.23 using the complementary Monte Carlo technique.

Solution

The ten complementary random numbers are 99, 10, 75, 71, 91, 63, 33, 93, 85, and 62. The ten new service times are: 1, 1/4, 1/2, 1/2, 1/2, 1/2, 1/4, 1/2, 1/2, and 1/2 minutes.

Combining these ten new service times with the original ten times gives an average of 0.4 minutes, which is closer to the true mean of 0.388.

21. DECISION THEORY

Decision Trees

A *tree diagram* is a graphical method of enumerating all of the possible outcomes of a sequence of actions. The total number of outcomes is given by the *fundamental principle of counting*.

$$N = \prod n_i \qquad 62.95$$

In Eq. 62.95, n_i is the number of elements in the ith set. The value calculated assumes that no outcomes are restricted and that all outcomes are distinctly different.

If a tree is used to model a decision process, it is called a *decision tree*. Usually decision trees have probabilities, losses, and rewards associated with the various possible outcomes. The tree is divided into segments called *generations* or *stages*. At the end of any given stage, the current status is described by the value of the *state variable*.

Example 62.25

Find the possible values of product $P = A \times B$, where A can take on the values (7, 8) and B can take on the values (1, 3, 5). There are no restricted combinations.

Solution

There are two elements in the first set and three in the second. From Eq. 62.95, the number of possible outcomes (combinations) is

$$N = \prod n_i = (2)(3) = 6$$

The tree diagram is

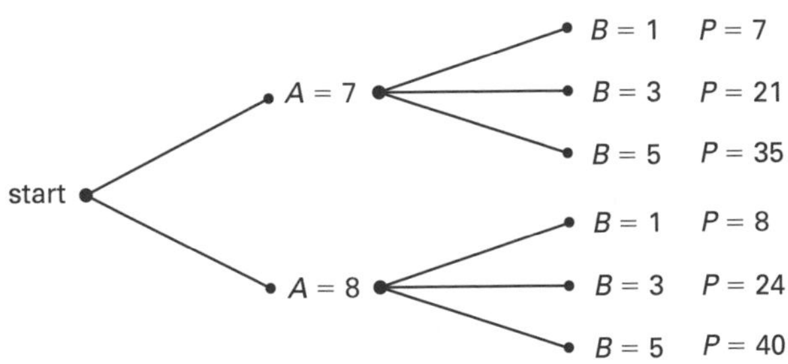

Therefore, the set of products is $P = (7, 8, 21, 24, 35, 40)$.

Decision Theory Problem Formulation

When decisions have to be made, there is usually a finite number of alternatives to be selected from. The set of alternatives is known as the *action space*. Each alternative is associated with a *benefit* or *efficiency*. Usually these efficiencies are expressed in monetary terms. If the benefit is negative, the term *loss* is used in place of efficiency. In general, it is necessary to choose from a set of alternatives, A.

$$A: \; (a_1, a_2, a_3, \ldots, a_n)$$

The state of nature set is Θ.

$$\Theta: \; (\theta_1, \theta_2, \theta_3, \ldots, \theta_m)$$

Decision theory commonly deals in *loss functions*, as opposed to *efficiency functions*. The difference is in sign only. The loss incurred depends only on the alternative chosen and the state of nature. The loss function is given by $L(a_i, \theta_j)$. If the loss function depends on a random variable, the expected loss should be used.

Decision-Making Criteria

The *minimax criterion* selects the alternative that minimizes the maximum loss. The criterion is very conservative and is seldom used since it assumes that nature is a conscious, malevolent opponent that inflicts maximum damage on the decision maker.

As an alternative to the minimax criterion, *Bayes' principle* requires knowledge of the probability distribution of the possible states of nature. Specifically, let $p\{\theta_j\}$ be the prior probability of the jth state. Then, the expected loss associated with the ith alternative is

$$E\{L(a_i)\} = p\{\theta_1\}L(a_i, \theta_1) + p\{\theta_2\}L(a_i, \theta_2) + \cdots + p\{\theta_m\}L(a_i, \theta_m) \qquad 62.96$$

Bayes' principle chooses the alternative that minimizes the expected loss.

22. MARKOV PROCESSES

Introduction

A *Markov process* is a random process in which the state variable may have one of a limited number of values at any given time. Since the process is random, it is not possible to predict with certainty which value the random variable will take in the next time increment. One way to describe the future or history of the process is the sequence of *state variables* $(T_0, T_1, T_2, \ldots, T_n)$.

All possible values of the state variable comprise the *state space*. Changes of state are called *transitions*. If the state actually changes, it is a *real transition*; otherwise, it is a *virtual transition*.

At any given time, the probability of moving to one of the other states depends on the current state and the current time. Thus, the T_j are not always identically distributed nor are they independent.

The process is known as a *Markov process* if the current value of the state variable, T_i, depends only on the previous value, T_{i-1}, and affects only the upcoming value, T_{i+1}. Such a linking of the values is sometimes called a *Markov chain*.

Plant Engineering

Transition and State Probabilities

The usual method of recording the transition probabilities is in a *transition matrix*, **P**. For a process with only three states,

$$\mathbf{P} = \begin{bmatrix} p_{11} & p_{12} & p_{13} \\ p_{21} & p_{22} & p_{23} \\ p_{31} & p_{32} & p_{33} \end{bmatrix}$$

In general, p_{jk} is the probability of making a move to state k given that the variable is currently in state j. Each row must sum to one, but the columns may not. Because the move from j to k is made in one step or jump, the above matrix is also called the *one-step transition matrix*, $\mathbf{P}^{(1)}$. If the matrix is the same from time increment to time increment, it is said to be a *stationary matrix*.

A *two-step transition matrix*, $\mathbf{P}^{(2)}$, contains values of p_{jk} that are probabilities of moving from state j to k in exactly two steps. The matrix $\mathbf{P}^{(2)}$ can be found by squaring **P** using matrix multiplication. The *n-step transition matrix* is found by raising the one-step matrix to the nth power. This is a specific case of the *Chapman-Kolmogorov equation*.

$$\mathbf{P}^{(n)} = (\mathbf{P})^n \qquad 62.97$$

If the value of T_0 is known, the probability of moving from state j to k in n steps can be found directly from the n-step transition matrix. However, often T_0 is unknown—given by its own probability distribution $\mathbf{P}_i^{(0)}$ where $\mathbf{P}_i^{(0)}$ is the probability of the state variable having the value i at time 0. Then, the probability of being in state k in n steps regardless of the initial state is given by Eq. 62.98.

$$p_k^{(n)} = p_1^{(0)} p_{1k}^{(n)} + p_2^{(0)} p_{2k}^{(n)} + p_3^{(0)} p_{3k}^{(n)} + \cdots \qquad 62.98$$

Notice that the state probabilities are the probabilities of *being* in a state after some number of steps, whereas the transition probabilities are the probabilities of moving *between* two states.

Steady-State Probabilities

The transition probabilities stabilize as the number of steps increases. That is, the starting position has increasingly less influence on the transition probabilities as n increases. The *steady-state transition probabilities* $(\pi_1, \pi_2, \pi_3, \ldots, \pi_n)$ can be found by solving the simultaneous equations specified by matrix equation Eq. 62.99 and the *normalizing equation*, Eq. 62.100.

$$[\pi_1 \ \pi_2 \ \pi_3] = [\pi_1 \ \pi_2 \ \pi_3]\mathbf{P} \qquad 62.99$$

$$\pi_1 + \pi_2 + \pi_3 = 1 \qquad 62.100$$

The reciprocals of the steady-state probabilities are the *average times between reassignments* to the ith state.

Special Cases

If the transition from state j to k is possible, we say that state k is *reachable* from state j. (See Fig. 62.9.) If j is also reachable from k, we say that the two states *communicate*. If all states in the process belong to a single communicating class, the process is said to be *irreducible*. Otherwise, it is *reducible*. A state is an *absorbing state* if a transition to another state is impossible, as in $p_{jj} = 1$.

A *closed set of states* is a set such that no state outside the set is reachable from a state within the set. If there are no absorbing states in the closed set, it is called a *minimal closed set*. States in a minimal closed set are called *recurrent states*. States outside of a closed set are called *transient states*.

If a recurrent state must be reached with a regular frequency due to the structure of the process, it is said to be *periodic*; otherwise, it is *aperiodic*.

A state that is not transient, not periodic, and has a finite mean recurrent time is called an *ergodic state*.

It is possible that the simultaneous equation method for finding steady-state probabilities will fail since these probabilities may not exist. However, no difficulties will arise if all states are ergodic.

Figure 62.9 *Types of Markov States*

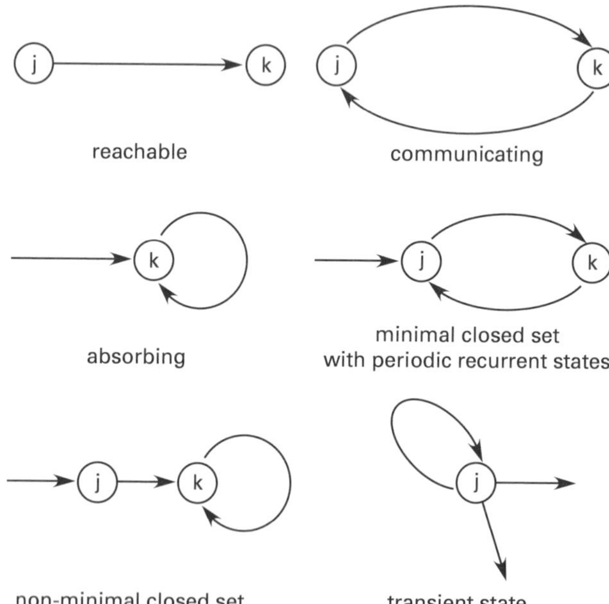

Example 62.26

A parrot in an oasis has three trees from which to choose. On any given day, the parrot can be in any tree, even the previous day's tree, but it may be in only one tree per day.

Once the parrot gets to tree 1, it will not move directly to tree 3. Moving to tree 2 or staying in tree 1 are equally likely. If the parrot gets to tree 2, it will not move directly to tree 1 but may stay in tree 2 or move to tree 3 with equal likelihood. If the parrot gets to tree 3, it will not stay in tree 3 but will move to tree 1 three times as often as to tree 2.

The one-step transition matrix is

$$\mathbf{P} = \begin{bmatrix} 0.50 & 0.50 & 0.00 \\ 0.00 & 0.50 & 0.50 \\ 0.75 & 0.25 & 0.00 \end{bmatrix}$$

(a) Given that the parrot is in tree 1, what is the probability that it will be in tree 2 after two moves? (b) Find the steady-state transition probabilities.

Solution

(a) The state variable T_j is a random variable representing the tree the parrot is in at time j. One way to describe the future or history of the parrot's wanderings is by the sequence $(T_0, T_1, T_2, \ldots, T_n)$.

Method 1: Enumeration

There are three paths the parrot can take to get from tree 1 to tree 2 in two days.

$$\begin{aligned} T_0 = 1,\ T_1 = 1,\ T_2 = 2;\ \text{probability} = (0.5)(0.5) &= 0.25 \\ T_0 = 1,\ T_1 = 2,\ T_2 = 2;\ \text{probability} = (0.5)(0.5) &= 0.25 \\ T_0 = 1,\ T_1 = 3,\ T_2 = 2;\ \text{probability} = (0)(0.25) &= 0.00 \\ p_{12}^{(2)} &= \overline{0.50} \end{aligned}$$

Method 2: Using $\mathbf{P}^{(2)}$

Squaring the one-step transition matrix gives

$$\mathbf{P}^{(2)} = (\mathbf{P})^2 = \begin{bmatrix} 0.250 & 0.500 & 0.250 \\ 0.375 & 0.375 & 0.250 \\ 0.375 & 0.500 & 0.125 \end{bmatrix}$$

The two-step transition probability of $p_{12}^{(2)}$ is 0.500.

(b) Solve the following four equations simultaneously.

$$\begin{aligned} \pi_1 &= 0.5\pi_1 + 0.75\pi_3 \\ \pi_2 &= 0.5\pi_1 + 0.5\pi_2 + 0.25\pi_3 \\ \pi_3 &= 0.5\pi_2 \\ 1 &= \pi_1 + \pi_2 + \pi_3 \end{aligned}$$

The solution is

$$\begin{aligned} \pi_1 &= 0.333\ldots \\ \pi_2 &= 0.444\ldots \\ \pi_3 &= 0.222\ldots \end{aligned}$$

23. NOMENCLATURE

BF	bonus factor	–
c	cycle time	various
c	number of failures (defects)	–
c_2	ratio of $\bar{\sigma}/\sigma'$	–
C	confidence level	–
C	cost	various
d_2	ratio of $\bar{R}/\sigma'$	–
D	duration	various
k	number	–
L	expected line length	–
L_q	expected queue length	–
M	number of consecutive samples	–
MTBF	mean time between failures	various
MTBFO	mean time before failure outage	various
MTTF	mean time to failure	various
MTTR	mean time to repair	various
n	actual number of observations	–
n	number	–
n	sample size	–
N	theoretical number of observations	–
p	observed fraction (n/N)	–
p	true fraction defective	–
$p\{x\}$	probability of event x	–
P	precision	–
Q	quantity	–
r	sample range	various
$\bar{R}$	average range	various
R	reliability	–
RB	relative bonus	–
RE	relative earnings	–
s	number of servers in the system	–
s	sample standard deviation	various
t	time	various
t_C	Student's t-distribution factor	–
W	waiting time in the system	various
W_q	waiting time in the queue	various
x	general variable	–
$\bar{X}$	sample average	–
$\bar{X}_W$	weighted average	–
z	standard normal variable	–
Z	objective function	–

Symbols

α	producer's risk	–
β	consumer's risk	–
ϵ	absolute error	various
λ	mean arrival rate	1/time
λ	mean failure rate	1/time
λ	mean of the Poisson distribution	–
μ	distribution mean	various

Plant Engineering

μ	mean service (repair) rate	1/time
ν	number of degrees of freedom	–
ρ	utilization factor, $\lambda/s\mu$	–
σ	standard deviation (of the sample)	various
σ'	standard deviation of the population	various

Subscripts

ave	average
C	at confidence level C
i	period i
max	maximum
min	minimum
p	fraction
q	queue
std	standard
t	total or at time t

63 Instrumentation and Measurements

Content in blue refers to the *NCEES Handbook*.

NCEES EXAM SPECIFICATIONS AND RELATED CONTENT

MACHINE DESIGN AND MATERIALS EXAM

- I.D.1. Strength of Materials: Stress/strain
  - 23. Stress Measurements in Known Directions
- I.D.5. Strength of Materials: Torsion
  - 23. Stress Measurements in Known Directions
- II.C.5. Testing and instrumentation
  - 2. Precision
  - 3. Stability
  - 9. Sensors

1. ACCURACY

A measurement is said to be *accurate* if it is substantially unaffected by (i.e., is insensitive to) all variation outside of the measurer's control.

For example, suppose a rifle is aimed at a point on a distant target and several shots are fired. The target point represents the "true value" of a measurement—the value that should be obtained. The impact points represent the measured values—what is obtained. The distance from the centroid of the points of impact to the target point is a measure of the alignment accuracy between the barrel and the sights. This difference between the true and measured values is known as the measurement *bias*.

2. PRECISION

Precision is not synonymous with *accuracy*. Precision is concerned with the repeatability of the measured results. If a measurement is repeated with identical results, the experiment is said to be precise. The average distance of each impact from the centroid of the impact group is a measure of precision. Therefore, it is possible to take highly precise measurements and still be inaccurate (i.e., have a large bias). **[Sensors and Transmitters]**

Most measurement techniques (e.g., taking multiple measurements and refining the measurement methods or procedures) that are intended to improve accuracy actually increase the precision.

Sometimes the term *reliability* is used with regard to the precision of a measurement. A *reliable measurement* is the same as a *precise estimate*.

3. STABILITY

Stability and *insensitivity* are synonymous terms. A stable measurement is insensitive to minor changes in the measurement process. Conversely, *instability* and *sensitivity* are synonymous. Sensitivity can be defined as the ratio of the change in the instrument output signal to the change in the measured system input variable. **[Sensors and Transmitters]**

Example 63.1

At 65°F (18°C), the centroid of an impact group on a rifle target is 2.1 in (5.3 cm) from the sight-in point. At 80°F (27°C), the distance is 2.3 in (5.8 cm). What is the sensitivity to temperature?

SI Solution

$$\begin{aligned}\text{sensitivity to temperature} &= \frac{\Delta\ \text{measurement}}{\Delta\ \text{temperature}}\\ &= \frac{5.8\ \text{cm} - 5.3\ \text{cm}}{27^\circ\text{C} - 18^\circ\text{C}}\\ &= 0.0556\ \text{cm}/^\circ\text{C}\end{aligned}$$

Customary U.S. Solution

$$\begin{aligned}\text{sensitivity to temperature} &= \frac{\Delta\ \text{measurement}}{\Delta\ \text{temperature}}\\ &= \frac{2.3\ \text{in} - 2.1\ \text{in}}{80^\circ\text{F} - 65^\circ\text{F}}\\ &= 0.0133\ \text{in}/^\circ\text{F}\end{aligned}$$

Plant Engineering

4. CALIBRATION

Calibration is used to determine or verify the scale of the measurement device. In order to calibrate a measurement device, one or more known values of the quantity to be measured (temperature, force, torque, etc.) are applied to the device and the behavior of the device is noted. (If the measurement device is linear, it may be adequate to use just a single calibration value. This is known as *single-point calibration.*)

Once a measurement device has been calibrated, the calibration signal should be reapplied as often as necessary to prove the reliability of the measurements. In some electronic measurement equipment, the calibration signal is applied continuously.

5. ERROR TYPES

Measurement errors can be categorized as *systematic (fixed) errors*, *random (accidental) errors*, *illegitimate errors*, and *chaotic errors.*

Systematic errors, such as improper calibration, use of the wrong scale, and incorrect (though consistent) technique, are essentially constant or similar in nature over time. *Loading errors* are systematic errors and occur when the act of measuring alters the true value.[1] Some *human errors*, if present in each repetition of the measurement, are also systematic. Systematic errors can be reduced or eliminated by refinement of the experimental method.

Random errors are caused by random and irregular influences generally outside the control of the measurer. Such errors are introduced by fluctuations in the environment, changes in the experimental method, and variations in materials and equipment operation. Since the occurrence of these errors is irregular, their effects can be reduced or eliminated by multiple repetitions of the experiment.

There is no reason to expect or tolerate *illegitimate errors* (e.g., errors in computations and other blunders). These are essentially mistakes that can be avoided through proper care and attention.

Chaotic errors, such as resonance, vibration, or experimental "noise," essentially mask or entirely invalidate the experimental results. Unlike the random errors previously mentioned, chaotic disturbances can be sufficiently large to reduce the experimental results to meaninglessness.[2] Chaotic errors must be eliminated.

6. ERROR MAGNITUDES

If a single measurement is taken of some quantity whose true value is known, the *error* is simply the difference between the true and measured values. However, the true value is never known in an experiment, and measurements are usually taken several times, not just once. Therefore, many conventions exist for estimating the unknown error.

When most experimental quantities are measured, the measurements tend to cluster around some "average value." The measurements will be distributed according to some distribution, such as linear, normal, Poisson, and so on. The measurements can be graphed in a *histogram* and the distribution inferred. Usually the data will be normally distributed.[3]

Certain error terms used with normally distributed data have been standardized. These are listed in Table 63.1.

[1]For example, inserting an air probe into a duct will change the flow pattern and velocity around the probe.

[2]Much has been written about *chaos theory*. This theory holds that, for many processes, the ending state is dependent on imperceptible differences in the starting state. Future weather conditions and the landing orientation of a finely balanced spinning top are often used as examples of states that are greatly affected by their starting conditions.

[3]The results of all numerical experiments are not automatically normally distributed. The throw of a die (one of two dice) is linearly distributed. Emissive power of a heated radiator is skewed with respect to wavelength. However, the means of sets of experimental data generally will be normally distributed, even if the raw measurements are not.

Table 63.1 Normal Distribution Error Terms

term	number of standard deviations	percent certainty	approximate odds of being incorrect
probable error	0.6745	50	1 in 2
mean deviation	0.6745	50	1 in 2
standard deviation	1.000	68.3	1 in 3
one-sigma error	1.000	68.3	1 in 3
90% error	1.6449	90	1 in 10
two-sigma error	2.000	95	1 in 20
three-sigma error	3.000	99.7	1 in 370
maximum error*	3.29	99.9+	1 in 1000

*The true maximum error is theoretically infinite.

7. POTENTIOMETERS

A *potentiometer* (*potentiometer transducer*, *variable resistor*) is a resistor with a sliding third contact. It converts linear or rotary motion into a variable resistance (voltage).[4] It consists of a resistance oriented in a linear or angular manner and a variable-position contact point known as the *tap*. A voltage is applied across the entire resistance, causing current to flow through the resistance. The voltage at the tap will vary with tap position. (See Fig. 63.1.)

Figure 63.1 Potentiometer Circuit Diagram

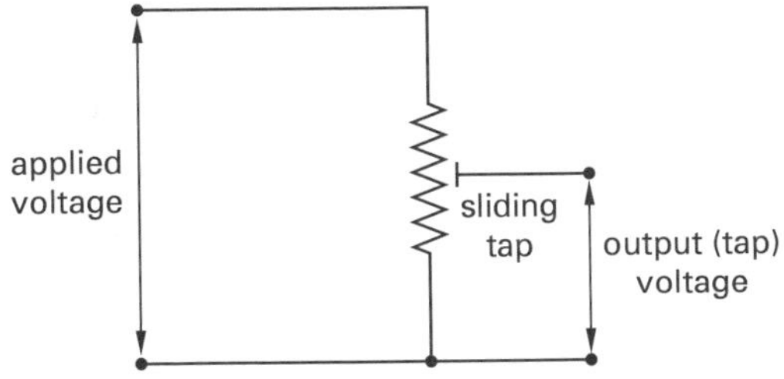

8. TRANSDUCERS

Physical quantities are often measured with transducers (*detector-transducers*). A *transducer* converts one variable into another. For example, a Bourdon tube pressure gauge converts pressure to angular displacement; a strain gauge converts stress to resistance change. Transducers are primarily mechanical in nature (e.g., pitot tube, spring devices, Bourdon tube pressure gauge) or electrical in nature (e.g., thermocouple, strain gauge, moving-core transformer).

9. SENSORS

While the term "transducer" is commonly used for devices that respond to mechanical input (force, pressure, torque, etc.), the term *sensor* is commonly applied to devices that respond to varying chemical conditions.[5] For example, an electrochemical sensor might respond to a specific gas, compound, or ion (known as a *target substance* or *species*). A sensor must provide enough variability in the output signal to effectively measure the entire range of the input state. Two types of electrochemical sensors are in use today: potentiometric and amperometric. These and their associated transmitters convert input variables to industry standard signals (e.g. 0-10 V and 4-20 mA). The time it takes to transmit the signal after the controlled variable change is known as the *sensor response time*. [**Sensors and Transmitters**]

Potentiometric sensors generate a measurable voltage at their terminals. In electrochemical sensors taking advantage of half-cell reactions at electrodes, the generated voltage is proportional to the absolute temperature, T, and is inversely proportional to the number of electrons, n, taking part in the chemical reaction at the half-cell. In Eq. 63.1, p_1 is the partial pressure of the target substance at the measurement electrode; p_2 is the partial pressure of the target substance at the reference electrode.

$$V \propto \left(\frac{T_{\text{absolute}}}{n}\right)\ln\frac{p_1}{p_2} \qquad 63.1$$

Amperometric sensors (also known as *voltammetric sensors*) generate a measurable current at their terminals. In the conventional electrochemical sensors known as *diffusion-controlled cells*, a high-conductivity acid or alkaline liquid electrolyte is used with a gas-permeable membrane that transmits ions from the outside to the inside of the sensor. A reference voltage is applied to two terminals within the electrolyte, and the current generated at a (third) sensing electrode is measured.

The maximum current generated is known as the *limiting current*. Current is proportional to the concentration, C, of the target substance; the permeability, P; the exposed sensor (membrane) area, A; and the number of electrons transferred per molecule detected, n. The current is inversely proportional to the membrane thickness, t.

$$I \propto \frac{nPCA}{t} \qquad 63.2$$

[4]There is a voltage-balancing device that shares the name *potentiometer* (*potentiometer circuit*). An unknown voltage source can be measured by adjusting a calibrated voltage until a null reading is obtained on a voltage meter. The applications are sufficiently different that no confusion occurs when the "pot is adjusted."

[5]The categorization is common but not universal. The terms "transducer," "sensor," "sending unit," and "pickup" are often used loosely.

10. VARIABLE-INDUCTANCE TRANSDUCERS

Inductive transducers contain a wire coil and a moving permeable *core*.[6] As the core moves, the flux linkage through the coil changes. The change in inductance affects the overall impedance of the detector circuit.

The *differential transformer* or *linear variable differential transformer* (LVDT) is an important type of *variable-inductance transducer*. (See Fig. 63.2.) It converts linear motion into a change in voltage. The transformer is supplied with a low AC voltage. When the core is centered between the two secondary windings, the LVDT is said to be in its *null position*.

Movement of the core changes the magnetic flux linkage between the primary and secondary windings. Over a reasonable displacement range, the output voltage is proportional to the displacement of the core from the null position, hence the "linear" designation. The voltage changes phase (by 180°) as the core passes through the null position.

Figure 63.2 *Linear Variable Differential Transformer Schematic and Performance Characteristic*

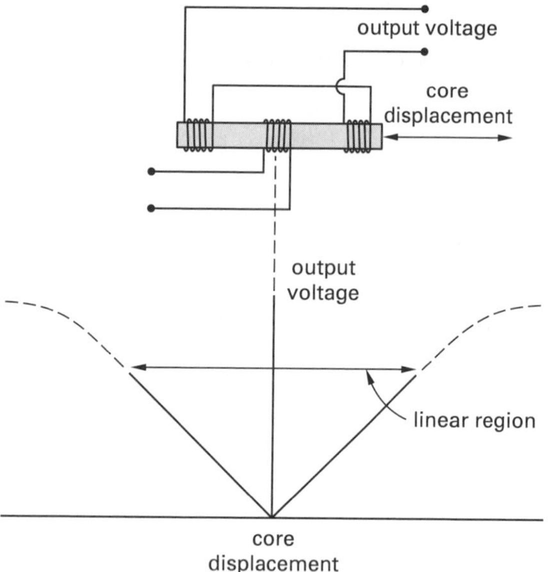

Sensitivity of an LVDT is measured in mV/in (mV/cm) of core movement. The sensitivity and output voltage depend on the frequency of the applied voltage (i.e., the *carrier frequency*) and are directly proportional to the magnitude of the applied voltage.

11. VARIABLE-RELUCTANCE TRANSDUCERS

A *variable-reluctance transducer* (*pickup*) is essentially a permanent magnet and a coil in the vicinity of the process being monitored.[7] There are no moving parts in this type of transducer. However, some of the magnet's magnetic flux passes through the surroundings, and the presence or absence of the process changes the coil voltage. Two typical applications of variable-reluctance pickups are measuring liquid levels and determining the rotational speed of a gear.

12. VARIABLE-CAPACITANCE TRANSDUCERS

In *variable-capacitance transducers*, the capacitance of a device can be modified by changing the plate separation, plate area, or dielectric constant of the medium separating the plates.

13. OTHER ELECTRICAL TRANSDUCERS

The *piezoelectric effect* is the name given to the generation of an electrical voltage when placed under stress.[8] *Piezoelectric transducers* generate a small voltage when stressed (strained). Since voltage is developed during the application of changing strain but not while strain is constant, piezoelectric transducers are limited to dynamic applications. Piezoelectric transducers may suffer from low voltage output, instability, and limited ranges in operating temperature and humidity.

The *photoelectric effect* is the generation of an electrical voltage when a material is exposed to light.[9] Devices using this effect are known as *photocells*, *photovoltaic cells*, *photosensors*, or *light-sensitive detectors*, depending on the applications. The sensitivity can be to radiation outside of the visible spectrum. Photoelectric detectors can be made that respond to infrared and ultraviolet radiation. The magnitude of the voltage (or of the current in an attached circuit) will depend on the amount of illumination. If the cell is reverse-biased by an external battery, its operation is similar to a constant-current source.[10] (See Fig. 63.3.)

[6]The term *core* is used even if the cross sections of the coil and core are not circular.
[7]*Reluctance* depends on the area, length, and permeability of the medium through which the magnetic flux passes.
[8]Quartz, table sugar, potassium sodium tartarate (Rochelle salt), and barium titanate are examples of piezoelectric materials. Quartz is commonly used to provide a stable frequency in electronic oscillators. Barium titanate is used in some ultrasonic cleaners and sonar-like equipment.
[9]While the *photogenerative* (*photovoltaic*) definition is the most common definition, the term "photoelectric" can also be used with *photoconductive devices* (those whose resistance changes with light) and *photoemissive devices* (those that emit light when a voltage is applied).
[10]A semiconductor device is *reverse-biased* when a negative battery terminal is connected to a p-type semiconductor material in the device, or when a positive battery terminal is connected to an n-type semiconductor material.

Figure 63.3 *Photovoltaic Device Characteristic Curves*

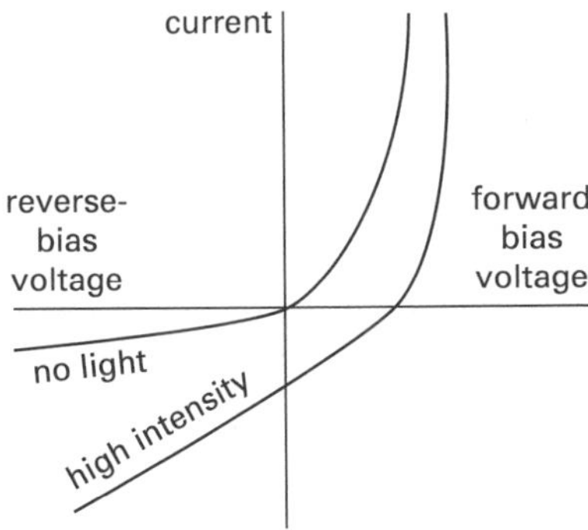

14. PHOTOSENSITIVE CONDUCTORS

Cadmium sulfide and *cadmium selenide* are two compounds that decrease in resistance when exposed to light. Cadmium sulfide is most sensitive to light in the 5000–6000 Å (0.5–0.6 μm) range, while cadmium selenide shows peak sensitivities in the 7000–8000 Å (0.7–0.8 μm) range. These compounds are used in *photosensitive conductors*. Due to hysteresis, photosensitive conductors do not react instantaneously to changes in intensity.[11] High-speed operation requires high light intensities and careful design.

15. RESISTANCE TEMPERATURE DETECTORS

Resistance temperature detectors (RTDs), also known as *resistance thermometers*, make use of changes in their resistance to determine changes in temperature. A fine wire is wrapped around a form and protected with glass or a ceramic coating. Nickel and copper are commonly used for industrial RTDs. Type D 100 Ω and type M 1000 Ω platinum are used when precision resistance thermometry is required. RTDs are connected through resistance bridges (see Sec. 63.19) to compensate for lead resistance. RTDs are categorized by their precision (tolerance). Class A RTDs are precise to within ±0.06% at 0°C, while class B RTDs are precise to within ±0.12% at 0°C. Tolerance varies with temperature, hence the need to specify a reference temperature.

Resistance in most conductors increases with temperature. The resistance at a given temperature can be calculated from the *coefficients of thermal resistance*, α and β.[12] Positive values of α are known as *positive temperature coefficients* (PTCs); negative values are known as *negative temperature coefficients* (NTCs). Although PTC platinum RTDs are most prevalent, RTDs can be constructed using materials with either positive or negative characteristics. The variation of resistance with temperature is nonlinear, though β is small and is often insignificant over short temperature ranges. Therefore, a linear relationship is often assumed and only α is used. In Eq. 63.3, R_{ref} is the resistance at the reference temperature, T_{ref}, usually 100 Ω at 32°F (0°C).

$$R_T \approx R_{\text{ref}}(1 + \alpha\Delta T + \beta\Delta T^2) \qquad 63.3$$

$$\Delta T = T - T_{\text{ref}} \qquad 63.4$$

In commercial RTDs, α is referred to by the literal term *alpha-value*. There are two applicable alpha values for platinum, depending on the purity. Commercial platinum RTDs produced in the United States generally have alpha values of 0.00391 1/°C, while RTDs produced in Europe and other countries generally have alpha values of 0.00385 1/°C.

Equation 63.3 and values of α and β are of academic interest. In practice, an equation similar in appearance using the actual temperature, T (not ΔT), is used. Equation 63.5 and Eq. 63.6 constitute a very accurate correlation of resistance and temperature for platinum. The correlation is known as the *Callendar-Van Dusen equation*. R_0 is the resistance of the RTD at 0°C, typically 100 Ω or 1000 Ω for commercial platinum RTDs, which are also referred to as *platinum resistance thermometers* (PRTs).

$$R_T = R_0\big(1 + AT + BT^2 + CT^3(T - 100°\text{C})\big) \quad [-200°\text{C} < T < 0°\text{C}] \qquad 63.5$$

$$R_T = R_0(1 + AT + BT^2) \quad [0°\text{C} < T < 850°\text{C}] \qquad 63.6$$

Values of A, B, and C can be calculated from the α and β values, if needed. For platinum RTDs with $\alpha = 0.00385$ 1/°C, these values are

- $A = 3.9083 \times 10^{-3}$ 1/°C
- $B = -5.775 \times 10^{-7}$ 1/°C^2
- $C = -4.183 \times 10^{-12}$ 1/°C^4

Table 63.2 gives resistances for standard Pt100 RTDs.

Plant Engineering

[11] *Hysteresis* is the tendency for the transducer to continue to respond (i.e., indicate) when the load is removed. Alternatively, hysteresis is the difference in transducer outputs when a specific load is approached from above and from below. Hysteresis is usually expressed in percent of the full-load reading during any single calibration cycle.

[12] Higher-order terms (third, fourth, etc.) are used when extreme accuracy is required.

Table 63.2 *Resistance of Standard Pt100 RTDs**

T (°C)	R (Ω)	T (°C)	R (Ω)	T (°C)	R (Ω)
−50	80.31	10	103.90	70	127.07
−45	82.29	15	105.85	75	128.98
−40	84.27	20	107.79	80	130.89
−35	86.25	25	109.73	85	132.80
−30	88.22	30	111.67	90	134.70
−25	90.19	35	113.61	95	136.60
−20	92.16	40	115.54	100	138.50
−15	94.12	45	117.47	105	140.39
−10	96.09	50	119.40	110	142.29
−5	98.04	55	121.32	150	157.31
0	100.00	60	123.24	200	175.84
5	101.95	65	125.16		

(Multiply tabulated resistances by 10 for Pt1000 RTDs.)

*Per *DIN International Electrotechnical Commission (IEC) Standard* 751

16. THERMISTORS

Thermistors are temperature-sensitive semiconductors constructed from oxides of manganese, nickel, and cobalt, and from sulfides of iron, aluminum, and copper. Thermistor materials are encapsulated in glass or ceramic materials to prevent penetration of moisture. Unlike RTDs, the resistance of thermistors decreases as the temperature increases.

Thermistor temperature-resistance characteristics are exponential. Depending on the brand, material, and construction, β typically varies between 3400K and 3900K.

$$R = R_o e^k \qquad 63.7$$

$$k = \beta\left(\frac{1}{T} - \frac{1}{T_o}\right) \quad [T \text{ in K}] \qquad 63.8$$

Thermistors can be connected to measurement circuits with copper wire and soldered connections. Compensation of lead wire effects is not required because resistance of thermistors is very large, far greater than the resistance of the leads. Since the negative temperature characteristic makes it difficult to design customized detection circuits, some thermistor and instrumentation standardization has occurred. The most common thermistors have resistances of 2252 Ω at 77°F (25°C), and most instrumentation is compatible with them. Other standardized resistances are 3000 Ω, 5000 Ω, 10,000 Ω, and 30,000 Ω at 77°F (25°C). (See Table 63.3.)

Thermistors typically are less precise and more unpredictable than metallic resistors. Since resistance varies exponentially, most thermistors are suitable for use only up to approximately 550°F (290°C).

17. THERMOCOUPLES

A *thermocouple* consists of two wires of dissimilar metals joined at both ends.[13] One set of ends, typically called a *junction*, is kept at a known *reference temperature* while the other junction is exposed to the unknown temperature.[14] (See Fig. 63.4.) In a laboratory, the reference junction is often maintained at the *ice point*, 32°F (0°C), in an ice/water bath for convenience in later analysis. In commercial applications, the reference temperature can be any value, with appropriate compensation being made.

Thermocouple materials, standard ANSI designations, and approximate useful temperature ranges are given in Table 63.4.[15] The "functional" temperature range can be much larger than the useful range. The most significant factors limiting the useful temperature range, sometimes referred to as the *error limits range*, are linearity, the rate at which the material will erode due to oxidation at higher temperatures, irreversible magnetic effects above magnetic critical points, and longer stabilization periods at higher temperatures.

Figure 63.4 *Thermocouple Circuits*

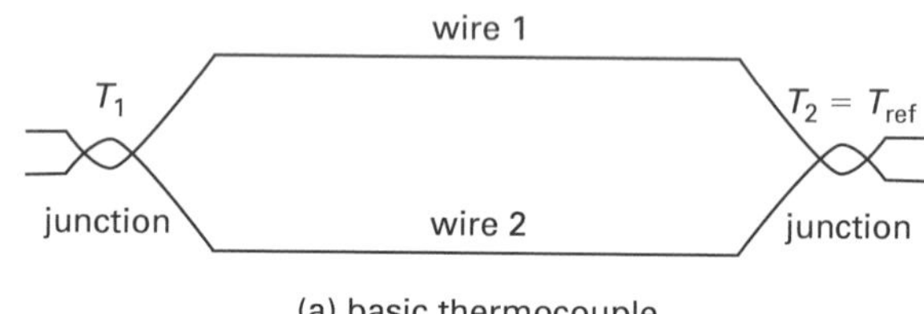

(a) basic thermocouple

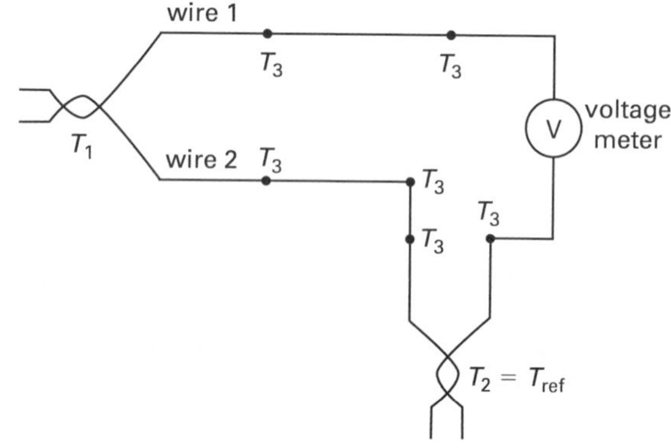

(b) thermocouple in measurement circuit

[13] The joint may be made by simply twisting the ends together. However, to achieve a higher mechanical strength and a better electrical connection, the ends should be soldered, brazed, or welded.

[14] The *ice point* is the temperature at which liquid water and ice are in equilibrium. Other standardized temperature references are: *oxygen point*, −297.346°F (90.19K); *steam point*, 212.0°F (373.16K); *sulfur point*, 832.28°F (717.76K); *silver point*, 1761.4°F (1233.96K); and *gold point*, 1945.4°F (1336.16K).

[15] It is not uncommon to list the two thermocouple materials with "vs." (as in *versus*). For example, a copper/constantan thermocouple might be designated as "copper vs. constantan."

Table 63.3 *Typical Resistivities and Coefficients of Thermal Resistance*[a]

conductor	resistivity[b] (Ω·cm)	α^c (1/°C)
alumel[d]	28.1×10^{-6}	0.0024 @ 212°F (100°C)
aluminum	2.82×10^{-6}	0.0039 @ 68°F (20°C)
		0.0040 @ 70°F (21°C)
brass	7×10^{-6}	0.002 @ 68°F (20°C)
constantan[e,f]	49×10^{-6}	0.00001 @ 68°F (20°C)
chromel[g]	0.706×10^{-6}	0.00032 @ 68°F (20°C)
copper, annealed	1.724×10^{-6}	0.0043 @ 32°F (0°C)
		0.0039 @ 70°F (21°C)
		0.0037 @ 100°F (38°C)
		0.0031 @ 200°F (93°C)
gold	2.44×10^{-6}	0.0034 @ 68°F (20°C)
iron (99.98% pure)	10×10^{-6}	0.005 @ 68°F (20°C)
isoelastic[h]	112×10^{-6}	0.00047
lead	22×10^{-6}	0.0039
magnesium	4.6×10^{-6}	0.004 @ 68°F (20°C)
manganin[i]	44×10^{-6}	0.0000 @ 68°F (20°C)
monel[j]	42×10^{-6}	0.002 @ 68°F (20°C)
nichrome[k]	100×10^{-6}	0.0004 @ 68°F (20°C)
nickel	7.8×10^{-6}	0.006 @ 68°F (20°C)
platinum	10×10^{-6}	0.0039 @ 32°F (0°C)
		0.0036 @ 70°F (21°C)
platinum-iridium[l]	24×10^{-6}	0.0013
platinum-rhodium[m]	18×10^{-6}	0.0017 @ 212°F (100°C)
silver	1.59×10^{-6}	0.004 @ 68°F (20°C)
tin	11.5×10^{-6}	0.0042 @ 68°F (20°C)
tungsten (drawn)	5.8×10^{-6}	0.0045 @ 70°F (21°C)

(Multiply 1/°C by 5/9 to obtain 1/°F.)

(Multiply ppm/°F by 1.8×10^{-6} to obtain 1/°C.)

[a]Compiled from various sources. Data is not to be taken too literally, as values depend on composition and cold working.
[b]At 20°C (68°F)
[c]Values vary with temperature. Common values given when no temperature is specified.
[d]Trade name for 94% Ni, 2.5% Mn, 2% Al, 1% Si, 0.5% Fe (TM of Hoskins Manufacturing Co.)
[e]60% Cu, 40% Ni, also known by trade names *Advance, Eureka,* and *Ideal.*
[f]Constantan is also the name given to the composition 55% Cu and 45% Ni, an alloy with slightly different properties.
[g]Trade name for 90% Ni, 10% Cr (TM of Hoskins Manufacturing Co.)
[h]36% Ni, 8% Cr, 0.5% Mo, remainder Fe
[i]9% to 18% Mn, 11% to 4% Ni, remainder Cu
[j]33% Cu, 67% Ni
[k]75% Ni, 12% Fe, 11% Cr, 2% Mn
[l]95% Pt, 5% Ir
[m]90% Pt, 10% Rh

A voltage is generated when the temperatures of the two junctions are different. This phenomenon is known as the *Seebeck effect.*[16] Referring to the polarities of the voltage generated, one metal is known as the *positive element* while the other is the *negative element.* The generated voltage is small, and thermocouples are calibrated in μV/°F or μV/°C. An amplifier may be required to provide usable signal levels, although thermocouples can be connected in series (a *thermopile*) to increase the value.[17] The accuracy (referred to as the *calibration*) of thermocouples is approximately ½–¾%, though manufacturers produce thermocouples with various guaranteed accuracies.

The voltage generated by a thermocouple is given by Eq. 63.9. Since the *thermoelectric constant*, k_T, varies with temperature, thermocouple problems can be solved with published tables of total generated voltage versus temperature. (See App. 63.A.)

$$V = k_T(T - T_{\text{ref}}) \qquad 63.9$$

Generation of thermocouple voltage in a measurement circuit is governed by three laws. The *law of homogeneous circuits* states that the temperature distribution along one or both of the thermocouple leads is irrelevant. Only the junction temperatures contribute to the generated voltage.

The *law of intermediate metals* states that an intermediate length of wire placed within one leg or at the junction of the thermocouple circuit will not affect the voltage generated as long as the two new junctions are at the same temperature. This law permits the use of a measuring device, soldered connections, and extension leads.

The *law of intermediate temperatures* states that if a thermocouple generates voltage V_1 when its junctions are at T_1 and T_2, and it generates voltage V_2 when its junctions are at T_2 and T_3, then it will generate voltage $V_1 + V_2$ when its junctions are at T_1 and T_3.

Example 63.2

A type-K (chromel-alumel) thermocouple produces a voltage of 10.87 mV. The "cold" junction is at 70°F. What is the temperature of the hot junction?

Solution

Use the law of intermediate temperatures. From App. 63.A, the thermoelectric constant for 70°F is 0.84 mV. If the cold junction had been at 32°C, the generated voltage would have been higher. The corrected reading is

$$10.87 \text{ mV} + 0.84 \text{ mV} = 11.71 \text{ mV}$$

The temperature corresponding to this voltage is 550°F.

Plant Engineering

Table 63.4 Typical Temperature Ranges of Thermocouple Materials[a]

materials	ANSI designation	useful range (°F (°C))
copper-constantan	T	–300 to 700 (–180 to 370)
chromel-constantan	E	32 to 1600 (0 to 870)
iron-constantan[b]	J	32 to 1400 (0 to 760)
chromel-alumel	K	32 to 2300 (0 to 1260)
platinum-10% rhodium	S	32 to 2700 (0 to 1480)
platinum-13% rhodium	R	32 to 2700 (0 to 1480)
Pt-6% Rh-Pt-30% Rh	B	1600 to 3100 (870 to 1700)
tungsten-Tu-25% rhenium	–	to 4200[c] (2320)
Tu-5% rhenium-Tu-26% rhenium	–	to 4200[c] (2320)
Tu-3% rhenium-Tu-25% rhenium	–	to 4200[c] (2320)
iridium-rhodium	–	to 3500[c] (1930)
nichrome-constantan	–	to 1600[c] (870)
nichrome-alumel	–	to 2200[c] (1200)

[a]Actual values will depend on wire gauge, atmosphere (oxidizing or reducing), use (continuous or intermittent), and manufacturer.
[b]Nonoxidizing atmospheres only.
[c]Approximate usable temperature range. Error limit range is less.

Example 63.3

A type-T (copper-constantan) thermocouple is connected directly to a voltage meter. The temperature of the meter's screw-terminals is measured by a nearby thermometer as 70°F. The thermocouple generates 5.262 mV. What is the temperature of its hot junction?

Solution

There are two connections at the meter. However, both connections are at the same temperature, so the meter can be considered to be a length of different wire, and the law of intermediate metals applies. Since the meter connections are not at 32°F, the law of intermediate temperatures applies. The corrected voltage is

$$5.262 \text{ mV} + 0.832 \text{ mV} = 6.094 \text{ mV}$$

From App. 63.A, 6.094 mV corresponds to 280°F.

18. STRAIN GAUGES

A *bonded strain gauge* is a metallic resistance device that is cemented to the surface of the unstressed member.[18] The gauge consists of a metallic conductor (known as the *grid*) on a backing (known as the *substrate*).[19] The substrate and grid experience the same strain as the surface of the member. The resistance of the gauge changes as the member is stressed due to changes in conductor cross section and intrinsic changes in resistivity with strain. Temperature effects must be compensated for by the circuitry or by using a second unstrained gauge as part of the bridge measurement system. (See Sec. 63.19.)

When simultaneous strain measurements in two or more directions are needed, it is convenient to use a commercial *rosette strain gauge*. A rosette consists of two or more *grids* properly oriented for application as a single unit. (See Fig. 63.5.)

***Figure 63.5** Strain Gauge*

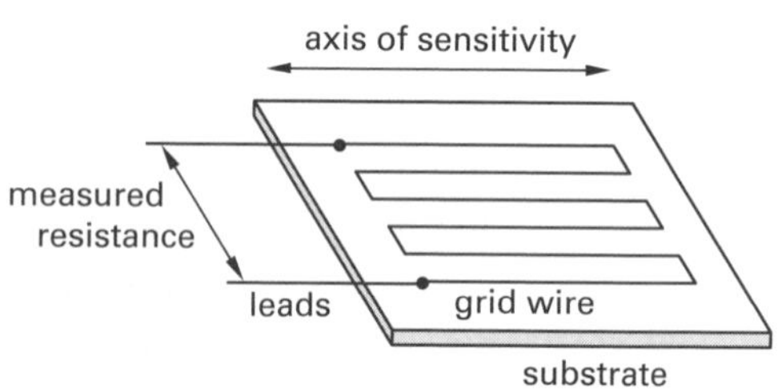

(a) folded-wire strain gauge

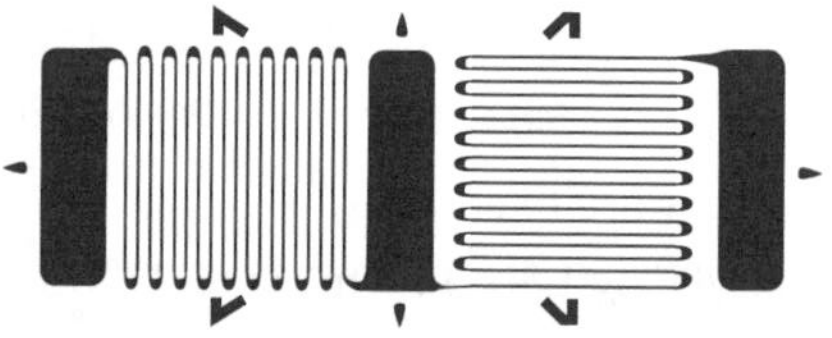

(b) commercial two-element rosette

The *gage factor* (*strain sensitivity factor*), GF, is the ratio of the fractional change in resistance to the fractional change in length (strain) along the detecting axis of the gauge. The higher the gage factor, the greater the

[16]The inverse of the Seebeck effect, that current flowing through a junction of dissimilar metals will cause either heating or cooling, is the *Peltier effect*, though the term is generally used in regard to cooling applications. An extension of the Peltier effect, known as the *Thompson effect*, is that heat will be carried along the conductor. Both the Peltier and Thompson effects occur simultaneously with the Seebeck effect. However, the Peltier and Thompson effects are so minuscule that they can be disregarded.
[17]There is no special name for a combination of thermocouples connected in parallel.
[18]A *bonded strain gauge* is constructed by bonding the conductor to the surface of the member. An *unbonded strain gauge* is constructed by wrapping the conductor tightly around the member or between two points on the member. Strain gauges on rotating shafts are usually connected through *slip rings* to the measurement circuitry.
[19]The grids of strain gauges were originally of the folded-wire variety. For example, nichrome wire with a total resistance under 1000 Ω was commonly used. Modern strain gauges are generally of the foil type manufactured by printed circuit techniques. Semiconductor gauges are also used when extreme sensitivity (i.e., gage factors in excess of 100) is required. However, semiconductor gauges are extremely temperature-sensitive.

sensitivity of the gauge. The gage factor is a function of the gauge material. It can be calculated from the grid material's properties and configuration. From a practical standpoint, however, the gage factor and gage resistance are provided by the gauge manufacturer. Only the change in resistance is measured. (See Table 63.5.)

$$\mathrm{GF} = 1 + 2\nu + \frac{\frac{\Delta\rho}{\rho_o}}{\frac{\Delta L}{L_o}} = \frac{\frac{\Delta R_g}{R_g}}{\frac{\Delta L}{L_o}} = \frac{\frac{\Delta R_g}{R_g}}{\epsilon} \quad \text{63.10}$$

***Table 63.5** Approximate Gage Factors*[a]

material	GF
constantan	2.0
iron, soft	4.2
isoelastic	3.5
manganin	0.47
monel	1.9
nichrome	2.0
nickel	–12[b]
platinum	4.8
platinum-iridium	5.1

[a]Other properties of strain gauge materials are listed in Table 63.3.
[b]Value depends on amount of preprocessing and cold working.

Constantan and isoelastic wires, along with metal foil with gage factors of approximately 2 and initial resistances of less than 1000 Ω (typically 120 Ω, 350 Ω, 600 Ω, and 700 Ω) are commonly used. In practice, the gage factor and initial gage resistance, R_g, are specified by the manufacturer of the gauge. Once the strain sensitivity factor is known, the strain, ϵ, can be determined from the change in resistance. Strain is often reported in units of μin/in (μm/m) and is given the name *microstrain*.

$$\epsilon = \frac{\Delta R_g}{(\mathrm{GF})R_g} \quad \text{63.11}$$

Theoretically, a strain gauge should not respond to strain in its transverse direction. However, the turn-around end-loops are also made of strain-sensitive material, and the end-loop material contributes to a nonzero sensitivity to strain in the transverse direction. Equation 63.12 defines the *transverse sensitivity factor*, K_t, which is of academic interest in most problems. The transverse sensitivity factor is seldom greater than 2%.

$$K_t = \frac{(\mathrm{GF})_{\mathrm{transverse}}}{(\mathrm{GF})_{\mathrm{longitudinal}}} \quad \text{63.12}$$

Example 63.4

A strain gauge with a nominal resistance of 120 Ω and gage factor of 2.0 is used to measure a strain of 1 μin/in. What is the change in resistance?

Solution

From Eq. 63.11,

$$\Delta R_g = (\mathrm{GF})R_g\epsilon = (2.0)(120\ \Omega)\left(1 \times 10^{-6}\ \frac{\mathrm{in}}{\mathrm{in}}\right)$$
$$= 2.4 \times 10^{-4}\ \Omega$$

19. WHEATSTONE BRIDGES

The *Wheatstone bridge*, shown in Fig. 63.6, is one type of *resistance bridge*.[20] The bridge can be used to determine the unknown resistance of a resistance transducer (e.g., thermistor or resistance-type strain gauge), say R_1 in Fig. 63.6. The potentiometer is adjusted (i.e., the bridge is "balanced") until no current flows through the meter or until there is no voltage across the meter (hence the name *null indicator*).[21,22] When the bridge is balanced and no current flows through the meter leg, Eq. 63.13 through Eq. 63.16 are applicable.

$$I_2 = I_4 \quad [\text{balanced}] \quad \text{63.13}$$

$$I_1 = I_3 \quad [\text{balanced}] \quad \text{63.14}$$

$$V_1 + V_3 = V_2 + V_4 \quad [\text{balanced}] \quad \text{63.15}$$

$$\frac{R_1}{R_2} = \frac{R_3}{R_4} \quad [\text{balanced}] \quad \text{63.16}$$

***Figure 63.6** Wheatstone Bridge*

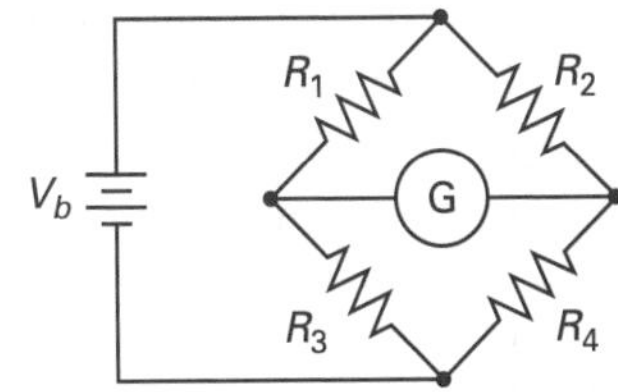

Since any one of the four resistances can be the unknown, up to three of the remaining resistances can be fixed or adjustable, and the battery and meter can be connected to either of two diagonal corners, it is sometimes confusing to apply Eq. 63.16 literally. However, the following bridge law statement can be used to help formulate the proper relationship: *When a series Wheatstone bridge is null-balanced, the ratio of resistance of any two adjacent arms equals the ratio of resistance of*

Plant Engineering

[20]Other types of resistance bridges are the *differential series balance bridge*, *shunt balance bridge*, and *differential shunt balance bridge*. These differ in the manner in which the adjustable resistor is incorporated into the circuit.
[21]This gives rise to the alternate names of *zero-indicating bridge* and *null-indicating bridge*.
[22]The unknown resistance can also be determined from the amount of voltage unbalance shown by the meter reading, in which case, the bridge is known as a *deflection bridge* rather than a null-indicating bridge. Deflection bridges are described in Sec. 63.20.

the remaining two arms, taken in the same sense. In this statement, "taken in the same sense" means that both ratios must be formed reading either left to right, right to left, top to bottom, or bottom to top.

20. STRAIN GAUGE DETECTION CIRCUITS

The resistance of a strain gauge can be measured by placing the gauge in either a ballast circuit or bridge circuit. A *ballast circuit* consists of a voltage source, V_b, of less than 10 V (typical); a current-limiting ballast resistance, R_b; and the strain gauge of known resistance, R_g, in series. (See Fig. 63.7.) This is essentially a voltage-divider circuit. The change in voltage, ΔV_g, across the strain gauge is measured. The strain, ϵ, can be determined from Eq. 63.17.

$$\Delta V_g = \frac{(\text{GF})\epsilon V_b R_b R_g}{(R_b + R_g)^2} \quad 63.17$$

Figure 63.7 Ballast Circuit

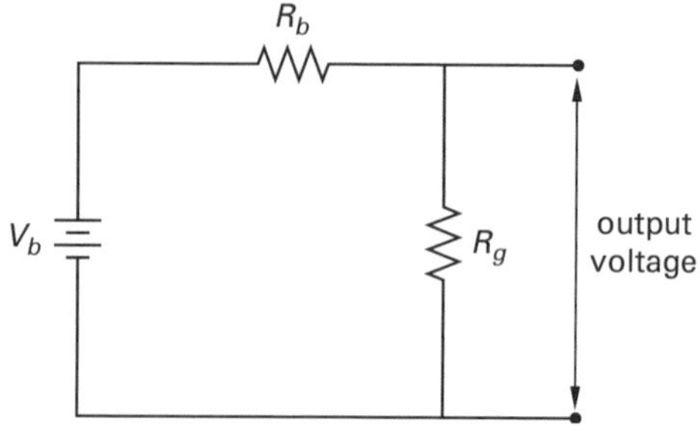

Ballast circuits do not provide temperature compensation, nor is their sensitivity adequate for measuring static strain. Ballast circuits, where used, are often limited to measurement of transient strains. A bridge detection circuit overcomes these limitations.

Figure 63.8 illustrates how a strain gauge can be used with a resistance bridge. Gauge 1 measures the strain, while *dummy gauge* 2 provides temperature compensation.[23] The meter voltage is a function of the input (battery) voltage and the resistors. (As with bridge circuits, the input voltage is typically less than 10 V.) The variable resistance is used for balancing the bridge prior to the strain. When the bridge is balanced, V_{meter} is zero.

Figure 63.8 Strain Gauge in Resistance Bridge

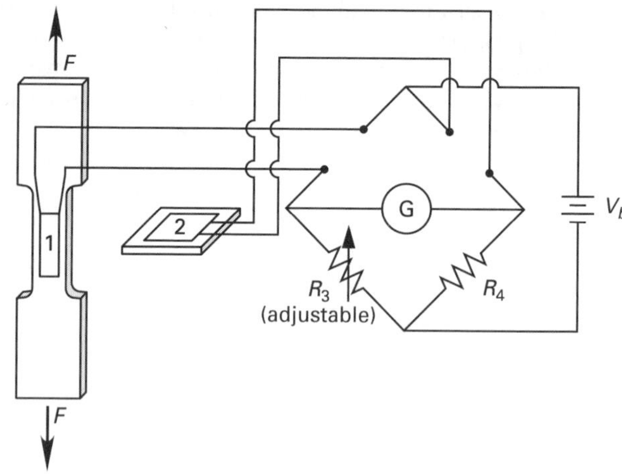

When the gauge is strained, the bridge becomes unbalanced. Assuming the bridge is initially balanced, the voltage at the meter (known as the *voltage deflection* from the null condition) will be[24]

$$V_{\text{meter}} = V_b\left(\frac{R_1}{R_1 + R_3} - \frac{R_2}{R_2 + R_4}\right) \quad [\tfrac{1}{4}\text{-bridge}] \quad 63.18$$

For a single strain gauge in a resistance bridge and neglecting lead resistance, the voltage deflection is related to the strain by Eq. 63.19.

$$V_{\text{meter}} = \frac{(\text{GF})\epsilon V_b}{4 + 2(\text{GF})\epsilon} \approx \tfrac{1}{4}(\text{GF})\epsilon V_b \quad [\tfrac{1}{4}\text{-bridge}] \quad 63.19$$

21. STRAIN GAUGE IN UNBALANCED RESISTANCE BRIDGE

A resistance bridge does not need to be balanced prior to use as long as an accurate digital voltmeter is used in the detection circuit. The voltage ratio difference, ΔVR, is defined as the fractional change in the output voltage from the unstrained to the strained condition.

$$\Delta\,\text{VR} = \left(\frac{V_{\text{meter}}}{V_b}\right)_{\text{strained}} - \left(\frac{V_{\text{meter}}}{V_b}\right)_{\text{unstrained}} \quad 63.20$$

[23]This is a "quarter-bridge" or "¼-bridge" configuration, as described in Sec. 63.23. The strain gauge used for temperature compensation is not active.
[24]Equation 63.18 applies to the unstrained condition as well. However, if the gauge is unstrained and the bridge is balanced, the resistance term in parentheses is zero.

If the only resistance change between the strained and unstrained conditions is in the strain gauge and lead resistance is disregarded, the fractional change in gage resistance for a single strain gauge in a resistance bridge is

$$\frac{\Delta R_g}{R_g} = \frac{-4\Delta(\text{VR})}{1+2\Delta(\text{VR})} \quad [\tfrac{1}{4}\text{-bridge}] \qquad 63.21$$

Since the fractional change in gage resistance also occurs in the definition of the gage factor (see Eq. 63.11), the strain is

$$\epsilon = \frac{-4\Delta\text{VR}}{(\text{GF})(1+2\Delta\text{VR})} \quad [\tfrac{1}{4}\text{-bridge}] \qquad 63.22$$

22. BRIDGE CONSTANT

The voltage deflection can be doubled (or quadrupled) by using two (or four) strain gauges in the bridge circuit. The larger voltage deflection is more easily detected, resulting in more accurate measurements.

Use of multiple strain gauges is generally limited to configurations where symmetrical strain is available on the member. For example, a beam in bending experiences the same strain on the top and bottom faces. Therefore, if the temperature-compensation strain gauge shown in Fig. 63.8 is bonded to the bottom of the beam, the resistance change would double.

The *bridge constant* (BC) is the ratio of the actual voltage deflection to the voltage deflection from a single gauge. Depending on the number and orientation of the gauges used, bridge constants of 1.0, 1.3, 2.0, 2.6, and 4.0 may be encountered (for materials with a Poisson's ratio of 0.3).

Figure 63.9 illustrates how (up to) four strain gauges can be connected in a Wheatstone bridge circuit. The total strain indicated will be the algebraic sum of the four strains detected. For example, if all four strains are equal in magnitude, ϵ_1 and ϵ_4 are tensile (i.e., positive), and ϵ_2 and ϵ_3 are compressive (i.e., negative), then the bridge constant would be 4.

$$\epsilon_t = \epsilon_1 - \epsilon_2 - \epsilon_3 + \epsilon_4 \qquad 63.23$$

Figure 63.9 *Wheatstone Bridge Strain Gauge Circuit**

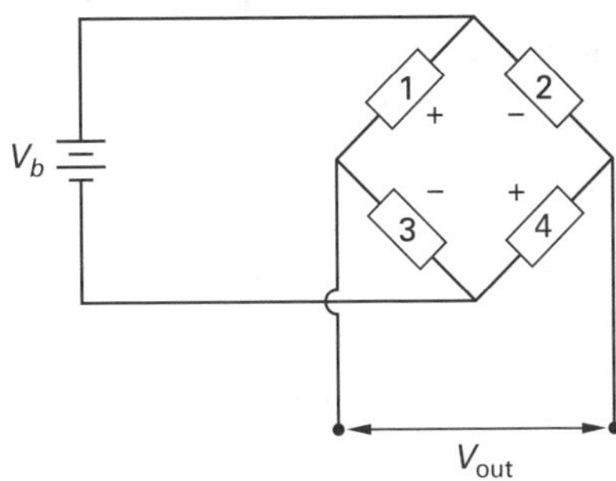

*The 45° orientations shown are figurative. Actual gauge orientation can be in any direction.

23. STRESS MEASUREMENTS IN KNOWN DIRECTIONS

Strain gauges are the most frequently used method of determining the stress in a member. Stress can be calculated from strain, or the measurement circuitry can be calibrated to give the stress directly.

For stress in only one direction (i.e., the *uniaxial stress* case), such as a simple bar in tension, only one strain gauge is required. The stress can be calculated from *Hooke's law.*

Hooke's Law

$$\sigma = E\epsilon \qquad 63.24$$

When a surface, such as that of a pressure vessel, experiences simultaneous stresses in two directions (the *biaxial stress* case), the strain in one direction affects the strain in the other direction.[25] Therefore, two strain gauges are required, even if the stress in only one direction is needed. The strains actually measured by the gauges are known as the *net strains.*

Hooke's Law

$$\epsilon_x = \frac{1}{E}(\sigma_x - \nu\sigma_y) \qquad 63.25$$

$$\epsilon_y = \frac{1}{E}(\sigma_y - \nu\sigma_x) \qquad 63.26$$

The stresses are determined by solving Eq. 63.25 and Eq. 63.26 simultaneously.

$$\sigma_x = \frac{E(\epsilon_x + \nu\epsilon_y)}{1-\nu^2} \qquad 63.27$$

$$\sigma_y = \frac{E(\epsilon_y + \nu\epsilon_x)}{1-\nu^2} \qquad 63.28$$

[25]Thin-wall pressure vessel theory shows that the circumferential (hoop) stress is twice the longitudinal stress. However, the ratio of circumferential to longitudinal strains is closer to 4:1 than to 2:1.

Figure 63.9 shows how four strain gauges can be interconnected in a bridge circuit. Figure 63.10 shows how (up to) four strain gauges would be physically oriented on a test specimen to measure different types of stress. Not all four gauges are needed in all cases. If four gauges are used, the arrangement is said to be a *full bridge*. If only one or two gauges are used, the terms *quarter-bridge* (¼-bridge) and *half-bridge* (½-bridge), respectively, apply.

In the case of up to four gauges applied to detect bending strain (see Fig. 63.10(a)), the bridge constant (BC) can be 1.0 (one gauge in position 1), 2.0 (two gauges in positions 1 and 2), or 4.0 (all four gauges). The relationships between the stress, strain, and applied force are

$$\sigma = E\epsilon = \frac{E\epsilon_t}{\text{BC}} \quad 63.29$$

$$\sigma = \frac{Mc}{I} = \frac{Mh}{2I} \quad 63.30$$

$$I = \frac{bh^3}{12} \quad \text{[rectangular section]} \quad 63.31$$

For axial strain (see Fig. 63.10(b)) and a material with a Poisson's ratio of 0.3, the bridge constant can be 1.0 (one gauge in position 1), 1.3 (two gauges in positions 1 and 2), 2.0 (two gauges in positions 1 and 3), or 2.6 (all four gauges).

$$\sigma = E\epsilon = \frac{E\epsilon_t}{\text{BC}} \quad 63.32$$

$$\sigma = \frac{F}{A} \quad 63.33$$

$$A = bh \quad \text{[rectangular section]} \quad 63.34$$

For shear strain (see Fig. 63.10(c)) and a material with a Poisson's ratio of 0.3, the bridge constant can be 2.0 (two gauges in positions 1 and 2) or 4.0 (all four gauges). The shear strain is twice the axial strain at 45°.

$$\tau = G\gamma = 2G\epsilon = \frac{2G\epsilon_t}{\text{BC}} \quad 63.35$$

$$\tau_{\text{max}} = \frac{FQ_{\text{max}}}{bI} \quad 63.36$$

$$\gamma = 2\epsilon \quad \text{[at 45°]} \quad 63.37$$

$$Q_{\text{max}} = \frac{bh^2}{8} \quad \text{[rectangular section]} \quad 63.38$$

$$G = \frac{E}{2(1+\nu)} \quad 63.39$$

Figure 63.10 *Orientation of Strain Gauges*

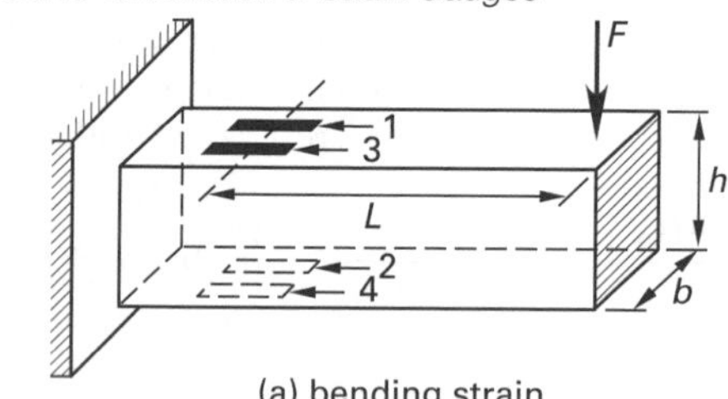

(a) bending strain

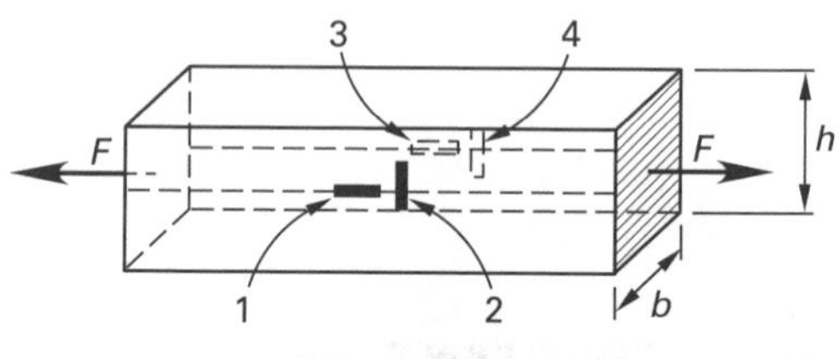

(b) axial strain

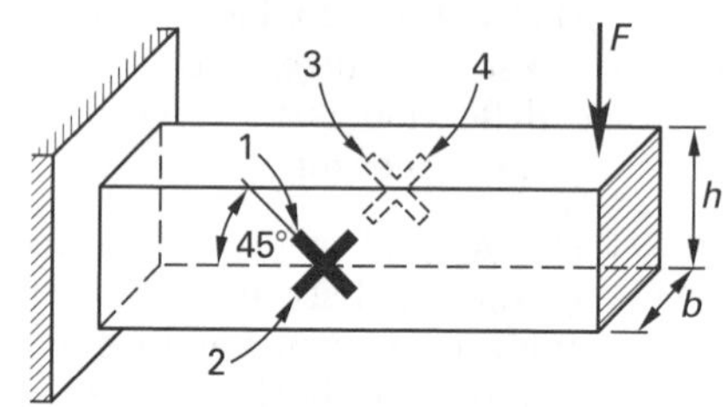

(c) shear strain

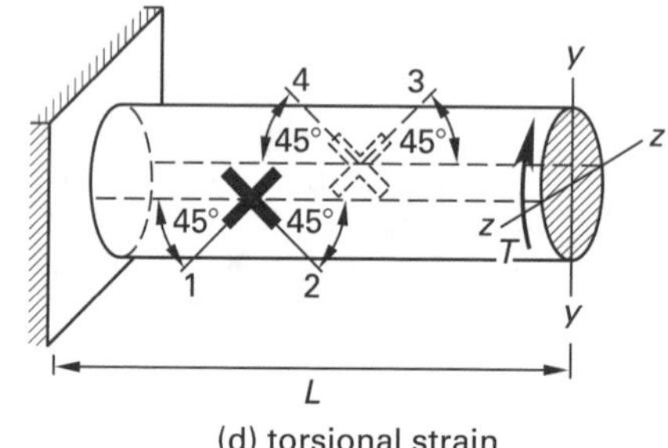

(d) torsional strain

For torsional strain (see Fig. 63.10(d)) and a material with a Poisson's ratio of 0.3, the bridge constant can be 2.0 (two gauges in positions 1 and 2) or 4.0 (all four gauges). The shear strain is twice the axial strain at 45°.

$$\tau = G\gamma = 2G\epsilon = \frac{2G\epsilon_t}{\text{BC}} \quad 63.40$$

Torsion

$$\tau = \frac{Tr}{J} = \frac{TD}{2J} \quad 63.41$$

$$\gamma = 2\epsilon \quad \text{[at 45°]} \quad 63.42$$

$$J = \frac{\pi D^4}{32} \quad \text{[solid circular section]} \quad 63.43$$

Torsional Strain

$$\phi = \frac{TL}{GJ} \quad 63.44$$

$$G = \frac{E}{2(1+\nu)} \quad 63.45$$

24. STRESS MEASUREMENTS IN UNKNOWN DIRECTIONS

In order to calculate the maximum stresses (i.e., the principal stresses) on the surface shown in Fig. 63.10, the gauges would be oriented in the known directions of the principal stresses.

In most cases, however, the directions of the principal stresses are not known. Therefore, rosettes of at least three gauges are used to obtain information in a third direction. Rosettes of three gauges (*rectangular* and *equiangular* (*delta*) *rosettes*) are used for this purpose. *T-delta rosettes* include a fourth strain gauge to refine and validate the results of the three primary gauges. Table 63.6 can be used for calculating principal stresses.

25. LOAD CELLS

Load cells are used to measure forces. A load cell is a transducer that converts a tensile or compressive force into an electrical signal. Though the details of the load cell will vary with the application, the basic elements are (a) a member that is strained by the force and (b) a strain detection system (e.g., strain gauge). The force is calculated from the observed deflection, y. In Eq. 63.46, the spring constant, k, is known as the load cell's *deflection constant.*

$$F = ky \quad 63.46$$

Because of their low cost and simple construction, *bending beam load cells* are the most common variety of load cells. Two strain gauges, one on the top and the other mounted on the bottom of a cantilever bar, are used. *Shear beam load cells* (which detect force by measuring the shear stress) can be used where the shear does not vary considerably with location, as in the web of an I-beam cross section.[26] The common S-shaped load cell constructed from a machined steel block can be instrumented as either a bending beam or shear beam load cell.

Load cell applications are categorized into grades or accuracy classes, with class III (500 to 10,000 scale divisions) being the most common. Commercial load cells meet standardized limits on errors due to temperature, nonlinearity, and hysteresis. The *temperature effect on output* (TEO) is typically stated in percentage change per 100°F (55.5°C) change in temperature.

Nonlinearity errors are reduced in proportion to the load cell's derating (i.e., using the load cell to measure forces less than its rated force). For example, a 2:1 derating will reduce the nonlinearity errors by 50%. Hysteresis is not normally reduced by derating.

The overall error of force measurement can be reduced by a factor of $1/\sqrt{n}$ (where n is the number of load cells that share the load equally) by using more than one load cell. Conversely, the applied force can vary by $\sqrt{n}$ times the known accuracy of a single load cell without decreasing the error.

26. DYNAMOMETERS

Torque from large motors and engines is measured by a *dynamometer. Absorption dynamometers* (e.g., the simple *friction brake*, *Prony brake*, *water brake*, and *fan brake*) dissipate energy as the torque is measured. Opposing torque in pumps and compressors must be supplied by a *driving dynamometer*, which has its own power input. *Transmission dynamometers* (e.g., *torque meters*, *torsion dynamometers*) use strain gauges to sense torque. They do not absorb or provide energy.

Using a brake dynamometer involves measuring a force, a moment arm, and the angular speed of rotation. The familiar torque-power-speed relationships are used with absorption dynamometers.

$$T = Fr \quad 63.47$$

$$P_{\text{ft-lbf/min}} = 2\pi T_{\text{ft-lbf}} N_{\text{rpm}} \quad 63.48$$

$$P_{\text{kW}} = \frac{T_{\text{N}\cdot\text{m}} N_{\text{rpm}}}{9549} \quad \text{[SI]} \quad 63.49(a)$$

$$P_{\text{hp}} = \frac{2\pi F_{\text{lbf}} r_{\text{ft}} N_{\text{rpm}}}{33{,}000} = \frac{2\pi T_{\text{ft-lbf}} N_{\text{rpm}}}{33{,}000} \quad \text{[U.S.]} \quad 63.49(b)$$

Some brakes and dynamometers are constructed with a "standard" brake arm whose length is 5.252 ft. In that case, the horsepower calculation conveniently reduces to

$$P_{\text{hp}} = \frac{F_{\text{lbf}} N_{\text{rpm}}}{1000} \quad \text{[standard brake arm]} \quad 63.50$$

[26]While shear in a rectangular beam varies parabolically with distance from the neutral axis, shear in the web of an I-beam is essentially constant at F/A. The flanges carry very little of the shear load. Other advantages of the shear beam load cell include protection from the load and environment, high side load rejection, lower creep, faster RTZ (return to zero) after load removal, and higher tolerance of vibration, dynamic forces, and noise.

Plant Engineering

Table 63.6 Stress-Strain Relationships for Strain Gauge Rosettes[a]

type of rosette	rectangular	equiangular (delta)	T-delta
principal strains, ϵ_p, ϵ_q	$\frac{1}{2}\left(\epsilon_a+\epsilon_c \pm\sqrt{2(\epsilon_a-\epsilon_b)^2+2(\epsilon_b-\epsilon_c)^2}\right)$	$\frac{1}{3}\left(\epsilon_a+\epsilon_b+\epsilon_c \pm\sqrt{2(\epsilon_a-\epsilon_b)^2+2(\epsilon_b-\epsilon_c)^2+2(\epsilon_c-\epsilon_a)^2}\right)$	$\frac{1}{2}\left(\epsilon_a+\epsilon_d \pm\sqrt{(\epsilon_a-\epsilon_d)^2+\frac{4}{3}(\epsilon_b-\epsilon_c)^2}\right)$
principal stresses, σ_1, σ_2	$\frac{E}{2}\left(\frac{\epsilon_a+\epsilon_c}{1-\nu}\pm\frac{1}{1+\nu}\times\sqrt{2(\epsilon_a-\epsilon_b)^2+2(\epsilon_b-\epsilon_c)^2}\right)$	$\frac{E}{3}\left(\frac{\epsilon_a+\epsilon_b+\epsilon_c}{1-\nu}\pm\frac{1}{1+\nu}\times\sqrt{2(\epsilon_a-\epsilon_b)^2+2(\epsilon_b-\epsilon_c)^2+2(\epsilon_c-\epsilon_a)^2}\right)$	$\frac{E}{2}\left(\frac{\epsilon_a+\epsilon_d}{1-\nu}\pm\frac{1}{1+\nu}\times\sqrt{(\epsilon_a-\epsilon_d)^2+\frac{4}{3}(\epsilon_b-\epsilon_c)^2}\right)$
maximum shear, τ_{max}	$\frac{E}{2(1+\nu)}\times\sqrt{2(\epsilon_a-\epsilon_b)^2+2(\epsilon_b-\epsilon_c)^2}$	$\frac{E}{3(1+\nu)}\times\sqrt{2(\epsilon_a-\epsilon_b)^2+2(\epsilon_b-\epsilon_c)^2+2(\epsilon_c-\epsilon_a)^2}$	$\frac{E}{2(1+\nu)}\times\sqrt{(\epsilon_a-\epsilon_d)^2+\frac{4}{3}(\epsilon_b-\epsilon_c)^2}$
$\tan 2\theta$[b]	$\frac{2\epsilon_b-\epsilon_a-\epsilon_c}{\epsilon_a-\epsilon_c}$	$\frac{\sqrt{3}(\epsilon_c-\epsilon_b)}{2\epsilon_a-\epsilon_b-\epsilon_c}$	$\frac{2}{\sqrt{3}}\left(\frac{\epsilon_c-\epsilon_b}{\epsilon_a-\epsilon_d}\right)$
$0<\theta<+90°$	$\epsilon_b>\frac{\epsilon_a+\epsilon_c}{2}$	$\epsilon_c>\epsilon_b$	$\epsilon_c>\epsilon_b$

[a]θ is measured in the counterclockwise direction from the a-axis of the rosette to the axis of the algebraically larger stress.
[b]θ is the angle from a-axis to axis of maximum normal stress.

Plant Engineering

If an absorption dynamometer uses a DC generator to dissipate energy, the generated voltage (V in volts) and line current (I in amps) are used to determine the power. Equation 63.49 and Eq. 63.50 are used to determine the torque.

$$P_{hp}=\frac{IV}{\eta\left(1000\ \frac{W}{kW}\right)\left(0.7457\ \frac{W}{hp}\right)}\quad\text{[absorption]}\qquad 63.51$$

For a driving dynamometer using a DC motor,

$$P_{hp}=\frac{IV\eta}{\left(1000\ \frac{W}{kW}\right)\left(0.7457\ \frac{W}{hp}\right)}\quad\text{[driving]}\qquad 63.52$$

Torque can be measured directly by a *torque meter* mounted to the power shaft. Either the angle of twist, ϕ, or the shear strain, τ/G, is measured. The torque in a solid shaft of diameter, D, and length, L, is

$$\begin{aligned}T&=\frac{JG\phi}{L}=\left(\frac{\pi}{32}\right)D^4\left(\frac{G\phi}{L}\right)\\&=\frac{\pi}{16}D^3\tau\quad\text{[solid round]}\end{aligned}\qquad 63.53$$

27. NOMENCLATURE

A	area	in^2	m^2
b	base length	in	m
BC	bridge constant	–	–
c	distance from neutral axis to extreme fiber	in	m
C	concentration	various	various
D	diameter	in	m
E	modulus of elasticity	lbf/in^2	Pa
F	force	lbf	N
G	shear modulus	lbf/in^2	Pa
GF	gage factor	–	–
h	height	in	m
I	current	A	A
I	moment of inertia	in^4	m^4
J	polar moment of inertia	in^4	m^4
k	constant	various	various
k	deflection constant	lbf/in	N/m
K	factor	–	–
L	shaft length	in	m
M	moment	in-lbf	N·m
n	number	–	–
N	rotational speed	rev/min	rev/min
p	pressure	lbf/in^2	Pa
P	permeability	various	various
P	power	hp	kW
Q	statical moment	in^3	m^3
r	radius	in	m
R	resistance	Ω	Ω

t	thickness	in	m
T	temperature	°R	K
T	torque	in-lbf	N·m
V	voltage	V	V
VR	voltage ratio	–	–
y	deflection	in	m

Symbols

α	temperature coefficient	1/°R	1/K
β	RTD temperature coefficient	$1/°R^2$	$1/K^2$
β	thermistor constant	°R	K
γ	shear strain	–	–
ϵ	strain	–	–
η	efficiency	–	–
θ	angle of twist	deg	deg
ν	Poisson's ratio	–	–
ρ	resistivity	Ω-in	Ω·cm
σ	stress	lbf/in^2	Pa
τ	shear stress	lbf/in^2	Pa
ϕ	angle of twist	rad	rad

Subscripts

b	battery
g	gage
o	original
r	ratio
ref	reference
t	transverse or total
T	at temperature T
x	in x-direction
y	in y-direction

Plant Engineering

64 Manufacturing Processes

Content in blue refers to the *NCEES Handbook*.

1. CHIP FORMATION

One of the most common ways that a workpiece can be shaped is by removing material through chip-forming operations such as turning, drilling, multitooth operations (e.g., milling, broaching, sawing, and filing), and specialized processes, such as thread and gear cutting.

The toughness of a workpiece can be determined from the nature of the chips produced. Brittle materials produce discrete fragments, known as *discontinuous chips*, *segmented chips*, or *type-one chips*. Ductile materials form long, helix-coiled string chips, known as *continuous chips* or *type-two chips*.[1] *Chip-breaker* grooves are often ground in the cutting tool face to cause long chips to break into shorter, more manageable pieces. Chip formation is optimum when chips are produced in the shapes of sixes and nines.

2. CUTTING TOOL SPEEDS AND FORCES

Figure 64.1 shows a chip produced during *orthogonal cutting* (i.e., two-dimensional cutting). Cutting involves both compressive and shear stresses. The chip expands from its undeformed thickness, t_o, to t_c (due to the release of compressive stress) as it slides over the cutting tool. The *chip thickness ratio*, $r = t_o/t_c$, is typically around 0.5.

Figure 64.1 Tool-Workpiece-Chip Geometry

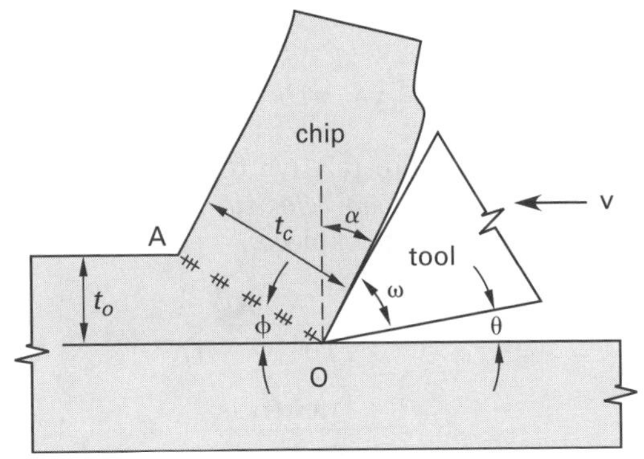

The cutting energy is minimum when the *shear angle*, ϕ, is approximately given by Eq. 64.1. (The angle β is shown in Fig. 64.2. It can be determined from F_n and F_t from toolpost dynamometer data.)

$$\phi_{\text{minimum energy}} = \frac{\pi}{4} + \frac{\alpha}{2} - \frac{\beta}{2} \qquad 64.1$$

The angle at which the tool meets the workpiece is characterized by the *true rake angle*, α. This angle has a major impact on chip formation, and it depends on the tool shape and set-up orientation of the tool. Small rake angles result in excessive compression, high tool forces, and excessive friction. Chips produced are thick, hot, and highly deformed. Conventional cutting tools are oriented with positive rake angles; sintered carbide and ceramic tools used for cutting steel are frequently designed to be oriented with negative rake angles to provide additional support of the cutting edge.

Figure 64.1 also shows the *clearance angle* (*relief angle*), θ, and the *wedge angle*, ω. The sum of the rake, clearance, and wedge angles is $\pi/2$ (90°).

$$\alpha + \theta + \omega = \frac{\pi}{2} \qquad 64.2$$

[1]There is also a *type-three chip*, a continuous chip with a *built-up edge* (BUE), which is not discussed here.

The rake angle, shear angle, and chip thickness ratio are related by Eq. 64.3 and Eq. 64.4.

From Fig. 64.1,

$$r = \frac{t_o}{t_c} = \frac{\sin\phi}{\cos(\phi-\alpha)} \qquad 64.3$$

Rearranging Eq. 64.3 and applying trigonometric identities,

$$\tan\phi = \frac{r\cos\alpha}{1-r\sin\alpha} \qquad 64.4$$

The relative velocity difference between the tool and the workpiece is the *cutting speed*, v. (See Table 64.1.) The cutting speed can be calculated from the diameter, D, of a rotating workpiece and the rotational speed, n. If D is in feet and n is in revolutions per minute, the cutting speed will be in traditional (in the United States) units of feet per minute.[2]

$$\text{v} = \pi D n \qquad 64.5$$

The velocity of the chip relative to the tool face is the *chip velocity*, v_c. The *shear velocity*, v_s, is the velocity of the chip relative to the workpiece.

$$\text{v}_c = r\text{v} = \frac{\text{v}\sin\phi}{\cos(\phi-\alpha)} \qquad 64.6$$

$$\text{v}_s = \frac{\text{v}\cos\alpha}{\cos(\phi-\alpha)} \qquad 64.7$$

Figure 64.2 illustrates that the resultant force, F_R, between the tool and the chip can be resolved into tangential force, F_t, and normal force, F_n, relative to the rake face of the tool, or alternatively, into a horizontal cutting force, F_h, and a vertical thrust force, F_v, relative to the cutting surface. The horizontal and vertical forces are commonly measured by a strain-gauge toolpost dynamometer. Equation 64.8 and Eq. 64.9 relate these two sets of forces.

$$F_t = F_h\sin\alpha + F_v\cos\alpha \qquad 64.8$$

$$F_n = F_h\cos\alpha + F_v\sin\alpha \qquad 64.9$$

A third set of axes can be chosen relative to the shear plane. The forces parallel, F_s, and normal, F_{ns}, to the shear plane can be derived from the horizontal and vertical forces. These are also shown in Fig. 64.2.

$$F_s = F_h\cos\phi + F_v\sin\phi \qquad 64.10$$

$$F_{ns} = F_h\sin\phi + F_v\cos\phi \qquad 64.11$$

Shear stress is the primary parameter affecting the cutting energy requirement. The average shear stress, τ, is F_s divided by the area of the shear plane, A_s, which depends on the chip width, b.

$$A_s = \frac{bt_o}{\sin\phi} \qquad 64.12$$

$$\tau = \frac{F_s}{A_s} = \frac{F_s\sin\phi}{bt_o} \qquad 64.13$$

The average normal stress is

$$\sigma = \frac{F_{ns}}{A_s} = \frac{F_{ns}\sin\phi}{bt_o} \qquad 64.14$$

The energy required per unit cutting time is the cutting power, P.

$$P = F_h\text{v} \qquad 64.15$$

If F_h is in pounds-force, and if v is in feet per minute, the horsepower requirement is

$$\text{cutting horsepower} = \frac{F_h\text{v}}{33{,}000} \qquad 64.16$$

The *metal removal rate*, Z_w, is

$$Z_w = bt_o\text{v} \qquad 64.17$$

The energy expended per unit volume removed, known as the *specific cutting energy*, is

$$U = \frac{P}{Z_w} = \frac{F_h\text{v}}{Z_w} = \frac{F_h}{bt_o} \qquad 64.18$$

For a simple lathe (turning) operation on a cylindrical workpiece of diameter D, the cutting time to make a cut of (horizontal) length L in a single pass is

$$t_{\text{min}} = \frac{L}{fn_{\text{rpm}}} = \frac{\pi LD}{f\text{v}} \qquad 64.19$$

Example 64.1

A cylindrical cast steel cylinder with a diameter of 6.00 in is faced in a lathe. The procedure removes the outer 0.40 in of the bar. The lathe develops a maximum power of 20 hp. The unit power limit for this material and process is 2.0 hp-min/in³. The cutting pressure is maintained at its maximum limit throughout. What is the approximate minimum time to face the bar?

Solution

The bar radius is

$$r = \frac{6.00 \text{ in}}{2} = 3.00 \text{ in}$$

[2]The symbol CS is also used for cutting speed.

Table 64.1 *Typical Cutting Speeds (ft/min)*

material	high-speed steel rough	high-speed steel finish	carbide rough	carbide finish
cast iron	50–60	80–110	120–200	350–400
semisteel*	40–50	65–90	140–160	250–300
malleable iron*	80–110	110–130	250–300	300–400
steel casting* (0.35C)	45–60	70–90	150–180	200–250
brass (85-5-5)	200–300	200–300	600–1000	600–1000
bronze (80-10-10)*	110–150	150–180	600	1000
aluminum	400	700	800	1000
SAE 1020*	80–100	100–120	300–400	300–400
SAE 1050*	60–80	100	200	200
stainless steel*	100–120	100–120	240–300	240–300

(Multiply ft/min by 5.08×10^{-3} to obtain m/s.)

*Appropriate lubricants are used to achieve listed speeds.

Used with permission from *Manufacturing Processes*, 5th Ed., by Myron L. Begeman and B. H. Amstead, published by John Wiley & Sons, Inc., copyright © 1963.

Figure 64.2 *Force Components in Orthogonal Cutting*

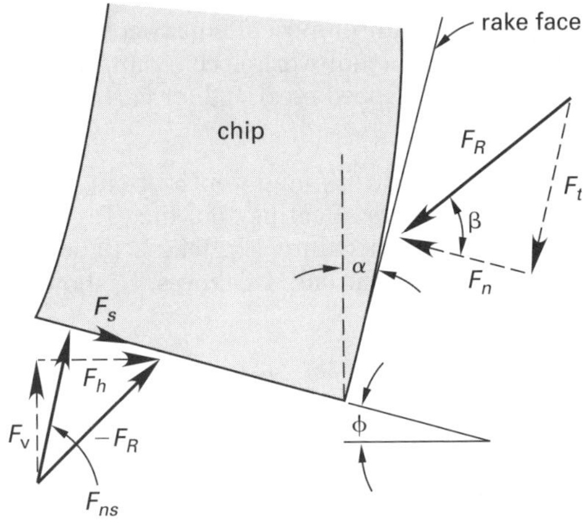

The volume of material to be removed is

$$V = AL = \pi r^2 L = \pi(3.00\,\text{in})^2(0.4\,\text{in})$$
$$= 11.31\,\text{in}^3$$

If the lathe develops the full 20 hp power throughout the operation, the minimum cutting time is

$$t = V\left(\frac{\text{unit power}}{\text{lathe power}}\right) = (11.31\ \text{in}^3)\left(\frac{2.0\ \frac{\text{hp-min}}{\text{in}^3}}{20\ \text{hp}}\right)$$
$$= 1.13\ \text{min}$$

Example 64.2

A cylindrical workpiece with a diameter of 4.00 in is turned on a lathe. The cutting speed is 200 ft/min. The depth of cut is 0.20 in. The feed is 0.010 ipr (inches per revolution). How long will it take to make a 14 in cut?

Solution

The minimum cutting time is

$$t_{\text{min}} = \frac{\pi LD}{f\text{v}} = \frac{\pi(14\ \text{in})(4.00\ \text{in})}{\left(12\ \frac{\text{in}}{\text{ft}}\right)\left(0.010\ \frac{\text{in}}{\text{rev}}\right)\left(200\ \frac{\text{ft}}{\text{min}}\right)}$$
$$= 7.33\ \text{min}$$

3. TOOL MATERIALS

Carbon tool steel is plain carbon steel with approximately 0.9% to 1.3% carbon, which has been hardened and tempered. It can be given a good edge but is restricted to use below 400°F to 600°F (200°C to 300°C) to prevent further tempering.

High-speed steel (HSS) contains tungsten or chromium and retains its hardness up to approximately 1100°F (600°C), a property known as *red hardness*. The common 18-4-1 formulation contains 18% tungsten, 4% chromium, and 1% vanadium. Other categories include *molybdenum high-speed steels* and *superhigh-speed steels*. Tools made with these steels can be run approximately twice as fast as carbon steel tools.

Cast nonferrous cutting tools have similar characteristics to carbides and are used in an as-cast condition. A common composition contains 45% cobalt, 34%

chromium, 18% tungsten, and 2% carbon. Cast nonferrous tools are brittle but can be used up to approximately 1700°F (925°C) and operate at speeds twice that of HSS tools.

Sintered carbides are produced through powder metallurgy from nonferrous metals (e.g., tungsten carbide and titanium carbide with some cobalt). Carbide tools are commonly of the throw-away type. They are very hard, can be used up to 2200°F (1200°C), and operate at cutting speeds two to five times as fast as HSS tools. However, they are less tough and cannot be used where impact forces are significant.

Ceramic tools manufactured from aluminum oxide have the same expected life as carbide tools but can operate at speeds from two to three times higher. They operate below 2000°F (1100°C).

Diamonds and diamond dust are used in specific cases, usually in finishing operations.

In addition to speed and temperature considerations, there should be no possibility of welding between the chip and tool material. Diamonds, for example, are soluble in the presence of high-temperature iron. Also, aluminum oxide tools are not satisfactory for machining aluminum.

4. TEMPERATURE AND COOLING FLUIDS

Friction is greatly reduced in *free-machining steels* that have had sulfur added as an alloying ingredient. However, only approximately 25% of the heat developed in cutting is due to friction between the tool and the workpiece. The remainder results from compression and shear stresses. Only 20% to 40% of this heat is removed by the tool and workpiece. The remainder must be removed by the chips and cooling fluids.

If all the heat generated goes into the chip (which does not actually occur), the *adiabatic chip temperature* is given by Eq. 64.20.

$$T_c = T_o + \frac{U}{\rho c_p} \quad \text{[SI]} \qquad 64.20(a)$$

$$T_c = T_o + \frac{U}{\rho c_p J} \quad \text{[U.S.]} \qquad 64.20(b)$$

Cutting fluids are used to reduce friction, remove heat, remove chips, and protect against corrosion. Gases, such as air, carbon dioxide, and water vapor, can be used, but they do not remove heat well, cannot be reused, and may require an exhaust system. Water is a good heat remover, but it promotes rust. (Addition of *sal soda* to water produces an efficient, inexpensive cutting fluid that does not promote rusting.)

Straight-cutting oils (i.e., petroleum-based nonsoluble oils) reduce friction and do not cause rust but are less efficient at heat removal than water. Therefore, emulsions of water and oil or water-miscible fluids (soluble oils) are often used with steel. Kerosene lubricants are commonly used with aluminum.

Chlorinated or sulfurized oils are used to decrease friction.[3] Other additives are used to inhibit rust, clean the workpiece, soften water, promote film formation, and inhibit bacterial growth.

5. TOOL LIFE

Tools wear and fail through abrasion, loss of hardness, and fracture. Three common types of failure are *flank wear*, *crater wear*, and *nose failure*. The life of a tool, T (expressed in minutes), is the length of time it will cut satisfactorily before requiring grinding and depends on the conditions of use. The *tool life equation*, also known as *Taylor's equation*, relates cutting speed, v, and tool life, T, for a particular combination of tool and workpiece.

$$\text{v}T^n = \text{constant} \qquad 64.21$$

The exponent n is an empirical constant that must be determined for each tool-workpiece setup. Typical values are 0.1 for high-speed steel, 0.2 for carbides, and 0.4 for ceramics.

Since the tool feed rate, f, and depth of cut, d, are also important parameters affecting tool life, Taylor's equation has been expanded into Eq. 64.22. (The depth of cut, d, is the same as the chip thickness, t_o, shown in Fig. 64.1.)

$$\text{v}T^n d^x f^y = \text{constant} \qquad 64.22$$

6. ABRASIVES AND GRINDING

Grinding (i.e., *abrasive machining*) is used as a finishing operation since very fine and dimensionally accurate surface finishes can be produced. However, grinding is also used for gross material removal. In fact, grinding is the only economical way to cut hardened steel.

Most modern grinding wheels are produced from aluminum oxide. However, grinding wheels can be produced from either *natural abrasives* or *synthetic abrasives*, as described in Table 64.2.

Abrasive grit size is measured by the smallest standard-size screen through which the grains will pass. *Coarse grits*, for example, will pass through no. 6 (i.e., having

[3]Chlorine and sulfur form metallic chlorides and sulfides at cutting temperatures. These compounds have low shear strength, and therefore, friction is reduced. Chlorinated oils work better at low speeds, whereas sulfurized oils work better under severe conditions.

Table 64.2 *Types of Grinding Wheel Abrasives*

natural abrasives
- sandstone
- solid quartz
- emery (50% to 60% Al_2O_3 plus iron oxide)
- corundum (75% to 90% Al_2O_3 plus iron oxide)
- garnet
- diamond

synthetic abrasives
- silicon carbide, SiC
- aluminum oxide, Al_2O_3
- boron carbide

six uniform openings per inch) to no. 24 screens, inclusive, but will be retained on any finer screen. Table 64.3 summarizes the abrasive size designations.

Table 64.3 *Abrasive Grit Sizes*

	screen sizes	
designations	English	metric (mm)
coarse	no. 6–no. 24	4.23–1.06
medium	no. 30–no. 60	0.847–0.423
fine*	no. 70–no. 600	0.363–0.042

(Multiply in by 25.4 to obtain mm.)

*Sizes no. 240 through no. 600 are also known as *flour grit.*

Snagging describes very rough grinding, such as that performed in foundries to remove gates, fins, and risers from castings.

Honing is grinding in which very little material, 0.001 in to 0.005 in (0.025 mm to 0.13 mm) is removed. Its purpose is to size the workpiece, to remove tool marks from a prior operation, and to produce very smooth surfaces. Coolants, such as sulfurized mineral-base oils and kerosene, are used to cool the workpiece and to flush away small chips. Because the stones are moved with an oscillatory pattern, honing leaves a characteristic cross-hatch pattern.

Lapping is used to produce dimensionally accurate surfaces by removing less than 0.001 in (0.025 mm). Parts are lapped to produce a close fit and to correct minor surface imperfections.

After any cutting or standard grinding operation, the surface of a workpiece will consist of *smear metal* (a fragmented, noncrystalline surface). *Superfinishing* or *ultrafinishing* using light pressure, short but fast oscillations of the stone, and copious amounts of lubricant-coolant, removes the smear metal and leaves a solid crystalline metal surface. The operation is similar to honing, but the stone moves with a different motion. There is essentially no dimensional change in the workpiece.

Other nonprecision methods of abrasion can be used to improve the surface finish and to remove burrs, scale, and oxides. Such methods include buffing, wire brushing, tumbling (i.e., barrel finishing), polishing, and vibratory finishing.

Centerless grinding is a method of grinding that does not require clamping, chucking, or holding round workpieces. The workpiece is supported between two abrasive wheels by a work-rest blade. One wheel rotates at the normal speed and does the actual grinding. The other wheel, the *regulating wheel,* is mounted at a slight offset angle and turns more slowly. Its purpose is to rotate and position the workpiece. (See Fig. 64.3.)

Figure 64.3 *Centerless Grinding*

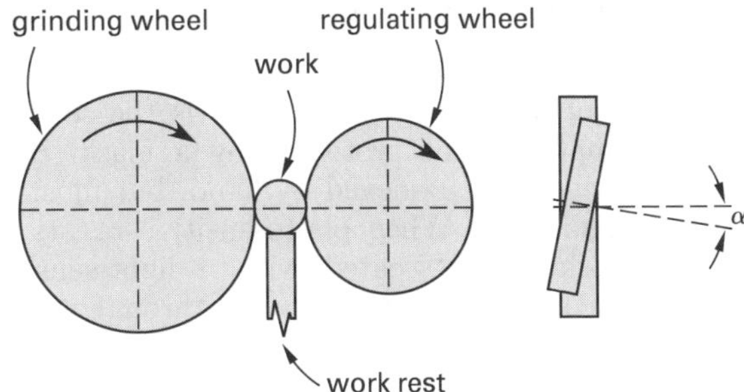

7. CHIPLESS (NONTRADITIONAL) MACHINING

Electrical discharge machining (EDM), also known as *electrodischarge machining, electrospark machining,* and *electronic erosion,* uses high-energy electrical discharges (i.e., sparks) to shape an electrically conducting workpiece. Thousands of controlled sparks are generated per second between a cutting head and the workpiece, while a servomechanism controls the separating gap. EDM requires the cutting to be performed in a dielectric liquid. The final cut surface consists of small craters melted by the arcs.

EDM can be used with all conductive metals, regardless of melting point, toughness, and hardness. Since there is no contact between the tool and the workpiece, delicate and intricate cutting is possible. However, the metal removal rate is low. Also, the tool material is lost much faster than the workpiece material. Wear ratios for the tool and workpiece vary between 20:1 for common brass tools to 4:1 for expensive tool materials.

Electrochemical machining (ECM) removes metal by electrolysis in a high-current deplating operation. Current densities of 1500 A/in^2 to 2000 A/in^2 (230 A/cm^2 to 310 A/cm^2) are common. A tool electrode (the cathode) with the approximate profile desired to be given to the workpiece (the anode) is brought close to the

Plant Engineering

workpiece. The separation is maintained by a servomechanism. A water-based electrolyte (e.g., sodium chlorate solution) is forced between the tool and workpiece. The electrolyte completes the circuit and removes the free ions.

ECM shapes and cuts metal of any hardness or toughness. Relatively high (compared with EDM) metal removal rates are possible. Unlike EDM, the tool is not consumed or changed in shape. The *current efficiency* is defined as the volume of metal removed per unit energy used. Typical units are cubic inches per 1000 ampere-minutes.

Electrochemical grinding (ECG), also known as *electrolytic grinding*, a variant of electrochemical machining, is used to shape and sharpen carbide cutting tools. It uses a rotating metal disk electrode with diamond dust (typically) bonded on the surface. Less than 1% of the workpiece material is removed by conventional grinding. The remainder is removed by electrolysis.

Chemical milling (*chem-milling*), typically used in the manufacture of printed circuit boards, is the selective removal of material not protected by a mask. Some masks are scribed and removed by hand, but most are *photosensitive resists*. When photosensitive resists are used, the workpiece is coated with a light-sensitive emulsion. The emulsion is then exposed through a negative and developed, which removes the unexposed emulsion. Finally, the workpiece is placed in a *reagent* (the *etchant*), which removes only unmasked workpiece metal.

Chemical milling works with almost any metal, such as copper, aluminum, magnesium, and steel. Although the removal rate is low, very large areas can be processed. For highly accurate work, the tendency of the etchant to undercut the mask must be known and compensated for. The *etching radius* (*etch factor*) is one method of quantifying this tendency.

Ultrasonic machining (USM) or *ultrasonic impact machining* works with metallic and nonmetallic materials of any hardness. USM can be used to shape hard and brittle materials such as glass, ceramics, crystals, and gemstones, as well as tool steel and other metals. Ultrasonic energy with a frequency between 15,000 Hz and 30,000 Hz is generated in a *transducer* through magnetostrictive and piezoelectric effects. Wear of the transducer is minimal.

The transducer is separated from the workpiece by a slurry of abrasive particles. The ultrasonic energy generated is used to hurl fine abrasive particles against the workpiece at ultrasonic velocities. The same abrasives used for grinding wheels are used with USM: aluminum oxide, silicon carbide, and boron carbide. Grit sizes of 280 mesh or finer are common.

Laser machining is used to cut or burn very small holes in the workpiece with high dimensional accuracy.

8. COLD- AND HOT-WORKING OPERATIONS

Whether a workpiece is considered cold worked or hot worked depends on whether the working temperature is below or above the recrystallization temperature, respectively. Table 64.4 categorizes most common forming operations.

Table 64.4 *Cold- and Hot-Working Operations*

cold working

- bending
- coining
- cold forging
- cold rolling
- cutting
- drawing
- drilling
- extruding
- grinding
- hobbing
- peening and burnishing
- riveting and staking
- rolling
- shearing, trimming, blanking, and piercing
- sizing
- spinning
- squeezing (e.g., swaging)
- thread rolling

hot working

- bending
- extruding and drawing (bar and wire)
- hot forging
- hot rolling
- piercing
- pipe welding
- spinning and shear forming
- swaging

9. PRESSWORK

Presswork is a general term used to denote the blanking, bending and forming, and shearing of thin-gage metals. Presses (also known as *brakes*) are used with dies and

Plant Engineering

punches to form the workpieces. Press forces are very high, and press capacities (known as *tonnage*) are often quoted in tons.

With *progressive dies*, the workpiece advances through a sequence of operations. Each of the press operations is performed at a *station*. Progressive dies can be of the *strip die* or *transfer die* varieties.

Shearing operations (blanking, punching, notching, etc.) cut pieces from flat plates, strips, and coil stock. Since the cutting is a shearing operation, the press force required to blank N items at one time is given by Eq. 64.23.

$$F = NS_{us}Lt \qquad \text{64.23}$$

The distinction between blanking and punching is relative. *Blanking* produces usable pieces (i.e., *blanks*), leaving the source piece behind as scrap. *Punching* is the operation of removing scrap blanks from the workpiece, leaving the source piece as the final product.

Bending and forming operations are often considered in the same category. *Bending dies* are used in press brakes to bend along a straight axis. *Forming dies* bend and form the blank along a curved axis and may incorporate other operations (e.g., notching, piercing, lancing, and cut-offs). There is little or no metal flow in a die-forming operation. The tension and compression on opposite surfaces of the blank are approximately equal.

Spring-back, *bend allowance*, and bending pressure can be calculated for bending and forming operations. However, it is generally necessary to make test runs to determine these values under realistic conditions.

Drawing is a cold-forming process that converts a flat blank into a hollow vessel (e.g., beverage cans). Drawing sheet metal blanks results in plastic metal flow along a curved axis. Double-acting presses may be required to accomplish deep draws.

Coining, as used in the production of coins, is a severe operation requiring high tonnage, due to the fact that the metal flow is completely confined within the die cavity. Because of this, coining is used mainly to form small parts.

Embossing forms shallow raised letters or other designs in relief on the surface of sheet metal blanks. It differs from coining in that the workpiece is not confined.

Swaging operations reduce the workpiece area by cold flowing the metal into a die cavity by a high compressive force or impact. It is applicable to small parts requiring close finishes. *Sizing* and *cold heading* of bolts and rivets are related operations.

10. FORGING

Forging is the repeated hammering of a workpiece to obtain the desired shape. Forging can be a cold-work process, but it is commonly considered to be a hot-work process when the term forging is used. Hot-work forging is carried out above the recrystallization temperature to produce a strain-free product. Table 64.5 lists the approximate temperatures for hot forging.

Table 64.5 *Approximate Hot-Forging Temperatures*

material	temperature	
	°F	°C
steel	2000–2300	1100–1250
copper alloys	1400–1700	750–925
magnesium alloys	600	300
aluminum alloys	700–850	375–450

The oldest form of forging is similar to what is done by blacksmiths. Commercial *hammer forging*, *smith forging*, or *open die forging* consists of repeatedly hammering the workpiece (known as the *stock*) in a powered forge. Accuracy is low since the shape is not defined by dies, and considerable operator skill is required.

Drop forging (*closed-die forming*) relies on closed-impression dies to produce the desired shape. One-half of the die set is stationary; the other half is attached to the hammer. The metal flows plastically into the die upon impact by the forge hammer. The forging blows are repeated at the rate of several times a minute (for *gravity drop hammers*, also known as *board hammers*) to more than 300 times a minute (for *powered hammers*).[4] Progressive forging operations are used to significantly change the shape of a part over several steps.

The total forming energy (work) supplied by a gravity drop hammer is the potential energy of the hammer.

$$E = mgh \qquad \text{[SI]} \qquad \text{64.24(a)}$$

$$E = \frac{mgh}{g_c} \qquad \text{[U.S.]} \qquad \text{64.24(b)}$$

The total forming energy supplied by a powered drop hammer falling from height h is

$$E = (mg + pA)h \qquad \text{[SI]} \qquad \text{64.25(a)}$$

$$E = \left(\frac{mg}{g_c} + pA\right)h \qquad \text{[U.S.]} \qquad \text{64.25(b)}$$

[4]Steam or compressed air can be used to lift the gravity drop hammer back into place, but the hammer is not powered during its downward travel.

Plant Engineering

The total forming energy supplied by an eccentric-crank press can be calculated from its rotational speeds before and after the impact.

$$E = \tfrac{1}{2}I(\omega_o^2 - \omega_f^2) = \tfrac{1}{2}I\left(\frac{\pi}{30}\right)^2(n_o^2 - n_f^2) \quad \text{[SI]} \quad 64.26(a)$$

$$E = \tfrac{1}{2}\left(\frac{I}{g_c}\right)(\omega_o^2 - \omega_f^2) = \tfrac{1}{2}\left(\frac{I}{g_c}\right)\left(\frac{\pi}{30}\right)^2(n_o^2 - n_f^2) \quad \text{[U.S.]} \quad 64.26(b)$$

In *impactor forging* or *counterblow forging*, the workpiece is held in position while the dies are hammered horizontally into it from both sides. *Upset forming* involves holding and applying pressure to round heated blanks. The part is progressively formed from one end to the other and, characteristic of upset forming, becomes shorter in length but larger in diameter.

With *press forging*, the part is shaped by a slow squeezing action, rather than rapid impacts. This allows more forging energy to be used in shaping rather than in transmission to the machine and foundation. The pressing action can be obtained through screw or hydraulic action.

Following forging, the part will have a thin projection of excess metal known as *flash* at the *parting line*. The flash is trimmed off by *trimmer dies* in a subsequent operation. Also, since the hot-worked part will be covered with scale, the part is cleaned in acid (an operation known as *pickling*). Additional processing may include shotpeening, tumbling, and heat treatment.

11. SAND MOLDING

With *sand molding*, a mold is produced by packing sand around a pattern. After the pattern is removed, the remaining cavity has the desired shape. To facilitiate the removal of the pattern, all surfaces parallel to the direction of withdrawal are slightly tapered. This taper is called *draft*. After molten metal is poured into the mold, the sand is removed, exposing the completed cast part.

Various types of foundry sand are used, including green (moist) sand, dry (baked) sand, and carbon dioxide process sands. Carbon dioxide process sands contain approximately 4% silicate of soda (Na_2SiO_3), which hardens upon exposure to carbon dioxide. Pure, dry silica sand has no binding capacity and is not suited for molding. Various types of clay can be added to dry sand to improve its bonding characteristics (*cohesiveness*).

Other additives permit gases to escape during casting (*permeability*) and enhance the sand's heat resistance (*refractoriness*). Small amounts of organic matter can be added to the sand to enhance its *collapsibility*. The organic matter burns out when exposed to the hot metal, permitting the mold to be easily removed.

The *gating system* shown in Fig. 64.4 is the set of openings and passages that brings molten metal to the mold cavity in a controlled manner. The metal is poured into a *sprue hole* and enters a vertical passage known as a *downgate*. The metal then passes directly into the distribution passageways, known as *runners*, before entering the cavity. An entrance to the cavity may be constricted

Figure 64.4 Sand-Casting Gating System

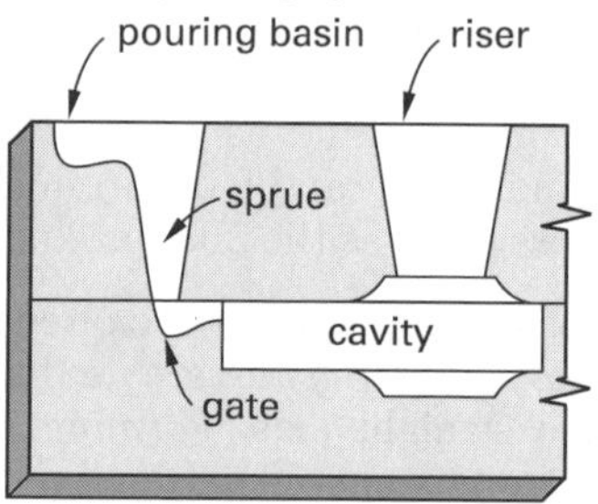

to control the rate of fill, and such constrictions are known as *gates*. *Risers* serve as accumulators to feed molten metal into the cavity during initial shrinkage. After hardening, the gates and risers can be broken off from iron castings, but a torch or cutting wheel is necessary with steel castings. Raw castings will be covered with sand and scale that must be removed by tumbling or sand blasting.

Figure 64.5 Green and Dry Sand Cores

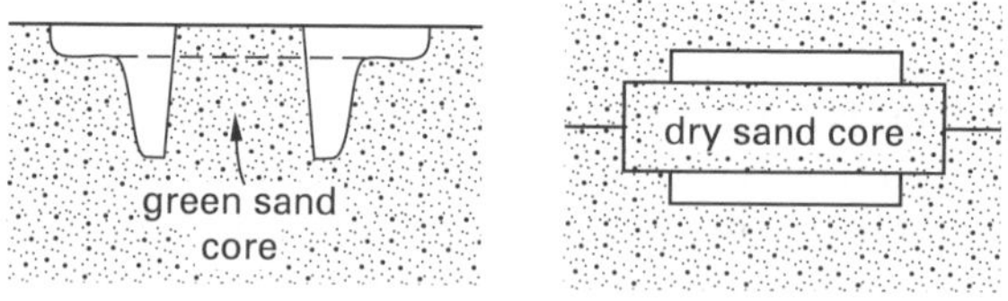

If a part has a recess or hole, a *core* must be placed in the mold. Cores come in *green sand core* and *dry sand core* varieties. (See Fig. 64.5.) Green sand cores are formed by the pattern itself. Dry sand cores are formed and baked in a separate operation and inserted into the sand mold.

12. GRAVITY MOLDING

With *gravity molding* (*tilt pouring, gravity die casting*, or *permanent molding*), molten metal is poured into a metal or graphite mold. Pressure is not used to fill the mold.[5] The mold may be coated prior to filling to prevent the casting from sticking to the mold's interior. Both ferrous and nonferrous metals (including magnesium, aluminum, and copper alloys) can be gravity molded. This method has the advantage (over sand casting) that a new mold is not required for each casting.

13. DIE CASTING

Die casting (*pressure die casting*) is suitable for creating parts of zinc, aluminum, copper, magnesium, and lead/tin alloys. (More than 75% of all die casting uses zinc alloys.) Molten metal is forced under pressure into a permanent metallic mold known as a *die*. Dies that produce one part per injection are known as *single cavity dies*. Dies that produce more than one part per injection are known as *multiple cavity dies*. The metal can be introduced by a plunger or compressed air, but never by gravity alone. The casting pressure is maintained until solidification is complete.

Dies for zinc, tin, and lead alloys are usually made from high-carbon and alloy steels, although low-carbon steel can be used for zinc casting dies. Dies for aluminum, magnesium, and copper alloys (which melt at higher temperatures) are made out of heat-resisting alloy steel.

Hot-chamber die casting and *cold-chamber die casting* are the two main variations of die casting. The hot-chamber method is limited to alloys (e.g., zinc, tin, and lead) that have melting temperatures below 1000°F (550°C) and that do not attack the injection apparatus. The injection apparatus (the *gooseneck*) is submerged in the molten metal, and low temperatures limit corrosion.

Brass (and other copper alloys), aluminum, and magnesium have high melting temperatures and require higher injection pressures. They also corrode ferrous machine parts and become contaminated by the iron they pick up. Brass and bronze, with their 1600°F to 1900°F (875°C to 1050°C) melting temperatures, particularly attack the steel in die casting machines. These alloys are usually melted in a separate furnace and ladled into the plunger cavity. This is the principle of the *cold-chamber method*.

After solidification, the sprues, gates, runners, and overflows are cut off in *trimming dies*. Die castings will typically be harder on the outside than on the inside due to the chilling action of the die. Also, gases may have been trapped inside the part, making the interior of the casting porous. Porous castings are brittle and subject to fracture.

14. CENTRIFUGAL CASTING

If the mold is rapidly rotated, the molten metal will be forced into the mold by centripetal action while the metal solidifies. This process is known as *centrifugal casting* or *centrifuging*. This method is particularly useful in producing objects with round and symmetrical (e.g., hexagonal) outer surfaces such as gun barrels and brake drums.

15. INVESTMENT CASTING

Casting methods that produce a molding cavity from a wax pattern are known as *investment casting, precision casting*, or the *lost-wax process*. These methods are suited to small, complex shapes and casting of precious metals.

A positive image of the part to be cast is created from wax.[6] The image is coated with the *investment material*, which can be finely ground refractory, plaster of paris or another ceramic material, or rubber, which becomes the mold. The mold is heated, and the liquid wax is poured out. The mold is then filled with molten metal. Centrifuging may be used to ensure complete filling. After the metal solidifies, the mold is broken off.

If more than one image is to be cast, a *master pattern* is made out of wood, steel, or plastic. The master pattern is used to make a *master die*. The wax patterns are then made in the master die.

16. CONTINUOUS CASTING

Any process in which molten material is continuously poured into a mold is known as *continuous casting*. This method can be used to produce sheets of glass, copper slabs, and brass or bronze bars. Continuous casting of bars is similar to extrusion, except that a cooling apparatus is included as part of the extrusion head.

17. PLASTIC MOLDING

Thermosetting compounds are purchased in liquid form, which makes them easy to combine with additives. Thermoplastic materials are commonly purchased in granular form. They are mixed with additives in a *muller* (i.e., a bulk mixer) before transfer to the feed hoppers. Thermoplastic materials can also be molded into

[5]There is limited application of strength-, strain-, and safety-critical components (e.g., the production of railway wheels and some steel ingots) of a process known as *pressure pouring*, in which the molten metal is forced into the mold by air pressure, or drawn into the mold by a vacuum, or both (as in the case of counter-pressure casting). Pressure pouring differs from die casting in that ferrous alloys are used, typically magnesium ferrosilicon-treated ductile iron (MgFeSi iron).

[6]Other variations of this process use plastics with low melting points, lead, and even frozen mercury (which freezes at 40°F (4°C)).

Plant Engineering

small pellets called *preforms* for easier handling in subsequent melting operations. Common additives for plastics are color pigments and tints, stabilizers, plasticizers, fillers, and resins.

The *hot compression molding* process is the oldest plastic forming process but is used extensively only for thermosetting polymers. (Compression molding is similar to the coining process for metals.) A measured amount of plastic material is placed in the open cavity of a heated mold. The mold is closed, and pressure is applied. The plastic flows into the mold, taking on its shape. Compression molding of thermoplastic resins is not very practical.[7]

Some plastic parts can be produced by *cold molding*. In this process, the part is simply cold pressed in a mold and then heated outside the press to fuse the particles.

The main method of forming thermoplastic resins is *injection molding*. The plastic molding compound is gravity fed into a heating chamber, where it is plasticized. Heating and metering is done by the *torpedo*, illustrated in Fig. 64.6. The molten plastic is injected under pressure into a water-cooled mold, where it solidifies almost immediately. (The mold temperature is constant.)

Figure 64.6 *Injection Molding Equipment*

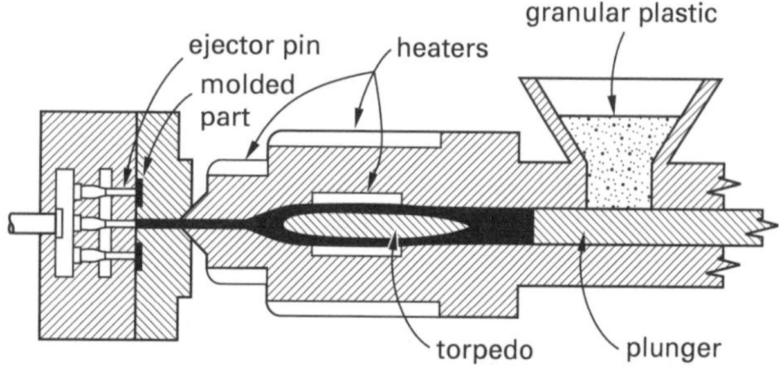

Transfer molding involves the heating of thermosetting plastic powder or preforms under pressure outside the mold cavity. The molten plastic is then forced from the transfer chamber through the gate and runner system into the mold cavity. The plastic is cured in the mold by maintaining the pressure and temperature. This method differs from injection molding in that the mold is kept heated and the plastic part is ejected while it is still hot.

Blow molding and *vacuum forming* both rely on an air-pressure differential to draw a heated thermoplastic sheet around a pattern or into a mold. The plastic retains the shape of the mold after cooling.

Most thermoplastics can be *extrusion-formed* into shapes (including sheets) of any length.[8] Solid plastic in granulated or powdered form is fed by a screw-feed mechanism into a heating chamber and then extruded through a die. The plastic is cooled by contact with air or water after extrusion. (See Fig. 64.7.)

Figure 64.7 *Extrusion Process*

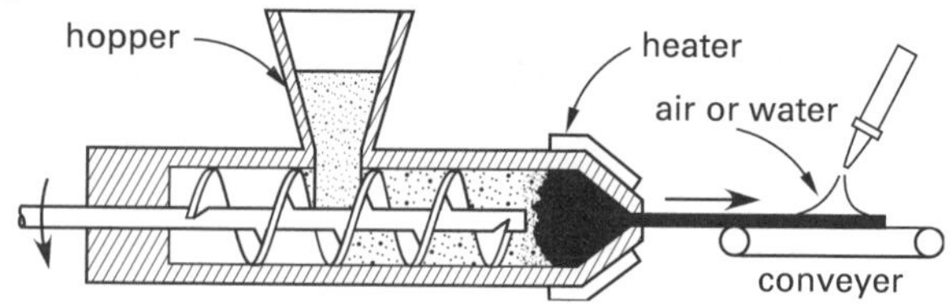

18. POWDER METALLURGY

Useful parts can be made by compressing a metal powder into shape and bonding the particles with heat—the principle of *powder metallurgy*. Typical powder metallurgy products include tungsten carbide cutting tools, copper motor brushes, bronze porous bearings, auto connecting rods and transmission parts, iron magnets, and filters.

Compared with machined products, parts made by powder metallurgy are relatively weak and expensive. However, the process is used for parts that cannot be easily produced in other manners. These are porous products, parts with complex shapes, items made from materials that are difficult to machine (e.g., tungsten carbide), and products that require the characteristics of two materials (e.g., copper/graphite electrical contacts).

The two most common types of powders are *iron-based powders* and *copper-based powders*. However, aluminum, nickel, tungsten, and other metals are also used. Bronze powders are regularly used to produce porous bearings; brass and iron are applicable to small machine parts where strength is important.

Metallic powders must be very fine. Most metal powders are produced by *atomizing* (i.e., using a jet of air to break up a fine stream of molten metal). Other methods include reduction of oxides, electrolytic deposition, and precipitation from a liquid or gas.

Pure metal powders are often mixed to improve manufacturing or performance characteristics. Graphite improves lubricating qualities and is added to powders used for bearings and electrical contacts. Cobalt and other metals are added to tungsten carbide to improve bonding.

Powders can be pressed into their final shape with a punch and die. The ejected shape is known as a *briquette* or *green compact*. With *isostatic molding*, hydraulic pressure is applied in all directions to the powder. Other

[7]Unless a mold is cooled before the part is removed, distortion of thermoplastics can result.
[8]Thermosetting plastics harden too quickly to be extruded.

methods of forming the green compact are centrifuging, *slip casting* (i.e., slurry casting), extrusion, and rolling. Due to internal friction, the density of a powder metallurgy part will not be consistent throughout but will be higher at the surface.

Sintering is heating to 70% to 90% of the melting point of the metal. The temperature is maintained for up to three hours, although the duration is commonly less than an hour. Typical sintering temperatures are 1600°F (875°C) for copper, 2050°F (1120°C) for iron, 2150°F (1175°C) for stainless steel, and 2700°F (1475°C) for tungsten carbide. To prevent the formation of metallic oxides, sintering must be performed in an inert or reducing gas (e.g., nitrogen) atmosphere.

19. HIGH ENERGY RATE FORMING

High energy rate forming (HERF), also known as *high velocity forming* (HVF), is the name given to several processes that plastically deform metals with blasts of high-pressure shock waves. Although these processes have traditionally been used to form thin metals, they are also applicable to other manufacturing needs such as powder metallurgy, forging, and welding.

The most common HERF method is *explosive forming*, as illustrated in Fig. 64.8. A small amount of low- or high-explosive is detonated. The resulting shock waves travel through the surrounding medium and force the metal into a shape determined by the dies. The medium can be either gas or liquid.

Figure 64.8 *Explosive Forming*

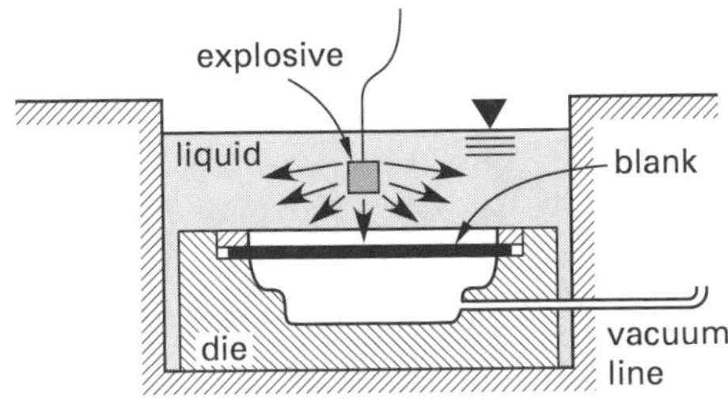

Another process using explosives is *explosive bonding*. A detonation is used to drive two similar metals together. When used with explosives in sheet form, this method has been successful in producing combinations of two metallic sheets (i.e., *cladding*). The resulting bond is almost metallurgically complete.

With *electro-hydraulic forming* (also known as *electrospark forming*), the forming pressure is obtained from the discharge of massive amounts of stored electricity. The electrical energy is built up in a capacitor bank. The energy used can be changed by adding or removing capacitors from the circuit. The discharge is across a spark gap between two electrodes in a nonconducting medium. Upon discharge, the electrical energy is converted directly into work. The usual medium is liquid.

Magnetic forming (*magnetic pulse forming*) is another example of the direct conversion of electrical energy into work. (See Fig. 64.9.) As with electrospark forming, a large amount of electrical energy is built up in a capacitor bank. A special expendable forming coil is placed around a part to be compressed or within a part to be expanded. (Magnetic forming can also be used for embossing if the workpiece is placed between the forming coil and the embossing die.) When the capacitor bank discharges, the current in the forming coil induces a current and a force in the workpiece. This force stresses the workpiece beyond its elastic limit.

Figure 64.9 *Magnetic Forming*

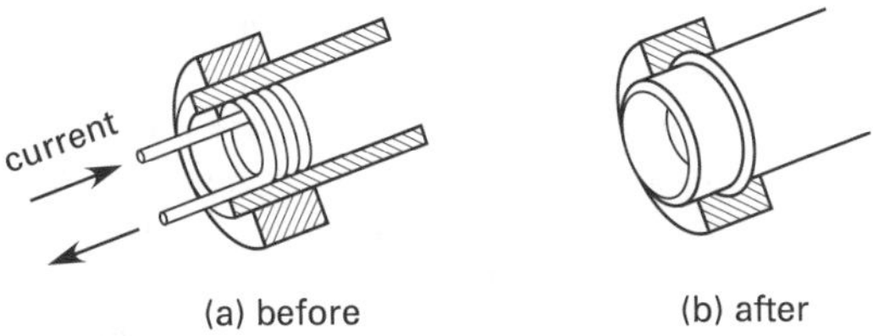

20. GAS WELDING

With *welding*, two metals are fused (i.e., melted) together by localized heat or pressure. This is known as *fusion* or *coalescence*. A welding rod of similar metal can be used to fill large voids between the two pieces but is not always necessary. The main types of welds are the *bead*, *groove*, *fillet*, and *plug* (*spot*) *welds* shown in Fig. 64.10.

Figure 64.10 *Types of Fusion Welds*

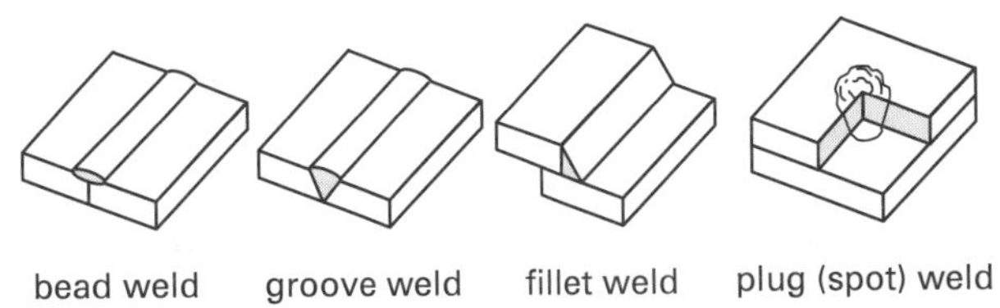

In *gas welding processes*, a combustible fuel and oxidizing gas are combined in a *torch* (*blowpipe*). Although natural gas and hydrogen can be used as fuels, *acetylene gas* (C_2H_2) is most common. Oxygen gas is the oxidizer, hence the names *oxyhydrogen welding* and *oxyacetylene welding*. The maximum welding temperature with oxyhydrogen welding is approximately 5100°F (2800°C), and with oxyacetylene welding, it is approximately 6000°F (3300°C).

MAPP gas (*methyacetylene propadiene*) is also extensively used. It is safer to store, is more dense, and provides more energy per unit volume than acetylene.

The proportions of oxygen and acetylene can be adjusted to obtain three different welding conditions: reducing, neutral, and oxidizing flames. Figure 64.11 illustrates the reactions that occur with a *neutral flame*, which is obtained when the oxygen:acetylene proportions are approximately 1:1 by volume. The inner luminous cone of the flame is distinctly blue

Plant Engineering

in color and is the hottest part of the flame. The outer envelope is only slightly luminous and may be difficult to see. Oxygen for the combustion of the outer envelope comes from the atmosphere.

Figure 64.11 *Neutral Oxyacetylene Flame*

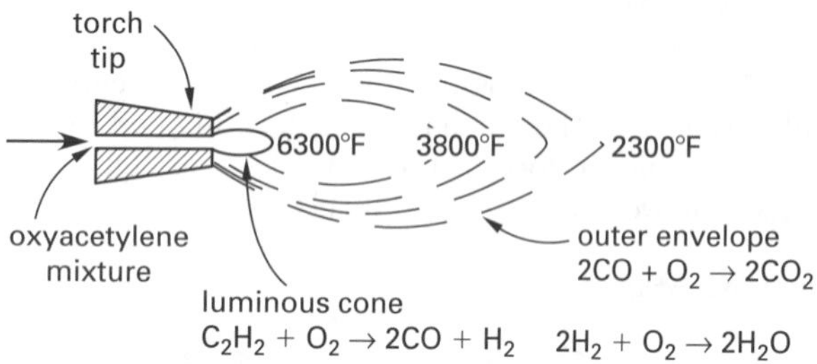

If the gas proportions are adjusted to an excess of acetylene, the flame is known as a *reducing flame* or *carburizing flame*. This process is used to weld many nonferrous metals (including Monel metal and nickel alloys as well as some alloy steels) and in applying several types of hard surfacing materials.

An *oxidizing flame* requires an excess of oxygen. Oxidizing fusion has some application to brass and bronze but is generally undesirable.[9]

21. ARC WELDING

Temperatures of up to 10,000°F (5550°C) can be obtained from *arc welding* using either DC or AC electrical current. The arc is created by first touching the electrode to the workpiece, establishing the current flow, and then moving the electrode slightly away. The current flow is maintained by the arc, and the electrical energy is converted to heat, which melts the metal.

If a *carbon electrode* is used to create an arc, a welding rod must be used to supply the filler material. Alternatively, in *metal electrode welding*, the electrode is itself melted by the arc and becomes the filler material.

Electrodes can be bare, but most have coatings known as *flux* that melt into slag and improve other welding characteristics.[10] The molten slag floats on and covers the molten metal, inhibiting the high temperature formation of oxides that weaken most welds. Welding with coated electrodes is known as *shielded metal arc welding* (SMAW). Most coatings have a significant amount of SiO_2 and/or TiO_2, plus small amounts of oxides of other metals. Hardened slag is chipped away after the weld has cooled.

In the United States, welding rods are classified according to the tensile strength (in ksi—thousands of lbf/in^2) of the deposited material. Thus, an E70 welding rod will produce a bead with a minimum nonstress relieved tensile strength of 70,000 lbf/in^2.

With *submerged-arc welding*, the flux is granular and is dispensed from a feed tube ahead of the welding process. The tip of the electrode and the arc are buried in the granular flux. The arc melts the granular flux, forming a coating that protects against oxidation.

Another method of protecting the molten weld metal from oxidation is by shielding with an inert gas. This is done when welding magnesium, aluminum, stainless steels, and some other steels. Argon gas is commonly used, although helium and argon-helium mixtures are also used. (Carbon dioxide gas can be used to shield the weld when working with plain-carbon and low-alloy steels.) This is the principle of *inert gas shielded arc welding*. With *TIG welding* (*tungsten inert gas*), the arc issues from an air- or water-cooled, nonconsumable tungsten electrode.

MIG welding (*metal inert gas*) is similar to TIG welding except that a consumable wire is used as the electrode. (See Fig. 64.12.) This method is also known as the *GMAW* (*gas metal arc welding*) process.

Figure 64.12 *TIG and MIG Welding Processes*

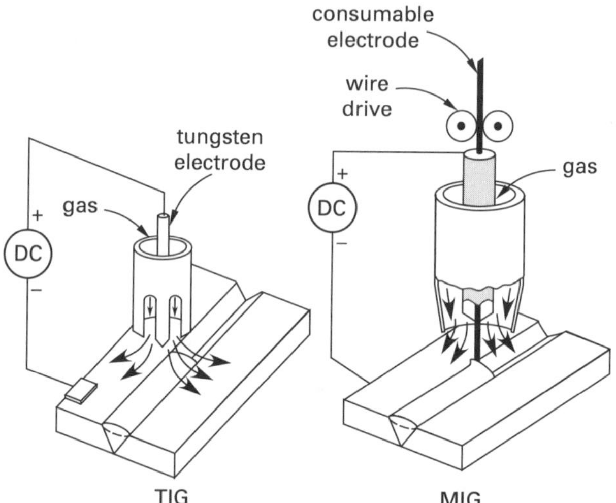

Plant Engineering

[9]An oxyacetylene *cutting torch* uses oxygen to cut steel, but the process is different from regular welding. A cutting torch uses a neutral flame to heat the steel. The oxygen used to oxidize (cut) the metal issues from a separate orifice in the torch and does not participate in the combustion of the acetylene.

[10]In addition to providing a protection from oxidation, the next most important function of the coating is to stabilize the arc (i.e., reduce the effect of variations in the separation of electrode and workpiece). Other functions include reducing weld metal splatter, adding alloying ingredients, changing the weld bead shape, improving overhead and vertical weldability, and providing additional filler material.

22. SOLDERING AND BRAZING

Soldering and *brazing* both use a molten dissimilar metal as glue between the two pieces. Generally, soldering uses a lead-tin filler with melting points below 800°F (425°C), whereas brazing uses copper-zinc or silver-based alloys with melting points above 800°F (425°C). Soldering is not a fusion process, since parts do not melt. Since some alloying occurs, brazing is considered to be a fusion process.

Usually, the solder or brazing material is drawn into the space separating the pieces by capillary action. However, a chemical flux can be used to remove oxides from the surfaces, inhibit additional oxidation, and improve cohesion of the filler material.

23. ADHESIVE BONDING

Both thermoplastic and thermosetting polymers are used as structural adhesives. Polymers of both types are sometimes combined to obtain the performance characteristics of both. A low-viscosity primer may be used to prepare the bonding surface for the adhesive. Almost all adhesive bonds are of the *lap-joint variety.*

Structural adhesives can be used to join any similar or dissimilar materials. However, even structural adhesives are limited to low-strength and low-temperature (e.g., less than 500°F or 250°C) applications.[11]

Thermoplastic adhesives, such as polyamides, vinyls, and nonvulcanized neoprene rubbers, soften when heated and cannot be used for elevated temperatures. They are generally used for nonstructural applications.

Thermosetting adhesives (e.g., epoxies, isocyanate, phenolic rubbers and vinyls, vulcanized rubbers, and neoprene) are used as structural adhesives. They must be used with elevated temperatures and where creep is unacceptable. Heat, pressure, ultraviolet radiation, and/or chemical hardeners must be used to activate the curing process.

The performance of an adhesive is determined by its toughness, tensile strength, peel strength, and temperature resistance. Adhesives with high tensile and shear strengths are usually hard and brittle and have low *peel strengths.* Adhesives with high peel strengths are usually more ductile and have fair tensile and shear strengths. These properties are typically provided in adhesive mechanical performance tables.

Example 64.3

A handle in the general configuration shown is to be bonded to either side of a container. The handle is expected to carry a load as depicted.

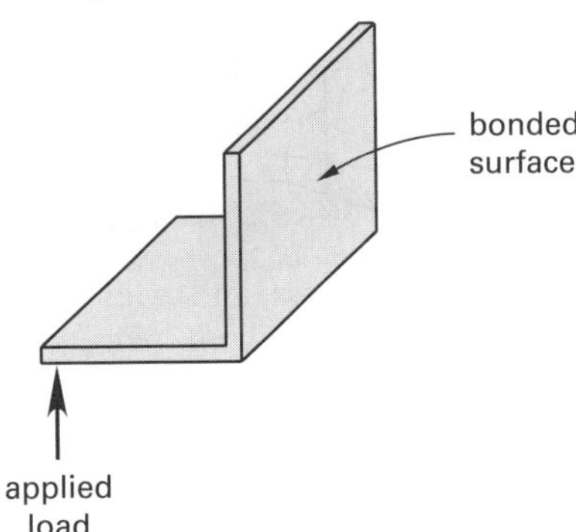

Four adhesives are available to bond the handle to the container. On initial evaluation, which would best work for the application?

(A) rubber-based adhesive

(B) cyanoacrylate

(C) urethane

(D) unmodified epoxy

Solution

To determine qualitatively which adhesive would perform the best in the given application, each adhesives lap shear strength and peel strength should be compared. The unmodified epoxy has the highest average shear strength, and therefore would be preferable. [**Mechanical Performance of Various Types of Adhesives**]

However, the joint configuration shown is prone to peeling failures, so peel strength must be considered. [**Design Practices That Improve Adhesive Bonding**]

Urethane has the second highest average shear strength and the highest average peel strength, so it would be the best fit for the application. [**Design Practices That Improve Adhesive Bonding**]

24. MANUFACTURE OF METAL PIPE

Pipes and tubes can be either seamed or seamless. Seamless pipe is made by piercing, whereas seamed pipe is made by forming and butt- or lap-welding the joined edges.

Thin-wall pipe can be formed by drawing a heated flat skelp through a welding bell. Prior to forming, the edges of the skelp are heated by flame or induction to the forging temperature, and joining occurs spontaneously in the welding bell, as shown in Fig. 64.13.

[11]Certain *ceramic adhesives* have useful ranges up to 1000°F (550°C).

Plant Engineering

Figure 64.13 Butt-Welding in a Welding Bell

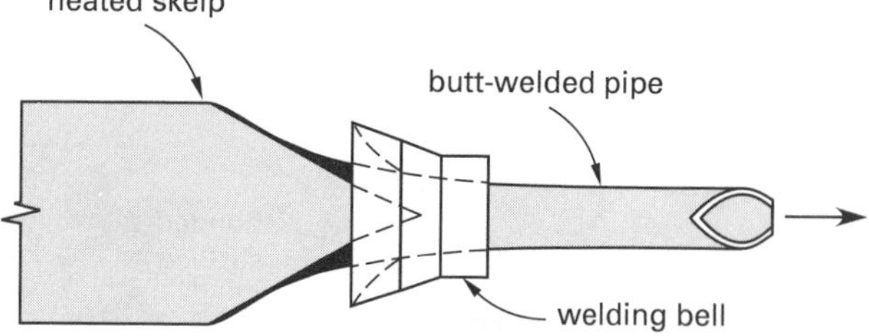

Pipes with larger diameters and wall thicknesses are formed with *roll forming* and *electric butt-welding*, in which the skelp is cold-rolled into circular shape by a series of roller pairs. One of the last rolling operations incorporates induction heating to bring the pipe edges up to forging temperature before they are pressed together. The flash is subsequently removed from the inside and outside of the pipe.

The skelp of *lap-welded pipe* passes through rollers that overlap the edges. Unlike roll forming, however, a fixed mandrel is placed inside the pipe. The heated skelp is rolled between the roller and mandrel at high pressure, which welds the heated edges together.

Seamless pipe is manufactured by heating a solid round bar known as a *billet* to forging temperature and piercing it with a mandrel. (See Fig. 64.14.) Subsequent operations with rollers strengthen, size, and finish the pipe.

Figure 64.14 Production of Seamless Pipe

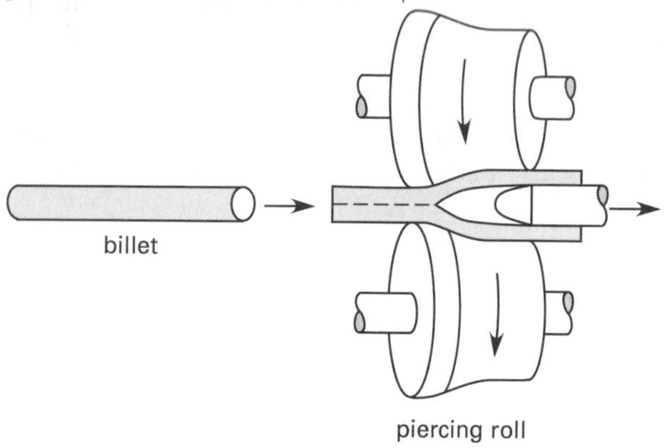

25. PIPE SUPPORT AND FLEXIBILITY

Pipe systems must be designed with adequate flexibility to control thermal expansion and contraction. Without adequate control, there is a risk of structural failure of pipes due to excessive stresses, equipment failure due to excessive forces and moments on attached equipment joints, and operational and safety impacts due to joint leakage.

Thermal expansion is a function of pressure and temperature for typical commercial pipe materials. Considerable linear expansion can be seen in high-temperature and high-pressure applications, up to 12 in per 100 ft of pipe.

There are several design methods for addressing thermal expansion, including L-bends, Z-bends, and U-bends. These methods use bends in the pipe to offset the thermal expansion.

L-bends are used when natural system routing related changes in direction are required, and Z-bends are used to route around pipe interferences (e.g., facilities related HVAC and electrical equipment and structural interferences). The ASHRAE L-bend and Z-bend equations find the bend length and associated anchor force as a function of the pipe diameter and anchor-to-anchor displacement. [**L-Bends**] [**Z-Bends**]

U-bends are used for long, unidirectional piping runs where system or facilities related directional changes are not a design option (e.g., long cross-country runs). To accommodate thermal expansion in these long pipe runs, the ASHRAE methodology uses the dimensions of the anchor-to-anchor expansion to determine the required bend length and anchor force.[**U-Bends and Pipe Loops**]

26. SURFACE FINISHING AND COATINGS

Finishes protect and improve the appearance of surfaces. Some processes involve material removal, and others involve material addition. Some of the common finishing methods and coating systems are listed as follows.

- *abrasive cleaning:* shooting sand (i.e., *sand blasting*), steel grit, or steel shot against workpieces to remove casting sand, scale, and oxidation.
- *anodizing:* an electroplating-acid bath oxidation process for aluminum and magnesium. The workpiece is the anode in the electrical circuit.
- *barrel finishing* (*tumbling*): rotating parts in a barrel filled with an abrasive or nonabrasive medium. Widely used to remove burrs, flash, scale, and oxides.
- *buffing:* a fine finishing operation, similar to polishing, using a very fine polishing compound (e.g., *rouge*).
- *burnishing:* a fine grinding or peening operation designed to leave a characteristic pattern on the surface of the workpiece.
- *calorizing:* the diffusing of aluminum into a steel surface, producing an aluminum oxide that protects the steel from high-temperature corrosion.
- *electroplating:* the electro-deposition of a coating onto the workpiece. Electrical current is used to drive ions in solution to the part. The workpiece is the cathode in the electrical circuit.
- *galvanizing:* a zinc coating applied to low-carbon steel to improve corrosion resistance. The coating

can be applied in a hot dip bath, by electroplating, or by dry tumbling (*sheradizing*).

- *hard surfacing:* the creation (by spraying, plating, fusion welding, or heat treatment) of a hard metal surface in a softer product.
- *honing:* a grinding operation using stones moving in a reciprocating pattern. Leaves a characteristic cross-hatch pattern.
- *lapping:* a fine grinding operation used to obtain exact fit and dimensional accuracy.
- *metal spraying:* the spraying of molten metal onto a product. Methods include *metallizing*, *metal powder spraying*, and *plasma flame spraying*.
- *organic finishes:* the covering of surfaces with an organic film of paint, enamel, or lacquer.
- *painting:* see *organic finishes*.
- *parkerizing:* application of a thin phosphate coating on steel to improve corrosion resistance. This process is known as *bonderizing* when used as a primer for paints.
- *pickling:* a process in which metal is dipped in dilute acid solutions to remove dirt, grease, and oxides.
- *polishing:* abrasion of parts against wheels or belts coated with polishing compounds.
- *sheradizing:* a specific method of zinc galvanizing in which parts are tumbled in zinc dust at high temperatures.
- *superfinishing:* a super-fine grinding operation used to expose nonfragmented, crystalline base metal.
- *tin-plating:* a hot-dip or electroplate application of tin to steel.

27. NOMENCLATURE

A	area	ft^2	m^2
b	chip width	ft	m
c	specific heat	Btu/lbm-°F	J/kg·°C
d	depth of cut	ft	m
D	diameter	ft	m
E	energy	ft-lbf	J
f	feed rate	ft/rev	m/rev
F	force	lbf	N
g	gravitational acceleration, 32.2 (9.81)	ft/sec^2	m/s^2
g_c	gravitational constant, 32.2	$lbm\text{-}ft/lbf\text{-}sec^2$	n.a.
h	height	ft	m
I	rotational moment of inertia	$lbm\text{-}ft^2$	$kg{\cdot}m^2$
J	Joule's constant, 778.17	ft-lbf/Btu	n.a.
L	length	ft	m
m	mass	lbm	kg
n	constant	–	–
n	rotational speed	rev/min	rev/min
N	number of items	–	–
p	pressure	lbf/ft^2	Pa
P	power	ft-lbf/min	W
r	chip thickness ratio	–	–
r	radius	ft	m
S	strength	lbf/ft^2	Pa
t	thickness	ft	m
t	time	sec	s
T	temperature	°F	°C
T	tool life	sec	s
U	specific cutting energy	$ft\text{-}lbf/ft^3$	J/m^3
v	cutting speed	ft/min	m/s
V	volume	ft^3	m^3
Z_w	metal removal rate	ft^3/min	m^3/s

Symbols

α	true rake angle	deg	deg
β	rake face resultant angle	deg	deg
θ	clearance angle	deg	deg
ρ	density	lbm/ft^3	kg/m^3
σ	normal stress	lbf/ft^2	Pa
τ	shear stress	lbf/ft^2	Pa
ϕ	shear angle	deg	deg
ω	rotational speed	rad/sec	rad/s
ω	wedge angle	deg	deg

Subscripts

c	chip
f	final
h	horizontal
n	normal
o	original (undeformed)
p	constant pressure
R	resultant
s	shear
t	tangent
u	ultimate
v	vertical

65 Materials Handling and Processing

Content in blue refers to the *NCEES Handbook*.

NCEES EXAM SPECIFICATIONS AND RELATED CONTENT

MACHINE DESIGN AND MATERIALS EXAM

II.A.6. Mechanical Components: Belt, pulley and chain drives

10. Belt Conveyors

1. WIRE ROPE

Wire rope is constructed by first winding individual *wires* into *strands* and then winding the strands into rope. (See Fig. 65.1.) Wire rope is specified by its diameter and numbers of strands and wires. The most common *hoisting cable* is 6 × 19, consisting of six strands of 19 wires each, wound around a core. This configuration is sometimes referred to as "standard wire rope." Other common configurations are 6 × 7 (stiff *transmission* or *haulage rope*), 8 × 19 (*extra-flexible hoisting rope*), and the abrasion-resistant 6 × 37.[1] The diameter and area of a wire rope are based on the circle that just encloses the rope.

Figure 65.1 *Wire Rope Cross Sections*

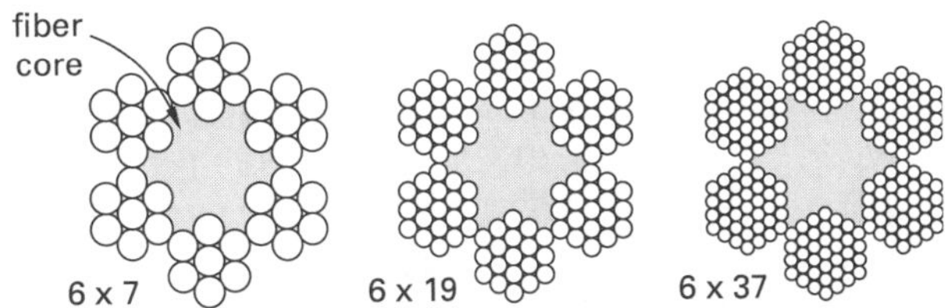

Wire rope can be obtained in a variety of materials and cross sections. In the past, wire ropes were available in iron, cast steel, traction steel (TS), mild plow steel (MPS), and plow steel (PS) grades. Modern wire ropes are generally available only in improved plow steel (IPS) and extra-improved plow steel (EIP) grades.[2] *Monitor* and *blue center steels* are essentially the same as improved plow steel. Table 65.1 gives minimum strengths of wire materials.

Table 65.1 *Minimum Strengths of Wire Materials*

material	ultimate strength, $S_{ut,w}$ (ksi)
iron (obsolete)	65
cast steel (obsolete)	140
extra-strong cast steel (obsolete)	160
plow steel (rare)	175–210
improved plow steel (IPS)	200–240
extra-improved plow steel (EIPS)	240–280
extra-extra improved plow steel (EEIPS)	280–310

(Multiply ksi by 6.8948 to obtain MPa.)

While manufacturer's data should be relied on whenever possible, general properties of 6 × 19 wire rope are given in Table 65.2.

In Table 65.2, the ultimate strength, $S_{ut,r}$, is the ultimate tensile load that the rope can carry without breaking. This is different from the ultimate strength, $S_{ut,w}$, of each wire given in Table 65.1. The rope's tensile strength will be only 80–95% of the combined tensile strengths of the individual wires. The modulus of elasticity for steel ropes is more a function of how the rope is constructed than the type of steel used. (See Table 65.3.)

[1]The designations 6 × 19 and 6 × 37 are only nominal designations. Improvements in wire rope design have resulted in the use of strands with widely varying numbers of wire. For example, none of the 6 × 37 ropes actually have 37 wires per strand. Typical 6 × 37 constructions include 6 × 33 for diameters under ½ in, 6 × 36 Warrington Seale (the most common 6 × 37 Class construction) offered in diameters ½ in through 1⅝ in, and 6 × 49 Filler Wire Seale over 1¾ in diameter.

[2]The term "plow steel" is somewhat traditional, as hard-drawn AISI 1070 or AISI 1080 might actually be used.

Table 65.2 *Properties of 6 x 19 Steel Wire Rope (improved plow steel, fiber core)*

diameter		mass		tensile strength[a,b] $S_{ut,r}$	
(in)	(mm)	(lbm/ft)	(kg/m)	(tons[c])	(tonnes)
1/4	(6.4)	0.11	(0.16)	2.74	(2.49)
3/8	(9.5)	0.24	(0.35)	6.10	(5.53)
1/2	(13)	0.42	(0.63)	10.7	(9.71)
5/8	(16)	0.66	(0.98)	16.7	(15.1)
7/8	(22)	1.29	(1.92)	32.2	(29.2)
1 1/8	(29)	2.13	(3.17)	52.6	(47.7)
1 3/8	(35)	3.18	(4.73)	77.7	(70.5)
1 5/8	(42)	4.44	(6.61)	107	(97.1)
1 7/8	(48)	5.91	(8.80)	141	(128)
2 1/8	(54)	7.59	(11.3)	179	(162)
2 3/8	(60)	9.48	(14.1)	222	(201)
2 5/8	(67)	11.6	(17.3)	268	(243)

(Multiply in by 25.4 to obtain mm.)
(Multiply lbm/ft by 1.488 to obtain kg/m.)
(Multiply tons by 0.9072 to obtain tonnes.)
[a]Add 7 1/2% for wire ropes with steel cores.
[b]Deduct 10% for galvanized wire ropes.
[c]tons of 2000 pounds

Table 65.3 *Typical Characteristics of Steel Wire Ropes*

configuration	mass (lbm/ft)	area (in²)	minimum sheave diameter	modulus of elasticity (psi)
6 × 7	$1.50d_r^2$	$0.380d_r^2$	42–72d_r	14×10^6
6 × 19	$1.60d_r^2$	$0.404d_r^2$	30–45d_r	12×10^6
6 × 37	$1.55d_r^2$	$0.404d_r^2$	18–27d_r	11×10^6
8 × 19	$1.45d_r^2$	$0.352d_r^2$	21–31d_r	10×10^6

(Multiply lbm/ft by 1.49 to obtain kg/m.)
(Multiply in² by 6.45 to obtain cm².)
(Multiply psi by 6.9×10^{-6} to obtain GPa.)

The central core can be of natural (e.g., hemp) or synthetic fibers or, for higher-temperature use, steel strands or cable. Core designations are FC for *fiber core*, IWRC for *independent wire rope core*, and WSC for *wire-strand core*. Wire rope is protected against corrosion by lubrication carried in the saturated fiber core. Steel-cored ropes are approximately 7.5% stronger than fiber-cored ropes. Figure 65.2 shows a 6 × 19 IWRC configuration.

Figure 65.2 *6 x 19 IWRC Configuration*

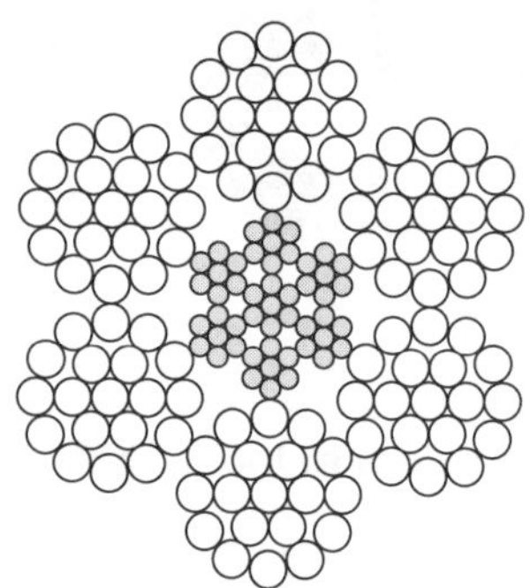

6 x 19 IWRC

Structural rope, *structural strand*, and *aircraft cabling* are similar in design to wire rope but are intended for permanent installation in bridges and aircraft, respectively. Structural rope and strand are galvanized to prevent corrosion, while aircraft cable is usually manufactured from corrosion-resistant steel.[3,4] Galvanized ropes should not be used for hoisting, as the galvanized coating will be worn off. Structural rope and strand have a nominal tensile strength of 220 ksi (1.5 GPa) and a modulus of elasticity of approximately 20,000 ksi (140 GPa) for diameters between 3/8 in and 4 in (0.95 cm and 10.2 cm).

The most common winding is *regular lay*, in which the wires are wound in one direction and the strands are wound around the core in the opposite direction. Regular lay ropes do not readily kink or unwind. Wires and strands in *lang lay* ropes are wound in the same direction, resulting in a wear-resistant rope that is more prone to unwinding. Lang lay ropes should not be used to support loads that are held in free suspension.

In addition to considering the primary tensile dead load, the significant effects of bending and sheave-bearing pressure must be considered when selecting wire rope. Self-weight may also be a factor for long cables. Appropriate dynamic factors should be applied to allow for acceleration, deceleration, and impacts. In general for hoisting and hauling, the working load should not exceed 20% of the breaking strength (i.e., a minimum factor of safety of 5 should be used).[5]

If D_w is the nominal wire strand diameter in inches,[6] D_r is the nominal wire rope diameter in inches, and D_{sh} is the sheave diameter in inches, the stress from bending around a drum or sheave is given by Eq. 65.1. E_w is the modulus of elasticity of the wire material

[3]Manufacture of structural rope and strand in the United States are in accordance with ASTM A603 and ASTM A586, respectively.
[4]Galvanizing usually reduces the strength of wire rope by approximately 10%.
[5]Factors of safety are much higher and may range as high as 8–12 for elevators and hoists carrying passengers.
[6]For 6 × 19 standard wire rope, the outer wire strands are typically 1/13–1/16 of the wire rope diameter. For 6 × 7 haulage rope, the ratio is approximately 1/9.

(approximately 3×10^7 psi (207 GPa for steel), not the rope's modulus of elasticity, though the latter is widely used in this calculation.[7]

$$\sigma_b = \frac{D_w E_w}{D_{\text{sh}}} \qquad 65.1$$

To reduce stress and eliminate permanent set in wire ropes, the diameter of the sheave should be kept as large as is practical, ideally 45–90 times the rope diameter. Alternatively, the minimum diameter of the sheave or drum may be stated as 400 times the diameter of the individual outer wires in the rope.[8] Table 65.3 lists minimum diameters for specific rope types.

For any allowable bending stress, the allowable load is calculated simply from the aggregate total area of all wire strands.

$$F_a = \sigma_{a,\text{bending}} \times \text{number of strands} \times A_{\text{strand}} \qquad 65.2$$

To prevent wear and fatigue of the sheave or drum, the radial bearing pressure should be kept as low as possible. Actual maximum bearing pressures are highly dependent on the sheave material, type of rope, and application. For 6×19 wire ropes, the acceptable bearing pressure can be as low as 500 psi (3.5 MPa) for cast-iron sheaves and as high as 2500 psi (17 MPa) for alloy steel sheaves. The approximate bearing pressure of the wire rope on the sheave or drum depends on the tensile force in the rope and is given by Eq. 65.3.

$$p_p = \frac{2F_t}{D_r D_{\text{sh}}} \qquad 65.3$$

Fatigue failure in wire rope can be avoided by keeping the ratio $p_p/S_{\text{ut},w}$ below approximately 0.014 for 6×19 wire rope.[9] ($S_{\text{ut},w}$ is the ultimate tensile strength of the wire material, not of the rope.)

2. ELEVATORS

Elevators classified for freight or passenger use are supported by counterweighted cables.[10] Both geared and gearless traction machines are used for power. Geared machines are suitable for elevator speeds up to approximately 300 ft/min (1.5 m/s); gearless machines (where the driving sheave is mounted directly to the motor shaft) should be used for higher speeds.

Elevator cable is a special category of wire rope with its own material grading categories. For satisfactory rope life, the sheave diameter is typically 40 to 50 times the rope diameter.

Elevator cabling passes over and is turned by a hoisting sheave. The required frictional force between cable and sheave is obtained by use of multiple cables and wraps. Roping is said to be *half-wrapped* if it passes over the sheave only once (or less) and *full-wrapped* if there are two successive half-wraps with an idler sheave in between.

The *roping* represents the ratios of load-to-cable tension and rope-to-car speeds. (See Fig. 65.3.) With 1:1 roping, the two ends of the rope are connected to the car and counterweight, and car and rope speeds are identical. 1:1 roping is suitable for light cars with speeds in excess of 600 ft/min (3 m/s). With 2:1 roping, the cables are attached at the penthouse level, and the car speed is half the rope speed. 3:1 roping, wherein the rope passes through deflector (idler) sheaves at the penthouse level before connecting to the car and counterweight, also is used occasionally for very heavy loads.

Figure 65.3 *Variations on Elevator Roping*

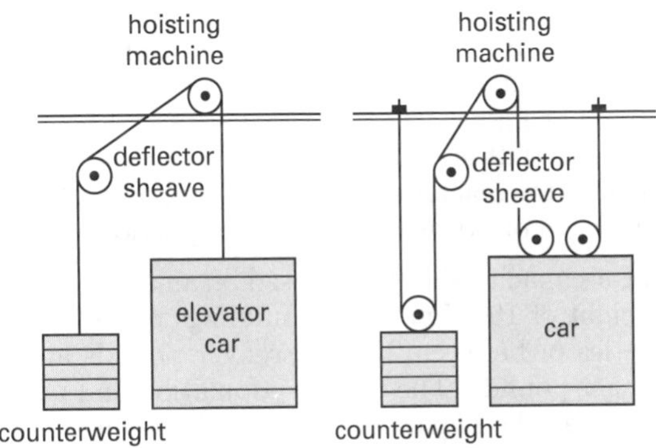

(a) 1:1 roping, single-wrap (b) 2:1 roping, single-wrap

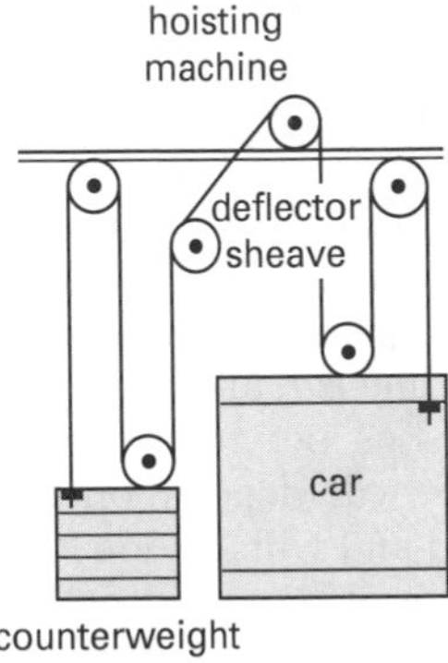

(c) 3:1 roping, single-wrap

Since the elevator weight is balanced by a counterweight, the motor only has to accelerate the unbalanced load. Passenger elevators are typically overbalanced (i.e., in addition to the elevator weight balanced to 40–50% of the fully loaded elevator capacity). With rises in excess of approximately 100 ft,

[7]Although E in Eq. 65.1 is often referred to as "the modulus of elasticity of the wire rope," it is understood that E is actually the modulus of elasticity of the wire rope material.
[8]For elevators and mine hoists, the sheave-to-wire diameter ratio may be as high as 1000.
[9]A maximum ratio of 0.001 is often quoted for wire rope regardless of configuration.
[10]Design, selection, and use of elevators are governed by local and national codes.

the weight of the cable between the traction device and counterweight should be counteracted by the weight of static cables or chains connected to the car bottom and shaft ceiling.

There are no height or speed limitations for the counterweighted design, and overtravel is inherently limited. If either the elevator or counterweight goes too high, one or the other bottoms out, eliminating the driving friction.[11]

The maximum travel rate (elevator speed) depends on the number of floors served. Commercial traction machines operate routinely at 1000 ft/min (5 m/s) and are capable of speeds between approximately 25 ft/min and 1600 ft/min (0.13 m/s and 8 m/s), with the highest speeds applicable only to the tallest buildings. Although local building codes may apply, a speed of 150 ft/min (0.75 m/s) is reasonable for 3 or fewer floors, 200 ft/min (1 m/s) for 4 floors, 300 ft/min (1.5 m/s) for 5 or 6 floors, 400 ft/min (2 m/s) for 7 to 9 floors, and 500 ft/min (2.5 m/s) for 10 to 13 floors.

For maximum comfort, acceleration and deceleration should be limited to $\frac{1}{8}$ g. For travel rates of 300 ft/min (1.5 m/s) or higher, the acceleration should be 3 ft/sec^2 to 5 ft/sec^2 (0.9 m/s^2 to 1.5 m/s^2). Acceleration should taper off after reaching 80% of the desired travel speed. Acceleration of slower elevators is proportionally lower.

Elevator capacity should be based on an average passenger weight of 150 lbm. Office buildings often have total capacities of between 2500 passenger pounds and 4000 passenger pounds. The service population and *peak load service* (i.e., the maximum number of passengers served in a 5 min period) may be estimated by rules of thumb or may be specified by contract or code.[12]

The average wait, known as the *interval*, for an elevator generally should not exceed 20 sec to 30 sec, although elevator costs must be balanced against the costs of waiting and other factors (crowding, safety, etc.). The theoretical number of elevators required should be calculated using queuing theory.[13] Equation 65.4 can be used if the 5 min peak service population is P, the average round-trip time is t, and the average peak period car loading (in persons per round trip) is e. The average round-trip time will depend on the average distance traveled per trip and will include allowances for elevator movement, acceleration and deceleration, door opening and closing, reaction time, and passenger movement.

$$\text{number of elevators} = \frac{P_{\text{peak,5 min}} t_{\text{min}}}{5e} \qquad 65.4$$

The *dispatch interval* is the average round trip time divided by the number of elevators.

3. BULK STORAGE PILES

Large quantities of free-flowing bulk solid materials are typically stored as mass piles that are conical or rectangular in shape. (Bins and hoppers are discussed in Sec. 65.4.) The angle formed between the surface of the pile and the horizontal, ϕ, is the *angle of repose* (also known as the *angle of natural slope*, *angle of natural friction*, and *angle of internal friction*).[14] Typical values for common industrial bulk materials are given in Table 65.4.[15]

Table 65.4 *Typical Properties of Bulk Materials*

material	angle of repose (degrees)	bulk density (lbm/ft^3)	bulk density (kg/m^3)
ash, anthracitic	45	40–45	640–720
ash, soft coal	40–45	35–40	560–640
cement, portland	40	90	1400
cinders, coal	25–50	40	640
clay, compact	20–25	110	1800
coal, anthracitic*	30–45	52–57	830–910
coal, bituminous*	35	44–52	700–830
coke, loose	30–45	23–35	370–560
earth, common	30–45	70–80	1100–1300
gravel	30–40	100	1600
lime, ground	43	60	960
sand, moist	15–30	110–130	1800–2100
slag, blast furnace	25	80–90	1300–1400
soda ash, light	37	20–35	320–560
sugar, granulated	30–45	50–55	800–880

(Multiply lbm/ft^3 by 16.0 to obtain kg/m^3.)

*Exercise care in storing coal. Spontaneous combustion is a problem when most coals, with the exception of anthracite, are stored for long periods. Accumulated coal gas represents an explosion hazard. Freshly pulverized coal is hygroscopic in nature.

Though the actual pile peak will be slightly rounded, the volume and height can be calculated assuming a conical shape. In Eq. 65.5 and Eq. 65.6, r is the radius at the base of the pile and h is the vertical height.

$$V = \tfrac{1}{3}\pi r^2 h = \tfrac{1}{3}\pi r^3 \tan\phi \qquad 65.5$$

$$h = \frac{3V}{\pi r^2} = \left(\frac{3V\tan^2\phi}{\pi}\right)^{1/3} \qquad 65.6$$

[11]Since there is no inherent safety against overtravel, drum-wound cabling is prohibited by code for passenger use. Similarly, almost all hydraulic elevator designs are obsolete.
[12]For example, the peak 5 min service load may be estimated as 10% to 15% of the building population.
[13]For continuity of service, at least two elevators should be installed in critical applications (e.g., in hospitals).
[14]The angle of a conical pile will be slightly less than the true angle of repose of the material due to the effect of impact.
[15]The actual angle of repose will depend on the moisture content and grading. Fines carry most of the moisture.

Rectangular piles can be assumed to have straight-sloped sides (inclined at the angle of repose) and four quarter-conical corner sections. Volumes of rectangular piles can be calculated from the dimensions of these individual sections.

The volumes of piles taking on the shapes of wedges and frusta of rectangular pyramids (see Fig. 65.4) can be calculated from basic shapes or from the following equations.

$$V = \left(\frac{h}{6}\right)b(2a + a_1) = \tfrac{1}{6}hb\left(3a - \frac{2h}{\tan\phi}\right) \qquad \text{65.7}$$

[wedge]

$$\begin{aligned} V &= \left(\frac{h}{6}\right)\left(ab + (a + a_1)(b + b_1) + a_1b_1\right) \\ &= \left(\frac{h}{6}\right)\left(ab + 4\left(a - \frac{h}{\tan\phi_1}\right)\left(b - \frac{h}{\tan\phi_2}\right) + \left(a - \frac{2h}{\tan\phi_1}\right)\left(b - \frac{2h}{\tan\phi_2}\right)\right) \end{aligned} \qquad \text{65.8}$$

[frustum of a rectangular pyramid]

$$a_1 = a - \frac{2h}{\tan\phi_1} \qquad \text{65.9}$$

$$b_1 = b - \frac{2h}{\tan\phi_2} \qquad \text{65.10}$$

A three-dimensional soil volume between two points is known as a soil *prismoid* or *prism*. (See Fig. 65.5.) The prismoid (prismatic) volume must be calculated in order to estimate hauling requirements. Such volume is generally expressed in units of cubic yards ("yards") or cubic meters. There are two methods of calculating the prismoid volume: the average end area method and the prismoidal formula method.

With the *average end area method*, the volume is calculated by averaging the two end areas and multiplying by the prism length. This disregards the slopes and orientations of the ends and sides, but is sufficiently accurate for most earthwork calculations. When the end area is complex, it may be necessary to use a planimeter or to plot the area on fine grid paper and simply count the squares. The average end area method usually over-estimates the actual soils volume, favoring the contractor in earthwork cost estimates.

$$V = \frac{L(A_1 + A_2)}{2} \qquad \text{65.11}$$

Figure 65.4 *Pile Shapes*

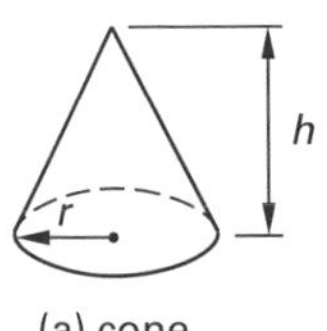

(a) cone

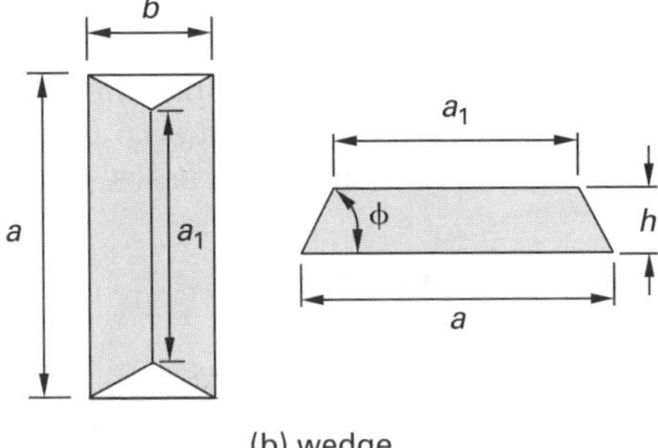

(b) wedge

(c) frustum of a rectangular pyramid

Figure 65.5 *Soil Prismoid*

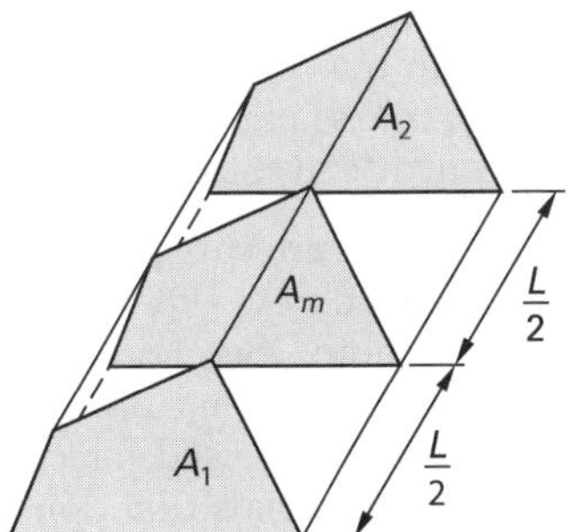

The precision obtained from the average end area method is generally sufficient unless one of the end areas is very small or zero. In that case, the volume should be computed as a pyramid or truncated pyramid.

$$V_{\text{pyramid}} = \frac{LA_{\text{base}}}{3} \qquad \text{65.12}$$

The *prismoidal formula* is preferred when the two end areas differ greatly or when the ground surface is irregular. It generally produces a smaller volume than the average end area method and thus favors the owner-developer in earthwork cost estimating.

Plant Engineering

The prismoidal formula uses the mean area, A_m, midway between the two end sections. In the absence of actual observed measurements, the dimensions of the middle area can be found by averaging the similar dimensions of the two end areas. The middle area is not found by averaging the two end areas.

$$V = \left(\frac{L}{6}\right)(A_1 + 4A_m + A_2) \qquad 65.13$$

When using the prismoidal formula, the volume is not found as LA_m, although that quantity is usually sufficiently accurate for estimating purposes.

4. BINS AND HOPPERS

Smaller quantities of dry bulk solids, including powders, pellets, flakes, and granules, are generally inventoried in conical bins, or wedge-shaped hoppers. Bins and hoppers must be carefully chosen to avoid the problems of arching, rat-holing, and segregation.

Arching or *bridging* occurs when the opening at the base of the bin is too small. The strength of the compressed or interlocked bulk solid is enough to form an arched cap above the bin opening. This stops flow. Arching is eliminated by either reducing cohesion in the material or increasing the opening size.

Rat-holing or *piping* occurs when the slope of the bin walls is not great enough for gravity to overcome the friction between the material and the walls. Material near the walls is trapped as dead (stagnant) storage, and a hole through the material appears from the opening to the top of the bin. *Segregation* occurs naturally during filling, when coarse particles fall to the sides and fine particles concentrate in the center.

Storage hoppers and bins are categorized as funnel-flow, mass-flow, and expanded-flow types, depending on the type of flow they promote. (See Fig. 65.6.) *Funnel-flow bins* have long, upright walls that narrow only near the bottom. They are commonly used for coarse (i.e., ¼ in (6 mm) or larger), dry materials that don't pack or deteriorate. Though the storage capacity of a funnel-flow bin is the largest of the three types, the effective capacity may be reduced by dead storage. Arching may also be a problem unless the opening is very wide. Since funnel-flow bins are particularly prone to segregation and formation of dead storage, they should not be used for materials that spoil or are damaged by long-term residence.

Figure 65.6 *Conical Bin Geometries*

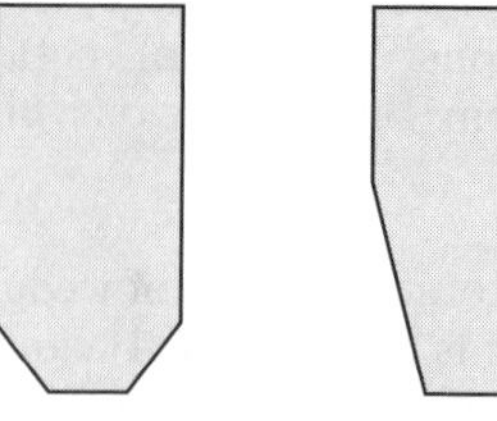

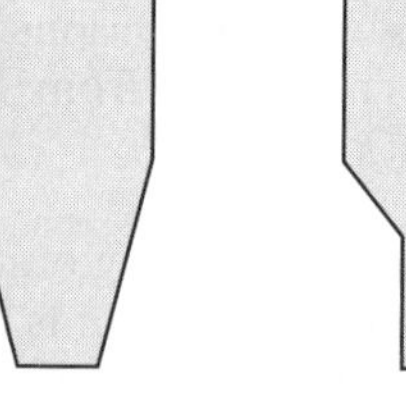

(a) funnel flow (b) mass flow (c) expanded flow

Mass-flow bins have steeper conical sections than funnel-flow bins. Though their theoretical storage capacity is the smallest of the three types of bins, mass-flow bins are less prone to arching and rat-holing. Material leaves the bin on a first-in, first-out basis, and segregated particles are efficiently remixed during discharge. Therefore, mass-flow bins are suitable for fine powders, cohesive bulk materials, and materials that degrade when stored for extended periods of time. *Expanded-flow* bins combine the designs of both funnel- and mass-flow bins. They have performance characteristics intermediate of the two types.

Wedge-shaped rectangular hoppers may offer advantages over conical bins. Typically, a wedge-shaped hopper can be 10° to 12° less steep without experiencing funnel flow. Since the diameter of a conical opening must be twice the width of a slot opening, hoppers can get by with smaller-width openings or provide higher flow rates. Rectangular hoppers use floor space more efficiently, and, for a given volume, hoppers require less headroom than conical bins.

In the past, the design of bins and hoppers was largely based on the desired volume and the angle of repose. Wall angles greater than the angle of repose were chosen. Modern bin and hopper design is based on the *angle of kinematic friction* (i.e., the angle of friction between the material and the storage unit wall), the stress in the material forming an arch, and numerous other factors. This constitutes an engineering economic storage analysis.[16]

Vertical pressure, p_v, at the bottom of a deep cylindrical hopper (e.g., bin, silo, etc.) with filled height, h, and diameter, D, is predicted by the *Janssen equation.* γ is the specific weight of the contents, and μ is the material's coefficient of internal friction. For a circular hopper, the *hydraulic radius*, R_h, is equal to $D/4$. k is the *pressure ratio* (*stress ratio*, *Janssen's constant*). Equation 65.16, derived from Mohr's circle, assumes frictionless contact between the stored material and the wall and is known in geomechanics as the *coefficient of active lateral earth pressure.* Various incompatible methods of incorporating wall friction have been proposed;

[16]This method of analysis was perfected by Jenike and Johanson in the early 1960s.

therefore, k should be determined experimentally. An approximate value of 0.4 can be used in the absence of other information.

$$R_h = \frac{\text{area of cross section}}{\text{length of cross-sectional perimeter}} \quad 65.14$$

$$\mu = \tan\phi \quad 65.15$$

$$k = \frac{1-\sin\phi}{1+\sin\phi} = \tan^2\left(45° - \frac{\phi}{2}\right) \quad \begin{bmatrix}\text{Rankine theory;}\\ \text{frictionless wall;}\\ \text{active case}\end{bmatrix} \quad 65.16$$

$$p_v = \frac{\rho g R_h}{g_c \mu k}\left(1 - \exp\left(\frac{-4h\mu k}{D}\right)\right) = \frac{\rho g D}{4 g_c \mu k}\left(1 - \exp\left(\frac{-4h\mu k}{D}\right)\right) \quad 65.17$$

Equation 65.18 gives the horizontal pressure against the hopper's wall.

$$p_h = k p_v = \frac{\rho g D}{4 g_c \mu}\left(1 - \exp\left(\frac{-4h\mu k}{D}\right)\right) \quad 65.18$$

The discharge rate of a free-flowing granular material experiencing mass flow from a circular hopper or bin with a circular outlet of diameter D is given by the *Johanson equation*, which is derived from basic principles. The discharge rate is a function of the outlet diameter and material characteristics. Discharge flow velocities are typically 10–150 ft/min (3–45 m/s). ρ_b is the materials bulk density, and θ is the angle of the hopper from the vertical.

$$\dot{m} = \rho_b A\sqrt{\frac{gD}{4\tan\theta}} = \frac{\rho_b \pi D^2}{4}\sqrt{\frac{gd}{4\tan\theta}} \quad 65.19$$

5. FEEDING DRY BULK SOLIDS

Depending on their characteristics, dry bulk powders, pellets, and flakes are categorized as being floodable, easy-flowing, difficult-flowing, or cohesive. *Floodable materials* (e.g., flour, gypsum, and diatomaceous earth) behave like liquids and require a positive sealing device such as intermeshed twin screws. *Easy-flowing materials* (e.g., sand and plastic pellets) can be transported by almost any system. *Difficult-flowing* materials (e.g., fiberglass and rubber particles) and *cohesive materials* (e.g., pigments and titanium dioxide) often require custom feeders and flow-aid devices.

Fully enclosed *single-screw feeders* (*screw conveyors*) are suitable for free-flowing powders and granular, pelletized, and flaked products (e.g., salt, sugar and plastic pellets). The path should be reasonably level and less than approximately 200 ft (60 m) long.[17] The screw may be of constant pitch or, in order to reduce the power required to shear material out of the overlying hopper load, of variable pitch with a tighter helix at the hopper.

Twin-screw feeders use two screws, either intermeshed or nonintermeshed. Intermeshed screws must turn in the same direction. Twin-screw feeders are suitable for almost all materials, including floodable, cohesive, and sticky materials. However, pelletized materials may jam or be crushed between intermeshed screws.

Volumetric flow rates and power requirements for screw feeders are proportional to length and depend on both the empty volume (for frictional effects) and the mass of material carried (for inertial effects). Power calculations are well documented by screw feeder manufacturers, each of which has its own unique set of coefficients and constants.

Rotary feeders are used for very fine, free-flowing powders. A rotor with radial vanes (blades) turns within a housing. A close tolerance between the vanes and housing keeps the material in the feed hopper from the discharge stream. Top-fed units are preferred for maximum throughput, as each cavity is completely filled during rotation of the rotor. Side-fed units are preferred for pelletized products since each cavity will be only partially filled, limiting shearing and tearing between the vanes and housing.

Belt feeders are used to supply high volumes of fragile materials, ores, and powders. Capacity is a function of the belt speed and width. Belt speeds up to 120 ft/min (0.6 m/s) are used, though 60 ft/min (0.3 m/s) is an upper limit for smaller feeders. *Vibratory feeders* are used to move abrasive or friable solids that could damage or be damaged by screw or rotary feeders.

For most noncyclical feeding devices, the feed rate is a function of the material bulk density, velocity, and flow cross section. The *turndown* of a feeder type refers to the ease and linearity of reducing the feed rate. Screw conveyors have large and linear turndowns; vibratory feeders have limited and nonlinear turndowns.

6. VENTURI FEEDERS

A *venturi feeder* (*venturi eductor*) is a method of conveying free-flowing materials in powder, pellet, flake, and particulate forms over relatively short distances without any moving parts. (See Fig. 65.7.) An air-entraining nozzle produces a jet of air that entrains the conveyed material, known as *dilute-phase conveying* and *suspension flow*. Control of the quantity of material conveyed must be provided separately (e.g., a flow-control butterfly valve inside the hopper, trickle valve, calibrated orifice plate, or gate/slide valve), as the venturi feeder has

[17]There is little loss in capacity for inclinations below 15°. However, an inclination of 20° can reduce the capacity of open screw conveyors by up to 50%.

Plant Engineering

no intrinsic flow control mechanism. In low-quantity feeders, throat pressures are essentially equal to the supply hopper pressure (i.e., 1 atm), allowing the material to feed in by gravity. For these feeders, a standard industrial (scroll) fan is adequate, and flow rates will be low. In high-capacity feeders, the pressure in the throat area is lower than atmospheric pressure, and material is drawn into the flow by suction. Relatively high supply pressures are used. The feeder design converts the supply pressure into a pressure sufficient to overcome the downstream friction. The *superficial velocity* is the velocity of the air alone, disregarding the material volume due to its higher specific gravity. The *loading ratio* (*material-air ratio*) is the ratio of the mass flow rate of conveyed material to the mass flow rate of air. The *specific air rate* (also known as *saturation*) is the ratio of volumetric air flow to the mass of solids conveyed.

Figure 65.7 *Venturi Feeder*

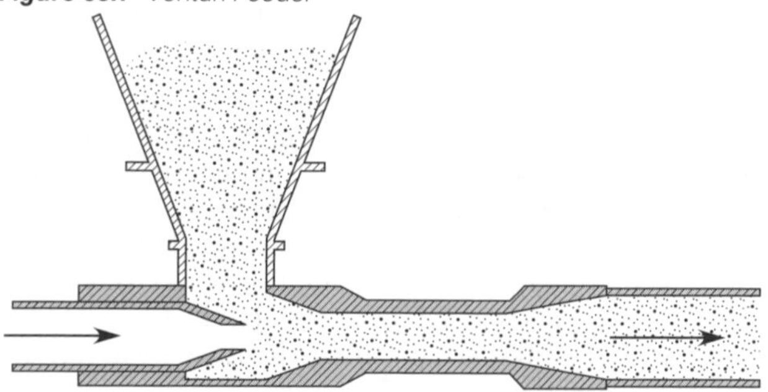

Frictional losses in the line are evaluated separately for the air and the solid. A conservative assumption is that the air and solids velocities are the same. Traditional methods can be used to calculate the air frictional loss. A friction factor of 0.015 is consistent with high Reynolds numbers. The contribution of the solid to friction loss is calculated from first principles as the work performed in increasing the solid's kinetic and potential energies, and in overcoming the sliding friction. The coefficient of sliding friction is the tangent of the angle of repose, typically 35–45°.

7. PNEUMATIC CONVEYORS

Pneumatic conveyors are used for dusts, coal, flyash, grains, granular material that is dry and free-flowing, and pellets less than approximately $^1/_4$ in (6 mm) in size. Material may be moved by positive pressure, vacuum, or a combination of the two. Positive-pressure systems are generally not used for multipoint feeding (i.e., from multiple bins or hoppers). Vacuum systems are generally not used for multipoint delivery.

Most pneumatic conveying systems operate with air speeds of 3000 ft/min to 7500 ft/min (15 m/s to 38 m/s). 3000 ft/min (15 m/s), the lower practical limit, is required for light materials such as flour and dusts, and 4000 ft/min to 6000 ft/min (20 m/s to 30 m/s) is required for heavier materials such as grains. Higher velocities, up to 20,000 ft/min (100 m/s), also may be used if economical.

The primary factor determining air velocity is the bulk density of the material. Equation 65.20 gives the theoretical air velocity to lift a smooth spherical particle of size *d*. Since most particles are neither spherical nor uniform, Eq. 65.20 should be used as a general check. Actual conveying velocities will be approximately 50% greater than calculated.

$$\mathrm{v}_{\mathrm{ft/min}} = 15{,}000\left(\frac{\mathrm{SG}}{\mathrm{SG}+1}\right)\sqrt{d_{\mathrm{in}}} \qquad 65.20$$

The required air volume must be determined by trial and error. Values range from 35 ft^3 to 40 ft^3 per pound for heavy, compact materials, to 90 ft^3 per pound for light, fluffy materials (e.g., cotton).

Rotary piston-type blowers are commonly used when the conveying distance is long and the flow rate is high. Their volumetric flow rates are largely insensitive to changes in the system pressure. The discharge:inlet pressure ratio for positive-displacement blowers is limited to approximately 2:1. This means that positive-pressure systems can operate up to two atmospheres; vacuum systems are limited to half an atmosphere.

Fans can be used as the motive power when the pressure differential required is not too great. Fans are ideal if the product can be conveyed through the fan itself, as is the case with shredded products (e.g., paper, sawdust, and polyethylene trimmings). Fan-driven systems generally operate at lower velocities.

Blow-tank systems are used for conveying materials long distances. A blast of air accelerates a plug of material through the pipeline. Delivery is cyclical, and a large volume of pressurized air is released with each plug. Lines must be sized to accommodate the instantaneous plug feed rate, not the average feed rate.

8. BRIQUETTING BULK SOLIDS

Briquetting and *pressure compaction* use pressure with or without a binder to form loose bulk solids into briquettes or pellets. Briquetting can improve the handling, storage, salability, and environmental acceptance (through dust reduction) of the product. Products commonly briquetted include charcoal, scrap metal, fertilizers, animal food, and salt for water softeners.

The process of producing enlarged particles is known as *agglomeration*. Confined compaction and extrusion are the two primary methods used. Roll briquetters, tabletizers, pelletizers, presses, screw extruders, and roll-extrusion pellet mills are typical agglomeration devices. A rotary knife device cuts extruded product into pellets of the desired length.

In roll machines, the *nip* is the gap between the rolling surfaces. The *angle of nip* defines the effective area of consolidation. (See Fig. 65.8.) Above the angle of nip, slip occurs within the material and between the material and the rollers, and the material is not captured by the rollers. Below the angle of nip, the material is caught and drawn into the rollers. The angle of nip is a primary factor in the compaction pressure. By using a screw conveyor to precompress the material, smaller-diameter rolls can be used to achieve the same briquetting pressure (compared to a gravity-fed system).

Figure 65.8 *Angle of Nip*

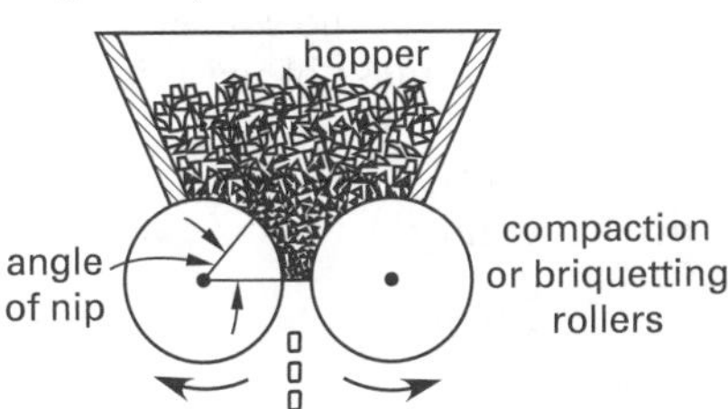

Briquetting pressures depend on the product. Coal, charcoal, coke, fertilizers, and animal feed can be briquetted at low pressures of 5 kips/in^2 to 20 kips/in^2 (35 MPa to 140 MPa). Medium pressures of 20 kips/in^2 to 50 kips/in^2 (140 MPa to 350 MPa) are needed for soft metal (e.g., brass and aluminum) turnings and borings and glass-making mixtures. High pressures of 50 kips/in^2 to 80 kips/in^2 (350 MPa to 550 MPa) are used for ductile metals, steel-chips and turnings, mill scale, and electric- and blast-furnace dust. Very high pressures, 80 kips/in^2 to more than 100 kips/in^2 (550 MPa to 700 MPa), are needed for metal powders and for titanium and stainless steel turnings.

Some materials are briquetted with a binder. *Matrix binders*, including wax, paraffin, clay, dry starch, and asphalt, fill the majority of the voids in the material. *Film binders* are added in liquid form.[18] As the solvent evaporates or cures, the binder remains to form solid bridges between particles. *Chemical-reaction binding* may be used in certain cases where a chemical reaction occurs between the binder and the briquetted material.

9. SIZE REDUCTION PROCESSES

Size reduction, also known as *comminution* and *pulverizing*, produces multiple pieces of smaller material from larger pieces. Large pieces are commonly crushed to size in primary operations, while smaller pieces are cut or ground in secondary operations. For example, freshly lined coal and friable ores are typically reduced in size in *roll crushers*, the rolls of which can be smooth, corrugated, or toothed. *Jaw* and *gyratory crushers*, which are more efficient and require less maintenance than roll crushers, are now generally used for harder ores. (See Fig. 65.9.)

Figure 65.9 *Jaw and Gyratory Crushers*

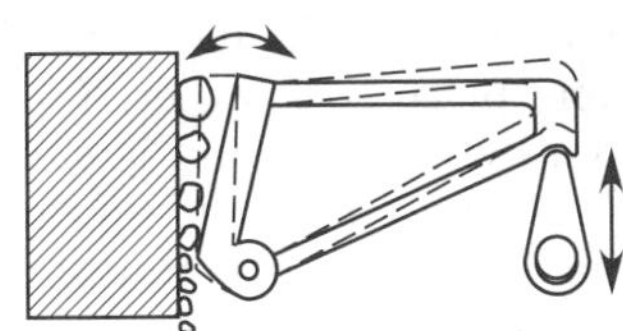

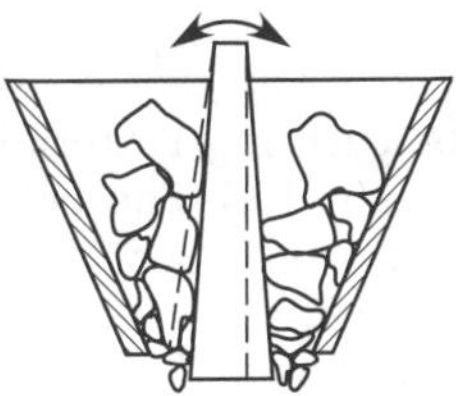

Rotary cutters, common in the recycling industry, are used for tough or fibrous materials such as pulp, wood, paper, fiberglass, plastics, and rags. Multiple cutting blades (known as *fly knives*) on a rotating rotor are juxtaposed to stationary blades (*bed knives*). Though there may be up to 16 fly knives per cutting head, typically only 2 to 4 bed knives are used. The softest, stickiest, and toughest materials require the fewest bed knives.

Disk mills using steel disks are similar in operation to old flour mills using millstones. They are used for grinding tough organic materials such as wood pulp.

Ring roller mills are used to size-reduce nonmetallic minerals (e.g., coal) and dry chemicals (limestone, starch, pigments, etc.) requiring a finished size of approximately 35 μm to 800 μm. Ring roller mills have rollers forced inside a cylindrical ring. Either the rollers or the ring can rotate. Operation of bowl- and ball-bearing mills is similar.

Agitator ball mills, also known as *tube-mills*, are commonly used for wet grinding. They use a cylindrical grinding bin partially filled with grinding balls. The grinding efficiency is low, and the ground product typically has a wide distribution of finished sizes.

In *mechanical impact mills*, including *hammer mills*, free-moving particles strike a single beating surface at high speed. Either the particles can be thrown outward by centrifugal force to a stationary beating surface, or the beaters can revolve at high speed.

Fine impact mills are mechanical impact devices that accept material that is smaller than approximately 4 in (100 mm). Only nonabrasive, brittle, and free-flowing materials are suitable for fine impact mills. Coal for direct firing is commonly pulverized in high-speed (225 rev/min and above) impact pulverizers. *Jet mills* accelerate the material by gas or steam jets, with impact between particles or solid surfaces reducing the particle size.

In *autogenous* and *semiautogenous* (SAG) grinding, the material being ground serves as the grinding media.[19] These methods have largely replaced conventional grinding methods for secondary grinding (i.e., following the primary gyratory crusher) in mineral processing plants for materials less than approximately 16 in (0.4 m) in initial size. Material on the belt is discharged at the head pulley.

[18]Plain water is the most common film binder.

[19]In SAG grinding, the material is supplemented by steel balls.

Plant Engineering

10. BELT CONVEYORS

A *belt conveyor* consists of an endless belt and is used for carrying material from one place to another. (See Fig. 65.10.) The belt is driven by a pulley and is supported on rollers (idlers) or (less often) on a flat runway. The belt may be flat or troughed. Carried material is placed on and removed from the belt along its run. Complete design of a belt conveyor will encompass selection of the belt material, the method of support, the driving method, and accessories to maintain belt tension and to load, unload, and clean the belt.

Figure 65.10 *Typical Belt Conveyor*

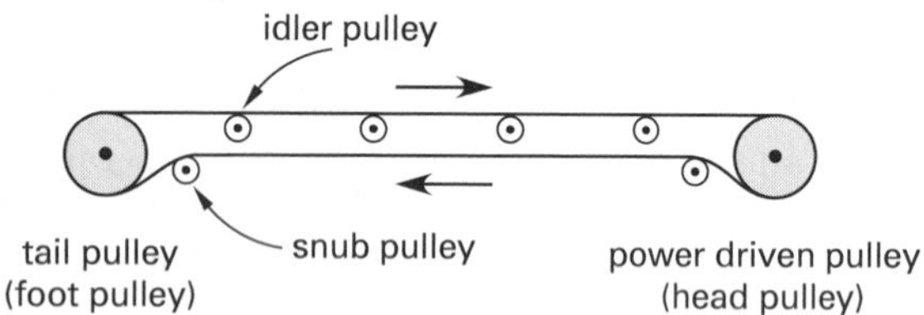

Power is required to (1) overcome friction in all of the moving parts, (2) move the load horizontally, (3) move the load vertically, and (4) drive any belt-powered accessories such as trippers.

Equation 65.21 calculates the power required to drive the belt in terms of the belt velocity, v, and net tension, F_{net}. The net tension, also known as the *effective tension*, is the difference between the tight- and loose-side belt tensions.[20] The required horsepower (kilowatts) can be calculated by dividing by the appropriate factor (e.g., 33,000 ft-lbf/hp-min).

$$P = (F_1 - F_2)\mathrm{v} = F_{net}\mathrm{v} \qquad 65.21$$

The force in the direction of impending motion is a function of the force applied to resist impending motion and the belt wrap angle.[21] The coefficient of friction, μ, between the belt and pulley is approximately 0.25 for dry, clean rubber belts on cast-iron pulleys and 0.35 when the pulleys are lagged or rubber-coated.[22] For dusty environments, the corresponding coefficients of friction are approximately 0.20 and 0.27. For wet work, the coefficient is approximately 0.20 for bare or lagged pulleys. Centrifugal force is usually not a factor in conveyor belt design.

Belt Friction

$$F_1 = F_2 e^{\mu\theta} \qquad 65.22$$

The belt tension in Eq. 65.21 results primarily from friction in the rotating idlers. Other frictional factors (i.e., friction of the carried material with the skirt boards) are considered to be minor in comparison and are accounted for by including an allowance with the idler friction. Modern designs use sealed antifriction bearings that do not require periodic lubrication.

Belts are seldom inclined more than 30° from the horizontal, and material slippage typically limits the slope to the 15° to 20° range.[23] If the conveyor is inclined, the power required to lift a mass of material per unit time, $\dot{m}$, through a height, h, is given by Eq. 65.23. Theoretically, an inclination of the belt does not affect the power required to drive it empty.[24]

$$P = \dot{m}gh \qquad \text{[SI]} \qquad 65.23(a)$$

$$P = \frac{\dot{m}gh}{g_c} \qquad \text{[U.S.]} \qquad 65.23(b)$$

In the absence of measurements of actual tension or specific manufacturer's data, little can be done to accurately predict the required driving power, though several quick approximations are widely used. A simple method for first approximations of horsepower for conveyors with modern antifriction idler bearings carrying materials with densities of 50 lbm/ft^3 to 100 lbm/ft^3 (800 kg/m^3 to 1600 kg/m^3) is given by Eq. 65.24.[25] In Eq. 65.24, L is the center-to-center length of the pulley, and h is the corresponding vertical rise.

$$P_{hp} = 0.4 + \left(\frac{\dot{m}_{lbm/hr}}{2000}\right)\left(\frac{0.00325L_{ft}}{100} + \frac{0.01h_{ft}}{10}\right) \qquad 65.24$$

Belt speed should be no faster than necessary to provide the capacity under conditions of full loading. However, it may be uneconomical (compared to other material handling methods) to run conveyor belts at much less than 150 ft/min (0.75 m/s). The main factor in selecting conveyor speed is belt wear due to contact with the carried material. It is more desirable to load a belt deep than shallow, as more material is carried for the same amount of belt damage. Because of this, wider belts can run faster than narrow belts because the loading can be deeper.

[20] There are cases—primarily when the conveyor is inclined, uses a tandem drive, or has a holdback—where the maximum tension is greater than F_1.

[21] It is seldom possible to get more than 180° with a plain pulley or more than 260° with a snub-nose pulley. However, with a *tandem-pulley drive* (i.e., a second drive pulley with a reverse bend in the belt), the combined angle of wrap can be up to approximately 410°. Tandem drives are seldom used in practice because of their expense.

[22] Ideally, the coefficient of friction between a rubber-coated pulley and a rubber belt is approximately 0.55. However, the value of 0.35 should be used to account for dirty conditions.

[23] Dry silica sand is limited to approximately 15°. Paper-wrapped packages are limited to 16°. Fine coal may be inclined up to about 22° if evenly loaded and 20° if lumpy. Run-of-mine coal and coke are limited to inclines of approximately 18°.

[24] There is some additional stress in the belt.

[25] This is a modification of the old rule of thumb calculating the horsepower as "2% of the tons per hour for every 100 feet or horizontal distance plus 1% of the tons per hour for every 10 feet of vertical distance." The 2% figure, applicable to early greased fittings, has been substantially reduced by use of low-friction bearings.

Approximate maximum speeds for common-design conveyors carrying coal, ore, and gravel vary linearly from 300 ft/min (1.5 m/s) for 12 in wide (0.3 m) belts to 600 ft/min (3 m/s) for 48 in wide (1.2 m) and wider belts. Approximate maximum speeds for conveyors carrying heavy grain (e.g., wheat and corn) vary linearly from 400 ft/min (2 m/s) for 12 in wide belts (0.3 m) to 700 ft/min (3.5 m/s) for 48 in wide (1.2 m) and wider belts. Lighter grains (e.g., bran, oats, and flour) can be carried at up to 500 ft/min (2.5 m/s) without being blown off the belt by air resistance.

Driving pulley diameter is not a factor in theoretical power calculation unless the belt is too stiff to bend around the pulley. Large pulleys reduce fatigue wear in the belt, but cost more and take up greater space. The driving pulley size is traditionally selected on the basis of 5 in (13 cm) of diameter for every belt ply. For foot and snub pulleys, use 4 in (10 cm); for tandem drives, use 6 in (15 cm).

Fabric belt materials are rated according to their maximum tension per ply per unit width. Calling this rating r and the width w, the required number of plies depends on the tight-side tension, F_1.

$$N = \frac{F_1}{rw} \quad 65.25$$

11. BUCKET ELEVATORS

A *bucket elevator* (*belt elevator*) for carrying bulk materials consists of buckets, a belt, a method of driving the belt, and accessories for loading and unloading the material, maintaining belt tension, and so on. Material may be carried at low speed and discharged by gravity, as in a *continuous bucket elevator*, or materials may be carried at high speed and discharged by centrifugal force, as in a *centrifugal discharge belt elevator*.

Figure 65.11 illustrates the *head pulley* of a typical belt elevator. The *foot pulley* is located in the *loading boot* at the bottom of the elevator. Loading is usually by the passage of the buckets through the material contained in the loading boot.

Since the carried material cascades over the previously emptied descending bucket, most continuous bucket elevators are inclined, and bucket spacing is closer than in centrifugal elevators. Buckets in continuous bucket elevators are generally angular, while buckets for centrifugal elevators have rounded bottoms.

Belt elevators are driven similarly to belt conveyors—using friction contact between a pulley and the belt. However, since snub and tandem pulleys cannot be used with bucketed belts, the contact angle cannot exceed

Figure 65.11 Belt Elevator

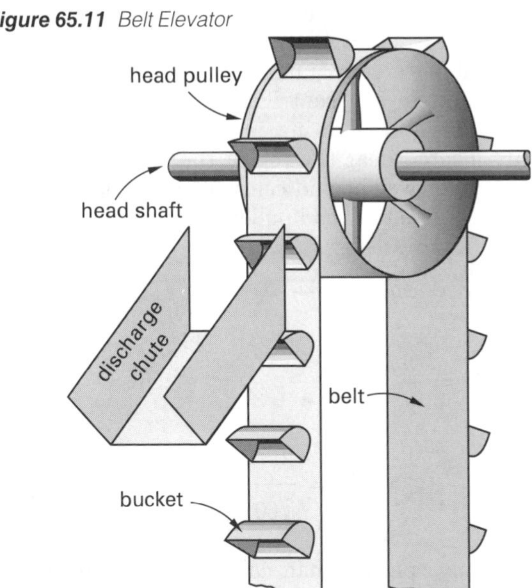

180°. Increased tension is used to obtain the desired speed and capacity. Coefficients of friction for belt elevators are essentially the same as for belt conveyors. Where the belt, buckets, and load are light, intrinsic belt tension may be sufficient. For heavier loads or wet or dusty environments, artificial tension may be needed.

Since bucket loading and delivery of material to the boot can vary over time, belt elevators and their motors should be sized for their expected peak minute loading, not for their average hourly loading. The theoretical capacity (mass per unit time) of a belt elevator is given by Eq. 65.26. However, in many situations, conditions for obtaining full bucket loads do not exist. Grain elevators typically run at 85% to 90% of their capacity; coal and ore elevators run at 75% or less.

$$\text{capacity} = \frac{(\text{bucket volume})(\text{belt speed})}{\text{bucket spacing}} \quad 65.26$$

Tension in the belt is the result of pulley and shaft friction, drag of the buckets through the loading boot, and weight of the material in the lifting buckets. The first two items are not easily estimated and must be derived indirectly from power tests of the elevator. The power required only to lift the material is given by Eq. 65.27. Appropriate conversion factors are required to convert to horsepower or kilowatts.[26]

$$\begin{aligned} P &= (\text{total material weight})(\text{elevator speed}) \\ &= \left(\frac{\text{weight}}{\text{unit time}}\right)(\text{elevator height}) \end{aligned} \quad 65.27$$

Location of discharge chutes for centrifugal elevators is somewhat heuristic. Some material is always scattered and spilled, but the amount will be small in good

[26]A rule of thumb is that the horsepower required to drive a boot-loaded bucket elevator is equal to the tons per hour multiplied by the lift in feet and divided by 500.

Plant Engineering

designs. The angle and location of the discharge chute are determined from an analysis of the centrifugal force on the carried material assuming frictionless discharge.[27] Discharge chutes for centrifugal elevators carrying liquids and dry, free-flowing materials like grain are commonly located near the top of the wheel where the bucket's content weight and centrifugal force are equal. In that case, where the head pulley has a radius r,

$$W = mg = \frac{m\mathrm{v}^2}{r} \quad \text{[SI]} \qquad 65.28(a)$$

$$W = \frac{mg}{g_c} = \frac{m\mathrm{v}^2}{g_c r} \quad \text{[U.S.]} \qquad 65.28(b)$$

$$\mathrm{v} = \sqrt{gr} = \frac{2\pi r n_{\mathrm{rev/min}}}{60\ \frac{\mathrm{sec}}{\mathrm{min}}} \qquad 65.29$$

Slower running speeds (than calculated by Eq. 65.29) are required when the material is hard, lumpy, dusty, or moist enough to stick to the buckets. Belt speeds for coal, ashes and cinders, coke, stone, ores, salt, and fertilizers are typically 80% or less of that calculated from Eq. 65.29.[28] Maximum speeds are typically 60 ft/min (0.3 m/s) when buckets are loaded by separate systems and less than 30 ft/min (0.15 m/s) when the buckets fill in a submerged boot.

12. CYCLONE SEPARATORS

Over the years, many theoretical and empirical models have been developed for *cyclone separator* (*vortex separator*) design. There is often little agreement among the models, and it is still not possible to predict the efficiency of collection from first principles. Instead, cyclone design typically starts with the geometry of a cyclone whose performance has already been tested. Cyclone specification and selection are mainly accomplished by reference to manufacturer's data and known designs. The two primary performance parameters are the pressure drop and collection efficiency.

The most common type of cyclone is the *reverse-flow cyclone separator*. (See Fig. 65.12.) Gas, which carries particles, is brought tangentially into the strong vortex that exists in the cylindrical section of diameter D. Since they have more mass than the gas molecules, the particles in the gas are subjected to greater centrifugal forces that move them radially outward, toward the inside surface of the cyclone. Near the bottom, the direction of the inner vortex gas flow reverses upward. Solids at the wall continue downward and out the bottom exit. Clean gas moving upward leaves the cyclone through the top outlet.

Figure 65.12 *Reverse-Flow Cyclone Separator*

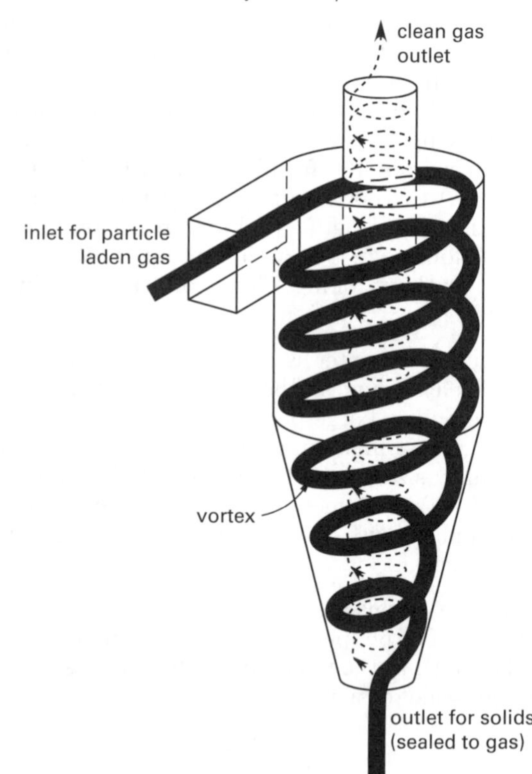

Since particles travel at the same speed, v_i, as the gas when they enter the cyclone, the centrifugal acceleration the particles experience is $m_p\mathrm{v}_i^2/r$. Since the centrifugal acceleration is so much greater than gravitational acceleration, gravity has little effect on cyclone performance.

Figure 65.13 illustrates the dimensions that can be varied in gas cyclone design. Some designs have become standard, including the 1D2D, 1D3D, and 2D2D designs. The "D" in the cyclone designation refers to the barrel diameter. The number preceding each "D" relates to the length of the barrel, L, and convergent section, Z, respectively. A 1D3D cyclone has a barrel length equal to the barrel diameter and a cone length that is three times the barrel diameter. Table 65.5 lists the dimensions for these standardized designs. D_e is the diameter of the top opening, where gas exits the cyclone.

[27] It is possible to proceed by assuming a coefficient of friction between the bucket and the contents, but the results are highly unreliable and generally do not lead to any practical result.

[28] Other factors, such as method of loading and bucket wear, are also considered when determining belt speed for hard-to-handle materials.

Figure 65.13 Dimensions of Gas Cyclones

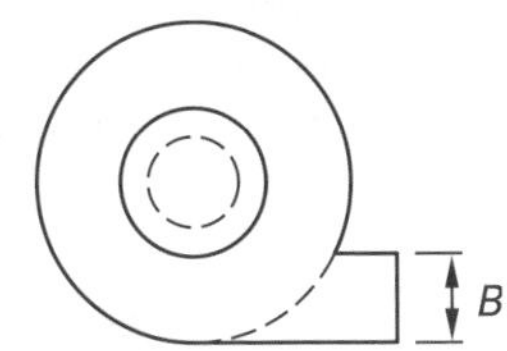

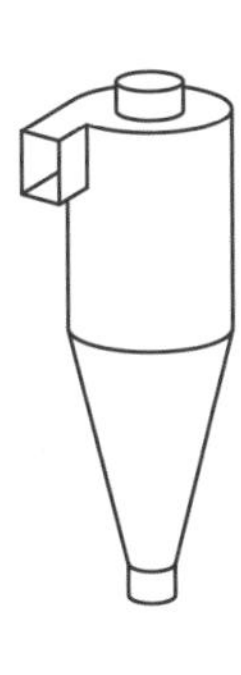

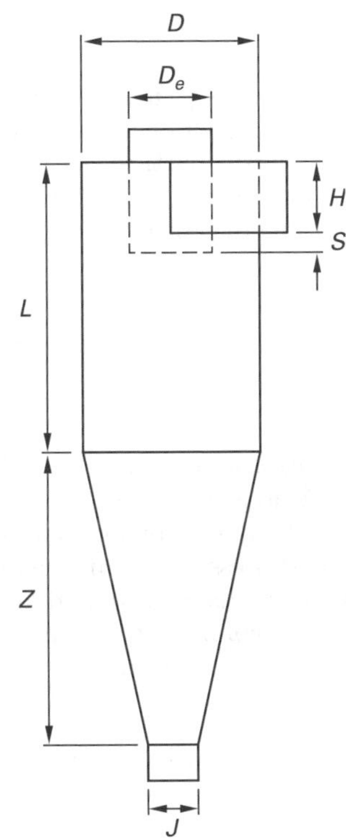

Table 65.5 Dimensions of Standardized Cyclones (multiples of D)

design	B	J	D_e	S	H	L	Z
1D2D	1/4	1/2	1/1.6	5/8	1/2	1	2
1D3D	1/8	1/4	1/2	1/8	1	1	3
2D2D	1/4	1/4	1/2	1/8	1/2	2	2

The *particle concentration*, C, in the gas entering the cyclone is

$$C = \frac{\dot{m}_s}{Q} \quad \text{65.30}$$

Performance can be correlated dimensionally with any of the velocities found in a cyclone. Therefore, the *characteristic velocity*, v_c, can be defined in several ways. The most logical definition is the average velocity in the cylindrical portion of the cyclone. This definition is often used in modern correlations and the calculation of dimensionless numbers.

$$v_c = \frac{4Q}{\pi D^2} \quad \text{65.31}$$

Historically, many researchers have used the *inlet velocity*, v_i, as the characteristic velocity in their formulas. Therefore, most formulas (for cut-size, efficiency, pressure drop, etc.) are based on the inlet velocity. Commercial inlet velocities are typically in the range of 2500–3500 ft/min (15–18 m/s).

$$v_i = \frac{Q}{A} = \frac{Q}{BH} \quad \text{65.32}$$

The *effective number of turns*, N_e, is the number of revolutions that the gas spins while passing downward through the cyclone outer vortex. Collection efficiency increases with more turns. In the standard (classical) method[29] of designing and analyzing gas cyclones, the effective number of turns is defined as

$$N_e = \frac{1}{H}\left(L + \frac{Z}{2}\right) \quad \text{65.33}$$

Scale-up of geometrically similar cyclones is often based on the dimensionless *Stokes number*. The *Stokes 50 number*, Stk_{50}, is the ratio of centrifugal force (less buoyancy) to the drag force for a particle size that is captured 50% of the time, d_{50}. (See Eq. 65.37 and Eq. 65.38.) v_c is the average velocity in the cylindrical portion of the cyclone.

$$\text{Stk}_{50} = \frac{\Delta p D}{\mu v_c} = \frac{d_{50}^2(\rho_s - \rho_g)v_c}{18\mu g_c D} \quad \text{65.34}$$

The *Euler number* (*resistance coefficient*), Eu, relates the pressure drop to a characteristic velocity. It is the ratio of the pressure drop between the gas inlet and the gas outlet to the dynamic pressure of flow within the cyclone body. v_c is the average velocity in the cylindrical portion of the cyclone.

$$\text{Eu} = \frac{\Delta p}{\frac{\rho_s v_c^2}{2g_c}} \approx \sqrt{\frac{12}{\text{Stk}_{50}}}\Bigg|_{\text{well-designed, well-known cyclones}} \quad \text{65.35}$$

The barrel diameter, D, required for a particular flow rate, Q, and pressure drop, Δp, is

$$D = 2\sqrt{\frac{Q}{\pi}\sqrt{\frac{\rho_s \text{Eu}}{2g_c \Delta p}}} \quad \text{65.36}$$

[29]The methods outlined in *Air Pollution Engineering Manual* (Davis) and *Air Pollution Control: A Design Approach* (Cooper and Alley) are considered the "standard method."

Plant Engineering

For the reverse-flow type of cyclone, the *cut-size* (*cut point*, *cut diameter*, *equiprobable size*, and *separation mesh*), d_{50}, is the particle size that is collected with 50% efficiency (i.e., 50% of the time). It is the size for which the centrifugal force is equal to the drag force. v_i is the inlet velocity into the cyclone. Since the gas density is much smaller than the solid's density, $\rho_s - \rho_g \approx \rho_s$.

$$d_{50} = \sqrt{\frac{18\mu_g g_c D(\text{Stk}_{50})}{v_c(\rho_s - \rho_g)}} = \sqrt{\frac{9\pi\mu_g g_c D^3(\text{Stk}_{50})}{2Q(\rho_s - \rho_g)}} = \sqrt{\frac{9\mu_g g_c B}{2\pi N_e v_i(\rho_s - \rho_g)}} \qquad 65.37$$

The theoretical (based on first principles) cut-size formula for a particular particle size, d_p, overestimates the collection efficiency due to assumptions about instantaneous complete mixing and the absence of reentrainment. The cut-size is also affected by the particle size distribution at the inlet. Therefore, an empirical factor is used to correct the single-particle size formula for use with material having a range of sizes. For the standardized designs, Lapple (1951) suggested Eq. 65.38, in which n_1 is 1.92 (1D2D), 1.61 (2D2D), and 1.62 (1D3D). The actual values depend on the particle size distribution shape, median size, and standard deviation.

$$d_{50} = \sqrt{\frac{9\mu_g g_c D}{8\pi v_i(\rho_s - \rho_g)n_1}} \qquad 65.38$$

The *critical particle size*, d_{100}, is the particle size that is collected with 100% efficiency. For the standardized designs, Shepard and Lapple (1940) suggested Eq. 65.39, in which n_2 is 0.84 (1D2D), 0.92 (2D2D), and 0.97 (1D3D).

$$d_{100} = \sqrt{\frac{9\mu_g g_c D_e}{2\pi v_i(\rho_s - \rho_g)n_2}} \qquad 65.39$$

The theoretical *cyclone efficiency* (*grade efficiency*, *collection efficiency*, etc.), η, for a particular particle size, d_p, is given by Eq. 65.40. Cyclones collect larger particles more easily than smaller ones. The efficiency of separation is the fraction by weight of particles collected compared to the total amount in the feed. Of particular interest is a cyclone's ability to remove PM-10,which is particulate matter 10 μm in size. In general, cyclones perform poorly with PM-10, and they are completely ineffective with PM-2.5 (2.5 μm in size).

$$\eta_{d_p} = 1 - \exp\left(\frac{-\pi d_p^2(\rho_s - \rho_g)N_e v_i}{9\mu_g g_c B}\right) \qquad 65.40$$

Equation 65.40 predicts actual performance poorly, and numerous studies have suggested increasingly complex replacements. The simple empirical *Lapple formula* has proven to be a better predictor of single-size particle collection efficiency than the theoretical formulas.

$$\eta_{d_p} = \frac{1}{1 + \left(\frac{d_{50}}{d_p}\right)^2} \qquad 65.41$$

A *fractional-efficiency curve* (*grade-efficiency curve*) graphs the fraction of particles collected (i.e., the efficiency) versus particle size. The *overall efficiency* (*absolute efficiency*) of a cyclone is a weighted average of the collection efficiencies for the various size ranges.

$$\eta_t = \frac{\sum_j m_j \eta_{d_p,j}}{\sum_j m_j} \qquad 65.42$$

The *pressure loss* across a cyclone is due to several factors: cyclone entry loss, kinetic energy loss, frictional loss in the outer vortex, kinetic energy loss due to rotation, and pressure loss in the inner vortex and exit tube. Researchers have evaluated all of these individually, but it is more common to use a simple equation that incorporates all factors. In Eq. 65.43, K is a constant that depends on the design and operating conditions and is 12 to 18 for standardized designs. $K = 16$ is used for tangential inlets without inlet vanes. $K = 7.5$ is used for cyclones with inlet vanes.

$$\Delta p = \frac{K\rho_g H B v_i^2}{2g_c D_e^2} \qquad 65.43$$

Most gas separators operate with a pressure loss of 0.07–0.18 lbf/in^2 (2–5 in of water; 0.5–1.3 kPa). In general, efficiency increases with increases in velocity. However, there is little value in designing or operating gas cyclones with pressure drops much greater than 0.2 lbf/in^2 (1.5 kPa). This is when the separation efficiency stops increasing with increases in pressure drop. There is also a common lower limit for pressure drop, arbitrarily set at 0.07 lbf/in^2 (0.5 kPa), below which cyclones are little more than settling chambers, operating at very low efficiencies.

Power increases with the square of the velocity. The power, P, required to overcome the pressure drop and keep the cyclone operating is

$$P = \Delta p Q = \frac{K\rho_g H B v_i^2 Q}{2g_c D_e^2} \qquad 65.44$$

13. HYDROCYCLONES

Hydrocyclones (also known as *hydraulic cyclones* and *hydroclones*) are used to separate solids suspended in liquid streams. They are common in ore processing plants because of their compactness and economy of operation.[30] (See Fig. 65.14.)

Figure 65.14 *Hydrocyclone*

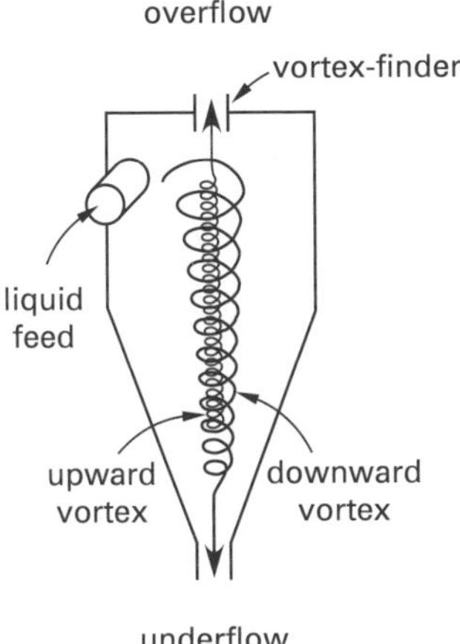

The primary cyclone design characteristic is the cyclone diameter, and the primary input variable is the *separation mesh* (or *cut point*), d_{50}—the particle diameter that is captured with an efficiency of 50%. Other factors affecting cyclone design and selection are feed flow rate, operating pressure, and pressure drop.

Like air cyclones, hydrocyclones consist of a cylindrical section joined to a conical section. The suspension is pumped tangentially into the upper part of the cylindrical section. Centrifugal force acts on solid particles, and the most coarse particles exit in the *underflow* at the bottom with a small amount of liquid. The *overflow* consists of most of the liquid and the finer particles. Half of the particles with diameters equal to d_{50} will appear in the overflow and half will appear in the underflow. The overflow is removed through a cylindrical pipe known as the *vortex finder*.

It is common for manufacturers of hydrocyclones to supply performance data based on separating sand from water. The separation efficiency for other solids and liquids can be calculated from Eq. 65.45. The density terms can be replaced by specific gravity terms (1.00 for water, 2.65 for sand, and 1.4 for coal).

$$d_{50,\text{actual}} = d_{50,\text{sand-water}}\sqrt{\frac{\rho_{\text{sand}} - \rho_{\text{water}}}{\rho_{\text{solid}} - \rho_{\text{liquid}}}} \qquad 65.45$$

The pressure needed to maintain a steady circulation (i.e., velocity head) inside the cyclone and to overcome the friction losses is normally provided by a centrifugal pump. Pressure drop is the only major operating cost, and it is probably not economical to exceed a loss of 5 atm. Smaller units typically operate with a loss of 4 atm to 5 atm, while the loss in larger units may be as low as 1 atm.

14. SIEVES AND SCREENS

Sieving is the process of separating different sizes of the same material. Vibratory shakers and air-jet fluidization are used to increase the performance of sieving. Sieving is practical for particles down to approximately 30 μm to 100 μm.

In the past, wet classification was accomplished with stationary and vibrating screens. This method has high operating costs, however, and has been essentially abandoned in favor of hydrocyclones.

15. AIR CLASSIFIERS

Classification is the process of separating different grades or sizes of the same material, and often follows a grinding process.[31] *Air classifier mills* combine the processes of classification and grinding, recycling the coarse material for further grinding. Particles and minerals most suitable for air classification are uniform, homogeneous, spherical, smooth, and dry.[32]

Air classifiers are primarily used to grade bulk powders. The *selectivity* or *efficiency* is the probability of a particle entering the coarse stream. The *selectivity curve* is a curve of particle size versus selectivity. (See Fig. 65.15.) The *cut point*, designated d_{50}, is the midpoint of the selectivity curve. This is the size of particles that appear in equal numbers in the fine and coarse streams. The *bandwidth* is the range of particle sizes captured.[33] The cutoff values are the top-size and bottom-size limits.[34]

There are many ways to describe the *sharpness*, a measure of how tight the particle-size distribution is around the cut point. The *yield* or *recovery* is the fraction by weight of the end product in the desired size range, compared to the total amount of material in the feed.

Plant Engineering

[30]Other methods for separating solids from liquids include settling with and without floc (sedimentation) and centrifuging. Centrifuging is more appropriate for particles in the micrometer size range.
[31]*Fractioning* is the process of separating a material into a variety of specifically sized groups.
[32]Grinding, drying, solvent removal, extraction, homogenizing, dispersion, or deagglomeration processes may be necessary before air classification.
[33]Some materials, such as chromatographic material and toners, have bandwidths of 1 μm to 5 μm.
[34]The *cut-off values* and *cut-size* are different quantities.

Figure 65.15 *Typical Selectivity Curve*

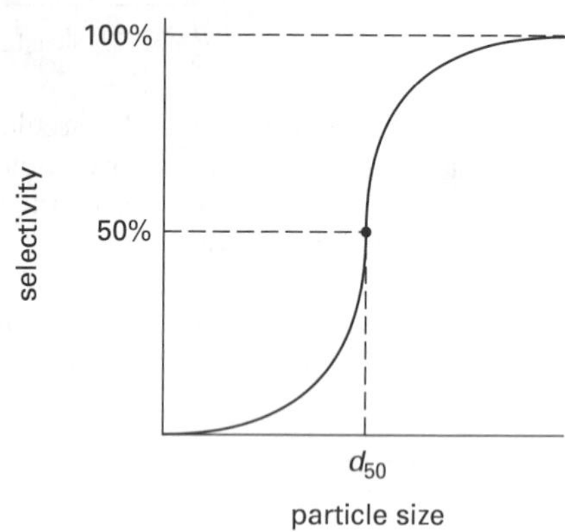

Air classification encompasses elutriation, free vortex classification, and forced vortex classification. *Elutriation* is a crude method of separating most of the fine and coarse particles. Feed is introduced into the airflow, which raises the fine particles but allows the coarse particles to fall. Elutriation is effective in the same range of sizes as sieving but has a greater throughput and is more flexible. The cut point depends on the air velocity.

Free vortex classification is the process that occurs in a traditional cyclone. *Forced vortex classification*, also known as *centrifugal classification* and *mechanical classification*, is used to obtain the most precise particle size control. The cut points of centrifugal classifiers are in the 1 μm to 100 μm range.

Depending on the classifier design, a rotating rotor (i.e., the *rejector* or *classifier wheel*) forces the feed stream into circular motion. (See Fig. 65.16.) The rotor is constructed from blades, rods, or spokes, or as a hollow perforated disk to allow smaller particles closer to the rotational axis to escape to the central outlet. Coarse particles are thrown to the outer housing and collected. *Sidedraft classifiers* perform better in all respects (i.e., yield, product recovery, and efficiency) than conventional gravity-fed *updraft classifiers*.

The separation can be adjusted by varying the rotor speed and air flow. Rejection increases with rotor speed. Rejection also can be increased by increasing the number of blades or rods in the rejector. Therefore, an increase in rejection increases fines but decreases the yield. Yield, in turn, can be increased by increasing the air velocity, but this makes the product more coarse and increases the pressure drop.[35]

Operating power requirements of disk rotor-type of classifiers are not easily modeled. For steady-state operation, the power calculations have been based on the rotational mass moment of inertia ($I = mr^2$) of the rotor

Figure 65.16 *Mechanical Air Classifier*

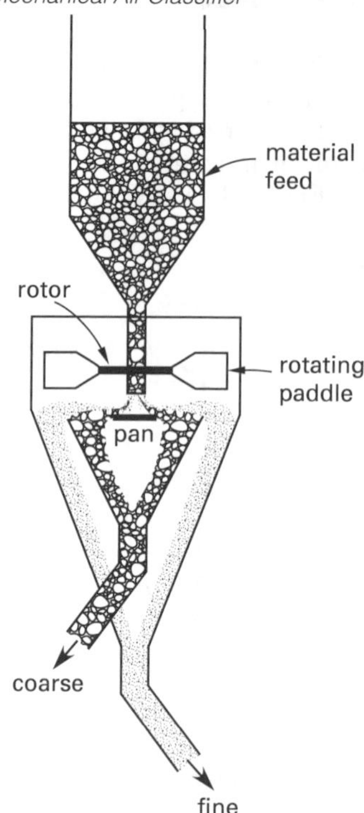

and the start-up time. Using a typical starting time, t_{start}, of 10 sec, the calculated steady-state power is conservative.

$$P = \frac{I\omega^2}{2t_{\text{start}}} \qquad \text{[SI]} \qquad 65.46(a)$$

$$P = \frac{I\omega^2}{2g_c t_{\text{start}}} \qquad \text{[U.S.]} \qquad 65.46(b)$$

Equation 65.46 is generally not applicable to higher-efficiency classifiers that draw more power while operating at higher speeds in more dense particle clouds. For these classifiers, the power requirement consists of two components: the power to distribute the feed and the power to overcome the drag force (i.e., friction between the rotor and the particle cloud).

$$P_t = P_{\text{feed}} + P_{\text{drag}} \qquad 65.47$$

The power to distribute the feed is approximated by calculating the kinetic energy imparted to the feed. Particle velocities cannot exceed the rotor's peripheral

[35]Pressure drop is proportional to the square of the air velocity.

velocity, so that value can be used to conservatively estimate power requirements if the distribution of particle velocities is not known. $\dot{m}$ is the material feed rate.

$$P_{\text{feed}} = \frac{\dot{m}v^2}{2} \quad \text{[SI]} \qquad 65.48(a)$$

$$P_{\text{feed}} = \frac{\dot{m}v^2}{2g_c} \quad \text{[U.S.]} \qquad 65.48(b)$$

The power dissipated in overcoming the drag force depends on the number of vanes or paddles in the rotor, N, and the effective peripheral velocity, v_e. The *effective peripheral velocity* is the resultant of the rotor's peripheral velocity and the air velocity of the feed. The drag force is calculated by traditional means using the drag coefficient of each vane or rod.

$$P_{\text{drag}} = NF_{\text{drag}}v_e \qquad 65.49$$

16. SUSPENSIONS OF SOLIDS IN LIQUIDS

Three states of suspensions are distinguished. (1) An *on-bottom suspension* is characterized by the complete motion of all the particles at the bottom of a vessel, although not all particles are suspended. (2) In an *off-bottom suspension*, all particles are in complete motion, and no particles remain on the base of the vessel for more than 1–2 sec. This is known as the *Zwietering criterion* (*condition* or *state*). The impeller speed that results in this condition is the *just-suspended speed*, n_{js}. (3) In a *uniform suspension*, the solid particles are evenly distributed throughout the vessel. An increase in agitation does not significantly change the distribution of the suspended solids.

The *solids loading* is the ratio of solid to liquid masses in the suspension.

$$X = \frac{m_s}{m_l} \qquad 65.50$$

For power calculations, the *suspension density* is equal to the pure liquid density until an off-bottom condition has been achieved. The density at full suspension is

$$\rho_{\text{suspension}} = \frac{1}{\dfrac{X}{\rho_s} + \dfrac{1-X}{\rho_l}} \quad \text{[full suspension]} \qquad 65.51$$

As Eq. 65.52 shows, the required mixing power increases significantly with increasing rotational speed and diameter. If the mixing objective is *mass transfer* (e.g., the transfer of oxygen in the water to the solid particles), operating at the minimum impeller speed that exposes all surface area of the particles will require the least power. The *just-suspended speed correlation*, n_{js}, was suggested by Zwietering (1958) and has since been adapted for practical use. The *Zwietering constant*, S, in Table 65.6 is a function of impeller and tank geometry.

$$n_{\text{js}} = \frac{S\nu^{0.1}\left(\dfrac{g(\rho_s - \rho_l)}{\rho_l}\right)^{0.45} X^{0.13}d_p^{0.2}}{D^{0.85}} \qquad 65.52$$

Table 65.6 *Typical Zwietering Constants*[a] *(flat-bottom tanks)*

impeller[b]	D/T	C/T			
		0.333	0.250	0.167	0.125
RDT (RT)	0.333	8.37	7.43	6.50	6.03
	0.500	4.72	4.20	3.67	3.41
PBT-45°	0.333	6.95	6.64	6.32	6.16
	0.500	4.97	4.74	4.52	4.40
Chemineer HE-3, hydrofoil, downward pumping	0.333	10.41	9.95	9.48	9.25
	0.500	7.14	6.82	6.50	6.35
FBT-90°	0.333	4.4			

[a]Representative values. Actual values depend on manufacturer and model.
[b]RDT or RT: Rushton disk turbine; PBT: pitched-blade turbine; HE: high-efficiency turbine; hydrofoils have tapered and shaped blades; FBT: flat-blade turbine

In stirred tanks with solid loadings, X, exceeding 10%, a clear interfacial layer may form between the solids-rich volume lower in the tank and the solids-free clear liquid volume near the liquid surface. The height of this distinctive cloud-like layer is the *cloud height*. Solid particles pass only infrequently into the upper clear liquid, and mixing time can be up to 20 times longer in the solids-free phase than in the solids-rich volume.

The required speed of a scaled-up, geometrically similar mixer will be less than the model speed and is predicted by Eq. 65.53. The exponent, k, depends on which characteristic remains constant. k takes on values of 0 (equal blend time), 2/3 (equal power per volume), 0.85 (Zwietering suspension), and 1.0 (equal tip speed and equal fluid velocity).

$$n_2 = n_1\left(\frac{d_1}{d_2}\right)^k \qquad 65.53$$

Plant Engineering

17. MIXING PHYSICS

The drag force on a paddle is given by the standard fluid drag force equation. For flat plates, the coefficient of drag, C_D, is approximately 1.8.

$$F_D = \frac{C_D A \rho \mathrm{v}^2_{\text{mixing}}}{2} \quad \text{[SI]} \quad 65.54(a)$$

$$F_D = \frac{C_D A \rho \mathrm{v}^2_{\text{mixing}}}{2g_c} = \frac{C_D A \gamma \mathrm{v}^2_{\text{mixing}}}{2g} \quad \text{[U.S.]} \quad 65.54(b)$$

The power required is calculated from the drag force and the mixing velocity. The average *mixing velocity*, $\mathrm{v}_{\text{mixing}}$, also known as the *relative paddle velocity*, is the difference in paddle and average water velocities. The mixing velocity is approximately 0.7–0.8 times the tip speed. R is the distance from the shaft to the paddle center in ft.

$$\mathrm{v}_{\text{paddle,ft/sec}} = \frac{2\pi R n_{\text{rpm}}}{60 \frac{\text{sec}}{\text{min}}} \quad 65.55$$

$$\mathrm{v}_{\text{mixing}} = \mathrm{v}_{\text{paddle}} - \mathrm{v}_{\text{water}} \quad 65.56$$

$$P_{\text{kW}} = \frac{F_D \mathrm{v}_{\text{mixing}}}{1000 \frac{\text{W}}{\text{kW}}} = \frac{C_D A \rho \mathrm{v}^3_{\text{mixing}}}{2\left(1000 \frac{\text{W}}{\text{kW}}\right)} \quad \text{[SI]} \quad 65.57(a)$$

$$P_{\text{hp}} = \frac{F_D \mathrm{v}_{\text{mixing}}}{550 \frac{\text{ft-lbf}}{\text{hp-sec}}} = \frac{C_D A \rho \mathrm{v}^3_{\text{mixing}}}{2g_c\left(550 \frac{\text{ft-lbf}}{\text{hp-sec}}\right)} = \frac{C_D A \gamma \mathrm{v}^3_{\text{mixing}}}{2g\left(550 \frac{\text{ft-lbf}}{\text{hp-sec}}\right)} \quad \text{[U.S.]} \quad 65.57(b)$$

For slow-moving paddle mixers, the *velocity gradient*, G, varies from 20 sec^{-1} to 75 sec^{-1} for a 15 min to 30 min mixing period. Typical units in Eq. 65.58 are ft-lbf/sec for power (multiply hp by 550 to obtain ft-lbf/sec), lbf-sec/ft^2 for μ, and ft^3 for volume. In the SI system, power in kW is multiplied by 1000 to obtain W, viscosity is in Pa·s, and volume is in m^3. (Multiply viscosity in cP by 0.001 to obtain Pa·s.)

$$G = \sqrt{\frac{P}{\mu V_{\text{tank}}}} \quad 65.58$$

Equation 65.58 can also be used for rapid mixers, in which case the mean velocity gradient is much higher: approximately 500–1000 sec^{-1} for 10–30 sec mixing period, or 3000–5000 sec^{-1} for a 0.5–1.0 sec mixing period in an in-line blender configuration.

Equation 65.58 can be rearranged to calculate the power requirement. Power is typically 0.5–1.5 hp/MGD for rapid mixers.

$$P = \mu G^2 V_{\text{tank}} \quad 65.59$$

The dimensionless product, Gt_d, of the velocity gradient and detention time is known as the *mixing opportunity parameter*. Typical values range from 10^4 to 10^5.

$$Gt_d = \frac{V_{\text{tank}}}{Q}\sqrt{\frac{P}{\mu V_{\text{tank}}}} = \frac{1}{Q}\sqrt{\frac{P V_{\text{tank}}}{\mu}} \quad 65.60$$

18. MIXERS AND IMPELLERS

Many industries use rotating impellers to keep solid material in suspension. The blades of *radial-flow impellers* (*paddle-type impellers*, *turbine impellers*, etc.) are parallel to the drive shaft. *Axial-flow impellers* (*propellers*, *pitched-blade impellers*, etc.) have blades inclined with respect to the drive shaft. (See Fig. 65.17.)

Figure 65.17 *Typical Axial Flow Mixing Impellers*

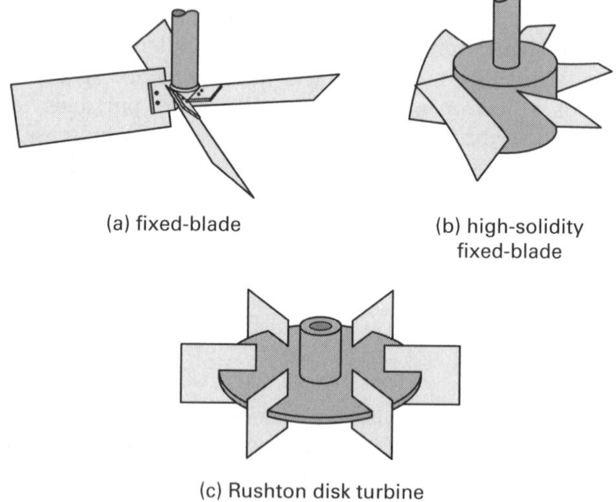

The *mixing Reynolds number*, Re, depends on the impeller diameter, d, suspension density, ρ, liquid viscosity, μ, and rotational speed, n (in rev/sec). Flow is in transition for $10 < \text{Re} < 10{,}000$, laminar below 10, and turbulent above 10,000.

$$\text{Re} = \frac{D^2 n \rho}{\mu} \quad \text{[SI]} \quad 65.61(a)$$

$$\text{Re} = \frac{D^2 n \rho}{g_c \mu} \quad \text{[U.S.]} \quad 65.61(b)$$

The dimensionless *power number*, N_P, of the impeller is related to the power required to drive the impeller. (For any given level of suspension, Eq. 65.62 through Eq. 65.71 apply to geometrically similar impellers and turbulent flow.[36])

$$P = N_P n^3 D^5 \rho \quad \text{[SI]} \qquad 65.62(a)$$

$$P = \frac{N_P n^3 D^5 \rho}{g_c} \quad \text{[U.S.]} \qquad 65.62(b)$$

$$P = \rho g Q h_v \quad \text{[SI]} \qquad 65.63(a)$$

$$P = \frac{\rho g Q h_v}{g_c} \quad \text{[U.S.]} \qquad 65.63(b)$$

The impeller's dimensionless *flow number*, N_Q, is defined by the flow rate equation.

$$Q = N_Q n D^3 \qquad 65.64$$

The velocity head can be calculated from a combination of the power and flow numbers.

$$h_v = \frac{N_P n^2 D^2}{N_Q g} \qquad 65.65$$

The *force number*, N_F, is defined by the force equation. F is the peak transverse force in normal operation. The upper shaft bending moment, M, is calculated by multiplying the peak force by the shaft length. (Units of area may be incorporated into N_F.)

$$F = \frac{N_F N_P \rho n^2 D^2}{g_c} \qquad 65.66$$

$$M = FL \qquad 65.67$$

A key equation for evaluating impeller design is the ratio of power to flow rate, obtained by dividing Eq. 65.63 by Eq. 65.64.

$$\frac{P}{Q} = \left(\frac{N_P}{N_Q}\right) n^2 D^2 \rho \quad \text{[SI]} \qquad 65.68(a)$$

$$\frac{P}{Q} = \frac{\left(\frac{N_P}{N_Q}\right) n^2 D^2 \rho}{g_c} \quad \text{[U.S.]} \qquad 65.68(b)$$

Equation 65.63 and Eq. 65.64 can be combined in other ways to express the P/Q ratio in terms of the fixed variables. If Q and d are fixed, rotational speed and power are variable. The ratio N_P/N_Q^3 characterizes the solids suspension performance of the impeller. Power is proportional to N_P/N_Q^3.

$$\frac{P}{Q} = \frac{\left(\frac{N_P}{N_Q^3}\right) Q^2 \rho}{D^4} \quad \text{[SI]} \qquad 65.69(a)$$

$$\frac{P}{Q} = \frac{\left(\frac{N_P}{N_Q^3}\right) Q^2 \rho}{g_c D^4} \quad \text{[U.S.]} \qquad 65.69(b)$$

For suspensions produced in small round tanks, the ratio of the impeller diameter to tank diameter, d/d_t, is a characteristic in many relationships. For common d/d_t ratios (e.g., 0.25 to 0.50) and a specific impeller design, Eq. 65.70 and Eq. 65.71 apply.[37]

$$P \propto \left(\frac{D}{D_t}\right)^{-1/5} \qquad 65.70$$

$$T \propto \left(\frac{D}{D_t}\right)^{2/3} \qquad 65.71$$

Vibration near the critical speed can be a major problem with low D/D_t ratios and with modern, high-efficiency, high-speed impellers. Mixing speed should be well below the first critical speed of the shaft. Other important design factors include tip speed and shaft bending moment.

The critical speed of a shaft corresponds to its natural harmonic frequency of vibration when loaded as a flexing beam. If the harmonic frequency, f_c, is in hertz, the shaft's critical rotational speed (in rev/min) will be $60f_c$. The critical speed depends on many factors, including the masses and locations of paddles and other shaft loads, shaft material, length, and end restraints. It is not important whether the shaft is vertical or horizontal, and the damping effects of the mixing fluid and any stabilizing devices are disregarded in calculations. For a lightweight (i.e., massless) mixing impeller that is on a suspended (cantilever) shaft of length L, has a uniform cross section, a mass per unit length of m_L, and is supported by a fixed upper bearing, the first critical speed, f_c, in hertz, will be the same as that of a cantilever beam and can be approximated from Eq. 65.72. Other configurations, including mixed shaft materials, stepped or

[36]Different levels of suspension range from being merely in motion (but not off the bottom) to being uniformly dispersed throughout the liquid.

[37]There is a risk of the impeller being "sanded in" if power to the mixer is lost. A considerably higher torque than predicted by Eq. 65.71 will be required to get the impeller moving again. The motor, shaft, and impeller design must take this into consideration.

Plant Engineering

varying shaft diameters, two or more support bearings, substantial impeller mass, and multiple impellers require different analysis methods.

$$f_c = 0.56\sqrt{\frac{EI}{L^4 m_L}} \quad \text{[SI]} \qquad 65.72(a)$$

$$f_c = 0.56\sqrt{\frac{EIg_c}{L^4 m_L}} \quad \text{[U.S.]} \qquad 65.72(b)$$

19. NOMENCLATURE

A	area	ft^2	m^2
B	inlet width	ft	m
C	clearance from tank bottom	ft	m
C	coefficient	–	–
C	concentration	lbm/ft^3	kg/m^3
d	particle size	ft	m
D	diameter	ft	m
D_e	exit diameter	ft	m
e	number of persons per trip	–	–
E	modulus of elasticity	lbf/ft^2	Pa
Eu	Euler number	–	–
f	coefficient of friction	–	–
f	frequency	Hz	Hz
F	force	lbf	N
g	acceleration of gravity, 32.2 (9.81)	ft/sec^2	m/s^2
g_c	gravitational constant, 32.2	lbm-ft/lbf-sec^2	n.a.
G	velocity gradient	1/sec	1/s
h	fluid head	ft	m
h	height	ft	m
H	gas entrance height	ft	m
I	moment of inertia	ft^4	m^4
I	rotational mass moment of inertia	lbm-ft^2	kg·m^2
J	solids exit diameter	ft	m
k	pressure ratio	–	–
K	cyclone constant	–	–
L	barrel length	ft	m
L	length	ft	m
m	mass	lbm	kg
$\dot{m}$	mass flow rate	lbm/sec	kg/s
M	moment	ft-lbf	N·m
n	rotational speed	rev/min	rev/min
N	dimensionless number	–	–
N	number (quantity)	–	–
N_e	effective number of turns	–	–
p	pressure	lbf/in^2	Pa
P	peak 5 min population	–	–
P	power	ft-lbf/sec	W
Q	flow rate	ft^3/sec	m^3/s
Q	gas flow	ft^3/sec	m^3/s
r	belt rating	lbf/ft	N/m
r	radius	ft	m
R	radius	ft	m
R_h	hydraulic radius	ft	m
Re	Reynolds number	–	–
S	exit tube extension length	ft	m
S	strength	lbf/in^2	Pa
S	Zwietering constant	–	–
SG	specific gravity	–	–
Stk	Stokes number	–	–
t	time	various	various
T	tank diameter	ft	m
T	torque	ft-lbf	N·m
v	velocity	ft/sec	m/s
V	volume	ft^3	m^3
w	width	ft	m
W	weight	lbf	N
X	mass fraction	–	–
Z	length of convergent section	ft	m

Symbols

γ	specific weight	lbf/ft^3	N/m^3
η	efficiency	–	–
θ	angle of wrap	rad	rad
θ	wall angle (from the vertical)	deg	deg
μ	coefficient of internal friction	–	–
μ	viscosity	lbf-sec/ft^2	Pa·s
ν	kinematic viscosity	ft^2/sec	m^2/s
ρ	density	lbm/ft^3	kg/m^3
σ	stress	lbf/in^2	Pa
ϕ	angle of internal friction	deg	deg
ω	rotational velocity	rad/sec	rad/s

Subscripts

b	bulk
c	characteristic, critical, or cylindrical
d	detention
D	drag
e	effective
F	force
g	gas
h	horizontal
i	inlet
js	just suspended
l	liquid
L	per unit length
p	particle
P	power
Q	flow or quantity
r	rope
s	solid
sh	sheave
t	tank, tensile, or total
u	ultimate
v	vertical
v	velocity
w	wire

Plant Engineering

66 Fire Protection Sprinkler Systems

Content in blue refers to the *NCEES Handbook*.

1. HAZARD CLASSIFICATION

The *hazard classification*[1,2] of a protected area will determine the characteristics (e.g., water supply rate, number of fire pumps, number of sprinklers) of the fire sprinkler design and installation. NFPA 13 differentiates fire hazards in storage facilities from non-storage facilities fire hazards.[3] Non-storage facilities are classified as light, ordinary, and extra hazard areas according to the combustibility, quantity of combustibles, rate of heat release, storage height, and quantity of flammable and/or combustible liquids.

Fire hazards in *storage space* are classified as Class I, II, III, IV, or plastics according to the commodity and, in some cases, the height of the storage. Class I commodities present the lowest severe fire hazard, and plastics represent the highest hazard. Four other storage commodities (rubber tires, baled cotton, rolled paper, and idle pallets) are given their own commodity classifications. Some storage and non-storage applications (e.g., laboratories using chemicals, and areas storing organic peroxide compounds) are classified as special hazards.

The annex (appendix) section of NFPA 13 defines the hazard classifications of *non-storage spaces* according to usage and amounts and nature of combustible materials contained within the spaces. (See Sec. 66.23, "Portable Fire Extinguishers," for classification of combustible materials.)

A *light (low) hazard occupancy* (LH) space contains only small amounts of Class A combustible materials, including furnishings, decorations, and contents. The contents are either noncombustible or are arranged such that a fire would not spread rapidly, and a relatively low heat release rate is anticipated. Small amounts of Class B flammable materials safely stored in closed containers may be present. Typical uses include offices, school classrooms, assembly halls, theaters, auditoriums, museums, churches, hospitals, restaurants, and library reading areas, as well as most residential uses.

An *ordinary (moderate) hazard occupancy* (OH) space contains greater amounts of Class A combustibles and Class B flammables are present in greater amounts than in a light hazard occupancy. Fires with moderate rates of heat release are expected. OH areas are subdivided into two groups.

Ordinary hazard group 1 (OH-1) occupancies contain materials whose combustibility is low, the quantity of combustibles is moderate, and combustibles do not exceed 8 ft (2.4 m) in storage height. Typical uses

[1]This chapter is based on Part 4, "Design of Water Sprinkler Systems," *Fire and Explosion Protection Systems* (Lindeburg), published by Professional Publications, Inc.

[2]The authoritative references for subjects in this chapter are NFPA 13 (*Standard for the Installation of Sprinkler Systems*) and NFPA 20 (*Standard for the Installation of Stationary Pumps for Fire Protection*), both of which are recommended.

[3]NPFA 13 doesn't include information about protecting storage areas containing flammable liquids, combustible liquids, or aerosol products. NFPA 30 (*Flammable and Combustible Liquids Code*) and NFPA 30B (*Code for the Manufacture and Storage of Aerosol Products*) apply to such areas.

include automobile parking garages, light electronic manufacturing, bakeries and canneries, laundries, and restaurant service areas.

Ordinary hazard group 2 (OH-2) occupancies contain combustible materials whose combustibility and quantity are moderate, and combustibles do not exceed 12 ft (3.7 m) in storage height. Typical uses include machine shops, mercantile, printing, repair garages, library stack rooms, and most manufacturing and mill operations.

Extra (high) hazard (EH) occupancies contain greater amounts of Class A combustibles and Class B flammables than ordinary hazards. Also included is warehousing of or in process storage of other than Class I and Class II commodities. EH areas are subdivided into two groups.

Extra hazard group 1 (EH-1) occupancies are those where large quantities of highly combustible content are processed. These occupancies have little or no flammable or combustible liquids but may have dust, lint, shavings, or other materials present which may produce severe fires. Typical usages are plywood and particle board manufacturing, textile manufacturing and processing, saw mills, rubber reclamation, and areas using combustible hydraulic fluids, cutting oils, and inks.

Extra hazard group 2 (EH-2) occupancies are those where moderate to substantial quantities of flammable or combustible liquids are stored and used. Typical usages include high-piled storage of high hazard materials, including flammable liquids spraying or coating, oil quenching, and large-scale use of solvents, varnishes, and paints.

2. TYPES OF SPRINKLER SYSTEMS

There are many ways to categorize sprinkler systems. *Wet pipe sprinkler systems* have sprinkler heads attached to a piping system that contains water and is connected to a water supply. Water discharges immediately from sprinklers opened by heat from a fire.

Dry pipe sprinkler systems have a piping system containing air or nitrogen under pressure. When a sprinkler opens and the pressurization drops, the supply water pressure opens a *dry pipe valve*, allowing water to flow into the piping system within 60 seconds. Dry pipe valves are designed so that a moderate air pressure will hold the valve clapper against a greater water pressure. Dry pipe sprinkler systems are used when wet pipe systems are inappropriate, such as where freezing temperatures are expected.

Preaction sprinkler systems are used where damage from an accidental water discharge cannot be tolerated. There is no water in preaction system piping. Unsupervised systems contain air at atmospheric pressure; supervised systems contain compressed (e.g., 10 psi (0.7 bar)) air or nitrogen.

Non-interlock preaction systems contain a control valve and an electric, pneumatic, or electronic detector. The valve is opened when either a sprinkler opens or the detector is triggered. If the detector alone is triggered, water will flow into the sprinkler piping in readiness for an open sprinkler. If a sprinkler opens first, the system will respond as a dry pipe system, providing water flow within 60 seconds.

Single interlock preaction systems are the most common types of preaction systems. An electrical detector must be activated, and the air pressure must be released from the sprinkler system before the *preaction valve* (also known as a *primary control valve*, *deluge valve*, or *dry pipe valve*) opens and sprinklers can discharge water. If the detector alone is triggered, the system will fill with water and an alarm signal will be generated, but no water will be discharged. (Opening the preaction valve transforms the system into a wet pipe system.) An open sprinkler alone will result in a low air pressure signal, but no water will be discharged. Once water is being discharged, the system must be manually turned off.

Double interlock preaction systems are designed to offer maximum protection against accidental water discharge.[4] Like single interlock preaction systems, activation of both the detection system and an open sprinkler are necessary for water to be discharged. However, the preaction valve does not trip and the piping does not fill with water until both conditions are met. The system must be turned off manually. Common applications are protection of freezers, ovens, and elevator shafts where water in the piping can freeze or vaporize.

Deluge systems use open sprinklers and a piping system connected to the water supply through a *deluge valve* to protect a two-dimensional floor area. Deluge systems are used where large volumes of water are needed quickly to control a fast-developing fire. Deluge valves can be operated electrically, pneumatically, or hydraulically. A hydraulically controlled deluge valve is held closed by the water pressure in a prime chamber that enters through a restricted orifice. When the valve opens, water flows into the piping system and discharges from all of the sprinklers. Once the deluge valve opens, it is held open by a *pressure operated relief valve* (PORV) reacting to the increased pressure in the sprinkler lines and venting water in the prime chamber.

Water spray systems are functionally identical to deluge systems. However, the piping and discharge nozzle spray patterns are specifically designed to protect well-defined, three-dimensional areas. Typical uses include

[4] In critical and high value facilities, such as electronic equipment rooms, art libraries, and museums, the fire and water damage associated with the delay in fire suppression may not be acceptable.

the protection of electrical transformers containing flammable oil and tanks containing flammable gas (e.g., hydrogen).

Piping in *foam-water systems* contain a source of foam concentrate in addition to a water supply. Foam-water systems can be wet pipe, dry pipe, deluge, or preaction systems. Typical uses include protection of flammable liquid hazards or airport hangars.

Water mist fire suppression is an alternative method applicable when it is critical that water damage or a high volume of water be avoided. Water mist is discharged at high velocity by high-pressure pumps or accumulators. Water mist fire suppression systems work by lowering the ambient temperature through evaporation. Accordingly, they may only slow the spread of, and not necessarily extinguish, a fire. Typical uses include protection of ships and some residential areas.

3. PIPING FOR FIRE PROTECTION SYSTEMS

During manufacturing of any pipe, either the outside or inside diameters (or both) are dimensionally controlled. The outside diameter is typically controlled when pipes are to be joined with threaded and/or slip fittings. Depending on the pipe material and application, *controlled outside diameter* is designated as either "DR" or "CTS." *Controlled inside diameter* is designated as either "PR" or "SIDR." Outside and inside diameter control is designated as "DR-PR."

Steel Pipe

The most common type of pipe used in nonresidential fire protection systems is *black steel pipe*. "Black steel" ("black iron") is a generic term that refers to the exterior color and has nothing to do with a given pipe's dimensions. When steel pipe is forged, a black oxide scale forms on its surface. Because steel is subject to rust and corrosion, the pipe is usually coated with protective oil, resulting in a pipe that will not rust and requires little maintenance.

Different standards specify such things as outside and inside diameters, wall thickness, and working pressure. The outer diameter of steel pipe is dimensionally controlled. The wall thickness and inner diameter are specified by a schedule designation. The most common standards for fire protection pipe are ASTM A795 (*Standard Specification for Black and Hot-Dipped Zinc-Coated (Galvanized) Welded and Seamless Steel Pipe for Fire Protection Use*); ANSI/ASTM A53 (*Standard Specification for Pipe, Steel, Black,* and *Hot-Dipped, Welded and Seamless*); ANSI/ASME B36.10 *(Welded and Seamless Wrought Steel Pipe)*; and ASTM A135 *(Standard Specification for Electric Resistance-Welded Steel Pipe)*. Appendix 16.B and App. 16.C give dimensions of B36.10 steel pipe.

Copper Pipe

Copper pipe types K, L, and M in standard sizes are designated by their outside diameters in accordance with ASTM B88, with the outside diameter always $^1/_8$ in (3 mm) larger than the designation. Each type represents a series of sizes with different wall thicknesses. For any given diameter, type K pipe has thicker walls than type L pipe, and type L pipe has thicker walls than type M pipe. All inside diameters depend on pipe size and wall thickness. (Appendix 16.J gives dimensions of copper pipe.) For pipe sizes larger than approximately $1^1/_4$ in (32 mm), there is virtually no difference in friction losses among the three types. To avoid system noise and erosion-corrosion, cold water flow velocities should be kept below 8 ft/sec (2.4 m/s). *Temper* designates the strength and hardness of the pipe. *Drawn temper pipe* is often referred to as *hard pipe*. *Annealed pipe* is referred to as *soft pipe*.

Types K, L, and M copper pipe can all be used for both pressure and non-pressure applications, and all three types are accepted by the NFPA. Type M hard is preferred for straight sections. Types K and L are preferred when bending is required. Grooved-end piping connections are common in fire sprinkler systems. The joints are sealed by a special clamp that incorporates an elastomeric gasket.

Plastic Pipe

Plastic pipe specifically listed for fire protection use is permitted for some applications, primarily residential. Chlorinated polyvinyl chloride (CPVC) pipe per ASTM F442 (*Standard Specification for Chlorinated Poly (Vinyl Chloride) (CPVC) Plastic Pipe (SDR–PR)*) and polybutylene (PB) pipe per ASTM D2241 (*Standard Specification for Poly (Vinyl Chloride) (PVC) Pressure-Rated Pipe (SDR Series)*) are both listed. Appendix 16.D lists dimensions for PVC and CPVC pipe.

Plastic pipe is classified in two ways: by *standard dimension ratio* (SDR) (or equivalently, *dimension ratio* (DR)); or by schedule XX pipe (e.g., schedule-40 or schedule-80). SDR, or DR, is the average outside diameter, D, divided by the minimum wall thickness, t. Since SDR is a diameter to thickness ratio, all pipes with the same SDR (and material) have the same pressure rating.

$$\text{SDR} = \frac{D}{t} \qquad 66.1$$

A plastic pipe's long-term *pressure rating* (*pressure class*) is calculated from the *hydrostatic design stress*, HDS, equal to the *hydrostatic design basis*, HDB, divided by a factor of safety, FS. The HDB is typically

Plant Engineering

4000 psi (27.6 GPa), and the FS is usually 2. In some instances, the pressure is derated by a *surge pressure allowance*, p_s.

$$\text{HDS} = \frac{\text{HDB}}{\text{FS}} \tag{66.2}$$

$$p_{\text{psi}} = \frac{2(\text{HDS})_{\text{psi}}}{\text{SDR} - 1} - p_s \quad \begin{bmatrix}\text{outside diameter}\\ \text{controlled}\end{bmatrix} \quad \text{[SI]} \tag{66.3(a)}$$

$$p_{\text{psi}} = \frac{2(\text{HDS})_{\text{psi}}}{\text{SDR} + 1} - p_s \quad \begin{bmatrix}\text{inside diameter}\\ \text{controlled}\end{bmatrix} \quad \text{[U.S.]} \tag{66.3(b)}$$

Chlorinated polyvinyl chloride (CPVC) is a thermoplastic pipe material. CPVC pipe is immune to galvanic corrosion and resists internal scale buildup. CPVC is fire resistant and will not burn without a flame source, making it suitable for light hazard areas and residential (single and multifamily) fire suppression systems. A *derating factor* is not needed for water temperatures below 80°F (27°C). For CPVC pipe carrying water at 90°F (32.2°C) and 100°F (37.8°C), the pressure derating factors, p_s, are 0.91 and 0.82, respectively.

Polybutylene (PB) *tubing* meeting ASTM D3309 (*Standard Specification for Polybutylene (PB) Plastic Hot- and Cold-Water Distribution Systems*) is permitted for use in residential applications. Although PB tubing may still be used for sprinkler systems, it is no longer manufactured.

Polyethylene (PE) *pipe* is available in standard and high-density (HDPE) varieties. Neither polyethylene nor *acrylonitrile-butadiene-styrene* (ABS) plastic pipe is approved for fire suppression use.

4. COMPONENTS OF SPRINKLER SYSTEMS

Most fire sprinkler systems include many common components and share similar architectures. Table 66.1 lists common components and relates some of them to Fig. 66.1.

Table 66.1 *Typical Fire Sprinkler System Components (See Fig. 66.1 for numbered component integration.)*

(1)	check valve trim*
(2)	drip check valve
(3)	fire department connection
(4)	retard chamber
(5)	alarm line strainer
(6)	alarm pressure switch
(7)	system-side pressure gauge
(8)	fire department hookup swing check valve
(9)	water-motor alarm (water-powered bell)
(10)	alarm check valve
(11)	main drain valve (normally closed)
(12)	alarm test shutoff valve (normally open)
(13)	main drain piping
(14)	water supply main lead-in
(15)	water supply side pressure gauge
(NS)	post-indicating valve (PIV) (water supply control valve)
(NS)	backflow preventer
(NS)	riser check valve
(NS)	restricted orifice
(NS)	drain check valve
(NS)	flow test valve (normally closed)

(NS) – Not shown in Fig. 66.1.

*The check valve is trimmed out with additional components, such as pressure gauges to monitor system pressure conditions, a bypass check valve, a main drain valve, and an alarm test valve.

A *post-indicating valve* (PIV) is a gate or butterfly valve installed in the main fire line serving a building that is used to turn the water supply to a fire protection system on and off. The PIV is commonly located in the yard near the building to provide a visual indication of a valve's position (open or closed), as well as the ability to access the valve from outside of a building when the building is not safe to enter. A special wrench is needed to change the valve's position.

Since the valve proper is inline with the water supply line, it is typically underground. The above-ground portion has a handle or a hand wheel to actuate the valve below. Depending on features and installation, this part is known by many names: *indicator post*, *I-post*, *IP*, *post indicator*, *PI*, *wall post*, *pit post*, *ground post*, and *pedestal post*. Most indicator posts have a window that indicates "OPEN" or "SHUT."

Some *indicating butterfly valves* (IBV) are listed for fire protection use. Due to their internal construction, an IBV's position is indicated by a directional arrow that is either aligned with the flow or perpendicular to it.

In order to prevent tampering or inadvertent cut-off, the physical position of post IBVs is electrically supervised by a *post control valve supervisory switch* (PCVS).

Usually, main water control valves for fire sprinklers must be *outside screw and yoke* (OS&Y) *valves* because, unlike gate and ball valves, it is always clear whether an OS&Y valve is open or closed. The valve handle and the stem are both threaded and they interact. When the

Figure 66.1 Simple Wet Pipe Sprinkler System Component Integration (See Table 66.1 for component names.)

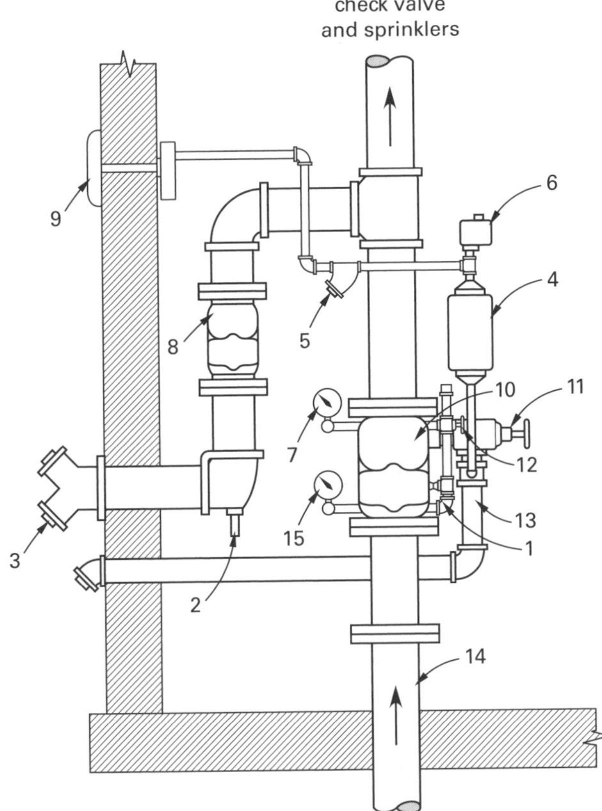

handle is turned, the stem raises or lowers outside the body of the valve, while the handle remains in a fixed position. As the stem rises, the gate inside the body of the valve rises in unison, letting water flow through the valve.

Alarm check valves are used in wet pipe sprinkler systems. They perform dual functions: preventing backflow that might contaminate the potable water supply, and providing a connection to a hydraulically-driven fire alarm (*water motor*) that is independent of the electrical power is also common.

NFPA 13 requires that all valves controlling water supplies be supervised in the open position. Normally, a control valve is required before and after each check valve in the water supply feed to permit each check valve to be isolated and serviced. However, control valves are not allowed in the fire department connection piping. Control valves can be installed downstream of the fire department connection piping, particularly in a multi-zone arrangement.

Virtually all sprinkler system piping contains confined air. If a water hammer or pressure surge occurs in the supply line, the increased pressure will compress the confined air and cause the clapper in the alarm valve to lift intermittently, sometimes enough to cause a false alarm. A hydraulic *retard chamber* introduces a time delay to eliminate false alarms caused by fluctuating water supply pressures. Where water pressure is constant, a retard chamber is not necessary.

Water from the alarm check valve enters the retard chamber through a *restricted orifice*. The retard chamber accumulates water and feeds it back into the main drain. When the drain capacity is exceeded, as when the alarm check valve clapper remains open, the chamber's capacity is reached. Then, water flows through the outlet to the water motor alarm and/or electric alarm switch. Additional electrical *vane switches* with pneumatic retards (VSR) can be added to trigger supplemental alarms and annunciators.

5. STANDPIPES

A *standpipe system* provides rapid access to water within a building interior. It is composed of pipes throughout the building that connect the water supply to hose connections within the building. They provide a pre-piped water system for building occupants or the fire department. Standpipe systems are common in tall buildings. Some older buildings have standpipe systems with hose connections only, while many newer buildings have a combined system supplying the sprinkler system and the hose connections. Figure 66.2 shows a typical standpipe system.

Figure 66.2 Typical Standpipe

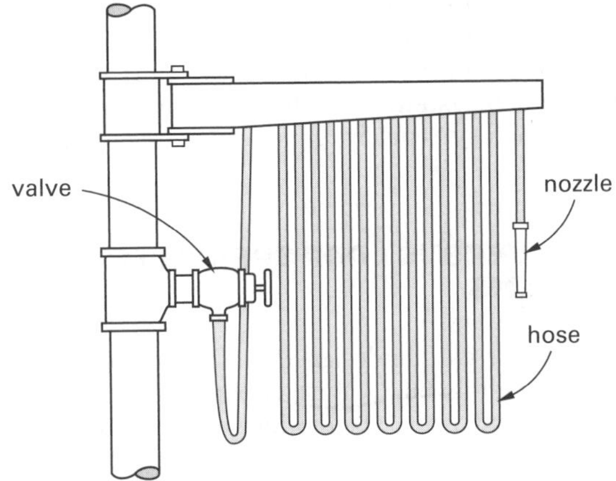

Standpipe systems may be automatic, semiautomatic, manual, wet, or dry. Dry standpipe systems (*dry risers*) bring water to various points in the building only when water is pumped in by the fire department; the pipes are normally empty. Wet systems are charged, meaning they are filled with water at all times. There are three classes of standpipe systems. A *class I standpipe system* has a design pressure of 100 psi (6.9 bar) and provides $2\frac{1}{2}$ in (63.5 mm) hose connections (but no hoses) for use by the fire department. A *class II standpipe system* provides $1\frac{1}{2}$ in (38.1 mm) hose connections, typically in cabinets with 100 ft (30.5 m) of hose attached. A *class III standpipe system* provides both $1\frac{1}{2}$ in (38.1 mm) and $2\frac{1}{2}$ in (63.5 mm) hose connections.

Plant Engineering

NFPA 14 (*Standard for the Installation of Standpipes and Hose Systems*) details the design, layout, and water flow requirements for standpipe systems. Class I and class III standpipes must use pipe diameters of at least 4 in (101.6 mm); pipe diameters in combined systems must be at least 6 in (152.4 mm). The maximum pressure at any point cannot exceed 350 psi (24.1 bar). Buildings with pressures over this maximum require the standpipe system to be split into different zones. A pressure regulator is required at hose connections if the pressure is greater than 175 psi (12.1 bar). If the hose is intended for occupant use, the hose pressure must be regulated to 100 psi (6.9 bar).

6. SCHEDULE-DESIGNED VERSUS HYDRAULICALLY DESIGNED SPRINKLER SYSTEMS

Pipes used in sprinkler systems can be sized in one of two ways. Before calculators and computers made sizing calculations simpler, the traditional manner was to use a *pipe schedule*—a table dictating the maximum number of fire sprinklers that can be served by any size of pipe. Different occupancy hazards call for different schedules. This method of design has been in use for more than a century.

Most contemporary sprinkler system designs (and all designs for deluge and water spray systems) are based on hydraulic calculations. The total water supply requirements for hydraulically designed systems are lower than those for schedule-designed systems. Also, pipe sizes can be reduced (down the run) in hydraulically designed systems. For these two reasons, hydraulically designed systems are generally more economical to install.

7. TREE VERSUS LOOP SPRINKLER SYSTEMS

Pipe networks can be designed as tree, gridded, or loop systems. (See Fig. 66.3.) With common *tree systems*, the first sprinkler to operate discharges at a greater-than-design rate due to the nature of the flow's declining pressure. *Loop and gridded systems* cannot eliminate the flow-related friction loss, but by providing multiple (parallel) paths to an open sprinkler, the friction loss can be reduced or minimized. Therefore, sprinkler pipe sizes can be smaller (or sprinkler spacing can be larger) with loop systems than with tree systems. Gridded systems are commonly used when the protected area is large and roughly rectangular.

Although loop and gridded designs have a slight economic benefit because they need fewer sprinklers or smaller piping, they require more piping and more complex design calculations. The cost of materials may be slightly greater for loop systems than for tree systems. With modern computer design methods, design costs are probably not an important factor.

Figure 66.3 *Tree, Loop, and Gridded Pipe Systems*

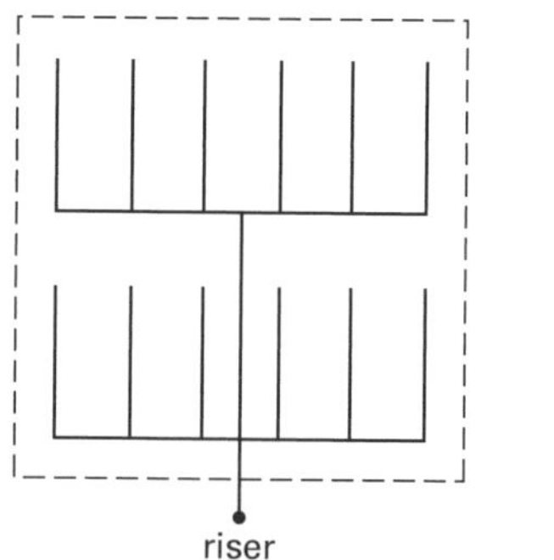

(a) tree system (side central feed shown)

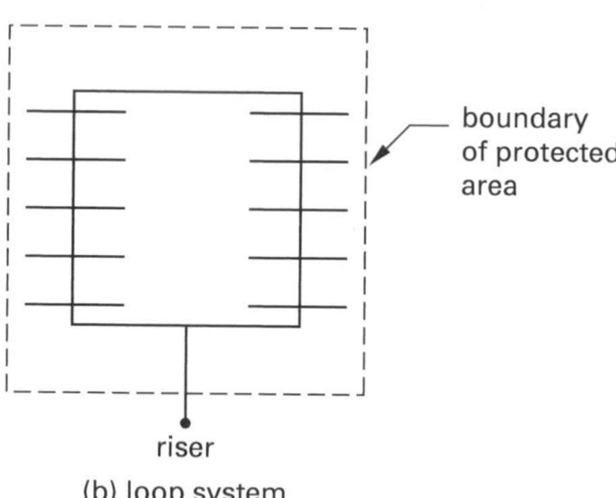

(b) loop system

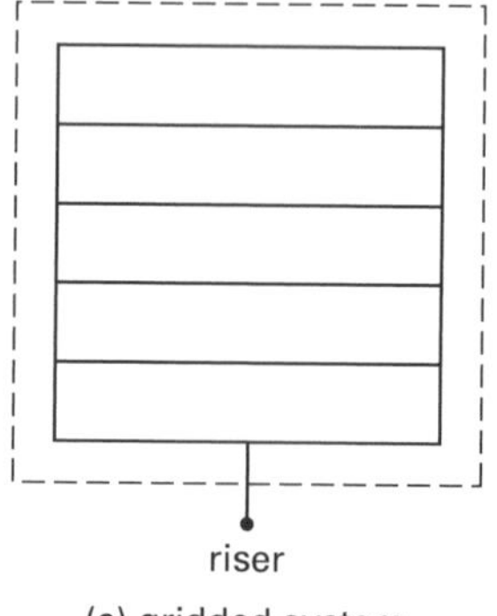

(c) gridded system

A *circulating closed-loop system*, which is a permitted application, is a wet pipe-looped sprinkler system in which the pipes have a second purpose—usually to carry heating or cooling water. Such systems are also known as *automatic sprinkler systems with nonfire protection connections*. Water is only circulated through the system; it is not used or removed from the system.

8. SPRINKLER SYSTEM DIAGRAMS

Descriptions of sprinkler systems make use of the following component terms. Other terms are illustrated in Fig. 66.4. Diagrams of sprinkler systems make use of the abbreviations listed in Table 66.2.

Figure 66.4 *Typical Sprinkler System Components*

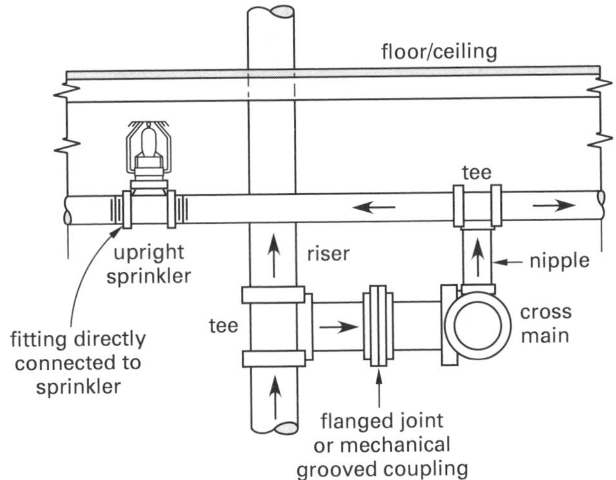

Table 66.2 *Standard Abbreviations for Sprinkler Diagrams*

ALV	alarm valve
BV	butterfly (wafer) check valve
Cr	cross
CV	swing check valve
Del V	deluge valve
DPV	dry-pipe valve
E	90° elbow
EE	45° elbow
GV	gate valve
Lt.E	long-turn elbow
OS&Y	outside stem (screw) and yoke manual control valve
PIV	post indicator valve (control valve)
St	strainer
T	tee, flow turned 90°
WCV	butterfly (wafer) check valve

- *branch line:* a pipe containing sprinklers crossing a cross main
- *cross main:* a pipe directly supplying cross lines containing sprinklers
- *feed main:* a supply riser or a cross main
- *riser:* a pipe, approximately vertical, usually extending one full story
- *system riser:* an above-ground pipe, approximately vertical, bringing water from the supply source

9. SUPPLY PUMPS FOR FIRE FIGHTING[5]

Water from tanks and reservoirs usually flows under gravity action to its fire protection systems. However, depending on the elevations, supply pumps may be required. In some cases, the municipal system can supply adequate volume but at too low a pressure. In such cases, booster pumps can be used to boost the pressure of the municipal water supply. All pumps used in fire protection systems must be approved and/or listed.

Approved fire pumps are almost always horizontal (centrifugal) pumps. However, vertical (centrifugal or turbine) pumps are also used depending on which is more economical and appropriate. Pumps may be single- or multiple-stage. Horizontal pumps must operate with positive suction head. This is particularly important with remote starting. Horizontal pumps must be split-case, end-suction, or in-line types. Single-stage, end-suction, and in-line pumps are limited to under 500 gpm.

To avoid loss of prime and other suction lift problems, *vertical submersible turbine pumps* (i.e., "sump pumps") with submerged impellers can be used when drawing water from a deep well or sump. Vertical pumps can operate without priming, and they are used in ponds, streams, and pits. However, the elevation difference between the source (when pumping at 150% of rated capacity) and the ground surface is limited to 200 ft.

Pumps may be driven by diesel engines, but electric drives are preferred because of their simplicity. Electric power must be available from a reliable source (i.e., one experiencing fewer than two outages per year) or from two independent sources. Spark-ignited gasoline-powered engines should not be used.

Pumps should start automatically unless other sources are capable of simultaneously supplying water for all firefighting and industrial demands. Each pump should have a manual shutdown switch.

A *pressure-maintenance pump* (also known as a *jockey pump* or a *makeup pump*) is usually a low-volume, electrically driven centrifugal pump. (See Fig. 66.5.) The impeller discharge is directly into the water sprinkler line. Operation is triggered by the initial water pressure drop resulting from the opening of a few sprinklers or other leakage. The automatic controller starts the primary fire pump when the capacity of the pressure-maintenance pump is exceeded.

There are three limiting points on the pump head-discharge curve:

- *churn* or *shutoff*, where the pump operates at rated speed with the discharge valve closed
- *rated (100%) capacity and rated (100%) head*
- *overload*, 65% of the rated head

[5]NFPA 20 (*Standard for the Installation of Stationary Pumps for Fire Protection*) is the accepted authority on this subject.

Figure 66.5 *Fire and Jockey Pump Integration*

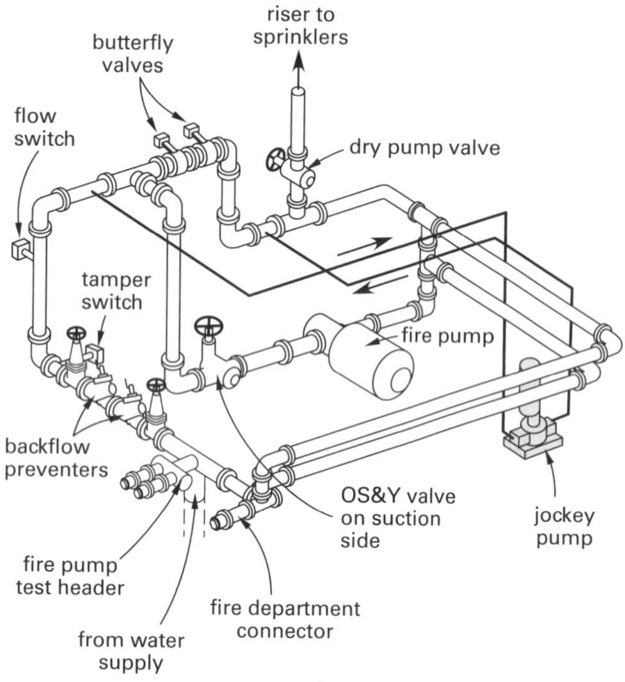

In selecting horizontal and vertical shaft turbine-type pumps for fire supply use, the churn (shutoff) head must not be greater than 140% of the rated head. This requirement eliminates pumps with high shutoff pressures and steep curves. The capacity at overload should not be less than 150% of the rated capacity. Figure 66.6 illustrates the pump characteristics that are described by these requirements.

Figure 66.6 *Standard Fire Protection Pump Curve*

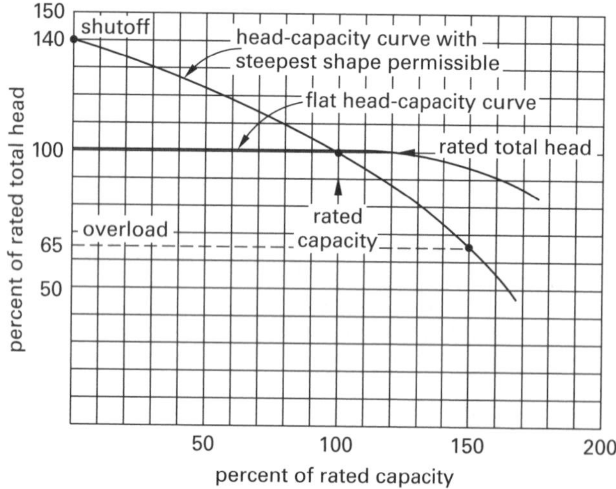

Standard rated capacities of approved horizontal and vertical fire pumps are 25 gpm, 50 gpm, 100 gpm, 150 gpm, 200 gpm, 250 gpm, 300 gpm, 400 gpm, 450 gpm, 500 gpm, 750 gpm, 1000 gpm, 1250 gpm, 1500 gpm, 2000 gpm, 2500 gpm, 3000 gpm, 3500 gpm, 4000 gpm, 4500 gpm, and 5000 gpm (95 L/min, 189 L/min, 379 L/min, 568 L/min, 757 L/min, 946 L/min, 1136 L/min, 1514 L/min, 1703 L/min, 1892 L/min, 2839 L/min, 3785 L/min, 4731 L/min, 5677 L/min, 7570 L/min, 9462 L/min, 11 355 L/min, 13 247 L/min, 15 140 L/min, 17 032 L/min, and 18 925 L/min).

Pressure ratings range from (approximately) 40 psi to 200 psi (280 kPa to 1.40 MPa; 2.8 bars to 14 bars) for horizontal pumps and from 75 psi to 280 psi (520 kPa to 1.9 MPa; 5.2 bars to 19 bars) for vertical pumps. Pressure ratings include the pressure boost of all stages in multistage pumps.

It is general practice to choose a pump for fire protection systems based on overload capacity. This is illustrated in Ex. 66.1. However, some local authorities may require a pump to be sized below overload (i.e., with a higher capacity).

Example 66.1

A fire protection system requires 1300 gpm (82 L/s), including hose streams of water at 80 psi (550 kPa). A suction tank is the source of the water. The equivalent pressure corresponding to the suction lift is 4 psi (28 kPa; 0.28 bars). What are the capacity and pressure ratings for an appropriate pump?

SI Solution

Meet the 82 L/s demand with the pump's overload (i.e., 150%) capacity. A trial rated capacity is

$$Q = \frac{82 \ \frac{\text{L}}{\text{s}}}{1.50} = 54.7 \text{ L/s}$$

The nearest standard pump rating is 63 L/s.

$$\frac{82 \ \frac{\text{L}}{\text{s}}}{63 \ \frac{\text{L}}{\text{s}}} = 1.30$$

Therefore, 82 L/s would be 130% of capacity.

From Fig. 66.6 at 130% of capacity, the total pressure is 80% of the rated pressure.

The total pump pressure is divided between the 550 kPa sprinkler and 28 kPa suction lift pressures. The total net pressure is

$$p_{\text{net}} = 550 \text{ kPa} + 28 \text{ kPa} = 578 \text{ kPa}$$

The rated pressure at 63 L/s is

$$p_{\text{rated}} = \frac{578 \text{ kPa}}{0.80} = 723 \text{ kPa}$$

The pump rating should be 63 L/s at 723 kPa.

Customary U.S. Solution

Meet the 1300 gpm demand with the pump's overload (i.e., 150%) capacity. A trial rated capacity is

$$Q = \frac{1300\ \frac{\text{gal}}{\text{min}}}{1.50} = 867\ \text{gpm}$$

The nearest standard pump rating is 1000 gpm. Therefore, 1300 gpm would be 130% of capacity.

From Fig. 66.6 at 130% of capacity, the total pressure is 80% of the rated pressure.

The total pump pressure is divided between the 80 psi sprinkler and 4 psi suction lift pressures. The total net pressure is

$$p_{\text{net}} = 80\ \frac{\text{lbf}}{\text{in}^2} + 4\ \frac{\text{lbf}}{\text{in}^2} = 84\ \text{psi}$$

The rated pressure at 1000 gpm is

$$p_{\text{rated}} = \frac{84\ \frac{\text{lbf}}{\text{in}^2}}{0.80} = 105\ \text{psi}$$

The pump rating should be 1000 gpm at 105 psi.

10. SPRINKLER HEAD CHARACTERISTICS

The most common sprinkler types are the *upright sprinkler* (which discharges water upward against the deflector), the *pendant sprinkler* (which discharges water downward against the deflector), and the *sidewall sprinkler* (which emits water in a one-quarter sphere spray away from the wall, with a small amount directed at the wall).[6] Figure 66.7 illustrates the uniform water distribution pattern that is characteristic of sprinklers in use since 1953.

Several types of sprinklers have been used in the past including fusible link, glass bulb, and soldier puck sprinklers. Modern sprinklers come in a variety of designs, but all sprinklers contain several important elements: a nozzle, a deflector, and a release mechanism. The release mechanism for the old-type *fusible-link* (also known as soldered *link-and-lever*) *sprinkler* is a soldered fusible link that separates at a specific temperature. Figure 66.8 illustrates a link-and-lever sprinkler.

Figure 66.7 *Distribution Pattern from Ceiling Sprinklers*

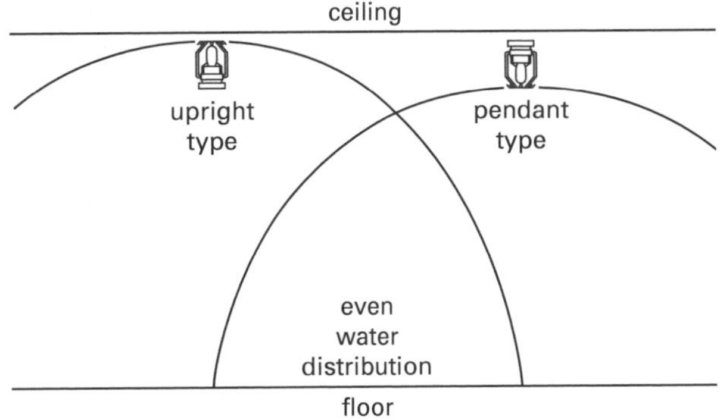

Figure 66.8 *Soldered Link-and-Lever Automatic Sprinkler*

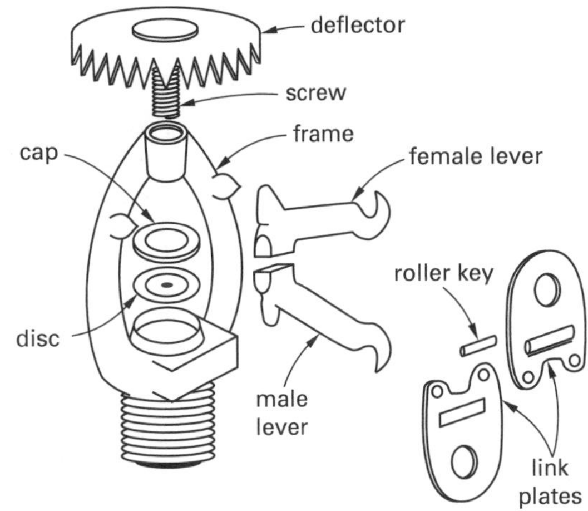

11. SPRINKLER DISCHARGE CHARACTERISTICS

The discharge from a sprinkler depends on the normal (i.e., static) pressure at the nozzle entrance. (See Table 66.3.) The minimum design pressure depends on the sprinkler head design and required coverage. Minimum pressure for any sprinkler is 7 psi (48 kPa; 0.48 bars), the pressure required to produce a 15 gpm flow (0.095 L/s; 57 L/min) in a standard ½ in (12.7 mm) orifice sprinkler.

If the *orifice constant* (also known as *nozzle constant*, *orifice discharge coefficient*, or *K-factor*), K, is known, Eq. 66.4 determines the relationship between the flow

[6]Older (i.e., pre-1953) sprinklers that discharge a significant fraction (i.e., 40%) of the water against the ceiling are no longer used except in special cases, such as wharves and other wood construction.

Table 66.3 Sprinkler Discharge Characteristics

nominal orifice size (in)	orifice type	K-factor	percent of nominal $^1/_2$ in discharge
$^1/_4$	small	1.3–1.5	25
$^5/_{16}$	small	1.8–2.0	33.3
$^3/_8$	small	2.6–2.9	50
$^7/_{16}$	small	4.0–4.4	75
$^1/_2$	standard	5.3–5.8	100
$^{17}/_{32}$	large	7.4–8.2	140
$^5/_8$	extra large	11.0–11.5	200
$^5/_8$	large drop	11.0–11.5	200
$^3/_4$	ESFR	13.5–14.5	250

rate, Q, in gpm (L/s), and normal pressure, p, in psi (kPa). The units of K, gpm/$\sqrt{\text{psi}}$ (L/s/$\sqrt{\text{kPa}}$), are usually not carried along in calculations.

$$Q_{\text{L/s}} = K_{\text{SI}}\sqrt{\frac{p}{\text{SG}}} \quad \text{[SI]} \qquad 66.4(a)$$

$$Q = K\sqrt{\frac{p}{\text{SG}}} \quad \text{[U.S.]} \qquad 66.4(b)$$

$$Q_{\text{L/min}} = K_{\text{bars}}\sqrt{\frac{p_{\text{bars}}}{\text{SG}}} \qquad 66.5$$

$$K_{\text{SI}} = 0.024K \quad \text{[SI]} \qquad 66.6(a)$$

$$K_{\text{bars}} = 14K \quad \text{[in bars]} \quad \text{[U.S.]} \qquad 66.6(b)$$

Although there are many sprinkler designs, most have similar operating characteristics. For example, the standard heads have a 0.5 in (12.7 mm) orifice, and they discharge 15 gpm (0.95 L/s; 57 L/min) at 7 psi (48 kPa; 0.48 bars). This is equivalent to a discharge constant, K, of approximately 5.6 to 5.8 (0.134 to 0.140). The coefficient of discharge, C_d, is taken as 0.75 to 0.78. Figure 66.9 illustrates the discharge characteristics for standard 0.5 in (12.7 mm) and $^{17}/_{32}$ in (13.5 mm) orifices.

Figure 66.9 *Orifice Discharge versus Static Pressure ($^1/_2$ in and $^{17}/_{32}$ in diameter orifices)*

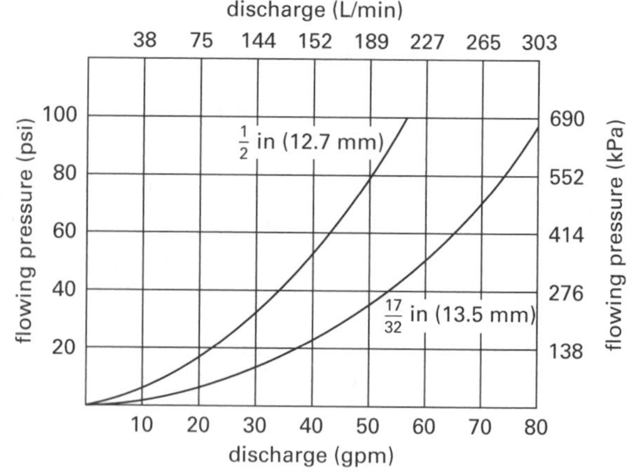

pressure at sprinkler psi (kPa)	approximate discharge gpm (L/min)	
	$\frac{1}{2}$ in (12.7 mm)	$\frac{17}{32}$ in (13.5 mm)
10 (69)	18 (68)	25 (95)
15 (103)	22 (83)	31 (117)
20 (138)	25 (95)	36 (136)
25 (172)	28 (106)	40 (151)
35 (241)	33 (125)	47 (178)
50 (345)	40 (151)	57 (216)
75 (517)	48 (182)	69 (261)
100 (690)	56 (212)	80 (303)

Heads with smaller orifices can be used for light hazard occupancies when hydraulically designed. Strainers are required for orifices less than $^3/_8$ in (9.5 mm) in diameter. Strainers are also a good idea with high-velocity flows. Strainers are often installed in risers.

Example 66.2

A sprinkler has a discharge constant of 5.6 (0.13 for liters and seconds) for pure water (SG = 1). What is the discharge from the sprinkler if the normal pressure is 23 psi (160 kPa)?

SI Solution

From Eq. 66.4(a),

$$\begin{aligned} Q &= K\sqrt{\frac{p}{\text{SG}}} = 0.13\sqrt{\frac{160 \text{ kPa}}{1}} \\ &= 1.64 \text{ L/s} \end{aligned}$$

Customary U.S. Solution

From Eq. 66.4(b),

$$Q = K\sqrt{\frac{p}{\text{SG}}} = 5.6\sqrt{\frac{23 \ \frac{\text{lbf}}{\text{in}^2}}{1}}$$
$$= 26.9 \text{ gpm}$$

12. SPRINKLER PROTECTION AREA

Theoretical coverage per sprinkler can vary from 60 ft^2 to 400 ft^2 (5.4 m^2 to 36 m^2) of floor area. However, the common operating range is more in the order of 80 ft^2 to 120 ft^2 (7.2 m^2 to 10.8 m^2) of floor area.

The spacing and layout design of sprinkler systems will be governed by the maximum floor area that one sprinkler can be expected to protect (i.e., the *protection area*). This, according to *Standard for the Installation of Sprinkler Systems* (NFPA 13), is in turn determined by two factors: the hazard classification and the type of ceiling construction. (Sidewall spray sprinklers have their own different area and spacing requirements.) Other sprinklers may be listed for larger coverage areas.

For example, light hazard areas as listed in Table 66.4 require one sprinkler for every 200 ft^2 (18 m^2) of floor area if the design is by schedule. The requirement is relaxed to one sprinkler for every 225 ft^2 (20.3 m^2) if the system is hydraulically calculated. For open-wood joist ceilings, one sprinkler is required for every 130 ft^2 (12 m^2).

Table 66.4 *Maximum Sprinkler Protection Areas*

	light hazard	ordinary hazard	extra hazard[a]	high-piled storage[b]
unobstructed ceiling[c]	225 ft^2[d]	130 ft^2	100 ft^2	100 ft^2
noncombustible obstructed ceiling	225 ft^2[d]	130 ft^2	100 ft^2	100 ft^2
combustible obstructed ceiling	168 ft^2[e,f]	130 ft^2	100 ft^2	100 ft^2

(Multiply ft^2 by 0.0929 to obtain m^2.)
(Multiply gpm/ft^2 by 40.75 to obtain L/min·m^2.)

[a]90 ft^2 when designed by schedule; 130 ft^2 (11.7 m^2) when hydraulically designed and density is less than 0.25 gpm/ft^2 (10.2 L/min·m^2).
[b]130 ft^2 (11.7 m^2) when hydraulically designed according to NFPA 231 and NFPA 231C and density is less than 0.25 gpm/ft^2 (10.2 L/min·m^2).
[c]Unobstructed construction excludes wood truss construction.
[d]200 ft^2 (18 m^2) when designed by schedule.
[e]130 ft^2 (11.7 m^2) for light combustible framing members spaced less than 3 ft on center.
[f]225 ft^2 (22.3 m^2) for heavy framing members spaced 3 ft or more on center.

When calculating numbers of sprinklers required for areas by dividing the floor area by the sprinkler coverage, fractional numbers are rounded up to the next higher integer. Similarly, when calculating numbers of sprinklers required for branches by dividing the design length by the sprinkler spacing, fractional numbers are rounded up to the next higher integer.

Even though the number of sprinklers is determined assuming that all sprinklers will discharge at the same rate, this does not usually occur in practice. The discharge rate depends on the pressure at the nozzle, which decreases toward the last open sprinkler. This introduces a conservative element in the design. The first (by time) sprinkler to open will discharge at a higher-than-average design rate until other sprinklers open.

13. SYSTEM WATER SUPPLY DEMAND

Supply lines are sized to provide water to all of the sprinklers in a particular area, as well as to meet a code-specified hose allowance. When using the *area-density method* to determine the total water demand, the *design area*, A_c, is the theoretical area of the building representing the worst case area where a fire could burn. The design area is not the same as the coverage per sprinkler, nor is it necessarily the total room/building area. The design area may be specified by local ordinances, insurance requirements, or a fire protection engineer. The *design density*, ρ_S, is a measurement of how much water per unit floor area should be applied to the design area. As Fig. 66.10 shows, density is a function of the hazard classification. The sprinkler water supply demand is calculated from the design density and design area.

$$Q = \rho_S A_c \qquad 66.7$$

Figure 66.10 *Fire Sprinkler Design Density*

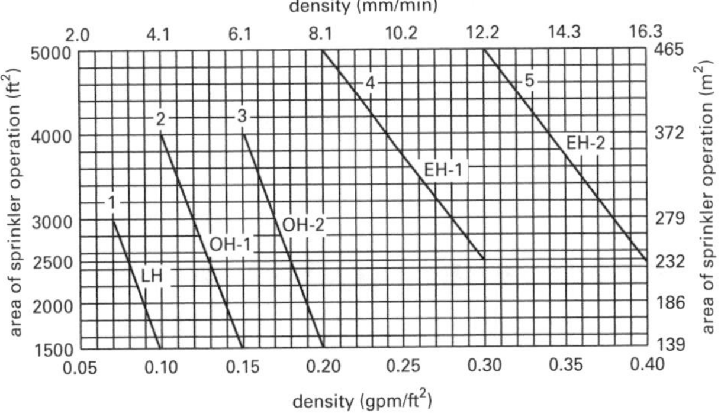

The minimum water supply requirements for a hydraulically designed sprinkler system must include a *hose stream allowance*, Q_{hose}, from Table 66.5. A hose stream allowance must be added to the sprinkler demand even if there are no hose connections in a building. Some fire sprinkler systems must provide their own water from an on-site tank to cover fire fighting needs until the local fire department arrives. Both the hose stream allowance and the duration of on-site storage is given in Table 66.5.

$$Q_{t,\mathrm{gpm}} = Q_{\mathrm{sprinkler,gpm}} + Q_{\mathrm{hose,gpm}} \qquad 66.8$$

$$\begin{aligned} V_{\mathrm{tank,gal}} &= V_{\mathrm{sprinkler,gal}} + V_{\mathrm{hose,gal}} \\ &= (Q_{\mathrm{sprinkler,gpm}} + Q_{\mathrm{hose,gpm}})t_{\mathrm{min}} \end{aligned} \qquad 66.9$$

Table 66.5 *Hose Stream Demand and Water Supply Duration Requirements*

occupancy hazard	inside hose stream allowance (gpm)	total combined inside and outside hose stream allowance (gpm)	duration (min)	
			monitored systems	unmonitored systems
light hazard (LH)	0, 50, 100	100	30	
ordinary hazard (OH)		250	60	90
extra hazard (EH)		500	90	120

(Multiply gpm by 3.785 to obtain L/min.)

14. DISTANCE BETWEEN SPRINKLERS

The distance between sprinklers on lines should be approximately the same as the distance between lines. The maximum spacing between adjacent sprinkler heads and branch lines is generally 15 ft (4.5 m) for light and ordinary hazard occupancies and 12 ft (3.6 m) for extra hazard and high-piled storage occupancies. There are two rare exceptions to these spacing requirements: bays and low discharge densities. (Sidewall spray sprinklers have their own, different area and spacing requirements.)

The maximum distance to the nearest wall is generally one-half of the required spacing but not less than 4 in (10.2 cm).

Sprinklers that are spaced too close together may interfere with other sprinklers' activations. For example, the water discharge from one sprinkler may cool and delay or prevent an adjacent sprinkler from opening. If sprinklers are spaced less than every 6 ft (1.8 m), baffles are required between them. This limitation might be increased to 8 ft by the local authority or the insurance underwriter. NFPA 13 should be consulted for baffle requirements for other types of sprinklers, such as large drop and early suppression fast response (ESFR) sprinklers.

Structural members (including joists and trusses) and nonflat ceiling construction greatly complicate spacing and placement standards.

15. FRICTION LOSSES

The Hazen-Williams equation (see Eq. 66.10) should be used to calculate the friction loss in hydraulically designed sprinkler systems. C is the Hazen-Williams coefficient, commonly known as the C-value. The actual (not nominal) pipe diameter should be used with Eq. 66.10. Use Eq. 66.10(c) to calculate a pressure drop in bars from a flow rate in L/min.

$$p_{f,\mathrm{kPag}} = \frac{(1.18 \times 10^8)L_{\mathrm{m}}Q_{\mathrm{L/s}}^{1.85}}{C^{1.85}d_{\mathrm{mm}}^{4.87}} \quad \text{[SI]} \qquad 66.10(a)$$

$$p_{f,\mathrm{psig}} = \frac{4.52L_{\mathrm{ft}}Q_{\mathrm{gpm}}^{1.85}}{C^{1.85}d_{\mathrm{in}}^{4.87}} \quad \text{[U.S.]} \qquad 66.10(b)$$

$$p_{f,\mathrm{barg}} = \frac{(6.05 \times 10^5)L_{\mathrm{m}}Q_{\mathrm{L/min}}^{1.85}}{C^{1.85}d_{\mathrm{mm}}^{4.87}} \quad \text{[in bars]} \qquad 66.10(c)$$

Table 66.6 gives the Hazen-Williams C-values to be used. The local building or fire officials, however, may specify other values. The C-value is assumed to be constant for a specific pipe roughness and is independent of velocity.

Table 66.6 *Hazen-Williams C-Values for Sprinkler System Design*

type of pipe	C
unlined cast or ductile iron	100
black steel (for dry and preaction systems)	100
black steel (for wet and deluge systems)	120
galvanized (all uses)	120
plastic (listed)	150
cement-lined cast or ductile iron	140
copper tube or stainless steel	150

New unlined steel pipe has a C-value of 140. However, as the pipe ages, the friction will increase. For that reason, 120 is the value used for wet systems. Dry and preaction systems have an even greater tendency to develop deposits and corrosion. Therefore, the C-value for dry systems is taken as 100. Both of these assumptions are conservative. When the pipe is new, greater-than-design flows will be achieved. Friction losses for other C-values can be calculated by multiplying the friction loss by the appropriate factor from Table 66.7.

Table 66.7 *Multipliers for Other C-Values*

	basis of graphical value	
actual C-value	$C = 100$	$C = 120$
150	0.472	0.662
140	0.537	0.752
130	0.615	0.862
120	0.714	1.00
110	0.838	1.18
100	1.00	1.40
90	1.22	1.70
80	1.51	2.12
70	1.93	2.71
60	2.57	3.61

Hydraulic calculations for new copper pipe and tube, plastic pipe and tube, and stainless steel pipe carrying water should use a C-value of 150. Copper pipe is subject to internal pitting and scaling which causes the pressure loss to increase over time (i.e., the C-value becomes smaller). Since plastic pipe is not subject to pitting or scaling, the C-value will remain constant as the system ages.

16. MINOR LOSSES

Ideally, all sources of friction, bends, valves, meters, and strainers should be recognized. However, in hydraulic calculations for sprinkler systems, only fittings involving changes in flow direction are included. Minor losses from straight-through run-of-tees and straight-across crosses are disregarded. Friction losses due to tapered reducers and reducers directly adjacent (attached) to spray nozzles are also disregarded.

Minor losses from tees at the top of risers are included with the cross mains, losses for tees at the bases of risers are included with the risers, and losses from crosses or tees at cross-main feed junctions are included with the cross mains.

For tees and reducing elbows, the equivalent length is based on the velocity and/or diameter of the smaller outlet.

Values for standard elbows are typically used for abrupt 90° turns constructed with threaded fittings. The long-turn elbow values are used with flanged, welded, or other mechanical connections.

Friction losses for fittings connected directly to sprinklers are omitted.

The equivalent lengths of valves and fittings for $C = 120$ are given in App. 66.A. Loss data for specialized elements (such as pressure-reducing valves, deluge valves, alarm valves, dry pipe valves, and strainers) must be obtained from the manufacturers.

Friction losses for other C-values can be calculated by dividing the equivalent lengths in App. 66.A by the appropriate factor in Table 66.8. (These factors assume that the friction loss through the fitting is independent of the piping C-value.)

Table 66.8 *Dividing Factors for Equivalent Lengths*

actual C-value	dividing factor
100	0.713
120	1.00
130	1.16
140	1.33
150	1.51

17. VELOCITY PRESSURE

Velocity pressure, since it is small, may be omitted at the discretion of the designer. This normally introduces a conservative error. However, if the velocity pressure is much more than 5% of the total pressure, it should be considered. If considered, velocity pressure must be included in calculations for both nonlooped branch lines and cross mains. NFPA 13 does not specify a maximum limit for water velocity.

$$p_{\text{v,kPa}} = \frac{9.81\text{v}^2}{2g} \quad \text{[SI]} \qquad \textit{66.11(a)}$$

$$p_{\text{v,psi}} = \frac{0.433\text{v}^2}{2g} \quad \text{[U.S.]} \qquad \textit{66.11(b)}$$

$$p_{\text{v,bars}} = \frac{0.0981\text{v}^2}{2g} \quad \text{[in bars]} \qquad \textit{66.11(c)}$$

Even when velocity pressure is considered, however, the method used to include it is peculiar to sprinkler design. Specifically, the velocity pressure downstream of a sprinkler is used for end outlets, while velocity pressure upstream (on the supply side) is used for other outlets.

Plant Engineering

End outlets include the last sprinkler on a dead-end branch, the final flowing branch on a dead-end cross main, any sprinkler with a flow split on a gridded branch line, and any branch line with a flow split on a loop system.

When velocity pressure is based on the upstream flow rate (which is initially unknown), the velocity pressure is determined by trial and error. This procedure starts with an estimated flow rate for the upstream side of the nozzle. This flow rate is used to determine the trial velocity pressure, and in turn, the normal pressure and a new flow rate. The procedure is repeated until the estimated and calculated flow rates converge sufficiently.

In analysis problems that consider velocity pressure, the discharge from the next-to-last sprinkler in a line may be lower than the last sprinkler. This condition may or may not require attention. If the discharge density from the next-to-last sprinkler exceeds the minimum and is less than approximately 3% of the total design flow, the design probably should be kept. Otherwise, the pipe supplying the end sprinkler should be increased in size.

Example 66.3

The flow through a 1 in (nominal) schedule-40 branch line is 36 gpm (2.3 L/s). What is the velocity pressure?

SI Solution

The internal cross-sectional area of a 1 in line is 5.574×10^{-10} m^2.

The velocity is

$$\text{v} = \frac{\dot{V}}{A} = \frac{\left(2.3\ \frac{\text{L}}{\text{s}}\right)\left(0.001\ \frac{\text{m}^3}{\text{L}}\right)}{5.574 \times 10^{-4}\ \text{m}^2} = 4.13\ \text{m/s}$$

From Eq. 66.11, the velocity pressure is

$$p_\text{v} = \frac{9.81\text{v}^2}{2g} = \frac{\left(9.81\ \frac{\text{kPa}}{\text{m}}\right)\left(4.13\ \frac{\text{m}}{\text{s}}\right)^2}{(2)\left(9.81\ \frac{\text{m}}{\text{s}^2}\right)} = 8.53\ \text{kPa}$$

Customary U.S. Solution

The internal cross-sectional area of a 1 in line is 0.0060 ft^2.

The velocity is

$$\text{v} = \frac{\dot{V}}{A} = \frac{\left(36\ \frac{\text{gal}}{\text{min}}\right)\left(0.002228\ \frac{\text{ft}^3}{\text{sec-}\frac{\text{gal}}{\text{min}}}\right)}{0.0060\ \text{ft}^2} = 13.37\ \text{ft/sec}$$

From Eq. 66.11, the velocity pressure is

$$p_\text{v} = \frac{0.433\text{v}^2}{2g} = \frac{\left(0.433\ \frac{\text{psi}}{\text{ft}}\right)\left(13.37\ \frac{\text{ft}}{\text{sec}}\right)^2}{(2)\left(32.2\ \frac{\text{ft}}{\text{sec}^2}\right)} = 1.20\ \text{psi}$$

18. NORMAL PRESSURE

The *normal pressure*, usually referred to in other subjects as the *static pressure*, is the difference between the total pressure and the velocity pressure.

$$p_n = p_t - p_\text{v}$$

19. HYDRAULIC DESIGN CONCEPTS

Hydraulic calculations, whether for analysis or design, can be performed by hand or by computer. However, to standardize designs, certain conventions (which follow) have been established. This includes standardized terminology (i.e., *normal pressure* instead of *static pressure*) and nomenclature.[7] Alternatively, more sophisticated methods may be used. However, it may be more difficult to justify or explain these methods to the local officials.

Analysis calculations of sprinkler systems start at the hydraulically most-remote nozzle using the minimum nozzle pressure (e.g., 20 psig). This most-remote nozzle is not necessarily the most-distant nozzle. Rather, it is the nozzle whose supply-to-nozzle path experiences the greatest friction loss. It may take several trial sets of calculations to verify which sprinkler is the most remote hydraulically.

To begin the design, either the pressure or the discharge quantity must be known for the most-remote nozzle. If the pressure is known, the discharge quantity is found from the pressure-volume curve for the sprinkler. If the density requirement is known, it can be used to calculate the discharge quantity. The pressure at the nozzle is

[7]The conventions and standardized nomenclature in this chapter are taken from *Standard for the Installation of Sprinkler Systems* (NFPA 13).

then found from the pressure-volume curve. Calculations of normal pressure then proceed, fitting by fitting, back to the point of water supply.

When there is only one sprinkler, as there is for typical protection of a closet or washroom, a reasonable assumption is that the sprinkler will be independent of the other sprinklers in the main room. As long as the sole sprinkler can discharge at the necessary rate, it can be omitted from the hydraulic calculations for main areas greater than 1500 ft^2 (135 m^3).

Branch calculations should be performed separately (i.e., on their own sheets). Calculations are normally omitted for identical and symmetrical branches.

Sprinkler systems supplied by loops are more difficult to analyze and design. Although looped systems can be designed by hand using the Hardy-Cross method, it is more expedient to use computerized methods.

20. HYDRAULIC DESIGN PROCEDURE: PRESSURE ALONG BRANCHES

The following analysis procedure starts at the hydraulically most-distant sprinkler. Note the difference in how velocity pressure is calculated for the third-to-last sprinkler in a branch and beyond.

step 1: Calculate the discharge rate (in gpm) for the last sprinkler from the area and density or from the normal pressure and discharge characteristics, whichever is known. Either Eq. 66.4 or graphical characteristics can be used. (No sprinkler may discharge at less than the minimum design rate.)

step 2: Using the flow rate, the pipe length to the next sprinkler, and the *C*-value, determine the Hazen-Williams friction pressure loss (in psi) from Eq. 66.10.

step 3: Add the friction pressure loss from step 2 to the normal pressure from step 1. This is the total pressure available at the next-to-last sprinkler.

step 4: If it is to be included in the calculations, calculate the velocity pressure (in psi) in the section of pipe used in step 2 from Eq. 66.11.

step 5: Subtract the velocity pressure from the total pressure. This is the normal pressure available at the next-to-last sprinkler. Notice that the downstream, not upstream, velocity pressure is used with the next-to-last sprinkler.

step 6: Calculate the discharge from the next-to-last sprinkler from the normal pressure and sprinkler discharge characteristics. (No sprinkler may discharge at less than the minimum design rate.)

step 7: Using the cumulative flow rate and pipe length to the next sprinkler, determine the Hazen-Williams friction pressure loss (in psi). Convert pressure loss to proper *C*-value if necessary.

step 8: If velocity pressure is to be considered, calculate it from the flow rate downstream of the next sprinkler.

step 9: Calculate the total pressure at the next upstream sprinkler by adding the normal pressure at the previous sprinkler, the friction pressure from step 7, and the velocity pressure from step 8.

step 10: Estimate the sprinkler discharge for the next upstream sprinkler. (Finding the discharge from the third-to-last and all subsequent sprinklers is an iterative process. The upstream velocity is used with all sprinklers except the next-to-last. Since the upstream velocity is unknown, it must be estimated. While the estimated variable could be either the upstream velocity or the upstream flow rate, it is common (and easier) to estimate the sprinkler discharge.)

step 11: Calculate the flow rate in the upstream pipeline by adding the estimated sprinkler discharge and the cumulative discharge from subsequent sprinklers.

step 12: Calculate the upstream velocity pressure from the flow rate determined in step 11.

step 13: Calculate the normal pressure by subtracting the velocity pressure calculated in step 12 from the total pressure calculated in step 9.

step 14: Determine a corrected discharge from the sprinkler using the normal pressure calculated in step 13. (No sprinkler may discharge at less than the minimum design rate.)

step 15: Compare the discharge assumed in step 10 with the corrected discharge calculated in step 14. If they are sufficiently the same, repeat steps 7 through 15 for all remaining sprinklers. If the two values are different, estimate a new discharge and repeat steps 10 through 15.

Example 66.4

The hydraulically most-remote branch of a sprinkler system consists of a 1 in (nominal) schedule-40 steel pipe and three sprinklers with standard ½ in diameter orifices. The last sprinkler is at the end of the branch, and the distance between sprinklers is 10 ft. The minimum pressure to any sprinkler is 10 psig. The discharge constant for the sprinklers is 5.6. What is the volume of water (SG = 1) discharged from each of the three open sprinklers?

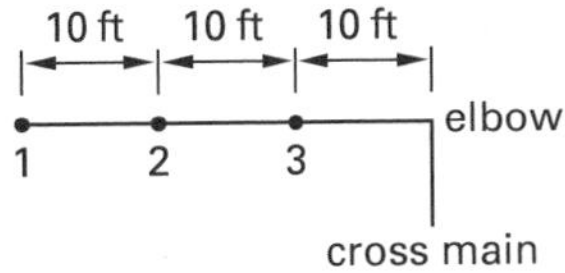

Plant Engineering

Solution

Follow the procedure in Sec. 66.20.

step 1: The normal (static) pressure at sprinkler 1 is 10 psig. From Eq. 66.4, the discharge from sprinkler 1 is

$$Q_1 = K\sqrt{\frac{p_{n,1}}{\text{SG}}} = 5.6\sqrt{\frac{10 \text{ psig}}{1}} = 17.71 \text{ gpm}$$

step 2: The Hazen-Williams coefficient for steel pipe is $C = 120$. Using Eq. 66.10, the friction loss between sprinklers 1 and 2 for a 1 in pipe with an inside diameter of 1.049 in and a flow rate of 17.7 gpm is approximately

$$\begin{aligned} p_{f,1-2} &= \frac{4.52LQ^{1.85}}{C^{1.85}d^{4.87}} \\ &= \frac{(4.52)(10 \text{ ft})\left(17.71 \frac{\text{gal}}{\text{min}}\right)^{1.85}}{(120)^{1.85}(1.049 \text{ in})^{4.87}} \\ &= 1.04 \text{ psig} \end{aligned}$$

step 3: The total pressure at sprinkler 2 (as illustrated) is

$$\begin{aligned} p_{t,2} &= p_{n,1} + p_{f,1-2} = 10 \text{ psig} + 1.04 \text{ psig} \\ &= 11.04 \text{ psig} \end{aligned}$$

step 4: The cross-sectional area of a 1 in diameter schedule-40 steel pipe is 0.0060 ft^2. The velocity in the pipe between sprinklers 1 and 2 is

$$\begin{aligned} \text{v} &= \frac{\dot{V}}{A} = \frac{\left(17.7 \frac{\text{gal}}{\text{min}}\right)\left(0.002228 \frac{\text{ft}^3}{\text{sec-}\frac{\text{gal}}{\text{min}}}\right)}{0.0060 \text{ ft}^2} \\ &= 6.57 \text{ ft/sec} \end{aligned}$$

From Eq. 66.11(b), the velocity pressure of the flow between sprinklers 1 and 2 is

$$\begin{aligned} p_{\text{v},1-2} &= \frac{0.433\text{v}^2}{2g} = \frac{\left(0.433 \frac{\text{psi}}{\text{ft}}\right)\left(6.57 \frac{\text{ft}}{\text{sec}}\right)^2}{(2)\left(32.2 \frac{\text{ft}}{\text{sec}^2}\right)} \\ &= 0.29 \text{ psig} \end{aligned}$$

step 5: The normal pressure to be used with sprinkler 2 is

$$\begin{aligned} p_{n,2} &= p_{t,2} - p_{\text{v},1-2} = 11.04 \text{ psig} - 0.29 \text{ psig} \\ &= 10.75 \text{ psig} \quad [> 10 \text{ psig, so OK}] \end{aligned}$$

step 6: The discharge from sprinkler 2 is

$$\begin{aligned} Q_2 &= K\sqrt{p_{n,2}} = 5.6\sqrt{10.75 \text{ psig}} \\ &= 18.36 \text{ gpm} \end{aligned}$$

step 7: The quantity flowing between sprinklers 2 and 3 is

$$\begin{aligned} Q_{2-3} &= Q_1 + Q_2 = 17.71 \frac{\text{gal}}{\text{min}} + 18.36 \frac{\text{gal}}{\text{min}} \\ &= 36.07 \text{ gpm} \end{aligned}$$

The friction loss between sprinklers 2 and 3 is

$$\begin{aligned} p_{f,2-3} &= (0.0051)\left(36.07 \frac{\text{gal}}{\text{min}}\right)^{1.85} \\ &= 3.88 \text{ psig} \end{aligned}$$

The velocity between sprinklers 2 and 3 is

$$\begin{aligned} \text{v} &= \left(36.07 \frac{\text{gal}}{\text{min}}\right)\left(0.3713 \frac{\text{ft}}{\text{sec-}\frac{\text{gal}}{\text{min}}}\right) \\ &= 13.4 \text{ ft/sec} \end{aligned}$$

The velocity pressure between sprinklers 2 and 3 is

$$\begin{aligned} p_{\text{v},2-3} &= (6.724 \times 10^{-3})\left(13.4 \frac{\text{ft}}{\text{sec}}\right)^2 \\ &= 1.21 \text{ psig} \end{aligned}$$

step 8: The total pressure at sprinkler 3 is

$$\begin{aligned} p_{t,3} &= p_{n,2} + p_{f,2-3} + p_{\text{v},2-3} \\ &= 10.75 \text{ psig} + 3.88 \text{ psig} + 1.21 \text{ psig} \\ &= 15.84 \text{ psig} \end{aligned}$$

step 9: Estimate the discharge from sprinkler 3 as 20 gpm. The normal pressure that would cause this discharge is not calculated.

step 10: The estimated flow rate from the cross-main elbow to sprinkler 3 is

$$\begin{aligned} Q_{3,\text{estimated}} &= Q_{2-3} + Q_{3,\text{estimated}} \\ &= 36.07 \frac{\text{gal}}{\text{min}} + 20 \frac{\text{gal}}{\text{min}} \\ &= 56.07 \text{ gpm} \end{aligned}$$

Illustration for Ex. 66.4

Contract No. ______ Sheet No. 1 of 1

Name and Location Example 66.4

nozzle type and location	flow gpm (L/min)	pipe size in	fitting and devices	pipe equivalent length	friction loss psi/ft (bars/m)	required pressure psi (bars)	normal pressure	$K = 5.6$ notes
1	q 17.71 Q 17.71		length 10 fittings 0 total 10		$C = 120$	p_t 10 p_f 1.04 p_e 0	p_t p_v p_n	$q = 5.6\sqrt{10} = 17.71$ steps 1–2
2	q 18.36 Q 36.07		length 10 fittings 0 total 10			p_t 11.04 p_f 3.88 p_e 0	p_t 11.0 p_v 0.29 p_n 10.75	$q = 5.6\sqrt{10.75} = 18.36$ steps 3–7
3	q 20.13 Q 56.20		length 10 fittings 0 total 10			p_t 15.84 p_f p_e	p_t 15.84 p_v 2.91 p_n 12.93	$p_t = 10.75 + 3.88 + 1.21$ $= 15.84$ $= 5.6\sqrt{12.93} = 20.13$ steps 8–14

step 11: The estimated velocity in the pipe from the cross-main elbow to sprinkler 3 is

$$\text{v}_{\text{elbow}-3,\text{estimated}} = \left(56.07\ \frac{\text{gal}}{\text{min}}\right)\left(0.3713\ \frac{\text{ft}}{\text{sec-}\frac{\text{gal}}{\text{min}}}\right)$$
$$= 20.82\ \text{ft/sec}$$

The estimated velocity pressure between the cross-main elbow and sprinkler 3 is

$$p_{\text{v,elbow}-3,\text{estimated}} = (6.724 \times 10^{-3})\left(20.82\ \frac{\text{ft}}{\text{sec}}\right)^2$$
$$= 2.91\ \text{psig}$$

step 12: The normal pressure at sprinkler 3 is

$$\begin{aligned} p_{n,3} &= p_{t,3} - p_{\text{v,elbow}-3,\text{estimated}} \\ &= 15.84\ \text{psig} - 2.91\ \text{psig} \\ &= 12.93\ \text{psig} \end{aligned}$$

step 13: The corrected discharge from sprinkler 3 is

$$Q_{3,\text{corrected}} = K\sqrt{\frac{p_{n,3}}{\text{SG}}} = 5.6\sqrt{\frac{12.93\ \text{psig}}{1}}$$
$$= 20.13\ \text{gpm}$$

step 14: The discharge from sprinkler 3 was assumed in step 10 to be 20 gpm. It was calculated as 20.13 gpm in step 13. Assuming this is sufficiently close, 20.13 gpm would be used in subsequent steps.

21. HYDRAULIC DESIGN PROCEDURE: PRESSURE AT CROSS MAINS

The total pressure at the cross-main connection to a branch line is the normal pressure at the nearest open sprinkler plus the friction loss and the velocity pressure in the pipe between the sprinkler and the cross-main connection. Thus, the connection is assumed to be an end outlet, and the velocity downstream of the cross-main connection is used to calculate the total pressure. Minor losses, such as tee and nipple losses, must be included in the friction loss.

The pressure at each subsequent upstream cross-main-to-branch connection is calculated similarly to that for branch lines, except that the velocity head (already included from the most-remote cross-main connection) is assumed to be unchanged. The normal pressure at each successive cross-main-to-branch line connection is assumed to be the normal pressure of the last cross-main connection plus the friction pressure loss between the two branch connections.[8]

[8]This is normally a valid assumption, since the velocities in the cross mains will be low. However, if necessary, a rigorous calculation based on actual velocities can be performed. This is seldom necessary.

The analysis procedure works back up the cross main from branch to subsequent upstream branch. While the normal pressure is known, the flow through each subsequent branch (not the hydraulically most-remote) is not. If the subsequent branch is identical to a previous branch, the subsequent branch can be treated as an orifice with the orifice constant, K, calculated from the flow rate and normal pressure at the cross main for the previous branch. All identical branches will have the same orifice constant. The discharge for the subsequent branch is simply calculated from Eq. 66.4.

If the subsequent branch is different from all previous branches, the new branch flow is initially calculated based on any conveniently assumed pressure at the end sprinkler in that branch, working up the branch back to the cross main. The final calculated normal pressure will not coincide with the normal pressure calculated by working up the cross main. This means that the calculated flow will also be incorrect.

Pressures at hydraulic junction points must balance within 0.5 psi (3.5 kPa; 0.035 bars). Pressure differences and their corresponding flows greater than this tolerance must be corrected. Orifice plates and sprinklers with mixed orifice sizes generally cannot be used to balance the system, although there are exceptions for small rooms.

The corrected flow quantity in the branch is calculated from Eq. 66.12.

$$\frac{Q_{\text{corrected}}}{Q_{\text{calculated}}} = \sqrt{\frac{p_{\text{cross main}}}{p_{\text{calculated}}}} \qquad 66.12$$

When two branches on opposite sides of a cross main have the same configuration, the flow calculated from one can be doubled. When the two opposite branches are different in configuration, the hydraulically shorter flow will have to be adjusted by using Eq. 66.12. The effect of this adjustment is to increase the flow in the shorter branch.

The corrected flow quantity is added to the cumulative flow in the cross main. Then, the next upstream branch is handled similarly until all branches in the design area are calculated. Once the flows from the design area have been calculated, the flow in the cross main is assumed constant all the way back to the supply valve. The distance from the design area to the supply valve is used to calculate the pressure friction loss based on the discharge from all sprinklers in the design area. However, branches outside the design area do not influence the flow rate in the cross main.

Example 66.5

The sprinkler branch from Ex. 66.4 is fed by a 1½ in (nominal) schedule-40 steel pipe cross main. What are the total and normal pressures in the cross main at the entrance to the elbow?

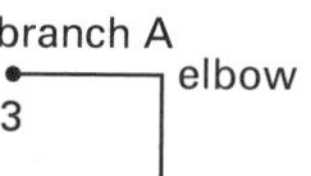

Solution

From Ex. 66.4, the flow rate between the elbow and sprinkler 3 is

$$Q_{\text{elbow}-3} = 36.07\ \frac{\text{gal}}{\text{min}} + 20.13\ \frac{\text{gal}}{\text{min}} = 56.20\ \text{gpm}$$

The velocity between sprinkler 3 and the elbow is

$$\begin{aligned} \text{v} &= \left(56.20\ \frac{\text{gal}}{\text{min}}\right)\left(0.3713\ \frac{\text{ft}}{\text{sec-}\frac{\text{gal}}{\text{min}}}\right) \\ &= 20.87\ \text{ft/sec} \end{aligned}$$

The velocity pressure between sprinkler 3 and the elbow is

$$\begin{aligned} p_{\text{v,elbow}-3} &= (6.724 \times 10^{-3})\left(20.87\ \frac{\text{ft}}{\text{sec}}\right)^2 \\ &= 2.93\ \text{psig} \end{aligned}$$

Since the flow is turned 90°, the minor loss of the elbow (based on the smaller diameter) must be included with the friction loss. From App. 66.A, the equivalent length of a 1 in elbow is 2 ft. The total friction loss between sprinkler 3 and the elbow is

$$\begin{aligned} p_{f,\text{elbow}-3} &= \frac{4.52LQ^{1.85}}{C^{1.85}d^{4.87}} \\ &= \frac{(4.52)(10\ \text{ft} + 2\ \text{ft})\left(56.20\ \frac{\text{gal}}{\text{min}}\right)^{1.85}}{(120)^{1.85}(1.049\ \text{in})^{4.87}} \\ &= 10.56\ \text{psig} \end{aligned}$$

The total pressure at the end of the cross main at the entrance to the elbow at branch A is

$$\begin{aligned} p_{t,\text{A}} &= p_{n,3} + p_{\text{v,elbow}-3} + p_{f,\text{elbow}-3} \\ &= 12.93\ \text{psig} + 2.93\ \text{psig} + 10.56\ \text{psig} \\ &= 26.42\ \text{psig} \end{aligned}$$

The normal pressure at the end of the cross main at the entrance to the elbow at branch A excludes the velocity pressure.

$$\begin{aligned} p_{n,\text{A}} &= p_{n,3} + p_{f,\text{elbow}-3} \\ &= 12.93 \text{ psig} + 10.56 \text{ psig} \\ &= 23.49 \text{ psig} \end{aligned}$$

Example 66.6

The 1½ in (nominal) cross main evaluated in Ex. 66.4 and Ex. 66.5 contains two additional opposing branches. Both branches use 1 in (nominal) schedule-40 steel pipe. The two opposing branches have different configurations. The branch flow rates and normal pressures at the cross connection based on an assumed minimum pressure of $p_{\text{min}} = 10$ psig are given. The distance along the cross main between branches A and B/C is 10 ft. Branch A, evaluated in Ex. 66.5, is the hydraulically most-remote. What is the total flow rate in the cross main?

branch A
elbow
cross
branch B
Q_B = 40 gpm
$p_{n,\text{B}}$ = 15 psig
branch C
Q_C = 35 gpm
$p_{n,\text{C}}$ = 13 psig

Solution

The diameter of 1½ in diameter pipe is 1.610 in.

The minor loss of the elbow has already been included. The cross is a straight-through connection, and its minor loss is disregarded. The friction pressure loss between the elbow and the cross is

$$\begin{aligned} p_{f,\text{elbow-cross}} &= \frac{4.52LQ^{1.85}}{C^{1.85}d^{4.87}} \\ &= \frac{(4.52)(10 \text{ ft})\left(56.20 \ \frac{\text{gal}}{\text{min}}\right)^{1.85}}{(120)^{1.85}(1.610 \text{ in})^{4.87}} \\ &= 1.09 \text{ psig} \end{aligned}$$

The normal pressure at the cross is

$$\begin{aligned} p_{n,\text{cross}} &= p_{n,\text{elbow}} + p_{f,\text{elbow-cross}} \\ &= 23.49 \text{ psig} + 1.09 \text{ psig} \\ &= 24.58 \text{ psig} \end{aligned}$$

The calculation for the normal pressure for branches B and C was not 24.58 psig. Therefore, the residual pressure at the ends of branches B and C will not be 10 psig —it will be higher. The flow rates are also incorrect. Equation 66.12 is used to correct the flows.

$$\begin{aligned} Q_{\text{B,corrected}} &= Q_{\text{B,calculated}}\sqrt{\frac{p_{\text{cross main}}}{p_{\text{calculated}}}} \\ &= \left(40 \ \frac{\text{gal}}{\text{min}}\right)\sqrt{\frac{24.58 \text{ psig}}{15 \text{ psig}}} \\ &= 51.2 \text{ gpm} \end{aligned}$$

$$Q_{\text{C,corrected}} = \left(35 \ \frac{\text{gal}}{\text{min}}\right)\sqrt{\frac{24.58 \text{ psig}}{13 \text{ psig}}} = 48.1 \text{ gpm}$$

When branches A, B, and C are all open, the total flow into the cross main will be

$$Q_{\text{total}} = 56.2 \ \frac{\text{gal}}{\text{min}} + 51.2 \ \frac{\text{gal}}{\text{min}} + 48.1 \ \frac{\text{gal}}{\text{min}} = 155.5 \text{ gpm}$$

22. HYDRAULIC DESIGN PROCEDURE: PRESSURE IN RISERS

The total pressure at the top of a riser is calculated by adding the normal pressure at the nearest (downstream) flowing branch, the total friction loss between the branch and the top of the riser (including the minor loss for the riser-to-cross-main fitting), and the velocity pressure in the cross main at the riser connection. Thus, the total pressure is based on the velocity downstream of the riser-to-cross-main connection.

The pressure at the bottom of a riser is the pressure at the top of the riser plus the friction loss in the riser plus the elevation (i.e., static) pressure corresponding to the change in elevation. (Each foot of water height corresponds to approximately 0.434 psi.)

Example 66.7

The 1½ in diameter cross main of Ex. 66.4 and Ex. 66.5 continues 10 ft and then connects to a 1½ in diameter, 12 ft high riser. The riser is fed by a 3 in feed main at the lower level. The riser-to-cross-main connection at the top of the riser is through a standard 90° elbow. The connection to the feed main at the base of the riser is through a tee. What are the normal pressures at the top and base of the riser?

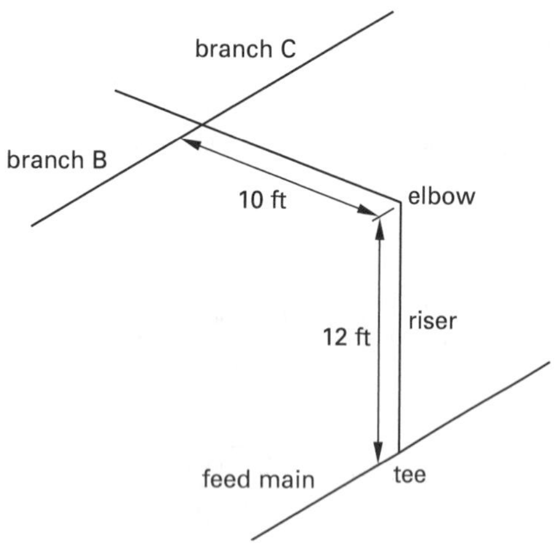

Solution

The equivalent length of a standard 1½ in 90° elbow is given in App. 66.A as 4 ft. The friction loss between branches B and C and the top of the riser is

$$p_{f,\text{cross-top}} = \frac{4.52LQ^{1.85}}{C^{1.85}d^{4.87}}$$

$$= \frac{(4.52)(10\text{ ft} + 4\text{ ft})\left(155.5\ \frac{\text{gal}}{\text{min}}\right)^{1.85}}{(120)^{1.85}(1.610\text{ in})^{4.87}}$$

$$= 10.05\text{ psig}$$

The normal pressure at the top of the riser is

$$p_{n,\text{top}} = p_{n,\text{cross}} + p_{f,\text{cross-top}} = 24.58\text{ psig} + 10.05\text{ psig} = 34.63\text{ psig}$$

The friction loss in the 12 ft of riser is

$$p_{f,\text{riser}} = \frac{4.52LQ^{1.85}}{C^{1.85}d^{4.87}} = \frac{(4.52)(12\text{ ft})\left(155.5\ \frac{\text{gal}}{\text{min}}\right)^{1.85}}{(120)^{1.85}(1.610\text{ in})^{4.87}} = 8.62\text{ psig}$$

The pressure at the base of the riser is the required feed main pressure. From App. 66.A, the equivalent length of a 1½ in tee is 8 ft. The friction loss in the tee is

$$p_{f,\text{tee}} = \frac{4.52LQ^{1.85}}{C^{1.85}d^{4.87}} = \frac{(4.52)(8\text{ ft})\left(155.5\ \frac{\text{gal}}{\text{min}}\right)^{1.85}}{(120)^{1.85}(1.610\text{ in})^{4.87}} = 5.74\text{ psig}$$

In addition, the elevation pressure head from 12 ft of water is

$$p_{\text{elevation}} = \gamma h = \frac{\left(62.4\ \frac{\text{lbf}}{\text{ft}^3}\right)(12\text{ ft})}{\left(12\ \frac{\text{in}}{\text{ft}}\right)^2} = 5.20\text{ psig}$$

The pressure at the base of the riser is

$$\begin{aligned} p_{n,\text{base}} &= p_{n,\text{top}} + p_{f,\text{riser}} + p_{f,\text{tee}} + p_{\text{elevation}} \\ &= 34.63\text{ psig} + 8.62\text{ psig} + 5.74\text{ psig} + 5.20\text{ psig} \\ &= 54.19\text{ psig} \end{aligned}$$

23. PORTABLE FIRE EXTINGUISHERS

Fires and *portable fire extinguishers* are both classified into five categories according to the nature of the substance that is burning. Each fire extinguisher has a geometric symbol that indicates the type of fire the extinguisher is intended for and, generally, a numerical rating that indicates the amount of fire the extinguisher can extinguish.

Class A extinguishers (geometric symbol: green triangle) are for ordinary combustible materials, such as paper, wood, cardboard, and most plastics. The numerical rating indicates the amount of water held.

Class B extinguishers (geometric symbol: red square) are for flammable or combustible liquids such as gasoline, grease, and oil. The numerical rating indicates the approximate coverage in square feet of fire.

Class C extinguishers (geometric symbol: blue circle) are for fires involving electrical equipment. Class C extinguishers do not have a numerical rating. The "C" classification indicates that the extinguishing agent is not conductive.

Class D fire extinguishers (geometric symbol: yellow decagon (star)) are for fires that involve combustible metals, such as magnesium, potassium, and sodium. They have no numerical rating, nor are they given a multi-purpose rating.

Class K fire extinguishers (geometric symbol: black hexagon) are for restaurant and cafeteria kitchen fires involving cooking oils and fats in cooking appliances.

Portable fire extinguishers are filled with compounds that are appropriate for their intended types of fires. Some extinguishers may be used on a combination of fire types.

Water extinguishers (air-pressurized water (APW)) are suitable for class A fires only. *Dry chemical extinguishers* are filled with foam or powder pressurized with

Plant Engineering

nitrogen. They leave a non-flammable substance on the extinguished material, reducing the likelihood of reignition.

Multipurpose ABC extinguishers are filled with monoammonium phosphate, a yellow powder that melts at combustion temperatures and leaves a sticky residue.[9]

BC dry chemical extinguishers are filled with sodium bicarbonate or potassium bicarbonate, both of which are mildly corrosive.

BC carbon dioxide (CO_2) *extinguishers* contain highly pressurized carbon dioxide. They are generally not able to displace enough oxygen to put out class A fires. However, unlike dry chemical extinguishers, they don't leave harmful residues, making them suitable for electrical fires.

24. FLAMMABLE AND COMBUSTIBLE LIQUIDS

Since it is the vapor of a liquid, not the liquid itself, that burns, vapor generation is the primary factor in determining the fire hazard of a flammable or combustible liquid. The *flash point* is the lowest temperature at which the liquid gives off enough vapor to be ignited at the surface.[10] Liquids with flash points below ambient storage temperatures generally exhibit a rapid rate of flame spread over their liquid surfaces, since it is not necessary for the liquid to be heated by the fire in order to generate more vapor. This leads to the expression "low flash–high hazard." The classification of flammable and combustible liquids is, therefore, based on flash point because this property is directly related to *volatility* (i.e., the liquid's ability to generate vapor). Sometimes more than one flash point is reported for a chemical. Blended liquids having several components (e.g., gasoline) may have a range of flash points.

A *flammable liquid* (*Class I liquid*) has a flash point below 100°F (37.8°C), with the exception of any mixture of components with flash points of 100°F (37.8°C) or higher, the total of which makes up 99% or more of the total mixture volume. Class I liquids are divided into three classes. *Class IA liquids* have flash points below 73°F (22.8°C) and boiling points below 100°F (37.8°C).[11] Class IA liquids (e.g., ethyl-ether, heptane, pentane, propylene oxide, and vinyl chloride) are the most dangerous of all flammable and combustible liquids. *Class IB liquids* have flash points below 73°F (22.8°C) and boiling points at or above 100°F (37.8°C). Examples include acetone, ethanol, gasoline, isopropyl alcohol, methanol, octane, and toluene. *Class IC liquids* have flash points at or above 73°F (22.8°C) but below 100°F (37.8°C). Examples include isobutyl alcohol, mineral spirits, turpentine, and xylene.

A *combustible liquid* has a flash point at or above 100°F (37.8°C). Combustible liquids are divided into two classes. *Class II liquids* have flash points at or above 100°F (37.8°C) but below 140°F (60°C), with the exception of mixtures of components with flash points of 200°F (93.3°C) or higher, the volume of which make up 99% or more of the total volume of the mixture. Diesel fuel, fuel oil, kerosene, and motor oil are all Class II liquids. *Class III liquids* have flash points at or above 140°F (60°C) and are subdivided into two subclasses. *Class IIIA liquids* have flash points at or above 140°F (60°C) but below 200°F (93.3°C), with the exception of mixture components with flash points of 200°F (93.3°C) or higher, the total volume of which make up 99% or more of the total volume of the mixture. Linseed oil, mineral oil, and oil-based paints are all Class IIIA liquids. *Class IIIB liquids* have flash points at or above 200°F (93.3°C). Ethylene glycol and glycerin are both Class IIIB liquids.

Classifications of flammable and combustible liquids are shown in Fig. 66.11.

Figure 66.11 *Classes of Flammable and Combustible Liquids*

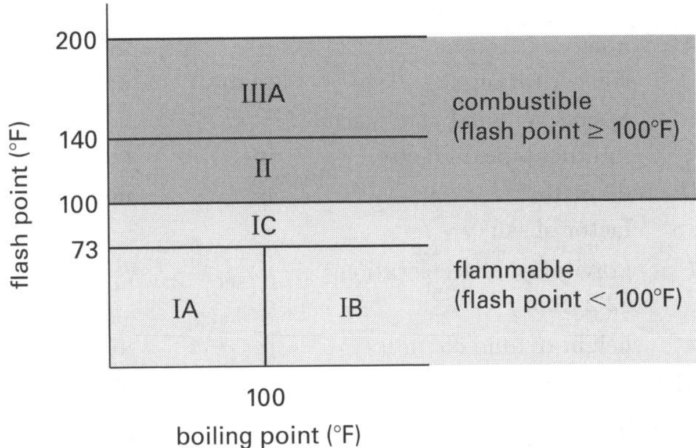

25. EXPLOSIVE LIMITS

When vapors of a flammable or combustible liquid are mixed with air in the proper proportions, ignition can cause rapid combustion or an explosion. There is a minimum concentration below which flame does not propagate, known as the *lower explosive limit* (LEL) or *lower flammable limit* (LFL). Below the LEL, the vapor/air mixture is too lean to burn or explode. There is also a maximum concentration above which flame does not propagate, known as the *upper explosive limit* (UEL) or

[9]Monoammonium phosphate is naturally clear in crystal form and white in powder form. Food dye is used to change its color.

[10]73°F (22.8°C) and 100°F (37.8°C) are the flash point demarcation temperatures. 73°F (22.8°C) is a normal room temperature, and 100°F (37.8°C) is the upper limit of outdoor ambient temperature in all but the hottest climates.

[11]The *boiling point* of a liquid is the temperature at which the vapor pressure of the liquid equals the atmospheric pressure. The lower the boiling point, the more volatile and hazardous the flammable liquid.

Plant Engineering

upper flammable limit (UFL). Above the UEL, the vapor/air mixture is too rich to burn or explode. The range between the LEL and UEL is known as the *flammability range* (*explosive range*).[12] The flammable range includes all concentrations of the flammable vapor (or gas) in air in which flame will propagate when the vapor/air mixture is ignited.

Gas concentrations can be expressed in parts per million by volume (ppm or ppmV), as a mass fraction, as a mole fraction, or as a volumetric percentage. Measurements can be converted between ppm and volumetric percentages with Eq. 66.13.

$$C_{\text{ppm}} = C_{V,\%} \times 10{,}000 \qquad \textit{66.13}$$

Explosive limits for mixtures of combustible gases are calculated using *Le Châtelier's mixing rule*, Eq. 66.14, which can be used to convert both LEL and UEL.

$$\text{LEL}_{\text{mixture,ppm}} = \frac{100}{\sum \frac{C_{V,i,\%}}{\text{LEL}_{i,\text{ppm}}}} \qquad \textit{66.14}$$

26. NOMENCLATURE

A	area	ft^2	m^2
C	concentration	ppm	ppm
C	Hazen-Williams coefficient	–	–
d	internal pipe diameter	in	mm
D	diameter	in	mm
FS	factor of safety	–	–
g	gravitational acceleration, 32.2 (9.81)	ft/sec^2	m/s^2
h	height of fluid column	ft	m
HDB	hydrostatic design basis	lbf/in^2	kPa
HDS	hydrostatic design stress	lbf/in^2	kPa
K	orifice constant	$\text{gpm}/\sqrt{\text{psi}}$	$\text{L/s}/\sqrt{\text{kPa}}$
L	length or distance	ft	m
LEL	lower explosive limit	ppm	ppm
m	mass	lbm	kg
$\dot{m}$	mass leak rate	lbm/sec	kg/s
n	number of sprinklers	–	–
N	number (quantity)	–	–
p	pressure	lbf/ft^2	kPa
q	water flow rate increment	gpm	L/s
Q	flow rate	gpm	L/s
SDR	standard dimension ratio	–	–
SG	specific gravity	–	–
t	time	sec	s
t	time (duration)	min	min
UEL	upper explosive limit	ppm	ppm
v	velocity	ft/sec	m/s
V	volume	gal	L
$\dot{V}$	volume flow rate	gal/sec	L/s

Most modern fire protection design is done in SI units. However, the major reference (*Standard for the Installation of Sprinkler Systems*, NFPA 13) uses bars as the unit of pressure, so this non-SI metric unit is included in this chapter.

Symbols

γ	specific weight	lbf/ft^3	n.a.
ρ	mass density	lbm/ft^3	kg/m^3
ρ_S	surface coverage density	gpm/ft^2	L/s·m^2

Subscripts

c	coverage
d	discharge
e	elevation difference
f	friction
n	normal
s	surge
S	surface (per unit area)
t	total
v	velocity
V	volumetric

[12]No attempt is made to differentiate between the terms *flammable* and *explosive* as applied to the limits of flammability.

67 Pollutants in the Environment

PART 1: GENERAL CONCEPTS

1. POLLUTION PREVENTION AND CONTROL

Pollution prevention (sometimes referred to as "P2") can be achieved in a number of ways. Table 67.1 lists some examples in order of desirability. The most desirable method is reduction at the source. Reduction is accomplished through process modifications, raw material quantity reduction, substitution of materials, improvements in material quality, and increased efficiencies. However, traditional *end-of-pipe* treatment and disposal processes after the pollution is generated remain the main focus. *Pollution control* is the limiting of pollutants in a planned and systematic manner.

Table 67.1 *Pollution Prevention Hierarchy**

source reduction
recycling
waste separation and concentration
waste exchange
energy/material recovery
waste treatment
disposal

*from most to least desirable

Pollution control, hazardous waste, and other environmental regulations vary from nation to nation and are constantly changing. Therefore, regulation-specific issues, including timetables, permit application processes, enforcement, and penalties for violations, are either omitted from this chapter or discussed in general terms. For a similar reason, few limits on pollutant emission rates and concentrations are given in this chapter.[1] Those that are given should be considered merely representative and typical of the general range of values.

PART 2: TYPES AND SOURCES OF POLLUTION

2. THE ENVIRONMENT

Specific regulations often deal with parts of the *environment*, such as the atmosphere (i.e., the "air"), oceans and other surface water, subsurface water, and the soil. *Nonattainment areas* are geographical areas identified by regulation that do not meet *national ambient air quality standards* (NAAQS). Nonattainment is usually the result of geography, concentrations of industrial facilities, and excessive vehicular travel, and can apply to any of the regulated substances (e.g., ozone, oxides of sulfur and nitrogen, and heavy metals).

3. POLLUTANTS

A *pollutant* is a material or substance that is accidentally or intentionally introduced to the environment in a quantity that exceeds what occurs naturally. Not all pollutants are toxic or hazardous, but the issue is moot when regulations limiting permissible concentrations are specific. As defined by regulations in the United States, *hazardous air pollutants* (HAPs), also known as *air toxics*, consist of trace metals (e.g., lead, beryllium, mercury, cadmium, nickel, and arsenic) and other substances for a total of approximately 200 "listed" substances.[2]

Another categorization defined by regulation in the United States separates pollutants into *criteria pollutants* (e.g., sulfur dioxide, nitrogen oxides, carbon monoxide, volatile organic compounds, particulate matter, and lead) and *noncriteria pollutants* (e.g., fluorides, sulfuric acid mist, reduced sulfur compounds, vinyl chloride, asbestos, beryllium, mercury, and other heavy metals).

4. WASTES

Process wastes are generated during manufacturing. *Intrinsic wastes* are part of the design of the product and manufacturing process. Examples of intrinsic wastes are impurities in the reactants and raw materials, by-products, residues, and spent materials. Reduction of intrinsic wastes usually means redesigning the product or manufacturing process. *Extrinsic wastes*, on the other hand, are usually reduced by administrative controls, maintenance, training, or recycling. Examples of extrinsic wastes are fugitive leaks and discharges during material handling, testing, or process shutdown.

Solid wastes are garbage, refuse, biosolids, and containerized solid, liquid, and gaseous wastes.[3] *Hazardous waste* is defined as solid waste, alone or in combination with other solids, that because of its quantity, concentration, or physical, chemical, or infectious characteristics may either (a) cause an increase in mortality or serious (irreversible or incapacitating) illness, or (b) pose a present or future hazard to health or the environment when improperly treated, stored, transported, disposed of, or managed.

A substance is *reactive* if it reacts violently with water; if its pH is less than 2 or greater than 12.5, it is *corrosive*; if its flash point is less than 140°F (60°C), it is *ignitable*. The *toxicity characteristic leaching procedure* (TCLP)[4] test is used to determine if the substance is *toxic*. The TCLP tests for the presence of eight metals (e.g., chromium, lead, and mercury) and 25 organic compounds (e.g., benzene and chlorinated hydrocarbons).

Hazardous wastes can be categorized by the nature of their sources. *F-wastes* (e.g., spent solvents and distillation residues) originate from nonspecific sources; *K-wastes* (e.g., separator sludge from petroleum refining) are generated from industry-specific sources; *P-wastes* (e.g., hydrogen cyanide) are acutely hazardous discarded commercial products, off-specification products, and spill residues; *U-wastes* (e.g., benzene and hydrogen sulfide) are other discarded commercial products, off-specification products, and spill residues.

Two notable "rules" pertain to hazardous wastes. The *mixture rule* states that any solid waste mixed with hazardous waste becomes hazardous. The *derived from rule* states that any waste derived from the treatment of a hazardous waste (e.g., ash from the incineration of hazardous waste) remains a hazardous waste.[5]

[1] Another factor complicating the publication of specific regulations is that the maximum permitted concentrations and emissions depend on the size and nature of the source.

[2] The most dangerous air toxics include asbestos, benzene, cadmium, carbon tetrachloride, chlorinated dioxins and dibenzofurans, chromium, ethylene dibromide, ethylene dichloride, ethylene oxide and methylene chloride, radionuclides, vinyl chloride, and emissions from coke ovens. Most of these substances are carcinogenic.

[3] The term *biosolids* is replacing the term *sludge* when it refers to organic waste produced from biological wastewater treatment processes. Sludge from industrial processes and flue gas cleanup (FGC) devices retains its name.

[4] Probably no subject in this book has more acronyms than environmental engineering. All the acronyms used in this chapter are in actual use; none was invented for the benefit of the chapter.

[5] Hazardous waste should not be incinerated with nonhazardous waste, as all of the ash would be considered hazardous by these rules.

5. POLLUTION SOURCES

A *pollution source* is any facility that produces pollution. A *generator* is any facility that generates hazardous waste. The term "major" (i.e., a *major source*) is defined differently for each class of nonattainment areas.[6]

The combustion of fossil fuels (e.g., coal, fuel oil, and natural gas) to produce steam in electrical generating plants is the most significant source of airborne pollution. For this reason, this industry is among the most highly regulated.

Specific regulations pertain to generators of hazardous waste. Generators of more than certain amounts (e.g., 100 kg/month) must be registered (i.e., with the Environmental Protection Agency (EPA)). Restrictions are placed on generators in the areas of storage, personnel training, shipping, treatment, and disposal.

6. ENVIRONMENTAL IMPACT REPORTS

New installations and large-scale construction projects must be assessed for potential environmental damage before being approved by state and local building officials. The assessment is documented in an *environmental impact report* (EIR). The EIR evaluates all of the potential ways that a project could affect the ecological balance and environment. A report that alleges the absence of any environmental impacts is known as a *negative declaration* ("negative dec").

The following issues are typically addressed in an EIR.

1. the nature of the proposed project
2. the nature of the project area, including distinguishing natural and human-made characteristics
3. current and proposed percentage uses by zoning: residential, commercial, industrial, public, and planned development
4. current and proposed percentage uses by application: built up, landscaped, agricultural, paved streets and highways, paved parking, surface and aerial utilities, railroad, and vacant and unimproved
5. the nature and degree that the earth will be altered: (a) changes in topology from excavation and earth movement to and from the site, (b) changes in slope, (c) changes in chemical composition of the soil, and (d) changes in structural capacity of the soil as a result of compaction, tilling, shoring, or changes in moisture content
6. the nature of known geologic hazards such as earthquake faults and soil instability (subsidence, landslides, and severe erosion)
7. increases in dust, smoke, or air pollutants generated by hauling during construction
8. changes in path, direction, and capacity of natural draining or tendency to flood
9. changes in erosion rates on and off the site
10. the extent to which the project will affect the potential use, extraction, or conservation of the earth's resources, such as crops, minerals, and groundwater
11. after construction is complete, the nature and extent to which the completed project's sewage, drainage, airborne particulate matter, and solid waste will affect the quality and characteristics of the soil, water, and air in the immediate project area and in the surrounding community
12. the quantity and source of fresh water consumed as a result of the project
13. effects on the plant and animal life currently on site
14. effects on any unique (i.e., not found anywhere else in the city, county, state, or nation) natural or human-made features
15. effects on any historically significant or archeological site
16. changes to the view of the site from points around the project
17. changes affecting wilderness use, open space, landscaping, recreational use, and other aesthetic considerations
18. effects on the health and safety of the people in the project area
19. changes to existing noise levels
20. changes in the number of people who will (a) live or (b) work in the project area
21. changes in the burden placed on roads, highways, intersections, railroads, mass transit, or other elements of the transportation system
22. changes in the burden placed on other municipal services, including sewage treatment; health, fire, and police services; utility companies; and so on.
23. the extent to which hazardous materials will be involved, generated, or disposed of during or after construction

[6]A nonattainment area is classified as marginal, moderate, serious, severe, or extreme based on the average pollution (e.g., ozone) level measured in the area.

PART 3: ENVIRONMENTAL ISSUES

7. INTRODUCTION

Engineers face many environmental issues. This part of the chapter discusses (in alphabetical order) some of them. In some cases, "listed substances" (e.g., those that are specifically regulated) are discussed. In other cases, environmental issues are discussed in general terms.

Environmental engineering covers an immense subject area, and this chapter is merely an introduction to some of the topics. Some subjects "belong" to other engineering disciplines. For example, coal-fired plants are typically designed by mechanical engineers. Other subjects, such as the storage and destruction of nuclear wastes, are specialized topics subject to changing politics, complex legislation, and sometimes-untested technologies. Finally, some wastes are considered nonhazardous industrial wastes and are virtually unregulated. They are not discussed in this chapter, either. Processing of medical wastes, personal safety, and the physiological effects of exposure, as important as they are, are beyond the scope of this chapter.

Other philosphical and political issues, such as nuclear fuel versus fossil fuel, plastic bags versus paper bags, and disposable diapers versus cloth diapers, are similarly not covered.

8. ACID GAS

Acid gas generally refers to *sulfur trioxide*, SO_3, in flue gas.[7,8] *Sulfuric acid*, H_2SO_4, formed when sulfur trioxide combines with water, has a low vapor pressure and, consequently, a high boiling point. Hydrochloric (HCl), hydrofluoric (HF), and nitric (HNO_3) acids also form in smaller quantities. However, unlike sulfuric acid, they do not lower the vapor pressure of water significantly. For this reason, any sulfuric acid present will control the dew point.

Sulfur trioxide has a large affinity for water, forming a strong acid even at very low concentrations. At the elevated temperatures in a stack, sulfuric acid attacks steel, almost all plastics, cement, and mortar. Sulfuric acid can be prevented from forming by keeping the temperature of the flue gas above the dew-point temperature. This may require preheating equipment prior to start-up and postheating during shutdown.

Stack dew points have been reported by various researchers to be in the range of 225°F to 300°F. However, the actual value is dependent on the amount of SO_3 in the flue gas, and therefore, on the amount of sulfur in the fuel. The theoretical equilibrium relationships are too complex to be useful, and empirical correlations or graphical methods are used. Equation 67.1 can be used to determine the dew point based on the partial pressures, p_w and p_s, of the water vapor and sulfur trioxide, respectively, in units of atmospheres.[9,10]

$$\begin{aligned}\frac{1000}{T_{\text{dp},K}} &= 1.7842 + 0.0269\log_{10}p_w \\ &\quad - 0.1029\log_{10}p_s \\ &\quad + 0.0329\log_{10}p_w\log_{10}p_s \\ &\quad [\text{error} < \pm 4\text{K}]\end{aligned} \qquad 67.1$$

Hydrochloric acid does not normally occur unless the fuel has a high chlorine content, as chlorinated solvents, municipal solid wastes (MSW), and refuse-derived fuels (RDF) all do. Hydrochloric acid formed during the combustion of MSW and RDF can be removed by semidry scrubbing. HCl removal efficiencies of 90–99% are common. (See also Sec. 67.9 and Sec. 67.47.)

Example 67.1

At a particular point in a stack, the flue gas has a total pressure of 30.2 in Hg. The flue gas is 8% water vapor by volume, and the sulfur trioxide concentration is 100 ppm. What is the approximate dew-point temperature at that point?

Solution

From Dalton's law, the partial pressures are volumetrically weighted.

$$p_w = B_w p_t = \frac{(0.08)(30.2\text{ in Hg})}{29.92\ \frac{\text{in Hg}}{\text{atm}}} = 0.0807\text{ atm}$$

$$\begin{aligned}p_s &= B_s p_t \\ &= \frac{(100\text{ ppm})(30.2\text{ in Hg})}{\left(29.92\ \frac{\text{in Hg}}{\text{atm}}\right)(10^6\text{ ppm})} \\ &= 1.01\times 10^{-4}\text{ atm}\end{aligned}$$

---

[7] The term *stack gas* is used interchangeably with *flue gas*.

[8] Sulfur dioxide normally is not a source of acidity in the flue gas.

[9] This correlation was reported by F. H. Verhoff and J. T. Banchero in *Chemical Engineering Progress*, 1974, Vol. 70, p. 71.

[10] *Dalton's law* states that the partial pressure of a component is volumetrically weighted. Therefore, knowing the volumetric flue gas concentration, C_A, in ppmv is equivalent to knowing the partial pressures.

$$p_A = p_t\left(\frac{C_A}{10^6}\right)$$

Use Eq. 67.1.

$$\begin{aligned}\frac{1000}{T_{\mathrm{dp},K}} &= 1.7842 + 0.0269\log_{10}(0.0807\ \mathrm{atm}) \\ &\quad -0.1029\log_{10}(1.01\times 10^{-4}\ \mathrm{atm}) \\ &\quad +0.0329\log_{10}(0.0807\ \mathrm{atm}) \\ &\quad \times\log_{10}(1.01\times 10^{-4}\ \mathrm{atm}) \\ &= 2.3097\end{aligned}$$

$$T_{\mathrm{dp}} = \frac{1000\mathrm{K}}{2.3097} = 433\mathrm{K} \quad (160°\mathrm{C},\ 320°\mathrm{F})$$

9. ACID RAIN

Acid rain consists of weak solutions of sulfuric, hydrochloric, and to a lesser extent, nitric acids. These acids are formed when emissions of sulfur oxides (SOx), hydrogen chloride (HCl), and nitrogen oxides (NOx) return to the ground in rain, fog, or snow, or as dry particles and gases. Acid rain affects lakes and streams, damages buildings and monuments, contributes to reduced visibility, and affects certain tree species. Acid rain may also represent a health hazard. (See also Sec. 67.8 and Sec. 67.47.)

10. ALLERGENS AND MICROORGANISMS

Allergens such as molds, viruses, bacteria, animal droppings, mites, cockroaches, and pollen can cause allergic reactions in humans. One form of bacteria, *legionella*, causes the potentially fatal *Legionnaires' disease*. Inside buildings, allergens and microorganisms become particularly concentrated in standing water, carpets, HVAC filters and humidifier pads, and in locations where birds and rodents have taken up residence. Legionella bacteria can also be spread by aerosol mists generated by cooling towers and evaporative condensers if the bacteria are present in the recirculating water.

Some buildings cause large numbers of people to simultaneously become sick, particularly after a major renovation or change has been made. This is known as *sick building syndrome* or *building-related illness*. This phenomenon can be averted by using building materials that do not release vapor over time (e.g., as does plywood impregnated with formaldehyde) or harbor other irritants. Carpets can accumulate dusts. Repainting, wallpapering, and installing new flooring can release new airborne chemicals. Areas must be flushed with fresh air until all noticeable effects have been eliminated.

Care must also be taken to ensure that filters in the HVAC system do not harbor microorganisms and are not contaminated by bird or rodent droppings. Air intakes must not be located near areas of chemical storage or parking garages.

11. ASBESTOS

Asbestos is a fibrous silicate mineral material that is inert, strong, and incombustible. Once released into the air, its fibers are light enough to stay airborne for a long time.

Asbestos has typically been used in woven and compressed forms in furnace insulation, gaskets, pipe coverings, boards, roofing felt, and shingles, and has been used as a filler and reinforcement in paint, asphalt, cement, and plastic. Asbestos is no longer banned outright in industrial products. However, regulations, well-publicized health risks associated with cancer and *asbestosis*, and potential liabilities have driven producers to investigate alternatives.

No single product has emerged as a suitable replacement for all asbestos applications. (Almost all replacements are more costly.) *Fiberglass* is an insulator with superior tensile strength but low heat resistance. Fiberglass has a melting temperature of approximately 1000°F (538°C). However, fiberglass treated with hydrochloric acid to leach out most of the silica (SiO_2) can withstand 2000–3500°F (1090–1930°C).

In typical static sealing applications, *aramid fibers* (known by the trade names Kevlar™ and Twaron™) are particularly useful up to approximately 800°F (427°C). However, aramid fibers cannot withstand the caustic, high-temperature environment encountered in curing concrete.

12. BOTTOM ASH

Ash is the residue left after combustion of a fossil fuel. *Bottom ash* (*bottoms ash* or *bottoms*) is the ash that is removed from the combustor after a fuel is burned. (The rest of the ash is fly ash.) The ash falls through the combustion grates into quenching troughs below. It may be continuously removed by *submerged scraper conveyors* (SSCs), screw-type devices, or ram dischargers. The ash can be dewatered to approximately 15% moisture content by compression or by being drawn up a dewatering slope. Bottom ash is combined with conditioned fly ash on the way to the ash storage bunker.[11] Most combined ash is eventually landfilled.

13. DISPOSAL OF ASH

Combined ash (bottom ash and fly ash) is usually landfilled. Other occasional uses for high-quality combustion ash (not ash from the incineration of municipal solid waste) include roadbed subgrades, road surfaces ("ash-phalt"), and building blocks.

[11]Approximately half of the electrical generating plants in the United States use wet fly ash handling.

Plant Engineering

14. CARBON DIOXIDE

Carbon dioxide, though an environmental issue, is not a hazardous material and is not regulated as a pollutant.[12] Carbon dioxide is not an environmental or human toxin, is not flammable, and is not explosive. Skin contact with solid or liquid carbon dioxide presents a freezing hazard. Other than the remote potential for causing frostbite, carbon dioxide has no long-term health effects. Its major hazard is that of asphyxiation, by excluding oxygen from the lungs.

As with other products of combustion, the fraction of carbon dioxide in a flue gas can be arbitrarily reduced by the introduction of dilution air into the flue stream. Therefore, carbon dioxide is reported on a standardized basis—typically as a dry volumetric fraction at some percentage (e.g., 3%) of oxygen.[13] The standardized value can be calculated from stoichiometric relationships. Since there is essentially no carbon dioxide in air (approximately 0.03% by volume), carbon dioxide in a flue gas has a unique source—carbon in the fuel. Knowing the theoretical flue gas composition is sufficient to calculate the standardized value. The analysis is independent of stack temperature.

For a standardized value of 3% oxygen, Table 67.2 gives the dry carbon dioxide volumetric fraction directly, based on the volumetric ratio of hydrogen to carbon in the fuel. For any other percentage of oxygen, the following procedure can be used.

step 1: Obtain the volumetric fuel composition. Gaseous fuel compositions are normally reported on a volumetric basis. Solid and liquid fuels are reported on a weight (gravimetric) basis. Convert weight basis analyses to a volumetric basis by dividing the gravimetric percentages by their respective atomic (or, molecular) weights. Combustion products are gaseous. (For gases, molar and volumetric ratios are the same.)

step 2: Write and balance the stoichiometric combustion equation using the volumetric fractions as coefficients for the fuel elements. Disregard trace emissions (NO, SO_2, CO, etc.) that contribute less than 1% to the flue gas volume, and disregard oxygen contributed by the fuel. Include a variable amount of excess air. Include nitrogen for the combustion and excess air at the ratio of 3.773 volumes of nitrogen for each volume of oxygen.

step 3: Divide the oxygen volume (i.e., the balanced reaction coefficient) by the sum of all flue gas volumes, excluding water vapor, and set this ratio equal to the standardized volumetric oxygen fraction. Solve for excess air.

step 4: Divide the carbon dioxide volume by the sum of all flue gas volumes, excluding water vapor.

(The fraction can be multiplied by 10^6 to obtain the volumetric fraction in ppm, though this is seldom done for carbon dioxide.)

Table 67.2 *Theoretical CO_2 Fraction at 3% Oxygen (dry volumetric basis)*

H/C ratio	CO_2	H/C ratio	CO_2
0	0.18	2.1	0.12723
0.1	0.17651	2.2	0.12548
0.2	0.17816	2.3	0.12378
0.3	0.16993	2.4	0.12212
0.4	0.16682	2.5	0.12050
0.5	0.16382	2.6	0.11893
0.6	0.16093	2.7	0.11740
0.7	0.15814	2.8	0.11590
0.8	0.15544	2.9	0.11445
0.9	0.15283	3.0	0.11303
1.0	0.15031	3.1	0.11165
1.1	0.14787	3.2	0.11029
1.2	0.14551	3.3	0.10898
1.3	0.14323	3.4	0.10768
1.4	0.14101	3.5	0.10643
1.5	0.13886	3.6	0.10520
1.6	0.13678	3.7	0.10400
1.7	0.13476	3.8	0.10283
1.8	0.13279	3.9	0.10168
1.9	0.13089	4.0	0.10056
2.0	0.12903		

The carbon dioxide concentration (mass emission per dry standard volume), C, can be calculated from the carbon dioxide's molecular weight ($M = 44$) and Eq. 67.17.

Example 67.2

Coal has the following gravimetric composition: carbon, 86.5%; hydrogen, 11.75%; nitrogen (N_2), 0.39%; sulfur, 0.40%; ash, 0.01%; andoxygen (O_2), 0.96%. Determine the theoretical carbon dioxide concentration in parts per million on a dry volumetric (ppmvd) basis at 3% oxygen by volume.

Solution

step 1: Disregarding the combustion of sulfur to SO_2 and other elements present in small quantities, flue gases will be products of carbon and hydrogen combustion and the excess air. The atomic weight of carbon is 12 lbm/lbmol. Since the hydrogen is present in elemental form, not as H_2 gas, the atomic weight is 1 lbm/lbmol. Consider 100 lbm of fuel. The carbon content will be

[12]Industrial exposure is regulated by the U.S. Occupational Safety and Health Administration (OSHA).
[13]The phrases "at 3% O_2" and "at 3% excess air" are not equivalent. The former means that oxygen comprises 3% of the gaseous reaction products by volume. The latter means that 3% more air (3% more oxygen) is provided in reactants than is needed.

(0.865)(100 lbm) = 86.5 lbm. The volumetric ratios (number of moles) are

$$\text{C:}\quad \frac{86.5\ \text{lbm}}{12\ \frac{\text{lbm}}{\text{lbmol}}} = 7.208\ \text{lbmol}$$

$$\text{H:}\quad \frac{11.75\ \text{lbm}}{1\ \frac{\text{lbm}}{\text{lbmol}}} = 11.75\ \text{lbmol}$$

step 2: The unbalanced combustion reaction is

$$7.208\,\text{C} + 11.75\,\text{H} + n_1\text{O}_2 + 3.773n_1\text{N}_2 \rightarrow n_2\text{CO}_2 + n_3\text{H}_2\text{O} + 3.773n_1\text{N}_2$$

The balanced combustion reaction is

$$7.208\,\text{C} + 11.75\,\text{H} + 10.146\text{O}_2 + 38.281\text{N}_2 \rightarrow 7.208\text{CO}_2 + 5.875\text{H}_2\text{O} + 38.281\text{N}_2$$

Let e represent the excess air fraction. All of the excess oxygen and nitrogen will appear in the flue gas.

$$\begin{aligned}7.208\,\text{C} &+ 11.75\,\text{H} + (1+e)10.146\text{O}_2 + (1+e)38.281\text{N}_2\\ &\rightarrow 7.208\text{CO}_2 + 5.875\text{H}_2\text{O} + (1+e)38.281\text{N}_2 + 10.146e\text{O}_2\end{aligned}$$

step 3: At 3% O_2,

$$0.03 = \frac{10.146e}{7.208 + (1+e)(38.281) + 10.146e}$$

$$e \approx 0.157 \quad (15.7\%\ \text{excess air})$$

The balanced combustion reaction at 3% oxygen is

$$\begin{aligned}7.208\,\text{C} &+ 11.75\,\text{H} + 11.739\text{O}_2 + 44.291\text{N}_2\\ &\rightarrow 7.208\text{CO}_2 + 5.875\text{H}_2\text{O} + 44.291\text{N}_2 + 1.593\text{O}_2\end{aligned}$$

step 4: The theoretical carbon dioxide fraction, on a dry volumetric basis at 3% oxygen, is

$$\frac{7.208}{7.208 + 44.291 + 1.593} = 0.1358 \quad (13.58\%)$$

(This is the same value as obtained from Table 67.2 for a volumetric fuel ratio of $H/C = 11.75/7.208 = 1.63$.)

Example 67.3

When the fuel described in Ex. 67.2 is burned, the carbon dioxide in the stack gas is measured on a wet, volumetric basis to be 10.4%. What is this value corrected to 3% O_2?

Solution

Since the theoretical carbon dioxide volumetric fraction can be calculated from the fuel composition, the actual carbon dioxide fraction in the stack is irrelevant. (The amount of excess air could be found from this value, however.) The theoretical carbon dioxide fraction, on a dry volumetric basis at 3% oxygen, is still 13.55%.

15. CARBON MONOXIDE

Carbon monoxide, CO, is formed during incomplete combustion of carbon in fuels. This is usually the result of an oxygen deficiency at lower temperatures. Carbon monoxide displaces oxygen in the bloodstream, so it represents an asphyxiation hazard. Carbon monoxide does not contribute to smog.

The generation of carbon monoxide can be minimized by furnace monitoring and control. For industrial sources, the American Boiler Manufacturers Association (ABMA) recommends limiting carbon monoxide to 400 ppm (corrected to 3% O_2) in oil- and gas-fired industrial boilers. This value can usually be met with reasonable ease. Local ordinances may be more limiting, however.

Most carbon monoxide released in highly populated areas comes from vehicles. Vehicular traffic may cause the CO concentration to exceed regulatory limits. For this reason, *oxygenated fuels* are required to be sold in those areas during certain parts of the year. Oxygenated gasoline has a minimum oxygen content of approximately 2.0%. Oxygen is increased in gasoline with additives such as ethanol (ethyl alcohol). Methyl tertiary butyl ether (MTBE) continues to be used as an oxygenate outside of the United States.

Minimization of carbon monoxide is compromised by efforts to minimize nitrogen oxides. Control of these pollutants is inversely related.

16. CHLOROFLUOROCARBONS

Most atmospheric oxygen is in the form of two-atom molecules, O_2. However, there is a thin layer in the stratosphere about 12 miles up where *ozone* molecules, O_3, are found in large quantities. Ozone filters out ultraviolet radiation that damages crops and causes skin cancer.

Chlorofluorocarbons (i.e., chlorinated fluorocarbons, such as Freon™) contribute to the deterioration of the earth's ozone layer. Ozone in the atmosphere is depleted in a complex process involving pollutants, wind patterns, and atmospheric ice. As chlorofluorocarbon molecules rise through the atmosphere, solar energy breaks the chlorine free. The chlorine molecules attach themselves to ozone molecules, and the new structure eventually decomposes into chlorine oxide and normal oxygen, O_2. The depletion process is particularly pronounced in the Antarctic because that continent's dry, cold air is

Plant Engineering

filled with ice crystals on whose surfaces the chlorine and ozone can combine. Also, the prevailing winter wind isolates and concentrates the chlorofluorocarbons.

The depletion is not limited to the Antarctic; it also occurs throughout the northern hemisphere, including virtually all of the United States.

The 1987 Montreal Protocol (conference) resulted in an international agreement to phase out worldwide production of chemical compounds that have ozone-depletion characteristics. Since 2000 (the peak of Antarctic ozone depletion), concentrations of atmospheric chlorine have decreased, and ozone layer recovery is increasing.

Over the years, the provisions of the 1987 Montreal Protocol have been modified numerous times. Substance lists have been amended, and action deadlines have been extended. In many cases, though production may have ceased in a particular country, significant recycling and stockpiling keeps chemicals in use. Voluntary compliance by some nations, particularly developing countries, is spotty or nonexistent. In the United States, provisions have been incorporated into the Clean Air Act (Title VI) and other legislation, but such provisions are subject to constant amendment.

Special allowances are made for aviation safety, national security, and fire suppression and explosion prevention if safe or effective substitutes are not available for those purposes. Excise taxes are used as interim disincentives for those who produce the compounds. Large reserves and recycling, however, probably ensure that chlorofluorocarbons and halons will be in use for many years after the deadlines have passed.

Replacements for chlorofluorocarbons (CFCs) include hydrochlorofluorocarbons (HCFCs) and hydrofluorocarbons (HFCs), both of which are environmentally more benign than CFCs, and blends of HCFCs and HFCs. (See Table 67.3.) The additional hydrogen atoms in the molecules make them less stable, allowing nearly all chlorine to dissipate in the lower atmosphere before reaching the ozone layer. The lifetime of HCFC molecules is 2 years to 25 years, compared with 100 years or longer for CFCs. The net result is that HCFCs have only 2–10% of the ozone-depletion ability of CFCs. HFCs have no chlorine and thus cannot deplete the ozone layer.

Most chemicals intended to replace CFCs still have chlorine, but at reduced levels. Additional studies are determining if HCFCs and HFCs accumulate in the atmosphere, how they decompose, and whether any byproducts could damage the environment.

Halon is a generic term used to refer to various liquefied, compressed gas *halomethanes* containing bromine, fluorine, and chlorine. Halons continue to be in widespread legal use for various specialty purposes known as *critical uses* (as defined by the EPA) because they are nonconducting, leave no residue upon evaporation, and are relatively safe for human exposure. For example, halons

Table 67.3 *Typical Replacement Compounds for Chlorofluorocarbons*

designation	applications
HCFC 22	low- and medium-temperature refrigerant; blowing agent; propellant
HCFC 123	replacement for CFC-11; industrial chillers and applications where potential for exposure is low; somewhat toxic; blowing agent; replacement for perchloroethylene (dry cleaning fluid)
HCFC 124	industrial chillers; blowing agent
HFC 134a	replacement for CFC-12; medium-temperature refrigeration systems; centrifugal and reciprocating chillers; propellant
HCFC 141b	replacement for CFC-11 as a blowing agent; solvent
HCFC 142b	replacement for CFC-12 as a blowing agent; propellant
IPC (isopropyl chloride)	replacement for CFC-11 as a blowing agent

are commonly found in aircraft, ship engine compartments, military vehicles, cleanrooms, and commercial kitchen fire suppression systems. Halon 1011 (bromochloromethane, CH_2BrCl) is used in portable fire extinguishers. Halon 1211 (bromochlorodifluoromethane, $CBrClF_2$) is used when the fire extinguishing application method is liquid streaming. Halon 1301 (bromotrifluoromethane, $CBrF_3$) is a gaseous flooding agent, as is required in aircraft suppression systems. Halons are rated for flammable liquids (class B fires) and electrical fires (class C fires), although they are also effective on common combustibles (class A fires). Halons are greenhouse gases, so at least in the United States, production of new halons has ceased. Existing inventory stockpiles, recycling, and importing are used to satisfy current demands.

17. COOLING TOWER BLOWDOWN

State-of-the art reuse programs in *cooling towers* (CTs) may recirculate water 15 to 20 times before it is removed through blowdown. Pollutants such as metals, herbicides, and pesticides originally in the makeup water are concentrated to five or six times their incoming concentrations. Most CTs are constructed with copper alloy condenser tubes, so the recirculated water becomes contaminated with copper ions as well. CT water also usually contains chlorine compounds or other biocides added to inhibit biofouling. Ideally, no water should leave the plant (i.e., a *zero-discharge facility*). If discharged, CT blowdown must be treated prior to disposal.

18. CONDENSER COOLING WATER

Approximately half of the electrical generation plants in the United States use *once-through* (OT) *cooling water*. The discharged cooling water may be a chemical or thermal pollutant. Since fouling in the main steam condenser significantly reduces performance, water can be treated by the intermittent addition of chlorine, chlorine dioxide, bromine, or ozone, and these chemicals may be present in residual form. *Total residual chlorine* (TRC) is regulated. Methods of chlorine control include *targeted chlorination* (the frequent application of small amounts of chlorine where needed) and *dechlorination*.

19. DIOXINS

Dioxins are a family of chlorinated dibenzo-*p*-dioxins (CDDs). The term *dioxin*, however, is commonly used to refer to the specific congener 2,3,7,8-tetrachlorodibenzo-*p*-dioxin (TCDD). Primary dioxin sources include herbicides containing 2,4-D, 2,4,5-trichlorophenol, and hexachlorophene. Other potential sources include incinerated municipal and industrial waste, leaded gasoline exhaust, chlorinated chemical wastes, incinerated polychlorinated biphenyls (PCBs), and any combustion in the presence of chlorine.

The exact mechanism of dioxin formation during incineration is complex but probably requires free chlorine (in the form of HCl vapor), heavy metal concentrations (often found in the ash), and a critical temperature window of 570–840°F (300–450°C). Dioxins in incinerators probably form near waste heat boilers, which operate in this temperature range.

Dioxin destruction is difficult because it is a large organic molecule with a high boiling point. Most destruction methods rely on high temperature since temperatures of 1550°F (850°C) denature the dioxins. Other methods include physical immobilization (i.e., vitrification), dehalogenation, oxidation, and catalytic cracking using catalysts such as platinum.

Dioxins liberated during the combustion of municipal solid waste (MSW) and refuse-derived fuel (RDF) can be controlled by the proper design and operation of the furnace combustion system. Once formed, they can be removed by end-of-pipe processes, including activated charcoal (AC) injection. Success has also been reported using the vanadium oxide catalyst used for NOx removal, as well as manganese oxide and tungsten oxide.

20. DUST, GENERAL

Dust or *fugitive dust* is any solid particulate matter (PM) that becomes airborne, with the exception of PM emitted from the exhaust stack of a combustion process. Nonhazardous fugitive dusts are commonly generated when a material (e.g., coal) is unloaded from trucks and railcars. Dusts are also generated by manufacturing, construction, earth-moving, sand blasting, demolition, and vehicle movement.

Dusts pose three types of hazards. (a) Inhalation of airborne dust or vapors, particularly those that carry hazardous compounds, is the major concern. Even without toxic compounds, odors can be objectionable. Dusts are easily observed and can cover cars and other objects left outside. (b) Dusts can transport hazardous materials, contaminating the environment far from the original source. (c) In closed environments, even nontoxic dusts can represent an explosion hazard.

Dusts are categorized by size according to how deep they can penetrate into the respiratory system. Particle size is based on *aerodynamic equivalent diameter* (AED), the diameter of a sphere with the density of water (62.4 lbm/ft^3 (1000 kg/m^3)) that would have the same settling velocity as the particle. The distribution of dust sizes is divided into three fractions, also known as *conventions*. The *inhalable fraction* (*inhalable convention*) (< 100 μm AED; $d_{50} = 100$ μm) can be breathed into the nose and mouth; the *thoracic fraction* (< 25 μm AED; $d_{50} = 10$ μm) can enter the larger lung airways; and the *respirable fraction* (< 10 μm AED; $d_{50} = 4$ μm) can penetrate beyond terminal bronchioles into the gas exchange regions. (See Fig. 67.1.)

Figure 67.1 *Airborne Particle Fractions*

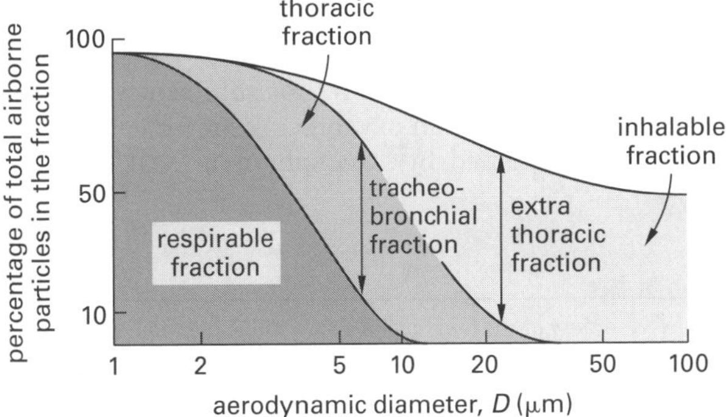

Plant Engineering

Dust emission reduction from *spot sources* (e.g., manufacturing processes such as grinders) is accomplished by *inertial separators* such as cyclone separators. Potential dust sources (e.g., truck loads and loose piles) can be covered, and dust generation can be reduced by spraying water mixed with other compounds.

There are three mechanisms of dust control by spraying. (a) In *particle capture* (as occurs in a spray curtain at a railcar unloading station), suspended particles are knocked down, wetted, and captured by liquid droplets. (b) In *bulk agglomeration* (as when a material being carried on a screw conveyor is sprayed), the moisture keeps the dust with the material being transported. (c) Spraying roads and coal piles to inhibit wind-blown dust is an example of *surface stabilization*.

Wetting agents are *surfactant* formulations added to water to improve water's ability to wet and agglomerate fine particles.[14] The resulting solution can be applied as liquid spray or as a foam.[15] *Humectant binders* (e.g., magnesium chloride and calcium chloride) and adhesive binders (e.g., waste oil) may also be added to the mixture to make the dust adhere to the contact surface if other water-based methods are ineffective.[16]

Surface stabilization of materials stored outside and exposed to wind, rain, freeze-thaw cycles, and ultraviolet radiation is enhanced by the addition of *crusting agents*.

21. DUST, COAL

Clean coals, western low-sulfur coal, eastern low-sulfur coal, eastern high-sulfur coal, low-rank lignite coal, and varieties in between have their own peculiar handling characteristics.[17] *Dry ultra-fine coal* (DUC) and coal slurries have their own special needs.

Western coals have a lower sulfur content, but because they are easily fractured, they generate more dust. Western coals also pose higher fire and explosion hazards than eastern coals. Water misting or foam must be applied to coal cars, storage piles, and conveyer transfer points. Adequate ventilation in storage silos and bunkers is also required, with the added benefit of reducing methane accumulation.

DUC is as fine as talcum (10 μm with less than 10% moisture).[18] It must be containerized for transport, because traditional railcars are not sufficiently airtight. Also, since the minimum oxygen content for combustion of 14–15% is satisfied by atmospheric air, DUC should be maintained in a pressurized, oxygen-depleted environment until used. Pneumatic flow systems are used to transport DUC through the combustion plant.

22. FUGITIVE EMISSIONS

Equipment leaks from plant equipment are called *fugitive emissions* (FEs). Leaks from pump and valve seals are common sources, though compressors, pressure-relief devices, connectors, sampling systems, closed vents, and storage vessels are also potential sources. FEs are reduced administratively by *leak detection and repair programs* (LDARs).

Some common causes of fugitive emissions include (a) equipment in poor condition, (b) off-design pump operation, (c) inadequate seal characteristics, and (d) inadequate boiling-point margin in the seal chamber. Other pump/shaft/seal problems that can increase emissions include improper seal-face material, excessive seal-face loading and seal-face deflections, and improper pressure balance ratio.

Inadequate *boiling-point margin* (BPM) results in poor seal performance and face damage. BPM is the difference between the seal chamber temperature and the initial boiling point temperature of the pumped product at the seal chamber pressure. When seals operate close to the boiling point, seal-generated heat can cause the pumped product between the seal faces to flash and the seal to run dry. A minimum BPM of 15°F (9°C) or a 25 psig (172 kPa) pressure margin is recommended to avoid flashing. Even greater margins will result in longer seal life and reduced emissions.

23. GASOLINE-RELATED POLLUTION

Gasoline-related pollution is primarily in the form of unburned hydrocarbons, nitrogen oxides, lead, and carbon monoxide. (The requirement for lead-free automobile gasoline has severely curtailed gasoline-related lead pollution. Low-lead aviation fuel remains in use.) Though a small reduction in gasoline-related pollution can be achieved by blending detergents into gasoline, reformulation is required for significant improvements. Table 67.4 lists typical characteristics for traditional and reformulated gasolines.

The potential for smog formation can be reduced by reformulating gasoline and/or reducing the summer *Reid vapor pressure* (RVP) of the blend. Regulatory requirements to reduce the RVP can be met by using less butane in the gasoline. However, reformulating to

[14]A *surface-acting agent* (*surfactant*) is a soluble compound that reduces a liquid's surface tension or reduces the interfacial tension between a liquid and a solid.

[15]Collapsible aqueous foam is an increasingly popular means of reducing the potential for explosions in secondary coal crushers.

[16]A *humectant* is a substance that absorbs or retains moisture.

[17]A valuable resource for this subject is NFPA 850, *Recommended Practice for Fire Protection for Electrical Generating Plants and High-Voltage Direct Current Converter Stations*, National Fire Protection Association, Quincy, MA.

[18]A μm is a *micrometer*, 10^{-6} m, and is commonly referred to as a *micron*.

remove pentanes is required when the RVP is required to achieve 7 psig (48 kPa) or below. Sulfur content can be reduced by pretreating the refinery feed in hydrodesulfurization (HDS) units. Heavier gasoline components must be removed to lower the 50% and 90% *distillation temperatures* (T50 and T90).

Table 67.4 *Typical Gasoline Characteristics*

fuel parameter	traditional	reformulated
sulfur	150 ppmw*	40 ppmw
benzene	2% by vol	1% by vol
olefins	9.9% by vol	4% by vol
aromatic hydrocarbons	32% by vol	25% by vol
oxygen	0	1.8 to 2.2% by wt
T90	330°F (166°C)	300°F (149°C)
T50	220°F (104°C)	210°F (99°C)
RVP	8.5 psig (59 kPa)	7 psig (48 kPa)

(Multiply psi by 6.89 to obtain kPa.)
*ppmw stands for parts per million by weight (same as ppmm, parts per million by mass).

24. GLOBAL WARMING

The *global warming* theory is that increased levels of atmospheric carbon dioxide, CO_2, from the combustion of carbon-rich fossil fuels and other *greenhouse gases* (e.g., water vapor, methane, nitrous oxide, and chlorofluorocarbons) trap an increasing amount of solar radiation in a *greenhouse effect*, gradually increasing the earth's temperature.

Recent studies have shown that atmospheric carbon dioxide has increased at the rate of about 0.4% per year since 1958. (For example, in one year carbon dioxide might increase from 390 ppmv to 392 ppmv. By comparison, oxygen is approximately 209,500 ppmv.) There has also been a 100% increase in atmospheric methane since the beginning of the industrial revolution. However, increases in carbon dioxide have not correlated conclusively with surface temperature change. According to most researchers, there has been a global temperature increase of approximately 1.6°F (0.91°C) during the past 50 years.[19] However, the 2014 NOAA/NASA *GISS Surface Temperature Analysis* data noted that the five-year mean surface temperature has not changed in more than ten years. Other studies have shown that temperatures since 2005 are not well-predicted by climate models, which produced higher results.

In addition to a temperature increase, other evidence of the global warming theory are record-breaking hot summers, widespread aberrations in the traditional seasonal weather patterns (e.g., hurricane-like storms in England), an increase in the length of the arctic melt season, melting of glaciers and the ice caps, and an average 7 in (18 cm) rise in sea level over the last century.[20] It has been predicted that there will be a temperature increase of 2–11.5°F (1.1–6.4°C) by the year 2100.

Although global warming is generally accepted, its anthropogenic (human-made) causes are not. The global warming theory is disputed by some scientists and has not been proven to be an absolute truth. Arguments against the theory center around the fact that manufactured carbon dioxide is a small fraction of what is naturally released (e.g., by wetlands, in rain forest fires, and during volcanic eruptions). It is argued that, in the face of such massive contributors, and since the earth's temperature has remained essentially constant for millenia, the earth already possesses some built-in mechanism, currently not perfectly understood, that reduces the earth's temperature swings.

Most major power generation industries have adopted goals of reducing carbon dioxide emissions. However, efforts to reduce carbon dioxide emissions by converting fossil fuel from one form to another are questioned by many engineers. Natural gas produces the least amount of carbon dioxide of any fossil fuel. Therefore, conversion of coal to a gas or liquid fuel in order to lower the carbon-to-hydrogen ratio would appear to lessen carbon dioxide emissions at the point of final use. However, the conversion processes consume energy derived from carbon-containing fuel. This additional consumption, taken over all sites, results in a net increase in carbon dioxide emission of 10–200%, depending on the process.

The use of ethanol as an alternative for gasoline is also problematic. Manufacturing processes that produce ethanol give off (at least) twice as much carbon dioxide as the gasoline being replaced produces during combustion.

Most synthetic fuels are intrinsically less efficient (based on their actual heating values compared with those theoretically obtainable from the fuels' components in elemental form). This results in an increase in the amount of fuel consumed. Thus, fossil fuels should be used primarily in their raw forms until cleaner sources of energy are available.

25. LANDFILL GAS

Most closed landfills are covered by a thick soil layer. As the covered refuse decays beneath the soil, the natural anaerobic biological reaction generates a low-Btu *landfill gas* (LFG) consisting of approximately 50%

[19]Other researchers detect no discernible upward trend.
[20]The rise in the level of the oceans is generally accepted though the causes are disputed. Satellite altimetry indicates a current rise rate of 12 in (30 cm) per century.

methane, carbon dioxide, and trace amounts of other gases. If uncontrolled, LFG migrates to the surface. If the LFG accumulates, an explosion hazard results. If it escapes, other environmental problems (including objectionable odors) occur. Therefore, synthetic and compacted clay liners and clay trenches are used to prevent gas from spreading laterally. Wells and collection pipes are used to collect and incinerate LFG in flares. However, emissions from flaring are also problematic.

Alternatives to flaring include using the LFG to produce hot water or steam for heating or electricity generation. During the 1980s, reciprocating engines and combustion turbines powered by LFG were tried. However, such engines generated relatively high emissions of their own due to impurities and composition variations in the fuel. True Rankine-cycle power plants (generally without reheat) avoid this problem, since boilers are less sensitive to impurities.

One problem with using LFG commercially is that LFG is withdrawn from landfills at less than atmospheric pressure. Conventional furnace burners need approximately 5 psig at the boiler front. Low-pressure burners requiring 2 psig are available, though expensive. Therefore, some of the plant power must be used to pressurize the LFG.

Although production is limited, LFG is produced for a long period after a landfill site is closed. Production slowly drops 3% to 5% annually to approximately 30% of its original value after about 20 to 25 years, which is considered to be the economical life of a gas-reclamation system. Approximately 40 ft^3 of gas will be produced per cubic yard (1.5 m^3 of gas per cubic meter) of landfill.

26. LANDFILL LEACHATE

Leachates are liquid wastes containing dissolved and finely suspended solid matter and microbial waste produced in landfills. Leachate becomes more concentrated as the landfill ages. Leachate forms from liquids brought into the landfill, water run-on, and precipitation. Leachate in a natural attenuation landfill will contaminate the surrounding soil and groundwater. In a lined containment landfill, leachate will percolate downward through the refuse and collect at the first landfill liner. Leachate must be removed to reduce hydraulic head on the liner and to reduce unacceptable concentrations of hazardous substances.

When a layer of liquid sludge is disposed of in a landfill, the consolidation of the sludge by higher layers will cause the water to be released. This released water is known as *pore-squeeze liquid.*

Some water will be absorbed by the MSW and will not percolate down. The quantity of water that can be held against the pull of gravity is referred to as the *field capacity*, FC. The potential quantity of leachate is the amount of moisture within the landfill in excess of the FC.

In general, the amount of leachate produced is directly related to the amount of external water entering the landfill. Theoretically, the leachate generation rate can be determined by writing a water mass balance on the landfill. This can be done on a preclosure and a postclosure basis. Typical units for all the terms are units of length (e.g., "1.2 in of rain") or mass per unit volume (e.g., "a field capacity of 4 lbm/yd^3").

The preclosure leachate generation rate is

$$\begin{aligned}&\text{preclosure leachate generation}\\&\quad= \text{moisture released by incoming waste,}\\&\qquad\quad \text{including pore-squeezed liquid}\\&\qquad + \text{precipitation}\\&\qquad - \text{moisture lost due to evaporation}\\&\qquad - \text{field capacity}\end{aligned} \tag{67.2}$$

The postclosure water balance is

$$\begin{aligned}&\text{postclosure leachate generation}\\&\quad= \text{precipitation}\\&\qquad - \text{surface runoff}\\&\qquad - \text{evapotranspiration}\\&\qquad - \text{moisture lost in formation}\\&\qquad\quad \text{of landfill gas and other}\\&\qquad\quad \text{chemical compounds}\\&\qquad - \text{water vapor removed along}\\&\qquad\quad \text{with landfill gas}\\&\qquad - \text{change in soil moisture storage}\end{aligned} \tag{67.3}$$

27. LEACHATE MIGRATION FROM LANDFILLS

From Darcy's law, migration of leachate contaminants that have passed through liners into aquifers or the groundwater table is proportional to the hydraulic conductivity, K, and the hydraulic gradient, i. Hydraulic conductivities of clay liners are 1.2×10^{-7} ft/hr to 1.2×10^{-5} ft/hr (10^{-9} cm/s to 10^{-7} cm/s). However, the properties of clay liners can change considerably over time due to interactions with materials in the landfill. If the clay dries out (desiccates), it will be much more permeable. For synthetic FMLs, hydraulic conductivities are 1.2×10^{-10} ft/hr to 1.2×10^{-7} ft/hr (10^{-12} cm/s to 10^{-9} cm/s). The average permeability of high-density polyethylene is approximately 1.2×10^{-11} ft/hr (1×10^{-13} cm/s).

$$Q = KiA \tag{67.4}$$

$$i = \frac{dH}{dL} \tag{67.5}$$

Plant Engineering

28. GROUNDWATER DEWATERING

It may be possible to prevent or reduce contaminant migration by reducing the elevation of the groundwater table (GWT). This is accomplished by dewatering the soil with relief-type *extraction drains* (*relief drains*). The *ellipse equation*, also known as the *Donnan formula* and the *Colding equation*, used for calculating pipe spacing, L, in draining agricultural fields, can be used to determine the spacing of groundwater dewatering systems. In Eq. 67.6, K is the hydraulic conductivity with units of length/time, a is the distance between the pipe and the impervious layer barrier (a is zero if the pipe is installed on the barrier), b is the maximum allowable table height above the barrier, and Q is the *recharge rate*, also known as the *drainage coefficient*, with dimensions of length/time. The units of K and Q must be on the same time basis.

$$L = 2\sqrt{\left(\frac{K}{Q}\right)(b^2 - a^2)} = 2\sqrt{\frac{b^2 - a^2}{i}} \qquad 67.6$$

Equation 67.6 is often used because of its simplicity, but the accuracy is only approximately ±20%. Therefore, the calculated spacing should be decreased by 10–20%.

For pipes above the impervious stratum, as illustrated in Fig. 67.2, the total discharge per unit length of each pipe (in ft^3/ft-sec or m^3/m·s) is given by Eq. 67.7. H is the maximum height of the water table above the pipe invert elevation. D is the average depth of flow.

$$Q_{\text{unit length}} = \frac{2\pi KHD}{L} \qquad 67.7$$

$$D = a + \frac{H}{2} \qquad 67.8$$

Figure 67.2 *Geometry for Groundwater Dewatering Systems*

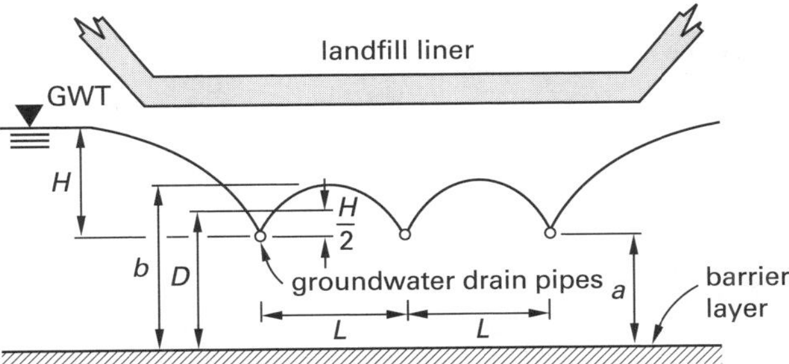

(a) pipes above impervious stratum

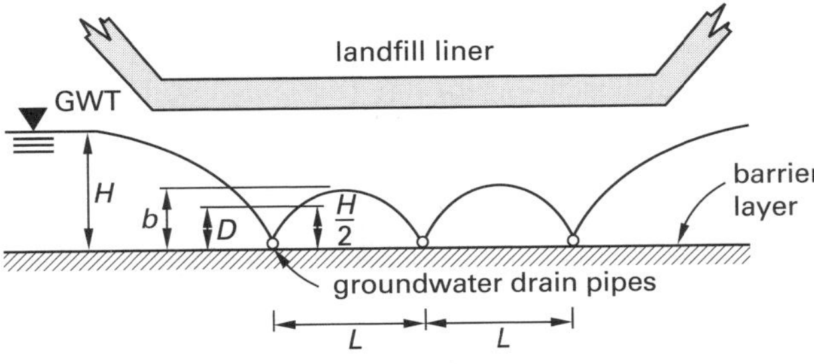

(b) pipes on impervious stratum

For pipes on the impervious stratum, the total discharge per unit length from the ends of each pipe (in ft^3/ft-sec or m^3/m·s) is

$$Q_{\text{unit length}} = \frac{4KH^2}{L} \qquad 67.9$$

Equation 67.10 gives the total discharge per pipe. If the pipe drains from both ends, the discharge per end would be half of that amount.

$$Q_{\text{pipe}} = LQ_{\text{unit length}} \qquad 67.10$$

Example 67.4

Subsurface leachate migration from a landfill is to be mitigated by maintaining the water table that surrounds the landfill site lower than the natural level. The surrounding area consists of 15 ft of saturated, homogeneous soil over an impervious rock layer. The water table, originally at the surface, has been lowered to a depth of 9 ft. Fully pervious parallel collector drains at a depth of 12 ft are present. The hydraulic conductivity of the soil is 0.23 ft/hr. The natural water table will recharge the site in 30 days if drainage stops. What collector drain separation is required?

Solution

Referring to Fig. 67.2(a),

$$a = 15\text{ ft} - 12\text{ ft} = 3\text{ ft}$$

$$b = 15\text{ ft} - 9\text{ ft} = 6\text{ ft}$$

$$K = \left(0.23\ \frac{\text{ft}}{\text{hr}}\right)\left(24\ \frac{\text{hr}}{\text{day}}\right) = 5.52\text{ ft/day}$$

To maintain the lowered water table level, the drainage rate must equal the recharge rate.

$$Q = \frac{9\text{ ft}}{30\text{ days}} = 0.3\text{ ft/day}$$

Use Eq. 67.6 to find the drain spacing.

$$\begin{aligned} L &= 2\sqrt{\left(\frac{K}{Q}\right)(b^2 - a^2)} \\ &= 2\sqrt{\left(\frac{5.52\ \frac{\text{ft}}{\text{day}}}{0.3\ \frac{\text{ft}}{\text{day}}}\right)\left((6\text{ ft})^2 - (3\text{ ft})^2\right)} \\ &= 44.57\text{ ft} \quad (45\text{ ft}) \end{aligned}$$

Plant Engineering

29. LEACHATE RECOVERY SYSTEMS

At least two distinct leachate recovery systems are required in landfills: one within the landfill to limit the hydraulic head of leachate that has reached the top liner, and another to catch the leachate that has passed through the top liner and drainage layer and has reached the bottom liner.

By removing leachate at the first liner, the hydrostatic pressure on the liner is reduced, minimizing the pressure gradient and hydraulic movement through the liner. A pump is used to raise the collected leachate to the surface once a predetermined level has been reached. Tracer compounds (e.g., lithium compounds or radioactive hydrogen) can be buried with the wastes to signal migration and leakage.

Leachate collection and recovery systems fail because of clogged drainage layers and pipe, crushed collection pipes due to waste load, pump failures, and faulty design.

30. LEACHATE TREATMENT

Leachate is essentially a very strong municipal wastewater, and it tends to become more concentrated as the landfill ages. Leachate from landfills contains extremely high concentrations of compounds and cannot be discharged directly into rivers or other water sources.

Leachate is treated with biological (i.e., trickling filter and activated sludge) and physical/chemical processes very similar to those used in wastewater treatment plants. For large landfills, these treatment facilities are located on the landfill site. A typical large landfill treatment facility would include an equalization tank, a primary clarifier, a first-stage activated sludge aerator and clarifier, a second-stage activated sludge aerator and clarifier, and a rapid sand filter. Additional equipment for sludge dewatering and digestion would also be required. Liquid effluent would be discharged to the municipal wastewater treatment plant.

31. LEAD

Lead, even in low concentrations, is toxic. Inhaled lead accumulates in the blood, bones, and vital organs. It can produce stomach cramps, fatigue, aches, and nausea. It causes irreparable damage to the brain, particularly in young and unborn children, and high blood pressure in adults. At high concentrations, lead damages the nervous system and can be fatal.

Lead was outlawed in paint in the late 1970s.[21] Lead has also been removed from most gasoline blends. Lead continues to be used in large quantities in automobile batteries and plating and metal-finishing operations. However, these manufacturing operations are tightly regulated. Lead enters the atmosphere during the combustion of fossil fuels and the smelting of sulfide ore. Lead enters lakes and streams primarily from acid mine drainage.

For lead in industrial and municipal wastewater, current remediation methods include pH adjustment with lime or alkali hydroxides, coagulation-sedimentation, reverse osmosis, and zeolite ion exchange.

32. MUNICIPAL SOLID WASTE

Municipal solid waste (MSW, previously known as *garbage*) has traditionally been disposed of in *municipal solid waste landfills* (MSWLs, previously referred to as *dumps*).[22] Waste is placed in layers, typically 2 ft to 3 ft thick (0.6 m to 0.9 m), and is compacted to the cell height before soil is added as a cover. Two to five passes by a tracked bulldozer are sufficient to compact the MSW to 800 lbm/yd^3 to 1500 lbm/yd^3 (470 kg/m^3 to 890 kg/m^3). (A density of 1000 lbm/ft^3 or 590 kg/m^3 is used in design studies.) The *lift* is the height of the covered layer, as shown in Fig. 67.3. When the landfill layer is full, the ratio of solid waste volume to soil cover volume will be approximately between 4:1 and 3:1.

Figure 67.3 *Landfill Cells*

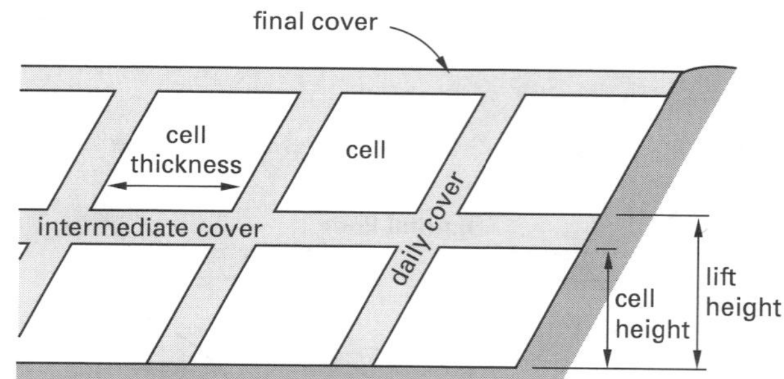

The *cell height* is typically taken as 8 ft (2.4 m) for design studies, although it can actually be much higher. The height should be chosen to minimize the cover material requirement consistent with the regulatory requirements. Cell slopes will be less than 40°, and typically less than 30°.

MSW is generated in the United States at the average rate of about 5 lbm/capita-day to 8 lbm/capita-day (2.3 kg/capita-day to 3.6 kg/capita-day).[23] (Characteristics

[21]Workers can be exposed to lead if they strip away lead-based paint, demolish old buildings, or weld or cut metals that have been coated with lead-based paint.

[22]In 1988, 73% of the nation's MSW was landfilled, 14% was incinerated, and 13% was recycled. In 2010, 54% was landfilled, 12% was incinerated with energy recovery, and 34% was recycled.

[23]In 2010, the EPA reported that the generation rate was 4.43 lbm/capita-day. 5 lbm/capita-day is a conservative estimate commonly used in design studies.

of MSW in the United States are given in Table 67.5.) The daily increase in landfill volume is predicted by Eq. 67.11.[24] N is the population size, G is the per-capita MSW generation rate, and γ is the landfill overall specific weight, calculated as a weighted average of soil and compacted MSW specific weights.[25]

$$\Delta V_{\text{day}} = \frac{NG(\text{LF})}{\gamma} \qquad 67.11$$

Table 67.5 Typical Nationwide Characteristics of MSW

component	percentage by weight
paper	40
yard waste	18
glass	8
plastic	7
steel	7
food waste	6
aluminum	2
other	12

The *loading factor*, LF, is 1.25 for a 4:1 volumetric ratio and is calculated from for other ratios.

$$\text{LF} = \frac{V_{\text{MSW}} + V_{\text{cover soil}}}{V_{\text{MSW}}} \qquad 67.12$$

Suitable (economical and safe) landfill sites are becoming difficult to find, and once identified, are objected to by residents near the site and in the MSW transport corridor. This is referred to as the *NIMBY* (not in my backyard) *syndrome*.[26]

33. NITROGEN OXIDES

Nitrogen oxides (NOx) are one of the primary causes of smog formation. NOx from the combustion of coal is primarily *nitric oxide* (NO) with small quantities of *nitrogen dioxide* (NO_2).[27] NO_2 can be a primary or a secondary pollutant. Although some NO_2 is emitted directly from combustion sources, it is also produced from the oxidation of nitric oxide, NO, in the atmosphere.

NOx is produced in two ways: (a) *thermal NOx* produced at high temperatures from free nitrogen and oxygen, and (b) *fuel NOx* (or *fuel-bound NOx*) formed from the decomposition/combustion of fuel-bound nitrogen.[28] When natural gas and light distillate oil are burned, almost all of the NOx produced is thermal. Residual fuel oil, however, can have a nitrogen content as high as 0.3%, and 50–60% of this can be converted to NOx. Coal has an even higher nitrogen content.[29]

Thermal NOx is usually produced in small (but significant) quantities when excess oxygen is present at the highest temperature point in a furnace, such that nitrogen (N_2) can dissociate.[30] Dissociation of N_2 and O_2 is negligible, and little or no thermal NOx is produced below approximately 3000°F (1650°C).

Formation of thermal NOx is approximately exponential with temperature. For this reason, many NOx-reduction techniques attempt to reduce the *peak flame temperature* (PFT). NOx formation can also be reduced by injecting urea or ammonia reagents directly into the furnace.[31] The relationship between NOx production and excess oxygen is complex, but NOx production appears to vary directly with the square root of the oxygen concentration.

In existing plants, retrofit NOx-reduction techniques include fuel-rich combustion (i.e., staged air burners), flue gas recirculation, changing to a low-nitrogen fuel, reduced air preheat, installing low-NOx burners, and using overfire air.[32] Low-NOx burners using controlled flow/split flame or internal fuel staging technology are essentially direct-replacement (i.e., "plug-in") units, differing from the original burners primarily in nozzle configuration. However, some fuel supply and air modifications may also be needed. Use of overfire air requires windbox modifications and separate ducts.

Plant Engineering

[24]Multiply cubic yards by 27/43,560 to get acre-feet (ac-ft).
[25]Soils have specific weights between 70 lbf/ft^3 and 130 lbf/ft^3 (densities between 1100 kg/w^3 and 2100 kg/m^3). A value of 100 lbf/ft^3 (1600 kg/m^3) is commonly used in design studies.
[26]In fact, NIMBY applies to landfills, incinerators, and any commercial or processing plant dealing with hazardous wastes.
[27]Other oxides are produced in insignificant amounts. These include *nitrous oxide* (N_2O), N_2O_4, N_2O_5, and NO_3, all of which are eventually oxidized to (and reported as) NO_2.
[28]Some engineers further divide the production of fuel-bound NOx into low- and high-temperature processes and declare a third fuel-related NOx-production method known as *prompt-NOx*. Prompt-NOx is the generation of the first 15–20 ppm of NOx from partial combustion of the fuel at lower temperatures.
[29]In order of increasing fuel-bound NOx production potential, common boiler fuels are: methanol, ethanol, natural gas, propane, butane, ultralow-nitrogen fuel oil, fuel oil No. 2, fuel oil No. 6, and coal.
[30]The *mean residence time* at the high temperature points of the combustion gases is also an important parameter. At temperatures below 2500°F (1370°C), several minutes of exposure may be required to generate any significant quantities of NOx. At temperatures above 3000°F (1650°C), dissociation can occur in less than a second. At the highest temperatures—3600°F (1980°C) and above—dissociation takes less than a tenth of a second.
[31]*Urea* (NH_2CONH_2), also known as *carbamide urea*, is a water-soluble organic compound prepared from ammonia. Urea has significant biological and industrial usefulness.
[32]Air is injected into the furnace at high velocity over the combustion bed to create turbulence and to provide oxygen.

Lime spray-dryer scrubbers of the type found in electrical generating plants do not remove all of the NOx. Further reduction requires that the remaining NOx be destroyed. Reburn, selective catalytic reduction (SCR), and selective noncatalytic methods are required.

Scrubbing, incineration, and other end-of-pipe methods typically have been used to reduce NOx emissions from stationary sources such as gas turbine-generators, although these methods are unwieldy for the smallest units. Water/steam injection, catalytic combustion, and *selective catalytic reduction* (SCR) are particularly suited for gas-turbine and combined-cycle installations. SCR is also effective in NOx reduction in all heater and boiler applications.

The volumetric fraction (in ppm) of NOx in the flue gas is calculated from the molecular weight and mass flow rates. The molecular weight of NOx is assumed to be 46.[33] The ratio of $\dot{m}_{NOx}/\dot{m}_{fluegas}$, percentage of water vapor, and flue gas molecular weight are derived from flue gas analysis.

$$B_{NOx} = \left(\frac{\dot{m}_{NOx}}{\dot{m}_{flue\,gas}}\right)(10^6)\left(\frac{M_{flue\,gas}}{M_{NOx}}\right) \times \left(\frac{100\%}{100\% - \%H_2O}\right) \quad 67.13$$

Since the apparent concentration of NOx could be arbitrarily decreased without reducing the NOx production rate by simply diluting the combustion gas with excess air, it is common to correct NOx readings to 3% O_2 by volume, dry basis. (This is indicated by the units *ppmvd*.) Standardizing is accomplished by multiplying the measured NOx reading (in ppmvd) by the O_2 correction factor from Eq. 67.14. (Corrections to 7%, 12%, and 15% oxygen content are also used. For a correction to any other percentage of O_2, replace the 3% in Eq. 67.14 with the new value.)

$$O_2\text{ correction factor} = \frac{21\% - 3\%}{21\% - \%O_2} = \frac{18\%}{21\% - \%O_2} \quad 67.14$$

If the flue gas analysis is on a wet basis, Eq. 67.15 can be used to calculate the multiplicative factor.

$$O_2\text{ correction factor} = \frac{21\% - 3\%}{21\% - \left(\frac{100\%}{100\% - \%H_2O}\right)(\%O_2)} = \frac{18\%}{21\% - \left(\frac{100\%}{100\% - \%H_2O}\right)(\%O_2)} \quad 67.15$$

Some regulations specify NOx limitations in terms of pounds per hour or in terms of pounds of NOx per million Btus (MMBtu) of gross heat released. The mass emission rate, E, in lbm/MMBtu, is calculated from the concentration in pounds per dry, standard cubic foot (dscf).

$$E_{lbm/MMBtu} = \frac{C_{lbm/dscf}V_{NOx,dscf/hr}}{q_{MMBtu/hr}} = \frac{C_{lbm/dscf}(100)\,V_{th,CO_2}}{C_{m,CO_2,\%}(HHV_{MMBtu/lbm})} \quad 67.16$$

The relationship between the NOx concentrations, C, expressed in pounds per dry standard volume and ppm can be calculated from Eq. 67.17.[34] Conversions between ppm and pounds are made assuming NOx has a molecular weight of 46.

$$C_{g/m^3} = (4.15 \times 10^{-5})C_{ppm}M \quad \text{[SI]} \quad 67.17(a)$$

$$C_{lbm/ft^3} = (2.59 \times 10^{-9})C_{ppm}M \quad \text{[U.S.]} \quad 67.17(b)$$

The theoretical volume of carbon dioxide, V_{th,CO_2}, produced can be determined stoichiometrically. However, Table 67.6 can be used for quick estimates with an accuracy of approximately ±5.9%.

Table 67.6 *Approximate CO_2 Production for Various Fuels (with 0% excess air)*

fuel	standard* ft³/10⁶ Btu	standard* m³/10⁶ cal
coal, anthracite	1980	0.222
coal, bituminous	1810	0.203
coal, lignite	1810	0.203
gas, butane	1260	0.412
gas, natural	1040	0.117
gas, propane	1200	0.135
oil	1430	0.161

*Standard conditions are 70°F (21°C) and 1 atm pressure.

Equation 67.16 is based on the theoretical volume of carbon dioxide gas produced per pound of fuel. Other approximations and correlations are based on the total dry volume of the flue gas in standard cubic feet per million Btu at 3% oxygen, $V_{t,dry}$. For quick estimates on

[33] 46 is the molecular weight of NO_2. The predominant oxide in the flue gas is NO, and NO_2 may be only 10% to 15% of the total NOx. Furthermore, NO is measured, not NO_2. However, NO has a short half-life and is quickly oxidized to NO_2 in the atmosphere.
[34] The constants in Eq. 67.17 are the same and can be used for any gas.

furnaces burning natural gas, propane, and butane, V_t is approximately 10,130 ft^3/MMBtu; for fuel oil, V_t is approximately 10,680 ft^3/MMBtu.

$$E_{\text{lbm/MMBtu}} = (C_{\text{ppmv at 3\% O}_2})(V_{t,\text{dry}}) \times \left(\frac{46 \ \frac{\text{lbm}}{\text{lbmol}}}{\left(379.3 \ \frac{\text{ft}^3}{\text{lbmol}}\right)(10^6)} \right) \quad \text{67.18}$$

As with all pollutants, the maximum allowable concentration or discharge of NOx is subject to continuous review and revision. Actual limits may depend on the type of geographical location, type of fuel, size of the facility, and so on. For steam/electrical (gas turbine) plants, the general target is 25 to 40 ppm. The lower values apply to combustion of natural gas, and the higher values apply to combustion of distillate oil. Even lower values (down to 9 ppm to 10 ppm) are imposed in some areas.

Example 67.5

A combustion turbine produces 25 lbm/hr of NOx while generating 550,000 lbm/hr of exhaust gases. The exhaust gas contains 10% water vapor and 11% oxygen by volume. Assume the molecular weight of NOx is 46 lbm/lbmol. What is the NOx concentration in ppm, dry volume basis, corrected to 3% oxygen?

Solution

Some of the gas analysis is missing, so the flue gas is assumed to be mostly nitrogen (the largest component of air) with a molecular weight of 28 lbm/lbmol.

The uncorrected, dry NOx volumetric fraction is

$$\begin{aligned} B_{\text{NOx}} &= \left(\frac{\dot{m}_{\text{NOx}}}{\dot{m}_{\text{flue gas}}}\right) \times \left(\frac{M_{\text{flue gas}}}{M_{\text{NOx}}}\right)(10^6)\left(\frac{100\%}{100\% - \%\text{H}_2\text{O}}\right) \\ &= \left(\frac{25 \ \frac{\text{lbm}}{\text{hr}}}{550{,}000 \ \frac{\text{lbm}}{\text{hr}}}\right)\left(\frac{28 \ \frac{\text{lbm}}{\text{lbmol}}}{46 \ \frac{\text{lbm}}{\text{lbmol}}}\right) \times (10^6)\left(\frac{100\%}{100\% - 10\%}\right) \\ &= 30.7 \text{ ppmvd} \quad [\text{uncorrected}] \end{aligned}$$

Equation 67.14 corrects the value to 3% oxygen.

$$\begin{aligned} \text{O}_2 \text{ correction factor} &= \frac{21\% - 3\%}{21\% - \%\text{O}_2} \\ &= \frac{18\%}{21\% - 11\%} \\ &= 1.8 \end{aligned}$$

The corrected value is

$$C = (1.8)(30.7 \text{ ppmvd}) = 55.3 \text{ ppmvd}$$

34. ODORS

Odors of unregulated substances can be eliminated at their source, contained by sealing and covering, diluted to unnoticeable levels with clean air, or removed by simple water washing, chemical scrubbing (using acid, alkali, or sodium hypochlorite), bioremediation, and activated carbon adsorption.

35. OIL SPILLS IN NAVIGABLE WATERS

Intentional and accidental releases of oil in navigable waters are prohibited. Deleterious effects of such spills include large-scale biological (i.e., sea life and wildlife) damage and destruction of scenic and recreational sites. Long-term toxicity can be harmful for microorganisms that normally live in the sediment.

36. OZONE

Ground-level ozone is a secondary pollutant. Ozone is not usually emitted directly, but is formed from hydrocarbons and nitrogen oxides (NO and NO_2) in the presence of sunlight. *Oxidants* are by-products of reactions between combustion products. Nitrogen oxides react with other organic substances (e.g., hydrocarbons) to form the oxidants ozone and peroxyacyl nitrates (PAN) in complex *photochemical reactions*. Ozone and PAN are usually considered to be the major components of smog.[35] (See also Sec. 67.44.)

37. PARTICULATE MATTER

Particulate matter (PM), also known as *aerosols*, is defined as all particles that are emitted by a combustion source. Particulate matter with aerodynamic diameters of less than or equal to a nominal 10 μm is known as *PM-10*. Particulate matter is generally inorganic in nature. It can be categorized into metals (or heavy metals), acids, bases, salts, and nonmetallic inorganics.

[35] The term *oxidant* sometimes is used to mean the original emission products of NOx and hydrocarbons.

Metallic inorganic PM from incinerators is controlled with baghouses or electrostatic precipitators (ESPs), while nonmetallic inorganics are removed by scrubbing (wet absorption). Flue gas PM, such as fly ash and lime particles from desulfurization processes, can be removed by fabric baghouses and electrostatic precipitators. These processes must be used with other processes that remove NOx and SO_2.

High temperatures cause the average flue gas particle to decrease in size toward or below 10 μm. With incineration, emission of trace metals into the atmosphere increases significantly. Because of this, incinerators should not operate above 1650°F (900°C).[36]

38. PCBs

Polychlorinated biphenyls (PCBs) are organic compounds (i.e., *chlorinated organics*) manufactured in oily liquid and solid forms through the late 1970s, and subsequently prohibited. PCBs are carcinogenic and can cause skin lesions and reproductive problems. PCBs build up, rather than dissipate, in the food chain, accumulating in fatty tissues. Most PCBs were used as dielectric and insulating liquids in large electrical transformers and capacitors, and in ballasts for fluorescent lights (which contain capacitors). PCBs were also used as heat transfer and hydraulic fluids, as dye carriers in carbonless copy paper, and as plasticizers in paints, adhesives, and caulking compounds.

Incineration of PCB liquids and PCB-contaminated materials (usually soil) has long been used as an effective mediation technique. Removal and landfilling of contaminated soil is expensive and regulated, but may be appropriate for quick cleanups. In addition to incineration and other thermal destruction processes, methods used to routinely remediate PCB-laden soils include biodegradation, ex situ thermal desorption and soil washing, in situ vaporization by heating or steam injection and subject vacuum extraction, and stabilization to prevent leaching. The application of quicklime (CaO) or high-calcium fly ash is now known to be ineffective.

Specialized PCB processes targeted at cleaning PCB from spent oil are also available. Final PCB concentrations are below detectable levels, and the cleaned oil can be recycled or used as fuel.

39. PESTICIDES

The term *traditional organochlorine pesticide* refers to a narrow group of persistent pesticides, including DDT, the "*drins*" (aldrin, endrin, and dieldrin), chlordane, endosulfan, hexachlorobenzene, lindane, mirex, and toxaphene. Traditional organochlorine insecticides used extensively between the 1950s and early 1970s have been widely banned because of their environmental persistence. There are, however, notable exceptions, and environmental levels of traditional organochlorine pesticides (especially DDT) are not necessarily declining throughout the world, especially in developing countries and countries with malaria. DDT, with its half-life of up to 60 years, does not always remain in the country where it is used. The semi-volatile nature of the chemicals means that at high temperatures they will tend to evaporate from the land, only to condense in cooler air. This *global distillation* is thought to be responsible for levels of organochlorines increasing in the Arctic.

The term *chlorinated pesticides* refers to a much wider group of insecticides, fungicides, and herbicides that contain organically bound chlorine. A major difference between traditional organochlorine and other chlorinated pesticides is that the former have been perceived to have high persistence and build up in the food chain, and the latter do not. However, even some chlorinated pesticides are persistent in the environment. There is little information available concerning the overall environmental impact of chlorinated pesticides. Far more studies exist concerning the effects of traditional organochlorines in the public domain than on chlorinated pesticides in general. Pesticides that have, for example, active organophosphate (OP), carbamate, or triazine parts of their molecules are chlorinated pesticides and may pose long-lasting environmental dangers.

In the United States, about 30–40% of pesticides are chlorinated. All the top five pesticides are chlorinated. Worldwide, half of the 10 top-selling herbicides are chlorinated (alachlor, metolachlor, 2,4-D, cyanazine, and atrazine). Four of the top 10 insecticides are chlorinated (chlorpyrifos, fenvalerate, endosulfan, and cypermethrin). Four (propiconazole, chlorothalonil, prochloraz, and triadimenol) of the 10 most popular fungicides are chlorinated.

40. PLASTICS

Plastics, of which there are six main chemical polymers as given in Table 67.7, generally do not degrade once disposed of and are considered a disposal issue. Disposal is not a problem per se, however, since plastics are lightweight, inert, and do not harm the environment when discarded.

A distinction is made between *biodegrading* and *recycling*. Most plastics, such as the polyethylene bags used to protect pressed shirts from the dry cleaner and to mail some magazines, are not biodegradable but are recyclable. Also, all plastics can be burned for their fuel value.

[36]Incineration of biosolids (sewage sludge) concentrates trace metals into the combustion ash, which is subsequently landfilled. Due to the high initial investment required, difficulty in using combustion energy, and increasingly stringent air quality and other environmental regulations, biosolid incineration is not widespread.

Table 67.7 Polymers

polymer	plastic ID number	common use
polyethylene terephthalate (PET)	1	clear beverage containers
high-density polyethylene (HDPE)	2	detergent; milk bottles; oil containers; toys
polyvinyl chloride (PVC)	3	clear bottles
low-density polyethylene (LDPE)	4	grocery bags; food wrap
polypropylene (PP)	5	labels; bottles; housewares
polystyrene (PS)	6	styrofoam cups; "clam shell" food containers

The collection and sorting problems often render low-volume plastic recycling efforts uneconomical. Complicating the drive toward recycling is the fact that many of the six different types cannot easily be distinguished visually, and they cannot be recycled successfully when intermixed. Also, some plastic products consist of layers of different polymers that cannot be separated mechanically. Some plastic products are marked with a plastic type identification number.

Sorting in low-volume applications is performed visually and manually. Commercial high-volume methods include hydrocycloning, flotation with flocculation (for all polymers), X-ray fluorescence (primarily for PVC detection), and near-infrared (NIR) spectroscopy (primarily for separating PVC, PET, PP, PE, and PS). Mass NIR spectroscopy is also promising, but has yet to be commercialized.

Unsorted plastics can be melted and reformed into some low-value products. This operation is known as *down-cycling*, since each successful cycle further degrades the material. This method is suitable only for a small fraction of the overall recyclable plastic.

Other operations that can reuse the compounds found in plastic products include hydrogenation, pyrolysis, and gasification. *Hydrogenation* is the conversion of mixed plastic scrap to "syncrude" (synthetic crude oil), in a high-temperature (i.e., 750–880°F (400–470°C)), high-pressure (i.e., 2200–4400 psig (15–30 MPa)), hydrogen-rich atmosphere. Since the end product is a crude oil substitute, hydrogenation operations must be integrated into refinery or petrochemical operations.

Gasification and pyrolysis are stand-alone operations that do not require integration with a refinery. *Pyrolysis* takes place in a fluidized bed between 750°F and 1475°F (400°C and 800°C). Cracked polymer gas or other inert gas fluidizes the sand bed, which promotes good mixing and heat transfer, resulting in liquid and gaseous petroleum products.

Gasification operates at higher temperatures, 1650–3600°F (900–2000°C), and lower pressures, around 870 psig (6 MPa). The waste stream is pyrolyzed at lower temperatures before being processed by the gasifier. The gas can be used on-site to generate steam. Gasification has the added advantage of being able to treat the entire municipal solid waste stream, avoiding the need for sorting plastics.

Biodegradable plastics have focused on polymers that are derived from agricultural sources (e.g., corn, potato, tapioca, and soybean starches), rather than from petroleum. *Bioplastics* have various degrees of *biodegradation* (i.e., breaking down into carbon dioxide, water, and biomass), *disintegration* (losing their shapes and identities and becoming invisible in the compost without needing to be screened out), and *eco-toxicity* (containing no toxic material that prevents plant growth in the compost). A bioplastic that satisfies all three characteristics is a *compostable plastic.* A *biodegradable plastic* will eventually be acted on by naturally occurring bacteria or fungi, but the time required is indeterminate. Also, biodegradable plastics may leave some toxic components. A *degradable plastic* will experience a significant change in its chemical structure and properties under specific environmental conditions, although it may not be affected by bacteria (i.e., be biodegradable) or satisfy any of the requirements for a compostable plastic.

Some engineers point out that biodegrading is not even a desirable characteristic for plastics and that being nonbiodegradable is not harmful. Biodegrading converts materials (such as the paper bags often preferred over plastic bags) to water and carbon dioxide, contributing to the greenhouse effect without even receiving the energy benefit of incineration. Biodegrading of most substances also results in gases and leachates that can be more harmful to the environment than the original substance. In a landfill, biodegrading serves no useful purpose, since the space occupied by the degraded plastic does not create additional useful space (volume).

41. RADON

Radon gas is a radioactive gas produced from the natural decay of radium within the rocks beneath a building. Radon accumulates in unventilated areas (e.g., basements), in stagnant water, and in air pockets formed when the ground settles beneath building slabs. Radon also can be brought into the home by radon-saturated well water used in baths and showers. The EPA's action level of 4 pCi/L for radon in air is contested by many as being too high.

Radon mitigation methods include (a) pressurizing to prevent the infiltration of radon, (b) installing depressurization systems to intercept radon below grade and vent it safely, (c) removing radon-producing soil, and (d) abandoning radon-producing sites.

42. RAINWATER RUNOFF

Rainwater percolating through piles of coal, fly ash, mine tailings, and other stored substances can absorb toxic compounds and eventually make its way into the earth, possibly contaminating soil and underground aquifers.

43. RAINWATER RUNOFF FROM HIGHWAYS

The *first flush* of a storm is generally considered to be the first half-inch of storm runoff or the runoff from the first 15 min of the storm. Along highways and other paved transportation corridors, the first flush contains potent pollutants such as petroleum products, asbestos fibers from brake pads, tire rubber, and fine metal dust from wearing parts. Under the National Pollutant Discharge Elimination System (NPDES), stormwater runoff in newly developed watersheds must be cleaned before it reaches existing drainage facilities, and runoff must be maintained at or below the present undeveloped runoff rate.

A good stormwater system design generally contains two separate basins or a single basin with two discrete compartments. The function of one compartment is water quality control, and the function of the other is peak runoff control.

The *water quality compartment* (WQC) should normally have sufficient volume and discharge rates to provide a minimum of one hour of detention time for 90–100% of the first flush volume. "Treatment" in the WQC consists of sedimentation of suspended solids and evaporation of volatiles. A removal goal of 75% of the suspended solids is reasonable in all but the most environmentally sensitive areas.

In environmentally sensitive areas, a filter berm of sand, a sand chamber, or a sand filter bed can further clarify the discharge from the WQC.

After the WQC becomes full from the first flush, subsequent runoff will be diverted to the *peak-discharge compartment* (PDC). This is done by designing a junction structure with an inlet for the incoming runoff and separate outlets from each compartment. If the elevation of the WQC outlet is lower than the inlet to the PDF, the first flush will be retained in the WQC.

Sediment from the WQC chamber and any filters should be removed every one to three years, or as required. The sediment must be properly handled, as it may be considered to be hazardous waste under the EPA's "mixture" and "derived-from" rules and its "contained-in" policy.

When designing chambers and filters, sizing should accommodate the first flush of a 100-year storm. The top of the berm between the WQC and PDC should include a minimum of 1 ft (0.3 m) of freeboard. In all cases, an emergency overflow weir should allow a storm greater than the design storm to discharge into the PDC or receiving water course.

Minimum chamber and berm width is approximately 8 ft (2.4 m). Optimum water depth in each chamber is approximately 2–5 ft (0.6–1.5 m). Each compartment can be sized by calculating the divided flows and staging each compartment. The outlet from the WQC should be sufficient to empty the compartment in approximately 24–28 hours after a 25-year storm.

A *filter berm* is essentially a sand layer between the WQC and the receiving chamber that filters water as it flows between the two compartments. The filter should be constructed as a layer of sand placed on geosynthetic fabric, protected with another sheet of geosynthetic fabric, and covered with coarse gravel, another geosynthetic cover, and finally a layer of medium stone. The sand, gravel, and stone layers should all have a minimum thickness of 1 ft (0.3 m). The rate of permeability is controlled by the sand size and front-of-fill material. Permeability calculations should assume that 50% of the filter fabric is clogged.

A *filter chamber* consists of a concrete structure with a removable filter pack. The filter pack consists of geosynthetic fabric wrapped around a plastic frame (core) that can be removed for backwashing and maintenance. The filter is supported on a metal screen mounted in the concrete chamber. The outlet of the chamber should be located at least 1 ft (0.3 m) behind the filter pack. The opening's size will determine the discharge rate through the filter, which should be designed as less than 2 ft/sec (60 cm/s) assuming that 50% of the filter area is clogged.

A *sand filter bed* is similar to the filter beds used for tertiary sewage treatment. The sand filter consists of a series of 4 in (100 mm) perforated PVC pipes in a gravel bed. The gravel bed is covered by geotextile fabric and 8 in (200 mm) of fine-to-medium sand. The perforated underdrains lead to the outlet channel or chamber.

44. SMOG

Photochemical smog (usually, just *smog*) consists of ground-level ozone and peroxyacyl nitrates (PAN). Smog is produced by the sunlight-induced reaction of ozone *precursors*, primarily nitrogen dioxide (NOx), hydrocarbons, and volatile organic compounds (VOCs). NOx and hydrocarbons are emitted by combustion sources such as automobiles, refineries, and industrial boilers. VOCs are emitted by manufacturing processes, dry cleaners, gasoline stations, print shops, painting operations, and municipal wastewater treatment plants. (See also Sec. 67.36.)

45. SMOKE

Smoke results from incomplete combustion and indicates unacceptable combustion conditions. In addition to being a nuisance problem, smoke contributes to air pollution and reduced visibility. Smoke generation can be minimized by proper furnace monitoring and control.

Opacity can be measured by a variety of informal and formal methods, including transmissometers mounted on the stack. The sum of the *opacity* (the fraction of light blocked) and the *transmittance* (the fraction of light transmitted) is 1.0.

Optical density is calculated from Eq. 67.19. The *smoke spot number* (SSN) can also be used to quantify smoke levels.

$$\text{optical density} = \log_{10}\frac{1}{1 - \text{opacity}} \qquad 67.19$$

Visible moisture plumes with opacities of 40% are common at large steam generators even when there are no unburned hydrocarbons emitted. High-sulfur fuels and the presence of ammonium chloride (a by-product of some ammonia-injection processes) seem to increase formation of visible plumes. Moisture plumes from saturated gas streams can be avoided by reheating prior to discharge to the atmosphere.

46. SPILLS, HAZARDOUS

Contamination by a hazardous material can occur accidentally (e.g., a spill) or intentionally (e.g., a previously used chemical-holding lagoon). Soil that has been contaminated with hazardous materials from spills or leaks from *underground storage tanks* (commonly known as *UST wastes*) is itself a hazardous waste.

The type of waste determines what laws are applicable, what permits are required, and what remediation methods are used. With contaminated soil, spilled substances can be (a) solid and nonhazardous or (b) nonhazardous liquid petroleum products (e.g., "UST nonhazardous"), and Resource Conservation and Recovery Act- (RCRA-) listed (c) hazardous substances, and (d) toxic substances.

Cleaning up a hazardous waste requires removing the waste from whatever air, soil, and water (lakes, rivers, and oceans) have been contaminated. The term *remediation* is used to mean the corrective steps taken to return the environment to its original condition. *Stabilization* refers to the act of reducing the waste concentrations to lower levels so that the waste can be transported, stored, or landfilled.

Remediation methods are classified as available or innovative. *Available methods* can be implemented immediately without being further tested. *Innovative methods* are new, unproven methods in various stages of study.

The remediation method used depends on the waste type. The two most common available methods are incineration and landfilling after stabilization.[37] Incineration can occur in rotary kilns, injection incinerators, infrared incineration, and fluidized-bed combustors. Landfilling requires the contaminated soil to be stabilized chemically or by other means prior to disposal. Technologies for general VOC-contaminated soil include vacuum extraction, bioremediation, thermal desorption, and soil washing.

47. SULFUR OXIDES

Sulfur oxides (SOx), consisting of *sulfur dioxide* (SO_2) and *sulfur trioxide* (SO_3), are the primary cause of acid rain. *Sulfurous acid* (H_2SO_3) and sulfuric acid (H_2SO_4) are produced when oxides of sulfur react with moisture in the flue gas. Both of these acids are corrosive.

$$SO_2 + H_2O \rightarrow H_2SO_3 \qquad 67.20$$

$$SO_3 + H_2O \rightarrow H_2SO_4 \qquad 67.21$$

Fuel switching (*coal substitution/blending* (CS/B)) is the burning of low-sulfur fuel. (Low-sulfur fuels are approximately 0.25–0.65% sulfur by weight, compared to high-sulfur coals with 2.4–3.5% sulfur.) However, unlike nitrogen oxides, which can be prevented during combustion, formation of sulfur oxides cannot be avoided when low-cost, high-sulfur fuels are burned.

Some air quality regulations regarding SOx production may be met by a combination of options. These options include fuel switching, flue gas scrubbing, derating, and allowance trading. The most economical blend of these options will vary from location to location.

In addition to fuel switching, available technology options for retrofitting existing coal-fired plants include wet scrubbing, dry scrubbing, sorbent injection, repowering with clean coal technology (CCT), and co-firing with natural gas.

Sulfur dioxide (like carbon dioxide and nitrogen oxides) emissions must be reported on a standardized basis. Equation 67.22 can be used to calculate the volumetric, dry basis concentration from a wet stack gas sample.

$$\frac{SO_{2,\text{dry, at }3\%\,O_2}}{SO_{2,\text{wet, at stack }O_2}} = \frac{CO_{2,\text{th, dry, at }3\%\,O_2}}{CO_{2,\text{wet, at stack }O_2}} \qquad 67.22$$

[37]Other technologies include in situ and ex situ bioremediation, chemical treatment, in situ flushing, in situ vitrification, soil vapor extraction, soil washing, solvent extraction, and thermal desorption.

Plant Engineering

The sulfur dioxide concentration (mass per dry standard volume, C) can be calculated from sulfur dioxide's molecular weight ($M = 64.07$) and Eq. 67.17.[38]

Example 67.6

A wet stack gas analysis shows 230 ppm of SO_2 and 10.5% CO_2. Based on stoichiometric combustion, the theoretical carbon dioxide percentage at 3% oxygen should have been 13.4%. What is the SO_2 concentration in ppm on a dry basis, standardized to 3% oxygen?

Solution

Use Eq. 67.22.

$$\begin{aligned} SO_{2,\text{dry,at }3\%\,O_2} &= SO_{2,\text{wet,at stack }O_2} \\ &\quad \times \frac{CO_{2,\text{th,dry,at }3\%\,O_2}}{CO_{2,\text{wet,at stack }O_2}} \\ &= \frac{(230 \text{ ppm})(0.134)}{0.105} \\ &= 294 \text{ ppm} \end{aligned}$$

48. TIRES

Discarded tires are more than a disposal problem. At 15,000 Btu/lbm (35 MJ/kg), their energy content is nearly 80% that of crude oil. Discarded tires are wasted energy resources. While tires can be incinerated, other processes can be used to gasify them to produce clean synthetic gas (i.e., *syngas*). The low-sulfur, hydrogen-rich syngas can subsequently be burned in combined-cycle plants or used as a feedstock for ammonia and methanol production, or the hydrogen can be recovered for separate use. Tires can also be converted in a non-chemical process into a strong asphalt-rubber pavement.

49. TRIHALOMETHANES

Trihalomethanes (THMs) are halogenated disinfection by-products (DBPs). These organic chemicals are produced during the disinfection of water and, therefore, constitute a drinking water quality problem. The chemically active elements of chlorine, iodine, and bromine react with various *organic precursors* to produce THMs.

Chlorine in various forms is widely used to disinfect water. Bromine can be present in gaseous chlorine as an impurity. Bromine also results from reacting chlorine with the bromide present in high-salinity water. Iodine is seldom used in disinfection. Therefore, only four THMs are found in significant quantities: trichloromethane, also known as chloroform ($CHCl_3$); tribromomethane, also known as bromoform ($CHBr_3$); bromodichloromethane ($CHBrCl_2$); and dibromochloromethane ($CHBr_2Cl$).

The *organic precursors* that react with chlorine to produce THMs are primarily naturally occurring (e.g., humic and fulvic acids from decaying vegetation). The precursors are not themselves harmful, but the THMs produced from them are presumed to be carcinogenic.[39]

Two main options exist for reducing THMs: formation avoidance and removal after formation. The first option includes (1) using ozone, chlorine dioxide, or potassium permanganate (60% to 90% THM formation reduction), (2) dechlorination after chlorination to prevent reaction of chlorine, and (3) adding ammonia to water prior to discharge to enhance chloramineformation, since chloramines suppress THM formation. The second option includes (1) using activated carbon to remove THMs (25% to 60% THM reduction), (2) using activated carbon, adsorbents, or zeolite filters to remove precursors, (3) moving the chlorination point to the end of the treatment process so that most precursors are removed in earlier processes prior to disinfection (70% to 75% THM reduction), (4) optimizing coagulation and settling water treatment processes to improve precursor removal, and (5) selecting water with fewer precursors.

Follow-up studies are necessary when changing to alternative disinfectants. An alternative disinfectant and its by-products should be evaluated to determine disinfecting power, residual power, toxicity, and other health effects. Costs of operation will increase when alternative disinfectants are used, although moving the point of application may not result in any significant increase in operating costs after a modest capital expenditure is made.

Ozone, frequently considered as an alternative disinfectant, is less expensive than chlorine dioxide, but it costs more than using chloramines or changing the point of chlorine application. An emerging consensus holds that ozonation alone is relatively ineffective (5% to 20% reduction at typical ozone doses) in controlling THM. Also, ozonation converts the precursors to *biodegradable organic matter* (BOM). Unless a biological process subsequently removes the BOM, aftergrowths in the distribution system may become a problem. Finally, ozonation does not leave any residual disinfectant in the water for use throughout the distribution system. Therefore, emphasis is on ozonation in combination with biological treatment to destroy precursors, followed by chlorine application for residual formation.

[38]See Ftn. 33.

[39]Tests have shown that chloroform in high doses is carcinogenic to rats. Other THMs are considered carcinogenic by association.

50. VOLATILE INORGANIC COMPOUNDS

Volatile inorganic compounds (VICs) include H_2S, NOx (except N_2O), SO_2, HCl, NH_3, and many other less common compounds.

51. VOLATILE ORGANIC COMPOUNDS

Volatile organic compounds (VOCs) (e.g., benzene, chloroform, formaldehyde, methylene chloride, napthalene, phenol, toluene, and trichloroethylene) are highly soluble in water. VOCs that have leaked from storage tanks or have been discharged often end up in groundwater and drinking supplies.

There are a large number of methods for removing VOCs, including incineration, chemical scrubbing with oxidants, water washing, air or steam stripping, activated charcoal adsorption processes, SCR (selective catalytic reduction), and bioremediation. Incineration of VOCs is fast and 99%+ effective, but incineration requires large amounts of fuel and produces NOx. Using SCR with heat recovery after incineration reduces the energy input but adds to the expense.

52. WATER VAPOR

Emitted water vapor is not generally considered to be a pollutant.

53. NOMENCLATURE

a	interfacial area per volume of packing	1/ft	1/m
A	area	ft^2	m^2
b	max table height above barrier	ft	m
B	volumetric fraction	–	–
C	concentration	various	various
D	flow depth	ft	μm
E	mass emission rate	lbm/MMBtu	kg/MJ
G	MSW generation rate per capita	lbm/day	kg/d
H	cyclone inlet height	ft	m
i	hydraulic gradient	ft/ft	m/m
K	hydraulic conductivity	ft/hr	cm/s
L	length	ft	m
LF	loading factor	–	–
$\dot{m}$	mass flow rate	lbm/hr	kg/h
M	molecular weight	lbm/lbmol	kg/kmol
N	population size	–	–
p	vapor pressure	lbf/in^2	kPa
q	heat generate rate or recharge rate	MMBtu/hr	kW
Q	total discharge per unit length of pipe	ft^3/ft-sec	m^3/m·s
Q	volumetric flow rate or recharge rate	ft^3/sec	m^3/s
T	temperature	°R	K
V	volume	ft^3	m^3

Symbols

γ	specific weight	lbf/ft^3	–

Subscripts

dp	dew point
s	sulfur
t	total
w	water

Plant Engineering

68 Storage and Disposition of Hazardous Materials

Content in blue refers to the *NCEES Handbook*.

1. GENERAL STORAGE

Storage of *hazardous materials* (*hazmats*) is often governed by local building codes in addition to state and federal regulations. Types of construction, maximum floor areas, and building layout may all be restricted.[1]

Good engineering judgment is called for in areas not specifically governed by the building code. Engineering consideration will need to be given to the following aspects of storage facility design: (1) spill containment provisions, (2) chemical resistance of construction and storage materials, (3) likelihood of and resistance to explosions, (4) exiting, (5) ventilation, (6) electrical design, (7) storage method, (8) personnel emergency equipment, (9) security, and (10) spill cleanup provisions.

2. STORAGE TANKS

Underground storage tanks (USTs) have traditionally been used to store bulk chemicals and petroleum products. Fire and explosion risks are low with USTs, but subsurface pollution is common since inspection is limited. Since 1988, the U.S. Environmental Protection Agency (EPA) has required USTs to have secondary containment, corrosion protection, and leak detection. UST operators also must carry insurance in an amount sufficient to clean up a tank failure.

Because of the cost of complying with UST legislation, above-ground storage tanks (ASTs) are becoming more popular. AST strengths and weaknesses are the reverse of USTs: ASTs reduce pollution caused by leaks, but the expected damage due to fire and explosion is greatly increased. Because of this, some local ordinances prohibit all ASTs for petroleum products.[2]

The following factors should be considered when deciding between USTs and ASTs: (1) space available, (2) zoning ordinances, (3) secondary containment, (4) leak-detection equipment, (5) operating limitations, and (6) economics.

Most ASTs are constructed of carbon or stainless steel. These provide better structural integrity and fire resistance than fiberglass-reinforced plastic and other composite tanks. Tanks can be either field-erected or factory-fabricated (capacities greater than approximately 50,000 gal (190 kL)). Factory-fabricated ASTs are usually designed according to UL-142 (Underwriters Laboratories *Standard for Safety*), which dictates steel type, wall thickness, and characteristics of compartments, bulkheads, and fittings. Most ASTs are not pres-surized, but those that are, and operate at pressures equal to or greater then 15 psig, must be designed in accordance with the ASME *Boiler and Pressure Vessel Code*, Section VIII.

NFPA 30 (*Flammable and Combustible Liquids Code*, National Fire Protection Association, Quincy, MA) specifies the minimum separation distances between ASTs, other tanks, structures, and public right-of-ways. The separation is a function of tank type, size, and con-tents. NFPA 30 also specifies installation, spill control, venting, and testing.

ASTs must be double-walled, concrete-encased, or contained in a dike or vault to prevent leaks and spills, and they must meet fire codes. Dikes should have a capacity in excess (e.g., 110% to 125%) of the tank volume. ASTs (as do USTs) must be equipped with overfill prevention systems. Piping should be above-ground wherever possi-ble. Reasonable protection against vandalism and hunters' bullets is also necessary.[3]

Though they are a good idea, leak-detection systems are not typically required for ASTs. Methodology for leak detection is evolving, but currently includes vacuum or pressure monitoring, electronic gauging, and optical and sniffing sensors. Double-walled tanks may also be fitted with sensors within the interstitial space.

Operationally, ASTs present special problems. In hot weather, volatile substances vaporize and represent an additional leak hazard. In cold weather, viscous content may need to be heated (often by steam tracing).

ASTs are not necessarily less expensive than USTs, but they are generally thought to be so. Additional hidden costs of regulatory compliance, secondary containment, fire protection, and land acquisition must also be considered.

[1]For example, flammable materials stored in rack systems are typically limited to heights of 25 ft (8.3 m).
[2]The American Society of Petroleum Operations Engineers (ASPOE) policy statement states, "Above-ground storage of liquid hydrocarbon motor fuels is inherently less safe than underground storage. Above-ground storage of Class 1 liquids (gasoline) should be prohibited at facilities open to the public."
[3]Approximately 20% of all spills from ASTs are caused by vandalism.

3. DISPOSITION OF HAZARDOUS WASTES

When a hazardous waste is disposed of, it must be taken to a registered *treatment*, *storage*, or *disposal facility* (TSDF). The EPA's *land ban* specifically prohibits the disposal of hazardous wastes on land prior to treatment. Incineration at sea is also prohibited. Waste must be treated to specific maximum concentration limits by specific technology prior to disposal in landfills.

Once treated to specific regulated concentrations, hazardous waste residues can be disposed of by incineration, by landfilling, or, less frequently, by deep-well injection. All disposal facilities must meet detailed design and operational standards.

Testing and Sampling

Content in blue refers to the *NCEES Handbook*.

1. EMISSIONS SAMPLING

There are numerous sampling procedures, as most pollutants have their own specific characteristics, nuances, and idiosyncracies.

Sampling of emissions in flue gases can either be continuous or by spot sampling using wet chemistry or whole air methods. With *wet chemistry sampling methods*, a sample is collected in a container, solid adsorbent, or liquid-impinger train. The sample is then moved from the field to the laboratory for analysis.

With *whole air sampling*, a volume of flue gas is collected in a bag-like container (e.g., a *Tedlar bag*), and the container is transported to the laboratory for analysis by gas chromatography (GC). Flame ionization detection (GC-FID) and a mass spectrophotometer (GC-MS) are both used. For some pollutants, residence time in the container is critical. Volatile organic compounds (VOCs) with short half-lives can be captured and tested in a sampling train (VOST), which allows for longer sample-holding times.

Particulates are detected and measured by EPA method 5, which, though complex, has become a de facto standard, or EPA method 17 (in-stack filtration). With method 5, the sampling device includes a series of absorbers connected in tandem. The absorption train is followed by a gas-drying tube and a vacuum pump. Several samples are usually taken. The amount of particulates is determined from the filter's weight increase. Particles in the sampling path are washed out, dried, and included in the weight. Stack gas moisture content is determined from the impinger train. Oxygen, carbon dioxide, and (by difference) nitrogen volumes are determined by analyzing a sample collected separately in a sample bag.

Tests of gas streams should be unbiased, which requires sampling to be isokinetic. The gas velocity and ratio of dry gas mass to pollutant mass are not disturbed by *isokinetic testing*.[1,2]

The number of samples taken depends on the proximity of the sampling point to upstream and downstream disturbances. The number of samples can be minimized by locating the sampling point such that flow is undisturbed by upstream and downstream changes of direction, changes in diameter, and equipment in the duct (i.e., after a "good straight run"). Sampling should be at least eight equivalent diameters downstream and two equivalent diameters upstream from disturbances in flow that would cause the flow pattern to be unsymmetrical (i.e., would cause swirling). For rectangular ducts, the equivalent diameter is

$$D_{\text{eq}} = \frac{(2)(\text{height})(\text{width})}{\text{height} + \text{width}} \qquad 69.1$$

Since a pollutant may not be dispersed evenly across the duct, the average concentration must be determined by taking multiple samples across the flow cross section. This process is known as *traversing* the area, and the sampling points are known as *traverse points*. For circular ducts, the cross section is divided into annular areas according to a standardized procedure. A minimum of six measurements should be taken along each perpendicular diameter. 24 is the usual maximum number of traverse points. (Figure 69.1 shows the locations for a 10-point traverse.) For rectangular ducts, the cross section is divided into a minimum of nine equal rectangular areas of the same shape and with an aspect ratio between 1:1 and 1:2. The traverse points are located at the centroids of these areas.

2. EMISSIONS MONITORING SYSTEMS

Continuous emissions monitoring systems (CEMs) are used in large installations, though smaller installations may also be subject to CEM requirements. CEM not only ensures compliance with regulatory limits, but also permits the installation to participate in any allowance trading components of the regulations. CEMs are meant as compliance monitors, not merely operational indicators. The name "continuous emissions monitor system"

[1]With *super-isokinetic* sampling, too little pollution is indicated. This can occur if the vacuum draws a sample such that sampling velocity is higher than the bulk flow velocity. Since particles are subject to inertial forces and gases are not, the particle count will be essentially unaffected, but the metered volume will be higher, resulting in a lower apparent concentration. Conversely, if the sample is drawn too slowly (i.e., *sub-isokinetic* sampling), the metered volume would be lower, and the apparent pollution concentration will be too high.

[2]Solid and liquid substances in flue gases require isokinetic sampling. Gaseous substances do not.

Figure 69.1 *Traverse Sampling Points*

(a) rectangular stack
(measure at center of at least 9 equal areas)

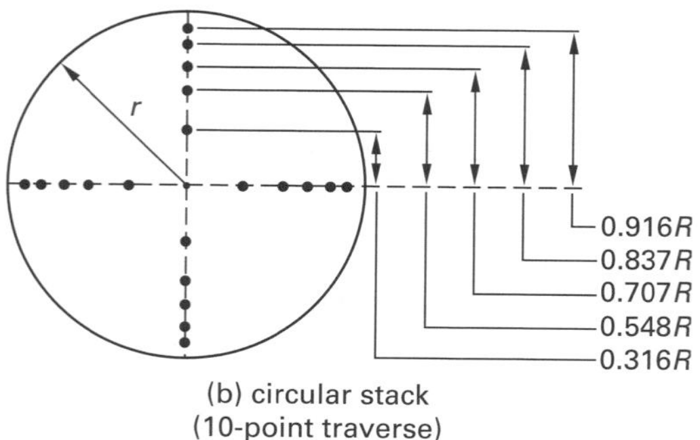

(b) circular stack
(10-point traverse)

has come to mean combined equipment for determining and recording SO_2 and NOx concentrations, volumetric flow, opacity, *diluent gases* (O_2 or CO_2), total hydrocarbons, and HCl.

CEMs consist of components that (1) obtain the sample, (2) analyze the sample, and (3) acquire and record analyses in real time.[3] With *straight extractive* CEMs, a sample is drawn from the stack by a vacuum pump and sent down a heated line to an analyzer at the base of the stack. This method must keep the stack gas heated above its 250°F to 300°F (120°C to 150°C) dew point to keep the moisture from condensing. Since samples are drawn by vacuum, leaks in the tube can lead to errors.

With *dilution extraction* CEMs, the gas is diluted with clean air or other inert carrier gas and passed to the analyzer located at a lower level. Dilution ratios are generally high—between 50:1 and 300:1, although they may be as low as approximately 12:1 at the outlet of an FGD unit. Since the dew point is lowered to below −10°F (−23°C) by the dilution, samples passed to the analyzer do not need to be heated. Samples are passed under positive pressure, ensuring that outside air and other gases cannot enter the tube. Dilution extraction analyses are on a wet basis.

With in situ CEM methods, direct measurements are taken by advanced electro-optical techniques. All instrumentation is located on the stack or duct. Data is sent by wire to remote monitors and the *data acquisition/reporting system* (DAS). The term "in situ" also implies that the flue gas analysis will be "wet" (i.e., the volumetric fraction will include water vapor).

There are three common methods of determining volumetric flow rates: (1) thermal sensing using a hotwire anemometer or thermal dispersion, (2) differential pressure using a single-point pitot tube or multipoint annubar, and (3) acoustic velocimetry using ultrasonic transducers.

With the advent of mass emission measurement requirements, pollutant (e.g., SO_2 and CO_2) concentration monitors and velocity monitors are being combined.

3. CENTRIFUGAL AIR SAMPLERS

A *centrifugal air sampler* (*aerosol centrifuge*, *cyclone sampler*) consists of a rotating shallow hollow drum, an impeller to draw air into the collection channel, and a collection lining (usually foil or a plastic strip covered with a thin layer of agar medium) on the inside of the channel. Samplers can be hand-held battery powered units or larger portable units. Centrifugal samplers impart rotational motion to the particles that are drawn into their sampling ports. The particles are acted upon by centrifugal forces, and those with sufficient mass move outward within the channel until they impact the collection medium at the inner wall of the channel. Discharged air may leave through the entrance port, although some units have separate discharge ports.

Unlike filters that collect particles of all sizes, the performance of inertial devices depends on the particle sizes. Centrifugal samplers are fairly efficient in collecting particles 15 μm and larger, but they are less capable with smaller particles. Less than 10% of particles under 2 μm may be captured. Collection efficiency is affected by the particle's *aerodynamic equivalent diameter*, D_o, which is the diameter of a smooth sphere with a density of 1 g/cm^3 and has the same settling velocity as the particle. For large spherical particles, the aerodynamic diameter is equal to the particle diameter, D_p. For very small or nonspherical particles, Eq. 69.2 should be used. $D_{p,\text{Stk}}$ is the *Stokes' particle diameter*, which is simply the diameter for smooth spherical particles. The reference density, ρ_o, is 1 g/cm^3. C_c is the *Cunningham slip factor* (*Cunningham correction factor*), which accounts for the reduction in drag on particles less than 1 μm in size when gas molecules slip past them. C_c approaches 1.0 for particles with $D_p > 5\mu\text{m}$. λ is the *mean free path length* of the gas. For air at 68°F (20°C) and 1 atm, λ is approximately 0.0665 μm.

$$D_o = D_{p,\text{Stk}}\sqrt{\frac{\rho_s C_c}{\rho_o C_o}} \approx D_{p,\text{Stk}}\sqrt{\frac{\rho_s}{\rho_o}} \quad [D_p \gg \lambda] \qquad 69.2$$

$$C_c \approx 1 + \frac{\lambda}{D_{p,\text{Stk}}}\left(2.514 + 0.80\exp\left(\frac{-0.55 D_{p,\text{Stk}}}{\lambda}\right)\right) \qquad 69.3$$

[3]The term "continuous" is somewhat of a misnomer. Though sampling is "automatic and frequent," data are rarely collected more frequently than once every 15 minutes.

$$\lambda_{\mu\mathrm{m}} \approx 0.0653\left(\frac{76}{p_{\mathrm{cm\,Hg}}}\right)\left(\frac{T_{\mathrm{K}}}{296.2}\right) \qquad 69.4$$

As shown in Fig. 69.2, particles enter the inlet a distance r_i from the rotational axis; they are collected at the outer wall, a distance r_o from the rotational axis. The *radial distance* (the *stopping distance*) they travel before being captured determines their free flight time and distance traveled. The *residence time* (*exposure time*), t_r, is defined more simplistically as the open (internal) volume, V, of the channel divided by the flow rate, Q.

$$t_r = \frac{V}{Q} = \frac{\pi b(r_o^2 - r_i^2)}{Q} \qquad 69.5$$

Figure 69.2 *Centrifugal Particle Sampler*

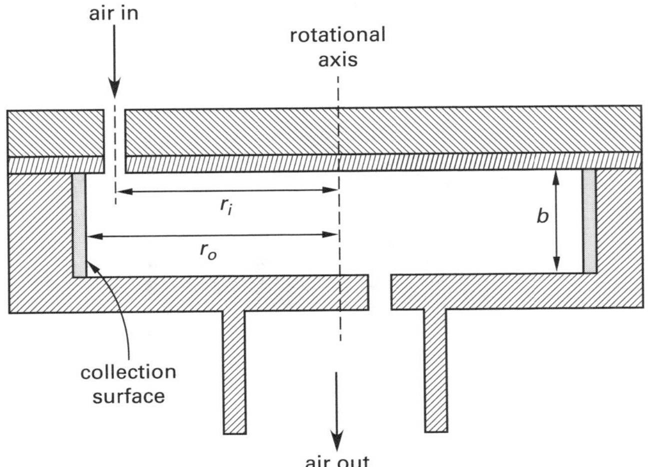

The instantaneous radial component of velocity of a particle's movement toward the inner wall, known as the *settling velocity*, *sedimentation rate*, and *terminal velocity*, is given by Eq. 69.6. The *hindrance factor*, $F_{\mathrm{hindrance}}$, is 1.0 for airflows with solids mass fractions less than 1%. The *shape factor*, F_{shape}, is 1.0 for spherical particles.

$$\mathrm{v}_{p,\mathrm{radial}} = \left(\frac{D_p^2(\rho_s - \rho_g)\omega^2 r}{18\mu g_c}\right)\left(\frac{F_{\mathrm{hindrance}}}{F_{\mathrm{shape}}}\right) \qquad 69.6$$

[Stokes' law range]

The time required for a particle to move radially a distance $r_1 - r_2$ away from the rotational axis is

$$t_{r_1 - r_2} = \left(\frac{18\mu g_c}{D_p^2(\rho_s - \rho_g)\omega^2}\right)\ln\frac{r_2}{r_1} \qquad 69.7$$

Small particles travel farther than large particles, and not all particles are captured. The smallest spherical particle diameter (*cut-point diameter* or *critical particle diameter*) that is captured 100% of the time is $D_{p,100\%}$. Particles larger than $D_{p,100\%}$ are captured; particles smaller than $D_{p,100\%}$ escape.

$$D_{p,100\%} = \sqrt{\frac{18\mu g_c Q}{\pi b(\rho_s - \rho_g)\omega^2(r_o^2 - r_i^2)}}\ \ln\frac{r_2}{r_1} \qquad 69.8$$

Equation 69.9 can be simplified by assuming the radial distance $r_o - r_i$ is much less than r_i and can be approximated by r_o.

$$D_{p,100\%} \approx \sqrt{\frac{9\mu g_c Q}{\pi b(\rho_s - \rho_g)\omega^2 r_o^2}} \qquad 69.9$$

$$[r_o - r_i \ll r_i \approx r_o]$$

Collection efficiency depends on the particle density, ρ_s, and diameter, D_p. Overall efficiency, η_{D_p}, as defined by the fraction of total particle mass entering the sampler, depends on the distribution of particle sizes.

$$\eta_{D_p} \approx 1 - \left(\frac{D_p}{D_{p,100\%}}\right)^2 \qquad 69.10$$

The *sigma factor* (also known as *capacity*), Σ, is a function of the centrifuge alone and is used to compare centrifuges of different sizes that are intended to perform the same function. It has units of length squared and represents the equivalent settling area of the centrifuge. $\mathrm{v}_{t,G}$ is the terminal velocity of a particle settling under the influence of gravity alone, as calculated from Stokes' law.

$$\Sigma = \frac{Q}{\mathrm{v}_{t,G}} = \frac{\pi\omega^2 b(r_o^2 - r_i^2)}{g\ln\frac{r_o}{r_i}} \qquad 69.11$$

The sigma factor is also defined from the parameters existing at the cut-point (i.e., at 50% capture), resulting in a different formula. For a constant rotational speed, the flow rate, Q, varies until the collection efficiency reaches 50%. Not only is the formula different, but the flow rate is different. As long as sigma is used consistently in ratios comparing two operating points, the ambiguity is harmless.

$$\Sigma_{1/2} = \frac{Q_{1/2}}{2\mathrm{v}_{t,G}} \qquad 69.12$$

$$\frac{Q_1}{\Sigma_1} = \frac{Q_2}{\Sigma_2} \qquad 69.13$$

Plant Engineering

4. PARTICLE DETECTORS, COUNTERS, AND SIZERS

Aspirating particle detectors (*automatic optical particle counters*, *particle sizers*) draw in gas samples and measure particle concentrations. Most are of the *photometer* variety, sensitive to light intensity. A particle detector is composed of a high-intensity light source (usually a laser producing monochromatic coherent light), a sample or particle delivery mechanism such as flowing air, a capillary viewing zone, optics that produce a shadow or scattering pattern, and a sensitive photodetector (photocell). (See Fig. 69.3.) When fully illuminated, the photodetector generates a reference voltage, E_o. When partially obscured by a particle shadow, the voltage drops. The voltage drop in the photocell is directly proportional to the size of the shadow, and hence, the size of the particle passing through.

$$\Delta E = \frac{A_{\text{shadow}} E_o}{A_{\text{detector}}} \qquad 69.14$$

Figure 69.3 *Optical Particle Counter*

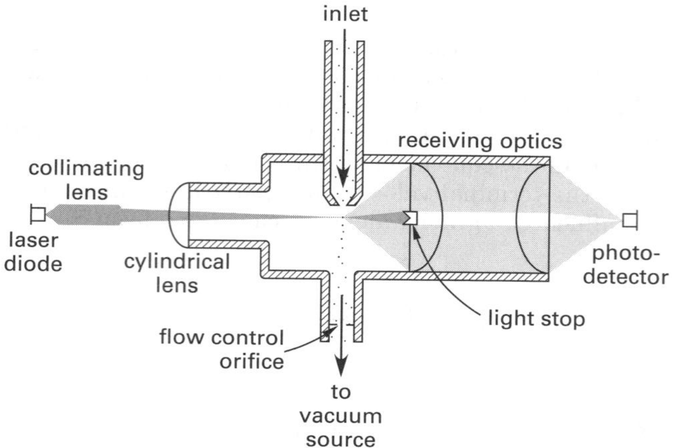

5. PARTICLE SIZE AND SHAPE

Particles or dust and other material are rarely perfect spheres. A *nonspherical particle* can be categorized in an optical detector by several different characteristics of its shadow. (See Fig. 69.4.) The *enclosing circle diameter* is the diameter of a circle that would contain the particle's shadow. The *sieve diameter* is the minimum width of a square aperture in a sieve screen through which the particle will pass.

The *perimeter diameter* is calculated from the shadow's perimeter length.

$$D_P = \frac{P_{\text{shadow}}}{\pi} \qquad 69.15$$

Figure 69.4 *Particle Parameters*

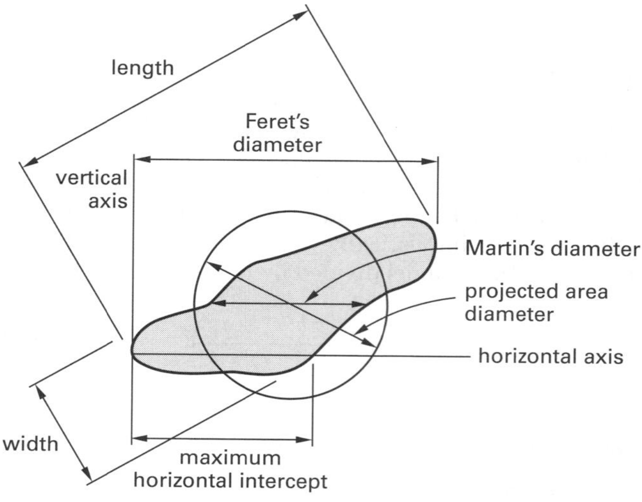

(a) dimensions

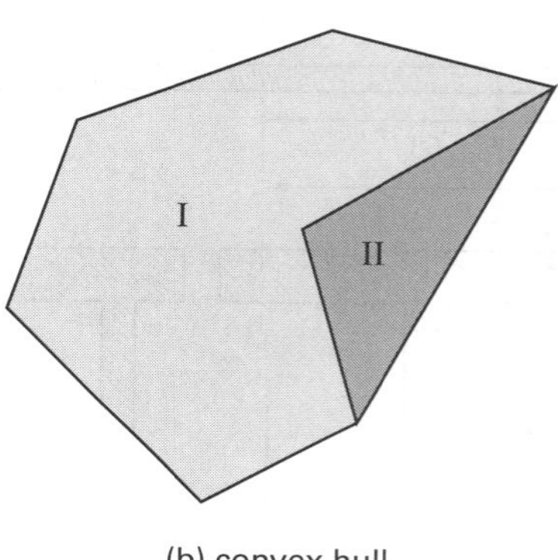

(b) convex hull

The *equivalent projected area diameter* (*circle equivalent diameter*, *equivalent area diameter*) is

$$D_A = \sqrt{\frac{4A_{\text{shadow}}}{\pi}} \qquad 69.16$$

The *equivalent surface area diameter* is the diameter of a sphere with the same surface area, S, as the particle.

$$D_S = \sqrt{\frac{S_{\text{particle}}}{\pi}} \qquad 69.17$$

The equivalent *volume diameter* is the diameter of a sphere with the same particle volume.

$$D_V = \sqrt[3]{\frac{6V_{\text{particle}}}{\pi}} \qquad 69.18$$

Feret's diameter (*maximum caliper*), D_F, is the perpendicular distance between parallel tangents touching opposite sides of the shadow. Feret's diameter depends on the shadow (particle, axis, etc.)

orientation, so $D_{F,\text{max}}$, $D_{F,\text{min}}$, and $D_{F,\text{mean}}$ values are recorded. The minimum value, $D_{F,\text{min}}$, is often used as the *sieve diameter*. *Martin's diameter*, D_M, is the length of a line that divides the shadow into two parts of equal areas.

Various inconsistent definitions of the *shape factor*, ψ, are used to characterize irregular particles without any axis of symmetry.[4] The *aspect ratio* (*anisotropy factor*, *elongation ratio*) is the ratio of two orthogonal maximum dimensions.

$$\psi_A = \frac{\text{length}}{\text{width}} = \frac{D_{F,\text{max}}}{D_{F,\text{min}}} \qquad 69.19$$

The *volume-surface ratio diameter* (*Sauter's diameter*) is

$$\psi_{SV} = \frac{D_V^3}{D_S^2} \qquad 69.20$$

"Circularity" and "sphericity" are often used (incorrectly) interchangeably. *Circularity* is a parameter associated with two-dimensional shapes, characterizing a particle's "closeness" to a perfect circle. Circularity has two definitions, the first of which is more common (Eq. 69.21(a)).

$$\psi_C = \frac{4\pi A_{\text{shadow}}}{P_{\text{shadow}}^2} \quad \text{[SI]} \qquad 69.21(a)$$

$$\psi_C = \frac{\pi D_A}{P_{\text{shadow}}} \quad \text{[U.S.]} \qquad 69.21(b)$$

Sphericity is a parameter associated with three-dimensional shapes. There are multiple definitions of sphericity, as given by Eq. 69.22.

$$\psi_S = \left.\frac{(36\pi V_{\text{particle}})^{1/3}}{S_{\text{particle}}}\right|_{\text{Wadell}} \qquad 69.22(a)$$

$$\psi_S = \left.\sqrt{\frac{D_{\text{inscribing,max}}}{D_{F,\text{max}}}}\right|_{\text{Riley}} \qquad 69.22(b)$$

$$\psi_S = \frac{S_{\text{particle}}}{\pi D_V^2} \qquad 69.22(c)$$

$$\psi_S = \frac{6V_{\text{particle}}}{D_V S_{\text{particle}}} \qquad 69.22(d)$$

There are several standard definitions of particle *roundness*, as well as at least another dozen definitions borne out of convenience.[5] *Wadell's roundness*, Eq. 69.23(b), is the ratio of the average radius of curvature of the corners to the radius of the largest inscribed circle.

$$\psi_R = \frac{4\pi A_{\text{shadow}}}{P_{\text{shadow}}^2} \qquad 69.23(a)$$

$$\psi_R = \left.\frac{\frac{1}{N}\sum_{i=1}^{N\text{ corners}} r_i}{r_{\text{inscribing,max}}}\right|_{\text{Wadell (1932)}} \qquad 69.23(b)$$

$$\psi_R = \left.\frac{r_{\text{sharpest convex corner}}}{r_{\text{inscribing,max}}}\right|_{\text{Dobkins and Folk (1970)}} \qquad 69.23(c)$$

$$\psi_R = \frac{r_{\text{sharpest convex corner}}}{r_{\text{inscribing}}} \qquad 69.23(d)$$

The *ratio of perimeter lengths* compares the actual perimeter to the perimeter of a circle with the same area.

$$\psi_{P/P} = \frac{P_{\text{shadow}}}{\pi D_A} \qquad 69.24$$

There are two definitions of *convexity* (also known as *solidity*), one based on perimeter, and the other based on area. Convexity is a measure of edge smoothness. Convexity is calculated by dividing the *convex hull perimeter* (area) by the actual particle perimeter (area).[6]

$$\psi_C = \frac{P_{\text{shadow}}}{P_{\text{convex hull}}} \quad \text{[SI]} \qquad 69.25(a)$$

$$\psi_C = \frac{A_{\text{shadow}}}{A_{\text{convex hull}}} \quad \text{[U.S.]} \qquad 69.25(b)$$

The *perimeter-free shape factor* uses the area of a circle with the maximum Feret's diameter, $D_{F,\text{max}}$.

$$\psi_{\text{PF}} = \frac{\pi D_{F,\text{max}}^2}{4A_{\text{shadow}}} \qquad 69.26$$

Plant Engineering

[4]There is frequent "mixing" of shape factor names and formulas in the industry. For example, a parameter may be defined as a ratio of perimeters, areas, or volume. Names (e.g., circularity, sphericity, solidarity, and roundness) may be used interchangeably. Inverses of all of the formulas are used with the same names. Some shape factors are fictitious, such as $4\pi A/P$, which has no physical significance.

[5]The term "roundness" should be reserved for a statistic analysis of the deviation traces when a shape is rotated about a specific axis.

[6]The convex hull perimeter can be visualized as the length of a rubber band placed around the shape.

The *specific surface area* (*specific area*) has two definitions.

$$\psi_{\text{SSA}} = \frac{S_{\text{particle}}}{V_{\text{particle}}} \quad \text{[SI]} \qquad 69.27(a)$$

$$\psi_{\text{SSA}} = \frac{S_{\text{particle}}}{m_{\text{particle}}} \quad \text{[U.S.]} \qquad 69.27(b)$$

6. NOMENCLATURE

A	area	ft^2	m^2
b	length of cylindrical centrifuge section	ft	m
C_c	Cunningham slip factor	–	–
D	diameter	ft	m
E	voltage	V	V
F	factor	–	–
g	acceleration of gravity, 32.2 (9.81)	ft/sec^2	m/s^2
g_c	gravitational constant, 32.2	$lbm\text{-}ft/lbf\text{-}sec^2$	n.a.
m	mass	lbm	kg
N	number (quantity)	–	–
p	pressure	lbf/ft^2	Pa
P	perimeter	ft	m
Q	flow rate	ft^3/sec	m^3/s
r	radius	ft	m
S	surface area	ft^2	m^2
T	temperature	°R	K
v	velocity	ft/sec	m/s
V	volume	ft^3	m^3

Symbols

Σ	sigma factor	ft^2	m^2
η	efficiency	–	–
λ	mean free path length	ft	m
μ	absolute viscosity	$lbf\text{-}sec/ft^2$	Pa·s
ρ	density	lbm/ft^3	kg/m^3
ψ	shape factor	–	–
ω	angular velocity	rad/sec	rad/s

Subscripts

A	area
eq	equivalent
F	Feret
g	gas
G	standard gravity
i	inner
M	Martin
o	outer or reference
p	particle
r	residence
s	solid
S	surface
Stk	Stokes
t	terminal
V	volume

70 Environmental Remediation

Content in blue refers to the *NCEES Handbook*.

1. INTRODUCTION

This chapter discusses (in alphabetical order) the methods and equipment that can be used to reduce or eliminate pollution. Legislation often requires use of the *best available control technology* (BACT—also known as the *best available technology*—BAT) and the *maximum achievable control technology* (MACT), *lowest achievable emission rate* (LAER), and *reasonably available control technology* (RACT) in the design of pollution-prevention systems.

2. ABSORPTION, GAS (GENERAL)

Gas *absorption processes* remove a gas (the *target substance*) from a gas stream by dissolving it in a liquid solvent.[1] Absorption can be used for flue gas cleanup (FGC) to remove sulfur dioxide, hydrogen sulfide, hydrogen chloride, chlorine, ammonia, nitrogen oxides, and light hydrocarbons. Gas absorption equipment includes packed towers, spray towers and chambers, and venturi absorbers.[2]

[1]Scrubbing, gas absorption, and stripping are distinguished by their target substances, the carrier flow phase, and the directions of flow. *Scrubbing* is the removal of particulate matter from a gas flow by exposing the flow to a liquid or slurry spray. *Gas absorption* is a countercurrent operation for the removal of a target gas from a gas mixture by exposing the mixture to a liquid bath or spray. *Stripping*, also known as *gas desorption* (also a countercurrent operation), is the removal of a dissolved gas or other volatile component from liquid by exposing the liquid to air or steam. Stripping is the reverse operation of gas absorption. Packed towers can be used for both gas absorption and stripping processes, and the processes look similar. The fundamental difference is that in gas absorption processes, the target substance (i.e., the substance to be removed) is in the gas flow moving up the tower, and in stripping processes, the target substance is in the liquid flow moving down the tower.

[2]Spray and packed towers, though they are capable, are not generally used for desulfurization of furnace or incineration combustion gases, as scrubbers are better suited for this task.

3. ABSORPTION, GAS (SPRAY TOWERS)

In a general spray tower or spray chamber (i.e., a *wet scrubber*), liquid and gas flow countercurrently or crosscurrently. The gas moves through a liquid spray that is carried downward by gravity. A mist eliminator removes entrained liquid from the gas flow. (See Fig. 70.1.) The liquid can be recirculated. The removal efficiency is moderate.

***Figure 70.1** Spray Tower Absorber*

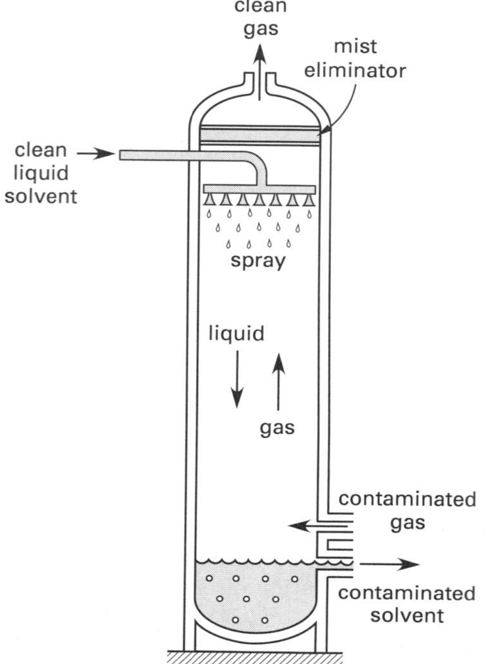

Spray towers are characterized by low pressure drops—typically 1 in wg to 2 in wg[3] (0.25 kPa to 0.5 kPa)—and their liquid-to-gas ratios—typically 20 gal/1000 ft^3 to 100 gal/1000 ft^3 (3 L/m^3 to 14 L/m^3). For self-contained units, fan power is also low—approximately 3×10^{-4} kW/ft^3 (0.01 kW/m^3) of gas moved.

Flooding of spray towers is an operational difficulty where the liquid spray is carried up the column by the gas stream. Flooding occurs when the gas stream velocity approaches the *flooding velocity.* To prevent this, the tower diameter is chosen as approximately 50% to 75% of the flooding velocity.

Flue gas desulfurization (FGD) wet scrubbers can remove approximately 90% to 95% of the sulfur with efficiencies of 98% claimed by some installations. Stack effluent leaves at approximately 150°F (65°C) and is saturated (or nearly saturated) with moisture. Dense steam plumes may be present unless the scrubbed stream is reheated.

Collected sludge waste and fly ash are removed within the scrubber by purely inertial means. After being dewatered, the sludge is landfilled.[4]

Spray towers, though simple in concept and requiring little energy to operate, are not simple to operate and maintain. Other disadvantages of spray scrubbers include high water usage, generation of wastewater or sludge, and low efficiency at removing particles smaller than 5 μm. Relative to dry scrubbing, the production of wet sludge and the requirement for a sludge-handling system is the major disadvantage of wet scrubbing.

4. ABSORPTION, GAS (IN PACKED TOWERS)

In a packed tower, clean liquid flows from top to bottom over the tower packing media, usually consisting of synthetic engineered shapes designed to maximize liquid-surface contact area. (See Fig. 70.2.) The contaminated gas flows countercurrently, from bottom to top, although crosscurrent designs also exist. As in spray towers, the liquid can be recirculated, and a mist eliminator is used.

***Figure 70.2** Packed Bed Spray Tower*

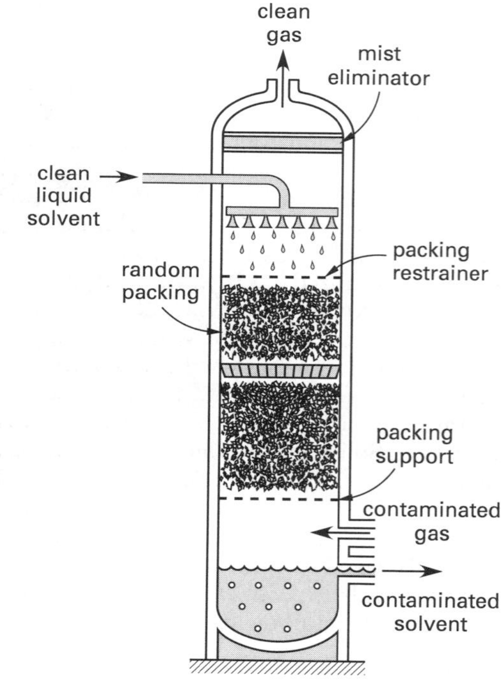

Pressure drops are in the 1 in wg to 8 in wg (0.25 kPa to 2.0 kPa) range. Typical liquid-to-gas ratios are 10 gal/1000 ft^3 to 20 gal/1000 ft^3 (1 L/m^3 to 3 L/m^3). Although the pressure drop is greater than in spray towers, the removal efficiency is much higher.

[3]"inwg" and "iwg" are abbreviations for "inches water gage," also referred to as "iwc" for "inches water column" and "inches of water."
[4]A typical 500 MW plant burning high-sulfur fuel can produce as much as 10^7 ft^3 (300,000 m^3) of dewatered sludge per year.

5. ADSORPTION (ACTIVATED CARBON)

Granular activated carbon (GAC), also known as *activated carbon* and *activated charcoal* (AC), processes are effective in removing a number of compounds, including volatile organic compounds (VOCs), heavy metals (e.g., lead and mercury), and dioxins.[5] AC is an effective adsorbent[6] for both air and water streams, and a VOC removal efficiency of 99% can be achieved. AC can be manufactured from almost any carbonaceous raw material, but wood, coal, and coconut shell are widely used.

Pollution control processes use AC in both solvent capture-recovery (i.e., recycling) and capture-destruction processes. AC is available in powder and granular form. Granules are preferred for use in recovery systems since the AC can be regenerated when *breakthrough*, also known as *breakpoint*, occurs. This is when the AC has become saturated with the solvent and traces of the solvent begin to appear in the exit air. Until breakthrough, removal efficiency is essentially constant. The *retentivity* of the AC is the ratio of adsorbed solvent mass to carbon mass.

6. ADSORPTION (SOLVENT RECOVERY)

AC for solvent recovery is used in a cyclic process where it is alternately exposed to the target substance and then regenerated by the removal of the target substance. For solvent recovery to be effective, the VOC inlet concentration should be at least 700 ppm. Regeneration of the AC is accomplished by heating, usually by passing low-pressure (e.g., 5 psig) steam over the AC to raise the temperature above the solvent-capture temperature. Figure 70.3 shows a stripping-AC process with recovery.

There are three main processes for solvent recovery. In a traditional *open-loop recovery system*, a countercurrent air stripper separates VOC from the incoming stream. The resulting VOC air/vapor stream passes through an AC bed. Periodically, the air/vapor stream is switched to an alternate bed, and the first bed is regenerated by passing steam through it. The VOC is recovered from the steam. The VOC-free air stream is freely discharged. Operation of a *closed-loop recovery system* is the same as an open loop system, except that the VOC-free air stream is returned to the air stripper.

For VOC-laden gas flows from 5000 SCFM to 100,000 SCFM (2.3 kL/s to 47 kL/s), a third type of solvent recovery system involves a *rotor concentrator*. VOCs are continuously adsorbed onto a multilayer, corrugated wheel whose honeycomb structure has been coated with powdered AC. The wheel area is divided into three zones: one for adsorption, one for desorption

Figure 70.3 *Stripping-AC Process with Recovery*

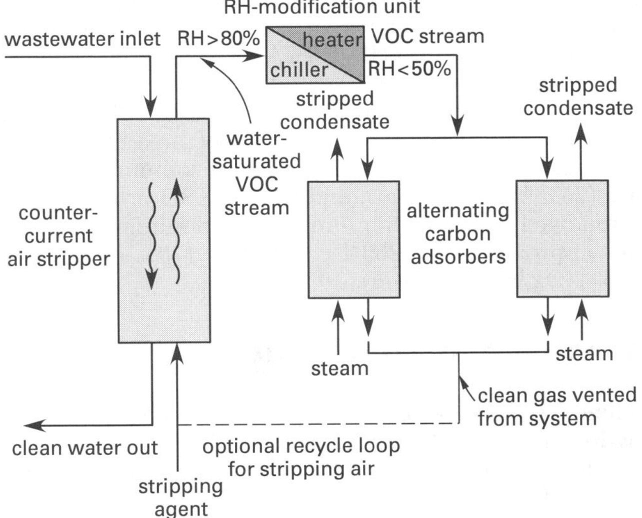

(i.e., regeneration), and one for cooling. Each of the zones is isolated from the other by tight sealing. The wheel rotates at low speed, continuously exposing new portions of the streams to each of the three zones. Though the equipment is expensive, operational efficiencies are high—95% to 98%.

7. ADSORPTION, HAZARDOUS WASTE

AC is particularly attractive for flue gas cleanup (FGC) at installations that burn spent oil, electrical cable, biosolids (i.e., sewage sludge), waste solvents, or tires as supplemental fuels. As with liquids, flue gases can be cleaned by passing them through fixed AC. Heavy metal dioxins are quickly adsorbed by the AC—in the first 8 in (20 cm) or so. AC injection (direct or spray) can remove 60% to 90% of the target substance present.

The spent AC creates its own waste disposal problem. For some substances (e.g., dioxins), AC can be incinerated. Heavy metals in AC must be removed by a wash process. Another process in use is vitrifying the AC and fly ash into a glassy, unleachable substance. *Vitrification* is a high-temperature process that turns incinerator ash into a safe, glass-like material. In some processes, heavy-metal salts are recovered separately. No gases and no hazardous wastes are formed.

8. ADVANCED FLUE GAS CLEANUP

Most air pollution control systems reduce NOx in the burner and remove SOx in the stack. *Advanced flue gas cleanup* (AFGC) methods combine processes to remove both NOx and SOx in the stack. Particulate matter is

[5]The term *activated* refers to the high-temperature removal of tarry substances from the interior of the carbon granule, leaving a highly porous structure.

[6]An *adsorbent* is a substance with high surface area per unit weight, an intricate pore structure, and a hydrophobic surface. An *adsorbent material* traps substances in fluid (liquid and gaseous) form on its exposed surfaces. In addition to activated carbon, other common industrial adsorbents include alumina, bauxite, bone char, Fuller's earth, magnesia, and silica gel.

Plant Engineering

removed by an electrostatic precipitator or baghouse as is typical. Promising AFGC methods include wet scrubbing with metal chelates such as ferrous ethylenediaminetetraacetate (Fe(II)-EDTA) (NOx/SOx removal efficiencies of 60%/90%), adding sodium hydroxide injection to dry scrubbing operations (35% NOx removal), in-duct sorbent injection of urea (80% NOx removal), in-duct sorbent injection of sodium bicarbonate (35% NOx removal), and the NOXSO process using a fluidized bed of sodium-impregnated alumina sorbent at approximately 250°F (120°C) (70% to 90%/90% NOx/SOx removal).

9. ADVANCED OXIDATION

The term *advanced oxidation* refers to the use of ozone, hydrogen peroxide, ultraviolet radiation, and other exotic methods that produce free hydroxyl radicals (OH^-). (See also Sec. 70.40.) Table 70.1 shows the relative oxidation powers of common oxidants.

Table 70.1 *Relative Oxidation Powers of Common Oxidants*

oxidant	oxidation power (relative to chlorine)
fluorine	2.25
hydroxyl radical (OH)	2.05
ozone (O_3)	1.52
permanganate radical (MnO_4)	1.23
chlorine dioxide (ClO_2)	1.10
hypochlorous acid (HClO)	1.10
chlorine	1.00
bromine	0.80
iodine	0.40
oxygen	0.29

10. BAGHOUSES

Baghouses have a reputation for excellent particulate removal, down to 0.005 grains/ft^3 (0.18 grains/m^3) (dry), with particulate emissions of 0.01 grains/ft^3 (0.35 grains/m^3) being routine. Removal efficiencies are in excess of 99% and are often as high as 99.99% (weight basis). Fabric filters have a high efficiency for removing particular matter less than 10 μm in size. Because of this, baghouses are effective at collecting air toxics, which preferentially condense on these particles at the baghouse operating temperature—less than 300°F (150°C). Gas temperatures can be up to 500°F (260°C) with short periods up to about 550°F (288°C), depending on the filter fabric. Most of the required operational power is to compensate for the system pressure drops due to bags, cake, and ducting.

Baghouse filter fabric has micro-sized holes but may be felted or woven, with the material depending on the nature of the flue gas and particulates. Woven fabrics are better suited for lower filtering velocities in the 1 ft/min to 2 ft/min (0.3 m/min to 0.7 m/min) range, while felted fabrics are better at 5 ft/min (1.7 m/min). The fabric is often used in tube configuration, but envelopes (i.e., flat bags) and pleated cartridges are also available.

The particulate matter collects on the outside of the bag, forming a cake-like coating. If lime has been introduced into the stream in a previous step, some of the cake will consist of unreacted lime. As the gases pass through this cake, additional neutralizing takes place.

When the pressure drop across the filter reaches a preset limit (usually 6 in wg to 8 in wg (1.5 kPa to 2 kPa)), the cake is dislodged by mechanical shaking, reverse-air cleaning, or pulse-jetting. The dislodged fly ash cake falls into collection hoppers below the bags. Fly ash is transported by pressure or vacuum conveying systems to the conditioning system (consisting of surge bins, rotary feeders, and pug mills). Figure 70.4 shows a typical baghouse.

Figure 70.4 *Typical Baghouse*

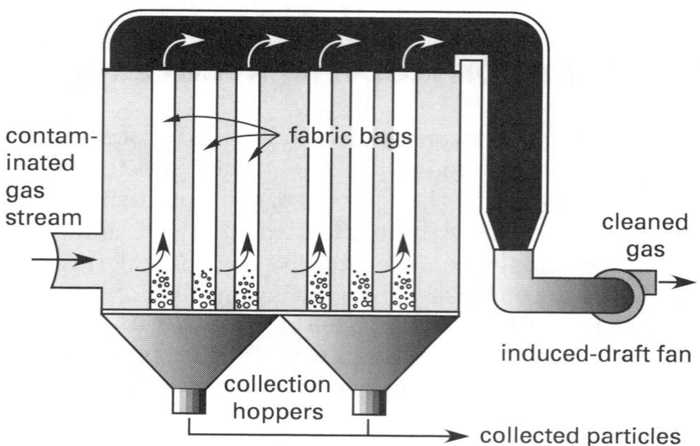

Baghouses are characterized by their air-to-cloth ratios and pressure drop. The *air-to-cloth ratio*, also known as *filter ratio*, *superficial face velocity*, and *filtering velocity*, is the ratio of the air volumetric flow rate in ft^3/min (m^3/s) to the exposed surface area in ft^2 (m^2). After canceling, the units are ft/min (m/s), hence the name *filtering velocity*. The higher the ratio, the smaller the baghouse and the higher the pressure drop. Shaker and reverse-air baghouses with woven fabrics have typical air-to-cloth ratios ranging from 1.0 ft/min to 6.0 ft/min (0.056 m/s to 0.34 m/s). Pulse-jet collectors with felted fabrics have higher ratios, ranging from 3.5 ft/min to 5.0 ft/min (0.0175 m/s to 0.025 m/s).

Advantages of baghouses include high efficiency and performance that is essentially independent of flow rate, particle size, and particle (electrical) resistivity. Also, baghouses produce the lowest opacity (generally less than 10, which is virtually invisible). Disadvantages include clogging, difficult cleaning, and bag breakage.

In a baghouse with N compartments, each with multiple bags, the *gross filtering velocity*, *gross air-to-cloth ratio*, and *gross filtering area*, $v_{f,N}$, refer to having all N compartments operating simultaneously. (See Eq. 70.1.) The *design filtering velocity*, *net filtering velocity*, *net air-to-cloth ratio*, and *net filtering area*, $v_{f,N-1}$, refer to one compartment taken offline for cleaning, meaning $N-1$ compartments are operating simultaneously. (See Eq. 70.2.) *Net net* refers to having $N-2$ compartments operating simultaneously.

$$v_{f,N} = \frac{Q_t}{A_{N\,\text{compartments}}} = \frac{Q_t}{NA_{\text{one compartment}}} \quad \text{70.1}$$

$$v_{f,N-1} = \frac{Q_t}{(N-1)A_{\text{one compartment}}} \quad \text{70.2}$$

The number of bags is

$$n_{\text{bags}} = \frac{A_N}{A_{\text{bag}}} = \frac{A_N}{\pi dh} \quad \text{70.3}$$

The *areal dust density*, W, is the mass of dust cake on the filter per unit area. The areal dust density can be calculated from the incoming particle concentration (*dust loading*, *fabric loading*), C, filtering velocity, and collection efficiency. Typical units are lbm/ft^2 (kg/m^2).

$$W = \eta C v_f t \approx C v_f t \quad \text{70.4}$$

In the *filter drag model*, the pressure drop through the baghouse is calculated from the *filter drag* (*filter resistance*), S, in units of inwg-min/ft (Pa·min/m), which depends on the permeabilities of the cloth, K_1, K_0, or K_e, and the particle cake, K_2 or K_s. (K_1 is also known as the *flow resistance* of the clean fabric; K_2 is the *specific resistance* of the cake.) Since Eq. 70.6 represents a straight line, K_e and K_s can be determined by plotting S versus W for a few test points. Maximum pressure drop is typically in the 5–20 in wg (1.2–5.0 kPa) range.

$$\Delta p = v_{\text{filtering}} S \quad \text{70.5(a)}$$

$$\Delta p_{\text{in wg}} = v_{f,\text{ft/min}} S_{\text{in wg-min/ft}} \quad \text{70.5(b)}$$

$$S = K_e + K_s W \quad \text{70.6}$$

Baghouses commonly have multiple compartments, and sufficient capacity must remain when one compartment is taken offline for cleaning. If there are N compartments, the time for N cycles of filtration (known as the *filtration time*) will be[7]

$$t_{\text{filtration}} = N(t_{\text{run}} + t_{\text{cleaning}}) - t_{\text{cleaning}} \quad \text{70.7}$$

Filtering velocity and pressure drop depend on the number of compartments that are online (i.e., they increase when one of the compartments is taken offline). The maximum pressure drop, Δp_m, occurs several times during the filtration time. The maximum pressure drop can be calculated from the maximum filter drag, which in turn, depends on the maximum areal density. The *actual filtering velocity*, $v_{f,\max\Delta p}$, is the velocity at the time the maximum pressure drop occurs.[8] c in Eq. 70.11 is a scaling constant that depends on the number of compartments. (See Table 70.2.)

$$\Delta p_{\max} = S_{\max\Delta p} v_{f,\max\Delta p} \quad \text{70.8}$$

$$S_{\max\Delta p} = K_e + K_s W_{\max\Delta p} \quad \text{70.9}$$

$$W_{\max\Delta p} = (N-1)C(v_{f,N} t_{\text{run}} + v_{f,N-1} t_{\text{cleaning}}) \quad \text{70.10}$$

$$v_{f,\max\Delta p} = c v_{f,N-1} \quad \text{70.11}$$

Table 70.2 *Values for Scaling Constant, c, Based on Number of Compartments, N*

number of compartments, N	c
3	0.87
4	0.80
5	0.76
7	0.71
10	0.67
12	0.65
15	0.64
20	0.62

The motor for the fan (blower) is sized based on the maximum pressure loss.

$$P = Q_t \Delta p_{\max} \quad \text{70.12(a)}$$

$$P_{\text{hp}} = \frac{Q_{t,\text{ft}^3/\text{min}} \Delta p_{\max,\text{in wg}}}{6356\eta_{\text{motor}}} \quad \text{[U.S. only]} \quad \text{70.12(b)}$$

There is no single formula derived from basic principles that predicts baghouse collection efficiency, η. Baghouse designs are normally based on experience, not on fractional efficiency curves. The instantaneous collection efficiency (as well as the pressure drop) of a baghouse increases with time as pores fill with particles. The empirical rate constant, k, is derived from a curve fit of actual performance.

$$\eta = 1 - e^{-kt} \quad \text{70.13}$$

[7]By convention, the filtration time includes only $N-1$ cleanings.
[8]The name "actual filtering velocity" is definitely confusing.

11. BIOREMEDIATION

Bioremediation encompasses the methods of biofiltration, bioreaction, bioreclamation, activated sludge, trickle filtration, fixed-film biological treatment, landfilling, and injection wells (for in situ treatment of soils and groundwater). Bioremediation relies on microorganisms in a moist, oxygen-rich environment to oxidize solid, liquid, or gaseous organic compounds, producing carbon dioxide and water. Bioremediation is effective for removing volatile organic compounds (VOCs) and easy-to-degrade organic compounds such as BTEXs (benzene, toluene, ethylbenzene, and xylene).[9] Wood-preserving wastes, such as creosote and other polynuclear aromatic hydrocarbons (PAHs), can also be treated.

Bioremediation can be carried out in open tanks, packed columns, beds of porous synthetic materials, composting piles, or soil. The effectiveness of bioremediation depends on the nature of the process, the time and physical space available, and the degradability of the substance. Table 70.3 categorizes gases according to their general degradabilities.

Table 70.3 Degradability of Volatile Organic and Inorganic Gases

rapidly degradable VOCs	rapidly reactive VICs	slowly degradable VOCs	very slowly degradable VOCs
alcohols	H_2S	hydrocarbons[a]	halogenated hydrocarbons[b]
aldehydes	NOx (but not N_2O)	phenols	polyaromatic hydrocarbons
ketones	SO_2	methylene chloride	CS_2
ethers	HCl		
esters	NH_3		
organic acids	PH_3		
amines	SiH_4		
thiols	HF		
other molecules containing O, N, or S functional groups			

[a]Aliphatics degrade faster than aromatics, such as xylene, toluene, benzene, and styrene.

[b]These include trichloroethylene, trichloroethane, carbon tetrachloride, and pentachlorophenol.

Though slow, limited by microorganisms with the specific affinity for the chemicals present, and susceptible to compounds that are toxic to the microorganisms, bioremediation has the advantage of destroying substances rather than merely concentrating them. Bioremediation is less effective when a variety of different compounds are present simultaneously.

12. BIOFILTRATION

The term *biofiltration* refers to the use of composting and soil beds. A *biofilter* is a bed of soil or compost through which runs a distribution system of perforated pipe. Contaminated air or liquid flows through the pipes and into the bed. Volatile organic compounds (VOCs) are oxidized to CO_2 by microorganisms. Volatile inorganic compounds (VICs) are oxidized to acids (e.g., HNO_3 and H_2SO_4) and salts.

Biofilters require no fuel or chemicals when processing VOCs, and the operational lifetime is essentially infinite. For VICs, the lifetime depends on the soil's capacity to neutralize acids. Though reaction times are long and absorption capacities are low (hence the large areas required), the oxidation continually regenerates (rather than depletes) the treatment capacity. Once operational, biofiltration is probably the least expensive method of eliminating VOCs and VICs.

The *removal efficiency* of a biofilter is given by Eq. 70.14. k is an empirical *reaction rate constant*, and t is the *bed residence time* of the carrier fluid (water or air) in the bed. The reaction rate constant depends on the temperature but is primarily a function of the biodegradability of the target substance. Biofiltration typically removes 80% to 99% of volatile organic and inorganic compounds (VOCs and VICs).

$$\eta = 1 - \frac{C_{\text{out}}}{C_{\text{in}}} = 1 - e^{-kt} \qquad 70.14$$

13. BIOREACTION

Bioreactors (*reactor tanks*) are open or closed tanks containing dozens or hundreds of slowly rotating disks covered with a biological film of microorganisms (i.e., colonies). Closed tanks can be used to maintain anaerobic or other design atmosphere conditions. For example, methanotrophic bacteria are useful in breaking down chlorinated hydrocarbons (e.g., trichlorethylene (TCE), dichloroethylene (DCE), and vinyl chloride (VC)) that would otherwise be considered nonbiodegradable. Methanotrophic bacteria derive their food from methane gas that is added to the bioreactor. An enzyme, known as MMO, secreted by the bacteria breaks down the chlorinated hydrocarbons.

14. BIOVENTING

Bioventing is the treatment of contaminated soil in a large plastic-covered tank. Clean air, water, and nutrients are continuously supplied to the tank while off-gases are suctioned off. The off-gas is cleaned with activated carbon (AC) adsorption or with thermal or

[9]BTEXs are common ingredients in gasoline.

catalytic oxidation prior to discharge. Bioventing has been used successfully to remove volatile hydrocarbon compounds (e.g., gasoline and BTEX compounds) from soil.

15. COAL CONDITIONING

Coal intended for electrical generating plants can be modified into "self-scrubbing" coal by conditioning prior to combustion. Conditioning consists of physical separation by size, by cleaning (e.g., cycloning to remove noncombustible material, including up to 90% of the pyritic sulfur), and by the optional addition of limestone and other additives to capture SO_2 during combustion. Small-sized material is pelletized to reduce loss of fines and particulate emissions (dust).

16. CYCLONE SEPARATORS

Inertial separators of the *double-vortex, single cyclone* variety are suitable for collecting medium- and large-sized (i.e., greater than 5 μm) particles from spot sources. During operation, particulate matter in the incoming gas stream spirals downward at the outside and upward at the inside.[10] The particles, because of their greater mass, move toward the outside wall, where they drop into a collection bin. Figure 70.5 shows a double-vortex, single cyclone separator.

Figure 70.5 *Double-Vortex, Single Cyclone*

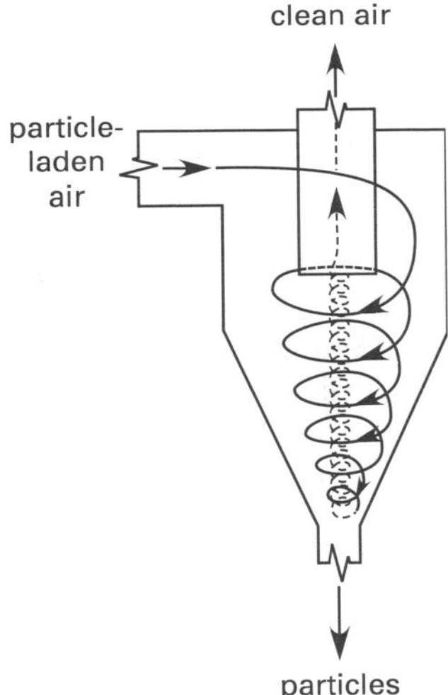

The *cut size* is the diameter of particles collected with a 50% efficiency. Separation efficiency varies directly with (1) particle diameter, (2) particle density, (3) inlet velocity, (4) cyclone body length (ratio of cyclone body diameter to outlet diameter), (5) smoothness of inner wall, (6) number of gas revolutions, and (7) amount of particles in the flow. Collection efficiency decreases with increases in (1) gas viscosity, (2) gas density, (3) cyclone diameter, (4) gas outlet diameter, (5) gas inlet duct width, and (6) inlet area.

Collection efficiencies are not particularly high, and dusts (5 μm to 10 μm) are too fine for most cyclones. For geometrically similar cyclones, the collection efficiency varies directly with the dimensionless *separation factor*, *S*.

$$S = \frac{v_{\text{inlet}}^2}{rg} \qquad 70.15$$

Pressure drop, h, in feet (meters) of air across a cyclone can be roughly (i.e., with an accuracy of only approximately ±30%) estimated by Eq. 70.16. H and B are the height and width of the rectangular cyclone inlet duct, respectively; D_e is the gas exit duct diameter; and K is an empirical constant that varies from approximately 7.5 to 18.4.

$$h = \frac{KBH v_{\text{inlet}}^2}{2gD_e^2} \qquad 70.16$$

17. DECHLORINATION

Industrial and municipal wastewaters containing excessive amounts of *total residual chlorine* (TRC) must be dechlorinated prior to discharge. Sulfur dioxide, sodium metabisulfate, and sulfite salts are effective dechlorinating agents. For sulfur dioxide, the dose is approximately 0.9 lbm/lbm (0.9 kg/kg) of chlorine to be removed. Reaction is essentially instantaneous, being completed within 10 pipe diameters at turbulent flow. For open channels, a submerged weir may be necessary to obtain the necessary turbulence. TRC is reduced to less than detectable levels.

18. ELECTROSTATIC PRECIPITATORS

Electrostatic precipitators (ESPs) are used to remove particulate matter from gas streams. Collection efficiency for particulate matter is usually in the 95% to 99% range. ESPs are preferred over scrubbers because they are more economical to operate, dependable, and predictable and because they don't produce a moisture plume.

ESPs used on steam generator/electrical utility units treat gas that is approximately 280°F to 300°F (140°C to 150°C) with moisture being superheated.[11] This high

[10]The number of gas revolutions varies approximately between 0.5 and 10.0, with averages of 1.5 revolutions for simple cyclones and 5 revolutions for high-efficiency cyclones.

[11]The flue gas, at approximately 1400°F (760°C), is cooled to this temperature range in a conditioning tower. Injected water increases the moisture content of the flue gas to approximately 25% by volume.

temperature enhances buoyancy and plume dissipation. However, ESPs generally cannot be used with moist flows, mists, or sticky or hygroscopic particles. Scrubbers should be used in those cases. Relatively humid flows can be treated, although entrained water droplets can insulate particles, lowering their resistivities. Table 70.4 gives typical design parameters for ESPs.

***Table 70.4** Typical Electrostatic Precipitator Design Parameters*

parameter	typical range U.S.	typical range SI
efficiency	90% to 98%	
gas velocity	2 ft/sec to 4 ft/sec	0.6 m/s to 1.2 m/s
gas temperature		
standard	≤ 700°F	≤ 370°C
high-temperature	≤ 1000°F	≤ 540°C
special	≤ 1300°F	≤ 700°C
drift velocity	0.1 ft/sec to 0.7 ft/sec	0.03 m/s to 0.21 m/s
treatment/residence time	2 sec to 10 sec	
draft pressure loss	0.1 iwg to 0.5 iwg	0.025 kPa to 0.125 kPa
plate spacing	12 in to 16 in	30 cm to 41 cm
plate height	30 ft to 50 ft	9 m to 15 m
plate length/height ratio	1.0 to 2.0	
applied voltage	30 kV to 75 kV	

In operation, the gas passes over negatively charged tungsten *corona wires* or grids. Particles are attracted to positively charged collection plates.[12] The speed at which the particles move toward the plate is known as the *drift velocity*, *w*. Drift velocity is approximately 0.20 ft/sec to 0.30 ft/sec (0.06 m/s to 0.09 m/s), with 0.25 ft/sec (0.075 m/s) being a reasonable design value. Periodically, *rappers* vibrate the collection plates and dislodge the particles, which drop into collection hoppers.

Advantages of ESPs include high reliability, low maintenance, low power requirements, and low pressure drop. Disadvantages are sensitivity to particle size and resistivity, and the need to heat ESPs during start-up and shutdown to avoid corrosion from acid gas condensation.

For a rectangular electrostatic precipitator (ESP) with interior dimensions $W \times H \times L$, the flow rate is

$$Q = \mathrm{v}_g WH \tag{70.17}$$

The *residence time* (*exposure time*), t_r, is

$$t_r = \frac{L}{\mathrm{v}_g} \tag{70.18}$$

Collection efficiency is affected by particle *resistivity*, which is divided into three categories: low (less than 1 × 108 Ω·cm); medium, moderate, or normal (1×10^8 Ω·cm to 2×10^{11} Ω·cm); and high (more than 2×10^{11} Ω·cm). Particles in the medium resistivity range are collected most easily. Particles with low resistivity are easily charged, but upon contact with the charged plate, they rapidly lose their charges and are re-entrained into the gas flow. Particles with high resistivity coat the collection plate with a layer of insulation, causing *flashovers* and *back corona*, a localized electrical discharge. High resistivity particles can be brought into the moderate range by *particle conditioning*. When the gas temperature is below 350°F (177°C), adding moisture to the gas stream, reducing the temperature, or adding SO_3 and ammonia will reduce resistivity into the moderate range. When the gas temperature is above 350°F (177°C), increasing the temperature lowers particle resistivity.

Collection efficiency is also affected by the particle size. The theoretical *drift velocity* (*migration velocity* or *precipitation rate*) is proportional to the particle diameter. The theoretical drift velocity, w, of a particle in an electric field can be calculated from basic principles. In Eq. 70.19, q is the charge on the particle, which can be many times the charge on an electron, q_e (1.602×10^{-19} C); E is the electric field strength in V/m; ϵ is the particle's *dielectric constant* (*relative permittivity*), approximately 1.5–2.6 for fly ash;[13] and ϵ_0 is the *permittivity of free space*, equal to 8.854×10^{-12} F/m (same as C/V, $C^2/N{\cdot}m^2$, and $A^2{\cdot}s^4/kg{\cdot}m^3$). C_c is the *Cunningham slip factor* (*Cunningham correction factor*), which accounts for the reduction in drag on particles less than 1 μm in size when gas molecules slip past them. C_c approaches 1.0 for particles with $d_p > 5$ μm. λ is the *mean free path length* of the gas. For air at 68°F (20°C) and 1 atm, λ is approximately 0.0665 μm. (See Eq. 70.20.)

$$w \approx \frac{qEC_c}{3\pi\mu_g d_p} = \frac{Nq_e EC_c}{3\pi\mu_g d_p} \approx \left.\frac{\epsilon\epsilon_0 E^2 d_p C_c}{(\epsilon+2)\mu_g}\right|_{\text{sat}} \tag{70.19}$$

$$C_c \approx 1 + \frac{\lambda}{d_{p,\text{Stk}}}\left(2.514 + 0.80\exp\left(\frac{-0.55 d_{p,\text{Stk}}}{\lambda}\right)\right) \tag{70.20}$$

[12]The *tubular ESP*, used for collecting moist or sticky particles, is a variation on this design.
[13]The dielectric constant of fly ash depends greatly on the temperature and amount of unburned carbon.

A particle will become saturated if it remains in a strong electric field sufficiently long. The *equilibrium charge* is found from the number of electron charges on a saturated spherical particle. The charging electric field is not necessarily the same as the drift electric field.

$$N_{\text{sat,spherical}} = \frac{3\epsilon\epsilon_0 \pi E_{\text{charging}} d_p^2}{(\epsilon+2)q_e} \qquad 70.21$$

In practice, the theoretical drift velocity is seldom used. Rather, an *effective drift velocity*, w_e, is used. Although it shares the "velocity" name and units, the effective drift velocity is not an actual velocity, but is an empirical design parameter derived from pilot tests of collection efficiency.

The *Deutsch-Anderson equation* predicts the single-particle fractional *collection efficiency* (*removal efficiency*), η, of an electrostatic precipitator with total collection area, A, of all plates. The exponent, y, is 1 for fly ash and for anything else in the absence of specific knowledge otherwise. *Penetration*, Pt, is the fraction of particles that pass through the ESP uncollected. Penetration is the complement of collection efficiency.

$$\begin{aligned}\eta &= \frac{C_{\text{in}} - C_{\text{out}}}{C_{\text{in}}} = \frac{\dot{m}_{\text{in}} - \dot{m}_{\text{out}}}{\dot{m}_{\text{in}}} \\ &= 1 - \exp\left(\frac{-Aw_e}{Q}\right)^y\end{aligned} \qquad 70.22$$

$$\text{Pt} = 1 - \eta \qquad 70.23$$

The effect of flow rate on the collection efficiency is predicted by Eq. 70.24.

$$\frac{Q_1}{Q_2} = \ln(\eta_1 - \eta_2) \qquad 70.24$$

Collection efficiency can also be expressed as a function of the *residence time*. K is the ratio of collection plate area to internal volume.

$$\begin{aligned}\eta &= 1 - \exp(-Kw_e t_r) \\ &= 1 - \exp\left(-\left(\frac{A}{WHL}\right)w_e t_r\right)\end{aligned} \qquad 70.25$$

One of the most important factors affecting the collection efficiency is the *specific collection area*, SCA, which is the ratio of the total collection surface area to the gas flow rate into the collector, usually reported in ft^2/1000 ft^3-min (m^2/1000 m^3·h). The higher the SCA, the higher the collection efficiency will be.

$$\text{SCA} = \frac{A}{Q} \qquad 70.26$$

Using units common to the industry, the instantaneous collection efficiency of an ESP collector is

$$\eta = 1 - \exp(-0.06 w_{e,\text{ft/sec}} \text{SCA}_{\text{ft}^2/1000\,\text{ACFM}}) \qquad 70.27$$

The *corona power ratio*, CPR, is the power consumed by the corona in developing the electric field divided by the airflow. Generally, collection efficiency increases with increased CPR. The *corona current ratio*, CCR, is the current per unit plate area. The *power density*, P_A, is the power per unit plate area.

$$\text{CPR}_{\text{W-min/ft}^3} = \frac{P_{c,\text{W}}}{Q_{\text{ft}^3/\text{min}}} = \frac{I_{c,\text{A}} V_{\text{V}}}{Q_{\text{ft}^3/\text{min}}} \qquad 70.28$$

$$\text{CCR}_{\mu\text{A/ft}^2} = \frac{I_{c,\mu\text{A}}}{A_{\text{ft}^2}} \qquad 70.29$$

$$P_{A,\text{W/ft}^2} = \frac{P}{A} \qquad 70.30$$

Typical values for ESP design parameters are given in Table 70.5.

***Table 70.5** Typical ESP Design Parameters*

design parameter	typical value
particle effective drift velocity	0.05–1.0 ft/sec (0.015–0.3 m/s)
gas velocity	2–5 ft/sec typical; 15 ft/sec max (0.6–1.5 m/s typical; 4.5 m/s max)
number of stages	1–7
total specific collection area	100–1000 ft^2/1000 ACFM (5.5–55 m^2/1000 m^3·h)
pressure drop	> 0.5 in wg (0.13 kPa)

Plant Engineering

19. FLUE GAS RECIRCULATION

NOx emissions can be reduced when thermal dissociation is the primary NOx source (as it is when low-nitrogen fuels such as natural gas are burned) by recirculating a portion (15% to 25%) of the flue gas back into the furnace. This process is known as *flue gas recirculation* (FGR). The recirculated gas absorbs heat energy from the flame and lowers the peak temperature. Thermal NOx formation can be reduced by up to 50%. The recirculated gas should not be more than 600°F (315°C).

20. FLUIDIZED-BED COMBUSTORS

Fluidized-bed combustors (FBCs) are increasingly being used in steam/electric generation systems and for destruction of hazardous wastes. A *bubbling bed FBC*, as shown in Fig. 70.6, consists of four major components: (1) a windbox (plenum) that receives the fluidizing/combustion air, (2) an air distribution plate that transmits the air at 10 ft/sec to 30 ft/sec (3 m/s to 9 m/s) from the windbox to the bed and prevents the bed material from sifting into the windbox, (3) the fluid bed of inert material (usually sand or product ash), and (4) the freeboard area above the bed.

Figure 70.6 Fluidized-Bed Combustor

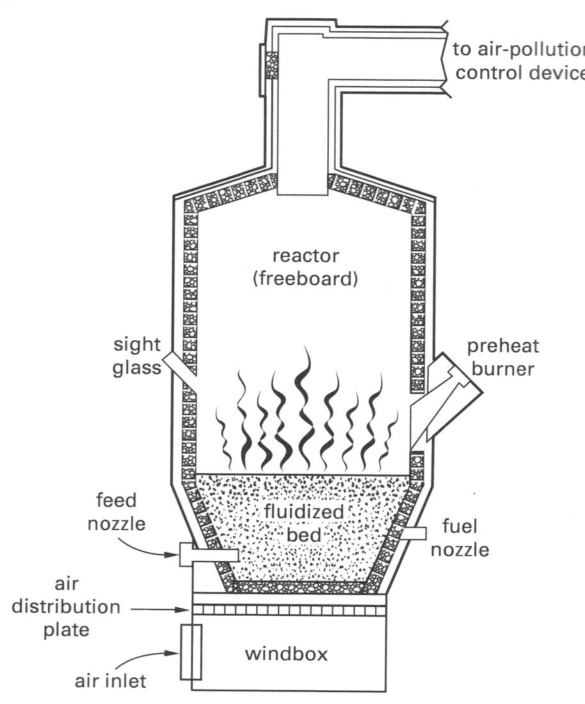

During the operation of a bubbling bed fluidized-bed combustor, the inert bed is levitated by the upcoming air, taking on many characteristics of a fluid (hence, the name FBC). The bed "boils" vigorously, offering an extremely large heat-transfer area, resulting in the thorough mixing of feed combustibles and air. Combustible material usually represents less than 1% of the bed mass, so the rest of the bed acts as a large thermal flywheel.

Combustion of volatile materials is completed in the freeboard area. Ash is generally reduced to small size so that it exits with the flue gas. In some cases (e.g., deliberate pelletization or wastes with high ash content), ash can accumulate in the bed. The fluid nature of the bed allows the ash to float on its surface, where it is removed through an overflow drain.

Most FBC systems use forced air with a single blower at the front end. If there are significant losses due to heat recovery or pollution control systems, an exhaust fan may also be used. In that case, the *null* (*balanced draft*) *point* should be in the freeboard area.

Large variations in the composition of the flue gases (known as *puffing*) is minimized by the long residence time and the large heat reservoir of the bed. Air pollution control equipment common to most boilers and incinerators is used with fluidized-bed combustors. Either wet scrubbers or baghouses can be used.

The temperature of the bed can be as high as approximately 1900°F (1040°C), though in most applications, temperatures this high are neither required nor desirable.[14] Most systems operate in the 1400°F to 1650°F (760°C to 900°C) range.

Three main options exist for reducing temperatures with overautogeneous fuels:[15] (1) water can be injected into the bed, which has the disadvantage of reducing downstream heat recovery; (2) excess air can be injected, requiring the entire system to be sized for the excess air, increasing its cost; (3) and heat-exchange coils can be placed within the bed itself, in which case, the name *fluidized-bed boiler* is applicable.

Air can be preheated to approximately 1000°F to 1250°F (540°C to 680°C) for use with subautogenous fuels, or auxiliary fuel can be used.

Contempory FBC boilers for steam/electricity generation are typically of the *circulating fluidized-bed boiler design*. An important aspect of FBC boiler operation is the in-bed gas desulfurization and dechlorination that occurs when limestone and other solid reagents are injected into the combustion area. In addition, circulating fluidized-bed boilers have very low NOx emissions (i.e., less than 200 ppm).

21. INJECTION WELLS

Properly treated and stabilized liquid and low-viscosity wastes can be injected under high pressure into appropriate strata 2000 ft to 6000 ft (600 m to 1800 m) below the surface. The wastes displace natural fluids, and the injection well is capped to maintain the pressure. Injection wells fail primarily by waste plumes through fractures, cracks, fault slips, and seepage around the well casing. Figure 70.7 shows a typical injection well installation.

22. INCINERATORS, FLUIDIZED-BED COMBUSTION

In an FBC, combustion is efficient and excess air required is low (e.g., 25% to 50%). Destruction is essentially complete due to the long residence time (5 sec to 8 sec for gases, and even longer for solids), high turbulence, and exposure to oxygen. Combustion control is simple, and combustion is often maintained within a 15°F (8°C) band. Both overautogeneous and subautogenous feeds can be handled. Due to the thermal mass of the bed, the FBC temperature drops slowly—at the rate of about 10°F/hr (5°C/h) after shutdown. Start-up is fast and operation can be intermittent.

[14]It is a common misconception that extremely high temperatures are required to destroy hazardous waste.

[15]A *subautogenous waste* has a heating value too low to sustain combustion and requires a supplemental fuel. Conversely, an *overautogeneous waste* has a heating value in excess of what is required to sustain combustion and requires temperature control.

Figure 70.7 Typical Injection Well Installation

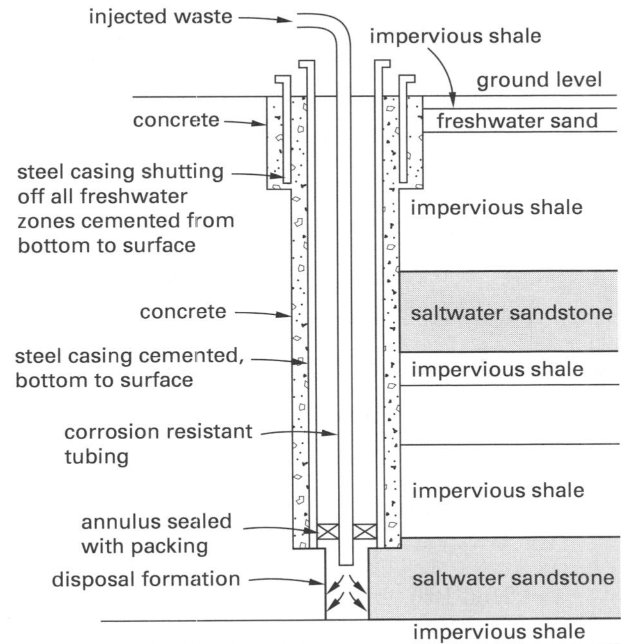

Relative to other hazardous waste disposal methods, NOx and metal emissions are low, and the organic content of the ash is very low (e.g., below 1%). Because of the long residence time, FBC systems do not usually require afterburners or secondary combustion chambers (SCCs). Due to the turbulent combustion process, the combustion temperature can be 200°F to 300°F (110°C to 170°C) lower than in rotary kilns. These factors translate into fuel savings, fewer or no NOx emissions, and lower metal emissions.

Most limitations of FBC systems relate to the feed. The feed must be of a form (small size and roughly regular in shape) that can be fluidized. This makes FBCs ideal for non-atomizable liquids, slurries, sludges, tars, and granular solids. It is more appropriate (and economical) to destroy atomizable liquid wastes in a boiler or special injection furnace and large bulky wastes in a rotary kiln or incinerator.

FBCs are usually inapplicable when the feed material or its ash melts below the bed temperature.[16] If the feed melts, it will agglomerate and defluidize the bed. When it is necessary to burn a feed with low melting temperature, two methods can be used. The bed can be operated as a *chemically active bed*, where operation is close to but below the ash melting point of 1350°F to 1450°F (730°C to 790°C). A small amount of controlled melting is permitted to occur, and the agglomerated ash is removed.

When a higher temperature is desirable in order to destroy hazardous materials, the melting point of feed can be increased by injecting low-cost additives (e.g., calcium hydroxide or kaolin clay). These combine with salts of alkali metals to form refractory-like materials with melting points in the 1950°F to 2350°F (1070°C to 1290°C) range.

The design of a fluidizing-bed incinerator starts with a heat balance. Energy enters the FBC during combustion of the feed and auxiliary fuel and from sensible heat contributions of the air and fuel. Some of the heat is recovered and reused. The remainder of the heat is lost through sensible heating of the combustion products, water vapor, excess air, and ash, and through radiation from the combustor vessel and associated equipment. Since these items may not all be known in advance, the following assumptions are reasonable.

- Radiation losses are approximately 5%.
- Sensible heat losses from the ash are minimal, particularly if the feed contains significant amounts of moisture.
- Excess air will be approximately 40%.
- Combustion temperature will be approximately 1400°F (760°C).
- If air is preheated, the preheat temperature will be approximately 1000°F (540°C).

The fuel's *specific feed characteristic* (SFC) is the ratio of the higher (gross) heating value to the moisture content.

$$\text{SFC} = \frac{\text{total heat of combustion}}{\text{mass of water}} = \frac{m_{\text{solid}}\text{HHV}}{m_w} \qquad 70.31$$

Typical values of the specific feed characteristic range from a low of approximately 1000 Btu/lbm (2300 kJ/kg) for wastewater treatment plant biosolids, to more than 60,000 Btu/lbm (140 MJ/kg) for barks, sawdust, and RDF.[17] Making the previous assumptions and using the SFC as an indicator, the following generalizations can be made.

- Fuels with SFCs that are less than 2600 Btu/lbm (6060 kJ/kg) require a 1000°F (540°C) hot windbox, and are autogenous at that value and subautogenous below that value. The SFC drops to 2400 Btu/lbm (5600 kJ/kg) with air preheated to 1200°F (650°C). Below the subautogenous SFC, water evaporation is the controlling design factor, and auxiliary fuel is required.
- Fuels with an SFC of 4000 Btu/lbm (9300 kJ/kg) are autogenous with a cold windbox and are overautogenous above that value. Combustion is the controlling design factor for overautogenous SFCs.

[16]The melting temperature of a eutectic mixture of two components may be lower than the individual melting temperatures.
[17]Although the units are the same, the specific feed characteristic is not the same as the heating value.

Example 70.1

A wastewater treatment sludge is dewatered to 75% water by weight. The solids have a heating value of 6500 Btu/lbm (15 000 kJ/kg). The dewatered sludge enters a fluidized-bed combustor with a 1000°F (540°C) windbox at the rate of 15,000 lbm/hr (1.9 kg/s). (a) What is the specific feed characteristic? (b) Approximately what energy (in Btu/hr) must the auxiliary fuel supply?

SI Solution

(a) The dewatered sludge consists of 25% combustible solids and 75% moisture. The total mass of sludge required to contain 1.0 kg water is

$$m_{\text{sludge}} = \frac{m_w}{x_w} = \frac{1 \text{ kg}}{0.75} = 1.333 \text{ kg}$$

The mass of combustible solids per kilogram of water is

$$\begin{aligned} m_{\text{solids}} &= 0.25 m_{\text{sludge}} = (0.25)(1.333 \text{ kg}) \\ &= 0.3333 \text{ kg} \end{aligned}$$

From Eq. 70.31, the specific feed characteristic is

$$\begin{aligned} \text{SFC} &= \frac{\text{total heat of combustion}}{m_w} = \frac{m_{\text{solids}}\text{HHV}}{m_w} \\ &= \frac{(0.3333 \text{ kg})\left(15\,000 \ \frac{\text{kJ}}{\text{kg}}\right)\left(1000 \ \frac{\text{J}}{\text{kJ}}\right)}{1 \text{ kg}} \\ &= 5.0 \times 10^6 \text{ J/kg} \end{aligned}$$

(b) Making the listed assumptions, operation with a 540°C hot windbox requires an SFC of approximately 6060 kJ/kg. Therefore, the energy per kilogram of water in the fuel that an auxiliary fuel must provide is

$$(6060 \text{ kJ})\left(1000 \ \frac{\text{J}}{\text{kJ}}\right) - 5 \times 10^6 \text{ J} = 1.06 \times 10^6 \text{ J}$$

The energy supplied by the auxiliary fuel is

$$\begin{aligned} &\dot{m}_{\text{fuel}} x_w \times \text{SFC deficit} \\ &\quad = \left(1.9 \ \frac{\text{kg}}{\text{s}}\right)(0.75)\left(1.06 \times 10^6 \ \frac{\text{J}}{\text{kg}}\right) \\ &\quad = 1.51 \times 10^6 \text{ J/s} \quad (1.5 \text{ MW}) \end{aligned}$$

Customary U.S. Solution

(a) The dewatered sludge consists of 25% combustible solids and 75% moisture. The total mass of sludge required to contain 1.0 lbm water is

$$m_{\text{sludge}} = \frac{m_w}{x_w} = \frac{1 \text{ lbm}}{0.75} = 1.333 \text{ lbm}$$

The mass of combustible solids per pound of water is

$$\begin{aligned} m_{\text{solids}} &= 0.25 m_{\text{sludge}} = (0.25)(1.333 \text{ lbm}) \\ &= 0.3333 \text{ lbm} \end{aligned}$$

From Eq. 70.31, the specific feed characteristic is

$$\begin{aligned} \text{SFC} &= \frac{\text{total heat of combustion}}{m_w} = \frac{m_{\text{solids}}\text{HHV}}{m_w} \\ &= \frac{(0.3333 \text{ lbm})\left(6500 \ \frac{\text{Btu}}{\text{lbm}}\right)}{1 \text{ lbm}} \\ &= 2166 \text{ Btu/lbm} \end{aligned}$$

(b) Making the listed assumptions, operation with a 1000°F hot windbox requires an autogenous SFC of approximately 2600 Btu/lbm. Therefore, the energy per pound of water in the fuel that an auxiliary fuel must provide is

$$2600 \text{ Btu} - 2166 \text{ Btu} = 434 \text{ Btu}$$

The energy supplied by the auxiliary fuel is

$$\begin{aligned} &\dot{m}_{\text{fuel}} x_w \times \text{SFC deficit} \\ &\quad = \left(15{,}000 \ \frac{\text{lbm}}{\text{hr}}\right)(0.75)\left(434 \ \frac{\text{Btu}}{\text{lbm}}\right) \\ &\quad = 4.88 \times 10^6 \text{ Btu/hr} \end{aligned}$$

23. INCINERATION, GENERAL

Most rotary kiln and liquid injection incinerators have *primary* and *secondary combustion chambers* (SCCs). Kiln temperatures are approximately 1200°F to 1400°F (650°C to 760°C) for soil incineration and up to 1700°F (930°C) for other waste types.[18] SCC temperatures are higher—1800°F (980°C) for most hazardous wastes and 2200°F (1200°C) for liquid PCBs. The waste heat may be recovered in a boiler, but the combustion gas must be cooled prior to further processing. *Thermal ballast* can be accomplished by injecting large amounts of excess air (typical when liquid fuels are burned) or by quenching with a water spray (typical in rotary kilns).

SCCs are necessary to destroy toxics in the off-gases. SCCs are vertical units with high-swirl, vortex-type burners. These produce high *destruction removal*

[18]Temperatures higher than 1400°F (760°C) may cause incinerated soil to vitrify and clog the incinerator.

efficiencies (DREs) with low retention times (e.g., 0.5 sec) and moderate-to-high temperatures, even for chlorinated compounds. When soil with fine clay is incinerated, fines can build up in the SCC, causing slagging and other problems. A refractory-lined cyclone located after the primary combustion chamber can be used to reduce particle carryover to the SCC.

Prior to full operation, incinerators must be tested in a trial burn with a *principal organic hazardous constituent* (POHC) that is in or has been added to the waste. The POHC must be destroyed with a DRE of at least 99.99% by weight.

For nontoxic organics, a DRE of 95% is a common requirement. Hazardous wastes require a DRE of 99.99%. Certain hazardous wastes, including PCBs and dioxins, require a 99.9999% DRE. This is known as the *six nines rule.*

Emission limitations of some pollutants depend on the incoming concentration and the height of the stack.[19] Thus, in certain circumstances, raising the stack is the most effective method of being in compliance. This is considered justified on the basis that ground-level concentrations will be lower with higher stacks.

Rules of thumb regarding incinerator performance are as follows.

- Stoichiometric combustion requires approximately 725 lbm (330 kg) of air for each million Btu of fuel or waste burned.
- 100% excess air is required.
- Stack gas dew point is approximately 180°F (80°C).
- Water-spray-quenched flue gas is approximately 40% moisture by weight.

Table 70.6 gives performance parameters typical of incinerators.

Table 70.6 *Representative Incinerator Performance*

	type of incinerator/use			
			soil incinerators	
	rotary kiln	liquid injection	hazardous	non-hazardous
waste heating value				
(Btu/lbm)	15,000	20,000	0	0
(MJ/kg)	35	46	0	0
kiln temp				
(°F)	1700	–	1650	850
(°C)	925	–	900	450
SCC temp				
(°F)	1800–2200	2000–2200	1800	1400
(°C)	980–1200	1100–1200	980	750
SCC mean residence time (sec)	2	2	2	1
O_2 in stack gas (%)	10%	12%	9%	6%

Example 70.2

If no additional fuel is added to the incinerator, what is the mass (in lbm/hr) of water required to spray-quench combustion gases from the incineration of hazardous waste with a heating rate of 50 MBtu/hr?

Solution

Use the rules of thumb. Assume 100% excess air is required. The total dry air required is

$$m_a = \frac{(2)\left(50\times 10^6\ \frac{\text{Btu}}{\text{hr}}\right)(725\ \text{lbm})}{10^6\ \frac{\text{Btu}}{\text{hr}}} = 72{,}500\ \text{lbm/hr}\quad[\text{dry}]$$

Since the quenched combustion gas is 40% water by weight, it is 60% dry air by weight. The total mass of wet combustion gas produced per hour is

$$m_t = \frac{m_a}{x_a} = \frac{72{,}500\ \frac{\text{lbm}}{\text{hr}}}{0.6} = 1.21\times 10^5\ \text{lbm/hr}$$

The required mass of quenching water is

$$m_w = x_w m_t = (0.4)\left(1.21\times 10^5\ \frac{\text{lbm}}{\text{hr}}\right) = 4.84\times 10^4\ \text{lbm/hr}$$

24. INCINERATION, HAZARDOUS WASTES

Most hazardous waste incinerators use rotary kilns. Waste in solid and paste form enters a rotating drum where it is burned at 1850°F to 2200°F (1000°C to 1200°C). (See Table 70.6 for other representative performance characteristics.) Slag is removed at the bottom, and toxic gases exit to a tall, vortex secondary combustion chamber (SCC). Gases remain in the SCC for 2 sec to 4 sec where they are completely burned at approximately 1850°F (1000°C). Liquid wastes are introduced into and destroyed by the SCC as well. Heat from off-gases may be recovered in a

[19]Some of the metallic pollutants treated this way include antimony, arsenic, barium, beryllium, cadmium, chromium, lead, mercury, silver, and thallium.

boiler. Typical flue gas cleaning processes include electrostatic precipitation, two-stage scrubbing, and NOx removal.

Common problems with hazardous waste incinerators include (1) inadequate combustion efficiency (easily caused by air leakage in the drum and uneven fuel loading) resulting in incomplete combustion of the primary organic hazardous component (POHC), emission of CO, NOx, and *products of incomplete combustion* (also known as *partially incinerated compounds* or PICS) and metals, (2) meeting low dioxin limits, and (3) minimizing the toxicity of slag and fly ash.

These problems are addressed by (1) reducing air leaks in the drum and (2) introducing air to the SCC through multiple sets of ports at specific levels. Gas is burned in the SCC in substoichiometric conditions, with the vortex ensuring adequate mixing to obtain complete combustion. Dioxin formation can be reduced by eliminating the waste-heat recovery process, since the lower temperatures present near waste-heat boilers are ideal for dioxin formation. Once formed, dioxin is removed by traditional end-of-pipe methods. Figure 70.8 shows a large-scale hazardous waste incinerator.

25. INCINERATION, INFRARED

Infrared incineration (II) is effective for reducing dioxins to undetectable levels. The basic II system consists of a waste feed conveyor, an electrical-heated primary chamber, a gas-fired afterburner, and a typical flue gas cleanup (FGC) system (i.e., scrubber, electrostatic precipitator, and/or baghouse). Electrical heating elements heat organic wastes to their combustion temperatures. Off-gas is burned in a secondary combustion chamber (SCC).

26. INCINERATION, LIQUIDS

Liquid-injection incinerators can be used for atomizable liquids. Such incinerators have burners that fire directly into a refractory-lined chamber. If the liquid waste contains salts or metals, a downfired liquid-injection incinerator is used with a submerged quench to capture the molten material. A typical flue gas cleanup (FGC) system (i.e., scrubber, electrostatic precipitator, and/or baghouse) completes the system.

Incineration of organic liquid wastes usually requires little external fuel, since the wastes are overautogenous and have good heating values. The heating value is approximately 20,000 Btu/lbm (47 MJ/kg) for solvents and approximately 8000 Btu/lbm to 18,000 Btu/lbm (19 MJ/kg to 42 MJ/kg) for chlorinated compounds.

27. INCINERATION, MUNICIPAL SOLID WASTE

Incinerator/generator facilities with separation capability are known as *resource-recovery plants*, commonly referred to as *waste-to-energy facilities*. Incineration of *municipal solid waste* (MSW) results in a 90% reduction in waste disposal volume and a mass reduction of 75%.

In *no-boiler incinerators*, MSW is efficiently burned without steam generation. However, incineration is often accompanied by electrical power generation. In *waste-to-energy* (WTE) plants, the combustion heat is used to generate steam and electrical power. *Mass burning* is the incineration of unprocessed MSW, typically on stoker grates or in rotary and waterwall combustors to generate steam. With mass burning, MSW is unloaded into a pit and then moved by crane to the furnace-conveying mechanism. (See Fig. 70.9.) Approximately 27% of the MSW remains as ash, which consists of glass, sand, stones, aluminum, and other noncombustible materials. It and the fly ash are usually collected and disposed of in municipal landfills.[20] Capacities of typical mass burn units vary from less than 400 tpd (360 Mg/d) to a high of 3000 tpd (2700 Mg/d), with the majority of units processing 1000 tpd to 2000 tpd (910 Mg/d to 1800 Mg/d).

MSW, as collected, has a heating value of approximately 4500 Btu/lbm (10 MJ/kg), though higher values have been reported. Per 1000 tpd (1000 Mg/d) of MSW incinerated, the yields are approximately 150,000 lbm/hr to 200,000 lbm/hr (75 Mg/h to 1000 Mg/h) of 650 psig to 750 psig (4.5 MPa to 5.2 MPa) steam at 700°F to 750°F (370°C to 400°C) and 25 MW to 30 MW (27.6 MW to 33.1 MW) of gross electrical power. Approximately 10% of the electrical power is used internally, and units generating much less than approximately 10 MW (gross) may use all of their generated electrical power internally.

The *burning rate* varies from approximately 40 lbm/hr-ft^2 to 60 lbm/hr-ft^2 (200 kg/h·m^2 to 300 kg/h·m^2) and is the fueling rate divided by the total effective grate area. Maximum heat release rates are approximately 300,000 Btu/hr-ft^2 (3.4 GJ/h·m^2). The *heat release rate*, HRR, is defined as

$$\text{HRR} = \frac{(\text{fueling rate})(\text{HHV})}{\text{total effective grate area}} \qquad 70.32$$

Refuse-derived fuel (RDF) is derived from MSW. First generation RDF plants used "crunch and burn" technology. In these plants, the MSW is shredded after ferrous metals are removed magnetically. First-generation RDF plants suffer from the same problems that have

[20]The U.S. Environmental Protection Agency (EPA) has ruled that ash from the incineration of MSW is not a hazardous waste. However, this is hotly contested and is subject to ongoing evaluation.

Figure 70.8 *Large-Scale Hazardous Waste Incinerator*

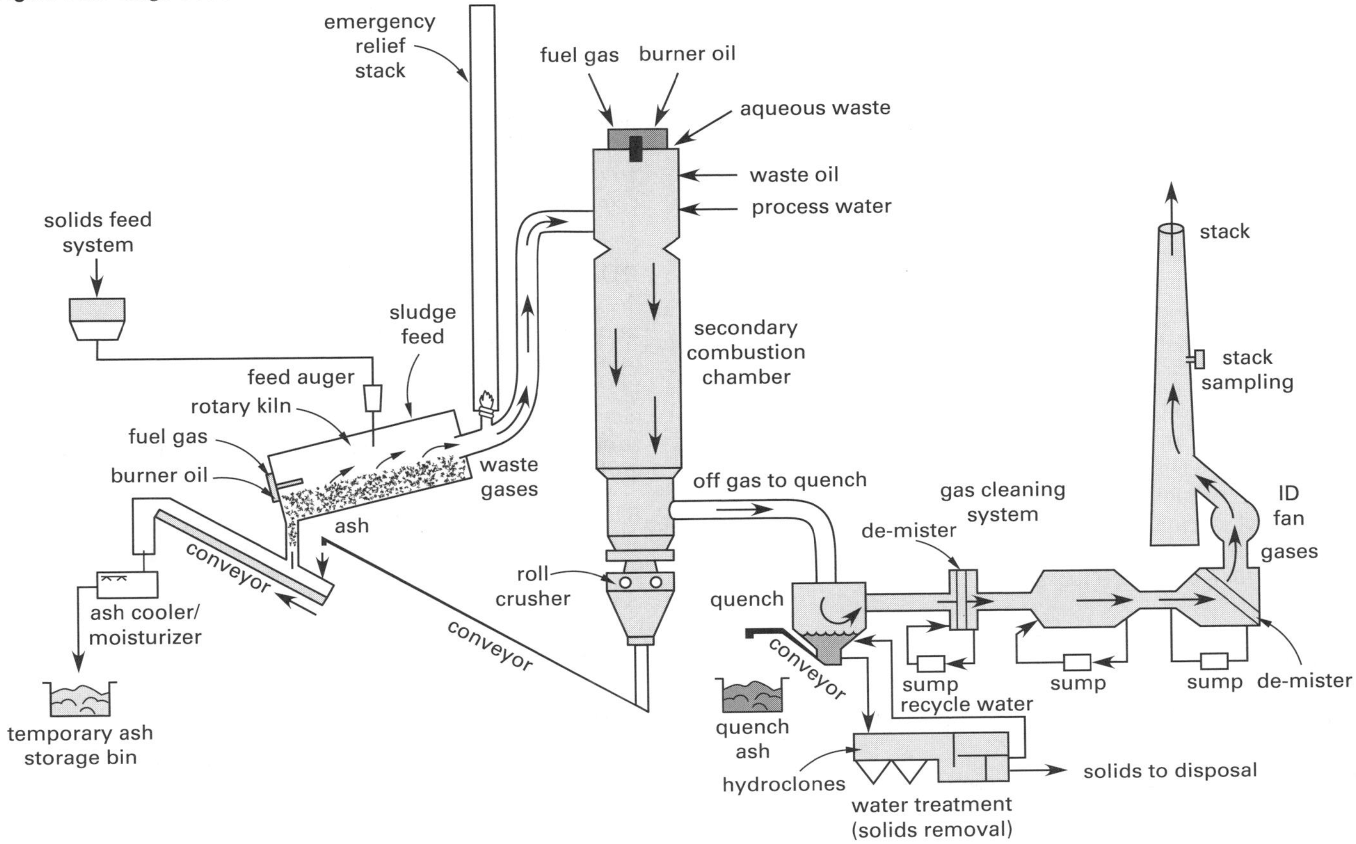

Figure 70.9 *Typical Mass-Burn Waste-to-Energy Plant*

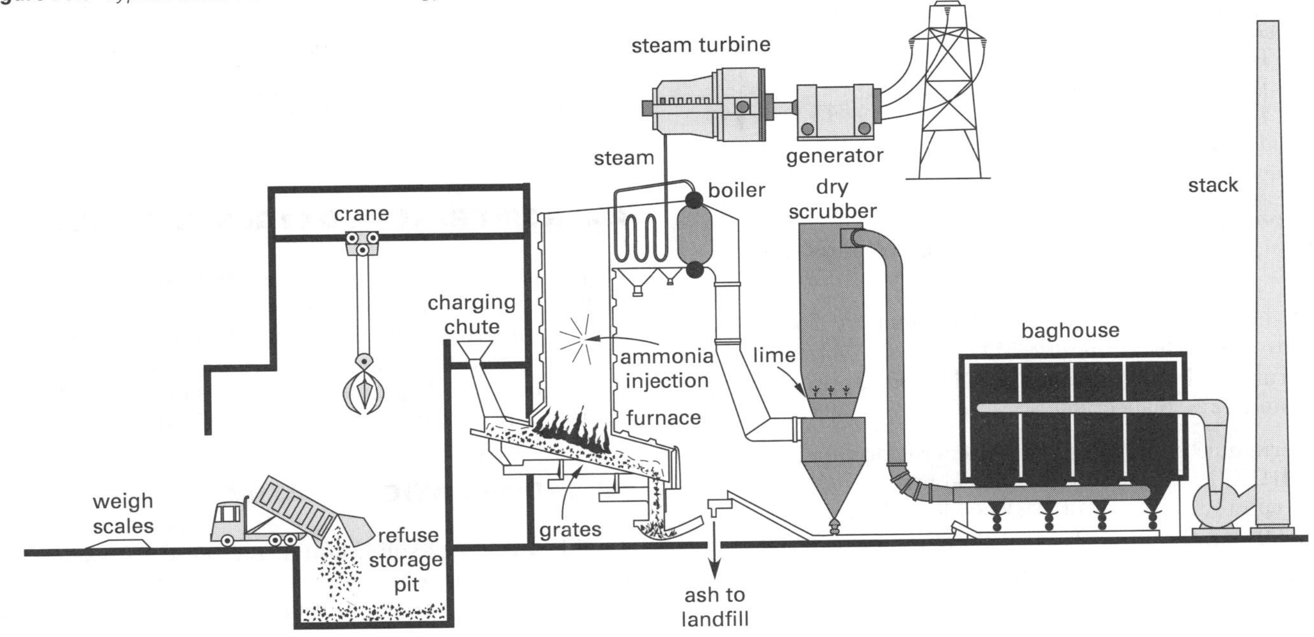

plagued early mass burn units, including ash with excessive quantities of noncombustible materials such as glass, grit and sand, and aluminum.

Second-generation plants incorporated screens and air classifiers to reduce noncombustible materials and to increase the recovery of some materials.[21] The ash content is reduced and the energy content of the RDF is increased. The MSW is converted into RDF pellets 2½ in to 6 in (6.4 cm to 15 cm) in size and introduced through feed ports above a traveling grate. Some of the fuel is burned in suspension, with the rest burned on the grate. Grate speed is varied so that the fuel is completely incinerated by the time it reaches the ash rejection ports at the front of the burner.

Large, 2000 tpd to 3000 tpd (1800 Mg/d to 2700 Mg/d), third-generation RDF plants, illustrated in Fig. 70.10, retain greater than 95% of the original MSW combustibles while reducing *mass yield* (i.e., the ratio of RDF mass to MSW mass) to below 85%.

Figure 70.10 *Typical RDF Processing*[a,b]

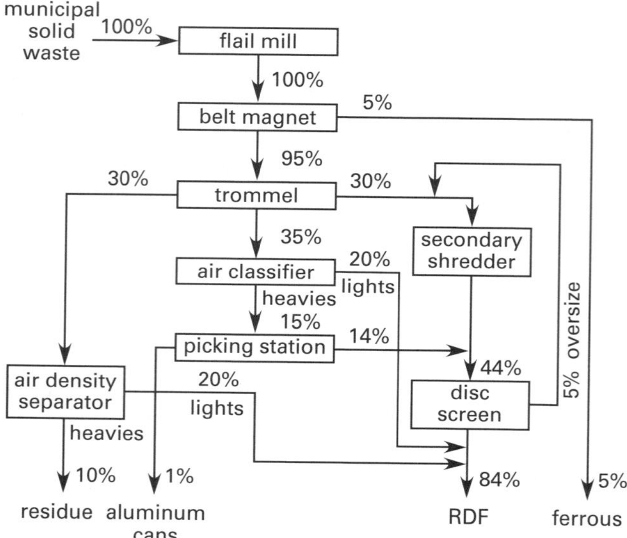

[a]Evaporative water losses not shown.
[b]Special lines to process bulky wastes and shred tires not shown.

RDF has a heating value of approximately 5500 Btu/lbm to 5900 Btu/lbm (12 MJ/kg to 14 MJ/kg). The moisture and ash contents of RDF are approximately 24% and 12%, respectively.

Performance of a typical third-generation facility burning RDF is similar to a mass-burn unit. For each 1000 tpd (1000 Mg/d) of MSW collected, approximately 150,000 lbm/hr to 250,000 lbm/hr (75 Mg/h to 125 Mg/h) of 750 psig to 850 psig (5.2 MPa to 5.9 MPa) steam at 750°F to 825°F (400°C to 440°C) can be generated, resulting in approximately 30 MW to 40 MW (33.1 MW to 44.1 MW) of electrical power.

Natural gas is introduced at start-up to bring the furnace up to the required 1800°F (1000°C) operating temperature. Natural gas can also be used upon demand, as when a load of particularly wet RDF lowers the furnace temperature.

Coal, oil, or natural gas can be used as a back-up fuel if RDF is unavailable or as an intended part of a *co-firing* design. Co-firing installations are relatively rare, and many have discontinued burning RDF due to poor economic performance. Typical problems with co-firing are (1) upper furnace wall slagging, (2) decreased efficiencies in electrostatic precipitators, (3) increased boiler tube corrosion, (4) excessive amounts of bottom ash, and (5) difficulties in receiving and handling RDF.

RDF is a low-sulfur fuel, but like MSW, RDF is high in ash and chlorine. Relative to coal, RDF produces less SO_2 but more hydrogen chloride. Bottom ash can also contain trace organics and lead, cadmium, and other heavy metals. Table 70.7 shows the typical composition of MSW and RDF.

Table 70.7 *Typical Ultimate Analyses of MSW and RDF*

	percentage by weight	
element	MSW	RDF
carbon	26.65	31.00
water	25.30	27.14
ash	23.65	13.63
oxygen	19.61	22.72
hydrogen	3.61	4.17
chlorine	0.55	0.66
nitrogen	0.46	0.49
sulfur	0.17	0.19
total	100.00	100.00

28. INCINERATION, OXYGEN-ENRICHED

Oxygen-enriched incineration is intended primarily for dioxin removal and is operationally similar to that of a rotary kiln. However, the burner includes oxidant jets. The jets aspirate furnace gases to provide more oxygen for combustion. Apparent advantages are low NOx production and increased incinerator feed rates.

29. INCINERATION, PLASMA

A wide variety of solid, liquid, and gaseous wastes can be treated in a *plasma incinerator*. Wastes are heated by an electric arc to higher than 5000°F (2760°C), dissociating them into component atoms. Upon cooling,

[21]*Material recovery facilities* (MRFs) specialize in sorting out recyclables from MSF.

atoms recombine into hydrogen gas, nitrogen gas, carbon monoxide, hydrochloric acid, and particulate carbon. The ash cools to a nonleachable, vitrified matrix. Off-gases pass through a normal train of cyclone, baghouse, and scrubbing operations. The process has a very high DRE. However, energy requirements are high.

30. INCINERATION, SOIL

Incineration can completely decontaminate soil. Incinerators are often thought of as being fixed facilities, as are cement kilns and special-use (e.g., Superfund) incinerators. However, mobile incinerators can be brought to sites when the soil quantities are large (e.g., 2000 tons to 100,000 tons (1800 Mg to 90,000 Mg)) and enough setup space is available.[22] (See also Sec. 70.31.)

31. INCINERATION, SOLIDS AND SLUDGES

Rotary kilns and fluidized-bed incinerators are commonly used to incinerate solids and sludges. The feed system depends on the waste's physical characteristics. Ram feeders are used for boxed or drummed solids. Bulk solids are fed via chutes or screw feeders. Sludges are fed via lances or by premixing with solids.

The constant rotation of the shell moves wastes through rotary kilns and promotes incineration. External fuel is required in rotary kilns if the heating value of the waste is below 1200 Btu/lbm (2800 kJ/mg) (i.e., is subautogenous). Additional fuel is required in the secondary chamber, as well.

Fluidized-bed incinerators work best when the waste is consistent in size and texture. An important benefit is the ability to introduce limestone and other solid reagents to the bed in order to remove HCl and SO_2.

32. INCINERATION, VAPORS

Vapor incinerators, also known as *afterburners* and *flares*, convert combustible materials (gases, vapors, and particulate matter) in the stack gas to carbon dioxide and water. Afterburners can be either direct-flame or catalytic in operation.

33. LANDFILLING

Since 1992, new and expanded municipal landfills in the United States must satisfy strict design regulations and are designated *Subtitle D landfills*.[23,24] The regulations apply to any landfill designed to hold *municipal solid waste* (MSW), biosolids, and ash from MSW combustion. Construction of landfills is prohibited near sensitive areas such as airports, floodplains, wetlands, earthquake zones, and geologically unstable terrain. Air quality control methods are required to control emission of dust, odors, and landfill gas. Runoff from storms must be controlled.

Subtitle D landfills have sophisticated liners and leachate collection systems. While states can specify greater protection, the basic bottom requirements are a 30 mil flexible PVC membrane liner (FML)[25] and at least 2 ft of compacted soil with a maximum hydraulic conductivity of 1.2×10^{-5} ft/hr (1×10^{-7} cm/s). A series of wells is required to detect high hydraulic heads and accumulation of heavy metals and volatile organic compounds (VOCs) in the leachate. The minimum thickness is 20 mils for the top FML, and the maximum hydraulic conductivity of the cover soil is 1.8×10^{-5} ft/hr (1.5×10^{-7} cm/s). Figure 70.11 shows a Subtitle D landfill in detail.

34. LOW EXCESS-AIR BURNERS

NOx formation in gas-fired boilers can be reduced by maintaining excess air below 5%.[26] *Low excess-air burners* use a forced-draft and self-recirculating combustion chamber configuration to approximate multistaged combustion.

35. LOW-NOX BURNERS

Low (or *ultra-low*) *NOx burners* (LNB) in gas-fired applications use a combination of staged-fuel burning and internal flue gas recirculation (FGR). Recirculation within a burner is induced by either the pressure of the fuel gas or other agents (e.g., medium-pressure steam or compressed air).

36. MECHANICAL SEALS

Fugitive emissions from pumps and other rotating equipment can be reduced or eliminated using current technology by the proper selection and installation of mechanical seals. The three major classes of mechanical seals are single seals, tandem seals (dual seals placed next to each other), and double seals (dual seals mounted back to back or face to face).

The most economical and reliable sealing device is a *single seal*. It has a minimum number of parts and requires no support devices. However, since the pumped product is usually the lubricant for the seal face, small amounts of the product escape into the environment. Emissions are generally below 1000 ppmv, and are often below 100 ppmv.

[22]Modified asphalt batch processing plants can be used to incinerate soils contaminated with low-heating value, nonchlorinated hydrocarbons.
[23]The reference is to Subtitle D of the RCRA.
[24]Closures are also regulated.
[25]If the membrane is high-density polyethylene (HDPE), the minimum thickness is 60 mils. Since 30 mil PVC costs less than 60 mil HDPE, bidded public projects may end up using the less expensive product, while contracted private projects will opt for the superior protection of the 60 mil product.
[26]Reducing excess air from 30% to 10%, for example, can reduce NOx emissions by 30%.

Figure 70.11 *Subtitle D Liner and Cover Detail*

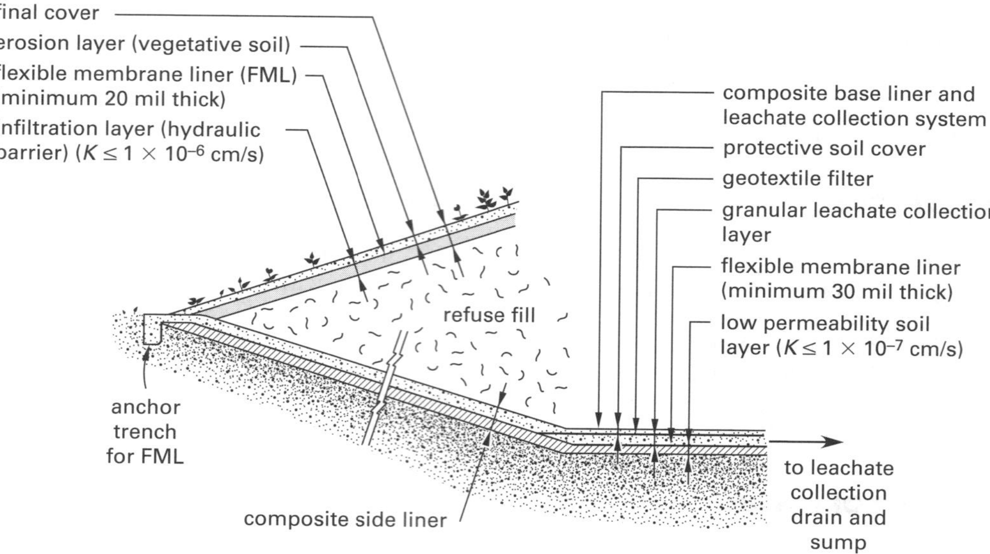

Tandem seals consist of two seal assemblies separated by a buffer fluid at a pressure lower than the seal-chamber liquid. The primary inner seal operates at full pump pressure. The outer seal operates in the buffer fluid. Tandem seals can achieve zero emissions when used with a vapor recovery system, provided that the pumped product's specific gravity is less than that of the buffer fluid and the product is immiscible with the buffer fluid.

If the buffer fluid used in a tandem dual seal is a controlled substance, emissions from the outer seal must also meet the emission limits. Seals using glycol as the buffer fluid typically achieve zero emissions. Seals using diesel oil or kerosene have emissions in the 25 ppmv to 100 ppmv range.

Double seals are recommended for hazardous fluids with specific gravities less than 0.4 and are a good choice when a vapor recovery system is not available. (Liquids with low specific gravities do not lubricate the seal well.) Double seals consist of two seal assemblies connected by a common collar; they operate with a barrier fluid kept at a pressure higher than that of the pumped product. Double seals can reduce the emission rate to zero.

Figure 70.12 shows the relationship between typical emissions and specific gravity for the different seal types.

Figure 70.12 *Representative Emissions to Atmosphere for Mechanical Seals* (1 cm from source)*

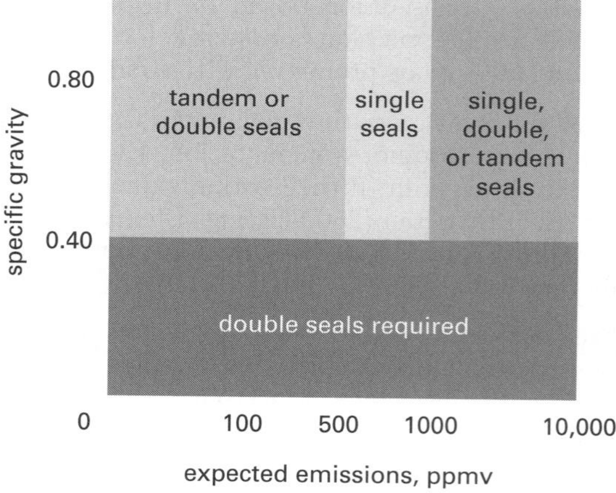

*maximum seal size, 6 in (153 mm); maximum pressure, 600 psig (40 bar); maximum speed, 3600 rpm.

37. MULTIPLE PIPE CONTAINMENT

The U.S. Environmental Protection Agency (EPA) requires *multiple pipe containment* (MPC) in the storage and transmission of hazardous fluids.[27] MPC systems consist of a carrier pipe or bundle of pipes, a common casing, and a leak detector consisting of redundant instrumentation with automatic shutdown features. Casings are generally not pressurized. If one of the pipes within the bundle develops a leak, the fluid will remain in the casing and the detector will signal an alarm and automatically shut off the liquid flow.

Some of the many factors that go into the design of an MPC system are (1) the number of pipes to be grouped together; (2) the weight and strength of the pipes, carrier, and supports; (3) the compatibility of pipe materials and fluids; (4) differential expansion of the container and carrier pipes; and (5) a method of leak detection.

Carrier pipes within the casing should be supported and separated from each other by internal supports (i.e., perforated baffles within the casing). Carrier pipes pass through oval-shaped holes in the internal supports. Holes are oval

[27]In addition to multiple pipe containment, other factors that contribute to reduction of fugitive emissions are (1) the use of tongue-in-groove flanges, and (2) the reduction of as many nozzles as possible from storage tanks and reactor vessels.

and oversized to allow flexing of the carrier pipes due to expansion. Other holes through the internal supports allow venting and draining of the casing in the event of a leak. If multiple detectors are used, isolation baffles can be included to separate the casing into smaller sections.

When a pipe bundle changes elevation as well as direction, the carrier pipes may change their positions relative to one another. A pipe on the outside (i.e., at 3 o'clock) may end up being the lower pipe (i.e., at 6 o'clock).[28] This change in relative positioning is known as *rotation*. Rotation complicates the accommodation of pipes that must enter and exit the bundle at specific locations.

Rotation occurs when a change in direction is located at the top or bottom of an elevation change. Therefore, unless rotation can be tolerated, changes in direction should not be combined with changes in elevation. In Fig. 70.13(a), pipe A remains to the left of pipe B. In Fig. 70.13(b), pipe A starts out to the left of pipe B and ends up above pipe B.

Figure 70.13 *Rotation in Pipe Bundles*

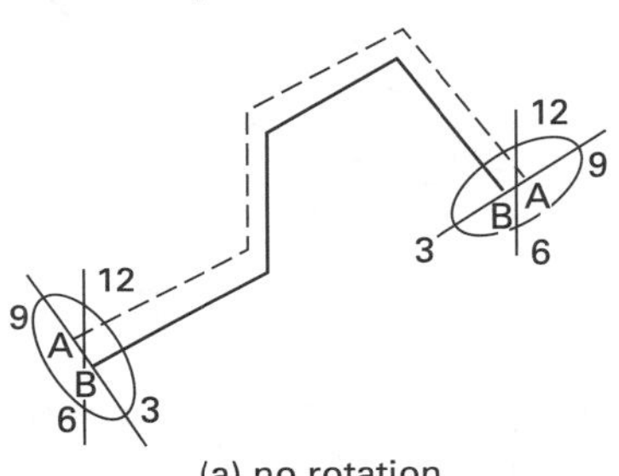

(a) no rotation

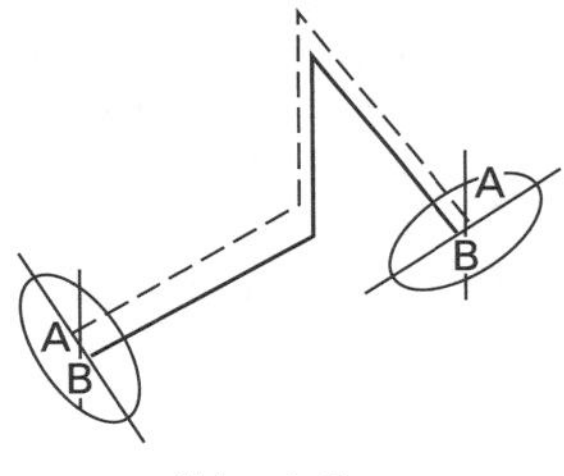

(b) rotation

Another problem associated with MPC systems is expansion of the containment casing and carrier pipes caused by differences in ambient and fluid temperatures and the use of different pipe materials. For pipes containing direct changes, there must be space at each elbow to permit the pipe expansion. If both ends of a pipe are fixed, expansion and contraction will cause the pipes to flex unless Z-bends or loops are included.

Differential expansion must be calculated from the largest expected temperature difference using Eq. 70.33.

α is the coefficient of thermal expansion found from Table 70.8.

$$\Delta L = \alpha L \Delta T \qquad 70.33$$

Computer programs are commonly used to design pipe networks. These programs are helpful in locating guides and anchors and in performing stress analyses on the final design. Drains, purge points, baffles, and joints are generally chosen by experience.

Table 70.8 *Approximate Coefficients of Thermal Expansion of Piping Materials (at 70°F (21°C))*

pipe material	U.S. (ft/ft-°F)	SI (m/m·°C)
carbon steel	6.1×10^{-6}	1.1×10^{-5}
chlorinated polyvinyl chloride (CPVC)	3.8×10^{-5}	6.8×10^{-5}
copper	9.5×10^{-6}	1.7×10^{-5}
fiberglass-reinforced polyethylene (FRP)	8.5×10^{-6}	1.5×10^{-5}
polyethylene (PE)	8.3×10^{-5}	1.5×10^{-4}
polyvinyl chloride (PVC)	3.0×10^{-5}	5.4×10^{-5}
stainless steel	9.1×10^{-6}	1.6×10^{-5}

(Multiply ft/ft-°F by 1.8 to get m/m·°C.)

38. OZONATION

Ozonation is one of several advanced oxidation methods capable of reducing pollutants from water. Ozone is a powerful oxidant produced by electric discharge through liquid oxygen. Ozone is routinely used in the water treatment industry for disinfection. Table 70.9 shows reactions used for oxidation of several common industrial wastewater pollutants.

Table 70.9 *Oxidation of Industrial Wastewater Pollutants**

	Cl_2	ClO_2	$KMnO_4$	O_3	H_2O_2	OH^-
amines	C	P	P	P	P	C
ammonia	C	N	P	N	N	N
bacteria	C	C	C	C	P	C
carbohydrates	P	P	P	P	N	C
chlorinated solvents	P	P	P	P	N	C
phenols	P	C	C	C	P	C
sulfides	C	C	C	C	C	C

*C = complete reaction; P = partially effective; N = not effective

[28]It is important to consistently "face" the same way (e.g., in the direction of flow).

Plant Engineering

39. SCRUBBING, GENERAL

Scrubbing is the act of removing particulate matter from a gaseous stream, although substances in gaseous and liquid forms may also be removed.[29]

40. SCRUBBING, CHEMICAL

Chemical scrubbing using oxidizing compounds such as chlorine, ozone, hypochlorite, or permanganate rapidly destroys volatile organic compounds (VOCs). Efficiencies are typically 95% for highly reactive substances, but are lower for hydrocarbons and substances with low reactivities.

41. SCRUBBING, DRY

Dry scrubbing, also known as *dry absorption* and *semi-dry scrubbing*, is one form of sorbent injection. It is commonly used to remove SO_2 from flue gas. In operation, flue gases pass through a scrubbing chamber where a slurry of lime and water is sprayed through them.[30] The slurry is produced by high-speed rotary atomizers. The sorbent (a reagent such as lime, limestone, hydrated lime, sodium bicarbonate, or trona, also called sodium sesquicarbonate) is injected into the flue gas in either dry or slurry form. The flue gas heat drives off the water. In either case, a dry powder is carried through the flue gas system. An electrostatic precipitator or baghouse captures the fly ash and calcium sulfate particulates.

Dry scrubbers do not saturate the gas stream, even when some moisture is used as a carrier. The moisture that is added is evaporated and does not condense, so unlike wet scrubbers, there is no visible moisture plume.

Dry scrubbing has an SO_2 removal efficiency of approximately 50% to 75%, with some installations reporting 90% efficiencies. NOx removal is not usually intended and is essentially zero. Though the removal efficiency is lower than with wet scrubbing, an advantage of dry scrubbing is that the waste is dry, requiring no sludge-handling equipment. The lower efficiency may be sufficient in older power plants with less stringent regulations. (See also Sec. 70.46.)

Particulate removal efficiencies of 90% can be achieved. Since wet scrubbers normally have a lower particulate removal efficiency than baghouses, they are usually combined with electrostatic precipitators.

42. SCRUBBING, VENTURI

Venturi scrubbers (also known as *atomizing scrubbers*) are effective with particle sizes greater than 0.2 μm. In a fixed-throat venturi scrubber, the gas stream enters a converging section and is accelerated toward the throat. In the throat section, the high velocity gas stream strikes liquid streams that are injected at right angles to the gas flow, shattering the liquid into small droplets. The particles impact the slower moving high-mass droplets and are collected in the droplets. Some particles are also collected in the diverging section where the gas stream slows and the droplets agglomerate. The droplet-entrained particles are collected in the flooded elbow and cyclonic separator. (See Fig. 70.14.)

Figure 70.14 *Wet Venturi Scrubber*

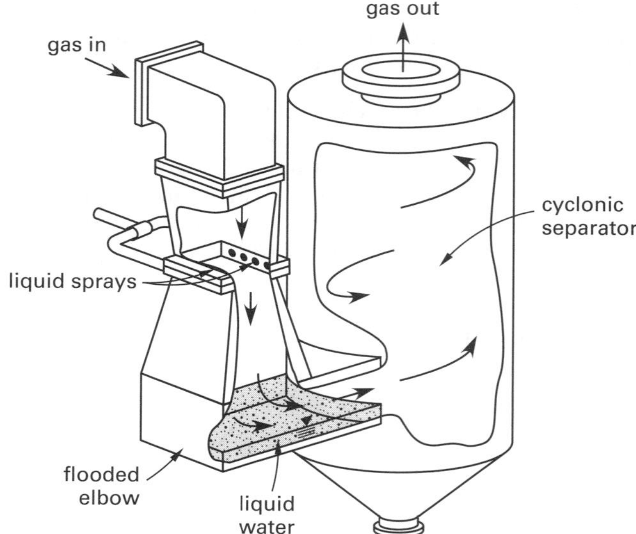

The pressure drop between the entrance of the converging section and the exit of the diverging section is proportional to the square of the gas velocity at the throat. Q_w/Q_g is known as the *liquid-gas ratio*, L/G. Typical operating ranges are: Δp, less than 60 inwg maximum, 5–25 inwg typical (less than 15 kPa maximum, 1.2–6.2 typical); $v_{g,\text{throat}}$, 18–45 ft/sec (60–150 m/s); Q_w/Q_g, 10–30 gal/1000 ACFM (80–240 L·h/1000 m^3). The high pressure drop results in a high operating cost.

$$\Delta p \propto \frac{Q_w v_{g,\text{throat}}^2}{Q_g} \qquad 70.34$$

The gas is considered incompressible and in a saturated condition at the throat. The throat velocity can be calculated from basic principles.

$$v_{g,\text{throat}} = \frac{Q_{\text{sat}}}{A_{\text{throat}}} = K_{\text{Calvert}}\sqrt{\frac{\Delta p}{\rho_{\text{sat}}}} \qquad 70.35$$

$$A_{\text{throat}} = \frac{A_{\text{inlet}} v_{g,\text{inlet}}}{v_{g,\text{throat}}} \qquad 70.36$$

[29]As it relates to air pollution control, the term *scrubbing* is loosely used. The term may be used in reference to any process that removes any substance from flue gas. In particular, any process that removes SO_2 by passing flue gas through lime or limestone is referred to as *scrubbing*.
[30]Limestone, sodium carbonate, and magnesium oxide can also be used instead of lime. Because they cost less, lime and limestone are the most common.

Plant Engineering

$$\rho_{\text{sat}} = \rho_{\text{dry air}} + \rho_w = \frac{p_{\text{air}}}{R_{\text{air}}T} + \frac{p_w}{R_wT} = \rho_{\text{dry air}}\left(\frac{1+x}{1+1.609x}\right) \quad \text{70.37}$$

For v in ft/sec, Δp in in wg, and ρ in lbm/ft^3, *Calvert's constant* is

$$K_{\text{Calvert}} = \sqrt{\frac{1850}{\frac{L}{G}}} \quad \text{70.38}$$

The derivation of an explicit equation for venturi scrubber efficiency is dependent on the assumptions made about the extent of liquid atomization, droplet sizes, acceleration and velocity of the droplets, coefficient of drag, and location of impaction. Accordingly, there are numerous models of varying complexities, none of which are generically applicable or particularly accurate. For example, the common *Calvert model* assumes an infinitely long throat section where all droplets travel at the throat velocity. The Calvert model incorporates an *impaction parameter*, $K_{\text{impaction}}$, defined as the ratio of particle stopping distance, $s_{p,\text{stopping}}$, to droplet diameter, d_d. The predicted efficiency is

$$\eta = \left(\frac{K_{\text{impaction}}}{K_{\text{impaction}} + 0.35}\right)^2 \quad \text{70.39}$$

$$K_{\text{impaction}} = \frac{s_{p,\text{stopping}}}{d_d} = \frac{C_c\rho_s d_p^2|\text{v}_g - \text{v}_d|}{18\mu_g g_c d_d} \quad \text{70.40}$$

Equation 70.41 is a generic equation for calculating collection efficiency for many collection system types, including scrubbers. K' is based on observation and is unique to the characteristics of the process and equipment. It is determined from pilot tests or actual performance measurements. L is a characteristic length, such as the height of a tower or length or width of a chamber, and may or may not be incorporated into K'.

$$\eta = 1 - \exp\left(\frac{K'L}{d_p}\right) \quad \text{70.41}$$

Collection efficiency may be expressed in terms of penetration. *Penetration*, Pt, is the fraction of particles that pass through the scrubber uncollected. Penetration is the complement of collection efficiency.

$$\text{Pt} = 1 - \eta \quad \text{70.42}$$

In order to compare relative (not absolute) changes in collection operations, penetration can be expressed in terms of the *number of transfer units*, NTU.

$$\text{NTU} = -\ln(\text{Pt}) \quad \text{70.43}$$

43. SELECTIVE CATALYTIC REDUCTION

Selective catalytic reduction (SCR) uses ammonia in combination with a catalyst to reduce nitrogen oxides (NOx) to nitrogen gas and water.[31] Vaporized ammonia is injected into the flue gas; the mixture passes through a catalytic reaction bed approximately 0.5 sec to 1.0 sec later. The reaction bed can be constructed from honeycomb plates or parallel-ridged plates. Alternatively, the reaction bed may consist of a packed bed of rings, pellets, or other shapes. The catalyst lowers the NOx decomposition activation energy. NOx and NH_3 combine on the catalyst's surface, producing nitrogen and water.

One mole of ammonia is required for every mole of NOx removed. However, in order to maximize the reduction efficiency, approximately 5 ppm to 10 ppm of unreacted ammonia is left behind.[32] The chemical reaction is

$$O_2 + 4NO + 4NH_3 \longrightarrow 4N_2 + 6H_2O$$

The optimum temperature for SCR is 600°F to 700°F (315°C to 370°C). Gas velocities are typically around 20 ft/sec (6 m/s). The pressure drop is 3 in wg to 4 in wg (0.75 kPa to 1 kPa). NOx removal efficiency is typically 90% but can range from 70% to 95% depending on the application.[33] SCR produces no liquid waste.

Several conditions can cause deactivation of the catalyst. *Poisoning* occurs when trace quantities of specific materials (e.g., arsenic, lead, phosphorous, other heavy metals, silicon, and sulfur from SO_2) react with the catalyst and lower its activity. Poisoning by SO_2 can be reduced by keeping the flue gas temperature above 608°F (320°C) and/or using poison-resisting compounds. *Masking* (or *plugging*) occurs when the catalytic surface becomes covered with fine particle dust, unburned solids, or ammonium salts. Internal cleaning

[31]Vanadium oxide, titanium, and platinum are metallic catalysts; zeolites and ceramic catalysts are also used.

[32]Leftover ammonia in the flue gas is referred to as *ammonia slip* and is usually measured in ppm. 50 ppm to 100 ppm would be considered an excessive ammonia slip. Since NOx and NH_3 react on a 1:1 molar basis, slip is easily calculated. In the following equations, the units can be either molar or volumetric (ppmvd, lbm/hr, mol/hr, scfm, etc.), but they must be consistent.

$$\begin{aligned} NH_{3,\text{slip}} &= NH_{3,\text{feed}} - NH_{3,\text{reacted}} \\ &= NH_{3,\text{feed}} - (NOx_{\text{in}} - NOx_{\text{out}}) \\ &= NH_{3,\text{feed}} - (NOx_{\text{in}})(\text{removal efficiency}) \end{aligned}$$

[33]Removal efficiencies of 95% to 99% are possible when SCR is used for VOC removal.

devices remove surface contaminants such as ash deposits and are used to increase the activity of the equipment.

44. SELECTIVE NONCATALYTIC REDUCTION

Selective noncatalytic reduction (SNCR) involves injecting ammonia or urea into the upper parts of the combustion chamber (or into a thermally favorable location downstream of the combustion chamber) to reduce NOx.[34] SNCR is effective when the oxygen content is low (e.g., 1%) and the combustion temperature is controlled. If the temperature is too high, the NH_3 will react more with oxygen than with NOx, forming even more NOx. If the temperature is too low, the reactions slow and unreacted ammonia enters the flue gas. The optimal temperature range is 1600°F to 1750°F (870°C to 950°C) for NH_3 injection and up to 1900°F (1040°C) for urea. (In general, the temperature ranges for NH_3 and urea are similar. However, various hydrocarbon additives can be used with urea to lower the temperature range.) The reactions are

$$6NO + 4NH_3 \longrightarrow 5N_2 + 6H_2O$$
$$6NO_2 + 8NH_3 \longrightarrow N_2 + 12H_2O$$

The NOx reduction efficiency is approximately 20% to 50% with an ammonia slip of 20 ppm to 30 ppm. The actual efficiency is highly dependent on the injection geometries and interrelations between ammonia slip, ash, and sulfur.

Since the NOx reduction efficiency is relatively low, the SNCR process cannot usually satisfy NOx regulations by itself. The use of urea must be balanced against the potential for ammonia slip and the conversion of NO to nitrous oxide. These and other problems relating to formation of ammonium salts have kept SNCR from gaining widespread popularity in small installations.

45. SOIL WASHING

Soil washing is effective in removing heavy metals, wood preserving wastes (PAHs), and BTEX compounds from contaminated soil. Soil washing is a two-step process. In the first step, soil is mixed with water to dissolve the contaminants. Additives are used as required to improve solubility. In the second step, additional water is used to flush the soil and to separate the fine soil from coarser particles. (Semi-volatile materials concentrate in the fines.) Metals are extracted by adding chelating agents to the wash water.[35] The contaminated wash water is subsequently treated.

46. SORBENT INJECTION

Sorbent injection (FSI) involves injecting a limestone slurry directly into the upper parts of the combustion chamber (or into a thermally favorable location downstream) to reduce SOx. Heat calcines the limestone into reactive lime. Fast drying prevents wet particles from building up in the duct. Lime particles are captured in a scrubber with or without an electrostatic precipitator (ESP). With only an ESP, the SOx removal efficiency is approximately 50%.

47. SPARGING

Sparging is the process of using air injection wells to bubble air through groundwater. The air pushes volatile contaminants into the overlying soil above the aquifer where they can be captured by vacuum extraction.

48. STAGED COMBUSTION

Staged combustion methods are primarily used with gas-fired burners to reduce formation of nitrogen oxides (NOx). Both staged-air burner and staged-fuel burner systems are used. *Staged-air burner systems* reduce NOx production 20% to 35% by admitting the combustion air through primary and secondary paths around each fuel nozzle. The fuel burns partially in the fuel-rich zone. Fuel-borne nitrogen is converted to compounds that are subsequently oxidized to nitrogen gas. Secondary air completes combustion and controls the flame size and shape. Combustion temperature is lowered by recirculation of combustion products within the burner. Staged burners have few disadvantages, the main one being longer flames.

In *staged-fuel burner systems*, a portion of the fuel gas is initially burned in a fuel-lean (air-rich) combustion. The peak flame temperature is reduced, with a corresponding reduction in thermal NOx production. The remainder of the fuel is injected through secondary nozzles. Combustion gases from the first stage dilute the combustion in the second stage, reducing peak temperature and oxygen content. Reductions in NOx formation of 50% to 60% are possible. Flame length is less than with staged-air burners, and the required excess air is lower.

49. STRIPPING, AIR

Air strippers are primarily used to remove volatile organic compounds (VOCs) or other target substances from water. In operation, contaminated water enters a stripping tower at the top, and fresh air enters at the bottom. The effectiveness of the process depends on the volatility of the compound, its temperature and concentration, and the liquid-air contact area. However, removal efficiencies of 80% to 90% (and above) are common for VOCs. Figure 70.15 shows a typical air stripping operation.

[34] *Urea*, which decomposes to NH_3 and carbon dioxide inside the combustion chamber, is safer and easier to handle.
[35] A *chelate* is a ring-like molecular structure formed by unshared electrons of neighboring atoms. A *chelating agent* (*chelant*) is an organic compound in which atoms form bonds with metals in solution. By combining with metal ions, chelates control the damaging effects of trace metal contamination. Ethylenediaminetetraacetic acid (EDTA) types are the leading industrial chelants.

Plant Engineering

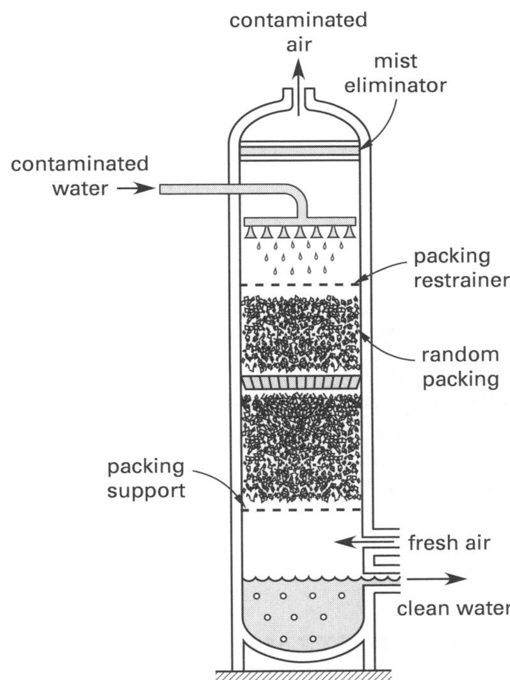

Figure 70.15 Schematic of Air Stripping Operation

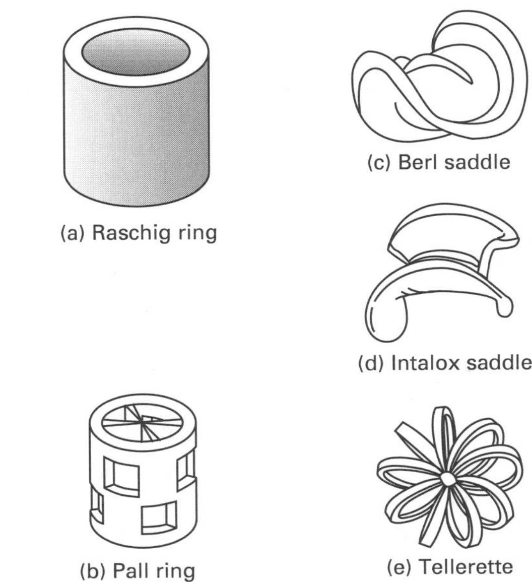

Figure 70.16 Packing Media Types

There are three types of stripping towers—*packed towers* filled with synthetic *packing media* (polypropylene balls, rings, saddles, etc.) like those shown in Fig. 70.16, *spray towers*, and (less frequently for VOC removal) *tray towers* with horizontal trays arranged in layers vertically. *Redistribution rings* (*wall wipers*) prevent channeling down the inside of the tower. (*Channeling* is the flow of liquid through a few narrow paths rather than an over-the-bed packing.) The stripping air is generated by a small blower at the column base. A mist eliminator at the top eliminates entrained water from the air.

As the contaminated water passes over packing media in a packed tower, the target substance leaves the liquid and enters the air where the concentration is lower. The mole fraction of the target substance in the water, x, decreases; the mole fraction of the target substance in the air, y, increases.

The discharged air, known as *off-gas*, is discharged to a process that destroys or recovers the target substance. (Since the quantities are small, recovery is rarer.) Destruction of the target substance can be accomplished by flaring, carbon absorption, and incineration. Since flaring is dangerous and carbon absorption creates a secondary waste if the AC is not regenerated, incineration is often preferred.

Henry's law, as it applies to water treatment, states that at equilibrium, the vapor pressure of target substance A is directly proportional to the target substance's mole fraction, x_A. (The maximum mole fraction is the substance's *solubility*.) Table 70.10 lists representative values of Henry's law constant, H.[36] When multiple compounds are present in the water, the *key component* is the one that is the most difficult to remove (i.e., the component with the highest concentration and lowest Henry's constant).

$$p_A = H_A x_A \qquad 70.44$$

The liquid *mass-transfer coefficient*, K_La, is the product of the *coefficient of liquid mass transfer*, K, and the *interfacial area* per volume of packing, a. The mass transfer coefficient is a measure of the efficiency of the air stripper as a whole. The higher the mass transfer coefficient, the higher the efficiency. The K_La value is largely a function of the size and shape of the packing material, and it is usually obtained from the packing manufacturer or theoretical correlations. For some types of packing, the gas-phase resistance is greater than the liquid-phase resistance, and the gas mass-transfer coefficient, K_Ga, may be given.

[36]Henry's law is fairly accurate for most gases when the partial pressure is less than 1 atm. If the partial pressure is greater than 1 atm, Henry's law constants will vary with partial pressure.

Table 70.10 *Typical Henry's Law Constants for Selected Volatile Organic Compounds (at low pressures and 25°C)*

VOC	Henry's law constant, H (atm)*
1,1,2,2-tetrachloroethane	24.02
1,1,2-trichloroethane	47.0
propylene dichloride	156.8
methylene chloride	177.4
chloroform	188.5
1,1,1-trichloroethane	273.56
1,2-dichloroethene	295.8
1,1-dichloroethane	303.0
hexachloroethane	547.7
hexachlorobutadiene	572.7
trichloroethylene	651.0
1,1-dichloroethene	834.03
perchloroethane	1596.0
carbon tetrachloride	1679.17

(Multiply atm by 101.3 to obtain kPa.)

*Henry's law indicates that the units of H are atmospheres per mole fraction. Mole fraction is dimensionless and does not appear in the units for H.

The *stripping factor*, R (the reciprocal of the *absorption factor*), is given by Eq. 70.45. Units for the gas and liquid flow rates, G and L, depend on the correlations used to solve the stripping equations. The air flow rate is limited by the acceptable pressure drop through the tower.

$$R = \frac{HG}{L} \qquad 70.45$$

The *transfer unit method* is a convenient way of designing and analyzing the performance of stripping towers. A *transfer unit* is a measure of the difficulty of the mass-transfer operation and depends on the solubility and concentrations. The overall number of transfer units is expressed as NTU_{OG} or NTU_{OL} (alternatively, N_{OG} or N_{OL}), depending on whether the gas or liquid resistance dominates.[37] The NTU value depends on the incoming and desired concentrations and the material flow rates.[38]

The *height of the packing*, z, is the effective height of the tower. It is calculated from the NTUs and the *height of a transfer unit*, HTU, using Eq. 70.46. The HTU value depends on the packing media.

$$z = (\text{HTU}_{\text{OG}})(\text{NTU}_{\text{OG}}) = (\text{HTU}_{\text{OL}})(\text{NTU}_{\text{OL}}) \qquad 70.46$$

10% of additional height should be added to the theoretical value as a safety factor. Packing heights greater than 30 ft to 40 ft (9.1 m to 12.2 m) are not recommended since the packing might be crushed, and greater heights produce little or no increase in removal efficiency.

The blower power depends on the packing shape and size and the air-to-water ratio. Typical pressure drops are 0.5 in wg to 1.0 in wg per vertical foot (0.38 kPa/m to 0.75 kPa/m) of packing. To minimize the blower power requirement, the ratio of tower diameter to gross packing dimension should be between 8 and 15. Blower fan power increases with increasing air-to-water ratios.

A significant problem for air strippers removing contaminants from water is fouling of the packing media through biological growth and solids (e.g., iron complexes) deposition. Biological growth can be inhibited by continuous or batch use of a *biocide* that does not interfere with the off-gas system.[39] Another problem is *flooding*, which occurs when excess liquid flow rates impede the flow of the air.

50. STRIPPING, STEAM

Steam stripping is more effective than air stripping for removing semi- and non-volatile compounds, such as diesel fuel, oil, and other organic compounds with boiling points up to approximately 400°F (200°C). Operation of a steam stripper is similar to that of an air stripper, except that steam is used in place of the air. Steam strippers can be operated at or below atmospheric pressure. Higher vacuums will remove greater amounts of the compound.

51. THERMAL DESORPTION

Thermal desorption is primarily used to remove volatile organic compounds (VOCs) from contaminated soil. The soil is heated directly or indirectly to approximately 1000°F (540°C) to evaporate the volatiles. This method differs from incineration in that the released gases are not burned but are captured in a subsequent process step (e.g., activated carbon filtration).

[37]The gas film resistance controls when the solubility of the target substance in the liquid is high. Conversely, when the solubility is low, the liquid film resistance controls. In flue gas cleanup (FGC) work, the gas film resistance usually controls.

[38]Stripping is one of many mass-transfer operations studied by chemical engineers. Determining the number of transfer units required and the height of a single transfer unit are typical chemical engineering calculations.

[39]A *biocide* is a chemical that kills living things, particularly microorganisms. Biocide categories include chlorinated isocyanurates (used in swimming pool disinfectants and dishwashing detergents), sodium bromide, inorganics (used in wood treatments), quaternaries ("quats") used in hard-surface cleaners and sanitizers, and iodophors (used in human-skin disinfectants). By comparison, *biostats* do not kill microorganisms already present, but they retard further growth of microorganisms from the moment they are incorporated. Biostats are organic acids and salts (e.g., sodium and potassium benzoate, sorbic acid, and potassium sorbate used "to preserve freshness" in foods).

52. VACUUM EXTRACTION

Vacuum extraction is used to remove many types of volatile organic compounds (VOCs) from soil. The VOCs are pulled from the soil through a well dug in the contaminated area. Air is withdrawn from the well, vacuuming volatile substances with it.

53. VAPOR CONDENSING

Some vapors can be removed simply by cooling them to below their dew points. Traditional contact (open) and surface (closed) condensers can be used.

54. VITRIFICATION

Vitrification melts and forms slag and ash wastes into glass-like pellets. Heavy metals and toxic compounds cannot leach out, and the pellets can be disposed of in hazardous waste landfills. Vitrification can occur in the incineration furnace or in a stand-alone process. Stand-alone vitrification occurs in an electrically heated vessel where the temperature is maintained at 2200°F to 2370°F (1200°C to 1300°C) for up to 20 hr or so. Since the electric heating is nonturbulent, flue gas cleaning systems are not needed.

55. WASTEWATER TREATMENT PROCESSES

Many industrial processes use water for cooling, rinsing, or mixing. Such water must be treated prior to being discharged to *publicly owned treatment works* (POTWs).[40] Table 70.11 lists some of the polluting characteristics of industrial wastewaters.

Most large-volume industrial wastewaters go through the following processes: (1) flow equalization, (2) neutralization, (3) oil and grease removal, (4) suspended solids removal, (5) metals removal, and (6) VOC removal. These processes are similar, in many cases, to processes with the same names used to treat municipal wastewater in a POTW.

Flow equalization reduces the chance of under- and over-loading a treatment process. Equalization of both hydraulic and chemical loading is required. *Hydraulic loading* is usually equalized by storing wastewater during high flow periods and discharging it during periods of low flow. *Chemical loading* is equalized by use of an equalization basin with mechanical mixing or air agitation. Mixing with air has the added advantages of oxidizing reducing compounds, stripping away volatiles, and eliminating odors.

There are two reasons for water *neutralization*. First, the pH of water is regulated and must be between 6.0 and 9.0 when discharged. Second, water that is too acidic or alkaline may not be properly processed by subsequent, particularly biological, processes. The optimum pH for biological processes is between 6.5 and 7.5. The effectiveness of the processes is greatly reduced with pHs below 4.0 and above 9.5.

Table 70.11 *Types of Pollution from Industrial Wastewater*

ammonia
biochemical oxygen demand (BOD)
carbon, total organic (TOC)
chemical oxygen demand (COD)
chloride
flow rate
metals, soluble
metals, nonsoluble
nitrate
nitrite
organic compounds, acid-extractable
organic compounds, base/neutral-extractable
organic compounds, volatile (VOC)
pH
phosphorus
sodium
solids, total suspended (TSS)
sulfate
surfactants
temperature
whole-effluent toxicity (LC_{50})

Acidic water is neutralized by adding lime (oxides and hydroxides of calcium and magnesium), limestone, or some other caustic solution. Alkaline waters are treated with sulfuric or hydrochloric acid or carbon dioxide gas.[41] Large volumes of oil and grease float to the surface and are skimmed off. Smaller volumes may require a dissolved-air process. Removing emulsified oil is more complex and may require use of chemical coagulants.

The method used to remove suspended solids depends on the solid size. Most industrial plants do not require strainers, bar screens, or fine screens to remove large solids (i.e., those larger than 1 in (25 mm)). *Grit* (i.e., sand and gravel) is removed in grit chambers by simple sedimentation. Fine solids are categorized into *settleable solids* (diameters more than 1 μm) and *colloids* (diameters between 0.001 μm and 1 μm). Settleable solids and colloids are removed in a sedimentation tank, with chemical coagulants or dissolved-air flotation being used to assist colloidal particles in settling out.

[40]Rainwater runoff from some industrial plants can also be a hazardous waste.
[41]In water, carbon dioxide gas forms *carbonic acid*.

Plant Engineering

Most metal (e.g., lead) removal occurs by precipitating its hydroxide, although precipitation as carbonates or sulfides is also used.[42] A caustic substance (e.g., lime) is added to the water to raise the pH below the solubility limit of the metal ion. When they come out of solution, the metallic compounds are *flocculated* into larger flakes that ultimately settle out. Floc is mechanically removed as inorganic heavy-metal sludge, which has a 96% to 99% water content by weight. The sludge is dewatered in a drying bed, vacuum filter, or filter press to 65% to 85% water. Depending on its composition, the sludge can be landfilled or treated as a hazardous waste by incineration or other methods.

VOCs such as benzene and toluene are removed by air stripping or adsorption in activated charcoal (AC) towers. Nonvolatile organic compounds (NVOCs) are removed by biological processes such as lagooning, trickle filtration, and activated sludge.

Although destruction of biological health hazards is not always needed for industrial wastewater, chemical oxidants are still needed to destroy odors, control bacterial growth downstream, and eliminate sulfur compounds.[43] (See Table 70.1.) Chemical oxidants can also reduce heavy metals (e.g., iron, manganese, silver, and lead) that were not removed in previous operations.

56. NOMENCLATURE

a	interfacial area per volume of packing	1/ft	1/m
A	area	ft^2	m^2
B	cyclone inlet width	ft	m
c	scaling constant	–	–
C	concentration	lbm/ft^3	kg/m^3
C_c	Cunningham slip factor	–	–
CCR	corona current ratio	A/ft^2	A/m^2
CPR	corona power ratio	W-min/ft^3	W·h/m^3
d	diameter	ft	m
D_e	cyclone exit diameter	ft	m
E	electric field strength	V/m	V/m
g	gravitational acceleration, 32.2	ft/sec^2	m/s^2
g_c	gravitational constant, 32.2	lbm-ft/lbf-sec^2	n.a.
h	height	ft	m
H	cyclone inlet height	ft	m
H	Henry's law constant	atm	atm
HHV	higher heating value	MMBtu/lbm	kJ/kg
HRR	heat release rate	Btu/hr-ft^2	W/m^2
HTU	height of a transfer unit	ft	m
I	current	A	A
k	reaction rate constant	1/sec	1/s
K	ESP constant	–	–
K	factor	various	various
K	ratio of area to volume	1/ft	1/m
L	length	ft	m
L	liquid loading rate	ft^3/min-ft^2	m^3/s·m^2
L/G	liquid-gas ratio	gal-min/1000 ft^3	L·h/m^3
m	mass	lbm	kg
$\dot{m}$	mass flow rate	lbm/hr	kg/h
N	number (quantity)	–	–
NTU	number of transfer units	–	–
p	pressure	lbf/ft^2	Pa
p	vapor pressure	lbf/in^2	kPa
P	power	ft-lbf/sec	W
Pt	penetration	–	–
q	charge	C	C
Q	flow rate	ft^3/sec	m^3/s
r	cyclone radius	ft	m
R	specific gas constant	ft-lbf/lbm-°R	J/kg·K
R	stripping factor	–	–
s	distance	ft	m
S	drag	lbf-min/ft^3	Pa·min/m
S	separation factor	–	–
SCA	specific collection area	ft^2-min/1000 ft^3	m^2·h/1000 m^3
SFC	specific feed characteristic	Btu/lbm	J/kg
t	time	sec	s
T	temperature	°R	K
v	velocity	ft/sec	m/s
V	voltage	V	V
w	drift velocity	ft/sec	m/s
W	areal dust density	lbm/ft^2	kg/m^2
x	fraction by weight	–	–
x	humidity ratio	lbm/lbm	kg/kg
y	exponent	–	–
z	packing height	ft	m

Symbols

α	coefficient of linear thermal expansion	ft/ft-°F	m/m·°C
ϵ	dielectric constant	–	–
ϵ_0	permittivity of free space, 8.854×10^{-12}	F/m	F/m
η	efficiency	%	%
μ	absolute viscosity	lbf-sec/ft^2	Pa·s
ρ	density	lbm/ft^3	kg/m^3

[42]Ion exchange, activated charcoal, and reverse osmosis methods can be used but may be more expensive.

[43]Meat packing and dairy (e.g., cheese) plants are examples of industrial processes that require chlorination to destroy bacteria in wastewater.

Subscripts

a	air
A	areal (per unit area)
d	droplet
e	electron or effective
f	filtering
g	gas
G	gas
L	liquid
o	reference
p	particle
r	residence
sat	saturation
t	total
w	water

71 Electricity and Electrical Equipment

Content in blue refers to the *NCEES Handbook*.

NCEES EXAM SPECIFICATIONS AND RELATED CONTENT

HVAC AND REFRIGERATION EXAM

I.A.3. Basic Engineering Practice: Electrical concepts

- 4. Resistivity and Resistance
- 7. Power in DC Circuits
- 10. Simple Series Circuits
- 19. Capacitors
- 25. Power Factor
- 33. Input Power in Three-Phase Systems
- 36. Torque and Power

MACHINE DESIGN AND MATERIALS EXAM

II.A.13. Mechanical Components: Motors and engines

- 25. Power Factor
- 33. Input Power in Three-Phase Systems
- 36. Torque and Power
- 41. Induction Motors
- 43. Synchronous Motors
- 47. Efficiency of Rotating Machines

1. ELECTRIC CHARGE

The charge on an electron is one *electrostatic unit* (esu). Since an esu is very small, electrostatic charges are more conveniently measured in *coulombs* (C). One coulomb is approximately 6.242×10^{18} esu. Another unit of charge, the *faraday*, is sometimes encountered in the description of ionic bonding and plating processes. One faraday is equal to one mole of electrons, approximately 96,485 C (96,485 A·s).

2. CURRENT

Current, I, is the movement of electrons. By historical convention, current moves in a direction opposite to the flow of electrons (i.e., current flows from the positive terminal to the negative terminal). Current is measured in *amperes* (A) and is the time rate change in charge.

$$I = \frac{dQ}{dt} \quad \text{[DC circuits]} \quad \text{[SI]} \qquad 71.1(a)$$

$$i(t) = \frac{dQ}{dt} \quad \text{[AC circuits]} \quad \text{[U.S.]} \qquad 71.1(b)$$

Plant Engineering

3. VOLTAGE SOURCES

A net *voltage*, V, causes electrons to move, hence the common synonym *electromotive force* (emf).[1] With *direct current* (DC) voltage sources, the voltage may vary in amplitude, but not in polarity. In simple problems where a battery serves as the voltage source, the magnitude is also constant.

With *alternating current* (AC) voltage sources, the magnitude and polarity both vary with time. Due to the method of generating electrical energy, AC voltages are typically sinusoidal.

4. RESISTIVITY AND RESISTANCE

Resistance, R, is the property of a *resistor* or resistive circuit to impede current flow.[2] A circuit with zero resistance is a *short circuit*, whereas an *open circuit* has infinite resistance. Adjustable resistors are known as *potentiometers*, *rheostats*, and *variable resistors*.

$$R = \frac{\rho L}{A} \quad \text{71.2}$$

Resistance of a circuit element depends on the *resistivity*, ρ (in Ω-in or Ω·cm), and length, L, of the material and the geometry of the current path through the element.

$$\rho = \frac{RA}{L} \quad \text{71.3}$$

The area, A, of circular conductors is often measured in *circular mils*, abbreviated *cmils*, the area of a 0.001 in diameter circle.

$$A_{\text{in}^2} = (7.854 \times 10^{-7})A_{\text{cmils}} \quad \text{71.4}$$

Resistivity depends on temperature. For most conductors, resistivity increases linearly with temperature. The resistivity of standard *International Annealed Copper Standard* (IACS) copper wire at 20°C is approximately

$$\begin{aligned} \rho_{\text{IACS Cu,20°C}} &= 1.7241 \times 10^{-6}\ \Omega\text{·cm} \\ &= 1.7241 \times 10^{-8}\ \text{m/S} \\ &= 0.67879 \times 10^{-6}\ \Omega\text{-in} \\ &= 10.371\ \Omega\text{-cmil/ft} \end{aligned} \quad \text{71.5}$$

Example 71.1

What is the resistance of a parallelepiped (1 cm × 1 cm × 1 m long) of IACS copper if current flows between the two smaller faces?

Solution

From Eq. 71.2 and Eq. 71.5, the resistance is

$$\begin{aligned} R = \frac{\rho L}{A} &= \frac{(1.7241 \times 10^{-6}\ \Omega\text{·cm})(1\ \text{m})\left(100\ \frac{\text{cm}}{\text{m}}\right)}{(1\ \text{cm})(1\ \text{cm})} \\ &= 1.724 \times 10^{-4}\ \Omega \end{aligned}$$

5. CONDUCTIVITY AND CONDUCTANCE

The reciprocals of resistivity and resistance are *conductivity*, σ, and *conductance*, G, respectively. The unit of conductance is the *siemens* (S).[3]

$$\sigma = \frac{1}{\rho} \quad \text{71.6}$$

$$G = \frac{1}{R} \quad \text{71.7}$$

Percent conductivity is the ratio of a substance's conductivity to the conductivity of standard IACS copper. (See Sec. 71.4.)

$$\begin{aligned} \%\ \text{conductivity} &= \frac{\sigma}{\sigma_{\text{Cu}}} \times 100\% \\ &= \frac{\rho_{\text{Cu}}}{\rho} \times 100\% \end{aligned} \quad \text{71.8}$$

6. OHM'S LAW

The *voltage drop*, also known as the *IR drop*, across a circuit or circuit element with resistance, R, is given by *Ohm's law*.[4]

$$V = IR \quad \text{[DC circuits]} \quad \text{71.9}$$

$$v(t) = i(t)R \quad \text{[AC circuits]} \quad \text{71.10}$$

7. POWER IN DC CIRCUITS

The *power*, P (in watts), dissipated across two terminals with resistance, R, and voltage drop, V, is given by *Joule's law*, Eq. 71.11.

[1]The symbol E (derived from the name electromotive force) has been commonly used to represent voltage induced by electromagnetic induction.
[2]*Resistance* is not the same as *inductance*, which is the property of a device to impede a *change* in current flow. (See Sec. 71.20.)
[3]The siemens is the inverse of an ohm and is the same as the obsolete unit, the *mho*.
[4]This book uses the convention that uppercase letters represent fixed, maximum, or effective values, and lowercase letters represent values that change with time.

Plant Engineering

Electrical Properties

Power

$$P = IV = \frac{V^2}{R} = I^2R$$
$$= V^2G \quad \text{[DC circuits]} \qquad 71.11$$

$$P(t) = i(t)v(t) = i(t)^2R = \frac{v(t)^2}{R}$$
$$= v(t)^2G \quad \text{[AC circuits]} \qquad 71.12$$

8. ELECTRICAL CIRCUIT SYMBOLS

Figure 71.1 illustrates symbols typically used to diagram electrical circuits in this book.

***Figure 71.1** Symbols for Electrical Circuit Elements*

(a) DC voltage source

(b) AC voltage source

I (c) current source

(d) resistor

(e) inductor

(f) capacitor

N_1 N_2 (g) transformer

9. RESISTORS IN COMBINATION

Resistors in series are added to obtain the total (equivalent) resistance of a circuit.

$$R_{\text{eq}} = R_1 + R_2 + R_3 + \cdots \quad \text{[series]} \qquad 71.13$$

Resistors in parallel are combined by adding their reciprocals. This is a direct result of the fact that conductances in parallel add.

$$G_{\text{eq}} = G_1 + G_2 + G_3 + \cdots \quad \text{[parallel]} \qquad 71.14$$

$$\frac{1}{R_{\text{eq}}} = \frac{1}{R_1} + \frac{1}{R_2} + \frac{1}{R_3} + \cdots \quad \text{[parallel]} \qquad 71.15$$

For two resistors in parallel, the equivalent resistance is

$$R_{\text{eq}} = \frac{R_1R_2}{R_1 + R_2} \quad \text{[two parallel resistors]} \qquad 71.16$$

10. SIMPLE SERIES CIRCUITS

Figure 71.2 illustrates a simple series DC circuit and its equivalent circuit.

- The current is the same through all circuit elements.

$$I = I_{R_1} = I_{R_2} = I_{R_3} \qquad 71.17$$

- The equivalent resistance is the sum of the individual resistances.

$$R_{\text{eq}} = R_1 + R_2 + R_3 \qquad 71.18$$

- The equivalent applied voltage is the sum of all voltage sources (polarities considered).

$$V_{\text{eq}} = \pm V_1 \pm V_2 \qquad 71.19$$

- The sum of all of the voltage drops across the components in the circuit (a *loop*) is equal to the equivalent applied voltage (i.e., $\sum V_{\text{rises}} = \sum V_{\text{drops}}$). This fact is known as *Kirchhoff's voltage law*. **[Kirchoff's Voltage Law for Closed Path (Loop)]**

$$V_{\text{eq}} = \sum IR_j = I\sum R_j = IR_{\text{eq}} \qquad 71.20$$

***Figure 71.2** Simple Series DC Circuit and Its Equivalent*

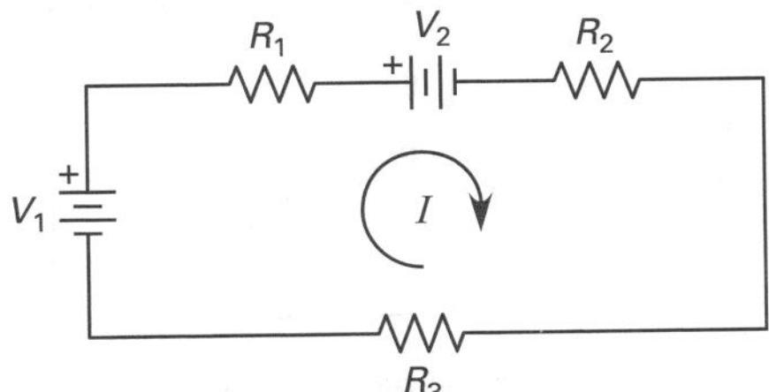

(a) original series circuit

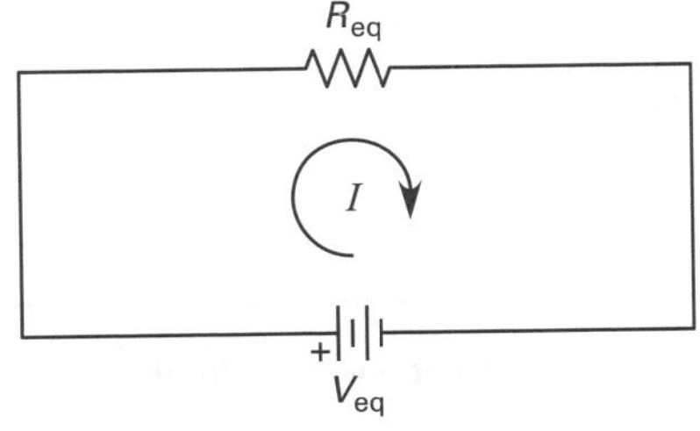

(b) equivalent circuit

11. SIMPLE PARALLEL CIRCUITS

Figure 71.3 illustrates a simple parallel DC circuit with one voltage source and its equivalent circuit.

- The voltage drop is the same across all legs.

$$\begin{aligned} V &= V_{R_1} = V_{R_2} = V_{R_3} \\ &= I_1R_1 = I_2R_2 = I_3R_3 \end{aligned} \qquad 71.21$$

- The reciprocal of the equivalent resistance is the sum of the reciprocals of the individual resistances.

$$\frac{1}{R_e} = \frac{1}{R_1} + \frac{1}{R_2} + \frac{1}{R_3} \qquad 71.22$$

$$G_e = G_1 + G_2 + G_3 \qquad 71.23$$

- The sum of all of the leg currents is equal to the total current. This fact is an extension of *Kirchhoff's current law:* The current flowing out of a connection (*node*) is equal to the current flowing into it.

$$\begin{aligned} I &= I_1 + I_2 + I_3 = \frac{V}{R_1} + \frac{V}{R_2} + \frac{V}{R_3} \\ &= V(G_1 + G_2 + G_3) \end{aligned} \qquad 71.24$$

Figure 71.3 *Simple Parallel DC Circuit and Its Equivalent*

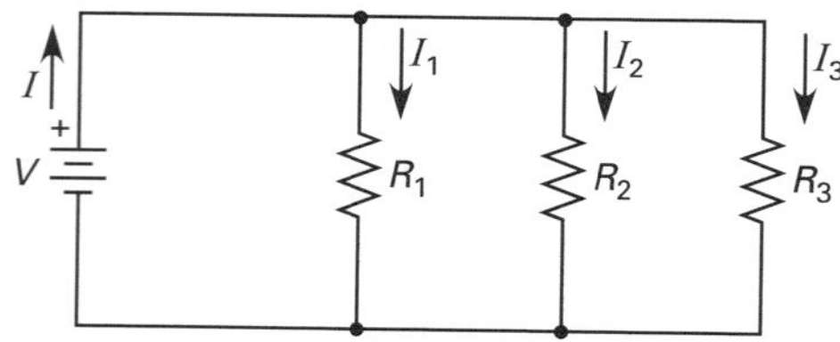

(a) original parallel circuit

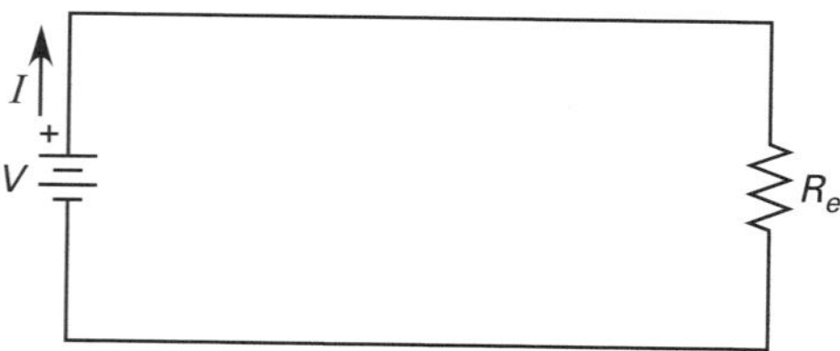

(b) equivalent circuit

12. VOLTAGE AND CURRENT DIVIDERS

Figure 71.4(a) illustrates a *voltage divider circuit.* The voltage across resistor 2 is

$$V_2 = V\left(\frac{R_2}{R_1 + R_2}\right) \qquad 71.25$$

Figure 71.4(b) illustrates a *current divider circuit.* The current through resistor 2 is

$$I_2 = I\left(\frac{R_1}{R_1 + R_2}\right) \qquad 71.26$$

Figure 71.4 *Divider Circuits*

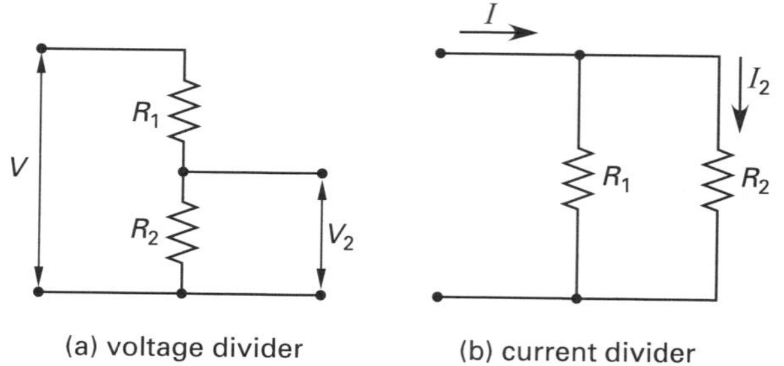

(a) voltage divider (b) current divider

13. AC VOLTAGE SOURCES

The term *alternating current* (AC) almost always means that the current is produced from the application of a voltage with sinusoidal waveform.[5] Sinusoidal variables can be specified without loss of generality as either sines or cosines. If a sine waveform is used, Eq. 71.27 gives the instantaneous voltage as a function of time. V_m is the *maximum value*, also known as the *amplitude*, of the sinusoid. If the time scale has been defined such that $v(t)$ is not zero at $t=0$, a *phase angle*, θ, must be included. In most cases, the voltage waveform is the reference, and $\theta = 0$.

$$v(t) = V_m \sin(\omega t + \theta) \text{ [trigonometric form]} \qquad 71.27$$

Figure 71.5 illustrates the *period*, T, of the waveform. The *frequency*, f (also known as *linear frequency*), of the sinusoid is the reciprocal of the period and is expressed in hertz (Hz).[6] *Angular frequency*, ω, in radians per second (rad/s), can also be specified.

$$f = \frac{1}{T} = \frac{\omega}{2\pi} \qquad 71.28$$

$$\omega = 2\pi f = \frac{2\pi}{T} \qquad 71.29$$

[5] Other alternating waveforms commonly encountered in commercial applications are the square, triangular, and sawtooth waveforms.

[6] In the United States, the standard frequency is 60 Hz. The standard frequency is 50 Hz in Japan; the British Isles and Commonwealth nations; continental Europe; and some Mediterranean, Near Eastern, African, and South American countries.

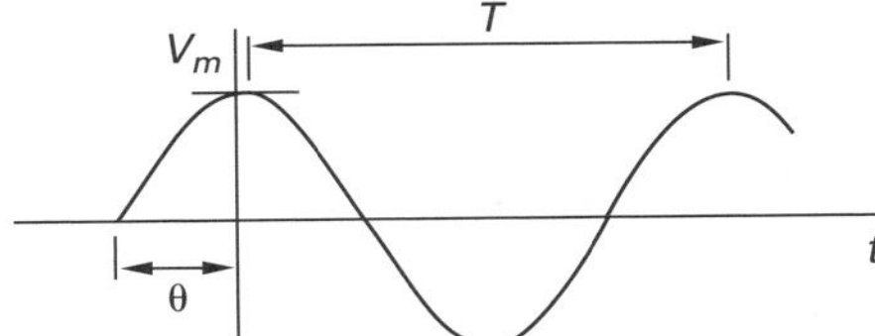

Figure 71.5 Sinusoidal Waveform

14. MAXIMUM, EFFECTIVE, AND AVERAGE VALUES

The *maximum value*, V_m, of a sinusoidal voltage is usually not specified in commercial and residential power systems. (See Fig. 71.5.) The *effective value*, also known as the *root-mean-square* (*rms*) *value*, is usually specified when referring to single- and three-phase voltages.[7] A DC current equal in magnitude to the effective value of a sinusoidal AC current produces the same heating effect as the sinusoid. The scale reading of a typical AC current meter is proportional to the effective current.

$$V_{\text{eff}} = \frac{V_m}{\sqrt{2}} \approx 0.707\,V_m \qquad 71.30$$

The *average value* of a symmetrical sinusoidal waveform is zero. However, the average value of a rectified sinusoid (or the average value of a sinusoid taken over half of the cycle) is $V_{\text{ave}} = 2V_m/\pi$. A DC current equal to the average value of a *rectified AC current* has the same electrolytic action (e.g., capacitor charging, plating, and ion formation) as the rectified sinusoid.[8]

15. IMPEDANCE

Simple alternating current circuits can be composed of three different types of passive circuit components—resistors, inductors, and capacitors. Each type of component affects both the magnitude of the current flowing as well as the phase angle of the current. (See Sec. 71.23.) For both individual components and combinations of components, these two effects are quantified by the *impedance*, **Z**. Impedance is a complex quantity with a magnitude (in ohms) and an associated *impedance angle*, ϕ. It is usually written in *phasor* (*polar*) *form* (e.g., $\mathbf{Z} \equiv Z\angle\phi$).

Multiple impedances in a circuit combine in the same manner as resistances: impedances in series add; reciprocals of impedances in parallel add. However, the addition must use complex (i.e., vector) arithmetic.

16. REACTANCE

Impedance, like any complex quantity, can also be written in rectangular form. In this case, impedance is written as the complex sum of the resistive, R, and reactive, X, components, both having units of ohms. The resistive and reactive components combine trigonometrically in the impedance triangle. (See Fig. 71.6.) The reactive component, X, is known as the *reactance*.

$$\mathbf{Z} \equiv R \pm jX \qquad 71.31$$

$$R = Z\cos\phi \quad \text{[resistive part]} \qquad 71.32$$

$$X = Z\sin\phi \quad \text{[reactive part]} \qquad 71.33$$

Figure 71.6 Impedance Triangle of a Complex Circuit

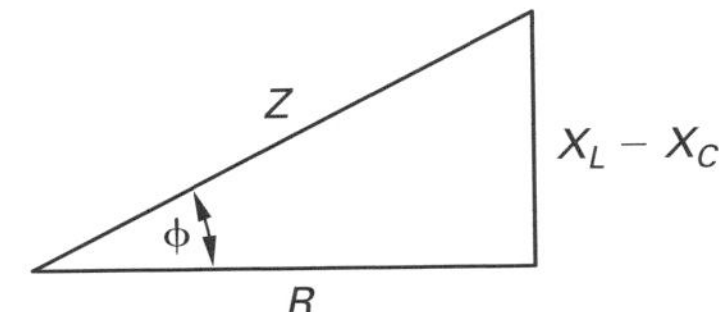

17. ADMITTANCE, CONDUCTANCE, AND SUSCEPTANCE

The reciprocal of impedance is the complex quantity *admittance*, **Y**. Admittance can be used to analyze parallel circuits, since admittances of parallel circuit elements add together.

$$\mathbf{Y} = \frac{1}{\mathbf{Z}} \equiv \frac{1}{Z}\angle -\phi \qquad 71.34$$

The reciprocal of the resistive part of the impedance is the *conductance*, G, with units of siemens (S). The reciprocal of the reactive part of impedance is the *susceptance*, B.

$$G = \frac{1}{R} \qquad 71.35$$

$$B = \frac{1}{X} \qquad 71.36$$

[7]The standard U.S. 110 V household voltage (also commonly referred to as 115 V, 117 V, and 120 V) is an effective value. This is sometimes referred to as the *nominal system voltage* or just *system voltage*. Other standard voltages used around the world, expressed as effective values, are 208 V, 220 V, 230 V, 240 V, 480 V, 550 V, 575 V, and 600 V.

[8]A *rectified waveform* has had all of its negative values converted to positive values of equal absolute value.

18. RESISTORS IN AC CIRCUITS

An ideal resistor, with resistance R, has no inductance or capacitance. The magnitude of the impedance is equal to the resistance, R (in ohms, Ω). The impedance angle is zero. Therefore, current and voltage are in phase in a purely resistive circuit.

$$\mathbf{Z}_R \equiv R\angle 0° \equiv R + j0 \qquad 71.37$$

19. CAPACITORS

A *capacitor* stores electrical charge. The charge on a capacitor is proportional to its *capacitance*, C (in farads, F), and voltage.[9]

Electrical Properties

$$q = CV \qquad 71.38$$

An ideal capacitor has no resistance or inductance. The magnitude of the impedance is the *capacitive reactance*, X_C, in ohms. The impedance angle is $-\pi/2$ ($-90°$). Therefore, current leads the voltage by 90° in a purely capacitive circuit.

$$\mathbf{Z}_C \equiv X_C\angle -90° \equiv 0 - jX_C \qquad 71.39$$

$$X_C = \frac{1}{\omega C} = \frac{1}{2\pi f C} \qquad 71.40$$

The capacitance of a parallel plate capacitor is calculated using Eq. 71.41 through Eq. 71.43.

Electrical Properties

$$C = \frac{\epsilon A}{d} \qquad 71.41$$

Permittivity, ϵ, is also expressed as the product of the dielectric constant, k, and the permittivity of free space, ϵ_0.

$$\epsilon = \epsilon_0 \kappa \qquad 71.42$$

$$\epsilon_0 = 8.85 \times 10^{12} \text{ F/m} \qquad 71.43$$

20. INDUCTORS

An ideal *inductor*, with an *inductance*, L (in units of the henry, H), has no resistance or capacitance. The magnitude of the impedance is the *inductive reactance*, X_L, in ohms. The impedance angle is $\pi/2$ (90°). Therefore, current lags the voltage by 90° in a purely inductive circuit.[10]

$$\mathbf{Z}_L \equiv X_L\angle 90° \equiv 0 + jX_L \qquad 71.44$$

$$X_L = \omega L = 2\pi f L \qquad 71.45$$

21. TRANSFORMERS

Transformers are used to change voltages, isolate circuits, and match impedances. Transformers usually consist of two coils of wire wound on magnetically permeable cores. One coil, designated as the *primary coil*, serves as the input; the other coil, the *secondary coil*, is the output. The primary current produces a magnetic flux in the core; the magnetic flux, in turn, induces a voltage in the secondary coil. In an *ideal transformer* (*lossless transformer* or *100% efficient transformer*), the coils have no electrical resistance, and all magnetic flux lines pass through both coils.

The ratio of the numbers, n, of primary to secondary coil windings is the *turns ratio* (*ratio of transformation*), a. If the turns ratio is greater than 1.0, the transformer decreases voltage and is a *step-down transformer*. If the turns ratio is less than 1.0, the transformer increases voltage and is a *step-up transformer*.

$$a = \frac{n_p}{n_s} \qquad 71.46$$

In an ideal transformer, the power transferred from the primary side equals the power received by the secondary side.

$$P = I_p V_p = I_s V_s \qquad 71.47$$

$$a = \frac{V_p}{V_s} = \frac{I_s}{I_p} = \sqrt{\frac{Z_p}{Z_s}} \qquad 71.48$$

22. OHM'S LAW FOR AC CIRCUITS

Ohm's law can be written in phasor (polar) form. (See Sec. 71.6.) Voltage and current can be represented by their maximum, effective (rms), or average values. However, both must be represented in the same manner. [Ohm's Law]

$$\mathbf{V} = \mathbf{IZ} \qquad 71.49$$

$$V = IZ \quad \text{[magnitudes only]} \qquad 71.50$$

$$\phi_V = \phi_I + \phi_Z \quad \text{[angles only]} \qquad 71.51$$

[9] Since a farad is a very large unit of capacitance, most capacitors are measured in microfarads, μF.

[10] Use the memory aid "ELI the ICE man" to remember that current (I) comes after voltage (electromagnetic force, E) in inductive (L) circuits. In capacitive (C) circuits, current leads voltage.

23. AC CURRENT AND PHASE ANGLE

The current and current phase angle of a circuit are determined by using Ohm's law in phasor form. The current *phase angle*, ϕ_I, is the angular difference between when the current and voltage waveforms peak.

$$\mathbf{I} = \frac{\mathbf{V}}{\mathbf{Z}} \quad 71.52$$

$$I = \frac{V}{Z} \quad \text{[magnitudes only]} \quad 71.53$$

$$\phi_I = \phi_V - \phi_Z \quad \text{[angles only]} \quad 71.54$$

Example 71.2

An inventor's black box is connected across standard household voltage (110 V rms). The current drawn is 1.7 A with a lagging phase angle of 14° with respect to the voltage. What is the impedance of the black box?

Solution

From Eq. 71.52,

$$\mathbf{Z} = \frac{\mathbf{V}}{\mathbf{I}} = \frac{110 \text{ V}\angle 0°}{1.7 \text{ A}\angle 14°} = 64.71\ \Omega\angle{-14°}$$

24. POWER IN AC CIRCUITS

In a purely resistive circuit, all of the current drawn contributes to dissipated energy. The flow of current causes resistors to increase in temperature, and heat is transferred to the environment.

In a typical AC circuit containing inductors and capacitors as well as resistors, some of the current drawn does not cause heating.[11] Rather, the current charges capacitors and creates magnetic fields in inductors. Since the voltage alternates in polarity, capacitors alternately charge and discharge. Energy is repeatedly drawn and returned by capacitors. Similarly, energy is repeatedly drawn and returned by inductors as their magnetic fields form and collapse.

Current through resistors in AC circuits causes heating, just as in DC circuits. The power dissipated is represented by the *real power* vector, **P**. Current through capacitors and inductors contributes to reactive power, represented by the *reactive power* vector, **Q**. Reactive power does not contribute to heating. For convenience, both real and reactive power are considered to be complex (vector) quantities, with magnitudes and associated angles. Real and reactive power combine as vectors into the *complex power* vector, **S**, as shown in Fig. 71.7. The angle, ϕ, is the *impedance angle*.

Figure 71.7 *Complex Power Triangle (lagging)*

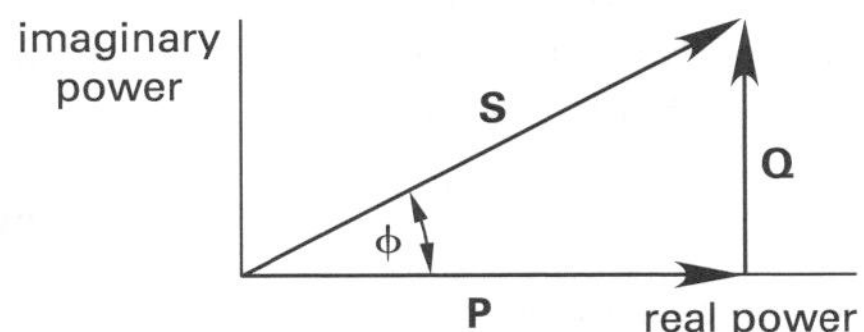

The magnitude of real power is known as the *average power*, P, and is measured in watts. The magnitude of the reactive power vector is known as *reactive power*, Q, and is measured in VARs (volt-amps reactive). The magnitude of the complex power vector is the *apparent power*, S, measured in VAs (volt-amps). The apparent power is easily calculated from measurements of the line current and line voltage.

$$S = I_{\text{eff}} V_{\text{eff}} = \tfrac{1}{2} I_m V_m \quad 71.55$$

25. POWER FACTOR

The complex power triangle shown in Fig. 71.7 is congruent to the impedance triangle, and the *power angle*, ϕ, is identical to the overall impedance angle. (See Fig. 71.6.)

$$S^2 = P^2 + Q^2 \quad 71.56$$

$$P = S \cos\phi \quad 71.57$$

$$Q = S \sin\phi \quad 71.58$$

$$\phi = \arctan\frac{Q}{P} \quad 71.59$$

The *power factor*, pf (also occasionally referred to as the *phase factor*), for a sinusoidal voltage is $\cos\phi$. By convention, the power factor is usually given in percent rather than by its equivalent decimal value. For a purely resistive circuit, pf = 100%; for a purely reactive circuit, pf = 0%.

$$\text{pf} = \frac{P}{S} \quad 71.60$$

Since the cosine is positive for both positive and negative angles, the terms "leading" and "lagging" are used to describe the nature of the circuit. In a circuit with a *leading power factor* (i.e., in a *leading circuit*), the load is primarily capacitive in nature. In a circuit with a *lagging power factor* (i.e., in a *lagging circuit*), the load is primarily inductive in nature.

[11]The notable exception is a *resonant circuit* in which the inductive and capacitive reactances are equal. In that case, the circuit is purely resistive in nature.

The power factor is also the ratio of kilowatt input to kilovolt-ampere input for single-, two-, and three-phase motors. The product of the power factor and motor efficiency is the apparent efficiency, E_a.

Power Factor

$$\text{kV-amp input} = \frac{\text{kW}}{\text{pf}} = \frac{(0.746)(\text{horsepower})}{E(\text{pf})} = \frac{(0.746)(\text{horsepower})}{E_a} \quad \text{71.61}$$

Example 71.3

A simple circuit contains a 208 V, 60 Hz power source powering a motor load with an inline ammeter reading of 10.850 A. The true power is 2.0 kW. What is the power factor of the circuit?

Solution

Find the apparent power using Eq. 71.55.

$$S = I_{\text{eff}}V_{\text{eff}} = \frac{(10.850 \text{ A})(208 \text{ V})}{1000 \ \frac{\text{VA}}{\text{kVA}}} = 2.257 \text{ kVA}$$

The true power, P, is 2.0 kW. From Eq. 71.60,

$$\text{pf} = \frac{P}{S} = \frac{2.0 \text{ kW}}{2.257 \text{ kVA}} = 0.886$$

26. COST OF ELECTRICAL ENERGY

Except for large industrial users, electrical meters at service locations usually measure and record real power only. Electrical utilities charge on the basis of the total energy used. Energy usage, commonly referred to as the *usage* or *demand*, is measured in kilowatt-hours, abbreviated kWh or kW-hr.

$$\text{cost} = (\text{cost per kW-hr}) \times E_{\text{kW-hr}} \quad \text{71.62}$$

The cost per kW-hr may not be a single value but may be tiered so that cost varies with cumulative usage. The lowest rate is the *baseline rate*.[12] To encourage conservation, the incremental cost of energy increases with increases in monthly usage.[13] To encourage cutbacks during the day, the cost may also increase during periods of peak demand.[14] The increase in cost for usage during peak demand may be accomplished by varying the rate, additively, or by use of a *peak demand multiplier*. There may also be different rates for summer and winter usage.

Although only real power is dissipated, reactive power contributes to total current. (Reactive power results from the current drawn in supplying the magnetization energy in motors and charges on capacitors.) The distribution system (wires, transformers, etc.) must be sized to carry the total current, not just the current supplying the heating effect. When real power alone is measured at the service location, the power factor is routinely monitored and its effect is built into the charge per kW-hr. This has the equivalent effect of charging the user for apparent power usage.

Example 71.4

A small office normally uses 700 kW-hr per month of electrical energy. The company adds a 1.5 kW heater (to be used at the rate of 1000 kW-hr/month) and a 5 hp motor with a mechanical efficiency of 90% (to be used at a rate of 240 hr/mo). What is the incremental cost of adding the heater and motor? The tiered rate structure is

electrical usage (kW-hr)	rate ($/kW-hr)
less than 350	0.1255
350–999	0.1427
1000–3999	0.1693

Solution

Motors are rated by their real power output, which is less than their real power demand. The incremental electrical usage per month is

$$1000 \text{ kW-hr} + \frac{(5 \text{ hp})\left(0.7457 \ \frac{\text{kW}}{\text{hp}}\right)(240 \text{ hr})}{0.90} = 1994 \text{ kW-hr}$$

The cumulative monthly electrical usage is

$$700 \text{ kW-hr} + 1994 \text{ kW-hr} = 2694 \text{ kW-hr}$$

The company was originally in the second tier. The new usage will be billed partially at the second and third tier rates.

[12]There might also be a *lifeline rate* for low-income individuals with very low usage.

[13]There are different rate structures for different categories of users. While increased use within certain categories of users (e.g., residential) results in a higher cost per kW-hr, larger users in another category may pay substantially less per kW-hr due to their volume "buying power."

[14]The day may be divided into *peak periods*, *partial-peak periods*, and *off-peak periods*.

The incremental cost is

$$(999 \text{ kW-hr} - 700 \text{ kW-hr})\left(0.1427 \frac{\$}{\text{kW-hr}}\right) + (2694 \text{ kW-hr} - 999 \text{ kW-hr})\left(0.1693 \frac{\$}{\text{kW-hr}}\right) = \$329.63$$

27. CIRCUIT BREAKER SIZING

The purpose of a circuit breaker ("breaker") is to protect the wiring insulation from overheating. Breakers operate as heat-activated switches. Overloads, short circuits, and faulty equipment result in large current draws. Extended currents in excess of a breaker's rating will cause it to trip.

A breaker should be sized based on the average expected electrical load for that circuit. *Dedicated circuits* are easily sized based on the amperage of the equipment served, taken either from the nameplate current or calculated from the nameplate power as $I = P/V$. For general purpose circuits, reasonable judgment is required to estimate how many of the outlets will be in use simultaneously, and what equipment will be plugged in. Although circuit breakers are designed to operate at their rated loads for at least three hours, breakers should be sized for about 1.25 times the expected maximum load, not to exceed the capacity of the wire.

28. NATIONAL ELECTRICAL CODE

While a circuit breaker can be simplistically sized by calculating the maximum current from the known loads, unless the practitioner is a licensed electrician or electrical power engineer, a little knowledge of circuit breaker operation and the *National Electrical Code* (NEC) can greatly complicate the analysis. For this reason, most engineers should stop after calculating the maximum current, and they should leave the actual device selection to another trade. Selecting protection for standby generator circuits is particularly complex.

In the United States, the NEC specifies circuit breaker size as a function of the wire size. The wire is measured in American wire gauge and is abbreviated AWG or just GA. (Numerical wire size gauge decreases as the physical size increases.) Typical wire sizes and maximum currents (*ampacity*) for copper wire are 14 AWG (used for lighting and light-duty circuits up to 15 A), 12 AWG (for power outlets, kitchen microwaves, and lighting circuits up to 20 A), 10 AWG (30 A), 8 AWG (40 A), 6 AWG (50 A or 60 A), 4 AWG (75 A), and 3 AWG (100 A). For aluminum wires, the wire size should be one gauge larger. Common household circuit breaker sizes are 15 A, 20 A, and 30 A, although specialty sizes are also available in 5 A increments. Industrial sizes are readily available through 800 A, and larger breakers can handle thousands of amps.

The NEC refers to circuit breakers as *overcurrent protection devices* (OCPD). The NEC permits a circuit breaker to run essentially continuously at 100% of its *rated current*, I_r. This ability is intrinsic to the circuit breaker design, as Fig. 71.8 shows. When the current is greater than the rated current, the circuit breaker will open (i.e., "clear the circuit") if the exposure duration is between the minimum and maximum curve times. The range of times between vertically adjacent points on the two curves represents the maximum delay for the breaker for that current, although exact values should not be expected. The lower minimum curve anticipates tripping due to thermal action; the upper maximum curve anticipates tripping due to magnetic action (i.e., short circuits).

Figure 71.8 shows that this particular curve allows circuit breakers to pass currents as large as 10 times their rated values, albeit for no more than a second, and twice their rated currents for up to 10 seconds. This overcurrent must be considered when selecting breakers with rated currents significantly in excess of the calculated maximum circuit current.

Figure 71.8 *Typical Thermal-Magnetic Breaker Trip Curve**

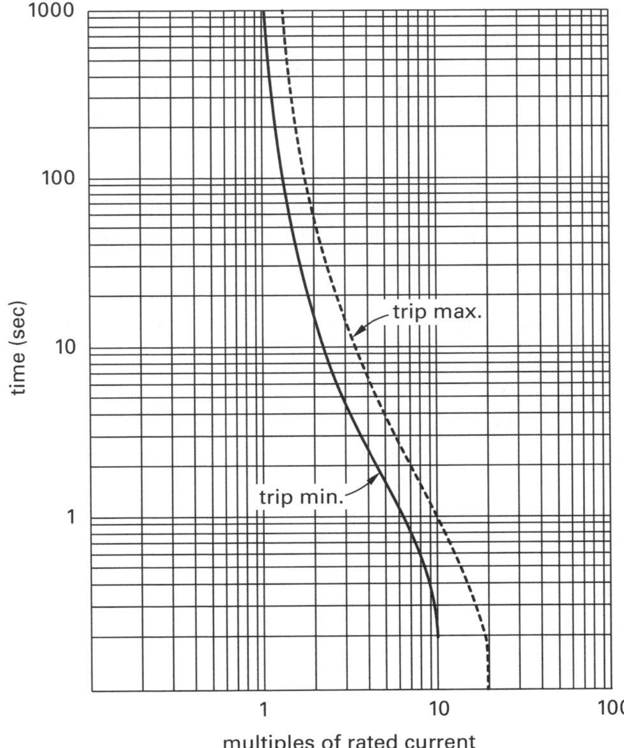

*Typical curve with Underwriters Laboratories (UL) specifications for use in the United States, 104°F (40°C).
International Electrotechnical Commission trip curves (IEC 947-2) for other countries are similar in shape but differ in specific values from UL curves.

Plant Engineering

Allowing a circuit breaker to run continuously at its rated value is based on ideal testing conditions, such as mounting in free air, no enclosure, and constant 104°F (40°C) air temperature. For typical installations with heating from adjacent breakers in an enclosure, the NEC requires continuously operated breakers to be derated to 80% of rated capacities. This derating should be applied with continuous operation, which NEC Art. 100 defines as operation that exceeds three hours in duration. Interior office lighting would be considered continuous, while electrical kitchen appliances normally would not be. Most electrical heating for vessels, pipes, and deicing is considered continuous. NEC Sec. 210.19(A)(1) and Sec. 210.20(A) specify the required OCPD (i.e., breaker) size as Eq. 71.63, which is sometimes referred to as the *80% rule*, since the reciprocal of 0.80 is 1.25.

$$\begin{array}{c}\text{required}\\ \text{rated capacity}\end{array} = \begin{array}{c}100\%\text{ of}\\ \text{noncontinuous load}\end{array} + \begin{array}{c}125\%\text{ of}\\ \text{continuous load}\end{array} \qquad 71.63$$

Circuit breakers are subject to heating from adjacent breakers in an enclosure regardless of whether their loads are continuous or noncontinuous. This does not mean that all circuit breakers in an enclosure must be derated. *Standard circuit breakers*, also known as *80% rated circuit breakers* and *non-100% rated circuit breakers*, have been tested and certified to function properly with noncontinuous loads in enclosures, so Eq. 71.63 can be used with confidence. Unfortunately, there is also a class of *100% rated circuit breakers* (including solid state devices that do not generate thermal energy), which have been tested and certified to function properly with continuous loads within enclosures. The 125% multiplier does not need to be applied to 100% rated circuit breakers. Clearly, the specification of actual OCPDs should be left to the experts.

Example 71.5

A 120 V (rms) circuit supplies 6 kW of office lighting and 125 A to one phase of a three-phase elevator induction motor. 175 A, 200 A, 225 A, and 250 A standard breakers are available. Which breaker should be specified?

Solution

The elevator motor is a noncontinuous load. The office lighting draws a continuous current of

$$I = \frac{P}{V} = \frac{(6\text{ kW})\left(1000\ \frac{\text{W}}{\text{kW}}\right)}{120\text{ V}} = 50\text{ A}$$

Since a standard (i.e., 80% rated) circuit breaker is called for, use Eq. 71.63.

$$\begin{aligned}\text{required rated capacity} &= \begin{array}{r}100\%\text{ of noncontinuous load}\\ +125\%\text{ of continuous load}\end{array}\\ &= (1.0)(125\text{ A}) + (1.25)(50\text{ A})\\ &= 187.5\text{ A}\end{aligned}$$

The 200 A circuit breaker should be specified.

29. POWER FACTOR CORRECTION

Inasmuch as apparent power is paid for but only real power is used to drive motors or provide light and heating, it may be possible to reduce electrical utility charges by reducing the power angle without changing the real power. This strategy, known as *power factor correction*, is routinely accomplished by changing the circuit reactance in order to reduce the reactive power. The change in reactive power needed to change the power angle from ϕ_1 to ϕ_2 is

$$\Delta Q = P(\tan\phi_1 - \tan\phi_2) \qquad 71.64$$

When a circuit is capacitive (i.e., leading), induction motors can be connected across the line to improve the power factor. (See Sec. 71.41.) In the more common situation, when a circuit is inductive (i.e., lagging), capacitors or synchronous capacitors can be added across the line. (See Sec. 71.43.) The size (in farads) of capacitor required is

$$C = \frac{\Delta Q}{\pi f V_m^2} \quad [V_m \text{ maximum}] \qquad 71.65$$

$$C = \frac{\Delta Q}{2\pi f V_{\text{eff}}^2} \quad [V_{\text{eff}} \text{ effective}] \qquad 71.66$$

Capacitors for power factor correction are generally rated in kVA. Equation 71.64 can be used to find that rating.

30. THREE-PHASE ELECTRICITY

Smaller electric loads, such as household loads, are served by single-phase power. The power company delivers a sinusoidal voltage of fixed frequency and amplitude connected between two wires—a phase wire and a neutral wire. Large electric loads, large buildings, and industrial plants are served by three-phase power. Three voltage signals are connected between three phase wires and a single neutral wire. The phases have equal frequency and amplitude, but they are out of phase by 120° (electrical) with each other. Such *three-phase systems* use smaller conductors to distribute electricity.[15] For the same delivered power, three-phase distribution systems have lower losses and are more efficient.

Three-phase motors provide a more uniform torque than do single-phase motors whose torque production pulsates.[16] Three-phase induction motors require no additional starting windings or associated switches. When rectified, three-phase voltage has a smoother waveform. (See Fig. 71.9.)

Figure 71.9 *Three-Phase Voltage*

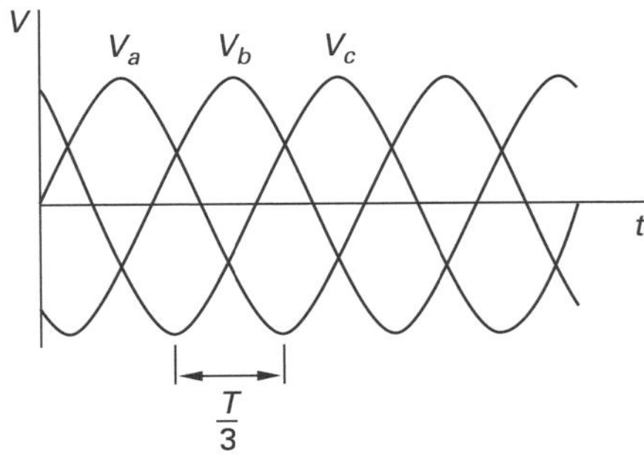

31. THREE-PHASE LOADS

Three impedances are required to load a three-phase voltage source fully. In three-phase motors and other devices, these impedances are three separate motor windings. Similarly, three-phase transformers have three separate sets of primary and secondary windings.

The impedances in a three-phase system are said to be *balanced loads* when they are identical in magnitude and angle. The voltages, line currents, and real, apparent, and reactive powers are all identical in a balanced system. Also, the power factor is the same for each phase. Therefore, only one phase of a balanced system needs to be analyzed (i.e., a *one-line analysis*).

32. LINE AND PHASE VALUES

The *line current*, I_l, is the current carried by the distribution lines (wires). The *phase current*, I_p, is the current flowing through each of the three separate loads (i.e., the phase) in the motor or device. Line and phase currents are both vector quantities.

Depending on how the motor or device is internally wired, the line and phase currents may or may not be the same. Figure 71.10 illustrates delta- and wye-connected loads. For balanced *wye-connected loads*, the line and phase currents are the same. For balanced *delta-connected loads*, they are not.

$$I_p = I_l \quad \text{[wye]} \qquad 71.67$$

$$I_p = \frac{I_l}{\sqrt{3}} \quad \text{[delta]} \qquad 71.68$$

Similarly, the *line voltage*, V_l (same as *line-to-line voltage*, commonly referred to as the *terminal voltage*), and *phase voltage*, V_p, may not be the same. With balanced delta-connected loads, the full line voltage appears across each phase. With balanced wye-connected loads, the line voltage appears across two loads.

$$V_p = \frac{V_l}{\sqrt{3}} \quad \text{[wye]} \qquad 71.69$$

$$V_p = V_l \quad \text{[delta]} \qquad 71.70$$

Figure 71.10 *Wye- and Delta-Connected Loads*

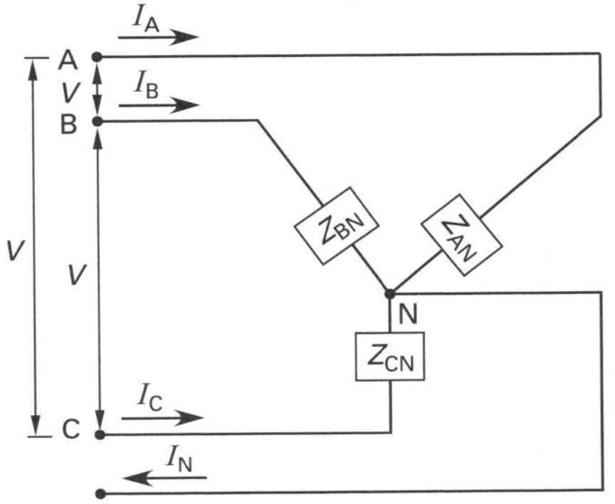

(a) wye-connected loads

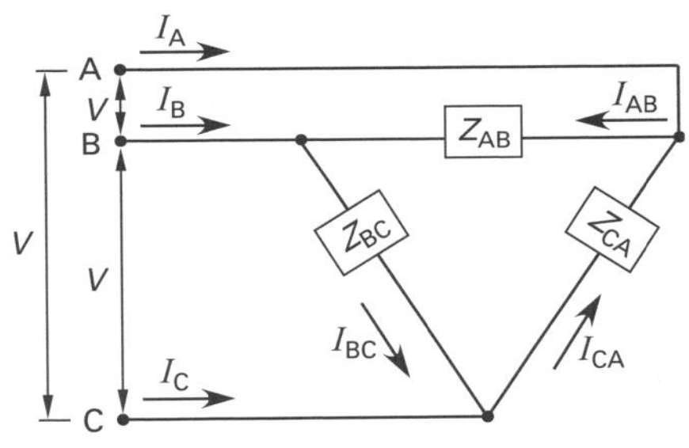

(b) delta-connected loads

33. INPUT POWER IN THREE-PHASE SYSTEMS

Each impedance in a balanced system dissipates the same real *phase power*, P_p. The power dissipated in a balanced three-phase system is three times the phase

[15]The uppercase Greek letter phi is often used as an abbreviation for the word "phase." For example, "3Φ" would be interpreted as "three-phase."
[16]Single-phase motors require auxiliary windings for starting, since one phase alone cannot get the magnetic field rotating.

power and is calculated in the same manner for both delta- and wye-connected loads. [**Power for Different Motor Phases**]

$$\begin{aligned} P_t &= 3P_p = 3V_pI_p\cos\phi \\ &= \sqrt{3}\,V_lI_l\cos\phi \end{aligned} \qquad \textit{71.71}$$

The real power component is sometimes referred to as "power in kW." Apparent power is sometimes referred to as "kVA value" or "power in kVA."

$$\begin{aligned} S_t &= 3S_p = 3V_pI_p \\ &= \sqrt{3}\,V_lI_l \end{aligned} \qquad \textit{71.72}$$

34. ROTATING MACHINES

Rotating machines are categorized as AC and DC machines. Both categories include machines that use electrical power (i.e., motors) and those that generate electrical power (alternators and generators).[17] Machines can be constructed in either single-phase or polyphase configurations, although single-phase machines may be inferior in terms of economics and efficiency. (See Sec. 71.30.)

Large AC motors are almost always three-phase. However, since the phases are balanced, it is necessary to analyze only one phase of the motor. Torque and power are divided evenly among the three phases.

35. REGULATION

Rotating machines (motors and alternators), as well as power supplies, are characterized by changes in voltage and speed under load. *Voltage regulation*, VR, is defined as

$$\text{VR} = \frac{\text{no-load voltage} - \text{full-load voltage}}{\text{full-load voltage}} \times 100\% \qquad \textit{71.73}$$

Speed regulation, SR, is defined as

$$\text{SR} = \frac{\text{no-load speed} - \text{full-load speed}}{\text{full-load speed}} \times 100\% \qquad \textit{71.74}$$

36. TORQUE AND POWER

For rotating machines, torque and power are basic operational parameters. It takes mechanical power to turn an alternator or generator. A motor converts electrical power into mechanical power. In SI units, power is given in watts (W) and kilowatts (kW). One horsepower (hp) is equivalent to 0.7457 kilowatts. The relationship between torque and power is

$$T_{\text{N·m}} = \frac{9549P_{\text{kW}}}{N_{\text{rpm}}} \qquad \text{[SI]} \qquad \textit{71.75(a)}$$

$$T_{\text{ft-lbf}} = \frac{5252P_{\text{horsepower}}}{N_{\text{rpm}}} \qquad \text{[U.S.]} \qquad \textit{71.75(b)}$$

There are many important torque parameters for motors. (See Fig. 71.11.) The *starting torque* (also known as *static torque*, *breakaway torque*, and *locked-rotor torque*) is the turning effort exerted by the motor when starting from rest. *Pull-up torque* (*acceleration torque*) is the minimum torque developed during the period of acceleration from rest to full speed. *Pull-in torque* (as developed in synchronous motors) is the maximum torque that brings the motor back to synchronous speed. (See Sec. 71.41.) *Nominal pull-in torque* is the torque that is developed at 95% of synchronous speed.

Figure 71.11 *Induction Motor Torque-Speed Characteristic (typical of design B frames)*

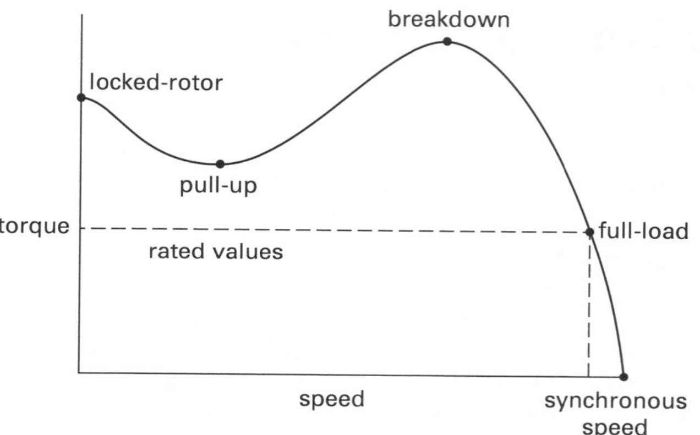

The *full-load torque* (*steady-state torque*) occurs at the rated speed and horsepower. Full-load torque is supplied to the load on a continuous basis. The full-load torque establishes the temperature increase that the motor must be able to withstand without deterioration. The *rated torque* is developed at rated speed and rated horsepower. The maximum torque that the motor can develop at its synchronous speed is the *pull-out torque*. *Breakdown torque* is the maximum torque that the motor can develop without stalling (i.e., without coming rapidly to a complete stop).

Motors are rated according to their output power (*rated power* or *brake power*). A 5 hp motor will normally deliver a full five horsepower when running at its rated speed. While the rated power is not affected by the motor's energy conversion efficiency, η, the electrical power input to the motor is.

[17] An *alternator* produces AC potential. A *generator* produces DC potential.

Mechanical Power

$$P_{\text{mechanical (rated)}} = \eta_{\text{motor}} P_{\text{electrical}} \qquad 71.76$$

Similarly, the electrical power from a generator converting mechanical power to electrical power is determined by Eq. 71.77. η_G is the electric generator efficiency.

Electrical Power

$$P_{\text{electrical (rated)}} = \eta_G P_{\text{mechanical}} \qquad 71.77$$

Table 71.1 lists standard motor sizes by rated horsepower.[18] The rated horsepower should be greater than the calculated brake power requirements. Since the rated power is the power actually produced, motors are not selected on the basis of their efficiency or electrical power input. The smaller motors listed in Table 71.1 are generally single-phase motors. The larger motors listed are three-phase motors.

Table 71.1 *Typical Standard Motor Sizes (horsepower)**

$\frac{1}{8}$	$\frac{1}{6}$	$\frac{1}{4}$	$\frac{1}{3}$	$\frac{1}{2}$	$\frac{3}{4}$
1	$1\frac{1}{2}$	2	3	5	$7\frac{1}{2}$
10	15	20	25	30	40
50	60	75	100	125	150
200	250				

(Multiply hp by 0.7457 to obtain kW.)

*1/8 hp and 1/6 hp motors are less common.

Example 71.6

A 60 Hz, 5 hp induction motor draws 53 A (rms) at 117 V (rms) with a 78.5% electrical-to-mechanical energy conversion efficiency. What capacitance should be connected across the line to increase the power factor to 92%?

Solution

The apparent power is found from the observed voltage and the current. Use Eq. 71.55.

$$S = IV = \frac{(53 \text{ A})(117 \text{ V})}{1000 \, \frac{\text{VA}}{\text{kVA}}} = 6.201 \text{ kVA}$$

The real power drawn from the line is calculated from the real work done by the motor. Use Eq. 71.76.

Mechanical Power

$$P_{\text{mechanical}} = \eta_{\text{motor}} P_{\text{electrical}}$$

$$P_{\text{electrical}} = \frac{P_{\text{mechanical}}}{\eta} = \frac{(5 \text{ hp})\left(0.7457 \, \frac{\text{kW}}{\text{hp}}\right)}{0.785} = 4.750 \text{ kW}$$

The reactive power and power angle are calculated from the real and apparent powers. From Eq. 71.56,

$$Q_1 = \sqrt{S^2 - P^2} = \sqrt{(6.201 \text{ kVA})^2 - (4.750 \text{ kW})^2} = 3.986 \text{ kVAR}$$

From Eq. 71.57,

$$\phi_1 = \arccos \frac{P}{S} = \arccos \frac{4.750 \text{ kW}}{6.201 \text{ kVA}} = 40.00°$$

The desired power factor angle is

$$\phi_2 = \arccos 0.92 = 23.07°$$

The reactive power after the capacitor is installed is given by Eq. 71.59.

$$Q_2 = P \tan \phi_2 = (4.750 \text{ kW})(\tan 23.07°) = 2.023 \text{ kVAR}$$

The required capacitance is found from Eq. 71.66.

$$C = \frac{\Delta Q}{2\pi f V_{\text{eff}}^2} = \frac{(3.986 \text{ kVAR} - 2.023 \text{ kVAR})\left(1000 \, \frac{\text{VAR}}{\text{kVAR}}\right)}{(2\pi)(60 \text{ Hz})(117 \text{ V})^2} = 3.8 \times 10^{-4} \text{ F} \quad (380 \; \mu\text{F})$$

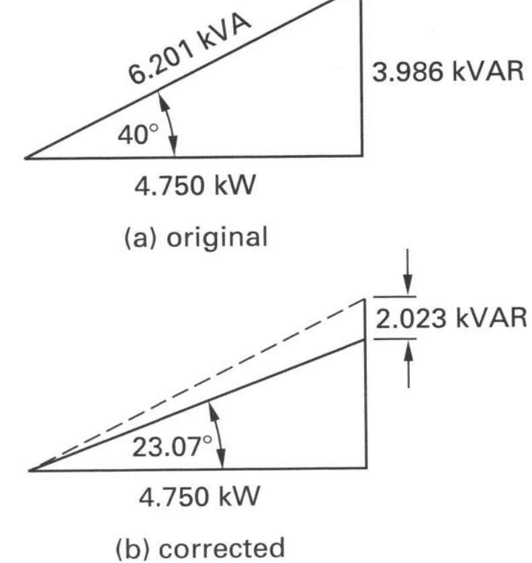

(a) original

(b) corrected

[18]For economics, standard motor sizes should be specified.

Plant Engineering

37. NEMA MOTOR CLASSIFICATIONS

Motors are classified by their NEMA design type—A, B, C, D, and F.[19,20] These designs are collectively referred to as *NEMA frame motors*. All NEMA motors of a given frame size are interchangeable as to bolt holes, shaft diameter, height, length, and various other dimensions.

- *Frame design A:* three-phase, squirrel-cage motor with high locked-rotor (starting) current but also higher breakdown torques; capable of handling intermittent overloads without stalling; 1–3% slip at full load.
- *Frame design B:* three-phase, squirrel-cage motor capable of withstanding full-voltage starting; locked-rotor and breakdown torques suitable for most applications; 1–3% slip at full load; often designated for "normal" usage; the most common design.[21]
- *Frame design C:* three-phase, squirrel-cage motor capable of withstanding full-voltage starting; high locked-rotor torque for special applications (e.g., conveyors and compressors); 1–3% slip at full load.
- *Frame design D:* three-phase, squirrel-cage motor capable of withstanding full-voltage starting; develops 275% locked-rotor torque for special applications; high breakdown torque; depending on design, 5–13% slip at full load; commonly known as a *high-slip motor*; used for cranes, hoists, oil well pump jacks, and valve actuators.
- *Frame design F:* three-phase, squirrel-cage motor with low starting current and low starting torque; used to meet applicable regulations regarding starting current; limited availability.

38. NAMEPLATE VALUES

A *nameplate* is permanently affixed to each motor's housing or frame. This nameplate is embossed or engraved with the *rated values* of the motor (*nameplate values* or *full-load values*). Nameplate information may include some or all of the following: voltage,[22] frequency,[23] number of phases, rated power (output power), running speed, duty cycle, locked-rotor and breakdown torques, starting current, current drawn at rated load (in kVA/hp), ambient temperature, temperature rise at rated load, temperature rise at service factor, insulation rating, power factor, efficiency, frame size, and enclosure type. (See Sec. 71.40 for more on duty cycle, and see Sec. 71.39 for more on temperature rise at service factor.)

It is important to recognize that the rated power of the motor is the actual output power (i.e., power delivered to the load), not the input power. Only the electrical power input is affected by the motor efficiency. (See Eq. 71.76.)

Nameplates are also provided on transformers. Transformer nameplate information includes the two winding voltages (either of which can be the primary winding), frequency, and kVA rating. Apparent power in kVA, not real power, is used in the rating because heating is proportional to the square of the supply current. As with motors, continuous operation at the rated values will not result in excessive heat buildup.

39. SERVICE FACTOR

The horsepower and torque ratings listed on the nameplate of an AC motor can be maintained on a continuous basis without overheating. However, motors can be operated at slightly higher loads without exceeding a safe temperature rise.[24] The ratio of the safe to rated loads (horsepower or torque) is the *service factor*, sf, usually expressed as a decimal.

$$\text{sf} = \frac{\text{maximum safe load}}{\text{nameplate load}} \qquad 71.78$$

Service factors vary from 1.15 to 1.4, with the lower values being applicable to the larger, more efficient motors. Typical values of service factor are 1.4 (up to $^1/_8$ hp motors), 1.35 ($^1/_6$–$^1/_3$ hp motors), 1.25 ($^1/_2$–1 hp motors), and 1.15 (1 or $1^1/_2$ to 200 hp).

When running above the rated load, the motor speed, temperature, power factor, full-load current, and efficiency will differ from the nameplate values. However, the locked-rotor and breakdown torques will remain the same, as will the starting current.

[19]NEMA stands for the National Electrical Manufacturers Association.

[20]There is no type E.

[21]Design B is estimated to be used in 90% of all applications.

[22]Standard NEMA nameplate voltages (effective) are 200 V, 230 V, 460 V, and 575 V. The NEMA motor voltage rating assumes that there will be a voltage drop from the network to the motor terminals. A 200 V motor is appropriate for a 208 V network, a 230 V motor on a 240 V network, and so on. NEMA motors are capable of operating in a range of only ±10% of their rated voltages. Thus, 230 V rated motors should not be used on 208 V systems.

[23]While some 60 Hz motors (notably those intended for 230 V operation) can be used at 50 Hz, most others (e.g., those intended for 200 V operation) are generally not suitable for use at 50 Hz.

[24]Higher temperatures have a deteriorating effect on the winding insulation. A general rule of thumb is that a motor loses two or three hours of useful life for each hour run at the service factor load.

Active current is proportional to torque and therefore, is proportional to the horsepower developed. (See Eq. 71.80.) The active current (line or phase) drawn is also proportional to the service factor. The current drawn per phase is given by Eq. 71.80.[25]

$$I_{\text{active}} = I_l(\text{pf}) \quad 71.79$$

$$I_{\text{actual},p} = \frac{(\text{sf})P_{\text{rated},p}}{V_p\eta(\text{pf})} = \frac{P_{\text{actual},p}}{V_p\eta(\text{pf})} \quad \text{[per phase; } P \text{ in watts]} \quad 71.80$$

Equation 71.80 can also be used when a motor is developing less than its rated power. In this case, the service factor can be considered as the fraction of the rated power being developed.

Example 71.7

What is the approximate phase current drawn by a 98% efficient, three-phase, 75 hp, 230 V (rms) motor that is running at 88% power factor and at its rated load?

Solution

The service factor is 1.00 because the motor is running at its rated load. From Eq. 71.80, the approximate phase current is

$$\begin{aligned} I_p &= \frac{(\text{sf})P_p}{V_p\eta(\text{pf})} \\ &= \frac{(1.00)\left(\frac{75 \text{ hp}}{3 \text{ phases}}\right)\left(0.7457 \frac{\text{kW}}{\text{hp}}\right)\left(1000 \frac{\text{W}}{\text{kW}}\right)}{(230 \text{ V})(0.98)(0.88)} \\ &= 93.99 \text{ A} \quad \text{[per phase]} \end{aligned}$$

40. DUTY CYCLE

Motors are categorized according to their *duty cycle: continuous duty* (24 hr/day); *intermittent* or *short-time duty* (15 min to 30 min); and *special duty* (application specific). Duty cycle is the amount of time a motor can be operated out of every hour without overheating the winding insulation. The idle time is required to allow the motor to cool.

41. INDUCTION MOTORS

The three-phase induction motor is by far the most frequently used motor in industry. In an induction motor, the magnetic field rotates at the synchronous speed. The *synchronous speed* can be calculated from the number of poles and frequency. The frequency, f, is either 60 Hz (in the United States) or 50 Hz (in Europe and other locations). The number of poles, p, must be even.[26] The most common motors have 2, 4, or 6 poles.

$$N_{\text{synchronous}} = \frac{120f}{p} \quad \text{[rpm]} \quad 71.81$$

For typical 60 Hz service, synchronous motor speeds for 2, 4, and 6 poles are 3600, 1800, and 1200 rpm, respectively. **[Synchronous Speed Motors]**

Due to rotor circuit resistance and other factors, rotors (and the motor shafts) in induction motors run slightly slower than their synchronous speeds. The percentage difference is known as the *slip*, s. Slip is seldom greater than 10%, and it is usually much less than that. 4% is a typical value.

$$s = \frac{N_{\text{synchronous}} - N_{\text{actual}}}{N_{\text{synchronous}}} \quad 71.82$$

The rotor's actual speed is[27]

$$N_{\text{actual}} = (1 - s)N_{\text{synchronous}} \quad 71.83$$

Induction motors are usually specified in terms of the *kVA ratings*. The kVA rating is not the same as the motor power in kilowatts, although one can be calculated from the other if the motor's power factor is known. The power factor generally varies from 0.8 to 0.9 depending on the motor size.

$$\text{kVA rating} = \frac{P_{\text{kW}}}{\text{pf}} \quad 71.84$$

$$P_{\text{kW}} = 0.7457P_{\text{mechanical,hp}} \quad 71.85$$

Induction motors can differ in the manner in which their rotors are constructed. A *wound rotor* is similar to an armature winding in a dynamo. Wound rotors have high-torque and soft-starting capabilities. There are no wire windings at all in a *squirrel-cage* rotor. Most motors use squirrel-cage rotors. Typical torque-speed characteristics of a design B induction motor are shown in Fig. 71.11.

Plant Engineering

[25]It is important to recognize the difference between the rated (i.e., nameplate) power and the actual power developed. The actual power should not be combined with the service factor, since the actual power developed is the product of the rated power and the service factor.

[26]There are various forms of Eq. 71.81. As written, the speed is given in rpm, and the number of poles, p, is twice the number of *pole pairs*. When the synchronous speed is specified as f/p, it is understood that the speed is in revolutions per second (rps) and p is the number of pole pairs.

[27]Some motors (i.e., *integral gear motors*) are manufactured with integral speed reducers. Common standard output speeds are 37 rpm, 45 rpm, 56 rpm, 68 rpm, 84 rpm, 100 rpm, 125 rpm, 155 rpm, 180 rpm, 230 rpm, 280 rpm, 350 rpm, 420 rpm, 520 rpm, and 640 rpm. While the integral gear motor is more compact, lower in initial cost, and easier to install than a separate motor with belt drive, coupling, and guard, the separate motor and reducer combination may nevertheless be preferred for its flexibility, especially in replacing and maintaining the motor.

Example 71.8

A pump is driven by a three-phase induction motor running at its rated values. The motor's nameplate lists the following rated values: 50 hp, 440 V, 92% lagging power factor, 90% efficiency, 60 Hz, and 4 poles. The motor's windings are delta-connected. The pump efficiency is 80%. When running under the pump load, the slip is 4%. What are the (a) total torque developed, (b) torque developed per phase, and (c) line current?

Solution

(a) The synchronous speed is given by Eq. 71.81.

$$\begin{aligned} N_{\text{synchronous}} &= \frac{120f}{p} = \frac{(120)(60\ \text{Hz})}{4} \\ &= 1800\ \text{rpm} \end{aligned}$$

From Eq. 71.83, the rotor speed is

$$\begin{aligned} N_{\text{actual}} &= (1-s)N_{\text{synchronous}} = (1-0.04)(1800\ \text{rpm}) \\ &= 1728\ \text{rpm} \end{aligned}$$

Since the motor is running at its rated values, the motor delivers 50 hp to the pump. The pump's efficiency is irrelevant. From Eq. 71.75, the total torque developed by all three phases is

$$\begin{aligned} T_t &= \frac{5252 P_{\text{horsepower}}}{N_{\text{rpm}}} = \frac{\left(5252\ \frac{\text{ft-lbf}}{\text{hp-min}}\right)(50\ \text{hp})}{1728\ \text{rpm}} \\ &= 152.0\ \text{ft-lbf} \quad [\text{total}] \end{aligned}$$

(b) The torque developed per phase is one-third of the total torque developed.

$$\begin{aligned} T_p &= \frac{T_t}{3} = \frac{152.0\ \text{ft-lbf}}{3\ \text{phases}} \\ &= 50.67\ \text{ft-lbf} \quad [\text{per phase}] \end{aligned}$$

(c) The total electrical input power is given by Eq. 71.76.

Mechanical Power

$$P_{\text{mechanical (rated)}} = \eta_{\text{motor}} P_{\text{electrical}}$$

$$\begin{aligned} P_{\text{electrical}} &= \frac{P_{\text{mechanical (rated)}}}{\eta} \\ &= \frac{(50\ \text{hp})\left(0.7457\ \frac{\text{kW}}{\text{hp}}\right)}{0.90} \\ &= 41.43\ \text{kW} \quad [\text{total}] \end{aligned}$$

Since the power factor is less than 1.0, more current is being drawn than is being converted into useful work. From Eq. 71.84, the apparent power in kVA per phase is

$$\begin{aligned} S_{\text{kVA}} &= \frac{P_{\text{kW}}}{\text{pf}} = \frac{41.43\ \text{kW}}{(3)(0.92)} \\ &= 15.01\ \text{kVA} \quad [\text{per phase}] \end{aligned}$$

The phase current is given by Eq. 71.55.

$$\begin{aligned} I_p &= \frac{S}{V} = \frac{(15.01\ \text{kVA})\left(1000\ \frac{\text{VA}}{\text{kVA}}\right)}{440\ \text{V}} \\ &= 34.11\ \text{A} \end{aligned}$$

Since the motor's windings are delta-connected across the three lines, the line current is found from Eq. 71.68.

$$\begin{aligned} I_l &= \sqrt{3}\, I_p = (\sqrt{3})(34.11\ \text{A}) \\ &= 59.08\ \text{A} \end{aligned}$$

42. INDUCTION MOTOR PERFORMANCE

The following rules of thumb can be used for initial estimates of induction motor performance.

- At 1800 rpm, a motor will develop a torque of 3 ft-lbf/hp.
- At 1200 rpm, a motor will develop a torque of 4.5 ft-lbf/hp.
- At 550 V, a three-phase motor will draw 1 A/hp.
- At 440 V, a three-phase motor will draw 1.25 A/hp.
- At 220 V, a three-phase motor will draw 2.5 A/hp.
- At 220 V, a single-phase motor will draw 5 A/hp.
- At 110 V, a single-phase motor will draw 10 A/hp.

43. SYNCHRONOUS MOTORS

Synchronous motors are essentially dynamo alternators operating in reverse. The stator field frequency is fixed, so regardless of load, the motor runs only at a single speed—the synchronous speed given by Eq. 71.81. Stalling occurs when the motor's counter-torque is exceeded. For some equipment that must be driven at constant speed, such as large air or gas compressors, the additional complexity of synchronous motors is justified. [Synchronous Speed Motors]

Power factor can be adjusted manually by varying the field current. With *normal excitation* field current, the power factor is 1.0. With *over-excitation*, the power factor is leading, and the field current is greater than normal. With *under-excitation*, the power factor is lagging, and the field current is less than normal.

Since a synchronous motor can be adjusted to draw leading current, it can be used for power factor correction. A synchronous motor used purely for power factor correction is referred to as a *synchronous capacitor* or *synchronous condenser*. A power factor of 80% is often specified or used with synchronous capacitors.

44. DC MACHINES

DC motors and generators can be wired in one of three ways: series, shunt, and compound. Operational characteristics are listed in Table 71.2. Equation 71.75 can be used to calculate torque and power.

45. CHOICE OF MOTOR TYPES

Squirrel-cage induction motors are commonly chosen because of their simple construction, low maintenance, and excellent efficiencies.[28] A wound-rotor induction motor should be used only if it is necessary to achieve a low starting kVA, controllable kVA, controllable torque, or variable speed.[29]

While induction motors are commonly used, synchronous motors are suitable for many applications normally handled by a NEMA design B squirrel-cage motor. They have adjustable power factors and higher efficiency. Their initial cost may also be less.

Selecting a motor type is greatly dependent on the power, torque, and speed requirements of the rotating load. Table 71.3, Fig. 71.12, App. 71.A, and App. 71.B can be used as starting points in the selection process.

46. LOSSES IN ROTATING MACHINES

Losses in rotating machines are typically divided into the following categories: armature copper losses, field copper losses, mechanical losses (including friction and windage), core losses (including hysteresis losses, eddy current losses, and brush resistance losses), and stray losses.

Copper losses (also known as *I^2R losses*, *impedance losses*, *heating losses*, and *real losses*) are real power losses due to wire and winding resistance heating.

$$P_{\text{Cu}} = \sum I^2R \tag{71.86}$$

Core losses (also known as *iron losses*) are constant losses that are independent of the load and, for that reason, are also known as *open-circuit* and *no-load losses*.

Mechanical losses (also known as *rotational losses*) include brush and bearing friction and *windage* (air friction). (Windage is a no-load loss but is not an electrical core loss.) Mechanical losses are determined by measuring the power input at the rated speed and with no load.

Stray losses are due to nonuniform current distribution in the conductors. Stray losses are approximately 1% for DC machines and zero for AC machines.

47. EFFICIENCY OF ROTATING MACHINES

Only real power is used to compute the efficiency of a rotating machine. This efficiency is sometimes referred to as *overall efficiency* and *commercial efficiency*.

$$\begin{aligned}\eta &= \frac{\text{output power}}{\text{input power}} \\ &= \frac{\text{output power}}{\text{output power} + \text{power losses}} \\ &= \frac{\text{input power} - \text{power losses}}{\text{input power}}\end{aligned} \tag{71.87}$$

The input power is typically given in kilowatts.

Electrical Concepts of Motors: Efficiency

$$\text{kilowatts input} = \frac{(0.746)(\text{horsepower})}{\text{efficiency}} \tag{71.88}$$

$$\text{efficiency} = \frac{(0.746)(\text{horsepower})}{\text{kilowatts input}} \tag{71.89}$$

Example 71.9

A DC shunt motor draws 40 A at 112 V when fully loaded. When running without a load at the same speed, it draws only 3 A at 106 V. The field resistance is 100 Ω, and the armature resistance is 0.125 Ω. (a) What is the efficiency of the motor? (b) What power (in hp) does the motor deliver at full load?

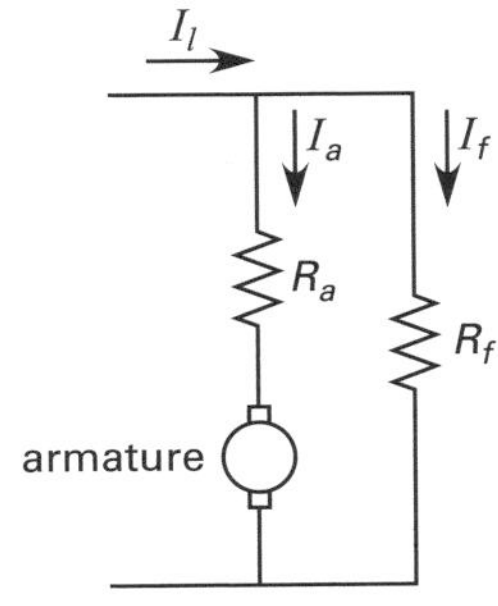

[28]The larger the motor and the higher the speed, the higher the efficiency. Large 3600 rpm induction motors have excellent performance.

[29]A constant-speed motor with a slip coupling could also be used.

Plant Engineering

Table 71.2 Operational Characteristics of DC and AC Machines

	motors			generators		
	shunt	series	compound	shunt	series	compound
equivalent circuit	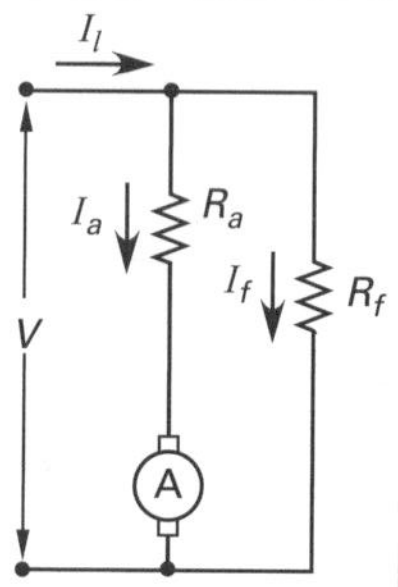	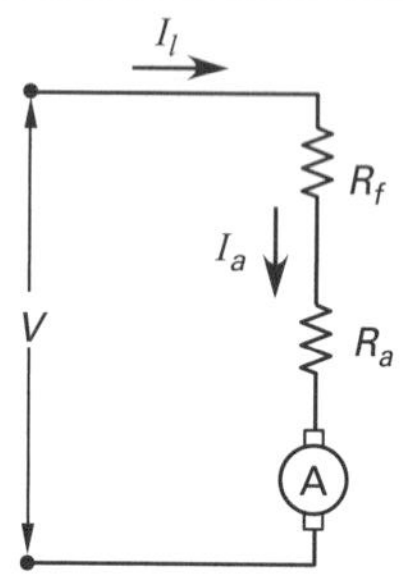	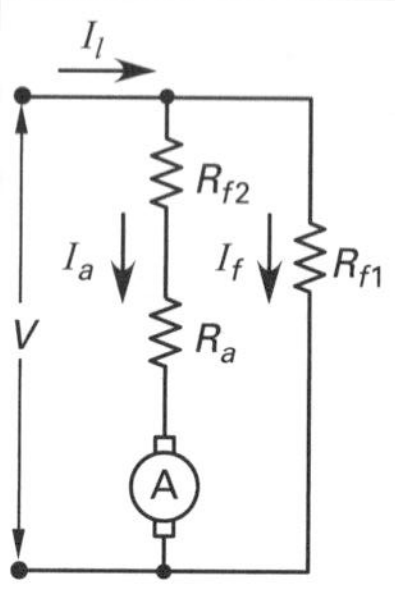	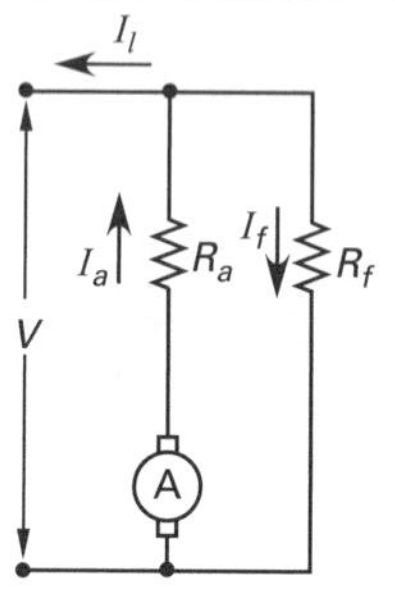	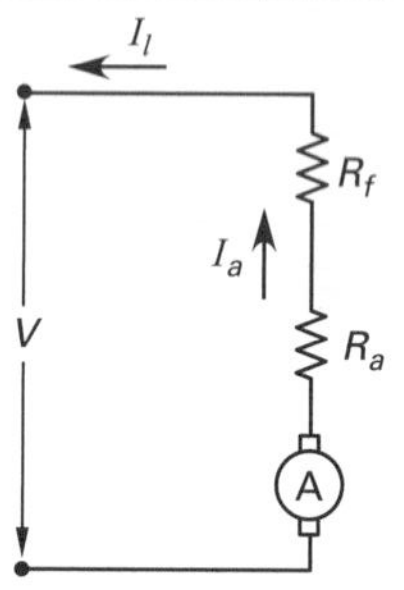	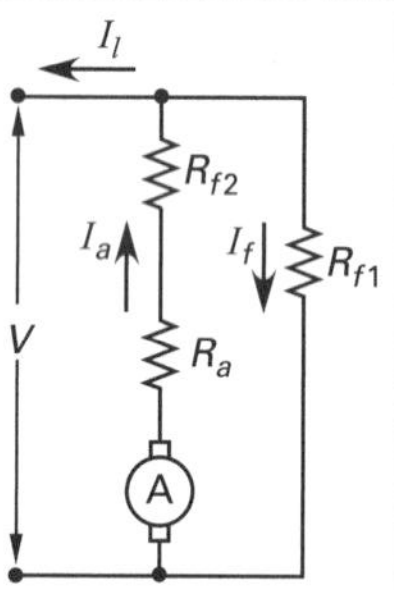
line voltage, V	V	V	V	V	V	V
line current, I_l	I_l	$I_l = I_a$	I_l	$I_l = I_a - I_f$	$I_l = I_a$	$I_l = I_a - I_f$
field current, I_f	$I_f = \frac{V}{R_f}$	$I_f = I_a$	$I_f = \frac{V}{R_{f1}}$	$I_f = \frac{V}{R_f}$	$I_f = I_a$	$I_f = \frac{V}{R_{f1}}$
armature current, I_a	$I_a = I_l - I_f$	$I_a = I_l$	$I_a = I_l - I_f$	$I_a = I_l + I_f$	$I_a = I_l$	$I_a = I_l + I_f$
armature circuit loss, V_a	$V_a = I_aR_a$	$V_a = I_a(R_a + R_f)$	$V_a = I_a(R_a + R_{f2})$	$V_a = I_aR_a$	$V_a = I_a(R_a + R_f)$	$V_a = I_a(R_a + R_{f2})$
counter emf, E_S	$E_S = V - V_a$ $= V - I_aR_a$	$E_S = V - V_a$ $= V - I_a(R_a + R_f)$	$E_S = V - I_a$ $\times (R_a + R_{f2})$	$E_S = V + V_a$ $= V + I_aR_a$	$E_S = V + V_a$ $= V + I_a(R_a + R_f)$	$E_S = V + V_a$ $= V + I_a(R_a + R_{f2})$
power in kW or hp [DC machines]	$P = VI_l$	$P = VI_l$	$P = VI_l$	$\text{hp} = \frac{2\pi NT}{33{,}000}$	$\text{hp} = \frac{2\pi NT}{33{,}000}$	$\text{hp} = \frac{2\pi NT}{33{,}000}$
power in kVA [AC machines]	$P = VI_l \cos \phi$	$P = VI_l \cos \phi$	$P = VI_l \cos \phi$	$\text{hp} = \frac{2\pi NT}{33{,}000}$	$\text{hp} = \frac{2\pi NT}{33{,}000}$	$\text{hp} = \frac{2\pi NT}{33{,}000}$
power out in hp or kW [DC machines]	$\text{hp} = \frac{2\pi NT}{33{,}000}$	$\text{hp} = \frac{2\pi NT}{33{,}000}$	$\text{hp} = \frac{2\pi NT}{33{,}000}$	$P = VI_l$	$P = VI_l$	$P = VI_l$
power out in kVA [AC machines]	$\text{hp} = \frac{2\pi NT}{33{,}000}$	$\text{hp} = \frac{2\pi NT}{33{,}000}$	$\text{hp} = \frac{2\pi NT}{33{,}000}$	$P = VI_l \cos \phi$	$P = VI_l \cos \phi$	$P = VI_l \cos \phi$

Table 71.3 Recommended Motor Voltage and Power Ranges

voltage	horsepower
direct current	
115	0–30 (max)
230	0–200 (max)
550 or 600	$\frac{1}{2}$ and upward
alternating current, one-phase	
110, 115, or 120	0–$1\frac{1}{2}$
220, 230, or 240	0–10
440 or 550	5–10*
alternating current, two- and three-phase	
110, 115, or 120	0–15
208, 220, 230, or 240	0–200
440 or 550	0–500
2200 or 2300	40 and upward
4000	75 and upward
6600	400 and upward

(Multiply hp by 0.7457 to obtain kW.)

*not recommended

Figure 71.12 Motor Rating According to Speed (general guidelines)

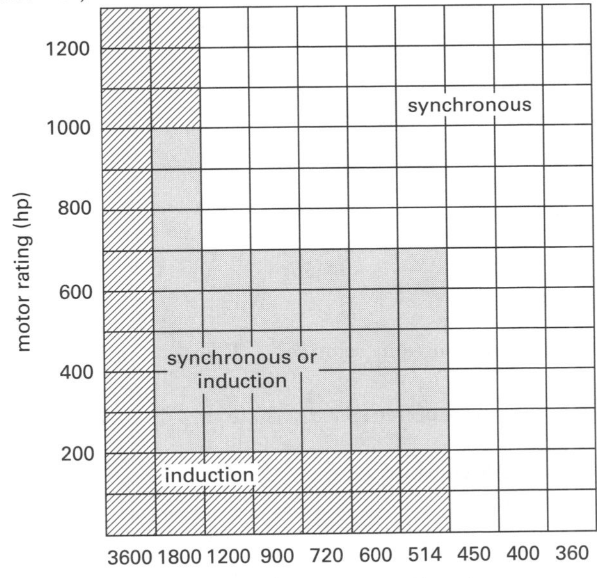

Adapted from *Mechanical Engineering, Design Manual NAVFAC DM-3*, Department of the Navy, © 1972.

Solution

(a) The field current is given by Eq. 71.9.

$$I_f = \frac{V}{R_f} = \frac{112 \text{ V}}{100 \ \Omega} = 1.12 \text{ A}$$

The total full-load line current is known to be 40 A. The full-load armature current is

$$I_a = I_l - I_f = 40 \text{ A} - 1.12 \text{ A} = 38.88 \text{ A}$$

The field copper loss is given by Eq. 71.11.

$$P_{\text{Cu},f} = I_f^2 R_f = (1.12 \text{ A})^2(100 \ \Omega) = 125.4 \text{ W}$$

The armature copper loss is given by Eq. 71.11.

$$P_{\text{Cu},a} = I_a^2 R_a = (38.88 \text{ A})^2(0.125 \ \Omega) = 189.0 \text{ W}$$

The total copper loss is

$$P_{\text{Cu},t} = P_{\text{Cu},f} + P_{\text{Cu},a} = 125.4 \text{ W} + 189.0 \text{ W} = 314.4 \text{ W}$$

Stray power is determined from the no-load conditions. At no load, the field current is given by Eq. 71.9.

$$I_f = \frac{V}{R_f} = \frac{106 \text{ V}}{100 \ \Omega} = 1.06 \text{ A}$$

At no load, the armature current is

$$I_a = I_l - I_f = 3 \text{ A} - 1.06 \text{ A} = 1.94 \text{ A}$$

The stray power loss is

$$\begin{aligned} P_{\text{stray}} &= V_l I_a - I_a^2 R_a \\ &= (106 \text{ V})(1.94 \text{ A}) - (1.94 \text{ A})^2(0.125 \ \Omega) \\ &= 205.2 \text{ W} \end{aligned}$$

The stray power loss is assumed to be independent of the load. The total losses at full load are

$$P_{\text{loss}} = P_{\text{Cu},t} + P_{\text{stray}} = 314.4 \text{ W} + 205.2 \text{ W} = 519.6 \text{ W}$$

The power input to the motor when fully loaded is

$$P_{\text{input}} = I_l V_l = (40 \text{ A})(112 \text{ V}) = 4480 \text{ W}$$

Plant Engineering

The efficiency is

$$\begin{aligned}\eta &= \frac{\text{input power} - \text{power losses}}{\text{input power}}\\ &= \frac{4480\ \text{W} - 519.6\ \text{W}}{4480\ \text{W}}\\ &= 0.884 \quad (88.4\%)\end{aligned}$$

(b) The real power (in hp) delivered at full load is

$$\begin{aligned}P_{\text{real}} &= \text{input power} - \text{power losses}\\ &= \frac{4480\ \text{W} - 519.6\ \text{W}}{\left(1000\ \frac{\text{W}}{\text{kW}}\right)\left(0.7457\ \frac{\text{kW}}{\text{hp}}\right)}\\ &= 5.3\ \text{hp}\end{aligned}$$

48. HIGH-EFFICIENCY MOTORS AND DRIVES

A premium, energy-efficient motor will have approximately 50% of the losses of a conventional motor. Due to the relatively high overall efficiency enjoyed by all motors, however, this translates into only a 5% increase in overall efficiency.

High-efficiency motors are often combined with *variable-frequency drives* (VFDs) to achieve continuous-variable speed control.[30] VFDs can substantially reduce the power drawn by the process. For example, with a motor-driven pump or fan, the motor can be slowed down instead of closing a valve or damper when the flow requirements decrease. Since the required motor horsepower varies with the cube root of the speed, for processes that do not always run at full-flow (fluid pumping, metering, flow control, etc.), the energy savings can be substantial.

A VFD uses an electronic controller to produce a variable-frequency signal that is not an ideal sine wave. This results in additional heating (e.g., an increase of 20–40%) from copper and core losses in the motor. In pumping applications, though, the process load drops off faster than additional heat is produced. Thus, the process power savings dominate.

Most low- and medium-power motors implement VFD with DC voltage intermediate circuits, a technique known as *voltage-source inversion*. Voltage-source inversion is further subdivided into *pulse width modulation* (PWM) and *pulse amplitude modulation* (PAM).

Under VFD control, the motor torque is approximately proportional to the applied voltage and drawn current but inversely proportional to the applied frequency.

$$T \propto \frac{VI}{f} \qquad 71.90$$

49. NOMENCLATURE

a	turns ratio	–	–
A	area	ft^2	m^2
B	susceptance	S	S
C	capacitance	F	F
d	diameter	ft	m
E	energy usage (demand)	kW-hr	kW·h
f	linear frequency	Hz	Hz
G	conductance	S	S
i	varying current	A	A
I	constant current	A	A
l	length	ft	m
L	inductance	H	H
N	speed	rev/min	rev/min
n	number of turns	–	–
p	number of poles	–	–
pf	power factor	–	–
P	power	W	W
q	charge	C	C
Q	reactive power	VAR	VAR
R	resistance	Ω	Ω
s	slip	–	–
sf	service factor	–	–
S	apparent power	VA	VA
SR	speed regulation	–	–
t	time	sec	s
T	period	sec	s
T	torque	ft-lbf	N·m
v	varying voltage	V	V
V	constant voltage	V	V
VR	voltage regulation	–	–
X	reactance	Ω	Ω
Y	admittance	S	S
Z	impedance	Ω	Ω

Symbols

ϵ	dielectric constant (relative permittivity)	F/m	F/m
ϵ_0	permittivity of free space, 8.85×10^{12}	F/m	F/m
κ	dielectric constant	–	–
η	efficiency	–	–
θ	phase angle	rad	rad
ρ	resistivity	Ω-ft	Ω·m
σ	conductivity	1/Ω-ft	1/Ω·m
ϕ	impedance angle	rad	rad
ω	angular frequency	rad/sec	rad/s

[30]Prior to VFDs, there were two common ways to change the speed of an induction motor: (a) increasing the slip, and (b) changing the number of pole pairs. Increasing the slip was accomplished by under-magnetizing the motor so that it received less input voltage than it was built for. Dropping or adding the number of active poles resulted in the speed changing by a factor of 2 (or ½).

Subscripts

a	armature
ave	average
C	capacitive
Cu	copper
eff	effective
eq	equivalent
f	field
G	generator
l	line
L	inductive
m	maximum
p	phase or primary
r	rated
R	resistive
s	secondary
t	total

72 Illumination and Sound

Content in blue refers to the *NCEES Handbook*.

NCEES EXAM SPECIFICATIONS AND RELATED CONTENT

HVAC AND REFRIGERATION EXAM

II.D.4. Supportive Knowledge: Acoustics
- 7. Sound and Noise
- 16. Sound Power Level
- 21. Combining Multiple Sources
- 24. Sound Rating Methods
- 29. Sound Transmission Class
- 30. Noise Reduction

1. LUMINOUS FLUX

The amount of visible light emitted from a source is the *luminous flux*, Φ, with units of lumens (lm).[1] Not all of the energy emitted will be in the visible region, though. Therefore, the total source power is not a good indicator of the visible lighting effect. The ratio of luminous flux to total radiant energy (also known as the *quantity of light*), Q, is the *luminous efficiency*, η_l.

$$\eta_l = \frac{\Phi}{Q} \tag{72.1}$$

Efficiency of a lamp is the ratio of light output in lumens to the lamp's input in lumens. *Efficacy* is the ratio of the lamp's output in lumens to its input power in watts. (See Table 72.1.)

$$\eta = \frac{\Phi}{P} \tag{72.2}$$

The *luminous emittance* (also known as *brightness*), M, of a source is the luminous flux per unit area of the source. (Compare this with *illumination* of a receiver covered in Sec. 72.3, which is the flux per unit of receiving area.) In the United States, the unit of brightness is the lm/ft² (foot-lambert).[2]

$$M = \frac{\Phi}{A_{\text{source}}} \tag{72.3}$$

[1] A lumen is approximately equal to 1.47×10^{-3} W of visible light with a wavelength of 5.55×10^{-7} m.
[2] The term *lambert* in the United States is a source of confusion. The old lambert unit is the same as lm/cm². A foot-lambert is the same as lm/ft².

Table 72.1 *Typical Values of Luminous Efficacy*

source		lm/W
	candle	0.1
25 W	tungsten lamp	10
100 W	tungsten lamp	16
1000 W	tungsten lamp	22
40 W	fluorescent lamp	80
400 W	mercury fluorescent lamp	58
1000 W	carbon arc	60
100 W	mercury arc	35
400 W	metal halide lamp	85
1000 W	metal halide lamp	100
400 W	high-pressure sodium lamp	125
1000 W	high-pressure sodium lamp	130
180 W	low-pressure sodium lamp	180

Example 72.1

A tungsten filament emitting 1600 lm has a surface area of 0.35 in^2 (2.26×10^{-4} m^2). What is its brightness?

SI Solution

$$M = \frac{\Phi}{A_{\text{surface}}} = \frac{1600 \text{ lm}}{2.26 \times 10^{-4} \text{ m}^2} = 7.08 \times 10^6 \text{ lm/m}^2$$

Customary U.S. Solution

$$M = \frac{\Phi}{A_{\text{surface}}} = \frac{(1600 \text{ lm})\left(12 \ \frac{\text{in}}{\text{ft}}\right)^2}{0.35 \text{ in}^2} = 6.58 \times 10^5 \text{ lm/ft}^2$$

2. LUMINOUS INTENSITY

The *luminous intensity*, I, of a source is a measure of flux per unit solid angle. Its units are lumens per steradian (lm/sr), referred to by the synonymous names of *candle*, *candela*, and *candlepower*.[3] Equation 72.4 relates flux from an omnidirectional point source to intensity measured on a spherical receptor of radius r.

$$I = \frac{\Phi}{\omega} = \frac{\Phi r^2}{A} = \frac{\Phi_t r^2}{4\pi r^2} = \frac{\Phi_t}{4\pi} \qquad 72.4$$

Example 72.2

Two lumens pass through a circular hole (diameter = 0.5 m) in a screen located 6.0 m from an omnidirectional source. (a) What is the luminous intensity of the source? (b) What luminous flux is emitted by the source?

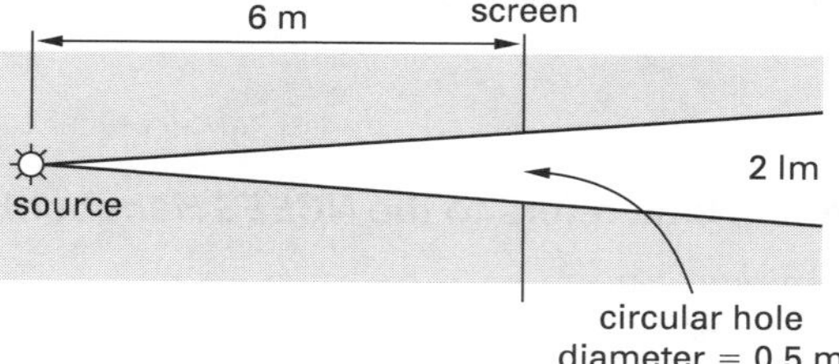

Solution

(a) The solid angle subtended by the hole is

$$\omega = \frac{A}{r^2} = \frac{\left(\frac{\pi}{4}\right)(0.5 \text{ m})^2}{(6.0 \text{ m})^2} = 0.005454 \text{ sr}$$

From Eq. 72.4, the intensity is

$$I = \frac{\Phi}{\omega} = \frac{2 \text{ lm}}{0.005454 \text{ sr}} = 366.7 \text{ lm/sr}$$

(b) From Eq. 72.4, the luminous flux is

$$\Phi_t = 4\pi I = (4\pi \text{ sr})\left(366.7 \ \frac{\text{lm}}{\text{sr}}\right) = 4608 \text{ lm}$$

3. ILLUMINANCE

The *illuminance* (*illumination*), E, is a measure of luminous flux per incident area. Typical units are lux (lm/m^2) and foot-candles (lm/ft^2). (See Table 72.2.) The name *irradiance* is used when the units are W/m^2.

Table 72.2 *Recommended Illuminance*

location	lm/ft^2 (fc)	lux
roadway	1–2	10–20
living room	5–15	50–150
library (reading)	30–70	300–700
evening sports	30–100	300–1000
factory (assembly)	100–200	1000–2000
office	100–200	1000–2000
factory (fine assembly)	500–1000	5000–10 000
hospital operating room	2000–2500	20 000–25 000

(Multiply foot-candles by 10.764 to obtain lux.)
(Multiply foot-candles by 0.10764 to obtain hectolux.)

[3]Intensity is not a measure of power, as the term *candlepower* may imply.

One foot-candle is the illumination from a one *candle-power* (cp) source at a distance of one foot.

$$E = \frac{\Phi}{A_{\text{receptor}}} \quad 72.5$$

For an omnidirectional source and a spherical receptor of radius r,

$$E = \frac{\Phi_t}{4\pi r^2} \quad 72.6$$

Equation 72.6 shows that illumination follows the *distance squared law.*

$$E_1 r_1^2 = E_2 r_2^2 \quad 72.7$$

Example 72.3

A lamp radiating hemispherically and rated at 2000 lm is positioned 20 ft (6 m) above the ground. What is the illumination on a walkway directly below the lamp?

SI Solution

From Eq. 72.5,

$$E = \frac{\Phi}{A} = \frac{\Phi}{\frac{1}{2}A_{\text{sphere}}} = \frac{2000 \text{ lm}}{\left(\frac{1}{2}\right)(4\pi)(6 \text{ m})^2}$$
$$= 8.84 \text{ lux}$$

Customary U.S. Solution

From Eq. 72.5,

$$E = \frac{\Phi}{A} = \frac{\Phi}{\frac{1}{2}A_{\text{sphere}}} = \frac{2000 \text{ lm}}{\left(\frac{1}{2}\right)(4\pi)(20 \text{ ft})^2}$$
$$= 0.796 \text{ lm/ft}^2 \text{ (fc)}$$

4. LUMINANCE

Illuminance of an object is the quantity of light from a source reaching the object. The amount of light reflected from the object's surface is referred to as the *luminance* or *brightness* of the object. Luminance depends on the *reflectance* of the surface and is measured in essentially the same units as illuminance.[4] For example, if an object has a reflectance of 40% and is illuminated at 150 fc, its brightness is (0.40)(150 fc) = 60 fL.

5. LIGHTING DESIGN

Design of lighting systems has many elements and is primarily performed by architects. A common requirement is to determine the number of lamps (luminaires) required. The *lumen method* (*total flux method*) can be used in rectangular rooms with a gridded lamp arrangement and where uniform illumination is needed. In Eq. 72.8, A is the area at the working plane level, CU is the *coefficient of utilization* (*utilization factor*), an allowance for the light distribution of the lamps and the room surfaces, and LLF is the *light loss factor* (*maintenance factor*), an allowance for reduced light output due to deterioration and dirt. The coefficient of utilization depends on wall color, lamp placement, aspect ratio of the room and lamp spacing, and other factors. In the absence of other information, the light loss factor is arbitrarily set to 0.80.

$$n = \frac{EA}{\Phi(\text{CU})(\text{LLF})} \quad 72.8$$

6. INTERACTION OF LIGHT WITH MATTER

Light travels through a vacuum as an electromagnetic wave. When the light makes contact with matter, some of the wave energy is absorbed by the matter, causing electrons to jump into higher energy states. (A polished metal surface, for example, will absorb only about 10% of the incident energy, reflecting the remaining 90% away.) Some of this absorbed energy is re-emitted when the electrons drop back to a lower energy level. Generally, the re-emitted light will not be at its original wavelength.

If the reflecting surface is smooth, the *reflection angle* for most of the light will be the same as the *incident angle*, and the light is said to be *regularly reflected* (i.e., the case of *specular reflection*). If the surface is rough, however, the light will be scattered and reflected randomly (the case of *diffuse reflection*). (See Fig. 72.1.)

Figure 72.1 *Specular and Diffuse Reflection*

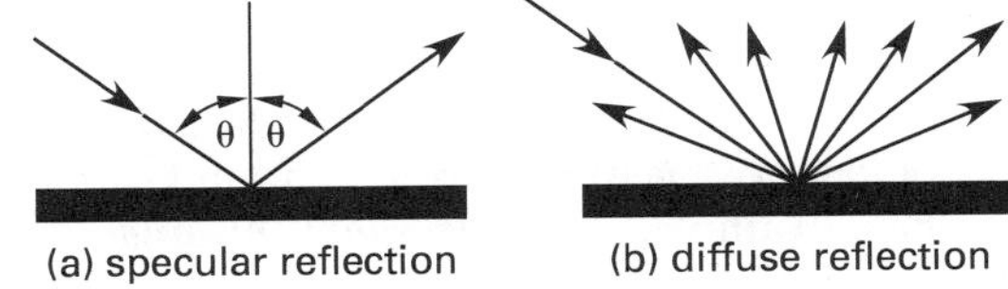

The energy that is absorbed is said to be *refracted.* In the case of an *opaque material,* the refracted energy is absorbed within a very thin layer and converted to heat. Light is able to pass through a *transparent material* without being absorbed. Light is partially absorbed in a *translucent material.*

7. SOUND AND NOISE

Sound is made up of the pressure waves and vibrations that are received from a vibrating source. *Noise* is unwanted sound. Noise can either be continuous or impulsive. For the purpose of classification, *impact*

[4]A *lambert* is equal to 929 foot-lamberts. A *millilambert* is equal to 0.929 foot-lamberts. Other units include the *stilb* (candle/cm^2) and *nit* (candle/m^2).

noise (*impulse noise*) is noise whose peak levels have a duration less than one second and occur at intervals greater than one second.

Background noise is the "normal" noise in the local environment. Background noise consists of environmental noises that are not components of the noise to be measured or studied. It should be measured separately and subtracted from any noise measurements made while equipment or other sources are operating. The *ASHRAE Handbook—HVAC Applications* provides typical design guidelines for HVAC-related background noise in various room types. [**Design Guidelines for HVAC-Related Background Sound in Rooms**]

A *pure tone* is one that consists essentially of a single frequency. Relatively pure tones can be generated at simple harmonic frequencies in rotating equipment. For example, a five-bladed fan turning at 50 rev/sec would generate harmonics with a fundamental frequency of 250 Hz. Pure tones and harmonics are of interest because they are difficult to attenuate. However, most noise is white noise, not pure tones. *White noise* is relatively evenly distributed over the frequency spectrum.

Structure-borne noise (also known as *mechanical noise*) is noise that is generated by mechanical elements in moving equipment. Panels, covers, housings, and other membrane-like elements are frequent sources of structure-borne noise. These elements produce noise at their own natural frequencies, not just at the cycling frequency of the equipment.

8. PHYSIOLOGICAL BASIS OF SOUND

Longitudinal sound waves in the 20 Hz to 20,000 Hz range can be heard by most people. The average human ear is most sensitive to frequencies around 3000 Hz. Even in this range, however, the intensity must be great enough for a sound to be detected. *Intensity* is the amount of energy passing through a unit area each second. For example, the *threshold of audibility* at 3000 Hz is approximately 10^{-12} W/m^2. If the sound is too intense (i.e., the intensity is above the *threshold of pain*, approximately 10 W/m^2), the sound will be experienced painfully. The thresholds of audibility and pain are both somewhat frequency dependent.

Loudness is a qualitative sensation that can be quantified approximately by comparing the sound intensity to a reference intensity (usually the assumed threshold of audibility, 10^{-12} W/m^2). Loudness is perceived by most individuals approximately logarithmically and is calculated in that manner.

$$\text{loudness} = 10\log\frac{I}{I_0} = 10\log\frac{I}{10^{-12}\ \frac{\text{W}}{\text{m}^2}} \qquad 72.9$$

9. LONGITUDINAL SOUND WAVES

Sound, like light and electromagnetic radiation, is a wave phenomenon. However, sound has two distinct differences from electromagnetic waves. First, sound is transmitted as *longitudinal waves*, also known as *compression waves*. Such waves are alternating compressions and expansions of the medium through which they travel. In comparison, electromagnetic waves are *transverse waves*. Second, longitudinal waves require a medium through which to travel—they cannot travel through a vacuum, as can electromagnetic waves.

For ease of visualization, longitudinal waves may be represented by transverse waves. The *wavelength* is the distance between successive compressions (or expansions), as shown in Fig. 72.2.

Figure 72.2 *Longitudinal Waves*

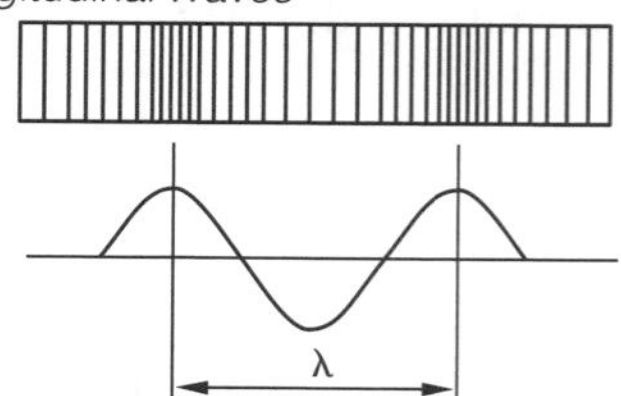

The usual relationships among wavelength, propagation velocity, period, and frequency are valid.

$$a = f\lambda \qquad 72.10$$

$$T = \frac{1}{f} \qquad 72.11$$

10. PROPAGATION VELOCITY OF LONGITUDINAL SOUND WAVES

The *propagation velocity* of longitudinal waves, commonly known as the *speed of sound* and *sonic velocity*, depends on the compressibility of the supporting medium. For solids, the compressibility is accounted for by the elastic modulus, E. For liquids, it is customary to refer to the elastic modulus as the *bulk modulus*.

$$c = \sqrt{\frac{E}{\rho}} \qquad \text{[SI]} \qquad 72.12(a)$$

$$c = \sqrt{\frac{Eg_c}{\rho}} \qquad \text{[U.S.]} \qquad 72.12(b)$$

For ideal gases, the propagation velocity is calculated from Eq. 72.13.

$$c = \sqrt{kRT} = \sqrt{\frac{kR^*T}{M}} \quad \text{[SI]} \qquad 72.13(a)$$

$$c = \sqrt{kg_cRT} = \sqrt{\frac{kg_cR^*T}{M}} \quad \text{[U.S.]} \qquad 72.13(b)$$

Table 72.3 lists approximate values for the speed of sound in various media.

***Table 72.3** Approximate Speeds of Sound (at one atmospheric pressure)*

	speed of sound	
material	(ft/sec)	(m/s)
air	1130 at 70°F	330 at 0°C
aluminum	16,400	4990
carbon dioxide	870 at 70°F	260 at 0°C
hydrogen	3310 at 70°F	1260 at 0°C
steel	16,900	5150
water	4880 at 70°F	1490 at 20°C

(Multiply ft/sec by 0.3048 to obtain m/s.)

11. DOPPLER EFFECT

If the distance between a sound source and a listener is changing, the frequency heard will differ from the frequency emitted. If the separation distance is decreasing, the frequency will be shifted higher; if the separation distance is increasing, the frequency will be shifted lower. This shifting is known as the *Doppler effect*. The ratio of observed frequency, f', to emitted frequency, f, depends on the local speed of sound, a, and the absolute velocities of the source and observer, v_s and v_o. In Eq. 72.14, v_s is positive if the source moves away from the observer and is negative otherwise; v_o is positive if the observer moves toward the source and negative otherwise.

$$\frac{f'}{f} = \frac{c+v_o}{c+v_s} \qquad 72.14$$

Example 72.4

On a particular day, the local speed of sound is 1130 ft/sec (344 m/s). What frequency is heard by a stationary pedestrian when a 1200 Hz siren on an emergency vehicle moves away at 45 mph (72 kph)?

SI Solution

The observer's velocity is zero. The source's velocity is positive because the vehicle is moving away. From Eq. 72.14, the frequency heard is

$$f' = f\left(\frac{c+v_o}{c+v_s}\right)$$

$$= (1200\text{ Hz})\left(\frac{344\ \frac{\text{m}}{\text{s}} + 0\ \frac{\text{m}}{\text{s}}}{344\ \frac{\text{m}}{\text{s}} + \frac{\left(72\ \frac{\text{km}}{\text{h}}\right)\left(1000\ \frac{\text{m}}{\text{km}}\right)}{3600\ \frac{\text{s}}{\text{h}}}}\right)$$

$$= 1134\text{ Hz}$$

Customary U.S. Solution

The observer's velocity is zero. The source velocity is positive because the vehicle is moving away. Convert the source velocity from miles per hour to feet per second.

$$v_s = \frac{\left(45\ \frac{\text{mi}}{\text{hr}}\right)\left(5280\ \frac{\text{ft}}{\text{mi}}\right)}{3600\ \frac{\text{sec}}{\text{hr}}} = 66\text{ ft/sec}$$

From Eq. 72.14, the frequency heard is

$$f' = f\left(\frac{c+v_o}{c+v_s}\right) = (1200\text{ Hz})\left(\frac{1130\ \frac{\text{ft}}{\text{sec}} + 0\ \frac{\text{ft}}{\text{sec}}}{1130\ \frac{\text{ft}}{\text{sec}} + 66\ \frac{\text{ft}}{\text{sec}}}\right)$$

$$= 1134\text{ Hz}$$

12. ACOUSTIC IMPEDANCE

The *acoustic impedance* (*sound impedance*), Z, is a function of the pressure, p, wave velocity, v, and the surface area, A, through which the sound travels. Common units are lbf-sec/ft^5 and N·s/m^5, equivalent to rayls/m^2. The quantity vA is known as the *volume velocity*.

$$Z = \frac{p}{vA} \qquad 72.15$$

The *specific acoustic impedance*, z, is

$$z = \frac{p}{v} = ZA \qquad 72.16$$

The *specific acoustic impedance* (*characteristic acoustic impedance*) *of the medium* (e.g., air, water, steel), Z_0, is given by Eq. 72.17. Common units are lbf-sec/ft^3 and

Plant Engineering

rayls, equivalent to N·s/m^3 and Pa·s/m (also known as an *acoustic ohm*). For 68°F (20°C) air at 1 atm, the specific acoustic impedance is between 410 rayls and 420 rayls.

$$Z_0 = \rho c \quad \text{[SI]} \qquad 72.17(a)$$

$$Z_0 = \frac{\rho c}{g_c} \quad \text{[U.S.]} \qquad 72.17(b)$$

Specific acoustic impedance is the vector resultant of two orthogonal vectors. The real, resistive part, R, represents energy lost due to friction and viscous effects when compression sound waves pass through the medium. The imaginary, reactive part, X, represents energy temporarily stored in and subsequently released from the compressed medium.

$$Z_0 = R + iX \qquad 72.18$$

13. SOUND METER MEASUREMENTS

Rather than attempt to describe sound in terms of subjective loudness, simple *sound meters* (*noise meters*) measure noise using various response curves. (A *response curve* is basically a set of weighting values that can be applied to each frequency detected.) Meter response curves have been standardized worldwide, and most industrial-quality meters respond accurately to three curves, designated as the A-, B-, and C-scales. The slow-response A-scale approximates human hearing the most accurately, so it is used the most frequently.[5,6]

When sound measurements are reported, a sound "level" always means a logarithmic value expressed in decibels. Sound level units are reported along with the scale used. For example, "15 dBA" (also written as "dB(A)" and "dB-A") indicates that the A-scale was used.

Some meters often also provide a fourth, unweighted reading using the basic frequency response of the microphone and meter circuitry. This reading is different for different meters. The terms *unweighted response*, *flat response*, *20 kHz response*, and *linear response* all refer to this case.

Peak levels from impulse sources are measured with special equipment (i.e., an impact meter or oscilloscope) that captures and holds the maximum instantaneous noise level.

14. SOUND FIELDS

Sound that is received directly from a source without any significant reflection (including that from the floor or ground) is known as *free-field* sound. Similarly, free-field measurements are those that do not include any contribution from reflections. Measurements made near walls or other objects where reflected sound is significant are called *reverberant field* measurements. Most in-room measurements are reverberant field measurements. In a reverberant field, sound pressure levels do not fall off with increases in distance from the source. Sound levels are essentially the same anywhere in a reverberant field.

Measurements made close to a noise source are called *near-field measurements.* They can be in error due to directionality effects and reflections from the source itself. The near-field effects will be significant if the measurement is taken closer than one wavelength (based on the lowest frequency of interest) from the source. Measurements made at large distances from a noise source are called *far-field measurements.* Far-field measurements can be in error due to background noise. In a true far-field environment, the sound pressure level will decay 6 dB for each doubling of the distance from the source.

Anechoic rooms (free-field throughout) and *reverberant rooms* (reverberant field throughout) represent extremes in measurement environments. Such rooms are routinely used for measuring directionality effects and absorption material performance.

15. SOUND PRESSURE LEVEL

Sound propagates as pressure fluctuates. The actual pressure at a particular point is the *sound pressure level*, L_p. (The symbol SPL is also used in some literature.) Sound pressure level is the quantity that is actually measured in sound meters.

Since the human ear responds to variations in air pressure over a range of more than a million to one, a logarithmic power ratio scale is used to measure sound in decibels. The minimum perceptible pressure amplitude (i.e., the approximate *threshold of hearing*) has been standardized as 20 μPa, and this value is used as a reference pressure.[7] The acoustic energy is proportional to the square of the sound pressure. For an rms (root-mean-squared or "effective") sound pressure, p, in pascals, the sound pressure level is

$$\begin{aligned} L_p &= 10\log\left(\frac{p}{p_{\text{ref}}}\right)^2 = 20\log\frac{p}{p_{\text{ref}}} \\ &= 20\log\frac{p}{20\times 10^{-6}\ \text{Pa}} \end{aligned} \qquad 72.19$$

Usually, the sound pressure level is sampled at several single frequencies over the audio spectrum. These samples are referred to as "narrow band" or "octave band"

[5] The A-curve discriminates against lower frequencies, the C-curve is nearly flat, and the B-curve falls in between.
[6] Refer to ANSI S1.4 for sound level meters.
[7] At one time, this was thought to be the threshold of hearing for an average young person at 1000 Hz.

measurements. The individual measurements can be combined into a *broad band* value by weighting each by standardized values.[8]

16. SOUND POWER LEVEL

The *sound power level* is used to measure the total acoustic power emitted by a source of sound. Sound power is, therefore, independent of the environment. Sound power is not measured by sound meters; it must be calculated. A standard reference level of 1×10^{-12} W is used.[9] The literature uses L_W and PWL as symbols for sound power level. For a source of sound emitting P watts, the sound power level in decibels is shown in Eq. 72.20. *ASHRAE Handbook—Fundamentals* provides several common examples of sound power outputs and levels. [**Examples of Sound Power Outputs and Sound Power Levels**]

$$L_W = 10\log\frac{P}{P_{\text{ref}}} = 10\log\frac{P}{1\times 10^{-12}\text{ W}} \qquad 72.20$$

17. DIRECTIVITY

The *directivity*, Q (also known as the *directivity factor*), affects the sound power that appears as sound pressure in a near-field measurement. (Directivity is not an issue in far-field measurements.) Directivity depends on the orientation of the source and the physical location of the source in relationship to walls and other solid surfaces. Directivity is 1 for an isotropic source radiating into free space, 2 for a source on the surface of an infinite flat plane, 4 for a source at the intersection of two perpendicular planes, and 8 for a source in the corner of three mutually perpendicular reflecting planes.

18. CONVERTING SOUND PRESSURE INTO SOUND POWER

Sound pressure level and sound power level are both reported in decibels. However, there is no simple conversion between them because sound pressure level is affected by directivity and separation distance from the source, while sound power level is not. However, the two scales have been established such that a decibel change in one quantity will produce the same decibel change in the other quantity. Therefore, a decibel decrease or increase in sound power will produce the same decibel decrease or increase in sound pressure.

The sound pressure at a distance r from a continuous noise source with known sound power, directivity, and room constant (which is covered in Sec. 72.27), in a confined space large enough to have both direct and reverberant sound, is

$$L_{p,\text{dBA}} = 0.2 + L_W + 10\log\left(\frac{Q}{4\pi r^2} + \frac{4}{R}\right) \quad \text{[SI]} \qquad 72.21(a)$$

$$L_{p,\text{dBA}} = 10.5 + L_W + 10\log\left(\frac{Q}{4\pi r^2} + \frac{4}{R}\right) \quad \text{[U.S.]} \qquad 72.21(b)$$

Example 72.5

The combined sound pressure level from four identical machines at a point equally distant from each is 100 dBA. Three machines are shut down. What is the new sound pressure level?

Solution

When three of the four machines are shut down, the total radiated sound power will be 25% of the original condition. The reduction in sound power level will be

$$\begin{aligned} L_{W,1} - L_{W,2} &= 10\log\frac{P_1}{P_2} = 10\log 4 \\ &= 6.02\text{ dB} \quad \text{[use 6 dB]} \end{aligned}$$

The change in sound pressure level is the same as the change in sound power level. Therefore, the new sound pressure level will be

$$L_p = 100\text{ dBA} - 6\text{ dB} = 94\text{ dBA}$$

19. FREQUENCY CONTENT OF NOISE

Detailed knowledge of the frequency content of noise is required for most noise control problems. *Frequency content* is described by dividing the sound spectrum into a series of frequency ranges called *bands*. Frequency bands are established by dividing the frequency range of interest into bands of either equal bandwidths or equal percentages.[10] The sound level in each frequency band can be reported or plotted against the midpoint of the frequency band.

20. OCTAVE BAND ANALYSIS

Percentage bandwidths and center frequencies have been standardized worldwide. The most common (and widest) standard bandwidth is 50%, which means that the ratio between the frequencies at the two ends of the band is 1:2. For example, the band whose center

Plant Engineering

[8]Weighting and combining are done automatically by the sound meter.

[9]Many years ago, a reference of 10^{-13} W was used. There may still be some references to this value in older literature.

[10]Equal bandwidths are generally used only when the frequency range of interest is small. There is little standardization of narrow bandwidths, primarily because narrow band analysis is used more for detailed investigation of noise sources than for reporting absolute levels.

frequency is 1000 Hz spans the range from 707 Hz to 1414 Hz. Since musical pitch frequency ratios of 1:2 are called *octaves*, the name *octave band* has been adopted.

Center frequencies for octave bands have been standardized at 63 Hz, 125 Hz, 250 Hz, 500 Hz, 1000 Hz, 2000 Hz, 4000 Hz, and 8000 Hz.[11,12] These bands are numbered 1, 2, 3, 4, 5, 6, 7, and 8, respectively. Octave bands can also be subdivided for better frequency detail. The most common subdivision is the *one-third octave band.*

21. COMBINING MULTIPLE SOURCES

Multiple sound sources combine to produce more sound. Sound pressures from multiple correlated sources combine linearly, while sound pressures combine as sums of squares. However, multiple sound pressure levels or sound power levels expressed in decibels cannot be directly added. Two 90 dB sources do not produce a 180 dB source. For multiple sound sources, the combined power is given by Eq. 72.22. This is known as an *unweighted sum* because the individual readings are used without modification. *ASHRAE Handbook—Fundamentals* provides a table indicating the number of decibels to add to the highest decibel level when combining two sound sources. [**Combining Multiple Sound Levels**]

$$L_p = 10\log\sum 10^{L_i/10} \qquad 72.22$$

Using Eq. 72.22 to add or subtract the sound from two or more sources is not the same as determining the *change* in sound level. As described in Sec. 72.30, Sec. 72.31, and Sec. 72.32, the relationship between the old, new, and change in sound levels is that of a simple linear combination (i.e., addition or subtraction).

Equation 72.22 illustrates an important noise reduction principle: The noisiest source must be identified and treated before significant overall noise reduction can be achieved. Sources with sound levels more than a few decibels lower than the noisiest source make a minor contribution to the total sound level.

Example 72.6

What is the unweighted combined sound pressure from two machines, one with a sound pressure of 89 dB and the other with a sound pressure of 94 dB?

Solution

Use Eq. 72.22.

$$\begin{aligned} L_p &= 10\log\sum 10^{L_i/10} \\ &= 10\log\left(10^{89\ \text{dB}/10} + 10^{94\ \text{dB}/10}\right) \\ &= 95.2\ \text{dB} \end{aligned}$$

22. BROAD BAND MEASUREMENTS

Octave band measurements contain enough information about the frequency content to permit calculation of the equivalent A-weighted level. The correction from Table 72.4 is made to each octave band measurement, and the corresponding levels are added using Eq. 72.22.

Table 72.4 *A-Weighting Corrections from Octave Band Analysis*

center frequency of band (Hz)	correction to be added (dB)*
31.5	–39.2
63	–26.2
125	–16.1
250	–8.6
500	–3.2
1000	0.0
2000	+1.2
4000	+1.0
8000	–1.1

*referenced to 20 μPa

23. LOUDNESS

Human hearing is not equally sensitive to pressure at all frequencies. In addition, some types of noise are more annoying than others. Subjective (i.e., perceived) noise level is called *loudness.*

There are two loudness scales: the *phon scale* and the *sone scale.* One *phon* is numerically equal to the sound pressure level in decibels at a frequency of 1000 Hz. At 1000 Hz, the values of loudness and L_p are the same.

The common loudness scale used by fan and duct manufacturers is the sone scale. By definition, one *sone* is the loudness of a 1000 Hz sound with a sound pressure of 0.02 μbar.[13] Sones are calculated from octave band sound pressure level measurements made through frequency filters in much the same way as are dBA measurements.[14]

[11] A now-obsolete series of octave bands was used until the late 1950s.

[12] Octave bands centered at 31.5 Hz and 16,000 Hz are also occasionally encountered. However, these two end bands are usually omitted, as they are at the extreme limits of most peoples' hearing abilities. For example, the A-weighted correction for 31.5 Hz is −39.2 dB. This is such a large reduction that the contribution from this band is negligible except for all but the most powerful or pure-tone sources.

[13] This is approximately the loudness of a quiet refrigerator in a quiet kitchen.

[14] However, sone values are combined and manipulated differently than decibel (dBA) values.

Sone values are easier to use than decibel values because the sone scale is linear, not logarithmic. A doubling of sones is perceived by most people to be a doubling of the loudness.

24. SOUND RATING METHODS

There are several methods used for rating sound as a function of its ability to interfere with speech. Some of these same methods can be used to evaluate sound quality and the impact of sound-generating components. *ASHRAE Handbook—HVAC Applications* provides a simple comparison of these methods. [**Comparison of Sound Rating Methods**]

One way of specifying limits on loudness is to require all octave band measurements to be below a certain value. However, since the human ear is not as sensitive to low frequencies as it is to high frequencies, the limiting value will depend on the frequency. *Noise criterion* (NC) *curves* (also known as *noise criteria curves*) are curves of acceptable sound pressure level plotted against frequency. There are several such curves, and these are designated NC-15, NC-20, and so on. The NC curve used depends on the type of location. NC-35 is commonly specified for such locations as apartment buildings, hotels, executive offices, and laboratories. NC-45 is commonly specified for washrooms, kitchens, lobbies, banking areas, and open offices. [**NC (Noise Criteria) Curves and Sample Spectrum (Curve with Symbols)**]

The method of determining the NC level or rating actually achieved consists of plotting the measured octave band sound pressure levels (the *noise spectrum*) on a set of NC curves. The NC rating is the lowest NC curve that is not exceeded by any measurement in the spectrum.

25. ACOUSTIC ABSORBER MATERIALS

Acoustic absorbers are materials that prevent reflection of incident sound. Fibrous glass, draperies, and open cell foams are acoustic absorbers. Absorbers generally are used in combination with barriers and dampers since absorbers do not block noise or reduce vibration when used alone. Absorbers work best at higher frequencies unless they are very thick or are tuned to low frequencies.

26. SOUND ABSORPTION COEFFICIENT

The *sound absorption coefficient*, α, is the ratio of sound energy absorbed by the surface compared to the sound energy striking the surface. Hard surfaces absorb essentially no sound energy and have sound absorption coefficients close to zero. Soft surfaces absorb most of the sound energy striking them. A value of 1.0 indicates a perfect absorber.

Since the sound absorption coefficient varies with frequency, the *noise reduction coefficient* rating (NRC) is used to simplify comparisons of materials. The NRC is the arithmetic average of the four absorption coefficients measured at frequencies of 250 Hz, 500 Hz, 1000 Hz, and 2000 Hz. Coefficients of common building materials and furnishings are listed in App. 72.A.

Absorptive products seldom can reduce ambient noise by more than 6 dB to 8 dB. They reduce noise reflection, but they do not stop sound transmission.

27. ROOM CONSTANT

If a room contains absorbing materials with known absorption coefficients, the average absorption coefficient can be calculated. For this method of determining the average absorption coefficient to be valid, the absorption materials must be well distributed in the room so that the value applies to the majority of locations. In Eq. 72.23, the product αA is known as the *surface absorption* or *sabin area*, S, with units of sabins.[15] One *sabin* (also known as an *open window unit*, OWU) is one unit area of perfectly absorbing surface. Generally, the surface absorption should be 20% to 50% of the room area. Otherwise, the room will be reverberatory.

$$S = \alpha A \qquad 72.23$$

$$\overline{\alpha} = \frac{\sum A_i \alpha_i}{\sum A_i} = \frac{\sum S_i}{\sum A_i} \qquad 72.24$$

The acoustic absorption characteristics of the room can be described by its *room constant*, R. A room with perfectly absorbing surfaces (i.e., an anechoic room) has an infinite room constant. A room with perfectly reflecting surfaces (i.e., a reverberant room) has a room constant of zero.

$$R = \frac{\overline{\alpha} A}{1 - \overline{\alpha}} = \frac{S}{1 - \overline{\alpha}} \qquad 72.25$$

28. ACOUSTIC BARRIER MATERIALS

Acoustic barriers block sound. The best performing *barrier materials* have high mass per unit area. Due to their low cost, gypsum board, plywood, and masonry are often used for this purpose.

Plant Engineering

[15]Derived from the early 1900s pioneer in room acoustics, Wallace Sabine, the spelling *sabine* is also encountered.

29. SOUND TRANSMISSION CLASS

The *sound transmission class* (STC) is a single-number rating that is based on a match of the barrier's transmission loss curve to one of several standard ASTM contours. STC is particularly useful in selecting noise barriers for speech isolation. However, STC ratings are not accurate enough for most industrial noise control work because they are functions of the material used, not the design or actual attenuation obtained, and the frequency of the noise to be attenuated is not considered. *ASHRAE Handbook—HVAC Applications* provides a table of STC values for mechanical equipment room wall, floor, and ceiling types. Additional typical STC values are given in Table 72.5. [**Sound Transmission Class (STC) and Transmission Loss Values of Typical Mechanical Equipment Room Wall, Floor, and Ceiling Types, dB**]

A sound barrier or partition is given an STC rating based on measured transmission losses at 16 different one-third octave frequencies between 125 Hz and 4000 Hz, which is consistent with the frequency range of speech. The STC rating does not assess sound transfer at lower or higher frequencies such as might be generated by mechanical equipment or musical instruments. Penetrations and openings through the wall, as well as *flanking paths* around the wall, can degrade the isolation performance of a barrier. Improvement in sound isolation is perceived as almost imperceptible at an STC rating of 1 dB, just barely perceptible at 3 dB, clearly perceptible at 5 dB, and half as loud at 10 dB.

The STC of a partition is determined by comparing the transmission loss spectrum with a standard reference contour. This contour consists of three segments with different vertical increments: 125–400 Hz (15 dB increase), 400–1250 Hz (5 dB increase), and 1250–4000 Hz (0 dB increase), as shown in Fig. 72.3. The STC rating is found by shifting the standard contour vertically until most of the measured values fall below the STC curve and two conditions are met: (1) the sum of all the deficiencies (i.e., measured values above the standard contour) do not exceed 32 dB, and (2) the maximum deficiency at any frequency does not exceed 8 dB. The contour is shifted vertically in 1 dB steps, and when the conforming position is found, the STC rating is read as the transmission loss corresponding to the reference contour at 500 Hz. For the transmission loss spectrum shown in Fig. 72.3, the STC rating is 31 dB.

30. NOISE REDUCTION

Received sound can be reduced by modifying either the sound source, the transmission path, or the area occupied by the source. The process of reducing the noise is

Table 72.5 *Typical Values of Sound Transmission Class (STC)*

barrier construction	STC (dB)
22-gauge steel	25
$^3/_8$ in plasterboard	26
1 lbm barrier or curtain	26
solid wood door	27
2 in fiberglass curtain with 1 lbm barrier	29
$^5/_8$ in gypsum wallboard	30
$^3/_{16}$ in steel	31
two sheets of $^1/_2$ in drywall on wood studs, uninsulated	33
2 in metal panel	35
two sheets of $^1/_2$ in drywall on wood studs, with fiberglass batt insulation	36–39
double $^1/_2$ in drywall on both sides, wood studs, with fiberglass batt insulation	45
$^1/_2$ in drywall on 6 in lightweight concrete block, painted both sides	46
4 in metal panel	41
concrete block, unpainted	44
12 in concrete	53
$^1/_2$ in drywall on 8 in dense concrete block, painted both sides	54
double $^1/_2$ in drywall on both sides, staggered wood studs, with fiberglass batt insulation	55
double $^1/_2$ in drywall on both sides, wood studs, resilient channels on one side, with fiberglass batt insulation	59

Figure 72.3 *STC Standard Contour and Sample Rating*

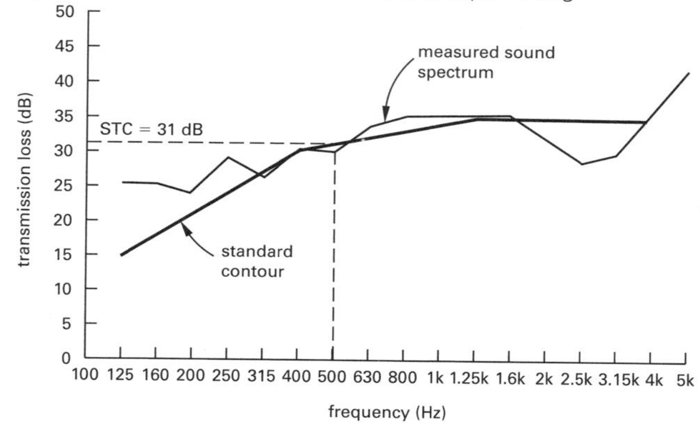

known as *noise reduction*, *sound abatement*, and *attenuation*. The amount of noise reduction, NR, is a decrease in sound pressure level due to such modifications.

$$\mathrm{NR} = L_{p,1} - L_{p,2} = L_{W,1} - L_{W,2} \quad 72.26$$

When adding noise reduction material to a room, the new surface absorption (sabin area) should be 3 to 10 times the original surface absorption. In a room with well-distributed absorption surfaces and at large distances from a single noise source, or for the case of many small noise sources distributed in the room, the noise reduction can be determined using Eq. 72.27.

$$\mathrm{NR} = 10\log\frac{\sum S_1}{\sum S_2} = 10\log\frac{R_1(1-\bar{\alpha}_1)}{R_2(1-\bar{\alpha}_2)} \quad 72.27$$

ASHRAE Handbook—HVAC Applications provides recommendations for maximum duct airflow velocities and transmission path adjustments, both of which are useful when working with specific acoustic design criteria. **[Maximum Recommended Duct Airflow Velocities to Achieve Specified Acoustic Design Criteria] [Duct Breakout Insertion Loss—Potential Low-Frequency Improvement over Bare Duct and Elbow]**

31. INSERTION LOSS

Insertion loss, IL, is the term used to describe the amount of noise reduction based on sound pressure levels at a particular place after the noise source has been isolated. The source is usually isolated by covering it with an enclosure or by constructing a wall between it and occupants. (The wall or enclosure has been "inserted" into the sound transmission path.)

$$\mathrm{IL} = L_{p,1} - L_{p,2} = 20\log\frac{p_1}{p_2} \quad 72.28$$

Insertion loss takes into account structure-borne sound and flanking and diffracted sound, which can bend around the edges of a barrier. Including these factors produces a better estimate of the noise reduction than transmission loss.

32. TRANSMISSION LOSS

The *transmission loss*, TL, (also known as the *sound reduction index*, SRI) is based on sound power levels and is the actual portion of the sound energy lost when sound passes through a wall or barrier. The fraction of the incident energy transmitted through a barrier is designated as the *transmission coefficient*, τ. Appendix 72.B lists the transmission loss of several common materials used for noise control.

$$\mathrm{TL} = L_{W,1} - L_{W,2} = 10\log\frac{P_1}{P_2} = 10\log\frac{1}{\tau} \quad 72.29$$

Unlike insertion loss, which is determined from measurements of actual sound pressure levels, with and without a sound barrier, transmission loss is solely related to a material's ability to reduce the flow of sound energy passing through it. Basically, transmission loss is a function of the materials used, while insertion loss is affected by materials, installation, and leaks. According to the *mass law* (i.e., a doubling of the mass of the wall will result in a 6 dB reduction in transmitted sound), transmission loss is greatly dependent on the mass per unit area, ρ_S, (also known as *surface density*) of the barrier. Transmission loss can be predicted at normal incidence for a simple, homogeneous panel (e.g., a concrete block wall or single layer of sheetrock on vertical wood studs) at a specific frequency, f, from the mathematical expression of the mass law, Eq. 72.30. Predicting transmission loss of walls receiving frequencies near their resonant frequencies, at other than normal incidence, and of more complex walls, including double panels with airspace and/or insulation between, requires more sophisticated expressions.

$$\mathrm{TL}_{\mathrm{dB}} = 20\log(f_{\mathrm{Hz}}\rho_{S,\mathrm{panel,kg/m^2}}) - 47 \quad \text{[SI]} \quad 72.30(a)$$

$$\mathrm{TL}_{\mathrm{dB}} = 20\log\left(4.88 f_{\mathrm{Hz}}\rho_{S,\mathrm{panel,lbm/ft^2}}\right) - 47 \quad \text{[U.S.]} \quad 72.30(b)$$

33. RESONANT FREQUENCY OF A PANEL

Most panels have fairly low resonant frequencies. Transmission loss is at a minimum for sound at the resonant frequency. The *fundamental resonant frequency* (*natural frequency, first harmonic frequency*), f_0, of a simple infinite panel (plate) of material with fixed edges, with a density ρ and a Poisson's ratio ν, is given by Eq. 72.31. The second harmonic frequency is found by multiplying the resonant frequency by 2, and so on.

$$f_{0,\mathrm{Hz}} = 2\pi a^2_{\mathrm{panel,m/s}}\sqrt{\frac{12\rho_{S,\mathrm{panel,kg/m^2}}(1-\nu^2_{\mathrm{panel}})}{E_{\mathrm{panel,Pa}}t^3_{\mathrm{m}}}}$$

$$= 2\pi\sqrt{\frac{12E_{\mathrm{panel,Pa}}}{\rho_{S,\mathrm{panel,kg/m^2}}t^3_{\mathrm{m}}(1-\nu^2_{\mathrm{panel}})}} \quad \text{[infinite single panel]} \quad 72.31$$

$$a_{\mathrm{panel,m/s}} = \sqrt{\frac{E_{\mathrm{panel,Pa}}}{\rho_{S,\mathrm{panel,kg/m^2}}(1-\nu^2_{\mathrm{panel}})}} \quad 72.32$$

Plant Engineering

The resonant frequency of a simple finite panel with fixed edges of length L, height h, and thickness t can be estimated from Eq. 72.33. B is the *bending stiffness per unit width*, with units of ft-lbf or N·m.

$$f_{0,\mathrm{Hz}} = \frac{\pi}{2}\sqrt{\frac{B_{\mathrm{N\cdot m}}}{\rho_{S,\mathrm{panel,kg/m^2}}}}\left(\frac{1}{L_{\mathrm{m}}^2}+\frac{1}{h_{\mathrm{m}}^2}\right) = 0.45a_{\mathrm{panel,m/s}}t_{\mathrm{m}}\left(\frac{1}{L_{\mathrm{m}}^2}+\frac{1}{h_{\mathrm{m}}^2}\right) \quad \text{[finite single panel]} \qquad 72.33$$

$$B = \frac{E_{\mathrm{panel,Pa}}t_{\mathrm{panel,m}}^3}{12(1-\nu_{\mathrm{panel}}^2)} \quad \text{[homogeneous panel]} \qquad 72.34$$

The resonant frequency of an infinite double panel with an air cavity of thickness t (e.g., a double-glazed window) depends on the two surface densities and the cavity thickness.

$$f_{0,\mathrm{Hz}} = 2\pi a_{\mathrm{air,m/s}}\sqrt{\left(\frac{\rho_{\mathrm{air,kg/m^3}}}{t_{\mathrm{air,m}}}\right)\times\left(\frac{1}{\rho_{S,\mathrm{panel\,1,kg/m^2}}}+\frac{1}{\rho_{S,\mathrm{panel\,2,kg/m^2}}}\right)} \qquad 72.35$$

34. ATTENUATION

Attenuation is the amount by which sound is decreased as it travels from the source to the receiver. This general term is used interchangeably with noise reduction, insertion loss, and transmission loss. An *attenuator* is any device (e.g., a muffler) that can be inserted into the sound path to reduce the noise received.

35. TIME WEIGHTED AVERAGE

In the United States, the parameter used in the Occupational Safety and Health Act (OSHA) regulations to assess a worker's exposure to noise is the *time weighted average*, TWA. TWA represents a worker's noise exposure normalized to an eight-hour day, taking into account the average levels of noise and the time spent in each area. Determining the TWA starts with calculating the *noise dose*, D. C_i is the time in hours spent at each noise exposure level, and t_i is the *reference duration* (*permitted duration*) in hours calculated from the sound pressure level in dB for each exposure level. Since they are different representations of the same level, the *OSHA action levels* are based on either TWA or dose percent. These action levels are 85 dB (or 50% dose) and 90 dB (or 100% dose). Equation 72.36 is used when the noise level varies during the measurement period. Preventative steps must be taken if the daily noise dose is more than 1.0.

$$D_{\%} = \sum\frac{C_i}{t_i}\times 100\% \qquad 72.36$$

$$t_i = \frac{8\ \mathrm{hr}}{2^{(L_i-90)/5}} \qquad 72.37$$

$$\mathrm{TWA} = 90 + 16.61\log_{10}\frac{D_{\%}}{100\%} \qquad 72.38$$

Unlike a traditional sound meter, which measures the instantaneous sound level, a *noise dosimeter* (*dose badge*) worn by an employee monitors and integrates the sound level over the entire work shift in order to determine the dose, D, time weighted average, TWA, average sound level, L_p, and other parameters directly.

36. ALLOWABLE EXPOSURE LIMITS

OSHA sets maximum limits on daily sound exposure. (See Table 72.6.) The "all-day" eight-hour noise level limit in the United States is 90 dBA. This is higher than the maximum level permitted in other countries.[16] In the United States, employees may not be exposed to steady sound levels above 115 dBA, regardless of the duration. Impact sound levels are limited to 140 dBA.

In the United States, hearing protection, educational programs, periodic examinations, and other actions are required for workers whose eight-hour exposure is more than 85 dBA or whose noise dose exceeds 50% of the action levels.

37. HEARING PROTECTIVE DEVICES

For continuous (as opposed to impulsive noise), when engineering or administrative controls are ineffective or not feasible, *hearing protective devices* (HPDs) (such as earmuffs and earplugs) can be used as a last resort. HPDs are labeled with a *noise reduction rating* (NRR) which is the manufacturer's claim of how much noise reduction (in dB) a hearing protective device provides. The NRR is essentially an average insertion loss. Theoretically, the entire NRR would be subtracted from the environmental noise level to obtain the noise level at the ear. However, since HPDs generally provide much less protection than their labels claim, OSHA (drawing on NIOSH standards) recommends in its field manual that inspectors calculate a more realistic measurement of effectiveness.

[16]The 2003 *European Union Noise Directive* (2003/10/EC) became effective in 2006. It sets the maximum "all day" eight-hour noise level exposure limit in EU countries at 87 dBA. The first action level, to train personnel and make hearing protection available, occurs at 80 dBA. The second action level, to implement a noise reduction program and require usage of hearing protection, occurs at 85 dBA.

Table 72.6 *Typical Permissible Noise Exposure Levels*[a,b]

sound level (dBA)	exposure (hours per day)
90	8
92	6
95	4
97	3
100	2
102	1½
105	1
110	½
115	¼ or less

[a]without hearing protection
[b]Federal Register, Title 29, Sec. 1910.95, Table G-16

Since sound level meters have multiple scales, the calculation method depends on the measurement scale used to measure the environmental noise. For formable earplugs, when noise is measured in dBA (i.e., on the A-weighted scale), the OSHA formula subtracts 7 dB from the NRR and divides the result by 2. (Values for earmuffs and nonformable earplugs are calculated differently.)

$$\Delta L_{\text{dBA}} = \frac{\text{NRR} - 7}{2} \quad \text{[formable earplugs; A-scale]} \qquad 72.39$$

When noise is measured in dBC (i.e., on the C-weighted scale), the OSHA formula subtracts 7 dB from the NRR.

$$\Delta L_{\text{dBC}} = \text{NRR} - 7 \quad \text{[formable earplugs; C-scale]} \qquad 72.40$$

For environments dominated by frequencies below 500 Hz, the C-weighted noise level should be used. With impulsive noise (e.g., gunfire), the NRR may not be an accurate indicator of hearing protection.

Example 72.7

A construction worker leaves an 86 dB environment after six hours and works in a 92 dB environment for three hours more. What are the (a) dose and (b) TWA?

Solution

Use Eq. 72.37.

$$t_1 = \frac{8\text{ hr}}{2^{(L_1-90)/5}} = \frac{8\text{ hr}}{2^{(86\text{ dB}-90\text{ dB})/5}} = 13.92\text{ hr}$$

$$t_2 = \frac{8\text{ hr}}{2^{(L_2-90)/5}} = \frac{8\text{ hr}}{2^{(92\text{ dB}-90\text{ dB})/5}} = 6.06\text{ hr}$$

From Eq. 72.36, the noise dose is

$$D_{\%} = \sum \frac{C_i}{t_i} \times 100\% = \left(\frac{6\text{ hr}}{13.92\text{ hr}} + \frac{3\text{ hr}}{6.06\text{ hr}}\right) \times 100\%$$
$$= 92.6\%$$

From Eq. 72.38, the time weighted average is

$$\begin{aligned}\text{TWA} &= 90 + 16.61\log_{10}\frac{D_{\%}}{100\%} \\ &= 90\text{ dB} + 16.61\log_{10}\frac{92.6\%}{100\%} \\ &= 89.46\text{ dB}\end{aligned}$$

38. NOMENCLATURE

A	area	ft²	m²
B	bending stiffness per unit width	ft-lbf	N·m
c	speed of light	ft/sec	m/s
c	speed of sound	ft/sec	m/s
C	exposure duration	hr	h
CU	coefficient of utilization	–	–
D	dose	–	–
E	illuminance	lm/ft²	lm/m²
E	modulus of elasticity	lbf/ft²	Pa
f	frequency	Hz	Hz
g	gravitational acceleration, 32.2 (9.81)	ft/sec²	m/s²
g_c	gravitational constant, 32.2	lbm-ft/lbf-sec²	n.a.
h	height	ft	m
I	light intensity	lm/sr	lm/sr
I	sound intensity	dB	dB
IL	insertion loss	dB	dB
k	ratio of specific heats	–	–
L	length	ft	m
L	sound level	dB	dB
LLF	light loss factor	–	–
m	mass	lbm	kg
M	brightness	lm/ft²	lm/m²
M	molecular weight	lbm/lbmol	kg/kmol
n	number	–	–
NR	noise reduction	dB	dB
NRR	noise reduction rating	dB	dB
p	pressure	lbf/ft²	Pa
P	power	W	W
Q	directivity	–	–
Q	quantity of light	lm	lm

Plant Engineering

r	distance	ft	m
R	acoustic resistance	lbf-sec/ft^5	N·s/m^5
R	room constant	ft^2	m^2
R	specific gas constant, air, 53.35 (287.03)	ft-lbf/lbm-°R	J/kg·K
R^*	universal gas constant, 1545.35 (8314.47)	ft-lbf/lbmol-°R	J/kmol·K
S	surface absorption	sabins	sabins
t	reference duration	hr	h
t	thickness	ft	m
t	time	sec	s
T	period	sec	s
T	temperature	°R	K
TL	transmission loss	dB	dB
TWA	time weighted average	dB	dB
v	velocity	ft/sec	m/s
X	acoustic reactance	lbf-sec/ft^5	N·s/m^5
z	specific acoustic impedance	lbf-sec/ft^3	N·s/m^3
Z	acoustic impedance	lbf-sec/ft^5	N·s/m^5
Z_0	specific acoustic impedance of the medium	lbf-sec/ft^3	N·s/m^3

Plant Engineering

Symbols

α	sound absorption coefficient	–	–
η	efficacy	lm/W	lm/W
η	efficiency	–	–
λ	wavelength	ft	m
ρ	density	lbm/ft^3	kg/m^3
τ	transmission coefficient	–	–
θ	angle	–	–
ν	Poisson's ratio	–	–
Φ	luminous flux	lm	lm
ω	solid angle	sr	sr

Subscripts

l	luminous
o	observer
p	pressure
ref	reference
s	source
S	surface per unit area
t	total
W	watts

Topic XIII: Economics

Chapter

73 Engineering Economic Analysis

Content in blue refers to the *NCEES Handbook*.

NCEES EXAM SPECIFICATIONS AND RELATED CONTENT

HVAC AND REFRIGERATION EXAM

I.A.2. Basic Engineering Practice: Economic analysis

6. Types of Cash Flows
10. Single-Payment Equivalence
11. Standard Cash Flow Factors and Symbols
12. Calculating Uniform Series Equivalence
13. Finding Past Values
16. The Meaning of Present Worth and i
22. Choice of Alternatives: Comparing One Alternative with Another Alternative
29. Economic Life: Retirement at Minimum Cost
34. Depreciation Methods
35. Accelerated Depreciation Methods
44. Rate and Period Changes
46. Probabilistic Problems
53. Break-Even Analysis

MACHINE DESIGN AND MATERIALS EXAM

I.A.4. Basic Engineering Practice: Project management and economic analysis

6. Types of Cash Flows
10. Single-Payment Equivalence
11. Standard Cash Flow Factors and Symbols
12. Calculating Uniform Series Equivalence
13. Finding Past Values

16. The Meaning of Present Worth and i
22. Choice of Alternatives: Comparing One Alternative with Another Alternative
29. Economic Life: Retirement at Minimum Cost
34. Depreciation Methods
35. Accelerated Depreciation Methods
44. Rate and Period Changes
46. Probabilistic Problems
53. Break-Even Analysis

THERMAL AND FLUID SYSTEMS EXAM

I.A.2. Basic Engineering Practice: Economic analysis

6. Types of Cash Flows
10. Single-Payment Equivalence
11. Standard Cash Flow Factors and Symbols
12. Calculating Uniform Series Equivalence
13. Finding Past Values
16. The Meaning of Present Worth and i
22. Choice of Alternatives: Comparing One Alternative with Another Alternative
29. Economic Life: Retirement at Minimum Cost
34. Depreciation Methods
35. Accelerated Depreciation Methods
44. Rate and Period Changes
46. Probabilistic Problems
53. Break-Even Analysis

1. INTRODUCTION

In its simplest form, an *engineering economic analysis* is a study of the desirability of making an investment.[1] The decision-making principles in this chapter can be applied by individuals as well as by companies. The nature of the spending opportunity or industry is not important. Farming equipment, personal investments, and multimillion dollar factory improvements can all be evaluated using the same principles.

Similarly, the applicable principles are insensitive to the monetary units. Although *dollars* are used in this chapter, it is equally convenient to use pounds, yen, or euros.

Finally, this chapter may give the impression that investment alternatives must be evaluated on a year-by-year basis. Actually, the *effective period* can be defined as a day, month, century, or any other convenient period of time.

2. MULTIPLICITY OF SOLUTION METHODS

Most economic conclusions can be reached in more than one manner. There are usually several different analyses that will eventually result in identical answers.[2] Other than the pursuit of elegant solutions in a timely manner, there is no reason to favor one procedural method over another.[3]

3. PRECISION AND SIGNIFICANT DIGITS

The full potential of electronic calculators will never be realized in engineering economic analyses. Considering that calculations are based on estimates of far-future cash flows and that unrealistic assumptions (no inflation, identical cost structures of replacement assets, etc.) are routinely made, it makes little sense to carry cents along in calculations.

The calculations in this chapter have been designed to illustrate and review the principles presented. Because of this, greater precision than is normally necessary in everyday problems may be used. Though used, such precision is not warranted.

Unless there is some compelling reason to strive for greater precision, the following rules are presented for use in reporting final answers to engineering economic analysis problems.

- Omit fractional parts of the dollar (i.e., cents).
- Report and record a number to a maximum of four significant digits unless the first digit of that number is 1, in which case, a maximum of five significant digits should be written. For example,

\$49	not	\$49.43
\$93,450	not	\$93,453
\$1,289,700	not	\$1,289,673

4. YEAR-END AND OTHER CONVENTIONS

Except in short-term transactions, it is simpler to assume that all receipts and disbursements (cash flows) take place at the end of the year in which they occur.[4] This is known as the *year-end convention*. The exceptions to the year-end convention are initial project cost (purchase cost), trade-in allowance, and other cash flows that are associated with the inception of the project at $t = 0$.

[1] This subject is also known as *engineering economics* and *engineering economy*. There is very little, if any, true economics in this subject.
[2] Because of round-off errors, particularly when factors are taken from tables, these different calculations will produce slightly different numerical results (e.g., \$49.49 versus \$49.50). However, this type of divergence is well known and accepted in engineering economic analysis.
[3] This does not imply that approximate methods, simplifications, and rules of thumb are acceptable.
[4] A *short-term transaction* typically has a lifetime of five years or less and has payments or compounding that are more frequent than once per year.

On the surface, such a convention appears grossly inappropriate since repair expenses, interest payments, corporate taxes, and so on seldom coincide with the end of a year. However, the convention greatly simplifies engineering economic analysis problems, and it is justifiable on the basis that the increased precision associated with a more rigorous analysis is not warranted (due to the numerous other simplifying assumptions and estimates initially made in the problem).

There are various established procedures, known as *rules* or *conventions*, imposed by the Internal Revenue Service on U.S. taxpayers. An example is the *half-year rule*, which permits only half of the first-year depreciation to be taken in the first year of an asset's life when certain methods of depreciation are used. These rules are subject to constantly changing legislation and are not covered in this book. The implementation of such rules is outside the scope of engineering practice and is best left to accounting professionals.

5. CASH FLOW DIAGRAMS

Although they are not always necessary in simple problems (and they are often unwieldy in very complex problems), *cash flow diagrams* can be drawn to help visualize and simplify problems having diverse receipts and disbursements.

The following conventions are used to standardize cash flow diagrams.

- The horizontal (time) axis is marked off in equal increments, one per period, up to the duration (or *horizon*) of the project.
- Two or more transfers in the same period are placed end to end, and these may be combined.
- Expenses incurred before $t=0$ are called *sunk costs.* Sunk costs are not relevant to the problem unless they have tax consequences in an after-tax analysis.
- *Receipts* are represented by arrows directed upward. *Disbursements* are represented by arrows directed downward. The arrow length is proportional to the magnitude of the cash flow.

Example 73.1

A mechanical device will cost $20,000 when purchased. Maintenance will cost $1000 each year. The device will generate revenues of $5000 each year for five years, after which the salvage value is expected to be $7000. Draw and simplify the cash flow diagram.

Solution

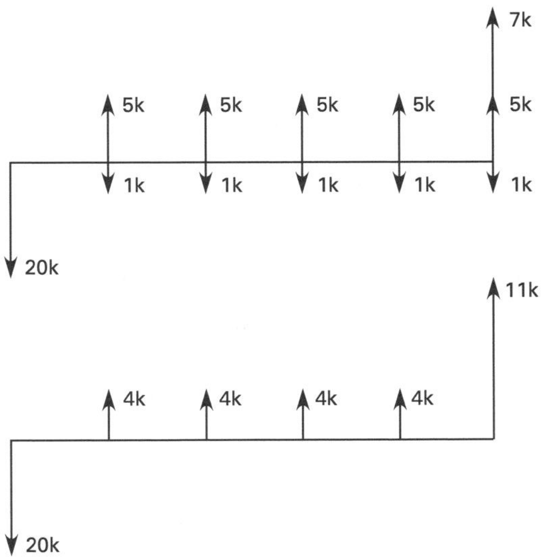

6. TYPES OF CASH FLOWS

To evaluate a real-world project, it is necessary to present the project's cash flows in terms of standard cash flows that can be handled by engineering economic analysis techniques. The standard cash flows are single payment cash flow, uniform series cash flow, gradient series cash flow, and the infrequently encountered exponential gradient series cash flow. (See Fig. 73.1.) **[Economic Analysis: Nomenclature and Definitions] [Economic Factor Conversions]**

A *single payment cash flow* can occur at the beginning of the time line (designated as $t = 0$), at the end of the time line (designated as $t = n$), or at any time in between.

The *uniform series cash flow* consists of a series of equal transactions starting at $t = 1$ and ending at $t = n$. The symbol A is typically given to the magnitude of each individual cash flow.[5]

The *gradient series cash flow* starts with a cash flow (typically given the symbol G) at $t = 2$ and increases by G each year until $t = n$, at which time the final cash flow is $(n-1)G$.

An *exponential gradient series cash flow* is based on a phantom value (typically given the symbol E_0) at $t = 0$ and grows or decays exponentially according to the following relationship.[6]

$$\text{amount at time } t = E_t = E_0(1+g)^t \quad [t = 1, 2, 3, \ldots, n] \qquad 73.1$$

[5]The cash flows do not begin at $t = 0$. This is an important concept with all of the series cash flows. This convention has been established to accommodate the timing of annual maintenance (and similar) cash flows for which the year-end convention is applicable.

[6]By convention, for an exponential cash flow series: The first cash flow, E_0, is at $t = 1$, as in the uniform annual series. However, the first cash flow is $E_0(1+g)$. The cash flow of E_0 at $t = 0$ is absent (i.e., is a *phantom cash flow*).

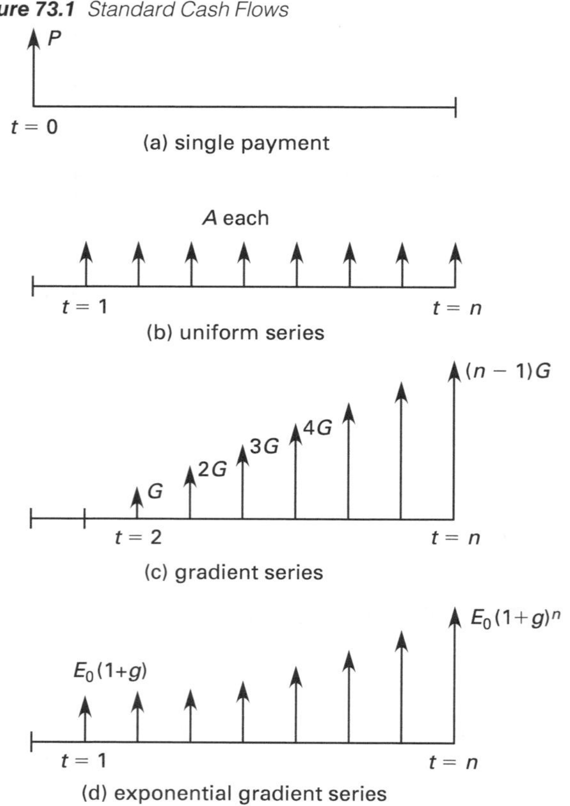

Figure 73.1 Standard Cash Flows

In Eq. 73.1, g is the *exponential growth rate*, which can be either positive or negative. Exponential gradient cash flows are rarely seen in economic justification projects assigned to engineers.[7]

7. TYPICAL PROBLEM TYPES

There is a wide variety of problem types that, collectively, are considered to be engineering economic analysis problems. By far, the majority of engineering economic analysis problems are *alternative comparisons*. With the exception of some investment and rate of return problems, the typical alternative comparison problem will have the following characteristics.

- An interest rate will be given.
- Two or more mutually exclusive investments will be competing for limited funding.
- Each alternative will have its own cash flows.
- It will be necessary to select the best alternative.

A variation of this is a *replacement/retirement analysis*, which is repeated each year to determine if an existing asset should be replaced. Finding the percentage return on an investment is a *rate of return problem*, one of the alternative comparison solution methods.

Investigating interest and principal amounts in loan payments is a *loan repayment problem*. An *economic life analysis* will determine when an asset should be retired. In addition, there are miscellaneous problems involving economic order quantity, learning curves, break-even points, product costs, and so on.

8. IMPLICIT ASSUMPTIONS

Several assumptions are implicitly made when solving engineering economic analysis problems. Some of these assumptions are made with the knowledge that they are or will be poor approximations of what really will happen. The assumptions are made, regardless, for the benefit of obtaining a solution.

The most common assumptions are the following.

- The year-end convention is applicable.
- There is no inflation now, nor will there be any during the lifetime of the project.
- Unless otherwise specifically called for, a before-tax analysis is needed.
- The effective interest rate used in the problem will be constant during the lifetime of the project.
- Nonquantifiable factors can be disregarded.
- Funds invested in a project are available and are not urgently needed elsewhere.
- Excess funds continue to earn interest at the effective rate used in the analysis.

This last assumption, like most of the assumptions listed, is almost never specifically mentioned in the body of a solution. However, it is a key assumption when comparing two alternatives that have different initial costs.

For example, suppose two investments, on costing \$10,000 and the other costing \$8000, are to be compared at 10%. It is obvious that \$10,000 in funds is available, otherwise the costlier investment would not be under consideration. If the smaller investment is chosen, what is done with the remaining \$2000? The last assumption yields the answer: The \$2000 is "put to work" in some investment earning (in this case) 10%.

9. EQUIVALENCE

Industrial decision makers using engineering economic analysis are concerned with the magnitude and timing of a project's cash flow as well as with the total profitability of that project. In this situation, a method is required to compare projects involving receipts and disbursements occurring at different times.

[7]For one of the few discussions on exponential cash flow, see *Capital Budgeting*, Robert V. Oakford, The Ronald Press Company, New York, 1970.

By way of illustration, consider $100 placed in a bank account that pays 5% effective annual interest at the end of each year. After the first year, the account will have grown to $105. After the second year, the account will have grown to $110.25.

Assume that you will have no need for money during the next two years, and any money received will immediately go into your 5% bank account. Then, which of the following options would be more desirable?

As illustrated, none of the options is superior under the assumptions given. If the first option is chosen, you will immediately place $100 into a 5% account, and in two years the account will have grown to $110.25. In fact, the account will contain $110.25 at the end of two years regardless of the option chosen. Therefore, these alternatives are said to be *equivalent.*

option A: $100 now

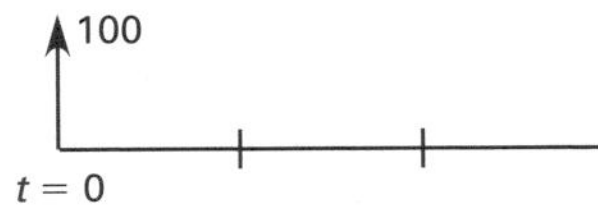

option B: $105 to be delivered in one year

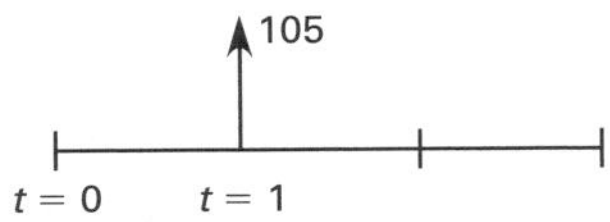

option C: $110.25 to be delivered in two years

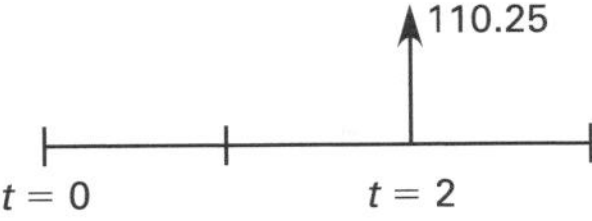

Equivalence may or may not be the case, depending on the interest rate, so an alternative that is acceptable to one decision maker may be unacceptable to another. The interest rate that is used in actual calculations is known as the *effective interest rate.*[8] If compounding is once a year, it is known as the *effective annual interest rate.* However, effective quarterly, monthly, daily, and so on, interest rates are also used.

The fact that $100 today grows to $105 in one year (at 5% annual interest) is an example of what is known as the *time value of money* principle. This principle simply articulates what is obvious: Funds placed in a secure investment will increase to an equivalent future amount.

The procedure for determining the present investment from the equivalent future amount is known as *discounting.*

10. SINGLE-PAYMENT EQUIVALENCE

The equivalence of any present amount, P, at $t = 0$, to any future amount, F, at $t = n$, is called the *future worth* and can be calculated from Eq. 73.2. **[Economic Analysis: Nomenclature and Definitions]** **[Economic Factor Conversions]**

$$F = P(1+i)^n \qquad 73.2$$

The factor $(1 + i)^n$ is known as the single payment *compound amount factor* and has been tabulated in economic factor tables for various combinations of i and n.

Similarly, the equivalence of any future amount to any present amount is called the *present worth* and can be calculated from Eq. 73.3. **[Economic Factor Conversions]**

$$P = F(1+i)^{-n} = \frac{F}{(1+i)^n} \qquad 73.3$$

The factor $(1 + i)^{-n}$ is known as the *single payment present worth factor.*[9]

The interest rate used in Eq. 73.2 and Eq. 73.3 must be the effective rate per period. Also, the basis of the rate (annually, monthly, etc.) must agree with the type of period used to count n. Therefore, it would be incorrect to use an effective annual interest rate if n was the number of compounding periods in months.

Example 73.2

How much should you put into a 10% (effective annual rate) savings account in order to have $10,000 in five years?

Solution

This problem could also be stated: What is the equivalent present worth of $10,000 five years from now if money is worth 10% per year? **[Economic Factor Conversions]**

$$\begin{aligned} P &= F(1+i)^{-n} = (\$10{,}000)(1+0.10)^{-5} \\ &= \$6209 \end{aligned}$$

[8]The adjective *effective* distinguishes this interest rate from other interest rates (e.g., nominal interest rates) that are not meant to be used directly in calculating equivalent amounts.

[9]The *present worth* is also called the *present value* and *net present value.* These terms are used interchangeably and no significance should be attached to the terms *value, worth,* and *net.*

Economics

The factor 0.6209 would usually be obtained from the tables.

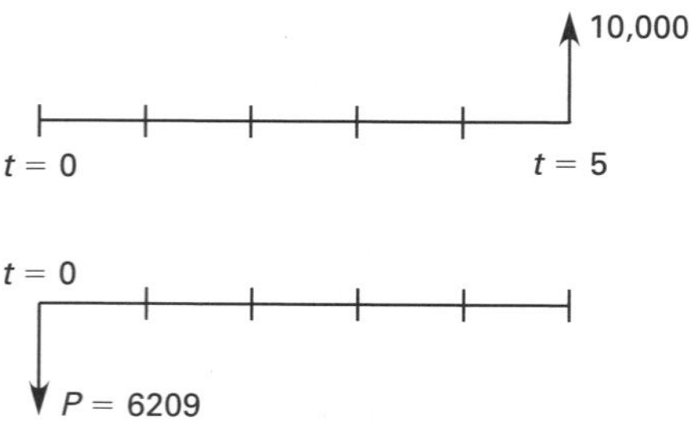

11. STANDARD CASH FLOW FACTORS AND SYMBOLS

Equation 73.2 and Eq. 73.3 may give the impression that solving engineering economic analysis problems involves a lot of calculator use, and, in particular, a lot of exponentiation. Such calculations may be necessary from time to time, but most problems are simplified by the use of tabulated values of the factors.

Rather than actually writing the formula for the compound amount factor (which converts a present amount to a future amount), it is common convention to substitute the standard functional notation of $(F/P,\ i\%,\ n)$. [**Economic Analysis: Nomenclature and Definitions**] [**Economic Factor Conversions**]

Therefore, the future value in n periods of a present amount would be symbolically written as

$$F = P(F/P, i\%, n) \qquad 73.4$$

Similarly, the present worth factor has a functional notation of $(P/F,\ i\%,\ n)$. Therefore, the present worth of a future amount n periods in the future would be symbolically written as

$$P = F(P/F, i\%, n) \qquad 73.5$$

Values of these *cash flow (discounting) factors* are tabulated in economic factor tables. There is often initial confusion about whether the (F/P) or (P/F) column should be used in a particular problem. There are several ways of remembering what the functional notations mean. [**Economic Analysis: Nomenclature and Definitions**]

One method of remembering which factor should be used is to think of the factors as conditional probabilities. The conditional probability of event **A** given that event **B** has occurred is written as $p\{\mathbf{A}|\mathbf{B}\}$, where the given event comes after the vertical bar. In the standard notational form of discounting factors, the given amount is similarly placed after the slash. What you want comes before the slash. (F/P) would be a factor to find F given P. [**Economic Factor Conversions**]

Another method of remembering the notation is to interpret the factors algebraically. The (F/P) factor could be thought of as the fraction F/P. Algebraically, Eq. 73.4 would be

$$F = P\left(\frac{F}{P}\right) \qquad 73.6$$

This algebraic approach is actually more than an interpretation. The numerical values of the discounting factors are consistent with this algebraic manipulation. The (F/A) factor could be calculated as $(F/P) \times (P/A)$. This consistent relationship can be used to calculate other factors that might be occasionally needed, such as (F/G) or (G/P). For instance, the annual cash flow that would be equivalent to a uniform gradient may be found from

$$A = G(P/G, i\%, n)(A/P, i\%, n) \qquad 73.7$$

Formulas for the compounding and discounting factors are contained in Table 73.1. Normally, it will not be necessary to calculate factors from the formulas. An economic factor table is adequate for solving most problems. [**Economic Analysis: Nomenclature and Definitions**]

Example 73.3

What factor will convert a gradient cash flow ending at $t = 8$ to a future value at $t = 8$? (That is, what is the $(F/G,\ i\%,\ n)$ factor?) The effective annual interest rate is 10%.

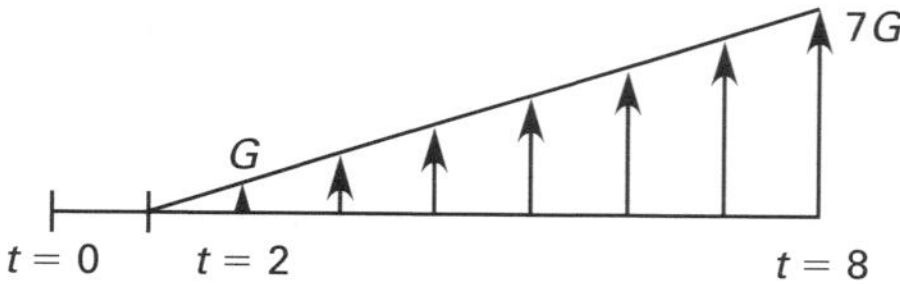

Solution

method 1:

Find the $(F/G,\ 10\%,\ 8)$ factor. [**Economic Factor Conversions**]

$$\begin{aligned}(F/G, 10\%, 8) &= \frac{(1+i)^n - 1}{i^2} - \left(\frac{n}{i}\right) \\ &= \frac{(1+0.10)^8 - 1}{(0.10)^2} - \frac{8}{0.10} \\ &= 34.3589\end{aligned}$$

Table 73.1 Discount Factors for Discrete Compounding

factor name	converts	symbol	formula
single payment compound amount	P to F	$(F/P, i\%, n)$	$(1+i)^n$
single payment present worth	F to P	$(P/F, i\%, n)$	$(1+i)^{-n}$
uniform series sinking fund	F to A	$(A/F, i\%, n)$	$\frac{i}{(1+i)^n - 1}$
capital recovery	P to A	$(A/P, i\%, n)$	$\frac{i(1+i)^n}{(1+i)^n - 1}$
uniform series compound amount	A to F	$(F/A, i\%, n)$	$\frac{(1+i)^n - 1}{i}$
uniform series present worth	A to P	$(P/A, i\%, n)$	$\frac{(1+i)^n - 1}{i(1+i)^n}$
uniform gradient present worth	G to P	$(P/G, i\%, n)$	$\frac{(1+i)^n - 1}{i^2(1+i)^n} - \frac{n}{i(1+i)^n}$
uniform gradient future worth	G to F	$(F/G, i\%, n)$	$\frac{(1+i)^n - 1}{i^2} - \frac{n}{i}$
uniform gradient uniform series	G to A	$(A/G, i\%, n)$	$\frac{1}{i} - \frac{n}{(1+i)^n - 1}$

method 2:

The tabulated values of (P/G) and (F/P) in economic factor tables can be used to calculate the factor. **[Economic Factor Tables]**

$$\begin{aligned}(F/G, 10\%, 8) &= (P/G, 10\%, 8)(F/P, 10\%, 8)\\ &= (16.0287)(2.1436)\\ &= 34.3591\end{aligned}$$

The (F/G) factor could also have been calculated as the product of the (A/G) and (F/A) factors. **[Economic Factor Conversions]**

$F = 34.3591G$

$t = 0$ $t = 8$

12. CALCULATING UNIFORM SERIES EQUIVALENCE

A cash flow that repeats each year for n years without change in amount is known as an *annual amount* and is given the symbol A. As an example, a piece of equipment may require annual maintenance, and the maintenance cost will be an annual amount. Although the equivalent value for each of the n annual amounts could be calculated and then summed, it is more expedient to use one of the uniform series factors. For example, it is possible to convert from an annual amount to a future amount by use of the (F/A) factor. **[Economic Factor Conversions]**

$$F = A(F/A, i\%, n) \tag{73.8}$$

A *sinking fund* is a fund or account into which annual deposits of A are made in order to accumulate F at $t = n$ in the future. Since the annual deposit is calculated as $A = F(A/F, i\%, n)$, the (A/F) factor is known as the *sinking fund factor*. An *annuity* is a series of equal payments (A) made over a period of time.[10] Usually, it is necessary to "buy into" an investment (a bond, an insurance policy, etc.) in order to ensure the annuity. **[Economic Factor Conversions]**

In the simplest case of an annuity that starts at the end of the first year and continues for n years, the purchase price or present worth P, is

$$P = A(P/A, i\%, n) \tag{73.9}$$

The present worth of an *infinite* (*perpetual*) *series* of annual amounts is known as a *capitalized cost*. There is no $(P/A, i\%, \infty)$ factor in the tables, but the capitalized cost can be calculated simply as

$$P = \frac{A}{i} \quad [i \text{ in decimal form}] \tag{73.10}$$

[10]An annuity may also consist of a lump sum payment made at some future time. However, this interpretation is not considered in this chapter.

Economics

Alternatives with different lives will generally be compared by way of *equivalent uniform annual cost* (EUAC). An EUAC is the annual amount that is equivalent to all of the cash flows in the alternative. The EUAC differs in sign from all of the other cash flows. Costs and expenses expressed as EUACs, which would normally be considered negative, are actually positive. The term *cost* in the designation EUAC serves to make clear the meaning of a positive number.

Example 73.4

Maintenance costs for a machine are $250 each year. What is the present worth of these maintenance costs over a 12-year period if the interest rate is 8%?

Solution

Find the present worth. [Economic Factor Conversions] [Economic Factor Tables]

$$P = A(P/A, 8\%, 12) = (-\$250)(7.5361)$$
$$= -\$1884$$

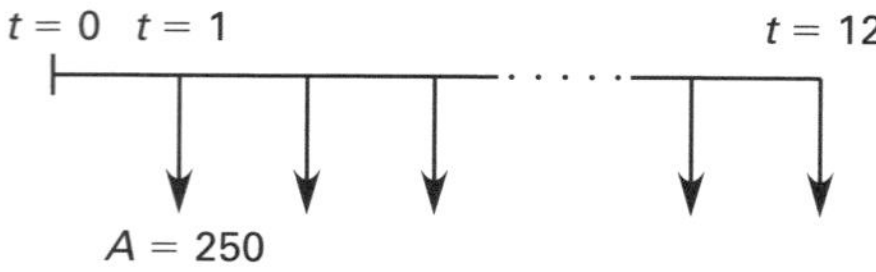

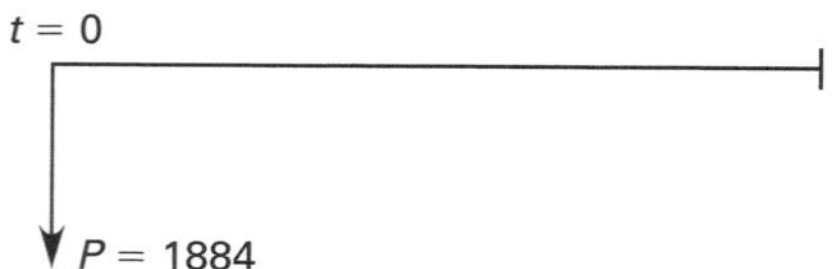

13. FINDING PAST VALUES

From time to time, it will be necessary to determine an amount in the past equivalent to some current (or future) amount. For example, you might have to calculate the original investment made 15 years ago given a current annuity payment.

Such problems are solved by placing the $t = 0$ point at the time of the original investment, and then calculating the past amount as a P value. For example, the original investment, P, can be extracted from the annuity, A, by using the standard cash flow factors. [Economic Factor Conversions]

$$P = A(P/A, i\%, n) \qquad 73.11$$

The choice of $t = 0$ is flexible. As a general rule, the $t = 0$ point should be selected for convenience in solving a problem.

Example 73.5

You currently pay $250 per month to lease your office phone equipment. You have three years (36 months) left on the five-year (60-month) lease. What would have been an equivalent purchase price two years ago? The effective interest rate per month is 1%.

Solution

The solution of this example is not affected by the fact that investigation is being performed in the middle of the horizon. This is a simple calculation of present worth. [Economic Factor Conversions] [Economic Factor Tables]

$$\begin{aligned} P &= A(P/A, 1\%, 60) \\ &= (-\$250)(44.9550) \\ &= -\$11{,}239 \end{aligned}$$

14. TIMES TO DOUBLE AND TRIPLE AN INVESTMENT

If an investment doubles in value (in n compounding periods and with $i\%$ effective interest), the ratio of current value to past investment will be 2.

$$F/P = (1 + i)^n = 2 \qquad 73.12$$

Similarly, the ratio of current value to past investment will be 3 if an investment triples in value. This can be written as

$$F/P = (1 + i)^n = 3 \qquad 73.13$$

It is a simple matter to extract the number of periods, n, from Eq. 73.12 and Eq. 73.13 to determine the *doubling time* and *tripling time*, respectively. For example, the doubling time is

$$n = \frac{\log 2}{\log(1 + i)} \qquad 73.14$$

When a quick estimate of the doubling time is needed, the *rule of 72* can be used. The doubling time is approximately $72/i$.

The tripling time is

$$n = \frac{\log 3}{\log(1 + i)} \qquad 73.15$$

Equation 73.14 and Eq. 73.15 form the basis of Table 73.2.

Economics

Table 73.2 Doubling and Tripling Times for Various Interest Rates

interest rate (%)	doubling time (periods)	tripling time (periods)
1	69.7	110.4
2	35.0	55.5
3	23.4	37.2
4	17.7	28.0
5	14.2	22.5
6	11.9	18.9
7	10.2	16.2
8	9.01	14.3
9	8.04	12.7
10	7.27	11.5
11	6.64	10.5
12	6.12	9.69
13	5.67	8.99
14	5.29	8.38
15	4.96	7.86
16	4.67	7.40
17	4.41	7.00
18	4.19	6.64
19	3.98	6.32
20	3.80	6.03

15. VARIED AND NONSTANDARD CASH FLOWS

Gradient Cash Flow

A common situation involves a uniformly increasing cash flow. If the cash flow has the proper form, its present worth can be determined by using the *uniform gradient factor*, $(P/G, i\%, n)$. The uniform gradient factor finds the present worth of a uniformly increasing cash flow that starts in year two (not in year one).

There are three common difficulties associated with the form of the uniform gradient. The first difficulty is that the initial cash flow occurs at $t = 2$. This convention recognizes that annual costs, if they increase uniformly, begin with some value at $t = 1$ (due to the year-end convention) but do not begin to increase until $t = 2$. The tabulated values of (P/G) have been calculated to find the present worth of only the increasing part of the annual expense. The present worth of the base expense incurred at $t = 1$ must be found separately with the (P/A) factor.

The second difficulty is that, even though the factor $(P/G, i\%, n)$ is used, there are only $n - 1$ actual cash flows. It is clear that n must be interpreted as the *period number* in which the last gradient cash flow occurs, not the number of gradient cash flows.

Finally, the sign convention used with gradient cash flows may seem confusing. If an expense increases each year (as in Ex. 73.6), the gradient will be negative, since it is an expense. If a revenue increases each year, the gradient will be positive. (See Fig. 73.2) In most cases, the sign of the gradient depends on whether the cash flow is an expense or a revenue.[11]

Figure 73.2 Positive and Negative Gradient Cash Flows

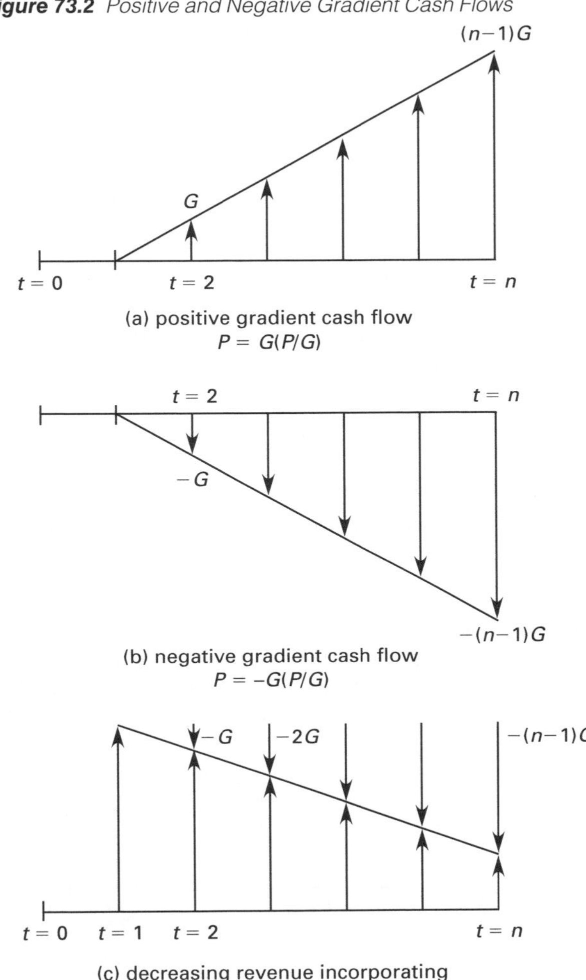

Example 73.6

Maintenance on an old machine is \$100 this year but is expected to increase by \$25 each year thereafter. What is the present worth of five years of the costs of maintenance? Use an interest rate of 10%.

[11]This is not a universal rule. It is possible to have a uniformly decreasing revenue as in Fig. 73.2(c). In this case, the gradient would be negative.

Solution

In this problem, the cash flow must be broken down into parts. (The five-year gradient factor is used even though there are only four nonzero gradient cash flows.) [**Economic Factor Conversions**] [**Economic Factor Tables**]

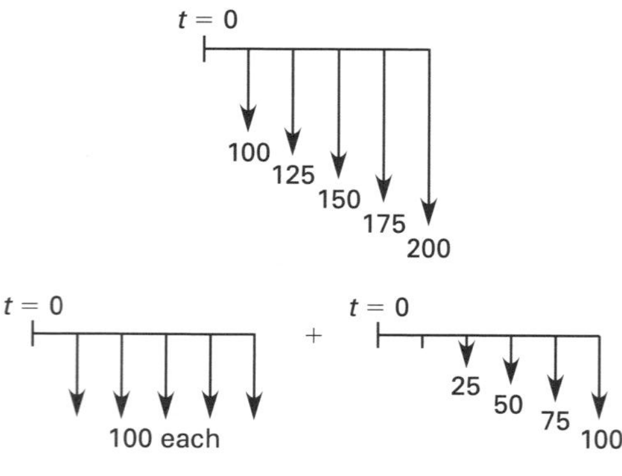

$$\begin{aligned} P &= A(P/A, 10\%, 5) + G(P/G, 10\%, 5) \\ &= (-\$100)(3.7908) - (\$25)(6.8618) \\ &= -\$551 \end{aligned}$$

Stepped Cash Flows

Stepped cash flows are easily handled by the technique of *superposition of cash flows*. This technique is used in Ex. 73.7.

Example 73.7

An investment costing \$1000 returns \$100 for the first five years and \$200 for the following five years. How would the present worth of this investment be calculated?

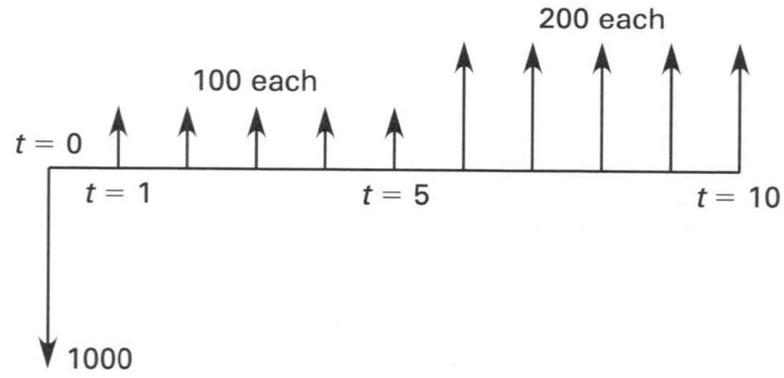

Solution

Using the principle of superposition, the revenue cash flow can be thought of as \$200 each year from $t = 1$ to $t = 10$, with a negative revenue of \$100 from $t = 1$ to $t = 5$. Superimposed, these two cash flows make up the actual performance cash flow. [**Economic Factor Conversions**]

$$P = -\$1000 + (\$200)(P/A, i\%, 10) - (\$100)(P/A, i\%, 5)$$

Missing and Extra Parts of Standard Cash Flows

A missing or extra part of a standard cash flow can also be handled by superposition. For example, suppose an annual expense is incurred each year for ten years, except in the ninth year. (The cash flow is illustrated in Fig. 73.3.) The present worth could be calculated as a subtractive process. [**Economic Factor Conversions**]

$$P = A(P/A, i\%, 10) - A(P/F, i\%, 9) \qquad 73.16$$

Figure 73.3 *Cash Flow with a Missing Part*

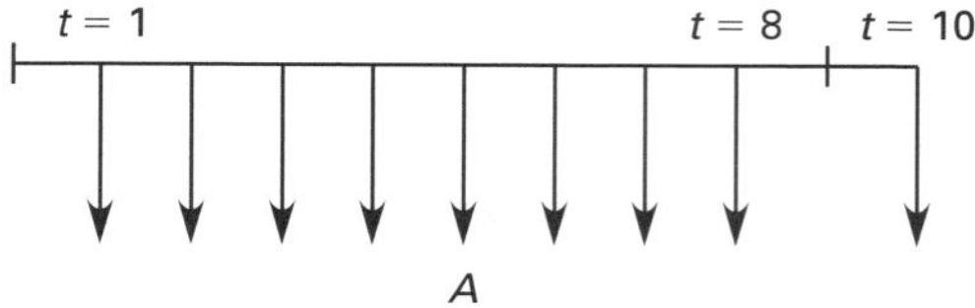

Alternatively, the present worth could be calculated as an additive process. [**Economic Factor Conversions**]

$$P = A(P/A, i\%, 8) + A(P/F, i\%, 10) \qquad 73.17$$

Delayed and Premature Cash Flows

There are cases when a cash flow matches a standard cash flow exactly, except that the cash flow is delayed or starts sooner than it should. Often, such cash flows can be handled with superposition. At other times, it may be more convenient to shift the time axis. This shift is known as the *projection method*. Example 73.8 demonstrates the projection method.

Example 73.8

An expense of \$75 is incurred starting at $t = 3$ and continues until $t = 9$. There are no expenses or receipts until $t = 3$. Use the projection method to determine the present worth of this stream of expenses.

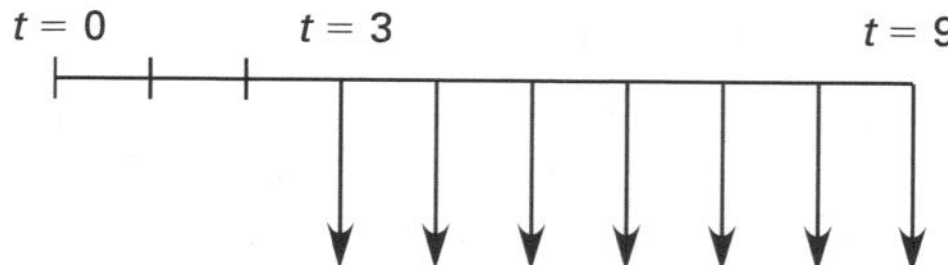

Solution

First, determine a cash flow at $t = 2$ that is equivalent to the entire expense stream. If $t = 0$ was where $t = 2$ actually is, the present worth of the expense stream would be

$$P' = (-\$75)(P/A, i\%, 7)$$

P' is a cash flow at $t = 2$. It is now simple to find the present worth (at $t = 0$) of this future amount. [Economic Factor Conversions]

$$P = P'(P/F, i\%, 2) = (-\$75)(P/A, i\%, 7)(P/F, i\%, 2)$$

Cash Flows at Beginnings of Years: The Christmas Club Problem

This type of problem is characterized by a stream of equal payments (or expenses) starting at $t = 0$ and ending at $t = n - 1$. (See Fig. 73.4.) It differs from the standard annual cash flow in the existence of a cash flow at $t = 0$ and the absence of a cash flow at $t = n$. This problem gets its name from the service provided by some savings institutions whereby money is automatically deposited each week or month (starting immediately, when the savings plan is opened) in order to accumulate money to purchase Christmas presents at the end of the year.

Figure 73.4 Cash Flows at Beginnings of Years

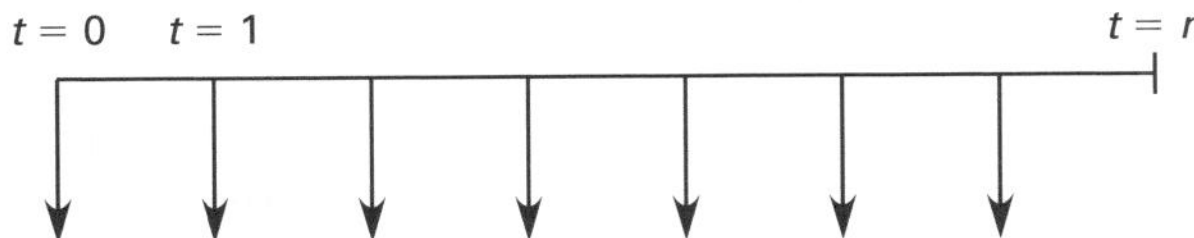

It may seem that the present worth of the savings stream can be determined by directly applying the (P/A) factor. However, this is not the case, since the Christmas Club cash flow and the standard annual cash flow differ. The Christmas Club problem is easily handled by superposition, as demonstrated by Ex. 73.9.

Example 73.9
How much can you expect to accumulate by $t = 10$ for a child's college education if you deposit \$300 at the beginning of each year for a total of ten payments?

Solution

The first payment is made at $t = 0$, and there is no payment at $t = 10$. The future worth of the first payment is calculated with the (F/P) factor. The absence of the payment at $t = 10$ is handled by superposition. This "correction" is not multiplied by a factor. [Economic Factor Conversions] [Economic Factor Tables]

$$\begin{aligned} F &= (\$300)(F/P, i\%, 10) + (\$300)(F/A, i\%, 10) - \$300 \\ &= (\$300)(F/A, i\%, 11) - \$300 \end{aligned}$$

16. THE MEANING OF PRESENT WORTH AND *i*

If \$100 is invested in a 5% bank account (using annual compounding), you can remove \$105 one year from now; if this investment is made, you will receive a *return on investment* (ROI) of \$5. [Economic Factor Tables]

The cash flow diagram (see Fig. 73.5) and the present worth of the two transactions are

$$\begin{aligned} P &= -\$100 + (\$105)(P/F, 5\%, 1) \\ &= -\$100 + (\$105)(0.9524) \\ &= 0 \end{aligned}$$

Figure 73.5 Cash Flow Diagram

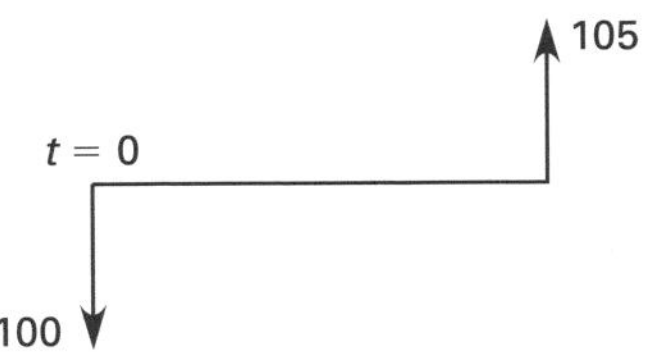

The present worth is zero even though you will receive a \$5 return on your investment.

However, if you are offered \$120 for the use of \$100 over a one-year period, the cash flow diagram (see Fig. 73.6) and present worth (at 5%) would be

$$\begin{aligned} P &= -\$100 + (\$120)(P/F, 5\%, 1) \\ &= -\$100 + (\$120)(0.9524) \\ &= \$14.29 \end{aligned}$$

Figure 73.6 Cash Flow Diagram

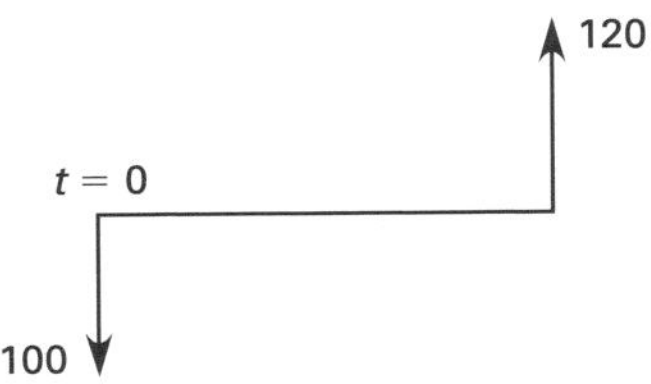

Therefore, the present worth of an alternative is seen to be equal to the equivalent value at $t = 0$ of the increase in return above that which you would be able to earn in an investment offering $i\%$ per period. In the previous case, \$14.29 is the present worth of (\$20 – \$5), the difference in the two ROIs.

The present worth is also the amount that you would have to be given to dissuade you from making an investment, since placing the initial investment amount along with the present worth into a bank account earning $i\%$ will yield the same eventual return on investment. Relating this to the previous paragraphs, you could be dissuaded from investing \$100 in an alternative that would return \$120 in one year by a $t = 0$ payment of \$14.29. Clearly, (\$100 + \$14.29) invested at $t = 0$ will also yield \$120 in one year at 5%.

Income-producing alternatives with negative present worths are undesirable, and alternatives with positive present worths are desirable because they increase the average earning power of invested capital. (In some

Economics

cases, such as municipal and public works projects, the present worths of all alternatives are negative, in which case, the least negative alternative is best.)

The selection of the interest rate is difficult in engineering economics problems. Usually, it is taken as the average rate of return that an individual or business organization has realized in past investments. Alternatively, the interest rate may be associated with a particular level of risk. Usually, i for individuals is the interest rate that can be earned in relatively *risk-free investments.*

17. SIMPLE AND COMPOUND INTEREST

If $100 is invested at 5%, it will grow to $105 in one year. During the second year, 5% interest continues to be accrued, but on $105, not on $100. This is the principle of *compound interest:* The interest accrues interest.[12]

If only the original principal accrues interest, the interest is said to be *simple interest.* Simple interest is rarely encountered in long-term engineering economic analyses, but the concept may be incorporated into short-term transactions.

18. EXTRACTING THE INTEREST RATE: RATE OF RETURN

An intuitive definition of the *rate of return* (ROR) is the effective annual interest rate at which an investment accrues income. That is, the rate of return of a project is the interest rate that would yield identical profits if all money were invested at that rate. Although this definition is correct, it does not provide a method of determining the rate of return.

It was previously seen that the present worth of a $100 investment invested at 5% is zero when $i = 5\%$ is used to determine equivalence. Therefore, a working definition of rate of return would be the effective annual interest rate that makes the present worth of the investment zero. Alternatively, rate of return could be defined as the effective annual interest rate that will discount all cash flows to a total present worth equal to the required initial investment.

It is tempting, but impractical, to determine a rate of return analytically. It is simply too difficult to extract the interest rate from the equivalence equation. For example, consider a $100 investment that pays back $75 at the end of each of the first two years. The present worth equivalence equation (set equal to zero in order to determine the rate of return) is

$$P = 0 = -\$100 + (\$75)(1+i)^{-1} + (\$75)(1+i)^{-2} \qquad 73.18$$

Solving Eq. 73.18 requires finding the roots of a quadratic equation. In general, for an investment or project spanning n years, the roots of an nth-order polynomial would have to be found. It should be obvious that an analytical solution would be essentially impossible for more complex cash flows. (The rate of return in this example is 31.87%.)

If the rate of return is needed, it can be found from a trial-and-error solution. To find the rate of return of an investment, proceed as follows.

step 1: Set up the problem as if to calculate the present worth.

step 2: Arbitrarily select a reasonable value for i. Calculate the present worth.

step 3: Choose another value of i (not too close to the original value), and again solve for the present worth.

step 4: Interpolate or extrapolate the value of i that gives a zero present worth.

step 5: For increased accuracy, repeat steps 2 and 3 with two more values that straddle the value found in step 4.

A common, although incorrect, method of calculating the rate of return involves dividing the annual receipts or returns by the initial investment. (See Sec. 73.54.) However, this technique ignores such items as salvage, depreciation, taxes, and the time value of money. This technique also is inadequate when the annual returns vary.

It is possible that more than one interest rate will satisfy the zero present worth criteria. This confusing situation occurs whenever there is more than one change in sign in the investment's cash flow.[13] Table 73.3 indicates the numbers of possible interest rates as a function of the number of sign reversals in the investment's cash flow.

[12]This assumes, of course, that the interest remains in the account. If the interest is removed and spent, only the remaining funds accumulate interest.

[13]There will always be at least one change of sign in the cash flow of a legitimate investment. (This excludes municipal and other tax-supported functions.) At $t = 0$, an investment is made (a negative cash flow). Hopefully, the investment will begin to return money (a positive cash flow) at $t = 1$ or shortly thereafter. Although it is possible to conceive of an investment in which all of the cash flows were negative, such an investment would probably be classified as a *hobby.*

Table 73.3 *Multiplicity of Rates of Return*

number of sign reversals	number of distinct rates of return
0	0
1	0 or 1
2	0, 1, or 2
3	0, 1, 2, or 3
4	0, 1, 2, 3, or 4
m	$0, 1, 2, 3, \ldots, m-1, m$

Difficulties associated with interpreting the meaning of multiple rates of return can be handled with the concepts of external investment and external rate of return. An *external investment* is an investment that is distinct from the investment being evaluated (which becomes known as the internal investment). The *external rate of return*, which is the rate of return earned by the external investment, does not need to be the same as the rate earned by the internal investment.

Generally, the multiple rates of return indicate that the analysis must proceed as though money will be invested outside of the project. The mechanics of how this is done are not covered here.

Example 73.10

A biomedical company is developing a new drug. A venture capital firm gives the company \$25 million initially and \$55 million more at the end of the first year. The drug patent will be sold at the end of year 5 to the highest bidder, and the biomedical company will receive \$80 million. (The venture capital firm will receive everything in excess of \$80 million.) The firm invests unused money in short-term commercial paper earning 10% effective interest per year through its bank. In the meantime, the biomedical company incurs development expenses of \$50 million annually for the first three years. The drug is to be evaluated by a government agency and there will be neither expenses nor revenues during the fourth year. What is the biomedical company's rate of return on this investment?

Solution

Normally, the rate of return is determined by setting up a present worth problem and varying the interest rate until the present worth is zero. Writing the cash flows, though, shows that there are two reversals of sign: one at $t = 2$ (positive to negative) and the other at $t = 5$ (negative to positive). Therefore, there could be two interest rates that produce a zero present worth. (In fact, there actually are two interest rates: 10.7% and 41.4%.)

time	cash flow (millions)
0	+25
1	$+55 - 50 = +5$
2	−50
3	−50
4	0
5	+80

However, this problem can be reduced to one with only one sign reversal in the cash flow series. The initial \$25 million is invested in commercial paper (an *external investment* having nothing to do with the drug development process) during the first year at 10%. The accumulation of interest and principal after one year is

$$(25)(1 + 0.10) = 27.5$$

This 27.5 is combined with the 5 (the money remaining after all expenses are paid at $t = 1$) and invested externally, again at 10%. The accumulation of interest and principal after one year (i.e., at $t = 2$) is

$$(27.5 + 5)(1 + 0.10) = 35.75$$

This 35.75 is combined with the development cost paid at $t = 2$.

The cash flow for the development project (the internal investment) is

time	cash flow (millions)
0	0
1	0
2	$35.75 - 50 = -14.25$
3	−50
4	0
5	+80

There is only one sign reversal in the cash flow series. The *internal rate of return* on this development project is found by the traditional method to be 10.3%. This is different from the rate the company can earn from investing externally in commercial paper.

19. RATE OF RETURN VERSUS RETURN ON INVESTMENT

Rate of return (ROR) is an effective annual interest rate, typically stated in percent per year. *Return on investment* (ROI) is a dollar amount. *Rate of return* and *return on investment* are not synonymous.

Return on investment can be calculated in two different ways. The accounting method is to subtract the total of all investment costs from the total of all net profits (i.e., revenues less expenses). The time value of money is not considered.

In engineering economic analysis, the return on investment is calculated from equivalent values. Specifically, the present worth (at $t = 0$) of all investment costs is subtracted from the future worth (at $t = n$) of all net profits.

When there are only two cash flows, a single investment amount and a single payback, the two definitions of return on investment yield the same numerical value. When there are more than two cash flows, the returns on investment will be different depending on which definition is used.

20. MINIMUM ATTRACTIVE RATE OF RETURN

A company may not know what effective interest rate, i, to use in engineering economic analysis. In such a case, the company can establish a minimum level of economic performance that it would like to realize on all investments. This criterion is known as the *minimum attractive rate of return* (MARR). Unlike the effective interest rate, i, the minimum attractive rate of return is not used in numerical calculations.[14] It is used only in comparisons with the rate of return.

Once a rate of return for an investment is known, it can be compared to the minimum attractive rate of return. To be a viable alternative, the rate of return must be greater than the minimum attractive rate of return.

The advantage of using comparisons to the minimum attractive rate of return is that an effective interest rate, i, never needs to be known. The minimum attractive rate of return becomes the correct interest rate for use in present worth and equivalent uniform annual cost calculations.

21. DURATIONS OF INVESTMENTS

Because they are handled differently, short-term investments and short-lived assets need to be distinguished from investments and assets that constitute an infinitely lived project. Short-term investments are easily identified: a drill press that is needed for three years or a temporary factory building that is being constructed to last five years.

Investments with perpetual cash flows are also (usually) easily identified: maintenance on a large flood control dam and revenues from a long-span toll bridge. Furthermore, some items with finite lives can expect renewal on a repeated basis.[15] For example, a major freeway with a pavement life of 20 years is unlikely to be abandoned; it will be resurfaced or replaced every 20 years.

Actually, if an investment's finite lifespan is long enough, it can be considered an infinite investment because money 50 or more years from now has little impact on current decisions. The $(P/F, 10\%, 50)$ factor, for example, is 0.0085. Therefore, one dollar at $t = 50$ has an equivalent present worth of less than one penny. Since these far-future cash flows are eclipsed by present cash flows, long-term investments can be considered finite or infinite without significant impact on the calculations.

22. CHOICE OF ALTERNATIVES: COMPARING ONE ALTERNATIVE WITH ANOTHER ALTERNATIVE

Several methods exist for selecting a superior alternative from among a group of proposals. Each method has its own merits and applications.

Present Worth Method

When two or more alternatives are capable of performing the same functions, the superior alternative will have the largest present worth. The *present worth method* is restricted to evaluating alternatives that are mutually exclusive and that have the same lives. This method is suitable for ranking the desirability of alternatives. [Economic Factor Conversions]

Example 73.11

Investment A costs $10,000 today and pays back $11,500 two years from now. Investment B costs $8000 today and pays back $4500 each year for two years. If an interest rate of 5% is used, which alternative is superior?

Solution

Find the superior alternative. [Economic Factor Conversions] [Economic Factor Tables]

$$\begin{aligned} P(\text{A}) &= -\$10{,}000 + (\$11{,}500)(P/F, 5\%, 2) \\ &= -\$10{,}000 + (\$11{,}500)(0.9070) \\ &= \$431 \\ P(\text{B}) &= -\$8000 + (\$4500)(P/A, 5\%, 2) \\ &= -\$8000 + (\$4500)(1.8594) \\ &= \$367 \end{aligned}$$

Alternative A is superior and should be chosen.

Capitalized Cost Method

The present worth of a project with an infinite life is known as the *capitalized cost* or *life-cycle cost*. Capitalized cost is the amount of money at $t = 0$ needed to

[14] Not everyone adheres to this rule. Some people use "minimum attractive rate of return" and "effective interest rate" interchangeably.
[15] The term *renewal* can be interpreted to mean replacement or repair.

perpetually support the project on the earned interest only. Capitalized cost is a positive number when expenses exceed income.

In comparing two alternatives, each of which is infinitely lived, the superior alternative will have the lowest capitalized cost.

Normally, it would be difficult to work with an infinite stream of cash flows since most economics tables do not list factors for periods in excess of 100 years. However, the (A/P) discounting factor approaches the interest rate as n becomes large. Since the (P/A) and (A/P) factors are reciprocals of each other, it is possible to divide an infinite series of equal cash flows by the interest rate in order to calculate the present worth of the infinite series. This is the basis of Eq. 73.19.

$$\text{capitalized cost} = \text{initial cost} + \frac{\text{annual costs}}{i} \qquad 73.19$$

Equation 73.19 can be used when the annual costs are equal in every year. If the operating and maintenance costs occur irregularly instead of annually, or if the costs vary from year to year, it will be necessary to somehow determine a cash flow of equal annual amounts (EAA) that is equivalent to the stream of original costs.

The equal annual amount may be calculated in the usual manner by first finding the present worth of all the actual costs and then multiplying the present worth by the interest rate (the (A/P) factor for an infinite series). However, it is not even necessary to convert the present worth to an equal annual amount since Eq. 73.20 will convert the equal amount back to the present worth.

$$\begin{aligned}\text{capitalized cost} &= \text{initial cost} + \frac{\text{EAA}}{i} \\ &= \text{initial cost} + \begin{array}{c}\text{present worth} \\ \text{of all expenses}\end{array}\end{aligned} \qquad 73.20$$

Example 73.12

What is the capitalized cost of a public works project that will cost \$25,000,000 now and will require \$2,000,000 in maintenance annually? The effective annual interest rate is 12%.

Solution

From Eq. 73.19, find the capitalized cost in millions of dollars. [**Economic Factor Conversions**]

$$\begin{aligned}\text{capitalized cost} &= 25 + (2)(P/A, 12\%, \infty) \\ &= 25 + \frac{2}{0.12} \\ &= 41.67\end{aligned}$$

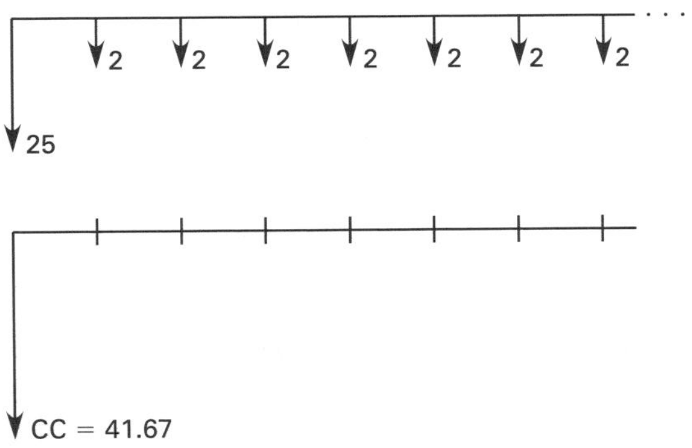

Annual Cost Method

Alternatives that accomplish the same purpose but that have unequal lives must be compared by the *annual cost method*.[16] The annual cost method assumes that each alternative will be replaced by an identical twin at the end of its useful life (infinite renewal). This method, which may also be used to rank alternatives according to their desirability, is also called the *annual return method* or *capital recovery method*.

Restrictions are that the alternatives must be mutually exclusive and repeatedly renewed up to the duration of the longest-lived alternative. The calculated annual cost is known as the *equivalent uniform annual cost* (EUAC) or just *equivalent annual cost*. Cost is a positive number when expenses exceed income.

Example 73.13

Which of the following alternatives is superior over a 30-year period if the interest rate is 7%?

	alternative A	alternative B
type	brick	wood
life	30 years	10 years
initial cost	\$1800	\$450
maintenance	\$5/year	\$20/year

[16]Of course, the annual cost method can be used to determine the superiority of assets with identical lives as well.

Solution

Determine the equivalent uniform annual cost. [Economic Factor Tables]

$$\begin{aligned} \text{EUAC(A)} &= (\$1800)(A/P, 7\%, 30) + \$5 \\ &= (\$1800)(0.0806) + \$5 \\ &= \$150 \\ \text{EUAC(B)} &= (\$450)(A/P, 7\%, 10) + \$20 \\ &= (\$450)(0.1424) + \$20 \\ &= \$84 \end{aligned}$$

Alternative B is superior since its annual cost of operation is the lowest. It is assumed that three wood facilities, each with a life of 10 years and a cost of $450, will be built to span the 30-year period.

23. CHOICE OF ALTERNATIVES: COMPARING AN ALTERNATIVE WITH A STANDARD

With specific economic performance criteria, it is possible to qualify an investment as acceptable or unacceptable without having to compare it with another investment. Two such performance criteria are the benefit-cost ratio and the minimum attractive rate of return.

Benefit-Cost Ratio Method

The *benefit-cost ratio method* is often used in municipal project evaluations where benefits and costs accrue to different segments of the community. With this method, the present worth of all benefits (irrespective of the beneficiaries) is divided by the present worth of all costs. The project is considered acceptable if the ratio equals or exceeds 1.0, that is, if $B/C \geq 1.0$.

When the benefit-cost ratio method is used, disbursements by the initiators or sponsors are *costs*. Disbursements by the users of the project are known as *disbenefits*. It is often difficult to determine whether a cash flow is a cost or a disbenefit (whether to place it in the numerator or denominator of the benefit-cost ratio calculation).

Regardless of where the cash flow is placed, an acceptable project will always have a benefit-cost ratio greater than or equal to 1.0, although the actual numerical result will depend on the placement. For this reason, the benefit-cost ratio method should not be used to rank competing projects.

The benefit-cost ratio method of comparing alternatives is used extensively in transportation engineering where the ratio is often (but not necessarily) written in terms of annual benefits and annual costs instead of present worths. Another characteristic of highway benefit-cost ratios is that the route (road, highway, etc.) is usually already in place and that various alternative upgrades are being considered. There will be existing benefits and costs associated with the current route. Therefore, the *change* (usually an increase) in benefits and costs is used to calculate the benefit-cost ratio.[17]

$$B/C = \frac{\Delta\,\text{user benefits}}{\Delta\begin{array}{c}\text{investment}\\ \text{cost}\end{array} + \Delta\,\text{maintenance} - \Delta\begin{array}{c}\text{residual}\\ \text{value}\end{array}} \quad 73.21$$

The change in *residual value* (*terminal value*) appears in the denominator as a negative item. An increase in the residual value would decrease the denominator.

Example 73.14

By building a bridge over a ravine, a state department of transportation can shorten the time it takes to drive through a mountainous area. Estimates of costs and benefits (due to decreased travel time, fewer accidents, reduced gas usage, etc.) have been prepared. Should the bridge be built? Use the benefit-cost ratio method of comparison.

	millions
initial cost	40
capitalized cost of perpetual annual maintenance	12
capitalized value of annual user benefits	49
residual value	0

Solution

If Eq. 73.21 is used, the benefit-cost ratio is

$$B/C = \frac{49}{40 + 12 - 0} = 0.942$$

Since the benefit-cost ratio is less than 1.00, the bridge should not be built.

[17]This discussion of highway benefit-cost ratios is not meant to imply that everyone agrees with Eq. 73.21. In *Economic Analysis for Highways* (International Textbook Company, Scranton, PA, 1969), author Robley Winfrey took a strong stand on one aspect of the benefits versus disbenefits issue: highway maintenance. According to Winfrey, regular highway maintenance costs should be placed in the numerator as a subtraction from the user benefits. Some have called this mandate the *Winfrey method.*

If the maintenance costs are placed in the numerator (per Ftn. 17), the benefit-cost ratio value will be different, but the conclusion will not change.

$$B/C_{\text{alternate method}} = \frac{49-12}{40} = 0.925$$

Rate of Return Method

The minimum attractive rate of return (MARR) has already been introduced as a standard of performance against which an investment's actual *rate of return* (ROR) is compared. If the rate of return is equal to or exceeds the minimum attractive rate of return, the investment is qualified. This is the basis for the *rate of return method* of alternative selection.

Finding the rate of return can be a long, iterative process. Usually, the actual numerical value of rate of return is not needed; it is sufficient to know whether or not the rate of return exceeds the minimum attractive rate of return. This *comparative analysis* can be accomplished without calculating the rate of return simply by finding the present worth of the investment using the minimum attractive rate of return as the effective interest rate (i.e., i = MARR). If the present worth is zero or positive, the investment is qualified. If the present worth is negative, the rate of return is less than the minimum attractive rate of return.

24. RANKING MUTUALLY EXCLUSIVE MULTIPLE PROJECTS

Ranking of multiple investment alternatives is required when there is sufficient funding for more than one investment. Since the best investments should be selected first, it is necessary to place all investments into an ordered list.

Ranking is relatively easy if the present worths, future worths, capitalized costs, or equivalent uniform annual costs have been calculated for all the investments. The highest ranked investment will be the one with the largest present or future worth, or the smallest capitalized or annual cost. Present worth, future worth, capitalized cost, and equivalent uniform annual cost can all be used to rank multiple investment alternatives.

However, neither rates of return nor benefit-cost ratios should be used to rank multiple investment alternatives. Specifically, if two alternatives both have rates of return exceeding the minimum acceptable rate of return, it is not sufficient to select the alternative with the highest rate of return.

An *incremental analysis*, also known as a *rate of return on added investment study*, should be performed if rate of return is used to select between investments. An incremental analysis starts by ranking the alternatives in order of increasing initial investment. Then, the cash flows for the investment with the lower initial cost are subtracted from the cash flows for the higher-priced alternative on a year-by-year basis. This produces, in effect, a third alternative representing the costs and benefits of the added investment. The added expense of the higher-priced investment is not warranted unless the rate of return of this third alternative exceeds the minimum attractive rate of return as well. The choice criterion is to select the alternative with the higher initial investment if the incremental rate of return exceeds the minimum attractive rate of return.

An incremental analysis is also required if ranking is to be done by the benefit-cost ratio method. The incremental analysis is accomplished by calculating the ratio of differences in benefits to differences in costs for each possible pair of alternatives. If the ratio exceeds 1.0, alternative 2 is superior to alternative 1. Otherwise, alternative 1 is superior.[18]

$$\frac{B_2 - B_1}{C_2 - C_1} \geq 1 \quad \text{[alternative 2 superior]} \qquad 73.22$$

25. ALTERNATIVES WITH DIFFERENT LIVES

Comparison of two alternatives is relatively simple when both alternatives have the same life. For example, a problem might be stated: "Which would you rather have: car A with a life of three years, or car B with a life of five years?"

However, care must be taken to understand what is going on when the two alternatives have different lives. If car A has a life of three years and car B has a life of five years, what happens at $t = 3$ if the five-year car is chosen? If a car is needed for five years, what happens at $t = 3$ if the three-year car is chosen?

In this type of situation, it is necessary to distinguish between the length of the need (the *analysis horizon*) and the lives of the alternatives or assets intended to meet that need. The lives do not have to be the same as the horizon.

Finite Horizon with Incomplete Asset Lives

If an asset with a five-year life is chosen for a three-year need, the disposition of the asset at $t = 3$ must be known in order to evaluate the alternative. If the asset is sold at $t = 3$, the salvage value is entered into the analysis (at $t = 3$) and the alternative is evaluated as a three-year investment. The fact that the asset is sold when it has some useful life remaining does not affect the analysis horizon.

[18]It goes without saying that the benefit-cost ratios for all investment alternatives by themselves must also be equal to or greater than 1.0.

Similarly, if a three-year asset is chosen for a five-year need, something about how the need is satisfied during the last two years must be known. Perhaps a rental asset will be used. Or, perhaps the function will be "farmed out" to an outside firm. In any case, the costs of satisfying the need during the last two years enter the analysis, and the alternative is evaluated as a five-year investment.

If both alternatives are "converted" to the same life, any of the alternative selection criteria (present worth method, annual cost method, etc.) can be used to determine which alternative is superior.

Finite Horizon with Integer Multiple Asset Lives

It is common to have a long-term horizon (need) that must be met with short-lived assets. In special instances, the horizon will be an integer number of asset lives. For example, a company may be making a 12-year transportation plan and may be evaluating two cars: one with a three-year life, and another with a four-year life.

In this example, four of the first car or three of the second car are needed to reach the end of the 12-year horizon.

If the horizon is an integer number of asset lives, any of the alternative selection criteria can be used to determine which is superior. If the present worth method is used, all alternatives must be evaluated over the entire horizon. (In this example, the present worth of 12 years of car purchases and use must be determined for both alternatives.)

If the equivalent uniform annual cost method is used, it may be possible to base the calculation of annual cost on one lifespan of each alternative only. It may not be necessary to incorporate all of the cash flows into the analysis. (In the running example, the annual cost over three years would be determined for the first car; the annual cost over four years would be determined for the second car.) This simplification is justified if the subsequent asset replacements (renewals) have the same cost and cash flow structure as the original asset. This assumption is typically made implicitly when the annual cost method of comparison is used.

Infinite Horizon

If the need horizon is infinite, it is not necessary to impose the restriction that asset lives of alternatives be integer multiples of the horizon. The superior alternative will be replaced (renewed) whenever it is necessary to do so, forever.

Infinite horizon problems are almost always solved with either the annual cost or capitalized cost method. It is common to (implicitly) assume that the cost and cash flow structure of the asset replacements (renewals) are the same as the original asset.

26. OPPORTUNITY COSTS

An *opportunity cost* is an imaginary cost representing what will not be received if a particular strategy is rejected. It is what you will lose if you do or do not do something. As an example, consider a growing company with an existing operational computer system. If the company trades in its existing computer as part of an upgrade plan, it will receive a *trade-in allowance*. (In other problems, a *salvage value* may be involved.)

If one of the alternatives being evaluated is not to upgrade the computer system at all, the trade-in allowance (or, salvage value in other problems) will not be realized. The amount of the trade-in allowance is an opportunity cost that must be included in the problem analysis.

Similarly, if one of the alternatives being evaluated is to wait one year before upgrading the computer, the *difference in trade-in allowances* is an opportunity cost that must be included in the problem analysis.

27. REPLACEMENT STUDIES

An investigation into the retirement of an existing process or piece of equipment is known as a *replacement study*. Replacement studies are similar in most respects to other alternative comparison problems: An interest rate is given, two alternatives exist, and one of the previously mentioned methods of comparing alternatives is used to choose the superior alternative. Usually, the annual cost method is used on a year-by-year basis.

In replacement studies, the existing process or piece of equipment is known as the *defender*. The new process or piece of equipment being considered for purchase is known as the *challenger*.

28. TREATMENT OF SALVAGE VALUE IN REPLACEMENT STUDIES

Since most defenders still have some market value when they are retired, the problem of what to do with the salvage arises. It seems logical to use the salvage value of the defender to reduce the initial purchase cost of the challenger. This is consistent with what would actually happen if the defender were to be retired.

By convention, however, the defender's salvage value is subtracted from the defender's present value. This does not seem logical, but it is done to keep all costs and benefits related to the defender with the defender. In this case, the salvage value is treated as an opportunity cost that would be incurred if the defender is not retired.

If the defender and the challenger have the same lives and a present worth study is used to choose the superior alternative, the placement of the salvage value will have no effect on the net difference between present worths for the challenger and defender. Although the values of the two present worths will be different depending on the placement, the difference in present worths will be the same.

If the defender and the challenger have different lives, an annual cost comparison must be made. Since the salvage value would be "spread over" a different number of years depending on its placement, it is important to abide by the conventions listed in this section.

There are a number of ways to handle salvage value in retirement studies. The best way is to calculate the cost of keeping the defender one more year. In addition to the usual operating and maintenance costs, that cost includes an opportunity interest cost incurred by not selling the defender, and also a drop in the salvage value if the defender is kept for one additional year. Specifically,

$$\begin{aligned}\underset{\text{(defender)}}{\text{EUAC}} &= \text{next year's maintenance costs} \\ &\quad + i(\text{current salvage value}) \\ &\quad + \text{current salvage} \\ &\quad - \text{next year's salvage}\end{aligned} \qquad 73.23$$

It is important in retirement studies not to double count the salvage value. That is, it would be incorrect to add the salvage value to the defender and at the same time subtract it from the challenger.

Equation 73.23 contains the difference in salvage value between two consecutive years. This calculation shows that the defender/challenger decision must be made on a year-by-year basis. One application of Eq. 73.23 will not usually answer the question of whether the defender should remain in service indefinitely. The calculation must be repeatedly made as long as there is a drop in salvage value from one year to the next.

29. ECONOMIC LIFE: RETIREMENT AT MINIMUM COST

As an asset grows older, its operating and maintenance costs typically increase. Eventually, the cost to keep the asset in operation becomes prohibitive, and the asset is retired or replaced. However, it is not always obvious when an asset should be retired or replaced.

As the asset's maintenance cost is increasing each year, the amortized cost of its initial purchase is decreasing. It is the sum of these two costs that should be evaluated to determine the point at which the asset should be retired or replaced. Since an asset's initial purchase price is likely to be high, the amortized cost will be the controlling factor in those years when the maintenance costs are low. Therefore, the EUAC of the asset will decrease in the initial part of its life.

However, as the asset grows older, the change in its amortized cost decreases while maintenance cost increases. Eventually, the sum of the two costs reaches a minimum and then starts to increase. The age of the asset at the minimum cost point is known as the *economic life* of the asset. The economic life generally is less than the length of need and the technological lifetime of the asset (see Fig. 73.7.)

Figure 73.7 *EUAC Versus Age at Retirement*

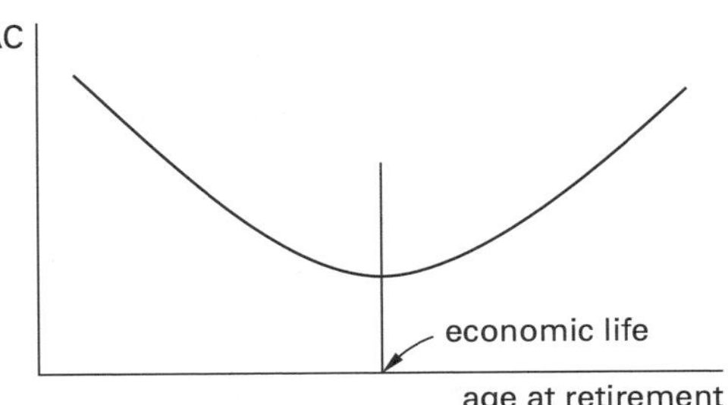

The determination of an asset's economic life is illustrated by Ex. 73.15.

Example 73.15

Buses in a municipal transit system have the characteristics listed. In order to minimize its annual operating expenses, when should the city replace its buses if money can be borrowed at 8%?

initial cost of bus: $120,000

year	maintenance cost	salvage value
1	35,000	60,000
2	38,000	55,000
3	43,000	45,000
4	50,000	25,000
5	65,000	15,000

Solution

The annual maintenance is different each year. Each maintenance cost must be spread over the life of the bus. This is done by first finding the present worth and then amortizing the maintenance costs. [**Economic Factor Conversions**] [**Economic Factor Tables**]

If a bus is kept for one year and then sold, the annual cost will be

$$\begin{aligned}\text{EUAC}(1) &= (\$120{,}000)(A/P, 8\%, 1) \\ &\quad + (\$35{,}000)(A/F, 8\%, 1) \\ &\quad - (\$60{,}000)(A/F, 8\%, 1) \\ &= (\$120{,}000)(1.0800) + (\$35{,}000)(1.000) \\ &\quad - (\$60{,}000)(1.000) \\ &= \$104{,}600\end{aligned}$$

Economics

If a bus is kept for two years and then sold, the annual cost will be

$$\begin{aligned}\text{EUAC}(2) &= \big(\$120{,}000 + (\$35{,}000)(P/F, 8\%, 1)\big) \\ &\quad \times (A/P, 8\%, 2) \\ &\quad + (\$38{,}000 - \$55{,}000)(A/F, 8\%, 2) \\ &= \big(\$120{,}000 + (\$35{,}000)(0.9259)\big)(0.5608) \\ &\quad + (\$38{,}000 - \$55{,}000)(0.4808) \\ &= \$77{,}296\end{aligned}$$

If a bus is kept for three years and then sold, the annual cost will be

$$\begin{aligned}\text{EUAC}(3) &= \big(\$120{,}000 + (\$35{,}000)(P/F, 8\%, 1) \\ &\quad + (\$38{,}000)(P/F, 8\%, 2)\big)(A/P, 8\%, 3) \\ &\quad + (\$43{,}000 - \$45{,}000)(A/F, 8\%, 3) \\ &= \big(\$120{,}000 + (\$35{,}000)(0.9259) \\ &\quad + (\$38{,}000)(0.8573)\big)(0.3880) \\ &\quad - (\$2000)(0.3080) \\ &= \$71{,}158\end{aligned}$$

This process is continued until the annual cost begins to increase. In this example, EUAC(4) is $71,700. Therefore, the buses should be retired after three years.

30. LIFE-CYCLE COST

The *life-cycle cost* of an alternative is the equivalent value (at $t = 0$) of the alternative's cash flow over the alternative's lifespan. Since the present worth is evaluated using an effective interest rate of i (which would be the interest rate used for all engineering economic analyses), the life-cycle cost is the same as the alternative's present worth. If the alternative has an infinite horizon, the life-cycle cost and capitalized cost will be identical.

31. CAPITALIZED ASSETS VERSUS EXPENSES

High expenses reduce profit, which in turn reduces income tax. It seems logical to label each and every expenditure, even an asset purchase, as an expense. As an alternative to this *expensing the asset*, it may be decided to capitalize the asset. *Capitalizing the asset* means that the cost of the asset is divided into equal or unequal parts, and only one of these parts is taken as an expense each year. Expensing is clearly the more desirable alternative, since the after-tax profit is increased early in the asset's life.

There are long-standing accounting conventions as to what can be expensed and what must be capitalized.[19] Some companies capitalize everything—regardless of cost—with expected lifetimes greater than one year. Most companies, however, expense items whose purchase costs are below a cutoff value. A cutoff value in the range of $250–500, depending on the size of the company, is chosen as the maximum purchase cost of an expensed asset. Assets costing more than this are capitalized.

It is not necessary for a large corporation to keep track of every lamp, desk, and chair for which the purchase price is greater than the cutoff value. Such assets, all of which have the same lives and have been purchased in the same year, can be placed into groups or *asset classes.* A group cost, equal to the sum total of the purchase costs of all items in the group, is capitalized as though the group was an identifiable and distinct asset itself.

32. PURPOSE OF DEPRECIATION

Depreciation is an *artificial expense* that spreads the purchase price of an asset or other property over a number of years.[20] Depreciating an asset is an example of capitalization, as previously defined. The inclusion of depreciation in engineering economic analysis problems will increase the after-tax present worth (profitability) of an asset. The larger the depreciation, the greater will be the profitability. Therefore, individuals and companies eligible to utilize depreciation want to maximize and accelerate the depreciation available to them.

Although the entire property purchase price is eventually recognized as an expense, the net recovery from the expense stream never equals the original cost of the asset. That is, depreciation cannot realistically be thought of as a fund (an annuity or sinking fund) that accumulates capital to purchase a replacement at the end of the asset's life. The primary reason for this is that the depreciation expense is reduced significantly by the impact of income taxes, as will be seen in later sections.

[19]For example, purchased vehicles must be capitalized; payments for leased vehicles can be expensed. Repainting a building with paint that will last five years is an expense, but the replacement cost of a leaking roof must be capitalized.

[20]In the United States, the tax regulations of Internal Revenue Service (IRS) allow depreciation on almost all forms of *business property* except land. The following types of property are distinguished: *real* (e.g., buildings used for business), *residential* (e.g., buildings used as rental property), and *personal* (e.g., equipment used for business). Personal property does *not* include items for personal use (such as a personal residence), despite its name. *Tangible personal property* is distinguished from *intangible property* (goodwill, copyrights, patents, trademarks, franchises, agreements not to compete, etc.).

33. DEPRECIATION BASIS OF AN ASSET

The *depreciation basis* of an asset is the part of the asset's purchase price that is spread over the *depreciation period*, also known as the *service life*.[21] Usually, the depreciation basis and the purchase price are not the same.

A common depreciation basis is the difference between the purchase price and the expected salvage value at the end of the depreciation period. That is,

$$\text{depreciation basis} = C - S_n \qquad 73.24$$

There are several methods of calculating the year-by-year depreciation of an asset. Equation 73.24 is not universally compatible with all depreciation methods. Some methods do not consider the salvage value. This is known as an *unadjusted basis*. When the depreciation method is known, the depreciation basis can be rigorously defined.[22]

34. DEPRECIATION METHODS

Generally, tax regulations do not allow the cost of an asset to be treated as a deductible expense in the year of purchase. Rather, portions of the depreciation basis must be allocated to each of the n years of the asset's depreciation period. The amount that is allocated each year is called the *depreciation*.

Various methods exist for calculating an asset's depreciation each year.[23] Although the depreciation calculations may be considered independently (for the purpose of determining book value or as an academic exercise), it is important to recognize that depreciation has no effect on engineering economic analyses unless income taxes are also considered.

Straight-Line Method

With the *straight-line* (SL) *method*, depreciation is the same each year. The depreciation basis ($C - S_n$) is allocated uniformly to all of the n years in the depreciation period. Each year, the depreciation will be

Depreciation: Straight Line

$$D_j = \frac{C - S_n}{n} \qquad 73.25$$

Constant Percentage Method

The *constant percentage method*[24] is similar to the straight-line method in that the depreciation is the same each year. If the fraction of the basis used as depreciation is $1/n$, there is no difference between the constant percentage and straight-line methods. The two methods differ only in what information is available. (With the straight-line method, the life is known. With the constant percentage method, the depreciation fraction is known.)

Each year, the depreciation will be

$$\begin{aligned} D &= (\text{depreciation fraction})(\text{depreciation basis}) \\ &= (\text{depreciation fraction})(C - S_n) \end{aligned} \qquad 73.26$$

Sum-of-the-Years' Digits Method

In *sum-of-the-years' digits* (SOYD) depreciation, the digits from 1 to n inclusive are summed. The total, T, can also be calculated from

$$T = \tfrac{1}{2}n(n+1) \qquad 73.27$$

The depreciation in year j can be found from Eq. 73.28. Notice that the depreciation in year j, D_j, decreases by a constant amount each year.

Depreciation: Sum-of-Years Digits Method

$$D_j = \frac{2(C - S_n)(n - j + 1)}{n(n+1)} \qquad 73.28$$

Double Declining Balance Method[25]

Double declining balance[26] (DDB) depreciation is independent of salvage value. Furthermore, the book value never stops decreasing, although the depreciation decreases in magnitude. Usually, any book value in excess of the salvage value is written off in the last year of the asset's depreciation period. Unlike any of the other depreciation methods, double declining balance depends on accumulated depreciation.

$$D_{\text{first year}} = \frac{2C}{n} \qquad 73.29$$

$$D_j = \frac{2\left(C - \sum_{m=1}^{j-1} D_m\right)}{n} \qquad 73.30$$

[21] The *depreciation period* is selected to be as short as possible within recognized limits. This depreciation will not normally coincide with the *economic life* or *useful life* of an asset. For example, a car may be capitalized over a depreciation period of three years. It may become uneconomical to maintain and use at the end of an economic life of nine years. However, the car may be capable of operation over a useful life of 25 years.

[22] For example, with the Accelerated Cost Recovery System (ACRS) the *depreciation basis* is the total purchase cost, regardless of the expected salvage value. With declining balance methods, the depreciation basis is the purchase cost less any previously taken depreciation.

[23] This discussion gives the impression that any form of depreciation may be chosen regardless of the nature and circumstances of the purchase. In reality, the IRS tax regulations place restrictions on the higher-rate (accelerated) methods, such as declining balance and sum-of-the-years' digits methods. Furthermore, the *Economic Recovery Act of 1981* and the *Tax Reform Act of 1986* substantially changed the laws relating to personal and corporate income taxes.

[24] The *constant percentage method* should not be confused with the declining balance method, which used to be known as the *fixed percentage on diminishing balance method*.

[25] In the past, the *declining balance method* has also been known as the *fixed percentage of book value* and *fixed percentage on diminishing balance method*.

[26] Double declining balance depreciation is a particular form of *declining balance depreciation*, as defined by the IRS tax regulations. Declining balance depreciation includes 125% declining balance and 150% declining balance depreciations that can be calculated by substituting 1.25 and 1.50, respectively, for the 2 in Eq. 73.29.

Calculating the depreciation in the middle of an asset's life appears particularly difficult with double declining balance, since all previous years' depreciation amounts seem to be required. It appears that the depreciation in the sixth year, for example, cannot be calculated unless the values of depreciation for the first five years are calculated. However, this is not true.

Depreciation in the middle of an asset's life can be found from the following equations. (d is known as the *depreciation rate.*)

$$d = \frac{2}{n} \quad \left[\begin{array}{c}\text{double declining}\\ \text{balance}\end{array}\right] \qquad 73.31$$

$$D_j = dC(1-d)^{j-1} \qquad 73.32$$

Statutory Depreciation Systems

In the United States, property placed into service in 1981 and thereafter must use the *Accelerated Cost Recovery System* (ACRS), and after 1986, the *Modified Accelerated Cost Recovery System* (MACRS) or other statutory method. Other methods (straight-line, declining balance, etc.) cannot be used except in special cases.

Property placed into service in 1980 or before must continue to be depreciated according to the method originally chosen (e.g., straight-line, declining balance, or sum-of-the-years' digits). ACRS and MACRS cannot be used.

Under ACRS and MACRS, the cost recovery amount in the jth year of an asset's cost recovery period is calculated by multiplying the initial cost by a factor.

Depreciation: Modified Accelerated Cost Recovery System (MACRS)

$$D_j = (\text{factor})C \qquad 73.33$$

The initial cost used is not reduced by the asset's salvage value for ACRS and MACRS calculations. The factor used depends on the asset's cost recovery period. (See Table 73.4.) Such factors are subject to continuing legislation changes. Current tax publications should be consulted before using this method. [**Depreciation: Modified Accelerated Cost Recovery System (MACRS)**]

Production or Service Output Method

If an asset has been purchased for a specific task and that task is associated with a specific lifetime amount of output or production, the depreciation may be calculated by the fraction of total production produced during the year. Under the *units of production* method, the depreciation is not expected to be the same each year.

$$D_j = (C - S_n)\left(\frac{\text{actual output in year } j}{\text{estimated lifetime output}}\right) \qquad 73.34$$

Table 73.4 *Representative MACRS Depreciation Factors**

year, j	depreciation rate for recovery period, n: 3 years	5 years	7 years	10 years
1	33.33%	20.00%	14.29%	10.00%
2	44.45%	32.00%	24.49%	18.00%
3	14.81%	19.20%	17.49%	14.40%
4	7.41%	11.52%	12.49%	11.52%
5		11.52%	8.93%	9.22%
6		5.76%	8.92%	7.37%
7			8.93%	6.55%
8			4.46%	6.55%
9				6.56%
10				6.55%
11				3.28%

*Values are for the "half-year" convention. This table gives typical values only. Since these factors are subject to continuing revision, they should not be used without consulting an accounting professional.

Sinking Fund Method

The *sinking fund method* is seldom used in industry because the initial depreciation is low. The formula for sinking fund depreciation (which increases each year) is

$$D_j = (C - S_n)(A/F, i\%, n)(F/P, i\%, j-1) \qquad 73.35$$

Example 73.16

An asset is purchased for $9000. Its estimated economic life is 10 years, after which it will be sold for $200. Find the depreciation in the first three years using straight-line, double declining balance, and sum-of-the-years' digits (SOYD) depreciation methods.

Solution

Use the straight-line method to find the depreciation in the first three years.

Depreciation: Straight Line

$$D = \frac{C - S_n}{n} = \frac{\$9000 - \$200}{10} = \$880$$

Use the double-declining balance method.

$$D_j = \frac{2C}{n}$$

$$D_1 = \frac{(2)(\$9000)}{10} = \$1800$$

$$D_2 = \frac{(2)(\$9000 - \$1800)}{10} = \$1440$$

$$D_3 = \frac{(2)(\$9000 - (\$1800 + \$1440))}{10} = \$1152$$

Use the SOYD depreciation method to find the depreciation in the first three years.

Depreciation: Sum-of-Years Digits Method

$$D_j = \frac{2(C - S_n)(n - j + 1)}{n(n+1)}$$

$$D_1 = \frac{(2)(\$9000 - \$200)(10 - 1 + 1)}{(10)(10+1)} = \$1600$$

$$D_2 = \frac{(2)(\$9000 - \$200)(10 - 2 + 1)}{(10)(10+1)} = \$1440$$

$$D_3 = \frac{(2)(\$9000 - \$200)(10 - 3 + 1)}{(10)(10+1)} = \$1280$$

or

$$\text{SOYD: } T = \left(\frac{1}{2}\right)(10)(11) = 55$$

$$D_1 = \left(\frac{10}{55}\right)(\$9000 - \$200) = \$1600 \text{ in year 1}$$

$$D_2 = \left(\frac{9}{55}\right)(\$8800) = \$1440 \text{ in year 2}$$

$$D_3 = \left(\frac{8}{55}\right)(\$8800) = \$1280 \text{ in year 3}$$

35. ACCELERATED DEPRECIATION METHODS

An *accelerated depreciation method* is one that calculates a depreciation amount greater than a straight-line amount. Double declining balance and sum-of-the-years' digits methods are accelerated methods. The ACRS and MACRS methods are explicitly accelerated methods. Straight-line and sinking fund methods are not accelerated methods.

Use of an accelerated depreciation method may result in unexpected tax consequences when the depreciated asset or property is disposed of. Professional tax advice should be obtained in this area.

Example 73.17

A business owner buys a laptop computer. The laptop computer has an anticipated life cycle of three years and the purchase price of the laptop computer is $4,000. Using the MACRS method, find the cost recovery amount in the second year of depreciation.

Solution

The depreciation basis is

$$\begin{aligned} \text{depreciation basis} &= C - S_n \\ &= \$4000 - \$0 \\ &= \$4000 \end{aligned}$$

The year two factor for a three-year recovery period is 44.45%.

Depreciation: Modified Accelerated Cost Recovery System (MACRS)

$$\begin{aligned} D_j &= (\text{factor})C \\ &= (0.4445)(\$4000) \\ &= \$1778 \end{aligned}$$

36. BONUS DEPRECIATION

Bonus depreciation is a special one-time depreciation authorized by legislation for specific types of equipment, to be taken in the first year, in addition to the standard depreciation normally available for the equipment. Bonus depreciation is usually enacted in order to stimulate investment in or economic recovery of specific industries (e.g., aircraft manufacturing).

37. BOOK VALUE

The difference between original purchase price and accumulated depreciation is known as *book value*.[27] At the end of each year, the book value (which is initially equal to the purchase price) is reduced by the depreciation in that year.

It is important to properly synchronize depreciation calculations. It is difficult to answer the question, "What is the book value in the fifth year?" unless the timing of the book value change is mutually agreed upon. It is better to be specific about an inquiry by identifying when the book value change occurs. For example, the following question is unambiguous: "What is the book value at the end of year 5, after subtracting depreciation in the fifth year?" or "What is the book value after five years?"

Unfortunately, this type of care is seldom taken in book value inquiries, and it is up to the respondent to exercise reasonable care in distinguishing between beginning-of-year book value and end-of-year book value. To be consistent, the book value equations in this chapter have been written in such a way that the year subscript, j, has the same meaning in book value and depreciation calculations. That is, BV_5 means the book value at the

[27]The balance sheet of a corporation usually has two asset accounts: the *equipment account* and the *accumulated depreciation account*. There is no book value account on this financial statement, other than the implicit value obtained from subtracting the accumulated depreciation account from the equipment account. The book values of various assets, as well as their original purchase cost, date of purchase, salvage value, and so on, and accumulated depreciation appear on detail sheets or other peripheral records for each asset.

end of the fifth year, after five years of depreciation, including D_5, have been subtracted from the original purchase price.

There can be a great difference between the book value of an asset and the *market value* of that asset. There is no legal requirement for the two values to coincide, and no intent for book value to be a reasonable measure of market value.[28] Therefore, it is apparent that book value is merely an accounting convention with little practical use. Even when a depreciated asset is disposed of, the book value is used to determine the consequences of disposal, not the price the asset should bring at sale.

The calculation of book value is relatively easy, even for the case of the declining balance depreciation method.

For the straight-line depreciation method, the book value at the end of the *j*th year, after the *j*th depreciation deduction has been made, is

$$\text{BV}_j = C - \frac{j(C - S_n)}{n} = C - jD \qquad 73.36$$

For the sum-of-the-years' digits method, the book value is

$$\text{BV}_j = (C - S_n)\left(1 - \frac{j(2n+1-j)}{n(n+1)}\right) + S_n \qquad 73.37$$

For the declining balance method, including double declining balance, the book value is

$$\text{BV}_j = C(1-d)^j \qquad 73.38$$

For the sinking fund method, the book value is calculated directly as

$$\text{BV}_j = C - (C - S_n)(A/F, i\%, n)(F/P, i\%, j) \qquad 73.39$$

Of course, the book value at the end of year j can always be calculated for any method by successive subtractions (i.e., subtraction of the accumulated depreciation), as Eq. 73.40 illustrates.

$$\text{BV}_j = C - \sum_{m=1}^{j} D_m \qquad 73.40$$

Figure 73.8 illustrates the book value of a hypothetical asset depreciated using several depreciation methods. Notice that the double declining balance method initially produces the fastest write-off, while the sinking fund method produces the slowest write-off. Also, the book value does not automatically equal the salvage value at the end of an asset's depreciation period with the double declining balance method.[29]

Figure 73.8 *Book Value with Different Depreciation Methods*

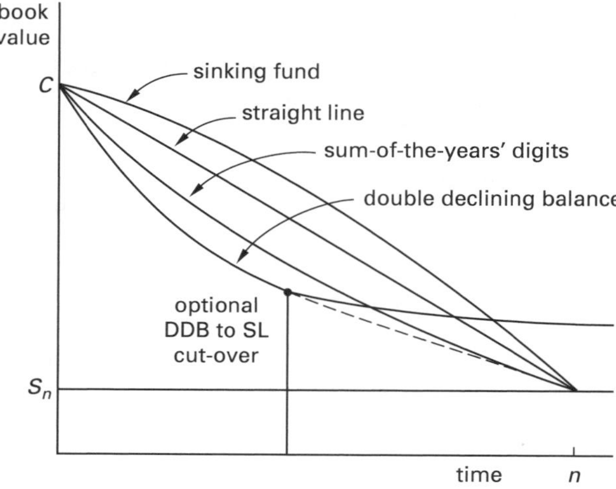

Example 73.18

For the asset described in Ex. 73.16, calculate the book value at the end of the first three years if sum-of-the-years' digits depreciation is used. The book value at the beginning of year 1 is \$9000.

Solution

From Eq. 73.40,

$$\text{BV}_1 = \$9000 - \$1600 = \$7400$$
$$\text{BV}_2 = \$7400 - \$1440 = \$5960$$
$$\text{BV}_3 = \$5960 - \$1280 = \$4680$$

38. AMORTIZATION

Amortization and depreciation are similar in that they both divide up the cost basis or value of an asset. In fact, in certain cases, the term "amortization" may be used in place of the term "depreciation." However, depreciation is a specific form of amortization.

Amortization spreads the cost basis or value of an asset over some base. The base can be time, units of production, number of customers, and so on. The asset can be tangible (e.g., a delivery truck or building) or intangible (e.g., goodwill or a patent).

[28]Common examples of assets with great divergences of book and market values are buildings (rental houses, apartment complexes, factories, etc.) and company luxury automobiles (Porsches, Mercedes, etc.) during periods of inflation. Book values decrease, but actual values increase.

[29]This means that the straight-line method of depreciation may result in a lower book value at some point in the depreciation period than if double declining balance is used. A *cut-over* from double declining balance to straight line may be permitted in certain cases. Finding the *cut-over point*, however, is usually done by comparing book values determined by both methods. The analytical method is complicated.

If the asset is tangible, if the base is time, and if the length of time is consistent with accounting standards and taxation guidelines, then the term "depreciation" is appropriate. However, if the asset is intangible, if the base is some other variable, or if some length of time other than the customary period is used, then the term "amortization" is more appropriate.[30]

Example 73.19

A company purchases complete and exclusive patent rights to an invention for \$1,200,000. It is estimated that once commercially produced, the invention will have a specific but limited market of 1200 units. For the purpose of allocating the patent right cost to production cost, what is the amortization rate in dollars per unit?

Solution

The patent should be amortized at the rate of

$$\frac{\$1,200,000}{1200 \text{ units}} = \$1000 \text{ per unit}$$

39. DEPLETION

Depletion is another artificial deductible operating expense, designed to compensate mining organizations for decreasing mineral reserves. Since original and remaining quantities of minerals are seldom known accurately, the *depletion allowance* is calculated as a fixed percentage of the organization's gross income. These percentages are usually in the 10–20% range and apply to such mineral deposits as oil, natural gas, coal, uranium, and most metal ores.

40. BASIC INCOME TAX CONSIDERATIONS

The issue of income taxes is often overlooked in academic engineering economic analysis exercises. Such a position is justifiable when an organization (e.g., a nonprofit school, a church, or the government) pays no income taxes. However, if an individual or organization is subject to income taxes, the income taxes must be included in an economic analysis of investment alternatives.

Assume that an organization pays a fraction, f, of its profits to the federal government as income taxes. If the organization also pays a fraction s of its profits as state income taxes and if state taxes paid are recognized by the federal government as tax-deductible expenses, then the composite tax rate is

$$t = s + f - sf \qquad 73.41$$

The basic principles used to incorporate taxation into engineering economic analyses are the following.

- Initial purchase expenditures are unaffected by income taxes.
- Salvage revenues are unaffected by income taxes.
- Deductible expenses, such as operating costs, maintenance costs, and interest payments, are reduced by the fraction t (i.e., multiplied by the quantity $(1 - t)$).
- Revenues are reduced by the fraction t (i.e., multiplied by the quantity $(1 - t)$).
- Since tax regulations allow the depreciation in any year to be handled as if it were an actual operating expense, and since operating expenses are deductible from the income base prior to taxation, the after-tax profits will be increased. If D is the depreciation, the net result to the after-tax cash flow will be the addition of tD. Depreciation is multiplied by t and added to the appropriate year's cash flow, increasing that year's present worth.

For simplicity, most engineering economics practice problems involving income taxes specify a single income tax rate. In practice, however, federal and most state tax rates depend on the income level. Each range of incomes and its associated tax rate are known as *income bracket* and *tax bracket*, respectively. For example, the state income tax rate might be 4% for incomes up to and including \$30,000, and 5% for incomes above \$30,000. The income tax for a taxpaying entity with an income of \$50,000 would have to be calculated in two parts.

$$\begin{aligned} \text{tax} &= (0.04)(\$30,000) + (0.05)(\$50,000 - \$30,000) \\ &= \$2200 \end{aligned}$$

Income taxes and depreciation have no bearing on municipal or governmental projects since municipalities, states, and the U.S. government pay no taxes.

Example 73.20

A corporation that pays 53% of its profit in income taxes invests \$10,000 in an asset that will produce a \$3000 annual revenue for eight years. If the annual expenses are \$700, salvage after eight years is \$500, and 9% interest is used, what is the after-tax present worth? Disregard depreciation.

Economics

[30]From time to time, the U.S. Congress has allowed certain types of facilities (e.g., emergency, grain storage, and pollution control) to be written off more rapidly than would otherwise be permitted in order to encourage investment in such facilities. The term "amortization" has been used with such write-off periods.

Solution

Find the after-tax present worth. [**Economic Factor Conversions**] [**Economic Factor Tables**]

$$\begin{aligned} P &= -\$10{,}000 + (\$3000)(P/A, 9\%, 8)(1 - 0.53) \\ &\quad - (\$700)(P/A, 9\%, 8)(1 - 0.53) \\ &\quad + (\$500)(P/F, 9\%, 8) \\ &= -\$10{,}000 + (\$3000)(5.5348)(0.47) \\ &\quad - (\$700)(5.5348)(0.47) + (\$500)(0.5019) \\ &= -\$3766 \end{aligned}$$

41. TAXATION AT THE TIMES OF ASSET PURCHASE AND SALE

There are numerous rules and conventions that governmental tax codes and the accounting profession impose on organizations. Engineers are not normally expected to be aware of most of the rules and conventions, but occasionally it may be necessary to incorporate their effects into an engineering economic analysis.

Tax Credit

A *tax credit* (also known as an *investment tax credit* or *investment credit*) is a one-time credit against income taxes.[31] Therefore, it is added to the after-tax present worth as a last step in an engineering economic analysis. Such tax credits may be allowed by the government from time to time for equipment purchases, employment of various classes of workers, rehabilitation of historic landmarks, and so on.

A tax credit is usually calculated as a fraction of the initial purchase price or cost of an asset or activity.

$$\text{TC} = \text{fraction} \times \text{initial cost} \qquad 73.42$$

When the tax credit is applicable, the fraction used is subject to legislation. A professional tax expert or accountant should be consulted prior to applying the tax credit concept to engineering economic analysis problems.

Since the investment tax credit reduces the buyer's tax liability, a tax credit should be included only in after-tax engineering economic analyses. The credit is assumed to be received at the end of the year.

Gain or Loss on the Sale of a Depreciated Asset

If an asset that has been depreciated over a number of prior years is sold for more than its current book value, the difference between the book value and selling price is taxable income in the year of the sale. Alternatively, if the asset is sold for less than its current book value, the difference between the selling price and book value is an expense in the year of the sale.

Example 73.21

One year, a company makes a \$5000 investment in a historic building. The investment is not depreciable, but it does qualify for a one-time 20% tax credit. In that same year, revenue is \$45,000 and expenses (exclusive of the \$5000 investment) are \$25,000. The company pays a total of 53% in income taxes. What is the after-tax present worth of this year's activities if the company's interest rate for investment is 10%?

Solution

The tax credit is

$$\text{TC} = (0.20)(\$5000) = \$1000$$

This tax credit is assumed to be received at the end of the year. Find the after-tax present worth. [**Economic Factor Conversions**] [**Economic Factor Tables**]

$$\begin{aligned} P &= -\$5000 + (\$45{,}000 - \$25{,}000)(1 - 0.53) \\ &\quad \times (P/F, 10\%, 1) + (\$1000)(P/F, 10\%, 1) \\ &= -\$5000 + (\$20{,}000)(0.47)(0.9091) \\ &\quad + (\$1000)(0.9091) \\ &= \$4455 \end{aligned}$$

42. DEPRECIATION RECOVERY

The economic effect of depreciation is to reduce the income tax in year j by tD_j. The present worth of the asset is also affected: The present worth is increased by $tD_j(P/F, i\%, j)$. The after-tax present worth of all depreciation effects over the depreciation period of the asset is called the *depreciation recovery* (DR).[32]

$$\text{DR} = t\sum_{j=1}^{n} D_j(P/F, i\%, j) \qquad 73.43$$

There are multiple ways depreciation can be calculated, as summarized in Table 73.5. *Straight-line* (SL) *depreciation recovery* from an asset is easily calculated, since the depreciation is the same each year. Assuming the asset has a constant depreciation of D and depreciation period of n years, the depreciation recovery is

$$\text{DR} = tD(P/A, i\%, n) \qquad 73.44$$

[31]Strictly, *tax credit* is the more general term, and applies to a credit for doing anything creditable. An *investment tax credit* requires an investment in something (usually real property or equipment).

[32]Since the depreciation benefit is reduced by taxation, depreciation cannot be thought of as an annuity to fund a replacement asset.

Table 73.5 Depreciation Calculation Summary

method	depreciation basis	depreciation in year j, D_j	book value after jth depreciation, BV_j	present worth of after-tax depreciation recovery (DR)	supplementary formulas
straight-line (SL)	$C-S_n$	$\frac{C-S_n}{n}$ [constant]	$C-jD$	$tD(P/A, i\%, n)$	
constant percentage	$C-S_n$	fraction × $(C-S_n)$ [constant]	$C-jD$	$tD(P/A, i\%, n)$	
sum-of-the-years' digits (SOYD)	$C-S_n$	$\frac{2(C-S_n)(n-j+1)}{n(n+1)}$	$(C-S_n)\times\left(1-\frac{j(2n+1-j)}{n(n+1)}\right)+S_n$	$\frac{t(C-S_n)}{T}\times\big(n(P/A, i\%, n)-(P/G, i\%, n)\big)$	$T=\frac{1}{2}n(n+1)$
double declining balance (DDB)	C	$dC(1-d)^{j-1}$	$C(1-d)^j$	$tC\left(\frac{d}{1-d}\right)\times(P/EG, z-1, n)$	$d=\frac{2}{n}$; $z=\frac{1+i}{1-d}$; $(P/EG, z-1, n)=\frac{z^n-1}{z^n(z-1)}$
sinking fund (SF)	$C-S_n$	$(C-S_n)\times(A/F, i\%, n)\times(F/P, i\%, j-1)$	$C-(C-S_n)\times(A/F, i\%, n)\times(F/A, i\%, j)$	$\frac{t(C-S_n)(A/F, i\%, n)}{1+i}$	
accelerated cost recovery system (ACRS/MACRS)	C	(factor) C	$C-\sum_{m=1}^{j}D_m$	$t\sum_{j=1}^{n}D_j(P/F, i\%, j)$	
units of production or service output	$C-S_n$	$(C-S_n)\times\left(\frac{\text{actual output in year } j}{\text{lifetime output}}\right)$	$C-\sum_{m=1}^{j}D_m$	$t\sum_{j=1}^{n}D_j(P/F, i\%, j)$	

Depreciation: Straight Line

$$D=\frac{C-S_n}{n} \quad \text{73.45}$$

Sum-of-the-years' digits (SOYD) *depreciation recovery* is also relatively easily calculated, since the depreciation decreases uniformly each year.

$$\text{DR}=\left(\frac{t(C-S_n)}{T}\right)\times\big(n(P/A, i\%, n)-(P/G, i\%, n)\big) \quad \text{73.46}$$

Finding *declining balance depreciation recovery* is more involved. There are three difficulties. The first (the apparent need to calculate all previous depreciations in order to determine the subsequent depreciation) has already been addressed by Eq. 73.32.

The second difficulty is that there is no way to ensure (that is, to force) the book value to be S_n at $t=n$. Therefore, it is common to write off the remaining book value (down to S_n) at $t=n$ in one lump sum. This assumes $\text{BV}_n \geq S_n$.

The third difficulty is that of finding the present worth of an *exponentially decreasing cash flow.* Although the proof is omitted here, such exponential cash flows can be handled with the *exponential gradient factor*, (P/EG).[33]

$$(P/EG, z-1, n)=\frac{z^n-1}{z^n(z-1)} \quad \text{73.47}$$

$$z=\frac{1+i}{1-d} \quad \text{73.48}$$

[33]The (P/A) columns in economic factor tables can be used for (P/EG) as long as the interest rate is assumed to be $z-1$.

Then, as long as $BV_n > S_n$, the declining balance depreciation recovery is

$$DR = tC\left(\frac{d}{1-d}\right)(P/EG, z-1, n) \qquad 73.49$$

Example 73.22

For the asset described in Ex. 73.16, calculate the after-tax depreciation recovery with straight-line and sum-of-the-years' digits depreciation methods. Use 6% interest with 48% income taxes.

Solution

Find the depreciation recovery using the straight-line method. [**Economic Factor Tables**]

$$\begin{aligned} DR &= (0.48)(\$880)(P/A, 6\%, 10) \\ &= (0.48)(\$880)(7.3601) \\ &= \$3109 \end{aligned}$$

Using SOYD, the depreciation series can be thought of as a constant $1600 term with a negative $160 gradient. [**Economic Factor Tables**]

$$\begin{aligned} DR &= (0.48)(\$1600)(P/A, 6\%, 10) \\ &\quad - (0.48)(\$160)(P/G, 6\%, 10) \\ &= (0.48)(\$1600)(7.3601) \\ &\quad - (0.48)(\$160)(29.6023) \\ &= \$3379 \end{aligned}$$

The ten-year (P/G) factor is used even though there are only nine years in which the gradient reduces the initial $1600 amount.

Example 73.23

What is the after-tax present worth of the asset described in Ex. 73.20 if straight-line, sum-of-the-years' digits, and double declining balance depreciation methods are used?

Solution

Using the straight-line method, find the depreciation recovery. [**Economic Factor Tables**]

$$\begin{aligned} DR &= (0.53)\left(\frac{\$10{,}000 - \$500}{8}\right)(P/A, 9\%, 8) \\ &= (0.53)\left(\frac{\$9500}{8}\right)(5.5348) \\ &= \$3483 \end{aligned}$$

$$T = \left(\frac{1}{2}\right)(8)(9) = 36$$

$$\text{depreciation base} = \$10{,}000 - \$500 = \$9500$$

$$D_1 = \left(\frac{8}{36}\right)(\$9500) = \$2111$$

$$G = \left(\frac{1}{36}\right)(\$9500) = \$264$$

$$\begin{aligned} DR &= (0.53)\big((\$2111)(P/A, 9\%, 8) \\ &\quad - (\$264)(P/G, 9\%, 8)\big) \\ &= (0.53)\big((\$2111)(5.5348) \\ &\quad - (\$264)(16.8877)\big) \\ &= \$3830 \end{aligned}$$

Using sum-of-the-years' digits (SOYD), find the depreciation.

Depreciation: Sum-of-Years Digits Method

$$\begin{aligned} D_1 &= \frac{2(C - S_n)(n - j + 1)}{n(n+1)} \\ &= \frac{(2)(\$10{,}000 - \$500)(8 - 1 + 1)}{(8)(8+1)} \\ &= \$2111 \end{aligned}$$

Determine the uniform gradient cash flow using T and the depreciation base.

$$G = \left(\frac{1}{36}\right)(\$9500) = \$264$$

Find the depreciation recovery. [**Economic Factor Tables**]

$$\begin{aligned} DR &= (0.53)((\$2111)(P/A, 9\%, 8) \\ &\quad - (\$264)(16.8877)) \\ &= (0.53)((\$2111)(5.5348) \\ &\quad - (\$264)(16.8877)) \\ &= \$3830 \end{aligned}$$

Using DDB, the depreciation recovery is calculated as follows.[34]

$$d = \frac{2}{8} = 0.25$$

$$z = \frac{1 + 0.09}{1 - 0.25} = 1.4533$$

$$(P/EG, z-1, n) = \frac{(1.4533)^8 - 1}{(1.4533)^8(0.4533)} = 2.095$$

[34] This method should start by checking that the book value at the end of the depreciation period is greater than the salvage value. In this example, such is the case. However, the step is not shown.

Economics

From Eq. 73.49,

$$\begin{aligned} \text{DR} &= (0.53)\left(\frac{(0.25)(\$10{,}000)}{0.75}\right)(2.095) \\ &= \$3701 \end{aligned}$$

The after-tax present worth, neglecting depreciation, was previously found to be −$3766.

The after-tax present worths, including depreciation recovery, are

$$\begin{aligned} \text{SL: } P &= -\$3766 + \$3483 = -\$283 \\ \text{SOYD: } P &= -\$3766 + \$3830 = \$64 \\ \text{DDB: } P &= -\$3766 + \$3701 = -\ \$65 \end{aligned}$$

43. OTHER INTEREST RATES

The *effective interest rate per period*, i_e (also called *yield* by banks), is the only interest rate that should be used in equivalence equations. The interest rates at the top of economic factor tables are implicitly all effective interest rates. Usually, the period will be one year, hence the name *effective annual interest rate*. However, there are other interest rates in use as well.

The term *nominal interest rate*, r (*rate per annum*), is encountered when compounding is more than once per year. The nominal rate does not include the effect of compounding and is not the same as the effective rate. And, since the effective interest rate can be calculated from the nominal rate only if the number of compounding periods per year is known, nominal rates cannot be compared unless the method of compounding is specified. The only practical use for a nominal rate per year is for calculating the effective rate per period.

44. RATE AND PERIOD CHANGES

If there are m compounding periods during the year (two for semiannual compounding, four for quarterly compounding, twelve for monthly compounding, etc.) and the nominal rate is r, the *effective rate per compounding period* is

$$\phi = \frac{r}{m} \qquad 73.50$$

The effective annual rate, i_e, can be calculated from the effective rate per period, ϕ, by using Eq. 73.51. **[Economic Analysis: Nomenclature and Definitions]**

$$\begin{aligned} i_e &= (1+\phi)^m - 1 \\ &= \left(1+\frac{r}{m}\right)^m - 1 \end{aligned} \qquad 73.51$$

Sometimes, only the effective rate per period (e.g., per month) is known. However, that will be a simple problem since compounding for n periods at an effective rate per period is not affected by the definition or length of the period.

The following rules may be used to determine which interest rate is given.

- Unless specifically qualified, the interest rate given is an annual rate.
- If the compounding is annual, the rate given is the effective rate. If compounding is other than annual, the rate given is the nominal rate.

The effective annual interest rate determined on a *daily compounding basis* will not be significantly different than if *continuous compounding* is assumed.[35] In the case of continuous (or daily) compounding, the discounting factors can be calculated directly from the nominal interest rate and number of years, without having to find the effective interest rate per period. Table 73.6 can be used to determine the discount factors for continuous compounding.

Table 73.6 *Discount Factors for Continuous Compounding (n is the number of years)*

symbol	formula
$(F/P, r\%, n)$	e^{rn}
$(P/F, r\%, n)$	e^{-rn}
$(A/F, r\%, n)$	$\dfrac{e^r - 1}{e^{rn} - 1}$
$(F/A, r\%, n)$	$\dfrac{e^{rn} - 1}{e^r - 1}$
$(A/P, r\%, n)$	$\dfrac{e^r - 1}{1 - e^{-rn}}$
$(P/A, r\%, n)$	$\dfrac{1 - e^{-rn}}{e^r - 1}$

Example 73.24

A savings and loan offers a nominal rate of 5.25% compounded daily over 365 days in a year. What is the effective annual rate?

Solution

method 1: Use Eq. 73.51.

$$r = 0.0525, \quad m = 365$$

$$i_e = \left(1+\frac{r}{m}\right)^m - 1 = \left(1+\frac{0.0525}{365}\right)^{365} - 1 = 0.0539$$

[35]The number of *banking days in a year* (250, 360, etc.) must be specifically known.

method 2: Assume daily compounding is the same as continuous compounding.

$$\begin{aligned} i_e &= (F/P,\ r\%,\ 1) - 1 \\ &= e^{0.0525} - 1 \\ &= 0.0539 \end{aligned}$$

Example 73.25

A real estate investment trust pays \$7,000,000 for an apartment complex with 100 units. The trust expects to sell the complex in 10 years for \$15,000,000. In the meantime, it expects to receive an average rent of \$900 per month from each apartment. Operating expenses are expected to be \$200 per month per occupied apartment. A 95% occupancy rate is predicted. In similar investments, the trust has realized a 15% effective annual return on its investment. Compare to those past investments the expected present worth of this investment when calculated assuming (a) annual compounding (i.e., the year-end convention), and (b) monthly compounding. Disregard taxes, depreciation, and all other factors.

Solution

(a) The net annual income will be

$$\begin{aligned} &(0.95)(100 \text{ units})\left(\frac{\$900}{\text{unit-mo}} - \frac{\$200}{\text{unit-mo}}\right)\left(12\ \frac{\text{mo}}{\text{yr}}\right) \\ &\quad = \$798{,}000/\text{yr} \end{aligned}$$

Find the present worth of ten years of operation. [Economic Factor Conversions]

$$\begin{aligned} P &= -\$7{,}000{,}000 + (\$798{,}000)(P/A, 15\%, 10) \\ &\quad + (\$15{,}000{,}000)(P/F, 15\%, 10) \\ &= -\$7{,}000{,}000 + (\$798{,}000)(5.0188) \\ &\quad + (\$15{,}000{,}000)(0.2472) \\ &= \$713{,}000 \end{aligned}$$

(b) The net monthly income is

$$\begin{aligned} &(0.95)(100 \text{ units})\left(\frac{\$900}{\text{unit-mo}} - \frac{\$200}{\text{unit-mo}}\right) \\ &\quad = \$66{,}500/\text{mo} \end{aligned}$$

Equation 73.51 is used to calculate the effective monthly rate, ϕ, from the effective annual rate, $i_e = 15\%$, and the number of compounding periods per year, $m = 12$.

$$\begin{aligned} \phi &= (1 + i_e)^{1/m} - 1 \\ &= (1 + 0.15)^{1/12} - 1 \\ &= 0.011715 \quad (1.1715\%) \end{aligned}$$

The number of compounding periods in ten years is

$$n = (10 \text{ yr})\left(12\ \frac{\text{mo}}{\text{yr}}\right) = 120 \text{ mo}$$

Find the present worth of 120 months of operation. [Economic Factor Conversions]

$$\begin{aligned} P &= -\$7{,}000{,}000 + (\$66{,}500)(P/A, 1.1715\%, 120) \\ &\quad + (\$15{,}000{,}000)(P/F, 1.1715\%, 120) \end{aligned}$$

Since table values for 1.1715% discounting factors are not available, the factors are calculated from Table 73.1. [Economic Factor Conversions]

$$\begin{aligned} (P/A,\ 1.1715\%,\ 120) &= \frac{(1 + i_e)^n - 1}{i(1 + i_e)^n} \\ &= \frac{(1 + 0.011715)^{120} - 1}{(0.011715)(1 + 0.011715)^{120}} \\ &= 64.261 \end{aligned}$$

$$\begin{aligned} (P/F,\ 1.1715\%,\ 120) &= (1 + i)^{-n} = (1 + 0.011715)^{-120} \\ &= 0.2472 \end{aligned}$$

Find the present worth over 120 monthly compounding periods. [Economic Factor Conversions]

$$\begin{aligned} P &= -\$7{,}000{,}000 + (\$66{,}500)(64.261) \\ &\quad + (\$15{,}000{,}000)(0.2472) \\ &= \$981{,}357 \end{aligned}$$

45. BONDS

A *bond* is a method of long-term financing commonly used by governments, states, municipalities, and very large corporations.[36] The bond represents a contract to pay the bondholder specific amounts of money at specific times. The holder purchases the bond in exchange for specific payments of interest and principal. Typical municipal bonds call for quarterly or semiannual interest payments and a payment of the *face value of the bond* on the *date of maturity* (end of the bond period).[37]

[36] In the past, 30-year bonds were typical. Shorter term 10-year, 15-year, 20-year, and 25-year bonds are also commonly issued.
[37] A *fully amortized bond* pays back interest and principal throughout the life of the bond. There is no balloon payment.

Due to the practice of discounting in the bond market, a bond's face value and its purchase price generally will not coincide.

In the past, a bondholder had to submit a coupon or ticket in order to receive an interim interest payment. This has given rise to the term *coupon rate*, which is the nominal annual interest rate on which the interest payments are made. Coupon books are seldom used with modern bonds, but the term survives. The coupon rate determines the magnitude of the semiannual (or otherwise) interest payments during the life of the bond. The bondholder's own effective interest rate should be used for economic decisions about the bond.

Actual *bond yield* is the bondholder's actual rate of return of the bond, considering the purchase price, interest payments, and face value payment (or, value realized if the bond is sold before it matures). By convention, bond yield is calculated as a nominal rate (rate per annum), not an effective rate per year. The bond yield should be determined by finding the effective rate of return per payment period (e.g., per semiannual interest payment) as a conventional rate of return problem. Then, the nominal rate can be found by multiplying the effective rate per period by the number of payments per year, as in Eq. 73.51.

46. PROBABILISTIC PROBLEMS

If an alternative's cash flows are specified by an implicit or explicit probability distribution rather than being known exactly, the problem is *probabilistic*.

Probabilistic problems typically possess the following characteristics.

- There is a chance of loss that must be minimized (or, rarely, a chance of gain that must be maximized) by selection of one of the alternatives.
- There are multiple alternatives. Each alternative offers a different degree of protection from the loss. Usually, the alternatives with the greatest protection will be the most expensive.
- The magnitude of loss or gain is independent of the alternative selected.

Probabilistic problems are typically solved using annual costs and expected values. An *expected value* is similar to an *average value* since it is calculated as the mean of the given probability distribution. If cost 1 has a probability of occurrence, p_1, cost 2 has a probability of occurrence, p_2, and so on, the expected value is

$$\mathcal{E}\{\text{cost}\} = p_1(\text{cost 1}) + p_2(\text{cost 2}) + \cdots \quad 73.52$$

Example 73.26

Flood damage in any year is given according to the following table. What is the present worth of flood damage for a ten-year period? Use 6% as the effective annual interest rate.

damage	probability
0	0.75
\$10,000	0.20
\$20,000	0.04
\$30,000	0.01

Solution

The expected value of flood damage in any given year is

$$\begin{aligned}\mathcal{E}\{\text{damage}\} &= (0)(0.75) + (\$10{,}000)(0.20) \\ &\quad + (\$20{,}000)(0.04) + (\$30{,}000)(0.01) \\ &= \$3100\end{aligned}$$

Find the present worth of ten years of expected flood damage. **[Economic Factor Conversions] [Economic Factor Tables]**

$$\begin{aligned}\text{present worth} &= (\$3100)(P/A, 6\%, 10) \\ &= (\$3100)(7.3601) \\ &= \$22{,}816\end{aligned}$$

Example 73.27

A dam is being considered on a river that periodically overflows and causes \$600,000 damage. The damage is essentially the same each time the river causes flooding. The project horizon is 40 years. A 10% interest rate is being used.

Three different designs are available, each with different costs and storage capacities.

design alternative	cost	maximum capacity
A	\$500,000	1 unit
B	\$625,000	1.5 units
C	\$900,000	2.0 units

The National Weather Service has provided a statistical analysis of annual rainfall runoff from the watershed draining into the river.

units annual rainfall	probability
0	0.10
0.1–0.5	0.60
0.6–1.0	0.15
1.1–1.5	0.10
1.6–2.0	0.04
2.1 or more	0.01

Economics

Which design alternative would you choose assuming the dam is essentially empty at the start of each rainfall season?

Solution

The sum of the construction cost and the expected damage should be minimized. If alternative A is chosen, it will have a capacity of 1 unit. Its capacity will be exceeded (causing \$600,000 damage) when the annual rainfall exceeds 1 unit. Present worth factors are determined from factor tables. [**Economic Factor Tables**]

$$\begin{aligned}\mathcal{E}\{\text{EUAC(A)}\} &= (\$500{,}000)(A/P, 10\%, 40) \\ &\quad + (\$600{,}000)(0.10 + 0.04 + 0.01) \\ &= (\$500{,}000)(0.1023) + (\$600{,}000)(0.15) \\ &= \$141{,}150\end{aligned}$$

Similarly,

$$\begin{aligned}\mathcal{E}\{\text{EUAC(B)}\} &= (\$625{,}000)(A/P, 10\%, 40) \\ &\quad + (\$600{,}000)(0.04 + 0.01) \\ &= (\$625{,}000)(0.1023) + (\$600{,}000)(0.05) \\ &= \$93{,}938\end{aligned}$$

$$\begin{aligned}\mathcal{E}\{\text{EUAC(C)}\} &= (\$900{,}000)(A/P, 10\%, 40) \\ &\quad + (\$600{,}000)(0.01) \\ &= (\$900{,}000)(0.1023) + (\$600{,}000)(0.01) \\ &= \$98{,}070\end{aligned}$$

Alternative B should be chosen.

47. WEIGHTED COSTS

The reliability of preliminary cost estimates can be increased by considering as much historical data as possible. For example, the cost of finishing a concrete slab when the contractor has not yet been selected should be estimated from actual recent costs from as many local jobs as is practical. Most jobs are not directly comparable, however, because they differ in size or in some other characteristic. A *weighted cost* (*weighted average cost*) is a cost that has been averaged over some rational basis. The weighted cost is calculated by weighting the individual cost elements, C_i, by their respective weights, w_i. Respective weights are usually *relative fractions* (*relative importance*) based on other characteristics, such as length, area, number of units, frequency of occurrence, points, etc.

$$C_{\text{weighted}} = w_1C_1 + w_2C_2 + \cdots + w_nC_n \qquad 73.53$$

$$w_j = \frac{A_j}{\sum_{i=1}^{n} A_i} \quad \text{[weighting by area]} \qquad 73.54$$

It is implicit in calculating weighted average costs that the costs with the largest weights are more important to the calculation. For example, costs from a job of 10,000 ft^2 are considered to be twice as reliable (important, relevant, etc.) as costs from a 5000 ft^2 job. This assumption must be carefully considered.

Determining a weighted average from Eq. 73.53 disregards the fixed and variable natures of costs. In effect, fixed costs are allocated over entire jobs, increasing the apparent variable costs. For that reason, it is important to include in the calculation only costs of similar cases. This requires common sense segregation of the initial cost data. For example, it is probably not appropriate to include very large (e.g., supermarket and mall) concrete finishing job costs in the calculation of a weighted average cost used to estimate residential slab finishing costs.

Example 73.28

Calculate a weighted average cost per unit length of concrete wall to be used for estimating future job costs. Data from three similar walls are available.

wall 1: length, 50 ft; actual cost, \$9000
wall 2: length, 90 ft; actual cost, \$15,000
wall 3: length, 65 ft; actual cost, \$11,000

Solution

Consider wall 1. The cost per foot of wall is \$9000/50 ft. According to Eq. 73.54, accuracy of cost data is proportional to wall length (i.e., cost per foot from a wall of length $2X$ is twice as reliable as cost per foot from a wall of length X). This cost per foot value is given a weight of 50 ft/(50 ft + 90 ft + 65 ft). So, the weighted cost per foot for wall 1 is

$$w_1C_{\text{ft},1} = \frac{(50\ \text{ft})\left(\frac{\$9000}{50\ \text{ft}}\right)}{50\ \text{ft} + 90\ \text{ft} + 65\ \text{ft}} = \frac{\$9000}{50\ \text{ft} + 90\ \text{ft} + 65\ \text{ft}}$$

Similarly, the weighted cost per foot for wall 2 is $w_2C_{\text{ft},2} = \$15{,}000/(50\ \text{ft} + 90\ \text{ft} + 65\ \text{ft})$, and the weighted cost per foot for wall 3 is $w_3C_{\text{ft},3} = \$11{,}000/(50\ \text{ft} + 90\ \text{ft} + 65\ \text{ft})$.

From Eq. 73.53, the total weighted cost per foot is

$$\begin{aligned}C_{\text{ft}} &= w_1C_{\text{ft},1} + w_2C_{\text{ft},2} + w_3C_{\text{ft},3} \\ &= \frac{\$9000 + \$15{,}000 + \$11{,}000}{50\ \text{ft} + 90\ \text{ft} + 65\ \text{ft}} \\ &= \$170.73\ \text{per foot}\end{aligned}$$

48. FIXED AND VARIABLE COSTS

The distinction between fixed and variable costs depends on how these costs vary when an independent variable changes. For example, factory or machine production is frequently the independent variable. However, it could just as easily be vehicle miles driven, hours of operation, or quantity (mass, volume, etc.). Examples of fixed and variable costs are given in Table 73.7.

Table 73.7 *Summary of Fixed and Variable Costs*

fixed costs
- rent
- property taxes
- interest on loans
- insurance
- janitorial service expense
- tooling expense
- setup, cleanup, and tear-down expenses
- depreciation expense
- marketing and selling costs
- cost of utilities
- general burden and overhead expense

variable costs
- direct material costs
- direct labor costs
- cost of miscellaneous supplies
- payroll benefit costs
- income taxes
- supervision costs

If a cost is a function of the independent variable, the cost is said to be a *variable cost*. The change in cost per unit variable change (i.e., what is usually called the *slope*) is known as the *incremental cost*. Material and labor costs are examples of variable costs. They increase in proportion to the number of product units manufactured.

If a cost is not a function of the independent variable, the cost is said to be a *fixed cost*. Rent and lease payments are typical fixed costs. These costs will be incurred regardless of production levels.

Some costs have both fixed and variable components, as Fig. 73.9 illustrates. The fixed portion can be determined by calculating the cost at zero production.

An additional category of cost is the *semivariable cost*. This type of cost increases stepwise. Semivariable cost structures are typical of situations where *excess capacity* exists. For example, supervisory cost is a stepwise function of the number of production shifts. Also, labor cost for truck drivers is a stepwise function of weight (volume) transported. As long as a truck has room left (i.e., excess capacity), no additional driver is needed. As soon as the truck is filled, labor cost will increase.

Figure 73.9 *Fixed and Variable Costs*

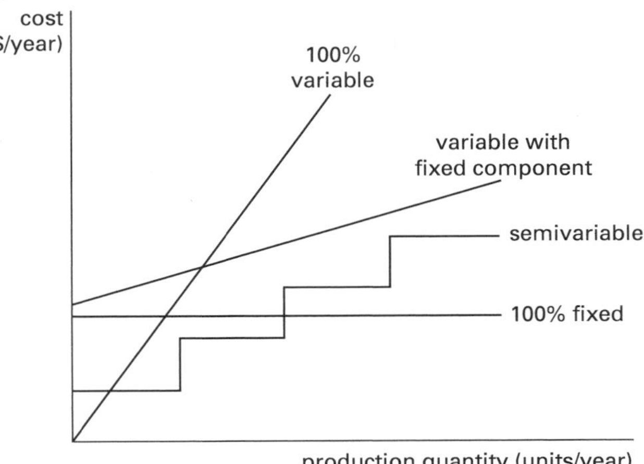

49. ACCOUNTING COSTS AND EXPENSE TERMS

The accounting profession has developed special terms for certain groups of costs. When annual costs are incurred due to the functioning of a piece of equipment, they are known as *operating and maintenance* (O&M) *costs*. The annual costs associated with operating a business (other than the costs directly attributable to production) are known as *general, selling, and administrative* (GS&A) *expenses*.

Direct labor costs are costs incurred in the factory, such as assembly, machining, and painting labor costs. *Direct material costs* are the costs of all materials that go into production.[38] Typically, both direct labor and direct material costs are given on a per-unit or per-item basis. The sum of the direct labor and direct material costs is known as the *prime cost*.

There are certain additional expenses incurred in the factory, such as the costs of factory supervision, stock-picking, quality control, factory utilities, and miscellaneous supplies (cleaning fluids, assembly lubricants, routing tags, etc.) that are not incorporated into the final product. Such costs are known as *indirect manufacturing expenses* (IME) or *indirect material and labor costs*.[39] The sum of the per-unit indirect manufacturing expense and prime cost is known as the *factory cost*.

[38]There may be problems with pricing the material when it is purchased from an outside vendor and the stock on hand derives from several shipments purchased at different prices.
[39]The *indirect material and labor costs* usually exclude costs incurred in the office area.

Economics

Research and development (R&D) *costs* and *administrative expenses* are added to the factory cost to give the *manufacturing cost* of the product.

Additional costs are incurred in marketing the product. Such costs are known as *selling expenses* or *marketing expenses*. The sum of the selling expenses and manufacturing cost is the *total cost* of the product. Figure 73.10 illustrates these terms.[40]

Figure 73.10 *Costs and Expenses Combined*

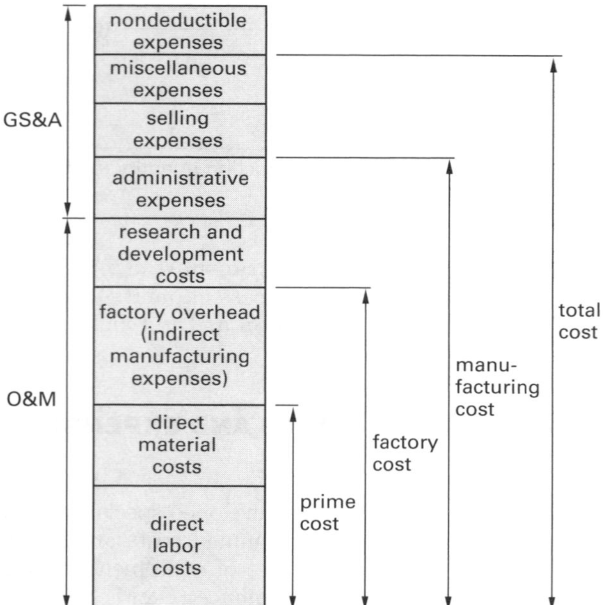

The distinctions among the various forms of cost (particularly with overhead costs) are not standardized. Each company must develop a classification system to deal with the various cost factors in a consistent manner. There are also other terms in use (e.g., *raw materials*, *operating supplies*, *general plant overhead*), but these terms must be interpreted within the framework of each company's classification system. Table 73.8 is typical of such classification systems.

Table 73.8 *Typical Classification of Expenses*

direct labor expenses
- machining and forming
- assembly
- finishing
- inspection
- testing

direct material expenses
- items purchased from other vendors
- manufactured assemblies

factory overhead expenses (indirect manufacturing expenses)
- supervision
- benefits (e.g., pension, medical insurance, vacations)
- wages overhead (e.g., unemployment comp., SS, & disability taxes)
- stock-picking
- quality control and inspection
- expediting
- rework
- maintenance
- miscellaneous supplies (e.g., routing tags, assembly lubricants, cleaning fluids, wiping cloths, janitorial supplies)
- packaging (materials and labor)
- factory utilities
- laboratory
- depreciation on factory equipment

research and development expenses
- engineering (labor)
- patents
- testing
- prototypes (material and labor)
- drafting
- O&M of R&D facility

administrative expenses
- corporate officers
- accounting
- secretarial/clerical/reception
- security (protection)
- medical (nurse)
- employment (personnel)
- reproduction
- data processing
- production control
- depreciation on nonfactory equipment
- office supplies
- office utilities
- O&M of offices

selling expenses
- marketing (labor)
- advertising
- transportation (if not paid by customer)
- outside sales force (labor and expenses)
- demonstration units
- commissions
- technical service and support
- order processing
- branch office expenses

miscellaneous expenses
- insurance
- property taxes
- interest on loans

nondeductible expenses
- federal income taxes
- fines and penalties

[40] *Total cost* does not include income taxes.

50. ACCOUNTING PRINCIPLES

Basic Bookkeeping

An accounting or *bookkeeping system* is used to record historical financial transactions. The resultant records are used for product costing, satisfaction of statutory requirements, reporting of profit for income tax purposes, and general company management.

Bookkeeping consists of two main steps: recording the transactions, followed by categorization of the transactions.[41] The transactions (receipts and disbursements) are recorded in a *journal* (*book of original entry*) to complete the first step. Such a journal is organized in a simple chronological and sequential manner. The transactions are then categorized (into interest income, advertising expense, etc.) and posted (i.e., entered or written) into the appropriate *ledger account*.[42]

The ledger accounts together constitute the *general ledger* or *ledger*. All ledger accounts can be classified into one of three types: *asset accounts*, *liability accounts*, and *owners' equity accounts*. Strictly speaking, income and expense accounts, kept in a separate journal, are included within the classification of owners' equity accounts.

Together, the journal and ledger are known simply as "the books" of the company, regardless of whether bound volumes of pages are actually involved.

Balancing the Books

In a business environment, *balancing the books* means more than reconciling the checkbook and bank statements. All accounting entries must be posted in such a way as to maintain the equality of the *basic accounting equation*,

$$\text{assets} = \text{liability} + \text{owner's equity} \qquad 73.55$$

In a *double-entry bookkeeping system*, the equality is maintained within the ledger system by entering each transaction into two balancing ledger accounts. For example, paying a utility bill would decrease the cash account (an asset account) and decrease the utility expense account (a liability account) by the same amount.

Transactions are either *debits* or *credits*, depending on their sign. Increases in asset accounts are debits; decreases are credits. For liability and equity accounts, the opposite is true: Increases are credits, and decreases are debits.[43]

Cash and Accrual Systems

The simplest form of bookkeeping is based on the *cash system*.[44] The only transactions that are entered into the journal are those that represent cash receipts and disbursements. In effect, a checkbook register or bank deposit book could serve as the journal.

During a given period (e.g., month or quarter), expense liabilities may be incurred even though the payments for those expenses have not been made. For example, an invoice (bill) may have been received but not paid. Under the *accrual system*, the obligation is posted into the appropriate expense account before it is paid.[45] Analogous to expenses, under the accrual system, income will be claimed before payment is received. Specifically, a sales transaction can be recorded as income when the customer's order is received, when the outgoing invoice is generated, or when the merchandise is shipped.

Financial Statements

Each period, two types of corporate financial statements are typically generated: the *balance sheet* and *profit and loss* (P&L) *statements worth*.[46] The profit and loss statement, also known as a *statement of income and retained earnings*, is a summary of sources of *income* or *revenue* (interest, sales, fees charged, etc.) and *expenses* (utilities, advertising, repairs, etc.) for the period. The expenses are subtracted from the revenues to give a *net income* (generally, before taxes).[47] Figure 73.11 illustrates a simple profit and loss statement.

Economics

[41]These two steps are not to be confused with the *double-entry bookkeeping method*.

[42]The two-step process is more typical of a *manual bookkeeping system* than a computerized *general ledger system*. However, even most computerized systems produce reports in journal entry order, as well as account summaries.

[43]There is a difference in sign between asset and liability accounts. An increase in an expense account is actually a decrease. The accounting profession, apparently, is comfortable with the common confusion that exists between debits and credits.

[44]There is also a distinction made between cash flows that are known and those that are expected. It is a *standard accounting principle* to record losses in full, at the time they are recognized, even before their occurrence. In the construction industry, for example, losses are recognized in full and projected to the end of a project as soon as they are foreseeable. Profits, on the other hand, are recognized only as they are realized (typically, as a percentage of project completion). The difference between cash and accrual systems is a matter of *bookkeeping*. The difference between loss and profit recognition is a matter of *accounting convention*. Engineers seldom need to be concerned with the accounting tradition.

[45]The expense for an item or service might be accrued even *before* the invoice is received. It might be recorded when the purchase order for the item or service is generated, or when the item or service is received.

[46]Other types of financial statements (*statements of changes in financial position*, *cost of sales statements*, inventory and asset reports, etc.) also will be generated, depending on the needs of the company.

[47]Financial statements also can be prepared with percentages (of total assets and net revenue) instead of dollars, in which case they are known as *common size financial statements*.

Figure 73.11 *Simplified Profit and Loss Statement*

revenue			
interest	2000		
sales	237,000		
returns	(23,000)		
net revenue		216,000	
expenses			
salaries	149,000		
utilities	6000		
advertising	28,000		
insurance	4000		
supplies	1000		
net expenses		188,000	
period net income			28,000	
beginning retained earnings			63,000	
net year-to-date earnings				91,000

The *balance sheet* presents the *basic accounting equation* in tabular form. The balance sheet lists the major categories of assets and outstanding liabilities. The difference between asset values and liabilities is the *equity*, as defined in Eq. 73.55. This equity represents what would be left over after satisfying all debts by liquidating the company.

There are several terms that appear regularly on balance sheets (see Fig. 73.12).

- *current assets:* cash and other assets that can be converted quickly into cash, such as accounts receivable, notes receivable, and merchandise (inventory). Also known as *liquid assets*.
- *fixed assets:* relatively permanent assets used in the operation of the business and relatively difficult to convert into cash. Examples are land, buildings, and equipment. Also known as *nonliquid assets*.
- *current liabilities:* liabilities due within a short period of time (e.g., within one year) and typically paid out of current assets. Examples are accounts payable, notes payable, and other accrued liabilities.
- *long-term liabilities:* obligations that are not totally payable within a short period of time (e.g., within one year).

Figure 73.12 *Simplified Balance Sheet*

ASSETS			
current assets			
cash	14,000		
accounts receivable	36,000		
notes receivable	20,000		
inventory	89,000		
prepaid expenses	3000		
total current assets		162,000	
plant, property, and equipment			
land and buildings	217,000		
motor vehicles	31,000		
equipment	94,000		
accumulated depreciation	(52,000)		
total fixed assets		290,000	
total assets			452,000
LIABILITIES AND OWNERS' EQUITY			
current liabilities			
accounts payable	66,000		
accrued income taxes	17,000		
accrued expenses	8000		
total current liabilities		91,000	
long-term debt			
notes payable	117,000		
mortgage	23,000		
total long-term debt		140,000	
owners' and stockholders' equity			
stock	130,000		
retained earnings	91,000		
total owners' equity		221,000	
total liabilities and owners' equity			452,000

Analysis of Financial Statements

Financial statements are evaluated by management, lenders, stockholders, potential investors, and many other groups for the purpose of determining the *health of the company*. The health can be measured in terms of *liquidity* (ability to convert assets to cash quickly), *solvency* (ability to meet debts as they become due), and *relative risk* (of which one measure is *leverage*—the portion of total capital contributed by owners).

The analysis of financial statements involves several common ratios, usually expressed as percentages. The following are some frequently encountered ratios.

- *current ratio:* an index of short-term paying ability.

$$\text{current ratio} = \frac{\text{current assets}}{\text{current liabilities}} \qquad 73.56$$

- *quick* (or *acid-test*) *ratio:* a more stringent measure of short-term debt-paying ability. The *quick assets* are defined to be current assets minus inventories and prepaid expenses.

$$\text{quick ratio} = \frac{\text{quick assets}}{\text{current liabilities}} \quad 73.57$$

- *receivable turnover:* a measure of the average speed with which accounts receivable are collected.

$$\text{receivable turnover} = \frac{\text{net credit sales}}{\text{average net receivables}} \quad 73.58$$

- *average age of receivables:* number of days, on the average, in which receivables are collected.

$$\text{average age of receivables} = \frac{365}{\text{receivable turnover}} \quad 73.59$$

- *inventory turnover:* a measure of the speed with which inventory is sold, on the average.

$$\text{inventory turnover} = \frac{\text{cost of goods sold}}{\text{average cost of inventory on hand}} \quad 73.60$$

- *days supply of inventory on hand:* number of days, on the average, that the current inventory would last.

$$\text{days supply of inventory on hand} = \frac{365}{\text{inventory turnover}} \quad 73.61$$

- *book value per share of common stock:* number of dollars represented by the balance sheet owners' equity for each share of common stock outstanding.

$$\text{book value per share of common stock} = \frac{\text{common shareholders' equity}}{\text{number of outstanding shares}} \quad 73.62$$

- *gross margin:* gross profit as a percentage of sales. (Gross profit is sales less cost of goods sold.)

$$\text{gross margin} = \frac{\text{gross profit}}{\text{net sales}} \quad 73.63$$

- *profit margin ratio:* percentage of each dollar of sales that is net income.

$$\text{profit margin} = \frac{\text{net income before taxes}}{\text{net sales}} \quad 73.64$$

- *return on investment ratio:* shows the percent return on owners' investment.

$$\text{return on investment} = \frac{\text{net income}}{\text{owners' equity}} \quad 73.65$$

- *price-earnings ratio:* indication of relationship between earnings and market price per share of common stock, useful in comparisons between alternative investments.

$$\text{price-earnings} = \frac{\text{market price per share}}{\text{earnings per share}} \quad 73.66$$

51. COST ACCOUNTING

Cost accounting is the system that determines the cost of manufactured products. Cost accounting is called *job cost accounting* if costs are accumulated by part number or contract. It is called *process cost accounting* if costs are accumulated by departments or manufacturing processes.

Cost accounting is dependent on historical and recorded data. The unit product cost is determined from actual expenses and numbers of units produced. Allowances (i.e., budgets) for future costs are based on these historical figures. Any deviation from historical figures is called a *variance*. Where adequate records are available, variances can be divided into *labor variance* and *material variance*.

When determining a unit product cost, the direct material and direct labor costs are generally clear-cut and easily determined. Furthermore, these costs are 100% variable costs. However, the indirect cost per unit of product is not as easily determined. Indirect costs (*burden*, *overhead*, etc.) can be fixed or semivariable costs. The amount of indirect cost allocated to a unit will depend on the unknown future overhead expense as well as the unknown future production (*vehicle size*).

A typical method of allocating indirect costs to a product is as follows.

step 1: Estimate the total expected indirect (and overhead) costs for the upcoming year.

step 2: Determine the most appropriate vehicle (basis) for allocating the overhead to production. Usually, this vehicle is either the number of units expected to be produced or the number of direct hours expected to be worked in the upcoming year.

step 3: Estimate the quantity or size of the overhead vehicle.

step 4: Divide expected overhead costs by the expected overhead vehicle to obtain the unit overhead.

step 5: Regardless of the true size of the overhead vehicle during the upcoming year, one unit of overhead cost is allocated per unit of overhead vehicle.

Once the prime cost has been determined and the indirect cost calculated based on projections, the two are combined into a *standard factory cost* or *standard cost*, which remains in effect until the next budgeting period (usually a year).

During the subsequent manufacturing year, the standard cost of a product is not generally changed merely because it is found that an error in projected indirect costs or production quantity (vehicle size) has been made. The allocation of indirect costs to a product is assumed to be independent of errors in forecasts. Rather, the difference between the expected and actual expenses, known as the *burden (overhead) variance*, experienced during the year is posted to one or more *variance accounts*.

Burden (overhead) variance is caused by errors in forecasting both the actual indirect expense for the upcoming year and the overhead vehicle size. In the former case, the variance is called *burden budget variance*; in the latter, it is called *burden capacity variance*.

Example 73.29

A company expects to produce 8000 items in the coming year. The current material cost is \$4.54 each. Sixteen minutes of direct labor are required per unit. Workers are paid \$7.50 per hour. 2133 direct labor hours are forecasted for the product. Miscellaneous overhead costs are estimated at \$45,000.

Find the per-unit (a) expected direct material cost, (b) direct labor cost, (c) prime cost, (d) burden as a function of production and direct labor, and (e) total cost.

Solution

(a) The direct material cost was given as \$4.54.

(b) The direct labor cost is

$$\left(\frac{16\text{ min}}{60\ \frac{\text{min}}{\text{hr}}}\right)\left(\frac{\$7.50}{\text{hr}}\right) = \$2.00$$

(c) The prime cost is

$$\$4.54 + \$2.00 = \$6.54$$

(d) If the burden vehicle is production, the burden rate is \$45,000/8000 = \$5.63 per item.

If the burden vehicle is direct labor hours, the burden rate is \$45,000/2133 = \$21.10 per hour.

(e) If the burden vehicle is production, the total cost is

$$\$4.54 + \$2.00 + \$5.63 = \$12.17$$

If the burden vehicle is direct labor hours, the total cost is

$$\$4.54 + \$2.00 + \left(\frac{16\text{ min}}{60\ \frac{\text{min}}{\text{hr}}}\right)\left(\frac{\$21.10}{\text{hr}}\right) = \$12.17$$

52. COST OF GOODS SOLD

Cost of goods sold (COGS) is an accounting term that represents an inventory account adjustment.[48] Cost of goods sold is the difference between the starting and ending inventory valuations. That is,

$$\begin{aligned}\text{COGS} = &\ \text{starting inventory valuation}\\ &-\text{ending inventory valuation}\end{aligned} \qquad \textit{73.67}$$

Cost of goods sold is subtracted from *gross profit* to determine the *net profit* of a company. Despite the fact that cost of goods sold can be a significant element in the profit equation, the inventory adjustment may not be made each accounting period (e.g., each month) due to the difficulty in obtaining an accurate inventory valuation.

With a *perpetual inventory system*, a company automatically maintains up-to-date inventory records, either through an efficient stocking and stock-releasing system or through a *point of sale* (POS) *system* integrated with the inventory records. If a company only counts its inventory (i.e., takes a *physical inventory*) at regular intervals (e.g., once a year), it is said to be operating on a *periodic inventory system*.

Inventory accounting is a source of many difficulties. The inventory value is calculated by multiplying the quantity on hand by the standard cost. In the case of completed items actually assembled or manufactured at the company, this standard cost usually is the manufacturing cost, although factory cost also can be used. In the case of purchased items, the standard cost will be the cost per item charged by the supplying vendor. In some cases, delivery and transportation costs will be included in this standard cost.

It is not unusual for the elements in an item's inventory to come from more than one vendor, or from one vendor in more than one order. Inventory valuation is more difficult if the price paid is different for these different purchases. There are four methods of determining the cost of elements in inventory. Any of these methods can be

[48]The cost of goods sold inventory adjustment is posted to the COGS *expense account*.

Economics

used (if applicable), but the method must be used consistently from year to year. The four methods are as follows.

- *specific identification method:* Each element can be uniquely associated with a cost. Inventory elements with serial numbers fit into this costing scheme. Stock, production, and sales records must include the serial number.
- *average cost method:* The standard cost of an item is the average of (recent or all) purchase costs for that item.
- *first-in, first-out* (FIFO) *method:* This method keeps track of how many of each item were purchased each time and the number remaining out of each purchase, as well as the price paid at each purchase. The inventory system assumes that the oldest elements are issued first.[49] Inventory value is a weighted average dependent on the number of elements from each purchase remaining. Items issued no longer contribute to the inventory value.
- *last-in, first-out* (LIFO) *method:* This method keeps track of how many of each item were purchased each time and the number remaining out of each purchase, as well as the price paid at each purchase.[50] The inventory value is a weighted average dependent on the number of elements from each purchase remaining. Items issued no longer contribute to the inventory value.

53. BREAK-EVEN ANALYSIS

Special Nomenclature

a	*incremental cost* to produce one additional item (also called *marginal cost* or *differential cost*)
C	total cost
f	fixed cost that does not vary with production
p	*incremental value* (price)
Q	quantity sold
$Q*$	quantity at break-even point
R	total revenue

Break-even analysis is a method of determining when the value of one alternative becomes equal to the value of another. A common application is that of determining when costs exactly equal revenue. If the manufactured quantity is less than the break-even quantity, a loss is incurred. If the manufactured quantity is greater than the break-even quantity, a profit is made. (See Fig. 73.13.) **[Economic Analysis: Nomenclature and Definitions]**

Assuming no change in the inventory, the *break-even point* can be found by setting costs equal to revenue ($C = R$).

$$C = f + aQ \tag{73.68}$$

$$R = pQ \tag{73.69}$$

$$Q^* = \frac{f}{p - a} \tag{73.70}$$

Figure 73.13 *Break-Even Quality*

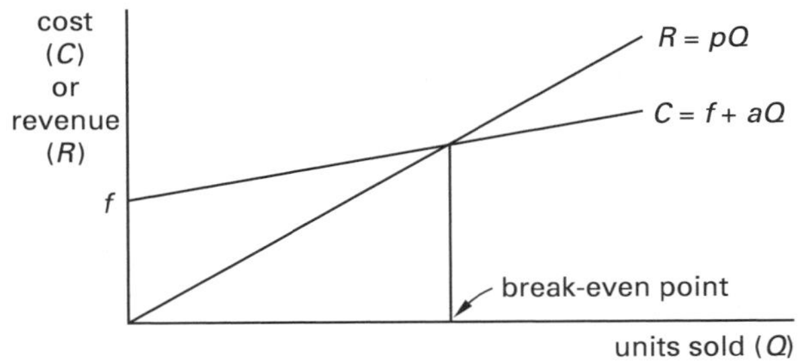

An alternative form of the break-even problem is to find the number of units per period for which two alternatives have the same total costs. Fixed costs are to be spread over a period longer than one year using the equivalent uniform annual cost (EUAC) concept. One of the alternatives will have a lower cost if production is less than the break-even point. The other will have a lower cost for production greater than the break-even point.

Example 73.30

Two plans are available for a company to obtain automobiles for its sales representatives. How many miles must the cars be driven each year for the two plans to have the same costs? Use an interest rate of 10%. (Use the year-end convention for all costs.)

plan A:	Lease the cars and pay \$0.15 per mile.
plan B:	Purchase the cars for \$5000. Each car has an economic life of three years, after which it can be sold for \$1200. Gas and oil cost \$0.04 per mile. Insurance is \$500 per year.

Economics

[49]If all elements in an item's inventory are identical, and if all shipments of that item are agglomerated, there will be no way to guarantee that the oldest element in inventory is issued first. But, unless *spoilage* is a problem, it really does not matter.
[50]See previous footnote.

Solution

Let x be the number of miles driven per year. Find the EUAC for both alternatives. [**Economic Factor Tables**]

$$\begin{aligned}\text{EUAC(A)} &= 0.15x \\ \text{EUAC(B)} &= 0.04x + \$500 + (\$5000)(A/P, 10\%, 3) \\ &\quad - (\$1200)(A/F, 10\%, 3) \\ &= 0.04x + \$500 + (\$5000)(0.4021) \\ &\quad - (\$1200)(0.3021) \\ &= 0.04x + 2148\end{aligned}$$

Setting EUAC(A) and EUAC(B) equal and solving for x yields 19,527 miles per year as the break-even point.

54. PAY-BACK PERIOD

The *pay-back period* is defined as the length of time, usually in years, for the cumulative net annual profit to equal the initial investment. It is tempting to introduce equivalence into pay-back period calculations, but by convention, this is generally not done.[51]

$$\text{pay-back period} = \frac{\text{initial investment}}{\text{net annual profit}} \qquad 73.71$$

Example 73.31

A ski resort installs two new ski lifts at a total cost of \$1,800,000. The resort expects the annual gross revenue to increase by \$500,000 while it incurs an annual expense of \$50,000 for lift operation and maintenance. What is the pay-back period?

Solution

From Eq. 73.71,

$$\text{pay-back period} = \frac{\$1{,}800{,}000}{\frac{\$500{,}000}{\text{yr}} - \frac{\$50{,}000}{\text{yr}}} = 4 \text{ yr}$$

55. MANAGEMENT GOALS

Depending on many factors (market position, age of the company, age of the industry, perceived marketing and sales windows, etc.), a company may select one of many production and marketing strategic goals. Three such strategic goals are

- maximization of product demand
- minimization of cost
- maximization of profit

Such goals require knowledge of how the dependent variable (e.g., demand quantity or quantity sold) varies as a function of the independent variable (e.g., price). Unfortunately, these three goals are not usually satisfied simultaneously. For example, minimization of product cost may require a large production run to realize economies of scale, while the actual demand is too small to take advantage of such economies of scale.

If sufficient data are available to plot the independent and dependent variables, it may be possible to optimize the dependent variable graphically. (See Fig. 73.14.) Of course, if the relationship between independent and dependent variables is known algebraically, the dependent variable can be optimized by taking derivatives or by use of other numerical methods.

Figure 73.14 *Graphs of Management Goal Functions*

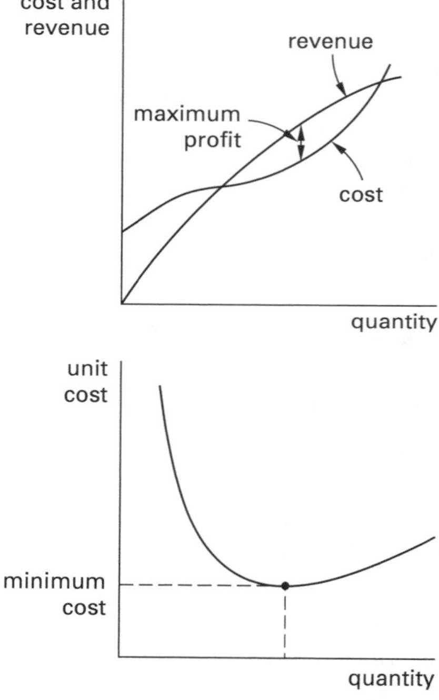

56. INFLATION

It is important to perform economic studies in terms of *constant value dollars*. One method of converting all cash flows to constant value dollars is to divide the flows by some annual *economic indicator* or price index.

[51]Equivalence (i.e., interest and compounding) generally is not considered when calculating the "pay-back period." However, if it is desirable to include equivalence, then the term *pay-back period* should not be used. Other terms, such as *cost recovery period* or *life of an equivalent investment*, should be used. Unfortunately, this convention is not always followed in practice.

If indicators are not available, cash flows can be adjusted by assuming that inflation is constant at a decimal rate (e), per year. Then, all cash flows can be converted to $t = 0$ dollars by dividing by $(1 + e)^n$, where n is the year of the cash flow.

An alternative is to replace the effective annual interest rate, i_e, with a value corrected for inflation. This corrected value, i', is

$$i' = i + e + ie \tag{73.72}$$

This method has the advantage of simplifying the calculations. However, precalculated factors are not available for the non-integer values of i'. Therefore, Table 73.1 must be used to calculate the factors.

Example 73.32

What is the uninflated present worth of a \$2000 future value in two years if the average inflation rate is 6% and i is 10%?

Solution

$$\begin{aligned} P &= \frac{F}{(1+i)^n(1+e)^n} \\ &= \frac{\$2000}{(1+0.10)^2(1+0.06)^2} \\ &= \$1471 \end{aligned}$$

Example 73.33

Repeat Ex. 73.32 using Eq. 73.72.

Solution

$$\begin{aligned} i' &= i + e + ie \\ &= 0.10 + 0.06 + (0.10)(0.06) \\ &= 0.166 \end{aligned}$$

$$P = \frac{F}{(1+i')^2} = \frac{\$2000}{(1+0.166)^2} = \$1471$$

57. CONSUMER LOANS

Special Nomenclature

BAL_j	balance after the jth payment
j	payment or period number
LV	principal total value loaned (cost minus down payment)
N	total number of payments to pay off the loan
PI_j	jth interest payment
PP_j	jth principal payment
PT_j	jth total payment
ϕ	effective rate per period (r/k)

Many different arrangements can be made between a borrower and a lender. With the advent of creative financing concepts, it often seems that there are as many variations of loans as there are loans made. Nevertheless, there are several traditional types of transactions. Real estate or investment texts, or a financial consultant, should be consulted for more complex problems.

Simple Interest

Interest due does not compound with a *simple interest loan.* The interest due is merely proportional to the length of time that the principal is outstanding. Because of this, simple interest loans are seldom made for long periods (e.g., more than one year). (For loans less than one year, it is commonly assumed that a year consists of 12 months of 30 days each.)

Example 73.34

A \$12,000 simple interest loan is taken out at 16% per annum interest rate. The loan matures in two years with no intermediate payments. How much will be due at the end of the second year?

Solution

The interest each year is

$$PI = (0.16)(\$12{,}000) = \$1920$$

The total amount due in two years is

$$PT = \$12{,}000 + (2)(\$1920) = \$15{,}840$$

Example 73.35

\$4000 is borrowed for 75 days at 16% per annum simple interest. How much will be due at the end of 75 days?

Solution

$$\begin{aligned} \text{amount due} &= \$4000 + (0.16)\left(\frac{75 \text{ days}}{360 \,\frac{\text{days}}{\text{bank yr}}}\right)(\$4000) \\ &= \$4133 \end{aligned}$$

Loans with Constant Amount Paid Toward Principal

With this loan type, the payment is not the same each period. The amount paid toward the principal is constant, but the interest varies from period to period. (See Fig. 73.15.) The equations that govern this type of loan are

$$BAL_j = LV - j(PP) \tag{73.73}$$

$$PI_j = \phi(BAL)_{j-1} \tag{73.74}$$

Economics

$$\text{PT}_j = \text{PP} + \text{PI}_j \qquad 73.75$$

$$\text{PP} = \frac{\text{LV}}{N} \qquad 73.76$$

$$N = \frac{\text{LV}}{\text{PP}} \qquad 73.77$$

$$\text{LV} = (\text{PP} + \text{PI}_1)(P/A, \phi, N) - \text{PI}_N(P/G, \phi, N) \qquad 73.78$$

$$1 = \left(\frac{1}{N} + \phi\right)(P/A, \phi, N) - \left(\frac{\phi}{N}\right)(P/G, \phi, N) \qquad 73.79$$

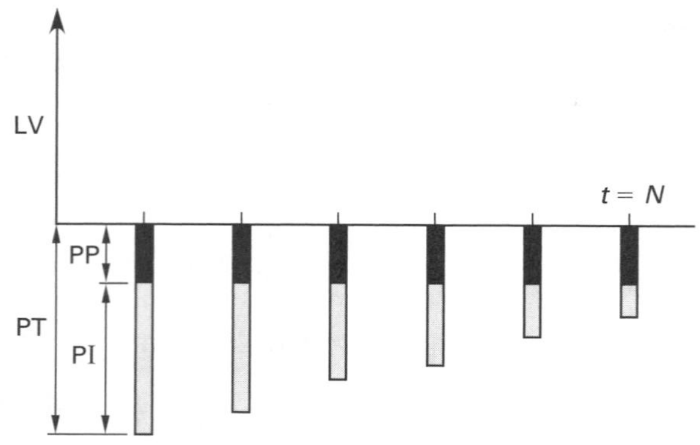

Figure 73.15 *Loan with Constant Amount Paid Toward Principal*

Example 73.36

A \$12,000 six-year loan is taken from a bank that charges 15% effective annual interest. Payments toward the principal are uniform, and repayments are made at the end of each year. Tabulate the interest, total payments, and the balance remaining after each payment is made.

Solution

The amount of each principal payment is

$$\text{PP} = \frac{\text{LV}}{N} = \frac{\$12{,}000}{6} = \$2000$$

At the end of the first year (before the first payment is made), the principal balance is \$12,000 (i.e., BAL_0 = \$12,000). From Eq. 73.74, the interest payment is

$$\text{PI}_1 = \phi(\text{BAL})_0 = (0.15)(\$12{,}000) = \$1800$$

The total first payment is

$$\text{PT}_1 = \text{PP} + \text{PI} = \$2000 + \$1800 = \$3800$$

The following table is similarly constructed.

j	BAL_j	PP_j	PI_j	PT_j
	(in dollars)			
0	12,000	–	–	–
1	10,000	2000	1800	3800
2	8000	2000	1500	3500
3	6000	2000	1200	3200
4	4000	2000	900	2900
5	2000	2000	600	2600
6	0	2000	300	2300

Direct Reduction Loans

This is the typical "interest paid on unpaid balance" loan. The amount of the periodic payment is constant, but the amounts paid toward the principal and interest both vary. (See Fig. 73.16.)

Figure 73.16 *Direct Reduction Loan*

$$\text{BAL}_{j-1} = \text{PT}\left(\frac{1-(1+\phi)^{j-1-N}}{\phi}\right) \qquad 73.80$$

$$\text{PI}_j = \phi(\text{BAL})_{j-1} \qquad 73.81$$

$$\text{PP}_j = \text{PT} - \text{PI}_j \qquad 73.82$$

$$\text{BAL}_j = \text{BAL}_{j-1} - \text{PP}_j \qquad 73.83$$

$$N = \frac{-\ln\left(1 - \frac{\phi(\text{LV})}{\text{PT}}\right)}{\ln(1+\phi)} \qquad 73.84$$

Equation 73.84 calculates the number of payments necessary to pay off a loan. This equation can be solved with effort for the total periodic payment (PT) or the initial value of the loan (LV). It is easier, however, to use the $(A/P, i\%, n)$ factor to find the payment and loan value.

$$\text{PT} = \text{LV}(A/P, \phi\%, N) \qquad 73.85$$

Economics

If the loan is repaid in yearly installments, then i is the effective annual rate. If the loan is paid off monthly, then i should be replaced by the effective rate per month (ϕ from Eq. 73.51). For monthly payments, N is the number of months in the loan period.

Example 73.37

A $45,000 loan is financed at 9.25% per annum. The monthly payment is $385. What are the amounts paid toward interest and principal in the 14th period? What is the remaining principal balance after the 14th payment has been made?

Solution

The effective rate per month is

$$\begin{aligned}\phi &= \frac{r}{m} = \frac{0.0925}{12} \\ &= 0.007708\end{aligned}$$

$$\begin{aligned}N &= \frac{-\ln\left(1-\dfrac{\phi(\text{LV})}{\text{PT}}\right)}{\ln(1+\phi)} \\ &= \frac{-\ln\left(1-\dfrac{(0.007708)(45{,}000)}{385}\right)}{\ln(1+0.007708)} \\ &= 301\end{aligned}$$

$$\begin{aligned}\text{BAL}_{14-1} &= \text{PT}\left(\frac{1-(1+\phi)^{14-1-N}}{\phi}\right) \\ &= (\$385)\left(\frac{1-(1+0.007708)^{14-1-301}}{0.007708}\right) \\ &= \$44{,}476.39\end{aligned}$$

$$\begin{aligned}\text{PI}_{14} &= \phi(\text{BAL})_{14-1} \\ &= (0.007708)(\$44{,}476.39) \\ &= \$342.82\end{aligned}$$

$$\text{PP}_{14} = \text{PT} - \text{PI}_{14} = \$385 - \$342.82 = \$42.18$$

Therefore, using Eq. 73.83, the remaining principal balance is

$$\begin{aligned}\text{BAL}_{14} &= \text{BAL}_{14-1} - \text{PP}_{14} \\ &= \$44{,}476.39 - \$42.18 \\ &= \$44{,}434.21\end{aligned}$$

Direct Reduction Loans with Balloon Payments

This type of loan has a constant periodic payment, but the duration of the loan is insufficient to completely pay back the principal (i.e., the loan is not fully amortized). Therefore, all remaining unpaid principal must be paid back in a lump sum when the loan matures. This large payment is known as a *balloon payment*.[52] (See Fig. 73.17.)

Equation 73.80 through Eq. 73.84 also can be used with this type of loan. The remaining balance after the last payment is the balloon payment. This balloon payment must be repaid along with the last regular payment calculated.

Figure 73.17 *Direct Reduction Loan with Balloon Payment*

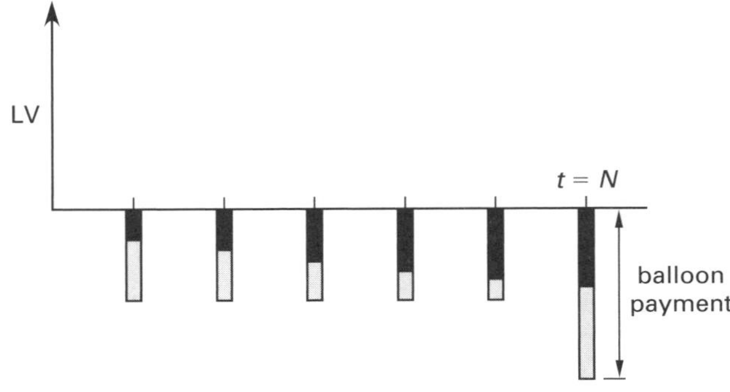

58. FORECASTING

There are many types of forecasting models, although most are variations of the basic types.[53] All models produce a *forecast*, F_{t+1}, of some quantity (*demand* is used in this section) in the next period based on actual measurements, D_j, in current and prior periods. All of the models also try to provide *smoothing* (or *damping*) of extreme data points.

Forecasts by Moving Averages

The method of *moving average forecasting* weights all previous demand data points equally and provides some smoothing of extreme data points. The amount of smoothing increases as the number of data points, n, increases.

$$F_{t+1} = \frac{1}{n}\sum_{m=t+1-n}^{t} D_m \qquad 73.86$$

[52]The term *balloon payment* may include the final interest payment as well. Generally, the problem statement will indicate whether the balloon payment is inclusive or exclusive of the regular payment made at the end of the loan period.
[53]For example, forecasting models that take into consideration steady (linear), cyclical, annual, and seasonal trends are typically variations of the exponentially weighted model. A truly different forecasting tool, however, is *Monte Carlo simulation*.

Economics

Forecasts by Exponentially Weighted Averages

With *exponentially weighted forecasts*, the more current (most recent) data points receive more weight. This method uses a *weighting factor*, α, also known as a *smoothing coefficient*, which typically varies between 0.01 and 0.30. An initial forecast is needed to start the method. Forecasts immediately following are sensitive to the accuracy of this first forecast. It is common to choose $F_0 = D_1$ to get started.

$$F_{t+1} = \alpha D_t + (1-\alpha)F_t \qquad 73.87$$

59. LEARNING CURVES

Special Nomenclature

b	learning curve constant
n	total number of items produced
R	decimal learning curve rate (2^{-b})
T_1	time or cost for the first item
T_n	time or cost for the nth item

The more products that are made, the more efficient the operation becomes due to experience gained. Therefore, direct labor costs decrease.[54] Usually, a *learning curve* is specified by the decrease in cost each time the cumulative quantity produced doubles. If there is a 20% decrease per doubling, the curve is said to be an 80% learning curve (i.e., the *learning curve rate*, R, is 80%).

Then, the time to produce the nth item is

$$T_n = T_1 n^{-b} \qquad 73.88$$

The total time to produce units from quantity n_1 to n_2 inclusive is approximately given by Eq. 73.89. T_1 is a constant, the time for item 1, and does not correspond to n unless $n_1 = 1$.

$$\int_{n_1}^{n_2} T_n d_n \approx \left(\frac{T_1}{1-b}\right)\left(\left(n_2+\tfrac{1}{2}\right)^{1-b} - \left(n_1-\tfrac{1}{2}\right)^{1-b}\right) \qquad 73.89$$

The *average time per unit* over the production from n_1 to n_2 is the above total time from Eq. 73.89 divided by the quantity produced, $(n_2 - n_1 + 1)$.

$$T_{\text{ave}} = \frac{\int_{n_1}^{n_2} T_n\,dn}{n_2 - n_1 + 1} \qquad 73.90$$

Table 73.9 lists representative values of the *learning curve constant*, b. For learning curve rates not listed in the table, Eq. 73.91 can be used to find b.

$$b = \frac{-\log_{10} R}{\log_{10}(2)} = \frac{-\log_{10} R}{0.301} \qquad 73.91$$

***Table 73.9** Learning Curve Constants*

learning curve rate, R	b
0.70 (70%)	0.515
0.75 (75%)	0.415
0.80 (80%)	0.322
0.85 (85%)	0.234
0.90 (90%)	0.152
0.95 (95%)	0.074

Example 73.38

A 70% learning curve is used with an item whose first production time is 1.47 hr. (a) How long will it take to produce the 11th item? (b) How long will it take to produce the 11th through 27th items?

Solution

(a) From Eq. 73.88,

$$T_{11} = T_1 n^{-b} = (1.47\ \text{hr})(11)^{-0.515} = 0.428\ \text{hr}$$

(b) The time to produce the 11th item through 27th item is given by Eq. 73.89.

$$\begin{aligned} T &\approx \left(\frac{T_1}{1-b}\right)\left(\left(n_{27}+\tfrac{1}{2}\right)^{1-b} - \left(n_{11}-\tfrac{1}{2}\right)^{1-b}\right) \\ &\approx \left(\frac{1.47\ \text{hr}}{1-0.515}\right)\left((27.5)^{1-0.515} - (10.5)^{1-0.515}\right) \\ &= 5.643\ \text{hr} \end{aligned}$$

60. ECONOMIC ORDER QUANTITY

Special Nomenclature

a	constant depletion rate (items/unit time)
h	inventory storage cost ($/item-unit time)
H	total inventory storage cost between orders ($)
K	fixed cost of placing an order ($)
Q	order quantity (original quantity on hand)
t^*	time at depletion

[54] Learning curve reductions apply only to direct labor costs. They are not applied to indirect labor or direct material costs.

The *economic order quantity* (EOQ) is the order quantity that minimizes the inventory costs per unit time. Although there are many different EOQ models, the simplest is based on the following assumptions.

- Reordering is instantaneous. The time between order placement and receipt is zero. (See Fig. 73.18.)
- Shortages are not allowed.
- Demand for the inventory item is deterministic (i.e., is not a random variable).
- Demand is constant with respect to time.
- An order is placed when the inventory is zero.

Figure 73.18 *Inventory with Instantaneous Reorder*

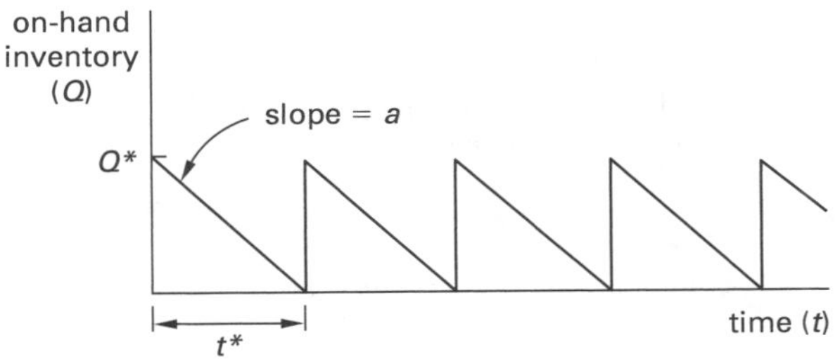

If the original quantity on hand is Q, the stock will be depleted at

$$t^* = \frac{Q}{a} \qquad 73.92$$

The total inventory storage cost between t_0 and t^* is

$$H = \tfrac{1}{2}Qht^* = \frac{Q^2h}{2a} \qquad 73.93$$

The total inventory and ordering cost per unit time is

$$C_t = \frac{aK}{Q} + \frac{hQ}{2} \qquad 73.94$$

C_t can be minimized with respect to Q. The economic order quantity and time between orders are

$$Q^* = \sqrt{\frac{2aK}{h}} \qquad 73.95$$

$$t^* = \frac{Q^*}{a} \qquad 73.96$$

61. SENSITIVITY ANALYSIS

Data analysis and forecasts in economic studies require estimates of costs that will occur in the future. There are always uncertainties about these costs. However, these uncertainties are insufficient reason not to make the best possible estimates of the costs. Nevertheless, a decision between alternatives often can be made more confidently if it is known whether or not the conclusion is sensitive to moderate changes in data forecasts. Sensitivity analysis provides this extra dimension to an economic analysis.

The sensitivity of a decision is determined by inserting a range of estimates for critical cash flows and other parameters. If radical changes can be made to a cash flow without changing the decision, the decision is said to be *insensitive* to uncertainties regarding that cash flow. However, if a small change in the estimate of a cash flow will alter the decision, that decision is said to be very *sensitive* to changes in the estimate. If the decision is sensitive only for a limited range of cash flow values, the term *variable sensitivity* is used. Figure 73.19 illustrates these terms.

Figure 73.19 *Types of Sensitivity*

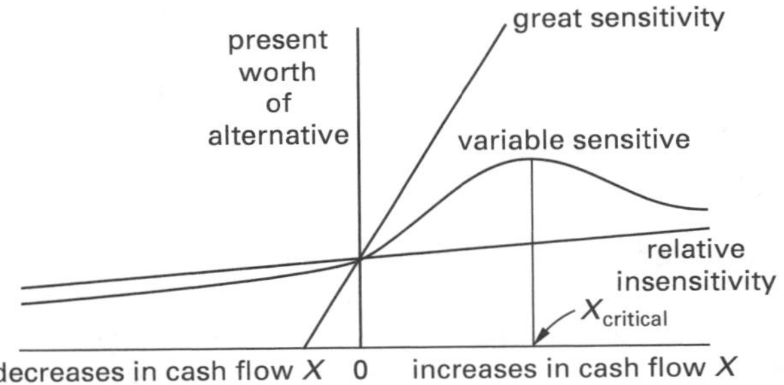

An established semantic tradition distinguishes between risk analysis and uncertainty analysis. *Risk analysis* addresses variables that have a known or estimated probability distribution. In this regard, statistics and probability theory can be used to determine the probability of a cash flow varying between given limits. On the other hand, *uncertainty analysis* is concerned with situations in which there is not enough information to determine the probability or frequency distribution for the variables involved.

As a first step, sensitivity analysis should be applied one at a time to the dominant factors. Dominant cost factors are those that have the most significant impact on the present value of the alternative.[55] If warranted, additional investigation can be used to determine the sensitivity to several cash flows varying simultaneously. Significant judgment is needed, however, to successfully determine the proper combinations of cash flows to

[55]In particular, engineering economic analysis problems are sensitive to the choice of effective interest rate, i_e, and to accuracy in cash flows at or near the beginning of the horizon. The problems will be less sensitive to accuracy in far-future cash flows, such as salvage value and subsequent generation replacement costs.

vary. It is common to plot the dependency of the present value on the cash flow being varied in a two-dimensional graph. Simple linear interpolation is used (within reason) to determine the critical value of the cash flow being varied.

62. VALUE ENGINEERING

The *value* of an investment is defined as the ratio of its return (performance or utility) to its cost (effort or investment). The basic object of *value engineering* (VE, also referred to as *value analysis*) is to obtain the maximum per-unit value.[56]

Value engineering concepts often are used to reduce the cost of mass-produced manufactured products. This is done by eliminating unnecessary, redundant, or superfluous features, by redesigning the product for a less expensive manufacturing method, and by including features for easier assembly without sacrificing utility and function.[57] However, the concepts are equally applicable to one-time investments, such as buildings, chemical processing plants, and space vehicles. In particular, value engineering has become an important element in all federally funded work.[58]

Typical examples of large-scale value engineering work are using stock-sized bearings and motors (instead of custom manufactured units), replacing rectangular concrete columns with round columns (which are easier to form), and substituting custom buildings with prefabricated structures.

Value engineering is usually a team effort. And, while the original designers may be on the team, usually outside consultants are utilized. The cost of value engineering is usually returned many times over through reduced construction and life-cycle costs.

63. NOMENCLATURE

A	annual amount	\$
B	present worth of all benefits	\$
BV_j	book value at end of the jth year	\$
C	cost or present worth of all costs	\$
d	declining balance depreciation rate	decimal
D	demand	various
D	depreciation	\$
DR	present worth of after-tax depreciation recovery	\$
e	constant inflation rate	decimal
$\mathcal{E}$	expected value	various
E_0	initial amount of an exponentially growing cash flow	\$
EAA	equivalent annual amount	\$
EUAC	equivalent uniform annual cost	\$
f	federal income tax rate	decimal
F	forecasted quantity	various
F	future worth	\$
g	exponential growth rate	decimal
G	uniform gradient amount	\$
i	interest rate per interest period	decimal
$i\%$	interest rate per interest period	%
i'	effective interest rate corrected for inflation	decimal
j	number of years	–
m	number of compounding periods per year	–
m	an integer	–
n	number of compounding periods or years in life of asset	–
p	probability	decimal
P	present worth	\$
r	nominal rate per year (rate per annum)	decimal per unit time
ROI	return on investment	\$
ROR	rate of return	decimal per unit time
s	state income tax rate	decimal
S_n	expected salvage value in year n	\$
t	composite tax rate	decimal
t	time	years (typical)
T	a quantity equal to $\frac{1}{2}n(n+1)$	–
TC	tax credit	\$
z	a quantity equal to $\frac{1+i}{1-d}$	decimal

Symbols

α	smoothing coefficient for forecasts	–
ϕ	effective rate per period	decimal

Subscripts

0	initial
e	effective
j	at time j
n	at time n
t	at time t

[56] Value analysis, the methodology that has become today's value engineering, was developed in the early 1950s by Lawrence D. Miles, an analyst at General Electric.

[57] Some people say that value engineering is the act of going over the plans and taking out everything that is interesting.

[58] U.S. Government Office of Management and Budget Circular A-131 outlines value engineering for federally funded construction projects.

Topic XIV: Law and Ethics

Chapter

74 Professional Services, Contracts, and Engineering Law

Content in blue refers to the *NCEES Handbook*.

1. FORMS OF COMPANY OWNERSHIP[1]

There are three basic forms of company ownership in the United States: (a) sole proprietorship, (b) partnership, and (c) corporation.[2] Each of these forms of ownership has advantages and disadvantages.

2. SOLE PROPRIETORSHIPS

A *sole proprietorship* (*single proprietorship*) is the easiest form of ownership to establish. Other than the necessary licenses, filings, and notices (which apply to all forms of ownership), no legal formalities are required to start business operations. A sole proprietor (the owner) has virtually total control of the business and makes all supervisory and management decisions.

Legally, there is no distinction between the sole proprietor and the sole proprietorship (the business). This is the greatest disadvantage of this form of business. The owner is solely responsible for the operation of the business, even if the owner hires others for assistance. The owner assumes personal, legal, and financial liability for all acts and debts of the company. If the company debts remain unpaid, or in the event there is a legal judgment against the company, the owner's personal assets (home, car, savings, etc.) can be seized or attached.

Another disadvantage of the sole proprietorship is the lack of significant organizational structure. In times of business crisis or trouble, there may be no one to share the responsibility or to help make decisions. When the owner is sick or dies, there may be no way to continue the business.

There is also no distinction between the incomes of the business and the owner. Therefore, the business income is taxed at the owner's income tax rate. Depending on the owner's financial position, the success of the business, and the tax structure, this can be an advantage or a disadvantage.[3]

3. PARTNERSHIPS

A *partnership* (also known as a *general partnership*) is ownership by two or more persons known as *general partners*. Legally, this form is very similar to a sole proprietorship, and the two forms of business have many of the same advantages and disadvantages. For example, with the exception of an optional *partnership agreement*, there are a minimum of formalities to setting up business. The partners make all business and management decisions themselves according to an agreed-upon process. The business income is split among the partners and taxed at the partners' individual tax rates.[4]

[1]This chapter is not intended to be a substitute for professional advice. Law is not always black and white. For every rule there are exceptions. For every legal principle, there are variations. For every type of injury, there are numerous legal precedents. This chapter covers the superficial basics of a small subset of U.S. law affecting engineers.
[2]The discussion of forms of company ownership in Sec. 74.2, Sec. 74.3, and Sec. 74.4 applies equally to service-oriented companies (e.g., consulting engineering firms) and product-oriented companies.
[3]To use a simplistic example, if the corporate tax rates are higher than the individual tax rates, it would be *financially* better to be a sole proprietor because the company income would be taxed at a lower rate.
[4]The percentage split is specified in the partnership agreement.

Continuity of the business is still a problem since most partnerships are automatically dissolved upon the withdrawal or death of one of the partners.[5]

One advantage of a partnership over a sole proprietorship is the increase in available funding. Not only do more partners bring in more start-up capital, but the resource pool may make business credit easier to obtain. Also, the partners bring a diversity of skills and talents.

Unless the partnership agreement states otherwise, each partner can individually obligate (i.e., *bind*) the partnership without the consent of the other partners. Similarly, each partner has personal responsibility and liability for the acts and debts of the partnership company, just as sole proprietors do. In fact, each partner assumes the *sole* responsibility, not just a proportionate share. If one or more partners are unable to pay, the remaining partners shoulder the entire debt. The possibility of one partner having to pay for the actions of another partner must be considered when choosing this form of business ownership.

A *limited partnership* differs from a general partnership in that one (or more) of the partners is silent. The *limited partners* make a financial contribution to the business and receive a share of the profit but do not participate in the management and cannot bind the partnership. While *general partners* have unlimited personal liabilities, limited partners are generally liable only to the extent of their investment.[6] A written partnership agreement is required, and the agreement must be filed with the proper authorities.

4. CORPORATIONS

A corporation is a legal entity (i.e., a legal person) distinct from the founders and owners. The separation of ownership and management makes the corporation a fundamentally different business form than a sole proprietorship or partnership, with very different advantages and disadvantages.

A corporation becomes legally distinct from its founders upon formation and proper registration. Ownership of the corporation is through shares of stock, distributed to the founders and investors according to some agreed-upon investment and distribution rule. The founders and investors become the stockholders (i.e., owners) of the corporation. A *closely held* (*private*) *corporation* is one in which all stock is owned by a family or small group of co-investors. A *public corporation* is one whose stock is available for the public-at-large to purchase.

There is no mandatory connection between ownership and management functions. The decision-making power is vested in the executive officers and a *board of directors* that governs by majority vote. The stockholders elect the board of directors which, in turn, hires the executive officers, management, and other employees. Employees of the corporation may or may not be stockholders.

Disadvantages (at least for a person or persons who could form a partnership or sole proprietorship) include the higher corporate tax rate, difficulty and complexity of formation (some states require a minimum number of persons on the board of directors), and additional legal and accounting paperwork.

However, since a corporation is distinctly separate from its founders and investors, those individuals are not liable for the acts and debts of the corporation. Debts are paid from the corporate assets. Income to the corporation is not taxable income to the owners. (Only the salaries, if any, paid to the employees by the corporation are taxable to the employees.) Even if the corporation were to go bankrupt, the assets of the owners would not ordinarily be subject to seizure or attachment.

A corporation offers the best guarantee of continuity of operation in the event of the death, incapacitation, or retirement of the founders since, as a legal entity, it is distinct from the founders and owners.

5. LIMITED LIABILITY ENTITIES

A variety of other legal entities have been established that blur the lines between the three traditional forms of business (i.e., proprietorship, partnership, and corporation). The *limited liability partnership*, LLP, extends a measure of corporate-like protection to professionals while permitting partnership-like personal participation in management decisions. Since LLPs are formed and operated under state laws, actual details vary from state to state. However, most LLPs allow the members to participate in management decisions, while not being responsible for the misdeeds of other partners. As in a corporation, the debts of the LLP do not become debts of the members.[7] The *double taxation* characteristic of traditional corporations is avoided, as profits to the LLP flow through to the members.

For engineers and architects (as well as doctors, lawyers, and accountants), the *professional corporation*, PC, offers protection from the actions (e.g., malpractice) of other professionals within a shared environment, such as a design firm. While a PC does not shield the individual from responsibility for personal negligence or malpractice, it does permit the professional to be associated with a larger entity, such as a partnership of other PCs, without accepting responsibility for the actions of the other members. In that sense, the protection is similar

[5]Some or all of the remaining partners may want to form a new partnership, but this is not always possible.
[6]That is, if the partnership fails or is liquidated to pay debts, the limited partners lose no more than their initial investments.
[7]Depending on the state, the shield may be complete or limited. It is common that the protection only applies to negligence-related claims, as opposed to intentional tort claims, contract-related obligations, and day-to-day operating expenses such as rent, utilities, and employees.

to that of an LLP. Unlike a traditional corporation, a PC may have a board of directors consisting of only a single individual, the professional.

The *limited liability company*,[8] LLC, also combines advantages from partnerships and corporations. In an LLC, the members are shielded from debts of the LLC while enjoying the pass-through of all profits.[9] Like a partnership or shareholder, a member's obligation is limited to the *membership interest* in (i.e., contribution to) the LLC. LLCs are directed and controlled by one or more managers who may also be members. A variation of the LLC specifically for design, medical, and other professionals is the *professional limited liability company*, PLLC.

The traditional corporation, as described in Sec. 74.4, is referred to as a *Subchapter C corporation* or "*C corp*."[10] A variant is the *S corporation* ("*S corp*") which combines characteristics of the C corporation with pass-through for taxation. S corporations can be limited or treated differently than C corporations by state and federal law.

6. PIERCING THE CORPORATE VEIL

An individual operating as a corporation, LLP, LLC, or PC entity may lose all protection if his or her actions are fraudulent, or if the court decides the business is an "alter ego" of the individual. Basically, this requires the business to be run as a business. Business and personal assets cannot be intermingled, and business decisions must be made and documented in a business-like manner. If operated fraudulently or loosely, a court may assign liability directly to an individual, an action known as *piercing the corporate veil*.

7. AGENCY

In some contracts, decision-making authority and right of action are transferred from one party (the owner, or *principal*) who would normally have that authority to another person (the *agent*). For example, in construction contracts, the engineer may be the agent of the owner for certain transactions. Agents are limited in what they can do by the scope of the agency agreement. Within that scope, however, an agent acts on behalf of the principal, and the principal is liable for the acts of the agent and is bound by contracts made in the principal's name by the agent.

Agents are required to execute their work with care, skill, and diligence. Specifically, agents have *fiduciary responsibility* toward their principal, meaning that agent must be honest and loyal. Agents are liable for damages resulting from a lack of diligence, loyalty, and/or honesty. If the agents misrepresented their skills when obtaining the agency, they can be liable for breach of contract or fraud.

8. GENERAL CONTRACTS

A *contract* is a legally binding agreement or promise to exchange goods or services.[11] A written contract is merely a documentation of the agreement. Some agreements must be in writing, but most agreements for engineering services can be verbal, particularly if the parties to the agreement know each other well.[12] Written contract documents do not need to contain intimidating legal language, but all agreements must satisfy three basic requirements to be enforceable (binding).

- There must be a clear, specific, and definite *offer* with no room for ambiguity or misunderstanding.
- There must be some form of conditional future *consideration* (i.e., payment).[13]
- There must be an *acceptance* of the offer.

There are other conditions that the agreement must meet to be enforceable. These conditions are not normally part of the explicit agreement but represent the conditions under which the agreement was made.

- The agreement must be *voluntary* for all parties.
- All parties must have *legal capacity* (i.e., be mentally competent, of legal age, and uninfluenced by drugs).
- The purpose of the agreement must be *legal*.

For small projects, a simple *letter of agreement* on one party's stationery may suffice. For larger, complex projects, a more formal document may be required. Some clients prefer to use a *purchase order*, which can function as a contract if all basic requirements are met.

Regardless of the format of the written document—letter of agreement, purchase order, or standard form—a contract should include the following features.[14]

- introduction, preamble, or preface indicating the purpose of the contract

Law and Ethics

[8]LLC does not mean *limited liability corporation*. LLCs are not corporations.
[9]LLCs enjoy *check the box taxation*, which means they can elect to be taxed as sole proprietorships, partnerships, or corporations.
[10]The reference is to subchapter C of the Internal Revenue Code.
[11]Not all agreements are legally binding (i.e., enforceable). Two parties may agree on something, but unless the agreement meets all of the requirements and conditions of a contract, the parties cannot hold each other to the agreement.
[12]All states have a *statute of frauds* that, among other things, specifies what types of contracts must be in writing to be enforceable. These include contracts for the sale of land, contracts requiring more than one year for performance, contracts for the sale of goods over $500 in value, contracts to satisfy the debts of another, and marriage contracts. Contracts to provide engineering services do not fall under the statute of frauds.
[13]Actions taken or payments made prior to the agreement are irrelevant. Also, it does not matter to the courts whether the exchange is based on equal value or not.
[14]*Construction contracts* are unique unto themselves. Items that might also be included as part of the *contract documents* are the agreement form, the general conditions, drawings, specifications, and addenda.

- name, address, and business forms of both contracting parties
- signature date of the agreement
- effective date of the agreement (if different from the signature date)
- duties and obligations of both parties
- deadlines and required service dates
- fee amount
- fee schedule and payment terms
- agreement expiration date
- standard boilerplate clauses
- signatures of parties or their agents
- declaration of authority of the signatories to bind the contracting parties
- supporting documents

9. STANDARD BOILERPLATE CLAUSES

It is common for full-length contract documents to include important *boilerplate clauses.* These clauses have specific wordings that should not normally be changed, hence the name "boilerplate." Some of the most common boilerplate clauses are paraphrased here.

- Delays and inadequate performance due to war, strikes, and acts of God and nature are forgiven (*force majeure*).
- The contract document is the complete agreement, superseding all prior verbal and written agreements.
- The contract can be modified or canceled only in writing.
- Parts of the contract that are determined to be void or unenforceable will not affect the enforceability of the remainder of the contract (*severability*). Alternatively, parts of the contract that are determined to be void or unenforceable will be rewritten to accomplish their intended purpose without affecting the remainder of the contract.
- None (or one, or both) of the parties can (or cannot) assign its (or their) rights and responsibilities under the contract (*assignment*).
- All notices provided for in the agreement must be in writing and sent to the address in the agreement.
- Time is of the essence.[15]
- The subject headings of the agreement paragraphs are for convenience only and do not control the meaning of the paragraphs.
- The laws of the state in which the contract is signed must be used to interpret and govern the contract.
- Disagreements shall be arbitrated according to the rules of the American Arbitration Association.
- Any lawsuits related to the contract must be filed in the county and state in which the contract is signed.
- Obligations under the agreement are unique, and in the event of a breach, the defaulting party waives the defense that the loss can be adequately compensated by monetary damages (*specific performance*).
- In the event of a lawsuit, the prevailing party is entitled to an award of reasonable attorneys' and court fees.[16]
- Consequential damages are not recoverable in a lawsuit.

10. SUBCONTRACTS

When a party to a contract engages a third party to perform the work in the original contract, the contract with the third party is known as a *subcontract.* Whether or not responsibilities can be subcontracted under the original contract depends on the content of the *assignment clause* in the original contract.

11. PARTIES TO A CONSTRUCTION CONTRACT

A specific set of terms has been developed for referring to parties in consulting and construction contracts. The *owner* of a construction project is the person, partnership, or corporation that actually owns the land, assumes the financial risk, and ends up with the completed project. The *developer* contracts with the architect and/or engineer for the design and with the contractors for the construction of the project. In some cases, the owner and developer are the same, in which case the term *owner-developer* can be used.

The *architect* designs the project according to established codes and guidelines but leaves most stress and capacity calculations to the *engineer.*[17] Depending on the construction contract, the engineer may work for the architect, or vice versa, or both may work for the developer.

[15]Without this clause in writing, damages for delay cannot be claimed.
[16]Without this clause in writing, attorneys' fees and court costs are rarely recoverable.
[17]On simple small projects, such as wood-framed residential units, the design may be developed by a *building designer.* The legal capacities of building designers vary from state to state.

Once there are approved plans, the developer hires *contractors* to do the construction. Usually, the entire construction project is awarded to a *general contractor*. Due to the nature of the construction industry, separate *subcontracts* are used for different tasks (electrical, plumbing, mechanical, framing, fire sprinkler installation, finishing, etc.). The general contractor who hires all of these different *subcontractors* is known as the *prime contractor* (or *prime*). (The subcontractors can also work directly for the owner-developer, although this is less common.) The prime contractor is responsible for all acts of the subcontractors and is liable for any damage suffered by the owner-developer due to those acts.

Construction is managed by an agent of the owner-developer known as the *construction manager*, who may be the engineer, the architect, or someone else.

12. STANDARD CONTRACTS FOR DESIGN PROFESSIONALS

Several professional organizations have produced standard agreement forms and other standard documents for design professionals.[18] Among other standard forms, notices, and agreements, the following standard contracts are available.[19]

- standard contract between engineer and client
- standard contract between engineer and architect
- standard contract between engineer and contractor
- standard contract between owner and construction manager

Besides completeness, the major advantage of a standard contract is that the meanings of the clauses are well established, not only among the design professionals and their clients but also in the courts. The clauses in these contracts have already been litigated many times. Where a clause has been found to be unclear or ambiguous, it has been rewritten to accomplish its intended purpose.

13. CONSULTING FEE STRUCTURE

Compensation for consulting engineering services can incorporate one or more of the following concepts.

- *lump-sum fee:* This is a predetermined fee agreed upon by client and engineer. This payment can be used for small projects where the scope of work is clearly defined.
- *cost plus fixed fee:* All costs (labor, material, travel, etc.) incurred by the engineer are paid by the client. The client also pays a predetermined fee as profit. This method has an advantage when the scope of services cannot be determined accurately in advance. Detailed records must be kept by the engineer in order to allocate costs among different clients.
- *per diem fee:* The engineer is paid a specific sum for each day spent on the job. Usually, certain direct expenses (e.g., travel and reproduction) are billed in addition to the per diem rate.
- *salary plus:* The client pays for the employees on an engineer's payroll (the salary) plus an additional percentage to cover indirect overhead and profit plus certain direct expenses.
- *retainer:* This is a minimum amount paid by the client, usually in total and in advance, for a normal amount of work expected during an agreed-upon period. None of the retainer is returned, regardless of how little work the engineer performs. The engineer can be paid for additional work beyond what is normal, however. Some direct costs, such as travel and reproduction expenses, may be billed directly to the client.
- *percentage of construction cost:* This method, which is widely used in construction design contracts, pays the architect and/or the engineer a percentage of the final total cost of the project. Costs of land, financing, and legal fees are generally not included in the construction cost, and other costs (plan revisions, project management labor, value engineering, etc.) are billed separately.

14. MECHANIC'S LIENS

For various reasons, providers and material, labor, and design services to construction sites may not be promptly paid or even paid at all. Such providers have, of course, the right to file a lawsuit demanding payment, but due to the nature of the construction industry, such

Law and Ethics

[18]There are two main sources of standardized construction and design agreements: EJCDC and AIA. Consensus documents, known as *ConsensusDOCS*, for every conceivable situation have been developed by the *Engineers Joint Contracts Documents Committee*, EJCDC. EJCDC includes the American Society of Civil Engineers (ASCE), the American Council of Engineering Companies (ACEC), National Society of Professional Engineers' (NSPE's) Professional Engineers in Private Practice Division, Associated General Contractors of America (AGC), and more than fifteen other participating professional engineering design, construction, owner, legal, and risk management organizations, including the Associated Builders and Contractors; American Subcontractors Association; Construction Users Roundtable; National Roofing Contractors Association; Mechanical Contractors Association of America; and National Plumbing, Heating-Cooling Contractors Association. The American Institute of Architects, AIA, has developed its own standardized agreements in a less collaborative manner. Though popular with architects, AIA provisions are considered less favorable to engineers, contractors, and subcontractors who believe the AIA documents assign too much authority to architects, too much risk and liability to contractors, and too little flexibility in how construction disputes are addressed and resolved.

[19]The Construction Specifications Institute (CSI) has produced standard specifications for materials. The standards have been organized according to a UNIFORMAT structure consistent with ASTM Standard E1557.

relief may be insufficient or untimely. Therefore, such providers have the right to file a *mechanic's lien* (also known as a *construction lien, materialman's lien, supplier's lien,* or *laborer's lien*) against the property. Although there are strict requirements for deadlines, filing, and notices, the procedure for obtaining (and removing) such a lien is simple. The lien establishes the supplier's security interest in the property. Although the details depend on the state, essentially the property owner is prevented from transferring title (i.e., selling) the property until the lien has been removed by the supplier. The act of filing a lawsuit to obtain payment is known as "perfecting the lien." Liens are perfected by forcing a judicial foreclosure sale. The court orders the property sold, and the proceeds are used to pay off any lien-holders.

15. DISCHARGE OF A CONTRACT

A contract is normally discharged when all parties have satisfied their obligations. However, a contract can also be terminated for the following reasons:

- mutual agreement of all parties to the contract
- impossibility of performance (e.g., death of a party to the contract)
- illegality of the contract
- material breach by one or more parties to the contract
- fraud on the part of one or more parties
- failure (i.e., loss or destruction) of consideration (e.g., the burning of a building one party expected to own or occupy upon satisfaction of the obligations)

Some contracts may be dissolved by actions of the court (e.g., bankruptcy), passage of new laws and public acts, or a declaration of war.

Extreme difficulty (including economic hardship) in satisfying the contract does not discharge it, even if it becomes more costly or less profitable than originally anticipated.

16. BREACH OF CONTRACT, NEGLIGENCE, MISREPRESENTATION, AND FRAUD

A *breach of contract* occurs when one of the parties fails to satisfy all of its obligations under a contract. The breach can be *willful* (as in a contractor walking off a construction job) or *unintentional* (as in providing less than adequate quality work or materials). A *material breach* is defined as nonperformance that results in the injured party receiving something substantially less than or different from what the contract intended.

Normally, the only redress that an *injured party* has through the courts in the event of a breach of contract is to force the breaching party to provide *specific performance*—that is, to satisfy all remaining contract provisions and to pay for any damage caused. Normally, *punitive damages* (to punish the breaching party) are unavailable.

Negligence is an action, willful or unwillful, taken without proper care or consideration for safety, resulting in damages to property or injury to persons. "Proper care" is a subjective term, but in general it is the diligence that would be exercised by a reasonably prudent person.[20] Damages sustained by a negligent act are recoverable in a tort action. (See Sec. 74.17.) If the plaintiff is partially at fault (as in the case of *comparative negligence*), the defendant will be liable only for the portion of the damage caused by the defendant.

Punitive damages are available, however, if the breaching party was fraudulent in obtaining the contract. In addition, the injured party has the right to void (nullify) the contract entirely. A *fraudulent act* is basically a special case of *misrepresentation* (i.e., an intentionally false statement known to be false at the time it is made). Misrepresentation that does not result in a contract is a tort. When a contract is involved, misrepresentation can be a breach of that contract (i.e., *fraud*).

Unfortunately, it is extremely difficult to prove *compensatory fraud* (i.e., fraud for which damages are available). Proving fraud requires showing *beyond a reasonable doubt* (a) a reckless or intentional misstatement of a material fact, (b) an intention to deceive, (c) it resulted in misleading the innocent party to contract, and (d) it was to the innocent party's detriment.

For example, if an engineer claims to have experience in designing steel buildings but actually has none, the court might consider the misrepresentation a fraudulent action. If, however, the engineer has some experience, but an insufficient amount to do an adequate job, the engineer probably will not be considered to have acted fraudulently.

17. TORTS

A *tort* is a civil wrong committed by one person causing damage to another person or person's property, emotional well-being, or reputation.[21] It is a breach of the rights of an individual to be secure in person or property. In order to correct the wrong, a civil lawsuit (*tort action* or *civil complaint*) is brought by the alleged

[20] Negligence of a design professional (e.g., an engineer or architect) is the absence of a *standard of care* (i.e., customary and normal care and attention) that would have been provided by other engineers. It is highly subjective.

[21] The difference between a *civil tort* (*lawsuit*) and a *criminal lawsuit* is the alleged injured party. A *crime* is a wrong against society. A criminal lawsuit is brought by the state against a defendant.

injured party (the *plaintiff*) against the *defendant*. To be a valid *tort action* (i.e., lawsuit), there must have been injury (i.e., damage). Generally, there will be no contract between the two parties, so the tort action cannot claim a breach of contract.[22]

Tort law is concerned with compensation for the injury, not punishment. Therefore, tort awards usually consist of general, compensatory, and special damages and rarely include punitive and exemplary damages. (See Sec. 74.20 for definitions of these damages.)

18. STRICT LIABILITY IN TORT

Strict liability in tort means that the injured party wins if the injury can be proven. It is not necessary to prove negligence, breach of explicit or implicit warranty, or the existence of a contract (*privity of contract*). Strict liability in tort is most commonly encountered in product liability cases. A defect in a product, regardless of how the defect got there, is sufficient to create strict liability in tort.

Case law surrounding defective products has developed and refined the following requirements for winning a strict liability in tort case. The following points must be proved.

- The product was defective in manufacture, design, labeling, and so on.
- The product was defective when used.
- The defect rendered the product unreasonably dangerous.
- The defect caused the injury.
- The specific use of the product that caused the damage was reasonably foreseeable.

19. MANUFACTURING AND DESIGN LIABILITY

Case law makes a distinction between *design professionals* (architects, structural engineers, building designers, etc.) and manufacturers of consumer products. Design professionals are generally consultants whose primary product is a design service sold to sophisticated clients. Consumer product manufacturers produce specific product lines sold through wholesalers and retailers to the unsophisticated public.

The law treats design professionals favorably. Such professionals are expected to meet a *standard of care* and skill that can be measured by comparison with the conduct of other professionals. However, professionals are not expected to be infallible. In the absence of a contract provision to the contrary, design professionals are not held to be guarantors of their work in the strict sense of legal liability. Damages incurred due to design errors are recoverable through tort actions, but proving a breach of contract requires showing negligence (i.e., not meeting the standard of care).

On the other hand, the law is much stricter with consumer product manufacturers, and perfection is (essentially) expected of them. They are held to the standard of strict liability in tort without regard to negligence. A manufacturer is held liable for all phases of the design and manufacturing of a product being marketed to the public.[23]

Prior to 1916, the court's position toward product defects was exemplified by the expression *caveat emptor* ("let the buyer beware").[24] Subsequent court rulings have clarified that "... a manufacturer is strictly liable in tort when an article [it] places on the market, knowing that it will be used without inspection, proves to have a defect that causes injury to a human being."[25]

Although all defectively designed products can be traced back to a design engineer or team, only the manufacturing company is usually held liable for injury caused by the product. This is more a matter of economics than justice. The company has liability insurance; the product design engineer (who is merely an employee of the company) probably does not. Unless the product design or manufacturing process is intentionally defective, or unless the defect is known in advance and covered up, the product design engineer will rarely be punished by the courts.[26]

20. DAMAGES

An injured party can sue for *damages* as well as for specific performance. Damages are the award made by the court for losses incurred by the injured party.

[22]It is possible for an injury to be both a breach of contract and a tort. Suppose an owner has an agreement with a contractor to construct a building, and the contract requires the contractor to comply with all state and federal safety regulations. If the owner is subsequently injured on a stairway because there was no guardrail, the injury could be recoverable both as a tort and as a breach of contract. If a third party unrelated to the contract was injured, however, that party could recover only through a tort action.

[23]The reason for this is that the public is not considered to be as sophisticated as a client who contracts with a design professional for building plans.

[24]1916, *MacPherson v. Buick*. MacPherson bought a Buick from a car dealer. The car had a defective wheel, and there was evidence that reasonable inspection would have uncovered the defect. MacPherson was injured when the wheel broke and the car collapsed, and he sued Buick. Buick defended itself under the ancient *prerequisite of privity* (i.e., the requirement of a face-to-face contractual relationship in order for liability to exist), since the dealer, not Buick, had sold the car to MacPherson, and no contract between Buick and MacPherson existed. The judge disagreed, thus establishing the concept of *third party liability* (i.e., manufacturers are responsible to consumers even though consumers do not buy directly from manufacturers).

[25]1963, *Greenman v. Yuba Power Products*. Greenman purchased and was injured by an electric power tool.

[26]The engineer can expect to be discharged from the company. However, for strategic reasons, this discharge probably will not occur until after the company loses the case.

- *General* or *compensatory damages* are awarded to make up for the injury that was sustained.
- *Special damages* are awarded for the direct financial loss due to the breach of contract.
- *Nominal damages* are awarded when responsibility has been established but the injury is so slight as to be inconsequential.
- *Liquidated damages* are amounts that are specified in the contract document itself for nonperformance.
- *Punitive* or *exemplary damages* are awarded, usually in tort and fraud cases, to punish and make an example of the defendant (i.e., to deter others from doing the same thing).
- *Consequential damages* provide compensation for indirect losses incurred by the injured party but not directly related to the contract.

21. INSURANCE

Most design firms and many independent design professionals carry *errors and omissions insurance* to protect them from claims due to their mistakes. Such policies are costly, and for that reason, some professionals choose to "go bare."[27] Policies protect against inadvertent mistakes only, not against willful, knowing, or conscious efforts to defraud or deceive.

[27] Going bare appears foolish at first glance, but there is a perverted logic behind the strategy. One-person consulting firms (and perhaps, firms that are not profitable) are "judgment-proof." Without insurance or other assets, these firms would be unable to pay any large judgments against them. When damage victims (and their lawyers) find this out in advance, they know that judgments will be uncollectable. So, often the lawsuit never makes its way to trial.

75 Engineering Ethics

Content in blue refers to the *NCEES Handbook*.

1. CREEDS, CODES, CANONS, STATUTES, AND RULES

It is generally conceded that an individual acting on his or her own cannot be counted on to always act in a proper and moral manner. Creeds, statutes, rules, and codes all attempt to complete the guidance needed for an engineer to do "... the correct thing."

A *creed* is a statement or oath, often religious in nature, taken or assented to by an individual in ceremonies. For example, the *Engineers' Creed* adopted by the National Society of Professional Engineers (NSPE) in 1954 is[1]

> As a professional engineer, I dedicate my professional knowledge and skill to the advancement and betterment of human welfare.
>
> I pledge ...
>
> ... to give the utmost of performance;
>
> ... to participate in none but honest enterprise;
>
> ... to live and work according to the laws of man and the highest standards of professional conduct;
>
> ... to place service before profit, the honor and standing of the profession before personal advantage, and the public welfare above all other considerations.
>
> In humility and with need for Divine Guidance, I make this pledge.

A *code* is a system of nonstatutory, nonmandatory canons of personal conduct. A *canon* is a fundamental belief that usually encompasses several rules. For example, the code of ethics of the American Society of Civil Engineers (ASCE) contains seven canons.

1. Engineers shall hold paramount the safety, health, and welfare of the public and shall strive to comply with the principles of sustainable development in the performance of their professional duties.

2. Engineers shall perform services only in areas of their competence.

3. Engineers shall issue public statements only in an objective and truthful manner.

4. Engineers shall act in professional matters for each employer or client as faithful agents or trustees and shall avoid conflicts of interest.

5. Engineers shall build their professional reputation on the merit of their service and shall not compete unfairly with others.

6. Engineers shall act in such a manner as to uphold and enhance the honor, integrity, and dignity of the engineering profession, and shall act with zero tolerance for bribery, fraud, and corruption.

7. Engineers shall continue their professional development throughout their careers and shall provide opportunities for the professional development of those engineers under their supervision.

A *rule* is a guide (principle, standard, or norm) for conduct and action in a certain situation. A *statutory rule* is enacted by the legislative branch of a state or federal government and carries the weight of law. Some U.S. engineering registration boards have statutory *rules of professional conduct*.

[1]The *Faith of an Engineer*, adopted by the Accreditation Board for Engineering and Technology (ABET), formerly the Engineer's Council for Professional Development (ECPD), is a similar but more detailed creed.

2. PURPOSE OF A CODE OF ETHICS

Many different sets of *codes of ethics* (*canons of ethics*, *rules of professional conduct*, etc.) have been produced by various engineering societies, registration boards, and other organizations.[2] The purpose of these ethical guidelines is to guide the conduct and decision making of engineers. Most codes are primarily educational. Nevertheless, from time to time they have been used by the societies and regulatory agencies as the basis for disciplinary actions.

Fundamental to ethical codes is the requirement that engineers render faithful, honest, professional service. In providing such service, engineers must represent the interests of their employers or clients and, at the same time, protect public health, safety, and welfare.

There is an important distinction between what is legal and what is ethical. Many actions that are legal can be violations of codes of ethical or professional behavior.[3] For example, an engineer's contract with a client may give the engineer the right to assign the engineer's responsibilities, but doing so without informing the client would be unethical.

Ethical guidelines can be categorized on the basis of who is affected by the engineer's actions—the client, vendors and suppliers, other engineers, or the public at large.[4]

3. ETHICAL PRIORITIES

There are frequently conflicting demands on engineers. While it is impossible to use a single decision-making process to solve every ethical dilemma, it is clear that ethical considerations will force engineers to subjugate their own self-interests. Specifically, the ethics of engineers dealing with others need to be considered in the following order, from highest to lowest priority.

- society and the public
- the law
- the engineering profession
- the engineer's client
- the engineer's firm
- other involved engineers
- the engineer personally

4. DEALING WITH CLIENTS AND EMPLOYERS

The most common ethical guidelines affecting engineers' interactions with their employer (the *client*) can be summarized as follows.[5]

- Engineers should not accept assignments for which he/she does not have the skill, knowledge, or time.
- Engineers must recognize his/her own limitations, They should use associates and other experts when the design requirements exceed their abilities.
- The client's interests must be protected. The extent of this protection exceeds normal business relationships and transcends the legal requirements of the engineer-client contract.
- Engineers must not be bound by what the client wants in instances where such desires would be unsuccessful, dishonest, unethical, unhealthy, or unsafe.
- Confidential client information remains the property of the client and must be kept confidential.
- Engineers must avoid conflicts of interest and should inform the client of any business connections or interests that might influence their judgment. Engineers should also avoid the *appearance* of a conflict of interest when such an appearance would be detrimental to the profession, their client, or themselves.
- The engineers' sole source of income for a particular project should be the fee paid by their client. Engineers should not accept compensation in any form from more than one party for the same services.
- If the client rejects the engineer's recommendations, the engineer should fully explain the consequences to the client.
- Engineers must freely and openly admit to the client any errors made.

All courts of law have required an engineer to perform in a manner consistent with normal professional standards. This is not the same as saying an engineer's work must be error-free. If an engineer completes a design,

[2]All of the major engineering technical and professional societies in the United States (ASCE, IEEE, ASME, AIChE, NSPE, etc.) and throughout the world have adopted codes of ethics. Most U.S. societies have endorsed the *Code of Ethics of Engineers* developed by the Accreditation Board for Engineering and Technology (ABET). The National Council of Examiners for Engineering and Surveying (NCEES) has developed its *Model Rules of Professional Conduct* as a guide for state registration boards in developing guidelines for the professional engineers in those states.

[3]Whether the guidelines emphasize ethical behavior or professional conduct is a matter of wording. The intention is the same: to provide guidelines that transcend the requirements of the law.

[4]Some authorities also include ethical guidelines for dealing with the employees of an engineer. However, these guidelines are no different for an engineering employer than they are for a supermarket, automobile assembly line, or airline employer. Ethics is not a unique issue when it comes to employees.

[5]These general guidelines contain references to contractors, plans, specifications, and contract documents. This language is common, though not unique, to the situation of an engineer supplying design services to an owner-developer or architect. However, most of the ethical guidelines are general enough to apply to engineers in industry as well.

has the design and calculations checked by another competent engineer, and an error is subsequently shown to have been made, the engineer may be held responsible, but the engineer will probably not be considered negligent.

5. DEALING WITH SUPPLIERS

Engineers routinely deal with manufacturers, contractors, and vendors (*suppliers*). In this regard, engineers have great responsibility and influence. Such a relationship requires that engineers deal justly with both clients and suppliers.

An engineer will often have an interest in maintaining good relationships with suppliers since this often leads to future work. Nevertheless, relationships with suppliers must remain highly ethical. Suppliers should not be encouraged to feel that they have any special favors coming to them because of a long-standing relationship with the engineer.

The ethical responsibilities relating to suppliers are listed as follows.

- The engineer must not accept or solicit gifts or other valuable considerations from a supplier during, prior to, or after any job. An engineer should not accept discounts, allowances, commissions, or any other indirect compensation from suppliers, contractors, or other engineers in connection with any work or recommendations.
- The engineer must enforce the plans and specifications (i.e., the *contract documents*), but must also interpret the contract documents fairly.
- Plans and specifications developed by an engineer on behalf of the client must be complete, definite, and specific.
- Suppliers should not be required to spend time or furnish materials that are not called for in the plans and contract documents.
- The engineer should not unduly delay the performance of suppliers.

6. DEALING WITH OTHER ENGINEERS

Engineers should try to protect the engineering profession as a whole, to strengthen it, and to enhance its public stature. The following ethical guidelines apply.

- An engineer should not attempt to maliciously injure the professional reputation, business practice, or employment position of another engineer. However, if there is proof that another engineer has acted unethically or illegally, the engineer should advise the proper authority.
- An engineer should not review someone else's work while the other engineer is still employed unless the other engineer is made aware of the review.
- An engineer should not try to replace another engineer once the other engineer has received employment.
- An engineer should not use the advantages of a salaried position to compete unfairly (i.e., moonlight) with other engineers who have to charge more for the same consulting services.
- Subject to legal and proprietary restraints, an engineer should freely report, publish, and distribute information that would be useful to other engineers.

7. DEALING WITH (AND AFFECTING) THE PUBLIC

In regard to the social consequences of engineering, the relationship between an engineer and the public is essentially straightforward. Responsibilities to the public demand that the engineer place service to humankind above personal gain. Furthermore, proper ethical behavior requires that an engineer avoid association with projects that are contrary to public health and welfare or that are of questionable legal character.

- Engineers must consider the safety, health, and welfare of the public in all work performed.
- Engineers must uphold the honor and dignity of his/her profession by refraining from self-laudatory advertising, by explaining (when required) their work to the public, and by expressing opinions only in areas of knowledge.
- When engineers issues a public statement, they must clearly indicate if the statement is being made on anyone's behalf (i.e., if anyone is benefitting from their position).
- Engineers must keep their skills at a state-of-the-art level.
- Engineers should develop public knowledge and appreciation of the engineering profession and its achievements.
- Engineers must notify the proper authorities when decisions adversely affecting public safety and welfare are made.[6]

[6]This practice has come to be known as *whistle-blowing*.

8. COMPETITIVE BIDDING

The ethical guidelines for dealing with other engineers presented here and in more detailed codes of ethics no longer include a prohibition on *competitive bidding.* Until 1971, most codes of ethics for engineers considered competitive bidding detrimental to public welfare, since cost-cutting normally results in a lower quality design.

However, in a 1971 case against NSPE that went all the way to the U.S. Supreme Court, the prohibition against competitive bidding was determined to be a violation of the Sherman Antitrust Act (i.e., it was an unreasonable restraint of trade).

The opinion of the Supreme Court does not *require* competitive bidding—it merely forbids a prohibition against competitive bidding in NSPE's code of ethics. The following points must be considered.

- Engineers and design firms may individually continue to refuse to bid competitively on engineering services.
- Clients are not required to seek competitive bids for design services.
- Federal, state, and local statutes governing the procedures for procuring engineering design services, even those statutes that prohibit competitive bidding, are not affected.
- Any prohibitions against competitive bidding in individual state engineering registration laws remain unaffected.
- Engineers and their societies may actively and aggressively lobby for legislation that would prohibit competitive bidding for design services by public agencies.

9. MODERN ETHICAL ISSUES

With few exceptions, ethical concepts have changed little for engineers since they were first developed by the fledgling engineering societies. Ethical rules covering such issues as service-before-self, conflicts of interest, bribery and kickbacks, moonlighting and unfair competition, respectable dealings with engineering competitors, and respect for the profession and its image may have seen wording changes, but their basic concepts have not changed. Even protection for whistle-blowers can be traced back to the early 1900s, although environmental whistle-blowing came into its own in the 1970s.

Modern engineers, however, have to deal with newer concepts such as environmental protection, sustainable (green) design, offshoring, energy efficiency and dependency, cultural diversity and tolerance, nutrition and safety of foodstuffs and drugs, national sufficiency, and issues affecting national security and personal privacy and security. Most engineers have little experience or training in these concepts, and for some issues, neither regulations nor technology exist to help engineers arrive at definitive decisions.

Environmental protection, from an ethical standpoint, places burdens on engineers that go far beyond adhering to EPA and U.S. Army Corps of Engineers regulations. Issues include waste, pollution, loss of biodiversity, introduction of invasive species, release of genetically modified organisms, genetically modified foodstuffs, and release of toxic substances.

In 2006, the American Society of Civil Engineers amended its first canon to include mention of *sustainable development*, which is subsequently defined as follows: "Sustainable development is the process of applying natural, human, and economic resources to enhance the safety, welfare, and quality of life for all of society while maintaining the availability of the remaining natural resources."

Offshoring broadly means the replacement of a service originally performed within an organization by a service located in a foreign country. Offshoring applies to designing facilities and processes sited in foreign countries, as well as the more common issue of sending jobs to foreign countries. Closer to home, the engineering professional itself has struggled with respecting and accepting offshore engineering services. The very cost-cutting moves that engineers have pursued in service to their employers have hurt engineers and their profession and has given rise to a form of nationalism founded on resentment.

Topic XV: Support Material

Appendices

Support Material

APPENDIX 7.A
Mensuration of Two-Dimensional Areas

Nomenclature

A	total surface area
b	base
c	chord length
d	distance
h	height
L	length
p	perimeter
r	radius
s	side (edge) length, arc length
θ	vertex angle, in radians
ϕ	central angle, in radians

Triangle

equilateral

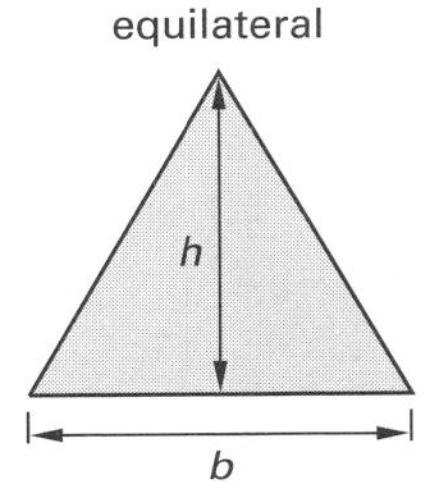

right

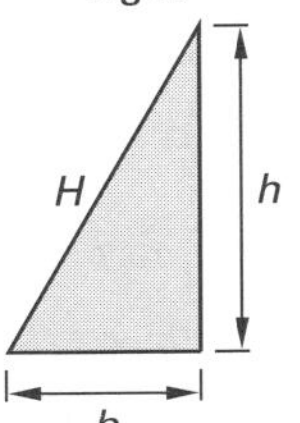

oblique

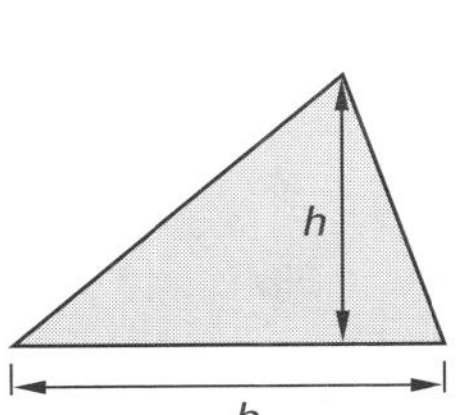

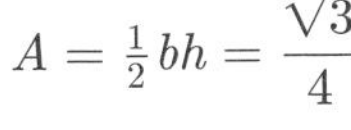

$$A = \tfrac{1}{2}bh = \frac{\sqrt{3}}{4}b^2$$

$$h = \frac{\sqrt{3}}{2}b$$

$$A = \tfrac{1}{2}bh$$

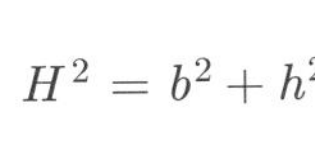

$$H^2 = b^2 + h^2$$

$$A = \tfrac{1}{2}bh$$

Circle

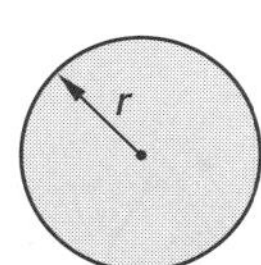

$$p = 2\pi r$$

$$A = \pi r^2 = \frac{p^2}{4\pi}$$

Circular Segment

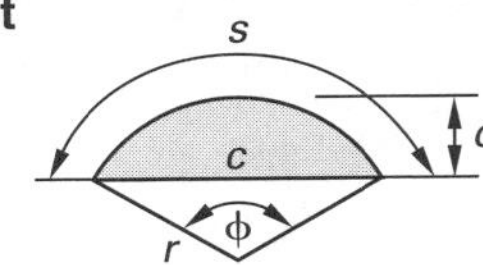

$$A = \tfrac{1}{2}r^2(\phi - \sin\phi)$$

$$\phi = \frac{s}{r} = 2\left(\arccos\frac{r-d}{r}\right)$$

$$c = 2r\sin\left(\frac{\phi}{2}\right)$$

Circular Sector

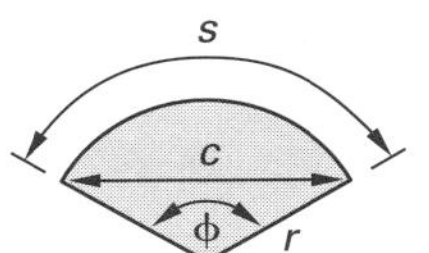

$$A = \tfrac{1}{2}\phi r^2 = \tfrac{1}{2}sr$$

$$\phi = \frac{s}{r}$$

$$s = r\phi$$

$$c = 2r\sin\left(\frac{\phi}{2}\right)$$

Parabola

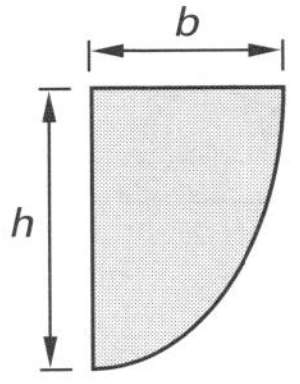

$$A = \tfrac{2}{3}bh$$

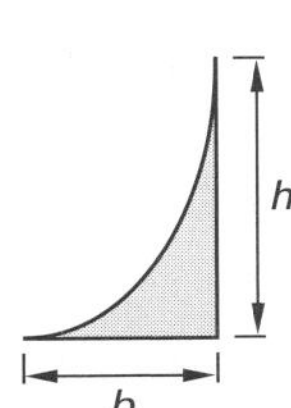

$$A = \tfrac{1}{3}bh$$

Ellipse

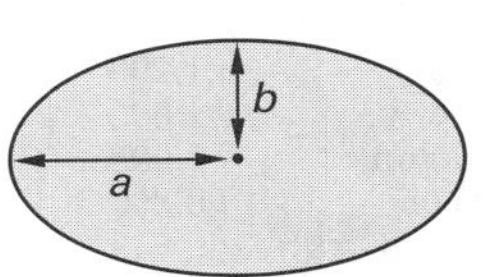

$$A = \pi ab$$

$$p \approx 2\pi\sqrt{\tfrac{1}{2}(a^2 + b^2)} \quad \left[\begin{array}{c}\text{Euler's} \\ \text{upper bound}\end{array}\right]$$

Support Material

APPENDIX 7.A *(continued)*
Mensuration of Two-Dimensional Areas

Trapezoid

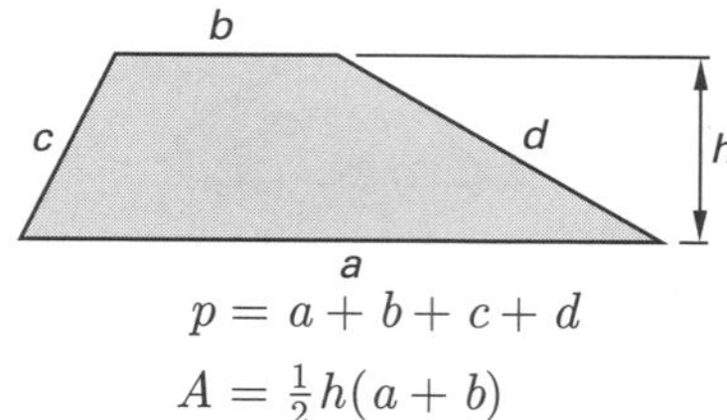

$$p = a + b + c + d$$

$$A = \tfrac{1}{2}h(a + b)$$

If $c = d$, the trapezoid is isosceles.

Parallelogram

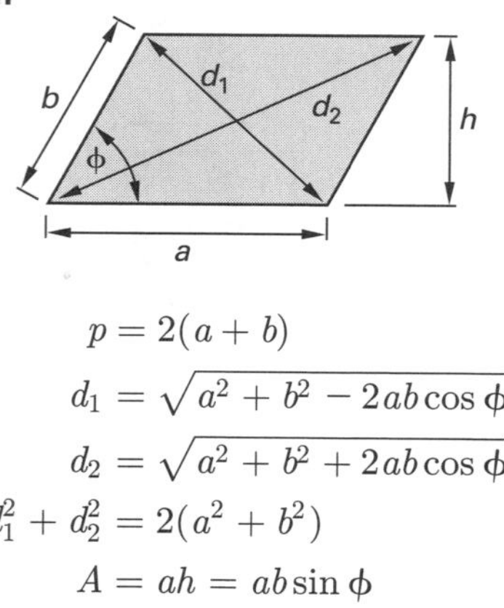

$$p = 2(a + b)$$

$$d_1 = \sqrt{a^2 + b^2 - 2ab\cos\phi}$$

$$d_2 = \sqrt{a^2 + b^2 + 2ab\cos\phi}$$

$$d_1^2 + d_2^2 = 2(a^2 + b^2)$$

$$A = ah = ab\sin\phi$$

If $a = b$, the parallelogram is a rhombus.

Regular Polygon (*n* equal sides)

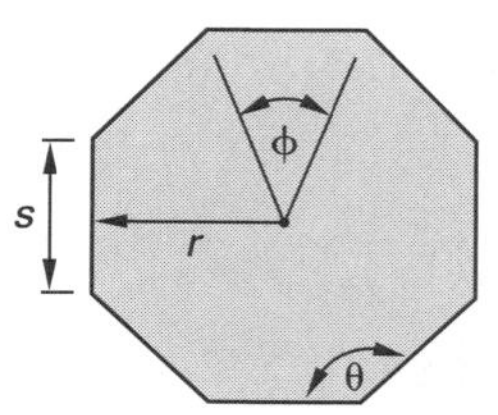

$$\phi = \frac{2\pi}{n}$$

$$\theta = \frac{\pi(n-2)}{n} = \pi - \phi$$

$$p = ns$$

$$s = 2r\tan\frac{\theta}{2}$$

$$A = \tfrac{1}{2}nsr$$

regular polygons		$d = 1$, A	$L = 1$, A	$L = 1$, r	L, $r = 1$	L, $p = 1$	$L = 1$, p
sides	name	area (A) when diameter of inscribed circle = 1	area (A) when side = 1	radius (r) of circumscribed circle when side = 1	length (L) of side when radius (r) of circumscribed circle = 1	length (L) of side when perpendicular to circle = 1	perpendicular (p) to center when side = 1
3	triangle	1.299	0.433	0.577	1.732	3.464	0.289
4	square	1.000	1.000	0.707	1.414	2.000	0.500
5	pentagon	0.908	1.720	0.851	1.176	1.453	0.688
6	hexagon	0.866	2.598	1.000	1.000	1.155	0.866
7	heptagon	0.843	3.634	1.152	0.868	0.963	1.038
8	octagon	0.828	4.828	1.307	0.765	0.828	1.207
9	nonagon	0.819	6.182	1.462	0.684	0.728	1.374
10	decagon	0.812	7.694	1.618	0.618	0.650	1.539
11	undecagon	0.807	9.366	1.775	0.563	0.587	1.703
12	dodecagon	0.804	11.196	1.932	0.518	0.536	1.866

APPENDIX 7.B
Mensuration of Three-Dimensional Volumes

Nomenclature

A	surface area
b	base
h	height
r	radius
R	radius
s	side (edge) length
V	internal volume

Sphere

$$V = \tfrac{4}{3}\pi r^3 = \tfrac{4}{3}\pi\left(\frac{d}{2}\right)^3 = \tfrac{1}{6}\pi d^3$$

$$A = 4\pi r^2$$

Right Circular Cone (excluding base area)

$$V = \tfrac{1}{3}\pi r^2 h = \tfrac{1}{3}\pi\left(\frac{d}{2}\right)^2 h = \tfrac{1}{12}\pi d^2 h$$

$$A = \pi r\sqrt{r^2 + h^2}$$

Right Circular Cylinder (excluding end areas)

$$V = \pi r^2 h$$

$$A = 2\pi r h$$

Spherical Segment (spherical cap)

Surface area of a spherical segment of radius r cut out by an angle θ_0 rotated from the center about a radius, r, is

$$A = 2\pi r^2(1 - \cos\theta_0)$$

$$\omega = \frac{A}{r^2} = 2\pi(1 - \cos\theta_0)$$

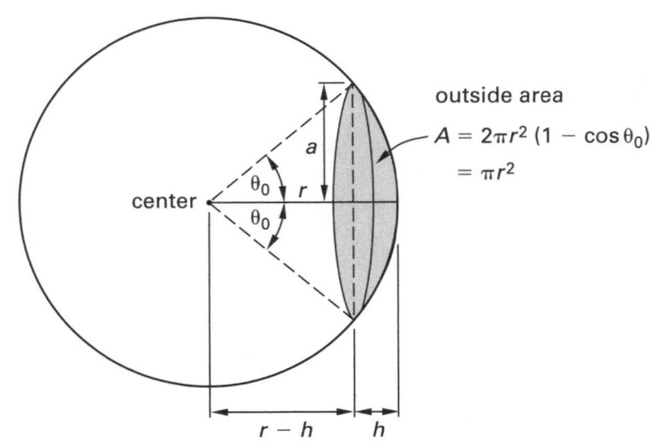

$$V_{cap} = \tfrac{1}{6}\pi h(3a^2 + h^2)$$

$$= \tfrac{1}{3}\pi h^2(3r - h)$$

$$a = \sqrt{h(2r - h)}$$

Paraboloid of Revolution

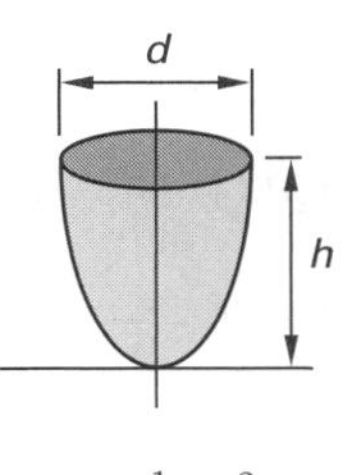

$$V = \tfrac{1}{8}\pi d^2 h$$

Torus

$$A = 4\pi^2 R r$$

$$V = 2\pi^2 R r^2$$

Regular Polyhedra (identical faces)

name	number of faces	form of faces	total surface area	volume
tetrahedron	4	equilateral triangle	$1.7321 s^2$	$0.1179 s^3$
cube	6	square	$6.0000 s^2$	$1.0000 s^3$
octahedron	8	equilateral triangle	$3.4641 s^2$	$0.4714 s^3$
dodecahedron	12	regular pentagon	$20.6457 s^2$	$7.6631 s^3$
icosahedron	20	equilateral triangle	$8.6603 s^2$	$2.1817 s^3$

The radius of a sphere inscribed within a regular polyhedron is

$$r = \frac{3V_{polyhedron}}{A_{polyhedron}}$$

Support Material

APPENDIX 9.A
Abbreviated Table of Indefinite Integrals
(In each case, add a constant of integration. All angles are measured in radians.)

General Formulas

1. $\int dx = x$

2. $\int c\,dx = c\int dx$

3. $\int (dx + dy) = \int dx + \int dy$

4. $\int u\,dv = uv - \int v\,du$ $\left[\begin{array}{c}\text{integration by parts; } u \text{ and } v \text{ are}\\ \text{functions of the same variable}\end{array}\right]$

Algebraic Forms

5. $\int x^n\,dx = \dfrac{x^{n+1}}{n+1} \quad [n \neq -1]$

6. $\int x^{-1}dx = \int \dfrac{dx}{x} = \ln|x|$

7. $\int (ax+b)^n\,dx = \dfrac{(ax+b)^{n+1}}{a(n+1)} \quad [n \neq -1]$

8. $\int \dfrac{dx}{ax+b} = \dfrac{1}{a}\ln(ax+b)$

9. $\int \dfrac{x\,dx}{ax+b} = \dfrac{1}{a^2}\big(ax+b-b\ln(ax+b)\big)$

10. $\int \dfrac{x\,dx}{(ax+b)^2} = \dfrac{1}{a^2}\left(\dfrac{b}{ax+b}+\ln(ax+b)\right)$

11. $\int \dfrac{dx}{x(ax+b)} = \dfrac{1}{b}\ln\left(\dfrac{x}{ax+b}\right)$

12. $\int \dfrac{dx}{x(ax+b)^2} = \dfrac{1}{b(ax+b)}+\dfrac{1}{b^2}\ln\left(\dfrac{x}{ax+b}\right)$

13. $\int \dfrac{dx}{x^2+a^2} = \dfrac{1}{a}\arctan\left(\dfrac{x}{a}\right)$

14. $\int \dfrac{dx}{a^2-x^2} = \dfrac{1}{a}\operatorname{arctanh}\left(\dfrac{x}{a}\right)$

15. $\int \dfrac{x\,dx}{ax^2+b} = \dfrac{1}{2a}\ln(ax^2+b)$

16. $\int \dfrac{dx}{x(ax^n+b)} = \dfrac{1}{bn}\ln\left(\dfrac{x^n}{ax^n+b}\right)$

17. $\int \dfrac{dx}{ax^2+bx+c} = \dfrac{1}{\sqrt{b^2-4ac}}\ln\left(\dfrac{2ax+b-\sqrt{b^2-4ac}}{2ax+b+\sqrt{b^2-4ac}}\right) \quad [b^2 > 4ac]$

18. $\int \dfrac{dx}{ax^2+bx+c} = \dfrac{2}{\sqrt{4ac-b^2}}\arctan\left(\dfrac{2ax+b}{\sqrt{4ac-b^2}}\right) \quad [b^2 < 4ac]$

19. $\int \sqrt{a^2-x^2}\,dx = \dfrac{x}{2}\sqrt{a^2-x^2}+\dfrac{a^2}{2}\arcsin\left(\dfrac{x}{a}\right)$

20. $\int x\sqrt{a^2-x^2}\,dx = -\tfrac{1}{3}(a^2-x^2)^{3/2}$

21. $\int \dfrac{dx}{\sqrt{a^2-x^2}} = \arcsin\left(\dfrac{x}{a}\right)$

22. $\int \dfrac{x\,dx}{\sqrt{a^2-x^2}} = -\sqrt{a^2-x^2}$

Support Material

APPENDIX 10.A
Laplace Transforms

$f(t)$	$\mathcal{L}(f(t))$	$f(t)$	$\mathcal{L}(f(t))$
$\delta(t)$ [unit impulse at $t=0$]	1	$1-\cos at$	$\frac{a^2}{s(s^2+a^2)}$
$\delta(t-c)$ [unit impulse at $t=c$]	e^{-cs}	$\cosh at$	$\frac{s}{s^2-a^2}$
1 or u_0 [unit step at $t=0$]	$\frac{1}{s}$	$t\cos at$	$\frac{s^2-a^2}{(s^2+a^2)^2}$
u_c [unit step at $t=c$]	$\frac{e^{-cs}}{s}$	t^n [n is a positive integer]	$\frac{n!}{s^{n+1}}$
t [unit ramp at $t=0$]	$\frac{1}{s^2}$	e^{at}	$\frac{1}{s-a}$
rectangular pulse, magnitude M, duration a	$\left(\frac{M}{s}\right)(1-e^{-as})$	$e^{at}\sin bt$	$\frac{b}{(s-a)^2+b^2}$
triangular pulse, magnitude M, duration $2a$	$\left(\frac{M}{as^2}\right)(1-e^{-as})^2$	$e^{at}\cos bt$	$\frac{s-a}{(s-a)^2+b^2}$
sawtooth pulse, magnitude M, duration a	$\left(\frac{M}{as^2}\right)\left(1-(as+1)e^{-as}\right)$	$e^{at}t^n$ [n is a positive integer]	$\frac{n!}{(s-a)^{n+1}}$
sinusoidal pulse, magnitude M, duration π/a	$\left(\frac{Ma}{s^2+a^2}\right)(1+e^{-\pi s/a})$	$1-e^{-at}$	$\frac{a}{s(s+a)}$
$\frac{t^{n-1}}{(n-1)!}$	$\frac{1}{s^n}$	$e^{-at}+at-1$	$\frac{a^2}{s^2(s+a)}$
$\sin at$	$\frac{a}{s^2+a^2}$	$\frac{e^{-at}-e^{-bt}}{b-a}$	$\frac{1}{(s+a)(s+b)}$
$at-\sin at$	$\frac{a^3}{s^2(s^2+a^2)}$	$\frac{(c-a)e^{-at}-(c-b)e^{-bt}}{b-a}$	$\frac{s+c}{(s+a)(s+b)}$
$\sinh at$	$\frac{a}{s^2-a^2}$	$\frac{1}{ab}+\frac{be^{-at}-ae^{-bt}}{ab(a-b)}$	$\frac{1}{s(s+a)(s+b)}$
$t\sin at$	$\frac{2as}{(s^2+a^2)^2}$	$t\sinh at$	$\frac{2as}{(s^2-a^2)^2}$
$\cos at$	$\frac{s}{s^2+a^2}$	$t\cosh at$	$\frac{s^2+a^2}{(s^2-a^2)^2}$

APPENDIX 11.A
Areas Under the Standard Normal Curve (0 to z)

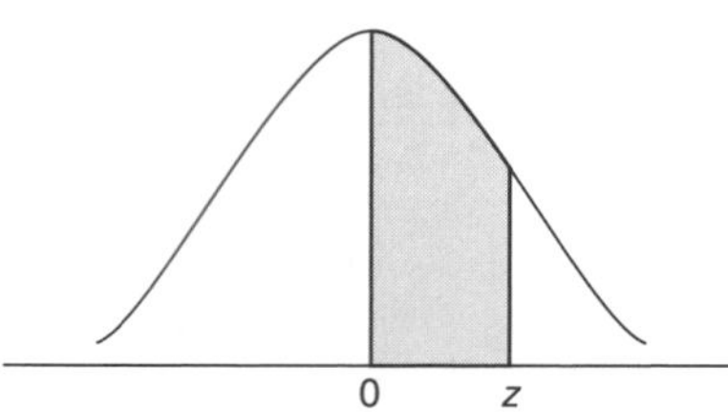

z	0	1	2	3	4	5	6	7	8	9
0.0	0.0000	0.0040	0.0080	0.0120	0.0160	0.0199	0.0239	0.0279	0.0319	0.0359
0.1	0.0398	0.0438	0.0478	0.0517	0.0557	0.0596	0.0636	0.0675	0.0714	0.0754
0.2	0.0793	0.0832	0.0871	0.0910	0.0948	0.0987	0.1026	0.1064	0.1103	0.1141
0.3	0.1179	0.1217	0.1255	0.1293	0.1331	0.1368	0.1406	0.1443	0.1480	0.1517
0.4	0.1554	0.1591	0.1628	0.1664	0.1700	0.1736	0.1772	0.1808	0.1844	0.1879
0.5	0.1915	0.1950	0.1985	0.2019	0.2054	0.2088	0.2123	0.2157	0.2190	0.2224
0.6	0.2258	0.2291	0.2324	0.2357	0.2389	0.2422	0.2454	0.2486	0.2518	0.2549
0.7	0.2580	0.2612	0.2642	0.2673	0.2704	0.2734	0.2764	0.2794	0.2823	0.2852
0.8	0.2881	0.2910	0.2939	0.2967	0.2996	0.3023	0.3051	0.3078	0.3106	0.3133
0.9	0.3159	0.3186	0.3212	0.3238	0.3264	0.3289	0.3315	0.3340	0.3365	0.3389
1.0	0.3413	0.3438	0.3461	0.3485	0.3508	0.3531	0.3554	0.3577	0.3599	0.3621
1.1	0.3643	0.3665	0.3686	0.3708	0.3729	0.3749	0.3770	0.3790	0.3810	0.3830
1.2	0.3849	0.3869	0.3888	0.3907	0.3925	0.3944	0.3962	0.3980	0.3997	0.4015
1.3	0.4032	0.4049	0.4066	0.4082	0.4099	0.4115	0.4131	0.4147	0.4162	0.4177
1.4	0.4192	0.4207	0.4222	0.4236	0.4251	0.4265	0.4279	0.4292	0.4306	0.4319
1.5	0.4332	0.4345	0.4357	0.4370	0.4382	0.4394	0.4406	0.4418	0.4429	0.4441
1.6	0.4452	0.4463	0.4474	0.4484	0.4495	0.4505	0.4515	0.4525	0.4535	0.4545
1.7	0.4554	0.4564	0.4573	0.4582	0.4591	0.4599	0.4608	0.4616	0.4625	0.4633
1.8	0.4641	0.4649	0.4656	0.4664	0.4671	0.4678	0.4686	0.4693	0.4699	0.4706
1.9	0.4713	0.4719	0.4726	0.4732	0.4738	0.4744	0.4750	0.4756	0.4761	0.4767
2.0	0.4772	0.4778	0.4783	0.4788	0.4793	0.4798	0.4803	0.4808	0.4812	0.4817
2.1	0.4821	0.4826	0.4830	0.4834	0.4838	0.4842	0.4846	0.4850	0.4854	0.4857
2.2	0.4861	0.4864	0.4868	0.4871	0.4875	0.4878	0.4881	0.4884	0.4887	0.4890
2.3	0.4893	0.4896	0.4898	0.4901	0.4904	0.4906	0.4909	0.4911	0.4913	0.4916
2.4	0.4918	0.4920	0.4922	0.4925	0.4927	0.4929	0.4931	0.4932	0.4934	0.4936
2.5	0.4938	0.4940	0.4941	0.4943	0.4945	0.4946	0.4948	0.4949	0.4951	0.4952
2.6	0.4953	0.4955	0.4956	0.4957	0.4959	0.4960	0.4961	0.4962	0.4963	0.4964
2.7	0.4965	0.4966	0.4967	0.4968	0.4969	0.4970	0.4971	0.4972	0.4973	0.4974
2.8	0.4974	0.4975	0.4976	0.4977	0.4977	0.4978	0.4979	0.4979	0.4980	0.4981
2.9	0.4981	0.4982	0.4982	0.4983	0.4984	0.4984	0.4985	0.4985	0.4986	0.4986
3.0	0.4987	0.4987	0.4987	0.4988	0.4988	0.4989	0.4989	0.4989	0.4990	0.4990
3.1	0.4990	0.4991	0.4991	0.4991	0.4992	0.4992	0.4992	0.4992	0.4993	0.4993
3.2	0.4993	0.4993	0.4994	0.4994	0.4994	0.4994	0.4994	0.4995	0.4995	0.4995
3.3	0.4995	0.4995	0.4996	0.4996	0.4996	0.4996	0.4996	0.4996	0.4996	0.4997
3.4	0.4997	0.4997	0.4997	0.4997	0.4997	0.4997	0.4997	0.4997	0.4997	0.4998
3.5	0.4998	0.4998	0.4998	0.4998	0.4998	0.4998	0.4998	0.4998	0.4998	0.4998
3.6	0.4998	0.4998	0.4999	0.4999	0.4999	0.4999	0.4999	0.4999	0.4999	0.4999
3.7	0.4999	0.4999	0.4999	0.4999	0.4999	0.4999	0.4999	0.4999	0.4999	0.4999
3.8	0.4999	0.4999	0.4999	0.4999	0.4999	0.4999	0.4999	0.4999	0.4999	0.4999
3.9	0.5000	0.5000	0.5000	0.5000	0.5000	0.5000	0.5000	0.5000	0.5000	0.5000

APPENDIX 11.B
Chi-Squared Distribution

degrees of	probability of exceeding the critical value, α				
freedom	0.10	0.05	0.025	0.01	0.001
1	2.706	3.841	5.024	6.635	10.828
2	4.605	5.991	7.378	9.210	13.816
3	6.251	7.815	9.348	11.345	16.266
4	7.779	9.488	11.143	13.277	18.467
5	9.236	11.070	12.833	15.086	20.515
6	10.645	12.592	14.449	16.812	22.458
7	12.017	14.067	16.013	18.475	24.322
8	13.362	15.507	17.535	20.090	26.125
9	14.684	16.919	19.023	21.666	27.877
10	15.987	18.307	20.483	23.209	29.588
11	17.275	19.675	21.920	24.725	31.264
12	18.549	21.026	23.337	26.217	32.910
13	19.812	22.362	24.736	27.688	34.528
14	21.064	23.685	26.119	29.141	36.123
15	22.307	24.996	27.488	30.578	37.697
16	23.542	26.296	28.845	32.000	39.252
17	24.769	27.587	30.191	33.409	40.790
18	25.989	28.869	31.526	34.805	42.312
19	27.204	30.144	32.852	36.191	43.820
20	28.412	31.410	34.170	37.566	45.315
21	29.615	32.671	35.479	38.932	46.797
22	30.813	33.924	36.781	40.289	48.268

APPENDIX 11.C

Values of t_C for Student's t-Distribution

(ν degrees of freedom; confidence level C; $\alpha = 1 - C$; shaded area = p)

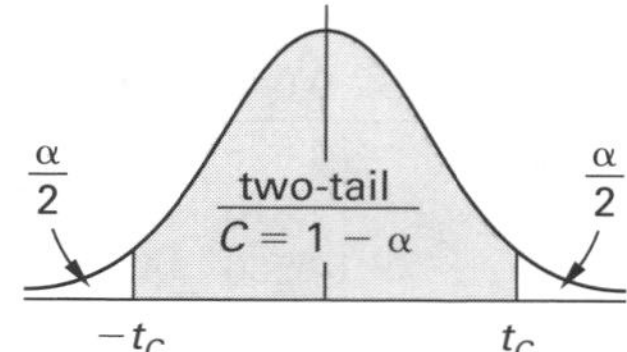

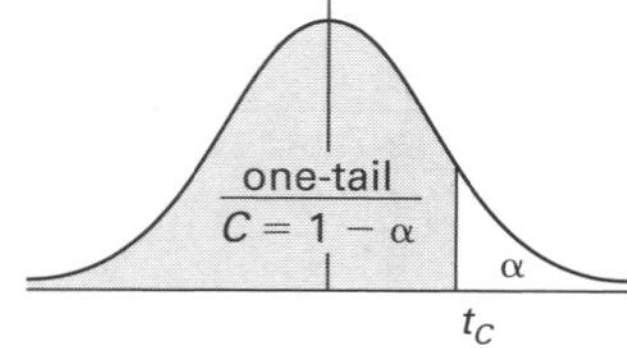

ν (df)										
two-tail, t_C	$t_{99\%}$	$t_{98\%}$	$t_{95\%}$	$t_{90\%}$	$t_{80\%}$	$t_{60\%}$	$t_{50\%}$	$t_{40\%}$	$t_{20\%}$	$t_{10\%}$
one-tail, t_C	$t_{99.5\%}$	$t_{99\%}$	$t_{97.5\%}$	$t_{95\%}$	$t_{90\%}$	$t_{80\%}$	$t_{75\%}$	$t_{70\%}$	$t_{60\%}$	$t_{55\%}$
1	63.657	31.821	12.706	6.314	3.078	1.376	1.000	0.727	0.325	0.158
2	9.925	6.965	4.303	2.920	1.886	1.061	0.816	0.617	0.289	0.142
3	5.841	4.541	3.182	2.353	1.638	0.978	0.765	0.584	0.277	0.137
4	4.604	3.747	2.776	2.132	1.533	0.941	0.741	0.569	0.271	0.134
5	4.032	3.365	2.571	2.015	1.476	0.920	0.727	0.559	0.267	0.132
6	3.707	3.143	2.447	1.943	1.440	0.906	0.718	0.553	0.265	0.131
7	3.499	2.998	2.365	1.895	1.415	0.896	0.711	0.549	0.263	0.130
8	3.355	2.896	2.306	1.860	1.397	0.889	0.706	0.546	0.262	0.130
9	3.250	2.821	2.262	1.833	1.383	0.883	0.703	0.543	0.261	0.129
10	3.169	2.764	2.228	1.812	1.372	0.879	0.700	0.542	0.260	0.129
11	3.106	2.718	2.201	1.796	1.363	0.876	0.697	0.540	0.260	0.129
12	3.055	2.681	2.179	1.782	1.356	0.873	0.695	0.539	0.259	0.128
13	3.012	2.650	2.160	1.771	1.350	0.870	0.694	0.538	0.259	0.128
14	2.977	2.624	2.145	1.761	1.345	0.868	0.692	0.537	0.258	0.128
15	2.947	2.602	2.131	1.753	1.341	0.866	0.691	0.536	0.258	0.128
16	2.921	2.583	2.120	1.746	1.337	0.865	0.690	0.535	0.258	0.128
17	2.898	2.567	2.110	1.740	1.333	0.863	0.689	0.534	0.257	0.128
18	2.878	2.552	2.101	1.734	1.330	0.862	0.688	0.534	0.257	0.127
19	2.861	2.539	2.093	1.729	1.328	0.861	0.688	0.533	0.257	0.127
20	2.845	2.528	2.086	1.725	1.325	0.860	0.687	0.533	0.257	0.127
21	2.831	2.518	2.080	1.721	1.323	0.859	0.686	0.532	0.257	0.127
22	2.819	2.508	2.074	1.717	1.321	0.858	0.686	0.532	0.256	0.127
23	2.807	2.500	2.069	1.714	1.319	0.858	0.685	0.532	0.256	0.127
24	2.797	2.492	2.064	1.711	1.318	0.857	0.685	0.531	0.256	0.127
25	2.787	2.485	2.060	1.708	1.316	0.856	0.684	0.531	0.256	0.127
26	2.779	2.479	2.056	1.706	1.315	0.856	0.684	0.531	0.256	0.127
27	2.771	2.473	2.052	1.703	1.314	0.855	0.684	0.531	0.256	0.127
28	2.763	2.467	2.048	1.701	1.313	0.855	0.683	0.530	0.256	0.127
29	2.756	2.462	2.045	1.699	1.311	0.854	0.683	0.530	0.256	0.127
30	2.750	2.457	2.042	1.697	1.310	0.854	0.683	0.530	0.256	0.127
40	2.705	2.423	2.021	1.684	1.303	0.851	0.681	0.529	0.255	0.126
60	2.660	2.390	2.000	1.671	1.296	0.848	0.679	0.527	0.254	0.126
120	2.617	2.358	1.980	1.658	1.289	0.845	0.677	0.526	0.254	0.126
∞	2.557	2.326	1.960	1.645	1.282	0.842	0.674	0.524	0.253	0.126

APPENDIX 11.D

Values of the Error Function and Complementary Error Function
(for positive values of x)

x	erf(x)	erfc(x)	x	erf(x)	erfc(x)	x	erf(x)	erfc(x)	x	erf(x)	erfc(x)	x	erf(x)	erfc(x)
0	0.0000	1.0000												
0.01	0.0113	0.9887	0.51	0.5292	0.4708	1.01	0.8468	0.1532	1.51	0.9673	0.0327	2.01	0.9955	0.0045
0.02	0.0226	0.9774	0.52	0.5379	0.4621	1.02	0.8508	0.1492	1.52	0.9684	0.0316	2.02	0.9957	0.0043
0.03	0.0338	0.9662	0.53	0.5465	0.4535	1.03	0.8548	0.1452	1.53	0.9695	0.0305	2.03	0.9959	0.0041
0.04	0.0451	0.9549	0.54	0.5549	0.4451	1.04	0.8586	0.1414	1.54	0.9706	0.0294	2.04	0.9961	0.0039
0.05	0.0564	0.9436	0.55	0.5633	0.4367	1.05	0.8624	0.1376	1.55	0.9716	0.0284	2.05	0.9963	0.0037
0.06	0.0676	0.9324	0.56	0.5716	0.4284	1.06	0.8661	0.1339	1.56	0.9726	0.0274	2.06	0.9964	0.0036
0.07	0.0789	0.9211	0.57	0.5798	0.4202	1.07	0.8698	0.1302	1.57	0.9736	0.0264	2.07	0.9966	0.0034
0.08	0.0901	0.9099	0.58	0.5879	0.4121	1.08	0.8733	0.1267	1.58	0.9745	0.0255	2.08	0.9967	0.0033
0.09	0.1013	0.8987	0.59	0.5959	0.4041	1.09	0.8768	0.1232	1.59	0.9755	0.0245	2.09	0.9969	0.0031
0.1	0.1125	0.8875	0.6	0.6039	0.3961	1.1	0.8802	0.1198	1.6	0.9763	0.0237	2.1	0.9970	0.0030
0.11	0.1236	0.8764	0.61	0.6117	0.3883	1.11	0.8835	0.1165	1.61	0.9772	0.0228	2.11	0.9972	0.0028
0.12	0.1348	0.8652	0.62	0.6194	0.3806	1.12	0.8868	0.1132	1.62	0.9780	0.0220	2.12	0.9973	0.0027
0.13	0.1459	0.8541	0.63	0.6270	0.3730	1.13	0.8900	0.1100	1.63	0.9788	0.0212	2.13	0.9974	0.0026
0.14	0.1569	0.8431	0.64	0.6346	0.3654	1.14	0.8931	0.1069	1.64	0.9796	0.0204	2.14	0.9975	0.0025
0.15	0.1680	0.8320	0.65	0.6420	0.3580	1.15	0.8961	0.1039	1.65	0.9804	0.0196	2.15	0.9976	0.0024
0.16	0.1790	0.8210	0.66	0.6494	0.3506	1.16	0.8991	0.1009	1.66	0.9811	0.0189	2.16	0.9977	0.0023
0.17	0.1900	0.8100	0.67	0.6566	0.3434	1.17	0.9020	0.0980	1.67	0.9818	0.0182	2.17	0.9979	0.0021
0.18	0.2009	0.7991	0.68	0.6638	0.3362	1.18	0.9048	0.0952	1.68	0.9825	0.0175	2.18	0.9980	0.0020
0.19	0.2118	0.7882	0.69	0.6708	0.3292	1.19	0.9076	0.0924	1.69	0.9832	0.0168	2.19	0.9980	0.0020
0.2	0.2227	0.7773	0.7	0.6778	0.3222	1.2	0.9103	0.0897	1.7	0.9838	0.0162	2.2	0.9981	0.0019
0.21	0.2335	0.7665	0.71	0.6847	0.3153	1.21	0.9130	0.0870	1.71	0.9844	0.0156	2.21	0.9982	0.0018
0.22	0.2443	0.7557	0.72	0.6914	0.3086	1.22	0.9155	0.0845	1.72	0.9850	0.0150	2.22	0.9983	0.0017
0.23	0.2550	0.7450	0.73	0.6981	0.3019	1.23	0.9181	0.0819	1.73	0.9856	0.0144	2.23	0.9984	0.0016
0.24	0.2657	0.7343	0.74	0.7047	0.2953	1.24	0.9205	0.0795	1.74	0.9861	0.0139	2.24	0.9985	0.0015
0.25	0.2763	0.7237	0.75	0.7112	0.2888	1.25	0.9229	0.0771	1.75	0.9867	0.0133	2.25	0.9985	0.0015
0.26	0.2869	0.7131	0.76	0.7175	0.2825	1.26	0.9252	0.0748	1.76	0.9872	0.0128	2.26	0.9986	0.0014
0.27	0.2974	0.7026	0.77	0.7238	0.2762	1.27	0.9275	0.0725	1.77	0.9877	0.0123	2.27	0.9987	0.0013
0.28	0.3079	0.6921	0.78	0.7300	0.2700	1.28	0.9297	0.0703	1.78	0.9882	0.0118	2.28	0.9987	0.0013
0.29	0.3183	0.6817	0.79	0.7361	0.2639	1.29	0.9319	0.0681	1.79	0.9886	0.0114	2.29	0.9988	0.0012
0.3	0.3286	0.6714	0.8	0.7421	0.2579	1.3	0.9340	0.0660	1.8	0.9891	0.0109	2.3	0.9989	0.0011
0.31	0.3389	0.6611	0.81	0.7480	0.2520	1.31	0.9361	0.0639	1.81	0.9895	0.0105	2.31	0.9989	0.0011
0.32	0.3491	0.6509	0.82	0.7538	0.2462	1.32	0.9381	0.0619	1.82	0.9899	0.0101	2.32	0.9990	0.0010
0.33	0.3593	0.6407	0.83	0.7595	0.2405	1.33	0.9400	0.0600	1.83	0.9903	0.0097	2.33	0.9990	0.0010
0.34	0.3694	0.6306	0.84	0.7651	0.2349	1.34	0.9419	0.0581	1.84	0.9907	0.0093	2.34	0.9991	0.0009
0.35	0.3794	0.6206	0.85	0.7707	0.2293	1.35	0.9438	0.0562	1.85	0.9911	0.0089	2.35	0.9991	0.0009
0.36	0.3893	0.6107	0.86	0.7761	0.2239	1.36	0.9456	0.0544	1.86	0.9915	0.0085	2.36	0.9992	0.0008
0.37	0.3992	0.6008	0.87	0.7814	0.2186	1.37	0.9473	0.0527	1.87	0.9918	0.0082	2.37	0.9992	0.0008
0.38	0.4090	0.5910	0.88	0.7867	0.2133	1.38	0.9490	0.0510	1.88	0.9922	0.0078	2.38	0.9992	0.0008
0.39	0.4187	0.5813	0.89	0.7918	0.2082	1.39	0.9507	0.0493	1.89	0.9925	0.0075	2.39	0.9993	0.0007
0.4	0.4284	0.5716	0.9	0.7969	0.2031	1.4	0.9523	0.0477	1.9	0.9928	0.0072	2.4	0.9993	0.0007
0.41	0.4380	0.5620	0.91	0.8019	0.1981	1.41	0.9539	0.0461	1.91	0.9931	0.0069	2.41	0.9993	0.0007
0.42	0.4475	0.5525	0.92	0.8068	0.1932	1.42	0.9554	0.0446	1.92	0.9934	0.0066	2.42	0.9994	0.0006
0.43	0.4569	0.5431	0.93	0.8116	0.1884	1.43	0.9569	0.0431	1.93	0.9937	0.0063	2.43	0.9994	0.0006
0.44	0.4662	0.5338	0.94	0.8163	0.1837	1.44	0.9583	0.0417	1.94	0.9939	0.0061	2.44	0.9994	0.0006
0.45	0.4755	0.5245	0.95	0.8209	0.1791	1.45	0.9597	0.0403	1.95	0.9942	0.0058	2.45	0.9995	0.0005
0.46	0.4847	0.5153	0.96	0.8254	0.1746	1.46	0.9611	0.0389	1.96	0.9944	0.0056	2.46	0.9995	0.0005
0.47	0.4937	0.5063	0.97	0.8299	0.1701	1.47	0.9624	0.0376	1.97	0.9947	0.0053	2.47	0.9995	0.0005
0.48	0.5027	0.4973	0.98	0.8342	0.1658	1.48	0.9637	0.0363	1.98	0.9949	0.0051	2.48	0.9995	0.0005
0.49	0.5117	0.4883	0.99	0.8385	0.1615	1.49	0.9649	0.0351	1.99	0.9951	0.0049	2.49	0.9996	0.0004
0.5	0.5205	0.4795	1	0.8427	0.1573	1.5	0.9661	0.0339	2	0.9953	0.0047	2.5	0.9996	0.0004

APPENDIX 14.A
Viscosity of Water in Other Units
(customary U.S. units)

temperature	absolute viscosity	kinematic viscosity	
(°F)	(cP)	(cSt)	(SSU)
32	1.79	1.79	33.0
50	1.31	1.31	31.6
60	1.12	1.12	31.2
70	0.98	0.98	30.9
80	0.86	0.86	30.6
85	0.81	0.81	30.4
100	0.68	0.69	30.2
120	0.56	0.57	30.0
140	0.47	0.48	29.7
160	0.40	0.41	29.6
180	0.35	0.36	29.5
212	0.28	0.29	29.3

Support Material

APPENDIX 14.B
Properties of Uncommon Fluids

Typical fluid viscosities are listed in the following table. The values given for thixotropic fluids are effective viscosities at normal pumping shear rates. Effective viscosity can vary greatly with changes in solids content, concentration, and so on.

Viscous Behavior Type: N—Newtonian T—Thixotropic

fluid	specific gravity	viscosity CPS	viscous behavior type
reference—water	1.0	1.0	N
adhesives			
"box" adhesives	1±	3000	T
PVA	1.3	100	T
rubber and solvents	1.0	15,000	N
bakery			
batter	1	2200	T
butter, melted	0.98	18 at 140°F	N
egg, whole	0.5	60 at 50°F	N
emulsifier		20	T
frosting	1	10,000	T
lecithin		3250 at 125°F	T
77% sweetened condensed milk	1.3	10,000 at 77°F	N
yeast slurry, 15%	1	180	T
beer, wine			
beer	1.0	1.1 at 40°F	N
brewers concentrated yeast, 80% solids		16,000 at 40°F	T
wine	1.0		
chemicals, miscellaneous			
glycols	1.1	35 at range	
confectionary			
caramel	1.2	400 at 140°F	
chocolate	1.1	17,000 at 120°F	T
fudge, hot	1.1	36,000	T
toffee	1.2	87,000	T
cosmetics, soaps			
face cream		10,000	T
gel, hair	1.4	5000	T
shampoo		5000	T
toothpaste		20,000	T
hand cleaner		2000	T
dairy			
cottage cheese	1.08	225	T
cream	1.02	20 at 40°F	N
milk	1.03	1.2 at 60°F	N
cheese, processed		30,000 at 160°F	T
yogurt		1100	T
detergents			
detergent concentrate		10	N
dyes and inks			
ink, printers	1–1.38	10,000	T
dye	1.1	10	N
gum		5000	T

fluid	specific gravity	viscosity CPS	viscous behavior type
meat products			
animal fat, melted	0.9	43 at 100°F	N
ground beef fat	0.9	11,000 at 60°F	T
meat emulsion	1.0	22,000 at 40°F	T
pet food	1.0	11,000 at 40°F	T
pork fat slurry	1.0	650 at 40°F	T
paint			
auto paint, metallic		200	T
solvents	0.8–0.9	0.5–10	N
titanium dioxide slurry		10,000	T
varnish	1.06	140 at 100°F	
turpentine	0.86	2 at 60°F	
paper and textile			
black liquor	1.3	1100 at 122°F	
black liquor soap		7000 at 122°F	
black liquor tar		2000 at 300°F	
paper coating, 35%		400	
sulfide, 6%		1600	
petroleum and petroleum products			
asphalt, unblended	1.3	500–2500	
gasoline	0.7	0.8 at 60°F	N
kerosene	0.8	3 at 68°F	N
fuel oil no. 6	0.9	660 at 122°F	N
auto lube oil SAE 40	0.9	200 at 100°F	N
auto trans oil SAE 90	0.9	320 at 100°F	N
propane	0.46	0.2 at 100°F	N
tars	1.2	wide range	
pharmaceuticals			
castor oil	0.96	350	N
cough syrup	1.0	190	N
"stomach" remedy slurries		1500	T
pill pastes		5000±	T
plastics*, resins			
butadiene	0.94	0.17 at 40°F	
polyester resin (typ)	1.4	3000	
PVA resin (typ)	1.3	65,000	T
starches, gums			
cornstarch sol 22°B	1.18	32	T
cornstarch sol 25°B	1.21	300	T

APPENDIX 14.B *(continued)*
Properties of Uncommon Fluids

Typical fluid viscosities are listed in the following table. The values given for thixotropic fluids are effective viscosities at normal pumping shear rates. Effective viscosity can vary greatly with changes in solids content, concentration, and so on.

Viscous Behavior Type: N—Newtonian T—Thixotropic

fluid	specific gravity	viscosity CPS	viscous behavior type
fats and oils			
corn oils	0.92	30	N
lard	0.96	60 at 100°F	N
linseed oil	0.93	30 at 100°F	N
peanut oil	0.92	42 at 100°F	N
soybean oil	0.95	36 at 100°F	N
vegetable oil	0.92	3 at 300°F	N
foods, miscellaneous			
black bean paste		10,000	T
cream style corn		130 at 190°F	T
catsup	1.11	560 at 145°F	T
pablum		4500	T
pear pulp		4000 at 160°F	T
potato, mashed	1.0	20,000	T
potato skins and caustic		20,000 at 100°F	T
prune juice	1.0	60 at 120°F	T
orange juice concentrate	1.1	5000 at 38°F	T
tapioca pudding	0.7	1000 at 235°F	T
mayonnaise	1.0	5000 at 75°F	T
tomato paste, 33%	1.14	7000	T
honey	1.5	1500 at 100°F	T
sugar, syrups, molasses			
corn syrup 41°Be	1.39	15,000 at 60°F	N
corn syrup 45°Be	1.45	12,000 at 130°F	N
glucose	1.42	10,000 at 100°F	
molasses			
A	1.4–1.47	280–5000 at 100°F	
B	1.43–1.48	1400–13,000 at 100°F	
C	1.46–1.49	2600–5000 at 100°F	
sugar syrups			
60 brix	1.29	75 at 60°F	N
68 brix	1.34	360 at 60°F	N
76 brix	1.39	4000 at 60°F	N
water and waste treatment			
clarified sewage sludge	1.1	2000 range	

*A wide variety of plastics can be pumped; viscosity varies greatly.

APPENDIX 14.C
Vapor Pressure of Various Hydrocarbons and Water
(Cox Chart)

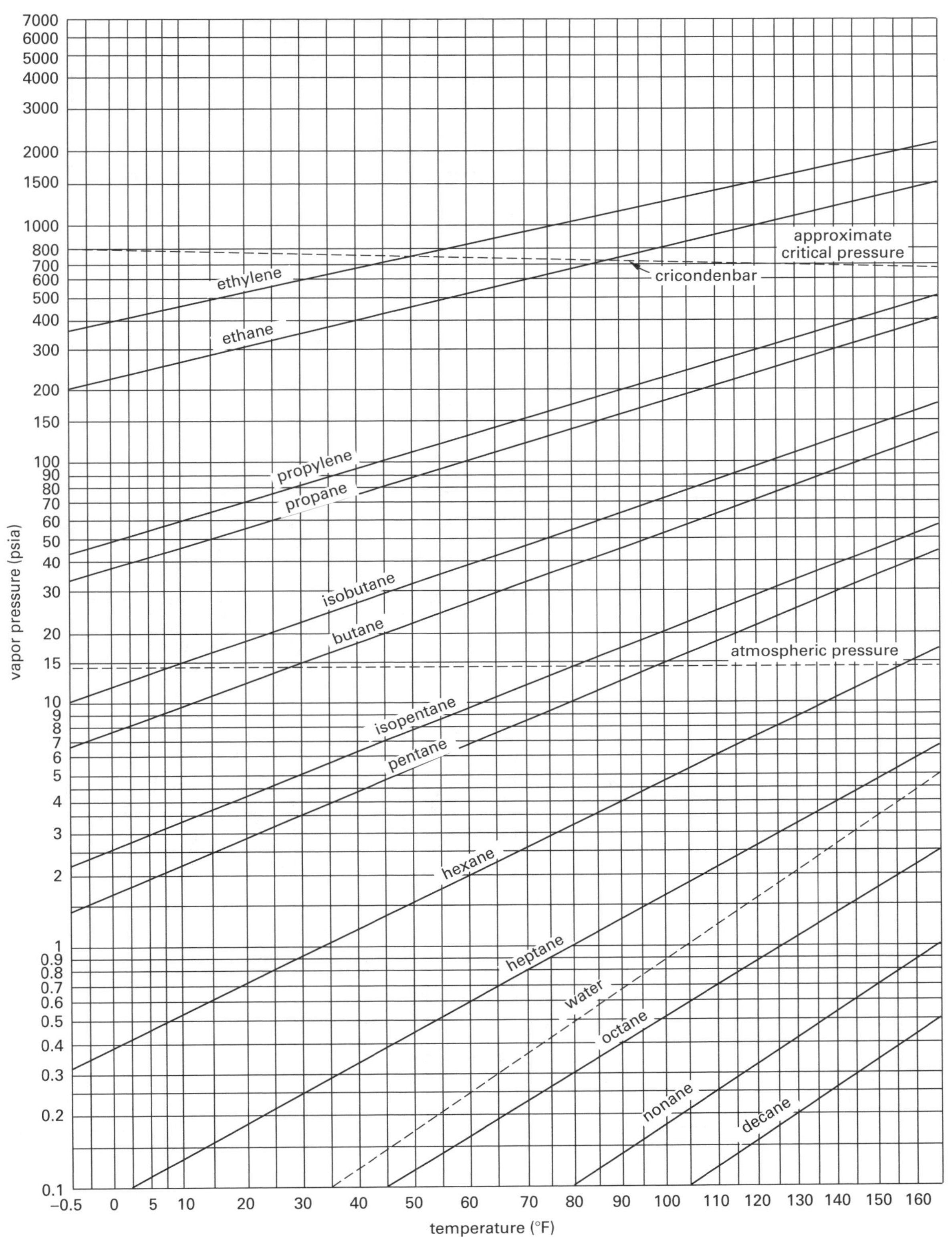

From *Hydraulic Handbook*, copyright © 1988, by Fairbanks Morse Pump. Reproduced with permission.

Support Material

APPENDIX 14.D
Specific Gravity of Hydrocarbons

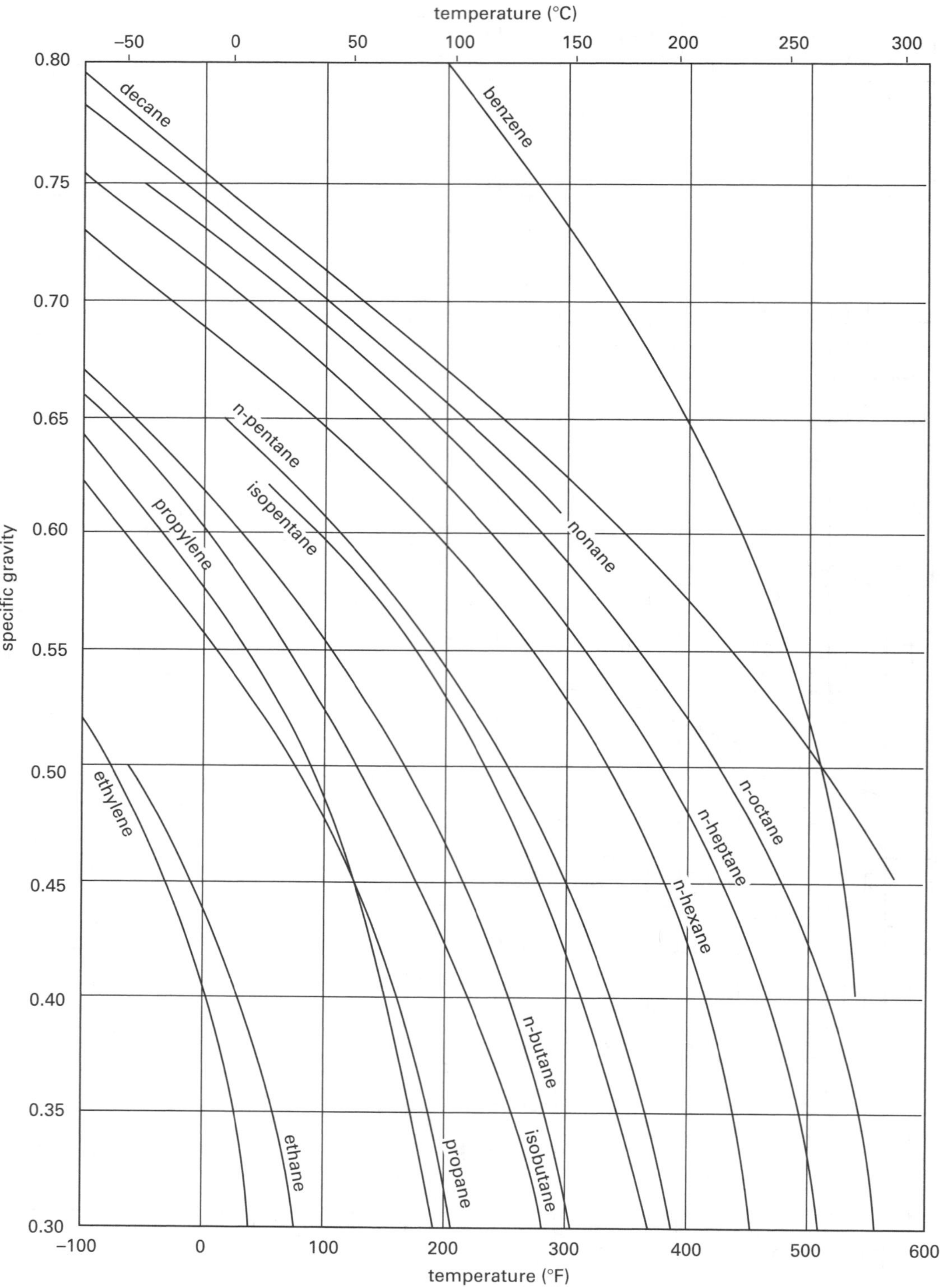

Table provided courtesy of Flowserve US, Inc. as successor in interest to Ingersoll-Dresser Pump Company, all rights reserved.

APPENDIX 14.E
Viscosity Conversion Chart
(Approximate Conversions for Newtonian Fluids)

kinematic viscosity scales	units	scale values
	centistokes	0, 100, 200, 300, 400, 500, 600, 700, 800, 900, 1000, 1100, 1200, 1300, 1400, 1500
Mobilometer 100 g 10 cm	seconds	10, 20, 30, 40, 50, 60, 70
Engler	degrees	25, 50, 75, 100, 125, 150, 175
Ford 4	seconds	25, 50, 75, 100, 125, 150, 175, 200, 225, 250, 275, 300, 325, 350, 375
Ford 3	seconds	25, 50, 75, 100, 125, 150, 175, 200, 225, 250, 275, 300, 325, 350, 375, 400, 425, 450, 475, 500, 550, 600, 650
Saybolt Universal	seconds	500, 1000, 1500, 2000, 2500, 3000, 3500, 4000, 4500, 5000, 5500, 6000, 6500
Saybolt Furol	seconds	50, 100, 150, 200, 250, 300, 350, 400, 450, 500, 550, 600, 650
Redwood 1 Standard	seconds	250, 500, 750, 1000, 1250, 1500, 2000, 2500, 3000, 3500, 4000, 4500, 5000, 5500, 6000
Ubbelohde	cSt	200, 300, 400, 500, 600, 700, 800, 900, 1000
Gardner Holdt		A2, A4, A, C, E, G, H, I, J, K, L, M, N, O, P, Q, R, S, T, U, V, W, X; A3, A1, B, D, F
Zahn 5	seconds	13, 20, 30, 40, 50, 60
Zahn 3	seconds	23, 30, 40, 50, 60

absolute viscosity scales	units	scale values
Kreb Stormer 200 g	K.U.	50, 60, 70, 80, 90
Stormer Cylinder 150 g	seconds	16, 27, 50, 115, 223
	centipoise	0, 100, 200, 300, 400, 500, 600, 700, 800, 900, 1000, 1100, 1200, 1300, 1400, 1500

(Multiply centistokes by 1.0764×10^{-5} to get ft^2/sec.)
(Multiply centistokes by 1.000×10^{-6} to get m^2/s.)

Support Material

APPENDIX 14.F
Viscosity Index Chart: 0–100 VI

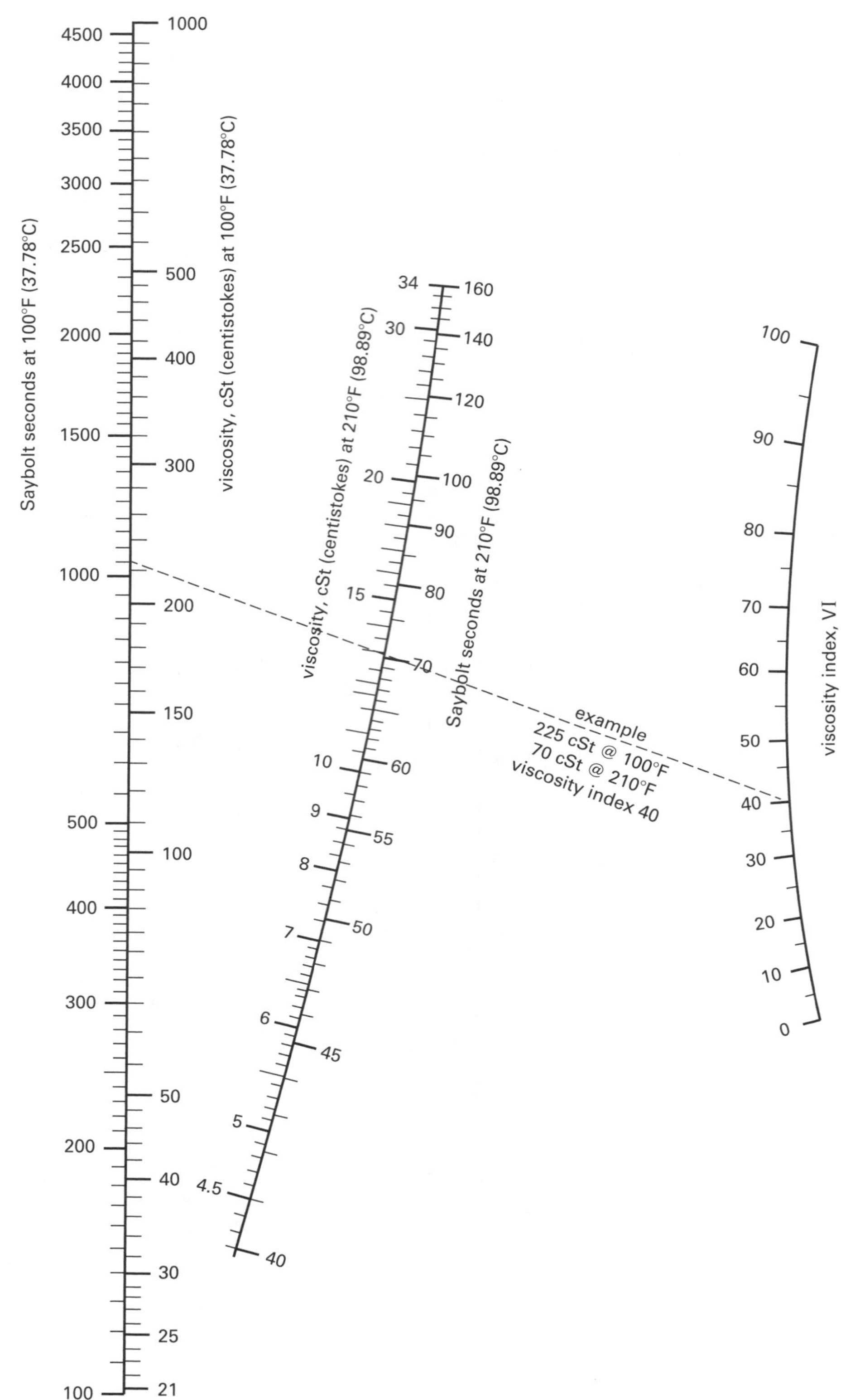

Based on correlations presented in ASTM D2270.

APPENDIX 16.A

Area, Wetted Perimeter, and Hydraulic Radius of Partially Filled Circular Pipes

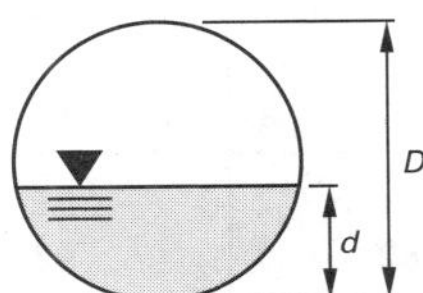

$\frac{d}{D}$	$\frac{\text{area}}{D^2}$	$\frac{\text{wetted perimeter}}{D}$	$\frac{r_h}{D}$	$\frac{d}{D}$	$\frac{\text{area}}{D^2}$	$\frac{\text{wetted perimeter}}{D}$	$\frac{r_h}{D}$
0.01	0.0013	0.2003	0.0066	0.51	0.4027	1.5908	0.2531
0.02	0.0037	0.2838	0.0132	0.52	0.4127	1.6108	0.2561
0.03	0.0069	0.3482	0.0197	0.53	0.4227	1.6308	0.2591
0.04	0.0105	0.4027	0.0262	0.54	0.4327	1.6509	0.2620
0.05	0.0147	0.4510	0.0326	0.55	0.4426	1.6710	0.2649
0.06	0.0192	0.4949	0.0389	0.56	0.4526	1.6911	0.2676
0.07	0.0242	0.5355	0.0451	0.57	0.4625	1.7113	0.2703
0.08	0.0294	0.5735	0.0513	0.58	0.4723	1.7315	0.2728
0.09	0.0350	0.6094	0.0574	0.59	0.4822	1.7518	0.2753
0.10	0.0409	0.6435	0.0635	0.60	0.4920	1.7722	0.2776
0.11	0.0470	0.6761	0.0695	0.61	0.5018	1.7926	0.2797
0.12	0.0534	0.7075	0.0754	0.62	0.5115	1.8132	0.2818
0.13	0.0600	0.7377	0.0813	0.63	0.5212	1.8338	0.2839
0.14	0.0688	0.7670	0.0871	0.64	0.5308	1.8546	0.2860
0.15	0.0739	0.7954	0.0929	0.65	0.5404	1.8755	0.2881
0.16	0.0811	0.8230	0.0986	0.66	0.5499	1.8965	0.2899
0.17	0.0885	0.8500	0.1042	0.67	0.5594	1.9177	0.2917
0.18	0.0961	0.8763	0.1097	0.68	0.5687	1.9391	0.2935
0.19	0.1039	0.9020	0.1152	0.69	0.5780	1.9606	0.2950
0.20	0.1118	0.9273	0.1206	0.70	0.5872	1.9823	0.2962
0.21	0.1199	0.9521	0.1259	0.71	0.5964	2.0042	0.2973
0.22	0.1281	0.9764	0.1312	0.72	0.6054	2.0264	0.2984
0.23	0.1365	1.0003	0.1364	0.73	0.6143	2.0488	0.2995
0.24	0.1449	1.0239	0.1416	0.74	0.6231	2.0714	0.3006
0.25	0.1535	1.0472	0.1466	0.75	0.6318	2.0944	0.3017
0.26	0.1623	1.0701	0.1516	0.76	0.6404	2.1176	0.3025
0.27	0.1711	1.0928	0.1566	0.77	0.6489	2.1412	0.3032
0.28	0.1800	1.1152	0.1614	0.78	0.6573	2.1652	0.3037
0.29	0.1890	1.1373	0.1662	0.79	0.6655	2.1895	0.3040
0.30	0.1982	1.1593	0.1709	0.80	0.6736	2.2143	0.3042
0.31	0.2074	1.1810	0.1755	0.81	0.6815	2.2395	0.3044
0.32	0.2167	1.2025	0.1801	0.82	0.6893	2.2653	0.3043
0.33	0.2260	1.2239	0.1848	0.83	0.6969	2.2916	0.3041
0.34	0.2355	1.2451	0.1891	0.84	0.7043	2.3186	0.3038
0.35	0.2450	1.2661	0.1935	0.85	0.7115	2.3462	0.3033
0.36	0.2546	1.2870	0.1978	0.86	0.7186	2.3746	0.3026
0.37	0.2642	1.3078	0.2020	0.87	0.7254	2.4038	0.3017
0.38	0.2739	1.3284	0.2061	0.88	0.7320	2.4341	0.3008
0.39	0.2836	1.3490	0.2102	0.89	0.7384	2.4655	0.2995
0.40	0.2934	1.3694	0.2142	0.90	0.7445	2.4981	0.2980
0.41	0.3032	1.3898	0.2181	0.91	0.7504	2.5322	0.2963

Support Material

APPENDIX 16.A *(continued)*
Area, Wetted Perimeter, and Hydraulic Radius of Partially Filled Circular Pipes

$\frac{d}{D}$	$\frac{\text{area}}{D^2}$	$\frac{\text{wetted perimeter}}{D}$	$\frac{r_h}{D}$	$\frac{d}{D}$	$\frac{\text{area}}{D^2}$	$\frac{\text{wetted perimeter}}{D}$	$\frac{r_h}{D}$
0.42	0.3130	1.4101	0.2220	0.92	0.7560	2.5681	0.2944
0.43	0.3229	1.4303	0.2257	0.93	0.7612	2.6061	0.2922
0.44	0.3328	1.4505	0.2294	0.94	0.7662	2.6467	0.2896
0.45	0.3428	1.4706	0.2331	0.95	0.7707	2.6906	0.2864
0.46	0.3527	1.4907	0.2366	0.96	0.7749	2.7389	0.2830
0.47	0.3627	1.5108	0.2400	0.97	0.7785	2.7934	0.2787
0.48	0.3727	1.5308	0.2434	0.98	0.7816	2.8578	0.2735
0.49	0.3827	1.5508	0.2467	0.99	0.7841	2.9412	0.2665
0.50	0.3927	1.5708	0.2500	1.00	0.7854	3.1416	0.2500

APPENDIX 16.B
Dimensions of Rigid PVC and CPVC Pipe
(customary U.S. units)

		schedule-40 ASTM D1785			schedule-80 ASTM D1785			class 200 ASTM D2241		
nominal diameter (in)	outside diameter (in)	internal diameter (in)	wall thickness (in)	pressure rating* (lbf/in²)	internal diameter (in)	wall thickness (in)	pressure rating* (lbf/in²)	internal diameter (in)	wall thickness (in)	pressure rating* (lbf/in²)
$\frac{1}{8}$	0.405	0.249	0.068	810	0.195	0.095	1230	–	–	–
$\frac{1}{4}$	0.540	0.344	0.088	780	0.282	0.119	1130	–	–	–
$\frac{3}{8}$	0.675	0.473	0.091	620	0.403	0.126	920	–	–	–
$\frac{1}{2}$	0.840	0.622	0.109	600	0.546	0.147	850	0.716	0.062	200
$\frac{3}{4}$	1.050	0.824	0.113	480	0.742	0.154	690	0.930	0.060	200
1	1.315	1.049	0.133	450	0.957	0.179	630	1.189	0.063	200
$1\frac{1}{4}$	1.660	1.380	0.140	370	1.278	0.191	520	1.502	0.079	200
$1\frac{1}{2}$	1.900	1.610	0.145	330	1.500	0.200	470	1.720	0.090	200
2	2.375	2.067	0.154	280	1.939	0.218	400	2.149	0.113	200
$2\frac{1}{2}$	2.875	2.469	0.203	300	2.323	0.276	420	2.601	0.137	200
3	3.500	3.068	0.216	260	2.900	0.300	370	3.166	0.167	200
4	4.500	4.026	0.237	220	3.826	0.337	320	4.072	0.214	200
5	5.563	5.047	0.258	190	4.768	0.375	290	–	–	–
6	6.625	6.065	0.280	180	5.761	0.432	280	5.993	0.316	200
8	8.625	7.961	0.332	160	7.565	0.500	250	7.740	0.410	200
10	10.750	9.976	0.365	140	9.492	0.593	230	9.650	0.511	200
12	12.750	11.890	0.406	130	11.294	0.687	230	11.450	0.606	200
14	14.000	13.073	0.447	130	12.410	0.750	220	–	–	–
16	16.000	14.940	0.500	130	14.213	0.843	220	–	–	–
18	18.000	16.809	0.562	130	16.014	0.937	220	–	–	–
20	20.000	18.743	0.5937	130	17.814	1.031	220	–	–	–
24	24.000	22.554	0.687	120	21.418	1.218	210	–	–	–

(Multiply in by 25.4 to obtain mm.)
(Multiply lbf/in² by 6.895 to obtain kPa.)
*Pressure ratings are for a pipe temperature of 68°F (20°C) and are subject to the following temperature derating factors. Operation above 140°F (60°C) is not permitted.

pipe operating temperature °F (°C)	73 (23)	80 (27)	90 (32)	100 (38)	110 (43)	120 (49)	130 (54)	140 (60)
derating factor	1.0	0.88	0.75	0.62	0.51	0.40	0.31	0.22

Support Material

APPENDIX 16.C
Dimensions of Large Diameter, Nonpressure, PVC Sewer and Water Pipe
(customary U.S. units)

nominal size (in)	nominal size (mm)	designations and dimensional controls	minimum wall thickness (in)	outside diameter (inside diameter for profile wall pipes) (in)
ASTM F679 – Gravity Sewer Pipe (solid wall)				
18	450	PS-46, T-2	0.499	18.701
		PS-46, T-1	0.536	
		PS-115, T-2	0.671	
		PS-115, T-1	0.720	
21	525	PS-46, T-2	0.588	22.047
		PS-46, T-1	0.632	
		PS-115, T-2	0.791	
		PS-115, T-1	0.849	
24	600	PS-46, T-2	0.661	24.803
		PS-46, T-1	0.709	
		PS-115, T-2	0.889	
		PS-115, T-1	0.954	
27	675	PS-46, T-2	0.745	27.953
		PS-46, T-1	0.745	
		PS-115, T-2	1.002	
		PS-115, T-1	1.075	
30	750	PS-46, T-2	0.853	32.00
		PS-46, T-1	0.914	
		PS-115, T-2	1.148	
		PS-115, T-1	1.231	
36	900	PS-46, T-2	1.021	38.30
		PS-46, T-1	1.094	
		PS-115, T-2	1.373	
		PS-115, T-1	1.473	
42	1050	PS-46, T-2	1.186	44.50
		PS-46, T-1	1.271	
		PS-115, T-2	1.596	
		PS-115, T-1	1.712	
48	1200	PS-46, T-2	1.354	50.80
		PS-46, T-1	1.451	
		PS-115, T-2	1.822	
		PS-115, T-1	1.954	
ASTM F758 – PVC Highway Underdrain Pipe (perforated)				
4	100		0.120	4.215
6	150		0.180	6.275
8	200		0.240	8.40
ASTM F789 – PVC Sewage Pipe (solid wall)				
available in sizes 4–18 in (100–450 mm) (obsolete)		available in two stiffnesses and three material grades: PS-46; T-1, T-2, T-3 PS-115; T-1, T-2, T-3		
ASTM F794 – Open Profile Wall Pipe				
available in sizes 4–48 in (100–1200 mm)		PS-46		internal diameter controlled

Support Material

APPENDIX 16.C *(continued)*
Dimensions of Large Diameter, Nonpressure, PVC Sewer and Water Pipe
(customary U.S. units)

nominal size (in)	nominal size (mm)	designations and dimensional controls	minimum wall thickness (in)	outside diameter (inside diameter for profile wall pipes) (in)
AWWA C900 – Water and Wastewater Pressure Pipe (solid wall)				
4	100	DR-25 (PC-100, PR-165, CL100); DR-18 (PC-150, PR-235, CL150); DR-14 (PC-200, PC-305, CL200)	$t_{\text{min}} = D_o/\text{DR}$	4.80
6	150			6.90
8	200			9.05
10	250			11.10
12	300			13.20
AWWA C905 – Water and Wastewater Pipe (solid wall)				
14	350	DR-51 (PR-80);	$t_{\text{min}} = D_o/\text{DR}$	15.30
16	400	DR-41 (PR-100);		17.40
18	450	DR-32.5 (PR-125);		19.50
20	500	DR-25 (PR-165);		21.60
24	600	DR-21 (PR-200);		25.8
30	750	DR-18 (PR-235);		32.0
36	900	DR-14 (PR-305)		38.3
ASTM F949 – Open Profile Wall Pipe				
available in sizes 4–36 in (100–900 mm)		PS-46		internal diameter controlled
ASTM D1785				
See App. 16.B.				
ASTM F1803 – Closed Profile Wall "Truss" Pipe (DWCP)				
available in sizes 18–60 in (450–1500 mm)		PS-46		internal diameter controlled
ASTM D2241 – PVC Sewer Pipe (solid wall)				
1	25	SDR-41 SDR-32.5 SDR-26 SDR-21 SDR-17 SDR-13.5	$t_{\text{min}} = D_o/\text{SDR}$	1.315
$1\frac{1}{4}$	31			1.660
$1\frac{1}{2}$	37			1.900
2	50			2.375
$2\frac{1}{2}$	62			2.875
3	75			3.500
$3\frac{1}{2}$	87			4.000
4	100			4.500
6	150			6.625
8	200			8.625
10	250			10.750
12	300			12.750
ASTM D2729 – PVC Drainage Pipe (solid wall; perforated)				
3			0.070	3.250
4			0.075	4.215
6			0.100	6.275

APPENDIX 16.C *(continued)*
Dimensions of Large Diameter, Nonpressure, PVC Sewer and Water Pipe
(customary U.S. units)

nominal size		designations and dimensional controls	minimum wall thickness (in)	outside diameter (inside diameter for profile wall pipes) (in)
(in)	(mm)			
ASTM D3034 – PVC Gravity Sewer Pipe and Perforated Drain Pipe (solid wall)				
4	100	SDR-35 & PS-46 (regular); SDR-26 & PS-115 (HW); SDR-23.5 & PS-135	$t_{min} = D_o/\text{SDR}$	4.215
5	135			5.640
6	150			6.275
8	200	SDR-41 & PS-28; SDR-35 & PS-46 (regular); SDR-26 & PS-115 (HW)		8.400
10	250			10.50
12	300			12.50
15	375			15.30

abbreviations

CL – pressure class (same as PC)
DR – dimensional ratio (constant ratio of outside diameter/wall thickness)
DWCP – double wall corrugated pipe
HW – heavy weight
PC – pressure class (same as CL)
PS – pipe stiffness (resistance to vertical diametral deflection due to compression from burial soil and surface loading, in lbf/in^2; calculated uncorrected as vertical load in lbf per inch of pipe length divided by deflection in inches)
SDR – standard dimension ratio (constant ratio of outside diameter/wall thickness)
(Multiply in by 25.4 to obtain mm.)
(Multiply lbf/in^2 by 6.895 to obtain kPa.)

APPENDIX 16.D
Dimensions and Weights of Concrete Sewer Pipe
(customary U.S. units)

ASTM C14 – Nonreinforced Round Sewer and Culvert Pipe for Nonpressure (Gravity Flow) Applications

nominal size and internal diameter		minimum wall thickness and required ASTM C14 $D_{0.01}$-load rating[a] (in and lbf/ft)		
(in)	(mm)	class 1[b]	class 2[b]	class 3[b]
4	100	0.625 (1500)	0.75 (2000)	0.875 (2400)
6	150	0.625 (1500)	0.75 (2000)	1.0 (2400)
8	200	0.75 (1500)	0.875 (2000)	1.125 (2400)
10	250	0.875 (1600)	1.0 (2000)	1.25 (2400)
12	300	1.0 (1800)	1.375 (2250)	1.75 (2500)
15	375	1.25 (2000)	1.625 (2500)	1.875 (2900)
18	450	1.5 (2200)	2.0 (3000)	2.25 (3300)
21	525	1.75 (2400)	2.25 (3300)	2.75 (3850)
24	600	2.125 (2600)	3.0 (3600)	3.375 (4400)
27	675	3.25 (2800)	3.75 (3950)	3.75 (4600)
30	750	3.5 (3000)	4.25 (4300)	4.25 (4750)
33	825	3.75 (3150)	4.5 (4400)	4.5 (4850)
36	900	4.0 (3300)	4.75 (4500)	4.75 (5000)

(Multiply in by 25.4 to obtain mm.)
(Multiply lbf/ft by 14.594 to obtain kN/m.)
[a]For C14 pipe, D-loads are stated in lbf/ft (pound per foot length of pipe).
[b]As defined by the strength (D-load) requirements of ASTM C14.

ASTM C76 – Reinforced Round Concrete Culvert, Storm Drain, and Sewer Pipe for Nonpressure (Gravity Flow) Applications

nominal size and internal diameter		minimum wall thickness and maximum ASTM C76 $D_{0.01}$-load rating (in and lbf/ft^2)		
(in)	(mm)	wall A[a]	wall B[a]	wall C[a]
8[b]	200		2.0	
10[b]	250		2.0	
12	300	1.75 (2000)	2.0 (3000)	2.75 (3000)
15	375	1.875 (2000)	2.25 (3000)	3.0 (3000)
18	450	2.0 (2000)	2.5 (3000)	3.25 (3000)
21	525	2.25 (2000)	2.75 (3000)	3.50 (3000)
24	600	2.5 (2000)	3.0 (3000)	3.75 (3000)
27	675	2.625 (2000)	3.25 (3000)	4.0 (3000)
30	750	2.75 (2000)	3.5 (3000)	4.25 (3000)
33	825	2.875 (1350)	3.75 (3000)	4.5 (3000)
36	900	3.0 (1350)	4.0 (3000)	4.75 (3000)
42	1050	3.5 (1350)	4.5 (3000)	5.25 (3000)
48	1200	4.0 (1350)	5.0 (3000)	5.75 (3000)
54	1350	4.5 (1350)	5.5 (2000)	6.25 (3000)
60	1500	5.0 (1350)	6.0 (2000)	6.75 (3000)
66	1650	5.5 (1350)	6.5 (2000)	7.25 (3000)
72	1800	6.0 (1350)	7.0 (2000)	7.75 (3000)
78	1950	6.5 (1350)	7.5 (1350)	8.25 (2000)
84	2100	7.0 (1350)	8.0 (1350)	8.75 (1350)
90	2250	7.5 (1350)	8.5 (1350)	9.25 (1350)
96	2400	8.0 (1350)	9.0 (1350)	9.75 (1350)
102	2550	8.5 (1350)	9.5 (1350)	10.25 (1350)
108	2700	9.0 (1350)	10.0 (1350)	10.75 (1350)

(Multiply in by 25.4 to obtain mm.)
(Multiply lbf/ft^2 by 0.04788 to obtain kN/m^2 (kPa).)
[a]wall A thickness in inches = diameter in feet; wall B thickness in inches = diameter in feet + 1 in; wall C thickness in inches = diameter in feet + 1.75 in
[b]Although not specifically called out in ASTM C76, 8 in and 10 in diameter circular concrete pipes are routinely manufactured.

APPENDIX 16.D *(continued)*
Dimensions and Weights of Concrete Sewer Pipe
(customary U.S. units)

ASTM C76 – Large Sizes of Pipe for Nonpressure (Gravity Flow) Applications

nominal size and internal diameter		minimum wall thickness (in)
(in)	(mm)	wall A*
114	2850	9.5
120	3000	10.0
126	3150	10.5
132	3300	11.0
138	3450	11.5
144	3600	12.0
150	3750	12.5
156	3900	13.0
162	4050	13.5
168	4200	14.0
174	4350	14.5
180	4500	15.0

(Multiply in by 25.4 to obtain mm.)

*wall A thickness in inches = diameter in feet

ASTM C361 – RCPP: Reinforced Concrete Low Head Pressure Pipe

"Low head" means 125 ft (54 psi; 375 kPa) or less. Standard pipe with 12 in (300 mm) to 108 in (2700 mm) diameters are available. ASTM C361 limits tensile stress (and, therefore, strain) and flexural deformation. Any reinforcement design that meets these limitations is permitted. Internal and external dimensions typically correspond to C76 (wall B) pipe, but wall C may also be used.

ASTM C655 – Reinforced Concrete D-Load Culvert, Storm Drain, and Sewer Pipe

Pipes smaller than 72 in (1800 mm) are often specified by D-load. Pipe manufactured to satisfy ASTM C655 will support a specified concentrated vertical load (applied in three-point loading) in pounds per ft of length per ft of diameter. Some standard C76 pipe meets this standard. In order to use C76 (and other) pipes, D-loads are mapped onto C76 pipe classes. In some cases, special pipe designs are required. Pipe selection by D-load is accomplished by first determining the size of pipe required to carry the flow, then choosing the appropriate class of pipe according to the required D-load.

$D_{0.01}$-load range	C76 pipe class
1–800	class 1 (I)
801–1000	class 2 (II)
1001–1350	class 3 (III)
1351–2000	class 4 (IV)
2000–3000	class 5 (V)
>3000	no equivalent class pipe

(Multiply in by 25.4 to obtain mm.)

(Multiply lbf/ft^2 by 0.04788 to obtain kN/m^2 (kPa).)

Support Material

APPENDIX 16.E
Dimensions of Cast-Iron and Ductile Iron Pipe Standard Pressure Classes
(customary U.S. units)

ANSI/AWWA C106 Gray Cast-Iron Pipe

The ANSI/AWWA C106 standard is obsolete. The outside diameters of 4–48 in (100–1200 mm) diameter gray cast-iron pipes are the same as for C150 pipes. Ductile iron pipe is interchangeable with respect to joining diameters, accessories, and fittings. However, inside diameters of cast-iron pipe are substantially greater than C150 pipe. Outside diameters and thicknesses of AWWA 1908 standard cast-iron pipe are substantially different from C150 values.

ANSI/AWWA C150/A21.50 and ANSI/AWWA C151/A21.51

		calculated minimum wall thickness[a], t (in)						
		pressure class and head (lbf/in^2 and ft)						
nominal size (in)	outside diameter, D_o (in)	150 (346)	200 (462)	250 (577)	300	350	casting tolerance (in)	inside diameter
3	3.96					0.25[b]	0.05[c]	
4	4.80					0.25[b]	0.05[c]	
6	6.90					0.25[b]	0.05[c]	
8	9.05					0.25[b]	0.05[c]	
10	11.10					0.26	0.06	
12	13.20					0.28	0.06	
14	15.30			0.28	0.30	0.31	0.07	
16	17.40			0.30	0.32	0.34	0.07	
18	19.50			0.31	0.34	0.36	0.07	$D_i = D_o - 2t$
20	21.60			0.33	0.36	0.38	0.07	
24	25.80		0.33	0.37	0.40	0.43	0.07	
30	32.00	0.34	0.38	0.42	0.45	0.49	0.07	
36	38.30	0.38	0.42	0.47	0.51	0.56	0.07	
42	44.50	0.41	0.47	0.52	0.57	0.63	0.07	
48	50.80	0.46	0.52	0.58	0.64	0.70	0.08	
54	57.56	0.51	0.58	0.65	0.72	0.79	0.09	
60	61.61	0.54	0.61	0.68	0.76	0.83	0.09	
64	65.67	0.56	0.64	0.72	0.80	0.87	0.09	

(Multiply in by 25.4 to obtain mm.)

(Multiply lbf/in^2 by 6.895 to obtain kPa.)

[a]Per ANSI/AWWA C150/A21.50, the tabulated minimum wall thicknesses include a 0.08 in (2 mm) service allowance and the appropriate casting tolerance. Listed thicknesses are adequate for the rated water working pressure plus a surge allowance of 100 lbf/in^2 (690 kPa). Values are based on a yield strength of 42,000 lbf/in^2 (290 MPa), the sum of the working pressure and 100 lbf/in^2 (690 kPa) surge allowance, and a safety factor of 2.0.

[b]Pressure class is defined as the rated gage water pressure of the pipe in lbf/in^2.

[c]Limited by manufacturing. Calculated required thickness is less.

Support Material

APPENDIX 16.F
Dimensions of Ductile Iron Pipe Special Pressure Classes
(customary U.S. units)

ANSI/AWWA C150/A21.50 and ANSI/AWWA C151/A21.51

nominal size[a] (in)	outside diameter (in)	wall thickness (in)						
		special pressure class[b]						
		50	51	52	53	54	55	56
4	4.80	–	0.26	0.29	0.32	0.35	0.38	0.41
6	6.90	0.25	0.28	0.31	0.34	0.37	0.40	0.43
8	9.05	0.27	0.30	0.33	0.36	0.39	0.42	0.45
10	11.10	0.29	0.32	0.35	0.38	0.41	0.44	0.47
12	13.20	0.31	0.34	0.37	0.40	0.43	0.46	0.49
14	15.30	0.33	0.36	0.39	0.42	0.45	0.48	0.51
16	17.40	0.34	0.37	0.40	0.43	0.46	0.49	0.52
18	19.50	0.35	0.38	0.41	0.44	0.47	0.50	0.53
20	21.60	0.36	0.39	0.42	0.45	0.48	0.51	0.54
24	25.80	0.38	0.41	0.44	0.47	0.50	0.53	0.56
30	32.00	0.39	0.43	0.47	0.51	0.55	0.59	0.63
36	38.30	0.43	0.48	0.53	0.58	0.63	0.68	0.73
42	44.50	0.47	0.53	0.59	0.65	0.71	0.77	0.83
48	50.80	0.51	0.58	0.65	0.72	0.79	0.86	0.93
54	57.56	0.57	0.65	0.73	0.81	0.89	0.97	1.05

(Multiply in by 25.4 to obtain mm.)

(Multiply lbf/in^2 by 6.895 to obtain kPa.)

[a]Formerly designated "standard thickness classes." These special pressure classes are as shown in AWWA C150 and C151. Special classes are most appropriate for some threaded, grooved, or ball and socket pipes, or for extraordinary design conditions. They are generally less available than standard pressure class pipe.

[b]60 in (1500 mm) and 64 in (1600 mm) pipe sizes are not available in special pressure classes.

ASTM A746 Cement-Lined Gravity Sewer Pipe

The thicknesses for cement-lined pipe are the same as those in AWWA C150 and C151.

ASTM A746 Flexible Lining Gravity Sewer Pipe

The thicknesses for flexible lining pipe are the same as those in AWWA C150 and C151.

APPENDIX 16.G
Dimensions of Copper Water Tubing
(customary U.S. units)

classification	nominal tube size (in)	outside diameter (in)	wall thickness (in)	inside diameter (in)	transverse area (in^2)	safe working pressure (psi)
hard	1/4	3/8	0.025	0.325	0.083	1000
	3/8	1/2	0.025	0.450	0.159	1000
	1/2	5/8	0.028	0.569	0.254	890
	3/4	7/8	0.032	0.811	0.516	710
	1	1 1/8	0.035	1.055	0.874	600
	1 1/4	1 3/8	0.042	1.291	1.309	590
type "M" 250 psi working pressure	1 1/2	1 5/8	0.049	1.527	1.831	580
	2	2 1/8	0.058	2.009	3.17	520
	2 1/2	2 5/8	0.065	2.495	4.89	470
	3	3 1/8	0.072	2.981	6.98	440
	3 1/2	3 5/8	0.083	3.459	9.40	430
	4	4 1/8	0.095	3.935	12.16	430
	5	5 1/8	0.109	4.907	18.91	400
	6	6 1/8	0.122	5.881	27.16	375
	8	8 1/8	0.170	7.785	47.6	375
hard	3/8	1/2	0.035	0.430	0.146	1000
	1/2	5/8	0.040	0.545	0.233	1000
	3/4	7/8	0.045	0.785	0.484	1000
	1	1 1/8	0.050	1.025	0.825	880
	1 1/4	1 3/8	0.055	1.265	1.256	780
type "L" 250 psi working pressure	1 1/2	1 5/8	0.060	1.505	1.78	720
	2	2 1/8	0.070	1.985	3.094	640
	2 1/2	2 5/8	0.080	2.465	4.77	580
	3	3 1/8	0.090	2.945	6.812	550
	3 1/2	3 5/8	0.100	3.425	9.213	530
	4	4 1/8	0.110	3.905	11.97	510
	5	5 1/8	0.125	4.875	18.67	460
	6	6 1/8	0.140	5.845	26.83	430
hard	1/4	3/8	0.032	0.311	0.076	1000
	3/8	1/2	0.049	0.402	0.127	1000
	1/2	5/8	0.049	0.527	0.218	1000
	3/4	7/8	0.065	0.745	0.436	1000
	1	1 1/8	0.065	0.995	0.778	780
type "K" 400 psi working pressure	1 1/4	1 3/8	0.065	1.245	1.217	630
	1 1/2	1 5/8	0.072	1.481	1.722	580
	2	2 1/8	0.083	1.959	3.014	510
	2 1/2	2 5/8	0.095	2.435	4.656	470
	3	3 1/8	0.109	2.907	6.637	450
	3 1/2	3 5/8	0.120	3.385	8.999	430
	4	4 1/8	0.134	3.857	11.68	420
	5	5 1/8	0.160	4.805	18.13	400
	6	6 1/8	0.192	5.741	25.88	400
soft	1/4	3/8	0.032	0.311	0.076	1000
	3/8	1/2	0.049	0.402	0.127	1000
	1/2	5/8	0.049	0.527	0.218	1000
	3/4	7/8	0.065	0.745	0.436	1000
	1	1 1/8	0.065	0.995	0.778	780
type "K" 250 psi working pressure	1 1/4	1 3/8	0.065	1.245	1.217	630
	1 1/2	1 5/8	0.072	1.481	1.722	580
	2	2 1/8	0.083	1.959	3.014	510
	2 1/2	2 5/8	0.095	2.435	4.656	470
	3	3 1/8	0.109	2.907	6.637	450
	1 1/2	2 5/8	0.120	3.385	8.999	430
	4	4 1/8	0.134	3.857	11.68	420
	5	5 2/8	0.160	4.805	18.13	400
	6	6 1/8	0.192	5.741	25.88	400

(Multiply in by 25.4 to obtain mm.)
(Multiply in^2 by 645 to obtain mm^2.)

APPENDIX 16.H
Dimensions of Brass and Copper Tubing
(customary U.S. units)

regular

pipe size (in)	nominal dimensions (in)			cross-sectional area of bore (in^2)	lbm/ft	
	O.D.	I.D.	wall		red brass	copper
1/8	0.405	0.281	0.062	0.062	0.253	0.259
1/4	0.540	0.376	0.082	0.110	0.447	0.457
3/8	0.675	0.495	0.090	0.192	0.627	0.641
1/2	0.840	0.626	0.107	0.307	0.934	0.955
3/4	1.050	0.822	0.114	0.531	1.270	1.300
1	1.315	1.063	0.126	0.887	1.780	1.820
1 1/4	1.660	1.368	0.146	1.470	2.630	2.690
1 1/2	1.900	1.600	0.150	2.010	3.130	3.200
2	2.375	2.063	0.156	3.340	4.120	4.220
2 1/2	2.875	2.501	0.187	4.910	5.990	6.120
3	3.500	3.062	0.219	7.370	8.560	8.750
3 1/2	4.000	3.500	0.250	9.620	11.200	11.400
4	4.500	4.000	0.250	12.600	12.700	12.900
5	5.562	5.062	0.250	20.100	15.800	16.200
6	6.625	6.125	0.250	29.500	19.000	19.400
8	8.625	8.001	0.312	50.300	30.900	31.600
10	10.750	10.020	0.365	78.800	45.200	46.200
12	12.750	12.000	0.375	113.000	55.300	56.500

extra strong

pipe size (in)	nominal dimensions (in)			cross-sectional area of bore (in^2)	lbm/ft	
	O.D.	I.D.	wall		red brass	copper
1/8	0.405	0.205	0.100	0.033	0.363	0.371
1/4	0.540	0.294	0.123	0.068	0.611	0.625
3/8	0.675	0.421	0.127	0.139	0.829	0.847
1/2	0.840	0.542	0.149	0.231	1.230	1.250
3/4	1.050	0.736	0.157	0.425	1.670	1.710
1	1.315	0.951	0.182	0.710	2.460	2.510
1 1/4	1.660	1.272	0.194	1.270	3.390	3.460
1 1/2	1.990	1.494	0.203	1.750	4.100	4.190
2	2.375	1.933	0.221	2.94	5.670	5.800
2 1/2	2.875	2.315	0.280	4.21	8.660	8.850
3	3.500	2.892	0.304	6.57	11.600	11.800
3 1/2	4.000	3.358	0.321	8.86	14.100	14.400
4	4.500	3.818	0.341	11.50	16.900	17.300
5	5.562	4.812	0.375	18.20	23.200	23.700
6	6.625	5.751	0.437	26.00	32.200	32.900
8	8.625	7.625	0.500	45.70	48.400	49.500
10	10.750	9.750	0.500	74.70	61.100	62.400

(Multiply in by 25.4 to obtain mm.)
(Multiply in^2 by 645 to obtain mm^2.)

APPENDIX 16.I

Dimensions of Seamless Steel Boiler (BWG) Tubing[a,b,c]
(customary U.S. units)

O.D. (in)	BWG	wall thickness (in)	O.D. (in)	BWG	wall thickness (in)
1	13	0.095	3	12	0.109
	12	0.109		11	0.120
	11	0.120		10	0.134
	10	0.134		9	0.148
1 1/4	13	0.095	3 1/4	11	0.120
	12	0.109		10	0.134
	11	0.120		9	0.148
	10	0.134		8	0.165
1 1/2	13	0.095	3 1/2	11	0.120
	12	0.109		10	0.134
	11	0.120		9	0.148
	10	0.134		8	0.165
1 3/4	13	0.095	4	10	0.134
	12	0.109		9	0.148
	11	0.120		8	0.165
	10	0.134		7	0.180
2	13	0.095	4 1/2	10	0.134
	12	0.109		9	0.148
	11	0.120		8	0.165
	10	0.134		7	0.180
2 1/4	13	0.095	5	9	0.148
	12	0.109		8	0.165
	11	0.120		7	0.180
	10	0.134		6	0.203
2 1/2	12	0.109	5 1/2	9	0.148
	11	0.120		8	0.165
	10	0.134		7	0.180
	9	0.148		6	0.203
2 3/4	12	0.109	6	7	0.180
	11	0.120		6	0.203
	10	0.134		5	0.220
	9	0.148		4	0.238

(Multiply in by 25.4 to obtain mm.)
(Multiply in^2 by 645 to obtain mm^2.)
[a]Abstracted from information provided by the United States Steel Corporation.
[b]Values in this table are not to be used for tubes in condensers and heat-exchangers unless those tubes are specified by BWG.
[c]Birmingham wire gauge, commonly used for ferrous tubing, is identical to Stubs iron-wire gauge.

APPENDIX 17.A
Darcy Friction Factors
(turbulent flow)

Reynolds no.	relative roughness, ϵ/D 0.00000	0.000001	0.0000015	0.00001	0.00002	0.00004	0.00005	0.00006	0.00008
2×10^3	0.0495	0.0495	0.0495	0.0495	0.0495	0.0495	0.0495	0.0495	0.0495
2.5×10^3	0.0461	0.0461	0.0461	0.0461	0.0461	0.0461	0.0461	0.0461	0.0461
3×10^3	0.0435	0.0435	0.0435	0.0435	0.0435	0.0436	0.0436	0.0436	0.0436
4×10^3	0.0399	0.0399	0.0399	0.0399	0.0399	0.0399	0.0400	0.0400	0.0400
5×10^3	0.0374	0.0374	0.0374	0.0374	0.0374	0.0374	0.0374	0.0375	0.0375
6×10^3	0.0355	0.0355	0.0355	0.0355	0.0355	0.0356	0.0356	0.0356	0.0356
7×10^3	0.0340	0.0340	0.0340	0.0340	0.0340	0.0341	0.0341	0.0341	0.0341
8×10^3	0.0328	0.0328	0.0328	0.0328	0.0328	0.0328	0.0329	0.0329	0.0329
9×10^3	0.0318	0.0318	0.0318	0.0318	0.0318	0.0318	0.0318	0.0319	0.0319
1×10^4	0.0309	0.0309	0.0309	0.0309	0.0309	0.0309	0.0310	0.0310	0.0310
1.5×10^4	0.0278	0.0278	0.0278	0.0278	0.0278	0.0279	0.0279	0.0279	0.0280
2×10^4	0.0259	0.0259	0.0259	0.0259	0.0259	0.0260	0.0260	0.0260	0.0261
2.5×10^4	0.0245	0.0245	0.0245	0.0245	0.0246	0.0246	0.0246	0.0247	0.0247
3×10^4	0.0235	0.0235	0.0235	0.0235	0.0235	0.0236	0.0236	0.0236	0.0237
4×10^4	0.0220	0.0220	0.0220	0.0220	0.0220	0.0221	0.0221	0.0222	0.0222
5×10^4	0.0209	0.0209	0.0209	0.0209	0.0210	0.0210	0.0211	0.0211	0.0212
6×10^4	0.0201	0.0201	0.0201	0.0201	0.0201	0.0202	0.0203	0.0203	0.0204
7×10^4	0.0194	0.0194	0.0194	0.0194	0.0195	0.0196	0.0196	0.0197	0.0197
8×10^4	0.0189	0.0189	0.0189	0.0189	0.0190	0.0190	0.0191	0.0191	0.0192
9×10^4	0.0184	0.0184	0.0184	0.0184	0.0185	0.0186	0.0186	0.0187	0.0188
1×10^5	0.0180	0.0180	0.0180	0.0180	0.0181	0.0182	0.0183	0.0183	0.0184
1.5×10^5	0.0166	0.0166	0.0166	0.0166	0.0167	0.0168	0.0169	0.0170	0.0171
2×10^5	0.0156	0.0156	0.0156	0.0157	0.0158	0.0160	0.0160	0.0161	0.0163
2.5×10^5	0.0150	0.0150	0.0150	0.0151	0.0152	0.0153	0.0154	0.0155	0.0157
3×10^5	0.0145	0.0145	0.0145	0.0146	0.0147	0.0149	0.0150	0.0151	0.0153
4×10^5	0.0137	0.0137	0.0137	0.0138	0.0140	0.0142	0.0143	0.0144	0.0146
5×10^5	0.0132	0.0132	0.0132	0.0133	0.0134	0.0137	0.0138	0.0140	0.0142
6×10^5	0.0127	0.0128	0.0128	0.0129	0.0131	0.0133	0.0135	0.0136	0.0139
7×10^5	0.0124	0.0124	0.0124	0.0126	0.0127	0.0131	0.0132	0.0134	0.0136
8×10^5	0.0121	0.0121	0.0121	0.0123	0.0125	0.0128	0.0130	0.0131	0.0134
9×10^5	0.0119	0.0119	0.0119	0.0121	0.0123	0.0126	0.0128	0.0130	0.0133
1×10^6	0.0116	0.0117	0.0117	0.0119	0.0121	0.0125	0.0126	0.0128	0.0131
1.5×10^6	0.0109	0.0109	0.0109	0.0112	0.0114	0.0119	0.0121	0.0123	0.0127
2×10^6	0.0104	0.0104	0.0104	0.0107	0.0110	0.0116	0.0118	0.0120	0.0124
2.5×10^6	0.0100	0.0100	0.0101	0.0104	0.0108	0.0113	0.0116	0.0118	0.0123
3×10^6	0.0097	0.0098	0.0098	0.0102	0.0105	0.0112	0.0115	0.0117	0.0122
4×10^6	0.0093	0.0094	0.0094	0.0098	0.0103	0.0110	0.0113	0.0115	0.0120
5×10^6	0.0090	0.0091	0.0091	0.0096	0.0101	0.0108	0.0111	0.0114	0.0119
6×10^6	0.0087	0.0088	0.0089	0.0094	0.0099	0.0107	0.0110	0.0113	0.0118
7×10^6	0.0085	0.0086	0.0087	0.0093	0.0098	0.0106	0.0110	0.0113	0.0118
8×10^6	0.0084	0.0085	0.0085	0.0092	0.0097	0.0106	0.0109	0.0112	0.0118
9×10^6	0.0082	0.0083	0.0084	0.0091	0.0097	0.0105	0.0109	0.0112	0.0117
1×10^7	0.0081	0.0082	0.0083	0.0090	0.0096	0.0105	0.0109	0.0112	0.0117
1.5×10^7	0.0076	0.0078	0.0079	0.0087	0.0094	0.0104	0.0108	0.0111	0.0116
2×10^7	0.0073	0.0075	0.0076	0.0086	0.0093	0.0103	0.0107	0.0110	0.0116
2.5×10^7	0.0071	0.0073	0.0074	0.0085	0.0093	0.0103	0.0107	0.0110	0.0116
3×10^7	0.0069	0.0072	0.0073	0.0084	0.0092	0.0103	0.0107	0.0110	0.0116
4×10^7	0.0067	0.0070	0.0071	0.0084	0.0092	0.0102	0.0106	0.0110	0.0115
5×10^7	0.0065	0.0068	0.0070	0.0083	0.0092	0.0102	0.0106	0.0110	0.0115

APPENDIX 17.A *(continued)*
Darcy Friction Factors
(turbulent flow)

	relative roughness, ϵ/D								
Reynolds no.	0.0001	0.00015	0.00020	0.00025	0.00030	0.00035	0.0004	0.0006	0.0008
2×10^3	0.0495	0.0496	0.0496	0.0496	0.0497	0.0497	0.0498	0.0499	0.0501
2.5×10^3	0.0461	0.0462	0.0462	0.0463	0.0463	0.0463	0.0464	0.0466	0.0467
3×10^3	0.0436	0.0437	0.0437	0.0437	0.0438	0.0438	0.0439	0.0441	0.0442
4×10^3	0.0400	0.0401	0.0401	0.0402	0.0402	0.0403	0.0403	0.0405	0.0407
5×10^3	0.0375	0.0376	0.0376	0.0377	0.0377	0.0378	0.0378	0.0381	0.0383
6×10^3	0.0356	0.0357	0.0357	0.0358	0.0359	0.0359	0.0360	0.0362	0.0365
7×10^3	0.0341	0.0342	0.0343	0.0343	0.0344	0.0345	0.0345	0.0348	0.0350
8×10^3	0.0329	0.0330	0.0331	0.0331	0.0332	0.0333	0.0333	0.0336	0.0339
9×10^3	0.0319	0.0320	0.0321	0.0321	0.0322	0.0323	0.0323	0.0326	0.0329
1×10^4	0.0310	0.0311	0.0312	0.0313	0.0313	0.0314	0.0315	0.0318	0.0321
1.5×10^4	0.0280	0.0281	0.0282	0.0283	0.0284	0.0285	0.0285	0.0289	0.0293
2×10^4	0.0261	0.0262	0.0263	0.0264	0.0265	0.0266	0.0267	0.0272	0.0276
2.5×10^4	0.0248	0.0249	0.0250	0.0251	0.0252	0.0254	0.0255	0.0259	0.0264
3×10^4	0.0238	0.0239	0.0240	0.0241	0.0243	0.0244	0.0245	0.0250	0.0255
4×10^4	0.0223	0.0224	0.0226	0.0227	0.0229	0.0230	0.0232	0.0237	0.0243
5×10^4	0.0212	0.0214	0.0216	0.0218	0.0219	0.0221	0.0223	0.0229	0.0235
6×10^4	0.0205	0.0207	0.0208	0.0210	0.0212	0.0214	0.0216	0.0222	0.0229
7×10^4	0.0198	0.0200	0.0202	0.0204	0.0206	0.0208	0.0210	0.0217	0.0224
8×10^4	0.0193	0.0195	0.0198	0.0200	0.0202	0.0204	0.0206	0.0213	0.0220
9×10^4	0.0189	0.0191	0.0194	0.0196	0.0198	0.0200	0.0202	0.0210	0.0217
1×10^5	0.0185	0.0188	0.0190	0.0192	0.0195	0.0197	0.0199	0.0207	0.0215
1.5×10^5	0.0172	0.0175	0.0178	0.0181	0.0184	0.0186	0.0189	0.0198	0.0207
2×10^5	0.0164	0.0168	0.0171	0.0174	0.0177	0.0180	0.0183	0.0193	0.0202
2.5×10^5	0.0158	0.0162	0.0166	0.0170	0.0173	0.0176	0.0179	0.0190	0.0199
3×10^5	0.0154	0.0159	0.0163	0.0166	0.0170	0.0173	0.0176	0.0188	0.0197
4×10^5	0.0148	0.0153	0.0158	0.0162	0.0166	0.0169	0.0172	0.0184	0.0195
5×10^5	0.0144	0.0150	0.0154	0.0159	0.0163	0.0167	0.0170	0.0183	0.0193
6×10^5	0.0141	0.0147	0.0152	0.0157	0.0161	0.0165	0.0168	0.0181	0.0192
7×10^5	0.0139	0.0145	0.0150	0.0155	0.0159	0.0163	0.0167	0.0180	0.0191
8×10^5	0.0137	0.0143	0.0149	0.0154	0.0158	0.0162	0.0166	0.0180	0.0191
9×10^5	0.0136	0.0142	0.0148	0.0153	0.0157	0.0162	0.0165	0.0179	0.0190
1×10^6	0.0134	0.0141	0.0147	0.0152	0.0157	0.0161	0.0165	0.0178	0.0190
1.5×10^6	0.0130	0.0138	0.0144	0.0149	0.0154	0.0159	0.0163	0.0177	0.0189
2×10^6	0.0128	0.0136	0.0142	0.0148	0.0153	0.0158	0.0162	0.0176	0.0188
2.5×10^6	0.0127	0.0135	0.0141	0.0147	0.0152	0.0157	0.0161	0.0176	0.0188
3×10^6	0.0126	0.0134	0.0141	0.0147	0.0152	0.0157	0.0161	0.0176	0.0187
4×10^6	0.0124	0.0133	0.0140	0.0146	0.0151	0.0156	0.0161	0.0175	0.0187
5×10^6	0.0123	0.0132	0.0139	0.0146	0.0151	0.0156	0.0160	0.0175	0.0187
6×10^6	0.0123	0.0132	0.0139	0.0145	0.0151	0.0156	0.0160	0.0175	0.0187
7×10^6	0.0122	0.0132	0.0139	0.0145	0.0151	0.0155	0.0160	0.0175	0.0187
8×10^6	0.0122	0.0131	0.0139	0.0145	0.0150	0.0155	0.0160	0.0175	0.0187
9×10^6	0.0122	0.0131	0.0139	0.0145	0.0150	0.0155	0.0160	0.0175	0.0187
1×10^7	0.0122	0.0131	0.0138	0.0145	0.0150	0.0155	0.0160	0.0175	0.0186
1.5×10^7	0.0121	0.0131	0.0138	0.0144	0.0150	0.0155	0.0159	0.0174	0.0186
2×10^7	0.0121	0.0130	0.0138	0.0144	0.0150	0.0155	0.0159	0.0174	0.0186
2.5×10^7	0.0121	0.0130	0.0138	0.0144	0.0150	0.0155	0.0159	0.0174	0.0186
3×10^7	0.0120	0.0130	0.0138	0.0144	0.0150	0.0155	0.0159	0.0174	0.0186
4×10^7	0.0120	0.0130	0.0138	0.0144	0.0150	0.0155	0.0159	0.0174	0.0186
5×10^7	0.0120	0.0130	0.0138	0.0144	0.0150	0.0155	0.0159	0.0174	0.0186

APPENDIX 17.A *(continued)*
Darcy Friction Factors
(turbulent flow)

	relative roughness, ϵ/D								
Reynolds no.	0.001	0.0015	0.002	0.0025	0.003	0.0035	0.004	0.006	0.008
2×10^3	0.0502	0.0506	0.0510	0.0513	0.0517	0.0521	0.0525	0.0539	0.0554
2.5×10^3	0.0469	0.0473	0.0477	0.0481	0.0485	0.0489	0.0493	0.0509	0.0524
3×10^3	0.0444	0.0449	0.0453	0.0457	0.0462	0.0466	0.0470	0.0487	0.0503
4×10^3	0.0409	0.0414	0.0419	0.0424	0.0429	0.0433	0.0438	0.0456	0.0474
5×10^3	0.0385	0.0390	0.0396	0.0401	0.0406	0.0411	0.0416	0.0436	0.0455
6×10^3	0.0367	0.0373	0.0378	0.0384	0.0390	0.0395	0.0400	0.0421	0.0441
7×10^3	0.0353	0.0359	0.0365	0.0371	0.0377	0.0383	0.0388	0.0410	0.0430
8×10^3	0.0341	0.0348	0.0354	0.0361	0.0367	0.0373	0.0379	0.0401	0.0422
9×10^3	0.0332	0.0339	0.0345	0.0352	0.0358	0.0365	0.0371	0.0394	0.0416
1×10^4	0.0324	0.0331	0.0338	0.0345	0.0351	0.0358	0.0364	0.0388	0.0410
1.5×10^4	0.0296	0.0305	0.0313	0.0320	0.0328	0.0335	0.0342	0.0369	0.0393
2×10^4	0.0279	0.0289	0.0298	0.0306	0.0315	0.0323	0.0330	0.0358	0.0384
2.5×10^4	0.0268	0.0278	0.0288	0.0297	0.0306	0.0314	0.0322	0.0352	0.0378
3×10^4	0.0260	0.0271	0.0281	0.0291	0.0300	0.0308	0.0317	0.0347	0.0374
4×10^4	0.0248	0.0260	0.0271	0.0282	0.0291	0.0301	0.0309	0.0341	0.0369
5×10^4	0.0240	0.0253	0.0265	0.0276	0.0286	0.0296	0.0305	0.0337	0.0365
6×10^4	0.0235	0.0248	0.0261	0.0272	0.0283	0.0292	0.0302	0.0335	0.0363
7×10^4	0.0230	0.0245	0.0257	0.0269	0.0280	0.0290	0.0299	0.0333	0.0362
8×10^4	0.0227	0.0242	0.0255	0.0267	0.0278	0.0288	0.0298	0.0331	0.0361
9×10^4	0.0224	0.0239	0.0253	0.0265	0.0276	0.0286	0.0296	0.0330	0.0360
1×10^5	0.0222	0.0237	0.0251	0.0263	0.0275	0.0285	0.0295	0.0329	0.0359
1.5×10^5	0.0214	0.0231	0.0246	0.0259	0.0271	0.0281	0.0292	0.0327	0.0357
2×10^5	0.0210	0.0228	0.0243	0.0256	0.0268	0.0279	0.0290	0.0325	0.0355
2.5×10^5	0.0208	0.0226	0.0241	0.0255	0.0267	0.0278	0.0289	0.0325	0.0355
3×10^5	0.0206	0.0225	0.0240	0.0254	0.0266	0.0277	0.0288	0.0324	0.0354
4×10^5	0.0204	0.0223	0.0239	0.0253	0.0265	0.0276	0.0287	0.0323	0.0354
5×10^5	0.0202	0.0222	0.0238	0.0252	0.0264	0.0276	0.0286	0.0323	0.0353
6×10^5	0.0201	0.0221	0.0237	0.0251	0.0264	0.0275	0.0286	0.0323	0.0353
7×10^5	0.0201	0.0221	0.0237	0.0251	0.0264	0.0275	0.0286	0.0322	0.0353
8×10^5	0.0200	0.0220	0.0237	0.0251	0.0263	0.0275	0.0286	0.0322	0.0353
9×10^5	0.0200	0.0220	0.0236	0.0251	0.0263	0.0275	0.0285	0.0322	0.0353
1×10^6	0.0199	0.0220	0.0236	0.0250	0.0263	0.0275	0.0285	0.0322	0.0353
1.5×10^6	0.0198	0.0219	0.0235	0.0250	0.0263	0.0274	0.0285	0.0322	0.0352
2×10^6	0.0198	0.0218	0.0235	0.0250	0.0262	0.0274	0.0285	0.0322	0.0352
2.5×10^6	0.0198	0.0218	0.0235	0.0249	0.0262	0.0274	0.0285	0.0322	0.0352
3×10^6	0.0197	0.0218	0.0235	0.0249	0.0262	0.0274	0.0285	0.0321	0.0352
4×10^6	0.0197	0.0218	0.0235	0.0249	0.0262	0.0274	0.0284	0.0321	0.0352
5×10^6	0.0197	0.0218	0.0235	0.0249	0.0262	0.0274	0.0284	0.0321	0.0352
6×10^6	0.0197	0.0218	0.0235	0.0249	0.0262	0.0274	0.0284	0.0321	0.0352
7×10^6	0.0197	0.0218	0.0234	0.0249	0.0262	0.0274	0.0284	0.0321	0.0352
8×10^6	0.0197	0.0218	0.0234	0.0249	0.0262	0.0274	0.0284	0.0321	0.0352
9×10^6	0.0197	0.0218	0.0234	0.0249	0.0262	0.0274	0.0284	0.0321	0.0352
1×10^7	0.0197	0.0218	0.0234	0.0249	0.0262	0.0273	0.0284	0.0321	0.0352
1.5×10^7	0.0197	0.0217	0.0234	0.0249	0.0262	0.0273	0.0284	0.0321	0.0352
2×10^7	0.0197	0.0217	0.0234	0.0249	0.0262	0.0273	0.0284	0.0321	0.0352
2.5×10^7	0.0196	0.0217	0.0234	0.0249	0.0262	0.0273	0.0284	0.0321	0.0352
3×10^7	0.0196	0.0217	0.0234	0.0249	0.0262	0.0273	0.0284	0.0321	0.0352
4×10^7	0.0196	0.0217	0.0234	0.0249	0.0262	0.0273	0.0284	0.0321	0.0352
5×10^7	0.0196	0.0217	0.0234	0.0249	0.0262	0.0273	0.0284	0.0321	0.0352

APPENDIX 17.A *(continued)*
Darcy Friction Factors
(turbulent flow)

Reynolds no.	relative roughness, ϵ/D 0.01	0.015	0.02	0.025	0.03	0.035	0.04	0.045	0.05
2×10^3	0.0568	0.0602	0.0635	0.0668	0.0699	0.0730	0.0760	0.0790	0.0819
2.5×10^3	0.0539	0.0576	0.0610	0.0644	0.0677	0.0709	0.0740	0.0770	0.0800
3×10^3	0.0519	0.0557	0.0593	0.0628	0.0661	0.0694	0.0725	0.0756	0.0787
4×10^3	0.0491	0.0531	0.0570	0.0606	0.0641	0.0674	0.0707	0.0739	0.0770
5×10^3	0.0473	0.0515	0.0555	0.0592	0.0628	0.0662	0.0696	0.0728	0.0759
6×10^3	0.0460	0.0504	0.0544	0.0583	0.0619	0.0654	0.0688	0.0721	0.0752
7×10^3	0.0450	0.0495	0.0537	0.0576	0.0613	0.0648	0.0682	0.0715	0.0747
8×10^3	0.0442	0.0489	0.0531	0.0571	0.0608	0.0644	0.0678	0.0711	0.0743
9×10^3	0.0436	0.0484	0.0526	0.0566	0.0604	0.0640	0.0675	0.0708	0.0740
1×10^4	0.0431	0.0479	0.0523	0.0563	0.0601	0.0637	0.0672	0.0705	0.0738
1.5×10^4	0.0415	0.0466	0.0511	0.0553	0.0592	0.0628	0.0664	0.0698	0.0731
2×10^4	0.0407	0.0459	0.0505	0.0547	0.0587	0.0624	0.0660	0.0694	0.0727
2.5×10^4	0.0402	0.0455	0.0502	0.0544	0.0584	0.0621	0.0657	0.0691	0.0725
3×10^4	0.0398	0.0452	0.0499	0.0542	0.0582	0.0619	0.0655	0.0690	0.0723
4×10^4	0.0394	0.0448	0.0496	0.0539	0.0579	0.0617	0.0653	0.0688	0.0721
5×10^4	0.0391	0.0446	0.0494	0.0538	0.0578	0.0616	0.0652	0.0687	0.0720
6×10^4	0.0389	0.0445	0.0493	0.0536	0.0577	0.0615	0.0651	0.0686	0.0719
7×10^4	0.0388	0.0443	0.0492	0.0536	0.0576	0.0614	0.0650	0.0685	0.0719
8×10^4	0.0387	0.0443	0.0491	0.0535	0.0576	0.0614	0.0650	0.0685	0.0718
9×10^4	0.0386	0.0442	0.0491	0.0535	0.0575	0.0613	0.0650	0.0684	0.0718
1×10^5	0.0385	0.0442	0.0490	0.0534	0.0575	0.0613	0.0649	0.0684	0.0718
1.5×10^5	0.0383	0.0440	0.0489	0.0533	0.0574	0.0612	0.0648	0.0683	0.0717
2×10^5	0.0382	0.0439	0.0488	0.0532	0.0573	0.0612	0.0648	0.0683	0.0717
2.5×10^5	0.0381	0.0439	0.0488	0.0532	0.0573	0.0611	0.0648	0.0683	0.0716
3×10^5	0.0381	0.0438	0.0488	0.0532	0.0573	0.0611	0.0648	0.0683	0.0716
4×10^5	0.0381	0.0438	0.0487	0.0532	0.0573	0.0611	0.0647	0.0682	0.0716
5×10^5	0.0380	0.0438	0.0487	0.0531	0.0572	0.0611	0.0647	0.0682	0.0716
6×10^5	0.0380	0.0438	0.0487	0.0531	0.0572	0.0611	0.0647	0.0682	0.0716
7×10^5	0.0380	0.0438	0.0487	0.0531	0.0572	0.0611	0.0647	0.0682	0.0716
8×10^5	0.0380	0.0437	0.0487	0.0531	0.0572	0.0611	0.0647	0.0682	0.0716
9×10^5	0.0380	0.0437	0.0487	0.0531	0.0572	0.0610	0.0647	0.0682	0.0716
1×10^6	0.0380	0.0437	0.0487	0.0531	0.0572	0.0610	0.0647	0.0682	0.0716
1.5×10^6	0.0379	0.0437	0.0487	0.0531	0.0572	0.0610	0.0647	0.0682	0.0716
2×10^6	0.0379	0.0437	0.0487	0.0531	0.0572	0.0610	0.0647	0.0682	0.0716
2.5×10^6	0.0379	0.0437	0.0487	0.0531	0.0572	0.0610	0.0647	0.0682	0.0716
3×10^6	0.0379	0.0437	0.0487	0.0531	0.0572	0.0610	0.0647	0.0682	0.0716
4×10^6	0.0379	0.0437	0.0486	0.0531	0.0572	0.0610	0.0647	0.0682	0.0716
5×10^6	0.0379	0.0437	0.0486	0.0531	0.0572	0.0610	0.0647	0.0682	0.0716
6×10^6	0.0379	0.0437	0.0486	0.0531	0.0572	0.0610	0.0647	0.0682	0.0716
7×10^6	0.0379	0.0437	0.0486	0.0531	0.0572	0.0610	0.0647	0.0682	0.0716
8×10^6	0.0379	0.0437	0.0486	0.0531	0.0572	0.0610	0.0647	0.0682	0.0716
9×10^6	0.0379	0.0437	0.0486	0.0531	0.0572	0.0610	0.0647	0.0682	0.0716
1×10^7	0.0379	0.0437	0.0486	0.0531	0.0572	0.0610	0.0647	0.0682	0.0716
1.5×10^7	0.0379	0.0437	0.0486	0.0531	0.0572	0.0610	0.0647	0.0682	0.0716
2×10^7	0.0379	0.0437	0.0486	0.0531	0.0572	0.0610	0.0647	0.0682	0.0716
2.5×10^7	0.0379	0.0437	0.0486	0.0531	0.0572	0.0610	0.0647	0.0682	0.0716
3×10^7	0.0379	0.0437	0.0486	0.0531	0.0572	0.0610	0.0647	0.0682	0.0716
4×10^7	0.0379	0.0437	0.0486	0.0531	0.0572	0.0610	0.0647	0.0682	0.0716
5×10^7	0.0379	0.0437	0.0486	0.0531	0.0572	0.0610	0.0647	0.0682	0.0716

APPENDIX 17.B
Water Pressure Drop in Schedule-40 Steel Pipe

pressure drop per 1000 ft of schedule-40 steel pipe (lbf/in^2)

discharge (gal/min)	velocity (ft/sec)	pressure drop	velocity (ft/sec)	pressure drop	velocity (ft/sec)	pressure drop	velocity (ft/sec)	pressure drop	velocity (ft/sec)	pressure drop	velocity (ft/sec)	pressure drop	velocity (ft/sec)	pressure drop	velocity (ft/sec)	pressure drop	velocity (ft/sec)	pressure drop
	1 in																	
1	0.37	0.49	1¼ in															
2	0.74	1.70	0.43	0.45	1½ in													
3	1.12	3.53	0.64	0.94	0.47	0.44												
4	1.49	5.94	0.86	1.55	0.63	0.74												
5	1.86	9.02	1.07	2.36	0.79	1.12	2 in											
6	2.24	12.25	1.28	3.30	0.95	1.53	0.57	0.46										
8	2.98	21.1	1.72	5.52	1.26	2.63	0.76	0.75	2½ in									
10	3.72	30.8	2.14	8.34	1.57	3.86	0.96	1.14	0.67	0.48								
15	5.60	64.6	3.21	17.6	2.36	8.13	1.43	2.33	1.00	0.99	3 in							
20	7.44	110.5	4.29	29.1	3.15	13.5	1.91	3.86	1.34	1.64	0.87	0.59	3½ in					
25			5.36	43.7	3.94	20.2	2.39	5.81	1.68	2.48	1.08	0.67	0.81	0.42				
30			6.43	62.9	4.72	29.1	2.87	8.04	2.01	3.43	1.30	1.21	0.97	0.60	4 in			
35			7.51	82.5	5.51	38.2	3.35	10.95	2.35	4.49	1.52	1.58	1.14	0.79	0.88	0.42		
40					6.30	47.8	3.82	13.7	2.68	5.88	1.74	2.06	1.30	1.00	1.01	0.53		
45					7.08	60.6	4.30	17.4	3.00	7.14	1.95	2.51	1.46	1.21	1.13	0.67		
50					7.87	74.7	4.78	20.6	3.35	8.82	2.17	3.10	1.62	1.44	1.26	0.80		
60							5.74	29.6	4.02	12.2	2.60	4.29	1.95	2.07	1.51	1.10	5 in	
70							6.69	38.6	4.69	15.3	3.04	5.84	2.27	2.71	1.76	1.50	1.12	0.48
80							7.65	50.3	5.37	21.7	3.48	7.62	2.59	3.53	2.01	1.87	1.28	0.63
90	6 in						8.60	63.6	6.04	26.1	3.91	9.22	2.92	4.46	2.26	2.37	1.44	0.80
100	1.11	0.39					9.56	75.1	6.71	32.3	4.34	11.4	3.24	5.27	2.52	2.81	1.60	0.95
125	1.39	0.56							8.38	48.2	5.42	17.1	4.05	7.86	3.15	4.38	2.00	1.48
150	1.67	0.78							10.06	60.4	6.51	23.5	4.86	11.3	3.78	6.02	2.41	2.04
175	1.94	1.06							11.73	90.0	7.59	32.0	5.67	14.7	4.41	8.20	2.81	2.78
200	2.22	1.32	8 in								8.68	39.7	6.48	19.2	5.04	10.2	3.21	3.46
225	2.50	1.66	1.44	0.44							9.77	50.2	7.29	23.1	5.67	12.9	3.61	4.37
250	2.78	2.05	1.60	0.55							10.85	61.9	8.10	28.5	6.30	15.9	4.01	5.14
275	3.06	2.36	1.76	0.63							11.94	75.0	8.91	34.4	6.93	18.3	4.41	6.22
300	3.33	2.80	1.92	0.75							13.02	84.7	9.72	40.9	7.56	21.8	4.81	7.41
325	3.61	3.29	2.08	0.88									10.53	45.5	8.18	25.5	5.21	8.25
350	3.89	3.62	2.24	0.97									11.35	52.7	8.82	29.7	5.61	9.57
375	4.16	4.16	2.40	1.11									12.17	60.7	9.45	32.3	6.01	11.0
400	4.44	4.72	2.56	1.27									12.97	68.9	10.08	36.7	6.41	12.5
425	4.72	5.34	2.72	1.43									13.78	77.8	10.70	41.5	6.82	14.1
450	5.00	5.96	2.88	1.60	10 in								14.59	87.3	11.33	46.5	7.22	15.0
475	5.27	6.66	3.04	1.69	1.93	0.30									11.96	51.7	7.62	16.7
500	5.55	7.39	3.20	1.87	2.04	0.63									12.59	57.3	8.02	18.5
550	6.11	8.94	3.53	2.26	2.24	0.70									13.84	69.3	8.82	22.4
600	6.66	10.6	3.85	2.70	2.44	0.86									15.10	82.5	9.62	26.7
650	7.21	11.8	4.17	3.16	2.65	1.01	12 in										10.42	31.3
700	7.77	13.7	4.49	3.69	2.85	1.18	2.01	0.48									11.22	36.3
750	8.32	15.7	4.81	4.21	3.05	1.35	2.15	0.55									12.02	41.6
800	8.88	17.8	5.13	4.79	3.26	1.54	2.29	0.62	14 in								12.82	44.7
850	9.44	20.2	5.45	5.11	3.46	1.74	2.44	0.70	2.02	0.43							13.62	50.5
900	10.00	22.6	5.77	5.73	3.66	1.94	2.58	0.79	2.14	0.48							14.42	56.6
950	10.55	23.7	6.09	6.38	3.87	2.23	2.72	0.88	2.25	0.53							15.22	63.1
1000	11.10	26.3	6.41	7.08	4.07	2.40	2.87	0.98	2.38	0.59							16.02	70.0
1100	12.22	31.8	7.05	8.56	4.48	2.74	3.16	1.18	2.61	0.68	16 in						17.63	84.6
1200	13.32	37.8	7.69	10.2	4.88	3.27	3.45	1.40	2.85	0.81	2.18	0.40						
1300	14.43	44.4	8.33	11.3	5.29	3.86	3.73	1.56	3.09	0.95	2.36	0.47						
1400	15.54	51.5	8.97	13.0	5.70	4.44	4.02	1.80	3.32	1.10	2.54	0.54						
1500	16.65	55.5	9.62	15.0	6.10	5.11	4.30	2.07	3.55	1.19	2.73	0.62						
1600	17.76	63.1	10.26	17.0	6.51	5.46	4.59	2.36	3.80	1.35	2.91	0.71	18 in					
1800	19.98	79.8	11.54	21.6	7.32	6.91	5.16	2.98	4.27	1.71	3.27	0.85	2.58	0.48				
2000	22.20	98.5	12.83	25.0	8.13	8.54	5.73	3.47	4.74	2.11	3.63	1.05	2.88	0.56				
2500			16.03	39.0	10.18	12.5	7.17	5.41	5.92	3.09	4.54	1.63	3.59	0.88	20 in			
3000			19.24	52.4	12.21	18.0	8.60	7.31	7.12	4.45	5.45	2.21	4.31	1.27	3.45	0.73		
3500			22.43	71.4	14.25	22.9	10.03	9.95	8.32	6.18	6.35	3.00	5.03	1.52	4.03	0.94	24 in	
4000			25.65	93.3	16.28	29.9	11.48	13.0	9.49	7.92	7.25	3.92	5.74	2.12	4.61	1.22	3.19	0.51
4500					18.31	37.8	12.90	15.4	10.67	9.36	8.17	4.97	6.47	2.50	5.19	1.55	3.59	0.60
5000					20.35	46.7	14.34	18.9	11.84	11.6	9.08	5.72	7.17	3.08	5.76	1.78	3.99	0.74
6000					24.42	67.2	17.21	27.3	14.32	15.4	10.88	8.24	8.62	4.45	6.92	2.57	4.80	1.00
7000					28.50	85.1	20.08	37.2	16.60	21.0	12.69	12.2	10.04	6.06	8.06	3.50	5.68	1.36
8000							22.95	45.1	18.98	27.4	14.52	13.6	11.48	7.34	9.23	4.57	6.38	1.78
9000							25.80	57.0	21.35	34.7	16.32	17.2	12.92	9.20	10.37	5.36	7.19	2.25
10,000							28.63	70.4	23.75	42.9	18.16	21.2	14.37	11.5	11.53	6.63	7.96	2.78
12,000							34.38	93.6	28.50	61.8	21.80	30.9	17.23	16.5	13.83	9.54	9.57	3.71
14,000									33.20	84.0	25.42	41.6	20.10	20.7	16.14	12.0	11.18	5.05
16,000											29.05	54.4	22.96	27.1	18.43	15.7	12.77	6.60

(Multiply gal/min by 0.0631 to obtain L/s.)
(Multiply ft/sec by 0.3048 to obtain m/s.)
(Multiply in by 25.4 to obtain mm.)
(Multiply lbf/in^2-1000 ft by 2.3 to obtain kPa/100 m.)

APPENDIX 17.C

Equivalent Length of Straight Pipe for Various (generic) Fittings
(in feet, turbulent flow only, for any fluid)

fittings			pipe size (in) 1/4	3/8	1/2	3/4	1	1 1/4	1 1/2	2	2 1/2	3	4	5	6	8	10	12	14	16	18	20	24
regular 90° ell	screwed	steel	2.3	3.1	3.6	4.4	5.2	6.6	7.4	8.5	9.3	11.0	13.0										
		cast iron										9.0	11.0										
	flanged	steel			0.92	1.2	1.6	2.1	2.4	3.1	3.6	4.4	5.9	7.3	8.9	12.0	14.0	17.0	18.0	21.0	23.0	25.0	30.0
		cast iron										3.6	4.8		7.2	9.8	12.0	15.0	17.0	19.0	22.0	24.0	28.0
long radius 90° ell	screwed	steel	1.5	2.0	2.2	2.3	2.7	3.2	3.4	3.6	3.6	4.0	4.6										
		cast iron										3.3	3.7										
	flanged	steel			1.1	1.3	1.6	2.0	2.3	2.7	2.9	3.4	4.2	5.0	5.7	7.0	8.0	9.0	9.4	10.0	11.0	12.0	14.0
		cast iron										2.8	3.4		4.7	5.7	6.8	7.8	8.6	9.6	11.0	11.0	13.0
regular 45° ell	screwed	steel	0.34	0.52	0.71	0.92	1.3	1.7	2.1	2.7	3.2	4.0	5.5										
		cast iron										3.3	4.5										
	flanged	steel			0.45	0.59	0.81	1.1	1.3	1.7	2.0	2.6	3.5	4.5	5.6	7.7	9.0	11.0	13.0	15.0	16.0	18.0	22.0
		cast iron										2.1	2.9		4.5	6.3	8.1	9.7	12.0	13.0	15.0	17.0	20.0
tee-line flow	screwed	steel	0.79	1.2	1.7	2.4	3.2	4.6	5.6	7.7	9.3	12.0	17.0										
		cast iron										9.9	14.0										
	flanged	steel			0.69	0.82	1.0	1.3	1.5	1.8	1.9	2.2	2.8	3.3	3.8	4.7	5.2	6.0	6.4	7.2	7.6	8.2	9.6
		cast iron										1.9	2.2		3.1	3.9	4.6	5.2	5.9	6.5	7.2	7.7	8.8
tee-branch flow	screwed	steel	2.4	3.5	4.2	5.3	6.6	8.7	9.9	12.0	13.0	17.0	21.0										
		cast iron										14.0	17.0										
	flanged	steel			2.0	2.6	3.3	4.4	5.2	6.6	7.5	9.4	12.0	15.0	18.0	24.0	30.0	34.0	37.0	43.0	47.0	52.0	62.0
		cast iron										7.7	10.0		15.0	20.0	25.0	30.0	35.0	39.0	44.0	49.0	57.0
180° return bend	screwed	steel	2.3	3.1	3.6	4.4	5.2	6.6	7.4	8.5	9.3	11.0	13.0										
		cast iron										9.0	11.0										
	regular flanged	steel			0.92	1.2	1.6	2.1	2.4	3.1	3.6	4.4	5.9	7.3	8.9	12.0	14.0	17.0	18.0	21.0	23.0	25.0	30.0
		cast iron										3.6	4.8		7.2	9.8	12.0	15.0	17.0	19.0	22.0	24.0	28.0
	long radius flanged	steel			1.1	1.3	1.6	2.0	2.3	2.7	2.9	3.4	4.2	5.0	5.7	7.0	8.0	9.0	9.4	10.0	11.0	12.0	14.0
		cast iron										2.8	3.4		4.7	5.7	6.8	7.8	8.6	9.6	11.0	11.0	13.0
globe valve	screwed	steel	21.0	22.0	22.0	24.0	29.0	37.0	42.0	54.0	62.0	79.0	110.0										
		cast iron										65.0	86.0										
	flanged	steel			38.0	40.0	45.0	54.0	59.0	70.0	77.0	94.0	120.0	150.0	190.0	260.0	310.0	390.0					
		cast iron										77.0	99.0		150.0	210.0	270.0	330.0					
gate valve	screwed	steel	0.32	0.45	0.56	0.67	0.84	1.1	1.2	1.5	1.7	1.9	2.5										
		cast iron										1.6	2.0										
	flanged	steel								2.6	2.7	2.8	2.9	3.1	3.2	3.2	3.2	3.2	3.2	3.2	3.2	3.2	3.2
		cast iron										2.3	2.4		2.6	2.7	2.8	2.9	2.9	3.0	3.0	3.0	3.0
angle valve	screwed	steel	12.8	15.0	15.0	15.0	17.0	18.0	18.0	18.0	18.0	18.0	18.0										
		cast iron										15.0	15.0										
	flanged	steel			15.0	15.0	17.0	18.0	18.0	21.0	22.0	28.0	38.0	50.0	63.0	90.0	120.0	140.0	160.0	190.0	210.0	240.0	300.0
		cast iron										23.0	31.0		52.0	74.0	98.0	120.0	150.0	170.0	200.0	230.0	280.0
swing check valve	screwed	steel	7.2	7.3	8.0	8.8	11.0	13.0	15.0	19.0	22.0	27.0	38.0										
		cast iron										22.0	31.0										
	flanged	steel			3.8	5.3	7.2	10.0	12.0	17.0	21.0	27.0	38.0	50.0	63.0	90.0	120.0	140.0					
		cast iron										22.0	31.0		52.0	74.0	98.0	120.0					
coupling or union	screwed	steel	0.14	0.18	0.21	0.24	0.29	0.36	0.39	0.45	0.47	0.53	0.65										
		cast iron										0.44	0.52										
inlet	bell mouth inlet	steel	0.04	0.07	0.10	0.13	0.18	0.26	0.31	0.43	0.52	0.67	0.95	1.3	1.6	2.3	2.9	3.5	4.0	4.7	5.3	6.1	7.6
		cast iron										0.55	0.77		1.3	1.9	2.4	3.0	3.6	4.3	5.0	5.7	7.0
	square mouth inlet	steel	0.44	0.68	0.96	1.3	1.8	2.6	3.1	4.3	5.2	6.7	9.5	13.0	16.0	23.0	29.0	35.0	40.0	47.0	53.0	61.0	76.0
		cast iron										5.5	7.7		13.0	19.0	24.0	30.0	36.0	43.0	50.0	57.0	70.0
	re-entrant pipe	steel	0.88	1.4	1.9	2.6	3.6	5.1	6.2	8.5	10.0	13.0	19.0	25.0	32.0	45.0	58.0	70.0	80.0	95.0	110.0	120.0	150.0
		cast iron										11.0	15.0		26.0	37.0	49.0	61.0	73.0	86.0	100.0	110.0	140.0

(Multiply in by 25.4 to obtain mm.)
(Multiply ft by 0.3048 to obtain m.)

APPENDIX 17.D
Hazen-Williams Nomograph
($C = 100$)

Quantity (i.e., flow rate) and velocity are proportional to the C-value. For values of C other than 100, the quantity and velocity must be converted according to $\dot{V}_{\text{actual}} = \dot{V}_{\text{chart}} C_{\text{actual}}/100$. When quantity is the unknown, use the chart with known values of diameter, slope, or velocity to find $\dot{V}_{\text{chart}}$, and then convert to $\dot{V}_{\text{actual}}$. When velocity is the unknown, use the chart with the known values of diameter, slope, or quantity to find $\dot{V}_{\text{chart}}$, then convert to $\dot{V}_{\text{actual}}$. If $\dot{V}_{\text{actual}}$ is known, it must be converted to $\dot{V}_{\text{chart}}$ before this nomograph can be used. In that case, the diameter, loss, and quantity are as read from this chart.

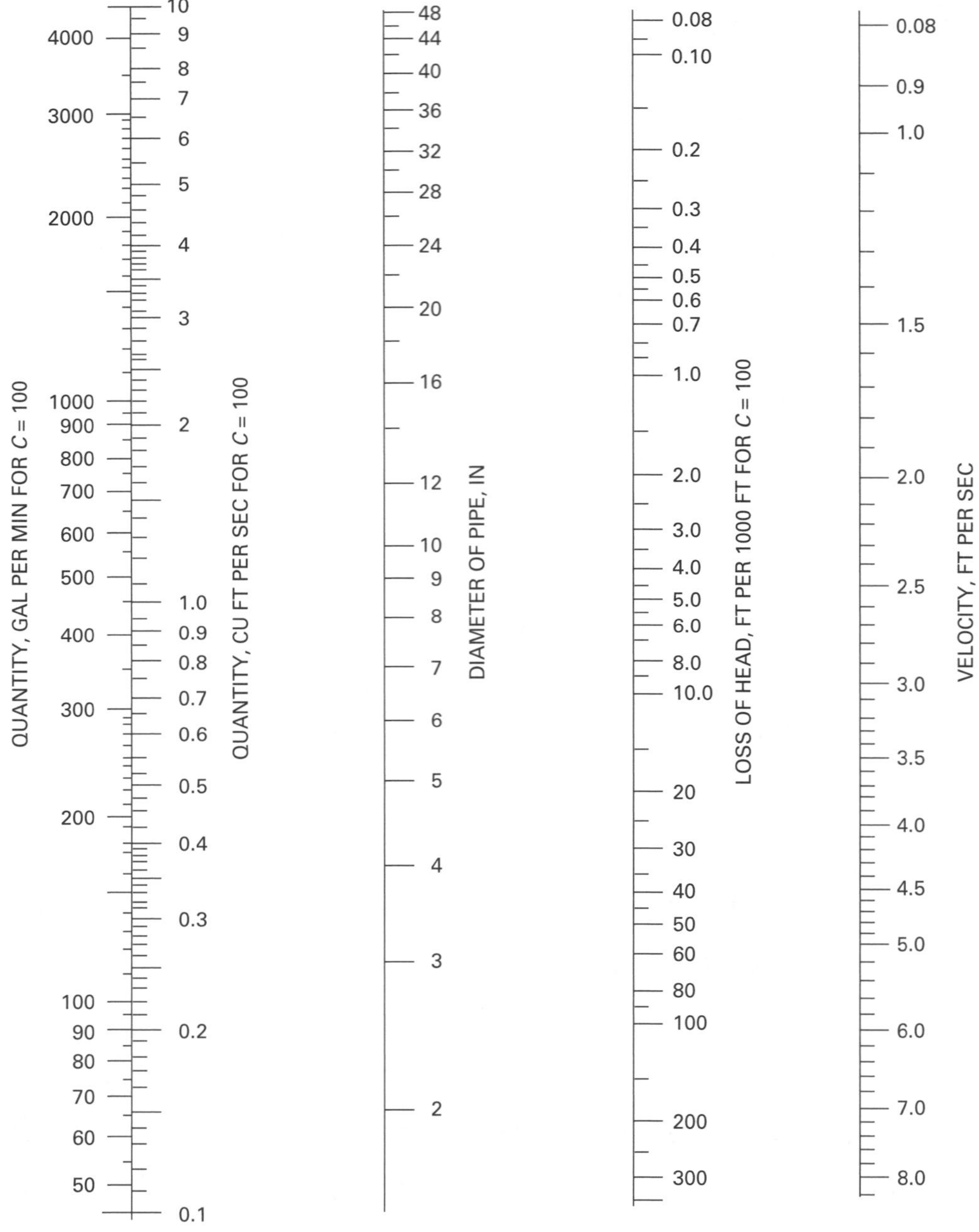

(Multiply gal/min by 0.0631 to obtain L/s.)
(Multiply ft^3/sec by 28.3 to obtain L/s.)
(Multiply in by 25.4 to obtain mm.)
(Multiply ft/1000 ft by 0.1 to obtain m/100 m.)
(Multiply ft/sec by 0.3048 to obtain m/s.)

APPENDIX 17.E
Corrugated Metal Pipe

Corrugated metal pipe (CMP, also known as *corrugated steel pipe*) is frequently used for culverts. Pipe is made from corrugated sheets of galvanized steel that are rolled and riveted together along a longitudinal seam. Aluminized steel may also be used in certain ranges of soil pH. Standard round pipe diameters range from 8 in to 96 in (200 mm to 2450 mm). Metric dimensions of standard diameters are usually rounded to the nearest 25 mm or 50 mm (e.g., a 42 in culvert would be specified as a 1050 mm culvert, not 1067 mm).

Larger and noncircular culverts can be created out of curved steel plate. Standard section lengths are 10 ft to 20 ft (3 m to 6 m). Though most corrugations are transverse (i.e., annular), helical corrugations are also used. Metal gages of 8, 10, 12, 14, and 16 are commonly used, depending on the depth of burial.

The most common corrugated steel pipe has transverse corrugations that are $\frac{1}{2}$ in (13 mm) deep and $2\frac{2}{3}$ in (68 mm) from crest to crest. These are referred to as "$2\frac{1}{2}$ inch" or "68 × 13" corrugations. For larger culverts, corrugations with a 2 in (25 mm) depth and 3, 5, and 6 in (76, 125, or 152 mm) pitches are used. Plate-based products using 6 in by 2 in (152 mm by 51 mm) corrugations are known as *structural plate corrugated steel pipe* (SPCSP) and *multiplate* after the trade-named product "Multi-Plate™."

The flow area for circular culverts is based on the nominal culvert diameter, regardless of the gage of the plate metal used to construct the pipe. Flow area is calculated to (at most) three significant digits.

A Hazen-Williams coefficient, C, of 60 is typically used with all sizes of corrugated pipe. Values of C and Manning's constant, n, for corrugated pipe are generally not affected by age. *Design Charts for Open Channel Flow* (U.S. Department of Transportation, 1979) recommends a Manning constant of $n = 0.024$ for all cases. The U.S. Department of the Interior recommends the following values. For standard ($2\frac{2}{3}$ in by $\frac{1}{2}$ in or 68 mm by 13 mm) corrugated pipe with the diameters given: 12 in (457 mm), 0.027; 24 in (610 mm), 0.025; 36 in to 48 in (914 mm to 1219 mm), 0.024; 60 in to 84 in (1524 mm to 2134 mm), 0.023; 96 in (2438 mm), 0.022. For (6 in by 2 in or 152 mm by 51 mm) multiplate construction with the diameters given: 5 ft to 6 ft (1.5 m to 1.8 m), 0.034; 7 ft to 8 ft (2.1 m to 2.4 m), 0.033; 9 ft to 11 ft (2.7 m to 3.3 m), 0.032; 12 ft to 13 ft (3.6 m to 3.9 m), 0.031; 14 ft to 15 ft (4.2 m to 4.5 m), 0.030; 16 ft to 18 ft (4.8 m to 5.4 m), 0.029; 19 ft to 20 ft (5.8 m to 6.0 m), 0.028; 21 ft to 22 ft (6.3 m to 6.6 m), 0.027.

If the inside of the corrugated pipe has been asphalted completely smooth 360° circumferentially, Manning's n ranges from 0.009 to 0.011. For culverts with 40% asphalted inverts, $n = 0.019$. For other percentages of paved invert, the resulting value is proportional to the percentage and the values normally corresponding to that diameter pipe. For field-bolted corrugated metal pipe arches, $n = 0.025$.

It is also possible to calculate the Darcy friction loss if the corrugation depth, 0.5 in (13 mm) for standard corrugations and 2.0 in (51 mm) for multiplate, is taken as the specific roughness.

APPENDIX 20.A
Atomic Numbers and Weights of the Elements
(referred to carbon-12)

name	symbol	atomic number	atomic weight	name	symbol	atomic number	atomic weight
actinium	Ac	89	–	meitnerium	Mt	109	–
aluminum	Al	13	26.9815	mendelevium	Md	101	–
americium	Am	95	–	mercury	Hg	80	200.59
antimony	Sb	51	121.760	molybdenum	Mo	42	95.96
argon	Ar	18	39.948	neodymium	Nd	60	144.242
arsenic	As	33	74.9216	neon	Ne	10	20.1797
astatine	At	85	–	neptunium	Np	93	237.048
barium	Ba	56	137.327	nickel	Ni	28	58.693
berkelium	Bk	97	–	niobium	Nb	41	92.906
beryllium	Be	4	9.0122	nitrogen	N	7	14.0067
bismuth	Bi	83	208.980	nobelium	No	102	–
bohrium	Bh	107	–	osmium	Os	76	190.23
boron	B	5	10.811	oxygen	O	8	15.9994
bromine	Br	35	79.904	palladium	Pd	46	106.42
cadmium	Cd	48	112.411	phosphorus	P	15	30.9738
calcium	Ca	20	40.078	platinum	Pt	78	195.084
californium	Cf	98	–	plutonium	Pu	94	–
carbon	C	6	12.0107	polonium	Po	84	–
cerium	Ce	58	140.116	potassium	K	19	39.0983
cesium	Cs	55	132.9054	praseodymium	Pr	59	140.9077
chlorine	Cl	17	35.453	promethium	Pm	61	–
chromium	Cr	24	51.996	protactinium	Pa	91	231.0359
cobalt	Co	27	58.9332	radium	Ra	88	–
copernicium	Cn	112	–	radon	Rn	86	226.025
copper	Cu	29	63.546	rhenium	Re	75	186.207
curium	Cm	96	–	rhodium	Rh	45	102.9055
darmstadtium	Ds	110	–	roentgenium	Rg	111	–
dubnium	Db	105	–	rubidium	Rb	37	85.4678
dysprosium	Dy	66	162.50	ruthenium	Ru	44	101.07
einsteinium	Es	99	–	rutherfordium	Rf	104	–
erbium	Er	68	167.259	samarium	Sm	62	150.36
europium	Eu	63	151.964	scandium	Sc	21	44.956
fermium	Fm	100	–	seaborgium	Sg	106	–
fluorine	F	9	18.9984	selenium	Se	34	78.96
francium	Fr	87	–	silicon	Si	14	28.0855
gadolinium	Gd	64	157.25	silver	Ag	47	107.868
gallium	Ga	31	69.723	sodium	Na	11	22.9898
germanium	Ge	32	72.64	strontium	Sr	38	87.62
gold	Au	79	196.9666	sulfur	S	16	32.065
hafnium	Hf	72	178.49	tantalum	Ta	73	180.94788
hassium	Hs	108	–	technetium	Tc	43	–
helium	He	2	4.0026	tellurium	Te	52	127.60
holmium	Ho	67	164.930	terbium	Tb	65	158.925
hydrogen	H	1	1.00794	thallium	Tl	81	204.383
indium	In	49	114.818	thorium	Th	90	232.038
iodine	I	53	126.90447	thulium	Tm	69	168.934
iridium	Ir	77	192.217	tin	Sn	50	118.710
iron	Fe	26	55.845	titanium	Ti	22	47.867
krypton	Kr	36	83.798	tungsten	W	74	183.84
lanthanum	La	57	138.9055	uranium	U	92	238.0289
lawrencium	Lr	103	–	vanadium	V	23	50.942
lead	Pb	82	207.2	xenon	Xe	54	131.293
lithium	Li	3	6.941	ytterbium	Yb	70	173.054
lutetium	Lu	71	174.9668	yttrium	Y	39	88.906
magnesium	Mg	12	24.305	zinc	Zn	30	65.38
manganese	Mn	25	54.9380	zirconium	Zr	40	91.224

APPENDIX 20.B

Saturation Concentrations of Dissolved Oxygen in Water[a]

temperature	chloride concentration in water (mg/L)			difference per 100 mg chloride	vapor pressure
	0[b]	5000	10,000		
(°C)	dissolved oxygen (mg/L)				(mm Hg)
0	14.60	13.79	12.97	0.0163	4.58
1	14.19	13.40	12.61	0.0158	4.93
2	13.81	13.05	12.28	0.0153	5.29
3	13.44	12.71	11.98	0.0146	5.69
4	13.09	12.39	11.69	0.0140	6.10
5	12.75	12.07	11.39	0.0136	6.54
6	12.43	11.78	11.12	0.0131	7.02
7	12.12	11.49	10.85	0.0127	7.52
8	11.83	11.22	10.61	0.0122	8.05
9	11.55	10.96	10.36	0.0119	8.61
10	11.27	10.70	10.13	0.0114	9.21
11	11.01	10.47	9.92	0.0109	9.85
12	10.76	10.24	9.72	0.0104	10.52
13	10.52	10.02	9.52	0.0100	11.24
14	10.29	9.81	9.32	0.0097	11.99
15	10.07	9.61	9.14	0.0093	12.79
16	9.85	9.41	8.96	0.0089	13.64
17	9.65	9.22	8.78	0.0087	14.54
18	9.45	9.04	8.62	0.0083	15.49
19	9.26	8.86	8.45	0.0081	16.49
20	9.07	8.69	8.30	0.0077	17.54
21	8.90	8.52	8.14	0.0076	18.66
22	8.72	8.36	7.99	0.0073	19.84
23	8.56	8.21	7.85	0.0071	21.08
24	8.40	8.06	7.71	0.0069	22.34
25	8.24	7.90	7.56	0.0068	23.77
26	8.09	7.76	7.42	0.0067	25.22
27	7.95	7.62	7.28	0.0067	26.75
28	7.81	7.48	7.14	0.0067	28.36
29	7.67	7.34	7.00	0.0067	30.05
30	7.54	7.20	6.86	0.0068	31.83

[a]For saturation at barometric pressures other than 760 mm Hg (29.92 in Hg), C_s' is related to the corresponding tabulated value, C_s, by the equation: $C_s' = C_s\left(\frac{P-p}{760-p}\right)$

C_s' = solubility at barometric pressure P and given temperature, mg/L

C_s = saturation solubility at given temperature from appendix, mg/L

P = barometric pressure, mm Hg

p = pressure of saturated water vapor at temperature of the water selected from appendix, mm Hg

[b]Zero-chloride values from *Volunteer Stream Monitoring: A Methods Manual* (EPA 841-B-97-003), Environmental Protection Agency, Office of Water, Sec. 5.2 "Dissolved Oxygen and Biochemical Oxygen Demand;" 1997.

APPENDIX 20.C
Water Chemistry $CaCO_3$ Equivalents

cations	formula	ionic weight	equivalent weight	substance to $CaCO_3$ factor
aluminum	Al^{+3}	27.0	9.0	5.56
ammonium	NH_4^+	18.0	18.0	2.78
calcium	Ca^{+2}	40.1	20.0	2.50
cupric copper	Cu^{+2}	63.6	31.8	1.57
cuprous copper	Cu^{+3}	63.6	21.2	2.36
ferric iron	Fe^{+3}	55.8	18.6	2.69
ferrous iron	Fe^{+2}	55.8	27.9	1.79
hydrogen	H^+	1.0	1.0	50.00
manganese	Mn^{+2}	54.9	27.5	1.82
magnesium	Mg^{+2}	24.3	12.2	4.10
potassium	K^+	39.1	39.1	1.28
sodium	Na^+	23.0	23.0	2.18

anions	formula	ionic weight	equivalent weight	substance to $CaCO_3$ factor
bicarbonate	HCO_3^-	61.0	61.0	0.82
carbonate	CO_3^{-2}	60.0	30.0	1.67
chloride	Cl^-	35.5	35.5	1.41
fluoride	F^-	19.0	19.0	2.66
hydroxide	OH^-	17.0	17.0	2.94
nitrate	NO_3^-	62.0	62.0	0.81
phosphate (tribasic)	PO_4^{-3}	95.0	31.7	1.58
phosphate (dibasic)	HPO_4^{-2}	96.0	48.0	1.04
phosphate (monobasic)	$H_2PO_4^-$	97.0	97.0	0.52
sulfate	SO_4^{-2}	96.1	48.0	1.04
sulfite	SO_3^{-2}	80.1	40.0	1.25

compounds	formula	molecular weight	equivalent weight	substance to $CaCO_3$ factor
aluminum hydroxide	$Al(OH)_3$	78.0	26.0	1.92
aluminum sulfate	$Al_2(SO_4)_3$	342.1	57.0	0.88
aluminum sulfate	$Al_2(SO_4)_3 \cdot 18H_2O$	666.1	111.0	0.45
alumina	Al_2O_3	102.0	17.0	2.94
sodium aluminate	$Na_2Al_2O_4$	164.0	27.3	1.83
calcium bicarbonate	$Ca(HCO_3)_2$	162.1	81.1	0.62
calcium carbonate	$CaCO_3$	100.1	50.1	1.00
calcium chloride	$CaCl_2$	111.0	55.5	0.90
calcium hydroxide (pure)	$Ca(OH)_2$	74.1	37.1	1.35
calcium hydroxide (90%)	$Ca(OH)_2$	–	41.1	1.22
calcium oxide (lime)	CaO	56.1	28.0	1.79
calcium sulfate (anhydrous)	$CaSO_4$	136.2	68.1	0.74
calcium sulfate (gypsum)	$CaSO_4 \cdot 2H_2O$	172.2	86.1	0.58

APPENDIX 20.C *(continued)*
Water Chemistry $CaCO_3$ Equivalents

compounds	formula	molecular weight	equivalent weight	substance to $CaCO_3$ factor
calcium phosphate	$Ca_3(PO_4)_2$	310.3	51.7	0.97
disodium phosphate	$Na_2HPO_4{\cdot}12H_2O$	358.2	119.4	0.42
disodium phosphate (anhydrous)	Na_2HPO_4	142.0	47.3	1.06
ferric oxide	Fe_2O_3	159.6	26.6	1.88
iron oxide (magnetic)	Fe_3O_4	321.4	–	–
ferrous sulfate (copperas)	$FeSO_4{\cdot}7H_2O$	278.0	139.0	0.36
magnesium oxide	MgO	40.3	20.2	2.48
magnesium bicarbonate	$Mg(HCO_3)_2$	146.3	73.2	0.68
magnesium carbonate	$MgCO_3$	84.3	42.2	1.19
magnesium chloride	$MgCl_2$	95.2	47.6	1.05
magnesium hydroxide	$Mg(OH)_2$	58.3	29.2	1.71
magnesium phosphate	$Mg_3(PO_4)_2$	263.0	43.8	1.14
magnesium sulfate	$MgSO_4$	120.4	60.2	0.83
monosodium phosphate	$NaH_2PO_4{\cdot}H_2O$	138.1	46.0	1.09
monosodium phosphate (anhydrous)	NaH_2PO_4	120.1	40.0	1.25
metaphosphate	$NaPO_3$	102.0	34.0	1.47
silica	SiO_2	60.1	30.0	1.67
sodium bicarbonate	$NaHCO_3$	84.0	84.0	0.60
sodium carbonate	Na_2CO_3	106.0	53.0	0.94
sodium chloride	$NaCl$	58.5	58.5	0.85
sodium hydroxide	$NaOH$	40.0	40.0	1.25
sodium nitrate	$NaNO_3$	85.0	85.0	0.59
sodium sulfate	Na_2SO_4	142.0	71.0	0.70
sodium sulfite	Na_2SO_3	126.1	63.0	0.79
tetrasodium EDTA	$(CH_2)_2N_2(CH_2COONa)_4$	380.2	95.1	0.53
trisodium phosphate	$Na_3PO_4{\cdot}12H_2O$	380.2	126.7	0.40
trisodium phosphate (anhydrous)	Na_3PO_4	164.0	54.7	0.91
trisodium NTA	$(CH_2)_3N(COONa)_3$	257.1	85.7	0.58

gases	formula	molecular weight	equivalent weight	substance to $CaCO_3$ factor
ammonia	NH_3	17	17	2.94
carbon dioxide	CO_2	44	22	2.27
hydrogen	H_2	2	1	50.00
hydrogen sulfide	H_2S	34	17	2.94
oxygen	O_2	32	8	6.25

Support Material

APPENDIX 20.C *(continued)*
Water Chemistry $CaCO_3$ Equivalents

acids	formula	molecular weight	equivalent weight	substance to $CaCO_3$ factor
carbonic	H_2CO_3	62.0	31.0	1.61
hydrochloric	HCl	36.5	36.5	1.37
phosphoric	H_3PO_4	98.0	32.7	1.53
sulfuric	H_2SO_4	98.1	49.1	1.02

(Multiply the concentration (in mg/L) of the substance by the corresponding factors to obtain the equivalent concentration in mg/L as $CaCO_3$. For example, 70 mg/L of Mg^{++} would be (70 mg/L)(4.1) = 287 mg/L as $CaCO_3$.)

Support Material

APPENDIX 20.D
Periodic Table of the Elements
(referred to carbon-12)

The Periodic Table of Elements (Long Form)

The number of electrons in filled shells is shown in the column at the extreme left; the remaining electrons for each element are shown immediately below the symbol for each element. Atomic numbers are enclosed in brackets. Atomic weights (rounded, based on carbon-12) are shown above the symbols. Atomic weight values in parentheses are those of the isotopes of longest half-life for certain radioactive elements whose atomic weights cannot be precisely quoted without knowledge of origin of the element.

	metals												nonmetals					
			transition metals															
periods	I A	II A	III B	IV B	V B	VI B	VII B	VIII	VIII	VIII	I B	II B	III A	IV A	V A	VI A	VII A	0
1 0	1.00794 H[1] 1																	4.00260 He[2] 2
2 2	6.941 Li[3] 1	9.01218 Be[4] 2											10.811 B[5] 3	12.0107 C[6] 4	14.0067 N[7] 5	15.9994 O[8] 6	18.9984 F[9] 7	20.1797 Ne[10] 8
3 2,8	22.9898 Na[11] 1	24.3050 Mg[12] 2											26.9815 Al[13] 3	28.0855 Si[14] 4	30.9738 P[15] 5	32.065 S[16] 6	35.453 Cl[17] 7	39.948 Ar[18] 8
4 2,8	39.0983 K[19] 8,1	40.078 Ca[20] 8,2	44.9559 Sc[21] 9,2	47.867 Ti[22] 10,2	50.9415 V[23] 11,2	51.9961 Cr[24] 13,1	54.9380 Mn[25] 13,2	55.845 Fe[26] 14,2	58.9332 Co[27] 15,2	58.6934 Ni[28] 16,2	63.546 Cu[29] 18,1	65.38 Zn[30] 18,2	69.723 Ga[31] 18,3	72.64 Ge[32] 18,4	74.9216 As[33] 18,5	78.96 Se[34] 18,6	79.904 Br[35] 18,7	83.798 Kr[36] 18,8
5 2,8,18	85.4678 Rb[37] 8,1	87.62 Sr[38] 8,2	88.9059 Y[39] 9,2	91.224 Zr[40] 10,2	92.9064 Nb[41] 12,1	95.96 Mo[42] 13,1	(98) Tc[43] 14,1	101.07 Ru[44] 15,1	102.906 Rh[45] 16,1	106.42 Pd[46] 18	107.868 Ag[47] 18,1	112.411 Cd[48] 18,2	114.818 In[49] 18,3	118.710 Sn[50] 18,4	121.760 Sb[51] 18,5	127.60 Te[52] 18,6	126.904 I[53] 18,7	131.293 Xe[54] 18,8
6 2,8,18	132.905 Cs[55] 18,8,1	137.327 Ba[56] 18,8,2	* (57-71)	178.49 Hf[72] 32,10,2	180.948 Ta[73] 32,11,2	183.84 W[74] 32,12,2	186.207 Re[75] 32,13,2	190.23 Os[76] 32,14,2	192.217 Ir[77] 32,15,2	195.084 Pt[78] 32,17,1	196.967 Au[79] 32,18,1	200.59 Hg[80] 32,18,2	204.383 Tl[81] 32,18,3	207.2 Pb[82] 32,18,4	208.980 Bi[83] 32,18,5	(209) Po[84] 32,18,6	(210) At[85] 32,18,7	(222) Rn[86] 32,18,8
7 2,8,18,32	(223) Fr[87] 18,8,1	(226) Ra[88] 18,8,2	† (89-103)	(265) Rf[104] 32,10,2	(268) Db[105] 32,11,2	(271) Sg[106] 32,12,2	(272) Bh[107] 32,13,2	(270) Hs[108] 32,14,2	(276) Mt[109] 32,15,2	(281) Ds[110] 32,17,1	(280) Rg[111] 32,18,1	(285) Cn[112] 32,18,2						

*lathanide series	138.905 La[57] 18,9,2	140.116 Ce[58] 20,8,2	140.908 Pr[59] 21,8,2	144.242 Nd[60] 22,8,2	(145) Pm[61] 23,8,2	150.36 Sm[62] 24,8,2	151.964 Eu[63] 25,8,2	157.25 Gd[64] 25,9,2	158.925 Tb[65] 27,8,2	162.500 Dy[66] 28,8,2	164.930 Ho[67] 29,8,2	167.259 Er[68] 30,8,2	168.934 Tm[69] 31,8,2	173.054 Yb[70] 32,8,2	174.967 Lu[71] 32,9,2
†actinide series	(227) Ac[89] 18,9,2	232.038 Th[90] 18,10,2	231.036 Pa[91] 20,9,2	238.029 U[92] 21,9,2	(237) Np[93] 23,8,2	(244) Pu[94] 24,8,2	(243) Am[95] 25,8,2	(247) Cm[96] 25,9,2	(247) Bk[97] 26,9,2	(251) Cf[98] 28,8,2	(252) Es[99] 29,8,2	(257) Fm[100] 30,8,2	(258) Md[101] 31,8,2	(259) No[102] 32,8,2	(262) Lr[103] 32,9,2

Support Material

APPENDIX 20.E
Approximate Solubility Product Constants at 25°C

substance	formula	K_{sp}
aluminum hydroxide	$Al(OH)_3$	1.3×10^{-33}
aluminum phosphate	$AlPO_4$	6.3×10^{-19}
barium carbonate	$BaCO_3$	5.1×10^{-9}
barium chromate	$BaCrO_4$	1.2×10^{-10}
barium fluoride	BaF_2	1.0×10^{-6}
barium hydroxide	$Ba(OH)_2$	5×10^{-3}
barium sulfate	$BaSO_4$	1.1×10^{-10}
barium sulfite	$BaSO_3$	8×10^{-7}
barium thiosulfate	BaS_2O_3	1.6×10^{-6}
bismuthyl chloride	$BiOCl$	1.8×10^{-31}
bismuthyl hydroxide	$BiOOH$	4×10^{-10}
cadmium carbonate	$CdCO_3$	5.2×10^{-12}
cadmium hydroxide	$Cd(OH)_2$	2.5×10^{-14}
cadmium oxalate	CdC_2O_4	1.5×10^{-8}
cadmium sulfide[a]	CdS	8×10^{-28}
calcium carbonate[b]	$CaCO_3$	2.8×10^{-9}
calcium chromate	$CaCrO_4$	7.1×10^{-4}
calcium fluoride	CaF_2	5.3×10^{-9}
calcium hydrogen phosphate	$CaHPO_4$	1×10^{-7}
calcium hydroxide	$Ca(OH)_2$	5.5×10^{-6}
calcium oxalate	CaC_2O_4	2.7×10^{-9}
calcium phosphate	$Ca_3(PO_4)_2$	2.0×10^{-29}
calcium sulfate	$CaSO_4$	9.1×10^{-6}
calcium sulfite	$CaSO_3$	6.8×10^{-8}
chromium (II) hydroxide	$Cr(OH)_2$	2×10^{-16}
chromium (III) hydroxide	$Cr(OH)_3$	6.3×10^{-31}
cobalt (II) carbonate	$CoCO_3$	1.4×10^{-13}
cobalt (II) hydroxide	$Co(OH)_2$	1.6×10^{-15}
cobalt (III) hydroxide	$Co(OH)_3$	1.6×10^{-44}
cobalt (II) sulfide[a]	CoS	4×10^{-21}
copper (I) chloride	$CuCl$	1.2×10^{-6}
copper (I) cyanide	$CuCN$	3.2×10^{-20}
copper (I) iodide	CuI	1.1×10^{-12}
copper (II) arsenate	$Cu_3(AsO_4)_2$	7.6×10^{-36}
copper (II) carbonate	$CuCO_3$	1.4×10^{-10}
copper (II) chromate	$CuCrO_4$	3.6×10^{-6}
copper (II) ferrocyanide	$Cu[Fe(CN)_6]$	1.3×10^{-16}
copper (II) hydroxide	$Cu(OH)_2$	2.2×10^{-20}
copper (II) sulfide[a]	CuS	6×10^{-37}
iron (II) carbonate	$FeCO_3$	3.2×10^{-11}
iron (II) hydroxide	$Fe(OH)_2$	8.0×10^{-16}
iron (II) sulfide[a]	FeS	6×10^{-19}
iron (III) arsenate	$FeAsO_4$	5.7×10^{-21}
iron (III) ferrocyanide	$Fe_4[Fe(CN)_6]_3$	3.3×10^{-41}
iron (III) hydroxide	$Fe(OH)_3$	4×10^{-38}
iron (III) phosphate	$FePO_4$	1.3×10^{-22}

APPENDIX 20.E *(continued)*
Approximate Solubility Product Constants at 25°C

substance	formula	K_{sp}
lead (II) arsenate	$Pb_3(AsO_4)_2$	4×10^{-36}
lead (II) azide	$Pb(N_3)_2$	2.5×10^{-9}
lead (II) bromide	$PbBr_2$	4.0×10^{-5}
lead (II) carbonate	$PbCO_3$	7.4×10^{-14}
lead (II) chloride	$PbCl_2$	1.6×10^{-5}
lead (II) chromate	$PbCrO_4$	2.8×10^{-13}
lead (II) fluoride	PbF_2	2.7×10^{-8}
lead (II) hydroxide	$Pb(OH)_2$	1.2×10^{-15}
lead (II) iodide	PbI_2	7.1×10^{-9}
lead (II) sulfate	$PbSO_4$	1.6×10^{-8}
lead (II) sulfide[a]	PbS	3×10^{-28}
lithium carbonate	Li_2CO_3	2.5×10^{-2}
lithium fluoride	LiF	3.8×10^{-3}
lithium phosphate	Li_3PO_4	3.2×10^{-9}
magnesium ammonium phosphate	$MgNH_4PO_4$	2.5×10^{-13}
magnesium arsenate	$Mg_3(AsO_4)_2$	2×10^{-20}
magnesium carbonate	$MgCO_3$	3.5×10^{-8}
magnesium fluoride	MgF_2	3.7×10^{-8}
magnesium hydroxide	$Mg(OH)_2$	1.8×10^{-11}
magnesium oxalate	MgC_2O_4	8.5×10^{-5}
magnesium phosphate	$Mg_3(PO_4)_2$	1×10^{-25}
manganese (II) carbonate	$MnCO_3$	1.8×10^{-11}
manganese (II) hydroxide	$Mn(OH)_2$	1.9×10^{-13}
manganese (II) sulfide[a]	MnS	3×10^{-14}
mercury (I) bromide	Hg_2Br_2	5.6×10^{-23}
mercury (I) chloride	Hg_2Cl_2	1.3×10^{-18}
mercury (I) iodide	Hg_2I_2	4.5×10^{-29}
mercury (II) sulfide[a]	HgS	2×10^{-53}
nickel (II) carbonate	$NiCO_3$	6.6×10^{-9}
nickel (II) hydroxide	$Ni(OH)_2$	2.0×10^{-15}
nickel (II) sulfide[a]	NiS	3×10^{-19}
scandium fluoride	ScF_3	4.2×10^{-18}
scandium hydroxide	$Sc(OH)_3$	8.0×10^{-31}
silver acetate	$AgC_2H_3O_2$	2.0×10^{-3}
silver arsenate	Ag_3AsO_4	1.0×10^{-22}
silver azide	AgN_3	2.8×10^{-9}
silver bromide	$AgBr$	5.0×10^{-13}
silver chloride	$AgCl$	1.8×10^{-10}
silver chromate	Ag_2CrO_4	1.1×10^{-12}
silver cyanide	$AgCN$	1.2×10^{-16}
silver iodate	$AgIO_3$	3.0×10^{-8}
silver iodide	AgI	8.5×10^{-17}
silver nitrite	$AgNO_2$	6.0×10^{-4}
silver sulfate	Ag_2SO_4	1.4×10^{-5}
silver sulfide[a]	Ag_2S	6×10^{-51}

Support Material

APPENDIX 20.E *(continued)*
Approximate Solubility Product Constants at 25°C

substance	formula	K_{sp}
silver sulfite	Ag_2SO_3	1.5×10^{-14}
silver thiocyanate	$AgSCN$	1.0×10^{-12}
strontium carbonate	$SrCO_3$	1.1×10^{-10}
strontium chromate	$SrCrO_4$	2.2×10^{-5}
strontium fluoride	SrF_2	2.5×10^{-9}
strontium sulfate	$SrSO_4$	3.2×10^{-7}
thallium (I) bromide	$TlBr$	3.4×10^{-6}
thallium (I) chloride	$TlCl$	1.7×10^{-4}
thallium (I) iodide	TlI	6.5×10^{-8}
thallium (III) hydroxide	$Tl(OH)_3$	6.3×10^{-46}
tin (II) hydroxide	$Sn(OH)_2$	1.4×10^{-28}
tin (II) sulfide[a]	SnS	1×10^{-26}
zinc carbonate	$ZnCO_3$	1.4×10^{-11}
zinc hydroxide	$Zn(OH)_2$	1.2×10^{-17}
zinc oxalate	ZnC_2O_4	2.7×10^{-8}
zinc phosphate	$Zn_3(PO_4)_2$	9.0×10^{-33}
zincsulfide[a]	ZnS	2×10^{-25}

[a]Sulfide equilibrium of the type:

$$MS(s) + H_2O(l) \rightleftharpoons M^{2+}(aq) + HS^-(aq) + OH^-(aq)$$

[b]Solubility product depends on mineral form.

Support Material

APPENDIX 20.F
Dissociation Constants of Acids at 25°C

acid		K_a
acetic	K_1	1.8×10^{-5}
arsenic	K_1	5.6×10^{-3}
	K_2	1.2×10^{-7}
	K_3	3.2×10^{-12}
arsenious	K_1	1.4×10^{-9}
benzoic	K_1	6.3×10^{-5}
boric	K_1	5.9×10^{-10}
carbonic	K_1^*	4.5×10^{-7}
	K_2	5.6×10^{-11}
chloroacetic	K_1	1.4×10^{-3}
chromic	K_2	3.2×10^{-7}
citric	K_1	7.4×10^{-4}
	K_2	1.7×10^{-5}
	K_3	3.9×10^{-7}
ethylenedinitrilotetracetic	K_1	1.0×10^{-2}
	K_2	2.1×10^{-3}
	K_3	6.9×10^{-7}
	K_4	7.4×10^{-11}
formic	K_1	1.8×10^{-4}
hydrocyanic	K_1	4.9×10^{-10}
hydrofluoric	K_1	6.8×10^{-4}
hydrogen sulfide	K_1	1.0×10^{-8}
	K_2	1.2×10^{-14}
hypochlorous	K_1	2.8×10^{-8}
iodic	K_1	1.8×10^{-1}
nitrous	K_1	4.5×10^{-4}
oxalic	K_1	5.4×10^{-2}
	K_2	5.1×10^{-5}
phenol	K_1	1.1×10^{-10}
phosphoric (ortho)	K_1	7.1×10^{-3}
	K_2	6.3×10^{-8}
	K_3	4.4×10^{-13}
o-phthalic	K_1	1.1×10^{-3}
	K_2	3.9×10^{-6}
salicylic	K_1	1.0×10^{-3}
	K_2	4.0×10^{-14}
sulfamic	K_1	1.0×10^{-1}
sulfuric	K_1	1.1×10^{-2}
sulfurous	K_1	1.7×10^{-2}
	K_2	6.3×10^{-8}
tartaric	K_1	9.2×10^{-4}
	K_2	4.3×10^{-5}
thiocyanic	K_1	1.4×10^{-1}

*apparent constant based on $C_{H_2CO_3} = [CO_2] + [H_2CO_3]$

APPENDIX 20.G
Dissociation Constants of Bases at 25°C

base		K_b
2-amino-2-(hydroxymethyl)-1,3-propanediol	K_1	1.2×10^{-6}
ammonia	K_1	1.8×10^{-5}
aniline	K_1	4.2×10^{-10}
diethylamine	K_1	1.3×10^{-3}
hexamethylenetetramine	K_1	1.0×10^{-9}
hydrazine	K_1	9.8×10^{-7}
hydroxylamine	K_1	9.6×10^{-9}
lead hydroxide	K_1	1.2×10^{-4}
piperidine	K_1	1.3×10^{-3}
pyridine	K_1	1.5×10^{-9}
silver hydroxide	K_1	6.0×10^{-5}

APPENDIX 20.H
Names and Formulas of Important Chemicals

common name	chemical name	chemical formula
acetone	acetone	$(CH_3)_2CO$
acetylene	acetylene	C_2H_2
ammonia	ammonia	NH_3
ammonium	ammonium hydroxide	NH_4OH
aniline	aniline	$C_6H_5NH_2$
bauxite	hydrated aluminum oxide	$Al_2O_3 \cdot 2H_2O$
bleach	calcium hypochlorite	$Ca(ClO)_2$
borax	sodium tetraborate	$Na_2B_4O_7 \cdot 10H_2O$
carbide	calcium carbide	CaC_2
carbolic acid	phenol	C_6H_5OH
carbon dioxide	carbon dioxide	CO_2
carborundum	silicon carbide	SiC
caustic potash	potassium hydroxide	KOH
caustic soda/lye	sodium hydroxide	$NaOH$
chalk	calcium carbonate	$CaCO_3$
cinnabar	mercuric sulfide	HgS
ether	diethyl ether	$(C_2H_5)_2O$
formic acid	methanoic acid	$HCOOH$
Glauber's salt	decahydrated sodium sulfate	$Na_2SO_4 \cdot 10H_2O$
glycerine	glycerine	$C_3H_5(OH)_3$
grain alcohol	ethanol	C_2H_5OH
graphite	crystalline carbon	C
gypsum	calcium sulfate	$CaSO_4 \cdot 2H_2O$
halite	sodium chloride	$NaCl$
iron chloride	ferrous chloride	$FeCl_2 \cdot 4H_2O$
laughing gas	nitrous oxide	N_2O
limestone	calcium carbonate	$CaCO_3$
magnesia	magnesium oxide	MgO
marsh gas	methane	CH_4
muriate of potash	potassium chloride	KCl
muriatic acid	hydrochloric acid	HCl
niter	sodium nitrate	$NaNO_3$
niter cake	sodium bisulfate	$NaHSO_4$
oleum	fuming sulfuric acid	SO_3 in H_2SO_4
potash	potassium carbonate	K_2CO_3
prussic acid	hydrogen cyanide	HCN
pyrites	ferrous sulfide	FeS
pyrolusite	manganese dioxide	MnO_2
quicklime	calcium oxide	CaO
sal soda	decahydrated sodium carbonate	$NaCO_3 \cdot 10H_2O$
salammoniac	ammonium chloride	NH_4Cl
sand or silica	silicon dioxide	SiO_2
salt cake	sodium sulfate (crude)	Na_2SO_4
slaked lime	calcium hydroxide	$Ca(OH)_2$

Support Material

APPENDIX 20.H *(continued)*
Names and Formulas of Important Chemicals

common name	chemical name	chemical formula
soda ash	sodium carbonate	Na_2CO_3
soot	amorphous carbon	C
stannous chloride	stannous chloride	$SnCl_2 \cdot 2H_2O$
superphosphate	monohydrated primary calcium phosphate	$Ca(H_2PO_4)_2 \cdot H_2O$
table salt	sodium chloride	NaCl
table sugar	sucrose	$C_{12}H_{22}O_{11}$
trilene	trichloroethylene	C_2HCl_3
urea	urea	$CO(NH_2)_2$
vinegar (acetic acid)	ethanoic acid	CH_3COOH
washing soda	decahydrated sodium carbonate	$Na_2CO_3 \cdot 10H_2O$
wood alcohol	methanol	CH_3OH
zinc blende	zinc sulfide	ZnS

APPENDIX 23.A
Properties of Compressed Water
(customary U.S. units)

T (°F)	p (psia)	ρ (lbm/ft³)	v (ft³/lbm)	x	h (Btu/lbm)	s (Btu/lbm-°R)	u (Btu/lbm)
32	200	62.46	0.01601	subcooled	0.5852	–0.00001534	–0.007334
100	200	62.03	0.01612	subcooled	68.56	0.1295	67.96
200	200	60.16	0.01662	subcooled	168.6	0.2939	167.9
300	200	57.34	0.01744	subcooled	270	0.437	269.3
381.8	200	54.39	0.01839	0	355.5	0.5438	354.8
381.8	200	0.437	2.288	1	1199	1.546	1114
32	400	62.5	0.016	subcooled	1.187	0.00000448	0.003032
100	400	62.07	0.01611	subcooled	69.09	0.1294	67.89
200	400	60.2	0.01661	subcooled	169	0.2936	167.8
300	400	57.39	0.01742	subcooled	270.3	0.4366	269.1
400	400	53.7	0.01862	subcooled	375.2	0.5662	373.8
444.6	400	51.7	0.01934	0	424.2	0.6217	422.7
444.6	400	0.8609	1.162	1	1205	1.485	1119
32	600	62.55	0.01599	subcooled	1.788	0.00002258	0.01296
100	600	62.1	0.0161	subcooled	69.61	0.1292	67.83
200	600	60.24	0.0166	subcooled	169.5	0.2934	167.6
300	600	57.44	0.01741	subcooled	270.7	0.4362	268.8
400	600	53.77	0.0186	subcooled	375.4	0.5657	373.4
486.2	600	49.65	0.02014	0	471.7	0.6724	469.5
486.2	600	1.298	0.7702	1	1204	1.446	1118
32	800	62.59	0.01598	subcooled	2.388	0.00003896	0.02244
100	800	62.14	0.01609	subcooled	70.14	0.1291	67.76
200	800	60.28	0.01659	subcooled	169.9	0.2931	167.5
300	800	57.49	0.01739	subcooled	271.1	0.4359	268.5
400	800	53.84	0.01857	subcooled	375.7	0.5652	372.9
500	800	48.99	0.02041	subcooled	487.9	0.6885	484.9
518.3	800	47.9	0.02088	0	509.9	0.7112	506.8
518.3	800	1.757	0.5692	1	1199	1.416	1115
32	1000	62.63	0.01597	subcooled	2.986	0.00005365	0.0315
100	1000	62.18	0.01608	subcooled	70.67	0.129	67.69
200	1000	60.31	0.01658	subcooled	170.4	0.2929	167.3
300	1000	57.54	0.01738	subcooled	271.5	0.4355	268.2
400	1000	53.9	0.01855	subcooled	375.9	0.5646	372.5
500	1000	49.1	0.02037	subcooled	487.8	0.6876	484
544.6	1000	46.3	0.0216	0	542.7	0.7435	538.7
544.6	1000	2.242	0.446	1	1193	1.391	1110
32	1500	62.74	0.01594	subcooled	4.477	0.00008309	0.05229
100	1500	62.27	0.01606	subcooled	71.98	0.1287	67.53
200	1500	60.41	0.01655	subcooled	171.5	0.2923	166.9
300	1500	57.65	0.01734	subcooled	272.4	0.4346	267.6
400	1500	54.06	0.0185	subcooled	376.5	0.5633	371.4
500	1500	49.36	0.02026	subcooled	487.6	0.6855	482
596.3	1500	42.62	0.02346	0	611.7	0.8085	605.2
596.3	1500	3.61	0.277	1	1169	1.336	1092

APPENDIX 23.A *(continued)*
Properties of Compressed Water
(customary U.S. units)

T (°F)	p (psia)	ρ (lbm/ft³)	v (ft³/lbm)	x	h (Btu/lbm)	s (Btu/lbm-°R)	u (Btu/lbm)
32	2000	62.85	0.01591	subcooled	5.959	0.0001023	0.0705
100	2000	62.36	0.01603	subcooled	73.3	0.1284	67.36
200	2000	60.51	0.01653	subcooled	172.7	0.2917	166.5
300	2000	57.77	0.01731	subcooled	273.3	0.4338	266.9
400	2000	54.22	0.01844	subcooled	377.1	0.5621	370.3
500	2000	49.62	0.02015	subcooled	487.5	0.6835	480.1
600	2000	42.89	0.02332	subcooled	614.4	0.809	605.8
635.8	2000	39.01	0.02564	0	671.8	0.8623	662.3
635.8	2000	5.316	0.1881	1	1136	1.286	1067
32	3000	63.06	0.01586	subcooled	8.903	0.0001113	0.09948
100	3000	62.55	0.01599	subcooled	75.91	0.1278	67.04
200	3000	60.7	0.01648	subcooled	174.9	0.2905	165.8
300	3000	58	0.01724	subcooled	275.2	0.4321	265.7
400	3000	54.53	0.01834	subcooled	378.4	0.5596	368.2
500	3000	50.1	0.01996	subcooled	487.5	0.6796	476.4
600	3000	43.94	0.02276	subcooled	610.1	0.8009	597.4
695.4	3000	29.12	0.03434	0	802.5	0.9733	783.4
695.4	3000	11.81	0.08466	1	1017	1.159	969.9
32	4000	63.26	0.01581	subcooled	11.82	0.00008277	0.119
100	4000	62.73	0.01594	subcooled	78.52	0.1271	66.72
200	4000	60.88	0.01642	subcooled	177.2	0.2894	165.1
300	4000	58.22	0.01718	subcooled	277.1	0.4304	264.4
400	4000	54.83	0.01824	subcooled	379.7	0.5572	366.2
500	4000	50.55	0.01978	subcooled	487.7	0.676	473.1
600	4000	44.81	0.02231	subcooled	607	0.794	590.5
700	4000	34.83	0.02871	subcooled	763.6	0.9347	742.3

Values in this table were calculated from *NIST Standard Reference Database 10*, "NIST/ASME Steam Properties," Ver. 2.11, National Institute of Standards and Technology, U.S. Department of Commerce, Gaithersburg, MD, 1997, which has been licensed to PPI.

APPENDIX 23.B
Enthalpy-Entropy (Mollier) Diagram for Steam
(customary U.S. units)

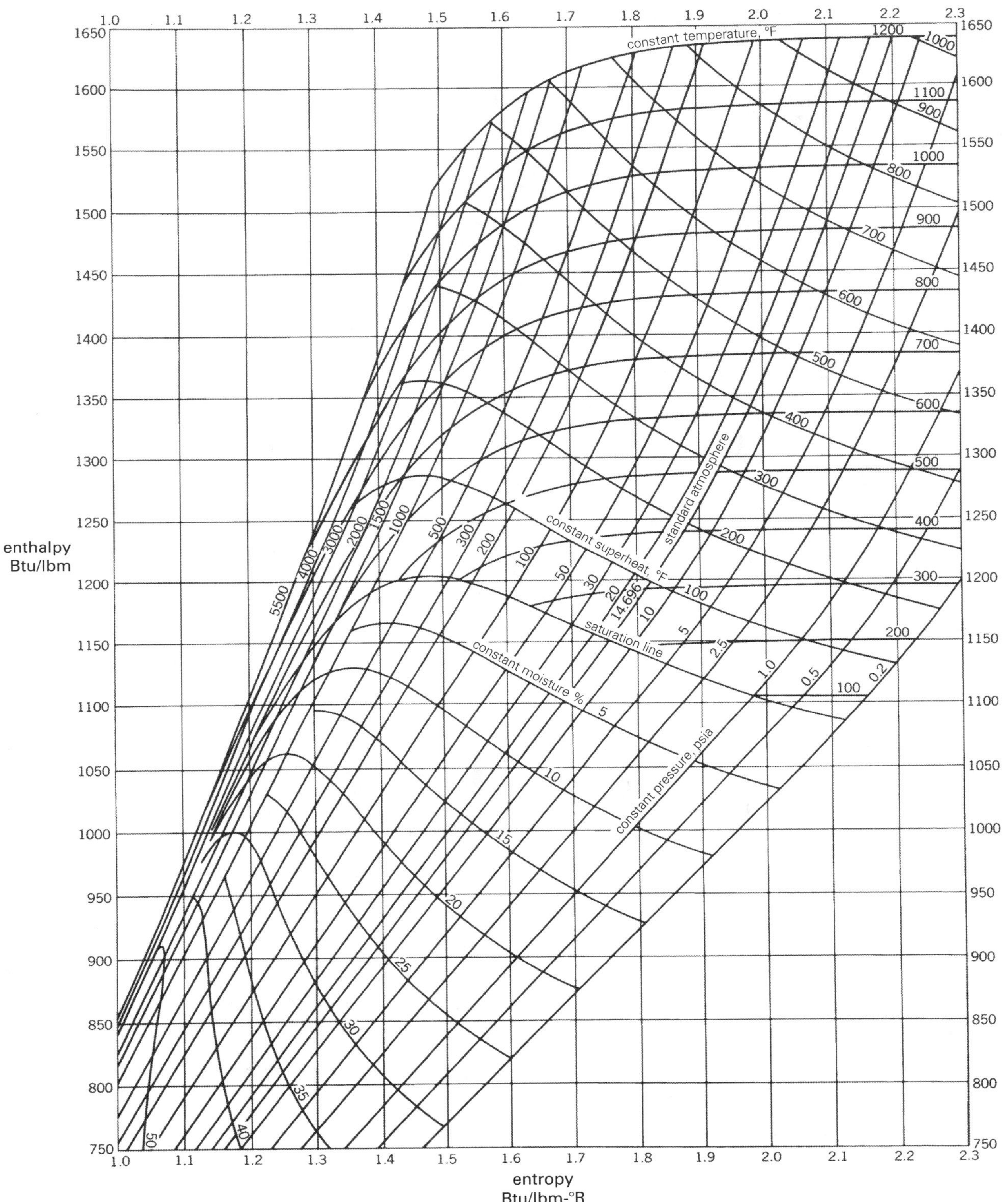

Used with permission from *Steam: Its Generation and Use*, 41st ed., edited by S. C. Stultz and J. B. Kitto, copyright © 2005, by The Babcock & Wilcox Company.

APPENDIX 23.C
Properties of Low-Pressure Air
(customary U.S. units)

T in °R; *h* and *u* in Btu/lbm; ϕ in Btu/lbm-°R

T	h	p_r	u	v_r	ϕ
200	47.67	0.04320	33.96	1714.9	0.36303
220	52.46	0.06026	37.38	1352.5	0.38584
240	57.25	0.08165	40.80	1088.8	0.40666
260	62.03	0.10797	44.21	892.0	0.42582
280	66.82	0.13986	47.63	741.6	0.44356
300	71.61	0.17795	51.04	624.5	0.46007
320	76.40	0.22290	54.46	531.8	0.47550
340	81.18	0.27545	57.87	457.2	0.49002
360	85.97	0.3363	61.29	396.6	0.50369
380	90.75	0.4061	64.70	346.6	0.51663
400	95.53	0.4858	68.11	305.0	0.52890
420	100.32	0.5760	71.52	270.1	0.54058
440	105.11	0.6776	74.93	240.6	0.55172
460	109.90	0.7913	78.36	215.33	0.56235
480	114.69	0.9182	81.77	193.65	0.57255
500	119.48	1.0590	85.20	174.90	0.58233
520	124.27	1.2147	88.62	158.58	0.59172
537	128.34	1.3593	91.53	146.34	0.59945
540	129.06	1.3860	92.04	144.32	0.60078
560	133.86	1.5742	95.47	131.78	0.60950
580	138.66	1.7800	98.90	120.70	0.61793
600	143.47	2.005	102.34	110.88	0.62607
620	148.28	2.249	105.78	102.12	0.63395
640	153.09	2.514	109.21	94.30	0.64159
660	157.92	2.801	112.67	87.27	0.64902
680	162.74	3.111	116.12	80.96	0.65621
700	167.56	3.446	119.58	75.25	0.66321
720	172.39	3.806	123.04	70.07	0.67002
740	177.23	4.193	126.51	65.38	0.67665
760	182.08	4.607	129.99	61.10	0.68312
780	186.94	5.051	133.47	57.20	0.68942
800	191.81	5.526	136.97	53.63	0.69558
820	196.69	6.033	140.47	50.35	0.70160
840	201.56	6.573	143.98	47.34	0.70747
860	206.46	7.149	147.50	44.57	0.71323
880	211.35	7.761	151.02	42.01	0.71886
900	216.26	8.411	154.57	39.64	0.72438
920	221.18	9.102	158.12	37.44	0.72979
940	226.11	9.834	161.68	35.41	0.73509
960	231.06	10.61	165.26	33.52	0.74030
980	236.02	11.43	168.83	31.76	0.74540
1000	240.98	12.30	172.43	30.12	0.75042
1040	250.95	14.18	179.66	27.17	0.76019
1080	260.97	16.28	186.93	24.58	0.76964
1120	271.03	18.60	194.25	22.30	0.77880
1160	281.14	21.18	201.63	20.29	0.78767

APPENDIX 23.C *(continued)*
Properties of Low-Pressure Air
(customary U.S. units)

T in °R; h and u in Btu/lbm; ϕ in Btu/lbm-°R

T	h	p_r	u	v_r	ϕ
1200	291.30	24.01	209.05	18.51	0.79628
1240	301.52	27.13	216.53	16.93	0.80466
1280	311.79	30.55	224.05	15.52	0.81280
1320	322.11	34.31	231.63	14.25	0.82075
1360	332.48	38.41	239.25	13.12	0.82848
1400	342.90	42.88	246.93	12.10	0.83604
1440	353.37	47.75	254.66	11.17	0.84341
1480	363.89	53.04	262.44	10.34	0.85062
1520	374.47	58.78	270.26	9.578	0.85767
1560	385.08	65.00	278.13	8.890	0.86456
1600	395.74	71.73	286.06	8.263	0.87130
1650	409.13	80.89	296.03	7.556	0.87954
1700	422.59	90.95	306.06	6.924	0.88758
1750	436.12	101.98	316.16	6.357	0.89542
1800	449.71	114.0	326.32	5.847	0.90308
1850	463.37	127.2	336.55	5.388	0.91056
1900	477.09	141.5	346.85	4.974	0.91788
1950	490.88	157.1	357.20	4.598	0.92504
2000	504.71	174.0	367.61	4.258	0.93205
2050	518.61	192.3	378.08	3.949	0.93891
2100	532.55	212.1	388.60	3.667	0.94564
2150	546.54	233.5	399.17	3.410	0.95222
2200	560.59	256.6	409.78	3.176	0.95868
2250	574.69	281.4	420.46	2.961	0.96501
2300	588.82	308.1	431.16	2.765	0.97123
2350	603.00	336.8	441.91	2.585	0.97732
2400	617.22	367.6	452.70	2.419	0.98331
2450	631.48	400.5	463.54	2.266	0.98919
2500	645.78	435.7	474.40	2.125	0.99497
2550	660.12	473.3	485.31	1.996	1.00064
2600	674.49	513.5	496.26	1.876	1.00623
2650	688.90	556.3	507.25	1.765	1.01172
2700	703.35	601.9	518.26	1.662	1.01712
2750	717.83	650.4	529.31	1.566	1.02244
2800	732.33	702.0	540.40	1.478	1.02767
2850	746.88	756.7	551.52	1.395	1.03282
2900	761.45	814.8	562.66	1.318	1.03788
2950	776.05	876.4	573.84	1.247	1.04288
3000	790.68	941.4	585.04	1.180	1.04779
3050	805.34	1011	596.28	1.118	1.05264
3100	820.03	1083	607.53	1.060	1.05741
3150	834.75	1161	618.82	1.006	1.06212
3200	849.48	1242	630.12	0.9546	1.06676
3250	864.24	1328	641.46	0.9069	1.07134
3300	879.02	1418	652.81	0.8621	1.07585

APPENDIX 23.C *(continued)*
Properties of Low-Pressure Air
(customary U.S. units)

T in °R; h and u in Btu/lbm; ϕ in Btu/lbm-°R

T	h	p_r	u	v_r	ϕ
3350	893.83	1513	664.20	0.8202	1.08031
3400	908.66	1613	675.60	0.7807	1.08470
3450	923.52	1719	687.04	0.7436	1.08904
3500	938.40	1829	698.48	0.7087	1.09332
3550	953.30	1946	709.95	0.6759	1.09755
3600	968.21	2068	721.44	0.6449	1.10172
3650	983.15	2196	732.95	0.6157	1.10584
3700	998.11	2330	744.48	0.5882	1.10991
3750	1013.1	2471	756.04	0.5621	1.11393
3800	1028.1	2618	767.60	0.5376	1.11791
3850	1043.1	2773	779.19	0.5143	1.12183
3900	1058.1	2934	790.80	0.4923	1.12571
3950	1073.2	3103	802.43	0.4715	1.12955
4000	1088.3	3280	814.06	0.4518	1.13334
4050	1103.4	3464	825.72	0.4331	1.13709
4100	1118.5	3656	837.40	0.4154	1.14079
4150	1133.6	3858	849.09	0.3985	1.14446
4200	1148.7	4067	860.81	0.3826	1.14809
4300	1179.0	4513	884.28	0.3529	1.15522
4400	1209.4	4997	907.81	0.3262	1.16221
4500	1239.9	5521	931.39	0.3019	1.16905
4600	1270.4	6089	955.04	0.2799	1.17575
4700	1300.9	6701	978.73	0.2598	1.18232
4800	1331.5	7362	1002.5	0.2415	1.18876
4900	1362.2	8073	1026.3	0.2248	1.19508
5000	1392.9	8837	1050.1	0.2096	1.20129
5100	1423.6	9658	1074.0	0.1956	1.20738
5200	1454.4	10539	1098.0	0.1828	1.21336
5300	1485.3	11481	1122.0	0.1710	1.21923

Gas Tables: Thermodynamic Properties of Air, Products of Combustion and Component Gases, Compressible Flow Functions, 2nd Edition, Joseph H. Keenan, Jing Chao, and Joseph Kaye, copyright © 1980. Reproduced with permission of John Wiley & Sons, Inc.

Support Material

APPENDIX 23.D
Properties of Compressed Water
(SI units)

T (°C)	p (bars)	ρ (kg/cm³)	v (cm³/kg)	x	h (kJ/kg)	s (kJ/kg·K)	u (kJ/kg)
0	25	0.0010011	998.94	subcooled	2.5	0.00000380	0.0026204
25	25	0.00099813	1001.9	subcooled	107.14	0.36658	104.63
50	25	0.00098908	1011	subcooled	211.49	0.70266	208.96
75	25	0.00097591	1024.7	subcooled	316.02	1.0142	313.45
100	25	0.00095947	1042.2	subcooled	420.97	1.3053	418.36
125	25	0.00094018	1063.6	subcooled	526.64	1.5794	523.98
150	25	0.00091815	1089.1	subcooled	633.43	1.8395	630.71
175	25	0.00089332	1119.4	subcooled	741.87	2.0885	739.07
200	25	0.00086538	1155.6	subcooled	852.65	2.329	849.76
223.95	25	0.00083512	1197.4	0	961.91	2.5543	958.91
223.95	25	0.000012508	79949	1	2801.9	6.2558	2602.1
0	50	0.0010023	997.68	subcooled	5.0325	0.0001383	0.044068
25	50	0.00099925	1000.8	subcooled	109.45	0.36592	104.44
50	50	0.00099016	1009.9	subcooled	213.64	0.7015	208.59
75	50	0.00097701	1023.5	subcooled	318.03	1.0127	312.92
100	50	0.00096063	1041	subcooled	422.85	1.3034	417.64
125	50	0.00094144	1062.2	subcooled	528.37	1.5771	523.06
150	50	0.00091956	1087.5	subcooled	634.98	1.8368	629.55
175	50	0.00089493	1117.4	subcooled	743.19	2.0852	737.61
200	50	0.00086726	1153.1	subcooled	853.68	2.3251	847.91
225	50	0.00083599	1196.2	subcooled	967.38	2.5592	961.4
250	50	0.00080009	1249.9	subcooled	1085.7	2.791	1079.5
263.94	50	0.00077737	1286.4	0	1154.6	2.921	1148.2
263.94	50	0.000025351	39446	1	2794.2	5.9737	2597
0	75	0.0010036	996.44	subcooled	7.5555	0.0002494	0.082204
25	75	0.0010004	999.64	subcooled	111.75	0.36526	104.25
50	75	0.00099124	1008.8	subcooled	215.79	0.70035	208.22
75	75	0.0009781	1022.4	subcooled	320.05	1.0111	312.38
100	75	0.00096179	1039.7	subcooled	424.73	1.3015	416.93
125	75	0.00094269	1060.8	subcooled	530.11	1.5748	522.15
150	75	0.00092095	1085.8	subcooled	636.54	1.8341	628.4
175	75	0.00089651	1115.4	subcooled	744.54	2.082	736.17
200	75	0.00086911	1150.6	subcooled	854.73	2.3212	846.1
225	75	0.00083824	1193	subcooled	968.01	2.5545	959.06
250	75	0.00080293	1245.4	subcooled	1085.7	2.7851	1076.4
275	75	0.0007614	1313.4	subcooled	1210.2	3.0174	1200.4
290.54	75	0.00073088	1368.2	0	1292.9	3.1662	1282.7
290.54	75	0.000039479	25330	1	2765.9	5.7793	2575.9
0	100	0.0010048	995.2	subcooled	10.069	0.00033757	0.1171
25	100	0.0010015	998.54	subcooled	114.05	0.3646	104.06
50	100	0.00099231	1007.8	subcooled	217.94	0.6992	207.86
75	100	0.00097919	1021.3	subcooled	322.07	1.0096	311.85
100	100	0.00096293	1038.5	subcooled	426.62	1.2996	416.23
125	100	0.00094393	1059.4	subcooled	531.84	1.5725	521.25
150	100	0.00092232	1084.2	subcooled	638.11	1.8313	627.27
175	100	0.00089807	1113.5	subcooled	745.89	2.0788	734.75
200	100	0.00087094	1148.2	subcooled	855.8	2.3174	844.31
225	100	0.00084044	1189.9	subcooled	968.68	2.5499	956.78
250	100	0.0008057	1241.2	subcooled	1085.8	2.7792	1073.4
275	100	0.00076513	1307	subcooled	1209.3	3.0097	1196.2
300	100	0.00071529	1398	subcooled	1343.3	3.2488	1329.4

APPENDIX 23.D *(continued)*
Properties of Compressed Water
(SI units)

T (°C)	p (bars)	ρ (kg/cm³)	ν (cm³/kg)	x	h (kJ/kg)	s (kJ/kg·K)	μ (kJ/kg)
0	150	0.0010073	992.76	subcooled	15.069	0.00044686	0.17746
25	150	0.0010037	996.35	subcooled	118.63	0.36325	103.69
50	150	0.00099443	1005.6	subcooled	222.23	0.6969	207.15
75	150	0.00098135	1019	subcooled	326.1	1.0065	310.81
100	150	0.0009652	1036.1	subcooled	430.39	1.2958	414.85
125	150	0.00094638	1056.7	subcooled	535.33	1.568	519.48
150	150	0.00092503	1081	subcooled	641.27	1.826	625.05
175	150	0.00090114	1109.7	subcooled	748.63	2.0725	731.98
200	150	0.0008745	1143.5	subcooled	857.99	2.31	840.84
225	150	0.00084471	1183.8	subcooled	970.12	2.5409	952.36
250	150	0.00081103	1233	subcooled	1086.1	2.768	1067.6
275	150	0.00077216	1295.1	subcooled	1207.8	2.9951	1188.3
300	150	0.00072555	1378.3	subcooled	1338.3	3.2279	1317.6
0	200	0.0010097	990.36	subcooled	20.033	0.00046962	0.22569
25	200	0.0010058	994.19	subcooled	123.2	0.36187	103.32
50	200	0.00099653	1003.5	subcooled	226.51	0.69461	206.44
75	200	0.00098348	1016.8	subcooled	330.13	1.0035	309.79
100	200	0.00096744	1033.7	subcooled	434.17	1.292	413.5
125	200	0.00094879	1054	subcooled	538.84	1.5635	517.76
150	200	0.00092769	1077.9	subcooled	644.45	1.8208	622.89
175	200	0.00090414	1106	subcooled	751.42	2.0664	729.3
200	200	0.00087797	1139	subcooled	860.27	2.3027	837.49
225	200	0.00084882	1178.1	subcooled	971.69	2.5322	948.13
250	200	0.00081609	1225.4	subcooled	1086.7	2.7573	1062.2
275	200	0.00077871	1284.2	subcooled	1206.7	2.9814	1181
300	200	0.00073471	1361.1	subcooled	1334.4	3.2091	1307.1
0	250	0.0010122	988	subcooled	24.962	0.00040919	0.26234
25	250	0.001008	992.07	subcooled	127.75	0.36047	102.95
50	250	0.00099861	1001.4	subcooled	230.79	0.69233	205.75
75	250	0.00098559	1014.6	subcooled	334.16	1.0004	308.79
100	250	0.00096965	1031.3	subcooled	437.95	1.2883	412.17
125	250	0.00095116	1051.3	subcooled	542.35	1.5591	516.07
150	250	0.0009303	1074.9	subcooled	647.66	1.8156	620.78
175	250	0.00090707	1102.5	subcooled	754.25	2.0604	726.69
200	250	0.00088133	1134.6	subcooled	862.61	2.2956	834.24
225	250	0.00085279	1172.6	subcooled	973.38	2.5237	944.06
250	250	0.00082092	1218.1	subcooled	1087.4	2.7471	1057
275	250	0.00078485	1274.1	subcooled	1206	2.9685	1174.2
300	250	0.00074302	1345.9	subcooled	1331.3	3.1919	1297.6
0	300	0.0010145	985.67	subcooled	29.858	0.00026879	0.28791
25	300	0.0010101	989.98	subcooled	132.28	0.35905	102.58
50	300	0.0010007	999.33	subcooled	235.05	0.69005	205.07
75	300	0.00098767	1012.5	subcooled	338.19	0.99746	307.81
100	300	0.00097182	1029	subcooled	441.74	1.2847	410.87
125	300	0.0009535	1048.8	subcooled	545.88	1.5548	514.42
150	300	0.00093286	1072	subcooled	650.89	1.8106	618.73
175	300	0.00090994	1099	subcooled	757.11	2.0545	724.15
200	300	0.00088462	1130.4	subcooled	865.02	2.2888	831.1
225	300	0.00085664	1167.4	subcooled	975.17	2.5156	940.15
250	300	0.00082556	1211.3	subcooled	1088.4	2.7373	1052
275	300	0.00079064	1264.8	subcooled	1205.7	2.9563	1167.7
300	300	0.00075066	1332.2	subcooled	1328.9	3.176	1288.9

(Multiply MPa by 10 to obtain bars.)

Values in this table were calculated from *NIST Standard Reference Database 10*, "NIST/ASME Steam Properties," Ver. 2.11, National Institute of Standards and Technology, U.S. Department of Commerce, Gaithersburg, MD, 1997, which has been licensed to PPI.

APPENDIX 23.E
Enthalpy-Entropy (Mollier) Diagram for Steam
(SI units)

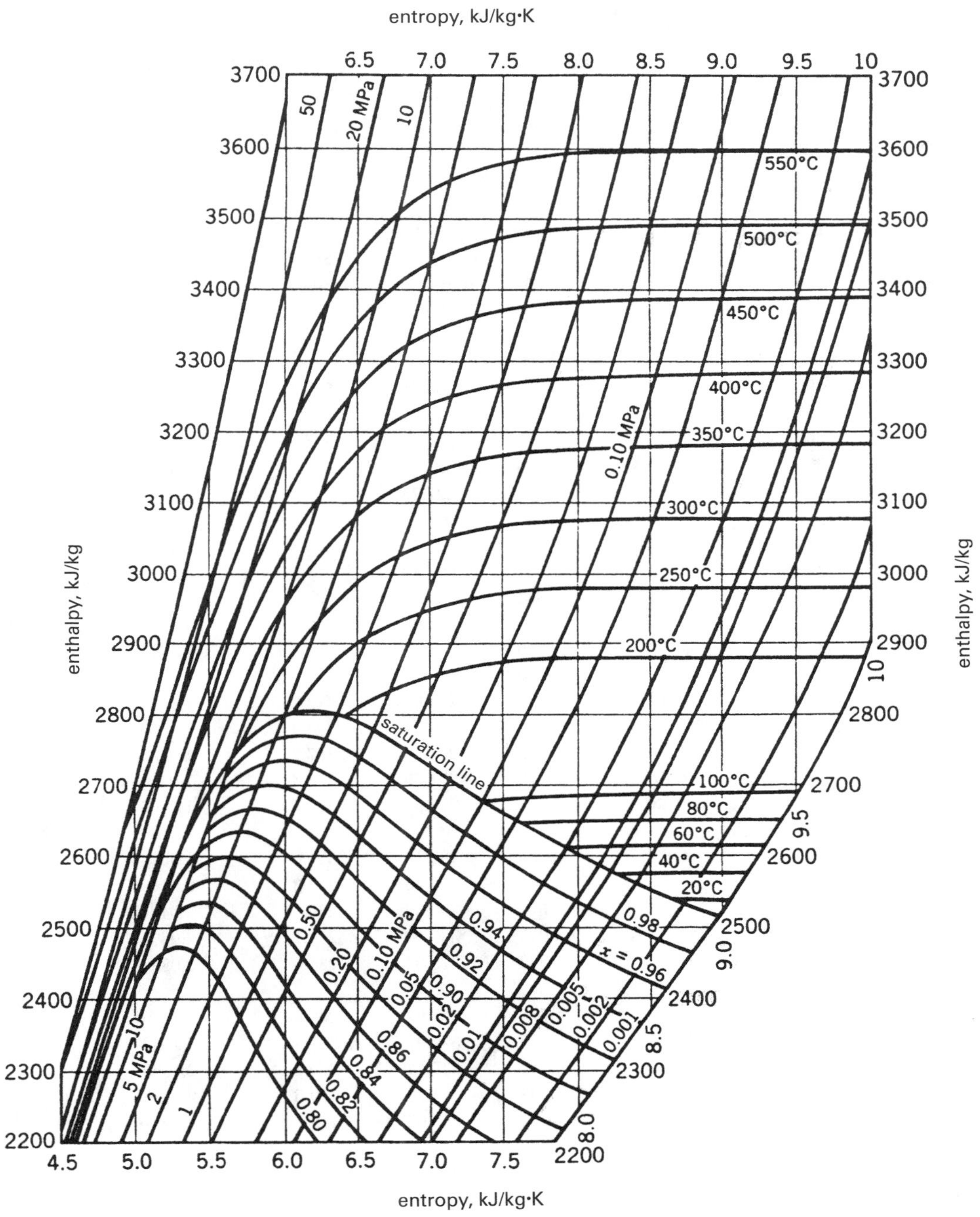

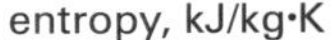

(Multiply MPa by 10 to obtain bars.)

Reproduced from *Steam Tables: Thermodynamic Properties of Water Including Vapor, Liquid, and Solid Phase* (SI version), by John H. Keenan, Frederick G. Keyes, Philip G. Hill, and Joan G. Moore, copyright © 1978. Reproduced with permission of John Wiley & Sons, Inc.

Support Material

APPENDIX 23.F
Properties of Low-Pressure Air
(SI units)

T in *K*; *h* and *u* in kJ/kg; ϕ in kJ/kg·K

T	h	p_r	u	v_r	ϕ
200	199.97	0.3363	142.56	1707	1.29559
210	209.97	0.3987	149.69	1512	1.34444
220	219.97	0.4690	156.82	1346	1.39105
230	230.02	0.5477	164.00	1205	1.43557
240	240.02	0.6355	171.13	1084	1.47824
250	250.05	0.7329	178.28	979	1.51917
260	260.09	0.8405	185.45	887.8	1.55848
270	270.11	0.9590	192.60	808.0	1.59634
280	280.13	1.0889	199.75	738.0	1.63279
285	285.14	1.1584	203.33	706.1	1.65055
290	290.16	1.2311	206.91	676.1	1.66802
295	295.17	1.3068	210.49	647.9	1.68515
300	300.19	1.3860	214.07	621.2	1.70203
305	305.22	1.4686	217.67	596.0	1.71865
310	310.24	1.5546	221.25	572.3	1.73498
315	315.27	1.6442	224.85	549.8	1.75106
320	320.29	1.7375	228.42	528.6	1.76690
325	325.31	1.8345	232.02	508.4	1.78249
330	330.34	1.9352	235.61	489.4	1.79783
340	340.42	2.149	242.82	454.1	1.82790
350	350.49	2.379	250.02	422.2	1.85708
360	360.58	2.626	257.24	393.4	1.88543
370	370.67	2.892	264.46	367.2	1.91313
380	380.77	3.176	271.69	343.4	1.94001
390	390.88	3.481	278.93	321.5	1.96633
400	400.98	3.806	286.16	301.6	1.99194
410	411.12	4.153	293.43	283.3	2.01699
420	421.26	4.522	300.69	266.6	2.04142
430	431.43	4.915	307.99	251.1	2.06533
440	441.61	5.332	315.30	236.8	2.08870
450	451.80	5.775	322.62	223.6	2.11161
460	462.02	6.245	329.97	211.4	2.13407
470	472.24	6.742	337.32	200.1	2.15604
480	482.49	7.268	344.70	189.5	2.17760
490	492.74	7.824	352.08	179.7	2.19876
500	503.02	8.411	359.49	170.6	2.21952
510	513.32	9.031	366.92	162.1	2.23993
520	523.63	9.684	374.36	154.1	2.25997
530	533.98	10.37	381.84	146.7	2.27967
540	544.35	11.10	389.34	139.7	2.29906
550	554.74	11.86	396.86	133.1	2.31809
560	565.17	12.66	404.42	127.0	2.33685
570	575.59	13.50	411.97	121.2	2.35531
580	586.04	14.38	419.55	115.7	2.37348
590	596.52	15.31	427.15	110.6	2.39140

APPENDIX 23.F *(continued)*
Properties of Low-Pressure Air
(SI units)

T in K; h and u in kJ/kg; ϕ in kJ/kg·K

T	h	p_r	u	v_r	ϕ
600	607.02	16.28	434.78	105.8	2.40902
610	617.53	17.30	442.42	101.2	2.42644
620	628.07	18.36	450.09	96.92	2.44356
630	638.63	19.84	457.78	92.84	2.46048
640	649.22	20.64	465.50	88.99	2.47716
650	659.84	21.86	473.25	85.34	2.49364
660	670.47	23.13	481.01	81.89	2.50985
670	681.14	24.46	488.81	78.61	2.52589
680	691.82	25.85	496.62	75.50	2.54175
690	702.52	27.29	504.45	72.56	2.55731
700	713.27	28.80	512.33	69.76	2.57277
710	724.04	30.38	520.23	67.07	2.58810
720	734.82	32.02	528.14	64.53	2.60319
730	745.62	33.72	536.07	62.13	2.61803
740	756.44	35.50	544.02	59.82	2.63280
750	767.29	37.35	551.99	57.63	2.64737
760	778.18	39.27	560.01	55.54	2.66176
770	789.11	41.31	568.07	53.39	2.67595
780	800.03	43.35	576.12	51.64	2.69013
790	810.99	45.55	584.21	49.86	2.70400
800	821.95	47.75	592.30	48.08	2.71787
820	843.98	52.59	608.59	44.84	2.74504
840	866.08	57.60	624.95	41.85	2.77170
860	888.27	63.09	641.40	39.12	2.79783
880	910.56	68.98	657.95	36.61	2.82344
900	932.93	75.29	674.58	34.31	2.84856
920	955.38	82.05	691.28	32.18	2.87324
940	977.92	89.28	708.08	30.22	2.89748
960	1000.55	97.00	725.02	28.40	2.92128
980	1023.25	105.2	741.98	26.73	2.94468
1000	1046.04	114.0	758.94	25.17	2.96770
1020	1068.89	123.4	776.10	23.72	2.99034
1040	1091.85	133.3	793.36	22.39	3.01260
1060	1114.86	143.9	810.62	21.14	3.03449
1080	1137.89	155.2	827.88	19.98	3.05608
1100	1161.07	167.1	845.33	18.896	3.07732
1120	1184.28	179.7	862.79	17.886	3.09825
1140	1207.57	193.1	880.35	16.946	3.11883
1160	1230.92	207.2	897.91	16.064	3.13916
1180	1254.34	222.2	915.57	15.241	3.15916
1200	1277.79	238.0	933.33	14.470	3.17888
1220	1301.31	254.7	951.09	13.747	3.19834
1240	1324.93	272.3	968.95	13.069	3.21751
1260	1348.55	290.8	986.90	12.435	3.23638
1280	1372.24	310.4	1004.76	11.835	3.25510

APPENDIX 23.F *(continued)*
Properties of Low-Pressure Air
(SI units)

T in *K*; *h* and *u* in kJ/kg; ϕ in kJ/kg·K

T	h	p_r	u	v_r	ϕ
1300	1395.97	330.9	1022.82	11.275	3.27345
1320	1419.76	352.5	1040.88	10.747	3.29160
1340	1443.60	375.3	1058.94	10.247	3.30959
1360	1467.49	399.1	1077.10	9.780	3.32724
1380	1491.44	424.2	1095.26	9.337	3.34474
1400	1515.42	450.5	1113.52	8.919	3.36200
1420	1539.44	478.0	1131.77	8.526	3.37901
1440	1563.51	506.9	1150.13	8.153	3.39586
1460	1587.63	537.1	1168.49	7.801	3.41247
1480	1611.79	568.8	1186.95	7.468	3.42892
1500	1635.97	601.9	1205.41	7.152	3.44516
1520	1660.23	636.5	1223.87	6.854	3.46120
1540	1684.51	672.8	1242.43	6.569	3.47712
1560	1708.82	710.5	1260.99	6.301	3.49276
1580	1733.17	750.0	1279.65	6.046	3.50829
1600	1757.57	791.2	1298.30	5.804	3.52364
1620	1782.00	834.1	1316.96	5.574	3.53879
1640	1806.46	878.9	1335.72	5.355	3.55381
1660	1830.96	925.6	1354.48	5.147	3.56867
1680	1855.50	974.2	1373.24	4.949	3.58335
1700	1880.1	1025	1392.7	4.761	3.5979
1750	1941.6	1161	1439.8	4.328	3.6336
1800	2003.3	1310	1487.2	3.944	3.6684
1850	2065.3	1475	1534.9	3.601	3.7023
1900	2127.4	1655	1582.6	3.295	3.7354
1950	2189.7	1852	1630.6	3.022	3.7677
2000	2252.1	2068	1678.7	2.776	3.7994
2050	2314.6	2303	1726.8	2.555	3.8303
2100	2377.4	2559	1775.3	2.356	3.8605
2150	2440.3	2837	1823.8	2.175	3.8901
2200	2503.2	3138	1872.4	2.012	3.9191
2250	2566.4	3464	1921.3	1.864	3.9474

Gas Tables: International Version—Thermodynamic Properties of Air, Products of Combustion and Component Gases, Compressible Flow Functions, Second Edition. Joseph H. Keenan, Jing Chao, and Joseph Kaye, copyright © 1983. Reproduced with permission of John Wiley & Sons, Inc.

APPENDIX 25.A
Fanno Flow Factors
($k = 1.4$)

M	p/p^*	$a/a^* = \rho^*/\rho$	T/T^*	p_0/p_0^*	$4fL/D$
0.00	∞	0.0000	1.200	∞	∞
0.05	21.903	0.0547	1.199	11.592	280.02
0.10	10.944	0.1094	1.197	5.822	66.922
0.12	9.116	0.131	1.1965	4.864	45.408
0.14	7.809	0.153	1.195	4.182	32.511
0.16	6.829	0.175	1.194	3.673	24.198
0.18	6.066	0.196	1.192	3.278	18.543
0.20	5.455	0.218	1.1905	2.963	14.533
0.25	4.355	0.272	1.185	2.403	8.483
0.30	3.619	0.3257	1.178	2.035	5.299
0.35	3.092	0.379	1.171	1.778	3.453
0.40	2.696	0.431	1.163	1.590	2.308
0.45	2.386	0.483	1.153	1.448	1.566
0.50	2.138	0.534	1.143	1.340	1.069
0.52	2.052	0.555	1.138	1.303	0.917
0.54	1.972	0.575	1.134	1.270	0.787
0.56	1.897	0.595	1.129	1.240	0.673
0.58	1.828	0.615	1.124	1.213	0.576
0.60	1.763	0.635	1.119	1.188	0.491
0.65	1.618	0.684	1.106	1.135	0.325
0.70	1.493	0.732	1.093	1.094	0.208
0.75	1.385	0.779	1.078	1.062	0.127
0.80	1.289	0.825	1.064	1.038	0.072
0.85	1.205	0.870	1.048	1.020	0.0363
0.90	1.129	0.914	1.0327	1.009	0.0145
0.95	1.061	0.958	1.0165	1.002	0.0033
1.00	1.000	1.000	1.000	1.000	0.000
1.20	0.804	1.158	0.932	1.030	0.0336
1.50	0.606	1.365	0.827	1.176	0.136
1.60	0.557	1.425	0.794	1.250	0.172
1.70	0.513	1.483	0.760	1.338	0.208
1.80	0.474	1.536	0.728	1.439	0.242
1.90	0.439	1.586	0.697	1.555	0.274
2.00	0.408	1.633	0.667	1.687	0.305
2.50	0.292	1.826	0.533	2.637	0.432
3.00	0.218	1.964	0.428	4.235	0.522
3.50	0.1685	2.064	0.348	6.789	0.586
4.00	0.134	2.138	0.286	10.719	0.633
4.50	0.108	2.194	0.237	16.562	0.667
5.00	0.0894	2.236	0.200	25.000	0.694

APPENDIX 25.B
Rayleigh Flow Factors
($k = 1.4$)

M	p/p^*	p_0/p_0^*	T/T^*	T_0/T_0^*	$a/a^* = \rho^*/\rho$
0.00	2.400	1.268	0.000	0.000	0.000
0.05	2.392	1.266	0.0143	0.0119	0.00598
0.10	2.367	1.259	0.056	0.0468	0.0237
0.12	2.353	1.255	0.079	0.0667	0.0339
0.14	2.336	1.251	0.107	0.089	0.0458
0.16	2.317	1.246	0.137	0.115	0.0593
0.18	2.296	1.241	0.1708	0.143	0.0744
0.20	2.273	1.235	0.2066	0.1735	0.091
0.25	2.207	1.218	0.304	0.257	0.138
0.30	2.131	1.198	0.409	0.3468	0.192
0.35	2.048	1.178	0.514	0.439	0.251
0.40	1.961	1.157	0.615	0.529	0.314
0.45	1.870	1.135	0.708	0.614	0.378
0.50	1.778	1.114	0.790	0.691	0.444
0.52	1.741	1.106	0.819	0.720	0.470
0.54	1.704	1.098	0.847	0.747	0.497
0.56	1.668	1.090	0.872	0.772	0.523
0.58	1.632	1.083	0.896	0.796	0.549
0.60	1.596	1.075	0.917	0.819	0.574
0.65	1.508	1.058	0.961	0.868	0.637
0.70	1.424	1.043	0.993	0.908	0.697
0.75	1.343	1.030	1.014	0.940	0.755
0.80	1.266	1.019	1.025	0.964	0.810
0.85	1.193	1.011	1.028	0.981	0.862
0.90	1.125	1.005	1.0245	0.992	0.911
0.95	1.060	1.001	1.0146	0.998	0.957
1.00	1.000	1.000	1.000	1.000	1.000
1.20	0.796	1.0194	0.912	0.978	1.146
1.50	0.578	1.122	0.753	0.909	1.301
1.60	0.523	1.176	0.702	0.884	1.340
1.70	0.475	1.240	0.654	0.859	1.375
1.80	0.433	1.316	0.609	0.836	1.405
1.90	0.396	1.403	0.567	0.814	1.431
2.00	0.363	1.503	0.529	0.794	1.455
2.50	0.246	2.222	0.378	0.710	1.538
3.00	0.176	3.424	0.280	0.654	1.588
3.50	0.132	5.328	0.214	0.616	1.619
4.00	0.1025	8.227	0.168	0.589	1.641
4.50	0.0818	12.502	0.135	0.569	1.656
5.00	0.0667	18.634	0.111	0.555	1.667

APPENDIX 25.C
International Standard Atmosphere

customary U.S. units

altitude (ft)	temperature (°R)	pressure (psia)	
0	518.7	14.696	troposphere
1000	515.1	14.175	
2000	511.6	13.664	
3000	508.0	13.168	
4000	504.4	12.692	
5000	500.9	12.225	
6000	497.3	11.778	
7000	493.7	11.341	
8000	490.2	10.914	
9000	486.6	10.501	
10,000	483.0	10.108	
11,000	479.5	9.720	
12,000	475.9	9.347	
13,000	472.3	8.983	
14,000	468.8	8.630	
15,000	465.2	8.291	
16,000	461.6	7.962	
17,000	458.1	7.642	
18,000	454.5	7.338	
19,000	450.9	7.038	
20,000	447.4	6.753	
21,000	443.8	6.473	
22,000	440.2	6.203	
23,000	436.7	5.943	
24,000	433.1	5.693	
25,000	429.5	5.452	
26,000	426.0	5.216	
27,000	422.4	4.990	
28,000	418.8	4.774	
29,000	415.3	4.563	
30,000	411.7	4.362	
31,000	408.1	4.165	
32,000	404.6	3.978	
33,000	401.0	3.797	
34,000	397.5	3.625	
35,000	393.9	3.458	tropopause
36,000	392.7	3.296	
37,000	392.7	3.143	
38,000	392.7	2.996	
39,000	392.7	2.854	
40,000	392.7	2.721	
41,000	392.7	2.593	
42,000	392.7	2.475	
43,000	392.7	2.358	
44,000	392.7	2.250	
45,000	392.7	2.141	
46,000	392.7	2.043	
47,000	392.7	1.950	
48,000	392.7	1.857	
49,000	392.7	1.768	
50,000	392.7	1.690	
51,000	392.7	1.611	
52,000	392.7	1.532	
53,000	392.7	1.464	
54,000	392.7	1.395	
55,000	392.7	1.331	
56,000	392.7	1.267	
57,000	392.7	1.208	
58,000	392.7	1.154	
59,000	392.7	1.100	
60,000	392.7	1.046	
61,000	392.7	0.997	
62,000	392.7	0.953	
63,000	392.7	0.909	
64,000	392.7	0.864	
65,000	392.7	0.825	

stratosphere (to approximately 160,000 ft)

SI units

altitude (m)	temperature (K)	pressure (bar)	
0	288.15	1.01325	troposphere
500	284.9	0.9546	
1000	281.7	0.8988	
1500	278.4	0.8456	
2000	275.2	0.7950	
2500	271.9	0.7469	
3000	268.7	0.7012	
3500	265.4	0.6578	
4000	262.2	0.6166	
4500	258.9	0.5775	
5000	255.7	0.5405	
5500	252.4	0.5054	
6000	249.2	0.4722	
6500	245.9	0.4408	
7000	242.7	0.4111	
7500	239.5	0.3830	
8000	236.2	0.3565	
8500	233.0	0.3315	
9000	229.7	0.3080	
9500	226.5	0.2858	
10 000	223.3	0.2650	
10 500	220.0	0.2454	tropopause
11 000	216.8	0.2270	
11 500	216.7	0.2098	
12 000	216.7	0.1940	
12 500	216.7	0.1793	
13 000	216.7	0.1658	
13 500	216.7	0.1533	
14 000	216.7	0.1417	
14 500	216.7	0.1310	
15 000	216.7	0.1211	
15 500	216.7	0.1120	
16 000	216.7	0.1035	
16 500	216.7	0.09572	
17 000	216.7	0.08850	
17 500	216.7	0.08182	
18 000	216.7	0.07565	
18 500	216.7	0.06995	
19 000	216.7	0.06467	
19 500	216.7	0.05980	
20 000	216.7	0.05529	
22 000	218.6	0.04047	stratosphere (to approximately 50 000 m)
24 000	220.6	0.02972	
26 000	222.5	0.02188	
28 000	224.5	0.01616	
30 000	226.5	0.01197	
32 000	228.5	0.00889	

Support Material

APPENDIX 33.A

Representative Thermal Conductivity[a,b]
(at 32°F (0°C) unless specified otherwise)

material	$\frac{\text{Btu-ft}}{\text{hr-ft}^2\text{-°F}}$	$\frac{\text{W}}{\text{m·K}}$
air	0.014	0.024
aluminum	117	202
asbestos	0.087	0.15
brass	56	97
brick, fire clay (400°F)	0.58	1.0
concrete	0.5	0.9
copper	224	388
cork	0.025	0.043
fiberglass	0.03	0.05
glass	0.63	1.1
glass, Pyrex™	0.68	1.2
gold (68°F)	169	292
hydrogen (100°F)	0.11	0.19
ice	1.3	2.2
iron, cast (4% C, 68°F)	30	52
iron, pure	36	62
lead (70°F)	20	35
mercury	4.83	8.36
nickel	34.4	59.5
oxygen	0.016	0.028
rubber, soft	0.10	0.17
silver	242	419
steel (1% C)[c]	24.8	42.9
tungsten	92	160
water	0.32	0.55
zinc	65	110

(Multiply Btu-ft/hr-ft²-°F by 12 to obtain Btu-in/hr-ft²-°F.)
(Multiply Btu-ft/hr-ft²-°F by 1.7307 to obtain W/m·K.)
(Multiply Btu-ft/hr-ft²-°F by 4.1365×10^{-3} to obtain cal·cm/s·cm²·°C.)

[a]Values of thermal conductivity are typically accurate to only ±20%, although in some cases the error may be as small as ±10%.
[b]Values are compiled from a variety of sources.
[c]Values depend greatly on alloy and carbon content, as well as temperature.

Support Material

APPENDIX 33.B
Properties of Metals and Alloys[a,b]

metal	thermal conductivity, k (Btu-ft/hr-ft^2-°F) 32°F (0°C)	212°F (100°C)	572°F (300°C)	932°F (500°C)	c_p (Btu/lbm-°F) 32°F (0°C)	ρ (lbm/ft^3) 32°F (0°C)
aluminum alloy	92	104	–	–	–	–
aluminum, pure	117	119	133	155	0.208	169
brass (70% Cu, 30% Zn)	58	60	66	–	0.092	532
bronze (75% Cu, 25% Sn)	15	–	–	–	0.082	540
copper, pure	224	218	212	207	0.091	558
iron, cast, alloy	30	28.3	27.0	–	0.10	455
iron, cast, plain	33	31.8	27.7	24.8	0.11	474
iron, pure	35.8	36.6	–	–	0.104	491
lead	20.1	19	18	–	0.030	705
magnesium	91	92	–	–	0.232	109
nickel/chrome	7.5	9.2	–	–	–	–
silver	242	238	–	–	0.056	655
steel, carbon (1% C)	26.5	26.0	25.0	22.0	0.11	490
steel, stainless	8.0	9.3	11.0	12.8	0.11	488
tin	36	34	–	–	0.054	456
zinc	65	64	59	–	0.091	446

(Multiply Btu-ft/hr-ft^2-°F by 12 to obtain Btu-in/hr-ft^2-°F.)
(Multiply Btu-ft/hr-ft^2-°F by 1.7307 to obtain W/m·K.)
(Multiply Btu-ft/hr-ft^2-°F by 4.1365×10^{-3} to obtain cal·cm/s·cm^2·°C.)
(Multiply Btu/lbm-°F by 4186.8 to obtain J/kg·K.)
(Multiply lbm/ft^3 by 16.0185 to obtain kg/m^3.)
[a]Values of thermal conductivity are typically accurate to only ±20%, although in some cases the error may be as small as ±10%.
[b]Values are compiled from a variety of sources.

Support Material

APPENDIX 33.C
Properties of Nonmetals[a,b]

material	average temperature (°F)	k $\left(\frac{\text{Btu-ft}}{\text{hr-ft}^2\text{-°F}}\right)$	c_p $\left(\frac{\text{Btu}}{\text{lbm-°F}}\right)$	ρ $\left(\frac{\text{lbm}}{\text{ft}^3}\right)$
asbestos	32	0.087	0.25	36
	392	0.12		
brick, building	70	0.38	0.20	106
brick, fire-clay	392	0.58	0.20	144
	1832	0.95		
brick, Kaolin				
insulating	932	0.15		27
	2102	0.26		
firebrick	392	0.05		19
	1400	0.11		
concrete, stone	70	0.54	0.20	144
with 10% moisture	70	0.70		140
diatomaceous earth				
powdered	100	0.030	0.21	14
	300	0.036		
	600	0.046		
glass, window	70	0.45	0.2	170
glass wool (fine)	20	0.022		
	100	0.031		1.5
	200	0.043		
glass wool (packed)	20	0.016		
	100	0.022		6.0
	200	0.029		
ice	32	1.28	0.46	57
magnesia, 85%, molded pipe	32	0.032		17
covering ($T < 600$°F)	200	0.037		
molded pipe covering,	400	0.051		26
diatomaceous silica	1600	0.088		
sand, dry	68	0.20		95
sand, with 10% water	68	0.60		100
soil, dry	70	0.20	0.44	
soil, wet	70	1.5		
wood, oak				
perpendicular to grain	70	0.12	0.57	51
parallel to grain	70	0.20	0.57	
wood, pine				
perpendicular to grain	70	0.06	0.67	31
parallel to grain	70	0.14	0.67	

(Multiply Btu-ft/hr-ft^2-°F by 12 to obtain Btu-in/hr-ft^2-°F.)
(Multiply Btu-ft/hr-ft^2-°F by 1.7307 to obtain W/m·K.)
(Multiply Btu-ft/hr-ft^2-°F by 4.1365×10^{-3} to obtain cal·cm/s·cm^2·°C.)
(Multiply Btu/lbm-°F by 4186.8 to obtain J/kg·K.)
(Multiply lbm/ft^3 by 16.0185 to obtain kg/m^3.)

[a]Values of thermal conductivity are typically accurate to only ±20%, although in some cases the error may be as small as ±10%.

[b]Values are compiled from a variety of sources.

Support Material

APPENDIX 33.D
Typical Thermal Properties of Foodstuffs[a]

item	density, ρ		thermal conductivity, k		diffusivity, α		size or characteristic length, L	average air velocity, v	convective heat transfer (film) coefficient in air, h, or Nusselt correlation[b]
	lbm/ft^3	kg/m^3	Btu/hr-ft-°F	W/m·°C	ft^2/sec	m^2/s	mm	m/s	W/m^2·°C
apples	52.4	837 (840)	0.323	0.559 (0.393–0.513)	1.18×10^{-6}	1.10×10^{-7}	52–62 (Jonathan)	0.0	11.1–11.4
								0.4	15.9–17.0
								0.9	26.1–27.3
								2.0	39.2–45.3
								5.1	50.5–54.5
							63–76 (Red Delicious)	1.5	14.2–27.3
								4.6	34.6–56.8
bananas (fresh)	61.2	980	0.278	0.481	1.51×10^{-6}	1.40×10^{-7}			
broccoli	35.0	561	0.223	0.386					
cherries	65.5	1050	0.315	0.545	1.42×10^{-6}	1.31×10^{-7}			
eggs							34 (Jifujitori)	2–8	$C = 0.46$; $a = 0$; $b = 0.56$
							44 (Leghorn)		$C = 0.71$; $a = 0$; $b = 0.55$
grapefruits							53–80	0.11–0.33	$C = 5.05$; $a = 0$; $b = 0.333$
oranges							53–80	0.11–0.33	$C = 5.05$; $a = 0$; $b = 0.333$ (16 in (41 cm) packing depth)
peaches	59.9	960 (930)	0.304	0.526 (0.581)	1.51×10^{-6}	1.40×10^{-7}			
plums	38.1	610	0.143	0.256					
potatoes	65.7	1040 (1070)	0.288	0.498 (0.648)	1.84×10^{-6}	1.71×10^{-7}	mixed, bulk packed	0.66	14.0
								4.0	19.1
								1.36	20.2
								1.73	24.4

(Multiply kg/m^3 by 0.06243 to obtain lbm/ft^3.)

(Multiply mm by 0.03937 to obtain in.)

(Multiply m/s by 196.85 to obtain ft/min.)

(Multiply m^2/s by 38,760 to obtain ft^2/hr.)

(Multiply m^2/s by 10.764 to obtain ft^2/sec.)

(Multiply W/m·°C by 0.5778 to obtain Btu-ft/hr-ft^2-°F.)

(Multiply W/m^2·°C by 0.1761 to obtain Btu/hr-ft^2-°F.)

[a]Values are typical for noncritical first estimates. Data for particular items were compiled from various international sources, so columns may be inconsistent. Some values are experimental, while others have been calculated from properties (e.g., water content), correlations, and/or definitions. Alternate values in parentheses are from secondary sources and are equally reliable.

[b]Correlations are of the form $\mathrm{Nu} = C\mathrm{Pr}^a\mathrm{Re}^b$. Any consistent set of units may be used. $\mathrm{Re} = \mathrm{v}L/\nu$, where L is the characteristic length of the food item. The thermal conductivity, k_{air}, and kinematic viscosity, ν, of air (the cooling medium) are given in App. 34.C and App. 34.D.

Support Material

APPENDIX 33.E

Transient Heat Flow Charts (solid spheres of radius r_o)

(The variable r_o used in this chart is the outside radius of the sphere, not the characteristic length.)

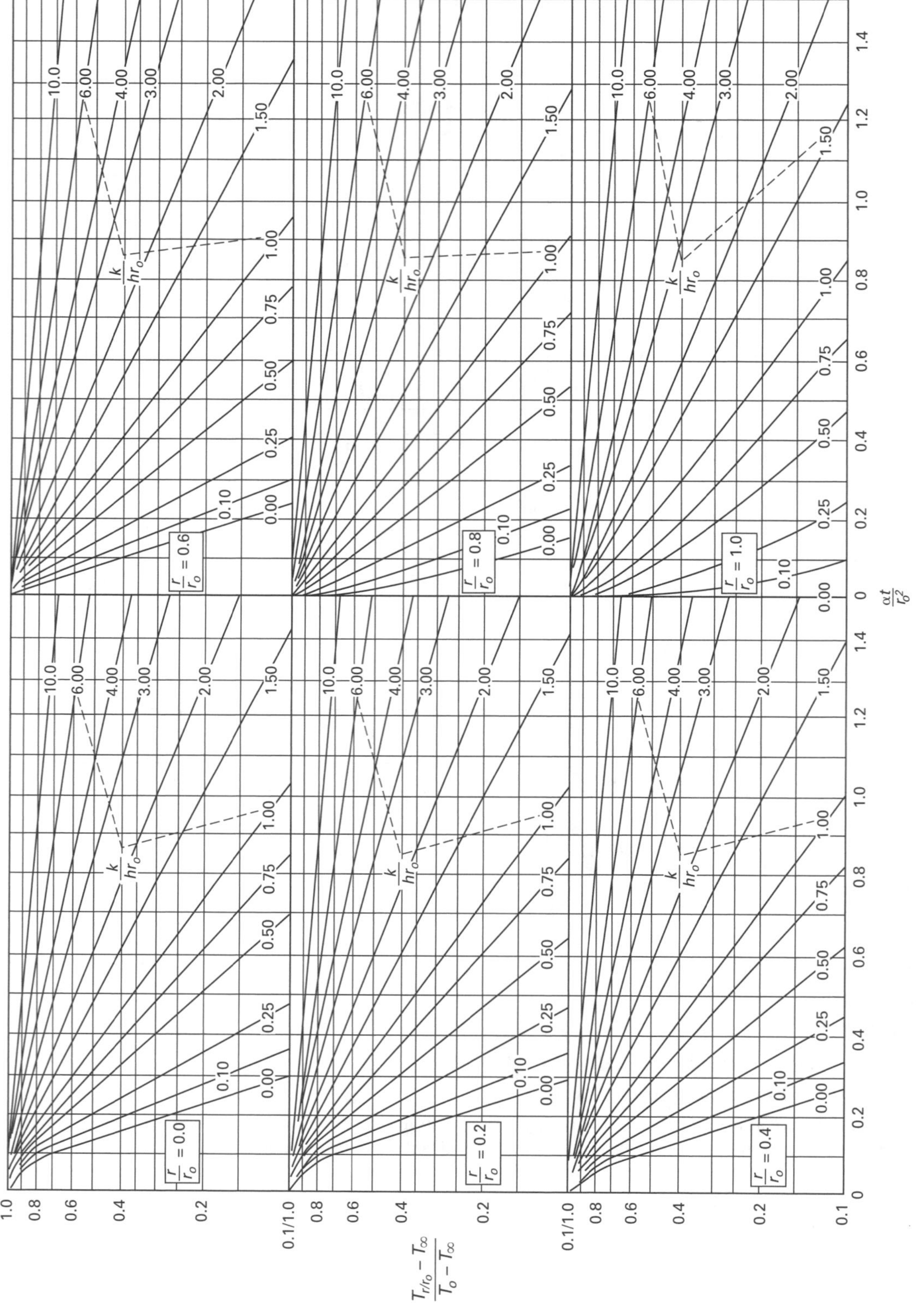

APPENDIX 33.F

Transient Heat Flow Charts (infinite solid circular cylinders of radius r_o)

(The variable r_o used in this chart is the outside radius of the cylinder, not the characteristic length.)

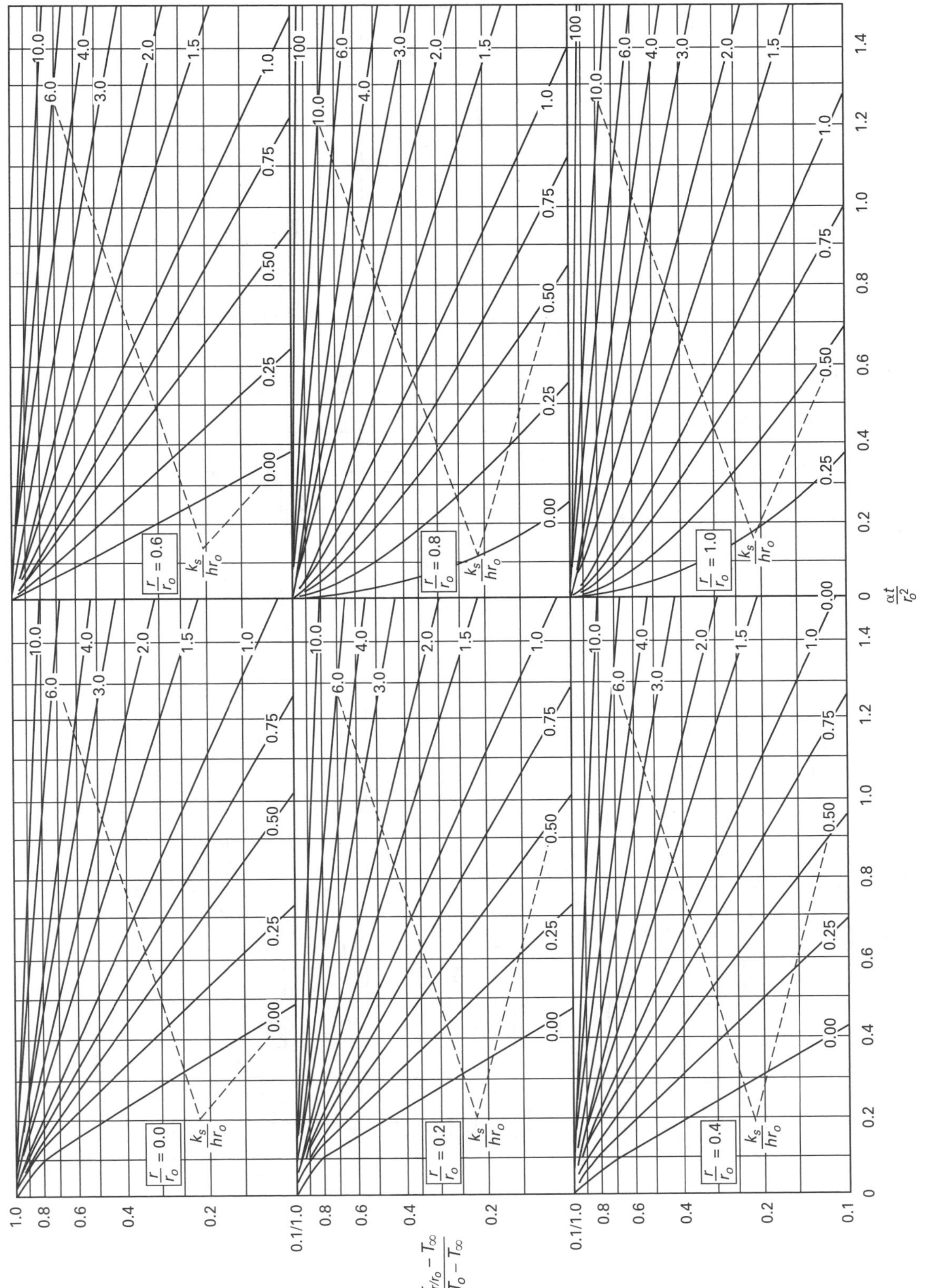

Support Material

APPENDIX 33.G
Transient Heat Flow Charts
(infinite flat slabs of thickness $2L$)

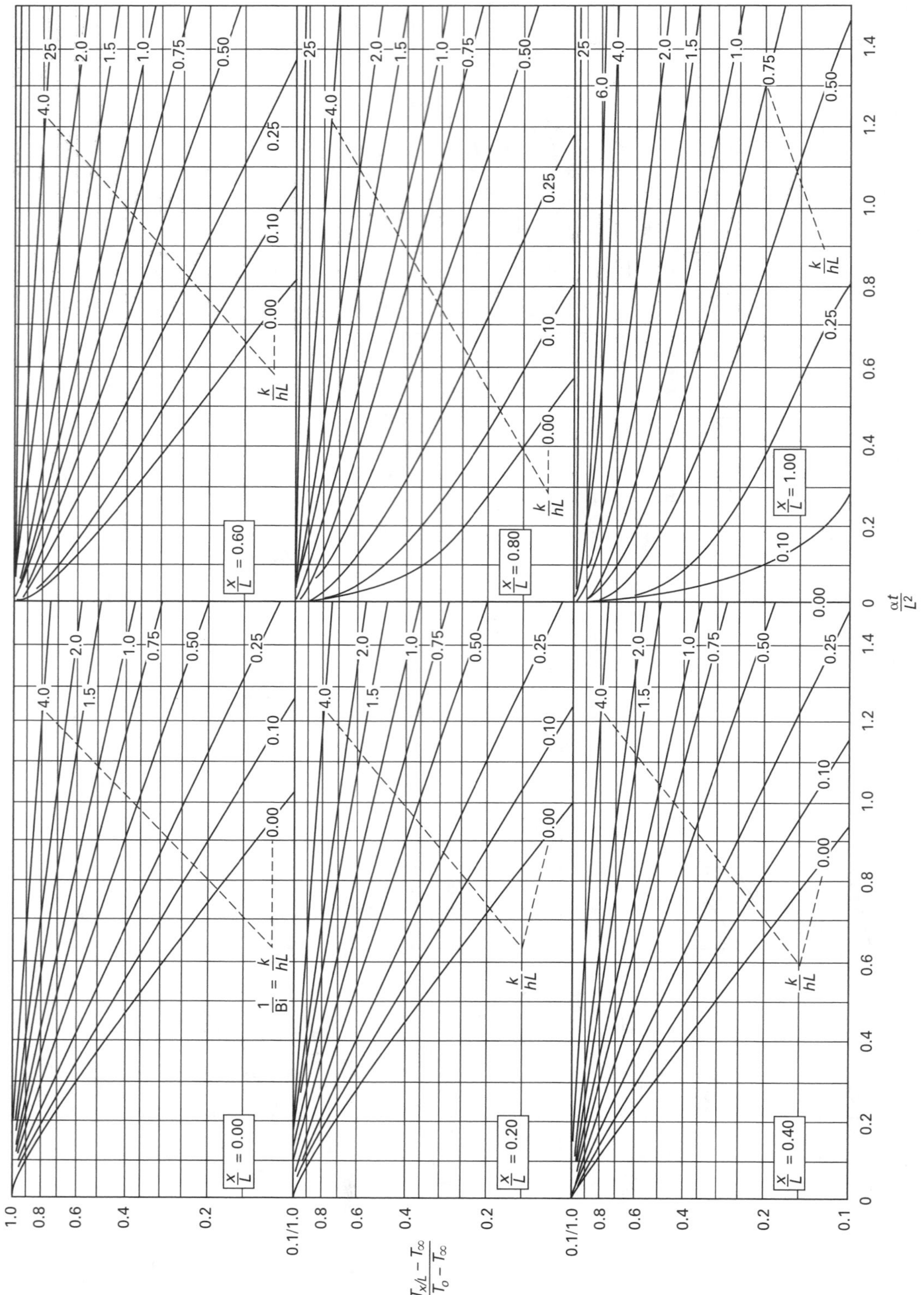

APPENDIX 33.H

Heisler Transient Heat Flow Chart
(temperature at center of a sphere of radius r_o)

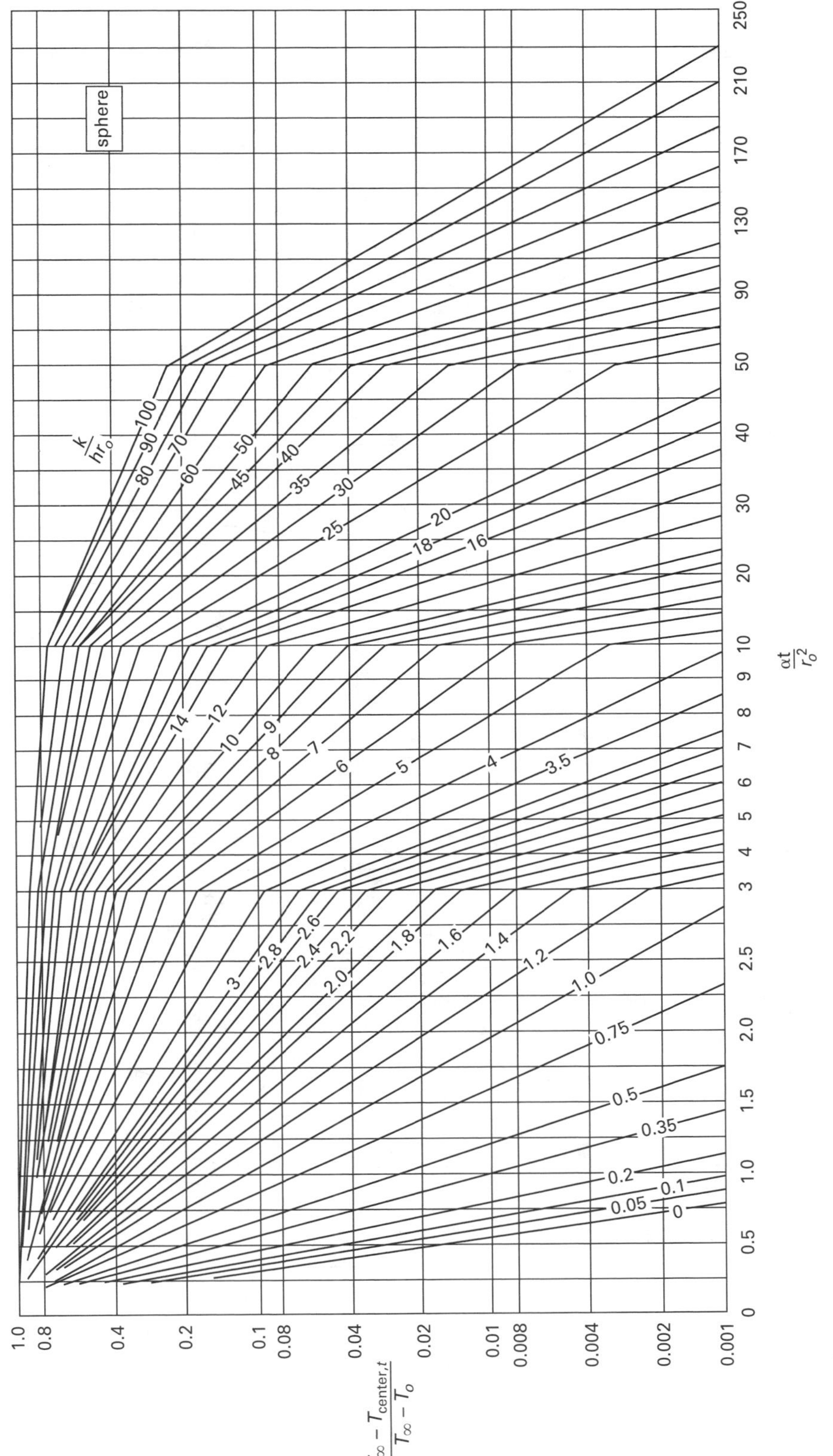

Support Material

APPENDIX 33.I

Heisler Transient Heat Flow Chart
(temperature at center of infinite cylinder of radius r_o)

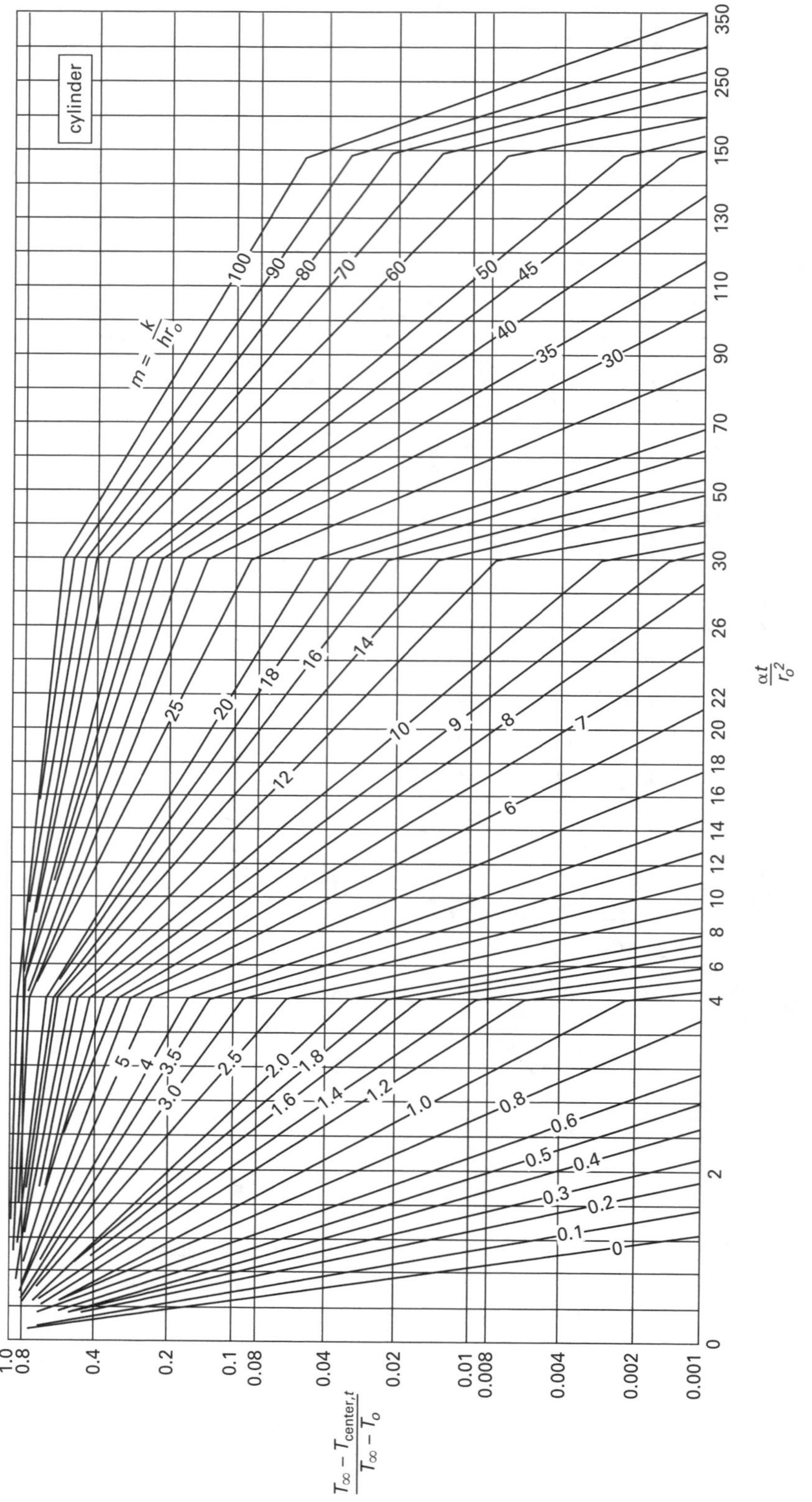

APPENDIX 33.J

Heisler Transient Heat Flow Chart
(temperature at center of infinite slab of thickness $2L$)

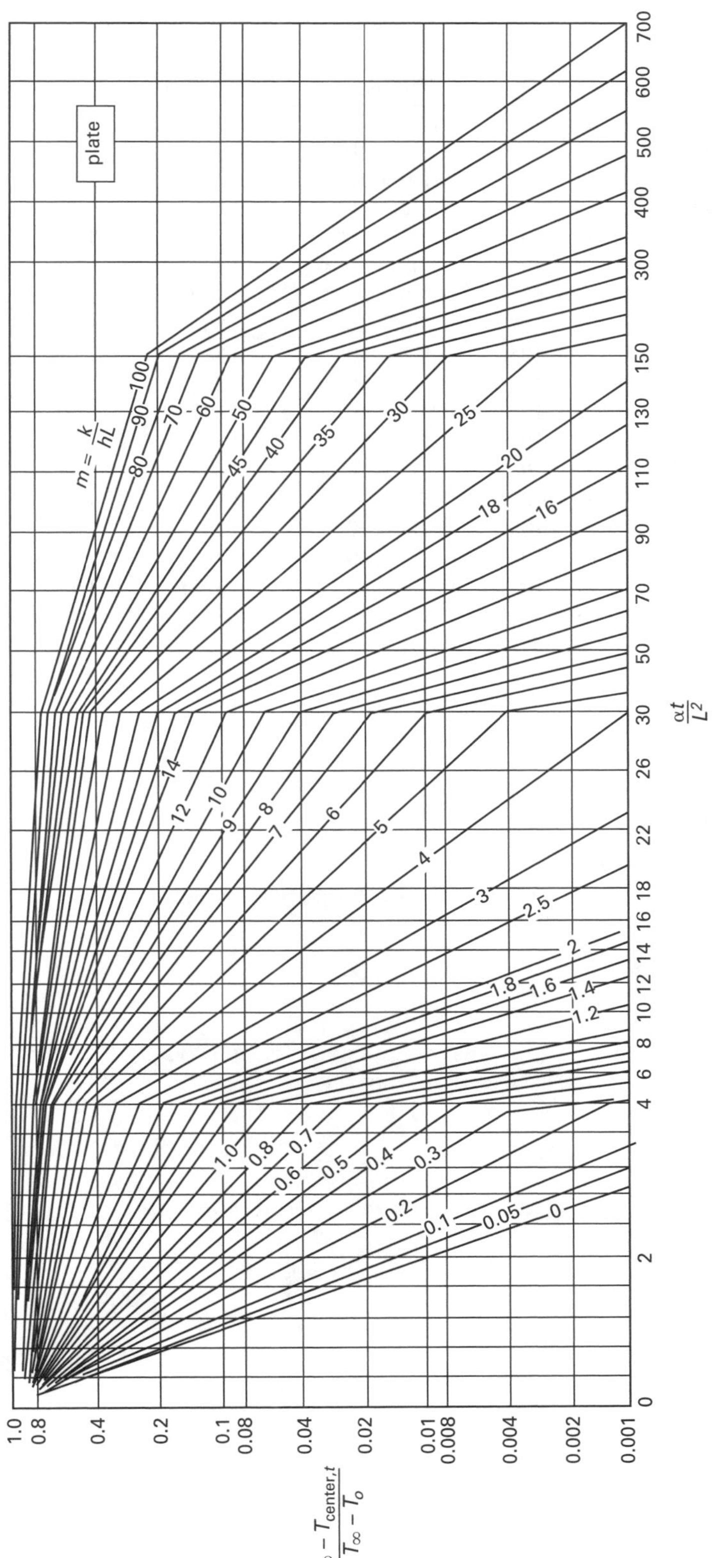

APPENDIX 34.A
Properties of Saturated Water
(customary U.S. units)

T (°F)	ρ (lbm/ft³)	c_p (Btu/lbm-°F)	μ (lbm/ft-sec)	ν (ft²/sec)	k (Btu/hr-ft-°F)	Pr	β (1/°F)	$\frac{g\beta\rho^2}{\mu^2}$ (1/ft³-°F)
32	62.4	1.01	1.20×10^{-3}	1.93×10^{-5}	0.319	13.7	-0.37×10^{-4}	
40	62.4	1.00	1.04×10^{-3}	1.67×10^{-5}	0.325	11.6	0.20×10^{-4}	2.3×10^{6}
50	62.4	1.00	0.88×10^{-3}	1.40×10^{-5}	0.332	9.55	0.49×10^{-4}	8.0×10^{6}
60	62.3	0.999	0.76×10^{-3}	1.22×10^{-5}	0.340	8.03	0.85×10^{-4}	18.4×10^{6}
70	62.3	0.998	0.658×10^{-3}	1.06×10^{-5}	0.347	6.82	1.2×10^{-4}	34.6×10^{6}
80	62.2	0.998	0.578×10^{-3}	0.93×10^{-5}	0.353	5.89	1.5×10^{-4}	56.0×10^{6}
90	62.1	0.997	0.514×10^{-3}	0.825×10^{-5}	0.359	5.13	1.8×10^{-4}	85.0×10^{6}
100	62.0	0.998	0.458×10^{-3}	0.740×10^{-5}	0.364	4.52	2.0×10^{-4}	118×10^{6}
150	61.2	1.00	0.292×10^{-3}	0.477×10^{-5}	0.384	2.74	3.1×10^{-4}	440×10^{6}
200	60.1	1.00	0.205×10^{-3}	0.341×10^{-5}	0.394	1.88	4.0×10^{-4}	1.11×10^{9}
250	58.8	1.01	0.158×10^{-3}	0.269×10^{-5}	0.396	1.45	4.8×10^{-4}	2.14×10^{9}
300	57.3	1.03	0.126×10^{-3}	0.220×10^{-5}	0.395	1.18	6.0×10^{-4}	4.00×10^{9}
350	55.6	1.05	0.105×10^{-3}	0.189×10^{-5}	0.391	1.02	6.9×10^{-4}	6.24×10^{9}
400	53.6	1.08	0.091×10^{-3}	0.170×10^{-5}	0.381	0.927	8.0×10^{-4}	8.95×10^{9}
450	51.6	1.12	0.080×10^{-3}	0.155×10^{-5}	0.367	0.876	9.0×10^{-4}	12.1×10^{9}
500	49.0	1.19	0.071×10^{-3}	0.145×10^{-5}	0.349	0.87	10.0×10^{-4}	15.3×10^{9}
550	45.9	1.31	0.064×10^{-3}	0.139×10^{-5}	0.325	0.93	11.0×10^{-4}	17.8×10^{9}
600	42.4	1.51	0.058×10^{-3}	0.137×10^{-5}	0.292	1.09	12.0×10^{-4}	20.6×10^{9}

(Multiply Btu/lbm-°F by 4.187 to obtain kJ/kg·K.)

(Multiply lbm/ft-sec by 3600 to obtain lbm/ft-hr.)

(Multiply lbm/ft-sec by 1.488 to obtain kg/m·s.)

(Multiply ft²/sec by 0.0929 to obtain m²/s.)

(Multiply Btu/hr-ft-°F by 1.730 to obtain W/m·K.)

(Multiply 1/°F by 5/9 to obtain 1/K.)

(Multiply 1/ft³-°F by 19.611 to obtain 1/m³·K.)

Source: *Handbook of Thermodynamic Tables and Charts* by K. Raznjevic, copyright © 1976.

Support Material

APPENDIX 34.B
Properties of Saturated Water (SI units)

T (°C)	T (K)	ρ (kg/m³)	c_p (kJ/kg·K)	μ (kg/m·s)	k (W/m·K)	Pr	β (1/K)	$\frac{g\beta\rho^2}{\mu^2}$ (1/m³·K)
0	273.2	999.6	4.229	1.786×10^{-3}	0.5694	13.3	-0.630×10^{-4}	
15.6	288.8	998.0	4.187	1.131×10^{-3}	0.5884	8.07	1.44×10^{-4}	10.93×10^{8}
26.7	299.9	996.4	4.183	0.860×10^{-3}	0.6109	5.89	2.34×10^{-4}	30.70×10^{8}
37.8	311.0	994.7	4.183	0.682×10^{-3}	0.6283	4.51	3.24×10^{-4}	68.0×10^{8}
65.6	338.8	981.9	4.187	0.432×10^{-3}	0.6629	2.72	5.04×10^{-4}	256.2×10^{8}
93.3	366.5	962.7	4.229	0.3066×10^{-3}	0.6802	1.91	6.66×10^{-4}	642×10^{8}
121.1	394.3	943.5	4.271	0.2381×10^{-3}	0.6836	1.49	8.46×10^{-4}	1300×10^{8}
148.9	422.1	917.9	4.312	0.1935×10^{-3}	0.6836	1.22	10.08×10^{-4}	2231×10^{8}
204.4	477.6	858.6	4.522	0.1384×10^{-3}	0.6611	0.950	14.04×10^{-4}	5308×10^{8}
260.0	533.2	784.9	4.982	0.1042×10^{-3}	0.6040	0.859	19.8×10^{-4}	$11\,030 \times 10^{8}$
315.6	588.8	679.4	6.322	0.0862×10^{-3}	0.5071	1.07	31.5×10^{-4}	$19\,260 \times 10^{8}$

(Divide kg/m³ by 16.0185 to obtain lbm/ft³.)

(Divide kJ/kg·K by 4.187 to obtain Btu/lbm-°F.)

(Divide kg/m·s by 1.488 to obtain lbm/ft-sec.)

(Divide W/m·K by 1.730 to obtain Btu/hr-ft-°F.)

(Divide 1/K by 5/9 to obtain 1/°F.)

(Divide 1/m³·K by 19.611 to obtain 1/ft³-°F.)

Support Material

APPENDIX 34.C
Properties of Atmospheric Air*
(customary U.S. units)

T (°F)	ρ (lbm/ft³)	c_p (Btu/lbm-°F)	μ (lbm/ft-sec)	ν (ft²/sec)	k (Btu/hr-ft-°F)	Pr	β (1/°F)	$\frac{g\beta\rho^2}{\mu^2}$ (1/ft³-°F)
0	0.086	0.239	1.110×10^{-5}	0.130×10^{-3}	0.0133	0.73	2.18×10^{-3}	4.2×10^{6}
32	0.081	0.240	1.165×10^{-5}	0.145×10^{-3}	0.0140	0.72	2.03×10^{-3}	3.16×10^{6}
100	0.071	0.240	1.285×10^{-5}	0.180×10^{-3}	0.0154	0.72	1.79×10^{-3}	1.76×10^{6}
200	0.060	0.241	1.440×10^{-5}	0.239×10^{-3}	0.0174	0.72	1.52×10^{-3}	0.850×10^{6}
300	0.052	0.243	1.610×10^{-5}	0.306×10^{-3}	0.0193	0.71	1.32×10^{-3}	0.444×10^{6}
400	0.046	0.245	1.75×10^{-5}	0.378×10^{-3}	0.0212	0.689	1.16×10^{-3}	0.258×10^{6}
500	0.0412	0.247	1.890×10^{-5}	0.455×10^{-3}	0.0231	0.683	1.04×10^{-3}	0.159×10^{6}
600	0.0373	0.250	2.000×10^{-5}	0.540×10^{-3}	0.0250	0.685	0.943×10^{-3}	0.106×10^{6}
700	0.0341	0.253	2.14×10^{-5}	0.625×10^{-3}	0.0268	0.690	0.862×10^{-3}	70.4×10^{3}
800	0.0314	0.256	2.25×10^{-5}	0.717×10^{-3}	0.0286	0.697	0.794×10^{-3}	49.8×10^{3}
900	0.0291	0.259	2.36×10^{-5}	0.815×10^{-3}	0.0303	0.705	0.735×10^{-3}	36.0×10^{3}
1000	0.0271	0.262	2.47×10^{-5}	0.917×10^{-3}	0.0319	0.713	0.685×10^{-3}	26.5×10^{3}
1500	0.0202	0.276	3.00×10^{-5}	1.47×10^{-3}	0.0400	0.739	0.510×10^{-3}	7.45×10^{3}
2000	0.0161	0.286	3.54×10^{-5}	2.14×10^{-3}	0.0471	0.753	0.406×10^{-3}	2.84×10^{3}
2500	0.0133	0.292	3.69×10^{-5}	2.80×10^{-3}	0.051	0.763	0.338×10^{-3}	1.41×10^{3}
3000	0.0114	0.297	3.85×10^{-5}	3.39×10^{-3}	0.054	0.765	0.289×10^{-3}	0.815×10^{3}

(Multiply lbm/ft³ by 16.0185 to obtain kg/m³.)

(Multiply Btu/lbm-°F by 4.187 to obtain kJ/kg·K.)

(Multiply lbm/ft-sec by 3600 to obtain lbm/ft-hr.)

(Multiply lbm/ft-sec by 1.488 to obtain kg/m·s.)

(Multiply Btu/hr-ft-°F by 1.730 to obtain W/m·K.)

(Multiply 1/°F by 5/9 to obtain 1/K.)

(Multiply 1/ft³-°F by 19.611 to obtain 1/m³·K.)

(Multiply ft²/sec by 0.0929 to obtain m²/s.)

*μ, k, c_p, and Pr do not greatly depend on pressure and may be used over a wide range of pressures.

Source: *Handbook of Thermodynamic Tables and Charts* by K. Raznjevic, copyright © 1976.

APPENDIX 34.D
Properties of Atmospheric Air*
(SI units)

T (°C)	T (K)	ρ (kg/m³)	c_p (kJ/kg·K)	μ (kg/m·s)	k (W/m·K)	Pr	β (1/K)	$\frac{g\beta\rho^2}{\mu^2}$ (1/m³·K)
−17.8	255.4	1.379	1.0048	1.62×10^{-5}	0.02250	0.720	3.92×10^{-3}	2.79×10^{8}
0	273.2	1.293	1.0048	1.72×10^{-5}	0.02423	0.715	3.65×10^{-3}	2.04×10^{8}
10.0	283.2	1.246	1.0048	1.78×10^{-5}	0.02492	0.713	3.53×10^{-3}	1.72×10^{8}
37.8	311.0	1.137	1.0048	1.90×10^{-5}	0.02700	0.705	3.22×10^{-3}	1.12×10^{8}
65.6	338.8	1.043	1.0090	2.03×10^{-5}	0.02925	0.702	2.95×10^{-3}	0.775×10^{8}
93.3	366.5	0.964	1.0090	2.15×10^{-5}	0.03115	0.694	2.74×10^{-3}	0.534×10^{8}
121.1	394.3	0.895	1.0132	2.27×10^{-5}	0.03323	0.692	2.54×10^{-3}	0.386×10^{8}
148.9	422.1	0.838	1.0174	2.37×10^{-5}	0.03531	0.689	2.38×10^{-3}	0.289×10^{8}
176.7	449.9	0.785	1.0216	2.50×10^{-5}	0.03721	0.687	2.21×10^{-3}	0.214×10^{8}
204.4	477.6	0.740	1.0258	2.60×10^{-5}	0.03894	0.686	2.09×10^{-3}	0.168×10^{8}
232.2	505.4	0.700	1.0300	2.71×10^{-5}	0.04084	0.684	1.98×10^{-3}	0.130×10^{8}
260.0	533.2	0.662	1.0341	2.80×10^{-5}	0.04258	0.680	1.87×10^{-3}	0.104×10^{8}

(Divide kg/m³ by 16.0185 to obtain lbm/ft³.)

(Divide kJ/kg·K by 4.187 to obtain Btu/lbm-°F.)

(Divide kg/m·s by 1.488 to obtain lbm/ft-sec.)

(Divide W/m·K by 1.730 to obtain Btu/hr-ft-°F.)

(Divide 1/K by 5/9 to obtain 1/°F.)

(Divide 1/m³·K by 19.611 to obtain 1/ft³-°F.)

Support Material

APPENDIX 34.E
Properties of Steam at One Atmosphere*
(customary U.S. units)

T (°F)	ρ (lbm/ft^3)	c_p (Btu/lbm-°F)	μ (lbm/ft-sec)	ν (ft^2/sec)	k (Btu/hr-ft-°F)	Pr	β (1/°F)	$\frac{g\beta\rho^2}{\mu^2}$ (1/ft^3-°F)
212	0.0372	0.451	0.870×10^{-5}	0.234×10^{-3}	0.0145	0.96	1.49×10^{-3}	0.877×10^6
300	0.0328	0.456	1.00×10^{-5}	0.303×10^{-3}	0.0171	0.95	1.32×10^{-3}	0.459×10^6
400	0.0288	0.462	1.13×10^{-5}	0.395×10^{-3}	0.0200	0.94	1.16×10^{-3}	0.243×10^6
500	0.0258	0.470	1.265×10^{-5}	0.490×10^{-3}	0.0228	0.94	1.04×10^{-3}	0.139×10^6
600	0.0233	0.477	1.420×10^{-5}	0.610×10^{-3}	0.0257	0.94	0.943×10^{-3}	82×10^3
700	0.0213	0.485	1.555×10^{-5}	0.725×10^{-3}	0.0288	0.93	0.862×10^{-3}	52.1×10^3
800	0.0196	0.494	1.70×10^{-5}	0.855×10^{-3}	0.0321	0.92	0.794×10^{-3}	34.0×10^3
900	0.0181	0.50	1.810×10^{-5}	0.987×10^{-3}	0.0355	0.91	0.735×10^{-3}	23.6×10^3
1000	0.0169	0.51	1.920×10^{-5}	1.13×10^{-3}	0.0388	0.91	0.685×10^{-3}	17.1×10^3
1200	0.0149	0.53	2.14×10^{-5}	1.44×10^{-3}	0.0457	0.88	0.603×10^{-3}	9.4×10^3
1400	0.0133	0.55	2.36×10^{-5}	1.78×10^{-3}	0.053	0.87	0.537×10^{-3}	5.49×10^3
1600	0.0120	0.56	2.58×10^{-5}	2.14×10^{-3}	0.061	0.87	0.485×10^{-3}	3.38×10^3
1800	0.0109	0.58	2.81×10^{-5}	2.58×10^{-3}	0.068	0.87	0.442×10^{-3}	2.14×10^3
2000	0.0100	0.60	3.03×10^{-5}	3.03×10^{-3}	0.076	0.86	0.406×10^{-3}	1.43×10^3
2500	0.0083	0.64	3.58×10^{-5}	4.30×10^{-3}	0.096	0.86	0.338×10^{-3}	0.603×10^3
3000	0.0071	0.67	4.00×10^{-5}	5.75×10^{-3}	0.114	0.86	0.289×10^{-3}	0.293×10^3

(Multiply lbm/ft^3 by 16.0185 to obtain kg/m^3.)

(Multiply Btu/lbm-°F by 4.187 to obtain kJ/kg·K.)

(Multiply lbm/ft-sec by 3600 to obtain lbm/ft-sec.)

(Multiply lbm/ft-sec by 1.488 to obtain kg/m·s.)

(Multiply ft^2/sec by 0.0929 to obtain m^2/s.)

(Multiply Btu/hr-ft-°F by 1.730 to obtain W/m·K.)

(Multiply 1/°F by 5/9 to obtain 1/K.)

(Multiply 1/ft^3-°F by 19.611 to obtain 1/m^3·K.)

*μ, k, c_p, and Pr do not greatly depend on pressure and may be used over a wide range of pressures.

Support Material

APPENDIX 34.F
Properties of Steam at One Atmosphere*
(SI units)

T (°C)	T (K)	ρ (kg/m³)	c_p (kJ/kg·K)	μ (kg/m·s)	k (W/m·K)	Pr	β (1/K)	$\frac{g\beta\rho^2}{\mu^2}$ (1/m³·K)
100.0	373.2	0.596	1.888	1.295×10^{-5}	0.02510	0.96	2.68×10^{-3}	0.557×10^{8}
148.9	422.1	0.525	1.909	1.488×10^{-5}	0.02960	0.95	2.38×10^{-3}	0.292×10^{8}
204.4	477.6	0.461	1.934	1.682×10^{-5}	0.03462	0.94	2.09×10^{-3}	0.154×10^{8}
260.0	533.2	0.413	1.968	1.883×10^{-5}	0.03946	0.94	1.87×10^{-3}	0.0883×10^{8}
315.6	588.8	0.373	1.997	2.113×10^{-5}	0.04448	0.94	1.70×10^{-3}	52.1×10^{5}
371.1	644.3	0.341	2.030	2.314×10^{-5}	0.04985	0.93	1.55×10^{-3}	33.1×10^{5}
426.7	699.9	0.314	2.068	2.529×10^{-5}	0.05556	0.92	1.43×10^{-3}	21.6×10^{5}

(Divide kg/m³ by 16.0185 to obtain lbm/ft³.)

(Divide kJ/kg·K by 4.187 to obtain Btu/lbm-°F.)

(Divide kg/m·s by 1.488 to obtain lbm/ft-sec.)

(Divide W/m·K by 1.730 to obtain Btu/hr-ft-°F.)

(Divide 1/K by 5/9 to obtain 1/°F.)

(Divide 1/m³·K by 19.611 to obtain 1/ft³-°F.)

*μ, k, c_p, and Pr do not greatly depend on pressure and may be used over a wide range of pressures.

Support Material

APPENDIX 35.A
Correction Factor, F_c, for the Logarithmic Mean Temperature Difference
(one shell pass, even number of tube passes)

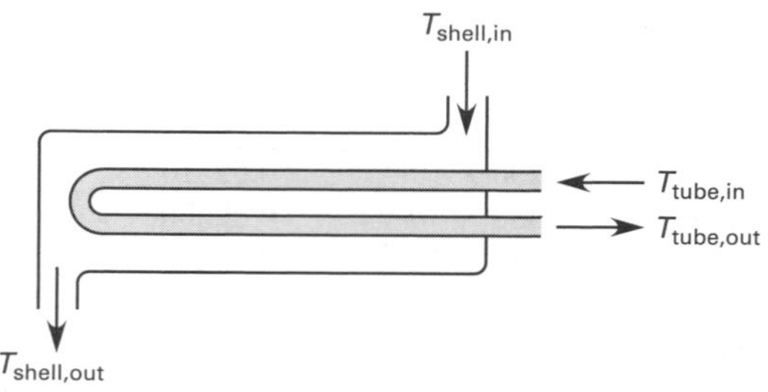

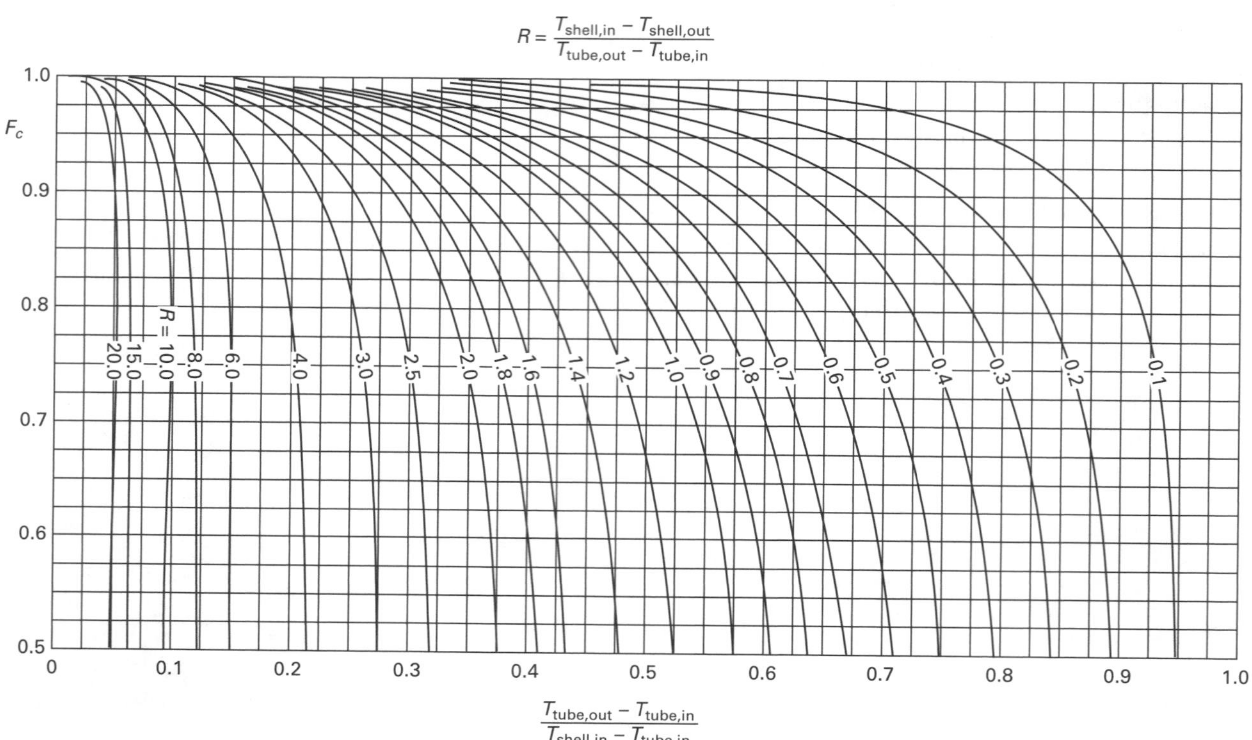

Support Material

APPENDIX 35.B

Correction Factor, F_c, for the Logarithmic Mean Temperature Difference
(two shell passes, multiple of four tube passes)

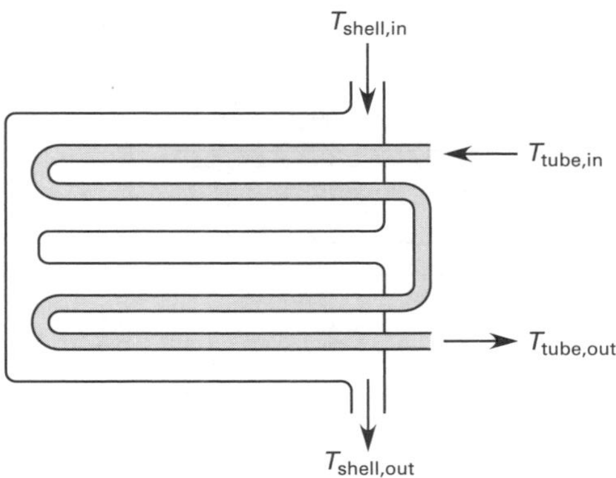

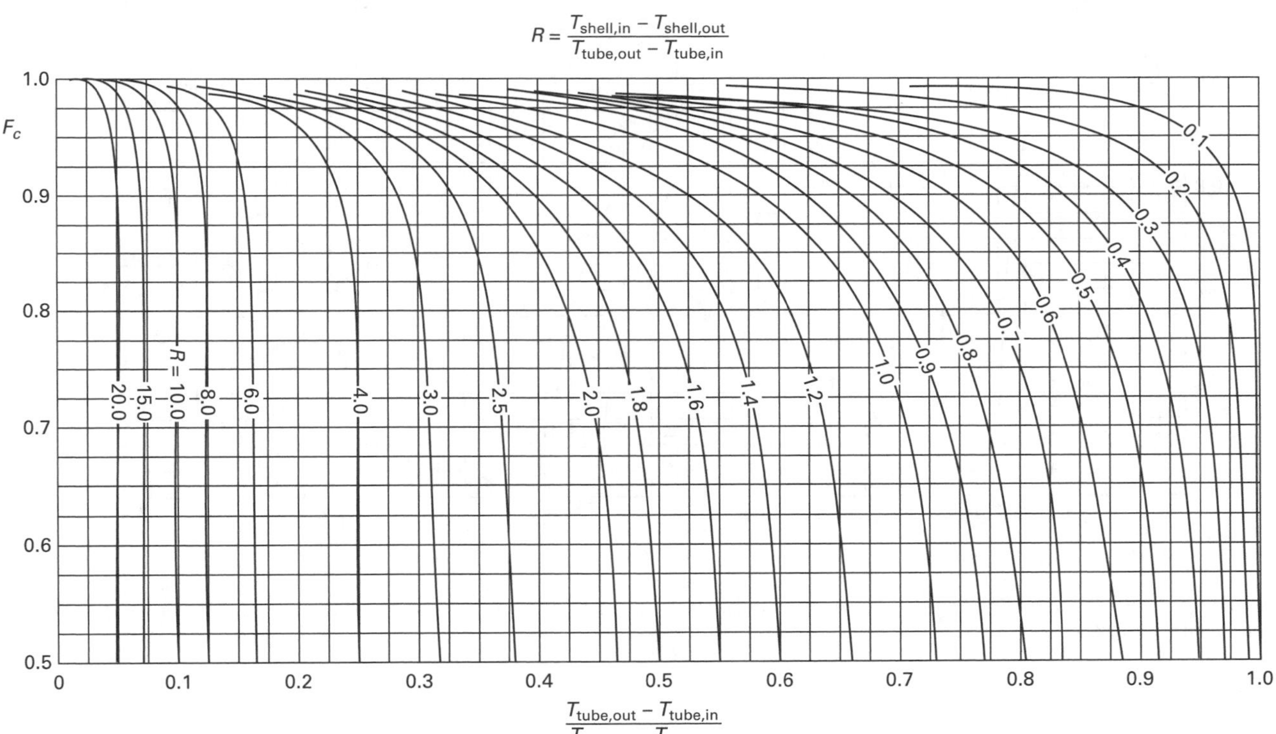

Support Material

APPENDIX 35.C

Correction Factor, F_c, for the Logarithmic Mean Temperature Difference
(one shell pass, three tube passes, two counter and one cocurrent)

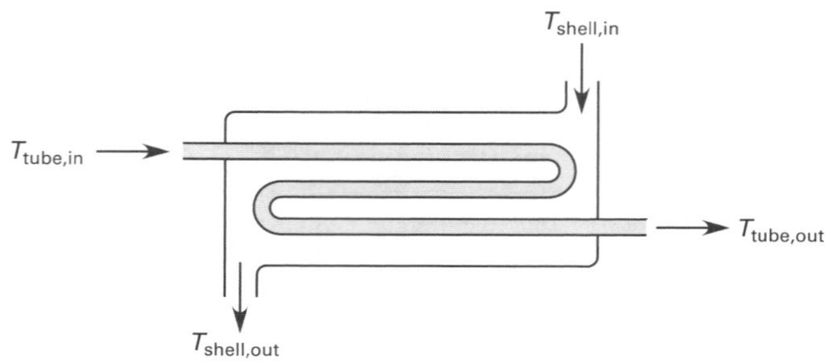

$$R = \frac{T_{shell,in} - T_{shell,out}}{T_{tube,out} - T_{tube,in}}$$

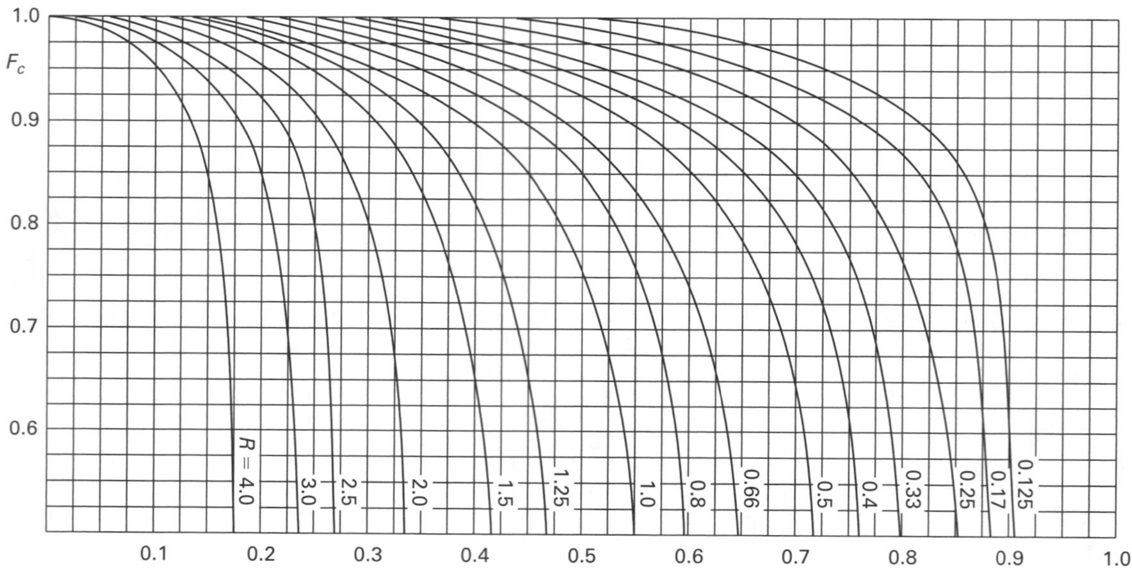

Support Material

APPENDIX 35.D

Correction Factor, F_c, for the Logarithmic Mean Temperature Difference
(crossflow shell, one tube pass, one fluid mixed)

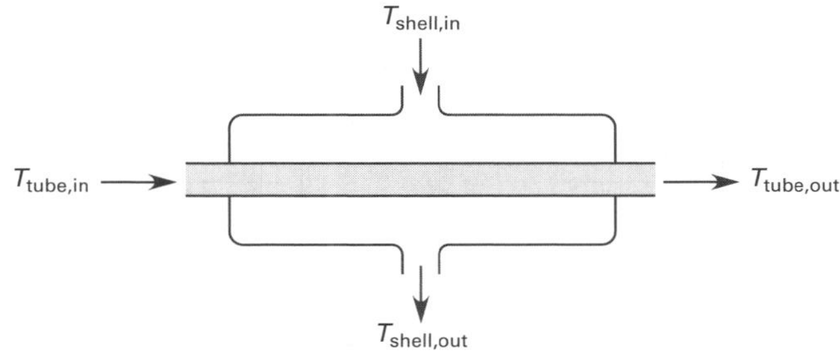

$$R = \frac{T_{shell,in} - T_{shell,out}}{T_{tube,out} - T_{tube,in}}$$

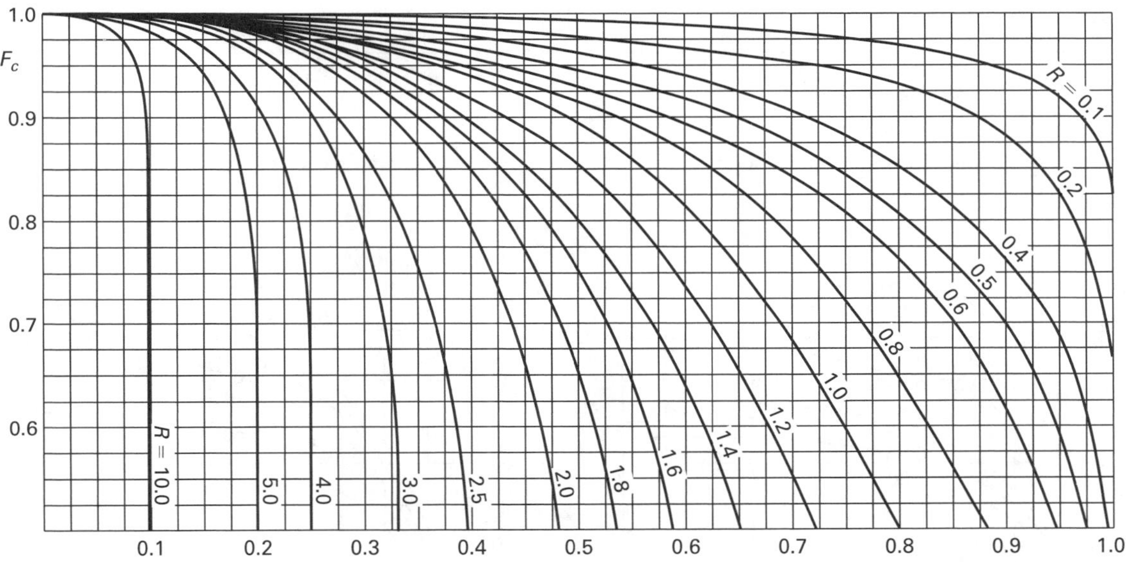

Support Material

APPENDIX 35.E
Heat Exchanger Effectiveness [a,b]
(%)

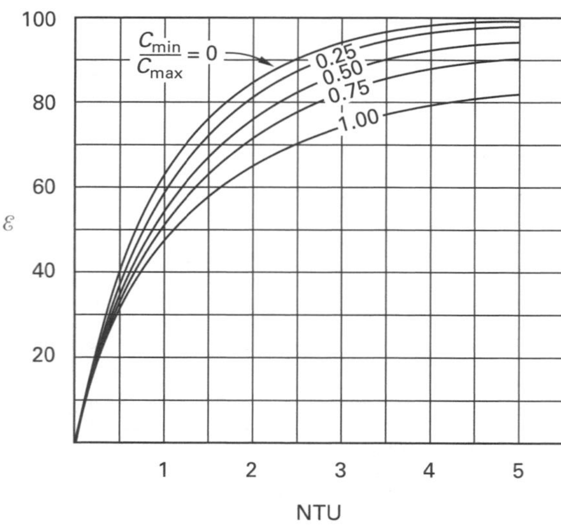

(a) single-pass, shell-and-tube, counterflow

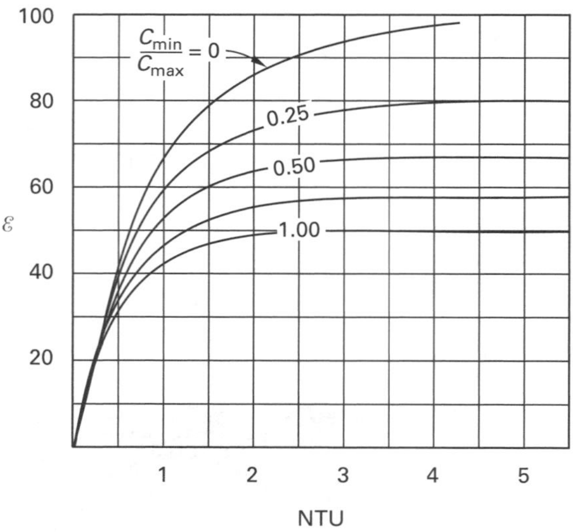

(b) single-pass, shell-and-tube, parallel flow

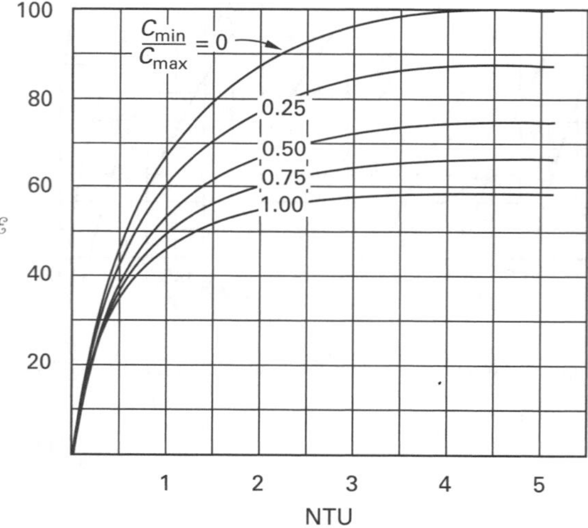

(c) one shell pass, two, four, six, eight . . . tube passes, parallel counterflow[a]

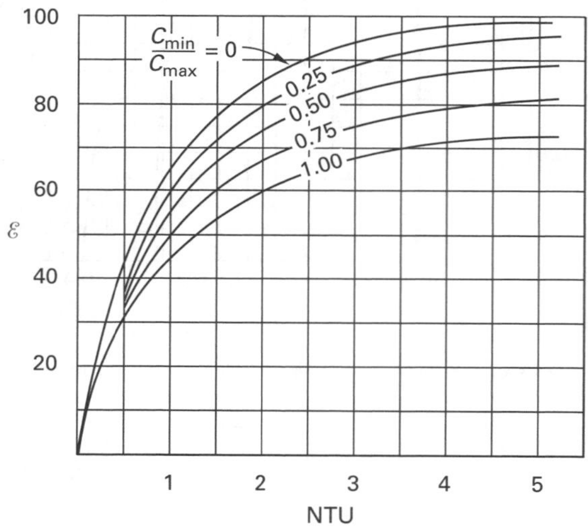

(d) two shell passes, four, eight, or twelve . . . tube passes, multipass counterflow

Support Material

APPENDIX 35.E *(continued)*
Heat Exchanger Effectiveness [a,b]
(%)

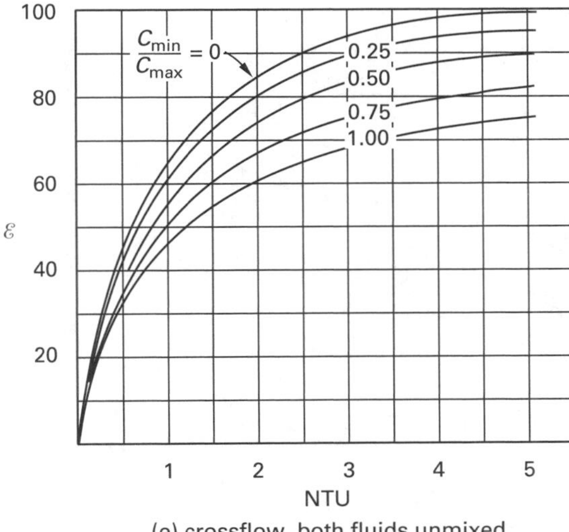

(e) crossflow, both fluids unmixed

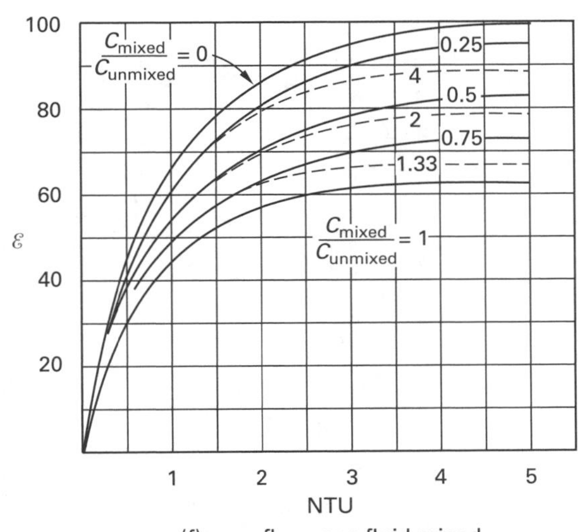

(f) crossflow, one fluid mixed

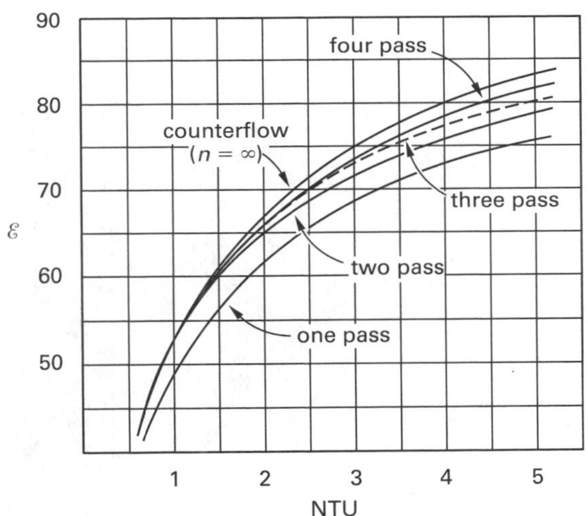

(g) crossflow, multiple passes, both fluids unmixed

[a]Formulas used to generate the curves for one shell pass are exact for two, four, and six tube passes, but produce a small error (approximately 1% to 5%) for higher multiples. The error associated with reading the graphs is generally higher than the error introduced by the formulas.
[b]Most formulas require effectiveness to be given as a decimal, not a percentage.

APPENDIX 37.A
ASHRAE Psychrometric Chart No. 1, Normal Temperature—Sea Level (32–120°F)
(customary U.S. units)

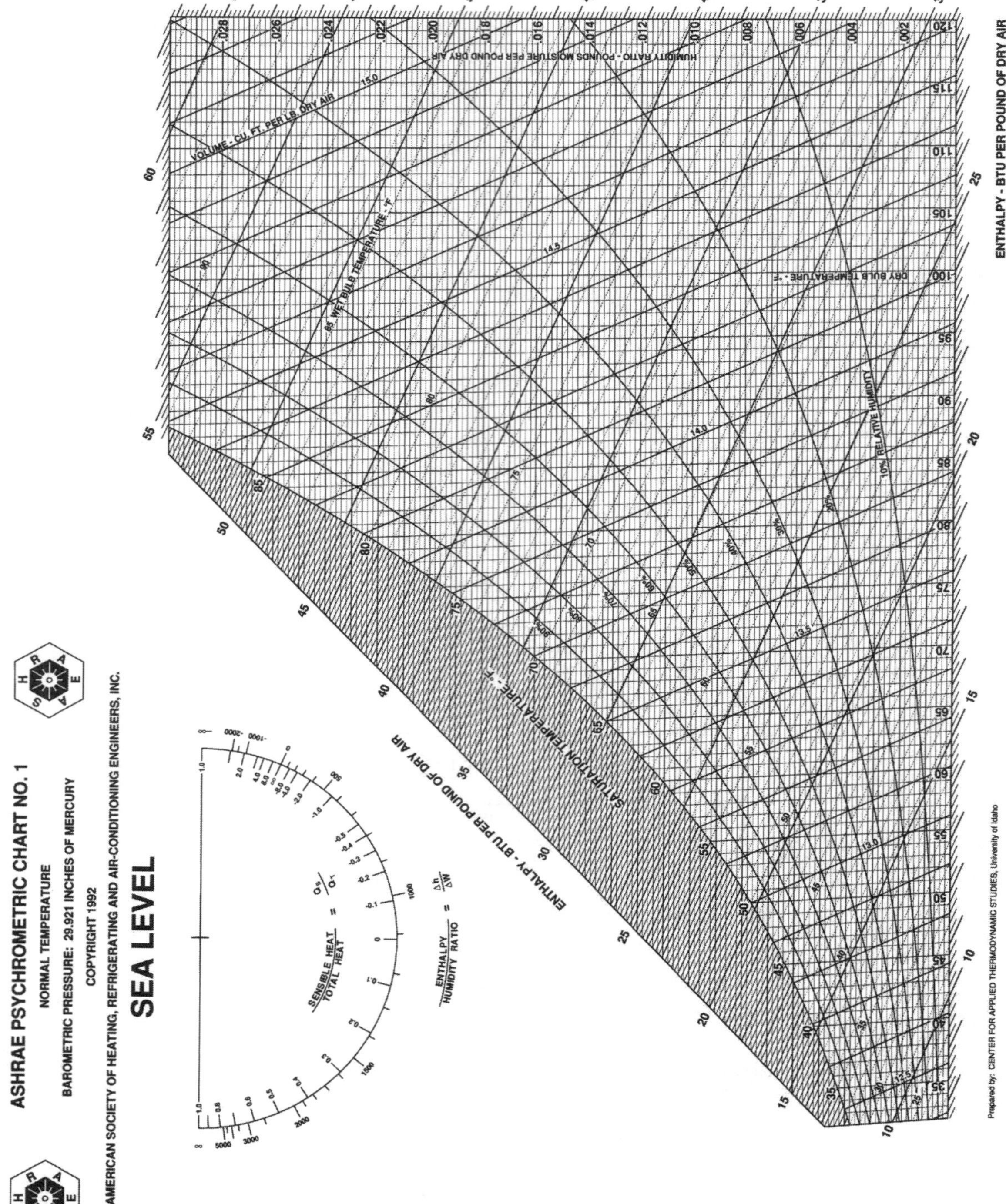

Support Material

APPENDIX 37.B

ASHRAE Psychrometric Chart No. 1, Normal Temperature—Sea Level (0–50°C)
(SI units)

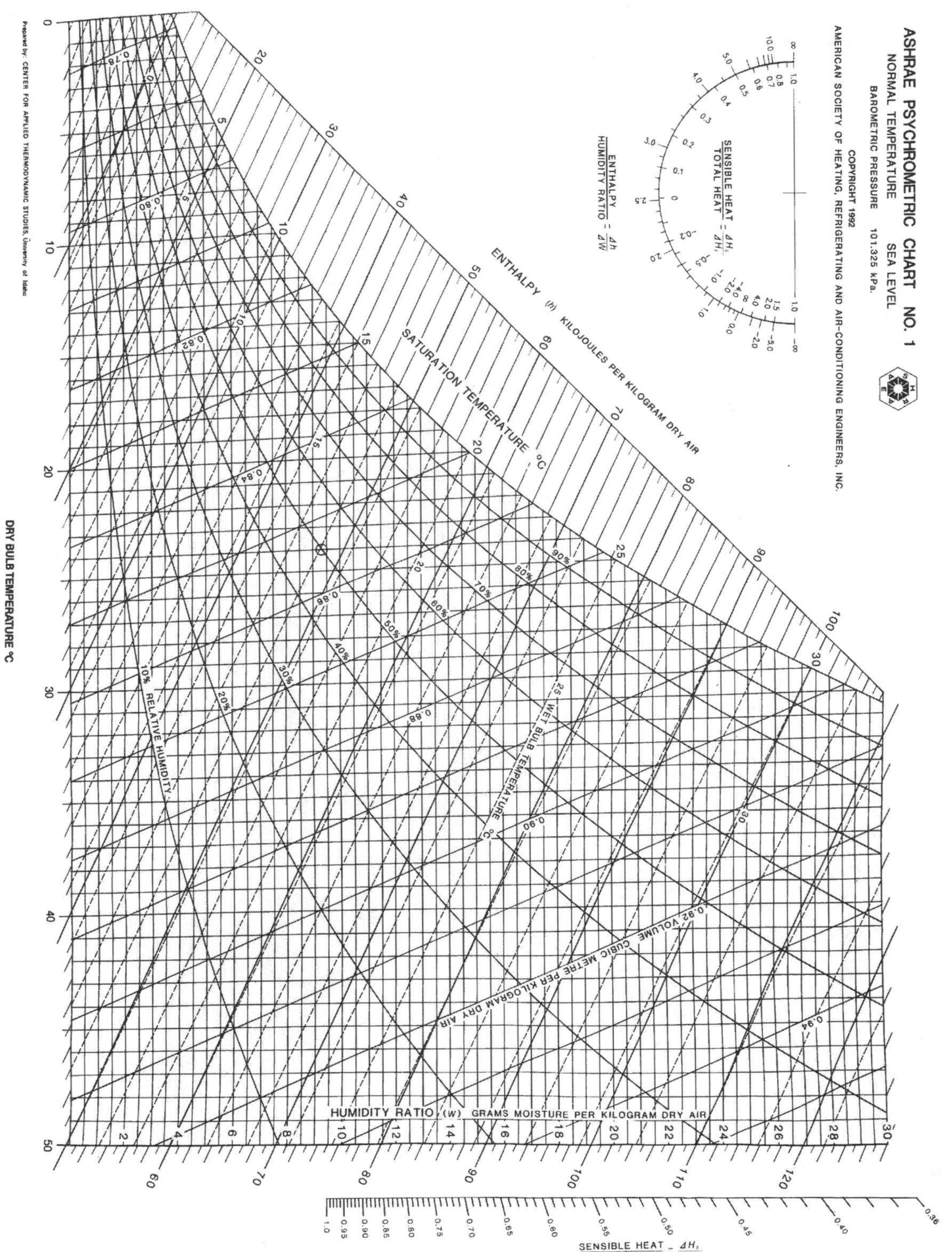

Support Material

APPENDIX 37.C

ASHRAE Psychrometric Chart No. 2, Low Temperature—Sea Level (–40–50°F)
(customary U.S. units)

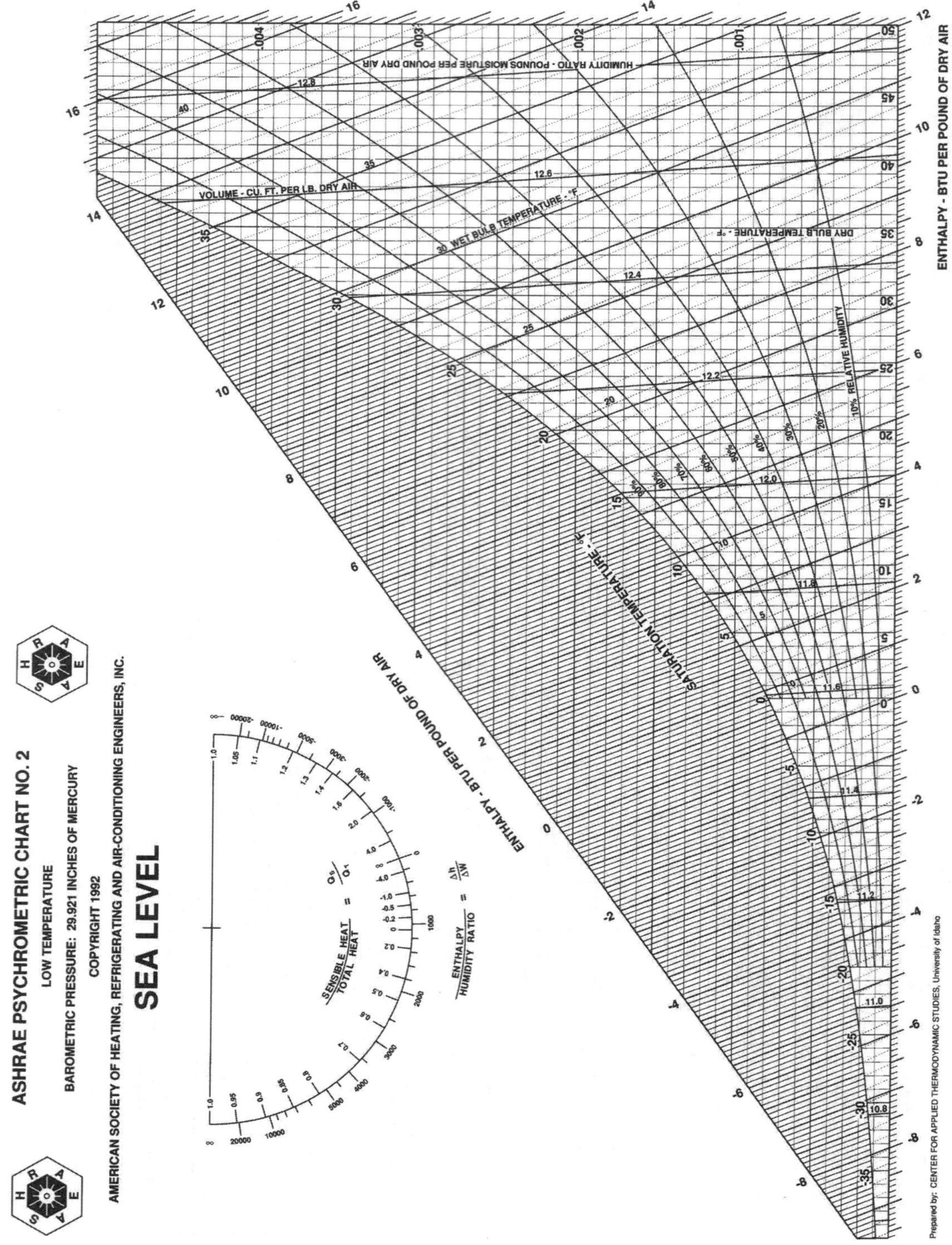

APPENDIX 37.D

ASHRAE Psychrometric Chart No. 3, High Temperature—Sea Level (60–250°F) (customary U.S. units)

ASHRAE PSYCHROMETRIC CHART NO. 3

HIGH TEMPERATURE

BAROMETRIC PRESSURE: 29.921 INCHES OF MERCURY

SEA LEVEL

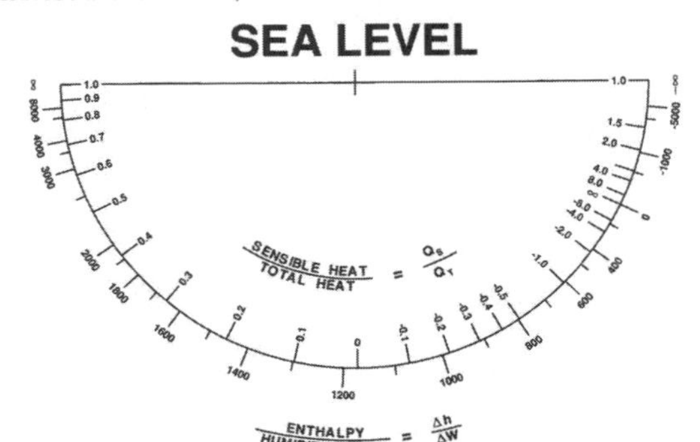

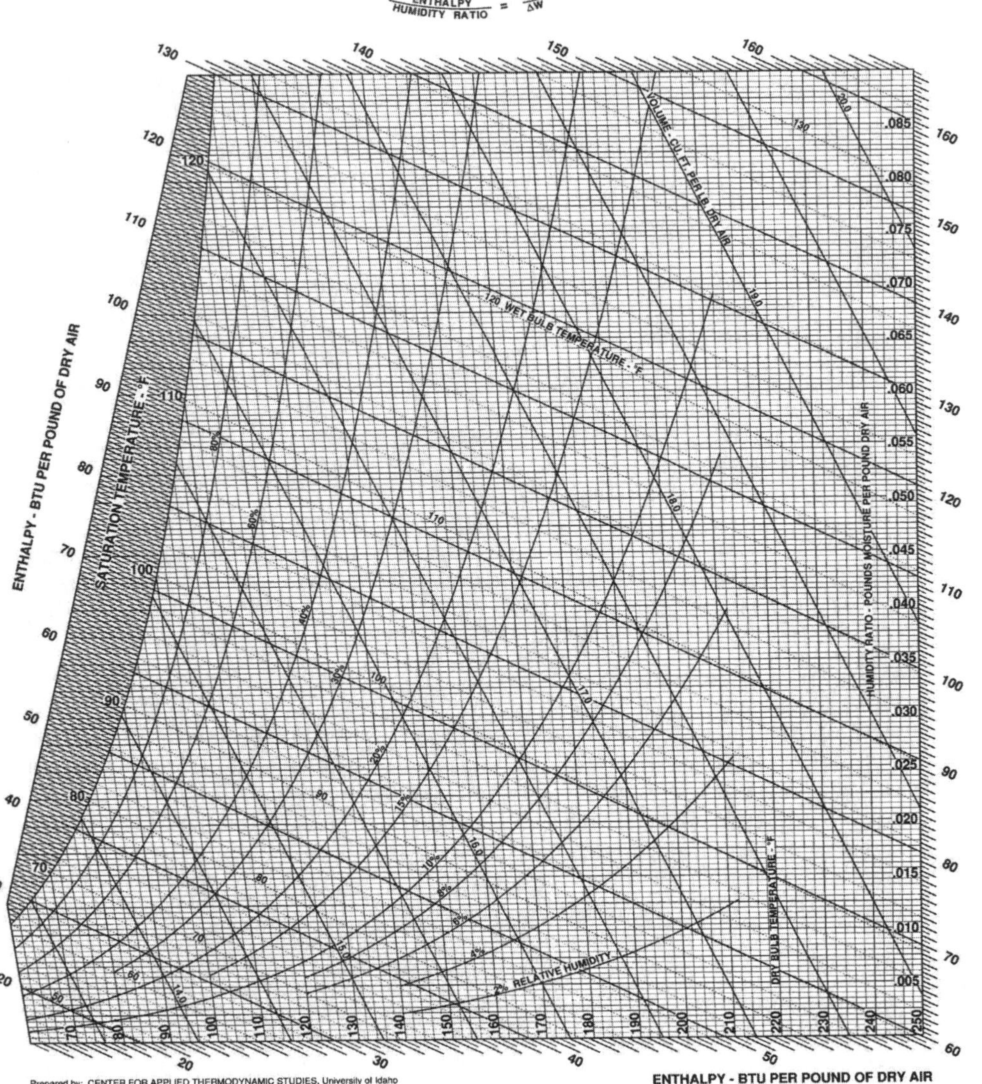

Support Material

APPENDIX 53.A
Spring Wire Diameters and Sheet Metal Gauges*
(inches)

no. of gauge	Washburn and Moen	American, or Brown and Sharpe	Birmingham or Stubs	U.S. standard for plate (iron and steel)	Stubs steel wire	British imperial wire gauge (S.W.G.)	Morse twist drill and steel wire	wood and machine screws	American S. & W. piano and music wire
7–0	0.4900	–	–	–	–	0.500	–	–	–
6–0	0.4615	0.5800	–	–	–	0.464	–	–	–
5–0	0.4305	0.5165	0.500	–	–	0.432	–	–	0.005
4–0	0.3938	0.4600	0.454	–	–	0.400	–	–	0.006
3–0	0.3625	0.4096	0.425	–	–	0.372	–	0.032	0.007
2–0	0.3310	0.3648	0.380	–	–	0.348	–	0.045	0.008
0	0.3065	0.3249	0.340	0.313	–	0.324	–	0.058	0.009
1	0.2830	0.2893	0.300	0.281	0.227	0.300	0.228	0.071	0.010
2	0.2625	0.2576	0.284	0.265	0.219	0.276	0.221	0.084	0.011
3	0.2437	0.2294	0.259	0.250	0.212	0.252	0.213	0.097	0.012
4	0.2253	0.2043	0.238	0.234	0.207	0.232	0.209	0.110	0.013
5	0.2070	0.1819	0.220	0.219	0.204	0.212	0.206	0.124	0.014
6	0.1920	0.1620	0.203	0.203	0.201	0.192	0.204	0.137	0.016
7	0.1770	0.1442	0.180	0.188	0.199	0.176	0.201	0.150	0.018
8	0.1620	0.1284	0.165	0.172	0.197	0.160	0.199	0.163	0.020
9	0.1483	0.1144	0.148	0.156	0.194	0.144	0.196	0.176	0.022
10	0.1350	0.1018	0.134	0.141	0.191	0.128	0.194	0.189	0.024
11	0.1205	0.0907	0.120	0.125	0.188	0.116	0.191	0.203	0.026
12	0.1055	0.0808	0.109	0.109	0.185	0.104	0.189	0.216	0.029
13	0.0915	0.0719	0.095	0.094	0.182	0.092	0.185	0.229	0.031
14	0.0800	0.0640	0.083	0.078	0.180	0.080	0.182	0.242	0.033
15	0.0720	0.0570	0.072	0.070	0.178	0.072	0.180	0.255	0.035
16	0.0625	0.0508	0.065	0.063	0.175	0.064	0.177	0.268	0.037
17	0.0540	0.0452	0.058	0.056	0.172	0.056	0.173	0.282	0.039
18	0.0475	0.0403	0.049	0.050	0.168	0.048	0.170	0.295	0.041
19	0.0410	0.0358	0.042	0.044	0.164	0.040	0.166	0.308	0.043
20	0.0348	0.0319	0.035	0.038	0.161	0.036	0.161	0.321	0.045
21	0.0317	0.0284	0.032	0.034	0.157	0.032	0.159	0.334	0.047
22	0.0286	0.0253	0.028	0.031	0.155	0.028	0.157	0.347	0.049
23	0.0258	0.0225	0.025	0.028	0.153	0.024	0.154	0.360	0.051
24	0.0230	0.0201	0.022	0.025	0.151	0.022	0.152	0.374	0.055
25	0.0204	0.0179	0.020	0.022	0.148	0.020	0.150	0.387	0.059
26	0.0181	0.0159	0.018	0.019	0.146	0.018	0.147	0.400	0.063
27	0.0173	0.0141	0.016	0.017	0.143	0.0164	0.144	0.413	0.067
28	0.0162	0.0126	0.014	0.016	0.139	0.0149	0.141	0.426	0.071
29	0.0150	0.0112	0.013	0.014	0.134	0.0136	0.136	0.439	0.075
30	0.0140	0.0100	0.012	0.013	0.127	0.0124	0.129	0.453	0.080
31	0.0132	0.0089	0.010	0.011	0.120	0.0116	0.120	0.466	0.085
32	0.0128	0.0079	0.009	0.010	0.115	0.0108	0.116	0.479	0.090
33	0.0118	0.0070	0.008	0.009	0.112	0.0100	0.133	0.492	0.095
34	0.0104	0.0063	0.007	0.0086	0.110	0.0092	0.111	0.505	0.100
35	0.0095	0.0056	0.005	0.0078	0.108	0.0084	0.110	0.518	0.106
36	0.0090	0.0050	0.004	0.007	0.106	0.0076	0.1065	0.532	0.112
37	0.0085	0.0044	–	0.0066	0.103	0.0068	0.104	0.545	0.118
38	0.0080	0.0039	–	0.0062	0.101	0.0060	0.1015	0.558	0.124
39	0.0075	0.0035	–	–	0.099	0.0052	0.0995	0.571	0.130
40	0.0070	0.0031	–	–	0.097	0.0048	0.098	0.584	0.138
41	0.0066	0.00280	–	–	0.095	0.0044	0.096	0.597	0.146
42	0.0062	0.00249	–	–	0.092	0.0040	0.094	0.611	0.154
43	0.0060	0.00222	–	–	0.088	0.0036	0.089	0.624	0.162
44	0.0058	0.00198	–	–	0.085	0.0032	0.086	0.637	0.170

APPENDIX 53.A *(continued)*
Spring Wire Diameters and Sheet Metal Gauges*
(inches)

no. of gauge	Washburn and Moen	American, or Brown and Sharpe	Birmingham or Stubs	U.S. standard for plate (iron and steel)	Stubs steel wire	British imperial wire gauge (S.W.G.)	Morse twist drill and steel wire	wood and machine screws	American S. & W. piano and music wire
45	0.0055	0.00176	–	–	0.081	0.0028	0.082	0.650	0.180
46	0.0052	0.00157	–	–	0.079	0.0024	0.081	0.663	–
47	0.0050	0.00140	–	–	0.077	0.0020	0.079	0.676	–
48	0.0048	0.00124	–	–	0.075	0.0016	0.076	0.690	–
49	0.0046	0.00111	–	–	0.072	0.0012	0.073	0.703	–
50	0.0044	0.00099	–	–	0.069	0.0010	0.070	0.716	–

(Multiply in by 25.4 to obtain mm.)

*Birmingham Wire Gauge (BWG), commonly used for ferrous tubing, is identical with Stubs iron-wire gauge.

Support Material

APPENDIX 53.B
Journal Bearing Correlations

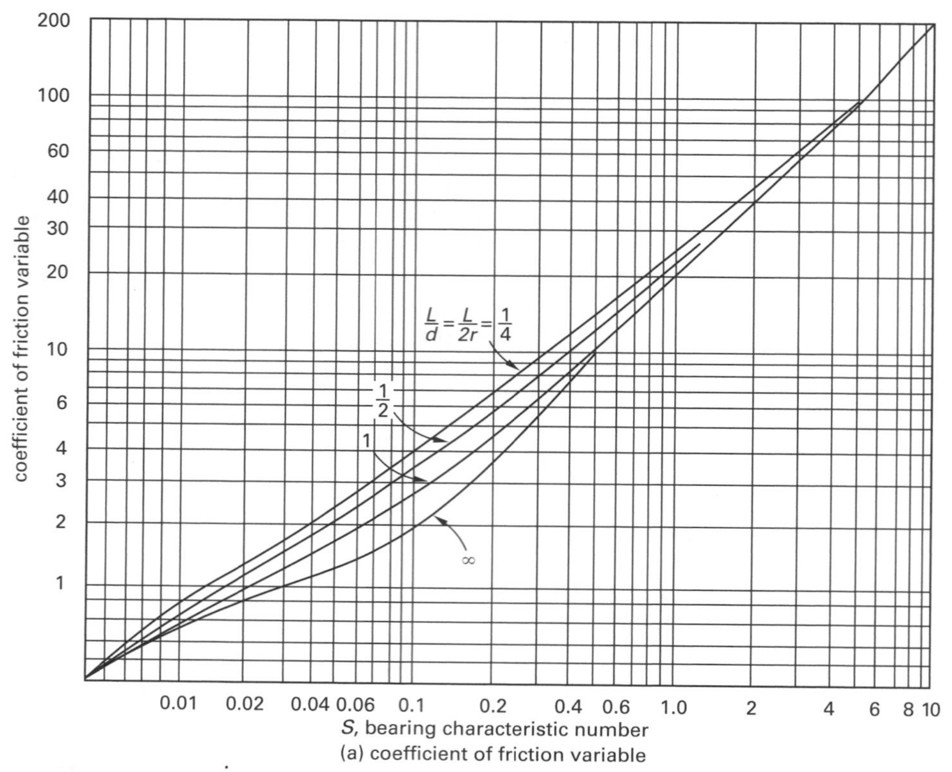

(a) coefficient of friction variable

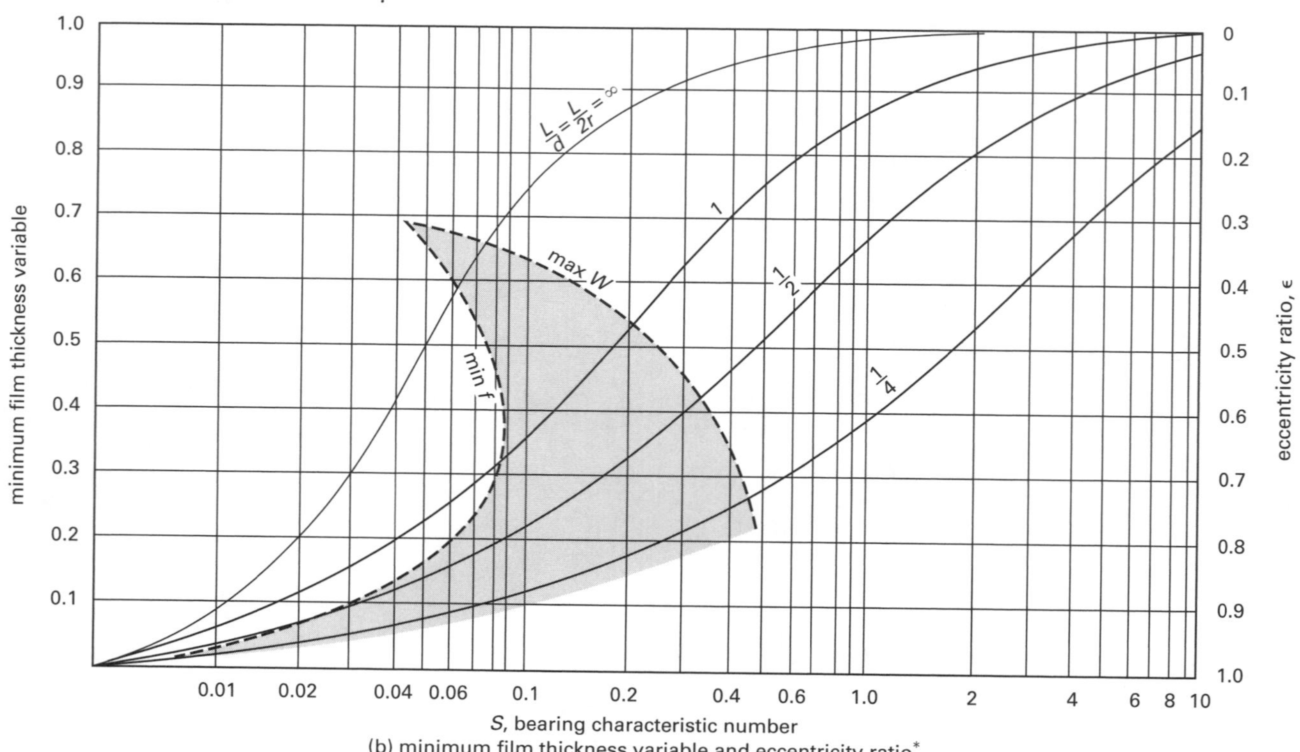

(b) minimum film thickness variable and eccentricity ratio*

*The left edge of the shaded area represents the optimum minimum thickness, resulting in the minimum friction. The right edge of the shaded area represents the optimum minimum thickness for maximum load-carrying ability.

Support Material

APPENDIX 55.A
Mass Moments of Inertia
(centroids at points labeled C)

slender rod	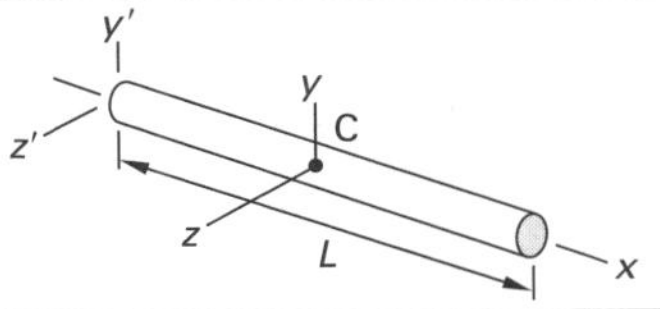	$I_y = I_z = \frac{mL^2}{12}$ $I_{y'} = I_{z'} = \frac{mL^2}{3}$
solid circular cylinder, radius r	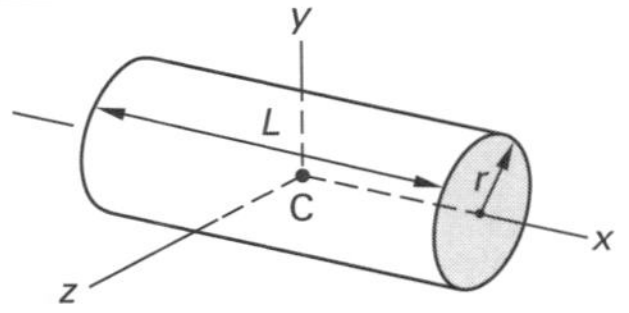	$I_x = \frac{mr^2}{2}$ $I_y = I_z = \frac{m(3r^2 + L^2)}{12}$
hollow circular cylinder, inner radius r_i, outer radius r_o	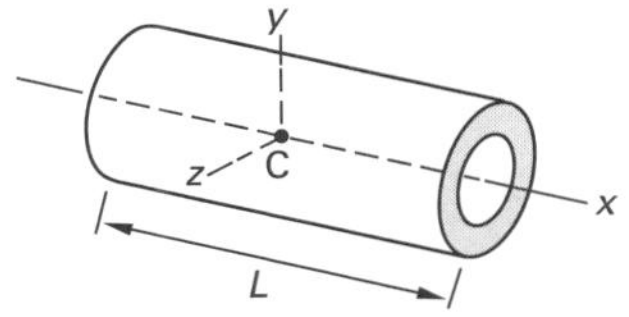	$I_x = \frac{m\left(r_o^2 + r_i^2\right)}{2}$ $= \left(\frac{\pi\rho L}{2}\right)\left(r_o^4 - r_i^4\right)$ $I_y = I_z = \frac{\pi\rho L}{12}\left(3(r_2^4 - r_1^4)\right.$ $\left.+L^2(r_2^2 - r_1^2)\right)$
thin disk, radius r	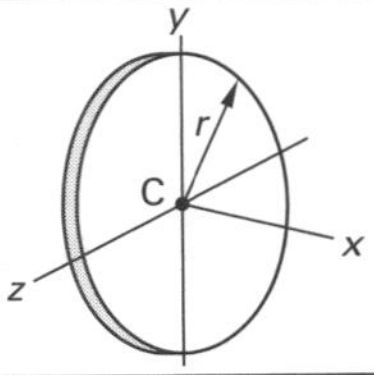	$I_x = \frac{mr^2}{2}$ $I_y = I_z = \frac{mr^2}{4}$
solid circular cone, base radius r	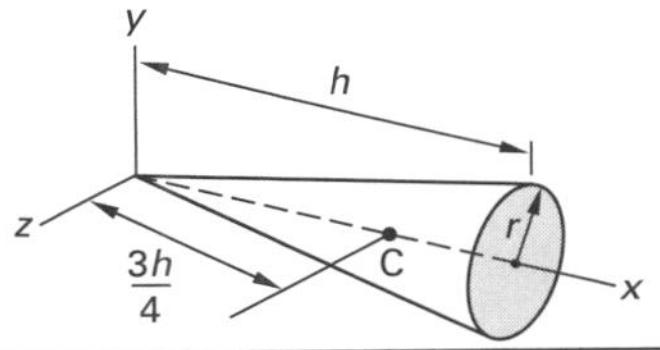	$I_x = \frac{3mr^2}{10}$ $I_y = I_z = \left(\frac{3m}{5}\right)\left(\frac{r^2}{4} + h^2\right)$
thin rectangular plate	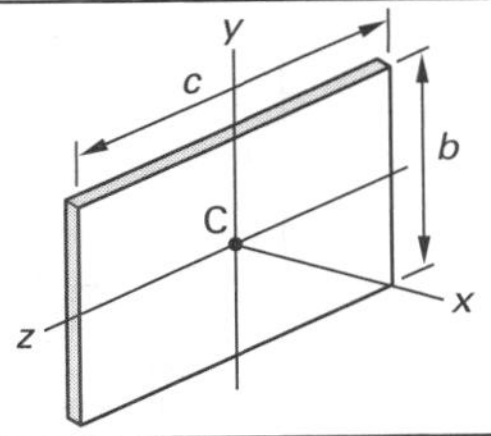	$I_x = \frac{m(b^2 + c^2)}{12}$ $I_y = \frac{mc^2}{12}$ $I_z = \frac{mb^2}{12}$
rectangular parallelepiped	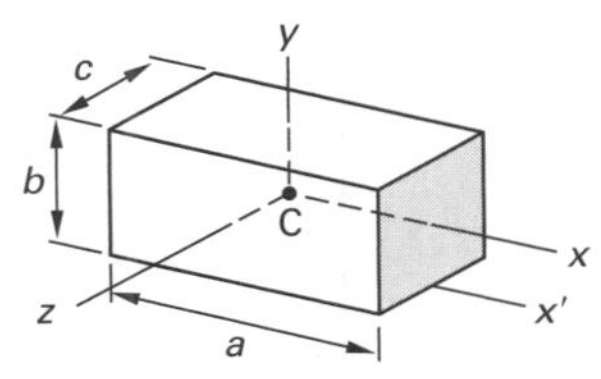	$I_x = \frac{m(b^2 + c^2)}{12}$ $I_y = \frac{m(c^2 + a^2)}{12}$ $I_z = \frac{m(a^2 + b^2)}{12}$ $I_{x'} = \frac{m(4b^2 + c^2)}{12}$
sphere, radius r	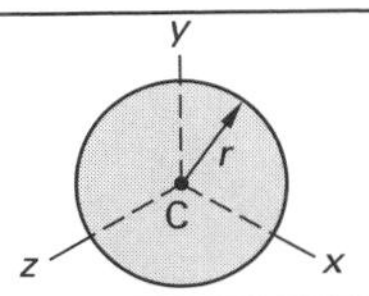	$I_x = I_y = I_z = \frac{2mr^2}{5}$

Support Material

APPENDIX 58.A

Standard Epicyclic Gear Train Ratios (for fixed gears, set $n = 0$)
(observe consistent sign convention for rotational speeds)

Double lines represent points at which two gears mesh. Vertical lines represent the web of the gear in profile view. Horizontal lines are shafts on which the gears are mounted, and bold bars are bearings which allow relative rotation between concentric shafts. Lines that extend from several gears downward and then horizontally identify the inputs (to the left) and output (to the right) for the gear train. All examples represent two inputs and one output.

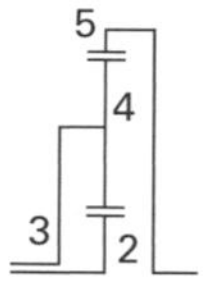

A

$$n_3\left(1+\frac{N_2}{N_5}\right) - n_2\left(\frac{N_2}{N_5}\right) = n_5$$

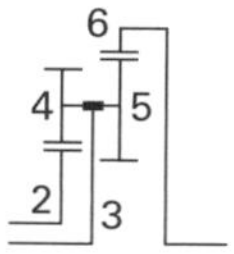

B

$$n_3\left(1+\frac{N_2N_5}{N_4N_6}\right) - n_2\left(\frac{N_2N_5}{N_4N_6}\right) = n_6$$

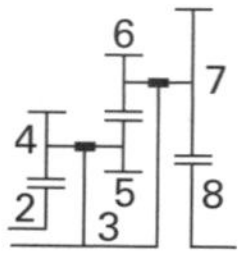

C

$$n_3\left(1+\frac{N_2N_5N_7}{N_4N_6N_8}\right) - n_2\left(\frac{N_2N_5N_7}{N_4N_6N_8}\right) = n_8$$

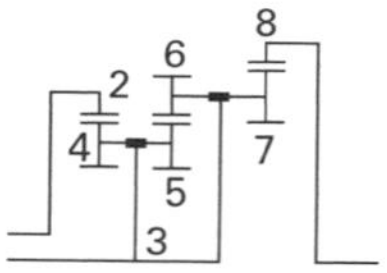

D

$$n_3\left(1+\frac{N_2N_5N_7}{N_4N_6N_8}\right) - n_2\left(\frac{N_2N_5N_7}{N_4N_6N_8}\right) = n_8$$

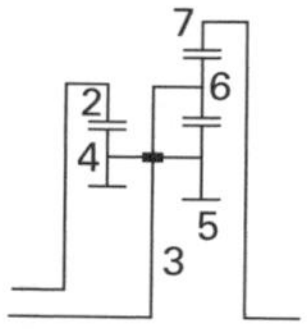

E

$$n_3\left(1+\frac{N_2N_5}{N_4N_7}\right) - n_2\left(\frac{N_2N_5}{N_4N_7}\right) = n_7$$

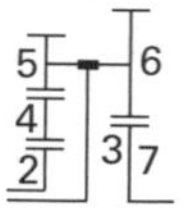

F

$$n_3\left(1+\frac{N_2N_6}{N_5N_7}\right) - n_2\left(\frac{N_2N_6}{N_5N_7}\right) = n_7$$

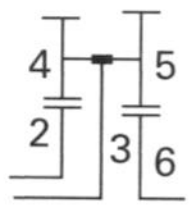

G

$$n_3\left(1-\frac{N_2N_5}{N_4N_6}\right) + n_2\left(\frac{N_2N_5}{N_4N_6}\right) = n_6$$

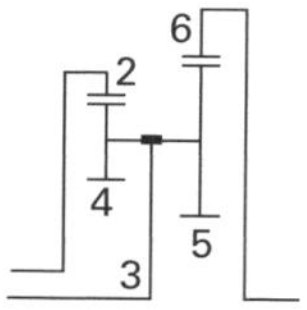

H

$$n_3\left(1-\frac{N_2N_5}{N_4N_6}\right) + n_2\left(\frac{N_2N_5}{N_4N_6}\right) = n_6$$

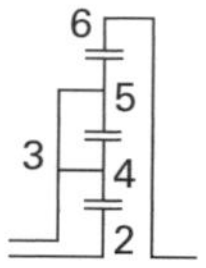

I

$$n_3\left(1-\frac{N_2}{N_6}\right) + n_2\left(\frac{N_2}{N_6}\right) = n_6$$

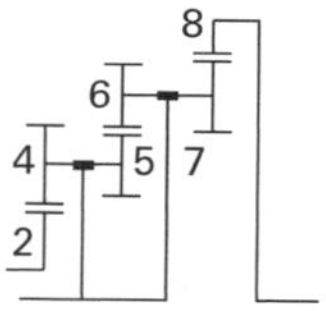

J

$$n_3\left(1-\frac{N_2N_5N_7}{N_4N_6N_8}\right) + n_2\left(\frac{N_2N_5N_7}{N_4N_6N_8}\right) = n_8$$

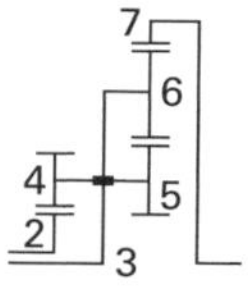

K

$$n_3\left(1-\frac{N_2N_5}{N_4N_7}\right) + n_2\left(\frac{N_2N_5}{N_4N_7}\right) = n_7$$

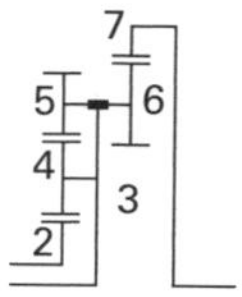

L

$$n_3\left(1-\frac{N_2N_6}{N_5N_7}\right) + n_2\left(\frac{N_2N_6}{N_5N_7}\right) = n_7$$

APPENDIX 63.A

Thermoelectric Constants for Thermocouples*(mV, reference 32°F (0°C))

(a) chromel-alumel (type K)

°F	0	10	20	30	40	50	60	70	80	90
					millivolts					
−300	−5.51	−5.60								
−200	−4.29	−4.44	−4.58	−4.71	−4.84	−4.96	−5.08	−5.20	−5.30	−5.41
−100	−2.65	−2.83	−3.01	−3.19	−3.36	−3.52	−3.69	−3.84	−4.00	−4.15
−0	−0.68	−0.89	−1.10	−1.30	−1.50	−1.70	−1.90	−2.09	−2.28	−2.47
+0	−0.68	−0.49	−0.26	−0.04	0.18	0.40	0.62	0.84	1.06	1.29
100	1.52	1.74	1.97	2.20	2.43	2.66	2.89	3.12	3.36	3.59
200	3.82	4.05	4.28	4.51	4.74	4.97	5.20	5.42	5.65	5.87
300	6.09	6.31	6.53	6.76	6.98	7.20	7.42	7.64	7.87	8.09
400	8.31	8.54	8.76	8.98	9.21	9.43	9.66	9.88	10.11	10.34
500	10.57	10.79	11.02	11.25	11.48	11.71	11.94	12.17	12.40	12.63
600	12.86	13.09	13.32	13.55	13.78	14.02	14.25	14.48	14.71	14.95
700	15.18	15.41	15.65	15.88	16.12	16.35	16.59	16.82	17.06	17.29
800	17.53	17.76	18.00	18.23	18.47	18.70	18.94	19.18	19.41	19.65
900	19.89	20.13	20.36	20.60	20.84	21.07	21.31	21.54	21.78	22.02
1000	22.26	22.49	22.73	22.97	23.20	23.44	23.68	23.91	24.15	24.39
1100	24.63	24.86	25.10	25.34	25.57	25.81	26.05	26.28	26.52	26.75
1200	26.98	27.22	27.45	27.69	27.92	28.15	28.39	28.62	28.86	29.09
1300	29.32	29.56	29.79	30.02	30.25	30.49	30.72	30.95	31.18	31.42
1400	31.65	31.88	32.11	32.34	32.57	32.80	33.02	33.25	33.48	33.71
1500	33.93	34.16	34.39	34.62	34.84	35.07	35.29	35.52	35.75	35.97
1600	36.19	36.42	36.64	36.87	37.09	37.31	37.54	37.76	37.98	38.20
1700	38.43	38.65	38.87	39.09	39.31	39.53	39.75	39.96	40.18	40.40
1800	40.62	40.84	41.05	41.27	41.49	41.70	41.92	42.14	42.35	42.57
1900	42.78	42.99	43.21	43.42	43.63	43.85	44.06	44.27	44.49	44.70
2000	44.91	45.12	45.33	45.54	45.75	45.96	46.17	46.38	46.58	46.79

(b) iron-constantan (type J)

°F	0	10	20	30	40	50	60	70	80	90
					millivolts					
−300	−7.52	−7.66								
−200	−5.76	−5.96	−6.16	−6.35	−6.53	−6.71	−6.89	−7.06	−7.22	−7.38
−100	−3.49	−3.73	−3.97	−4.21	−4.44	−4.68	−4.90	−5.12	−5.34	−5.55
−0	−0.89	−1.16	−1.43	−1.70	−1.96	−2.22	−2.48	−2.74	−2.99	−3.24
+0	−0.89	−0.61	−0.34	−0.06	0.22	0.50	0.79	1.07	1.36	1.65
100	1.94	2.23	2.52	2.82	3.11	3.41	3.71	4.01	4.31	4.61
200	4.91	5.21	5.51	5.81	6.11	6.42	6.72	7.03	7.33	7.64
300	7.94	8.25	8.56	8.87	9.17	9.48	9.79	10.10	10.41	10.72
400	11.03	11.34	11.65	11.96	12.26	12.57	12.88	13.19	13.50	13.81
500	14.12	14.42	14.73	15.04	15.34	15.65	15.96	16.26	16.57	16.88
600	17.18	17.49	17.80	18.11	18.41	18.72	19.03	19.34	19.64	19.95
700	20.26	20.56	20.87	21.18	21.48	21.79	22.10	22.40	22.71	23.01
800	23.32	23.63	23.93	24.24	24.55	24.85	25.16	25.47	25.78	26.09
900	26.40	26.70	27.02	27.33	27.64	27.95	28.26	28.58	28.89	29.21
1000	29.52	29.84	30.16	30.48	30.80	31.12	31.44	31.76	32.08	32.40
1100	32.72	33.05	33.37	33.70	34.03	34.36	34.68	35.01	35.35	35.68
1200	36.01	36.35	36.69	37.02	37.36	37.71	38.05	38.39	38.74	39.08
1300	39.43	39.78	40.13	40.48	40.83	41.19	41.54	41.90	42.25	42.61

APPENDIX 63.A *(continued)*
Thermoelectric Constants for Thermocouples*(mV, reference 32°F (0°C))

(c) copper-constantan (type T)

°F	0	10	20	30	40	50	60	70	80	90
−300	−5.284	−5.379			millivolts					
−200	−4.111	−4.246	−4.377	−4.504	−4.627	−4.747	−4.863	−4.974	−5.081	−5.185
−100	−2.559	−2.730	−2.897	−3.062	−3.223	−3.380	−3.533	−3.684	−3.829	−3.972
−0	−0.670	−0.872	−1.072	−1.270	−1.463	−1.654	−1.842	−2.026	−2.207	−2.385
+0	−0.670	−0.463	−0.254	−0.042	0.171	0.389	0.609	0.832	1.057	1.286
100	1.517	1.751	1.987	2.226	2.467	2.711	2.958	3.207	3.458	3.712
200	3.967	4.225	4.486	4.749	5.014	5.280	5.550	5.821	6.094	6.370
300	6.647	6.926	7.208	7.491	7.776	8.064	8.352	8.642	8.935	9.229
400	9.525	9.823	10.123	10.423	10.726	11.030	11.336	11.643	11.953	12.263
500	12.575	12.888	13.203	13.520	13.838	14.157	14.477	14.799	15.122	15.447

*This appendix is included to support general study and noncritical applications. *NBS Circular 561* has been superseded by *NBS Monograph 125, Thermocouple Reference Tables Based on the IPTS-68: Reference Tables in degrees Fahrenheit for Thermoelements versus Platinum*, presenting the same information as British Standards Institution standard *B.S. 4937*. In the spirit of globalization, these have been superseded in other countries by *IEC 584-1 (60584-1) Thermocouples Part 1: Reference Tables*, published in 1995. In addition to being based on modern correlations and the 1990 International Temperature Scale (ITS), the 160-page IEC 60584-1 duplicates all text in English and French, refers only to the Celsius temperature scale, and is available only by purchase or licensing.

APPENDIX 66.A

Typical Equivalent Lengths of Valves and Fittings for Fire Protection Systems

(Hazen-Williams $C = 120$. See Table 66.7 multipliers for other C-values.)
(all types of approved pipe)
(fittings and valves expressed in equivalent ft (m) of pipe)

	3/4 in		1 in		1 1/4 in		1 1/2 in		2 in		2 1/2 in		3 in	
45° elbow	1	(0.3)	1	(0.3)	1	(0.3)	2	(0.6)	2	(0.6)	3	(0.9)	3	(0.9)
90° standard elbow	2	(0.6)	2	(0.6)	3	(0.9)	4	(1.2)	5	(1.5)	6	(1.8)	7	(2.1)
90° long turn elbow	1	(0.3)	2	(0.6)	2	(0.6)	2	(0.6)	3	(0.9)	4	(1.2)	5	(1.5)
tee or cross (flow turned 90°)	4	(1.2)	5	(1.5)	6	(1.8)	8	(2.4)	10	(3.1)	12	(3.7)	15	(4.6)
gate valve	–		–		–		–		1	(0.3)	1	(0.3)	1	(0.3)
butterfly valve	–		–		–		–		6	(1.8)	7	(2.1)	10	(3.1)
swing check valve*	4	(1.2)	5	(1.5)	7	(2.1)	9	(2.7)	11	(3.4)	14	(4.3)	16	(4.9)

	3 1/2 in		4 in		5 in		6 in		8 in		10 in		12 in	
45° elbow	3	(0.9)	4	(1.2)	5	(1.5)	7	(2.1)	9	(2.7)	11	(3.4)	13	(4.0)
90° standard elbow	8	(2.4)	10	(3.1)	12	(3.7)	14	(4.3)	18	(5.5)	22	(6.7)	27	(8.2)
90° long turn elbow	5	(1.5)	6	(1.8)	8	(2.4)	9	(2.7)	13	(4.0)	16	(4.9)	18	(5.5)
tee or cross (flow turned 90°)	17	(5.2)	20	(6.1)	25	(7.6)	30	(9.2)	35	(10.7)	50	(15.3)	60	(18.3)
gate valve	1	(0.3)	2	(0.6)	2	(0.6)	3	(0.9)	4	(1.2)	5	(1.5)	6	(1.8)
butterfly valve	–		12	(3.7)	9	(2.7)	10	(3.1)	12	(3.7)	19	(5.8)	21	(6.4)
swing check valve*	19	(5.8)	22	(6.7)	27	(8.2)	32	(9.8)	15	(13.7)	55	(16.8)	65	(19.8)

*average values

Reprinted with permission from NFPA 15, *Standard for Water Spray Fixed Systems for Fire Protection*, copyright © 1996, National Fire Protection Association, Quincy, MA 02269. This reprinted material is not the complete and official position of the National Fire Protection Association on the referred subject which is represented only by the standard in entirety.

Support Material

APPENDIX 71.A
Polyphase Motor Classifications and Characteristics

speed regulations	speed control	starting torque	breakdown torque	application
general-purpose squirrel cage (NEMA design B)				
Drops about 3% for large to 5% for small sizes.	None, except multispeed types, designed for two to four fixed speeds.	100% for large; 275% for 1 hp 4 pole unit.	200% of full load.	Constant-speed service where starting is not excessive. Fans, blowers, rotary compressors, and centrifugal pumps.
high-torque squirrel cage (NEMA design C)				
Drops about 3% for large to 6% for small sizes.	None, except multispeed types, designed for two and four fixed speeds.	250% of full load for high-speed to 200% for low-speed designs.	200% of full load.	Constant-speed where fairly high starting torque is required infrequently with starting current about 550% of full load. Reciprocating pumps and compressors, crushers, etc.
high-slip squirrel cage (NEMA design D)				
Drops about 10% to 15% from no load to full load.	None, except multispeed types, designed for two to four fixed speeds.	225% to 300% full load, depending on speed with rotor resistance.	200%. Will usually not stall until loaded to maximum torque, which occurs at standstill.	Constant-speed and high starting torque, if starting is not too frequent, and for high-peak loads with or without flywheels. Punch presses, shears, elevators, etc.
low-torque squirrel cage (NEMA design F)				
Drops about 3% for large to 5% for small sizes.	None, except multispeed types, designed for two to four fixed speeds.	50% of full load for high-speed to 90% for low-speed designs.	135% to 170% of full load.	Constant-speed service where starting duty is light. Fans, blowers, centrifugal pumps, and similar loads.
wound rotor				
With rotor rings short circuited, drops about 3% for large to 5% for small sizes.	Speed can be reduced to 50% by rotor resistance. Speed varies inversely as load.	Up to 300% depending on external resistance in rotor circuit and how distributed.	300% when rotor slip rings are short circuited.	Where high starting torque with low starting current or where limited speed control is required. Fans, centrifugal and plunger pumps, compressors, conveyors, hoists, cranes, etc.
synchronous				
Constant.	None, except special motors designed for two fixed speeds.	40% for slow to 160% for medium-speed 80% pf. Specials develop higher.	Unity-pf motors 170%, 80%-pf motors 225%. Specials, up to 300%	For constant-speed service, direct connection to slow-speed machines and where pf correction is required.

Adapted from *Mechanical Engineering, Design Manual*, NAVFAC DM-3, Department of the Navy, copyright © 1972.

Support Material

APPENDIX 71.B
DC and Single-Phase Motor Classifications and Characteristics

speed regulations	speed control	starting torque	breakdown torque	application
		series		
Varies inversely as load. Races on light loads and full voltage.	Zero to maximum depending on control and load.	High. Varies as square of voltage. Limited by commutation, heating, and line capacity.	High. Limited by commutation, heating, and line capacity.	Where high starting torque is required and speed can be regulated. Traction, bridges, hoists, gates, car dumpers, car retarders.
		shunt		
Drops 3% to 5% from no load to full load.	Any desired range depending on design, type of system.	Good. With constant field, varies directly as voltage applied to armature.	High. Limited by commutation, heating, and line capacity.	Where constant or adjustable speed is required and starting conditions are not severe. Fans, blowers, centrifugal pumps, conveyors, wood and metal-working machines, elevators.
		compound		
Drops 7% to 20% from no load to full load depending on amount of compounding.	Any desired range, depending on design, type of control.	Higher than for shunt, depending on amount of compounding.	High. Limited by commutation, heating and line capacity.	Where high starting torque and fairly constant speed is required. Plunger pumps, punch presses, shears, bending rolls, geared elevators, conveyors, hoists.
		split-phase		
Drops about 10% from no load to full load.	None.	75% for large to 175% for small sizes.	150% for large to 200% for small sizes.	Constant-speed service where starting is easy. Small fans, centrifugal pumps and light-running machines, where polyphase is not available.
		capacitor start		
Drops about 5% for large to 10% for small sizes.	None.	150% to 350% of full load depending on design and size.	50% for large to 200% for small sizes.	Constant-speed service for any starting duty and quiet operation, where polyphase current cannot be used.
		commutator type		
Drops about 5% for large to 10% for small sizes.	Repulsion induction, none. Brush-shifting types, four to one at full load.	250% for large to 350% for small sizes.	150% for large to 250%.	Constant-speed service for any starting duty where speed control is required and polyphase current cannot be used.

Adapted from *Mechanical Engineering, Design Manual*, NAVFAC DM-3, Department of the Navy, copyright © 1972.

Support Material

APPENDIX 72.A
Noise Reduction and Absorption Coefficients

	noise absorption coefficients						
material	125 Hz	250 Hz	500 Hz	1000 Hz	2000 Hz	4000 Hz	NRC
brick, unglazed	0.03	0.03	0.03	0.04	0.05	0.07	0.04
brick, unglazed, painted	0.01	0.01	0.02	0.02	0.02	0.03	0.02
carpet							
1/8 in pile height	0.05	0.05	0.10	0.20	0.30	0.40	0.16
1/4 in pile height	0.05	0.10	0.15	0.30	0.50	0.55	0.26
3/16 in combined pile and foam	0.05	0.10	0.10	0.30	0.40	0.50	0.23
5/16 in combined pile and foam	0.05	0.15	0.30	0.40	0.50	0.60	0.34
concrete block, painted	0.10	0.05	0.06	0.07	0.09	0.08	0.07
fabrics							
light velour, 10 oz per sq yd, hung straight, in contact with wall	0.03	0.04	0.11	0.17	0.24	0.35	0.14
medium velour, 14 oz per sq yd, draped to half area	0.07	0.31	0.49	0.75	0.70	0.60	0.56
heavy velour, 18 oz per sq yd, draped to half area	0.14	0.35	0.55	0.72	0.70	0.65	0.62
floors							
concrete or terrazzo	0.01	0.01	0.01	0.02	0.02	0.02	0.02
linoleum, asphalt, rubber or cork tile on concrete	0.02	0.03	0.03	0.03	0.03	0.02	0.03
wood	0.15	0.11	0.10	0.07	0.06	0.07	0.09
wood parquet in asphalt on concrete	0.04	0.04	0.07	0.06	0.06	0.07	0.06
glass							
1/4 in, sealed, large panes	0.05	0.03	0.02	0.02	0.03	0.02	0.03
24 oz, operable windows (in closed condition)	0.10	0.05	0.04	0.03	0.03	0.03	0.04
gypsum board, 1/2 in nailed to 2 × 4's 16 in o.c., painted	0.10	0.08	0.05	0.03	0.03	0.03	0.05
marble or glazed tile	0.01	0.01	0.01	0.01	0.02	0.02	0.01
plaster, gypsum or lime,							
rough finish or lath	0.02	0.03	0.04	0.05	0.04	0.03	0.04
same, with smooth finish	0.02	0.02	0.03	0.04	0.04	0.03	0.03
hardwood plywood paneling 1/4 in thick, wood frame	0.58	0.22	0.07	0.04	0.03	0.07	0.09
water surface, as in a swimming pool	0.01	0.01	0.01	0.01	0.02	0.03	0.01
wood roof decking, tongue-and-groove cedar	0.24	0.19	0.14	0.08	0.13	0.10	0.14
air, sabins per 1000 cubic ft at 50% RH				0.9	2.3	7.2	
audience, seated, depending on spacing and upholstery of seats*	2.5–4.0	3.5–5.0	4.0–5.5	4.5–6.5	5.0–7.0	4.5–7.0	
seats, heavily upholstered with fabric*	1.5–3.5	3.5–4.5	4.0–5.0	4.0–5.5	3.5–5.5	3.5–4.5	
seats, heavily upholstered with leather, plastic, etc.*	2.5–3.5	3.0–4.5	3.0–4.0	2.0–4.0	1.5–3.5	1.0–3.0	
seats, lightly upholstered with leather, plastic, etc.*			1.5–2.0				
seats, wood veneer, no upholstery*	0.15	0.20	0.25	0.30	0.50	0.50	

*Values given are in sabins per person or unit of seating at the indicated frequency.

Derived from *Performance Data for Acoustical Materials Bulletin*, Acoustical and Board Products Association, 1975.

Support Material

APPENDIX 72.B
Transmission Loss Through Common Materials
(decibels)

material	density per unit area (lbm/in^2)	125 Hz	250 Hz	500 Hz	1000 Hz	2000 Hz	4000 Hz	8000 Hz
lead								
1/32 in thick	2	22	24	29	33	40	43	49
1/64 in thick	1	19	20	24	27	33	39	43
plywood								
3/4 in thick	2	24	22	27	28	25	27	35
1/4 in thick	0.7	17	15	20	24	28	27	25
lead vinyl	0.5	11	12	15	20	26	32	37
lead vinyl	1.0	15	17	21	28	33	37	43
steel								
18-gauge	2.0	15	19	31	32	35	48	53
16-gauge	2.5	21	30	34	37	40	47	52
sheet metal (viscoelastic laminate core)	2	15	25	28	32	39	42	47
plexiglass								
1/4 in thick	1.45	16	17	22	28	33	35	35
1/2 in thick	2.9	21	23	26	32	32	37	37
1 in thick	5.8	25	28	32	32	34	46	46
glass								
1/8 in thick	1.5	11	17	23	25	26	27	28
1/4 in thick	3	17	23	25	27	28	29	30
double glass								
1/4 × 1/2 × 1/4 in		23	24	24	27	28	30	36
1/4 × 6 × 1/4 in		25	28	31	37	40	43	47
5/8 in gypsum								
on 2 × 2 in stud		23	28	33	43	50	49	50
on staggered stud		26	35	42	52	57	55	57
concrete, 4 in thick	48	29	35	37	43	44	50	55
concrete block, 6 in	36	33	34	35	38	46	52	65
panels of 16 gauge steel, 4 in absorbent, 20 gauge steel		25	35	43	48	52	55	56

(Multiply lbm/ft^2 by 4.882 to obtain kg/m^2.)

Reprinted from *Industrial Noise Control Manual*, HEW Publication NIOSH 75-183, 1975.

Support Material

APPENDIX 73.A
Standard Cash Flow Factors

	multiply	by	to obtain	
	F	$P = F(1+i)^{-n}$ $(P/F, i\%, n)$	P	
	P	$F = P(1+i)^{n}$ $(F/P, i\%, n)$	F	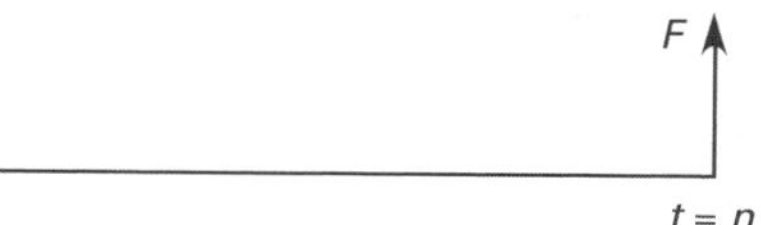
	A	$P = A\left(\frac{(1+i)^n - 1}{i(1+i)^n}\right)$ $(P/A, i\%, n)$	P	
	P	$A = P\left(\frac{i(1+i)^n}{(1+i)^n - 1}\right)$ $(A/P, i\%, n)$	A	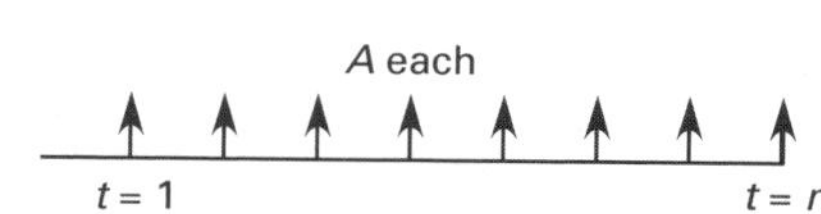
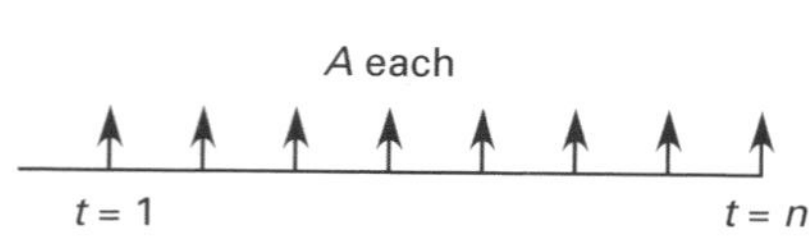	A	$F = A\left(\frac{(1+i)^n - 1}{i}\right)$ $(F/A, i\%, n)$	F	
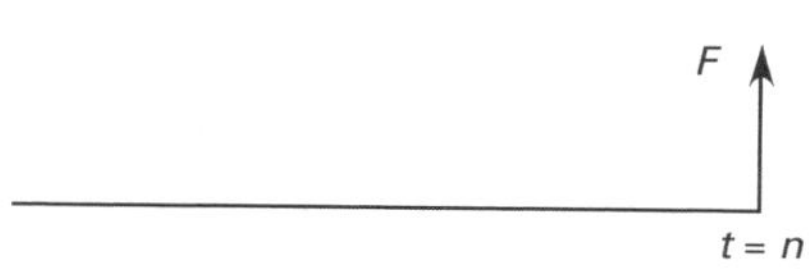	F	$A = F\left(\frac{i}{(1+i)^n - 1}\right)$ $(A/F, i\%, n)$	A	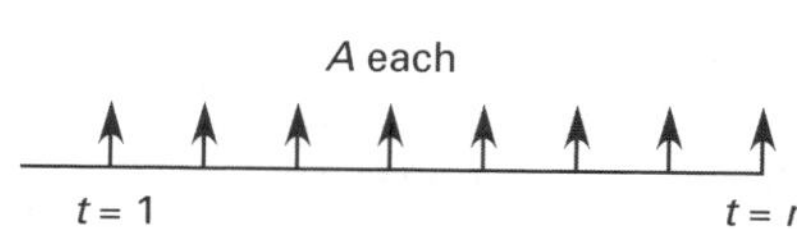
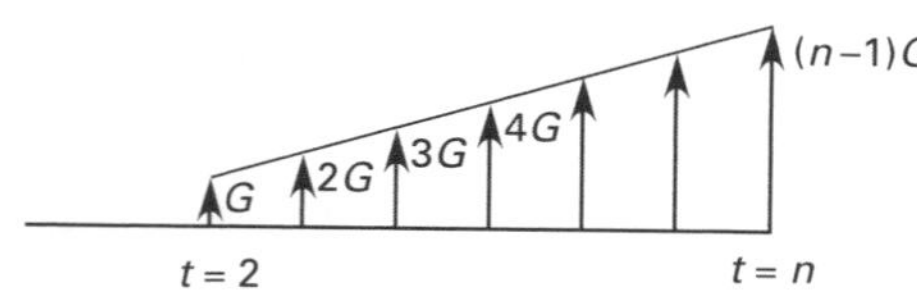	G	$P = G\left(\frac{(1+i)^n - 1}{i^2(1+i)^n} - \frac{n}{i(1+i)^n}\right)$ $(P/G, i\%, n)$	P	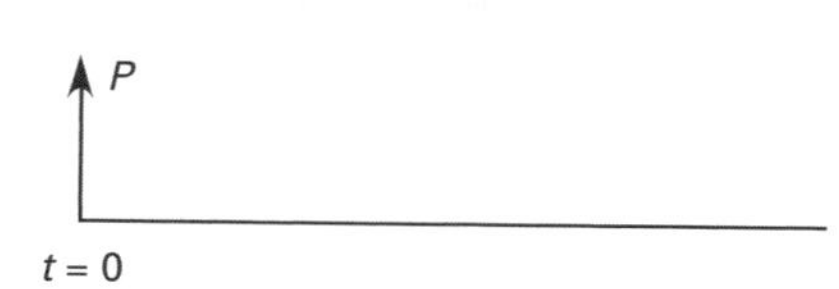
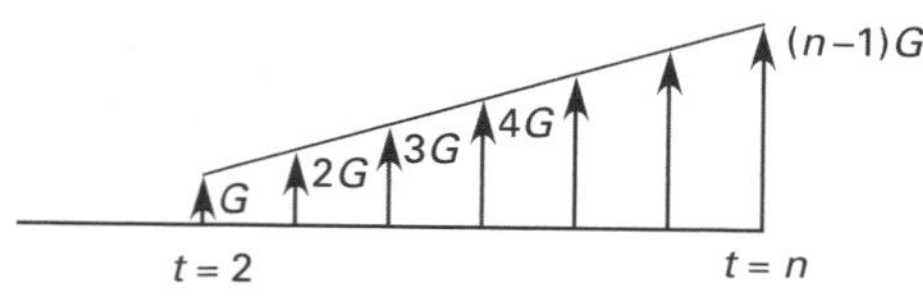	G	$A = G\left(\frac{1}{i} - \frac{n}{(1+i)^n - 1}\right)$ $(A/G, i\%, n)$	A	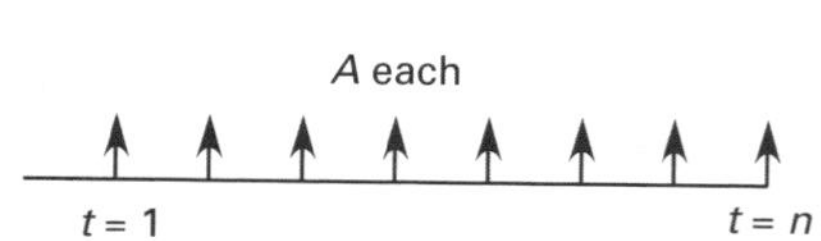

Index

B

C

D

INDEX - D

INDEX - D

E

INDEX - F

G

INDEX - F

I

INDEX - H

INDEX - L

INDEX - L

M

INDEX - M

INDEX - M

N

INDEX - M

O

P

INDEX - P

INDEX - P

INDEX - P

Q

R

INDEX - R

INDEX - R

S

INDEX - R

INDEX - S

INDEX - S

INDEX - S

INDEX - S

T

INDEX - S

INDEX - T

INDEX - T

INDEX - T

X

Y

Z